Contents

...us

University of Florida

with the assistance of

David C. Falvo

The Pennsylvania State University
The Behrend College

BROOKS/COLE
CENGAGE Learning

Australia • Brazil • Japa... United States

Calculus I with Precalculus
A One-Year Course, Third Edition
Larson/Edwards

Editor in Chief: Michelle Julet
Executive Editor: Liz Covello
Assistant Editor: Liza Neustaetter
Editorial Assistant: Jennifer Staller
Media Editor: Lynh Pham
Senior Marketing Manager: Jennifer Pursley Jones
Marketing Coordinator: Michael Ledesma
Marketing Communications Manager: Mary Anne Payumo
Content Project Manager: Jill Clark
Senior Art Director: Jill Ort
Senior Manufacturing Buyer: Diane Gibbons
Rights Acquisition Specialist, Image: Mandy Groszko
Senior Rights Acquisition Specialist, Text: Katie Huha
Cover Designer: Jill Ort
Cover Image: istockphoto.com © Raycat
Compositor: Larson Texts, Inc.

International Edition:
ISBN-13: 978-0-8400-6920-7
ISBN-10: 0-8400-6920-0

Cengage Learning International Offices

Asia
www.cengageasia.com
tel: (65) 6410 1200

Australia/New Zealand
www.cengage.com.au
tel: (61) 3 9685 4111

Brazil
www.cengage.com.br
tel: (55) 11 3665 9900

India
www.cengage.co.in
tel: (91) 1 4364 1111

Latin America
www.cengage.com.mx
tel: (52) 55 1500 6000

UK/Europe/Middle East/Africa
www.cengage.co.uk
tel: (44) 0 1264 332 424

Represented in Canada by Nelson Education, Ltd.
www.nelson.com
tel: (416) 752 9100 / (800) 668 0671

Cengage Learning is a leading provider of customized learning solutions with office locations around the globe, including Singapore, the United Kingdom, Australia, Mexico, Brazil, and Japan. Locate your local office at: **www.cengage.com/global**.

For product information: **www.cengage.com/international**
Visit your local office: **www.cengage.com/global**
Visit our corporate website: **www.cengage.com**

Printed in the United States of America
3 4 5 6 7 14 13 12

CHAPTER 6 Integration 397

CHAPTER 7 Exponential and Logarithmic Functions 471

CHAPTER 8 Exponential and Logarithmic Functions and Calculus 527

ADDITIONAL APPENDICES

A Word from the Authors

Integrating Precalculus and Calculus I

As its title suggests, *Calculus I with Precalculus: A One-Year Course*, Third Edition, is comprised of both precalculus topics and Calculus I topics. Rather than simply presenting all of the precalculus topics in the first half of the book, the precalculus topics are integrated throughout the text, according to function type—like, algebraic functions, exponential and logarithmic functions, and trigonometric and inverse trigonometric functions. This *function-driven* approach—covering precalculus topics before covering calculus topics—is repeated throughout the text, as illustrated below.

Function-Driven Approach			
Function Type	Precalculus	Calculus	Semester
Algebraic Functions	Chapters P–3	Chapters 4–6	I
Exponential and Logarithmic Functions	Chapter 7	Chapter 8	II
Trigonometric and Inverse Trigonometric Functions	Chapters 9, 10, 12	Chapters 11, 12	II
Additional Topics in Trigonometry and Analytic Geometry	Chapters 12, 13	Chapter 13	II

Additional precalculus topics are covered in Chapters P and 13. Chapter P offers a review of basic algebra, which can be covered quickly or assigned as outside reading. Chapter 13 can be covered at almost any point in the course.

Function-Driven Approach

Schools that offer a course combining precalculus and Calculus I have reported several advantages to the function-driven approach over the traditional precalculus-calculus sequence.

1. Students are motivated because they study calculus early in the semester as do their peers in the regular calculus sequence.
2. Students are asking calculus questions early in their study of precalculus.
3. Instructors have the opportunity to incorporate calculus examples and exercises into the later chapters that cover additional topics in trigonometry and analytic geometry, including parametric and polar equations.

Full Preparation for Calculus II

With the integration of precalculus, *Calculus I with Precalculus,* Third Edition, is intended for a slower-paced calculus course. This in turn makes the course more manageable, especially for students who have already struggled through a calculus course. Despite the slower pace, students will enter a Calculus II course as prepared and on the same level as their peers.

Calculus courses have been evolving and changing since we first began teaching and writing calculus. With these changes, we have made every effort to continue to provide instructors and students with quality textbooks and resources to accommodate their instructional and educational needs. We are excited about the opportunity to offer a textbook in a newly emerging market. We hope you enjoy this third edition of an innovative text.

If you have any suggestions for improving the text, please feel free to write us.

Ron Larson Bruce H. Edwards

Acknowledgments

We would like to thank the following people who have reviewed and provided feedback for the content of this text. Their suggestions, criticisms, and encouragement have been invaluable to us.

Reviewers of the Third Edition

Michael Axtell, *University of St. Thomas*
Patrick Bibby, *University of Miami*
Ekemezie Joseph Emeka, *Quincy University*
Marion Graziano, *Montgomery County Community College*
Benny Lo, *DeVry University –Freemont*
Lew Ludwig, *Denison University*
Sudeepa Pathak, *Williamston High School*
Thomas W. Simpson, *University of South Carolina–Union*

Reviewers of the Previous Editions

James Alsobrook, *Southern Union State Community College*; Anthony Austin, *Sherman High School, TX*; Raymond Badalian, *Los Angeles City College*; Virginia Bale, *Skyline Hight School, CA*; Carlos Barron, *Mountain View College, TX*; Rudranath Beharrysingh, *Southwestern Community College, NC*; John Berger, *Medina High School, OH*; Sharry Biggers, *Clemson University*; Charles Biles, *Humboldt State University*; Randall Boan, *Aims Community College*; John Burnette, *Savannah Country Day School, GA*; Christopher Butler, *Case Western Reserve University*; Dane R. Camp, *New Trier High School, IL*; Jeremy Carr, *Pensacola Junior College*; D.J. Clark, *Portland Community College*; Donald Clayton, *Madisonville Community College*; Barbara Cortzen, *DePaul University*; Linda Crabtree, *Metropolitan Community College*; David DeLatte, *University of North Texas*; Catherine DiChiaro, *Lincoln School, R*; Gregory Dlabach, *Northeastern Oklahoma A&M College*; Sadeq Elbaneh, *Sweet Home High School, NY*; Christian Eriksen, *Alameda Senior High School, CO*; Duane Frankiewicz, *Spooner High School, WI*; Nicholas Gorgievski, *Nichols College, MA*; Steve Gottlieb, *Albany High School, CA*; Dave Grim, *Liberty Center High School, OH*; Allen Grommet, *East Arkansas Community College, AR*; Joseph Lloyd Harris, *Gulf Coast Community College*; Jeff Heiking, *St. Petersburg Junior College*; Linda Henderson, *Ursuline Academy Upper School, DE*; Eugene A. Herman, *Grinnell College*; Celeste Hernandez, *Richland College*; Kathy Hoke, *University of Richmond*; Heidi Howard, *Florida Community College at Jacksonville*; Tami Jenkins, *Colorado Mountain College, CO*; Clay Laughary, *Forest Ridge School, WA*; Beth Long, *Pellissippi State Technical College*; Wanda Long, *St. Charles Community College*; John McDermott, *Bogan Technical High School, IL*; Wayne F. Mackey, *University of Arkansas*; Rhonda MacLeod, *Florida State University*; M. Maheswaran, *University Wisconsin–Marathon County*; Diane Maltby, *Westminster Christian School, FL*; Arda Melkonian, *Victor Valley College, CA*; Gordon Melrose, *Old Dominion University*; Robert Milano, *Notre Dame High School, CT*; Valerie Miller, *Georgia State University*; Katharine Muller, *Cisco Junior College*; Larry Norris, *North Carolina State University*; Bonnie Oppenheimer, *Mississippi University for Women*; Eleanor Palais, *Belmont High School, MA*; James Pohl, *Florida Atlantic University*; Hari Pulapaka, *Valdosta State University*; Lila Roberts, *Georgia Southern University*; Alma Runey, *Bishop England High School, SC*; Michael Russo, *Suffolk County Community College*;

Doreen Sabella, *County College of Morris, NJ*; John Santomas, *Villanova University*; Susan Schindler, *Baruch College–CUNY*; Cynthia Floyd Sikes, *Georgia Southern University*; Thomas Simpson, *University of South Carolina–Union SC*; Lynn Smith, *Gloucester County College*; Stanley Smith, *Black Hills State University*; Anthony Thomas, *University of Wisconsin–Platteville*; Nora Thornber, *Raritan Valley Community College, NJ*; Barry Trippett, *St. Clair County Community College, MI*; David Weinreich, *Gettsburg College, PA*; Charles Wheeler, *Montgomery College*.

Many thanks to Robert Hostetler, The Beherend College, The Pennsylvania State University, and David Heyd, The Behernd College, The Pennsylvania State University, for their significant contributions to previous editions of this text.

We would also like to thank the staff at Larson Texts, Inc., who assisted in preparing the manuscript, rendering the art package, typesetting, and proofreading the pages and supplements.

On a personal level, we are grateful to our wives, Deanna Gilbert Larson, and Consuelo Edwards, for their love, patience, and support. Also, a special note of thanks goes out to R. Scott O'Neil.

If you have suggestions for improving this text, please feel free to write to us. Over the years we have received many useful comments from both instructors and students, and we value these very much.

Ron Larson

Bruce H. Edwards

our Course. Your Way.

Calculus Textbook Options

The traditional calculus course is available in a variety of textbook configurations to address the different ways instructors teach—and students take—their classes.

The book can be customized to meet your individual needs and is available through iChapters—www.ichapters.com.

TOPICS COVERED	APPROACH			
	Integrated coverage	**Late Transcendental Functions**	**Early Transcendental Functions**	**Late Trigonometry**
Single Variable Only	Calculus I with Precalculus 3e	Calculus 9e Single Variable	Calculus: Early Transcendental Functions 5e Single Variable	
3-semester		Calculus 9e	Calculus: Early Transcendental Functions 5e	Calculus with Late Trigonometry
Multivariable		Calculus 9e Multivariable	Calculus 9e Multivariable	
Custom All of these textbook choices can be customized to fit the individual needs of your course.	Calculus I with Precalculus 3e	Calculus 9e	Calculus: Early Transcendental Functions 5e	Calculus with Late Trigonometry

Textbook Features

Tools to Build Mastery

CAPSTONES

NEW! Capstone exercises now appear in every section. These exercises synthesize the main concepts of each section and show students how the topics relate. They are often multipart problems that contain conceptual and noncomputational parts, and can be used for classroom discussion or test prep.

CAPSTONE

58. Use the graph of f' shown in the figure to answer the following, given that $f(0) = -4$.

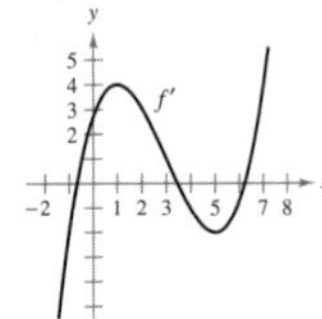

(a) Approximate the slope of f at $x = 4$. Explain.

(b) Is it possible that $f(2) = -1$? Explain.

(c) Is $f(5) - f(4) > 0$? Explain.

(d) Approximate the value of x where f is maximum. Explain.

(e) Approximate any intervals in which the graph of f is concave upward and any intervals in which it is concave downward. Approximate the x-coordinates of any points of inflection.

(f) Approximate the x-coordinate of the minimum of $f''(x)$.

(g) Sketch an approximate graph of f. To print an enlarged copy of the graph, go to the website *www.mathgraphs.com*.

WRITING ABOUT CONCEPTS

51. State the Fundamental Theorem of Calculus.

52. The graph of f is shown in the figure.

(a) Evaluate $\int_1^7 f(x)\,dx$.

(b) Determine the average value of f on the interval $[1, 7]$.

(c) Determine the answers to parts (a) and (b) if the graph is translated two units upward.

53. If $r'(t)$ represents the rate of growth of a dog in pounds per year, what does $r(t)$ represent? What does

$$\int_2^6 r'(t)\,dt$$

represent about the dog?

WRITING ABOUT CONCEPTS

These writing exercises are questions designed to test students' understanding of basic concepts in each section. The exercises encourage students to verbalize and write answers, promoting technical communication skills that will be invaluable in their future careers.

STUDY TIPS

The devil is in the details. Study Tips help point out some of the troublesome common mistakes, indicate special cases that can cause confusion, or expand on important concepts. These tips provide students with valuable information, similar to what an instructor might comment on in class.

STUDY TIP Because integration is usually more difficult than differentiation, you should always check your answer to an integration problem by differentiating. For instance, in Example 3 you should differentiate $\frac{1}{3}(2x$ ... that you obtain the

STUDY TIP Later in this chapter, you will learn convenient methods for calculating $\int_a^b f(x)\,dx$ for continuous ... For now, you must use the ...ion.

STUDY TIP Remember that you can check your answer by differentiating.

EXAMPLE 6 Evaluation of a Definite Integral

Evaluate $\int_1^3 (-x^2 + 4x - 3)\,dx$ using each of the following values.

$$\int_1^3 x^2\,dx = \frac{26}{3}, \qquad \int_1^3 x\,dx = 4, \qquad \int_1^3 dx = 2$$

Solution

$$\begin{aligned}\int_1^3 (-x^2 + 4x - 3)\,dx &= \int_1^3 (-x^2)\,dx + \int_1^3 4x\,dx + \int_1^3 (-3)\,dx \\ &= -\int_1^3 x^2\,dx + 4\int_1^3 x\,dx - 3\int_1^3 dx \\ &= -\left(\frac{26}{3}\right) + 4(4) - 3(2) = \frac{4}{3}\end{aligned}$$

EXAMPLES

Throughout the text, examples are worked out step-by-step. These worked examples demonstrate the procedures and techniques for solving problems, and give students an increased understanding of the concepts of calculus. Many examples are presented in a side-by-side format to help students see that a problem can be solved in more than one way.

EXERCISES

Practice makes perfect. Exercises are often the first place students turn to in a textbook. The authors have spent a great deal of time analyzing and revising the exercises, and the result is a comprehensive and robust set of exercises at the end of every section. A variety of exercise types and levels of difficulty are included to accommodate students with all learning styles.

In addition to the exercises in the book, 3,000 algorithmic exercises appear in the WebAssign® course that accompanies Calculus.

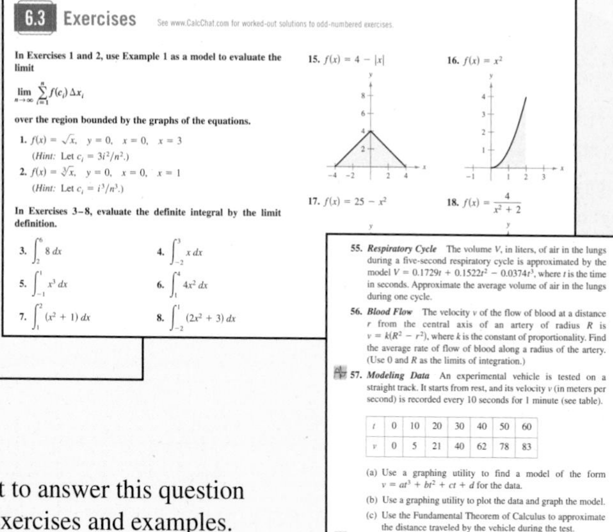

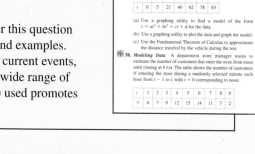

APPLICATIONS

"When will I use this?" The authors attempt to answer this question for students with carefully chosen applied exercises and examples. Applications are pulled from diverse sources, such as current events, world data, industry trends, and more, and relate to a wide range of interests. Understanding where calculus is (or can be) used promotes fuller understanding of the material.

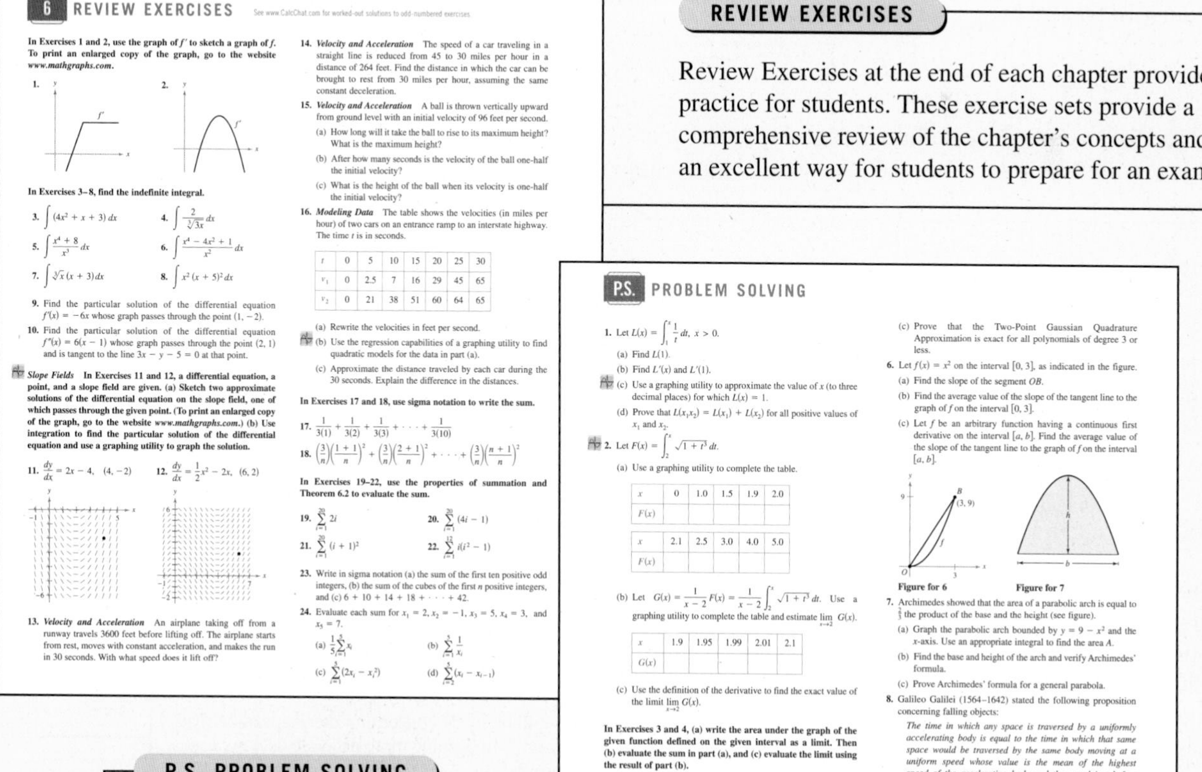

REVIEW EXERCISES

Review Exercises at the end of each chapter provide more practice for students. These exercise sets provide a comprehensive review of the chapter's concepts and are an excellent way for students to prepare for an exam.

P.S. PROBLEM SOLVING

These sets of exercises at the end of each chapter test students' abilities with challenging, thought-provoking questions.

Classic Calculus with Contemporary Relevance

THEOREMS

Theorems provide the conceptual framework for calculus. Theorems are clearly stated and separated from the rest of the text by boxes for quick visual reference. Key proofs often follow the theorem, and other proofs are provided in an in-text appendix.

THEOREM 6.9 THE FUNDAMENTAL THEOREM OF CALCULUS

If a function f is continuous on the closed interval $[a, b]$ and F is an antiderivative of f on the interval $[a, b]$, then

$$\int_a^b f(x)\,dx = F(b) - F(a).$$

DEFINITIONS

As with the theorems, definitions are clearly stated using precise, formal wording and are separated from the text by boxes for quick visual reference.

DEFINITION OF DEFINITE INTEGRAL

If f is defined on the closed interval $[a, b]$ and the limit of Riemann sums over partitions Δ

$$\lim_{\|\Delta\|\to 0} \sum_{i=1}^{n} f(c_i)\,\Delta x_i$$

exists (as described above), then f is said to be **integrable** on $[a, b]$ and the limit is denoted by

$$\lim_{\|\Delta\|\to 0} \sum_{i=1}^{n} f(c_i)\,\Delta x_i = \int_a^b f(x)\,dx.$$

The limit is called the **definite integral** of f from a to b. ' **lower limit** of integration, and the number b is the **uppe**

PROCEDURES

Formal procedures are set apart from the text for easy reference. The procedures provide students with step-by-step instructions that will help them solve problems quickly and efficiently.

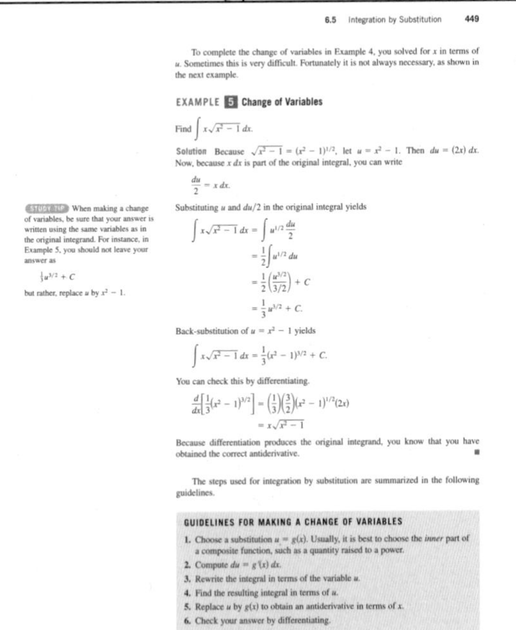

6.5 Integration by Substitution 449

To complete the change of variables in Example 4, you solved for x in terms of u. Sometimes this is very difficult. Fortunately it is not always necessary, as shown in the next example.

EXAMPLE 5 Change of Variables

Find $\int x\sqrt{x^2 - 1}\,dx$.

Solution Because $\sqrt{x^2 - 1} = (x^2 - 1)^{1/2}$, let $u = x^2 - 1$. Then $du = (2x)\,dx$. Now, because $x\,dx$ is part of the original integral, you can write

$$\frac{du}{2} = x\,dx.$$

Substituting u and $du/2$ in the original integral yields

$$\begin{aligned}\int x\sqrt{x^2 - 1}\,dx &= \int u^{1/2}\frac{du}{2}\\ &= \frac{1}{2}\int u^{1/2}\,du\\ &= \frac{1}{2}\left(\frac{u^{3/2}}{3/2}\right) + C\\ &= \frac{1}{3}u^{3/2} + C.\end{aligned}$$

Back-substitution of $u = x^2 - 1$ yields

$$\int x\sqrt{x^2 - 1}\,dx = \frac{1}{3}(x^2 - 1)^{3/2} + C.$$

You can check this by differentiating.

$$\begin{aligned}\frac{d}{dx}\left[\frac{1}{3}(x^2 - 1)^{3/2}\right] &= \left(\frac{1}{3}\right)\left(\frac{3}{2}\right)(x^2 - 1)^{1/2}(2x)\\ &= x\sqrt{x^2 - 1}\end{aligned}$$

Because differentiation produces the original integrand, you know that you have obtained the correct antiderivative. ■

STUDY TIP When making a change of variables, be sure that your answer is written using the same variables as in the original integrand. For instance, in Example 5, you should not leave your answer as

$$\tfrac{1}{3}u^{3/2} + C$$

but rather, replace u by $x^2 - 1$.

The steps used for integration by substitution are summarized in the following guidelines.

GUIDELINES FOR MAKING A CHANGE OF VARIABLES

1. Choose a substitution $u = g(x)$. Usually, it is best to choose the *inner* part of a composite function, such as a quantity raised to a power.
2. Compute $du = g'(x)\,dx$.
3. Rewrite the integral in terms of the variable u.
4. Find the resulting integral in terms of u.
5. Replace u by $g(x)$ to obtain an antiderivative in terms of x.
6. Check your answer by differentiating.

NOTES

Notes provide additional details about theorems, definitions, and examples. They offer additional insight, or important generalizations that students might not immediately see. Like the study tips, notes can be invaluable to students.

NOTE There are two important points that should be made concerning the Trapezoidal Rule (or the Midpoint Rule). First, the approximation tends to become more accurate as n increases. For instance, in Example 1, if $n = 16$, the Trapezoidal Rule yields an approximation of 1.2189. Second, although you could have used the Fundamental Theorem to evaluate the integral in Example 1, this theorem cannot be used to evaluate an integral as simple as $\int_0^1 \sqrt{x^3 + 1}\,dx$ because $\sqrt{x^3 + 1}$ has no elementary antiderivative. Yet, the Trapezoidal Rule can be applied easily to estimate this integral. ■

Expanding the Experience of Calculus

CHAPTER OPENERS

Chapter Openers provide initial motivation for the upcoming chapter material. Along with a map of the chapter objectives, an important concept in the chapter is related to an application of the topic in the real world. Students are encouraged to see the real-life relevance of calculus.

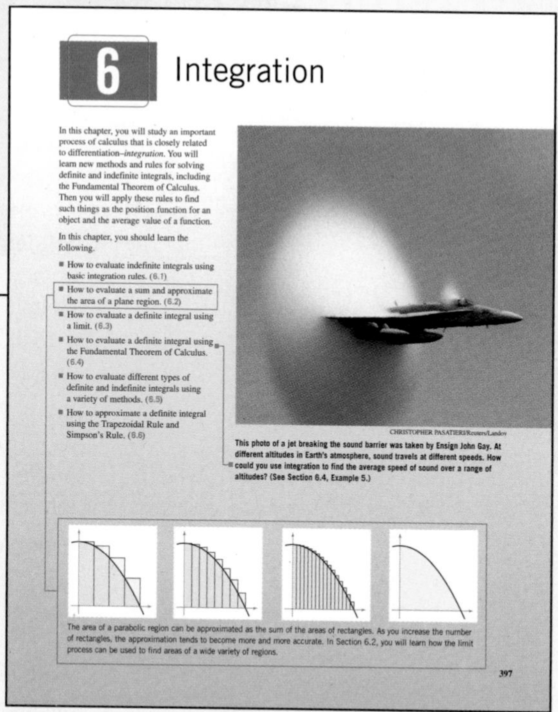

6 Integration

In this chapter, you will study an important process of calculus that is closely related to differentiation–*integration*. You will learn new methods and rules for solving definite and indefinite integrals, including the Fundamental Theorem of Calculus. Then you will apply these rules to find such things as the position function for an object and the average value of a function.

In this chapter, you should learn the following.

- How to evaluate indefinite integrals using basic integration rules. (6.1)
- How to evaluate a sum and approximate the area of a plane region. (6.2)
- How to evaluate a definite integral using a limit. (6.3)
- How to evaluate a definite integral using the Fundamental Theorem of Calculus. (6.4)
- How to evaluate different types of definite and indefinite integrals using a variety of methods. (6.5)
- How to approximate a definite integral using the Trapezoidal Rule and Simpson's Rule. (6.6)

CHRISTOPHER PASATIERI/Reuters/Landov

This photo of a jet breaking the sound barrier was taken by Ensign John Gay. At different altitudes in Earth's atmosphere, sound travels at different speeds. How could you use integration to find the average speed of sound over a range of altitudes? (See Section 6.4, Example 5.)

The area of a parabolic region can be approximated as the sum of the areas of rectangles. As you increase the number of rectangles, the approximation tends to become more and more accurate. In Section 6.2, you will learn how the limit process can be used to find areas of a wide variety of regions.

397

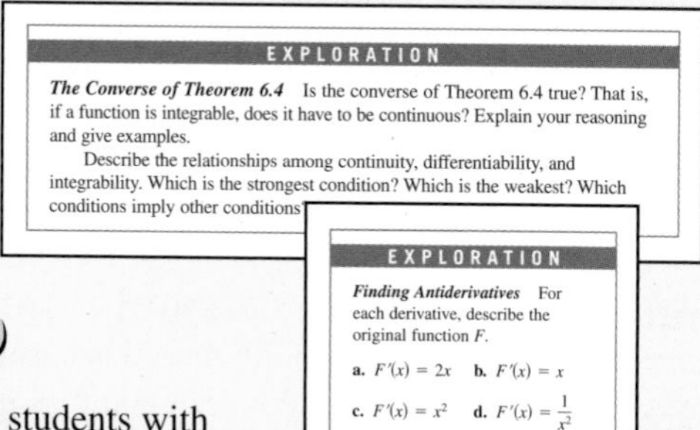

EXPLORATION

The Converse of Theorem 6.4 Is the converse of Theorem 6.4 true? That is, if a function is integrable, does it have to be continuous? Explain your reasoning and give examples.

Describe the relationships among continuity, differentiability, and integrability. Which is the strongest condition? Which is the weakest? Which conditions imply other conditions

EXPLORATION

Finding Antiderivatives For each derivative, describe the original function F.

a. $F'(x) = 2x$ b. $F'(x) = x$

c. $F'(x) = x^2$ d. $F'(x) = \frac{1}{x^2}$

e. $F'(x) = \frac{1}{x^3}$

What strategy did you use to find F?

EXPLORATIONS

Explorations provide students with unique challenges to study concepts that have not yet been formally covered. They allow students to learn by discovery and introduce topics related to ones they are presently studying. By exploring topics in this way, students are encouraged to think outside the box.

HISTORICAL NOTES AND BIOGRAPHIES

Historical Notes provide students with background information on the foundations of calculus, and Biographies help humanize calculus and teach students about the people who contributed to its formal creation.

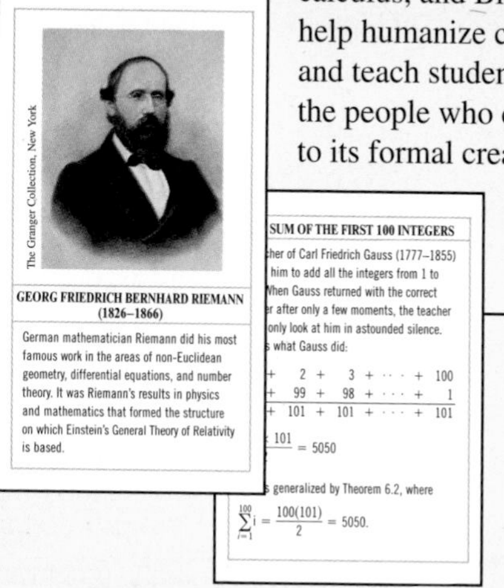

The Granger Collection, New York

GEORG FRIEDRICH BERNHARD RIEMANN (1826–1866)

German mathematician Riemann did his most famous work in the areas of non-Euclidean geometry, differential equations, and number theory. It was Riemann's results in physics and mathematics that formed the structure on which Einstein's General Theory of Relativity is based.

SUM OF THE FIRST 100 INTEGERS

her of Carl Friedrich Gauss (1777–1855) him to add all the integers from 1 to When Gauss returned with the correct r after only a few moments, the teacher only look at him in astounded silence. what Gauss did:

$$\begin{array}{rrrrr} + & 2 + & 3 + & \cdots + & 100 \\ + & 99 + & 98 + & \cdots + & 1 \\ \hline + & 101 + & 101 + & \cdots + & 101 \end{array}$$

$\frac{101}{} = 5050$

generalized by Theorem 6.2, where

$$\sum_{i=1}^{100} i = \frac{100(101)}{2} = 5050.$$

TECHNOLOGY

Throughout the book, technology boxes give students a glimpse of how technology may be used to help solve problems and explore the concepts of calculus. They provide discussions of not only where technology succeeds, but also where it may fail.

TECHNOLOGY Most graphing utilities and computer algebra systems have built-in programs that can be used to approximate the value of a definite integral. Try using such a program to approximate the integral in Example 1.

When you use such a program, you need to be aware of its limitations. Often, you are given no indication of the degree of accuracy of the approximation. Other times, you may be given an approximation that is completely wrong. For instance, try using a built-in numerical integration program to evaluate

$$\int_{-1}^{2} \frac{1}{x}\, dx.$$

Your calculator should give an error message. Does yours?

SECTION PROJECTS

Projects appear in selected sections and more deeply explore applications related to the topics being studied. They provide an interesting and engaging way for students to work and investigate ideas collaboratively.

SECTION PROJECT

Demonstrating the Fundamental Theorem

Use a graphing utility to graph the function

$$y_1 = \frac{t}{\sqrt{1+t}}$$

on the interval $2 \le t \le 5$. Let $F(x)$ be the following function of x.

$$F(x) = \int_2^x \frac{t}{\sqrt{1+t}}\, dt$$

(a) Complete the table. Explain why the values of F are increasing.

x	2	2.5	3	3.5	4	4.5	5
$F(x)$							

(b) Use the integration capabilities of a graphing utility to graph F.

(c) Use the differentiation capabilities of a graphing utility to graph $F'(x)$. How is this graph related to the graph in part (b)?

(d) Verify that the derivative of

$$y = \frac{2}{3}(t-2)\sqrt{1+t}$$

is $t/\sqrt{1+t}$. Graph y and write a short paragraph about how this graph is related to those in parts (b) and (c).

Additional Resources

Student Resources

- **Student Solutions Manual** (ISBN 0-8400-6912-X)—Need a leg up on your homework or help to prepare for an exam? The *Student Solutions Manual* contains worked-out solutions for all odd-numbered exercises in the text. It is a great resource to help you understand how to solve those tough problems.
- **CalcLabs with Maple® and Mathematica®** (CalcLabs with Maple for Single Variable Calculus: ISBN 0-8400-5811-X; CalcLabs with Mathematica for Single Variable Calculus: ISBN 0-8400-5814-4)—Working with *Maple* or *Mathematica* in class? Be sure to pick up one these comprehensive manuals that will help you use each program efficiently.
- **Enhanced WebAssign®** (ISBN 0-538-73810-3)—Enhanced WebAssign is designed for you to do your homework online. This proven and reliable system uses pedagogy and content found in this text, and then enhances it to help you learn Calculus I with Precalculus more effectively. Automatically graded homework allows you to focus on your learning and get interactive study assistance outside of class.
- **CourseMate**—The more you study, the better the results. Make the most of your study time by accessing everything you need in one place. Read your textbook, take notes, review flashcards, watch videos, and take practice quizzes—online with CourseMate.
- **CengageBrain.com**—To access additional course materials including CourseMate, please visit www.cengagebrain.com. At the CengageBrain.com home page, search for the ISBN of your title (from the back cover of your book) using the search box at the top of the page. This will take you to the product page where these resources can be found.

Instructor Resources

- **Enhanced WebAssign®** (ISBN 0-538-73810-3)—Exclusively from Cengage Learning, Enhanced WebAssign® offers an extensive online program for Calculus I with Precalculus to encourage the practice that is so critical for concept mastery. The meticulously crafted pedagogy and exercises in our proven texts become even more effective in Enhanced WebAssign, supplemented by multimedia tutorial support and immediate feedback as students complete their assignments. Key features include:
  - Read It eBook pages, Watch It videos, Master It tutorials, and Chat About It links
  - As many as 3000 homework problems that match your textbook's end-of-section exercises

Instructor Resources
(continued)

- New! Premium eBook with highlighting, note-taking, and search features, as well as links to multimedia resources
- New! Personal Study Plans (based on diagnostic quizzing) that identify chapter topics that students still need to master
- Algorithmic problems, allowing you to assign unique versions to each student
- Practice Another Version feature (activated at the instructor's discretion) allows students to attempt the questions with new sets of values until they feel confident enough to work the original problem
- GraphPad enables students to graph lines, segments, parabolas, and circles as they answer questions
- MathPad simplifies the input of mathematical symbols
- New! WebAssign Answer Evaluator recognizes and accepts equivalent mathematical responses in the same way an instructor grades. Student responses are analyzed for correctness and intent so students are not penalized for mathematically equivalent responses
- New! A *Show Your Work* feature gives instructors the option of seeing students' detailed solutions

- **Instructor's Complete Solutions Manual** (ISBN 0-8400-6911-1)—This manual contains worked-out solutions for all exercises in the text.
- **Solution Builder** (www.cengage.com/solutionbuilder)—This online instructor database offers complete worked-out solutions to all exercises in the text, allowing you to create customized, secure solutions printouts (in PDF format) matched exactly to the problems you assign in class.
- **PowerLecture** (ISBN 0-8400-6913-8)—This comprehensive CD-ROM includes instructor resources such as PowerPoint Slides® and Diploma Computerized Testing featuring algorithmically created questions that can be used to create, deliver, and customize tests.
- **Diploma Computerized Testing**—Diploma testing software allows instructors to quickly create, deliver, and customize tests for class in print and online formats, and features automatic grading. This software includes a test bank with hundreds of questions customized directly to the text. Diploma Testing is available within the PowerLecture CD-ROM.
- **CourseMate**—Cengage Learning's CourseMate bring concepts to life with interactive learning, study, and exam preparation tools that support the printed textbook. Watch student comprehension soar as your class works with the printed textbook and the textbook-specific website. CourseMate goes beyond the book to deliver what you need!
- **CengageBrain.com**—To access additional course materials including CourseMate, please visit http://login.cengage.com. At the CengageBrain.com home page, search for the ISBN of your title (from the back cover of your book) using the search box at the top of the page. This will take you to the product page where these resources can be found.

Prerequisites

In this chapter, you should learn the following.

- How to solve equations, including linear, quadratic, and higher-degree polynomial equations, as well as equations involving radicals and absolute values. (P.1)
- How to solve inequalities, including linear, absolute value, polynomial, and rational inequalities. (P.2)
- How to represent data graphically, find the distance between two points, and find the midpoint of a line segment. (P.3)
- How to identify the characteristics of equations and sketch their graphs, including equations and graphs of circles. (P.4)
- How to find and graph equations of lines, including parallel and perpendicular lines, using the concept of slope. (P.5)

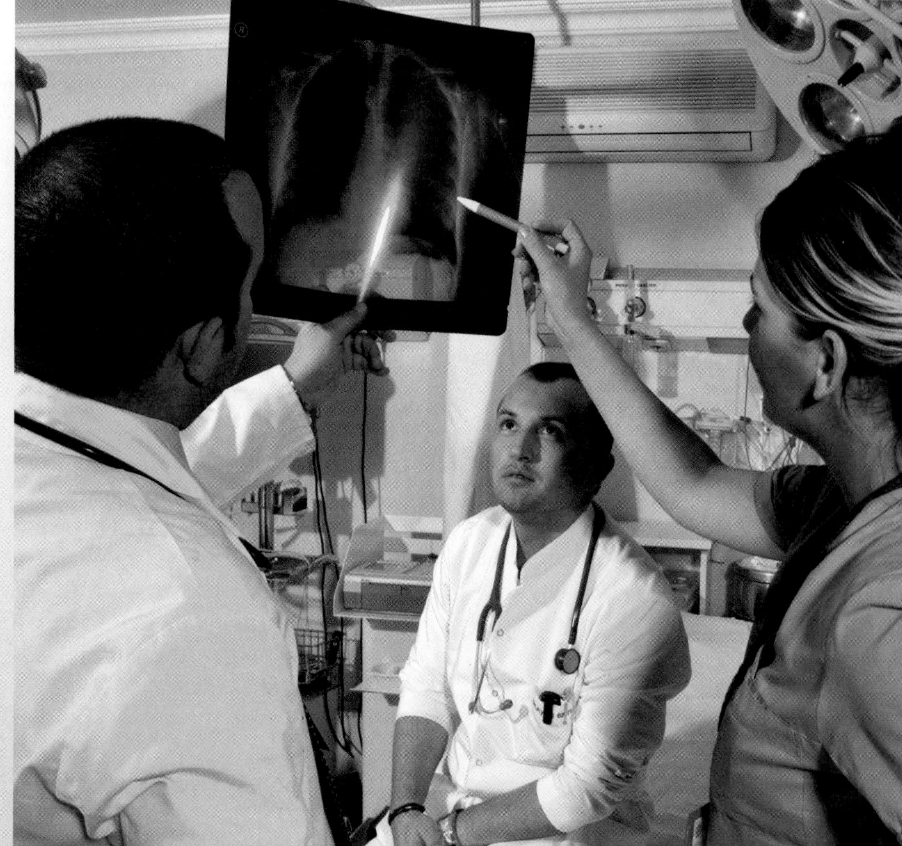

Levent Konuk, 2010/Used under license from Shutterstock.com

The numbers of doctors of osteopathic medicine in the United States increased each year from 2000 through 2008. How can you use this information to estimate the number of doctors of osteopathic medicine in 2012? (See Section P.5, Exercise 135.)

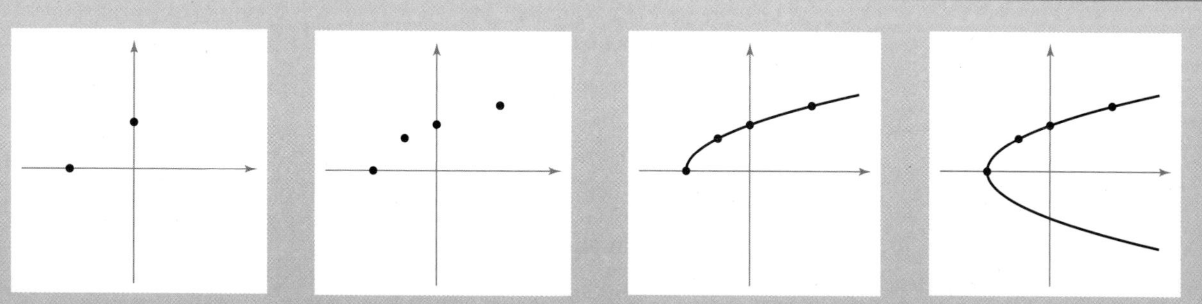

You can represent the solutions of an equation in two variables visually by making a graph in a rectangular coordinate system. (See Section P.4.)

P.1 Solving Equations

- Identify different types of equations.
- Solve linear equations in one variable and equations that lead to linear equations.
- Solve quadratic equations by factoring, extracting square roots, completing the square, and using the Quadratic Formula.
- Solve polynomial equations of degree three or greater.
- Solve equations involving radicals.
- Solve equations with absolute values.

Equations and Solutions of Equations

NOTE Recall that the set of real numbers is made up of rational numbers (integers and fractions) and irrational numbers such as $\sqrt{2}$, $\sqrt{3}$, π, and so on. Graphically the real numbers are represented by a number line with zero as its origin.

Negative direction | Origin | $\sqrt{2}$ | Positive direction | −3 −2 −1 0 1 2 3 | x

The set of real numbers for which an algebraic expression is defined is the **domain** of the expression.

An **equation** in x is a statement that two algebraic expressions are equal. For example,

$$3x - 5 = 7, x^2 - x - 6 = 0, \text{ and } \sqrt{2x} = 4$$

are equations. To **solve** an equation in x means to find all values of x for which the equation is true. Such values are **solutions.** For instance, $x = 4$ is a solution of the equation

$$3x - 5 = 7$$

because $3(4) - 5 = 7$ is a true statement.

The solutions of an equation depend on the kinds of numbers being considered. For instance, in the set of rational numbers, $x^2 = 10$ has no solution because there is no rational number whose square is 10. However, in the set of real numbers, the equation has the two solutions $x = \sqrt{10}$ and $x = -\sqrt{10}$.

An equation that is true for *every* real number in the *domain* of the variable is called an **identity.** The domain is the set of all real numbers for which the equation is defined. For example,

$$x^2 - 9 = (x + 3)(x - 3) \qquad \text{Identity}$$

is an identity because it is a true statement for any real value of x. The equation

$$\frac{x}{3x^2} = \frac{1}{3x} \qquad \text{Identity}$$

where $x \neq 0$, is an identity because it is true for any nonzero real value of x.

An equation that is true for just *some* (or even none) of the real numbers in the domain of the variable is called a **conditional equation.** For example, the equation

$$x^2 - 9 = 0 \qquad \text{Conditional equation}$$

is conditional because $x = 3$ and $x = -3$ are the only values in the domain that satisfy the equation. The equation $2x - 4 = 2x + 1$ is conditional because there are no real values of x for which the equation is true.

Linear Equations in One Variable

DEFINITION OF A LINEAR EQUATION

A **linear equation in one variable** x is an equation that can be written in the standard form

$$ax + b = 0$$

where a and b are real numbers with $a \neq 0$.

A linear equation in one variable, written in standard form, always has *exactly one* solution. To see this, consider the following steps.

$ax + b = 0$ Original equation, with $a \neq 0$

$ax = -b$ Subtract b from each side.

$x = -\frac{b}{a}$ Divide each side by a.

To solve a conditional equation in x, isolate x on one side of the equation by a sequence of **equivalent** (and usually simpler) **equations,** each having the same solution(s) as the original equation.

STUDY TIP Note that some linear equations in nonstandard form have *no solution* or *infinitely many solutions.* For instance,

$$x = x + 1$$

has no solution because it is not true for any value of x. Because

$$5x + 10 = 5(x + 2)$$

is true for any value of x, the equation has infinitely many solutions.

GENERATING EQUIVALENT EQUATIONS: PROPERTIES OF EQUALITY

An equation can be transformed into an *equivalent equation* by one or more of the following steps.

	Given Equation	*Equivalent Equation*
1. Remove symbols of grouping, combine like terms, or simplify fractions on one or both sides of the equation.	$2x - x = 4$	$x = 4$
2. Add (or subtract) the same quantity to (from) *each* side of the equation.	$x + 1 = 6$	$x = 5$
3. Multiply (or divide) *each* side of the equation by the same *nonzero* quantity.	$2x = 6$	$x = 3$
4. Interchange the two sides of the equation.	$2 = x$	$x = 2$

EXAMPLE 1 Solving a Linear Equation

Solve $3x - 6 = 0$.

Solution

$3x - 6 = 0$ Write original equation.

$3x = 6$ Add 6 to each side.

$x = 2$ Divide each side by 3.

Check After solving an equation, you should check each solution in the original equation.

$3x - 6 = 0$ Write original equation.

$3(2) - 6 \stackrel{?}{=} 0$ Substitute 2 for x.

$0 = 0$ Solution checks. ✓

So, $x = 2$ is a solution. ■

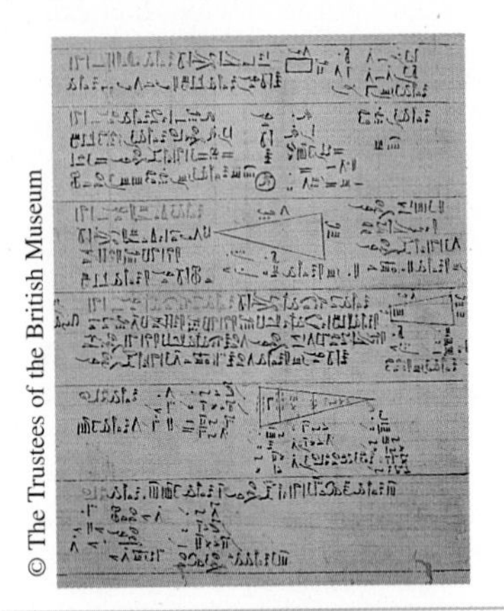

EGYPTIAN PAPYRUS (1650 B.C.)

This ancient Egyptian papyrus, discovered in 1858, contains one of the earliest examples of mathematical writing in existence. The papyrus itself dates back to around 1650 B.C, but it is actually a copy of writings from two centuries earlier. The algebraic equations on the papyrus were written in words. Diophantus, a Greek who lived around A.D. 250, is often called the Father of Algebra. He was the first to use abbreviated word forms in equations.

To solve an equation involving fractional expressions, find the least common denominator (LCD) of all terms and multiply every term by this LCD.

EXAMPLE 2 An Equation Involving Fractional Expressions

Solve $\dfrac{x}{3} + \dfrac{3x}{4} = 2$.

Solution

$$\frac{x}{3} + \frac{3x}{4} = 2 \qquad \text{Write original equation.}$$

$$(12)\frac{x}{3} + (12)\frac{3x}{4} = (12)2 \qquad \text{Multiply each term by the LCD of 12.}$$

$$4x + 9x = 24 \qquad \text{Divide out and multiply.}$$

$$x = \frac{24}{13} \qquad \text{Combine like terms and divide each side by 13.}$$

Check

$$\frac{x}{3} + \frac{3x}{4} = 2 \qquad \text{Write original equation.}$$

$$\frac{24/13}{3} + \frac{3(24/13)}{4} \stackrel{?}{=} 2 \qquad \text{Substitute } \tfrac{24}{13} \text{ for } x.$$

$$\frac{8}{13} + \frac{18}{13} \stackrel{?}{=} 2 \qquad \text{Simplify.}$$

$$\frac{26}{13} = 2 \qquad \text{Solution checks. ✓}$$

So, the solution is $x = \frac{24}{13}$. ■

Multiplying or dividing an equation by a *variable* quantity may introduce an extraneous solution. An **extraneous solution** does not satisfy the original equation.

EXAMPLE 3 An Equation with an Extraneous Solution

Solve $\dfrac{1}{x-2} = \dfrac{3}{x+2} - \dfrac{6x}{x^2-4}$.

Solution The LCD is $x^2 - 4$, or $(x + 2)(x - 2)$. Multiply each term by this LCD.

$$\frac{1}{x-2}(x+2)(x-2) = \frac{3}{x+2}(x+2)(x-2) - \frac{6x}{x^2-4}(x+2)(x-2)$$

$$x + 2 = 3(x - 2) - 6x, \quad x \neq \pm 2$$

$$x + 2 = 3x - 6 - 6x$$

$$x + 2 = -3x - 6$$

$$4x = -8 \implies x = -2 \qquad \text{Extraneous solution}$$

In the original equation, $x = -2$ yields a denominator of zero. So, $x = -2$ is an extraneous solution, and the original equation has *no solution*. ■

Quadratic Equations

A **quadratic equation** in x is an equation that can be written in the general form

$$ax^2 + bx + c = 0$$

where a, b, and c are real numbers, with $a \neq 0$. A quadratic equation in x is also known as a **second-degree polynomial equation** in x.

You should be familiar with the following four methods of solving quadratic equations.

NOTE The **Zero-Factor Property** states that if the product of two factors is zero, then one (or both) of the factors must be zero.

STUDY TIP The Square Root Principle is also referred to as *extracting square roots.*

SOLVING A QUADRATIC EQUATION

Factoring: If $ab = 0$, then $a = 0$ or $b = 0$. Zero Factor Property

Example:

$$x^2 - x - 6 = 0$$
$$(x - 3)(x + 2) = 0$$
$$x - 3 = 0 \implies x = 3$$
$$x + 2 = 0 \implies x = -2$$

Square Root Principle: If $u^2 = c$, where $c > 0$, then $u = \pm\sqrt{c}$.

Example:

$$(x + 3)^2 = 16$$
$$x + 3 = \pm 4$$
$$x = -3 \pm 4$$
$$x = 1 \quad \text{or} \quad x = -7$$

Completing the Square: If $x^2 + bx = c$, then

$$x^2 + bx + \left(\frac{b}{2}\right)^2 = c + \left(\frac{b}{2}\right)^2 \qquad \text{Add } \left(\frac{b}{2}\right)^2 \text{ to each side.}$$
$$\left(x + \frac{b}{2}\right)^2 = c + \frac{b^2}{4}.$$

Example:

$$x^2 + 6x = 5$$
$$x^2 + 6x + 3^2 = 5 + 3^2 \qquad \text{Add } \left(\frac{6}{2}\right)^2 \text{ to each side.}$$
$$(x + 3)^2 = 14$$
$$x + 3 = \pm\sqrt{14}$$
$$x = -3 \pm \sqrt{14}$$

Quadratic Formula: If $ax^2 + bx + c = 0$, then $x = \dfrac{-b \pm \sqrt{b^2 - 4ac}}{2a}$.

Example: $2x^2 + 3x - 1 = 0$

$$x = \frac{-3 \pm \sqrt{3^2 - 4(2)(-1)}}{2(2)}$$
$$x = \frac{-3 \pm \sqrt{17}}{4}$$

NOTE The Quadratic Formula can be derived by completing the square with the general form $ax^2 + bx + c = 0$. ■

EXAMPLE 4 Solving a Quadratic Equation by Factoring

Solve each equation by factoring.

a. $2x^2 + 9x + 7 = 3$

b. $6x^2 - 3x = 0$

Solution

a. $2x^2 + 9x + 7 = 3$ Original equation

$2x^2 + 9x + 4 = 0$ Write in general form.

$(2x + 1)(x + 4) = 0$ Factor.

$2x + 1 = 0 \Rightarrow x = -\frac{1}{2}$ Set 1st factor equal to 0.

$x + 4 = 0 \Rightarrow x = -4$ Set 2nd factor equal to 0.

The solutions are $x = -\frac{1}{2}$ and $x = -4$. Check these in the original equation.

b. $6x^2 - 3x = 0$ Original equation

$3x(2x - 1) = 0$ Factor.

$3x = 0 \Rightarrow x = 0$ Set 1st factor equal to 0.

$2x - 1 = 0 \Rightarrow x = \frac{1}{2}$ Set 2nd factor equal to 0.

The solutions are $x = 0$ and $x = \frac{1}{2}$. Check these in the original equation. ■

Note that the method of solution in Example 4 is based on the Zero-Factor Property. Be sure you see that this property works *only* for equations written in general form (in which the right side of the equation is zero). So, all terms must be collected on one side *before* factoring. For instance, in the equation

$$(x - 5)(x + 2) = 8$$

it is *incorrect* to set each factor equal to 8. Try to solve this equation correctly.

EXAMPLE 5 Extracting Square Roots

Solve each equation by extracting square roots.

a. $4x^2 = 12$

b. $(x - 3)^2 = 7$

Solution

a. $4x^2 = 12$ Write original equation.

$x^2 = 3$ Divide each side by 4.

$x = \pm\sqrt{3}$ Extract square roots.

When you take the square root of a variable expression, you must account for both positive and negative solutions. So, the solutions are $x = \sqrt{3}$ and $x = -\sqrt{3}$. Check these in the original equation.

b. $(x - 3)^2 = 7$ Write original equation.

$x - 3 = \pm\sqrt{7}$ Extract square roots.

$x = 3 \pm \sqrt{7}$ Add 3 to each side.

The solutions are $x = 3 \pm \sqrt{7}$. Check these in the original equation. ■

To solve quadratic equations by completing the square, you must add $(b/2)^2$ to *each side* in order to maintain equality. When the leading coefficient is *not* 1, you must divide each side of the equation by the leading coefficient *before* completing the square, as shown in Example 7.

EXAMPLE 6 Completing the Square: Leading Coefficient is 1

Solve $x^2 + 2x - 6 = 0$ by completing the square.

Solution

$$x^2 + 2x - 6 = 0 \qquad \text{Write original equation.}$$

$$x^2 + 2x = 6 \qquad \text{Add 6 to each side.}$$

$$x^2 + 2x + 1^2 = 6 + 1^2 \qquad \text{Add } 1^2 \text{ to each side.}$$

$(\text{half of } 2)^2$

$$(x + 1)^2 = 7 \qquad \text{Simplify.}$$

$$x + 1 = \pm\sqrt{7} \qquad \text{Take square root of each side.}$$

$$x = -1 \pm \sqrt{7} \qquad \text{Subtract 1 from each side.}$$

The solutions are $x = -1 \pm \sqrt{7}$. Check these in the original equation.

EXAMPLE 7 Completing the Square: Leading Coefficient is Not 1

Solve $3x^2 - 4x - 5 = 0$ by completing the square.

Solution

$$3x^2 - 4x - 5 = 0 \qquad \text{Original equation}$$

$$3x^2 - 4x = 5 \qquad \text{Add 5 to each side.}$$

$$x^2 - \frac{4}{3}x = \frac{5}{3} \qquad \text{Divide each side by 3.}$$

$$x^2 - \frac{4}{3}x + \left(-\frac{2}{3}\right)^2 = \frac{5}{3} + \left(-\frac{2}{3}\right)^2 \qquad \text{Add } \left(-\tfrac{2}{3}\right)^2 \text{ to each side.}$$

$\left(\text{half of } -\frac{4}{3}\right)^2$

$$x^2 - \frac{4}{3}x + \frac{4}{9} = \frac{19}{9} \qquad \text{Simplify.}$$

$$\left(x - \frac{2}{3}\right)^2 = \frac{19}{9} \qquad \text{Perfect square trinomial}$$

$$x - \frac{2}{3} = \pm\frac{\sqrt{19}}{3} \qquad \text{Extract square roots.}$$

$$x = \frac{2}{3} \pm \frac{\sqrt{19}}{3} \qquad \text{Solutions}$$

The solutions are $x = \frac{2}{3} \pm \frac{\sqrt{19}}{3}$. Check these in the original equation. ■

STUDY TIP When using the Quadratic Formula, *before* the formula can be applied, you must first write the quadratic equation in general form.

EXAMPLE 8 The Quadratic Formula: Two Distinct Solutions

Use the Quadratic Formula to solve $x^2 + 3x = 9$.

Solution

$$x^2 + 3x = 9 \qquad \text{Write original equation.}$$

$$x^2 + 3x - 9 = 0 \qquad \text{Write in general form.}$$

$$x = \frac{-b \pm \sqrt{b^2 - 4ac}}{2a} \qquad \text{Quadratic Formula}$$

$$x = \frac{-3 \pm \sqrt{(3)^2 - 4(1)(-9)}}{2(1)} \qquad \text{Substitute } a = 1,\ b = 3, \text{ and } c = -9.$$

$$x = \frac{-3 \pm \sqrt{45}}{2} \qquad \text{Simplify.}$$

$$x = \frac{-3 \pm 3\sqrt{5}}{2} \qquad \text{Simplify.}$$

The equation has two solutions:

$$x = \frac{-3 + 3\sqrt{5}}{2} \quad \text{and} \quad x = \frac{-3 - 3\sqrt{5}}{2}.$$

Check these in the original equation.

EXAMPLE 9 The Quadratic Formula: One Solution

Use the Quadratic Formula to solve $8x^2 - 24x + 18 = 0$.

Solution

$$8x^2 - 24x + 18 = 0 \qquad \text{Write original equation.}$$

$$4x^2 - 12x + 9 = 0 \qquad \text{Divide out common factor of 2.}$$

$$x = \frac{-b \pm \sqrt{b^2 - 4ac}}{2a} \qquad \text{Quadratic Formula}$$

$$x = \frac{-(-12) \pm \sqrt{(-12)^2 - 4(4)(9)}}{2(4)} \qquad \text{Substitute } a = 4,\ b = -12, \text{ and } c = 9.$$

$$x = \frac{12 \pm \sqrt{0}}{8} = \frac{3}{2} \qquad \text{Simplify.}$$

This quadratic equation has only one solution: $x = \frac{3}{2}$. Check this in the original equation as shown below.

Check

$$8x^2 - 24x + 18 = 0 \qquad \text{Write original equation.}$$

$$8\left(\frac{3}{2}\right)^2 - 24\left(\frac{3}{2}\right) + 18 \stackrel{?}{=} 0 \qquad \text{Substitute } \tfrac{3}{2} \text{ for } x.$$

$$18 - 36 + 18 = 0 \qquad \text{Solution checks. } \checkmark$$

■

Polynomial Equations of Higher Degree

The methods used to solve quadratic equations can sometimes be extended to solve polynomial equations of higher degree.

EXAMPLE 10 Solving a Polynomial Equation by Factoring

Solve $3x^4 = 48x^2$.

Solution First write the polynomial equation in general form with zero on one side, factor the other side, and then set each factor equal to zero and solve.

$$3x^4 = 48x^2 \quad \text{Write original equation.}$$
$$3x^4 - 48x^2 = 0 \quad \text{Write in general form.}$$
$$3x^2(x^2 - 16) = 0 \quad \text{Factor.}$$
$$3x^2(x + 4)(x - 4) = 0 \quad \text{Factor completely.}$$
$$3x^2 = 0 \Rightarrow x = 0 \quad \text{Set 1st factor equal to 0.}$$
$$x + 4 = 0 \Rightarrow x = -4 \quad \text{Set 2nd factor equal to 0.}$$
$$x - 4 = 0 \Rightarrow x = 4 \quad \text{Set 3rd factor equal to 0.}$$

You can check these solutions by substituting in the original equation as shown.

Check

$$3(0)^4 = 48(0)^2 \quad \text{0 checks. ✓}$$
$$3(-4)^4 = 48(-4)^2 \quad \text{−4 checks. ✓}$$
$$3(4)^4 = 48(4)^2 \quad \text{4 checks. ✓}$$

So, you can conclude that the solutions are $x = 0$, $x = -4$, and $x = 4$. ■

A common mistake that is made in solving an equation such as that in Example 10 is to divide each side of the equation by the variable factor x^2. This loses the solution $x = 0$. When solving an equation, be sure to write the equation in general form, then factor the equation and set *each* factor equal to zero. Don't divide each side of an equation by a variable factor in an attempt to simplify the equation.

EXAMPLE 11 Solving a Polynomial Equation by Factoring

Solve $x^3 - 3x^2 - 3x + 9 = 0$.

Solution

$$x^3 - 3x^2 - 3x + 9 = 0 \quad \text{Write original equation.}$$
$$(x^3 - 3x^2) + (-3x + 9) = 0 \quad \text{Group terms.}$$
$$x^2(x - 3) - 3(x - 3) = 0 \quad \text{Factor by grouping.}$$
$$(x - 3)(x^2 - 3) = 0 \quad \text{Distributive Property}$$
$$x - 3 = 0 \Rightarrow x = 3 \quad \text{Set 1st factor equal to 0.}$$
$$x^2 - 3 = 0 \Rightarrow x = \pm\sqrt{3} \quad \text{Set 2nd factor equal to 0.}$$

The solutions are $x = 3$, $x = \sqrt{3}$, and $x = -\sqrt{3}$. Check these in the original equation. ■

Equations Involving Radicals

Operations such as squaring each side of an equation, raising each side of an equation to a rational power, and multiplying each side of an equation by a variable quantity all can introduce extraneous solutions. So, when you use any of these operations, checking your solutions is crucial.

NOTE The essential operations in Example 12 are isolating the radical and squaring each side. In Example 13, this is equivalent to isolating the factor with the rational exponent and raising each side to the *reciprocal power*.

EXAMPLE 12 Solving Equations Involving Radicals

Solve each equation.

a. $\sqrt{2x+7} - x = 2$

b. $\sqrt{2x-5} - \sqrt{x-3} = 1$

Solution

a.
$$\sqrt{2x+7} - x = 2 \qquad \text{Original equation}$$
$$\sqrt{2x+7} = x + 2 \qquad \text{Isolate radical.}$$
$$2x + 7 = x^2 + 4x + 4 \qquad \text{Square each side.}$$
$$0 = x^2 + 2x - 3 \qquad \text{Write in general form.}$$
$$0 = (x+3)(x-1) \qquad \text{Factor.}$$
$$x + 3 = 0 \Rightarrow x = -3 \qquad \text{Set 1st factor equal to 0.}$$
$$x - 1 = 0 \Rightarrow x = 1 \qquad \text{Set 2nd factor equal to 0.}$$

By checking these values, you can determine that the only solution is $x = 1$.

b.
$$\sqrt{2x-5} - \sqrt{x-3} = 1 \qquad \text{Original equation}$$
$$\sqrt{2x-5} = \sqrt{x-3} + 1 \qquad \text{Isolate } \sqrt{2x-5}.$$
$$2x - 5 = x - 3 + 2\sqrt{x-3} + 1 \qquad \text{Square each side.}$$
$$2x - 5 = x - 2 + 2\sqrt{x-3} \qquad \text{Combine like terms.}$$
$$x - 3 = 2\sqrt{x-3} \qquad \text{Isolate } 2\sqrt{x-3}.$$
$$x^2 - 6x + 9 = 4(x-3) \qquad \text{Square each side.}$$
$$x^2 - 10x + 21 = 0 \qquad \text{Write in general form.}$$
$$(x-3)(x-7) = 0 \qquad \text{Factor.}$$
$$x - 3 = 0 \Rightarrow x = 3 \qquad \text{Set 1st factor equal to 0.}$$
$$x - 7 = 0 \Rightarrow x = 7 \qquad \text{Set 2nd factor equal to 0.}$$

The solutions are $x = 3$ and $x = 7$. Check these in the original equation.

STUDY TIP When an equation contains two radicals, it may not be possible to isolate both. In such cases, you may have to raise each side of the equation to a power at two different stages in the solution, as shown in Example 12(b).

EXAMPLE 13 Solving an Equation Involving a Rational Exponent

Solve $(x-4)^{2/3} = 25$.

Solution

$$(x-4)^{2/3} = 25 \qquad \text{Original equation}$$
$$\sqrt[3]{(x-4)^2} = 25 \qquad \text{Rewrite in radical form.}$$
$$(x-4)^2 = 15{,}625 \qquad \text{Cube each side.}$$
$$x - 4 = \pm 125 \qquad \text{Take square root of each side.}$$
$$x = 129, \quad x = -121 \qquad \text{Add 4 to each side.}$$

The solutions are $x = 129$ and $x = -121$. Check these in the original equation. ■

Equations with Absolute Values

To solve an equation involving an absolute value, remember that the expression inside the absolute value signs can be positive or negative. This results in two separate equations, each of which must be solved. For instance, the equation

$$|x - 2| = 3$$

results in the two equations

$$x - 2 = 3 \text{ and } -(x - 2) = 3$$

which implies that the equation has two solutions: $x = 5$ and $x = -1$.

EXAMPLE 14 Solving an Equation Involving Absolute Value

Solve $|x^2 - 3x| = -4x + 6$.

Solution Because the variable expression inside the absolute value signs can be positive or negative, you must solve the following two equations.

First Equation

$$x^2 - 3x = -4x + 6 \qquad \text{Use positive expression.}$$

$$x^2 + x - 6 = 0 \qquad \text{Write in general form.}$$

$$(x + 3)(x - 2) = 0 \qquad \text{Factor.}$$

$$x + 3 = 0 \Rightarrow x = -3 \qquad \text{Set 1st factor equal to 0.}$$

$$x - 2 = 0 \Rightarrow x = 2 \qquad \text{Set 2nd factor equal to 0.}$$

Second Equation

$$-(x^2 - 3x) = -4x + 6 \qquad \text{Use negative expression.}$$

$$x^2 - 7x + 6 = 0 \qquad \text{Write in general form.}$$

$$(x - 1)(x - 6) = 0 \qquad \text{Factor.}$$

$$x - 1 = 0 \Rightarrow x = 1 \qquad \text{Set 1st factor equal to 0.}$$

$$x - 6 = 0 \Rightarrow x = 6 \qquad \text{Set 2nd factor equal to 0.}$$

Check

$$|(-3)^2 - 3(-3)| \stackrel{?}{=} -4(-3) + 6 \qquad \text{Substitute } -3 \text{ for } x.$$

$$18 = 18 \qquad -3 \text{ checks. } \checkmark$$

$$|(2)^2 - 3(2)| \stackrel{?}{=} -4(2) + 6 \qquad \text{Substitute 2 for } x.$$

$$2 \neq -2 \qquad \text{2 does not check.}$$

$$|(1)^2 - 3(1)| \stackrel{?}{=} -4(1) + 6 \qquad \text{Substitute 1 for } x.$$

$$2 = 2 \qquad \text{1 checks. } \checkmark$$

$$|(6)^2 - 3(6)| \stackrel{?}{=} -4(6) + 6 \qquad \text{Substitute 6 for } x.$$

$$18 \neq -18 \qquad \text{6 does not check.}$$

The solutions are $x = -3$ and $x = 1$. ■

P.1 Exercises

See www.CalcChat.com for worked-out solutions to odd-numbered exercises.

In Exercises 1–4, fill in the blanks.

1. A(n) ________ is a statement that equates two algebraic expressions.
2. A linear equation in one variable is an equation that can be written in the standard form ________.
3. When solving an equation, it is possible to introduce an ________ solution, which is a value that does not satisfy the original equation.
4. The four methods that can be used to solve a quadratic equation are ________, ________, ________, and the ________.

In Exercises 5–10, determine whether the equation is an identity or a conditional equation.

5. $4(x + 1) = 4x + 4$
6. $-6(x - 3) + 5 = -2x + 10$
7. $4(x + 1) - 2x = 2(x + 2)$
8. $x^2 + 2(3x - 2) = x^2 + 6x - 4$
9. $3 + \dfrac{1}{x + 1} = \dfrac{4x}{x + 1}$
10. $\dfrac{5}{x} + \dfrac{3}{x} = 24$

In Exercises 11–24, solve the equation and check your solution.

11. $x + 11 = 15$
12. $7 - x = 19$
13. $7 - 2x = 25$
14. $7x + 2 = 23$
15. $8x - 5 = 3x + 20$
16. $7x + 3 = 3x - 17$
17. $4y + 2 - 5y = 7 - 6y$
18. $3(x + 3) = 5(1 - x) - 1$
19. $x - 3(2x + 3) = 8 - 5x$
20. $9x - 10 = 5x + 2(2x - 5)$
21. $\dfrac{3x}{8} - \dfrac{4x}{3} = 4$
22. $\dfrac{x}{5} - \dfrac{x}{2} = 3 + \dfrac{3x}{10}$
23. $\frac{3}{2}(z + 5) - \frac{1}{4}(z + 24) = 0$
24. $0.60x + 0.40(100 - x) = 50$

In Exercises 25–38, solve the equation and check your solution. (If not possible, explain why.)

25. $x + 8 = 2(x - 2) - x$
26. $8(x + 2) - 3(2x + 1) = 2(x + 5)$
27. $\dfrac{100 - 4x}{3} = \dfrac{5x + 6}{4} + 6$
28. $\dfrac{17 + y}{y} + \dfrac{32 + y}{y} = 100$
29. $\dfrac{5x - 4}{5x + 4} = \dfrac{2}{3}$
30. $\dfrac{15}{x} - 4 = \dfrac{6}{x} + 3$
31. $3 = 2 + \dfrac{2}{z + 2}$
32. $\dfrac{1}{x} + \dfrac{2}{x - 5} = 0$
33. $\dfrac{x}{x + 4} + \dfrac{4}{x + 4} + 2 = 0$
34. $\dfrac{7}{2x + 1} - \dfrac{8x}{2x - 1} = -4$
35. $\dfrac{3}{x^2 - 3x} + \dfrac{4}{x} = \dfrac{1}{x - 3}$
36. $\dfrac{6}{x} - \dfrac{2}{x + 3} = \dfrac{3(x + 5)}{x^2 + 3x}$
37. $(x + 2)^2 + 5 = (x + 3)^2$
38. $(2x + 1)^2 = 4(x^2 + x + 1)$

In Exercises 39–42, write the quadratic equation in general form.

39. $2x^2 = 3 - 8x$
40. $13 - 3(x + 7)^2 = 0$
41. $\frac{1}{5}(3x^2 - 10) = 18x$
42. $x(x + 2) = 5x^2 + 1$

In Exercises 43–54, solve the quadratic equation by factoring.

43. $6x^2 + 3x = 0$
44. $9x^2 - 4 = 0$
45. $x^2 - 2x - 8 = 0$
46. $x^2 - 10x + 9 = 0$
47. $x^2 - 12x + 35 = 0$
48. $4x^2 + 12x + 9 = 0$
49. $3 + 5x - 2x^2 = 0$
50. $2x^2 = 19x + 33$
51. $x^2 + 4x = 12$
52. $\frac{1}{8}x^2 - x - 16 = 0$
53. $x^2 + 2ax + a^2 = 0$, a is a real number
54. $(x + a)^2 - b^2 = 0$, a and b are real numbers

In Exercises 55–66, solve the equation by extracting square roots.

55. $x^2 = 49$
56. $x^2 = 32$
57. $3x^2 = 81$
58. $9x^2 = 36$
59. $(x - 12)^2 = 16$
60. $(x + 13)^2 = 25$
61. $(x + 2)^2 = 14$
62. $(x - 5)^2 = 30$
63. $(2x - 1)^2 = 18$
64. $(2x + 3)^2 - 27 = 0$
65. $(x - 7)^2 = (x + 3)^2$
66. $(x + 5)^2 = (x + 4)^2$

In Exercises 67–76, solve the quadratic equation by completing the square.

67. $x^2 + 4x - 32 = 0$
68. $x^2 + 6x + 2 = 0$
69. $x^2 + 12x + 25 = 0$
70. $x^2 + 8x + 14 = 0$
71. $8 + 4x - x^2 = 0$
72. $9x^2 - 12x = 14$
73. $2x^2 + 5x - 8 = 0$
74. $4x^2 - 4x - 99 = 0$
75. $5x^2 - 15x + 7 = 0$
76. $3x^2 + 9x + 5 = 0$

In Exercises 77–92, use the Quadratic Formula to solve the equation.

77. $2x^2 + x - 1 = 0$
78. $25x^2 - 20x + 3 = 0$
79. $2 + 2x - x^2 = 0$
80. $x^2 - 10x + 22 = 0$
81. $x^2 + 14x + 44 = 0$
82. $6x = 4 - x^2$
83. $12x - 9x^2 = -3$
84. $4x^2 - 4x - 4 = 0$
85. $9x^2 + 24x + 16 = 0$
86. $16x^2 - 40x + 5 = 0$
87. $28x - 49x^2 = 4$
88. $3x + x^2 - 1 = 0$
89. $8t = 5 + 2t^2$
90. $25h^2 + 80h + 61 = 0$
91. $(y - 5)^2 = 2y$
92. $(\frac{5}{7}x - 14)^2 = 8x$

In Exercises 93–96, use the Quadratic Formula to solve the equation. (Round your answer to three decimal places.)

93. $0.1x^2 + 0.2x - 0.5 = 0$
94. $-0.005x^2 + 0.101x - 0.193 = 0$
95. $422x^2 - 506x - 347 = 0$
96. $-3.22x^2 - 0.08x + 28.651 = 0$

In Exercises 97–104, solve the equation using any convenient method.

97. $x^2 - 2x - 1 = 0$
98. $11x^2 + 33x = 0$
99. $(x + 3)^2 = 81$
100. $x^2 - 14x + 49 = 0$
101. $x^2 - x - \frac{11}{4} = 0$
102. $3x + 4 = 2x^2 - 7$
103. $4x^2 + 2x + 4 = 2x + 8$
104. $a^2x^2 - b^2 = 0$, a and b are real numbers, $a \neq 0$

In Exercises 105–118, find all real solutions of the equation. Check your solutions in the original equation.

105. $2x^4 - 50x^2 = 0$
106. $20x^3 - 125x = 0$
107. $x^4 - 81 = 0$
108. $x^6 - 64 = 0$
109. $x^3 + 216 = 0$
110. $9x^4 - 24x^3 + 16x^2 = 0$
111. $x^3 - 3x^2 - x + 3 = 0$
112. $x^3 + 2x^2 + 3x + 6 = 0$
113. $x^4 + x = x^3 + 1$
114. $x^4 - 2x^3 = 16 + 8x - 4x^3$
115. $x^4 - 4x^2 + 3 = 0$
116. $36t^4 + 29t^2 - 7 = 0$
117. $x^6 + 7x^3 - 8 = 0$
118. $x^6 + 3x^3 + 2 = 0$

In Exercises 119–144, find all solutions of the equation. Check your solutions in the original equation.

119. $\sqrt{2x} - 10 = 0$
120. $7\sqrt{x} - 6 = 0$
121. $\sqrt{x - 10} - 4 = 0$
122. $\sqrt{5 - x} - 3 = 0$
123. $\sqrt{2x + 5} + 3 = 0$
124. $\sqrt{3 - 2x} - 2 = 0$
125. $\sqrt[3]{2x + 1} + 8 = 0$
126. $\sqrt[3]{4x - 3} + 2 = 0$
127. $\sqrt{5x - 26} + 4 = x$
128. $\sqrt{x + 5} = \sqrt{2x - 5}$
129. $(x - 6)^{3/2} = 8$
130. $(x + 3)^{3/2} = 8$
131. $(x + 3)^{2/3} = 5$
132. $(x^2 - x - 22)^{4/3} = 16$
133. $3x(x - 1)^{1/2} + 2(x - 1)^{3/2} = 0$
134. $4x^2(x - 1)^{1/3} + 6x(x - 1)^{4/3} = 0$
135. $x = \frac{3}{x} + \frac{1}{2}$
136. $\frac{4}{x + 1} - \frac{3}{x + 2} = 1$
137. $\frac{20 - x}{x} = x$
138. $4x + 1 = \frac{3}{x}$
139. $|2x - 1| = 5$
140. $|13x + 1| = 12$
141. $|x| = x^2 + x - 3$
142. $|x^2 + 6x| = 3x + 18$
143. $|x + 1| = x^2 - 5$
144. $|x - 10| = x^2 - 10x$

WRITING ABOUT CONCEPTS

145. To solve the equation $2x^2 + 3x = 15x$, a student divides each side by x and solves the equation $2x + 3 = 15$. The resulting solution is 6. Is the student correct? Explain your reasoning.

146. To solve the equation $4x^2 + 4x = 15$, a student factors $4x$ from the left side of the equation, sets each factor equal to 15, and solves the equations $4x = 15$ and $x + 1 = 15$. The resulting solutions are $x = \frac{15}{4}$ and $x = 14$. Is the student correct? Explain your reasoning.

147. What is meant by *equivalent equations*? Give an example of two equivalent equations.

148. In your own words, describe the steps used to transform an equation into an equivalent equation.

Anthropology **In Exercises 149 and 150, use the following information. The relationship between the length of an adult's femur (thigh bone) and the height of the adult can be approximated by the linear equations**

$y = 0.432x - 10.44$ **Female**

$y = 0.449x - 12.15$ **Male**

where y is the length of the femur in inches and x is the height of the adult in inches (see figure).

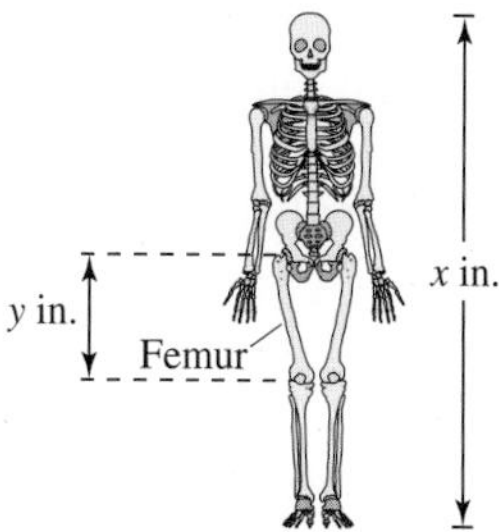

149. An anthropologist discovers a femur belonging to an adult human female. The bone is 16 inches long. Estimate the height of the female.

150. From the foot bones of an adult human male, an anthropologist estimates that the person's height was 69 inches. A few feet away from the site where the foot bones were discovered, the anthropologist discovers a male adult femur that is 19 inches long. Is it likely that both the foot bones and the thigh bone came from the same person?

151. ***Voting Population*** The total voting-age population P (in millions) in the United States from 1990 through 2006 can be modeled by

$$P = \frac{182.17 - 1.542t}{1 - 0.018t}, \quad 0 \le t \le 16$$

where t represents the year, with $t = 0$ corresponding to 1990. *(Source: U.S. Census Bureau)*

(a) In which year did the total voting-age population reach 200 million?

(b) Use the model to predict the year in which the total voting-age population will reach 241 million. Is this prediction reasonable? Explain.

152. ***Airline Passengers*** An airline offers daily flights between Chicago and Denver. The total monthly cost C (in millions of dollars) of these flights is $C = \sqrt{0.2x + 1}$, where x is the number of passengers (in thousands). The total cost of the flights for June is 2.5 million dollars. How many passengers flew in June?

True or False? **In Exercises 153 and 154, determine whether the statement is true or false. Justify your answer.**

153. An equation can never have more than one extraneous solution.

154. When solving an absolute value equation, you will always have to check more than one solution.

Think About It **In Exercises 155–158, write a quadratic equation that has the given solutions. (There are many correct answers.)**

155. -3 and 6

156. -4 and -11

157. $1 + \sqrt{2}$ and $1 - \sqrt{2}$

158. $-3 + \sqrt{5}$ and $-3 - \sqrt{5}$

In Exercises 159 and 160, consider an equation of the form $x + |x - a| = b$, where a and b are constants.

159. Find a and b when the solution of the equation is $x = 9$. (There are many correct answers.)

160. ***Writing*** Write a short paragraph listing the steps required to solve this equation involving absolute values, and explain why it is important to check your solutions.

161. Solve each equation, given that a and b are not zero.

(a) $ax^2 + bx = 0$ (b) $ax^2 - ax = 0$

CAPSTONE

162. (a) Explain the difference between a conditional equation and an identity.

(b) Give an example of an absolute value equation that has only one solution.

(c) State the Quadratic Formula in words.

(d) Does raising each side of an equation to the nth power always yield an equivalent equation? Explain.

SECTION PROJECT

Projectile Motion

An object is projected straight upward from an initial height of s_0 (in feet) with initial velocity v_0 (in feet per second). The object's height s (in feet) is given by $s = -16t^2 + v_0t + s_0$, where t is the elapsed time (in seconds).

(a) An object is projected upward with an initial velocity of 251 feet per second from a height of 32 feet (see figure). During what time period will its height exceed 91 feet?

(b) You have thrown a baseball straight upward from a height of about 6 feet. A friend has used a stopwatch to record the time the ball is in the air and determines that it takes approximately 6.5 seconds for the ball to strike the ground (see figure). Explain how you can find the ball's initial velocity.

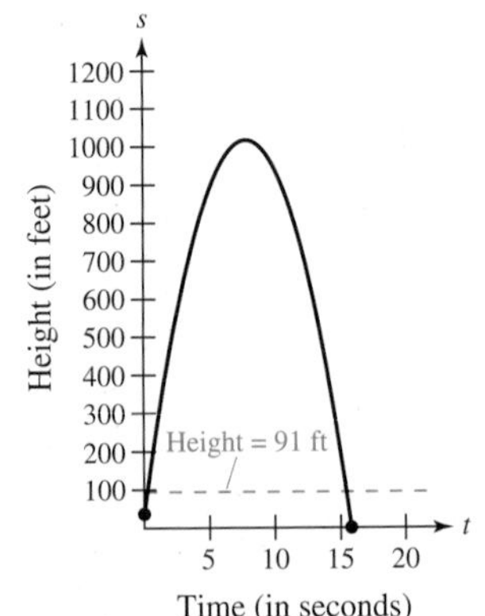

(a)

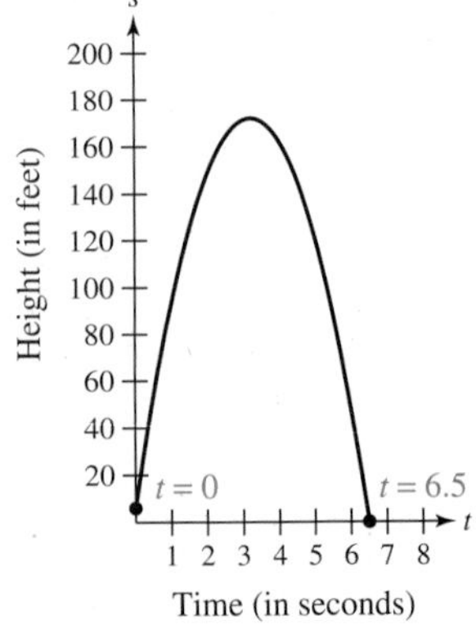

(b)

P.2 Solving Inequalities

- Represent solutions of linear inequalities in one variable.
- Use properties of inequalities to create equivalent inequalities.
- Solve linear inequalities in one variable.
- Solve inequalities involving absolute values.
- Solve polynomial and rational inequalities.

Introduction

In a previous course, you learned to use the inequality symbols $<$, $\leq$, $>$, and $\geq$ to compare two numbers and to denote subsets of real numbers. For instance, the simple inequality

$$x \geq 3$$

denotes all real numbers x that are greater than or equal to 3.

Now, you will expand your work with inequalities to include more involved statements such as

$$5x - 7 < 3x + 9$$

and

$$-3 \leq 6x - 1 < 3.$$

As with an equation, you **solve an inequality** in the variable x by finding all values of x for which the inequality is true. Such values are **solutions** and are said to **satisfy** the inequality. The set of all real numbers that are solutions of an inequality is the **solution set** of the inequality. For instance, the solution set of

$$x + 1 < 4$$

is all real numbers that are less than 3.

The set of all points on the real number line that represents the solution set is the **graph of the inequality.** Graphs of many types of inequalities consist of intervals on the real number line. Note that each type of interval can be classified as *bounded* or *unbounded.* **Bounded** intervals are of the form $[a, b]$, (a, b), $[a, b)$, and $(a, b]$. **Unbounded** intervals are of the form $(-\infty, b)$, $(-\infty, b]$, (a, ∞), $[a, \infty)$, and $(-\infty, \infty)$.

NOTE The intervals (a, b), $(-\infty, b)$, and (a, ∞) are *open*. The intervals $[a, b]$, $(-\infty, b]$, and $[a, \infty)$ are *closed*. The interval $(-\infty, \infty)$ is considered open and closed. The intervals $(a, b]$ and $[a, b)$ are neither open nor closed. ■

EXAMPLE 1 Intervals and Inequalities

Write an inequality to represent each interval, and state whether the interval is bounded or unbounded.

a. $(-3, 5]$ **b.** $(-3, \infty)$ **c.** $[0, 2]$ **d.** $(-\infty, \infty)$

Solution

a. $(-3, 5]$ corresponds to $-3 < x \leq 5$. Bounded

b. $(-3, \infty)$ corresponds to $-3 < x$. Unbounded

c. $[0, 2]$ corresponds to $0 \leq x \leq 2$. Bounded

d. $(-\infty, \infty)$ corresponds to $-\infty < x < \infty$. Unbounded ■

Properties of Inequalities

The procedures for solving linear inequalities in one variable are much like those for solving linear equations. To isolate the variable, you can make use of the **Properties of Inequalities.** These properties are similar to the properties of equality, but there are two important exceptions. When each side of an inequality is multiplied or divided by a negative number, the direction of the inequality symbol must be reversed. Here is an example.

$$-2 < 5 \qquad \text{Original inequality}$$

$$(-3)(-2) > (-3)(5) \qquad \text{Multiply each side by } -3 \text{ and reverse inequality.}$$

$$6 > -15 \qquad \text{Simplify.}$$

Notice that if the inequality was not reversed, you would obtain the false statement $6 < -15$.

Two inequalities that have the same solution set are **equivalent.** For instance, the inequalities

$$x + 2 < 5$$

and

$$x < 3$$

are equivalent. To obtain the second inequality from the first, you can subtract 2 from each side of the inequality. The following list describes the operations that can be used to create equivalent inequalities.

PROPERTIES OF INEQUALITIES

Let a, b, c, and d be real numbers.

1. Transitive Property

$$a < b \text{ and } b < c \implies a < c$$

2. Addition of Inequalities

$$a < b \text{ and } c < d \implies a + c < b + d$$

3. Addition of a Constant

$$a < b \implies a + c < b + c$$

4. Multiplication by a Constant

$$\text{For } c > 0,\ a < b \implies ac < bc$$

$$\text{For } c < 0,\ a < b \implies ac > bc \qquad \text{Reverse the inequality.}$$

NOTE Each of the properties above is true if the symbol $<$ is replaced by $\le$ and the symbol $>$ is replaced by $\ge$. For instance, another form of the multiplication property would be as follows.

$$\text{For } c > 0,\ a \le b \implies ac \le bc$$

$$\text{For } c < 0,\ a \le b \implies ac \ge bc \qquad \text{Reverse the inequality.}$$

■

Solving a Linear Inequality in One Variable

The simplest type of inequality is a **linear inequality** in one variable. For instance, $2x + 3 > 4$ is a linear inequality in x.

In the following examples, pay special attention to the steps in which the inequality symbol is reversed. Remember that when you multiply or divide by a negative number, you must reverse the inequality symbol.

EXAMPLE 2 Solving a Linear Inequality

Solve $5x - 7 > 3x + 9$.

Solution

$5x - 7 > 3x + 9$	Write original inequality.
$2x - 7 > 9$	Subtract $3x$ from each side.
$2x > 16$	Add 7 to each side.
$x > 8$	Divide each side by 2.

STUDY TIP Checking the solution set of an inequality is not as simple as checking the solutions of an equation. You can, however, get an indication of the validity of a solution set by substituting a few convenient values of x. For instance, in Example 2, try substituting $x = 5$ and $x = 10$ into the original inequality.

The solution set is all real numbers that are greater than 8, which is denoted by $(8, \infty)$. The graph of this solution set is shown in Figure P.1. Note that a parenthesis at 8 on the real number line indicates that 8 *is not* part of the solution set.

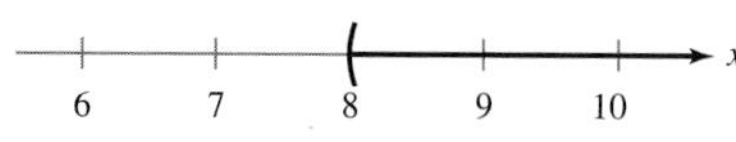

Solution interval: $(8, \infty)$
Figure P.1

EXAMPLE 3 Solving a Linear Inequality

Solve $1 - \frac{3}{2}x \geq x - 4$.

Algebraic Solution

$1 - \frac{3x}{2} \geq x - 4$	Write original inequality.
$2 - 3x \geq 2x - 8$	Multiply each side by 2.
$2 - 5x \geq -8$	Subtract $2x$ from each side.
$-5x \geq -10$	Subtract 2 from each side.
$x \leq 2$	Divide each side by -5 and reverse the inequality.

The solution set is all real numbers that are less than or equal to 2, which is denoted by $(-\infty, 2]$. The graph of this solution set is shown in Figure P.2. Note that a bracket at 2 on the real number line indicates that 2 *is* part of the solution set.

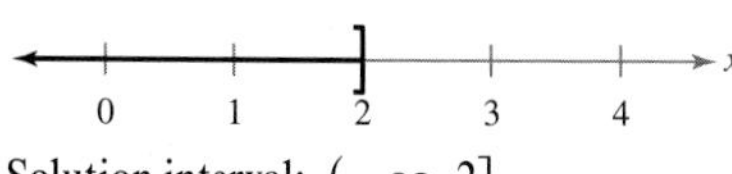

Solution interval: $(-\infty, 2]$
Figure P.2

Graphical Solution

Use a graphing utility to graph $y_1 = 1 - \frac{3}{2}x$ and $y_2 = x - 4$ in the same viewing window. In Figure P.3, you can see that the graphs appear to intersect at the point $(2, -2)$. Use the *intersect* feature of the graphing utility to confirm this. The graph of y_1 lies above the graph of y_2 to the left of their point of intersection, which implies that $y_1 \geq y_2$ for all $x \leq 2$.

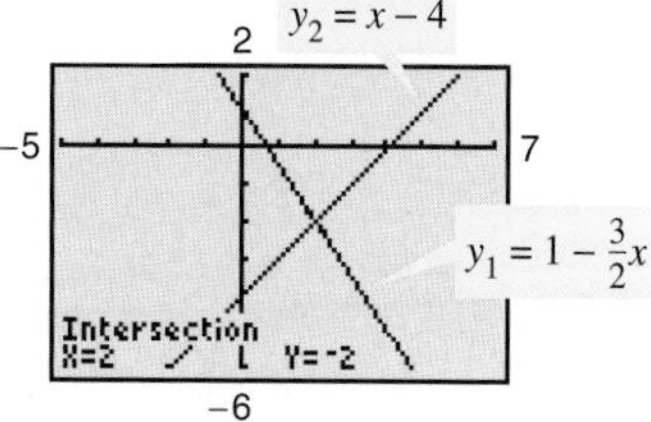

Figure P.3

Sometimes it is possible to write two inequalities as a **double inequality.** For instance, you can write the two inequalities $-4 \le 5x - 2$ and $5x - 2 < 7$ more simply as

$$-4 \le 5x - 2 < 7. \qquad \text{Double inequality}$$

This form allows you to solve the two inequalities together, as demonstrated in Example 4.

EXAMPLE 4 Solving a Double Inequality

Solve the inequality.

$$-3 \le 6x - 1 < 3$$

Solution To solve a double inequality, you can isolate x as the middle term.

$$-3 \le 6x - 1 < 3 \qquad \text{Original inequality}$$

$$-3 + 1 \le 6x - 1 + 1 < 3 + 1 \qquad \text{Add 1 to each part.}$$

$$-2 \le 6x < 4 \qquad \text{Simplify.}$$

$$\frac{-2}{6} \le \frac{6x}{6} < \frac{4}{6} \qquad \text{Divide each part by 6.}$$

$$-\frac{1}{3} \le x < \frac{2}{3} \qquad \text{Simplify.}$$

The solution set is all real numbers that are greater than or equal to $-\frac{1}{3}$ and less than $\frac{2}{3}$, which is denoted by $\left[-\frac{1}{3}, \frac{2}{3}\right)$. The graph of this solution set is shown in Figure P.4.

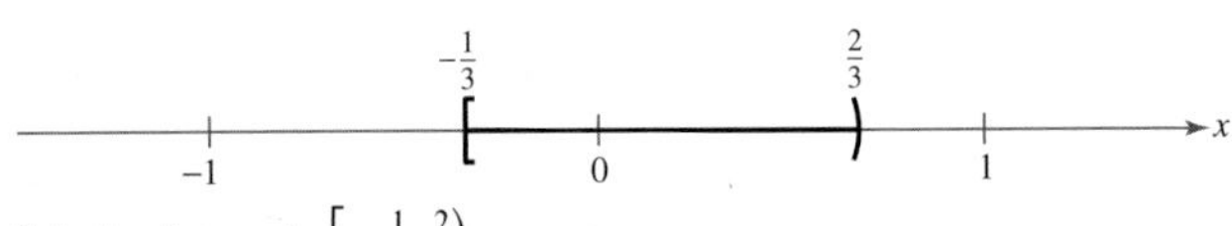

Solution interval: $\left[-\frac{1}{3}, \frac{2}{3}\right)$
Figure P.4

■

The double inequality in Example 4 could have been solved in two parts, as follows.

$$-3 \le 6x - 1 \quad \text{and} \quad 6x - 1 < 3$$

$$-2 \le 6x \qquad\qquad 6x < 4$$

$$-\frac{1}{3} \le x \qquad\qquad x < \frac{2}{3}$$

The solution set consists of all real numbers that satisfy both inequalities. In other words, the solution set is the set of all values of x for which

$$-\frac{1}{3} \le x < \frac{2}{3}.$$

When combining two inequalities to form a double inequality, be sure that the inequalities satisfy the Transitive Property. For instance, it is *incorrect* to combine the inequalities $3 < x$ and $x \le -1$ as $3 < x \le -1$. This "inequality" is wrong because 3 is not less than -1.

TECHNOLOGY A graphing utility can be used to identify the solution set of the graph of an inequality. For instance, to find the solution set of $|x - 5| < 2$ (see Example 5a), rewrite the inequality as $|x - 5| - 2 < 0$, enter

Y1 = abs (X − 5) − 2

and press the *graph* key. The graph should look like the one shown in Figure P.5. Notice that the graph is below the x-axis on the interval $(3, 7)$.

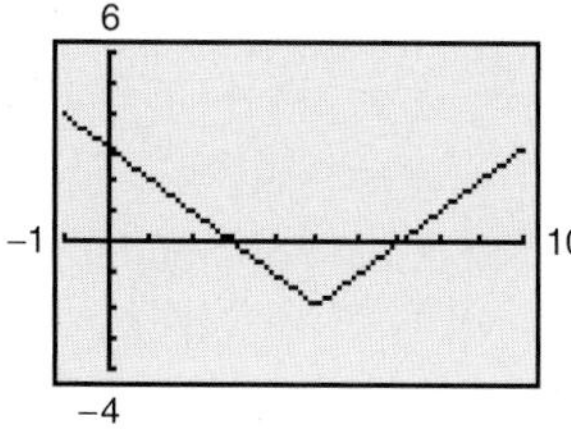

Figure P.5

Inequalities Involving Absolute Values

SOLVING AN ABSOLUTE VALUE INEQUALITY

Let x be a variable or an algebraic expression and let a be a real number such that $a \geq 0$.

1. The solutions of $|x| < a$ are all values of x that lie between $-a$ and a.

   $|x| < a$ if and only if $-a < x < a$. Double inequality

2. The solutions of $|x| > a$ are all values of x that are less than $-a$ or greater than a.

   $|x| > a$ if and only if $x < -a$ or $x > a$. Compound inequality

These rules are also valid when $<$ is replaced by $\leq$ and $>$ is replaced by $\geq$.

EXAMPLE 5 Solving an Absolute Value Inequality

Solve each inequality.

a. $|x - 5| < 2$

b. $|x + 3| \geq 7$

Solution

a.

$|x - 5| < 2$ Write original inequality.

$-2 < x - 5 < 2$ Write equivalent inequalities.

$-2 + 5 < x - 5 + 5 < 2 + 5$ Add 5 to each part.

$3 < x < 7$ Simplify.

The solution set is all real numbers that are greater than 3 and less than 7, which is denoted by $(3, 7)$. The graph of this solution set is shown in Figure P.6.

STUDY TIP When working with absolute value inequalities, a "less than" inequality can be solved as a double inequality and the solution lies between two numbers. A "greater than" inequality must be solved in two parts and the solution set is disjoint.

b.

$|x + 3| \geq 7$ Write original inequality.

$x + 3 \leq -7$ or $x + 3 \geq 7$ Write equivalent inequalities.

$x + 3 - 3 \leq -7 - 3$ $\quad$ $x + 3 - 3 \geq 7 - 3$ Subtract 3 from each side.

$x \leq -10$ $\quad$ $x \geq 4$ Simplify.

The solution set is all real numbers that are less than or equal to -10 *or* greater than or equal to 4. The interval notation for this solution set is $(-\infty, -10] \cup [4, \infty)$. The symbol $\cup$ is called a *union* symbol and is used to denote the combining of two sets. The graph of this solution set is shown in Figure P.7.

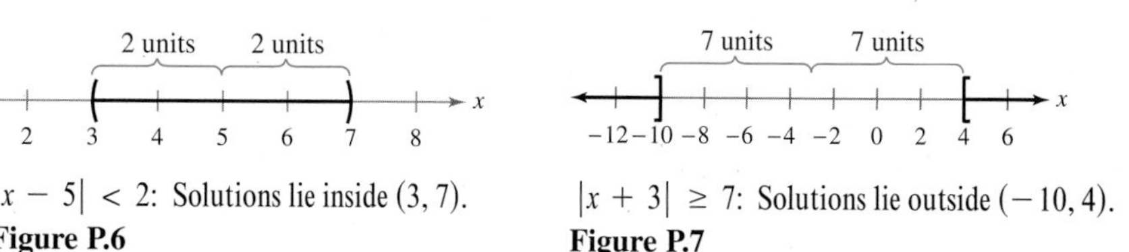

$|x - 5| < 2$: Solutions lie inside $(3, 7)$.
Figure P.6

$|x + 3| \geq 7$: Solutions lie outside $(-10, 4)$.
Figure P.7

NOTE The graph of the inequality $|x - 5| < 2$ can be described as all real numbers *within* two units of 5, as shown in Figure P.5.

Other Types of Inequalities

To solve a polynomial inequality, you can use the fact that a polynomial can change signs only at its zeros (the x-values that make the polynomial equal to zero). Between two consecutive zeros, a polynomial must be entirely positive or entirely negative. This means that when the real zeros of a polynomial are put in order, they divide the real number line into intervals in which the polynomial has no sign changes. These zeros are the **key numbers** of the inequality, and the resulting intervals are the **test intervals** for the inequality.

EXAMPLE 6 Solving a Polynomial Inequality

Solve $x^2 - x - 6 < 0$.

Solution By factoring the polynomial as

$$x^2 - x - 6 = (x + 2)(x - 3)$$

you can see that the key numbers are $x = -2$ and $x = 3$. So, the polynomial's test intervals are

$$(-\infty, -2),\quad (-2, 3),\quad \text{and} \quad (3, \infty). \qquad \text{Test intervals}$$

In each test interval, choose a representative x-value and evaluate the polynomial.

Test Interval	*x-Value*	*Polynomial Value*	*Conclusion*
$(-\infty, -2)$	$x = -3$	$(-3)^2 - (-3) - 6 = 6$	Positive
$(-2, 3)$	$x = 0$	$(0)^2 - (0) - 6 = -6$	Negative
$(3, \infty)$	$x = 4$	$(4)^2 - (4) - 6 = 6$	Positive

From this you can conclude that the inequality is satisfied for all x-values in $(-2, 3)$. This implies that the solution of the inequality $x^2 - x - 6 < 0$ is the interval $(-2, 3)$, as shown in Figure P.8. Note that the original inequality contains a "less than" symbol. This means that the solution set does not contain the endpoints of the test interval $(-2, 3)$.

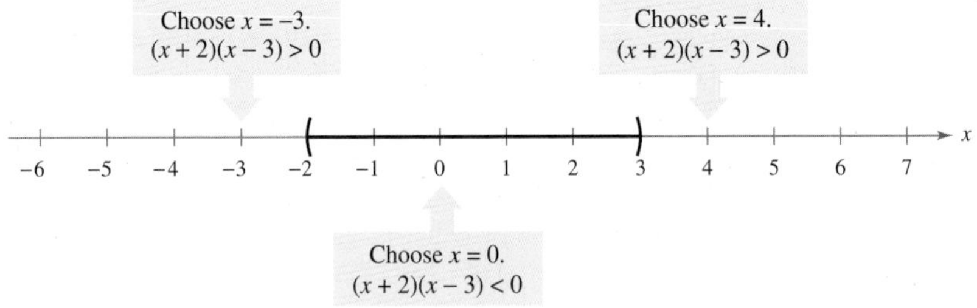

Solution interval: $(-2, 3)$
Figure P.8

As with linear inequalities, you can check the reasonableness of a solution by substituting x-values into the original inequality. For instance, to check the solution found in Example 6, try substituting several x-values from the interval $(-2, 3)$ into the inequality $x^2 - x - 6 < 0$. Regardless of which x-values you choose, the inequality should be satisfied.

In Example 6, the polynomial inequality was given in general form (with the polynomial on one side and zero on the other). Whenever this is not the case, you should begin the solution process by writing the inequality in general form.

The concepts of key numbers and test intervals can be extended to rational inequalities. To do this, use the fact that the value of a rational expression can change sign only at its *zeros* (the x-values for which its numerator is zero) and its *undefined values* (the x-values for which its denominator is zero). These two types of numbers make up the *key numbers* of a rational inequality. When solving a rational inequality, begin by writing the inequality in general form with the rational expression on the left and zero on the right.

STUDY TIP In Example 7, if you write 3 as $\frac{3}{1}$, you should be able to see that the LCD (least common denominator) is $(x - 5)(1) = x - 5$. So, you can rewrite the general form as

$$\frac{2x - 7}{x - 5} - \frac{3(x - 5)}{x - 5} \leq 0$$

which simplifies as shown.

EXAMPLE 7 Solving a Rational Inequality

$$\frac{2x - 7}{x - 5} \leq 3 \qquad \text{Original inequality}$$

$$\frac{2x - 7}{x - 5} - 3 \leq 0 \qquad \text{Write in general form.}$$

$$\frac{2x - 7 - 3x + 15}{x - 5} \leq 0 \qquad \text{Find the LCD and subtract fractions.}$$

$$\frac{-x + 8}{x - 5} \leq 0 \qquad \text{Simplify.}$$

Key numbers: $x = 5, x = 8$ Zeros and undefined values of rational expression

Test intervals: $(-\infty, 5), (5, 8), (8, \infty)$

Test: Is $\dfrac{-x + 8}{x - 5} \leq 0$?

Interval	*x-Value*	*Expression Value*	*Conclusion*
$(-\infty, 5)$	$x = 4$	$\frac{-4 + 8}{4 - 5} = -4$	Negative
$(5, 8)$	$x = 6$	$\frac{-6 + 8}{6 - 5} = 2$	Positive
$(8, \infty)$	$x = 9$	$\frac{-9 + 8}{9 - 5} = -\frac{1}{4}$	Negative

You can see that the inequality is satisfied on the open intervals $(-\infty, 5)$ and $(8, \infty)$. Moreover, because $\dfrac{-x + 8}{x - 5} = 0$ when $x = 8$, you can conclude that the solution set consists of all real numbers in the intervals $(-\infty, 5) \cup [8, \infty)$, as shown in Figure P.9. (Be sure to use a closed interval to indicate that x can equal 8.)

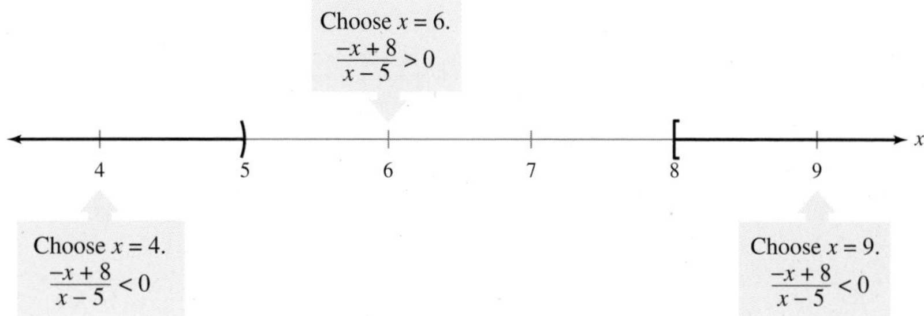

Solution interval: $(-\infty, 5) \cup [8, \infty)$
Figure P.9

■

A common application of inequalities is finding the domain of an expression that involves a square root, as shown in Example 8.

EXAMPLE 8 Finding the Domain of an Expression

Find the domain of

$\sqrt{64 - 4x^2}$.

Solution Remember that the domain of an expression is the set of all x-values for which the expression is defined. Because $\sqrt{64 - 4x^2}$ is defined (has real values) only if $64 - 4x^2$ is nonnegative, the domain is given by $64 - 4x^2 \geq 0$.

$$64 - 4x^2 \geq 0 \qquad \text{Write in general form.}$$

$$16 - x^2 \geq 0 \qquad \text{Divide each side by 4.}$$

$$(4 - x)(4 + x) \geq 0 \qquad \text{Write in factored form.}$$

So, the inequality has two key numbers:

$x = -4$ and $x = 4$.

You can use these two numbers to test the inequality as follows.

Key numbers: $x = -4, x = 4$

Test intervals: $(-\infty, -4), (-4, 4), (4, \infty)$

Test: For what values of x is $\sqrt{64 - 4x^2} \geq 0$?

Interval	*x-Value*	*Expression Value*	*Conclusion*
$(-\infty, -4)$	$x = -5$	$\sqrt{64 - 4(-5)^2} = \sqrt{-36}$	Undefined
$(-4, 4)$	$x = 0$	$\sqrt{64 - 4(0)^2} = \sqrt{64}$	Positive
$(4, \infty)$	$x = 5$	$\sqrt{64 - 4(5)^2} = \sqrt{-36}$	Undefined

From the test, you can see that the inequality is satisfied on the open interval $(-4, 4)$. Also, because $\sqrt{64 - 4x^2} = 0$ when $x = -4$ and $x = 4$, you can conclude that the solution set consists of all real numbers in the *closed interval* $[-4, 4]$. So, the domain of the expression $\sqrt{64 - 4x^2}$ is the interval $[-4, 4]$, as shown in Figure P.10.

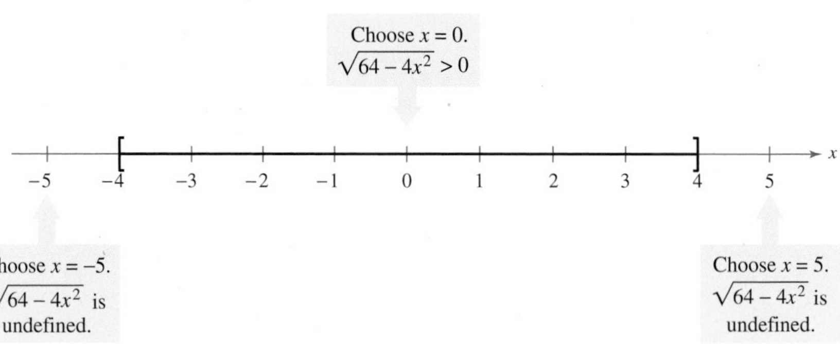

Solution interval: $[-4, 4]$

Figure P.10

■

P.2 Exercises

See www.CalcChat.com for worked-out solutions to odd-numbered exercises.

In Exercises 1–6, fill in the blanks.

1. The set of all real numbers that are solutions of an inequality is the ________ ________ of the inequality.
2. The set of all points on the real number line that represents the solution set of an inequality is the ________ of the inequality.
3. To solve a linear inequality in one variable, you can use the properties of inequalities, which are identical to those used to solve equations, with the exception of multiplying or dividing each side by a ________ number.
4. The symbol $\cup$ is called a ________ symbol and is used to denote the combining of two sets.
5. To solve a polynomial inequality, find the ________ numbers of the polynomial, and use these numbers to create ________ ________ for the inequality.
6. The key numbers of a rational expression are its ________ and its ________ ________.

In Exercises 7–14, (a) write an inequality that represents the interval and (b) state whether the interval is bounded or unbounded.

7. $[0, 9)$
8. $(-7, 4)$
9. $[-1, 5]$
10. $(2, 10]$
11. $(11, \infty)$
12. $[-5, \infty)$
13. $(-\infty, -2)$
14. $(-\infty, 7]$

In Exercises 15–22, match the inequality with its graph. [The graphs are labeled (a)–(h).]

(a)
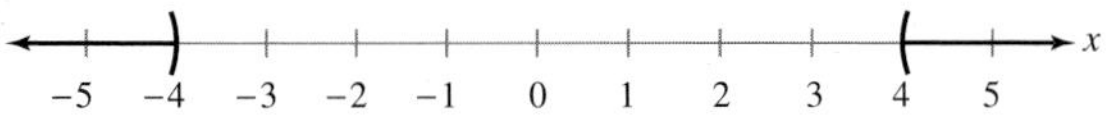

(b)
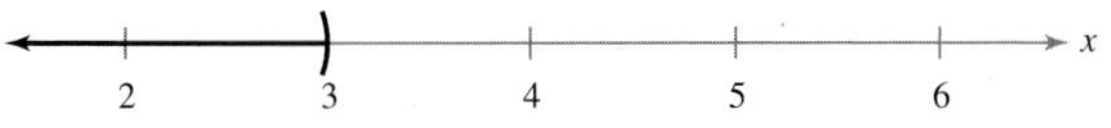

(c)
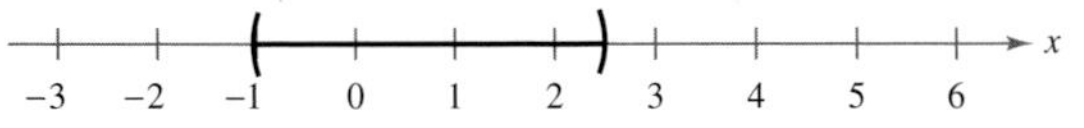

(d)
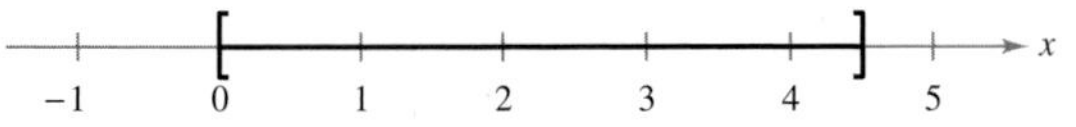

(e)
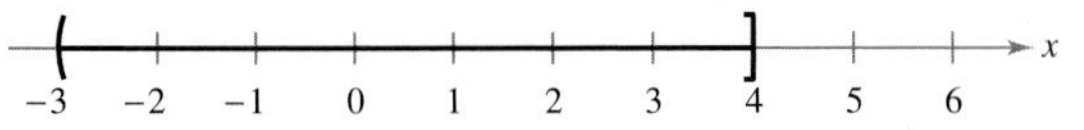

(f)
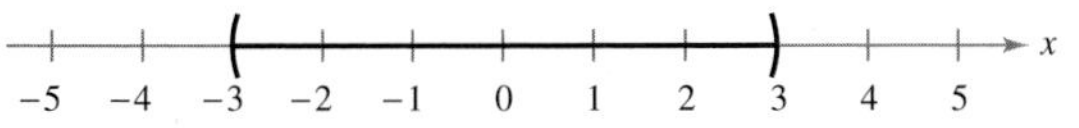

(g)
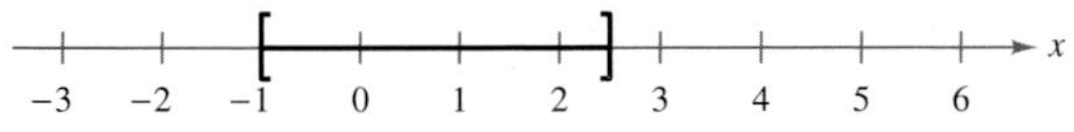

(h)
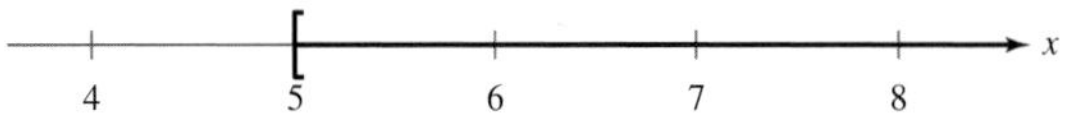

15. $x < 3$
16. $x \geq 5$
17. $-3 < x \leq 4$
18. $0 \leq x \leq \frac{9}{2}$
19. $|x| < 3$
20. $|x| > 4$
21. $-1 \leq x \leq \frac{5}{2}$
22. $-1 < x < \frac{5}{2}$

In Exercises 23–28, determine whether each value of x is a solution of the inequality.

Inequality	*Values*	
23. $5x - 12 > 0$	(a) $x = 3$	(b) $x = -3$
	(c) $x = \frac{5}{2}$	(d) $x = \frac{3}{2}$
24. $2x + 1 < -3$	(a) $x = 0$	(b) $x = -\frac{1}{4}$
	(c) $x = -4$	(d) $x = -\frac{3}{2}$
25. $0 < \dfrac{x-2}{4} < 2$	(a) $x = 4$	(b) $x = 10$
	(c) $x = 0$	(d) $x = \frac{7}{2}$
26. $-5 < 2x - 1 \leq 1$	(a) $x = -\frac{1}{2}$	(b) $x = -\frac{5}{2}$
	(c) $x = \frac{4}{3}$	(d) $x = 0$
27. $\lvert x - 10 \rvert \geq 3$	(a) $x = 13$	(b) $x = -1$
	(c) $x = 14$	(d) $x = 9$
28. $\lvert 2x - 3 \rvert < 15$	(a) $x = -6$	(b) $x = 0$
	(c) $x = 12$	(d) $x = 7$

In Exercises 29–56, solve the inequality and sketch the solution on the real number line. (Some inequalities have no solutions.)

29. $4x < 12$
30. $10x < -40$
31. $-2x > -3$
32. $-6x > 15$
33. $x - 5 \geq 7$
34. $x + 7 \leq 12$
35. $2x + 7 < 3 + 4x$
36. $3x + 1 \geq 2 + x$
37. $2x - 1 \geq 1 - 5x$
38. $6x - 4 \leq 2 + 8x$
39. $4 - 2x < 3(3 - x)$
40. $4(x + 1) < 2x + 3$
41. $\frac{3}{4}x - 6 \leq x - 7$
42. $3 + \frac{2}{7}x > x - 2$
43. $\frac{1}{2}(8x + 1) \geq 3x + \frac{5}{2}$
44. $9x - 1 < \frac{3}{4}(16x - 2)$
45. $3.6x + 11 \geq -3.4$
46. $15.6 - 1.3x < -5.2$
47. $1 < 2x + 3 < 9$
48. $-8 \leq -(3x + 5) < 13$
49. $-8 \leq 1 - 3(x - 2) < 13$
50. $0 \leq 2 - 3(x + 1) < 20$

51. $-4 < \frac{2x-3}{3} < 4$

52. $0 \le \frac{x+3}{2} < 5$

53. $\frac{3}{4} > x + 1 > \frac{1}{4}$

54. $-1 < 2 - \frac{x}{3} < 1$

55. $3.2 \le 0.4x - 1 \le 4.4$

56. $4.5 > \frac{1.5x+6}{2} > 10.5$

In Exercises 57–72, solve the inequality and sketch the solution on the real number line. (Some inequalities have no solution.)

57. $|x| < 5$

58. $|x| \ge 8$

59. $\left|\frac{x}{2}\right| > 1$

60. $\left|\frac{x}{5}\right| > 3$

61. $|x - 5| < -1$

62. $|x - 7| < -5$

63. $|x - 20| \le 6$

64. $|x - 8| \ge 0$

65. $|3 - 4x| \ge 9$

66. $|1 - 2x| < 5$

67. $\left|\frac{x-3}{2}\right| \ge 4$

68. $\left|1 - \frac{2x}{3}\right| < 1$

69. $|9 - 2x| - 2 < -1$

70. $|x + 14| + 3 > 17$

71. $2|x + 10| \ge 9$

72. $3|4 - 5x| \le 9$

Graphical Analysis **In Exercises 73–82, use a graphing utility to graph the inequality and identify the solution set.**

73. $6x > 12$

74. $3x - 1 \le 5$

75. $5 - 2x \ge 1$

76. $20 < 6x - 1$

77. $4(x - 3) \le 8 - x$

78. $3(x + 1) < x + 7$

79. $|x - 8| \le 14$

80. $|2x + 9| > 13$

81. $2|x + 7| \ge 13$

82. $\frac{1}{2}|x + 1| \le 3$

In Exercises 83–88, find the interval(s) on the real number line for which the radicand is nonnegative.

83. $\sqrt{x - 5}$

84. $\sqrt{x - 10}$

85. $\sqrt{x + 3}$

86. $\sqrt{3 - x}$

87. $\sqrt[4]{7 - 2x}$

88. $\sqrt[4]{6x + 15}$

89. ***Think About It*** The graph of $|x - 5| < 3$ can be described as all real numbers within three units of 5. Give a similar description of $|x - 10| < 8$.

90. ***Think About It*** The graph of $|x - 2| > 5$ can be described as all real numbers more than five units from 2. Give a similar description of $|x - 8| > 4$.

In Exercises 91–98, use absolute value notation to define the interval (or pair of intervals) on the real number line.

91. 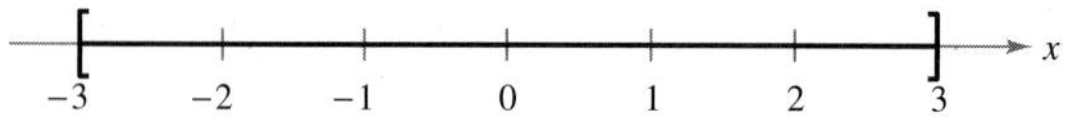

92. 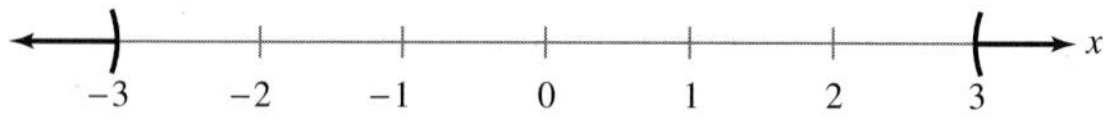

93. 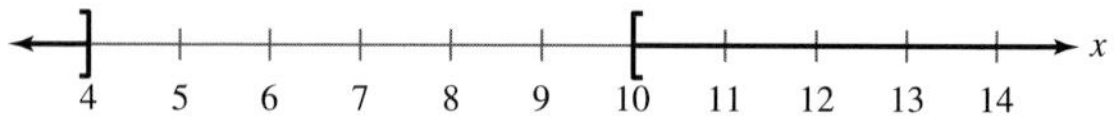

94. 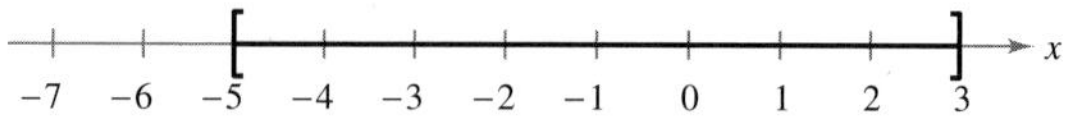

95. All real numbers within 10 units of 12

96. All real numbers at least five units from 8

97. All real numbers more than four units from -3

98. All real numbers no more than seven units from -6

In Exercises 99–102, determine whether each value of *x* is a solution of the inequality.

Inequality	*Values*	
99. $x^2 - 3 < 0$	(a) $x = 3$	(b) $x = 0$
	(c) $x = \frac{3}{2}$	(d) $x = -5$
100. $x^2 - x - 12 \ge 0$	(a) $x = 5$	(b) $x = 0$
	(c) $x = -4$	(d) $x = -3$
101. $\frac{x+2}{x-4} \ge 3$	(a) $x = 5$	(b) $x = 4$
	(c) $x = -\frac{9}{2}$	(d) $x = \frac{9}{2}$
102. $\frac{3x^2}{x^2+4} < 1$	(a) $x = -2$	(b) $x = -1$
	(c) $x = 0$	(d) $x = 3$

In Exercises 103–106, find the key numbers of the expression.

103. $3x^2 - x - 2$

104. $9x^3 - 25x^2$

105. $\frac{1}{x-5} + 1$

106. $\frac{x}{x+2} - \frac{2}{x-1}$

In Exercises 107–124, solve the inequality and graph the solution on the real number line.

107. $x^2 < 9$

108. $x^2 \le 16$

109. $(x + 2)^2 \le 25$

110. $(x - 3)^2 \ge 1$

111. $x^2 + 4x + 4 \ge 9$

112. $x^2 - 6x + 9 < 16$

113. $x^2 + x < 6$

114. $x^2 + 2x > 3$

The symbol indicates an exercise in which you are instructed to use graphing technology or a symbolic computer algebra system. The solutions of other exercises may also be facilitated by use of appropriate technology.

115. $x^2 + 2x - 3 < 0$
116. $x^2 > 2x + 8$
117. $3x^2 - 11x > 20$
118. $-2x^2 + 6x + 15 \le 0$
119. $x^2 - 3x - 18 > 0$
120. $x^3 + 2x^2 - 4x - 8 \le 0$
121. $x^3 - 3x^2 - x > -3$
122. $2x^3 + 13x^2 - 8x - 46 \ge 6$
123. $4x^2 - 4x + 1 \le 0$
124. $x^2 + 3x + 8 > 0$

In Exercises 125–130, solve the inequality and write the solution set in interval notation.

125. $4x^3 - 6x^2 < 0$
126. $4x^3 - 12x^2 > 0$
127. $x^3 - 4x \ge 0$
128. $2x^3 - x^4 \le 0$
129. $(x - 1)^2(x + 2)^3 \ge 0$
130. $x^4(x - 3) \le 0$

In Exercises 131–144, solve the inequality and graph the solution on the real number line.

131. $\dfrac{4x - 1}{x} > 0$
132. $\dfrac{x^2 - 1}{x} < 0$
133. $\dfrac{3x - 5}{x - 5} \ge 0$
134. $\dfrac{5 + 7x}{1 + 2x} \le 4$
135. $\dfrac{x + 6}{x + 1} - 2 < 0$
136. $\dfrac{x + 12}{x + 2} - 3 \ge 0$
137. $\dfrac{2}{x + 5} > \dfrac{1}{x - 3}$
138. $\dfrac{5}{x - 6} > \dfrac{3}{x + 2}$
139. $\dfrac{1}{x - 3} \le \dfrac{9}{4x + 3}$
140. $\dfrac{1}{x} \ge \dfrac{1}{x + 3}$
141. $\dfrac{x^2 + 2x}{x^2 - 9} \le 0$
142. $\dfrac{x^2 + x - 6}{x} \ge 0$
143. $\dfrac{3}{x - 1} + \dfrac{2x}{x + 1} > -1$
144. $\dfrac{3x}{x - 1} \le \dfrac{x}{x + 4} + 3$

In Exercises 145–150, find the domain of x in the expression.

145. $\sqrt{4 - x^2}$
146. $\sqrt{x^2 - 4}$
147. $\sqrt{x^2 - 9x + 20}$
148. $\sqrt{81 - 4x^2}$
149. $\sqrt{\dfrac{x}{x^2 - 2x - 35}}$
150. $\sqrt{\dfrac{x}{x^2 - 9}}$

In Exercises 151–156, solve the inequality. (Round your answers to two decimal places.)

151. $0.4x^2 + 5.26 < 10.2$
152. $-1.3x^2 + 3.78 > 2.12$
153. $-0.5x^2 + 12.5x + 1.6 > 0$
154. $1.2x^2 + 4.8x + 3.1 < 5.3$
155. $\dfrac{1}{2.3x - 5.2} > 3.4$
156. $\dfrac{2}{3.1x - 3.7} > 5.8$

WRITING ABOUT CONCEPTS

157. Identify the graph of the inequality $|x - a| \ge 2$.

(a) open interval from $a - 2$ to $a + 2$; labels: $a-2$, a, $a+2$, x
(b) $a-2$, a, $a+2$, x
(c) $2-a$, 2, $2+a$, x
(d) $2-a$, 2, $2+a$, x

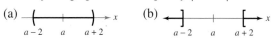

158. Identify the graph of the inequality $|x - b| < 4$.

(a) $b-4$, b, $b+4$, x
(b) $b-4$, b, $b+4$, x
(c) $b-4$, b, $b+4$, x
(d) $b-4$, b, $b+4$, x

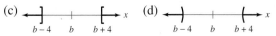

159. Find sets of values for $a, b,$ and c such that $0 \le x \le 10$ is a solution of the inequality $|ax - b| \le c, a \ne 0$.

160. Consider the polynomial $(x - a)(x - b)$ and the real number line shown below.

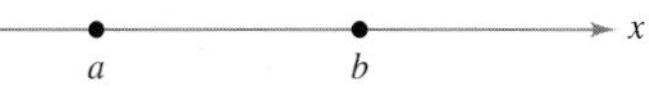

(a) Identify the points on the line at which the polynomial is zero.

(b) In each of the three subintervals of the line, write the sign of each factor and the sign of the product.

(c) At what x-values does the polynomial change signs?

161. ***Job Offers*** You are considering two job offers. The first job pays \$13.50 per hour. The second job pays \$9.00 per hour plus \$0.75 per unit produced per hour. Write an inequality yielding the number of units x that must be produced per hour to make the second job pay the greater hourly wage. Solve the inequality.

162. ***Job Offers*** You are considering two job offers. The first job pays \$3000 per month. The second job pays \$1000 per month plus a commission of 4% of your gross sales. Write an inequality yielding the gross sales x per month for which the second job will pay the greater monthly wage. Solve the inequality.

163. ***Investment*** In order for an investment of \$1000 to grow to more than \$1062.50 in 2 years, what must the annual interest rate be? $[A = P(1 + rt)]$

164. ***Investment*** In order for an investment of \$750 to grow to more than \$825 in 2 years, what must the annual interest rate be? $[A = P(1 + rt)]$

165. ***Egg Production*** The numbers of eggs E (in billions) produced in the United States from 1990 through 2006 can be modeled by

$$E = 1.52t + 68.0, \quad 0 \le t \le 16$$

where t represents the year, with $t = 0$ corresponding to 1990. *(Source: U.S. Department of Agriculture)*

(a) According to this model, when was the annual egg production 70 billion, but no more than 80 billion?

(b) According to this model, when will the annual egg production exceed 100 billion?

166. ***Daily Sales*** A doughnut shop sells a dozen doughnuts for \$4.50. Beyond the fixed costs (rent, utilities, and insurance) of \$220 per day, it costs \$2.75 for enough materials (flour, sugar, and so on) and labor to produce a dozen doughnuts. The daily profit from doughnut sales varies from \$60 to \$270. Between what levels (in dozens) do the daily sales vary?

167. ***Height*** The heights h of two-thirds of the members of a population satisfy the inequality

$$\left|\frac{h - 68.5}{2.7}\right| \le 1$$

where h is measured in inches. Determine the interval on the real number line in which these heights lie.

168. ***Meteorology*** An electronic device is to be operated in an environment with relative humidity h in the interval defined by $|h - 50| \le 30$. What are the minimum and maximum relative humidities for the operation of this device?

169. ***Geometry*** A rectangular playing field with a perimeter of 100 meters is to have an area of at least 500 square meters. Within what bounds must the length of the rectangle lie?

170. ***Geometry*** A rectangular parking lot with a perimeter of 440 feet is to have an area of at least 8000 square feet. Within what bounds must the length of the rectangle lie?

171. ***Investment*** P dollars, invested at interest rate r compounded annually, increases to an amount

$$A = P(1 + r)^2$$

in 2 years. An investment of \$1000 is to increase to an amount greater than \$1100 in 2 years. The interest rate must be greater than what percent?

172. ***Cost, Revenue, and Profit*** The revenue and cost equations for a product are

$$R = x(50 - 0.0002x) \quad \text{and} \quad C = 12x + 150{,}000$$

where R and C are measured in dollars and x represents the number of units sold. How many units must be sold to obtain a profit of at least \$1,650,000?

Height of a Projectile **In Exercises 173 and 174, use the position equation $s = -16t^2 + v_0t + s_0$, where s represents the height of an object (in feet), v_0 represents the initial velocity of the object (in feet per second), s_0 represents the initial height of the object (in feet), and t represents the time (in seconds).**

173. A projectile is fired straight upward from ground level $(s_0 = 0)$ with an initial velocity of 160 feet per second.

(a) At what instant will it be back at ground level?

(b) When will the height exceed 384 feet?

174. A projectile is fired straight upward from ground level $(s_0 = 0)$ with an initial velocity of 128 feet per second.

(a) At what instant will it be back at ground level?

(b) When will the height be less than 128 feet?

175. ***Resistors*** When two resistors of resistances R_1 and R_2 are connected in parallel (see figure), the total resistance R satisfies the equation

$$\frac{1}{R} = \frac{1}{R_1} + \frac{1}{R_2}.$$

Find R_1 for a parallel circuit in which $R_2 = 2$ ohms and R must be at least 1 ohm.

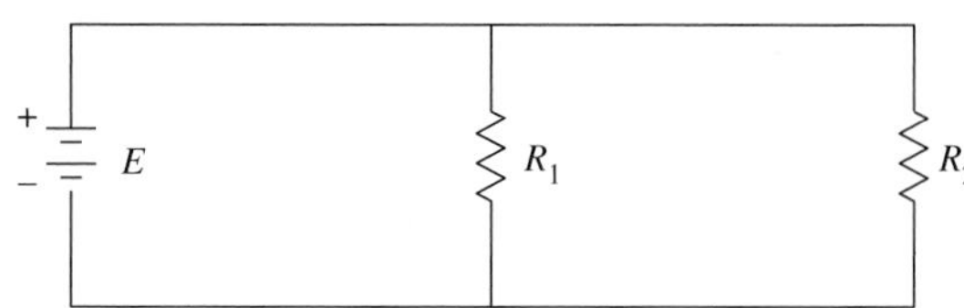

176. ***Safe Load*** The maximum safe load uniformly distributed over a one-foot section of a two-inch-wide wooden beam is approximated by the model Load $= 168.5d^2 - 472.1$, where d is the depth of the beam.

(a) Evaluate the model for $d = 4, 6, 8, 10$, and 12. Use the results to create a bar graph.

(b) Determine the minimum depth of the beam that will safely support a load of 2000 pounds.

True or False? **In Exercises 177–179, determine whether the statement is true or false. Justify your answer.**

177. If a, b, and c are real numbers, and $a \le b$, then $ac \le bc$.

178. If $-10 \le x \le 8$, then $-10 \ge -x$ and $-x \ge -8$.

179. The solution set of the inequality $\frac{3}{2}x^2 + 3x + 6 \ge 0$ is the entire set of real numbers.

CAPSTONE

180. Describe any differences between properties of equalities and properties of inequalities.

P.3 Graphical Representation of Data

- **Plot points in the Cartesian plane.**
- **Use the Distance Formula to find the distance between two points.**
- **Use the Midpoint Formula to find the midpoint of a line segment.**
- **Use a coordinate plane to model and solve real-life problems.**

RENÉ DESCARTES (1596–1650)

The Cartesian coordinate plane named after René Descartes was developed independently by another French mathematician, Pierre de Fermat. Fermat's *Introduction to Loci,* written about 1629, was clearer and more systematic than Descartes's *La géométrié.* However, Fermat's work was not published during his lifetime. Consequently, Descartes received the credit for the development of the coordinate plane with the now familiar x- and y-axes.

The Cartesian Plane

Just as you can represent real numbers by points on a real number line, you can represent ordered pairs of real numbers by points in a plane called the **rectangular coordinate system,** or the **Cartesian plane,** named after the French mathematician René Descartes (1596–1650).

The Cartesian plane is formed by using two real number lines intersecting at right angles, as shown in Figure P.11. The horizontal real number line is usually called the **x-axis,** and the vertical real number line is usually called the **y-axis.** The point of intersection of these two axes is the **origin,** and the two axes divide the plane into four parts called **quadrants.**

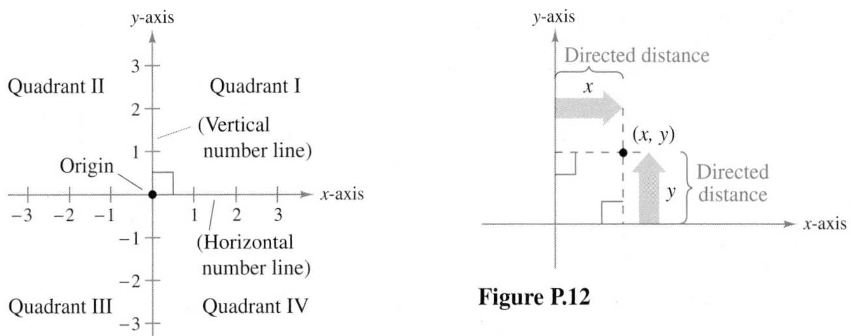

The Cartesian Plane
Figure P.11

Figure P.12

Each point in the plane corresponds to an **ordered pair** (x, y) of real numbers x and y, called **coordinates** of the point. The **x-coordinate** represents the directed distance from the y-axis to the point, and the **y-coordinate** represents the directed distance from the x-axis to the point, as shown in Figure P.12.

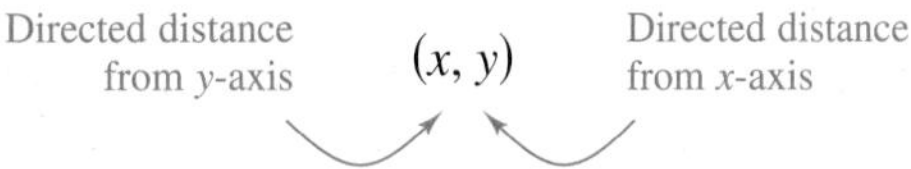

NOTE The notation (x, y) denotes both a point in the plane and an open interval on the real number line. The context will tell you which meaning is intended. ■

EXAMPLE 1 Plotting Points in the Cartesian Plane

Plot the points $(-1, 2)$, $(3, 4)$, $(0, 0)$, $(3, 0)$, and $(-2, -3)$.

Solution To plot the point $(-1, 2)$, imagine a vertical line through -1 on the x-axis and a horizontal line through 2 on the y-axis. The intersection of these two lines is the point $(-1, 2)$. The other four points can be plotted in a similar way, as shown in Figure P.13. ■

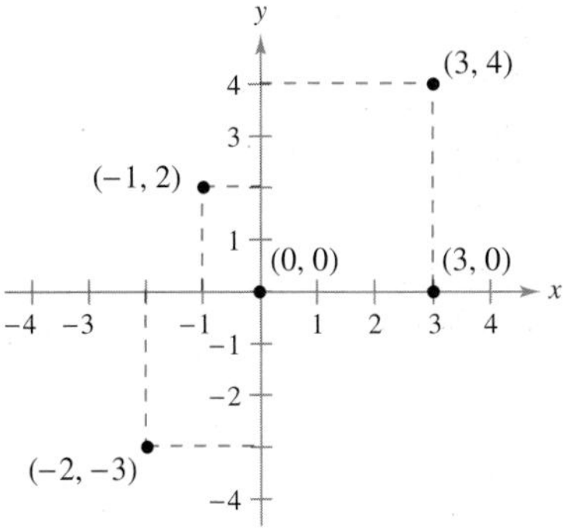

Figure P.13

The beauty of a rectangular coordinate system is that it allows you to *see* relationships between two variables. It would be difficult to overestimate the importance of Descartes's introduction of coordinates in the plane. Today, his ideas are in common use in virtually every scientific and business-related field.

EXAMPLE 2 Sketching a Scatter Plot

From 1994 through 2007, the numbers N (in millions) of subscribers to a cellular telecommunication service in the United States are shown in the table, where t represents the year. Sketch a scatter plot of the data. *(Source: CTIA-The Wireless Association)*

Year, t	1994	1995	1996	1997	1998	1999	2000	2001
Subscribers, N	24.1	33.8	44.0	55.3	69.2	86.0	109.5	128.4

Year, t	2002	2003	2004	2005	2006	2007
Subscribers, N	140.8	158.7	182.1	207.9	233.0	255.4

Solution To sketch a *scatter plot* of the data shown in the table, you simply represent each pair of values by an ordered pair (t, N) and plot the resulting points, as shown in Figure P.14. For instance, the first pair of values is represented by the ordered pair (1994, 24.1). Note that the break in the t-axis indicates that the numbers between 0 and 1994 have been omitted.

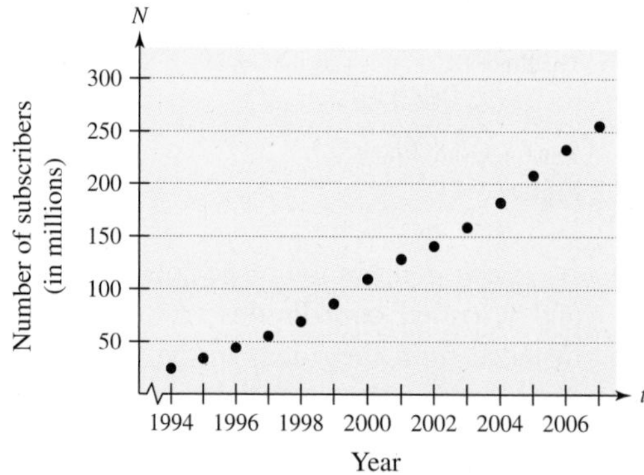

Figure P.14

In Example 2, you could have let $t = 1$ represent the year 1994. In that case, the horizontal axis would not have been broken, and the tick marks would have been labeled 1 through 14 (instead of 1994 through 2007).

TECHNOLOGY The scatter plot in Example 2 is only one way to represent the data graphically. You could also represent the data using a bar graph or a line graph. If you have access to a graphing utility, try using it to represent graphically the data given in Example 2.

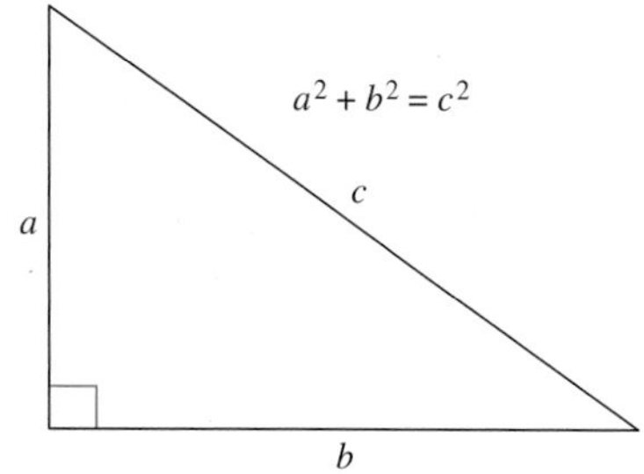

Figure P.15

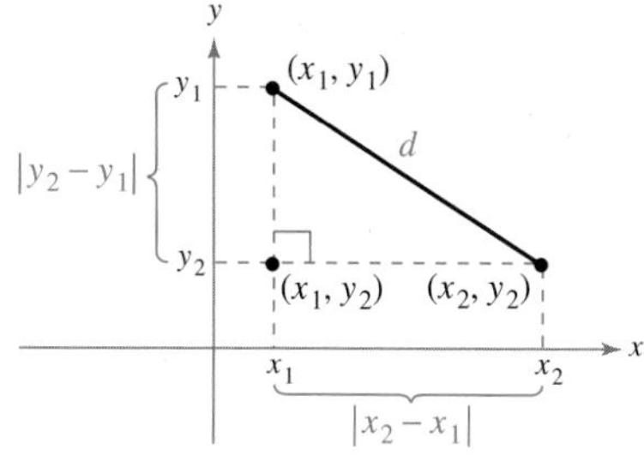

Figure P.16

The Distance Formula

Recall from the Pythagorean Theorem that, for a right triangle with hypotenuse of length c and sides of lengths a and b, you have

$$a^2 + b^2 = c^2 \qquad \text{Pythagorean Theorem}$$

as shown in Figure P.15. (The converse is also true. That is, if $a^2 + b^2 = c^2$, then the triangle is a right triangle.)

Suppose you want to determine the distance d between two points (x_1, y_1) and (x_2, y_2) in the plane. With these two points, a right triangle can be formed, as shown in Figure P.16. The length of the vertical side of the triangle is $|y_2 - y_1|$, and the length of the horizontal side is $|x_2 - x_1|$. By the Pythagorean Theorem, you can write

$$d^2 = |x_2 - x_1|^2 + |y_2 - y_1|^2$$

$$d = \sqrt{|x_2 - x_1|^2 + |y_2 - y_1|^2}$$

$$= \sqrt{(x_2 - x_1)^2 + (y_2 - y_1)^2}.$$

This result is the **Distance Formula.**

THE DISTANCE FORMULA

The distance d between the points (x_1, y_1) and (x_2, y_2) in the plane is

$$d = \sqrt{(x_2 - x_1)^2 + (y_2 - y_1)^2}.$$

EXAMPLE 3 Finding a Distance

Find the distance between the points $(-2, 1)$ and $(3, 4)$.

Algebraic Solution

Let $(x_1, y_1) = (-2, 1)$ and $(x_2, y_2) = (3, 4)$. Then apply the Distance Formula.

$$d = \sqrt{(x_2 - x_1)^2 + (y_2 - y_1)^2} \qquad \text{Distance Formula}$$

$$= \sqrt{[3 - (-2)]^2 + (4 - 1)^2} \qquad \text{Substitute for } x_1, y_1, x_2, \text{ and } y_2.$$

$$= \sqrt{(5)^2 + (3)^2} \qquad \text{Simplify.}$$

$$= \sqrt{34} \qquad \text{Simplify.}$$

$$\approx 5.83 \qquad \text{Use a calculator.}$$

So, the distance between the points is about 5.83 units. You can use the Pythagorean Theorem to check that the distance is correct.

$$d^2 \stackrel{?}{=} 3^2 + 5^2 \qquad \text{Pythagorean Theorem}$$

$$\left(\sqrt{34}\right)^2 \stackrel{?}{=} 3^2 + 5^2 \qquad \text{Substitute for } d.$$

$$34 = 34 \qquad \text{Distance checks. ✓}$$

Graphical Solution

Use centimeter graph paper to plot the points $A(-2, 1)$ and $B(3, 4)$. Carefully sketch the line segment from A to B. Then use a centimeter ruler to measure the length of the segment.

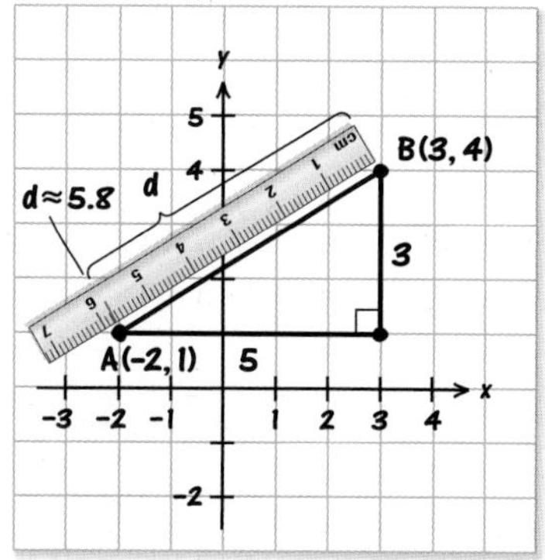

Figure P.17

The line segment measures about 5.8 centimeters, as shown in Figure P.17. So, the distance between the points is about 5.8 units. ■

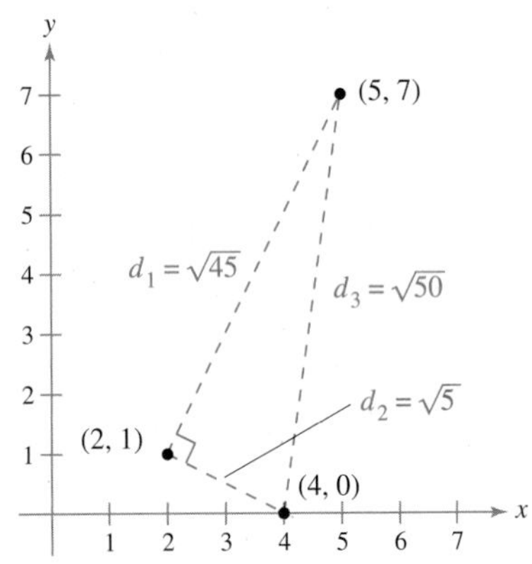

Figure P.18

EXAMPLE 4 Verifying a Right Triangle

Show that the points (2, 1), (4, 0), and (5, 7) are vertices of a right triangle.

Solution The three points are plotted in Figure P.18. Using the Distance Formula, you can find the lengths of the three sides as follows.

$$d_1 = \sqrt{(5-2)^2 + (7-1)^2} = \sqrt{9+36} = \sqrt{45}$$
$$d_2 = \sqrt{(4-2)^2 + (0-1)^2} = \sqrt{4+1} = \sqrt{5}$$
$$d_3 = \sqrt{(5-4)^2 + (7-0)^2} = \sqrt{1+49} = \sqrt{50}$$

Because

$$(d_1)^2 + (d_2)^2 = 45 + 5 = 50 = (d_3)^2$$

you can conclude by the Pythagorean Theorem that the triangle must be a right triangle. ■

The Midpoint Formula

To find the **midpoint** of the line segment that joins two points in a coordinate plane, you can simply find the average values of the respective coordinates of the two endpoints using the **Midpoint Formula.**

THE MIDPOINT FORMULA

The midpoint of the line segment joining the points (x_1, y_1) and (x_2, y_2) is given by the Midpoint Formula

$$\text{Midpoint} = \left(\frac{x_1 + x_2}{2}, \frac{y_1 + y_2}{2}\right).$$

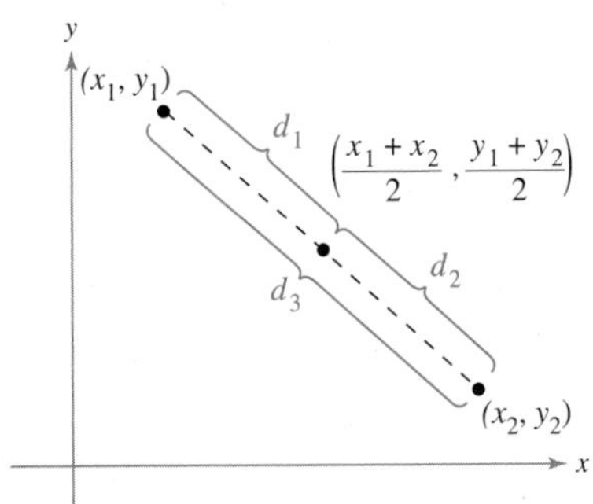

Figure P.19

PROOF Using Figure P.19, you must show that $d_1 = d_2$ and $d_1 + d_2 = d_3$. By the Distance Formula, you obtain

$$d_1 = \sqrt{\left(\frac{x_1 + x_2}{2} - x_1\right)^2 + \left(\frac{y_1 + y_2}{2} - y_1\right)^2} = \frac{1}{2}\sqrt{(x_2 - x_1)^2 + (y_2 - y_1)^2}$$

$$d_2 = \sqrt{\left(x_2 - \frac{x_1 + x_2}{2}\right)^2 + \left(y_2 - \frac{y_1 + y_2}{2}\right)^2} = \frac{1}{2}\sqrt{(x_2 - x_1)^2 + (y_2 - y_1)^2}$$

$$d_3 = \sqrt{(x_2 - x_1)^2 + (y_2 - y_1)^2}.$$

So, it follows that $d_1 = d_2$ and $d_1 + d_2 = d_3$.

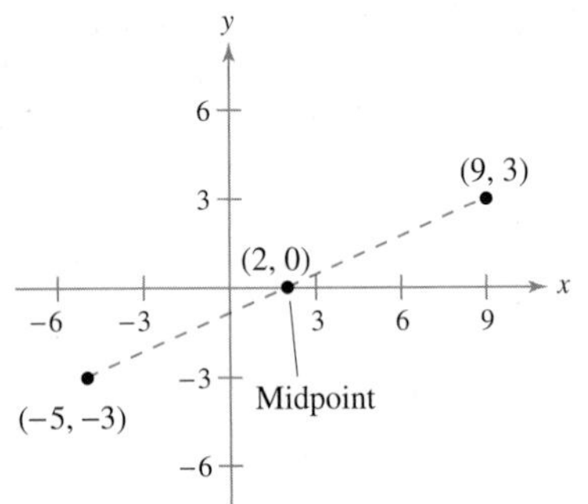

Figure P.20

EXAMPLE 5 Finding a Line Segment's Midpoint

Find the midpoint of the line segment joining the points (−5, −3) and (9, 3).

Solution Let $(x_1, y_1) = (-5, -3)$ and $(x_2, y_2) = (9, 3)$.

$$\text{Midpoint} = \left(\frac{x_1 + x_2}{2}, \frac{y_1 + y_2}{2}\right) = \left(\frac{-5 + 9}{2}, \frac{-3 + 3}{2}\right) = (2, 0)$$

The midpoint of the line segment is (2, 0), as shown in Figure P.20. ■

Applications

EXAMPLE 6 Finding the Length of a Pass

A football quarterback throws a pass from the 28-yard line, 40 yards from the sideline. The pass is caught by a wide receiver on the 5-yard line, 20 yards from the same sideline, as shown in Figure P.21. How long is the pass?

Solution You can find the length of the pass by finding the distance between the points (40, 28) and (20, 5).

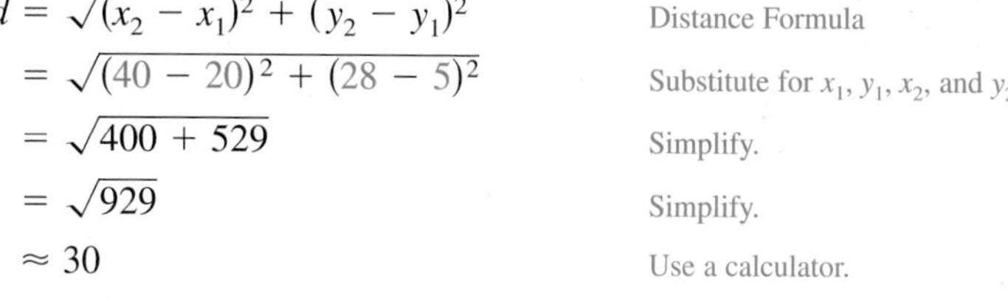

$$
\begin{aligned}
d &= \sqrt{(x_2 - x_1)^2 + (y_2 - y_1)^2} && \text{Distance Formula} \\
&= \sqrt{(40 - 20)^2 + (28 - 5)^2} && \text{Substitute for } x_1, y_1, x_2, \text{ and } y_2. \\
&= \sqrt{400 + 529} && \text{Simplify.} \\
&= \sqrt{929} && \text{Simplify.} \\
&\approx 30 && \text{Use a calculator.}
\end{aligned}
$$

So, the pass is about 30 yards long.

NOTE In Example 6, the scale along the goal line does not normally appear on a football field. However, when you use coordinate geometry to solve real-life problems, you are free to place the coordinate system in any way that is convenient for the solution of the problem.

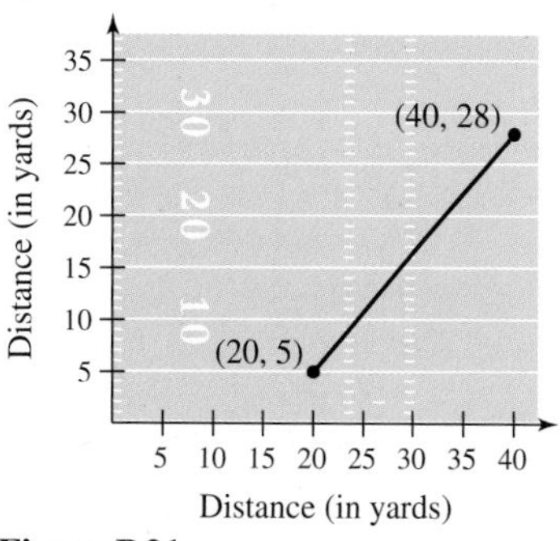

Figure P.21

EXAMPLE 7 Estimating Annual Revenue

Barnes & Noble had annual sales of approximately \$5.1 billion in 2005, and \$5.4 billion in 2007. Without knowing any additional information, what would you estimate the 2006 sales to have been? *(Source: Barnes & Noble, Inc.)*

Solution One solution to the problem is to assume that sales followed a linear pattern. With this assumption, you can estimate the 2006 sales by finding the midpoint of the line segment connecting the points (2005, 5.1) and (2007, 5.4).

$$
\begin{aligned}
\text{Midpoint} &= \left(\frac{x_1 + x_2}{2}, \frac{y_1 + y_2}{2}\right) && \text{Midpoint Formula} \\
&= \left(\frac{2005 + 2007}{2}, \frac{5.1 + 5.4}{2}\right) && \text{Substitute for } x_1, x_2, y_1, \text{ and } y_2. \\
&= (2006, 5.25) && \text{Simplify.}
\end{aligned}
$$

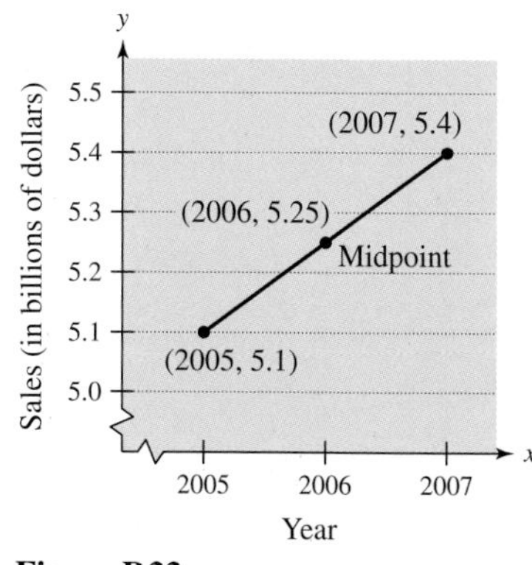

Figure P.22

So, you would estimate the 2006 sales to have been about \$5.25 billion, as shown in Figure P.22. (The actual 2006 sales were about \$5.26 billion.) ■

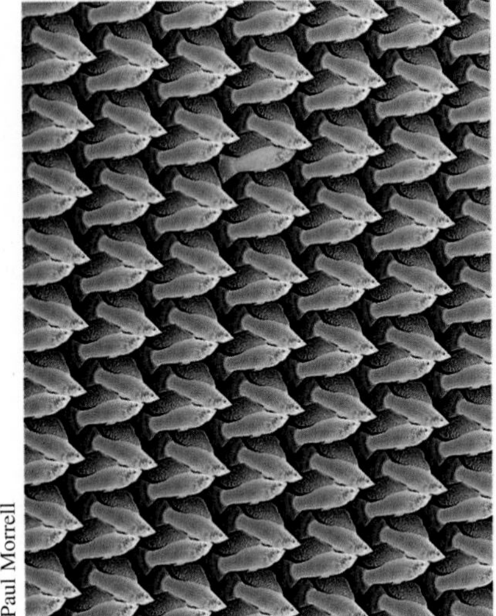

Paul Morrell

Much of computer graphics, including this computer-generated goldfish tessellation, consists of transformations of points in a coordinate plane. One type of transformation, a translation, is illustrated in Example 8. Other types include reflections (as illustrated in Example 9), rotations, and stretches.

EXAMPLE 8 Translating Points in the Plane

The triangle in Figure P.23(a) has vertices at the points $(-1, 2)$, $(1, -4)$, and $(2, 3)$. Shift the triangle three units to the right and two units upward.

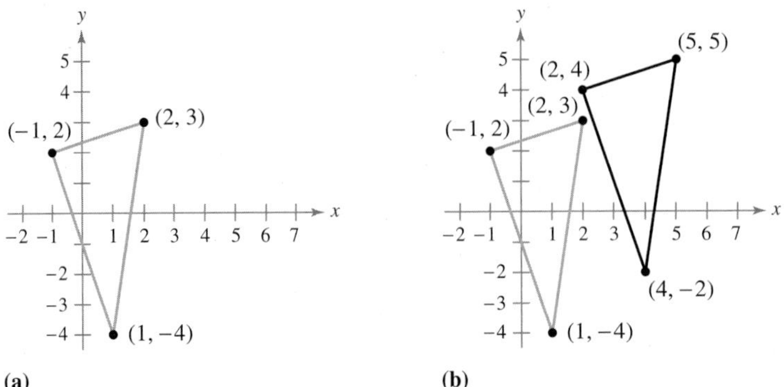

Figure P.23

Solution To shift the vertices three units to the right, add 3 to each of the x-coordinates. To shift the vertices two units upward, add 2 to each of the y-coordinates. The result is shown in Figure P.23(b).

Original Point	*Translated Point*
$(-1, 2)$	$(-1 + 3, 2 + 2) = (2, 4)$
$(1, -4)$	$(1 + 3, -4 + 2) = (4, -2)$
$(2, 3)$	$(2 + 3, 3 + 2) = (5, 5)$

EXAMPLE 9 Reflecting Points in the Plane

The triangle in Figure P.24(a) has vertices at the points $(1, 1)$, $(4, 2)$, and $(2, 4)$. Reflect the triangle in the y-axis.

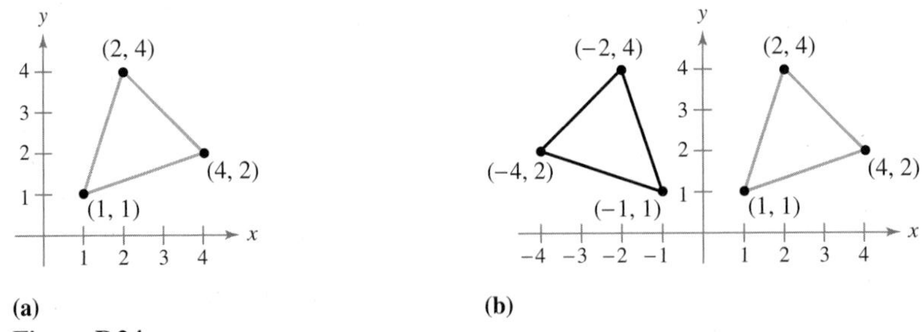

Figure P.24

Solution To reflect the vertices in the y-axis, negate each x-coordinate. The result is shown in Figure P.24(b).

Original Point	*Reflected Point*
$(1, 1)$	$(-1, 1)$
$(4, 2)$	$(-4, 2)$
$(2, 4)$	$(-2, 4)$

■

The figures provided with Examples 8 and 9 were not really essential to the solutions. Nevertheless, it is strongly recommended that you develop the habit of including sketches with your solutions—even if they are not required.

P.3 Exercises

See www.CalcChat.com for worked-out solutions to odd-numbered exercises.

1. Match each term with its definition.
   (a) x-axis
   (b) y-axis
   (c) origin
   (d) quadrants
   (e) x-coordinate
   (f) y-coordinate
   (i) point of intersection of vertical axis and horizontal axis
   (ii) directed distance from the x-axis
   (iii) directed distance from the y-axis
   (iv) four regions of the coordinate plane
   (v) horizontal real number line
   (vi) vertical real number line

In Exercises 2–4, fill in the blanks.

2. An ordered pair of real numbers can be represented in a plane called the rectangular coordinate system or the ________ plane.
3. The ________ ________ is a result derived from the Pythagorean Theorem.
4. Finding the average values of the representative coordinates of the two endpoints of a line segment in a coordinate plane is also known as using the ________ ________.

In Exercises 5 and 6, approximate the coordinates of the points.

5. 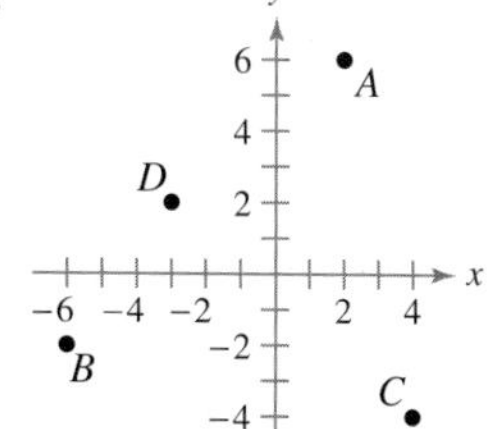

6. 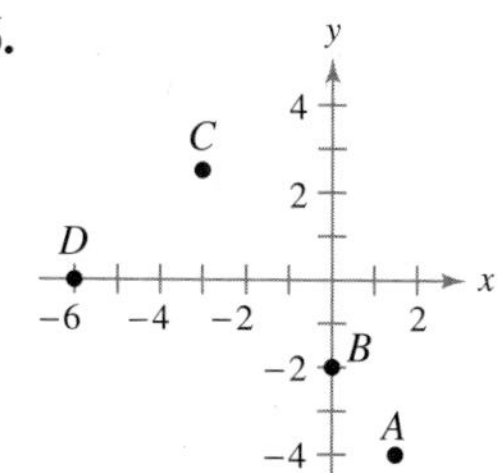

In Exercises 7–10, plot the points in the Cartesian plane.

7. $(-4, 2), (-3, -6), (0, 5), (1, -4)$
8. $(0, 0), (3, 1), (-2, 4), (1, -1)$
9. $(3, 8), (0.5, -1), (5, -6), (-2, 2.5)$
10. $\left(1, -\frac{1}{3}\right), \left(\frac{3}{4}, 3\right), (-3, 4), \left(-\frac{4}{3}, -\frac{3}{2}\right)$

In Exercises 11–14, find the coordinates of the point.

11. The point is located three units to the left of the y-axis and four units above the x-axis.
12. The point is located eight units below the x-axis and four units to the right of the y-axis.
13. The point is located five units below the x-axis and the coordinates of the point are equal.
14. The point is on the x-axis and 12 units to the left of the y-axis.

In Exercises 15–24, determine the quadrant(s) in which (x, y) is located so that the condition(s) is (are) satisfied.

15. $x > 0$ and $y < 0$
16. $x < 0$ and $y < 0$
17. $x = -4$ and $y > 0$
18. $x > 2$ and $y = 3$
19. $y < -5$
20. $x > 4$
21. $x < 0$ and $-y > 0$
22. $-x > 0$ and $y < 0$
23. $xy > 0$
24. $xy < 0$

In Exercises 25 and 26, sketch a scatter plot of the data shown in the table.

25. ***Number of Stores*** The table shows the number y of Wal-Mart stores for each year x from 2000 through 2007. *(Source: Wal-Mart Stores, Inc.)*

Year, x	2000	2001	2002	2003
Number of stores, y	4189	4414	4688	4906

Year, x	2004	2005	2006	2007
Number of stores, y	5289	6141	6779	7262

26. ***Meteorology*** The table shows the lowest temperature on record y (in degrees Fahrenheit) in Duluth, Minnesota for each month x, where $x = 1$ represents January. *(Source: NOAA)*

Month, x	1	2	3	4
Temperature, y	−39	−39	−29	−5

Month, x	5	6	7	8
Temperature, y	17	27	35	32

Month, x	9	10	11	12
Temperature, y	22	8	−23	−34

In Exercises 27–38, find the distance between the points.

27. $(6, -3), (6, 5)$

28. $(1, 4), (8, 4)$

29. $(-3, -1), (2, -1)$

30. $(-3, -4), (-3, 6)$

31. $(-2, 6), (3, -6)$

32. $(8, 5), (0, 20)$

33. $(1, 4), (-5, -1)$

34. $(1, 3), (3, -2)$

35. $\left(\frac{1}{2}, \frac{4}{3}\right), (2, -1)$

36. $\left(-\frac{2}{3}, 3\right), \left(-1, \frac{5}{4}\right)$

37. $(-4.2, 3.1), (-12.5, 4.8)$

38. $(9.5, -2.6), (-3.9, 8.2)$

In Exercises 39–42, (a) find the length of each side of the right triangle, and (b) show that these lengths satisfy the Pythagorean Theorem.

39.

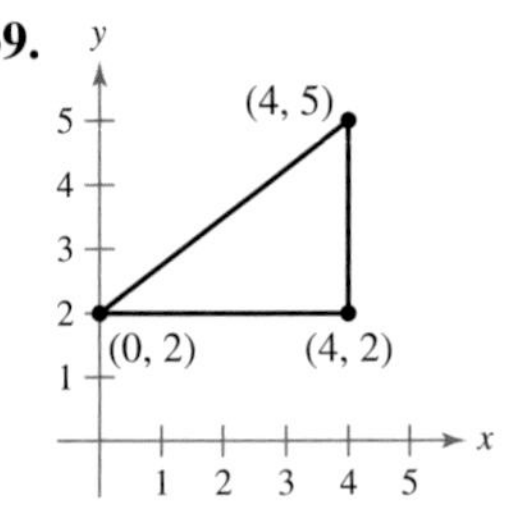

40.

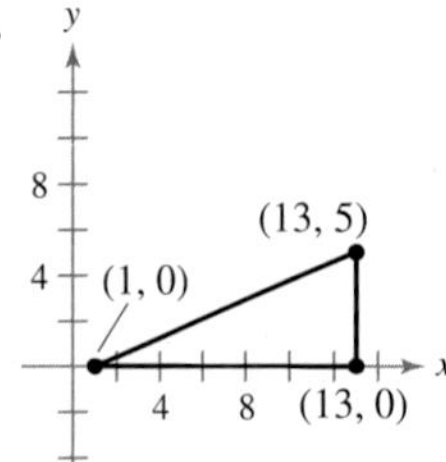

41.

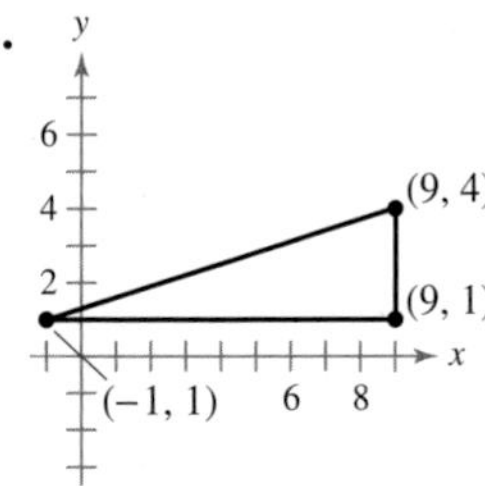

42.

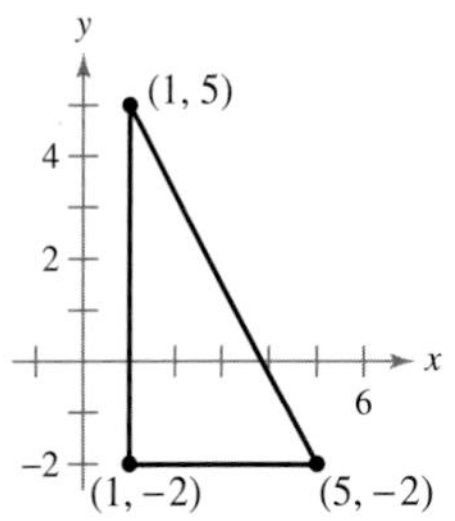

In Exercises 43–46, show that the points form the vertices of the indicated polygon.

43. Right triangle: $(4, 0), (2, 1), (-1, -5)$

44. Right triangle: $(-1, 3), (3, 5), (5, 1)$

45. Isosceles triangle: $(1, -3), (3, 2), (-2, 4)$

46. Isosceles triangle: $(2, 3), (4, 9), (-2, 7)$

In Exercises 47–56, (a) plot the points, (b) find the distance between the points, and (c) find the midpoint of the line segment joining the points.

47. $(1, 1), (9, 7)$

48. $(1, 12), (6, 0)$

49. $(-4, 10), (4, -5)$

50. $(-7, -4), (2, 8)$

51. $(-1, 2), (5, 4)$

52. $(2, 10), (10, 2)$

53. $\left(\frac{1}{2}, 1\right), \left(-\frac{5}{2}, \frac{4}{3}\right)$

54. $\left(-\frac{1}{3}, -\frac{1}{3}\right), \left(-\frac{1}{6}, -\frac{1}{2}\right)$

55. $(6.2, 5.4), (-3.7, 1.8)$

56. $(-16.8, 12.3), (5.6, 4.9)$

57. ***Flying Distance*** An airplane flies from Naples, Italy in a straight line to Rome, Italy, which is 120 kilometers north and 150 kilometers west of Naples. How far does the plane fly?

58. ***Sports*** A soccer player passes the ball from a point that is 18 yards from the endline and 12 yards from the sideline. The pass is received by a teammate who is 42 yards from the same endline and 50 yards from the same sideline, as shown in the figure. How long is the pass?

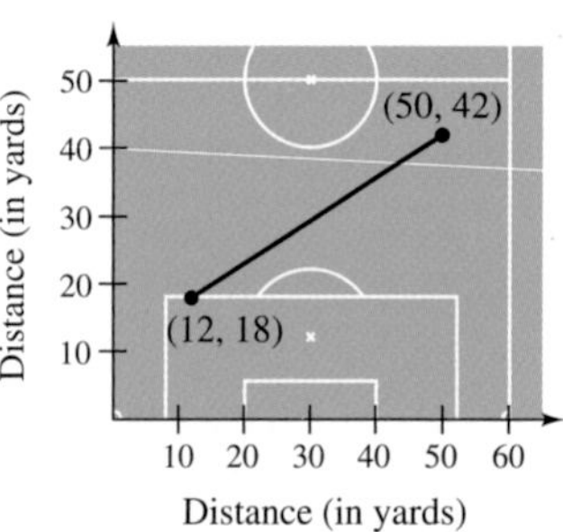

Sales **In Exercises 59 and 60, use the Midpoint Formula to estimate the sales of Big Lots, Inc. and Dollar Tree Stores, Inc. in 2005, given the sales in 2003 and 2007. Assume that the sales followed a linear pattern.** ***(Source: Big Lots, Inc.; Dollar Tree Stores, Inc.)***

59. Big Lots

Year	2003	2007
Sales (in millions)	\$4174	\$4656

60. Dollar Tree

Year	2003	2007
Sales (in millions)	\$2800	\$4243

In Exercises 61–64, the polygon is shifted to a new position in the plane. Find the coordinates of the vertices of the polygon in its new position.

61. 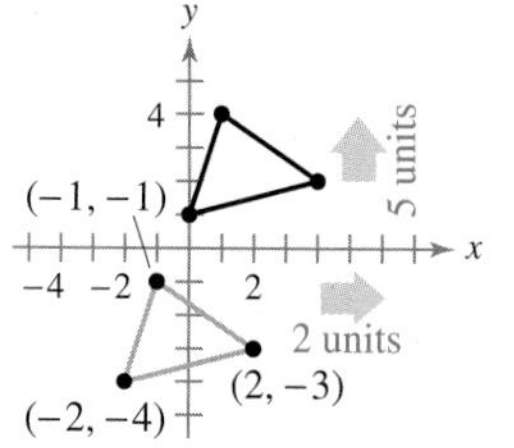

62. 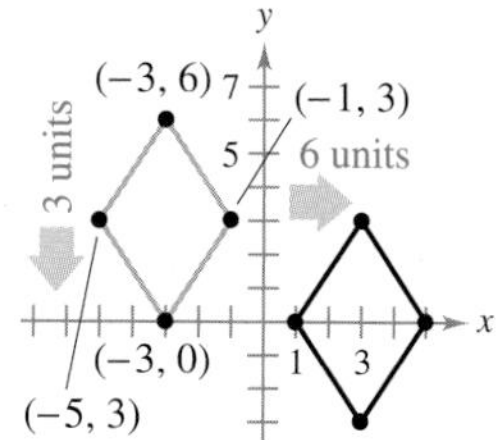

63. Original coordinates of vertices: $(-7, -2), (-2, 2), (-2, -4), (-7, -4)$

Shift: eight units upward, four units to the right

64. Original coordinates of vertices: $(5, 8)$, $(3, 6)$, $(7, 6)$, $(5, 2)$

Shift: 6 units downward, 10 units to the left

In Exercises 65–68, the vertices of a polygon are given. Find the coordinates of the vertices when the polygon is reflected in the *y*-axis.

65. 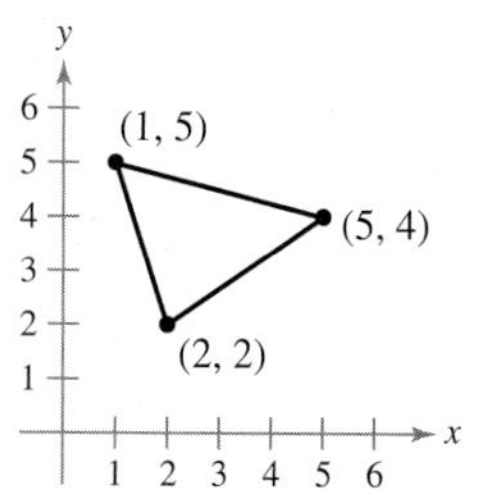

66. 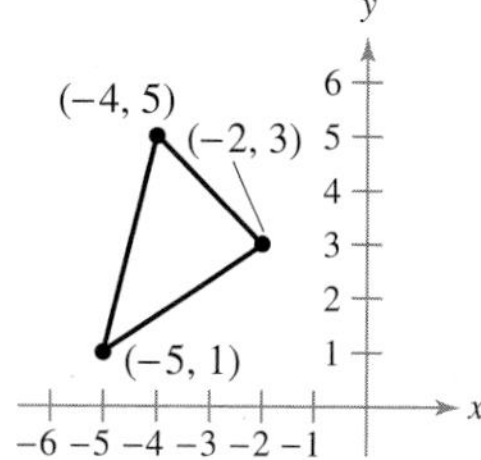

67. Quadrilateral: $(0, 3)$, $(3, -2)$, $(6, 3)$, $(3, 8)$

68. Quadrilateral: $(-7, 1)$, $(-5, 4)$, $(-1, 4)$, $(-3, 1)$

WRITING ABOUT CONCEPTS

In Exercises 69 and 70, find the length of each side of the right triangle and show that the lengths satisfy the Pythagorean Theorem.

69. 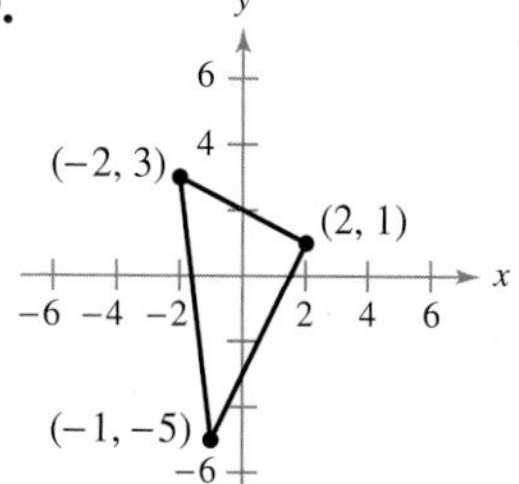

70. 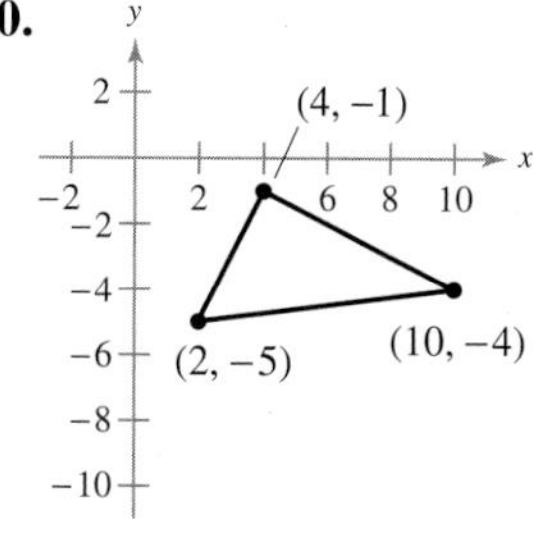

71. What is the y-coordinate of any point on the x-axis? What is the x-coordinate of any point on the y-axis?

WRITING ABOUT CONCEPTS (continued)

72. Plot the points $(2, 1)$, $(-3, 5)$, and $(7, -3)$ on a rectangular coordinate system. Then change the sign of the x-coordinate of each point and plot the three new points on the same rectangular coordinate system. Make a conjecture about the location of a point when each of the following occurs.

(a) The sign of the x-coordinate is changed.

(b) The sign of the y-coordinate is changed.

(c) The signs of both coordinates are changed.

Retail Price **In Exercises 73 and 74, use the graph which shows the average retail prices of 1 gallon of whole milk from 1996 to 2007.** ***(Source: U.S. Bureau of Labor Statistics)***

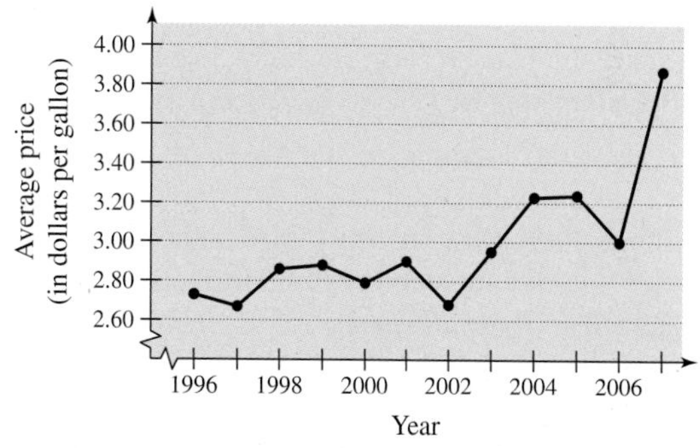

73. Approximate the highest price of a gallon of whole milk shown in the graph. When did this occur?

74. Approximate the percent change in the price of milk from the price in 1996 to the highest price shown in the graph.

75. ***Advertising*** The graph shows the average costs of a 30-second television spot (in thousands of dollars) during the Super Bowl from 2000 to 2008. *(Source: Nielsen Media and TNS Media Intelligence)*

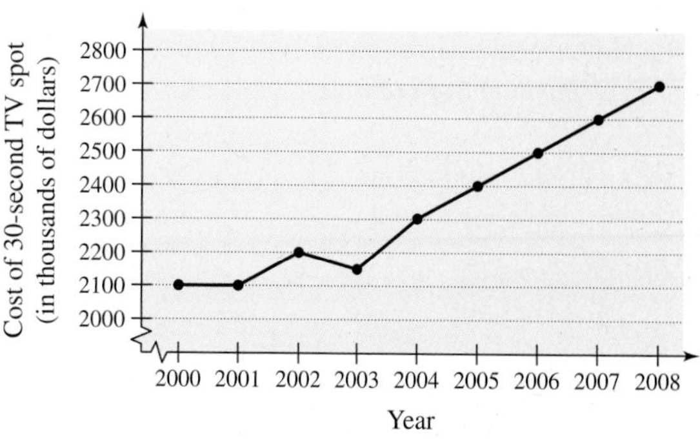

(a) Estimate the percent increase in the average cost of a 30-second spot from Super Bowl XXXIV in 2000 to Super Bowl XXXVIII in 2004.

(b) Estimate the percent increase in the average cost of a 30-second spot from Super Bowl XXXIV in 2000 to Super Bowl XLII in 2008.

76. ***Advertising*** The graph shows the average costs of a 30-second television spot (in thousands of dollars) during the Academy Awards from 1995 to 2007. *(Source: Nielsen Monitor-Plus)*

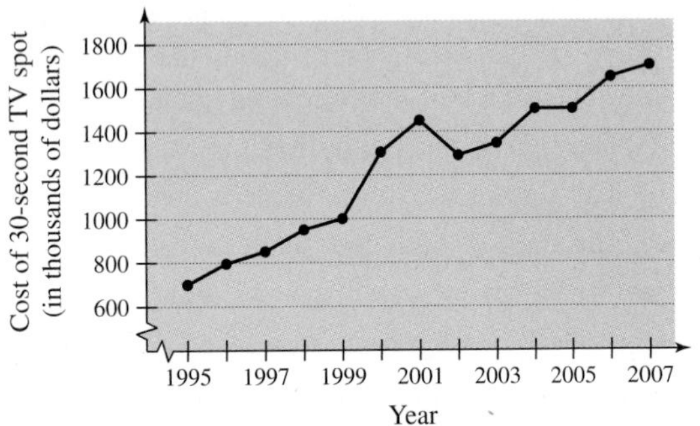

(a) Estimate the percent increase in the average cost of a 30-second spot in 1996 to the cost in 2002.

(b) Estimate the percent increase in the average cost of a 30-second spot in 1996 to the cost in 2007.

77. ***Music*** The graph shows the numbers of performers who were elected to the Rock and Roll Hall of Fame from 1991 through 2008. Describe any trends in the data. From these trends, predict the number of performers elected in 2010. *(Source: rockhall.com)*

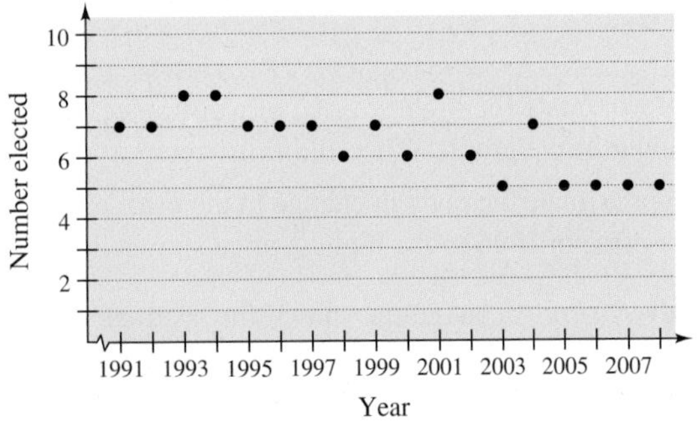

78. ***Labor Force*** The graph shows the minimum wage in the United States (in dollars) from 1950 to 2009. *(Source: U.S. Department of Labor)*

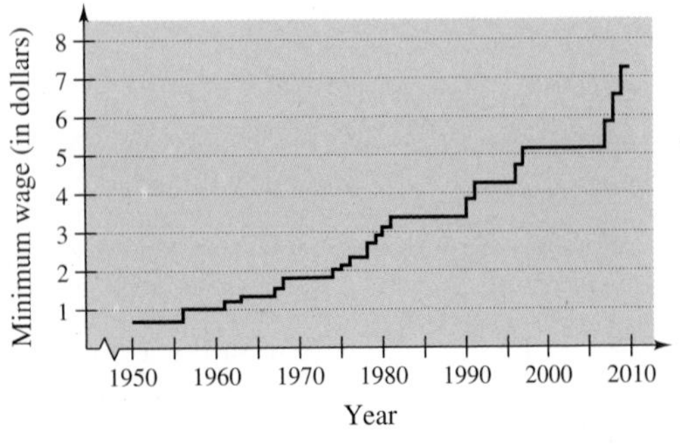

(a) Which decade shows the greatest increase in minimum wage?

(b) Approximate the percent increases in the minimum wage from 1990 to 1995 and from 1995 to 2009.

(c) Use the percent increase from 1995 to 2009 to predict the minimum wage in 2013.

(d) Do you believe that your prediction in part (c) is reasonable? Explain.

79. ***Sales*** The Coca-Cola Company had sales of \$19,805 million in 1999 and \$28,857 million in 2007. Use the Midpoint Formula to estimate the sales in 2003. Assume that the sales followed a linear pattern. *(Source: The Coca-Cola Company)*

80. ***Data Analysis: Exam Scores*** The table shows the mathematics entrance test scores x and the final examination scores y in an algebra course for a sample of 10 students.

x	22	29	35	40	44	48	53	58	65	76
y	53	74	57	66	79	90	76	93	83	99

(a) Sketch a scatter plot of the data.

(b) Find the entrance test score of any student with a final exam score in the 80s.

(c) Does a higher entrance test score imply a higher final exam score? Explain.

81. ***Data Analysis: Mail*** The table shows the number y of pieces of mail handled (in billions) by the U.S. Postal Service for each year x from 1996 through 2008. *(Source: U.S. Postal Service)*

Year, x	1996	1997	1998	1999	2000
Pieces of mail, y	183	191	197	202	208

Year, x	2001	2002	2003	2004	2005
Pieces of mail, y	207	203	202	206	212

Year, x	2006	2007	2008
Pieces of mail, y	213	212	203

(a) Sketch a scatter plot of the data.

(b) Approximate the year in which there was the greatest decrease in the number of pieces of mail handled.

(c) Why do you think the number of pieces of mail handled decreased?

82. ***Data Analysis: Athletics*** The table shows the numbers of men's M and women's W college basketball teams for each year x from 1994 through 2007. *(Source: National Collegiate Athletic Association)*

Year, x	1994	1995	1996	1997	1998
Men's teams, M	858	868	866	865	895
Women's teams, W	859	864	874	879	911

Year, x	1999	2000	2001	2002	2003
Men's teams, M	926	932	937	936	967
Women's teams, W	940	956	958	975	1009

Year, x	2004	2005	2006	2007
Men's teams, M	981	983	984	982
Women's teams, W	1008	1036	1018	1003

(a) Sketch scatter plots of these two sets of data on the same set of coordinate axes.

(b) Find the year in which the numbers of men's and women's teams were nearly equal.

(c) Find the year in which the difference between the numbers of men's and women's teams was the greatest. What was this difference?

83. A line segment has (x_1, y_1) as one endpoint and (x_m, y_m) as its midpoint. Find the other endpoint (x_2, y_2) of the line segment in terms of x_1, y_1, x_m, and y_m.

84. Use the result of Exercise 83 to find the coordinates of the endpoint of a line segment if the coordinates of the other endpoint and midpoint are, respectively,

(a) $(1, -2), (4, -1)$ and (b) $(-5, 11), (2, 4)$.

85. Use the Midpoint Formula three times to find the three points that divide the line segment joining (x_1, y_1) and (x_2, y_2) into four parts.

86. Use the result of Exercise 85 to find the points that divide the line segment joining the points (a) $(1, -2), (4, -1)$ and (b) $(-2, -3), (0, 0)$ into four equal parts.

87. ***Think About It*** When plotting points on the rectangular coordinate system, is it true that the scales on the x- and y-axes must be the same? Explain.

88. ***Collinear Points*** Three or more points are *collinear* if they all lie on the same line. Use the steps below to determine if the set of points $\{A(2, 3), B(2, 6), C(6, 3)\}$ and the set of points $\{A(8, 3), B(5, 2), C(2, 1)\}$ are collinear.

(a) For each set of points, use the Distance Formula to find the distances from A to B, from B to C, and from A to C. What relationship exists among these distances for each set of points?

(b) Plot each set of points in the Cartesian plane. Do all the points of either set appear to lie on the same line?

(c) Compare your conclusions from part (a) with the conclusions you made from the graphs in part (b). Make a general statement about how to use the Distance Formula to determine collinearity.

True or False? **In Exercises 89 and 90, determine whether the statement is true or false. Justify your answer.**

89. In order to divide a line segment into 16 equal parts, you would have to use the Midpoint Formula 16 times.

90. The points $(-8, 4)$, $(2, 11)$, and $(-5, 1)$ represent the vertices of an isosceles triangle.

91. ***Proof*** Prove that the diagonals of the parallelogram in the figure intersect at their midpoints.

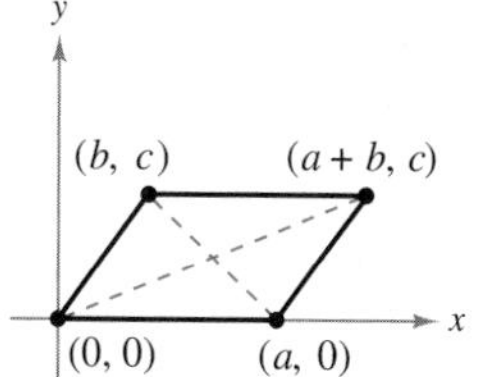

CAPSTONE

92. Use the plot of the point (x_0, y_0) in the figure. Match the transformation of the point with the correct plot. Explain your reasoning. [The plots are labeled (i), (ii), (iii), and (iv).]

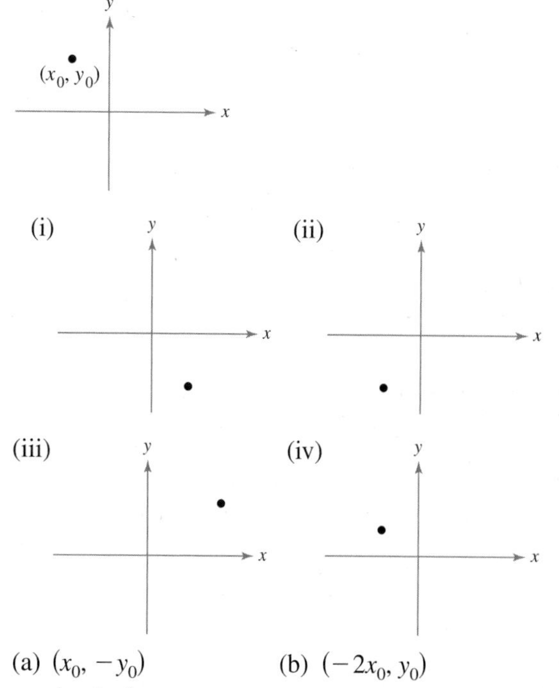

(a) $(x_0, -y_0)$ (b) $(-2x_0, y_0)$

(c) $\left(x_0, \frac{1}{2}y_0\right)$ (d) $(-x_0, -y_0)$

P.4 Graphs of Equations

- Sketch graphs of equations.
- Find x- and y-intercepts of graphs of equations.
- Use symmetry to sketch graphs of equations.
- Find equations of and sketch graphs of circles.
- Use graphs of equations in solving real-life problems.

The Graph of an Equation

In Section P.3, you used a coordinate system to represent graphically the relationship between two quantities. There, the graphical picture consisted of a collection of points in a coordinate plane.

Frequently, a relationship between two quantities is expressed as an **equation in two variables.** For instance, $y = 7 - 3x$ is an equation in x and y. An ordered pair (a, b) is a **solution** or **solution point** of an equation in x and y if the equation is true when a is substituted for x and b is substituted for y. For instance, $(1, 4)$ is a solution of $y = 7 - 3x$ because $4 = 7 - 3(1)$ is a true statement.

In this section you will review some basic procedures for sketching the graph of an equation in two variables. The **graph of an equation** is the set of all points that are solutions of the equation.

EXAMPLE 1 Determining Solution Points

Determine whether each point lies on the graph of $y = 10x - 7$.

a. $(2, 13)$ **b.** $(-1, -3)$

Solution

a.

$y = 10x - 7$ Write original equation.

$13 \stackrel{?}{=} 10(2) - 7$ Substitute 2 for x and 13 for y.

$13 = 13$ $(2, 13)$ is a solution. ✓

The point $(2, 13)$ *does* lie on the graph of $y = 10x - 7$ because it is a solution point of the equation.

b.

$y = 10x - 7$ Write original equation.

$-3 \stackrel{?}{=} 10(-1) - 7$ Substitute -1 for x and -3 for y.

$-3 \neq -17$ $(-1, -3)$ is not a solution.

The point $(-1, -3)$ *does not* lie on the graph of $y = 10x - 7$ because it is *not* a solution point of the equation. ■

The basic technique used for sketching the graph of an equation is the **point-plotting method.**

SKETCHING THE GRAPH OF AN EQUATION BY POINT PLOTTING

1. If possible, rewrite the equation so that one of the variables is isolated on one side of the equation.
2. Make a table of values showing several solution points.
3. Plot these points on a rectangular coordinate system.
4. Connect the points with a smooth curve or line.

EXAMPLE 2 Sketching the Graph of an Equation

Sketch the graph of

$$y = 7 - 3x.$$

Solution Because the equation is already solved for y, construct a table of values that consists of several solution points of the equation. For instance, when $x = -1$,

$$\begin{aligned} y &= 7 - 3(-1) \\ &= 10 \end{aligned}$$

which implies that $(-1, 10)$ is a solution point of the graph.

x	$y = 7 - 3x$	(x, y)
-1	10	$(-1, 10)$
0	7	$(0, 7)$
1	4	$(1, 4)$
2	1	$(2, 1)$
3	-2	$(3, -2)$
4	-5	$(4, -5)$

From the table, it follows that

$$(-1, 10),\ (0, 7),\ (1, 4),\ (2, 1),\ (3, -2),\ \text{and } (4, -5)$$

are solution points of the equation. After plotting these points, you can see that they appear to lie on a line, as shown in Figure P.25. The graph of the equation is the line that passes through the six plotted points.

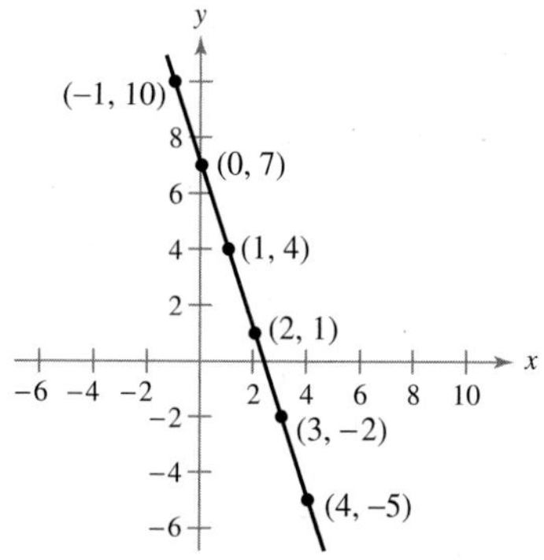

Figure P.25

■

EXAMPLE 3 Sketching the Graph of an Equation

Sketch the graph of

$$y = x^2 - 2.$$

Solution Because the equation is already solved for y, begin by constructing a table of values.

x	-2	-1	0	1	2	3
$y = x^2 - 2$	2	-1	-2	-1	2	7
(x, y)	$(-2, 2)$	$(-1, -1)$	$(0, -2)$	$(1, -1)$	$(2, 2)$	$(3, 7)$

NOTE One of your goals in this course is to learn to classify the basic shape of a graph from its equation. For instance, you will learn that the *linear equation* in Example 2 has the form

$$y = mx + b$$

and its graph is a line. Similarly, the *quadratic equation* in Example 3 has the form

$$y = ax^2 + bx + c$$

and its graph is a parabola.

Next, plot the points given in the table, as shown in Figure P.26(a). Finally, connect the points with a smooth curve, as shown in Figure P.26(b).

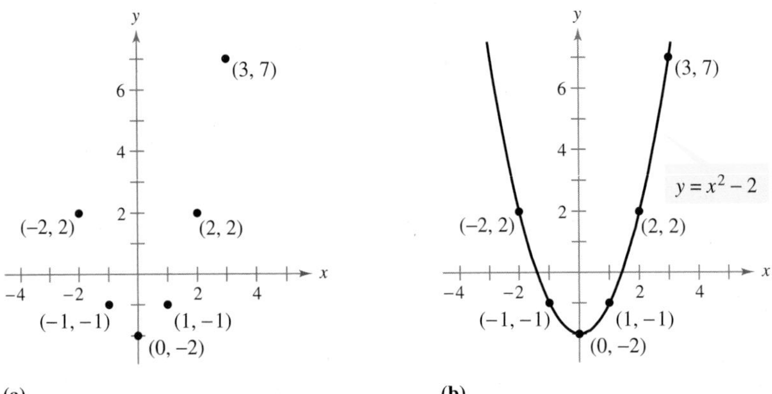

Figure P.26

The point-plotting method demonstrated in Examples 2 and 3 is easy to use, but it has some shortcomings. With too few solution points, you can misrepresent the graph of an equation. For instance, if only the four points

$$(-2, 2), (-1, -1), (1, -1), \text{ and } (2, 2)$$

in Figure P.26(a) were plotted, any one of the three graphs in Figure P.27 would be reasonable.

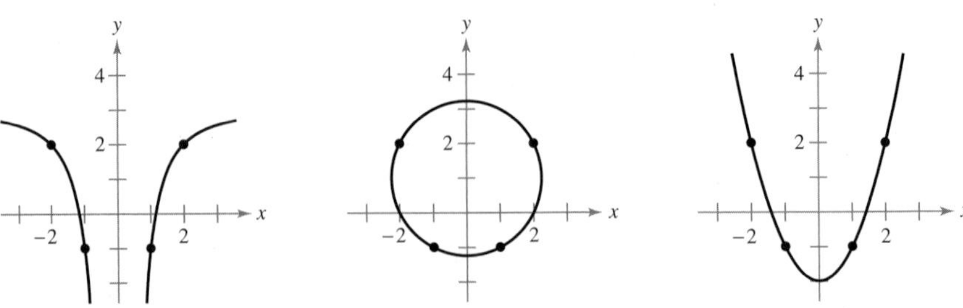

Figure P.27

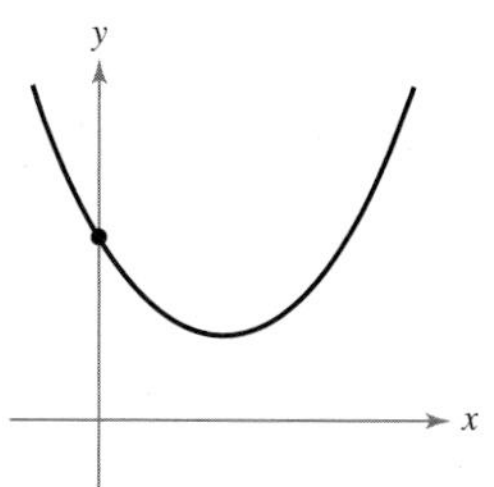

No x-intercepts; one y-intercept

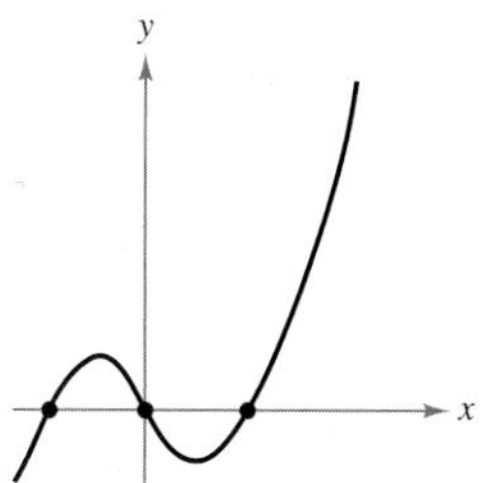

Three x-intercepts; one y-intercept

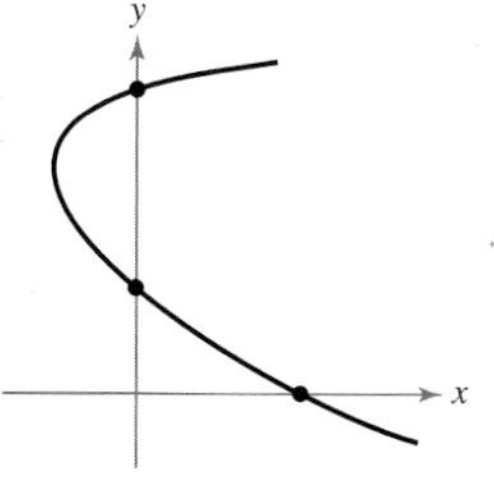

One x-intercept; two y-intercepts

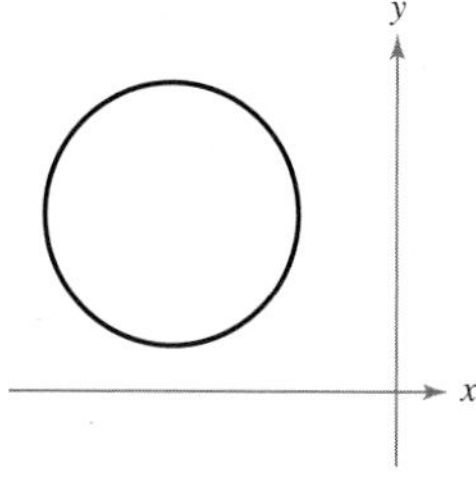

No intercepts

Figure P.28

TECHNOLOGY To graph an equation involving x and y on a graphing utility, use the following procedure.

1. Rewrite the equation so that y is isolated on the left side.
2. Enter the equation into the graphing utility.
3. Determine a *viewing window* that shows all important features of the graph.
4. Graph the equation.

Intercepts of a Graph

It is often easy to determine the solution points that have zero as either the x-coordinate or the y-coordinate. These points are called **intercepts** because they are the points at which the graph intersects or touches the x- or y-axis. It is possible for a graph to have no intercepts, one intercept, or several intercepts, as shown in Figure P.28.

Note that an x-intercept can be written as the ordered pair $(x, 0)$ and a y-intercept can be written as the ordered pair $(0, y)$. Some texts denote the x-intercept as the x-coordinate of the point $(a, 0)$ [and the y-intercept as the y-coordinate of the point $(0, b)$] rather than the point itself. Unless it is necessary to make a distinction, we will use the term *intercept* to mean either the point or the coordinate.

FINDING INTERCEPTS

1. To find x-intercepts, let y be zero and solve the equation for x.
2. To find y-intercepts, let x be zero and solve the equation for y.

EXAMPLE 4 Finding x- and y-Intercepts

Find the x- and y-intercepts of the graph of $y = x^3 - 4x$.

Solution Let $y = 0$. Then

$$0 = x^3 - 4x = x(x^2 - 4)$$

has solutions $x = 0$ and $x = \pm 2$.

$$x\text{-intercepts: } (0, 0), (2, 0), (-2, 0)$$

Let $x = 0$. Then

$$y = (0)^3 - 4(0)$$

has one solution, $y = 0$.

$$y\text{-intercept: } (0, 0)$$ See Figure P.29.

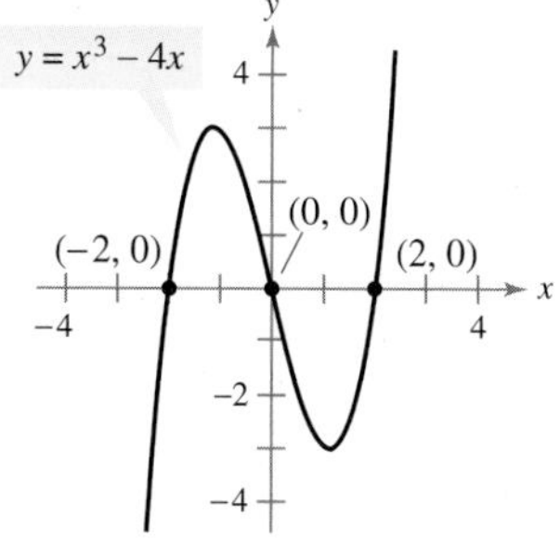

Figure P.29

■

Symmetry

Graphs of equations can have **symmetry** with respect to one of the coordinate axes or with respect to the origin. Symmetry with respect to the x-axis means that if the Cartesian plane were folded along the x-axis, the portion of the graph above the x-axis would coincide with the portion below the x-axis. Symmetry with respect to the y-axis or the origin can be described in a similar manner, as shown in Figure P.30.

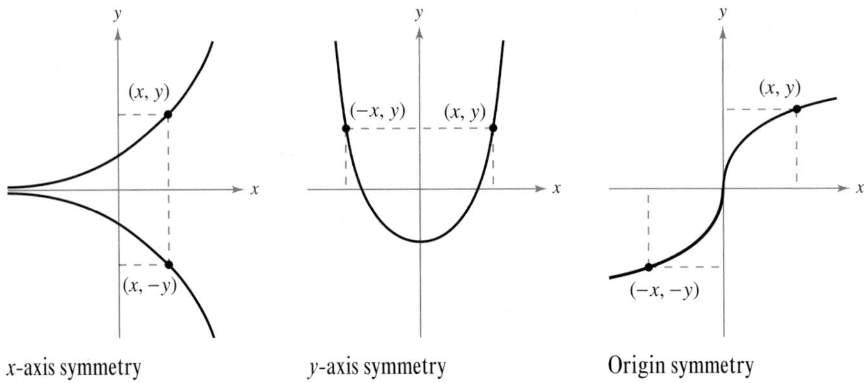

Figure P.30

Knowing the symmetry of a graph *before* attempting to sketch it is helpful, because then you need only half as many solution points to sketch the graph. There are three basic types of symmetry, described as follows.

GRAPHICAL TESTS FOR SYMMETRY

1. A graph is **symmetric with respect to the x-axis** if, whenever (x, y) is on the graph, $(x, -y)$ is also on the graph.
2. A graph is **symmetric with respect to the y-axis** if, whenever (x, y) is on the graph, $(-x, y)$ is also on the graph.
3. A graph is **symmetric with respect to the origin** if, whenever (x, y) is on the graph, $(-x, -y)$ is also on the graph.

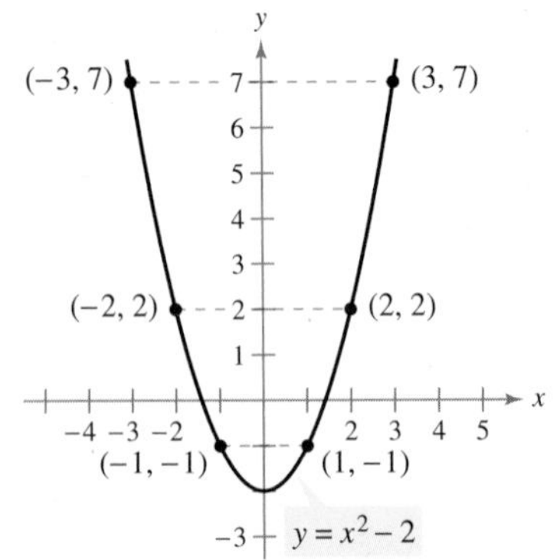

y-axis symmetry
Figure P.31

You can conclude that the graph of $y = x^2 - 2$ is symmetric with respect to the y-axis because the point $(-x, y)$ is also on the graph of $y = x^2 - 2$. (See the table below and Figure P.31.)

x	-3	-2	-1	1	2	3
y	7	2	-1	-1	2	7
(x, y)	$(-3, 7)$	$(-2, 2)$	$(-1, -1)$	$(1, -1)$	$(2, 2)$	$(3, 7)$

ALGEBRAIC TESTS FOR SYMMETRY

1. The graph of an equation is symmetric with respect to the x-axis if replacing y with $-y$ yields an equivalent equation.
2. The graph of an equation is symmetric with respect to the y-axis if replacing x with $-x$ yields an equivalent equation.
3. The graph of an equation is symmetric with respect to the origin if replacing x with $-x$ and y with $-y$ yields an equivalent equation.

EXAMPLE 5 Testing for Symmetry

Test $y = 2x^3$ for symmetry with respect to both axes and the origin.

Solution

x-axis:	$y = 2x^3$	Write original equation.
	$-y = 2x^3$	Replace y with $-y$. Result is *not* an equivalent equation.
y-axis:	$y = 2x^3$	Write original equation.
	$y = 2(-x)^3$	Replace x with $-x$.
	$y = -2x^3$	Simplify. Result is *not* an equivalent equation.
Origin:	$y = 2x^3$	Write original equation.
	$-y = 2(-x)^3$	Replace y with $-y$ and x with $-x$.
	$-y = -2x^3$	Simplify.
	$y = 2x^3$	Equivalent equation

Of the three tests for symmetry, the only one that is satisfied is the test for origin symmetry (see Figure P.32).

Figure P.32

EXAMPLE 6 Using Symmetry as a Sketching Aid

Use symmetry to sketch the graph of $x - y^2 = 1$.

Solution Of the three tests for symmetry, the only one that is satisfied is the test for x-axis symmetry because $x - (-y)^2 = 1$ is equivalent to $x - y^2 = 1$. So, the graph is symmetric with respect to the x-axis. Using symmetry, you only need to find the solution points above the x-axis and then reflect them to obtain the graph, as shown in Figure P.33.

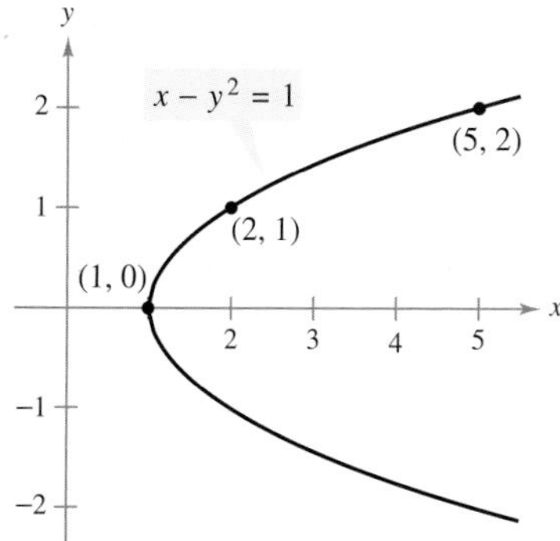

Figure P.33

y	0	1	2
$x = y^2 + 1$	1	2	5
(x, y)	(1, 0)	(2, 1)	(5, 2)

EXAMPLE 7 Sketching the Graph of an Equation

Sketch the graph of $y = |x - 1|$.

Solution This equation fails all three tests for symmetry and consequently its graph is not symmetric with respect to either axis or to the origin. The absolute value sign indicates that y is always nonnegative. Create a table of values and plot the points, as shown in Figure P.34. From the table, you can see that $x = 0$ when $y = 1$. So, the y-intercept is $(0, 1)$. Similarly, $y = 0$ when $x = 1$. So, the x-intercept is $(1, 0)$.

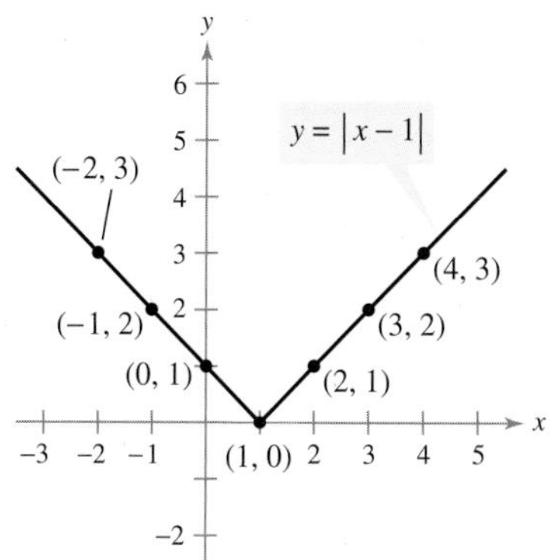

Figure P.34

x	−2	−1	0	1	2	3	4
$y = \|x - 1\|$	3	2	1	0	1	2	3
(x, y)	(−2, 3)	(−1, 2)	(0, 1)	(1, 0)	(2, 1)	(3, 2)	(4, 3)

■

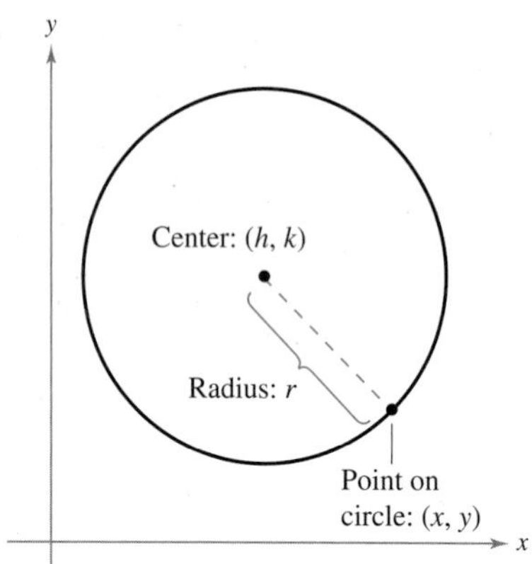

The standard form of the equation of a circle is $(x - h)^2 + (y - k)^2 = r^2$.
Figure P.35

Throughout this course, you will learn to recognize several types of graphs from their equations. For instance, you will learn to recognize that the graph of a second-degree equation of the form

$$y = ax^2 + bx + c, \quad a \neq 0$$

is a parabola (see Example 3). The graph of a **circle** is also easy to recognize.

Circles

Consider the circle shown in Figure P.35. A point (x, y) is on the circle if and only if its distance from the center (h, k) is r. By the Distance Formula,

$$\sqrt{(x - h)^2 + (y - k)^2} = r.$$

By squaring each side of this equation, you obtain the **standard form of the equation of a circle.**

STANDARD FORM OF THE EQUATION OF A CIRCLE

The point (x, y) lies on the circle of **radius** r and **center** (h, k) if and only if

$$(x - h)^2 + (y - k)^2 = r^2.$$

From this result, you can see that the standard form of the equation of a *circle with its center at the origin,* $(h, k) = (0, 0)$, is simply

$$x^2 + y^2 = r^2. \qquad \text{Circle with center at origin}$$

EXAMPLE 8 Finding the Equation of a Circle

The point $(3, 4)$ lies on a circle whose center is at $(-1, 2)$, as shown in Figure P.36. Write the standard form of the equation of this circle.

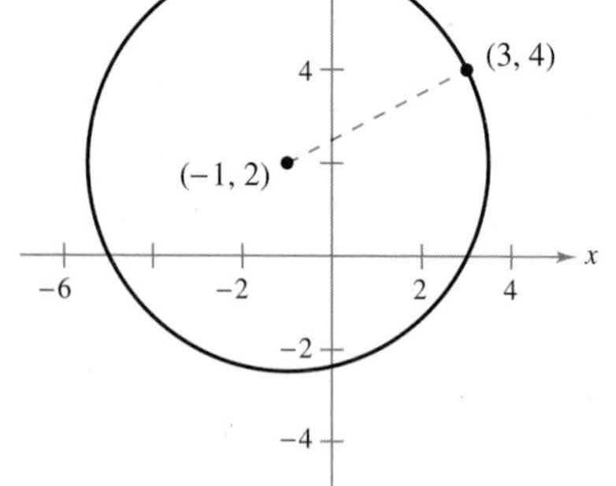

Figure P.36

Solution The radius of the circle is the distance between $(-1, 2)$ and $(3, 4)$.

$$\begin{aligned} r &= \sqrt{(x - h)^2 + (y - k)^2} && \text{Distance Formula} \\ &= \sqrt{[3 - (-1)]^2 + (4 - 2)^2} && \text{Substitute for } x, y, h, \text{ and } k. \\ &= \sqrt{4^2 + 2^2} && \text{Simplify.} \\ &= \sqrt{16 + 4} && \text{Simplify.} \\ &= \sqrt{20} && \text{Radius} \end{aligned}$$

Using $(h, k) = (-1, 2)$ and $r = \sqrt{20}$, the equation of the circle is

$$\begin{aligned} (x - h)^2 + (y - k)^2 &= r^2 && \text{Equation of circle} \\ [x - (-1)]^2 + (y - 2)^2 &= \left(\sqrt{20}\right)^2 && \text{Substitute for } h, k, \text{ and } r. \\ (x + 1)^2 + (y - 2)^2 &= 20. && \text{Standard form} \end{aligned}$$

■

NOTE In Example 8, to find the correct h and k from the equation of the circle, it may be helpful to rewrite the quantities $(x + 1)^2$ and $(y - 2)^2$ using subtraction.

$$(x + 1)^2 = [x - (-1)]^2,$$
$$(y - 2)^2 = [y - (2)]^2$$

So, $h = -1$ and $k = 2$.

■

Application

STUDY TIP You should develop the habit of using at least two approaches to solve every problem. This helps build your intuition and helps you check that your answers are reasonable.

In this course, you will learn that there are many ways to approach a problem. Three common approaches are illustrated in Example 9.

A *Numerical Approach:* Construct and use a table.

A *Graphical Approach:* Draw and use a graph.

An *Algebraic Approach:* Use the rules of algebra.

EXAMPLE 9 Recommended Weight

The median recommended weight y (in pounds) for men of medium frame who are 25 to 59 years old can be approximated by the mathematical model

$$y = 0.073x^2 - 6.99x + 289.0, \quad 62 \le x \le 76$$

where x is the man's height (in inches). *(Source: Metropolitan Life Insurance Company)*

a. Construct a table of values that shows the median recommended weights for men with heights of 62, 64, 66, 68, 70, 72, 74, and 76 inches. Then use the table to estimate *numerically* the median recommended weight for a man whose height is 71 inches.

b. Use the table of values to sketch a graph of the model. Then use the graph to estimate *graphically* the median recommended weight for a man whose height is 71 inches.

c. Use the model to confirm *algebraically* the estimates you found in parts (a) and (b).

Solution

a. You can use a calculator to complete the table, as shown below.

x	62	64	66	68	70	72	74	76
y	136.2	140.6	145.6	151.2	157.4	164.2	171.5	179.4

When $x = 71$, $y \approx 161$.

When $x = 71$, you can estimate that $y \approx 161$ pounds.

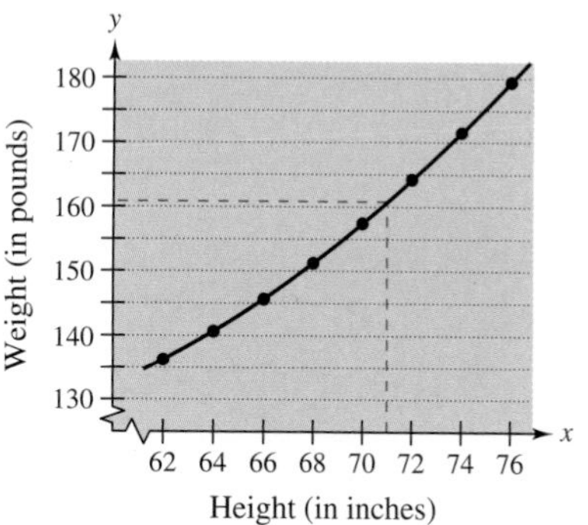

Figure P.37

b. The table of values can be used to sketch the graph of the equation, as shown in Figure P.37. From the graph, you can estimate that a height of 71 inches corresponds to a weight of about 161 pounds.

c. To confirm algebraically the estimate found in parts (a) and (b), you can substitute 71 for x in the model.

$$y = 0.073x^2 - 6.99x + 289.0 \qquad \text{Write original model.}$$
$$= 0.073(71)^2 - 6.99(71) + 289.0 \qquad \text{Substitute 71 for } x.$$
$$\approx 160.70 \qquad \text{Use a calculator.}$$

So, the estimate of 161 pounds is fairly good. ■

P.4 Exercises

See www.CalcChat.com for worked-out solutions to odd-numbered exercises.

In Exercises 1–6, fill in the blanks.

1. An ordered pair (a, b) is a ________ of an equation in x and y if the equation is true when a is substituted for x, and b is substituted for y.
2. The set of all solution points of an equation is the ________ of the equation.
3. The points at which a graph intersects or touches an axis are called the ________ of the graph.
4. A graph is symmetric with respect to the ________ if, whenever (x, y) is on the graph, $(-x, y)$ is also on the graph.
5. The equation $(x - h)^2 + (y - k)^2 = r^2$ is the standard form of the equation of a ________ with center ________ and radius ________.
6. When you construct and use a table to solve a problem, you are using a ________ approach.

In Exercises 7–14, determine whether each point lies on the graph of the equation.

	Equation	*Points*	
7.	$y = \sqrt{x + 4}$	(a) $(0, 2)$	(b) $(5, 3)$
8.	$y = \sqrt{5 - x}$	(a) $(1, 2)$	(b) $(5, 0)$
9.	$y = x^2 - 3x + 2$	(a) $(2, 0)$	(b) $(-2, 8)$
10.	$y = 4 - \|x - 2\|$	(a) $(1, 5)$	(b) $(6, 0)$
11.	$y = \|x - 1\| + 2$	(a) $(2, 3)$	(b) $(-1, 0)$
12.	$2x - y - 3 = 0$	(a) $(1, 2)$	(b) $(1, -1)$
13.	$x^2 + y^2 = 20$	(a) $(3, -2)$	(b) $(-4, 2)$
14.	$y = \frac{1}{3}x^3 - 2x^2$	(a) $\left(2, -\frac{16}{3}\right)$	(b) $(-3, 9)$

In Exercises 15–18, complete the table. Use the resulting solution points to sketch the graph of the equation.

15. $y = -2x + 5$

x	-1	0	1	2	$\frac{5}{2}$
y					
(x, y)					

16. $y = \frac{3}{4}x - 1$

x	-2	0	1	$\frac{4}{3}$	2
y					
(x, y)					

17. $y = x^2 - 3x$

x	-1	0	1	2	3
y					
(x, y)					

18. $y = 5 - x^2$

x	-2	-1	0	1	2
y					
(x, y)					

In Exercises 19–22, graphically estimate the x- and y-intercepts of the graph. Verify your results algebraically.

19. $y = (x - 3)^2$

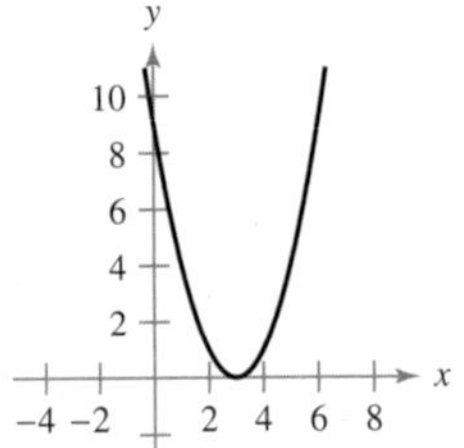

20. $y = 16 - 4x^2$

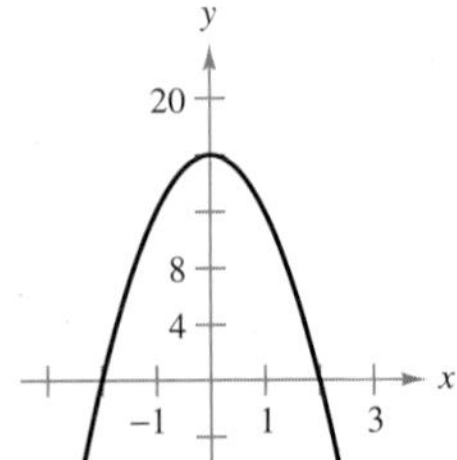

21. $y = |x + 2|$

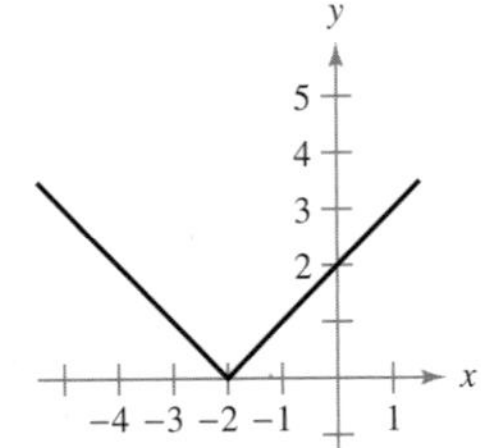

22. $y^2 = 4 - x$

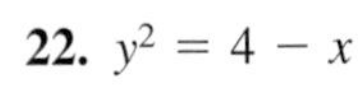

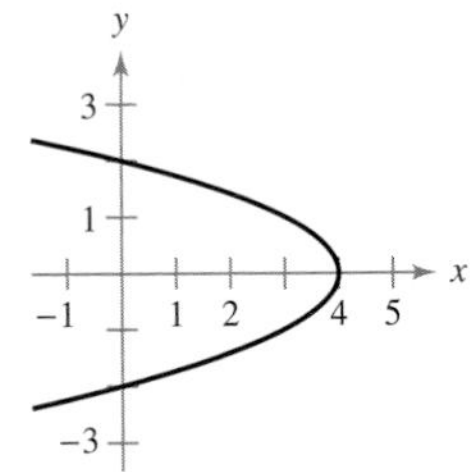

In Exercises 23–32, find the x- and y-intercepts of the graph of the equation.

23. $y = 5x - 6$
24. $y = 8 - 3x$
25. $y = \sqrt{x + 4}$
26. $y = \sqrt{2x - 1}$
27. $y = |3x - 7|$
28. $y = -|x + 10|$
29. $y = 2x^3 - 4x^2$
30. $y = x^4 - 25$
31. $y^2 = 6 - x$
32. $y^2 = x + 1$

In Exercises 33–40, use the algebraic tests to check for symmetry with respect to both axes and the origin.

33. $x^2 - y = 0$

34. $x - y^2 = 0$

35. $y = x^3$

36. $y = x^4 - x^2 + 3$

37. $y = \dfrac{x}{x^2 + 1}$

38. $y = \dfrac{1}{x^2 + 1}$

39. $xy^2 + 10 = 0$

40. $xy = 4$

In Exercises 41–52, identify any intercepts and test for symmetry. Then sketch the graph of the equation.

41. $y = -3x + 1$

42. $y = 2x - 3$

43. $y = x^2 - 2x$

44. $y = -x^2 - 2x$

45. $y = x^3 + 3$

46. $y = x^3 - 1$

47. $y = \sqrt{x - 3}$

48. $y = \sqrt{1 - x}$

49. $y = |x - 6|$

50. $y = 1 - |x|$

51. $x = y^2 - 1$

52. $x = y^2 - 5$

In Exercises 53–64, use a graphing utility to graph the equation. Use a standard setting. Approximate any intercepts.

53. $y = 3 - \frac{1}{2}x$

54. $y = \frac{2}{3}x - 1$

55. $y = x^2 - 4x + 3$

56. $y = x^2 + x - 2$

57. $y = \dfrac{2x}{x - 1}$

58. $y = \dfrac{4}{x^2 + 1}$

59. $y = \sqrt[3]{x} + 2$

60. $y = \sqrt[3]{x + 1}$

61. $y = x\sqrt{x + 6}$

62. $y = (6 - x)\sqrt{x}$

63. $y = |x + 3|$

64. $y = 2 - |x|$

In Exercises 65–72, write the standard form of the equation of the circle with the given characteristics.

65. Center: $(0, 0)$; Radius: 4

66. Center: $(0, 0)$; Radius: 5

67. Center: $(2, -1)$; Radius: 4

68. Center: $(-7, -4)$; Radius: 7

69. Center: $(-1, 2)$; Solution point: $(0, 0)$

70. Center: $(3, -2)$; Solution point: $(-1, 1)$

71. Endpoints of a diameter: $(0, 0)$, $(6, 8)$

72. Endpoints of a diameter: $(-4, -1)$, $(4, 1)$

In Exercises 73–78, find the center and radius of the circle, and sketch its graph.

73. $x^2 + y^2 = 25$

74. $x^2 + y^2 = 16$

75. $(x - 1)^2 + (y + 3)^2 = 9$

76. $x^2 + (y - 1)^2 = 1$

77. $\left(x - \frac{1}{2}\right)^2 + \left(y - \frac{1}{2}\right)^2 = \frac{9}{4}$

78. $(x - 2)^2 + (y + 3)^2 = \frac{16}{9}$

WRITING ABOUT CONCEPTS

In Exercises 79–82, assume that the graph has the indicated type of symmetry. Sketch the complete graph of the equation. To print an enlarged copy of the graph, go to the website *www.mathgraphs.com*.

79.

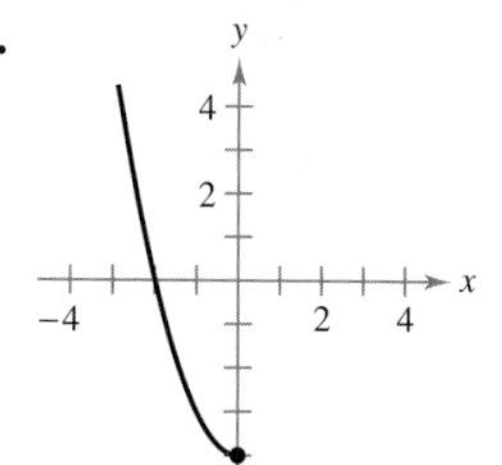

y-Axis symmetry

80.

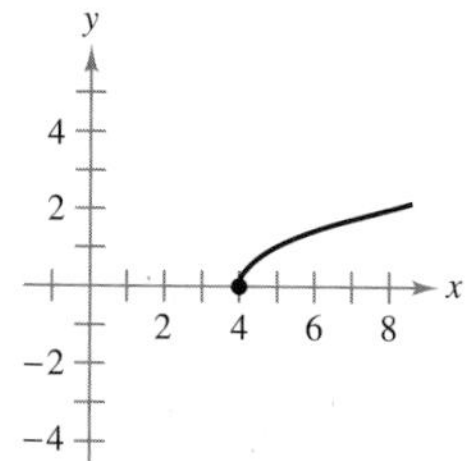

x-Axis symmetry

81.

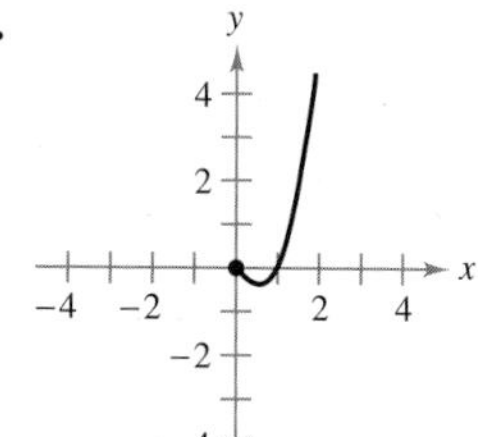

Origin symmetry

82.

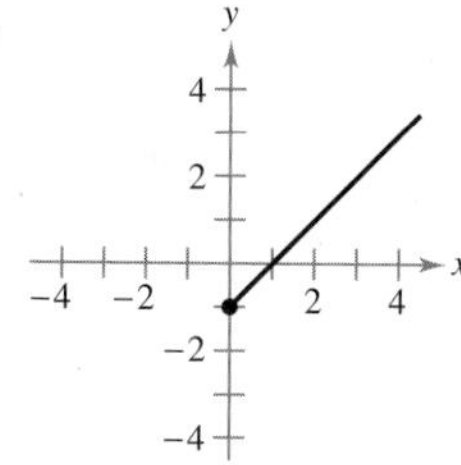

y-Axis symmetry

In Exercises 83 and 84, write an equation whose graph has the given property. (There is more than one correct answer.)

83. The graph has intercepts at $x = -2$, $x = 4$, and $x = 6$.

84. The graph has intercepts at $x = -\frac{5}{2}$, $x = 2$, and $x = \frac{3}{2}$.

85. ***Geometry*** A regulation NFL playing field (including the end zones) of length x and width y has a perimeter of $346\frac{2}{3}$ or $\frac{1040}{3}$ yards.

(a) Draw a rectangle that gives a visual representation of the problem. Use the specified variables to label the sides of the rectangle.

(b) Show that the width of the rectangle is

$$y = \frac{520}{3} - x \text{ and its area is } A = x\left(\frac{520}{3} - x\right).$$

(c) Use a graphing utility to graph the area equation. Be sure to adjust your window settings.

(d) From the graph in part (c), estimate the dimensions of the rectangle that yield a maximum area.

(e) Use your school's library, the Internet, or some other reference source to find the actual dimensions and area of a regulation NFL playing field and compare your findings with the results of part (d).

86. *Geometry* A soccer playing field of length x and width y has a perimeter of 360 meters.

(a) Draw a rectangle that gives a visual representation of the problem. Use the specified variables to label the sides of the rectangle.

(b) Show that the width of the rectangle is $y = 180 - x$ and its area is $A = x(180 - x)$.

(c) Use a graphing utility to graph the area equation. Be sure to adjust your window settings.

(d) From the graph in part (c), estimate the dimensions of the rectangle that yield a maximum area.

(e) Use your school's library, the Internet, or some other reference source to find the actual dimensions and area of a regulation Major League Soccer field and compare your findings with the results of part (d).

87. *Population Statistics* The table shows the life expectancies of a child (at birth) in the United States for selected years from 1920 to 2000. *(Source: U.S. National Center for Health Statistics)*

Year	1920	1930	1940	1950	1960
Life Expectancy, y	54.1	59.7	62.9	68.2	69.7

Year	1970	1980	1990	2000
Life Expectancy, y	70.8	73.7	75.4	77.0

A model for the life expectancy during this period is

$$y = -0.0025t^2 + 0.574t + 44.25, \quad 20 \le t \le 100$$

where y represents the life expectancy and t is the time in years, with $t = 20$ corresponding to 1920.

(a) Use a graphing utility to graph the data from the table and the model in the same viewing window. How well does the model fit the data? Explain.

(b) Determine the life expectancy in 1990 both graphically and algebraically.

(c) Use the graph to determine the year when life expectancy was approximately 76.0. Verify your answer algebraically.

(d) One projection for the life expectancy of a child born in 2015 is 78.9. How does this compare with the projection given by the model?

(e) Do you think this model can be used to predict the life expectancy of a child 50 years from now? Explain.

88. *Electronics* The resistance y (in ohms) of 1000 feet of solid copper wire at 68 degrees Fahrenheit can be approximated by the model

$$y = \frac{10{,}770}{x^2} - 0.37, \quad 5 \le x \le 100$$

where x is the diameter of the wire in mils (0.001 inch). *(Source: American Wire Gage)*

(a) Complete the table.

x	5	10	20	30	40	50
y						

x	60	70	80	90	100
y					

(b) Use the table of values in part (a) to sketch a graph of the model. Then use your graph to estimate the resistance when $x = 85.5$.

(c) Use the model to confirm algebraically the estimate you found in part (b).

(d) What can you conclude in general about the relationship between the diameter of the copper wire and the resistance?

89. *Think About It* Find a and b if the graph of $y = ax^2 + bx^3$ is symmetric with respect to (a) the y-axis and (b) the origin. (There are many correct answers.)

CAPSTONE

90. Match the equation or equations with the given characteristic.

(i) $y = 3x^3 - 3x$ (ii) $y = (x + 3)^2$

(iii) $y = 3x - 3$ (iv) $y = \sqrt[3]{x}$

(v) $y = 3x^2 + 3$ (vi) $y = \sqrt{x + 3}$

(a) Symmetric with respect to the y-axis

(b) Three x-intercepts

(c) Symmetric with respect to the x-axis

(d) $(-2, 1)$ is a point on the graph

(e) Symmetric with respect to the origin

(f) Graph passes through the origin

91. *Writing* In your own words, explain how the display of a graphing utility changes if the maximum setting for x is changed from 10 to 20.

P.5 Linear Equations in Two Variables

- **Use slope to graph linear equations in two variables.**
- **Find the slope of a line given two points on the line.**
- **Write linear equations in two variables.**
- **Use slope to identify parallel and perpendicular lines.**
- **Use slope and linear equations in two variables to model and solve real-life problems.**

Using Slope

The simplest mathematical model for relating two variables is the **linear equation in two variables** $y = mx + b$. The equation is called *linear* because its graph is a line. (In mathematics, the term *line* means *straight line.*) By letting $x = 0$, you obtain

$$\begin{aligned} y &= m(0) + b && \text{Substitute 0 for } x. \\ &= b. \end{aligned}$$

So, the line crosses the y-axis at $y = b$, as shown in Figure P.38. In other words, the y-intercept is $(0, b)$. The steepness or slope of the line is m.

$$y = mx + b$$

Slope (points to m) — y-Intercept (points to b)

The **slope** of a nonvertical line is the number of units the line rises (or falls) vertically for each unit of horizontal change from left to right, as shown in Figure P.38.

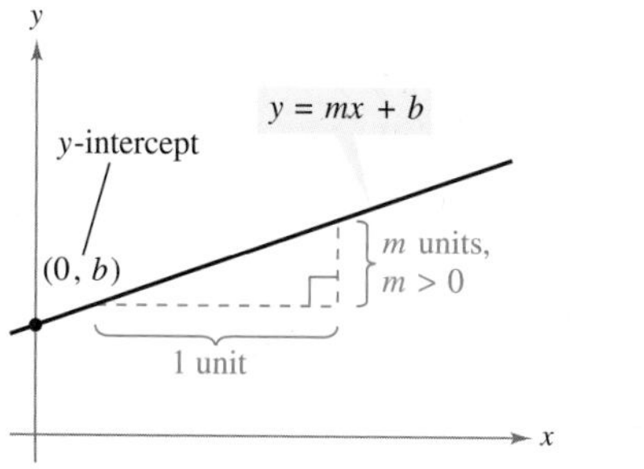

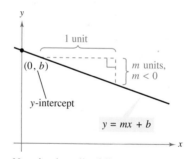

Positive slope, line rises. Negative slope, line falls.
Figure P.38

A linear equation that is written in the form $y = mx + b$ is said to be written in **slope-intercept form.**

> **THE SLOPE-INTERCEPT FORM OF THE EQUATION OF A LINE**
>
> The graph of the equation
>
> $$y = mx + b$$
>
> is a line whose slope is m and whose y-intercept is $(0, b)$.

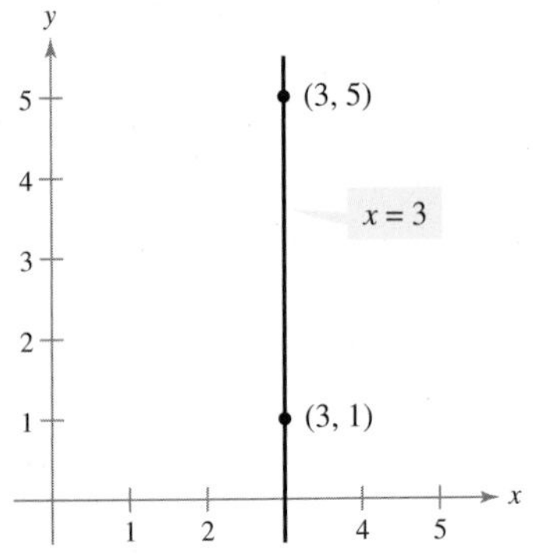

Figure P.39 Slope is undefined.

Once you have determined the slope and the y-intercept of a line, it is a relatively simple matter to sketch its graph. In the next example, note that none of the lines is vertical. A vertical line has an equation of the form

$$x = a. \qquad \text{Vertical line}$$

The equation of a vertical line cannot be written in the form $y = mx + b$ because the slope of a vertical line is undefined, as indicated in Figure P.39. Later in this section you will see that the undefined slope of a vertical line derives algebraically from division by zero.

EXAMPLE 1 Graphing a Linear Equation

Sketch the graph of each linear equation.

a. $y = 2x + 1$

b. $y = 2$

c. $x + y = 2$

Solution

a. Because $b = 1$, the y-intercept is $(0, 1)$. Moreover, because the slope is $m = 2$, the line rises two units for each unit the line moves to the right, as shown in Figure P.40(a).

b. By writing this equation in the form $y = (0)x + 2$, you can see that the y-intercept is $(0, 2)$ and the slope is zero. A zero slope implies that the line is horizontal—that is, it doesn't rise or fall, as shown in Figure P.40(b).

c. By writing this equation in slope-intercept form

$$x + y = 2 \qquad \text{Write original equation.}$$
$$y = -x + 2 \qquad \text{Subtract } x \text{ from each side.}$$
$$y = (-1)x + 2 \qquad \text{Write in slope-intercept form.}$$

you can see that the y-intercept is $(0, 2)$. Moreover, because the slope is $m = -1$, the line *falls* one unit for each unit the line moves to the right, as shown in Figure P.40(c).

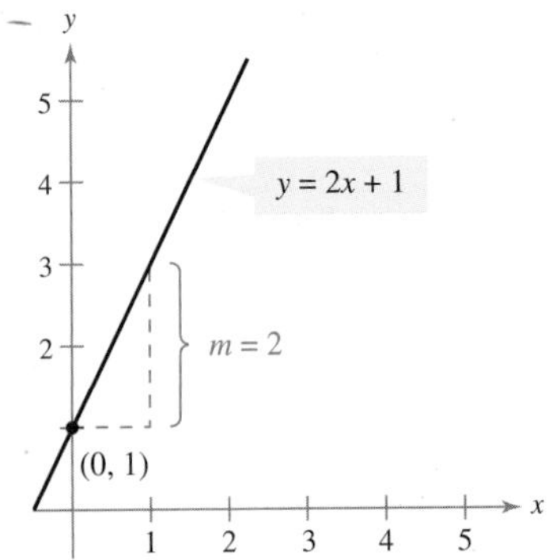

(a) When m is positive, the line rises.

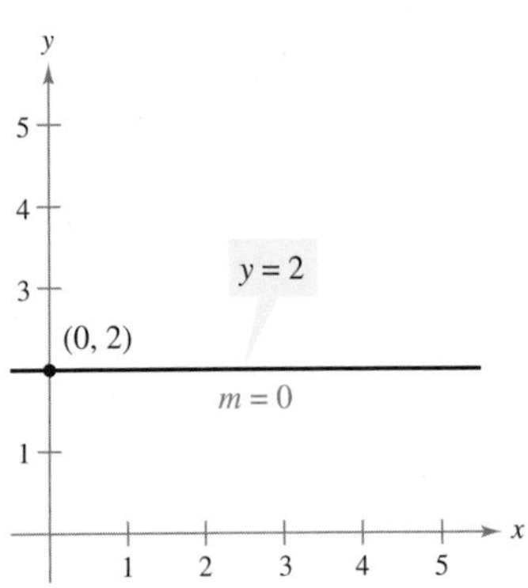

(b) When m is 0, the line is horizontal.

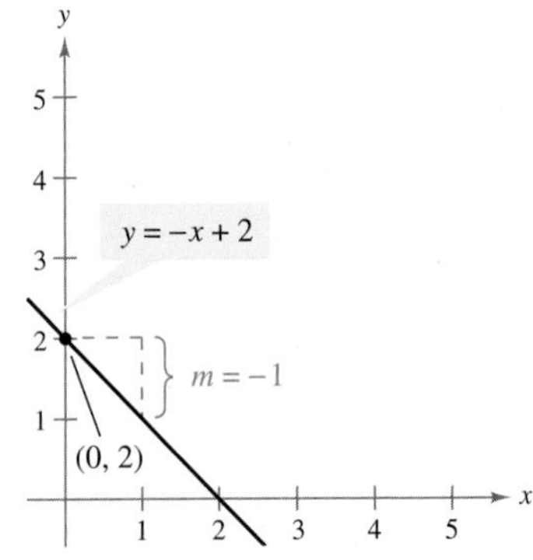

(c) When m is negative, the line falls.

Figure P.40

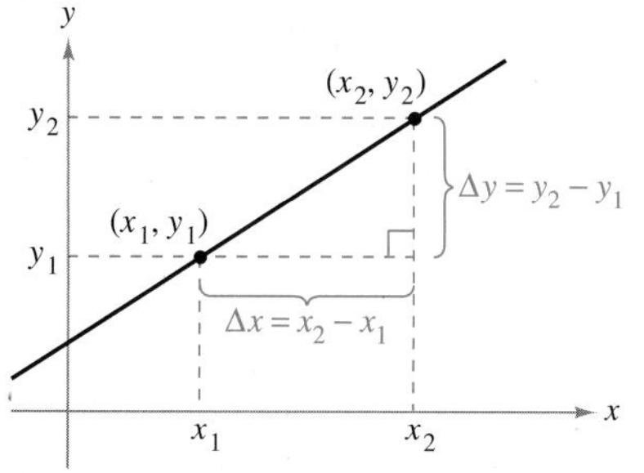

Figure P.41

Finding the Slope of a Line

Given an equation of a nonvertical line, you can find its slope by writing the equation in slope-intercept form. If you are not given an equation, you can still find the slope of a line. For instance, suppose you want to find the slope of the line passing through the points (x_1, y_1) and (x_2, y_2), as shown in Figure P.41. As you move from left to right along this line, a change of $(y_2 - y_1)$ units in the vertical direction corresponds to a change of $(x_2 - x_1)$ units in the horizontal direction.

$$\begin{aligned}\Delta y &= y_2 - y_1\\ &= \text{the change in } y = \text{rise}\end{aligned}$$

and

$$\begin{aligned}\Delta x &= x_2 - x_1\\ &= \text{the change in } x = \text{run}\end{aligned}$$

The ratio of $(y_2 - y_1)$ to $(x_2 - x_1)$ represents the slope of the line that passes through the points (x_1, y_1) and (x_2, y_2).

$$\begin{aligned}\text{Slope} &= \frac{\text{change in } y}{\text{change in } x}\\ &= \frac{\Delta y}{\Delta x}\\ &= \frac{y_2 - y_1}{x_2 - x_1}\end{aligned}$$

NOTE The symbol Δ is the Greek letter *delta*, and the symbols Δy and Δx are read "delta y" and "delta x." This notation is used frequently in calculus.

THE SLOPE OF A LINE PASSING THROUGH TWO POINTS

The **slope** m of the nonvertical line through (x_1, y_1) and (x_2, y_2) is

$$m = \frac{\Delta y}{\Delta x} = \frac{y_2 - y_1}{x_2 - x_1}$$

where $x_1 \neq x_2$.

When this formula is used for slope, the *order of subtraction* is important. Given two points on a line, you are free to label either one of them as (x_1, y_1) and the other as (x_2, y_2). However, once you have done this, you must form the numerator and denominator using the same order of subtraction.

$$\underbrace{m = \frac{y_2 - y_1}{x_2 - x_1}}_{\text{Correct}} \qquad \underbrace{m = \frac{y_1 - y_2}{x_1 - x_2}}_{\text{Correct}} \qquad \underbrace{\cancel{m = \frac{y_2 - y_1}{x_1 - x_2}}}_{\text{Incorrect}}$$

For instance, the slope of the line passing through the points $(3, 4)$ and $(5, 7)$ can be calculated as

$$m = \frac{7 - 4}{5 - 3} = \frac{3}{2}$$

or, reversing the subtraction order in both the numerator and denominator, as

$$m = \frac{4 - 7}{3 - 5} = \frac{-3}{-2} = \frac{3}{2}.$$

EXAMPLE 2 Finding the Slope of a Line Through Two Points

Find the slope of the line passing through each pair of points.

a. $(-2, 0)$ and $(3, 1)$

b. $(-1, 2)$ and $(2, 2)$

c. $(0, 4)$ and $(1, -1)$

d. $(3, 4)$ and $(3, 1)$

Solution

a. Letting $(x_1, y_1) = (-2, 0)$ and $(x_2, y_2) = (3, 1)$, you obtain a slope of

$$m = \frac{y_2 - y_1}{x_2 - x_1} = \frac{1 - 0}{3 - (-2)} = \frac{1}{5}.$$ See Figure P.42(a).

b. The slope of the line passing through $(-1, 2)$ and $(2, 2)$ is

$$m = \frac{2 - 2}{2 - (-1)} = \frac{0}{3} = 0.$$ See Figure P.42(b).

c. The slope of the line passing through $(0, 4)$ and $(1, -1)$ is

$$m = \frac{-1 - 4}{1 - 0} = \frac{-5}{1} = -5.$$ See Figure P.42(c).

d. The slope of the line passing through $(3, 4)$ and $(3, 1)$ is

$$m = \frac{1 - 4}{3 - 3} = \cancel{\frac{-3}{0}}.$$ See Figure P.42(d).

Because division by 0 is undefined, the slope is undefined and the line is vertical.

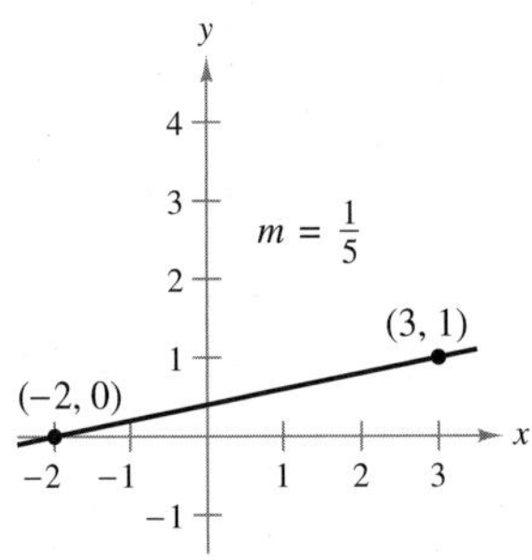

(a)

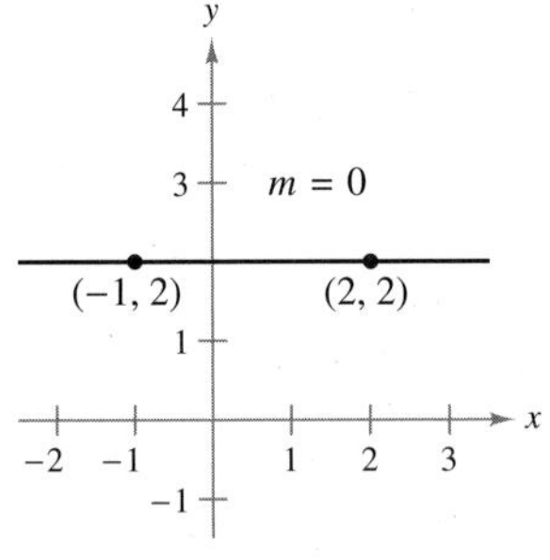

(b)

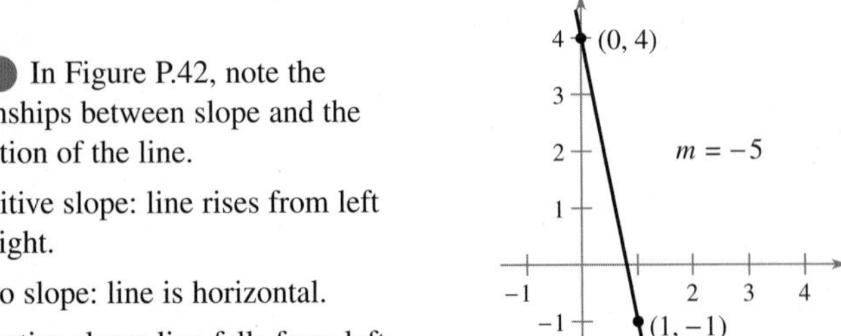

(c)

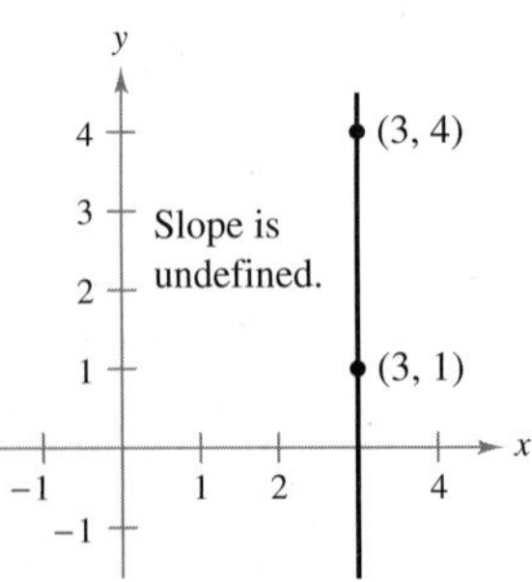

(d)

Figure P.42

NOTE In Figure P.42, note the relationships between slope and the orientation of the line.

a. Positive slope: line rises from left to right.

b. Zero slope: line is horizontal.

c. Negative slope: line falls from left to right.

d. Undefined slope: line is vertical.

Writing Linear Equations in Two Variables

If (x_1, y_1) is a point on a nonvertical line of slope m and (x, y) is *any other* point on the line, then

$$\frac{y - y_1}{x - x_1} = m.$$

This equation, involving the variables x and y, can be rewritten in the form

$$y - y_1 = m(x - x_1)$$

which is the **point-slope form** of the equation of a line.

NOTE Remember that only nonvertical lines have slopes. Consequently, vertical lines cannot be written in point-slope form. For instance, the equation of the vertical line passing through the point $(1, -2)$ is $x = 1$.

POINT-SLOPE FORM OF THE EQUATION OF A LINE

The equation of the nonvertical line with slope m passing through the point (x_1, y_1) is $y - y_1 = m(x - x_1)$.

The point-slope form is most useful for *finding* the equation of a nonvertical line. You should remember this form.

EXAMPLE 3 Using the Point-Slope Form

Find the slope-intercept form of the equation of the line that has a slope of 3 and passes through the point $(1, -2)$.

Solution Use the point-slope form with $m = 3$ and $(x_1, y_1) = (1, -2)$.

$$y - y_1 = m(x - x_1) \quad \text{Point-slope form}$$

$$y - (-2) = 3(x - 1) \quad \text{Substitute for } m, x_1, \text{ and } y_1.$$

$$y + 2 = 3x - 3 \quad \text{Simplify.}$$

$$y = 3x - 5 \quad \text{Write in slope-intercept form.}$$

The slope-intercept form of the equation of the line is $y = 3x - 5$. The graph of this line is shown in Figure P.43. ■

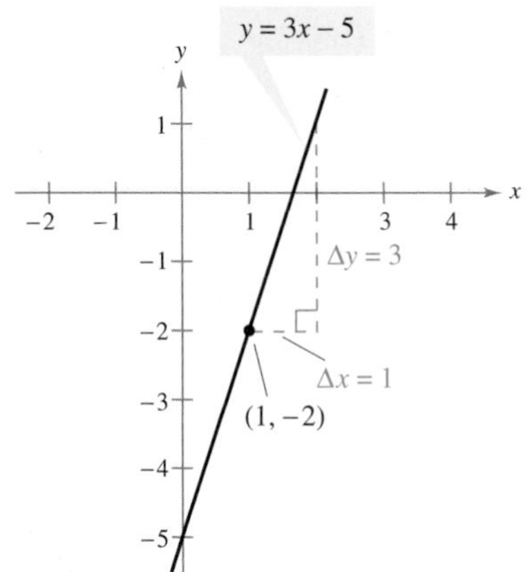

Figure P.43

The point-slope form can be used to find an equation of the nonvertical line passing through two points (x_1, y_1) and (x_2, y_2). To do this, first find the slope of the line

$$m = \frac{y_2 - y_1}{x_2 - x_1}, \quad x_1 \neq x_2$$

and then use the point-slope form to obtain the equation

$$y - y_1 = \frac{y_2 - y_1}{x_2 - x_1}(x - x_1). \quad \text{Two-point form}$$

This is sometimes called the **two-point form** of the equation of a line. Here is an example. The line passing through $(1, 3)$ and $(2, 5)$ is given by

$$y - 3 = \frac{5 - 3}{2 - 1}(x - 1)$$

$$y - 3 = 2(x - 1)$$

$$y = 2x + 1.$$

STUDY TIP When you find an equation of the line that passes through two given points, you only need to substitute the coordinates of one of the points in the point-slope form. It does not matter which point you choose because both points will yield the same result.

Parallel and Perpendicular Lines

Slope can be used to decide whether two nonvertical lines in a plane are parallel, perpendicular, or neither.

> **PARALLEL AND PERPENDICULAR LINES**
>
> 1. Two distinct nonvertical lines are **parallel** if and only if their slopes are equal. That is, $m_1 = m_2$.
> 2. Two nonvertical lines are **perpendicular** if and only if their slopes are negative reciprocals of each other. That is, $m_1 = -1/m_2$.

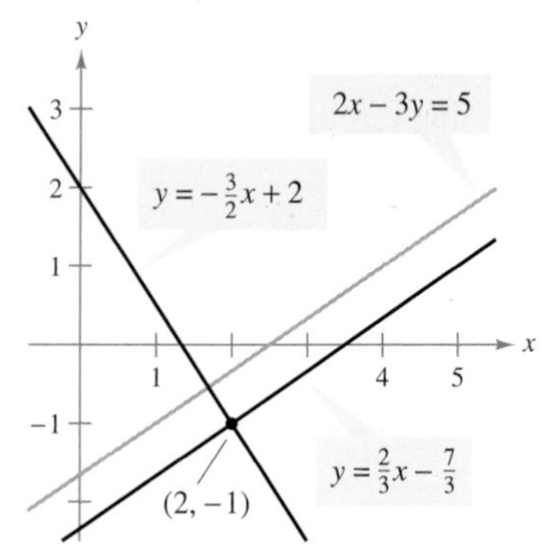

Figure P.44

EXAMPLE 4 Finding Parallel and Perpendicular Lines

Find the slope-intercept forms of the equations of the lines that pass through the point $(2, -1)$ and are (a) parallel to and (b) perpendicular to the line $2x - 3y = 5$.

Solution By writing the equation of the given line in slope-intercept form

$2x - 3y = 5$	Write original equation.
$-3y = -2x + 5$	Subtract $2x$ from each side.
$y = \frac{2}{3}x - \frac{5}{3}$	Write in slope-intercept form.

you can see that it has a slope of $m = \frac{2}{3}$, as shown in Figure P.44.

a. Any line parallel to the given line must also have a slope of $\frac{2}{3}$. So, the line through $(2, -1)$ that is parallel to the given line has the following equation.

$y - (-1) = \frac{2}{3}(x - 2)$	Write in point-slope form.
$3(y + 1) = 2(x - 2)$	Multiply each side by 3.
$3y + 3 = 2x - 4$	Distributive Property
$y = \frac{2}{3}x - \frac{7}{3}$	Write in slope-intercept form.

b. Any line perpendicular to the given line must have a slope of $-\frac{3}{2}$ $\left(\text{because } -\frac{3}{2} \text{ is the negative reciprocal of } \frac{2}{3}\right)$. So, the line through $(2, -1)$ that is perpendicular to the given line has the following equation.

$y - (-1) = -\frac{3}{2}(x - 2)$	Write in point-slope form.
$2(y + 1) = -3(x - 2)$	Multiply each side by 2.
$2y + 2 = -3x + 6$	Distributive Property
$y = -\frac{3}{2}x + 2$	Write in slope-intercept form.

■

Notice in Example 4 how the slope-intercept form is used to obtain information about the graph of a line, whereas the point-slope form is used to write the equation of a line.

TECHNOLOGY On a graphing utility, lines will not appear to have the correct slope unless you use a viewing window that has a square setting. For instance, try graphing the lines in Example 4 using the standard setting $-10 \le x \le 10$ and $-10 \le y \le 10$. Then reset the viewing window with the square setting $-9 \le x \le 9$ and $-6 \le y \le 6$. On which setting do the lines $y = \frac{2}{3}x - \frac{5}{3}$ and $y = -\frac{3}{2}x + 2$ appear to be perpendicular?

Applications

In real-life problems, the slope of a line can be interpreted as either a *ratio* or a *rate*. If the x-axis and y-axis have the same unit of measure, then the slope has no units and is a **ratio.** If the x-axis and y-axis have different units of measure, then the slope is a **rate** or **rate of change.**

EXAMPLE 5 Using Slope as a Ratio

The maximum recommended slope of a wheelchair ramp is $\frac{1}{12}$. A business is installing a wheelchair ramp that rises 22 inches over a horizontal length of 24 feet. Is the ramp steeper than recommended? *(Source: Americans with Disabilities Act Handbook)*

Solution The horizontal length of the ramp is 24 feet or $12(24) = 288$ inches, as shown in Figure P.45. So, the slope of the ramp is

$$\text{Slope} = \frac{\text{vertical change}}{\text{horizontal change}}$$

$$= \frac{22 \text{ in.}}{288 \text{ in.}} \approx 0.076.$$

Because $\frac{1}{12} \approx 0.083$, the slope of the ramp is not steeper than recommended.

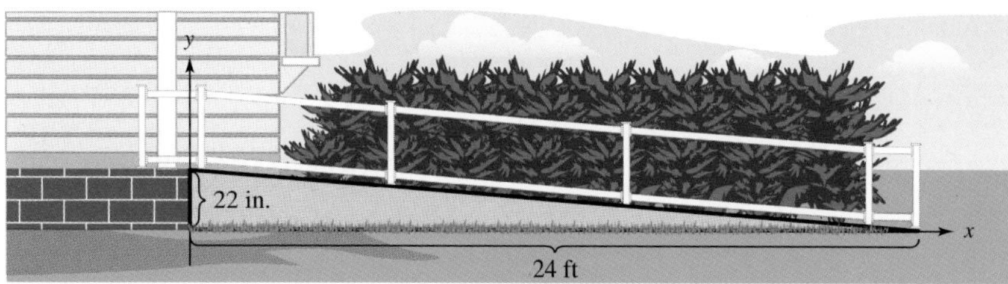

Figure P.45

EXAMPLE 6 Using Slope as a Rate of Change

The population of Kentucky was 3,961,000 in 1999 and 4,314,000 in 2009. Over this 10-year period, the average rate of change of the population was

$$\text{Rate of change} = \frac{\text{change in population}}{\text{change in years}}$$

$$= \frac{4{,}314{,}000 - 3{,}961{,}000}{2009 - 1999}$$

$$= \frac{353{,}000}{10}$$

$$= 35{,}300 \text{ people per year.}$$

If Kentucky's population continues to increase at this same rate for the next 10 years, it will have a 2019 population of 4,667,000 (see Figure P.46). *(Source: U.S. Census Bureau)* ■

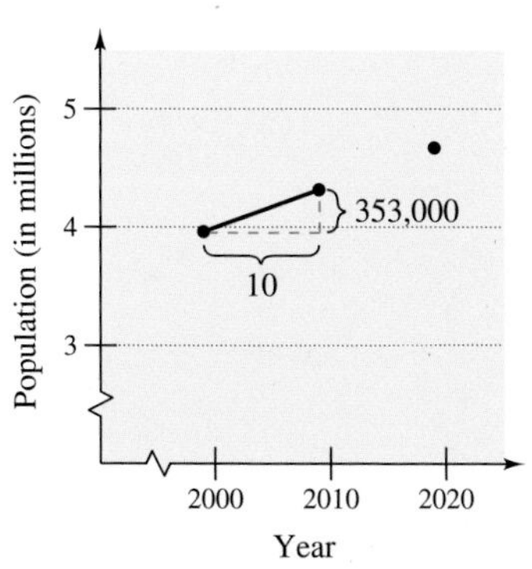

Population of Kentucky in census years
Figure P.46

The rate of change found in Example 6 is an average rate of change. An average rate of change is always calculated over an interval. In this case, the interval is [1999, 2009]. In Chapter 4, you will study another type of rate of change called an *instantaneous rate of change.*

Most business expenses can be deducted in the same year they occur. One exception is the cost of property that has a useful life of more than 1 year. Such costs must be *depreciated* (decreased in value) over the useful life of the property. If the *same amount* is depreciated each year, the procedure is called *linear* or *straight-line depreciation.* The *book value* is the difference between the original value and the total amount of depreciation accumulated to date.

EXAMPLE 7 Straight-Line Depreciation

A college purchased exercise equipment worth \$12,000 for the new campus fitness center. The equipment has a useful life of 8 years. The salvage value at the end of 8 years is \$2000. Write a linear equation that describes the book value of the equipment each year.

Solution Let V represent the value of the equipment at the end of year t. You can represent the initial value of the equipment by the data point (0, 12,000) and the salvage value of the equipment by the data point (8, 2000). The slope of the line is

$$m = \frac{2000 - 12{,}000}{8 - 0} = -\$1250$$

which represents the annual depreciation in *dollars per year.* Using the point-slope form, you can write the equation of the line as follows.

$$V - 12{,}000 = -1250(t - 0) \qquad \text{Write in point-slope form.}$$

$$V = -1250t + 12{,}000 \qquad \text{Write in slope-intercept form.}$$

The table shows the book value at the end of each year, and the graph of the equation is shown in Figure P.47.

Year, t	0	1	2	3	4	5	6	7	8
Value, V	12,000	10,750	9500	8250	7000	5750	4500	3250	2000

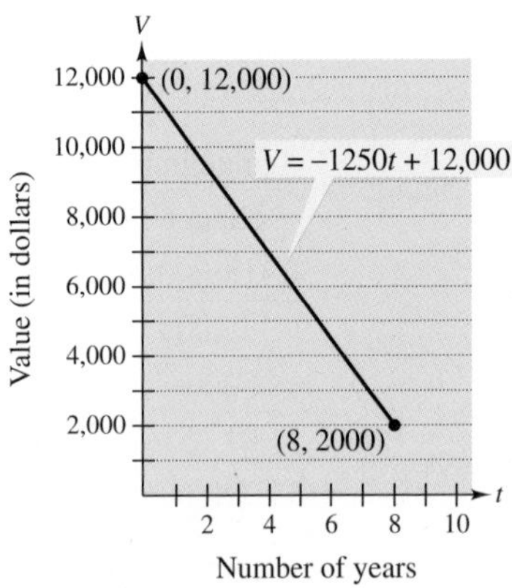

Straight-line depreciation
Figure P.47

In many real-life applications, the two data points that determine the line are often given in a disguised form. Note how the data points are described in Example 7.

EXAMPLE 8 Predicting Sales

The sales for Best Buy were approximately \$35.9 billion in 2006 and \$40.0 billion in 2007. Using only this information, write a linear equation that gives the sales (in billions of dollars) in terms of the year. Then predict the sales for 2012. *(Source: Best Buy Company, Inc.)*

Solution Let $t = 6$ represent 2006. Then the two given values are represented by the data points $(6, 35.9)$ and $(7, 40.0)$. The slope of the line through these points is

$$m = \frac{40.0 - 35.9}{7 - 6}$$

$$= 4.1.$$

Using the point-slope form, you can find the equation that relates the sales y and the year t to be

$$y - 35.9 = 4.1(t - 6) \qquad \text{Write in point-slope form.}$$

$$y = 4.1t + 11.3. \qquad \text{Write in slope-intercept form.}$$

According to this equation, the sales for 2012 will be

$$y = 4.1(12) + 11.3$$

$$= 49.2 + 11.3$$

$$= \$60.5 \text{ billion.} \quad \text{(See Figure P.48.)}$$

■

Sales (in billions of dollars)

Year (6 ↔ 2006)

Figure P.48

The prediction method illustrated in Example 8 is called **linear extrapolation.** Note in Figure P.49(a) that an extrapolated point does not lie between the given points. When the estimated point lies between two given points, as shown in Figure P.49(b), the procedure is called **linear interpolation.**

Because the slope of a vertical line is not defined, its equation cannot be written in slope-intercept form. However, every line has an equation that can be written in the **general form**

$$Ax + By + C = 0 \qquad \text{General form}$$

where A and B are not both zero. For instance, the vertical line given by $x = a$ can be represented by the general form $x - a = 0$.

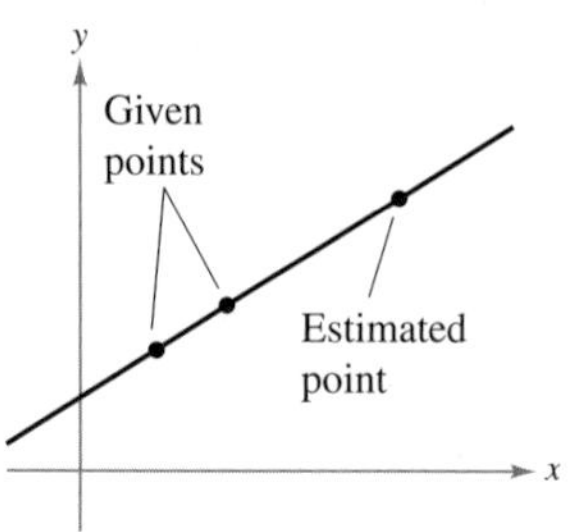

(a) Linear extrapolation

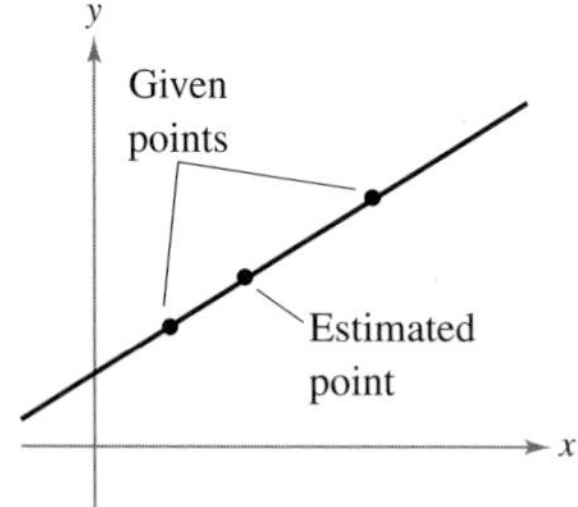

(b) Linear interpolation

Figure P.49

SUMMARY OF EQUATIONS OF LINES

1. General form: $Ax + By + C = 0$
2. Vertical line: $x = a$
3. Horizontal line: $y = b$
4. Slope-intercept form: $y = mx + b$
5. Point-slope form: $y - y_1 = m(x - x_1)$
6. Two-point form: $y - y_1 = \dfrac{y_2 - y_1}{x_2 - x_1}(x - x_1)$

P.5 Exercises

See www.CalcChat.com for worked-out solutions to odd-numbered exercises.

In Exercises 1–7, fill in the blanks.

1. The simplest mathematical model for relating two variables is the ________ equation in two variables $y = mx + b$.
2. For a line, the ratio of the change in y to the change in x is called the ________ of the line.
3. Two lines are ________ if and only if their slopes are equal.
4. Two lines are ________ if and only if their slopes are negative reciprocals of each other.
5. When the x-axis and y-axis have different units of measure, the slope can be interpreted as a ________.
6. The prediction method ________ ________ is the method used to estimate a point on a line when the point does not lie between the given points.
7. Every line has an equation that can be written in ________ form.

8. Match each equation of a line with its form.

(a) $Ax + By + C = 0$	(i) Vertical line
(b) $x = a$	(ii) Slope-intercept form
(c) $y = b$	(iii) General form
(d) $y = mx + b$	(iv) Point-slope form
(e) $y - y_1 = m(x - x_1)$	(v) Horizontal line

In Exercises 9 and 10, sketch the lines through the point with the indicated slopes on the same set of coordinate axes.

	Point	*Slopes*
9.	$(2, 3)$	(a) 0 (b) 1 (c) 2 (d) -3
10.	$(-4, 1)$	(a) 3 (b) -3 (c) $\frac{1}{2}$ (d) Undefined

In Exercises 11–14, estimate the slope of the line.

11.

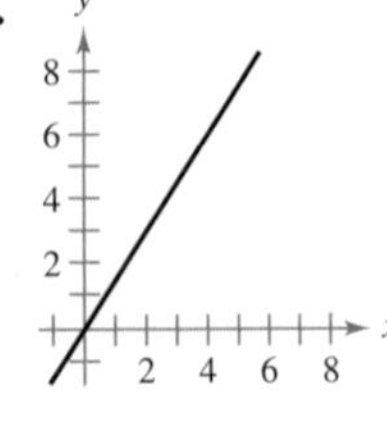

12.

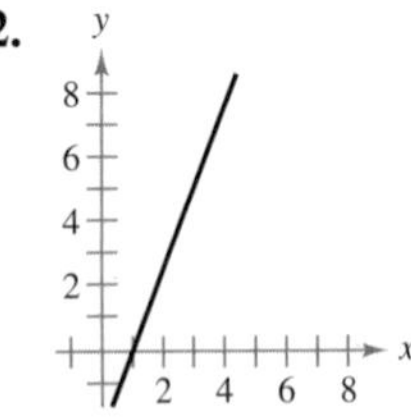

13.

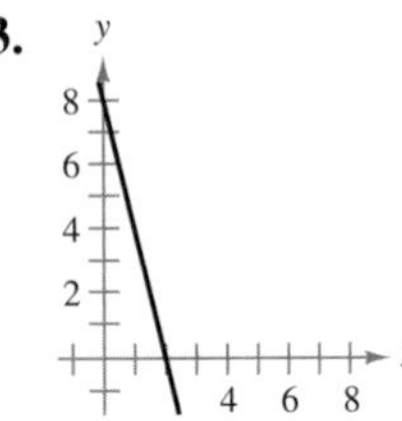

14.

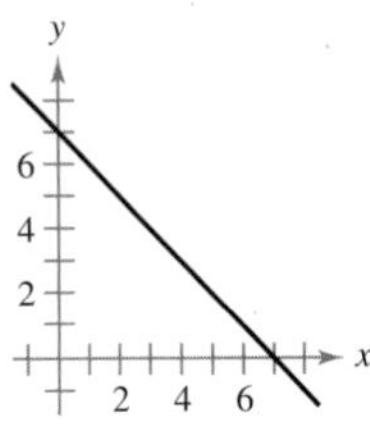

In Exercises 15–26, find the slope and y-intercept (if possible) of the equation of the line. Sketch the line.

15. $y = 5x + 3$
16. $y = x - 10$
17. $y = -\frac{1}{2}x + 4$
18. $y = -\frac{3}{2}x + 6$
19. $5x - 2 = 0$
20. $3y + 5 = 0$
21. $7x + 6y = 30$
22. $2x + 3y = 9$
23. $y - 3 = 0$
24. $y + 4 = 0$
25. $x + 5 = 0$
26. $x - 2 = 0$

In Exercises 27–38, plot the points and find the slope of the line passing through the pair of points.

27. $(0, 9), (6, 0)$
28. $(12, 0), (0, -8)$
29. $(-3, -2), (1, 6)$
30. $(2, 4), (4, -4)$
31. $(5, -7), (8, -7)$
32. $(-2, 1), (-4, -5)$
33. $(-6, -1), (-6, 4)$
34. $(0, -10), (-4, 0)$
35. $\left(\frac{11}{2}, -\frac{4}{3}\right), \left(-\frac{3}{2}, -\frac{1}{3}\right)$
36. $\left(\frac{7}{8}, \frac{3}{4}\right), \left(\frac{5}{4}, -\frac{1}{4}\right)$
37. $(4.8, 3.1), (-5.2, 1.6)$
38. $(-1.75, -8.3), (2.25, -2.6)$

WRITING ABOUT CONCEPTS

In Exercises 39 and 40, identify the line that has each slope.

39. (a) $m = \frac{2}{3}$
(b) m is undefined.
(c) $m = -2$

40. (a) $m = 0$
(b) $m = -\frac{3}{4}$
(c) $m = 1$

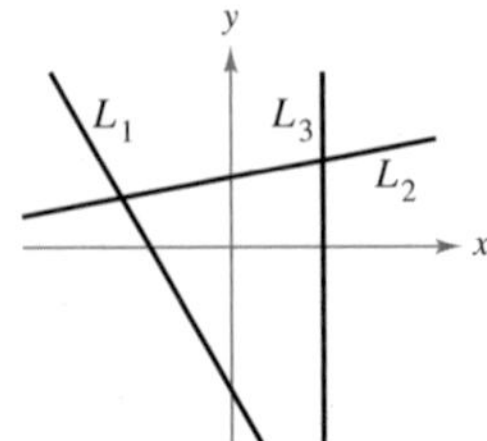

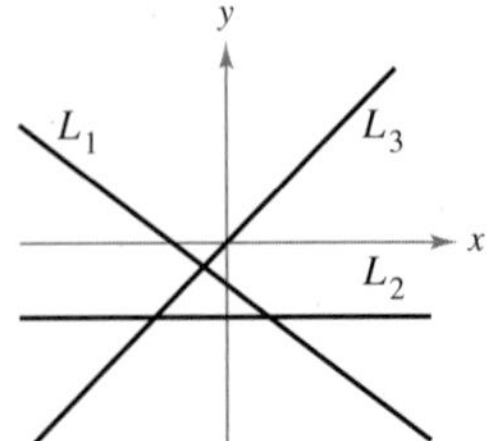

In Exercises 41–50, use the point on the line and the slope m of the line to find three additional points through which the line passes. (There are many correct answers.)

41. $(2, 1)$, $m = 0$
42. $(3, -2)$, $m = 0$
43. $(5, -6)$, $m = 1$
44. $(10, -6)$, $m = -1$
45. $(-8, 1)$, m is undefined.
46. $(1, 5)$, m is undefined.
47. $(-5, 4)$, $m = 2$
48. $(0, -9)$, $m = -2$
49. $(7, -2)$, $m = \frac{1}{2}$
50. $(-1, -6)$, $m = -\frac{1}{2}$

In Exercises 51–64, find the slope-intercept form of the equation of the line that passes through the given point and has the indicated slope m. Sketch the line.

51. $(0, -2)$, $m = 3$
52. $(0, 10)$, $m = -1$
53. $(-3, 6)$, $m = -2$
54. $(0, 0)$, $m = 4$
55. $(4, 0)$, $m = -\frac{1}{3}$
56. $(8, 2)$, $m = \frac{1}{4}$
57. $(2, -3)$, $m = -\frac{1}{2}$
58. $(-2, -5)$, $m = \frac{3}{4}$
59. $(6, -1)$, m is undefined.
60. $(-10, 4)$, m is undefined.
61. $\left(4, \frac{5}{2}\right)$, $m = 0$
62. $\left(-\frac{1}{2}, \frac{3}{2}\right)$, $m = 0$
63. $(-5.1, 1.8)$, $m = 5$
64. $(2.3, -8.5)$, $m = -2.5$

In Exercises 65–78, find the slope-intercept form of the equation of the line passing through the points. Sketch the line.

65. $(5, -1), (-5, 5)$
66. $(4, 3), (-4, -4)$
67. $(-8, 1), (-8, 7)$
68. $(-1, 4), (6, 4)$
69. $\left(2, \frac{1}{2}\right), \left(\frac{1}{2}, \frac{5}{4}\right)$
70. $(1, 1), \left(6, -\frac{2}{3}\right)$
71. $\left(-\frac{1}{10}, -\frac{3}{5}\right), \left(\frac{9}{10}, -\frac{9}{5}\right)$
72. $\left(\frac{3}{4}, \frac{3}{2}\right), \left(-\frac{4}{3}, \frac{7}{4}\right)$
73. $(1, 0.6), (-2, -0.6)$
74. $(-8, 0.6), (2, -2.4)$
75. $(2, -1), \left(\frac{1}{3}, -1\right)$
76. $\left(\frac{1}{5}, -2\right), (-6, -2)$
77. $\left(\frac{7}{3}, -8\right), \left(\frac{7}{3}, 1\right)$
78. $(1.5, -2), (1.5, 0.2)$

In Exercises 79–82, determine whether the lines are parallel, perpendicular, or neither.

79. L_1: $y = \frac{1}{3}x - 2$
L_2: $y = \frac{1}{3}x + 3$
80. L_1: $y = 4x - 1$
L_2: $y = 4x + 7$
81. L_1: $y = \frac{1}{2}x - 3$
L_2: $y = -\frac{1}{2}x + 1$
82. L_1: $y = -\frac{4}{5}x - 5$
L_2: $y = \frac{5}{4}x + 1$

In Exercises 83–86, determine whether the lines L_1 and L_2 passing through the pairs of points are parallel, perpendicular, or neither.

83. L_1: $(0, -1), (5, 9)$
L_2: $(0, 3), (4, 1)$
84. L_1: $(-2, -1), (1, 5)$
L_2: $(1, 3), (5, -5)$
85. L_1: $(3, 6), (-6, 0)$
L_2: $(0, -1), \left(5, \frac{7}{3}\right)$
86. L_1: $(4, 8), (-4, 2)$
L_2: $(3, -5), \left(-1, \frac{1}{3}\right)$

In Exercises 87–96, write the slope-intercept forms of the equations of the lines through the given point (a) parallel to the given line and (b) perpendicular to the given line.

87. $4x - 2y = 3$, $(2, 1)$
88. $x + y = 7$, $(-3, 2)$
89. $3x + 4y = 7$, $\left(-\frac{2}{3}, \frac{7}{8}\right)$
90. $5x + 3y = 0$, $\left(\frac{7}{8}, \frac{3}{4}\right)$
91. $y + 3 = 0$, $(-1, 0)$
92. $y - 2 = 0$, $(-4, 1)$
93. $x - 4 = 0$, $(3, -2)$
94. $x + 2 = 0$, $(-5, 1)$
95. $x - y = 4$, $(2.5, 6.8)$
96. $6x + 2y = 9$, $(-3.9, -1.4)$

In Exercises 97–102, use the *intercept form* to find the equation of the line with the given intercepts. The intercept form of the equation of a line with intercepts $(a, 0)$ and $(0, b)$ is $x/a + y/b = 1$, $a \neq 0$, $b \neq 0$.

97. x-intercept: $(2, 0)$
y-intercept: $(0, 3)$
98. x-intercept: $(-3, 0)$
y-intercept: $(0, 4)$
99. x-intercept: $\left(-\frac{1}{6}, 0\right)$
y-intercept: $\left(0, -\frac{2}{3}\right)$
100. x-intercept: $\left(\frac{2}{3}, 0\right)$
y-intercept: $(0, -2)$
101. Point on line: $(1, 2)$
x-intercept: $(c, 0)$
y-intercept: $(0, c)$, $c \neq 0$
102. Point on line: $(-3, 4)$
x-intercept: $(d, 0)$
y-intercept: $(0, d)$, $d \neq 0$

Graphical Analysis **In Exercises 103–106, identify any relationships that exist among the lines, and then use a graphing utility to graph the three equations in the same viewing window. Adjust the viewing window so that the slope appears visually correct—that is, so that parallel lines appear parallel and perpendicular lines appear to intersect at right angles.**

103. (a) $y = 2x$ (b) $y = -2x$ (c) $y = \frac{1}{2}x$
104. (a) $y = \frac{2}{3}x$ (b) $y = -\frac{3}{2}x$ (c) $y = \frac{2}{3}x + 2$
105. (a) $y = -\frac{1}{2}x$ (b) $y = -\frac{1}{2}x + 3$ (c) $y = 2x - 4$
106. (a) $y = x - 8$ (b) $y = x + 1$ (c) $y = -x + 3$

In Exercises 107–110, find a relationship between x and y such that (x, y) is equidistant (the same distance) from the two points.

107. $(4, -1), (-2, 3)$
108. $(6, 5), (1, -8)$
109. $\left(3, \frac{5}{2}\right), (-7, 1)$
110. $\left(-\frac{1}{2}, -4\right), \left(\frac{7}{2}, \frac{5}{4}\right)$

111. ***Sales*** The following are the slopes of lines representing annual sales y in terms of time x in years. Use the slopes to interpret any change in annual sales for a one-year increase in time.

(a) The line has a slope of $m = 135$.

(b) The line has a slope of $m = 0$.

(c) The line has a slope of $m = -40$.

112. ***Revenue*** The following are the slopes of lines representing daily revenues y in terms of time x in days. Use the slopes to interpret any change in daily revenues for a one-day increase in time.

(a) The line has a slope of $m = 400$.

(b) The line has a slope of $m = 100$.

(c) The line has a slope of $m = 0$.

113. ***Average Salary*** The graph shows the average salaries for senior high school principals from 1996 through 2008. *(Source: Educational Research Service)*

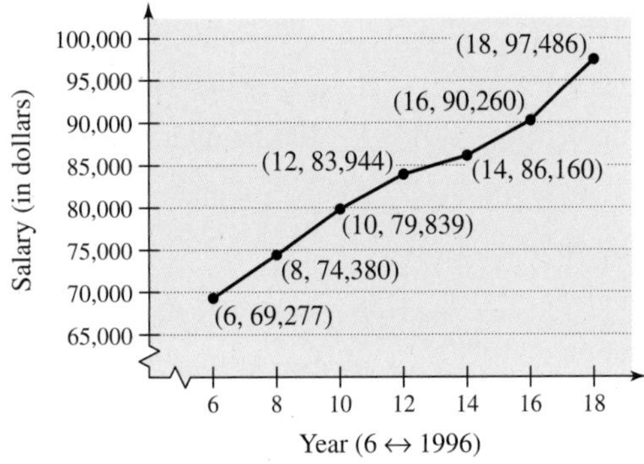

(a) Use the slopes of the line segments to determine the time periods in which the average salary increased the greatest and the least.

(b) Find the slope of the line segment connecting the points for the years 1996 and 2008.

(c) Interpret the meaning of the slope in part (b) in the context of the problem.

114. ***Sales*** The graph shows the sales (in billions of dollars) for Apple Inc. for the years 2001 through 2007. *(Source: Apple Inc.)*

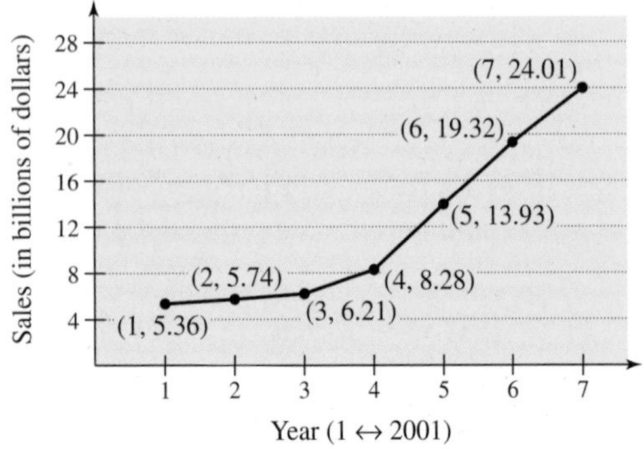

(a) Use the slopes of the line segments to determine the years in which the sales showed the greatest increase and the least increase.

(b) Find the slope of the line segment connecting the points for the years 2001 and 2007.

(c) Interpret the meaning of the slope in part (b) in the context of the problem.

115. ***Road Grade*** You are driving on a road that has a 6% uphill grade (see figure). This means that the slope of the road is $\frac{6}{100}$. Approximate the amount of vertical change in your position if you drive 200 feet.

116. ***Road Grade*** From the top of a mountain road, a surveyor takes several horizontal measurements x and several vertical measurements y, as shown in the table (x and y are measured in feet).

x	300	600	900	1200
y	-25	-50	-75	-100

x	1500	1800	2100
y	-125	-150	-175

(a) Sketch a scatter plot of the data.

(b) Use a straightedge to sketch the line that you think best fits the data.

(c) Find an equation for the line you sketched in part (b).

(d) Interpret the meaning of the slope of the line in part (c) in the context of the problem.

(e) The surveyor needs to put up a road sign that indicates the steepness of the road. For instance, a surveyor would put up a sign that states "8% grade" on a road with a downhill grade that has a slope of $-\frac{8}{100}$. What should the sign state for the road in this problem?

Rate of Change **In Exercises 117 and 118, you are given the dollar value of a product in 2010 and the rate at which the value of the product is expected to change during the next 5 years. Use this information to write a linear equation that gives the dollar value V of the product in terms of the year t. (Let $t = 10$ represent 2010.)**

	2010 *Value*	*Rate*
117.	\$2540	\$125 decrease per year
118.	\$156	\$4.50 increase per year

119. ***Depreciation*** The value V of a molding machine t years after it is purchased is $V = -4000t + 58{,}500$, $0 \le t \le 5$. Explain what the V-intercept and the slope measure.

120. ***Cost*** The cost C of producing n computer laptop bags is given by $C = 1.25n + 15{,}750$, $0 < n$. Explain what the C-intercept and the slope measure.

121. ***Depreciation*** A sub shop purchases a used pizza oven for \$875. After 5 years, the oven will have to be replaced. Write a linear equation giving the value V of the equipment during the 5 years it will be in use.

122. ***Depreciation*** A school district purchases a high-volume printer, copier, and scanner for \$25,000. After 10 years, the equipment will have to be replaced. Its value at that time is expected to be \$2000. Write a linear equation giving the value V of the equipment during the 10 years it will be in use.

123. ***Sales*** A discount outlet is offering a 20% discount on all items. Write a linear equation giving the sale price S for an item with a list price L.

124. ***Hourly Wage*** A microchip manufacturer pays its assembly line workers \$12.25 per hour. In addition, workers receive a piecework rate of \$0.75 per unit produced. Write a linear equation for the hourly wage W in terms of the number of units x produced per hour.

125. ***Monthly Salary*** A pharmaceutical salesperson receives a monthly salary of \$2500 plus a commission of 7% of sales. Write a linear equation for the salesperson's monthly wage W in terms of monthly sales S.

126. ***Business Costs*** A sales representative of a company using a personal car receives \$120 per day for lodging and meals plus \$0.55 per mile driven. Write a linear equation giving the daily cost C to the company in terms of x, the number of miles driven.

127. ***Cash Flow per Share*** The cash flow per share for the Timberland Co. was \$1.21 in 1999 and \$1.46 in 2007. Write a linear equation that gives the cash flow per share in terms of the year. Let $t = 9$ represent 1999. Then predict the cash flows for the years 2012 and 2014. *(Source: The Timberland Co.)*

128. ***Number of Stores*** In 2003 there were 1078 J.C. Penney stores and in 2007 there were 1067 stores. Write a linear equation that gives the number of stores in terms of the year. Let $t = 3$ represent 2003. Then predict the numbers of stores for the years 2012 and 2014. Are your answers reasonable? Explain. *(Source: J.C. Penney Co.)*

129. ***College Enrollment*** The Pennsylvania State University had enrollments of 40,571 students in 2000 and 44,112 students in 2008 at its main campus in University Park, Pennsylvania. *(Source: Penn State Fact Book)*

(a) Assuming the enrollment growth is linear, find a linear model that gives the enrollment in terms of the year t, where $t = 0$ corresponds to 2000.

(b) Use your model from part (a) to predict the enrollments in 2010 and 2015.

(c) What is the slope of your model? Explain its meaning in the context of the situation.

130. ***College Enrollment*** The University of Florida had enrollments of 46,107 students in 2000 and 51,413 students in 2008. *(Source: University of Florida)*

(a) What was the average annual change in enrollment from 2000 to 2008?

(b) Use the average annual change in enrollment to estimate the enrollments in 2002, 2004, and 2006.

(c) Write the equation of a line that represents the given data in terms of the year t, where $t = 0$ corresponds to 2000. What is its slope? Interpret the slope in the context of the problem.

(d) Using the results of parts (a)–(c), write a short paragraph discussing the concepts of *slope* and *average rate of change.*

131. ***Cost, Revenue, and Profit*** A roofing contractor purchases a shingle delivery truck with a shingle elevator for \$42,000. The vehicle requires an average expenditure of \$6.50 per hour for fuel and maintenance, and the operator is paid \$11.50 per hour.

(a) Write a linear equation giving the total cost C of operating this equipment for t hours. (Include the purchase cost of the equipment.)

(b) Assuming that customers are charged \$30 per hour of machine use, write an equation for the revenue R derived from t hours of use.

(c) Use the formula for profit $P = R - C$ to write an equation for the profit derived from t hours of use.

(d) Use the result of part (c) to find the break-even point—that is, the number of hours this equipment must be used to yield a profit of 0 dollars.

132. *Rental Demand* A real estate office handles an apartment complex with 50 units. When the rent per unit is \$580 per month, all 50 units are occupied. However, when the rent is \$625 per month, the average number of occupied units drops to 47. Assume that the relationship between the monthly rent p and the demand x is linear.

(a) Write the equation of the line giving the demand x in terms of the rent p.

(b) Use this equation to predict the number of units occupied when the rent is \$655.

(c) Predict the number of units occupied when the rent is \$595.

133. *Geometry* The length and width of a rectangular garden are 15 meters and 10 meters, respectively. A walkway of width x surrounds the garden.

(a) Draw a diagram that gives a visual representation of the problem.

(b) Write the equation for the perimeter y of the walkway in terms of x.

(c) Use a graphing utility to graph the equation for the perimeter.

(d) Determine the slope of the graph in part (c). For each additional one-meter increase in the width of the walkway, determine the increase in its perimeter.

134. *Average Annual Salary* The average salaries (in millions of dollars) of Major League Baseball players from 2000 through 2007 are shown in the scatter plot. Find the equation of the line that you think best fits these data. (Let y represent the average salary and let t represent the year, with $t = 0$ corresponding to 2000.) *(Source: Major League Baseball Players Association)*

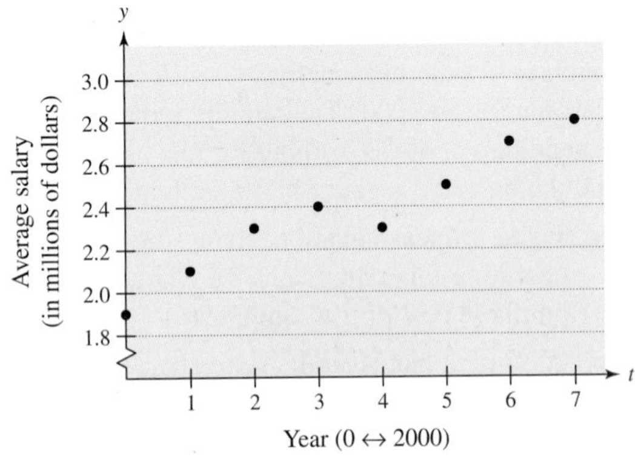

135. *Data Analysis: Number of Doctors* The numbers of doctors of osteopathic medicine y (in thousands) in the United States from 2000 through 2008, where x is the year, are shown as data points (x, y). *(Source: American Osteopathic Association)*

$(2000, 44.9)$, $(2001, 47.0)$, $(2002, 49.2)$, $(2003, 51.7)$, $(2004, 54.1)$, $(2005, 56.5)$, $(2006, 58.9)$, $(2007, 61.4)$, $(2008, 64.0)$

(a) Sketch a scatter plot of the data. Let $x = 0$ correspond to 2000.

(b) Use a straightedge to sketch the line that you think best fits the data.

(c) Find the equation of the line from part (b). Explain the procedure you used.

(d) Write a short paragraph explaining the meanings of the slope and y-intercept of the line in terms of the data.

(e) Compare the values obtained using your model with the actual values.

(f) Use your model to estimate the number of doctors of osteopathic medicine in 2012.

136. *Data Analysis: Average Scores* An instructor gives regular 20-point quizzes and 100-point exams in an algebra course. Average scores for six students, given as data points (x, y), where x is the average quiz score and y is the average test score, are $(18, 87)$, $(10, 55)$, $(19, 96)$, $(16, 79)$, $(13, 76)$, and $(15, 82)$. [Note: There are many correct answers for parts (b)–(d).]

(a) Sketch a scatter plot of the data.

(b) Use a straightedge to sketch the line that you think best fits the data.

(c) Find an equation for the line you sketched in part (b).

(d) Use the equation in part (c) to estimate the average test score for a person with an average quiz score of 17.

(e) The instructor adds 4 points to the average test score of each student in the class. Describe the changes in the positions of the plotted points and the change in the equation of the line.

True or False? **In Exercises 137 and 138, determine whether the statement is true or false. Justify your answer.**

137. A line with a slope of $-\frac{5}{7}$ is steeper than a line with a slope of $-\frac{6}{7}$.

138. The line through $(-8, 2)$ and $(-1, 4)$ and the line through $(0, -4)$ and $(-7, 7)$ are parallel.

139. Explain how you could show that the points $A(2, 3)$, $B(2, 9)$, and $C(4, 3)$ are the vertices of a right triangle.

140. Explain why the slope of a vertical line is said to be undefined.

141. With the information shown in the graphs, is it possible to determine the slope of each line? Is it possible that the lines could have the same slope? Explain.

(a)

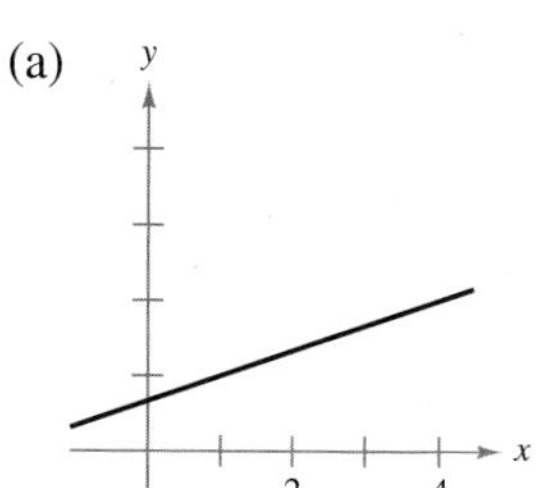

(b)

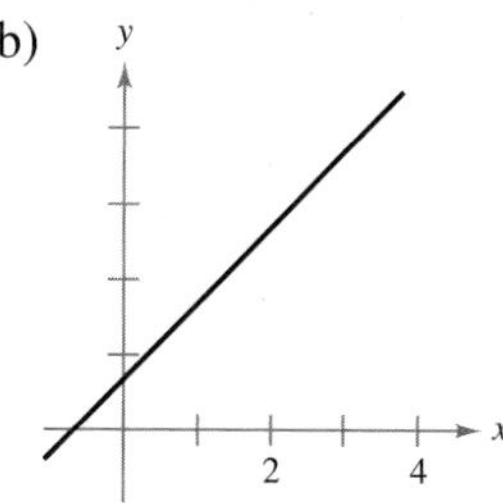

142. The slopes of two lines are -4 and $\frac{5}{2}$. Which is steeper? Explain.

Think About It **In Exercises 143 and 144, determine which pair of equations may be represented by the graphs shown.**

143.

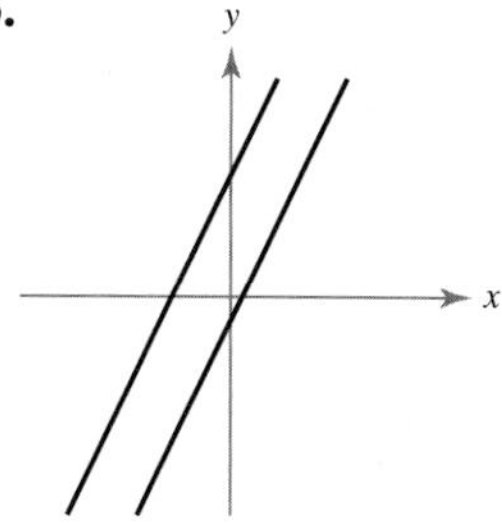

(a) $2x - y = 5$
$2x - y = 1$

(b) $2x + y = -5$
$2x + y = 1$

(c) $2x - y = -5$
$2x - y = 1$

(d) $x - 2y = -5$
$x - 2y = -1$

144.

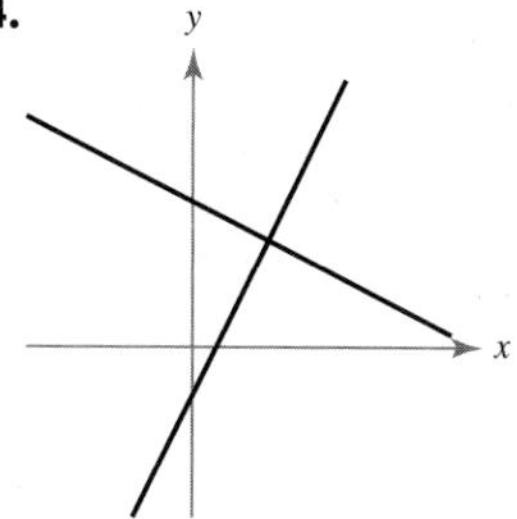

(a) $2x - y = 2$
$x + 2y = 12$

(b) $x - y = 1$
$x + y = 6$

(c) $2x + y = 2$
$x - 2y = 12$

(d) $x - 2y = 2$
$x + 2y = 12$

145. Use a graphing utility to compare the slopes of the lines $y = mx$, where $m = 0.5$, 1, 2, and 4. Which line rises most quickly? Now, let $m = -0.5$, -1, -2, and -4. Which line falls most quickly? Use a square setting to obtain a true geometric perspective. What can you conclude about the slope and the "rate" at which the line rises or falls?

146. Find d_1 and d_2 in terms of m_1 and m_2, respectively (see figure). Then use the Pythagorean Theorem to find a relationship between m_1 and m_2.

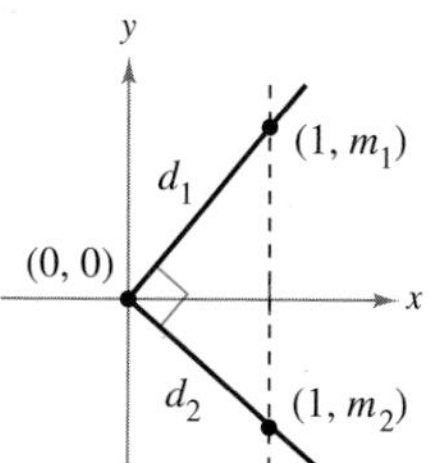

147. ***Think About It*** Is it possible for two lines with positive slopes to be perpendicular? Explain.

CAPSTONE

148. Match the description of the situation with its graph. Also determine the slope and y-intercept of each graph and interpret the slope and y-intercept in the context of the situation. [The graphs are labeled (i), (ii), (iii), and (iv).]

(i)

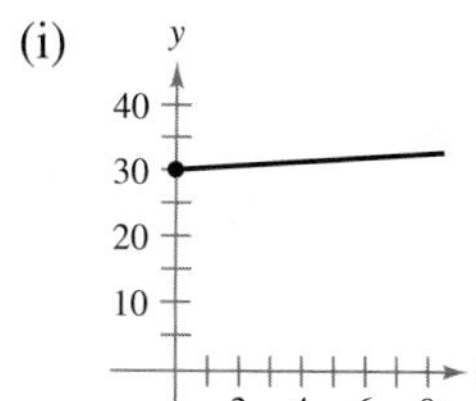

(ii)

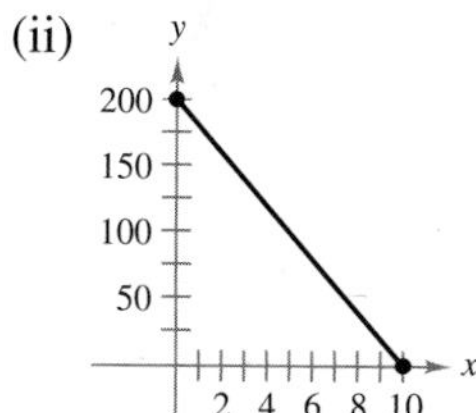

(iii)

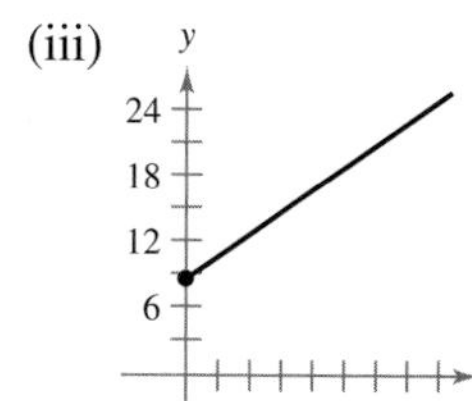

(iv)

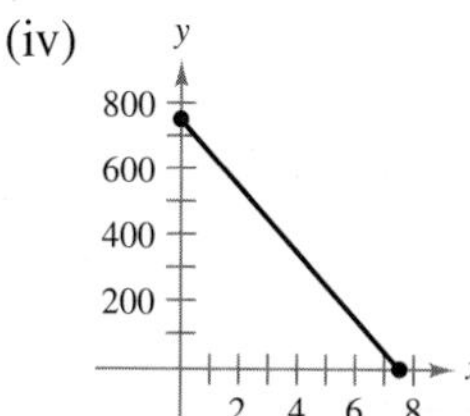

(a) A person is paying \$20 per week to a friend to repay a \$200 loan.

(b) An employee is paid \$8.50 per hour plus \$2 for each unit produced per hour.

(c) A sales representative receives \$30 per day for food plus \$0.50 for each mile traveled.

(d) A computer that was purchased for \$750 depreciates \$100 per year.

P CHAPTER SUMMARY

Section P.1	Review Exercises
■ Identify different types of equations (*p. 2*).	*1–4*
■ Solve linear equations in one variable and equations that lead to linear equations (*p. 2*).	*5–8, 35*
■ Solve quadratic equations by factoring, extracting square roots, completing the square, and using the Quadratic Formula (*p. 5*).	*9–18*
■ Solve polynomial equations of degree three or greater (*p. 9*).	*19–22*
■ Solve equations involving radicals (*p. 10*).	*23–30, 36*
■ Solve equations with absolute values (*p. 11*).	*31–34*
Section P.2	
■ Represent solutions of linear inequalities in one variable (*p. 15*).	*37, 38*
■ Use properties of inequalities to create equivalent inequalities (*p. 16*) and solve linear inequalities in one variable (*p. 17*).	*39–44, 49*
■ Solve inequalities involving absolute values (*p. 19*).	*45–48, 50*
■ Solve polynomial inequalities (*p. 20*).	*51–54, 59*
■ Solve rational inequalities (*p. 21*).	*55–58, 60*
Section P.3	
■ Plot points in the Cartesian plane (*p. 27*).	*61–64*
■ Use the Distance Formula to find the distance between two points (*p. 29*) and use the Midpoint Formula to find the midpoint of a line segment (*p. 30*).	*65–68*
■ Use a coordinate plane to model and solve real-life problems (*p. 31*).	*69–72*
Section P.4	
■ Sketch graphs of equations (*p. 38*).	*73–82*
■ Find x- and y-intercepts of graphs of equations (*p. 41*).	*83–86*
■ Use symmetry to sketch graphs of equations (*p. 42*).	*87–94*
■ Find equations of and sketch graphs of circles (*p. 44*).	*95–102*
■ Use graphs of equations in solving real-life problems (*p. 45*).	*103, 104*
Section P.5	
■ Use slope to graph linear equations in two variables (*p. 49*).	*105–110*
■ Find the slope of a line given two points on the line (*p. 51*).	*111–114*
■ Write linear equations in two variables (*p. 53*).	*115–124*
■ Use slope to identify parallel and perpendicular lines (*p. 54*).	*125–130*
■ Use slope and linear equations in two variables to model and solve real-life problems (*p. 55*).	*131, 132*

P REVIEW EXERCISES

See www.CalcChat.com for worked-out solutions to odd-numbered exercises.

In Exercises 1–4, determine whether the equation is an identity or a conditional equation.

1. $6 - (x - 2)^2 = 2 + 4x - x^2$
2. $3(x - 2) + 2x = 2(x + 3)$
3. $-x^3 + x(7 - x) + 3 = x(-x^2 - x) + 7(x + 1) - 4$
4. $3(x^2 - 4x + 8) = -10(x + 2) - 3x^2 + 6$

In Exercises 5–8, solve the equation (if possible) and check your solution.

5. $3x - 2(x + 5) = 10$
6. $4x + 2(7 - x) = 5$
7. $4(x + 3) - 3 = 2(4 - 3x) - 4$
8. $\frac{1}{2}(x - 3) - 2(x + 1) = 5$

In Exercises 9–18, use any method to solve the equation.

9. $2x^2 - x - 28 = 0$
10. $15 + x - 2x^2 = 0$
11. $16x^2 = 25$
12. $6 = 3x^2$
13. $(x - 8)^2 = 15$
14. $(x + 4)^2 = 18$
15. $x^2 + 6x - 3 = 0$
16. $x^2 - 12x + 30 = 0$
17. $-20 - 3x + 3x^2 = 0$
18. $-2x^2 - 5x + 27 = 0$

In Exercises 19–34, find all solutions of the equation. Check your solutions in the original equation.

19. $4x^3 - 6x^2 = 0$
20. $5x^4 - 12x^3 = 0$
21. $9x^4 + 27x^3 - 4x^2 - 12x = 0$
22. $x^4 - 5x^2 + 6 = 0$
23. $\sqrt{x - 2} - 8 = 0$
24. $\sqrt{x + 4} = 3$
25. $\sqrt{3x - 2} = 4 - x$
26. $2\sqrt{x} - 5 = x$
27. $(x + 2)^{3/4} = 27$
28. $(x - 1)^{2/3} - 25 = 0$
29. $8x^2(x^2 - 4)^{1/3} + (x^2 - 4)^{4/3} = 0$
30. $(x + 4)^{1/2} + 5x(x + 4)^{3/2} = 0$
31. $|2x + 3| = 7$
32. $|x - 5| = 10$
33. $|x^2 - 6| = x$
34. $|x^2 - 3| = 2x$

35. ***Mixture Problem*** A car radiator contains 10 liters of a 30% antifreeze solution. How many liters will have to be replaced with pure antifreeze if the resulting solution is to be 50% antifreeze?

36. ***Demand*** The demand equation for a product is

$$p = 42 - \sqrt{0.001x + 2}$$

where x is the number of units demanded per day and p is the price per unit (in dollars). Find the demand if the price is set at \$29.95.

In Exercises 37 and 38, determine whether each value of x is a solution of the inequality.

37. $6x - 17 > 0$ (a) $x = 3$ (b) $x = -4$
38. $-3 \le \frac{x - 3}{5} < 2$ (a) $x = 3$ (b) $x = -12$

In Exercises 39–48, solve the inequality.

39. $9x - 8 \le 7x + 16$
40. $4(5 - 2x) \le \frac{1}{2}(8 - x)$
41. $\frac{15}{2}x + 4 > 3x - 5$
42. $\frac{1}{2}(3 - x) > \frac{1}{3}(2 - 3x)$
43. $-19 < \frac{3x - 17}{2} \le 34$
44. $-3 \le \frac{2x - 5}{3} < 5$
45. $|x + 1| \le 5$
46. $|x - 2| < 1$
47. $|x - 3| > 4$
48. $\left|x - \frac{3}{2}\right| \ge \frac{3}{2}$

49. ***Cost, Revenue, and Profit*** The revenue for selling x units of a product is $R = 125.33x$. The cost of producing x units is $C = 92x + 1200$. To obtain a profit, the revenue must be greater than the cost. Determine the smallest value of x for which this product returns a profit.

50. ***Geometry*** The side of a square stained glass window is measured as 19.3 centimeters with a possible error of 0.5 centimeter. Using these measurements, determine the interval containing the area of the glass.

In Exercises 51–58, solve the inequality.

51. $x^2 - 6x - 27 < 0$
52. $x^2 - 2x \ge 3$
53. $6x^2 + 5x < 4$
54. $2x^2 + x \ge 15$
55. $\frac{2}{x + 1} \le \frac{3}{x - 1}$
56. $\frac{x - 5}{3 - x} < 0$
57. $\frac{x^2 + 7x + 12}{x} \ge 0$
58. $\frac{1}{x - 2} > \frac{1}{x}$

59. ***Investment*** P dollars invested at interest rate r compounded annually increases to an amount $A = P(1 + r)^2$ in 2 years. An investment of \$5000 is to increase to an amount greater than \$5500 in 2 years. The interest rate must be greater than what percent?

60. ***Population of Ladybugs*** A biologist introduces 200 ladybugs into a crop field. The population P of the ladybugs is approximated by the model

$$P = \frac{1000(1 + 3t)}{5 + t}$$

where t is the time in days. Find the time required for the population to increase to at least 2000 ladybugs.

In Exercises 61 and 62, plot the points in the Cartesian plane.

61. $(5, 5), (-2, 0), (-3, 6), (-1, -7)$

62. $(0, 6), (8, 1), (4, -2), (-3, -3)$

In Exercises 63 and 64, determine the quadrant(s) in which (x, y) is located so that the condition(s) is (are) satisfied.

63. $x > 0$ and $y = -2$

64. $xy = 4$

In Exercises 65–68, (a) plot the points, (b) find the distance between the points, and (c) find the midpoint of the line segment joining the points.

65. $(-3, 8), (1, 5)$

66. $(-2, 6), (4, -3)$

67. $(5.6, 0), (0, 8.2)$

68. $(1.8, 7.4), (-0.6, -14.5)$

In Exercises 69 and 70, the polygon is shifted to a new position in the plane. Find the coordinates of the vertices of the polygon in its new position.

69. Original coordinates of vertices:

$(4, 8), (6, 8), (4, 3), (6, 3)$

Shift: eight units downward, four units to the left

70. Original coordinates of vertices:

$(0, 1), (3, 3), (0, 5), (-3, 3)$

Shift: three units upward, two units to the left

71. ***Sales*** Starbucks had annual sales of $2.17 billion in 2000 and $10.38 billion in 2008. Use the Midpoint Formula to estimate the sales in 2004. *(Source: Starbucks Corp.)*

72. ***Meteorology*** The apparent temperature is a measure of relative discomfort to a person from heat and high humidity. The table shows the actual temperatures x (in degrees Fahrenheit) versus the apparent temperatures y (in degrees Fahrenheit) for a relative humidity of 75%.

x	70	75	80	85	90	95	100
y	70	77	85	95	109	130	150

(a) Sketch a scatter plot of the data shown in the table.

(b) Find the change in the apparent temperature when the actual temperature changes from 70°F to 100°F.

In Exercises 73–76, complete a table of values. Use the solution points to sketch the graph of the equation.

73. $y = 3x - 5$

74. $y = -\frac{1}{2}x + 2$

75. $y = x^2 - 3x$

76. $y = 2x^2 - x - 9$

In Exercises 77–82, sketch the graph by hand.

77. $y - 2x - 3 = 0$

78. $3x + 2y + 6 = 0$

79. $y = \sqrt{5 - x}$

80. $y = \sqrt{x + 2}$

81. $y + 2x^2 = 0$

82. $y = x^2 - 4x$

In Exercises 83–86, find the x- and y-intercepts of the graph of the equation.

83. $y = 2x + 7$

84. $y = |x + 1| - 3$

85. $y = (x - 3)^2 - 4$

86. $y = x\sqrt{4 - x^2}$

In Exercises 87–94, identify any intercepts and test for symmetry. Then sketch the graph of the equation.

87. $y = -4x + 1$

88. $y = 5x - 6$

89. $y = 5 - x^2$

90. $y = x^2 - 10$

91. $y = x^3 + 3$

92. $y = -6 - x^3$

93. $y = \sqrt{x + 5}$

94. $y = |x| + 9$

In Exercises 95–100, find the center and radius of the circle and sketch its graph.

95. $x^2 + y^2 = 9$

96. $x^2 + y^2 = 4$

97. $(x + 2)^2 + y^2 = 16$

98. $x^2 + (y - 8)^2 = 81$

99. $\left(x - \frac{1}{2}\right)^2 + (y + 1)^2 = 36$

100. $(x + 4)^2 + \left(y - \frac{3}{2}\right)^2 = 100$

101. Find the standard form of the equation of the circle for which the endpoints of a diameter are $(0, 0)$ and $(4, -6)$.

102. Find the standard form of the equation of the circle for which the endpoints of a diameter are $(-2, -3)$ and $(4, -10)$.

103. ***Number of Stores*** The numbers N of Walgreen stores for the years 2000 through 2008 can be approximated by the model

$$N = 439.9t + 2987, \quad 0 \le t \le 8$$

where t represents the year, with $t = 0$ corresponding to 2000. *(Source: Walgreen Co.)*

(a) Sketch a graph of the model.

(b) Use the graph to estimate the year in which the number of stores was 6500.

104. ***Physics*** The force F (in pounds) required to stretch a spring x inches from its natural length (see figure) is

$F = \frac{5}{4}x, \ 0 \le x \le 20.$

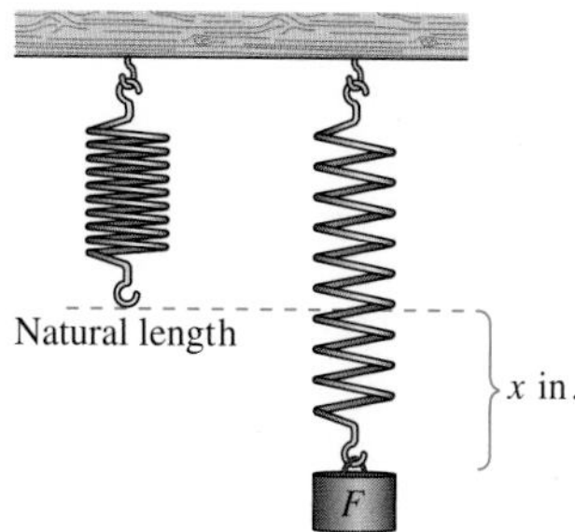

(a) Use the model to complete the table.

x	0	4	8	12	16	20
Force, F						

(b) Sketch a graph of the model.

(c) Use the graph to estimate the force necessary to stretch the spring 10 inches.

In Exercises 105–108, find the slope and y-intercept (if possible) of the equation of the line. Sketch the line.

105. $y = 6$ **106.** $x = -3$

107. $y = 3x + 13$ **108.** $y = -10x + 9$

In Exercises 109 and 110, match each value of slope m with the corresponding line in the figure.

109. (a) $m = \frac{3}{2}$

(b) $m = 0$

(c) $m = -3$

(d) $m = -\frac{1}{5}$

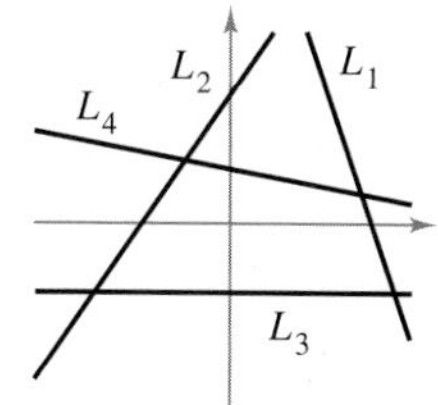

110. (a) $m = -\frac{5}{2}$

(b) m is undefined.

(c) $m = 0$

(d) $m = \frac{1}{2}$

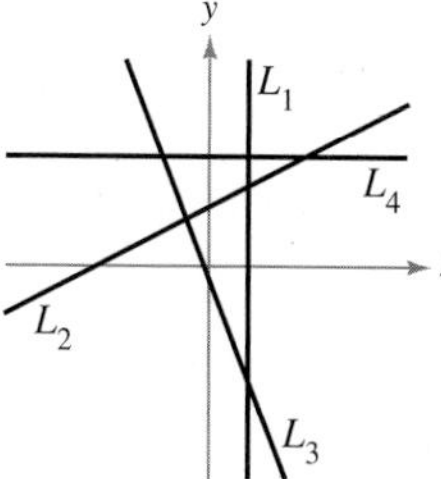

In Exercises 111–114, plot the points and find the slope of the line passing through the pair of points.

111. $(6, 4), (-3, -4)$ **112.** $\left(\frac{3}{2}, 1\right), \left(5, \frac{5}{2}\right)$

113. $(-4.5, 6), (2.1, 3)$ **114.** $(-3, 2), (8, 2)$

In Exercises 115–120, find the slope-intercept form of the equation of the line that passes through the given point and has the indicated slope. Sketch the line.

	Point	*Slope*
115.	$(3, 0)$	$m = \frac{2}{3}$
116.	$(10, -3)$	$m = -\frac{1}{2}$
117.	$(-2, 6)$	$m = 0$
118.	$(-3, 1)$	$m = 0$
119.	$(-8, 5)$	m is undefined.
120.	$(12, -6)$	m is undefined.

In Exercises 121–124, find the slope-intercept form of the equation of the line passing through the points.

121. $(0, 0), (0, 10)$ **122.** $(2, -1), (4, -1)$

123. $(-1, 0), (6, 2)$ **124.** $(11, -2), (6, -1)$

In Exercises 125–130, write the slope-intercept forms of the equations of the lines through the given point (a) parallel to the given line and (b) perpendicular to the given line.

	Point	*Line*
125.	$(2, -1)$	$x - 5 = 0$
126.	$(3, 2)$	$x + 4 = 0$
127.	$(-2, 1)$	$y + 6 = 0$
128.	$(3, 4)$	$y - 1 = 0$
129.	$(3, -2)$	$5x - 4y = 8$
130.	$(-8, 3)$	$2x + 3y = 5$

Rate of Change **In Exercises 131 and 132, you are given the dollar value of a product in 2010 and the rate at which the value of the product is expected to change during the next 5 years. Use this information to write a linear equation that gives the dollar value V of the product in terms of the year t. (Let $t = 10$ represent 2010.)**

	2010 Value	*Rate*
131.	$12,500	$850 decrease per year
132.	$72.95	$5.15 increase per year

In Exercises 133 and 134, consider an equation of the form $x + \sqrt{x - a} = b$, where a and b are constants.

133. Find a and b when the solution of the equation is $x = 20$. (There are many correct answers.)

134. ***Writing*** Write a short paragraph listing the steps required to solve this equation involving radicals, and explain why it is important to check your solutions.

P CHAPTER TEST

Take this test as you would take a test in class. When you are finished, check your work against the answers given in the back of the book.

In Exercises 1–6, solve the equation. (If not possible, explain why.)

1. $\frac{2}{3}(x - 1) + \frac{1}{4}x = 10$

2. $\frac{x - 2}{x + 2} + \frac{4}{x + 2} + 4 = 0$

3. $(x - 3)(x + 2) = 14$

4. $x^4 + x^2 - 6 = 0$

5. $x - \sqrt{2x + 1} = 1$

6. $|3x - 1| = 7$

In Exercises 7–10, solve the inequality and sketch the solution on the real number line.

7. $-3 \leq 2(x + 4) < 14$

8. $|x - 15| \geq 5$

9. $2x^2 + 5x > 12$

10. $\frac{2}{x} > \frac{5}{x + 6}$

11. Plot the points $(-2, 5)$ and $(6, 0)$. Find the coordinates of the midpoint of the line segment joining the points and the distance between the points.

12. A triangle has vertices at the points $(-2, 1)$, $(4, -1)$, and $(5, 2)$. Shift the triangle three units downward and two units to the left and find the vertices of the shifted triangle.

In Exercises 13–15, use intercepts and symmetry to sketch the graph of the equation.

13. $y = 3 - 5x$

14. $y = 4 - |x|$

15. $y = x^2 - 1$

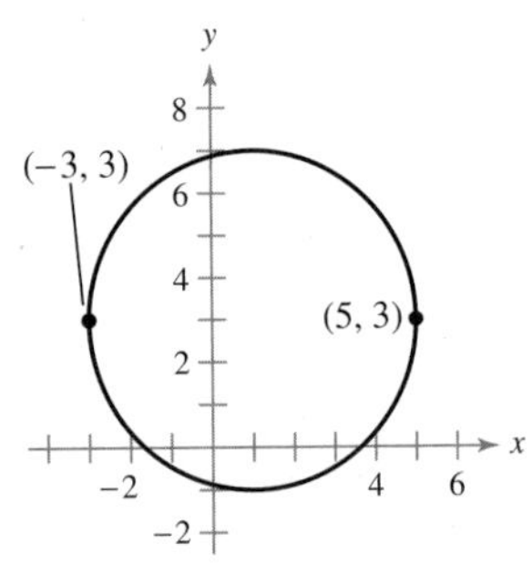

Figure for 16

16. Write the standard form of the equation of the circle shown at the left.

In Exercises 17 and 18, find the slope-intercept form of the equation of the line passing through the points.

17. $(2, -3), (-4, 9)$

18. $(3, 0.8), (7, -6)$

19. Find equations of the lines that pass through the point $(0, 4)$ and are (a) parallel to and (b) perpendicular to the line $5x + 2y = 3$.

20. The admissions office of a college wants to determine whether there is a relationship between IQ scores x and grade-point averages y after the first year of school. An equation that models the data the admissions office obtained is $y = 0.067x - 5.638$.

(a) Use a graphing utility to graph the model.

(b) Use the graph to estimate the values of x that predict a grade-point average of at least 3.0.

21. The maximum heart rate of a person in normal health is related to the person's age by the equation $r = 220 - A$, where r is the maximum heart rate in beats per minute and A is the person's age in years. Some physiologists recommend that during physical activity a sedentary person should strive to increase his or her heart rate to at least 50% of the maximum heart rate, and a highly fit person should strive to increase his or her heart rate to at most 85% of the maximum heart rate. Express as an interval the range of the target heart rate for a 20-year-old. *(Source: American Heart Association)*

P.S. PROBLEM SOLVING

1. Solve $3(x + 4)^2 + (x + 4) - 2 = 0$ in two ways.

(a) Let $u = x + 4$, and solve the resulting equation for u. Then solve the u-solution for x.

(b) Expand and collect like terms in the equation, and solve the resulting equation for x.

(c) Which method is easier? Explain your reasoning.

2. Solve the equations, given that a and b are not zero.

(a) $ax^2 + bx = 0$

(b) $ax^2 - (a - b)x - b = 0$

3. In parts (a)–(d), find the interval for b such that the equation has at least one real solution.

(a) $x^2 + bx + 4 = 0$

(b) $x^2 + bx - 4 = 0$

(c) $3x^2 + bx + 10 = 0$

(d) $2x^2 + bx + 5 = 0$

(e) Write a conjecture about the interval for b in parts (a)–(d). Explain your reasoning.

(f) What is the center of the interval for b in parts (a)–(d)?

4. Michael Kasha of Florida State University used physics and mathematics to design a new classical guitar. The model he used for the frequency of the vibrations on a circular plate was $v = (2.6t/d^2)\sqrt{E/\rho}$, where v is the frequency (in vibrations per second), t is the plate thickness (in millimeters), d is the diameter of the plate, E is the elasticity of the plate material, and ρ is the density of the plate material. For fixed values of d, E, and ρ, the graph of the equation is a line (see figure).

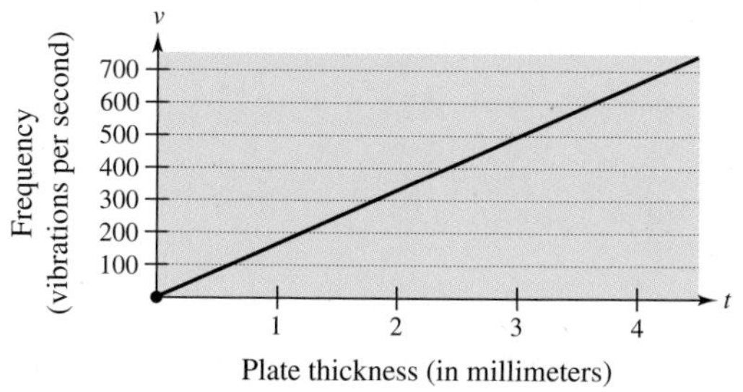

(a) Estimate the frequency when the plate thickness is 2 millimeters.

(b) Estimate the plate thickness when the frequency is 600 vibrations per second.

(c) Approximate the interval for the plate thickness when the frequency is between 200 and 400 vibrations per second.

(d) Approximate the interval for the frequency when the plate thickness is less than 3 millimeters.

5. The graphs show the solutions of equations plotted on the real number line. In each case, determine whether the solution(s) is (are) for a linear equation, a quadratic equation, both, or neither. Explain your reasoning.

(a) 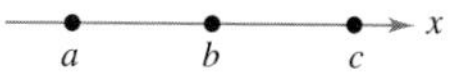

(b) 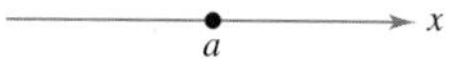

(c) 

(d) 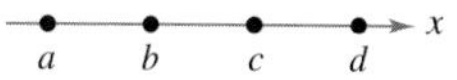

6. Consider the circle $x^2 + y^2 - 6x - 8y = 0$ shown in the figure.

(a) Find the center and radius of the circle.

(b) Find an equation of the tangent line to the circle at the point $(0, 0)$. A tangent line contains exactly one point of the circle.

(c) Find an equation of the tangent line to the circle at the point $(6, 0)$.

(d) Where do the two tangent lines intersect?

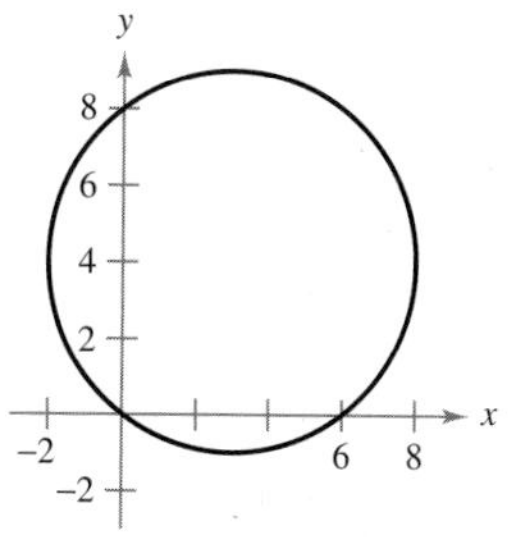

7. Let d_1 and d_2 be the distances from the point (x, y) to the points $(-1, 0)$ and $(1, 0)$, respectively, as shown in the figure. Show that the equation of the graph of all points (x, y) satisfying $d_1 d_2 = 1$ is $(x^2 + y^2)^2 = 2(x^2 - y^2)$. This curve is called a **lemniscate.** Sketch the lemniscate and identify three points on the graph.

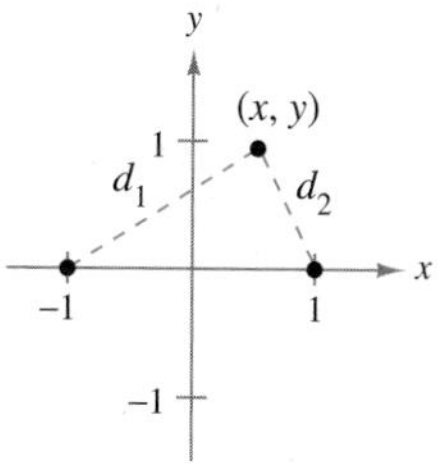

8. Write a paragraph describing how each of the following transformed points is related to the original point.

Original Point	*Transformed Point*
(a) (x, y)	$(-x, y)$
(b) (x, y)	$(x, -y)$
(c) (x, y)	$(-x, -y)$

9. The 2000 and 2010 enrollments at a college are shown in the bar graph.

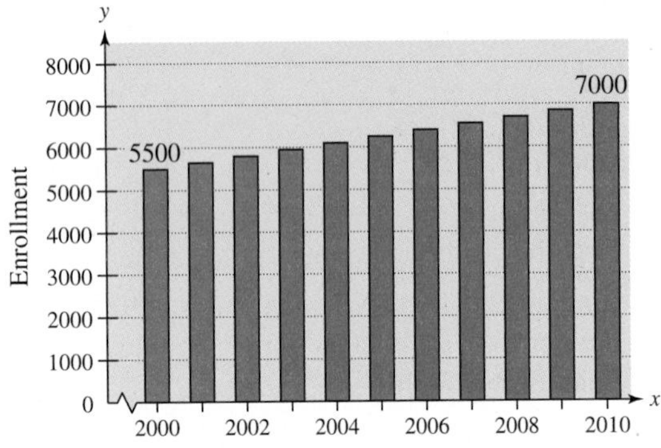

(a) Determine the average annual change in enrollment from 2000 to 2010.

(b) Use the average annual change in enrollment to estimate the enrollments in 2003, 2007, and 2009.

(c) Write an equation of the line that represents the data in part (b). What is the slope? Interpret the slope in the context of the real-life setting.

10. The per capita consumptions (in gallons) of milk M and bottled water B from 2002 through 2007 can be modeled by

$$M = -0.23t + 22.3$$

and

$$B = 1.87t + 16.1$$

where $t = 2$ represents 2002. *(Source: U.S. Dept. of Agriculture)*

(a) Find the point of intersection of these graphs algebraically.

(b) Use a graphing utility to graph the equations in the same viewing window. Explain why you chose the viewing window settings that you used.

(c) Verify your answer to part (a) using either the *zoom* and *trace* features or the *intersect* feature of your graphing utility.

(d) Explain what the point of intersection of these equations represents.

11. You want to determine whether there is a linear relationship between an athlete's body weight x (in pounds) and the athlete's maximum bench-press weight y (in pounds). The table shows a sample of data from 12 athletes.

Athlete's weight, x	165	184	150	210	196	240
Bench-press weight, y	170	185	200	255	205	295

Athlete's weight, x	202	170	185	190	230	160
Bench-press weight, y	190	175	195	185	250	155

(a) Use a graphing utility to plot the data.

(b) A model for the data is $y = 1.3x - 36$. Use a graphing utility to graph the model in the same viewing window used in part (a).

(c) Use the graph to estimate the values of x that predict a maximum bench-press weight of at least 200 pounds.

(d) Verify your estimate from part (c) algebraically.

(e) Use the graph to write a statement about the accuracy of the model. If you think the graph indicates that an athlete's weight is not a particularly good indicator of the athlete's maximum bench-press weight, list other factors that might influence an individual's maximum bench-press weight.

12. The table shows the numbers S (in millions) of cellular telephone subscribers in the United States from 2002 to 2008, where $t = 2$ represents 2002. Use the regression capabilities of a graphing utility to find a linear model for the data. Determine both analytically and graphically when the total number of subscribers exceeded 300 million. *(Source: Cellular Telecommunications and Internet Association)*

t	2	3	4	5
S	140.8	158.7	182.1	207.9

t	6	7	8
S	233.0	255.4	270.3

13. Your employer offers you a choice of wage scales: a monthly salary of \$3000 plus commission of 7% of sales or a salary of \$3400 plus a 5% commission of sales.

(a) Write a linear equation representing your wages W in terms of the sales s for both offers.

(b) At what sales level would both options yield the same wage?

(c) Write a paragraph discussing how you would choose your option.

1 Functions and Their Graphs

In this chapter, you will study several concepts that will help you prepare for your study of calculus. These concepts include sketching the graphs of equations and functions, and fitting mathematical models to data. It is important to know these concepts before moving on to calculus.

In this chapter, you should learn the following.

- How to recognize, represent, and evaluate functions. (1.1)
- How to analyze graphs of functions. (1.2)
- How to use transformations to sketch graphs of functions. (1.3)
- How to form combinations of functions. (1.4)
- How to find inverse functions. (1.5)
- How to use and write mathematical models. (1.6)

Andy Z., 2010/Used under license from Shutterstock.com

Given a function that estimates the force of water against the face of a dam in terms of the depth of the water, how can you determine the depth at which the force against the dam is 1,000,000 tons? (See Section 1.1, Exercise 106.)

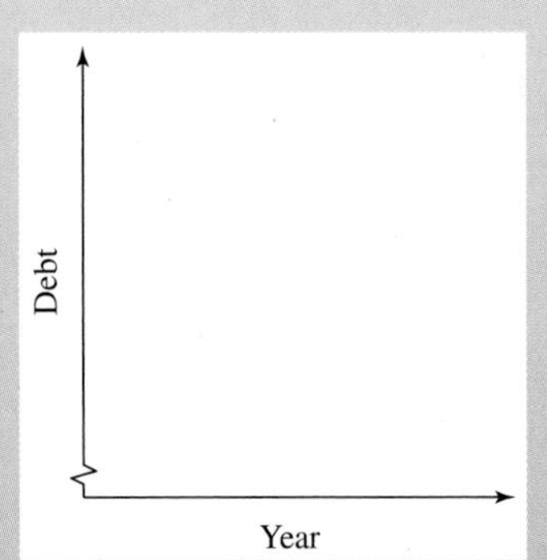

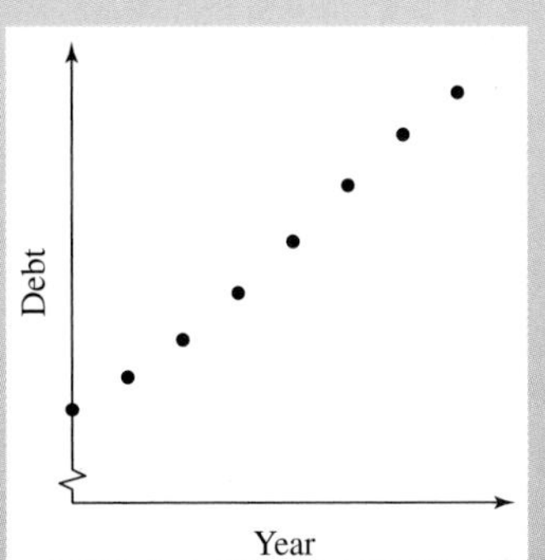

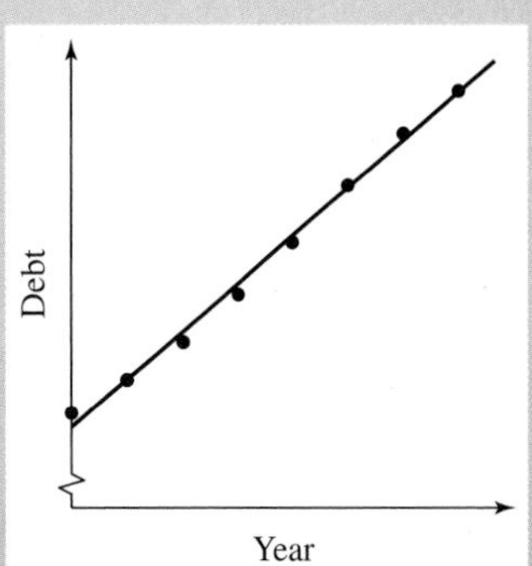

Mathematical models are commonly used to describe data sets. The best-fitting *linear* model is called the least squares regression line. (See Section 1.6.)

1.1 Functions

- **Determine whether relations between two variables are functions.**
- **Use function notation and evaluate functions.**
- **Find the domains of functions.**
- **Use functions to model and solve real-life problems.**

Introduction to Functions

Many everyday phenomena involve two quantities that are related to each other by some rule of correspondence. The mathematical term for such a rule of correspondence is a **relation.** In mathematics, relations are often represented by mathematical equations and formulas. For instance, the simple interest I earned on \$1000 for 1 year is related to the annual interest rate r by the formula $I = 1000r$.

The formula $I = 1000r$ represents a special kind of relation that matches each item from one set with *exactly one* item from a different set. Such a relation is called a **function.**

DEFINITION OF FUNCTION

A **function** f from a set A to a set B is a relation that assigns to each element x in the set A exactly one element y in the set B. The set A is the **domain** (or set of inputs) of the function f, and the set B contains the **range** (or set of outputs).

To help understand this definition, look at the function that relates the time of day to the temperature in Figure 1.1.

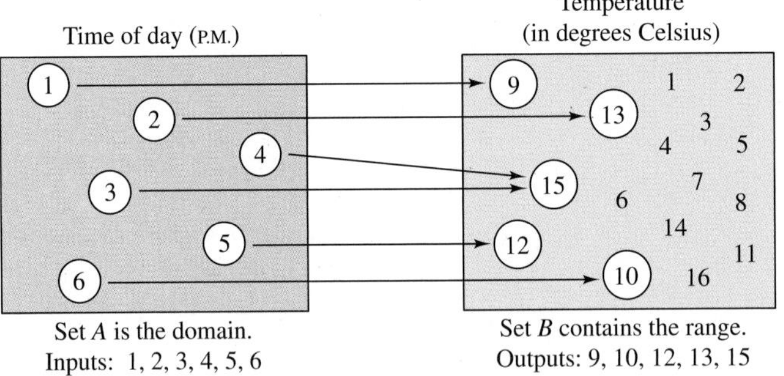

Figure 1.1

This function can be represented by the following set of ordered pairs, in which the first coordinate (x-value) is the input and the second coordinate (y-value) is the output.

$$\{(1, 9^\circ), (2, 13^\circ), (3, 15^\circ), (4, 15^\circ), (5, 12^\circ), (6, 10^\circ)\}$$

CHARACTERISTICS OF A FUNCTION FROM SET A TO SET B

1. Each element in A must be matched with an element in B.
2. Some elements in B may not be matched with any element in A.
3. Two or more elements in A may be matched with the same element in B.
4. An element in A (the domain) cannot be matched with two different elements in B.

Functions are commonly represented in four ways.

FOUR WAYS TO REPRESENT A FUNCTION

1. *Verbally* by a sentence that describes how the input variable is related to the output variable
2. *Numerically* by a table or a list of ordered pairs that matches input values with output values
3. *Graphically* by points on a graph in a coordinate plane in which the input values are represented by the horizontal axis and the output values are represented by the vertical axis
4. *Analytically* by an equation in two variables

To determine whether or not a relation is a function, you must decide whether each input value is matched with exactly one output value. When any input value is matched with two or more output values, the relation is not a function.

EXAMPLE 1 Testing for Functions

Determine whether the relation represents y as a function of x.

a. The input value x is the number of representatives from a state, and the output value y is the number of senators.

b.

Input, x	Output, y
2	11
2	10
3	8
4	5
5	1

c.

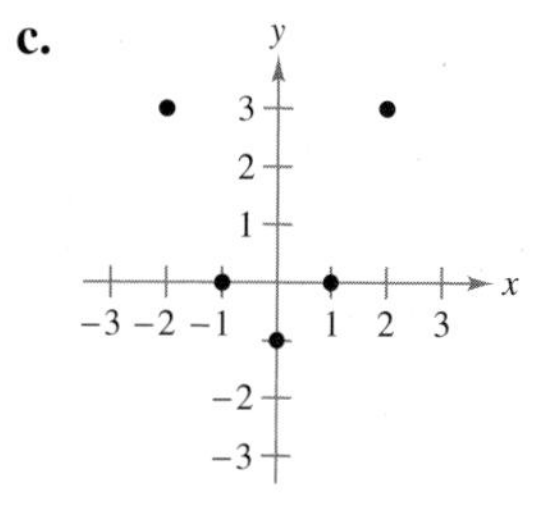

Figure 1.2

Solution

a. This verbal description *does* describe y as a function of x. Regardless of the value of x, the value of y is always 2. Such functions are called *constant functions.*

b. This table *does not* describe y as a function of x. The input value 2 is matched with two different y-values.

c. The graph in Figure 1.2 *does* describe y as a function of x. Each input value is matched with exactly one output value. ■

Representing functions by sets of ordered pairs is common in *discrete mathematics.* In algebra and calculus, however, it is more common to represent functions by equations or formulas involving two variables. For instance, the equation

$$y = x^2 \qquad y \text{ is a function of } x.$$

represents the variable y as a function of the variable x. In this equation, x is the **independent variable** and y is the **dependent variable.** The domain of the function is the set of all values taken on by the independent variable x, and the range of the function is the set of all values taken on by the dependent variable y.

Michael Nicholson/CORBIS

LEONHARD EULER (1707–1783)

Leonhard Euler, a Swiss mathematician, is considered to have been the most prolific and productive mathematician in history. One of his greatest influences on mathematics was his use of symbols, or notation. The function notation $y = f(x)$ was introduced by Euler.

EXAMPLE 2 Testing for Functions Analytically

Does the equation represent y as a function of x?

a. $x^2 + y = 1$

b. $-x + y^2 = 1$

c. $y - 2 = 0$

Solution To determine whether y is a function of x, try to solve for y in terms of x.

a. Solving for y yields

$$x^2 + y = 1 \qquad \text{Write original equation.}$$

$$y = 1 - x^2. \qquad \text{Solve for } y.$$

To each value of x there corresponds exactly one value of y. So, y is a function of x.

b. Solving for y yields

$$-x + y^2 = 1 \qquad \text{Write original equation.}$$

$$y^2 = 1 + x \qquad \text{Add } x \text{ to each side.}$$

$$y = \pm\sqrt{1 + x}. \qquad \text{Solve for } y.$$

The $\pm$ indicates that to a given value of x there correspond two values of y. So, y is not a function of x.

c. Solving for y yields

$$y - 2 = 0 \qquad \text{Write original equation.}$$

$$y = 2. \qquad \text{Solve for } y.$$

To each value of x there corresponds exactly one value of y, which is $y = 2$. So, y is a function of x. ■

Function Notation

When an equation is used to represent a function, it is convenient to name the function so that it can be referenced easily. For example, you know that the equation $y = 1 - x^2$ describes y as a function of x. Suppose you give this function the name "f." Then you can use the following **function notation.**

Input	*Output*	*Equation*
x	$f(x)$	$f(x) = 1 - x^2$

The symbol $f(x)$ is read as *the value of f at x* or simply *f of x*. The symbol $f(x)$ corresponds to the y-value for a given x. So, you can write $y = f(x)$. Keep in mind that f is the *name* of the function, whereas $f(x)$ is the *value* of the function at x. For instance, the function given by

$$f(x) = 3 - 2x$$

has *function values* denoted by $f(-1)$, $f(0)$, $f(2)$, and so on. To find these values, substitute the specified input values into the given equation.

For $x = -1$, $\quad f(-1) = 3 - 2(-1) = 3 + 2 = 5.$

For $x = 0$, $\quad f(0) = 3 - 2(0) = 3 - 0 = 3.$

For $x = 2$, $\quad f(2) = 3 - 2(2) = 3 - 4 = -1.$

Although f is often used as a convenient function name and x is often used as the independent variable, you can use other letters. For instance,

$$f(x) = x^2 - 4x + 7, \quad f(t) = t^2 - 4t + 7, \quad \text{and} \quad g(s) = s^2 - 4s + 7$$

all define the same function. In fact, the role of the independent variable is that of a "placeholder." Consequently, the function could be described by

$$f(\square) = (\square)^2 - 4(\square) + 7$$

where any real number or algebraic expression can be put in the box.

NOTE In Example 3, note that $g(x + 2)$ is not equal to $g(x) + g(2)$. In general, $g(u + v) \neq g(u) + g(v)$.

EXAMPLE 3 Evaluating a Function

Let $g(x) = -x^2 + 4x + 1$. Find each function value.

a. $g(2)$

b. $g(t)$

c. $g(x + 2)$

Solution

a. Replacing x with 2 in $g(x) = -x^2 + 4x + 1$ yields the following.

$$\begin{aligned} g(2) &= -(2)^2 + 4(2) + 1 \\ &= -4 + 8 + 1 \\ &= 5 \end{aligned}$$

b. Replacing x with t yields the following.

$$\begin{aligned} g(t) &= -(t)^2 + 4(t) + 1 \\ &= -t^2 + 4t + 1 \end{aligned}$$

c. Replacing x with $x + 2$ yields the following.

$$\begin{aligned} g(x + 2) &= -(x + 2)^2 + 4(x + 2) + 1 \\ &= -(x^2 + 4x + 4) + 4x + 8 + 1 \\ &= -x^2 - 4x - 4 + 4x + 8 + 1 \\ &= -x^2 + 5 \end{aligned}$$

■

A function defined by two or more equations over a specified domain is called a **piecewise-defined function.**

EXAMPLE 4 A Piecewise-Defined Function

Evaluate the function when $x = -1, 0,$ and 1.

$$f(x) = \begin{cases} x^2 + 1, & x < 0 \\ x - 1, & x \geq 0 \end{cases}$$

Solution Because $x = -1$ is less than 0, use $f(x) = x^2 + 1$ to obtain

$$f(-1) = (-1)^2 + 1 = 2.$$

For $x = 0$, use $f(x) = x - 1$ to obtain

$$f(0) = (0) - 1 = -1.$$

For $x = 1$, use $f(x) = x - 1$ to obtain

$$f(1) = (1) - 1 = 0.$$

■

EXAMPLE 5 Finding Values for Which $f(x) = 0$

Find all real values of x such that $f(x) = 0$.

a. $f(x) = -2x + 10$

b. $f(x) = x^2 - 5x + 6$

Solution For each function, set $f(x) = 0$ and solve for x.

a. $-2x + 10 = 0$ — Set $f(x)$ equal to 0.

$-2x = -10$ — Subtract 10 from each side.

$x = 5$ — Divide each side by -2.

So, $f(x) = 0$ when $x = 5$.

b. $x^2 - 5x + 6 = 0$ — Set $f(x)$ equal to 0.

$(x - 2)(x - 3) = 0$ — Factor.

$x - 2 = 0 \Rightarrow x = 2$ — Set 1st factor equal to 0.

$x - 3 = 0 \Rightarrow x = 3$ — Set 2nd factor equal to 0.

So, $f(x) = 0$ when $x = 2$ or $x = 3$.

EXAMPLE 6 Finding Values for Which $f(x) = g(x)$

Find the values of x for which $f(x) = g(x)$.

a. $f(x) = x^2 + 1$

and

$g(x) = 3x - x^2$

b. $f(x) = x^2 - 1$

and

$g(x) = -x^2 + x + 2$

Solution

a. $x^2 + 1 = 3x - x^2$ — Set $f(x)$ equal to $g(x)$.

$2x^2 - 3x + 1 = 0$ — Write in general form.

$(2x - 1)(x - 1) = 0$ — Factor.

$2x - 1 = 0 \Rightarrow x = \frac{1}{2}$ — Set 1st factor equal to 0.

$x - 1 = 0 \Rightarrow x = 1$ — Set 2nd factor equal to 0.

So, $f(x) = g(x)$ when $x = \dfrac{1}{2}$ or $x = 1$.

b. $x^2 - 1 = -x^2 + x + 2$ — Set $f(x)$ equal to $g(x)$.

$2x^2 - x - 3 = 0$ — Write in general form.

$(2x - 3)(x + 1) = 0$ — Factor.

$2x - 3 = 0 \Rightarrow x = \frac{3}{2}$ — Set 1st factor equal to 0.

$x + 1 = 0 \Rightarrow x = -1$ — Set 2nd factor equal to 0.

So, $f(x) = g(x)$ when $x = \dfrac{3}{2}$ or $x = -1$. ■

EXPLORATION

Use a graphing utility to graph the functions given by $y = \sqrt{9 - x^2}$ and $y = \sqrt{x^2 - 9}$. What is the domain of each function? Do the domains of these two functions overlap? If so, for what values do the domains overlap?

The Domain of a Function

The domain of a function can be described explicitly or it can be *implied* by the expression used to define the function. The **implied domain** is the set of all real numbers for which the expression is defined. For instance, the function given by

$$f(x) = \frac{1}{x^2 - 4}$$ Domain excludes x-values that result in division by zero.

has an implied domain that consists of all real x other than $x = \pm 2$. These two values are excluded from the domain because division by zero is undefined. Another common type of implied domain is that used to avoid even roots of negative numbers. For example, the function given by

$$f(x) = \sqrt{x}$$ Domain excludes x-values that result in even roots of negative numbers.

is defined only for $x \geq 0$. So, its implied domain is the interval $[0, \infty)$. In general, the domain of a function *excludes* values that would cause division by zero *or* that would result in the even root of a negative number.

EXAMPLE 7 Finding the Domain of a Function

Find the domain of each function.

a. f: $\{(-3, 0), (-1, 4), (0, 2), (2, 2), (4, -1)\}$

b. $g(x) = \dfrac{1}{x + 5}$

c. Volume of a sphere: $V = \frac{4}{3}\pi r^3$

d. $h(x) = \sqrt{4 - x^2}$

e. $y = x^2 + 3x + 4$

Solution

a. The domain of f consists of all first coordinates in the set of ordered pairs.

$$\text{Domain} = \{-3, -1, 0, 2, 4\}$$

b. Excluding x-values that yield zero in the denominator, the domain of g is the set of all real numbers x except $x = -5$.

c. Because this function represents the volume of a sphere, the values of the radius r must be positive. So, the domain is the set of all real numbers r such that $r > 0$.

d. This function is defined only for x-values for which

$$4 - x^2 \geq 0.$$

Using the methods described in Section P.2, you can conclude that $-2 \leq x \leq 2$. So, the domain is the interval $[-2, 2]$.

e. This function is defined for all values of x. So, the domain is the set of all real numbers. ■

In Example 7(c), note that the domain of a function may be implied by the physical context. For instance, from the equation

$$V = \frac{4}{3}\pi r^3$$

you would have no reason to restrict r to positive values, but the physical context implies that a sphere cannot have a negative or zero radius.

Figure 1.3

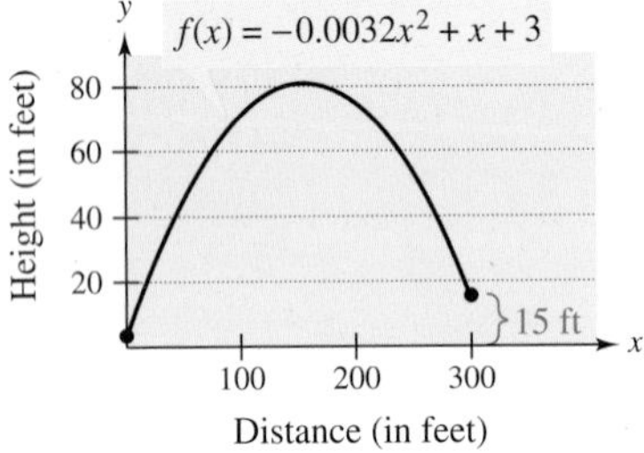

Figure 1.4

Applications

EXAMPLE 8 The Dimensions of a Container

You work in the marketing department of a soft-drink company and are experimenting with a new can for iced tea that is slightly narrower and taller than a standard can. For your experimental can, the ratio of the height to the radius is 4, as shown in Figure 1.3.

a. Write the volume of the can as a function of the radius r.

b. Write the volume of the can as a function of the height h.

Solution

a. $V(r) = \pi r^2 h = \pi r^2(4r) = 4\pi r^3$ Write V as a function of r.

b. $V(h) = \pi\left(\frac{h}{4}\right)^2 h = \frac{\pi h^3}{16}$ Write V as a function of h.

EXAMPLE 9 The Path of a Baseball

A baseball is hit at a point 3 feet above ground at a velocity of 100 feet per second and an angle of 45°. The path of the baseball is given by the function $f(x) = -0.0032x^2 + x + 3$, where x and $f(x)$ are measured in feet, as shown in Figure 1.4. Will the baseball clear a 10-foot fence located 300 feet from home plate?

Solution When $x = 300$, the height of the baseball is

$$f(300) = -0.0032(300)^2 + 300 + 3 = 15 \text{ feet.}$$

So, the ball will clear the fence. ■

One of the basic definitions in calculus employs the ratio

$$\frac{f(x + \Delta x) - f(x)}{\Delta x}, \quad \Delta x \neq 0.$$

This ratio is called a **difference quotient,** as illustrated in Example 10.

EXAMPLE 10 Evaluating a Difference Quotient

For $f(x) = x^2 - 4x + 7$, find $\dfrac{f(x + \Delta x) - f(x)}{\Delta x}$.

Solution

$$\begin{aligned}\frac{f(x + \Delta x) - f(x)}{\Delta x} &= \frac{[(x + \Delta x)^2 - 4(x + \Delta x) + 7] - (x^2 - 4x + 7)}{\Delta x} \\ &= \frac{x^2 + 2x(\Delta x) + (\Delta x)^2 - 4x - 4\Delta x + 7 - x^2 + 4x - 7}{\Delta x} \\ &= \frac{2x(\Delta x) + (\Delta x)^2 - 4\Delta x}{\Delta x} \\ &= \frac{\Delta x(2x + \Delta x - 4)}{\Delta x} \\ &= 2x + \Delta x - 4, \quad \Delta x \neq 0\end{aligned}$$

■

1.1 Exercises

See www.CalcChat.com for worked-out solutions to odd-numbered exercises.

In Exercises 1–6, fill in the blanks.

1. A relation that assigns to each element x from a set of inputs, or ________, exactly one element y in a set of outputs, or ________, is called a ________.
2. Functions are commonly represented in four different ways, ________, ________, ________, and ________.
3. For an equation that represents y as a function of x, the set of all values taken on by the ________ variable x is the domain, and the set of all values taken on by the ________ variable y is the range.
4. The function given by

$$f(x) = \begin{cases} 2x - 1, & x < 0 \\ x^2 + 4, & x \geq 0 \end{cases}$$

is an example of a ________ function.

5. If the domain of the function f is not given, then the set of values of the independent variable for which the expression is defined is called the ________ ________.
6. In calculus, one of the basic definitions is that of a ________ ________, given by

$$\frac{f(x + h) - f(x)}{h}, \quad h \neq 0.$$

In Exercises 7–12, is the relationship a function?

7. *Domain* *Range*

Domain: -2, -1, 0, 1, 2
Range: 5, 6, 7, 8

8. *Domain* *Range*

Domain: -2, -1, 0, 1, 2
Range: 3, 4, 5

9. *Domain* *Range*

Domain: -5, -4, -3, -2
Range: 1, 2, 3, 4, 5

10. *Domain* *Range*

Domain: 1, 2, 3, 4, 5
Range: -4, -2, 0

11. *Domain* *Range*

National League: Cubs, Pirates, Dodgers

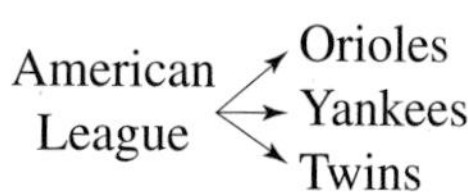

12. *Domain* *Range*

(Year) (Number of North Atlantic tropical storms and hurricanes)

Domain: 1999, 2000, 2001, 2002, 2003, 2004, 2005, 2006, 2007, 2008
Range: 10, 12, 15, 16, 21, 27

In Exercises 13–16, determine whether the relation represents y as a function of x.

13.

Input, x	-2	-1	0	1	2
Output, y	-8	-1	0	1	8

14.

Input, x	0	1	2	1	0
Output, y	-4	-2	0	2	4

15.

Input, x	10	7	4	7	10
Output, y	3	6	9	12	15

16.

Input, x	0	3	9	12	15
Output, y	3	3	3	3	3

In Exercises 17 and 18, which sets of ordered pairs represent functions from A to B? Explain.

17. $A = \{0, 1, 2, 3\}$ and $B = \{-2, -1, 0, 1, 2\}$
    (a) $\{(0, 1), (1, -2), (2, 0), (3, 2)\}$
    (b) $\{(0, -1), (2, 2), (1, -2), (3, 0), (1, 1)\}$
    (c) $\{(0, 0), (1, 0), (2, 0), (3, 0)\}$
    (d) $\{(0, 2), (3, 0), (1, 1)\}$
18. $A = \{a, b, c\}$ and $B = \{0, 1, 2, 3\}$
    (a) $\{(a, 1), (c, 2), (c, 3), (b, 3)\}$
    (b) $\{(a, 1), (b, 2), (c, 3)\}$
    (c) $\{(1, a), (0, a), (2, c), (3, b)\}$
    (d) $\{(c, 0), (b, 0), (a, 3)\}$

Circulation of Newspapers **In Exercises 19 and 20, use the graph, which shows the circulation (in millions) of daily newspapers in the United States.** ***(Source: Editor & Publisher Company)***

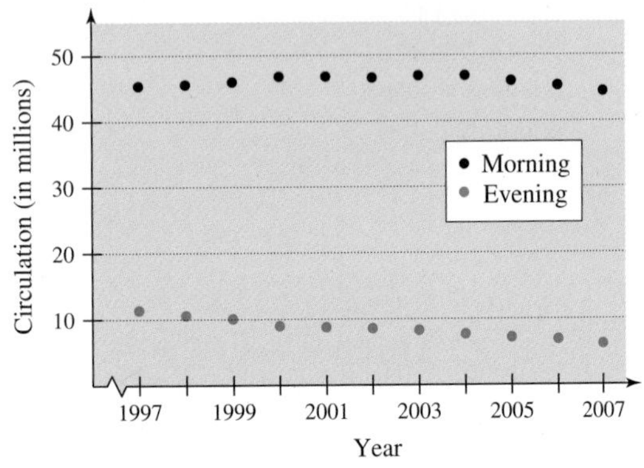

19. Is the circulation of morning newspapers a function of the year? Is the circulation of evening newspapers a function of the year? Explain.

20. Let $f(x)$ represent the circulation of evening newspapers in year x. Find $f(2002)$.

In Exercises 21–38, determine whether the equation represents *y* as a function of *x*.

21. $x^2 + y^2 = 4$ **22.** $x^2 - y^2 = 16$

23. $x^2 + y = 4$ **24.** $y - 4x^2 = 36$

25. $2x + 3y = 4$ **26.** $2x + 5y = 10$

27. $(x + 2)^2 + (y - 1)^2 = 25$

28. $(x - 2)^2 + y^2 = 4$

29. $y^2 = x^2 - 1$ **30.** $x + y^2 = 4$

31. $y = \sqrt{16 - x^2}$ **32.** $y = \sqrt{x + 5}$

33. $y = |4 - x|$ **34.** $|y| = 4 - x$

35. $x = 14$ **36.** $y = -75$

37. $y + 5 = 0$ **38.** $x - 1 = 0$

In Exercises 39–54, evaluate the function at each specified value of the independent variable and simplify.

39. $f(x) = 2x - 3$
(a) $f(1)$ (b) $f(-3)$ (c) $f(x - 1)$

40. $g(y) = 7 - 3y$
(a) $g(0)$ (b) $g\left(\frac{7}{3}\right)$ (c) $g(s + 2)$

41. $V(r) = \frac{4}{3}\pi r^3$
(a) $V(3)$ (b) $V\left(\frac{3}{2}\right)$ (c) $V(2r)$

42. $S(r) = 4\pi r^2$
(a) $S(2)$ (b) $S\left(\frac{1}{2}\right)$ (c) $S(3r)$

43. $g(t) = 4t^2 - 3t + 5$
(a) $g(2)$ (b) $g(t - 2)$ (c) $g(t) - g(2)$

44. $h(t) = t^2 - 2t$
(a) $h(2)$ (b) $h(1.5)$ (c) $h(x + 2)$

45. $f(y) = 3 - \sqrt{y}$
(a) $f(4)$ (b) $f(0.25)$ (c) $f(4x^2)$

46. $f(x) = \sqrt{x + 8} + 2$
(a) $f(-8)$ (b) $f(1)$ (c) $f(x - 8)$

47. $q(x) = 1/(x^2 - 9)$
(a) $q(0)$ (b) $q(3)$ (c) $q(y + 3)$

48. $q(t) = (2t^2 + 3)/t^2$
(a) $q(2)$ (b) $q(0)$ (c) $q(-x)$

49. $f(x) = |x|/x$
(a) $f(2)$ (b) $f(-2)$ (c) $f(x - 1)$

50. $f(x) = |x| + 4$
(a) $f(2)$ (b) $f(-2)$ (c) $f(x^2)$

51. $f(x) = \begin{cases} 2x + 1, & x < 0 \\ 2x + 2, & x \geq 0 \end{cases}$
(a) $f(-1)$ (b) $f(0)$ (c) $f(2)$

52. $f(x) = \begin{cases} x^2 + 2, & x \leq 1 \\ 2x^2 + 2, & x > 1 \end{cases}$
(a) $f(-2)$ (b) $f(1)$ (c) $f(2)$

53. $f(x) = \begin{cases} 3x - 1, & x < -1 \\ 4, & -1 \leq x \leq 1 \\ x^2, & x > 1 \end{cases}$
(a) $f(-2)$ (b) $f\left(-\frac{1}{2}\right)$ (c) $f(3)$

54. $f(x) = \begin{cases} 4 - 5x, & x \leq -2 \\ 0, & -2 < x < 2 \\ x^2 + 1, & x \geq 2 \end{cases}$
(a) $f(-3)$ (b) $f(4)$ (c) $f(-1)$

In Exercises 55–60, complete the table.

55. $f(x) = x^2 - 3$

x	-2	-1	0	1	2
$f(x)$					

56. $g(x) = \sqrt{x - 3}$

x	3	4	5	6	7
$g(x)$					

57. $h(t) = \frac{1}{2}|t + 3|$

t	-5	-4	-3	-2	-1
$h(t)$					

58. $f(s) = \dfrac{|s-2|}{s-2}$

s	0	1	$\frac{3}{2}$	$\frac{5}{2}$	4
$f(s)$					

59. $f(x) = \begin{cases} -\frac{1}{2}x + 4, & x \le 0 \\ (x-2)^2, & x > 0 \end{cases}$

x	-2	-1	0	1	2
$f(x)$					

60. $f(x) = \begin{cases} 9 - x^2, & x < 3 \\ x - 3, & x \ge 3 \end{cases}$

x	1	2	3	4	5
$f(x)$					

In Exercises 61–68, find all real values of x such that $f(x) = 0$.

61. $f(x) = 15 - 3x$
62. $f(x) = 5x + 1$
63. $f(x) = \dfrac{3x - 4}{5}$
64. $f(x) = \dfrac{12 - x^2}{5}$
65. $f(x) = x^2 - 9$
66. $f(x) = x^2 - 8x + 15$
67. $f(x) = x^3 - x$
68. $f(x) = x^3 - x^2 - 4x + 4$

In Exercises 69–72, find the value(s) of x for which $f(x) = g(x)$.

69. $f(x) = x^2, \quad g(x) = x + 2$
70. $f(x) = x^2 + 2x + 1, \quad g(x) = 7x - 5$
71. $f(x) = x^4 - 2x^2, \quad g(x) = 2x^2$
72. $f(x) = \sqrt{x} - 4, \quad g(x) = 2 - x$

In Exercises 73–84, find the domain of the function.

73. $f(x) = 5x^2 + 2x - 1$
74. $g(x) = 1 - 2x^2$
75. $h(t) = \dfrac{4}{t}$
76. $s(y) = \dfrac{3y}{y + 5}$
77. $g(y) = \sqrt{y - 10}$
78. $f(t) = \sqrt[3]{t + 4}$
79. $g(x) = \dfrac{1}{x} - \dfrac{3}{x + 2}$
80. $h(x) = \dfrac{10}{x^2 - 2x}$
81. $f(s) = \dfrac{\sqrt{s - 1}}{s - 4}$
82. $f(x) = \dfrac{\sqrt{x + 6}}{6 + x}$
83. $f(x) = \dfrac{x - 4}{\sqrt{x}}$
84. $f(x) = \dfrac{x + 2}{\sqrt{x - 10}}$

In Exercises 85–88, assume that the domain of f is the set $A = \{-2, -1, 0, 1, 2\}$. Determine the set of ordered pairs that represents the function f.

85. $f(x) = x^2$
86. $f(x) = (x - 3)^2$
87. $f(x) = |x| + 2$
88. $f(x) = |x + 1|$

WRITING ABOUT CONCEPTS

89. Does the relationship shown in the figure represent a function from set A to set B? Explain.

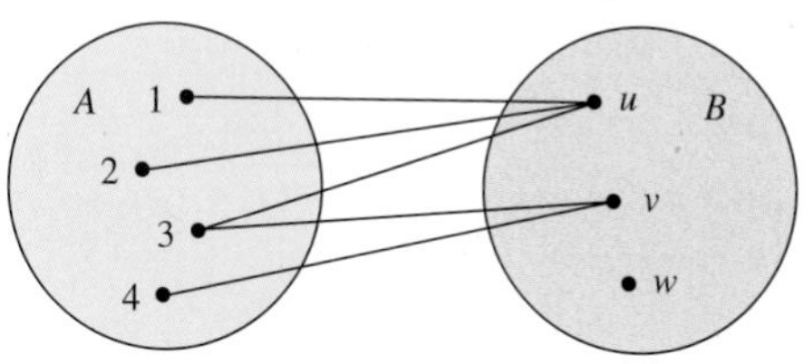

90. Describe an advantage of function notation.

91. ***Geometry*** Write the area A of a square as a function of its perimeter P.

92. ***Geometry*** Write the area A of a circle as a function of its circumference C.

93. ***Geometry*** Write the area A of an isosceles triangle with a height of 8 inches and a base of b inches as a function of the length s of one of its two equal sides.

94. ***Geometry*** Write the area A of an equilateral triangle as a function of the length s of its sides.

95. ***Maximum Volume*** An open box of maximum volume is to be made from a square piece of material 24 centimeters on a side by cutting equal squares from the corners and turning up the sides (see figure).

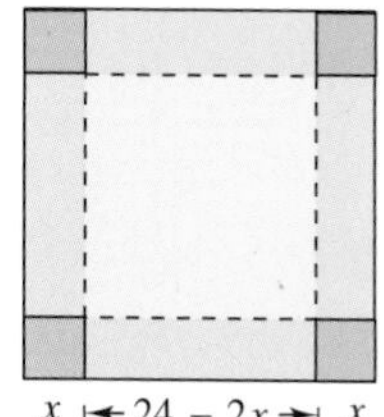

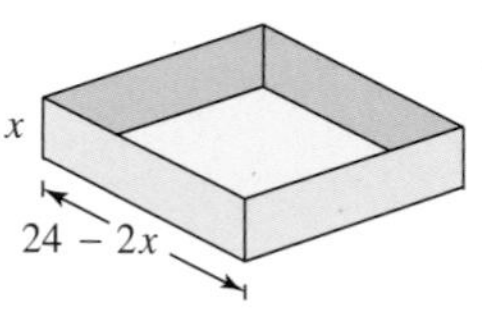

(a) The table shows the volumes V (in cubic centimeters) of the box for various heights x (in centimeters). Use the table to estimate the maximum volume.

Height, x	1	2	3	4	5	6
Volume, V	484	800	972	1024	980	864

(b) Plot the points (x, V) from the table in part (a). Does the relation defined by the ordered pairs represent V as a function of x?

(c) If V is a function of x, write the function and determine its domain.

96. ***Maximum Profit*** The cost per unit in the production of an MP3 player is \$60. The manufacturer charges \$90 per unit for orders of 100 or less. To encourage large orders, the manufacturer reduces the charge by \$0.15 per MP3 player for each unit ordered in excess of 100 (for example, there would be a charge of \$87 per MP3 player for an order size of 120).

(a) The table shows the profits P (in dollars) for various numbers of units ordered, x. Use the table to estimate the maximum profit.

Units, x	110	120	130	140
Profit, P	3135	3240	3315	3360

Units, x	150	160	170
Profit, P	3375	3360	3315

(b) Plot the points (x, P) from the table in part (a). Does the relation defined by the ordered pairs represent P as a function of x?

(c) If P is a function of x, write the function and determine its domain.

97. ***Geometry*** A right triangle is formed in the first quadrant by the x- and y-axes and a line through the point $(2, 1)$ (see figure). Write the area A of the triangle as a function of x, and determine the domain of the function.

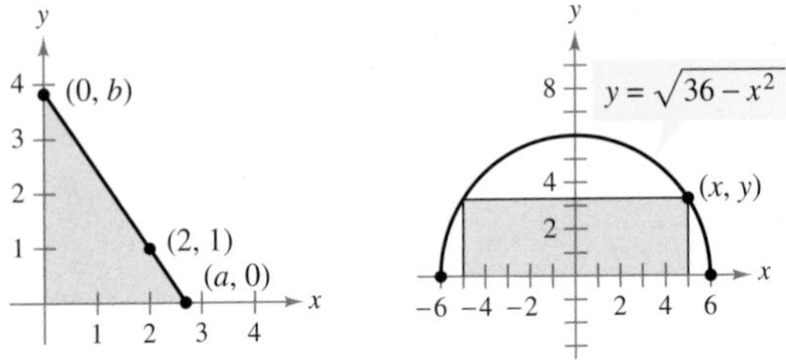

Figure for 97 **Figure for 98**

98. ***Geometry*** A rectangle is bounded by the x-axis and the semicircle $y = \sqrt{36 - x^2}$ (see figure). Write the area A of the rectangle as a function of x, and graphically determine the domain of the function.

99. ***Path of a Ball*** The height y (in feet) of a baseball thrown by a child is

$$y = -\frac{1}{10}x^2 + 3x + 6$$

where x is the horizontal distance (in feet) from where the ball was thrown. Will the ball fly over the head of another child 30 feet away trying to catch the ball? (Assume that the child who is trying to catch the ball holds a baseball glove at a height of 5 feet.)

100. ***Prescription Drugs*** The numbers d (in millions) of drug prescriptions filled by independent outlets in the United States from 2000 through 2007 (see figure) can be approximated by the model

$$d(t) = \begin{cases} 10.6t + 699, & 0 \le t \le 4 \\ 15.5t + 637, & 5 \le t \le 7 \end{cases}$$

where t represents the year, with $t = 0$ corresponding to 2000. Use this model to find the number of drug prescriptions filled by independent outlets in each year from 2000 through 2007. *(Source: National Association of Chain Drug Stores)*

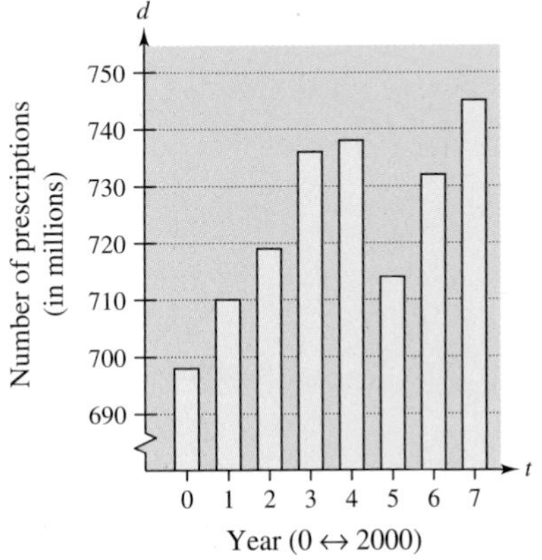

101. ***Median Sales Price*** The median sale prices p (in thousands of dollars) of an existing one-family home in the United States from 1998 through 2007 (see figure) can be approximated by the model

$$p(t) = \begin{cases} 1.011t^2 - 12.38t + 170.5, & 8 \le t \le 13 \\ -6.950t^2 + 222.55t - 1557.6, & 14 \le t \le 17 \end{cases}$$

where t represents the year, with $t = 8$ corresponding to 1998. Use this model to find the median sale price of an existing one-family home in each year from 1998 through 2007. *(Source: National Association of Realtors)*

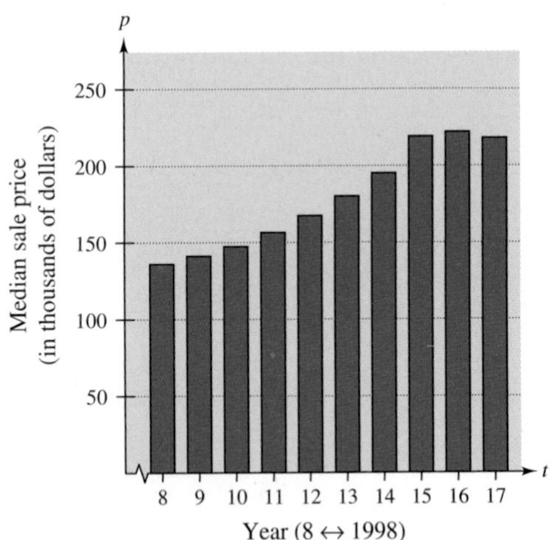

102. ***Postal Regulations*** A rectangular package to be sent by the U.S. Postal Service can have a maximum combined length and girth (perimeter of a cross section) of 108 inches (see figure).

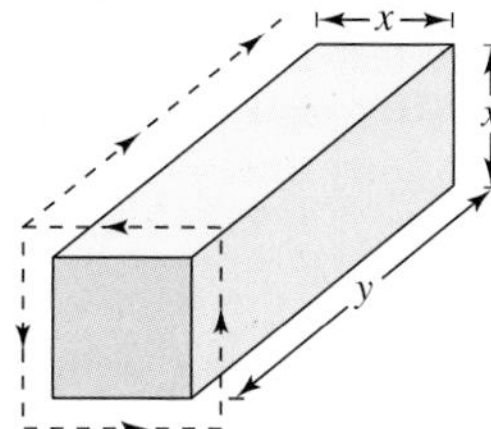

(a) Write the volume V of the package as a function of x. What is the domain of the function?

(b) Use a graphing utility to graph your function. Be sure to use an appropriate window setting.

(c) What dimensions will maximize the volume of the package? Explain your answer.

103. ***Cost, Revenue, and Profit*** A company produces a product for which the variable cost is $12.30 per unit and the fixed costs are $98,000. The product sells for $17.98. Let x be the number of units produced and sold.

(a) The total cost for a business is the sum of the variable cost and the fixed costs. Write the total cost C as a function of the number of units produced.

(b) Write the revenue R as a function of the number of units sold.

(c) Write the profit P as a function of the number of units sold. (Note: $P = R - C$)

104. ***Average Cost*** The inventor of a new game believes that the variable cost for producing the game is $0.95 per unit and the fixed costs are $6000. The inventor sells each game for $1.69. Let x be the number of games sold.

(a) The total cost for a business is the sum of the variable cost and the fixed costs. Write the total cost C as a function of the number of games sold.

(b) Write the average cost per unit $\overline{C} = C/x$ as a function of x.

105. ***Transportation*** For groups of 80 or more people, a charter bus company determines the rate per person according to the formula

$$\text{Rate} = 8 - 0.05(n - 80), \quad n \geq 80$$

where the rate is given in dollars and n is the number of people.

(a) Write the revenue R for the bus company as a function of n.

(b) Use the function in part (a) to complete the table. What can you conclude?

n	90	100	110	120	130	140	150
$R(n)$							

106. ***Physics*** The force F (in tons) of water against the face of a dam is estimated by the function $F(y) = 149.76\sqrt{10}y^{5/2}$, where y is the depth of the water (in feet).

(a) Complete the table. What can you conclude from the table?

y	5	10	20	30	40
$F(y)$					

(b) Use the table to approximate the depth at which the force against the dam is 1,000,000 tons.

(c) Find the depth at which the force against the dam is 1,000,000 tons analytically.

107. ***E-Filing*** The table shows the numbers of tax returns (in millions) made through e-file from 2000 through 2007. Let $f(t)$ represent the number of tax returns made through e-file in the year t. *(Source: Internal Revenue Service)*

Year	2000	2001	2002	2003
Number (in millions)	35.4	40.2	46.9	52.9

Year	2004	2005	2006	2007
Number (in millions)	61.5	68.5	73.3	80.0

(a) Find $\dfrac{f(2007) - f(2000)}{2007 - 2000}$ and interpret the result in the context of the problem.

(b) Make a scatter plot of the data.

(c) Find a linear model for the data analytically. Let N represent the number of tax returns made through e-file and let $t = 0$ correspond to 2000.

(d) Use the model found in part (c) to complete the table. Compare your results with the actual data.

t	0	1	2	3	4	5	6	7
N								

(e) Use a graphing utility to find a linear model for the data. Let $x = 0$ correspond to 2000. How does the model you found in part (c) compare with the model given by the graphing utility?

108. ***Height of a Balloon*** A balloon carrying a transmitter ascends vertically from a point 3000 feet from the receiving station.

(a) Draw a diagram that gives a visual representation of the problem. Let h represent the height of the balloon and let d represent the distance between the balloon and the receiving station.

(b) Write the height of the balloon as a function of d. What is the domain of the function?

In Exercises 109–116, find the difference quotient and simplify your answer.

109. $f(x) = x^2 - x + 1, \quad \dfrac{f(2 + \Delta x) - f(2)}{\Delta x}, \quad \Delta x \neq 0$

110. $f(x) = 5x - x^2, \quad \dfrac{f(5 + \Delta x) - f(5)}{\Delta x}, \quad \Delta x \neq 0$

111. $f(x) = x^3 + 2x - 1, \quad \dfrac{f(x + c) - f(x)}{c}, \quad c \neq 0$

112. $f(x) = x^3 - x + 1, \quad \dfrac{f(x + c) - f(x)}{c}, \quad c \neq 0$

113. $g(x) = 3x - 1, \quad \dfrac{g(x) - g(3)}{x - 3}, \quad x \neq 3$

114. $f(t) = \dfrac{1}{t}, \quad \dfrac{f(t) - f(1)}{t - 1}, \quad t \neq 1$

115. $f(x) = \sqrt{5x}, \quad \dfrac{f(x) - f(5)}{x - 5}, \quad x \neq 5$

116. $f(x) = x^{2/3} + 1, \quad \dfrac{f(x) - f(8)}{x - 8}, \quad x \neq 8$

In Exercises 117–120, match the data with one of the following functions

$$f(x) = cx,\ g(x) = cx^2,\ h(x) = c\sqrt{|x|},\ \text{and}\ r(x) = \frac{c}{x}$$

and determine the value of the constant c that will make the function fit the data in the table.

117.

x	-4	-1	0	1	4
y	-32	-2	0	-2	-32

118.

x	-4	-1	0	1	4
y	-1	$-\frac{1}{4}$	0	$\frac{1}{4}$	1

119.

x	-4	-1	0	1	4
y	-8	-32	Undefined	32	8

120.

x	-4	-1	0	1	4
y	6	3	0	3	6

True or False? **In Exercises 121–126, determine whether the statement is true or false. Justify your answer.**

121. Every relation is a function.

122. Every function is a relation.

123. A function can assign all elements in the domain to a single element in the range.

124. A function can assign one element from the domain to two or more elements in the range.

125. The domain of the function given by $f(x) = x^4 - 1$ is $(-\infty, \infty)$, and the range of $f(x)$ is $(0, \infty)$.

126. The set of ordered pairs $\{(-8, -2), (-6, 0), (-4, 0), (-2, 2), (0, 4), (2, -2)\}$ represents a function.

127. ***Think About It*** Consider

$$f(x) = \sqrt{x - 1} \quad \text{and} \quad g(x) = \frac{1}{\sqrt{x - 1}}.$$

Why are the domains of f and g different?

128. ***Think About It*** Consider

$$f(x) = \sqrt{x - 2} \quad \text{and} \quad g(x) = \sqrt[3]{x - 2}.$$

Why are the domains of f and g different?

129. ***Think About It*** Given $f(x) = x^2$, is f the independent variable? Why or why not?

CAPSTONE

130. (a) Describe any differences between a *relation* and a *function.*

(b) In your own words, explain the meanings of *domain* and *range.*

In Exercises 131 and 132, determine whether the statements use the word *function* in ways that are mathematically correct. Explain your reasoning.

131. (a) The sales tax on a purchased item is a function of the selling price.

(b) Your score on the next algebra exam is a function of the number of hours you study the night before the exam.

132. (a) The amount in your savings account is a function of your salary.

(b) The speed at which a free-falling baseball strikes the ground is a function of the height from which it was dropped.

1.2 Analyzing Graphs of Functions

- **Use the Vertical Line Test for functions.**
- **Find the zeros of functions.**
- **Determine intervals on which functions are increasing or decreasing and determine relative maximum and relative minimum values of functions.**
- **Identify and graph linear functions.**
- **Identify and graph step and other piecewise-defined functions.**
- **Identify even and odd functions.**

The Graph of a Function

In Section 1.1, you studied functions from an analytic point of view. In this section, you will study functions from a graphical perspective.

The **graph of a function** f is the collection of ordered pairs $(x, f(x))$ such that x is in the domain of f. As you study this section, remember that

$x =$ the directed distance from the y-axis

$y = f(x) =$ the directed distance from the x-axis

as shown in Figure 1.5.

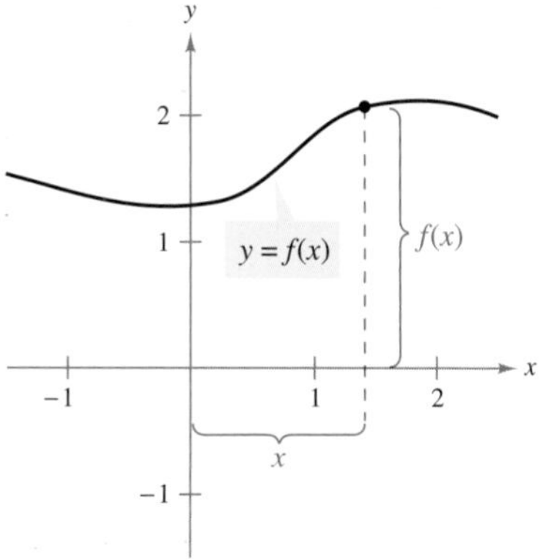

Figure 1.5

EXAMPLE 1 Finding the Domain and Range of a Function

Use the graph of the function f, shown in Figure 1.6, to find (a) the domain of f, (b) the function values $f(-1)$ and $f(2)$, and (c) the range of f.

Solution

a. The closed dot at $(-1, 1)$ indicates that $x = -1$ is in the domain of f, whereas the open dot at $(5, 2)$ indicates that $x = 5$ is not in the domain. So, the domain of f is all x in the interval $[-1, 5)$.

b. Because $(-1, 1)$ is a point on the graph of f, it follows that $f(-1) = 1$. Similarly, because $(2, -3)$ is a point on the graph of f, it follows that $f(2) = -3$.

c. Because the graph does not extend below $f(2) = -3$ or above $f(0) = 3$, the range of f is the interval $[-3, 3]$. ■

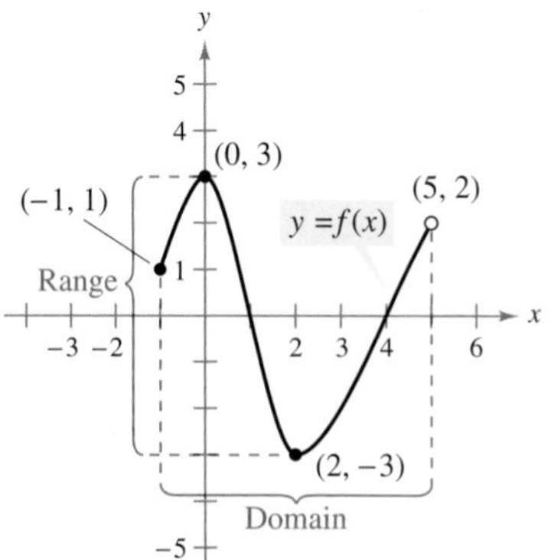

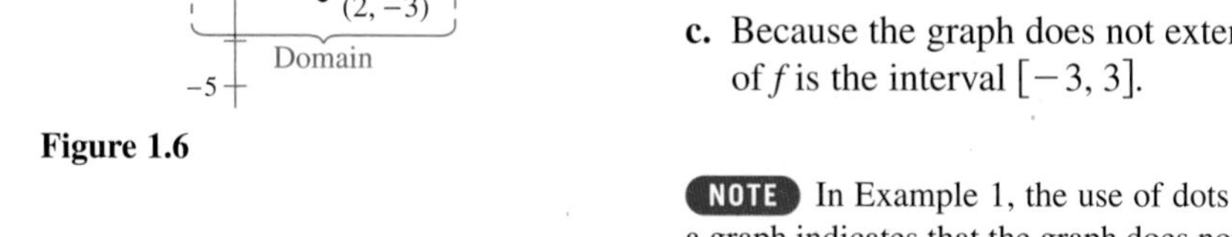

Figure 1.6

NOTE In Example 1, the use of dots (open or closed) at the extreme left and right points of a graph indicates that the graph does not extend beyond these points. If no such dots are shown, assume that the graph extends beyond these points. ■

By the definition of a function, at most one y-value corresponds to a given x-value. This means that the graph of a function cannot have two or more different points with the same x-coordinate, and no two points on the graph of a function can be vertically above or below each other. It follows, then, that a vertical line can intersect the graph of a function at most once. This observation provides a convenient visual test called the **Vertical Line Test** for functions.

VERTICAL LINE TEST FOR FUNCTIONS

A set of points in a coordinate plane is the graph of y as a function of x if and only if no *vertical* line intersects the graph at more than one point.

EXAMPLE 2 Vertical Line Test for Functions

Use the Vertical Line Test to decide whether the graphs in Figure 1.7 represent y as a function of x.

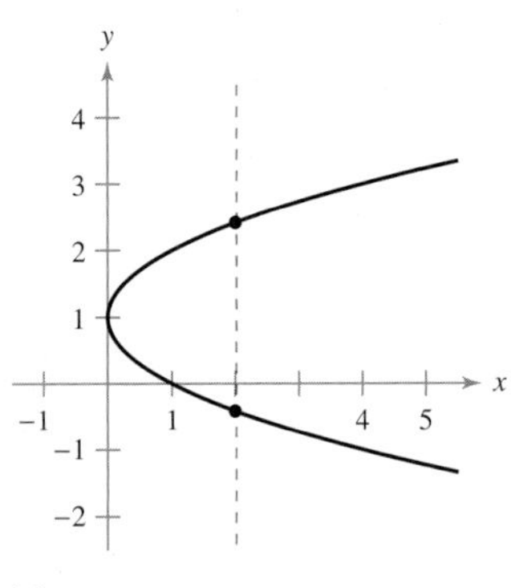

(a)

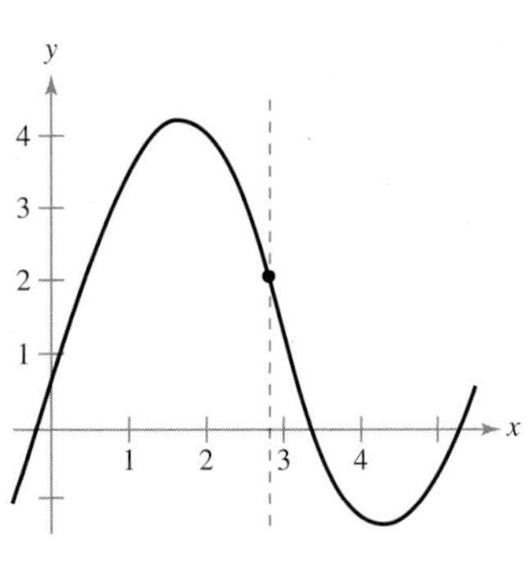

(b)

TECHNOLOGY PITFALL

Most graphing utilities are designed to graph functions of x more easily than other types of equations. For instance, the graph shown in Figure 1.7(a) represents the equation $x - (y - 1)^2 = 0$. To use a graphing utility to duplicate this graph, you must first solve the equation for y to obtain $y = 1 \pm \sqrt{x}$, and then graph the two equations $y_1 = 1 + \sqrt{x}$ and $y_2 = 1 - \sqrt{x}$ in the same viewing window.

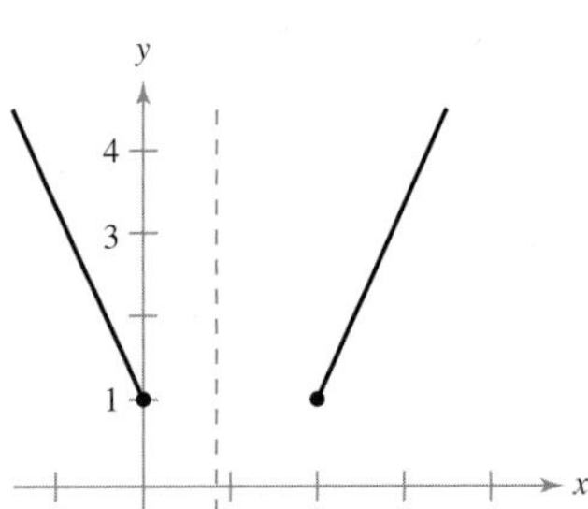

(c)

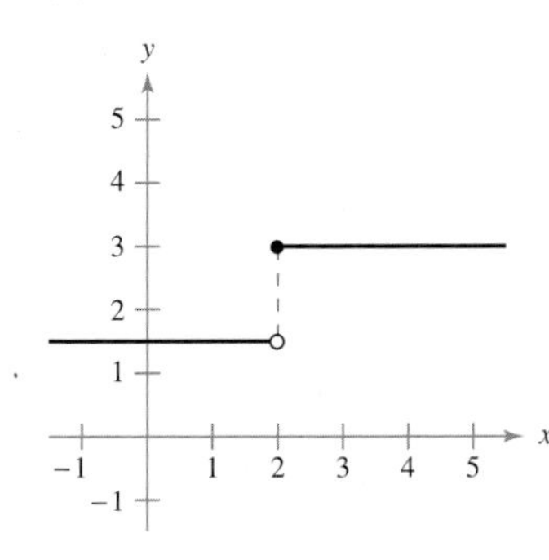

(d)

Figure 1.7

Solution

a. This *is not* a graph of y as a function of x, because you can find a vertical line that intersects the graph twice. That is, for a particular input x, there is more than one output y.

b. This *is* a graph of y as a function of x, because every vertical line intersects the graph at most once. That is, for a particular input x, there is at most one output y.

c. This *is* a graph of y as a function of x. That is, for a particular input x, there is at most one output y.

d. This *is* a graph of y as a function of x. Note that $f(2) = 3$, not 1.5. ■

NOTE In Example 2(c), notice that if a vertical line does not intersect the graph, it simply means that the function is undefined for that particular value of x.

Zeros of a Function

If the graph of a function of x has an x-intercept at $(a, 0)$, then a is a *zero* of the function.

ZEROS OF A FUNCTION

The **zeros of a function** f of x are the x-values for which $f(x) = 0$.

To find the zeros of a function, set the function equal to zero and solve for the independent variable.

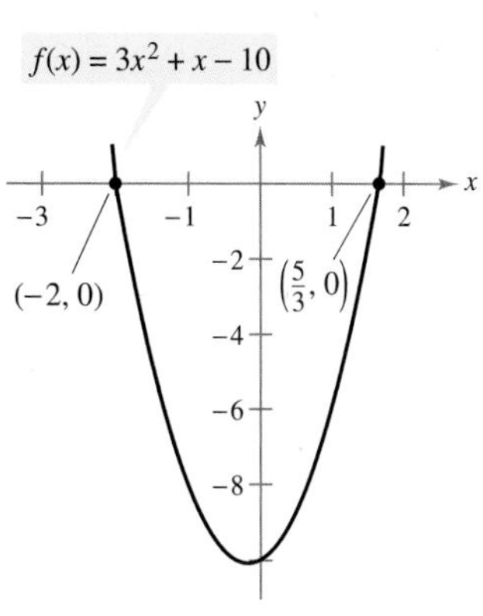

Zeros of f: $x = -2, x = \frac{5}{3}$
(a)

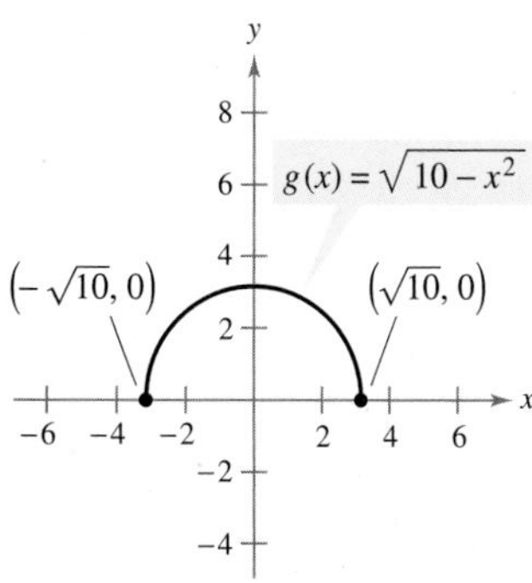

Zeros of g: $x = \pm\sqrt{10}$
(b)

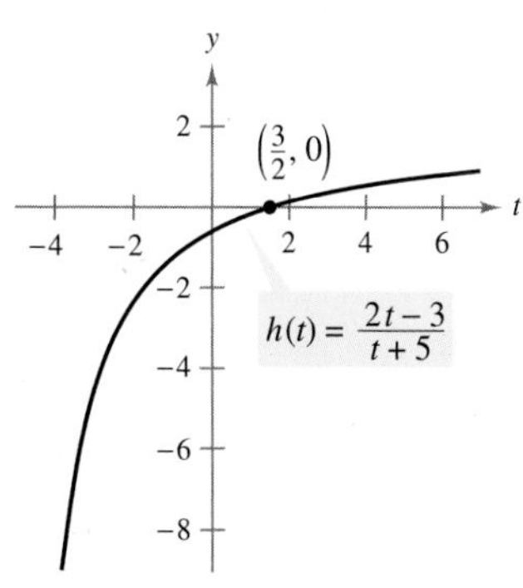

Zeros of h: $t = \frac{3}{2}$
(c)
Figure 1.8

EXAMPLE 3 Finding the Zeros of a Function

Find the zeros of each function.

a. $f(x) = 3x^2 + x - 10$ **b.** $g(x) = \sqrt{10 - x^2}$ **c.** $h(t) = \dfrac{2t - 3}{t + 5}$

Solution

a.

$3x^2 + x - 10 = 0$		Set $f(x)$ equal to 0.
$(3x - 5)(x + 2) = 0$		Factor.
$3x - 5 = 0$	$\Rightarrow$ $x = \frac{5}{3}$	Set 1st factor equal to 0.
$x + 2 = 0$	$\Rightarrow$ $x = -2$	Set 2nd factor equal to 0.

The zeros of f are $x = \frac{5}{3}$ and $x = -2$. In Figure 1.8(a), note that the graph of f has $\left(\frac{5}{3}, 0\right)$ and $(-2, 0)$ as its x-intercepts.

b.

$\sqrt{10 - x^2} = 0$	Set $g(x)$ equal to 0.
$10 - x^2 = 0$	Square each side.
$10 = x^2$	Add x^2 to each side.
$\pm\sqrt{10} = x$	Extract square roots.

The zeros of g are $x = -\sqrt{10}$ and $x = \sqrt{10}$. In Figure 1.8(b), note that the graph of g has $\left(-\sqrt{10}, 0\right)$ and $\left(\sqrt{10}, 0\right)$ as its x-intercepts.

c.

$\dfrac{2t - 3}{t + 5} = 0$	Set $h(t)$ equal to 0.
$2t - 3 = 0$	Multiply each side by $t + 5$.
$2t = 3$	Add 3 to each side.
$t = \dfrac{3}{2}$	Divide each side by 2.

The zero of h is $t = \frac{3}{2}$. In Figure 1.8(c), note that the graph of h has $\left(\frac{3}{2}, 0\right)$ as its t-intercept. ■

You can check that an x-value is a zero of a function by substituting into the original function. For instance, in Example 3(a), you can check that $x = \frac{5}{3}$ is a zero as shown.

$$f\left(\tfrac{5}{3}\right) = 3\left(\tfrac{5}{3}\right)^2 + \tfrac{5}{3} - 10$$
$$= \tfrac{25}{3} + \tfrac{5}{3} - 10 = 0 \checkmark$$

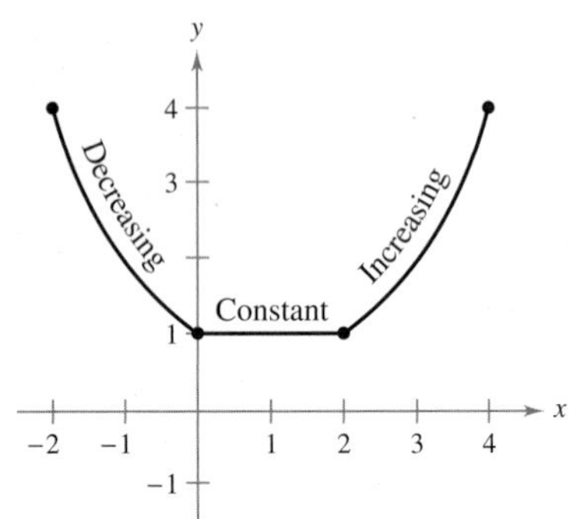

Figure 1.9

Increasing and Decreasing Functions

The more you know about the graph of a function, the more you know about the function itself. Consider the graph shown in Figure 1.9. As you move from *left to right,* this graph falls from $x = -2$ to $x = 0$, is constant from $x = 0$ to $x = 2$, and rises from $x = 2$ to $x = 4$.

> **INCREASING, DECREASING, AND CONSTANT FUNCTIONS**
>
> A function f is **increasing** on an interval if, for any x_1 and x_2 in the interval, $x_1 < x_2$ implies $f(x_1) < f(x_2)$.
>
> A function f is **decreasing** on an interval if, for any x_1 and x_2 in the interval, $x_1 < x_2$ implies $f(x_1) > f(x_2)$.
>
> A function f is **constant** on an interval if, for any x_1 and x_2 in the interval, $f(x_1) = f(x_2)$.

EXAMPLE 4 Increasing and Decreasing Functions

Use the graphs in Figure 1.10 to describe the increasing or decreasing behavior of each function.

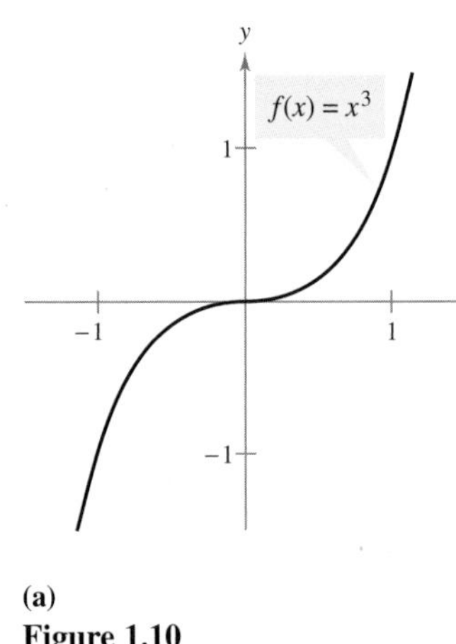

(a)

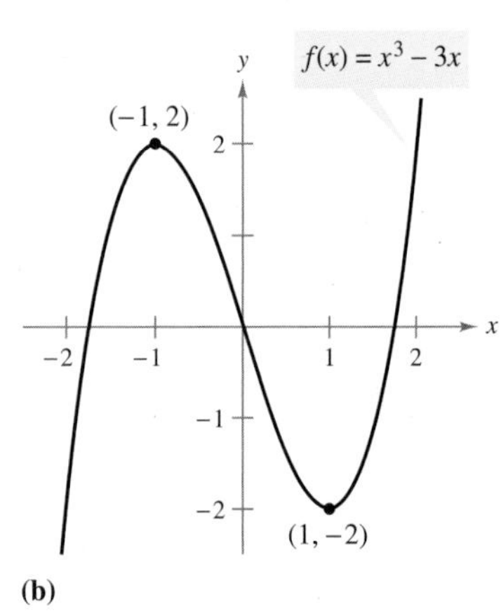
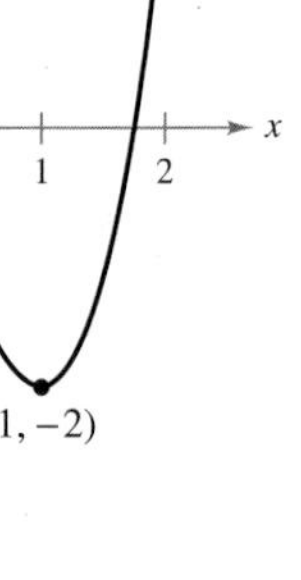

(b)

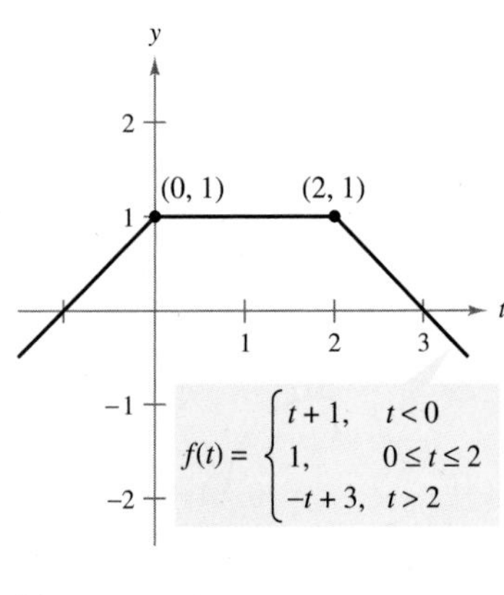

(c)

Figure 1.10

Solution

a. This function is increasing over the entire real line.

b. This function is increasing on the interval $(-\infty, -1)$, decreasing on the interval $(-1, 1)$, and increasing on the interval $(1, \infty)$.

c. This function is increasing on the interval $(-\infty, 0)$, constant on the interval $(0, 2)$, and decreasing on the interval $(2, \infty)$. ■

To help you decide whether a function is increasing, decreasing, or constant on an interval, you can evaluate the function for several values of x. However, calculus is needed to determine, for certain, all intervals on which a function is increasing, decreasing, or constant.

STUDY TIP A relative minimum or relative maximum is also referred to as a *local* minimum or *local* maximum.

The points at which a function changes its increasing, decreasing, or constant behavior are helpful in determining the **relative minimum** or **relative maximum** values of the function.

DEFINITIONS OF RELATIVE MINIMUM AND RELATIVE MAXIMUM

A function value $f(a)$ is called a **relative minimum** of f if there exists an interval (x_1, x_2) that contains a such that

$$x_1 < x < x_2 \quad \text{implies} \quad f(a) \le f(x).$$

A function value $f(a)$ is called a **relative maximum** of f if there exists an interval (x_1, x_2) that contains a such that

$$x_1 < x < x_2 \quad \text{implies} \quad f(a) \ge f(x).$$

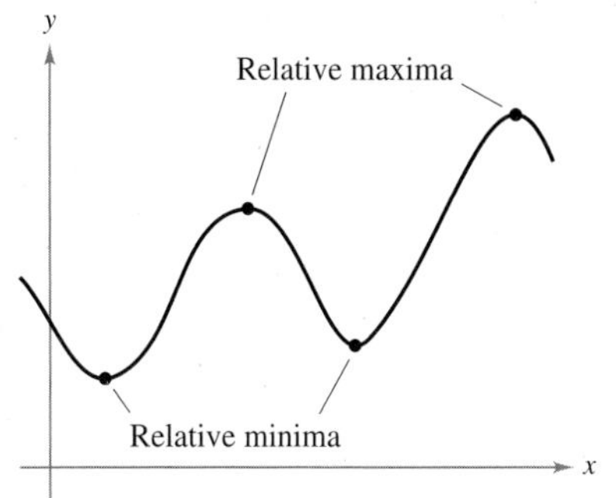

Figure 1.11

Figure 1.11 shows several different examples of relative minima and relative maxima. In Section 2.1, you will study a technique for finding the *exact point* at which a second-degree polynomial function has a relative minimum or relative maximum. For the time being, however, you can use a graphing utility to find reasonable approximations of these points.

EXAMPLE 5 Approximating a Relative Minimum

Use a graphing utility to approximate the relative minimum of the function given by $f(x) = 3x^2 - 4x - 2$.

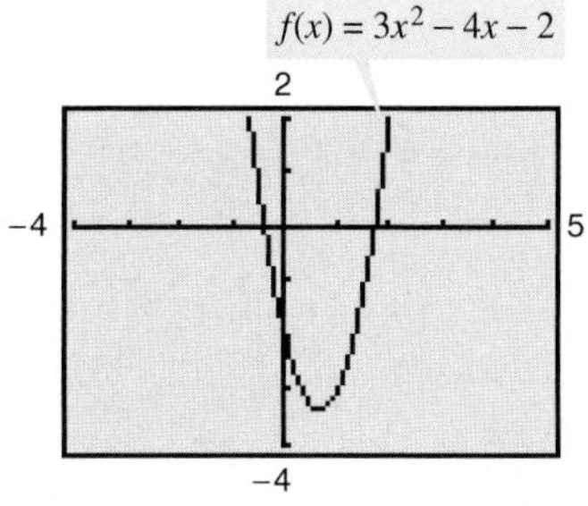

Figure 1.12

Solution The graph of f is shown in Figure 1.12. By using the *zoom* and *trace* features or the *minimum* feature of a graphing utility, you can estimate that the function has a relative minimum at the point

$(0.67, -3.33)$. Relative minimum

Later, in Section 2.1, you will be able to determine that the exact point at which the relative minimum occurs is $\left(\frac{2}{3}, -\frac{10}{3}\right)$. ■

You can also use the *table* feature of a graphing utility to approximate numerically the relative minimum of the function in Example 5. Using a table that begins at 0.6 and increments the value of x by 0.01, you can approximate the relative minimum of $f(x) = 3x^2 - 4x - 2$ to be -3.33, which occurs at the point $(0.67, -3.33)$.

x	0.60	0.61	0.62	0.63	0.64	0.65
$f(x)$	−3.32	−3.3237	−3.3268	−3.3293	−3.3312	−3.3325

x	0.66	0.67	0.68	0.69	0.70
$f(x)$	−3.3332	−3.3333	−3.3328	−3.3317	−3.33

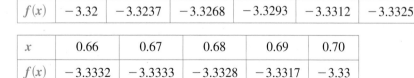

TECHNOLOGY When you use a graphing utility to estimate the x- and y-values of a relative minimum or relative maximum, the *zoom* feature will often produce graphs that are nearly flat. To overcome this problem, you can manually change the vertical setting of the viewing window. The graph will stretch vertically if the values of Ymin and Ymax are closer together.

Linear Functions

A **linear function** of x is a function of the form

$$f(x) = mx + b. \qquad \text{Linear function}$$

In Section P.5, you learned that the graph of such a function is a line that has a slope of m and a y-intercept at $(0, b)$.

EXAMPLE 6 Graphing a Linear Function

Sketch the graph of the linear function given by $f(x) = -\frac{1}{2}x + 3$.

Solution The graph of this function is a line that has a slope of $m = -\frac{1}{2}$ and a y-intercept at $(0, 3)$. To sketch the line, plot the y-intercept. Then, because the slope is $-\frac{1}{2}$, move two units to the right and one unit downward and plot a second point, as shown in Figure 1.13(a). Finally, draw the line that passes through these two points, as shown in Figure 1.13(b).

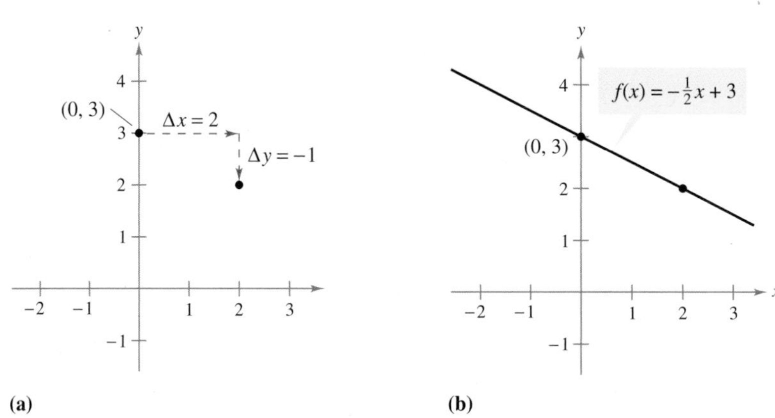

Figure 1.13

EXAMPLE 7 Writing a Linear Function

Write the linear function f for which $f(1) = 3$ and $f(4) = 0$.

Solution To find the equation of the line that passes through $(x_1, y_1) = (1, 3)$ and $(x_2, y_2) = (4, 0)$, first find the slope of the line.

$$m = \frac{\Delta y}{\Delta x} = \frac{y_2 - y_1}{x_2 - x_1} = \frac{0 - 3}{4 - 1} = -1$$

Next, use the point-slope form of the equation of a line.

$$y - y_1 = m(x - x_1) \qquad \text{Point-slope form, Section P.5}$$

$$y - 3 = -1(x - 1) \qquad \text{Substitute.}$$

$$y = -x + 4 \qquad \text{Simplify.}$$

$$f(x) = -x + 4 \qquad \text{Function notation}$$

You can check this result as shown.

$$f(1) = -(1) + 4 = 3 \checkmark$$

$$f(4) = -(4) + 4 = 0 \checkmark$$

The graph of f is shown in Figure 1.14. ■

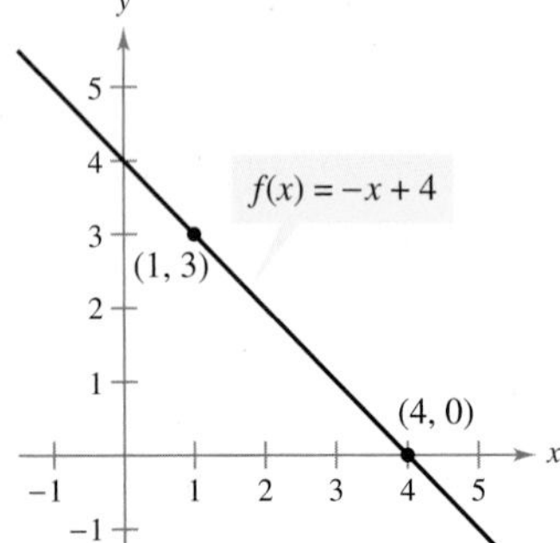

Figure 1.14

Step and Piecewise-Defined Functions

Functions whose graphs resemble sets of stairsteps are known as **step functions.** The most famous of the step functions is the **greatest integer function,** which is denoted by $[\![x]\!]$ and defined as

$$f(x) = [\![x]\!] = \textit{the greatest integer less than or equal to } x.$$

Some values of the greatest integer function are as follows.

$$[\![-1]\!] = -1 \qquad [\![-0.5]\!] = -1 \qquad [\![0]\!] = 0$$
$$[\![0.5]\!] = 0 \qquad [\![1]\!] = 1 \qquad [\![1.5]\!] = 1$$

The graph of the greatest integer function

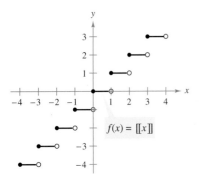

$$f(x) = [\![x]\!]$$

has the following characteristics, as shown in Figure 1.15.

Figure 1.15

- The domain of the function is the set of all real numbers.
- The range of the function is the set of all integers.
- The graph has a y-intercept at $(0, 0)$ and x-intercepts in the interval $[0, 1)$.
- The graph is constant between each pair of consecutive integers.
- The graph jumps vertically one unit at each integer value.

TECHNOLOGY When graphing a step function, you should set your graphing utility to *dot* mode.

Recall from Section 1.1 that a piecewise-defined function is defined by two or more equations over a specified domain. To graph a piecewise-defined function, graph each equation separately over the specified domain, as shown in Example 8.

EXAMPLE 8 Graphing a Piecewise-Defined Function

Sketch the graph of

$$f(x) = \begin{cases} 2x + 3, & x \le 1 \\ -x + 4, & x > 1 \end{cases}.$$

Solution This piecewise-defined function is composed of two linear functions. At $x = 1$ and to the left of $x = 1$ the graph is the line $y = 2x + 3$, and to the right of $x = 1$ the graph is the line $y = -x + 4$, as shown in Figure 1.16. Notice that the point $(1, 5)$ is a solid dot and the point $(1, 3)$ is an open dot. This is because $f(1) = 2(1) + 3 = 5$.

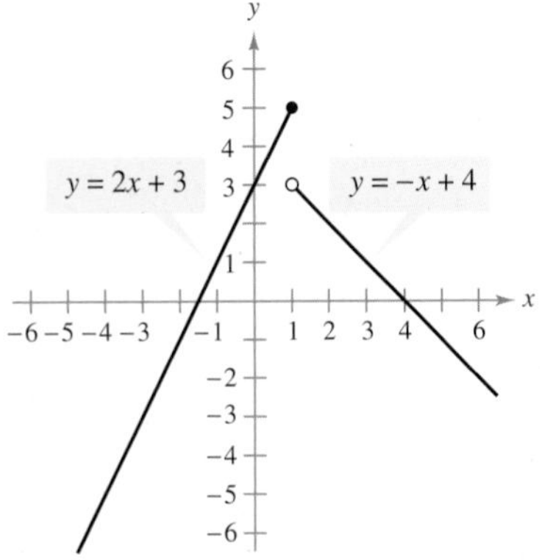

Figure 1.16

■

Even and Odd Functions

In Section P.4, you studied different types of symmetry of a graph. In the terminology of functions, a function is said to be **even** if its graph is symmetric with respect to the y-axis and to be **odd** if its graph is symmetric with respect to the origin. The symmetry tests in Section P.4 yield the following tests for even and odd functions.

> **EXPLORATION**
>
> Graph each function with a graphing utility. Determine whether the function is *even*, *odd*, or *neither*.
>
> $f(x) = x^2 - x^4$
>
> $g(x) = 2x^3 + 1$
>
> $h(x) = x^5 - 2x^3 + x$
>
> $j(x) = 2 - x^6 - x^8$
>
> $k(x) = x^5 - 2x^4 + x - 2$
>
> $p(x) = x^9 + 3x^5 - x^3 + x$
>
> What do you notice about the equations of functions that are odd? What do you notice about the equations of functions that are even? Can you describe a way to identify a function as odd or even by inspecting the equation? Can you describe a way to identify a function as neither odd nor even by inspecting the equation?

TESTS FOR EVEN AND ODD FUNCTIONS

A function $y = f(x)$ is **even** if, for each x in the domain of f,

$f(-x) = f(x)$. Symmetric to y-axis

A function $y = f(x)$ is **odd** if, for each x in the domain of f,

$f(-x) = -f(x)$. Symmetric to origin

EXAMPLE 9 Even and Odd Functions

Determine whether each function is even, odd, or neither.

a. $g(x) = x^3 - x$ **b.** $h(x) = x^2 + 1$

Solution

a. The function $g(x) = x^3 - x$ is odd because $g(-x) = -g(x)$, as follows.

$$\begin{aligned} g(-x) &= (-x)^3 - (-x) && \text{Substitute } -x \text{ for } x. \\ &= -x^3 + x && \text{Simplify.} \\ &= -(x^3 - x) && \text{Distributive Property} \\ &= -g(x) && \text{Test for odd function} \end{aligned}$$

b. The function $h(x) = x^2 + 1$ is even because $h(-x) = h(x)$, as follows.

$$\begin{aligned} h(-x) &= (-x)^2 + 1 && \text{Substitute } -x \text{ for } x. \\ &= x^2 + 1 && \text{Simplify.} \\ &= h(x) && \text{Test for even function} \end{aligned}$$

The graphs and symmetry of these two functions are shown in Figure 1.17.

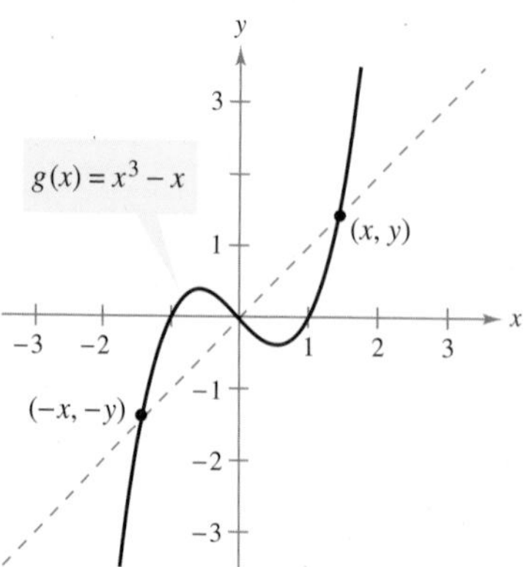

(a) Symmetric to origin: Odd Function

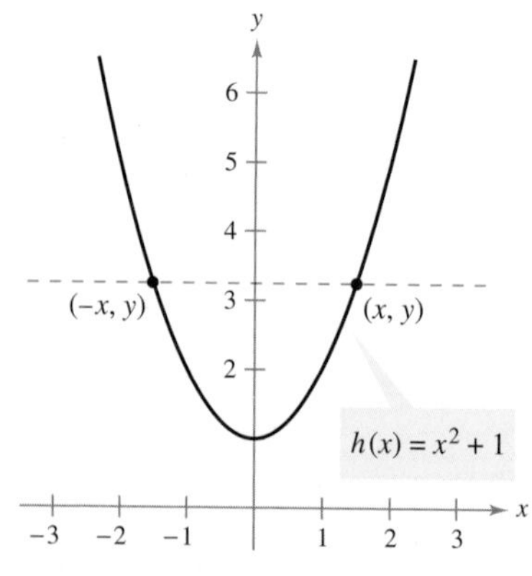

(b) Symmetric to y-axis: Even Function

Figure 1.17

1.2 Exercises

See www.CalcChat.com for worked-out solutions to odd-numbered exercises.

In Exercises 1–8, fill in the blanks.

1. The graph of a function f is the collection of ________ ________ $(x, f(x))$ such that x is in the domain of f.
2. The ________ ________ ________ is used to determine whether the graph of an equation is a function of y in terms of x.
3. The ________ of a function f are the values of x for which $f(x) = 0$.
4. A function f is ________ on an interval if, for any x_1 and x_2 in the interval, $x_1 < x_2$ implies $f(x_1) > f(x_2)$.
5. A function value $f(a)$ is a relative ________ of f if there exists an interval (x_1, x_2) containing a such that $x_1 < x < x_2$ implies $f(a) \geq f(x)$.
6. Functions whose graphs resemble sets of stairsteps are known as ________ functions, the most famous being the ________ ________ function.
7. A function f is ________ if, for each x in the domain of f, $f(-x) = -f(x)$.
8. A function f is ________ if its graph is symmetric with respect to the y-axis.

In Exercises 9–12, use the graph of the function to find the domain and range of f.

9.

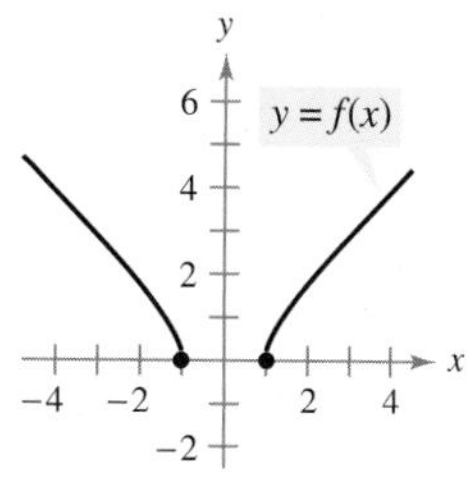

10.

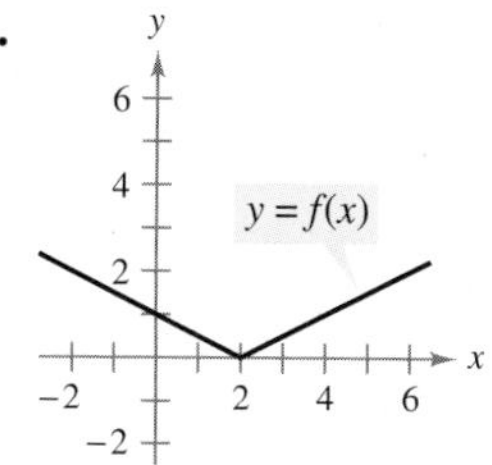

11.

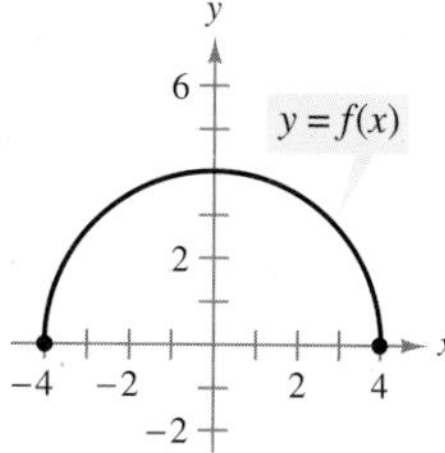

12.

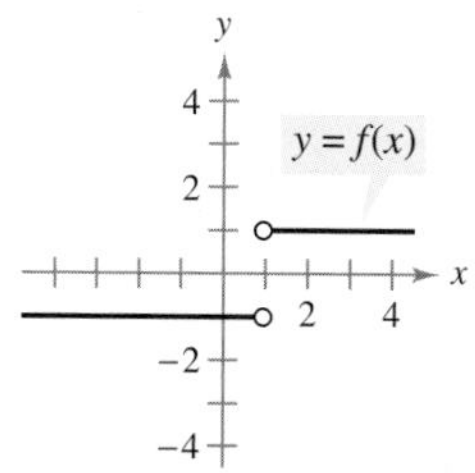

In Exercises 13–16, use the graph of the function to find the domain and range of f and the indicated function values.

13. (a) $f(-2)$ (b) $f(-1)$ (c) $f\left(\frac{1}{2}\right)$ (d) $f(1)$

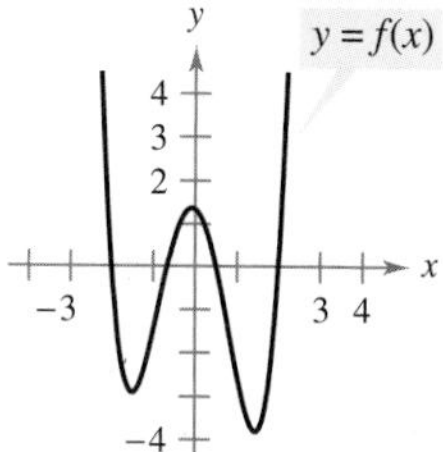

14. (a) $f(-1)$ (b) $f(2)$ (c) $f(0)$ (d) $f(1)$

15. (a) $f(2)$ (b) $f(1)$ (c) $f(3)$ (d) $f(-1)$

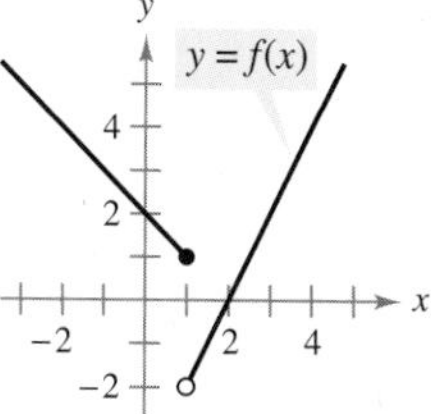

16. (a) $f(-2)$ (b) $f(1)$ (c) $f(0)$ (d) $f(2)$

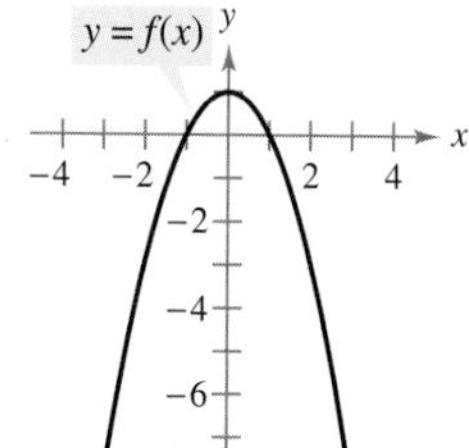

In Exercises 17–22, use the Vertical Line Test to determine whether y is a function of x. To print an enlarged copy of the graph, go to the website *www.mathgraphs.com*.

17. $y = \frac{1}{2}x^2$

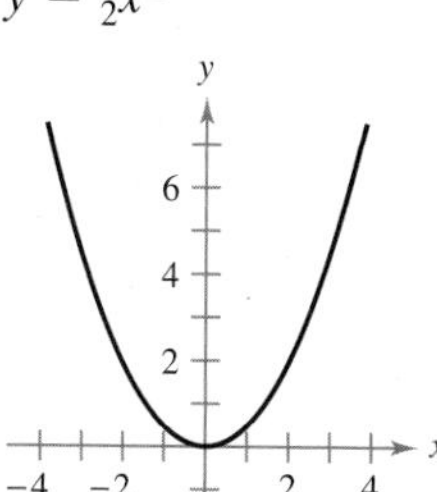

18. $y = \frac{1}{4}x^3$

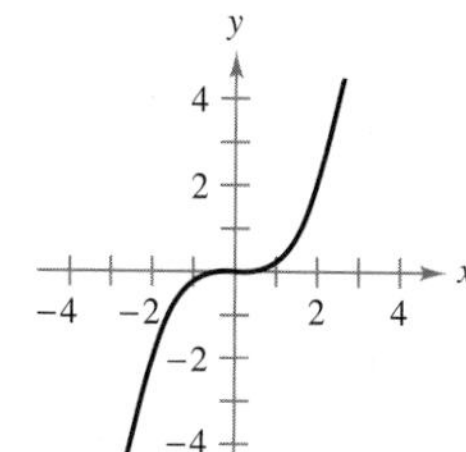

19. $x - y^2 = 1$

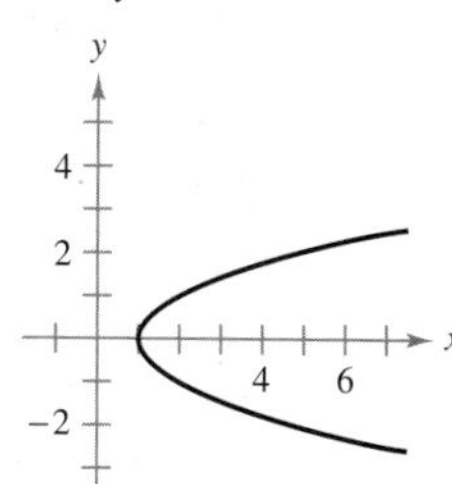

20. $x^2 + y^2 = 25$

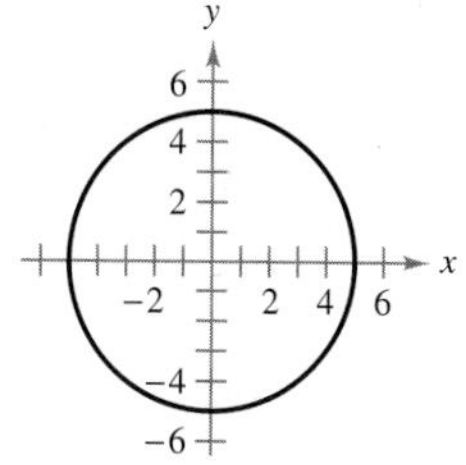

21. $x^2 = 2xy - 1$

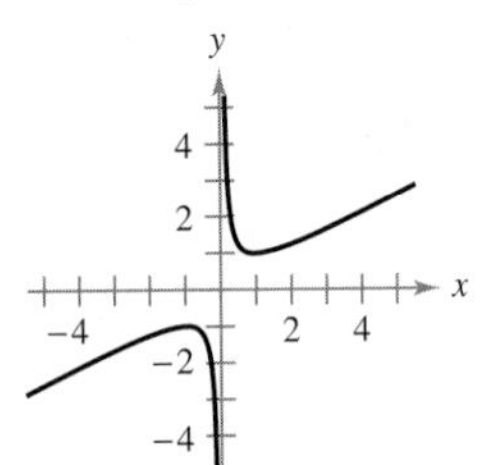

22. $x = |y + 2|$

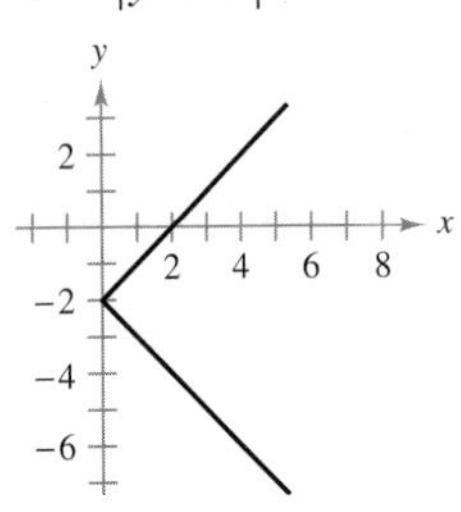

In Exercises 23–32, find the zeros of the function analytically.

23. $f(x) = 2x^2 - 7x - 30$
24. $f(x) = 3x^2 + 22x - 16$
25. $f(x) = \dfrac{x}{9x^2 - 4}$
26. $f(x) = \dfrac{x^2 - 9x + 14}{4x}$
27. $f(x) = \frac{1}{2}x^3 - x$
28. $f(x) = x^3 - 4x^2 - 9x + 36$
29. $f(x) = 4x^3 - 24x^2 - x + 6$
30. $f(x) = 9x^4 - 25x^2$
31. $f(x) = \sqrt{2x} - 1$
32. $f(x) = \sqrt{3x + 2}$

In Exercises 33–38, (a) use a graphing utility to graph the function and find the zeros of the function and (b) verify your results from part (a) analytically.

33. $f(x) = 3 + \dfrac{5}{x}$
34. $f(x) = x(x - 7)$
35. $f(x) = \sqrt{2x + 11}$
36. $f(x) = \sqrt{3x - 14} - 8$
37. $f(x) = \dfrac{3x - 1}{x - 6}$
38. $f(x) = \dfrac{2x^2 - 9}{3 - x}$

In Exercises 39–42, determine the intervals over which the function is increasing, decreasing, or constant.

39. $f(x) = \frac{3}{2}x$

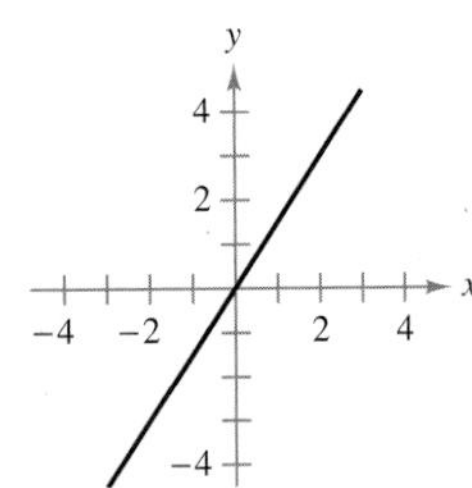

40. $f(x) = x^2 - 4x$

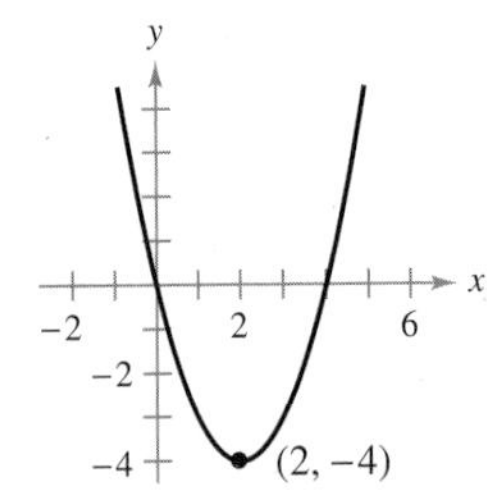

41. $f(x) = x^3 - 3x^2 + 2$

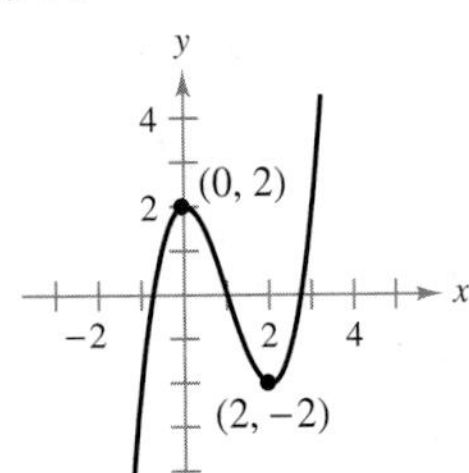

42. $f(x) = \sqrt{x^2 - 1}$

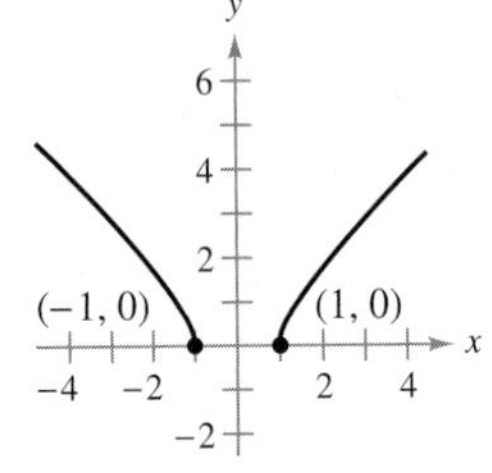

In Exercises 43–58, (a) use a graphing utility to graph the function and visually determine the intervals over which the function is increasing, decreasing, or constant, and (b) make a table of values to verify whether the function is increasing, decreasing, or constant over the intervals you identified in part (a).

43. $f(x) = 3$
44. $g(x) = x$
45. $g(s) = \dfrac{s^2}{4}$
46. $h(x) = x^2 - 4$
47. $f(t) = -t^4$
48. $f(x) = 3x^4 - 6x^2$
49. $f(x) = \sqrt{1 - x}$
50. $f(x) = x\sqrt{x + 3}$
51. $f(x) = x^{3/2}$
52. $f(x) = x^{2/3}$
53. $g(t) = \sqrt[3]{t - 1}$
54. $f(x) = \sqrt[3]{x + 5}$
55. $f(x) = |x + 2| - |x - 2|$
56. $f(x) = |x + 1| + |x - 1|$
57. $f(x) = \begin{cases} x + 3, & x \le 0 \\ 3, & 0 < x \le 2 \\ 2x - 1, & x > 2 \end{cases}$
58. $f(x) = \begin{cases} 2x + 1, & x \le -1 \\ x^2 - 2, & x > -1 \end{cases}$

In Exercises 59–68, use a graphing utility to graph the function and approximate (to two decimal places) any relative minimum or relative maximum values.

59. $f(x) = (x - 4)(x + 2)$
60. $f(x) = 3x^2 - 2x - 5$
61. $f(x) = -x^2 + 3x - 2$
62. $f(x) = -2x^2 + 9x$
63. $f(x) = x(x - 2)(x + 3)$
64. $f(x) = x^3 - 3x^2 - x + 1$
65. $g(x) = 2x^3 + 3x^2 - 12x$
66. $h(x) = x^3 - 6x^2 + 15$
67. $h(x) = (x - 1)\sqrt{x}$
68. $g(x) = x\sqrt{4 - x}$

In Exercises 69–76, sketch the graph of the linear function. Label the y-intercept.

69. $f(x) = 1 - 2x$
70. $f(x) = 3x - 11$
71. $f(x) = -x - \frac{3}{4}$
72. $f(x) = 3x - \frac{5}{2}$
73. $f(x) = -\frac{1}{6}x - \frac{5}{2}$
74. $f(x) = \frac{5}{6} - \frac{2}{3}x$
75. $f(x) = -1.8 + 2.5x$
76. $f(x) = 10.2 + 3.1x$

In Exercises 77–82, (a) write the linear function f such that it has the indicated function values and (b) sketch the graph of the function.

77. $f(1) = 4, f(0) = 6$ **78.** $f(-3) = -8, f(1) = 2$

79. $f(5) = -4, f(-2) = 17$

80. $f(3) = 9, f(-1) = -11$

81. $f(-5) = -1, f(5) = -1$

82. $f\left(\frac{2}{3}\right) = -\frac{15}{2}, f(-4) = -11$

In Exercises 83–88, sketch the graph of the function.

83. $g(x) = -[\![x]\!]$ **84.** $g(x) = 4[\![x]\!]$

85. $g(x) = [\![x]\!] - 2$ **86.** $g(x) = [\![x]\!] - 1$

87. $g(x) = [\![x + 1]\!]$ **88.** $g(x) = [\![x - 3]\!]$

In Exercises 89–96, graph the function.

89. $f(x) = \begin{cases} 2x + 3, & x < 0 \\ 3 - x, & x \geq 0 \end{cases}$

90. $g(x) = \begin{cases} x + 6, & x \leq -4 \\ \frac{1}{2}x - 4, & x > -4 \end{cases}$

91. $f(x) = \begin{cases} \sqrt{4 + x}, & x < 0 \\ \sqrt{4 - x}, & x \geq 0 \end{cases}$

92. $f(x) = \begin{cases} 1 - (x - 1)^2, & x \leq 2 \\ \sqrt{x - 2}, & x > 2 \end{cases}$

93. $f(x) = \begin{cases} x^2 + 5, & x \leq 1 \\ -x^2 + 4x + 3, & x > 1 \end{cases}$

94. $h(x) = \begin{cases} 3 - x^2, & x < 0 \\ x^2 + 2, & x \geq 0 \end{cases}$

95. $h(x) = \begin{cases} 4 - x^2, & x < -2 \\ 3 + x, & -2 \leq x < 0 \\ x^2 + 1, & x \geq 0 \end{cases}$

96. $k(x) = \begin{cases} 2x + 1, & x \leq -1 \\ 2x^2 - 1, & -1 < x \leq 1 \\ 1 - x^2, & x > 1 \end{cases}$

In Exercises 97–106, graph the function and determine the interval(s) for which $f(x) \geq 0$.

97. $f(x) = 4 - x$ **98.** $f(x) = 4x + 2$

99. $f(x) = 9 - x^2$ **100.** $f(x) = x^2 - 4x$

101. $f(x) = \sqrt{x - 1}$ **102.** $f(x) = \sqrt{x + 2}$

103. $f(x) = -(1 + |x|)$ **104.** $f(x) = \frac{1}{2}(2 + |x|)$

105. $f(x) = \begin{cases} 1 - 2x^2, & x \leq -2 \\ -x + 8, & x > -2 \end{cases}$

106. $f(x) = \begin{cases} \sqrt{x - 5}, & x > 5 \\ x^2 + x - 1, & x \leq 5 \end{cases}$

In Exercises 107–110, (a) use a graphing utility to graph the function, (b) state the domain and range of the function, and (c) describe the pattern of the graph.

107. $s(x) = 2\left(\frac{1}{4}x - [\![\frac{1}{4}x]\!]\right)$ **108.** $g(x) = 2\left(\frac{1}{4}x - [\![\frac{1}{4}x]\!]\right)^2$

109. $h(x) = 4\left(\frac{1}{2}x - [\![\frac{1}{2}x]\!]\right)$ **110.** $k(x) = 4\left(\frac{1}{2}x - [\![\frac{1}{2}x]\!]\right)^2$

WRITING ABOUT CONCEPTS

In Exercises 111–114, use the graph to determine (a) the domain, (b) the range, and (c) the intervals over which the function is increasing, decreasing, and constant.

111.

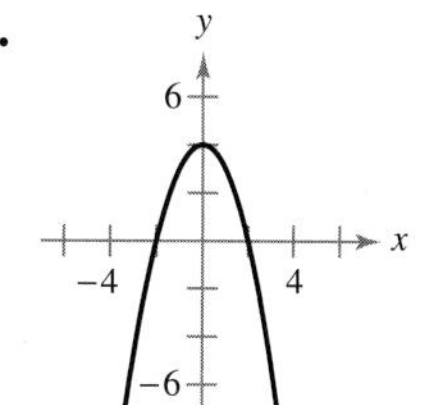

112.

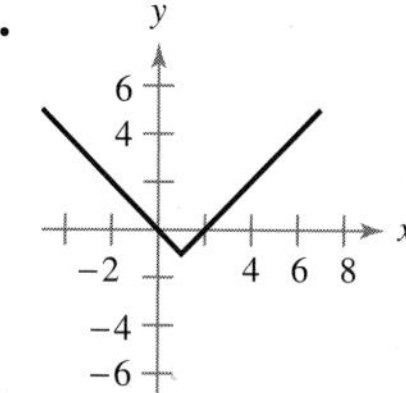

113.

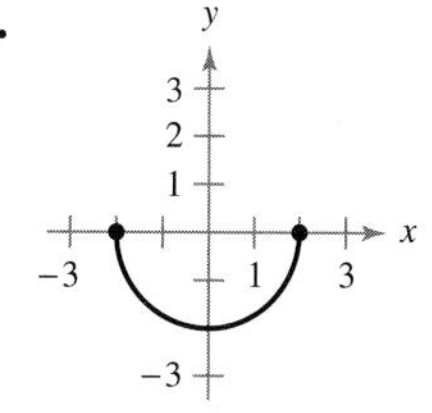

114.

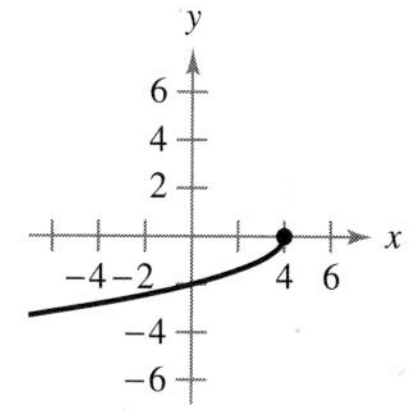

115. Use the graph of $y = f(x)$.

(a) Evaluate $f(-1)$.

(b) Evaluate $f(1)$.

(c) Approximate the intervals over which f is increasing and decreasing.

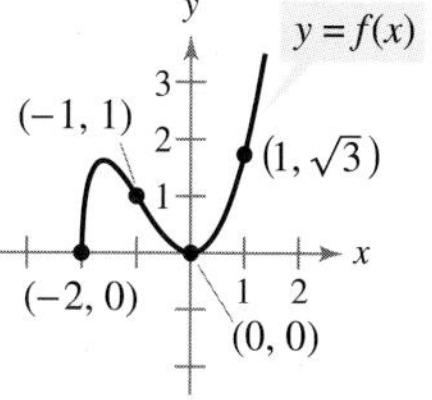

Figure for 115

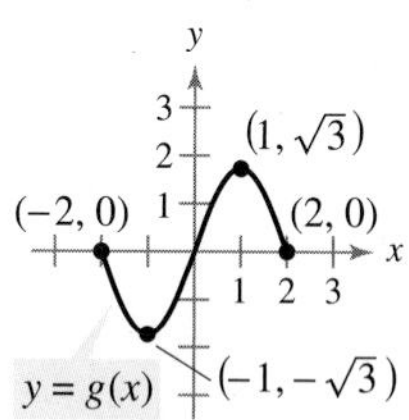

Figure for 116

116. Use the graph of $y = g(x)$.

(a) Evaluate $g(-1)$.

(b) Evaluate $g(1)$.

(c) Determine the intervals over which g is increasing and decreasing.

In Exercises 117–124, determine whether the function is even, odd, or neither. Then describe the symmetry.

117. $f(x) = x^6 - 2x^2 + 3$ **118.** $h(x) = x^3 - 5$

119. $g(x) = x^3 - 5x$ **120.** $f(t) = t^2 + 2t - 3$

121. $h(x) = x\sqrt{x + 5}$ **122.** $f(x) = x\sqrt{1 - x^2}$

123. $f(s) = 4s^{3/2}$ **124.** $g(s) = 4s^{2/3}$

In Exercises 125–134, sketch a graph of the function and determine whether it is even, odd, or neither. Verify your answers analytically.

125. $f(x) = 5$ **126.** $f(x) = -9$

127. $f(x) = 3x - 2$ **128.** $f(x) = 5 - 3x$

129. $h(x) = x^2 - 4$ **130.** $f(x) = -x^2 - 8$

131. $f(x) = \sqrt{1 - x}$ **132.** $g(t) = \sqrt[3]{t - 1}$

133. $f(x) = |x + 2|$ **134.** $f(x) = -|x - 5|$

In Exercises 135–138, write the height *h* of the rectangle as a function of *x*.

135.

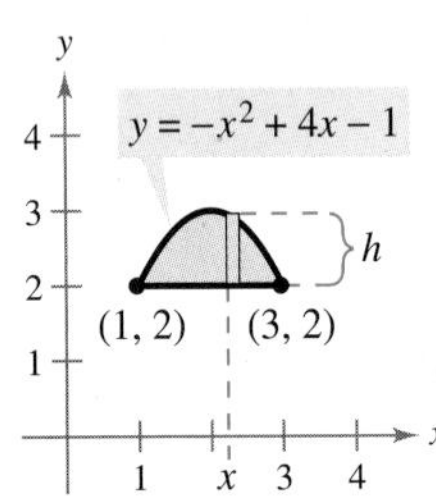

136.

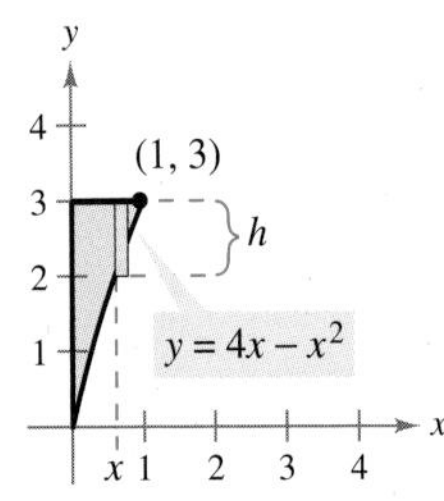

137.

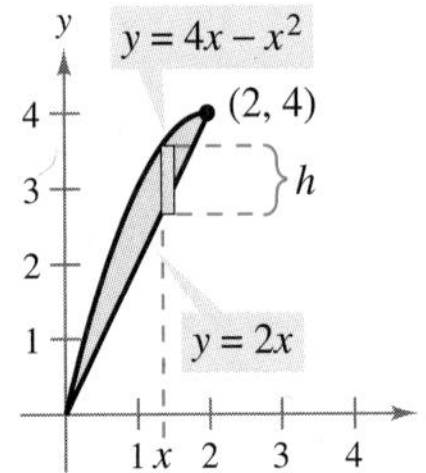

138.

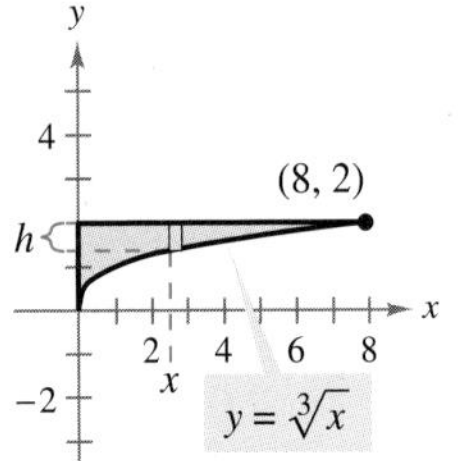

In Exercises 139–142, write the length *L* of the rectangle as a function of *y*.

139.

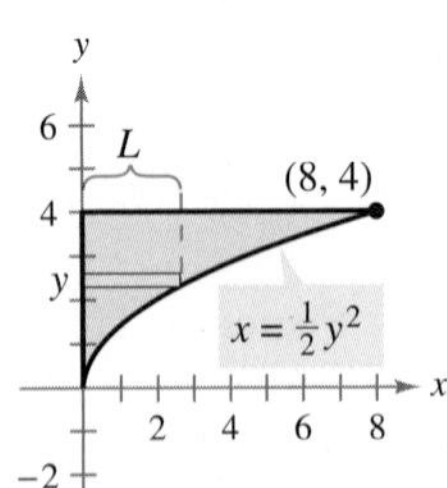

140.

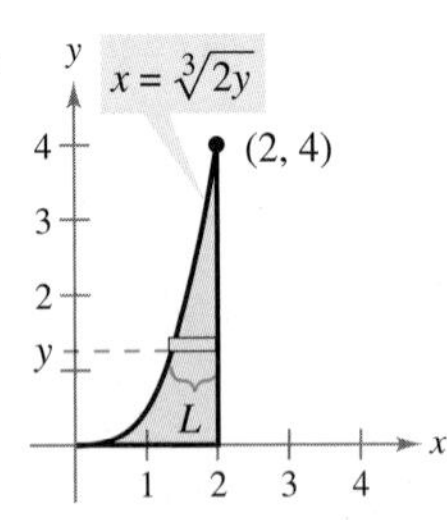

141.

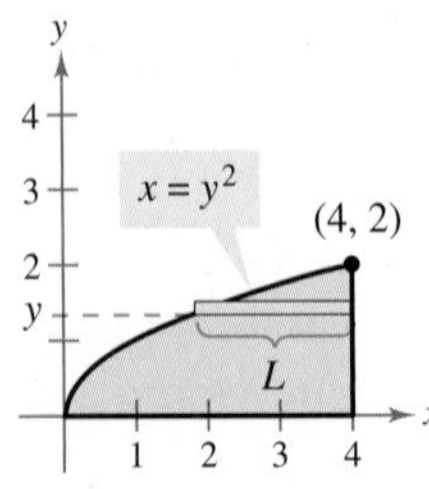

142.

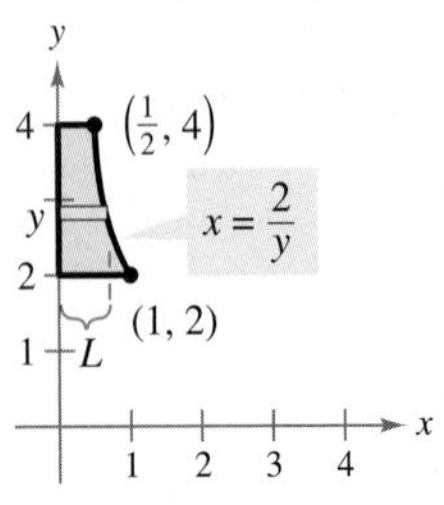

143. ***Electronics*** The number of lumens (time rate of flow of light) L from a fluorescent lamp can be approximated by the model

$$L = -0.294x^2 + 97.744x - 664.875, \quad 20 \le x \le 90$$

where x is the wattage of the lamp.

(a) Use a graphing utility to graph the function.

(b) Use the graph from part (a) to estimate the wattage necessary to obtain 2000 lumens.

144. ***Data Analysis: Temperature*** The table shows the temperatures y (in degrees Fahrenheit) in a certain city over a 24-hour period. Let x represent the time of day, where $x = 0$ corresponds to 6 A.M.

Time, x	Temperature, y
0	34
2	50
4	60
6	64
8	63
10	59
12	53
14	46
16	40
18	36
20	34
22	37
24	45

A model that represents these data is given by

$$y = 0.026x^3 - 1.03x^2 + 10.2x + 34, \quad 0 \le x \le 24.$$

(a) Use a graphing utility to create a scatter plot of the data. Then graph the model in the same viewing window.

(b) How well does the model fit the data?

(c) Use the graph to approximate the times when the temperature was increasing and decreasing.

(d) Use the graph to approximate the maximum and minimum temperatures during this 24-hour period.

(e) Could this model be used to predict the temperatures in the city during the next 24-hour period? Why or why not?

145. ***Delivery Charges*** The cost of sending an overnight package from Los Angeles to Miami is \$23.40 for a package weighing up to but not including 1 pound and \$3.75 for each additional pound or portion of a pound. A model for the total cost C (in dollars) of sending the package is $C = 23.40 + 3.75[\![x]\!],\ x > 0$, where x is the weight in pounds.

(a) Sketch a graph of the model.

(b) Determine the cost of sending a package that weighs 9.25 pounds.

146. ***Delivery Charges*** The cost of sending an overnight package from New York to Atlanta is \$22.65 for a package weighing up to but not including 1 pound and \$3.70 for each additional pound or portion of a pound.

(a) Use the greatest integer function to create a model for the cost C of overnight delivery of a package weighing x pounds, $x > 0$.

(b) Sketch the graph of the function.

147. ***Coordinate Axis Scale*** Each function described below models the specified data for the years 1998 through 2008, with $t = 8$ corresponding to 1998. Estimate a reasonable scale for the vertical axis (e.g., hundreds, thousands, millions, etc.) of the graph and justify your answer. (There are many correct answers.)

(a) $f(t)$ represents the average salary of college professors.

(b) $f(t)$ represents the U.S. population.

(c) $f(t)$ represents the percent of the civilian work force that is unemployed.

148. ***Geometry*** Corners of equal size are cut from a square with sides of length 8 meters (see figure).

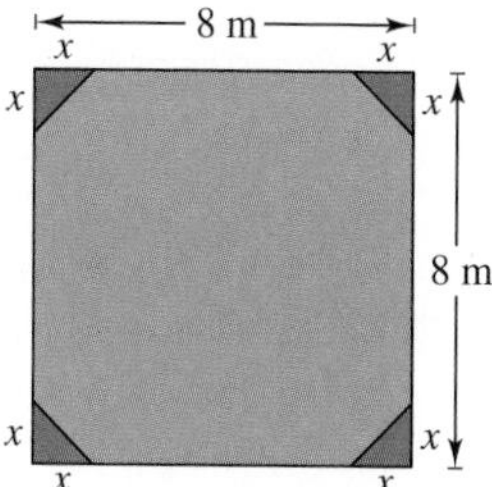

(a) Write the area A of the resulting figure as a function of x. Determine the domain of the function.

(b) Use a graphing utility to graph the area function over its domain. Use the graph to find the range of the function.

(c) Identify the figure that would result if x were chosen to be the maximum value in the domain of the function. What would be the length of each side of the figure?

True or False? **In Exercises 149 and 150, determine whether the statement is true or false. Justify your answer.**

149. A function with a square root cannot have a domain that is the set of real numbers.

150. A piecewise-defined function will always have at least one x-intercept or at least one y-intercept.

151. If f is an even function, determine whether g is even, odd, or neither. Explain.

(a) $g(x) = -f(x)$ (b) $g(x) = f(-x)$

(c) $g(x) = f(x) - 2$ (d) $g(x) = f(x - 2)$

152. ***Think About It*** Does the graph in Exercise 19 represent x as a function of y? Explain.

Think About It **In Exercises 153–158, find the coordinates of a second point on the graph of a function f if the given point is on the graph and the function is (a) even and (b) odd.**

153. $\left(-\frac{3}{2}, 4\right)$ **154.** $\left(-\frac{5}{3}, -7\right)$

155. $(4, 9)$ **156.** $(5, -1)$

157. $(x, -y)$ **158.** $(2a, 2c)$

159. Find the values of a and b so that the function

$$f(x) = \begin{cases} x + 2, & x < -2 \\ 0, & -2 \le x \le 2 \\ ax + b, & x > 2 \end{cases}$$

(a) is an odd function.

(b) is an even function.

CAPSTONE

160. Use the graph of the function to answer (a)–(e).

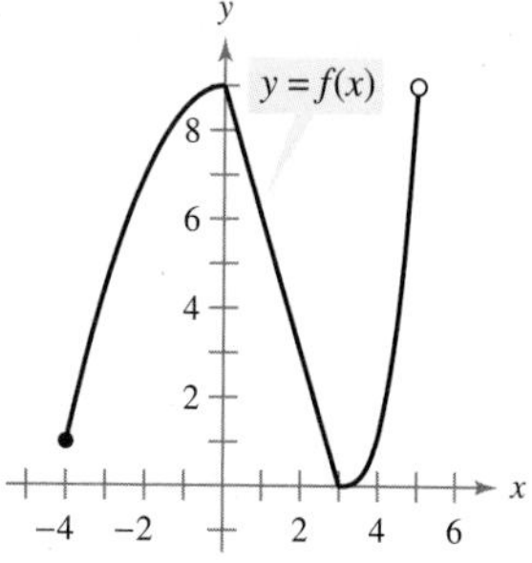

(a) Find the domain and range of f.

(b) Find the zero(s) of f.

(c) Determine the intervals over which f is increasing, decreasing, or constant.

(d) Approximate any relative minimum or relative maximum values of f.

(e) Is f even, odd, or neither?

1.3 Transformations of Functions

- Recognize graphs of common functions.
- Use vertical and horizontal shifts to sketch graphs of functions.
- Use reflections to sketch graphs of functions.
- Use nonrigid transformations to sketch graphs of functions.

Summary of Graphs of Common Functions

One of the goals of this text is to enable you to recognize the basic shapes of the graphs of different types of functions. For instance, from your study of lines in Section P.5, you can determine the basic shape of the graph of the linear function $f(x) = mx + b$. Specifically, you know that the graph of this function is a line whose slope is m and whose y-intercept is b.

The six graphs shown in Figure 1.18 represent the most commonly used functions in algebra and calculus. Familiarity with the basic characteristics of these simple graphs will help you analyze the shapes of more complicated graphs.

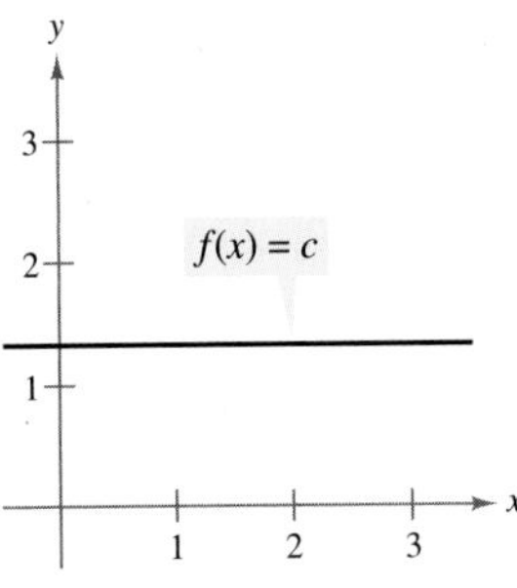

(a) Constant function

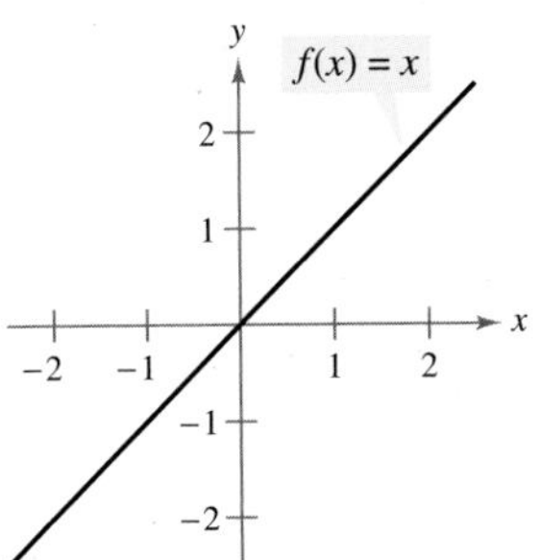

(b) Identity function

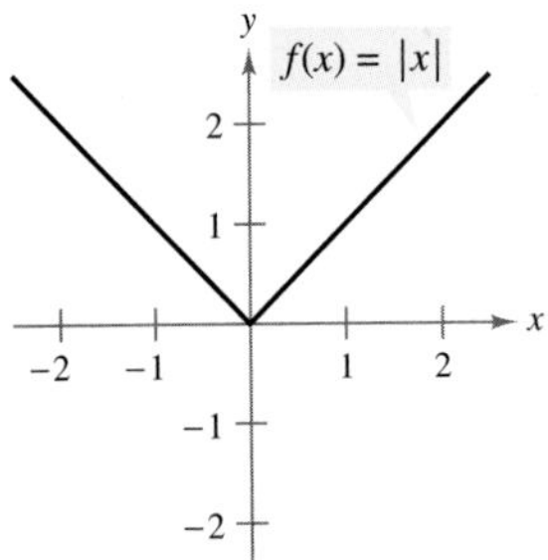

(c) Absolute value function

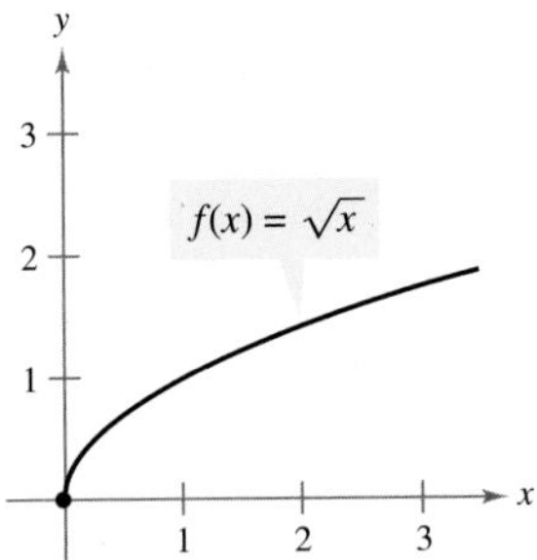

(d) Square root function

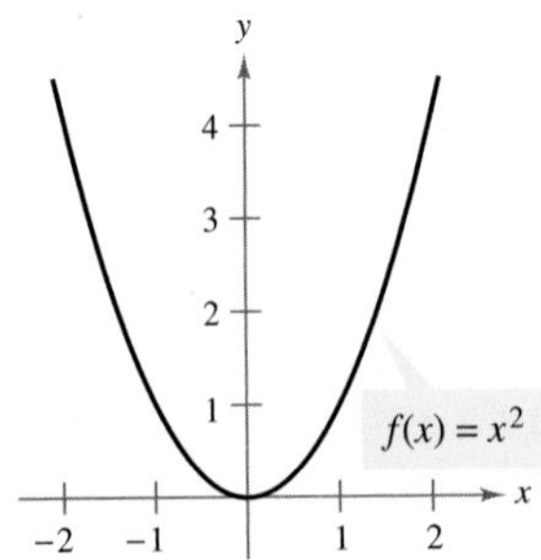

(e) Squaring function

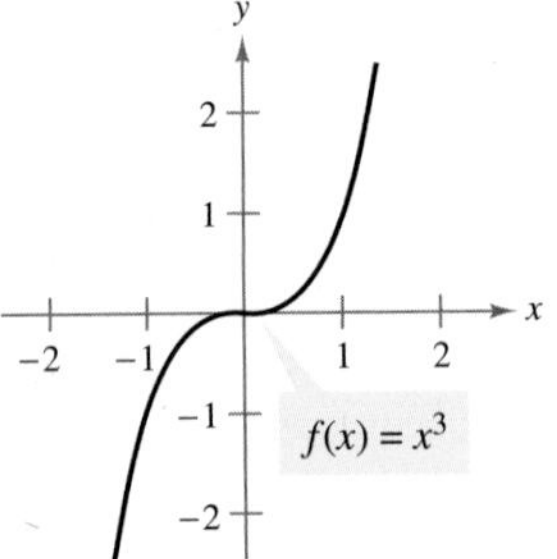

(f) Cubing function

Figure 1.18

Shifting Graphs

Many functions have graphs that are simple transformations of the common graphs summarized on page 98. For example, you can obtain the graph of

$$h(x) = x^2 + 2$$

by shifting the graph of $f(x) = x^2$ *upward* two units, as shown in Figure 1.19. In function notation, h and f are related as follows.

$$h(x) = x^2 + 2 = f(x) + 2 \qquad \text{Upward shift of two units}$$

Similarly, you can obtain the graph of

$$g(x) = (x - 2)^2$$

by shifting the graph of $f(x) = x^2$ to the *right* two units, as shown in Figure 1.20. In this case, the functions g and f have the following relationship.

$$g(x) = (x - 2)^2 = f(x - 2) \qquad \text{Right shift of two units}$$

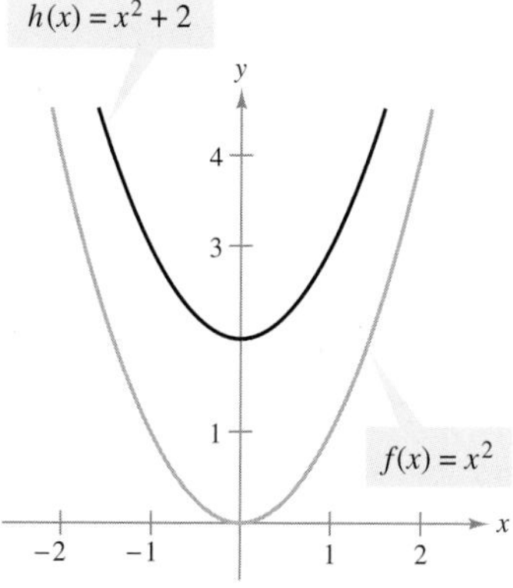

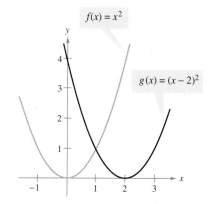

Figure 1.19 **Figure 1.20**

The following list summarizes this discussion about horizontal and vertical shifts.

VERTICAL AND HORIZONTAL SHIFTS

Let c be a positive real number. **Vertical and horizontal shifts** in the graph of $y = f(x)$ are represented as follows.

1. Vertical shift c units *upward:* $h(x) = f(x) + c$
2. Vertical shift c units *downward:* $h(x) = f(x) - c$
3. Horizontal shift c units to the *right:* $h(x) = f(x - c)$
4. Horizontal shift c units to the *left:* $h(x) = f(x + c)$

NOTE In items 3 and 4, be sure you see that $h(x) = f(x - c)$ corresponds to a *right* shift and $h(x) = f(x + c)$ corresponds to a *left* shift for $c > 0$. ■

Some graphs can be obtained from combinations of vertical and horizontal shifts, as demonstrated in Example 1(b). Vertical and horizontal shifts generate a *family of graphs,* each with the same shape but at different locations in the plane.

EXAMPLE 1 Shifts in the Graphs of a Function

Use the graph of $f(x) = x^3$ to sketch the graph of each function.

a. $g(x) = x^3 - 1$

b. $h(x) = (x + 2)^3 + 1$

Solution

a. Relative to the graph of $f(x) = x^3$, the graph of

$$g(x) = x^3 - 1$$

is a downward shift of one unit, as shown in Figure 1.21.

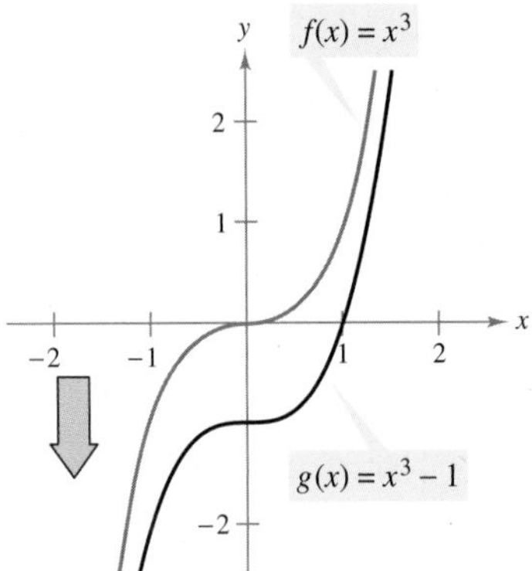

Figure 1.21

STUDY TIP In Example 1(a), note that $g(x) = f(x) - 1$. In Example 1(b), note that $h(x) = f(x + 2) + 1$.

b. Relative to the graph of $f(x) = x^3$, the graph of

$$h(x) = (x + 2)^3 + 1$$

involves a left shift of two units and an upward shift of one unit, as shown in Figure 1.22.

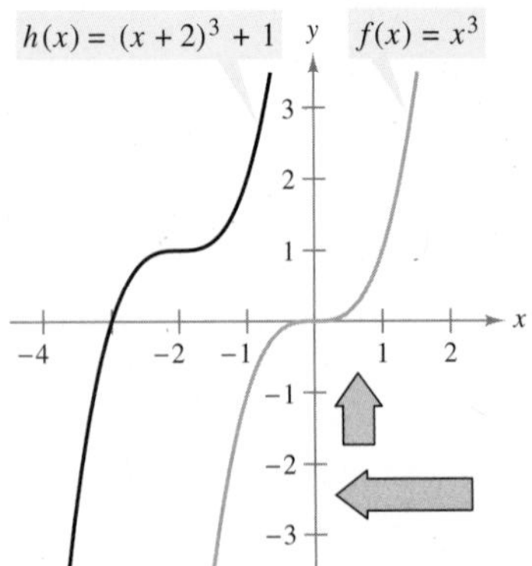

Figure 1.22 ■

NOTE In Figure 1.22, notice that the same result is obtained if the vertical shift precedes the horizontal shift *or* if the horizontal shift precedes the vertical shift. ■

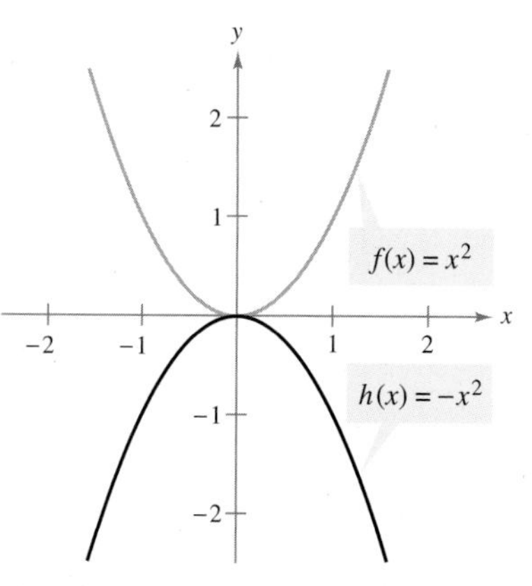

The graph of h is a reflection of the graph of f in the x-axis.
Figure 1.23

Reflecting Graphs

The second common type of transformation is a **reflection.** For instance, if you consider the x-axis to be a mirror, the graph of

$$h(x) = -x^2$$

is the mirror image (or reflection) of the graph of

$$f(x) = x^2$$

as shown in Figure 1.23.

> **REFLECTIONS IN THE COORDINATE AXES**
>
> **Reflections** in the coordinate axes of the graph of $y = f(x)$ are represented as follows.
>
> **1.** Reflection in the x-axis: $h(x) = -f(x)$
>
> **2.** Reflection in the y-axis: $h(x) = f(-x)$

EXAMPLE 2 Reflections and Shifts

Compare the graph of each function with the graph of $f(x) = \sqrt{x}$.

a. $g(x) = -\sqrt{x}$ **b.** $h(x) = \sqrt{-x}$ **c.** $k(x) = -\sqrt{x+2}$

Solution

a. The graph of g is a reflection of the graph of f in the x-axis because

$$g(x) = -\sqrt{x} = -f(x).$$

The graph of g compared with f is shown in Figure 1.24(a).

b. The graph of h is a reflection of the graph of f in the y-axis because

$$h(x) = \sqrt{-x} = f(-x).$$

The graph of h compared with f is shown in Figure 1.24(b).

c. The graph of k is a left shift of two units followed by a reflection in the x-axis because

$$k(x) = -\sqrt{x+2} = -f(x+2).$$

The graph of k compared with f is shown in Figure 1.24(c).

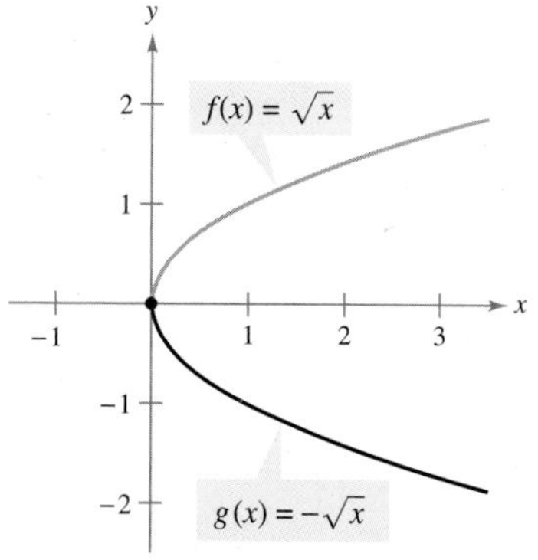

(a) Reflection in x-axis

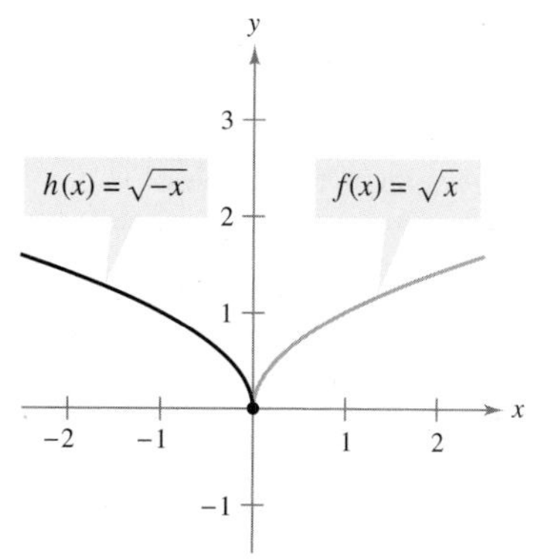

(b) Reflection in y-axis

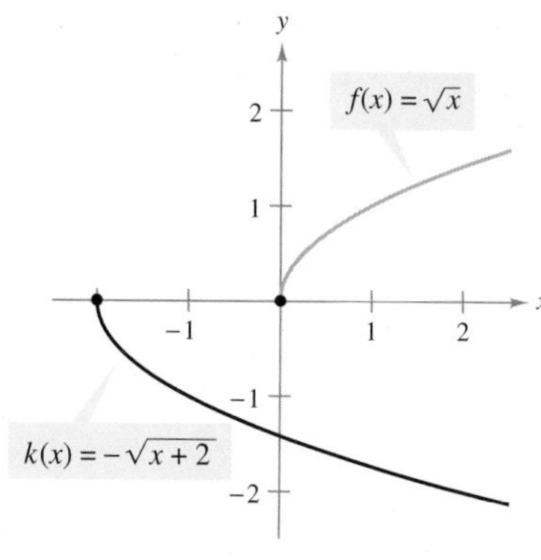

(c) Left shift and reflection in x-axis

Figure 1.24

■

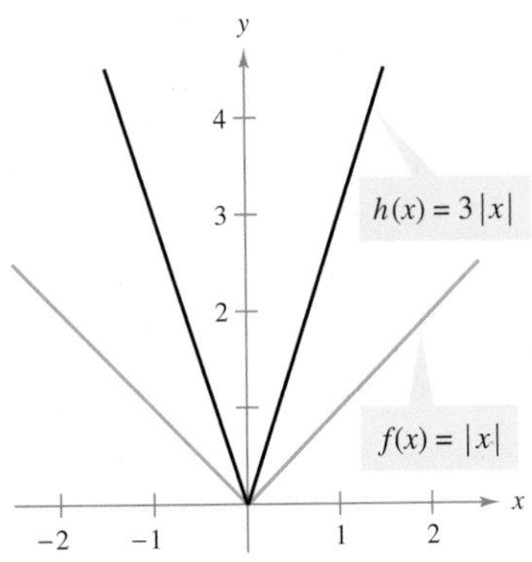

(a) Vertical stretch

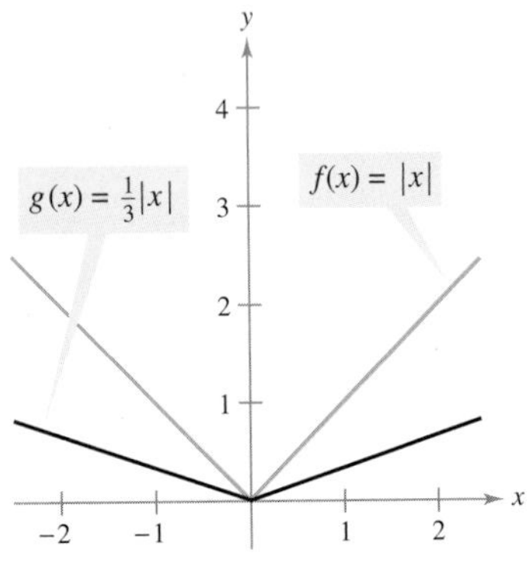

(b) Vertical shrink

Figure 1.25

Nonrigid Transformations

Horizontal shifts, vertical shifts, and reflections are **rigid transformations** because the basic shape of the graph is unchanged. These transformations change only the *position* of the graph in the coordinate plane. **Nonrigid transformations** are those that cause a *distortion*—a change in the shape of the original graph. For instance, a nonrigid transformation of the graph of $y = f(x)$ is represented by $g(x) = cf(x)$, where the transformation is a **vertical stretch** if $c > 1$ and a **vertical shrink** if $0 < c < 1$. Another nonrigid transformation of the graph of $y = f(x)$ is represented by $h(x) = f(cx)$, where the transformation is a **horizontal shrink** if $c > 1$ and a **horizontal stretch** if $0 < c < 1$.

EXAMPLE 3 Nonrigid Transformations

Compare the graph of each function with the graph of $f(x) = |x|$.

a. $h(x) = 3|x|$

b. $g(x) = \frac{1}{3}|x|$

Solution

a. Relative to the graph of $f(x) = |x|$, the graph of

$$h(x) = 3|x| = 3f(x)$$

is a vertical stretch (each y-value is multiplied by 3) of the graph of f. (See Figure 1.25(a).)

b. Similarly, the graph of

$$g(x) = \frac{1}{3}|x| = \frac{1}{3}f(x)$$

is a vertical shrink $\left(\text{each } y\text{-value is multiplied by } \frac{1}{3}\right)$ of the graph of f. (See Figure 1.25(b).)

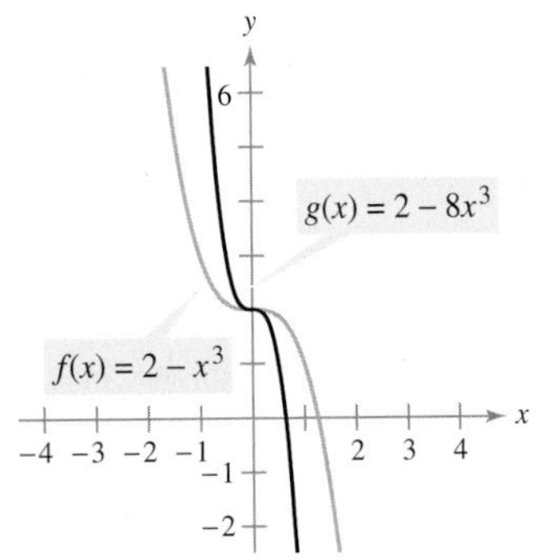

(a) Horizontal shrink

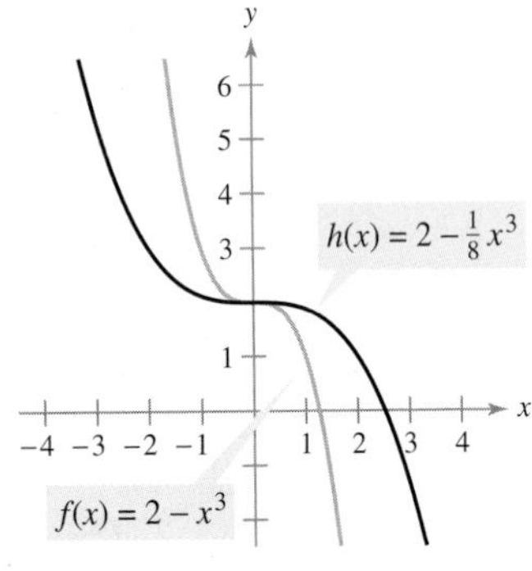

(b) Horizontal stretch

Figure 1.26

EXAMPLE 4 Nonrigid Transformations

Compare the graph of each function with the graph of $f(x) = 2 - x^3$.

a. $g(x) = f(2x)$

b. $h(x) = f\left(\frac{1}{2}x\right)$

Solution

a. Relative to the graph of $f(x) = 2 - x^3$, the graph of

$$g(x) = f(2x) = 2 - (2x)^3 = 2 - 8x^3$$

is a horizontal shrink $(c > 1)$ of the graph of f. (See Figure 1.26(a).)

b. Similarly, the graph of

$$h(x) = f\left(\frac{1}{2}x\right) = 2 - \left(\frac{1}{2}x\right)^3 = 2 - \frac{1}{8}x^3$$

is a horizontal stretch $(0 < c < 1)$ of the graph of f. (See Figure 1.26(b).)

■

1.3 Exercises

See www.CalcChat.com for worked-out solutions to odd-numbered exercises.

In Exercises 1–5, fill in the blanks.

1. Horizontal shifts, vertical shifts, and reflections are called ________ transformations.
2. A reflection in the x-axis of $y = f(x)$ is represented by $h(x) =$ ________, while a reflection in the y-axis of $y = f(x)$ is represented by $h(x) =$ ________.
3. Transformations that cause a distortion in the shape of the graph of $y = f(x)$ are called ________ transformations.
4. A nonrigid transformation of $y = f(x)$ represented by $h(x) = f(cx)$ is a ________ ________ if $c > 1$ and a ________ ________ if $0 < c < 1$.
5. A nonrigid transformation of $y = f(x)$ represented by $g(x) = cf(x)$ is a ________ ________ if $c > 1$ and a ________ ________ if $0 < c < 1$.

6. Match the rigid transformation of $y = f(x)$ with the correct representation of the graph of h, where $c > 0$.
   (a) $h(x) = f(x) + c$
   (b) $h(x) = f(x) - c$
   (c) $h(x) = f(x + c)$
   (d) $h(x) = f(x - c)$
   (i) A horizontal shift of f, c units to the right
   (ii) A vertical shift of f, c units downward
   (iii) A horizontal shift of f, c units to the left
   (iv) A vertical shift of f, c units upward
7. For each function, sketch (on the same set of coordinate axes) a graph for $c = -1, 1$, and 3.
   (a) $f(x) = |x| + c$
   (b) $f(x) = |x - c|$
   (c) $f(x) = |x + 4| + c$
8. For each function, sketch (on the same set of coordinate axes) a graph for $c = -3, -1, 1$, and 3.
   (a) $f(x) = \sqrt{x} + c$
   (b) $f(x) = \sqrt{x - c}$
   (c) $f(x) = \sqrt{x - 3} + c$
9. For each function, sketch (on the same set of coordinate axes) a graph for $c = -2, 0$, and 2.
   (a) $f(x) = [\![x]\!] + c$
   (b) $f(x) = [\![x + c]\!]$
   (c) $f(x) = [\![x - 1]\!] + c$
10. For each function, sketch (on the same set of coordinate axes) a graph for $c = -3, -1, 1$, and 3.
   (a) $f(x) = \begin{cases} x^2 + c, & x < 0 \\ -x^2 + c, & x \ge 0 \end{cases}$
   (b) $f(x) = \begin{cases} (x + c)^2, & x < 0 \\ -(x + c)^2, & x \ge 0 \end{cases}$
   (c) $f(x) = \begin{cases} (x + 1)^2 + c, & x < 0 \\ -(x + 1)^2 + c, & x \ge 0 \end{cases}$
11. Use the graph of $f(x) = x^2$ to write an equation for each function whose graph is shown.

(a)

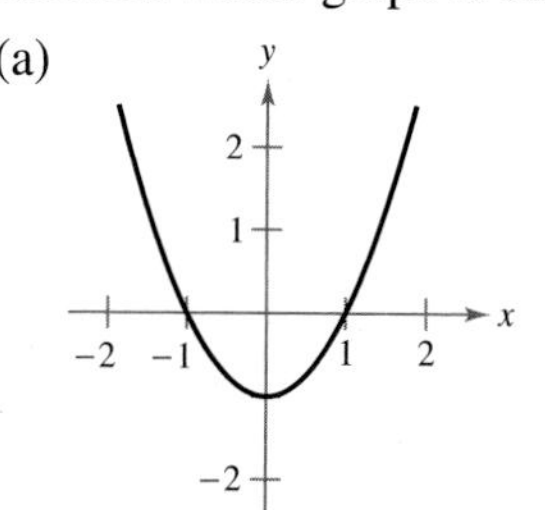

(b)

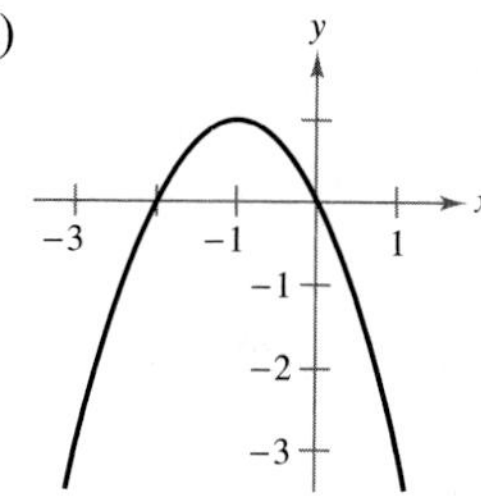

(c)

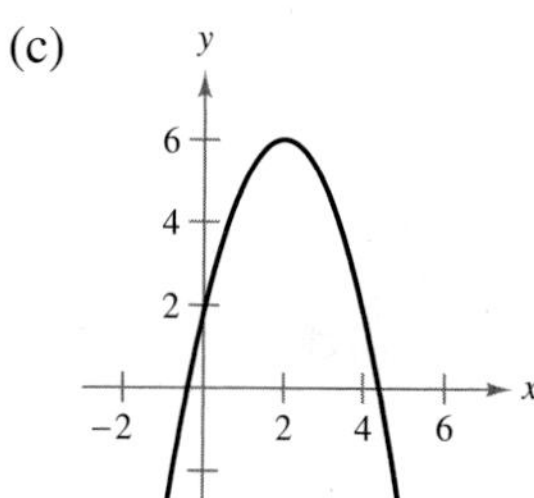

(d)

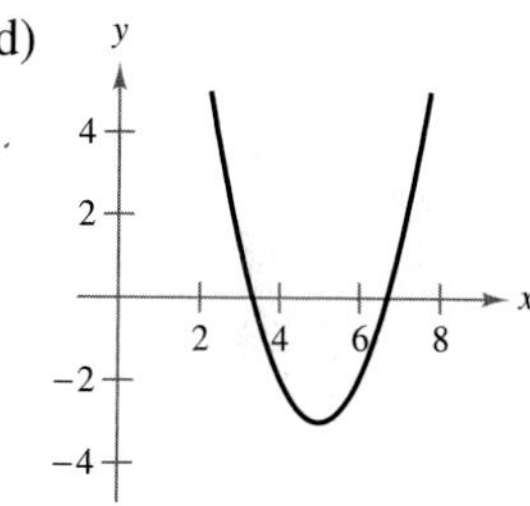

12. Use the graph of $f(x) = x^3$ to write an equation for each function whose graph is shown.

(a)

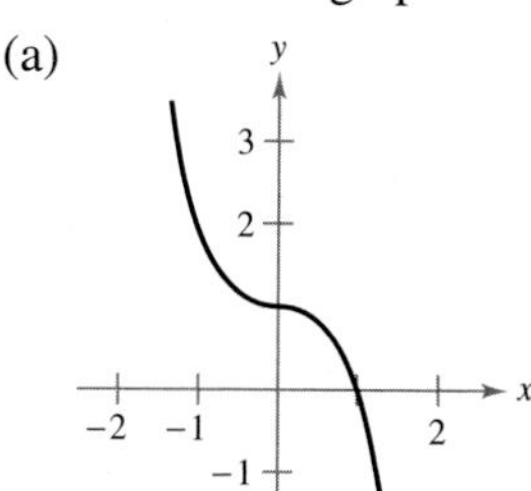

(b)

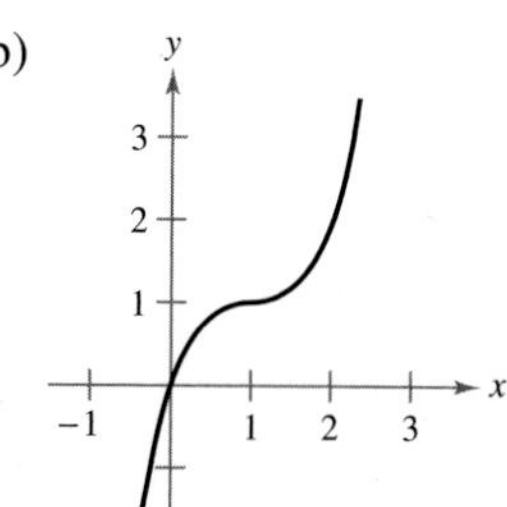

(c)

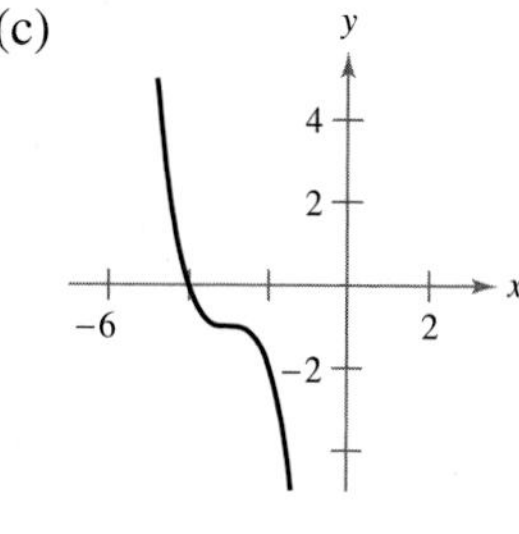

(d)

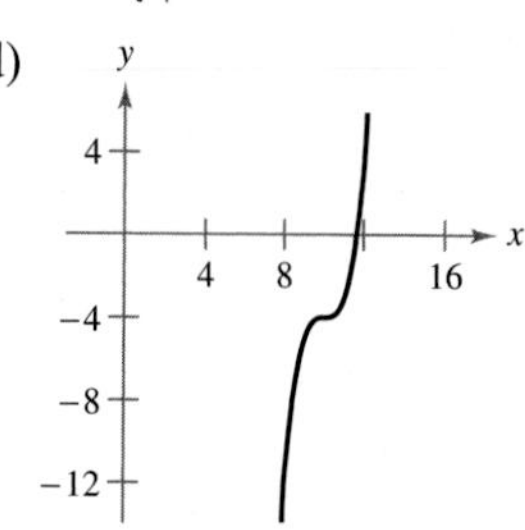

13. Use the graph of $f(x) = |x|$ to write an equation for each function whose graph is shown.

(a)

(b)

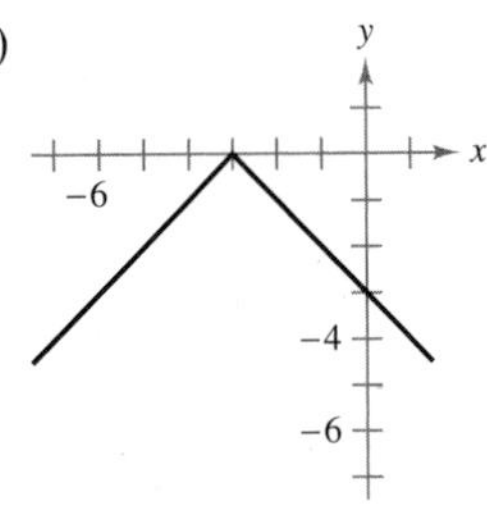

(c)

(d)

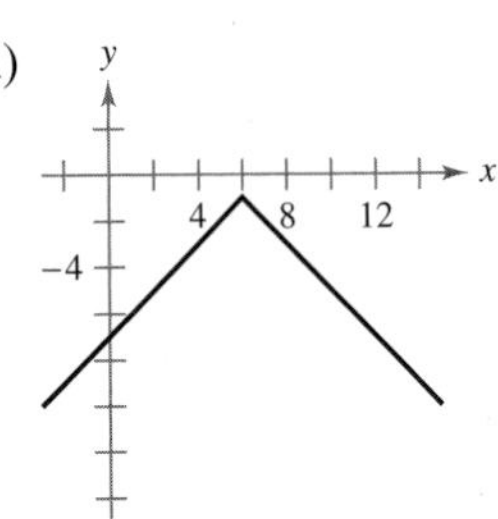

14. Use the graph of $f(x) = \sqrt{x}$ to write an equation for each function whose graph is shown.

(a)

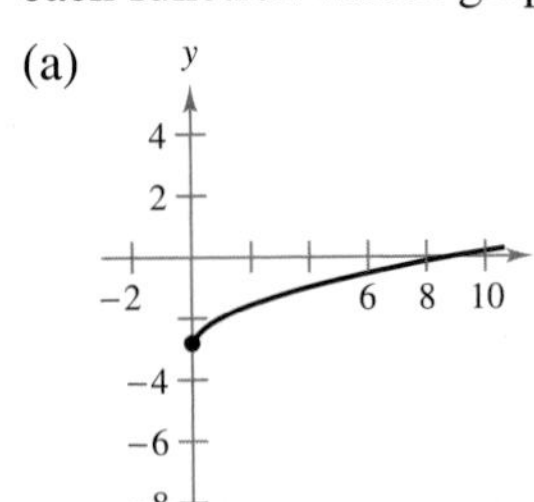

(b)

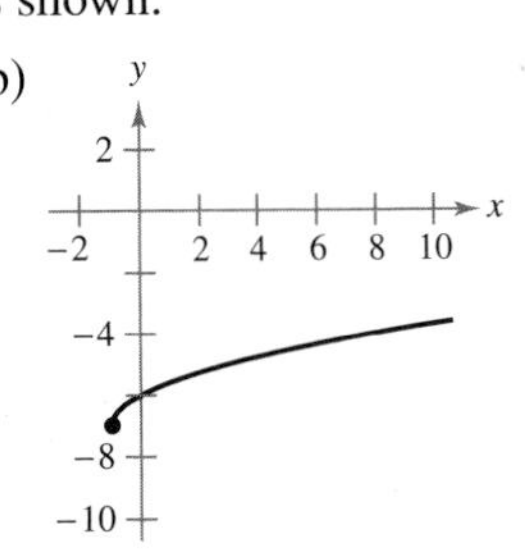

(c)

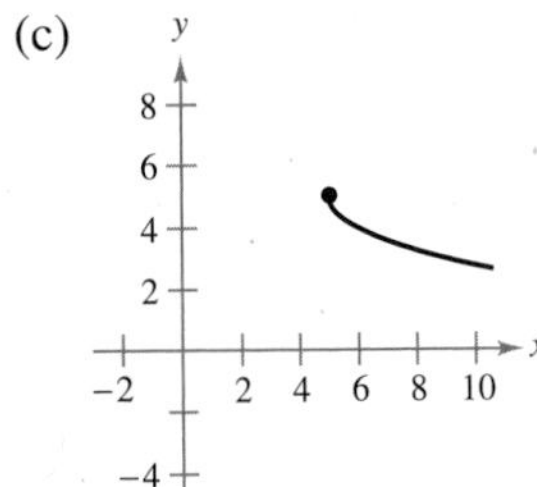

(d)

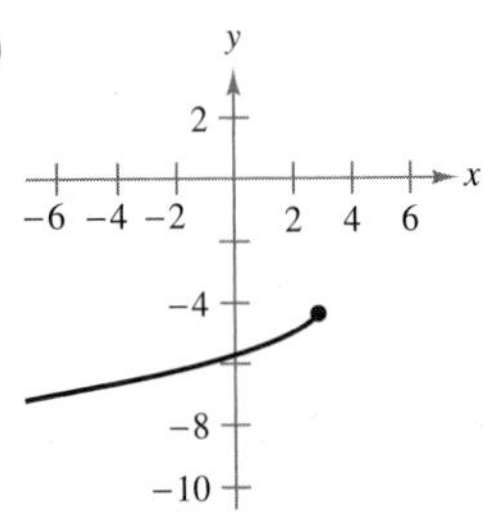

In Exercises 15–20, identify the common function and the transformation shown in the graph. Write an equation for the function shown in the graph.

15.

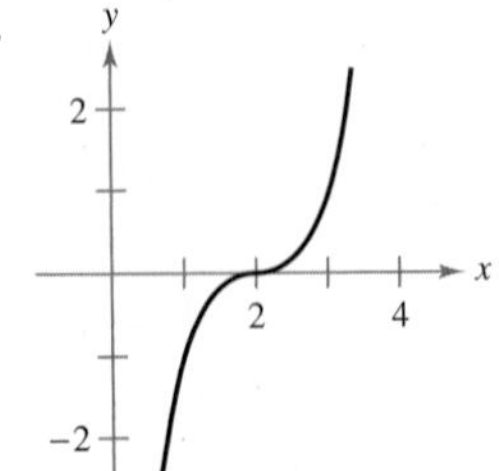

16.

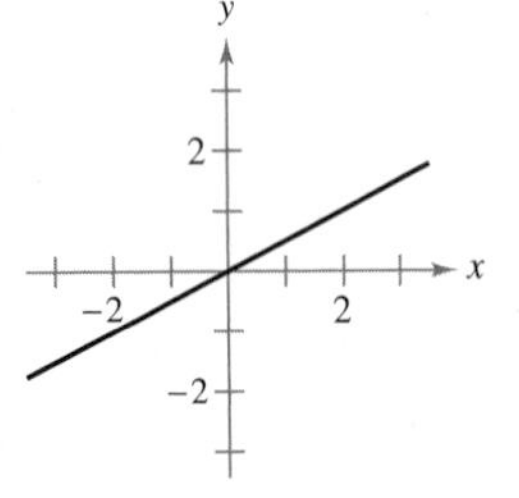

17.

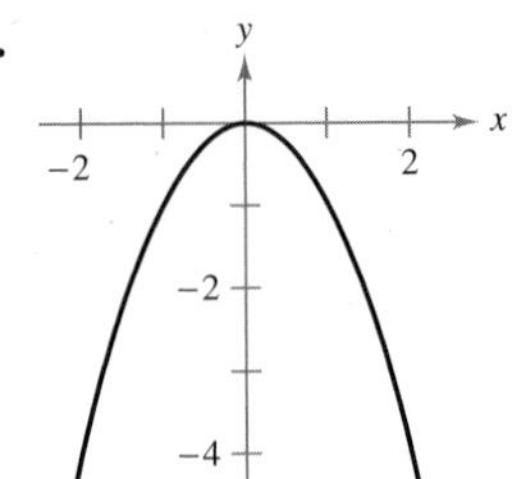

18.

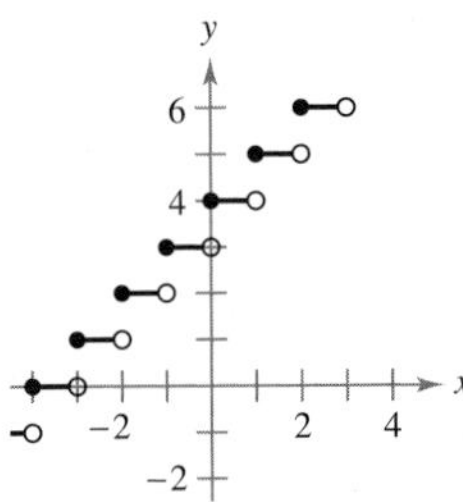

19.

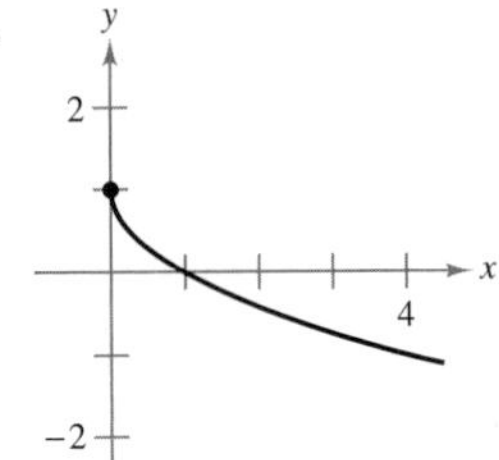

20.

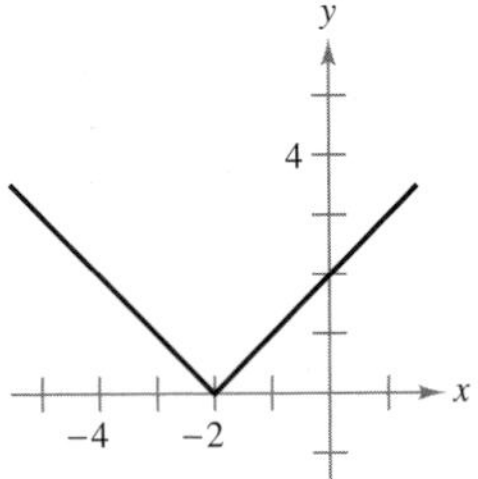

In Exercises 21–36, g is related to one of the common functions described on page 98. (a) Identify the common function f. (b) Describe the sequence of transformations from f to g. (c) Sketch the graph of g. (d) Use function notation to write g in terms of f.

21. $g(x) = 12 - x^2$

22. $g(x) = (x - 8)^2$

23. $g(x) = x^3 + 7$

24. $g(x) = -x^3 - 1$

25. $g(x) = 2 - (x + 5)^2$

26. $g(x) = -(x + 10)^2 + 5$

27. $g(x) = (x - 1)^3 + 2$

28. $g(x) = (x + 3)^3 - 10$

29. $g(x) = -|x| - 2$

30. $g(x) = 6 - |x + 5|$

31. $g(x) = -|x + 4| + 8$

32. $g(x) = |-x + 3| + 9$

33. $g(x) = \sqrt{x - 9}$

34. $g(x) = \sqrt{x + 4} + 8$

35. $g(x) = \sqrt{7 - x} - 2$

36. $g(x) = -\frac{1}{2}\sqrt{x + 3} - 1$

In Exercises 37–44, write an equation for the function that is described by the given characteristics.

37. The shape of $f(x) = x^2$, but shifted three units to the right and seven units downward

38. The shape of $f(x) = x^2$, but shifted two units to the left and nine units upward, and reflected in the x-axis

39. The shape of $f(x) = x^3$, but shifted 13 units to the right

40. The shape of $f(x) = x^3$, but shifted six units to the left and six units downward, and reflected in the y-axis

41. The shape of $f(x) = |x|$, but shifted 12 units upward and reflected in the x-axis

42. The shape of $f(x) = |x|$, but shifted four units to the left and eight units downward

43. The shape of $f(x) = \sqrt{x}$, but shifted six units to the left and reflected in both the x-axis and the y-axis

44. The shape of $f(x) = \sqrt{x}$, but shifted nine units downward and reflected in both the x-axis and the y-axis

In Exercises 45–48, use the graph of f to write an equation for each function whose graph is shown.

45. $f(x) = x^2$

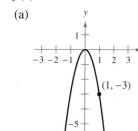

(a) (b)

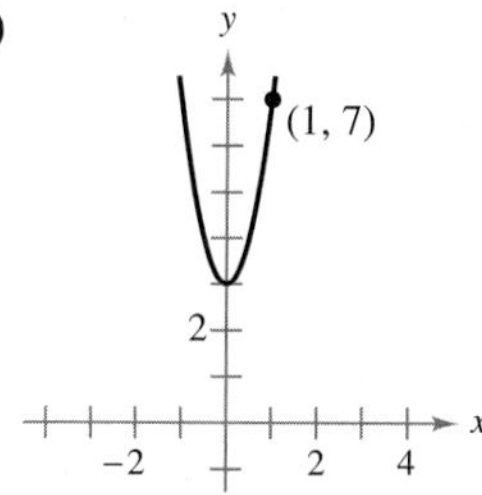

46. $f(x) = x^3$

(a)

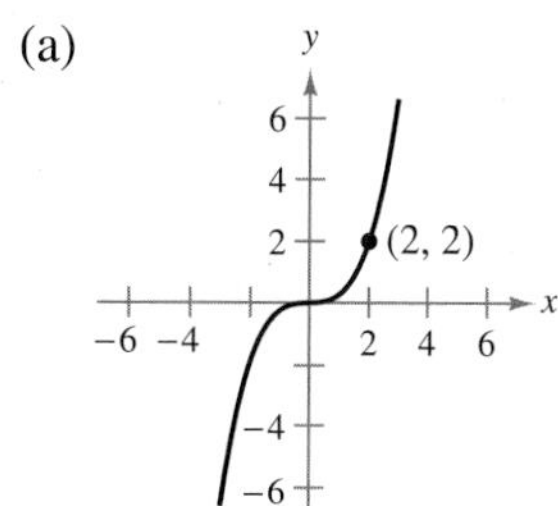

(b)

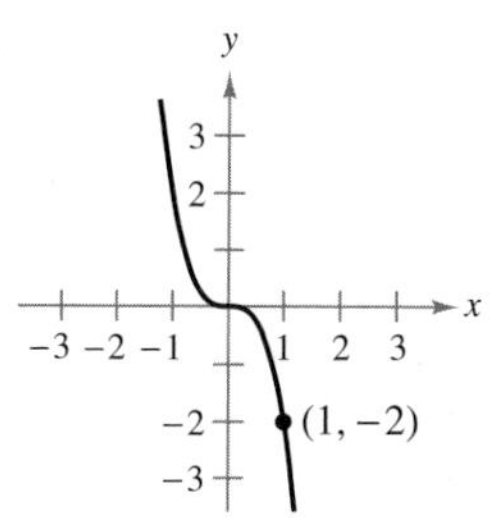

47. $f(x) = |x|$

(a)

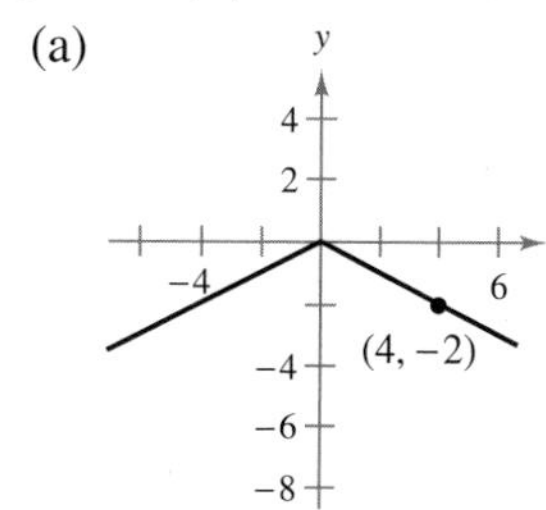

(b)

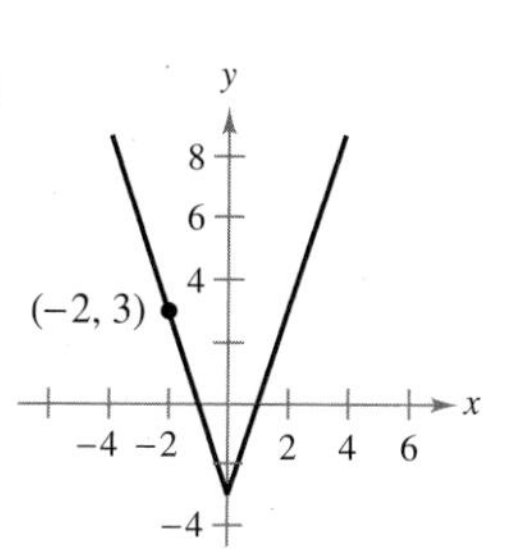

48. $f(x) = \sqrt{x}$

(a)

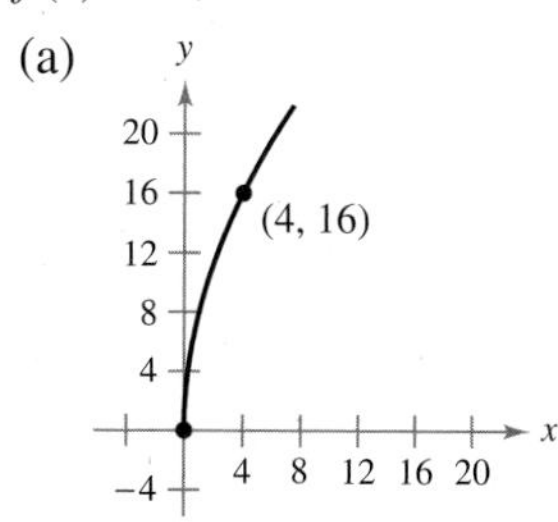

(b)

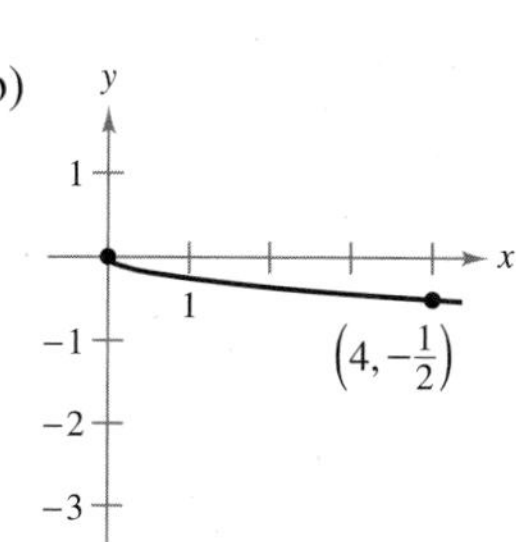

In Exercises 49–52, identify the common function and the transformation shown in the graph. Write an equation for the function shown in the graph. Then use a graphing utility to verify your answer.

49.

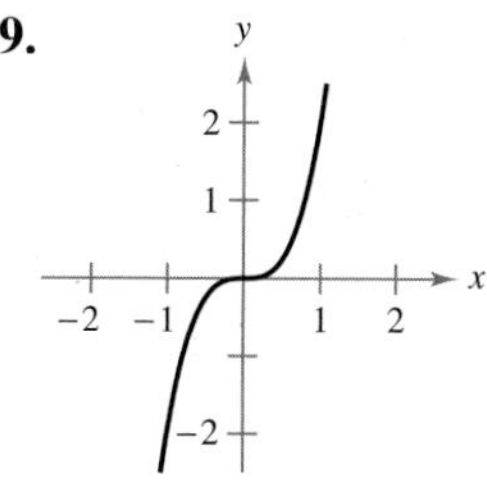

50.

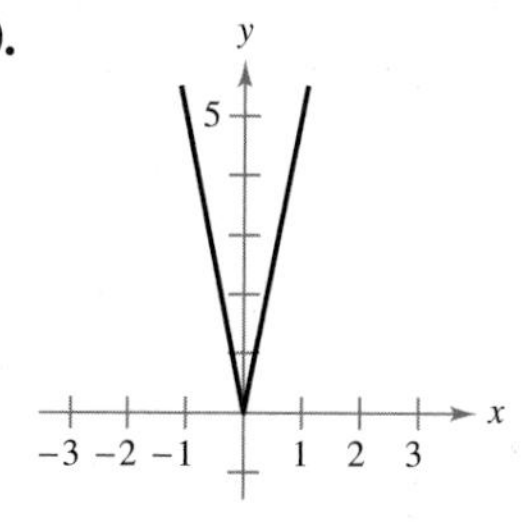

51.

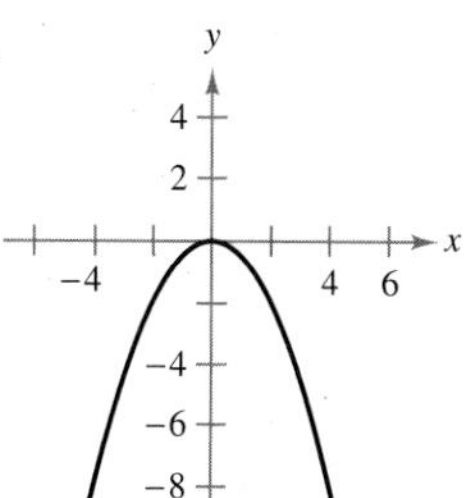

52.

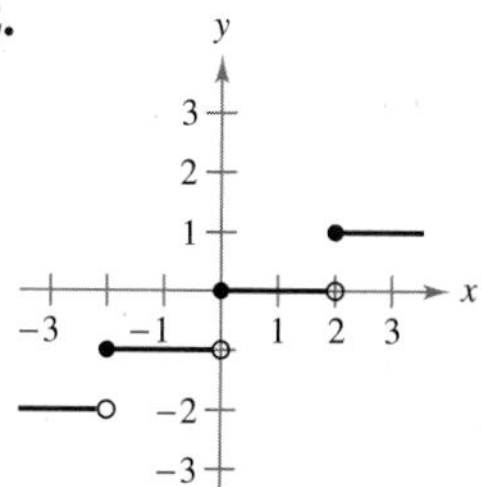

Graphical Analysis **In Exercises 53–56, use the viewing window shown to write a possible equation for the transformation of the common function.**

53.

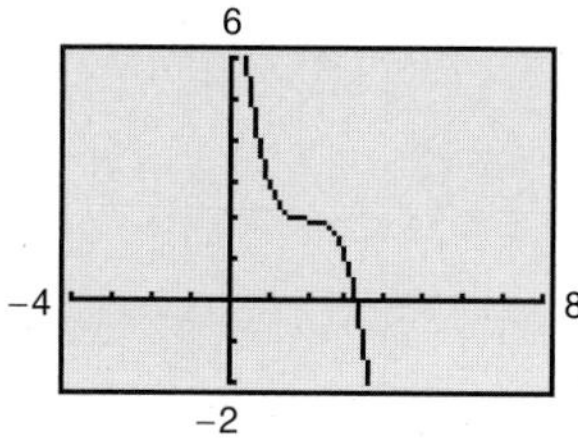

54.

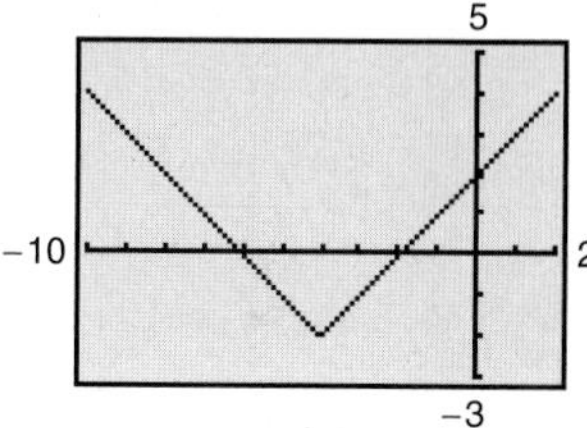

55.

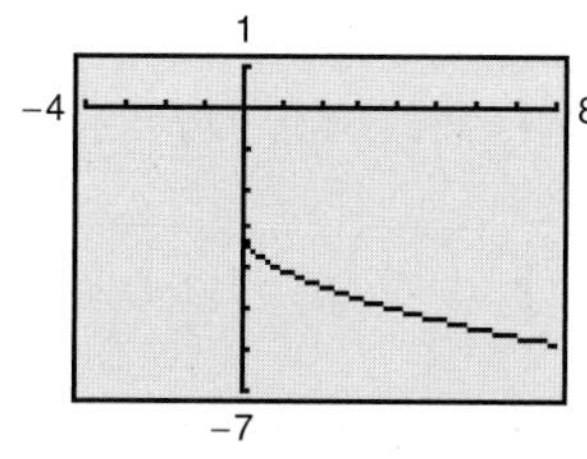

56.

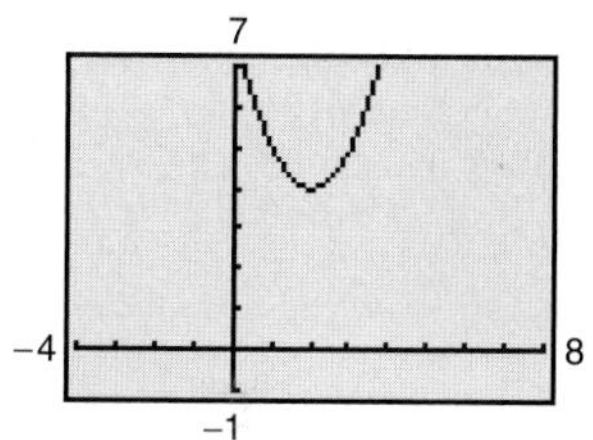

WRITING ABOUT CONCEPTS

In Exercises 57–60, use the graph of f to sketch each graph. To print an enlarged copy of the graph, go to the website *www.mathgraphs.com.*

57. (a) $y = f(x) + 2$
(b) $y = f(x - 2)$
(c) $y = 2f(x)$
(d) $y = -f(x)$
(e) $y = f(x + 3)$
(f) $y = f(-x)$
(g) $y = f\left(\frac{1}{2}x\right)$

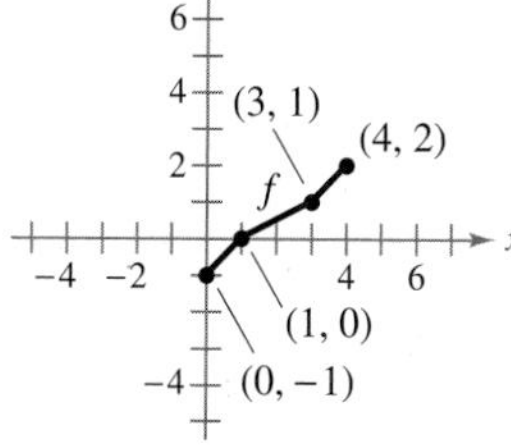

58. (a) $y = f(-x)$
(b) $y = f(x) + 4$
(c) $y = 2f(x)$
(d) $y = -f(x - 4)$
(e) $y = f(x) - 3$
(f) $y = -f(x) - 1$
(g) $y = f(2x)$

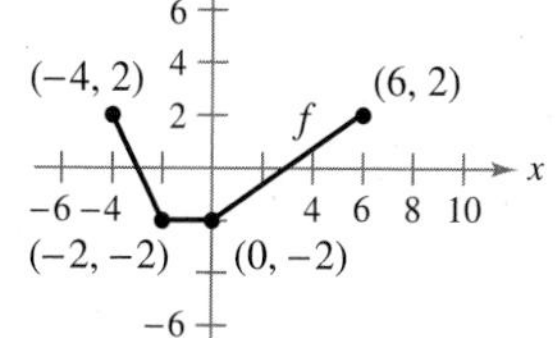

WRITING ABOUT CONCEPTS (continued)

59. (a) $y = f(x) - 1$
(b) $y = f(x - 1)$
(c) $y = f(-x)$
(d) $y = f(x + 1)$
(e) $y = -f(x - 2)$
(f) $y = \frac{1}{2}f(x)$
(g) $y = f(2x)$

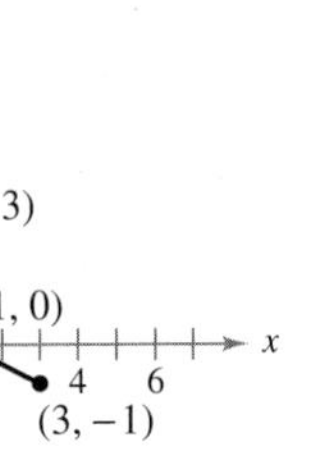

60. (a) $y = f(x - 5)$
(b) $y = -f(x) + 3$
(c) $y = \frac{1}{3}f(x)$
(d) $y = -f(x + 1)$
(e) $y = f(-x)$
(f) $y = f(x) - 10$
(g) $y = f\left(\frac{1}{3}x\right)$

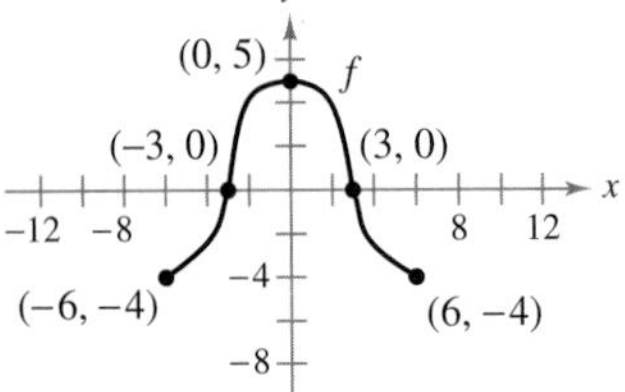

In Exercises 61 and 62, use the graph of f to sketch the graph of g. To print an enlarged copy of the graph, go to the website *www.mathgraphs.com*.

61.

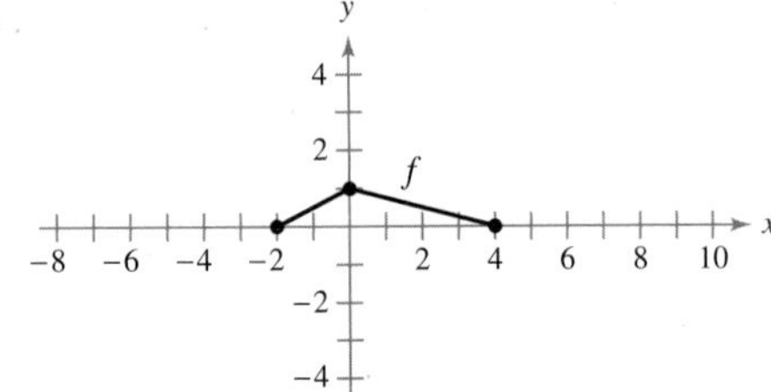

(a) $g(x) = f(x) + 2$
(b) $g(x) = f(x) - 1$
(c) $g(x) = f(-x)$
(d) $g(x) = -2f(x)$
(e) $g(x) = f(4x)$
(f) $g(x) = f\left(\frac{1}{2}x\right)$

62.

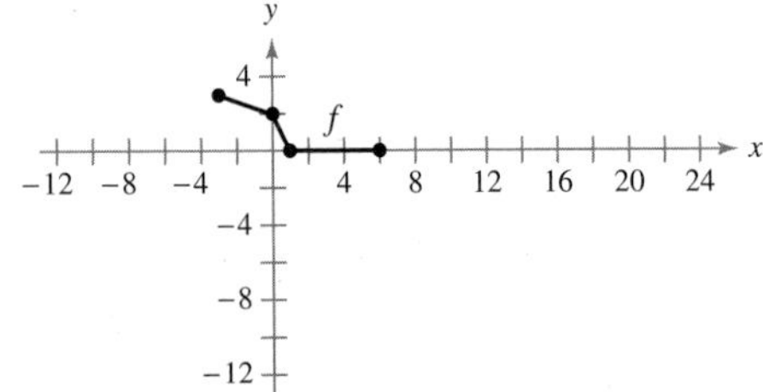

(a) $g(x) = f(x) - 5$
(b) $g(x) = f(x) + \frac{1}{2}$
(c) $g(x) = f(-x)$
(d) $g(x) = -4f(x)$
(e) $g(x) = f(2x) + 1$
(f) $g(x) = f\left(\frac{1}{4}x\right) - 2$

63. ***Miles Driven*** The total numbers of miles M (in billions) driven by vans, pickups, and SUVs (sport utility vehicles) in the United States from 1990 through 2006 can be approximated by the function

$$M = 527 + 128.0\sqrt{t}, \quad 0 \le t \le 16$$

where t represents the year, with $t = 0$ corresponding to 1990. *(Source: U.S. Federal Highway Administration)*

(a) Describe the transformation of the common function $f(x) = \sqrt{x}$. Then use a graphing utility to graph the function over the specified domain.

(b) Rewrite the function so that $t = 0$ represents 2000. Explain how you got your answer.

64. ***Married Couples*** The numbers N (in thousands) of married couples with stay-at-home mothers from 2000 through 2007 can be approximated by the function

$$N = -24.70(t - 5.99)^2 + 5617, \quad 0 \le t \le 7$$

where t represents the year, with $t = 0$ corresponding to 2000. *(Source: U.S. Census Bureau)*

(a) Describe the transformation of the common function $f(x) = x^2$. Then use a graphing utility to graph the function over the specified domain.

(b) Use the model to predict the number of married couples with stay-at-home mothers in 2015. Does your answer seem reasonable? Explain.

True or False? **In Exercises 65–67, determine whether the statement is true or false. Justify your answer.**

65. The graphs of $f(x) = |x| + 6$ and $f(x) = |-x| + 6$ are identical.

66. If the graph of the common function $f(x) = x^2$ is shifted six units to the right and three units upward, and reflected in the x-axis, then the point $(-2, 19)$ will lie on the graph of the transformation.

67. If f is an even function, then $y = f(x) + c$ is also even for any value of c.

CAPSTONE

68. Use the fact that the graph of $y = f(x)$ is increasing on the intervals $(-\infty, -1)$ and $(2, \infty)$ and decreasing on the interval $(-1, 2)$ to find the intervals on which the graph is increasing and decreasing. If not possible, state the reason.

(a) $y = f(-x)$
(b) $y = -f(x)$
(c) $y = \frac{1}{2}f(x)$
(d) $y = -f(x - 1)$
(e) $y = f(x - 2) + 1$

1.4 Combinations of Functions

- **Add, subtract, multiply, and divide functions.**
- **Find the composition of one function with another function.**
- **Use combinations and compositions of functions to model and solve real-life problems.**

Arithmetic Combinations of Functions

Just as two real numbers can be combined by the operations of addition, subtraction, multiplication, and division to form other real numbers, two *functions* can be combined to create new functions. For example, the functions given by $f(x) = 2x - 3$ and $g(x) = x^2 - 1$ can be combined to form the sum, difference, product, and quotient of f and g.

$$f(x) + g(x) = (2x - 3) + (x^2 - 1)$$
$$= x^2 + 2x - 4 \qquad \text{Sum}$$
$$f(x) - g(x) = (2x - 3) - (x^2 - 1)$$
$$= -x^2 + 2x - 2 \qquad \text{Difference}$$
$$f(x)g(x) = (2x - 3)(x^2 - 1)$$
$$= 2x^3 - 3x^2 - 2x + 3 \qquad \text{Product}$$
$$\frac{f(x)}{g(x)} = \frac{2x - 3}{x^2 - 1}, \quad x \neq \pm 1 \qquad \text{Quotient}$$

The domain of an **arithmetic combination** of functions f and g consists of all real numbers that are common to the domains of f and g. In the case of the quotient $f(x)/g(x)$, there is the further restriction that $g(x) \neq 0$.

SUM, DIFFERENCE, PRODUCT, AND QUOTIENT OF FUNCTIONS

Let f and g be two functions with overlapping domains. Then, for all x common to both domains, the *sum, difference, product,* and *quotient* of f and g are defined as follows.

1. *Sum:* $(f + g)(x) = f(x) + g(x)$
2. *Difference:* $(f - g)(x) = f(x) - g(x)$
3. *Product:* $(fg)(x) = f(x) \cdot g(x)$
4. *Quotient:* $\left(\frac{f}{g}\right)(x) = \frac{f(x)}{g(x)}, \quad g(x) \neq 0$

EXAMPLE 1 Finding the Sum of Two Functions

Given $f(x) = 2x + 1$ and $g(x) = x^2 + 2x - 1$, find $(f + g)(x)$. Then evaluate the sum when $x = 3$.

Solution

$$(f + g)(x) = f(x) + g(x) = (2x + 1) + (x^2 + 2x - 1) = x^2 + 4x$$

When $x = 3$, the value of this sum is

$$(f + g)(3) = 3^2 + 4(3) = 21.$$

■

EXAMPLE 2 Finding the Difference of Two Functions

Given $f(x) = 2x + 1$ and $g(x) = x^2 + 2x - 1$, find $(f - g)(x)$. Then evaluate the difference when $x = 2$.

Solution The difference of f and g is

$$\begin{aligned}(f - g)(x) &= f(x) - g(x) && \text{Definition of } (f-g)(x)\\ &= (2x + 1) - (x^2 + 2x - 1) && \text{Substitute.}\\ &= -x^2 + 2. && \text{Simplify.}\end{aligned}$$

When $x = 2$, the value of the difference is

$$\begin{aligned}(f - g)(2) &= -(2)^2 + 2\\ &= -2.\end{aligned}$$

EXAMPLE 3 Finding the Product of Two Functions

Given $f(x) = x^2$ and $g(x) = x - 3$, find $(fg)(x)$. Then evaluate the product when $x = 4$.

Solution The product of f and g is

$$\begin{aligned}(fg)(x) &= f(x)g(x) && \text{Definition of } (fg)(x)\\ &= (x^2)(x - 3) && \text{Substitute.}\\ &= x^3 - 3x^2 && \text{Simplify.}\end{aligned}$$

When $x = 4$, the value of this product is

$$\begin{aligned}(fg)(4) &= 4^3 - 3(4)^2\\ &= 16.\end{aligned}$$

■

NOTE In Examples 1–3, both f and g have domains that consist of all real numbers. So, the domains of $f + g$, $f - g$, and fg are also the set of all real numbers. Remember that any restrictions on the domains of f and g must be considered when forming the sum, difference, product, or quotient of f and g. ■

EXAMPLE 4 Finding the Quotients of Two Functions

Find $(f/g)(x)$ and $(g/f)(x)$ for the functions given by $f(x) = \sqrt{x}$ and $g(x) = \sqrt{4 - x^2}$. Then find the domains of f/g and g/f.

Solution The quotient of f and g is

$$\left(\frac{f}{g}\right)(x) = \frac{f(x)}{g(x)} = \frac{\sqrt{x}}{\sqrt{4 - x^2}}$$

and the quotient of g and f is

$$\left(\frac{g}{f}\right)(x) = \frac{g(x)}{f(x)} = \frac{\sqrt{4 - x^2}}{\sqrt{x}}.$$

The domain of f is $[0, \infty)$ and the domain of g is $[-2, 2]$. The intersection of these domains is $[0, 2]$. So, the domains of f/g and g/f are as follows.

Domain of $\frac{f}{g}$: $[0, 2)$ Domain of $\frac{g}{f}$: $(0, 2]$ ■

STUDY TIP Note that the domain of f/g includes $x = 0$, but not $x = 2$, because $x = 2$ yields a zero in the denominator, whereas the domain of g/f includes $x = 2$, but not $x = 0$, because $x = 0$ yields a zero in the denominator.

Composition of Functions

Another way of combining two functions is to form the **composition** of one with the other. For instance, if $f(x) = x^2$ and $g(x) = x + 1$, the composition of f with g is

$$\begin{aligned} f(g(x)) &= f(x + 1) \\ &= (x + 1)^2. \end{aligned}$$

This composition is denoted as $f \circ g$ and reads as "f composed with g."

DEFINITION OF COMPOSITION OF TWO FUNCTIONS

The **composition** of the function f with the function g is

$$(f \circ g)(x) = f(g(x)).$$

The domain of $f \circ g$ is the set of all x in the domain of g such that $g(x)$ is in the domain of f. (See Figure 1.27.)

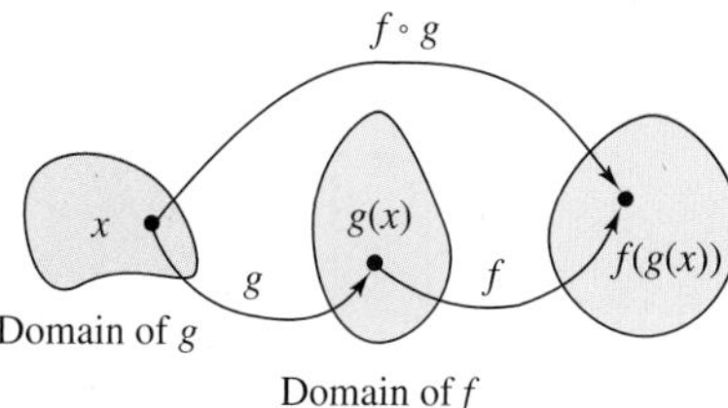

Figure 1.27

EXAMPLE 5 Composition of Functions

Given $f(x) = x + 2$ and $g(x) = 4 - x^2$, find the following.

a. $(f \circ g)(x)$ **b.** $(g \circ f)(x)$ **c.** $(g \circ f)(-2)$

Solution

a. The composition of f with g is as follows.

$$\begin{aligned} (f \circ g)(x) &= f(g(x)) && \text{Definition of } f \circ g \\ &= f(4 - x^2) && \text{Definition of } g(x) \\ &= (4 - x^2) + 2 && \text{Definition of } f(x) \\ &= -x^2 + 6 && \text{Simplify.} \end{aligned}$$

b. The composition of g with f is as follows.

$$\begin{aligned} (g \circ f)(x) &= g(f(x)) && \text{Definition of } g \circ f \\ &= g(x + 2) && \text{Definition of } f(x) \\ &= 4 - (x + 2)^2 && \text{Definition of } g(x) \\ &= -x^2 - 4x && \text{Simplify.} \end{aligned}$$

Note that, in this case, $(f \circ g)(x) \neq (g \circ f)(x)$.

c. Using the result of part (b), you can write the following.

$$\begin{aligned} (g \circ f)(-2) &= -(-2)^2 - 4(-2) && \text{Substitute.} \\ &= -4 + 8 && \text{Simplify.} \\ &= 4 && \text{Simplify.} \end{aligned}$$

■

NOTE The following tables of values help illustrate the composition of $(f \circ g)(x)$ given in Example 5.

x	0	1	2	3
$g(x)$	4	3	0	-5

$g(x)$	4	3	0	-5
$f(g(x))$	6	5	2	-3

x	0	1	2	3
$f(g(x))$	6	5	2	-3

Note that the first two tables can be combined (or "composed") to produce the values given in the third table.

EXAMPLE 6 Finding the Domain of a Composite Function

Find the domain of $(f \circ g)(x)$ for the functions given by

$$f(x) = x^2 - 9 \quad \text{and} \quad g(x) = \sqrt{9 - x^2}.$$

Algebraic Solution

The composition of the functions is as follows.

$$\begin{aligned}(f \circ g)(x) &= f(g(x)) \\ &= f\left(\sqrt{9 - x^2}\right) \\ &= \left(\sqrt{9 - x^2}\right)^2 - 9 \\ &= 9 - x^2 - 9 \\ &= -x^2\end{aligned}$$

From this, it might appear that the domain of the composition is the set of all real numbers. This, however, is not true. Because the domain of f is the set of all real numbers and the domain of g is $[-3, 3]$, the domain of $f \circ g$ is $[-3, 3]$.

Graphical Solution

You can use a graphing utility to graph the composition of the functions $(f \circ g)(x)$ as $y = \left(\sqrt{9 - x^2}\right)^2 - 9$. Enter the functions as follows.

$$y_1 = \sqrt{9 - x^2} \qquad y_2 = y_1^2 - 9$$

Graph y_2, as shown in Figure 1.28. Use the *trace* feature to determine that the x-coordinates of points on the graph extend from -3 to 3. So, you can graphically estimate the domain of $f \circ g$ to be $[-3, 3]$.

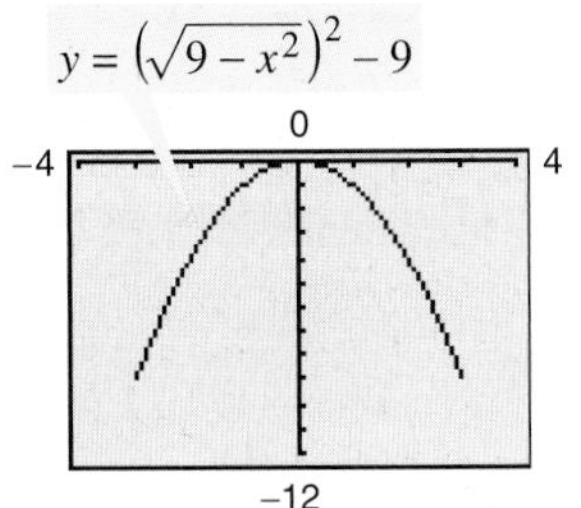

Figure 1.28

In Examples 5 and 6, you formed the composition of two given functions. In calculus, it is also important to be able to identify two functions that make up a given composite function. For instance, the function h given by $h(x) = (3x - 5)^3$ is the composition of f with g, where $f(x) = x^3$ and $g(x) = 3x - 5$. That is,

$$h(x) = (3x - 5)^3 = f(3x - 5) = f(g(x)).$$

Basically, to "decompose" a composite function, look for an "inner" function and an "outer" function. In the function h above, $g(x) = 3x - 5$ is the inner function and $f(x) = x^3$ is the outer function.

STUDY TIP For the composition $(f \circ g)(x) = f(g(x))$, consider f as the *outer* function and g as the *inner* function. For either $f(g(x))$ or $g(f(x))$, the domain of the composite function is either *equal to* or *a restriction of the domain of the inner function.*

EXAMPLE 7 Decomposing a Composite Function

Write the function given by $h(x) = \dfrac{1}{(x - 2)^2}$ as a composition of two functions.

Solution One way to write h as a composition of two functions is to take the inner function to be $g(x) = x - 2$ and the outer function to be

$$f(x) = \frac{1}{x^2}.$$

Then you can write

$$h(x) = \frac{1}{(x - 2)^2} = f(x - 2) = f(g(x)).$$

NOTE There are other correct answers to Example 7. For instance, let $g(x) = (x - 2)^2$ and let $f(x) = \dfrac{1}{x}$. Then $f(g(x)) = f([x - 2]^2) = \dfrac{1}{(x - 2)^2} = h(x)$.

Application

EXAMPLE 8 Bacteria Count

The number N of bacteria in a refrigerated food is given by

$$N(T) = 20T^2 - 80T + 500, \quad 2 \leq T \leq 14$$

where T is the temperature of the food in degrees Celsius. When the food is removed from refrigeration, the temperature of the food is given by

$$T(t) = 4t + 2, \quad 0 \leq t \leq 3$$

where t is the time in hours.

a. Find the composition $N(T(t))$ and interpret its meaning in context.

b. Find the time when the bacteria count reaches 2000.

Solution

a.
$$\begin{aligned} N(T(t)) &= 20(4t + 2)^2 - 80(4t + 2) + 500 \\ &= 20(16t^2 + 16t + 4) - 320t - 160 + 500 \\ &= 320t^2 + 320t + 80 - 320t - 160 + 500 \\ &= 320t^2 + 420 \end{aligned}$$

The composite function $N(T(t))$ represents the number of bacteria in the food as a function of the amount of time the food has been out of refrigeration.

b. The bacteria count will reach 2000 when $320t^2 + 420 = 2000$. Solve this equation for t as shown.

$$\begin{aligned} 320t^2 + 420 &= 2000 \\ 320t^2 &= 1580 \\ t^2 &= \frac{79}{16} \\ t &= \frac{\sqrt{79}}{4} \\ t &\approx 2.2 \end{aligned}$$

So, the count will reach 2000 when $t \approx 2.2$ hours.

When you solve this equation, note that the negative value is rejected because it is not in the domain of the composite function. ■

EXPLORATION

You are buying an automobile that costs $18,500. Which of the following options would you choose? Explain your reasoning.

a. You are given a factory rebate of $2000, followed by a dealer discount of 10%.

b. You are given a dealer discount of 10%, followed by a factory rebate of $2000.

Let $f(x) = x - 2000$ and let $g(x) = 0.9x$. Which option is represented by the composite $f(g(x))$? Which is represented by the composite $g(f(x))$?

1.4 Exercises

See www.CalcChat.com for worked-out solutions to odd-numbered exercises.

In Exercises 1–4, fill in the blanks.

1. Two functions f and g can be combined by the arithmetic operations of ________, ________, ________, and ________ to create new functions.
2. The ________ of the function f with g is $(f \circ g)(x) = f(g(x))$.
3. The domain of $(f \circ g)$ is all x in the domain of g such that ________ is in the domain of f.
4. To decompose a composite function, look for an ________ function and an ________ function.

In Exercises 5–12, find (a) $(f + g)(x)$, (b) $(f - g)(x)$, (c) $(fg)(x)$, and (d) $(f/g)(x)$. What is the domain of f/g?

5. $f(x) = x + 2, \quad g(x) = x - 2$
6. $f(x) = 2x - 5, \quad g(x) = 2 - x$
7. $f(x) = x^2, \quad g(x) = 4x - 5$
8. $f(x) = 3x + 1, \quad g(x) = 5x - 4$
9. $f(x) = x^2 + 6, \quad g(x) = \sqrt{1 - x}$
10. $f(x) = \sqrt{x^2 - 4}, \quad g(x) = \dfrac{x^2}{x^2 + 1}$
11. $f(x) = \dfrac{1}{x}, \quad g(x) = \dfrac{1}{x^2}$
12. $f(x) = \dfrac{x}{x + 1}, \quad g(x) = x^3$

In Exercises 13–24, evaluate the indicated function for $f(x) = x^2 + 1$ and $g(x) = x - 4$.

13. $(f + g)(2)$
14. $(f - g)(-1)$
15. $(f - g)(0)$
16. $(f + g)(1)$
17. $(f - g)(3t)$
18. $(f + g)(t - 2)$
19. $(fg)(6)$
20. $(fg)(-6)$
21. $(f/g)(5)$
22. $(f/g)(0)$
23. $(f/g)(-1) - g(3)$
24. $(fg)(5) + f(4)$

In Exercises 25–28, graph the functions f, g, and $f + g$ on the same set of coordinate axes.

25. $f(x) = \frac{1}{2}x, \quad g(x) = x - 1$
26. $f(x) = \frac{1}{3}x, \quad g(x) = -x + 4$
27. $f(x) = x^2, \quad g(x) = -2x$
28. $f(x) = 4 - x^2, \quad g(x) = x$

Graphical Reasoning **In Exercises 29–32, use a graphing utility to graph f, g, and $f + g$ in the same viewing window. Which function contributes most to the magnitude of the sum when $0 \le x \le 2$? Which function contributes most to the magnitude of the sum when $x > 6$?**

29. $f(x) = 3x, \quad g(x) = -\dfrac{x^3}{10}$
30. $f(x) = \dfrac{x}{2}, \quad g(x) = \sqrt{x}$
31. $f(x) = 3x + 2, \quad g(x) = -\sqrt{x + 5}$
32. $f(x) = x^2 - \frac{1}{2}, \quad g(x) = -3x^2 - 1$

In Exercises 33–36, find (a) $f \circ g$, (b) $g \circ f$, and (c) $g \circ g$.

33. $f(x) = x^2, \quad g(x) = x - 1$
34. $f(x) = 3x + 5, \quad g(x) = 5 - x$
35. $f(x) = \sqrt[3]{x - 1}, \quad g(x) = x^3 + 1$
36. $f(x) = x^3, \quad g(x) = \dfrac{1}{x}$

In Exercises 37–44, find (a) $f \circ g$ and (b) $g \circ f$. Find the domain of each function and each composite function.

37. $f(x) = \sqrt{x + 4}, \quad g(x) = x^2$
38. $f(x) = \sqrt[3]{x - 5}, \quad g(x) = x^3 + 1$
39. $f(x) = x^2 + 1, \quad g(x) = \sqrt{x}$
40. $f(x) = x^{2/3}, \quad g(x) = x^6$
41. $f(x) = |x|, \quad g(x) = x + 6$
42. $f(x) = |x - 4|, \quad g(x) = 3 - x$
43. $f(x) = \dfrac{1}{x}, \quad g(x) = x + 3$
44. $f(x) = \dfrac{3}{x^2 - 1}, \quad g(x) = x + 1$

In Exercises 45–52, find two functions f and g such that $(f \circ g)(x) = h(x)$. (There are many correct answers.)

45. $h(x) = (2x + 1)^2$
46. $h(x) = (1 - x)^3$
47. $h(x) = \sqrt[3]{x^2 - 4}$
48. $h(x) = \sqrt{9 - x}$
49. $h(x) = \dfrac{1}{x + 2}$
50. $h(x) = \dfrac{4}{(5x + 2)^2}$
51. $h(x) = \dfrac{-x^2 + 3}{4 - x^2}$
52. $h(x) = \dfrac{27x^3 + 6x}{10 - 27x^3}$

WRITING ABOUT CONCEPTS

In Exercises 53–56, use the graphs of f and g to graph $h(x) = (f + g)(x)$. To print an enlarged copy of the graph, go to the website *www.mathgraphs.com.*

53.

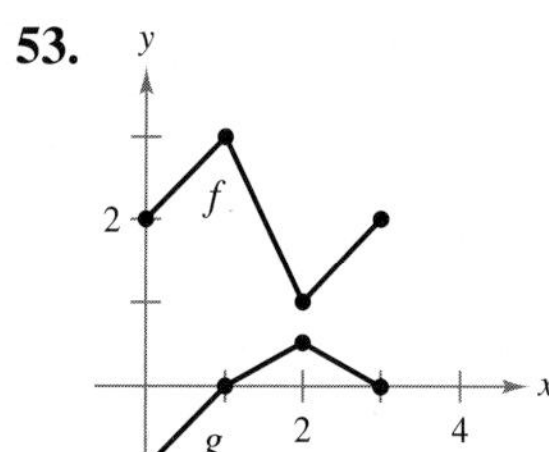

54.

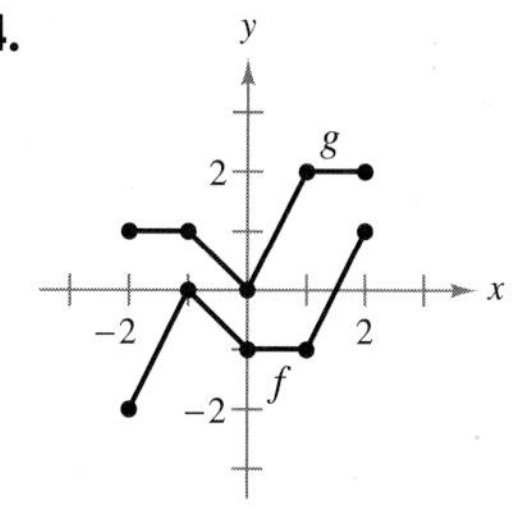

55.

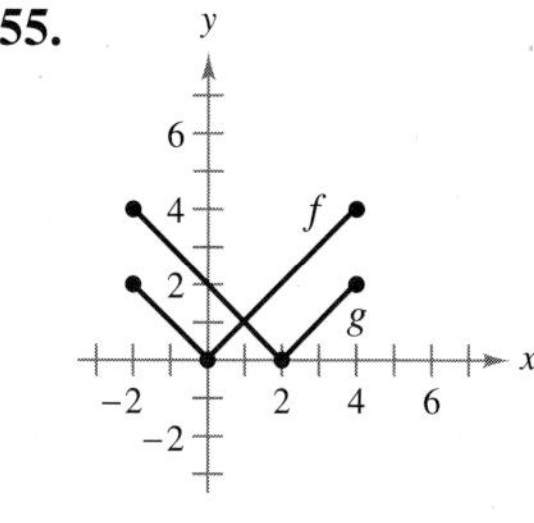

56.

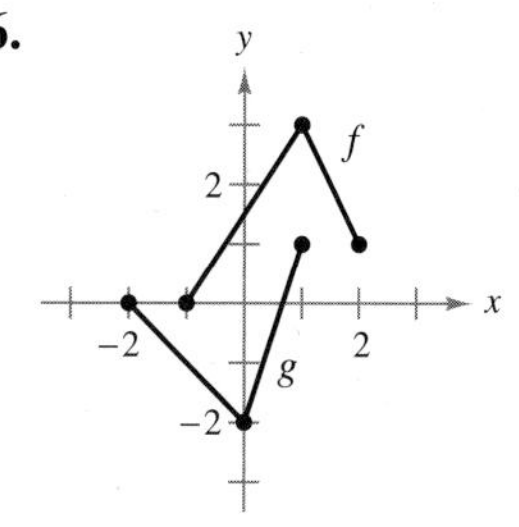

In Exercises 57–60, use the graphs of f and g to evaluate the functions.

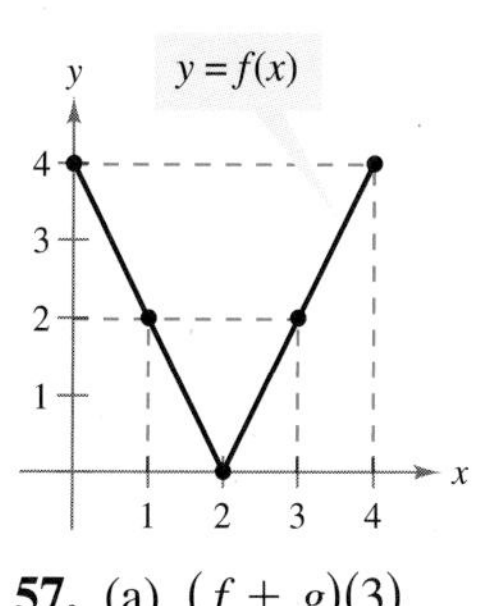

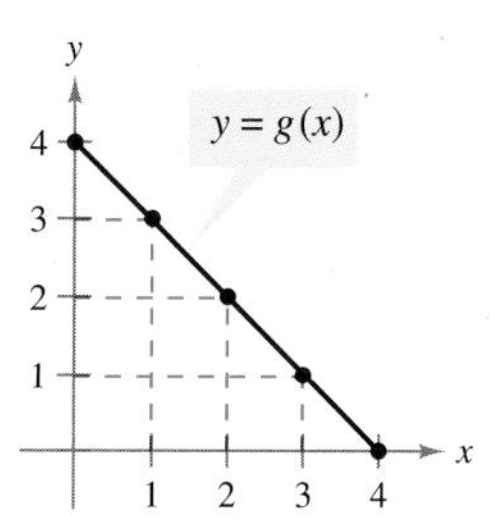

57. (a) $(f + g)(3)$ (b) $(f/g)(2)$

58. (a) $(f - g)(1)$ (b) $(fg)(4)$

59. (a) $(f \circ g)(2)$ (b) $(g \circ f)(2)$

60. (a) $(f \circ g)(1)$ (b) $(g \circ f)(3)$

61. ***Stopping Distance*** The research and development department of an automobile manufacturer has determined that when a driver is required to stop quickly to avoid an accident, the distance (in feet) the car travels during the driver's reaction time is given by $R(x) = \frac{3}{4}x$, where x is the speed of the car in miles per hour. The distance (in feet) traveled while the driver is braking is given by $B(x) = \frac{1}{15}x^2$.

(a) Find the function that represents the total stopping distance T.

(b) Graph the functions R, B, and T on the same set of coordinate axes for $0 \le x \le 60$.

(c) Which function contributes most to the magnitude of the sum at higher speeds? Explain.

62. ***Sales*** From 2003 through 2008, the sales R_1 (in thousands of dollars) for one of two restaurants owned by the same parent company can be modeled by

$$R_1 = 480 - 8t - 0.8t^2, \quad t = 3, 4, 5, 6, 7, 8$$

where $t = 3$ represents 2003. During the same six-year period, the sales R_2 (in thousands of dollars) for the second restaurant can be modeled by

$$R_2 = 254 + 0.78t, \quad t = 3, 4, 5, 6, 7, 8.$$

(a) Write a function R_3 that represents the total sales of the two restaurants owned by the same parent company.

(b) Use a graphing utility to graph R_1, R_2, and R_3 in the same viewing window.

Births and Deaths **In Exercises 63 and 64, use the data, which shows the total numbers of births B (in thousands) and deaths D (in thousands) in the United States from 1990 through 2006.** ***(Source: U.S. Census Bureau)***

Year, t	*Births, B*	*Deaths, D*
1990	4158	2148
1991	4111	2170
1992	4065	2176
1993	4000	2269
1994	3953	2279
1995	3900	2312
1996	3891	2315
1997	3881	2314
1998	3942	2337
1999	3959	2391
2000	4059	2403
2001	4026	2416
2002	4022	2443
2003	4090	2448
2004	4112	2398
2005	4138	2448
2006	4266	2426

The models for these data are

$$B(t) = -0.197t^3 + 8.96t^2 - 90.0t + 4180$$

and

$$D(t) = -1.21t^2 + 38.0t + 2137$$

where t represents the year, with $t = 0$ corresponding to 1990.

63. Find and interpret $(B - D)(t)$.

64. Evaluate $B(t)$, $D(t)$, and $(B - D)(t)$ for the years 2010 and 2012. What does each function value represent?

65. ***Sports*** The numbers of people playing tennis T (in millions) in the United States from 2000 through 2007 can be approximated by the function

$$T(t) = 0.0233t^4 - 0.3408t^3 + 1.556t^2 - 1.86t + 22.8$$

and the U.S. population P (in millions) from 2000 through 2007 can be approximated by the function $P(t) = 2.78t + 282.5$, where t represents the year, with $t = 0$ corresponding to 2000. *(Source: Tennis Industry Association, U.S. Census Bureau)*

(a) Find and interpret $h(t) = T(t)/P(t)$.

(b) Evaluate the function in part (a) for $t = 0$, 3, and 6.

66. ***Graphical Reasoning*** An electronically controlled thermostat in a home is programmed to lower the temperature automatically during the night. The temperature in the house T (in degrees Fahrenheit) is given in terms of t, the time in hours on a 24-hour clock (see figure).

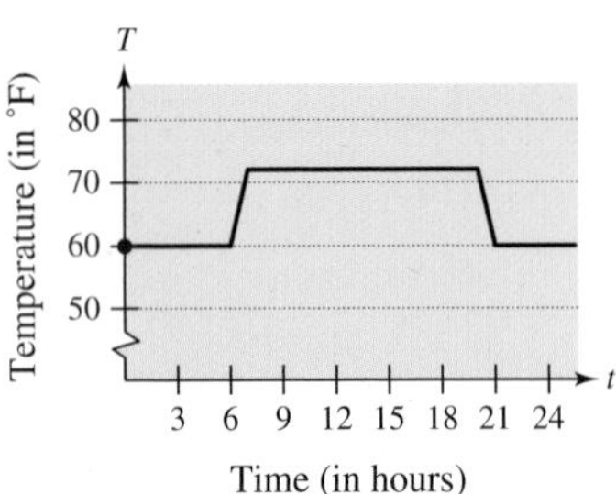

(a) Explain why T is a function of t.

(b) Approximate $T(4)$ and $T(15)$.

(c) The thermostat is reprogrammed to produce a temperature H for which $H(t) = T(t - 1)$. How does this change the temperature?

(d) The thermostat is reprogrammed to produce a temperature H for which $H(t) = T(t) - 1$. How does this change the temperature?

(e) Write a piecewise-defined function that represents the graph.

67. ***Geometry*** A square concrete foundation is prepared as a base for a cylindrical tank (see figure).

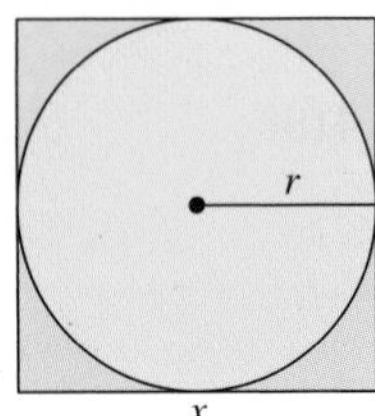

(a) Write the radius r of the tank as a function of the length x of the sides of the square.

(b) Write the area A of the circular base of the tank as a function of the radius r.

(c) Find and interpret $(A \circ r)(x)$.

68. ***Ripples*** A pebble is dropped into a calm pond, causing ripples in the form of concentric circles. The radius r (in feet) of the outer ripple is $r(t) = 0.6t$, where t is the time in seconds after the pebble strikes the water. The area A of the circle is given by the function $A(r) = \pi r^2$. Find and interpret $(A \circ r)(t)$.

69. ***Cost*** The weekly cost C of producing x units in a manufacturing process is given by $C(x) = 60x + 750$. The number of units x produced in t hours is given by $x(t) = 50t$.

(a) Find and interpret $(C \circ x)(t)$.

(b) Find the cost of the units produced in 4 hours.

(c) Find the time that must elapse in order for the cost to increase to $15,000.

70. ***Salary*** You are a sales representative for a clothing manufacturer. You are paid an annual salary, plus a bonus of 3% of your sales over $500,000. Consider the two functions given by $f(x) = x - 500{,}000$ and $g(x) = 0.03x$. If x is greater than $500,000, which of the following represents your bonus? Explain your reasoning.

(a) $f(g(x))$ (b) $g(f(x))$

True or False? **In Exercises 71 and 72, determine whether the statement is true or false. Justify your answer.**

71. If $f(x) = x + 1$ and $g(x) = 6x$, then

$(f \circ g)(x) = (g \circ f)(x)$.

72. If you are given two functions $f(x)$ and $g(x)$, you can calculate $(f \circ g)(x)$ if and only if the range of g is a subset of the domain of f.

73. ***Proof***

(a) Given a function f, prove that $g(x)$ is even and $h(x)$ is odd, where $g(x) = \frac{1}{2}[f(x) + f(-x)]$ and $h(x) = \frac{1}{2}[f(x) - f(-x)]$.

(b) Use the result of part (a) to prove that any function can be written as a sum of even and odd functions. [*Hint:* Add the two equations in part (a).]

(c) Use the result of part (b) to write each function as a sum of even and odd functions.

$$f(x) = x^2 - 2x + 1, \quad k(x) = \frac{1}{x + 1}$$

CAPSTONE

74. Consider the functions $f(x) = x^2$ and $g(x) = \sqrt{x}$.

(a) Find f/g and its domain.

(b) Find $f \circ g$ and $g \circ f$. Find the domain of each composite function. Are they the same? Explain.

1.5 Inverse Functions

- Find inverse functions informally and verify that two functions are inverse functions of each other.
- Use graphs of functions to determine whether functions have inverse functions.
- Use the Horizontal Line Test to determine if functions are one-to-one.
- Find inverse functions analytically.

Inverse Functions

Recall from Section 1.1 that a function can be represented by a set of ordered pairs. For instance, the function $f(x) = x + 4$ from the set $A = \{1, 2, 3, 4\}$ to the set $B = \{5, 6, 7, 8\}$ can be written as follows.

$$f(x) = x + 4: \ \{(1, 5), (2, 6), (3, 7), (4, 8)\}$$

In this case, by interchanging the first and second coordinates of each of these ordered pairs, you can form the **inverse function** of f, which is denoted by f^{-1}. It is a function from the set B to the set A, and can be written as follows.

$$f^{-1}(x) = x - 4: \ \{(5, 1), (6, 2), (7, 3), (8, 4)\}$$

Note that the domain of f is equal to the range of f^{-1}, and vice versa, as shown in Figure 1.29. Also note that the functions f and f^{-1} have the effect of "undoing" each other. In other words, when you form the composition of f with f^{-1} or the composition of f^{-1} with f, you obtain the identity function.

$$f(f^{-1}(x)) = f(x - 4) = (x - 4) + 4 = x$$
$$f^{-1}(f(x)) = f^{-1}(x + 4) = (x + 4) - 4 = x$$

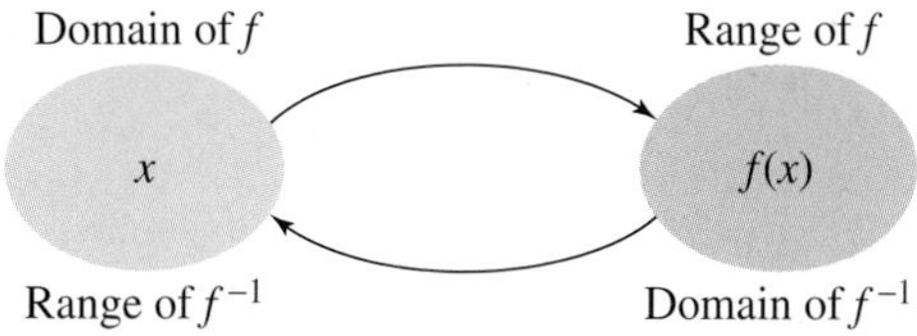

Figure 1.29

EXAMPLE 1 Finding Inverse Functions Informally

Find the inverse function of $f(x) = 4x$. Then verify that both $f(f^{-1}(x))$ and $f^{-1}(f(x))$ are equal to the identity function.

Solution The function f *multiplies* each input by 4. To "undo" this function, you need to *divide* each input by 4. So, the inverse function of $f(x) = 4x$ is

$$f^{-1}(x) = \frac{x}{4}.$$

You can verify that both $f(f^{-1}(x)) = x$ and $f^{-1}(f(x)) = x$ as follows.

$$f(f^{-1}(x)) = f\left(\frac{x}{4}\right) = 4\left(\frac{x}{4}\right) = x$$

$$f^{-1}(f(x)) = f^{-1}(4x) = \frac{4x}{4} = x$$

■

NOTE Do not be confused by the use of -1 to denote the inverse function f^{-1}. In this text, whenever f^{-1} is written, it *always* refers to the inverse function of the function f and *not* to the reciprocal of f.

DEFINITION OF INVERSE FUNCTION

Let f and g be two functions such that

$$f(g(x)) = x \qquad \text{for every } x \text{ in the domain of } g$$

and

$$g(f(x)) = x \qquad \text{for every } x \text{ in the domain of } f.$$

Under these conditions, the function g is the **inverse function** of the function f. The function g is denoted by f^{-1} (read "f-inverse"). So,

$$f(f^{-1}(x)) = x \qquad \text{and} \qquad f^{-1}(f(x)) = x.$$

The domain of f must be equal to the range of f^{-1}, and the range of f must be equal to the domain of f^{-1}.

EXPLORATION

Consider the functions given by

$$f(x) = 2x - 1$$

and

$$g(x) = \frac{x+1}{2}.$$

Complete the table.

x	-1	0	1	2
$f(x)$				
$g(x)$				
$f(g(x))$				
$g(f(x))$				

What can you conclude about the functions f and g?

If the function g is the inverse function of the function f, it must also be true that the function f is the inverse function of the function g. For this reason, you can say that the functions f and g are *inverse functions of each other.*

EXAMPLE 2 Verifying Inverse Functions

Which of the functions is the inverse function of $f(x) = \dfrac{5}{x-2}$?

$$g(x) = \frac{x-2}{5} \qquad h(x) = \frac{5}{x} + 2$$

Solution By forming the composition of f with g, you have

$$f(g(x)) = f\left(\frac{x-2}{5}\right)$$

$$= \frac{5}{\left(\frac{x-2}{5}\right) - 2} \qquad \text{Substitute } \frac{x-2}{5} \text{ for } x.$$

$$= \frac{25}{x-12} \neq x.$$

Because this composition is not equal to the identity function x, it follows that g *is not* the inverse function of f. By forming the composition of f with h, you have

$$f(h(x)) = f\left(\frac{5}{x} + 2\right)$$

$$= \frac{5}{\left(\frac{5}{x} + 2\right) - 2}$$

$$= \frac{5}{\left(\frac{5}{x}\right)} = x.$$

So, it appears that h *is* the inverse function of f. You can confirm this by showing that the composition of h with f is also equal to the identity function, as shown below.

$$h(f(x)) = h\left(\frac{5}{x-2}\right) = \frac{5}{\left(\frac{5}{x-2}\right)} + 2 = x - 2 + 2 = x$$

■

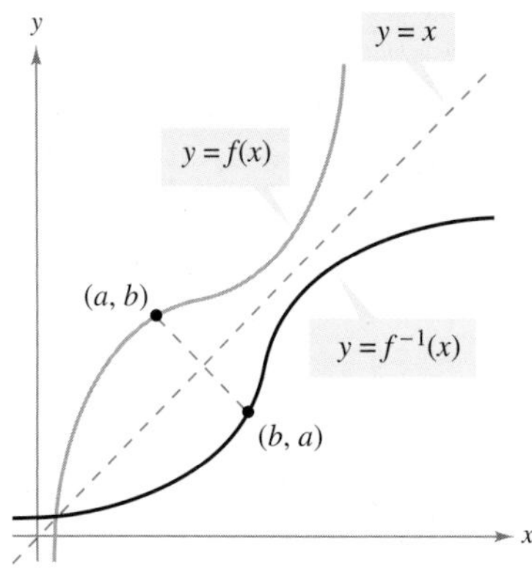

Figure 1.30

The Graph of an Inverse Function

The graphs of a function f and its inverse function f^{-1} are related to each other in the following way. If the point (a, b) lies on the graph of f, then the point (b, a) must lie on the graph of f^{-1}, and vice versa. This means that the graph of f^{-1} is a *reflection* of the graph of f in the line $y = x$, as shown in Figure 1.30.

EXAMPLE 3 Finding Inverse Functions Graphically

Sketch the graphs of the inverse functions

$$f(x) = 2x - 3$$

and

$$f^{-1}(x) = \tfrac{1}{2}(x + 3)$$

on the same rectangular coordinate system and show that the graphs are reflections of each other in the line $y = x$.

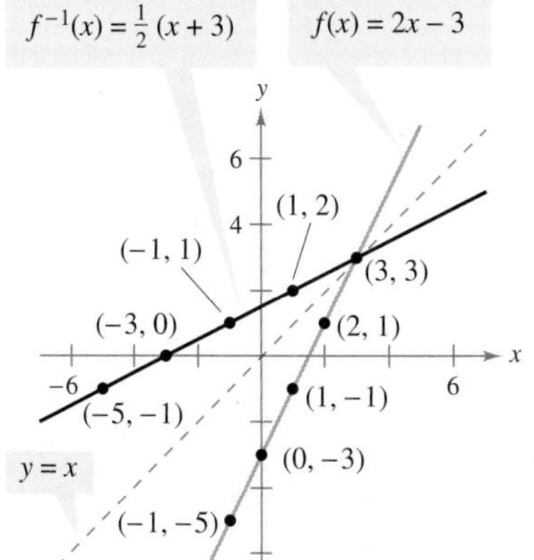

Figure 1.31

Solution The graphs of f and f^{-1} are shown in Figure 1.31. It appears that the graphs are reflections of each other in the line $y = x$. You can further verify this reflective property by testing a few points on each graph. Note in the following list that if the point (a, b) is on the graph of f, the point (b, a) is on the graph of f^{-1}.

Graph of $f(x) = 2x - 3$	*Graph of* $f^{-1}(x) = \frac{1}{2}(x + 3)$
$(-1, -5)$	$(-5, -1)$
$(0, -3)$	$(-3, 0)$
$(1, -1)$	$(-1, 1)$
$(2, 1)$	$(1, 2)$
$(3, 3)$	$(3, 3)$

EXAMPLE 4 Finding Inverse Functions Graphically

Sketch the graphs of the inverse functions

$$f(x) = x^2 \ (x \geq 0)$$

and

$$f^{-1}(x) = \sqrt{x}$$

on the same rectangular coordinate system and show that the graphs are reflections of each other in the line $y = x$.

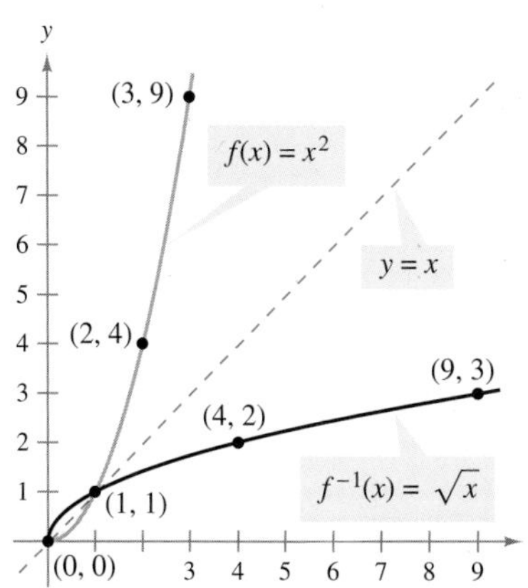

Figure 1.32

Solution The graphs of f and f^{-1} are shown in Figure 1.32. It appears that the graphs are reflections of each other in the line $y = x$. You can further verify this reflective property by testing a few points on each graph. Note in the following list that if the point (a, b) is on the graph of f, the point (b, a) is on the graph of f^{-1}.

Graph of $f(x) = x^2, \quad x \geq 0$	*Graph of* $f^{-1}(x) = \sqrt{x}$
$(0, 0)$	$(0, 0)$
$(1, 1)$	$(1, 1)$
$(2, 4)$	$(4, 2)$
$(3, 9)$	$(9, 3)$

Try showing that $f(f^{-1}(x)) = x$ and $f^{-1}(f(x)) = x$. ■

One-to-One Functions

The reflective property of the graphs of inverse functions gives you a nice *geometric* test for determining whether a function has an inverse function. This test is called the **Horizontal Line Test** for inverse functions.

HORIZONTAL LINE TEST FOR INVERSE FUNCTIONS
A function f has an inverse function if and only if no *horizontal* line intersects the graph of f at more than one point.

If no horizontal line intersects the graph of f at more than one point, then no y-value is matched with more than one x-value. This is the essential characteristic of what are called **one-to-one functions.**

NOTE The domain of the function given by $f(x) = x^2$ can be restricted so that the function does have an inverse function. For instance, if the domain is restricted as follows

$$f(x) = x^2, \quad x \geq 0$$

the function has an inverse function, as shown in Example 4.

ONE-TO-ONE FUNCTIONS
A function f is **one-to-one** if each value of the dependent variable corresponds to exactly one value of the independent variable. A function f has an inverse function if and only if f is one-to-one.

Consider the function given by $f(x) = x^2$. The first table is a table of values for $f(x) = x^2$. The second table of values is made up by interchanging the rows of the first table. The second table does not represent a function because the input $x = 4$ is matched with two different outputs: $y = -2$ and $y = 2$. So, $f(x) = x^2$ is not one-to-one and does not have an inverse function.

x	-2	-1	0	1	2	3
$f(x) = x^2$	4	1	0	1	4	9

x	4	1	0	1	4	9
y	-2	-1	0	1	2	3

$f(x) = x^3 - 1$

(a)

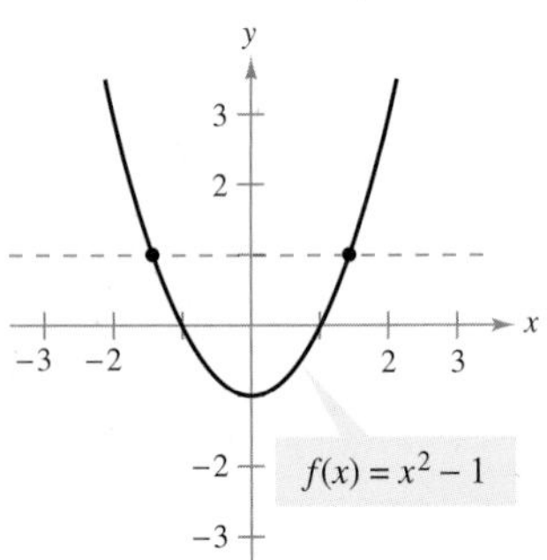

(b)

Figure 1.33

EXAMPLE 5 Applying the Horizontal Line Test

Use the Horizontal Line Test to determine whether each function has an inverse function.

a. $f(x) = x^3 - 1$ **b.** $f(x) = x^2 - 1$

Solution

a. The graph of the function given by $f(x) = x^3 - 1$ is shown in Figure 1.33(a). Because no horizontal line intersects the graph of f at more than one point, you can conclude that f *is* a one-to-one function and *does* have an inverse function.

b. The graph of the function given by $f(x) = x^2 - 1$ is shown in Figure 1.33(b). Because it is possible to find a horizontal line that intersects the graph of f at more than one point, you can conclude that f *is not* a one-to-one function and *does not* have an inverse function. ■

STUDY TIP Note what happens when you try to find the inverse function of a function that is not one-to-one.

$f(x) = x^2 + 1$	Original function
$y = x^2 + 1$	Replace $f(x)$ by y.
$x = y^2 + 1$	Interchange x and y.
$x - 1 = y^2$	Isolate y-term.
$y = \pm\sqrt{x - 1}$	Solve for y.

You obtain two y-values for each x.

Finding Inverse Functions Analytically

For simple functions (such as the one in Example 1), you can find inverse functions by inspection. For more complicated functions, however, it is best to use the following guidelines. The key step in these guidelines is Step 3—interchanging the roles of x and y. This step corresponds to the fact that inverse functions have ordered pairs with the coordinates reversed.

GUIDELINES FOR FINDING AN INVERSE FUNCTION

1. Use the Horizontal Line Test to decide whether f has an inverse function.
2. In the equation for $f(x)$, replace $f(x)$ by y.
3. Interchange the roles of x and y, and solve for y.
4. Replace y by $f^{-1}(x)$ in the new equation.
5. Verify that f and f^{-1} are inverse functions of each other by showing that the domain of f is equal to the range of f^{-1}, the range of f is equal to the domain of f^{-1}, and $f(f^{-1}(x)) = x$ and $f^{-1}(f(x)) = x$.

EXAMPLE 6 Finding an Inverse Function Analytically

Find the inverse function of $f(x) = \dfrac{5 - 3x}{2}$.

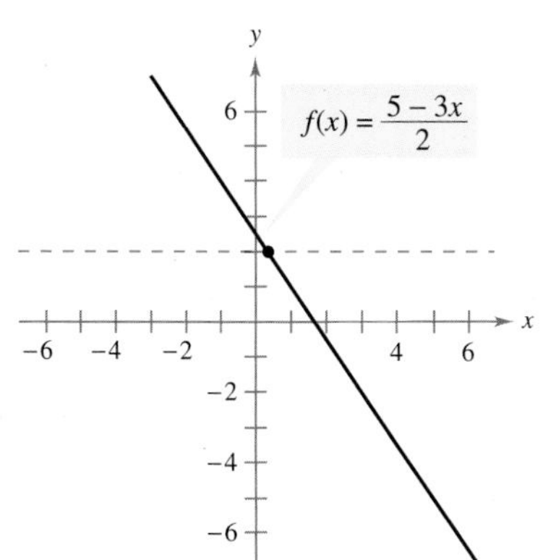

Figure 1.34

Solution The graph of f is a line, as shown in Figure 1.34. This graph passes the Horizontal Line Test. So, you know that f is one-to-one and has an inverse function.

$f(x) = \dfrac{5 - 3x}{2}$	Write original function.
$y = \dfrac{5 - 3x}{2}$	Replace $f(x)$ by y.
$x = \dfrac{5 - 3y}{2}$	Interchange x and y.
$2x = 5 - 3y$	Multiply each side by 2.
$3y = 5 - 2x$	Isolate the y-term.
$y = \dfrac{5 - 2x}{3}$	Solve for y.
$f^{-1}(x) = \dfrac{5 - 2x}{3}$	Replace y by $f^{-1}(x)$.

Note that both f and f^{-1} have domains and ranges that consist of the entire set of real numbers.

Check

$$f(f^{-1}(x)) = f\left(\frac{5 - 2x}{3}\right) = \frac{5 - 3\left(\frac{5 - 2x}{3}\right)}{2} = \frac{5 - (5 - 2x)}{2} = x \checkmark$$

$$f^{-1}(f(x)) = f^{-1}\left(\frac{5 - 3x}{2}\right) = \frac{5 - 2\left(\frac{5 - 3x}{2}\right)}{3} = \frac{5 - (5 - 3x)}{3} = x \checkmark$$

■

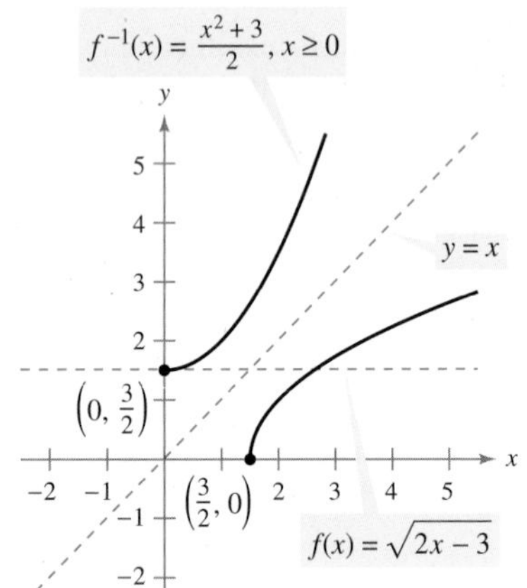

Figure 1.35

EXAMPLE 7 Finding an Inverse Function

Find the inverse function of

$$f(x) = \sqrt{2x - 3}.$$

Solution The graph of f is a curve, as shown in Figure 1.35. Because this graph passes the Horizontal Line Test, you know that f is one-to-one and has an inverse function.

$f(x) = \sqrt{2x - 3}$	Write original function.
$y = \sqrt{2x - 3}$	Replace $f(x)$ by y.
$x = \sqrt{2y - 3}$	Interchange x and y.
$x^2 = 2y - 3$	Square each side.
$2y = x^2 + 3$	Isolate y.
$y = \dfrac{x^2 + 3}{2}$	Solve for y.
$f^{-1}(x) = \dfrac{x^2 + 3}{2}, \quad x \geq 0$	Replace y by $f^{-1}(x)$.

The graph of f^{-1} in Figure 1.35 is the reflection of the graph of f in the line $y = x$. Note that the range of f is the interval $[0, \infty)$, which implies that the domain of f^{-1} is the interval $[0, \infty)$. Moreover, the domain of f is the interval $\left[\frac{3}{2}, \infty\right)$, which implies that the range of f^{-1} is the interval $\left[\frac{3}{2}, \infty\right)$. Verify that $f(f^{-1}(x)) = x$ and $f^{-1}(f(x)) = x$.

1.5 Exercises

See www.CalcChat.com for worked-out solutions to odd-numbered exercises.

In Exercises 1–6, fill in the blanks.

1. If the composite functions $f(g(x))$ and $g(f(x))$ both equal x, then the function g is the ________ function of f.
2. The inverse function of f is denoted by ________.
3. The domain of f is the ________ of f^{-1}, and the ________ of f^{-1} is the range of f.
4. The graphs of f and f^{-1} are reflections of each other in the line ________.
5. A function f is ________ if each value of the dependent variable corresponds to exactly one value of the independent variable.
6. A graphical test for the existence of an inverse function of f is called the ________ Line Test.

In Exercises 7–16, find the inverse function of f informally. Verify that $f(f^{-1}(x)) = x$ and $f^{-1}(f(x)) = x$.

7. $f(x) = 6x$
8. $f(x) = \frac{1}{3}x$
9. $f(x) = x + 9$
10. $f(x) = x - 4$
11. $f(x) = 3x + 1$
12. $f(x) = -2x - 9$
13. $f(x) = \dfrac{x - 1}{5}$
14. $f(x) = \dfrac{4x + 7}{2}$
15. $f(x) = \sqrt[3]{x}$
16. $f(x) = x^5$

In Exercises 17–28, show that f and g are inverse functions (a) analytically and (b) graphically.

17. $f(x) = 2x, \quad g(x) = \dfrac{x}{2}$
18. $f(x) = x - 5, \quad g(x) = x + 5$
19. $f(x) = 7x + 1, \quad g(x) = \dfrac{x - 1}{7}$
20. $f(x) = 3 - 4x, \quad g(x) = \dfrac{3 - x}{4}$
21. $f(x) = \dfrac{x^3}{8}, \quad g(x) = \sqrt[3]{8x}$
22. $f(x) = \dfrac{1}{x}, \quad g(x) = \dfrac{1}{x}$
23. $f(x) = \sqrt{x - 4}, \quad g(x) = x^2 + 4, \quad x \geq 0$
24. $f(x) = 1 - x^3, \quad g(x) = \sqrt[3]{1 - x}$
25. $f(x) = 9 - x^2, \quad x \geq 0, \quad g(x) = \sqrt{9 - x}, \quad x \leq 9$
26. $f(x) = \dfrac{1}{1 + x}, \quad x \geq 0, \quad g(x) = \dfrac{1 - x}{x}, \quad 0 < x \leq 1$
27. $f(x) = \dfrac{x - 1}{x + 5}, \quad g(x) = -\dfrac{5x + 1}{x - 1}$
28. $f(x) = \dfrac{x + 3}{x - 2}, \quad g(x) = \dfrac{2x + 3}{x - 1}$

In Exercises 29 and 30, does the function have an inverse function?

29.

x	-1	0	1	2	3	4
$f(x)$	-2	1	2	1	-2	-6

30.

x	-3	-2	-1	0	2	3
$f(x)$	10	6	4	1	-3	-10

In Exercises 31–34, does the function have an inverse function?

31.

32.

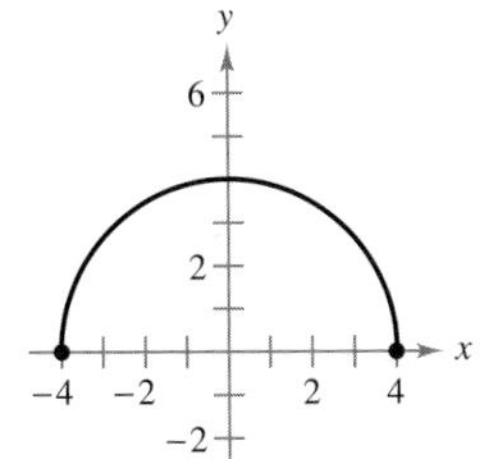

33.

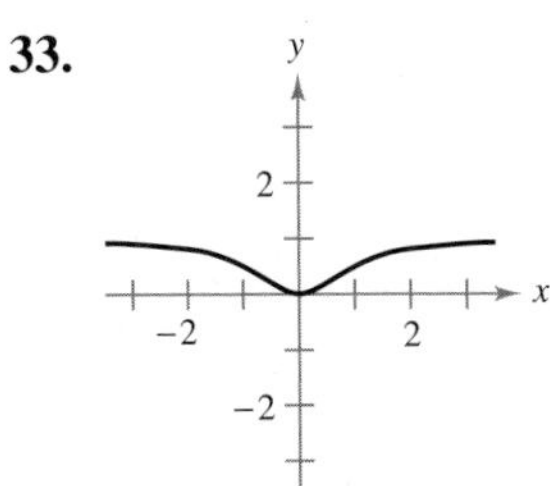

34.

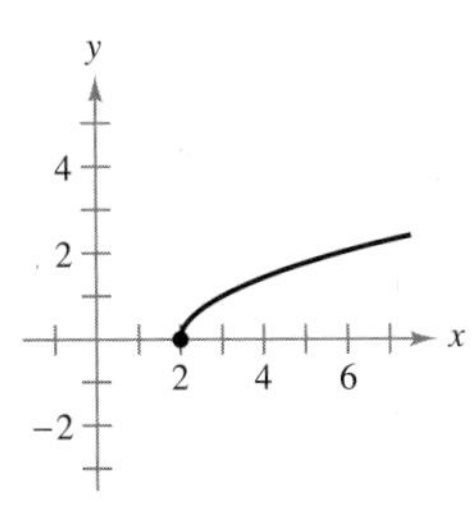

In Exercises 35–40, use a graphing utility to graph the function, and use the Horizontal Line Test to determine whether the function is one-to-one and so has an inverse function.

35. $g(x) = \dfrac{4 - x}{6}$

36. $f(x) = 10$

37. $h(x) = |x + 4| - |x - 4|$

38. $g(x) = (x + 5)^3$

39. $f(x) = -2x\sqrt{16 - x^2}$

40. $f(x) = \frac{1}{8}(x + 2)^2 - 1$

In Exercises 41–54, (a) find the inverse function of f, (b) graph both f and f^{-1} on the same set of coordinate axes, (c) describe the relationship between the graphs of f and f^{-1}, and (d) state the domain and range of f and f^{-1}.

41. $f(x) = 2x - 3$

42. $f(x) = 3x + 1$

43. $f(x) = x^5 - 2$

44. $f(x) = x^3 + 1$

45. $f(x) = \sqrt{4 - x^2}, \quad 0 \le x \le 2$

46. $f(x) = x^2 - 2, \quad x \le 0$

47. $f(x) = \dfrac{4}{x}$

48. $f(x) = -\dfrac{2}{x}$

49. $f(x) = \dfrac{x + 1}{x - 2}$

50. $f(x) = \dfrac{x - 3}{x + 2}$

51. $f(x) = \sqrt[3]{x - 1}$

52. $f(x) = x^{3/5}$

53. $f(x) = \dfrac{6x + 4}{4x + 5}$

54. $f(x) = \dfrac{8x - 4}{2x + 6}$

In Exercises 55–68, determine whether the function has an inverse function. If it does, find the inverse function.

55. $f(x) = x^4$

56. $f(x) = \dfrac{1}{x^2}$

57. $g(x) = \dfrac{x}{8}$

58. $f(x) = 3x + 5$

59. $p(x) = -4$

60. $f(x) = \dfrac{3x + 4}{5}$

61. $f(x) = (x + 3)^2, \quad x \ge -3$

62. $q(x) = (x - 5)^2$

63. $f(x) = \begin{cases} x + 3, & x < 0 \\ 6 - x, & x \ge 0 \end{cases}$

64. $f(x) = \begin{cases} -x, & x \le 0 \\ x^2 - 3x, & x > 0 \end{cases}$

65. $h(x) = -\dfrac{4}{x^2}$

66. $f(x) = |x - 2|, \quad x \le 2$

67. $f(x) = \sqrt{2x + 3}$

68. $f(x) = \sqrt{x - 2}$

In Exercises 69–74, use the functions given by $f(x) = \frac{1}{8}x - 3$ and $g(x) = x^3$ to find the indicated value or function.

69. $(f^{-1} \circ g^{-1})(1)$

70. $(g^{-1} \circ f^{-1})(-3)$

71. $(f^{-1} \circ f^{-1})(6)$

72. $(g^{-1} \circ g^{-1})(-4)$

73. $(f \circ g)^{-1}$

74. $g^{-1} \circ f^{-1}$

In Exercises 75–78, use the functions given by $f(x) = x + 4$ and $g(x) = 2x - 5$ to find the specified function.

75. $g^{-1} \circ f^{-1}$

76. $f^{-1} \circ g^{-1}$

77. $(f \circ g)^{-1}$

78. $(g \circ f)^{-1}$

WRITING ABOUT CONCEPTS

In Exercises 79–82, match the graph of the function with the graph of its inverse function. [The graphs of the inverse functions are labeled (a), (b), (c), and (d).]

(a)
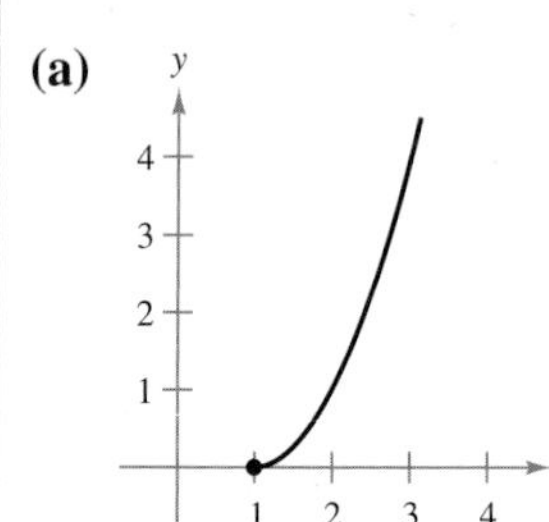

(b)
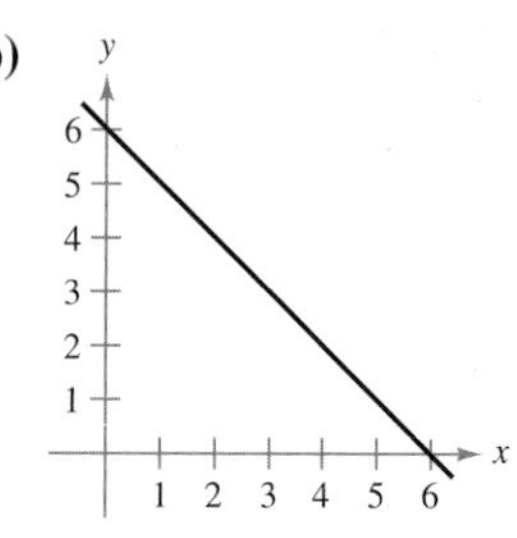

(c)
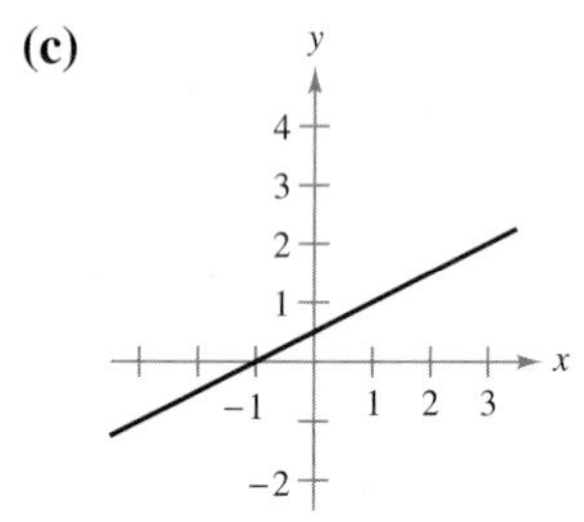

(d)
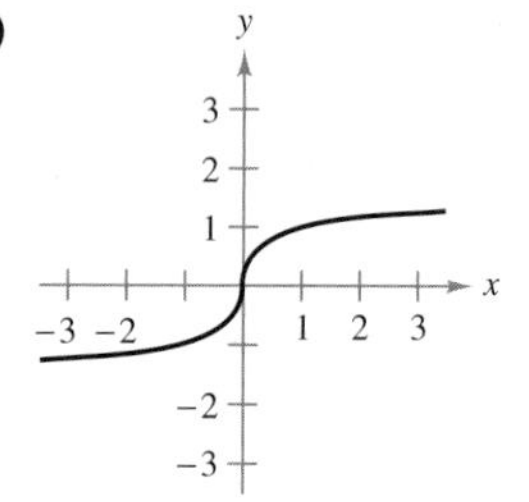

79.
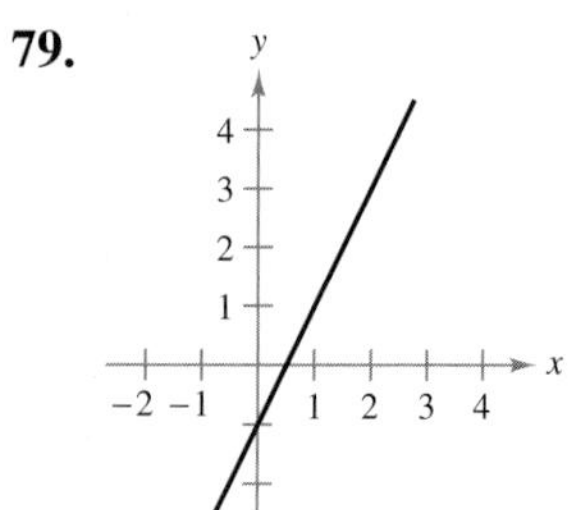

80.
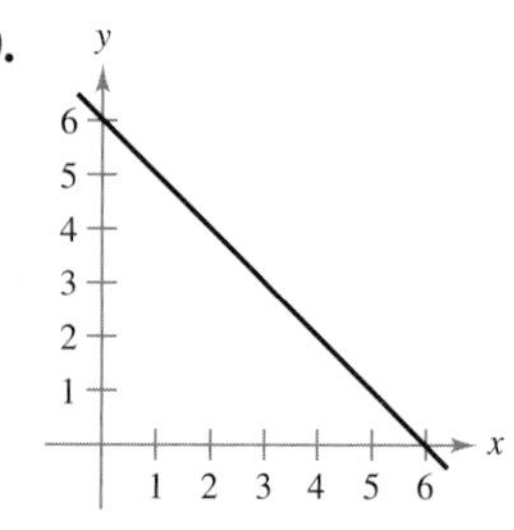

81.
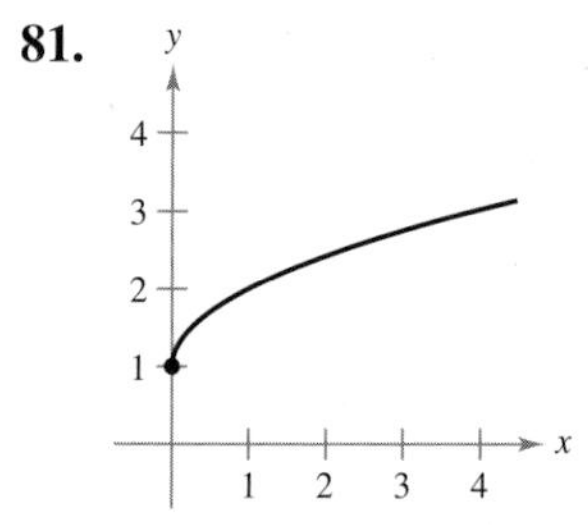

82.
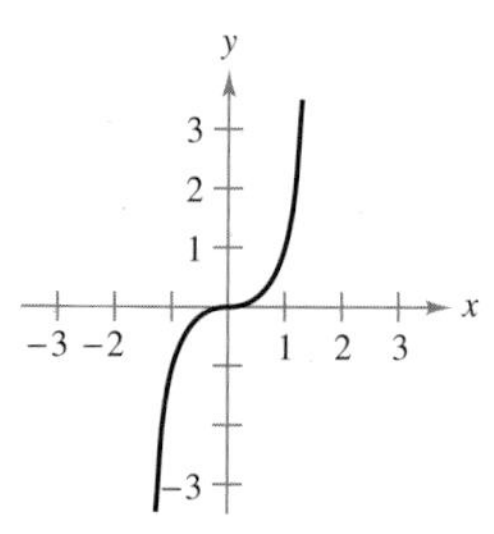

In Exercises 83 and 84, use the table of values for $y = f(x)$ to complete a table for $y = f^{-1}(x)$.

83.

x	-2	-1	0	1	2	3
$f(x)$	-2	0	2	4	6	8

84.

x	-3	-2	-1	0	1	2
$f(x)$	-10	-7	-4	-1	2	5

85. ***Hourly Wage*** Your wage is \$10.00 per hour plus \$0.75 for each unit produced per hour. So, your hourly wage y in terms of the number of units produced x is $y = 10 + 0.75x$.

(a) Find the inverse function. What does each variable represent in the inverse function?

(b) Determine the number of units produced when your hourly wage is \$24.25.

86. ***Cost*** You need 50 pounds of two commodities costing \$1.25 and \$1.60 per pound, respectively.

(a) Verify that the total cost is

$$y = 1.25x + 1.60(50 - x)$$

where x is the number of pounds of the less expensive commodity.

(b) Find the inverse function of the cost function. What does each variable represent in the inverse function?

(c) Use the context of the problem to determine the domain of the inverse function.

(d) Determine the number of pounds of the less expensive commodity purchased when the total cost is \$73.

87. ***Diesel Mechanics*** The function given by

$$y = 0.03x^2 + 245.50, \quad 0 < x < 100$$

approximates the exhaust temperature y in degrees Fahrenheit, where x is the percent load for a diesel engine.

(a) Find the inverse function. What does each variable represent in the inverse function?

(b) Use a graphing utility to graph the inverse function.

(c) The exhaust temperature of the engine must not exceed 500 degrees Fahrenheit. What is the percent load interval?

88. ***Population*** The projected populations P (in millions of people) in the United States for 2015 through 2040 are shown in the table. The time (in years) is given by t, with $t = 15$ corresponding to 2015. *(Source: U.S. Census Bureau)*

t	15	20	25	30	35	40
$P(t)$	325.5	341.4	357.5	373.5	389.5	405.7

(a) Does P^{-1} exist?

(b) If P^{-1} exists, what does it represent in the context of the problem?

(c) If P^{-1} exists, find $P^{-1}(357.5)$.

(d) If the table was extended to 2050 and if the projected population of the U.S. for that year was 373.5 million, would P^{-1} exist? Explain.

89. *Telecommunications* The amounts f (in billions of dollars) of cellular telecommunication service revenue in the United States from 2002 to 2009 are shown in the table and in the bar graph. The time (in years) is given by t, with $t = 2$ corresponding to 2002. *(Source: Cellular Telecommunications and Internet Association)*

Year, t	2	3	4	5
Amount, $f(t)$	76.5	87.6	102.1	113.5

Year, t	6	7	8	9
Amount, $f(t)$	125.5	138.9	148.1	152.6

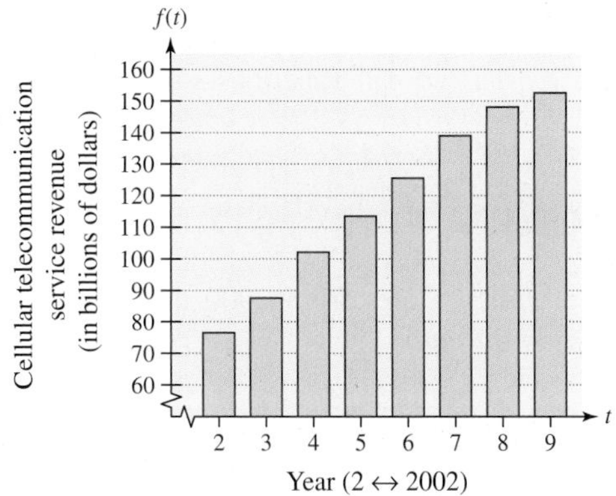

(a) Find $f^{-1}(113.5)$.

(b) What does f^{-1} mean in the context of the problem?

(c) Use the *regression* feature of a graphing utility to find a linear model $y = mx + b$ for the data.

(d) Analytically find the inverse function of the linear model in part (c).

(e) Use the inverse function of the linear model you found in part (d) to approximate $f^{-1}(226.4)$.

90. *U.S. Households* The numbers of households f (in thousands) in the United States from 2002 through 2009 are shown in the table and in the bar graph. The time (in years) is given by t, with $t = 2$ corresponding to 2002. *(Source: U.S. Census Bureau)*

Year, t	2	3	4
Households, $f(t)$	109,297	111,278	112,000

Year, t	5	6	7
Households, $f(t)$	113,343	114,384	116,011

Year, t	8	9
Households, $f(t)$	116,783	117,181

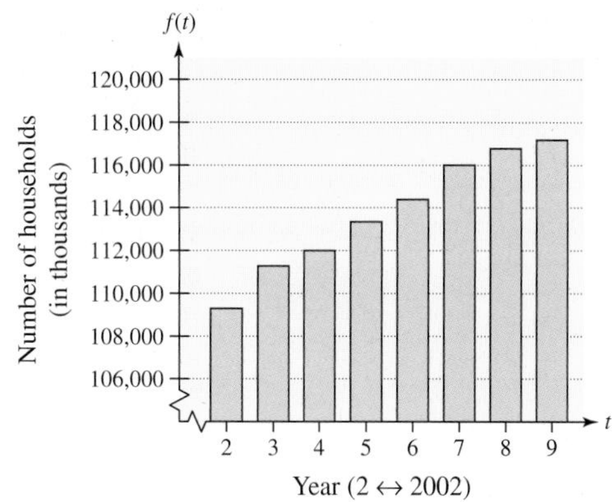

Figure for 90

(a) Find $f^{-1}(116{,}011)$.

(b) What does f^{-1} mean in the context of the problem?

(c) Use the *regression* feature of a graphing utility to find a linear model $y = mx + b$ for the data. (Round m and b to two decimal places.)

(d) Analytically find the inverse function of the linear model in part (c).

(e) Use the inverse function of the linear model you found in part (d) to approximate $f^{-1}(123{,}477)$.

True or False? **In Exercises 91 and 92, determine whether the statement is true or false. Justify your answer.**

91. If f is an even function, then f^{-1} exists.

92. If the inverse function of f exists and the graph of f has a y-intercept, then the y-intercept of f is an x-intercept of f^{-1}.

93. *Proof* Prove that if f and g are one-to-one functions, then $(f \circ g)^{-1}(x) = (g^{-1} \circ f^{-1})(x)$.

CAPSTONE

94. Describe and correct the error.

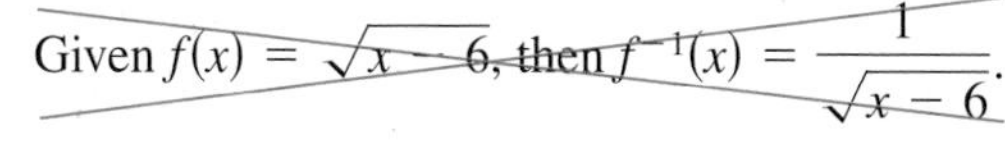

In Exercises 95–98, determine if the situation could be represented by a one-to-one function. If so, write a statement that describes the inverse function.

95. The number of miles n a marathon runner has completed in terms of the time t in hours

96. The population p of South Carolina in terms of the year t from 1960 through 2011

97. The depth of the tide d at a beach in terms of the time t over a 24-hour period

98. The height h in inches of a human born in the year 2000 in terms of his or her age n in years

1.6 Mathematical Modeling and Variation

- **Use mathematical models to approximate sets of data points.**
- **Use the *regression* feature of a graphing utility to find the equation of a least squares regression line.**
- **Write mathematical models for direct variation.**
- **Write mathematical models for direct variation as an *n*th power.**
- **Write mathematical models for inverse variation.**
- **Write mathematical models for joint variation.**

Introduction

You have already studied some techniques for fitting models to data. For instance, in Section P.5, you learned how to find the equation of a line that passes through two points. In this section, you will study other techniques for fitting models to data: *least squares regression* and *direct and inverse variation.* The resulting models are either polynomial functions or rational functions. (Rational functions will be studied in Chapter 2.)

EXAMPLE 1 A Mathematical Model

The populations y (in millions) of the United States from 2000 through 2007 are shown in the table. *(Source: U.S. Census Bureau)*

Year	2000	2001	2002	2003	2004	2005	2006	2007
Population, y	282.4	285.3	288.2	290.9	293.6	296.3	299.2	302.0

A linear model that approximates the data is

$$y = 2.78t + 282.5, \quad 0 \le t \le 7$$

where t is the year, with $t = 0$ corresponding to 2000. Plot the actual data *and* the model on the same graph. How closely does the model represent the data?

Solution The actual data are plotted in Figure 1.36, along with the graph of the linear model. From the graph, it appears that the model is a "good fit" for the actual data. You can see how well the model fits by comparing the actual values of y with the values of y given by the model. The values given by the model are labeled $y*$ in the table below.

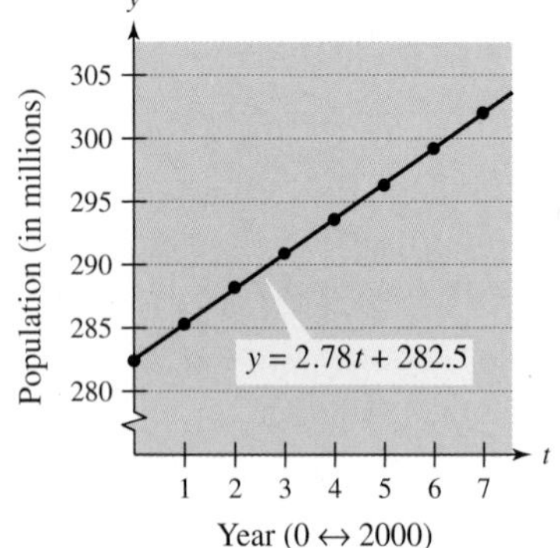

Figure 1.36

t	0	1	2	3	4	5	6	7
y	282.4	285.3	288.2	290.9	293.6	296.3	299.2	302.0
$y*$	282.5	285.3	288.1	290.8	293.6	296.4	299.2	302.0

■

NOTE In Example 1, you could have chosen any two points to find a line that fits the data. However, the linear model above was found using the *regression* feature of a graphing utility and is the line that *best* fits the data. This concept of a "best-fitting" line is discussed on the next page. ■

Least Squares Regression

So far in this text, you have worked with many different types of mathematical models that approximate real-life data. In some instances the model was given (as in Example 1), whereas in other instances you were asked to find the model using simple algebraic techniques or a graphing utility.

To find a model that approximates the data most accurately, statisticians use a measure called the **sum of square differences,** which is the sum of the squares of the differences between actual data values and model values. The "best-fitting" linear model, called the **least squares regression line,** is the one with the least sum of square differences. Recall that you can approximate this line visually by plotting the data points and drawing the line that appears to fit best—or you can enter the data points into a calculator or computer and use the *linear regression* feature of the calculator or computer.

EXAMPLE 2 Fitting a Linear Model to Data

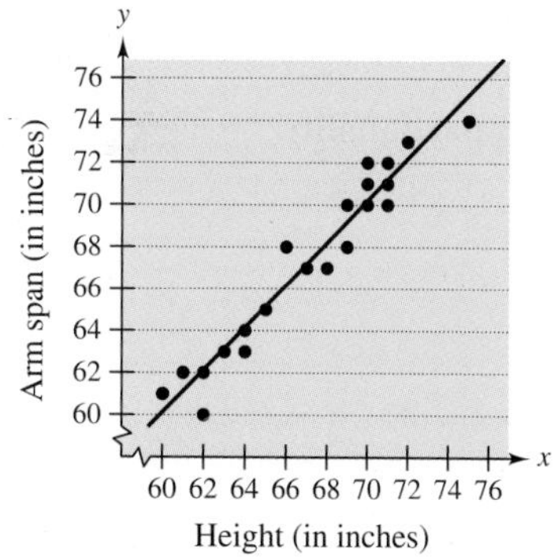

Linear model and data
Figure 1.37

A class of 28 people collected the following data, which represents their heights x and arm spans y (rounded to the nearest inch).

$(60, 61)$, $(65, 65)$, $(68, 67)$, $(72, 73)$, $(61, 62)$, $(63, 63)$, $(70, 71)$,
$(75, 74)$, $(71, 72)$, $(62, 60)$, $(65, 65)$, $(66, 68)$, $(62, 62)$, $(72, 73)$,
$(70, 70)$, $(69, 68)$, $(69, 70)$, $(60, 61)$, $(63, 63)$, $(64, 64)$, $(71, 71)$,
$(68, 67)$, $(69, 70)$, $(70, 72)$, $(65, 65)$, $(64, 63)$, $(71, 70)$, $(67, 67)$

Find a linear model to represent these data.

Solution There are different ways to model these data with an equation. The simplest would be to observe from a table of values that x and y are about the same and list the model as simply $y = x$. A more careful analysis would be to use a procedure from statistics called linear regression. The least squares regression line for these data is

$$y = 1.006x - 0.23. \quad \text{Least squares regression line}$$

The graph of the model and the data are shown in Figure 1.37. From this model, you can see that a person's arm span tends to be about the same as his or her height. ■

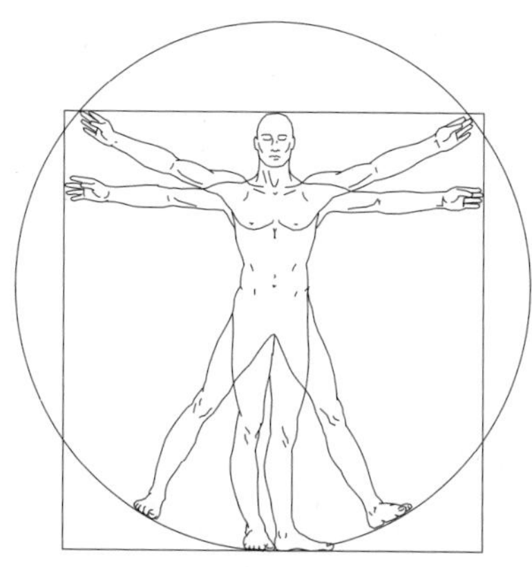

A computer graphics drawing based on the pen and ink drawing of Leonardo da Vinci's famous study of human proportions, called *Vitruvian Man*

NOTE One basic technique of modern science is gathering data and then describing the data with a mathematical model. For instance, the data given in Example 2 are inspired by Leonardo da Vinci's famous drawing that indicates that a person's height and arm span are equal. ■

TECHNOLOGY Many scientific and graphing calculators have built-in least squares regression programs. Typically, you enter the data into the calculator and then run the linear regression program. The program usually displays the slope and y-intercept of the best-fitting line and the *correlation coefficient* r. The correlation coefficient gives a measure of how well the model fits the data. The closer $|r|$ is to 1, the better the model fits the data. For instance, the correlation coefficient for the model in Example 2 is $r \approx 0.97$, which indicates that the model is a good fit for the data. If the r-value is positive, the variables have a positive correlation, as in Example 2. If the r-value is negative, the variables have a negative correlation.

Direct Variation

There are two basic types of linear models. The more general model has a y-intercept that is nonzero.

$$y = mx + b, \quad b \neq 0$$

The simpler model

$$y = kx$$

has a y-intercept that is zero. In the simpler model, y is said to **vary directly** as x, or to be **directly proportional** to x.

> **DIRECT VARIATION**
>
> The following statements are equivalent.
>
> **1.** y **varies directly** as x.
>
> **2.** y is **directly proportional** to x.
>
> **3.** $y = kx$ for some nonzero constant k.
>
> k is the **constant of variation** or the **constant of proportionality.**

EXAMPLE 3 Direct Variation

In Pennsylvania, the state income tax is directly proportional to *gross income.* You are working in Pennsylvania and your state income tax deduction is \$46.05 for a gross monthly income of \$1500. Find a mathematical model that gives the Pennsylvania state income tax in terms of gross income.

Solution

Verbal Model: State income tax $= k \cdot$ Gross income

Labels:

State income tax $= y$ (dollars)

Gross income $= x$ (dollars)

Income tax rate $= k$ (percent in decimal form)

Equation: $y = kx$

To solve for k, substitute the given information into the equation $y = kx$, and then solve for k.

$y = kx$ Write direct variation model.

$46.05 = k(1500)$ Substitute for y and x.

$0.0307 = k$ Simplify.

So, the equation (or model) for state income tax in Pennsylvania is

$$y = 0.0307x.$$

In other words, Pennsylvania has a state income tax rate of 3.07% of gross income. The graph of this equation is shown in Figure 1.38. ■

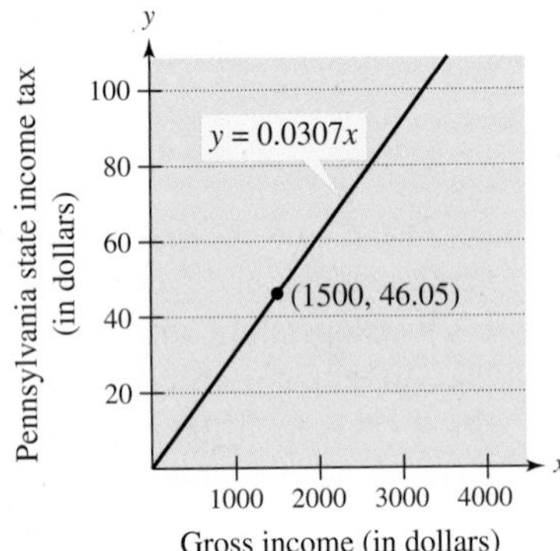

Figure 1.38

Direct Variation as an nth Power

Another type of direct variation relates one variable to a *power* of another variable. For example, in the formula for the area of a circle

$$A = \pi r^2$$

the area A is directly proportional to the square of the radius r. In this formula, π is the constant of proportionality.

DIRECT VARIATION AS AN nTH POWER

The following statements are equivalent.

1. y **varies directly as the nth power** of x.
2. y is **directly proportional to the nth power** of x.
3. $y = kx^n$ for some constant k.

Note that the direct variation model $y = kx$ is a special case of $y = kx^n$ with $n = 1$.

EXAMPLE 4 Direct Variation as an nth Power

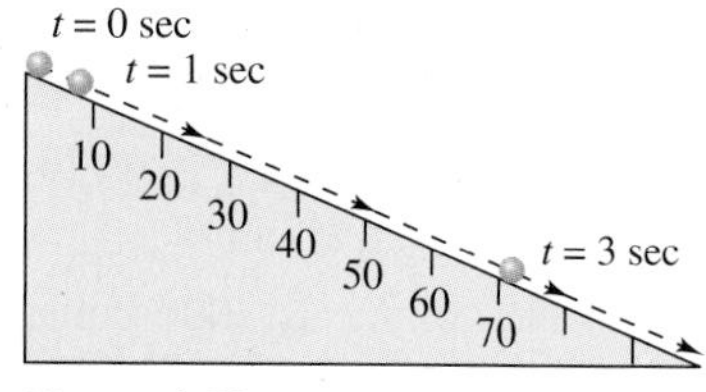

Figure 1.39

The distance a ball rolls down an inclined plane is directly proportional to the square of the time it rolls. During the first second, the ball rolls 8 feet. (See Figure 1.39.)

a. Find a mathematical model that relates the distance traveled to the time.

b. How far will the ball roll during the first 3 seconds?

Solution

a. Letting d be the distance (in feet) the ball rolls and letting t be the time (in seconds), you have

$$d = kt^2.$$

Now, because $d = 8$ when $t = 1$, you can see that $k = 8$, as follows.

$$d = kt^2$$
$$8 = k(1)^2$$
$$8 = k$$

So, the equation relating distance to time is

$$d = 8t^2.$$

b. When $t = 3$, the distance traveled is $d = 8(3)^2 = 8(9) = 72$ feet. ■

In Examples 3 and 4, the direct variations are such that an *increase* in one variable corresponds to an *increase* in the other variable. This is also true in the model $d = \frac{1}{5}F$, $F > 0$, where an increase in F results in an increase in d. You should not, however, assume that this always occurs with direct variation. For example, in the model $y = -3x$, an increase in x results in a *decrease* in y, and yet y is said to vary directly as x.

Inverse Variation

INVERSE VARIATION

The following statements are equivalent.

1. y **varies inversely** as x.
2. y is **inversely proportional** to x.
3. $y = \frac{k}{x}$ for some constant k.

If x and y are related by an equation of the form $y = k/x^n$, then y varies inversely as the nth power of x (or y is inversely proportional to the nth power of x).

Some applications of variation involve problems with both direct and inverse variation in the same model. These types of models are said to have **combined variation.**

EXAMPLE 5 Direct and Inverse Variation

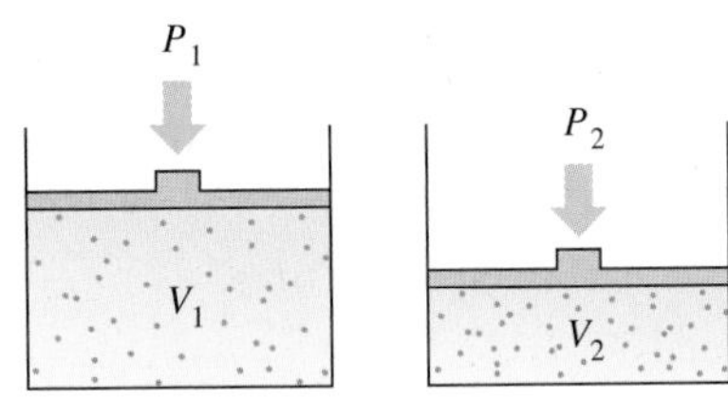

$P_2 > P_1$ then $V_2 < V_1$

If the temperature is held constant and pressure increases, volume decreases.
Figure 1.40

A gas law states that the volume of an enclosed gas varies directly as the temperature *and* inversely as the pressure, as shown in Figure 1.40. The pressure of a gas is 0.75 kilogram per square centimeter when the temperature is 294 K and the volume is 8000 cubic centimeters.

a. Find a mathematical model that relates pressure, temperature, and volume.

b. Find the pressure when the temperature is 300 K and the volume is 7000 cubic centimeters.

Solution

a. Let V be volume (in cubic centimeters), let P be pressure (in kilograms per square centimeter), and let T be temperature (in Kelvin). Because V varies directly as T and inversely as P, you have

$$V = \frac{kT}{P}.$$

Now, because $P = 0.75$ when $T = 294$ and $V = 8000$, you have

$$8000 = \frac{k(294)}{0.75}$$

$$\frac{8000(0.75)}{294} = k$$

$$k = \frac{6000}{294} = \frac{1000}{49}.$$

So, the mathematical model that relates pressure, temperature, and volume is

$$V = \frac{1000}{49}\left(\frac{T}{P}\right).$$

b. When $T = 300$ and $V = 7000$, the pressure is

$$P = \frac{1000}{49}\left(\frac{300}{7000}\right) = \frac{300}{343} \approx 0.87 \text{ kilogram per square centimeter.}$$ ■

Joint Variation

In Example 5, note that when a direct variation and an inverse variation occur in the same statement, they are coupled with the word "and." To describe two different *direct* variations in the same statement, the word **jointly** is used.

JOINT VARIATION

The following statements are equivalent.

1. z **varies jointly** as x and y.
2. z is **jointly proportional** to x and y.
3. $z = kxy$ for some nonzero constant k.

If x, y, and z are related by an equation of the form

$$z = kx^n y^m$$

then z varies jointly as the nth power of x and the mth power of y.

EXAMPLE 6 Joint Variation

The *simple* interest for a certain savings account is jointly proportional to the time and the principal. After one quarter (3 months), the interest on a principal of \$5000 is \$43.75.

a. Find a mathematical model that relates the interest, principal, and time.

b. Find the interest after three quarters.

Solution

a. Let I = interest (in dollars), P = principal (in dollars), and t = time (in years). Because I is jointly proportional to P and t, you have

$$I = kPt.$$

For $I = 43.75$, $P = 5000$, and $t = \frac{1}{4}$, you have

$$43.75 = k(5000)\left(\frac{1}{4}\right)$$

$$\frac{43.75(4)}{5000} = k$$

$$k = \frac{175}{5000}$$

$$= 0.035$$

So, the mathematical model that relates interest, principal, and time is

$$I = 0.035Pt$$

which is the familiar equation for simple interest where the constant of proportionality, 0.035, represents an annual interest rate of 3.5%.

b. When $P = \$5000$ and $t = \frac{3}{4}$, the interest is

$$I = (0.035)(5000)\left(\frac{3}{4}\right)$$

$$= \$131.25.$$

■

1.6 Exercises

See www.CalcChat.com for worked-out solutions to odd-numbered exercises.

In Exercises 1–10, fill in the blanks.

1. Two techniques for fitting models to data are called direct ________ and least squares ________.
2. Statisticians use a measure called ________ of ________ ________ to find a model that approximates a set of data most accurately.
3. The linear model with the least sum of square differences is called the ________ ________ ________ line.
4. An r-value of a set of data, also called a ________ ________, gives a measure of how well a model fits a set of data.
5. Direct variation models can be described as "y varies directly as x," or "y is ________ ________ to x."
6. In direct variation models of the form $y = kx$, k is called the ________ of ________.
7. The direct variation model $y = kx^n$ can be described as "y varies directly as the nth power of x," or "y is ________ ________ to the nth power of x."
8. The mathematical model $y = \frac{k}{x}$ is an example of ________ variation.
9. Mathematical models that involve both direct and inverse variation are said to have ________ variation.
10. The joint variation model $z = kxy$ can be described as "z varies jointly as x and y," or "z is ________ ________ to x and y."

11. ***Employment*** The total numbers of people (in thousands) in the U.S. civilian labor force from 1992 through 2007 are given by the following ordered pairs.

(1992, 128,105) (2000, 142,583)
(1993, 129,200) (2001, 143,734)
(1994, 131,056) (2002, 144,863)
(1995, 132,304) (2003, 146,510)
(1996, 133,943) (2004, 147,401)
(1997, 136,297) (2005, 149,320)
(1998, 137,673) (2006, 151,428)
(1999, 139,368) (2007, 153,124)

A linear model that approximates the data is $y = 1695.9t + 124{,}320$, where y represents the number of employees (in thousands) and $t = 2$ represents 1992. Plot the actual data and the model on the same set of coordinate axes. How closely does the model represent the data? *(Source: U.S. Bureau of Labor Statistics)*

12. ***Sports*** The winning times (in minutes) in the women's 400-meter freestyle swimming event in the Olympics from 1948 through 2008 are given by the following ordered pairs.

(1948, 5.30) (1972, 4.32) (1996, 4.12)
(1952, 5.20) (1976, 4.16) (2000, 4.10)
(1956, 4.91) (1980, 4.15) (2004, 4.09)
(1960, 4.84) (1984, 4.12) (2008, 4.05)
(1964, 4.72) (1988, 4.06)
(1968, 4.53) (1992, 4.12)

A linear model that approximates the data is $y = -0.020t + 5.00$, where y represents the winning time (in minutes) and $t = 0$ represents 1950. Plot the actual data and the model on the same set of coordinate axes. How closely does the model represent the data? Does it appear that another type of model may be a better fit? Explain. *(Source: International Olympic Committee)*

13. ***Sports*** The lengths (in feet) of the winning men's discus throws in the Olympics from 1920 through 2008 are listed below. *(Source: International Olympic Committee)*

Year	Length	Year	Length	Year	Length
1920	146.6	1956	184.9	1984	218.5
1924	151.3	1960	194.2	1988	225.8
1928	155.3	1964	200.1	1992	213.7
1932	162.3	1968	212.5	1996	227.7
1936	165.6	1972	211.3	2000	227.3
1948	173.2	1976	221.5	2004	229.3
1952	180.5	1980	218.7	2008	225.8

(a) Sketch a scatter plot of the data. Let y represent the length of the winning discus throw (in feet) and let $t = 20$ represent 1920.

(b) Use a straightedge to sketch the best-fitting line through the points and find an equation of the line.

(c) Use the *regression* feature of a graphing utility to find the least squares regression line that fits the data.

(d) Compare the linear model you found in part (b) with the linear model given by the graphing utility in part (c).

(e) Use the models from parts (b) and (c) to estimate the winning men's discus throw in the year 2012.

14. ***Sales*** The total sales (in billions of dollars) for Coca-Cola Enterprises from 2000 through 2007 are listed below. *(Source: Coca-Cola Enterprises, Inc.)*

2000	14.750	2004	18.185
2001	15.700	2005	18.706
2002	16.899	2006	19.804
2003	17.330	2007	20.936

(a) Sketch a scatter plot of the data. Let y represent the total revenue (in billions of dollars) and let $t = 0$ represent 2000.

(b) Use a straightedge to sketch the best-fitting line through the points and find an equation of the line.

(c) Use the *regression* feature of a graphing utility to find the least squares regression line that fits the data.

(d) Compare the linear model you found in part (b) with the linear model given by the graphing utility in part (c).

(e) Use the models from parts (b) and (c) to estimate the sales of Coca-Cola Enterprises in 2008.

(f) Use your school's library, the Internet, or some other reference source to analyze the accuracy of the estimate in part (e).

15. ***Data Analysis: Broadway Shows*** The table shows the annual gross ticket sales S (in millions of dollars) for Broadway shows in New York City from 1995 through 2006. *(Source: The League of American Theatres and Producers, Inc.)*

Year	1995	1996	1997	1998	1999	2000
Sales, S	406	436	499	558	588	603

Year	2001	2002	2003	2004	2005	2006
Sales, S	666	643	721	771	769	862

(a) Use a graphing utility to create a scatter plot of the data. Let $t = 5$ represent 1995.

(b) Use the *regression* feature of a graphing utility to find the equation of the least squares regression line that fits the data.

(c) Use the graphing utility to graph the scatter plot you created in part (a) and the model you found in part (b) in the same viewing window. How closely does the model represent the data?

(d) Use the model to estimate the annual gross ticket sales in 2007 and 2009.

(e) Interpret the meaning of the slope of the linear model in the context of the problem.

16. ***Data Analysis: Television Sets*** The table shows the numbers N (in millions) of television sets in U.S. households from 2000 through 2006. *(Source: Television Bureau of Advertising, Inc.)*

Year	2000	2001	2002	2003
Television sets, N	245	248	254	260

Year	2004	2005	2006
Television sets, N	268	287	301

(a) Use the *regression* feature of a graphing utility to find the equation of the least squares regression line that fits the data. Let $t = 0$ represent 2000.

(b) Use the graphing utility to create a scatter plot of the data. Then graph the model you found in part (a) and the scatter plot in the same viewing window. How closely does the model represent the data?

(c) Use the model to estimate the number of television sets in U.S. households in 2008.

(d) Use your school's library, the Internet, or some other reference source to analyze the accuracy of the estimate in part (c).

Think About It **In Exercises 17 and 18, use the graph to determine whether y varies directly as some power of x or inversely as some power of x. Explain.**

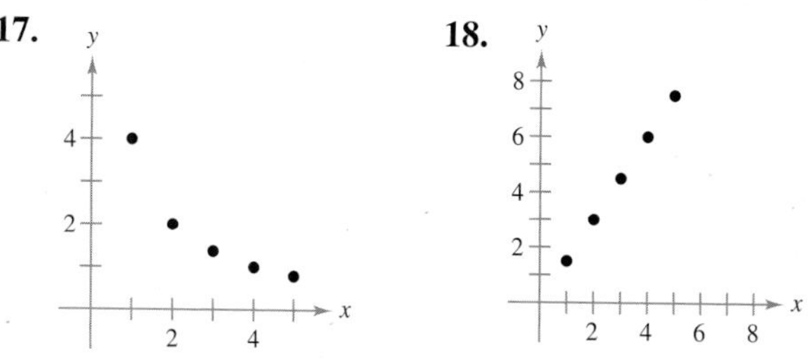

In Exercises 19–22, use the given value of k to complete the table for the direct variation model

$$y = kx^2.$$

Plot the points on a rectangular coordinate system.

x	2	4	6	8	10
$y = kx^2$					

19. $k = 1$

20. $k = 2$

21. $k = \frac{1}{2}$

22. $k = \frac{1}{4}$

In Exercises 23–26, use the given value of k to complete the table for the inverse variation model

$$y = \frac{k}{x^2}.$$

Plot the points on a rectangular coordinate system.

x	2	4	6	8	10
$y = \frac{k}{x^2}$					

23. $k = 2$ **24.** $k = 5$ **25.** $k = 10$ **26.** $k = 20$

In Exercises 27–30, determine whether the variation model is of the form $y = kx$ or $y = k/x$, and find k. Then write a model that relates y and x.

27.

x	5	10	15	20	25
y	1	$\frac{1}{2}$	$\frac{1}{3}$	$\frac{1}{4}$	$\frac{1}{5}$

28.

x	5	10	15	20	25
y	2	4	6	8	10

29.

x	5	10	15	20	25
y	-3.5	-7	-10.5	-14	-17.5

30.

x	5	10	15	20	25
y	24	12	8	6	$\frac{24}{5}$

Direct Variation **In Exercises 31–34, assume that y is directly proportional to x. Use the given x-value and y-value to find a linear model that relates y and x.**

31. $x = 5, y = 12$ **32.** $x = 2, y = 14$

33. $x = 10, y = 2050$ **34.** $x = 6, y = 580$

WRITING ABOUT CONCEPTS

In Exercises 35–40, find a mathematical model for the verbal statement.

35. A varies directly as the square of r.

36. V varies directly as the cube of e.

37. y varies inversely as the square of x.

38. h varies inversely as the square root of s.

39. F varies directly as g and inversely as r^2.

40. z is jointly proportional to the square of x and the cube of y.

WRITING ABOUT CONCEPTS (continued)

In Exercises 41–46, write a sentence using the variation terminology of this section to describe the formula.

41. Area of a triangle: $A = \frac{1}{2}bh$

42. Surface area of a sphere: $S = 4\pi r^2$

43. Volume of a sphere: $V = \frac{4}{3}\pi r^3$

44. Volume of a right circular cylinder: $V = \pi r^2 h$

45. Average speed: $r = d/t$

46. Free vibrations: $\omega = \sqrt{(kg)/W}$

In Exercises 47–50, discuss how well the data shown in the scatter plot can be approximated by a linear model.

47.

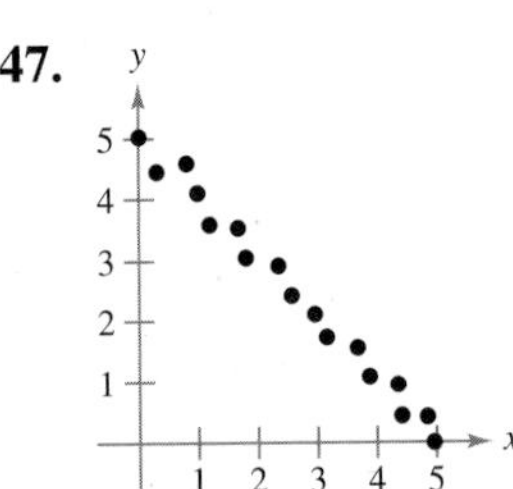

48.

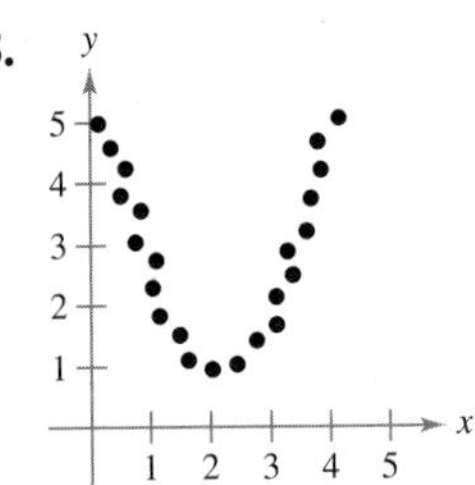

49.

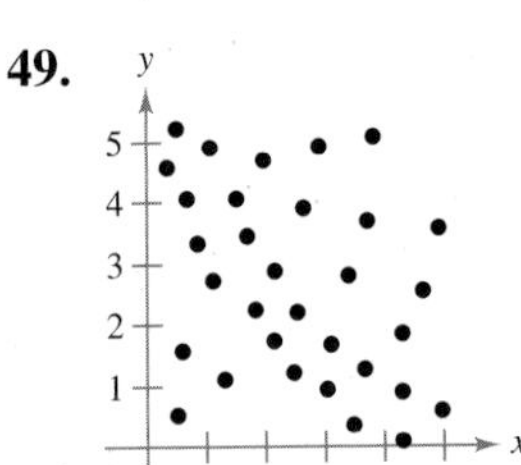

50.

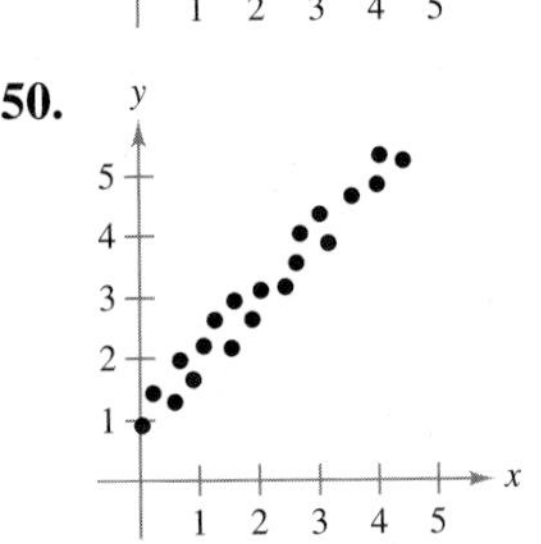

In Exercises 51–54, sketch the line that you think best approximates the data in the scatter plot. Then find an equation of the line. To print an enlarged copy of the graph, go to the website *www.mathgraphs.com*.

51.

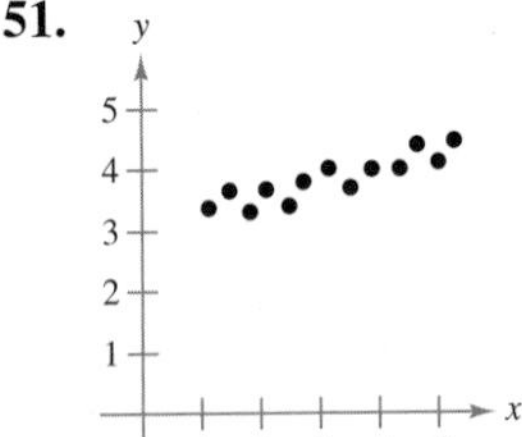

52.

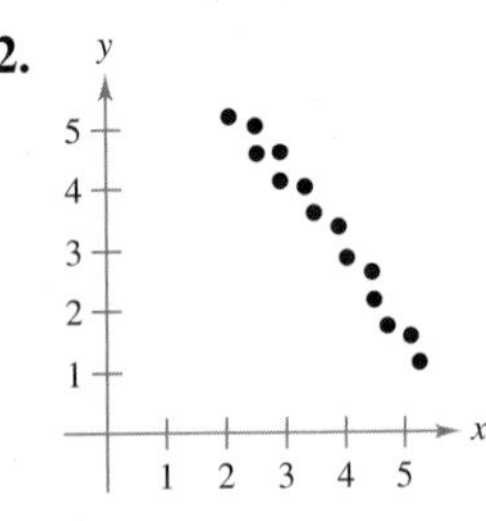

53.

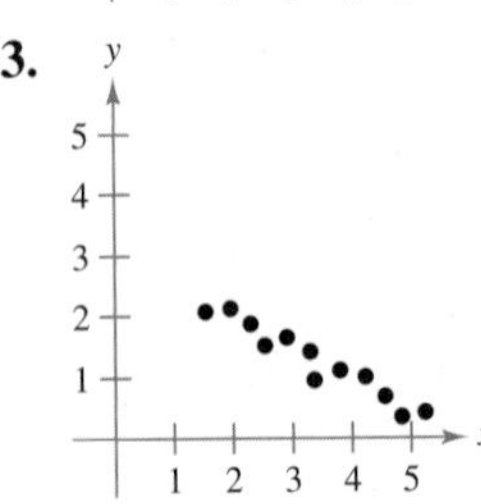

54.

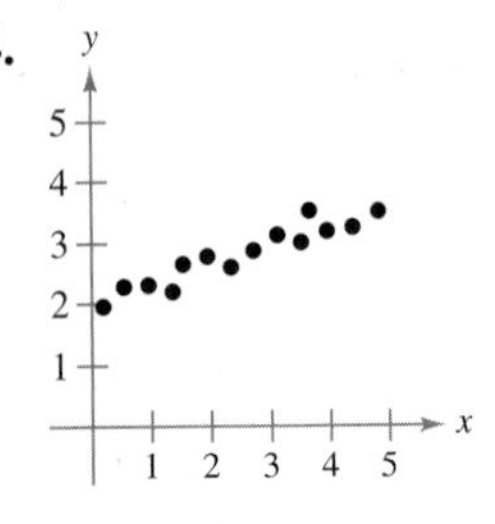

55. ***Simple Interest*** The simple interest on an investment is directly proportional to the amount of the investment. By investing \$3250 in a certain bond issue, you obtained an interest payment of \$113.75 after 1 year. Find a mathematical model that gives the interest I for this bond issue after 1 year in terms of the amount invested P.

56. ***Simple Interest*** The simple interest on an investment is directly proportional to the amount of the investment. By investing \$6500 in a municipal bond, you obtained an interest payment of \$211.25 after 1 year. Find a mathematical model that gives the interest I for this municipal bond after 1 year in terms of the amount invested P.

57. ***Measurement*** On a yardstick with scales in inches and centimeters, you notice that 13 inches is approximately the same length as 33 centimeters. Use this information to find a mathematical model that relates centimeters y to inches x. Then use the model to find the numbers of centimeters in 10 inches and 20 inches.

58. ***Measurement*** When buying gasoline, you notice that 14 gallons of gasoline is approximately the same amount of gasoline as 53 liters. Use this information to find a linear model that relates liters y to gallons x. Then use the model to find the numbers of liters in 5 gallons and 25 gallons.

59. ***Taxes*** Property tax is based on the assessed value of a property. A house that has an assessed value of \$150,000 has a property tax of \$5520. Find a mathematical model that gives the amount of property tax y in terms of the assessed value x of the property. Use the model to find the property tax on a house that has an assessed value of \$225,000.

60. ***Taxes*** State sales tax is based on retail price. An item that sells for \$189.99 has a sales tax of \$11.40. Find a mathematical model that gives the amount of sales tax y in terms of the retail price x. Use the model to find the sales tax on a \$639.99 purchase.

In Exercises 61–64, find a mathematical model for the verbal statement.

61. ***Boyle's Law:*** For a constant temperature, the pressure P of a gas is inversely proportional to the volume V of the gas.

62. ***Newton's Law of Cooling:*** The rate of change R of the temperature of an object is proportional to the difference between the temperature T of the object and the temperature T_e of the environment in which the object is placed.

63. ***Newton's Law of Universal Gravitation:*** The gravitational attraction F between two objects of masses m_1 and m_2 is proportional to the product of the masses and inversely proportional to the square of the distance r between the objects.

64. ***Logistic Growth:*** The rate of growth R of a population is jointly proportional to the size S of the population and the difference between S and the maximum population size L that the environment can support.

Hooke's Law **In Exercises 65–68, use Hooke's Law for springs, which states that the distance a spring is stretched (or compressed) varies directly as the force on the spring.**

65. A force of 265 newtons stretches a spring 0.15 meter (see figure).

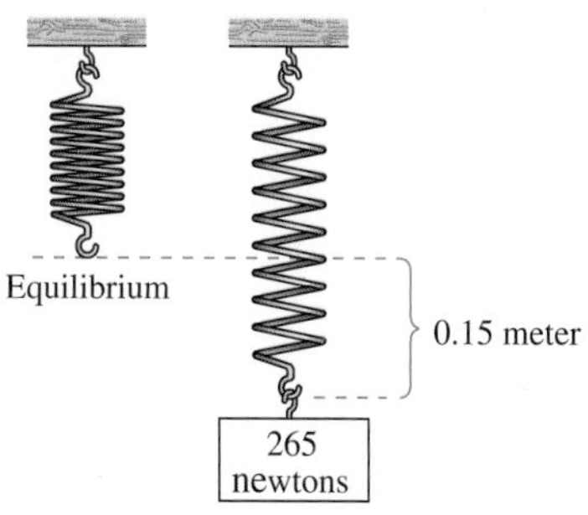

(a) How far will a force of 90 newtons stretch the spring?

(b) What force is required to stretch the spring 0.1 meter?

66. A force of 220 newtons stretches a spring 0.12 meter. What force is required to stretch the spring 0.16 meter?

67. The coiled spring of a toy supports the weight of a child. The spring is compressed a distance of 1.9 inches by the weight of a 25-pound child. The toy will not work properly if its spring is compressed more than 3 inches. What is the weight of the heaviest child who should be allowed to use the toy?

68. An overhead garage door has two springs, one on each side of the door (see figure). A force of 15 pounds is required to stretch each spring 1 foot. Because of a pulley system, the springs stretch only one-half the distance the door travels. The door moves a total of 8 feet, and the springs are at their natural length when the door is open. Find the combined lifting force applied to the door by the springs when the door is closed.

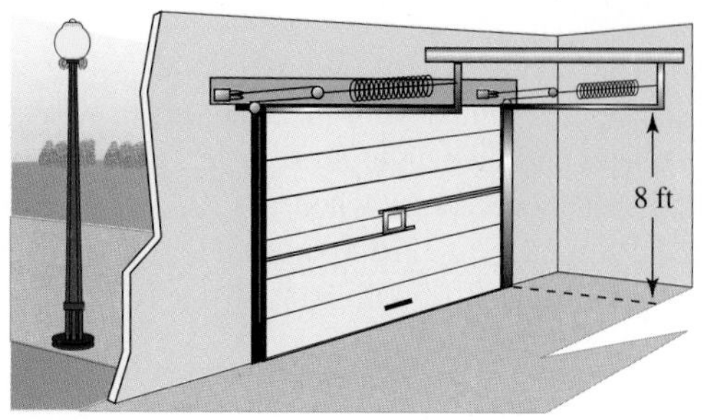

In Exercises 69–76, find a mathematical model representing the statement. (In each case, determine the constant of proportionality.)

69. A varies directly as r^2. ($A = 9\pi$ when $r = 3$.)
70. y varies inversely as x. ($y = 3$ when $x = 25$.)
71. y is inversely proportional to x. ($y = 7$ when $x = 4$.)
72. z varies jointly as x and y. ($z = 64$ when $x = 4$ and $y = 8$.)
73. F is jointly proportional to r and the third power of s. ($F = 4158$ when $r = 11$ and $s = 3$.)
74. P varies directly as x and inversely as the square of y. $\left(P = \frac{28}{3}\text{ when } x = 42 \text{ and } y = 9.\right)$
75. z varies directly as the square of x and inversely as y. ($z = 6$ when $x = 6$ and $y = 4$.)
76. v varies jointly as p and q and inversely as the square of s. ($v = 1.5$ when $p = 4.1$, $q = 6.3$, and $s = 1.2$.)

Ecology **In Exercises 77 and 78, use the fact that the diameter of the largest particle that can be moved by a stream varies approximately directly as the square of the velocity of the stream.**

77. A stream with a velocity of $\frac{1}{4}$ mile per hour can move coarse sand particles about 0.02 inch in diameter. Approximate the velocity required to carry particles 0.12 inch in diameter.
78. A stream of velocity v can move particles of diameter d or less. By what factor does d increase when the velocity is doubled?

Resistance **In Exercises 79 and 80, use the fact that the resistance of a wire carrying an electrical current is directly proportional to its length and inversely proportional to its cross-sectional area.**

79. If #28 copper wire (which has a diameter of 0.0126 inch) has a resistance of 66.17 ohms per thousand feet, what length of #28 copper wire will produce a resistance of 33.5 ohms?
80. A 14-foot piece of copper wire produces a resistance of 0.05 ohm. Use the constant of proportionality from Exercise 79 to find the diameter of the wire.
81. ***Work*** The work W (in joules) done when lifting an object varies jointly with the mass m (in kilograms) of the object and the height h (in meters) that the object is lifted. The work done when a 120-kilogram object is lifted 1.8 meters is 2116.8 joules. How much work is done when lifting a 100-kilogram object 1.5 meters?
82. ***Music*** The frequency of vibrations of a piano string varies directly as the square root of the tension on the string and inversely as the length of the string. The middle A string has a frequency of 440 vibrations per second. Find the frequency of a string that has 1.25 times as much tension and is 1.2 times as long.
83. ***Fluid Flow*** The velocity v of a fluid flowing in a conduit is inversely proportional to the cross-sectional area of the conduit. (Assume that the volume of the flow per unit of time is held constant.)
    (a) Determine the change in the velocity of water flowing from a hose when a person places a finger over the end of the hose to decrease its cross-sectional area by 25%.
    (b) Use the fluid velocity model in part (a) to determine the effect on the velocity of a stream when it is dredged to increase its cross-sectional area by one-third.
84. ***Beam Load*** The maximum load that can be safely supported by a horizontal beam varies jointly as the width of the beam and the square of its depth, and inversely as the length of the beam. Determine the changes in the maximum safe load under the following conditions.
    (a) The width and length of the beam are doubled.
    (b) The width and depth of the beam are doubled.
    (c) All three of the dimensions are doubled.
    (d) The depth of the beam is halved.
85. ***Data Analysis: Ocean Temperatures*** An oceanographer took readings of the water temperatures C (in degrees Celsius) at several depths d (in meters). The data collected are shown in the table.

Depth, d	1000	2000	3000	4000	5000
Temperature, C	4.2°	1.9°	1.4°	1.2°	0.9°

    (a) Sketch a scatter plot of the data.
    (b) Does it appear that the data can be modeled by the inverse variation model $C = k/d$? If so, find k for each pair of coordinates.
    (c) Determine the mean value of k from part (b) to find the inverse variation model $C = k/d$.
    (d) Use a graphing utility to plot the data points and the inverse model from part (c).
    (e) Use the model to approximate the depth at which the water temperature is 3°C.

86. ***Data Analysis: Physics Experiment*** An experiment in a physics lab requires a student to measure the compressed lengths y (in centimeters) of a spring when various forces of F pounds are applied. The data are shown in the table.

Force, F	0	2	4	6	8	10	12
Length, y	0	1.15	2.3	3.45	4.6	5.75	6.9

(a) Sketch a scatter plot of the data.

(b) Does it appear that the data can be modeled by Hooke's Law? If so, estimate k. (See Exercises 65–68.)

(c) Use the model in part (b) to approximate the force required to compress the spring 9 centimeters.

87. ***Data Analysis: Light Intensity*** A light probe is located x centimeters from a light source, and the intensity y (in microwatts per square centimeter) of the light is measured. The results are shown in the table.

x	30	34	38
y	0.1881	0.1543	0.1172

x	42	46	50
y	0.0998	0.0775	0.0645

A model for the data is $y = 262.76/x^{2.12}$.

(a) Use a graphing utility to plot the data points and the model in the same viewing window.

(b) Use the model to approximate the light intensity 25 centimeters from the light source.

88. ***Illumination*** The illumination from a light source varies inversely as the square of the distance from the light source. When the distance from a light source is doubled, how does the illumination change? Discuss this model in terms of the data given in Exercise 87. Give a possible explanation of the difference.

True or False? **In Exercises 89–92, decide whether the statement is true or false. Justify your answer.**

89. The statements "y varies directly as x" and "y is inversely proportional to x" are equivalent.

90. A mathematical equation for "a is jointly proportional to y and z with the constant of proportionality k" can be written as

$$a = k\frac{y}{z}.$$

91. In the equation for kinetic energy, $E = \frac{1}{2}mv^2$, the amount of kinetic energy E is directly proportional to the mass m of an object and the square of its velocity v.

92. If the correlation coefficient for a least squares regression line is close to -1, the regression line cannot be used to describe the data.

93. ***Writing*** A linear model for predicting prize winnings at a race is based on data for 3 years. Write a paragraph discussing the potential accuracy or inaccuracy of such a model.

94. ***Writing*** Suppose the constant of proportionality is positive and y varies directly as x. When one of the variables increases, how will the other change? Explain your reasoning.

95. ***Writing***

(a) Given that y varies inversely as the square of x and x is doubled, how will y change? Explain.

(b) Given that y varies directly as the square of x and x is doubled, how will y change? Explain.

CAPSTONE

96. The prices of three sizes of pizza at a pizza shop are as follows.

9-inch: \$8.78, 12-inch: \$11.78, 15-inch: \$14.18

You would expect that the price of a certain size of pizza would be directly proportional to its surface area. Is that the case for this pizza shop? If not, which size of pizza is the best buy?

SECTION PROJECT

Hooke's Law

In physics, Hooke's Law for springs states that the distance a spring is stretched or compressed from its natural or equilibrium length varies directly as the force on the spring. Distance is measured in inches (or meters) and force is measured in pounds (or newtons). One newton is approximately equivalent to 0.225 pound.

(a) Use direct variation to find an equation relating the distance stretched (or compressed) to the force applied.

(b) If a force of 100 newtons stretches a spring 0.75 meter, how far will a force of 80 newtons stretch the spring?

(c) Conduct your own experiment, and record your results.

(d) Write a brief summary comparing the theoretical result with your experimental results.

1 CHAPTER SUMMARY

Section 1.1	Review Exercises
■ Determine whether relations between two variables are functions *(p. 72).*	*1–4*
■ Use function notation and evaluate functions *(p. 74).*	*5, 6*
■ Find the domains of functions *(p. 77).*	*7–10*
■ Use functions to model and solve real-life problems *(p. 78).*	*11–14*
Section 1.2	
■ Use the Vertical Line Test for functions *(p. 86).*	*15, 16*
■ Find the zeros of functions *(p. 87).*	*17–20*
■ Determine intervals on which functions are increasing or decreasing *(p. 88)* and determine relative maximum and relative minimum values of functions *(p. 89).*	*21–26*
■ Identify and graph linear functions *(p. 90).*	*27, 28*
■ Identify and graph step and other piecewise-defined functions *(p. 91).*	*29–32*
■ Identify even and odd functions *(p. 92).*	*33–36*
Section 1.3	
■ Recognize graphs of common functions *(p. 98),* and use vertical and horizontal shifts *(p. 99),* reflections *(p. 101),* and nonrigid transformations *(p. 102)* to sketch graphs of functions.	*37–52*
Section 1.4	
■ Add, subtract, multiply, and divide functions *(p. 107).*	*53, 54*
■ Find the composition of one function with another function *(p. 109).*	*55–58*
■ Use combinations and compositions of functions to model and solve real-life problems *(p. 111).*	*59, 60*
Section 1.5	
■ Find inverse functions informally and verify that two functions are inverse functions of each other *(p. 115).*	*61, 62*
■ Use graphs of functions to determine whether functions have inverse functions *(p. 117).*	*63, 64*
■ Use the Horizontal Line Test to determine if functions are one-to-one *(p. 118).*	*65–68*
■ Find inverse functions analytically *(p. 119).*	*69–74*
Section 1.6	
■ Use mathematical models to approximate sets of data points *(p. 124),* and use the *regression* feature of a graphing utility to find the equation of a least squares regression line *(p. 125).*	*75, 76*
■ Write mathematical models for direct variation, direct variation as an nth power, inverse variation, and joint variation *(pp. 126–129).*	*77–84*

1 REVIEW EXERCISES

See www.CalcChat.com for worked-out solutions to odd-numbered exercises.

In Exercises 1–4, determine whether the equation represents y as a function of x.

1. $16x - y^4 = 0$
2. $2x - y - 3 = 0$
3. $y = \sqrt{1 - x}$
4. $|y| = x + 2$

In Exercises 5 and 6, evaluate the function at each specified value of the independent variable and simplify.

5. $f(x) = x^2 + 1$

   (a) $f(2)$ (b) $f(-4)$ (c) $f(t^2)$ (d) $f(t + 1)$

6. $h(x) = \begin{cases} 2x + 1, & x \le -1 \\ x^2 + 2, & x > -1 \end{cases}$

   (a) $h(-2)$ (b) $h(-1)$ (c) $h(0)$ (d) $h(2)$

In Exercises 7–10, find the domain of the function. Verify your result with a graph.

7. $f(x) = \sqrt{25 - x^2}$
8. $g(s) = \dfrac{5s + 5}{3s - 9}$
9. $h(x) = \dfrac{x}{x^2 - x - 6}$
10. $h(t) = |t + 1|$

11. ***Physics*** The velocity of a ball projected upward from ground level is given by $v(t) = -32t + 48$, where t is the time in seconds and v is the velocity in feet per second.

    (a) Find the velocity when $t = 1$.

    (b) Find the time when the ball reaches its maximum height. [*Hint:* Find the time when $v(t) = 0$.]

    (c) Find the velocity when $t = 2$.

12. ***Mixture Problem*** From a full 50-liter container of a 40% concentration of acid, x liters is removed and replaced with 100% acid.

    (a) Write the amount of acid in the final mixture as a function of x.

    (b) Determine the domain and range of the function.

    (c) Determine x if the final mixture is 50% acid.

In Exercises 13 and 14, find the difference quotient and simplify your answer.

13. $f(x) = 2x^2 + 3x - 1, \quad \dfrac{f(x + h) - f(x)}{h}, \quad h \ne 0$

14. $f(x) = x^3 - 5x^2 + x, \quad \dfrac{f(x + h) - f(x)}{h}, \quad h \ne 0$

In Exercises 15 and 16, use the Vertical Line Test to determine whether y is a function of x. To print an enlarged copy of the graph, go to the website *www.mathgraphs.com*.

15. $y = (x - 3)^2$

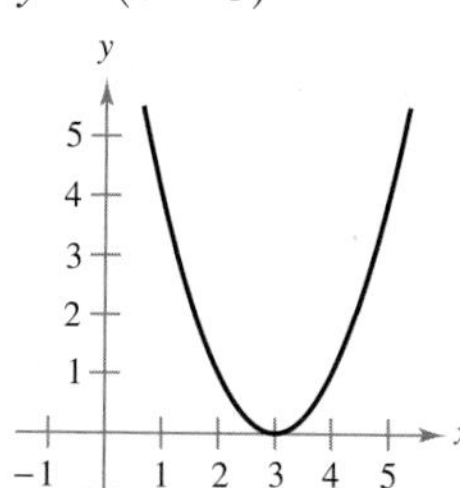

16. $x = -|4 - y|$

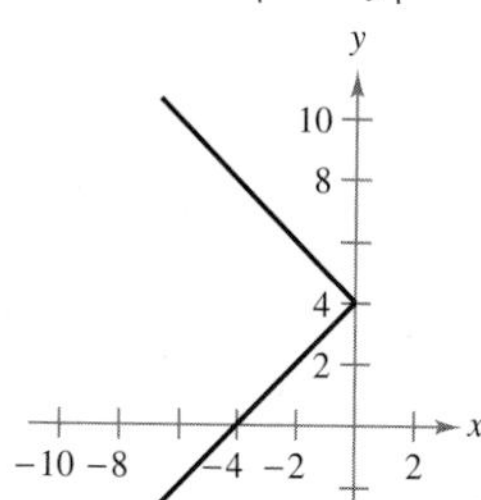

In Exercises 17–20, find the zeros of the function.

17. $f(x) = 3x^2 - 16x + 21$
18. $f(x) = 5x^2 + 4x - 1$
19. $f(x) = \dfrac{8x + 3}{11 - x}$
20. $f(x) = x^3 - x^2 - 25x + 25$

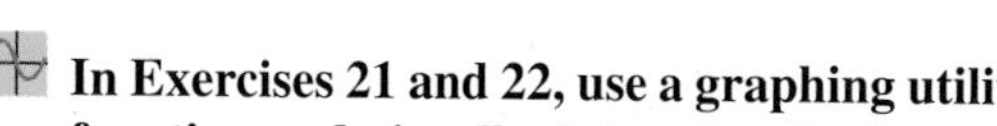

In Exercises 21 and 22, use a graphing utility to graph the function and visually determine the intervals over which the function is increasing, decreasing, or constant.

21. $f(x) = |x| + |x + 1|$
22. $f(x) = (x^2 - 4)^2$

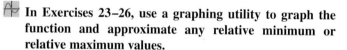

In Exercises 23–26, use a graphing utility to graph the function and approximate any relative minimum or relative maximum values.

23. $f(x) = -x^2 + 2x + 1$
24. $f(x) = x^4 - 4x^2 - 2$
25. $f(x) = x^3 - 6x^4$
26. $f(x) = x^3 - 4x^2 - 1$

In Exercises 27 and 28, write the linear function f such that it has the indicated function values. Then sketch the graph of the function.

27. $f(2) = -6, \ f(-1) = 3$
28. $f(0) = -5, \ f(4) = -8$

In Exercises 29–32, graph the function.

29. $f(x) = [\![x]\!] + 2$
30. $g(x) = [\![x + 4]\!]$
31. $f(x) = \begin{cases} 5x - 3, & x \ge -1 \\ -4x + 5, & x < -1 \end{cases}$
32. $f(x) = \begin{cases} x^2 - 2, & x < -2 \\ 5, & -2 \le x \le 0 \\ 8x - 5, & x > 0 \end{cases}$

In Exercises 33–36, determine whether the function is even, odd, or neither.

33. $f(x) = x^5 + 4x - 7$
34. $f(x) = x^4 - 20x^2$
35. $f(x) = 2x\sqrt{x^2 + 3}$
36. $f(x) = \sqrt[5]{6x^2}$

In Exercises 37 and 38, identify the common function and describe the transformation shown in the graph.

37.

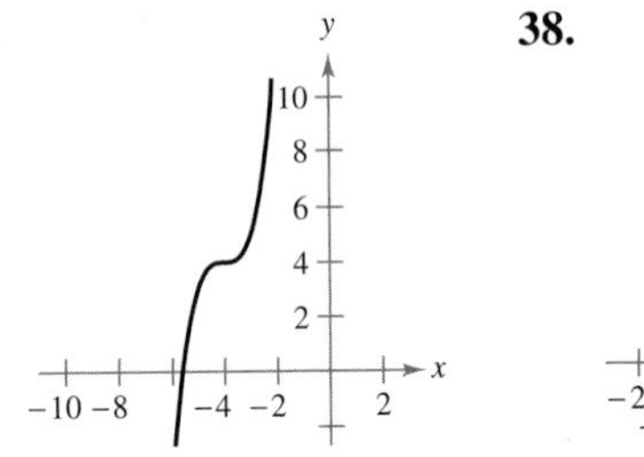

38.

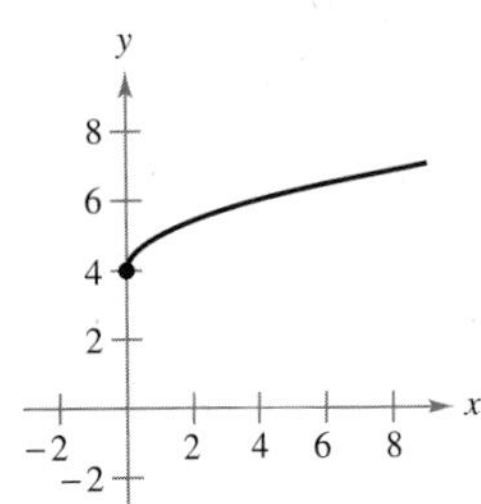

In Exercises 39–52, h is related to one of the common functions described in this chapter. (a) Identify the common function f. (b) Describe the sequence of transformations from f to h. (c) Sketch the graph of h. (d) Use function notation to write h in terms of f.

39. $h(x) = x^2 - 9$ **40.** $h(x) = (x - 2)^3 + 2$

41. $h(x) = -\sqrt{x} + 4$ **42.** $h(x) = |x + 3| - 5$

43. $h(x) = -(x + 2)^2 + 3$ **44.** $h(x) = \frac{1}{2}(x - 1)^2 - 2$

45. $h(x) = -[\![x]\!] + 6$ **46.** $h(x) = -\sqrt{x + 1} + 9$

47. $h(x) = -|-x + 4| + 6$ **48.** $h(x) = -(x + 1)^2 - 3$

49. $h(x) = 5[\![x - 9]\!]$ **50.** $h(x) = -\frac{1}{3}x^3$

51. $h(x) = -2\sqrt{x - 4}$ **52.** $h(x) = \frac{1}{2}|x| - 1$

In Exercises 53 and 54, find (a) $(f + g)(x)$, (b) $(f - g)(x)$, (c) $(fg)(x)$, and (d) $(f/g)(x)$. What is the domain of f/g?

53. $f(x) = x^2 + 3,\quad g(x) = 2x - 1$

54. $f(x) = x^2 - 4,\quad g(x) = \sqrt{3 - x}$

In Exercises 55 and 56, find (a) $f \circ g$ and (b) $g \circ f$. Find the domain of each function and each composite function.

55. $f(x) = \frac{1}{3}x - 3,\quad g(x) = 3x + 1$

56. $f(x) = x^3 - 4,\quad g(x) = \sqrt[3]{x + 7}$

In Exercises 57 and 58, find two functions f and g such that $(f \circ g)(x) = h(x)$. (There are many correct answers.)

57. $h(x) = (1 - 2x)^3$ **58.** $h(x) = \sqrt[3]{x + 2}$

59. ***Phone Expenditures*** The average annual expenditures (in dollars) for residential $r(t)$ and cellular $c(t)$ phone services from 2001 through 2006 can be approximated by the functions $r(t) = 27.5t + 705$ and $c(t) = 151.3t + 151$, where t represents the year, with $t = 1$ corresponding to 2001. *(Source: Bureau of Labor Statistics)*

(a) Find and interpret $(r + c)(t)$.

(b) Use a graphing utility to graph $r(t)$, $c(t)$, and $(r + c)(t)$ in the same viewing window.

(c) Find $(r + c)(13)$. Use the graph in part (b) to verify your result.

60. ***Bacteria Count*** The number N of bacteria in a refrigerated food is given by

$$N(T) = 25T^2 - 50T + 300, \quad 2 \le T \le 20$$

where T is the temperature of the food in degrees Celsius. When the food is removed from refrigeration, the temperature of the food is given by

$$T(t) = 2t + 1, \quad 0 \le t \le 9$$

where t is the time in hours. (a) Find the composition $N(T(t))$ and interpret its meaning in context, and (b) find the time when the bacteria count reaches 750.

In Exercises 61 and 62, find the inverse function of f informally. Verify that $f(f^{-1}(x)) = x$ and $f^{-1}(f(x)) = x$.

61. $f(x) = 3x + 8$ **62.** $f(x) = \dfrac{x - 4}{5}$

In Exercises 63 and 64, determine whether the function has an inverse function.

63.

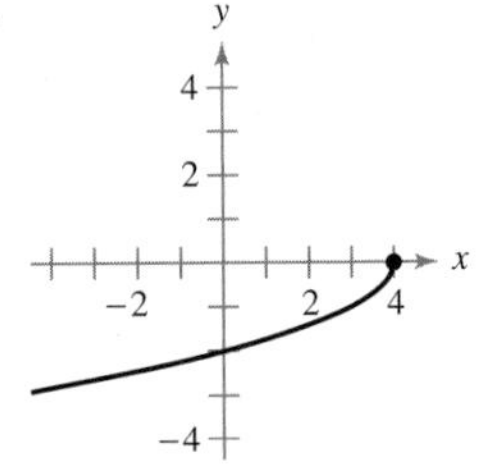

64.

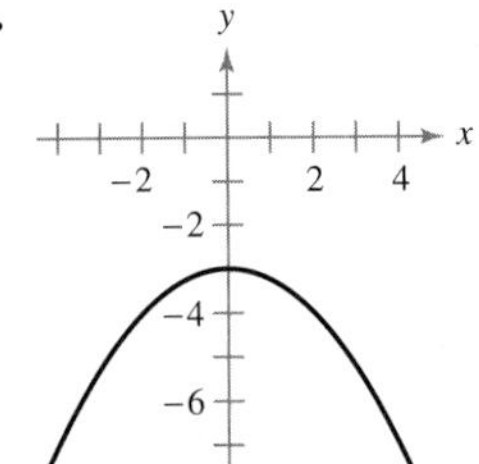

In Exercises 65–68, use a graphing utility to graph the function, and use the Horizontal Line Test to determine whether the function is one-to-one and so has an inverse function.

65. $f(x) = 4 - \frac{1}{3}x$ **66.** $f(x) = (x - 1)^2$

67. $h(t) = \dfrac{2}{t - 3}$ **68.** $g(x) = \sqrt{x + 6}$

In Exercises 69–72, (a) find the inverse function of f, (b) graph both f and f^{-1} on the same set of coordinate axes, (c) describe the relationship between the graphs of f and f^{-1}, and (d) state the domains and ranges of f and f^{-1}.

69. $f(x) = \frac{1}{2}x - 3$ **70.** $f(x) = 5x - 7$

71. $f(x) = \sqrt{x + 1}$ **72.** $f(x) = x^3 + 2$

In Exercises 73 and 74, restrict the domain of the function f to an interval over which the function is increasing and determine f^{-1} over that interval.

73. $f(x) = 2(x - 4)^2$ **74.** $f(x) = |x - 2|$

75. ***Compact Discs*** The values V (in billions of dollars) of shipments of compact discs in the United States from 2000 through 2007 are shown in the table. A linear model that approximates these data is

$$V = -0.742t + 13.62$$

where t represents the year, with $t = 0$ corresponding to 2000. *(Source: Recording Industry Association of America)*

Year	2000	2001	2002	2003
Value, V	13.21	12.91	12.04	11.23

Year	2004	2005	2006	2007
Value, V	11.45	10.52	9.37	7.45

(a) Plot the actual data and the model on the same set of coordinate axes.

(b) How closely does the model represent the data?

76. ***Data Analysis: TV Usage*** The table shows the projected numbers of hours H of television usage in the United States from 2003 through 2011. *(Source: Communications Industry Forecast and Report)*

Year	2003	2004	2005	2006	2007
Hours, H	1615	1620	1659	1673	1686

Year	2008	2009	2010	2011
Hours, H	1704	1714	1728	1742

(a) Use a graphing utility to create a scatter plot of the data. Let t represent the year, with $t = 3$ corresponding to 2003.

(b) Use the *regression* feature of the graphing utility to find the equation of the least squares regression line that fits the data. Then graph the model and the scatter plot you found in part (a) in the same viewing window. How closely does the model represent the data?

(c) Use the model to estimate the projected number of hours of television usage in 2020.

(d) Interpret the meaning of the slope of the linear model in the context of the problem.

77. ***Measurement*** You notice a billboard indicating that it is 2.5 miles or 4 kilometers to the next restaurant of a national fast-food chain. Use this information to find a mathematical model that relates miles to kilometers. Then use the model to find the numbers of kilometers in 2 miles and 10 miles.

78. ***Energy*** The power P produced by a wind turbine is proportional to the cube of the wind speed S. A wind speed of 27 miles per hour produces a power output of 750 kilowatts. Find the output for a wind speed of 40 miles per hour.

79. ***Frictional Force*** The frictional force F between the tires and the road required to keep a car on a curved section of a highway is directly proportional to the square of the speed s of the car. If the speed of the car is doubled, the force will change by what factor?

80. ***Demand*** A company has found that the daily demand x for its boxes of chocolates is inversely proportional to the price p. When the price is \$5, the demand is 800 boxes. Approximate the demand when the price is increased to \$6.

81. ***Travel Time*** The travel time between two cities is inversely proportional to the average speed. A train travels between the cities in 3 hours at an average speed of 65 miles per hour. How long would it take to travel between the cities at an average speed of 80 miles per hour?

82. ***Cost*** The cost of constructing a wooden box with a square base varies jointly as the height of the box and the square of the width of the box. A box of height 16 inches and width 6 inches costs \$28.80. How much would a box of height 14 inches and width 8 inches cost?

In Exercises 83 and 84, find a mathematical model representing the statement. (In each case, determine the constant of proportionality.)

83. y is inversely proportional to x. ($y = 9$ when $x = 5.5$.)

84. F is jointly proportional to x and to the square root of y. ($F = 6$ when $x = 9$ and $y = 4$.)

True or False? **In Exercises 85 and 86, determine whether the statement is true or false. Justify your answer.**

85. Relative to the graph of $f(x) = \sqrt{x}$, the function given by $h(x) = -\sqrt{x + 9} - 13$ is shifted 9 units to the left and 13 units downward, then reflected in the x-axis.

86. If f and g are two inverse functions, then the domain of g is equal to the range of f.

87. ***Writing*** Explain the difference between the Vertical Line Test and the Horizontal Line Test.

88. ***Writing*** Explain how to tell whether a relation between two variables is a function.

89. ***Think About It*** If y is directly proportional to x for a particular linear model, what is the y-intercept of the graph of the model?

1 CHAPTER TEST

Take this test as you would take a test in class. When you are finished, check your work against the answers given in the back of the book.

1. Evaluate $f(x) = \dfrac{\sqrt{x+9}}{x^2 - 81}$ at each value: (a) $f(7)$ (b) $f(-5)$ (c) $f(x-9)$.

2. Find the domain of $f(x) = 10 - \sqrt{3-x}$.

In Exercises 3–5, (a) find the zeros of the function, (b) use a graphing utility to graph the function, (c) approximate the intervals over which the function is increasing, decreasing, or constant, and (d) determine whether the function is even, odd, or neither.

3. $f(x) = 2x^6 + 5x^4 - x^2$ **4.** $f(x) = 4x\sqrt{3-x}$ **5.** $f(x) = |x+5|$

In Exercises 6 and 7, (a) write the linear function *f* such that it has the indicated function values and (b) sketch the graph of the function.

6. $f(-10) = 12,\ f(16) = -1$ **7.** $f\left(\frac{1}{2}\right) = -6,\ f(4) = -3$

8. Sketch the graph of $f(x) = \begin{cases} 3x + 7, & x \le -3 \\ 4x^2 - 1, & x > -3 \end{cases}$.

In Exercises 9–11, identify the common function in the transformation. Then sketch a graph of the function.

9. $h(x) = -[\![x]\!]$ **10.** $h(x) = -\sqrt{x+5} + 8$ **11.** $h(x) = -2(x-5)^3 + 3$

In Exercises 12 and 13, find (a) $(f+g)(x)$, (b) $(f-g)(x)$, (c) $(fg)(x)$, (d) $(f/g)(x)$, (e) $(f \circ g)(x)$, and (f) $(g \circ f)(x)$.

12. $f(x) = 3x^2 - 7,\quad g(x) = -x^2 - 4x + 5$ **13.** $f(x) = 1/x,\quad g(x) = 2\sqrt{x}$

In Exercises 14–16, determine whether or not the function has an inverse function, and if so, find the inverse function.

14. $f(x) = x^3 + 8$ **15.** $f(x) = |x^2 - 3| + 6$ **16.** $f(x) = 3x\sqrt{x}$

In Exercises 17–19, find a mathematical model representing the statement. (In each case, determine the constant of proportionality.)

17. v varies directly as the square root of s. ($v = 24$ when $s = 16$.)

18. A varies jointly as x and y. ($A = 500$ when $x = 15$ and $y = 8$.)

19. b varies inversely as a. ($b = 32$ when $a = 1.5$.)

20. The sales S (in billions of dollars) of lottery tickets in the United States from 2000 through 2007 are shown in the table. *(Source: TLF Publications, Inc.)*

(a) Use a graphing utility to create a scatter plot of the data. Let t represent the year, with $t = 0$ corresponding to 2000.

(b) Use the *regression* feature of the graphing utility to find the equation of the least squares regression line that fits the data.

(c) Use the graphing utility to graph the model in the same viewing window used for the scatter plot. How well does the model fit the data?

(d) Use the model to predict the sales of lottery tickets in 2015. Does your answer seem reasonable? Explain.

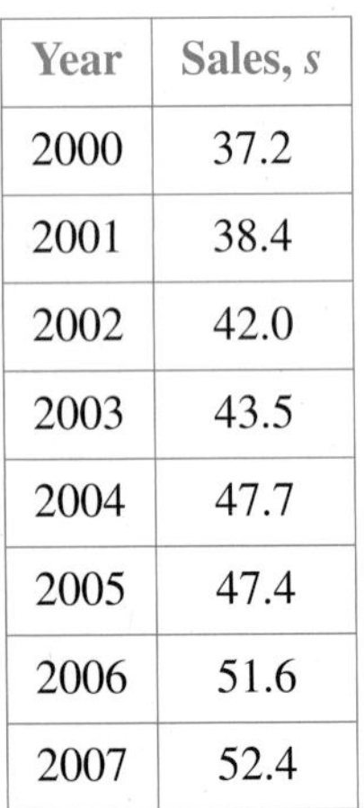

Year	Sales, s
2000	37.2
2001	38.4
2002	42.0
2003	43.5
2004	47.7
2005	47.4
2006	51.6
2007	52.4

Table for 20

P.S. PROBLEM SOLVING

1. For the numbers 2 through 9 on a telephone keypad (see figure), create two relations: one mapping numbers onto letters, and the other mapping letters onto numbers. Are both relations functions? Explain.

2. What can be said about the sum and difference of each of the following?

(a) Two even functions

(b) Two odd functions

(c) An odd function and an even function

3. Prove that a function of the following form is odd.

$$y = a_{2n+1}x^{2n+1} + a_{2n-1}x^{2n-1} + \cdot \cdot \cdot + a_3x^3 + a_1x$$

4. Prove that a function of the following form is even.

$$y = a_{2n}x^{2n} + a_{2n-2}x^{2n-2} + \cdot \cdot \cdot + a_2x^2 + a_0$$

5. Use a graphing utility to graph each function in parts (a)–(f). Write a paragraph describing any similarities and differences you observe among the graphs.

(a) $y = x$ (b) $y = x^2$ (c) $y = x^3$

(d) $y = x^4$ (e) $y = x^5$ (f) $y = x^6$

(g) Use the results of parts (a)–(f) to make a conjecture about the graphs of $y = x^7$ and $y = x^8$. Use a graphing utility to graph the functions and compare the results with your conjecture.

6. Explain why the graph of $y = -f(x)$ is a reflection of the graph of $y = f(x)$ about the x-axis.

7. The graph of $y = f(x)$ passes through the points $(0, 1)$, $(1, 2)$, and $(2, 3)$. Find the corresponding points on the graph of $y = f(x + 2) - 1$.

8. Prove that the product of two odd functions is an even function, and that the product of two even functions is an even function.

9. Use examples to hypothesize whether the product of an odd function and an even function is even or odd. Then prove your hypothesis.

10. Management originally predicted that the profits from the sales of a new product would be approximated by the graph of the function f shown. The actual profits are shown by the function g along with a verbal description. Use the concepts of transformations of graphs to write g in terms of f.

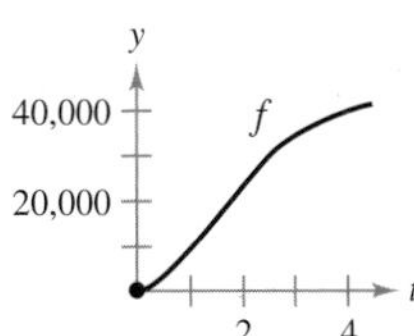

(a) The profits were only three-fourths as large as expected.

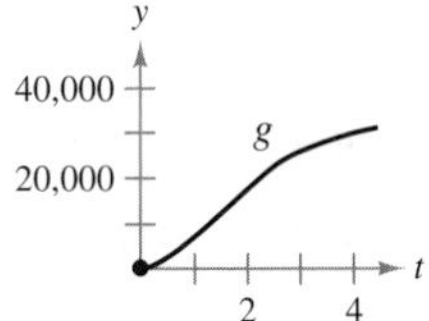

(b) The profits were consistently $10,000 greater than predicted.

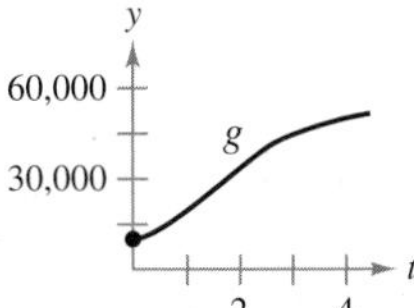

(c) There was a two-year delay in the introduction of the product. After sales began, profits grew as expected.

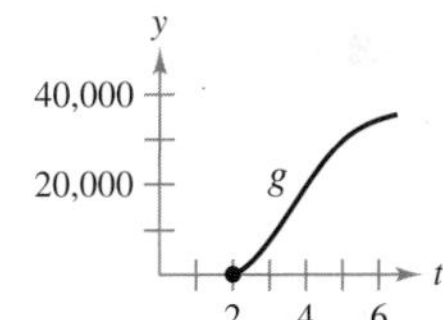

In Exercises 11–14, use the graph of the function f to create a table of values for the given points. Then create a second table that can be used to find f^{-1} and sketch the graph of f^{-1} if possible.

11.

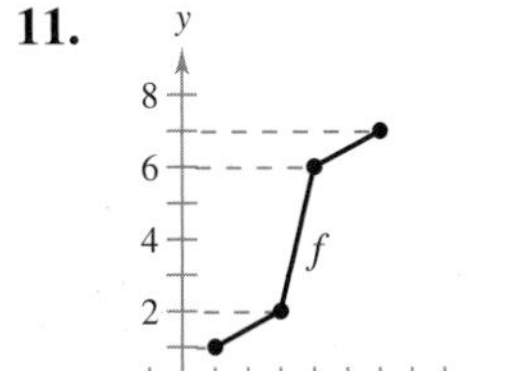

12.

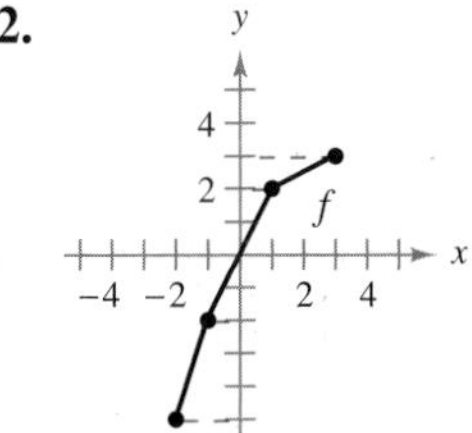

13.

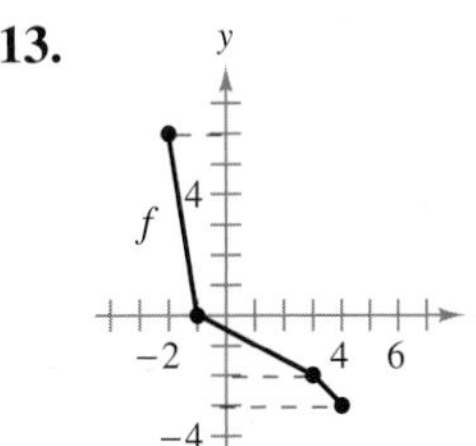

14.

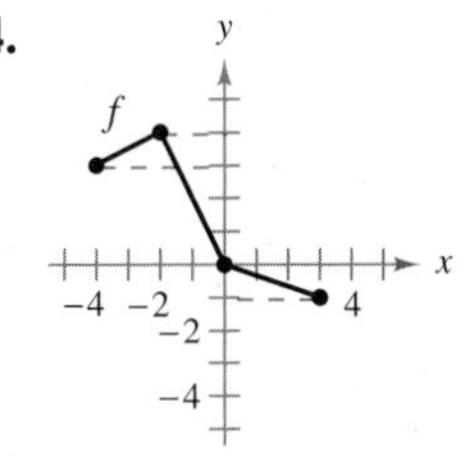

15. The function given by

$$f(x) = k(2 - x - x^3)$$

has an inverse function, and $f^{-1}(3) = -2$. Find k.

16. You are in a boat 2 miles from the nearest point on the coast. You are to travel to a point Q, 3 miles down the coast and 1 mile inland (see figure). You can row at 2 miles per hour and you can walk at 4 miles per hour.

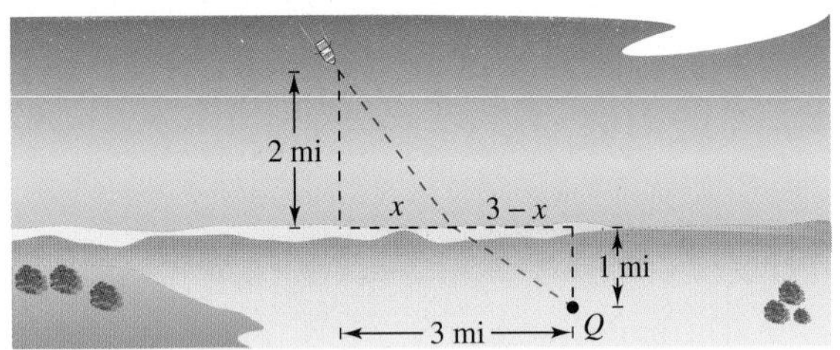

(a) Write the total time T of the trip as a function of x.

(b) Determine the domain of the function.

(c) Use a graphing utility to graph the function. Be sure to choose an appropriate viewing window.

(d) Use the *zoom* and *trace* features to find the value of x that minimizes T.

(e) Write a brief paragraph interpreting these values.

17. The **Heaviside function** $H(x)$ is widely used in engineering applications. (See figure.) To print an enlarged copy of the graph, go to the website *www.mathgraphs.com.*

$$H(x) = \begin{cases} 1, & x \geq 0 \\ 0, & x < 0 \end{cases}$$

Sketch the graph of each function by hand.

(a) $H(x) - 2$ (b) $H(x - 2)$

(c) $-H(x)$ (d) $H(-x)$

(e) $\frac{1}{2}H(x)$ (f) $-H(x - 2) + 2$

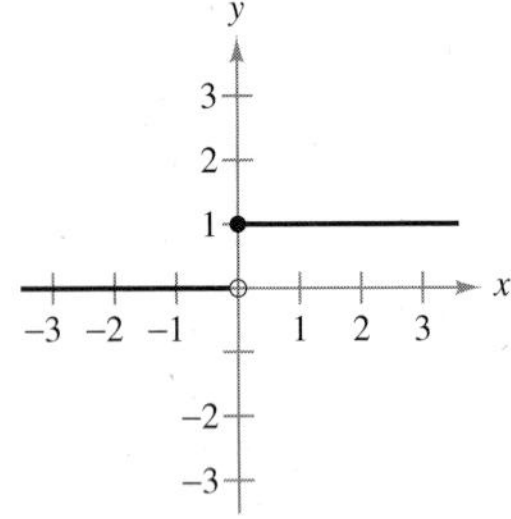

18. Let $f(x) = \dfrac{1}{1 - x}$.

(a) What are the domain and range of f?

(b) Find $f(f(x))$. What is the domain of this function?

(c) Find $f(f(f(x)))$. Is the graph a line? Why or why not?

19. Show that the Associative Property holds for compositions of functions—that is,

$$(f \circ (g \circ h))(x) = ((f \circ g) \circ h)(x).$$

20. Consider the graph of the function f shown in the figure. Use this graph to sketch the graph of each function. To print an enlarged copy of the graph, go to the website *www.mathgraphs.com.*

(a) $f(x + 1)$ (b) $f(x) + 1$ (c) $2f(x)$

(d) $f(-x)$ (e) $-f(x)$

(f) $|f(x)|$ (g) $f(|x|)$

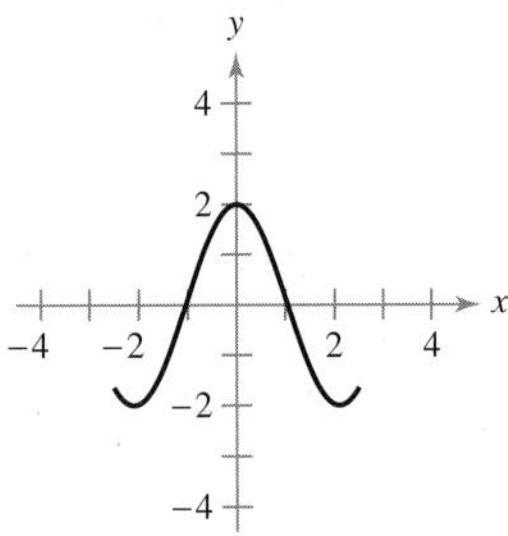

21. Use the graphs of f and f^{-1} to complete each table of function values.

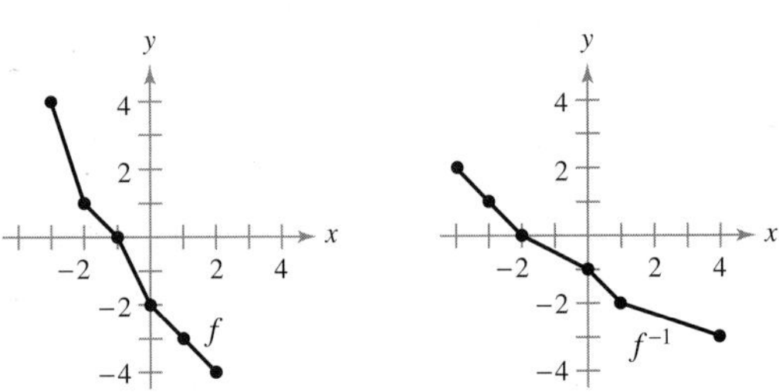

(a)

x	-4	-2	0	4
$f(f^{-1}(x))$				

(b)

x	-3	-2	0	1
$(f + f^{-1})(x)$				

(c)

x	-3	-2	0	1
$(f \cdot f^{-1})(x)$				

(d)

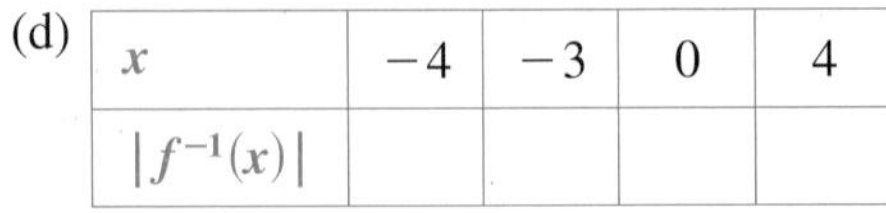

x	-4	-3	0	4
$\lvert f^{-1}(x) \rvert$				

2 Polynomial and Rational Functions

In this chapter, you will continue to study concepts that will help you prepare for your study of calculus. These concepts include analyzing and sketching graphs of polynomial and rational functions. It is important to know these concepts before moving on to calculus.

In this chapter, you should learn the following.

- How to analyze and sketch graphs of quadratic functions. (2.1)
- How to analyze and sketch graphs of polynomial functions of higher degree. (2.2)
- How to divide polynomials. (2.3)
- How to perform operations with complex numbers and find complex solutions of quadratic equations. (2.4)
- How to find zeros of polynomial functions. (2.5)
- How to analyze and sketch graphs of rational functions. (2.6)

Michael Newman / PhotoEdit

Given a polynomial function that models the per capita cigarette consumption in the United States, how can you determine whether the addition of cigarette warnings affected consumption? (See Section 2.1, Exercise 91.)

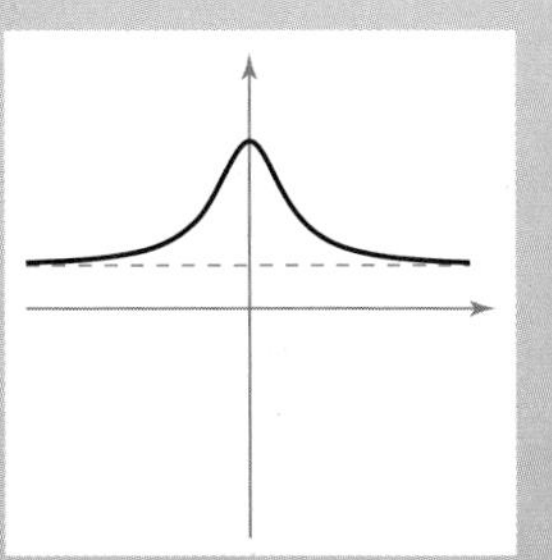

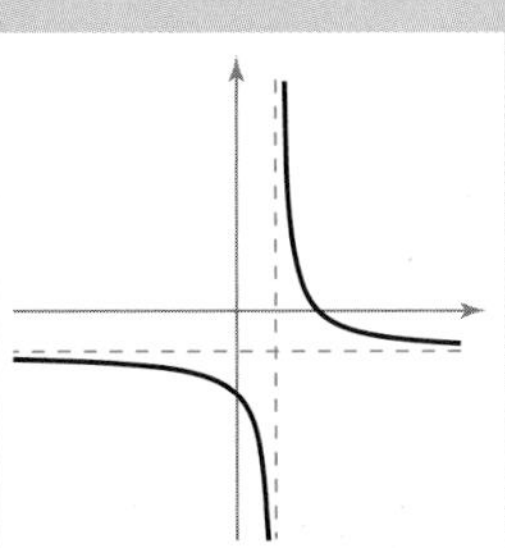

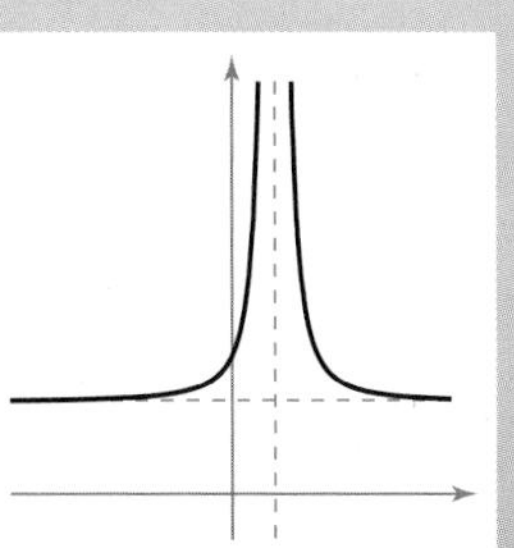

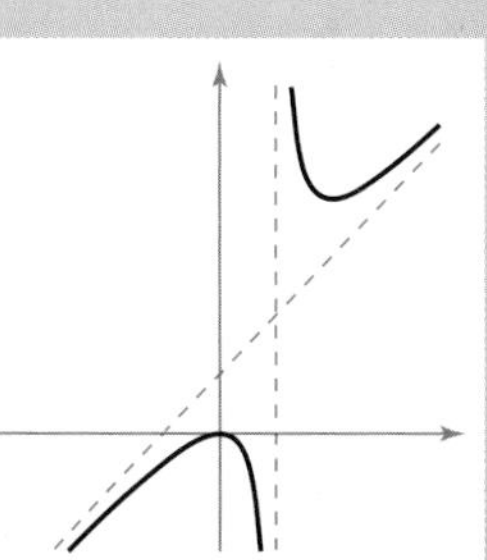

If you move far enough along a curve of the graph of a rational function, there is a straight line that you will increasingly approach but never cross or touch. This line is called an asymptote. (See Section 2.6.)

2.1 Quadratic Functions and Models

- Analyze graphs of quadratic functions.
- Write quadratic functions in standard form and use the results to sketch graphs of quadratic functions.
- Find minimum and maximum values of quadratic functions in real-life applications.

The Graph of a Quadratic Function

In this and the next section, you will study the graphs of polynomial functions. In Chapter 1, you were introduced to the following basic functions.

$f(x) = ax + b$ Linear function

$f(x) = c$ Constant function

$f(x) = x^2$ Squaring function

These functions are examples of **polynomial functions.**

DEFINITION OF POLYNOMIAL FUNCTION

Let n be a nonnegative integer and let $a_n, a_{n-1}, \ldots, a_2, a_1, a_0$ be real numbers with $a_n \neq 0$. The function given by

$$f(x) = a_n x^n + a_{n-1}x^{n-1} + \cdots + a_2x^2 + a_1x + a_0$$

is called a **polynomial function of x with degree n.**

Polynomial functions are classified by degree. For instance, a constant function $f(x) = c$ with $c \neq 0$ has degree 0, and a linear function $f(x) = ax + b$ with $a \neq 0$ has degree 1. In this section, you will study second-degree polynomial functions, which are called **quadratic functions.**

For instance, each of the following functions is a quadratic function.

$$f(x) = x^2 + 6x + 2$$
$$g(x) = 2(x + 1)^2 - 3$$
$$h(x) = 9 + \tfrac{1}{4}x^2$$
$$k(x) = -3x^2 + 4$$
$$m(x) = (x - 2)(x + 1)$$

Note that the squaring function is a simple quadratic function that has degree 2.

DEFINITION OF QUADRATIC FUNCTION

Let a, b, and c be real numbers with $a \neq 0$. The function given by

$$f(x) = ax^2 + bx + c$$

is called a **quadratic function.**

The graph of a quadratic function is a special type of "U"-shaped curve called a **parabola.** Parabolas occur in many real-life applications—especially those involving reflective properties of satellite dishes and flashlight reflectors. You will study these properties in Section 12.1.

All parabolas are symmetric with respect to a line called the **axis of symmetry,** or simply the **axis** of the parabola. The point where the axis intersects the parabola is the **vertex** of the parabola, as shown in Figure 2.1. When $a > 0$, the graph of

$$f(x) = ax^2 + bx + c$$

is a parabola that opens upward. When $a < 0$, the graph of

$$f(x) = ax^2 + bx + c$$

is a parabola that opens downward.

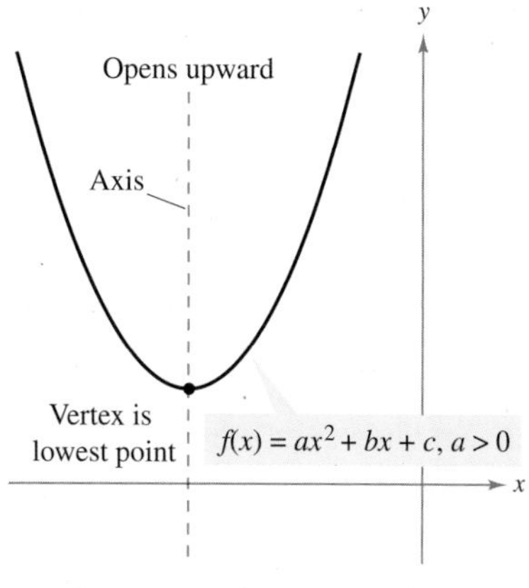

$a > 0$

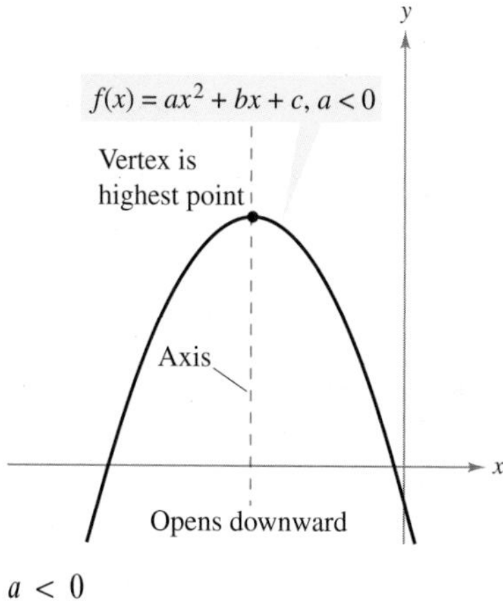

$a < 0$

Figure 2.1

The simplest type of quadratic function is

$$f(x) = ax^2.$$

Its graph is a parabola whose vertex is $(0, 0)$. When $a > 0$, the vertex is the point with the *minimum* y-value on the graph, and when $a < 0$, the vertex is the point with the *maximum* y-value on the graph, as shown in Figure 2.2.

NOTE A precise definition of the terms *minimum* and *maximum* will be given in Section 5.1.

EXPLORATION

Graph $y = ax^2$ for $a = -2, -1, -0.5, 0.5, 1,$ and 2. How does changing the value of a affect the graph?

Graph $y = (x - h)^2$ for $h = -4, -2, 2,$ and 4. How does changing the value of h affect the graph?

Graph $y = x^2 + k$ for $k = -4, -2, 2,$ and 4. How does changing the value of k affect the graph?

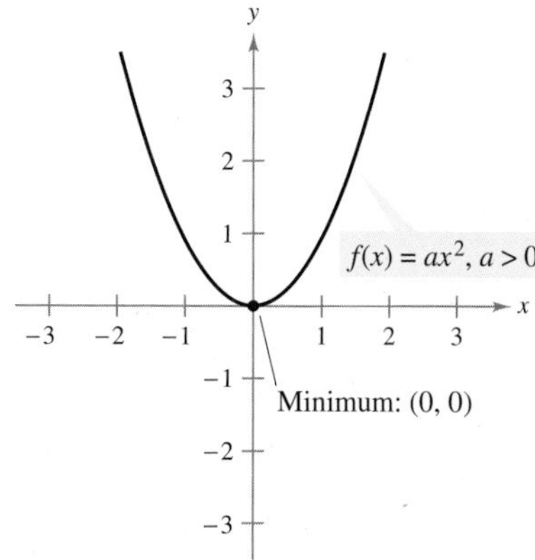

Minimum occurs at vertex.

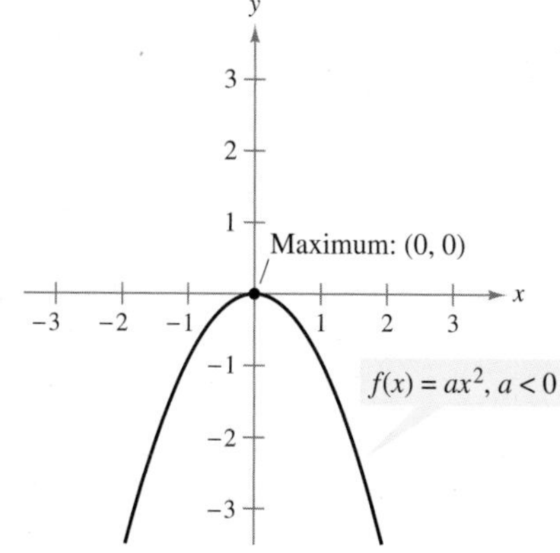

Maximum occurs at vertex.

Figure 2.2

Recall from Section 1.3 that the graphs of $y = f(x \pm c)$, $y = f(x) \pm c$, $y = f(-x)$, and $y = -f(x)$ are rigid transformations of the graph of $y = f(x)$ because they do not change the basic shape of the graph. The graph of $y = af(x)$ is a nonrigid transformation, provided $a \neq \pm 1$.

EXAMPLE 1 Sketching Graphs of Quadratic Functions

Sketch the graph of each function and compare it with the graph of $y = x^2$.

a. $f(x) = -x^2 + 1$

b. $g(x) = (x + 2)^2 - 3$

c. $h(x) = \frac{1}{3}x^2$

d. $k(x) = 2x^2$

Solution

a. To obtain the graph of $f(x) = -x^2 + 1$, reflect the graph of $y = x^2$ in the x-axis. Then shift the graph upward one unit, as shown in Figure 2.3(a).

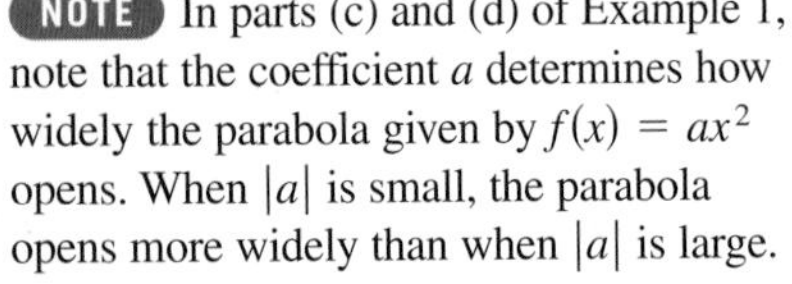
NOTE In parts (c) and (d) of Example 1, note that the coefficient a determines how widely the parabola given by $f(x) = ax^2$ opens. When $|a|$ is small, the parabola opens more widely than when $|a|$ is large.

b. To obtain the graph of $g(x) = (x + 2)^2 - 3$, shift the graph of $y = x^2$ two units to the left and three units downward, as shown in Figure 2.3(b).

c. Compared with $y = x^2$, each output of $h(x) = \frac{1}{3}x^2$ "shrinks" by a factor of $\frac{1}{3}$, creating the broader parabola shown in Figure 2.3(c).

d. Compared with $y = x^2$, each output of $k(x) = 2x^2$ "stretches" by a factor of 2, creating the narrower parabola shown in Figure 2.3(d).

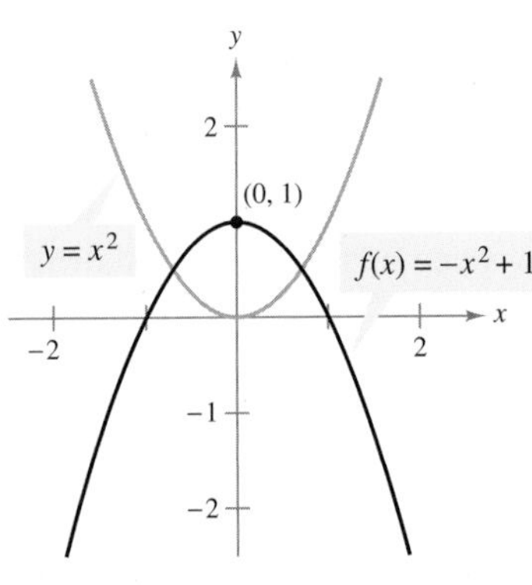

(a)

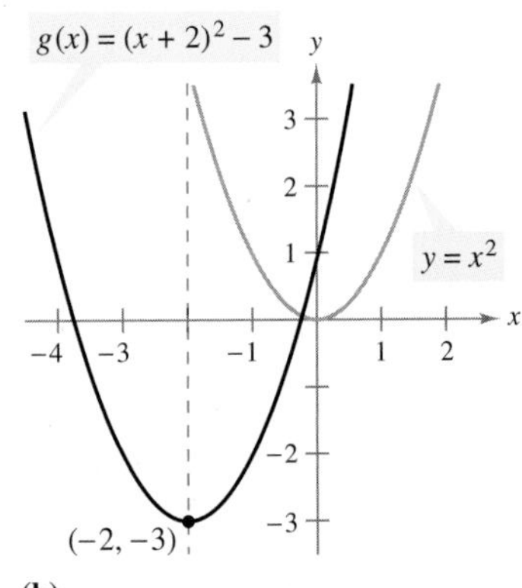

(b)

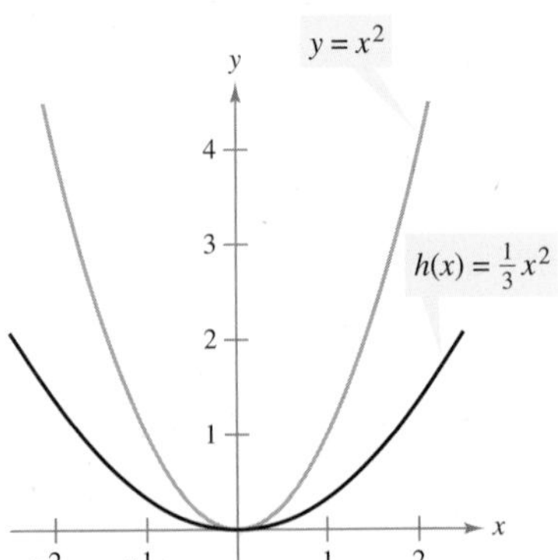

(c)

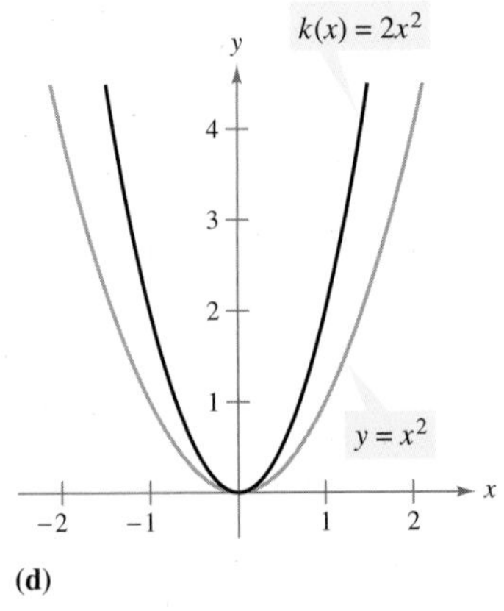

(d)

Figure 2.3

■

The Standard Form of a Quadratic Function

STUDY TIP The standard form of a quadratic function identifies four basic transformations of the graph of $y = x^2$.

a. The factor $|a|$ produces a vertical stretch or shrink.

b. When $a < 0$, the graph is reflected in the x-axis.

c. The factor $(x - h)^2$ represents a horizontal shift of h units.

d. The term k represents a vertical shift of k units.

The **standard form** of a quadratic function is $f(x) = a(x - h)^2 + k$. This form is especially convenient because it identifies the vertex of the parabola as (h, k).

> **STANDARD FORM OF A QUADRATIC FUNCTION**
>
> The quadratic function given by
>
> $$f(x) = a(x - h)^2 + k, \quad a \neq 0$$
>
> is in **standard form.** The graph of f is a parabola whose axis is the vertical line $x = h$ and whose vertex is the point (h, k). When $a > 0$, the parabola opens upward, and when $a < 0$, the parabola opens downward.

To graph a parabola, it is helpful to begin by writing the quadratic function in standard form using the process of completing the square, as illustrated in Example 2. In this example, notice that to complete the square, you *add and subtract* the square of half the coefficient of x within the parentheses instead of adding the value to each side of the equation as is done in Section P.1.

EXAMPLE 2 Using Standard Form to Graph a Parabola

Sketch the graph of

$$f(x) = 2x^2 + 8x + 7$$

and identify the vertex and the axis of the parabola.

Solution Begin by writing the quadratic function in standard form. Notice that the first step in completing the square is to factor out any coefficient of x^2 that is not 1.

$$\begin{aligned} f(x) &= 2x^2 + 8x + 7 && \text{Write original function.} \\ &= 2(x^2 + 4x) + 7 && \text{Factor 2 out of } x\text{-terms.} \\ &= 2(x^2 + 4x + 4 - 4) + 7 && \text{Add and subtract 4 within parentheses.} \end{aligned}$$

$(4/2)^2$

After adding and subtracting 4 within the parentheses, you must now regroup the terms to form a perfect square trinomial. The -4 can be removed from inside the parentheses; however, because of the 2 outside of the parentheses, you must multiply -4 by 2, as shown below.

$$\begin{aligned} f(x) &= 2(x^2 + 4x + 4) - 2(4) + 7 && \text{Regroup terms.} \\ &= 2(x^2 + 4x + 4) - 8 + 7 && \text{Simplify.} \\ &= 2(x + 2)^2 - 1 && \text{Write in standard form.} \end{aligned}$$

From this form, you can see that the graph of f is a parabola that opens upward and has its vertex at

$$(h, k) = (-2, -1).$$

This corresponds to a left shift of two units and a downward shift of one unit relative to the graph of $y = 2x^2$, as shown in Figure 2.4. In the figure, you can see that the axis of the parabola is the vertical line through the vertex, $x = -2$. ■

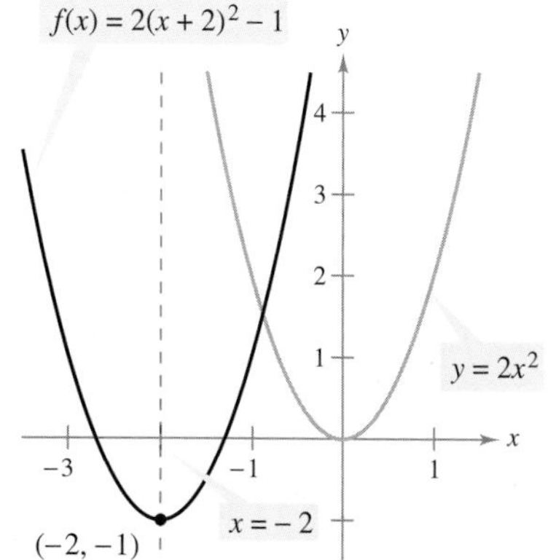

Figure 2.4

To find the x-intercepts of the graph of $f(x) = ax^2 + bx + c$, you can solve the equation $ax^2 + bx + c = 0$. When $ax^2 + bx + c$ does not factor, you can use the Quadratic Formula to find the x-intercepts. Remember, however, that a parabola may not have x-intercepts.

EXAMPLE 3 Finding the Vertex and x-Intercepts of a Parabola

Sketch the graph of

$$f(x) = -x^2 + 6x - 8$$

and identify the vertex and x-intercepts.

Solution

$$\begin{aligned} f(x) &= -x^2 + 6x - 8 && \text{Write original function.} \\ &= -(x^2 - 6x) - 8 && \text{Factor } -1 \text{ out of } x\text{-terms.} \\ &= -(x^2 - 6x + 9 - 9) - 8 && \text{Add and subtract 9 within parentheses.} \\ & \qquad\qquad (-6/2)^2 \\ &= -(x^2 - 6x + 9) - (-9) - 8 && \text{Regroup terms.} \\ &= -(x - 3)^2 + 1 && \text{Write in standard form.} \end{aligned}$$

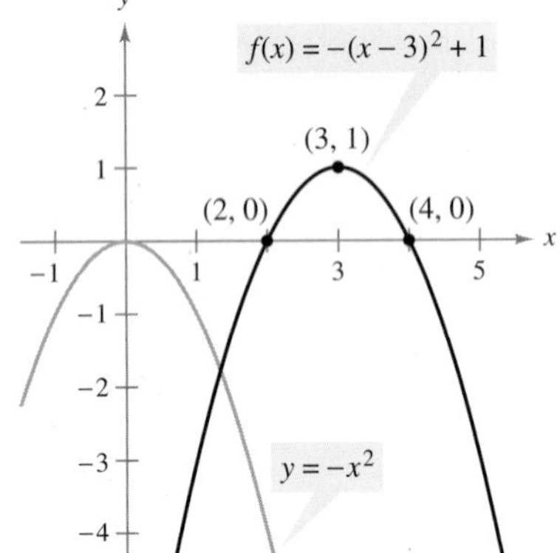

Figure 2.5

From this form, you can see that f is a parabola that opens downward with vertex $(3, 1)$. The x-intercepts of the graph are determined as follows.

$$\begin{aligned} -x^2 + 6x - 8 &= 0 && \text{Write original equation.} \\ -(x^2 - 6x + 8) &= 0 && \text{Factor out } -1. \\ -(x - 2)(x - 4) &= 0 && \text{Factor.} \\ x - 2 &= 0 \;\Rightarrow\; x = 2 && \text{Set 1st factor equal to 0.} \\ x - 4 &= 0 \;\Rightarrow\; x = 4 && \text{Set 2nd factor equal to 0.} \end{aligned}$$

So, the x-intercepts are $(2, 0)$ and $(4, 0)$, as shown in Figure 2.5.

EXAMPLE 4 Writing the Equation of a Parabola

Write the standard form of the equation of the parabola whose vertex is $(1, 2)$ and that passes through the point $(3, -6)$.

Solution Because the vertex of the parabola is at $(h, k) = (1, 2)$, the equation has the form

$$f(x) = a(x - 1)^2 + 2. \qquad \text{Substitute for } h \text{ and } k \text{ in standard form.}$$

Because the parabola passes through the point $(3, -6)$, it follows that $f(3) = -6$. So,

$$\begin{aligned} f(x) &= a(x - 1)^2 + 2 && \text{Write in standard form.} \\ -6 &= a(3 - 1)^2 + 2 && \text{Substitute 3 for } x \text{ and } -6 \text{ for } f(x). \\ -6 &= 4a + 2 && \text{Simplify.} \\ -8 &= 4a && \text{Subtract 2 from each side.} \\ -2 &= a. && \text{Divide each side by 4.} \end{aligned}$$

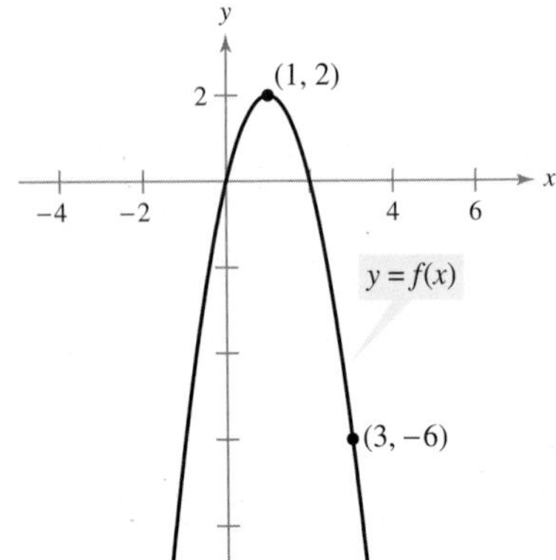

Figure 2.6

The equation in standard form is

$$f(x) = -2(x - 1)^2 + 2.$$

The graph of f is shown in Figure 2.6. ■

Finding Minimum and Maximum Values

Many applications involve finding the maximum or minimum value of a quadratic function. By completing the square of the quadratic function $f(x) = ax^2 + bx + c$, you can rewrite the function in standard form (see Exercise 73).

$$f(x) = a\left(x + \frac{b}{2a}\right)^2 + \left(c - \frac{b^2}{4a}\right) \qquad \text{Standard form}$$

So, the vertex of the graph of f is $\left(-\frac{b}{2a}, f\left(-\frac{b}{2a}\right)\right)$, which implies the following.

MINIMUM AND MAXIMUM VALUES OF QUADRATIC FUNCTIONS

Consider the function $f(x) = ax^2 + bx + c$ with vertex $\left(-\frac{b}{2a}, f\left(-\frac{b}{2a}\right)\right)$.

1. When $a > 0$, f has a *minimum* at $x = -\frac{b}{2a}$. The minimum value is $f\left(-\frac{b}{2a}\right)$.

2. When $a < 0$, f has a *maximum* at $x = -\frac{b}{2a}$. The maximum value is $f\left(-\frac{b}{2a}\right)$.

EXAMPLE 5 The Maximum Height of a Baseball

A baseball is hit at a point 3 feet above the ground at a velocity of 100 feet per second and at an angle of 45° with respect to the ground. The path of the baseball is given by the function $f(x) = -0.0032x^2 + x + 3$, where $f(x)$ is the height of the baseball (in feet) and x is the horizontal distance from home plate (in feet). What is the maximum height reached by the baseball?

Algebraic Solution

For this quadratic function, you have

$$\begin{aligned} f(x) &= ax^2 + bx + c \\ &= -0.0032x^2 + x + 3 \end{aligned}$$

which implies that $a = -0.0032$ and $b = 1$. Because $a < 0$, the function has a maximum when $x = -b/(2a)$. So, you can conclude that the baseball reaches its maximum height when it is x feet from home plate, where x is

$$\begin{aligned} x &= -\frac{b}{2a} \\ &= -\frac{1}{2(-0.0032)} \\ &= 156.25 \text{ feet.} \end{aligned}$$

At this distance, the maximum height is

$$\begin{aligned} f(156.25) &= -0.0032(156.25)^2 + 156.25 + 3 \\ &= 81.125 \text{ feet.} \end{aligned}$$

Graphical Solution

Use a graphing utility to graph

$$y = -0.0032x^2 + x + 3$$

so that you can see the important features of the parabola. Use the *maximum* feature (see Figure 2.7) or the *zoom* and *trace* features (see Figure 2.8) of the graphing utility to approximate the maximum height on the graph to be $y \approx 81.125$ feet at $x \approx 156.25$.

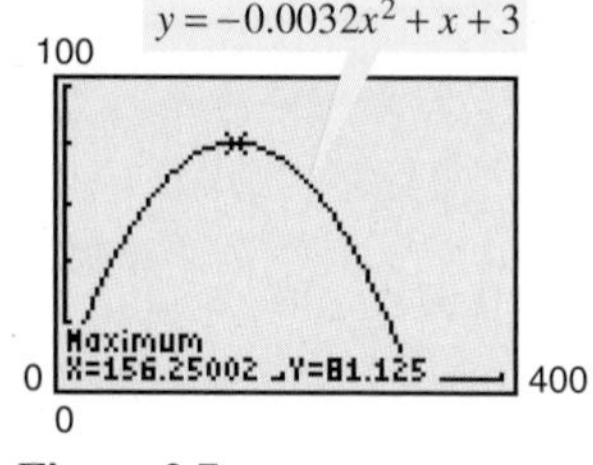

Figure 2.7

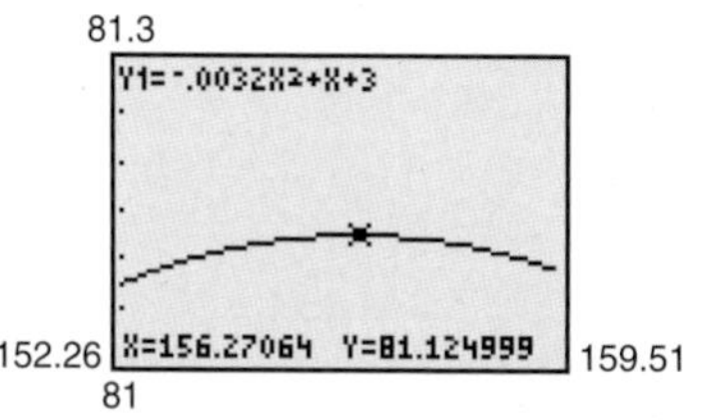

Figure 2.8

■

2.1 Exercises

See www.CalcChat.com for worked-out solutions to odd-numbered exercises.

In Exercises 1–6, fill in the blanks.

1. Linear, constant, and squaring functions are examples of ________ functions.
2. A polynomial function of degree n and leading coefficient a_n is a function of the form
$f(x) = a_nx^n + a_{n-1}x^{n-1} + \cdot \cdot \cdot + a_1x + a_0 \quad (a_n \neq 0)$
where n is a ________ ________ and $a_n, a_{n-1}, \ldots, a_1, a_0$ are ________ numbers.
3. A ________ function is a second-degree polynomial function, and its graph is called a ________.
4. The graph of a quadratic function is symmetric about its ________.
5. If the graph of a quadratic function opens upward, then its leading coefficient is ________ and the vertex of the graph is a ________.
6. If the graph of a quadratic function opens downward, then its leading coefficient is ________ and the vertex of the graph is a ________.

In Exercises 7–12, match the quadratic function with its graph. [The graphs are labeled (a), (b), (c), (d), (e), and (f).]

(a)
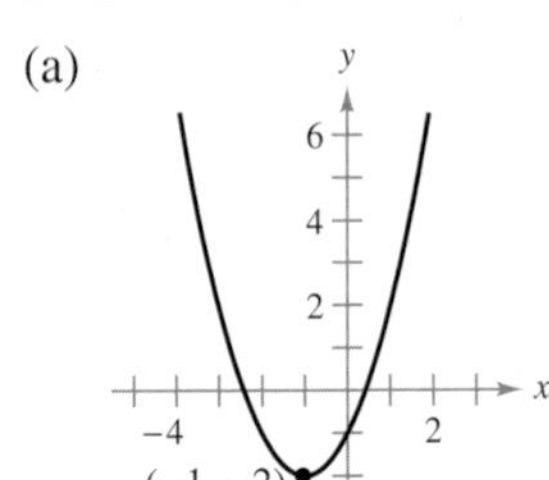

(b)
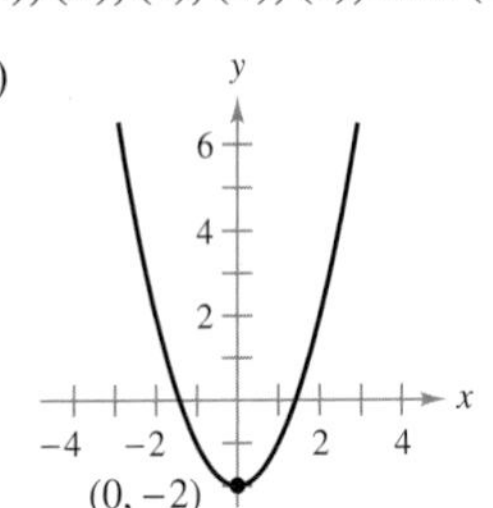

(c)
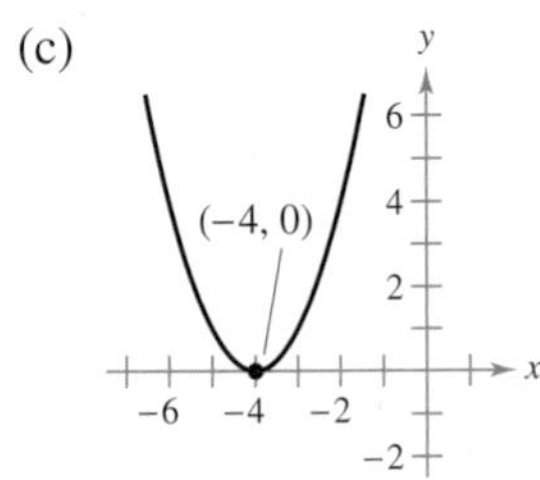

(d)
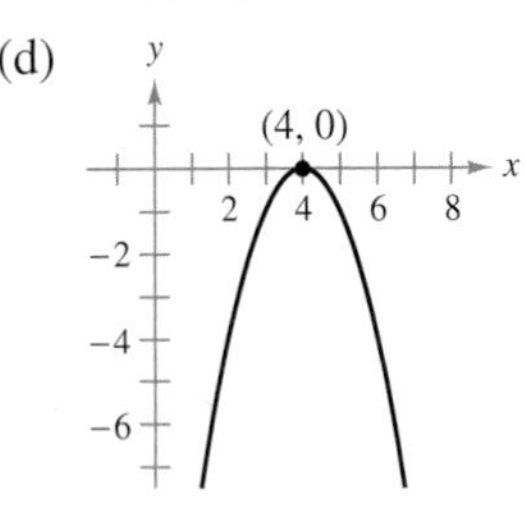

(e)
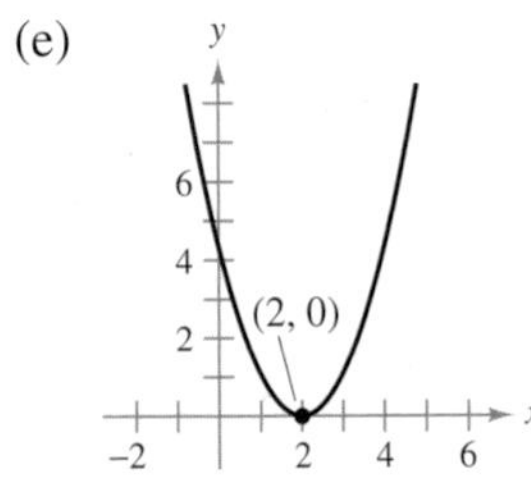

(f)
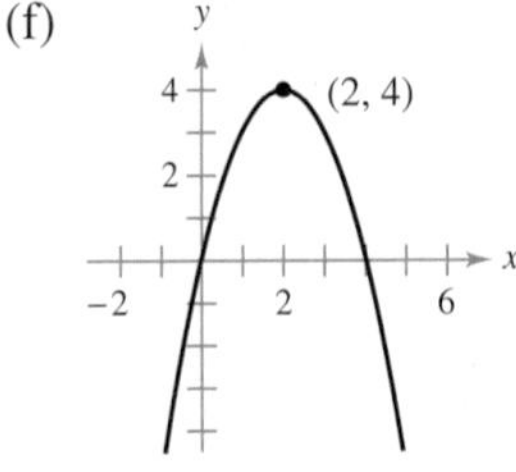

7. $f(x) = (x - 2)^2$
8. $f(x) = (x + 4)^2$
9. $f(x) = x^2 - 2$
10. $f(x) = (x + 1)^2 - 2$
11. $f(x) = 4 - (x - 2)^2$
12. $f(x) = -(x - 4)^2$

In Exercises 13–16, graph each function. Compare the graph of each function with the graph of $y = x^2$.

13. (a) $f(x) = \frac{1}{2}x^2$ (b) $g(x) = -\frac{1}{8}x^2$
(c) $h(x) = \frac{3}{2}x^2$ (d) $k(x) = -3x^2$
14. (a) $f(x) = x^2 + 1$ (b) $g(x) = x^2 - 1$
(c) $h(x) = x^2 + 3$ (d) $k(x) = x^2 - 3$
15. (a) $f(x) = (x - 1)^2$
(b) $g(x) = (3x)^2 + 1$
(c) $h(x) = \left(\frac{1}{3}x\right)^2 - 3$
(d) $k(x) = (x + 3)^2$
16. (a) $f(x) = -\frac{1}{2}(x - 2)^2 + 1$
(b) $g(x) = \left[\frac{1}{2}(x - 1)\right]^2 - 3$
(c) $h(x) = -\frac{1}{2}(x + 2)^2 - 1$
(d) $k(x) = [2(x + 1)]^2 + 4$

In Exercises 17–34, sketch the graph of the quadratic function without using a graphing utility. Identify the vertex, axis of symmetry, and x-intercept(s).

17. $f(x) = 1 - x^2$
18. $g(x) = x^2 - 8$
19. $f(x) = x^2 + 7$
20. $h(x) = 12 - x^2$
21. $f(x) = \frac{1}{2}x^2 - 4$
22. $f(x) = 16 - \frac{1}{4}x^2$
23. $f(x) = (x + 4)^2 - 3$
24. $f(x) = (x - 6)^2 + 8$
25. $h(x) = x^2 - 8x + 16$
26. $g(x) = x^2 + 2x + 1$
27. $f(x) = x^2 - x + \frac{5}{4}$
28. $f(x) = x^2 + 3x + \frac{1}{4}$
29. $f(x) = -x^2 + 2x + 5$
30. $f(x) = -x^2 - 4x + 1$
31. $h(x) = 4x^2 - 4x + 21$
32. $f(x) = 2x^2 - x + 1$
33. $f(x) = \frac{1}{4}x^2 - 2x - 12$
34. $f(x) = -\frac{1}{3}x^2 + 3x - 6$

In Exercises 35–42, use a graphing utility to graph the quadratic function. Identify the vertex, axis of symmetry, and x-intercepts. Then check your results analytically by writing the quadratic function in standard form.

35. $f(x) = -(x^2 + 2x - 3)$
36. $f(x) = -(x^2 + x - 30)$
37. $g(x) = x^2 + 8x + 11$
38. $f(x) = x^2 + 10x + 14$
39. $f(x) = 2x^2 - 16x + 31$
40. $f(x) = -4x^2 + 24x - 41$
41. $g(x) = \frac{1}{2}(x^2 + 4x - 2)$
42. $f(x) = \frac{3}{5}(x^2 + 6x - 5)$

In Exercises 43–46, write an equation for the parabola in standard form.

43.

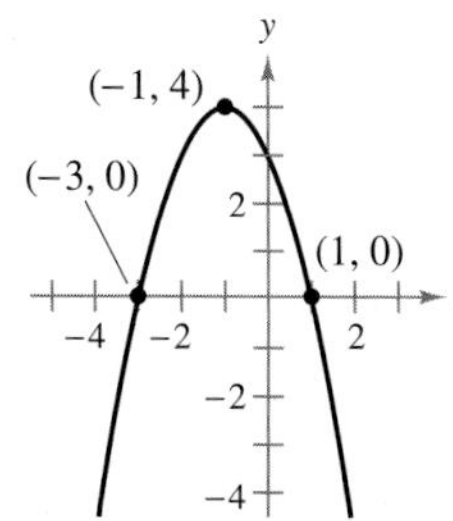

44.

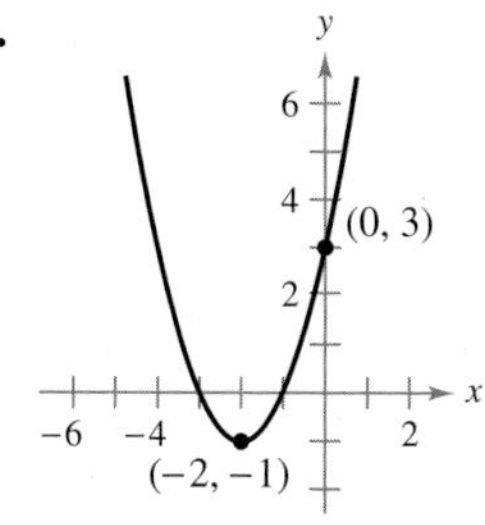

45.

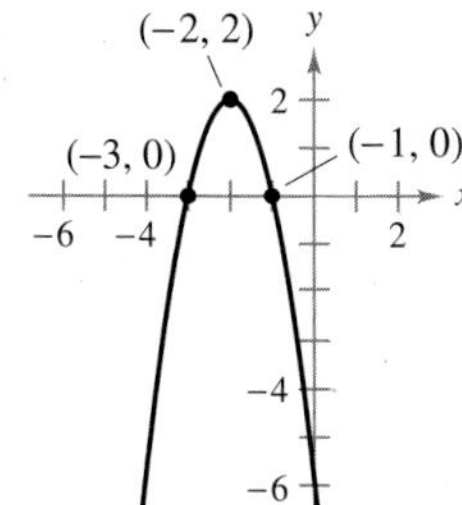

46.

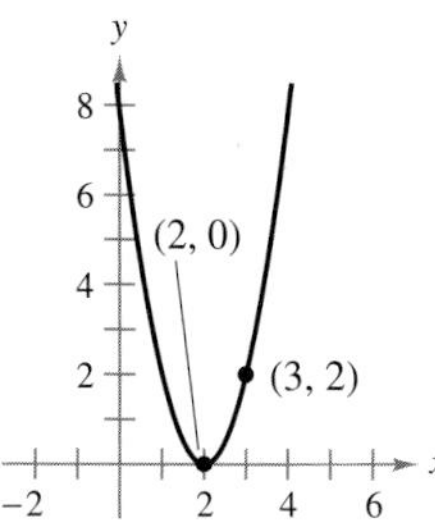

In Exercises 47–56, write the standard form of the equation of the parabola that has the indicated vertex and whose graph passes through the given point.

47. Vertex: $(-2, 5)$; point: $(0, 9)$

48. Vertex: $(4, -1)$; point: $(2, 3)$

49. Vertex: $(1, -2)$; point: $(-1, 14)$

50. Vertex: $(2, 3)$; point: $(0, 2)$

51. Vertex: $(5, 12)$; point: $(7, 15)$

52. Vertex: $(-2, -2)$; point: $(-1, 0)$

53. Vertex: $\left(-\frac{1}{4}, \frac{3}{2}\right)$; point: $(-2, 0)$

54. Vertex: $\left(\frac{5}{2}, -\frac{3}{4}\right)$; point: $(-2, 4)$

55. Vertex: $\left(-\frac{5}{2}, 0\right)$; point: $\left(-\frac{7}{2}, -\frac{16}{3}\right)$

56. Vertex: $(6, 6)$; point: $\left(\frac{61}{10}, \frac{3}{2}\right)$

In Exercises 57–62, use a graphing utility to graph the quadratic function. Find the x-intercepts of the graph and compare them with the solutions of the corresponding quadratic equation when $f(x) = 0$.

57. $f(x) = x^2 - 4x$

58. $f(x) = -2x^2 + 10x$

59. $f(x) = x^2 - 9x + 18$

60. $f(x) = x^2 - 8x - 20$

61. $f(x) = 2x^2 - 7x - 30$

62. $f(x) = \frac{7}{10}(x^2 + 12x - 45)$

In Exercises 63–68, find two quadratic functions, one that opens upward and one that opens downward, whose graphs have the given x-intercepts. (There are many correct answers.)

63. $(-1, 0), (3, 0)$

64. $(-5, 0), (5, 0)$

65. $(0, 0), (10, 0)$

66. $(4, 0), (8, 0)$

67. $(-3, 0), \left(-\frac{1}{2}, 0\right)$

68. $\left(-\frac{5}{2}, 0\right), (2, 0)$

WRITING ABOUT CONCEPTS

In Exercises 69–72, (a) determine the x-intercepts, if any, of the graph visually, (b) explain how the x-intercepts relate to the solutions of the quadratic equation when $f(x) = 0$, and (c) find the x-intercepts analytically to confirm your results.

69. $y = x^2 - 4x - 5$

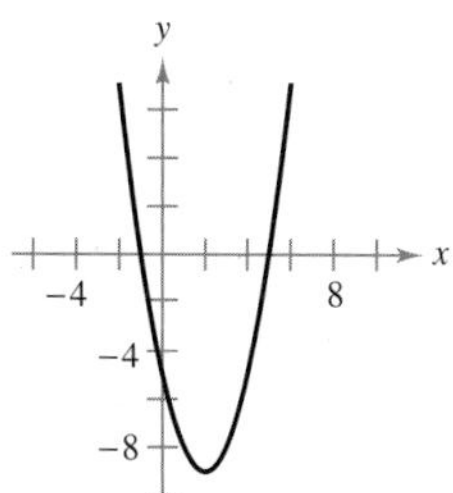

70. $y = 2x^2 + 5x - 3$

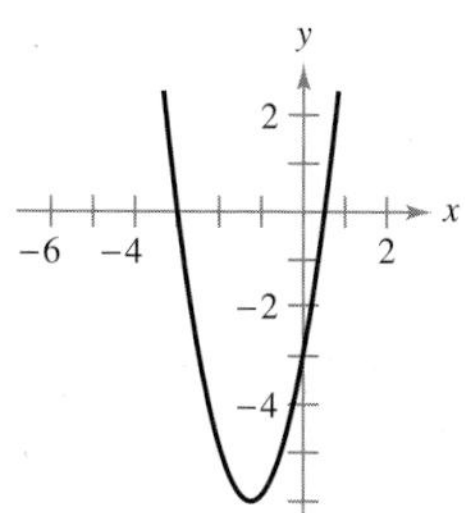

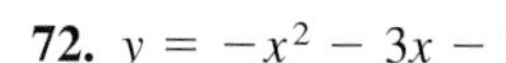

71. $y = -x^2 - 2x - 1$

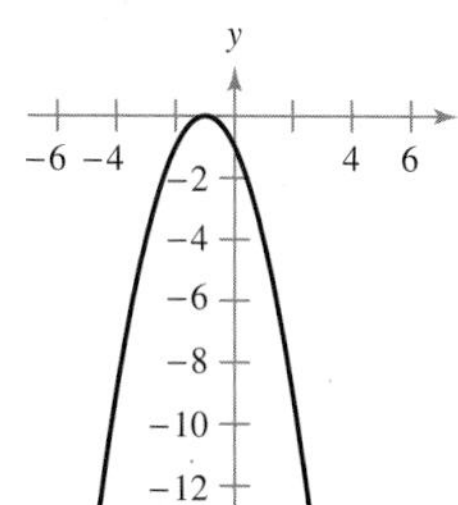

72. $y = -x^2 - 3x - 3$

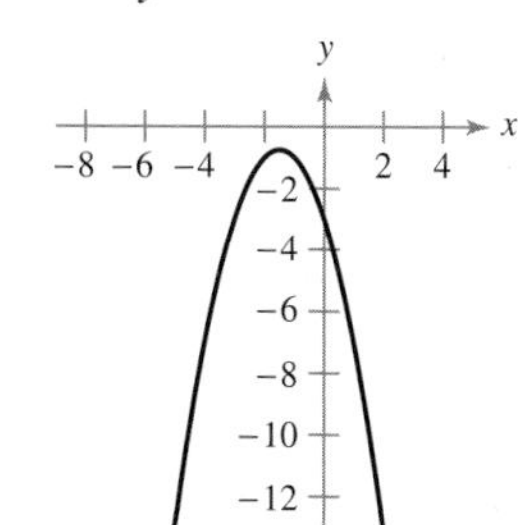

73. Write the quadratic function $f(x) = ax^2 + bx + c$ in standard form to verify that the vertex occurs at

$$\left(-\frac{b}{2a}, f\left(-\frac{b}{2a}\right)\right).$$

74. (a) Is it possible for the graph of a quadratic equation to have only one x-intercept? Explain.

(b) Is it possible for the graph of a quadratic equation to have no x-intercepts? Explain.

In Exercises 75–78, find two positive real numbers whose product is a maximum.

75. The sum is 110.

76. The sum is S.

77. The sum of the first and twice the second is 24.

78. The sum of the first and three times the second is 42.

Geometry **In Exercises 79 and 80, consider a rectangle of length x and perimeter P. (a) Write the area A as a function of x and determine the domain of the function. (b) Graph the area function. (c) Find the length and width of the rectangle of maximum area.**

79. $P = 100$ feet

80. $P = 36$ meters

81. ***Path of a Diver*** The path of a diver is given by

$$y = -\frac{4}{9}x^2 + \frac{24}{9}x + 12$$

where y is the height (in feet) and x is the horizontal distance from the end of the diving board (in feet). What is the maximum height of the diver?

82. ***Height of a Ball*** The height y (in feet) of a punted football is given by

$$y = -\frac{16}{2025}x^2 + \frac{9}{5}x + 1.5$$

where x is the horizontal distance (in feet) from the point at which the ball is punted.

(a) How high is the ball when it is punted?

(b) What is the maximum height of the punt?

(c) How long is the punt?

83. ***Minimum Cost*** A manufacturer of lighting fixtures has daily production costs of

$$C = 800 - 10x + 0.25x^2$$

where C is the total cost (in dollars) and x is the number of units produced. How many fixtures should be produced each day to yield a minimum cost?

84. ***Maximum Profit*** The profit P (in hundreds of dollars) that a company makes depends on the amount x (in hundreds of dollars) the company spends on advertising according to the model $P = 230 + 20x - 0.5x^2$. What expenditure for advertising will yield a maximum profit?

85. ***Maximum Revenue*** The total revenue R earned (in thousands of dollars) from manufacturing handheld video games is given by

$$R(p) = -25p^2 + 1200p$$

where p is the price per unit (in dollars).

(a) Find the revenues when the price per unit is \$20, \$25, and \$30.

(b) Find the unit price that will yield a maximum revenue. What is the maximum revenue? Explain your results.

86. ***Maximum Revenue*** The total revenue R earned per day (in dollars) from a pet-sitting service is given by

$$R(p) = -12p^2 + 150p$$

where p is the price charged per pet (in dollars).

(a) Find the revenues when the price per pet is \$4, \$6, and \$8.

(b) Find the price that will yield a maximum revenue. What is the maximum revenue? Explain your results.

87. ***Numerical, Graphical, and Analytical Analysis*** A rancher has 200 feet of fencing to enclose two adjacent rectangular corrals (see figure).

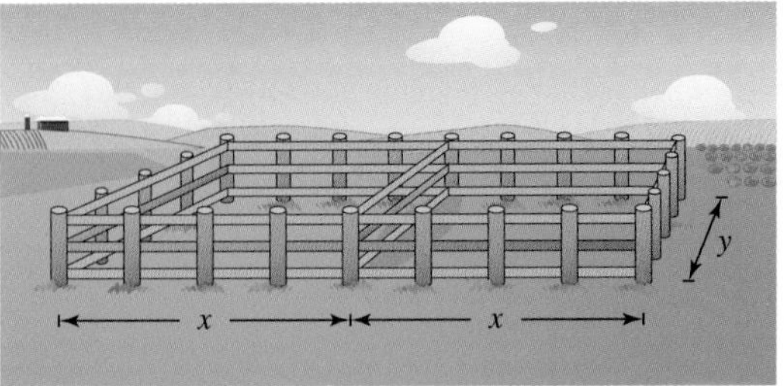

(a) Write the area A of the corrals as a function of x.

(b) Create a table showing possible values of x and the corresponding areas of the corral. Use the table to estimate the dimensions that will produce the maximum enclosed area.

(c) Use a graphing utility to graph the area function. Use the graph to approximate the dimensions that will produce the maximum enclosed area.

(d) Write the area function in standard form to find analytically the dimensions that will produce the maximum area.

(e) Compare your results from parts (b), (c), and (d).

88. ***Geometry*** An indoor physical fitness room consists of a rectangular region with a semicircle on each end. The perimeter of the room is to be a 200-meter single-lane running track.

(a) Draw a diagram that illustrates the problem. Let x and y represent the length and width of the rectangular region, respectively.

(b) Determine the radius of each semicircular end of the room. Determine the distance, in terms of y, around the inside edge of each semicircular part of the track.

(c) Use the result of part (b) to write an equation, in terms of x and y, for the distance traveled in one lap around the track. Solve for y.

(d) Use the result of part (c) to write the area A of the rectangular region as a function of x. What dimensions will produce a rectangle of maximum area?

89. ***Maximum Revenue*** A small theater has a seating capacity of 2000. When the ticket price is \$20, attendance is 1500. For each \$1 decrease in price, attendance increases by 100.

(a) Write the revenue R of the theater as a function of ticket price x.

(b) What ticket price will yield a maximum revenue? What is the maximum revenue?

90. ***Maximum Area*** A Norman window is constructed by adjoining a semicircle to the top of an ordinary rectangular window (see figure). The perimeter of the window is 16 feet.

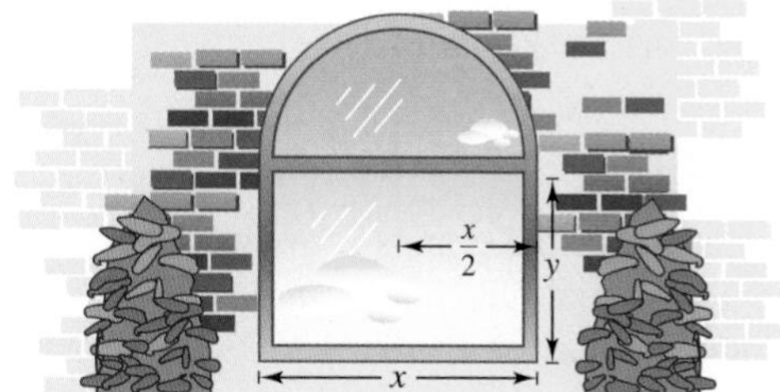

(a) Write the area A of the window as a function of x.

(b) What dimensions will produce a window of maximum area?

91. ***Graphical Analysis*** From 1950 through 2005, the per capita consumption C of cigarettes by Americans (age 18 and older) can be modeled by $C = 3565.0 + 60.30t - 1.783t^2$, $0 \leq t \leq 55$, where t is the year, with $t = 0$ corresponding to 1950. *(Source: Tobacco Outlook Report)*

(a) Use a graphing utility to graph the model.

(b) Use the graph of the model to approximate the maximum average annual consumption. Beginning in 1966, all cigarette packages were required by law to carry a health warning. Do you think the warning had any effect? Explain.

(c) In 2005, the U.S. population (age 18 and over) was 296,329,000. Of those, about 59,858,458 were smokers. What was the average annual cigarette consumption *per smoker* in 2005? What was the average daily cigarette consumption *per smoker*?

92. ***Data Analysis: Sales*** The sales y (in billions of dollars) for Harley-Davidson from 2000 through 2007 are shown in the table. *(Source: U.S. Harley-Davidson, Inc.)*

Year	2000	2001	2002	2003
Sales, S	2.91	3.36	4.09	4.62

Year	2004	2005	2006	2007
Sales, S	5.02	5.34	5.80	5.73

(a) Use a graphing utility to create a scatter plot of the data. Let x represent the year, with $x = 0$ corresponding to 2000.

(b) Use the *regression* feature of the graphing utility to find a quadratic model for the data.

(c) Use the graphing utility to graph the model in the same viewing window as the scatter plot. How well does the model fit the data?

(d) Use the *trace* feature of the graphing utility to approximate the year in which the sales for Harley-Davidson were the greatest.

(e) Verify your answer to part (d) algebraically.

(f) Use the model to predict the sales for Harley-Davidson in 2011.

True or False? **In Exercises 93–96, determine whether the statement is true or false. Justify your answer.**

93. The function given by $f(x) = -12x^2 - 1$ has no x-intercepts.

94. The graphs of $f(x) = -4x^2 - 10x + 7$ and $g(x) = 12x^2 + 30x + 1$ have the same axis of symmetry.

95. The graph of a quadratic function with a negative leading coefficient will have a maximum value at its vertex.

96. The graph of a quadratic function with a positive leading coefficient will have a minimum value at its vertex.

Think About It **In Exercises 97–100, find the values of b such that the function has the given maximum or minimum value.**

97. $f(x) = -x^2 + bx - 75$; Maximum value: 25

98. $f(x) = -x^2 + bx - 16$; Maximum value: 48

99. $f(x) = x^2 + bx + 26$; Minimum value: 10

100. $f(x) = x^2 + bx - 25$; Minimum value: -50

101. Describe the sequence of transformations from f to g given that $f(x) = x^2$ and $g(x) = a(x - h)^2 + k$. (Assume a, h, and k are positive.)

CAPSTONE

102. The profit P (in millions of dollars) for a recreational vehicle retailer is modeled by a quadratic function of the form

$$P = at^2 + bt + c$$

where t represents the year. If you were president of the company, which of the models below would you prefer? Explain your reasoning.

(a) a is positive and $-b/(2a) \leq t$.

(b) a is positive and $t \leq -b/(2a)$.

(c) a is negative and $-b/(2a) \leq t$.

(d) a is negative and $t \leq -b/(2a)$.

103. Assume that the function given by $f(x) = ax^2 + bx + c$, $a \neq 0$ has two real zeros. Show that the x-coordinate of the vertex of the graph is the average of the zeros of f. (*Hint:* Use the Quadratic Formula.)

2.2 Polynomial Functions of Higher Degree

- **Use transformations to sketch graphs of polynomial functions.**
- **Use the Leading Coefficient Test to determine the end behavior of graphs of polynomial functions.**
- **Find and use zeros of polynomial functions as sketching aids.**

Graphs of Polynomial Functions

NOTE A precise definition of the term *continuous* is given in Section 3.4.

In this section, you will study basic features of the graphs of polynomial functions. The first feature is that the graph of a polynomial function is *continuous*. Essentially, this means that the graph of a polynomial function has no breaks, holes, or gaps, as shown in Figure 2.9(a). The graph shown in Figure 2.9(b) is an example of a piecewise-defined function that is not continuous.

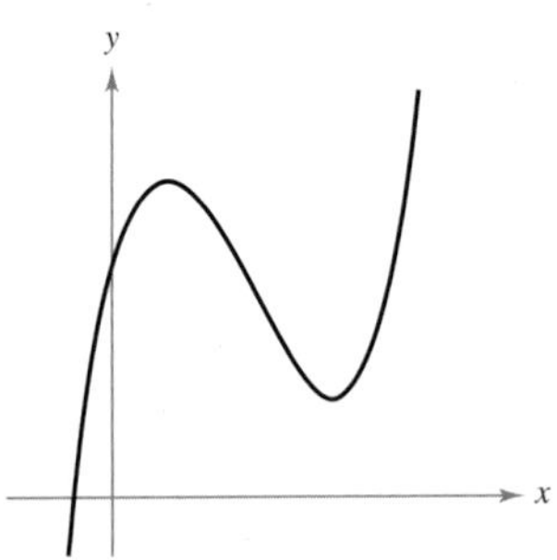

(a) Polynomial functions have continuous graphs.

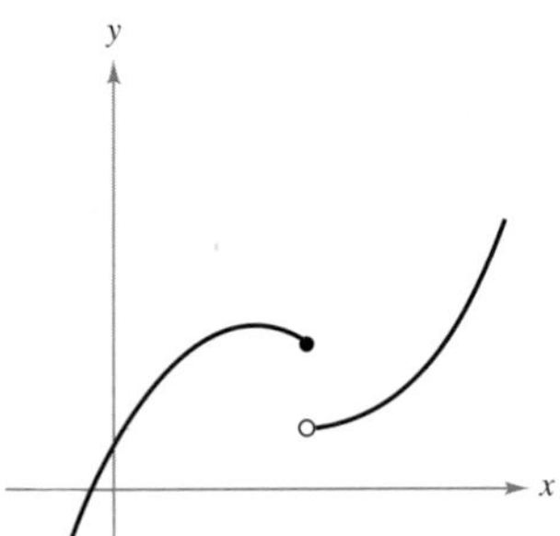

(b) Functions with graphs that are not continuous are not polynomial functions.

Figure 2.9

The second feature is that the graph of a polynomial function has only smooth, rounded turns, as shown in Figure 2.10(a). A polynomial function cannot have a sharp turn. For instance, the function given by $f(x) = |x|$, which has a sharp turn at the point $(0, 0)$, as shown in Figure 2.10(b), is not a polynomial function.

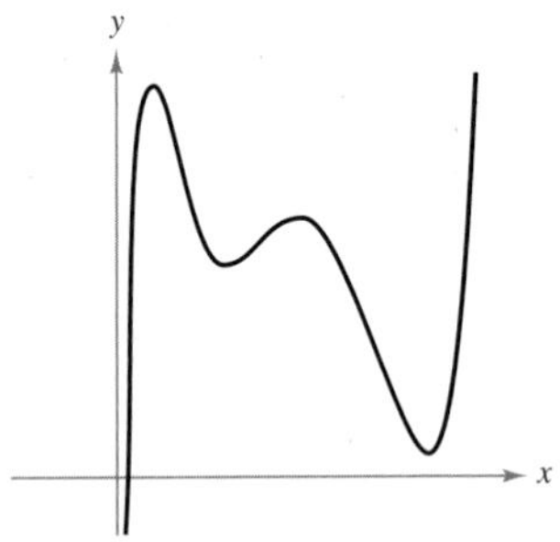

(a) Polynomial functions have graphs with smooth, rounded turns.

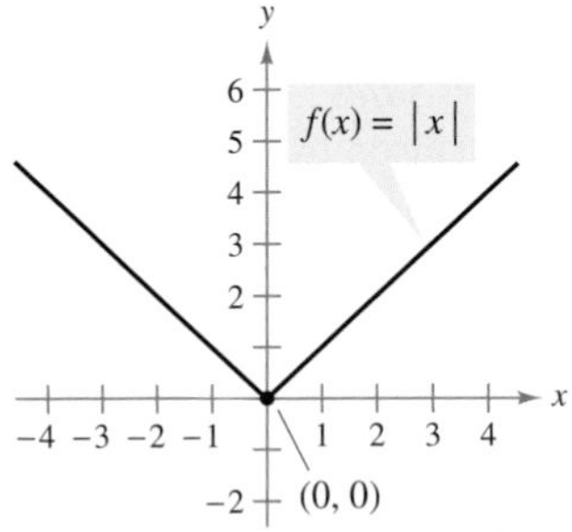

(b) Graphs of polynomial functions cannot have sharp turns.

Figure 2.10

The graphs of polynomial functions of degree greater than 2 are more difficult to analyze than the graphs of polynomials of degree 0, 1, or 2. However, using the features presented in this section, coupled with your knowledge of point plotting, intercepts, and symmetry, you should be able to make reasonably accurate sketches *by hand*. In Chapter 5, you will learn more techniques for analyzing the graphs of polynomial functions.

STUDY TIP For power functions given by $f(x) = x^n$, if n is even, then the graph of the function is symmetric with respect to the y-axis, and if n is odd, then the graph of the function is symmetric with respect to the origin.

The polynomial functions that have the simplest graphs are monomials of the form

$$f(x) = x^n$$

where n is an integer greater than zero. From Figure 2.11, you can see that when n is *even,* the graph is similar to the graph of $f(x) = x^2$, and when n is *odd,* the graph is similar to the graph of $f(x) = x^3$. Moreover, the greater the value of n, the flatter the graph near the origin. Polynomial functions of the form $f(x) = x^n$ are often referred to as **power functions.**

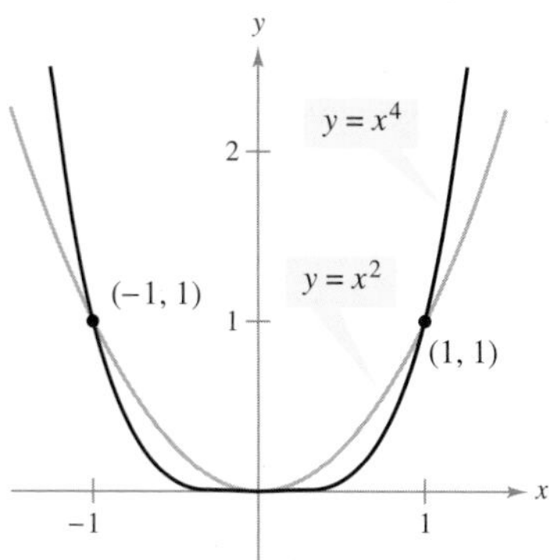

(a) When n is even, the graph of $y = x^n$ touches the axis at the x-intercept.

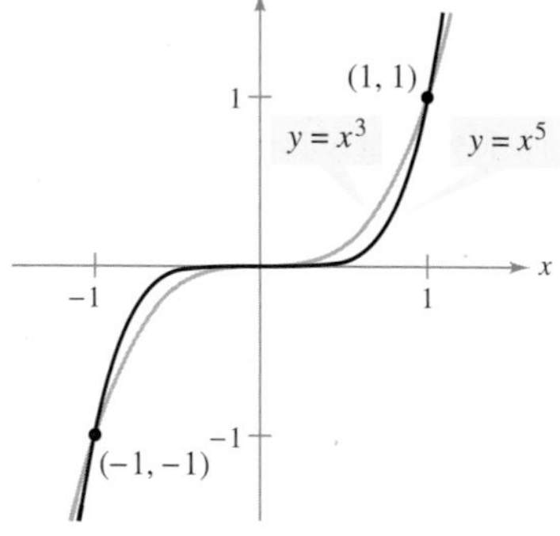

(b) When n is odd, the graph of $y = x^n$ crosses the axis at the x-intercept.

Figure 2.11

EXAMPLE 1 Sketching Transformations of Power Functions

Sketch the graph of each function.

a. $f(x) = -x^5$

b. $h(x) = (x + 1)^4$

Solution

a. Because the degree of $f(x) = -x^5$ is odd, its graph is similar to the graph of $y = x^3$. As shown in Figure 2.12(a), the graph of $f(x) = -x^5$ is a reflection in the x-axis of the graph of $y = x^5$.

b. The graph of $h(x) = (x + 1)^4$, as shown in Figure 2.12(b), is a left shift by one unit of the graph of $y = x^4$.

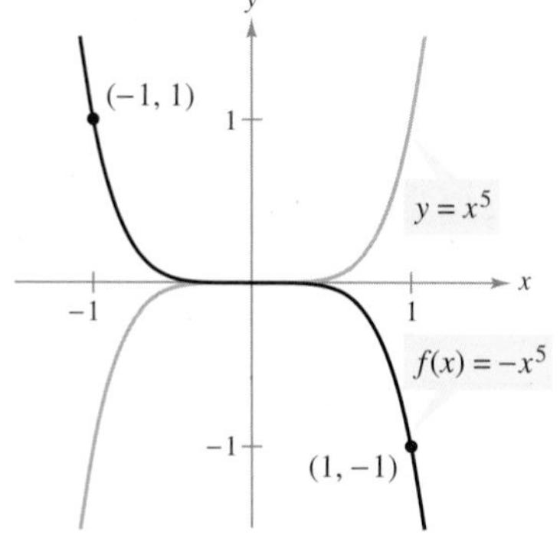

(a)

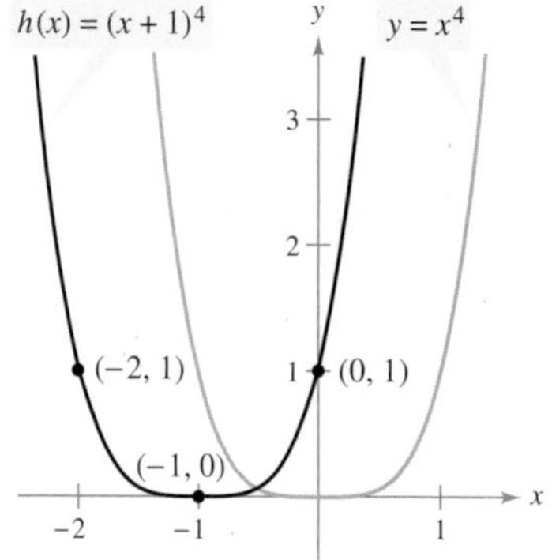

(b)

Figure 2.12

The Leading Coefficient Test

In Example 1, note that both graphs eventually rise or fall without bound as x moves to the right. Whether the graph of a polynomial function eventually rises or falls can be determined by the function's degree (even or odd) and by its leading coefficient, as indicated in the **Leading Coefficient Test.**

LEADING COEFFICIENT TEST

As x moves without bound to the left or to the right, the graph of the polynomial function $f(x) = a_n x^n + \cdot \cdot \cdot + a_1 x + a_0$ eventually rises or falls in the following manner.

1. When n is *odd:*

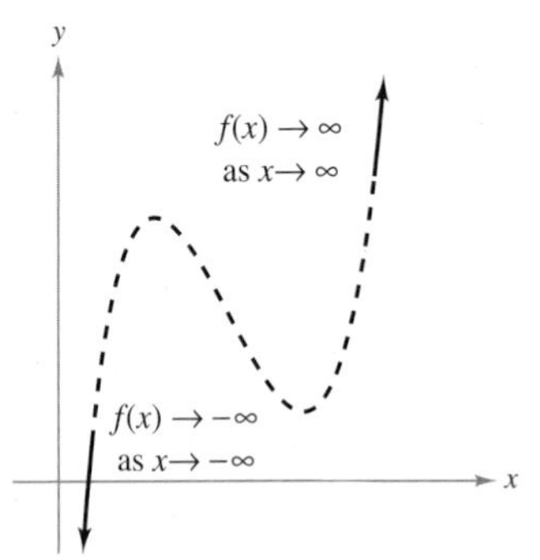

If the leading coefficient is positive ($a_n > 0$), the graph falls to the left and rises to the right.

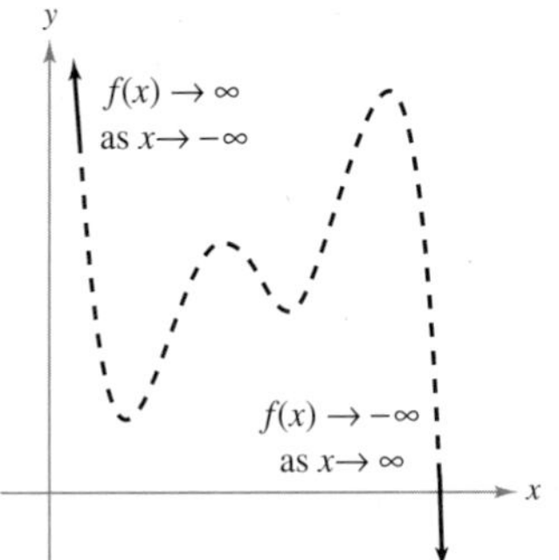

If the leading coefficient is negative ($a_n < 0$), the graph rises to the left and falls to the right.

2. When n is *even:*

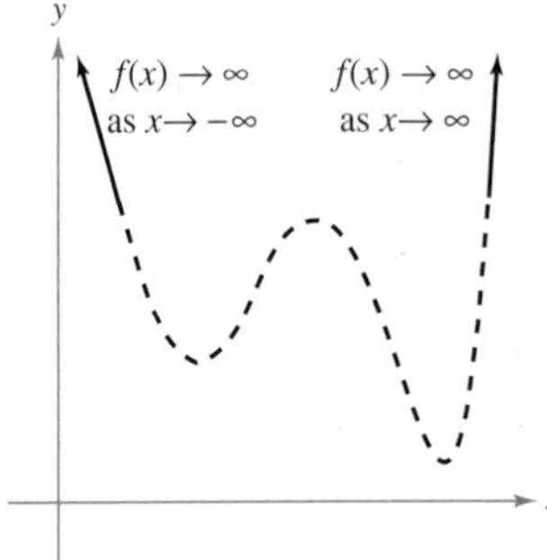

If the leading coefficient is positive ($a_n > 0$), the graph rises to the left and right.

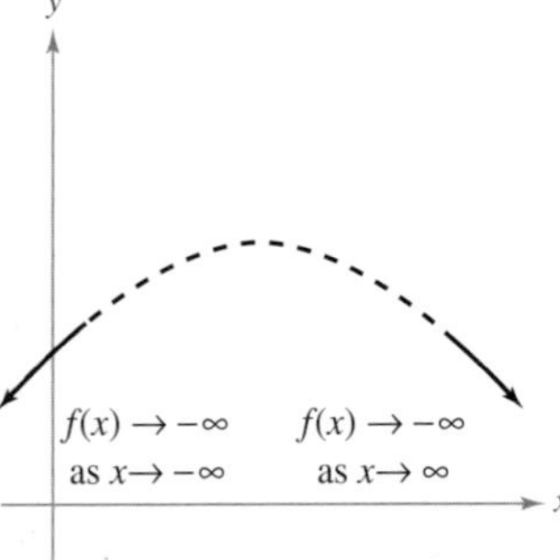

If the leading coefficient is negative ($a_n < 0$), the graph falls to the left and right

The dashed portions of the graphs indicate that the test determines *only* the right-hand and left-hand behavior of the graph.

EXPLORATION

For each polynomial function, identify the degree of the function and whether the degree of the function is even or odd. Identify the leading coefficient and whether it is positive or negative. Use a graphing utility to graph each function. Describe the relationship between the degree of the function and the sign of the leading coefficient of the polynomial function and the right-hand and left-hand behaviors of the graph of the function.

a. $f(x) = x^3 - 2x^2 - x + 1$

b. $f(x) = 2x^5 + 2x^2 - 5x + 1$

c. $f(x) = -2x^5 - x^2 + 5x + 3$

d. $f(x) = -x^3 + 5x - 2$

e. $f(x) = 2x^2 + 3x - 4$

f. $f(x) = x^4 - 3x^2 + 2x - 1$

g. $f(x) = x^2 + 3x + 2$

The notation "$f(x) \to \infty$ as $x \to \infty$" indicates that the graph rises without bound to the right. The notations "$f(x) \to \infty$ as $x \to -\infty$," "$f(x) \to -\infty$ as $x \to \infty$," and "$f(x) \to -\infty$ as $x \to -\infty$" have similar meanings. You will study precise definitions of these concepts in Section 5.5.

STUDY TIP A polynomial function is written in **standard form** when its terms are written in descending order of exponents from left to right. Before applying the Leading Coefficient Test to a polynomial function, it is a good idea to make sure that the polynomial function is written in standard form.

EXAMPLE 2 Applying the Leading Coefficient Test

Describe the right-hand and left-hand behaviors of the graph of each function.

a. $f(x) = -x^3 + 4x$

b. $f(x) = x^4 - 5x^2 + 4$

c. $f(x) = x^5 - x$

Solution

a. Because the degree is odd and the leading coefficient is negative, the graph rises to the left and falls to the right, as shown in Figure 2.13(a).

b. Because the degree is even and the leading coefficient is positive, the graph rises to the left and right, as shown in Figure 2.13(b).

c. Because the degree is odd and the leading coefficient is positive, the graph falls to the left and rises to the right, as shown in Figure 2.13(c).

EXPLORATION

For each of the graphs in Example 2, count the number of zeros of the polynomial function and the number of relative minima and relative maxima. Compare these numbers with the degree of the polynomial. What do you observe?

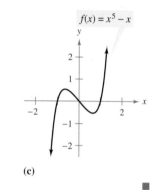

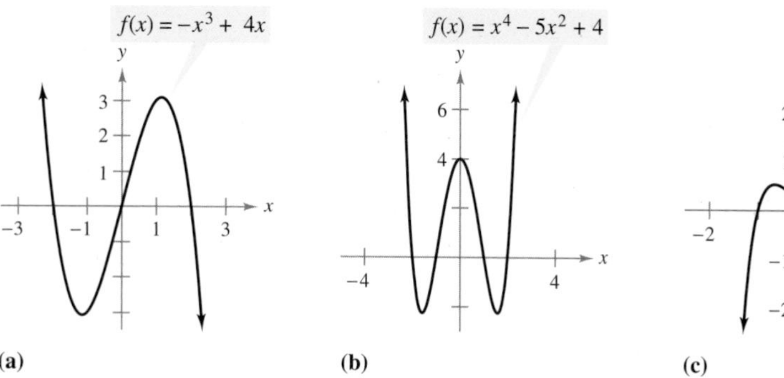

Figure 2.13

In Example 2, note that the Leading Coefficient Test tells you only whether the graph *eventually* rises or falls to the right or left. Other characteristics of the graph, such as intercepts and minimum and maximum points, must be determined by using other tests.

Real Zeros of Polynomial Functions

It can be shown that for a polynomial function f of degree n, the following statements are true. (Remember that the *zeros* of a function of x are the x-values for which the function is zero.)

1. The function f has, at most, n real zeros. (You will study this result in detail in the discussion of the Fundamental Theorem of Algebra in Section 2.5.)
2. The graph of f has, at most, $n - 1$ turning points. (Turning points, also called relative minima or relative maxima, are points at which the graph changes from increasing to decreasing or vice versa.)

Finding the zeros of polynomial functions is one of the most important problems in algebra. There is a strong interplay between graphical and analytic approaches to this problem. Sometimes you can use information about the graph of a function to help find its zeros, and in other cases you can use information about the zeros of a function to help sketch its graph.

REAL ZEROS OF POLYNOMIAL FUNCTIONS

When f is a polynomial function and a is a real number, the following statements are equivalent.

1. $x = a$ is a *zero* of the function f.
2. $x = a$ is a *solution* of the polynomial equation $f(x) = 0$.
3. $(x - a)$ is a *factor* of the polynomial $f(x)$.
4. $(a, 0)$ is an *x-intercept* of the graph of f.

NOTE In the equivalent statements above, notice that finding real zeros of polynomial functions is closely related to factoring and finding x-intercepts. ■

EXAMPLE 3 Find the Zeros of a Polynomial Function

Find all real zeros of

$$f(x) = -2x^4 + 2x^2.$$

The determine the number of turning points of the graph of the function.

Algebraic Solution

To find the real zeros of the function, set $f(x)$ equal to zero and solve for x.

$$-2x^4 + 2x^2 = 0 \quad \text{Set } f(x) \text{ equal to 0.}$$

$$-2x^2(x^2 - 1) = 0 \quad \text{Remove common monomial factor.}$$

$$-2x^2(x - 1)(x + 1) = 0 \quad \text{Factor completely.}$$

So, the real zeros are $x = 0$, $x = 1$, and $x = -1$. Because the function is a fourth-degree polynomial, the graph of f can have at most $4 - 1 = 3$ turning points.

Graphical Solution

Use a graphing utility to graph $y = -2x^4 + 2x^2$. In Figure 2.14, the graph appears to have zeros at $(0, 0)$, $(1, 0)$, and $(-1, 0)$. Use the *zero* or *root* feature, or the *zoom* and *trace* features, of the graphing utility to verify these zeros. So, the real zeros are $x = 0$, $x = 1$, and $x = -1$. From the figure, you can see that the graph has three turning points. This is consistent with the fact that a fourth-degree polynomial can have at most three turning points.

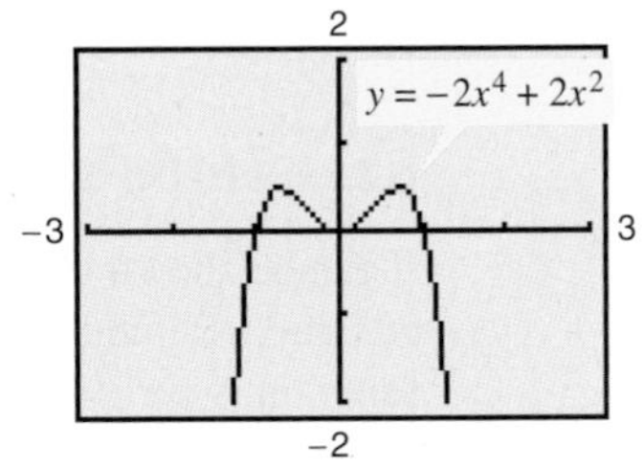

Figure 2.14 ■

NOTE In Example 3, note that because the exponent is greater than 1, the factor $-2x^2$ yields the *repeated* zero $x = 0$. Because the exponent is even, the graph touches the x-axis at $x = 0$, as shown in Figure 2.14.

REPEATED ZEROS

A factor $(x - a)^k$, $k > 1$, yields a **repeated zero** $x = a$ of **multiplicity** k.

1. When k is odd, the graph *crosses* the x-axis at $x = a$.
2. When k is even, the graph *touches* the x-axis (but does not cross the x-axis) at $x = a$.

TECHNOLOGY Example 4 uses an *analytic approach* to describe the graph of the function. A graphing utility is a complement to this approach. Remember that an important aspect of using a graphing utility is to find a viewing window that shows all significant features of the graph. For instance, the viewing window in Figure 2.16(a) illustrates all of the significant features of the function in Example 4 while the viewing window in Figure 2.16(b) does not.

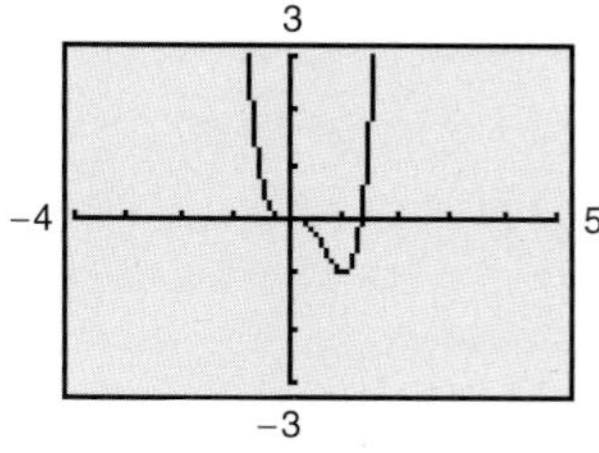

(a)

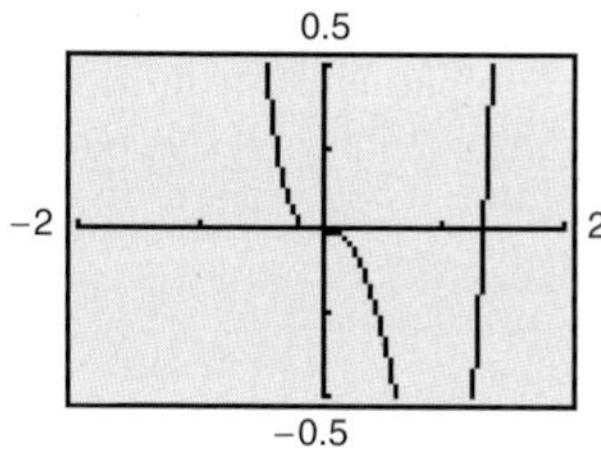

(b)

Figure 2.16

EXAMPLE 4 Sketching the Graph of a Polynomial Function

Sketch the graph of

$$f(x) = 3x^4 - 4x^3.$$

Solution

1. *Apply the Leading Coefficient Test.* Because the leading coefficient is positive and the degree is even, you know that the graph eventually rises to the left and to the right [see Figure 2.15(a)].
2. *Find the Real Zeros of the Polynomial.* By factoring $f(x) = 3x^4 - 4x^3$ as $f(x) = x^3(3x - 4)$, you can see that the zeros of f are $x = 0$ and $x = \frac{4}{3}$ (both of odd multiplicity). So, the x-intercepts occur at $(0, 0)$ and $\left(\frac{4}{3}, 0\right)$. Add these points to your graph, as shown in Figure 2.15(a).
3. *Plot a Few Additional Points.* Use the zeros of the polynomial to find the test intervals. In each test interval, choose a representative x-value and evaluate the polynomial function, as shown in the table.

Test interval	Representative x-value	Value of f	Sign	Point on graph
$(-\infty, 0)$	-1	$f(-1) = 7$	Positive	$(-1, 7)$
$\left(0, \frac{4}{3}\right)$	1	$f(1) = -1$	Negative	$(1, -1)$
$\left(\frac{4}{3}, \infty\right)$	1.5	$f(1.5) = 1.6875$	Positive	$(1.5, 1.6875)$

4. *Draw the Graph.* Draw a continuous curve through the points, as shown in Figure 2.15(b). Because both zeros are of odd multiplicity, you know that the graph should cross the x-axis at $x = 0$ and $x = \frac{4}{3}$.

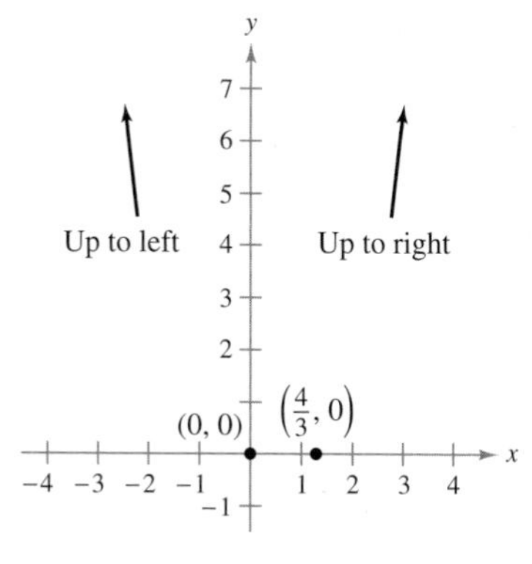

(a)

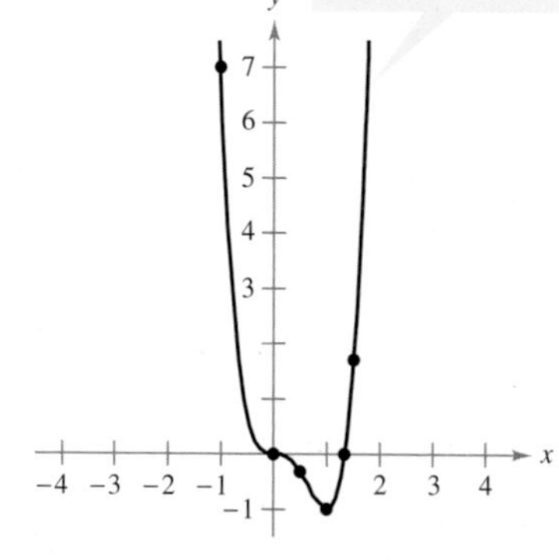

(b)

Figure 2.15

NOTE If you are unsure of the shape of a portion of the graph of a polynomial function, plot some additional points, such as the point $(0.5, -0.3125)$, as shown in Figure 2.15(b).

Before applying the Leading Coefficient Test to a polynomial function, it is a good idea to check that the polynomial function is written in *standard form.*

EXAMPLE 5 Sketching the Graph of a Polynomial Function

Sketch the graph of

$$f(x) = -2x^3 + 6x^2 - \tfrac{9}{2}x.$$

Solution

1. *Apply the Leading Coefficient Test.* Because the leading coefficient is negative and the degree is odd, you know that the graph eventually rises to the left and falls to the right [see Figure 2.17(a)].
2. *Find the Real Zeros of the Polynomial.* By factoring

$$\begin{aligned} f(x) &= -2x^3 + 6x^2 - \tfrac{9}{2}x \\ &= -\tfrac{1}{2}x(4x^2 - 12x + 9) \\ &= -\tfrac{1}{2}x(2x - 3)^2 \end{aligned}$$

   you can see that the zeros of f are $x = 0$ (odd multiplicity) and $x = \frac{3}{2}$ (even multiplicity). So, the x-intercepts occur at $(0, 0)$ and $\left(\frac{3}{2}, 0\right)$. Add these points to your graph, as shown in Figure 2.17(a).
3. *Plot a Few Additional Points.* Use the zeros of the polynomial to find the test intervals. In each test interval, choose a representative x-value and evaluate the polynomial function, as shown in the table.

Test interval	Representative x-value	Value of f	Sign	Point on graph
$(-\infty, 0)$	-0.5	$f(-0.5) = 4$	Positive	$(-0.5, 4)$
$\left(0, \frac{3}{2}\right)$	0.5	$f(0.5) = -1$	Negative	$(0.5, -1)$
$\left(\frac{3}{2}, \infty\right)$	2	$f(2) = -1$	Negative	$(2, -1)$

4. *Draw the Graph.* Draw a continuous curve through the points, as shown in Figure 2.17(b). As indicated by the multiplicities of the zeros, the graph crosses the x-axis at $(0, 0)$ but does not cross the x-axis at $\left(\frac{3}{2}, 0\right)$.

NOTE Observe in Example 5 that the sign of $f(x)$ is positive to the left of and negative to the right of the zero $x = 0$. Similarly, the sign of $f(x)$ is negative to the left and to the right of the zero $x = \frac{3}{2}$. This suggests that if the zero of a polynomial function is of *odd* multiplicity, then the sign of $f(x)$ changes from one side of the zero to the other side. If the zero is of *even* multiplicity, then the sign of $f(x)$ does not change from one side of the zero to the other side.

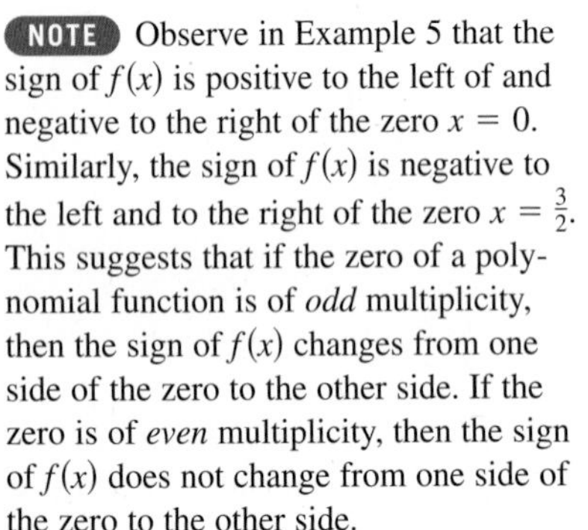

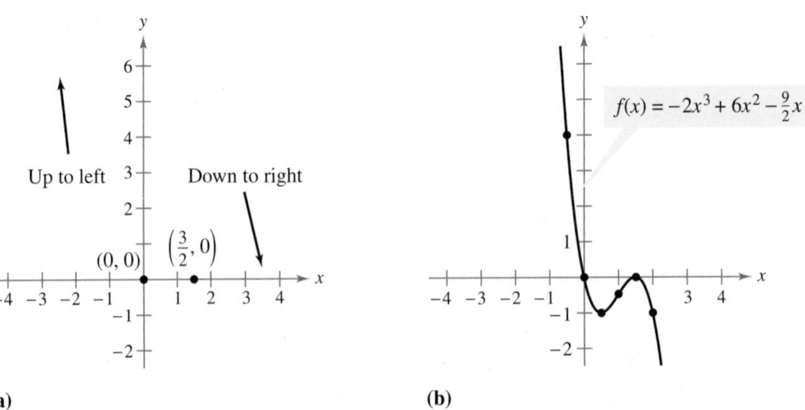

(a) (b)

Figure 2.17

■

2.2 Exercises

See www.CalcChat.com for worked-out solutions to odd-numbered exercises.

In Exercises 1–8, fill in the blanks.

1. The graphs of all polynomial functions are ________, which means that the graphs have no breaks, holes, or gaps.
2. The ________ ________ ________ is used to determine the left-hand and right-hand behavior of the graph of a polynomial function.
3. Polynomial functions of the form $f(x) =$ ________ are often referred to as power functions.
4. A polynomial function of degree n has at most ________ real zeros and at most ________ turning points.
5. If $x = a$ is a zero of a polynomial function f, then the following three statements are true.
   (a) $x = a$ is a ________ of the polynomial equation $f(x) = 0$.
   (b) ________ is a factor of the polynomial $f(x)$.
   (c) $(a, 0)$ is an ________ of the graph of f.
6. A factor $(x - a)^k$, $k > 1$, yields a ________ zero $x = a$ of ________ k.
7. If a real zero of a polynomial function is of even multiplicity, then the graph of f ________ the x-axis at $x = a$, and if it is of odd multiplicity, then the graph of f ________ the x-axis at $x = a$.
8. A polynomial function is written in ________ form when its terms are written in descending order of exponents from left to right.

In Exercises 9–16, match the polynomial function with its graph. [The graphs are labeled (a), (b), (c), (d), (e), (f), (g), and (h).]

(a)

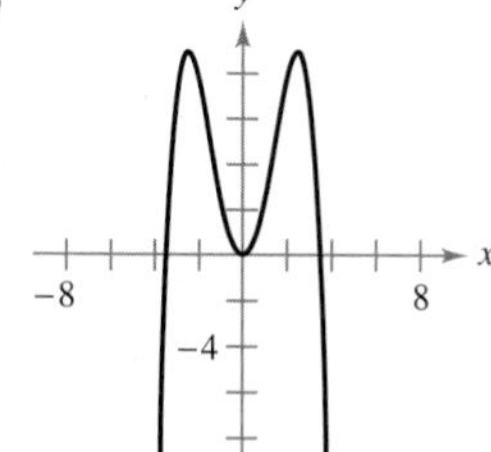

(b)

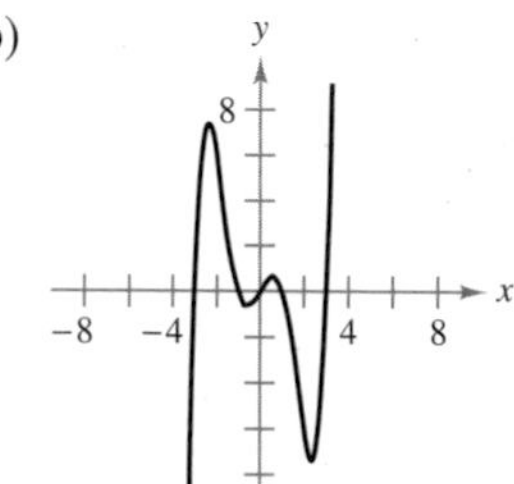

(c)

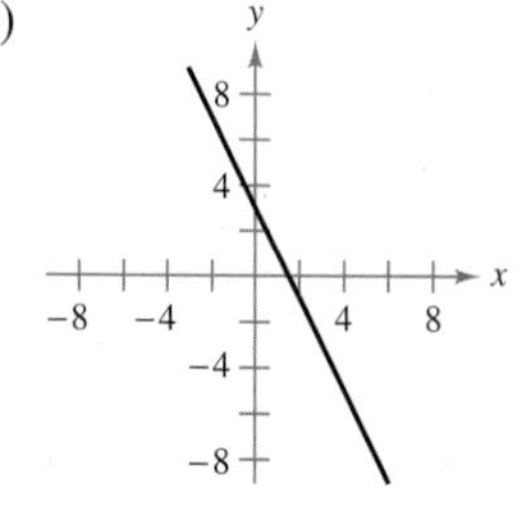

(d)

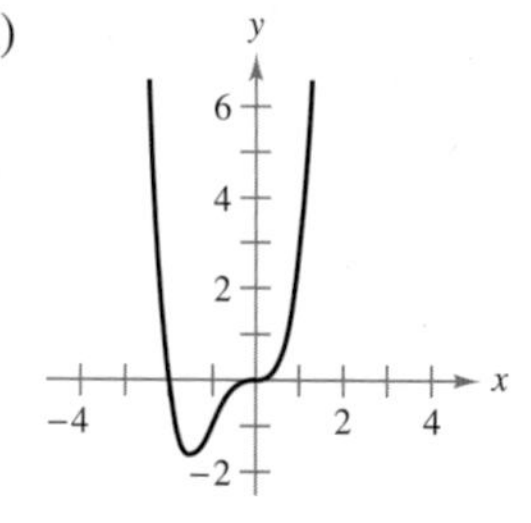

(e)

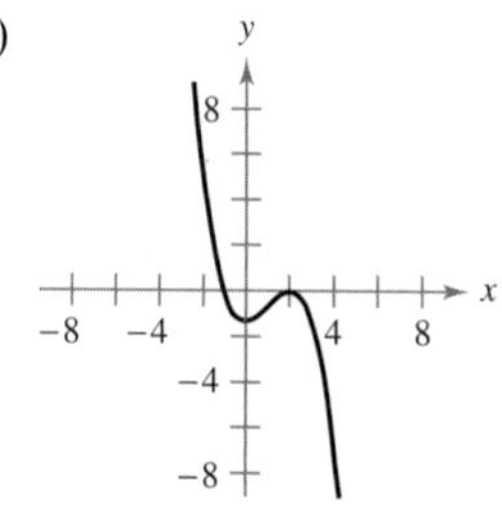

(f)

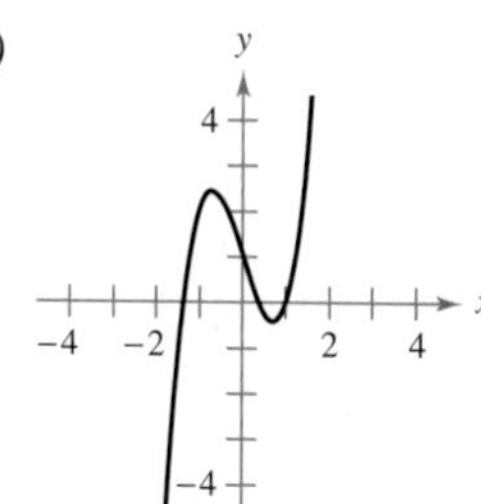

(g)

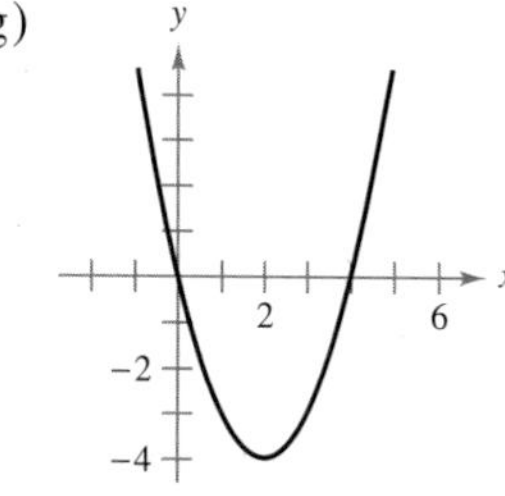

(h)

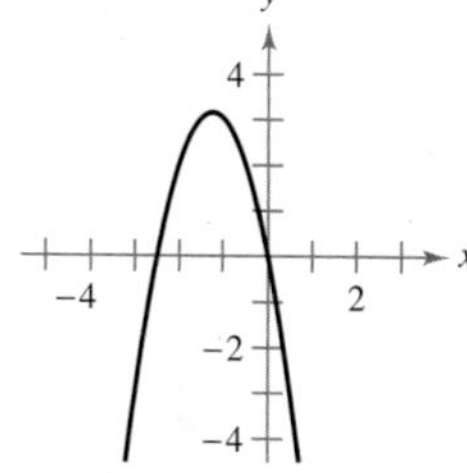

9. $f(x) = -2x + 3$
10. $f(x) = x^2 - 4x$
11. $f(x) = -2x^2 - 5x$
12. $f(x) = 2x^3 - 3x + 1$
13. $f(x) = -\frac{1}{4}x^4 + 3x^2$
14. $f(x) = -\frac{1}{3}x^3 + x^2 - \frac{4}{3}$
15. $f(x) = x^4 + 2x^3$
16. $f(x) = \frac{1}{5}x^5 - 2x^3 + \frac{9}{5}x$

In Exercises 17–20, sketch the graph of $y = x^n$ and each transformation.

17. $y = x^3$
   (a) $f(x) = (x - 4)^3$ (b) $f(x) = x^3 - 4$
   (c) $f(x) = -\frac{1}{4}x^3$ (d) $f(x) = (x - 4)^3 - 4$
18. $y = x^5$
   (a) $f(x) = (x + 1)^5$ (b) $f(x) = x^5 + 1$
   (c) $f(x) = 1 - \frac{1}{2}x^5$ (d) $f(x) = -\frac{1}{2}(x + 1)^5$
19. $y = x^4$
   (a) $f(x) = (x + 3)^4$ (b) $f(x) = x^4 - 3$
   (c) $f(x) = 4 - x^4$ (d) $f(x) = \frac{1}{2}(x - 1)^4$
   (e) $f(x) = (2x)^4 + 1$ (f) $f(x) = \left(\frac{1}{2}x\right)^4 - 2$
20. $y = x^6$
   (a) $f(x) = -\frac{1}{8}x^6$ (b) $f(x) = (x + 2)^6 - 4$
   (c) $f(x) = x^6 - 5$ (d) $f(x) = -\frac{1}{4}x^6 + 1$
   (e) $f(x) = \left(\frac{1}{4}x\right)^6 - 2$ (f) $f(x) = (2x)^6 - 1$

In Exercises 21–30, describe the right-hand and left-hand behaviors of the graph of the polynomial function.

21. $f(x) = \frac{1}{5}x^3 + 4x$
22. $f(x) = 2x^2 - 3x + 1$
23. $g(x) = 5 - \frac{7}{2}x - 3x^2$
24. $h(x) = 1 - x^6$
25. $f(x) = -2.1x^5 + 4x^3 - 2$

26. $f(x) = 4x^5 - 7x + 6.5$

27. $f(x) = 6 - 2x + 4x^2 - 5x^3$

28. $f(x) = (3x^4 - 2x + 5)/4$

29. $h(t) = -\frac{3}{4}(t^2 - 3t + 6)$

30. $f(s) = -\frac{7}{8}(s^3 + 5s^2 - 7s + 1)$

Graphical Analysis **In Exercises 31–34, use a graphing utility to graph the functions f and g in the same viewing window. Zoom out sufficiently far to show that the right-hand and left-hand behaviors of f and g appear identical.**

31. $f(x) = 3x^3 - 9x + 1, \quad g(x) = 3x^3$

32. $f(x) = -\frac{1}{3}(x^3 - 3x + 2), \quad g(x) = -\frac{1}{3}x^3$

33. $f(x) = -(x^4 - 4x^3 + 16x), \quad g(x) = -x^4$

34. $f(x) = 3x^4 - 6x^2, \quad g(x) = 3x^4$

In Exercises 35–50, (a) find all the real zeros of the polynomial function, (b) determine the multiplicity of each zero and the number of turning points of the graph of the function, and (c) use a graphing utility to graph the function and verify your answers.

35. $f(x) = x^2 - 36$

36. $f(x) = 81 - x^2$

37. $h(t) = t^2 - 6t + 9$

38. $f(x) = x^2 + 10x + 25$

39. $f(x) = \frac{1}{3}x^2 + \frac{1}{3}x - \frac{2}{3}$

40. $f(x) = \frac{1}{2}x^2 + \frac{5}{2}x - \frac{3}{2}$

41. $f(x) = 3x^3 - 12x^2 + 3x$

42. $g(x) = 5x(x^2 - 2x - 1)$

43. $f(t) = t^3 - 8t^2 + 16t$

44. $f(x) = x^4 - x^3 - 30x^2$

45. $g(t) = t^5 - 6t^3 + 9t$

46. $f(x) = x^5 + x^3 - 6x$

47. $f(x) = 3x^4 + 9x^2 + 6$

48. $f(x) = 2x^4 - 2x^2 - 40$

49. $g(x) = x^3 + 3x^2 - 4x - 12$

50. $f(x) = x^3 - 4x^2 - 25x + 100$

Graphical Analysis **In Exercises 51–54, (a) use a graphing utility to graph the function, (b) use the graph to approximate any x-intercepts of the graph, (c) set $y = 0$ and solve the resulting equation, and (d) compare the results of part (c) with any x-intercepts of the graph.**

51. $y = 4x^3 - 20x^2 + 25x$

52. $y = 4x^3 + 4x^2 - 8x - 8$

53. $y = x^5 - 5x^3 + 4x$

54. $y = \frac{1}{4}x^3(x^2 - 9)$

In Exercises 55–64, find a polynomial function that has the given zeros. (There are many correct answers.)

55. $0, 8$

56. $0, -7$

57. $2, -6$

58. $-4, 5$

59. $0, -4, -5$

60. $0, 1, 10$

61. $4, -3, 3, 0$

62. $-2, -1, 0, 1, 2$

63. $1 + \sqrt{3}, 1 - \sqrt{3}$

64. $2, 4 + \sqrt{5}, 4 - \sqrt{5}$

In Exercises 65–74, find a polynomial of degree n that has the given zero(s). (There are many correct answers.)

	Zero(s)	*Degree*
65.	$x = -3$	$n = 2$
66.	$x = -12, -6$	$n = 2$
67.	$x = -5, 0, 1$	$n = 3$
68.	$x = -2, 4, 7$	$n = 3$
69.	$x = 0, \sqrt{3}, -\sqrt{3}$	$n = 3$
70.	$x = 0, 2\sqrt{2}, -2\sqrt{2}$	$n = 3$
71.	$x = 1, -2, 1 \pm \sqrt{3}$	$n = 4$
72.	$x = 3, -2, 2 \pm \sqrt{5}$	$n = 4$
73.	$x = 0, -4$	$n = 5$
74.	$x = -1, 4, 7, 8$	$n = 5$

In Exercises 75–88, sketch the graph of the function by (a) applying the Leading Coefficient Test, (b) finding the zeros of the polynomial, (c) plotting sufficient solution points, and (d) drawing a continuous curve through the points.

75. $f(x) = x^3 - 25x$

76. $g(x) = x^4 - 9x^2$

77. $f(t) = \frac{1}{4}(t^2 - 2t + 15)$

78. $g(x) = -x^2 + 10x - 16$

79. $f(x) = x^3 - 2x^2$

80. $f(x) = 8 - x^3$

81. $f(x) = 3x^3 - 15x^2 + 18x$

82. $f(x) = -4x^3 + 4x^2 + 15x$

83. $f(x) = -5x^2 - x^3$

84. $f(x) = -48x^2 + 3x^4$

85. $f(x) = x^2(x - 4)$

86. $h(x) = \frac{1}{3}x^3(x - 4)^2$

87. $g(t) = -\frac{1}{4}(t - 2)^2(t + 2)^2$

88. $g(x) = \frac{1}{10}(x + 1)^2(x - 3)^3$

In Exercises 89–92, use a graphing utility to graph the function. Use the *zero* or *root* feature to approximate the real zeros of the function. Then determine the multiplicity of each zero.

89. $f(x) = x^3 - 16x$

90. $f(x) = \frac{1}{4}x^4 - 2x^2$

91. $g(x) = \frac{1}{5}(x + 1)^2(x - 3)(2x - 9)$

92. $h(x) = \frac{1}{5}(x + 2)^2(3x - 5)^2$

WRITING ABOUT CONCEPTS

93. Sketch a graph of the function given by $f(x) = x^4$. Explain how the graph of g differs (if it does) from the graph of f. Determine whether g is odd, even, or neither.

(a) $g(x) = f(x) + 2$ (b) $g(x) = f(x + 2)$

(c) $g(x) = f(-x)$ (d) $g(x) = -f(x)$

(e) $g(x) = f\left(\frac{1}{2}x\right)$ (f) $g(x) = \frac{1}{2}f(x)$

(g) $g(x) = f\left(x^{3/4}\right)$ (h) $g(x) = (f \circ f)(x)$

94. ***Revenue*** The total revenue R (in millions of dollars) for a company is related to its advertising expense by the function

$$R = \frac{1}{100{,}000}(-x^3 + 600x^2), \quad 0 \leq x \leq 400$$

where x is the amount spent on advertising (in tens of thousands of dollars). Use the graph of this function, shown in the figure, to estimate the point on the graph at which the function is increasing most rapidly. This point is called the *point of diminishing returns* because any expense above this amount will yield less return per dollar invested in advertising.

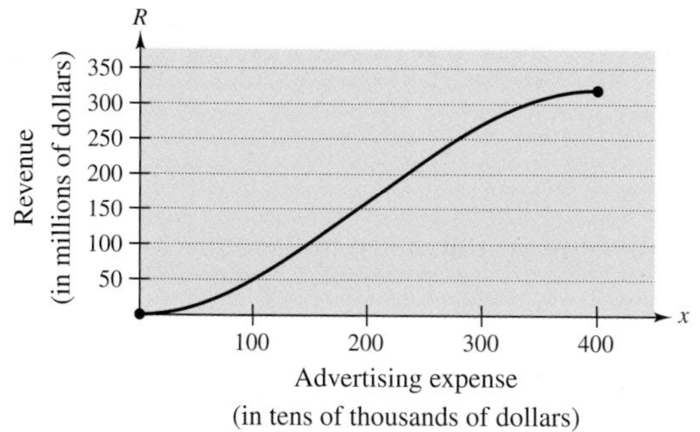

95. ***Numerical and Graphical Analysis*** An open box is to be made from a square piece of material, 36 inches on a side, by cutting equal squares with sides of length x from the corners and turning up the sides (see figure).

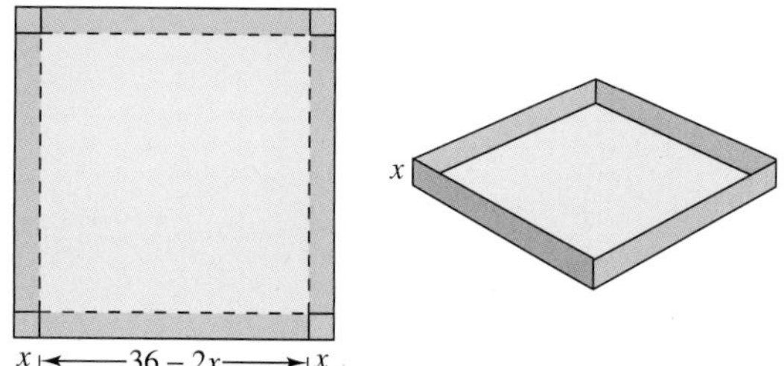

(a) Write a function $V(x)$ that represents the volume of the box.

(b) Determine the domain of the function.

(c) Use a graphing utility to create a table that shows box heights x and the corresponding volumes V. Use the table to estimate the dimensions that will produce a maximum volume.

(d) Use a graphing utility to graph V and use the graph to estimate the value of x for which $V(x)$ is maximum. Compare your result with that of part (c).

96. ***Maximum Volume*** An open box with locking tabs is to be made from a square piece of material 24 inches on a side. This is to be done by cutting equal squares from the corners and folding along the dashed lines shown in the figure.

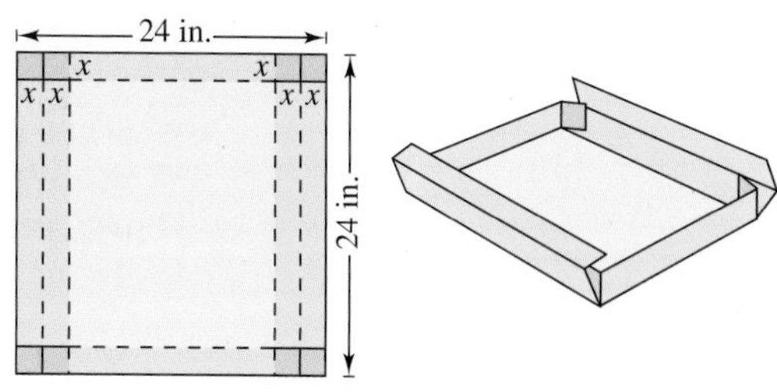

(a) Write a function $V(x)$ that represents the volume of the box.

(b) Determine the domain of the function V.

(c) Sketch a graph of the function and estimate the value of x for which $V(x)$ is maximum.

True or False? **In Exercises 97–99, determine whether the statement is true or false. Justify your answer.**

97. A fifth-degree polynomial can have five turning points in its graph.

98. It is possible for a sixth-degree polynomial to have only one solution.

99. The graph of the function given by

$$f(x) = 2 + x - x^2 + x^3 - x^4 + x^5 + x^6 - x^7$$

rises to the left and falls to the right.

CAPSTONE

100. For each graph, describe a polynomial function that could represent the graph. (Indicate the degree of the function and the sign of its leading coefficient.)

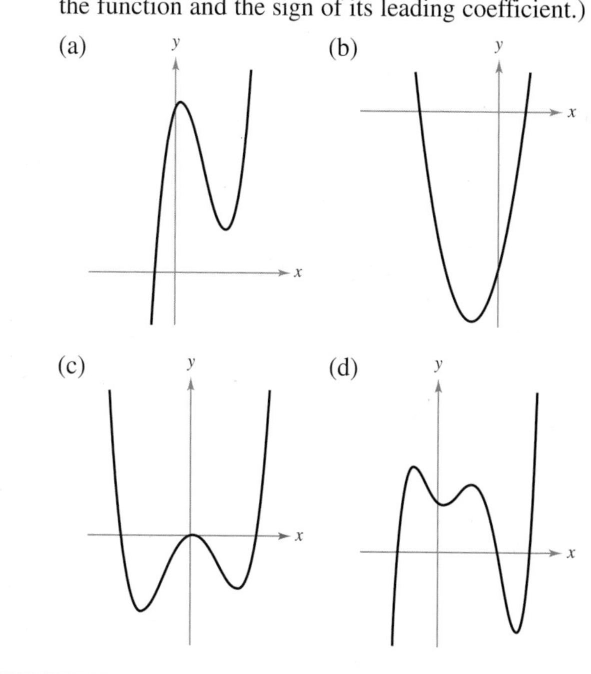

2.3 Polynomial and Synthetic Division

- **Divide polynomials using long division.**
- **Use synthetic division to divide polynomials by binomials of the form $(x - k)$.**
- **Use the Remainder Theorem and the Factor Theorem.**
- **Use polynomial division to answer questions about real-life problems.**

Long Division of Polynomials

In this section, you will study two procedures for *dividing* polynomials. These procedures are especially valuable in factoring and finding the zeros of polynomial functions. To begin, suppose you are given the graph of

$$f(x) = 6x^3 - 19x^2 + 16x - 4.$$

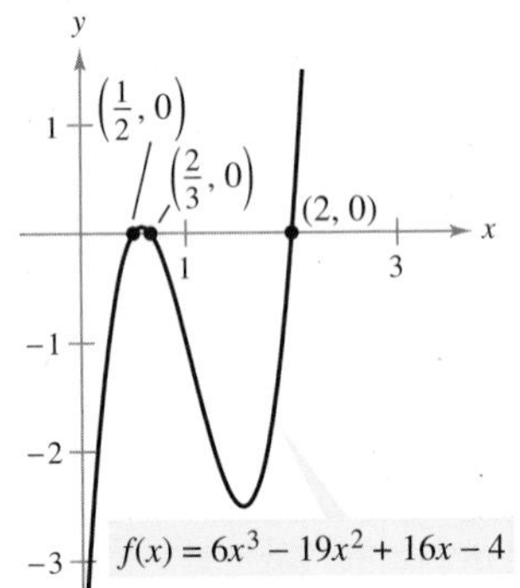

Figure 2.18

Notice that a zero of f occurs at $x = 2$, as shown in Figure 2.18. Because $x = 2$ is a zero of f, you know that $(x - 2)$ is a factor of $f(x)$. This means that there exists a second-degree polynomial $q(x)$ such that

$$f(x) = (x - 2) \cdot q(x).$$

To find $q(x)$, you can use **long division,** as illustrated in Example 1.

EXAMPLE 1 Long Division of Polynomials

Divide $6x^3 - 19x^2 + 16x - 4$ by $x - 2$, and use the result to factor the polynomial completely.

Solution

Think $\dfrac{6x^3}{x} = 6x^2$.

Think $\dfrac{-7x^2}{x} = -7x$.

Think $\dfrac{2x}{x} = 2$.

$$\begin{array}{r} 6x^2 - 7x + 2 \\ x - 2 \overline{)\, 6x^3 - 19x^2 + 16x - 4} \\ \underline{6x^3 - 12x^2} \qquad\qquad \\ -7x^2 + 16x \qquad \\ \underline{-7x^2 + 14x} \qquad \\ 2x - 4 \\ \underline{2x - 4} \\ 0 \end{array}$$

$6x^3 - 12x^2$	Multiply: $6x^2(x - 2)$.
$-7x^2 + 16x$	Subtract.
$-7x^2 + 14x$	Multiply: $-7x(x - 2)$.
$2x - 4$	Subtract.
$2x - 4$	Multiply: $2(x - 2)$.
0	Subtract.

From this division, you can conclude that

$$6x^3 - 19x^2 + 16x - 4 = (x - 2)(6x^2 - 7x + 2)$$

and by factoring the quadratic $6x^2 - 7x + 2$, you have

$$6x^3 - 19x^2 + 16x - 4 = (x - 2)(2x - 1)(3x - 2).$$ ■

NOTE Note that this factorization agrees with the graph shown in Figure 2.18 in that the three x-intercepts occur at $x = 2$, $x = \frac{1}{2}$, and $x = \frac{2}{3}$. ■

In Example 1, $x - 2$ is a factor of the polynomial $6x^3 - 19x^2 + 16x - 4$, and the long division process produces a remainder of zero. Often, long division will produce a nonzero remainder. For instance, when you divide $x^2 + 3x + 5$ by $x + 1$, you obtain the following.

$$\begin{array}{rrl} & x + 2 & \leftarrow \text{Quotient} \\ \text{Divisor} \rightarrow \quad x + 1 \,\big)\!\!\! & \overline{x^2 + 3x + 5} & \leftarrow \text{Dividend} \\ & \underline{x^2 + x} & \\ & 2x + 5 & \\ & \underline{2x + 2} & \\ & 3 & \leftarrow \text{Remainder} \end{array}$$

In fractional form, you can write this result as shown.

$$\overbrace{\frac{x^2 + 3x + 5}{\underbrace{x + 1}_{\text{Divisor}}}}^{\text{Dividend}} = \overbrace{x + 2}^{\text{Quotient}} + \frac{\overset{\text{Remainder}}{3}}{\underbrace{x + 1}_{\text{Divisor}}}$$

This implies that

$$x^2 + 3x + 5 = (x + 1)(x + 2) + 3 \qquad \text{Multiply each side by } (x + 1).$$

which illustrates the following theorem, called the **Division Algorithm.**

THE DIVISION ALGORITHM

When $f(x)$ and $d(x)$ are polynomials such that $d(x) \neq 0$, and the degree of $d(x)$ is less than or equal to the degree of $f(x)$, there exist unique polynomials $q(x)$ and $r(x)$ such that

$$\underset{\text{Dividend}}{f(x)} = \underset{\text{Divisor}}{d(x)}\underset{\text{Quotient}}{q(x)} + \underset{\text{Remainder}}{r(x)}$$

where $r(x) = 0$ *or* the degree of $r(x)$ is less than the degree of $d(x)$. When the remainder $r(x)$ is zero, $d(x)$ *divides evenly* into $f(x)$.

The Division Algorithm can also be written as

$$\frac{f(x)}{d(x)} = q(x) + \frac{r(x)}{d(x)}.$$

In the Division Algorithm, the rational expression $f(x)/d(x)$ is **improper** because the degree of $f(x)$ is greater than or equal to the degree of $d(x)$. On the other hand, the rational expression $r(x)/d(x)$ is **proper** because the degree of $r(x)$ is less than the degree of $d(x)$. Here are some examples.

$$\frac{x^2 + 3x + 5}{x + 1} \qquad \text{Improper rational expression}$$

$$\frac{3}{x + 1} \qquad \text{Proper rational expression}$$

EXAMPLE 2 Long Division of Polynomials

Divide $x^3 - 1$ by $x - 1$.

Solution Because there is no x^2-term or x-term in the dividend, you need to line up the subtraction by using zero coefficients (or leaving spaces) for the missing terms.

$$
\begin{array}{r}
x^2 + x + 1 \\
x - 1 \,\overline{)\, x^3 + 0x^2 + 0x - 1} \\
\underline{x^3 - x^2 } \\
x^2 + 0x \\
\underline{x^2 - x } \\
x - 1 \\
\underline{x - 1} \\
0
\end{array}
$$

So, $x - 1$ divides evenly into $x^3 - 1$, and you can write

$$\frac{x^3 - 1}{x - 1} = x^2 + x + 1, \quad x \neq 1.$$

Check You can check the result of a division problem by multiplying.

$$
\begin{aligned}
(x - 1)(x^2 + x + 1) &= x^3 + x^2 + x - x^2 - x - 1 \\
&= x^3 - 1 \checkmark
\end{aligned}
$$

EXAMPLE 3 Long Division of Polynomials

Divide $-5x^2 - 2 + 3x + 2x^4 + 4x^3$ by $2x - 3 + x^2$.

Solution Begin by writing the dividend and divisor in descending powers of x.

$$
\begin{array}{r}
2x^2 + 1 \\
x^2 + 2x - 3 \,\overline{)\, 2x^4 + 4x^3 - 5x^2 + 3x - 2} \\
\underline{2x^4 + 4x^3 - 6x^2 } \\
x^2 + 3x - 2 \\
\underline{x^2 + 2x - 3} \\
x + 1
\end{array}
$$

Note that the first subtraction eliminated two terms from the dividend. When this happens, the quotient skips a term. You can write the result as

$$\frac{2x^4 + 4x^3 - 5x^2 + 3x - 2}{x^2 + 2x - 3} = 2x^2 + 1 + \frac{x + 1}{x^2 + 2x - 3}.$$

Check

$$
\begin{aligned}
(x^2 + 2x - 3)\left(2x^2 + 1 + \frac{x + 1}{x^2 + 2x - 3}\right) &= (x^2 + 2x - 3)(2x^2 + 1) + (x + 1) \\
&= (2x^4 + 4x^3 - 5x^2 + 2x - 3) + (x + 1) \\
&= 2x^4 + 4x^3 - 5x^2 + 3x - 2 \checkmark
\end{aligned}
$$

■

NOTE Synthetic division works *only* for divisors of the form $x - k$. You cannot use synthetic division to divide a polynomial by a quadratic such as $x^2 - 3$.

Synthetic Division

There is a nice shortcut for long division of polynomials by divisors of the form $x - k$. This shortcut is called **synthetic division.** The pattern for synthetic division of a cubic polynomial is summarized below. (The pattern for higher-degree polynomials is similar.)

SYNTHETIC DIVISION (FOR A CUBIC POLYNOMIAL)

To divide $ax^3 + bx^2 + cx + d$ by $x - k$, use the following pattern.

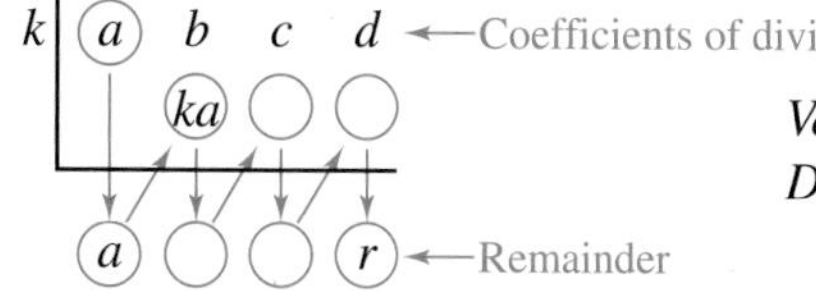

Vertical pattern: Add terms.
Diagonal pattern: Multiply by k.

EXAMPLE 4 Using Synthetic Division

Use synthetic division to divide $x^4 - 10x^2 - 2x + 4$ by $x + 3$.

Solution You should set up the array as shown below. Note that a zero is included for each missing term in the dividend.

$$\begin{array}{r|rrrrr} -3 & 1 & 0 & -10 & -2 & 4 \\ & & \bigcirc & \bigcirc & \bigcirc & \bigcirc \\ \hline & \bigcirc & \bigcirc & \bigcirc & \bigcirc & \bigcirc \end{array}$$

Then, use the synthetic division pattern by adding terms in columns and multiplying the results by -3.

Divisor: $x + 3$ Dividend: $x^4 - 10x^2 - 2x + 4$

$$\begin{array}{r|rrrrr} -3 & 1 & 0 & -10 & -2 & 4 \\ & & -3 & 9 & 3 & -3 \\ \hline & 1 & -3 & -1 & 1 & \boxed{1} \end{array}$$

← Remainder: 1

Quotient: $x^3 - 3x^2 - x + 1$

So, you have

$$\frac{x^4 - 10x^2 - 2x + 4}{x + 3} = x^3 - 3x^2 - x + 1 + \frac{1}{x + 3}.$$

Check

$$\begin{aligned} (x + 3)\left(x^3 - 3x^2 - x + 1 + \frac{1}{x + 3}\right) &= (x + 3)(x^3 - 3x^2 - x + 1) + 1 \\ &= (x^4 - 10x^2 - 2x + 3) + 1 \\ &= x^4 - 10x^2 - 2x + 4 \checkmark \end{aligned}$$

■

The Remainder and Factor Theorems

The remainder obtained in the synthetic division process has an important interpretation, as described in the **Remainder Theorem.**

THEOREM 2.1 THE REMAINDER THEOREM

When a polynomial $f(x)$ is divided by $x - k$, the remainder is

$$r = f(k).$$

PROOF From the Division Algorithm, you have

$$f(x) = (x - k)q(x) + r(x)$$

and because either $r(x) = 0$ or the degree of $r(x)$ is less than the degree of $x - k$, you know that $r(x)$ must be a constant. That is, $r(x) = r$. Now, by evaluating $f(x)$ at $x = k$, you have

$$f(k) = (k - k)q(k) + r = (0)q(k) + r = r.$$ ■

The Remainder Theorem tells you that synthetic division can be used to evaluate a polynomial function. That is, to evaluate a polynomial function $f(x)$ when $x = k$, divide $f(x)$ by $x - k$. The remainder will be $f(k)$, as shown in Example 5.

EXAMPLE 5 Using the Remainder Theorem

Use the Remainder Theorem to evaluate the following function at $x = -2$.

$$f(x) = 3x^3 + 8x^2 + 5x - 7$$

Solution Using synthetic division, you obtain the following.

$$\begin{array}{r|rrrr} -2 & 3 & 8 & 5 & -7 \\ & & -6 & -4 & -2 \\ \hline & 3 & 2 & 1 & -9 \end{array}$$

Because the remainder is $r = -9$, you can conclude that

$$f(-2) = -9.$$

This means that $(-2, -9)$ is a point on the graph of f. You can check this by substituting $x = -2$ in the original function. ■

Another important theorem is the **Factor Theorem,** which is stated below. This theorem states that you can test to see whether a polynomial has $(x - k)$ as a factor by evaluating the polynomial at $x = k$. When the result is 0, $(x - k)$ is a factor.

THEOREM 2.2 THE FACTOR THEOREM

A polynomial $f(x)$ has a factor $(x - k)$ if and only if $f(k) = 0$.

PROOF Using the Division Algorithm with the factor $(x - k)$, you have

$$f(x) = (x - k)q(x) + r(x).$$

By the Remainder Theorem, $r(x) = r = f(k)$, and you have

$$f(x) = (x - k)q(x) + f(k)$$

where $q(x)$ is a polynomial of lesser degree than $f(x)$. If $f(k) = 0$, then

$$f(x) = (x - k)q(x)$$

and you see that $(x - k)$ is a factor of $f(x)$. Conversely, if $(x - k)$ is a factor of $f(x)$, division of $f(x)$ by $(x - k)$ yields a remainder of 0. So, by the Remainder Theorem, you have $f(k) = 0$. ■

EXAMPLE 6 Factoring a Polynomial: Repeated Division

Show that $(x - 2)$ and $(x + 3)$ are factors of $f(x) = 2x^4 + 7x^3 - 4x^2 - 27x - 18$. Then find the remaining factors of $f(x)$.

Algebraic Solution

Use synthetic division with the factor $(x - 2)$.

$$\begin{array}{r|rrrrr} 2 & 2 & 7 & -4 & -27 & -18 \\ & & 4 & 22 & 36 & 18 \\ \hline & 2 & 11 & 18 & 9 & 0 \end{array}$$

← 0 remainder, so $f(2) = 0$ and $(x - 2)$ is a factor.

Use the result of this division to perform synthetic division again with the factor $(x + 3)$.

$$\begin{array}{r|rrrr} -3 & 2 & 11 & 18 & 9 \\ & & -6 & -15 & -9 \\ \hline & 2 & 5 & 3 & 0 \end{array}$$

← 0 remainder, so $f(-3) = 0$ and $(x + 3)$ is a factor.

$2x^2 + 5x + 3$

Because the resulting quadratic expression factors as

$$2x^2 + 5x + 3 = (2x + 3)(x + 1)$$

the complete factorization of $f(x)$ is

$$f(x) = (x - 2)(x + 3)(2x + 3)(x + 1).$$

Graphical Solution

From the graph of $f(x) = 2x^4 + 7x^3 - 4x^2 - 27x - 18$, you can see that there are four x-intercepts (see Figure 2.19). These occur at $x = -3$, $x = -\frac{3}{2}$, $x = -1$, and $x = 2$. (Check this algebraically.) This implies that $(x + 3)$, $\left(x + \frac{3}{2}\right)$, $(x + 1)$, and $(x - 2)$ are factors of $f(x)$. $\left[\text{Note that } \left(x + \frac{3}{2}\right) \text{ and } (2x + 3) \text{ are equivalent factors because they both yield the same zero, } x = -\frac{3}{2}.\right]$

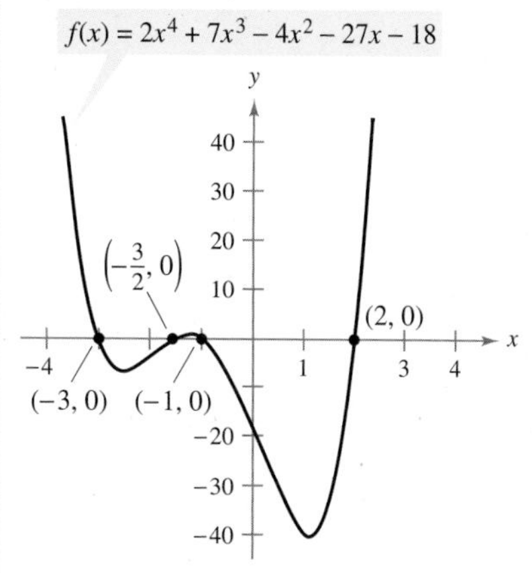

Figure 2.19 ■

STUDY TIP Note in Example 6 that the complete factorization of $f(x)$ implies that f has four real zeros: $x = 2$, $x = -3$, $x = -\frac{3}{2}$, and $x = -1$. This is confirmed by the graph of f, which is shown in Figure 2.19.

USES OF THE REMAINDER IN SYNTHETIC DIVISION

The remainder r, obtained in the synthetic division of $f(x)$ by $x - k$, provides the following information.

1. The remainder r gives the value of f at $x = k$. That is, $r = f(k)$.
2. When $r = 0$, $(x - k)$ is a factor of $f(x)$.
3. When $r = 0$, $(k, 0)$ is an x-intercept of the graph of f.

Application

EXAMPLE 7 Take-Home Pay

The 2010 monthly take-home pay for an employee who is single and claimed one deduction is given by the function

$$y = -0.00002320x^2 + 0.95189x + 37.564, \quad 500 \le x \le 5000$$

where y represents the take-home pay (in dollars) and x represents the gross monthly salary (in dollars). Find a function that gives the take-home pay as a *percent* of the gross monthly salary.

Solution Because the gross monthly salary is given by x and the take-home pay is given by y, the percent P of gross monthly salary that the person takes home is

$$\begin{aligned} P &= \frac{y}{x} \\ &= \frac{-0.00002320x^2 + 0.95189x + 37.564}{x} \\ &= -0.00002320x + 0.95189 + \frac{37.564}{x}. \end{aligned}$$

The graphs of y and P are shown in Figures 2.20(a) and (b), respectively. Note in Figure 2.20(b) that as a person's gross monthly salary increases, the *percent* that he or she takes home decreases.

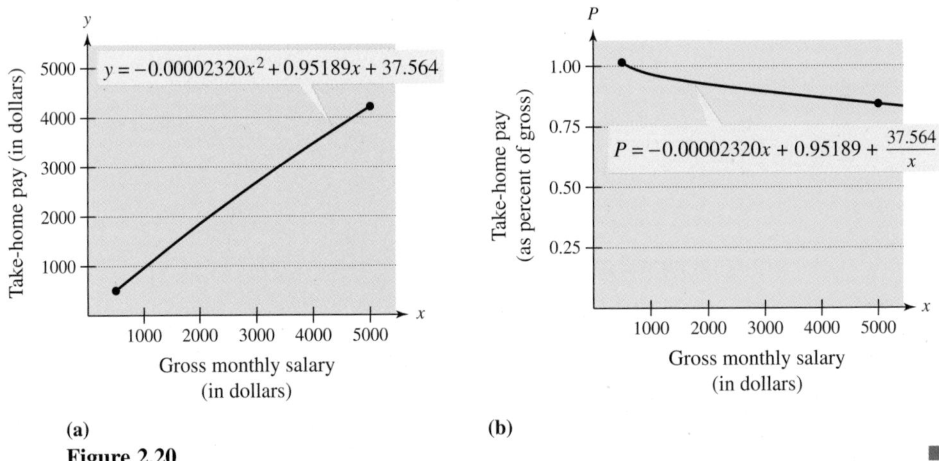

(a) (b)

Figure 2.20

Throughout this text, the importance of developing several problem-solving strategies is emphasized. In the exercises for this section, try using more than one strategy to solve several of the exercises. For instance, if you find that $x - k$ divides evenly into $f(x)$ (with no remainder), try sketching the graph of f. You should find that $(k, 0)$ is an x-intercept of the graph. Your problem-solving skills will be enhanced, too, by using a graphing utility to verify algebraic calculations, and conversely, to verify graphing utility results by analytic methods.

2.3 Exercises

See www.CalcChat.com for worked-out solutions to odd-numbered exercises.

1. Two forms of the Division Algorithm are shown below. Identify and label each term or function.

$$f(x) = d(x)q(x) + r(x) \qquad \frac{f(x)}{d(x)} = q(x) + \frac{r(x)}{d(x)}$$

In Exercises 2–6, fill in the blanks.

2. The rational expression $p(x)/q(x)$ is called ________ when the degree of the numerator is greater than or equal to that of the denominator, and is called ________ when the degree of the numerator is less than that of the denominator.

3. In the Division Algorithm, the rational expression $f(x)/d(x)$ is ________ because the degree of $f(x)$ is greater than or equal to the degree of $d(x)$.

4. An alternative method to long division of polynomials is called ________ ________, in which the divisor must be of the form $x - k$.

5. The ________ Theorem states that a polynomial $f(x)$ has a factor $(x - k)$ if and only if $f(k) = 0$.

6. The ________ Theorem states that when a polynomial $f(x)$ is divided by $x - k$, the remainder is $r = f(k)$.

Analytical Analysis **In Exercises 7 and 8, use long division to verify that $y_1 = y_2$.**

7. $y_1 = \dfrac{x^2}{x+2}$, $y_2 = x - 2 + \dfrac{4}{x+2}$

8. $y_1 = \dfrac{x^4 - 3x^2 - 1}{x^2 + 5}$, $y_2 = x^2 - 8 + \dfrac{39}{x^2 + 5}$

Graphical Analysis **In Exercises 9 and 10, (a) use a graphing utility to graph the two equations in the same viewing window, (b) use the graphs to verify that the expressions are equivalent, and (c) use long division to verify the results analytically.**

9. $y_1 = \dfrac{x^2 + 2x - 1}{x+3}$, $y_2 = x - 1 + \dfrac{2}{x+3}$

10. $y_1 = \dfrac{x^4 + x^2 - 1}{x^2 + 1}$, $y_2 = x^2 - \dfrac{1}{x^2 + 1}$

In Exercises 11–26, use long division to divide.

11. $(2x^2 + 10x + 12) \div (x + 3)$

12. $(5x^2 - 17x - 12) \div (x - 4)$

13. $(4x^3 - 7x^2 - 11x + 5) \div (4x + 5)$

14. $(6x^3 - 16x^2 + 17x - 6) \div (3x - 2)$

15. $(x^4 + 5x^3 + 6x^2 - x - 2) \div (x + 2)$

16. $(x^3 + 4x^2 - 3x - 12) \div (x - 3)$

17. $(x^3 - 27) \div (x - 3)$

18. $(x^3 + 125) \div (x + 5)$

19. $(7x + 3) \div (x + 2)$

20. $(8x - 5) \div (2x + 1)$

21. $(x^3 - 9) \div (x^2 + 1)$

22. $(x^5 + 7) \div (x^3 - 1)$

23. $(3x + 2x^3 - 9 - 8x^2) \div (x^2 + 1)$

24. $(5x^3 - 16 - 20x + x^4) \div (x^2 - x - 3)$

25. $\dfrac{x^4}{(x-1)^3}$

26. $\dfrac{2x^3 - 4x^2 - 15x + 5}{(x-1)^2}$

In Exercises 27–46, use synthetic division to divide.

27. $(3x^3 - 17x^2 + 15x - 25) \div (x - 5)$

28. $(5x^3 + 18x^2 + 7x - 6) \div (x + 3)$

29. $(6x^3 + 7x^2 - x + 26) \div (x - 3)$

30. $(2x^3 + 14x^2 - 20x + 7) \div (x + 6)$

31. $(4x^3 - 9x + 8x^2 - 18) \div (x + 2)$

32. $(9x^3 - 16x - 18x^2 + 32) \div (x - 2)$

33. $(-x^3 + 75x - 250) \div (x + 10)$

34. $(3x^3 - 16x^2 - 72) \div (x - 6)$

35. $(5x^3 - 6x^2 + 8) \div (x - 4)$

36. $(5x^3 + 6x + 8) \div (x + 2)$

37. $\dfrac{10x^4 - 50x^3 - 800}{x - 6}$

38. $\dfrac{x^5 - 13x^4 - 120x + 80}{x + 3}$

39. $\dfrac{x^3 + 512}{x + 8}$

40. $\dfrac{x^3 - 729}{x - 9}$

41. $\dfrac{-3x^4}{x - 2}$

42. $\dfrac{-3x^4}{x + 2}$

43. $\dfrac{180x - x^4}{x - 6}$

44. $\dfrac{5 - 3x + 2x^2 - x^3}{x + 1}$

45. $\dfrac{4x^3 + 16x^2 - 23x - 15}{x + \frac{1}{2}}$

46. $\dfrac{3x^3 - 4x^2 + 5}{x - \frac{3}{2}}$

In Exercises 47–54, write the function in the form $f(x) = (x - k)q(x) + r$ for the given value of k, and demonstrate that $f(k) = r$.

47. $f(x) = x^3 - x^2 - 14x + 11$, $k = 4$

48. $f(x) = x^3 - 5x^2 - 11x + 8$, $k = -2$

49. $f(x) = 15x^4 + 10x^3 - 6x^2 + 14$, $k = -\frac{2}{3}$

50. $f(x) = 10x^3 - 22x^2 - 3x + 4$, $k = \frac{1}{5}$

51. $f(x) = x^3 + 3x^2 - 2x - 14$, $k = \sqrt{2}$

52. $f(x) = x^3 + 2x^2 - 5x - 4, \quad k = -\sqrt{5}$

53. $f(x) = -4x^3 + 6x^2 + 12x + 4, \quad k = 1 - \sqrt{3}$

54. $f(x) = -3x^3 + 8x^2 + 10x - 8, \quad k = 2 + \sqrt{2}$

In Exercises 55–58, use the Remainder Theorem and synthetic division to find each function value. Verify your answers using another method.

55. $f(x) = 2x^3 - 7x + 3$
   (a) $f(1)$ (b) $f(-2)$ (c) $f(\frac{1}{2})$ (d) $f(2)$

56. $g(x) = 2x^6 + 3x^4 - x^2 + 3$
   (a) $g(2)$ (b) $g(1)$ (c) $g(3)$ (d) $g(-1)$

57. $h(x) = x^3 - 5x^2 - 7x + 4$
   (a) $h(3)$ (b) $h(2)$ (c) $h(-2)$ (d) $h(-5)$

58. $f(x) = 4x^4 - 16x^3 + 7x^2 + 20$
   (a) $f(1)$ (b) $f(-2)$ (c) $f(5)$ (d) $f(-10)$

In Exercises 59–66, use synthetic division to show that the given value of x is a solution of the third-degree polynomial equation, and use the result to factor the polynomial completely. List all real solutions of the equation.

59. $x^3 - 7x + 6 = 0, \quad x = 2$

60. $x^3 - 28x - 48 = 0, \quad x = -4$

61. $2x^3 - 15x^2 + 27x - 10 = 0, \quad x = \frac{1}{2}$

62. $48x^3 - 80x^2 + 41x - 6 = 0, \quad x = \frac{2}{3}$

63. $x^3 + 2x^2 - 3x - 6 = 0, \quad x = \sqrt{3}$

64. $x^3 + 2x^2 - 2x - 4 = 0, \quad x = \sqrt{2}$

65. $x^3 - 3x^2 + 2 = 0, \quad x = 1 + \sqrt{3}$

66. $x^3 - x^2 - 13x - 3 = 0, \quad x = 2 - \sqrt{5}$

In Exercises 67–74, (a) verify the given factors of the function f, (b) find the remaining factor(s) of f, (c) use your results to write the complete factorization of f, (d) list all real zeros of f, and (e) confirm your results by using a graphing utility to graph the function.

Function	*Factors*
67. $f(x) = 2x^3 + x^2 - 5x + 2$	$(x + 2), (x - 1)$
68. $f(x) = 3x^3 + 2x^2 - 19x + 6$	$(x + 3), (x - 2)$
69. $f(x) = x^4 - 4x^3 - 15x^2 + 58x - 40$	$(x - 5), (x + 4)$
70. $f(x) = 8x^4 - 14x^3 - 71x^2 - 10x + 24$	$(x + 2), (x - 4)$
71. $f(x) = 6x^3 + 41x^2 - 9x - 14$	$(2x + 1), (3x - 2)$
72. $f(x) = 10x^3 - 11x^2 - 72x + 45$	$(2x + 5), (5x - 3)$
73. $f(x) = 2x^3 - x^2 - 10x + 5$	$(2x - 1), (x + \sqrt{5})$
74. $f(x) = x^3 + 3x^2 - 48x - 144$	$(x + 4\sqrt{3}), (x + 3)$

Graphical Analysis **In Exercises 75–80, (a) use the *zero* or *root* feature of a graphing utility to approximate the zeros of the function accurate to three decimal places, (b) determine one of the exact zeros, and (c) use synthetic division to verify your result from part (b), and then factor the polynomial completely.**

75. $f(x) = x^3 - 2x^2 - 5x + 10$

76. $g(x) = x^3 - 4x^2 - 2x + 8$

77. $h(t) = t^3 - 2t^2 - 7t + 2$

78. $f(s) = s^3 - 12s^2 + 40s - 24$

79. $h(x) = x^5 - 7x^4 + 10x^3 + 14x^2 - 24x$

80. $g(x) = 6x^4 - 11x^3 - 51x^2 + 99x - 27$

In Exercises 81–84, simplify the rational expression by using long division or synthetic division.

81. $\dfrac{4x^3 - 8x^2 + x + 3}{2x - 3}$

82. $\dfrac{x^3 + x^2 - 64x - 64}{x + 8}$

83. $\dfrac{x^4 + 6x^3 + 11x^2 + 6x}{x^2 + 3x + 2}$

84. $\dfrac{x^4 + 9x^3 - 5x^2 - 36x + 4}{x^2 - 4}$

WRITING ABOUT CONCEPTS

In Exercises 85 and 86, perform the division by assuming that n is a positive integer.

85. $\dfrac{x^{3n} + 9x^{2n} + 27x^n + 27}{x^n + 3}$

86. $\dfrac{x^{3n} - 3x^{2n} + 5x^n - 6}{x^n - 2}$

87. Briefly explain what it means for a divisor to divide evenly into a dividend.

88. Briefly explain how to check polynomial division, and justify your reasoning. Give an example.

In Exercises 89 and 90, find the constant c such that the denominator will divide evenly into the numerator.

89. $\dfrac{x^3 + 4x^2 - 3x + c}{x - 5}$

90. $\dfrac{x^5 - 2x^2 + x + c}{x + 2}$

91. *Data Analysis: Higher Education* The amounts A (in billions of dollars) donated to support higher education in the United States from 2000 through 2007 are shown in the table, where t represents the year, with $t = 0$ corresponding to 2000.

Year, t	0	1	2	3
Amount, A	23.2	24.2	23.9	23.9

Year, t	4	5	6	7
Amount, A	24.4	25.6	28.0	29.8

(a) Use a graphing utility to create a scatter plot of the data.

(b) Use the *regression* feature of the graphing utility to find a cubic model for the data. Graph the model in the same viewing window as the scatter plot.

(c) Use the model to create a table of estimated values of A. Compare the model with the original data.

(d) Use synthetic division to evaluate the model for the year 2010. Even though the model is relatively accurate for estimating the given data, would you use this model to predict the amount donated to higher education in the future? Explain.

92. *Data Analysis: Health Care* The amounts A (in billions of dollars) of national health care expenditures in the United States from 2000 through 2007 are shown in the table, where t represents the year, with $t = 0$ corresponding to 2000.

Year, t	0	1	2	3
Amount, A	30.5	32.2	34.2	38.0

Year, t	4	5	6	7
Amount, A	42.7	47.9	52.7	57.6

(a) Use a graphing utility to create a scatter plot of the data.

(b) Use the *regression* feature of the graphing utility to find a cubic model for the data. Graph the model in the same viewing window as the scatter plot.

(c) Use the model to create a table of estimated values of A. Compare the model with the original data.

(d) Use synthetic division to evaluate the model for the year 2010.

True or False? **In Exercises 93–97, determine whether the statement is true or false. Justify your answer.**

93. If $(7x + 4)$ is a factor of some polynomial function f, then $\frac{4}{7}$ is a zero of f.

94. $(2x - 1)$ is a factor of the polynomial

$$6x^6 + x^5 - 92x^4 + 45x^3 + 184x^2 + 4x - 48.$$

95. The rational expression

$$\frac{x^3 + 2x^2 - 13x + 10}{x^2 - 4x - 12}$$

is improper.

96. If $x = k$ is a zero of a function f, then $f(k) = 0$.

97. To divide $x^4 - 3x^2 + 4x - 1$ by $x + 2$ using synthetic division, the setup would appear as shown.

$$\begin{array}{r|rrrr} -2 & 1 & -3 & 4 & -1 \\ & & & & \\ \hline \end{array}$$

98. Use the form $f(x) = (x - k)q(x) + r$ to create a cubic function that (a) passes through the point $(2, 5)$ and rises to the right, and (b) passes through the point $(-3, 1)$ and falls to the right. (There are many correct answers.)

99. *Think About It* Find the value of k such that $x - 4$ is a factor of $x^3 - kx^2 + 2kx - 8$.

100. *Think About It* Find the value of k such that $x - 3$ is a factor of $x^3 - kx^2 + 2kx - 12$.

101. *Writing* Complete each polynomial division. Write a brief description of the pattern that you obtain, and use your result to find a formula for the polynomial division $(x^n - 1)/(x - 1)$. Create a numerical example to test your formula.

(a) $\dfrac{x^2 - 1}{x - 1} = \square$

(b) $\dfrac{x^3 - 1}{x - 1} = \square$

(c) $\dfrac{x^4 - 1}{x - 1} = \square$

CAPSTONE

102. Consider the division

$$f(x) \div (x - k)$$

where

$$f(x) = (x + 3)^2(x - 3)(x + 1)^3.$$

(a) What is the remainder when $k = -3$? Explain.

(b) If it is necessary to find $f(2)$, is it easier to evaluate the function directly or to use synthetic division? Explain.

2.4 Complex Numbers

- **Use the imaginary unit *i* to write complex numbers.**
- **Add, subtract, and multiply complex numbers.**
- **Use complex conjugates to write the quotient of two complex numbers in standard form.**
- **Find complex solutions of quadratic equations.**

The Imaginary Unit *i*

You have learned that some quadratic equations have no real solutions. For instance, the quadratic equation $x^2 + 1 = 0$ has no real solution because there is no real number x that can be squared to produce -1. To overcome this deficiency, mathematicians created an expanded system of numbers using the **imaginary unit *i*,** defined as

$$i = \sqrt{-1} \qquad \text{Imaginary unit}$$

where $i^2 = -1$. By adding real numbers to real multiples of this imaginary unit, the set of **complex numbers** is obtained. Each complex number can be written in the **standard form $a + bi$.** For instance, the standard form of the complex number $-5 + \sqrt{-9}$ is $-5 + 3i$ because

$$-5 + \sqrt{-9} = -5 + \sqrt{3^2(-1)} = -5 + 3\sqrt{-1} = -5 + 3i.$$

In the standard form $a + bi$, the real number a is called the **real part** of the complex number $a + bi$, and the number bi (where b is a real number) is called the **imaginary part** of the complex number.

DEFINITION OF A COMPLEX NUMBER

When a and b are real numbers, the number $a + bi$ is a **complex number,** and it is said to be written in **standard form.** When $b = 0$, the number $a + bi = a$ is a real number. When $b \neq 0$, the number $a + bi$ is called an **imaginary number.** A number of the form bi, where $b \neq 0$, is called a **pure imaginary number.**

The set of real numbers is a subset of the set of complex numbers, as shown in Figure 2.21. This is true because every real number a can be written as a complex number using $b = 0$. That is, for every real number a, you can write $a = a + 0i$.

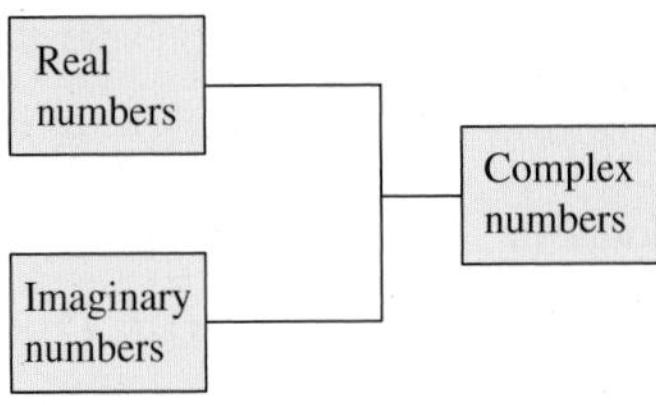

Figure 2.21

EQUALITY OF COMPLEX NUMBERS

Two complex numbers $a + bi$ and $c + di$, written in standard form, are equal to each other

$$a + bi = c + di \qquad \text{Equality of two complex numbers}$$

if and only if $a = c$ and $b = d$.

Operations with Complex Numbers

To add (or subtract) two complex numbers, you add (or subtract) the real and imaginary parts of the numbers separately.

ADDITION AND SUBTRACTION OF COMPLEX NUMBERS

When $a + bi$ and $c + di$ are two complex numbers written in standard form, their sum and difference are defined as follows.

Sum: $(a + bi) + (c + di) = (a + c) + (b + d)i$

Difference: $(a + bi) - (c + di) = (a - c) + (b - d)i$

The **additive identity** in the complex number system is zero (the same as in the real number system). Furthermore, the **additive inverse** of the complex number $a + bi$ is

$-(a + bi) = -a - bi.$ Additive inverse

So, you have

$(a + bi) + (-a - bi) = 0 + 0i = 0.$

EXAMPLE 1 Adding and Subtracting Complex Numbers

Perform the operations on the complex numbers.

a. $(3 - i) + (2 + 3i)$

b. $2i + (-4 - 2i)$

c. $3 - (-2 + 3i) + (-5 + i)$

d. $(3 + 2i) + (4 - i) - (7 + i)$

Solution

a. $(3 - i) + (2 + 3i) = 3 - i + 2 + 3i$ Remove parentheses.

$= 3 + 2 - i + 3i$ Group like terms.

$= (3 + 2) + (-1 + 3)i$

$= 5 + 2i$ Write in standard form.

b. $2i + (-4 - 2i) = 2i - 4 - 2i$ Remove parentheses.

$= -4 + 2i - 2i$ Group like terms.

$= -4$ Write in standard form.

c. $3 - (-2 + 3i) + (-5 + i) = 3 + 2 - 3i - 5 + i$

$= 3 + 2 - 5 - 3i + i$

$= 0 - 2i$

$= -2i$

d. $(3 + 2i) + (4 - i) - (7 + i) = 3 + 2i + 4 - i - 7 - i$

$= 3 + 4 - 7 + 2i - i - i$

$= 0 + 0i$

$= 0$ ■

Note in Example 1(b) that the sum of two complex numbers can be a real number.

EXPLORATION

Complete the following.

$i^1 = i$	$i^7 =$
$i^2 = -1$	$i^8 =$
$i^3 = -i$	$i^9 =$
$i^4 = 1$	$i^{10} =$
$i^5 =$	$i^{11} =$
$i^6 =$	$i^{12} =$

What pattern do you see? Write a brief description of how you would find i raised to any positive integer power.

Many of the properties of real numbers are valid for complex numbers as well. Here are some examples.

Associative Properties of Addition and Multiplication

Commutative Properties of Addition and Multiplication

Distributive Property of Multiplication Over Addition

Notice below how these properties are used when two complex numbers are multiplied.

$$\begin{aligned}(a + bi)(c + di) &= a(c + di) + bi(c + di) && \text{Distributive Property}\\ &= ac + (ad)i + (bc)i + (bd)i^2 && \text{Distributive Property}\\ &= ac + (ad)i + (bc)i + (bd)(-1) && i^2 = -1\\ &= ac - bd + (ad)i + (bc)i && \text{Commutative Property}\\ &= (ac - bd) + (ad + bc)i && \text{Associative Property}\end{aligned}$$

Rather than trying to memorize this multiplication rule, you should simply remember how the Distributive Property is used to multiply two complex numbers. The procedure is similar to multiplying two polynomials and combining like terms.

EXAMPLE 2 Multiplying Complex Numbers

Multiply the complex numbers.

a. $4(-2 + 3i)$

b. $(i)(-3i)$

c. $(2 - i)(4 + 3i)$

d. $(3 + 2i)(3 - 2i)$

e. $(3 + 2i)^2$

Solution

a. $$\begin{aligned}4(-2 + 3i) &= 4(-2) + 4(3i) && \text{Distributive Property}\\ &= -8 + 12i && \text{Simplify.}\end{aligned}$$

b. $$\begin{aligned}(i)(-3i) &= -3i^2 && \text{Multiply.}\\ &= -3(-1) && i^2 = -1\\ &= 3 && \text{Simplify.}\end{aligned}$$

c. $$\begin{aligned}(2 - i)(4 + 3i) &= 8 + 6i - 4i - 3i^2 && \text{Product of binomials}\\ &= 8 + 6i - 4i - 3(-1) && i^2 = -1\\ &= (8 + 3) + (6i - 4i) && \text{Group like terms.}\\ &= 11 + 2i && \text{Write in standard form.}\end{aligned}$$

d. $$\begin{aligned}(3 + 2i)(3 - 2i) &= 9 - 6i + 6i - 4i^2 && \text{Product of binomials}\\ &= 9 - 6i + 6i - 4(-1) && i^2 = -1\\ &= 9 + 4 && \text{Simplify.}\\ &= 13 && \text{Write in standard form.}\end{aligned}$$

e. $$\begin{aligned}(3 + 2i)^2 &= 9 + 6i + 6i + 4i^2 && \text{Product of binomials}\\ &= 9 + 6i + 6i + 4(-1) && i^2 = -1\\ &= 9 + 12i - 4 && \text{Simplify.}\\ &= 5 + 12i && \text{Write in standard form.}\end{aligned}$$

■

Complex Conjugates

Notice in Example 2(d) that the product of two complex numbers can be a real number. This occurs with pairs of complex numbers of the form $a + bi$ and $a - bi$, called **complex conjugates.**

$$\begin{aligned}(a + bi)(a - bi) &= a^2 - abi + abi - b^2i^2\\ &= a^2 - b^2(-1)\\ &= a^2 + b^2\end{aligned}$$

EXAMPLE 3 Multiplying Conjugates

Multiply each complex number by its complex conjugate.

a. $1 + i$

b. $4 - 3i$

Solution

a. The complex conjugate of $1 + i$ is $1 - i$.

$$(1 + i)(1 - i) = 1^2 - i^2 = 1 - (-1) = 2$$

b. The complex conjugate of $4 - 3i$ is $4 + 3i$.

$$(4 - 3i)(4 + 3i) = 4^2 - (3i)^2 = 16 - 9i^2 = 16 - 9(-1) = 25$$

■

STUDY TIP Note that when you multiply the numerator and denominator of a quotient of complex numbers by

$$\frac{c - di}{c - di}$$

you are actually multiplying the quotient by a form of 1. You are not changing the original expression, you are only creating an expression that is equivalent to the original expression.

To write the quotient of $a + bi$ and $c + di$ in standard form, where c and d are not both zero, multiply the numerator and denominator by the complex conjugate of the *denominator* to obtain

$$\begin{aligned}\frac{a + bi}{c + di} &= \frac{a + bi}{c + di}\left(\frac{c - di}{c - di}\right)\\ &= \frac{(ac + bd) + (bc - ad)i}{c^2 + d^2}\\ &= \frac{ac + bd}{c^2 + d^2} + \frac{(bc - ad)i}{c^2 + d^2}. \end{aligned}$$

Standard form

EXAMPLE 4 Writing a Complex Number in Standard Form

Write the complex number $\dfrac{2 + 3i}{4 - 2i}$ in standard form.

Solution

$$\frac{2 + 3i}{4 - 2i} = \frac{2 + 3i}{4 - 2i}\left(\frac{4 + 2i}{4 + 2i}\right)$$

Multiply numerator and denominator by complex conjugate of denominator.

$$= \frac{8 + 4i + 12i + 6i^2}{16 - 4i^2}$$

Expand.

$$= \frac{8 + 4i + 12i - 6}{16 + 4}$$

$i^2 = -1$

$$= \frac{2 + 16i}{20} = \frac{1}{10} + \frac{4}{5}i$$

Simplify and write in standard form.

■

Complex Solutions of Quadratic Equations

When using the Quadratic Formula to solve a quadratic equation, you often obtain a result such as $\sqrt{-3}$, which you know is not a real number. By factoring out $i = \sqrt{-1}$, you can write this number in standard form.

$$\sqrt{-3} = \sqrt{3(-1)} = \sqrt{3}\sqrt{-1} = \sqrt{3}i$$

The number $\sqrt{3}i$ is called the **principal square root** of -3.

STUDY TIP The definition of principal square root uses the rule

$$\sqrt{ab} = \sqrt{a}\sqrt{b}$$

for $a > 0$ and $b < 0$. This rule is not valid when *both* a and b are negative. For example,

$$\begin{aligned}\sqrt{-5}\sqrt{-5} &= \sqrt{5(-1)}\sqrt{5(-1)}\\ &= \sqrt{5}i\sqrt{5}i\\ &= \sqrt{25}i^2\\ &= 5i^2 = -5\end{aligned}$$

whereas

$$\sqrt{(-5)(-5)} = \sqrt{25} = 5.$$

To avoid problems with multiplying square roots of negative numbers, be sure to convert complex numbers to standard form *before* multiplying.

PRINCIPAL SQUARE ROOT OF A NEGATIVE NUMBER

When a is a positive number, the **principal square root** of the negative number $-a$ is defined as

$$\sqrt{-a} = \sqrt{a}i.$$

EXAMPLE 5 Writing Complex Numbers in Standard Form

Write each complex number in standard form and simplify.

a. $\sqrt{-3}\sqrt{-12}$ **b.** $\sqrt{-48} - \sqrt{-27}$ **c.** $\left(-1 + \sqrt{-3}\right)^2$

Solution

a. $\sqrt{-3}\sqrt{-12} = \sqrt{3}i\sqrt{12}i = \sqrt{36}i^2 = 6(-1) = -6$

b. $\sqrt{-48} - \sqrt{-27} = \sqrt{48}i - \sqrt{27}i = 4\sqrt{3}i - 3\sqrt{3}i = \sqrt{3}i$

c.
$$\begin{aligned}\left(-1 + \sqrt{-3}\right)^2 &= \left(-1 + \sqrt{3}i\right)^2\\ &= (-1)^2 - 2\sqrt{3}i + \left(\sqrt{3}\right)^2(i^2)\\ &= 1 - 2\sqrt{3}i + 3(-1)\\ &= -2 - 2\sqrt{3}i\end{aligned}$$

EXAMPLE 6 Complex Solutions of a Quadratic Equation

Solve each quadratic equation.

a. $x^2 + 4 = 0$ **b.** $3x^2 - 2x + 5 = 0$

Solution

a. $x^2 + 4 = 0$ Write original equation.

$x^2 = -4$ Subtract 4 from each side.

$x = \pm 2i$ Extract square roots.

b. $3x^2 - 2x + 5 = 0$ Write original equation.

$x = \dfrac{-(-2) \pm \sqrt{(-2)^2 - 4(3)(5)}}{2(3)}$ Quadratic Formula

$= \dfrac{2 \pm \sqrt{-56}}{6}$ Simplify.

$= \dfrac{2 \pm 2\sqrt{14}i}{6}$ Write $\sqrt{-56}$ in standard form.

$= \dfrac{1}{3} \pm \dfrac{\sqrt{14}}{3}i$ Write in standard form. ■

2.4 Exercises

See www.CalcChat.com for worked-out solutions to odd-numbered exercises.

1. Match the type of complex number $a + bi$ with its definition.

(a) Real number (i) $a \neq 0, \quad b \neq 0$

(b) Imaginary number (ii) $a = 0, \quad b \neq 0$

(c) Pure imaginary number (iii) $b = 0$

In Exercises 2–4, fill in the blanks.

2. The imaginary unit i is defined as $i =$ ________, where $i^2 =$ ________.

3. When a is a positive number, the ________ ________ root of the negative number $-a$ is defined as $\sqrt{-a} = \sqrt{a}\,i$.

4. The numbers $a + bi$ and $a - bi$ are called ________ ________, and their product is a real number $a^2 + b^2$.

In Exercises 5–8, find real numbers *a* and *b* such that the equation is true.

5. $a + bi = -12 + 7i$

6. $a + bi = 13 + 4i$

7. $(a - 1) + (b + 3)i = 5 + 8i$

8. $(a + 6) + 2bi = 6 - 5i$

In Exercises 9–16, write the complex number in standard form.

9. $8 + \sqrt{-25}$

10. $2 - \sqrt{-27}$

11. $\sqrt{-80}$

12. $\sqrt{-4}$

13. $\sqrt{-0.09}$

14. 14

15. $-10i + i^2$

16. $-4i^2 + 2i$

In Exercises 17–24, perform the addition or subtraction and write the result in standard form.

17. $(7 + i) + (3 - 4i)$

18. $(13 - 2i) + (-5 + 6i)$

19. $(9 - i) - (8 - i)$

20. $(3 + 2i) - (6 + 13i)$

21. $\left(-2 + \sqrt{-8}\right) + \left(5 - \sqrt{-50}\right)$

22. $\left(8 + \sqrt{-18}\right) - \left(4 + 3\sqrt{2}i\right)$

23. $13i - (14 - 7i)$

24. $-\left(\frac{3}{2} + \frac{5}{2}i\right) + \left(\frac{5}{3} + \frac{11}{3}i\right)$

In Exercises 25–34, perform the operation and write the result in standard form.

25. $\sqrt{-5} \cdot \sqrt{-10}$

26. $\left(\sqrt{-75}\right)^2$

27. $(1 + i)(3 - 2i)$

28. $(7 - 2i)(3 - 5i)$

29. $12i(1 - 9i)$

30. $-8i(9 + 4i)$

31. $\left(\sqrt{14} + \sqrt{10}i\right)\left(\sqrt{14} - \sqrt{10}i\right)$

32. $\left(3 + \sqrt{-5}\right)\left(7 - \sqrt{-10}\right)$

33. $(6 + 7i)^2$

34. $(5 - 4i)^2$

In Exercises 35–38, write the complex conjugate of the complex number. Then multiply the number by its complex conjugate.

35. $9 + 2i$

36. $-1 - \sqrt{5}i$

37. $\sqrt{-20}$

38. $\sqrt{6}$

In Exercises 39–44, write the quotient in standard form.

39. $3/i$

40. $-14/(2i)$

41. $\dfrac{13}{1 - i}$

42. $\dfrac{6 - 7i}{1 - 2i}$

43. $\dfrac{8 + 16i}{2i}$

44. $\dfrac{3i}{(4 - 5i)^2}$

In Exercises 45–48, perform the operation and write the result in standard form.

45. $\dfrac{2}{1 + i} - \dfrac{3}{1 - i}$

46. $\dfrac{2i}{2 + i} + \dfrac{5}{2 - i}$

47. $\dfrac{i}{3 - 2i} + \dfrac{2i}{3 + 8i}$

48. $\dfrac{1 + i}{i} - \dfrac{3}{4 - i}$

In Exercises 49–54, use the Quadratic Formula to solve the quadratic equation.

49. $x^2 - 2x + 2 = 0$

50. $4x^2 + 16x + 17 = 0$

51. $9x^2 - 6x + 37 = 0$

52. $16t^2 - 4t + 3 = 0$

53. $1.4x^2 - 2x - 10 = 0$

54. $\frac{3}{2}x^2 - 6x + 9 = 0$

In Exercises 55–60, simplify the complex number and write it in standard form.

55. $-6i^3 + i^2$

56. $4i^2 - 2i^3$

57. $(-i)^3$

58. $\left(\sqrt{-2}\right)^6$

59. $1/i^3$

60. $1/(2i)^3$

WRITING ABOUT CONCEPTS

61. Show that the product of a complex number $a + bi$ and its complex conjugate is a real number.

62. Describe the error.

$$\sqrt{-6}\sqrt{-6} = \sqrt{(-6)(-6)} = \sqrt{36} = 6$$

63. Show that the complex conjugate of the sum of two complex numbers $a_1 + b_1 i$ and $a_2 + b_2 i$ is the sum of their complex conjugates.

64. Raise each complex number to the fourth power.

(a) 2 (b) -2 (c) $2i$ (d) $-2i$

65. Write each of the powers of i as i, $-i$, 1, or -1.

(a) i^{40} (b) i^{25} (c) i^{50} (d) i^{67}

CAPSTONE

66. Consider the functions

$$f(x) = 2(x - 3)^2 - 4 \text{ and } g(x) = -2(x - 3)^2 - 4.$$

(a) Without graphing either function, determine whether the graph of f and the graph of g have x-intercepts. Explain your reasoning.

(b) Solve $f(x) = 0$ and $g(x) = 0$.

(c) Explain how the zeros of f and g are related to whether their graphs have x-intercepts.

(d) For the function $f(x) = a(x - h)^2 + k$, make a general statement about how a, h, and k affect whether the graph of f has x-intercepts, and whether the zeros of f are real or complex.

True or False? **In Exercises 67–69, determine whether the statement is true or false. Justify your answer.**

67. There is no complex number that is equal to its complex conjugate.

68. $-i\sqrt{6}$ is a solution of $x^4 - x^2 + 14 = 56$.

69. $i^{44} + i^{150} - i^{74} - i^{109} + i^{61} = -1$

70. ***Impedance*** The opposition to current in an electrical circuit is called its impedance. The impedance z in a parallel circuit with two pathways satisfies the equation

$$\frac{1}{z} = \frac{1}{z_1} + \frac{1}{z_2}$$

where z_1 is the impedance (in ohms) of pathway 1 and z_2 is the impedance of pathway 2.

(a) The impedance of each pathway in a parallel circuit is found by adding the impedances of all components in the pathway. Use the table to find z_1 and z_2.

(b) Find the impedance z.

	Resistor	Inductor	Capacitor
Symbol	$a\Omega$	$b\Omega$	$c\Omega$
Impedance	a	bi	$-ci$

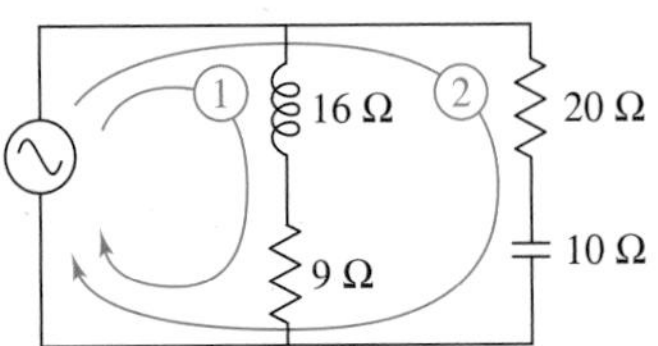

SECTION PROJECT

The Mandelbrot Set

Graphing utilities can be used to draw pictures of fractals in the complex plane. The most famous fractal is called the **Mandelbrot Set,** after the Polish-born mathematician Benoit Mandelbrot. To construct the Mandelbrot Set, consider the following sequence of numbers.

$$c, c^2 + c, (c^2 + c)^2 + c, [(c^2 + c)^2 + c]^2 + c, . . .$$

The behavior of this sequence depends on the value of the complex number c. For some values of c this sequence is **bounded,** which means that the absolute value of each number $\left(|a + bi| = \sqrt{a^2 + b^2}\right)$ in the sequence is less than some fixed number N. For other values of c the sequence is **unbounded,** which means that the absolute values of the terms of the sequence become infinitely large. When the sequence is bounded, the complex number c is in the Mandelbrot Set. When the sequence is unbounded, the complex number c is not in the Mandelbrot Set.

(a) The pseudo code below can be translated into a program for a graphing utility. (Programs for several models of graphing calculators can be found at our website *academic.cengage.com.*) The program determines whether the complex number c is in the Mandelbrot Set. To run the program for $c = -1 + 0.2i$, enter -1 for A and 0.2 for B. Press ENTER to see the first term of the sequence. Press ENTER again to see the second term of the sequence. Continue pressing ENTER. When the terms become large, the sequence is unbounded. For the number $c = -1 + 0.2i$, the terms are $-1 + 0.2i$, $-0.04 - 0.2i$, $-1.038 + 0.216i$, $0.032 - 0.249i, . . .$, and so the sequence is bounded. So, $c = -1 + 0.2i$ is in the Mandelbrot Set.

Program

1. Enter the real part A.

2. Enter the imaginary part B.

3. Store A in C. **4.** Store B in D.

5. Store 0 in N (number of term).

6. Label 1.

7. Increment N. **8.** Display N.

9. Display A. **10.** Display B.

11. Store A in F. **12.** Store B in G.

13. Store $F^2 - G^2 + C$ in A.

14. Store $2FG + D$ in B. **15.** Go to Label 1.

(b) Use a graphing calculator program or a computer program to determine whether the complex numbers $c = 1$, $c = -1 + 0.5i$, and $c = 0.1 + 0.1i$ are in the Mandelbrot Set.

2.5 The Fundamental Theorem of Algebra

- **Understand and use the Fundamental Theorem of Algebra.**
- **Find all the zeros of a polynomial function.**
- **Write a polynomial function with real coefficients, given its zeros.**

The Fundamental Theorem of Algebra

You know that an nth-degree polynomial can have *at most* n real zeros. In the complex number system, this statement can be improved. That is, in the complex number system, every nth-degree polynomial function has *precisely* n zeros. This important result is derived from the **Fundamental Theorem of Algebra,** first proved by the German mathematician Carl Friedrich Gauss (1777–1855).

THEOREM 2.3 THE FUNDAMENTAL THEOREM OF ALGEBRA

If $f(x)$ is a polynomial of degree n, where $n > 0$, then f has at least one zero in the complex number system.

Using the Fundamental Theorem of Algebra and the equivalence of zeros and factors, you obtain the **Linear Factorization Theorem.** (A proof is given in Appendix A.)

NOTE The Fundamental Theorem of Algebra and the Linear Factorization Theorem tell you only that the zeros or factors of a polynomial exist, not how to find them. Such theorems are called *existence theorems*. Remember that the n zeros of a polynomial function can be real or complex, and they may be repeated.

THEOREM 2.4 LINEAR FACTORIZATION THEOREM

If $f(x)$ is a polynomial of degree n, where $n > 0$, then f has precisely n linear factors

$$f(x) = a_n(x - c_1)(x - c_2) \cdot \cdot \cdot (x - c_n)$$

where $c_1, c_2, . . . , c_n$ are complex numbers.

EXAMPLE 1 Zeros of Polynomial Functions

STUDY TIP Recall that in order to find the zeros of a function $f(x)$, set $f(x)$ equal to 0 and solve the resulting equation for x. For instance, the function in Example 1(a) has a zero at $x = 2$ because

$$x - 2 = 0$$
$$x = 2.$$

Find the zeros of (a) $f(x) = x - 2$, (b) $f(x) = x^2 - 6x + 9$, (c) $f(x) = x^3 + 4x$, and (d) $f(x) = x^4 - 1$.

Solution

a. The first-degree polynomial $f(x) = x - 2$ has exactly *one* zero: $x = 2$.

b. Counting multiplicity, the second-degree polynomial function

$$f(x) = x^2 - 6x + 9 = (x - 3)(x - 3)$$

has exactly *two* zeros: $x = 3$ and $x = 3$. (This is called a *repeated zero*.)

c. The third-degree polynomial function

$$f(x) = x^3 + 4x = x(x^2 + 4) = x(x - 2i)(x + 2i)$$

has exactly *three* zeros: $x = 0, x = 2i$, and $x = -2i$.

d. The fourth-degree polynomial function

$$f(x) = x^4 - 1 = (x - 1)(x + 1)(x - i)(x + i)$$

has exactly *four* zeros: $x = 1, x = -1, x = i$, and $x = -i$. ■

Fogg Art Museum/Harvard University

JEAN LE ROND D'ALEMBERT (1717–1783)

D'Alembert worked independently of Carl Gauss in trying to prove the Fundamental Theorem of Algebra. His efforts were such that, in France, the Fundamental Theorem of Algebra is frequently known as the Theorem of d'Alembert.

The Rational Zero Test

The **Rational Zero Test** relates the possible rational zeros of a polynomial (having integer coefficients) to the leading coefficient and to the constant term of the polynomial. Recall that a rational number is any real number that can be written as the ratio of two integers.

THE RATIONAL ZERO TEST

When the polynomial $f(x) = a_nx^n + a_{n-1}x^{n-1} + \cdot \cdot \cdot + a_2x^2 + a_1x + a_0$ has *integer* coefficients, every rational zero of f has the form

$$\text{Rational zero} = \frac{p}{q}$$

where p and q have no common factors other than 1, and

$p =$ a factor of the constant term a_0

$q =$ a factor of the leading coefficient a_n.

To use the Rational Zero Test, you should first list all rational numbers whose numerators are factors of the constant term and whose denominators are factors of the leading coefficient.

$$\text{Possible rational zeros} = \frac{\text{factors of constant term}}{\text{factors of leading coefficient}}$$

Having formed this list of *possible rational zeros,* use a trial-and-error method to determine which, if any, are actual zeros of the polynomial.

NOTE When the leading coefficient is 1, the possible rational zeros are simply the factors of the constant term. ■

EXAMPLE 2 Rational Zero Test with Leading Coefficient of 1

Find the rational zeros of

$$f(x) = x^3 + x + 1.$$

Solution Because the leading coefficient is 1, the possible rational zeros are ± 1, the factors of the constant term.

Possible rational zeros: ± 1

By testing these possible zeros, you can see that neither works.

$f(1) = 1^3 + 1 + 1 = 3$ 1 is *not* a zero.

$f(-1) = (-1)^3 + (-1) + 1 = -1$ -1 is *not* a zero.

So, you can conclude that the given polynomial has no *rational* zeros. Note from the graph of f in Figure 2.22 that f does have one real zero between -1 and 0. However, by the Rational Zero Test, you know that this real zero is *not* a rational number. ■

Figure 2.22

The next few examples show how synthetic division can be used to test for rational zeros.

STUDY TIP When the list of possible rational zeros is small, as in Example 2, it may be quicker to test the zeros by evaluating the function. When the list of possible rational zeros is large, as in Example 3, it may be quicker to use a different approach to test the zeros, such as using synthetic division or sketching a graph.

EXAMPLE 3 Rational Zero Test with Leading Coefficient of 1

Find the rational zeros of

$$f(x) = x^4 - x^3 + x^2 - 3x - 6.$$

Solution Because the leading coefficient is 1, the possible rational zeros are the factors of the constant term.

Possible rational zeros: $\pm1, \pm2, \pm3, \pm6$

A test of these possible zeros shows that $x = -1$ and $x = 2$ are the only two that work. To test that $x = -1$ and $x = 2$ are zeros of f, you can apply synthetic division, as shown.

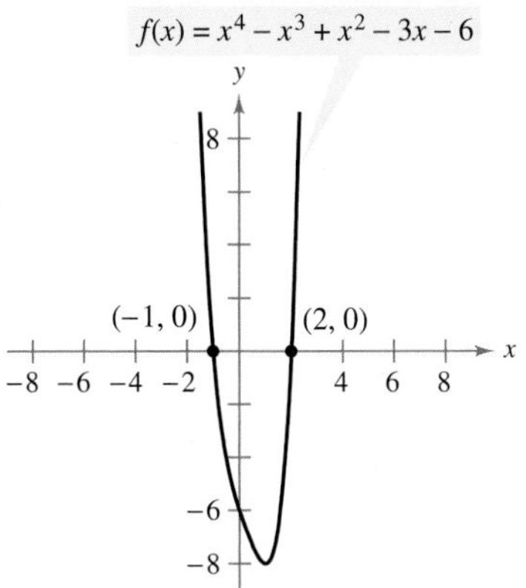

Figure 2.23

$$\begin{array}{r|rrrrr} -1 & 1 & -1 & 1 & -3 & -6 \\ & & -1 & 2 & -3 & 6 \\ \hline & 1 & -2 & 3 & -6 & 0 \end{array}$$

← 0 remainder, so $x = -1$ is a zero.

$$\begin{array}{r|rrrr} 2 & 1 & -2 & 3 & -6 \\ & & 2 & 0 & 6 \\ \hline & 1 & 0 & 3 & 0 \end{array}$$

← 0 remainder, so $x = 2$ is a zero.

So, $f(x)$ factors as

$$f(x) = (x + 1)(x - 2)(x^2 + 3).$$

Because the factor $(x^2 + 3)$ produces no real zeros, you can conclude that $x = -1$ and $x = 2$ are the only *real* zeros of f, which is verified in Figure 2.23. ■

When the leading coefficient of a polynomial is not 1, the list of possible rational zeros can increase dramatically. In such cases, the search can be shortened in several ways: (1) a programmable graphing utility can be used to speed up the calculations; (2) a graph, drawn either by hand or with a graphing utility, can give a good estimate of the locations of the zeros; and (3) synthetic division can be used to test the possible rational zeros and to assist in factoring the polynomial.

EXAMPLE 4 Using the Rational Zero Test

Find the rational zeros of

$$f(x) = 2x^3 + 3x^2 - 8x + 3.$$

Solution The leading coefficient is 2 and the constant term is 3.

$$\textit{Possible rational zeros: } \frac{\text{Factors of } 3}{\text{Factors of } 2} = \frac{\pm1, \pm3}{\pm1, \pm2} = \pm1, \pm3, \pm\frac{1}{2}, \pm\frac{3}{2}$$

By synthetic division, you can determine that $x = 1$ is a zero.

$$\begin{array}{r|rrrr} 1 & 2 & 3 & -8 & 3 \\ & & 2 & 5 & -3 \\ \hline & 2 & 5 & -3 & 0 \end{array}$$

So, $f(x)$ factors as

$$f(x) = (x - 1)(2x^2 + 5x - 3) = (x - 1)(2x - 1)(x + 3)$$

and you can conclude that the rational zeros of f are $x = 1$, $x = \frac{1}{2}$, and $x = -3$. ■

Conjugate Pairs

In Examples 1(c) and 1(d), note that the pairs of complex zeros are **conjugates.** That is, they are of the form $a + bi$ and $a - bi$.

NOTE Be sure you see that this result is true only if the polynomial function has *real coefficients.* For instance, the result applies to the function given by $f(x) = x^2 + 1$ but not to the function given by $g(x) = x - i$.

> **THEOREM 2.5 COMPLEX ZEROS OCCUR IN CONJUGATE PAIRS**
>
> Let $f(x)$ be a polynomial function that has *real coefficients.* When $a + bi$, where $b \neq 0$, is a zero of the function, the conjugate $a - bi$ is also a zero of the function.

EXAMPLE 5 Finding a Polynomial with Given Zeros

Find a fourth-degree polynomial function with real coefficients that has -1, -1, and $3i$ as zeros.

Solution Because $3i$ is a zero *and* the polynomial is stated to have real coefficients, you know that the conjugate $-3i$ must also be a zero. So, from the Linear Factorization Theorem, $f(x)$ can be written as

$$f(x) = a(x + 1)(x + 1)(x - 3i)(x + 3i).$$

For simplicity, let $a = 1$ to obtain

$$\begin{aligned} f(x) &= (x^2 + 2x + 1)(x^2 + 9) \\ &= x^4 + 2x^3 + 10x^2 + 18x + 9. \end{aligned}$$

■

Factoring a Polynomial

The Linear Factorization Theorem shows that you can write any nth-degree polynomial as the product of n linear factors.

$$f(x) = a_n(x - c_1)(x - c_2)(x - c_3) \cdot \cdot \cdot (x - c_n)$$

However, this result includes the possibility that some of the values of c_i are complex. The following theorem states that even when you do not want to get involved with "complex factors," you can still write $f(x)$ as the product of linear and/or quadratic factors.

> **THEOREM 2.6 FACTORS OF A POLYNOMIAL**
>
> Every polynomial of degree $n > 0$ with real coefficients can be written as the product of linear and quadratic factors with real coefficients, where the quadratic factors have no real zeros.

PROOF To begin, use the Linear Factorization Theorem to conclude that $f(x)$ can be *completely* factored in the form

$$f(x) = d(x - c_1)(x - c_2)(x - c_3) \cdot \cdot \cdot (x - c_n).$$

When each c_k is real, there is nothing more to prove. If any c_k is complex ($c_k = a + bi$, $b \neq 0$), then, because the coefficients of $f(x)$ are real, you know that the conjugate $c_j = a - bi$ is also a zero. By multiplying the corresponding factors, you obtain

$$(x - c_k)(x - c_j) = [x - (a + bi)][x - (a - bi)] = x^2 - 2ax + (a^2 + b^2)$$

where each coefficient of the quadratic expression is real. ■

A quadratic factor with no real zeros is said to be *prime* or **irreducible over the reals.** Be sure you see that this is not the same as being *irreducible over the rationals.* For example, the quadratic $x^2 + 1 = (x - i)(x + i)$ is irreducible over the reals (and therefore over the rationals). On the other hand, the quadratic $x^2 - 2 = \left(x - \sqrt{2}\right)\left(x + \sqrt{2}\right)$ is irreducible over the rationals, but *reducible* over the reals.

EXAMPLE 6 Finding the Zeros of a Polynomial Function

Find all the zeros of

$$f(x) = x^4 - 3x^3 + 6x^2 + 2x - 60$$

given that $1 + 3i$ is a zero of f.

Algebraic Solution

Because complex zeros occur in conjugate pairs, you know that $1 - 3i$ is also a zero of f. This means that both

$$[x - (1 + 3i)] \quad \text{and} \quad [x - (1 - 3i)]$$

are factors of f. Multiplying these two factors produces

$$\begin{aligned}[x - (1 + 3i)][x - (1 - 3i)] &= [(x - 1) - 3i][(x - 1) + 3i] \\ &= (x - 1)^2 - 9i^2 \\ &= x^2 - 2x + 1 - 9(-1) \\ &= x^2 - 2x + 10.\end{aligned}$$

Using long division, you can divide $x^2 - 2x + 10$ into f to obtain the following.

$$\begin{array}{r} x^2 - x - 6 \\ x^2 - 2x + 10 \overline{\smash{)}\, x^4 - 3x^3 + 6x^2 + 2x - 60} \\ \underline{x^4 - 2x^3 + 10x^2} \\ -x^3 - 4x^2 + 2x \\ \underline{-x^3 + 2x^2 - 10x} \\ -6x^2 + 12x - 60 \\ \underline{-6x^2 + 12x - 60} \\ 0 \end{array}$$

So, you have

$$\begin{aligned}f(x) &= (x^2 - 2x + 10)(x^2 - x - 6) \\ &= (x^2 - 2x + 10)(x - 3)(x + 2)\end{aligned}$$

and you can conclude that the zeros of f are $x = 1 + 3i$, $x = 1 - 3i$, $x = 3$, and $x = -2$.

Graphical Solution

Because complex zeros always occur in conjugate pairs, you know that $1 - 3i$ is also a zero of f. Because the polynomial is a fourth-degree polynomial, you know that there are at most two other zeros of the function. Use a graphing utility to graph

$$y = x^4 - 3x^3 + 6x^2 + 2x - 60$$

as shown in Figure 2.24.

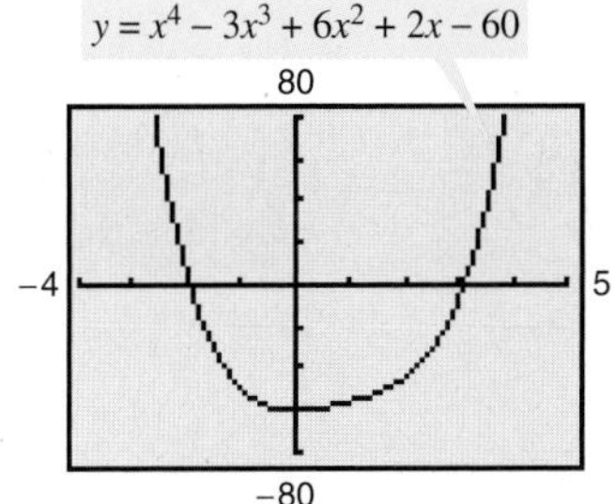

Figure 2.24

You can see that -2 and 3 appear to be zeros of the graph of the function. Use the *zero* or *root* feature or the *zoom* and *trace* features of the graphing utility to confirm that $x = -2$ and $x = 3$ are zeros of the graph. So, you can conclude that the zeros of f are $x = 1 + 3i$, $x = 1 - 3i$, $x = 3$, and $x = -2$.

■

In Example 6, if you were not told that $1 + 3i$ is a zero of f, you could still find all zeros of the function by using synthetic division to find the real zeros -2 and 3. Then you could factor the polynomial as $(x + 2)(x - 3)(x^2 - 2x + 10)$. Finally, by using the Quadratic Formula, you could determine that the zeros are $x = -2$, $x = 3$, $x = 1 + 3i$, and $x = 1 - 3i$.

Example 7 shows how to find all the zeros of a polynomial function, including complex zeros.

EXAMPLE 7 Finding the Zeros of a Polynomial Function

Write

$$f(x) = x^5 + x^3 + 2x^2 - 12x + 8$$

as the product of linear factors, and list all of its zeros.

Solution The possible rational zeros are $\pm 1, \pm 2, \pm 4$, and ± 8. Synthetic division produces the following.

$$\begin{array}{r|rrrrrr} 1 & 1 & 0 & 1 & 2 & -12 & 8 \\ & & 1 & 1 & 2 & 4 & -8 \\ \hline & 1 & 1 & 2 & 4 & -8 & 0 \end{array} \quad \longleftarrow \text{ 1 is a zero.}$$

$$\begin{array}{r|rrrrr} 1 & 1 & 1 & 2 & 4 & -8 \\ & & 1 & 2 & 4 & 8 \\ \hline & 1 & 2 & 4 & 8 & 0 \end{array} \quad \longleftarrow \text{ 1 is a repeated zero.}$$

$$\begin{array}{r|rrrr} -2 & 1 & 2 & 4 & 8 \\ & & -2 & 0 & -8 \\ \hline & 1 & 0 & 4 & 0 \end{array} \quad \longleftarrow \text{ } -2 \text{ is a zero.}$$

So, you have

$$f(x) = x^5 + x^3 + 2x^2 - 12x + 8 = (x - 1)(x - 1)(x + 2)(x^2 + 4).$$

By factoring $x^2 + 4$ as

$$\begin{aligned} x^2 - (-4) &= \left(x - \sqrt{-4}\right)\left(x + \sqrt{-4}\right) \\ &= (x - 2i)(x + 2i) \end{aligned}$$

you obtain

$$f(x) = (x - 1)(x - 1)(x + 2)(x - 2i)(x + 2i)$$

which gives the following five zeros of f.

$$x = 1, \quad x = 1, \quad x = -2, \quad x = 2i, \quad \text{and} \quad x = -2i$$

Note from the graph of f shown in Figure 2.25 that the *real* zeros are the only ones that appear as x-intercepts. ■

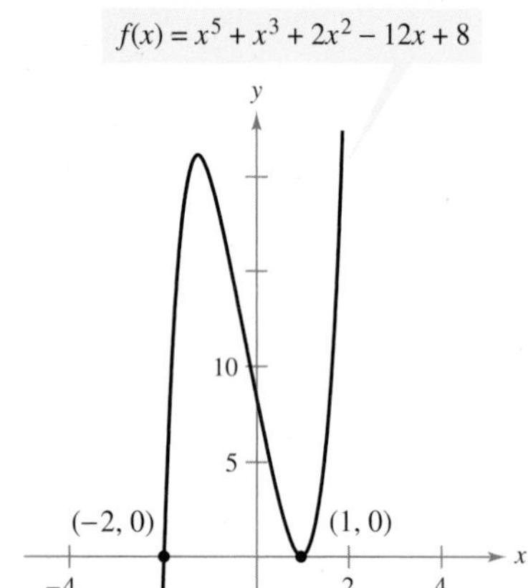

Figure 2.25

STUDY TIP In Example 7, the fifth-degree polynomial function has three real zeros. In such cases, you can use the *zoom* and *trace* features or the *zero* or *root* feature of a graphing utility to approximate the real zeros. You can then use these real zeros to determine the complex zeros analytically.

TECHNOLOGY You can use the *table* feature of a graphing utility to help you determine which of the possible rational zeros are zeros of the polynomial in Example 7. The table should be set to *ask* mode. Then enter each of the possible rational zeros in the table. When you do this, you will see that there are two rational zeros, -2 and 1, as shown in the table below.

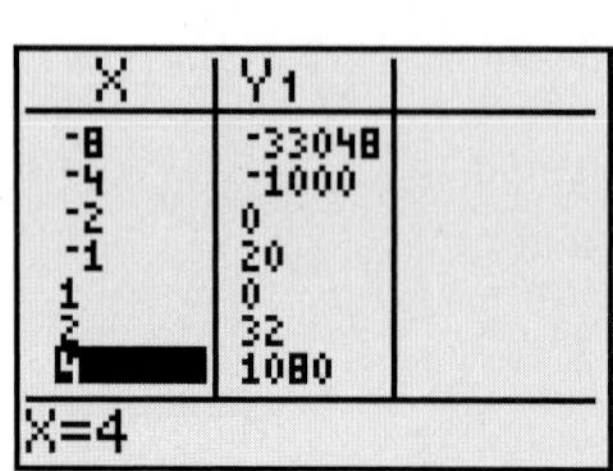

X	Y1
-8	-33048
-4	-1000
-2	0
-1	20
1	0
2	32
4	1080

X=4

Before concluding this section, here are two additional hints that can help you find the real zeros of a polynomial.

1. When the terms of $f(x)$ have a common monomial factor, it should be factored out before applying the tests in this section. For instance, by writing

$$f(x) = x^4 - 5x^3 + 3x^2 + x$$
$$= x(x^3 - 5x^2 + 3x + 1)$$

you can see that $x = 0$ is a zero of f and that the remaining zeros can be obtained by analyzing the cubic factor.

2. When you are able to find all but two zeros of $f(x)$, you can always use the Quadratic Formula on the remaining quadratic factor. For instance, if you succeeded in writing

$$f(x) = x^4 - 5x^3 + 3x^2 + x$$
$$= x(x - 1)(x^2 - 4x - 1)$$

you can apply the Quadratic Formula to $x^2 - 4x - 1$ to conclude that the two remaining zeros are $x = 2 + \sqrt{5}$ and $x = 2 - \sqrt{5}$.

EXAMPLE 8 Using a Polynomial Model

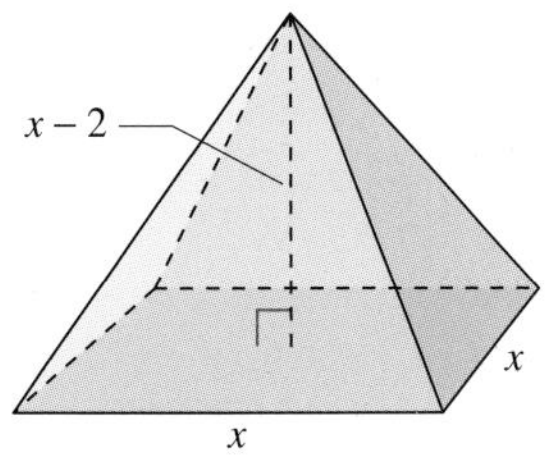

Figure 2.26

You are designing candle-making kits. Each kit contains 25 cubic inches of candle wax and a mold for making a pyramid-shaped candle. You want the height of the candle to be 2 inches less than the length of each side of the candle's square base, as shown in Figure 2.26. What should the dimensions of your candle mold be?

Solution The volume of a pyramid is given by $V = \frac{1}{3}Bh$, where B is the area of the base and h is the height. The area of the base is x^2 and the height is $(x - 2)$. So, the volume of the pyramid is

$$V = \frac{1}{3}Bh$$
$$= \frac{1}{3}x^2(x - 2).$$

Substituting 25 for the volume yields

$$25 = \frac{1}{3}x^2(x - 2) \qquad \text{Substitute 25 for } V.$$
$$75 = x^3 - 2x^2 \qquad \text{Multiply each side by 3.}$$
$$0 = x^3 - 2x^2 - 75 \qquad \text{Write in general form.}$$

The possible rational solutions are

$$x = \pm 1, \pm 3, \pm 5, \pm 15, \pm 25, \text{ and } \pm 75.$$

Use synthetic division to test some of the possible solutions. Note that in this case, it makes sense to test only positive x-values. Using synthetic division, you can determine that $x = 5$ is a solution and you have $0 = (x - 5)(x^2 + 3x + 15)$. The two solutions of the quadratic factor are imaginary and can be discarded. You can conclude that the base of the candle mold should be 5 inches by 5 inches and the height of the mold should be $5 - 2 = 3$ inches. ■

2.5 Exercises

See www.CalcChat.com for worked-out solutions to odd-numbered exercises.

In Exercises 1–6, fill in the blanks.

1. The ________ ________ of ________ states that if $f(x)$ is a polynomial of degree n $(n > 0)$, then f has at least one zero in the complex number system.
2. The ________ ________ ________ states that if $f(x)$ is a polynomial of degree n $(n > 0)$, then f has precisely n linear factors, $f(x) = a_n(x - c_1)(x - c_2) \cdot \cdot \cdot (x - c_n)$, where $c_1, c_2, \ldots, c_n$ are complex numbers.
3. The test that gives a list of the possible rational zeros of a polynomial function is called the ________ ________ Test.
4. If $a + bi$ is a complex zero of a polynomial with real coefficients, then so is its ________, $a - bi$.
5. Every polynomial of degree $n > 0$ with real coefficients can be written as the product of ________ and ________ factors with real coefficients, where the ________ factors have no real zeros.
6. A quadratic factor that cannot be factored further as a product of linear factors containing real numbers is said to be ________ over the ________.

In Exercises 7–12, find all the zeros of the function.

7. $f(x) = x(x - 6)^2$
8. $f(x) = x^2(x + 3)(x^2 - 1)$
9. $g(x) = (x - 2)(x + 4)^3$
10. $f(x) = (x + 5)(x - 8)^2$
11. $f(x) = (x + 6)(x + i)(x - i)$
12. $h(t) = (t - 3)(t - 2)(t - 3i)(t + 3i)$

In Exercises 13–16, use the Rational Zero Test to list all possible rational zeros of f. Verify that the zeros of f shown on the graph are contained in the list.

13. $f(x) = x^3 + 2x^2 - x - 2$

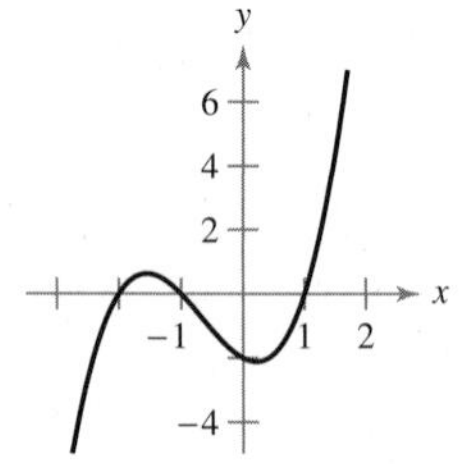

14. $f(x) = x^3 - 4x^2 - 4x + 16$

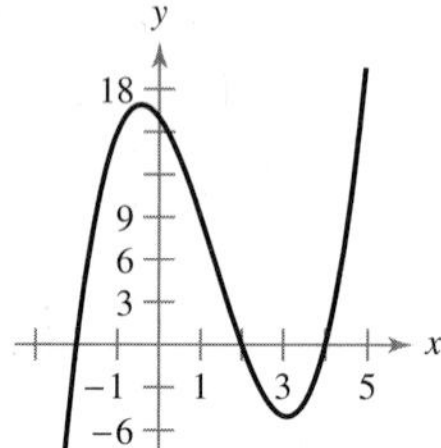

15. $f(x) = 2x^4 - 17x^3 + 35x^2 + 9x - 45$

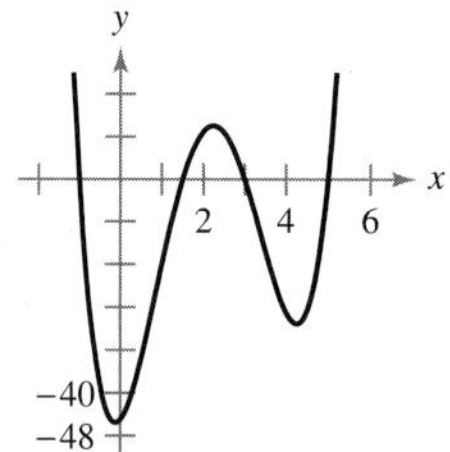

16. $f(x) = 4x^5 - 8x^4 - 5x^3 + 10x^2 + x - 2$

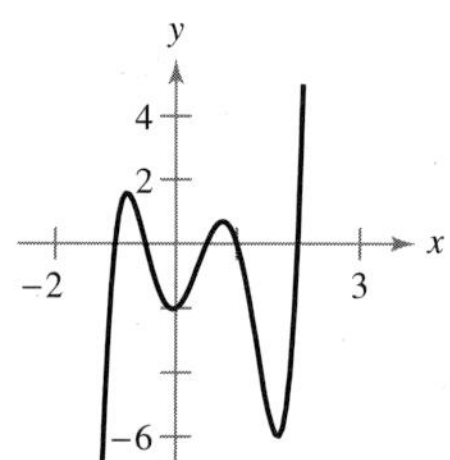

In Exercises 17–26, find all the rational zeros of the function.

17. $f(x) = x^3 - 6x^2 + 11x - 6$
18. $f(x) = x^3 - 7x - 6$
19. $g(x) = x^3 - 4x^2 - x + 4$
20. $h(x) = x^3 - 9x^2 + 20x - 12$
21. $h(t) = t^3 + 8t^2 + 13t + 6$
22. $p(x) = x^3 - 9x^2 + 27x - 27$
23. $C(x) = 2x^3 + 3x^2 - 1$
24. $f(x) = 3x^3 - 19x^2 + 33x - 9$
25. $f(x) = 9x^4 - 9x^3 - 58x^2 + 4x + 24$
26. $f(x) = 2x^4 - 15x^3 + 23x^2 + 15x - 25$

In Exercises 27–30, find all real solutions of the polynomial equation.

27. $z^4 + z^3 + z^2 + 3z - 6 = 0$
28. $x^4 - 13x^2 - 12x = 0$
29. $2y^4 + 3y^3 - 16y^2 + 15y - 4 = 0$
30. $x^5 - x^4 - 3x^3 + 5x^2 - 2x = 0$

In Exercises 31–34, (a) list the possible rational zeros of f, (b) sketch the graph of f so that some of the possible zeros in part (a) can be disregarded, and then (c) determine all real zeros of f.

31. $f(x) = x^3 + x^2 - 4x - 4$

32. $f(x) = -3x^3 + 20x^2 - 36x + 16$

33. $f(x) = -4x^3 + 15x^2 - 8x - 3$

34. $f(x) = 4x^3 - 12x^2 - x + 15$

In Exercises 35–38, (a) list the possible rational zeros of f, (b) use a graphing utility to graph f so that some of the possible zeros in part (a) can be disregarded, and then (c) determine all real zeros of f.

35. $f(x) = -2x^4 + 13x^3 - 21x^2 + 2x + 8$

36. $f(x) = 4x^4 - 17x^2 + 4$

37. $f(x) = 32x^3 - 52x^2 + 17x + 3$

38. $f(x) = 4x^3 + 7x^2 - 11x - 18$

Graphical Analysis **In Exercises 39–42, (a) use the *zero* or *root* feature of a graphing utility to approximate the zeros of the function accurate to three decimal places, (b) determine one of the exact zeros (use synthetic division to verify your result), and (c) factor the polynomial completely.**

39. $f(x) = x^4 - 3x^2 + 2$

40. $P(t) = t^4 - 7t^2 + 12$

41. $h(x) = x^5 - 7x^4 + 10x^3 + 14x^2 - 24x$

42. $g(x) = 6x^4 - 11x^3 - 51x^2 + 99x - 27$

In Exercises 43–48, find a polynomial function with real coefficients that has the given zeros. (There are many correct answers.)

43. $1, 5i$

44. $4, -3i$

45. $2, 5 + i$

46. $5, 3 - 2i$

47. $\frac{2}{3}, -1, 3 + \sqrt{2}i$

48. $-5, -5, 1 + \sqrt{3}i$

In Exercises 49–52, write the polynomial (a) as the product of factors that are irreducible over the *rationals*, (b) as the product of linear and quadratic factors that are irreducible over the *reals*, and (c) in completely factored form.

49. $f(x) = x^4 + 6x^2 - 27$

50. $f(x) = x^4 - 2x^3 - 3x^2 + 12x - 18$

(*Hint:* One factor is $x^2 - 6$.)

51. $f(x) = x^4 - 4x^3 + 5x^2 - 2x - 6$

(*Hint:* One factor is $x^2 - 2x - 2$.)

52. $f(x) = x^4 - 3x^3 - x^2 - 12x - 20$

(*Hint:* One factor is $x^2 + 4$.)

In Exercises 53–60, use the given zero to find all the zeros of the function.

	Function	*Zero*
53.	$f(x) = x^3 - x^2 + 4x - 4$	$2i$
54.	$f(x) = 2x^3 + 3x^2 + 18x + 27$	$3i$
55.	$f(x) = 2x^4 - x^3 + 49x^2 - 25x - 25$	$5i$
56.	$g(x) = x^3 - 7x^2 - x + 87$	$5 + 2i$
57.	$g(x) = 4x^3 + 23x^2 + 34x - 10$	$-3 + i$
58.	$h(x) = 3x^3 - 4x^2 + 8x + 8$	$1 - \sqrt{3}i$
59.	$f(x) = x^4 + 3x^3 - 5x^2 - 21x + 22$	$-3 + \sqrt{2}i$
60.	$f(x) = x^3 + 4x^2 + 14x + 20$	$-1 - 3i$

In Exercises 61–78, find all the zeros of the function and write the polynomial as a product of linear factors.

61. $f(x) = x^2 + 36$

62. $f(x) = x^2 - x + 56$

63. $h(x) = x^2 - 2x + 17$

64. $g(x) = x^2 + 10x + 17$

65. $f(x) = x^4 - 16$

66. $f(y) = y^4 - 256$

67. $f(z) = z^2 - 2z + 2$

68. $h(x) = x^3 - 3x^2 + 4x - 2$

69. $g(x) = x^3 - 3x^2 + x + 5$

70. $f(x) = x^3 - x^2 + x + 39$

71. $h(x) = x^3 - x + 6$

72. $h(x) = x^3 + 9x^2 + 27x + 35$

73. $f(x) = 5x^3 - 9x^2 + 28x + 6$

74. $g(x) = 2x^3 - x^2 + 8x + 21$

75. $g(x) = x^4 - 4x^3 + 8x^2 - 16x + 16$

76. $h(x) = x^4 + 6x^3 + 10x^2 + 6x + 9$

77. $f(x) = x^4 + 10x^2 + 9$

78. $f(x) = x^4 + 29x^2 + 100$

In Exercises 79–84, find all the zeros of the function. When there is an extended list of possible rational zeros, use a graphing utility to graph the function in order to discard any rational zeros that are obviously not zeros of the function.

79. $f(x) = x^3 + 24x^2 + 214x + 740$

80. $f(s) = 2s^3 - 5s^2 + 12s - 5$

81. $f(x) = 16x^3 - 20x^2 - 4x + 15$

82. $f(x) = 9x^3 - 15x^2 + 11x - 5$

83. $f(x) = 2x^4 + 5x^3 + 4x^2 + 5x + 2$

84. $g(x) = x^5 - 8x^4 + 28x^3 - 56x^2 + 64x - 32$

In Exercises 85–88, find all the real zeros of the function.

85. $f(x) = 4x^3 - 3x - 1$

86. $f(z) = 12z^3 - 4z^2 - 27z + 9$

87. $f(y) = 4y^3 + 3y^2 + 8y + 6$

88. $g(x) = 3x^3 - 2x^2 + 15x - 10$

In Exercises 89–92, find all the rational zeros of the polynomial function.

89. $P(x) = x^4 - \frac{25}{4}x^2 + 9 = \frac{1}{4}(4x^4 - 25x^2 + 36)$

90. $f(x) = x^3 - \frac{3}{2}x^2 - \frac{23}{2}x + 6 = \frac{1}{2}(2x^3 - 3x^2 - 23x + 12)$

91. $f(x) = x^3 - \frac{1}{4}x^2 - x + \frac{1}{4} = \frac{1}{4}(4x^3 - x^2 - 4x + 1)$

92. $f(z) = z^3 + \frac{11}{6}z^2 - \frac{1}{2}z - \frac{1}{3} = \frac{1}{6}(6z^3 + 11z^2 - 3z - 2)$

In Exercises 93–96, match the cubic function with the numbers of rational and irrational zeros.

(a) Rational zeros: 0; irrational zeros: 1

(b) Rational zeros: 3; irrational zeros: 0

(c) Rational zeros: 1; irrational zeros: 2

(d) Rational zeros: 1; irrational zeros: 0

93. $f(x) = x^3 - 1$

94. $f(x) = x^3 - 2$

95. $f(x) = x^3 - x$

96. $f(x) = x^3 - 2x$

WRITING ABOUT CONCEPTS

97. A third-degree polynomial function f has real zeros $-2, \frac{1}{2}$, and 3, and its leading coefficient is negative. Write an equation for f. Sketch the graph of f. How many polynomial functions are possible for f?

98. Sketch the graph of a fifth-degree polynomial function whose leading coefficient is positive and that has a zero at $x = 3$ of multiplicity 2.

99. Use the information in the table to answer each question.

Interval	Value of $f(x)$
$(-\infty, -2)$	Positive
$(-2, 1)$	Negative
$(1, 4)$	Negative
$(4, \infty)$	Positive

(a) What are the real zeros of the polynomial function f?

(b) What can be said about the behavior of the graph of f at $x = 1$?

(c) What is the least possible degree of f? Explain. Can the degree of f ever be odd? Explain.

WRITING ABOUT CONCEPTS (continued)

(d) Is the leading coefficient of f positive or negative? Explain.

(e) Write an equation for f. (There are many correct answers.)

(f) Sketch a graph of the equation you wrote in part (e).

100. Use the information in the table to answer each question.

Interval	Value of $f(x)$
$(-\infty, -2)$	Negative
$(-2, 0)$	Positive
$(0, 2)$	Positive
$(2, \infty)$	Positive

(a) What are the real zeros of the polynomial function f?

(b) What can be said about the behavior of the graph of f at $x = 0$ and $x = 2$?

(c) What is the least possible degree of f? Explain. Can the degree of f ever be even? Explain.

(d) Is the leading coefficient of f positive or negative? Explain.

(e) Write an equation for f. (There are many correct answers.)

(f) Sketch a graph of the equation you wrote in part (e).

101. ***Geometry*** A rectangular package to be sent by a delivery service (see figure) can have a maximum combined length and girth (perimeter of a cross section) of 120 inches.

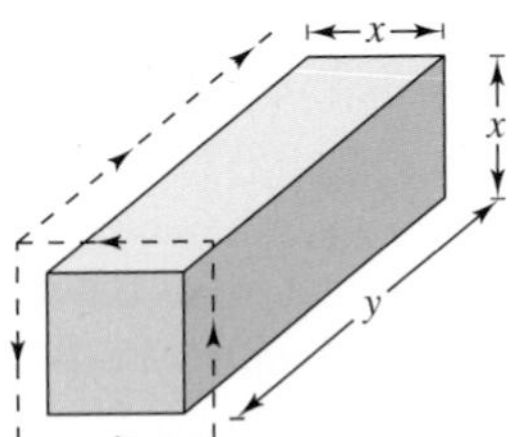

(a) Write a function $V(x)$ that represents the volume of the package.

(b) Use a graphing utility to graph the function and approximate the dimensions of the package that will yield a maximum volume.

(c) Find values of x such that $V = 13{,}500$. Which of these values is a physical impossibility in the construction of the package? Explain.

102. ***Geometry*** An open box is to be made from a rectangular piece of material, 15 centimeters by 9 centimeters, by cutting equal squares from the corners and turning up the sides.

(a) Let x represent the length of the sides of the squares removed. Draw a diagram showing the squares removed from the original piece of material and the resulting dimensions of the open box.

(b) Use the diagram to write the volume V of the box as a function of x. Determine the domain of the function.

(c) Sketch a graph of the function and approximate the dimensions of the box that will yield a maximum volume.

(d) Find values of x such that $V = 56$. Which of these values is a physical impossibility in the construction of the box? Explain.

103. ***Advertising Cost*** A company that produces MP3 players estimates that the profit P (in dollars) for selling a particular model is given by

$$P = -76x^3 + 4830x^2 - 320{,}000, \quad 0 \le x \le 60$$

where x is the advertising expense (in tens of thousands of dollars). Using this model, find the smaller of two advertising amounts that will yield a profit of \$2,500,000.

104. ***Advertising Cost*** A company that manufactures bicycles estimates that the profit P (in dollars) for selling a particular model is given by

$$P = -45x^3 + 2500x^2 - 275{,}000, \quad 0 \le x \le 50$$

where x is the advertising expense (in tens of thousands of dollars). Using this model, find the smaller of two advertising amounts that will yield a profit of \$800,000.

105. ***Geometry*** A bulk food storage bin with dimensions 2 feet by 3 feet by 4 feet needs to be increased in size to hold five times as much food as the current bin. (Assume each dimension is increased by the same amount.)

(a) Write a function that represents the volume V of the new bin.

(b) Find the dimensions of the new bin.

106. ***Geometry*** A manufacturer wants to enlarge an existing manufacturing facility such that the total floor area is 1.5 times that of the current facility. The floor area of the current facility is rectangular and measures 250 feet by 160 feet. The manufacturer wants to increase each dimension by the same amount.

(a) Write a function that represents the new floor area A.

(b) Find the dimensions of the new floor.

(c) Another alternative is to increase the current floor's length by an amount that is twice an increase in the floor's width. The total floor area is 1.5 times that of the current facility. Repeat parts (a) and (b) using these criteria.

107. ***Cost*** The ordering and transportation cost C (in thousands of dollars) for the components used in manufacturing a product is given by

$$C = 100\left(\frac{200}{x^2} + \frac{x}{x + 30}\right), \quad x \ge 1$$

where x is the order size (in hundreds). In Section 5.1, you will learn that the cost is a minimum when

$$3x^3 - 40x^2 - 2400x - 36{,}000 = 0.$$

Use a calculator to approximate the optimal order size to the nearest hundred units.

108. ***Athletics*** The attendance A (in millions) at NCAA women's college basketball games for the years 2000 through 2007 is shown in the table. *(Source: National Collegiate Athletic Association, Indianapolis, IN)*

Year	2000	2001	2002	2003
Attendance, A	8.7	8.8	9.5	10.2

Year	2004	2005	2006	2007
Attendance, A	10.0	9.9	9.9	10.9

(a) Use a graphing utility to create a scatter plot of the data. Let t represent the year, with $t = 0$ corresponding to 2000.

(b) Use the *regression* feature of the graphing utility to find a quartic model for the data.

(c) Graph the model and the scatter plot in the same viewing window. How well does the model fit the data?

(d) According to the model in part (b), in what year(s) was the attendance at least 10 million?

(e) According to the model, will the attendance continue to increase? Explain.

109. ***Height of a Baseball*** A baseball is thrown upward from a height of 6 feet with an initial velocity of 48 feet per second, and its height h (in feet) is

$$h(t) = -16t^2 + 48t + 6, \quad 0 \le t \le 3$$

where t is the time (in seconds). You are told the ball reaches a height of 64 feet. Is this possible?

110. ***Profit*** The demand equation for a certain product is $p = 140 - 0.0001x$, where p is the unit price (in dollars) of the product and x is the number of units produced and sold. The cost equation for the product is $C = 80x + 150{,}000$, where C is the total cost (in dollars) and x is the number of units produced. The total profit obtained by producing and selling x units is $P = R - C = xp - C$. You are working in the marketing department of the company that produces this product, and you are asked to determine a price p that will yield a profit of 9 million dollars. Is this possible? Explain.

True or False? **In Exercises 111 and 112, decide whether the statement is true or false. Justify your answer.**

111. It is possible for a third-degree polynomial function with integer coefficients to have no real zeros.

112. If $x = -i$ is a zero of the function given by

$$f(x) = x^3 + ix^2 + ix - 1$$

then $x = i$ must also be a zero of f.

Think About It **In Exercises 113–118, determine (if possible) the zeros of the function g if the function f has zeros at $x = r_1$, $x = r_2$, and $x = r_3$.**

113. $g(x) = -f(x)$

114. $g(x) = 3f(x)$

115. $g(x) = f(x - 5)$

116. $g(x) = f(2x)$

117. $g(x) = 3 + f(x)$

118. $g(x) = f(-x)$

In Exercises 119 and 120, the graph of a cubic polynomial function $y = f(x)$ is shown. It is known that one of the zeros is $1 + i$. Write an equation for f.

119.

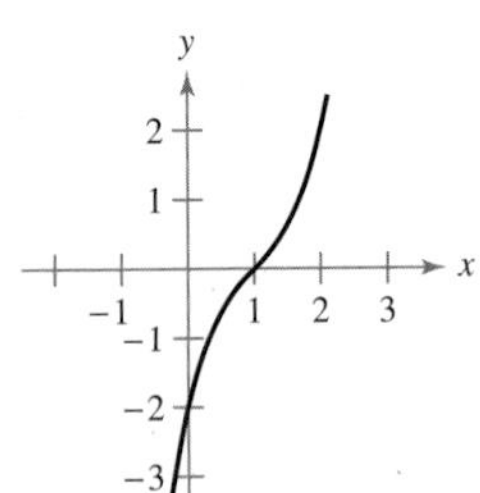

120.

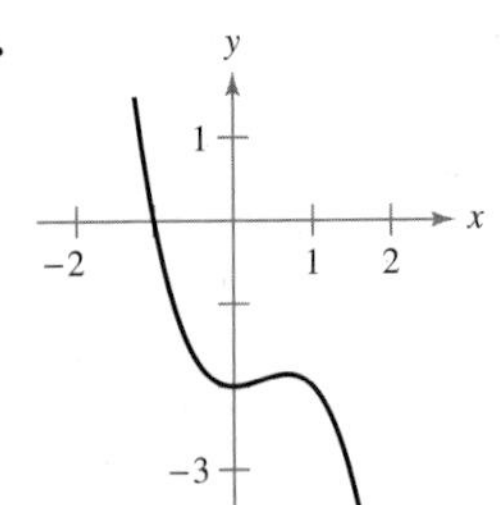

121. ***Think About It*** Let $y = f(x)$ be a quartic polynomial with leading coefficient $a = 1$ and $f(i) = f(2i) = 0$. Write an equation for f.

122. ***Think About It*** Let $y = f(x)$ be a cubic polynomial with leading coefficient $a = -1$ and $f(2) = f(i) = 0$. Write an equation for f.

123. ***Writing*** Compile a list of all the various techniques for factoring a polynomial that have been covered so far in the text. Give an example illustrating each technique, and write a paragraph discussing when the use of each technique is appropriate.

CAPSTONE

124. Use a graphing utility to graph the function given by

$$f(x) = x^4 - 4x^2 + k$$

for different values of k. Find values of k such that the zeros of f satisfy the specified characteristics. (Some parts do not have unique answers.)

(a) Four real zeros

(b) Two real zeros, each of multiplicity 2

(c) Two real zeros and two complex zeros

(d) Four complex zeros

(e) Will the answers to parts (a) through (d) change for the function g, where $g(x) = f(x - 2)$?

(f) Will the answers to parts (a) through (d) change for the function g, where $g(x) = f(2x)$?

125. (a) Find a quadratic function f (with integer coefficients) that has $\pm\sqrt{b}i$ as zeros. Assume that b is a positive integer.

(b) Find a quadratic function f (with integer coefficients) that has $a \pm bi$ as zeros. Assume that b is a positive integer.

126. ***Graphical Reasoning*** The graph of one of the following functions is shown below. Identify the function shown in the graph. Explain why each of the others is not the correct function. Use a graphing utility to verify your result.

(a) $f(x) = x^2(x + 2)(x - 3.5)$

(b) $g(x) = (x + 2)(x - 3.5)$

(c) $h(x) = (x + 2)(x - 3.5)(x^2 + 1)$

(d) $k(x) = (x + 1)(x + 2)(x - 3.5)$

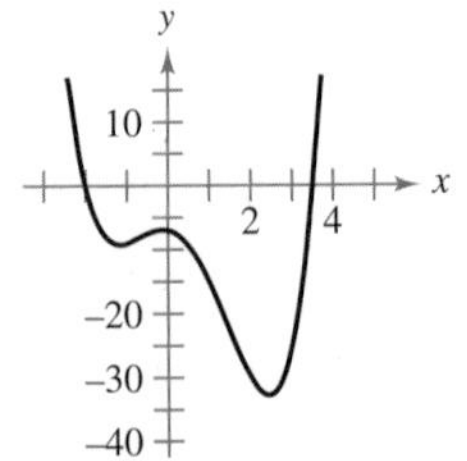

2.6 Rational Functions

- Find the domains of rational functions.
- Find the vertical and horizontal asymptotes of graphs of rational functions.
- Analyze and sketch graphs of rational functions.
- Sketch graphs of rational functions that have slant asymptotes.
- Use rational functions to model and solve real-life problems.

Introduction

A **rational function** is a quotient of polynomial functions. It can be written in the form

$$f(x) = \frac{N(x)}{D(x)}$$

where $N(x)$ and $D(x)$ are polynomials and $D(x)$ is not the zero polynomial.

In general, the *domain* of a rational function of x includes all real numbers except x-values that make the denominator zero. Much of the discussion of rational functions will focus on their graphical behavior near these x-values excluded from the domain.

EXAMPLE 1 Finding the Domain of a Rational Function

Find the domain of the reciprocal function

$$f(x) = \frac{1}{x}$$

and discuss the behavior of f near any excluded x-values.

Solution Because the denominator is zero when $x = 0$, the domain of f is all real numbers except $x = 0$. To determine the behavior of f near this excluded value, evaluate $f(x)$ to the left and right of $x = 0$, as indicated in the tables below.

x approaches 0 from the left. →

x	-1	-0.5	-0.1	-0.01	-0.001	→	0
$f(x)$	-1	-2	-10	-100	-1000	→	$-\infty$

x	0	←	0.001	0.01	0.1	0.5	1
$f(x)$	∞	←	1000	100	10	2	1

← x approaches 0 from the left.

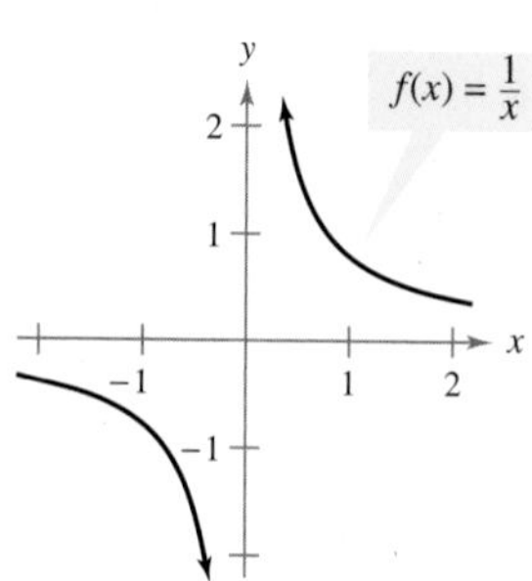

Figure 2.27

Note that as x approaches 0 *from the left,* $f(x)$ decreases without bound. In contrast, as x approaches 0 *from the right,* $f(x)$ increases without bound. The graph of f is shown in Figure 2.27. ■

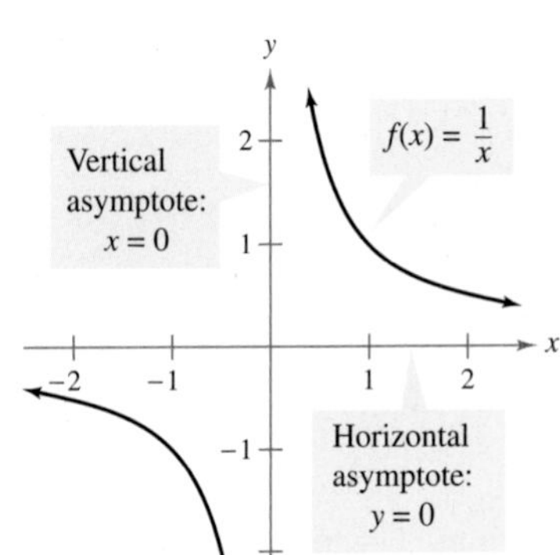

Figure 2.28

Vertical and Horizontal Asymptotes

In Example 1, the behavior of f near $x = 0$ is denoted as follows.

$$\underbrace{f(x) \longrightarrow -\infty \text{ as } x \longrightarrow 0^-}_{f(x) \text{ decreases without bound as } x \text{ approaches 0 from the left.}} \qquad \underbrace{f(x) \longrightarrow \infty \text{ as } x \longrightarrow 0^+}_{f(x) \text{ increases without bound as } x \text{ approaches 0 from the right.}}$$

The line $x = 0$ is a **vertical asymptote** of the graph of f, as shown in Figure 2.28. From this figure, you can see that the graph of f also has a **horizontal asymptote**—the line $y = 0$. This means that the values of $f(x) = \frac{1}{x}$ approach zero as x increases or decreases without bound.

$$\underbrace{f(x) \longrightarrow 0 \text{ as } x \longrightarrow -\infty}_{f(x) \text{ approaches 0 as } x \text{ decreases without bound.}} \qquad \underbrace{f(x) \longrightarrow 0 \text{ as } x \longrightarrow \infty}_{f(x) \text{ approaches 0 as } x \text{ increases without bound.}}$$

VERTICAL AND HORIZONTAL ASYMPTOTES

1. The line $x = a$ is a **vertical asymptote** of the graph of f when

$$f(x) \longrightarrow \infty \quad \text{or} \quad f(x) \longrightarrow -\infty$$

as $x \longrightarrow a$, either from the right or from the left.

2. The line $y = b$ is a **horizontal asymptote** of the graph of f when

$$f(x) \longrightarrow b$$

as $x \longrightarrow \infty$ or $x \longrightarrow -\infty$.

NOTE A more precise discussion of a *vertical asymptote* is given in Section 3.5. A more precise discussion of *horizontal asymptote* is given in Section 5.5. ■

Eventually (as $x \longrightarrow \infty$ or $x \longrightarrow -\infty$), the distance between the horizontal asymptote and the points on the graph must approach zero. Figure 2.29 shows the vertical and horizontal asymptotes of the graphs of three rational functions.

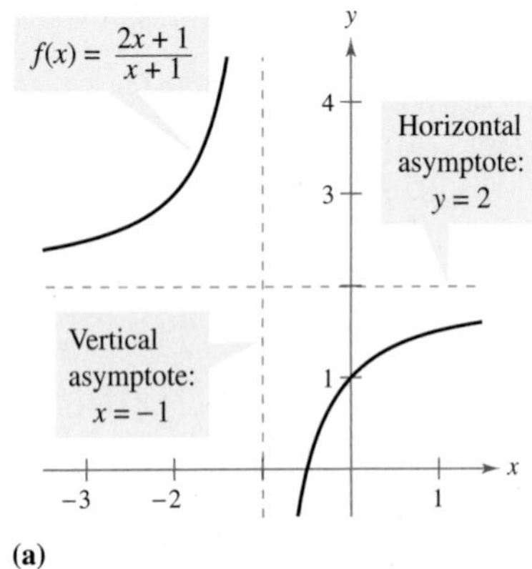

(a)

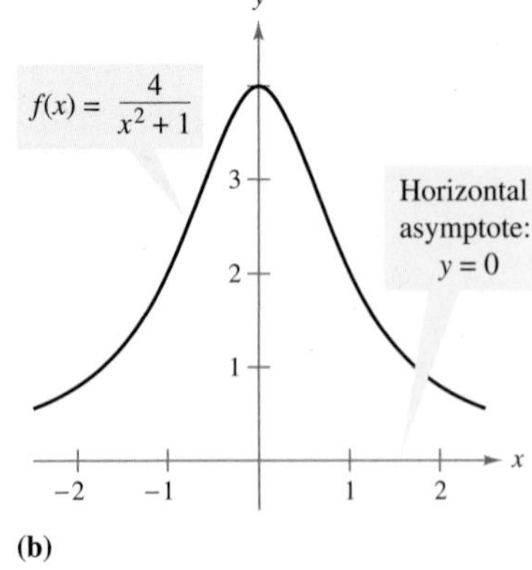

(b)

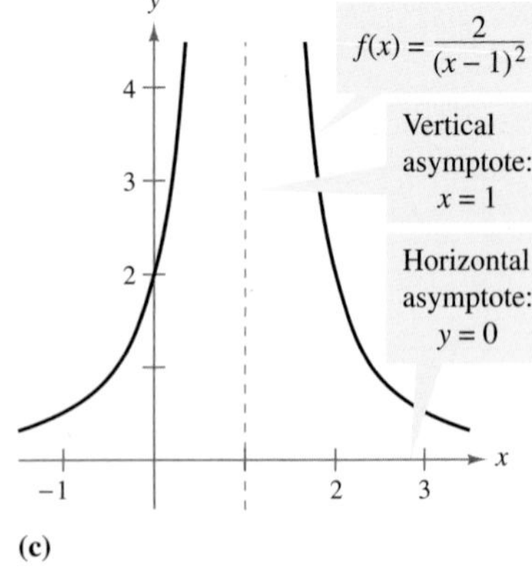

(c)

Figure 2.29

The graphs of $f(x) = \frac{1}{x}$ in Figure 2.28 and $f(x) = \frac{2x + 1}{x + 1}$ in Figure 2.29(a) are *hyperbolas*. You will study hyperbolas in Section 12.3.

VERTICAL AND HORIZONTAL ASYMPTOTES OF A RATIONAL FUNCTION

Let f be the rational function given by

$$f(x) = \frac{N(x)}{D(x)} = \frac{a_n x^n + a_{n-1}x^{n-1} + \cdots + a_1 x + a_0}{b_m x^m + b_{m-1}x^{m-1} + \cdots + b_1 x + b_0}$$

where $N(x)$ and $D(x)$ have no common factors.

1. The graph of f has *vertical* asymptotes at the zeros of $D(x)$.
2. The graph of f has one or no *horizontal* asymptote determined by comparing the degrees of $N(x)$ and $D(x)$.
   a. When $n < m$, the graph of f has the line $y = 0$ (the x-axis) as a horizontal asymptote.
   b. When $n = m$, the graph of f has the line $y = \frac{a_n}{b_m}$ (ratio of the leading coefficients) as a horizontal asymptote.
   c. When $n > m$, the graph of f has no horizontal asymptote.

EXAMPLE 2 Finding Vertical and Horizontal Asymptotes

Find all vertical and horizontal asymptotes of the graph of each rational function.

a. $f(x) = \dfrac{2x^2}{x^2 - 1}$

b. $f(x) = \dfrac{x^2 + x - 2}{x^2 - x - 6}$

Solution

a. For this rational function, the degree of the numerator is *equal to* the degree of the denominator. The leading coefficient of the numerator is 2 and the leading coefficient of the denominator is 1, so the graph has the line $y = 2$ as a horizontal asymptote. To find any vertical asymptotes, set the denominator equal to zero and solve the resulting equation for x.

$$x^2 - 1 = 0 \qquad \text{Set denominator equal to zero.}$$

$$(x + 1)(x - 1) = 0 \qquad \text{Factor.}$$

$$x + 1 = 0 \implies x = -1 \qquad \text{Set 1st factor equal to 0.}$$

$$x - 1 = 0 \implies x = 1 \qquad \text{Set 2nd factor equal to 0.}$$

This equation has two real solutions, $x = -1$ and $x = 1$, so the graph has the lines $x = -1$ and $x = 1$ as vertical asymptotes. The graph of the function is shown in Figure 2.30.

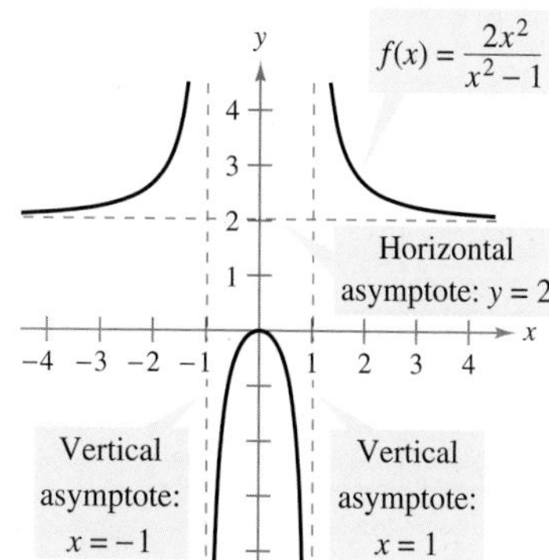

Figure 2.30

b. For this rational function, the degree of the numerator is equal to the degree of the denominator. The leading coefficient of both the numerator and denominator is 1, so the graph has the line $y = 1$ as a horizontal asymptote. To find any vertical asymptotes, first factor the numerator and denominator as follows.

$$f(x) = \frac{x^2 + x - 2}{x^2 - x - 6} = \frac{(x - 1)\cancel{(x + 2)}}{\cancel{(x + 2)}(x - 3)} = \frac{x - 1}{x - 3}, \quad x \neq -2$$

By setting the denominator $x - 3$ (of the simplified function) equal to zero, you can determine that the graph has the line $x = 3$ as a vertical asymptote. ■

STUDY TIP You may also want to test for symmetry when graphing rational functions, especially for simple rational functions. For example, the graph of

$$f(x) = \frac{1}{x}$$

is symmetric with respect to the origin, and the graph of

$$g(x) = \frac{1}{x^2}$$

is symmetric with respect to the y-axis.

Analyzing Graphs of Rational Functions

To sketch the graph of a rational function, use the following guidelines.

GUIDELINES FOR ANALYZING GRAPHS OF RATIONAL FUNCTIONS

Let $f(x) = \dfrac{N(x)}{D(x)}$, where $N(x)$ and $D(x)$ are polynomials.

1. Simplify f, if possible.
2. Find and plot the y-intercept (if any) by evaluating $f(0)$.
3. Find the zeros of the numerator (if any) by solving the equation $N(x) = 0$. Then plot the corresponding x-intercepts.
4. Find the zeros of the denominator (if any) by solving the equation $D(x) = 0$. Then sketch the corresponding vertical asymptotes.
5. Find and sketch the horizontal asymptote (if any) by using the rule for finding the horizontal asymptote of a rational function.
6. Plot at least one point *between* and one point *beyond* each x-intercept and vertical asymptote.
7. Use smooth curves to complete the graph between and beyond the vertical asymptotes.

TECHNOLOGY PITFALL Some graphing utilities have difficulty graphing rational functions that have vertical asymptotes. Often, the utility will connect parts of the graph that are not supposed to be connected. For instance, Figure 2.31(a) shows the graph of $f(x) = 1/(x - 2)$. Notice that the graph should consist of two unconnected portions—one to the left of $x = 2$ and the other to the right of $x = 2$. To eliminate this problem, you can try changing the mode of the graphing utility to *dot mode*. The problem with this mode is that the graph is then represented as a collection of dots [as shown in Figure 2.31(b)] rather than as a smooth curve.

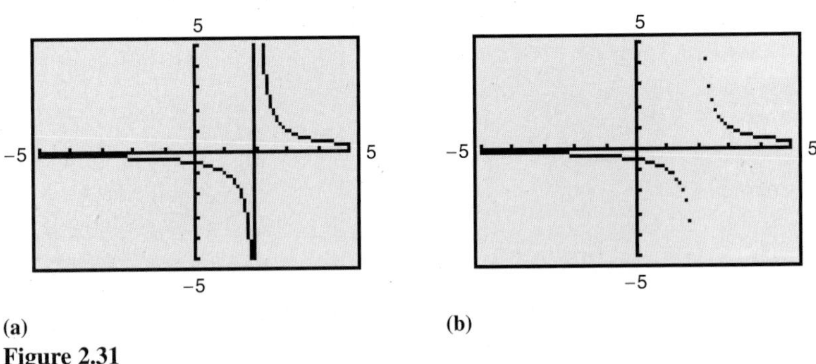

(a) (b)
Figure 2.31

The concept of *test intervals* from Section 2.2 can be extended to graphing of rational functions. To do this, use the fact that a rational function can change signs only at its zeros and its undefined values (the x-values for which its denominator is zero). Between two consecutive zeros of the numerator and the denominator, a rational function must be entirely positive or entirely negative. This means that when the zeros of the numerator and the denominator of a rational function are put in order, they divide the real number line into test intervals in which the function has no sign changes. A representative x-value is chosen to determine if the value of the rational function is positive (the graph lies above the x-axis) or negative (the graph lies below the x-axis).

EXAMPLE 3 Sketching the Graph of a Rational Function

Sketch the graph of $g(x) = \dfrac{3}{x - 2}$ and state its domain.

Solution

y-intercept: $\left(0, -\frac{3}{2}\right)$, because $g(0) = -\frac{3}{2}$

x-intercept: None, because $3 \neq 0$

Vertical asymptote: $x = 2$, zero of denominator

Horizontal asymptote: $y = 0$, because degree of $N(x) <$ degree of $D(x)$

Additional points:

Test interval	Representative x-value	Value of g	Sign	Point on graph
$(-\infty, 2)$	-4	$g(-4) = -0.5$	Negative	$(-4, -0.5)$
$(2, \infty)$	3	$g(3) = 3$	Positive	$(3, 3)$

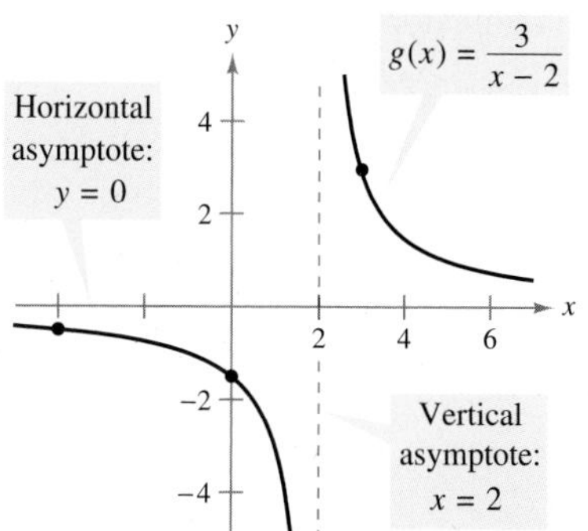

Figure 2.32

By plotting the intercepts, asymptotes, and a few additional points, you can obtain the graph shown in Figure 2.32. The domain of g is all real numbers x except $x = 2$. ■

NOTE The graph of g in Example 3 is a vertical stretch and a right shift of the graph of $f(x) = 1/x$ because $g(x) = 3/(x - 2) = 3[1/(x - 2)] = 3f(x - 2)$. ■

EXAMPLE 4 Sketching the Graph of a Rational Function

Sketch the graph of

$$f(x) = \frac{2x - 1}{x}$$

and state its domain.

Solution

y-intercept: None, because $x = 0$ is not in the domain

x-intercept: $\left(\frac{1}{2}, 0\right)$, because $2x - 1 = 0$

Vertical asymptote: $x = 0$, zero of denominator

Horizontal asymptote: $y = 2$, because degree of $N(x) =$ degree of $D(x)$

Additional points:

Test interval	Representative x-value	Value of f	Sign	Point on graph
$(-\infty, 0)$	-1	$f(-1) = 3$	Positive	$(-1, 3)$
$\left(0, \frac{1}{2}\right)$	$\frac{1}{4}$	$f\left(\frac{1}{4}\right) = -2$	Negative	$\left(\frac{1}{4}, -2\right)$
$\left(\frac{1}{2}, \infty\right)$	4	$f(4) = 1.75$	Positive	$(4, 1.75)$

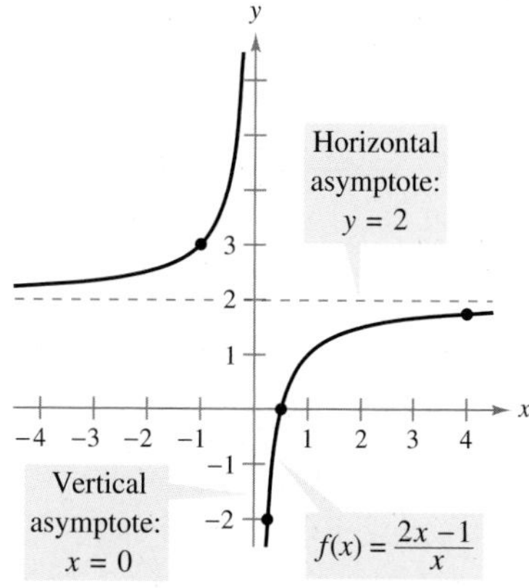

Figure 2.33

By plotting the intercepts, asymptotes, and a few additional points, you can obtain the graph shown in Figure 2.33. The domain of f is all real numbers x except $x = 0$. ■

EXAMPLE 5 Sketching the Graph of a Rational Function

Sketch the graph of $f(x) = \dfrac{x}{x^2 - x - 2}$.

Solution Factoring the denominator, you have $f(x) = \dfrac{x}{(x+1)(x-2)}$.

y-intercept: $(0, 0)$, because $f(0) = 0$

x-intercept: $(0, 0)$

Vertical asymptotes: $x = -1$, $x = 2$, zeros of denominator

Horizontal asymptote: $y = 0$, because degree of $N(x) <$ degree of $D(x)$

Additional points:

Test interval	Representative x-value	Value of f	Sign	Point on graph
$(-\infty, -1)$	-3	$f(-3) = -0.3$	Negative	$(-3, -0.3)$
$(-1, 0)$	-0.5	$f(-0.5) = 0.4$	Positive	$(-0.5, 0.4)$
$(0, 2)$	1	$f(1) = -0.5$	Negative	$(1, -0.5)$
$(2, \infty)$	3	$f(3) = 0.75$	Positive	$(3, 0.75)$

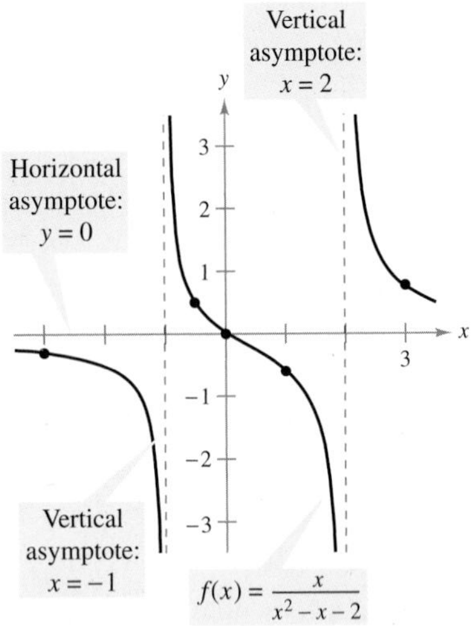

Figure 2.34

The graph is shown in Figure 2.34.

EXAMPLE 6 A Rational Function with Common Factors

Sketch the graph of $f(x) = \dfrac{x^2 - 9}{x^2 - 2x - 3}$.

Solution By factoring the numerator and denominator, you have

$$f(x) = \frac{x^2 - 9}{x^2 - 2x - 3} = \frac{(x-3)(x+3)}{(x-3)(x+1)} = \frac{x+3}{x+1}, \quad x \neq 3.$$

y-intercept: $(0, 3)$, because $f(0) = 3$

x-intercept: $(-3, 0)$, because $f(-3) = 0$

Vertical asymptote: $x = -1$, zero of (simplified) denominator

Horizontal asymptote: $y = 1$, because degree of $N(x) =$ degree of $D(x)$

Additional points:

Test interval	Representative x-value	Value of f	Sign	Point on graph
$(-\infty, -3)$	-4	$f(-4) = 0.33$	Positive	$(-4, 0.33)$
$(-3, -1)$	-2	$f(-2) = -1$	Negative	$(-2, -1)$
$(-1, \infty)$	2	$f(2) = 1.67$	Positive	$(2, 1.67)$

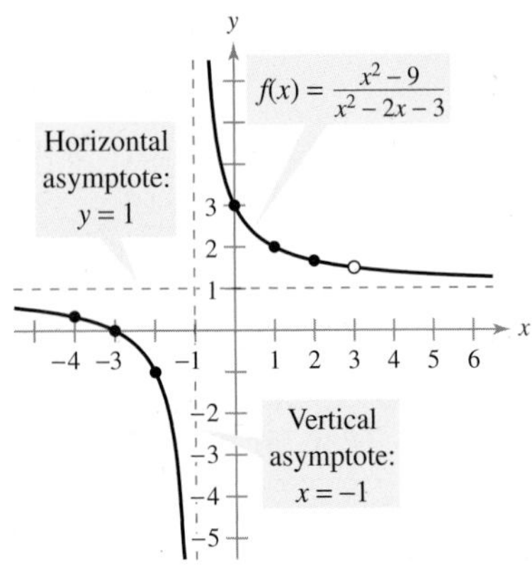

Hole at $x = 3$

Figure 2.35

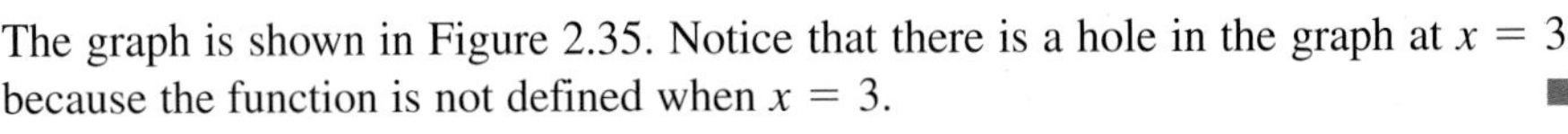

The graph is shown in Figure 2.35. Notice that there is a hole in the graph at $x = 3$, because the function is not defined when $x = 3$. ■

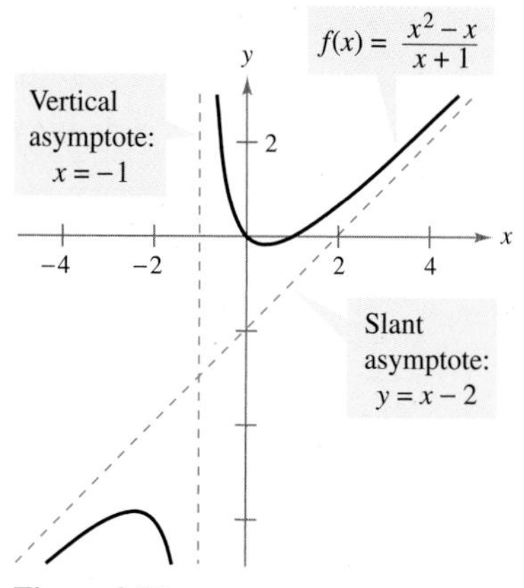

Figure 2.36

NOTE A more detailed explanation of the term *slant asymptote* is given in Section 5.6.

Slant Asymptotes

Consider a rational function whose denominator is of degree 1 or greater. When the degree of the numerator is exactly one more than the degree of the denominator, the graph of the function has a **slant** (or **oblique**) **asymptote.** For example, the graph of

$$f(x) = \frac{x^2 - x}{x + 1}$$

has a slant asymptote, as shown in Figure 2.36. To find the equation of a slant asymptote, use long division. For instance, by dividing $x + 1$ into $x^2 - x$, you obtain

$$f(x) = \frac{x^2 - x}{x + 1} = \underbrace{x - 2}_{\substack{\text{Slant asymptote} \\ (y = x - 2)}} + \frac{2}{x + 1}.$$

As x increases or decreases without bound, the remainder term $2/(x + 1)$ approaches 0, so the graph of f approaches the line $y = x - 2$, as shown in Figure 2.36.

EXAMPLE 7 A Rational Function with a Slant Asymptote

Sketch the graph of

$$f(x) = \frac{x^2 - x - 2}{x - 1}.$$

Solution Factoring the numerator as $(x - 2)(x + 1)$ allows you to recognize the x-intercepts. Using long division

$$f(x) = \frac{x^2 - x - 2}{x - 1}$$

$$= x - \frac{2}{x - 1}$$

allows you to recognize that the line $y = x$ is a slant asymptote of the graph.

y-intercept: $(0, 2)$, because $f(0) = 2$

x-intercepts: $(-1, 0)$ and $(2, 0)$

Vertical asymptote: $x = 1$, zero of denominator

Slant asymptote: $y = x$

Additional points:

Test interval	Representative x-value	Value of f	Sign	Point on graph
$(-\infty, -1)$	-2	$f(-2) = -1.3\overline{3}$	Negative	$(-2, -1.3\overline{3})$
$(-1, 1)$	0.5	$f(0.5) = 4.5$	Positive	$(0.5, 4.5)$
$(1, 2)$	1.5	$f(1.5) = -2.5$	Negative	$(1.5, -2.5)$
$(2, \infty)$	3	$f(3) = 2$	Positive	$(3, 2)$

The graph is shown in Figure 2.37. ■

Figure 2.37

Applications

There are many examples of asymptotic behavior in real life. For instance, Example 8 shows how a vertical asymptote can be used to analyze the cost of removing pollutants from smokestack emissions.

EXAMPLE 8 Cost-Benefit Model

A utility company burns coal to generate electricity. The cost of removing a certain *percent* of the pollutants from smokestack emissions is typically not a linear function. That is, if it costs C dollars to remove 25% of the pollutants, it would cost more than $2C$ dollars to remove 50% of the pollutants. As the percent of removed pollutants approaches 100%, the cost tends to increase without bound, becoming prohibitive. The cost C (in dollars) of removing $p\%$ of the smokestack pollutants is given by

$$C = \frac{80{,}000p}{100 - p}, \quad 0 \le p < 100.$$

You are a member of a state legislature considering a law that would require utility companies to remove 90% of the pollutants from their smokestack emissions. The current law requires 85% removal. How much additional cost would the utility company incur as a result of the new law?

Algebraic Solution

Because the current law requires 85% removal, the current cost to the utility company is

$$C = \frac{80{,}000(85)}{100 - 85} \approx \$453{,}333. \qquad \text{Evaluate } C \text{ when } p = 85.$$

When the new law increases the percent removal to 90%, the cost to the utility company will be

$$C = \frac{80{,}000(90)}{100 - 90} = \$720{,}000. \qquad \text{Evaluate } C \text{ when } p = 90.$$

So, the new law would require the utility company to spend an additional

$$720{,}000 - 453{,}333 = \$266{,}667. \qquad \text{Subtract 85\% removal cost from 90\% removal cost.}$$

Graphical Solution

Use a graphing utility to graph the function

$$y_1 = \frac{80{,}000x}{100 - x}$$

using a viewing window similar to that shown in Figure 2.38. Note that the graph has a vertical asymptote at $x = 100$. Then use the *trace* or *value* feature to approximate the values of y_1 when $x = 85$ and $x = 90$. You should obtain the following values.

When $x = 85$, $y_1 \approx 453{,}333$.

When $x = 90$, $y_1 = 720{,}000$.

So, the new law would require the utility company to spend an additional

$$720{,}000 - 453{,}333 = \$266{,}667.$$

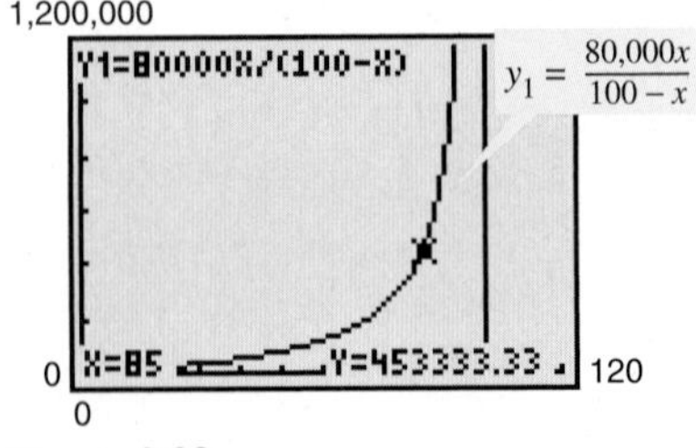

Figure 2.38

■

EXAMPLE 9 Finding a Minimum Area

A rectangular page is designed to contain 48 square inches of print. The margins at the top and bottom of the page are each 1 inch deep. The margins on each side are $1\frac{1}{2}$ inches wide. What should the dimensions of the page be so that the least amount of paper is used?

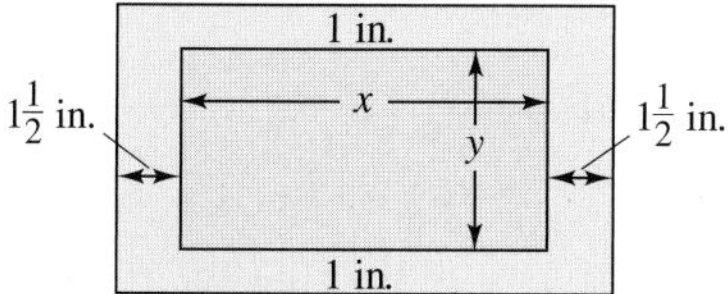

Figure 2.39

Graphical Solution

Let A be the area to be minimized. From Figure 2.39, you can write

$$A = (x + 3)(y + 2).$$

The printed area inside the margins is modeled by $48 = xy$ or $y = 48/x$. To find the minimum area, rewrite the equation for A in terms of just one variable by substituting $48/x$ for y.

$$A = (x + 3)\left(\frac{48}{x} + 2\right)$$

$$= \frac{(x + 3)(48 + 2x)}{x}, \quad x > 0$$

The graph of this rational function is shown in Figure 2.40. Because x represents the width of the printed area, you need consider only the portion of the graph for which x is positive. Using a graphing utility, you can approximate the minimum value of A to occur when $x \approx 8.5$ inches. The corresponding value of y is $48/8.5 \approx 5.6$ inches. So, the dimensions should be

$$x + 3 \approx 11.5 \text{ inches} \quad \text{by} \quad y + 2 \approx 7.6 \text{ inches}.$$

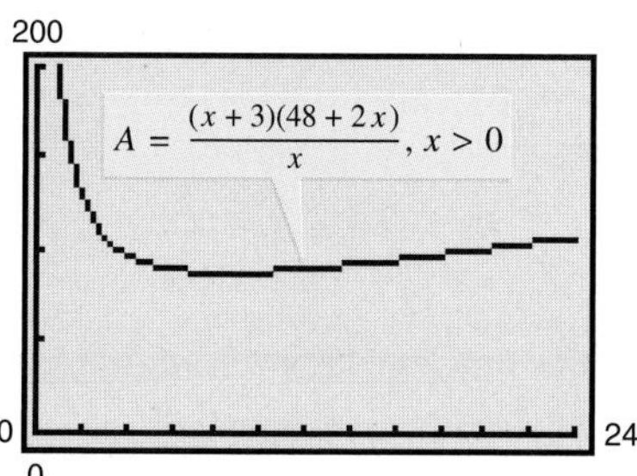

Figure 2.40

Numerical Solution

Let A be the area to be minimized. From Figure 2.39, you can write

$$A = (x + 3)(y + 2).$$

The printed area inside the margins is modeled by $48 = xy$ or $y = 48/x$. To find the minimum area, rewrite the equation for A in terms of just one variable by substituting $48/x$ for y.

$$A = (x + 3)\left(\frac{48}{x} + 2\right)$$

$$= \frac{(x + 3)(48 + 2x)}{x}, \quad x > 0$$

Use the *table* feature of a graphing utility to create a table of values for the function

$$y_1 = \frac{(x + 3)(48 + 2x)}{x}$$

beginning at $x = 1$. From the table, you can see that the minimum value of y_1 occurs when x is somewhere between 8 and 9, as shown in Figure 2.41. To approximate the minimum value of y_1 to one decimal place, change the table so that it starts at $x = 8$ and increases by 0.1. The minimum value of y_1 occurs when $x \approx 8.5$, as shown in Figure 2.42. The corresponding value of y is $48/8.5 \approx 5.6$ inches. So, the dimensions should be $x + 3 \approx 11.5$ inches by $y + 2 \approx 7.6$ inches.

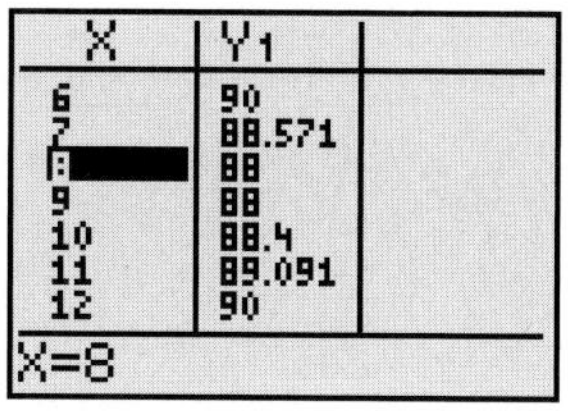

X	Y1
6	90
7	88.571
8	88
9	88
10	88.4
11	89.091
12	90

X=8

Figure 2.41

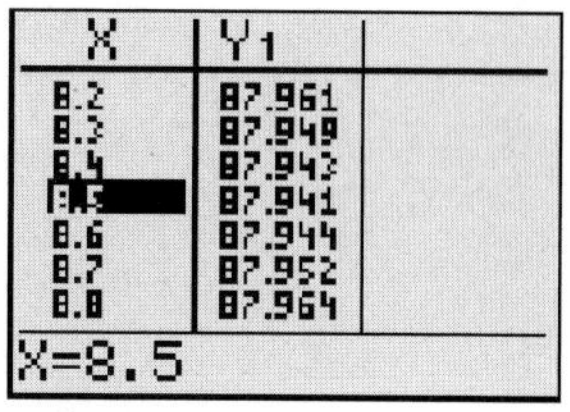

X	Y1
8.2	87.961
8.3	87.949
8.4	87.943
8.5	87.941
8.6	87.944
8.7	87.952
8.8	87.964

X=8.5

Figure 2.42

In Chapter 5, you will learn an analytic technique for finding the exact value of x that produces a minimum area. In this case, that value is $x = 6\sqrt{2} \approx 8.485$.

2.6 Exercises

See www.CalcChat.com for worked-out solutions to odd-numbered exercises.

In Exercises 1–4, fill in the blanks.

1. Functions of the form $f(x) = N(x)/D(x)$, where $N(x)$ and $D(x)$ are polynomials and $D(x)$ is not the zero polynomial, are called ________ ________.
2. If $f(x) \to \pm\infty$ as $x \to a$ from the left or the right, then $x = a$ is a ________ ________ of the graph of f.
3. If $f(x) \to b$ as $x \to \pm\infty$, then $y = b$ is a ________ ________ of the graph of f.
4. For the rational function given by $f(x) = N(x)/D(x)$, if the degree of $N(x)$ is exactly one more than the degree of $D(x)$, then the graph of f has a ________ (or oblique) ________.

In Exercises 5–8, (a) complete each table for the function, (b) determine the vertical and horizontal asymptotes of the graph of the function, and (c) find the domain of the function.

x	$f(x)$
0.5	
0.9	
0.99	
0.999	

x	$f(x)$
1.5	
1.1	
1.01	
1.001	

x	$f(x)$
5	
10	
100	
1000	

5. $f(x) = 1/(x - 1)$

6. $f(x) = 5x/(x - 1)$

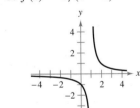

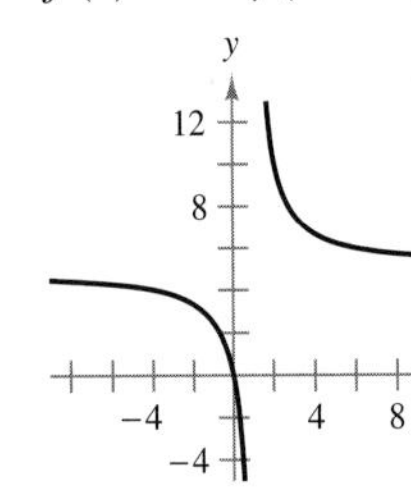

7. $f(x) = 3x^2/(x^2 - 1)$

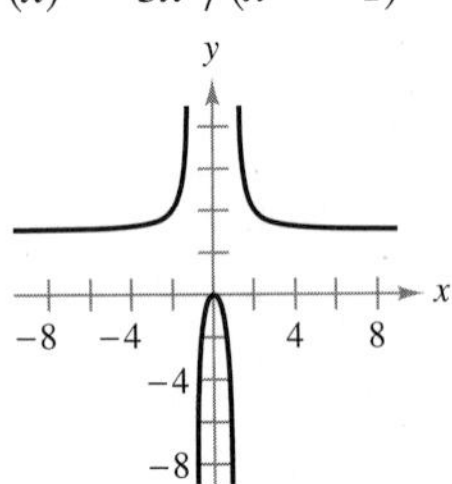

8. $f(x) = 4x/(x^2 - 1)$

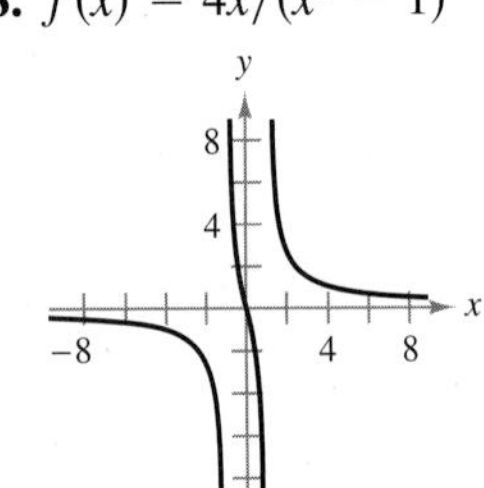

In Exercises 9–16, find the domain of the function and identify any vertical and horizontal asymptotes.

9. $f(x) = \dfrac{4}{x^2}$

10. $f(x) = \dfrac{4}{(x - 2)^3}$

11. $f(x) = \dfrac{5 + x}{5 - x}$

12. $f(x) = \dfrac{3 - 7x}{3 + 2x}$

13. $f(x) = \dfrac{x^3}{x^2 - 1}$

14. $f(x) = \dfrac{4x^2}{x + 2}$

15. $f(x) = \dfrac{3x^2 + 1}{x^2 + x + 9}$

16. $f(x) = \dfrac{3x^2 + x - 5}{x^2 + 1}$

In Exercises 17–20, match the rational function with its graph. [The graphs are labeled (a), (b), (c), and (d).]

(a)

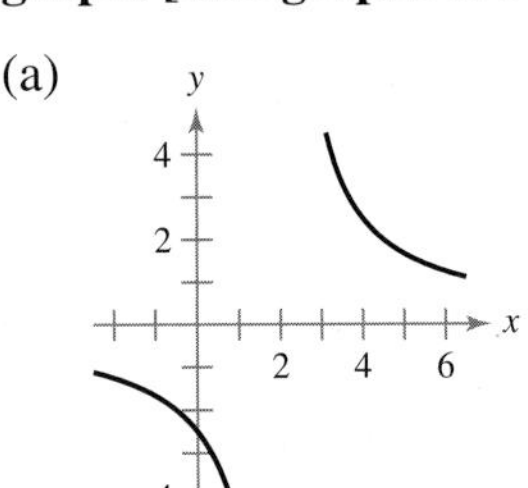

(b)

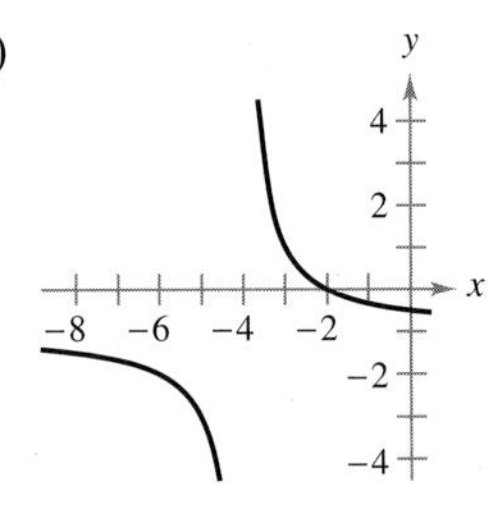

(c)

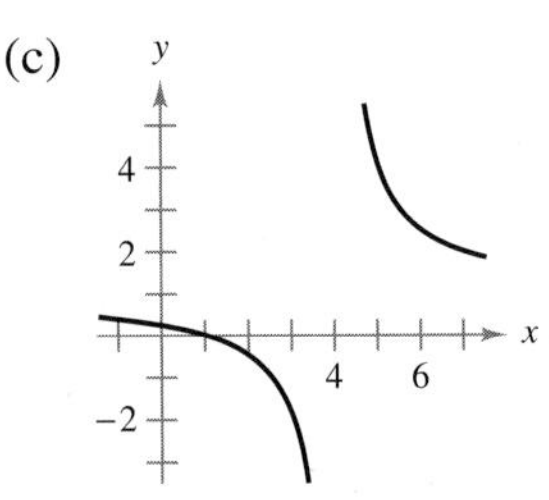

(d)

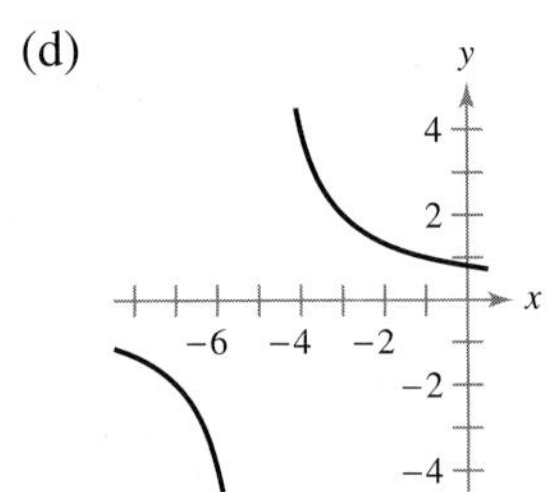

17. $f(x) = \dfrac{4}{x + 5}$

18. $f(x) = \dfrac{5}{x - 2}$

19. $f(x) = \dfrac{x - 1}{x - 4}$

20. $f(x) = -\dfrac{x + 2}{x + 4}$

In Exercises 21–24, find the zeros (if any) of the rational function.

21. $g(x) = \dfrac{x^2 - 9}{x + 3}$

22. $h(x) = 4 + \dfrac{10}{x^2 + 5}$

23. $f(x) = 1 - \dfrac{2}{x - 7}$

24. $g(x) = \dfrac{x^3 - 8}{x^2 + 1}$

In Exercises 25–30, find the domain of the function and identify any vertical and horizontal asymptotes.

25. $f(x) = \dfrac{x - 4}{x^2 - 16}$

26. $f(x) = \dfrac{x + 1}{x^2 - 1}$

27. $f(x) = \dfrac{x^2 - 25}{x^2 - 4x - 5}$

28. $f(x) = \dfrac{x^2 - 4}{x^2 - 3x + 2}$

29. $f(x) = \dfrac{x^2 - 3x - 4}{2x^2 + x - 1}$

30. $f(x) = \dfrac{6x^2 - 11x + 3}{6x^2 - 7x - 3}$

In Exercises 31–50, (a) state the domain of the function, (b) identify all intercepts, (c) find any vertical and horizontal asymptotes, and (d) plot additional solution points as needed to sketch the graph of the rational function.

31. $f(x) = \dfrac{1}{x+2}$

32. $f(x) = \dfrac{1}{x-3}$

33. $h(x) = \dfrac{-1}{x+4}$

34. $g(x) = \dfrac{1}{6-x}$

35. $C(x) = \dfrac{7+2x}{2+x}$

36. $P(x) = \dfrac{1-3x}{1-x}$

37. $f(x) = \dfrac{x^2}{x^2+9}$

38. $f(t) = \dfrac{1-2t}{t}$

39. $g(s) = \dfrac{4s}{s^2+4}$

40. $f(x) = -\dfrac{1}{(x-2)^2}$

41. $h(x) = \dfrac{x^2-5x+4}{x^2-4}$

42. $g(x) = \dfrac{x^2-2x-8}{x^2-9}$

43. $f(x) = \dfrac{2x^2-5x-3}{x^3-2x^2-x+2}$

44. $f(x) = \dfrac{x^2-x-2}{x^3-2x^2-5x+6}$

45. $f(x) = \dfrac{x^2+3x}{x^2+x-6}$

46. $f(x) = \dfrac{5(x+4)}{x^2+x-12}$

47. $f(x) = \dfrac{2x^2-5x+2}{2x^2-x-6}$

48. $f(x) = \dfrac{3x^2-8x+4}{2x^2-3x-2}$

49. $f(t) = \dfrac{t^2-1}{t-1}$

50. $f(x) = \dfrac{x^2-36}{x+6}$

Analytical, Numerical, and Graphical Analysis **In Exercises 51–54, do the following.**

(a) Determine the domains of *f* and *g*.

(b) Simplify *f* and find any vertical asymptotes of the graph of *f*.

(c) Compare the functions by completing the table.

(d) Use a graphing utility to graph *f* and *g* in the same viewing window.

(e) Explain why the graphing utility may not show the difference in the domains of *f* and *g*.

51. $f(x) = \dfrac{x^2-1}{x+1}$, $g(x) = x - 1$

x	-3	-2	-1.5	-1	-0.5	0	1
$f(x)$							
$g(x)$							

52. $f(x) = \dfrac{x^2(x-2)}{x^2-2x}$, $g(x) = x$

x	-1	0	1	1.5	2	2.5	3
$f(x)$							
$g(x)$							

53. $f(x) = \dfrac{x-2}{x^2-2x}$, $g(x) = \dfrac{1}{x}$

x	-0.5	0	0.5	1	1.5	2	3
$f(x)$							
$g(x)$							

54. $f(x) = \dfrac{2x-6}{x^2-7x+12}$, $g(x) = \dfrac{2}{x-4}$

x	0	1	2	3	4	5	6
$f(x)$							
$g(x)$							

In Exercises 55–68, (a) state the domain of the function, (b) identify all intercepts, (c) identify any vertical and slant asymptotes, and (d) plot additional solution points as needed to sketch the graph of the rational function.

55. $h(x) = \dfrac{x^2-9}{x}$

56. $g(x) = \dfrac{x^2+5}{x}$

57. $f(x) = \dfrac{2x^2+1}{x}$

58. $f(x) = \dfrac{1-x^2}{x}$

59. $g(x) = \dfrac{x^2+1}{x}$

60. $h(x) = \dfrac{x^2}{x-1}$

61. $f(t) = -\dfrac{t^2+1}{t+5}$

62. $f(x) = \dfrac{x^2}{3x+1}$

63. $f(x) = \dfrac{x^3}{x^2-4}$

64. $g(x) = \dfrac{x^3}{2x^2-8}$

65. $f(x) = \dfrac{x^2-x+1}{x-1}$

66. $f(x) = \dfrac{2x^2-5x+5}{x-2}$

67. $f(x) = \dfrac{2x^3-x^2-2x+1}{x^2+3x+2}$

68. $f(x) = \dfrac{2x^3+x^2-8x-4}{x^2-3x+2}$

WRITING ABOUT CONCEPTS

69. Give an example of a rational function whose domain is the set of all real numbers. Give an example of a rational function whose domain is the set of all real numbers except $x = 2$.

70. Describe what is meant by an asymptote of a graph.

In Exercises 71–74, use a graphing utility to graph the rational function. Give the domain of the function and identify any asymptotes. Then zoom out sufficiently far so that the graph appears as a line. Identify the line.

71. $f(x) = \dfrac{x^2 + 5x + 8}{x + 3}$

72. $f(x) = \dfrac{2x^2 + x}{x + 1}$

73. $g(x) = \dfrac{1 + 3x^2 - x^3}{x^2}$

74. $h(x) = \dfrac{12 - 2x - x^2}{2(4 + x)}$

Graphical Reasoning **In Exercises 75–78, (a) use the graph to determine any x-intercepts of the graph of the rational function and (b) set $y = 0$ and solve the resulting equation to confirm your result in part (a).**

75. $y = \dfrac{x + 1}{x - 3}$

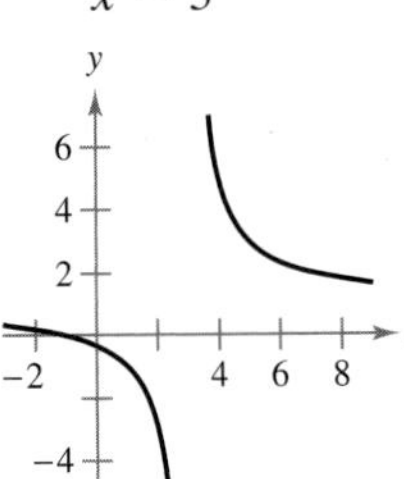

76. $y = \dfrac{2x}{x - 3}$

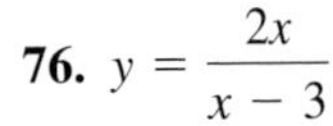

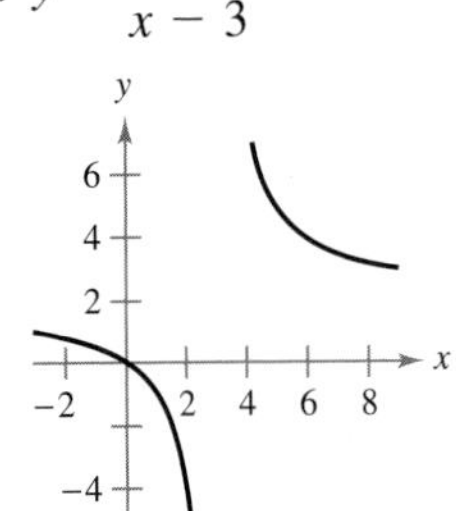

77. $y = \dfrac{1}{x} - x$

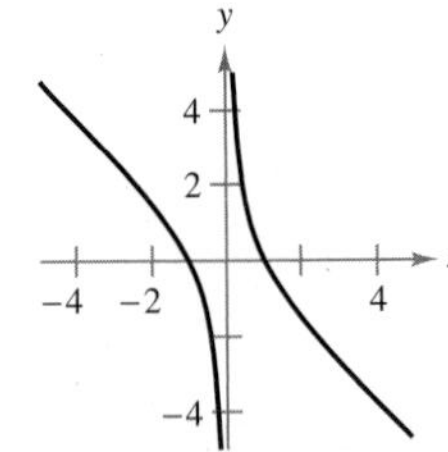

78. $y = x - 3 + \dfrac{2}{x}$

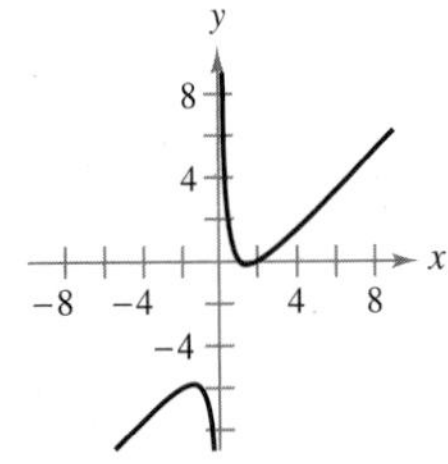

79. ***Pollution*** The cost C (in millions of dollars) of removing $p\%$ of the industrial and municipal pollutants discharged into a river is given by

$$C = \frac{255p}{100 - p}, \quad 0 \le p < 100.$$

(a) Use a graphing utility to graph the cost function.

(b) Find the costs of removing 10%, 40%, and 75% of the pollutants.

(c) According to this model, would it be possible to remove 100% of the pollutants? Explain.

80. ***Recycling*** In a pilot project, a rural township is given recycling bins for separating and storing recyclable products. The cost C (in dollars) of supplying bins to $p\%$ of the population is given by

$$C = \frac{25{,}000p}{100 - p}, \quad 0 \le p < 100.$$

(a) Use a graphing utility to graph the cost function.

(b) Find the costs of supplying bins to 15%, 50%, and 90% of the population.

(c) According to this model, would it be possible to supply bins to 100% of the residents? Explain.

81. ***Population Growth*** The game commission introduces 100 deer into newly acquired state game lands. The population N of the herd is modeled by

$$N = \frac{20(5 + 3t)}{1 + 0.04t}, \quad t \ge 0$$

where t is the time in years.

(a) Find the populations when $t = 5$, $t = 10$, and $t = 25$.

(b) What is the limiting size of the herd as time increases?

82. ***Concentration of a Mixture*** A 1000-liter tank contains 50 liters of a 25% brine solution. You add x liters of a 75% brine solution to the tank.

(a) Show that the concentration C, the proportion of brine to total solution, in the final mixture is

$$C = \frac{3x + 50}{4(x + 50)}.$$

(b) Determine the domain of the function based on the physical constraints of the problem.

(c) Sketch a graph of the concentration function.

(d) As the tank is filled, what happens to the rate at which the concentration of brine is increasing? What percent does the concentration of brine appear to approach?

83. ***Average Speed*** A driver averaged 50 miles per hour on the round trip between Akron, Ohio, and Columbus, Ohio, 100 miles away. The average speeds for going and returning were x and y miles per hour, respectively.

(a) Show that $y = \dfrac{25x}{x - 25}$.

(b) Determine the vertical and horizontal asymptotes of the graph of the function.

(c) Use a graphing utility to graph the function.

(d) Complete the table.

x	30	35	40	45	50	55	60
y							

(e) Are the results in the table what you expected? Explain.

(f) Is it possible to average 20 miles per hour in one direction and still average 50 miles per hour on the round trip? Explain.

84. ***Page Design*** A page that is x inches wide and y inches high contains 30 square inches of print. The top and bottom margins are 1 inch deep, and the margins on each side are 2 inches wide (see figure).

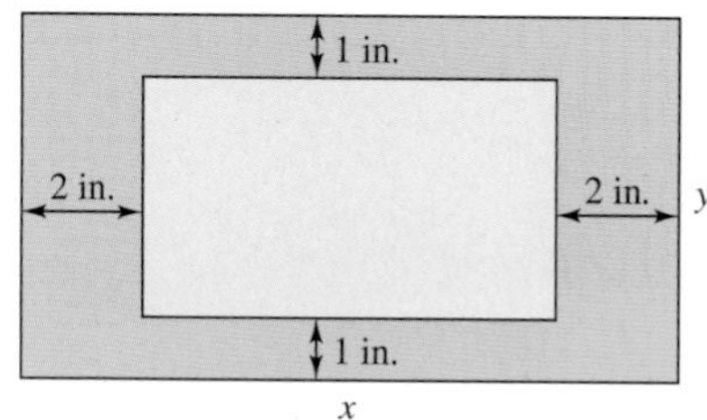

(a) Write a function for the total area A of the page in terms of x.

(b) Determine the domain of the function based on the physical constraints of the problem.

(c) Use a graphing utility to graph the area function and approximate the page size for which the least amount of paper will be used. Verify your answer numerically using the *table* feature of the graphing utility.

True or False? **In Exercises 85–87, determine whether the statement is true or false. Justify your answer.**

85. A polynomial can have infinitely many vertical asymptotes.

86. The graph of a rational function can never cross one of its asymptotes.

87. The graph of a function can have a vertical asymptote, a horizontal asymptote, and a slant asymptote.

CAPSTONE

88. Write a rational function f that has the specified characteristics. (There are many correct answers.)

(a) Vertical asymptote: $x = 2$
Horizontal asymptote: $y = 0$
Zero: $x = 1$

(b) Vertical asymptote: $x = -1$
Horizontal asymptote: $y = 0$
Zero: $x = 2$

(c) Vertical asymptotes: $x = -2, x = 1$
Horizontal asymptote: $y = 2$
Zeros: $x = 3, x = -3$

(d) Vertical asymptotes: $x = -1, x = 2$
Horizontal asymptote: $y = -2$
Zeros: $x = -2, x = 3$

89. ***Writing*** Is every rational function a polynomial function? Is every polynomial function a rational function? Explain.

SECTION PROJECT

Rational Functions

The numbers N (in thousands) of insured commercial banks in the United States for the years 1998 through 2007 are shown in the table. *(Source: U.S. Federal Deposit Insurance Corporation)*

Year	1998	1999	2000	2001	2002
Banks, N	8.8	8.6	8.3	8.1	7.9

Year	2003	2004	2005	2006	2007
Banks, N	7.8	7.6	7.5	7.4	7.3

For each of the following, let $t = 8$ represent 1998.

(a) Use the *regression* feature of a graphing utility to find a linear model for the data. Use a graphing utility to plot the data points and graph the linear model in the same viewing window.

(b) In order to find a rational model to fit the data, use the following steps. Add a third row to the table with entries $1/N$. Again use a graphing utility to find a linear model to fit the new set of data. Use t for the independent variable and $1/N$ for the dependent variable. The resulting linear model has the form $1/N = at + b$. Solve this equation for N. This is your rational model.

(c) Use a graphing utility to plot the original data (t, N) and graph your rational model in the same viewing window.

(d) Use the *table* feature of a graphing utility to show the actual data and the predicted number of banks based on each model for each of the years in the given table. Which model do you prefer? Explain why you chose the model you did.

2 CHAPTER SUMMARY

Section 2.1	Review Exercises
■ Analyze graphs of quadratic functions (*p. 144*).	*1, 2*
■ Write quadratic functions in standard form and use the results to sketch graphs of quadratic functions (*p. 147*).	*3–18*
■ Find minimum and maximum values of quadratic functions in real-life applications (*p. 149*).	*19–22*
Section 2.2	
■ Use transformations to sketch graphs of polynomial functions (*p. 154*).	*23–28*
■ Use the Leading Coefficient Test to determine the end behavior of graphs of polynomial functions (*p. 156*).	*29–32*
■ Find and use zeros of polynomial functions as sketching aids (*p. 157*).	*33–42*
Section 2.3	
■ Divide polynomials using long division (*p. 164*).	*43–48*
■ Use synthetic division to divide polynomials by binomials of the form $(x - k)$ (*p. 167*).	*49–54*
■ Use the Remainder Theorem and the Factor Theorem (*p. 168*).	*55–60*
■ Use polynomial division to answer questions about real-life problems (*p. 170*).	*61–64*
Section 2.4	
■ Use the imaginary unit i to write complex numbers (*p. 174*).	*65–68*
■ Add, subtract, and multiply complex numbers (*p. 175*).	*69–76*
■ Use complex conjugates to write the quotient of two complex numbers in standard form (*p. 177*).	*77–80*
■ Find complex solutions of quadratic equations (*p. 178*).	*81–84*
Section 2.5	
■ Understand and use the Fundamental Theorem of Algebra (*p. 181*).	*85–90*
■ Find all the zeros of a polynomial function (*p. 182*).	*91–106*
■ Write a polynomial function with real coefficients, given its zeros (*p. 184*).	*107, 108*
Section 2.6	
■ Find the domains of rational functions (*p. 193*).	*109–112*
■ Find the vertical and horizontal asymptotes of graphs of rational functions (*p. 194*).	*113–116*
■ Analyze and sketch graphs of rational functions (*p. 196*).	*117–128*
■ Sketch graphs of rational functions that have slant asymptotes (*p. 199*).	*129–132*
■ Use rational functions to model and solve real-life problems (*p. 200*).	*133, 134*

2 REVIEW EXERCISES

See www.CalcChat.com for worked-out solutions to odd-numbered exercises.

In Exercises 1 and 2, graph each function. Compare the graph of each function with the graph of $y = x^2$.

1. (a) $f(x) = 2x^2$
   (b) $g(x) = -2x^2$
   (c) $h(x) = x^2 + 2$
   (d) $k(x) = (x + 2)^2$

2. (a) $f(x) = x^2 - 4$
   (b) $g(x) = 4 - x^2$
   (c) $h(x) = (x - 3)^2$
   (d) $k(x) = \frac{1}{2}x^2 - 1$

In Exercises 3–14, write the quadratic function in standard form and sketch its graph. Identify the vertex, axis of symmetry, and x-intercept(s).

3. $g(x) = x^2 - 2x$
4. $f(x) = 6x - x^2$
5. $f(x) = x^2 + 8x + 10$
6. $h(x) = 3 + 4x - x^2$
7. $f(t) = -2t^2 + 4t + 1$
8. $f(x) = x^2 - 8x + 12$
9. $h(x) = 4x^2 + 4x + 13$
10. $f(x) = x^2 - 6x + 1$
11. $h(x) = x^2 + 5x - 4$
12. $f(x) = 4x^2 + 4x + 5$
13. $f(x) = \frac{1}{3}(x^2 + 5x - 4)$
14. $f(x) = \frac{1}{2}(6x^2 - 24x + 22)$

In Exercises 15–18, write the standard form of the equation of the parabola that has the indicated vertex and whose graph passes through the given point.

15. Vertex: $(4, 1)$; point: $(2, -1)$
16. Vertex: $(2, 2)$; point: $(0, 3)$
17. Vertex: $(1, -4)$; point: $(2, -3)$
18. Vertex: $(2, 3)$; point: $(-1, 6)$

19. ***Geometry*** The perimeter of a rectangle is 1000 meters.
    (a) Draw a diagram that gives a visual representation of the problem. Label the length and width as x and y, respectively.
    (b) Write y as a function of x. Use the result to write the area as a function of x.
    (c) Of all possible rectangles with perimeters of 1000 meters, find the dimensions of the one with the maximum area.

20. ***Maximum Revenue*** The total revenue R earned (in dollars) from producing a gift box of candles is given by $R(p) = -10p^2 + 800p$, where p is the price per unit (in dollars).
    (a) Find the revenues when the prices per box are \$20, \$25, and \$30.
    (b) Find the unit price that will yield a maximum revenue. What is the maximum revenue? Explain your results.

21. ***Minimum Cost*** A soft-drink manufacturer has daily production costs of $C = 70{,}000 - 120x + 0.055x^2$, where C is the total cost (in dollars) and x is the number of units produced. How many units should be produced each day to yield a minimum cost?

22. ***Sociology*** The average age of the groom at a first marriage for a given age of the bride can be approximated by the model

$$y = -0.107x^2 + 5.68x - 48.5, \quad 20 \le x \le 25$$

where y is the age of the groom and x is the age of the bride. Sketch a graph of the model. For what age of the bride is the average age of the groom 26? *(Source: U.S. Census Bureau)*

In Exercises 23–28, sketch the graphs of $y = x^n$ and the transformation.

23. $y = x^3$, $f(x) = -(x - 2)^3$
24. $y = x^3$, $f(x) = -4x^3$
25. $y = x^4$, $f(x) = 6 - x^4$
26. $y = x^4$, $f(x) = 2(x - 8)^4$
27. $y = x^5$, $f(x) = (x - 5)^5$
28. $y = x^5$, $f(x) = \frac{1}{2}x^5 + 3$

In Exercises 29–32, describe the right-hand and left-hand behaviors of the graph of the polynomial function.

29. $f(x) = -2x^2 - 5x + 12$
30. $f(x) = \frac{1}{2}x^3 + 2x$
31. $g(x) = \frac{3}{4}(x^4 + 3x^2 + 2)$
32. $h(x) = -x^7 + 8x^2 - 8x$

In Exercises 33–38, find all the real zeros of the polynomial function. Determine the multiplicity of each zero and the number of turning points of the graph of the function. Use a graphing utility to verify your answers.

33. $f(x) = 3x^2 + 20x - 32$
34. $f(x) = x(x + 3)^2$
35. $f(t) = t^3 - 3t$
36. $f(x) = x^3 - 8x^2$
37. $f(x) = -18x^3 + 12x^2$
38. $g(x) = x^4 + x^3 - 12x^2$

In Exercises 39–42, sketch the graph of the function by (a) applying the Leading Coefficient Test, (b) finding the zeros of the polynomial, (c) plotting sufficient solution points, and (d) drawing a continuous curve through the points.

39. $f(x) = -x^3 + x^2 - 2$
40. $g(x) = 2x^3 + 4x^2$
41. $f(x) = x(x^3 + x^2 - 5x + 3)$
42. $h(x) = 3x^2 - x^4$

In Exercises 43–48, use long division to divide.

43. $\dfrac{30x^2 - 3x + 8}{5x - 3}$

44. $\dfrac{4x + 7}{3x - 2}$

45. $\dfrac{5x^3 - 21x^2 - 25x - 4}{x^2 - 5x - 1}$

46. $\dfrac{3x^4}{x^2 - 1}$

47. $\dfrac{x^4 - 3x^3 + 4x^2 - 6x + 3}{x^2 + 2}$

48. $\dfrac{6x^4 + 10x^3 + 13x^2 - 5x + 2}{2x^2 - 1}$

In Exercises 49–52, use synthetic division to divide.

49. $\dfrac{6x^4 - 4x^3 - 27x^2 + 18x}{x - 2}$

50. $\dfrac{0.1x^3 + 0.3x^2 - 0.5}{x - 5}$

51. $\dfrac{2x^3 - 25x^2 + 66x + 48}{x - 8}$

52. $\dfrac{5x^3 + 33x^2 + 50x - 8}{x + 4}$

In Exercises 53 and 54, use synthetic division to determine whether the given values of x are zeros of the function.

53. $f(x) = 20x^4 + 9x^3 - 14x^2 - 3x$
(a) $x = -1$ (b) $x = \frac{3}{4}$ (c) $x = 0$ (d) $x = 1$

54. $f(x) = 3x^3 - 8x^2 - 20x + 16$
(a) $x = 4$ (b) $x = -4$ (c) $x = \frac{2}{3}$ (d) $x = -1$

In Exercises 55 and 56, use the Remainder Theorem and synthetic division to find each function value.

55. $f(x) = x^4 + 10x^3 - 24x^2 + 20x + 44$
(a) $f(-3)$ (b) $f(-1)$

56. $g(t) = 2t^5 - 5t^4 - 8t + 20$
(a) $g(-4)$ (b) $g(\sqrt{2})$

In Exercises 57–60, (a) verify the given factor(s) of the function f, (b) find the remaining factors of f, (c) use your results to write the complete factorization of f, (d) list all real zeros of f, and (e) confirm your results by using a graphing utility to graph the function.

Function	*Factor(s)*
57. $f(x) = x^3 + 4x^2 - 25x - 28$	$(x - 4)$
58. $f(x) = 2x^3 + 11x^2 - 21x - 90$	$(x + 6)$
59. $f(x) = x^4 - 4x^3 - 7x^2 + 22x + 24$	$(x + 2)(x - 3)$
60. $f(x) = x^4 - 11x^3 + 41x^2 - 61x + 30$	$(x - 2)(x - 5)$

Data Analysis **In Exercises 61–64, use the following information. The total annual attendance A (in millions) at women's Division I basketball games for the years 1997 through 2009 is shown in the table. The variable t represents the year, with $t = 7$ corresponding to 1997.** ***(Source: NCAA)***

Year, t	7	8	9	10	11	12	13
Attendance, A	4.9	5.4	5.8	6.4	6.5	6.9	7.4

Year, t	14	15	16	17	18	19
Attendance, A	7.2	7.1	7.1	7.9	8.1	8.0

61. Use the *regression* feature of a graphing utility to find a cubic model for the data.

62. Use a graphing utility to plot the data and graph the model in the same viewing window. Compare the model with the data.

63. Use the model to create a table of estimated values of A. Compare the estimated values with the actual data.

64. Use synthetic division to evaluate the model for the year 2014. Do you think the model is accurate in predicting the future attendance? Explain your reasoning.

In Exercises 65–68, write the complex number in standard form.

65. $8 + \sqrt{-100}$

66. $5 - \sqrt{-49}$

67. $i^2 + 3i$

68. $-5i + i^2$

In Exercises 69–80, perform the operation and write the result in standard form.

69. $(7 + 5i) + (-4 + 2i)$

70. $\left(\frac{\sqrt{2}}{2} - \frac{\sqrt{2}}{2}i\right) - \left(\frac{\sqrt{2}}{2} + \frac{\sqrt{2}}{2}i\right)$

71. $7i(11 - 9i)$

72. $(1 + 6i)(5 - 2i)$

73. $(10 - 8i)(2 - 3i)$

74. $i(6 + i)(3 - 2i)$

75. $(8 - 5i)^2$

76. $(4 + 7i)^2 + (4 - 7i)^2$

77. $\dfrac{6 + i}{4 - i}$

78. $\dfrac{8 - 5i}{i}$

79. $\dfrac{4}{2 - 3i} + \dfrac{2}{1 + i}$

80. $\dfrac{1}{2 + i} - \dfrac{5}{1 + 4i}$

In Exercises 81–84, find all solutions of the equation.

81. $5x^2 + 2 = 0$

82. $2 + 8x^2 = 0$

83. $x^2 - 2x + 10 = 0$

84. $6x^2 + 3x + 27 = 0$

In Exercises 85–90, find all the zeros of the function.

85. $f(x) = 4x(x - 3)^2$

86. $f(x) = (x - 4)(x + 9)^2$

87. $f(x) = x^2 - 11x + 18$

88. $f(x) = x^3 + 10x$

89. $f(x) = (x + 4)(x - 6)(x - 2i)(x + 2i)$

90. $f(x) = (x - 8)(x - 5)^2(x - 3 + i)(x - 3 - i)$

In Exercises 91 and 92, use the Rational Zero Test to list all possible rational zeros of f.

91. $f(x) = -4x^3 + 8x^2 - 3x + 15$

92. $f(x) = 3x^4 + 4x^3 - 5x^2 - 8$

In Exercises 93–98, find all the rational zeros of the function.

93. $f(x) = x^3 + 3x^2 - 28x - 60$

94. $f(x) = 4x^3 - 27x^2 + 11x + 42$

95. $f(x) = x^3 - 10x^2 + 17x - 8$

96. $f(x) = x^3 + 9x^2 + 24x + 20$

97. $f(x) = x^4 + x^3 - 11x^2 + x - 12$

98. $f(x) = 25x^4 + 25x^3 - 154x^2 - 4x + 24$

In Exercises 99–102, use the given zero to find all the zeros of the function.

	Function	*Zero*
99.	$f(x) = x^3 - 4x^2 + x - 4$	i
100.	$h(x) = -x^3 + 2x^2 - 16x + 32$	$-4i$
101.	$g(x) = 2x^4 - 3x^3 - 13x^2 + 37x - 15$	$2 + i$
102.	$f(x) = 4x^4 - 11x^3 + 14x^2 - 6x$	$1 - i$

In Exercises 103–106, find all the zeros of the function and write the polynomial as a product of linear factors.

103. $f(x) = x^3 + 4x^2 - 5x$ **104.** $g(x) = x^3 - 7x^2 + 36$

105. $g(x) = x^4 + 4x^3 - 3x^2 + 40x + 208$

106. $f(x) = x^4 + 8x^3 + 8x^2 - 72x - 153$

In Exercises 107 and 108, find a polynomial function with real coefficients that has the given zeros. (There are many correct answers.)

107. $\frac{2}{3}, 4, \sqrt{3}i$ **108.** $2, -3, 1 - 2i$

In Exercises 109–112, find the domain of the rational function.

109. $f(x) = \dfrac{3x}{x + 10}$ **110.** $f(x) = \dfrac{4x^3}{2 + 5x}$

111. $f(x) = \dfrac{8}{x^2 - 10x + 24}$ **112.** $f(x) = \dfrac{x^2 + x - 2}{x^2 + 4}$

In Exercises 113–116, identify any vertical or horizontal asymptotes.

113. $f(x) = \dfrac{4}{x + 3}$ **114.** $f(x) = \dfrac{2x^2 + 5x - 3}{x^2 + 2}$

115. $h(x) = \dfrac{5x + 20}{x^2 - 2x - 24}$ **116.** $h(x) = \dfrac{x^3 - 4x^2}{x^2 + 3x + 2}$

In Exercises 117–128, (a) state the domain of the function, (b) identify all intercepts, (c) find any vertical and horizontal asymptotes, and (d) plot additional solution points as needed to sketch the graph of the rational function.

117. $f(x) = \dfrac{-3}{2x^2}$ **118.** $f(x) = \dfrac{4}{x}$

119. $g(x) = \dfrac{2 + x}{1 - x}$ **120.** $h(x) = \dfrac{x - 4}{x - 7}$

121. $p(x) = \dfrac{5x^2}{4x^2 + 1}$ **122.** $f(x) = \dfrac{2x}{x^2 + 4}$

123. $f(x) = \dfrac{x}{x^2 + 1}$ **124.** $h(x) = \dfrac{9}{(x - 3)^2}$

125. $f(x) = \dfrac{-6x^2}{x^2 + 1}$ **126.** $f(x) = \dfrac{2x^2}{x^2 - 4}$

127. $f(x) = \dfrac{6x^2 - 11x + 3}{3x^2 - x}$ **128.** $f(x) = \dfrac{6x^2 - 7x + 2}{4x^2 - 1}$

In Exercises 129–132, (a) state the domain of the function, (b) identify all intercepts, (c) identify any vertical and slant asymptotes, and (d) plot additional solution points as needed to sketch the graph of the rational function.

129. $f(x) = \dfrac{2x^3}{x^2 + 1}$

130. $f(x) = \dfrac{x^2 + 1}{x + 1}$

131. $f(x) = \dfrac{3x^3 - 2x^2 - 3x + 2}{3x^2 - x - 4}$

132. $f(x) = \dfrac{3x^3 - 4x^2 - 12x + 16}{3x^2 + 5x - 2}$

133. ***Average Cost*** A business has a production cost of $C = 0.5x + 500$ for producing x units of a product. The average cost per unit, $\overline{C}$, is given by

$$\overline{C} = \frac{C}{x} = \frac{0.5x + 500}{x}, \quad x > 0.$$

Determine the average cost per unit as x increases without bound. (Find the horizontal asymptote.)

134. ***Seizure of Illegal Drugs*** The cost C (in millions of dollars) for the federal government to seize $p\%$ of an illegal drug as it enters the country is given by

$$C = \frac{528p}{100 - p}, \quad 0 \le p < 100.$$

(a) Use a graphing utility to graph the cost function.

(b) Find the costs of seizing 25%, 50%, and 75% of the drug.

(c) According to this model, would it be possible to seize 100% of the drug?

2 CHAPTER TEST

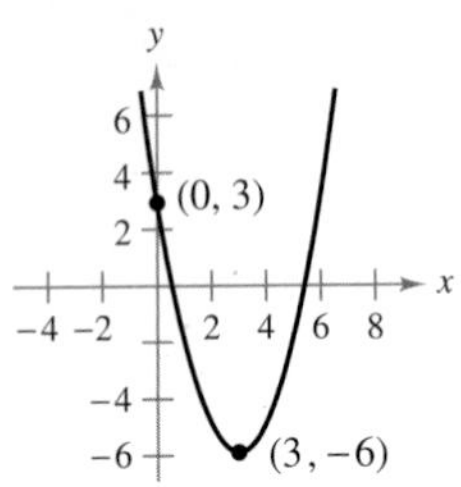

Figure for 2

Take this test as you would take a test in class. When you are finished, check your work against the answers given in the back of the book.

1. Describe how the graph of g differs from the graph of $f(x) = x^2$.

   (a) $g(x) = 2 - x^2$ (b) $g(x) = \left(x - \frac{3}{2}\right)^2$

2. Find an equation of the parabola shown in the figure at the left.
3. The path of a ball is given by $y = -\frac{1}{20}x^2 + 3x + 5$, where y is the height (in feet) of the ball and x is the horizontal distance (in feet) from where the ball was thrown.

   (a) Find the maximum height of the ball.

   (b) Which number determines the height at which the ball was thrown? Does changing this value change the coordinates of the maximum height of the ball? Explain.

4. Determine the right-hand and left-hand behavior of the graph of the function $h(t) = -\frac{3}{4}t^5 + 2t^2$. Then sketch its graph.
5. Divide using long division.

   $$\frac{3x^3 + 4x - 1}{x^2 + 1}$$

6. Divide using synthetic division.

   $$\frac{2x^4 - 5x^2 - 3}{x - 2}$$

7. Use synthetic division to show that $x = \frac{5}{2}$ is a zero of the function given by $f(x) = 2x^3 - 5x^2 - 6x + 15$. Use the result to factor the polynomial function completely and list all the real zeros of the function.
8. Perform each operation and write the result in standard form.

   (a) $10i - \left(3 + \sqrt{-25}\right)$ (b) $\left(2 + \sqrt{3}i\right)\left(2 - \sqrt{3}i\right)$

9. Write the quotient in standard form: $\dfrac{5}{2 + i}$.

In Exercises 10 and 11, find a polynomial function with real coefficients that has the given zeros. (There are many correct answers.)

10. $0, 3, 2 + i$
11. $1 - \sqrt{3}i, 2, 2$

In Exercises 12 and 13, find all the zeros of the function.

12. $f(x) = 3x^3 + 14x^2 - 7x - 10$
13. $f(x) = x^4 - 9x^2 - 22x - 24$

In Exercises 14–16, identify any intercepts and asymptotes of the graph of the function. Then sketch a graph of the function.

14. $h(x) = \dfrac{4}{x^2} - 1$
15. $f(x) = \dfrac{2x^2 - 5x - 12}{x^2 - 16}$
16. $g(x) = \dfrac{x^2 + 2}{x - 1}$

17. The amount y of CO_2 uptake (in milligrams per square decimeter per hour) at optimal temperatures and with the natural supply of CO_2 is approximated by the model

    $$y = \frac{18.47x - 2.96}{0.23x + 1}, \quad x > 0$$

    where x is the light intensity (in watts per square meter). Use a graphing utility to graph the function and determine the limiting amount of CO_2 uptake.

P.S. PROBLEM SOLVING

1. At a glassware factory, molten cobalt glass is poured into molds to make paperweights. Each mold is a rectangular prism whose height is 3 inches greater than the length of each side of the square base. A machine pours 20 cubic inches of liquid glass into each mold. What are the dimensions of the mold?

2. Determine whether the statement is true or false. If false, provide one or more reasons why the statement is false and correct the statement. Let $f(x) = ax^3 + bx^2 + cx + d$, $a \neq 0$, and let $f(2) = -1$. Then

$$\frac{f(x)}{x+1} = q(x) + \frac{2}{x+1}$$

where $q(x)$ is a second-degree polynomial.

3. Given the function $f(x) = a(x - h)^2 + k$, state the values of a, h, and k that give a reflection in the x-axis with either a shrink or a stretch of the graph of the function $f(x) = x^2$.

4. Explore the transformations of the form

$$g(x) = a(x - h)^5 + k.$$

(a) Use a graphing utility to graph the functions

$$y_1 = -\tfrac{1}{3}(x - 2)^5 + 1$$

and

$$y_2 = \tfrac{3}{5}(x + 2)^5 - 3.$$

Determine whether the graphs are increasing or decreasing. Explain.

(b) Will the graph of g always be increasing or decreasing? If so, is this behavior determined by a, h, or k? Explain.

(c) Use a graphing utility to graph the function given by

$$H(x) = x^5 - 3x^3 + 2x + 1.$$

Use the graph and the result of part (b) to determine whether H can be written in the form

$$H(x) = a(x - h)^5 + k.$$

Explain.

5. Consider the function given by

$$f(x) = \frac{ax}{(x - b)^2}.$$

(a) Determine the effect on the graph of f if $b \neq 0$ and a is varied. Consider cases in which a is positive and a is negative.

(b) Determine the effect on the graph of f if $a \neq 0$ and b is varied.

6. The growth of a red oak tree is approximated by the function

$$G = -0.003t^3 + 0.137t^2 + 0.458t - 0.839$$

where G is the height of the tree (in feet) and t $(2 \leq t \leq 34)$ is its age (in years).

(a) Use a graphing utility to graph the function. (*Hint:* Use a viewing window in which $-10 \leq x \leq 45$ and $-5 \leq y \leq 60$.)

(b) Estimate the age of the tree when it is growing most rapidly. This point is called the *point of diminishing returns* because the increase in size will be less with each additional year.

(c) Using calculus, the point of diminishing returns can also be found by finding the vertex of the parabola given by

$$y = -0.009t^2 + 0.274t + 0.458.$$

Find the vertex of this parabola.

(d) Compare your results from parts (b) and (c).

7. Consider the function given by

$$f(x) = (2x^2 + x - 1)/(x + 1).$$

(a) Use a graphing utility to graph the function. Does the graph have a vertical asymptote at $x = -1$?

(b) Rewrite the function in simplified form.

(c) Use the *zoom* and *trace* features to determine the value of the graph near $x = -1$.

8. A wire 100 centimeters in length is cut into two pieces. One piece is bent to form a square and the other to form a circle. Let x equal the length of the wire used to form the square.

(a) Write the function that represents the combined area of the two figures.

(b) Determine the domain of the function.

(c) Find the value(s) of x that yield a maximum and minimum area.

(d) Explain your reasoning.

9. The multiplicative inverse of z is a complex number z_m such that $z \cdot z_m = 1$. Find the multiplicative inverse of each complex number.

(a) $z = 1 + i$ (b) $z = 3 - i$ (c) $z = -2 + 8i$

10. The parabola shown in the figure has an equation of the form $y = ax^2 + bx + c$. Find the equation for this parabola by the following methods. (a) Find the equation analytically. (b) Use the *regression* feature of a graphing utility to find the equation.

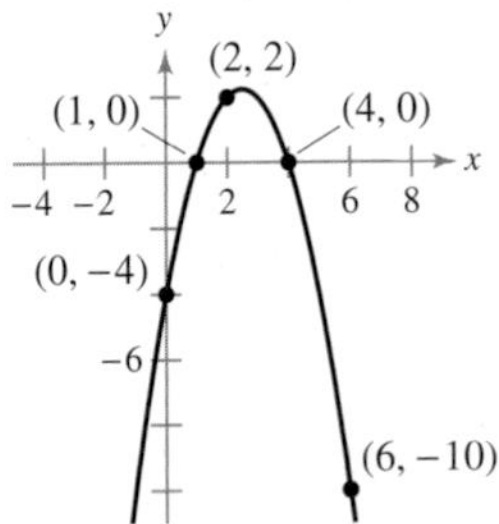

11. One of the fundamental themes of calculus is to find the slope of the tangent line to a curve at a point. To see how this can be done, consider the point (2, 4) on the graph of the quadratic function $f(x) = x^2$.

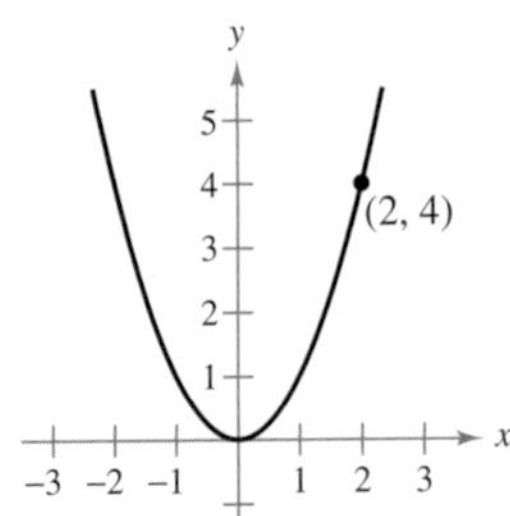

(a) Find the slope of the line joining (2, 4) and (3, 9). Is the slope of the tangent line at (2, 4) greater than or less than the slope of the line through (2, 4) and (3, 9)?

(b) Find the slope of the line joining (2, 4) and (1, 1). Is the slope of the tangent line at (2, 4) greater than or less than the slope of the line through (2, 4) and (1, 1)?

(c) Find the slope of the line joining (2, 4) and (2.1, 4.41). Is the slope of the tangent line at (2, 4) greater than or less than the slope of the line through (2, 4) and (2.1, 4.41)?

(d) Find the slope of the line joining (2, 4) and $(2 + h, f(2 + h))$ in terms of the nonzero number h.

(e) Evaluate the slope formula from part (d) for $h = -1$, 1, and 0.1. Compare these values with those in parts (a)–(c).

(f) What can you conclude the slope of the tangent line at (2, 4) to be? Explain.

12. A rancher plans to fence a rectangular pasture adjacent to a river. The rancher has 100 meters of fence, and no fencing is needed along the river.

(a) Write the area A as a function of x, the length of the side of the pasture parallel to the river. What is the feasible domain of A?

(b) Graph the function A and estimate the dimensions that yield the maximum area for the pasture.

(c) Find the exact dimensions that yield the maximum area for the pasture by writing the quadratic function in standard form.

13. Match the graph of the rational function

$$f(x) = \frac{ax + b}{cx + d}$$

with the given conditions.

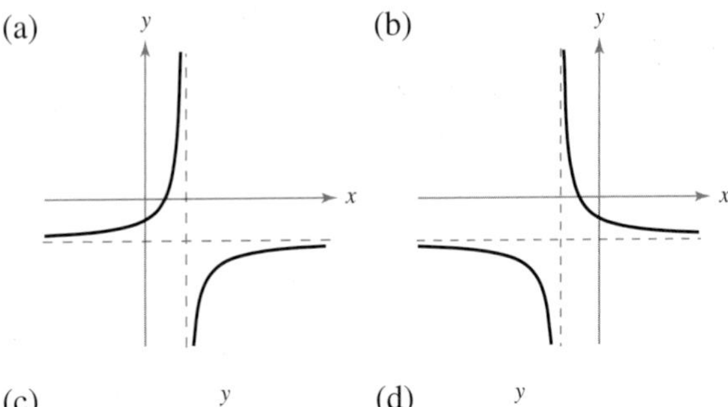

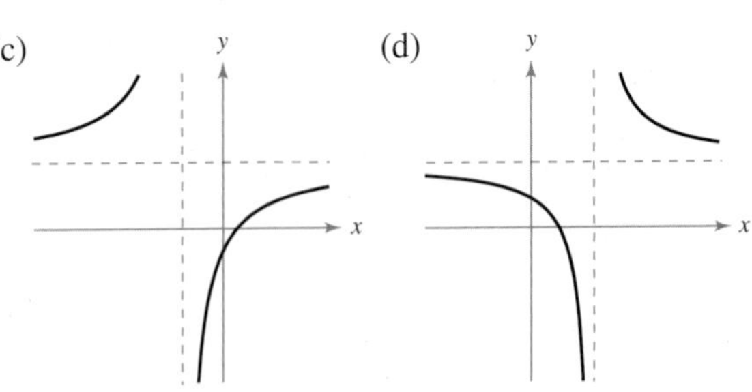

(i) $a > 0$	(ii) $a > 0$	(iii) $a < 0$	(iv) $a > 0$
$b < 0$	$b > 0$	$b > 0$	$b < 0$
$c > 0$	$c < 0$	$c > 0$	$c > 0$
$d < 0$	$d < 0$	$d < 0$	$d > 0$

3 Limits and Their Properties

The limit of a function is the primary concept that distinguishes calculus from algebra and analytic geometry. The notion of a limit is fundamental to the study of calculus. Thus, it is important to acquire a good working knowledge of limits before moving on to other topics in calculus.

In this chapter, you should learn the following.

- How calculus compares with precalculus. (3.1)
- How to find limits graphically and numerically. (3.2)
- How to evaluate limits analytically. (3.3)
- How to determine continuity at a point and on an open interval, and how to determine one-sided limits. (3.4)
- How to determine infinite limits and find vertical asymptotes. (3.5)

European Space Agency, NASA

According to NASA, the coldest place in the known universe is the Boomerang nebula. The nebula is five thousand light years from Earth and has a temperature of −272°C. That is only 1° warmer than absolute zero, the coldest possible temperature. How did scientists determine that absolute zero is the "lower limit" of the temperature of matter? (See Section 3.4, Example 6.)

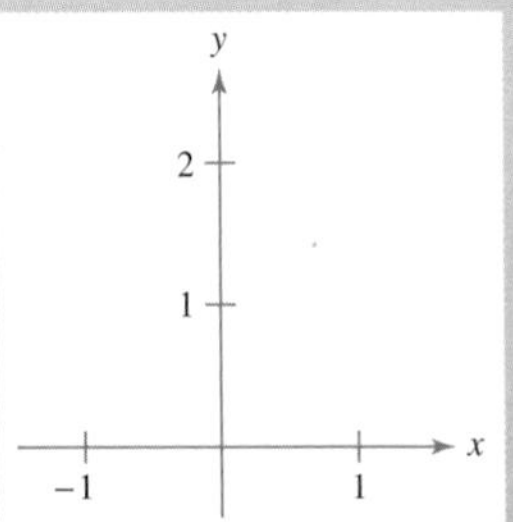

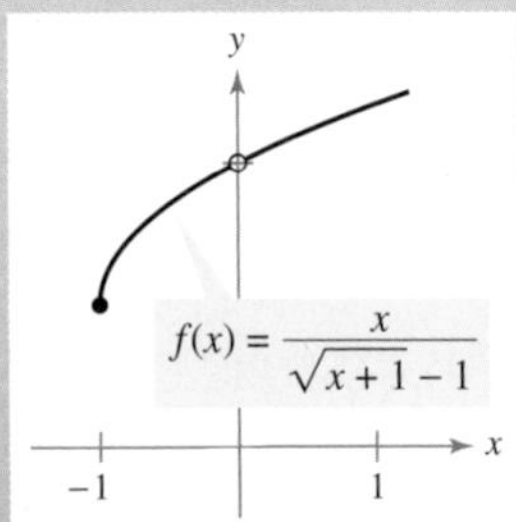

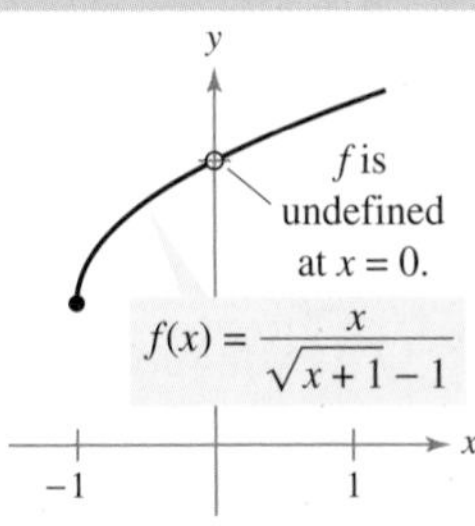

The limit process is a fundamental concept of calculus. One technique you can use to estimate a limit is to graph the function and then determine the behavior of the graph as the independent variable approaches a specific value. (See Section 3.2.)

3.1 A Preview of Calculus

- **Understand what calculus is and how it compares with precalculus.**
- **Understand that the tangent line problem is basic to calculus.**
- **Understand that the area problem is also basic to calculus.**

What Is Calculus?

Calculus is the mathematics of change. For instance, calculus is the mathematics of velocities, accelerations, tangent lines, slopes, areas, volumes, arc lengths, centroids, curvatures, and a variety of other concepts that have enabled scientists, engineers, and economists to model real-life situations.

Although precalculus mathematics also deals with velocities, accelerations, tangent lines, slopes, and so on, there is a fundamental difference between precalculus mathematics and calculus. Precalculus mathematics is more static, whereas calculus is more dynamic. Here are some examples.

- An object traveling at a constant velocity can be analyzed with precalculus mathematics. To analyze the velocity of an accelerating object, you need calculus.
- The slope of a line can be analyzed with precalculus mathematics. To analyze the slope of a curve, you need calculus.
- The curvature of a circle is constant and can be analyzed with precalculus mathematics. To analyze the variable curvature of a general curve, you need calculus.
- The area of a rectangle can be analyzed with precalculus mathematics. To analyze the area under a general curve, you need calculus.

Each of these situations involves the same general strategy—the reformulation of precalculus mathematics through the use of a limit process. So, one way to answer the question "What is calculus?" is to say that calculus is a "limit machine" that involves three stages. The first stage is precalculus mathematics, such as the slope of a line or the area of a rectangle. The second stage is the limit process, and the third stage is a new calculus formulation, such as a derivative or integral.

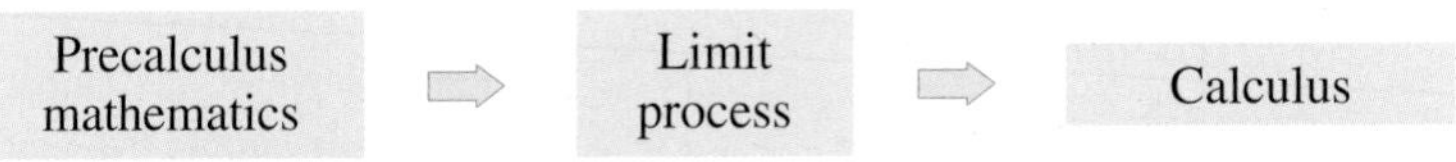

Some students try to learn calculus as if it were simply a collection of new formulas. This is unfortunate. If you reduce calculus to the memorization of differentiation and integration formulas, you will miss a great deal of understanding, self-confidence, and satisfaction.

On the following two pages are listed some familiar precalculus concepts coupled with their calculus counterparts. Throughout the text, your goal should be to learn how precalculus formulas and techniques are used as building blocks to produce the more general calculus formulas and techniques. Don't worry if you are unfamiliar with some of the concepts listed on the following two pages—you will be reviewing all of them.

As you proceed through this text, come back to this discussion repeatedly. Try to keep track of where you are relative to the three stages involved in the study of calculus. For example, the first five chapters break down as follows.

Chapters P, 1, 2:	Preparation for Calculus	Precalculus
Chapter 3:	Limits and Their Properties	Limit process
Chapter 4:	Differentiation	Calculus

STUDY TIP As you progress through this course, remember that learning calculus is just one of your goals. Your most important goal is to learn how to use calculus to model and solve real-life problems. Here are a few problem-solving strategies that may help you.

- Be sure you understand the question. What is given? What are you asked to find?
- Outline a plan. There are many approaches you could use: look for a pattern, solve a simpler problem, work backwards, draw a diagram, use technology, or any of several other approaches.
- Complete your plan. Be sure to answer the question. Verbalize your answer. For example, rather than writing the answer as $x = 4.6$, it would be better to write the answer as "The area of the region is 4.6 square meters."
- Look back at your work. Does your answer make sense? Is there a way you can check the reasonableness of your answer?

Without Calculus		With Differential Calculus	
Value of $f(x)$ when $x = c$	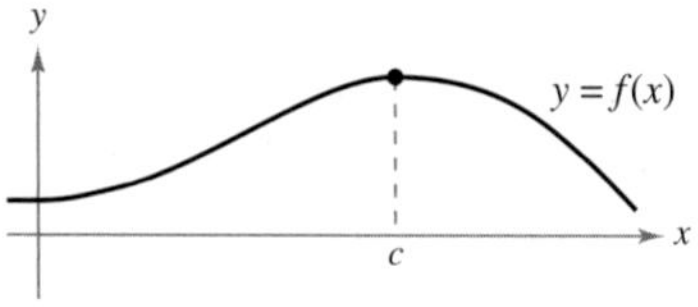	Limit of $f(x)$ as x approaches c	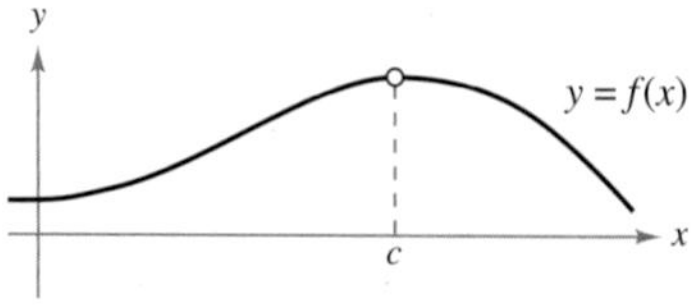
Slope of a line	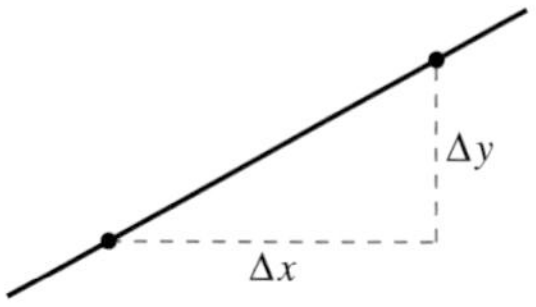	Slope of a curve	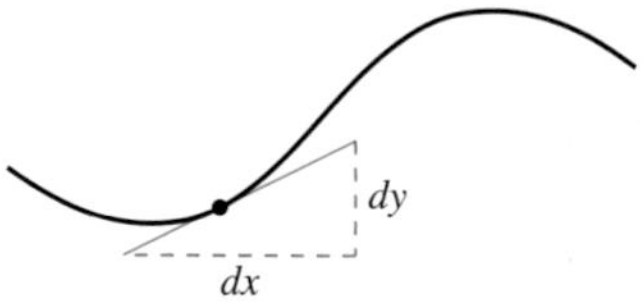
Secant line to a curve	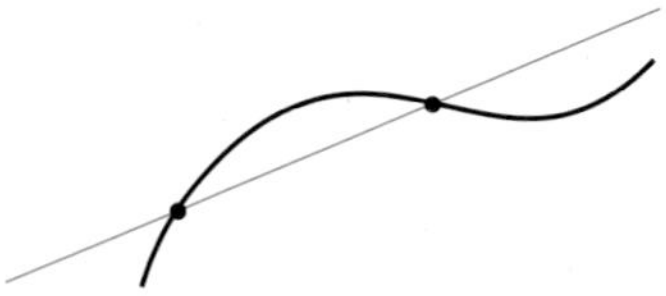	Tangent line to a curve	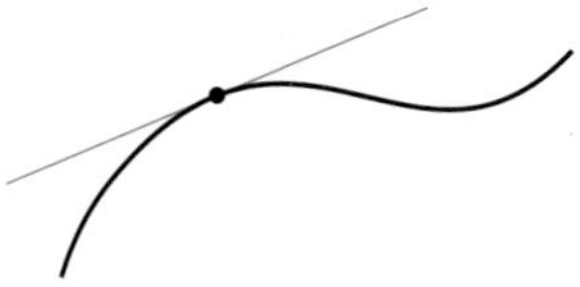
Average rate of change between $t = a$ and $t = b$		Instantaneous rate of change at $t = c$	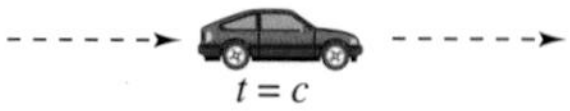
Curvature of a circle	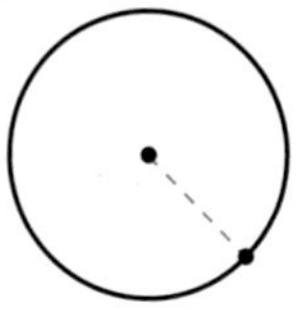	Curvature of a curve	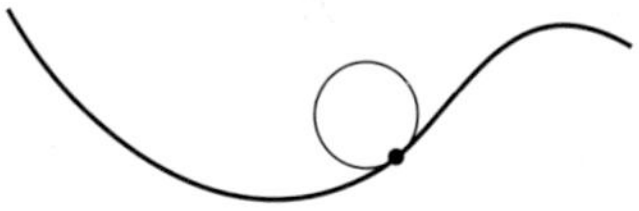
Height of a curve when $x = c$	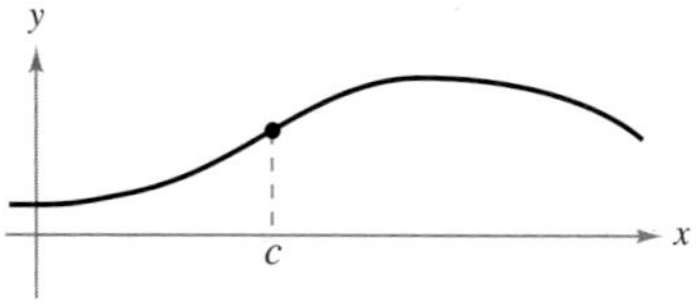	Maximum height of a curve on an interval	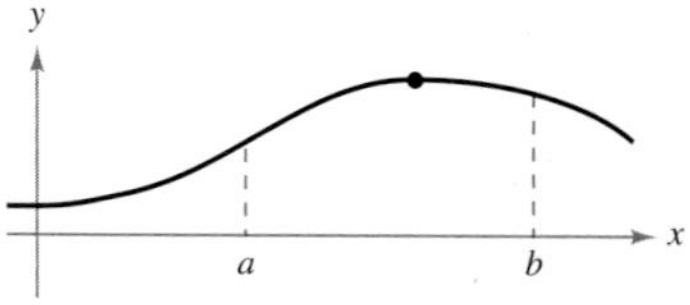
Tangent plane to a sphere	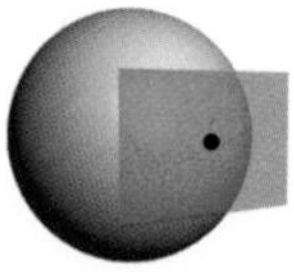	Tangent plane to a surface	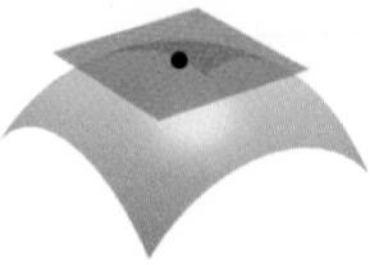
Direction of motion along a line	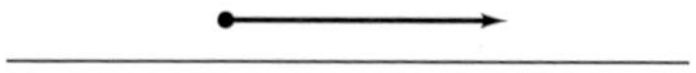	Direction of motion along a curve	

Without Calculus		With Integral Calculus	
Area of a rectangle		Area under a curve	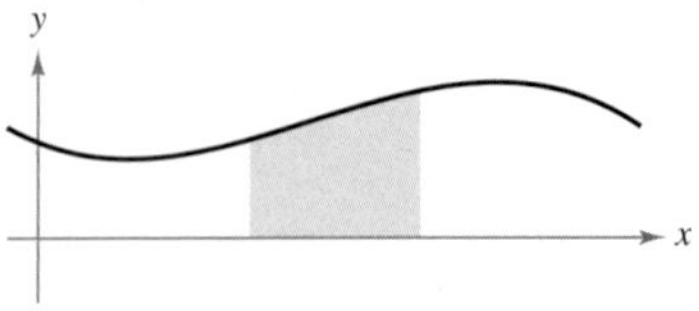
Work done by a constant force	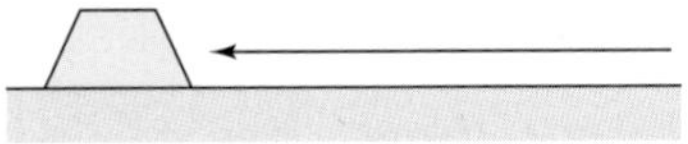	Work done by a variable force	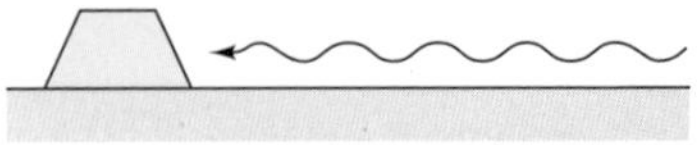
Center of a rectangle	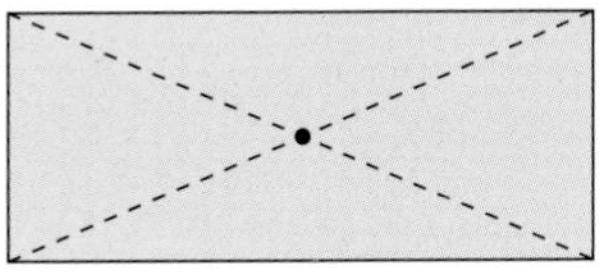	Centroid of a region	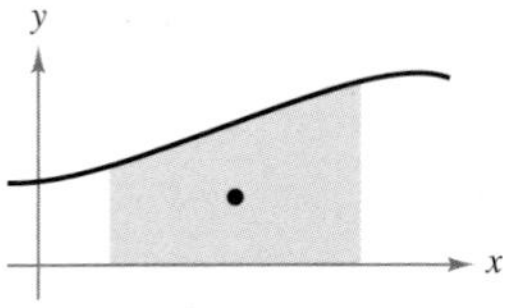
Length of a line segment	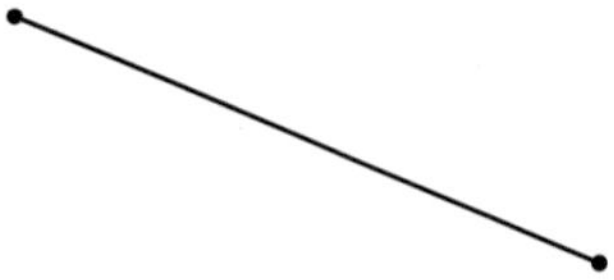	Length of an arc	
Surface area of a cylinder		Surface area of a solid of revolution	
Mass of a solid of constant density		Mass of a solid of variable density	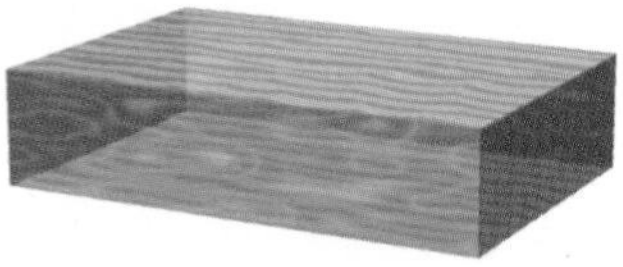
Volume of a rectangular solid		Volume of a region under a surface	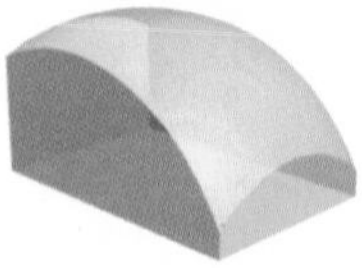
Sum of a finite number of terms	$a_1 + a_2 + \cdots + a_n = S$	Sum of an infinite number of terms	$a_1 + a_2 + a_3 + \cdots = S$

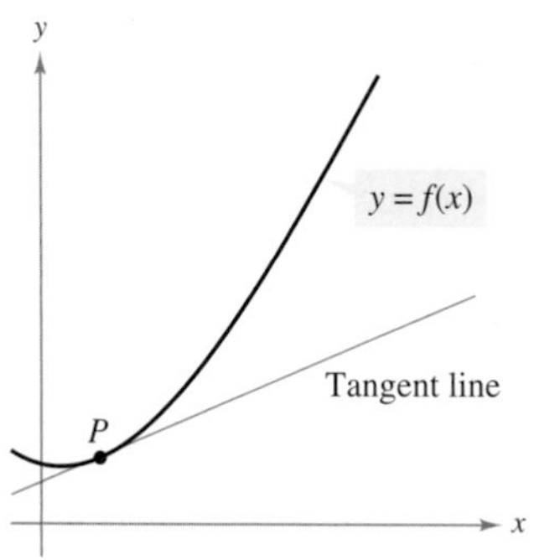

The tangent line to the graph of f at P
Figure 3.1

The Tangent Line Problem

The notion of a limit is fundamental to the study of calculus. The following brief descriptions of two classic problems in calculus—*the tangent line problem* and *the area problem*—should give you some idea of the way limits are used in calculus.

In the tangent line problem, you are given a function f and a point P on its graph and are asked to find an equation of the tangent line to the graph at point P, as shown in Figure 3.1.

Except for cases involving a vertical tangent line, the problem of finding the **tangent line** at a point P is equivalent to finding the *slope* of the tangent line at P. You can approximate this slope by using a line through the point of tangency and a second point on the curve, as shown in Figure 3.2(a). Such a line is called a **secant line.** If $P(c, f(c))$ is the point of tangency and

$$Q(c + \Delta x, f(c + \Delta x))$$

is a second point on the graph of f, the slope of the secant line through these two points can be found using precalculus and is given by

$$m_{sec} = \frac{f(c + \Delta x) - f(c)}{c + \Delta x - c} = \frac{f(c + \Delta x) - f(c)}{\Delta x}.$$

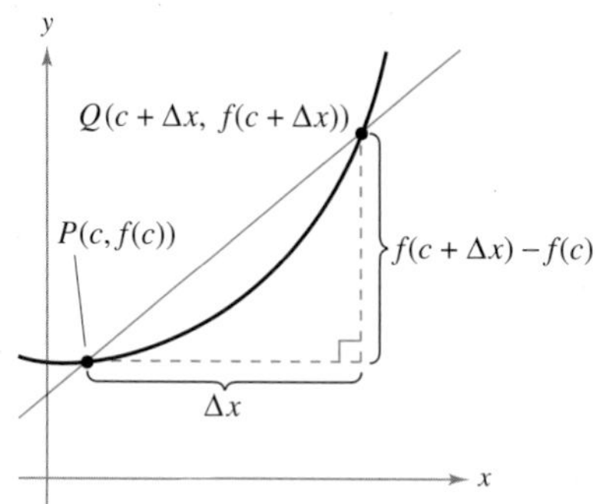

(a) The secant line through $(c, f(c))$ and $(c + \Delta x, f(c + \Delta x))$

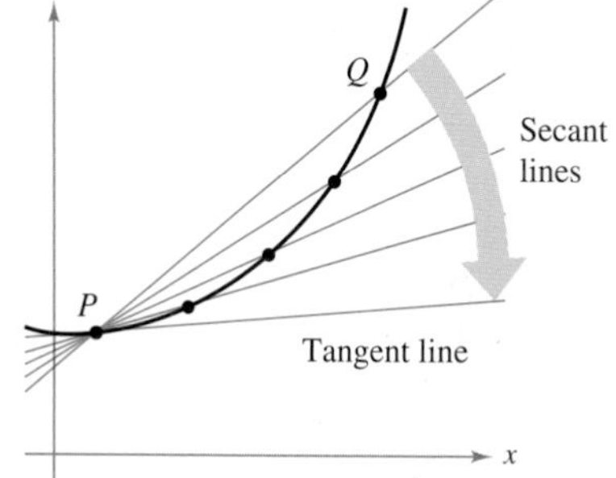

(b) As Q approaches P, the secant lines approach the tangent line.

Figure 3.2

As point Q approaches point P, the slopes of the secant lines approach the slope of the tangent line, as shown in Figure 3.2(b). When such a "limiting position" exists, the slope of the tangent line is said to be the **limit** of the slopes of the secant lines. (Much more will be said about this important calculus concept in Chapter 4.)

Girton College

GRACE CHISHOLM YOUNG (1868–1944)

Grace Chisholm Young received her degree in mathematics from Girton College in Cambridge, England. Her early work was published under the name of William Young, her husband. Between 1914 and 1916, Grace Young published work on the foundations of calculus that won her the Gamble Prize from Girton College.

EXPLORATION

The following points lie on the graph of $f(x) = x^2$.

$$Q_1(1.5, f(1.5)), \quad Q_2(1.1, f(1.1)), \quad Q_3(1.01, f(1.01)),$$
$$Q_4(1.001, f(1.001)), \quad Q_5(1.0001, f(1.0001))$$

Each successive point gets closer to the point $P(1, 1)$. Find the slopes of the secant lines through Q_1 and P, Q_2 and P, and so on. Graph these secant lines on a graphing utility. Then use your results to estimate the slope of the tangent line to the graph of f at the point P.

The Area Problem

In the tangent line problem, you saw how the limit process can be applied to the slope of a line to find the slope of a general curve. A second classic problem in calculus is finding the area of a plane region that is bounded by the graphs of functions. This problem can also be solved with a limit process. In this case, the limit process is applied to the area of a rectangle to find the area of a general region.

As a simple example, consider the region bounded by the graph of the function $y = f(x)$, the x-axis, and the vertical lines $x = a$ and $x = b$, as shown in Figure 3.3. You can approximate the area of the region with several rectangular regions, as shown in Figure 3.4. As you increase the number of rectangles, the approximation tends to become better and better because the amount of area missed by the rectangles decreases. Your goal is to determine the limit of the sum of the areas of the rectangles as the number of rectangles increases without bound.

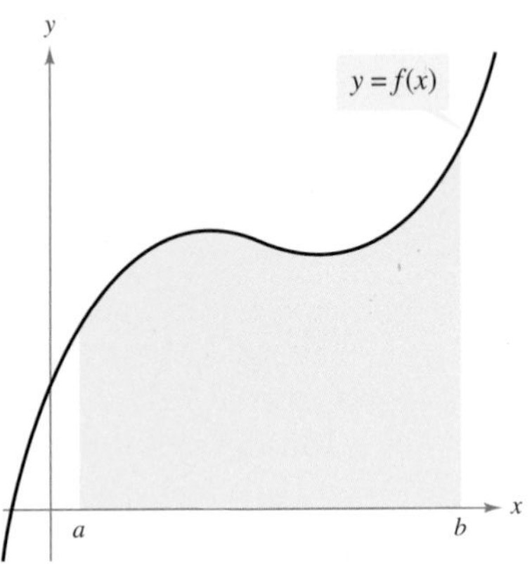

Area under a curve
Figure 3.3

HISTORICAL NOTE

In one of the most astounding events ever to occur in mathematics, it was discovered that the tangent line problem and the area problem are closely related. This discovery led to the birth of calculus. You will learn about the relationship between these two problems when you study the Fundamental Theorem of Calculus in Chapter 6.

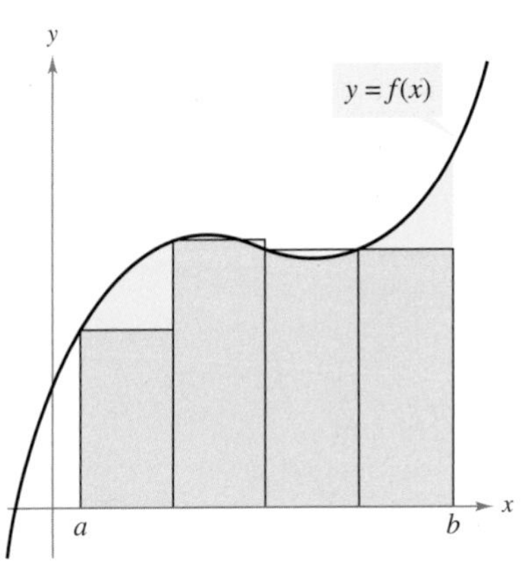

Approximation using four rectangles

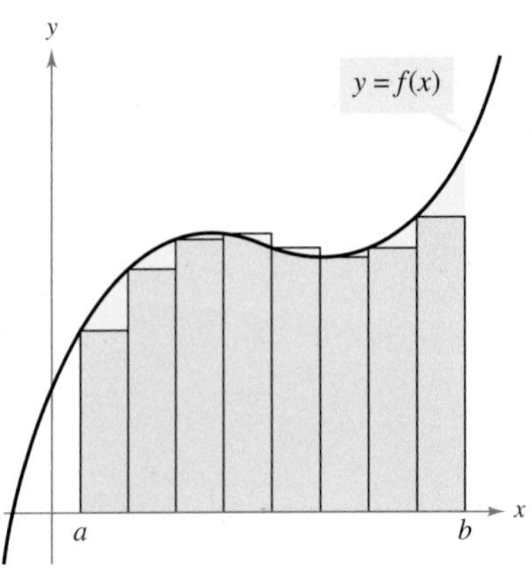

Approximation using eight rectangles
Figure 3.4

Consider the region bounded by the graphs of $f(x) = x^2$, $y = 0$, and $x = 1$, as shown in part (a) of the figure. The area of the region can be approximated by two sets of rectangles—one set inscribed within the region and the other set circumscribed over the region, as shown in parts (b) and (c). Find the sum of the areas of each set of rectangles. Then use your results to approximate the area of the region.

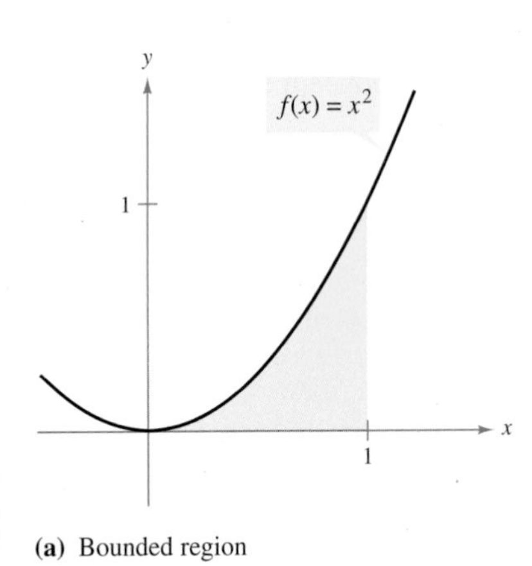

(a) Bounded region

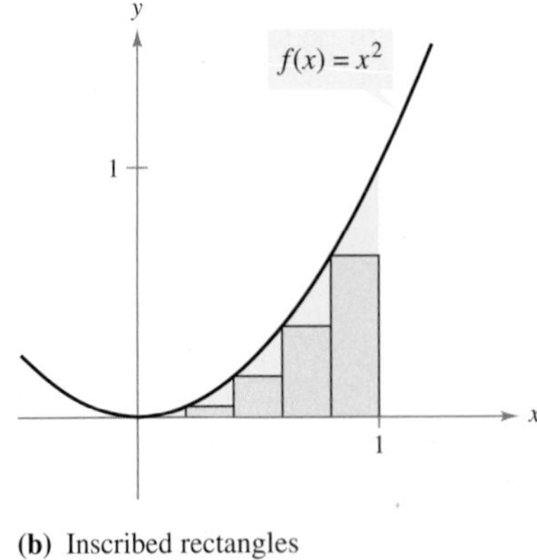

(b) Inscribed rectangles

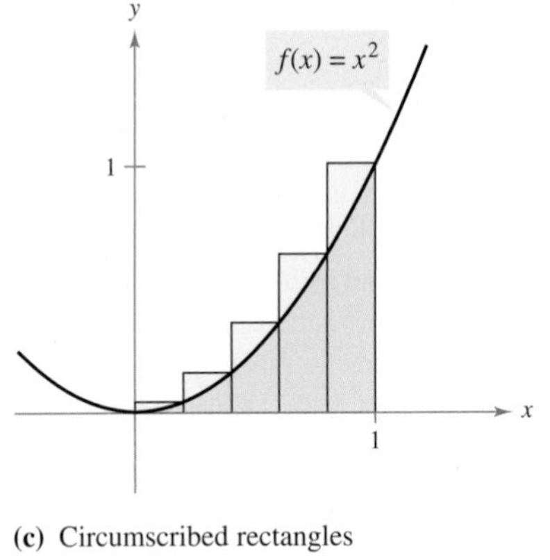

(c) Circumscribed rectangles

3.1 Exercises

See www.CalcChat.com for worked-out solutions to odd-numbered exercises.

In Exercises 1–4, decide whether the problem can be solved using precalculus or whether calculus is required. If the problem can be solved using precalculus, solve it. If the problem seems to require calculus, explain your reasoning and use a graphical or numerical approach to estimate the solution.

1. Find the distance traveled in 15 seconds by an object traveling at a constant velocity of 20 feet per second.

2. Find the distance traveled in 15 seconds by an object moving with a velocity of $v(t) = 20 + 3t$ feet per second, where t is the time in seconds.

3. A bicyclist is riding on a path modeled by the function $f(x) = 0.04(8x - x^2)$, (see figure), where x and $f(x)$ are measured in miles. Find the rate of change of elevation at $x = 2$.

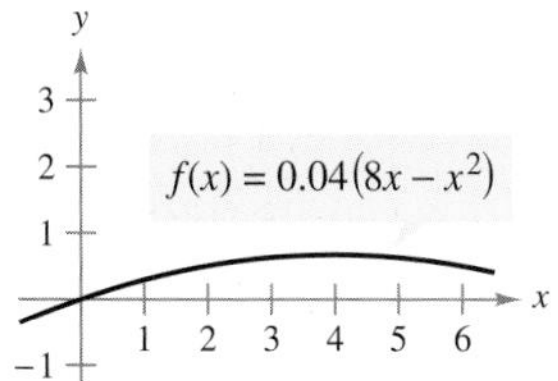

Figure for 3

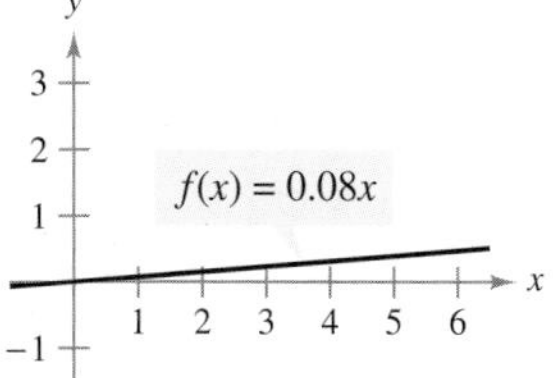

Figure for 4

4. A bicyclist is riding on a path modeled by the function $f(x) = 0.08x$, (see figure), where x and $f(x)$ are measured in miles. Find the rate of change of elevation at $x = 2$.

5. Find the area of the shaded region.

(a)

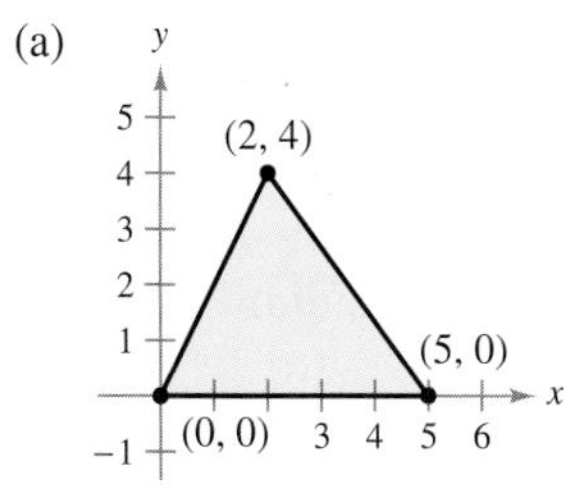

(b)

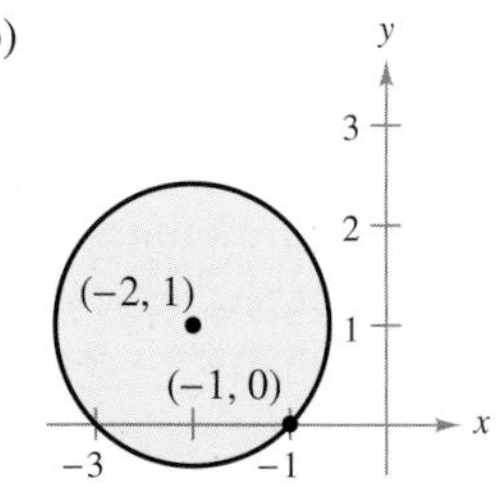

6. ***Secant Lines*** Consider the function $f(x) = \sqrt{x}$ and the point $P(4, 2)$ on the graph of f.

(a) Graph f and the secant lines passing through $P(4, 2)$ and $Q(x, f(x))$ for x-values of 1, 3, and 5.

(b) Find the slope of each secant line.

(c) Use the results of part (b) to estimate the slope of the tangent line to the graph of f at $P(4, 2)$. Describe how to improve your approximation of the slope.

7. ***Secant Lines*** Consider the function $f(x) = 6x - x^2$ and the point $P(2, 8)$ on the graph of f.

(a) Graph f and the secant lines passing through $P(2, 8)$ and $Q(x, f(x))$ for x-values of 3, 2.5, and 1.5.

(b) Find the slope of each secant line.

(c) Use the results of part (b) to estimate the slope of the tangent line to the graph of f at $P(2, 8)$. Describe how to improve your approximation of the slope.

8. Use the rectangles in each graph to approximate the area of the region bounded by $y = 4x - x^2$, $y = 0$, $x = 0$, and $x = 4$.

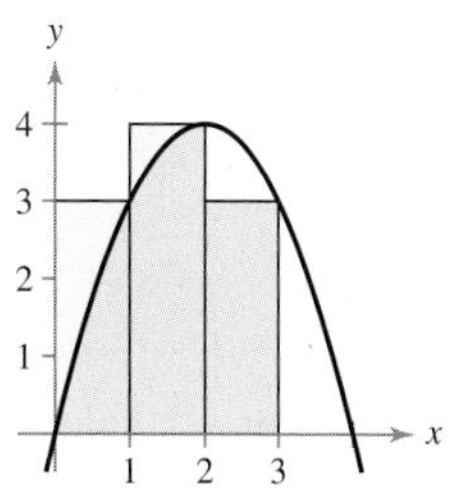

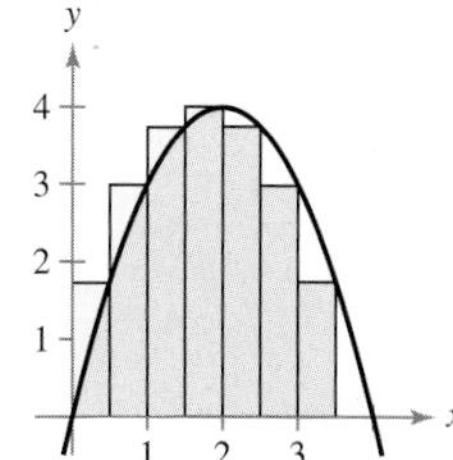

9. Use the rectangles in each graph to approximate the area of the region bounded by $y = 5/x$, $y = 0$, $x = 1$, and $x = 5$.

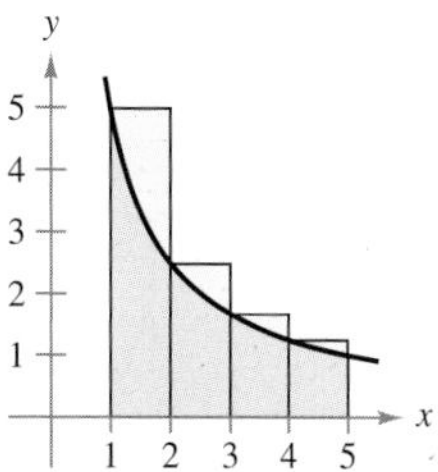

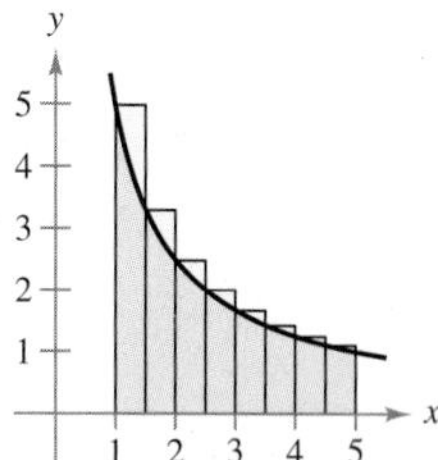

CAPSTONE

10. How would you describe the instantaneous rate of change of an automobile's position on the highway?

WRITING ABOUT CONCEPTS

11. Consider the length of the graph of

$f(x) = 5/x$

from $(1, 5)$ to $(5, 1)$.

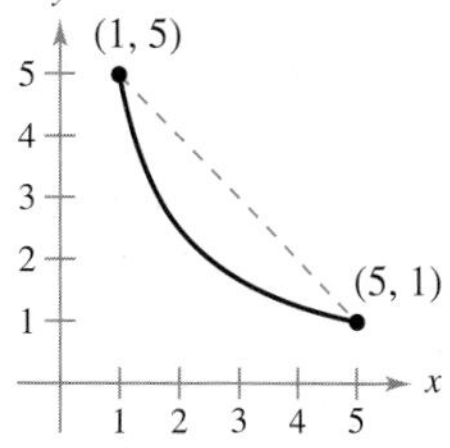

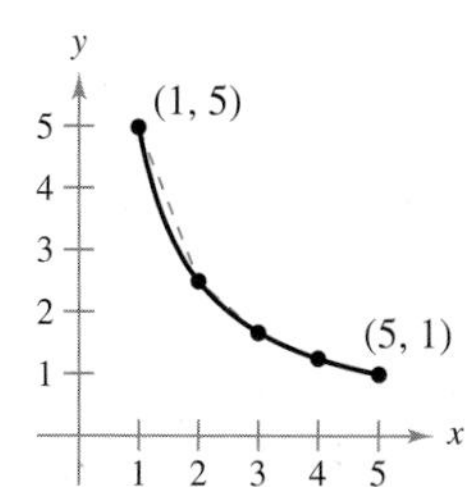

(a) Approximate the length of the curve by finding the distance between its two endpoints, as shown in the first figure.

(b) Approximate the length of the curve by finding the sum of the lengths of four line segments, as shown in the second figure.

(c) Describe how you could continue this process to obtain a more accurate approximation of the length of the curve.

3.2 Finding Limits Graphically and Numerically

- **Estimate a limit using a numerical or graphical approach.**
- **Learn different ways that a limit can fail to exist.**
- **Study and use a formal definition of limit.**

An Introduction to Limits

Suppose you are asked to sketch the graph of the function f given by

$$f(x) = \frac{x^3 - 1}{x - 1}, \quad x \neq 1.$$

For all values other than $x = 1$, you can use standard curve-sketching techniques. However, at $x = 1$, it is not clear what to expect. To get an idea of the behavior of the graph of f near $x = 1$, you can use two sets of x-values—one set that approaches 1 from the left and one set that approaches 1 from the right, as shown in the table.

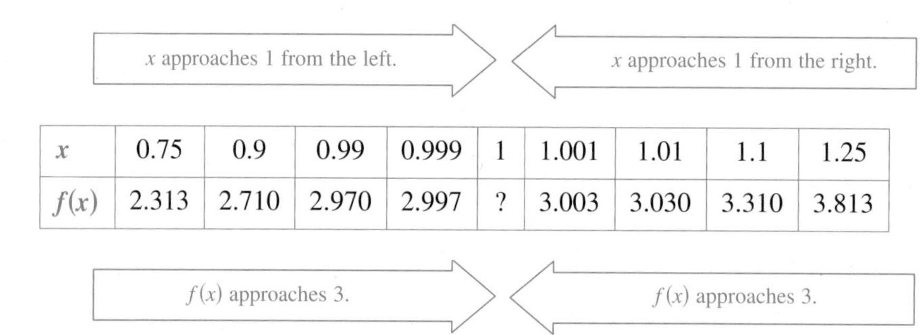

x	0.75	0.9	0.99	0.999	1	1.001	1.01	1.1	1.25
$f(x)$	2.313	2.710	2.970	2.997	?	3.003	3.030	3.310	3.813

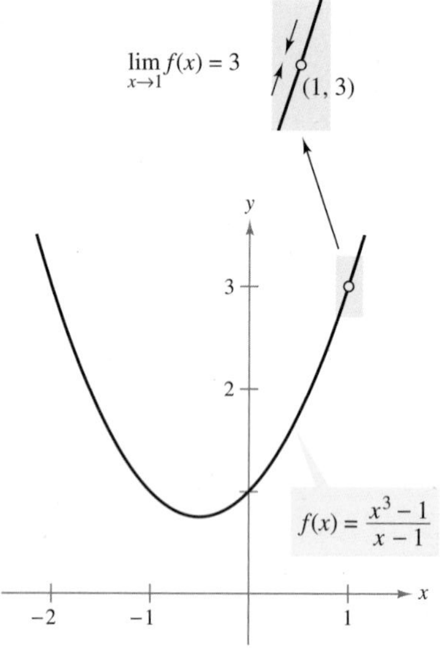

The limit of $f(x)$ as x approaches 1 is 3.
Figure 3.5

The graph of f is a parabola that has a gap at the point $(1, 3)$, as shown in Figure 3.5. Although x cannot equal 1, you can move arbitrarily close to 1, and as a result $f(x)$ moves arbitrarily close to 3. Using limit notation, you can write

$$\lim_{x \to 1} f(x) = 3. \qquad \text{This is read as "the limit of } f(x) \text{ as } x \text{ approaches 1 is 3."}$$

This discussion leads to an informal definition of limit. If $f(x)$ becomes arbitrarily close to a single number L as x approaches c from either side, the **limit** of $f(x)$, as x approaches c, is L. This limit is written as

$$\lim_{x \to c} f(x) = L.$$

EXPLORATION

The discussion above gives an example of how you can estimate a limit *numerically* by constructing a table and *graphically* by drawing a graph. Estimate the following limit numerically by completing the table.

$$\lim_{x \to 2} \frac{x^2 - 3x + 2}{x - 2}$$

x	1.75	1.9	1.99	1.999	2	2.001	2.01	2.1	2.25
$f(x)$	?	?	?	?	?	?	?	?	?

Then use a graphing utility to estimate the limit graphically.

EXAMPLE 1 Estimating a Limit Numerically

Evaluate the function $f(x) = x/\left(\sqrt{x+1} - 1\right)$ at several points near $x = 0$ and use the results to estimate the limit

$$\lim_{x \to 0} \frac{x}{\sqrt{x+1} - 1}.$$

Solution The table lists the values of $f(x)$ for several x-values near 0.

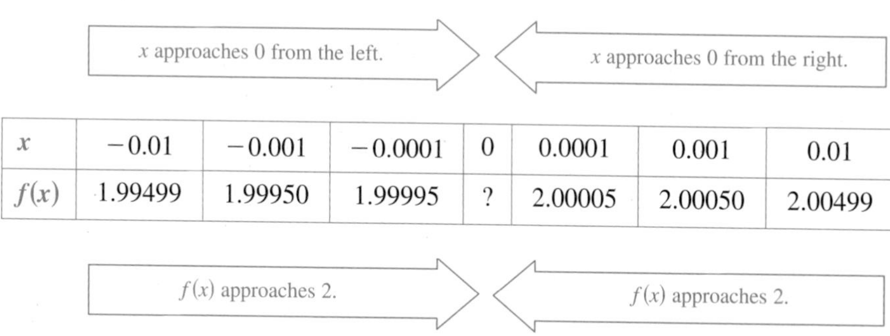

x	-0.01	-0.001	-0.0001	0	0.0001	0.001	0.01
$f(x)$	1.99499	1.99950	1.99995	?	2.00005	2.00050	2.00499

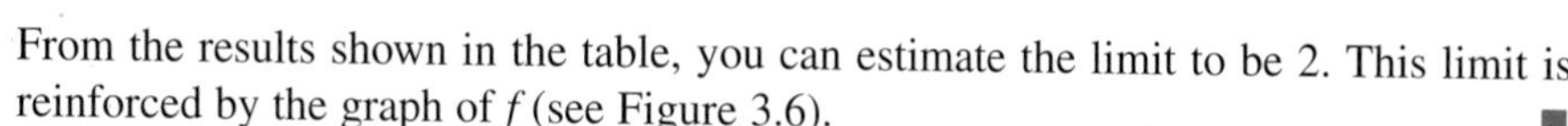

The limit of $f(x)$ as x approaches 0 is 2.
Figure 3.6

From the results shown in the table, you can estimate the limit to be 2. This limit is reinforced by the graph of f (see Figure 3.6). ■

In Example 1, note that the function is undefined at $x = 0$ and yet $f(x)$ appears to be approaching a limit as x approaches 0. This often happens, and it is important to realize that *the existence or nonexistence of $f(x)$ at $x = c$ has no bearing on the existence of the limit of $f(x)$ as x approaches c.*

EXAMPLE 2 Finding a Limit

Find the limit of $f(x)$ as x approaches 2, where f is defined as

$$f(x) = \begin{cases} 1, & x \neq 2 \\ 0, & x = 2 \end{cases}.$$

Solution Because $f(x) = 1$ for all x other than $x = 2$, you can conclude that the limit is 1, as shown in Figure 3.7. So, you can write

$$\lim_{x \to 2} f(x) = 1.$$

The fact that $f(2) = 0$ has no bearing on the existence or value of the limit as x approaches 2. For instance, if the function were defined as

$$f(x) = \begin{cases} 1, & x \neq 2 \\ 2, & x = 2 \end{cases}$$

the limit would be the same. ■

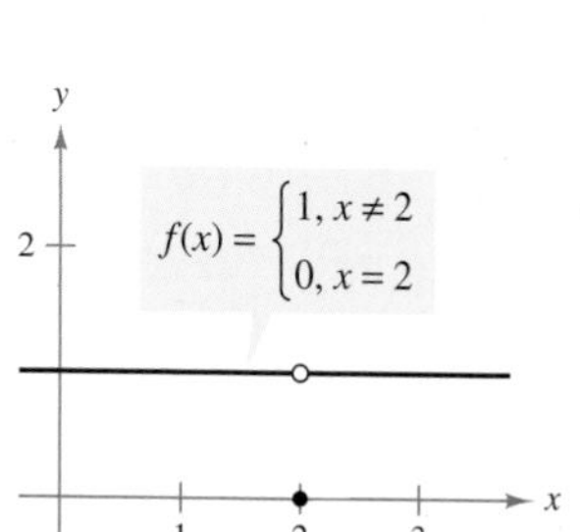

The limit of $f(x)$ as x approaches 2 is 1.
Figure 3.7

So far in this section, you have been estimating limits numerically and graphically. Each of these approaches produces an estimate of the limit. In Section 3.3, you will study analytic techniques for evaluating limits. Throughout the course, try to develop a habit of using this three-pronged approach to problem solving.

1. Numerical approach — Construct a table of values.
2. Graphical approach — Draw a graph by hand or using technology.
3. Analytic approach — Use algebra or calculus.

Limits That Fail to Exist

In the next two examples you will examine some limits that fail to exist.

EXAMPLE 3 Behavior That Differs from the Right and from the Left

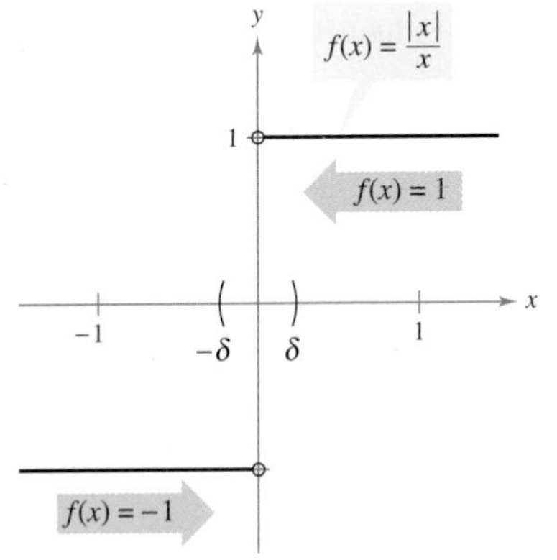

$\lim_{x \to 0} f(x)$ does not exist.

Figure 3.8

Show that the limit

$$\lim_{x \to 0} \frac{|x|}{x}$$

does not exist.

Solution Consider the graph of the function $f(x) = |x|/x$. From Figure 3.8 and the definition of absolute value

$$|x| = \begin{cases} x, & \text{if } x \geq 0 \\ -x, & \text{if } x < 0 \end{cases} \qquad \text{Definition of absolute value}$$

you can see that

$$\frac{|x|}{x} = \begin{cases} 1, & \text{if } x > 0 \\ -1, & \text{if } x < 0 \end{cases}.$$

This means that no matter how close x gets to 0, there will be both positive and negative x-values that yield $f(x) = 1$ or $f(x) = -1$. Specifically, if δ (the lowercase Greek letter *delta*) is a positive number, then for x-values satisfying the inequality $0 < |x| < \delta$, you can classify the values of $|x|/x$ as follows.

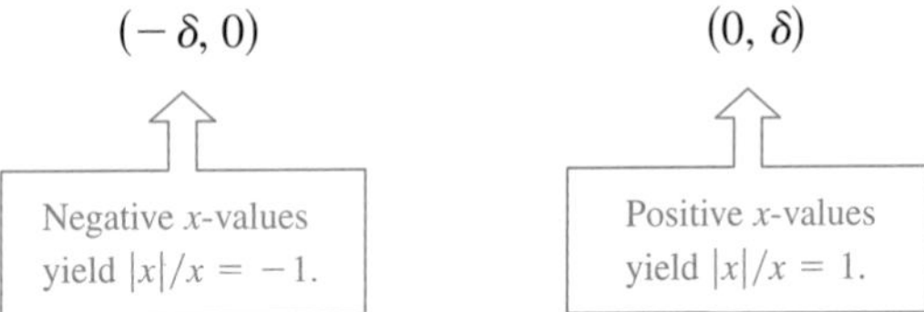

Because $|x|/x$ approaches a different number from the right side of 0 than it approaches from the left side, the limit $\lim_{x \to 0} (|x|/x)$ does not exist.

EXAMPLE 4 Unbounded Behavior

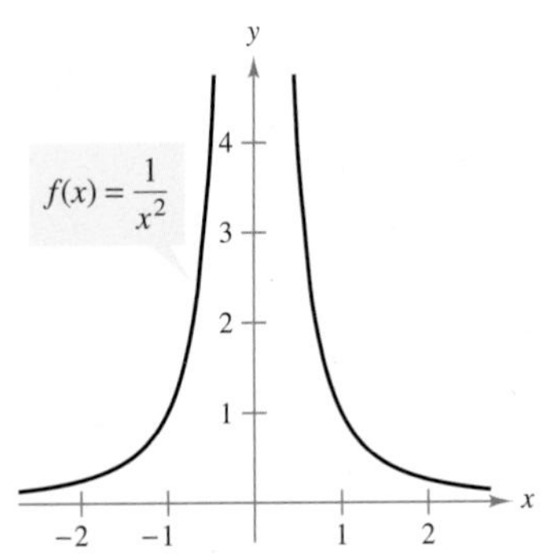

$\lim_{x \to 0} f(x)$ does not exist.

Figure 3.9

Discuss the existence of the limit $\lim_{x \to 0} \frac{1}{x^2}$.

Solution Let $f(x) = 1/x^2$. In Figure 3.9, you can see that as x approaches 0 from either the right or the left, $f(x)$ increases without bound. This means that by choosing x close enough to 0, you can force $f(x)$ to be as large as you want. For instance, $f(x)$ will be larger than 100 if you choose x that is within $\frac{1}{10}$ of 0. That is,

$$0 < |x| < \frac{1}{10} \Rightarrow f(x) = \frac{1}{x^2} > 100.$$

Similarly, you can force $f(x)$ to be larger than 1,000,000, as follows.

$$0 < |x| < \frac{1}{1000} \Rightarrow f(x) = \frac{1}{x^2} > 1{,}000{,}000$$

Because $f(x)$ is not approaching a real number L as x approaches 0, you can conclude that the limit does not exist. ■

A Formal Definition of Limit

Let's take another look at the informal definition of limit. If $f(x)$ becomes arbitrarily close to a single number L as x approaches c from either side, then the limit of $f(x)$ as x approaches c is L, written as

$$\lim_{x \to c} f(x) = L.$$

At first glance, this definition looks fairly technical. Even so, it is informal because exact meanings have not yet been given to the two phrases

"$f(x)$ becomes arbitrarily close to L"

and

"x approaches c."

The first person to assign mathematically rigorous meanings to these two phrases was Augustin-Louis Cauchy. His ε-δ **definition of limit** is the standard used today.

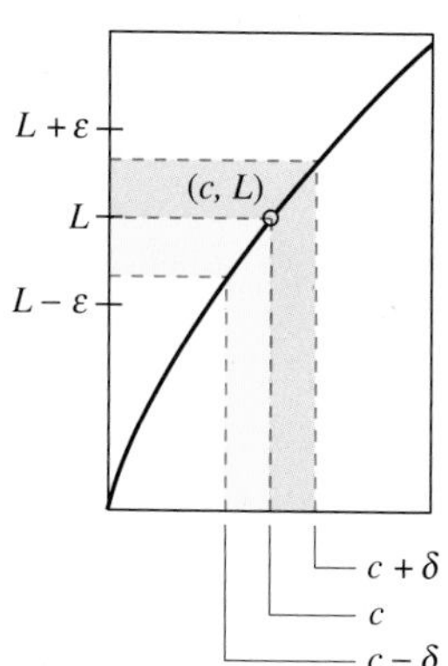

The ε-δ definition of the limit of $f(x)$ as x approaches c
Figure 3.10

In Figure 3.10, let ε (the lowercase Greek letter *epsilon*) represent a (small) positive number. Then the phrase "$f(x)$ becomes arbitrarily close to L" means that $f(x)$ lies in the interval $(L - \varepsilon, L + \varepsilon)$. Using absolute value, you can write this as

$$|f(x) - L| < \varepsilon. \qquad L - \varepsilon < f(x) < L + \varepsilon \text{ is equivalent.}$$

Similarly, the phrase "x approaches c" means that there exists a positive number δ such that x lies in either the interval $(c - \delta, c)$ or the interval $(c, c + \delta)$. This fact can be concisely expressed by the double inequality

$$0 < |x - c| < \delta. \qquad c - \delta < x < c + \delta \text{ is equivalent.}$$

The first inequality

$$0 < |x - c| \qquad \text{The distance between } x \text{ and } c \text{ is more than 0.}$$

expresses the fact that $x \neq c$. The second inequality

$$|x - c| < \delta \qquad x \text{ is within } \delta \text{ units of } c.$$

states that x is within a distance δ of c.

DEFINITION OF LIMIT

Let f be a function defined on an open interval containing c (except possibly at c) and let L be a real number. The statement

$$\lim_{x \to c} f(x) = L$$

means that for each $\varepsilon > 0$ there exists a $\delta > 0$ such that if

$$0 < |x - c| < \delta, \quad \text{then} \quad |f(x) - L| < \varepsilon.$$

NOTE Throughout this text, the expression

$$\lim_{x \to c} f(x) = L$$

implies two statements—the limit exists *and* the limit is L. ■

■ **FOR FURTHER INFORMATION** For more on the introduction of rigor to calculus, see "Who Gave You the Epsilon? Cauchy and the Origins of Rigorous Calculus" by Judith V. Grabiner in *The American Mathematical Monthly*. To view this article, go to the website *www.matharticles.com.*

Some functions do not have limits as $x \to c$, but those that do cannot have two different limits as $x \to c$. That is, *if the limit of a function exists, it is unique* (see Exercise 59).

The next three examples should help you develop a better understanding of the ε-δ definition of limit.

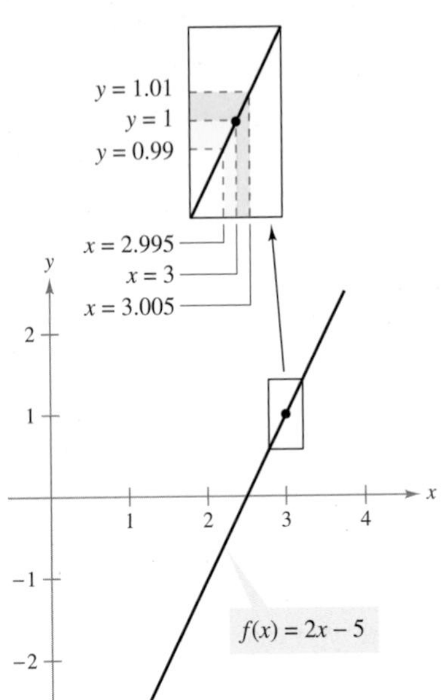

The limit of $f(x)$ as x approaches 3 is 1.
Figure 3.11

EXAMPLE 5 Finding a δ for a Given ε

Given

$$\lim_{x\to 3} (2x - 5) = 1$$

find δ such that $|(2x - 5) - 1| < 0.01$ whenever $0 < |x - 3| < \delta$.

Solution In this problem, you are working with a given value of ε—namely, $\varepsilon = 0.01$. To find an appropriate δ, notice that

$$|(2x - 5) - 1| = |2x - 6| = 2|x - 3|.$$

Because the inequality $|(2x - 5) - 1| < 0.01$ is equivalent to $2|x - 3| < 0.01$, you can choose $\delta = \frac{1}{2}(0.01) = 0.005$. This choice works because

$$0 < |x - 3| < 0.005$$

implies that

$$|(2x - 5) - 1| = 2|x - 3| < 2(0.005) = 0.01$$

as shown in Figure 3.11. ■

NOTE In Example 5, note that 0.005 is the *largest* value of δ that will guarantee $|(2x - 5) - 1| < 0.01$ whenever $0 < |x - 3| < \delta$. Any *smaller* positive value of δ would also work. ■

In Example 5, you found a δ-value for a *given* ε. This does not prove the existence of the limit. To do that, you must prove that you can find a δ for *any* ε, as shown in the next example.

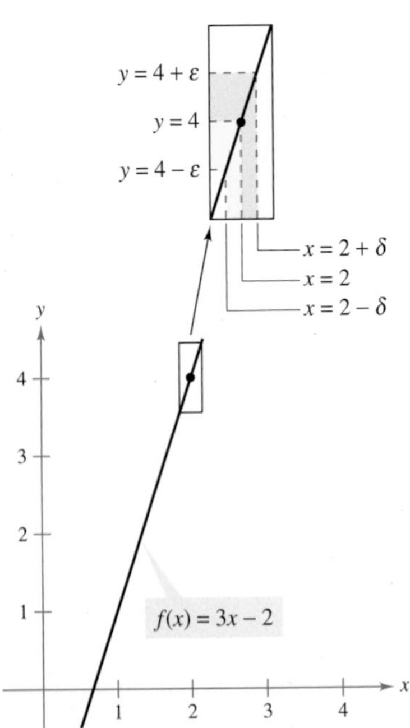

The limit of $f(x)$ as x approaches 2 is 4.
Figure 3.12

EXAMPLE 6 Using the ε-δ Definition of Limit

Use the ε-δ definition of limit to prove that

$$\lim_{x\to 2} (3x - 2) = 4.$$

Solution You must show that for each $\varepsilon > 0$, there exists a $\delta > 0$ such that $|(3x - 2) - 4| < \varepsilon$ whenever $0 < |x - 2| < \delta$. Because your choice of δ depends on ε, you need to establish a connection between the absolute values $|(3x - 2) - 4|$ and $|x - 2|$.

$$|(3x - 2) - 4| = |3x - 6| = 3|x - 2|$$

So, for a given $\varepsilon > 0$ you can choose $\delta = \varepsilon/3$. This choice works because

$$0 < |x - 2| < \delta = \frac{\varepsilon}{3}$$

implies that

$$|(3x - 2) - 4| = 3|x - 2| < 3\delta = 3\left(\frac{\varepsilon}{3}\right) = \varepsilon$$

as shown in Figure 3.12. ■

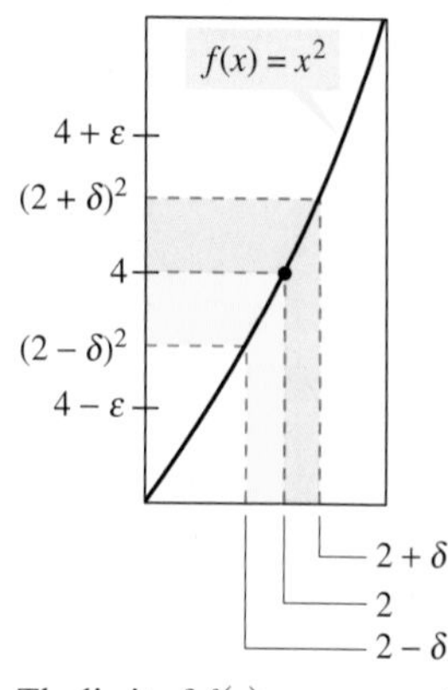

The limit of $f(x)$ as x approaches 2 is 4.
Figure 3.13

NOTE In Example 7, for $|x - 2| < \delta$ you want

$$|x^2 - 4| = |x - 2||x + 2| < \delta(\text{number}) = \varepsilon.$$

On $(1, 3)$, $|x + 2| < 5$, so you have $\delta \cdot 5 = \varepsilon$ or $\delta = \varepsilon/5$ as your choice for δ.

EXAMPLE 7 Using the ε-δ Definition of Limit

Use the ε-δ definition of limit to prove that

$$\lim_{x \to 2} x^2 = 4.$$

Solution You must show that for each $\varepsilon > 0$, there exists a $\delta > 0$ such that

$$|x^2 - 4| < \varepsilon \text{ whenever } 0 < |x - 2| < \delta.$$

To find an appropriate δ, begin by writing $|x^2 - 4| = |x - 2||x + 2|$. For all x in the interval $(1, 3)$, $x + 2 < 5$ and thus $|x + 2| < 5$. So, letting δ be the minimum of $\varepsilon/5$ and 1, it follows that, whenever $0 < |x - 2| < \delta$, you have

$$|x^2 - 4| = |x - 2||x + 2| < \left(\frac{\varepsilon}{5}\right)(5) = \varepsilon$$

as shown in Figure 3.13. ■

Throughout this chapter you will use the ε-δ definition of limit primarily to prove theorems about limits and to establish the existence or nonexistence of particular types of limits. For *finding* limits, you will learn techniques that are easier to use than the ε-δ definition of limit.

3.2 Exercises

See www.CalcChat.com for worked-out solutions to odd-numbered exercises.

In Exercises 1–6, complete the table and use the result to estimate the limit. Use a graphing utility to graph the function to confirm your result.

1. $\lim_{x \to 4} \dfrac{x - 4}{x^2 - 3x - 4}$

x	3.9	3.99	3.999	4.001	4.01	4.1
$f(x)$						

2. $\lim_{x \to 2} \dfrac{x - 2}{x^2 - 4}$

x	1.9	1.99	1.999	2.001	2.01	2.1
$f(x)$						

3. $\lim_{x \to 0} \dfrac{\sqrt{x + 6} - \sqrt{6}}{x}$

x	-0.1	-0.01	-0.001	0.001	0.01	0.1
$f(x)$						

4. $\lim_{x \to -5} \dfrac{\sqrt{4 - x} - 3}{x + 5}$

x	-5.1	-5.01	-5.001	-4.999	-4.99	-4.9
$f(x)$						

5. $\lim_{x \to 3} \dfrac{[1/(x + 1)] - (1/4)}{x - 3}$

x	2.9	2.99	2.999	3.001	3.01	3.1
$f(x)$						

6. $\lim_{x \to 4} \dfrac{[x/(x + 1)] - (4/5)}{x - 4}$

x	3.9	3.99	3.999	4.001	4.01	4.1
$f(x)$						

In Exercises 7–10, create a table of values for the function and use the result to estimate the limit. Use a graphing utility to graph the function to confirm your result.

7. $\lim_{x \to 1} \dfrac{x - 2}{x^2 + x - 6}$

8. $\lim_{x \to -3} \dfrac{x + 3}{x^2 + 7x + 12}$

9. $\lim_{x \to 1} \dfrac{x^4 - 1}{x^6 - 1}$

10. $\lim_{x \to -2} \dfrac{x^3 + 8}{x + 2}$

In Exercises 11–16, use the graph to find the limit (if it exists). If the limit does not exist, explain why.

11. $\lim_{x\to 3} (4 - x)$

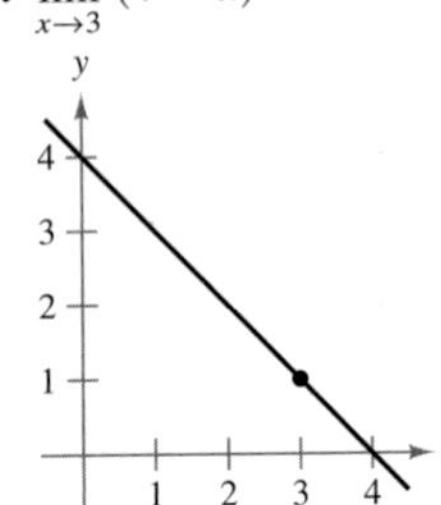

12. $\lim_{x\to 1} (x^2 + 3)$

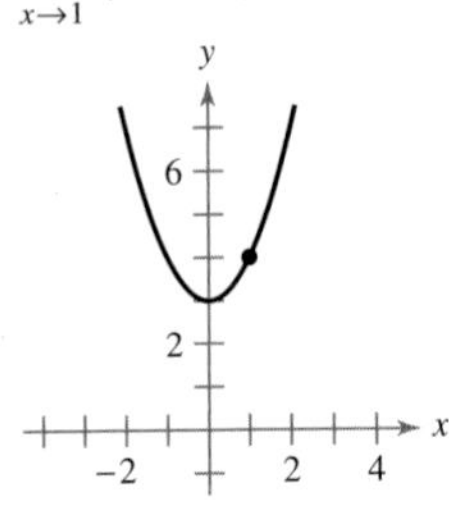

13. $\lim_{x\to 2} f(x)$

$$f(x) = \begin{cases} 4 - x, & x \neq 2 \\ 0, & x = 2 \end{cases}$$

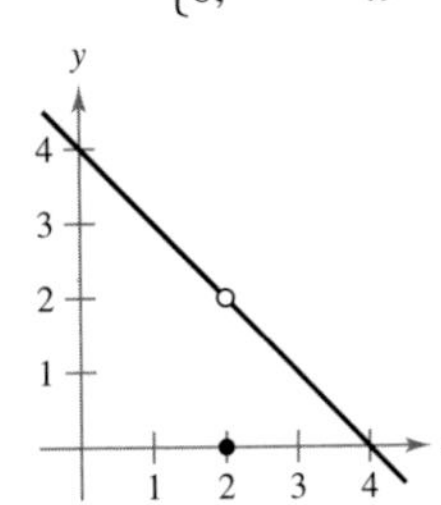

14. $\lim_{x\to 1} f(x)$

$$f(x) = \begin{cases} x^2 + 3, & x \neq 1 \\ 2, & x = 1 \end{cases}$$

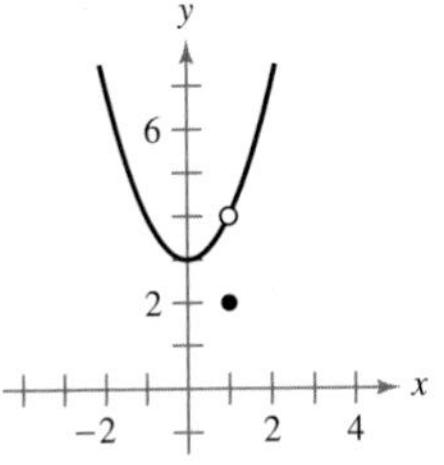

15. $\lim_{x\to 2} \dfrac{|x - 2|}{x - 2}$

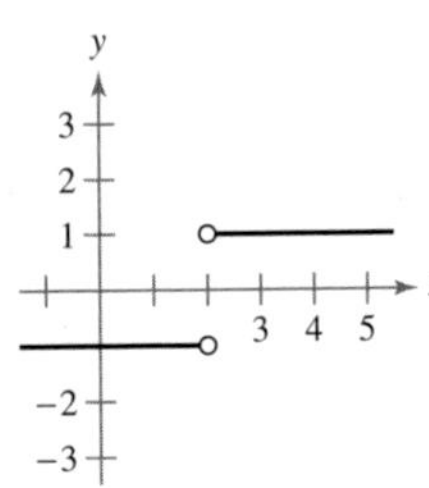

16. $\lim_{x\to 5} \dfrac{2}{x - 5}$

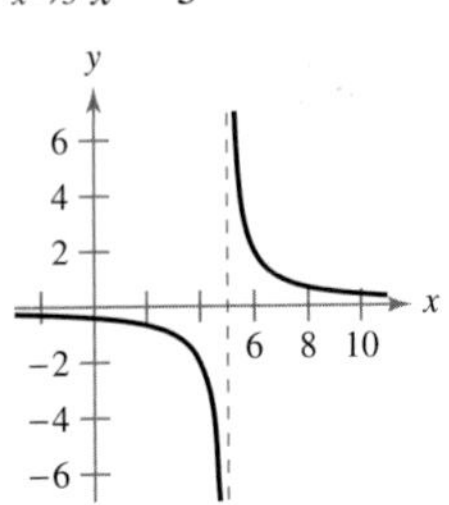

In Exercises 17 and 18, use the graph of the function *f* to decide whether the value of the given quantity exists. If it does, find it. If not, explain why.

17. (a) $f(1)$ (b) $\lim_{x\to 1} f(x)$

(c) $f(4)$ (d) $\lim_{x\to 4} f(x)$

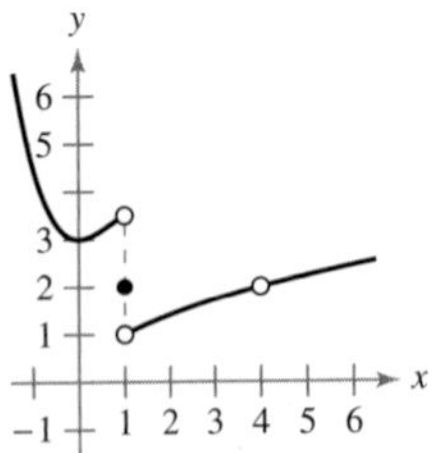

18. (a) $f(-2)$ (b) $\lim_{x\to -2} f(x)$

(c) $f(0)$ (d) $\lim_{x\to 0} f(x)$

(e) $f(2)$ (f) $\lim_{x\to 2} f(x)$

(g) $f(4)$ (h) $\lim_{x\to 4} f(x)$

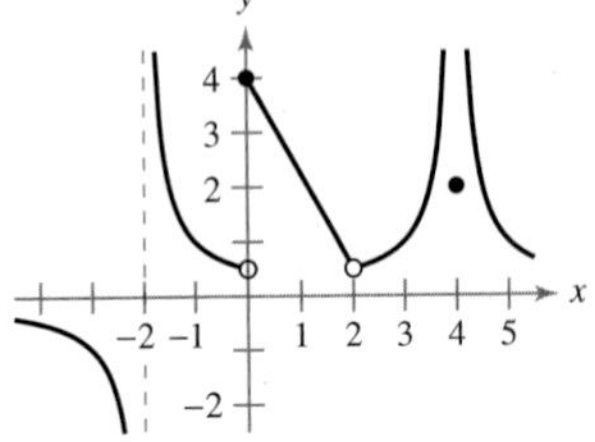

In Exercises 19 and 20, use the graph of *f* to identify the values of *c* for which $\lim_{x\to c} f(x)$ exists.

19.

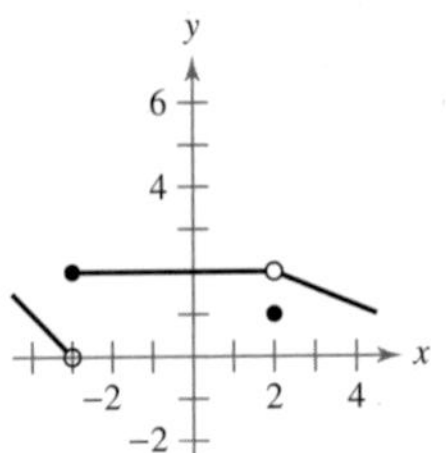

20.

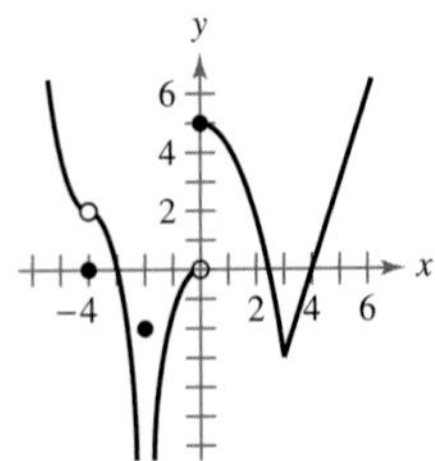

21. The graph of

$$f(x) = 2 - \frac{1}{x}$$

is shown in the figure. Find δ such that if $0 < |x - 1| < \delta$ then $|f(x) - 1| < 0.1$.

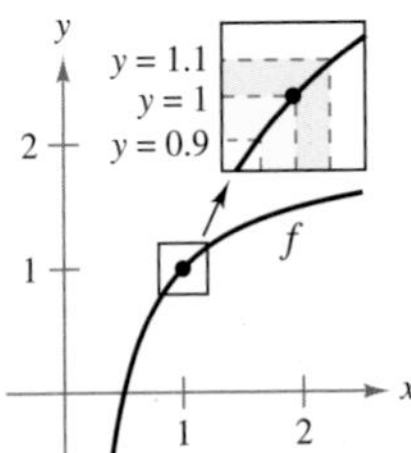

22. The graph of $f(x) = x^2 - 1$ is shown in the figure. Find δ such that if $0 < |x - 2| < \delta$ then $|f(x) - 3| < 0.2$.

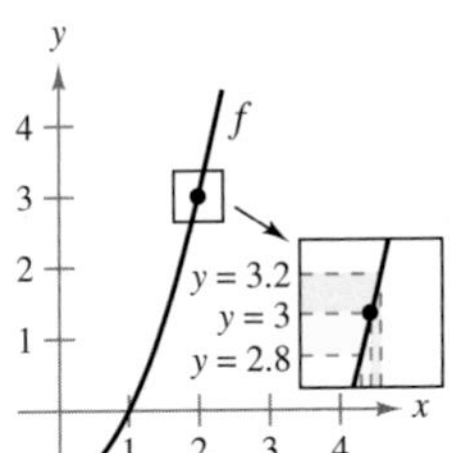

In Exercises 23–26, find the limit *L*. Then find $\delta > 0$ such that $|f(x) - L| < 0.01$ whenever $0 < |x - c| < \delta$.

23. $\lim_{x\to 2} (3x + 2)$

24. $\lim_{x\to 4} \left(4 - \dfrac{x}{2}\right)$

25. $\lim_{x\to 2} (x^2 - 3)$

26. $\lim_{x\to 5} (x^2 + 4)$

In Exercises 27–40, find the limit *L*. Then use the ε-δ definition to prove that the limit is *L*.

27. $\lim_{x\to 4} (x + 2)$

28. $\lim_{x\to -3} (2x + 5)$

29. $\lim_{x\to -4} \left(\frac{1}{2}x - 1\right)$

30. $\lim_{x\to 1} \left(\frac{2}{5}x + 7\right)$

31. $\lim_{x\to 6} 3$

32. $\lim_{x\to 2} (-1)$

33. $\lim_{x\to 0} \sqrt[3]{x}$

34. $\lim_{x\to 4} \sqrt{x}$

35. $\lim_{x\to -5} |x - 5|$

36. $\lim_{x\to 6} |x - 6|$

37. $\lim_{x\to 1} (x^2 + 1)$

38. $\lim_{x\to -3} (x^2 + 3x)$

39. What is the limit of $f(x) = 4$ as x approaches π?

40. What is the limit of $g(x) = x$ as x approaches π?

Writing **In Exercises 41–44, use a graphing utility to graph the function and estimate the limit (if it exists). What is the domain of the function? Can you detect a possible error in determining the domain of a function solely by analyzing the graph generated by a graphing utility? Write a short paragraph about the importance of examining a function analytically as well as graphically.**

41. $f(x) = \dfrac{\sqrt{x+5} - 3}{x - 4}$

$\lim_{x \to 4} f(x)$

42. $f(x) = \dfrac{x - 3}{x^2 - 4x + 3}$

$\lim_{x \to 3} f(x)$

43. $f(x) = \dfrac{x - 9}{\sqrt{x} - 3}$

$\lim_{x \to 9} f(x)$

44. $f(x) = \dfrac{x - 3}{x^2 - 9}$

$\lim_{x \to 3} f(x)$

WRITING ABOUT CONCEPTS

45. Write a brief description of the meaning of the notation $\lim_{x \to 8} f(x) = 25$.

46. Identify two types of behavior associated with the nonexistence of a limit. Illustrate each type with a graph of a function.

47. Determine the limit of the function describing the atmospheric pressure on a plane as it descends from 32,000 feet to land at Honolulu, located at sea level. (The atmospheric pressure at sea level is 14.7 pounds per square inch.)

CAPSTONE

48. (a) If $f(2) = 4$, can you conclude anything about the limit of $f(x)$ as x approaches 2? Explain your reasoning.

(b) If the limit of $f(x)$ as x approaches 2 is 4, can you conclude anything about $f(2)$? Explain your reasoning.

49. ***Jewelry*** A jeweler resizes a ring so that its inner circumference is 6 centimeters.

(a) What is the radius of the ring?

(b) If the ring's inner circumference can vary between 5.5 centimeters and 6.5 centimeters, how can the radius vary?

(c) Use the ε-δ definition of limit to describe this situation. Identify ε and δ.

50. ***Sports*** A sporting goods manufacturer designs a golf ball having a volume of 2.48 cubic inches.

(a) What is the radius of the golf ball?

(b) If the ball's volume can vary between 2.45 cubic inches and 2.51 cubic inches, how can the radius vary?

(c) Use the ε-δ definition of limit to describe this situation. Identify ε and δ.

51. Consider the function $f(x) = (1 + x)^{1/x}$. Estimate the limit

$$\lim_{x \to 0} (1 + x)^{1/x}$$

by evaluating f at x-values near 0. Sketch the graph of f.

52. Find two functions f and g such that $\lim_{x \to 0} f(x)$ and $\lim_{x \to 0} g(x)$ do not exist, but $\lim_{x \to 0} [f(x) + g(x)]$ does exist.

True or False? **In Exercises 53–56, determine whether the statement is true or false. If it is false, explain why or give an example that shows it is false.**

53. If f is undefined at $x = c$, then the limit of $f(x)$ as x approaches c does not exist.

54. If the limit of $f(x)$ as x approaches c is 0, then there must exist a number k such that $f(k) < 0.001$.

55. If $f(c) = L$, then $\lim_{x \to c} f(x) = L$.

56. If $\lim_{x \to c} f(x) = L$, then $f(c) = L$.

In Exercises 57 and 58, consider the function $f(x) = \sqrt{x}$.

57. Is $\lim_{x \to 0.25} \sqrt{x} = 0.5$ a true statement? Explain.

58. Is $\lim_{x \to 0} \sqrt{x} = 0$ a true statement? Explain.

59. Prove that if the limit of $f(x)$ as $x \to c$ exists, then the limit must be unique. [*Hint:* Let

$$\lim_{x \to c} f(x) = L_1 \quad \text{and} \quad \lim_{x \to c} f(x) = L_2$$

and prove that $L_1 = L_2$.]

60. Consider the line $f(x) = mx + b$, where $m \neq 0$. Use the ε-δ definition of limit to prove that $\lim_{x \to c} f(x) = mc + b$.

61. Prove that $\lim_{x \to c} f(x) = L$ is equivalent to $\lim_{x \to c} [f(x) - L] = 0$.

62. (a) Given that

$$\lim_{x \to 0} (3x + 1)(3x - 1)x^2 + 0.01 = 0.01$$

prove that there exists an open interval (a, b) containing 0 such that $(3x + 1)(3x - 1)x^2 + 0.01 > 0$ for all $x \neq 0$ in (a, b).

(b) Given that $\lim_{x \to c} g(x) = L$, where $L > 0$, prove that there exists an open interval (a, b) containing c such that $g(x) > 0$ for all $x \neq c$ in (a, b).

63. ***Writing*** The definition of limit on page 223 requires that f is a function defined on an open interval containing c, except possibly at c. Why is this requirement necessary?

3.3 Evaluating Limits Analytically

- Evaluate a limit using properties of limits.
- Develop and use a strategy for finding limits.
- Evaluate a limit using dividing out and rationalizing techniques.
- Evaluate a limit using the Squeeze Theorem.

Properties of Limits

In Section 3.2, you learned that the limit of $f(x)$ as x approaches c does not depend on the value of f at $x = c$. It may happen, however, that the limit is precisely $f(c)$. In such cases, the limit can be evaluated by **direct substitution.** That is,

$$\lim_{x \to c} f(x) = f(c). \qquad \text{Substitute } c \text{ for } x.$$

Such *well-behaved* functions are **continuous at c.** You will examine this concept more closely in Section 3.4.

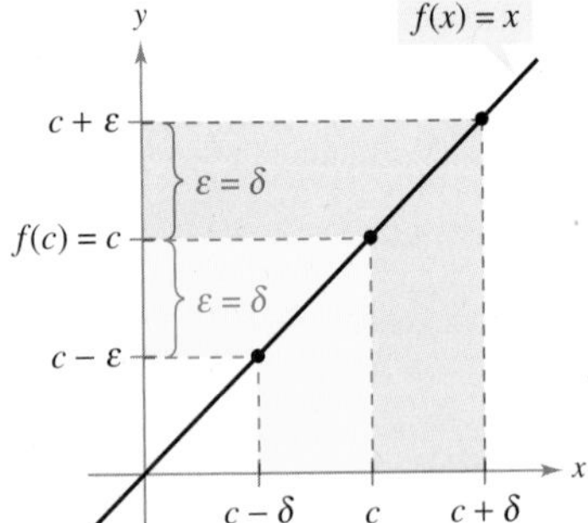

Figure 3.14

THEOREM 3.1 SOME BASIC LIMITS

Let b and c be real numbers and let n be a positive integer.

1. $\lim_{x \to c} b = b$ **2.** $\lim_{x \to c} x = c$ **3.** $\lim_{x \to c} x^n = c^n$

PROOF To prove Property 2 of Theorem 3.1, you need to show that for each $\varepsilon > 0$ there exists a $\delta > 0$ such that $|x - c| < \varepsilon$ whenever $0 < |x - c| < \delta$. To do this, choose $\delta = \varepsilon$. The second inequality then implies the first, as shown in Figure 3.14. This completes the proof. (Proofs of the other properties of limits in this section are listed in Appendix A or are discussed in the exercises.) ■

NOTE When you encounter new notations or symbols in mathematics, be sure you know how the notations are read. For instance, the limit in Example 1(c) is read as "the limit of x^2 as x approaches 2 is 4."

EXAMPLE 1 Evaluating Basic Limits

a. $\lim_{x \to 2} 3 = 3$ **b.** $\lim_{x \to -4} x = -4$ **c.** $\lim_{x \to 2} x^2 = 2^2 = 4$ ■

THEOREM 3.2 PROPERTIES OF LIMITS

Let b and c be real numbers, let n be a positive integer, and let f and g be functions with the following limits.

$$\lim_{x \to c} f(x) = L \quad \text{and} \quad \lim_{x \to c} g(x) = K$$

1. Scalar multiple: $\lim_{x \to c} [b f(x)] = bL$

2. Sum or difference: $\lim_{x \to c} [f(x) \pm g(x)] = L \pm K$

3. Product: $\lim_{x \to c} [f(x)g(x)] = LK$

4. Quotient: $\lim_{x \to c} \frac{f(x)}{g(x)} = \frac{L}{K}$, provided $K \neq 0$

5. Power: $\lim_{x \to c} [f(x)]^n = L^n$

EXAMPLE 2 The Limit of a Polynomial

Find the limit.

$$\lim_{x \to 2} (4x^2 + 3)$$

Solution

$$\begin{aligned}
\lim_{x \to 2} (4x^2 + 3) &= \lim_{x \to 2} 4x^2 + \lim_{x \to 2} 3 && \text{Property 2} \\
&= 4\left(\lim_{x \to 2} x^2\right) + \lim_{x \to 2} 3 && \text{Property 1} \\
&= 4(2^2) + 3 && \text{Theorem 3.1} \\
&= 19 && \text{Simplify.}
\end{aligned}$$

■

In Example 2, note that the limit (as $x \to 2$) of the *polynomial function* $p(x) = 4x^2 + 3$ is simply the value of p at $x = 2$.

$$\begin{aligned}
\lim_{x \to 2} p(x) &= p(2) \\
&= 4(2^2) + 3 \\
&= 19
\end{aligned}$$

This *direct substitution* property is valid for all polynomial and rational functions with nonzero denominators.

THEOREM 3.3 LIMITS OF POLYNOMIAL AND RATIONAL FUNCTIONS

If p is a polynomial function and c is a real number, then

$$\lim_{x \to c} p(x) = p(c).$$

If r is a rational function given by

$$r(x) = p(x)/q(x)$$

and c is a real number such that $q(c) \neq 0$, then

$$\lim_{x \to c} r(x) = r(c) = \frac{p(c)}{q(c)}.$$

EXAMPLE 3 The Limit of a Rational Function

Find the limit.

$$\lim_{x \to 1} \frac{x^2 + x + 2}{x + 1}$$

Solution Because the denominator is not 0 when $x = 1$, you can apply Theorem 3.3 to obtain

$$\begin{aligned}
\lim_{x \to 1} \frac{x^2 + x + 2}{x + 1} &= \frac{1^2 + 1 + 2}{1 + 1} && \text{Apply Theorem 3.3.} \\
&= \frac{4}{2} && \text{Simplify.} \\
&= 2. && \text{Simplify.}
\end{aligned}$$

■

THE SQUARE ROOT SYMBOL

The first use of a symbol to denote the square root can be traced to the sixteenth century. Mathematicians first used the symbol $\surd$, which had only two strokes. This symbol was chosen because it resembled a lowercase *r*, to stand for the Latin word *radix*, meaning root.

Polynomial functions and rational functions are two of the three basic types of algebraic functions. The following theorem deals with the limit of the third type of algebraic function—one that involves a radical. See Appendix A for a proof of this theorem.

THEOREM 3.4 THE LIMIT OF A FUNCTION INVOLVING A RADICAL

Let n be a positive integer. The following limit is valid for all c if n is odd, and is valid for $c > 0$ if n is even.

$$\lim_{x\to c} \sqrt[n]{x} = \sqrt[n]{c}$$

The following theorem greatly expands your ability to evaluate limits because it shows how to analyze the limit of a composite function by direct substitution. See Appendix A for a proof of this theorem.

THEOREM 3.5 THE LIMIT OF A COMPOSITE FUNCTION

If f and g are functions such that $\lim\limits_{x\to c} g(x) = L$ and $\lim\limits_{x\to L} f(x) = f(L)$, then

$$\lim_{x\to c} f(g(x)) = f\left(\lim_{x\to c} g(x)\right) = f(L).$$

EXAMPLE 4 The Limit of a Composite Function

Find each limit.

a. $\lim\limits_{x\to 0} \sqrt{x^2+4}$

b. $\lim\limits_{x\to 3} \sqrt[3]{2x^2-10}$

Solution

a. Let $g(x) = x^2 + 4$ and let $f(x) = \sqrt{x}$. Because

$$\begin{aligned}\lim_{x\to 0} g(x) &= \lim_{x\to 0} (x^2+4) \\ &= 0^2 + 4 \\ &= 4\end{aligned} \qquad \text{and} \qquad \begin{aligned}\lim_{x\to 4} f(x) &= \lim_{x\to 4} \sqrt{x} \\ &= \sqrt{4} \\ &= 2\end{aligned}$$

it follows from Theorem 3.5 that

$$\lim_{x\to 0} \sqrt{x^2+4} = \sqrt{\lim_{x\to 0} (x^2+4)} = \sqrt{4} = 2.$$

b. Let $g(x) = 2x^2 - 10$ and let $f(x) = \sqrt[3]{x}$. Because

$$\begin{aligned}\lim_{x\to 3} g(x) &= \lim_{x\to 3} (2x^2-10) \\ &= 2(3^2) - 10 \\ &= 8\end{aligned} \qquad \text{and} \qquad \begin{aligned}\lim_{x\to 8} f(x) &= \lim_{x\to 8} \sqrt[3]{x} \\ &= \sqrt[3]{8} \\ &= 2\end{aligned}$$

it follows from Theorem 3.5 that

$$\lim_{x\to 3} \sqrt[3]{2x^2-10} = \sqrt[3]{\lim_{x\to 3} (2x^2-10)} = \sqrt[3]{8} = 2.$$

■

A Strategy for Finding Limits

On the previous three pages, you studied several types of functions whose limits can be evaluated by direct substitution. This knowledge, together with the following theorem, can be used to develop a strategy for finding limits. A proof of this theorem is given in Appendix A.

> **THEOREM 3.6 FUNCTIONS THAT AGREE AT ALL BUT ONE POINT**
>
> Let c be a real number and let $f(x) = g(x)$ for all $x \neq c$ in an open interval containing c. If the limit of $g(x)$ as x approaches c exists, then the limit of $f(x)$ also exists and
>
> $$\lim_{x \to c} f(x) = \lim_{x \to c} g(x).$$

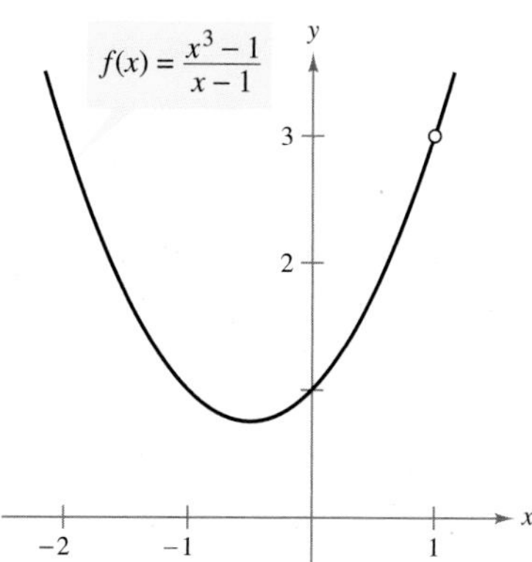

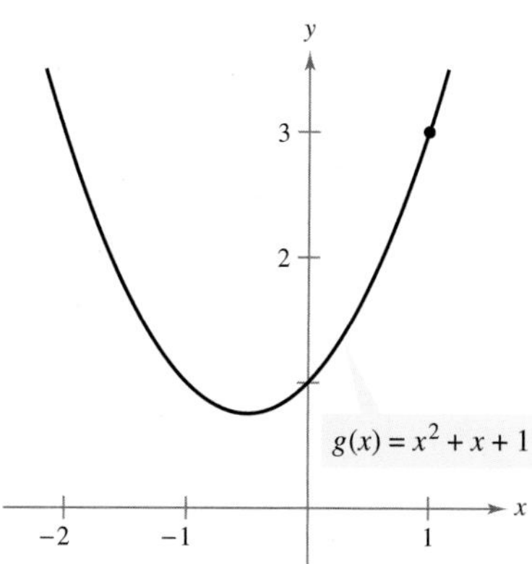

f and *g* agree at all but one point.
Figure 3.15

EXAMPLE 5 Finding the Limit of a Function

Find the limit: $\lim_{x \to 1} \dfrac{x^3 - 1}{x - 1}$.

Solution Let $f(x) = (x^3 - 1)/(x - 1)$. By factoring and dividing out like factors, you can rewrite f as

$$f(x) = \frac{\cancel{(x - 1)}(x^2 + x + 1)}{\cancel{(x - 1)}} = x^2 + x + 1 = g(x), \quad x \neq 1.$$

So, for all x-values other than $x = 1$, the functions f and g agree, as shown in Figure 3.15. Because $\lim_{x \to 1} g(x)$ exists, you can apply Theorem 3.6 to conclude that f and g have the same limit at $x = 1$.

$$\lim_{x \to 1} \frac{x^3 - 1}{x - 1} = \lim_{x \to 1} \frac{(x - 1)(x^2 + x + 1)}{x - 1} \qquad \text{Factor.}$$

$$= \lim_{x \to 1} \frac{\cancel{(x - 1)}(x^2 + x + 1)}{\cancel{x - 1}} \qquad \text{Divide out like factors.}$$

$$= \lim_{x \to 1} (x^2 + x + 1) \qquad \text{Apply Theorem 3.6.}$$

$$= 1^2 + 1 + 1 \qquad \text{Use direct substitution.}$$

$$= 3 \qquad \text{Simplify.}$$

■

STUDY TIP When applying this strategy for finding a limit, remember that some functions do not have a limit (as x approaches c). For instance, the following limit does not exist.

$$\lim_{x \to 1} \frac{x^3 + 1}{x - 1}$$

> **A STRATEGY FOR FINDING LIMITS**
>
> 1. Learn to recognize which limits can be evaluated by direct substitution. (These limits are listed in Theorems 3.1 through 3.5.)
> 2. If the limit of $f(x)$ as x approaches c *cannot* be evaluated by direct substitution, try to find a function g that agrees with f for all x other than $x = c$. [Choose g such that the limit of $g(x)$ *can* be evaluated by direct substitution.]
> 3. Apply Theorem 3.6 to conclude *analytically* that
>
> $$\lim_{x \to c} f(x) = \lim_{x \to c} g(x) = g(c).$$
>
> 4. Use a *graph* or *table* to reinforce your conclusion.

Dividing Out and Rationalizing Techniques

Two techniques for finding limits analytically are shown in Examples 6 and 7. The dividing out technique involves dividing out common factors, and the rationalizing technique involves rationalizing the numerator of a fractional expression.

EXAMPLE 6 Dividing Out Technique

Find the limit: $\lim\limits_{x \to -3} \dfrac{x^2 + x - 6}{x + 3}$.

Solution Although you are taking the limit of a rational function, you *cannot* apply Theorem 3.3 because the limit of the denominator is 0.

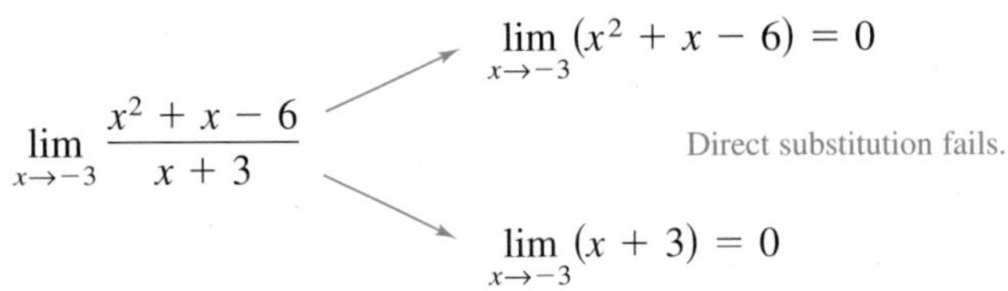

$\lim\limits_{x \to -3} (x^2 + x - 6) = 0$
$\lim\limits_{x \to -3} \dfrac{x^2 + x - 6}{x + 3}$
Direct substitution fails.
$\lim\limits_{x \to -3} (x + 3) = 0$

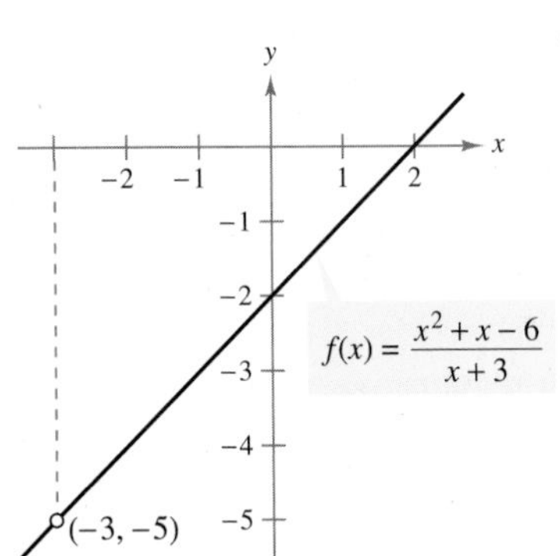

f is undefined when $x = -3$.
Figure 3.16

Because the limit of the numerator is also 0, the numerator and denominator have a *common factor* of $(x + 3)$. So, for all $x \neq -3$, you can divide out this factor to obtain

$$f(x) = \frac{x^2 + x - 6}{x + 3} = \frac{(x + 3)(x - 2)}{x + 3} = x - 2 = g(x), \quad x \neq -3.$$

Using Theorem 3.6, it follows that

$$\lim_{x \to -3} \frac{x^2 + x - 6}{x + 3} = \lim_{x \to -3} (x - 2) \qquad \text{Apply Theorem 3.6.}$$
$$= -5. \qquad \text{Use direct substitution.}$$

This result is shown graphically in Figure 3.16. Note that the graph of the function f coincides with the graph of the function $g(x) = x - 2$, except that the graph of f has a gap at the point $(-3, -5)$. ■

STUDY TIP In the solution of Example 6, notice that the Factor Theorem as discussed in Section 2.3 is applied. From the theorem you know that when c is a zero of a polynomial function, $(x - c)$ is a factor of the polynomial. So, when you apply direct substitution to a rational function and obtain

$$r(c) = \frac{p(c)}{q(c)} = \frac{0}{0}$$

you can conclude that $(x - c)$ must be a common factor of both $p(x)$ and $q(x)$.

In Example 6, direct substitution produced the meaningless fractional form $0/0$. An expression such as $0/0$ is called an **indeterminate form** because you cannot (from the form alone) determine the limit. When you try to evaluate a limit and encounter this form, remember that you must rewrite the fraction so that the new denominator does not have 0 as its limit. One way to do this is to *divide out like factors*, as shown in Example 6. A second way is to *rationalize the numerator*, as shown in Example 7.

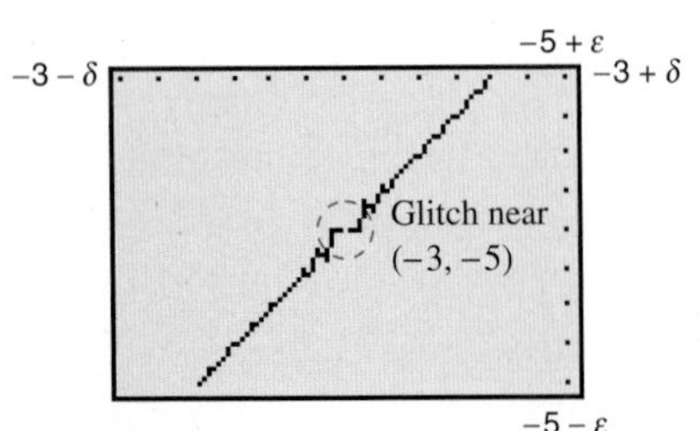

Incorrect graph of f
Figure 3.17

TECHNOLOGY PITFALL Because the graphs of

$$f(x) = \frac{x^2 + x - 6}{x + 3} \quad \text{and} \quad g(x) = x - 2$$

differ only at the point $(-3, -5)$, a standard graphing utility setting may not distinguish clearly between these graphs. However, because of the pixel configuration and rounding error of a graphing utility, it may be possible to find screen settings that distinguish between the graphs. Specifically, by repeatedly zooming in near the point $(-3, -5)$ on the graph of f, your graphing utility may show glitches or irregularities that do not exist on the actual graph. (See Figure 3.17.) By changing the screen settings on your graphing utility, you may obtain the correct graph of f.

EXAMPLE 7 Rationalizing Technique

Find the limit: $\displaystyle\lim_{x\to 0} \frac{\sqrt{x+1}-1}{x}$.

Solution By direct substitution, you obtain the indeterminate form 0/0.

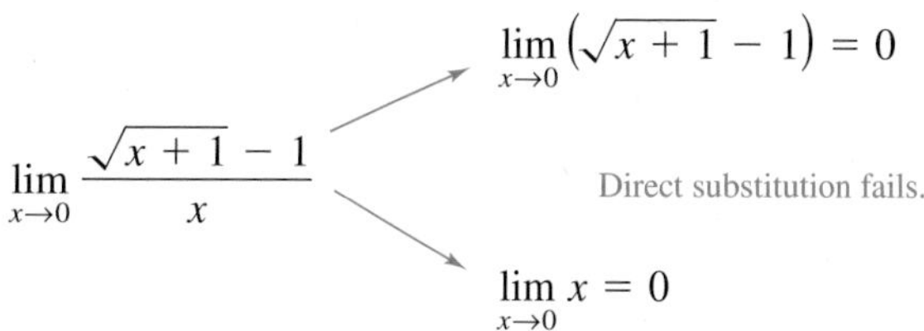

In this case, you can rewrite the fraction by rationalizing the numerator.

$$\begin{aligned}\frac{\sqrt{x+1}-1}{x} &= \left(\frac{\sqrt{x+1}-1}{x}\right)\left(\frac{\sqrt{x+1}+1}{\sqrt{x+1}+1}\right)\\ &= \frac{(x+1)-1}{x\left(\sqrt{x+1}+1\right)}\\ &= \frac{1}{\sqrt{x+1}+1}, \quad x \neq 0\end{aligned}$$

Now, using Theorem 3.6, you can evaluate the limit as shown.

$$\lim_{x\to 0} \frac{\sqrt{x+1}-1}{x} = \lim_{x\to 0} \frac{1}{\sqrt{x+1}+1} = \frac{1}{1+1} = \frac{1}{2}$$

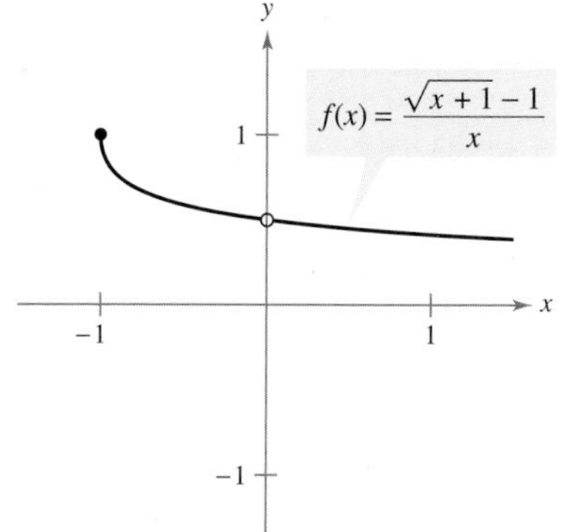

The limit of $f(x)$ as x approaches 0 is $\frac{1}{2}$.
Figure 3.18

A table or a graph can reinforce your conclusion that the limit is $\frac{1}{2}$. (See Figure 3.18.)

x approaches 0 from the left. → ← x approaches 0 from the right.

x	-0.25	-0.1	-0.01	-0.001	0	0.001	0.01	0.1	0.25
$f(x)$	0.5359	0.5132	0.5013	0.5001	?	0.4999	0.4988	0.4881	0.4721

■

NOTE The rationalizing technique for evaluating limits is based on multiplication by a convenient form of 1. In Example 7, the convenient form is

$$1 = \frac{\sqrt{x+1}+1}{\sqrt{x+1}+1}.$$

The Squeeze Theorem

The next theorem concerns the limit of a function that is squeezed between two other functions, each of which has the same limit at a given x-value, as shown in Figure 3.19. (The proof of this theorem is given in Appendix A.)

$h(x) \le f(x) \le g(x)$

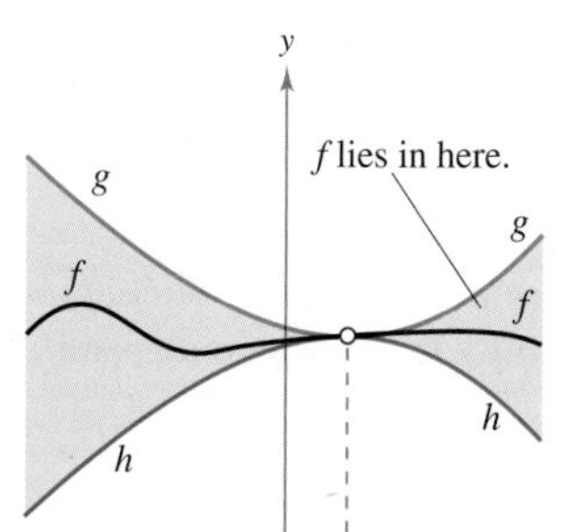

The Squeeze Theorem
Figure 3.19

> **THEOREM 3.7 THE SQUEEZE THEOREM**
>
> If $h(x) \le f(x) \le g(x)$ for all x in an open interval containing c, except possibly at c itself, and if $\displaystyle\lim_{x\to c} h(x) = L = \lim_{x\to c} g(x)$, then $\displaystyle\lim_{x\to c} f(x)$ exists and is equal to L.

3.3 Exercises

See www.CalcChat.com for worked-out solutions to odd-numbered exercises.

In Exercises 1–4, use a graphing utility to graph the function and visually estimate the limits.

1. $h(x) = -x^2 + 4x$
   (a) $\lim_{x\to 4} h(x)$
   (b) $\lim_{x\to -1} h(x)$

2. $g(x) = \dfrac{12(\sqrt{x} - 3)}{x - 9}$
   (a) $\lim_{x\to 4} g(x)$
   (b) $\lim_{x\to 0} g(x)$

3. $f(x) = x\sqrt{6 - x}$
   (a) $\lim_{x\to 0} f(x)$
   (b) $\lim_{x\to 2} f(x)$

4. $f(t) = t|t - 4|$
   (a) $\lim_{t\to 4} f(t)$
   (b) $\lim_{t\to -1} f(t)$

In Exercises 5–22, find the limit.

5. $\lim_{x\to 2} x^3$
6. $\lim_{x\to -2} x^4$
7. $\lim_{x\to 0} (2x - 1)$
8. $\lim_{x\to -3} (3x + 2)$
9. $\lim_{x\to -3} (x^2 + 3x)$
10. $\lim_{x\to 1} (-x^2 + 1)$
11. $\lim_{x\to -3} (2x^2 + 4x + 1)$
12. $\lim_{x\to 1} (3x^3 - 2x^2 + 4)$
13. $\lim_{x\to 3} \sqrt{x + 1}$
14. $\lim_{x\to 4} \sqrt[3]{x + 4}$
15. $\lim_{x\to -4} (x + 3)^2$
16. $\lim_{x\to 0} (2x - 1)^3$
17. $\lim_{x\to 2} (1/x)$
18. $\lim_{x\to -3} [2/(x + 2)]$
19. $\lim_{x\to 1} [x/(x^2 + 4)]$
20. $\lim_{x\to 1} [(2x - 3)/(x + 5)]$
21. $\lim_{x\to 7} \dfrac{3x}{\sqrt{x + 2}}$
22. $\lim_{x\to 2} \dfrac{\sqrt{x + 2}}{x - 4}$

In Exercises 23–26, find the limits.

23. $f(x) = 5 - x,\ g(x) = x^3$
   (a) $\lim_{x\to 1} f(x)$ (b) $\lim_{x\to 4} g(x)$ (c) $\lim_{x\to 1} g(f(x))$

24. $f(x) = x + 7,\ g(x) = x^2$
   (a) $\lim_{x\to -3} f(x)$ (b) $\lim_{x\to 4} g(x)$ (c) $\lim_{x\to -3} g(f(x))$

25. $f(x) = 4 - x^2,\ g(x) = \sqrt{x + 1}$
   (a) $\lim_{x\to 1} f(x)$ (b) $\lim_{x\to 3} g(x)$ (c) $\lim_{x\to 1} g(f(x))$

26. $f(x) = 2x^2 - 3x + 1,\ g(x) = \sqrt[3]{x + 6}$
   (a) $\lim_{x\to 4} f(x)$ (b) $\lim_{x\to 21} g(x)$ (c) $\lim_{x\to 4} g(f(x))$

In Exercises 27–30, use the information to evaluate the limits.

27. $\lim_{x\to c} f(x) = 3$
   $\lim_{x\to c} g(x) = 2$
   (a) $\lim_{x\to c} [5g(x)]$
   (b) $\lim_{x\to c} [f(x) + g(x)]$
   (c) $\lim_{x\to c} [f(x)g(x)]$
   (d) $\lim_{x\to c} [f(x)/g(x)]$

28. $\lim_{x\to c} f(x) = \frac{3}{2}$
   $\lim_{x\to c} g(x) = \frac{1}{2}$
   (a) $\lim_{x\to c} [4f(x)]$
   (b) $\lim_{x\to c} [f(x) + g(x)]$
   (c) $\lim_{x\to c} [f(x)g(x)]$
   (d) $\lim_{x\to c} [f(x)/g(x)]$

29. $\lim_{x\to c} f(x) = 4$
   (a) $\lim_{x\to c} [f(x)]^3$
   (b) $\lim_{x\to c} \sqrt{f(x)}$
   (c) $\lim_{x\to c} [3f(x)]$
   (d) $\lim_{x\to c} [f(x)]^{3/2}$

30. $\lim_{x\to c} f(x) = 27$
   (a) $\lim_{x\to c} \sqrt[3]{f(x)}$
   (b) $\lim_{x\to c} \dfrac{f(x)}{18}$
   (c) $\lim_{x\to c} [f(x)]^2$
   (d) $\lim_{x\to c} [f(x)]^{2/3}$

In Exercises 31–34, use the graph to determine the limit visually (if it exists). Write a simpler function that agrees with the given function at all but one point.

31. $g(x) = \dfrac{x^2 - x}{x}$

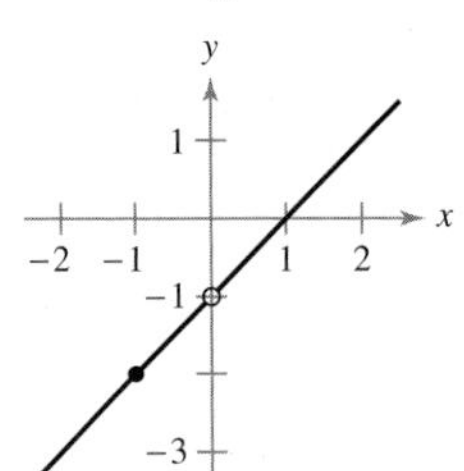

   (a) $\lim_{x\to 0} g(x)$
   (b) $\lim_{x\to -1} g(x)$

32. 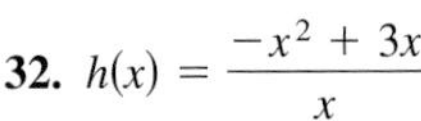 $h(x) = \dfrac{-x^2 + 3x}{x}$

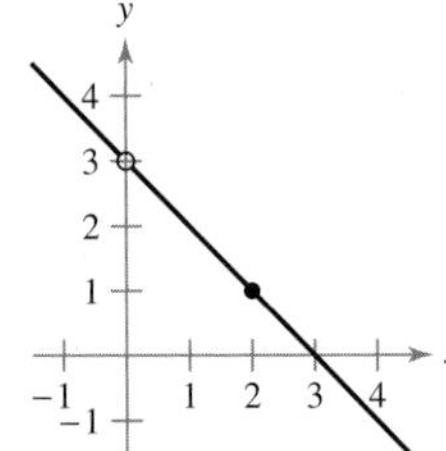

   (a) $\lim_{x\to 2} h(x)$
   (b) $\lim_{x\to 0} h(x)$

33. $g(x) = \dfrac{x^3 - x}{x - 1}$

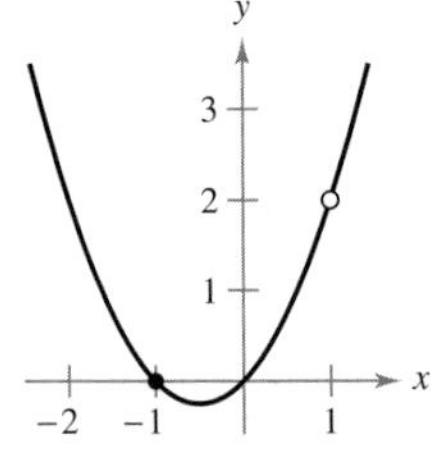

   (a) $\lim_{x\to 1} g(x)$
   (b) $\lim_{x\to -1} g(x)$

34. 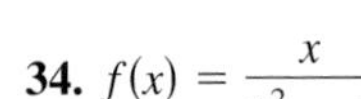 $f(x) = \dfrac{x}{x^2 - x}$

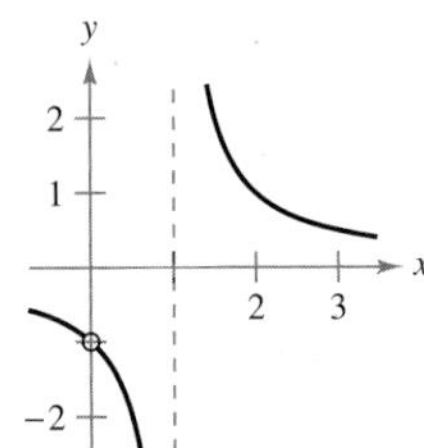

   (a) $\lim_{x\to 1} f(x)$
   (b) $\lim_{x\to 0} f(x)$

In Exercises 35–38, find the limit of the function (if it exists). Write a simpler function that agrees with the given function at all but one point. Use a graphing utility to confirm your result.

35. $\lim_{x\to -1} \dfrac{x^2 - 1}{x + 1}$
36. $\lim_{x\to -1} \dfrac{2x^2 - x - 3}{x + 1}$
37. $\lim_{x\to 2} \dfrac{x^3 - 8}{x - 2}$
38. $\lim_{x\to -1} \dfrac{x^3 + 1}{x + 1}$

In Exercises 39–52, find the limit (if it exists).

39. $\lim_{x\to 0} \dfrac{x}{x^2 - x}$
40. $\lim_{x\to 0} \dfrac{3x}{x^2 + 2x}$

41. $\lim_{x\to4} \frac{x-4}{x^2-16}$

42. $\lim_{x\to3} \frac{3-x}{x^2-9}$

43. $\lim_{x\to-3} \frac{x^2+x-6}{x^2-9}$

44. $\lim_{x\to4} \frac{x^2-5x+4}{x^2-2x-8}$

45. $\lim_{x\to4} \frac{\sqrt{x+5}-3}{x-4}$

46. $\lim_{x\to0} \frac{\sqrt{2+x}-\sqrt{2}}{x}$

47. $\lim_{x\to0} \frac{[1/(3+x)]-(1/3)}{x}$

48. $\lim_{x\to0} \frac{[1/(x+4)]-(1/4)}{x}$

49. $\lim_{\Delta x\to0} \frac{2(x+\Delta x)-2x}{\Delta x}$

50. $\lim_{\Delta x\to0} \frac{(x+\Delta x)^2-x^2}{\Delta x}$

51. $\lim_{\Delta x\to0} \frac{(x+\Delta x)^2-2(x+\Delta x)+1-(x^2-2x+1)}{\Delta x}$

52. $\lim_{\Delta x\to0} \frac{(x+\Delta x)^3-x^3}{\Delta x}$

Graphical, Numerical, and Analytic Analysis **In Exercises 53–56, use a graphing utility to graph the function and estimate the limit. Use a table to reinforce your conclusion. Then find the limit by analytic methods.**

53. $\lim_{x\to0} \frac{\sqrt{x+2}-\sqrt{2}}{x}$

54. $\lim_{x\to16} \frac{4-\sqrt{x}}{x-16}$

55. $\lim_{x\to0} \frac{[1/(2+x)]-(1/2)}{x}$

56. $\lim_{x\to2} \frac{x^5-32}{x-2}$

In Exercises 57–60, find $\lim_{\Delta x\to0} \frac{f(x+\Delta x)-f(x)}{\Delta x}$.

57. $f(x) = 3x - 2$

58. $f(x) = \sqrt{x}$

59. $f(x) = \frac{1}{x+3}$

60. $f(x) = x^2 - 4x$

In Exercises 61 and 62, use the Squeeze Theorem to find $\lim_{x\to c} f(x)$.

61. $c = 0;\ 4 - x^2 \le f(x) \le 4 + x^2$

62. $c = a;\ b - |x-a| \le f(x) \le b + |x-a|$

WRITING ABOUT CONCEPTS

63. In the context of finding limits, discuss what is meant by two functions that agree at all but one point.

64. Give an example of two functions that agree at all but one point.

65. What is meant by an indeterminate form?

66. In your own words, explain the Squeeze Theorem.

Free-Falling Object **In Exercises 67 and 68, use the position function $s(t) = -16t^2 + 500$, which gives the height (in feet) of an object that has fallen for t seconds from a height of 500 feet. The velocity at time $t = a$ seconds is given by**

$$\lim_{t\to a} \frac{s(a)-s(t)}{a-t}.$$

67. If a construction worker drops a wrench from a height of 500 feet, how fast will the wrench be falling after 2 seconds?

68. If a construction worker drops a wrench from a height of 500 feet, when will the wrench hit the ground? At what velocity will the wrench impact the ground?

Free-Falling Object **In Exercises 69 and 70, use the position function $s(t) = -4.9t^2 + 200$, which gives the height (in meters) of an object that has fallen from a height of 200 meters. The velocity at time $t = a$ seconds is given by**

$$\lim_{t\to a} \frac{s(a)-s(t)}{a-t}.$$

69. Find the velocity of the object when $t = 3$.

70. At what velocity will the object impact the ground?

71. Find two functions f and g such that $\lim_{x\to0} f(x)$ and $\lim_{x\to0} g(x)$ do not exist, but that $\lim_{x\to0} [f(x)/g(x)]$ does exist.

72. Prove that if $\lim_{x\to c} f(x)$ exists and $\lim_{x\to c} [f(x) + g(x)]$ does not exist, then $\lim_{x\to c} g(x)$ does not exist.

73. Prove Property 1 of Theorem 3.1.

74. Prove Property 3 of Theorem 3.1. (You may use Property 3 of Theorem 3.2.)

75. Prove Property 1 of Theorem 3.2.

CAPSTONE

76. Let $f(x) = \begin{cases} 3, & x \ne 2 \\ 5, & x = 2 \end{cases}$. Find $\lim_{x\to2} f(x)$.

True or False? **In Exercises 77–82, determine whether the statement is true or false. If it is false, explain why or give an example that shows it is false.**

77. $\lim_{x\to0} \frac{|x|}{x} = 1$

78. $\lim_{x\to0} x^3 = 0$

79. If $f(x) = g(x)$ for all real numbers other than $x = 0$, and $\lim_{x\to0} f(x) = L$, then $\lim_{x\to0} g(x) = L$.

80. If $\lim_{x\to c} f(x) = L$, then $f(c) = L$.

81. $\lim_{x\to2} f(x) = 3$, where $f(x) = \begin{cases} 3, & x \le 2 \\ 0, & x > 2 \end{cases}$

82. If $f(x) < g(x)$ for all $x \ne a$, then $\lim_{x\to a} f(x) < \lim_{x\to a} g(x)$.

83. Let $f(x) = \begin{cases} 0, & \text{if } x \text{ is rational} \\ 1, & \text{if } x \text{ is irrational} \end{cases}$

and

$g(x) = \begin{cases} 0, & \text{if } x \text{ is rational} \\ x, & \text{if } x \text{ is irrational.} \end{cases}$

Find (if possible) $\lim_{x\to0} f(x)$ and $\lim_{x\to0} g(x)$.

3.4 Continuity and One-Sided Limits

- **Determine continuity at a point and continuity on an open interval.**
- **Determine one-sided limits and continuity on a closed interval.**
- **Use properties of continuity.**
- **Understand and use the Intermediate Value Theorem.**

Continuity at a Point and on an Open Interval

In Section 2.2, you learned about the continuity of a polynomial function. In this section, you will add to your understanding of continuity by studying continuity at a point c and on an open interval (a, b). Informally, to say that a function f is continuous at $x = c$ means that there is no interruption in the graph of f at c. That is, its graph is unbroken at c and there are no holes, jumps, or gaps. Figure 3.20 identifies three values of x at which the graph of f is *not* continuous. At all other points in the interval (a, b), the graph of f is uninterrupted and **continuous.**

EXPLORATION

Informally, you might say that a function is *continuous* on an open interval if its graph can be drawn with a pencil without lifting the pencil from the paper. Use a graphing utility to graph each function on the given interval. From the graphs, which functions would you say are continuous on the interval? Do you think you can trust the results you obtained graphically? Explain your reasoning.

	Function	*Interval*
a.	$y = x^2 + 1$	$(-3, 3)$
b.	$y = \dfrac{1}{x - 2}$	$(-3, 3)$
c.	$y = \dfrac{x^2 - 4}{x + 2}$	$(-3, 3)$
d.	$y = \begin{cases} 2x - 4, & x \le 0 \\ x + 1, & x > 0 \end{cases}$	$(-3, 3)$

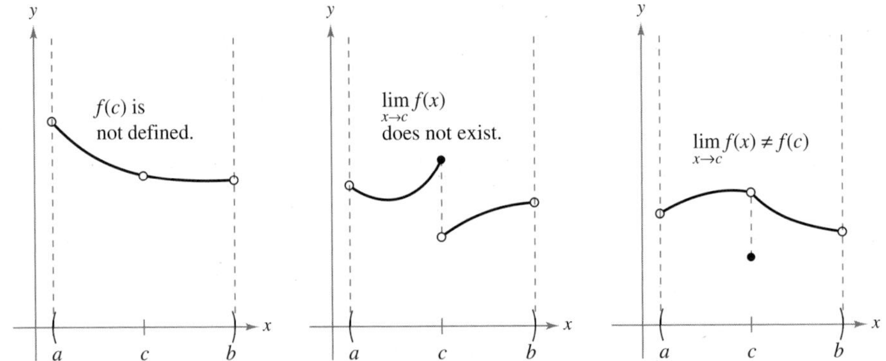

Three conditions exist for which the graph of f is not continuous at $x = c$.
Figure 3.20

In Figure 3.20, it appears that continuity at $x = c$ can be destroyed by any one of the following conditions.

1. The function is not defined at $x = c$.
2. The limit of $f(x)$ does not exist at $x = c$.
3. The limit of $f(x)$ exists at $x = c$, but it is not equal to $f(c)$.

If *none* of the three conditions above is true, the function f is called **continuous at c,** as indicated in the following important definition.

DEFINITION OF CONTINUITY

Continuity at a Point: A function f is **continuous at c** if the following three conditions are met.

1. $f(c)$ is defined.
2. $\lim\limits_{x \to c} f(x)$ exists.
3. $\lim\limits_{x \to c} f(x) = f(c)$

Continuity on an Open Interval: A function is **continuous on an open interval (a, b)** if it is continuous at each point in the interval. A function that is continuous on the entire real line $(-\infty, \infty)$ is **everywhere continuous.**

■ **FOR FURTHER INFORMATION** For more information on the concept of continuity, see the article "Leibniz and the Spell of the Continuous" by Hardy Grant in *The College Mathematics Journal*. To view this article, go to the website *www.matharticles.com*.

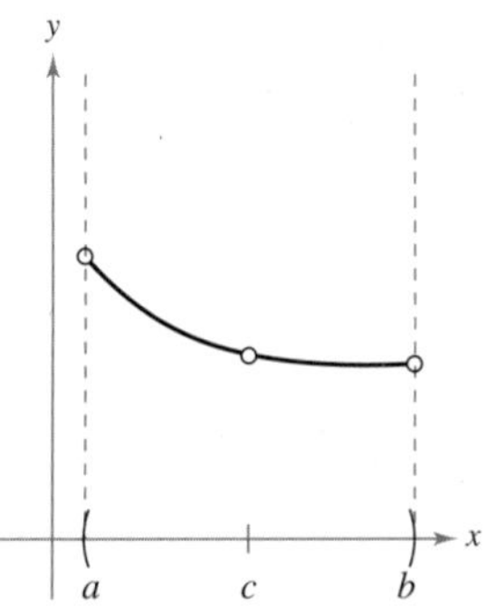

(a) Removable discontinuity

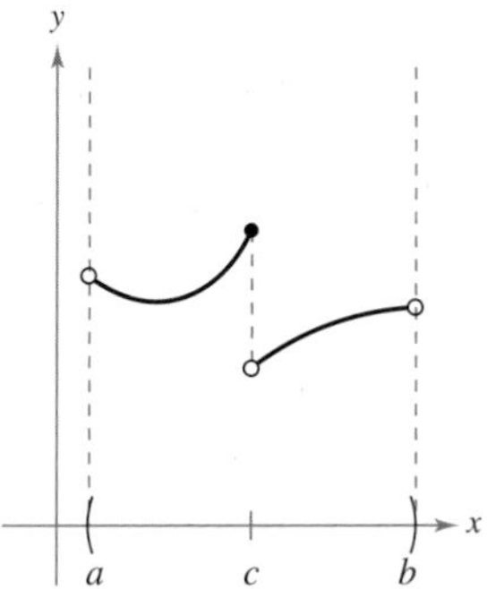

(b) Nonremovable discontinuity

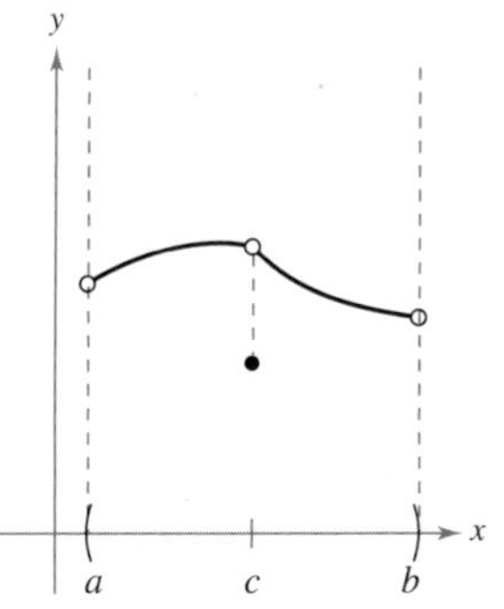

(c) Removable discontinuity
Figure 3.21

Consider an open interval I that contains a real number c. If a function f is defined on I (except possibly at c), and f is not continuous at c, then f is said to have a **discontinuity** at c. Discontinuities fall into two categories: **removable** and **nonremovable.** A discontinuity at c is called removable if f can be made continuous by appropriately defining (or redefining) $f(c)$. For instance, the functions shown in Figures 3.21(a) and (c) have removable discontinuities at c and the function shown in Figure 3.21(b) has a nonremovable discontinuity at c.

EXAMPLE 1 Continuity of a Function

Discuss the continuity of each function.

a. $f(x) = \dfrac{1}{x}$ **b.** $g(x) = \dfrac{x^2 - 1}{x - 1}$ **c.** $h(x) = \begin{cases} x + 1, & x \le 0 \\ x^2 + 1, & x > 0 \end{cases}$ **d.** $k(x) = x^2$

Solution

a. The domain of f is all nonzero real numbers. From Theorem 3.3, you can conclude that f is continuous at every x-value in its domain. At $x = 0$, f has a nonremovable discontinuity, as shown in Figure 3.22(a). In other words, there is no way to define $f(0)$ so as to make the function continuous at $x = 0$.

b. The domain of g is all real numbers except $x = 1$. From Theorem 3.3, you can conclude that g is continuous at every x-value in its domain. At $x = 1$, the function has a removable discontinuity, as shown in Figure 3.22(b). If $g(1)$ is defined as 2, the "newly defined" function is continuous for all real numbers.

c. The domain of h is all real numbers. The function h is continuous on $(-\infty, 0)$ and $(0, \infty)$, and, because $\lim\limits_{x \to 0} h(x) = 1$, h is continuous on the entire real line, as shown in Figure 3.22(c).

d. The domain of k is all real numbers. From Theorem 3.3, you can conclude that the function is continuous on its entire domain, $(-\infty, \infty)$, as shown in Figure 3.22(d).

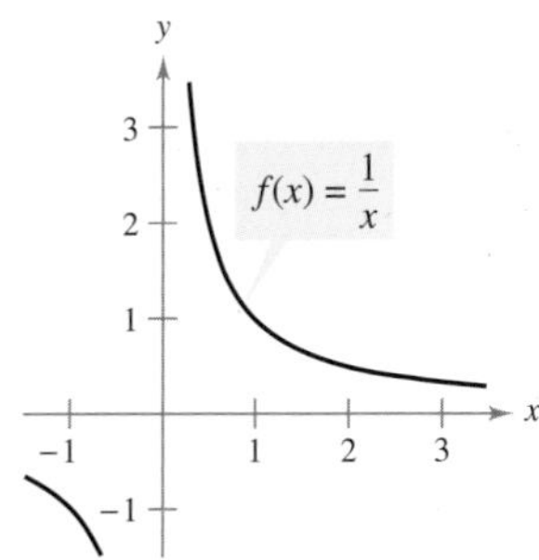

(a) Nonremovable discontinuity at $x = 0$

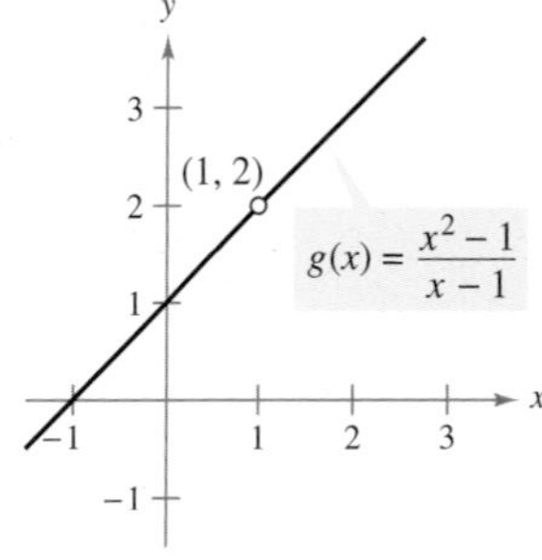

(b) Removable discontinuity at $x = 1$

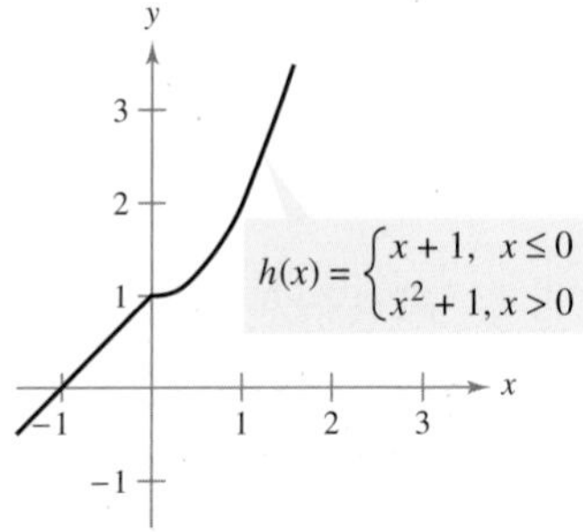

(c) Continuous on entire real line
Figure 3.22

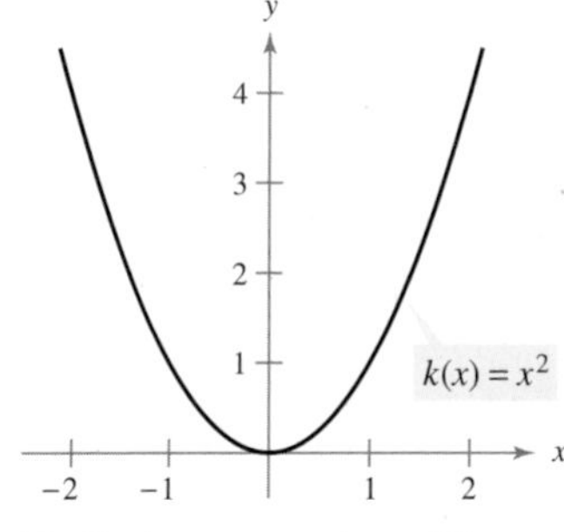

(d) Continuous on entire real line

■

STUDY TIP Some people may refer to the function in Example 1(a) as "discontinuous." We have found that this terminology can be confusing. Rather than saying that the function is discontinuous, we prefer to say that it has a discontinuity at $x = 0$.

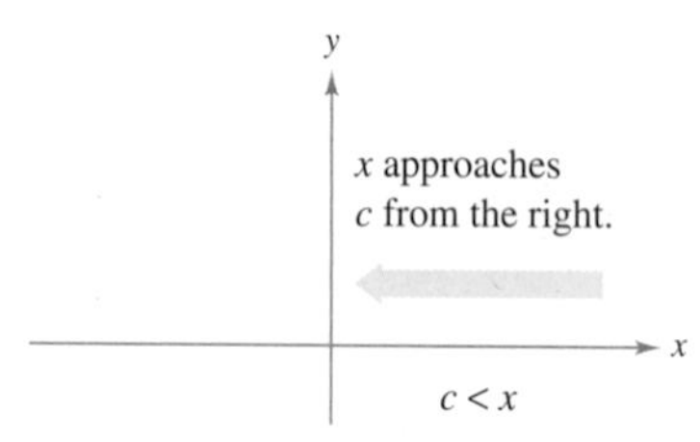

(a) Limit from right

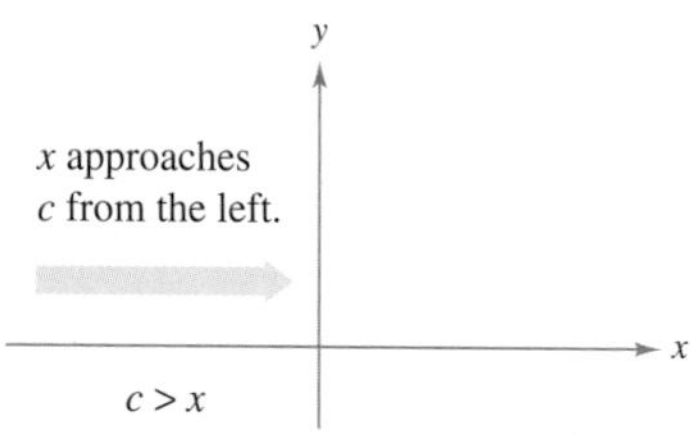

(b) Limit from left
Figure 3.23

One-Sided Limits and Continuity on a Closed Interval

To understand continuity on a closed interval, you first need to look at a different type of limit called a **one-sided limit.** For example, the **limit from the right** (or right-hand limit) means that x approaches c from values greater than c [see Figure 3.23(a)]. This limit is denoted as

$$\lim_{x\to c^+} f(x) = L. \qquad \text{Limit from the right}$$

Similarly, the **limit from the left** (or left-hand limit) means that x approaches c from values less than c [see Figure 3.23(b)]. This limit is denoted as

$$\lim_{x\to c^-} f(x) = L. \qquad \text{Limit from the left}$$

One-sided limits are useful in taking limits of functions involving radicals. For instance, if n is an even integer,

$$\lim_{x\to 0^+} \sqrt[n]{x} = 0.$$

EXAMPLE 2 A One-Sided Limit

Find the limit of $f(x) = \sqrt{4 - x^2}$ as x approaches -2 from the right.

Solution As shown in Figure 3.24, the limit as x approaches -2 from the right is

$$\lim_{x\to -2^+} \sqrt{4 - x^2} = \sqrt{4 - 4}$$
$$= 0.$$

■

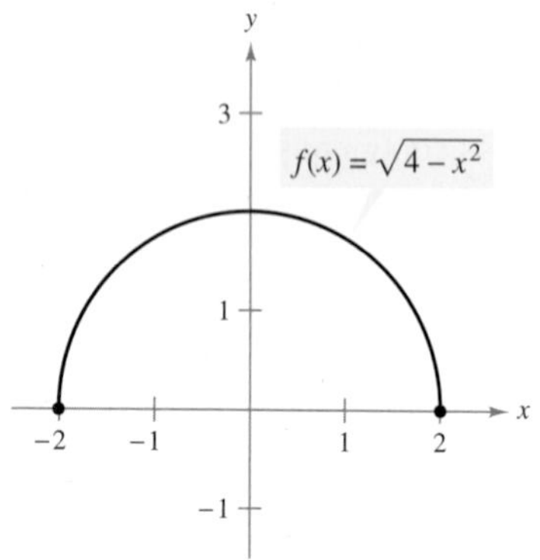

The limit of $f(x)$ as x approaches -2 from the right is 0.
Figure 3.24

One-sided limits can be used to investigate the behavior of *step functions.* Recall from Section 1.2 that one common type of step function is the *greatest integer function* $[\![x]\!]$, defined by

$$[\![x]\!] = \text{greatest integer } n \text{ such that } n \le x. \qquad \text{Greatest integer function}$$

EXAMPLE 3 The Greatest Integer Function

Find the limit of the greatest integer function $f(x) = [\![x]\!]$ as x approaches 0 from the left and from the right.

Solution As shown in Figure 3.25, the limit as x approaches 0 *from the left* is given by

$$\lim_{x\to 0^-} [\![x]\!] = -1$$

and the limit as x approaches 0 *from the right* is given by

$$\lim_{x\to 0^+} [\![x]\!] = 0.$$

The greatest integer function has a discontinuity at zero because the left and right limits at zero are different. By similar reasoning, you can see that the greatest integer function has a discontinuity at any integer n.

■

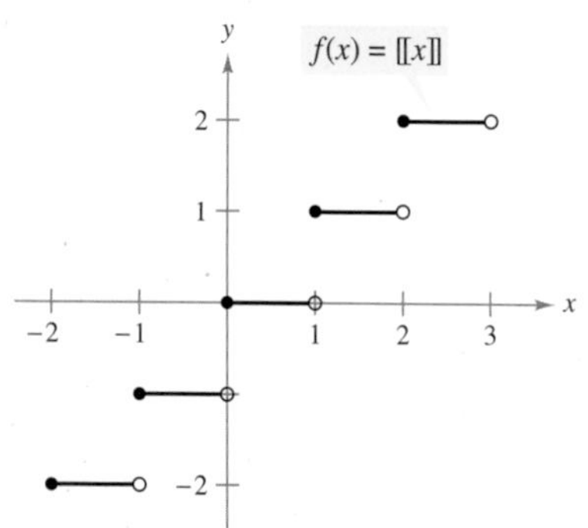

Greatest integer function
Figure 3.25

When the limit from the left is not equal to the limit from the right, the (two-sided) limit *does not exist*. The next theorem makes this more explicit. The proof of this theorem follows directly from the definition of a one-sided limit.

THEOREM 3.8 THE EXISTENCE OF A LIMIT

Let f be a function and let c and L be real numbers. The limit of $f(x)$ as x approaches c is L if and only if

$$\lim_{x\to c^-} f(x) = L \quad \text{and} \quad \lim_{x\to c^+} f(x) = L.$$

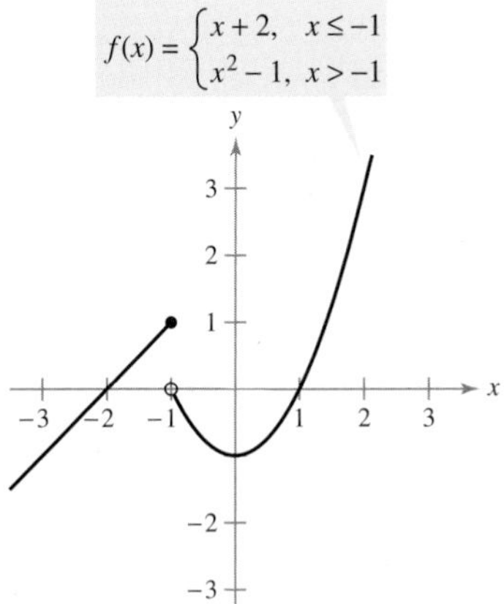

Figure 3.26

EXAMPLE 4 Limit of a Piecewise-Defined Function

Discuss the continuity of

$$f(x) = \begin{cases} x+2, & x \le -1 \\ x^2 - 1, & x > -1 \end{cases}.$$

Solution Because f is a polynomial for $x < -1$ and for $x > -1$, it is continuous everywhere except at $x = -1$. The one-sided limits

$$\lim_{x\to -1^-} f(x) = \lim_{x\to -1^-} (x + 2) = 1 \qquad \text{Limit from left of } x = -1$$

$$\lim_{x\to -1^+} f(x) = \lim_{x\to -1^+} (x^2 - 1) = 0 \qquad \text{Limit from right of } x = -1$$

show that $\lim_{x\to -1} f(x)$ does not exist and that f has a discontinuity at $x = -1$. The graph of f is shown in Figure 3.26. ■

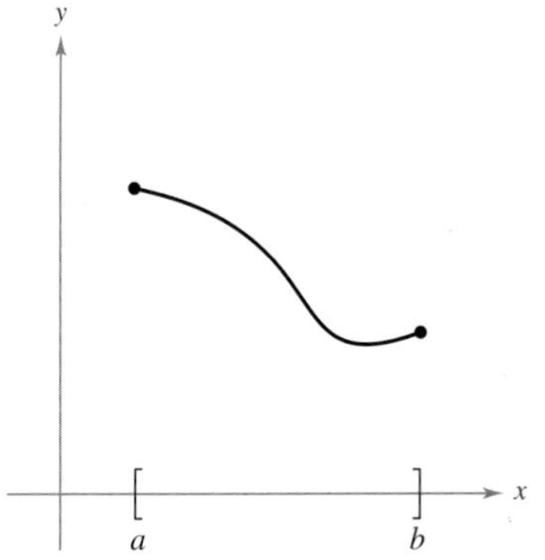

Continuous function on a closed interval
Figure 3.27

The concept of a one-sided limit allows you to extend the definition of continuity to closed intervals. Basically, a function is continuous on a closed interval if it is continuous in the interior of the interval and exhibits one-sided continuity at the endpoints. This is stated formally as follows.

DEFINITION OF CONTINUITY ON A CLOSED INTERVAL

A function f is **continuous on the closed interval $[a, b]$** if it is continuous on the open interval (a, b) and

$$\lim_{x\to a^+} f(x) = f(a) \quad \text{and} \quad \lim_{x\to b^-} f(x) = f(b).$$

The function f is **continuous from the right** at a and **continuous from the left** at b (see Figure 3.27).

Similar definitions can be made to cover continuity on intervals of the form $(a, b]$ and $[a, b)$ that are neither open nor closed, or on infinite intervals. For example, the function

$$f(x) = \sqrt{x}$$

is continuous on the infinite interval $[0, \infty)$, and the function

$$g(x) = \sqrt{2 - x}$$

is continuous on the infinite interval $(-\infty, 2]$.

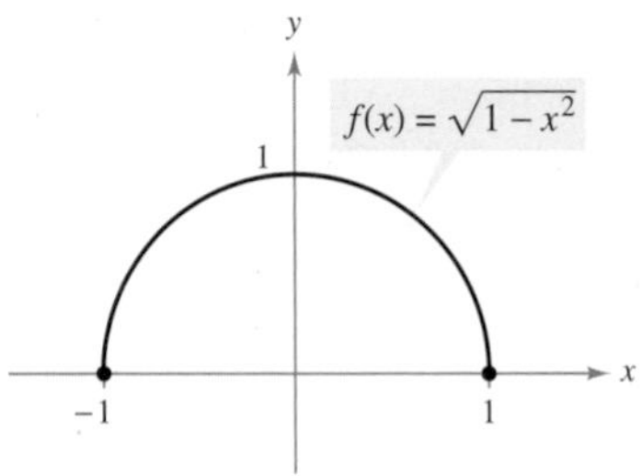

f is continuous on $[-1, 1]$.
Figure 3.28

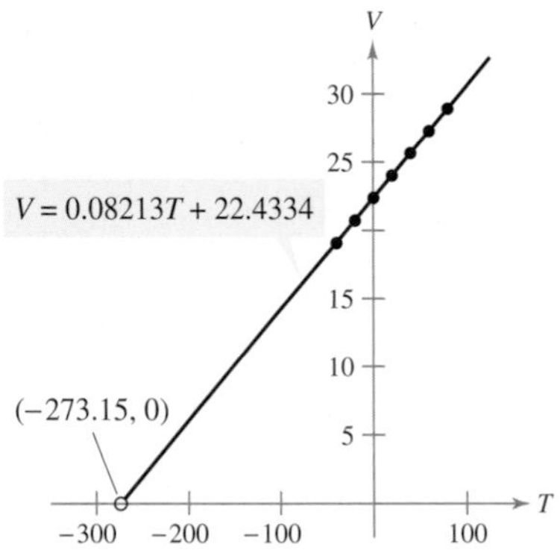

The volume of hydrogen gas depends on its temperature.
Figure 3.29

EXAMPLE 5 Continuity on a Closed Interval

Discuss the continuity of

$$f(x) = \sqrt{1 - x^2}.$$

Solution The domain of f is the closed interval $[-1, 1]$. At all points in the open interval $(-1, 1)$, the continuity of f follows from Theorems 3.4 and 3.5. Moreover, because

$$\lim_{x \to -1^+} \sqrt{1 - x^2} = 0 = f(-1) \qquad \text{Continuous from the right}$$

and

$$\lim_{x \to 1^-} \sqrt{1 - x^2} = 0 = f(1) \qquad \text{Continuous from the left}$$

you can conclude that f is continuous on the closed interval $[-1, 1]$, as shown in Figure 3.28.

EXAMPLE 6 Charles's Law and Absolute Zero

On the Kelvin scale, *absolute zero* is the temperature 0 K. Although temperatures very close to 0 K have been produced in laboratories, absolute zero has never been attained. In fact, evidence suggests that absolute zero *cannot* be attained. How did scientists determine that 0 K is the "lower limit" of the temperature of matter? What is absolute zero on the Celsius scale?

Solution The determination of absolute zero stems from the work of the French physicist Jacques Charles (1746–1823). Charles discovered that the volume of gas at a constant pressure increases linearly with the temperature of the gas. The table illustrates this relationship between volume and temperature. To generate the values in the table, one mole of hydrogen is held at a constant pressure of one atmosphere. The volume V is approximated and is measured in liters, and the temperature T is measured in degrees Celsius.

T	-40	-20	0	20	40	60	80
V	19.1482	20.7908	22.4334	24.0760	25.7186	27.3612	29.0038

The points represented by the table are shown in Figure 3.29. Moreover, by using the points in the table, you can determine that T and V are related by the linear equation

$$V = 0.08213T + 22.4334 \qquad \text{or} \qquad T = \frac{V - 22.4334}{0.08213}.$$

By reasoning that the volume of the gas can approach 0 (but can never equal or go below 0), you can determine that the "least possible temperature" is given by

$$\lim_{V \to 0^+} T = \lim_{V \to 0^+} \frac{V - 22.4334}{0.08213}$$

$$= \frac{0 - 22.4334}{0.08213} \qquad \text{Use direct substitution.}$$

$$\approx -273.15.$$

So, absolute zero on the Kelvin scale (0 K) is approximately $-273.15°$ on the Celsius scale. ■

In 2003, researchers at the Massachusetts Institute of Technology used lasers and evaporation to produce a supercold gas in which atoms overlap. This gas is called a Bose-Einstein condensate. They measured a temperature of about 450 pK (picokelvin), or approximately -273.14999999955°C. (*Source:* Science *magazine, September 12, 2003*)

AUGUSTIN-LOUIS CAUCHY (1789–1857)

The concept of a continuous function was first introduced by Augustin-Louis Cauchy in 1821. The definition given in his text *Cours d'Analyse* stated that indefinite small changes in y were the result of indefinite small changes in x. "... $f(x)$ will be called a *continuous* function if ... the numerical values of the difference $f(x + \alpha) - f(x)$ decrease indefinitely with those of α ...."

Properties of Continuity

In Section 3.3, you studied several properties of limits. Each of those properties yields a corresponding property pertaining to the continuity of a function. For instance, Theorem 3.9 follows directly from Theorem 3.2. (A proof of Theorem 3.9 is given in Appendix A.)

THEOREM 3.9 PROPERTIES OF CONTINUITY

If b is a real number and f and g are continuous at $x = c$, then the following functions are also continuous at c.

1. Scalar multiple: bf
2. Sum or difference: $f \pm g$
3. Product: fg
4. Quotient: $\dfrac{f}{g}$, if $g(c) \neq 0$

The following types of functions are continuous at every point in their domains.

1. Polynomial: $p(x) = a_n x^n + a_{n-1}x^{n-1} + \cdot \cdot \cdot + a_1 x + a_0$
2. Rational: $r(x) = \dfrac{p(x)}{q(x)}$, $q(x) \neq 0$
3. Radical: $f(x) = \sqrt[n]{x}$

By combining Theorem 3.9 with this summary, you can conclude that a wide variety of elementary functions are continuous at every point in their domains.

EXAMPLE 7 Applying Properties of Continuity

By Theorem 3.9, it follows that each of the functions below is continuous at every point in its domain.

$$f(x) = x + \sqrt{x} \qquad f(x) = 3\sqrt{x} \qquad f(x) = \frac{x^2 + 1}{\sqrt{x}}$$ ■

The next theorem, which is a consequence of Theorem 3.5, allows you to determine the continuity of *composite* functions such as

$$f(x) = \sqrt{x^2 + 1}$$

and

$$f(x) = \sqrt[3]{2x + 1}.$$

THEOREM 3.10 CONTINUITY OF A COMPOSITE FUNCTION

If g is continuous at c and f is continuous at $g(c)$, then the composite function given by $(f \circ g)(x) = f(g(x))$ is continuous at c.

NOTE One consequence of Theorem 3.10 is that if f and g satisfy the given conditions, you can determine the limit of $f(g(x))$ as x approaches c to be

$$\lim_{x \to c} f(g(x)) = f(g(c)).$$

PROOF By the definition of continuity, $\lim_{x \to c} g(x) = g(c)$ and $\lim_{x \to g(c)} f(x) = f(g(c))$. Apply Theorem 3.5 with $L = g(c)$ to obtain $\lim_{x \to c} f(g(x)) = f\left(\lim_{x \to c} g(x)\right) = f(g(c))$. So, $(f \circ g) = f(g(x))$ is continuous at c. ■

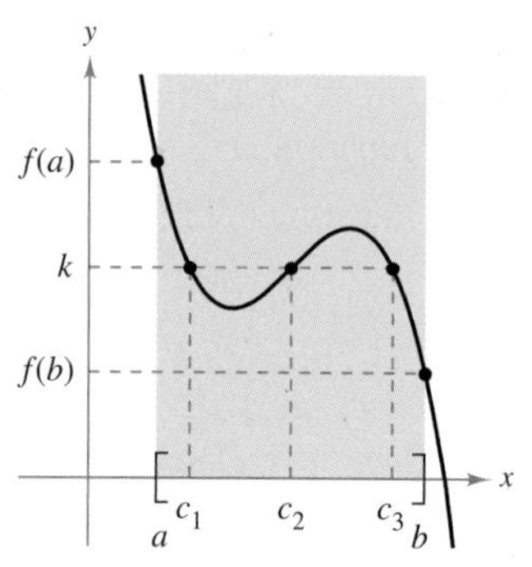

f is continuous on $[a, b]$.
[There exist three *c*'s such that $f(c) = k$.]
Figure 3.30

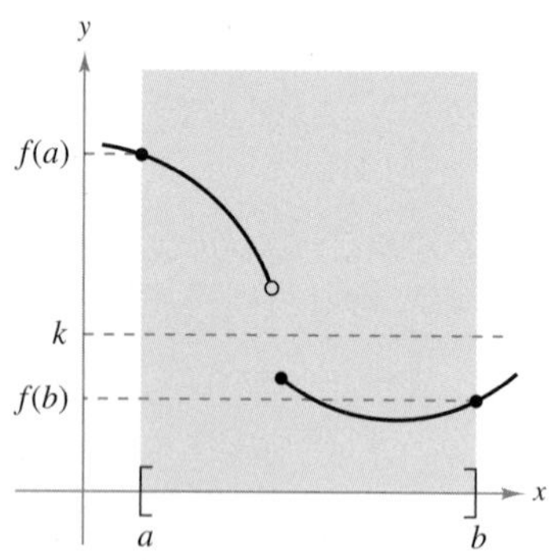

f is not continuous on $[a, b]$.
[There are no *c*'s such that $f(c) = k$.]
Figure 3.31

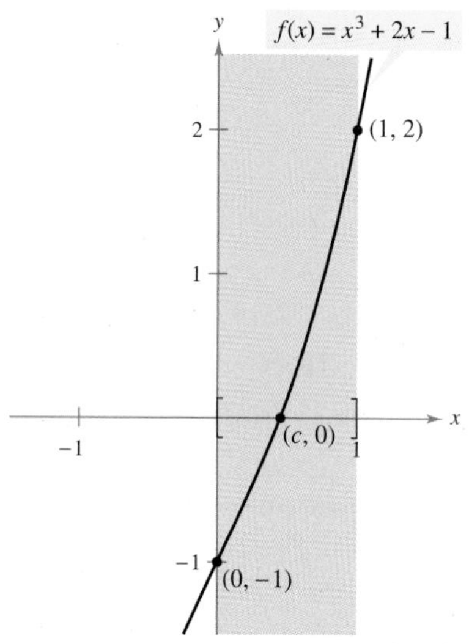

f is continuous on $[0, 1]$ with $f(0) < 0$ and $f(1) > 0$.
Figure 3.32

The Intermediate Value Theorem

Theorem 3.11 is an important theorem concerning the behavior of functions that are continuous on a closed interval.

> **THEOREM 3.11 INTERMEDIATE VALUE THEOREM**
>
> If *f* is continuous on the closed interval $[a, b]$, $f(a) \neq f(b)$, and k is any number between $f(a)$ and $f(b)$, then there is at least one number c in $[a, b]$ such that
>
> $$f(c) = k.$$

NOTE The Intermediate Value Theorem tells you that at least one number c exists, but it does not provide a method for finding c. Such theorems are called **existence theorems.** By referring to a text on advanced calculus, you will find that a proof of this theorem is based on a property of real numbers called *completeness*. The Intermediate Value Theorem states that for a continuous function f, if x takes on all values between a and b, $f(x)$ must take on all values between $f(a)$ and $f(b)$. ■

As a simple example of the application of this theorem, consider a person's height. Suppose that a girl is 5 feet tall on her thirteenth birthday and 5 feet 7 inches tall on her fourteenth birthday. Then, for any height h between 5 feet and 5 feet 7 inches, there must have been a time t when her height was exactly h. This seems reasonable because human growth is continuous and a person's height does not abruptly change from one value to another.

The Intermediate Value Theorem guarantees the existence of *at least one* number c in the closed interval $[a, b]$. There may, of course, be more than one number c such that $f(c) = k$, as shown in Figure 3.30. A function that is not continuous does not necessarily exhibit the intermediate value property. For example, the graph of the function shown in Figure 3.31 jumps over the horizontal line given by $y = k$, and for this function there is no value of c in $[a, b]$ such that $f(c) = k$.

The Intermediate Value Theorem often can be used to locate the zeros of a function that is continuous on a closed interval. Specifically, if *f* is continuous on $[a, b]$ and $f(a)$ and $f(b)$ differ in sign, the Intermediate Value Theorem guarantees the existence of at least one zero of *f* in the closed interval $[a, b]$.

EXAMPLE 8 An Application of the Intermediate Value Theorem

Use the Intermediate Value Theorem to show that the polynomial function

$$f(x) = x^3 + 2x - 1$$

has a zero in the interval $[0, 1]$.

Solution Note that *f* is continuous on the closed interval $[0, 1]$. Because

$$f(0) = 0^3 + 2(0) - 1 = -1$$

and

$$f(1) = 1^3 + 2(1) - 1 = 2$$

it follows that $f(0) < 0$ and $f(1) > 0$. You can therefore apply the Intermediate Value Theorem to conclude that there must be some c in $[0, 1]$ such that

$$f(c) = 0 \qquad f \text{ has a zero in the closed interval } [0, 1].$$

as shown in Figure 3.32. ■

The **bisection method** for approximating the real zeros of a continuous function is similar to the method used in Example 8. If you know that a zero exists in the closed interval $[a, b]$, the zero must lie in the interval $[a, (a + b)/2]$ or $[(a + b)/2, b]$. From the sign of $f([a + b]/2)$, you can determine which interval contains the zero. By repeatedly bisecting the interval, you can "close in" on the zero of the function.

TECHNOLOGY You can also use the *zoom* feature of a graphing utility to approximate the real zeros of a continuous function. By repeatedly zooming in on the point where the graph crosses the x-axis, and adjusting the x-axis scale, you can approximate the zero of the function to any desired accuracy. The zero of $x^3 + 2x - 1$ is approximately 0.453, as shown in Figure 3.33.

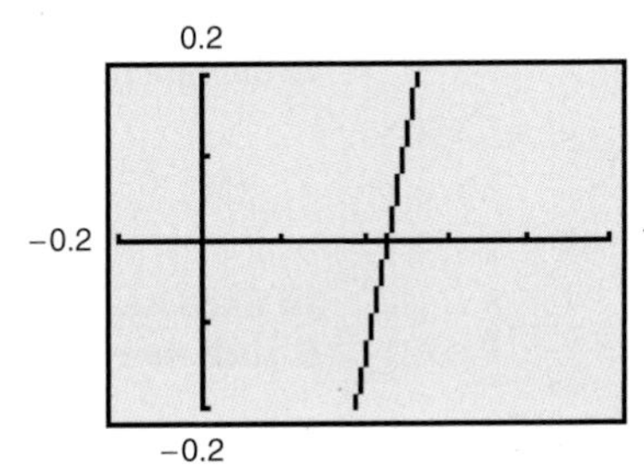

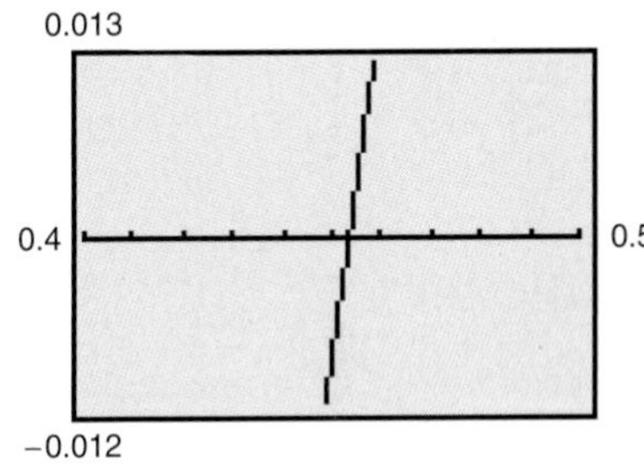

Figure 3.33 Zooming in on the zero of $f(x) = x^3 + 2x - 1$

3.4 Exercises

See www.CalcChat.com for worked-out solutions to odd-numbered exercises.

In Exercises 1–6, use the graph to determine the limit, and discuss the continuity of the function.

(a) $\lim_{x \to c^+} f(x)$ (b) $\lim_{x \to c^-} f(x)$ (c) $\lim_{x \to c} f(x)$

1\.

(4, 3)

$c = 4$

2\.

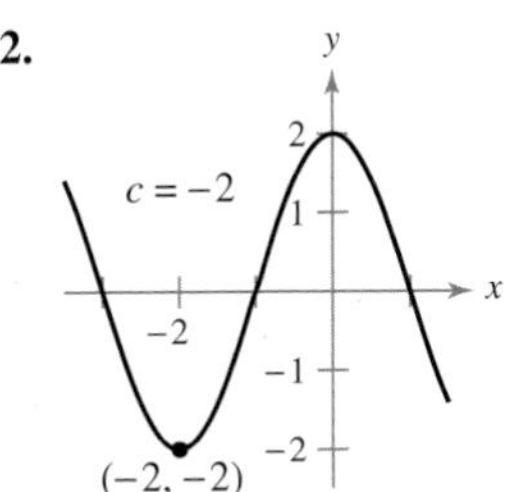

3\.

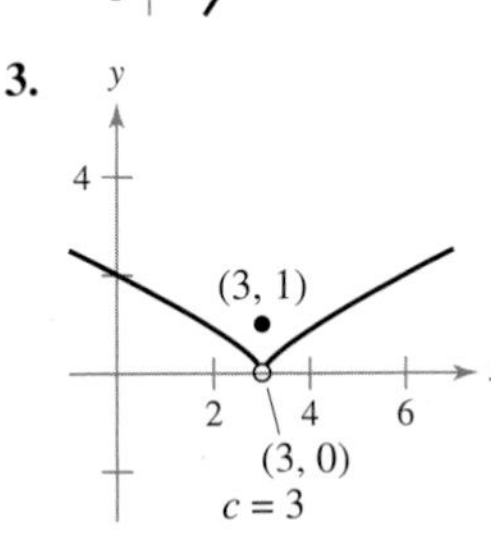

4\.

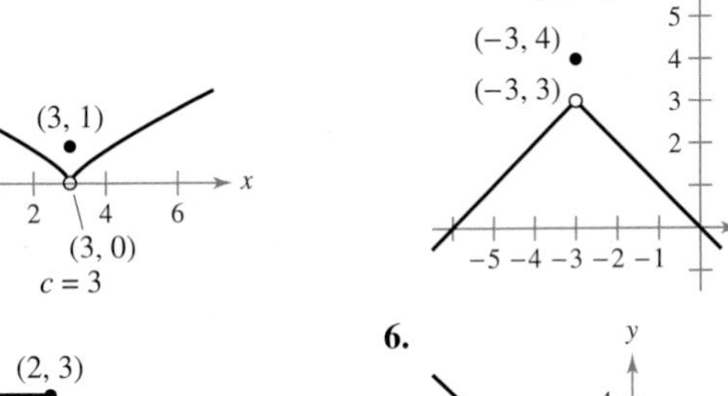

5\.

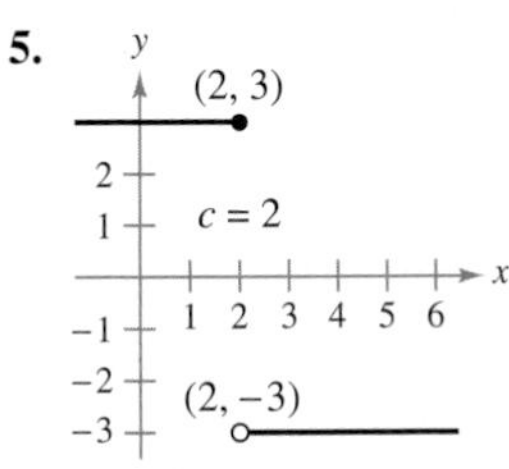

6\.

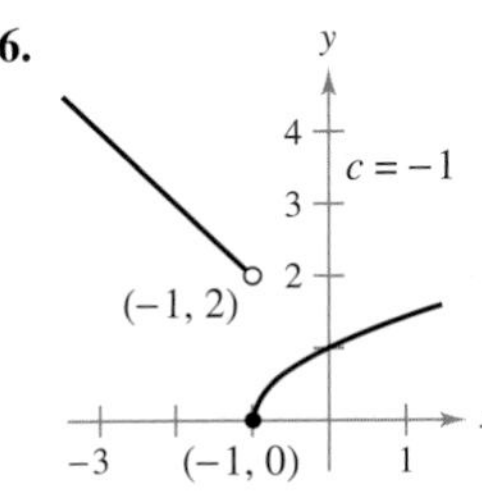

In Exercises 7–22, find the limit (if it exists). If it does not exist, explain why.

7\. $\lim_{x \to 5^+} \dfrac{x - 5}{x^2 - 25}$

8\. $\lim_{x \to 2^+} \dfrac{2 - x}{x^2 - 4}$

9\. $\lim_{x \to -3^-} \dfrac{x}{\sqrt{x^2 - 9}}$

10\. $\lim_{x \to 9^-} \dfrac{\sqrt{x} - 3}{x - 9}$

11\. $\lim_{x \to 0^-} \dfrac{|x|}{x}$

12\. $\lim_{x \to 10^+} \dfrac{|x - 10|}{x - 10}$

13\. $\lim_{\Delta x \to 0^-} \dfrac{\dfrac{1}{x + \Delta x} - \dfrac{1}{x}}{\Delta x}$

14\. $\lim_{\Delta x \to 0^+} \dfrac{(x + \Delta x)^2 + x + \Delta x - (x^2 + x)}{\Delta x}$

15\. $\lim_{x \to 3^-} f(x)$, where $f(x) = \begin{cases} \dfrac{x + 2}{2}, & x \le 3 \\ \dfrac{12 - 2x}{3}, & x > 3 \end{cases}$

16\. $\lim_{x \to 2} f(x)$, where $f(x) = \begin{cases} x^2 - 4x + 6, & x < 2 \\ -x^2 + 4x - 2, & x \ge 2 \end{cases}$

17\. $\lim_{x \to 1} f(x)$, where $f(x) = \begin{cases} x^3 + 1, & x < 1 \\ x + 1, & x \ge 1 \end{cases}$

18\. $\lim_{x \to 1^+} f(x)$, where $f(x) = \begin{cases} x, & x \le 1 \\ 1 - x, & x > 1 \end{cases}$

19. $\lim_{x\to 4^-} (5[\![x]\!] - 7)$

20. $\lim_{x\to 2^+} (2x - [\![x]\!])$

21. $\lim_{x\to 3} (2 - [\![-x]\!])$

22. $\lim_{x\to 1}\left(1 - \left[\!\!\left[-\dfrac{x}{2}\right]\!\!\right]\right)$

In Exercises 23–26, discuss the continuity of each function.

23. $f(x) = \dfrac{1}{x^2 - 4}$

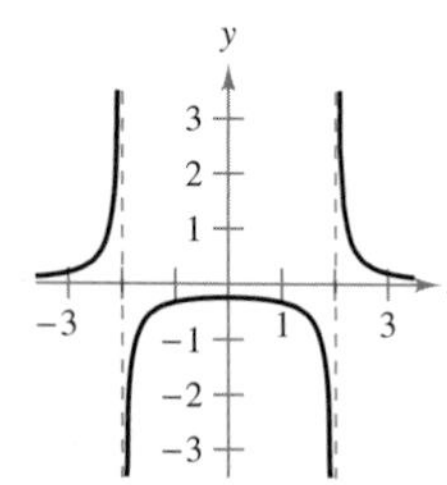

24. $f(x) = \dfrac{x^2 - 1}{x + 1}$

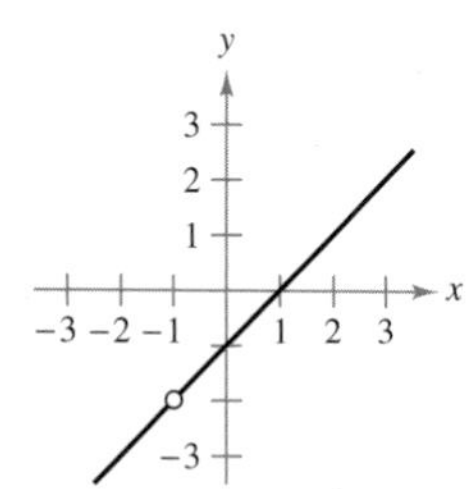

25. $f(x) = \frac{1}{2}[\![x]\!] + x$

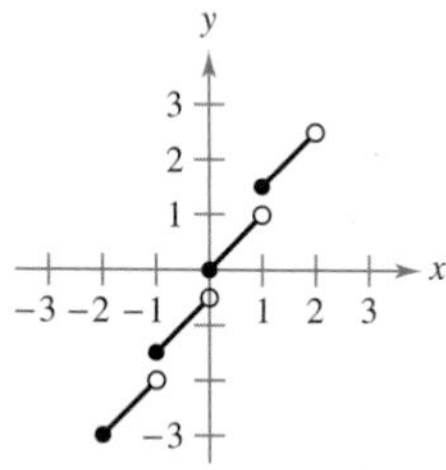

26. $f(x) = \begin{cases} x, & x < 1 \\ 2, & x = 1 \\ 2x - 1, & x > 1 \end{cases}$

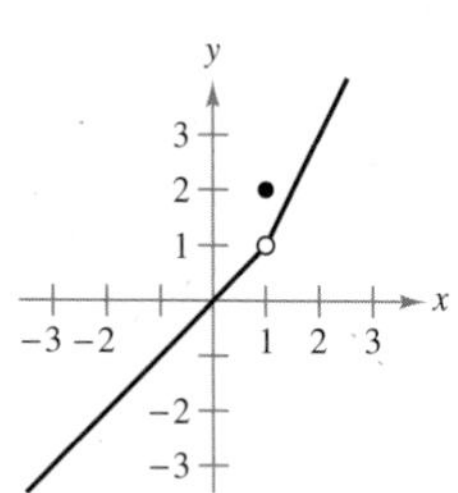

In Exercises 27–30, discuss the continuity of the function on the closed interval.

Function	*Interval*
27. $g(x) = \sqrt{49 - x^2}$	$[-7, 7]$
28. $f(t) = 3 - \sqrt{9 - t^2}$	$[-3, 3]$
29. $f(x) = \begin{cases} 3 - x, & x \le 0 \\ 3 + \frac{1}{2}x, & x > 0 \end{cases}$	$[-1, 4]$
30. $g(x) = \dfrac{1}{x^2 - 4}$	$[-1, 2]$

In Exercises 31–46, find the x-values (if any) at which f is not continuous. Which of the discontinuities are removable?

31. $f(x) = x^2 - 2x + 1$

32. $f(x) = \dfrac{1}{x^2 + 1}$

33. $f(x) = \dfrac{x}{x^2 - x}$

34. $f(x) = \dfrac{x}{x^2 - 1}$

35. $f(x) = \dfrac{x}{x^2 + 1}$

36. $f(x) = \dfrac{x - 6}{x^2 - 36}$

37. $f(x) = \dfrac{x + 2}{x^2 - 3x - 10}$

38. $f(x) = \dfrac{x - 1}{x^2 + x - 2}$

39. $f(x) = \dfrac{|x + 7|}{x + 7}$

40. $f(x) = \dfrac{|x - 8|}{x - 8}$

41. $f(x) = \begin{cases} x, & x \le 1 \\ x^2, & x > 1 \end{cases}$

42. $f(x) = \begin{cases} -2x + 3, & x < 1 \\ x^2, & x \ge 1 \end{cases}$

43. $f(x) = \begin{cases} \frac{1}{2}x + 1, & x \le 2 \\ 3 - x, & x > 2 \end{cases}$

44. $f(x) = \begin{cases} -2x, & x \le 2 \\ x^2 - 4x + 1, & x > 2 \end{cases}$

45. $f(x) = [\![x - 8]\!]$

46. $f(x) = 5 - [\![x]\!]$

In Exercises 47 and 48, use a graphing utility to graph the function. From the graph, estimate

$$\lim_{x\to 0^+} f(x) \quad \text{and} \quad \lim_{x\to 0^-} f(x).$$

Is the function continuous on the entire real line? Explain.

47. $f(x) = \dfrac{|x^2 - 4|x}{x + 2}$

48. $f(x) = \dfrac{|x^2 + 4x|(x + 2)}{x + 4}$

In Exercises 49–52, find the constant a, or the constants a and b, such that the function is continuous on the entire real line.

49. $f(x) = \begin{cases} 3x^2, & x \ge 1 \\ ax - 4, & x < 1 \end{cases}$

50. $f(x) = \begin{cases} 3x^3, & x \le 1 \\ ax + 5, & x > 1 \end{cases}$

51. $f(x) = \begin{cases} 2, & x \le -1 \\ ax + b, & -1 < x < 3 \\ -2, & x \ge 3 \end{cases}$

52. $g(x) = \begin{cases} \dfrac{x^2 - a^2}{x - a}, & x \ne a \\ 8, & x = a \end{cases}$

In Exercises 53–58, discuss the continuity of the composite function $h(x) = f(g(x))$.

53. $f(x) = x^2$
$g(x) = x - 1$

54. $f(x) = \sqrt{x}$
$g(x) = x^2$

55. $f(x) = \dfrac{1}{\sqrt{x}}$
$g(x) = \dfrac{1}{x}$

56. $f(x) = \dfrac{1}{\sqrt{x}}$
$g(x) = x - 1$

57. $f(x) = \dfrac{1}{x - 6}$
$g(x) = x^2 + 5$

58. $f(x) = \dfrac{1}{x}$
$g(x) = \dfrac{1}{x - 1}$

In Exercises 59–62, use a graphing utility to graph the function. Use the graph to determine any x-values at which the function is not continuous.

59. $f(x) = [\![x]\!] - x$

60. $h(x) = \dfrac{1}{x^2 - x - 2}$

61. $g(x) = \begin{cases} x^2 - 3x, & x > 4 \\ 2x - 5, & x \le 4 \end{cases}$

62. $f(x) = \begin{cases} x^2 - 2x + 2, & x < 2 \\ -x^2 + 6x - 6, & x \ge 2 \end{cases}$

In Exercises 63–66, describe the interval(s) on which the function is continuous.

63. $f(x) = \dfrac{x}{x^2 + x + 2}$

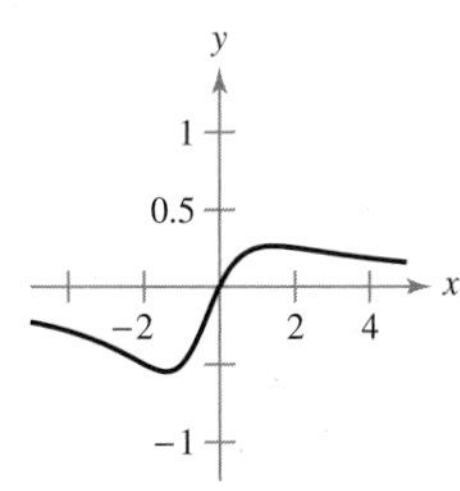

64. $f(x) = x\sqrt{x + 3}$

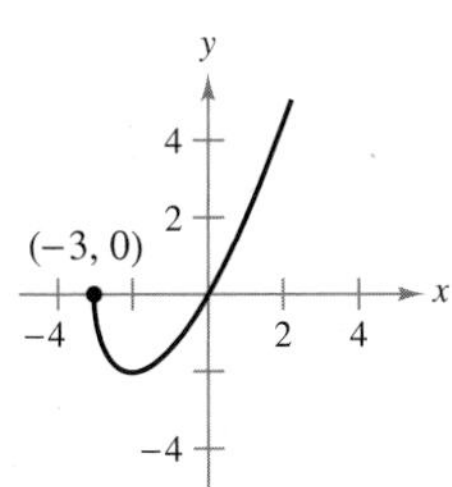

65. $f(x) = \dfrac{x^2}{x^2 - 36}$

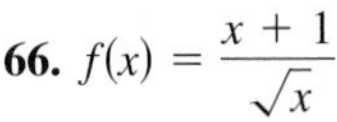

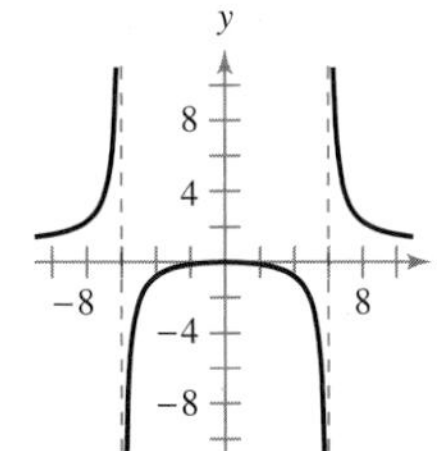

66. $f(x) = \dfrac{x + 1}{\sqrt{x}}$

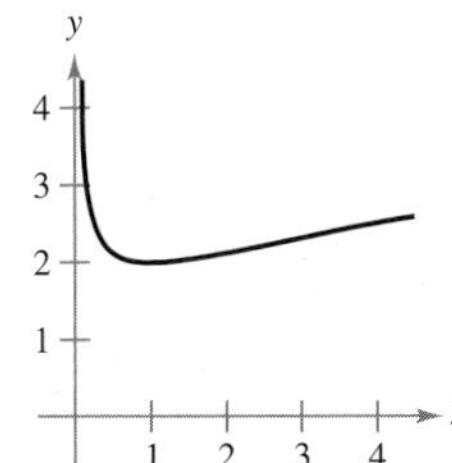

Writing **In Exercises 67 and 68, use a graphing utility to graph the function on the interval $[-4, 4]$. Does the graph of the function appear to be continuous on this interval? Is the function continuous on $[-4, 4]$? Write a short paragraph about the importance of examining a function analytically as well as graphically.**

67. $f(x) = \dfrac{x^2 - x - 2}{x + 1}$

68. $f(x) = \dfrac{x^3 - 8}{x - 2}$

Writing **In Exercises 69 and 70, explain why the function has a zero in the given interval.**

	Function	*Interval*
69.	$f(x) = \frac{1}{12}x^4 - x^3 + 4$	$[1, 2]$
70.	$f(x) = x^3 + 5x - 3$	$[0, 1]$

In Exercises 71–74, use the Intermediate Value Theorem and a graphing utility to approximate the zero of the function in the interval [0, 1]. Repeatedly "zoom in" on the graph of the function to approximate the zero accurate to two decimal places. Use the *zero* or *root* feature of the graphing utility to approximate the zero accurate to four decimal places.

71. $f(x) = x^3 + x - 1$

72. $f(x) = x^3 + 5x - 3$

73. $g(t) = 3\sqrt{t^2 + 1} - 4$

74. $h(s) = 5 - \dfrac{2}{s^3}$

In Exercises 75–78, verify that the Intermediate Value Theorem applies to the indicated interval and find the value of c guaranteed by the theorem.

75. $f(x) = x^2 + x - 1, \quad [0, 5], \quad f(c) = 11$

76. $f(x) = x^2 - 6x + 8, \quad [0, 3], \quad f(c) = 0$

77. $f(x) = x^3 - x^2 + x - 2, \quad [0, 3], \quad f(c) = 4$

78. $f(x) = \dfrac{x^2 + x}{x - 1}, \quad \left[\dfrac{5}{2}, 4\right], \quad f(c) = 6$

WRITING ABOUT CONCEPTS

79. State how continuity is destroyed at $x = c$ for each of the following graphs.

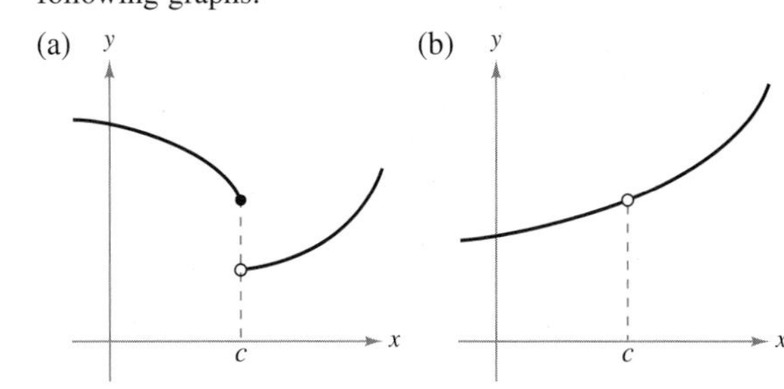

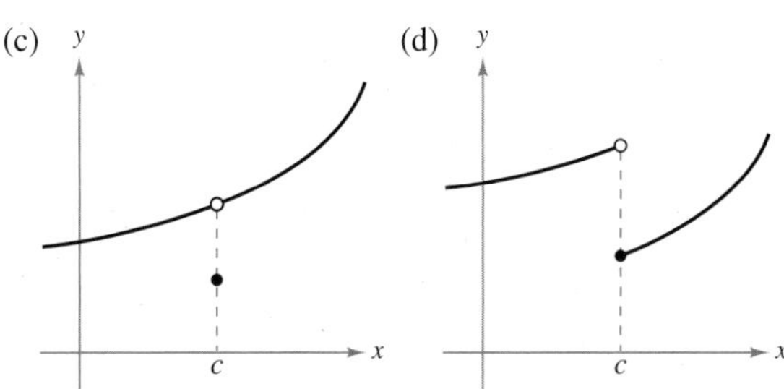

80. Sketch the graph of any function f such that

$$\lim_{x\to 3^+} f(x) = 1$$

and

$$\lim_{x\to 3^-} f(x) = 0.$$

Is the function continuous at $x = 3$? Explain.

81. If the functions f and g are continuous for all real x, is $f + g$ always continuous for all real x? Is f/g always continuous for all real x? If either is not continuous, give an example to verify your conclusion.

CAPSTONE

82. Describe the difference between a discontinuity that is removable and one that is nonremovable. In your explanation, give examples of the following descriptions.

(a) A function with a nonremovable discontinuity at $x = 4$

(b) A function with a removable discontinuity at $x = -4$

(c) A function that has both of the characteristics described in parts (a) and (b)

True or False? **In Exercises 83–86, determine whether the statement is true or false. If it is false, explain why or give an example that shows it is false.**

83. If $\lim_{x \to c} f(x) = L$ and $f(c) = L$, then f is continuous at c.

84. If $f(x) = g(x)$ for $x \neq c$ and $f(c) \neq g(c)$, then either f or g is not continuous at c.

85. A rational function can have infinitely many x-values at which it is not continuous.

86. The function $f(x) = |x - 1|/(x - 1)$ is continuous on $(-\infty, \infty)$.

87. ***Think About It*** Describe how the functions

$$f(x) = 3 + [\![x]\!]$$

and

$$g(x) = 3 - [\![-x]\!]$$

differ.

88. ***Telephone Charges*** A long distance phone service charges \$0.40 for the first 10 minutes and \$0.05 for each additional minute or fraction thereof. Use the greatest integer function to write the cost C of a call in terms of time t (in minutes). Sketch the graph of this function and discuss its continuity.

89. ***Inventory Management*** The number of units in inventory in a small company is given by

$$N(t) = 25\left(2\left[\!\!\left[\frac{t+2}{2}\right]\!\!\right] - t\right)$$

where t is the time in months. Sketch the graph of this function and discuss its continuity. How often must this company replenish its inventory?

90. ***Déjà Vu*** At 8:00 A.M. on Saturday, a man begins running up the side of a mountain to his weekend campsite (see figure). On Sunday morning at 8:00 A.M., he runs back down the mountain. It takes him 20 minutes to run up, but only 10 minutes to run down. At some point on the way down, he realizes that he passed the same place at exactly the same time on Saturday. Prove that he is correct. [*Hint*: Let $s(t)$ and $r(t)$ be the position functions for the runs up and down, and apply the Intermediate Value Theorem to the function $f(t) = s(t) - r(t)$.]

Saturday 8:00 A.M. Sunday 8:00 A.M.

91. ***Volume*** Use the Intermediate Value Theorem to show that for all spheres with radii in the interval $[5, 8]$, there is one with a volume of 1500 cubic centimeters.

92. Prove that if f is continuous and has no zeros on $[a, b]$, then either $f(x) > 0$ for all x in $[a, b]$ or $f(x) < 0$ for all x in $[a, b]$.

93. Show that the **Dirichlet function**

$$f(x) = \begin{cases} 0, & \text{if } x \text{ is rational} \\ 1, & \text{if } x \text{ is irrational} \end{cases}$$

is not continuous at any real number.

94. ***Modeling Data*** The table lists the speeds S (in feet per second) of a falling object at various times t (in seconds).

t	0	5	10	15	20	25	30
S	0	48.2	53.5	55.2	55.9	56.2	56.3

(a) Create a line graph of the data.

(b) Does there appear to be a limiting speed of the object? If there is a limiting speed, identify a possible cause.

95. ***Creating Models*** A swimmer crosses a pool of width b by swimming in a straight line from $(0, 0)$ to $(2b, b)$. (See figure.)

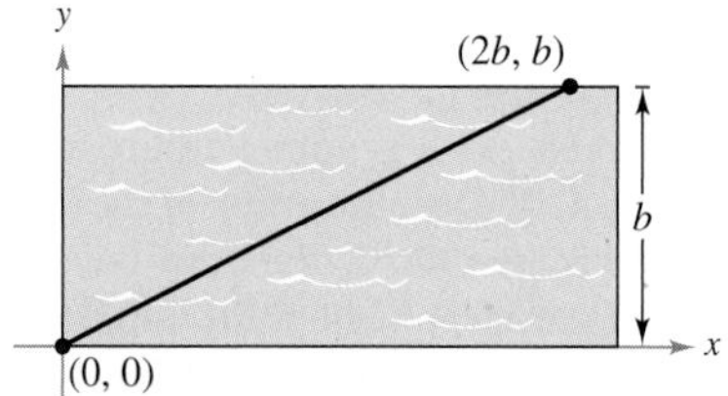

(a) Let f be a function defined as the y-coordinate of the point on the long side of the pool that is nearest the swimmer at any given time during the swimmer's crossing of the pool. Determine the function f and sketch its graph. Is f continuous? Explain.

(b) Let g be the minimum distance between the swimmer and the long sides of the pool. Determine the function g and sketch its graph. Is g continuous? Explain.

96. Discuss the continuity of the function $h(x) = x[\![x]\!]$.

97. Let

$$f(x) = \left(\sqrt{x + c^2} - c\right)/x,\ c > 0.$$

What is the domain of f? How can you define f at $x = 0$ in order for f to be continuous there?

98. Let $f_1(x)$ and $f_2(x)$ be continuous on the closed interval $[a, b]$. If $f_1(a) < f_2(a)$ and $f_1(b) > f_2(b)$, prove that there exists c between a and b such that $f_1(c) = f_2(c)$.

3.5 Infinite Limits

- **Determine infinite limits from the left and from the right.**
- **Find and sketch the vertical asymptotes of the graph of a function.**

Infinite Limits

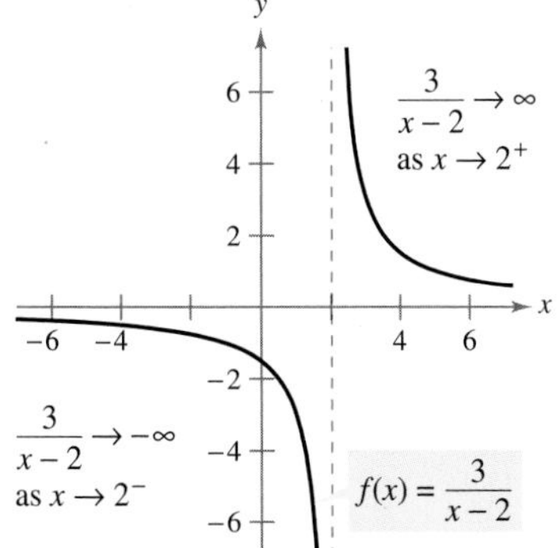

$f(x)$ increases and decreases without bound as x approaches 2.
Figure 3.34

Let f be the function given by $3/(x - 2)$. From Figure 3.34 and the table, you can see that $f(x)$ *decreases without bound* as x approaches 2 from the left, and $f(x)$ *increases without bound* as x approaches 2 from the right. This behavior is denoted as

$$\lim_{x\to 2^-} \frac{3}{x-2} = -\infty \qquad f(x) \text{ decreases without bound as } x \text{ approaches 2 from the left.}$$

and

$$\lim_{x\to 2^+} \frac{3}{x-2} = \infty \qquad f(x) \text{ increases without bound as } x \text{ approaches 2 from the right.}$$

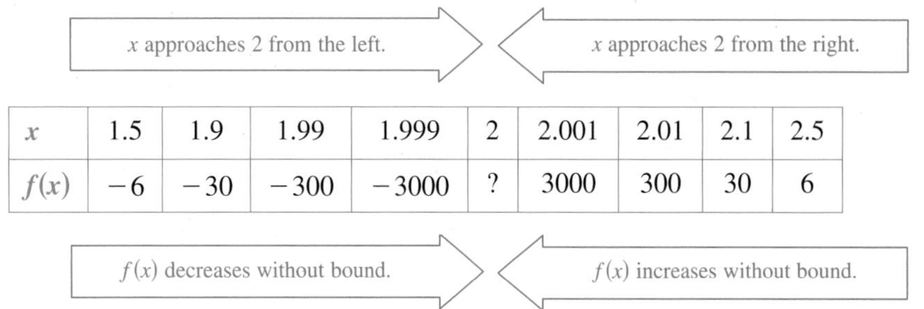

x	1.5	1.9	1.99	1.999	2	2.001	2.01	2.1	2.5
$f(x)$	−6	−30	−300	−3000	?	3000	300	30	6

$f(x)$ decreases without bound. $f(x)$ increases without bound.

A limit in which $f(x)$ increases or decreases without bound as x approaches c is called an **infinite limit.**

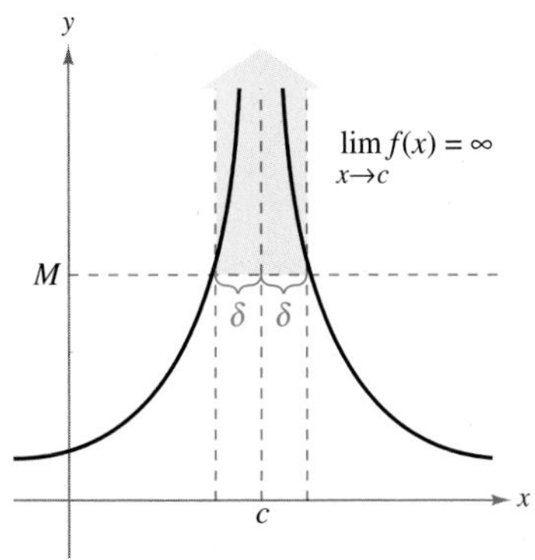

Infinite limits
Figure 3.35

DEFINITION OF INFINITE LIMITS

Let f be a function that is defined at every real number in some open interval containing c (except possibly at c itself). The statement

$$\lim_{x\to c} f(x) = \infty$$

means that for each $M > 0$ there exists a $\delta > 0$ such that $f(x) > M$ whenever $0 < |x - c| < \delta$ (see Figure 3.35). Similarly, the statement

$$\lim_{x\to c} f(x) = -\infty$$

means that for each $N < 0$ there exists a $\delta > 0$ such that $f(x) < N$ whenever $0 < |x - c| < \delta$.

To define the **infinite limit from the left,** replace $0 < |x - c| < \delta$ by $c - \delta < x < c$. To define the **infinite limit from the right,** replace $0 < |x - c| < \delta$ by $c < x < c + \delta$.

NOTE Be sure you see that the equal sign in the statement $\lim_{x\to c} f(x) = \infty$ does not mean that the limit exists! On the contrary, it tells you how the limit *fails to exist* by denoting the unbounded behavior of $f(x)$ as x approaches c. ■

EXPLORATION

Use a graphing utility to graph each function. For each function, analytically find the single real number c that is not in the domain. Then graphically find the limit (if it exists) of $f(x)$ as x approaches c from the left and from the right.

a. $f(x) = \dfrac{3}{x - 4}$

b. $f(x) = \dfrac{1}{2 - x}$

c. $f(x) = \dfrac{2}{(x - 3)^2}$

d. $f(x) = \dfrac{-3}{(x + 2)^2}$

EXAMPLE 1 Determining Infinite Limits from a Graph

Determine the limit of each function shown in Figure 3.36 as x approaches 1 from the left and from the right.

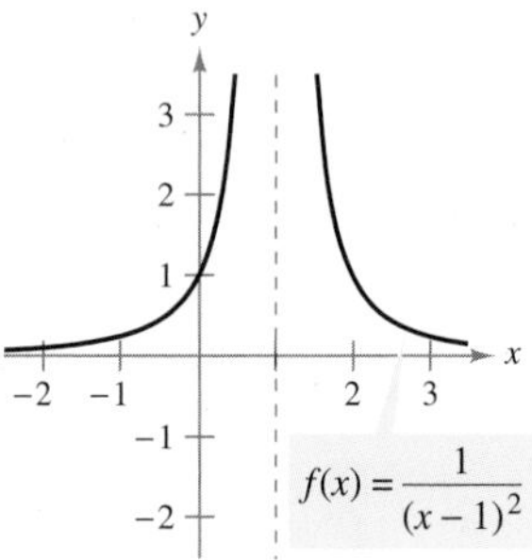

(a)

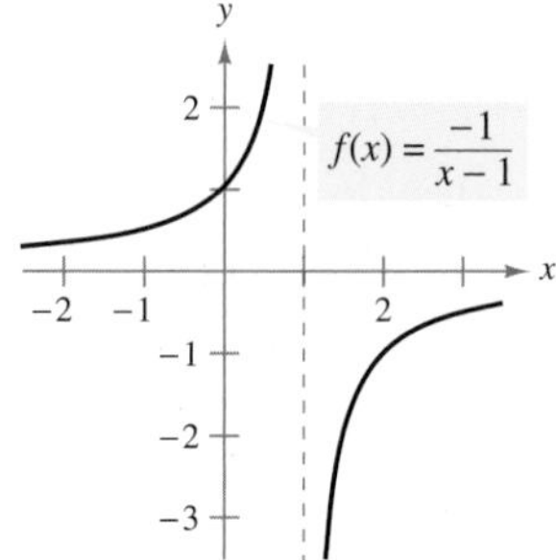

(b)

Each graph has an asymptote at $x = 1$.

Figure 3.36

Solution

a. When x approaches 1 from the left or the right, $(x - 1)^2$ is a small positive number. Thus, the quotient $1/(x - 1)^2$ is a large positive number and $f(x)$ approaches infinity from each side of $x = 1$. So, you can conclude that

$$\lim_{x \to 1} \frac{1}{(x - 1)^2} = \infty. \qquad \text{Limit from each side is infinity.}$$

Figure 3.36(a) confirms this analysis.

b. When x approaches 1 from the left, $x - 1$ is a small negative number. Thus, the quotient $-1/(x - 1)$ is a large positive number and $f(x)$ approaches infinity from the left of $x = 1$. So, you can conclude that

$$\lim_{x \to 1^-} \frac{-1}{x - 1} = \infty. \qquad \text{Limit from the left side is infinity.}$$

When x approaches 1 from the right, $x - 1$ is a small positive number. Thus, the quotient $-1/(x - 1)$ is a large negative number and $f(x)$ approaches negative infinity from the right of $x = 1$. So, you can conclude that

$$\lim_{x \to 1^+} \frac{-1}{x - 1} = -\infty. \qquad \text{Limit from the right side is negative infinity.}$$

Figure 3.36(b) confirms this analysis. ■

Vertical Asymptotes

In Section 2.6, you studied **vertical asymptotes** of graphs of rational functions. The definition of a vertical asymptote is reviewed below.

DEFINITION OF VERTICAL ASYMPTOTE

If $f(x)$ approaches infinity (or negative infinity) as x approaches c from the right or the left, then the line

$$x = c$$

is a **vertical asymptote** of the graph of f.

NOTE If the graph of a function f has a vertical asymptote at $x = c$, then f is *not continuous* at c.

In Example 1, note that each of the functions is a *quotient* and that the vertical asymptote occurs at a number at which the denominator is 0 (and the numerator is not 0). The next theorem generalizes this observation. (A proof of this theorem is given in Appendix A.)

THEOREM 3.12 VERTICAL ASYMPTOTES

Let f and g be continuous on an open interval containing c. If $f(c) \neq 0$, $g(c) = 0$, and there exists an open interval containing c such that $g(x) \neq 0$ for all $x \neq c$ in the interval, then the graph of the function given by

$$h(x) = \frac{f(x)}{g(x)}$$

has a vertical asymptote at $x = c$.

EXAMPLE 2 Finding Vertical Asymptotes

Determine all vertical asymptotes of the graph of each function.

a. $f(x) = \dfrac{1}{2(x+1)}$

b. $f(x) = \dfrac{x^2+1}{x^2-1}$

c. $f(x) = \dfrac{x^2-1}{x-2}$

Solution

a. When $x = -1$, the denominator of

$$f(x) = \frac{1}{2(x+1)}$$

is 0 and the numerator is not 0. So, by Theorem 3.12, you can conclude that $x = -1$ is a vertical asymptote, as shown in Figure 3.37(a).

b. By factoring the denominator as

$$f(x) = \frac{x^2+1}{x^2-1} = \frac{x^2+1}{(x-1)(x+1)}$$

you can see that the denominator is 0 at $x = -1$ and $x = 1$. Moreover, because the numerator is not 0 at these two points, you can apply Theorem 3.12 to conclude that the graph of f has two vertical asymptotes, as shown in Figure 3.37(b).

c. When $x = 2$, the denominator of

$$f(x) = \frac{x^2-1}{x-2}$$

is 0 and the numerator is not 0. So, by Theorem 3.12, you can conclude that $x = 2$ is a vertical asymptote, as shown in Figure 3.37(c). ■

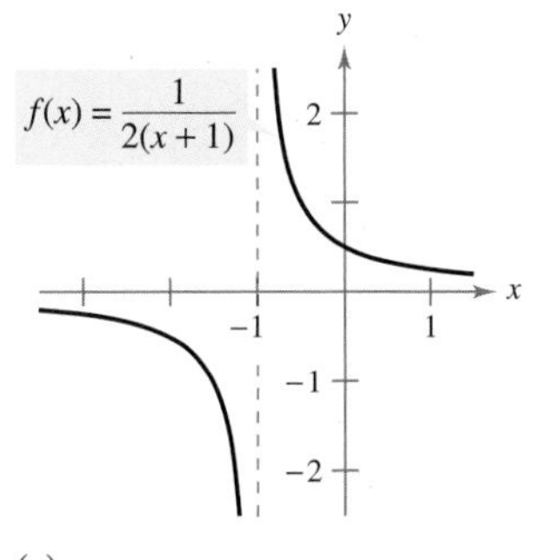

(a)

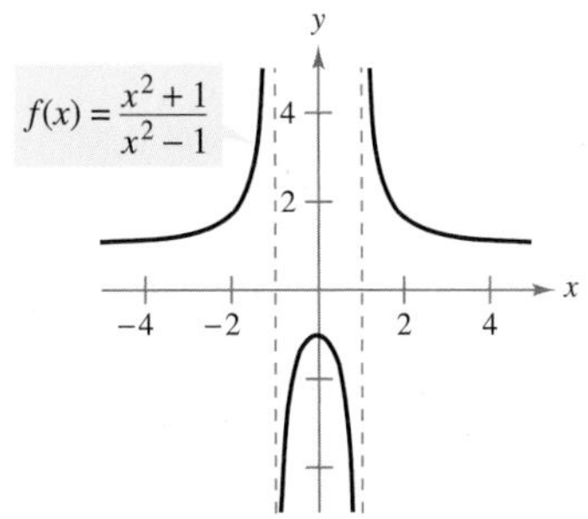

(b)

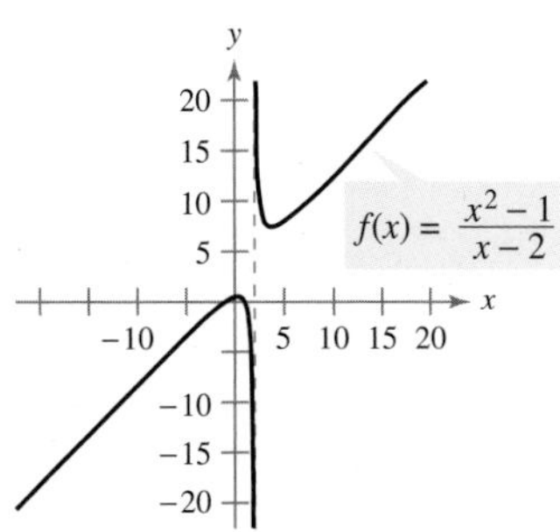

(c)
Functions with vertical asymptotes
Figure 3.37

Theorem 3.12 requires that the value of the numerator at $x = c$ be nonzero. If both the numerator and the denominator are 0 at $x = c$, you obtain the *indeterminate form* $0/0$, and you cannot determine the limit behavior at $x = c$ without further investigation, as illustrated in Example 3. Refer to Example 6 in Section 3.3 to review how to evaluate this indeterminate form.

EXAMPLE 3 A Rational Function with Common Factors

Determine all vertical asymptotes of the graph of

$$f(x) = \frac{x^2 + 2x - 8}{x^2 - 4}.$$

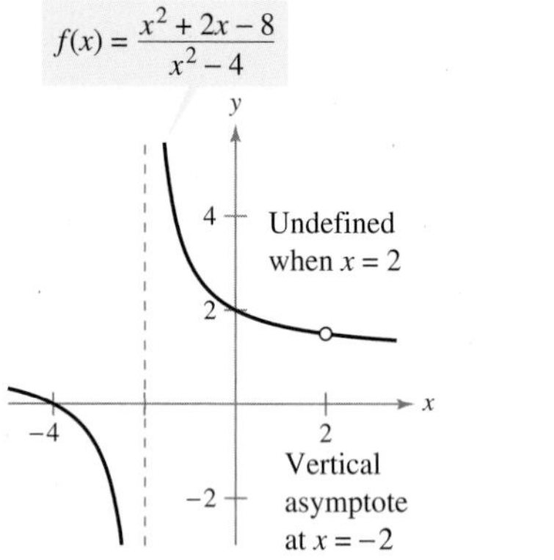

Solution Begin by simplifying the expression, as shown.

$$\begin{aligned} f(x) &= \frac{x^2 + 2x - 8}{x^2 - 4} \\ &= \frac{(x + 4)\cancel{(x - 2)}}{(x + 2)\cancel{(x - 2)}} \\ &= \frac{x + 4}{x + 2}, \quad x \neq 2 \end{aligned}$$

$f(x) = \dfrac{x^2 + 2x - 8}{x^2 - 4}$

Undefined when $x = 2$

Vertical asymptote at $x = -2$

$f(x)$ increases and decreases without bound as x approaches -2.
Figure 3.38

At all x-values other than $x = 2$, the graph of f coincides with the graph of $g(x) = (x + 4)/(x + 2)$. So, you can apply Theorem 3.12 to g to conclude that there is a vertical asymptote at $x = -2$, as shown in Figure 3.38. From the graph, you can see that

$$\lim_{x \to -2^-} \frac{x^2 + 2x - 8}{x^2 - 4} = -\infty \quad \text{and} \quad \lim_{x \to -2^+} \frac{x^2 + 2x - 8}{x^2 - 4} = \infty.$$

Note that $x = 2$ is *not* a vertical asymptote.

EXAMPLE 4 Determining Infinite Limits

Find each limit.

$$\lim_{x \to 1^-} \frac{x^2 - 3x}{x - 1} \quad \text{and} \quad \lim_{x \to 1^+} \frac{x^2 - 3x}{x - 1}$$

Solution Because the denominator is 0 when $x = 1$ (and the numerator is not zero), you know that the graph of

$$f(x) = \frac{x^2 - 3x}{x - 1}$$

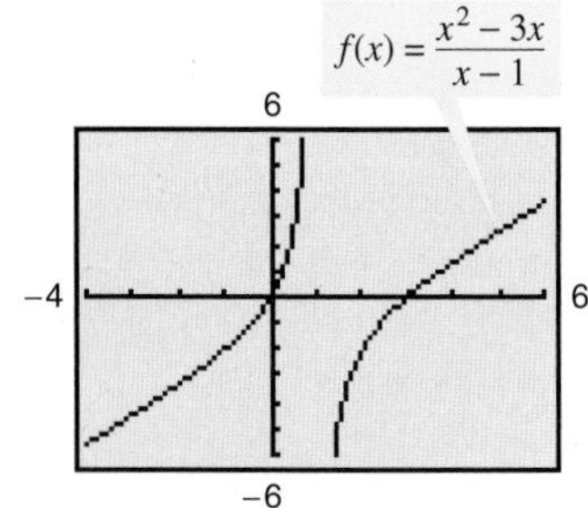

f has a vertical asymptote at $x = 1$.
Figure 3.39

has a vertical asymptote at $x = 1$. This means that each of the given limits is either ∞ or $-\infty$. You can determine the result by analyzing f at values of x close to 1, or by using a graphing utility. From the graph of f shown in Figure 3.39, you can see that the graph approaches ∞ from the left of $x = 1$ and approaches $-\infty$ from the right of $x = 1$. So, you can conclude that

$$\lim_{x \to 1^-} \frac{x^2 - 3x}{x - 1} = \infty \qquad \text{The limit from the left is infinity.}$$

and

$$\lim_{x \to 1^+} \frac{x^2 - 3x}{x - 1} = -\infty. \qquad \text{The limit from the right is negative infinity.}$$

■

TECHNOLOGY PITFALL When using a graphing calculator or graphing software, be careful to interpret correctly the graph of a function with a vertical asymptote—graphing utilities often have difficulty drawing this type of graph.

THEOREM 3.13 PROPERTIES OF INFINITE LIMITS

Let c and L be real numbers and let f and g be functions such that

$$\lim_{x \to c} f(x) = \infty \quad \text{and} \quad \lim_{x \to c} g(x) = L.$$

1. Sum or difference: $\lim_{x \to c} [f(x) \pm g(x)] = \infty$

2. Product: $\lim_{x \to c} [f(x)g(x)] = \infty, \quad L > 0$

$\lim_{x \to c} [f(x)g(x)] = -\infty, \quad L < 0$

3. Quotient: $\lim_{x \to c} \dfrac{g(x)}{f(x)} = 0$

Similar properties hold for one-sided limits and for functions for which the limit of $f(x)$ as x approaches c is $-\infty$.

PROOF To show that the limit of $f(x) + g(x)$ is infinite, choose $M > 0$. You then need to find $\delta > 0$ such that

$$[f(x) + g(x)] > M$$

whenever $0 < |x - c| < \delta$. For simplicity's sake, you can assume L is positive. Let $M_1 = M + 1$. Because the limit of $f(x)$ is infinite, there exists δ_1 such that $f(x) > M_1$ whenever $0 < |x - c| < \delta_1$. Also, because the limit of $g(x)$ is L, there exists δ_2 such that $|g(x) - L| < 1$ whenever $0 < |x - c| < \delta_2$. By letting δ be the smaller of δ_1 and δ_2, you can conclude that $0 < |x - c| < \delta$ implies $f(x) > M + 1$ and $|g(x) - L| < 1$. The second of these two inequalities implies that $g(x) > L - 1$, and, adding this to the first inequality, you can write

$$f(x) + g(x) > (M + 1) + (L - 1) = M + L > M.$$

So, you can conclude that

$$\lim_{x \to c} [f(x) + g(x)] = \infty.$$

The proofs of the remaining properties are left as exercises (see Exercise 72). ■

EXAMPLE 5 Determining Limits

Find each limit.

a. $\displaystyle \lim_{x \to 0} \left(1 + \frac{1}{x^2}\right)$ **b.** $\displaystyle \lim_{x \to 1^-} \frac{x^2 + 1}{1/(x - 1)}$

Solution

a. Because $\lim_{x \to 0} 1 = 1$ and $\lim_{x \to 0} \dfrac{1}{x^2} = \infty$, you can write

$$\lim_{x \to 0} \left(1 + \frac{1}{x^2}\right) = \infty. \qquad \text{Property 1, Theorem 3.13}$$

b. Because $\lim_{x \to 1^-} (x^2 + 1) = 2$ and $\lim_{x \to 1^-} [1/(x - 1)] = -\infty$, you can write

$$\lim_{x \to 1^-} \frac{x^2 + 1}{1/(x - 1)} = 0. \qquad \text{Property 3, Theorem 3.13}$$

■

3.5 Exercises

See www.CalcChat.com for worked-out solutions to odd-numbered exercises.

In Exercises 1–4, determine whether $f(x)$ approaches ∞ or $-\infty$ as x approaches 4 from the left and from the right.

1. $f(x) = \dfrac{1}{x-4}$

2. $f(x) = \dfrac{-1}{x-4}$

3. $f(x) = \dfrac{1}{(x-4)^2}$

4. $f(x) = \dfrac{-1}{(x-4)^2}$

In Exercises 5 and 6, determine whether $f(x)$ approaches ∞ or $-\infty$ as x approaches -2 from the left and from the right.

5. $f(x) = 2\left|\dfrac{x}{x^2-4}\right|$

6. 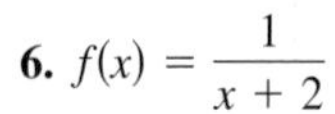 $f(x) = \dfrac{1}{x+2}$

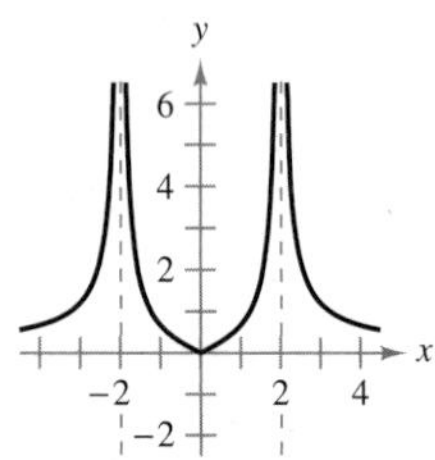

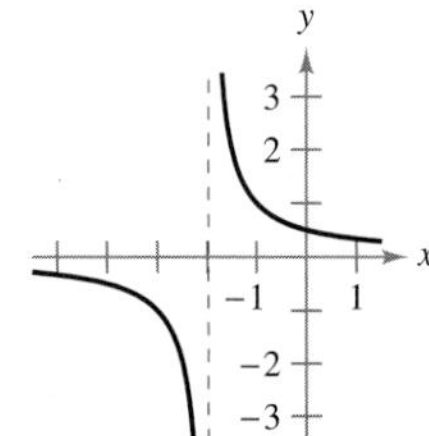

Numerical and Graphical Analysis **In Exercises 7–10, determine whether $f(x)$ approaches ∞ or $-\infty$ as x approaches -3 from the left and from the right by completing the table. Use a graphing utility to graph the function to confirm your answer.**

x	-3.5	-3.1	-3.01	-3.001
$f(x)$				

x	-2.999	-2.99	-2.9	-2.5
$f(x)$				

7. $f(x) = \dfrac{1}{x^2-9}$

8. $f(x) = \dfrac{x}{x^2-9}$

9. $f(x) = \dfrac{x^2}{x^2-9}$

10. $f(x) = \dfrac{x^3}{x^2-9}$

In Exercises 11 and 12, find the vertical asymptotes of the graph of the function.

11. $f(x) = \dfrac{x^2-2}{x^2-x-2}$

12. 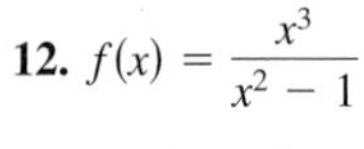 $f(x) = \dfrac{x^3}{x^2-1}$

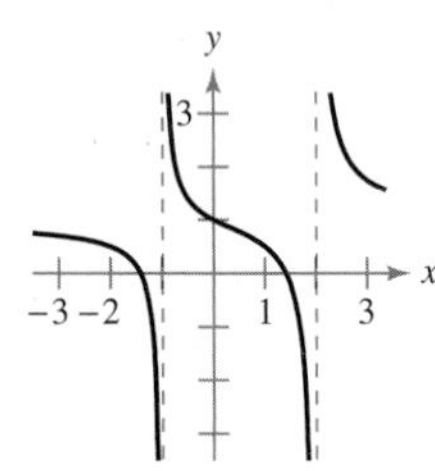

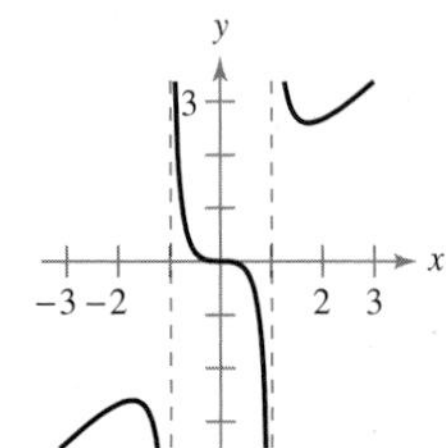

In Exercises 13–28, find the vertical asymptotes (if any) of the graph of the function.

13. $f(x) = \dfrac{1}{x^2}$

14. $f(x) = \dfrac{4}{(x-2)^3}$

15. $f(x) = \dfrac{x^2}{x^2-4}$

16. $f(x) = \dfrac{-4x}{x^2+4}$

17. $g(t) = \dfrac{t-1}{t^2+1}$

18. $h(s) = \dfrac{2s-3}{s^2-25}$

19. $h(x) = \dfrac{x^2-2}{x^2-x-6}$

20. $g(x) = \dfrac{2+x}{x^2(1-x)}$

21. $T(t) = 1 - \dfrac{4}{t^2}$

22. $g(x) = \dfrac{\frac{1}{2}x^3 - x^2 - 4x}{3x^2-6x-24}$

23. $f(x) = \dfrac{3}{x^2+x-2}$

24. $f(x) = \dfrac{4x^2+4x-24}{x^4-2x^3-9x^2+18x}$

25. $g(x) = \dfrac{x^3+1}{x+1}$

26. $h(x) = \dfrac{x^2-4}{x^3+2x^2+x+2}$

27. $f(x) = \dfrac{x^2-2x-15}{x^3-5x^2+x-5}$

28. $h(t) = \dfrac{t^2-2t}{t^4-16}$

In Exercises 29–32, determine whether the graph of the function has a vertical asymptote or a removable discontinuity at $x = -1$. Graph the function using a graphing utility to confirm your answer.

29. $f(x) = \dfrac{x^2-1}{x+1}$

30. $f(x) = \dfrac{x^2-6x-7}{x+1}$

31. $f(x) = \dfrac{x^2+1}{x+1}$

32. $f(x) = \dfrac{x-1}{x+1}$

In Exercises 33–44, find the limit (if it exists).

33. $\lim_{x\to -1^+} \dfrac{1}{x+1}$

34. $\lim_{x\to 1^-} \dfrac{-1}{(x-1)^2}$

35. $\lim_{x\to 2^+} \dfrac{x}{x-2}$

36. $\lim_{x\to 1^+} \dfrac{2+x}{1-x}$

37. $\lim_{x\to 1^+} \dfrac{x^2}{(x-1)^2}$

38. $\lim_{x\to 4^-} \dfrac{x^2}{x^2+16}$

39. $\lim_{x\to -3^-} \dfrac{x+3}{x^2+x-6}$

40. $\lim_{x\to(-1/2)^+} \dfrac{6x^2+x-1}{4x^2-4x-3}$

41. $\lim_{x\to 1} \dfrac{x-1}{(x^2+1)(x-1)}$

42. $\lim_{x\to 3} \dfrac{x-2}{x^2}$

43. $\lim_{x\to 0^-} \left(1+\dfrac{1}{x}\right)$

44. $\lim_{x\to 0^-} \left(x^2-\dfrac{1}{x}\right)$

In Exercises 45–50, find the indicated limit (if it exists), given that

$$f(x) = \frac{1}{(x-4)^2}$$

and

$$g(x) = x^2 - 5x.$$

45. $\lim_{x\to 4} f(x)$

46. $\lim_{x\to 4} g(x)$

47. $\lim_{x\to 4} [f(x)+g(x)]$

48. $\lim_{x\to 4} [f(x)g(x)]$

49. $\lim_{x\to 4} \left[\dfrac{f(x)}{g(x)}\right]$

50. $\lim_{x\to 4} \left[\dfrac{g(x)}{f(x)}\right]$

In Exercises 51–54, use a graphing utility to graph the function and determine the one-sided limit.

51. $f(x) = \dfrac{x^2+x+1}{x^3-1}$

$\lim_{x\to 1^+} f(x)$

52. $f(x) = \dfrac{x^3-1}{x^2+x+1}$

$\lim_{x\to 1^-} f(x)$

53. $f(x) = \dfrac{1}{x^2-25}$

$\lim_{x\to 5^-} f(x)$

54. $f(x) = \dfrac{6-x}{\sqrt{x-3}}$

$\lim_{x\to 3^+} f(x)$

WRITING ABOUT CONCEPTS

55. In your own words, describe the meaning of an infinite limit. Is ∞ a real number?

56. In your own words, describe what is meant by an asymptote of a graph.

57. Write a rational function with vertical asymptotes at $x = 6$ and $x = -2$, and with a zero at $x = 3$.

58. Does the graph of every rational function have a vertical asymptote? Explain.

WRITING ABOUT CONCEPTS (continued)

59. Use the graph of the function f (see figure) to sketch the graph of $g(x) = 1/f(x)$ on the interval $[-2, 3]$. To print an enlarged copy of the graph, go to the website *www.mathgraphs.com.*

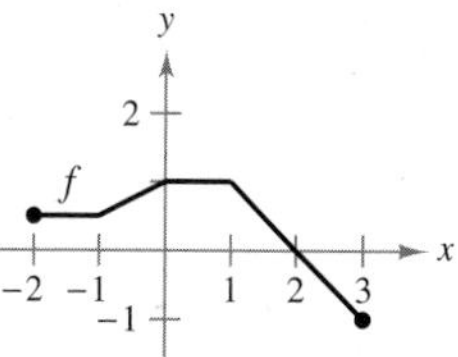

CAPSTONE

60. Given a polynomial $p(x)$, is it true that the graph of the function given by $f(x) = \dfrac{p(x)}{x-1}$ has a vertical asymptote at $x = 1$? Why or why not?

61. ***Boyle's Law*** For a quantity of gas at a constant temperature, the pressure P is inversely proportional to the volume V. Find the limit of P as $V \to 0^+$.

62. A given sum S is inversely proportional to $1 - r$, where $0 < |r| < 1$. Find the limit as $r \to 1^-$.

63. ***Pollution*** The cost C in dollars of removing p percent of the air pollutants from the stack emission of a utility company that burns coal to generate electricity is given by

$$C = \frac{80{,}000p}{100-p}, \quad 0 \le p < 100.$$

(a) Find the cost of removing 15 percent.

(b) Find the cost of removing 50 percent.

(c) Find the cost of removing 90 percent.

(d) Find the limit of C as $p \to 100^-$ and interpret its meaning.

64. ***Illegal Drugs*** The cost C in millions of dollars for a government agency to seize $x\%$ of an illegal drug is given by

$$C = \frac{528x}{100-x}, \quad 0 \le x < 100.$$

(a) Find the cost of seizing 25% of the drug.

(b) Find the cost of seizing 50% of the drug.

(c) Find the cost of seizing 75% of the drug.

(d) Find the limit of C as $x \to 100^-$ and interpret its meaning.

65. ***Relativity*** According to the theory of relativity, the mass m of a particle depends on its velocity v. That is,

$$m = \frac{m_0}{\sqrt{1-(v^2/c^2)}}$$

where m_0 is the mass when the particle is at rest and c is the speed of light. Find the limit of the mass as v approaches c^-.

66. ***Rate of Change*** A 25-foot ladder is leaning against a house (see figure). If the base of the ladder is pulled away from the house at a rate of 2 feet per second, the top will move down the wall at a rate of

$$r = \frac{2x}{\sqrt{625 - x^2}} \text{ ft/sec}$$

where x is the distance between the base of the ladder and the house.

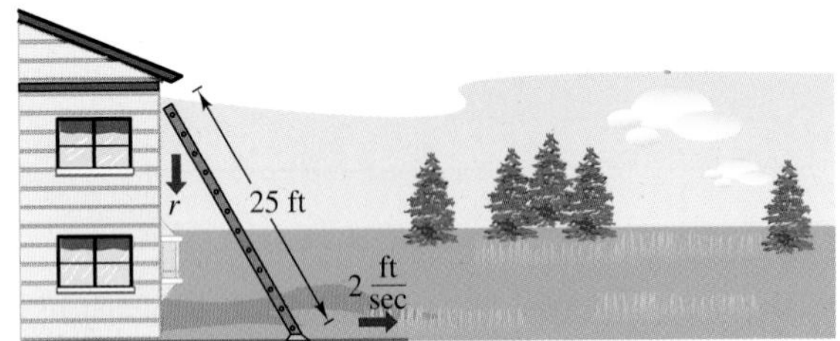

(a) Find the rate r when x is 7 feet.

(b) Find the rate r when x is 15 feet.

(c) Find the limit of r as $x \to 25^-$.

67. ***Average Speed*** On a trip of d miles to another city, a truck driver's average speed was x miles per hour. On the return trip the average speed was y miles per hour. The average speed for the round trip was 50 miles per hour.

(a) Verify that $y = \dfrac{25x}{x - 25}$. What is the domain?

(b) Complete the table.

x	30	40	50	60
y				

Are the values of y different than you expected? Explain.

(c) Find the limit of y as $x \to 25^+$ and interpret its meaning.

68. ***Average Speed*** On the first 150 miles of a 300-mile trip, your average speed is x miles per hour and on the second 150 miles, your average speed is y miles per hour. The average speed for the entire trip is 60 miles per hour.

(a) Write y as a function of x.

(b) If the average speed for the second half of the trip cannot exceed 65 miles per hour, what is the minimum possible average speed for the first half of the trip?

(c) Find the limit of y as $x \to 30^+$.

True or False? **In Exercises 69–71, determine whether the statement is true or false. If it is false, explain why or give an example that shows it is false.**

69. The graph of a rational function has at least one vertical asymptote.

70. The graphs of polynomial functions have no vertical asymptotes.

71. If f has a vertical asymptote at $x = 0$, then f is undefined at $x = 0$.

72. Prove the difference, product, and quotient properties in Theorem 3.13.

73. Prove that if $\lim_{x \to c} f(x) = \infty$, then $\lim_{x \to c} \dfrac{1}{f(x)} = 0$.

74. Prove that if $\lim_{x \to c} \dfrac{1}{f(x)} = 0$, then $\lim_{x \to c} f(x)$ does not exist.

Infinite Limits **In Exercises 75 and 76, use the ε-δ definition of infinite limits to prove the statement.**

75. $\lim_{x \to 3^+} \dfrac{1}{x - 3} = \infty$

76. $\lim_{x \to 5^-} \dfrac{1}{x - 5} = -\infty$

SECTION PROJECT

Graphs and Limits of Functions

Consider the functions given by

$$f(x) = \frac{\sqrt{x^3 - 2x^2 + x}}{|x - 1|} \quad \text{and} \quad g(x) = \frac{\sqrt{x^3 - 2x^2 + x}}{x - 1}.$$

(a) Determine the domains of the functions f and g.

(b) Use a graphing utility to graph the function f on the interval $[0, 9]$. Use the graph to determine if $\lim_{x \to 1} f(x)$ exists. Estimate the limit if it exists.

(c) Explain how you could use a table of values to confirm the value of the limit in part (b) numerically.

(d) Use a graphing utility to graph the function g on the interval $[0, 9]$. Determine if $\lim_{x \to 1} g(x)$ exists. Explain your reasoning.

(e) Verify that $h(x) = \sqrt{x}$ agrees with f for all x except $x = 1$.

(f) Graph the function h by hand. Sketch the tangent line at the point $(1, 1)$ and visually estimate its slope.

(g) Let $(x, \sqrt{x})$ be a point on the graph of h near the point $(1, 1)$, and write a formula for the slope of the secant line joining $(x, \sqrt{x})$ and $(1, 1)$. Evaluate the formula for $x = 1.1$ and $x = 1.01$. Then use limits to determine the exact slope of the tangent line to h at the point $(1, 1)$.

3 CHAPTER SUMMARY

	Review Exercises
Section 3.1	
■ Understand what calculus is and how it compares with precalculus (*p. 214*), understand that the tangent line problem is basic to calculus (*p. 217*), and understand that the area problem is also basic to calculus (*p. 218*).	*1, 2*
Section 3.2	
■ Estimate a limit using a numerical or graphical approach (*p. 220*), and learn different ways that a limit can fail to exist (*p. 222*).	*3–6*
■ Study and use a formal definition of limit (*p. 223*).	*7–10*
Section 3.3	
■ Evaluate a limit using properties of limits (*p. 228*).	*11–14*
■ Develop and use a strategy for finding limits (*p. 231*), and evaluate a limit using dividing out and rationalizing techniques (*p. 232*).	*15–30*
■ Evaluate a limit using the Squeeze Theorem (*p. 233*).	*31, 32*
Section 3.4	
■ Determine continuity at a point and continuity on an open interval (*p. 236*), determine one-sided limits and continuity on a closed interval (*p. 238*), and use properties of continuity (*p. 241*).	*33–53*
■ Understand and use the Intermediate Value Theorem (*p. 242*).	*54*
Section 3.5	
■ Determine infinite limits from the left and from the right (*p. 247*).	*55–62*
■ Find and sketch the vertical asymptotes of the graph of a function (*p. 248*).	*63–67*

3 REVIEW EXERCISES

See www.CalcChat.com for worked-out solutions to odd-numbered exercises.

In Exercises 1 and 2, determine whether the problem can be solved using precalculus or if calculus is required. If the problem can be solved using precalculus, solve it. If the problem seems to require calculus, explain your reasoning. Use a graphical or numerical approach to estimate the solution.

1. Find the distance between the points $(1, 1)$ and $(3, 9)$ along the curve $y = x^2$.
2. Find the distance between the points $(1, 1)$ and $(3, 9)$ along the line $y = 4x - 3$.

In Exercises 3 and 4, complete the table and use the result to estimate the limit. Use a graphing utility to graph the function to confirm your result.

x	-0.1	-0.01	-0.001	0.001	0.01	0.1
$f(x)$						

3. $\lim\limits_{x\to 0} \dfrac{[4/(x+2)] - 2}{x}$

4. $\lim\limits_{x\to 0} \dfrac{4(\sqrt{x+2} - \sqrt{2})}{x}$

In Exercises 5 and 6, use the graph to determine each limit.

5. $h(x) = \dfrac{4x - x^2}{x}$

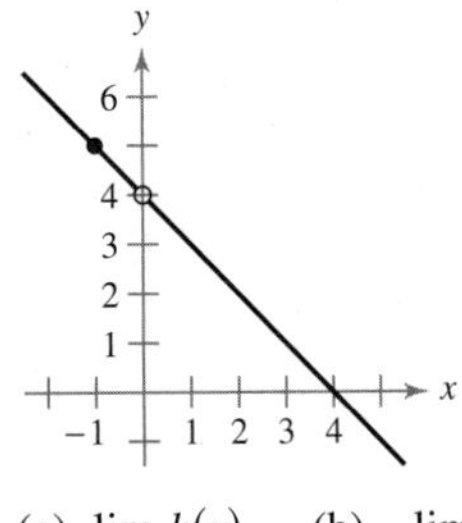

(a) $\lim\limits_{x\to 0} h(x)$ (b) $\lim\limits_{x\to -1} h(x)$

6. $g(x) = \dfrac{-2x}{x - 3}$

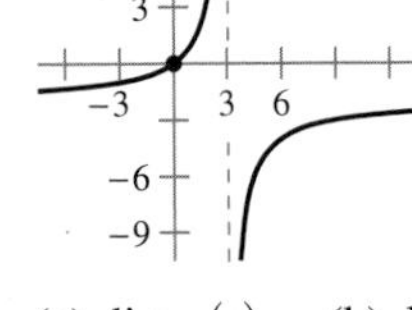

(a) $\lim\limits_{x\to 3} g(x)$ (b) $\lim\limits_{x\to 0} g(x)$

In Exercises 7–10, find the limit L. Then use the ε-δ definition to prove that the limit is L.

7. $\lim\limits_{x\to 1} (x + 4)$
8. $\lim\limits_{x\to 9} \sqrt{x}$
9. $\lim\limits_{x\to 2} (1 - x^2)$
10. $\lim\limits_{x\to 5} 9$

In Exercises 11–14, evaluate the limit given $\lim\limits_{x\to c} f(x) = -\frac{3}{4}$ and $\lim\limits_{x\to c} g(x) = \frac{2}{3}$.

11. $\lim\limits_{x\to c} [f(x)g(x)]$
12. $\lim\limits_{x\to c} \dfrac{f(x)}{g(x)}$
13. $\lim\limits_{x\to c} [f(x) + 2g(x)]$
14. $\lim\limits_{x\to c} [f(x)]^2$

In Exercises 15–26, find the limit (if it exists).

15. $\lim\limits_{x\to 6} (x - 2)^2$
16. $\lim\limits_{x\to 7} (10 - x)^4$
17. $\lim\limits_{t\to 4} \sqrt{t + 2}$
18. $\lim\limits_{y\to 4} 3|y - 1|$
19. $\lim\limits_{t\to -2} \dfrac{t + 2}{t^2 - 4}$
20. $\lim\limits_{t\to 3} \dfrac{t^2 - 9}{t - 3}$
21. $\lim\limits_{x\to 4} \dfrac{\sqrt{x - 3} - 1}{x - 4}$
22. $\lim\limits_{x\to 0} \dfrac{\sqrt{4 + x} - 2}{x}$
23. $\lim\limits_{x\to 0} \dfrac{[1/(x + 1)] - 1}{x}$
24. $\lim\limits_{s\to 0} \dfrac{(1/\sqrt{1 + s}) - 1}{s}$
25. $\lim\limits_{x\to -5} \dfrac{x^3 + 125}{x + 5}$
26. $\lim\limits_{x\to -2} \dfrac{x^2 - 4}{x^3 + 8}$

Numerical, Graphical, and Analytic Analysis **In Exercises 27 and 28, consider**

$$\lim_{x\to 1^+} f(x).$$

(a) Complete the table to estimate the limit.

(b) Use a graphing utility to graph the function and use the graph to estimate the limit.

(c) Rationalize the numerator to find the exact value of the limit analytically.

x	1.1	1.01	1.001	1.0001
$f(x)$				

27. $f(x) = \dfrac{\sqrt{2x + 1} - \sqrt{3}}{x - 1}$

28. $f(x) = \dfrac{1 - \sqrt[3]{x}}{x - 1}$

[*Hint:* $a^3 - b^3 = (a - b)(a^2 + ab + b^2)$]

Free-Falling Object **In Exercises 29 and 30, use the position function $s(t) = -4.9t^2 + 250$, which gives the height (in meters) of an object that has fallen from a height of 250 meters. The velocity at time $t = a$ seconds is given by**

$$\lim_{t\to a} \frac{s(a) - s(t)}{a - t}.$$

29. Find the velocity of the object when $t = 4$.
30. At what velocity will the object impact the ground?

In Exercises 31 and 32, use the Squeeze Theorem to find $\lim\limits_{x\to c} f(x)$.

31. $c = -1;\ 3 - |x + 1| \le f(x) \le 3 + |x + 1|$
32. $c = 0;\ a - x^2 \le f(x) \le a + x^2$

In Exercises 33–38, find the limit (if it exists). If the limit does not exist, explain why.

33. $\lim_{x\to 3^-} \frac{|x-3|}{x-3}$

34. $\lim_{x\to 4} [\![x-1]\!]$

35. $\lim_{x\to 2} f(x)$, where $f(x) = \begin{cases} (x-2)^2, & x \le 2 \\ 2-x, & x > 2 \end{cases}$

36. $\lim_{x\to 1^+} g(x)$, where $g(x) = \begin{cases} \sqrt{1-x}, & x \le 1 \\ x+1, & x > 1 \end{cases}$

37. $\lim_{t\to 1} h(t)$, where $h(t) = \begin{cases} t^3+1, & t < 1 \\ \frac{1}{2}(t+1), & t \ge 1 \end{cases}$

38. $\lim_{s\to -2} f(s)$, where $f(s) = \begin{cases} -s^2-4s-2, & s \le -2 \\ s^2+4s+6, & s > -2 \end{cases}$

In Exercises 39–48, determine the intervals on which the function is continuous.

39. $f(x) = -3x^2 + 7$

40. $f(x) = x^2 - \frac{2}{x}$

41. $f(x) = [\![x+3]\!]$

42. $f(x) = \frac{3x^2-x-2}{x-1}$

43. $f(x) = \begin{cases} \frac{3x^2-x-2}{x-1}, & x \ne 1 \\ 0, & x = 1 \end{cases}$

44. $f(x) = \begin{cases} 5-x, & x \le 2 \\ 2x-3, & x > 2 \end{cases}$

45. $f(x) = \frac{1}{(x-2)^2}$

46. $f(x) = \sqrt{\frac{x+1}{x}}$

47. $f(x) = \frac{3}{x+1}$

48. $f(x) = \frac{x+1}{2x+2}$

49. Determine the value of c such that the function is continuous on the entire real line.

$$f(x) = \begin{cases} x+3, & x \le 2 \\ cx+6, & x > 2 \end{cases}$$

50. Determine the values of b and c such that the function is continuous on the entire real line.

$$f(x) = \begin{cases} x+1, & 1 < x < 3 \\ x^2+bx+c, & |x-2| \ge 1 \end{cases}$$

51. Let $f(x) = \frac{x^2-4}{|x-2|}$. Find each limit (if possible).

(a) $\lim_{x\to 2^-} f(x)$

(b) $\lim_{x\to 2^+} f(x)$

(c) $\lim_{x\to 2} f(x)$

52. Let $f(x) = \sqrt{x(x-1)}$.

(a) Find the domain of f.

(b) Find $\lim_{x\to 0^-} f(x)$.

(c) Find $\lim_{x\to 1^+} f(x)$.

53. ***Delivery Charges*** The cost of sending an overnight package from New York to Atlanta is \$12.80 for the first pound and \$2.50 for each additional pound or fraction thereof. Use the greatest integer function to create a model for the cost C of overnight delivery of a package weighing x pounds. Use a graphing utility to graph the function and discuss its continuity.

54. Use the Intermediate Value Theorem to show that

$$f(x) = 2x^3 - 3$$

has a zero in the interval $[1, 2]$.

In Exercises 55–62, find the one-sided limit (if it exists).

55. $\lim_{x\to -2^-} \frac{2x^2+x+1}{x+2}$

56. $\lim_{x\to (1/2)^+} \frac{x}{2x-1}$

57. $\lim_{x\to -1^+} \frac{x+1}{x^3+1}$

58. $\lim_{x\to -1^-} \frac{x+1}{x^4-1}$

59. $\lim_{x\to 1^-} \frac{x^2+2x+1}{x-1}$

60. $\lim_{x\to -1^+} \frac{x^2-2x+1}{x+1}$

61. $\lim_{x\to 0^+} \left(x - \frac{1}{x^3}\right)$

62. $\lim_{x\to 2^-} \frac{1}{\sqrt[3]{x^2-4}}$

In Exercises 63–66, find the vertical asymptotes (if any) of the graph of the function.

63. $g(x) = 1 + \frac{2}{x}$

64. $h(x) = \frac{4x}{4-x^2}$

65. $f(x) = \frac{8}{(x-10)^2}$

66. $f(x) = \frac{x+3}{x(x^2+1)}$

67. ***Boating*** A boat is pulled into a dock by means of a winch 12 feet above the deck of the boat (see figure). The winch pulls in rope at the rate of 2 feet per second. The rate r at which the boat is moving is given by

$$r = \frac{2L}{\sqrt{L^2-144}}$$

where L is the length of the rope between the winch and the boat.

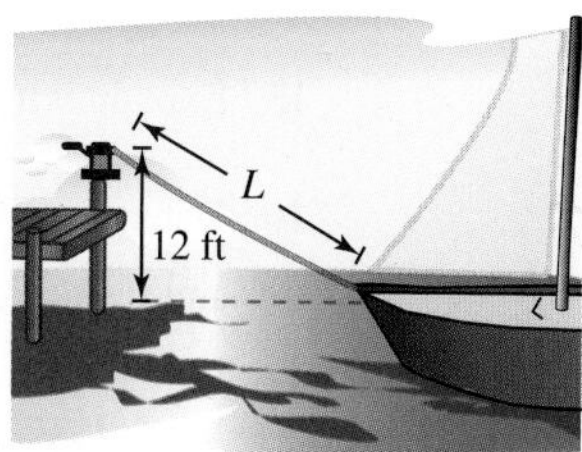

Not drawn to scale

(a) Find r when L is 25 feet.

(b) Find r when L is 13 feet.

(c) Find the limit of r as $L \to 12^+$.

3 CHAPTER TEST

Take this test as you would take a test in class. When you are finished, check your work against the answers given in the back of the book.

x	$f(x)$
1.9	
1.99	
1.999	
2.001	
2.01	
2.1	

Table for 1

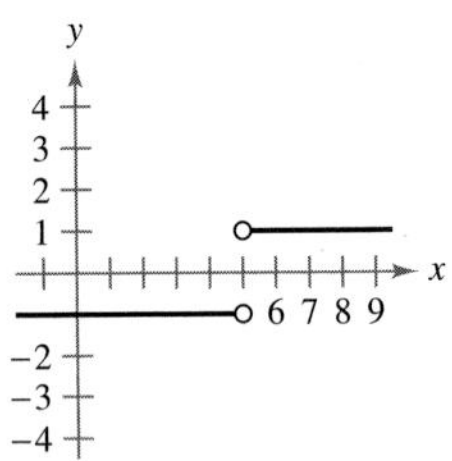

Figure for 2

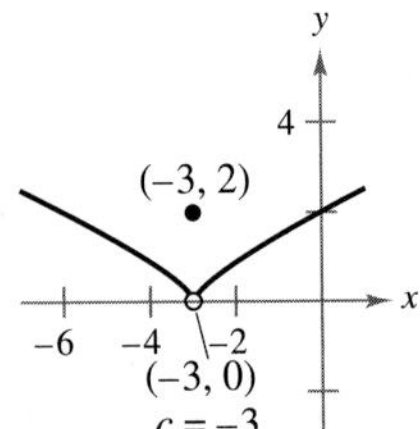

Figure for 11

1. Complete the table and use the result to estimate $\lim_{x\to 2} \dfrac{x-2}{x^2-x-2}$. Use a graphing utility to graph the function to confirm your result.

2. Use the graph at the left to find $\lim_{x\to 5} \dfrac{|x-5|}{x-5}$ (if it exists). If the limit does not exist, explain why.

3. Sketch the graph of $f(x) = \begin{cases} x^2, & x \le 2 \\ 8-2x, & 2 < x < 4 \\ 4, & x \ge 4 \end{cases}$. Then identify the values of c for which $\lim_{x\to c} f(x)$ exists.

4. Find $\lim_{x\to 2} (x+3)$. Then use the ε-δ definition of limit to prove your result.

In Exercises 5–8, find the limit.

5. $\lim_{x\to 5} \sqrt{x+4}$

6. $\lim_{x\to 13} \dfrac{\sqrt{x+3}-4}{x-13}$

7. $\lim_{x\to 4^+} \dfrac{x-4}{x^2-16}$

8. $\lim_{x\to 2} \dfrac{x-2}{(x^2+4)(x-2)}$

9. For the functions $f(x) = 12 - x$ and $g(x) = x^3$, find the limits.

(a) $\lim_{x\to 9} f(x)$ (b) $\lim_{x\to 3} g(x)$ (c) $\lim_{x\to 9} g(f(x))$

10. Use $\lim_{x\to c} f(x) = 2$ and $\lim_{x\to c} g(x) = 5$ to evaluate the limits.

(a) $\lim_{x\to c} [3g(x)]$ (b) $\lim_{x\to c} [f(x) - g(x)]$

(c) $\lim_{x\to c} [f(x)g(x)]$ (d) $\lim_{x\to c} \dfrac{f(x)}{g(x)}$

11. Use the graph at the left to determine the limits, and discuss the continuity of the function.

(a) $\lim_{x\to c^+} f(x)$ (b) $\lim_{x\to c^-} f(x)$ (c) $\lim_{x\to c} f(x)$

12. Discuss the continuity of $f(x) = \begin{cases} 3x-2, & x < 1 \\ 0, & x = 1. \\ x, & x > 1 \end{cases}$

13. Find the x-value at which $f(x) = \dfrac{|x+2|}{x+2}$ is not continuous. Is the discontinuity removable?

14. For the functions $f(x) = \dfrac{1}{x-8}$ and $g(x) = x^2 + 4$, discuss the continuity of the composite function $h(x) = f(g(x))$.

15. Explain why $f(x) = \frac{1}{16}x^4 - x^3 + 3$ has a zero in the interval $[1, 2]$.

16. Determine whether $f(x) = \dfrac{1}{(x-2)^2}$ approaches ∞ or $-\infty$ as x approaches 2 from the left and from the right.

17. Find the vertical asymptotes of the graph of $f(x) = \dfrac{x^2}{x^2+x-6}$.

18. Determine whether the graph of $f(x) = \dfrac{x^2-9}{x+3}$ has a vertical asymptote or a removable discontinuity at $x = -3$. Graph the function using a graphing utility to confirm your answer.

19. Use a graphing utility to graph $f(x) = \dfrac{x^2+2x+4}{x^3-8}$ and determine $\lim_{x\to 2^+} f(x)$.

P.S. PROBLEM SOLVING

1. Let $P(x, y)$ be a point on the parabola $y = x^2$ in the first quadrant. Consider the triangle $\triangle PAO$ formed by P, $A(0, 1)$, and the origin $O(0, 0)$, and the triangle $\triangle PBO$ formed by P, $B(1, 0)$, and the origin.

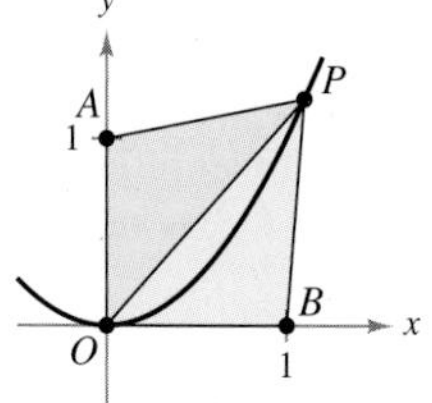

(a) Write the perimeter of each triangle in terms of x.

(b) Let $r(x)$ be the ratio of the perimeters of the two triangles,

$r(x) = \dfrac{\text{Perimeter } \triangle PAO}{\text{Perimeter } \triangle PBO}$. Complete the table.

x	4	2	1	0.1	0.01
Perimeter $\triangle PAO$					
Perimeter $\triangle PBO$					
$r(x)$					

(c) Calculate $\lim_{x \to 0^+} r(x)$.

2. Let $P(x, y)$ be a point on the parabola $y = x^2$ in the first quadrant. Consider the triangle $\triangle PAO$ formed by P, $A(0, 1)$, and the origin $O(0, 0)$, and the triangle $\triangle PBO$ formed by P, $B(1, 0)$, and the origin.

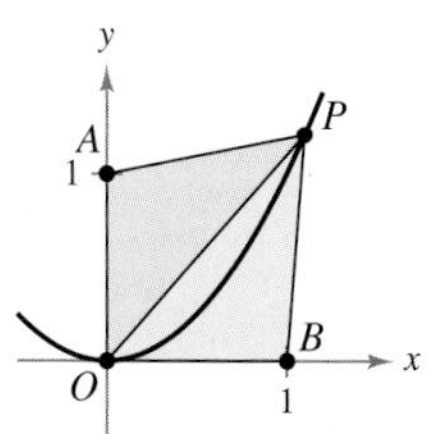

(a) Write the area of each triangle in terms of x.

(b) Let $a(x)$ be the ratio of the areas of the two triangles,

$a(x) = \dfrac{\text{Area } \triangle PBO}{\text{Area } \triangle PAO}$. Complete the table.

x	4	2	1	0.1	0.01
Area $\triangle PAO$					
Area $\triangle PBO$					
$a(x)$					

(c) Calculate $\lim_{x \to 0^+} a(x)$.

3. Let $P(3, 4)$ be a point on the circle $x^2 + y^2 = 25$.

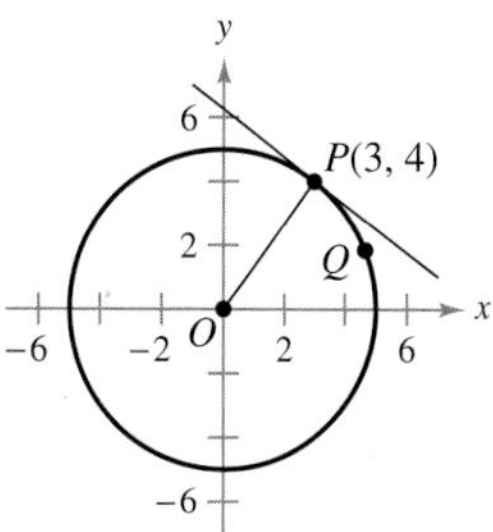

(a) What is the slope of the line joining P and $O(0, 0)$?

(b) Find an equation of the tangent line to the circle at P.

(c) Let $Q(x, y)$ be another point on the circle in the first quadrant. Find the slope m_x of the line joining P and Q in terms of x.

(d) Calculate $\lim_{x \to 3} m_x$. How does this number relate to your answer in part (b)?

4. Let $P(5, -12)$ be a point on the circle $x^2 + y^2 = 169$.

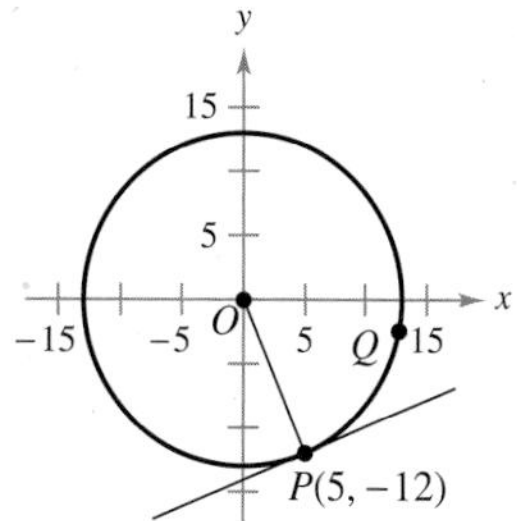

(a) What is the slope of the line joining P and $O(0, 0)$?

(b) Find an equation of the tangent line to the circle at P.

(c) Let $Q(x, y)$ be another point on the circle in the fourth quadrant. Find the slope m_x of the line joining P and Q in terms of x.

(d) Calculate $\lim_{x \to 5} m_x$. How does this number relate to your answer in part (b)?

5. Find the values of the constants a and b such that

$$\lim_{x \to 0} \frac{\sqrt{a + bx} - \sqrt{3}}{x} = \sqrt{3}.$$

6. Consider the function $f(x) = \dfrac{\sqrt{3 + x^{1/3}} - 2}{x - 1}$.

(a) Find the domain of f.

(b) Use a graphing utility to graph the function.

(c) Calculate $\lim_{x \to -27^+} f(x)$.

(d) Calculate $\lim_{x \to 1} f(x)$.

7. Consider the graphs of the four functions g_1, g_2, g_3, and g_4.

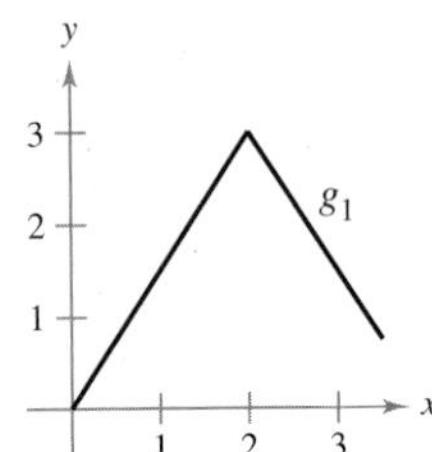

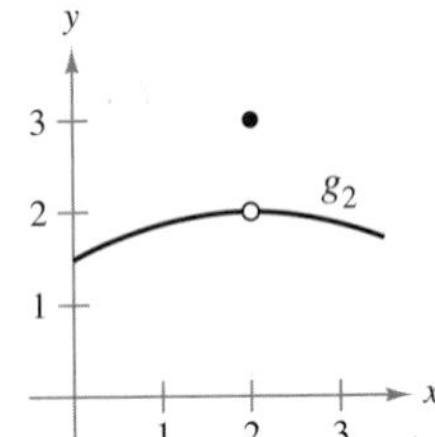

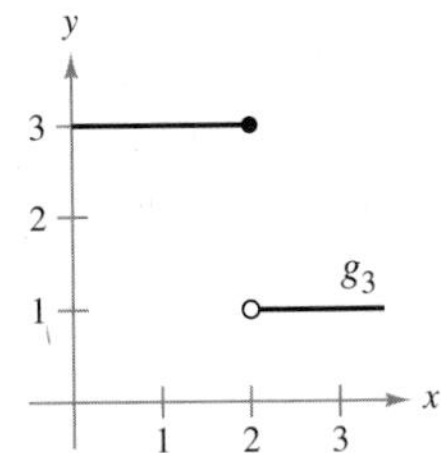

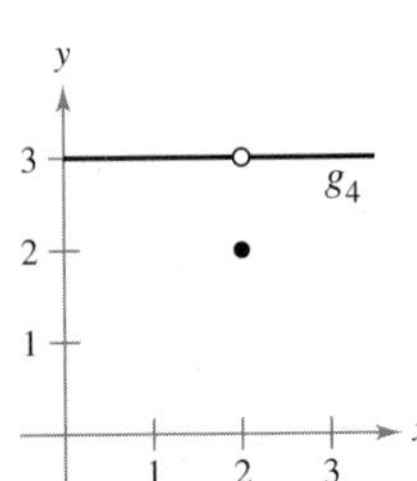

For each given condition of the function f, which of the graphs could be the graph of f?

(a) $\lim_{x\to 2} f(x) = 3$

(b) f is continuous at 2.

(c) $\lim_{x\to 2^-} f(x) = 3$

8. Sketch the graph of the function $f(x) = \left[\!\left[\frac{1}{x}\right]\!\right]$.

(a) Evaluate $f(\frac{1}{4})$, $f(3)$, and $f(1)$.

(b) Evaluate the limits $\lim_{x\to 1^-} f(x)$, $\lim_{x\to 1^+} f(x)$, $\lim_{x\to 0^-} f(x)$, and $\lim_{x\to 0^+} f(x)$.

(c) Discuss the continuity of the function.

9. Sketch the graph of the function $f(x) = [\![x]\!] + [\![-x]\!]$.

(a) Evaluate $f(1)$, $f(0)$, $f(\frac{1}{2})$, and $f(-2.7)$.

(b) Evaluate the limits $\lim_{x\to 1^-} f(x)$, $\lim_{x\to 1^+} f(x)$, and $\lim_{x\to \frac{1}{2}} f(x)$.

(c) Discuss the continuity of the function.

10. To escape Earth's gravitational field, a rocket must be launched with an initial velocity called the **escape velocity.** A rocket launched from the surface of Earth has velocity v (in miles per second) given by

$$v = \sqrt{\frac{2GM}{r} + v_0^2 - \frac{2GM}{R}} \approx \sqrt{\frac{192{,}000}{r} + v_0^2 - 48}$$

where v_0 is the initial velocity, r is the distance from the rocket to the center of Earth, G is the gravitational constant, M is the mass of Earth, and R is the radius of Earth (approximately 4000 miles).

(a) Find the value of v_0 for which you obtain an infinite limit for r as v approaches zero. This value of v_0 is the escape velocity for Earth.

(b) A rocket launched from the surface of the moon has velocity v (in miles per second) given by

$$v = \sqrt{\frac{1920}{r} + v_0^2 - 2.17}.$$

Find the escape velocity for the moon.

(c) A rocket launched from the surface of a planet has velocity v (in miles per second) given by

$$v = \sqrt{\frac{10{,}600}{r} + v_0^2 - 6.99}.$$

Find the escape velocity for this planet. Is the mass of this planet larger or smaller than that of Earth? (Assume that the mean density of this planet is the same as that of Earth.)

11. For positive numbers $a < b$, the **pulse function** is defined as

$$P_{a,b}(x) = H(x-a) - H(x-b) = \begin{cases} 0, & x < a \\ 1, & a \le x < b \\ 0, & x \ge b \end{cases}$$

where $H(x) = \begin{cases} 1, & x \ge 0 \\ 0, & x < 0 \end{cases}$ is the Heaviside function.

(a) Sketch the graph of the pulse function.

(b) Find the following limits:

(i) $\lim_{x\to a^+} P_{a,b}(x)$ (ii) $\lim_{x\to a^-} P_{a,b}(x)$

(iii) $\lim_{x\to b^+} P_{a,b}(x)$ (iv) $\lim_{x\to b^-} P_{a,b}(x)$

(c) Discuss the continuity of the pulse function.

(d) Why is

$$U(x) = \frac{1}{b-a}P_{a,b}(x)$$

called the **unit** pulse function?

12. Let a be a nonzero constant. Prove that if $\lim_{x\to 0} f(x) = L$, then $\lim_{x\to 0} f(ax) = L$. Show by means of an example that a must be nonzero.

4 Differentiation

In this chapter you will study one of the most important processes of calculus–*differentiation*. In each section, you will learn new methods and rules for finding derivatives of functions. Then you will apply these rules to find such things as velocity, acceleration, and the rates of change of two or more related variables.

In this chapter, you should learn the following.

- How to find the derivative of a function using the limit definition and understand the relationship between differentiability and continuity. (4.1)
- How to find the derivative of a function using basic differentiation rules. (4.2)
- How to find the derivative of a function using the Product Rule and the Quotient Rule. (4.3)
- How to find the derivative of a function using the Chain Rule and the General Power Rule. (4.4)
- How to find the derivative of a function using implicit differentiation. (4.5)
- How to find a related rate. (4.6)

Al Bello/Getty Images

When jumping from a platform, a diver's velocity is briefly positive because of the upward movement, but then becomes negative when falling. How can you use calculus to determine the velocity of a diver at impact? (See Section 4.2, Example 9.)

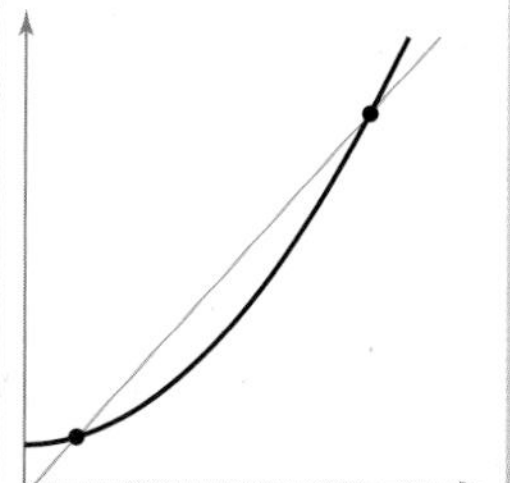

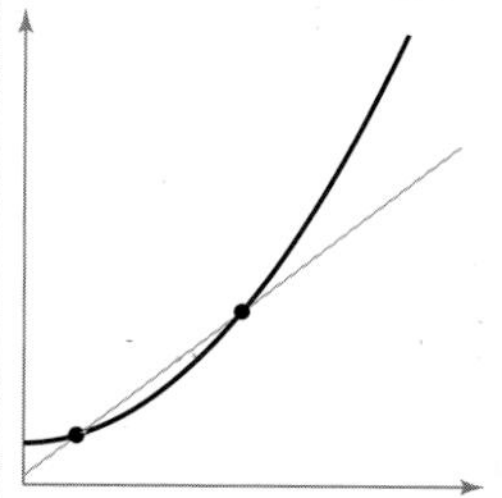

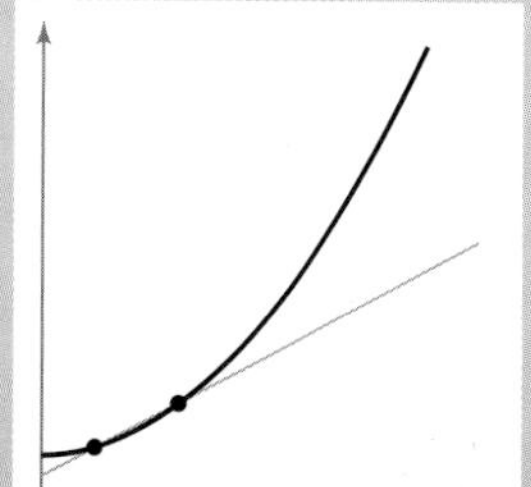

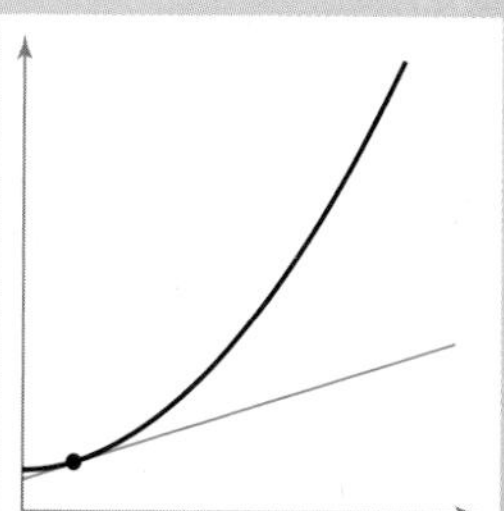

To approximate the slope of a tangent line to a graph at a given point, find the slope of the secant line through the given point and a second point on the graph. As the second point approaches the given point, the approximation tends to become more accurate. (See Section 4.1.)

4.1 The Derivative and the Tangent Line Problem

- **Find the slope of the tangent line to a curve at a point.**
- **Use the limit definition to find the derivative of a function.**
- **Understand the relationship between differentiability and continuity.**

The Tangent Line Problem

ISAAC NEWTON (1642–1727)

In addition to his work in calculus, Newton made revolutionary contributions to physics, including the Law of Universal Gravitation and his three laws of motion.

Calculus grew out of four major problems that European mathematicians were working on during the seventeenth century.

1. The tangent line problem (Section 3.1 and this section)
2. The velocity and acceleration problem (Sections 4.2 and 4.3)
3. The minimum and maximum problem (Section 5.1)
4. The area problem (Sections 3.1 and 6.2)

Each problem involves the notion of a limit, and calculus can be introduced with any of the four problems.

A brief introduction to the tangent line problem is given in Section 3.1. Although partial solutions to this problem were given by Pierre de Fermat (1601–1665), René Descartes (1596–1650), Christian Huygens (1629–1695), and Isaac Barrow (1630–1677), credit for the first general solution is usually given to Isaac Newton (1642–1727) and Gottfried Leibniz (1646–1716). Newton's work on this problem stemmed from his interest in optics and light refraction.

What does it mean to say that a line is tangent to a curve at a point? For a circle, the tangent line at a point P is the line that is perpendicular to the radial line at point P, as shown in Figure 4.1.

For a general curve, however, the problem is more difficult. For example, how would you define the tangent lines shown in Figure 4.2? You might say that a line is tangent to a curve at a point P if it touches, but does not cross, the curve at point P. This definition would work for the first curve shown in Figure 4.2, but not for the second. *Or* you might say that a line is tangent to a curve if the line touches or intersects the curve at exactly one point. This definition would work for a circle but not for more general curves, as the third curve in Figure 4.2 shows.

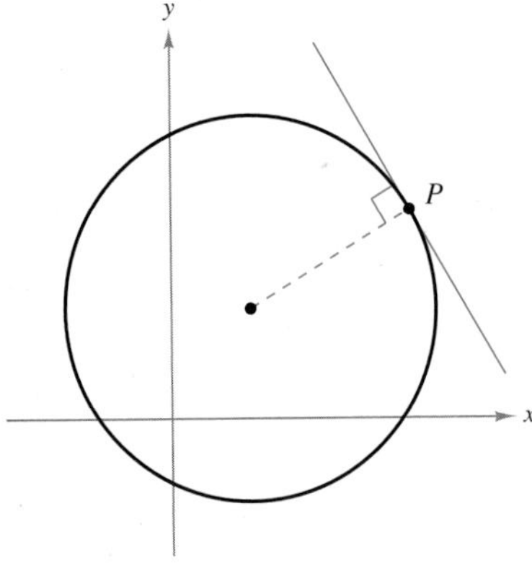

Tangent line to a circle
Figure 4.1

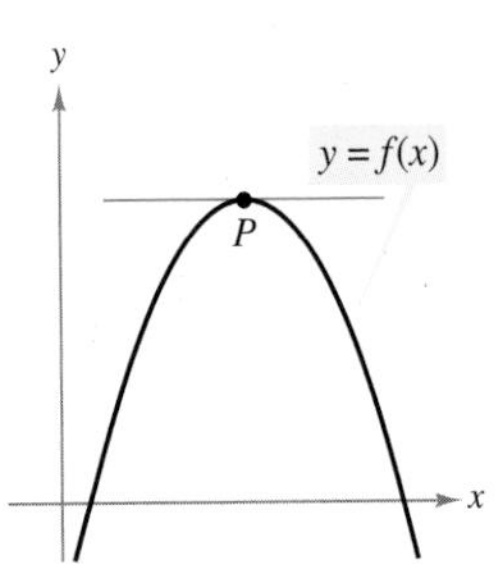

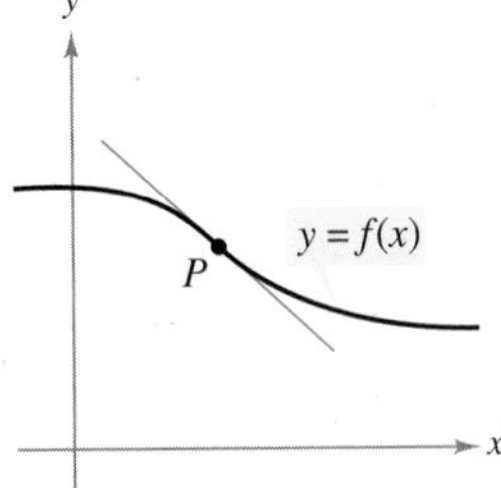

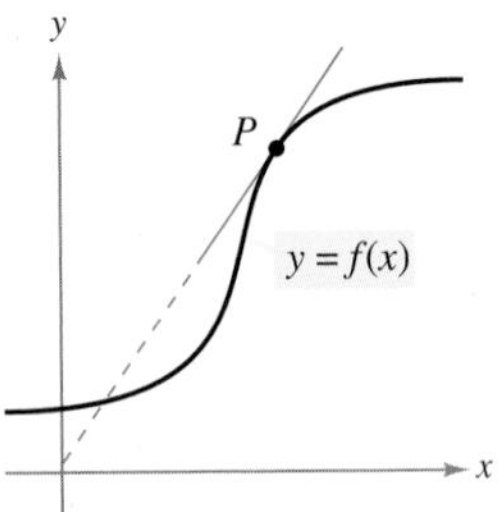

Tangent line to a curve at a point
Figure 4.2

EXPLORATION

Identifying a Tangent Line Use a graphing utility to graph the function $f(x) = 2x^3 - 4x^2 + 3x - 5$. On the same screen, graph $y = x - 5$, $y = 2x - 5$, and $y = 3x - 5$. Which of these lines, if any, appears to be tangent to the graph of f at the point $(0, -5)$? Explain your reasoning.

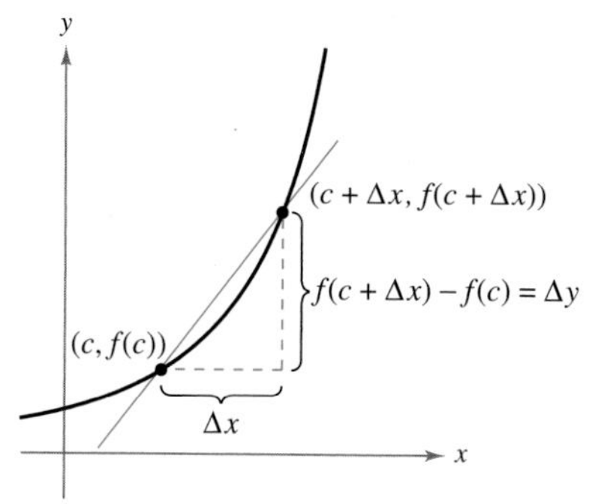

The secant line through $(c, f(c))$ and $(c + \Delta x, f(c + \Delta x))$
Figure 4.3

Essentially, the problem of finding the tangent line at a point P boils down to the problem of finding the *slope* of the tangent line at point P. You can approximate this slope using a **secant line*** through the point of tangency and a second point on the curve, as shown in Figure 4.3. If $(c, f(c))$ is the point of tangency and $(c + \Delta x, f(c + \Delta x))$ is a second point on the graph of f, the slope of the secant line through the two points is given by substitution into the slope formula

$$m = \frac{y_2 - y_1}{x_2 - x_1} \qquad \text{Slope formula}$$

$$m_{\text{sec}} = \frac{f(c + \Delta x) - f(c)}{(c + \Delta x) - c} \qquad \frac{\text{Change in } y}{\text{Change in } x}$$

$$m_{\text{sec}} = \frac{f(c + \Delta x) - f(c)}{\Delta x}. \qquad \text{Slope of secant line}$$

The right-hand side of this equation is a **difference quotient.** The denominator Δx is the **change in *x*,** and the numerator $\Delta y = f(c + \Delta x) - f(c)$ is the **change in *y*.**

The beauty of this procedure is that you can obtain more and more accurate approximations of the slope of the tangent line by choosing points closer and closer to the point of tangency, as shown in Figure 4.4.

> **THE TANGENT LINE PROBLEM**
>
> In 1637, mathematician René Descartes stated this about the tangent line problem:
>
> "And I dare say that this is not only the most useful and general problem in geometry that I know, but even that I ever desire to know."

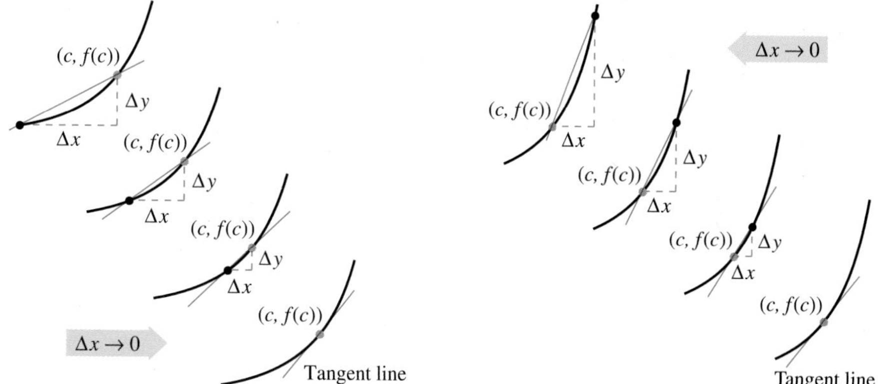

Tangent line approximations
Figure 4.4

> **DEFINITION OF TANGENT LINE WITH SLOPE *m***
>
> If f is defined on an open interval containing c, and if the limit
>
> $$\lim_{\Delta x \to 0} \frac{\Delta y}{\Delta x} = \lim_{\Delta x \to 0} \frac{f(c + \Delta x) - f(c)}{\Delta x} = m$$
>
> exists, then the line passing through $(c, f(c))$ with slope m is the **tangent line** to the graph of f at the point $(c, f(c))$.

The slope of the tangent line to the graph of f at the point $(c, f(c))$ is also called the **slope of the graph of *f* at $x = c$.**

---

* *This use of the word* secant *comes from the Latin* secare, *meaning to cut, and is not a reference to the trigonometric function of the same name.*

EXAMPLE 1 The Slope of the Graph of a Linear Function

Find the slope of the graph of

$$f(x) = 2x - 3$$

at the point (2, 1).

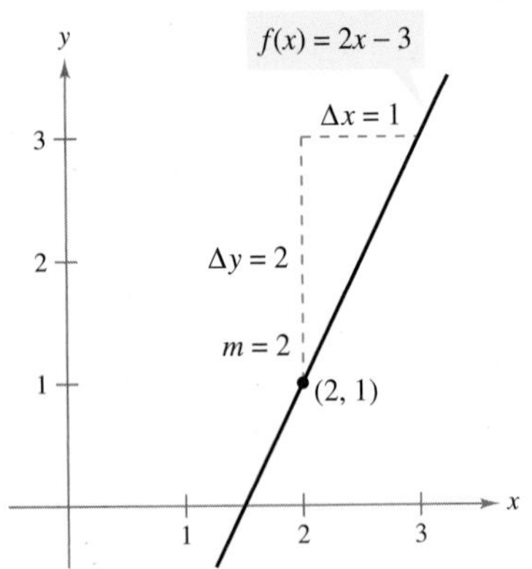

The slope of f at (2, 1) is $m = 2$.
Figure 4.5

Solution To find the slope of the graph of f when $c = 2$, you can apply the definition of the slope of a tangent line, as shown.

$$\begin{aligned}\lim_{\Delta x \to 0} \frac{f(2 + \Delta x) - f(2)}{\Delta x} &= \lim_{\Delta x \to 0} \frac{[2(2 + \Delta x) - 3] - [2(2) - 3]}{\Delta x} \\ &= \lim_{\Delta x \to 0} \frac{4 + 2\Delta x - 3 - 4 + 3}{\Delta x} \\ &= \lim_{\Delta x \to 0} \frac{2\cancel{\Delta x}}{\cancel{\Delta x}} \\ &= \lim_{\Delta x \to 0} 2 \\ &= 2\end{aligned}$$

The slope of f at $(c, f(c)) = (2, 1)$ is $m = 2$, as shown in Figure 4.5. ■

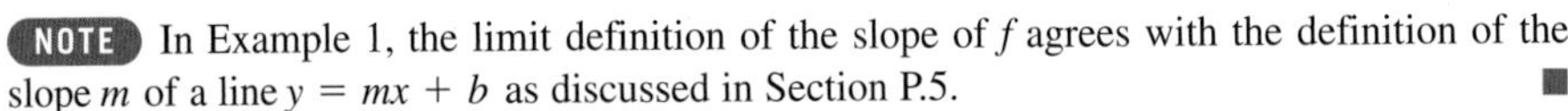

NOTE In Example 1, the limit definition of the slope of f agrees with the definition of the slope m of a line $y = mx + b$ as discussed in Section P.5. ■

The graph of a linear function has the same slope at any point. This is not true of nonlinear functions, as shown in the following example.

EXAMPLE 2 Tangent Lines to the Graph of a Nonlinear Function

Find the slopes of the tangent lines to the graph of

$$f(x) = x^2 + 1$$

at the points (0, 1) and $(-1, 2)$, as shown in Figure 4.6.

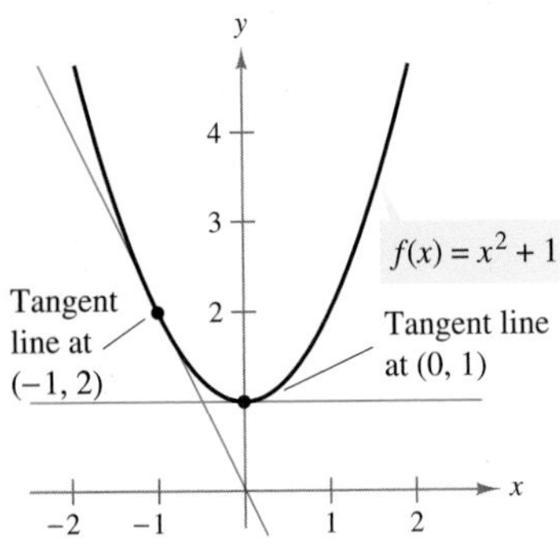

The slope of f at any point $(c, f(c))$ is $m = 2c$.
Figure 4.6

Solution Let $(c, f(c))$ represent an arbitrary point on the graph of f. Then the slope of the tangent line at $(c, f(c))$ is given by

$$\begin{aligned}\lim_{\Delta x \to 0} \frac{f(c + \Delta x) - f(c)}{\Delta x} &= \lim_{\Delta x \to 0} \frac{[(c + \Delta x)^2 + 1] - (c^2 + 1)}{\Delta x} \\ &= \lim_{\Delta x \to 0} \frac{c^2 + 2c(\Delta x) + (\Delta x)^2 + 1 - c^2 - 1}{\Delta x} \\ &= \lim_{\Delta x \to 0} \frac{2c(\Delta x) + (\Delta x)^2}{\Delta x} \\ &= \lim_{\Delta x \to 0} \frac{\cancel{\Delta x}(2c + \Delta x)}{\cancel{\Delta x}} \\ &= \lim_{\Delta x \to 0} (2c + \Delta x) \\ &= 2c.\end{aligned}$$

So, the slope at *any* point $(c, f(c))$ on the graph of f is $m = 2c$. At the point (0, 1), the slope is $m = 2(0) = 0$, and at $(-1, 2)$, the slope is $m = 2(-1) = -2$. ■

NOTE In Example 2, note that c is held constant in the limit process (as $\Delta x \to 0$). ■

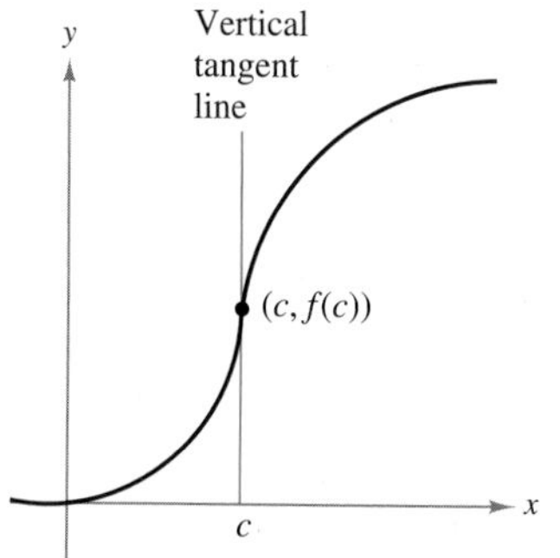

The graph of f has a vertical tangent line at $(c, f(c))$.
Figure 4.7

The definition of a tangent line to a curve does not cover the possibility of a vertical tangent line. For vertical tangent lines, you can use the following definition. If f is continuous at c and

$$\lim_{\Delta x \to 0} \frac{f(c + \Delta x) - f(c)}{\Delta x} = \infty \quad \text{or} \quad \lim_{\Delta x \to 0} \frac{f(c + \Delta x) - f(c)}{\Delta x} = -\infty$$

the vertical line $x = c$ passing through $(c, f(c))$ is a **vertical tangent line** to the graph of f. For example, the function shown in Figure 4.7 has a vertical tangent line at $(c, f(c))$. If the domain of f is the closed interval $[a, b]$, you can extend the definition of a vertical tangent line to include the endpoints by considering continuity and limits from the right (for $x = a$) and from the left (for $x = b$).

The Derivative of a Function

You have now arrived at a crucial point in the study of calculus. The limit used to define the slope of a tangent line is also used to define one of the two fundamental operations of calculus—**differentiation.**

DEFINITION OF THE DERIVATIVE OF A FUNCTION

The **derivative** of f at x is given by

$$f'(x) = \lim_{\Delta x \to 0} \frac{f(x + \Delta x) - f(x)}{\Delta x}$$

provided the limit exists. For all x for which this limit exists, f' is a function of x.

Be sure you see that the derivative of a function of x is also a function of x. This "new" function gives the slope of the tangent line to the graph of f at the point $(x, f(x))$, provided that the graph has a tangent line at this point.

The process of finding the derivative of a function is called **differentiation.** A function is **differentiable** at x if its derivative exists at x and is **differentiable on an open interval (a, b)** if it is differentiable at every point in the interval.

In addition to $f'(x)$, which is read as "f prime of x," other notations are used to denote the derivative of $y = f(x)$. The most common are

$$f'(x), \quad \frac{dy}{dx}, \quad y', \quad \frac{d}{dx}[f(x)], \quad D_x[y].$$

Notation for derivatives

The notation dy/dx is read as "the derivative of y *with respect to* x" or simply "dy, dx." Using limit notation, you can write

$$\begin{aligned} \frac{dy}{dx} &= \lim_{\Delta x \to 0} \frac{\Delta y}{\Delta x} \\ &= \lim_{\Delta x \to 0} \frac{f(x + \Delta x) - f(x)}{\Delta x} \\ &= f'(x). \end{aligned}$$

FOR FURTHER INFORMATION For more information on the crediting of mathematical discoveries to the first "discoverers," see the article "Mathematical Firsts—Who Done It?" by Richard H. Williams and Roy D. Mazzagatti in *Mathematics Teacher*. To view this article, go to the website *www.matharticles.com*.

EXAMPLE 3 Finding the Derivative by the Limit Process

Find the derivative of $f(x) = x^3 + 2x$.

Solution

$$f'(x) = \lim_{\Delta x \to 0} \frac{f(x + \Delta x) - f(x)}{\Delta x}$$ Definition of derivative

$$= \lim_{\Delta x \to 0} \frac{\overbrace{(x + \Delta x)^3 + 2(x + \Delta x)}^{f(x+\Delta x)} - \overbrace{(x^3 + 2x)}^{f(x)}}{\Delta x}$$ Substitute.

$$= \lim_{\Delta x \to 0} \frac{x^3 + 3x^2\Delta x + 3x(\Delta x)^2 + (\Delta x)^3 + 2x + 2\Delta x - x^3 - 2x}{\Delta x}$$

$$= \lim_{\Delta x \to 0} \frac{3x^2\Delta x + 3x(\Delta x)^2 + (\Delta x)^3 + 2\Delta x}{\Delta x}$$

$$= \lim_{\Delta x \to 0} \frac{\cancel{\Delta x}[3x^2 + 3x\Delta x + (\Delta x)^2 + 2]}{\cancel{\Delta x}}$$

$$= \lim_{\Delta x \to 0} [3x^2 + 3x\Delta x + (\Delta x)^2 + 2]$$

$$= 3x^2 + 2$$ ■

STUDY TIP When using the definition to find a derivative of a function, the key is to rewrite the difference quotient so that Δx can be divided out of the denominator.

Remember that the derivative of a function f is itself a function, which can be used to find the slope of the tangent line at the point $(x, f(x))$ on the graph of f.

EXAMPLE 4 Using the Derivative to Find the Slope at a Point

Find $f'(x)$ for $f(x) = \sqrt{x}$. Then find the slopes of the graph of f at the points $(1, 1)$ and $(4, 2)$. Discuss the behavior of f at $(0, 0)$.

Solution Use the procedure for rationalizing numerators, as discussed in Section 3.3.

$$f'(x) = \lim_{\Delta x \to 0} \frac{f(x + \Delta x) - f(x)}{\Delta x}$$ Definition of derivative

$$= \lim_{\Delta x \to 0} \frac{\overbrace{\sqrt{x + \Delta x}}^{f(x+\Delta x)} - \overbrace{\sqrt{x}}^{f(x)}}{\Delta x}$$ Substitute.

$$= \lim_{\Delta x \to 0} \left(\frac{\sqrt{x + \Delta x} - \sqrt{x}}{\Delta x}\right)\left(\frac{\sqrt{x + \Delta x} + \sqrt{x}}{\sqrt{x + \Delta x} + \sqrt{x}}\right)$$ Rationalize numerator.

$$= \lim_{\Delta x \to 0} \frac{(x + \Delta x) - x}{\Delta x\left(\sqrt{x + \Delta x} + \sqrt{x}\right)}$$

$$= \lim_{\Delta x \to 0} \frac{\cancel{\Delta x}}{\cancel{\Delta x}\left(\sqrt{x + \Delta x} + \sqrt{x}\right)}$$

$$= \lim_{\Delta x \to 0} \frac{1}{\sqrt{x + \Delta x} + \sqrt{x}}$$

$$= \frac{1}{2\sqrt{x}}, \quad x > 0$$

At the point $(1, 1)$, the slope is $f'(1) = \frac{1}{2}$. At the point $(4, 2)$, the slope is $f'(4) = \frac{1}{4}$. See Figure 4.8. At the point $(0, 0)$, the slope is undefined. Moreover, the graph of f has a vertical tangent line at $(0, 0)$. ■

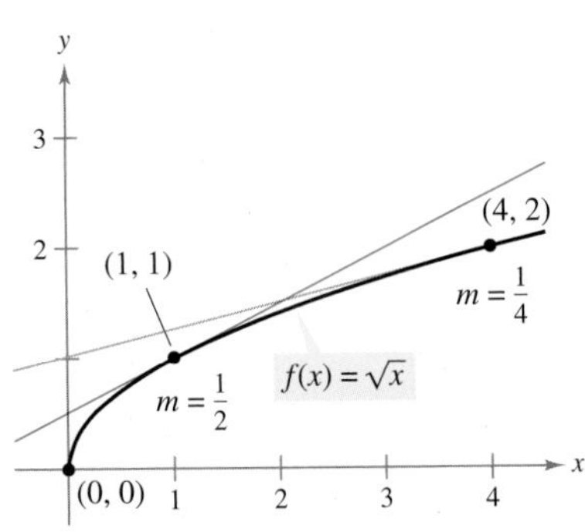

The slope of f at $(x, f(x))$, $x > 0$, is $m = 1/\left(2\sqrt{x}\right)$.
Figure 4.8

In many applications, it is convenient to use a variable other than x as the independent variable, as shown in Example 5.

EXAMPLE 5 Finding the Derivative of a Function

Find the derivative with respect to t for the function $y = 2/t$.

Solution Considering $y = f(t)$, you obtain

$$\frac{dy}{dt} = \lim_{\Delta t \to 0} \frac{f(t + \Delta t) - f(t)}{\Delta t}$$ Definition of derivative

$$= \lim_{\Delta t \to 0} \frac{\dfrac{2}{t + \Delta t} - \dfrac{2}{t}}{\Delta t}$$ $f(t + \Delta t) = 2/(t + \Delta t)$ and $f(t) = 2/t$

$$= \lim_{\Delta t \to 0} \frac{\dfrac{2t - 2(t + \Delta t)}{t(t + \Delta t)}}{\Delta t}$$ Combine fractions in numerator.

$$= \lim_{\Delta t \to 0} \frac{-2\Delta t}{\Delta t(t)(t + \Delta t)}$$ Divide out common factor of Δt.

$$= \lim_{\Delta t \to 0} \frac{-2}{t(t + \Delta t)}$$ Simplify.

$$= -\frac{2}{t^2}.$$ Evaluate limit as $\Delta t \to 0$. ■

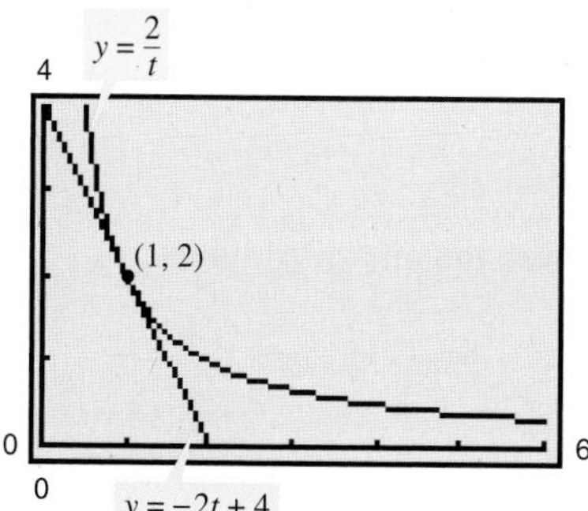

At the point $(1, 2)$, the line $y = -2t + 4$ is tangent to the graph of $y = 2/t$.
Figure 4.9

TECHNOLOGY A graphing utility can be used to reinforce the result given in Example 5. For instance, using the formula $dy/dt = -2/t^2$, you know that the slope of the graph of $y = 2/t$ at the point $(1, 2)$ is $m = -2$. Using the point-slope form, you can find that the equation of the tangent line to the graph at $(1, 2)$ is

$$y - 2 = -2(t - 1) \quad \text{or} \quad y = -2t + 4$$

as shown in Figure 4.9.

Differentiability and Continuity

The following alternative limit form of the derivative is useful in investigating the relationship between differentiability and continuity. The derivative of f at c is

$$f'(c) = \lim_{x \to c} \frac{f(x) - f(c)}{x - c}$$ Alternative form of derivative

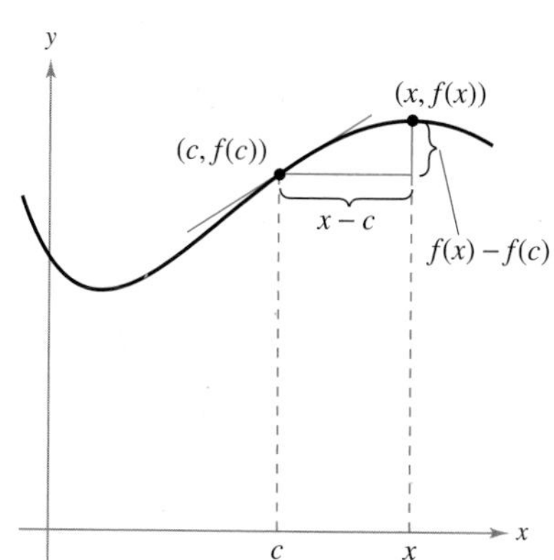

As x approaches c, the secant line approaches the tangent line.
Figure 4.10

provided this limit exists (see Figure 4.10). (A proof of the equivalence of this form is given in Appendix A.) Note that the existence of the limit in this alternative form requires that the one-sided limits

$$\lim_{x \to c^-} \frac{f(x) - f(c)}{x - c} \quad \text{and} \quad \lim_{x \to c^+} \frac{f(x) - f(c)}{x - c}$$

exist and are equal. These one-sided limits are called the **derivatives from the left and from the right,** respectively. It follows that f is **differentiable on the closed interval** $[a, b]$ if it is differentiable on (a, b) and if the derivative from the right at a and the derivative from the left at b both exist.

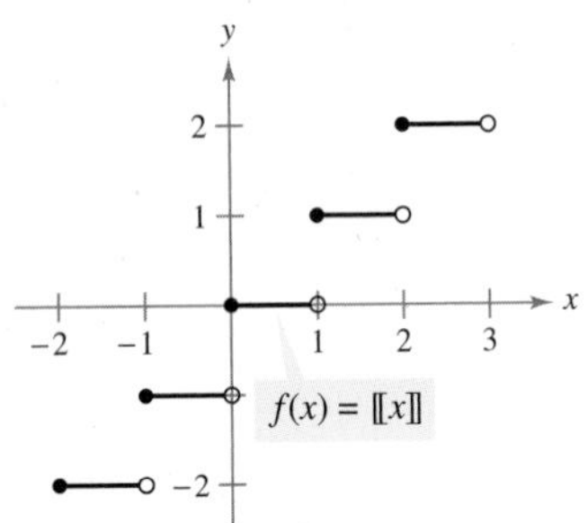

The greatest integer function is not differentiable at $x = 0$, because it is not continuous at $x = 0$.
Figure 4.11

If a function is not continuous at $x = c$, it is also not differentiable at $x = c$. For instance, the greatest integer function

$$f(x) = [\![x]\!]$$

is not continuous at $x = 0$, and so it is not differentiable at $x = 0$ (see Figure 4.11). You can verify this by observing that

$$\lim_{x \to 0^-} \frac{f(x) - f(0)}{x - 0} = \lim_{x \to 0^-} \frac{[\![x]\!] - 0}{x} = \infty \qquad \text{Derivative from the left}$$

and

$$\lim_{x \to 0^+} \frac{f(x) - f(0)}{x - 0} = \lim_{x \to 0^+} \frac{[\![x]\!] - 0}{x} = 0. \qquad \text{Derivative from the right}$$

Although it is true that differentiability implies continuity (as shown in Theorem 4.1 on the next page), the converse is not true. That is, it is possible for a function to be continuous at $x = c$ and *not* differentiable at $x = c$. Examples 6 and 7 illustrate this possibility.

EXAMPLE 6 A Graph with a Sharp Turn

The function

$$f(x) = |x - 2|$$

shown in Figure 4.12 is continuous at $x = 2$. However, the one-sided limits

$$\lim_{x \to 2^-} \frac{f(x) - f(2)}{x - 2} = \lim_{x \to 2^-} \frac{|x - 2| - 0}{x - 2} = -1 \qquad \text{Derivative from the left}$$

and

$$\lim_{x \to 2^+} \frac{f(x) - f(2)}{x - 2} = \lim_{x \to 2^+} \frac{|x - 2| - 0}{x - 2} = 1 \qquad \text{Derivative from the right}$$

are not equal. So, f is not differentiable at $x = 2$ and the graph of f does not have a tangent line at the point $(2, 0)$.

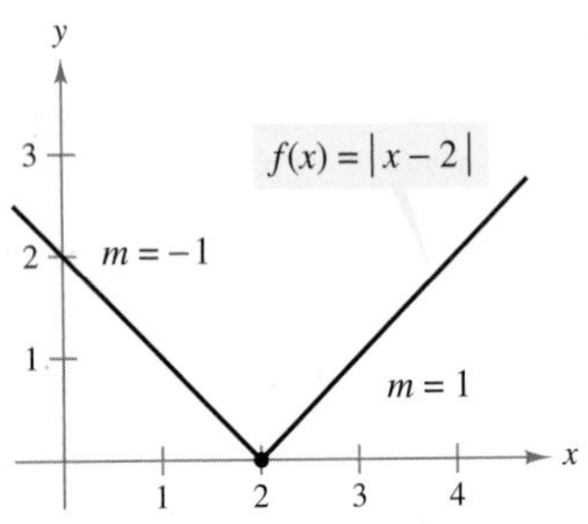

f is not differentiable at $x = 2$, because the derivatives from the left and from the right are not equal.
Figure 4.12

EXAMPLE 7 A Graph with a Vertical Tangent Line

The function

$$f(x) = x^{1/3}$$

is continuous at $x = 0$, as shown in Figure 4.13. However, because the limit

$$\begin{aligned}\lim_{x \to 0} \frac{f(x) - f(0)}{x - 0} &= \lim_{x \to 0} \frac{x^{1/3} - 0}{x} \\ &= \lim_{x \to 0} \frac{1}{x^{2/3}} \\ &= \infty\end{aligned}$$

is infinite, you can conclude that the tangent line is vertical at $x = 0$. So, f is not differentiable at $x = 0$. ■

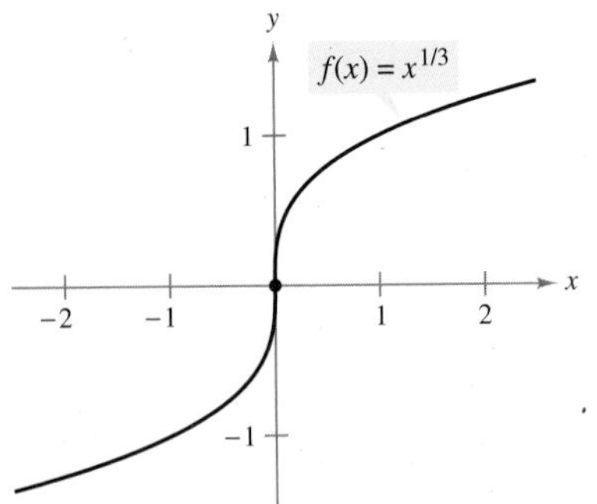

f is not differentiable at $x = 0$, because f has a vertical tangent line at $x = 0$.
Figure 4.13

From Examples 6 and 7, you can see that a function is not differentiable at a point at which its graph has a sharp turn *or* a vertical tangent line.

TECHNOLOGY Some graphing utilities, such as *Maple*, *Mathematica*, and the *TI-89*, perform symbolic differentiation. Others perform *numerical differentiation* by finding values of derivatives using the formula

$$f'(x) \approx \frac{f(x + \Delta x) - f(x - \Delta x)}{2\Delta x}$$

where Δx is a small number such as 0.001. Can you see any problems with this definition? For instance, using this definition, what is the value of the derivative of $f(x) = |x|$ when $x = 0$?

THEOREM 4.1 DIFFERENTIABILITY IMPLIES CONTINUITY

If f is differentiable at $x = c$, then f is continuous at $x = c$.

PROOF You can prove that f is continuous at $x = c$ by showing that $f(x)$ approaches $f(c)$ as $x \to c$. To do this, use the differentiability of f at $x = c$ and consider the following limit.

$$\begin{aligned}\lim_{x\to c}[f(x) - f(c)] &= \lim_{x\to c}\left[(x - c)\left(\frac{f(x) - f(c)}{x - c}\right)\right]\\ &= \left[\lim_{x\to c}(x - c)\right]\left[\lim_{x\to c}\frac{f(x) - f(c)}{x - c}\right]\\ &= (0)[f'(c)]\\ &= 0\end{aligned}$$

Because the difference $f(x) - f(c)$ approaches zero as $x \to c$, you can conclude that $\lim_{x\to c} f(x) = f(c)$. So, f is continuous at $x = c$. ■

The following statements summarize the relationship between continuity and differentiability.

1. If a function is differentiable at $x = c$, then it is continuous at $x = c$. So, differentiability implies continuity.
2. It is possible for a function to be continuous at $x = c$ and not be differentiable at $x = c$. So, continuity does not imply differentiability (see Example 6).

4.1 Exercises

See www.CalcChat.com for worked-out solutions to odd-numbered exercises.

In Exercises 1 and 2, estimate the slope of the graph at the points (x_1, y_1) and (x_2, y_2).

1. (a)

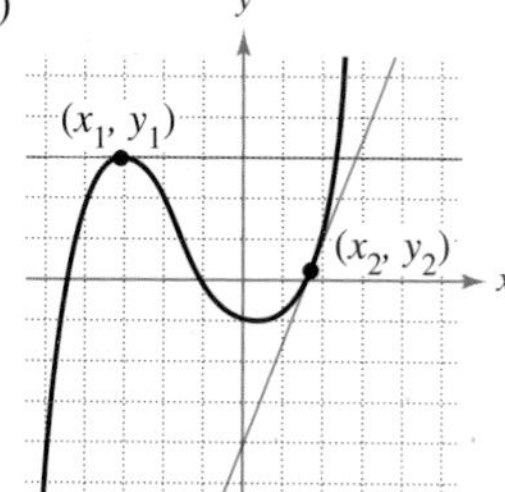

(b)

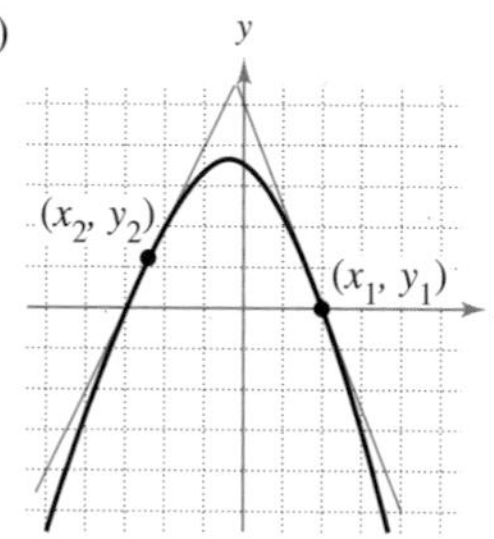

2. (a)

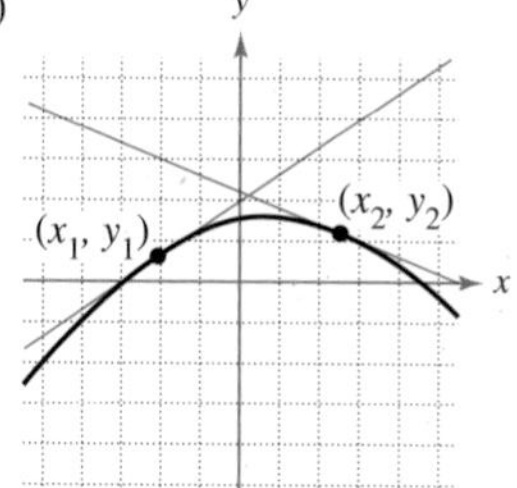

(b)

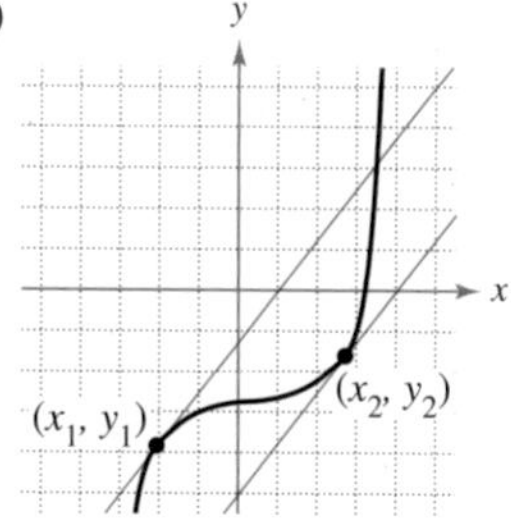

In Exercises 3 and 4, use the graph shown in the figure. To print an enlarged copy of the graph, go to the website *www.mathgraphs.com*.

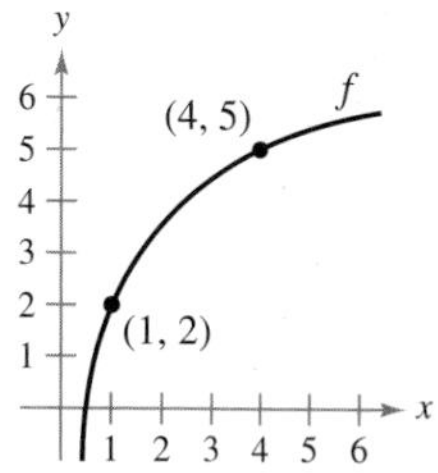

3. Identify or sketch each of the quantities on the figure.
   (a) $f(1)$ and $f(4)$ (b) $f(4) - f(1)$
   (c) $y = \dfrac{f(4) - f(1)}{4 - 1}(x - 1) + f(1)$
4. Insert the proper inequality symbol (< or >) between the given quantities.
   (a) $\dfrac{f(4) - f(1)}{4 - 1}$ ▭ $\dfrac{f(4) - f(3)}{4 - 3}$
   (b) $\dfrac{f(4) - f(1)}{4 - 1}$ ▭ $f'(1)$

In Exercises 5–10, find the slope of the tangent line to the graph of the function at the given point.

5. $f(x) = 3 - 5x, \quad (-1, 8)$
6. $g(x) = \frac{3}{2}x + 1, \quad (-2, -2)$
7. $g(x) = x^2 - 9, \quad (2, -5)$
8. $g(x) = 6 - x^2, \quad (1, 5)$
9. $f(t) = 3t - t^2, \quad (0, 0)$
10. $h(t) = t^2 + 3, \quad (-2, 7)$

In Exercises 11–24, find the derivative by the limit process.

11. $f(x) = 7$
12. $g(x) = -3$
13. $f(x) = -10x$
14. $f(x) = 3x + 2$
15. $h(s) = 3 + \frac{2}{3}s$
16. $f(x) = 8 - \frac{1}{5}x$
17. $f(x) = x^2 + x - 3$
18. $f(x) = 2 - x^2$
19. $f(x) = x^3 - 12x$
20. $f(x) = x^3 + x^2$
21. $f(x) = \dfrac{1}{x - 1}$
22. $f(x) = \dfrac{1}{x^2}$
23. $f(x) = \sqrt{x + 4}$
24. $f(x) = \dfrac{4}{\sqrt{x}}$

In Exercises 25–32, (a) find an equation of the tangent line to the graph of f at the given point, (b) use a graphing utility to graph the function and its tangent line at the point, and (c) use the *derivative* feature of a graphing utility to confirm your results.

25. $f(x) = x^2 + 3, \quad (1, 4)$
26. $f(x) = x^2 + 3x + 4, \quad (-2, 2)$
27. $f(x) = x^3, \quad (2, 8)$
28. $f(x) = x^3 + 1, \quad (1, 2)$
29. $f(x) = \sqrt{x}, \quad (1, 1)$
30. $f(x) = \sqrt{x - 1}, \quad (5, 2)$
31. $f(x) = x + \dfrac{4}{x}, \quad (4, 5)$
32. $f(x) = \dfrac{1}{x + 1}, \quad (0, 1)$

In Exercises 33–38, find an equation of the line that is tangent to the graph of f *and* parallel to the given line.

Function	*Line*
33. $f(x) = x^2$	$2x - y + 1 = 0$
34. $f(x) = 2x^2$	$4x + y + 3 = 0$
35. $f(x) = x^3$	$3x - y + 1 = 0$
36. $f(x) = x^3 + 2$	$3x - y - 4 = 0$
37. $f(x) = \dfrac{1}{\sqrt{x}}$	$x + 2y - 6 = 0$
38. $f(x) = \dfrac{1}{\sqrt{x - 1}}$	$x + 2y + 7 = 0$

In Exercises 39–42, the graph of f is given. Select the graph of f'.

39.

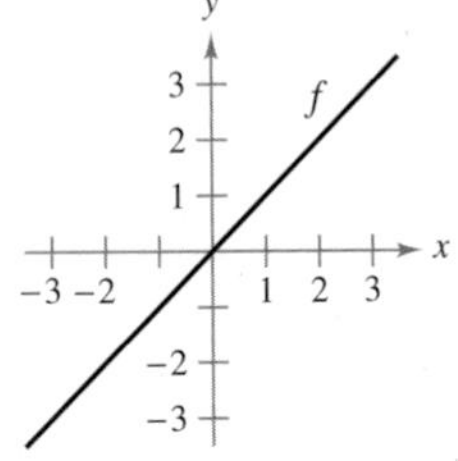

40.

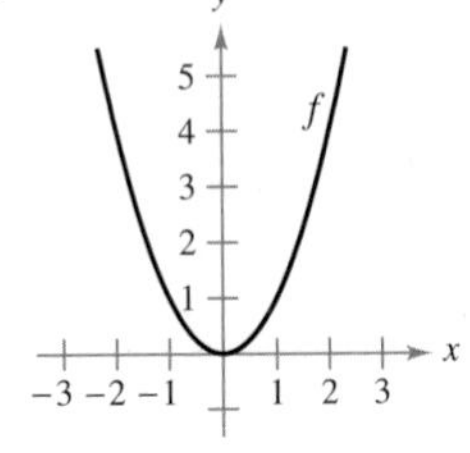

41.

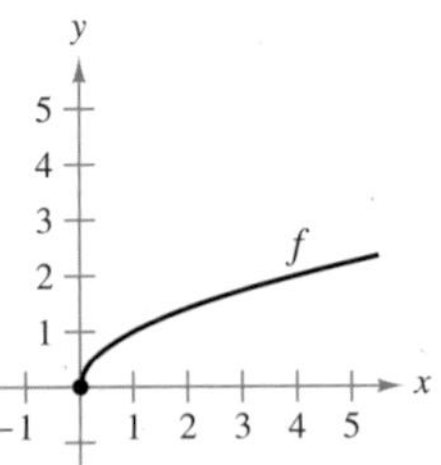

42.

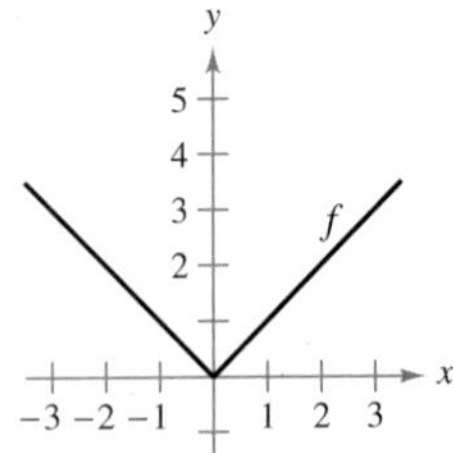

(a)

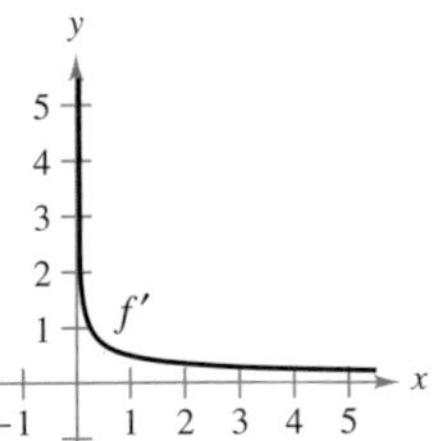

(b)

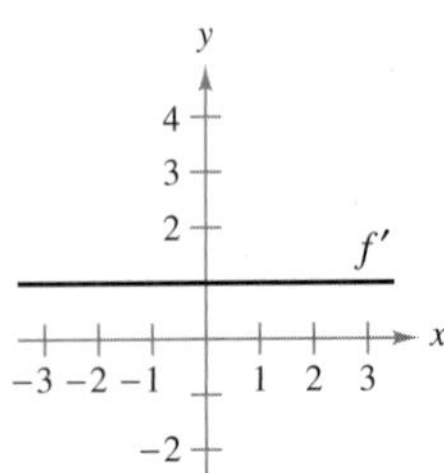

(c)

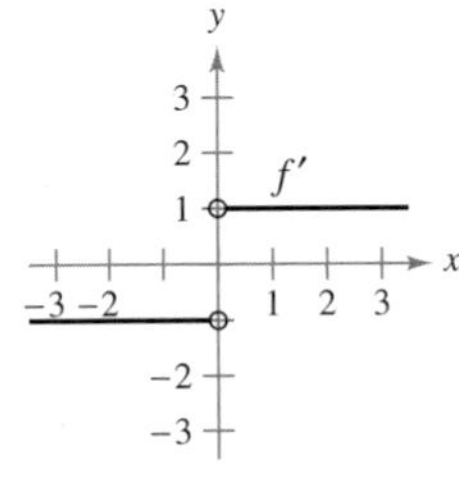

(d)

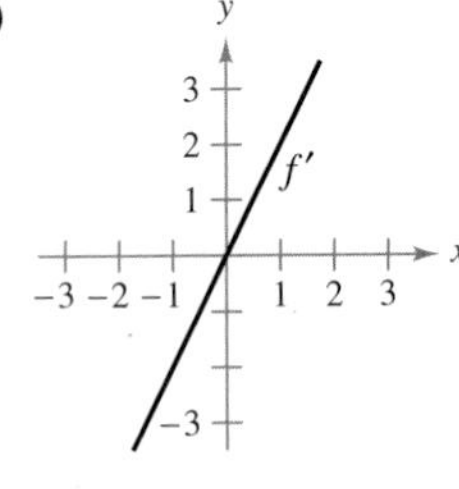

43. The tangent line to the graph of $y = g(x)$ at the point $(4, 5)$ passes through the point $(7, 0)$. Find $g(4)$ and $g'(4)$.

44. The tangent line to the graph of $y = h(x)$ at the point $(-1, 4)$ passes through the point $(3, 6)$. Find $h(-1)$ and $h'(-1)$.

WRITING ABOUT CONCEPTS

In Exercises 45–50, sketch the graph of f'. Explain how you found your answer.

45.

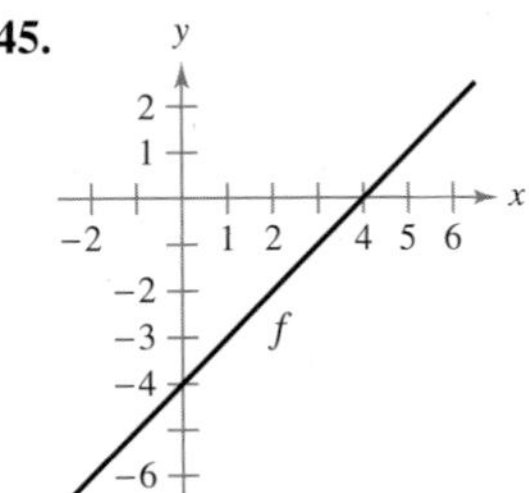

46.

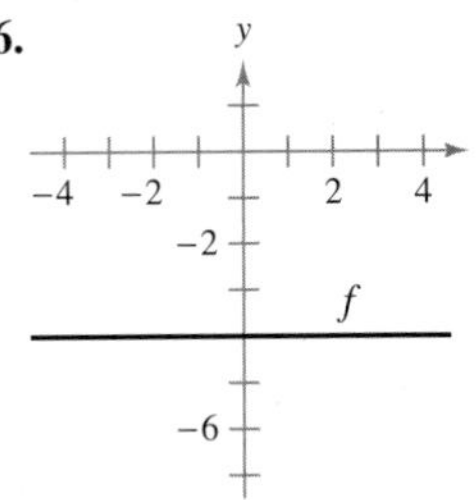

47.

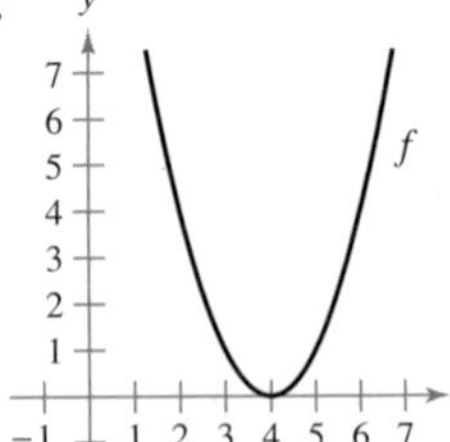

48.

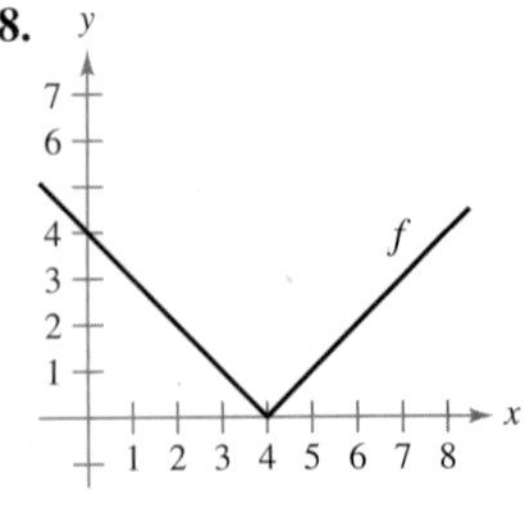

WRITING ABOUT CONCEPTS (continued)

49.

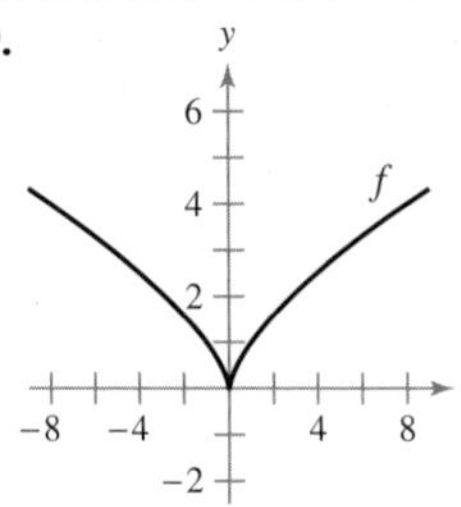

50.

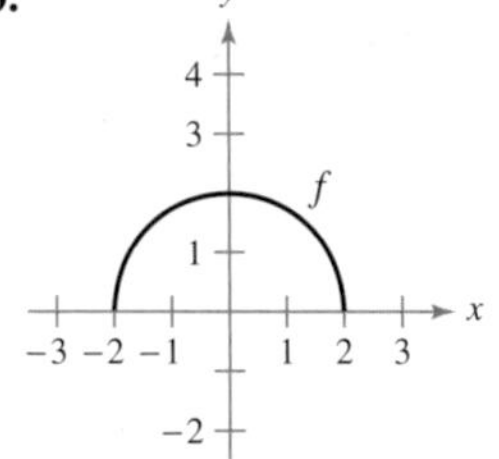

51. Sketch a graph of a function whose derivative is always negative. Explain how you found your answer.

52. Sketch a graph of a function whose derivative is always positive. Explain how you found your answer.

In Exercises 53–56, the limit represents $f'(c)$ for a function f and a number c. Find f and c.

53. $\lim_{\Delta x \to 0} \frac{[5 - 3(1 + \Delta x)] - 2}{\Delta x}$

54. $\lim_{\Delta x \to 0} \frac{(-2 + \Delta x)^3 + 8}{\Delta x}$

55. $\lim_{x \to 6} \frac{-x^2 + 36}{x - 6}$

56. $\lim_{x \to 9} \frac{2\sqrt{x} - 6}{x - 9}$

In Exercises 57–59, identify a function f that has the given characteristics. Then sketch the function.

57. $f(0) = 2;$ $f'(x) = -3, -\infty < x < \infty$

58. $f(0) = 4; f'(0) = 0;$ $f'(x) < 0$ for $x < 0;$ $f'(x) > 0$ for $x > 0$

59. $f(0) = 0; f'(0) = 0; f'(x) > 0$ for $x \neq 0$

60. Assume that $f'(c) = 3$. Find $f'(-c)$ if (a) f is an odd function and if (b) f is an even function.

In Exercises 61 and 62, find equations of the two tangent lines to the graph of f that pass through the indicated point.

61. $f(x) = 4x - x^2$

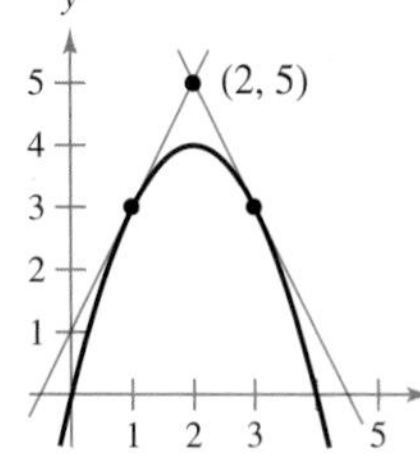

62. $f(x) = x^2$

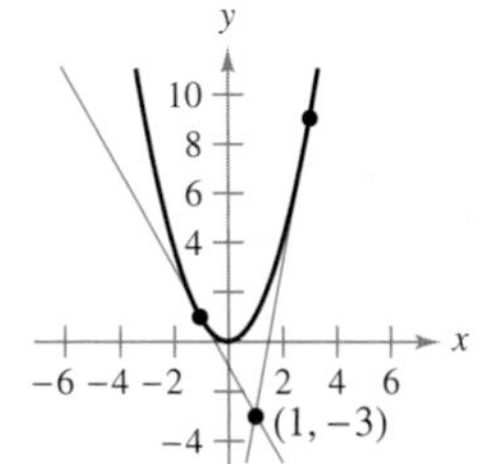

63. ***Graphical Reasoning*** Use a graphing utility to graph each function and its tangent lines at $x = -1$, $x = 0$, and $x = 1$. Based on the results, determine whether the slopes of tangent lines to the graph of a function at different values of x are always distinct.

(a) $f(x) = x^2$ (b) $g(x) = x^3$

CAPSTONE

64. The figure shows the graph of g'.

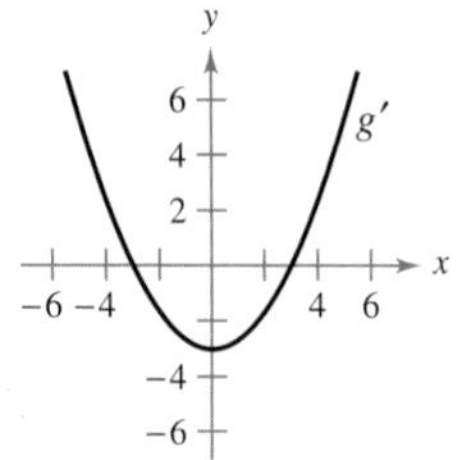

(a) $g'(0) = \square$ (b) $g'(3) = \square$

(c) What can you conclude about the graph of g knowing that $g'(1) = -\frac{8}{3}$?

(d) What can you conclude about the graph of g knowing that $g'(-4) = \frac{7}{3}$?

(e) Is $g(6) - g(4)$ positive or negative? Explain.

(f) Is it possible to find $g(2)$ from the graph? Explain.

65. ***Graphical Analysis*** Consider the function $f(x) = \frac{1}{2}x^2$.

(a) Use a graphing utility to graph the function and estimate the values of $f'(0)$, $f'(\frac{1}{2})$, $f'(1)$, and $f'(2)$.

(b) Use your results from part (a) to determine the values of $f'(-\frac{1}{2})$, $f'(-1)$, and $f'(-2)$.

(c) Sketch a possible graph of f'.

(d) Use the definition of derivative to find $f'(x)$.

66. ***Graphical Analysis*** Consider the function $f(x) = \frac{1}{3}x^3$.

(a) Use a graphing utility to graph the function and estimate the values of $f'(0)$, $f'(\frac{1}{2})$, $f'(1)$, $f'(2)$, and $f'(3)$.

(b) Use your results from part (a) to determine the values of $f'(-\frac{1}{2})$, $f'(-1)$, $f'(-2)$, and $f'(-3)$.

(c) Sketch a possible graph of f'.

(d) Use the definition of derivative to find $f'(x)$.

Graphical Reasoning **In Exercises 67 and 68, use a graphing utility to graph the functions f and g in the same viewing window, where**

$$g(x) = \frac{f(x + 0.01) - f(x)}{0.01}.$$

Label the graphs and describe the relationship between them.

67. $f(x) = 2x - x^2$

68. $f(x) = 3\sqrt{x}$

In Exercises 69 and 70, evaluate $f(2)$ and $f(2.1)$ and use the results to approximate $f'(2)$.

69. $f(x) = x(4 - x)$

70. $f(x) = \frac{1}{4}x^3$

Graphical Reasoning **In Exercises 71 and 72, use a graphing utility to graph the function and its derivative in the same viewing window. Label the graphs and describe the relationship between them.**

71. $f(x) = \frac{1}{\sqrt{x}}$

72. $f(x) = \frac{x^3}{4} - 3x$

In Exercises 73–82, use the alternative form of the derivative to find the derivative at $x = c$ (if it exists).

73. $f(x) = x^2 - 5, \quad c = 3$

74. $g(x) = x(x - 1), \quad c = 1$

75. $f(x) = x^3 + 2x^2 + 1, \quad c = -2$

76. $f(x) = x^3 + 6x, \quad c = 2$

77. $g(x) = \sqrt{|x|}, \quad c = 0$

78. $f(x) = 2/x, \quad c = 5$

79. $f(x) = (x - 6)^{2/3}, \quad c = 6$

80. $g(x) = (x + 3)^{1/3}, \quad c = -3$

81. $h(x) = |x + 7|, \quad c = -7$

82. $f(x) = |x - 6|, \quad c = 6$

In Exercises 83–88, describe the x-values at which f is differentiable.

83. $f(x) = \dfrac{2}{x - 3}$

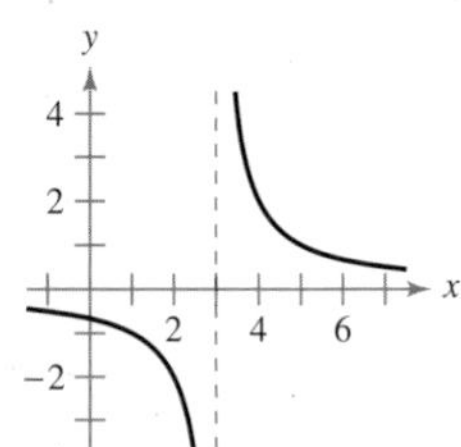

84. $f(x) = |x^2 - 9|$

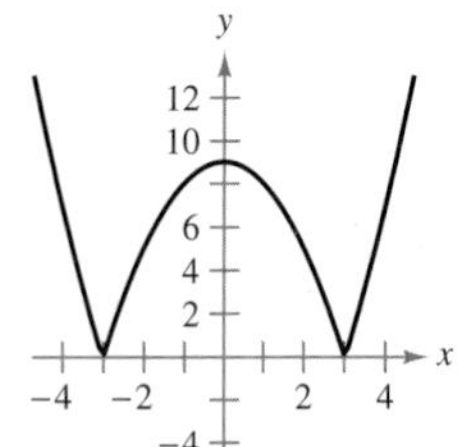

85. $f(x) = (x + 4)^{2/3}$

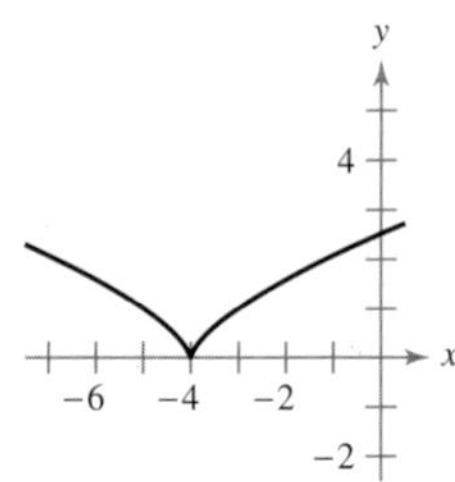

86. $f(x) = \dfrac{x^2}{x^2 - 4}$

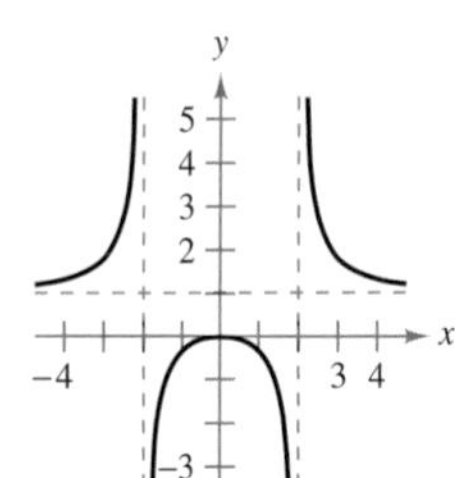

87. $f(x) = \sqrt{x - 1}$

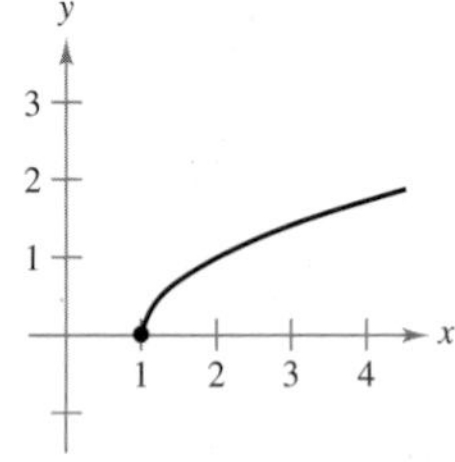

88. $f(x) = \begin{cases} x^2 - 4, & x \le 0 \\ 4 - x^2, & x > 0 \end{cases}$

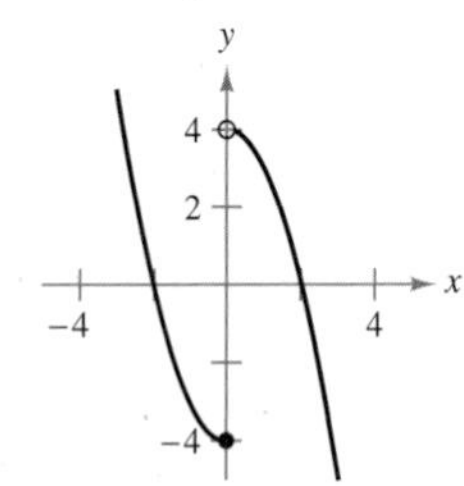

Graphical Analysis **In Exercises 89–92, use a graphing utility to graph the function and find the x-values at which f is differentiable.**

89. $f(x) = |x - 5|$

90. $f(x) = \dfrac{4x}{x - 3}$

91. $f(x) = x^{2/5}$

92. $f(x) = \begin{cases} x^3 - 3x^2 + 3x, & x \le 1 \\ x^2 - 2x, & x > 1 \end{cases}$

In Exercises 93–96, find the derivatives from the left and from the right at $x = 1$ (if they exist). Is the function differentiable at $x = 1$?

93. $f(x) = |x - 1|$

94. $f(x) = \sqrt{1 - x^2}$

95. $f(x) = \begin{cases} (x - 1)^3, & x \le 1 \\ (x - 1)^2, & x > 1 \end{cases}$

96. $f(x) = \begin{cases} x, & x \le 1 \\ x^2, & x > 1 \end{cases}$

In Exercises 97 and 98, determine whether the function is differentiable at $x = 2$.

97. $f(x) = \begin{cases} x^2 + 1, & x \le 2 \\ 4x - 3, & x > 2 \end{cases}$

98. $f(x) = \begin{cases} \frac{1}{2}x + 1, & x < 2 \\ \sqrt{2x}, & x \ge 2 \end{cases}$

99. ***Graphical Reasoning*** A line with slope m passes through the point $(0, 4)$ and has the equation $y = mx + 4$.
   (a) Write the distance d between the line and the point $(3, 1)$ as a function of m.
   (b) Use a graphing utility to graph the function d in part (a). Based on the graph, is the function differentiable at every value of m? If not, where is it not differentiable?

100. ***Conjecture*** Consider the functions $f(x) = x^2$ and $g(x) = x^3$.
   (a) Graph f and f' on the same set of axes.
   (b) Graph g and g' on the same set of axes.
   (c) Identify a pattern between f and g and their respective derivatives. Use the pattern to make a conjecture about $h'(x)$ if $h(x) = x^n$, where n is an integer and $n \ge 2$.
   (d) Find $f'(x)$ if $f(x) = x^4$. Compare the result with the conjecture in part (c). Is this a proof of your conjecture? Explain.

True or False? **In Exercises 101–104, determine whether the statement is true or false. If it is false, explain why or give an example that shows it is false.**

101. The slope of the tangent line to the differentiable function f at the point $(2, f(2))$ is $\dfrac{f(2 + \Delta x) - f(2)}{\Delta x}$.

102. If a function is continuous at a point, then it is differentiable at that point.

103. If a function has derivatives from both the right and the left at a point, then it is differentiable at that point.

104. If a function is differentiable at a point, then it is continuous at that point.

105. Determine whether the limit yields the derivative of a differentiable function f. Explain.
   (a) $\displaystyle\lim_{\Delta x \to 0} \frac{f(x + 2\Delta x) - f(x)}{2\Delta x}$
   (b) $\displaystyle\lim_{\Delta x \to 0} \frac{f(x + 2) - f(x)}{\Delta x}$

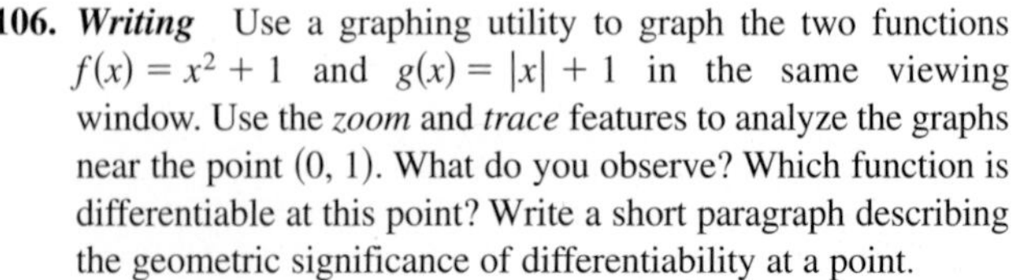

106. ***Writing*** Use a graphing utility to graph the two functions $f(x) = x^2 + 1$ and $g(x) = |x| + 1$ in the same viewing window. Use the *zoom* and *trace* features to analyze the graphs near the point $(0, 1)$. What do you observe? Which function is differentiable at this point? Write a short paragraph describing the geometric significance of differentiability at a point.

4.2 Basic Differentiation Rules and Rates of Change

- **Find the derivative of a function using the Constant Rule.**
- **Find the derivative of a function using the Power Rule.**
- **Find the derivative of a function using the Constant Multiple Rule.**
- **Find the derivative of a function using the Sum and Difference Rules.**
- **Use derivatives to find rates of change.**

The Constant Rule

In Section 4.1, you used the limit definition to find derivatives. In this and the next two sections, you will be introduced to several "differentiation rules" that allow you to find derivatives without the *direct* use of the limit definition.

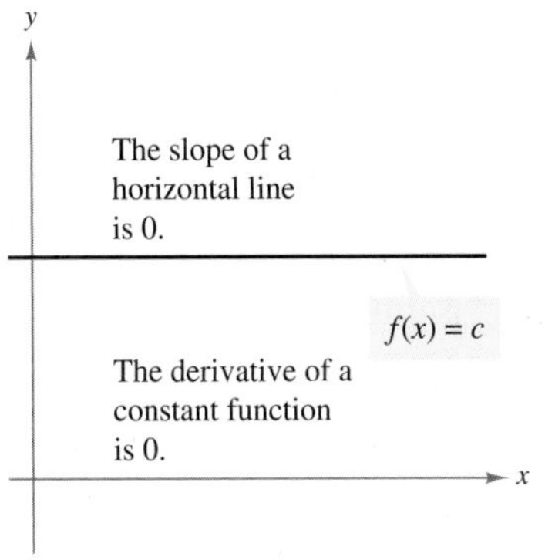

y

The slope of a horizontal line is 0.

$f(x) = c$

The derivative of a constant function is 0.

x

Notice that the Constant Rule is equivalent to saying that the slope of a horizontal line is 0. This demonstrates the relationship between slope and derivative.
Figure 4.14

THEOREM 4.2 THE CONSTANT RULE

The derivative of a constant function is 0. That is, if c is a real number, then

$$\frac{d}{dx}[c] = 0.$$

(See Figure 4.14.)

PROOF Let $f(x) = c$. Then, by the limit definition of the derivative,

$$\begin{aligned}\frac{d}{dx}[c] &= f'(x)\\ &= \lim_{\Delta x \to 0} \frac{f(x + \Delta x) - f(x)}{\Delta x}\\ &= \lim_{\Delta x \to 0} \frac{c - c}{\Delta x}\\ &= \lim_{\Delta x \to 0} 0 = 0.\end{aligned}$$ ■

EXAMPLE 1 Using the Constant Rule

	Function	*Derivative*
a.	$y = 7$	$dy/dx = 0$
b.	$f(x) = 0$	$f'(x) = 0$
c.	$s(t) = -3$	$s'(t) = 0$
d.	$y = k\pi^2$, k is constant	$y' = 0$

■

EXPLORATION

Writing a Conjecture Use the definition of the derivative given in Section 4.1 to find the derivative of each function. What patterns do you see? Use your results to write a conjecture about the derivative of $f(x) = x^n$.

a. $f(x) = x^1$ **b.** $f(x) = x^2$ **c.** $f(x) = x^3$
d. $f(x) = x^4$ **e.** $f(x) = x^{1/2}$ **f.** $f(x) = x^{-1}$

The Power Rule

Before proving the next rule, it is important to review the procedure for expanding a binomial.

$$(x + \Delta x)^2 = x^2 + 2x\Delta x + (\Delta x)^2$$
$$(x + \Delta x)^3 = x^3 + 3x^2\Delta x + 3x(\Delta x)^2 + (\Delta x)^3$$

The general binomial expansion for a positive integer n is

$$(x + \Delta x)^n = x^n + nx^{n-1}(\Delta x) + \underbrace{\frac{n(n-1)x^{n-2}}{2}(\Delta x)^2 + \cdots + (\Delta x)^n}_{(\Delta x)^2 \text{ is a factor of these terms.}}.$$

This binomial expansion is used in proving a special case of the Power Rule.

NOTE From Example 7 in Section 4.1, you know that the function $f(x) = x^{1/3}$ is defined at $x = 0$, but is not differentiable at $x = 0$. This is because $x^{-2/3}$ is not defined on an interval containing 0.

> **THEOREM 4.3 THE POWER RULE**
>
> If n is a rational number, then the function $f(x) = x^n$ is differentiable and
>
> $$\frac{d}{dx}[x^n] = nx^{n-1}.$$
>
> For f to be differentiable at $x = 0$, n must be a number such that x^{n-1} is defined on an interval containing 0.

PROOF If n is a positive integer greater than 1, then the binomial expansion produces

$$\begin{aligned}\frac{d}{dx}[x^n] &= \lim_{\Delta x \to 0} \frac{(x + \Delta x)^n - x^n}{\Delta x}\\ &= \lim_{\Delta x \to 0} \frac{x^n + nx^{n-1}(\Delta x) + \dfrac{n(n-1)x^{n-2}}{2}(\Delta x)^2 + \cdots + (\Delta x)^n - x^n}{\Delta x}\\ &= \lim_{\Delta x \to 0} \left[nx^{n-1} + \frac{n(n-1)x^{n-2}}{2}(\Delta x) + \cdots + (\Delta x)^{n-1}\right]\\ &= nx^{n-1} + 0 + \cdots + 0\\ &= nx^{n-1}.\end{aligned}$$

This proves the case for which n is a positive integer greater than 1. It is left to you to prove the case for $n = 1$. Example 6 in Section 4.3 proves the case for which n is a negative integer. In Exercise 64 in Section 4.5, you are asked to prove the case for which n is rational. (In Section 8.2, the Power Rule will be extended to cover irrational values of n.) ■

When using the Power Rule, the case for which $n = 1$ is best thought of as a separate differentiation rule. That is,

$$\frac{d}{dx}[x] = 1. \qquad \text{Power Rule when } n = 1$$

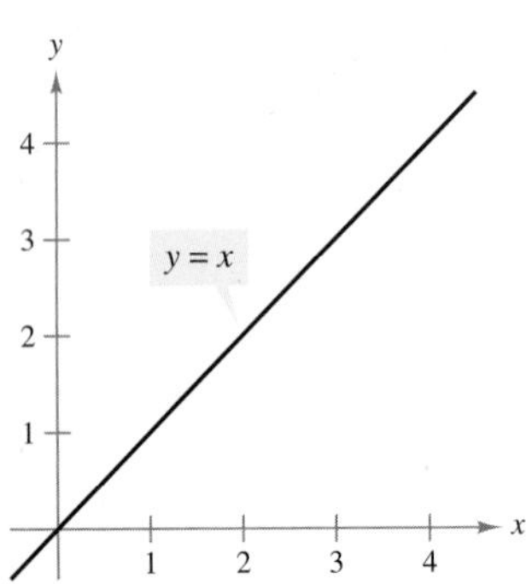

The slope of the line $y = x$ is 1.
Figure 4.15

This rule is consistent with the fact that the slope of the line $y = x$ is 1, as shown in Figure 4.15.

EXAMPLE 2 Using the Power Rule

Function	*Derivative*
a. $f(x) = x^3, n = 3$	$f'(x) = 3x^2$
b. $g(x) = \sqrt[3]{x}, n = \dfrac{1}{3}$	$g'(x) = \dfrac{d}{dx}[x^{1/3}] = \dfrac{1}{3}x^{-2/3} = \dfrac{1}{3x^{2/3}}$
c. $y = \dfrac{1}{x^2}, n = -2$	$\dfrac{dy}{dx} = \dfrac{d}{dx}[x^{-2}] = (-2)x^{-3} = -\dfrac{2}{x^3}$

■

In Example 2(c), note that *before* differentiating, $1/x^2$ was rewritten as x^{-2}. Rewriting is the first step in *many* differentiation problems.

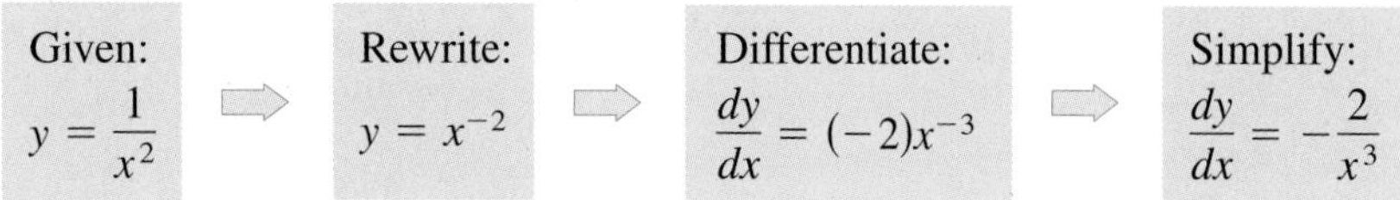

Given: $y = \dfrac{1}{x^2}$
Rewrite: $y = x^{-2}$
Differentiate: $\dfrac{dy}{dx} = (-2)x^{-3}$
Simplify: $\dfrac{dy}{dx} = -\dfrac{2}{x^3}$

EXAMPLE 3 Finding the Slope of a Graph

Find the slope of the graph of $f(x) = x^4$ when

a. $x = -1$

b. $x = 0$

c. $x = 1$.

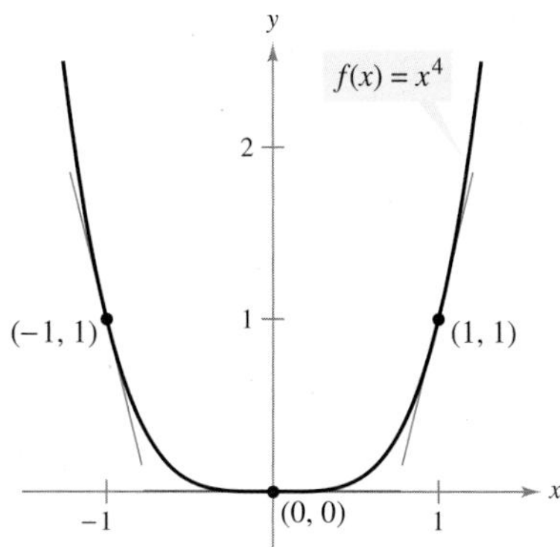

Figure 4.16

Solution The slope of a graph at a point is the value of the derivative at that point. The derivative of f is $f'(x) = 4x^3$.

a. When $x = -1$, the slope is $f'(-1) = 4(-1)^3 = -4$. Slope is negative.

b. When $x = 0$, the slope is $f'(0) = 4(0)^3 = 0$. Slope is zero.

c. When $x = 1$, the slope is $f'(1) = 4(1)^3 = 4$. Slope is positive.

Note that the slope of the graph is negative at the point $(-1, 1)$, the slope is zero at the point $(0, 0)$, and the slope is positive at the point $(1, 1)$. See Figure 4.16.

EXAMPLE 4 Finding an Equation of a Tangent Line

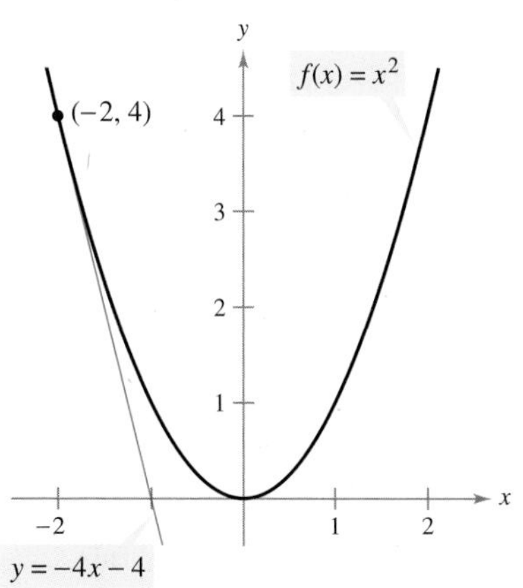

The line $y = -4x - 4$ is tangent to the graph of $f(x) = x^2$ at the point $(-2, 4)$.
Figure 4.17

Find an equation of the tangent line to the graph of $f(x) = x^2$ when $x = -2$.

Solution To find the *point* on the graph of f, evaluate the original function at $x = -2$.

$$(-2, f(-2)) = (-2, 4) \qquad \text{Point on graph}$$

To find the *slope* of the graph when $x = -2$, evaluate the derivative, $f'(x) = 2x$, at $x = -2$.

$$m = f'(-2) = -4 \qquad \text{Slope of graph at } (-2, 4)$$

Now, using the point-slope form of the equation of a line, you can write

$$y - y_1 = m(x - x_1) \qquad \text{Point-slope form}$$
$$y - 4 = -4[x - (-2)] \qquad \text{Substitute for } y_1, m, \text{ and } x_1.$$
$$y = -4x - 4. \qquad \text{Simplify.}$$

See Figure 4.17. ■

The Constant Multiple Rule

THEOREM 4.4 THE CONSTANT MULTIPLE RULE

If f is a differentiable function and c is a real number, then cf is also differentiable and $\frac{d}{dx}[cf(x)] = cf'(x)$.

PROOF

$$\begin{aligned}
\frac{d}{dx}[cf(x)] &= \lim_{\Delta x \to 0} \frac{cf(x + \Delta x) - cf(x)}{\Delta x} && \text{Definition of derivative}\\
&= \lim_{\Delta x \to 0} c\left[\frac{f(x + \Delta x) - f(x)}{\Delta x}\right]\\
&= c\left[\lim_{\Delta x \to 0} \frac{f(x + \Delta x) - f(x)}{\Delta x}\right] && \text{Apply Theorem 3.2.}\\
&= cf'(x)
\end{aligned}$$

■

Informally, the Constant Multiple Rule states that constants can be factored out of the differentiation process, even if the constants appear in the denominator.

$$\frac{d}{dx}[cf(x)] = c\frac{d}{dx}[\;f(x)] = cf'(x)$$

$$\begin{aligned}
\frac{d}{dx}\left[\frac{f(x)}{c}\right] &= \frac{d}{dx}\left[\left(\frac{1}{c}\right)f(x)\right]\\
&= \left(\frac{1}{c}\right)\frac{d}{dx}[\;f(x)] = \left(\frac{1}{c}\right)f'(x)
\end{aligned}$$

EXAMPLE 5 Using the Constant Multiple Rule

Function	*Derivative*
a. $y = \frac{2}{x}$	$\frac{dy}{dx} = \frac{d}{dx}[2x^{-1}] = 2\frac{d}{dx}[x^{-1}] = 2(-1)x^{-2} = -\frac{2}{x^2}$
b. $f(t) = \frac{4t^2}{5}$	$f'(t) = \frac{d}{dt}\left[\frac{4}{5}t^2\right] = \frac{4}{5}\frac{d}{dt}[t^2] = \frac{4}{5}(2t) = \frac{8}{5}t$
c. $y = 2\sqrt{x}$	$\frac{dy}{dx} = \frac{d}{dx}[2x^{1/2}] = 2\left(\frac{1}{2}x^{-1/2}\right) = x^{-1/2} = \frac{1}{\sqrt{x}}$
d. $y = \frac{1}{2\sqrt[3]{x^2}}$	$\frac{dy}{dx} = \frac{d}{dx}\left[\frac{1}{2}x^{-2/3}\right] = \frac{1}{2}\left(-\frac{2}{3}\right)x^{-5/3} = -\frac{1}{3x^{5/3}}$
e. $y = -\frac{3x}{2}$	$y' = \frac{d}{dx}\left[-\frac{3}{2}x\right] = -\frac{3}{2}(1) = -\frac{3}{2}$

■

The Constant Multiple Rule and the Power Rule can be combined into one rule. The combination rule is

$$\frac{d}{dx}[cx^n] = cnx^{n-1}.$$

STUDY TIP In Example 6, pay special attention to the numerical factor in the denominator. If its exponent is positive, keep it in the denominator. Only in part (d) is the numerical factor moved to the numerator.

EXAMPLE 6 Using Parentheses When Differentiating

Original Function	*Rewrite*	*Differentiate*	*Simplify*
a. $y = \frac{5}{2x^3}$	$y = \frac{5}{2}(x^{-3})$	$y' = \frac{5}{2}(-3x^{-4})$	$y' = -\frac{15}{2x^4}$
b. $y = \frac{5}{(2x)^3}$	$y = \frac{5}{8}(x^{-3})$	$y' = \frac{5}{8}(-3x^{-4})$	$y' = -\frac{15}{8x^4}$
c. $y = \frac{7}{3x^{-2}}$	$y = \frac{7}{3}(x^2)$	$y' = \frac{7}{3}(2x)$	$y' = \frac{14x}{3}$
d. $y = \frac{7}{(3x)^{-2}}$	$y = 63(x^2)$	$y' = 63(2x)$	$y' = 126x$

■

The Sum and Difference Rules

THEOREM 4.5 THE SUM AND DIFFERENCE RULES

The sum (or difference) of two differentiable functions f and g is itself differentiable. Moreover, the derivative of $f + g$ (or $f - g$) is the sum (or difference) of the derivatives of f and g.

$$\frac{d}{dx}[f(x) + g(x)] = f'(x) + g'(x) \qquad \text{Sum Rule}$$

$$\frac{d}{dx}[f(x) - g(x)] = f'(x) - g'(x) \qquad \text{Difference Rule}$$

PROOF A proof of the Sum Rule follows from Theorem 3.2. (The Difference Rule can be proved in a similar way.)

$$\begin{aligned}
\frac{d}{dx}[f(x) + g(x)] &= \lim_{\Delta x \to 0} \frac{[f(x + \Delta x) + g(x + \Delta x)] - [f(x) + g(x)]}{\Delta x} \\
&= \lim_{\Delta x \to 0} \frac{f(x + \Delta x) + g(x + \Delta x) - f(x) - g(x)}{\Delta x} \\
&= \lim_{\Delta x \to 0} \left[\frac{f(x + \Delta x) - f(x)}{\Delta x} + \frac{g(x + \Delta x) - g(x)}{\Delta x}\right] \\
&= \lim_{\Delta x \to 0} \frac{f(x + \Delta x) - f(x)}{\Delta x} + \lim_{\Delta x \to 0} \frac{g(x + \Delta x) - g(x)}{\Delta x} \\
&= f'(x) + g'(x)
\end{aligned}$$

■

The Sum and Difference Rules can be extended to any finite number of functions. For instance, if $F(x) = f(x) + g(x) - h(x) - k(x)$, then $F'(x) = f'(x) + g'(x) - h'(x) - k'(x)$.

EXAMPLE 7 Using the Sum and Difference Rules

Function	*Derivative*
a. $f(x) = x^3 - 4x + 5$	$f'(x) = 3x^2 - 4$
b. $g(x) = -\frac{x^4}{2} + 3x^3 - 2x$	$g'(x) = -2x^3 + 9x^2 - 2$

■

Rates of Change

You have seen how the derivative is used to determine slope. The derivative can also be used to determine the rate of change of one variable with respect to another. Applications involving rates of change occur in a wide variety of fields. A few examples are population growth rates, production rates, water flow rates, velocity, and acceleration.

A common use for rate of change is to describe the motion of an object moving in a straight line. In such problems, it is customary to use either a horizontal or a vertical line with a designated origin to represent the line of motion. On such lines, movement to the right (or upward) is considered to be in the positive direction, and movement to the left (or downward) is considered to be in the negative direction.

The function s that gives the position (relative to the origin) of an object as a function of time t is called a **position function.** If, over a period of time Δt, the object changes its position by the amount $\Delta s = s(t + \Delta t) - s(t)$, then, by the familiar formula

$$\text{Rate} = \frac{\text{distance}}{\text{time}}$$

the **average velocity** is

$$\frac{\text{Change in distance}}{\text{Change in time}} = \frac{\Delta s}{\Delta t}.$$

EXAMPLE 8 Finding Average Velocity of a Falling Object

If a billiard ball is dropped from a height of 100 feet, its height s at time t is given by the position function

$$s = -16t^2 + 100 \qquad \text{Position function}$$

where s is measured in feet and t is measured in seconds. Find the average velocity over each of the following time intervals.

a. $[1, 2]$ **b.** $[1, 1.5]$ **c.** $[1, 1.1]$

Solution

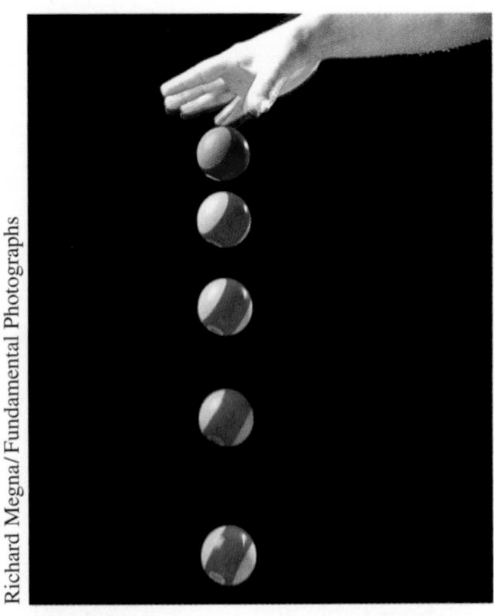
Richard Megna/Fundamental Photographs

Time-lapse photograph of a free-falling billiard ball

a. For the interval $[1, 2]$, the object falls from a height of $s(1) = -16(1)^2 + 100 = 84$ feet to a height of $s(2) = -16(2)^2 + 100 = 36$ feet. The average velocity is

$$\frac{\Delta s}{\Delta t} = \frac{s(2) - s(1)}{2 - 1} = \frac{36 - 84}{2 - 1} = \frac{-48}{1} = -48 \text{ feet per second.}$$

b. For the interval $[1, 1.5]$, the object falls from a height of 84 feet to a height of 64 feet. The average velocity is

$$\frac{\Delta s}{\Delta t} = \frac{s(1.5) - s(1)}{1.5 - 1} = \frac{64 - 84}{1.5 - 1} = \frac{-20}{0.5} = -40 \text{ feet per second.}$$

c. For the interval $[1, 1.1]$, the object falls from a height of 84 feet to a height of 80.64 feet. The average velocity is

$$\frac{\Delta s}{\Delta t} = \frac{s(1.1) - s(1)}{1.1 - 1} = \frac{80.64 - 84}{1.1 - 1} = \frac{-3.36}{0.1} = -33.6 \text{ feet per second.}$$

Note that the average velocities are *negative*, indicating that the object is moving downward. ■

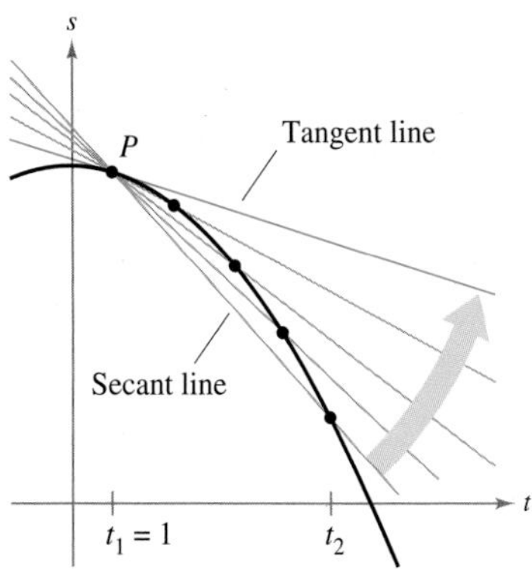

The average velocity between t_1 and t_2 is the slope of the secant line, and the instantaneous velocity at t_1 is the slope of the tangent line.
Figure 4.18

Suppose that in Example 8 you wanted to find the *instantaneous* velocity (or simply the velocity) of the object when $t = 1$. Just as you can approximate the slope of the tangent line by calculating the slope of the secant line, you can approximate the velocity at $t = 1$ by calculating the average velocity over a small interval $[1, 1 + \Delta t]$ (see Figure 4.18). By taking the limit as Δt approaches zero, you obtain the velocity when $t = 1$. Try doing this—you will find that the velocity when $t = 1$ is -32 feet per second.

In general, if $s = s(t)$ is the position function for an object moving along a straight line, the **velocity** of the object at time t is

$$v(t) = \lim_{\Delta t \to 0} \frac{s(t + \Delta t) - s(t)}{\Delta t} = s'(t).$$ Velocity function

In other words, the velocity function is the *derivative* of the position function. Velocity can be negative, zero, or positive. The **speed** of an object is the absolute value of its velocity. Speed cannot be negative.

The position of a free-falling object (neglecting air resistance) under the influence of gravity can be represented by the equation

$$s(t) = \frac{1}{2}gt^2 + v_0 t + s_0$$ Position function

where s_0 is the initial height of the object, v_0 is the initial velocity of the object, and g is the acceleration due to gravity. On Earth, the value of g is approximately -32 feet per second per second or -9.8 meters per second per second.

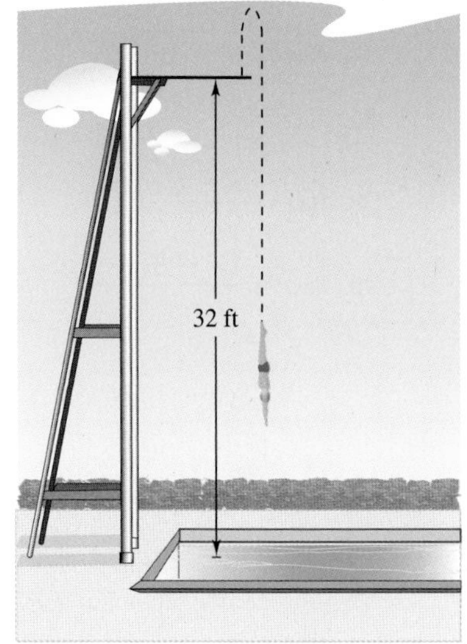

Velocity is positive when an object is rising, and is negative when an object is falling. Notice that the diver moves upward for the first half-second because the velocity is positive for $0 < t < \frac{1}{2}$. When the velocity is 0, the diver has reached the maximum height of the dive.
Figure 4.19

EXAMPLE 9 Using the Derivative to Find Velocity

At time $t = 0$, a diver jumps from a platform diving board that is 32 feet above the water (see Figure 4.19). The position of the diver is given by

$$s(t) = -16t^2 + 16t + 32$$ Position function

where s is measured in feet and t is measured in seconds.

a. When does the diver hit the water?

b. What is the diver's velocity at impact?

Solution

a. To find the time t when the diver hits the water, let $s = 0$ and solve for t.

$$-16t^2 + 16t + 32 = 0$$ Set position function equal to 0.

$$-16(t + 1)(t - 2) = 0$$ Factor.

$$t = -1 \text{ or } 2$$ Solve for t.

Because $t \geq 0$, choose the positive value to conclude that the diver hits the water at $t = 2$ seconds.

b. The velocity at time t is given by the derivative $s'(t) = -32t + 16$. So, the velocity at time $t = 2$ is

$$s'(2) = -32(2) + 16 = -48 \text{ feet per second.}$$ ■

4.2 Exercises

See www.CalcChat.com for worked-out solutions to odd-numbered exercises.

In Exercises 1 and 2, use the graph to estimate the slope of the tangent line to $y = x^n$ at the point $(1, 1)$. Verify your answer analytically. To print an enlarged copy of the graph, go to the website *www.mathgraphs.com*.

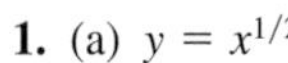
1. (a) $y = x^{1/2}$

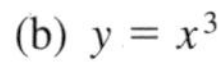
(b) $y = x^3$

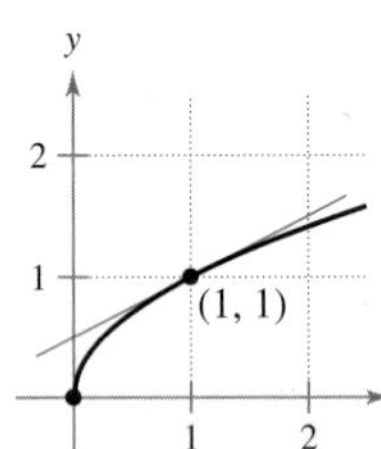

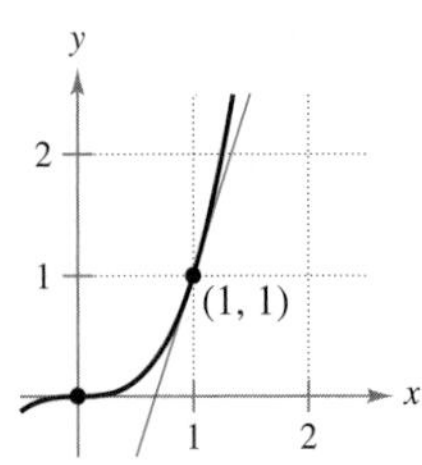

2. (a) $y = x^{-1/2}$ (b) $y = x^{-1}$

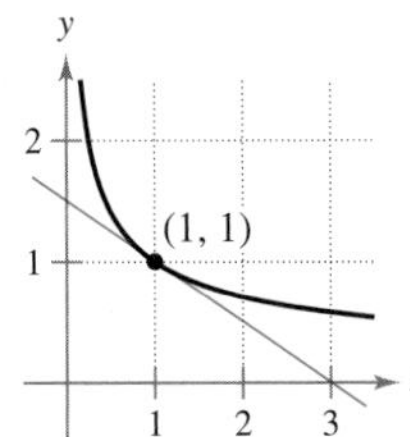

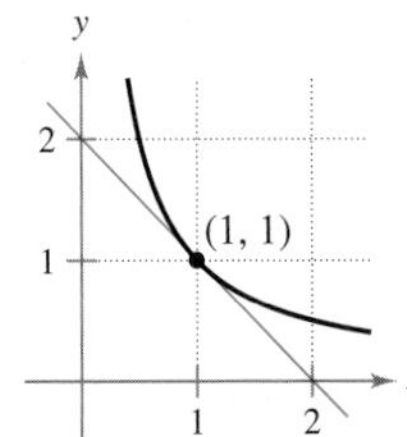

In Exercises 3–20, use the rules of differentiation to find the derivative of the function.

3. $y = 12$
4. $f(x) = -9$
5. $y = x^7$
6. $y = x^{16}$
7. $y = \dfrac{1}{x^5}$
8. $y = \dfrac{1}{x^8}$
9. $f(x) = \sqrt[5]{x}$
10. $g(x) = \sqrt[4]{x}$
11. $f(x) = x + 11$
12. $g(x) = 3x - 1$
13. $f(t) = -2t^2 + 3t - 6$
14. $y = t^2 + 2t - 3$
15. $y = 16 - 3x - \frac{1}{2}x^2$
16. $h(s) = 480 + 64s - 16s^2$
17. $g(x) = x^2 + 4x^3$
18. $y = 8 - x^3$
19. $s(t) = t^3 + 5t^2 - 3t + 8$
20. $f(x) = 2x^3 - x^2 + 3x$

In Exercises 21–26, complete the table, using Example 6 as a model.

Original Function	*Rewrite*	*Differentiate*	*Simplify*
21. $y = \dfrac{5}{2x^2}$			
22. $y = \dfrac{2}{3x^2}$			
23. $y = \dfrac{6}{(5x)^3}$			
24. $y = \dfrac{\pi}{(3x)^2}$			
25. $y = \dfrac{\sqrt{x}}{x}$			
26. $y = \dfrac{4}{x^{-3}}$			

In Exercises 27–32, find the slope of the graph of the function at the given point. Use the *derivative* feature of a graphing utility to confirm your results.

Function	*Point*
27. $f(x) = \dfrac{8}{x^2}$	$(2, 2)$
28. $f(t) = 3 - \dfrac{3}{5t}$	$\left(\frac{3}{5}, 2\right)$
29. $f(x) = -\frac{1}{2} + \frac{7}{5}x^3$	$\left(0, -\frac{1}{2}\right)$
30. $y = 3x^3 - 10$	$(2, 14)$
31. $y = (4x + 1)^2$	$(0, 1)$
32. $f(x) = 3(5 - x)^2$	$(5, 0)$

In Exercises 33–46, find the derivative of the function.

33. $f(x) = x^2 + 5 - 3x^{-2}$
34. $f(x) = x^2 - 3x - 3x^{-2}$
35. $g(t) = t^2 - \dfrac{4}{t^3}$
36. $f(x) = x + \dfrac{1}{x^2}$
37. $f(x) = \dfrac{4x^3 + 3x^2}{x}$
38. $f(x) = \dfrac{x^3 - 6}{x^2}$
39. $f(x) = \dfrac{x^3 - 3x^2 + 4}{x^2}$
40. $h(x) = \dfrac{2x^2 - 3x + 1}{x}$
41. $y = x(x^2 + 1)$
42. $y = 3x(6x - 5x^2)$
43. $f(x) = \sqrt{x} - 6\sqrt[3]{x}$
44. $f(x) = \sqrt[3]{x} + \sqrt[5]{x}$
45. $h(s) = s^{4/5} - s^{2/3}$
46. $f(t) = t^{2/3} - t^{1/3} + 4$

In Exercises 47–50, (a) find an equation of the tangent line to the graph of f at the given point, (b) use a graphing utility to graph the function and its tangent line at the point, and (c) use the *derivative* feature of a graphing utility to confirm your results.

Function	*Point*
47. $y = x^4 - 3x^2 + 2$	$(1, 0)$
48. $y = x^3 + x$	$(-1, -2)$
49. $f(x) = \dfrac{2}{\sqrt[4]{x^3}}$	$(1, 2)$
50. $y = (x^2 + 2x)(x + 1)$	$(1, 6)$

In Exercises 51–54, determine the point(s) (if any) at which the graph of the function has a horizontal tangent line.

51. $y = x^4 - 2x^2 + 3$

52. $y = x^3 + x$

53. $y = \frac{1}{x^2}$

54. $y = x^2 + 9$

In Exercises 55–60, find k such that the line is tangent to the graph of the function.

	Function	*Line*
55.	$f(x) = x^2 - kx$	$y = 5x - 4$
56.	$f(x) = k - x^2$	$y = -6x + 1$
57.	$f(x) = \frac{k}{x}$	$y = -\frac{3}{4}x + 3$
58.	$f(x) = k\sqrt{x}$	$y = x + 4$
59.	$f(x) = kx^3$	$y = x + 1$
60.	$f(x) = kx^4$	$y = 4x - 1$

61. Sketch the graph of a function f such that $f' > 0$ for all x and the rate of change of the function is decreasing.

CAPSTONE

62. Use the graph of f to answer each question. To print an enlarged copy of the graph, go to the website *www.mathgraphs.com.*

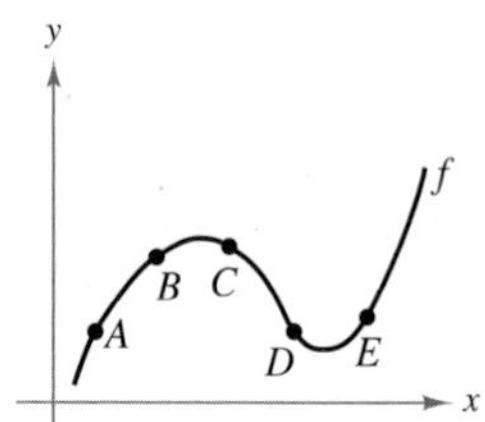

(a) Between which two consecutive points is the average rate of change of the function greatest?

(b) Is the average rate of change of the function between A and B greater than or less than the instantaneous rate of change at B?

(c) Sketch a tangent line to the graph between C and D such that the slope of the tangent line is the same as the average rate of change of the function between C and D.

WRITING ABOUT CONCEPTS

In Exercises 63 and 64, the relationship between f and g is given. Explain the relationship between f' and g'.

63. $g(x) = f(x) + 6$

64. $g(x) = -5f(x)$

WRITING ABOUT CONCEPTS (continued)

In Exercises 65 and 66, the graphs of a function f and its derivative f' are shown on the same set of coordinate axes. Label the graphs as f or f' and write a short paragraph stating the criteria you used in making your selection. To print an enlarged copy of the graph, go to the website *www.mathgraphs.com.*

65.

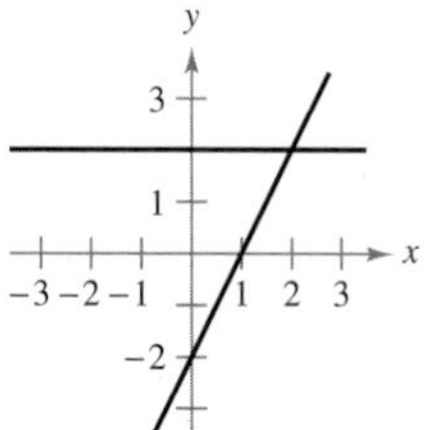

66.

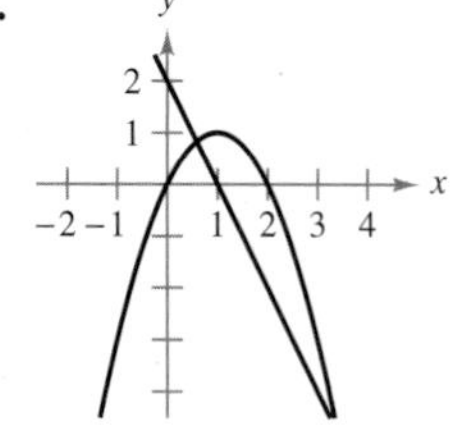

67. Sketch the graphs of $y = x^2$ and $y = -x^2 + 6x - 5$, and sketch the two lines that are tangent to both graphs. Find equations of these lines.

68. Show that the graphs of the two equations $y = x$ and $y = 1/x$ have tangent lines that are perpendicular to each other at their point of intersection.

In Exercises 69 and 70, find an equation of the tangent line to the graph of the function f through the point (x_0, y_0) not on the graph. To find the point of tangency (x, y) on the graph of f, solve the equation

$$f'(x) = \frac{y_0 - y}{x_0 - x}.$$

69. $f(x) = \sqrt{x}$

$(x_0, y_0) = (-4, 0)$

70. $f(x) = \frac{2}{x}$

$(x_0, y_0) = (5, 0)$

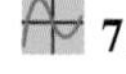

71. ***Linear Approximation*** Use a graphing utility, with a square window setting, to zoom in on the graph of

$$f(x) = 4 - \tfrac{1}{2}x^2$$

to approximate $f'(1)$. Use the derivative to find $f'(1)$.

72. ***Linear Approximation*** Use a graphing utility, with a square window setting, to zoom in on the graph of

$$f(x) = 4\sqrt{x} + 1$$

to approximate $f'(4)$. Use the derivative to find $f'(4)$.

73. ***Linear Approximation*** Consider the function $f(x) = x^{3/2}$ with the solution point $(4, 8)$.

(a) Use a graphing utility to graph f. Use the *zoom* feature to obtain successive magnifications of the graph in the neighborhood of the point $(4, 8)$. After zooming in a few times, the graph should appear nearly linear. Use the *trace* feature to determine the coordinates of a point near $(4, 8)$. Find an equation of the secant line $S(x)$ through the two points.

(b) Find the equation of the line

$$T(x) = f'(4)(x - 4) + f(4)$$

tangent to the graph of f passing through the given point. Why are the linear functions S and T nearly the same?

(c) Use a graphing utility to graph f and T in the same viewing window. Note that T is a good approximation of f when x is close to 4. What happens to the accuracy of the approximation as you move farther away from the point of tangency?

(d) Demonstrate the conclusion in part (c) by completing the table.

Δx	-3	-2	-1	-0.5	-0.1	0
$f(4 + \Delta x)$						
$T(4 + \Delta x)$						

Δx	0.1	0.5	1	2	3
$f(4 + \Delta x)$					
$T(4 + \Delta x)$					

74. ***Linear Approximation*** Repeat Exercise 73 for the function $f(x) = x^3$, where $T(x)$ is the line tangent to the graph at the point $(1, 1)$. Explain why the accuracy of the linear approximation decreases more rapidly than in Exercise 73.

True or False? **In Exercises 75–80, determine whether the statement is true or false. If it is false, explain why or give an example that shows it is false.**

75. If $f'(x) = g'(x)$, then $f(x) = g(x)$.

76. If $f(x) = g(x) + c$, then $f'(x) = g'(x)$.

77. If $y = \pi^2$, then $dy/dx = 2\pi$.

78. If $y = x/\pi$, then $dy/dx = 1/\pi$.

79. If $g(x) = 3f(x)$, then $g'(x) = 3f'(x)$.

80. If $f(x) = 1/x^n$, then $f'(x) = 1/(nx^{n-1})$.

In Exercises 81–84, find the average rate of change of the function over the given interval. Compare this average rate of change with the instantaneous rates of change at the endpoints of the interval.

81. $f(t) = 4t + 5$, $[1, 2]$

82. $f(t) = t^2 - 7$, $[3, 3.1]$

83. $f(x) = \dfrac{-1}{x}$, $[1, 2]$

84. $f(x) = \dfrac{1}{x + 1}$, $[0, 3]$

Vertical Motion **In Exercises 85 and 86, use the position function $s(t) = -16t^2 + v_0 t + s_0$ for free-falling objects.**

85. A silver dollar is dropped from the top of a building that is 1362 feet tall.

(a) Determine the position and velocity functions for the coin.

(b) Determine the average velocity on the interval $[1, 2]$.

(c) Find the instantaneous velocities when $t = 1$ second and $t = 2$ seconds.

(d) Find the time required for the coin to reach ground level.

(e) Find the velocity of the coin at impact.

86. A ball is thrown straight down from the top of a 220-foot building with an initial velocity of -22 feet per second. What is its velocity after 3 seconds? What is its velocity after falling 108 feet?

Vertical Motion **In Exercises 87 and 88, use the position function $s(t) = -4.9t^2 + v_0 t + s_0$ for free-falling objects.**

87. A projectile is shot upward from the surface of Earth with an initial velocity of 120 meters per second. What is its velocity after 5 seconds? After 10 seconds?

88. To estimate the height of a building, a stone is dropped from the top of the building into a pool of water at ground level. How high is the building if the splash is seen 5.6 seconds after the stone is dropped?

Think About It **In Exercises 89 and 90, the graph of a position function is shown. It represents the distance in miles that a person drives during a 10-minute trip to work. Make a sketch of the corresponding velocity function.**

89.

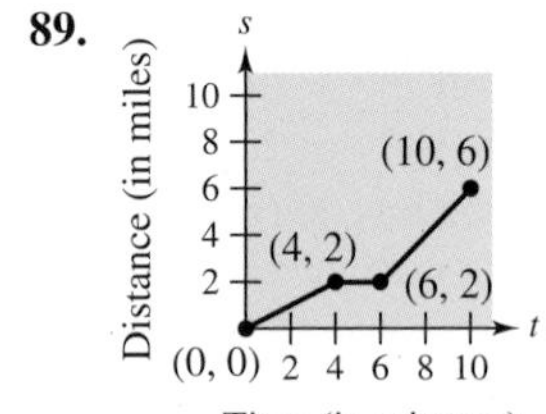

90.

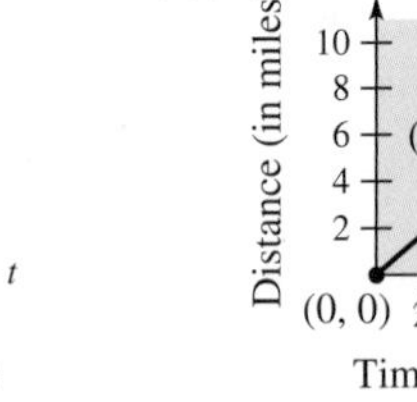

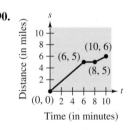

Think About It **In Exercises 91 and 92, the graph of a velocity function is shown. It represents the velocity in miles per hour during a 10-minute drive to work. Make a sketch of the corresponding position function.**

91.

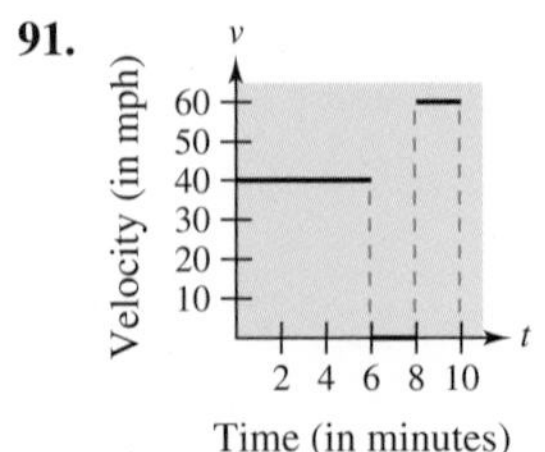

92.

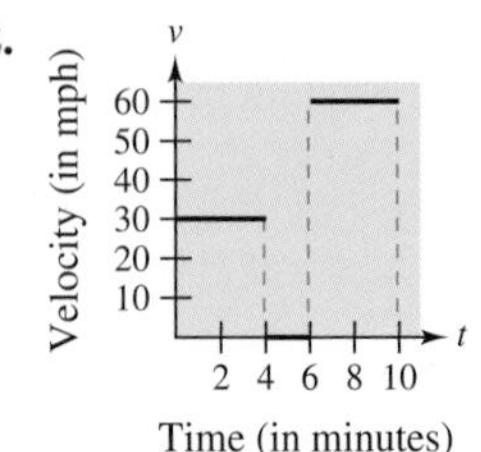

93. ***Modeling Data*** The stopping distance of an automobile, on dry, level pavement, traveling at a speed v (kilometers per hour), is the distance R (meters) the car travels during the reaction time of the driver plus the distance B (meters) the car travels after the brakes are applied (see figure). The table shows the results of an experiment.

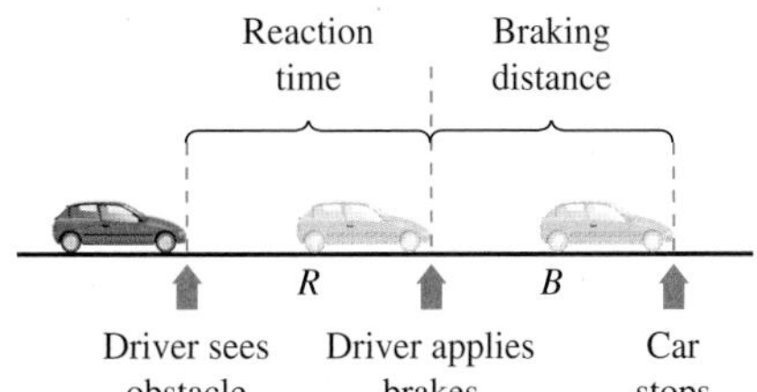

Speed, v	20	40	60	80	100
Reaction Time Distance, R	8.3	16.7	25.0	33.3	41.7
Braking Time Distance, B	2.3	9.0	20.2	35.8	55.9

(a) Use the regression capabilities of a graphing utility to find a linear model for reaction time distance.

(b) Use the regression capabilities of a graphing utility to find a quadratic model for braking distance.

(c) Determine the polynomial giving the total stopping distance T.

(d) Use a graphing utility to graph the functions R, B, and T in the same viewing window.

(e) Find the derivative of T and the rates of change of the total stopping distance for $v = 40$, $v = 80$, and $v = 100$.

(f) Use the results of this exercise to draw conclusions about the total stopping distance as speed increases.

94. ***Fuel Cost*** A car is driven 15,000 miles a year and gets x miles per gallon. Assume that the average fuel cost is \$2.76 per gallon. Find the annual cost of fuel C as a function of x and use this function to complete the table.

x	10	15	20	25	30	35	40
C							
dC/dx							

Who would benefit more from a one-mile-per-gallon increase in fuel efficiency—the driver of a car that gets 15 miles per gallon or the driver of a car that gets 35 miles per gallon? Explain.

95. ***Volume*** The volume of a cube with sides of length s is given by $V = s^3$. Find the rate of change of the volume with respect to s when $s = 6$ centimeters.

96. ***Area*** The area of a square with sides of length s is given by $A = s^2$. Find the rate of change of the area with respect to s when $s = 6$ meters.

97. ***Velocity*** Verify that the average velocity over the time interval $[t_0 - \Delta t, t_0 + \Delta t]$ is the same as the instantaneous velocity at $t = t_0$ for the position function

$$s(t) = -\tfrac{1}{2}at^2 + c.$$

98. ***Inventory Management*** The annual inventory cost C for a manufacturer is

$$C = \frac{1{,}008{,}000}{Q} + 6.3Q$$

where Q is the order size when the inventory is replenished. Find the change in annual cost when Q is increased from 350 to 351, and compare this with the instantaneous rate of change when $Q = 350$.

99. ***Writing*** The number of gallons N of regular unleaded gasoline sold by a gasoline station at a price of p dollars per gallon is given by

$$N = f(p).$$

(a) Describe the meaning of $f'(2.979)$.

(b) Is $f'(2.979)$ usually positive or negative? Explain.

100. ***Newton's Law of Cooling*** This law states that the rate of change of the temperature of an object is proportional to the difference between the object's temperature T and the temperature T_a of the surrounding medium. Write an equation for this law.

101. Find an equation of the parabola

$$y = ax^2 + bx + c$$

that passes through $(0, 1)$ and is tangent to the line $y = x - 1$ at $(1, 0)$.

102. Let (a, b) be an arbitrary point on the graph of

$$y = \frac{1}{x}, \quad x > 0.$$

Prove that the area of the triangle formed by the tangent line through (a, b) and the coordinate axes is 2.

103. Find the tangent line(s) to the curve

$$y = x^3 - 9x$$

through the point $(1, -9)$.

104. Find the equation(s) of the tangent line(s) to the parabola $y = x^2$ through the given point.

(a) $(0, a)$ (b) $(a, 0)$

Are there any restrictions on the constant a?

105. Find a and b such that

$$f(x) = \begin{cases} ax^3, & x \le 2 \\ x^2 + b, & x > 2 \end{cases}$$

is differentiable everywhere.

106. Show that the graph of the function given by

$$f(x) = x^5 + 3x^3 + 5x$$

does not have a tangent line with a slope of 3.

4.3 Product and Quotient Rules and Higher-Order Derivatives

- **Find the derivative of a function using the Product Rule.**
- **Find the derivative of a function using the Quotient Rule.**
- **Find a higher-order derivative of a function.**

The Product Rule

In Section 4.2 you learned that the derivative of the sum of two functions is simply the sum of their derivatives. The rules for the derivatives of the product and quotient of two functions are not as simple.

THEOREM 4.6 THE PRODUCT RULE

The product of two differentiable functions f and g is itself differentiable. Moreover, the derivative of fg is the first function times the derivative of the second, plus the second function times the derivative of the first.

$$\frac{d}{dx}[f(x)g(x)] = f(x)g'(x) + g(x)f'(x)$$

NOTE A version of the Product Rule that some people prefer is

$$\frac{d}{dx}[f(x)g(x)] = f'(x)g(x) + f(x)g'(x).$$

The advantage of this form is that it generalizes easily to products of three or more factors.

PROOF Some mathematical proofs, such as the proof of the Sum Rule, are straightforward. Others involve clever steps that may appear unmotivated to a reader. This proof involves such a step—subtracting and adding the same quantity—which is shown in color.

$$\begin{aligned}
\frac{d}{dx}[f(x)g(x)] &= \lim_{\Delta x\to 0}\frac{f(x+\Delta x)g(x+\Delta x) - f(x)g(x)}{\Delta x}\\
&= \lim_{\Delta x\to 0}\frac{f(x+\Delta x)g(x+\Delta x) - f(x+\Delta x)g(x) + f(x+\Delta x)g(x) - f(x)g(x)}{\Delta x}\\
&= \lim_{\Delta x\to 0}\left[f(x+\Delta x)\frac{g(x+\Delta x)-g(x)}{\Delta x} + g(x)\frac{f(x+\Delta x)-f(x)}{\Delta x}\right]\\
&= \lim_{\Delta x\to 0}\left[f(x+\Delta x)\frac{g(x+\Delta x)-g(x)}{\Delta x}\right] + \lim_{\Delta x\to 0}\left[g(x)\frac{f(x+\Delta x)-f(x)}{\Delta x}\right]\\
&= \lim_{\Delta x\to 0} f(x+\Delta x)\cdot\lim_{\Delta x\to 0}\frac{g(x+\Delta x)-g(x)}{\Delta x} + \lim_{\Delta x\to 0} g(x)\cdot\lim_{\Delta x\to 0}\frac{f(x+\Delta x)-f(x)}{\Delta x}\\
&= f(x)g'(x) + g(x)f'(x)
\end{aligned}$$

■

Note that $\lim_{\Delta x\to 0} f(x+\Delta x) = f(x)$ because f is given to be differentiable and therefore is continuous.

The Product Rule can be extended to cover products involving more than two factors. For example, if f, g, and h are differentiable functions of x, then

$$\frac{d}{dx}[f(x)g(x)h(x)] = f'(x)g(x)h(x) + f(x)g'(x)h(x) + f(x)g(x)h'(x).$$

For instance, the derivative of $y = x^2(x+1)(2x-3)$ is

$$\begin{aligned}
\frac{dy}{dx} &= 2x(x+1)(2x-3) + x^2(1)(2x-3) + x^2(x+1)2\\
&= x(4x^2 - 2x - 6 + 2x^2 - 3x + 2x^2 + 2x)\\
&= x(8x^2 - 3x - 6).
\end{aligned}$$

THE PRODUCT RULE

When Leibniz originally wrote a formula for the Product Rule, he was motivated by the expression

$$(x + dx)(y + dy) - xy$$

from which he subtracted $dx\,dy$ (as being negligible) and obtained the differential form $x\,dy + y\,dx$. This derivation resulted in the traditional form of the Product Rule. (*Source: The History of Mathematics by David M. Burton*)

The derivative of a product of two functions is not (in general) given by the product of the derivatives of the two functions. To see this, try comparing the product of the derivatives of $f(x) = 3x - 2x^2$ and $g(x) = 5 + 4x$ with the derivative in Example 1.

EXAMPLE 1 Using the Product Rule

Find the derivative of

$$h(x) = (3x - 2x^2)(5 + 4x).$$

Solution

$$\begin{aligned} h'(x) &= \underbrace{(3x - 2x^2)}_{\text{First}}\underbrace{\frac{d}{dx}[5 + 4x]}_{\text{Derivative of second}} + \underbrace{(5 + 4x)}_{\text{Second}}\underbrace{\frac{d}{dx}[3x - 2x^2]}_{\text{Derivative of first}} && \text{Apply Product Rule.} \\ &= (3x - 2x^2)(4) + (5 + 4x)(3 - 4x) && \text{Differentiate.} \\ &= (12x - 8x^2) + (15 - 8x - 16x^2) \\ &= -24x^2 + 4x + 15 && \text{Simplify.} \end{aligned}$$

■

In Example 1, you have the option of finding the derivative with or without the Product Rule. To find the derivative without the Product Rule, you can write

$$\begin{aligned} D_x[(3x - 2x^2)(5 + 4x)] &= D_x[-8x^3 + 2x^2 + 15x] && \text{Multiply binomials.} \\ &= -24x^2 + 4x + 15. \end{aligned}$$

EXAMPLE 2 Product Rule Versus Constant Multiple Rule

Find the derivative of each function.

a. $y = \sqrt{x}g(x)$

b. $y = \sqrt{2}g(x)$

Solution

a. Using the Product Rule, you obtain

$$\begin{aligned} \frac{dy}{dx} &= \sqrt{x}\left(\frac{d}{dx}[g(x)]\right) + g(x)\left(\frac{d}{dx}[\sqrt{x}]\right) && \text{Apply Product Rule.} \\ &= \sqrt{x}g'(x) + g(x)\left(\frac{1}{2}x^{-1/2}\right) && \text{Differentiate.} \\ &= \sqrt{x}g'(x) + g(x)\frac{1}{2\sqrt{x}}. && \text{Simplify.} \end{aligned}$$

b. Using the Constant Multiple Rule, you obtain

$$\frac{dy}{dx} = \sqrt{2}\frac{d}{dx}g(x) = \sqrt{2}g'(x).$$

■

In Example 2, notice that the Product Rule is used when both factors of the product are variable, and the Constant Multiple Rule is used when one of the two factors is a constant. The Constant Multiple Rule also applies to fractions with a constant denominator, as shown below.

$$y = \frac{2x^3 + 5x}{7} \quad \Longrightarrow \quad \frac{dy}{dx} = \frac{1}{7}\left[\frac{d}{dx}(2x^3 + 5x)\right] = \frac{1}{7}(6x^2 + 5)$$

The Quotient Rule

THEOREM 4.7 THE QUOTIENT RULE

The quotient f/g of two differentiable functions f and g is itself differentiable at all values of x for which $g(x) \neq 0$. Moreover, the derivative of f/g is given by *the denominator times the derivative of the numerator minus the numerator times the derivative of the denominator, all divided by the square of the denominator.*

$$\frac{d}{dx}\left[\frac{f(x)}{g(x)}\right] = \frac{g(x)f'(x) - f(x)g'(x)}{[g(x)]^2}, \quad g(x) \neq 0$$

STUDY TIP It is useful to learn the *verbal* version of the Quotient Rule. This is given in italics in Theorem 4.7.

NOTE $\lim_{\Delta x \to 0} g(x + \Delta x) = g(x)$ because g is given to be differentiable and therefore is continuous.

PROOF As with the proof of Theorem 4.6, the key to this proof is subtracting and adding the same quantity.

$$\frac{d}{dx}\left[\frac{f(x)}{g(x)}\right] = \lim_{\Delta x \to 0} \frac{\dfrac{f(x + \Delta x)}{g(x + \Delta x)} - \dfrac{f(x)}{g(x)}}{\Delta x} \qquad \text{Definition of derivative}$$

$$= \lim_{\Delta x \to 0} \frac{g(x)f(x + \Delta x) - f(x)g(x + \Delta x)}{\Delta x g(x)g(x + \Delta x)}$$

$$= \lim_{\Delta x \to 0} \frac{g(x)f(x + \Delta x) - f(x)g(x) + f(x)g(x) - f(x)g(x + \Delta x)}{\Delta x g(x)g(x + \Delta x)}$$

$$= \frac{\lim\limits_{\Delta x \to 0} \dfrac{g(x)[f(x + \Delta x) - f(x)]}{\Delta x} - \lim\limits_{\Delta x \to 0} \dfrac{f(x)[g(x + \Delta x) - g(x)]}{\Delta x}}{\lim\limits_{\Delta x \to 0} [g(x)g(x + \Delta x)]}$$

$$= \frac{g(x)\left[\lim\limits_{\Delta x \to 0} \dfrac{f(x + \Delta x) - f(x)}{\Delta x}\right] - f(x)\left[\lim\limits_{\Delta x \to 0} \dfrac{g(x + \Delta x) - g(x)}{\Delta x}\right]}{\lim\limits_{\Delta x \to 0} [g(x)g(x + \Delta x)]}$$

$$= \frac{g(x)f'(x) - f(x)g'(x)}{[g(x)]^2}$$

■

TECHNOLOGY A graphing utility can be used to compare the graph of a function with the graph of its derivative. For instance, in Figure 4.20, the graph of the function in Example 3 appears to have two points that have horizontal tangent lines. What are the values of y' at these two points?

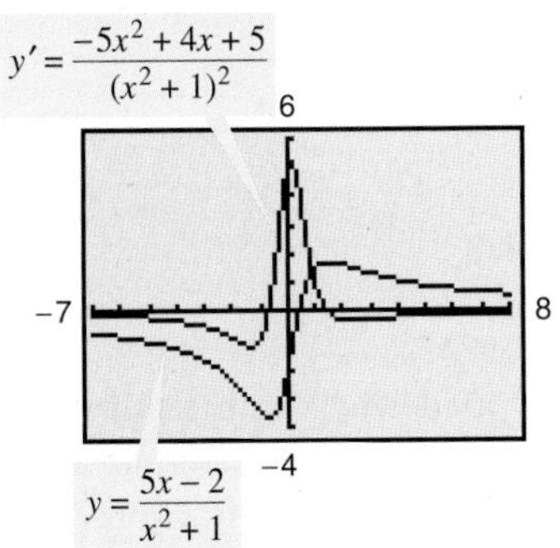

Graphical comparison of a function and its derivative
Figure 4.20

EXAMPLE 3 Using the Quotient Rule

Find the derivative of

$$y = \frac{5x - 2}{x^2 + 1}.$$

Solution

$$\frac{d}{dx}\left[\frac{5x - 2}{x^2 + 1}\right] = \frac{(x^2 + 1)\frac{d}{dx}[5x - 2] - (5x - 2)\frac{d}{dx}[x^2 + 1]}{(x^2 + 1)^2} \qquad \text{Apply Quotient Rule.}$$

$$= \frac{(x^2 + 1)(5) - (5x - 2)(2x)}{(x^2 + 1)^2} \qquad \text{Differentiate.}$$

$$= \frac{(5x^2 + 5) - (10x^2 - 4x)}{(x^2 + 1)^2}$$

$$= \frac{-5x^2 + 4x + 5}{(x^2 + 1)^2} \qquad \text{Simplify.}$$

■

Note the use of parentheses in Example 3. A liberal use of parentheses is recommended for *all* types of differentiation problems. For instance, with the Quotient Rule, it is a good idea to enclose all factors and derivatives in parentheses, and to pay special attention to the subtraction required in the numerator.

When differentiation rules were introduced in the preceding section, the need for rewriting *before* differentiating was emphasized. The next example illustrates this point with the Quotient Rule.

EXAMPLE 4 Rewriting Before Differentiating

Find an equation of the tangent line to the graph of $f(x) = \dfrac{3 - (1/x)}{x + 5}$ at $(-1, 1)$.

Solution Begin by rewriting the function.

$$f(x) = \frac{3 - (1/x)}{x + 5} \qquad \text{Write original function.}$$

$$= \frac{x\left(3 - \frac{1}{x}\right)}{x(x + 5)} \qquad \text{Multiply numerator and denominator by } x.$$

$$= \frac{3x - 1}{x^2 + 5x} \qquad \text{Rewrite.}$$

$$f'(x) = \frac{(x^2 + 5x)(3) - (3x - 1)(2x + 5)}{(x^2 + 5x)^2} \qquad \text{Quotient Rule}$$

$$= \frac{(3x^2 + 15x) - (6x^2 + 13x - 5)}{(x^2 + 5x)^2}$$

$$= \frac{-3x^2 + 2x + 5}{(x^2 + 5x)^2} \qquad \text{Simplify.}$$

To find the slope at $(-1, 1)$, evaluate $f'(-1)$.

$$f'(-1) = 0 \qquad \text{Slope of graph at } (-1, 1)$$

Then, using the point-slope form of the equation of a line, you can determine that the equation of the tangent line at $(-1, 1)$ is $y = 1$. See Figure 4.21. ■

$f(x) = \dfrac{3 - \frac{1}{x}}{x + 5}$

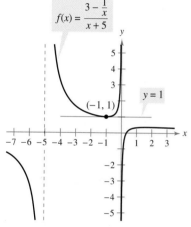

The line $y = 1$ is tangent to the graph of $f(x)$ at the point $(-1, 1)$.
Figure 4.21

Not every quotient needs to be differentiated by the Quotient Rule. For example, each quotient in the next example can be considered as the product of a constant times a function of x. In such cases it is more convenient to use the Constant Multiple Rule.

EXAMPLE 5 Using the Constant Multiple Rule

	Original Function	*Rewrite*	*Differentiate*	*Simplify*
a.	$y = \dfrac{x^2 + 3x}{6}$	$y = \dfrac{1}{6}(x^2 + 3x)$	$y' = \dfrac{1}{6}(2x + 3)$	$y' = \dfrac{2x + 3}{6}$
b.	$y = \dfrac{5x^4}{8}$	$y = \dfrac{5}{8}x^4$	$y' = \dfrac{5}{8}(4x^3)$	$y' = \dfrac{5}{2}x^3$
c.	$y = \dfrac{-3(3x - 2x^2)}{7x}$	$y = -\dfrac{3}{7}(3 - 2x)$	$y' = -\dfrac{3}{7}(-2)$	$y' = \dfrac{6}{7}$
d.	$y = \dfrac{9}{5x^2}$	$y = \dfrac{9}{5}(x^{-2})$	$y' = \dfrac{9}{5}(-2x^{-3})$	$y' = -\dfrac{18}{5x^3}$

■

NOTE To see the benefit of using the Constant Multiple Rule for some quotients, try using the Quotient Rule to differentiate the functions in Example 5—you should obtain the same results, but with more work.

In Section 4.2, the Power Rule was proved only for the case in which the exponent n is a positive integer greater than 1. The next example extends the proof to include negative integer exponents.

EXAMPLE 6 Proof of the Power Rule (Negative Integer Exponents)

Use the Quotient Rule to prove the Power Rule for the case when n is a negative integer.

Solution If n is a negative integer, there exists a positive integer k such that $n = -k$. So, by the Quotient Rule, you can write

$$\begin{aligned}\frac{d}{dx}[x^n] &= \frac{d}{dx}\left[\frac{1}{x^k}\right]\\ &= \frac{x^k(0) - (1)(kx^{k-1})}{(x^k)^2} && \text{Quotient Rule and Power Rule}\\ &= \frac{0 - kx^{k-1}}{x^{2k}}\\ &= -kx^{-k-1}\\ &= nx^{n-1}. && \text{Substitute } n \text{ for } -k.\end{aligned}$$

So, the Power Rule

$$\frac{d}{dx}[x^n] = nx^{n-1} \qquad \text{Power Rule}$$

is valid for any integer. In Exercise 64 in Section 4.5, you are asked to prove the case for which n is any rational number. ■

The summary below shows that much of the work in obtaining a simplified form of a derivative occurs *after* differentiating. Note that two characteristics of a simplified form are the absence of negative exponents and the combining of like terms.

STUDY TIP Initially, the exercise answers for the Product and Quotient Rules will be given in both forms—unsimplified and simplified.

	$f'(x)$ After Differentiating	$f'(x)$ After Simplifying
Example 1	$(3x - 2x^2)(4) + (5 + 4x)(3 - 4x)$	$-24x^2 + 4x + 15$
Example 3	$\dfrac{(x^2 + 1)(5) - (5x - 2)(2x)}{(x^2 + 1)^2}$	$\dfrac{-5x^2 + 4x + 5}{(x^2 + 1)^2}$
Example 4	$\dfrac{(x^2 + 5x)(3) - (3x - 1)(2x + 5)}{(x^2 + 5x)^2}$	$\dfrac{-3x^2 + 2x + 5}{(x^2 + 5x)^2}$

For a quotient with a monomial denominator, it may be advantageous to factor and reduce or rewrite the quotient as a sum or difference before differentiating, as shown below.

	Quotient	*Rewrite*	*Differentiate*	*Simplify*
1.	$\dfrac{12x^3 - 3x^2}{6x^2}$	$\dfrac{1}{2}(4x - 1)$	$\dfrac{1}{2}(4)$	2
2.	$\dfrac{8x^3 + 5x}{4x^2}$	$2x + \dfrac{5}{4}x^{-1}$	$2 - \dfrac{5}{4}x^{-2}$	$2 - \dfrac{5}{4x^2}$

Use the Quotient Rule with the problems above and compare methods.

Higher-Order Derivatives

Just as you can obtain a velocity function by differentiating a position function, you can obtain an **acceleration function** by differentiating a velocity function. Another way of looking at this is that you can obtain an acceleration function by differentiating a position function *twice*.

$$\begin{aligned} &s(t) && \text{Position function} \\ v(t) = {} &s'(t) && \text{Velocity function = rate of change in position} \\ a(t) = v'(t) = {} &s''(t) && \text{Acceleration function = rate of change in velocity} \end{aligned}$$

NOTE The second derivative of f is the derivative of the first derivative of f.

The function given by $a(t)$ is the **second derivative** of $s(t)$ and is denoted by $s''(t)$.

The second derivative is an example of a **higher-order derivative.** You can define derivatives of any positive integer order. For instance, the **third derivative** is the derivative of the second derivative. Higher-order derivatives are denoted as follows.

First derivative:	y',	$f'(x)$,	$\dfrac{dy}{dx}$,	$\dfrac{d}{dx}[f(x)]$,	$D_x[y]$
Second derivative:	y'',	$f''(x)$,	$\dfrac{d^2y}{dx^2}$,	$\dfrac{d^2}{dx^2}[f(x)]$,	$D_x^2[y]$
Third derivative:	y''',	$f'''(x)$,	$\dfrac{d^3y}{dx^3}$,	$\dfrac{d^3}{dx^3}[f(x)]$,	$D_x^3[y]$
Fourth derivative:	$y^{(4)}$,	$f^{(4)}(x)$,	$\dfrac{d^4y}{dx^4}$,	$\dfrac{d^4}{dx^4}[f(x)]$,	$D_x^4[y]$
$\vdots$					
nth derivative:	$y^{(n)}$,	$f^{(n)}(x)$,	$\dfrac{d^ny}{dx^n}$,	$\dfrac{d^n}{dx^n}[f(x)]$,	$D_x^n[y]$

EXAMPLE 7 Finding the Acceleration Due to Gravity

Because the moon has no atmosphere, a falling object on the moon encounters no air resistance. In 1971, astronaut David Scott demonstrated that a feather and a hammer fall at the same rate on the moon. The position function for each of these falling objects is given by

$$s(t) = -0.81t^2 + 2$$

where $s(t)$ is the height in meters and t is the time in seconds. What is the ratio of Earth's gravitational force to the moon's?

NASA

THE MOON

The moon's mass is 7.349×10^{22} kilograms, and Earth's mass is 5.976×10^{24} kilograms. The moon's radius is 1737 kilometers, and Earth's radius is 6378 kilometers. Because the gravitational force on the surface of a planet is directly proportional to its mass and inversely proportional to the square of its radius, the ratio of the gravitational force on Earth to the gravitational force on the moon is

$$\frac{(5.976 \times 10^{24})/6378^2}{(7.349 \times 10^{22})/1737^2} \approx 6.0.$$

Solution To find the acceleration, differentiate the position function twice.

$$\begin{aligned} s(t) &= -0.81t^2 + 2 && \text{Position function} \\ s'(t) &= -1.62t && \text{Velocity function} \\ s''(t) &= -1.62 && \text{Acceleration function} \end{aligned}$$

So, the acceleration due to gravity on the moon is -1.62 meters per second per second. Because the acceleration due to gravity on Earth is -9.8 meters per second per second, the ratio of Earth's gravitational force to the moon's is

$$\begin{aligned} \frac{\text{Earth's gravitational force}}{\text{Moon's gravitational force}} &= \frac{-9.8}{-1.62} \\ &\approx 6.0. \end{aligned}$$

■

4.3 Exercises

See www.CalcChat.com for worked-out solutions to odd-numbered exercises.

In Exercises 1–6, use the Product Rule to differentiate the function.

1. $g(x) = (x^2 + 3)(x^2 - 4x)$
2. $f(x) = (6x + 5)(x^3 - 2)$
3. $h(t) = \sqrt{t}(1 - t^2)$
4. $g(s) = \sqrt{s}(s^2 + 8)$
5. $g(t) = (2t^2 - 3)(4 - t^2 - t^4)$
6. $h(t) = (t^5 - 1)(4t^2 - 7t - 3)$

In Exercises 7–12, use the Quotient Rule to differentiate the function.

7. $f(x) = \dfrac{x}{x^2 + 1}$
8. $g(t) = \dfrac{t^2 + 4}{5t - 3}$
9. $h(x) = \dfrac{\sqrt{x}}{x^3 + 1}$
10. $h(s) = \dfrac{s}{\sqrt{s} - 1}$
11. $f(x) = \dfrac{x^3 + 3x + 2}{x^2 - 1}$
12. $g(x) = \dfrac{3 - 2x - x^2}{x^2 - 1}$

In Exercises 13–20, find $f'(x)$ and $f'(c)$.

Function	*Value of c*
13. $f(x) = \dfrac{5}{x^2}(x + 3)$	$c = 1$
14. $f(x) = \frac{1}{7}(5 - 6x^2)$	$c = 1$
15. $f(x) = (x^3 + 4x)(3x^2 + 2x - 5)$	$c = 0$
16. $f(x) = (x^2 - 2x + 1)(x^3 - 1)$	$c = 1$
17. $f(x) = \dfrac{x^2 - 4}{x - 3}$	$c = 1$
18. $f(x) = \dfrac{x + 5}{x - 5}$	$c = 4$
19. $f(x) = (x - 1)(x^2 - 3x + 2)$	$c = 0$
20. $f(x) = (x^5 - 3x)\left(\dfrac{1}{x^2}\right)$	$c = -1$

In Exercises 21–26, complete the table without using the Quotient Rule (see Example 5).

Function	*Rewrite*	*Differentiate*	*Simplify*
21. $y = \dfrac{x^2 + 3x}{7}$			
22. $y = \dfrac{5x^2 - 3}{4}$			
23. $y = \dfrac{6}{7x^2}$			
24. $y = \dfrac{10}{3x^3}$			
25. $y = \dfrac{4x^{3/2}}{x}$			
26. $y = \dfrac{5x^2 - 8}{11}$			

In Exercises 27–40, find the derivative of the function.

27. $f(x) = \dfrac{4 - 3x - x^2}{x^2 - 1}$
28. $f(x) = \dfrac{x^3 + 5x + 3}{x^2 - 1}$
29. $f(x) = x\left(1 - \dfrac{4}{x + 3}\right)$
30. $f(x) = x^4\left(1 - \dfrac{2}{x + 1}\right)$
31. $f(x) = \dfrac{3x - 1}{\sqrt{x}}$
32. $f(x) = \sqrt[3]{x}(\sqrt{x} + 3)$
33. $h(s) = (s^3 - 2)^2$
34. $h(x) = (x^2 - 1)^2$
35. $f(x) = \dfrac{2 - \dfrac{1}{x}}{x - 3}$
36. $g(x) = x^2\left(\dfrac{2}{x} - \dfrac{1}{x + 1}\right)$
37. $f(x) = (2x^3 + 5x)(x - 3)(x + 2)$
38. $f(x) = (x^3 - x)(x^2 + 2)(x^2 + x - 1)$
39. $f(x) = \dfrac{x^2 + c^2}{x^2 - c^2}$, c is a constant
40. $f(x) = \dfrac{c^2 - x^2}{c^2 + x^2}$, c is a constant

In Exercises 41 and 42, use a computer algebra system to differentiate the function.

41. $g(x) = \left(\dfrac{x + 1}{x + 2}\right)(2x - 5)$
42. $f(x) = \left(\dfrac{x^2 - x - 3}{x^2 + 1}\right)(x^2 + x + 1)$

In Exercises 43–46, (a) find an equation of the tangent line to the graph of f at the given point, (b) use a graphing utility to graph the function and its tangent line at the point, and (c) use the *derivative* feature of a graphing utility to confirm your results.

43. $f(x) = (x^3 + 4x - 1)(x - 2)$, $(1, -4)$
44. $f(x) = (x + 3)(x^2 - 2)$, $(-2, 2)$
45. $f(x) = \dfrac{x}{x + 4}$, $(-5, 5)$
46. $f(x) = \dfrac{(x - 1)}{(x + 1)}$, $\left(2, \dfrac{1}{3}\right)$

Famous Curves **In Exercises 47–50, find an equation of the tangent line to the graph at the given point. (The graphs in Exercises 47 and 48 are called *Witches of Agnesi*. The graphs in Exercises 49 and 50 are called *serpentines*.)**

47.

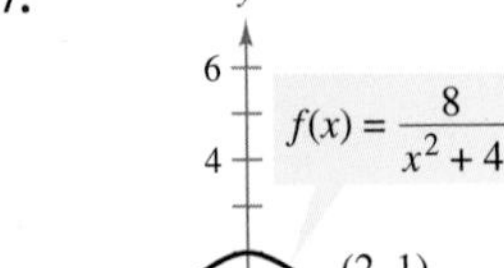
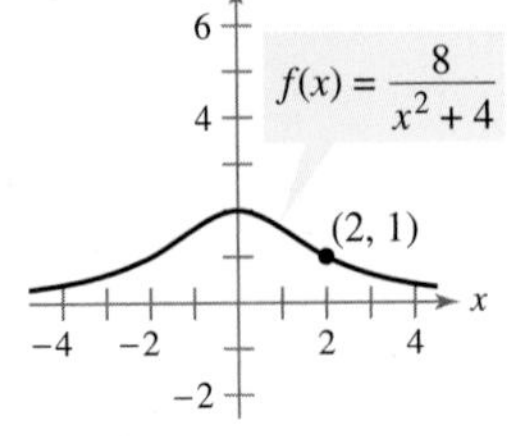

48.

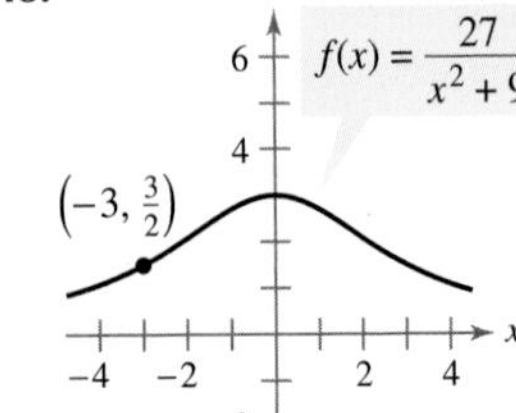

49.

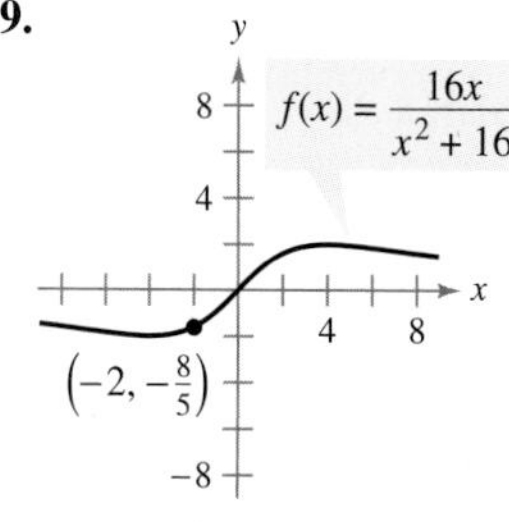

50.

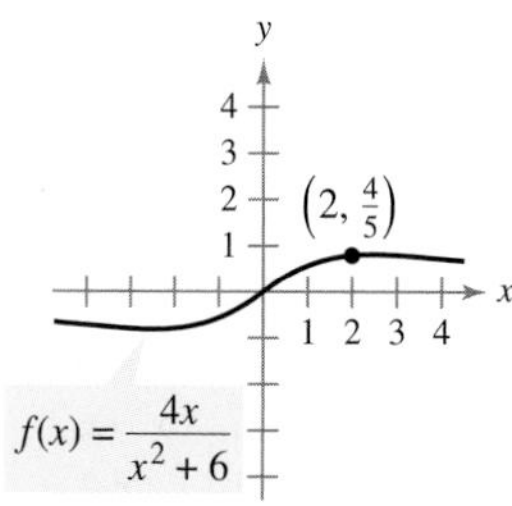

In Exercises 51–54, determine the point(s) at which the graph of the function has a horizontal tangent line.

51. $f(x) = \dfrac{2x - 1}{x^2}$

52. $f(x) = \dfrac{x^2}{x^2 + 1}$

53. $f(x) = \dfrac{x^2}{x - 1}$

54. $f(x) = \dfrac{x - 4}{x^2 - 7}$

55. ***Tangent Lines*** Find equations of the tangent lines to the graph of $f(x) = (x + 1)/(x - 1)$ that are parallel to the line $2y + x = 6$. Then graph the function and the tangent lines.

56. ***Tangent Lines*** Find equations of the tangent lines to the graph of $f(x) = x/(x - 1)$ that pass through the point $(-1, 5)$. Then graph the function and the tangent lines.

In Exercises 57 and 58, verify that $f'(x) = g'(x)$, and explain the relationship between f and g. [*Hint:* Use long division.]

57. $f(x) = \dfrac{3x}{x + 2}, \quad g(x) = \dfrac{5x + 4}{x + 2}$

58. $f(x) = \dfrac{5}{x - 3}, \quad g(x) = \dfrac{x + 2}{x - 3}$

In Exercises 59 and 60, use the graphs of f and g. Let $p(x) = f(x)g(x)$ and $q(x) = f(x)/g(x)$.

59. (a) Find $p'(1)$.
(b) Find $q'(4)$.

60. (a) Find $p'(4)$.
(b) Find $q'(7)$.

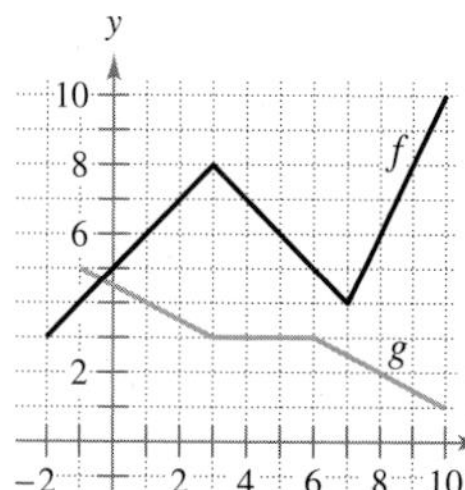

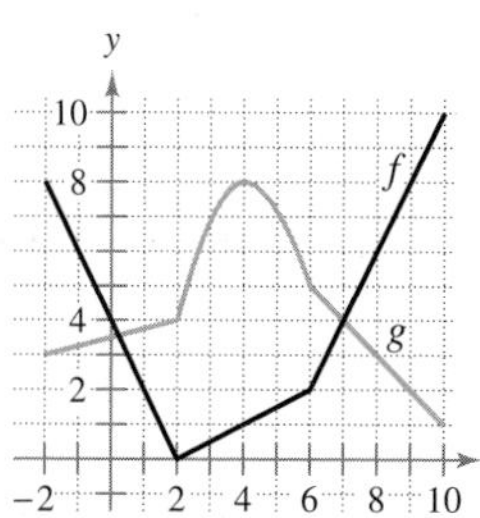

61. ***Area*** The length of a rectangle is given by $6t + 5$ and its height is $\sqrt{t}$, where t is time in seconds and the dimensions are in centimeters. Find the rate of change of the area with respect to time.

62. ***Boyle's Law*** This law states that if the temperature of a gas remains constant, its pressure is inversely proportional to its volume. Use the derivative to show that the rate of change of the pressure is inversely proportional to the square of the volume.

63. ***Population Growth*** A population of 500 bacteria is introduced into a culture and grows in number according to the equation

$$P(t) = 500\left(1 + \frac{4t}{50 + t^2}\right)$$

where t is measured in hours. Find the rate at which the population is growing when $t = 2$.

64. ***Modeling Data*** The table shows the quantities q (in millions) of personal computers shipped in the United States and the values v (in billions of dollars) of these shipments for the years 1999 through 2004. The year is represented by t, with $t = 9$ corresponding to 1999. (*Source: U.S. Census Bureau*)

Year, t	9	10	11	12	13	14
q	19.6	15.9	14.6	12.9	15.0	15.8
v	26.8	22.6	18.9	16.2	14.7	15.3

(a) Use a graphing utility to find cubic models for the quantity of personal computers shipped $q(t)$ and the value $v(t)$ of the personal computers.

(b) Graph each model found in part (a).

(c) Find $A = v(t)/q(t)$, then graph A. What does this function represent?

(d) Interpret $A'(t)$ in the context of these data.

In Exercises 65–70, find the second derivative of the function.

65. $f(x) = x^4 + 2x^3 - 3x^2 - x$

66. $f(x) = 8x^6 - 10x^5 + 5x^3$

67. $f(x) = 4x^{3/2}$

68. $f(x) = x + 32x^{-2}$

69. $f(x) = \dfrac{x}{x - 1}$

70. $f(x) = \dfrac{x^2 + 2x - 1}{x}$

In Exercises 71–74, find the given higher-order derivative.

71. $f'(x) = x^2, \quad f''(x)$

72. $f''(x) = 2 - \dfrac{2}{x}, \quad f'''(x)$

73. $f'''(x) = 2\sqrt{x}, \quad f^{(4)}(x)$

74. $f^{(4)}(x) = 2x + 1, \quad f^{(6)}(x)$

In Exercises 75–78, use the given information to find $f'(2)$.

$g(2) = 3$ and $g'(2) = -2$

$h(2) = -1$ and $h'(2) = 4$

75. $f(x) = 2g(x) + h(x)$

76. $f(x) = 4 - h(x)$

77. $f(x) = \dfrac{g(x)}{h(x)}$

78. $f(x) = g(x)h(x)$

WRITING ABOUT CONCEPTS

79. Sketch the graph of a differentiable function f such that $f(2) = 0$, $f' < 0$ for $-\infty < x < 2$, and $f' > 0$ for $2 < x < \infty$. Explain how you found your answer.

80. Sketch the graph of a differentiable function f such that $f > 0$ and $f' < 0$ for all real numbers x. Explain how you found your answer.

WRITING ABOUT CONCEPTS (continued)

In Exercises 81 and 82, the graphs of f, f', and f'' are shown on the same set of coordinate axes. Identify each graph. Explain your reasoning. To print an enlarged copy of the graph, go to the website *www.mathgraphs.com*.

81.

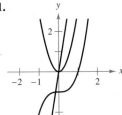

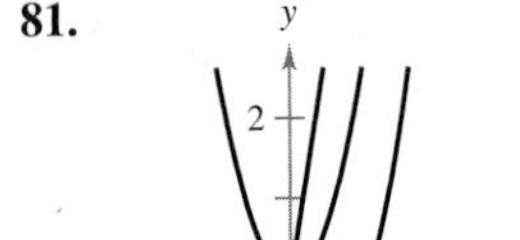

82.

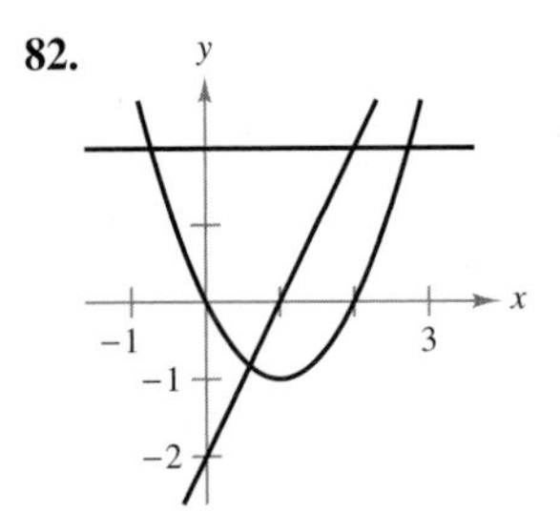

CAPSTONE

88. ***Particle Motion*** The figure shows the graphs of the position, velocity, and acceleration functions of a particle.

(a) Copy the graphs of the functions shown. Identify each graph. Explain your reasoning. To print an enlarged copy of the graph, go to the website *www.mathgraphs.com*.

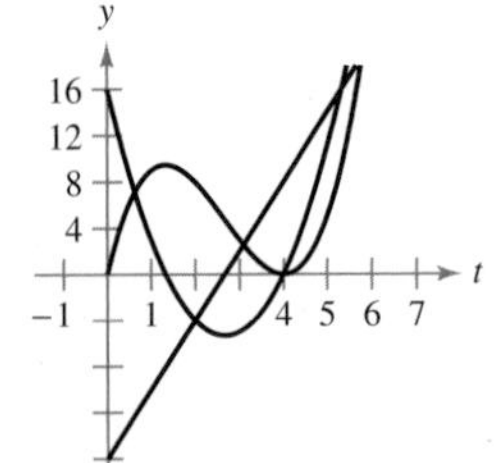

(b) On your sketch, identify when the particle speeds up and when it slows down. Explain your reasoning.

In Exercises 83 and 84, the graph of f is shown. Sketch the graphs of f' and f''. To print an enlarged copy of the graph, go to the website *www.mathgraphs.com*.

83.

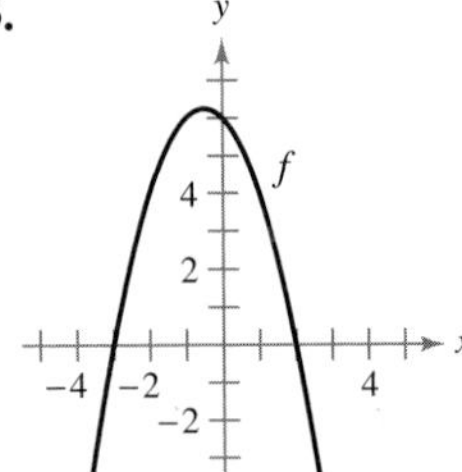

84.

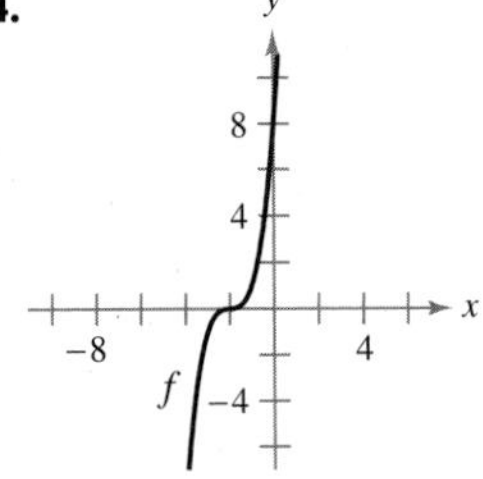

85. ***Acceleration*** The velocity of an object in meters per second is

$v(t) = 36 - t^2, 0 \le t \le 6.$

Find the velocity and acceleration of the object when $t = 3$. What can be said about the speed of the object when the velocity and acceleration have opposite signs?

86. ***Acceleration*** An automobile's velocity starting from rest is

$$v(t) = \frac{100t}{2t + 15}$$

where v is measured in feet per second. Find the acceleration at (a) 5 seconds, (b) 10 seconds, and (c) 20 seconds.

87. ***Stopping Distance*** A car is traveling at a rate of 66 feet per second (45 miles per hour) when the brakes are applied. The position function for the car is $s(t) = -8.25t^2 + 66t$, where s is measured in feet and t is measured in seconds. Use this function to complete the table, and find the average velocity during each time interval.

t	0	1	2	3	4
$s(t)$					
$v(t)$					
$a(t)$					

Finding a Pattern **In Exercises 89 and 90, develop a general rule for $f^{(n)}(x)$ given $f(x)$.**

89. $f(x) = x^n$

90. $f(x) = \dfrac{1}{x}$

91. ***Finding a Pattern*** Consider the function $f(x) = g(x)h(x)$.

(a) Use the Product Rule to generate rules for finding $f''(x)$, $f'''(x)$, and $f^{(4)}(x)$.

(b) Use the results of part (a) to write a general rule for $f^{(n)}(x)$.

92. ***Finding a Pattern*** Develop a general rule for $[xf(x)]^{(n)}$, where f is a differentiable function of x.

Differential Equations **In Exercises 93 and 94, verify that the function satisfies the differential equation.**

	Function	*Differential Equation*
93.	$y = \dfrac{1}{x}, x > 0$	$x^3y'' + 2x^2y' = 0$
94.	$y = 2x^3 - 6x + 10$	$-y''' - xy'' - 2y' = -24x^2$

True or False? **In Exercises 95–100, determine whether the statement is true or false. If it is false, explain why or give an example that shows it is false.**

95. If $y = f(x)g(x)$, then $dy/dx = f'(x)g'(x)$.

96. If $y = (x + 1)(x + 2)(x + 3)(x + 4)$, then $d^5y/dx^5 = 0$.

97. If $f'(c)$ and $g'(c)$ are zero and $h(x) = f(x)g(x)$, then $h'(c) = 0$.

98. If $f(x)$ is an nth-degree polynomial, then $f^{(n+1)}(x) = 0$.

99. The second derivative represents the rate of change of the first derivative.

100. If the velocity of an object is constant, then its acceleration is zero.

101. Find the derivative of $f(x) = x|x|$. Does $f''(0)$ exist?

102. ***Think About It*** Let f and g be functions whose first and second derivatives exist on an interval I. Which of the following formulas is (are) true?

(a) $fg'' - f''g = (fg' - f'g)'$

(b) $fg'' + f''g = (fg)''$

4.4 The Chain Rule

- **Find the derivative of a composite function using the Chain Rule.**
- **Find the derivative of a function using the General Power Rule.**
- **Simplify the derivative of a function using algebra.**

The Chain Rule

This text has yet to discuss one of the most powerful differentiation rules—the **Chain Rule.** This rule deals with composite functions and adds a surprising versatility to the rules discussed in the two previous sections. For example, compare the functions shown below. Those on the left can be differentiated without the Chain Rule, and those on the right are best differentiated with the Chain Rule.

Without the Chain Rule	*With the Chain Rule*
$y = x^2 + 1$	$y = \sqrt{x^2 + 1}$
$y = 3x + 2$	$y = (3x + 2)^5$
$y = x + 2$	$y = \sqrt[3]{x + 2}$

Basically, the Chain Rule states that if y changes dy/du times as fast as u, and u changes du/dx times as fast as x, then y changes $(dy/du)(du/dx)$ times as fast as x.

EXAMPLE 1 The Derivative of a Composite Function

A set of gears is constructed, as shown in Figure 4.22, such that the second and third gears are on the same axle. As the first axle revolves, it drives the second axle, which in turn drives the third axle. Let y, u, and x represent the numbers of revolutions per minute of the first, second, and third axles, respectively. Find dy/du, du/dx, and dy/dx, and show that

$$\frac{dy}{dx} = \frac{dy}{du} \cdot \frac{du}{dx}.$$

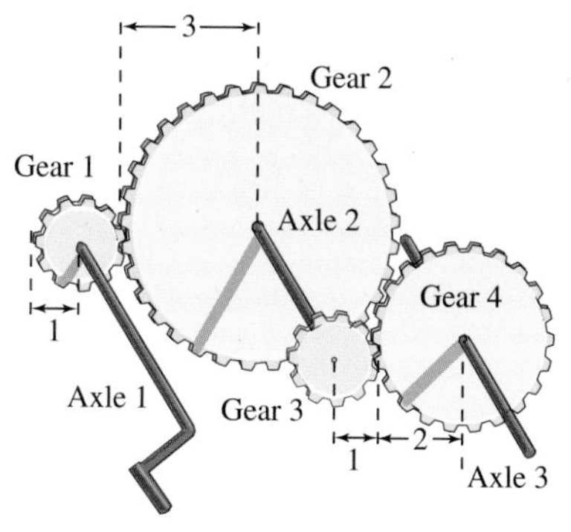

Axle 1: y revolutions per minute
Axle 2: u revolutions per minute
Axle 3: x revolutions per minute
Figure 4.22

Solution Because the circumference of the second gear is three times that of the first, the first axle must make three revolutions to turn the second axle once. Similarly, the second axle must make two revolutions to turn the third axle once, and you can write

$$\frac{dy}{du} = 3 \quad \text{and} \quad \frac{du}{dx} = 2.$$

Combining these two results, you know that the first axle must make six revolutions to turn the third axle once. So, you can write

$$\begin{aligned}\frac{dy}{dx} &= \boxed{\text{Rate of change of first axle with respect to second axle}} \cdot \boxed{\text{Rate of change of second axle with respect to third axle}} \\ &= \frac{dy}{du} \cdot \frac{du}{dx} = 3 \cdot 2 = 6 \\ &= \boxed{\text{Rate of change of first axle with respect to third axle}}.\end{aligned}$$

In other words, the rate of change of y with respect to x is the product of the rate of change of y with respect to u and the rate of change of u with respect to x. ■

EXPLORATION

Using the Chain Rule Each of the following functions can be differentiated using rules that you studied in Sections 4.2 and 4.3. For each function, find the derivative using those rules. Then find the derivative using the Chain Rule. Compare your results. Which method is simpler?

a. $\dfrac{2}{3x + 1}$

b. $(x + 2)^3$

Example 1 illustrates a simple case of the Chain Rule. The general rule is stated below.

THEOREM 4.8 THE CHAIN RULE

If $y = f(u)$ is a differentiable function of u and $u = g(x)$ is a differentiable function of x, then $y = f(g(x))$ is a differentiable function of x and

$$\frac{dy}{dx} = \frac{dy}{du} \cdot \frac{du}{dx}$$

or, equivalently,

$$\frac{d}{dx}[f(g(x))] = f'(g(x))g'(x).$$

PROOF Let $h(x) = f(g(x))$. Then, using the alternative form of the derivative, you need to show that, for $x = c$,

$$h'(c) = f'(g(c))g'(c).$$

An important consideration in this proof is the behavior of g as x approaches c. A problem occurs if there are values of x, other than c, such that $g(x) = g(c)$. Appendix A shows how to use the differentiability of f and g to overcome this problem. For now, assume that $g(x) \neq g(c)$ for values of x other than c. In the proofs of the Product Rule and the Quotient Rule, the same quantity was added and subtracted to obtain the desired form. This proof uses a similar technique—multiplying and dividing by the same (nonzero) quantity. Note that because g is differentiable, it is also continuous, and it follows that $g(x) \to g(c)$ as $x \to c$.

$$\begin{aligned} h'(c) &= \lim_{x\to c} \frac{f(g(x)) - f(g(c))}{x - c} \\ &= \lim_{x\to c} \left[\frac{f(g(x)) - f(g(c))}{g(x) - g(c)} \cdot \frac{g(x) - g(c)}{x - c}\right], \quad g(x) \neq g(c) \\ &= \left[\lim_{x\to c} \frac{f(g(x)) - f(g(c))}{g(x) - g(c)}\right]\left[\lim_{x\to c} \frac{g(x) - g(c)}{x - c}\right] \\ &= f'(g(c))g'(c) \end{aligned}$$

■

When applying the Chain Rule, it is helpful to think of the composite function $f \circ g$ as having two parts—an inner function and an outer function, as discussed in Section 1.4.

Outer function

$$y = f(g(x)) = f(u)$$

Inner function

The derivative of $y = f(u)$ is the derivative of the outer function (at the inner function u) *times* the derivative of the inner function.

$$y' = f'(u) \cdot u'$$

The next example is a review of the decomposition skills you learned in Section 1.4.

EXAMPLE 2 Decomposition of a Composite Function

Write each function as a composition of two simpler functions.

a. $y = \dfrac{1}{x + 1}$ **b.** $y = \sqrt{3x^2 - x + 1}$

Solution

$y = f(g(x))$	$u = g(x)$	$y = f(u)$
a. $y = \dfrac{1}{x + 1}$	$u = x + 1$	$y = \dfrac{1}{u}$
b. $y = \sqrt{3x^2 - x + 1}$	$u = 3x^2 - x + 1$	$y = \sqrt{u}$

EXAMPLE 3 Using the Chain Rule

Find dy/dx for $y = (x^2 + 1)^3$.

Solution For this function, you can consider the inside function to be $u = x^2 + 1$. By the Chain Rule, you obtain

$$\frac{dy}{dx} = \underbrace{3(x^2 + 1)^2}_{\frac{dy}{du}}\underbrace{(2x)}_{\frac{du}{dx}} = 6x(x^2 + 1)^2.$$

■

STUDY TIP You could also solve the problem in Example 3 without using the Chain Rule by observing that

$$y = (x^2 + 1)^3$$
$$= x^6 + 3x^4 + 3x^2 + 1$$

and

$$y' = 6x^5 + 12x^3 + 6x.$$

Verify that this is the same as the derivative in Example 3. Which method would you use to find

$$\frac{d}{dx}(x^2 + 1)^{50}?$$

The General Power Rule

The function in Example 3 is an example of one of the most common types of composite functions, $y = [u(x)]^n$. The rule for differentiating such functions is called the **General Power Rule,** and it is a special case of the Chain Rule.

> **THEOREM 4.9 THE GENERAL POWER RULE**
>
> If $y = [u(x)]^n$, where u is a differentiable function of x and n is a rational number, then
>
> $$\frac{dy}{dx} = n[u(x)]^{n-1}\frac{du}{dx}$$
>
> or, equivalently,
>
> $$\frac{d}{dx}[u^n] = nu^{n-1}u'.$$

PROOF Because $y = u^n$, you apply the Chain Rule to obtain

$$\frac{dy}{dx} = \left(\frac{dy}{du}\right)\left(\frac{du}{dx}\right) = \frac{d}{du}[u^n]\frac{du}{dx}.$$

By the (Simple) Power Rule in Section 4.2, you have $D_u[u^n] = nu^{n-1}$, and it follows that

$$\frac{dy}{dx} = n[u(x)]^{n-1}\frac{du}{dx}.$$

■

EXAMPLE 4 Applying the General Power Rule

Find the derivative of $f(x) = (3x - 2x^2)^3$.

Solution Let $u = 3x - 2x^2$. Then

$$f(x) = (3x - 2x^2)^3 = u^3$$

and, by the General Power Rule, the derivative is

$$\begin{aligned} f'(x) &= \underset{n}{3}\underbrace{(3x - 2x^2)^2}_{u^{n-1}}\underbrace{\frac{d}{dx}[3x - 2x^2]}_{u'} && \text{Apply General Power Rule.} \\ &= 3(3x - 2x^2)^2(3 - 4x). && \text{Differentiate } 3x - 2x^2. \end{aligned}$$

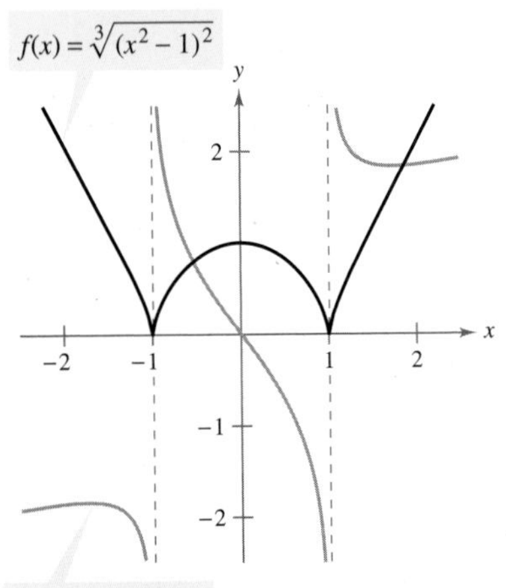

The derivative of f is 0 at $x = 0$ and is undefined at $x = \pm 1$.
Figure 4.23

EXAMPLE 5 Differentiating Functions Involving Radicals

Find all points on the graph of $f(x) = \sqrt[3]{(x^2 - 1)^2}$ for which $f'(x) = 0$ and those for which $f'(x)$ does not exist.

Solution Begin by rewriting the function as

$$f(x) = (x^2 - 1)^{2/3}.$$

Then, applying the General Power Rule (with $u = x^2 - 1$) produces

$$\begin{aligned} f'(x) &= \underset{n}{\frac{2}{3}}\underbrace{(x^2 - 1)^{-1/3}}_{u^{n-1}}\underbrace{(2x)}_{u'} && \text{Apply General Power Rule.} \\ &= \frac{4x}{3\sqrt[3]{x^2 - 1}}. && \text{Write in radical form.} \end{aligned}$$

So, $f'(x) = 0$ when $x = 0$ and $f'(x)$ does not exist when $x = \pm 1$, as shown in Figure 4.23.

EXAMPLE 6 Differentiating Quotients with Constant Numerators

Differentiate $g(t) = \dfrac{-7}{(2t - 3)^2}$.

Solution Begin by rewriting the function as

$$g(t) = -7(2t - 3)^{-2}.$$

Then, applying the General Power Rule produces

$$\begin{aligned} g'(t) &= \underbrace{(-7)}_{\text{Constant Multiple Rule}}\underbrace{(-2)}_{n}\underbrace{(2t - 3)^{-3}}_{u^{n-1}}\underbrace{(2)}_{u'} && \text{Apply General Power Rule.} \\ &= 28(2t - 3)^{-3} && \text{Simplify.} \\ &= \frac{28}{(2t - 3)^3}. && \text{Write with positive exponent.} \end{aligned}$$

■

NOTE Try differentiating the function in Example 6 using the Quotient Rule. You should obtain the same result, but using the Quotient Rule is less efficient than using the General Power Rule.

Simplifying Derivatives

The next three examples illustrate some techniques for simplifying the "raw derivatives" of functions involving products, quotients, and composites.

STUDY TIP You can also simplify raw derivatives by a rationalizing technique that removes negative exponents. In Example 7, multiply and divide the "raw derivative" by $(1 - x^2)^{1/2}$ to obtain

$$\frac{-x^3(1-x^2)^0 + 2x(1-x^2)^1}{(1-x^2)^{1/2}} = \frac{-x^3(1) + 2x - 2x^3}{\sqrt{1-x^2}} = \frac{x(2-3x^2)}{\sqrt{1-x^2}}.$$

For Example 8, the rationalizing factor would be $(x^2 + 4)^{2/3}$. Try it and compare with the given method.

TECHNOLOGY Symbolic differentiation utilities are capable of differentiating very complicated functions. Often, however, the result is given in unsimplified form. If you have access to such a utility, use it to find the derivatives of the functions given in Examples 7, 8, and 9. Then compare the results with those given in these examples.

EXAMPLE 7 Simplifying by Factoring Out the Least Powers

$$f(x) = x^2\sqrt{1-x^2} \qquad \text{Original function}$$
$$= x^2(1-x^2)^{1/2} \qquad \text{Rewrite.}$$
$$f'(x) = x^2\frac{d}{dx}[(1-x^2)^{1/2}] + (1-x^2)^{1/2}\frac{d}{dx}[x^2] \qquad \text{Product Rule}$$
$$= x^2\left[\frac{1}{2}(1-x^2)^{-1/2}(-2x)\right] + (1-x^2)^{1/2}(2x) \qquad \text{General Power Rule}$$
$$= -x^3(1-x^2)^{-1/2} + 2x(1-x^2)^{1/2} \qquad \text{Simplify.}$$
$$= x(1-x^2)^{-1/2}[-x^2(1) + 2(1-x^2)] \qquad \text{Factor.}$$
$$= \frac{x(2-3x^2)}{\sqrt{1-x^2}} \qquad \text{Simplify.}$$

EXAMPLE 8 Simplifying the Derivative of a Quotient

$$f(x) = \frac{x}{\sqrt[3]{x^2+4}} \qquad \text{Original function}$$
$$= \frac{x}{(x^2+4)^{1/3}} \qquad \text{Rewrite.}$$
$$f'(x) = \frac{(x^2+4)^{1/3}(1) - x(1/3)(x^2+4)^{-2/3}(2x)}{(x^2+4)^{2/3}} \qquad \text{Quotient Rule}$$
$$= \frac{1}{3}(x^2+4)^{-2/3}\left[\frac{3(x^2+4) - (2x^2)(1)}{(x^2+4)^{2/3}}\right] \qquad \text{Factor.}$$
$$= \frac{x^2+12}{3(x^2+4)^{4/3}} \qquad \text{Simplify.}$$

EXAMPLE 9 Simplifying the Derivative of a Power

$$y = \left(\frac{3x-1}{x^2+3}\right)^2 \qquad \text{Original function}$$
$$y' = \underbrace{2}_{n}\underbrace{\left(\frac{3x-1}{x^2+3}\right)}_{u^{n-1}}\underbrace{\frac{d}{dx}\left[\frac{3x-1}{x^2+3}\right]}_{u'} \qquad \text{General Power Rule}$$
$$= \left[\frac{2(3x-1)}{x^2+3}\right]\left[\frac{(x^2+3)(3) - (3x-1)(2x)}{(x^2+3)^2}\right] \qquad \text{Quotient Rule}$$
$$= \frac{2(3x-1)(3x^2+9-6x^2+2x)}{(x^2+3)^3} \qquad \text{Multiply.}$$
$$= \frac{2(3x-1)(-3x^2+2x+9)}{(x^2+3)^3} \qquad \text{Simplify.}$$

■

This section concludes with a summary of the differentiation rules studied so far. To become skilled at differentiation, you should memorize each rule.

SUMMARY OF DIFFERENTIATION RULES

General Differentiation Rules Let f, g, and u be differentiable functions of x.

Constant Multiple Rule:
$$\frac{d}{dx}[cf] = cf'$$

Sum or Difference Rule:
$$\frac{d}{dx}[f \pm g] = f' \pm g'$$

Product Rule:
$$\frac{d}{dx}[fg] = fg' + gf'$$

Quotient Rule:
$$\frac{d}{dx}\left[\frac{f}{g}\right] = \frac{gf' - fg'}{g^2}$$

Derivatives of Algebraic Functions

Constant Rule:
$$\frac{d}{dx}[c] = 0$$

(Simple) Power Rule:
$$\frac{d}{dx}[x^n] = nx^{n-1}, \quad \frac{d}{dx}[x] = 1$$

Chain Rule

Chain Rule:
$$\frac{d}{dx}[f(u)] = f'(u)\, u'$$

General Power Rule:
$$\frac{d}{dx}[u^n] = nu^{n-1}\, u'$$

4.4 Exercises

See www.CalcChat.com for worked-out solutions to odd-numbered exercises.

In Exercises 1–6, complete the table using Example 2 as a model.

$y = f(g(x))$	$u = g(x)$	$y = f(u)$
1. $y = (5x - 8)^4$		
2. $y = (5x - 2)^{3/2}$		
3. $y = (x^2 - 3x + 4)^6$		
4. $y = \dfrac{3}{x + 2}$		
5. $y = \dfrac{1}{\sqrt{x + 1}}$		
6. $y = \sqrt{x^3 - 7}$		

In Exercises 7–36, find the derivative of the function.

7. $y = (4x - 1)^3$

8. $y = 2(6 - x^2)^5$

9. $g(x) = 3(4 - 9x)^4$

10. $f(t) = (9t + 2)^{2/3}$

11. $f(t) = \sqrt{5 - t}$

12. $g(x) = \sqrt{9 - 4x}$

13. $y = \sqrt[3]{6x^2 + 1}$

14. $g(x) = \sqrt{x^2 - 2x + 1}$

15. $y = 2\sqrt[4]{9 - x^2}$

16. $f(x) = -3\sqrt[4]{2 - 9x}$

17. $y = \dfrac{1}{x - 2}$

18. $s(t) = \dfrac{1}{t^2 + 3t - 1}$

19. $f(t) = \left(\dfrac{1}{t - 3}\right)^2$

20. $y = -\dfrac{5}{(t + 3)^3}$

21. $y = \dfrac{1}{\sqrt{x + 2}}$

22. $g(t) = \sqrt{\dfrac{1}{t^2 - 2}}$

23. $f(x) = x^2(x - 2)^4$

24. $f(x) = x(3x - 9)^3$

25. $y = x\sqrt{1 - x^2}$

26. $y = \frac{1}{2}x^2\sqrt{16 - x^2}$

27. $y = \dfrac{x}{\sqrt{x^2 + 1}}$

28. $y = \dfrac{x}{\sqrt{x^4 + 4}}$

29. $g(x) = \left(\dfrac{x + 5}{x^2 + 2}\right)^2$

30. $h(t) = \left(\dfrac{t^2}{t^3 + 2}\right)^2$

31. $f(v) = \left(\dfrac{1 - 2v}{1 + v}\right)^3$

32. $g(x) = \left(\dfrac{3x^2 - 2}{2x + 3}\right)^3$

33. $f(x) = ((x^2 + 3)^5 + x)^2$

34. $g(x) = (2 + (x^2 + 1)^4)^3$

35. $f(x) = \sqrt{2 + \sqrt{2 + \sqrt{x}}}$

36. $g(t) = \sqrt{\sqrt{t + 1} + 1}$

In Exercises 37–40, use a computer algebra system to find the derivative of the function. Then use the utility to graph the function and its derivative on the same set of coordinate axes. Describe the behavior of the function that corresponds to any zeros of the graph of the derivative.

37. $y = \dfrac{\sqrt{x} + 1}{x^2 + 1}$

38. $y = \sqrt{\dfrac{2x}{x + 1}}$

39. $y = \sqrt{\dfrac{x + 1}{x}}$

40. $g(x) = \sqrt{x - 1} + \sqrt{x + 1}$

In Exercises 41–46, evaluate the derivative of the function at the given point. Use a graphing utility to verify your result.

41. $s(t) = \sqrt{t^2 + 6t - 2}, \quad (3, 5)$

42. $y = \sqrt[5]{3x^3 + 4x}, \quad (2, 2)$

43. $f(x) = \dfrac{5}{x^3 - 2}, \quad \left(-2, -\dfrac{1}{2}\right)$

44. $f(x) = \dfrac{1}{(x^2 - 3x)^2}, \quad \left(4, \dfrac{1}{16}\right)$

45. $f(t) = \dfrac{3t + 2}{t - 1}, \quad (0, -2)$

46. $f(x) = \dfrac{x + 1}{2x - 3}, \quad (2, 3)$

In Exercises 47–50, (a) find an equation of the tangent line to the graph of *f* at the given point, (b) use a graphing utility to graph the function and its tangent line at the point, and (c) use the *derivative* feature of the graphing utility to confirm your results.

Function	*Point*
47. $f(x) = \sqrt{2x^2 - 7}$	$(4, 5)$
48. $f(x) = \frac{1}{3}x\sqrt{x^2 + 5}$	$(2, 2)$
49. $y = (4x^3 + 3)^2$	$(-1, 1)$
50. $f(x) = (9 - x^2)^{2/3}$	$(1, 4)$

In Exercises 51–54, (a) use a graphing utility to find the derivative of the function at the given point, (b) find an equation of the tangent line to the graph of the function at the given point, and (c) use the utility to graph the function and its tangent line in the same viewing window.

51. $g(t) = \dfrac{3t^2}{\sqrt{t^2 + 2t - 1}}, \quad \left(\dfrac{1}{2}, \dfrac{3}{2}\right)$

52. $f(x) = \sqrt{x}(2 - x)^2, \quad (4, 8)$

53. $s(t) = \dfrac{(4 - 2t)\sqrt{1 + t}}{3}, \quad \left(0, \dfrac{4}{3}\right)$

54. $y = (t^2 - 9)\sqrt{t + 2}, \quad (2, -10)$

Famous Curves **In Exercises 55 and 56, find an equation of the tangent line to the graph at the given point. Then use a graphing utility to graph the function and its tangent line in the same viewing window.**

55. Top half of circle

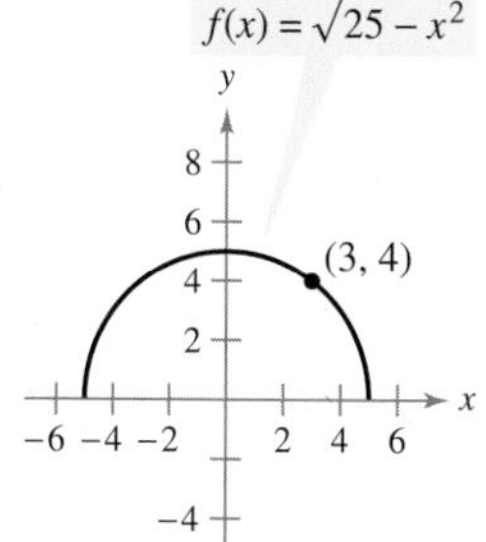

56. Bullet-nose curve

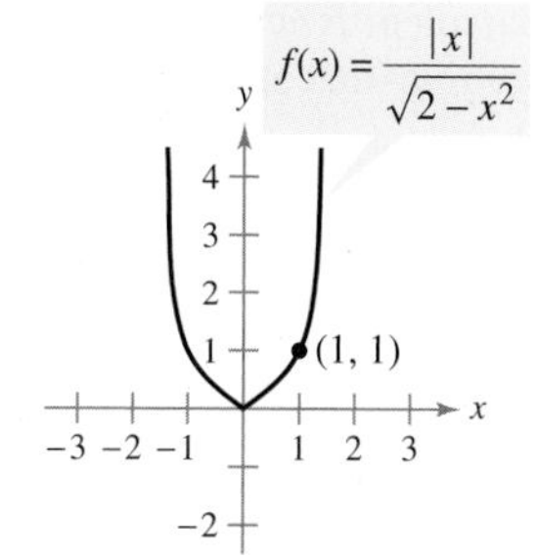

In Exercises 57–60, find the second derivative of the function.

57. $f(x) = 4(x^2 - 2)^3$

58. $f(t) = \dfrac{\sqrt{t^2 + 1}}{t}$

59. $f(x) = \sqrt{x^2 + x + 1}$

60. $f(x) = \dfrac{1}{x - 6}$

In Exercises 61 and 62, evaluate the second derivative of the function at the given point. Use a computer algebra system to verify your result.

61. $h(x) = \frac{1}{9}(3x + 1)^3, \quad \left(1, \frac{64}{9}\right)$

62. $f(x) = \dfrac{1}{\sqrt{x + 4}}, \quad \left(0, \dfrac{1}{2}\right)$

WRITING ABOUT CONCEPTS

In Exercises 63 and 64, the graphs of a function *f* and its derivative *f′* are shown. Label the graphs as *f* or *f′* and write a short paragraph stating the criteria you used in making your selection. To print an enlarged copy of the graph, go to the website *www.mathgraphs.com*.

63.

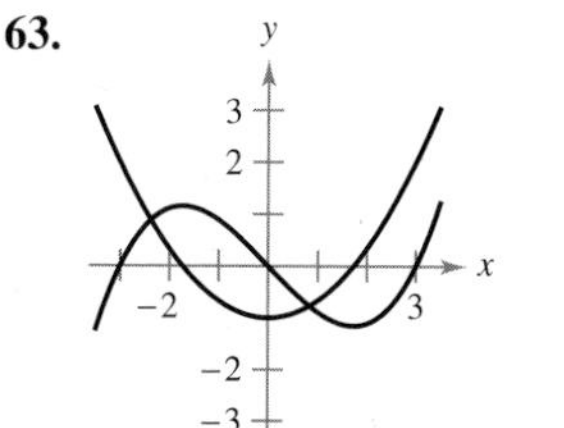

64.

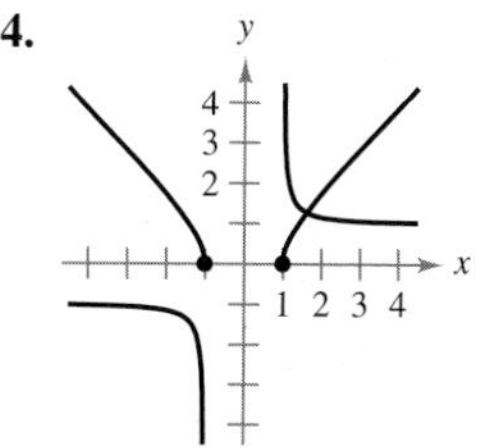

In Exercises 65 and 66, the relationship between *f* and *g* is given. Explain the relationship between *f′* and *g′*.

65. $g(x) = f(3x)$

66. $g(x) = f(x^2)$

67. ***Think About It*** The table shows some values of the derivative of an unknown function *f*. Use the table to find (if possible) the derivative of each transformation of *f*.

(a) $g(x) = f(x) - 2$

(b) $h(x) = 2f(x)$

(c) $r(x) = f(-3x)$

(d) $s(x) = f(x + 2)$

x	-2	-1	0	1	2	3
$f'(x)$	4	$\frac{2}{3}$	$-\frac{1}{3}$	-1	-2	-4

CAPSTONE

68. Given that $g(5) = -3$, $g'(5) = 6$, $h(5) = 3$, and $h'(5) = -2$, find $f'(5)$ (if possible) for each of the following. If it is not possible, state what additional information is required.

(a) $f(x) = g(x)h(x)$

(b) $f(x) = g(h(x))$

(c) $f(x) = \dfrac{g(x)}{h(x)}$

(d) $f(x) = [g(x)]^3$

In Exercises 69 and 70, the graphs of f and g are shown. Let $h(x) = f(g(x))$ and $s(x) = g(f(x))$. Find each derivative, if it exists. If the derivative does not exist, explain why.

69. (a) Find $h'(1)$.

(b) Find $s'(5)$.

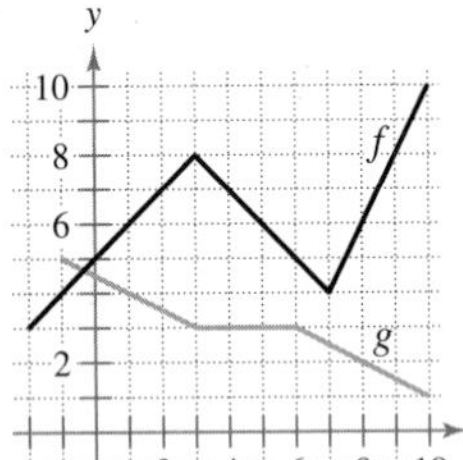

70. (a) Find $h'(3)$.

(b) Find $s'(9)$.

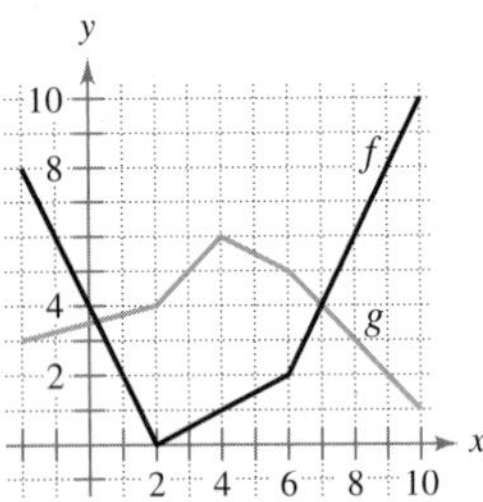

71. ***Doppler Effect*** The frequency F of a fire truck siren heard by a stationary observer is $F = 132{,}400/(331 \pm v)$, where $\pm v$ represents the velocity of the accelerating fire truck in meters per second (see figure). Find the rate of change of F with respect to v when

(a) the fire truck is approaching at a velocity of 30 meters per second (use $-v$).

(b) the fire truck is moving away at a velocity of 30 meters per second (use $+v$).

$$F = \frac{132{,}400}{331 + v} \qquad F = \frac{132{,}400}{331 - v}$$

72. ***Circulatory System*** The speed S of blood that is r centimeters from the center of an artery is $S = C(R^2 - r^2)$, where C is a constant, R is the radius of the artery, and S is measured in centimeters per second. Suppose a drug is administered and the artery begins to dilate at a rate of dR/dt. At a constant distance r, find the rate at which S changes with respect to t for $C = 1.76 \times 10^5$, $R = 1.2 \times 10^{-2}$, and $dR/dt = 10^{-5}$.

73. ***Modeling Data*** The cost of producing x units of a product is $C = 60x + 1350$. For one week management determined the number of units produced at the end of t hours during an eight-hour shift. The average values of x for the week are shown in the table.

t	0	1	2	3	4	5	6	7	8
x	0	16	60	130	205	271	336	384	392

(a) Use a graphing utility to fit a cubic model to the data.

(b) Use the Chain Rule to find dC/dt.

(c) Explain why the cost function is not increasing at a constant rate during the eight-hour shift.

74. ***Think About It*** Let $r(x) = f(g(x))$ and $s(x) = g(f(x))$, where f and g are shown in the figure. Find (a) $r'(1)$ and (b) $s'(4)$.

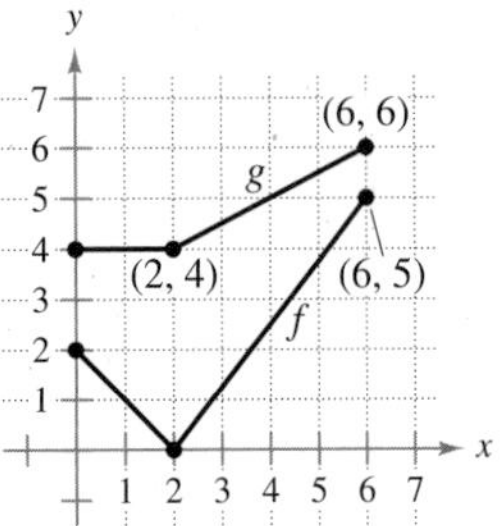

75. (a) Show that the derivative of an odd function is even. That is, if $f(-x) = -f(x)$, then $f'(-x) = f'(x)$.

(b) Show that the derivative of an even function is odd. That is, if $f(-x) = f(x)$, then $f'(-x) = -f'(x)$.

76. Let u be a differentiable function of x. Use the fact that $|u| = \sqrt{u^2}$ to prove that

$$\frac{d}{dx}[|u|] = u'\frac{u}{|u|}, \quad u \neq 0.$$

In Exercises 77 and 78, use the result of Exercise 76 to find the derivative of the function.

77. $g(x) = |3x - 5|$

78. $f(x) = |x^2 - 9|$

Linear and Quadratic Approximations **The linear and quadratic approximations of a function f at $x = a$ are**

$$P_1(x) = f'(a)(x - a) + f(a) \text{ and}$$

$$P_2(x) = \tfrac{1}{2}f''(a)(x - a)^2 + f'(a)(x - a) + f(a).$$

In Exercises 79 and 80, (a) find the specified linear and quadratic approximations of f, (b) use a graphing utility to graph f and the approximations, (c) determine whether P_1 or P_2 is the better approximation, and (d) state how the accuracy changes as you move farther from $x = a$.

79. $f(x) = \dfrac{1}{\sqrt{x^2 - 3}}$

$a = 2$

80. $f(x) = \sqrt{x^2 - 3}$

$a = 2$

True or False? **In Exercises 81 and 82, determine whether the statement is true or false. If it is false, explain why or give an example that shows it is false.**

81. If $y = (1 - x)^{1/2}$, then $y' = \frac{1}{2}(1 - x)^{-1/2}$.

82. If y is a differentiable function of u, u is a differentiable function of v, and v is a differentiable function of x, then

$$\frac{dy}{dx} = \frac{dy}{du}\frac{du}{dv}\frac{dv}{dx}.$$

4.5 Implicit Differentiation

- **Distinguish between functions written in implicit form and explicit form.**
- **Use implicit differentiation to find the derivative of a function.**

Implicit and Explicit Functions

EXPLORATION

Graphing an Implicit Equation How could you use a graphing utility to sketch the graph of the equation

$$x^2 - 2y^3 + 4y = 2?$$

Here are two possible approaches.

a. Solve the equation for x. Switch the roles of x and y and graph the two resulting equations. The combined graphs will show a 90° rotation of the graph of the original equation.

b. Set the graphing utility to *parametric* mode and graph the equations

$$x = -\sqrt{2t^3 - 4t + 2}$$
$$y = t$$

and

$$x = \sqrt{2t^3 - 4t + 2}$$
$$y = t.$$

From either of these two approaches, can you decide whether the graph has a tangent line at the point (0, 1)? Explain your reasoning.

Up to this point in the text, most functions have been expressed in **explicit form.** For example, in the equation

$$y = 3x^2 - 5 \qquad \text{Explicit form}$$

the variable y is explicitly written as a function of x. Some functions, however, are only implied by an equation. For instance, the function $y = 1/x$ is defined **implicitly** by the equation $xy = 1$. Suppose you were asked to find dy/dx for this equation. You could begin by writing y explicitly as a function of x and then differentiating.

Implicit Form	*Explicit Form*	*Derivative*
$xy = 1$	$y = \frac{1}{x} = x^{-1}$	$\frac{dy}{dx} = -x^{-2} = -\frac{1}{x^2}$

This strategy works whenever you can solve for the function explicitly. You cannot, however, use this procedure when you are unable to solve for y as a function of x. For instance, how would you find dy/dx for the equation

$$x^2 - 2y^3 + 4y = 2$$

where it is very difficult to express y as a function of x explicitly? To do this, you can use **implicit differentiation.**

To understand how to find dy/dx implicitly, you must realize that the differentiation is taking place *with respect to x*. This means that when you differentiate terms involving x alone, you can differentiate as usual. However, when you differentiate terms involving y, you must apply the Chain Rule, because you are assuming that y is defined implicitly as a differentiable function of x.

EXAMPLE 1 Differentiating with Respect to *x*

a. $\frac{d}{dx}[x^3] = 3x^2$ Variables agree: use Simple Power Rule.

Variables agree

b. $\frac{d}{dx}[\underbrace{y^3}_{u^n}] = \underbrace{3y^2}_{nu^{n-1}}\underbrace{\frac{dy}{dx}}_{u'}$ Variables disagree: use Chain Rule.

Variables disagree

c. $\frac{d}{dx}[x + 3y] = 1 + 3\frac{dy}{dx}$ Chain Rule: $\frac{d}{dx}[3y] = 3y'$

d. $\frac{d}{dx}[xy^2] = x\frac{d}{dx}[y^2] + y^2\frac{d}{dx}[x]$ Product Rule

$= x\left(2y\frac{dy}{dx}\right) + y^2(1)$ Chain Rule

$= 2xy\frac{dy}{dx} + y^2$ Simplify. ■

Implicit Differentiation

GUIDELINES FOR IMPLICIT DIFFERENTIATION

1. Differentiate both sides of the equation *with respect to x*.
2. Collect all terms involving dy/dx on the left side of the equation and move all other terms to the right side of the equation.
3. Factor dy/dx out of the left side of the equation.
4. Solve for dy/dx.

In Example 2, note that implicit differentiation can produce an expression for dy/dx that contains both x and y.

EXAMPLE 2 Implicit Differentiation

Find dy/dx given that $y^3 + y^2 - 5y - x^2 = -4$.

Solution

1. Differentiate both sides of the equation with respect to x.

$$\frac{d}{dx}[y^3 + y^2 - 5y - x^2] = \frac{d}{dx}[-4]$$

$$\frac{d}{dx}[y^3] + \frac{d}{dx}[y^2] - \frac{d}{dx}[5y] - \frac{d}{dx}[x^2] = \frac{d}{dx}[-4]$$

$$3y^2\frac{dy}{dx} + 2y\frac{dy}{dx} - 5\frac{dy}{dx} - 2x = 0$$

2. Collect the dy/dx terms on the left side of the equation and move all other terms to the right side of the equation.

$$3y^2\frac{dy}{dx} + 2y\frac{dy}{dx} - 5\frac{dy}{dx} = 2x$$

3. Factor dy/dx out of the left side of the equation.

$$\frac{dy}{dx}(3y^2 + 2y - 5) = 2x$$

4. Solve for dy/dx by dividing by $(3y^2 + 2y - 5)$.

$$\frac{dy}{dx} = \frac{2x}{3y^2 + 2y - 5}$$ ■

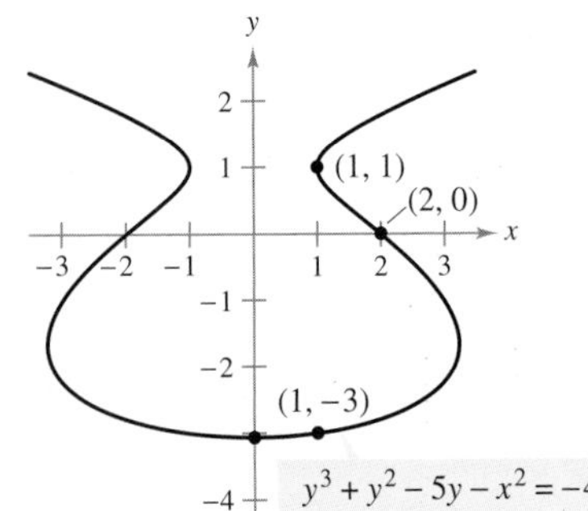

Point on Graph	Slope of Graph
(2, 0)	$-\frac{4}{5}$
(1, −3)	$\frac{1}{8}$
$x = 0$	0
(1, 1)	Undefined

Figure 4.24

To see how you can use an *implicit derivative*, consider the graph shown in Figure 4.24. From the graph, you can see that y is not a function of x. Even so, the derivative found in Example 2 gives a formula for the slope of the tangent line at a point on this graph. The slopes at several points on the graph are shown below the graph.

TECHNOLOGY With most graphing utilities, it is easy to graph an equation that explicitly represents y as a function of x. Graphing other equations, however, can require some ingenuity. For instance, to graph the equation given in Example 2, use a graphing utility, set in *parametric* mode, to graph the parametric representations $x = \sqrt{t^3 + t^2 - 5t + 4}$, $y = t$, and $x = -\sqrt{t^3 + t^2 - 5t + 4}$, $y = t$, for $-5 \le t \le 5$. How does the result compare with the graph shown in Figure 4.24?

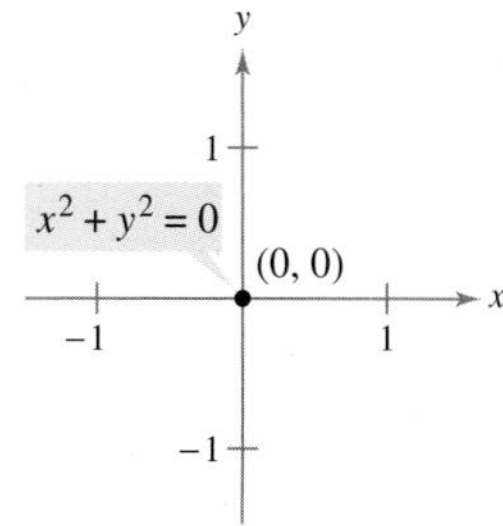

(a)

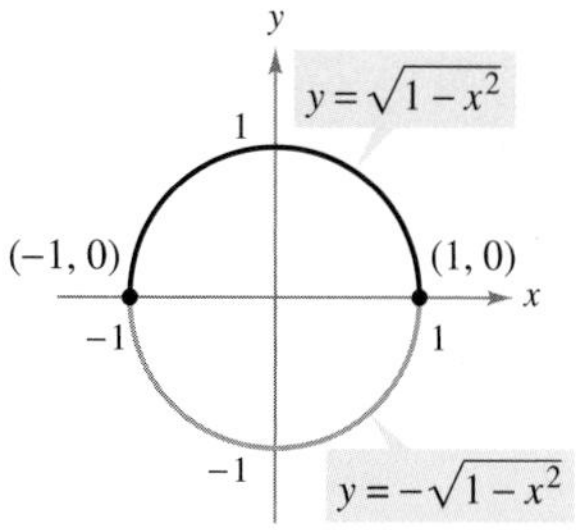

(b)

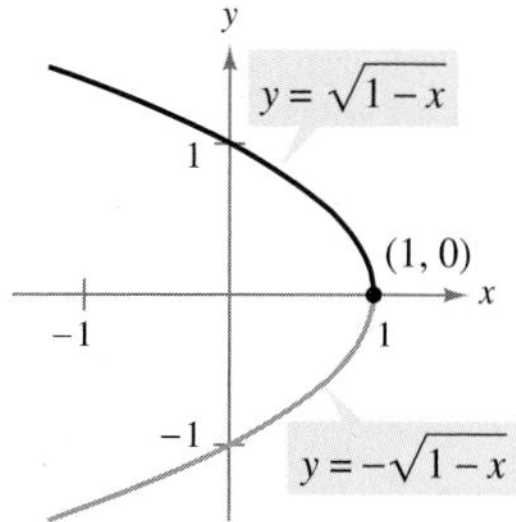

(c)
Some graph segments can be represented by differentiable functions.
Figure 4.25

It is meaningless to solve for dy/dx in an equation that has no solution points. (For example, $x^2 + y^2 = -4$ has no solution points.) If, however, a segment of a graph can be represented by a differentiable function, dy/dx will have meaning as the slope at each point on the segment. Recall that a function is not differentiable at (a) points with vertical tangents and (b) points at which the function is not continuous.

EXAMPLE 3 Representing a Graph by Differentiable Functions

If possible, represent y as a differentiable function of x.

a. $x^2 + y^2 = 0$ **b.** $x^2 + y^2 = 1$ **c.** $x + y^2 = 1$

Solution

a. The graph of this equation is a single point. So, it does not define y as a differentiable function of x. See Figure 4.25(a).

b. The graph of this equation is the unit circle, centered at $(0, 0)$. The upper semicircle is given by the differentiable function

$$y = \sqrt{1 - x^2}, \quad -1 < x < 1$$

and the lower semicircle is given by the differentiable function

$$y = -\sqrt{1 - x^2}, \quad -1 < x < 1.$$

At the points $(-1, 0)$ and $(1, 0)$, the slope of the graph is undefined. See Figure 4.25(b).

c. The upper half of this parabola is given by the differentiable function

$$y = \sqrt{1 - x}, \quad x < 1$$

and the lower half of this parabola is given by the differentiable function

$$y = -\sqrt{1 - x}, \quad x < 1.$$

At the point $(1, 0)$, the slope of the graph is undefined. See Figure 4.25(c).

EXAMPLE 4 Finding the Slope of a Graph Implicitly

Determine the slope of the tangent line to the graph of

$$x^2 + 4y^2 = 4$$

at the point $\left(\sqrt{2}, -1/\sqrt{2}\right)$. See Figure 4.26.

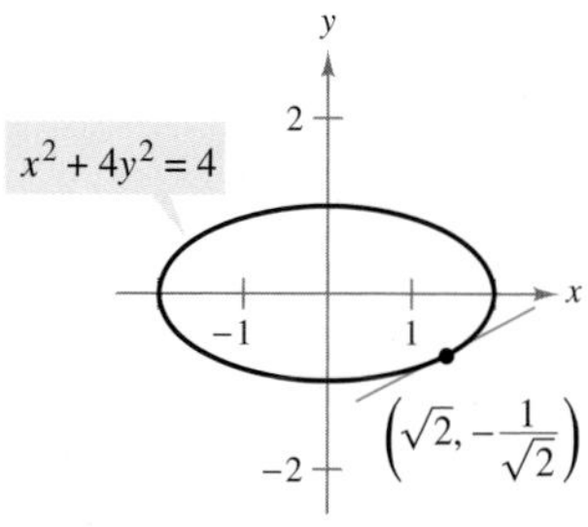

Figure 4.26

Solution

$$x^2 + 4y^2 = 4 \qquad \text{Write original equation.}$$

$$2x + 8y\frac{dy}{dx} = 0 \qquad \text{Differentiate with respect to } x.$$

$$\frac{dy}{dx} = \frac{-2x}{8y} = \frac{-x}{4y} \qquad \text{Solve for } \frac{dy}{dx}.$$

So, at $\left(\sqrt{2}, -1/\sqrt{2}\right)$, the slope is

$$\frac{dy}{dx} = \frac{-\sqrt{2}}{-4/\sqrt{2}} = \frac{1}{2}. \qquad \text{Evaluate } \frac{dy}{dx} \text{ when } x = \sqrt{2} \text{ and } y = -\frac{1}{\sqrt{2}}.$$

■

NOTE To see the benefit of implicit differentiation, try doing Example 4 using the explicit function $y = -\frac{1}{2}\sqrt{4 - x^2}$. ■

EXAMPLE 5 Finding the Slope of a Graph Implicitly

Find the slope of the tangent line to the graph of $3(x^2 + y^2)^2 = 100xy$ at the point $(3, 1)$.

Solution

$$3(x^2 + y^2)^2 = 100xy$$

$$\frac{d}{dx}[3(x^2 + y^2)^2] = \frac{d}{dx}[100xy]$$

$$3(2)(x^2 + y^2)\left(2x + 2y\frac{dy}{dx}\right) = 100\left[x\frac{dy}{dx} + y(1)\right]$$

$$12y(x^2 + y^2)\frac{dy}{dx} - 100x\frac{dy}{dx} = 100y - 12x(x^2 + y^2)$$

$$[12y(x^2 + y^2) - 100x]\frac{dy}{dx} = 100y - 12x(x^2 + y^2)$$

$$\frac{dy}{dx} = \frac{100y - 12x(x^2 + y^2)}{-100x + 12y(x^2 + y^2)}$$

$$= \frac{25y - 3x(x^2 + y^2)}{-25x + 3y(x^2 + y^2)}$$

$3(x^2 + y^2)^2 = 100xy$

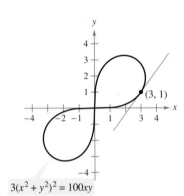

Lemniscate
Figure 4.27

At the point $(3, 1)$, the slope of the graph is

$$\frac{dy}{dx} = \frac{25(1) - 3(3)(3^2 + 1^2)}{-25(3) + 3(1)(3^2 + 1^2)} = \frac{25 - 90}{-75 + 30} = \frac{-65}{-45} = \frac{13}{9}$$

as shown in Figure 4.27. This graph is called a **lemniscate.**

EXAMPLE 6 Determining a Differentiable Function

Find dy/dx implicitly for the equation

$$4x - y^3 + 12y = 0$$

and use Figure 4.28 to find the largest interval of the form $-a < y < a$ on which y is a differentiable function of x.

Solution

$$4x - y^3 + 12y = 0 \qquad \text{Write original equation.}$$

$$\frac{d}{dx}[4x - y^3 + 12y] = \frac{d}{dx}[0] \qquad \text{Differentiate with respect to } x.$$

$$4 - 3y^2\frac{dy}{dx} + 12\frac{dy}{dx} = 0$$

$$\frac{dy}{dx}(-3y^2 + 12) = -4 \qquad \text{Factor and simplify.}$$

$$\frac{dy}{dx} = \frac{4}{3(y^2 - 4)} \qquad \text{Divide each side by } -3(y^2 - 4).$$

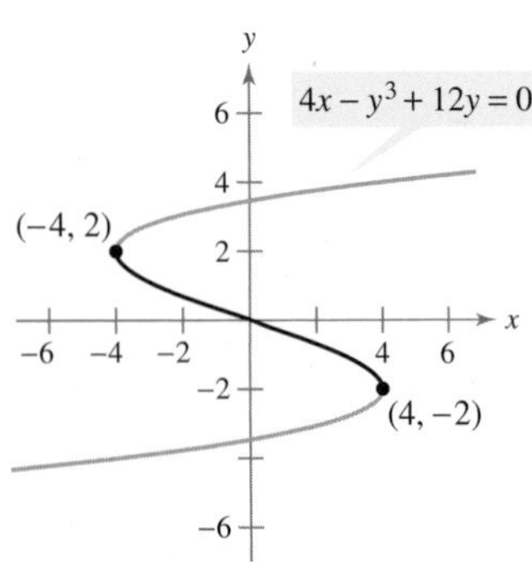

Figure 4.28

From Figure 4.28 you can see that the largest interval about the origin for which y is a differentiable function of x is $-2 < y < 2$. ■

The Granger Collection, New York

ISAAC BARROW (1630–1677)

The graph in Figure 4.29 is called the **kappa curve** because it resembles the Greek letter kappa, κ. The general solution for the tangent line to this curve was discovered by the English mathematician Isaac Barrow. Newton was Barrow's student, and they corresponded frequently regarding their work in the early development of calculus.

With implicit differentiation, the form of the derivative often can be simplified by an appropriate use of the *original* equation. A similar technique can be used to find and simplify higher-order derivatives obtained implicitly.

EXAMPLE 7 Finding the Second Derivative Implicitly

Given $x^2 + y^2 = 25$, find $\dfrac{d^2y}{dx^2}$. Evaluate the first and second derivatives at the point $(-3, 4)$.

Solution Differentiating each term with respect to x produces

$$2x + 2y\frac{dy}{dx} = 0$$

$$2y\frac{dy}{dx} = -2x$$

$$\frac{dy}{dx} = \frac{-2x}{2y} = -\frac{x}{y}.$$

At $(-3, 4)$: $\dfrac{dy}{dx} = -\dfrac{(-3)}{4} = \dfrac{3}{4}$.

Differentiating a second time with respect to x yields

$$\frac{d^2y}{dx^2} = -\frac{(y)(1) - (x)(dy/dx)}{y^2} \qquad \text{Quotient Rule}$$

$$= -\frac{y - (x)(-x/y)}{y^2} = -\frac{y^2 + x^2}{y^3} = -\frac{25}{y^3}.$$

At $(-3, 4)$: $\dfrac{d^2y}{dx^2} = -\dfrac{25}{4^3} = -\dfrac{25}{64}$.

EXAMPLE 8 Finding a Tangent Line to a Graph

Find the tangent line to the graph given by $x^2(x^2 + y^2) = y^2$ at the point $\left(\sqrt{2}/2, \sqrt{2}/2\right)$, as shown in Figure 4.29.

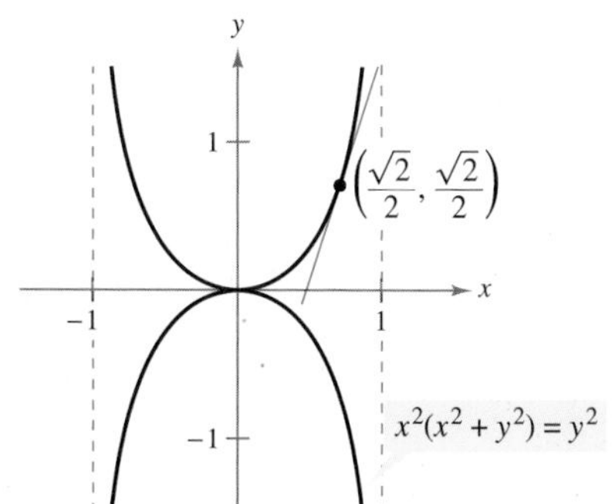

The kappa curve
Figure 4.29

Solution By rewriting and differentiating implicitly, you obtain

$$x^4 + x^2y^2 - y^2 = 0$$

$$4x^3 + \left[x^2\left(2y\frac{dy}{dx}\right) + 2xy^2\right] - 2y\frac{dy}{dx} = 0 \qquad \text{Product Rule}$$

$$2y(x^2 - 1)\frac{dy}{dx} = -2x(2x^2 + y^2) \qquad \text{Collect like terms.}$$

$$\frac{dy}{dx} = \frac{x(2x^2 + y^2)}{y(1 - x^2)}.$$

At the point $\left(\sqrt{2}/2, \sqrt{2}/2\right)$, the slope is

$$\frac{dy}{dx} = \frac{\left(\sqrt{2}/2\right)[2(1/2) + (1/2)]}{\left(\sqrt{2}/2\right)[1 - (1/2)]} = \frac{3/2}{1/2} = 3$$

and the equation of the tangent line at this point is

$$y - \frac{\sqrt{2}}{2} = 3\left(x - \frac{\sqrt{2}}{2}\right)$$

$$y = 3x - \sqrt{2}.$$

■

4.5 Exercises

See www.CalcChat.com for worked-out solutions to odd-numbered exercises.

In Exercises 1–10, find dy/dx by implicit differentiation.

1. $x^2 + y^2 = 9$
2. $x^2 - y^2 = 25$
3. $x^{1/2} + y^{1/2} = 16$
4. $x^3 + y^3 = 64$
5. $x^3 - xy + y^2 = 7$
6. $x^2y + y^2x = -2$
7. $x^3y^3 - y = x$
8. $\sqrt{xy} = x^2y + 1$
9. $x^3 - 3x^2y + 2xy^2 = 12$
10. $x^3 - 2x^2y + 3xy^2 = 38$

In Exercises 11–14, (a) find two explicit functions by solving the equation for y in terms of x, (b) sketch the graph of the equation and label the parts given by the corresponding explicit functions, (c) differentiate the explicit functions, and (d) find dy/dx implicitly and show that the result is equivalent to that of part (c).

11. $x^2 + y^2 = 64$
12. $x^2 + y^2 - 4x + 6y + 9 = 0$
13. $16x^2 + 25y^2 = 400$
14. $16y^2 - x^2 = 16$

In Exercises 15–20, find dy/dx by implicit differentiation and evaluate the derivative at the given point.

15. $xy = 6, \quad (-6, -1)$
16. $x^2 - y^3 = 0, \quad (1, 1)$
17. $y^2 = \dfrac{x^2 - 49}{x^2 + 49}, \quad (7, 0)$
18. $(x + y)^3 = x^3 + y^3, \quad (-1, 1)$
19. $x^{2/3} + y^{2/3} = 5, \quad (8, 1)$
20. $x^3 + y^3 = 6xy - 1, \quad (2, 3)$

Famous Curves **In Exercises 21–24, find the slope of the tangent line to the graph at the given point.**

21. Witch of Agnesi:
$(x^2 + 4)y = 8$
Point: $(2, 1)$

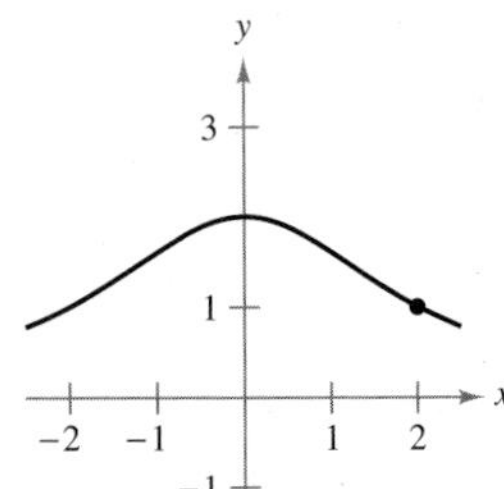

22. Cissoid:
$(4 - x)y^2 = x^3$
Point: $(2, 2)$

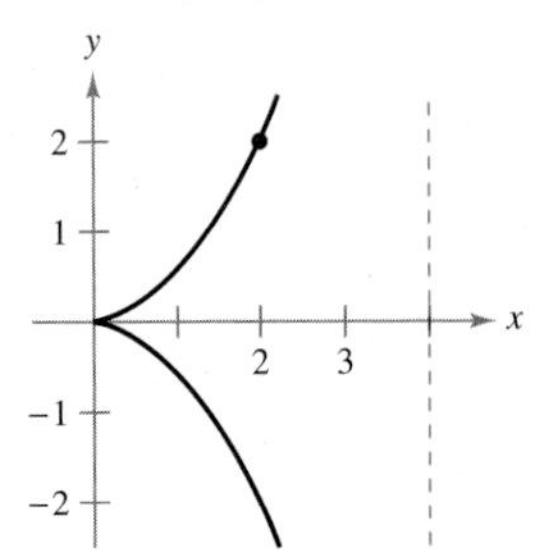

23. Bifolium:
$(x^2 + y^2)^2 = 4x^2y$
Point: $(1, 1)$

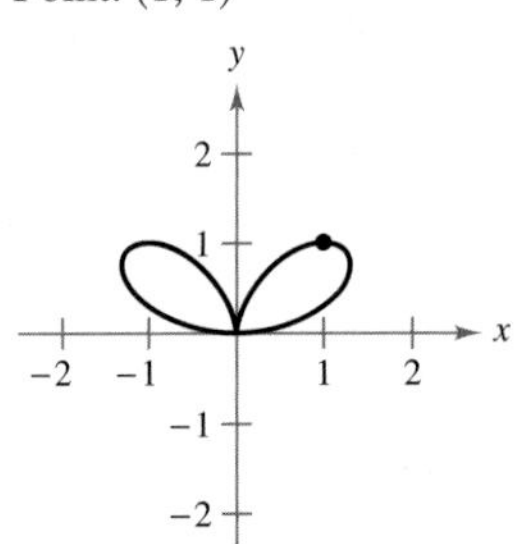

24. Folium of Descartes:
$x^3 + y^3 - 6xy = 0$
Point: $\left(\frac{4}{3}, \frac{8}{3}\right)$

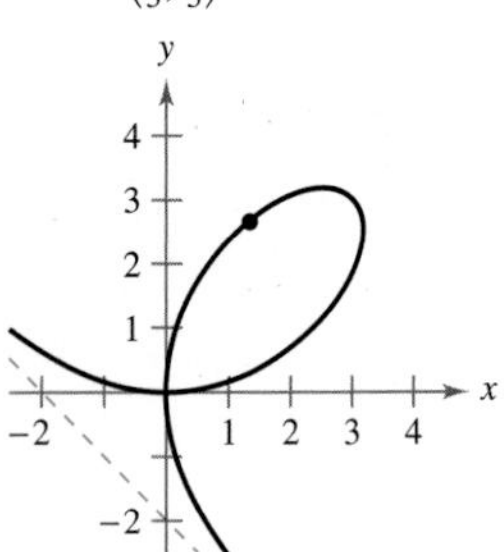

Famous Curves **In Exercises 25–32, find an equation of the tangent line to the graph at the given point. To print an enlarged copy of the graph, go to the website *www.mathgraphs.com*.**

25. Parabola

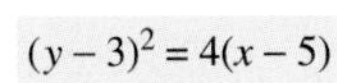

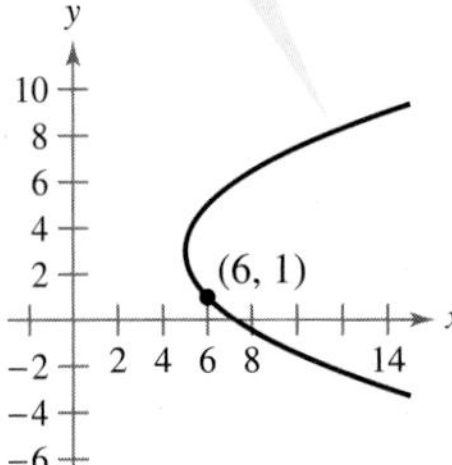

26. Circle

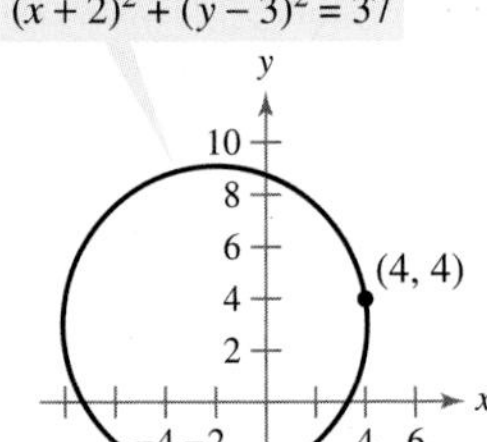

27. Rotated hyperbola

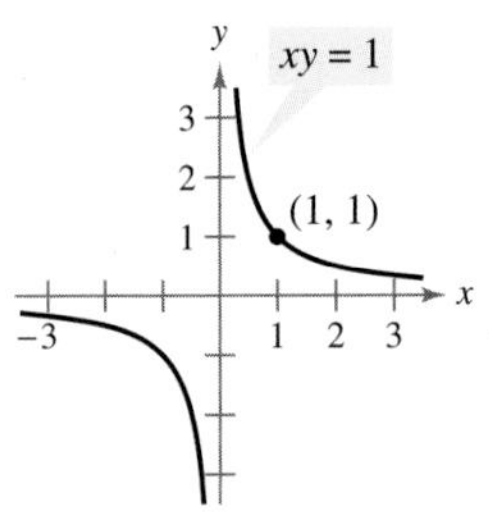

28. Rotated ellipse

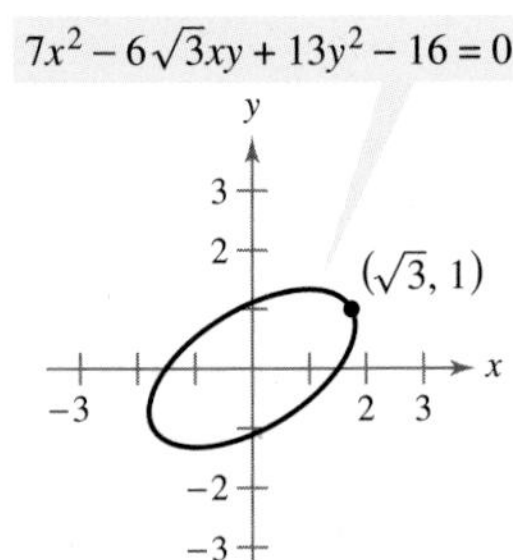

29. Cruciform

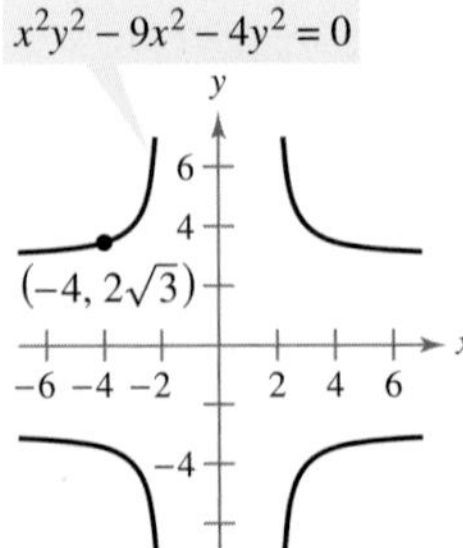

30. Astroid

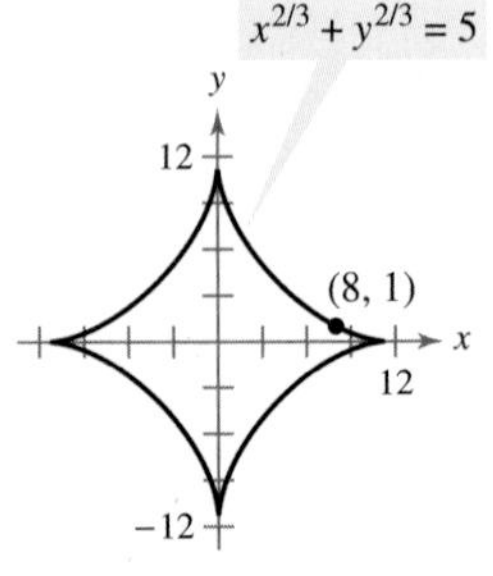

31. Lemniscate

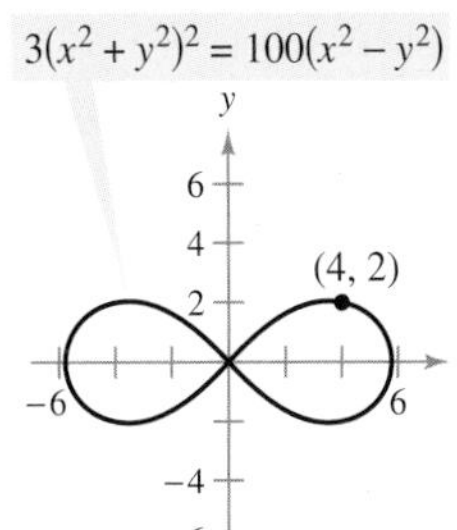

32. Kappa curve

$y^2(x^2+y^2) = 2x^2$

(1, 1)

33. (a) Use implicit differentiation to find an equation of the tangent line to the ellipse $\frac{x^2}{2} + \frac{y^2}{8} = 1$ at $(1, 2)$.

(b) Show that the equation of the tangent line to the ellipse $\frac{x^2}{a^2} + \frac{y^2}{b^2} = 1$ at (x_0, y_0) is $\frac{x_0x}{a^2} + \frac{y_0y}{b^2} = 1$.

34. (a) Use implicit differentiation to find an equation of the tangent line to the hyperbola $\frac{x^2}{6} - \frac{y^2}{8} = 1$ at $(3, -2)$.

(b) Show that the equation of the tangent line to the hyperbola $\frac{x^2}{a^2} - \frac{y^2}{b^2} = 1$ at (x_0, y_0) is $\frac{x_0x}{a^2} - \frac{y_0y}{b^2} = 1$.

In Exercises 35–40, find d^2y/dx^2 in terms of x and y.

35. $x^2 + y^2 = 4$

36. $x^2y^2 - 2x = 3$

37. $x^2 - y^2 = 36$

38. $1 - xy = x - y$

39. $y^2 = x^3$

40. $y^2 = 10x$

In Exercises 41 and 42, use a graphing utility to graph the equation. Find an equation of the tangent line to the graph at the given point and graph the tangent line in the same viewing window.

41. $\sqrt{x} + \sqrt{y} = 5, \quad (9, 4)$

42. $y^2 = \frac{x-1}{x^2+1}, \quad \left(2, \frac{\sqrt{5}}{5}\right)$

In Exercises 43 and 44, find equations for the tangent line and normal line to the circle at each given point. (The *normal line* at a point is perpendicular to the tangent line at the point.) Use a graphing utility to graph the equation, tangent line, and normal line.

43. $x^2 + y^2 = 25$
$(4, 3), (-3, 4)$

44. $x^2 + y^2 = 36$
$(6, 0), \left(5, \sqrt{11}\right)$

45. Show that the normal line at any point on the circle $x^2 + y^2 = r^2$ passes through the origin.

46. Two circles of radius 4 are tangent to the graph of $y^2 = 4x$ at the point $(1, 2)$. Find equations of these two circles.

In Exercises 47 and 48, find the points at which the graph of the equation has a vertical or horizontal tangent line.

47. $25x^2 + 16y^2 + 200x - 160y + 400 = 0$

48. $4x^2 + y^2 - 8x + 4y + 4 = 0$

Orthogonal Trajectories **In Exercises 49–52, use a graphing utility to sketch the intersecting graphs of the equations and show that they are orthogonal. [Two graphs are *orthogonal* if at their point(s) of intersection their tangent lines are perpendicular to each other.]**

49. $2x^2 + y^2 = 6$
$y^2 = 4x$

50. $y^2 = x^3$
$2x^2 + 3y^2 = 5$

51. $x + y = 0$
$x^2 + y^2 = 4$

52. $x^3 = 3(y - 1)$
$x(3y - 29) = 3$

Orthogonal Trajectories **In Exercises 53 and 54, verify that the two families of curves are orthogonal, where C and K are real numbers. Use a graphing utility to graph the two families for two values of C and two values of K.**

53. $xy = C, \quad x^2 - y^2 = K$

54. $x^2 + y^2 = C^2, \quad y = Kx$

In Exercises 55 and 56, differentiate (a) with respect to x (y is a function of x) and (b) with respect to t (x and y are functions of t).

55. $2y^2 - 3x^4 = 0$

56. $x^2 - 3xy^2 + y^3 = 10$

WRITING ABOUT CONCEPTS

57. Describe the difference between the explicit form of a function and an implicit equation. Give an example of each.

58. In your own words, state the guidelines for implicit differentiation.

59. ***Orthogonal Trajectories*** The figure below shows the topographic map carried by a group of hikers. The hikers are in a wooded area on top of the hill shown on the map and they decide to follow a path of steepest descent (orthogonal trajectories to the contours on the map). Draw their routes if they start from point A and if they start from point B. If their goal is to reach the road along the top of the map, which starting point should they use? To print an enlarged copy of the graph, go to the website *www.mathgraphs.com*.

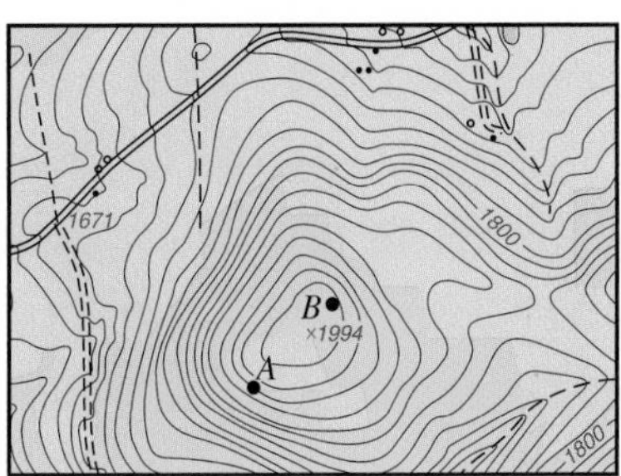

60. ***Weather Map*** The weather map shows several *isobars*—curves that represent areas of constant air pressure. Three high pressures H and one low pressure L are shown on the map. Given that wind speed is greatest along the orthogonal trajectories of the isobars, use the map to determine the areas having high wind speed.

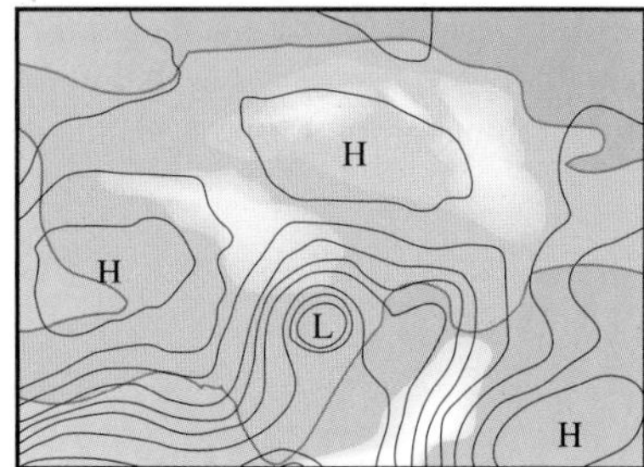

61. Consider the equation $x^4 = 4(4x^2 - y^2)$.

(a) Use a graphing utility to graph the equation.

(b) Find and graph the four tangent lines to the curve for $y = 3$.

(c) Find the exact coordinates of the point of intersection of the two tangent lines in the first quadrant.

CAPSTONE

62. Determine if the statement is true. If it is false, explain why and correct it. For each statement, assume y is a function of x.

(a) $\frac{d}{dx}x^2 = 2x$ (b) $\frac{d}{dy}x^2 = 2x$

(c) $\frac{d}{dx}y^2 = 2y$ (d) $\frac{d}{dy}y^2 = 2y$

63. Let L be any tangent line to the curve $\sqrt{x} + \sqrt{y} = \sqrt{c}$. Show that the sum of the x- and y-intercepts of L is c.

64. Prove (Theorem 4.3) that $d/dx[x^n] = nx^{n-1}$ for the case in which n is a rational number. (*Hint:* Write $y = x^{p/q}$ in the form $y^q = x^p$ and differentiate implicitly. Assume that p and q are integers, where $q > 0$.)

65. ***Horizontal Tangent*** Determine the point(s) at which the graph of $y^4 = y^2 - x^2$ has a horizontal tangent.

66. ***Tangent Lines*** Find equations of both tangent lines to the ellipse $\frac{x^2}{4} + \frac{y^2}{9} = 1$ that passes through the point $(4, 0)$.

67. ***Normals to a Parabola*** The graph shows the normal lines from the point $(2, 0)$ to the graph of the parabola $x = y^2$. How many normal lines are there from the point $(x_0, 0)$ to the graph of the parabola if (a) $x_0 = \frac{1}{4}$, (b) $x_0 = \frac{1}{2}$, and (c) $x_0 = 1$? For what value of x_0 are two of the normal lines perpendicular to each other?

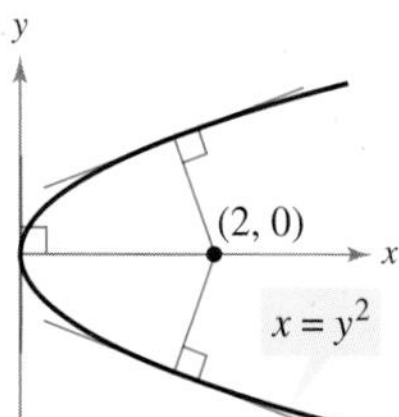

68. ***Normal Lines*** (a) Find an equation of the normal line to the ellipse $\frac{x^2}{32} + \frac{y^2}{8} = 1$ at the point $(4, 2)$. (b) Use a graphing utility to graph the ellipse and the normal line. (c) At what other point does the normal line intersect the ellipse?

SECTION PROJECT

Optical Illusions

In each graph below, an optical illusion is created by having lines intersect a family of curves. In each case, the lines appear to be curved. Find the value of dy/dx for the given values of x and y.

(a) Circles: $x^2 + y^2 = C^2$

$x = 3, y = 4, C = 5$

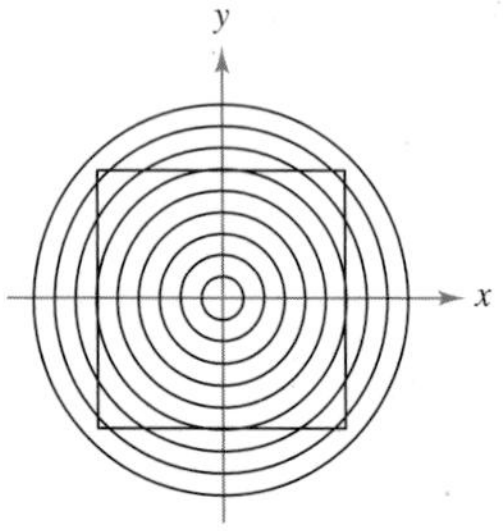

(b) Hyperbolas: $xy = C$

$x = 1, y = 4, C = 4$

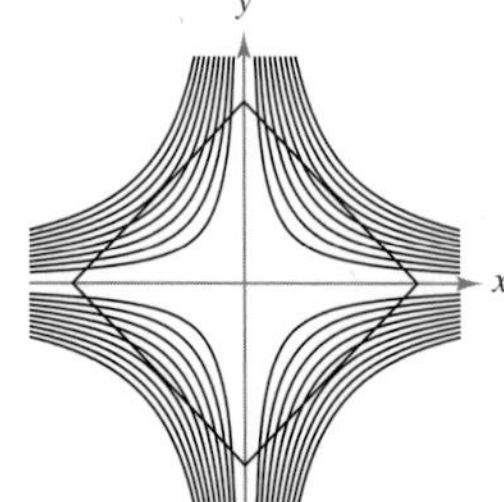

(c) Lines: $ax = by$

$x = \sqrt{3}, y = 3,$

$a = \sqrt{3}, b = 1$

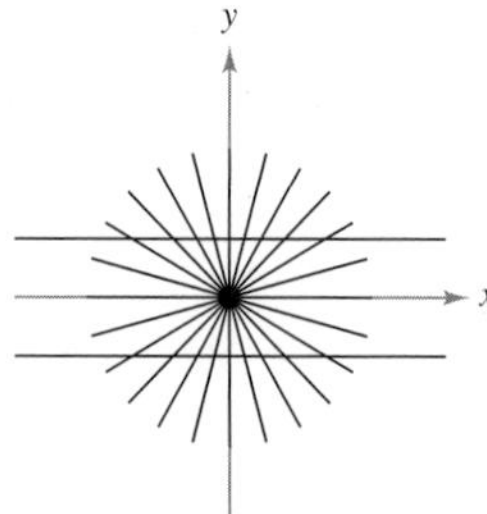

FOR FURTHER INFORMATION For more information on the mathematics of optical illusions, see the article "Descriptive Models for Perception of Optical Illusions" by David A. Smith in *The UMAP Journal*.

4.6 Related Rates

- **Find a related rate.**
- **Use related rates to solve real-life problems.**

Finding Related Rates

You have seen how the Chain Rule can be used to find dy/dx implicitly. Another important use of the Chain Rule is to find the rates of change of two or more related variables that are changing with respect to *time*.

For example, when water is drained out of a conical tank (see Figure 4.30), the volume V, the radius r, and the height h of the water level are all functions of time t. Knowing that these variables are related by the equation

$$V = \frac{\pi}{3} r^2 h \qquad \text{Original equation}$$

you can differentiate implicitly with respect to t to obtain the **related-rate** equation

$$\frac{d}{dt}(V) = \frac{d}{dt}\left(\frac{\pi}{3} r^2 h\right) = \frac{\pi}{3}\left[\frac{d}{dt}(r^2 h)\right] \qquad \text{Constant Multiple Rule}$$

$$\frac{dV}{dt} = \frac{\pi}{3}\left[r^2 \frac{dh}{dt} + h\left(2r\frac{dr}{dt}\right)\right] \qquad \text{Differentiate with respect to } t, \text{ using the Product and Chain Rules.}$$

$$= \frac{\pi}{3}\left(r^2 \frac{dh}{dt} + 2rh\frac{dr}{dt}\right).$$

From this equation you can see that the rate of change of V is related to the rates of change of both h and r.

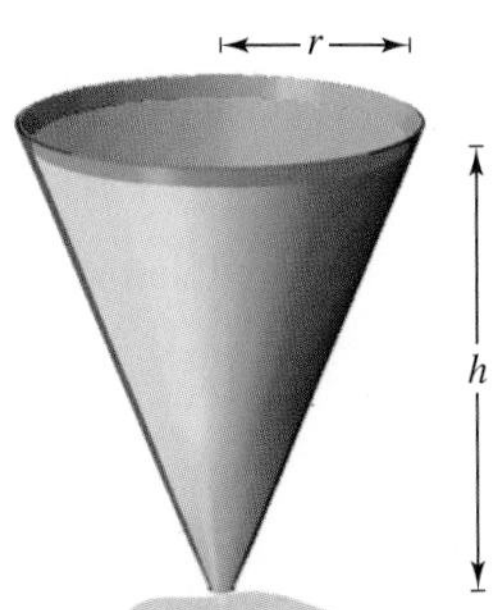

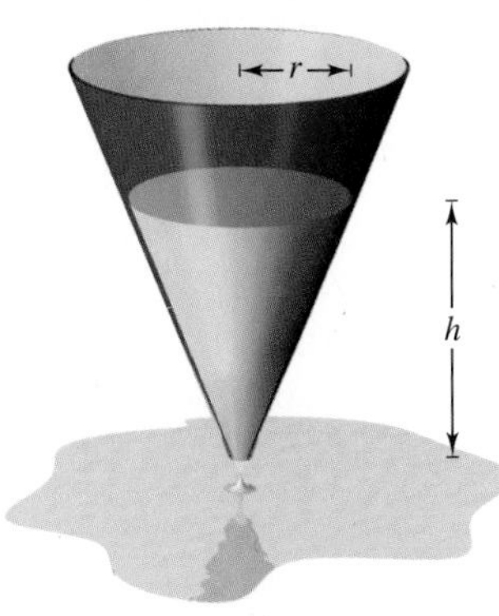

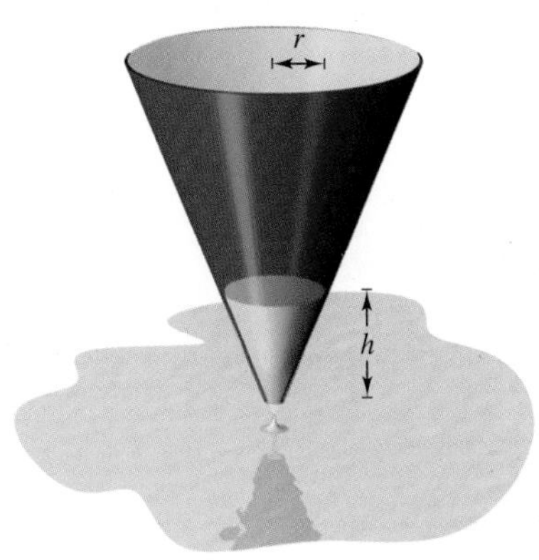

Volume is related to radius and height.
Figure 4.30

EXPLORATION

Finding a Related Rate In the conical tank shown in Figure 4.30, suppose that the height of the water level is changing at a rate of -0.2 foot per minute and the radius is changing at a rate of -0.1 foot per minute. What is the rate of change in the volume when the radius is $r = 1$ foot and the height is $h = 2$ feet? Does the rate of change in the volume depend on the values of r and h? Explain.

■ **FOR FURTHER INFORMATION** To learn more about the history of related-rate problems, see the article "The Lengthening Shadow: The Story of Related Rates" by Bill Austin, Don Barry, and David Berman in *Mathematics Magazine*. To view this article, go to the website *www.matharticles.com*.

EXAMPLE 1 Two Rates That Are Related

Suppose x and y are both differentiable functions of t and are related by the equation $y = x^2 + 3$. Find dy/dt when $x = 1$, given that $dx/dt = 2$ when $x = 1$.

Solution Using the Chain Rule, you can differentiate both sides of the equation *with respect to t*.

$$y = x^2 + 3 \qquad \text{Write original equation.}$$

$$\frac{d}{dt}[y] = \frac{d}{dt}[x^2 + 3] \qquad \text{Differentiate with respect to } t.$$

$$\frac{dy}{dt} = 2x\frac{dx}{dt} \qquad \text{Chain Rule}$$

When $x = 1$ and $dx/dt = 2$, you have

$$\frac{dy}{dt} = 2(1)(2) = 4.$$

■

Problem Solving with Related Rates

In Example 1, you were *given* an equation that related the variables x and y and were asked to find the rate of change of y when $x = 1$.

Equation: $y = x^2 + 3$

Given rate: $\frac{dx}{dt} = 2$ when $x = 1$

Find: $\frac{dy}{dt}$ when $x = 1$

In each of the remaining examples in this section, you must *create* a mathematical model from a verbal description.

EXAMPLE 2 Ripples in a Pond

A pebble is dropped into a calm pond, causing ripples in the form of concentric circles, as shown in Figure 4.31. The radius r of the outer ripple is increasing at a constant rate of 1 foot per second. When the radius is 4 feet, at what rate is the total area A of the disturbed water changing?

Russ Bishop / Alamy

Total area increases as the outer radius increases.
Figure 4.31

Solution The variables r and A are related by $A = \pi r^2$. The rate of change of the radius r is $dr/dt = 1$.

Equation: $A = \pi r^2$

Given rate: $\frac{dr}{dt} = 1$

Find: $\frac{dA}{dt}$ when $r = 4$

With this information, you can proceed as in Example 1.

$$\frac{d}{dt}[A] = \frac{d}{dt}[\pi r^2] \qquad \text{Differentiate with respect to } t.$$

$$\frac{dA}{dt} = 2\pi r\frac{dr}{dt} \qquad \text{Chain Rule}$$

$$\frac{dA}{dt} = 2\pi(4)(1) = 8\pi \qquad \text{Substitute 4 for } r \text{ and 1 for } dr/dt.$$

When the radius is 4 feet, the area is changing at a rate of 8π square feet per second. ■

GUIDELINES FOR SOLVING RELATED-RATE PROBLEMS

1. Identify all *given* quantities and quantities *to be determined.* Make a sketch and label the quantities.
2. Write an equation involving the variables whose rates of change either are given or are to be determined.
3. Using the Chain Rule, implicitly differentiate both sides of the equation *with respect to time t.*
4. *After* completing Step 3, substitute into the resulting equation all known values for the variables and their rates of change. Then solve for the required rate of change.

NOTE When using these guidelines, be sure you perform Step 3 before Step 4. Substituting the known values of the variables before differentiating will produce an inappropriate derivative.

The table below lists examples of mathematical models involving rates of change. For instance, the rate of change in the first example is the velocity of a car.

Verbal Statement	Mathematical Model
The velocity of a car after traveling for 1 hour is 50 miles per hour.	$x =$ distance traveled $\frac{dx}{dt} = 50$ when $t = 1$
Water is being pumped into a swimming pool at a rate of 10 cubic meters per hour.	$V =$ volume of water in pool $\frac{dV}{dt} = 10\ \text{m}^3/\text{hr}$
A population of bacteria is increasing at the rate of 2000 per hour.	$x =$ number in population $\frac{dx}{dt} = 2000$ bacteria per hour

EXAMPLE 3 An Inflating Balloon

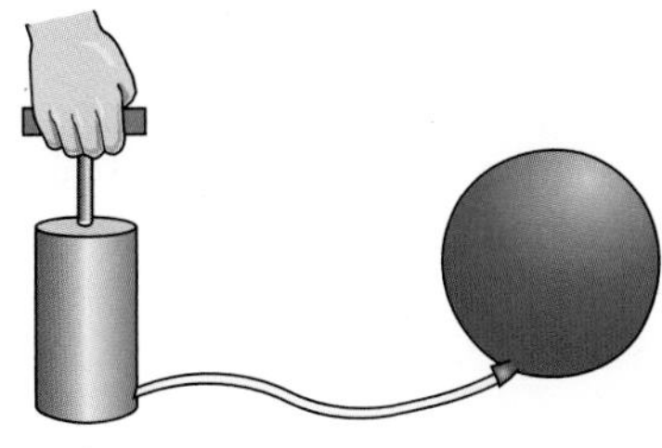

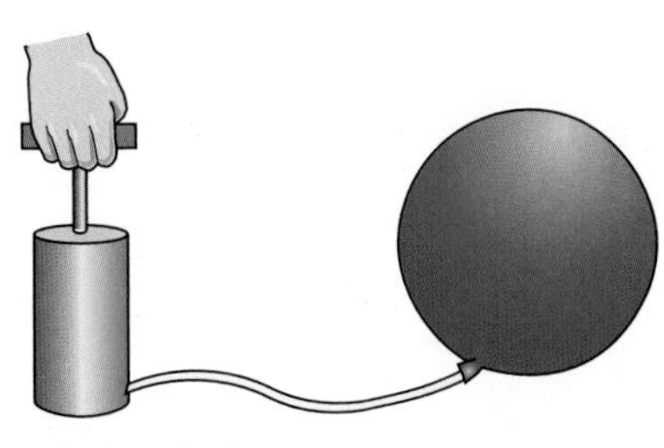

Inflating a balloon
Figure 4.32

Air is being pumped into a spherical balloon (see Figure 4.32) at a rate of 4.5 cubic feet per minute. Find the rate of change of the radius when the radius is 2 feet.

Solution Let V be the volume of the balloon and let r be its radius. Because the volume is increasing at a rate of 4.5 cubic feet per minute, you know that at time t the rate of change of the volume is $dV/dt = \frac{9}{2}$. So, the problem can be stated as shown.

Given rate: $\frac{dV}{dt} = \frac{9}{2}$ (constant rate)

Find: $\frac{dr}{dt}$ when $r = 2$

To find the rate of change of the radius, you must find an equation that relates the radius r to the volume V.

Equation: $V = \frac{4}{3}\pi r^3$ Volume of a sphere

Differentiating both sides of the equation with respect to t produces

$$\frac{dV}{dt} = 4\pi r^2 \frac{dr}{dt}$$ Differentiate with respect to t.

$$\frac{dr}{dt} = \frac{1}{4\pi r^2}\left(\frac{dV}{dt}\right).$$ Solve for dr/dt.

Finally, when $r = 2$, the rate of change of the radius is

$$\frac{dr}{dt} = \frac{1}{16\pi}\left(\frac{9}{2}\right) \approx 0.09 \text{ foot per minute.}$$ ■

In Example 3, note that the volume is increasing at a *constant* rate but the radius is increasing at a *variable* rate. Just because two rates are related does not mean that they are proportional. In this particular case, the radius is growing more and more slowly as t increases. Do you see why?

EXAMPLE 4 The Speed of an Airplane Tracked by Radar

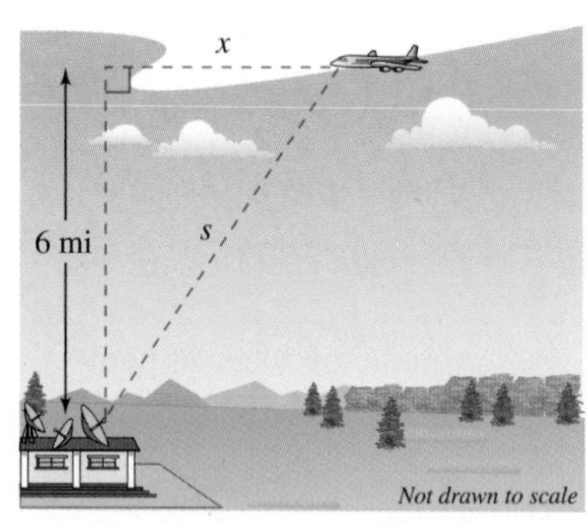

An airplane is flying at an altitude of 6 miles, s miles from the station.
Figure 4.33

An airplane is flying on a flight path that will take it directly over a radar tracking station, as shown in Figure 4.33. If s is decreasing at a rate of 400 miles per hour when $s = 10$ miles, what is the speed of the plane?

Solution Let x be the horizontal distance from the station, as shown in Figure 4.33. Notice that when $s = 10$, $x = \sqrt{10^2 - 36} = 8$.

Given rate: $ds/dt = -400$ when $s = 10$

Find: dx/dt when $s = 10$ and $x = 8$

You can find the velocity of the plane as shown.

Equation: $x^2 + 6^2 = s^2$ — Pythagorean Theorem

$$2x\frac{dx}{dt} = 2s\frac{ds}{dt}$$ Differentiate with respect to t.

$$\frac{dx}{dt} = \frac{s}{x}\left(\frac{ds}{dt}\right)$$ Solve for dx/dt.

$$\frac{dx}{dt} = \frac{10}{8}(-400)$$ Substitute for s, x, and ds/dt.

$$= -500 \text{ miles per hour}$$ Simplify.

Because the velocity is -500 miles per hour, the *speed* is 500 miles per hour.

NOTE Note that the velocity in Example 4 is negative because x represents a distance that is decreasing.

EXAMPLE 5 Tracking an Accelerating Object

Find the rate of change in the distance between the camera shown in Figure 4.34 and the base of the shuttle 10 seconds after lift-off. Assume that the camera and the base of the shuttle are level with each other when $t = 0$.

Solution Let r be the distance between the camera and the base of the shuttle (see Figure 4.34). Find the velocity of the rocket by differentiating s with respect to t.

Given rate: $\frac{ds}{dt} = 100t =$ velocity of rocket — From $s = 50t^2$

Find: $\frac{dr}{dt}$ when $t = 10$

Using Figure 4.34, you can relate s and r by the equation $r^2 = 2000^2 + s^2$.

Equation: $r^2 = 2000^2 + s^2$ — Pythagorean Theorem

$$2r\frac{dr}{dt} = 2s\frac{ds}{dt}$$ Differentiate with respect to t.

$$\frac{dr}{dt} = \frac{s}{r} \cdot \frac{ds}{dt} = \frac{s}{r}(100t)$$ Substitute $100t$ for ds/dt.

Now, when $t = 10$, you know that $s = 50(10)^2 = 5000$, and you obtain

$$r = \sqrt{2000^2 + 5000^2} = 1000\sqrt{29}.$$

Finally, the rate of change of r when $t = 10$ is

$$\frac{dr}{dt} = \frac{5000}{1000\sqrt{29}}(100)(10) \approx 928.48 \text{ feet per second.}$$

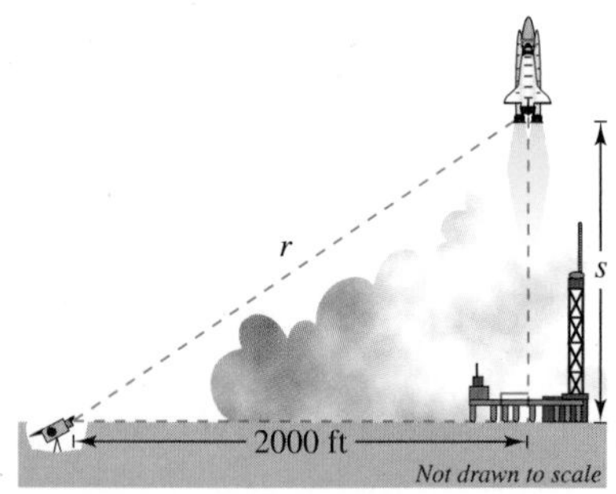

A television camera at ground level is filming the lift-off of a space shuttle that is rising vertically according to the position equation $s = 50t^2$, where s is measured in feet and t is measured in seconds. The camera is 2000 feet from the launch pad.
Figure 4.34

4.6 Exercises

See www.CalcChat.com for worked-out solutions to odd-numbered exercises.

In Exercises 1–4, assume that x and y are both differentiable functions of t and find the required values of dy/dt and dx/dt.

Equation	*Find*	*Given*
1. $y = \sqrt{x}$	(a) $\dfrac{dy}{dt}$ when $x = 4$	$\dfrac{dx}{dt} = 3$
	(b) $\dfrac{dx}{dt}$ when $x = 25$	$\dfrac{dy}{dt} = 2$
2. $y = 4(x^2 - 5x)$	(a) $\dfrac{dy}{dt}$ when $x = 3$	$\dfrac{dx}{dt} = 2$
	(b) $\dfrac{dx}{dt}$ when $x = 1$	$\dfrac{dy}{dt} = 5$
3. $xy = 4$	(a) $\dfrac{dy}{dt}$ when $x = 8$	$\dfrac{dx}{dt} = 10$
	(b) $\dfrac{dx}{dt}$ when $x = 1$	$\dfrac{dy}{dt} = -6$
4. $x^2 + y^2 = 25$	(a) $\dfrac{dy}{dt}$ when $x = 3, y = 4$	$\dfrac{dx}{dt} = 8$
	(b) $\dfrac{dx}{dt}$ when $x = 4, y = 3$	$\dfrac{dy}{dt} = -2$

In Exercises 5–8, a point is moving along the graph of the given function such that dx/dt is 2 centimeters per second. Find dy/dt for the given values of x.

5. $y = 2x^2 + 1$	(a) $x = -1$	(b) $x = 0$	(c) $x = 1$
6. $y = \dfrac{1}{1 + x^2}$	(a) $x = -2$	(b) $x = 0$	(c) $x = 2$
7. $\sqrt{x} + \sqrt{y} = 4$	(a) $x = 1$	(b) $x = 4$	(c) $x = 9$
8. $x^2 + \sqrt{y} + y = 3$	(a) $x = -1$	(b) $x = 0$	(c) $x = 1$

WRITING ABOUT CONCEPTS

9. Consider the linear function $y = ax + b$. If x changes at a constant rate, does y change at a constant rate? If so, does it change at the same rate as x? Explain.

10. In your own words, state the guidelines for solving related-rate problems.

11. Find the rate of change of the distance between the origin and a moving point on the graph of $y = x^2 + 1$ if $dx/dt = 2$ centimeters per second.

12. Find the rate of change of the distance between the origin and a moving point on the graph of $y = \sqrt{x}$ if $dx/dt = 2$ centimeters per second.

13. ***Area*** The radius r of a circle is increasing at a rate of 4 centimeters per minute. Find the rates of change of the area when (a) $r = 8$ centimeters and (b) $r = 32$ centimeters.

14. ***Area*** Let A be the area of a circle of radius r that is changing with respect to time. If dr/dt is constant, is dA/dt constant? Explain.

15. ***Area*** The base of an isosceles triangle has length b and the two sides of equal length each measure 30 centimeters.

(a) Find the area A of the triangle as a function of b.

(b) If b is increasing at a rate of 3 centimeters per second, find the rate of change of the area when $b = 20$ centimeters and $b = 56$ centimeters.

(c) Explain why the rate of change of the area of the triangle is not constant even though db/dt is constant.

16. ***Volume*** The radius r of a sphere is increasing at a rate of 3 inches per minute.

(a) Find the rates of change of the volume when $r = 9$ inches and $r = 36$ inches.

(b) Explain why the rate of change of the volume of the sphere is not constant even though dr/dt is constant.

17. ***Volume*** A spherical balloon is inflated with gas at the rate of 800 cubic centimeters per minute. How fast is the radius of the balloon increasing at the instant the radius is (a) 30 centimeters and (b) 60 centimeters?

18. ***Volume*** All edges of a cube are expanding at a rate of 6 centimeters per second. How fast is the volume changing when each edge is (a) 2 centimeters and (b) 10 centimeters?

19. ***Surface Area*** The conditions are the same as in Exercise 18. Determine how fast the *surface area* is changing when each edge is (a) 2 centimeters and (b) 10 centimeters.

20. ***Volume*** The formula for the volume of a cone is $V = \frac{1}{3}\pi r^2 h$. Find the rates of change of the volume if dr/dt is 2 inches per minute and $h = 3r$ when (a) $r = 6$ inches and (b) $r = 24$ inches.

21. ***Volume*** At a sand and gravel plant, sand is falling off a conveyor and onto a conical pile at a rate of 10 cubic feet per minute. The diameter of the base of the cone is approximately three times the altitude. At what rate is the height of the pile changing when the pile is 15 feet high?

22. ***Depth*** A conical tank (with vertex down) is 10 feet across the top and 12 feet deep. If water is flowing into the tank at a rate of 10 cubic feet per minute, find the rate of change of the depth of the water when the water is 8 feet deep.

23. ***Depth*** A swimming pool is 12 meters long, 6 meters wide, 1 meter deep at the shallow end, and 3 meters deep at the deep end (see figure). Water is being pumped into the pool at $\frac{1}{4}$ cubic meter per minute, and there is 1 meter of water at the deep end.

(a) What percent of the pool is filled?

(b) At what rate is the water level rising?

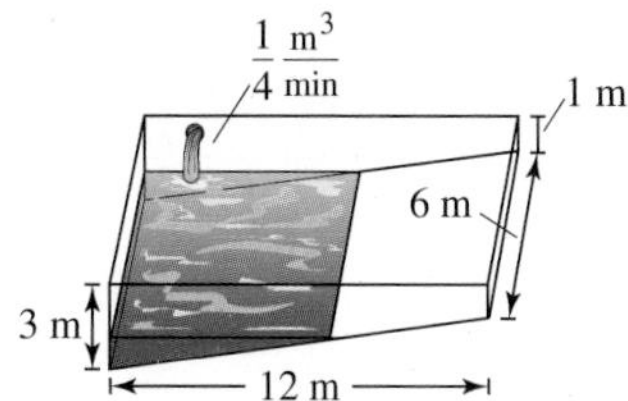

24. ***Depth*** A trough is 12 feet long and 3 feet across the top (see figure). Its ends are isosceles triangles with altitudes of 3 feet.

(a) If water is being pumped into the trough at 2 cubic feet per minute, how fast is the water level rising when the depth h is 1 foot?

(b) If the water is rising at a rate of $\frac{3}{8}$ inch per minute when $h = 2$, determine the rate at which water is being pumped into the trough.

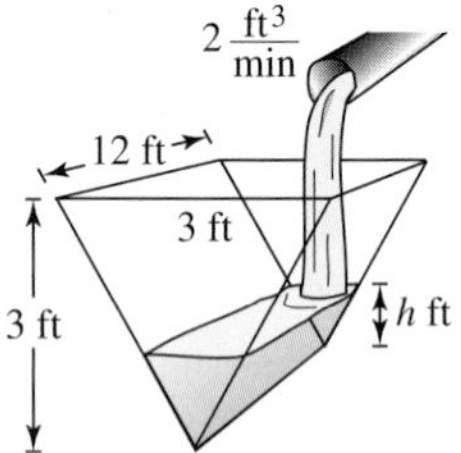

25. ***Moving Ladder*** A ladder 25 feet long is leaning against the wall of a house (see figure). The base of the ladder is pulled away from the wall at a rate of 2 feet per second.

(a) How fast is the top of the ladder moving down the wall when its base is 7 feet, 15 feet, and 24 feet from the wall?

(b) Consider the triangle formed by the side of the house, the ladder, and the ground. Find the rate at which the area of the triangle is changing when the base of the ladder is 7 feet from the wall.

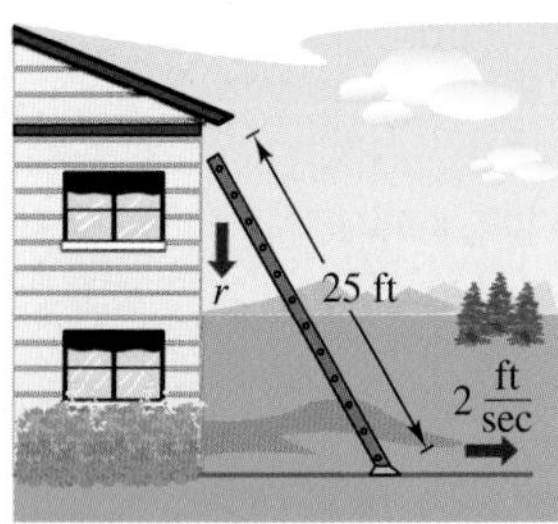

Figure for 25

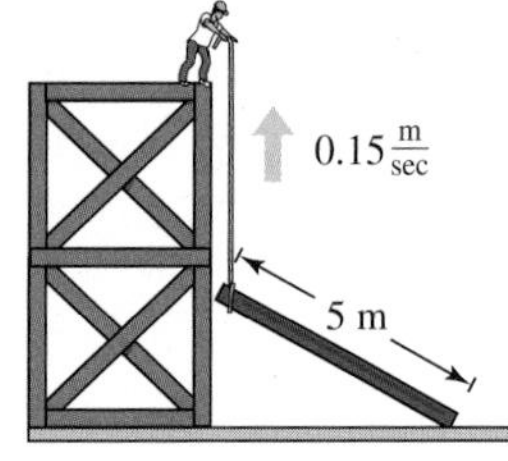

Figure for 26

■ **FOR FURTHER INFORMATION** For more information on the mathematics of moving ladders, see the article "The Falling Ladder Paradox" by Paul Scholten and Andrew Simoson in *The College Mathematics Journal*. To view this article, go to the website *www.matharticles.com*.

26. ***Construction*** A construction worker pulls a five-meter plank up the side of a building under construction by means of a rope tied to one end of the plank (see figure). Assume the opposite end of the plank follows a path perpendicular to the wall of the building and the worker pulls the rope at a rate of 0.15 meter per second. How fast is the end of the plank sliding along the ground when it is 2.5 meters from the wall of the building?

27. ***Construction*** A winch at the top of a 12-meter building pulls a pipe of the same length to a vertical position, as shown in the figure. The winch pulls in rope at a rate of -0.2 meter per second. Find the rate of vertical change and the rate of horizontal change at the end of the pipe when $y = 6$.

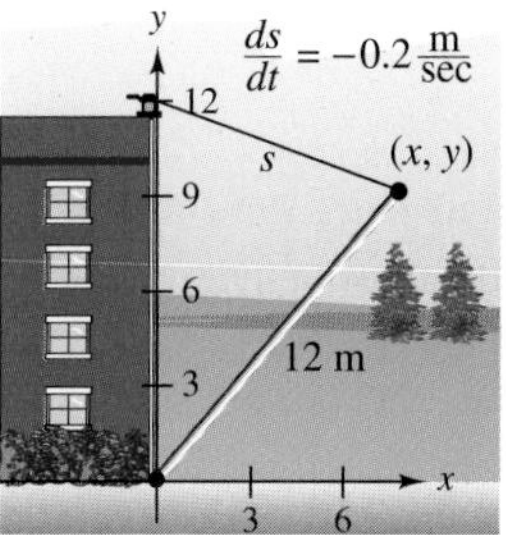

Figure for 27

28. ***Boating*** A boat is pulled into a dock by means of a winch 12 feet above the deck of the boat (see figure).

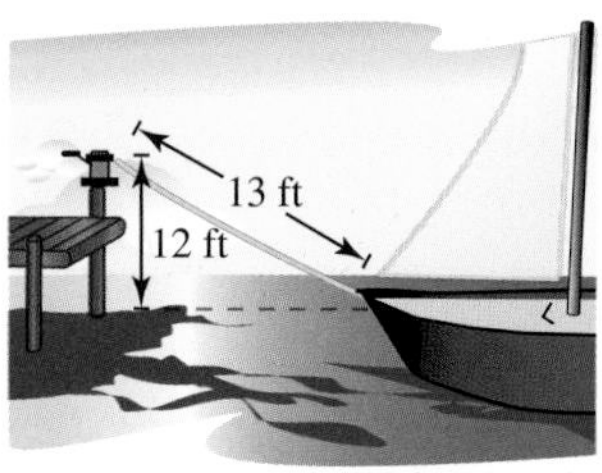

(a) The winch pulls in rope at a rate of 4 feet per second. Determine the speed of the boat when there is 13 feet of rope out. What happens to the speed of the boat as it gets closer to the dock?

(b) Suppose the boat is moving at a constant rate of 4 feet per second. Determine the speed at which the winch pulls in rope when there is a total of 13 feet of rope out. What happens to the speed at which the winch pulls in rope as the boat gets closer to the dock?

29. ***Air Traffic Control*** An air traffic controller spots two planes at the same altitude converging on a point as they fly at right angles to each other (see figure). One plane is 225 miles from the point moving at 450 miles per hour. The other plane is 300 miles from the point moving at 600 miles per hour.

(a) At what rate is the distance between the planes decreasing?

(b) How much time does the air traffic controller have to get one of the planes on a different flight path?

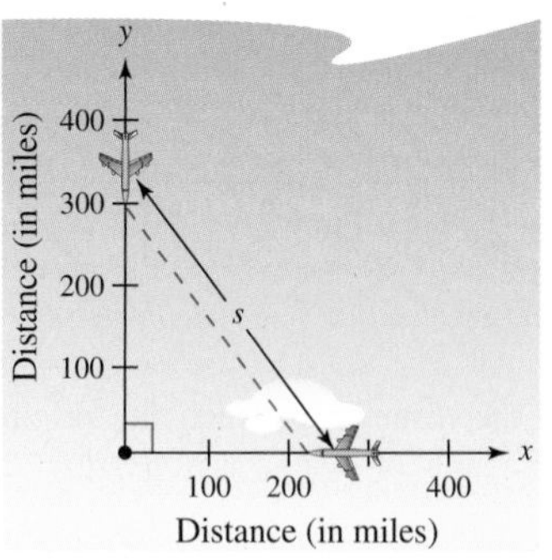

Figure for 29

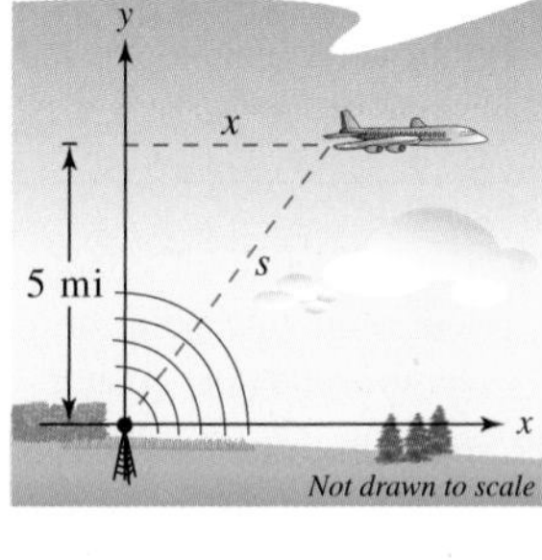

Figure for 30

30. ***Air Traffic Control*** An airplane is flying at an altitude of 5 miles and passes directly over a radar antenna (see figure). When the plane is 10 miles away ($s = 10$), the radar detects that the distance s is changing at a rate of 240 miles per hour. What is the speed of the plane?

31. ***Sports*** A baseball diamond has the shape of a square with sides 90 feet long (see figure). A player running from second base to third base at a speed of 25 feet per second is 20 feet from third base. At what rate is the player's distance s from home plate changing?

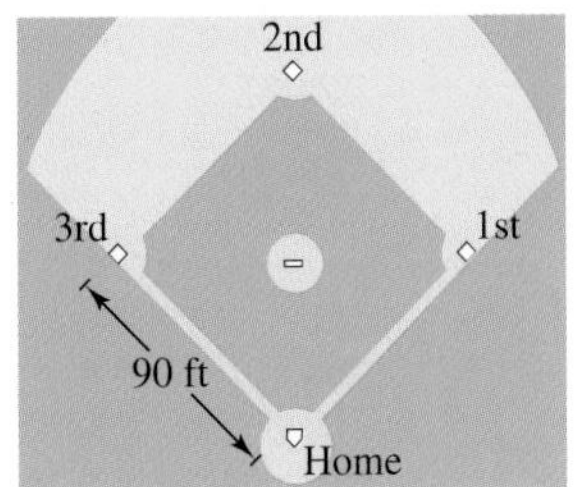

Figure for 31 and 32

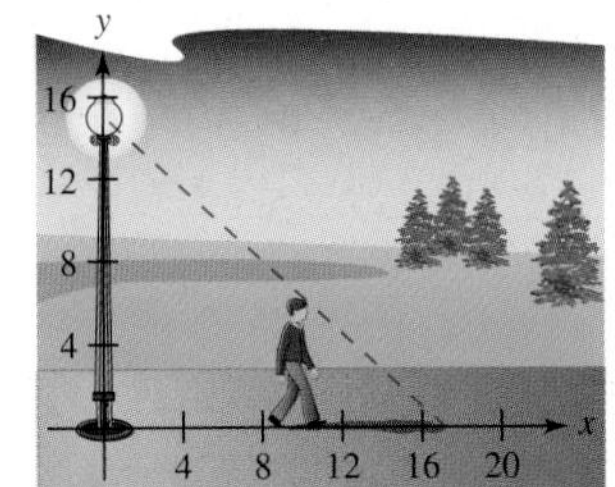

Figure for 33

32. ***Sports*** For the baseball diamond in Exercise 31, suppose the player is running from first to second at a speed of 25 feet per second. Find the rate at which the distance from home plate is changing when the player is 20 feet from second base.

33. ***Shadow Length*** A man 6 feet tall walks at a rate of 5 feet per second away from a light that is 15 feet above the ground (see figure). When he is 10 feet from the base of the light,

(a) at what rate is the tip of his shadow moving?

(b) at what rate is the length of his shadow changing?

34. ***Shadow Length*** Repeat Exercise 33 for a man 6 feet tall walking at a rate of 5 feet per second *toward* a light that is 20 feet above the ground.

35. ***Evaporation*** As a spherical raindrop falls, it reaches a layer of dry air and begins to evaporate at a rate that is proportional to its surface area $(S = 4\pi r^2)$. Show that the radius of the raindrop decreases at a constant rate.

CAPSTONE

36. Using the graph of f, (a) determine whether dy/dt is positive or negative given that dx/dt is negative, and (b) determine whether dx/dt is positive or negative given that dy/dt is positive.

(i)

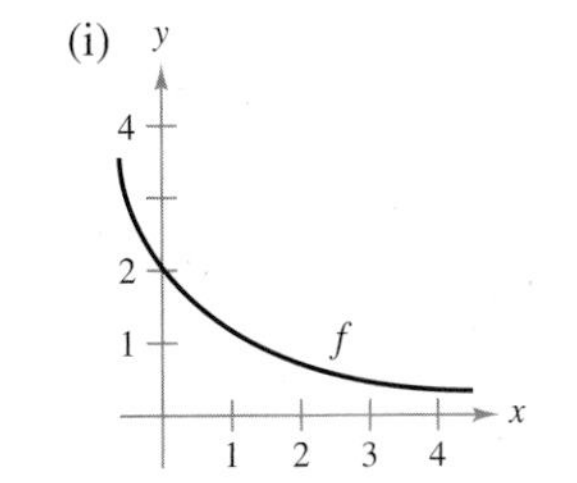

(ii)

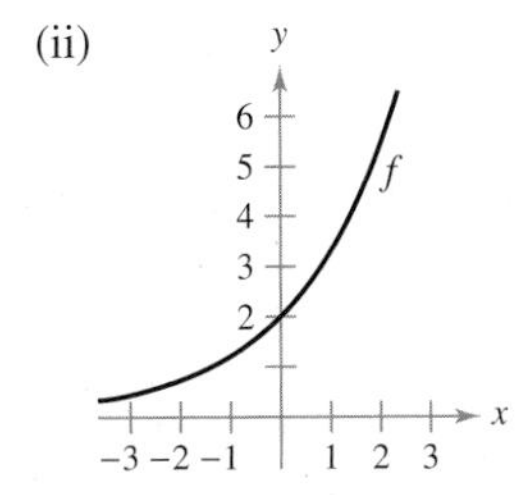

37. ***Electricity*** The combined electrical resistance R of R_1 and R_2, connected in parallel, is given by

$$\frac{1}{R} = \frac{1}{R_1} + \frac{1}{R_2}$$

where R, R_1, and R_2 are measured in ohms. R_1 and R_2 are increasing at rates of 1 and 1.5 ohms per second, respectively. At what rate is R changing when $R_1 = 50$ ohms and $R_2 = 75$ ohms?

38. ***Adiabatic Expansion*** When a certain polyatomic gas undergoes adiabatic expansion, its pressure p and volume V satisfy the equation

$$pV^{1.3} = k$$

where k is a constant. Find the relationship between the related rates dp/dt and dV/dt.

Acceleration **In Exercises 39 and 40, find the acceleration of the specified object. (*Hint:* Recall that if a variable is changing at a constant rate, its acceleration is zero.)**

39. Find the acceleration of the top of the ladder described in Exercise 25 when the base of the ladder is 7 feet from the wall.

40. Find the acceleration of the boat in Exercise 28(a) when there is a total of 13 feet of rope out.

41. ***Think About It*** Describe the relationship between the rate of change of y and the rate of change of x in each expression. Assume all variables and derivatives are positive.

(a) $\dfrac{dy}{dt} = 3\dfrac{dx}{dt}$ (b) $\dfrac{dy}{dt} = x(L - x)\dfrac{dx}{dt}, \quad 0 \le x \le L$

42. ***Modeling Data*** The table shows the numbers (in millions) of single women (never married) s and married women m in the civilian work force in the United States for the years 1997 through 2005. *(Source: U.S. Bureau of Labor Statistics)*

Year	1997	1998	1999	2000	2001	2002	2003	2004	2005
s	16.5	17.1	17.6	17.8	18.0	18.2	18.4	18.6	19.2
m	33.8	33.9	34.4	35.1	35.2	35.5	36.0	35.8	35.9

(a) Use the regression capabilities of a graphing utility to find a model of the form $m(s) = as^3 + bs^2 + cs + d$ for the data, where t is the time in years, with $t = 7$ corresponding to 1997.

(b) Find dm/dt. Then use the model to estimate dm/dt for $t = 10$ if it is predicted that the number of single women in the work force will increase at the rate of 0.75 million per year.

43. ***Moving Shadow*** A ball is dropped from a height of 20 meters, 12 meters away from the top of a 20-meter lamppost (see figure). The ball's shadow, caused by the light at the top of the lamppost, is moving along the level ground. How fast is the shadow moving 1 second after the ball is released? *(Submitted by Dennis Gittinger, St. Philips College, San Antonio, TX)*

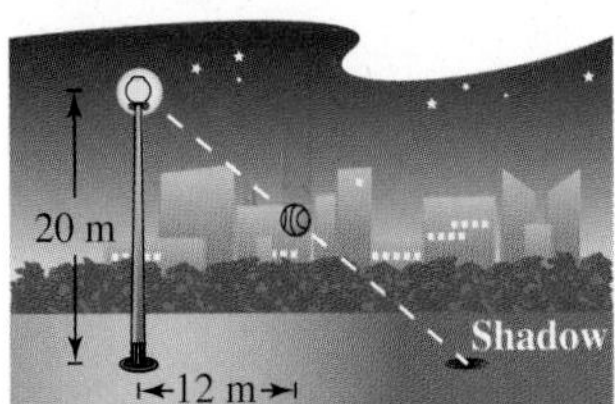

4 CHAPTER SUMMARY

	Review Exercises
Section 4.1	
■ Find the slope of the tangent line to a curve at a point *(p. 262).*	*1–6*
■ Use the limit definition to find the derivative of a function *(p. 265).*	*7–12*
■ Understand the relationship between differentiability and continuity *(p. 267).*	*13–16*
Section 4.2	
■ Find the derivative of a function using the Constant Rule *(p. 273).*	*17, 18*
■ Find the derivative of a function using the Power Rule *(p. 274).*	*19, 20*
■ Find the derivative of a function using the Constant Multiple Rule *(p. 276).*	*21–24*
■ Find the derivative of a function using the Sum and Difference Rules *(p. 277).*	*25–28*
■ Use derivatives to find rates of change *(p. 278).*	*29–36*
Section 4.3	
■ Find the derivative of a function using the Product Rule *(p. 284).*	*37–40*
■ Find the derivative of a function using the Quotient Rule *(p. 286).*	*41–46*
■ Find a higher-order derivative of a function *(p. 289).*	*47–54*
Section 4.4	
■ Find the derivative of a composite function using the Chain Rule *(p. 293),* find the derivative of a function using the General Power Rule *(p. 295),* and simplify the derivative of a function using algebra *(p. 297).*	*55–86*
Section 4.5	
■ Distinguish between functions written in implicit form and explicit form *(p. 301),* and use implicit differentiation to find the derivative of a function *(p. 302).*	*87–92*
Section 4.6	
■ Find a related rate *(p. 309),* and use related rates to solve real-life problems *(p. 310).*	*93–99*

4 REVIEW EXERCISES

See www.CalcChat.com for worked-out solutions to odd-numbered exercises.

In Exercises 1 and 2, find the slope of the tangent line to the graph of the function at the given point.

1. $g(x) = \frac{2}{3}x^2 - \frac{x}{6}, \quad \left(-1, \frac{5}{6}\right)$

2. $h(x) = \frac{3x}{8} - 2x^2, \quad \left(-2, -\frac{35}{4}\right)$

 **In Exercises 3 and 4, (a) find an equation of the tangent line to the graph of f at the given point, (b) use a graphing utility to graph the function and its tangent line at the point, and (c) use the *derivative* feature of the graphing utility to confirm your results.**

3. $f(x) = x^3 - 1, \quad (-1, -2)$

4. $f(x) = \frac{2}{x + 1}, \quad (0, 2)$

Writing **In Exercises 5 and 6, the figure shows the graphs of a function and its derivative. Label the graphs as f or f' and write a short paragraph stating the criteria you used in making your selection. To print an enlarged copy of the graph, go to the website *www.mathgraphs.com*.**

5. 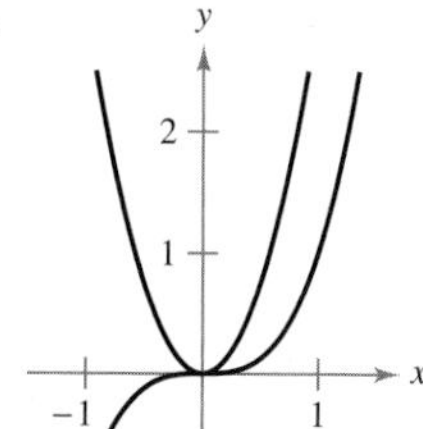

6. 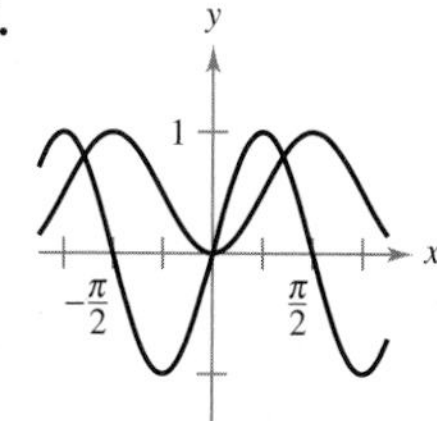

In Exercises 7–10, find the derivative of the function by using the definition of the derivative.

7. $f(x) = x^2 - 4x + 5$

8. $f(x) = \sqrt{x} + 1$

9. $f(x) = \frac{x + 1}{x - 1}$

10. $f(x) = \frac{6}{x}$

In Exercises 11 and 12, use the alternative form of the derivative to find the derivative at $x = c$ (if it exists).

11. $g(x) = x^2(x - 1), \quad c = 2$

12. $f(x) = \frac{1}{x + 4}, \quad c = 3$

In Exercises 13 and 14, describe the x-values at which f is differentiable.

13. $f(x) = (x - 3)^{2/5}$

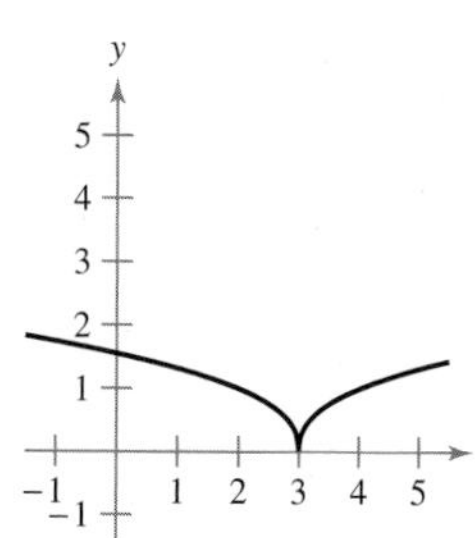

14. $f(x) = \frac{3x}{x + 1}$

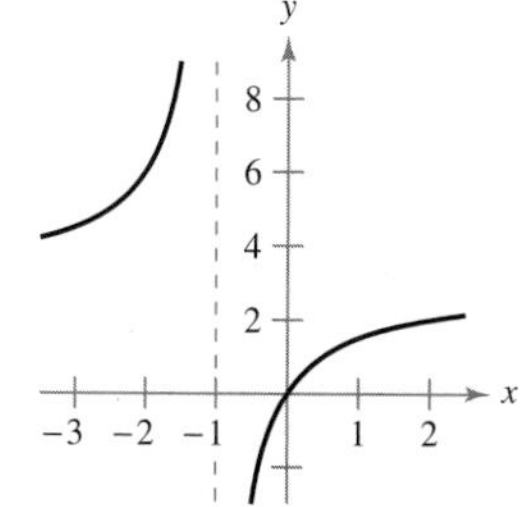

15. Sketch the graph of $f(x) = 4 - |x - 2|$.
    (a) Is f continuous at $x = 2$?
    (b) Is f differentiable at $x = 2$? Explain.

16. Sketch the graph of $f(x) = \begin{cases} x^2 + 4x + 2, & x < -2 \\ 1 - 4x - x^2, & x \geq -2. \end{cases}$
    (a) Is f continuous at $x = -2$?
    (b) Is f differentiable at $x = -2$? Explain.

In Exercises 17–28, use the rules of differentiation to find the derivative of the function.

17. $y = 25$

18. $y = -30$

19. $f(x) = x^8$

20. $g(x) = x^{20}$

21. $h(t) = 13t^4$

22. $f(t) = -8t^5$

23. $g(t) = \frac{2}{3t^2}$

24. $h(x) = \frac{10}{(7x)^2}$

25. $f(x) = x^3 - 11x^2$

26. $g(s) = 4s^4 - 5s^2$

27. $h(x) = 6\sqrt{x} + 3\sqrt[3]{x}$

28. $f(x) = x^{1/2} - x^{-1/2}$

29. ***Vibrating String*** When a guitar string is plucked, it vibrates with a frequency of $F = 200\sqrt{T}$, where F is measured in vibrations per second and the tension T is measured in pounds. Find the rates of change of F when (a) $T = 4$ and (b) $T = 9$.

30. ***Vertical Motion*** A ball is dropped from a height of 100 feet. One second later, another ball is dropped from a height of 75 feet. Which ball hits the ground first?

31. ***Vertical Motion*** To estimate the height of a building, a weight is dropped from the top of the building into a pool at ground level. How high is the building if the splash is seen 9.2 seconds after the weight is dropped?

32. ***Vertical Motion*** A bomb is dropped from an airplane at an altitude of 14,400 feet. How long will it take for the bomb to reach the ground? (Because of the motion of the plane, the fall will not be vertical, but the time will be the same as that for a vertical fall.) The plane is moving at 600 miles per hour. How far will the bomb move horizontally after it is released from the plane?

33. ***Projectile Motion*** A thrown ball follows a path described by $y = x - 0.02x^2$.
    (a) Sketch a graph of the path.
    (b) Find the total horizontal distance the ball is thrown.
    (c) At what x-value does the ball reach its maximum height? (Use the symmetry of the path.)
    (d) Find an equation that gives the instantaneous rate of change of the height of the ball with respect to the horizontal change. Evaluate the equation at $x = 0, 10, 25, 30,$ and 50.
    (e) What is the instantaneous rate of change of the height when the ball reaches its maximum height?

34. ***Projectile Motion*** The path of a projectile thrown at an angle of 45° with level ground is

$$y = x - \frac{32}{v_0^2}(x^2)$$

where the initial velocity is v_0 feet per second.

(a) Find the x-coordinate of the point where the projectile strikes the ground. Use the symmetry of the path of the projectile to locate the x-coordinate of the point where the projectile reaches its maximum height.

(b) What is the instantaneous rate of change of the height when the projectile is at its maximum height?

(c) Show that doubling the initial velocity of the projectile multiplies both the maximum height and the range by a factor of 4.

(d) Find the maximum height and range of a projectile thrown with an initial velocity of 70 feet per second. Use a graphing utility to graph the path of the projectile.

35. ***Horizontal Motion*** The position function of a particle moving along the x-axis is

$$x(t) = t^2 - 3t + 2 \quad \text{for} \quad -\infty < t < \infty.$$

(a) Find the velocity of the particle.

(b) Find the open t-interval(s) in which the particle is moving to the left.

(c) Find the position of the particle when the velocity is 0.

(d) Find the speed of the particle when the position is 0.

36. ***Modeling Data*** The speed of a car in miles per hour and the stopping distance in feet are recorded in the table.

Speed, x	20	30	40	50	60
Stopping Distance, y	25	55	105	188	300

(a) Use the regression capabilities of a graphing utility to find a quadratic model for the data.

(b) Use a graphing utility to plot the data and graph the model.

(c) Use a graphing utility to graph dy/dx.

(d) Use the model to approximate the stopping distance at a speed of 65 miles per hour.

(e) Use the graphs in parts (b) and (c) to explain the change in stopping distance as the speed increases.

In Exercises 37–44, find the derivative of the function.

37. $f(x) = (5x^2 + 8)(x^2 - 4x - 6)$

38. $g(x) = (x^3 + 7x)(x + 3)$

39. $h(t) = \sqrt{t}(t^3 + 4t - 1)$

40. $f(z) = \sqrt[3]{z}(z^2 + 5z)$

41. $f(x) = \dfrac{x^2 + x - 1}{x^2 - 1}$ **42.** $f(x) = \dfrac{6x - 5}{x^2 + 1}$

43. $f(x) = \dfrac{1}{9 - 4x^2}$ **44.** $f(x) = \dfrac{9}{3x^2 - 2x}$

In Exercises 45 and 46, find an equation of the tangent line to the graph of f at the given point.

45. $f(x) = \dfrac{2x^3 - 1}{x^2}$, $(1, 1)$ **46.** $f(x) = \dfrac{x + 1}{x - 1}$, $\left(\dfrac{1}{2}, -3\right)$

47. ***Acceleration*** The velocity of an object in meters per second is $v(t) = 36 - t^2$, $0 \le t \le 6$. Find the velocity and acceleration of the object when $t = 4$.

48. ***Acceleration*** The velocity of an automobile starting from rest is

$$v(t) = \frac{90t}{4t + 10}$$

where v is measured in feet per second. Find the vehicle's velocity and acceleration at each of the following times.

(a) 1 second (b) 5 seconds (c) 10 seconds

In Exercises 49–54, find the second derivative of the function.

49. $g(t) = -8t^3 - 5t + 12$ **50.** $h(x) = 21x^{-3} + 3x$

51. $f(x) = 15x^{5/2}$ **52.** $f(x) = 20\sqrt[5]{x}$

53. $f(x) = (3x^2 + 7)(x^2 - 2x + 3)$

54. $g(t) = \dfrac{t + 3}{t - 4}$

In Exercises 55–62, find the derivative of the function.

55. $h(x) = (x^2 - 4x)^3$

56. $g(x) = (5 - 3x)^4$

57. $f(x) = -2(1 - 4x^2)^2$

58. $f(x) = \left[\sqrt{x}(x - 3)\right]^2$

59. $h(x) = \left(\dfrac{x + 5}{x^2 + 3}\right)^2$ **60.** $f(x) = \left(x^2 + \dfrac{1}{x}\right)^5$

61. $f(s) = (s^2 - 1)^{5/2}(s^3 + 5)$ **62.** $h(t) = \dfrac{t}{(1 - t)^3}$

In Exercises 63 and 64, find the derivative of the function at the given point.

63. $f(x) = \sqrt{1 - x^3}$, $(-2, 3)$

64. $f(x) = \sqrt[3]{x^2 - 1}$, $(3, 2)$

In Exercises 65–70, use a computer algebra system to find the derivative of the function. Use the utility to graph the function and its derivative on the same set of coordinate axes. Describe the behavior of the function that corresponds to any zeros of the graph of the derivative.

65. $f(t) = t^2(t - 1)^5$

66. $f(x) = [(x - 2)(x + 4)]^2$

67. $g(x) = \dfrac{2x}{\sqrt{x + 1}}$ **68.** $g(x) = x\sqrt{x^2 + 1}$

69. $f(t) = \sqrt{t + 1}\,\sqrt[3]{t + 1}$ **70.** $y = \sqrt{3x}\,(x + 2)^3$

In Exercises 71 and 72, (a) use a computer algebra system to find the derivative of the function at the given point, (b) find an equation of the tangent line to the graph of the function at the point, and (c) graph the function and its tangent line on the same set of coordinate axes.

71. $f(t) = t^2(t - 1)^5, \quad (2, 4)$
72. $g(x) = x\sqrt{x^2 + 1}, \quad \left(3, 3\sqrt{10}\right)$

In Exercises 73–76, find the second derivative of the function.

73. $f(x) = \sqrt{x^2 + 9}$
74. $g(x) = \sqrt{5 - 2x}$
75. $y = \dfrac{4}{(x - 2)^2}$
76. $y = \dfrac{3}{\sqrt{x - 1}}$

In Exercises 77–80, use a computer algebra system to find the second derivative of the function.

77. $f(t) = \dfrac{4t^2}{(1 - t)^2}$
78. $g(x) = \dfrac{6x - 5}{x^2 + 1}$
79. $g(x) = \dfrac{x}{\sqrt{x + 3}}$
80. $h(x) = 5x\sqrt{x^2 - 16}$

In Exercises 81–84, (a) find an equation of the tangent line to the graph of the equation at the given point, (b) use a graphing utility to graph the equation and its tangent line at the point, and (c) use the *derivative* feature of the graphing utility to confirm your results.

81. $y = (x + 3)^3, \quad (-2, 1)$
82. $y = (x - 2)^2, \quad (2, 0)$
83. $y = \sqrt[3]{(x - 2)^2}, \quad (3, 1)$
84. $y = \dfrac{2x}{1 - x^2}, \quad (0, 0)$

85. ***Refrigeration*** The temperature T (in degrees Fahrenheit) of food in a freezer is

$$T = \frac{700}{t^2 + 4t + 10}$$

where t is the time in hours. Find the rate of change of T with respect to t at each of the following times.

(a) $t = 1$ (b) $t = 3$ (c) $t = 5$ (d) $t = 10$

86. ***Fluid Flow*** The emergent velocity v of a liquid flowing from a hole in the bottom of a tank is given by $v = \sqrt{2gh}$, where g is the acceleration due to gravity (32 feet per second per second) and h is the depth of the liquid in the tank. Find the rates of change of v with respect to h when (a) $h = 9$ and (b) $h = 4$. (Note that $g = +32$ feet per second per second. The sign of g depends on how a problem is modeled. In this case, letting g be negative would produce an imaginary value for v.)

In Exercises 87–90, find *dy/dx* by implicit differentiation.

87. $x^2 + 3xy + y^3 = 10$
88. $y^2 = (x - y)(x^2 + y)$
89. $\sqrt{xy} = x - 4y$
90. $y\sqrt{x} - x\sqrt{y} = 25$

In Exercises 91 and 92, find the equations of the tangent line and the normal line to the graph of the equation at the given point. Use a graphing utility to graph the equation, the tangent line, and the normal line.

91. $x^2 + y^2 = 10, \quad (3, 1)$
92. $x^2 - y^2 = 20, \quad (6, 4)$

93. A point moves along the curve $y = \sqrt{x}$ in such a way that the y-value is increasing at a rate of 2 units per second. At what rate is x changing for each of the following values?

(a) $x = \frac{1}{2}$
(b) $x = 1$
(c) $x = 4$

94. The same conditions exist as in Exercise 93. Find the rate of change of the distance between the origin and a point (x, y) on the graph for each of the following.

(a) $x = \frac{1}{2}$
(b) $x = 1$
(c) $x = 4$

95. ***Surface Area*** The edges of a cube are expanding at a rate of 8 centimeters per second. How fast is the surface area changing when each edge is 6.5 centimeters?

96. ***Depth*** The cross section of a five-meter trough is an isosceles trapezoid with a two-meter lower base, a three-meter upper base, and an altitude of 2 meters. Water is running into the trough at a rate of 1 cubic meter per minute. How fast is the water level rising when the water is 1 meter deep?

97. ***Distance*** Two cars start at the same point and at the same time. The first car travels west at 65 miles per hour and the second car travels north at 50 miles per hour. At what rate is the distance between the two cars changing after $\frac{1}{2}$ hour?

98. ***Environment*** An accident at an oil drilling platform is causing a circular oil slick. The slick is 0.08 foot thick, and when the radius is 750 feet, the radius of the slick is increasing at the rate of 0.5 foot per minute. At what rate (in cubic feet per minute) is oil flowing from the site of the accident?

99. ***Depth*** A swimming pool is 40 feet long, 20 feet wide, 4 feet deep at the shallow end, and 9 feet deep at the deep end (see figure). Water is being pumped into the pool at the rate of 10 cubic feet per minute. How fast is the water level rising when there is 4 feet of water in the deep end?

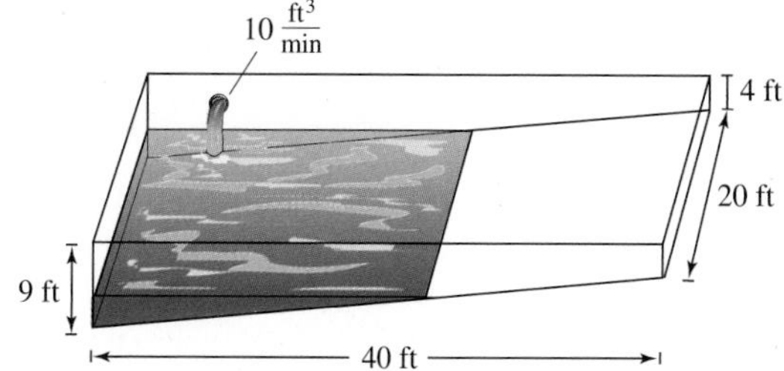

4 CHAPTER TEST

Take this test as you would take a test in class. When you are finished, check your work against the answers given in the back of the book.

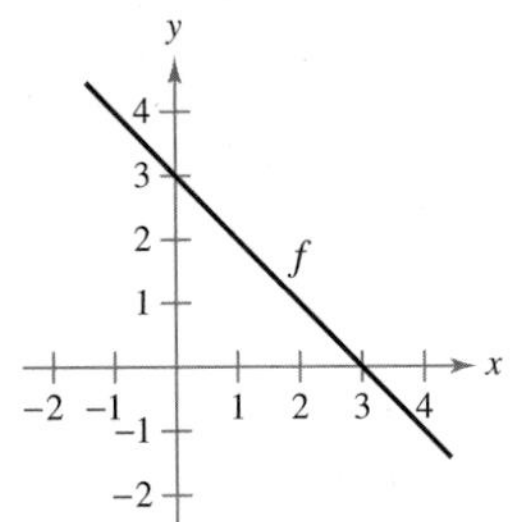

Figure for 2

1. Use the limit process to find the derivative of $f(x) = x^2 - x + 4$.
2. Use the graph of f at the left to sketch the graph of f'. Explain how you found your answer.

In Exercises 3–8, find the derivative of the function.

3. $s(t) = t^3 - 3t^2 + 2t + 6$
4. $g(x) = (x^2 + 1)(x^2 - 2x)$
5. $h(x) = \dfrac{\sqrt[3]{x}}{x^2 + 1}$
6. $f(x) = \dfrac{3 - 2x - x^2}{x^2 - 1}$
7. $f(t) = \left(\dfrac{1}{t + 2}\right)^3$
8. $y = \dfrac{x}{\sqrt{x^2 + 3}}$
9. Find the slope of the graph of $f(x) = -\frac{1}{8} + \frac{3}{4}x^3$ at the point $\left(0, -\frac{1}{8}\right)$. Use the *derivative* feature of a graphing utility to confirm your result.
10. Evaluate the derivative of $f(x) = \dfrac{x + 2}{3x - 2}$ at the point $(2, 1)$. Use a graphing utility to verify your result.

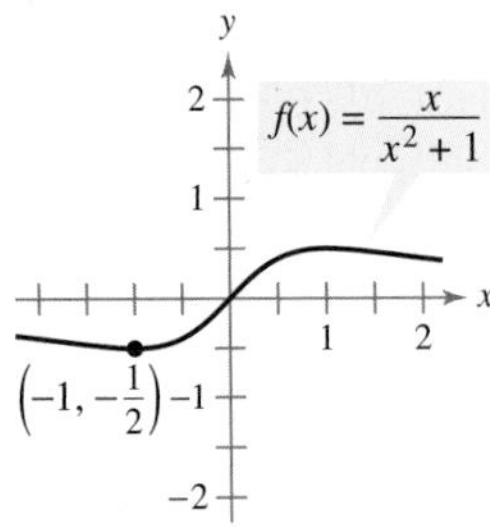

Figure for 13

In Exercises 11 and 12, (a) find an equation of the tangent line to the graph of f at the given point, (b) use a graphing utility to graph the function and its tangent line at the point, and (c) use the *derivative* feature of the graphing utility to confirm your results.

11. $f(x) = \sqrt{x}, \quad (4, 2)$
12. $f(x) = \sqrt{3x^2 - 2}, \quad (3, 5)$
13. For the graph at the left, find an equation of the tangent line at the given point.
14. Determine the point(s) at which the graph of $f(x) = \frac{1}{3}x^3 - x$ has a horizontal tangent line.

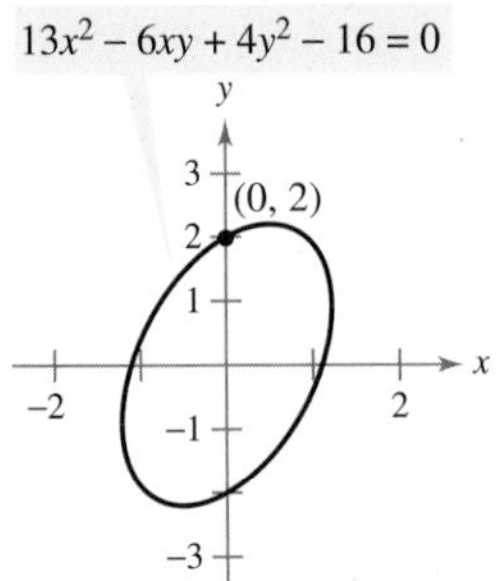

Figure for 18

15. Find the second derivative of $f(x) = \dfrac{x}{x - 3}$.
16. Find $\dfrac{dy}{dx}$ by implicit differentiation given that $x^3 + y^3 = 8$.
17. Find $\dfrac{dy}{dx}$ by implicit differentiation given that $(x - y)^3 = x^3 - y^3$. Evaluate the derivative at the point $(-1, -1)$.
18. For the graph at the left, find an equation of the tangent line at the given point.
19. The radius r of a circle is increasing at a rate of 3 centimeters per minute. Find the rate of change of the area when (a) $r = 6$ centimeters and (b) $r = 24$ centimeters.
20. A ladder 15 feet long is leaning against the wall of a house (see figure). The base of the ladder is pulled away from the wall at a rate of 1 foot per second.

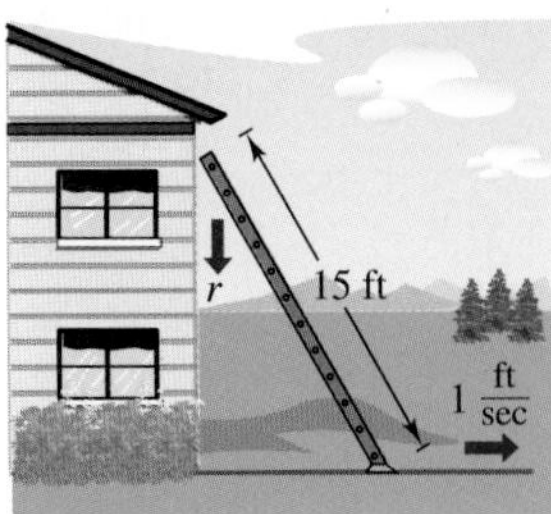

(a) How fast is the top of the ladder moving down the wall when its base is 9 feet, 12 feet, and 14 feet from the wall?

(b) Consider the triangle formed by the side of the house, the ladder, and the ground. Find the rate at which the area of the triangle is changing when the base of the ladder is 9 feet from the wall.

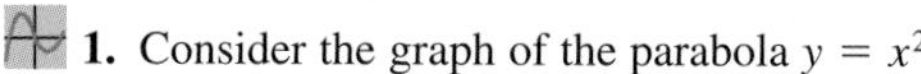

P.S. PROBLEM SOLVING

1. Consider the graph of the parabola $y = x^2$.
   (a) Find the radius r of the largest possible circle centered on the y-axis that is tangent to the parabola at the origin, as shown in the figure. This circle is called the **circle of curvature.** Find the equation of this circle. Use a graphing utility to graph the circle and parabola in the same viewing window to verify your answer.
   (b) Find the center $(0, b)$ of the circle of radius 1 centered on the y-axis that is tangent to the parabola at two points, as shown in the figure. Find the equation of this circle. Use a graphing utility to graph the circle and parabola in the same viewing window to verify your answer.

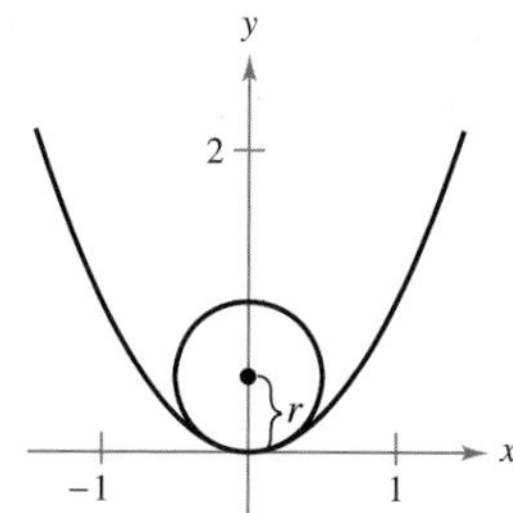

Figure for 1(a) **Figure for 1(b)**

2. Graph the two parabolas $y = x^2$ and $y = -x^2 + 2x - 5$ in the same coordinate plane. Find equations of the two lines simultaneously tangent to both parabolas.
3. (a) Find the polynomial $P_1(x) = a_0 + a_1x$ whose value and slope agree with the value and slope of $f(x) = \sqrt{x+1}$ at the point $x = 0$.
   (b) Find the polynomial $P_2(x) = a_0 + a_1x + a_2x^2$ whose value and first two derivatives agree with the value and first two derivatives of $f(x) = \sqrt{x+1}$ at the point $x = 0$. This polynomial is called the second-degree **Taylor polynomial** of $f(x) = \sqrt{x+1}$ at $x = 0$.
   (c) Complete the table comparing the values of $f(x) = \sqrt{x+1}$ and $P_2(x)$. What do you observe?

x	-1.0	-0.1	-0.001	0	0.001	0.1	1.0
$\sqrt{x+1}$							
$P_2(x)$							

   (d) Use a graphing utility to graph the polynomial $P_2(x)$ together with $f(x) = \sqrt{x+1}$ in the same viewing window. What do you observe?
4. (a) Find an equation of the tangent line to the parabola $y = x^2$ at the point $(2, 4)$.
   (b) Find an equation of the normal line to $y = x^2$ at the point $(2, 4)$. (The normal line is perpendicular to the tangent line.) Where does this line intersect the parabola a second time?
   (c) Find equations of the tangent line and normal line to $y = x^2$ at the point $(0, 0)$.
   (d) Prove that for any point $(a, b) \neq (0, 0)$ on the parabola $y = x^2$, the normal line intersects the graph a second time.
5. Find a third-degree polynomial $p(x)$ that is tangent to the line $y = 14x - 13$ at the point $(1, 1)$, and tangent to the line $y = -2x - 5$ at the point $(-1, -3)$.
6. Find a function of the form $f(x) = a + b\sqrt{x}$ that is tangent to the line $2y - 3x = 5$ at the point $(1, 4)$.
7. The graph of the **eight curve**

$$x^4 = a^2(x^2 - y^2), \quad a \neq 0$$

is shown below.

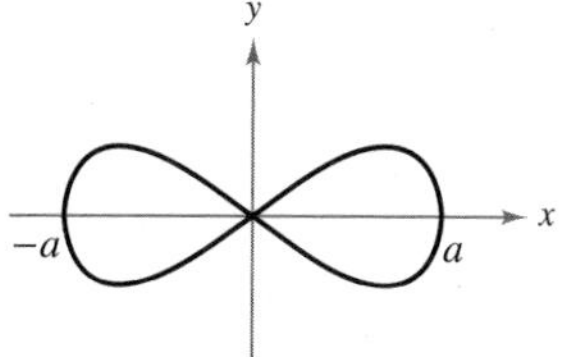

   (a) Explain how you could use a graphing utility to graph this curve.
   (b) Use a graphing utility to graph the curve for various values of the constant a. Describe how a affects the shape of the curve.
   (c) Determine the points on the curve at which the tangent line is horizontal.
8. The graph of the **pear-shaped quartic**

$$b^2y^2 = x^3(a - x), \quad a, b > 0$$

is shown below.

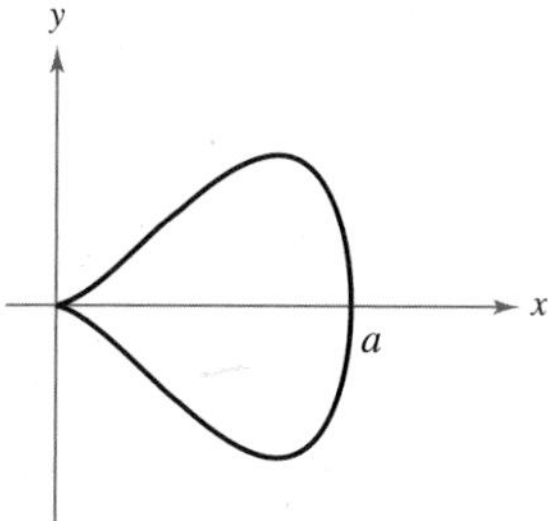

   (a) Explain how you could use a graphing utility to graph this curve.
   (b) Use a graphing utility to graph the curve for various values of the constants a and b. Describe how a and b affect the shape of the curve.
   (c) Determine the points on the curve at which the tangent line is horizontal.

9. A man 6 feet tall walks at a rate of 5 feet per second toward a streetlight that is 30 feet high (see figure). The man's 3-foot-tall child follows at the same speed, but 10 feet behind the man. At times, the shadow behind the child is caused by the man, and at other times, by the child.

(a) Suppose the man is 90 feet from the streetlight. Show that the man's shadow extends beyond the child's shadow.

(b) Suppose the man is 60 feet from the streetlight. Show that the child's shadow extends beyond the man's shadow.

(c) Determine the distance d from the man to the streetlight at which the tips of the two shadows are exactly the same distance from the streetlight.

(d) Determine how fast the tip of the man's shadow is moving as a function of x, the distance between the man and the streetlight. Discuss the continuity of this shadow speed function.

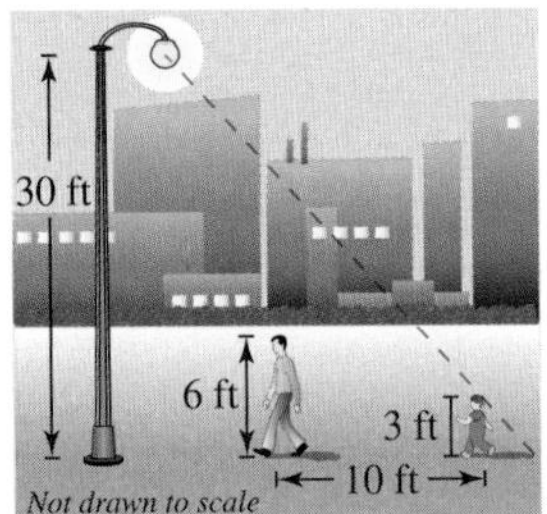

10. A particle is moving along the graph of

$$y = \sqrt[3]{x}$$

(see figure). When $x = 8$, the y-component of the position of the particle is increasing at the rate of 1 centimeter per second.

(a) How fast is the x-component changing at this moment?

(b) How fast is the distance from the origin changing at this moment?

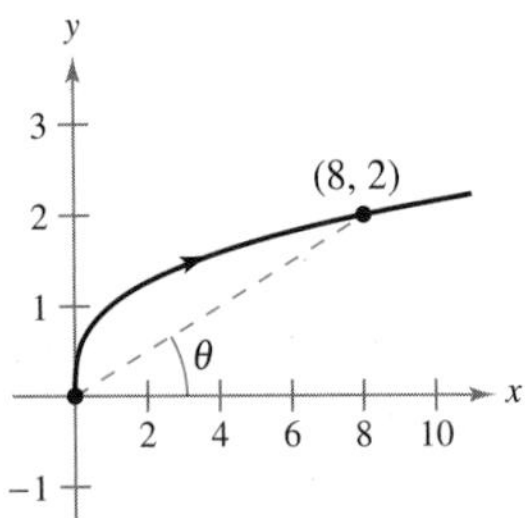

11. Let L be a differentiable function for all x. Prove that if $L(a + b) = L(a) + L(b)$ for all a and b, then $L'(x) = L'(0)$ for all x. Sketch the graph of L.

12. Let E be a function satisfying $E(0) = E'(0) = 1$. Prove that if $E(a + b) = E(a)E(b)$ for all a and b, then E is differentiable and $E'(x) = E(x)$ for all x. Find an example of a function satisfying $E(a + b) = E(a)E(b)$.

13. Consider the hyperbola $y = 1/x$ and its tangent line at $P = (1, 1)$, as shown in the figure. The tangent line intersects the x- and y-axes at A and B, respectively.

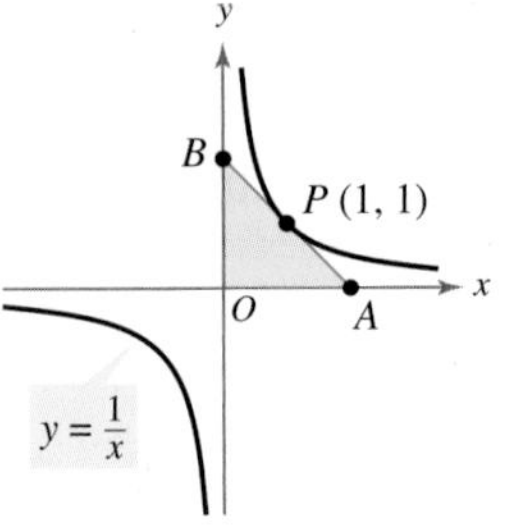

(a) Show that P is the midpoint of the line segment AB.

(b) Find the area of the triangle $\triangle OAB$.

(c) Let $P = \left(2, \frac{1}{2}\right)$. Show that P is the midpoint of the line segment AB and that the area of triangle $\triangle OAB$ is the same as in part (b).

(d) Let P be an arbitrary point on the hyperbola $y = a/x$. The tangent line at P intersects the x- and y-axes at A and B, respectively. Show that P is the midpoint of the line segment AB and that the area of triangle $\triangle OAB$ is not dependent on the location of the point P.

14. An astronaut standing on the moon throws a rock upward. The height of the rock is

$$s = -\frac{27}{10}t^2 + 27t + 6$$

where s is measured in feet and t is measured in seconds.

(a) Find expressions for the velocity and acceleration of the rock.

(b) Find the time when the rock is at its highest point by finding the time when the velocity is zero. What is the height of the rock at this time?

(c) How does the acceleration of the rock compare with the acceleration due to gravity on Earth?

15. If a is the acceleration of an object, the *jerk* j is defined by $j = a'(t)$.

(a) Use this definition to give a physical interpretation of j.

(b) Find j for the slowing vehicle in Exercise 87 in Section 4.3 and interpret the result.

(c) The figure shows the graphs of the position, velocity, acceleration, and jerk functions of a vehicle. Identify each graph and explain your reasoning.

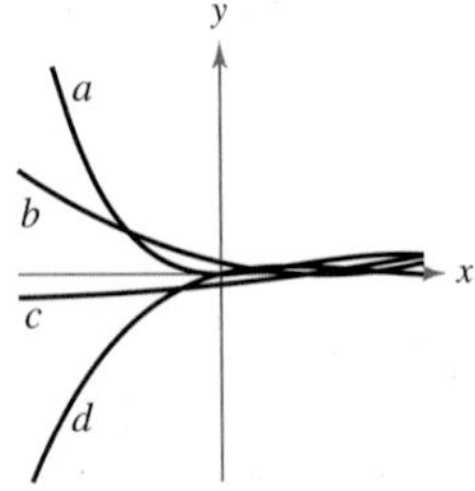

5 Applications of Differentiation

This chapter discusses several applications of the derivative of a function. These applications fall into three basic categories—curve sketching, optimization, and approximation techniques.

In this chapter, you should learn the following.

- How to use a derivative to locate the minimum and maximum values of a function on a closed interval. (5.1)
- How numerous results in this chapter depend on two important theorems called *Rolle's Theorem* and the *Mean Value Theorem*. (5.2)
- How to use the first derivative to determine whether a function is increasing or decreasing. (5.3)
- How to use the second derivative to determine whether the graph of a function is concave upward or concave downward. (5.4)
- How to find horizontal asymptotes of the graph of a function. (5.5)
- How to graph a function using the techniques from Chapters P–5. (5.6)
- How to solve optimization problems. (5.7)
- How to use approximation techniques to solve problems. (5.8)

E.J. Baumeister Jr. / Alamy

A small aircraft starts its descent from an altitude of 1 mile, 4 miles west of the runway. Given a function that models the glide path of the plane, when would the plane be descending at the greatest rate? (See Section 5.4, Exercise 65.)

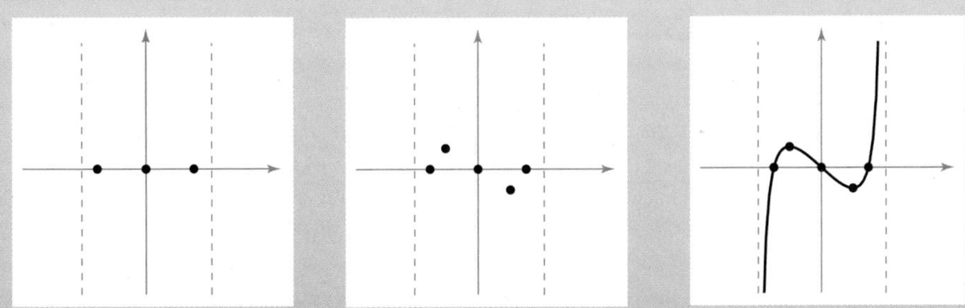

In Chapter 5, you will use calculus to analyze graphs of functions. For example, you can use the derivative of a function to determine the function's maximum and minimum values. You can use limits to identify any asymptotes of the function's graph. In Section 5.6, you will combine these techniques to sketch the graph of a function.

5.1 Extrema on an Interval

- **Understand the definition of extrema of a function on an interval.**
- **Understand the definition of relative extrema of a function on an open interval.**
- **Find extrema on a closed interval.**

Extrema of a Function

In calculus, much effort is devoted to determining the behavior of a function f on an interval I. Does f have a maximum value on I? Does it have a minimum value? Where is the function increasing? Where is it decreasing? In Section 1.2, you answered these questions using graphical and numerical analysis. In this chapter, you will learn how derivatives can be used to answer these questions. You will also see why these questions are important in real-life applications.

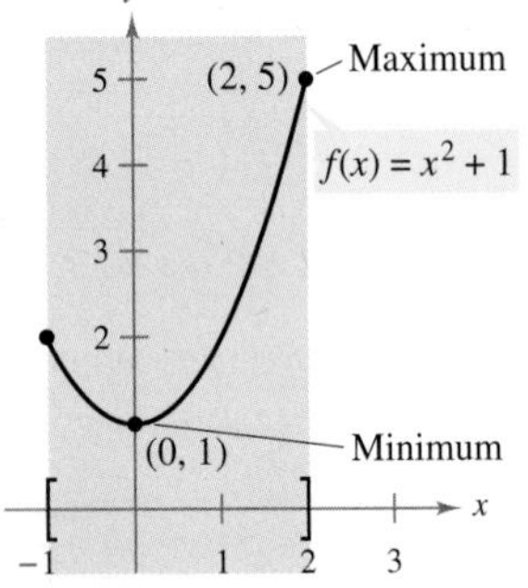

(a) f is continuous, $[-1, 2]$ is closed.

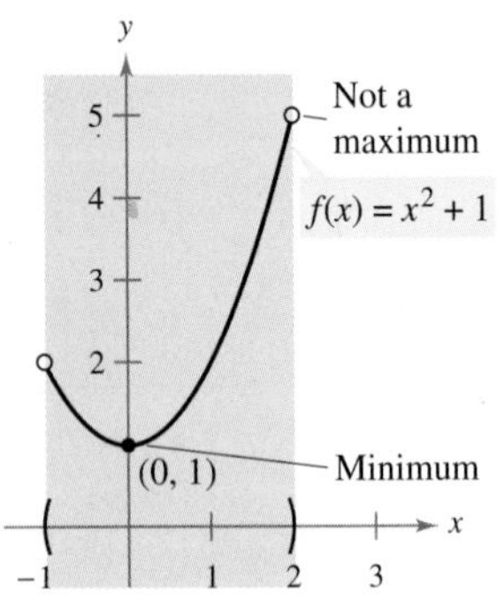

(b) f is continuous, $(-1, 2)$ is open.

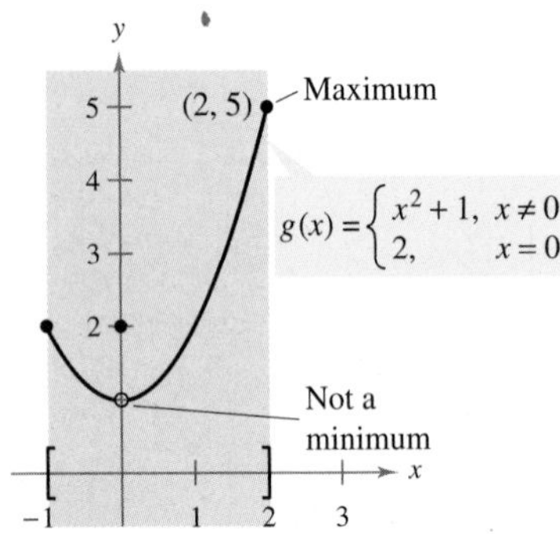

(c) g is not continuous, $[-1, 2]$ is closed.
Extrema can occur at interior points or endpoints of an interval. Extrema that occur at the endpoints are called **endpoint extrema.**
Figure 5.1

DEFINITION OF EXTREMA

Let f be defined on an interval I containing c.

1. $f(c)$ is the **minimum of f on I** if $f(c) \leq f(x)$ for all x in I.
2. $f(c)$ is the **maximum of f on I** if $f(c) \geq f(x)$ for all x in I.

The minimum and maximum of a function on an interval are the **extreme values,** or **extrema** (the singular form of extrema is extremum), of the function on the interval. The minimum and maximum of a function on an interval are also called the **absolute minimum** and **absolute maximum,** or the **global minimum** and **global maximum,** on the interval.

A function need not have a minimum or a maximum on an interval. For instance, in Figure 5.1(a) and (b), you can see that the function $f(x) = x^2 + 1$ has both a minimum and a maximum on the closed interval $[-1, 2]$, but does not have a maximum on the open interval $(-1, 2)$. Moreover, in Figure 5.1(c), you can see that continuity (or the lack of it) can affect the existence of an extremum on the interval. This suggests the theorem below. (Although the Extreme Value Theorem is intuitively plausible, a proof of this theorem is not within the scope of this text.)

THEOREM 5.1 THE EXTREME VALUE THEOREM

If f is continuous on a closed interval $[a, b]$, then f has both a minimum and a maximum on the interval.

EXPLORATION

Finding Minimum and Maximum Values The Extreme Value Theorem (like the Intermediate Value Theorem) is an *existence theorem* because it tells of the existence of minimum and maximum values but does not show how to find these values. Use the extreme-value capability of a graphing utility to find the minimum and maximum values of each of the following functions. In each case, do you think the x-values are exact or approximate? Explain your reasoning.

a. $f(x) = x^2 - 4x + 5$ on the closed interval $[-1, 3]$

b. $f(x) = x^3 - 2x^2 - 3x - 2$ on the closed interval $[-1, 3]$

Relative Extrema and Critical Numbers

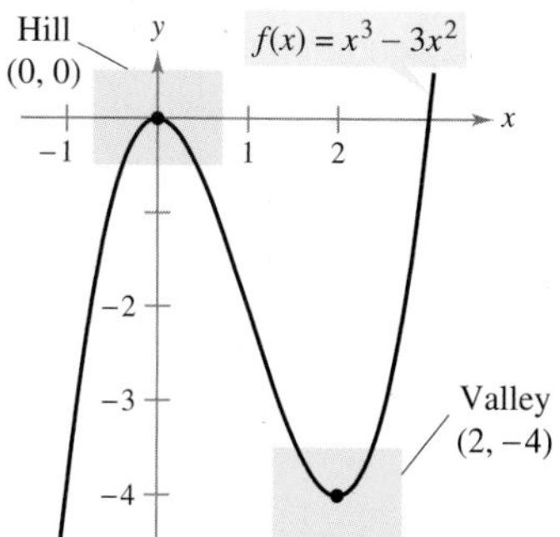

f has a relative maximum at $(0, 0)$ and a relative minimum at $(2, -4)$.
Figure 5.2

In Figure 5.2, the graph of $f(x) = x^3 - 3x^2$ has a **relative maximum** at the point $(0, 0)$ and a **relative minimum** at the point $(2, -4)$. Informally, for a continuous function, you can think of a relative maximum as occurring on a "hill" on the graph, and a relative minimum as occurring in a "valley" on the graph. Such a hill and valley can occur in two ways. If the hill (or valley) is smooth and rounded, the graph has a horizontal tangent line at the high point (or low point). If the hill (or valley) is sharp and peaked, the graph represents a function that is not differentiable at the high point (or low point).

DEFINITION OF RELATIVE EXTREMA

1. If there is an open interval containing c on which $f(c)$ is a maximum, then $f(c)$ is called a **relative maximum** of f, or you can say that f has a **relative maximum at $(c, f(c))$.**
2. If there is an open interval containing c on which $f(c)$ is a minimum, then $f(c)$ is called a **relative minimum** of f, or you can say that f has a **relative minimum at $(c, f(c))$.**

The plural of relative maximum is relative maxima, and the plural of relative minimum is relative minima. Relative maximum and relative minimum are sometimes called **local maximum** and **local minimum,** respectively.

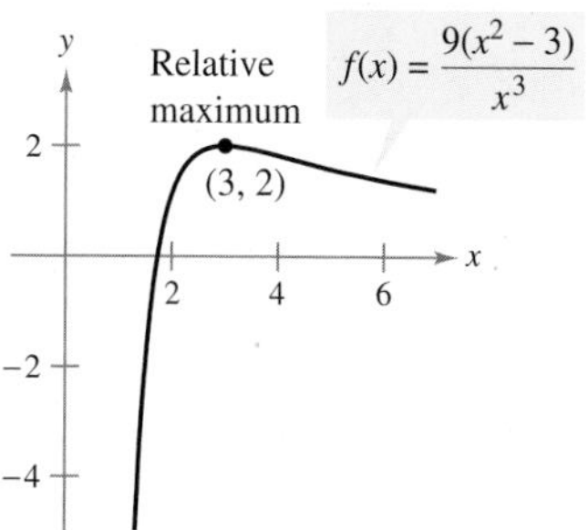

(a) $f'(3) = 0$

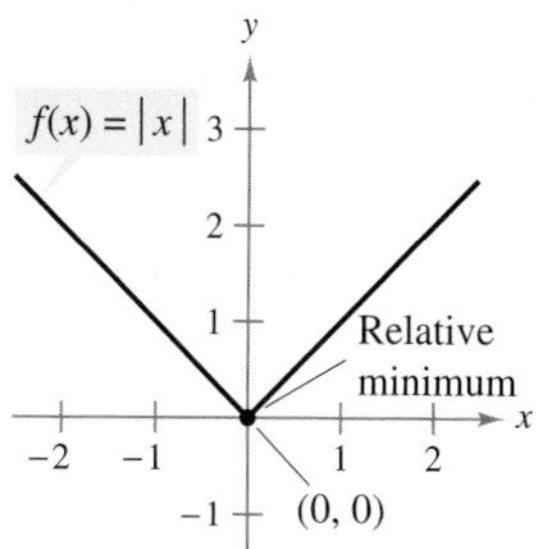

(b) $f'(0)$ does not exist.

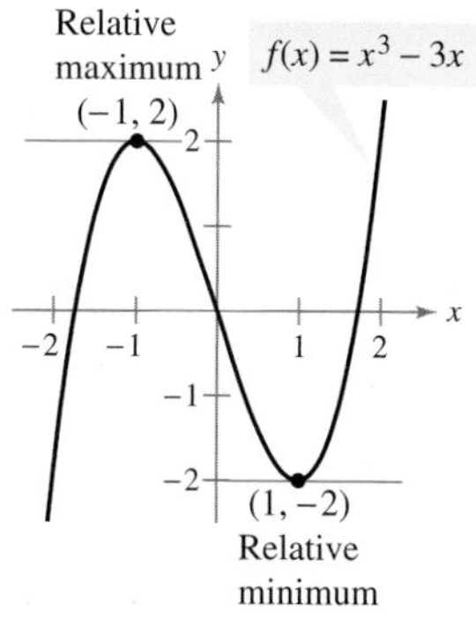

(c) $f(-1) = 0; f(1) = 0$
Figure 5.3

Example 1 examines the derivatives of functions at *given* relative extrema. (Much more is said about *finding* the relative extrema of a function in Section 5.3.)

EXAMPLE 1 The Value of the Derivative at Relative Extrema

Find the value of the derivative at each relative extremum shown in Figure 5.3.

Solution

a. The derivative of $f(x) = \dfrac{9(x^2 - 3)}{x^3}$ is

$$f'(x) = \frac{x^3(18x) - (9)(x^2 - 3)(3x^2)}{(x^3)^2} \qquad \text{Differentiate using Quotient Rule.}$$

$$= \frac{9(9 - x^2)}{x^4}. \qquad \text{Simplify.}$$

At the point $(3, 2)$, the value of the derivative is $f'(3) = 0$ [see Figure 5.3(a)].

b. At $x = 0$, the derivative of $f(x) = |x|$ *does not exist* because the following one-sided limits differ [see Figure 5.3(b)].

$$\lim_{x \to 0^-} \frac{f(x) - f(0)}{x - 0} = \lim_{x \to 0^-} \frac{|x|}{x} = -1 \qquad \text{Limit from the left}$$

$$\lim_{x \to 0^+} \frac{f(x) - f(0)}{x - 0} = \lim_{x \to 0^+} \frac{|x|}{x} = 1 \qquad \text{Limit from the right}$$

c. The derivative of $f(x) = x^3 - 3x$ is $f'(x) = 3x^2 - 3$. At the point $(-1, 2)$, the value of the derivative is $f'(-1) = 3(-1)^2 - 3 = 0$, and at the point $(1, -2)$ the value of the derivative is $f'(1) = 3(1)^2 - 3 = 0$. See Figure 5.3(c). ■

Note in Example 1 that at each relative extremum, the derivative either is zero or does not exist. The x-values at these special points are called **critical numbers.** Figure 5.4 illustrates the two types of critical numbers. Notice in the definition that the critical number c has to be in the domain of f, but c does not have to be in the domain of f'.

DEFINITION OF A CRITICAL NUMBER

Let f be defined at c. If $f'(c) = 0$ or if f is not differentiable at c, then c is a **critical number** of f.

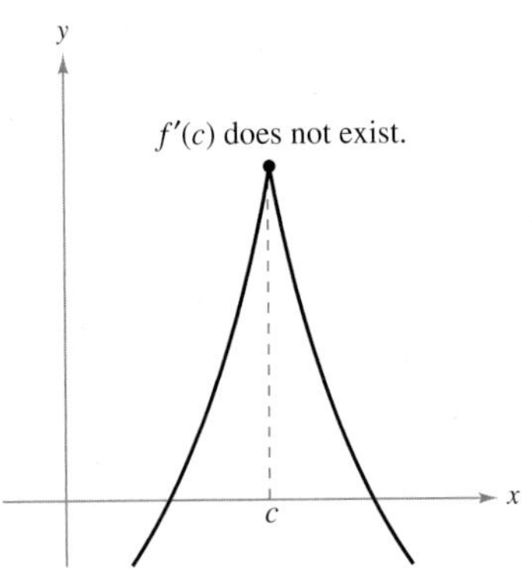

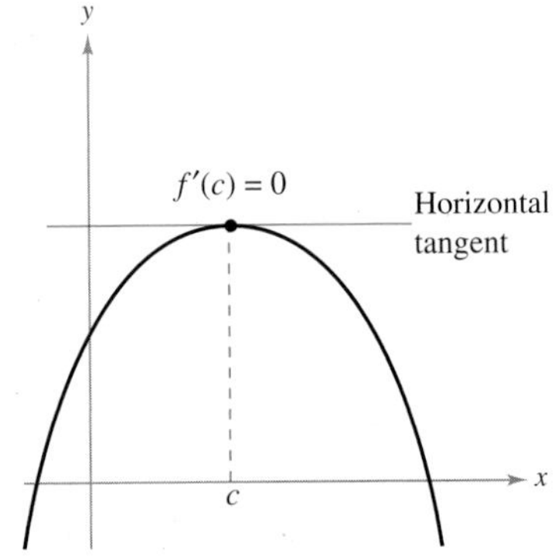

c is a critical number of f.
Figure 5.4

THEOREM 5.2 RELATIVE EXTREMA OCCUR ONLY AT CRITICAL NUMBERS

If f has a relative minimum or relative maximum at $x = c$, then c is a critical number of f.

PIERRE DE FERMAT (1601–1665)

For Fermat, who was trained as a lawyer, mathematics was more of a hobby than a profession. Nevertheless, Fermat made many contributions to analytic geometry, number theory, calculus, and probability. In letters to friends, he wrote of many of the fundamental ideas of calculus, long before Newton or Leibniz. For instance, Theorem 5.2 is sometimes attributed to Fermat.

PROOF

Case 1: If f is *not* differentiable at $x = c$, then, by definition, c is a critical number of f and the theorem is valid.

Case 2: If f is differentiable at $x = c$, then $f'(c)$ must be positive, negative, or 0. Suppose $f'(c)$ is positive. Then

$$f'(c) = \lim_{x \to c} \frac{f(x) - f(c)}{x - c} > 0$$

which implies that there exists an interval (a, b) containing c such that

$$\frac{f(x) - f(c)}{x - c} > 0, \text{ for all } x \neq c \text{ in } (a, b).$$

[See Exercise 62(b), Section 3.2.]

Because this quotient is positive, the signs of the denominator and numerator must agree. This produces the following inequalities for x-values in the interval (a, b).

Left of c: $x < c$ and $f(x) < f(c)$ ⇨ $f(c)$ is not a relative minimum.

Right of c: $x > c$ and $f(x) > f(c)$ ⇨ $f(c)$ is not a relative maximum.

So, the assumption that $f'(c) > 0$ contradicts the hypothesis that $f(c)$ is a relative extremum. Assuming that $f'(c) < 0$ produces a similar contradiction, you are left with only one possibility—namely, $f'(c) = 0$. So, by definition, c is a critical number of f and the theorem is valid. ■

Finding Extrema on a Closed Interval

Theorem 5.2 states that the relative extrema of a function can occur *only* at the critical numbers of the function. Knowing this, you can use the following guidelines to find extrema on a closed interval.

STUDY TIP Be aware that the critical numbers arise from the zeros of the derivative as well as from points in the domain where the derivative is undefined.

GUIDELINES FOR FINDING EXTREMA ON A CLOSED INTERVAL

To find the extrema of a continuous function f on a closed interval $[a, b]$, use the following steps.

1. Find the critical numbers of f in (a, b).
2. Evaluate f at each critical number in (a, b).
3. Evaluate f at each endpoint of $[a, b]$.
4. The least of these values is the minimum. The greatest is the maximum.

The next three examples show how to apply these guidelines. Be sure you see that finding the critical numbers of the function is only part of the procedure. Evaluating the function at the critical numbers *and* the endpoints is the other part.

EXAMPLE 2 Finding Extrema on a Closed Interval

Find the extrema of $f(x) = 3x^4 - 4x^3$ on the interval $[-1, 2]$.

Solution Begin by differentiating the function.

$$f(x) = 3x^4 - 4x^3 \qquad \text{Write original function.}$$
$$f'(x) = 12x^3 - 12x^2 \qquad \text{Differentiate.}$$

To find the critical numbers of f, you must find all x-values for which $f'(x) = 0$ and all x-values for which $f'(x)$ does not exist.

$$f'(x) = 12x^3 - 12x^2 = 0 \qquad \text{Set } f'(x) \text{ equal to 0.}$$
$$12x^2(x - 1) = 0 \qquad \text{Factor.}$$
$$x = 0, 1 \qquad \text{Critical numbers}$$

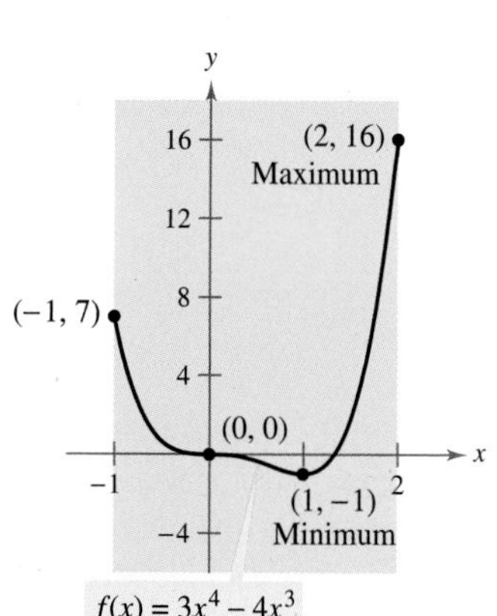

On the closed interval $[-1, 2]$, f has a minimum at $(1, -1)$ and a maximum at $(2, 16)$.
Figure 5.5

Because f' is defined for all x, you can conclude that these are the only critical numbers of f. By evaluating f at these two critical numbers and at the endpoints of $[-1, 2]$, you can determine that the maximum is $f(2) = 16$ and the minimum is $f(1) = -1$, as shown in the table. The graph of f is shown in Figure 5.5.

Left Endpoint	Critical Number	Critical Number	Right Endpoint
$f(-1) = 7$	$f(0) = 0$	$f(1) = -1$ Minimum	$f(2) = 16$ Maximum

■

In Figure 5.5, note that the critical number $x = 0$ does not yield a relative minimum or a relative maximum. This tells you that the converse of Theorem 5.2 is not true. In other words, *the critical numbers of a function need not produce relative extrema.*

EXAMPLE 3 Finding Extrema on a Closed Interval

Find the extrema of

$$f(x) = 2x - 3x^{2/3}$$

on the interval $[-1, 3]$.

Solution Begin by differentiating the function.

$$f(x) = 2x - 3x^{2/3} \qquad \text{Write original function.}$$

$$f'(x) = 2 - \frac{2}{x^{1/3}} = 2\left(\frac{x^{1/3} - 1}{x^{1/3}}\right) \qquad \text{Differentiate.}$$

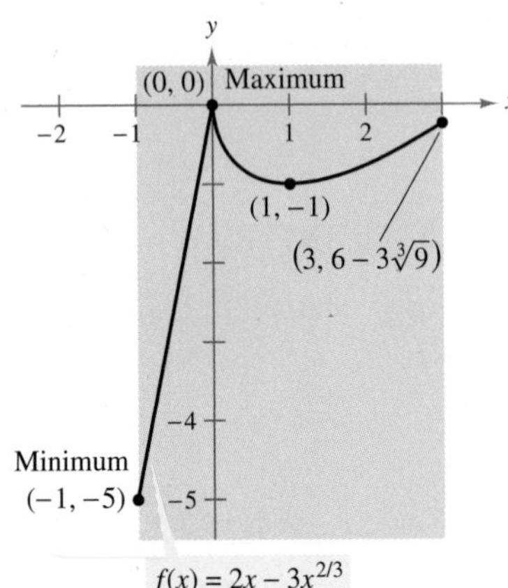

On the closed interval $[-1, 3]$, f has a minimum at $(-1, -5)$ and a maximum at $(0, 0)$.
Figure 5.6

From this derivative, you can see that the function has two critical numbers in the interval $[-1, 3]$. The number 1 is a critical number because $f'(1) = 0$, and the number 0 is a critical number because $f'(0)$ does not exist. By evaluating f at these two numbers and at the endpoints of the interval, you can conclude that the minimum is $f(-1) = -5$ and the maximum is $f(0) = 0$, as shown in the table. The graph of f is shown in Figure 5.6.

Left Endpoint	Critical Number	Critical Number	Right Endpoint
$f(-1) = -5$ Minimum	$f(0) = 0$ Maximum	$f(1) = -1$	$f(3) = 6 - 3\sqrt[3]{9} \approx -0.24$

NOTE To see how to determine the formula for the derivative of the absolute value function, see Section 4.4, Exercise 76.

EXAMPLE 4 Finding Extrema on a Closed Interval

Find the extrema of

$$f(x) = |1 - x^2|$$

on the interval $[-2, 2]$.

Solution Begin by differentiating the function.

$$f(x) = |1 - x^2| \qquad \text{Write original function.}$$

$$f'(x) = (-2x)\frac{1 - x^2}{|1 - x^2|} \qquad \text{Differentiate.}$$

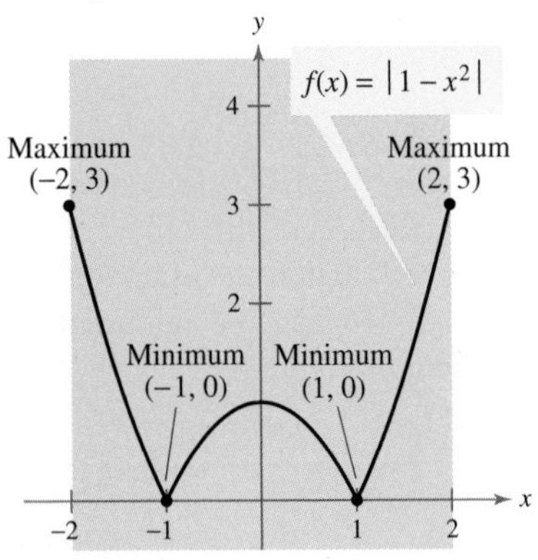

Figure 5.7

From this derivative, you can see that the function has three critical numbers in the interval $[-2, 2]$. The number 0 is a critical number because $f'(0) = 0$, and the numbers ± 1 are critical numbers because $f'(\pm 1)$ does not exist. By evaluating f at these three numbers and the endpoints of the interval, you can conclude that the maximum occurs at *two* points, $f(-2) = 3$ and $f(2) = 3$, and the minimum occurs at *two* points, $f(-1) = 0$ and $f(1) = 0$, as shown in the table. The graph of f is shown in Figure 5.7.

Left Endpoint	Critical Number	Critical Number	Critical Number	Right Endpoint
$f(-2) = 3$ Maximum	$f(-1) = 0$ Minimum	$f(0) = 1$	$f(1) = 0$ Minimum	$f(2) = 3$ Maximum

■

5.1 Exercises

See www.CalcChat.com for worked-out solutions to odd-numbered exercises.

In Exercises 1–6, find the value of the derivative (if it exists) at each indicated extremum.

1. $f(x) = \dfrac{x^2}{x^2 + 4}$

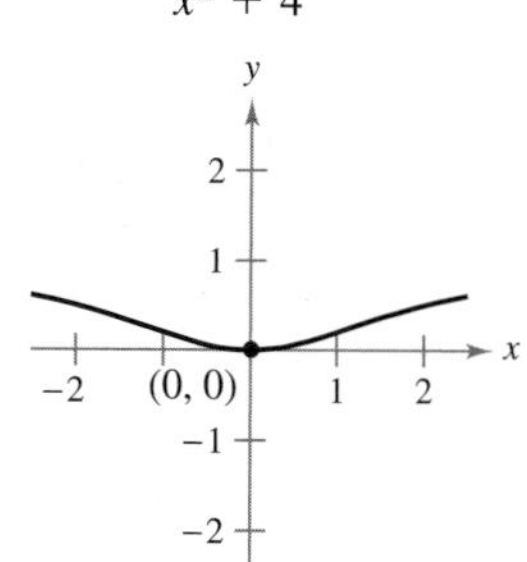

2. $f(x) = \dfrac{1}{2}x^3 - \dfrac{3}{2}x^2 + 1$

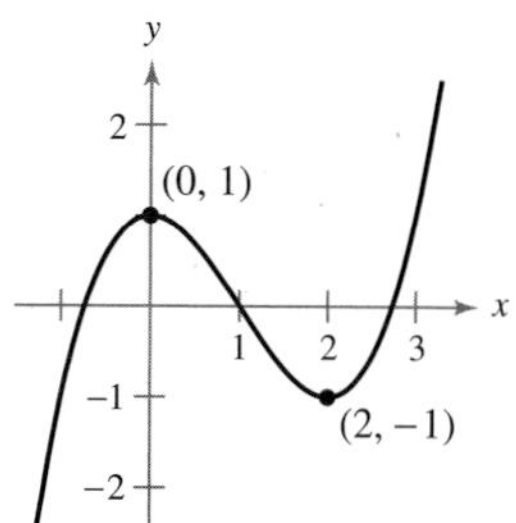

3. $g(x) = x + \dfrac{4}{x^2}$

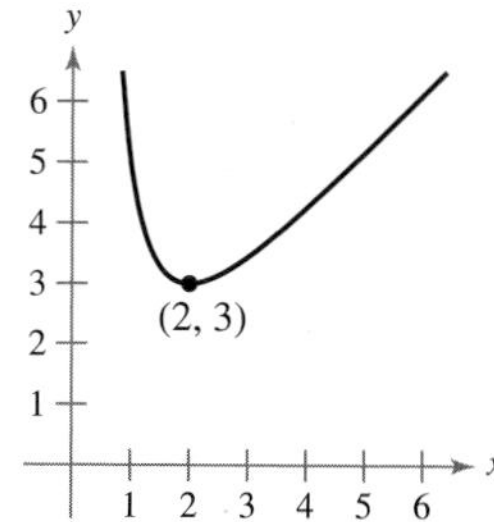

4. $f(x) = -3x\sqrt{x + 1}$

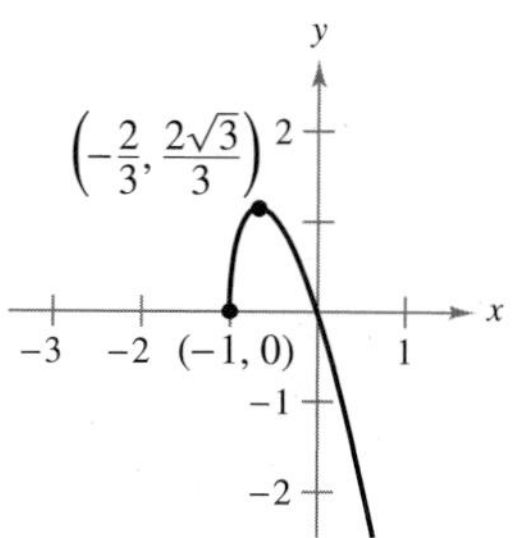

5. $f(x) = (x + 2)^{2/3}$

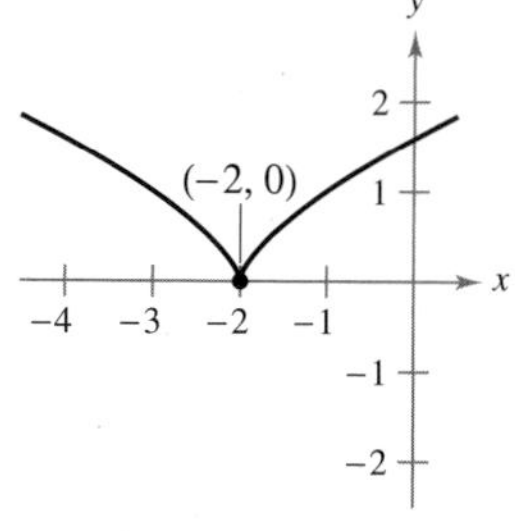

6. $f(x) = 4 - |x|$

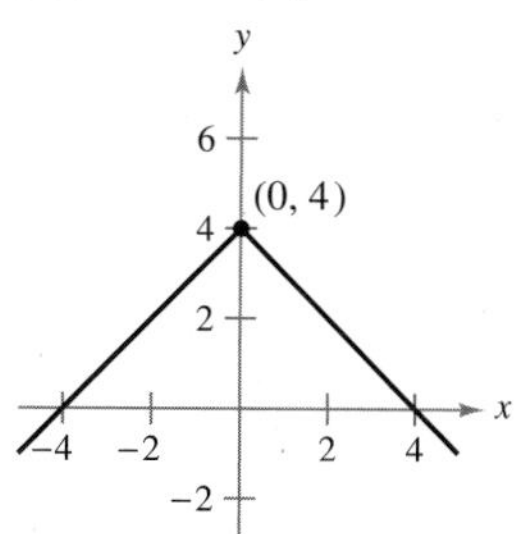

In Exercises 7–10, approximate the critical numbers of the function shown in the graph. Determine whether the function has a relative maximum, a relative minimum, an absolute maximum, an absolute minimum, or none of these at each critical number on the interval shown.

7.

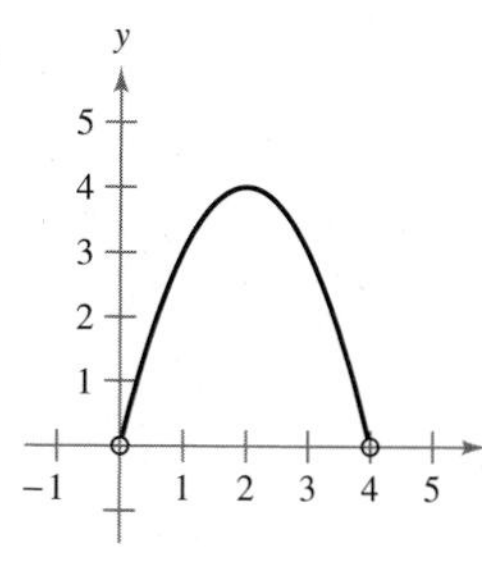

8.

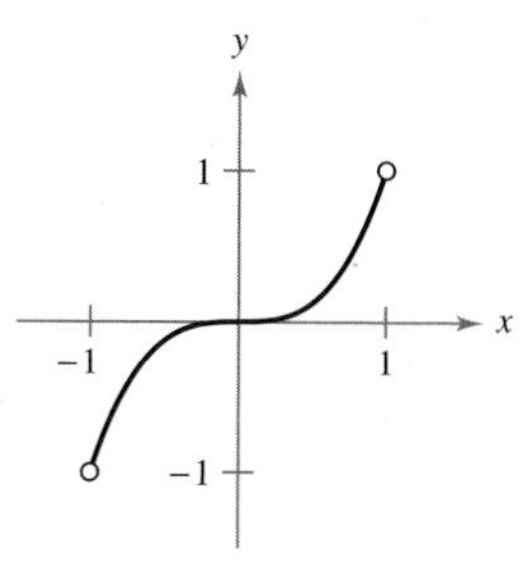

9.

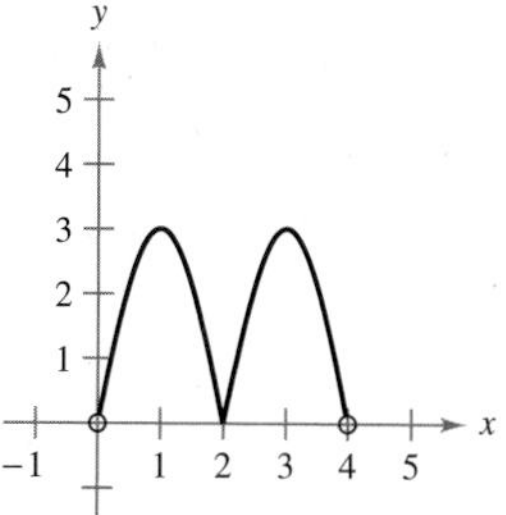

10.

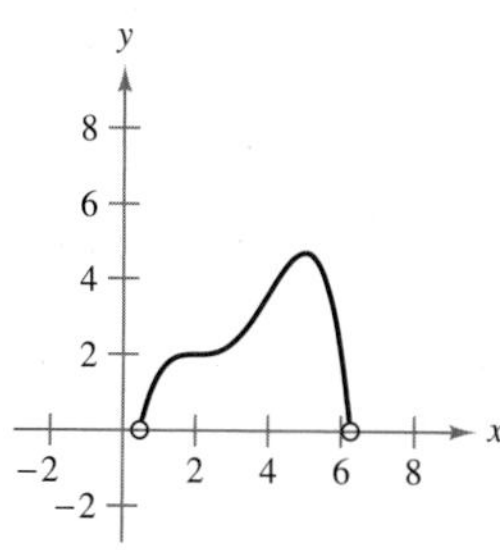

In Exercises 11–14, find any critical numbers of the function.

11. $f(x) = x^3 - 3x^2$

12. $g(x) = x^4 - 4x^2$

13. $g(t) = t\sqrt{4 - t}, \ t < 3$

14. $f(x) = \dfrac{4x}{x^2 + 1}$

In Exercises 15–30, locate the absolute extrema of the function on the closed interval.

15. $f(x) = 3 - x, \ [-1, 2]$

16. $f(x) = \dfrac{2x + 5}{3}, \ [0, 5]$

17. $g(x) = x^2 - 2x, \ [0, 4]$

18. $h(x) = -x^2 + 3x - 5, \ [-2, 1]$

19. $f(x) = x^3 - \dfrac{3}{2}x^2, \ [-1, 2]$

20. $f(x) = x^3 - 12x, \ [0, 4]$

21. $y = 3x^{2/3} - 2x, \ [-1, 1]$

22. $g(x) = \sqrt[3]{x}, \ [-1, 1]$

23. $g(t) = \dfrac{t^2}{t^2 + 3}, \ [-1, 1]$

24. $f(x) = \dfrac{2x}{x^2 + 1}, \ [-2, 2]$

25. $h(s) = \dfrac{1}{s - 2}, \ [0, 1]$

26. $h(t) = \dfrac{t}{t - 2}, \ [3, 5]$

27. $y = 3 - |t - 3|, \ [-1, 5]$

28. $g(x) = \dfrac{1}{1 + |x + 1|}, \ [-3, 3]$

29. $f(x) = [\![x]\!], \ [-2, 2]$

30. $h(x) = [\![2 - x]\!], \ [-2, 2]$

In Exercises 31–34, locate the absolute extrema of the function (if any exist) over each interval.

31. $f(x) = 2x - 3$

(a) $[0, 2]$ (b) $[0, 2)$
(c) $(0, 2]$ (d) $(0, 2)$

32. $f(x) = 5 - x$

(a) $[1, 4]$ (b) $[1, 4)$
(c) $(1, 4]$ (d) $(1, 4)$

33. $f(x) = x^2 - 2x$

(a) $[-1, 2]$ (b) $(1, 3]$
(c) $(0, 2)$ (d) $[1, 4)$

34. $f(x) = \sqrt{4 - x^2}$

(a) $[-2, 2]$ (b) $[-2, 0)$
(c) $(-2, 2)$ (d) $[1, 2)$

In Exercises 35–38, use a graphing utility to graph the function. Locate the absolute extrema of the function on the given interval.

35. $f(x) = \begin{cases} 2x + 2, & 0 \le x \le 1 \\ 4x^2, & 1 < x \le 3 \end{cases}, \quad [0, 3]$

36. $f(x) = \begin{cases} 2 - x^2, & 1 \le x < 3 \\ 2 - 3x, & 3 \le x \le 5 \end{cases}, \quad [1, 5]$

37. $f(x) = \dfrac{3}{x - 1}, \quad (1, 4]$

38. $f(x) = \dfrac{2}{2 - x}, \quad [0, 2)$

In Exercises 39 and 40, (a) use a computer algebra system to graph the function and approximate any absolute extrema on the given interval. (b) Use the utility to find any critical numbers, and use them to find any absolute extrema not located at the endpoints. Compare the results with those in part (a).

39. $f(x) = 3.2x^5 + 5x^3 - 3.5x, \quad [0, 1]$

40. $f(x) = \dfrac{4}{3}x\sqrt{3 - x}, \quad [0, 3]$

In Exercises 41 and 42, use a computer algebra system to find the maximum value of $|f''(x)|$ on the closed interval. (This value is used in the error estimate for the Trapezoidal Rule, as discussed in Section 6.6.)

41. $f(x) = \sqrt{1 + x^3}, \quad [0, 2]$

42. $f(x) = \dfrac{1}{x^2 + 1}, \quad \left[\dfrac{1}{2}, 3\right]$

In Exercises 43 and 44, use a computer algebra system to find the maximum value of $|f^{(4)}(x)|$ on the closed interval. (This value is used in the error estimate for Simpson's Rule, as discussed in Section 6.6.)

43. $f(x) = (x + 1)^{2/3}, \quad [0, 2]$

44. $f(x) = \dfrac{1}{x^2 + 1}, \quad [-1, 1]$

45. ***Writing*** Write a short paragraph explaining why a continuous function on an open interval may not have a maximum or minimum. Illustrate your explanation with a sketch of the graph of such a function.

CAPSTONE

46. Decide whether each labeled point is an absolute maximum or minimum, a relative maximum or minimum, or neither.

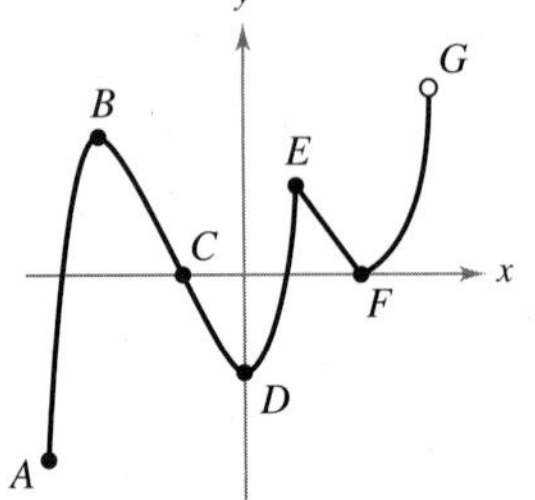

WRITING ABOUT CONCEPTS

In Exercises 47 and 48, graph a function on the interval $[-2, 5]$ having the given characteristics.

47. Absolute maximum at $x = -2$, absolute minimum at $x = 1$, relative maximum at $x = 3$

48. Relative minimum at $x = -1$, critical number (but no extremum) at $x = 0$, absolute maximum at $x = 2$, absolute minimum at $x = 5$

In Exercises 49–52, determine from the graph whether f has a minimum in the open interval (a, b).

49. (a) (b)

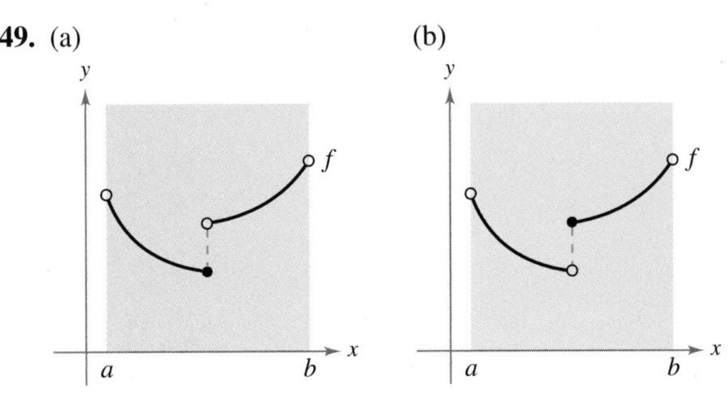

50. (a) (b)

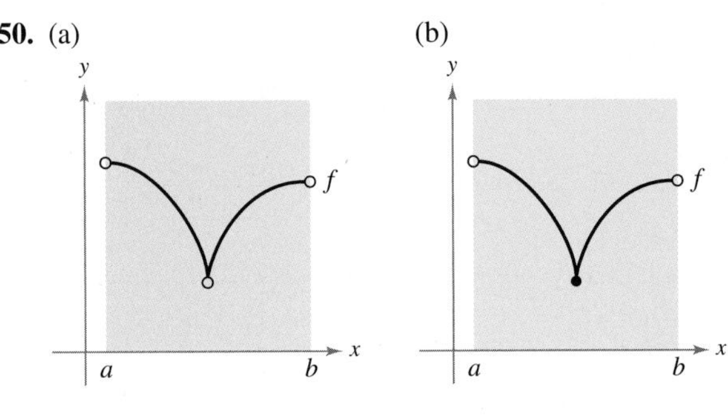

51. (a) (b)

52. (a) (b)

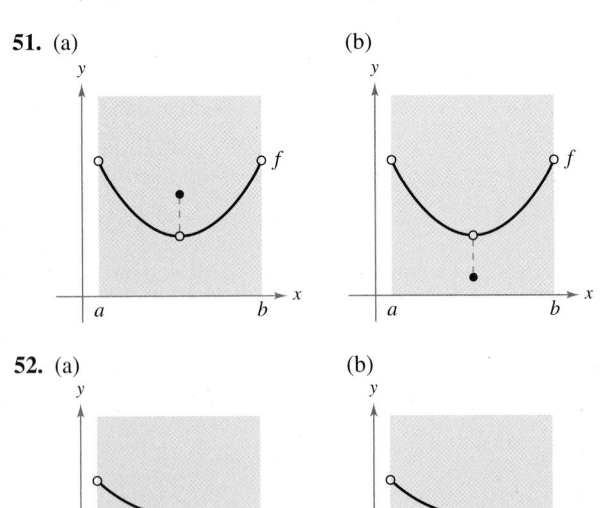

53. ***Power*** The formula for the power output P of a battery is $P = VI - RI^2$, where V is the electromotive force in volts, R is the resistance in ohms, and I is the current in amperes. Find the current that corresponds to a maximum value of P in a battery for which $V = 12$ volts and $R = 0.5$ ohm. Assume that a 15-ampere fuse bounds the output in the interval $0 \le I \le 15$. Could the power output be increased by replacing the 15-ampere fuse with a 20-ampere fuse? Explain.

54. ***Inventory Cost*** A retailer has determined that the cost C of ordering and storing x units of a product is

$$C = 2x + \frac{300{,}000}{x}, \quad 1 \le x \le 300.$$

The delivery truck can bring at most 300 units per order. Find the order size that will minimize cost. Could the cost be decreased if the truck were replaced with one that could bring at most 400 units? Explain your reasoning.

55. ***Fertility Rates*** The graph of the United States fertility rate shows the number of births per 1000 women in their lifetime according to the birth rate in that particular year. *(Source: U.S. National Center for Health Statistics)*

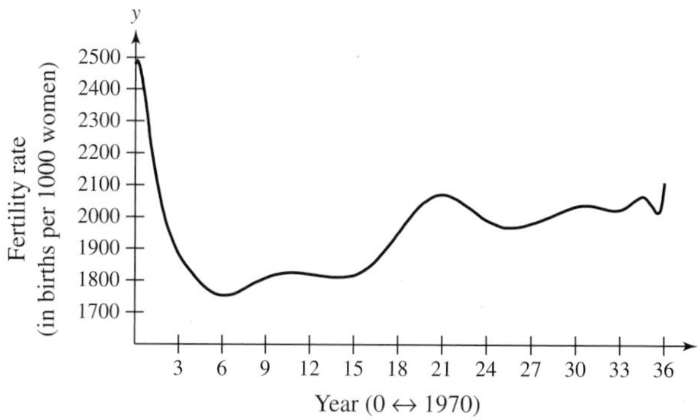

(a) Around what year was the fertility rate the highest, and to how many births per 1000 women did this rate correspond?

(b) During which time periods was the fertility rate increasing most rapidly? Most slowly?

(c) During which time periods was the fertility rate decreasing most rapidly? Most slowly?

(d) Give some possible real-life reasons for fluctuations in the fertility rate.

56. ***Population*** The resident population P (in millions) of the United States from 1790 through 2000 can be modeled by

$$P = 0.00000583t^3 + 0.005003t^2 + 0.13776t + 4.658$$

$-10 \le t \le 200$, where $t = 0$ corresponds to 1800. *(Source: U.S. Census Bureau)*

(a) Make a conjecture about the maximum and minimum populations in the U.S. from 1790 to 2000.

(b) Analytically find the maximum and minimum populations over the interval.

(c) Write a brief paragraph comparing your conjecture with your results in part (b).

57. ***Highway Design*** In order to build a highway, it is necessary to fill a section of a valley where the grades (slopes) of the sides are 9% and 6% (see figure). The top of the filled region will have the shape of a parabolic arc that is tangent to the two slopes at the points A and B. The horizontal distances from A to the y-axis and from B to the y-axis are both 500 feet.

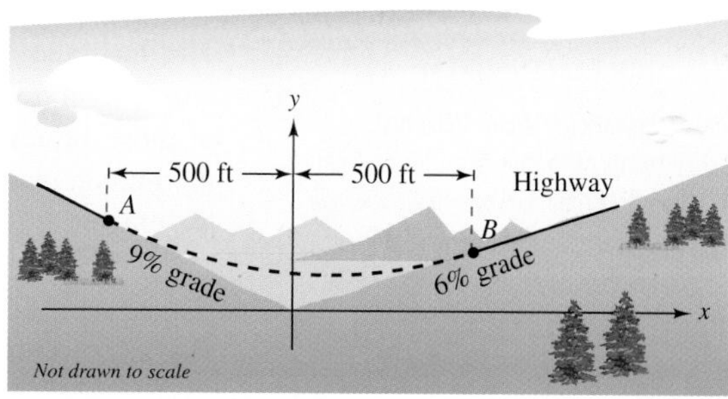

(a) Find the coordinates of A and B.

(b) Find a quadratic function $y = ax^2 + bx + c$, $-500 \le x \le 500$, that describes the top of the filled region.

(c) Construct a table giving the depths d of the fill for $x = -500, -400, -300, -200, -100, 0, 100, 200, 300, 400$, and 500.

(d) What will be the lowest point on the completed highway? Will it be directly over the point where the two hillsides come together?

58. Find all critical numbers of the greatest integer function $f(x) = [\![x]\!]$.

True or False? **In Exercises 59–62, determine whether the statement is true or false. If it is false, explain why or give an example that shows it is false.**

59. The maximum of a function that is continuous on a closed interval can occur at two different values in the interval.

60. If a function is continuous on a closed interval, then it must have a minimum on the interval.

61. If $x = c$ is a critical number of the function f, then it is also a critical number of the function $g(x) = f(x) + k$, where k is a constant.

62. If $x = c$ is a critical number of the function f, then it is also a critical number of the function $g(x) = f(x - k)$, where k is a constant.

63. Let the function f be differentiable on an interval I containing c. If f has a maximum value at $x = c$, show that $-f$ has a minimum value at $x = c$.

64. Consider the cubic function $f(x) = ax^3 + bx^2 + cx + d$, where $a \ne 0$. Show that f can have zero, one, or two critical numbers and give an example of each case.

65. Explain why the function given by $f(x) = 3/(x - 2)$ has a maximum on $[3, 5]$ but not on $[1, 3]$.

5.2 Rolle's Theorem and the Mean Value Theorem

- **Understand and use Rolle's Theorem.**
- **Understand and use the Mean Value Theorem.**

Rolle's Theorem

The Extreme Value Theorem (Section 5.1) states that a continuous function on a closed interval $[a, b]$ must have both a minimum and a maximum on the interval. Both of these values, however, can occur at the endpoints. **Rolle's Theorem,** named after the French mathematician Michel Rolle (1652–1719), gives conditions that guarantee the existence of an extreme value in the *interior* of a closed interval.

ROLLE'S THEOREM

French mathematician Michel Rolle first published the theorem that bears his name in 1691. Before this time, however, Rolle was one of the most vocal critics of calculus, stating that it gave erroneous results and was based on unsound reasoning. Later in life, Rolle came to see the usefulness of calculus.

EXPLORATION

Extreme Values in a Closed Interval Sketch a rectangular coordinate plane on a piece of paper. Label the points $(1, 3)$ and $(5, 3)$. Using a pencil or pen, draw the graph of a differentiable function f that starts at $(1, 3)$ and ends at $(5, 3)$. Is there at least one point on the graph for which the derivative is zero? Would it be possible to draw the graph so that there *is not* a point for which the derivative is zero? Explain your reasoning.

THEOREM 5.3 ROLLE'S THEOREM

Let f be continuous on the closed interval $[a, b]$ and differentiable on the open interval (a, b). If

$$f(a) = f(b)$$

then there is at least one number c in (a, b) such that $f'(c) = 0$.

PROOF Let $f(a) = d = f(b)$.

Case 1: If $f(x) = d$ for all x in $[a, b]$, f is constant on the interval and, by Theorem 4.2, $f'(x) = 0$ for all x in (a, b).

Case 2: Suppose $f(x) > d$ for some x in (a, b). By the Extreme Value Theorem, you know that f has a maximum at some c in the interval. Moreover, because $f(c) > d$, this maximum does not occur at either endpoint. So, f has a maximum in the *open* interval (a, b). This implies that $f(c)$ is a *relative* maximum and, by Theorem 5.2, c is a critical number of f. Finally, because f is differentiable at c, you can conclude that $f'(c) = 0$.

Case 3: If $f(x) < d$ for some x in (a, b), you can use an argument similar to that in Case 2, but involving the minimum instead of the maximum. ■

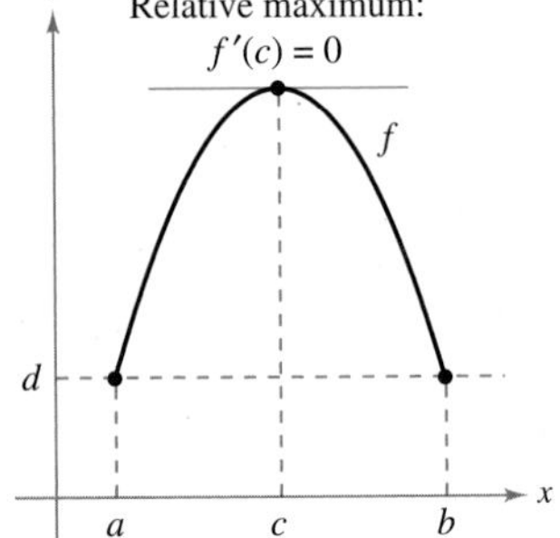

(a) f is continuous on $[a, b]$ and differentiable on (a, b).

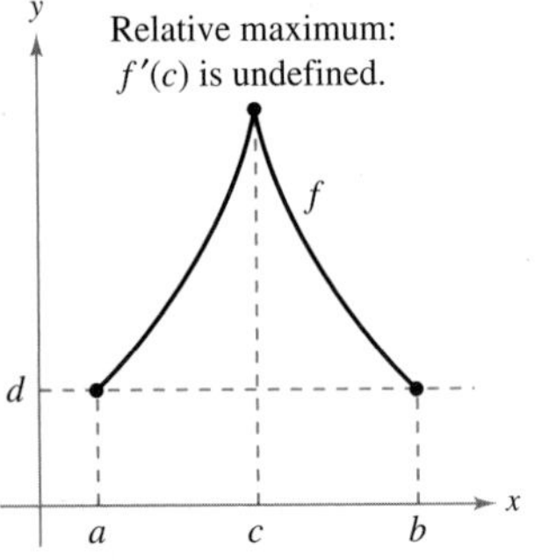

(b) f is continuous on $[a, b]$, but not differentiable on (a, b).

Figure 5.8

From Rolle's Theorem, you can see that if a function f is continuous on $[a, b]$ and differentiable on (a, b), and if $f(a) = f(b)$, there must be at least one x-value between a and b at which the graph of f has a horizontal tangent, as shown in Figure 5.8(a). If the differentiability requirement is dropped from Rolle's Theorem, f will still have a critical number in (a, b), but it may not yield a horizontal tangent. Such a case is shown in Figure 5.8(b).

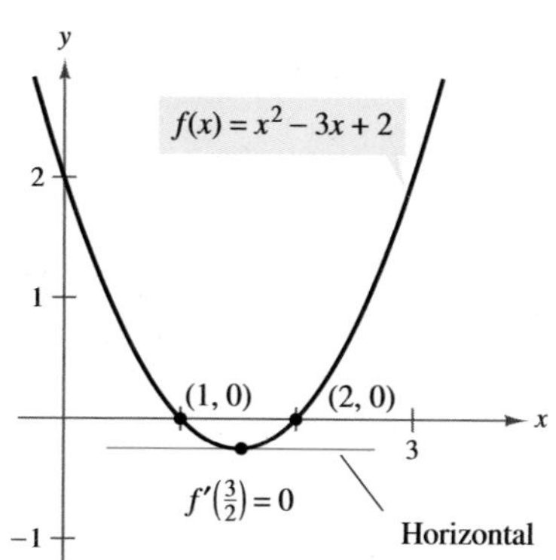

The x-value for which $f'(x) = 0$ is between the two x-intercepts.
Figure 5.9

EXAMPLE 1 Illustrating Rolle's Theorem

Find the two x-intercepts of

$$f(x) = x^2 - 3x + 2$$

and show that $f'(x) = 0$ at some point between the two x-intercepts.

Solution Note that f is differentiable on the entire real line. Setting $f(x)$ equal to 0 produces

$$x^2 - 3x + 2 = 0 \qquad \text{Set } f(x) \text{ equal to 0.}$$

$$(x - 1)(x - 2) = 0. \qquad \text{Factor.}$$

So, $f(1) = f(2) = 0$, and from Rolle's Theorem you know that there *exists* at least one c in the interval $(1, 2)$ such that $f'(c) = 0$. To *find* such a c, you can solve the equation

$$f'(x) = 2x - 3 = 0 \qquad \text{Set } f'(x) \text{ equal to 0.}$$

and determine that $f'(x) = 0$ when $x = \frac{3}{2}$. Note that this x-value lies in the open interval $(1, 2)$, as shown in Figure 5.9. ■

Rolle's Theorem states that if f satisfies the conditions of the theorem, there must be *at least* one point between a and b at which the derivative is 0. There may of course be more than one such point, as shown in the next example.

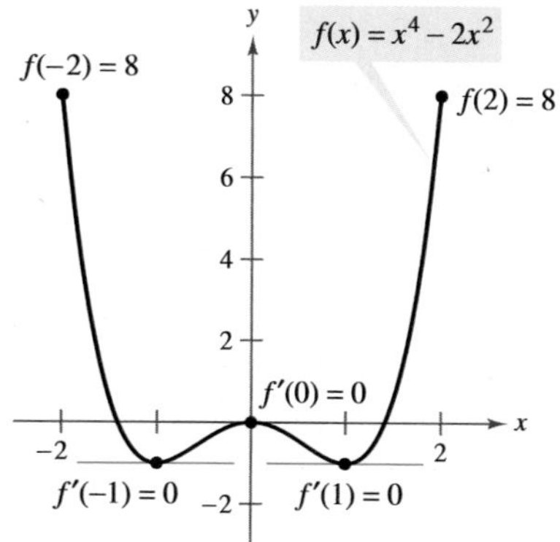

$f'(x) = 0$ for more than one x-value in the interval $(-2, 2)$.
Figure 5.10

EXAMPLE 2 Illustrating Rolle's Theorem

Let $f(x) = x^4 - 2x^2$. Find all values of c in the interval $(-2, 2)$ such that $f'(c) = 0$.

Solution To begin, note that the function satisfies the conditions of Rolle's Theorem. That is, f is continuous on the interval $[-2, 2]$ and differentiable on the interval $(-2, 2)$. Moreover, because $f(-2) = f(2) = 8$, you can conclude that there exists at least one c in $(-2, 2)$ such that $f'(c) = 0$. Setting the derivative equal to 0 produces

$$f'(x) = 4x^3 - 4x = 0 \qquad \text{Set } f'(x) \text{ equal to 0.}$$

$$4x(x - 1)(x + 1) = 0 \qquad \text{Factor.}$$

$$x = 0, 1, -1. \qquad x\text{-values for which } f'(x) = 0$$

So, in the interval $(-2, 2)$, the derivative is zero at three different values of x, as shown in Figure 5.10. ■

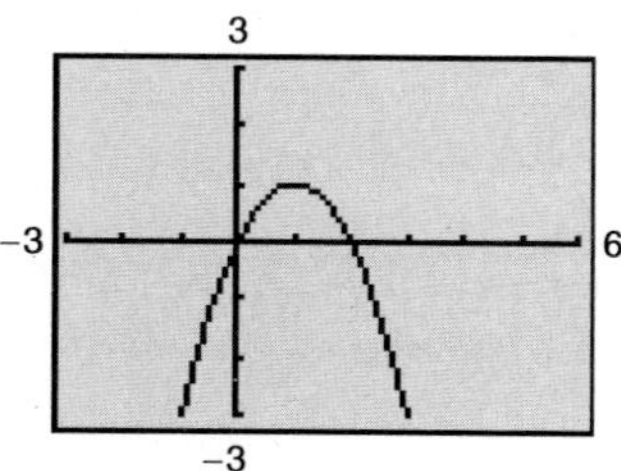

Figure 5.11

TECHNOLOGY PITFALL A graphing utility can be used to indicate whether the points on the graphs in Examples 1 and 2 are relative minima or relative maxima of the functions. When using a graphing utility, however, you should keep in mind that it can give misleading pictures of graphs. For example, use a graphing utility to graph

$$f(x) = 1 - (x - 1)^2 - \frac{1}{1000(x - 1)^{1/7} + 1}.$$

With most viewing windows, it appears that the function has a maximum of 1 when $x = 1$ (see Figure 5.11). By evaluating the function at $x = 1$, however, you can see that $f(1) = 0$. To determine the behavior of this function near $x = 1$, you need to examine the graph analytically to get the complete picture.

The Mean Value Theorem

Rolle's Theorem can be used to prove another theorem—the **Mean Value Theorem.**

> **THEOREM 5.4 THE MEAN VALUE THEOREM**
>
> If f is continuous on the closed interval $[a, b]$ and differentiable on the open interval (a, b), then there exists a number c in (a, b) such that
>
> $$f'(c) = \frac{f(b) - f(a)}{b - a}.$$

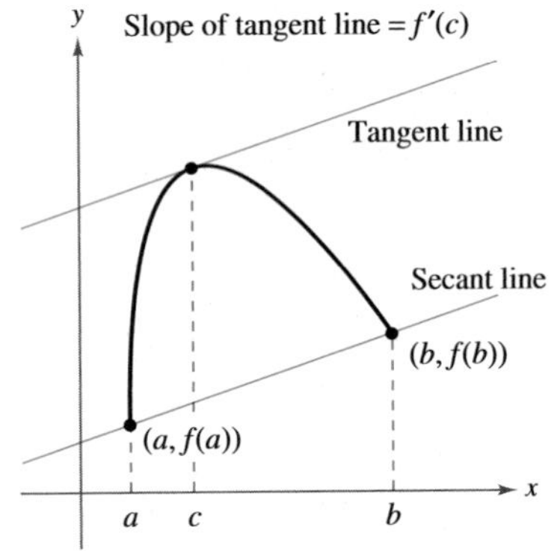

Figure 5.12

PROOF Refer to Figure 5.12. The equation of the secant line that passes through the points $(a, f(a))$ and $(b, f(b))$ is

$$y = \left[\frac{f(b) - f(a)}{b - a}\right](x - a) + f(a).$$

Let $g(x)$ be the difference between $f(x)$ and y. Then

$$\begin{aligned} g(x) &= f(x) - y \\ &= f(x) - \left[\frac{f(b) - f(a)}{b - a}\right](x - a) - f(a). \end{aligned}$$

By evaluating g at a and b, you can see that $g(a) = 0 = g(b)$. Because f is continuous on $[a, b]$, it follows that g is also continuous on $[a, b]$. Furthermore, because f is differentiable, g is also differentiable, and you can apply Rolle's Theorem to the function g. So, there exists a number c in (a, b) such that $g'(c) = 0$, which implies that

$$\begin{aligned} 0 &= g'(c) \\ &= f'(c) - \frac{f(b) - f(a)}{b - a}. \end{aligned}$$

So, there exists a number c in (a, b) such that

$$f'(c) = \frac{f(b) - f(a)}{b - a}.$$

■

NOTE The "mean" in the Mean Value Theorem refers to the mean (or average) rate of change of f in the interval $[a, b]$. ■

JOSEPH-LOUIS LAGRANGE (1736–1813)

The Mean Value Theorem was first proved by the famous mathematician Joseph-Louis Lagrange. Born in Italy, Lagrange held a position in the court of Frederick the Great in Berlin for 20 years. Afterward, he moved to France, where he met emperor Napoleon Bonaparte, who is quoted as saying, "Lagrange is the lofty pyramid of the mathematical sciences."

Although the Mean Value Theorem can be used directly in problem solving, it is used more often to prove other theorems. In fact, some people consider this to be the most important theorem in calculus—it is closely related to the Fundamental Theorem of Calculus discussed in Section 6.4. For now, you can get an idea of the versatility of the Mean Value Theorem by looking at the results stated in Exercises 65–70 in this section.

The Mean Value Theorem has implications for both basic interpretations of the derivative. Geometrically, the theorem guarantees the existence of a tangent line that is parallel to the secant line through the points $(a, f(a))$ and $(b, f(b))$, as shown in Figure 5.12. Example 3 illustrates this geometric interpretation of the Mean Value Theorem. In terms of rates of change, the Mean Value Theorem implies that there must be a point in the open interval (a, b) at which the instantaneous rate of change is equal to the average rate of change over the interval $[a, b]$. This is illustrated in Example 4.

EXAMPLE 3 Finding a Tangent Line

Given $f(x) = 5 - (4/x)$, find all values of c in the open interval $(1, 4)$ such that

$$f'(c) = \frac{f(4) - f(1)}{4 - 1}.$$

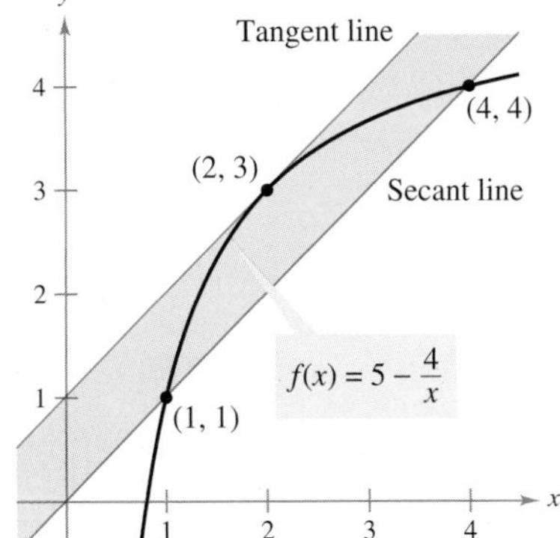

The tangent line at $(2, 3)$ is parallel to the secant line through $(1, 1)$ and $(4, 4)$.
Figure 5.13

Solution The slope of the secant line through $(1, f(1))$ and $(4, f(4))$ is

$$\frac{f(4) - f(1)}{4 - 1} = \frac{4 - 1}{4 - 1} = 1.$$

Note that the function satisfies the conditions of the Mean Value Theorem. That is, f is continuous on the interval $[1, 4]$ and differentiable on the interval $(1, 4)$. So, there exists at least one number c in $(1, 4)$ such that $f'(c) = 1$. Solving the equation $f'(x) = 1$ yields

$$f'(x) = \frac{4}{x^2} = 1$$

which implies that $x = \pm 2$. So, in the interval $(1, 4)$, you can conclude that $c = 2$, as shown in Figure 5.13.

EXAMPLE 4 Finding an Instantaneous Rate of Change

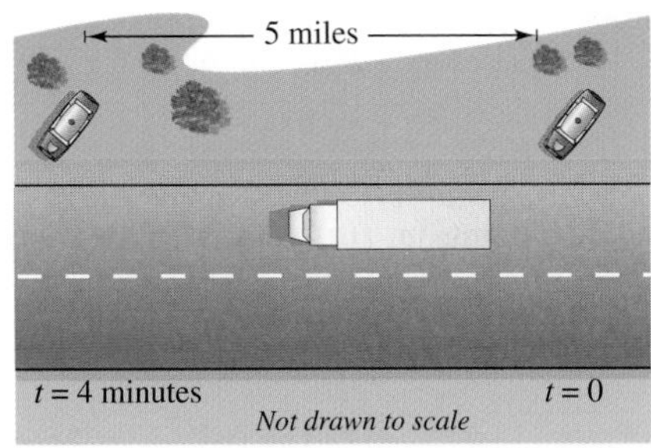

At some time t, the instantaneous velocity is equal to the average velocity over 4 minutes.
Figure 5.14

Two stationary patrol cars equipped with radar are 5 miles apart on a highway, as shown in Figure 5.14. As a truck passes the first patrol car, its speed is clocked at 55 miles per hour. Four minutes later, when the truck passes the second patrol car, its speed is clocked at 50 miles per hour. Prove that the truck must have exceeded the speed limit (of 55 miles per hour) at some time during the 4 minutes.

Solution Let $t = 0$ be the time (in hours) when the truck passes the first patrol car. The time when the truck passes the second patrol car is

$$t = \frac{4}{60} = \frac{1}{15} \text{ hour.}$$

By letting $s(t)$ represent the distance (in miles) traveled by the truck, you have $s(0) = 0$ and $s(\frac{1}{15}) = 5$. So, the average velocity of the truck over the five-mile stretch of highway is

$$\text{Average velocity} = \frac{s(1/15) - s(0)}{(1/15) - 0} = \frac{5}{1/15} = 75 \text{ miles per hour.}$$

Assuming that the position function is differentiable, you can apply the Mean Value Theorem to conclude that the truck must have been traveling at a rate of 75 miles per hour sometime during the 4 minutes. ■

A useful alternative form of the Mean Value Theorem is as follows: If f is continuous on $[a, b]$ and differentiable on (a, b), then there exists a number c in (a, b) such that

$$f(b) = f(a) + (b - a)f'(c).$$ Alternative form of Mean Value Theorem

NOTE When doing the exercises for this section, keep in mind that polynomial functions and rational functions are differentiable at all points in their domains. ■

5.2 Exercises

See www.CalcChat.com for worked-out solutions to odd-numbered exercises.

In Exercises 1–4, explain why Rolle's Theorem does not apply to the function even though there exist a and b such that $f(a) = f(b)$.

1. $f(x) = \left|\dfrac{1}{x}\right|$

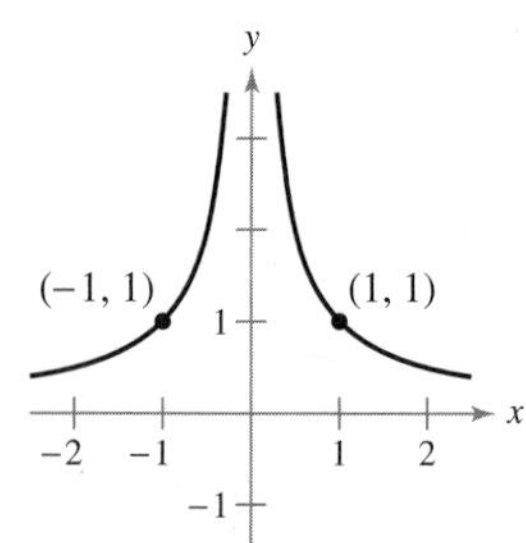

2. $f(x) = \dfrac{x^2 - 4}{x^2}$

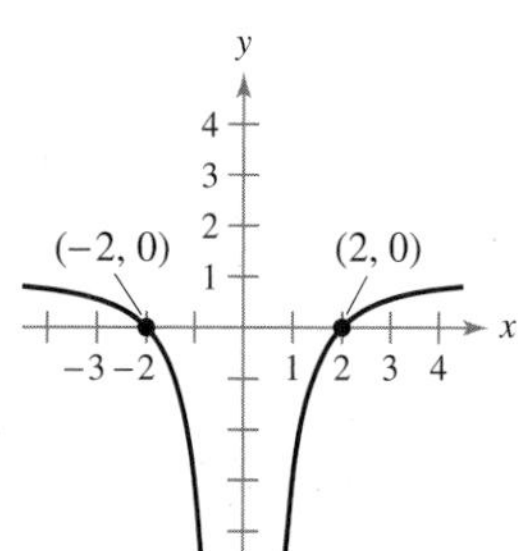

3. $f(x) = 1 - |x - 1|$

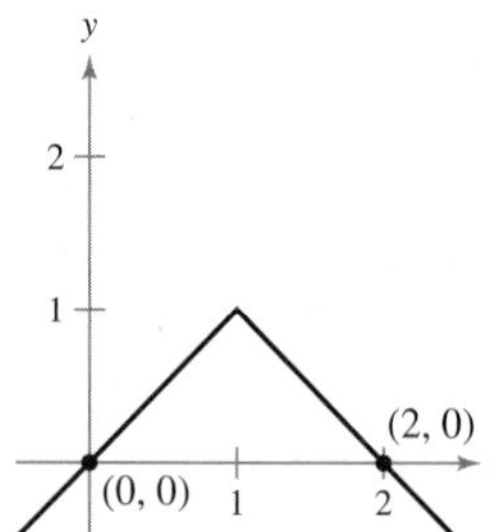

4. $f(x) = \sqrt{(2 - x^{2/3})^3}$

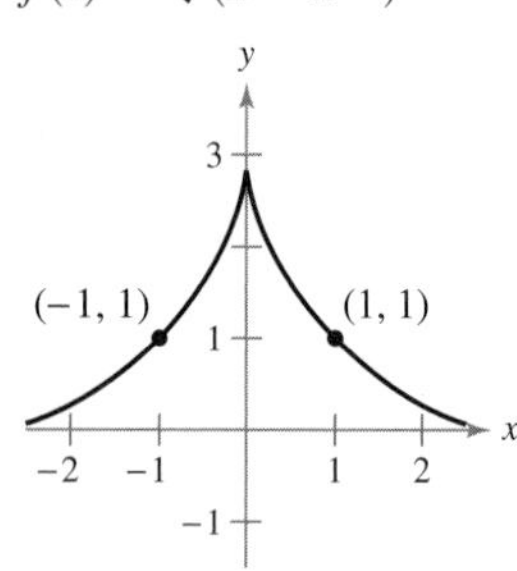

In Exercises 5–8, find the two x-intercepts of the function f and show that $f'(x) = 0$ at some point between the two x-intercepts.

5. $f(x) = x^2 - x - 2$
6. $f(x) = x(x - 3)$
7. $f(x) = x\sqrt{x + 4}$
8. $f(x) = -3x\sqrt{x + 1}$

In Exercises 9–16, determine whether Rolle's Theorem can be applied to f on the closed interval $[a, b]$. If Rolle's Theorem can be applied, find all values of c in the open interval (a, b) such that $f'(c) = 0$. If Rolle's Theorem cannot be applied, explain why not.

9. $f(x) = -x^2 + 3x,\quad [0, 3]$
10. $f(x) = x^2 - 5x + 4,\quad [1, 4]$
11. $f(x) = (x - 1)(x - 2)(x - 3),\quad [1, 3]$
12. $f(x) = (x - 3)(x + 1)^2,\quad [-1, 3]$
13. $f(x) = x^{2/3} - 1,\quad [-8, 8]$
14. $f(x) = 3 - |x - 3|,\quad [0, 6]$
15. $f(x) = \dfrac{x^2 - 2x - 3}{x + 2},\quad [-1, 3]$
16. $f(x) = \dfrac{x^2 - 1}{x},\quad [-1, 1]$

In Exercises 17 and 18, use a graphing utility to graph the function on the closed interval $[a, b]$. Determine whether Rolle's Theorem can be applied to f on the interval and, if so, find all values of c in the open interval (a, b) such that $f'(c) = 0$.

17. $f(x) = |x| - 1,\quad [-1, 1]$
18. $f(x) = x - x^{1/3},\quad [0, 1]$

19. ***Vertical Motion*** The height of a ball t seconds after it is thrown upward from a height of 6 feet and with an initial velocity of 48 feet per second is $f(t) = -16t^2 + 48t + 6$.
    (a) Verify that $f(1) = f(2)$.
    (b) According to Rolle's Theorem, what must the velocity be at some time in the interval $(1, 2)$? Find that time.

20. ***Reorder Costs*** The ordering and transportation cost C for components used in a manufacturing process is approximated by $C(x) = 10\left(\dfrac{1}{x} + \dfrac{x}{x + 3}\right)$, where C is measured in thousands of dollars and x is the order size in hundreds.
    (a) Verify that $C(3) = C(6)$.
    (b) According to Rolle's Theorem, the rate of change of the cost must be 0 for some order size in the interval $(3, 6)$. Find that order size.

In Exercises 21 and 22, copy the graph and sketch the secant line to the graph through the points $(a, f(a))$ and $(b, f(b))$. Then sketch any tangent lines to the graph for each value of c guaranteed by the Mean Value Theorem. To print an enlarged copy of the graph, go to the website *www.mathgraphs.com*.

21.

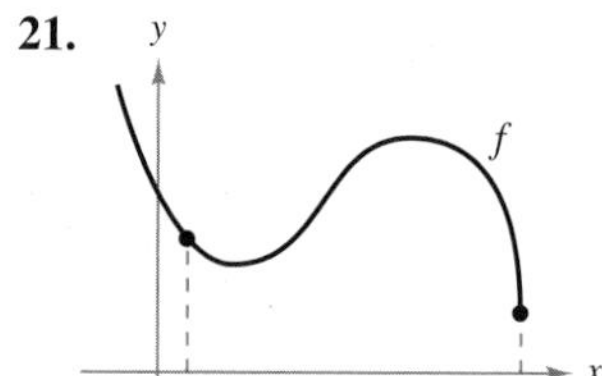

22.

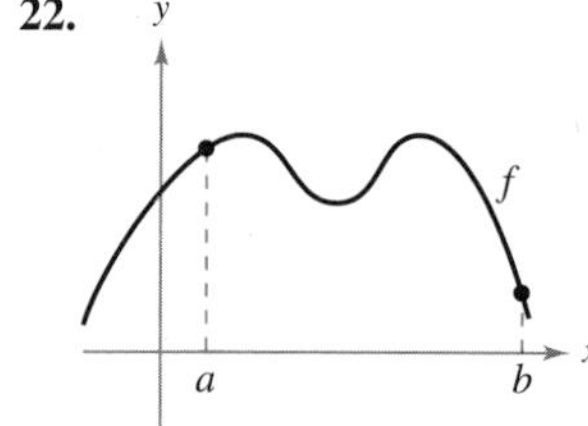

Writing **In Exercises 23–26, explain why the Mean Value Theorem does not apply to the function f on the interval $[0, 6]$.**

23.

24.

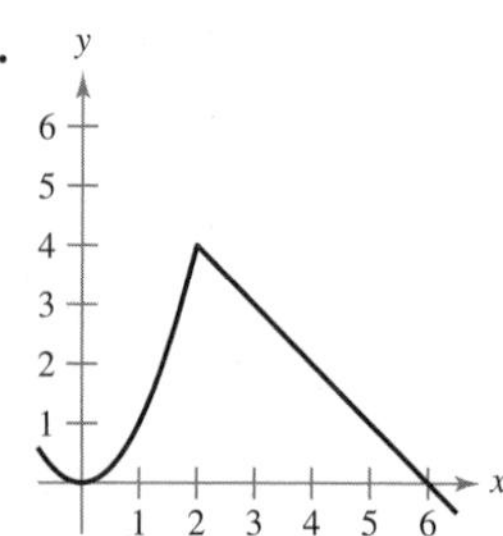

25. $f(x) = \dfrac{1}{x - 3}$

26. $f(x) = |x - 3|$

27. ***Mean Value Theorem*** Consider the graph of the function $f(x) = -x^2 + 5$. (a) Find the equation of the secant line joining the points $(-1, 4)$ and $(2, 1)$. (b) Use the Mean Value Theorem to determine a point c in the interval $(-1, 2)$ such that the tangent line at c is parallel to the secant line. (c) Find the equation of the tangent line through c. (d) Then use a graphing utility to graph f, the secant line, and the tangent line.

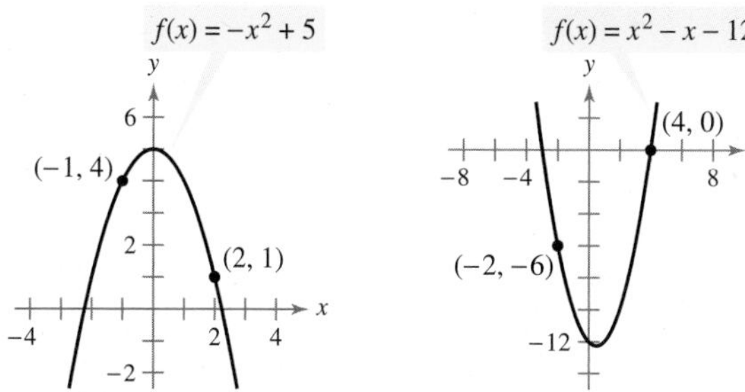

Figure for 27 **Figure for 28**

28. ***Mean Value Theorem*** Consider the graph of the function $f(x) = x^2 - x - 12$. (a) Find the equation of the secant line joining the points $(-2, -6)$ and $(4, 0)$. (b) Use the Mean Value Theorem to determine a point c in the interval $(-2, 4)$ such that the tangent line at c is parallel to the secant line. (c) Find the equation of the tangent line through c. (d) Then use a graphing utility to graph f, the secant line, and the tangent line.

In Exercises 29–36, determine whether the Mean Value Theorem can be applied to f on the closed interval $[a, b]$. If the Mean Value Theorem can be applied, find all values of c in the open interval (a, b) such that $f'(c) = \dfrac{f(b) - f(a)}{b - a}$. If the Mean Value Theorem cannot be applied, explain why not.

29. $f(x) = x^2, \quad [-2, 1]$

30. $f(x) = x^3, \quad [0, 1]$

31. $f(x) = x^3 + 2x, \quad [-1, 1]$

32. $f(x) = x^4 - 8x, \quad [0, 2]$

33. $f(x) = x^{2/3}, \quad [0, 1]$

34. $f(x) = \dfrac{x + 1}{x}, \quad [-1, 2]$

35. $f(x) = |2x + 1|, \quad [-1, 3]$

36. $f(x) = \sqrt{2 - x}, \quad [-7, 2]$

In Exercises 37–40, use a graphing utility to (a) graph the function f on the given interval, (b) find and graph the secant line through points on the graph of f at the endpoints of the given interval, and (c) find and graph any tangent lines to the graph of f that are parallel to the secant line.

37. $f(x) = \dfrac{x}{x + 1}, \quad \left[-\frac{1}{2}, 2\right]$

38. $f(x) = \sqrt{x}, \quad [1, 9]$

39. $f(x) = -x^2 + x^3, \quad [0, 1]$

40. $f(x) = x^4 - 2x^3 + x^2, \quad [0, 6]$

41. ***Vertical Motion*** The height of an object t seconds after it is dropped from a height of 300 meters is $s(t) = -4.9t^2 + 300$.

(a) Find the average velocity of the object during the first 3 seconds.

(b) Use the Mean Value Theorem to verify that at some time during the first 3 seconds of fall the instantaneous velocity equals the average velocity. Find that time.

42. ***Sales*** A company introduces a new product for which the number of units sold S is

$$S(t) = 200\left(5 - \frac{9}{2 + t}\right)$$

where t is the time in months.

(a) Find the average rate of change of $S(t)$ during the first year.

(b) During what month of the first year does $S'(t)$ equal the average rate of change?

WRITING ABOUT CONCEPTS

43. Let f be continuous on $[a, b]$ and differentiable on (a, b). If there exists c in (a, b) such that $f'(c) = 0$, does it follow that $f(a) = f(b)$? Explain.

44. Let f be continuous on the closed interval $[a, b]$ and differentiable on the open interval (a, b). Also, suppose that $f(a) = f(b)$ and that c is a real number in the interval such that $f'(c) = 0$. Find an interval for the function g over which Rolle's Theorem can be applied, and find the corresponding critical number of g (k is a constant).

(a) $g(x) = f(x) + k$ (b) $g(x) = f(x - k)$

(c) $g(x) = f(kx)$

45. The function

$$f(x) = \begin{cases} 0, & x = 0 \\ 1 - x, & 0 < x \le 1 \end{cases}$$

is differentiable on $(0, 1)$ and satisfies $f(0) = f(1)$. However, its derivative is never zero on $(0, 1)$. Does this contradict Rolle's Theorem? Explain.

46. Can you find a function f such that $f(-2) = -2$, $f(2) = 6$, and $f'(x) < 1$ for all x? Why or why not?

47. ***Speed*** A plane begins its takeoff at 2:00 P.M. on a 2500-mile flight. After 5.5 hours, the plane arrives at its destination. Explain why there are at least two times during the flight when the speed of the plane is 400 miles per hour.

48. ***Temperature*** When an object is removed from a furnace and placed in an environment with a constant temperature of 90°F, its core temperature is 1500°F. Five hours later the core temperature is 390°F. Explain why there must exist a time in the interval when the temperature is decreasing at a rate of 222°F per hour.

49. ***Velocity*** Two bicyclists begin a race at 8:00 A.M. They both finish the race 2 hours and 15 minutes later. Prove that at some time during the race, the bicyclists are traveling at the same velocity.

50. ***Acceleration*** At 9:13 A.M., a sports car is traveling 35 miles per hour. Two minutes later, the car is traveling 85 miles per hour. Prove that at some time during this two-minute interval, the car's acceleration is exactly 1500 miles per hour squared.

51. Consider the function $f(x) = |9 - x^2|$.

(a) Use a graphing utility to graph f and f'.

(b) Is f a continuous function? Is f' a continuous function?

(c) Does Rolle's Theorem apply on the interval $[-1, 1]$? Does it apply on the interval $[2, 4]$? Explain.

(d) Evaluate, if possible, $\lim_{x\to 3^-} f'(x)$ and $\lim_{x\to 3^+} f'(x)$.

CAPSTONE

52. ***Graphical Reasoning*** The figure shows two parts of the graph of a continuous differentiable function f on $[-10, 4]$. The derivative f' is also continuous. To print an enlarged copy of the graph, go to the website *www.mathgraphs.com.*

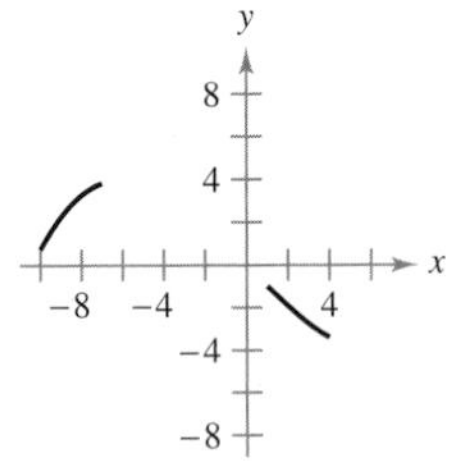

(a) Explain why f must have at least one zero in $[-10, 4]$.

(b) Explain why f' must also have at least one zero in the interval $[-10, 4]$. What are these zeros called?

(c) Make a possible sketch of the function with one zero of f' on the interval $[-10, 4]$.

(d) Make a possible sketch of the function with two zeros of f' on the interval $[-10, 4]$.

(e) Were the conditions of continuity of f and f' necessary to do parts (a) through (d)? Explain.

Think About It **In Exercises 53 and 54, sketch the graph of an arbitrary function f that satisfies the given condition but does not satisfy the conditions of the Mean Value Theorem on the interval $[-5, 5]$.**

53. f is continuous on $[-5, 5]$.

54. f is not continuous on $[-5, 5]$.

55. Determine the values a, b, and c such that the function f satisfies the hypotheses of the Mean Value Theorem on the interval $[0, 3]$.

$$f(x) = \begin{cases} 1, & x = 0 \\ ax + b, & 0 < x \le 1 \\ x^2 + 4x + c, & 1 < x \le 3 \end{cases}$$

56. Determine the values a, b, c, and d such that the function f satisfies the hypotheses of the Mean Value Theorem on the interval $[-1, 2]$.

$$f(x) = \begin{cases} a, & x = -1 \\ 2, & -1 < x \le 0 \\ bx^2 + c, & 0 < x \le 1 \\ dx + 4, & 1 < x \le 2 \end{cases}$$

Differential Equations **In Exercises 57–60, find a function f that has the derivative $f'(x)$ and whose graph passes through the given point. Explain your reasoning.**

57. $f'(x) = 0, \quad (2, 5)$

58. $f'(x) = 4, \quad (0, 1)$

59. $f'(x) = 2x, \quad (1, 0)$

60. $f'(x) = 2x + 3, \quad (1, 0)$

True or False? **In Exercises 61–64, determine whether the statement is true or false. If it is false, explain why or give an example that shows it is false.**

61. The Mean Value Theorem can be applied to $f(x) = 1/x$ on the interval $[-1, 1]$.

62. If the graph of a function has three x-intercepts, then it must have at least two points at which its tangent line is horizontal.

63. If the graph of a polynomial function has three x-intercepts, then it must have at least two points at which its tangent line is horizontal.

64. If $f'(x) = 0$ for all x in the domain of f, then f is a constant function.

65. Prove that if $a > 0$ and n is any positive integer, then the polynomial function $p(x) = x^{2n+1} + ax + b$ cannot have two real roots.

66. Prove that if $f'(x) = 0$ for all x in an interval (a, b), then f is constant on (a, b).

67. Let $p(x) = Ax^2 + Bx + C$. Prove that for any interval $[a, b]$, the value c guaranteed by the Mean Value Theorem is the midpoint of the interval.

68. (a) Let $f(x) = x^2$

and

$g(x) = -x^3 + x^2 + 3x + 2.$

Then

$f(-1) = g(-1)$

and

$f(2) = g(2).$

Show that there is at least one value c in the interval $(-1, 2)$ where the tangent line to f at $(c, f(c))$ is parallel to the tangent line to g at $(c, g(c))$. Identify c.

(b) Let f and g be differentiable functions on $[a, b]$ where $f(a) = g(a)$ and $f(b) = g(b)$. Show that there is at least one value c in the interval (a, b) where the tangent line to f at $(c, f(c))$ is parallel to the tangent line to g at $(c, g(c))$.

69. Prove that if f is differentiable on $(-\infty, \infty)$ and $f'(x) < 1$ for all real numbers, then f has at most one fixed point. A fixed point of a function f is a real number c such that $f(c) = c$.

70. Let $0 < a < b$. Use the Mean Value Theorem to show that

$$\sqrt{b} - \sqrt{a} < \frac{b - a}{2\sqrt{a}}.$$

5.3 Increasing and Decreasing Functions and the First Derivative Test

- **Determine intervals on which a function is increasing or decreasing.**
- **Apply the First Derivative Test to find relative extrema of a function.**

Increasing and Decreasing Functions

In this section, you will learn how derivatives can be used to *classify* relative extrema as either relative minima or relative maxima. First, it is important to review the definitions of increasing and decreasing functions.

DEFINITIONS OF INCREASING AND DECREASING FUNCTIONS

A function f is **increasing** on an interval if for any two numbers x_1 and x_2 in the interval, $x_1 < x_2$ implies $f(x_1) < f(x_2)$.

A function f is **decreasing** on an interval if for any two numbers x_1 and x_2 in the interval, $x_1 < x_2$ implies $f(x_1) > f(x_2)$.

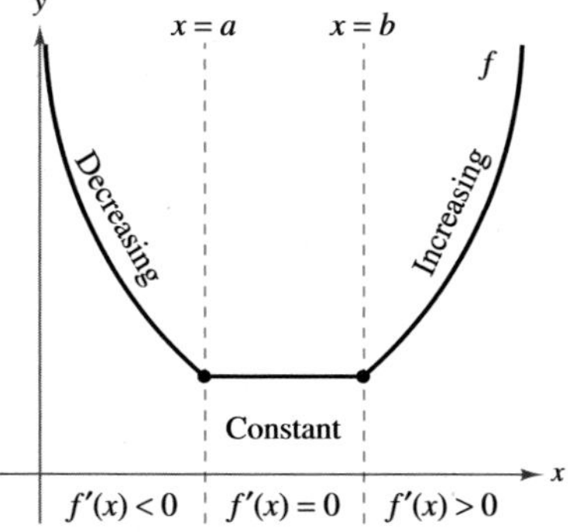

The derivative is related to the slope of a function.
Figure 5.15

A function is increasing if, *as x moves to the right*, its graph moves up, and is decreasing if its graph moves down. For example, the function in Figure 5.15 is decreasing on the interval $(-\infty, a)$, is constant on the interval (a, b), and is increasing on the interval (b, ∞). As shown in Theorem 5.5 below, a positive derivative implies that the function is increasing; a negative derivative implies that the function is decreasing; and a zero derivative on an entire interval implies that the function is constant on that interval.

THEOREM 5.5 TEST FOR INCREASING AND DECREASING FUNCTIONS

Let f be a function that is continuous on the closed interval $[a, b]$ and differentiable on the open interval (a, b).

1. If $f'(x) > 0$ for all x in (a, b), then f is increasing on $[a, b]$.
2. If $f'(x) < 0$ for all x in (a, b), then f is decreasing on $[a, b]$.
3. If $f'(x) = 0$ for all x in (a, b), then f is constant on $[a, b]$.

PROOF To prove the first case, assume that $f'(x) > 0$ for all x in the interval (a, b) and let $x_1 < x_2$ be any two points in the interval. By the Mean Value Theorem, you know that there exists a number c such that $x_1 < c < x_2$, and

$$f'(c) = \frac{f(x_2) - f(x_1)}{x_2 - x_1}.$$

Because $f'(c) > 0$ and $x_2 - x_1 > 0$, you know that

$$f(x_2) - f(x_1) > 0$$

which implies that $f(x_1) < f(x_2)$. So, f is increasing on the interval. The second case has a similar proof (see Exercise 68), and the third case is a consequence of Exercise 66 in Section 5.2. ■

NOTE The conclusions in the first two cases of Theorem 5.5 are valid even if $f'(x) = 0$ at a finite number of x-values in (a, b). ■

EXAMPLE 1 Intervals on Which f Is Increasing or Decreasing

Find the open intervals on which $f(x) = x^3 - \frac{3}{2}x^2$ is increasing or decreasing.

Solution Note that f is differentiable on the entire real number line. To determine the critical numbers of f, set $f'(x)$ equal to zero.

$$f(x) = x^3 - \frac{3}{2}x^2 \quad \text{Write original function.}$$

$$f'(x) = 3x^2 - 3x = 0 \quad \text{Differentiate and set } f'(x) \text{ equal to 0.}$$

$$3(x)(x - 1) = 0 \quad \text{Factor.}$$

$$x = 0, 1 \quad \text{Critical numbers}$$

Because there are no points for which f' does not exist, you can conclude that $x = 0$ and $x = 1$ are the only critical numbers. The table summarizes the testing of the three intervals determined by these two critical numbers.

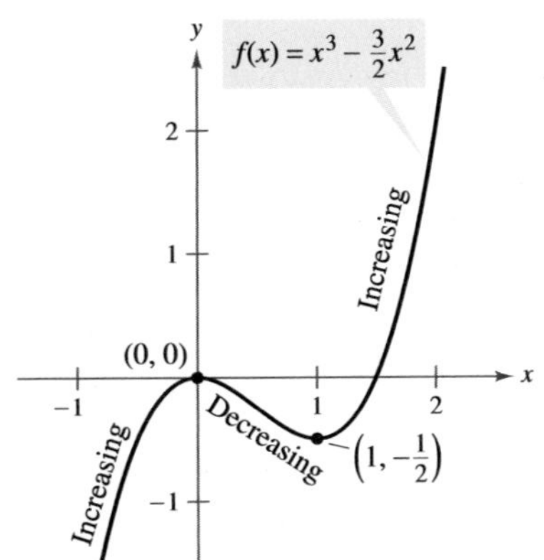

Figure 5.16

Interval	$-\infty < x < 0$	$0 < x < 1$	$1 < x < \infty$
Test Value	$x = -1$	$x = \frac{1}{2}$	$x = 2$
Sign of $f'(x)$	$f'(-1) = 6 > 0$	$f'(\frac{1}{2}) = -\frac{3}{4} < 0$	$f'(2) = 6 > 0$
Conclusion	Increasing	Decreasing	Increasing

So, f is increasing on the intervals $(-\infty, 0)$ and $(1, \infty)$ and decreasing on the interval $(0, 1)$, as shown in Figure 5.16. ■

Example 1 gives you one example of how to find intervals on which a function is increasing or decreasing. The guidelines below summarize the steps followed in that example.

GUIDELINES FOR FINDING INTERVALS ON WHICH A FUNCTION IS INCREASING OR DECREASING

Let f be continuous on the interval (a, b). To find the open intervals on which f is increasing or decreasing, use the following steps.

1. Locate the critical numbers of f in (a, b), and use these numbers to determine test intervals.
2. Determine the sign of $f'(x)$ at one test value in each of the intervals.
3. Use Theorem 5.5 to determine whether f is increasing or decreasing on each interval.

These guidelines are also valid if the interval (a, b) is replaced by an interval of the form $(-\infty, b)$, (a, ∞), or $(-\infty, \infty)$.

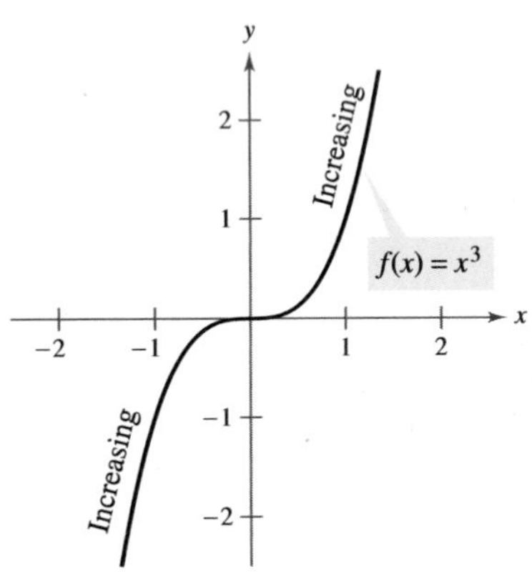

(a) Strictly monotonic function

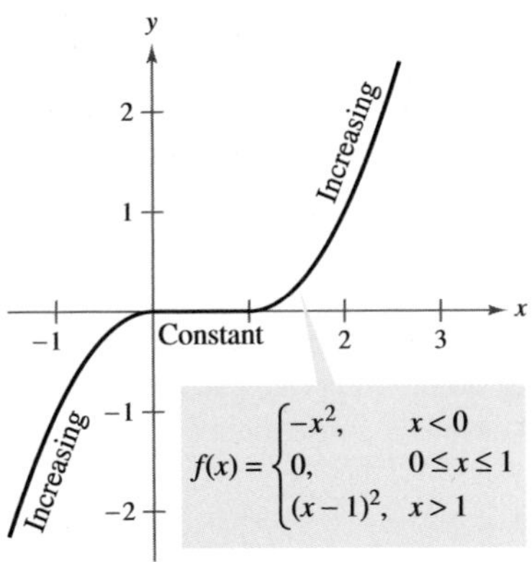

(b) Not strictly monotonic

Figure 5.17

A function is **strictly monotonic** on an interval if it is either increasing on the entire interval or decreasing on the entire interval. For instance, the function $f(x) = x^3$ is strictly monotonic on the entire real number line because it is increasing on the entire real number line, as shown in Figure 5.17(a). The function shown in Figure 5.17(b) is not strictly monotonic on the entire real number line because it is constant on the interval $[0, 1]$.

The First Derivative Test

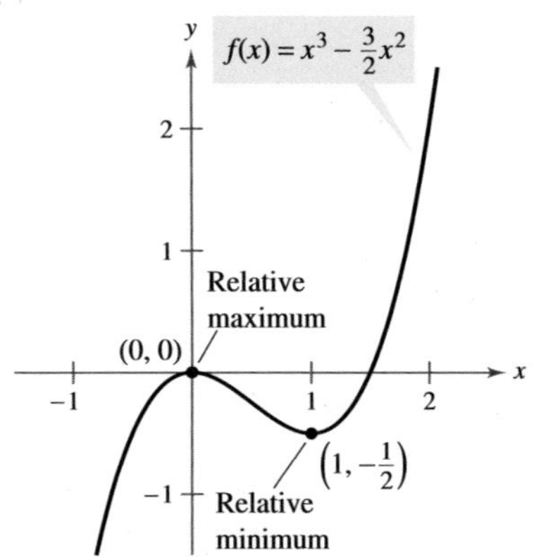

Relative extrema of f
Figure 5.18

After you have determined the intervals on which a function is increasing or decreasing, it is not difficult to locate the relative extrema of the function. For instance, in Figure 5.18 (from Example 1), the function

$$f(x) = x^3 - \frac{3}{2}x^2$$

has a relative maximum at the point $(0, 0)$ because f is increasing immediately to the left of $x = 0$ and decreasing immediately to the right of $x = 0$. Similarly, f has a relative minimum at the point $\left(1, -\frac{1}{2}\right)$ because f is decreasing immediately to the left of $x = 1$ and increasing immediately to the right of $x = 1$. The following theorem, called the First Derivative Test, makes this more explicit.

THEOREM 5.6 THE FIRST DERIVATIVE TEST

Let c be a critical number of a function f that is continuous on an open interval I containing c. If f is differentiable on the interval, except possibly at c, then $f(c)$ can be classified as follows.

1. If $f'(x)$ changes from negative to positive at c, then f has a *relative minimum* at $(c, f(c))$.
2. If $f'(x)$ changes from positive to negative at c, then f has a *relative maximum* at $(c, f(c))$.
3. If $f'(x)$ is positive on both sides of c or negative on both sides of c, then $f(c)$ is neither a relative minimum nor a relative maximum.

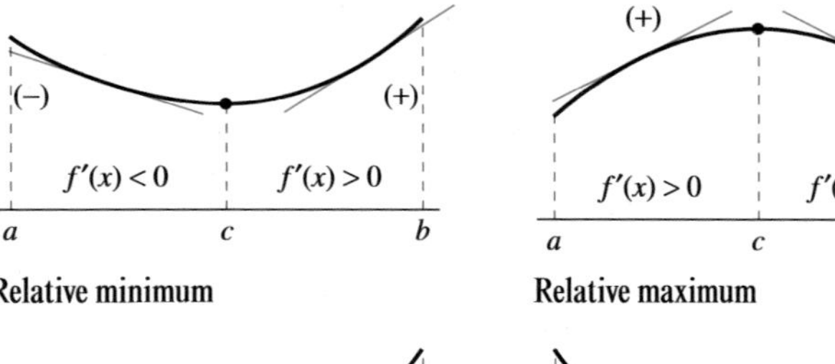

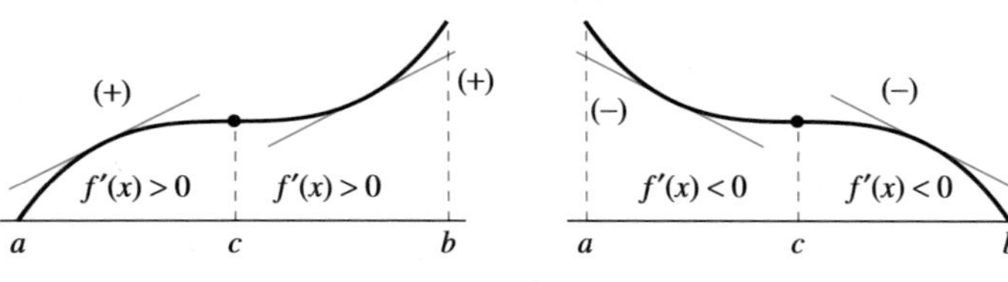

Neither relative minimum nor relative maximum

PROOF Assume that $f'(x)$ changes from negative to positive at c. Then there exist a and b in I such that

$$f'(x) < 0 \text{ for all } x \text{ in } (a, c)$$

and

$$f'(x) > 0 \text{ for all } x \text{ in } (c, b).$$

By Theorem 5.5, f is decreasing on $[a, c]$ and increasing on $[c, b]$. So, $f(c)$ is a minimum of f on the open interval (a, b) and, consequently, a relative minimum of f. This proves the first case of the theorem. The second case can be proved in a similar way (see Exercise 69). ■

EXAMPLE 2 Applying the First Derivative Test

STUDY TIP The only places at which a continuous function can change from increasing to decreasing or vice versa are at its critical numbers.

Find the relative extrema of the function given by

$$f(x) = 2x^3 - 3x^2 - 36x + 14.$$

Solution Note that f is continuous on the entire real line. To determine the critical numbers of f, set $f'(x)$ equal to 0.

$$f'(x) = 6x^2 - 6x - 36 = 0 \quad \text{Set } f'(x) \text{ equal to 0.}$$
$$6(x^2 - x - 6) = 0$$
$$6(x + 2)(x - 3) = 0$$
$$x = -2, 3 \quad \text{Critical numbers}$$

Because there are no points for which f' does not exist, you can conclude that $x = -2$ and $x = 3$ are the only critical numbers. The table summarizes the testing of the three intervals determined by these two critical numbers.

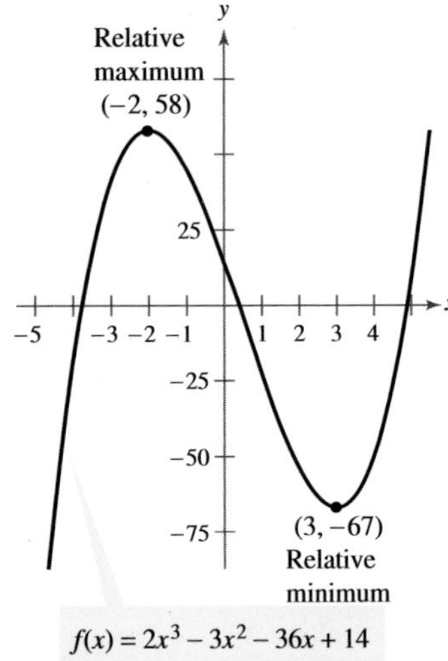

Figure 5.19

Interval	$-\infty < x < -2$	$-2 < x < 3$	$3 < x < \infty$
Test Value	$x = -3$	$x = 0$	$x = 4$
Sign of $f'(x)$	$f'(-3) > 0$	$f'(0) < 0$	$f'(4) > 0$
Conclusion	Increasing	Decreasing	Increasing

By applying the First Derivative Test, you can conclude that f has a relative minimum at the point where

$$x = 3 \quad x\text{-value where relative minimum occurs}$$

and a relative maximum at the point where

$$x = -2 \quad x\text{-value where relative maximum occurs}$$

as shown in Figure 5.19. ■

NOTE A relative minimum occurs when f changes from decreasing to increasing, and a relative maximum occurs when f changes from increasing to decreasing. ■

EXPLORATION

Comparing Graphical and Analytic Approaches From Section 5.2, you know that, *by itself*, a graphing utility can give misleading information about the relative extrema of a graph. *Used in conjunction with an analytic approach*, however, a graphing utility can provide a good way to reinforce your conclusions. Use a graphing utility to graph the function in Example 2. Then use the *zoom* and *trace* features to estimate the relative extrema. How close are your graphical approximations?

Note that in Examples 1 and 2, the given functions are differentiable on the entire real number line. For such functions, the only critical numbers are those for which $f'(x) = 0$. Example 3 concerns a function that has two types of critical numbers—those for which $f'(x) = 0$ and those for which f is not differentiable.

EXAMPLE 3 Applying the First Derivative Test

Find the relative extrema of

$$f(x) = (x^2 - 4)^{2/3}.$$

Solution Begin by noting that f is continuous on the entire real number line. The derivative of f

$$f'(x) = \frac{2}{3}(x^2 - 4)^{-1/3}(2x) \qquad \text{General Power Rule}$$

$$= \frac{4x}{3(x^2 - 4)^{1/3}} \qquad \text{Simplify.}$$

is 0 when $x = 0$ and does not exist when $x = \pm 2$. So, the critical numbers are $x = -2$, $x = 0$, and $x = 2$. The table summarizes the testing of the four intervals determined by these three critical numbers.

Interval	$-\infty < x < -2$	$-2 < x < 0$	$0 < x < 2$	$2 < x < \infty$
Test Value	$x = -3$	$x = -1$	$x = 1$	$x = 3$
Sign of $f'(x)$	$f'(-3) < 0$	$f'(-1) > 0$	$f'(1) < 0$	$f'(3) > 0$
Conclusion	Decreasing	Increasing	Decreasing	Increasing

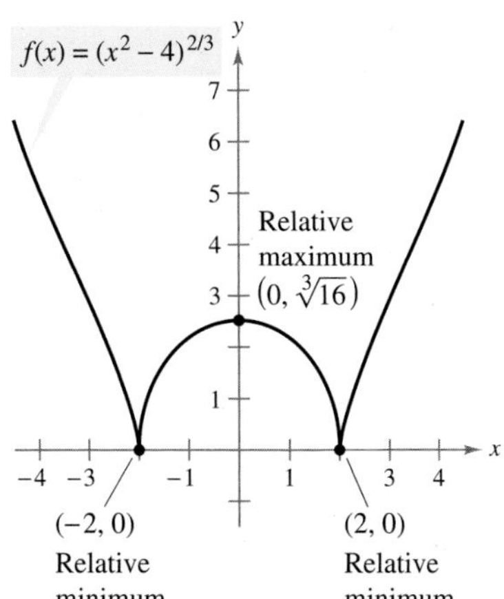

You can apply the First Derivative Test to find relative extrema.
Figure 5.20

By applying the First Derivative Test, you can conclude that f has a relative minimum at the point $(-2, 0)$, a relative maximum at the point $(0, \sqrt[3]{16})$, and another relative minimum at the point $(2, 0)$, as shown in Figure 5.20. ■

TECHNOLOGY PITFALL When using a graphing utility to graph a function involving radicals or rational exponents, be sure you understand the way the utility evaluates radical expressions. For instance, even though

$$f(x) = (x^2 - 4)^{2/3}$$

and

$$g(x) = [(x^2 - 4)^2]^{1/3}$$

are the same algebraically, some graphing utilities distinguish between these two functions. Which of the graphs shown in Figure 5.21 is incorrect? Why did the graphing utility produce an incorrect graph?

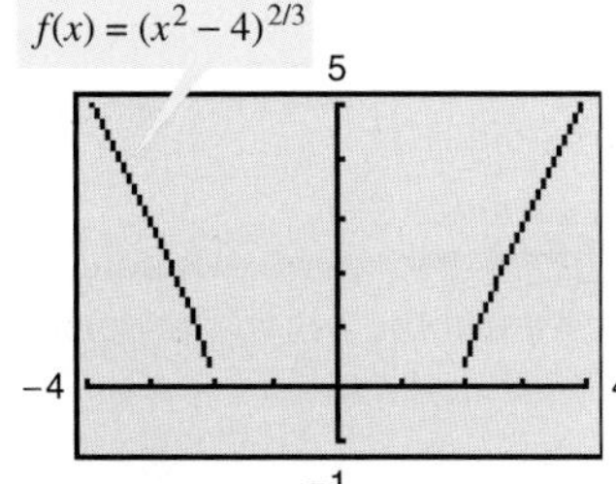

Which graph is incorrect?
Figure 5.21

When using the First Derivative Test, be sure to consider the domain of the function. For instance, in the next example, the function

$$f(x) = \frac{x^4 + 1}{x^2}$$

is not defined when $x = 0$. This x-value must be used with the critical numbers to determine the test intervals.

EXAMPLE 4 Applying the First Derivative Test

Find the relative extrema of $f(x) = \dfrac{x^4 + 1}{x^2}$.

Solution

$f(x) = x^2 + x^{-2}$ — Rewrite original function.

$f'(x) = 2x - 2x^{-3}$ — Differentiate.

$= 2x - \dfrac{2}{x^3}$ — Rewrite with positive exponent.

$= \dfrac{2(x^4 - 1)}{x^3}$ — Simplify.

$= \dfrac{2(x^2 + 1)(x - 1)(x + 1)}{x^3}$ — Factor.

So, $f'(x)$ is zero at $x = \pm 1$. Moreover, because $x = 0$ is not in the domain of f, you should use this x-value along with the critical numbers to determine the test intervals.

$x = \pm 1$ — Critical numbers, $f'(\pm 1) = 0$

$x = 0$ — 0 is not in the domain of f.

The table summarizes the testing of the four intervals determined by these three x-values.

Interval	$-\infty < x < -1$	$-1 < x < 0$	$0 < x < 1$	$1 < x < \infty$
Test Value	$x = -2$	$x = -\frac{1}{2}$	$x = \frac{1}{2}$	$x = 2$
Sign of $f'(x)$	$f'(-2) < 0$	$f'(-\frac{1}{2}) > 0$	$f'(\frac{1}{2}) < 0$	$f'(2) > 0$
Conclusion	Decreasing	Increasing	Decreasing	Increasing

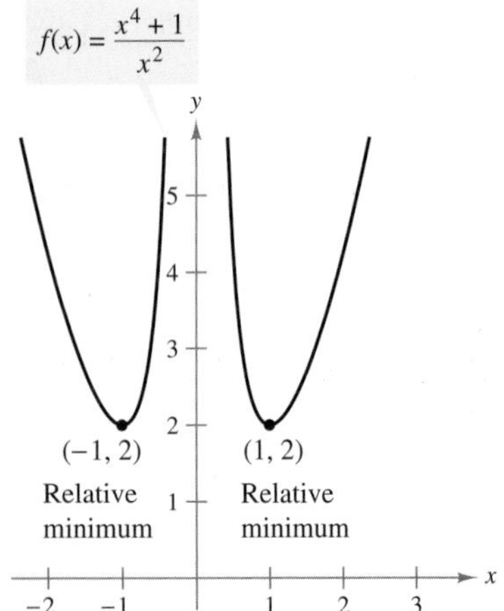

x-values that are not in the domain of f, as well as critical numbers, determine test intervals for f'.
Figure 5.22

By applying the First Derivative Test, you can conclude that f has one relative minimum at the point $(-1, 2)$ and another at the point $(1, 2)$, as shown in Figure 5.22. ■

TECHNOLOGY The most difficult step in applying the First Derivative Test is finding the values for which the derivative is equal to 0. For instance, the values of x for which the derivative of

$$f(x) = \frac{x^4 + 1}{x^2 + 1}$$

is equal to zero are $x = 0$ and $x = \pm\sqrt{\sqrt{2} - 1}$. If you have access to technology that can perform symbolic differentiation and solve equations, use it to apply the First Derivative Test to this function.

5.3 Exercises

See www.CalcChat.com for worked-out solutions to odd-numbered exercises.

In Exercises 1 and 2, use the graph of f to find (a) the largest open interval on which f is increasing, and (b) the largest open interval on which f is decreasing.

1.

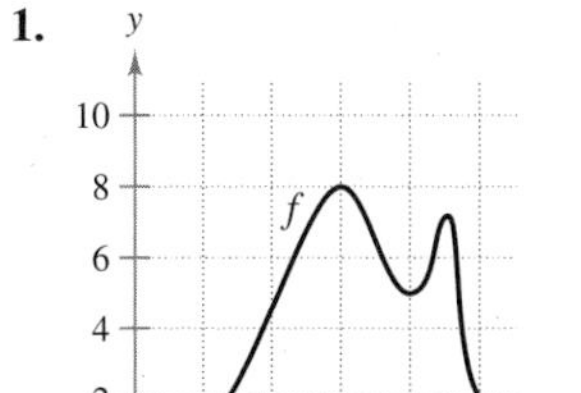

2.

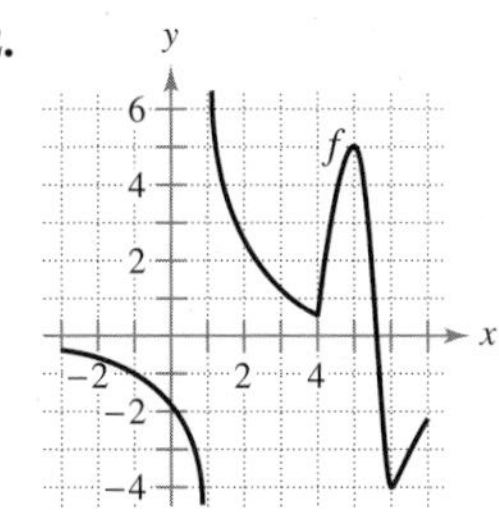

In Exercises 3–8, use the graph to estimate the open intervals on which the function is increasing or decreasing. Then find the open intervals analytically.

3. $f(x) = x^2 - 6x + 8$

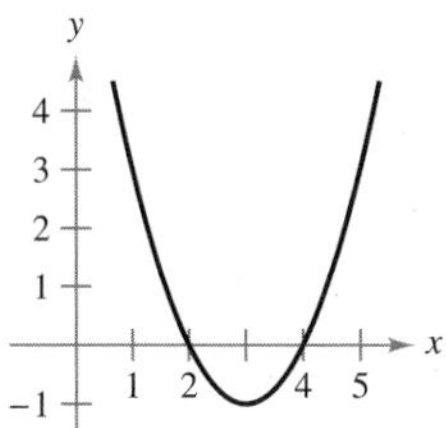

4. $y = -(x + 1)^2$

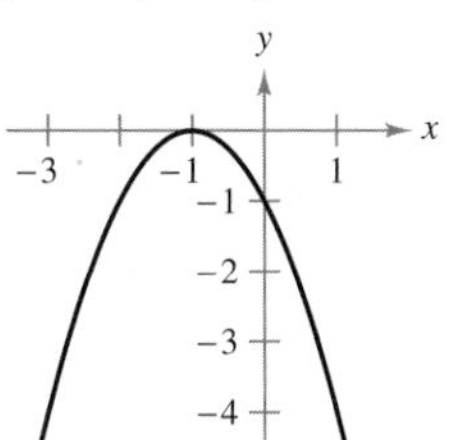

5. $y = \dfrac{x^3}{4} - 3x$

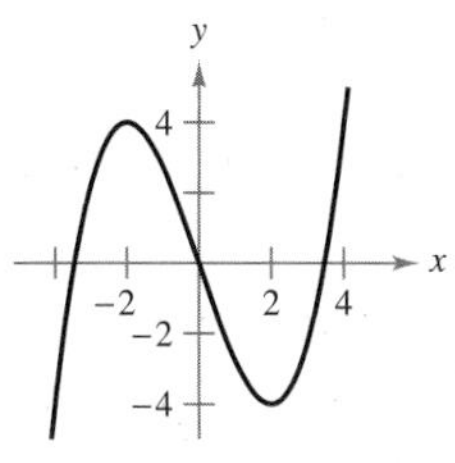

6. $f(x) = x^4 - 2x^2$

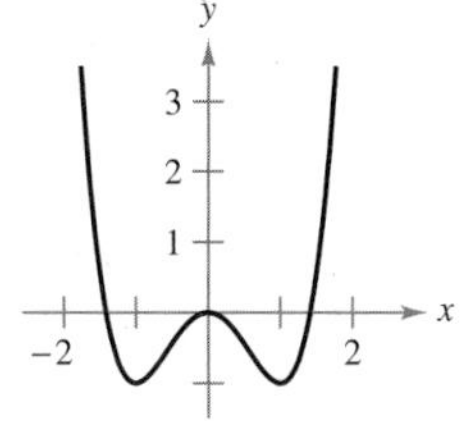

7. $f(x) = \dfrac{1}{(x + 1)^2}$

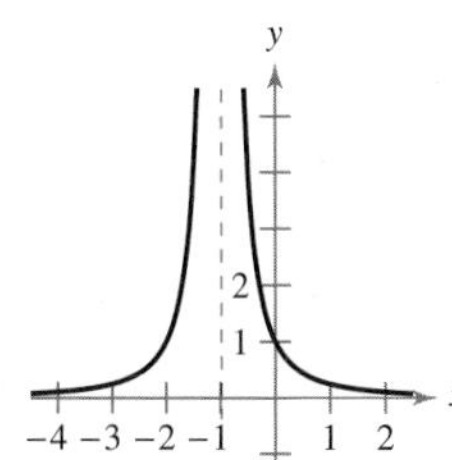

8. $y = \dfrac{x^2}{2x - 1}$

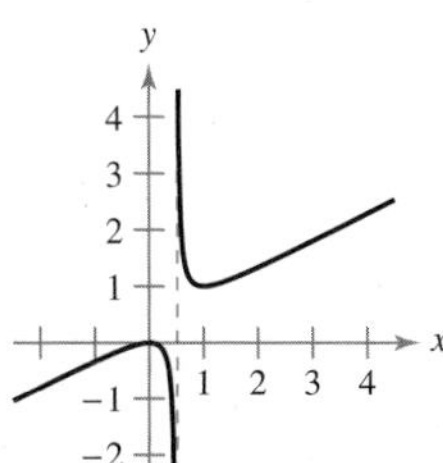

In Exercises 9–12, identify the open intervals on which the function is increasing or decreasing.

9. $g(x) = x^2 - 2x - 8$

10. $h(x) = 27x - x^3$

11. $y = x\sqrt{16 - x^2}$

12. $y = x + \dfrac{4}{x}$

In Exercises 13–38, (a) find the critical numbers of f (if any), (b) find the open interval(s) on which the function is increasing or decreasing, (c) apply the First Derivative Test to identify all relative extrema, and (d) use a graphing utility to confirm your results.

13. $f(x) = x^2 - 4x$

14. $f(x) = x^2 + 6x + 10$

15. $f(x) = -2x^2 + 4x + 3$

16. $f(x) = -(x^2 + 8x + 12)$

17. $f(x) = 2x^3 + 3x^2 - 12x$

18. $f(x) = x^3 - 6x^2 + 15$

19. $f(x) = (x - 1)^2(x + 3)$

20. $f(x) = (x + 2)^2(x - 1)$

21. $f(x) = \dfrac{x^5 - 5x}{5}$

22. $f(x) = x^4 - 32x + 4$

23. $f(x) = x^{1/3} + 1$

24. $f(x) = x^{2/3} - 4$

25. $f(x) = (x + 2)^{2/3}$

26. $f(x) = (x - 3)^{1/3}$

27. $f(x) = 5 - |x - 5|$

28. $f(x) = |x + 3| - 1$

29. $f(x) = 2x + \dfrac{1}{x}$

30. $f(x) = \dfrac{x}{x + 3}$

31. $f(x) = \dfrac{x^2}{x^2 - 9}$

32. $f(x) = \dfrac{x + 4}{x^2}$

33. $f(x) = \dfrac{x^2 - 2x + 1}{x + 1}$

34. $f(x) = \dfrac{x^2 - 3x - 4}{x - 2}$

35. $f(x) = \begin{cases} 4 - x^2, & x \le 0 \\ -2x, & x > 0 \end{cases}$

36. $f(x) = \begin{cases} 2x + 1, & x \le -1 \\ x^2 - 2, & x > -1 \end{cases}$

37. $f(x) = \begin{cases} 3x + 1, & x \le 1 \\ 5 - x^2, & x > 1 \end{cases}$

38. $f(x) = \begin{cases} -x^3 + 1, & x \le 0 \\ -x^2 + 2x, & x > 0 \end{cases}$

In Exercises 39 and 40, (a) use a computer algebra system to differentiate the function, (b) sketch the graphs of f and f' on the same set of coordinate axes over the given interval, (c) find the critical numbers of f in the open interval, and (d) find the interval(s) on which f' is positive and the interval(s) on which it is negative. Compare the behavior of f and the sign of f'.

39. $f(x) = 2x\sqrt{9 - x^2}, \quad [-3, 3]$

40. $f(x) = 10\left(5 - \sqrt{x^2 - 3x + 16}\right), \quad [0, 5]$

In Exercises 41 and 42, use symmetry, extrema, and zeros to sketch the graph of *f*. How do the functions *f* and *g* differ?

41. $f(x) = \dfrac{x^5 - 4x^3 + 3x}{x^2 - 1}$, $g(x) = x(x^2 - 3)$

42. $f(x) = \dfrac{x^6 - 5x^4 + 6x^2}{x^2 - 2}$, $g(x) = x^2(x^2 - 3)$

Think About It **In Exercises 43–48, the graph of *f* is shown in the figure. Sketch a graph of the derivative of *f*. To print an enlarged copy of the graph, go to the website *www.mathgraphs.com*.**

43.

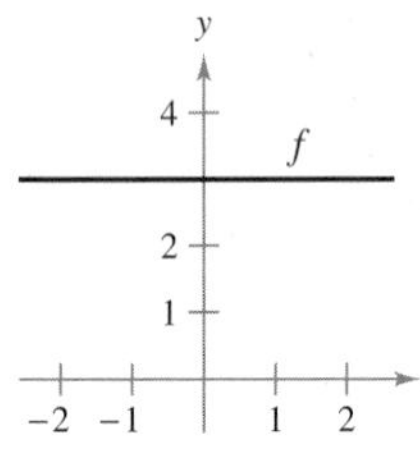

44.

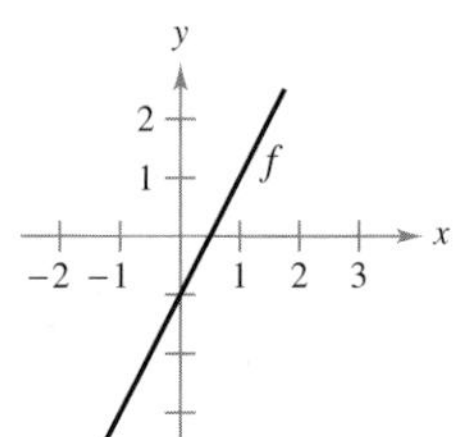

45.

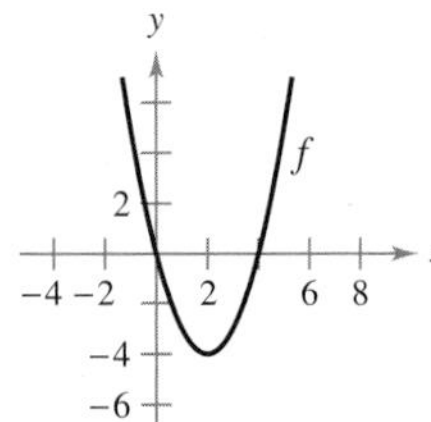

46.

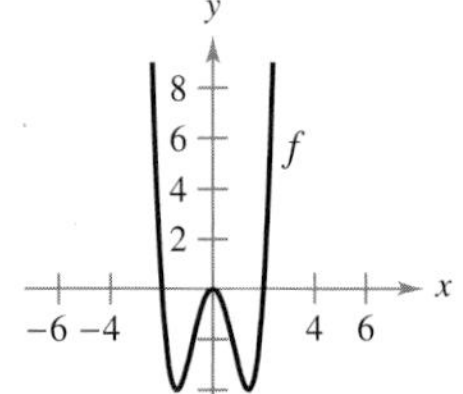

47.

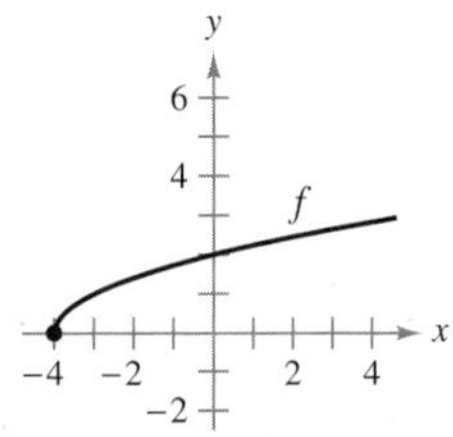

48.

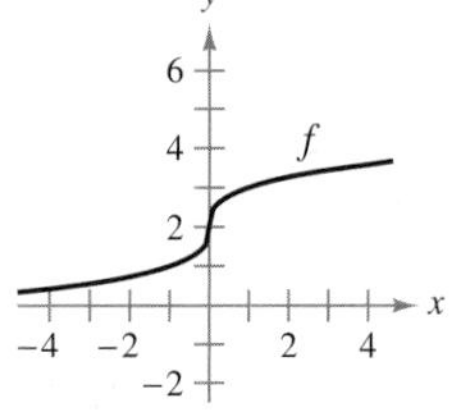

In Exercises 49–52, use the graph of f' to (a) identify the interval(s) on which *f* is increasing or decreasing, and (b) estimate the value(s) of *x* at which *f* has a relative maximum or minimum.

49.

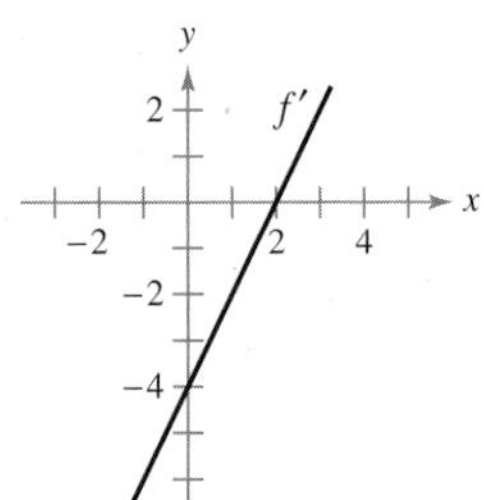

50.

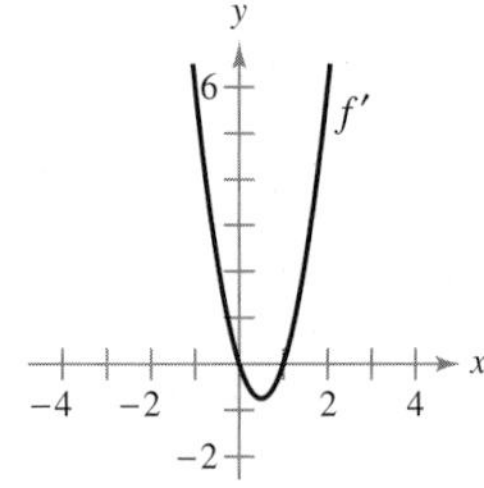

51.

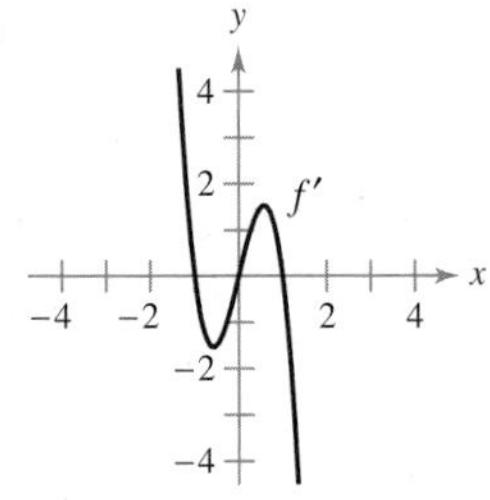

52.

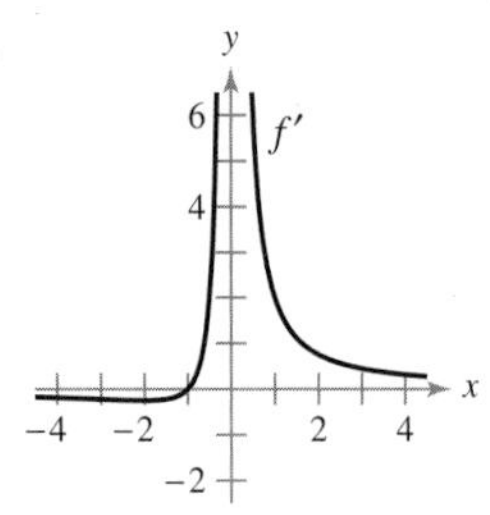

WRITING ABOUT CONCEPTS

In Exercises 53–58, assume that *f* is differentiable for all *x*. The signs of f' are as follows.

$f'(x) > 0$ **on** $(-\infty, -4)$

$f'(x) < 0$ **on** $(-4, 6)$

$f'(x) > 0$ **on** $(6, \infty)$

Supply the appropriate inequality sign for the indicated value of *c*.

	Function	*Sign of $g'(c)$*
53.	$g(x) = f(x) + 5$	$g'(0)$ ▢ 0
54.	$g(x) = 3f(x) - 3$	$g'(-5)$ ▢ 0
55.	$g(x) = -f(x)$	$g'(-6)$ ▢ 0
56.	$g(x) = -f(x)$	$g'(0)$ ▢ 0
57.	$g(x) = f(x - 10)$	$g'(0)$ ▢ 0
58.	$g(x) = f(x - 10)$	$g'(8)$ ▢ 0

59. Sketch the graph of the arbitrary function *f* such that

$$f'(x)\begin{cases} > 0, & x < 4 \\ \text{undefined}, & x = 4. \\ < 0, & x > 4 \end{cases}$$

CAPSTONE

60. A differentiable function *f* has one critical number at $x = 5$. Identify the relative extrema of *f* at the critical number if $f'(4) = -2.5$ and $f'(6) = 3$.

61. ***Think About It*** The function *f* is differentiable on the interval $[-1, 1]$. The table shows the values of f' for selected values of *x*. Sketch the graph of *f*, approximate the critical numbers, and identify the relative extrema.

x	-1	-0.75	-0.50	-0.25	0
$f'(x)$	-10	-3.2	-0.5	0.8	5.6

x	0.25	0.50	0.75	1
$f'(x)$	3.6	-0.2	-6.7	-20.1

62. ***Profit*** The profit *P* (in dollars) made by a fast-food restaurant selling *x* hamburgers is

$$P = 2.44x - \frac{x^2}{20{,}000} - 5000, \quad 0 \le x \le 35{,}000.$$

Find the open intervals on which *P* is increasing or decreasing.

63. ***Trachea Contraction*** Coughing forces the trachea (windpipe) to contract, which affects the velocity v of the air passing through the trachea. The velocity of the air during coughing is $v = k(R - r)r^2$, $0 \le r < R$, where k is a constant, R is the normal radius of the trachea, and r is the radius during coughing. What radius will produce the maximum air velocity?

64. ***Numerical, Graphical, and Analytic Analysis*** The concentration C of a chemical in the bloodstream t hours after injection into muscle tissue is

$$C(t) = \frac{3t}{27 + t^3}, \quad t \ge 0.$$

(a) Complete the table and use it to approximate the time when the concentration is greatest.

t	0	0.5	1	1.5	2	2.5	3
$C(t)$							

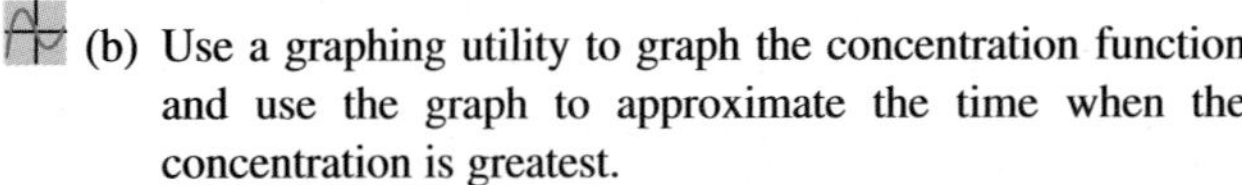

(b) Use a graphing utility to graph the concentration function and use the graph to approximate the time when the concentration is greatest.

(c) Use calculus to determine analytically the time when the concentration is greatest.

65. ***Inventory Cost*** The cost of inventory depends on the ordering and storage costs according to the inventory model

$$C = \left(\frac{Q}{x}\right)s + \left(\frac{x}{2}\right)r.$$

Determine the order size that will minimize the cost, assuming that sales occur at a constant rate, Q is the number of units sold per year, r is the cost of storing one unit for 1 year, s is the cost of placing an order, and x is the number of units per order.

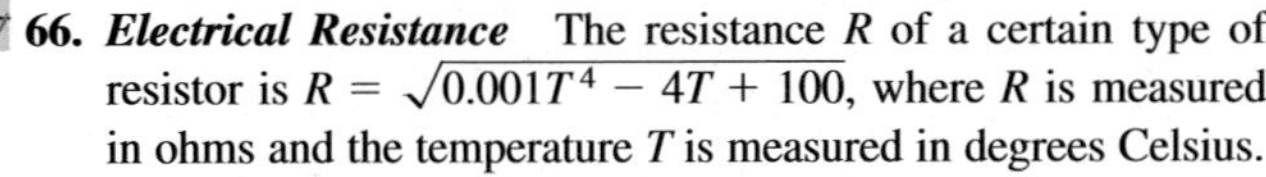

66. ***Electrical Resistance*** The resistance R of a certain type of resistor is $R = \sqrt{0.001T^4 - 4T + 100}$, where R is measured in ohms and the temperature T is measured in degrees Celsius.

(a) Use a computer algebra system to find dR/dT and the critical number of the function. Determine the minimum resistance for this type of resistor.

(b) Use a graphing utility to graph the function R and use the graph to approximate the minimum resistance for this type of resistor.

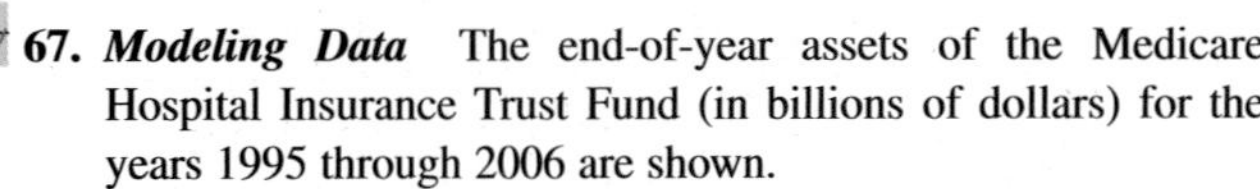

67. ***Modeling Data*** The end-of-year assets of the Medicare Hospital Insurance Trust Fund (in billions of dollars) for the years 1995 through 2006 are shown.

1995: 130.3; 1996: 124.9; 1997: 115.6; 1998: 120.4;

1999: 141.4; 2000: 177.5; 2001: 208.7; 2002: 234.8;

2003: 256.0; 2004: 269.3; 2005: 285.8; 2006: 305.4

(Source: U.S. Centers for Medicare and Medicaid Services)

(a) Use the regression capabilities of a graphing utility to find a model of the form $M = at^4 + bt^3 + ct^2 + dt + e$ for the data. (Let $t = 5$ represent 1995.)

(b) Use a graphing utility to plot the data and graph the model.

(c) Find the minimum value of the model and compare the result with the actual data.

68. Prove the second case of Theorem 5.5.

69. Prove the second case of Theorem 5.6.

70. Let $x > 0$ and $n > 1$ be real numbers. Prove that $(1 + x)^n > 1 + nx$.

71. Use the definitions of increasing and decreasing functions to prove that $f(x) = x^3$ is increasing on $(-\infty, \infty)$.

72. Use the definitions of increasing and decreasing functions to prove that $f(x) = 1/x$ is decreasing on $(0, \infty)$.

Motion Along a Line **In Exercises 73–76, the function $s(t)$ describes the motion of a particle along a line. For each function, (a) find the velocity function of the particle at any time $t \ge 0$, (b) identify the time interval(s) in which the particle is moving in a positive direction, (c) identify the time interval(s) in which the particle is moving in a negative direction, and (d) identify the time(s) at which the particle changes direction.**

73. $s(t) = 6t - t^2$

74. $s(t) = t^2 - 7t + 10$

75. $s(t) = t^3 - 5t^2 + 4t$

76. $s(t) = t^3 - 20t^2 + 128t - 280$

Creating Polynomial Functions **In Exercises 77–80, find a polynomial function**

$$f(x) = a_n x^n + a_{n-1}x^{n-1} + \cdots + a_2x^2 + a_1x + a_0$$

that has only the specified extrema. (a) Determine the minimum degree of the function and give the criteria you used in determining the degree. (b) Using the fact that the coordinates of the extrema are solution points of the function, and that the x-coordinates are critical numbers, determine a system of linear equations whose solution yields the coefficients of the required function. (c) Use a graphing utility to solve the system of equations and determine the function. (d) Use a graphing utility to confirm your result graphically.

77. Relative minimum: $(0, 0)$; Relative maximum: $(2, 2)$

78. Relative minimum: $(0, 0)$; Relative maximum: $(4, 1000)$

79. Relative minima: $(0, 0)$, $(4, 0)$; Relative maximum: $(2, 4)$

80. Relative minimum: $(1, 2)$; Relative maxima: $(-1, 4)$, $(3, 4)$

True or False? **In Exercises 81–86, determine whether the statement is true or false. If it is false, explain why or give an example that shows it is false.**

81. The sum of two increasing functions is increasing.

82. The product of two increasing functions is increasing.

83. Every nth-degree polynomial has $(n - 1)$ critical numbers.

84. An nth-degree polynomial has at most $(n - 1)$ critical numbers.

85. There is a relative maximum or minimum at each critical number.

86. The relative maxima of the function f are $f(1) = 4$ and $f(3) = 10$. So, f has at least one minimum for some x in the interval $(1, 3)$.

5.4 Concavity and the Second Derivative Test

- **Determine intervals on which a function is concave upward or concave downward.**
- **Find any points of inflection of the graph of a function.**
- **Apply the Second Derivative Test to find relative extrema of a function.**

Concavity

You have already seen that locating the intervals in which a function f increases or decreases helps to describe its graph. In this section, you will see how locating the intervals in which f' increases or decreases can be used to determine where the graph of f is *curving upward* or *curving downward.*

DEFINITION OF CONCAVITY

Let f be differentiable on an open interval I. The graph of f is **concave upward** on I if f' is increasing on the interval and **concave downward** on I if f' is decreasing on the interval.

The following graphical interpretation of concavity is useful. (See Appendix A for a proof of these results.)

1. Let f be differentiable on an open interval I. If the graph of f is concave *upward* on I, then the graph of f lies *above* all of its tangent lines on I. [See Figure 5.23(a).]
2. Let f be differentiable on an open interval I. If the graph of f is concave *downward* on I, then the graph of f lies *below* all of its tangent lines on I. [See Figure 5.23(b).]

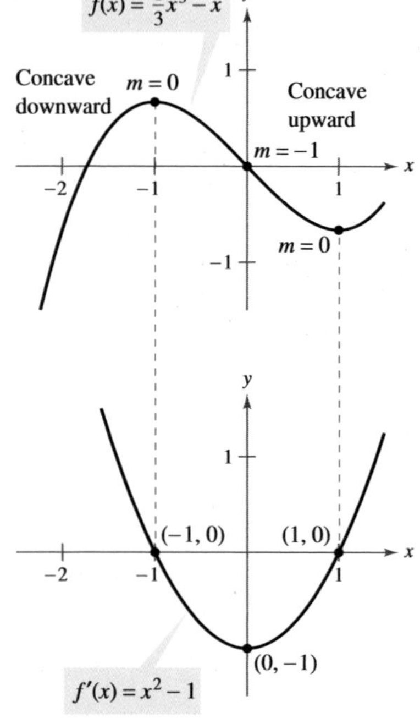

The concavity of f is related to the slope of the derivative.
Figure 5.24

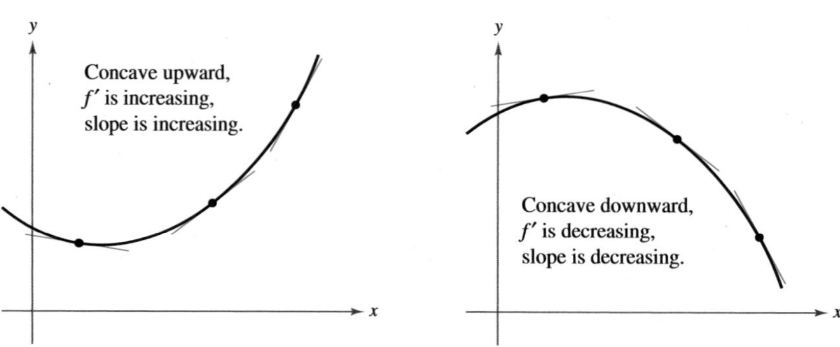

(a) The graph of f lies above its tangent lines. **(b)** The graph of f lies below its tangent lines.
Figure 5.23

To find the open intervals on which the graph of a function f is concave upward or concave downward, you need to find the intervals on which f' is increasing or decreasing. For instance, the graph of

$$f(x) = \tfrac{1}{3}x^3 - x$$

is concave downward on the open interval $(-\infty, 0)$ because $f'(x) = x^2 - 1$ is decreasing there. (See Figure 5.24.) Similarly, the graph of f is concave upward on the interval $(0, \infty)$ because f' is increasing on $(0, \infty)$.

STUDY TIP Theorem 5.7 is a parallel of Theorem 5.5. This means that the second derivative determines the concavity of the graph in the same way that the first derivative determines whether the graph is increasing or decreasing. Moreover, the only places that the concavity of the graph may change is at the values for which $f''(x)$ is zero or is undefined.

The following theorem shows how to use the *second* derivative of a function f to determine intervals on which the graph of f is concave upward or concave downward. A proof of this theorem (see Appendix A) follows directly from Theorem 5.5 and the definition of concavity.

THEOREM 5.7 TEST FOR CONCAVITY

Let f be a function whose second derivative exists on an open interval I.

1. If $f''(x) > 0$ for all x in I, then the graph of f is concave upward on I.
2. If $f''(x) < 0$ for all x in I, then the graph of f is concave downward on I.

NOTE A third case of Theorem 5.7 could be that if $f''(x) = 0$ for all x in I, then f is linear. Note, however, that concavity is not defined for a line. In other words, a straight line is neither concave upward nor concave downward.

To apply Theorem 5.7, locate the x-values at which $f''(x) = 0$ or f'' does not exist. Second, use these x-values to determine test intervals. Finally, test the sign of $f''(x)$ in each of the test intervals.

EXAMPLE 1 Determining Concavity

Determine the open intervals on which the graph of

$$f(x) = \frac{6}{x^2 + 3}$$

is concave upward or downward.

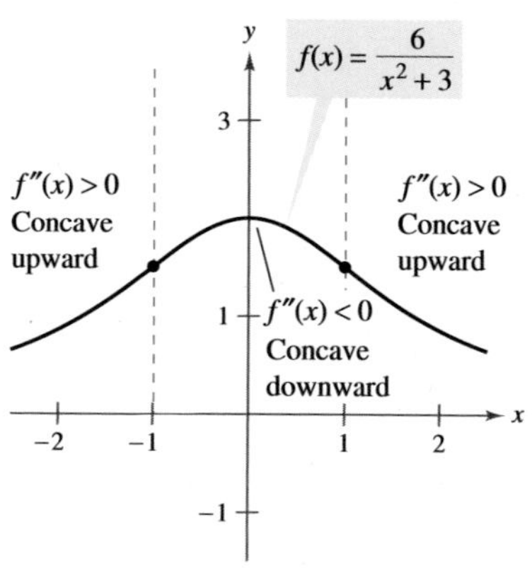

From the sign of f'' you can determine the concavity of the graph of f.
Figure 5.25

Solution Begin by observing that f is continuous on the entire real line. Next, find the second derivative of f.

$$f(x) = 6(x^2 + 3)^{-1} \qquad \text{Rewrite original function.}$$

$$f'(x) = (-6)(x^2 + 3)^{-2}(2x) \qquad \text{Differentiate.}$$

$$= \frac{-12x}{(x^2 + 3)^2} \qquad \text{First derivative}$$

$$f''(x) = \frac{(x^2 + 3)^2(-12) - (-12x)(2)(x^2 + 3)(2x)}{(x^2 + 3)^4} \qquad \text{Differentiate.}$$

$$= \frac{36(x^2 - 1)}{(x^2 + 3)^3} \qquad \text{Second derivative}$$

Because $f''(x) = 0$ when $x = \pm 1$ and f is differentiable on the entire real line, you should test f'' in the intervals $(-\infty, -1)$, $(-1, 1)$, and $(1, \infty)$. The results are shown in the table and in Figure 5.25.

Interval	$-\infty < x < -1$	$-1 < x < 1$	$1 < x < \infty$
Test Value	$x = -2$	$x = 0$	$x = 2$
Sign of $f''(x)$	$f''(-2) > 0$	$f''(0) < 0$	$f''(2) > 0$
Conclusion	Concave upward	Concave downward	Concave upward

■

The function given in Example 1 is continuous on the entire real line. If there are x-values at which the function is not continuous, these values should be used, along with the points at which $f''(x) = 0$ or $f''(x)$ does not exist, to form the test intervals.

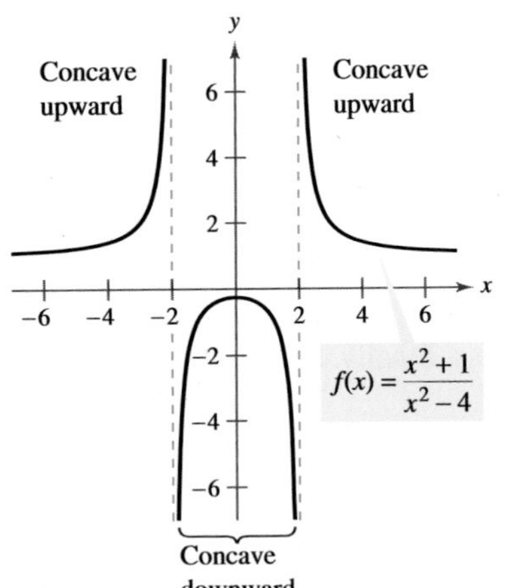

Figure 5.26

EXAMPLE 2 Determining Concavity

Determine the open intervals on which the graph of $f(x) = \frac{x^2 + 1}{x^2 - 4}$ is concave upward or concave downward.

Solution Differentiate the function twice.

$$f(x) = \frac{x^2 + 1}{x^2 - 4} \qquad \text{Write original function.}$$

$$f'(x) = \frac{(x^2 - 4)(2x) - (x^2 + 1)(2x)}{(x^2 - 4)^2} \qquad \text{Differentiate.}$$

$$= \frac{-10x}{(x^2 - 4)^2} \qquad \text{First derivative}$$

$$f''(x) = \frac{(x^2 - 4)^2(-10) - (-10x)(2)(x^2 - 4)(2x)}{(x^2 - 4)^4} \qquad \text{Differentiate.}$$

$$= \frac{10(3x^2 + 4)}{(x^2 - 4)^3} \qquad \text{Second derivative}$$

There are no points at which $f''(x) = 0$, but at $x = \pm 2$ the function f is not continuous, so test for concavity in the intervals $(-\infty, -2)$, $(-2, 2)$, and $(2, \infty)$, as shown in the table. The graph of f is shown in Figure 5.26.

Interval	$-\infty < x < -2$	$-2 < x < 2$	$2 < x < \infty$
Test Value	$x = -3$	$x = 0$	$x = 3$
Sign of $f''(x)$	$f''(-3) > 0$	$f''(0) < 0$	$f''(3) > 0$
Conclusion	Concave upward	Concave downward	Concave upward

■

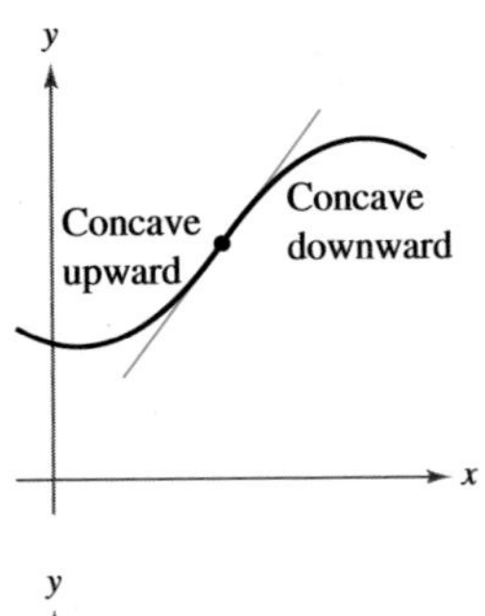

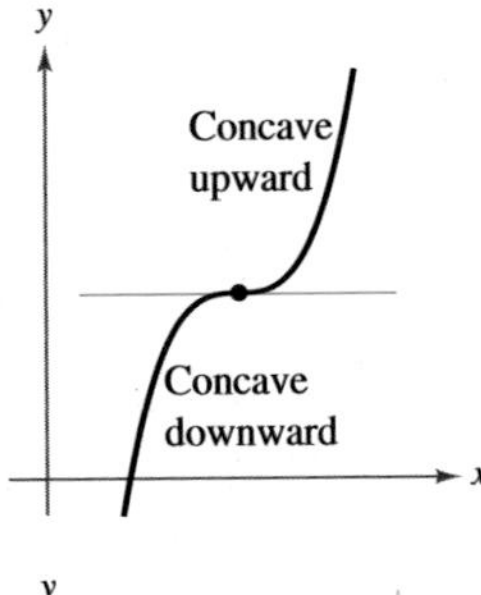

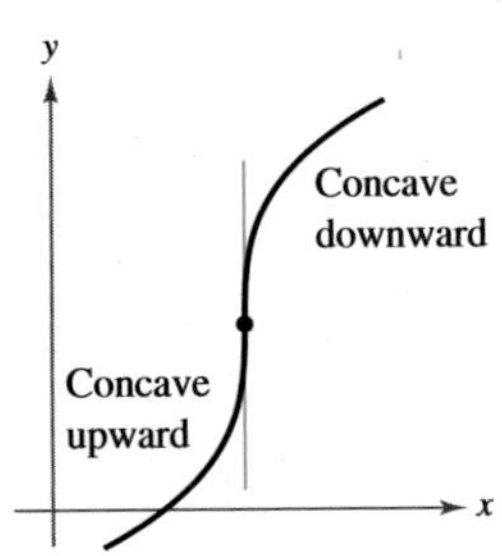

The concavity of f changes at a point of inflection. Note that the graph crosses its tangent line at a point of inflection.
Figure 5.27

Points of Inflection

The graph in Figure 5.25 has two points at which the concavity changes. If the tangent line to the graph exists at such a point, that point is a **point of inflection.** Three types of points of inflection are shown in Figure 5.27.

DEFINITION OF POINT OF INFLECTION

Let f be a function that is continuous on an open interval and let c be a point in the interval. If the graph of f has a tangent line at this point $(c, f(c))$, then this point is a **point of inflection** of the graph of f if the concavity of f changes from upward to downward (or downward to upward) at the point.

NOTE The definition of *point of inflection* given above requires that the tangent line exists at the point of inflection. Some books do not require this. For instance, we do not consider the function

$$f(x) = \begin{cases} x^3, & x < 0 \\ x^2 + 2x, & x \ge 0 \end{cases}$$

to have a point of inflection at the origin, even though the concavity of the graph changes from concave downward to concave upward. ■

To locate *possible* points of inflection, you can determine the values of x for which $f''(x) = 0$ or $f''(x)$ does not exist. You then use the second derivative to test for a point of inflection in the same way you use the first derivative to test for relative extrema.

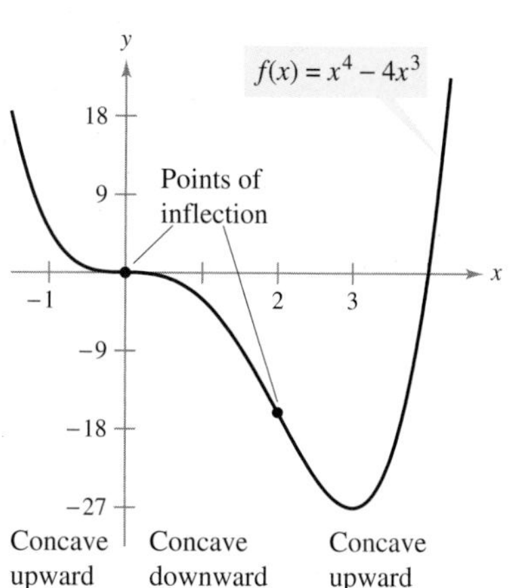

Points of inflection can occur where $f''(x) = 0$ or f'' does not exist.
Figure 5.28

THEOREM 5.8 POINTS OF INFLECTION

If $(c, f(c))$ is a point of inflection of the graph of f, then either $f''(c) = 0$ or f'' does not exist at $x = c$.

EXAMPLE 3 Finding Points of Inflection

Determine the points of inflection and discuss the concavity of the graph of

$$f(x) = x^4 - 4x^3.$$

Solution Differentiate the function twice.

$$f(x) = x^4 - 4x^3 \qquad \text{Write original function.}$$
$$f'(x) = 4x^3 - 12x^2 \qquad \text{Find first derivative.}$$
$$f''(x) = 12x^2 - 24x = 12x(x - 2) \qquad \text{Find second derivative.}$$

Setting $f''(x) = 0$, you can determine that the possible points of inflection occur at $x = 0$ and $x = 2$. By testing the intervals determined by these x-values, you can conclude that they both yield points of inflection. A summary of this testing is shown in the table, and the graph of f is shown in Figure 5.28.

Interval	$-\infty < x < 0$	$0 < x < 2$	$2 < x < \infty$
Test Value	$x = -1$	$x = 1$	$x = 3$
Sign of $f''(x)$	$f''(-1) > 0$	$f''(1) < 0$	$f''(3) > 0$
Conclusion	Concave upward	Concave downward	Concave upward

■

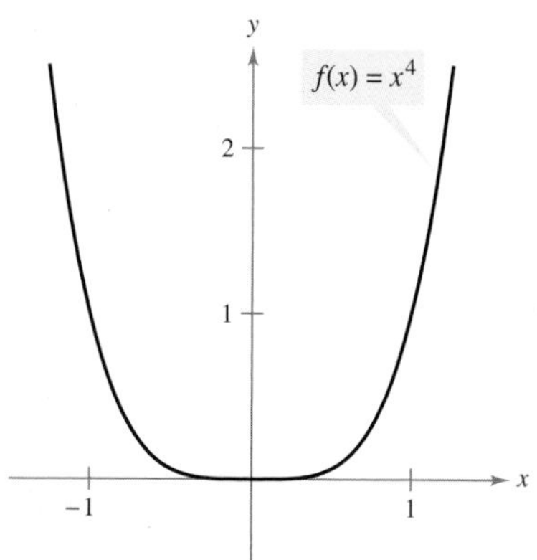

$f''(x) = 0$, but $(0, 0)$ is not a point of inflection.
Figure 5.29

The converse of Theorem 5.8 is not generally true. That is, it is possible for the second derivative to be 0 at a point that is *not* a point of inflection. For instance, the graph of $f(x) = x^4$ is shown in Figure 5.29. The second derivative is 0 when $x = 0$, but the point $(0, 0)$ is not a point of inflection because the graph of f is concave upward in both intervals $-\infty < x < 0$ and $0 < x < \infty$.

EXPLORATION

Consider a general cubic function of the form

$$f(x) = ax^3 + bx^2 + cx + d.$$

You know that the value of d has a bearing on the location of the graph but has no bearing on the value of the first derivative at given values of x. Graphically, this is true because changes in the value of d shift the graph up or down but do not change its basic shape. Use a graphing utility to graph several cubics with different values of c. Then give a graphical explanation of why changes in c do not affect the values of the second derivative.

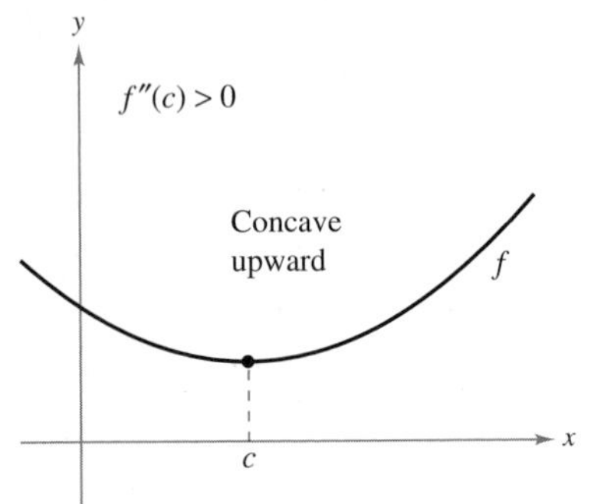

If $f'(c) = 0$ and $f''(c) > 0$, $f(c)$ is a relative minimum.

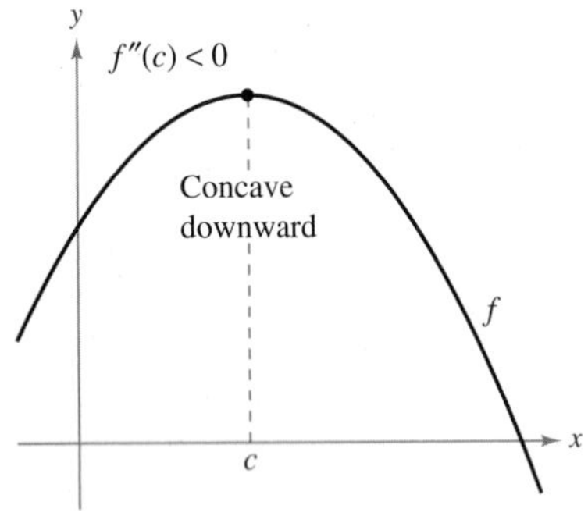

If $f'(c) = 0$ and $f''(c) < 0$, $f(c)$ is a relative maximum.

Figure 5.30

The Second Derivative Test

In addition to testing for concavity, the second derivative can be used to perform a simple test for relative maxima and minima. The test is based on the fact that if the graph of a function f is concave upward on an open interval containing c, and $f'(c) = 0$, $f(c)$ must be a relative minimum of f. Similarly, if the graph of a function f is concave downward on an open interval containing c, and $f'(c) = 0$, $f(c)$ must be a relative maximum of f (see Figure 5.30).

THEOREM 5.9 SECOND DERIVATIVE TEST

Let f be a function such that $f'(c) = 0$ and the second derivative of f exists on an open interval containing c.

1. If $f''(c) > 0$, then f has a relative minimum at $(c, f(c))$.
2. If $f''(c) < 0$, then f has a relative maximum at $(c, f(c))$.

If $f''(c) = 0$, the test fails. That is, f may have a relative maximum at $(c, f(c))$, a relative minimum at $(c, f(c))$, or neither. In such cases, you can use the First Derivative Test.

PROOF If $f'(c) = 0$ and $f''(c) > 0$, there exists an open interval I containing c for which

$$\frac{f'(x) - f'(c)}{x - c} = \frac{f'(x)}{x - c} > 0$$

for all $x \neq c$ in I. If $x < c$, then $x - c < 0$ and $f'(x) < 0$. Also, if $x > c$, then $x - c > 0$ and $f'(x) > 0$. So, $f'(x)$ changes from negative to positive at c, and the First Derivative Test implies that $f(c)$ is a relative minimum. A proof of the second case is left to you. ■

EXAMPLE 4 Using the Second Derivative Test

Find the relative extrema for $f(x) = -3x^5 + 5x^3$.

Solution Begin by finding the critical numbers of f.

$$f'(x) = -15x^4 + 15x^2 = 15x^2(1 - x^2) = 0 \qquad \text{Set } f'(x) \text{ equal to 0.}$$

$$x = -1, 0, 1 \qquad \text{Critical numbers}$$

Using

$$f''(x) = -60x^3 + 30x = 30(-2x^3 + x)$$

you can apply the Second Derivative Test as shown below.

Point	$(-1, -2)$	$(1, 2)$	$(0, 0)$
Sign of $f''(x)$	$f''(-1) > 0$	$f''(1) < 0$	$f''(0) = 0$
Conclusion	Relative minimum	Relative maximum	Test fails

Because the Second Derivative Test fails at $(0, 0)$, you can use the First Derivative Test and observe that f increases to the left and right of $x = 0$. So, $(0, 0)$ is neither a relative minimum nor a relative maximum (even though the graph has a horizontal tangent line at this point). The graph of f is shown in Figure 5.31. ■

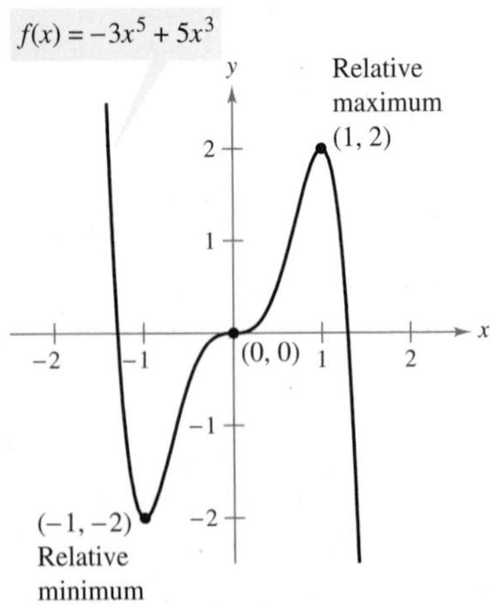

(0, 0) is neither a relative minimum nor a relative maximum.

Figure 5.31

5.4 Exercises

See www.CalcChat.com for worked-out solutions to odd-numbered exercises.

In Exercises 1–4, the graph of f is shown. State the signs of f' and f'' on the interval $(0, 2)$.

1. 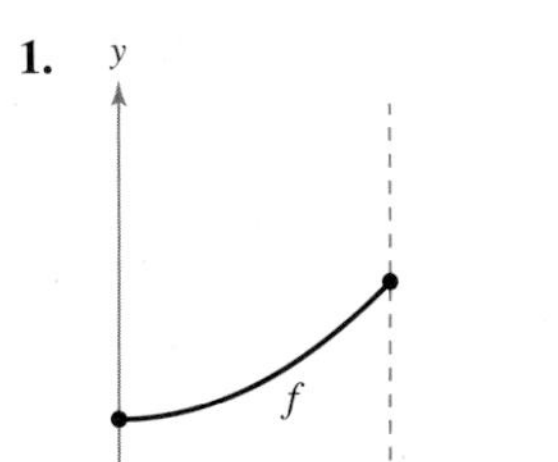

2. 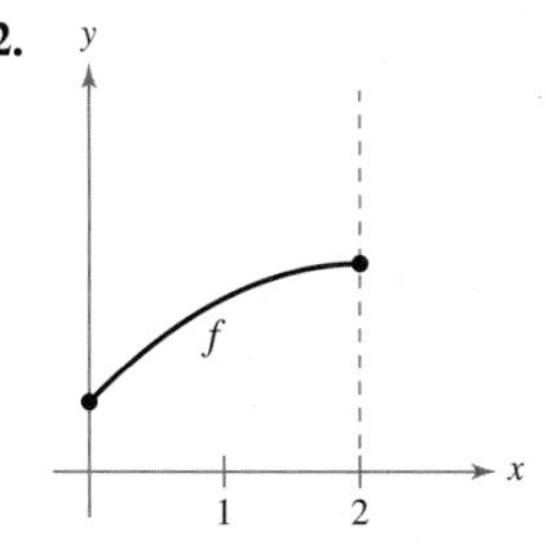

3. 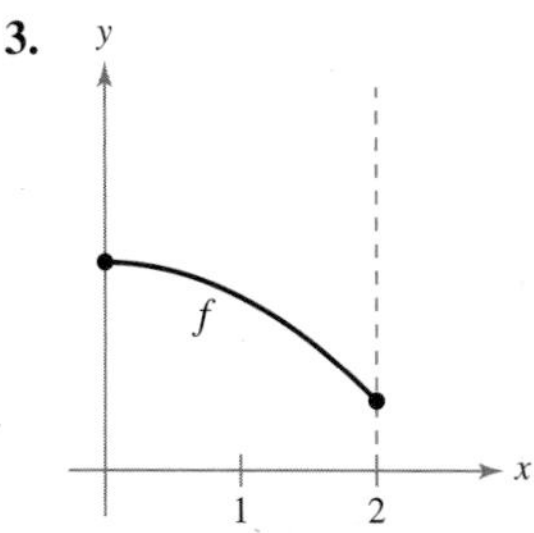

4. 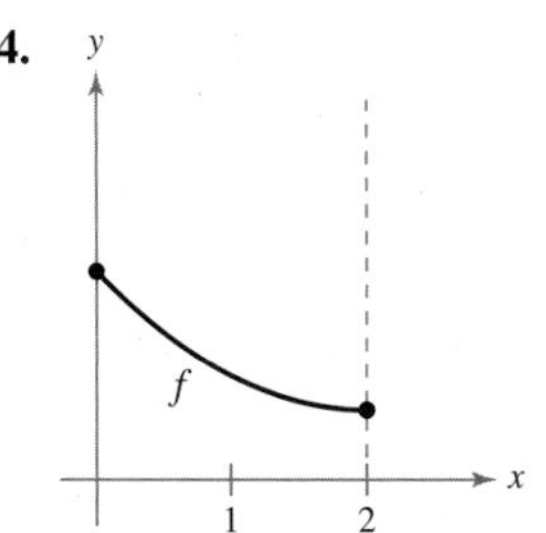

In Exercises 5–18, determine the open intervals on which the graph is concave upward or concave downward.

5. $y = x^2 - x - 2$
6. $y = -x^3 + 3x^2 - 2$
7. $g(x) = 3x^2 - x^3$
8. $h(x) = x^5 - 5x + 2$
9. $f(x) = -x^3 + 6x^2 - 9x - 1$
10. $f(x) = x^5 + 5x^4 - 40x^2$
11. $f(x) = \dfrac{24}{x^2 + 12}$
12. $f(x) = \dfrac{x^2}{x^2 + 1}$
13. $f(x) = \dfrac{x^2 + 1}{x^2 - 1}$
14. $y = \dfrac{-3x^5 + 40x^3 + 135x}{270}$
15. $g(x) = \dfrac{x^2 + 4}{4 - x^2}$
16. $h(x) = \dfrac{x^2 - 1}{2x - 1}$
17. $y = 2x$
18. $y = x + \dfrac{2}{x}$

In Exercises 19–30, find the points of inflection and discuss the concavity of the graph of the function.

19. $f(x) = \frac{1}{2}x^4 + 2x^3$
20. $f(x) = -x^4 + 24x^2$
21. $f(x) = x^3 - 6x^2 + 12x$
22. $f(x) = 2x^3 - 3x^2 - 12x + 5$
23. $f(x) = \frac{1}{4}x^4 - 2x^2$
24. $f(x) = 2x^4 - 8x + 3$
25. $f(x) = x(x - 4)^3$
26. $f(x) = (x - 2)^3(x - 1)$
27. $f(x) = x\sqrt{x + 3}$
28. $f(x) = x\sqrt{9 - x}$
29. $f(x) = \dfrac{4}{x^2 + 1}$
30. $f(x) = \dfrac{x + 1}{\sqrt{x}}$

In Exercises 31–44, find all relative extrema. Use the Second Derivative Test where applicable.

31. $f(x) = (x - 5)^2$
32. $f(x) = -(x - 5)^2$
33. $f(x) = 6x - x^2$
34. $f(x) = x^2 + 3x - 8$
35. $f(x) = x^3 - 3x^2 + 3$
36. $f(x) = x^3 - 5x^2 + 7x$
37. $f(x) = x^4 - 4x^3 + 2$
38. $f(x) = -x^4 + 4x^3 + 8x^2$
39. $g(x) = x^2(6 - x)^3$
40. $g(x) = -\frac{1}{8}(x + 2)^2(x - 4)^2$
41. $f(x) = x^{2/3} - 3$
42. $f(x) = \sqrt{x^2 + 1}$
43. $f(x) = x + \dfrac{4}{x}$
44. $f(x) = \dfrac{x}{x - 1}$

In Exercises 45 and 46, use a computer algebra system to analyze the function over the given interval. (a) Find the first and second derivatives of the function. (b) Find any relative extrema and points of inflection. (c) Graph f, f', and f'' on the same set of coordinate axes and state the relationship between the behavior of f and the signs of f' and f''.

45. $f(x) = 0.2x^2(x - 3)^3$, $[-1, 4]$
46. $f(x) = x^2\sqrt{6 - x^2}$, $[-\sqrt{6}, \sqrt{6}]$

WRITING ABOUT CONCEPTS

47. Consider a function f such that f' is increasing. Sketch graphs of f for (a) $f' < 0$ and (b) $f' > 0$.
48. Consider a function f such that f' is decreasing. Sketch graphs of f for (a) $f' < 0$ and (b) $f' > 0$.
49. Sketch the graph of a function f that does *not* have a point of inflection at $(c, f(c))$ even though $f''(c) = 0$.
50. S represents weekly sales of a product. What can be said of S' and S'' for each of the following statements?
    (a) The rate of change of sales is increasing.
    (b) Sales are increasing at a slower rate.
    (c) The rate of change of sales is constant.
    (d) Sales are steady.
    (e) Sales are declining, but at a slower rate.
    (f) Sales have bottomed out and have started to rise.

In Exercises 51–54, the graph of f is shown. Graph f, f', and f'' on the same set of coordinate axes. To print an enlarged copy of the graph, go to the website *www.mathgraphs.com*.

51.

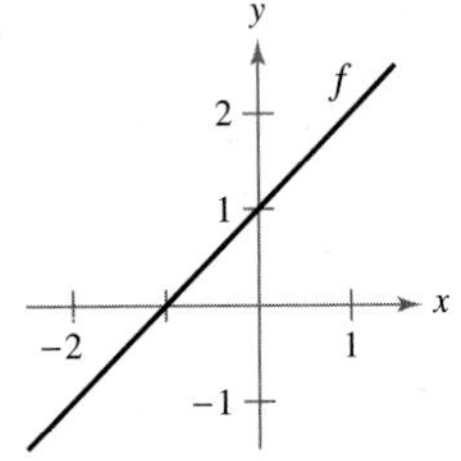

52.

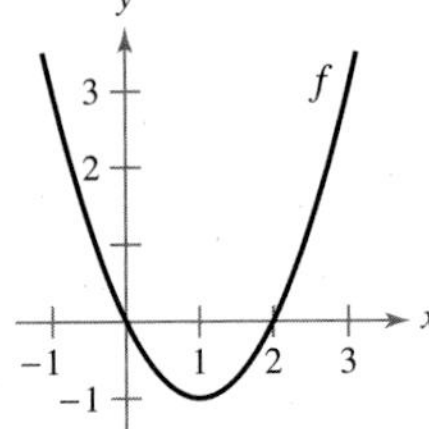

53.

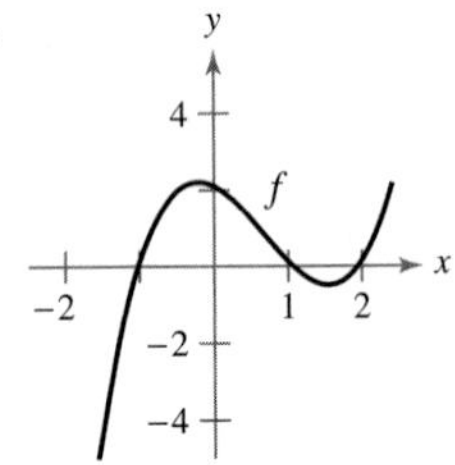

54.

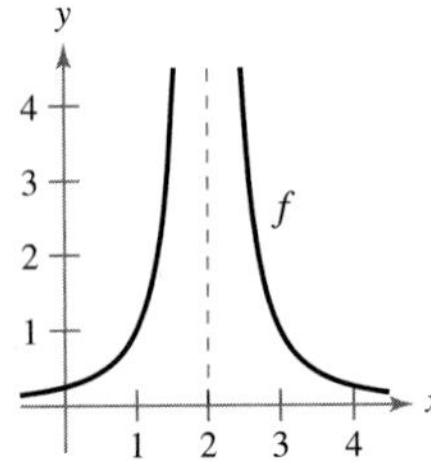

Think About It **In Exercises 55–58, sketch the graph of a function f having the given characteristics.**

55. $f(2) = f(4) = 0$
$f'(x) < 0$ if $x < 3$
$f'(3)$ does not exist.
$f'(x) > 0$ if $x > 3$
$f''(x) < 0, x \neq 3$

56. $f(0) = f(2) = 0$
$f'(x) > 0$ if $x < 1$
$f'(1) = 0$
$f'(x) < 0$ if $x > 1$
$f''(x) < 0$

57. $f(2) = f(4) = 0$
$f'(x) > 0$ if $x < 3$
$f'(3)$ does not exist.
$f'(x) < 0$ if $x > 3$
$f''(x) > 0, x \neq 3$

58. $f(0) = f(2) = 0$
$f'(x) < 0$ if $x < 1$
$f'(1) = 0$
$f'(x) > 0$ if $x > 1$
$f''(x) > 0$

59. ***Think About It*** The figure shows the graph of f''. Sketch a graph of f. (The answer is not unique.) To print an enlarged copy of the graph, go to the website *www.mathgraphs.com*.

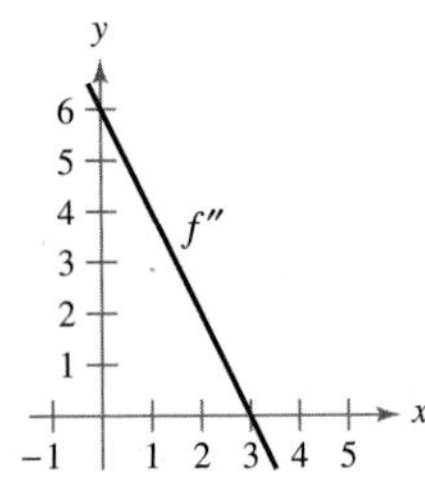

Figure for 59

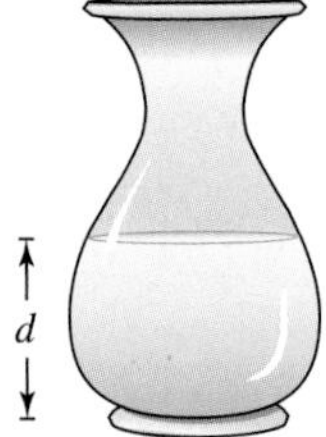

Figure for 60

CAPSTONE

60. ***Think About It*** Water is running into the vase shown in the figure at a constant rate.
(a) Graph the depth d of water in the vase as a function of time.
(b) Does the function have any extrema? Explain.
(c) Interpret the inflection points of the graph of d.

61. ***Conjecture*** Consider the function $f(x) = (x - 2)^n$.
(a) Use a graphing utility to graph f for $n = 1, 2, 3$, and 4. Use the graphs to make a conjecture about the relationship between n and any inflection points of the graph of f.
(b) Verify your conjecture in part (a).

62. (a) Graph $f(x) = \sqrt[3]{x}$ and identify the inflection point.
(b) Does $f''(x)$ exist at the inflection point? Explain.

In Exercises 63 and 64, find a, b, c, and d such that the cubic $f(x) = ax^3 + bx^2 + cx + d$ satisfies the given conditions.

63. Relative maximum: $(3, 3)$
Relative minimum: $(5, 1)$
Inflection point: $(4, 2)$

64. Relative maximum: $(2, 4)$
Relative minimum: $(4, 2)$
Inflection point: $(3, 3)$

65. ***Aircraft Glide Path*** A small aircraft starts its descent from an altitude of 1 mile, 4 miles west of the runway (see figure).

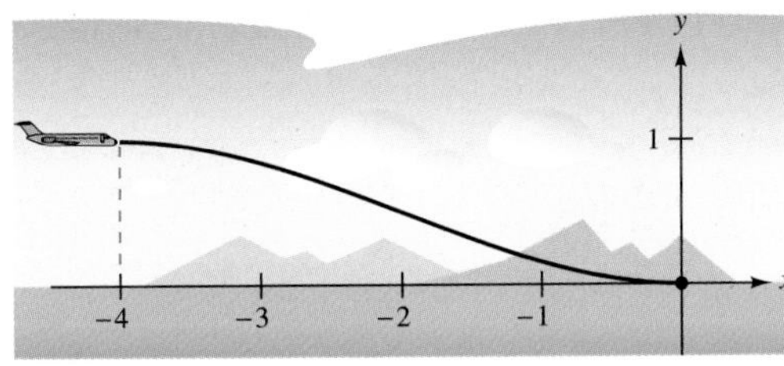

(a) Find the cubic $f(x) = ax^3 + bx^2 + cx + d$ on the interval $[-4, 0]$ that describes a smooth glide path for the landing.
(b) The function in part (a) models the glide path of the plane. When would the plane be descending at the greatest rate?

FOR FURTHER INFORMATION For more information on this type of modeling, see the article "How Not to Land at Lake Tahoe!" by Richard Barshinger in *The American Mathematical Monthly*. To view this article, go to the website *www.matharticles.com*.

66. ***Highway Design*** A section of highway connecting two hillsides with grades of 6% and 4% is to be built between two points that are separated by a horizontal distance of 2000 feet (see figure). At the point where the two hillsides come together, there is a 50-foot difference in elevation.

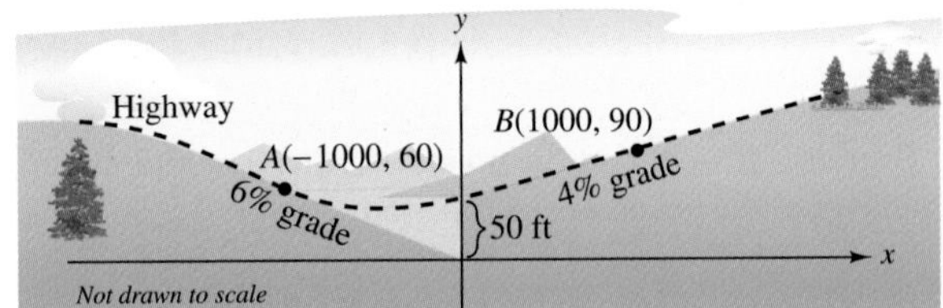

(a) Design a section of highway connecting the hillsides modeled by the function $f(x) = ax^3 + bx^2 + cx + d$ $(-1000 \leq x \leq 1000)$. At the points A and B, the slope of the model must match the grade of the hillside.
(b) Use a graphing utility to graph the model.
(c) Use a graphing utility to graph the derivative of the model.
(d) Determine the grade at the steepest part of the transitional section of the highway.

67. ***Beam Deflection*** The deflection D of a beam of length L is $D = 2x^4 - 5Lx^3 + 3L^2x^2$, where x is the distance from one end of the beam. Find the value of x that yields the maximum deflection.

68. ***Specific Gravity*** A model for the specific gravity of water S is

$$S = \frac{5.755}{10^8}T^3 - \frac{8.521}{10^6}T^2 + \frac{6.540}{10^5}T + 0.99987, \; 0 < T < 25$$

where T is the water temperature in degrees Celsius.

(a) Use a computer algebra system to find the coordinates of the maximum value of the function.

(b) Sketch a graph of the function over the specified domain. (Use a setting in which $0.996 \le S \le 1.001$.)

(c) Estimate the specific gravity of water when $T = 20°$.

69. ***Average Cost*** A manufacturer has determined that the total cost C of operating a factory is $C = 0.5x^2 + 15x + 5000$, where x is the number of units produced. At what level of production will the average cost per unit be minimized? (The average cost per unit is C/x.)

70. ***Inventory Cost*** The total cost C of ordering and storing x units is $C = 2x + (300{,}000/x)$. What order size will produce a minimum cost?

71. ***Sales Growth*** The annual sales S of a new product are given by $S = \dfrac{5000t^2}{8 + t^2}$, $0 \le t \le 3$, where t is time in years.

(a) Complete the table. Then use it to estimate when the annual sales are increasing at the greatest rate.

t	0.5	1	1.5	2	2.5	3
S						

(b) Use a graphing utility to graph the function S. Then use the graph to estimate when the annual sales are increasing at the greatest rate.

(c) Find the exact time when the annual sales are increasing at the greatest rate.

72. ***Modeling Data*** The average typing speed S (in words per minute) of a typing student after t weeks of lessons is shown in the table.

t	5	10	15	20	25	30
S	38	56	79	90	93	94

A model for the data is $S = \dfrac{100t^2}{65 + t^2}$, $t > 0$.

(a) Use a graphing utility to plot the data and graph the model.

(b) Use the second derivative to determine the concavity of S. Compare the result with the graph in part (a).

(c) What is the sign of the first derivative for $t > 0$? By combining this information with the concavity of the model, what inferences can be made about the typing speed as t increases?

Linear and Quadratic Approximations **In Exercises 73 and 74, use a graphing utility to graph the function. Then graph the linear and quadratic approximations**

$$P_1(x) = f(a) + f'(a)(x - a)$$

and

$$P_2(x) = f(a) + f'(a)(x - a) + \tfrac{1}{2}f''(a)(x - a)^2$$

in the same viewing window. Compare the values of f, P_1, and P_2 and their first derivatives at $x = a$. How do the approximations change as you move farther away from $x = a$?

	Function	*Value of a*
73.	$f(x) = \sqrt{1 - x}$	$a = 0$
74.	$f(x) = \dfrac{\sqrt{x}}{x - 1}$	$a = 2$

True or False? **In Exercises 75–78, determine whether the statement is true or false. If it is false, explain why or give an example that shows it is false.**

75. The graph of every cubic polynomial has precisely one point of inflection.

76. The graph of $f(x) = 1/x$ is concave downward for $x < 0$ and concave upward for $x > 0$, and thus it has a point of inflection at $x = 0$.

77. If $f'(c) > 0$, then f is concave upward at $x = c$.

78. If $f''(2) = 0$, then the graph of f must have a point of inflection at $x = 2$.

In Exercises 79 and 80, let f and g represent differentiable functions such that $f'' \neq 0$ and $g'' \neq 0$.

79. Show that if f and g are concave upward on the interval (a, b), then $f + g$ is also concave upward on (a, b).

80. Prove that if f and g are positive, increasing, and concave upward on the interval (a, b), then fg is also concave upward on (a, b).

81. Show that the point of inflection of

$$f(x) = x(x - 6)^2$$

lies midway between the relative extrema of f.

82. Prove that every cubic function with three distinct real zeros has a point of inflection whose x-coordinate is the average of the three zeros.

83. Show that the cubic polynomial

$$p(x) = ax^3 + bx^2 + cx + d$$

has exactly one point of inflection (x_0, y_0), where

$$x_0 = \frac{-b}{3a} \quad \text{and} \quad y_0 = \frac{2b^3}{27a^2} - \frac{bc}{3a} + d.$$

Use this formula to find the point of inflection of $p(x) = x^3 - 3x^2 + 2$.

5.5 Limits at Infinity

- **Determine (finite) limits at infinity.**
- **Determine the horizontal asymptotes, if any, of the graph of a function.**
- **Determine infinite limits at infinity.**

Limits at Infinity

So far, your primary focus on graphs has been their behavior at certain points or on finite intervals. This section discusses the "end behavior" of a function on an *infinite* interval. Consider the graph of

$$f(x) = \frac{3x^2}{x^2 + 1}$$

as shown in Figure 5.32. Graphically, you can see that the values of $f(x)$ appear to approach 3 as x increases without bound or decreases without bound. You can come to the same conclusions numerically, as shown in the table.

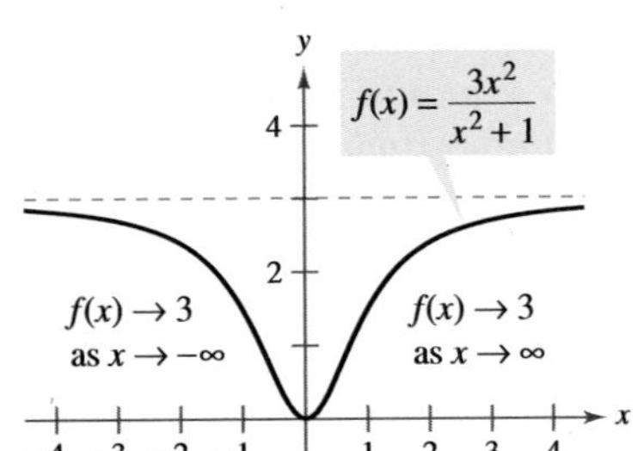

The limit of $f(x)$ as x approaches $-\infty$ or ∞ is 3.
Figure 5.32

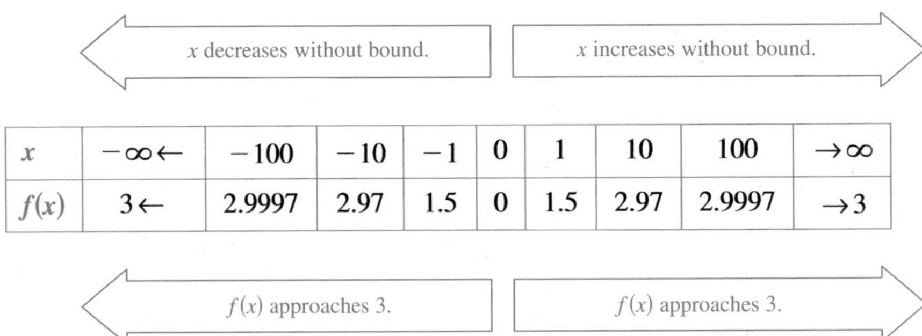

x	$-\infty \leftarrow$	-100	-10	-1	0	1	10	100	$\to \infty$
$f(x)$	$3 \leftarrow$	2.9997	2.97	1.5	0	1.5	2.97	2.9997	$\to 3$

The table suggests that the value of $f(x)$ approaches 3 as x increases without bound $(x \to \infty)$. Similarly, $f(x)$ approaches 3 as x decreases without bound $(x \to -\infty)$. These **limits at infinity** are denoted by

$$\lim_{x \to -\infty} f(x) = 3 \qquad \text{Limit at negative infinity}$$

and

$$\lim_{x \to \infty} f(x) = 3. \qquad \text{Limit at positive infinity}$$

NOTE The statement $\lim_{x \to -\infty} f(x) = L$ or $\lim_{x \to \infty} f(x) = L$ means that the limit exists *and* the limit is equal to L.

To say that a statement is true as x increases *without bound* means that for some (large) real number M, the statement is true for *all* x in the interval $\{x: x > M\}$. The following definition uses this concept.

DEFINITION OF LIMITS AT INFINITY

Let L be a real number.

1. The statement $\lim_{x \to \infty} f(x) = L$ means that for each $\varepsilon > 0$ there exists an $M > 0$ such that $|f(x) - L| < \varepsilon$ whenever $x > M$.
2. The statement $\lim_{x \to -\infty} f(x) = L$ means that for each $\varepsilon > 0$ there exists an $N < 0$ such that $|f(x) - L| < \varepsilon$ whenever $x < N$.

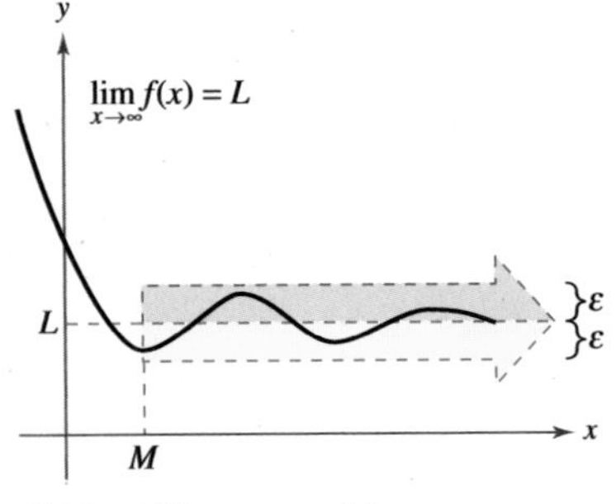

$f(x)$ is within ε units of L as $x \to \infty$.
Figure 5.33

The definition of a limit at infinity is shown in Figure 5.33. In this figure, note that for a given positive number ε there exists a positive number M such that, for $x > M$, the graph of f will lie between the horizontal lines given by $y = L + \varepsilon$ and $y = L - \varepsilon$.

EXPLORATION

Use a graphing utility to graph

$$f(x) = \frac{2x^2 + 4x - 6}{3x^2 + 2x - 16}.$$

Describe all the important features of the graph. Can you find a single viewing window that shows all of these features clearly? Explain your reasoning.

What are the horizontal asymptotes of the graph? How far to the right do you have to move on the graph so that the graph is within 0.001 unit of its horizontal asymptote? Explain your reasoning.

Horizontal Asymptotes

In Figure 5.33, the graph of f approaches the line $y = L$ as x increases without bound. As you learned in Section 2.6, the line $y = L$ is called a **horizontal asymptote** of the graph of f.

DEFINITION OF A HORIZONTAL ASYMPTOTE

The line $y = L$ is a **horizontal asymptote** of the graph of f if

$$\lim_{x\to-\infty} f(x) = L \quad \text{or} \quad \lim_{x\to\infty} f(x) = L.$$

Note that from this definition, it follows that the graph of a *function* of x can have at most two horizontal asymptotes—one to the right and one to the left.

Limits at infinity have many of the same properties of limits discussed in Section 3.3. For example, if $\lim_{x\to\infty} f(x)$ and $\lim_{x\to\infty} g(x)$ both exist, then

$$\lim_{x\to\infty} [f(x) + g(x)] = \lim_{x\to\infty} f(x) + \lim_{x\to\infty} g(x)$$

and

$$\lim_{x\to\infty} [f(x)g(x)] = \left[\lim_{x\to\infty} f(x)\right]\left[\lim_{x\to\infty} g(x)\right].$$

Similar properties hold for limits at $-\infty$.

When evaluating limits at infinity, the following theorem is helpful. (A proof of this theorem is given in Appendix A.)

THEOREM 5.10 LIMITS AT INFINITY

If r is a positive rational number and c is any real number, then $\lim_{x\to\infty} \frac{c}{x^r} = 0.$

Furthermore, if x^r is defined when $x < 0$, then $\lim_{x\to-\infty} \frac{c}{x^r} = 0.$

EXAMPLE 1 Finding a Limit at Infinity

Find the limit: $\lim_{x\to\infty} \left(4 - \frac{3}{x^2}\right).$

Algebraic Solution

Using Theorem 5.10, you can write

$$\begin{aligned}\lim_{x\to\infty} \left(4 - \frac{3}{x^2}\right) &= \lim_{x\to\infty} 4 - \lim_{x\to\infty} \frac{3}{x^2}\\ &= 4 - 0\\ &= 4.\end{aligned}$$

So, the limit of $f(x) = 4 - \frac{3}{x^2}$ as x approaches ∞ is 4.

Graphical Solution

Use a graphing utility to graph $y = 4 - 3/x^2$. Then use the *trace* feature to determine that as x gets larger and larger, y gets closer and closer to 4, as shown in Figure 5.34. Note that the line $y = 4$ is a horizontal asymptote to the right.

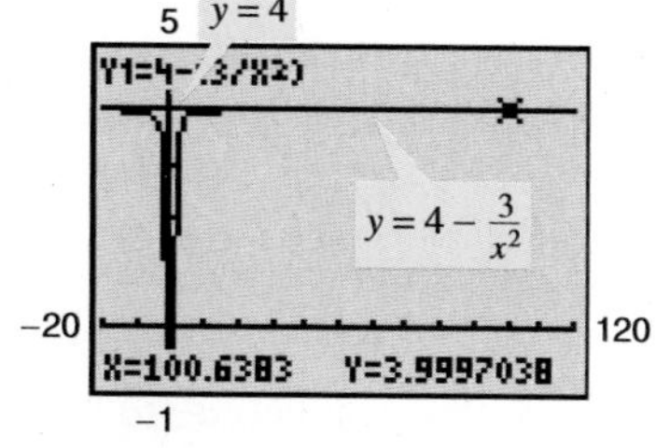

Figure 5.34

EXAMPLE 2 Finding a Limit at Infinity

Find the limit:

$$\lim_{x\to\infty} \frac{2x - 1}{x + 1}.$$

Solution Note that both the numerator and the denominator approach infinity as x approaches infinity.

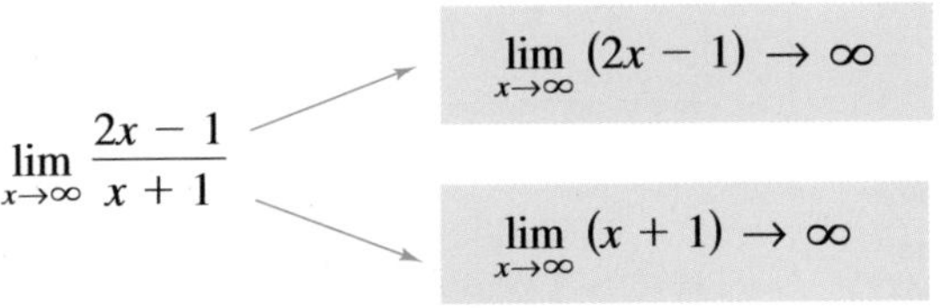

This results in $\frac{\infty}{\infty}$, an **indeterminate form.** To resolve this problem, you can divide both the numerator and the denominator by x. After dividing, the limit may be evaluated as shown.

NOTE When you encounter an indeterminate form such as the one in Example 2, you should divide the numerator and denominator by the highest power of x in the *denominator.*

$$\lim_{x\to\infty} \frac{2x - 1}{x + 1} = \lim_{x\to\infty} \frac{\frac{2x-1}{x}}{\frac{x+1}{x}} \qquad \text{Divide numerator and denominator by } x.$$

$$= \lim_{x\to\infty} \frac{2 - \frac{1}{x}}{1 + \frac{1}{x}} \qquad \text{Simplify.}$$

$$= \frac{\lim_{x\to\infty} 2 - \lim_{x\to\infty} \frac{1}{x}}{\lim_{x\to\infty} 1 + \lim_{x\to\infty} \frac{1}{x}} \qquad \text{Take limits of numerator and denominator.}$$

$$= \frac{2 - 0}{1 + 0} \qquad \text{Apply Theorem 5.10.}$$

$$= 2$$

So, the line $y = 2$ is a horizontal asymptote to the right. By taking the limit as $x \to -\infty$, you can see that $y = 2$ is also a horizontal asymptote to the left. The graph of the function is shown in Figure 5.35. ■

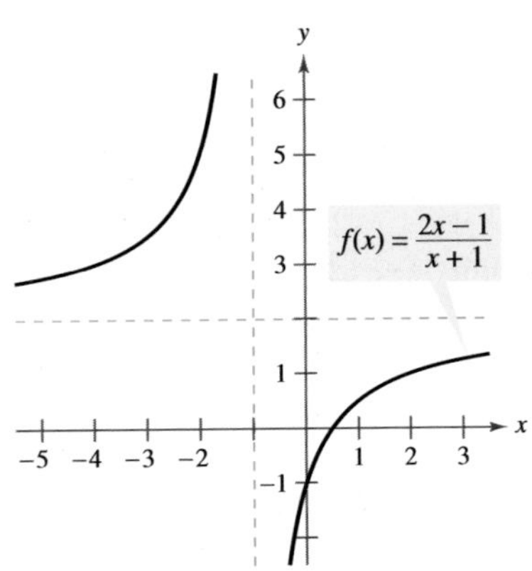

$y = 2$ is a horizontal asymptote.
Figure 5.35

TECHNOLOGY You can test the reasonableness of the limit found in Example 2 by evaluating $f(x)$ for a few large positive values of x. For instance,

$$f(100) \approx 1.9703, \quad f(1000) \approx 1.9970, \quad \text{and} \quad f(10{,}000) \approx 1.9997.$$

Another way to test the reasonableness of the limit is to use a graphing utility. For instance, in Figure 5.36, the graph of

$$f(x) = \frac{2x - 1}{x + 1}$$

is shown with the horizontal line $y = 2$. Note that as x increases, the graph of f moves closer and closer to its horizontal asymptote.

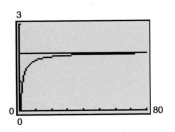

As x increases, the graph of f moves closer and closer to the line $y = 2$.
Figure 5.36

EXAMPLE 3 A Comparison of Three Rational Functions

MARIA GAETANA AGNESI (1718–1799)

Agnesi was one of a handful of women to receive credit for significant contributions to mathematics before the twentieth century. In her early twenties, she wrote the first text that included both differential and integral calculus. By age 30, she was an honorary member of the faculty at the University of Bologna.

For more information on the contributions of women to mathematics, see the article "Why Women Succeed in Mathematics" by Mona Fabricant, Sylvia Svitak, and Patricia Clark Kenschaft in *Mathematics Teacher.* To view this article, go to the website *www.matharticles.com.*

Find each limit.

a. $\lim_{x\to\infty} \dfrac{2x+5}{3x^2+1}$ **b.** $\lim_{x\to\infty} \dfrac{2x^2+5}{3x^2+1}$ **c.** $\lim_{x\to\infty} \dfrac{2x^3+5}{3x^2+1}$

Solution In each case, attempting to evaluate the limit produces the indeterminate form ∞/∞.

a. Divide both the numerator and the denominator by x^2.

$$\lim_{x\to\infty} \frac{2x+5}{3x^2+1} = \lim_{x\to\infty} \frac{(2/x)+(5/x^2)}{3+(1/x^2)} = \frac{0+0}{3+0} = \frac{0}{3} = 0$$

b. Divide both the numerator and the denominator by x^2.

$$\lim_{x\to\infty} \frac{2x^2+5}{3x^2+1} = \lim_{x\to\infty} \frac{2+(5/x^2)}{3+(1/x^2)} = \frac{2+0}{3+0} = \frac{2}{3}$$

c. Divide both the numerator and the denominator by x^2.

$$\lim_{x\to\infty} \frac{2x^3+5}{3x^2+1} = \lim_{x\to\infty} \frac{2x+(5/x^2)}{3+(1/x^2)} = \frac{\infty}{3}$$

You can conclude that the limit *does not exist* because the numerator increases without bound while the denominator approaches 3. ■

GUIDELINES FOR FINDING LIMITS AT $\pm\infty$ OF RATIONAL FUNCTIONS

1. If the degree of the numerator is *less than* the degree of the denominator, then the limit of the rational function is 0.
2. If the degree of the numerator is *equal to* the degree of the denominator, then the limit of the rational function is the ratio of the leading coefficients.
3. If the degree of the numerator is *greater than* the degree of the denominator, then the limit of the rational function does not exist.

Use these guidelines to check the results in Example 3. These limits seem reasonable when you consider that for large values of x, the highest-power term of the rational function is the most "influential" in determining the limit. For instance, the limit as x approaches infinity of the function

$$f(x) = \frac{1}{x^2+1}$$

is 0 because the denominator overpowers the numerator as x increases or decreases without bound, as shown in Figure 5.37.

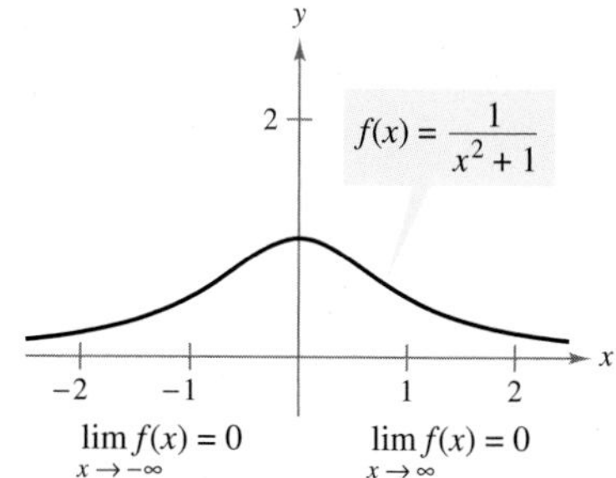

f has a horizontal asymptote at $y = 0$.
Figure 5.37

The function shown in Figure 5.37 is a special case of a type of curve studied by the Italian mathematician Maria Gaetana Agnesi. The general form of this function is

$$f(x) = \frac{8a^3}{x^2+4a^2} \qquad \text{Witch of Agnesi}$$

and, through a mistranslation of the Italian word *vertéré*, the curve has come to be known as the Witch of Agnesi. Agnesi's work with this curve first appeared in a comprehensive text on calculus that was published in 1748.

In Figure 5.37, you can see that the function $f(x) = 1/(x^2 + 1)$ approaches the same horizontal asymptote to the right and to the left. This is always true of rational functions. Functions that are not rational, however, may approach different horizontal asymptotes to the right and to the left. This is demonstrated in Example 4.

EXAMPLE 4 A Function with Two Horizontal Asymptotes

Find each limit.

a. $\displaystyle\lim_{x\to\infty} \frac{3x - 2}{\sqrt{2x^2 + 1}}$ **b.** $\displaystyle\lim_{x\to-\infty} \frac{3x - 2}{\sqrt{2x^2 + 1}}$

Solution

a. For $x > 0$, you can write $x = \sqrt{x^2}$. So, dividing both the numerator and the denominator by x produces

$$\frac{3x - 2}{\sqrt{2x^2 + 1}} = \frac{\frac{3x - 2}{x}}{\frac{\sqrt{2x^2 + 1}}{\sqrt{x^2}}} = \frac{3 - \frac{2}{x}}{\sqrt{\frac{2x^2 + 1}{x^2}}} = \frac{3 - \frac{2}{x}}{\sqrt{2 + \frac{1}{x^2}}}$$

and you can take the limit as follows.

$$\lim_{x\to\infty} \frac{3x - 2}{\sqrt{2x^2 + 1}} = \lim_{x\to\infty} \frac{3 - \frac{2}{x}}{\sqrt{2 + \frac{1}{x^2}}} = \frac{3 - 0}{\sqrt{2 + 0}} = \frac{3}{\sqrt{2}}$$

b. For $x < 0$, you can write $x = -\sqrt{x^2}$. So, dividing both the numerator and the denominator by x produces

$$\frac{3x - 2}{\sqrt{2x^2 + 1}} = \frac{\frac{3x - 2}{x}}{\frac{\sqrt{2x^2 + 1}}{-\sqrt{x^2}}} = \frac{3 - \frac{2}{x}}{-\sqrt{\frac{2x^2 + 1}{x^2}}} = \frac{3 - \frac{2}{x}}{-\sqrt{2 + \frac{1}{x^2}}}$$

and you can take the limit as follows.

$$\lim_{x\to-\infty} \frac{3x - 2}{\sqrt{2x^2 + 1}} = \lim_{x\to-\infty} \frac{3 - \frac{2}{x}}{-\sqrt{2 + \frac{1}{x^2}}} = \frac{3 - 0}{-\sqrt{2 + 0}} = -\frac{3}{\sqrt{2}}$$

The graph of $f(x) = \dfrac{3x - 2}{\sqrt{2x^2 + 1}}$ is shown in Figure 5.38. ■

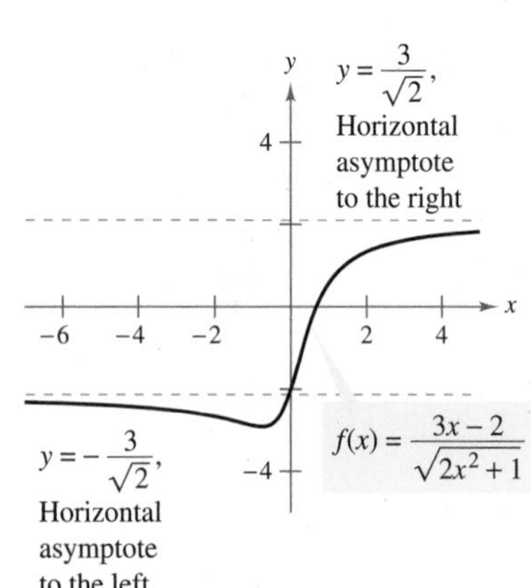

Functions that are not rational may have different right and left horizontal asymptotes.
Figure 5.38

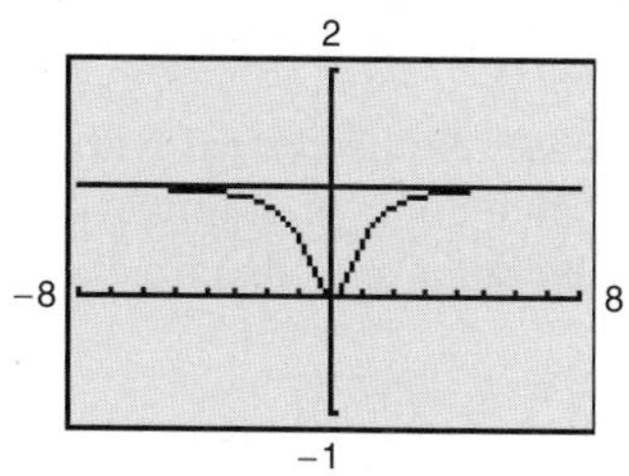

The horizontal asymptote appears to be the line $y = 1$ but it is actually the line $y = 2$.
Figure 5.39

TECHNOLOGY PITFALL If you use a graphing utility to help estimate a limit, be sure that you also confirm the estimate analytically—the pictures shown by a graphing utility can be misleading. For instance, Figure 5.39 shows one view of the graph of

$$y = \frac{2x^3 + 1000x^2 + x}{x^3 + 1000x^2 + x + 1000}.$$

From this view, one could be convinced that the graph has $y = 1$ as a horizontal asymptote. An analytical approach shows that the horizontal asymptote is actually $y = 2$. Confirm this by enlarging the viewing window on the graphing utility.

EXAMPLE 5 Oxygen Level in a Pond

Suppose that $f(t)$ measures the level of oxygen in a pond, where $f(t) = 1$ is the normal (unpolluted) level and the time t is measured in weeks. When $t = 0$, organic waste is dumped into the pond, and as the waste material oxidizes, the level of oxygen in the pond is

$$f(t) = \frac{t^2 - t + 1}{t^2 + 1}.$$

What percent of the normal level of oxygen exists in the pond after 1 week? After 2 weeks? After 10 weeks? What is the limit as t approaches infinity?

Solution When $t = 1$, 2, and 10, the levels of oxygen are as shown.

$$f(1) = \frac{1^2 - 1 + 1}{1^2 + 1} = \frac{1}{2} = 50\% \qquad \text{1 week}$$

$$f(2) = \frac{2^2 - 2 + 1}{2^2 + 1} = \frac{3}{5} = 60\% \qquad \text{2 weeks}$$

$$f(10) = \frac{10^2 - 10 + 1}{10^2 + 1} = \frac{91}{101} \approx 90.1\% \qquad \text{10 weeks}$$

To find the limit as t approaches infinity, divide the numerator and the denominator by t^2 to obtain

$$\lim_{t \to \infty} \frac{t^2 - t + 1}{t^2 + 1} = \lim_{t \to \infty} \frac{1 - (1/t) + (1/t^2)}{1 + (1/t^2)} = \frac{1 - 0 + 0}{1 + 0} = 1 = 100\%.$$

See Figure 5.40.

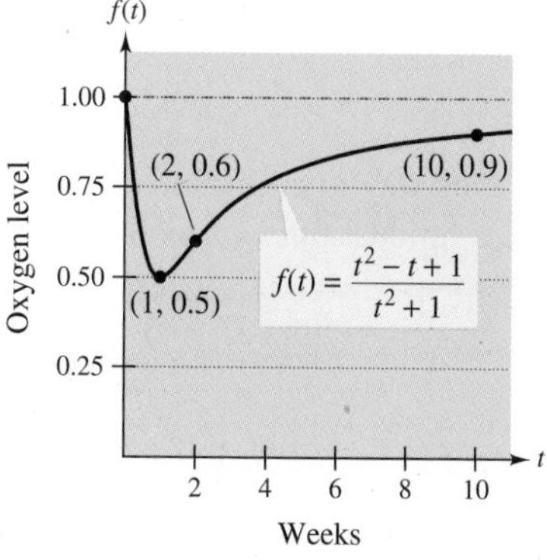

The level of oxygen in a pond approaches the normal level of 1 as t approaches ∞.
Figure 5.40

■

Infinite Limits at Infinity

Many functions do not approach a finite limit as x increases (or decreases) without bound. For instance, no polynomial function has a finite limit at infinity. The following definition is used to describe the behavior of polynomial and other functions at infinity.

NOTE Determining whether a function has an infinite limit at infinity is useful in analyzing the "end behavior" of its graph. You will see examples of this in Section 5.6 on curve sketching.

DEFINITION OF INFINITE LIMITS AT INFINITY

Let f be a function defined on the interval (a, ∞).

1. The statement $\lim_{x\to\infty} f(x) = \infty$ means that for each positive number M, there is a corresponding number $N > 0$ such that $f(x) > M$ whenever $x > N$.
2. The statement $\lim_{x\to\infty} f(x) = -\infty$ means that for each negative number M, there is a corresponding number $N > 0$ such that $f(x) < M$ whenever $x > N$.

Similar definitions can be given for the statements $\lim_{x\to-\infty} f(x) = \infty$ and $\lim_{x\to-\infty} f(x) = -\infty$.

EXAMPLE 6 Finding Infinite Limits at Infinity

Find each limit.

a. $\lim_{x\to\infty} x^3$ **b.** $\lim_{x\to-\infty} x^3$

Solution

a. As x increases without bound, x^3 also increases without bound. So, you can write $\lim_{x\to\infty} x^3 = \infty$.

b. As x decreases without bound, x^3 also decreases without bound. So, you can write $\lim_{x\to-\infty} x^3 = -\infty$.

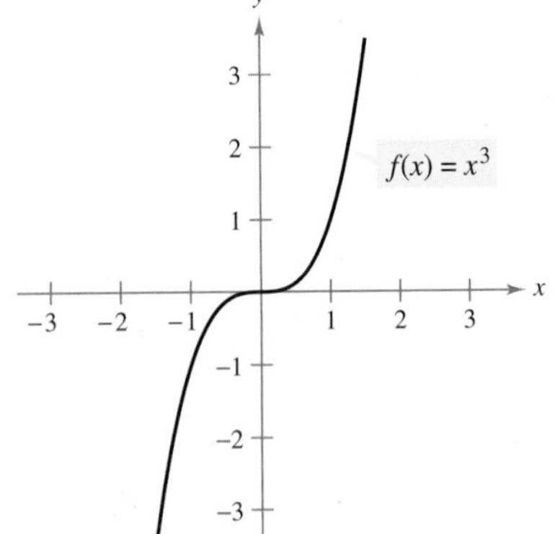

Figure 5.41

The graph of $f(x) = x^3$ in Figure 5.41 illustrates these two results. These results agree with the Leading Coefficient Test for polynomial functions as described in Section 2.2.

EXAMPLE 7 Finding Infinite Limits at Infinity

Find each limit.

a. $\lim_{x\to\infty} \dfrac{2x^2 - 4x}{x + 1}$ **b.** $\lim_{x\to-\infty} \dfrac{2x^2 - 4x}{x + 1}$

Solution One way to evaluate each of these limits is to use long division to rewrite the improper rational function as the sum of a polynomial and a rational function.

$$\textbf{a.}\ \lim_{x\to\infty} \frac{2x^2 - 4x}{x + 1} = \lim_{x\to\infty}\left(2x - 6 + \frac{6}{x + 1}\right) = \infty$$

$$\textbf{b.}\ \lim_{x\to-\infty} \frac{2x^2 - 4x}{x + 1} = \lim_{x\to-\infty}\left(2x - 6 + \frac{6}{x + 1}\right) = -\infty$$

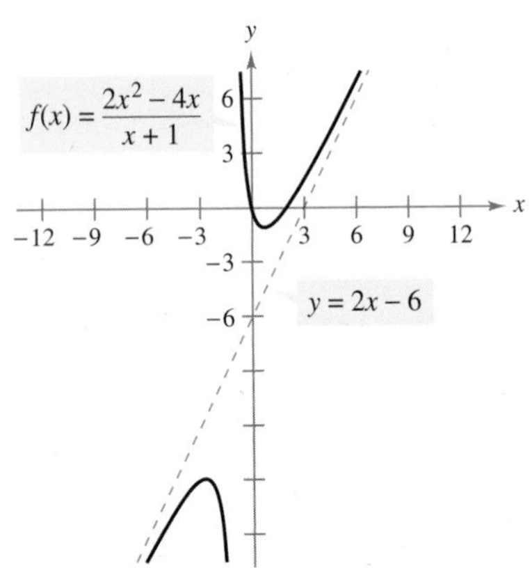

Figure 5.42

The statements above can be interpreted as saying that as x approaches $\pm\infty$, the function $f(x) = (2x^2 - 4x)/(x + 1)$ behaves like the function $g(x) = 2x - 6$. In Section 5.6, you will see that this is graphically described by saying that the line $y = 2x - 6$ is a slant asymptote of the graph of f, as shown in Figure 5.42. ■

5.5 Exercises

See www.CalcChat.com for worked-out solutions to odd-numbered exercises.

In Exercises 1–6, match the function with one of the graphs [(a), (b), (c), (d), (e), or (f)] using horizontal asymptotes as an aid.

(a)

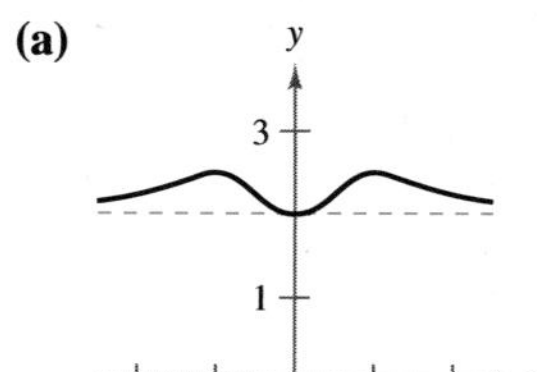

(b)

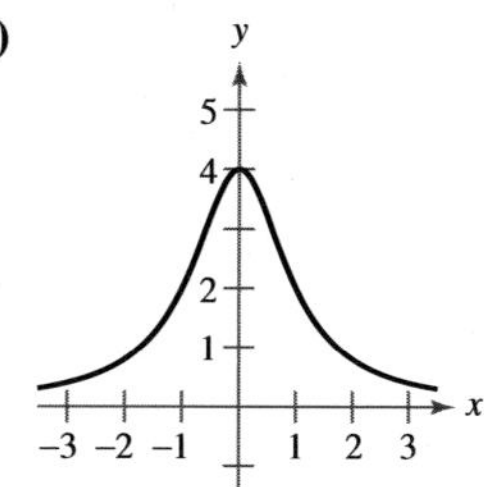

(c)

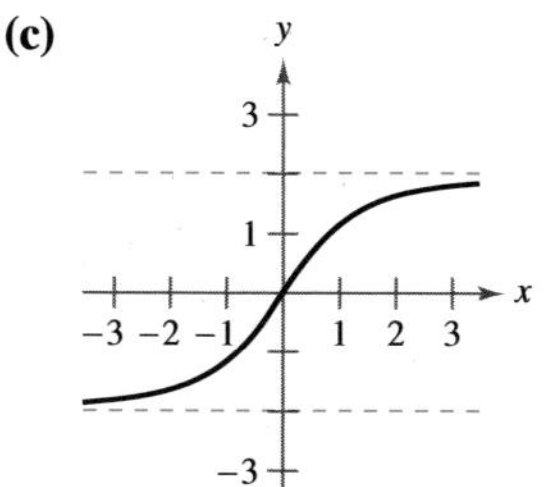

(d)

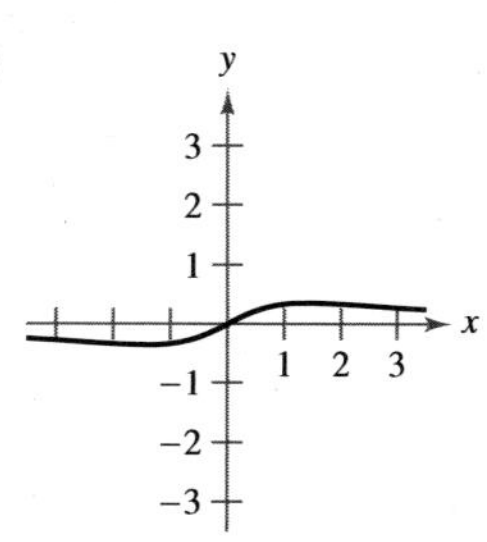

(e)

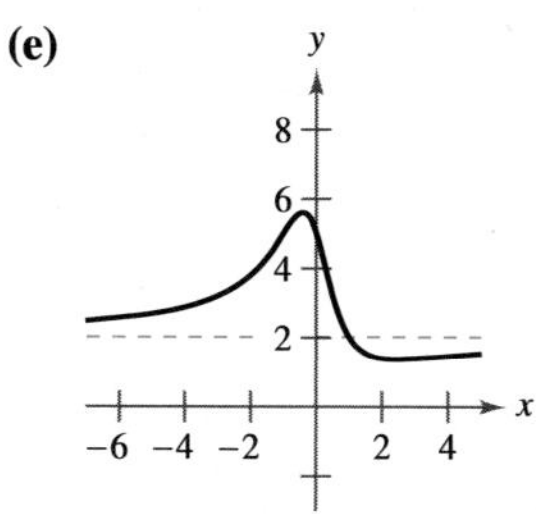

(f)

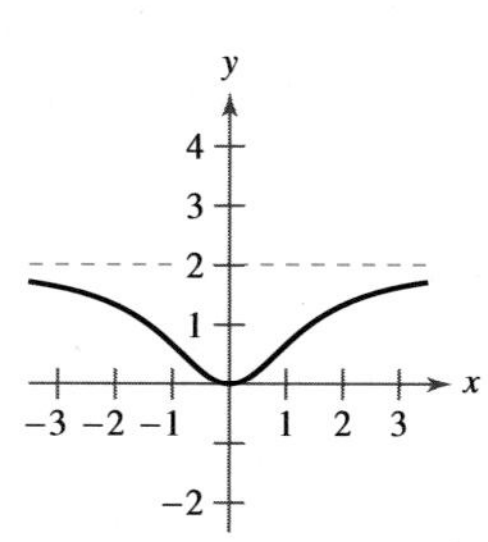

1. $f(x) = \frac{2x^2}{x^2 + 2}$

2. $f(x) = \frac{2x}{\sqrt{x^2 + 2}}$

3. $f(x) = \frac{x}{x^2 + 2}$

4. $f(x) = 2 + \frac{x^2}{x^4 + 1}$

5. $f(x) = \frac{4}{x^2 + 1}$

6. $f(x) = \frac{2x^2 - 3x + 5}{x^2 + 1}$

Numerical and Graphical Analysis **In Exercises 7–12, use a graphing utility to complete the table and estimate the limit as x approaches infinity. Then use a graphing utility to graph the function and estimate the limit graphically.**

x	10^0	10^1	10^2	10^3	10^4	10^5	10^6
$f(x)$							

7. $f(x) = \frac{4x + 3}{2x - 1}$

8. $f(x) = \frac{2x^2}{x + 1}$

9. $f(x) = \frac{-6x}{\sqrt{4x^2 + 5}}$

10. $f(x) = \frac{20x}{\sqrt{9x^2 - 1}}$

11. $f(x) = 5 - \frac{1}{x^2 + 1}$

12. $f(x) = 4 + \frac{3}{x^2 + 2}$

In Exercises 13 and 14, find $\lim_{x\to\infty} h(x)$, if possible.

13. $f(x) = 5x^3 - 3x^2 + 10x$

(a) $h(x) = \frac{f(x)}{x^2}$

(b) $h(x) = \frac{f(x)}{x^3}$

(c) $h(x) = \frac{f(x)}{x^4}$

14. $f(x) = -4x^2 + 2x - 5$

(a) $h(x) = \frac{f(x)}{x}$

(b) $h(x) = \frac{f(x)}{x^2}$

(c) $h(x) = \frac{f(x)}{x^3}$

In Exercises 15–18, find each limit, if possible.

15. (a) $\lim_{x\to\infty} \frac{x^2 + 2}{x^3 - 1}$

(b) $\lim_{x\to\infty} \frac{x^2 + 2}{x^2 - 1}$

(c) $\lim_{x\to\infty} \frac{x^2 + 2}{x - 1}$

16. (a) $\lim_{x\to\infty} \frac{3 - 2x}{3x^3 - 1}$

(b) $\lim_{x\to\infty} \frac{3 - 2x}{3x - 1}$

(c) $\lim_{x\to\infty} \frac{3 - 2x^2}{3x - 1}$

17. (a) $\lim_{x\to\infty} \frac{5 - 2x^{3/2}}{3x^2 - 4}$

(b) $\lim_{x\to\infty} \frac{5 - 2x^{3/2}}{3x^{3/2} - 4}$

(c) $\lim_{x\to\infty} \frac{5 - 2x^{3/2}}{3x - 4}$

18. (a) $\lim_{x\to\infty} \frac{5x^{3/2}}{4x^2 + 1}$

(b) $\lim_{x\to\infty} \frac{5x^{3/2}}{4x^{3/2} + 1}$

(c) $\lim_{x\to\infty} \frac{5x^{3/2}}{4\sqrt{x} + 1}$

In Exercises 19–34, find the limit.

19. $\lim_{x\to\infty} \left(4 + \frac{3}{x}\right)$

20. $\lim_{x\to-\infty} \left(\frac{5}{x} - \frac{x}{3}\right)$

21. $\lim_{x\to\infty} \frac{2x - 1}{3x + 2}$

22. $\lim_{x\to\infty} \frac{x^2 + 3}{2x^2 - 1}$

23. $\lim_{x\to\infty} \frac{x}{x^2 - 1}$

24. $\lim_{x\to\infty} \frac{5x^3 + 1}{10x^3 - 3x^2 + 7}$

25. $\lim_{x\to-\infty} \frac{5x^2}{x + 3}$

26. $\lim_{x\to-\infty} \left(\frac{1}{2}x - \frac{4}{x^2}\right)$

27. $\lim_{x\to-\infty} \frac{x}{\sqrt{x^2 - x}}$

28. $\lim_{x\to-\infty} \frac{x}{\sqrt{x^2 + 1}}$

29. $\lim_{x\to-\infty} \frac{2x + 1}{\sqrt{x^2 - x}}$

30. $\lim_{x\to-\infty} \frac{-3x + 1}{\sqrt{x^2 + x}}$

31. $\lim_{x\to\infty} \frac{\sqrt{x^2 - 1}}{2x - 1}$

32. $\lim_{x\to-\infty} \frac{\sqrt{x^4 - 1}}{x^3 - 1}$

33. $\lim_{x\to\infty} \frac{x + 1}{(x^2 + 1)^{1/3}}$

34. $\lim_{x\to-\infty} \frac{2x}{(x^6 - 1)^{1/3}}$

In Exercises 35–38, use a graphing utility to graph the function and identify any horizontal asymptotes.

35. $f(x) = \frac{|x|}{x + 1}$

36. $f(x) = \frac{|3x + 2|}{x - 2}$

37. $f(x) = \frac{3x}{\sqrt{x^2 + 2}}$

38. $f(x) = \frac{\sqrt{9x^2 - 2}}{2x + 1}$

In Exercises 39–42, find the limit. (*Hint:* Treat the expression as a fraction whose denominator is 1, and rationalize the numerator.) Use a graphing utility to verify your result.

39. $\lim_{x \to -\infty} \left(x + \sqrt{x^2 + 3}\right)$ **40.** $\lim_{x \to \infty} \left(x - \sqrt{x^2 + x}\right)$

41. $\lim_{x \to -\infty} \left(3x + \sqrt{9x^2 - x}\right)$ **42.** $\lim_{x \to \infty} \left(4x - \sqrt{16x^2 - x}\right)$

Numerical, Graphical, and Analytic Analysis **In Exercises 43–46, use a graphing utility to complete the table and estimate the limit as x approaches infinity. Then use a graphing utility to graph the function and estimate the limit. Finally, find the limit analytically and compare your results with the estimates.**

x	10^0	10^1	10^2	10^3	10^4	10^5	10^6
$f(x)$							

43. $f(x) = x - \sqrt{x(x - 1)}$

44. $f(x) = x^2 - x\sqrt{x(x - 1)}$

45. $f(x) = 2x - \sqrt{4x^2 + 1}$

46. $f(x) = \dfrac{x + 1}{x\sqrt{x}}$

WRITING ABOUT CONCEPTS

47. Sketch a graph of a differentiable function f that satisfies the following conditions and has $x = 2$ as its only critical number.

$f'(x) < 0$ for $x < 2$ $f'(x) > 0$ for $x > 2$

$\lim_{x \to -\infty} f(x) = \lim_{x \to \infty} f(x) = 6$

48. Is it possible to sketch a graph of a function that satisfies the conditions of Exercise 47 and has *no* points of inflection? Explain.

49. If f is a continuous function such that $\lim_{x \to \infty} f(x) = 5$, find, if possible, $\lim_{x \to -\infty} f(x)$ for each specified condition.

(a) The graph of f is symmetric with respect to the y-axis.

(b) The graph of f is symmetric with respect to the origin.

CAPSTONE

50. The graph of a function f is shown below. To print an enlarged copy of the graph, go to the website *www.mathgraphs.com*.

(a) Sketch f'.

(b) Use the graphs to estimate $\lim_{x \to \infty} f(x)$ and $\lim_{x \to \infty} f'(x)$.

(c) Explain the answers you gave in part (b).

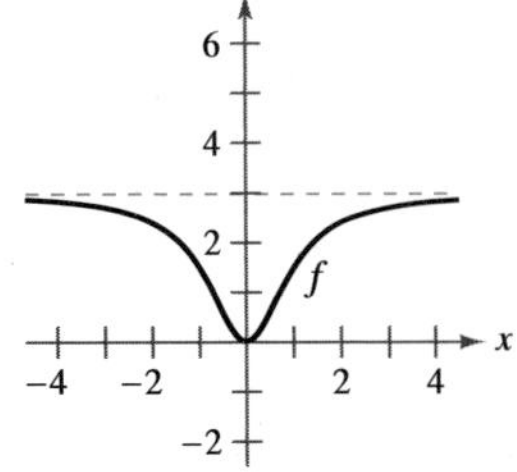

In Exercises 51–68, sketch the graph of the equation using extrema, intercepts, symmetry, and asymptotes. Then use a graphing utility to verify your result.

51. $y = \dfrac{x}{1 - x}$ **52.** $y = \dfrac{x - 4}{x - 3}$

53. $y = \dfrac{x + 1}{x^2 - 4}$ **54.** $y = \dfrac{2x}{9 - x^2}$

55. $y = \dfrac{x^2}{x^2 + 16}$ **56.** $y = \dfrac{x^2}{x^2 - 16}$

57. $y = \dfrac{2x^2}{x^2 - 4}$ **58.** $y = \dfrac{2x^2}{x^2 + 4}$

59. $xy^2 = 9$ **60.** $x^2y = 9$

61. $y = \dfrac{3x}{1 - x}$ **62.** $y = \dfrac{3x}{1 - x^2}$

63. $y = 2 - \dfrac{3}{x^2}$ **64.** $y = 1 + \dfrac{1}{x}$

65. $y = 3 + \dfrac{2}{x}$ **66.** $y = 4\left(1 - \dfrac{1}{x^2}\right)$

67. $y = \dfrac{x^3}{\sqrt{x^2 - 4}}$ **68.** $y = \dfrac{x}{\sqrt{x^2 - 4}}$

In Exercises 69–76, use a computer algebra system to analyze the graph of the function. Label any extrema and/or asymptotes that exist.

69. $f(x) = 9 - \dfrac{5}{x^2}$ **70.** $f(x) = \dfrac{x^2}{x^2 - 1}$

71. $f(x) = \dfrac{x}{x^2 - 4}$ **72.** $f(x) = \dfrac{1}{x^2 - x - 2}$

73. $f(x) = \dfrac{x - 2}{x^2 - 4x + 3}$ **74.** $f(x) = \dfrac{x + 1}{x^2 + x + 1}$

75. $f(x) = \dfrac{3x}{\sqrt{4x^2 + 1}}$ **76.** $g(x) = \dfrac{2x}{\sqrt{3x^2 + 1}}$

In Exercises 77 and 78, (a) use a graphing utility to graph f and g in the same viewing window, (b) verify analytically that f and g represent the same function, and (c) zoom out sufficiently far so that the graph appears as a line. What equation does this line appear to have? (Note that the points at which the function is not continuous are not readily seen when you zoom out.)

77. $f(x) = \dfrac{x^3 - 3x^2 + 2}{x(x - 3)}$ **78.** $f(x) = -\dfrac{x^3 - 2x^2 + 2}{2x^2}$

$g(x) = x + \dfrac{2}{x(x - 3)}$ $g(x) = -\dfrac{1}{2}x + 1 - \dfrac{1}{x^2}$

79. ***Engine Efficiency*** The efficiency of an internal combustion engine is

$$\text{Efficiency (\%)} = 100\left[1 - \frac{1}{(v_1/v_2)^c}\right]$$

where v_1/v_2 is the ratio of the uncompressed gas to the compressed gas and c is a positive constant dependent on the engine design. Find the limit of the efficiency as the compression ratio approaches infinity.

80. *Average Cost* A business has a cost of $C = 0.5x + 500$ for producing x units. The average cost per unit is

$$\overline{C} = \frac{C}{x}.$$

Find the limit of $\overline{C}$ as x approaches infinity.

81. *Temperature* The graph shows the temperature T, in degrees Fahrenheit, of molten glass t seconds after it is removed from a kiln.

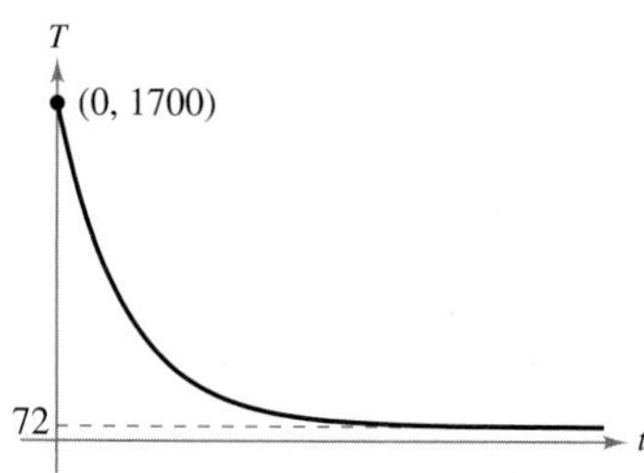

(a) Find $\lim_{t\to 0^+} T$. What does this limit represent?

(b) Find $\lim_{t\to\infty} T$. What does this limit represent?

(c) Will the temperature of the glass ever actually reach room temperature? Why?

82. The graph of $f(x) = \dfrac{2x^2}{x^2 + 2}$ is shown.

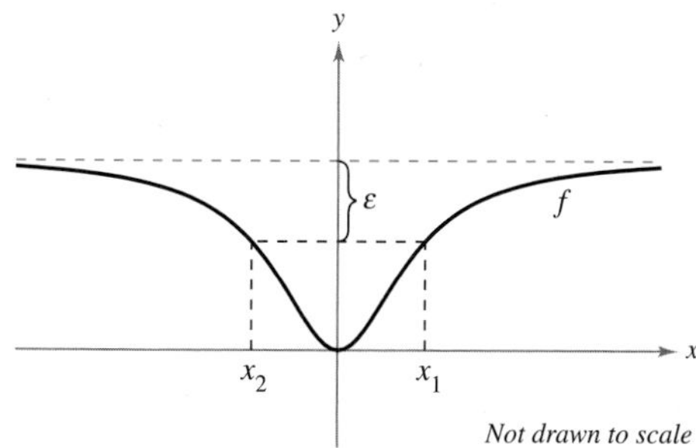

(a) Find $L = \lim_{x\to\infty} f(x)$.

(b) Determine x_1 and x_2 in terms of ε.

(c) Determine M, where $M > 0$, such that $|f(x) - L| < \varepsilon$ for $x > M$.

(d) Determine N, where $N < 0$, such that $|f(x) - L| < \varepsilon$ for $x < N$.

83. *Modeling Data* The average typing speeds S (in words per minute) of a typing student after t weeks of lessons are shown in the table.

t	5	10	15	20	25	30
S	28	56	79	90	93	94

A model for the data is $S = \dfrac{100t^2}{65 + t^2}$, $t > 0$.

(a) Use a graphing utility to plot the data and graph the model.

(b) Does there appear to be a limiting typing speed? Explain.

84. *Modeling Data* A heat probe is attached to the heat exchanger of a heating system. The temperature T (in degrees Celsius) is recorded t seconds after the furnace is started. The results for the first 2 minutes are recorded in the table.

t	0	15	30	45	60
T	25.2°	36.9°	45.5°	51.4°	56.0°

t	75	90	105	120
T	59.6°	62.0°	64.0°	65.2°

(a) Use the regression capabilities of a graphing utility to find a model of the form $T_1 = at^2 + bt + c$ for the data.

(b) Use a graphing utility to graph T_1.

(c) A rational model for the data is $T_2 = \dfrac{1451 + 86t}{58 + t}$. Use a graphing utility to graph T_2.

(d) Find $T_1(0)$ and $T_2(0)$.

(e) Find $\lim_{t\to\infty} T_2$.

(f) Interpret the result in part (e) in the context of the problem. Is it possible to do this type of analysis using T_1? Explain.

85. A line with slope m passes through the point $(0, 4)$.

(a) Write the distance d between the line and the point $(3, 1)$ as a function of m.

(b) Use a graphing utility to graph the equation in part (a).

(c) Find $\lim_{m\to\infty} d(m)$ and $\lim_{m\to-\infty} d(m)$. Interpret the results geometrically.

86. A line with slope m passes through the point $(0, -2)$.

(a) Write the distance d between the line and the point $(4, 2)$ as a function of m.

(b) Use a graphing utility to graph the equation in part (a).

(c) Find $\lim_{m\to\infty} d(m)$ and $\lim_{m\to-\infty} d(m)$. Interpret the results geometrically.

True or False? **In Exercises 87 and 88, determine whether the statement is true or false. If it is false, explain why or give an example that shows it is false.**

87. If $f'(x) > 0$ for all real numbers x, then f increases without bound.

88. If $f''(x) < 0$ for all real numbers x, then f decreases without bound.

89. Prove that if $p(x) = a_nx^n + \cdots + a_1x + a_0$ and $q(x) = b_mx^m + \cdots + b_1x + b_0$ ($a_n \neq 0$, $b_m \neq 0$), then

$$\lim_{x\to\infty} \frac{p(x)}{q(x)} = \begin{cases} 0, & n < m \\ \dfrac{a_n}{b_m}, & n = m. \\ \pm\infty, & n > m \end{cases}$$

5.6 A Summary of Curve Sketching

- **Analyze and sketch the graph of a function.**

Analyzing the Graph of a Function

It would be difficult to overstate the importance of using graphs in mathematics. Descartes's introduction of analytic geometry contributed significantly to the rapid advances in calculus that began during the mid-seventeenth century. In the words of Lagrange, "As long as algebra and geometry traveled separate paths their advance was slow and their applications limited. But when these two sciences joined company, they drew from each other fresh vitality and thenceforth marched on at a rapid pace toward perfection."

So far, you have studied several concepts that are useful in analyzing the graph of a function.

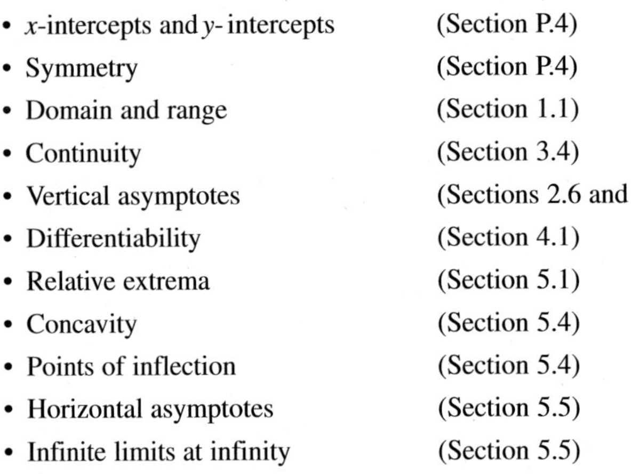

- x-intercepts and y-intercepts (Section P.4)
- Symmetry (Section P.4)
- Domain and range (Section 1.1)
- Continuity (Section 3.4)
- Vertical asymptotes (Sections 2.6 and 3.5)
- Differentiability (Section 4.1)
- Relative extrema (Section 5.1)
- Concavity (Section 5.4)
- Points of inflection (Section 5.4)
- Horizontal asymptotes (Section 5.5)
- Infinite limits at infinity (Section 5.5)

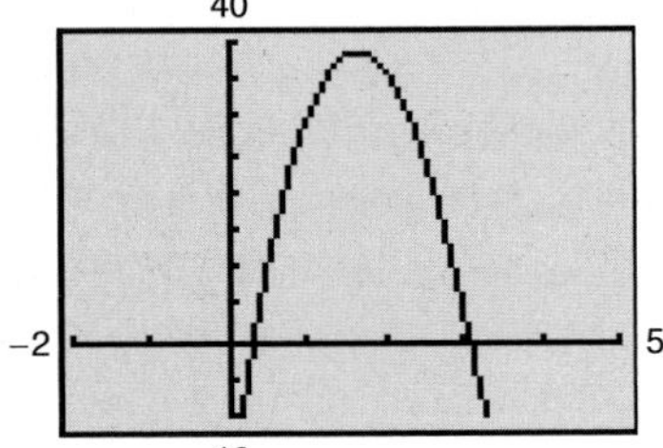

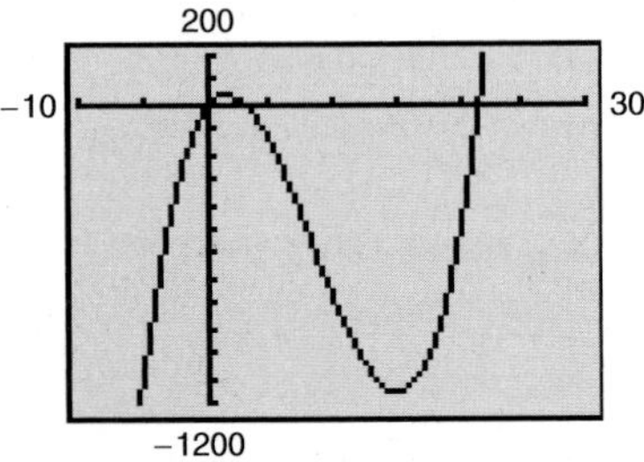

Different viewing windows for the graph of $f(x) = x^3 - 25x^2 + 74x - 20$
Figure 5.43

When you are sketching the graph of a function, either by hand or with a graphing utility, remember that normally you cannot show the *entire* graph. The decision as to which part of the graph you choose to show is often crucial. For instance, which of the viewing windows in Figure 5.43 better represents the graph of

$$f(x) = x^3 - 25x^2 + 74x - 20?$$

By seeing both views, it is clear that the second viewing window gives a more complete representation of the graph. But would a third viewing window reveal other interesting portions of the graph? To answer this, you need to use calculus to interpret the first and second derivatives. Here are some guidelines for determining a good viewing window for the graph of a function.

GUIDELINES FOR ANALYZING THE GRAPH OF A FUNCTION

1. Determine the domain and range of the function.
2. Determine the intercepts, asymptotes, and symmetry of the graph.
3. Locate the x-values for which $f'(x)$ and $f''(x)$ either are zero or do not exist. Use the results to determine relative extrema and points of inflection.

NOTE In these guidelines, note the importance of *algebra* (as well as calculus) for solving the equations $f(x) = 0$, $f'(x) = 0$, and $f''(x) = 0$. ■

EXAMPLE 1 Sketching the Graph of a Rational Function

Analyze and sketch the graph of $f(x) = \dfrac{2(x^2 - 9)}{x^2 - 4}$.

Solution

First derivative: $f'(x) = \dfrac{20x}{(x^2 - 4)^2}$

Second derivative: $f''(x) = \dfrac{-20(3x^2 + 4)}{(x^2 - 4)^3}$

x-intercepts: $(-3, 0), (3, 0)$

y-intercept: $\left(0, \frac{9}{2}\right)$

Vertical asymptotes: $x = -2, x = 2$

Horizontal asymptote: $y = 2$

Critical number: $x = 0$

Possible points of inflection: None

Domain: All real numbers except $x = \pm 2$

Symmetry: With respect to y-axis

Test intervals: $(-\infty, -2), (-2, 0), (0, 2), (2, \infty)$

Using calculus, you can be certain that you have determined all characteristics of the graph of f.
Figure 5.44

The table shows how the test intervals are used to determine several characteristics of the graph. The graph of f is shown in Figure 5.44.

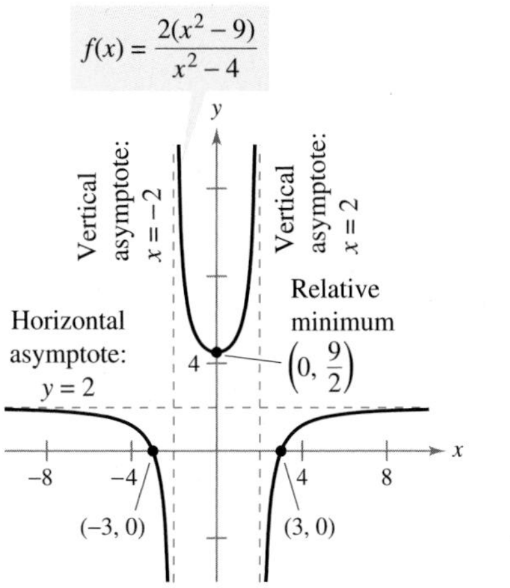

	$f(x)$	$f'(x)$	$f''(x)$	Characteristic of Graph
$-\infty < x < -2$		$-$	$-$	Decreasing, concave downward
$x = -2$	Undef.	Undef.	Undef.	Vertical asymptote
$-2 < x < 0$		$-$	$+$	Decreasing, concave upward
$x = 0$	$\frac{9}{2}$	0	$+$	Relative minimum
$0 < x < 2$		$+$	$+$	Increasing, concave upward
$x = 2$	Undef.	Undef.	Undef.	Vertical asymptote
$2 < x < \infty$		$+$	$-$	Increasing, concave downward

■ **FOR FURTHER INFORMATION** For more information on the use of technology to graph rational functions, see the article "Graphs of Rational Functions for Computer Assisted Calculus" by Stan Byrd and Terry Walters in *The College Mathematics Journal*. To view this article, go to the website *www.matharticles.com*.

Be sure you understand all of the implications of creating a table such as that shown in Example 1. By using calculus, you can *be sure* that the graph has no relative extrema or points of inflection other than those shown in Figure 5.44.

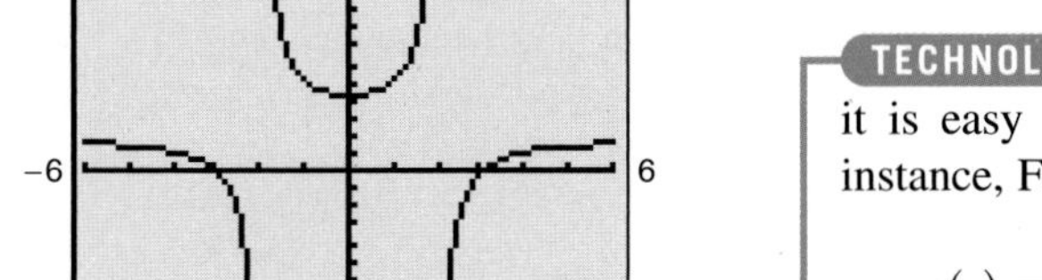

By not using calculus, you may overlook important characteristics of the graph of g.
Figure 5.45

TECHNOLOGY PITFALL Without using the type of analysis outlined in Example 1, it is easy to obtain an incomplete view of a graph's basic characteristics. For instance, Figure 5.45 shows a view of the graph of

$$g(x) = \frac{2(x^2 - 9)(x - 20)}{(x^2 - 4)(x - 21)}.$$

From this view, it appears that the graph of g is about the same as the graph of f shown in Figure 5.44. The graphs of these two functions, however, differ significantly. Try enlarging the viewing window to see the differences.

EXAMPLE 2 Sketching the Graph of a Rational Function

Analyze and sketch the graph of $f(x) = \dfrac{x^2 - 2x + 4}{x - 2}$.

Solution

First derivative: $f'(x) = \dfrac{x(x-4)}{(x-2)^2}$

Second derivative: $f''(x) = \dfrac{8}{(x-2)^3}$

x-intercepts: None

y-intercept: $(0, -2)$

Vertical asymptote: $x = 2$

Horizontal asymptotes: None

End behavior: $\lim_{x\to-\infty} f(x) = -\infty, \ \lim_{x\to\infty} f(x) = \infty$

Critical numbers: $x = 0, x = 4$

Possible points of inflection: None

Domain: All real numbers except $x = 2$

Test intervals: $(-\infty, 0), (0, 2), (2, 4), (4, \infty)$

The analysis of the graph of f is shown in the table, and the graph is shown in Figure 5.46.

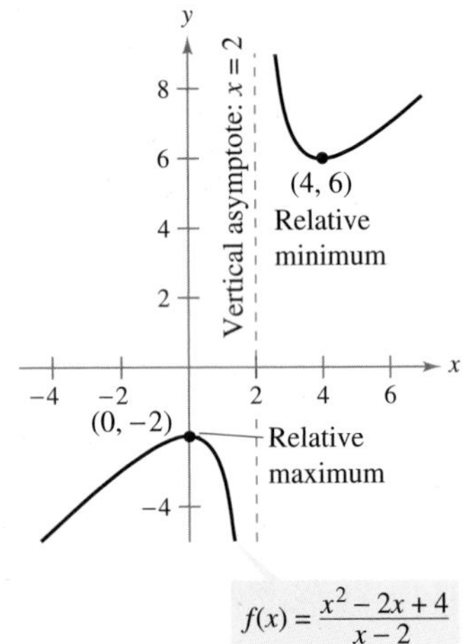

Figure 5.46

	$f(x)$	$f'(x)$	$f''(x)$	Characteristic of Graph
$-\infty < x < 0$		$+$	$-$	Increasing, concave downward
$x = 0$	-2	0	$-$	Relative maximum
$0 < x < 2$		$-$	$-$	Decreasing, concave downward
$x = 2$	Undef.	Undef.	Undef.	Vertical asymptote
$2 < x < 4$		$-$	$+$	Decreasing, concave upward
$x = 4$	6	0	$+$	Relative minimum
$4 < x < \infty$		$+$	$+$	Increasing, concave upward

■

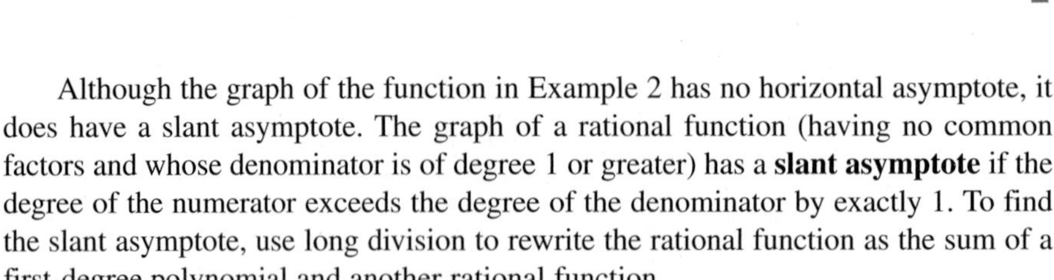

Although the graph of the function in Example 2 has no horizontal asymptote, it does have a slant asymptote. The graph of a rational function (having no common factors and whose denominator is of degree 1 or greater) has a **slant asymptote** if the degree of the numerator exceeds the degree of the denominator by exactly 1. To find the slant asymptote, use long division to rewrite the rational function as the sum of a first-degree polynomial and another rational function.

$$f(x) = \frac{x^2 - 2x + 4}{x - 2} \quad \text{Write original equation.}$$

$$= x + \frac{4}{x - 2} \quad \text{Rewrite using long division.}$$

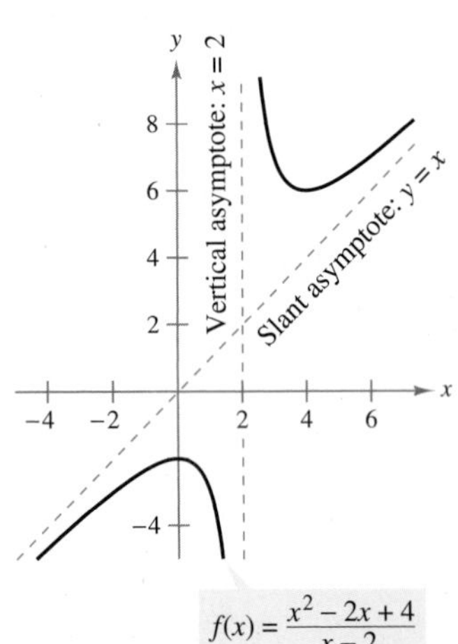

A slant asymptote
Figure 5.47

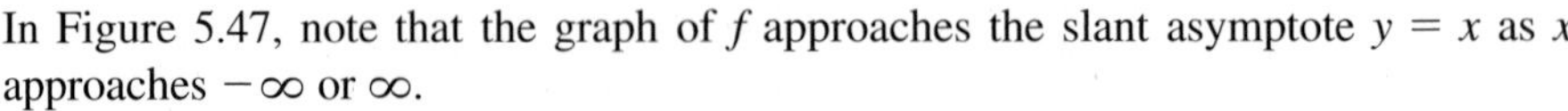

In Figure 5.47, note that the graph of f approaches the slant asymptote $y = x$ as x approaches $-\infty$ or ∞.

EXAMPLE 3 Sketching the Graph of a Radical Function

Analyze and sketch the graph of

$$f(x) = \frac{x}{\sqrt{x^2 + 2}}.$$

Solution

$$f'(x) = \frac{2}{(x^2 + 2)^{3/2}} \qquad f''(x) = -\frac{6x}{(x^2 + 2)^{5/2}}$$

The graph has only one intercept, $(0, 0)$. It has no vertical asymptotes, but it has two horizontal asymptotes: $y = 1$ (to the right) and $y = -1$ (to the left). The function has no critical numbers and one possible point of inflection (at $x = 0$). The domain of the function is all real numbers, and the graph is symmetric with respect to the origin. The analysis of the graph of f is shown in the table, and the graph is shown in Figure 5.48.

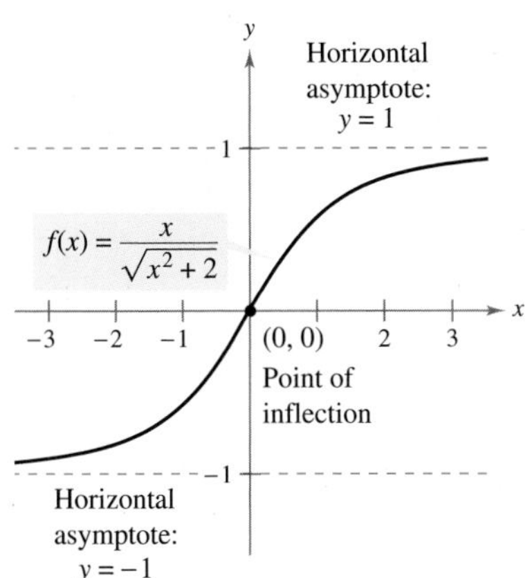

Figure 5.48

	$f(x)$	$f'(x)$	$f''(x)$	Characteristic of Graph
$-\infty < x < 0$		+	+	Increasing, concave upward
$x = 0$	0	$\frac{1}{\sqrt{2}}$	0	Point of inflection
$0 < x < \infty$		+	−	Increasing, concave downward

EXAMPLE 4 Sketching the Graph of a Radical Function

Analyze and sketch the graph of

$$f(x) = 2x^{5/3} - 5x^{4/3}.$$

Solution

$$f'(x) = \frac{10}{3}x^{1/3}(x^{1/3} - 2) \qquad f''(x) = \frac{20(x^{1/3} - 1)}{9x^{2/3}}$$

The function has two intercepts: $(0, 0)$ and $\left(\frac{125}{8}, 0\right)$. There are no horizontal or vertical asymptotes. The function has two critical numbers ($x = 0$ and $x = 8$) and two possible points of inflection ($x = 0$ and $x = 1$). The domain is all real numbers. The analysis of the graph of f is shown in the table, and the graph is shown in Figure 5.49.

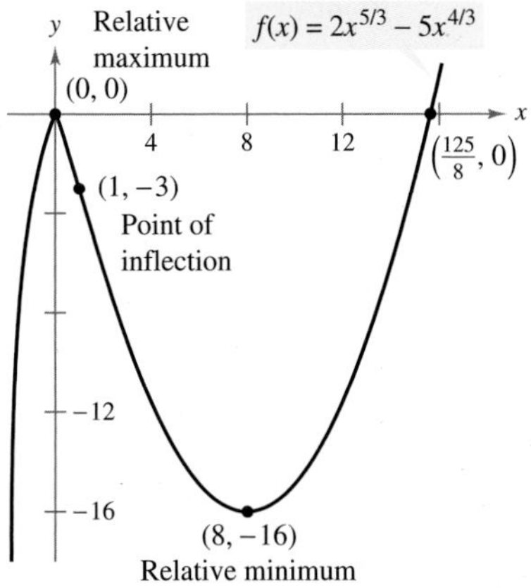

Figure 5.49

	$f(x)$	$f'(x)$	$f''(x)$	Characteristic of Graph
$-\infty < x < 0$		+	−	Increasing, concave downward
$x = 0$	0	0	Undef.	Relative maximum
$0 < x < 1$		−	−	Decreasing, concave downward
$x = 1$	−3	−	0	Point of inflection
$1 < x < 8$		−	+	Decreasing, concave upward
$x = 8$	−16	0	+	Relative minimum
$8 < x < \infty$		+	+	Increasing, concave upward

■

(a)

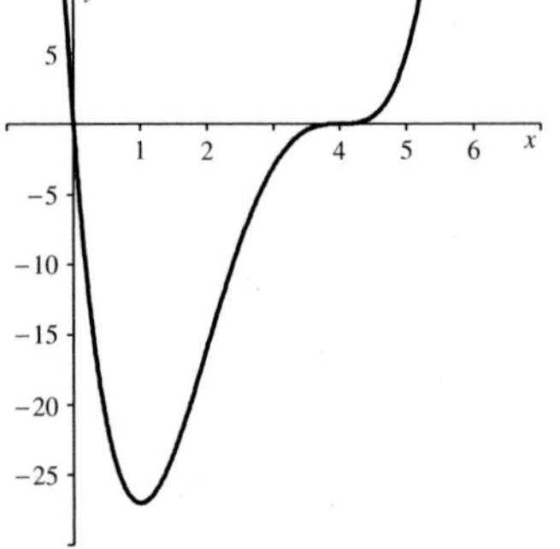

(b)
A polynomial function of even degree must have at least one relative extremum.
Figure 5.50

EXAMPLE 5 Sketching the Graph of a Polynomial Function

Analyze and sketch the graph of

$$f(x) = x^4 - 12x^3 + 48x^2 - 64x.$$

Solution Begin by factoring to obtain

$$f(x) = x^4 - 12x^3 + 48x^2 - 64x$$
$$= x(x - 4)^3.$$

Then, using the factored form of $f(x)$, you can perform the following analysis.

First derivative: $f'(x) = 4(x - 1)(x - 4)^2$
Second derivative: $f''(x) = 12(x - 4)(x - 2)$
x-intercepts: (0, 0), (4, 0)
y-intercept: (0, 0)
Vertical asymptotes: None
Horizontal asymptotes: None
End behavior: $\lim_{x\to-\infty} f(x) = \infty, \ \lim_{x\to\infty} f(x) = \infty$
Critical numbers: $x = 1, x = 4$
Possible points of inflection: $x = 2, x = 4$
Domain: All real numbers
Test intervals: $(-\infty, 1), (1, 2), (2, 4), (4, \infty)$

The analysis of the graph of f is shown in the table, and the graph is shown in Figure 5.50(a). Using a computer algebra system such as *Maple* [see Figure 5.50(b)] can help you verify your analysis.

	$f(x)$	$f'(x)$	$f''(x)$	Characteristic of Graph
$-\infty < x < 1$		−	+	Decreasing, concave upward
$x = 1$	−27	0	+	Relative minimum
$1 < x < 2$		+	+	Increasing, concave upward
$x = 2$	−16	+	0	Point of inflection
$2 < x < 4$		+	−	Increasing, concave downward
$x = 4$	0	0	0	Point of inflection
$4 < x < \infty$		+	+	Increasing, concave upward

■

The fourth-degree polynomial function in Example 5 has one relative minimum and no relative maxima. In general, a polynomial function of degree n can have *at most* $n - 1$ relative extrema, and *at most* $n - 2$ points of inflection. Moreover, polynomial functions of even degree must have *at least* one relative extremum.

Remember from the Leading Coefficient Test described in Section 2.2 that the "end behavior" of the graph of a polynomial function is determined by its leading coefficient and its degree. For instance, because the polynomial in Example 5 has a positive leading coefficient, the graph rises to the right. Moreover, because the degree is even, the graph also rises to the left.

5.6 Exercises

See www.CalcChat.com for worked-out solutions to odd-numbered exercises.

In Exercises 1–4, match the graph of f in the left column with that of its derivative in the right column.

Graph of f | *Graph of f′*

1. 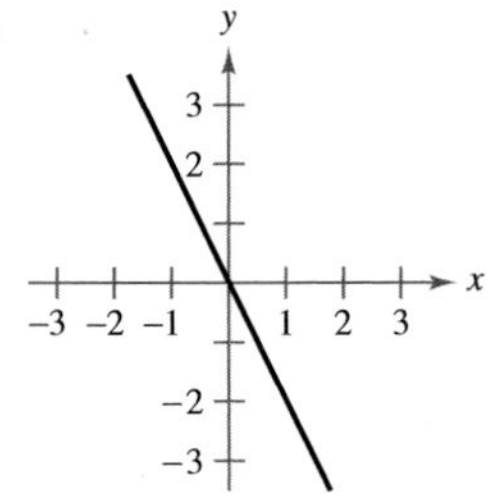
(a) 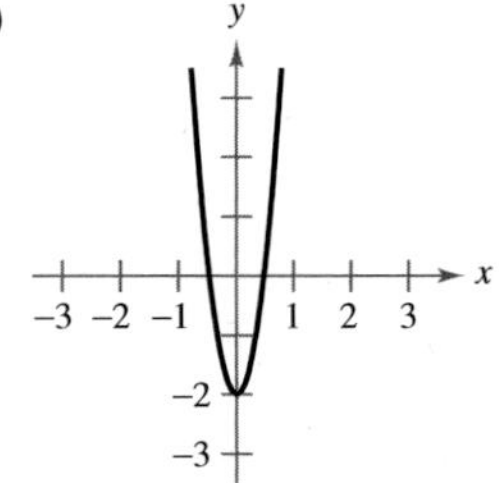

2. 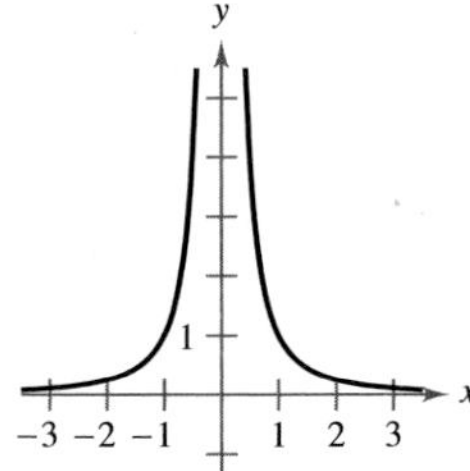
(b) 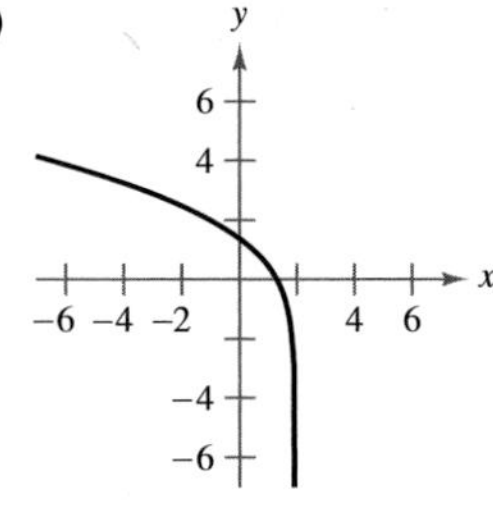

3. 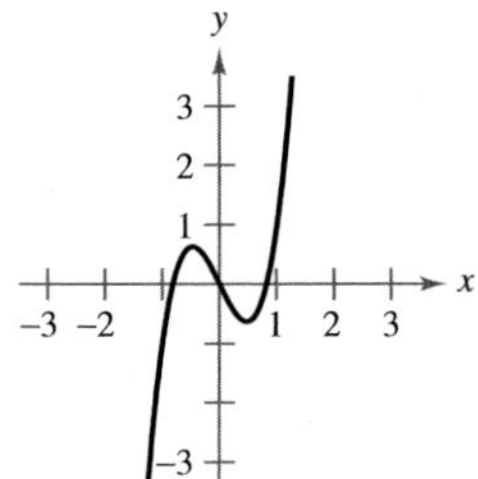
(c) 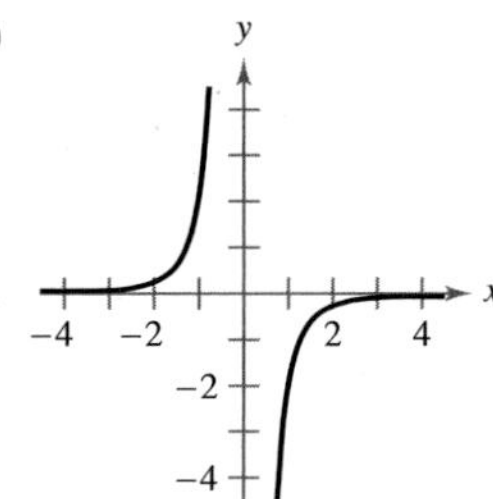

4. 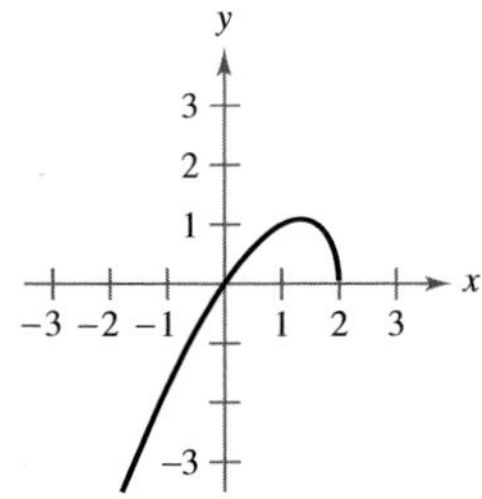
(d) 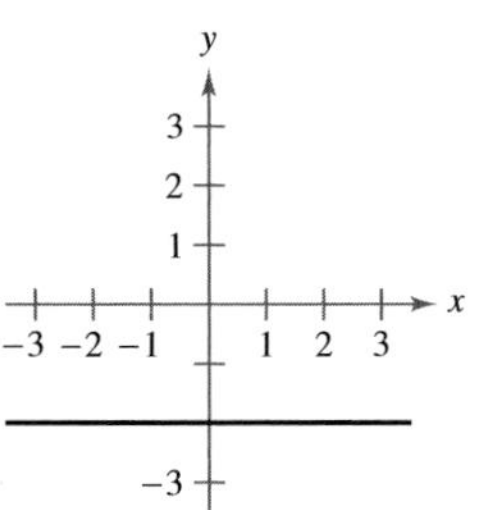

In Exercises 5–36, analyze and sketch a graph of the function. Label any intercepts, relative extrema, points of inflection, and asymptotes. Use a graphing utility to verify your results.

5. $y = \dfrac{1}{x-2} - 3$

6. $y = \dfrac{x}{x^2+1}$

7. $y = \dfrac{x^2}{x^2+3}$

8. $y = \dfrac{x^2+1}{x^2-4}$

9. $y = \dfrac{3x}{x^2-1}$

10. $f(x) = \dfrac{x-3}{x}$

11. $g(x) = x - \dfrac{8}{x^2}$

12. $f(x) = x + \dfrac{32}{x^2}$

13. $f(x) = \dfrac{x^2+1}{x}$

14. $f(x) = \dfrac{x^3}{x^2-9}$

15. $y = \dfrac{x^2-6x+12}{x-4}$

16. $y = \dfrac{2x^2-5x+5}{x-2}$

17. $y = x\sqrt{4-x}$

18. $g(x) = x\sqrt{9-x}$

19. $h(x) = x\sqrt{4-x^2}$

20. $g(x) = x\sqrt{9-x^2}$

21. $y = 3x^{2/3} - 2x$

22. $y = 3(x-1)^{2/3} - (x-1)^2$

23. $y = x^3 - 3x^2 + 3$

24. $y = -\frac{1}{3}(x^3 - 3x + 2)$

25. $y = 2 - x - x^3$

26. $f(x) = \frac{1}{3}(x-1)^3 + 2$

27. $f(x) = 3x^3 - 9x + 1$

28. $f(x) = (x+1)(x-2)(x-5)$

29. $y = 3x^4 + 4x^3$

30. $y = 3x^4 - 6x^2 + \frac{5}{3}$

31. $f(x) = x^4 - 4x^3 + 16x$

32. $f(x) = x^4 - 8x^3 + 18x^2 - 16x + 5$

33. $y = x^5 - 5x$

34. $y = (x-1)^5$

35. $y = |2x - 3|$

36. $y = |x^2 - 6x + 5|$

In Exercises 37–40, use a computer algebra system to analyze and graph the function. Identify any relative extrema, points of inflection, and asymptotes.

37. $f(x) = \dfrac{20x}{x^2+1} - \dfrac{1}{x}$

38. $f(x) = x + \dfrac{4}{x^2+1}$

39. $f(x) = \dfrac{-2x}{\sqrt{x^2+7}}$

40. $f(x) = \dfrac{4x}{\sqrt{x^2+15}}$

WRITING ABOUT CONCEPTS

41. Suppose $f'(t) < 0$ for all t in the interval $(2, 8)$. Explain why $f(3) > f(5)$.

42. Suppose $f(0) = 3$ and $2 \le f'(x) \le 4$ for all x in the interval $[-5, 5]$. Determine the greatest and least possible values of $f(2)$.

In Exercises 43 and 44, the graphs of f, f', and f'' are shown on the same set of coordinate axes. Which is which? Explain your reasoning. To print an enlarged copy of the graph, go to the website *www.mathgraphs.com*.

43. 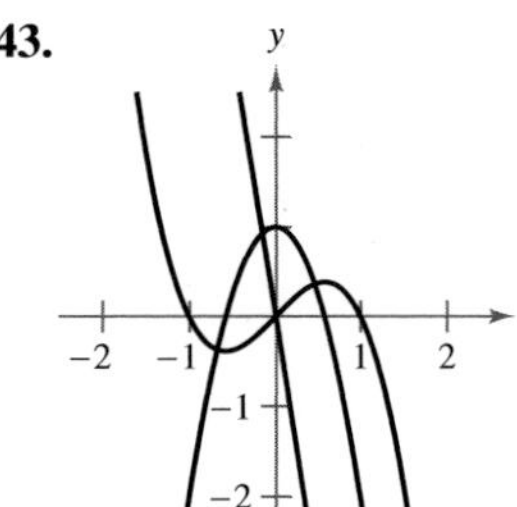

44. 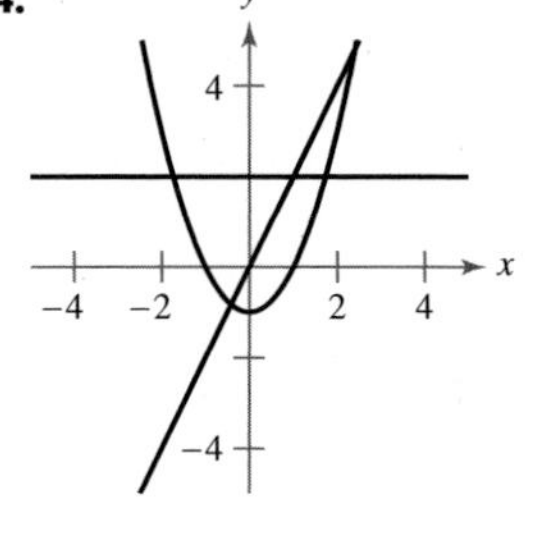

WRITING ABOUT CONCEPTS (continued)

In Exercises 45 and 46, use a graphing utility to graph the function. Use the graph to determine whether it is possible for the graph of a function to cross its horizontal asymptote. Do you think it is possible for the graph of a function to cross its vertical asymptote? Why or why not?

45. $f(x) = \dfrac{4(x-1)^2}{x^2 - 4x + 5}$

46. $g(x) = \dfrac{3x^4 - 5x + 3}{x^4 + 1}$

In Exercises 47 and 48, use a graphing utility to graph the function. Explain why there is no vertical asymptote when a superficial examination of the function may indicate that there should be one.

47. $h(x) = \dfrac{6 - 2x}{3 - x}$ **48.** $g(x) = \dfrac{x^2 + x - 2}{x - 1}$

In Exercises 49–52, use a graphing utility to graph the function and determine the slant asymptote of the graph. Zoom out repeatedly and describe how the graph on the display appears to change. Why does this occur?

49. $f(x) = -\dfrac{x^2 - 3x - 1}{x - 2}$ **50.** $g(x) = \dfrac{2x^2 - 8x - 15}{x - 5}$

51. $f(x) = \dfrac{2x^3}{x^2 + 1}$ **52.** $h(x) = \dfrac{-x^3 + x^2 + 4}{x^2}$

Graphical Reasoning **In Exercises 53–56, use the graph of f' to sketch a graph of f and the graph of f''. To print an enlarged copy of the graph, go to the website *www.mathgraphs.com*.**

53.

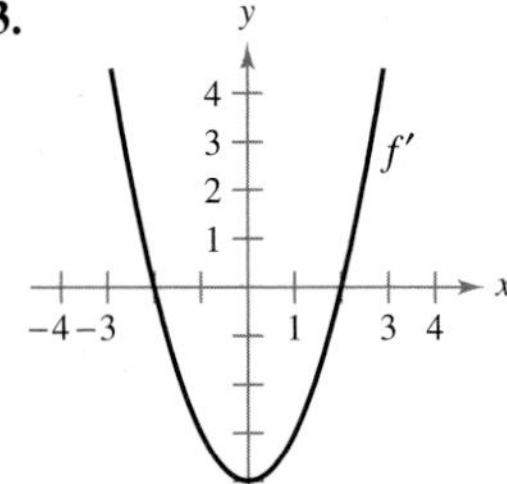

54.

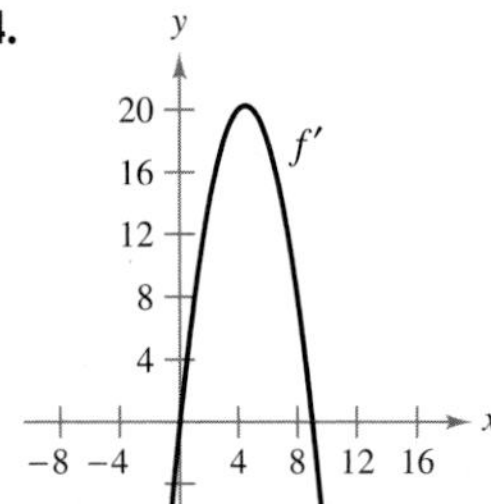

55.

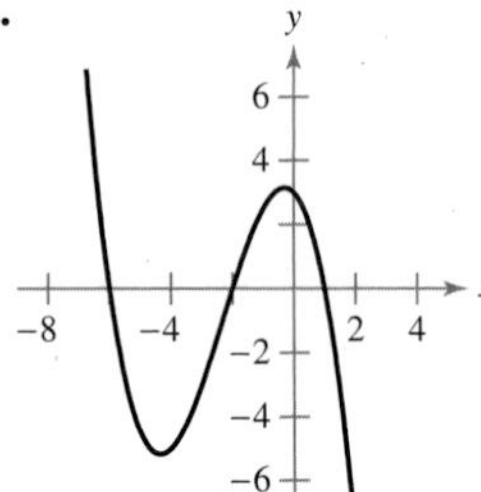

56.

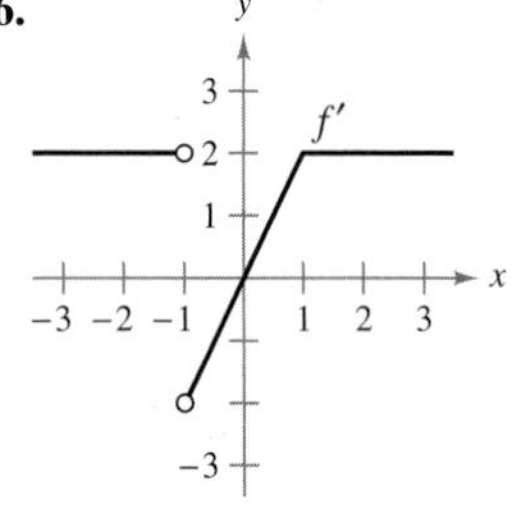

(Submitted by Bill Fox, Moberly Area Community College, Moberly, MO)

Think About It **In Exercises 57–60, create a function whose graph has the given characteristics. (There is more than one correct answer.)**

57. Vertical asymptote: $x = 3$
Horizontal asymptote: $y = 0$

58. Vertical asymptote: $x = -5$
Horizontal asymptote: None

59. Vertical asymptote: $x = 3$
Slant asymptote: $y = 3x + 2$

60. Vertical asymptote: $x = 2$
Slant asymptote: $y = -x$

61. ***Graphical Reasoning*** The graph of f is shown in the figure.

(a) For which values of x is $f'(x)$ zero? Positive? Negative?

(b) For which values of x is $f''(x)$ zero? Positive? Negative?

(c) On what interval is f' an increasing function?

(d) For which value of x is $f'(x)$ minimum? For this value of x, how does the rate of change of f compare with the rates of change of f for other values of x? Explain.

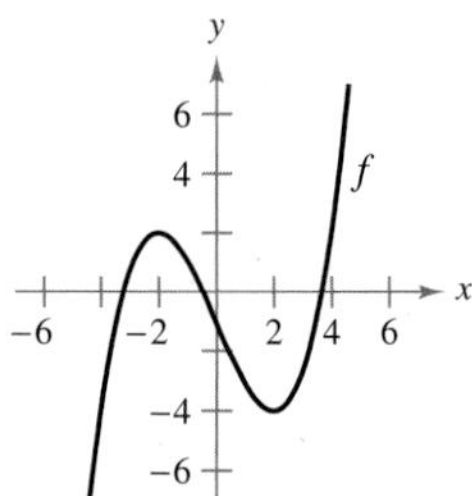

CAPSTONE

62. ***Graphical Reasoning*** Identify the real numbers $x_0, x_1, x_2, x_3,$ and x_4 in the figure such that each of the following is true.

(a) $f'(x) = 0$

(b) $f''(x) = 0$

(c) $f'(x)$ does not exist.

(d) f has a relative maximum.

(e) f has a point of inflection.

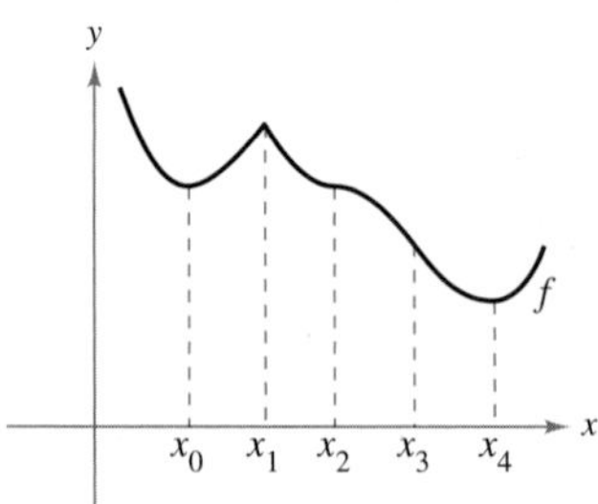

63. ***Graphical Reasoning*** Consider the function

$$f(x) = \frac{ax}{(x-b)^2}.$$

Determine the effect on the graph of f as a and b are changed. Consider cases where a and b are both positive or both negative, and cases where a and b have opposite signs.

64. Consider the function $f(x) = \frac{1}{2}(ax)^2 - ax,\ a \neq 0.$

(a) Determine the changes (if any) in the intercepts, extrema, and concavity of the graph of f when a is varied.

(b) In the same viewing window, use a graphing utility to graph the function for four different values of a.

65. ***Investigation*** Consider the function

$$f(x) = \frac{2x^n}{x^4 + 1}$$

for nonnegative integer values of n.

(a) Discuss the relationship between the value of n and the symmetry of the graph.

(b) For which values of n will the x-axis be the horizontal asymptote?

(c) For which value of n will $y = 2$ be the horizontal asymptote?

(d) What is the asymptote of the graph when $n = 5$?

(e) Use a graphing utility to graph f for the indicated values of n in the table. Use the graph to determine the number of extrema M and the number of inflection points N of the graph.

n	0	1	2	3	4	5
M						
N						

66. ***Investigation*** Let $P(x_0, y_0)$ be an arbitrary point on the graph of f such that $f'(x_0) \neq 0$, as shown in the figure. Verify each statement.

(a) The x-intercept of the tangent line is $\left(x_0 - \dfrac{f(x_0)}{f'(x_0)}, 0\right).$

(b) The y-intercept of the tangent line is $(0, f(x_0) - x_0 f'(x_0)).$

(c) The x-intercept of the normal line is $(x_0 + f(x_0)f'(x_0), 0).$

(d) The y-intercept of the normal line is $\left(0, y_0 + \dfrac{x_0}{f'(x_0)}\right).$

(e) $|BC| = \left|\dfrac{f(x_0)}{f'(x_0)}\right|$

(f) $|PC| = \left|\dfrac{f(x_0)\sqrt{1 + [f'(x_0)]^2}}{f'(x_0)}\right|$

(g) $|AB| = |f(x_0)f'(x_0)|$

(h) $|AP| = |f(x_0)|\sqrt{1 + [f'(x_0)]^2}$

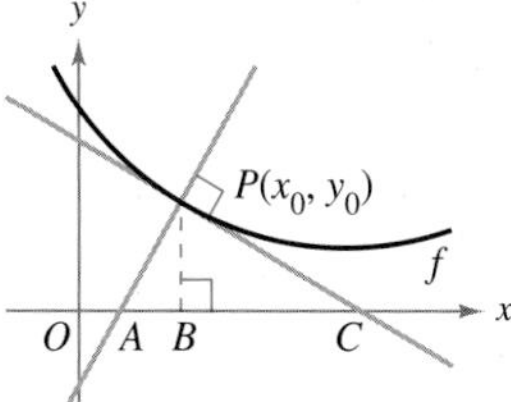

67. ***Profit*** The management of a company is considering three possible models for predicting the company's profits from 2006 through 2011. Model I gives the expected annual profits if the current trends continue. Models II and III give the expected annual profits for various combinations of increased labor and energy costs. In each model, p is the profit (in billions of dollars) and $t = 0$ corresponds to 2006.

Model I: $p = 0.03t^2 - 0.01t + 3.39$

Model II: $p = 0.08t + 3.36$

Model III: $p = -0.07t^2 + 0.05t + 3.38$

(a) Use a graphing utility to graph all three models in the same viewing window.

(b) For which models are profits increasing during the interval from 2006 through 2011?

(c) Which model is the most optimistic? Which is the most pessimistic? Which model would you choose? Explain.

68. ***Modeling Data*** The data in the table show the number N of bacteria in a culture at time t, where t is measured in days.

t	1	2	3	4	5	6	7	8
N	25	200	804	1756	2296	2434	2467	2473

A model for these data is given by

$$N = \frac{24{,}670 - 35{,}153t + 13{,}250t^2}{100 - 39t + 7t^2}, \quad 1 \leq t \leq 8.$$

(a) Use a graphing utility to plot the data and graph the model.

(b) Use the model to estimate the number of bacteria when $t = 10$.

(c) Approximate the day when the number of bacteria is greatest.

(d) Use a computer algebra system to determine the time when the rate of increase in the number of bacteria is greatest.

(e) Find $\lim_{t\to\infty} N(t)$.

Slant Asymptotes **In Exercises 69 and 70, the graph of the function has two slant asymptotes. Identify each slant asymptote. Then graph the function and its asymptotes.**

69. $y = \sqrt{4 + 16x^2}$

70. $y = \sqrt{x^2 + 6x}$

5.7 Optimization Problems

■ **Use calculus to solve applied minimum and maximum problems.**

Applied Minimum and Maximum Problems

One of the most common applications of calculus involves the determination of minimum and maximum values. Consider how frequently you hear or read terms such as greatest profit, least cost, least time, greatest voltage, optimum size, least size, greatest strength, and greatest distance. Before outlining a general problem-solving strategy for such problems, let's look at an example.

EXAMPLE 1 Finding Maximum Volume

A manufacturer wants to design an open box having a square base and a surface area of 108 square inches, as shown in Figure 5.51. What dimensions will produce a box with maximum volume?

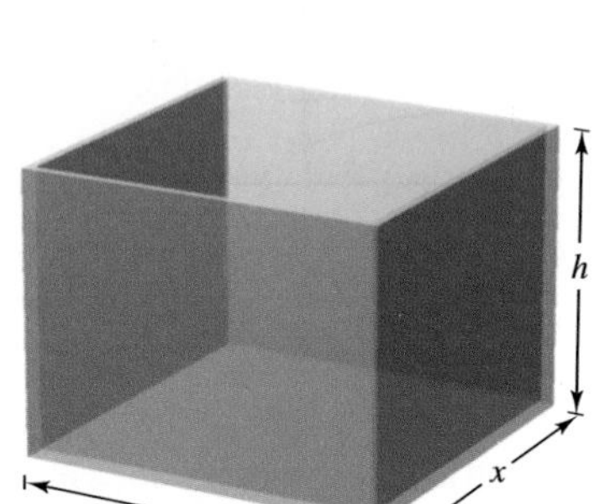

Open box with square base:
$S = x^2 + 4xh = 108$
Figure 5.51

Solution Because the box has a square base, its volume is

$$V = x^2h. \qquad \text{Primary equation}$$

This equation is called the **primary equation** because it gives a formula for the quantity to be optimized. The surface area of the box is

$$S = (\text{area of base}) + (\text{area of four sides})$$
$$S = x^2 + 4xh = 108. \qquad \text{Secondary equation}$$

Because V is to be maximized, you want to write V as a function of just one variable. To do this, you can solve the secondary equation $x^2 + 4xh = 108$ for h in terms of x to obtain $h = (108 - x^2)/(4x)$. Substituting into the primary equation produces

$$\begin{aligned} V &= x^2h && \text{Function of two variables} \\ &= x^2\left(\frac{108 - x^2}{4x}\right) && \text{Substitute for } h. \\ &= 27x - \frac{x^3}{4}. && \text{Function of one variable} \end{aligned}$$

Before finding which x-value will yield a maximum value of V, you should determine the *feasible domain*. That is, what values of x make sense in this problem? You know that $V \geq 0$. You also know that x must be nonnegative and that the area of the base $(A = x^2)$ is at most 108. So, the feasible domain is

$$0 \leq x \leq \sqrt{108}. \qquad \text{Feasible domain}$$

To maximize V, find the critical numbers of the volume function on the interval $(0, \sqrt{108})$.

$$\begin{aligned} \frac{dV}{dx} = 27 - \frac{3x^2}{4} &= 0 && \text{Set derivative equal to 0.} \\ 3x^2 &= 108 && \text{Simplify.} \\ x &= \pm 6 && \text{Critical numbers} \end{aligned}$$

So, the critical numbers are $x = \pm 6$. You do not need to consider $x = -6$ because it is not in the domain. Evaluating V at the critical number $x = 6$ and at the endpoints of the domain produces $V(0) = 0$, $V(6) = 108$, and $V(\sqrt{108}) = 0$. So, V is maximum when $x = 6$ and the dimensions of the box are $6 \times 6 \times 3$ inches. ■

TECHNOLOGY You can verify your answer in Example 1 by using a graphing utility to graph the volume function

$$V = 27x - \frac{x^3}{4}.$$

Use a viewing window in which $0 \leq x \leq \sqrt{108} \approx 10.4$ and $0 \leq y \leq 120$, and use the *trace* feature to determine the maximum value of V.

In Example 1, you should realize that there are infinitely many open boxes having 108 square inches of surface area. To begin solving the problem, you might ask yourself which basic shape would seem to yield a maximum volume. Should the box be tall, squat, or nearly cubical?

You might even try calculating a few volumes, as shown in Figure 5.52, to see if you can get a better feeling for what the optimum dimensions should be. Remember that you are not ready to begin solving a problem until you have clearly identified what the problem is.

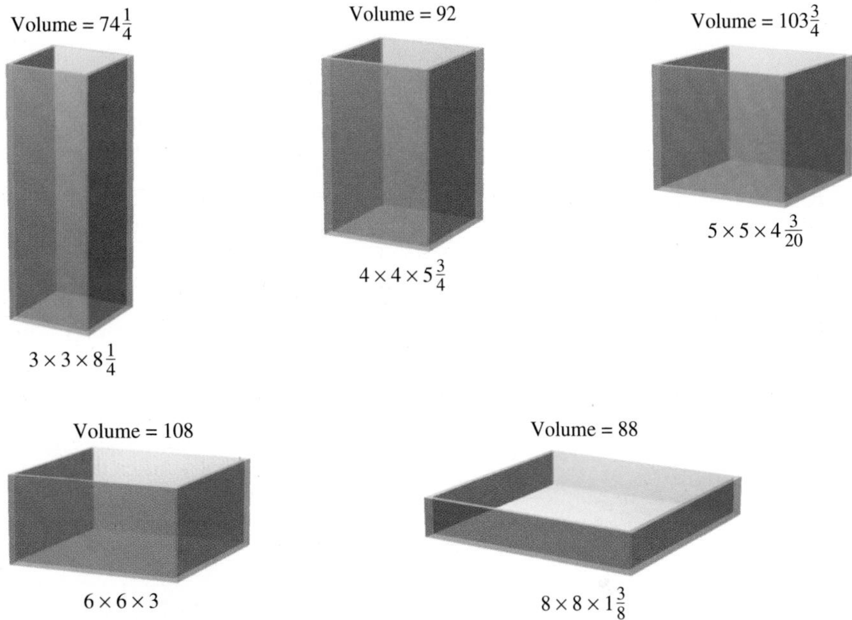

Which box has the greatest volume?
Figure 5.52

Example 1 illustrates the following guidelines for solving applied minimum and maximum problems.

GUIDELINES FOR SOLVING APPLIED MINIMUM AND MAXIMUM PROBLEMS

1. Identify all *given* quantities and all quantities *to be determined.* If possible, make a sketch and label it with any relevant measurements.
2. Write a **primary equation** for the quantity that is to be maximized or minimized. (A review of several useful formulas from geometry is presented inside the back cover.)
3. Reduce the primary equation to one having a *single independent variable.* This may involve the use of **secondary equations** relating the independent variables of the primary equation.
4. Determine the feasible domain of the primary equation. That is, determine the values for which the stated problem makes sense.
5. Determine the desired maximum or minimum value by the calculus techniques discussed in Sections 5.1 through 5.4.

NOTE When performing Step 5, recall that to determine the maximum or minimum value of a continuous function f on a closed interval, you should compare the values of f at its critical numbers with the values of f at the endpoints of the interval.

EXAMPLE 2 Finding Minimum Distance

Which points on the graph of

$$y = 4 - x^2$$

are closest to the point $(0, 2)$?

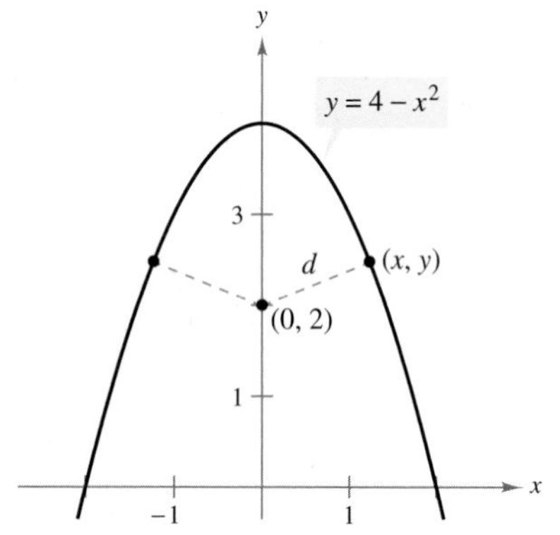

The quantity to be minimized is distance: $d = \sqrt{(x - 0)^2 + (y - 2)^2}$.
Figure 5.53

Solution Figure 5.53 shows that there are two points at a minimum distance from the point $(0, 2)$. The distance between the point $(0, 2)$ and a point (x, y) on the graph of $y = 4 - x^2$ is given by

$$d = \sqrt{(x - 0)^2 + (y - 2)^2}. \qquad \text{Primary equation}$$

Using the secondary equation $y = 4 - x^2$, you can rewrite the primary equation as

$$d = \sqrt{x^2 + (4 - x^2 - 2)^2} = \sqrt{x^4 - 3x^2 + 4}.$$

Because d is smallest when the expression inside the radical is smallest, you need only find the critical numbers of $f(x) = x^4 - 3x^2 + 4$. Note that the domain of f is the entire real line. So, there are no endpoints of the domain to consider. Moreover, setting $f'(x)$ equal to 0 yields

$$f'(x) = 4x^3 - 6x = 2x(2x^2 - 3) = 0$$

$$x = 0, \sqrt{\frac{3}{2}}, -\sqrt{\frac{3}{2}}.$$

The First Derivative Test verifies that $x = 0$ yields a relative maximum, whereas both $x = \sqrt{3/2}$ and $x = -\sqrt{3/2}$ yield a minimum distance. So, the closest points are $\left(\sqrt{3/2}, 5/2\right)$ and $\left(-\sqrt{3/2}, 5/2\right)$.

EXAMPLE 3 Finding Minimum Area

A rectangular page is to contain 24 square inches of print. The margins at the top and bottom of the page are to be $1\frac{1}{2}$ inches, and the margins on the left and right are to be 1 inch (see Figure 5.54). What should the dimensions of the page be so that the least amount of paper is used?

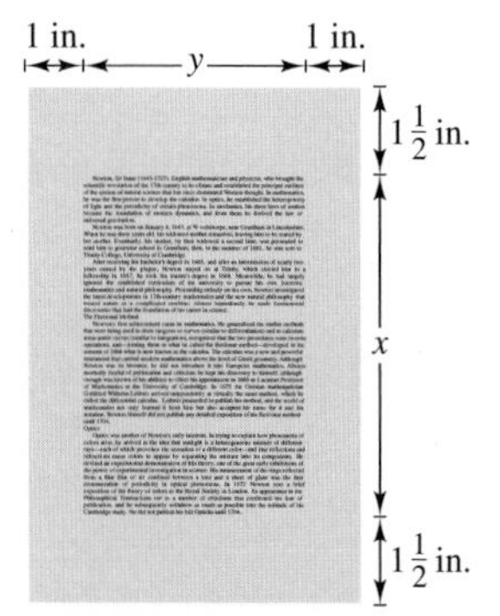

The quantity to be minimized is area: $A = (x + 3)(y + 2)$.
Figure 5.54

Solution Let A be the area to be minimized.

$$A = (x + 3)(y + 2) \qquad \text{Primary equation}$$

The printed area inside the margins is given by

$$24 = xy. \qquad \text{Secondary equation}$$

Solving this equation for y produces $y = 24/x$. Substitution into the primary equation produces

$$A = (x + 3)\left(\frac{24}{x} + 2\right) = 30 + 2x + \frac{72}{x}. \qquad \text{Function of one variable}$$

Because x must be positive, you are interested only in values of A for $x > 0$. To find the critical numbers, differentiate with respect to x.

$$\frac{dA}{dx} = 2 - \frac{72}{x^2} = 0 \implies x^2 = 36$$

So, the critical numbers are $x = \pm 6$. You do not have to consider $x = -6$ because it is outside the domain. The First Derivative Test confirms that A is a minimum when $x = 6$. So, $y = \frac{24}{6} = 4$ and the dimensions of the page should be $x + 3 = 9$ inches by $y + 2 = 6$ inches. ■

EXAMPLE 4 Finding Minimum Length

Two posts, one 12 feet high and the other 28 feet high, stand 30 feet apart. They are to be stayed by two wires, attached to a single stake, running from ground level to the top of each post. Where should the stake be placed to use the least amount of wire?

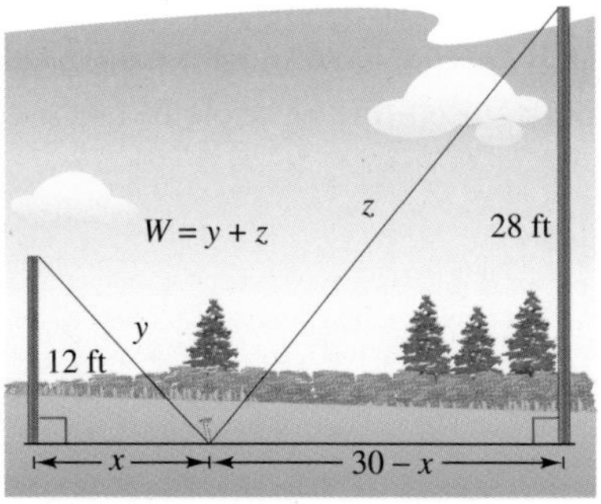

The quantity to be minimized is length. From the diagram, you can see that x varies between 0 and 30.
Figure 5.55

Solution Let W be the wire length to be minimized. Using Figure 5.55, you can write

$$W = y + z. \qquad \text{Primary equation}$$

In this problem, rather than solving for y in terms of z (or vice versa), you can solve for both y and z in terms of a third variable x, as shown in Figure 5.55. From the Pythagorean Theorem, you obtain

$$x^2 + 12^2 = y^2$$
$$(30 - x)^2 + 28^2 = z^2$$

which implies that

$$y = \sqrt{x^2 + 144}$$
$$z = \sqrt{x^2 - 60x + 1684}.$$

So, W is given by

$$\begin{aligned} W &= y + z \\ &= \sqrt{x^2 + 144} + \sqrt{x^2 - 60x + 1684}, \quad 0 \le x \le 30. \end{aligned}$$

Differentiating W with respect to x yields

$$\frac{dW}{dx} = \frac{x}{\sqrt{x^2 + 144}} + \frac{x - 30}{\sqrt{x^2 - 60x + 1684}}.$$

By letting $dW/dx = 0$, you obtain

$$\begin{aligned} \frac{x}{\sqrt{x^2 + 144}} + \frac{x - 30}{\sqrt{x^2 - 60x + 1684}} &= 0 \\ x\sqrt{x^2 - 60x + 1684} &= (30 - x)\sqrt{x^2 + 144} \\ x^2(x^2 - 60x + 1684) &= (30 - x)^2(x^2 + 144) \\ x^4 - 60x^3 + 1684x^2 &= x^4 - 60x^3 + 1044x^2 - 8640x + 129{,}600 \\ 640x^2 + 8640x - 129{,}600 &= 0 \\ 320(x - 9)(2x + 45) &= 0 \\ x &= 9, -22.5. \end{aligned}$$

Because $x = -22.5$ is not in the domain and

$$W(0) \approx 53.04, \quad W(9) = 50, \quad \text{and} \quad W(30) \approx 60.31$$

you can conclude that the wire should be staked at 9 feet from the 12-foot pole. ■

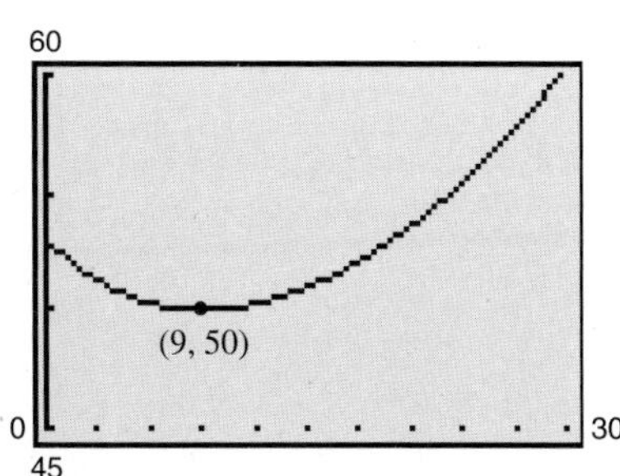

You can confirm the minimum value of W with a graphing utility.
Figure 5.56

TECHNOLOGY From Example 4, you can see that applied optimization problems can involve a lot of algebra. If you have access to a graphing utility, you can confirm that $x = 9$ yields a minimum value of W by graphing

$$W = \sqrt{x^2 + 144} + \sqrt{x^2 - 60x + 1684}$$

as shown in Figure 5.56.

In each of the first four examples, the extreme value occurred at a critical number. Although this happens often, remember that an extreme value can also occur at an endpoint of an interval, as shown in Example 5.

EXAMPLE 5 An Endpoint Maximum

Four feet of wire is to be used to form a square and a circle. How much of the wire should be used for the square and how much should be used for the circle to enclose the maximum total area?

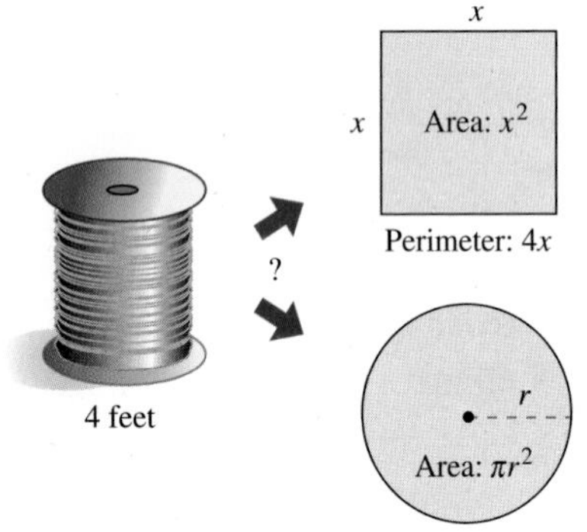

The quantity to be maximized is area: $A = x^2 + \pi r^2$.
Figure 5.57

Solution The total area (see Figure 5.57) is given by

$$A = (\text{area of square}) + (\text{area of circle})$$

$$A = x^2 + \pi r^2. \qquad \text{Primary equation}$$

Because the total length of wire is 4 feet, you obtain

$$4 = (\text{perimeter of square}) + (\text{circumference of circle})$$

$$4 = 4x + 2\pi r. \qquad \text{Secondary equation}$$

So, $r = 2(1 - x)/\pi$, and by substituting into the primary equation you have

$$\begin{aligned} A &= x^2 + \pi\left[\frac{2(1-x)}{\pi}\right]^2 \\ &= x^2 + \frac{4(1-x)^2}{\pi} \\ &= \frac{1}{\pi}[(\pi + 4)x^2 - 8x + 4]. \end{aligned}$$

The feasible domain is $0 \le x \le 1$ restricted by the square's perimeter. Because

$$\frac{dA}{dx} = \frac{2(\pi + 4)x - 8}{\pi}$$

the only critical number in $(0, 1)$ is $x = 4/(\pi + 4) \approx 0.56$. So, using

$$A(0) \approx 1.273, \quad A(0.56) \approx 0.56, \quad \text{and} \quad A(1) = 1$$

you can conclude that the maximum area occurs when $x = 0$. That is, *all* the wire is used for the circle. ■

> **EXPLORATION**
>
> What would the answer be if Example 5 asked for the dimensions needed to enclose the *minimum* total area?

Let's review the primary equations developed in the first five examples. As applications go, these five examples are fairly simple, and yet the resulting primary equations are quite complicated.

$$V = 27x - \frac{x^3}{4} \qquad W = \sqrt{x^2 + 144} + \sqrt{x^2 - 60x + 1684}$$

$$d = \sqrt{x^4 - 3x^2 + 4} \qquad A = \frac{1}{\pi}[(\pi + 4)x^2 - 8x + 4]$$

$$A = 30 + 2x + \frac{72}{x}$$

You must expect that real-life applications often involve equations that are *at least as complicated* as these five. Remember that one of the main goals of this course is to learn to use calculus to analyze equations that initially seem formidable.

5.7 Exercises

See www.CalcChat.com for worked-out solutions to odd-numbered exercises.

1. ***Numerical, Graphical, and Analytic Analysis*** Find two positive numbers whose sum is 110 and whose product is a maximum.

(a) Analytically complete six rows of a table such as the one below. (The first two rows are shown.)

First Number x	Second Number	Product P
10	$110 - 10$	$10(110 - 10) = 1000$
20	$110 - 20$	$20(110 - 20) = 1800$

(b) Use a graphing utility to generate additional rows of the table. Use the table to estimate the solution. (*Hint:* Use the *table* feature of the graphing utility.)

(c) Write the product P as a function of x.

(d) Use a graphing utility to graph the function in part (c) and estimate the solution from the graph.

(e) Use calculus to find the critical number of the function in part (c). Then find the two numbers.

2. ***Numerical, Graphical, and Analytic Analysis*** An open box of maximum volume is to be made from a square piece of material, 24 inches on a side, by cutting equal squares from the corners and turning up the sides (see figure).

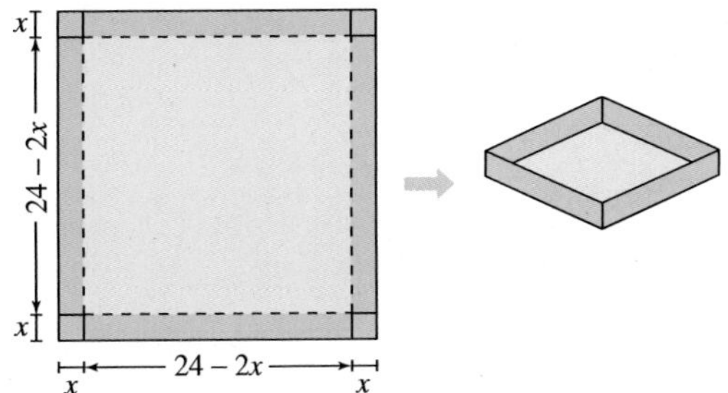

(a) Analytically complete six rows of a table such as the one below. (The first two rows are shown.) Use the table to guess the maximum volume.

Height x	Length and Width	Volume V
1	$24 - 2(1)$	$1[24 - 2(1)]^2 = 484$
2	$24 - 2(2)$	$2[24 - 2(2)]^2 = 800$

(b) Write the volume V as a function of x.

(c) Use calculus to find the critical number of the function in part (b) and find the maximum value.

(d) Use a graphing utility to graph the function in part (b) and verify the maximum volume from the graph.

In Exercises 3–8, find two positive numbers that satisfy the given requirements.

3. The sum is S and the product is a maximum.
4. The product is 185 and the sum is a minimum.
5. The product is 147 and the sum of the first number plus three times the second number is a minimum.
6. The second number is the reciprocal of the first number and the sum is a minimum.
7. The sum of the first number and twice the second number is 108 and the product is a maximum.
8. The sum of the first number squared and the second number is 54 and the product is a maximum.

In Exercises 9 and 10, find the length and width of a rectangle that has the given perimeter and a maximum area.

9. Perimeter: 80 meters
10. Perimeter: P units

In Exercises 11 and 12, find the length and width of a rectangle that has the given area and a minimum perimeter.

11. Area: 32 square feet
12. Area: A square centimeters

In Exercises 13–16, find the point on the graph of the function that is closest to the given point.

	Function	*Point*		*Function*	*Point*
13.	$f(x) = x^2$	$\left(2, \frac{1}{2}\right)$	14.	$f(x) = (x - 1)^2$	$(-5, 3)$
15.	$f(x) = \sqrt{x}$	$(4, 0)$	16.	$f(x) = \sqrt{x - 8}$	$(12, 0)$

17. ***Area*** A rectangular page is to contain 30 square inches of print. The margins on each side are 1 inch. Find the dimensions of the page such that the least amount of paper is used.

18. ***Area*** A rectangular page is to contain 36 square inches of print. The margins on each side are $1\frac{1}{2}$ inches. Find the dimensions of the page such that the least amount of paper is used.

19. ***Chemical Reaction*** In an autocatalytic chemical reaction, the product formed is a catalyst for the reaction. If Q_0 is the amount of the original substance and x is the amount of catalyst formed, the rate of chemical reaction is

$$\frac{dQ}{dx} = kx(Q_0 - x).$$

For what value of x will the rate of chemical reaction be greatest?

20. ***Traffic Control*** On a given day, the flow rate F (cars per hour) on a congested roadway is

$$F = \frac{v}{22 + 0.02v^2}$$

where v is the speed of the traffic in miles per hour. What speed will maximize the flow rate on the road?

21. ***Area*** A farmer plans to fence a rectangular pasture adjacent to a river (see figure). The pasture must contain 245,000 square meters in order to provide enough grass for the herd. What dimensions will require the least amount of fencing if no fencing is needed along the river?

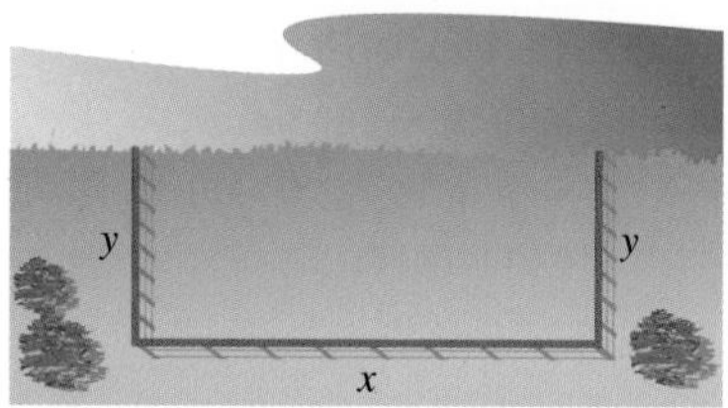

22. ***Maximum Area*** A rancher has 200 feet of fencing with which to enclose two adjacent rectangular corrals (see figure).

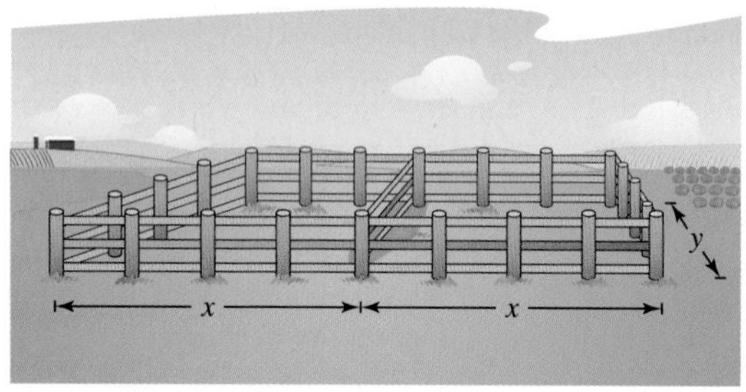

In Exercise 87 in Section 2.1, you numerically and graphically found the dimensions that would produce a maximum enclosed area. Use calculus to determine analytically the dimensions that should be used so that the enclosed area will be a maximum.

23. ***Maximum Volume***

(a) Verify that each of the rectangular solids shown in the figure has a surface area of 150 square inches.

(b) Find the volume of each solid.

(c) Determine the dimensions of a rectangular solid (with a square base) of maximum volume if its surface area is 150 square inches.

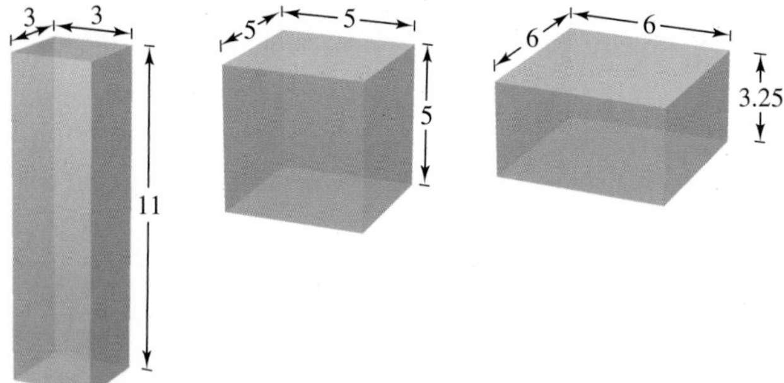

24. ***Maximum Volume*** Determine the dimensions of a rectangular solid (with a square base) with maximum volume if its surface area is 337.5 square centimeters.

25. ***Maximum Area*** A Norman window is constructed by adjoining a semicircle to the top of an ordinary rectangular window (see figure). Find the dimensions of a Norman window of maximum area if the total perimeter is 16 feet.

Figure for 25

26. ***Maximum Area*** A rectangle is bounded by the x- and y-axes and the graph of $y = (6 - x)/2$ (see figure). What length and width should the rectangle have so that its area is a maximum?

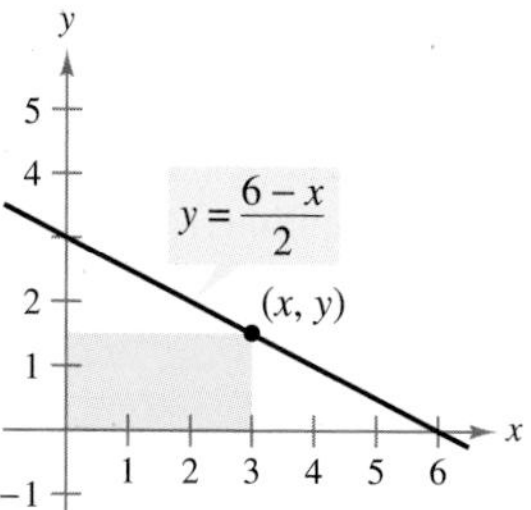

27. ***Minimum Length*** A right triangle is formed in the first quadrant by the x- and y-axes and a line through the point $(1, 2)$ (see figure).

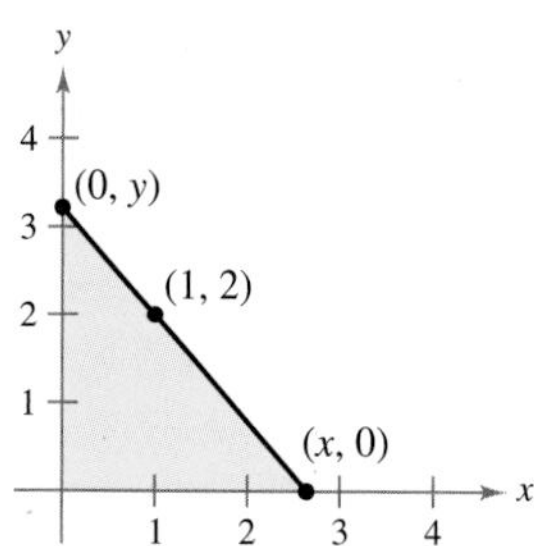

(a) Write the length L of the hypotenuse as a function of x.

 (b) Use a graphing utility to approximate x graphically such that the length of the hypotenuse is a minimum.

(c) Find the vertices of the triangle such that its area is a minimum.

28. ***Maximum Area*** Find the area of the largest isosceles triangle that can be inscribed in a circle of radius 6 (see figure).

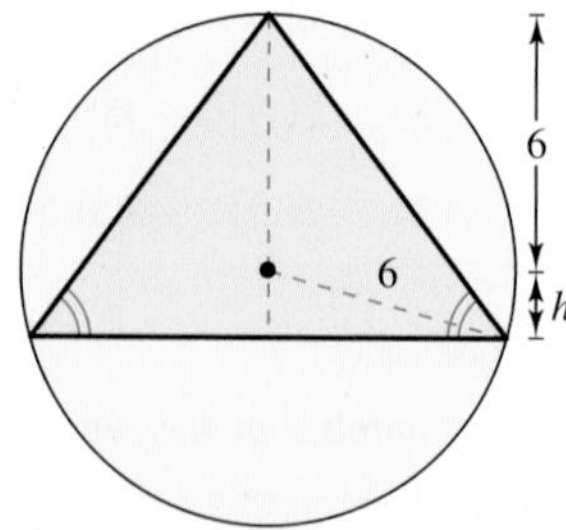

(a) Solve by writing the area as a function of h.

(b) Identify the type of triangle of maximum area.

29. *Maximum Area* A rectangle is bounded by the x-axis and the semicircle $y = \sqrt{25 - x^2}$ (see figure). What length and width should the rectangle have so that its area is a maximum?

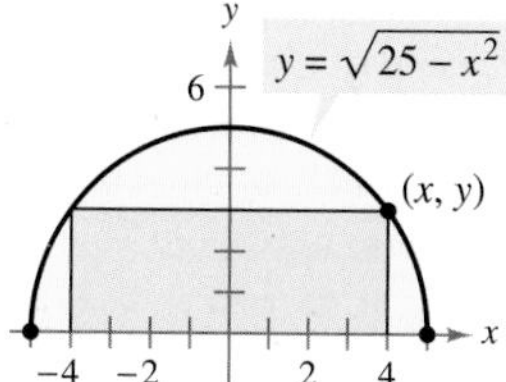

30. *Area* Find the dimensions of the largest rectangle that can be inscribed in a semicircle of radius r (see Exercise 29).

31. *Numerical, Graphical, and Analytic Analysis* An exercise room consists of a rectangle with a semicircle on each end. A 200-meter running track runs around the outside of the room.

(a) Draw a figure to represent the problem. Let x and y represent the length and width of the rectangle.

(b) Analytically complete six rows of a table such as the one below. (The first two rows are shown.) Use the table to guess the maximum area of the rectangular region.

Length x	Width y	Area xy
10	$\frac{2}{\pi}(100 - 10)$	$(10)\frac{2}{\pi}(100 - 10) \approx 573$
20	$\frac{2}{\pi}(100 - 20)$	$(20)\frac{2}{\pi}(100 - 20) \approx 1019$

(c) Write the area A as a function of x.

(d) Use calculus to find the critical number of the function in part (c) and find the maximum value.

(e) Use a graphing utility to graph the function in part (c) and verify the maximum area from the graph.

32. *Numerical, Graphical, and Analytic Analysis* A right circular cylinder is to be designed to hold 22 cubic inches of a soft drink (approximately 12 fluid ounces).

(a) Analytically complete six rows of a table such as the one below. (The first two rows are shown.)

Radius r	Height	Surface Area S
0.2	$\frac{22}{\pi(0.2)^2}$	$2\pi(0.2)\left[0.2 + \frac{22}{\pi(0.2)^2}\right] \approx 220.3$
0.4	$\frac{22}{\pi(0.4)^2}$	$2\pi(0.4)\left[0.4 + \frac{22}{\pi(0.4)^2}\right] \approx 111.0$

(b) Use a graphing utility to generate additional rows of the table. Use the table to estimate the minimum surface area. (*Hint:* Use the *table* feature of the graphing utility.)

(c) Write the surface area S as a function of r.

(d) Use a graphing utility to graph the function in part (c) and estimate the minimum surface area from the graph.

(e) Use calculus to find the critical number of the function in part (c) and find dimensions that will yield the minimum surface area.

33. *Maximum Volume* A rectangular package to be sent by a postal service can have a maximum combined length and girth (perimeter of a cross section) of 108 inches (see figure).

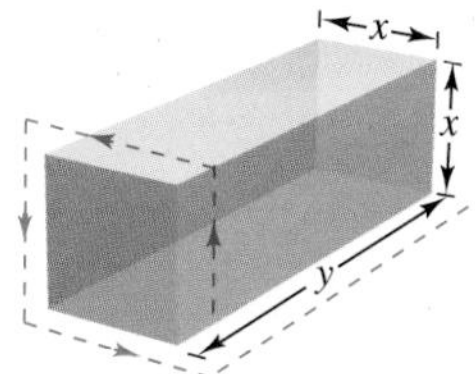

In Exercise 102 in Section 1.1, you graphically found the dimensions of the package of maximum volume. Use calculus to find analytically the dimensions of the package of maximum volume that can be sent. (Assume the cross section is square.)

34. *Maximum Volume* Rework Exercise 33 for a cylindrical package. (The cross section is circular.)

35. *Maximum Volume* Find the volume of the largest right circular cone that can be inscribed in a sphere of radius r.

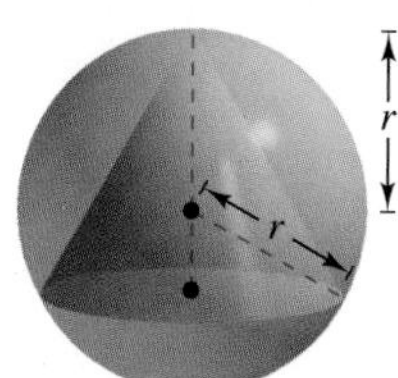

36. *Maximum Volume* Find the volume of the largest right circular cylinder that can be inscribed in a sphere of radius r.

WRITING ABOUT CONCEPTS

37. A shampoo bottle is a right circular cylinder. Because the surface area of the bottle does not change when it is squeezed, is it true that the volume remains the same? Explain.

CAPSTONE

38. The perimeter of a rectangle is 20 feet. Of all possible dimensions, the maximum area is 25 square feet when its length and width are both 5 feet. Are there dimensions that yield a minimum area? Explain.

39. *Minimum Surface Area* A solid is formed by adjoining two hemispheres to the ends of a right circular cylinder. The total volume of the solid is 14 cubic centimeters. Find the radius of the cylinder that produces the minimum surface area.

40. *Minimum Cost* An industrial tank of the shape described in Exercise 39 must have a volume of 4000 cubic feet. The hemispherical ends cost twice as much per square foot of surface area as the sides. Find the dimensions that will minimize cost.

41. *Minimum Area* The sum of the perimeters of an equilateral triangle and a square is 10. Find the dimensions of the triangle and the square that produce a minimum total area.

42. *Maximum Area* Twenty feet of wire is to be used to form an equilateral triangle and a square. How much wire should be used for each figure so that the total enclosed area is maximum?

43. ***Beam Strength*** A wooden beam has a rectangular cross section of height h and width w (see figure). The strength S of the beam is directly proportional to the width and the square of the height. What are the dimensions of the strongest beam that can be cut from a round log of diameter 20 inches? (*Hint*: $S = kh^2w$, where k is the proportionality constant.)

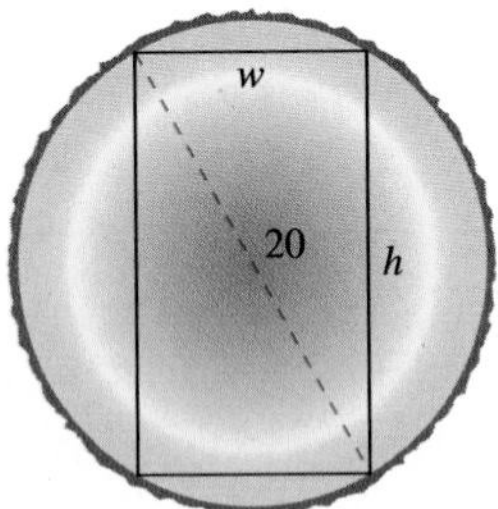

44. ***Minimum Length*** Two factories are located at the coordinates $(-x, 0)$ and $(x, 0)$, and their power supply is at $(0, h)$ (see figure). Find y such that the total length of power line from the power supply to the factories is a minimum.

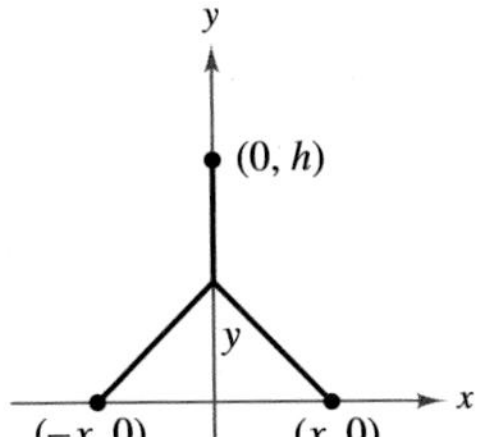

45. ***Conjecture*** Consider the functions $f(x) = \frac{1}{2}x^2$ and $g(x) = \frac{1}{16}x^4 - \frac{1}{2}x^2$ on the domain $[0, 4]$.
   (a) Use a graphing utility to graph the functions on the specified domain.
   (b) Write the vertical distance d between the functions as a function of x and use calculus to find the value of x for which d is maximum.
   (c) Find the equations of the tangent lines to the graphs of f and g at the critical number found in part (b). Graph the tangent lines. What is the relationship between the lines?
   (d) Make a conjecture about the relationship between tangent lines to the graphs of two functions at the value of x at which the vertical distance between the functions is greatest, and prove your conjecture.

46. ***Illumination*** The illumination from a light source is directly proportional to the strength of the source and inversely proportional to the square of the distance from the source. Two light sources of intensities I_1 and I_2 are d units apart. What point on the line segment joining the two sources has the least illumination?

47. ***Maximum Yield*** A home gardener estimates that 16 apple trees will have an average yield of 80 apples per tree. But because of the size of the garden, for each additional tree planted the yield will decrease by four apples per tree. How many trees should be planted to maximize the total yield of apples? What is the maximum yield?

48. ***Farming*** A strawberry farmer will receive \$30 per bushel of strawberries during the first week of harvesting. Each week after that, the value will drop \$0.80 per bushel. The farmer estimates that there are approximately 120 bushels of strawberries in the fields, and that the crop is increasing at a rate of four bushels per week. When should the farmer harvest the strawberries to maximize their value? How many bushels of strawberries will yield the maximum value? What is the maximum value of the strawberries?

49. ***Minimum Time*** A man is in a boat 2 miles from the nearest point on the coast. He is to go to a point Q, located 3 miles down the coast and 1 mile inland (see figure). He can row at 2 miles per hour and walk at 4 miles per hour. Toward what point on the coast should he row in order to reach point Q in the least time?

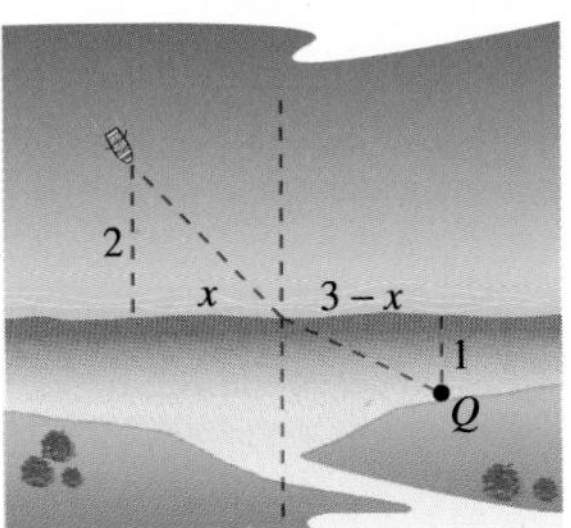

50. ***Minimum Cost*** An offshore oil well is 2 kilometers off the coast. The refinery is 4 kilometers down the coast. Laying pipe in the ocean is twice as expensive as on land. What path should the pipe follow in order to minimize the cost?

51. ***Maximum Profit*** Assume that the amount of money deposited in a bank is proportional to the square of the interest rate the bank pays on this money. Furthermore, the bank can reinvest this money at 12%. Find the interest rate the bank should pay to maximize profit. (Use the simple interest formula.)

52. ***Minimum Cost*** The ordering and transportation cost C of the components used in manufacturing a product is

$$C = 100\left(\frac{200}{x^2} + \frac{x}{x + 30}\right), \quad x \geq 1$$

where C is measured in thousands of dollars and x is the order size in hundreds. Find the order size that minimizes the cost. (*Hint:* Use the *root* feature of a graphing utility.)

Minimum Distance **In Exercises 53–55, consider a fuel distribution center located at the origin of the rectangular coordinate system (units in miles; see figures on next page). The center supplies three factories with coordinates (4, 1), (5, 6), and (10, 3). A trunk line will run from the distribution center along the line $y = mx$, and feeder lines will run to the three factories. The objective is to find m such that the lengths of the feeder lines are minimized.**

53. Minimize the sum of the squares of the lengths of the vertical feeder lines (see figure) given by

$$S_1 = (4m - 1)^2 + (5m - 6)^2 + (10m - 3)^2.$$

Find the equation of the trunk line by this method and then determine the sum of the lengths of the feeder lines.

54. Minimize the sum of the absolute values of the lengths of the vertical feeder lines (see figure) given by

$$S_2 = |4m - 1| + |5m - 6| + |10m - 3|.$$

Find the equation of the trunk line by this method and then determine the sum of the lengths of the feeder lines. (*Hint:* Use a graphing utility to graph the function S_2 and approximate the required critical number.)

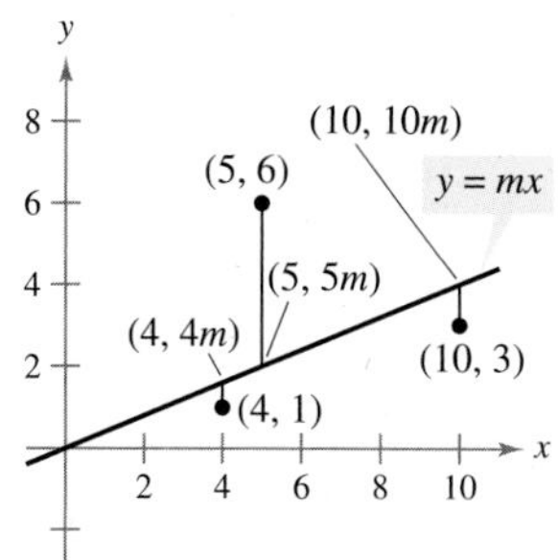

Figure for 53 and 54

55. Minimize the sum of the perpendicular distances (see figure) from the trunk line to the factories given by

$$S_3 = \frac{|4m - 1|}{\sqrt{m^2 + 1}} + \frac{|5m - 6|}{\sqrt{m^2 + 1}} + \frac{|10m - 3|}{\sqrt{m^2 + 1}}.$$

Find the equation of the trunk line by this method and then determine the sum of the lengths of the feeder lines. (*Hint:* Use a graphing utility to graph the function S_3 and approximate the required critical number.)

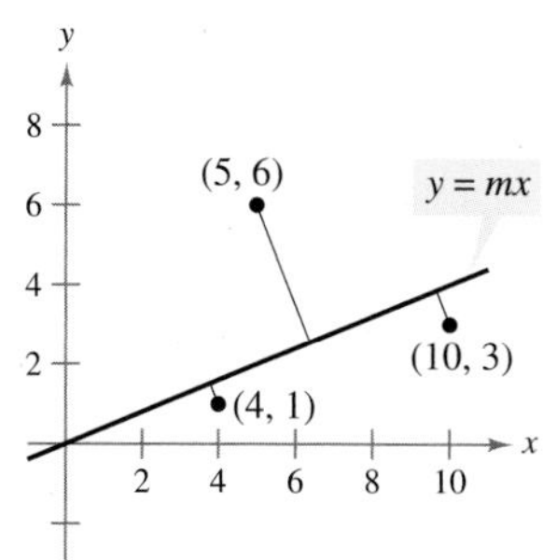

56. ***Maximum Area*** Consider a symmetric cross inscribed in a circle of radius r (see figure). Write the area A of the cross as a function of x and find the value of x that maximizes the area.

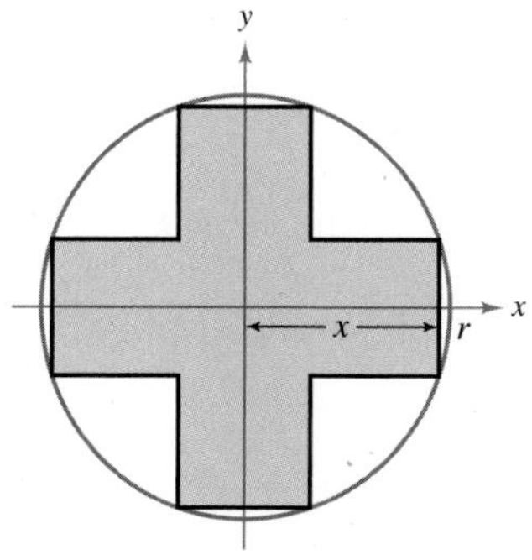

57. ***Diminishing Returns*** The profit P (in thousands of dollars) for a company spending an amount s (in thousands of dollars) on advertising is

$$P = -\frac{1}{10}s^3 + 6s^2 + 400.$$

(a) Find the amount of money the company should spend on advertising in order to yield a maximum profit.

(b) The *point of diminishing returns* is the point at which the rate of growth of the profit function begins to decline. Find the point of diminishing returns.

SECTION PROJECT

Connecticut River

Whenever the Connecticut River reaches a level of 105 feet above sea level, two Northampton, Massachusetts flood control station operators begin a round-the-clock river watch. Every 2 hours, they check the height of the river, using a scale marked off in tenths of a foot, and record the data in a log book. In the spring of 1996, the flood watch lasted from April 4, when the river reached 105 feet and was rising at 0.2 foot per hour, until April 25, when the level subsided again to 105 feet. Between those dates, their log shows that the river rose and fell several times, at one point coming close to the 115-foot mark. If the river had reached 115 feet, the city would have closed down Mount Tom Road (Route 5, south of Northampton).

The graph below shows the rate of change of the level of the river during one portion of the flood watch. Use the graph to answer each question.

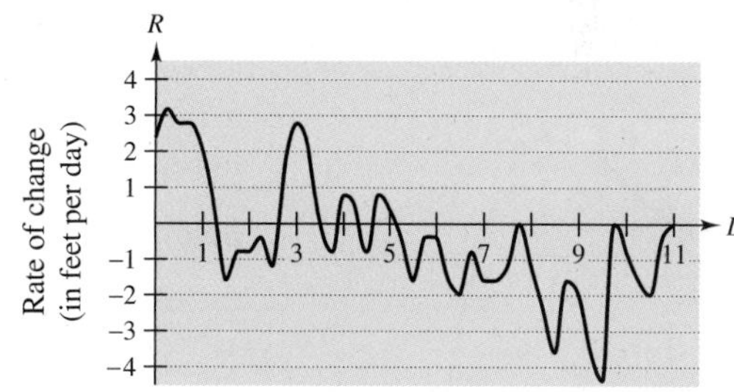

(a) On what date was the river rising most rapidly? How do you know?

(b) On what date was the river falling most rapidly? How do you know?

(c) There were two dates in a row on which the river rose, then fell, then rose again during the course of the day. On which days did this occur, and how do you know?

(d) At 1 minute past midnight, April 14, the river level was 111.0 feet. Estimate its height 24 hours later and 48 hours later. Explain how you made your estimates.

(e) The river crested at 114.4 feet. On what date do you think this occurred?

(Submitted by Mary Murphy, Smith College, Northampton, MA)

5.8 Differentials

- **Understand the concept of a tangent line approximation.**
- **Compare the value of the differential, *dy*, with the actual change in *y*, Δy.**
- **Estimate a propagated error using a differential.**
- **Find the differential of a function using differentiation formulas.**

> **EXPLORATION**
>
> ***Tangent Line Approximation***
> Use a graphing utility to graph
>
> $$f(x) = x^2.$$
>
> In the same viewing window, graph the tangent line to the graph of f at the point $(1, 1)$. Zoom in twice on the point of tangency. Does your graphing utility distinguish between the two graphs? Use the *trace* feature to compare the two graphs. As the x-values get closer to 1, what can you say about the y-values?

Tangent Line Approximations

In this section, you will look at examples in which the graph of a function can be approximated by a straight line.

To begin, consider a function f that is differentiable at c. The equation for the tangent line at the point $(c, f(c))$ is given by

$$y - f(c) = f'(c)(x - c)$$

$$y = f(c) + f'(c)(x - c)$$

and is called the **tangent line approximation** (or **linear approximation**) **of f at c.** Because c is a constant, y is a linear function of x. Moreover, by restricting the values of x to those sufficiently close to c, the values of y can be used as approximations (to any desired degree of accuracy) of the values of the function f. In other words, as $x \to c$, the limit of y is $f(c)$.

EXAMPLE 1 Using a Tangent Line Approximation

Find the tangent line approximation of

$$f(x) = 1 + x - \frac{x^3}{3}$$

at the point $(0, 1)$. Then use a table to compare the y-values of the linear function with those of $f(x)$ on an open interval containing $x = 0$.

Solution The derivative of f is

$$f'(x) = 1 - x^2. \quad \text{First derivative}$$

So, the equation of the tangent line to the graph of f at the point $(0, 1)$ is

$$y - f(0) = f'(0)(x - 0)$$
$$y - 1 = (1)(x - 0)$$
$$y = 1 + x. \quad \text{Tangent line approximation}$$

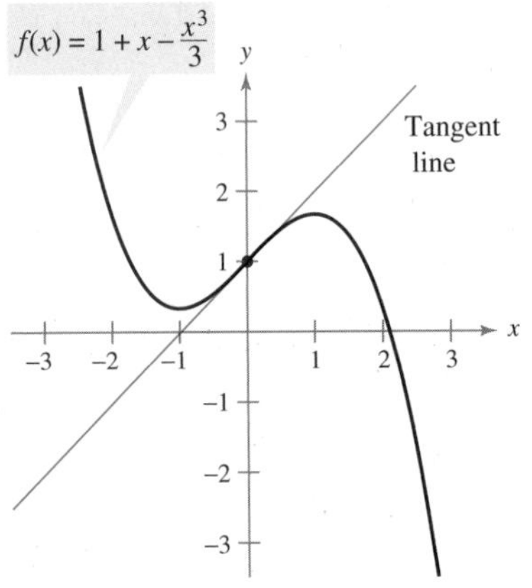

The tangent line approximation of f at the point $(0, 1)$
Figure 5.58

The table compares the values of y given by this linear approximation with the values of $f(x)$ near $x = 0$. Notice that the closer x is to 0, the better the approximation is. This conclusion is reinforced by the graph shown in Figure 5.58.

x	-0.5	-0.1	-0.01	0	0.01	0.1	0.5
$f(x) = 1 + x - \frac{x^3}{3}$	0.542	0.9003	0.9900003	1	1.0099997	1.0997	1.458
$y = 1 + x$	0.5	0.9	0.99	1	1.01	1.1	1.5

■

NOTE Be sure you see that this linear approximation of

$$f(x) = 1 + x - \frac{x^3}{3}$$

depends on the point of tangency. At a different point on the graph of f, you would obtain a different tangent line approximation.

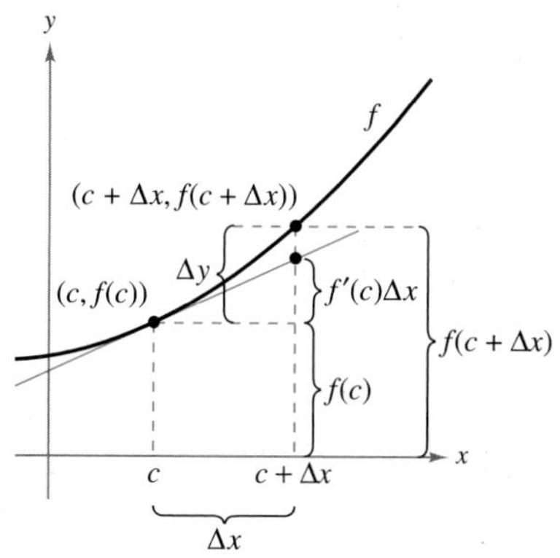

When Δx is small, $\Delta y = f(c + \Delta x) - f(c)$ is approximated by $f'(c)\Delta x$.
Figure 5.59

Differentials

When the tangent line to the graph of f at the point $(c, f(c))$

$$y = f(c) + f'(c)(x - c) \qquad \text{Tangent line at } (c, f(c))$$

is used as an approximation of the graph of f, the quantity $x - c$ is called the change in x, and is denoted by Δx, as shown in Figure 5.59. When Δx is small, the change in y (denoted by Δy) can be approximated as shown.

$$\Delta y = f(c + \Delta x) - f(c) \qquad \text{Actual change in } y$$

$$\approx f'(c)\Delta x \qquad \text{Approximate change in } y$$

For such an approximation, the quantity Δx is traditionally denoted by dx, and is called the **differential of x.** The expression $f'(x)\,dx$ is denoted by dy, and is called the **differential of y.**

DEFINITION OF DIFFERENTIALS

Let $y = f(x)$ represent a function that is differentiable on an open interval containing x. The **differential of x** (denoted by dx) is any nonzero real number. The **differential of y** (denoted by dy) is

$$dy = f'(x)\,dx.$$

STUDY TIP The differential of y can be described from another viewpoint. In the derivative notation

$$\frac{dy}{dx} = f'(x)$$

it is appropriate to think of dx and dy as small real numbers with $dx \neq 0$. So, if both sides are multiplied by dx, you obtain the differential of y as

$$\frac{dy}{dx} \cdot dx = f'(x) \cdot dx$$

$$dy = f'(x)\,dx.$$

In many types of applications, the differential of y can be used as an approximation of the change in y. That is,

$$\Delta y \approx dy \qquad \text{or} \qquad \Delta y \approx f'(x)dx.$$

EXAMPLE 2 Comparing Δy and dy

Let $y = x^2$. Find dy when $x = 1$ and $dx = 0.01$. Compare this value with Δy for $x = 1$ and $\Delta x = 0.01$.

Solution Because $y = f(x) = x^2$, you have $f'(x) = 2x$, and the differential dy is given by

$$dy = f'(x)\,dx = f'(1)(0.01) = 2(0.01) = 0.02. \qquad \text{Differential of } y$$

Now, using $\Delta x = 0.01$, the change in y is

$$\Delta y = f(x + \Delta x) - f(x) = f(1.01) - f(1) = (1.01)^2 - 1^2 = 0.0201. \qquad \text{Change in } y$$

Figure 5.60 shows the geometric comparison of dy and Δy. Try comparing other values of dy and Δy. You will see that the values become closer to each other as dx (or Δx) approaches 0. ■

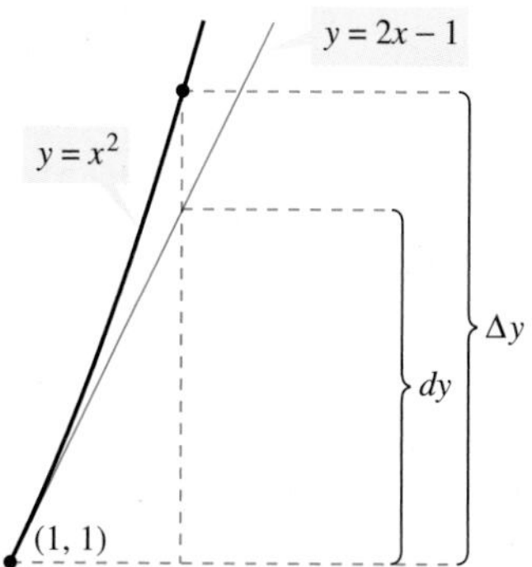

The change in y, Δy, is approximated by the differential of y, dy.
Figure 5.60

In Example 2, the tangent line to the graph of $f(x) = x^2$ at $x = 1$ is

$$y = 2x - 1 \qquad \text{or} \qquad g(x) = 2x - 1. \qquad \text{Tangent line to the graph of } f \text{ at } x = 1.$$

For x-values near 1, this line is close to the graph of f, as shown in Figure 5.60. For instance,

$$f(1.01) = 1.01^2 = 1.0201 \qquad \text{and} \qquad g(1.01) = 2(1.01) - 1 = 1.02.$$

Error Propagation

Physicists and engineers tend to make liberal use of the approximation of Δy by dy. One way this occurs in practice is in the estimation of errors propagated by physical measuring devices. For example, if you let x represent the measured value of a variable and let $x + \Delta x$ represent the exact value, then Δx is the *error in measurement*. Finally, if the measured value x is used to compute another value $f(x)$, the difference between $f(x + \Delta x)$ and $f(x)$ is the **propagated error.**

$$\underbrace{f(x + \overbrace{\Delta x}^{\text{Measurement error}})}_{\text{Exact value}} - \underbrace{f(x)}_{\text{Measured value}} = \overbrace{\Delta y}^{\text{Propagated error}}$$

EXAMPLE 3 Estimation of Error

The measured radius of a ball bearing is 0.7 inch, as shown in Figure 5.61. If the measurement is correct to within 0.01 inch, estimate the propagated error in the volume V of the ball bearing.

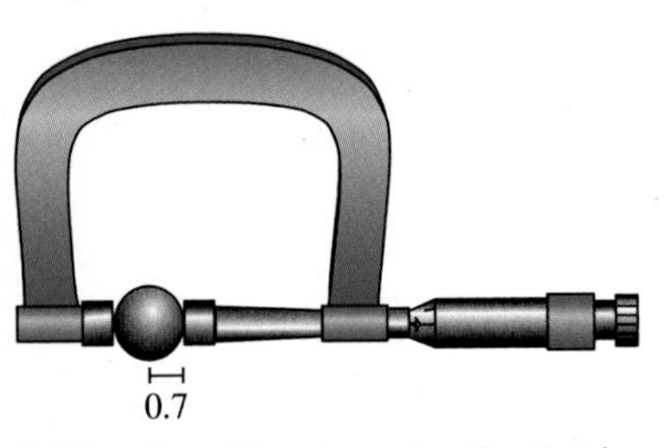

Ball bearing with measured radius that is correct to within 0.01 inch
Figure 5.61

Solution The formula for the volume of a sphere is

$$V = \frac{4}{3}\pi r^3$$

where r is the radius of the sphere. So, you can write

$$r = 0.7 \qquad \text{Measured radius}$$

and

$$-0.01 \le \Delta r \le 0.01. \qquad \text{Possible error}$$

To approximate the propagated error in the volume, differentiate V to obtain $dV/dr = 4\pi r^2$ and write

$$\begin{aligned} \Delta V &\approx dV && \text{Approximate } \Delta V \text{ by } dV. \\ &= 4\pi r^2\, dr \\ &= 4\pi(0.7)^2(\pm 0.01) && \text{Substitute for } r \text{ and } dr. \\ &\approx \pm 0.06158 \text{ cubic inch.} \end{aligned}$$

So, the volume has a propagated error of about 0.06 cubic inch. ■

Would you say that the propagated error in Example 3 is large or small? The answer is best given in *relative* terms by comparing dV with V. The ratio

$$\begin{aligned} \frac{dV}{V} &= \frac{4\pi r^2\, dr}{\frac{4}{3}\pi r^3} && \text{Ratio of } dV \text{ to } V \\ &= \frac{3\, dr}{r} && \text{Simplify.} \\ &\approx \frac{3}{0.7}(\pm 0.01) && \text{Substitute for } dr \text{ and } r. \\ &\approx \pm 0.0429 \end{aligned}$$

is called the **relative error.** The corresponding **percent error** is approximately 4.29%.

Calculating Differentials

Each of the differentiation rules that you studied in Chapter 4 can be written in **differential form.** For example, suppose u and v are differentiable functions of x. By the definition of differentials, you have

$$du = u'\,dx \quad \text{and} \quad dv = v'\,dx.$$

So, you can write the differential form of the Product Rule as shown below.

$$\begin{aligned} d[uv] &= \frac{d}{dx}[uv]\,dx && \text{Differential of } uv \\ &= [uv' + vu']\,dx && \text{Product Rule} \\ &= uv'\,dx + vu'\,dx \\ &= u\,dv + v\,du \end{aligned}$$

DIFFERENTIAL FORMULAS

Let u and v be differentiable functions of x.

Constant multiple: $d[cu] = c\,du$

Sum or difference: $d[u \pm v] = du \pm dv$

Product: $d[uv] = u\,dv + v\,du$

Quotient: $d\left[\frac{u}{v}\right] = \frac{v\,du - u\,dv}{v^2}$

EXAMPLE 4 Finding Differentials

Function	*Derivative*	*Differential*
a. $y = x^2$	$\frac{dy}{dx} = 2x$	$dy = 2x\,dx$
b. $y = 2x^3 + x$	$\frac{dy}{dx} = 6x^2 + 1$	$dy = (6x^2 + 1)\,dx$
c. $y = \sqrt{2x + 1}$	$\frac{dy}{dx} = \frac{1}{\sqrt{2x + 1}}$	$dy = \left(\frac{1}{\sqrt{2x + 1}}\right) dx$
d. $y = \frac{1}{x}$	$\frac{dy}{dx} = -\frac{1}{x^2}$	$dy = -\frac{dx}{x^2}$

■

Mary Evans Picture Library

GOTTFRIED WILHELM LEIBNIZ (1646–1716)

Both Leibniz and Newton are credited with creating calculus. It was Leibniz, however, who tried to broaden calculus by developing rules and formal notation. He often spent days choosing an appropriate notation for a new concept.

The notation in Example 4 is called the **Leibniz notation** for derivatives and differentials, named after the German mathematician Gottfried Wilhelm Leibniz. The beauty of this notation is that it provides an easy way to remember several important calculus formulas by making it seem as though the formulas were derived from algebraic manipulations of differentials. For instance, in Leibniz notation, the *Chain Rule*

$$\frac{dy}{dx} = \frac{dy}{du}\frac{du}{dx}$$

would appear to be true because the du's divide out. Even though this reasoning is *incorrect*, the notation does help one remember the Chain Rule.

EXAMPLE 5 Finding the Differential of a Composite Function

Use the Chain Rule to find dy.

a. $y = (3x^2 + x)^4$ **b.** $y = (x^2 + 1)^{1/2}$

Solution

a.

$$y = f(x) = (3x^2 + x)^4 \qquad \text{Write original function.}$$

$$f'(x) = 4(3x^2 + x)^3(6x + 1) \qquad \text{Apply Chain Rule.}$$

$$dy = f'(x)\,dx = 4(6x + 1)(3x^2 + x)^3\,dx \qquad \text{Differential form}$$

b.

$$y = f(x) = (x^2 + 1)^{1/2} \qquad \text{Write original function.}$$

$$f'(x) = \frac{1}{2}(x^2 + 1)^{-1/2}(2x) = \frac{x}{\sqrt{x^2 + 1}} \qquad \text{Apply Chain Rule.}$$

$$dy = f'(x)\,dx = \frac{x}{\sqrt{x^2 + 1}}\,dx \qquad \text{Differential form}$$

■

Differentials can be used to approximate function values. To do this for the function given by $y = f(x)$, use the formula

$$f(x + \Delta x) \approx f(x) + dy = f(x) + f'(x)\,dx$$

which is derived from the approximation $\Delta y = f(x + \Delta x) - f(x) \approx dy$. The key to using this formula is to choose a value for x that makes the calculations easier, as shown in Example 6. (This formula is equivalent to the tangent line approximation given earlier in this section.)

EXAMPLE 6 Approximating Function Values

Use differentials to approximate $\sqrt{16.5}$.

Solution Using $f(x) = \sqrt{x}$, you can write

$$f(x + \Delta x) \approx f(x) + f'(x)\,dx = \sqrt{x} + \frac{1}{2\sqrt{x}}\,dx.$$

Now, choosing $x = 16$ and $dx = 0.5$, you obtain the following approximation.

$$f(x + \Delta x) = \sqrt{16.5} \approx \sqrt{16} + \frac{1}{2\sqrt{16}}(0.5) = 4 + \left(\frac{1}{8}\right)\left(\frac{1}{2}\right) = 4.0625$$

■

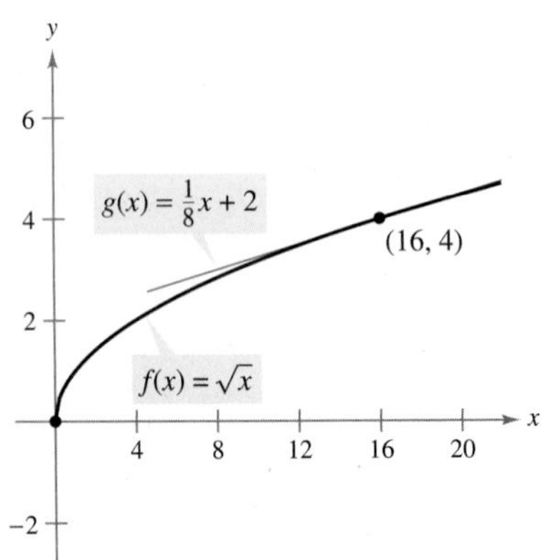

Figure 5.62

The tangent line approximation to $f(x) = \sqrt{x}$ at $x = 16$ is the line $g(x) = \frac{1}{8}x + 2$. For x-values near 16, the graphs of f and g are close together, as shown in Figure 5.62. For instance,

$$f(16.5) = \sqrt{16.5} \approx 4.0620 \quad \text{and} \quad g(16.5) = \frac{1}{8}(16.5) + 2 = 4.0625.$$

In fact, if you use a graphing utility to zoom in near the point of tangency (16, 4), you will see that the two graphs appear to coincide. Notice also that as you move farther away from the point of tangency, the linear approximation becomes less accurate.

5.8 Exercises

See www.CalcChat.com for worked-out solutions to odd-numbered exercises.

In Exercises 1–4, find the equation of the tangent line T to the graph of f at the given point. Use this linear approximation to complete the table.

x	1.9	1.99	2	2.01	2.1
$f(x)$					
$T(x)$					

1. $f(x) = x^2$, $(2, 4)$

2. $f(x) = \dfrac{6}{x^2}$, $\left(2, \dfrac{3}{2}\right)$

3. $f(x) = x^5$, $(2, 32)$

4. $f(x) = \sqrt{x}$, $\left(2, \sqrt{2}\right)$

In Exercises 5–8, use the information to evaluate and compare Δy and dy.

5. $y = x^3$ $\quad x = 1 \quad \Delta x = dx = 0.1$

6. $y = 1 - 2x^2$ $\quad x = 0 \quad \Delta x = dx = -0.1$

7. $y = x^4 + 1$ $\quad x = -1 \quad \Delta x = dx = 0.01$

8. $y = 2 - x^4$ $\quad x = 2 \quad \Delta x = dx = 0.01$

In Exercises 9–18, find the differential dy of the given function.

9. $y = 3x^2 - 4$

10. $y = 3x^{2/3}$

11. $y = 4x^3$

12. $y = 3$

13. $y = \dfrac{x + 1}{2x - 1}$

14. $y = \dfrac{x}{x + 5}$

15. $y = \sqrt{x}$

16. $y = \sqrt{9 - x^2}$

17. $y = x\sqrt{1 - x^2}$

18. $y = \sqrt{x} + \dfrac{1}{\sqrt{x}}$

In Exercises 19–22, use differentials and the graph of f to approximate (a) $f(1.9)$ and (b) $f(2.04)$. To print an enlarged copy of the graph, go to the website *www.mathgraphs.com*.

19.

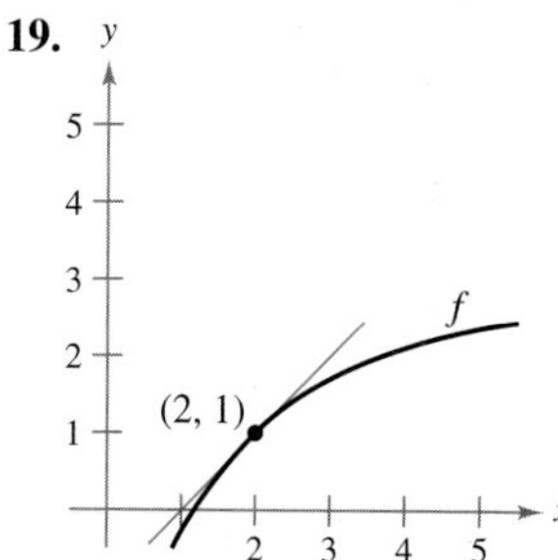

20.

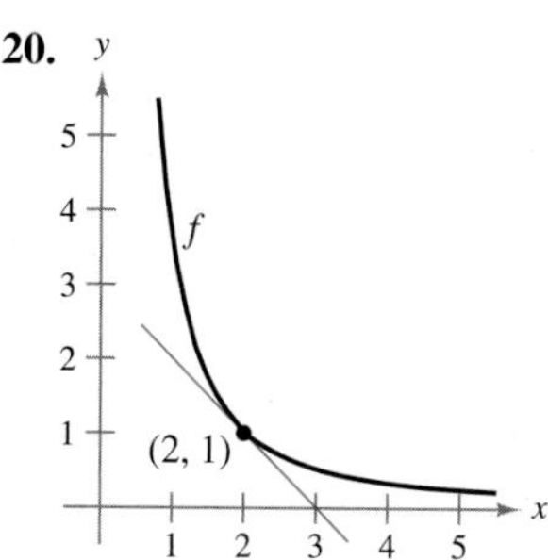

21.

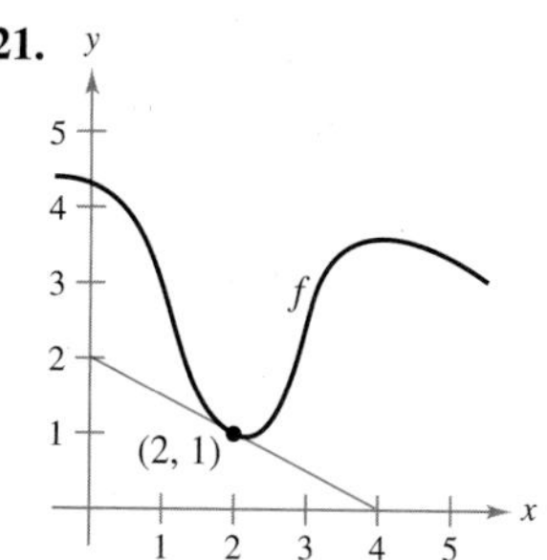

22.

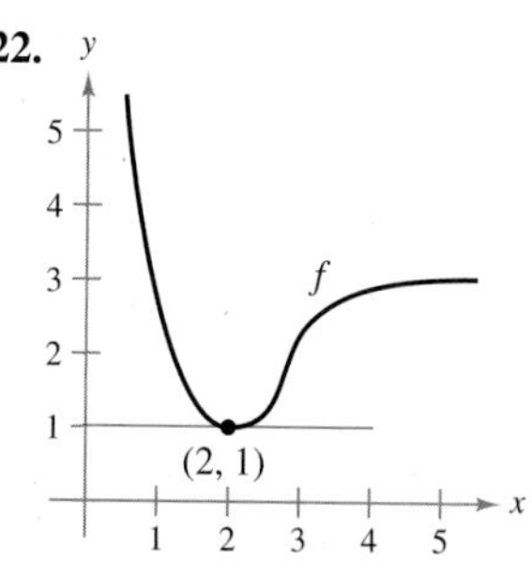

In Exercises 23 and 24, use differentials and the graph of g' to approximate (a) $g(2.93)$ and (b) $g(3.1)$ given that $g(3) = 8$.

23.

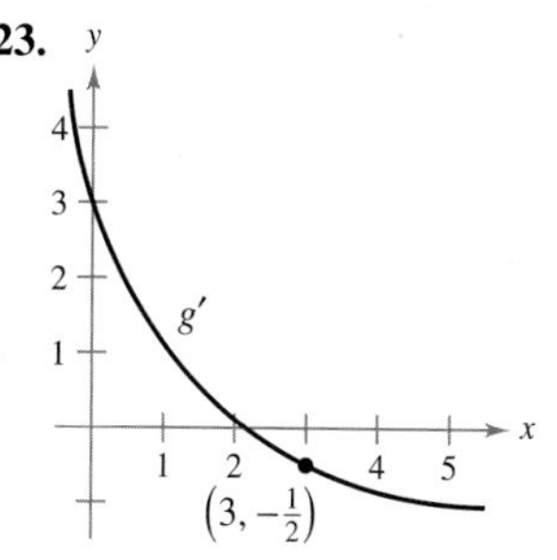

24.

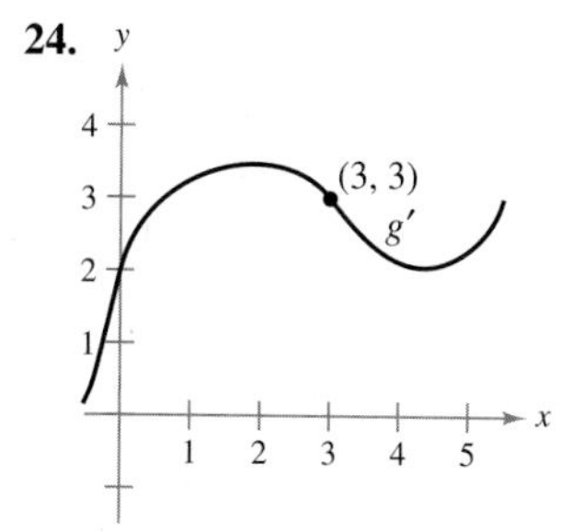

25. ***Area*** The measurement of the side of a square floor tile is 10 inches, with a possible error of $\frac{1}{32}$ inch. Use differentials to approximate the possible propagated error in computing the area of the square.

26. ***Area*** The measurements of the base and altitude of a triangle are found to be 36 and 50 centimeters, respectively. The possible error in each measurement is 0.25 centimeter. Use differentials to approximate the possible propagated error in computing the area of the triangle.

27. ***Area*** The measurement of the radius of the end of a log is found to be 16 inches, with a possible error of $\frac{1}{4}$ inch. Use differentials to approximate the possible propagated error in computing the area of the end of the log.

28. ***Volume and Surface Area*** The measurement of the edge of a cube is found to be 15 inches, with a possible error of 0.03 inch. Use differentials to approximate the maximum possible propagated error in computing (a) the volume of the cube and (b) the surface area of the cube.

29. ***Area*** The measurement of a side of a square is found to be 12 centimeters, with a possible error of 0.05 centimeter.

(a) Approximate the percent error in computing the area of the square.

(b) Estimate the maximum allowable percent error in measuring the side if the error in computing the area cannot exceed 2.5%.

30. ***Volume and Surface Area*** The radius of a spherical balloon is measured as 8 inches, with a possible error of 0.02 inch. Use differentials to approximate the maximum possible error in calculating (a) the volume of the sphere, (b) the surface area of the sphere, and (c) the relative errors in parts (a) and (b).

31. ***Circumference*** The measurement of the circumference of a circle is found to be 64 centimeters, with a possible error of 0.9 centimeter.

(a) Approximate the percent error in computing the area of the circle.

(b) Estimate the maximum allowable percent error in measuring the circumference if the error in computing the area cannot exceed 3%.

32. ***Stopping Distance*** The total stopping distance T of a vehicle is

$$T = 2.5x + 0.5x^2$$

where T is in feet and x is the speed in miles per hour. Approximate the change and percent change in total stopping distance as speed changes from $x = 25$ to $x = 26$ miles per hour.

33. ***Profit*** The profit P for a company producing x units is

$$P = (500x - x^2) - \left(\frac{1}{2}x^2 - 77x + 3000\right).$$

Approximate the change and percent change in profit as production changes from $x = 115$ to $x = 120$ units.

34. ***Revenue*** The revenue R for a company selling x units is

$$R = 900x - 0.1x^2.$$

Use differentials to approximate the change in revenue if sales increase from $x = 3000$ to $x = 3100$ units.

35. ***Demand*** The demand function for a product is modeled by

$$p = 75 - 0.25x.$$

(a) If x changes from 7 to 8, what is the corresponding change in p? Compare the values of Δp and dp.

(b) Repeat part (a) when x changes from 70 to 71 units.

36. ***Biology: Wildlife Management*** A state game commission introduces 50 deer into newly acquired state game lands. The population N of the herd can be modeled by

$$N = \frac{10(5 + 3t)}{1 + 0.04t}$$

where t is the time in years. Use differentials to approximate the change in the herd size from $t = 5$ to $t = 6$.

Volume **In Exercises 37 and 38, the thickness of each shell is 0.2 centimeter. Use differentials to approximate the volume of each shell.**

37.

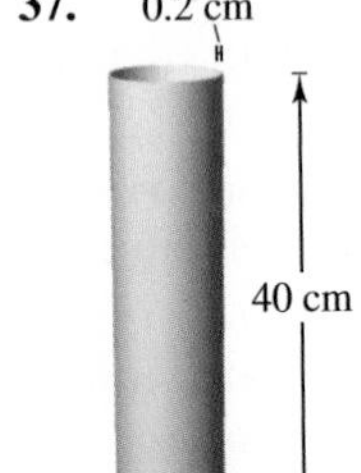

38.

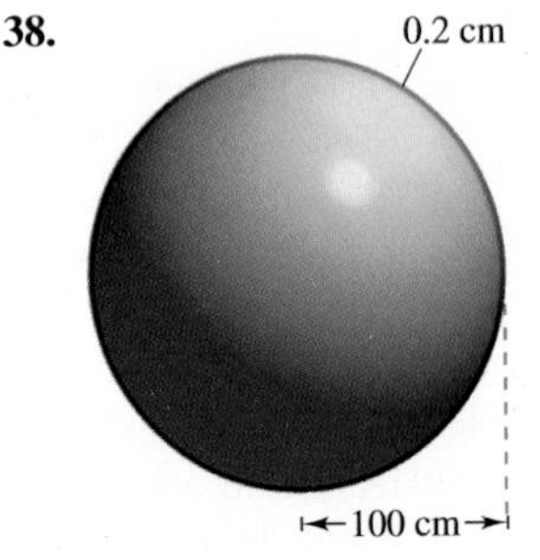

39. ***Pendulum*** The period of a pendulum is given by

$$T = 2\pi\sqrt{\frac{L}{g}}$$

where L is the length of the pendulum in feet, g is the acceleration due to gravity, and T is the time in seconds. The pendulum has been subjected to an increase in temperature such that the length has increased by $\frac{1}{2}\%$.

(a) Find the approximate percent change in the period.

(b) Using the result in part (a), find the approximate error in this pendulum clock in 1 day.

40. ***Ohm's Law*** A current of I amperes passes through a resistor of R ohms. **Ohm's Law** states that the voltage E applied to the resistor is $E = IR$. If the voltage is constant, show that the magnitude of the relative error in R caused by a change in I is equal in magnitude to the relative error in I.

In Exercises 41–44, use differentials to approximate the value of the expression. Compare your answer with that of a calculator.

41. $\sqrt{99.4}$

42. $\sqrt[3]{26}$

43. $\sqrt[4]{624}$

44. $(2.99)^3$

In Exercises 45 and 46, verify the tangent line approximation of the function at the given point. Then use a graphing utility to graph the function and its approximation in the same viewing window.

	Function	*Approximation*	*Point*
45.	$f(x) = \sqrt{x + 4}$	$y = 2 + \dfrac{x}{4}$	$(0, 2)$
46.	$f(x) = \dfrac{1}{1 - x}$	$y = 1 + x$	$(0, 1)$

WRITING ABOUT CONCEPTS

47. Describe the change in accuracy of dy as an approximation for Δy when Δx is decreased.

48. When using differentials, what is meant by the terms *propagated error*, *relative error*, and *percent error*?

49. Give a short explanation of why the approximation is valid.

(a) $\sqrt{4.02} \approx 2 + \frac{1}{4}(0.02)$ (b) $\frac{1}{2.04} \approx \frac{1}{2} - \frac{1}{4}(0.04)$

CAPSTONE

50. Would you use $y = x$ to approximate $f(x) = x^2 + x$ near $x = 0$? Why or why not?

True or False? **In Exercises 51–54, determine whether the statement is true or false. If it is false, explain why or give an example that shows it is false.**

51. If $y = x + c$, then $dy = dx$.

52. If $y = ax + b$, then $\Delta y/\Delta x = dy/dx$.

53. If y is differentiable, then $\lim_{\Delta x \to 0} (\Delta y - dy) = 0$.

54. If $y = f(x)$, f is increasing and differentiable, and $\Delta x > 0$, then $\Delta y \geq dy$.

5 CHAPTER SUMMARY

	Review Exercises
Section 5.1	
■ Understand the definition of extrema of a function on an interval *(p. 324),* understand the definition of relative extrema of a function on an open interval *(p. 325),* and find extrema on a closed interval *(p. 327).*	*1–6*
Section 5.2	
■ Understand and use Rolle's Theorem *(p. 332).*	*7–11*
■ Understand and use the Mean Value Theorem *(p. 334).*	*12–20*
Section 5.3	
■ Determine intervals on which a function is increasing or decreasing *(p. 339).*	*21–26*
■ Apply the First Derivative Test to find relative extrema of a function *(p. 341).*	*27–30*
Section 5.4	
■ Determine intervals on which a function is concave upward or concave downward *(p. 348),* and find any points of inflection of the graph of a function *(p. 350).*	*31–34*
■ Apply the Second Derivative Test to find relative extrema of a function *(p. 352).*	*35–40*
Section 5.5	
■ Determine (finite) limits at infinity *(p. 356).*	*41–46*
■ Determine the horizontal asymptotes, if any, of the graph of a function *(p. 357).*	*47–50*
■ Determine infinite limits at infinity *(p. 362).*	*51–54*
Section 5.6	
■ Analyze and sketch the graph of a function *(p. 366).*	*55–72*
Section 5.7	
■ Use calculus to solve applied minimum and maximum problems *(p. 374).*	*73–80*
Section 5.8	
■ Understand the concept of a tangent line approximation *(p. 384),* compare the value of the differential, dy, with the actual change in y, Δy *(p. 385),* estimate a propagated error using a differential *(p. 386),* and find the differential of a function using differentiation formulas *(p. 387).*	*81–84*

5 REVIEW EXERCISES

See www.CalcChat.com for worked-out solutions to odd-numbered exercises.

1. Give the definition of a critical number, and graph a function f showing the different types of critical numbers.

2. Consider the odd function f that is continuous and differentiable and has the functional values shown in the table.

x	-5	-4	-1	0	2	3	6
$f(x)$	1	3	2	0	-1	-4	0

(a) Determine $f(4)$.

(b) Determine $f(-3)$.

(c) Plot the points and make a possible sketch of the graph of f on the interval $[-6, 6]$. What is the smallest number of critical points in the interval? Explain.

(d) Does there exist at least one real number c in the interval $(-6, 6)$ where $f'(c) = -1$? Explain.

(e) Is it possible that $\lim_{x\to 0} f(x)$ does not exist? Explain.

(f) Is it necessary that $f'(x)$ exists at $x = 2$? Explain.

In Exercises 3–6, find the absolute extrema of the function on the closed interval. Use a graphing utility to graph the function over the given interval to confirm your results.

3. $f(x) = x^2 + 5x,\ [-4, 0]$

4. $h(x) = 3\sqrt{x} - x,\ [0, 9]$

5. $f(x) = x\sqrt{5 - x},\ [0, 4]$

6. $f(x) = \dfrac{x}{\sqrt{x^2 + 1}},\ [0, 2]$

In Exercises 7–10, determine whether Rolle's Theorem can be applied to f on the closed interval $[a, b]$. If Rolle's Theorem can be applied, find all values of c in the open interval (a, b) such that $f'(c) = 0$. If Rolle's Theorem cannot be applied, explain why not.

7. $f(x) = 2x^2 - 7,\ [0, 4]$

8. $f(x) = (x - 2)(x + 3)^2,\ [-3, 2]$

9. $f(x) = \dfrac{x^2}{1 - x^2},\ [-2, 2]$

10. $f(x) = |x - 2| - 2,\ [0, 4]$

11. Consider the function $f(x) = 3 - |x - 4|$.

(a) Graph the function and verify that $f(1) = f(7)$.

(b) Note that $f'(x)$ is not equal to zero for any x in $[1, 7]$. Explain why this does not contradict Rolle's Theorem.

12. Can the Mean Value Theorem be applied to the function $f(x) = 1/x^2$ on the interval $[-2, 1]$? Explain.

In Exercises 13–18, determine whether the Mean Value Theorem can be applied to f on the closed interval $[a, b]$. If the Mean Value Theorem can be applied, find all values of c in the open interval (a, b) such that $f'(c) = \dfrac{f(b) - f(a)}{b - a}$. If the Mean Value Theorem cannot be applied, explain why not.

13. $f(x) = x^{2/3},\ [1, 8]$

14. $f(x) = \dfrac{1}{x},\ [1, 4]$

15. $f(x) = |5 - x|,\ [2, 6]$

16. $f(x) = |x^2 - 9|,\ [0, 2]$

17. $f(x) = 2x - 3\sqrt{x},\ [-1, 1]$

18. $f(x) = \sqrt{x} - 2x,\ [0, 4]$

19. For the function $f(x) = Ax^2 + Bx + C$, determine the value of c guaranteed by the Mean Value Theorem on the interval $[x_1, x_2]$.

20. Demonstrate the result of Exercise 19 for $f(x) = 2x^2 - 3x + 1$ on the interval $[0, 4]$.

In Exercises 21–26, find the critical numbers (if any) and the open intervals on which the function is increasing or decreasing.

21. $f(x) = x^2 + 3x - 12$

22. $h(x) = (x + 2)^{1/3} + 8$

23. $f(x) = (x - 1)^2(x - 3)$

24. $g(x) = (x + 1)^3$

25. $h(x) = \sqrt{x}(x - 3),\ x > 0$

26. $y = 3(x - 2)^{2/3} - 2x + 4$

In Exercises 27–30, use the First Derivative Test to find any relative extrema of the function. Use a graphing utility to confirm your results.

27. $f(x) = 4x^3 - 5x$

28. $g(x) = \dfrac{x^3 - 8x}{4}$

29. $h(t) = \dfrac{1}{4}t^4 - 8t$

30. $g(x) = \dfrac{1}{27}(x^4 + 4x^3)$

In Exercises 31–34, determine the points of inflection and discuss the concavity of the graph of the function.

31. $f(x) = x^3 - 9x^2$

32. $g(x) = x\sqrt{x + 5}$

33. $f(x) = (x + 2)^2(x - 4)$

34. $f(x) = -x^3 + 6x^2 - 9x + 1$

In Exercises 35 and 36, use the Second Derivative Test to find all relative extrema.

35. $g(x) = 2x^2(1 - x^2)$

36. $h(t) = t - 4\sqrt{t + 1}$

Think About It **In Exercises 37 and 38, sketch the graph of a function f having the given characteristics.**

37. $f(0) = f(6) = 0$
$f'(3) = f'(5) = 0$
$f'(x) > 0$ if $x < 3$
$f'(x) > 0$ if $3 < x < 5$
$f'(x) < 0$ if $x > 5$
$f''(x) < 0$ if $x < 3$ or $x > 4$
$f''(x) > 0$ if $3 < x < 4$

38. $f(0) = 4,\ f(6) = 0$
$f'(x) < 0$ if $x < 2$ or $x > 4$
$f'(2)$ does not exist.
$f'(4) = 0$
$f'(x) > 0$ if $2 < x < 4$
$f''(x) < 0$ if $x \neq 2$

39. ***Writing*** A newspaper headline states that "The rate of growth of the national deficit is decreasing." What does this mean? What does it imply about the graph of the deficit as a function of time?

40. *Modeling Data* The manager of a store recorded the annual sales S (in thousands of dollars) of a product over a period of 7 years, as shown in the table, where t is the time in years, with $t = 1$ corresponding to 2001.

t	1	2	3	4	5	6	7
S	5.4	6.9	11.5	15.5	19.0	22.0	23.6

(a) Use the regression capabilities of a graphing utility to find a model of the form $S = at^3 + bt^2 + ct + d$ for the data.

(b) Use a graphing utility to plot the data and graph the model.

(c) Use calculus and the model to find the time t when sales were increasing at the greatest rate.

(d) Do you think the model would be accurate for predicting future sales? Explain.

In Exercises 41–46, find the limit.

41. $\lim_{x\to\infty}\left(8+\frac{1}{x}\right)$ **42.** $\lim_{x\to\infty}\frac{2x}{3x^2+5}$

43. $\lim_{x\to\infty}\frac{2x^2}{3x^2+5}$ **44.** $\lim_{x\to-\infty}\frac{3x^2}{x+5}$

45. $\lim_{x\to-\infty}\frac{\sqrt{x^2+x}}{-2x}$ **46.** $\lim_{x\to\infty}\frac{3x}{\sqrt{x^2+4}}$

In Exercises 47–50, find any vertical and horizontal asymptotes of the graph of the function. Use a graphing utility to verify your results.

47. $f(x) = \frac{3}{x} - 2$ **48.** $g(x) = \frac{5x^2}{x^2+2}$

49. $h(x) = \frac{2x+3}{x-4}$ **50.** $f(x) = \frac{3x}{\sqrt{x^2+2}}$

In Exercises 51–54, use a graphing utility to graph the function. Use the graph to approximate any relative extrema or asymptotes.

51. $f(x) = x^3 + \frac{243}{x}$ **52.** $f(x) = |x^3 - 3x^2 + 2x|$

53. $f(x) = \frac{x-1}{1+3x^2}$ **54.** $g(x) = \frac{2(x^2+1)}{x^2-4}$

In Exercises 55–70, analyze and sketch the graph of the function.

55. $f(x) = 4x - x^2$ **56.** $f(x) = 4x^3 - x^4$

57. $f(x) = x\sqrt{16-x^2}$ **58.** $f(x) = (x^2-4)^2$

59. $f(x) = (x-1)^3(x-3)^2$ **60.** $f(x) = (x-3)(x+2)^3$

61. $f(x) = x^{1/3}(x+3)^{2/3}$ **62.** $f(x) = (x-2)^{1/3}(x+1)^{2/3}$

63. $f(x) = \frac{5-3x}{x-2}$ **64.** $f(x) = \frac{2x}{1+x^2}$

65. $f(x) = \frac{4}{1+x^2}$ **66.** $f(x) = \frac{x^2}{1+x^4}$

67. $f(x) = x^3 + x + \frac{4}{x}$ **68.** $f(x) = x^2 + \frac{1}{x}$

69. $f(x) = |x^2 - 9|$ **70.** $f(x) = |x-1| + |x-3|$

71. Find the maximum and minimum points on the graph of

$$x^2 + 4y^2 - 2x - 16y + 13 = 0$$

(a) without using calculus.

(b) using calculus.

72. Consider the function $f(x) = x^n$ for positive integer values of n.

(a) For what values of n does the function have a relative minimum at the origin?

(b) For what values of n does the function have a point of inflection at the origin?

73. *Distance* At noon, ship A is 100 kilometers due east of ship B. Ship A is sailing west at 12 kilometers per hour, and ship B is sailing south at 10 kilometers per hour. At what time will the ships be nearest to each other, and what will this distance be?

74. *Maximum Area* Find the dimensions of the rectangle of maximum area, with sides parallel to the coordinate axes, that can be inscribed in the ellipse given by

$$\frac{x^2}{144} + \frac{y^2}{16} = 1.$$

75. *Minimum Length* A right triangle in the first quadrant has the coordinate axes as sides, and the hypotenuse passes through the point $(1, 8)$. Find the vertices of the triangle such that the length of the hypotenuse is minimum.

76. *Minimum Length* The wall of a building is to be braced by a beam that must pass over a parallel fence 5 feet high and 4 feet from the building. Find the length of the shortest beam that can be used.

77. *Maximum Area* Three sides of a trapezoid have the same length s. Of all such possible trapezoids, show that the one of maximum area has a fourth side of length $2s$.

78. *Maximum Area* Show that the greatest area of any rectangle inscribed in a triangle is one-half the area of the triangle.

Minimum Cost **In Exercises 79 and 80, find the speed v, in miles per hour, that will minimize costs on a 110-mile delivery trip. The cost per hour for fuel is C dollars, and the driver is paid W dollars per hour. (Assume there are no costs other than wages and fuel.)**

79. Fuel cost: $C = \frac{v^2}{600}$

Driver: $W = \$5$

80. Fuel cost: $C = \frac{v^2}{500}$

Driver: $W = \$7.50$

In Exercises 81 and 82, find the differential dy.

81. $y = (3x^2 - 2)^3$ **82.** $y = \sqrt{36 - x^2}$

83. *Surface Area and Volume* The diameter of a sphere is measured as 18 centimeters, with a maximum possible error of 0.05 centimeter. Use differentials to approximate the possible propagated error and percent error in calculating the surface area and the volume of the sphere.

84. *Demand Function* A company finds that the demand for its commodity is $p = 75 - (1/4)x$. If x changes from 7 to 8, find and compare the values of Δp and dp.

5 CHAPTER TEST

Take this test as you would take a test in class. When you are finished, check your work against the answers given in the back of the book.

1. Find the absolute extrema of $f(x) = \frac{4}{3}x^3 - 2x^2$ on the interval $[-3, 3]$.
2. Find the absolute extrema of $f(x) = 3x - 4$ (if any exist) over each interval.
   (a) $[0, 3]$ (b) $[0, 3)$ (c) $(0, 3]$ (d) $(0, 3)$
3. Determine whether Rolle's Theorem can be applied to $f(x) = \dfrac{x^2 - 9}{x + 5}$ on the closed interval $[-3, 3]$. If Rolle's Theorem can be applied, find all value(s) of c in the open interval $(-3, 3)$ such that $f'(c) = 0$. If Rolle's Theorem cannot be applied, explain why not.
4. Determine whether the Mean Value Theorem can be applied to $f(x) = x^4 + 8x$ on the closed interval $[-2, 0]$. If the Mean Value Theorem can be applied, find all value(s) of c in the open interval $(-2, 0)$ such that $f'(c) = \dfrac{f(b) - f(a)}{b - a}$. If the Mean Value Theorem cannot be applied, explain why not.
5. Identify the open intervals on which $y = 2x\sqrt{4 - x^2}$ is increasing or decreasing.
6. For $f(x) = x^3 - 3x^2 - 9x$, (a) find the critical numbers of f, (b) find the open interval(s) on which the function is increasing or decreasing, (c) apply the First Derivative Test to identify all relative extrema, and (d) use a graphing utility to confirm your results.
7. Determine the open intervals on which the graph of $f(x) = \dfrac{1}{x^2 + 1}$ is concave upward or concave downward.
8. For $f(x) = 3x^3 - 18x^2 - 10x + 4$, find the point(s) of inflection and discuss the concavity of the graph of the function.

In Exercises 9–12, find the limit.

9. $\displaystyle\lim_{x\to\infty} \frac{x^2 - 1}{x^3 + 2}$
10. $\displaystyle\lim_{x\to\infty} \frac{x^2 - 1}{x^2 + 2}$
11. $\displaystyle\lim_{x\to\infty} \frac{x^2 - 1}{x + 2}$
12. $\displaystyle\lim_{x\to\infty} \frac{\sqrt{x^2 - 1}}{x + 2}$

13. Sketch the graph of $y = \dfrac{3x^2}{x^2 - 9}$ using extrema, intercepts, symmetry, and asymptotes. Then use a graphing utility to verify your results.
14. Analyze and sketch a graph of $y = 6x^{2/3} + 2x$. Label any intercepts, relative extrema, and points of inflection. Use a graphing utility to verify your results.
15. Find the point on the graph of $f(x) = x^2$ that is closest to the point $\left(16, \frac{1}{2}\right)$.

Figure for 16

16. A farmer plans to fence a rectangular pasture adjacent to a river (see figure). The pasture must contain 180,000 square meters in order to provide enough grass for the herd. What dimensions will require the least amount of fencing if no fencing is needed along the river?

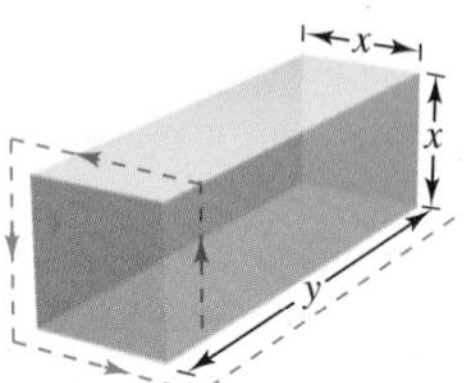

Figure for 17

17. A rectangular package to be sent by a shipping service can have a maximum combined length and girth (perimeter of a cross section) of 165 inches (see figure). Find the dimensions of the package of maximum volume that can be sent. (Assume the cross section is a square.)
18. Let $y = x^4 - 1$. Find dy when $x = -1$ and $dx = 0.01$. Compare this value with Δy for $x = -1$ and $\Delta x = 0.01$.
19. The measurement of the radius of the end of a log is found to be 14 inches, with a possible error of $\frac{1}{4}$ inch. Use differentials to approximate the possible propagated error in computing the area of the end of the log.

P.S. PROBLEM SOLVING

1. Graph the fourth-degree polynomial $p(x) = x^4 + ax^2 + 1$ for various values of the constant a.
   (a) Determine the values of a for which p has exactly one relative minimum.
   (b) Determine the values of a for which p has exactly one relative maximum.
   (c) Determine the values of a for which p has exactly two relative minima.
   (d) Show that the graph of p cannot have exactly two relative extrema.

2. (a) Graph the fourth-degree polynomial $p(x) = ax^4 - 6x^2$ for $a = -3, -2, -1, 0, 1, 2,$ and 3. For what values of the constant a does p have a relative minimum or relative maximum?
   (b) Show that p has a relative maximum for all values of the constant a.
   (c) Determine analytically the values of a for which p has a relative minimum.
   (d) Let $(x, y) = (x, p(x))$ be a relative extremum of p. Show that (x, y) lies on the graph of $y = -3x^2$. Verify this result graphically by graphing $y = -3x^2$ together with the seven curves from part (a).

3. Let $f(x) = \frac{c}{x} + x^2$. Determine all values of the constant c such that f has a relative minimum, but no relative maximum.

4. (a) Let $f(x) = ax^2 + bx + c$, $a \neq 0$, be a quadratic polynomial. How many points of inflection does the graph of f have?
   (b) Let $f(x) = ax^3 + bx^2 + cx + d$, $a \neq 0$, be a cubic polynomial. How many points of inflection does the graph of f have?
   (c) Suppose the function $y = f(x)$ satisfies the equation $\frac{dy}{dx} = ky\left(1 - \frac{y}{L}\right)$, where k and L are positive constants. Show that the graph of f has a point of inflection at the point where $y = \frac{L}{2}$. (This equation is called the **logistic differential equation.**)

5. Prove **Darboux's Theorem:** Let f be differentiable on the closed interval $[a, b]$ such that $f'(a) = y_1$ and $f'(b) = y_2$. If d lies between y_1 and y_2, then there exists c in (a, b) such that $f'(c) = d$.

6. Let f and g be functions that are continuous on $[a, b]$ and differentiable on (a, b). Prove that if $f(a) = g(a)$ and $g'(x) > f'(x)$ for all x in (a, b), then $g(b) > f(b)$.

7. Prove the following **Extended Mean Value Theorem.** If f and f' are continuous on the closed interval $[a, b]$, and if f'' exists in the open interval (a, b), then there exists a number c in (a, b) such that

$$f(b) = f(a) + f'(a)(b - a) + \frac{1}{2}f''(c)(b - a)^2.$$

8. (a) Let $V = x^3$. Find dV and ΔV. Show that for small values of x, the difference $\Delta V - dV$ is very small in the sense that there exists ε such that $\Delta V - dV = \varepsilon\Delta x$, where $\varepsilon \to 0$ as $\Delta x \to 0$.
   (b) Generalize this result by showing that if $y = f(x)$ is a differentiable function, then $\Delta y - dy = \varepsilon\Delta x$, where $\varepsilon \to 0$ as $\Delta x \to 0$.

9. Let L be a line through the point (p, q), intersecting the coordinate axes at the points A and B, as shown in the figure.

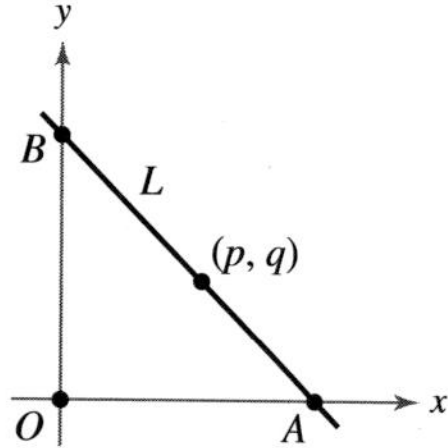

   (a) Find the minimum value of $OA + OB$ in terms of p and q.
   (b) Find the minimum value of $OA \cdot OB$ in terms of p and q.
   (c) Find the minimum value of AB in terms of p and q.

10. Consider a room in the shape of a cube, 4 meters on each side. A bug at point P wants to walk to point Q at the opposite corner, as shown in the figure. Use calculus to determine the shortest path. Can you solve the problem without calculus?

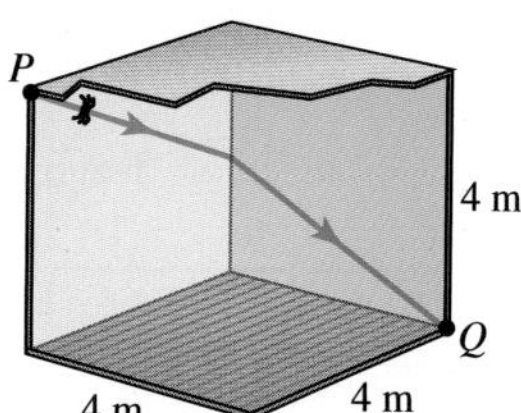

11. The line joining P and Q crosses the two parallel lines, as shown in the figure. The point R is d units from P. How far from Q should the point S be positioned so that the sum of the areas of the two shaded triangles is a minimum? So that the sum is a maximum?

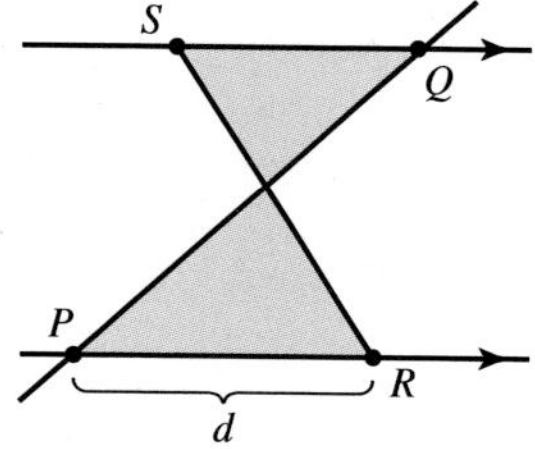

12. The figures show a rectangle, a circle, and a semicircle inscribed in a triangle bounded by the coordinate axes and the first-quadrant portion of the line with intercepts $(3, 0)$ and $(0, 4)$. Find the dimensions of each inscribed figure such that its area is maximum. State whether calculus was helpful in finding the required dimensions. Explain your reasoning.

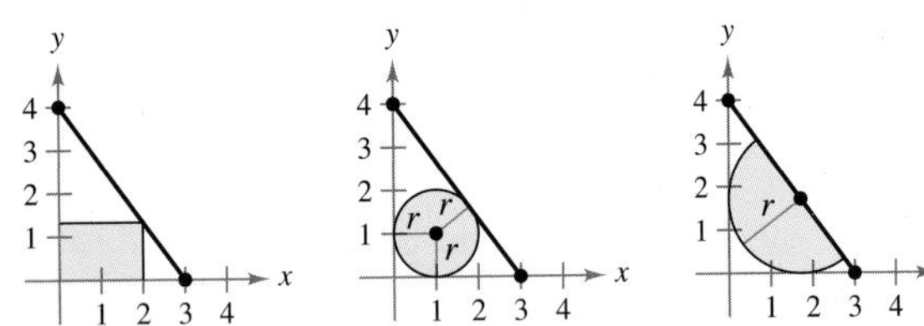

13. (a) Prove that $\lim_{x\to\infty} x^2 = \infty$.

(b) Prove that $\lim_{x\to\infty} \left(\frac{1}{x^2}\right) = 0$.

(c) Let L be a real number. Prove that if $\lim_{x\to\infty} f(x) = L$, then $\lim_{y\to 0^+} f\left(\frac{1}{y}\right) = L$.

14. Find the point on the graph of $y = \dfrac{1}{1 + x^2}$ (see figure) where the tangent line has the greatest slope, and the point where the tangent line has the least slope.

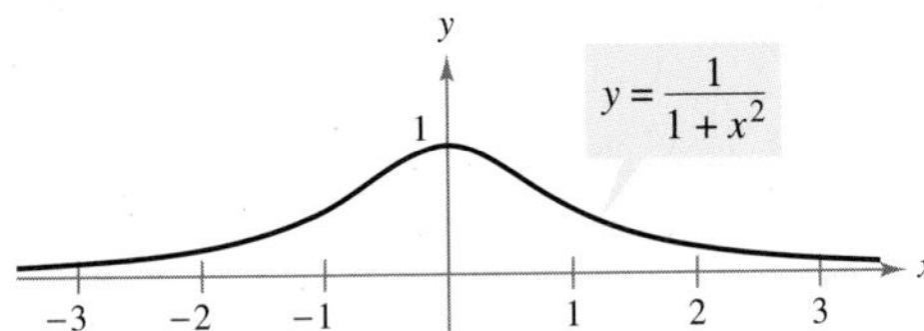

15. (a) Let x be a positive number. Use the *table* feature of a graphing utility to verify that $\sqrt{1 + x} < \frac{1}{2}x + 1$.

(b) Use the Mean Value Theorem to prove that $\sqrt{1 + x} < \frac{1}{2}x + 1$ for all positive real numbers x.

16. The police department must determine the speed limit on a bridge such that the flow rate of cars is maximum per unit time. The greater the speed limit, the farther apart the cars must be in order to keep a safe stopping distance. Experimental data on the stopping distances d (in meters) for various speeds v (in kilometers per hour) are shown in the table.

v	20	40	60	80	100
d	5.1	13.7	27.2	44.2	66.4

(a) Convert the speeds v in the table to speeds s in meters per second. Use the regression capabilities of a graphing utility to find a model of the form $d(s) = as^2 + bs + c$ for the data.

(b) Consider two consecutive vehicles of average length 5.5 meters, traveling at a safe speed on the bridge. Let T be the difference between the times (in seconds) when the front bumpers of the vehicles pass a given point on the bridge. Verify that this difference in times is given by

$$T = \frac{d(s)}{s} + \frac{5.5}{s}.$$

(c) Use a graphing utility to graph the function T and estimate the speed s that minimizes the time between vehicles.

(d) Use calculus to determine the speed that minimizes T. What is the minimum value of T? Convert the required speed to kilometers per hour.

(e) Find the optimal distance between vehicles for the posted speed limit determined in part (d).

17. Let R be the area of $\triangle ABC$. Use calculus to determine the area of the largest possible inscribed parallelogram, as shown in the figure. Can you solve the problem without calculus?

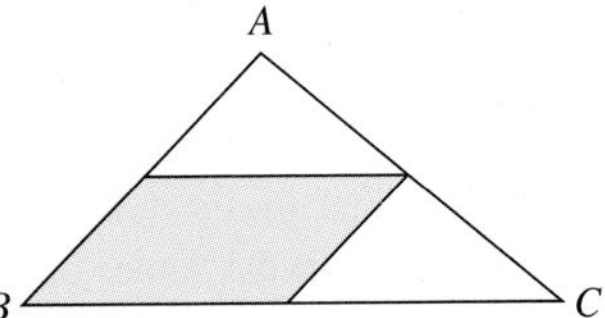

18. Graph the function given by $f(x) = |8 - x^3|$.

(a) Rewrite the function without using the absolute value notation.

(b) Find a formula for $f'(x)$.

(c) Determine the open intervals on which the graph of f is increasing and those on which the graph of f is decreasing.

(d) Determine the open intervals on which the graph of f is concave upward and those on which the graph of f is concave downward.

(e) Find all points of inflection of f.

19. A legal-sized sheet of paper (8.5 inches by 14 inches) is folded so that corner P touches the opposite 14-inch edge at R (see figure). (*Note:* $PQ = \sqrt{C^2 - x^2}$.)

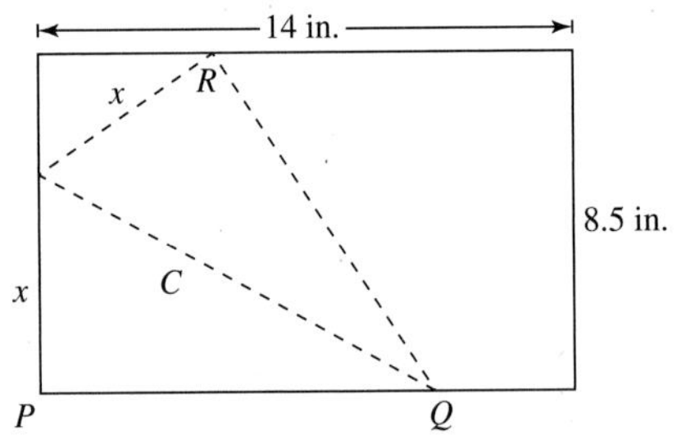

(a) Show that $C^2 = \dfrac{2x^3}{2x - 8.5}$.

(b) What is the domain of C?

(c) Determine the x-value that minimizes C.

(d) Determine the minimum length C.

20. The polynomial $P(x) = c_0 + c_1(x - a) + c_2(x - a)^2$ is the quadratic approximation of the function f at $(a, f(a))$ if $P(a) = f(a)$, $P'(a) = f'(a)$, and $P''(a) = f''(a)$.

(a) Find the quadratic approximation of

$$f(x) = \frac{x}{x + 1}$$

at $(0, 0)$.

(b) Use a graphing utility to graph $P(x)$ and $f(x)$ in the same viewing window.

6 Integration

In this chapter, you will study an important process of calculus that is closely related to differentiation–*integration*. You will learn new methods and rules for solving definite and indefinite integrals, including the Fundamental Theorem of Calculus. Then you will apply these rules to find such things as the position function for an object and the average value of a function.

In this chapter, you should learn the following.

- How to evaluate indefinite integrals using basic integration rules. (6.1)
- How to evaluate a sum and approximate the area of a plane region. (6.2)
- How to evaluate a definite integral using a limit. (6.3)
- How to evaluate a definite integral using the Fundamental Theorem of Calculus. (6.4)
- How to evaluate different types of definite and indefinite integrals using a variety of methods. (6.5)
- How to approximate a definite integral using the Trapezoidal Rule and Simpson's Rule. (6.6)

CHRISTOPHER PASATIERI/Reuters /Landov

This photo of a jet breaking the sound barrier was taken by Ensign John Gay. At different altitudes in Earth's atmosphere, sound travels at different speeds. How could you use integration to find the average speed of sound over a range of altitudes? (See Section 6.4, Example 5.)

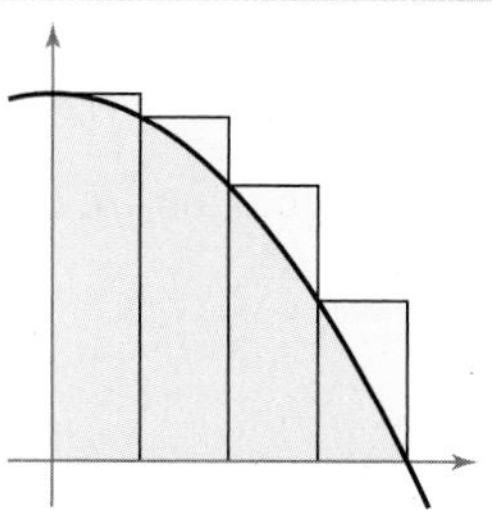

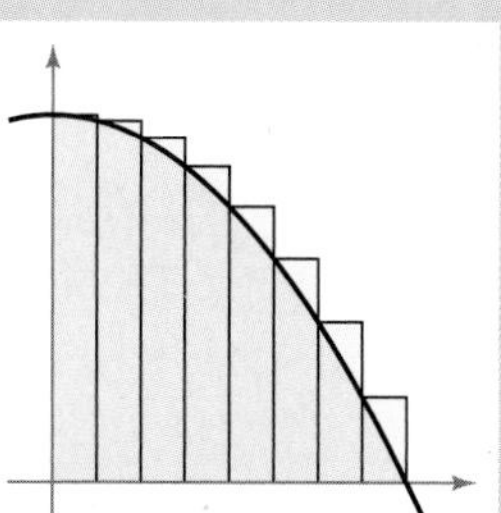

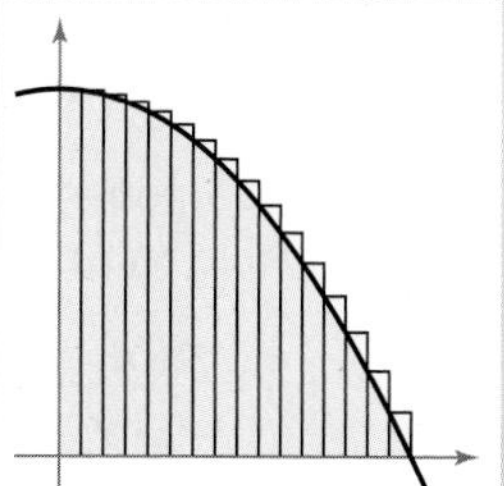

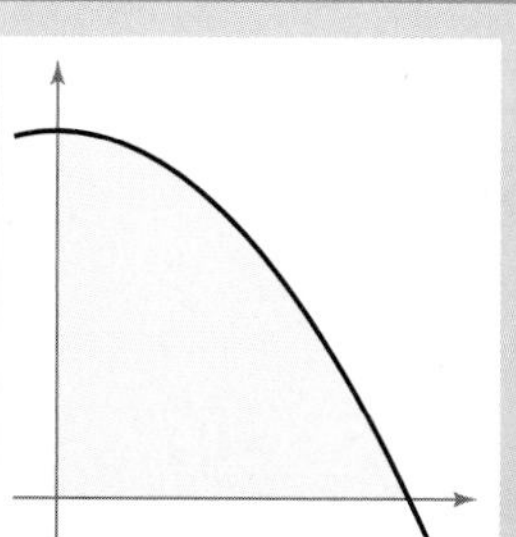

The area of a parabolic region can be approximated as the sum of the areas of rectangles. As you increase the number of rectangles, the approximation tends to become more and more accurate. In Section 6.2, you will learn how the limit process can be used to find areas of a wide variety of regions.

6.1 Antiderivatives and Indefinite Integration

- Write the general solution of a differential equation.
- Use indefinite integral notation for antiderivatives.
- Use basic integration rules to find antiderivatives.
- Find a particular solution of a differential equation.

Antiderivatives

Suppose you were asked to find a function F whose derivative is $f(x) = 3x^2$. From your knowledge of derivatives, you would probably say that

$$F(x) = x^3 \text{ because } \frac{d}{dx}[x^3] = 3x^2.$$

The function F is an *antiderivative* of f.

DEFINITION OF ANTIDERIVATIVE

A function F is an **antiderivative** of f on an interval I if $F'(x) = f(x)$ for all x in I.

EXPLORATION

Finding Antiderivatives For each derivative, describe the original function F.

a. $F'(x) = 2x$ **b.** $F'(x) = x$

c. $F'(x) = x^2$ **d.** $F'(x) = \dfrac{1}{x^2}$

e. $F'(x) = \dfrac{1}{x^3}$

What strategy did you use to find F?

Note that F is called *an* antiderivative of f, rather than *the* antiderivative of f. To see why, observe that

$$F_1(x) = x^3, \quad F_2(x) = x^3 - 5, \quad \text{and} \quad F_3(x) = x^3 + 97$$

are all antiderivatives of $f(x) = 3x^2$. In fact, for any constant C, the function given by $F(x) = x^3 + C$ is an antiderivative of f.

THEOREM 6.1 REPRESENTATION OF ANTIDERIVATIVES

If F is an antiderivative of f on an interval I, then G is an antiderivative of f on the interval I if and only if G is of the form $G(x) = F(x) + C$, for all x in I where C is a constant.

STUDY TIP Up to this point, you have been doing differential calculus:

Given f, find f'.

Here you begin work with the inverse process:

Given f', find f.

PROOF The proof of Theorem 6.1 in one direction is straightforward. That is, if $G(x) = F(x) + C$, $F'(x) = f(x)$, and C is a constant, then

$$G'(x) = \frac{d}{dx}[F(x) + C] = F'(x) + 0 = f(x).$$

To prove this theorem in the other direction, assume that G is an antiderivative of f. Define a function H such that

$$H(x) = G(x) - F(x).$$

For any two points a and b $(a < b)$ in the interval, H is continuous on $[a, b]$ and differentiable on (a, b). By the Mean Value Theorem,

$$H'(c) = \frac{H(b) - H(a)}{b - a}$$

for some c in (a, b). However, $H'(c) = 0$, so $H(a) = H(b)$. Because a and b are arbitrary points in the interval, you know that H is a constant function C. So, $G(x) - F(x) = C$ and it follows that $G(x) = F(x) + C$. ■

Using Theorem 6.1, you can represent the entire family of antiderivatives of a function by adding a constant to a *known* antiderivative. For example, knowing that $D_x[x^2] = 2x$, you can represent the family of *all* antiderivatives of $f(x) = 2x$ by

$$G(x) = x^2 + C \qquad \text{Family of all antiderivatives of } f(x) = 2x$$

where C is a constant. The constant C is called the **constant of integration.** The family of functions represented by G is the **general antiderivative** of f, and $G(x) = x^2 + C$ is the **general solution** of the *differential equation*

$$G'(x) = 2x. \qquad \text{Differential equation}$$

A **differential equation** in x and y is an equation that involves x, y, and derivatives of y. For instance, $y' = 3x$ and $y' = x^2 + 1$ are examples of differential equations.

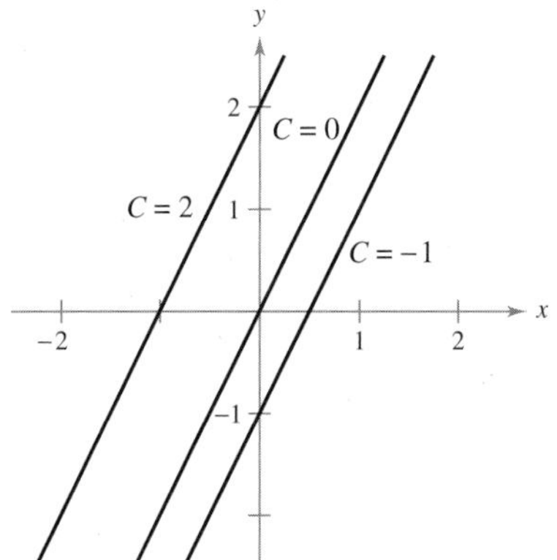

Functions of the form $y = 2x + C$
Figure 6.1

EXAMPLE 1 Solving a Differential Equation

Find the general solution of the differential equation $y' = 2$.

Solution To begin, you need to find a function whose derivative is 2. One such function is

$$y = 2x. \qquad 2x \text{ is } an \text{ antiderivative of 2.}$$

Now, you can use Theorem 6.1 to conclude that the general solution of the differential equation is

$$y = 2x + C. \qquad \text{General solution}$$

The graphs of several functions of the form $y = 2x + C$ are shown in Figure 6.1. ■

Notation for Antiderivatives

When solving a differential equation of the form

$$\frac{dy}{dx} = f(x)$$

it is convenient to write it in the equivalent differential form

$$dy = f(x)\,dx.$$

The operation of finding all solutions of this equation is called **antidifferentiation** (or **indefinite integration**) and is denoted by an integral sign $\int$. The general solution is denoted by

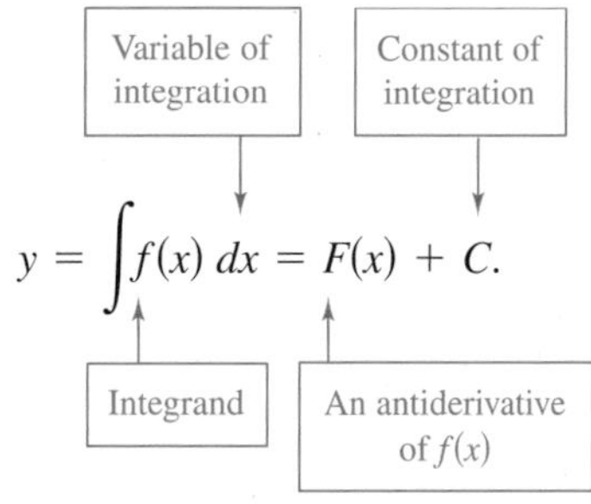

$$y = \int f(x)\,dx = F(x) + C.$$

The expression $\int f(x)dx$ is read as the *antiderivative of f with respect to x.* So, the differential dx serves to identify x as the variable of integration. The term **indefinite integral** is a synonym for antiderivative.

NOTE In this text, the notation $\int f(x)\,dx = F(x) + C$ means that F is an antiderivative of f *on an interval.*

Basic Integration Rules

The inverse nature of integration and differentiation can be verified by substituting $F'(x)$ for $f(x)$ in the indefinite integration definition to obtain

$$\int F'(x)\,dx = F(x) + C. \qquad \text{Integration is the "inverse" of differentiation.}$$

Moreover, if $\int f(x)\,dx = F(x) + C$, then differentiating both sides yields

$$\frac{d}{dx}\left[\int f(x)\,dx\right] = f(x). \qquad \text{Differentiation is the "inverse" of integration.}$$

These two equations allow you to obtain integration formulas directly from differentiation formulas, as shown in the following summary.

BASIC INTEGRATION RULES

Differentiation Formula	*Integration Formula*
$\frac{d}{dx}[C] = 0$	$\int 0\,dx = C$
$\frac{d}{dx}[kx] = k$	$\int k\,dx = kx + C$
$\frac{d}{dx}[kf(x)] = kf'(x)$	$\int kf(x)\,dx = k\int f(x)\,dx$
$\frac{d}{dx}[f(x) \pm g(x)] = f'(x) \pm g'(x)$	$\int [f(x) \pm g(x)]\,dx = \int f(x)\,dx \pm \int g(x)\,dx$
$\frac{d}{dx}[x^n] = nx^{n-1}$	$\int x^n\,dx = \frac{x^{n+1}}{n+1} + C,\ n \neq -1$ Power Rule

NOTE Note that the Power Rule for Integration has the restriction that $n \neq -1$. The evaluation of $\int 1/x\,dx$ must wait until the introduction of the natural logarithmic function in Chapter 8.

EXAMPLE 2 Applying the Basic Integration Rules

Describe the antiderivatives of $3x$.

Solution

$$\int 3x\,dx = 3\int x^1\,dx \qquad \text{Constant Multiple Rule and rewrite } x \text{ as } x^1.$$

$$= 3\left(\frac{x^2}{2}\right) + C \qquad \text{Power Rule } (n = 1)$$

$$= \frac{3}{2}x^2 + C \qquad \text{Simplify.}$$

■

When indefinite integrals are evaluated, a strict application of the basic integration rules tends to produce complicated constants of integration. For instance, in Example 2, you could have written

$$\int 3x\,dx = 3\int x\,dx = 3\left(\frac{x^2}{2} + C\right) = \frac{3}{2}x^2 + 3C.$$

However, because C represents *any* constant, it is both cumbersome and unnecessary to write $3C$ as the constant of integration. So, $\frac{3}{2}x^2 + 3C$ is written in the simpler form, $\frac{3}{2}x^2 + C$.

In Example 2, note that the general pattern of integration is similar to that of differentiation.

Original integral ⇨ Rewrite ⇨ Integrate ⇨ Simplify

TECHNOLOGY Some software programs, such as *Maple*, *Mathematica*, and the *TI-89*, are capable of performing integration symbolically. If you have access to such a symbolic integration utility, try using it to evaluate the indefinite integrals in Example 3.

EXAMPLE 3 Rewriting Before Integrating

	Original Integral	*Rewrite*	*Integrate*	*Simplify*
a.	$\int \frac{1}{x^3}\,dx$	$\int x^{-3}\,dx$	$\frac{x^{-2}}{-2} + C$	$-\frac{1}{2x^2} + C$
b.	$\int \sqrt{x}\,dx$	$\int x^{1/2}\,dx$	$\frac{x^{3/2}}{3/2} + C$	$\frac{2}{3}x^{3/2} + C$

■

The basic integration rules listed on page 400 allow you to integrate any polynomial function, as demonstrated in Example 4.

EXAMPLE 4 Integrating Polynomial Functions

a. $\int dx = \int 1\,dx = x + C$ Integrand is understood to be 1.

b. $\int (x+2)\,dx = \int x\,dx + \int 2\,dx$

$= \frac{x^2}{2} + C_1 + 2x + C_2$ Integrate.

$= \frac{x^2}{2} + 2x + C$ $C = C_1 + C_2$

NOTE The second line in the solution of Example 4(b) is usually omitted.

c. $\int (3x^4 - 5x^2 + x)\,dx = 3\left(\frac{x^5}{5}\right) - 5\left(\frac{x^3}{3}\right) + \frac{x^2}{2} + C$ Integrate.

$= \frac{3}{5}x^5 - \frac{5}{3}x^3 + \frac{1}{2}x^2 + C$ Simplify.

EXAMPLE 5 Rewriting Before Integrating

$\int \frac{x+1}{\sqrt{x}}\,dx = \int \left(\frac{x}{\sqrt{x}} + \frac{1}{\sqrt{x}}\right)dx$ Rewrite as two fractions.

$= \int (x^{1/2} + x^{-1/2})\,dx$ Rewrite with fractional exponents.

$= \frac{x^{3/2}}{3/2} + \frac{x^{1/2}}{1/2} + C$ Integrate.

$= \frac{2}{3}x^{3/2} + 2x^{1/2} + C$ Simplify. ■

STUDY TIP Remember that you can check your answer by differentiating.

NOTE When integrating quotients, do not integrate the numerator and denominator separately. This is no more valid in integration than it is in differentiation. For instance, in Example 5, be sure you understand that

$$\int \frac{x+1}{\sqrt{x}}\,dx \neq \frac{\int (x+1)\,dx}{\int \sqrt{x}\,dx}.$$

■

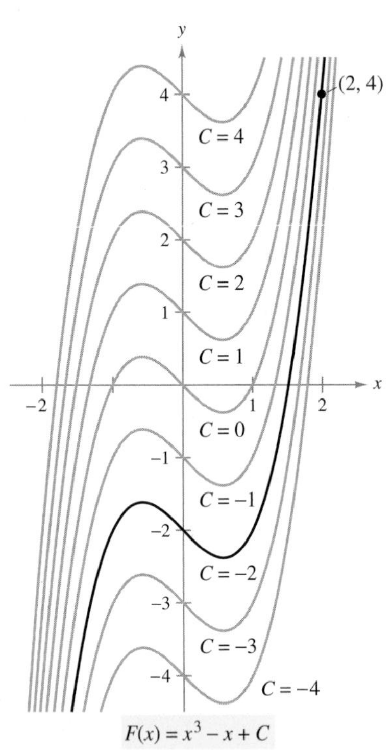

The particular solution that satisfies the initial condition $F(2) = 4$ is $F(x) = x^3 - x - 2$.
Figure 6.2

Initial Conditions and Particular Solutions

You have already seen that the equation $y = \int f(x)\,dx$ has many solutions (each differing from the others by a constant). This means that the graphs of any two antiderivatives of f are vertical translations of each other. For example, Figure 6.2 shows the graphs of several antiderivatives of the form

$$y = \int (3x^2 - 1)\,dx = x^3 - x + C \qquad \text{General solution}$$

for various integer values of C. Each of these antiderivatives is a solution of the differential equation

$$\frac{dy}{dx} = 3x^2 - 1.$$

In many applications of integration, you are given enough information to determine a **particular solution.** To do this, you need only know the value of $y = F(x)$ for one value of x. This information is called an **initial condition.** For example, in Figure 6.2, only one curve passes through the point $(2, 4)$. To find this curve, you can use the following information.

$$F(x) = x^3 - x + C \qquad \text{General solution}$$

$$F(2) = 4 \qquad \text{Initial condition}$$

By using the initial condition in the general solution, you can determine that $F(2) = 8 - 2 + C = 4$, which implies that $C = -2$. So, you obtain

$$F(x) = x^3 - x - 2. \qquad \text{Particular solution}$$

EXAMPLE 6 Finding a Particular Solution

Find the general solution of

$$F'(x) = \frac{1}{x^2}, \quad x > 0$$

and find the particular solution that satisfies the initial condition $F(1) = 0$.

Solution To find the general solution, integrate to obtain

$$F(x) = \int \frac{1}{x^2}\,dx \qquad F(x) = \int F'(x)\,dx$$

$$= \int x^{-2}\,dx \qquad \text{Rewrite as a power.}$$

$$= \frac{x^{-1}}{-1} + C \qquad \text{Integrate.}$$

$$= -\frac{1}{x} + C, \quad x > 0. \qquad \text{General solution}$$

Using the initial condition $F(1) = 0$, you can solve for C as follows.

$$F(1) = -\frac{1}{1} + C = 0 \implies C = 1$$

So, the particular solution, as shown in Figure 6.3, is

$$F(x) = -\frac{1}{x} + 1, \quad x > 0. \qquad \text{Particular solution}$$

■

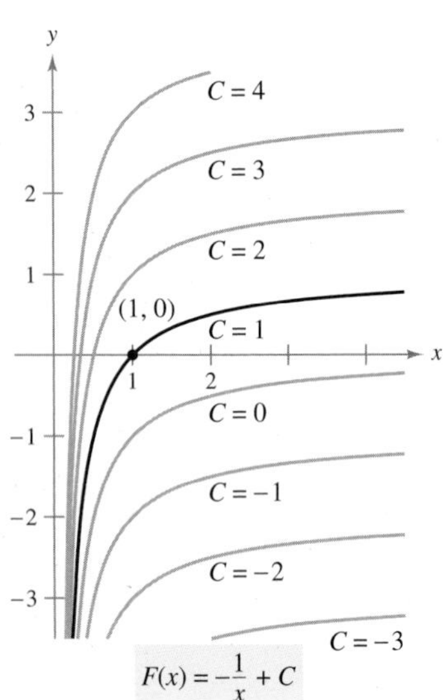

The particular solution that satisfies the initial condition $F(1) = 0$ is $F(x) = -(1/x) + 1$, $x > 0$.
Figure 6.3

So far in this section you have been using x as the variable of integration. In applications, it is often convenient to use a different variable. For instance, in the following example involving *time*, the variable of integration is t.

EXAMPLE 7 Solving a Vertical Motion Problem

A ball is thrown upward with an initial velocity of 64 feet per second from an initial height of 80 feet.

a. Find the position function giving the height s as a function of the time t.

b. When does the ball hit the ground?

Solution

a. Let $t = 0$ represent the initial time. The two given initial conditions can be written as follows.

$$s(0) = 80 \qquad \text{Initial height is 80 feet.}$$

$$s'(0) = 64 \qquad \text{Initial velocity is 64 feet per second.}$$

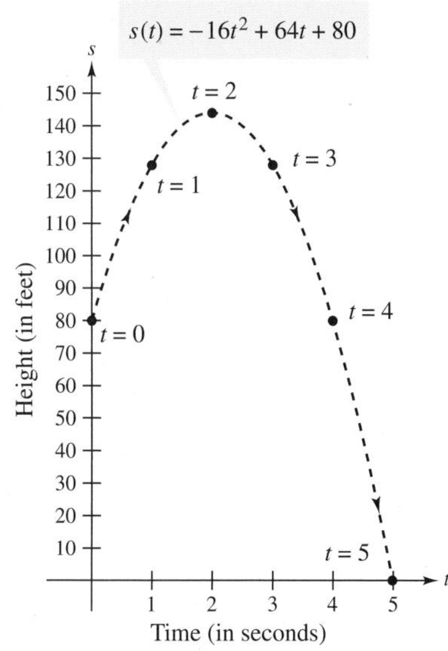

Height of a ball at time t
Figure 6.4

Using -32 feet per second per second as the acceleration due to gravity, you can write

$$s''(t) = -32$$

$$s'(t) = \int s''(t)\, dt = \int -32\, dt = -32t + C_1.$$

Using the initial velocity, you obtain $s'(0) = 64 = -32(0) + C_1$, which implies that $C_1 = 64$. Next, by integrating $s'(t)$, you obtain

$$s(t) = \int s'(t)\, dt = \int (-32t + 64)\, dt = -16t^2 + 64t + C_2.$$

Using the initial height, you obtain

$$s(0) = 80 = -16(0)^2 + 64(0) + C_2$$

which implies that $C_2 = 80$. So, the position function is

$$s(t) = -16t^2 + 64t + 80. \qquad \text{See Figure 6.4.}$$

b. Using the position function found in part (a), you can find the time at which the ball hits the ground by solving the equation $s(t) = 0$.

$$s(t) = -16t^2 + 64t + 80 = 0$$

$$-16(t + 1)(t - 5) = 0$$

$$t = -1, 5$$

Because t must be positive, you can conclude that the ball hits the ground 5 seconds after it was thrown. ■

NOTE In Example 7, note that the position function has the form

$$s(t) = \tfrac{1}{2}gt^2 + v_0t + s_0$$

where $g = -32$, v_0 is the initial velocity, and s_0 is the initial height, as presented in Section 4.2.

Example 7 shows how to use calculus to analyze vertical motion problems in which the acceleration is determined by a gravitational force. You can use a similar strategy to analyze other linear motion problems (vertical or horizontal) in which the acceleration (or deceleration) is the result of some other force, as you will see in Exercises 69–77.

Before you begin the exercise set, be sure you realize that one of the most important steps in integration is *rewriting the integrand* in a form that fits the basic integration rules. To illustrate this point further, here are some additional examples.

Original Integral	*Rewrite*	*Integrate*	*Simplify*
$\int \frac{2}{\sqrt{x}}\,dx$	$2\int x^{-1/2}\,dx$	$2\left(\frac{x^{1/2}}{1/2}\right) + C$	$4x^{1/2} + C$
$\int (t^2+1)^2\,dt$	$\int (t^4 + 2t^2 + 1)\,dt$	$\frac{t^5}{5} + 2\left(\frac{t^3}{3}\right) + t + C$	$\frac{1}{5}t^5 + \frac{2}{3}t^3 + t + C$
$\int \frac{x^3+3}{x^2}\,dx$	$\int (x + 3x^{-2})\,dx$	$\frac{x^2}{2} + 3\left(\frac{x^{-1}}{-1}\right) + C$	$\frac{1}{2}x^2 - \frac{3}{x} + C$
$\int \sqrt[3]{x}(x-4)\,dx$	$\int (x^{4/3} - 4x^{1/3})\,dx$	$\frac{x^{7/3}}{7/3} - 4\left(\frac{x^{4/3}}{4/3}\right) + C$	$\frac{3}{7}x^{7/3} - 3x^{4/3} + C$

6.1 Exercises

See www.CalcChat.com for worked-out solutions to odd-numbered exercises.

In Exercises 1–4, verify the statement by showing that the derivative of the right side equals the integrand of the left side.

1. $\int \left(-\frac{6}{x^4}\right) dx = \frac{2}{x^3} + C$

2. $\int \left(8x^3 + \frac{1}{2x^2}\right) dx = 2x^4 - \frac{1}{2x} + C$

3. $\int (x-4)(x+4)\,dx = \frac{1}{3}x^3 - 16x + C$

4. $\int \frac{x^2-1}{x^{3/2}}\,dx = \frac{2(x^2+3)}{3\sqrt{x}} + C$

In Exercises 5–8, find the general solution of the differential equation and check the result by differentiation.

5. $\frac{dy}{dt} = 9t^2$

6. $\frac{dr}{d\theta} = \pi$

7. $\frac{dy}{dx} = x^{3/2}$

8. $\frac{dy}{dx} = 2x^{-3}$

In Exercises 9–14, complete the table.

	Original Integral	*Rewrite*	*Integrate*	*Simplify*
9.	$\int \sqrt[3]{x}\,dx$			
10.	$\int \frac{1}{4x^2}\,dx$			
11.	$\int \frac{1}{x\sqrt{x}}\,dx$			
12.	$\int x(x^3+1)\,dx$			
13.	$\int \frac{1}{2x^3}\,dx$			
14.	$\int \frac{1}{(3x)^2}\,dx$			

In Exercises 15–34, find the indefinite integral and check the result by differentiation.

15. $\int (x+7)\,dx$

16. $\int (13-x)\,dx$

17. $\int (2x - 3x^2)\,dx$

18. $\int (8x^3 - 9x^2 + 4)\,dx$

19. $\int (x^5 + 1)\,dx$

20. $\int (x^3 - 10x - 3)\,dx$

21. $\int (x^{3/2} + 2x + 1)\,dx$

22. $\int \left(\sqrt{x} + \frac{1}{2\sqrt{x}}\right) dx$

23. $\int \sqrt[3]{x^2}\,dx$

24. $\int \left(\sqrt[4]{x^3} + 1\right) dx$

25. $\int \frac{1}{x^5}\,dx$

26. $\int \frac{1}{x^6}\,dx$

27. $\int \frac{x+6}{\sqrt{x}}\,dx$

28. $\int \frac{x^2 + 2x - 3}{x^4}\,dx$

29. $\int (x+1)(3x-2)\,dx$

30. $\int (2t^2 - 1)^2\,dt$

31. $\int y^2\sqrt{y}\,dy$

32. $\int (1+3t)t^2\,dt$

33. $\int dx$

34. $\int 14\,dt$

In Exercises 35 and 36, sketch the graphs of the function $g(x) = f(x) + C$ for $C = -2$, $C = 0$, and $C = 3$ on the same set of coordinate axes.

35. $f(x) = \dfrac{1}{x}$

36. $f(x) = \sqrt{x}$

In Exercises 37–40, the graph of the derivative of a function is given. Sketch the graphs of two functions that have the given derivative. (There is more than one correct answer.) To print an enlarged copy of the graph, go to the website *www.mathgraphs.com*.

37.

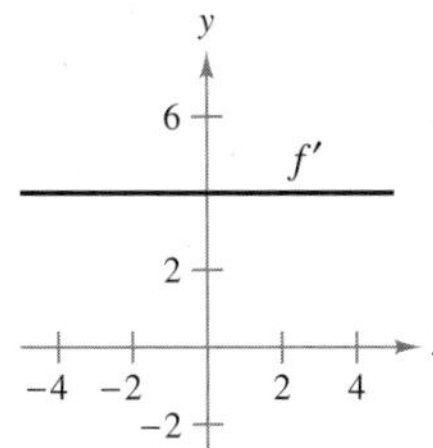

38.

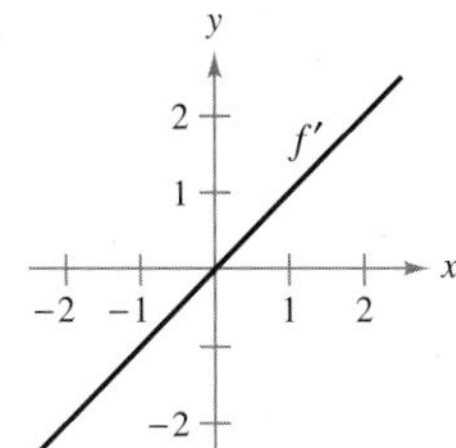

39.

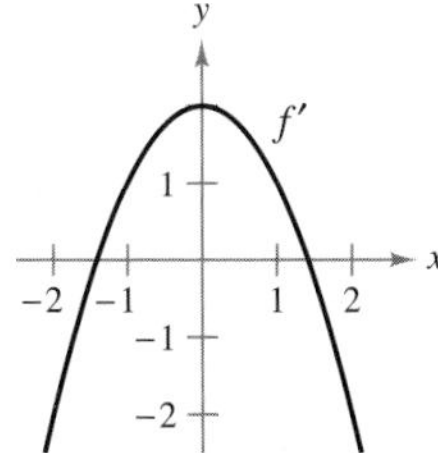

40.

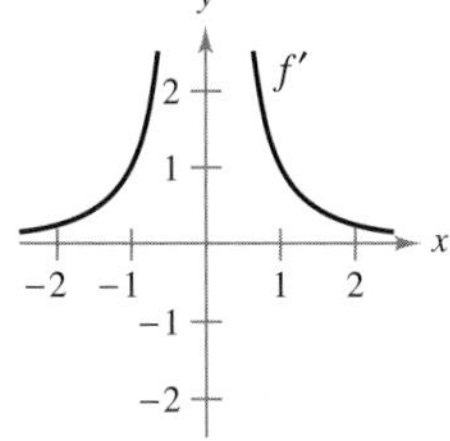

In Exercises 41–44, find the equation of y, given the derivative and the indicated point on the curve.

41. $\dfrac{dy}{dx} = 2x - 1$

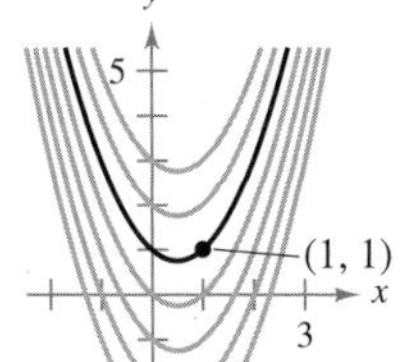

42. $\dfrac{dy}{dx} = 2(x - 1)$

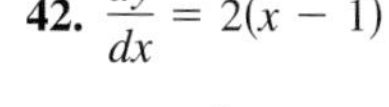

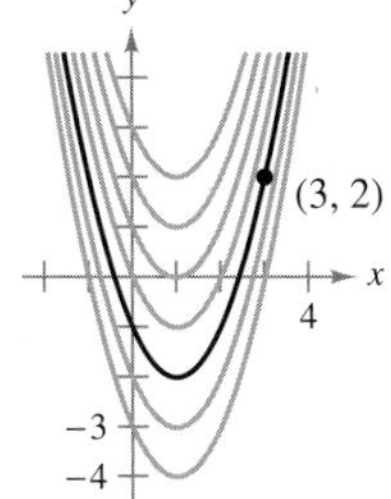

43. $\dfrac{dy}{dx} = 3x^2 - 1$

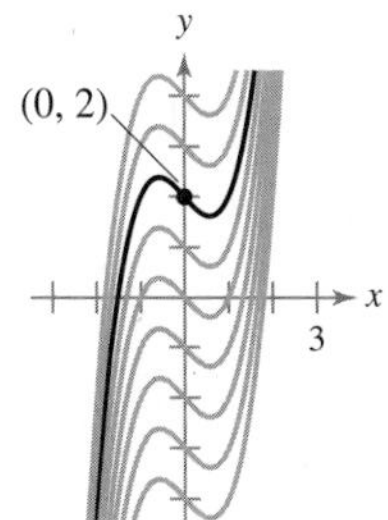

44. $\dfrac{dy}{dx} = -\dfrac{1}{x^2}, \quad x > 0$

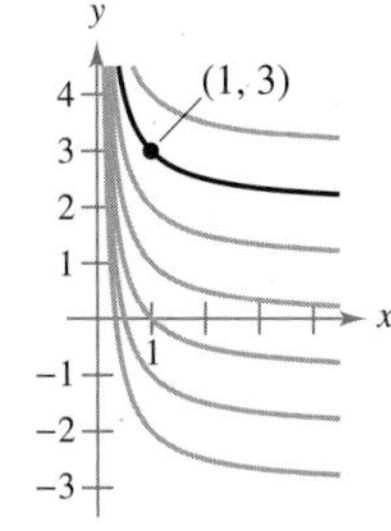

Slope Fields **In Exercises 45 and 46, a differential equation, a point, and a slope field are given. A *slope field* (or *direction field*) consists of line segments with slopes given by the differential equation. These line segments give a visual perspective of the slopes of the solutions of the differential equation. (a) Sketch two approximate solutions of the differential equation on the slope field, one of which passes through the indicated point. (To print an enlarged copy of the graph, go to the website *www.mathgraphs.com*.) (b) Use integration to find the particular solution of the differential equation and use a graphing utility to graph the solution. Compare the result with the sketches in part (a).**

45. $\dfrac{dy}{dx} = \dfrac{1}{2}x - 1, \ (4, 2)$

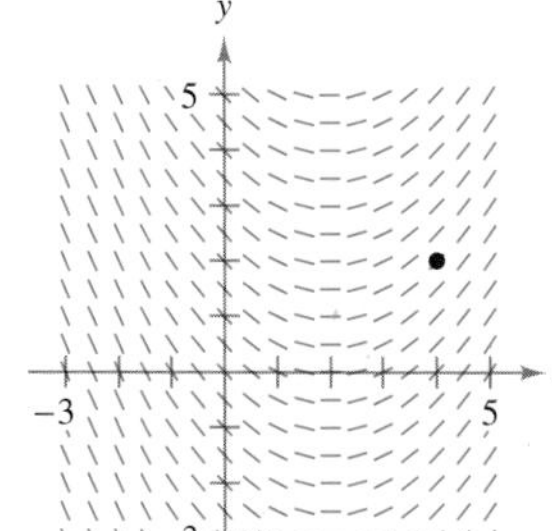

46. $\dfrac{dy}{dx} = x^2 - 1, \ (-1, 3)$

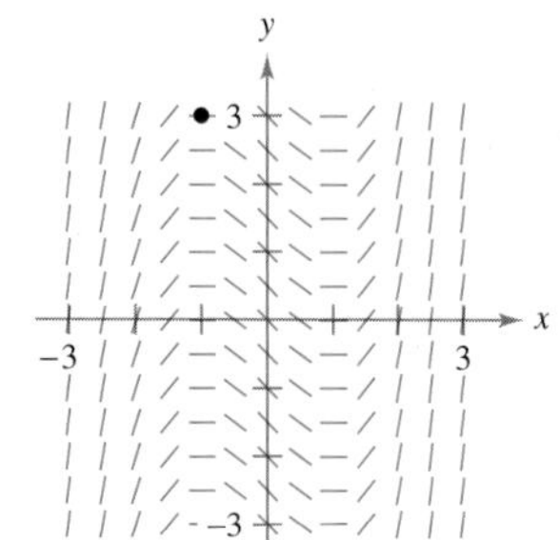

Slope Fields **In Exercises 47 and 48, (a) use a graphing utility to graph a slope field for the differential equation, (b) use integration and the given point to find the particular solution of the differential equation, and (c) graph the solution and the slope field in the same viewing window.**

47. $\dfrac{dy}{dx} = 2x, \ (-2, -2)$

48. $\dfrac{dy}{dx} = 2\sqrt{x}, \ (4, 12)$

In Exercises 49–54, solve the differential equation.

49. $f'(x) = 6x, \ f(0) = 8$

50. $g'(x) = 6x^2, \ g(0) = -1$

51. $h'(t) = 8t^3 + 5, \ h(1) = -4$

52. $f'(s) = 10s - 12s^3, \ f(3) = 2$

53. $f''(x) = 2, \ f'(2) = 5, \ f(2) = 10$

54. $f''(x) = x^2, \ f'(0) = 8, \ f(0) = 4$

55. ***Tree Growth*** An evergreen nursery usually sells a certain type of shrub after 6 years of growth and shaping. The growth rate during those 6 years is approximated by

$$\frac{dh}{dt} = 1.5t + 5$$

where t is the time in years and h is the height in centimeters. The seedlings are 12 centimeters tall when planted ($t = 0$).

(a) Find the height after t years.

(b) How tall are the shrubs when they are sold?

56. ***Population Growth*** The rate of growth dP/dt of a population of bacteria is proportional to the square root of t, where P is the population size and t is the time in days $(0 \leq t \leq 10)$. That is, $dP/dt = k\sqrt{t}$. The initial size of the population is 500. After 1 day the population has grown to 600. Estimate the population after 7 days.

WRITING ABOUT CONCEPTS

57. The graphs of f and f' each pass through the origin. Use the graph of f'' shown in the figure to sketch the graphs of f and f'. To print an enlarged copy of the graph, go to the website *www.mathgraphs.com.*

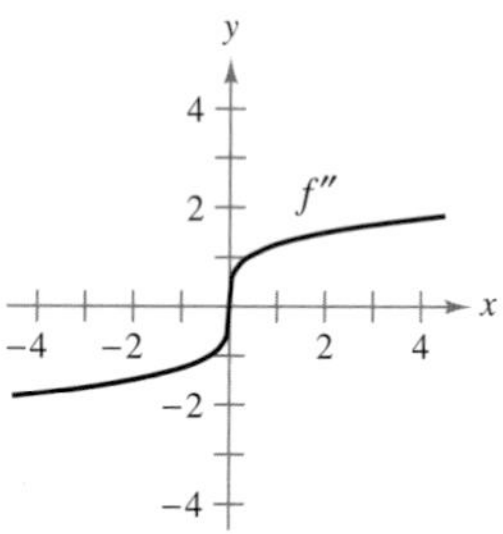

CAPSTONE

58. Use the graph of f' shown in the figure to answer the following, given that $f(0) = -4$.

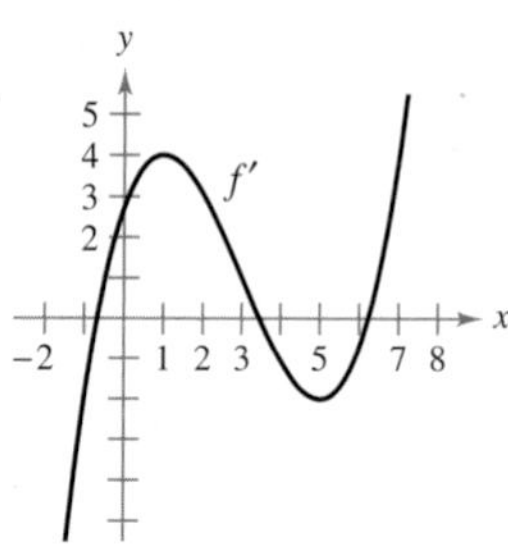

(a) Approximate the slope of f at $x = 4$. Explain.

(b) Is it possible that $f(2) = -1$? Explain.

(c) Is $f(5) - f(4) > 0$? Explain.

(d) Approximate the value of x where f is maximum. Explain.

(e) Approximate any intervals in which the graph of f is concave upward and any intervals in which it is concave downward. Approximate the x-coordinates of any points of inflection.

(f) Approximate the x-coordinate of the minimum of $f''(x)$.

(g) Sketch an approximate graph of f. To print an enlarged copy of the graph, go to the website *www.mathgraphs.com.*

Vertical Motion **In Exercises 59–62, use $a(t) = -32$ feet per second per second as the acceleration due to gravity. (Neglect air resistance.)**

59. A ball is thrown vertically upward from a height of 6 feet with an initial velocity of 60 feet per second. How high will the ball go?

60. Show that the height above the ground of an object thrown upward from a point s_0 feet above the ground with an initial velocity of v_0 feet per second is given by the function

$$f(t) = -16t^2 + v_0t + s_0.$$

61. With what initial velocity must an object be thrown upward (from ground level) to reach the top of the Washington Monument (approximately 550 feet)?

62. A balloon, rising vertically with a velocity of 16 feet per second, releases a sandbag at the instant it is 64 feet above the ground.

(a) How many seconds after its release will the bag strike the ground?

(b) At what velocity will it hit the ground?

Vertical Motion **In Exercises 63–66, use $a(t) = -9.8$ meters per second per second as the acceleration due to gravity. (Neglect air resistance.)**

63. Show that the height above the ground of an object thrown upward from a point s_0 meters above the ground with an initial velocity of v_0 meters per second is given by the function

$$f(t) = -4.9t^2 + v_0t + s_0.$$

64. The Grand Canyon is 1800 meters deep at its deepest point. A rock is dropped from the rim above this point. Write the height of the rock as a function of the time t in seconds. How long will it take the rock to hit the canyon floor?

65. A baseball is thrown upward from a height of 2 meters with an initial velocity of 10 meters per second. Determine its maximum height.

66. With what initial velocity must an object be thrown upward (from a height of 2 meters) to reach a maximum height of 200 meters?

67. ***Lunar Gravity*** On the moon, the acceleration due to gravity is -1.6 meters per second per second. A stone is dropped from a cliff on the moon and hits the surface of the moon 20 seconds later. How far did it fall? What was its velocity at impact?

68. ***Escape Velocity*** The minimum velocity required for an object to escape Earth's gravitational pull is obtained from the solution of the equation

$$\int v\,dv = -GM\int \frac{1}{y^2}\,dy$$

where v is the velocity of the object projected from Earth, y is the distance from the center of Earth, G is the gravitational constant, and M is the mass of Earth. Show that v and y are related by the equation

$$v^2 = v_0^2 + 2GM\left(\frac{1}{y} - \frac{1}{R}\right)$$

where v_0 is the initial velocity of the object and R is the radius of Earth.

Rectilinear Motion **In Exercises 69–71, consider a particle moving along the x-axis where $x(t)$ is the position of the particle at time t, $x'(t)$ is its velocity, and $x''(t)$ is its acceleration.**

69. $x(t) = t^3 - 6t^2 + 9t - 2, \quad 0 \le t \le 5$

(a) Find the velocity and acceleration of the particle.

(b) Find the open t-intervals on which the particle is moving to the right.

(c) Find the velocity of the particle when the acceleration is 0.

70. Repeat Exercise 69 for the position function

$x(t) = (t - 1)(t - 3)^2, \quad 0 \le t \le 5$

71. A particle moves along the x-axis at a velocity of $v(t) = 1/\sqrt{t}$, $t > 0$. At time $t = 1$, its position is $x = 4$. Find the acceleration and position functions for the particle.

72. ***Acceleration*** The maker of an automobile advertises that it takes 13 seconds to accelerate from 25 kilometers per hour to 80 kilometers per hour. Assuming constant acceleration, compute the following.

(a) The acceleration in meters per second per second

(b) The distance the car travels during the 13 seconds

73. ***Deceleration*** A car traveling at 45 miles per hour is brought to a stop, at constant deceleration, 132 feet from where the brakes are applied.

(a) How far has the car moved when its speed has been reduced to 30 miles per hour?

(b) How far has the car moved when its speed has been reduced to 15 miles per hour?

(c) Draw the real number line from 0 to 132, and plot the points found in parts (a) and (b). What can you conclude?

74. ***Acceleration*** At the instant the traffic light turns green, a car that has been waiting at an intersection starts with a constant acceleration of 6 feet per second per second. At the same instant, a truck traveling with a constant velocity of 30 feet per second passes the car.

(a) How far beyond its starting point will the car pass the truck?

(b) How fast will the car be traveling when it passes the truck?

75. ***Acceleration*** Assume that a fully loaded plane starting from rest has a constant acceleration while moving down a runway. The plane requires 0.7 mile of runway and a speed of 160 miles per hour in order to lift off. What is the plane's acceleration?

76. ***Airplane Separation*** Two airplanes are in a straight-line landing pattern and, according to FAA regulations, must keep at least a three-mile separation. Airplane A is 10 miles from touchdown and is gradually decreasing its speed from 150 miles per hour to a landing speed of 100 miles per hour. Airplane B is 17 miles from touchdown and is gradually decreasing its speed from 250 miles per hour to a landing speed of 115 miles per hour.

(a) Assuming the deceleration of each airplane is constant, find the position functions s_A and s_B for airplane A and airplane B. Let $t = 0$ represent the times when the airplanes are 10 and 17 miles from the airport.

(b) Use a graphing utility to graph the position functions.

(c) Find a formula for the magnitude of the distance d between the two airplanes as a function of t. Use a graphing utility to graph d. Is $d < 3$ for some time prior to the landing of airplane A? If so, find that time.

77. ***Data Analysis*** A vehicle slows to a stop from 45 miles per hour in 6 seconds. The table shows the velocities in feet per second.

t	0	1	2	3	4	5	6
v	66.0	61.1	48.9	33.0	17.1	4.8	0

(a) Use the *regression* feature of a graphing utility to fit a cubic model to the data.

(b) Approximate the distance traveled by the car during the 6 seconds.

78. Find a function f such that the graph of f has a horizontal tangent at $(2, 0)$ and $f''(x) = 2x$.

True or False? **In Exercises 79–84, determine whether the statement is true or false. If it is false, explain why or give an example that shows it is false.**

79. Each antiderivative of an nth-degree polynomial function is an $(n + 1)$th-degree polynomial function.

80. If $p(x)$ is a polynomial function, then p has exactly one antiderivative whose graph contains the origin.

81. If $F(x)$ and $G(x)$ are antiderivatives of $f(x)$, then $F(x) = G(x) + C$.

82. If $f'(x) = g(x)$, then $\int g(x)\,dx = f(x) + C$.

83. $\int f(x)g(x)\,dx = \int f(x)\,dx \int g(x)\,dx$

84. The antiderivative of $f(x)$ is unique.

85. The graph of f' is shown. Sketch the graph of f given that f is continuous and $f(0) = 1$.

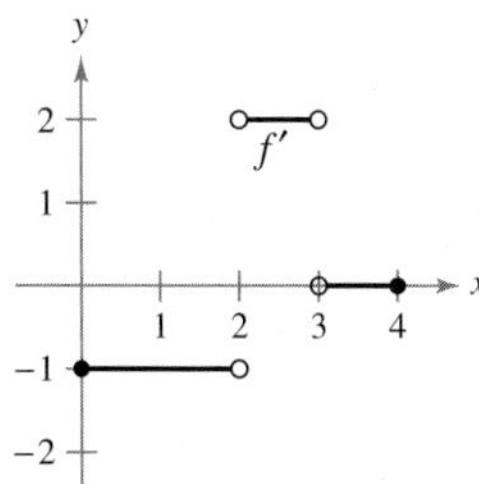

86. If $f'(x) = \begin{cases} 1, & 0 \le x < 2 \\ 3x, & 2 \le x \le 5 \end{cases}$, f is continuous, and $f(1) = 3$, find f. Is f differentiable at $x = 2$?

87. Let $s(x)$ and $c(x)$ be two functions satisfying $s'(x) = c(x)$ and $c'(x) = -s(x)$ for all x. If $s(0) = 0$ and $c(0) = 1$, prove that $[s(x)]^2 + [c(x)]^2 = 1$.

6.2 Area

- **Use sigma notation to write and evaluate a sum.**
- **Understand the concept of area.**
- **Use rectangles to approximate the area of a plane region.**
- **Find the area of a plane region using limits.**

Sigma Notation

In the preceding section, you studied antidifferentiation. In this section, you will look further into a problem introduced in Section 3.1—that of finding the area of a region in the plane. At first glance, these two ideas may seem unrelated, but you will discover in Section 6.4 that they are closely related by an extremely important theorem called the Fundamental Theorem of Calculus.

This section begins by introducing a concise notation for sums. This notation is called **sigma notation** because it uses the uppercase Greek letter sigma, written as Σ.

SIGMA NOTATION

The sum of n terms $a_1, a_2, a_3, \ldots, a_n$ is written as

$$\sum_{i=1}^{n} a_i = a_1 + a_2 + a_3 + \cdots + a_n$$

where i is the **index of summation,** a_i is the ***i*th term** of the sum, and the **upper and lower bounds of summation** are n and 1.

NOTE The upper and lower bounds must be constant with respect to the index of summation. However, the lower bound doesn't have to be 1. Any integer less than or equal to the upper bound is legitimate. ■

EXAMPLE 1 Examples of Sigma Notation

a. $\displaystyle\sum_{i=1}^{6} i = 1 + 2 + 3 + 4 + 5 + 6$

b. $\displaystyle\sum_{i=0}^{5} (i + 1) = 1 + 2 + 3 + 4 + 5 + 6$

c. $\displaystyle\sum_{j=3}^{7} j^2 = 3^2 + 4^2 + 5^2 + 6^2 + 7^2$

d. $\displaystyle\sum_{k=1}^{n} \frac{1}{n}(k^2 + 1) = \frac{1}{n}(1^2 + 1) + \frac{1}{n}(2^2 + 1) + \cdots + \frac{1}{n}(n^2 + 1)$

e. $\displaystyle\sum_{i=1}^{n} f(x_i)\,\Delta x = f(x_1)\,\Delta x + f(x_2)\,\Delta x + \cdots + f(x_n)\,\Delta x$

From parts (a) and (b), notice that the same sum can be represented in different ways using sigma notation. ■

Although any variable can be used as the index of summation, i, j, and k are often used. Notice in Example 1 that the index of summation does not appear in the terms of the expanded sum.

■ **FOR FURTHER INFORMATION** For a geometric interpretation of summation formulas, see the article "Looking at $\sum_{k=1}^{n} k$ and $\sum_{k=1}^{n} k^2$ Geometrically" by Eric Hegblom in *Mathematics Teacher*. To view this article, go to the website *www.matharticles.com*.

THE SUM OF THE FIRST 100 INTEGERS

A teacher of Carl Friedrich Gauss (1777–1855) asked him to add all the integers from 1 to 100. When Gauss returned with the correct answer after only a few moments, the teacher could only look at him in astounded silence. This is what Gauss did:

$$\begin{array}{rcrcrcccr} 1 & + & 2 & + & 3 & + & \cdots & + & 100 \\ 100 & + & 99 & + & 98 & + & \cdots & + & 1 \\ \hline 101 & + & 101 & + & 101 & + & \cdots & + & 101 \end{array}$$

$$\frac{100 \times 101}{2} = 5050$$

This is generalized by Theorem 6.2, where

$$\sum_{i=1}^{100} i = \frac{100(101)}{2} = 5050.$$

The following properties of summation can be derived using the associative and commutative properties of addition and the distributive property of addition over multiplication. (In the first property, k is a constant.)

1. $\displaystyle\sum_{i=1}^{n} ka_i = k\sum_{i=1}^{n} a_i$

2. $\displaystyle\sum_{i=1}^{n} (a_i \pm b_i) = \sum_{i=1}^{n} a_i \pm \sum_{i=1}^{n} b_i$

The next theorem lists some useful formulas for sums of powers. A proof of this theorem is given in Appendix A.

THEOREM 6.2 SUMMATION FORMULAS

1. $\displaystyle\sum_{i=1}^{n} c = cn$

2. $\displaystyle\sum_{i=1}^{n} i = \frac{n(n+1)}{2}$

3. $\displaystyle\sum_{i=1}^{n} i^2 = \frac{n(n+1)(2n+1)}{6}$

4. $\displaystyle\sum_{i=1}^{n} i^3 = \frac{n^2(n+1)^2}{4}$

EXAMPLE 2 Evaluating a Sum

Evaluate $\displaystyle\sum_{i=1}^{n} \frac{i+1}{n^2}$ for $n = 10$, 100, 1000, and 10,000.

Solution Applying Theorem 6.2, you can write

$$\begin{aligned} \sum_{i=1}^{n} \frac{i+1}{n^2} &= \frac{1}{n^2}\sum_{i=1}^{n}(i+1) && \text{Factor the constant } 1/n^2 \text{ out of sum.} \\ &= \frac{1}{n^2}\left(\sum_{i=1}^{n} i + \sum_{i=1}^{n} 1\right) && \text{Write as two sums.} \\ &= \frac{1}{n^2}\left[\frac{n(n+1)}{2} + n\right] && \text{Apply Theorem 6.2.} \\ &= \frac{1}{n^2}\left[\frac{n^2+3n}{2}\right] && \text{Simplify.} \\ &= \frac{n+3}{2n}. && \text{Simplify.} \end{aligned}$$

Now you can evaluate the sum by substituting the appropriate values of n, as shown in the table at the left. ■

n	$\displaystyle\sum_{i=1}^{n} \frac{i+1}{n^2} = \frac{n+3}{2n}$
10	0.65000
100	0.51500
1,000	0.50150
10,000	0.50015

In the table, note that the sum appears to approach a limit as n increases. Although the discussion of limits at infinity in Section 5.5 applies to a variable x, where x can be any real number, many of the same results hold true for limits involving the variable n, where n is restricted to positive integer values. So, to find the limit of

$$\frac{n+3}{2n}$$

as n approaches infinity, you can write

$$\lim_{n\to\infty} \frac{n+3}{2n} = \lim_{n\to\infty}\left(\frac{n}{2n} + \frac{3}{2n}\right) = \lim_{n\to\infty}\left(\frac{1}{2} + \frac{3}{2n}\right) = \frac{1}{2} + 0 = \frac{1}{2}.$$

Area

In Euclidean geometry, the simplest type of plane region is a rectangle. Although people often say that the *formula* for the area of a rectangle is $A = bh$, it is actually more proper to say that this is the *definition* of the **area of a rectangle.**

From this definition, you can develop formulas for the areas of many other plane regions. For example, to determine the area of a triangle, you can form a rectangle whose area is twice that of the triangle, as shown in Figure 6.5. Once you know how to find the area of a triangle, you can determine the area of any polygon by subdividing the polygon into triangular regions, as shown in Figure 6.6.

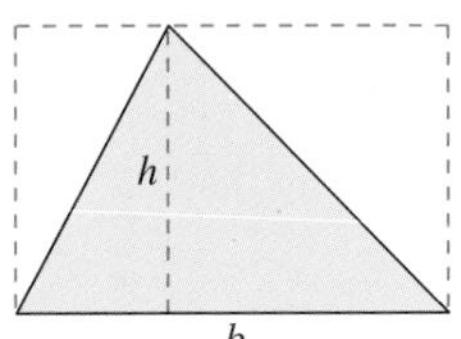

Triangle: $A = \frac{1}{2}bh$
Figure 6.5

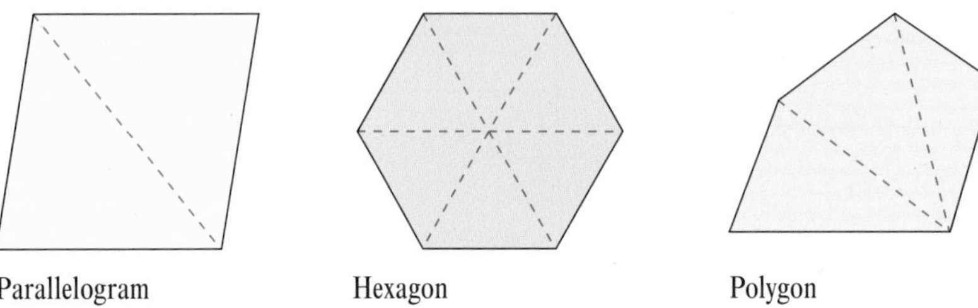

Parallelogram Hexagon Polygon
Figure 6.6

ARCHIMEDES (287–212 B.C.)

Archimedes used the method of exhaustion to derive formulas for the areas of ellipses, parabolic segments, and sectors of a spiral. He is considered to have been the greatest applied mathematician of antiquity.

Finding the areas of regions other than polygons is more difficult. The ancient Greeks were able to determine formulas for the areas of some general regions (principally those bounded by conics) by the *exhaustion* method. The clearest description of this method was given by Archimedes. Essentially, the method is a limiting process in which the area is squeezed between two polygons—one inscribed in the region and one circumscribed about the region.

For instance, in Figure 6.7 the area of a circular region is approximated by an n-sided inscribed polygon and an n-sided circumscribed polygon. For each value of n, the area of the inscribed polygon is less than the area of the circle, and the area of the circumscribed polygon is greater than the area of the circle. Moreover, as n increases, the areas of both polygons become better and better approximations of the area of the circle.

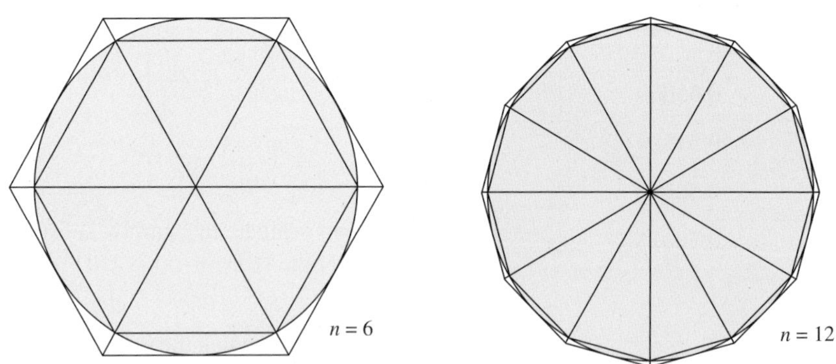

The exhaustion method for finding the area of a circular region
Figure 6.7

■ **FOR FURTHER INFORMATION** For an alternative development of the formula for the area of a circle, see the article "Proof Without Words: Area of a Disk is πR^2" by Russell Jay Hendel in *Mathematics Magazine*. To view this article, go to the website *www.matharticles.com.*

A process that is similar to that used by Archimedes to determine the area of a plane region is used in the remaining examples in this section.

The Area of a Plane Region

Recall from Section 3.1 that the origins of calculus are connected to two classic problems: the tangent line problem and the area problem. Example 3 begins the investigation of the area problem.

EXAMPLE 3 Approximating the Area of a Plane Region

Use the five rectangles in Figure 6.8(a) and (b) to find *two* approximations of the area of the region lying between the graph of

$$f(x) = -x^2 + 5$$

and the x-axis between $x = 0$ and $x = 2$.

Solution

a. The right endpoints of the five intervals are $\frac{2}{5}i$, where $i = 1, 2, 3, 4, 5$. The width of each rectangle is $\frac{2}{5}$, and the height of each rectangle can be obtained by evaluating f at the right endpoint of each interval.

$$\left[0, \frac{2}{5}\right], \left[\frac{2}{5}, \frac{4}{5}\right], \left[\frac{4}{5}, \frac{6}{5}\right], \left[\frac{6}{5}, \frac{8}{5}\right], \left[\frac{8}{5}, \frac{10}{5}\right]$$

Evaluate f at the right endpoints of these intervals.

The sum of the areas of the five rectangles is

$$\sum_{i=1}^{5} \underbrace{f\left(\frac{2i}{5}\right)}_{\text{Height}} \underbrace{\left(\frac{2}{5}\right)}_{\text{Width}} = \sum_{i=1}^{5}\left[-\left(\frac{2i}{5}\right)^2 + 5\right]\left(\frac{2}{5}\right) = \frac{162}{25} = 6.48.$$

Because each of the five rectangles lies inside the parabolic region, you can conclude that the area of the parabolic region is greater than 6.48.

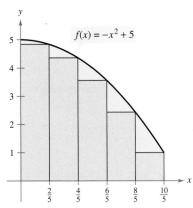

(a) The area of the parabolic region is greater than the area of the rectangles.

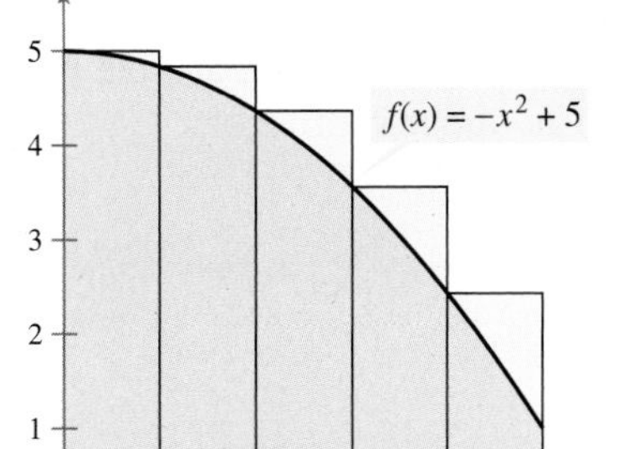

(b) The area of the parabolic region is less than the area of the rectangles.

Figure 6.8

b. The left endpoints of the five intervals are $\frac{2}{5}(i - 1)$, where $i = 1, 2, 3, 4, 5$. The width of each rectangle is $\frac{2}{5}$, and the height of each rectangle can be obtained by evaluating f at the left endpoint of each interval. So, the sum is

$$\sum_{i=1}^{5} \underbrace{f\left(\frac{2i-2}{5}\right)}_{\text{Height}} \underbrace{\left(\frac{2}{5}\right)}_{\text{Width}} = \sum_{i=1}^{5}\left[-\left(\frac{2i-2}{5}\right)^2 + 5\right]\left(\frac{2}{5}\right) = \frac{202}{25} = 8.08.$$

Because the parabolic region lies within the union of the five rectangular regions, you can conclude that the area of the parabolic region is less than 8.08.

By combining the results in parts (a) and (b), you can conclude that

$$6.48 < (\text{Area of region}) < 8.08.$$

■

NOTE By increasing the number of rectangles used in Example 3, you can obtain closer and closer approximations of the area of the region. For instance, using 25 rectangles of width $\frac{2}{25}$ each, you can conclude that

$$7.17 < (\text{Area of region}) < 7.49.$$

■

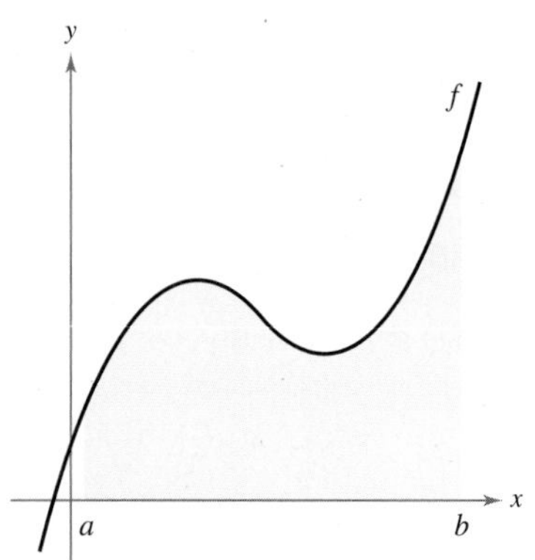

The region under a curve
Figure 6.9

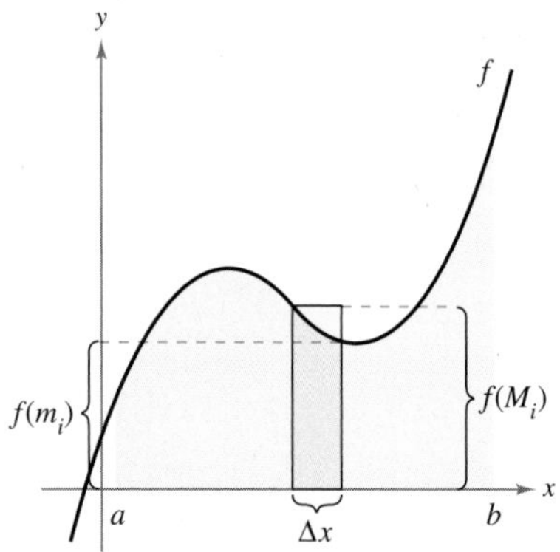

The interval $[a, b]$ is divided into n subintervals of width $\Delta x = \dfrac{b - a}{n}$.
Figure 6.10

Upper and Lower Sums

The procedure used in Example 3 can be generalized as follows. Consider a plane region bounded above by the graph of a nonnegative, continuous function $y = f(x)$, as shown in Figure 6.9. The region is bounded below by the x-axis, and the left and right boundaries of the region are the vertical lines $x = a$ and $x = b$.

To approximate the area of the region, begin by subdividing the interval $[a, b]$ into n subintervals, each of width $\Delta x = (b - a)/n$, as shown in Figure 6.10. The endpoints of the intervals are as follows.

$$\overbrace{a + 0(\Delta x)}^{a = x_0} < \overbrace{a + 1(\Delta x)}^{x_1} < \overbrace{a + 2(\Delta x)}^{x_2} < \cdot \cdot \cdot < \overbrace{a + n(\Delta x)}^{x_n = b}$$

Because f is continuous, the Extreme Value Theorem guarantees the existence of a minimum and a maximum value of $f(x)$ in *each* subinterval.

$f(m_i)$ = Minimum value of $f(x)$ in ith subinterval

$f(M_i)$ = Maximum value of $f(x)$ in ith subinterval

Next, define an **inscribed rectangle** lying *inside* the ith subregion and a **circumscribed rectangle** extending *outside* the ith subregion. The height of the ith inscribed rectangle is $f(m_i)$ and the height of the ith circumscribed rectangle is $f(M_i)$. For *each* i, the area of the inscribed rectangle is less than or equal to the area of the circumscribed rectangle.

$$\begin{pmatrix}\text{Area of inscribed}\\ \text{rectangle}\end{pmatrix} = f(m_i)\,\Delta x \le f(M_i)\,\Delta x = \begin{pmatrix}\text{Area of circumscribed}\\ \text{rectangle}\end{pmatrix}$$

The sum of the areas of the inscribed rectangles is called a **lower sum,** and the sum of the areas of the circumscribed rectangles is called an **upper sum.**

$$\text{Lower sum} = s(n) = \sum_{i=1}^{n} f(m_i)\,\Delta x \qquad \text{Area of inscribed rectangles}$$

$$\text{Upper sum} = S(n) = \sum_{i=1}^{n} f(M_i)\,\Delta x \qquad \text{Area of circumscribed rectangles}$$

From Figure 6.11, you can see that the lower sum $s(n)$ is less than or equal to the upper sum $S(n)$. Moreover, the actual area of the region lies between these two sums.

$$s(n) \le (\text{Area of region}) \le S(n)$$

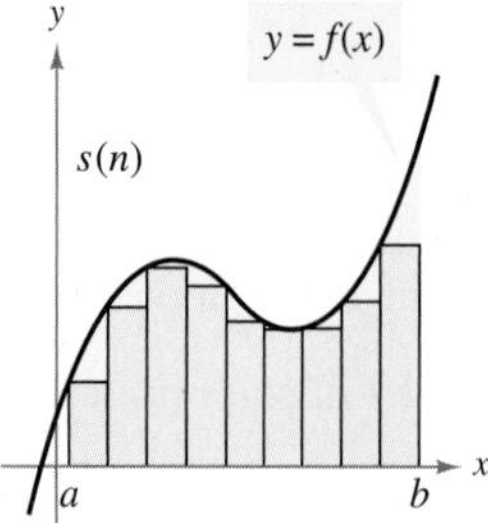

Area of inscribed rectangles is less than area of region.

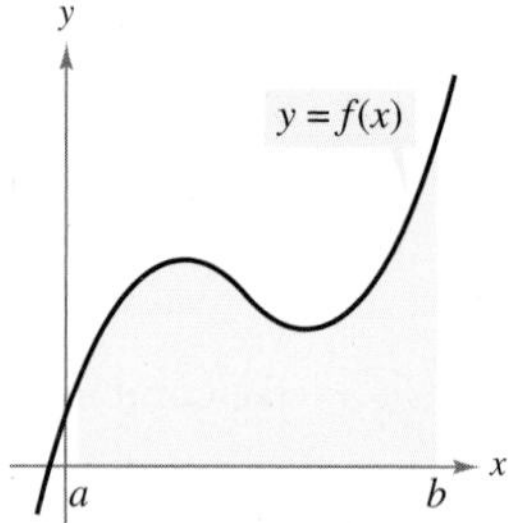

Area of region

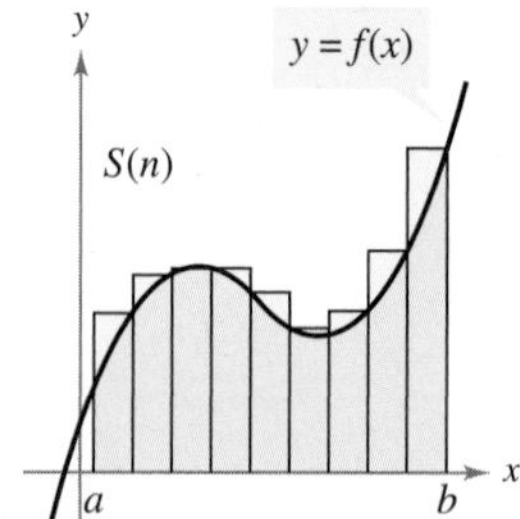

Area of circumscribed rectangles is greater than area of region.

Figure 6.11

EXAMPLE 4 Finding Upper and Lower Sums for a Region

Find the upper and lower sums for the region bounded by the graph of $f(x) = x^2$ and the x-axis between $x = 0$ and $x = 2$.

Solution To begin, partition the interval $[0, 2]$ into n subintervals, each of width

$$\Delta x = \frac{b-a}{n} = \frac{2-0}{n} = \frac{2}{n}.$$

Figure 6.12 shows the endpoints of the subintervals and several inscribed and circumscribed rectangles. Because f is increasing on the interval $[0, 2]$, the minimum value on each subinterval occurs at the left endpoint, and the maximum value occurs at the right endpoint.

Left Endpoints	*Right Endpoints*
$m_i = 0 + (i-1)\left(\frac{2}{n}\right) = \frac{2(i-1)}{n}$	$M_i = 0 + i\left(\frac{2}{n}\right) = \frac{2i}{n}$

Using the left endpoints, the lower sum is

$$\begin{aligned} s(n) = \sum_{i=1}^{n} f(m_i)\,\Delta x &= \sum_{i=1}^{n} f\left[\frac{2(i-1)}{n}\right]\left(\frac{2}{n}\right) \\ &= \sum_{i=1}^{n} \left[\frac{2(i-1)}{n}\right]^2\left(\frac{2}{n}\right) \\ &= \sum_{i=1}^{n} \left(\frac{8}{n^3}\right)(i^2 - 2i + 1) \\ &= \frac{8}{n^3}\left(\sum_{i=1}^{n} i^2 - 2\sum_{i=1}^{n} i + \sum_{i=1}^{n} 1\right) \\ &= \frac{8}{n^3}\left\{\frac{n(n+1)(2n+1)}{6} - 2\left[\frac{n(n+1)}{2}\right] + n\right\} \\ &= \frac{4}{3n^3}(2n^3 - 3n^2 + n) \\ &= \frac{8}{3} - \frac{4}{n} + \frac{4}{3n^2}. \qquad \text{Lower sum} \end{aligned}$$

Using the right endpoints, the upper sum is

$$\begin{aligned} S(n) = \sum_{i=1}^{n} f(M_i)\,\Delta x &= \sum_{i=1}^{n} f\left(\frac{2i}{n}\right)\left(\frac{2}{n}\right) \\ &= \sum_{i=1}^{n} \left(\frac{2i}{n}\right)^2\left(\frac{2}{n}\right) \\ &= \sum_{i=1}^{n} \left(\frac{8}{n^3}\right) i^2 \\ &= \frac{8}{n^3}\left[\frac{n(n+1)(2n+1)}{6}\right] \\ &= \frac{4}{3n^3}(2n^3 + 3n^2 + n) \\ &= \frac{8}{3} + \frac{4}{n} + \frac{4}{3n^2}. \qquad \text{Upper sum} \end{aligned}$$

■

Inscribed rectangles

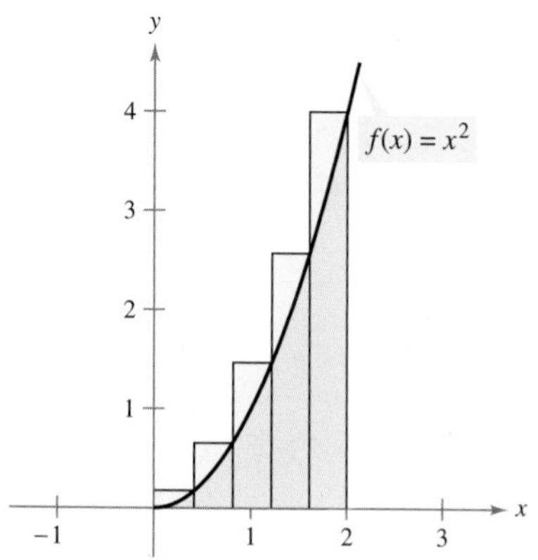

Circumscribed rectangles
Figure 6.12

EXPLORATION

For the region given in Example 4, evaluate the lower sum

$$s(n) = \frac{8}{3} - \frac{4}{n} + \frac{4}{3n^2}$$

and the upper sum

$$S(n) = \frac{8}{3} + \frac{4}{n} + \frac{4}{3n^2}$$

for $n = 10$, 100, and 1000. Use your results to determine the area of the region.

Example 4 illustrates some important things about lower and upper sums. First, notice that for any value of n, the lower sum is less than (or equal to) the upper sum.

$$s(n) = \frac{8}{3} - \frac{4}{n} + \frac{4}{3n^2} < \frac{8}{3} + \frac{4}{n} + \frac{4}{3n^2} = S(n)$$

Second, the difference between these two sums lessens as n increases. In fact, if you take the limits as $n \to \infty$, both the upper sum and the lower sum approach $\frac{8}{3}$.

$$\lim_{n\to\infty} s(n) = \lim_{n\to\infty} \left(\frac{8}{3} - \frac{4}{n} + \frac{4}{3n^2}\right) = \frac{8}{3} \qquad \text{Lower sum limit}$$

$$\lim_{n\to\infty} S(n) = \lim_{n\to\infty} \left(\frac{8}{3} + \frac{4}{n} + \frac{4}{3n^2}\right) = \frac{8}{3} \qquad \text{Upper sum limit}$$

The next theorem shows that the equivalence of the limits (as $n \to \infty$) of the upper and lower sums is not mere coincidence. It is true for all functions that are continuous and nonnegative on the closed interval $[a, b]$. The proof of this theorem is best left to a course in advanced calculus.

NOTE Refer to Section 5.5 to review the rule for finding limits at infinity of rational functions.

THEOREM 6.3 LIMITS OF THE LOWER AND UPPER SUMS

Let f be continuous and nonnegative on the interval $[a, b]$. The limits as $n \to \infty$ of both the lower and upper sums exist and are equal to each other. That is,

$$\begin{aligned}\lim_{n\to\infty} s(n) &= \lim_{n\to\infty} \sum_{i=1}^{n} f(m_i)\,\Delta x \\ &= \lim_{n\to\infty} \sum_{i=1}^{n} f(M_i)\,\Delta x \\ &= \lim_{n\to\infty} S(n)\end{aligned}$$

where $\Delta x = (b - a)/n$ and $f(m_i)$ and $f(M_i)$ are the minimum and maximum values of f on the subinterval.

Because the same limit is attained for both the minimum value $f(m_i)$ and the maximum value $f(M_i)$, it follows from the Squeeze Theorem (Theorem 3.7) that the choice of x in the ith subinterval does not affect the limit. This means that you are free to choose an *arbitrary* x-value in the ith subinterval, as in the following *definition of the area of a region in the plane.*

DEFINITION OF THE AREA OF A REGION IN THE PLANE

Let f be continuous and nonnegative on the interval $[a, b]$. The area of the region bounded by the graph of f, the x-axis, and the vertical lines $x = a$ and $x = b$ is

$$\text{Area} = \lim_{n\to\infty} \sum_{i=1}^{n} f(c_i)\,\Delta x, \quad x_{i-1} \le c_i \le x_i$$

where $\Delta x = \dfrac{b - a}{n}$ (see Figure 6.13).

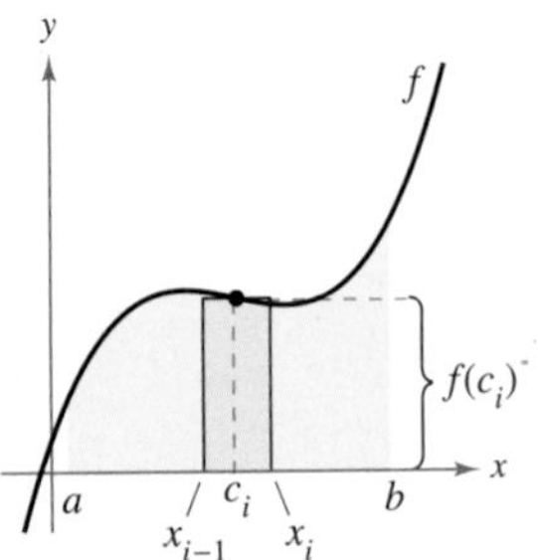

The width of the ith subinterval is $\Delta x = x_i - x_{i-1}$.
Figure 6.13

EXAMPLE 5 Finding Area by the Limit Definition

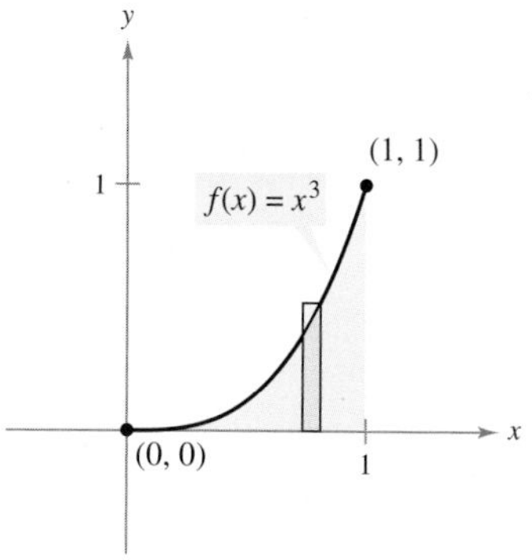

Find the area of the region bounded by the graph $f(x) = x^3$, the x-axis, and the vertical lines $x = 0$ and $x = 1$, as shown in Figure 6.14.

The area of the region bounded by the graph of f, the x-axis, $x = 0$, and $x = 1$ is $\frac{1}{4}$.
Figure 6.14

Solution Begin by noting that f is continuous and nonnegative on the interval $[0, 1]$. Next, partition the interval $[0, 1]$ into n subintervals, each of width $\Delta x = 1/n$. According to the definition of area, you can choose any x-value in the ith subinterval. For this example, the right endpoints $c_i = i/n$ are convenient.

$$\text{Area} = \lim_{n\to\infty} \sum_{i=1}^{n} f(c_i)\,\Delta x = \lim_{n\to\infty} \sum_{i=1}^{n} \left(\frac{i}{n}\right)^3\left(\frac{1}{n}\right) \qquad \text{Right endpoints: } c_i = a + i(\Delta x) = \frac{i}{n}$$

$$= \lim_{n\to\infty} \frac{1}{n^4}\sum_{i=1}^{n} i^3$$

$$= \lim_{n\to\infty} \frac{1}{n^4}\left[\frac{n^2(n+1)^2}{4}\right]$$

$$= \lim_{n\to\infty} \left(\frac{1}{4} + \frac{1}{2n} + \frac{1}{4n^2}\right)$$

$$= \frac{1}{4}$$

The area of the region is $\frac{1}{4}$.

EXAMPLE 6 Finding Area by the Limit Definition

Find the area of the region bounded by the graph of $f(x) = 4 - x^2$, the x-axis, and the vertical lines $x = 1$ and $x = 2$, as shown in Figure 6.15.

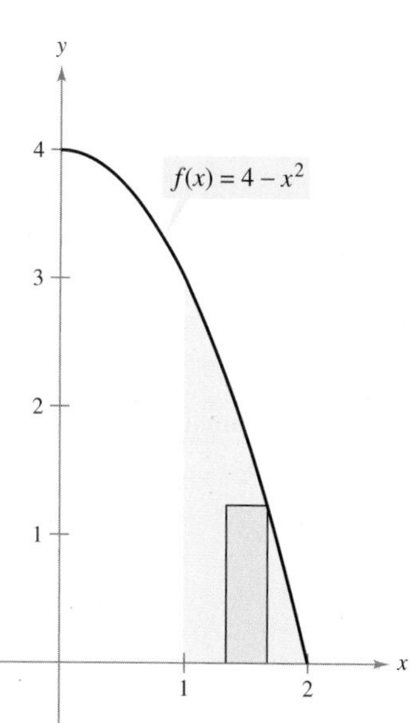

The area of the region bounded by the graph of f, the x-axis, $x = 1$, and $x = 2$ is $\frac{5}{3}$.
Figure 6.15

Solution The function f is continuous and nonnegative on the interval $[1, 2]$, and so begin by partitioning the interval into n subintervals, each of width $\Delta x = 1/n$. Choosing the right endpoint

$$c_i = a + i(\Delta x) = 1 + \frac{i}{n} \qquad \text{Right endpoints}$$

of each subinterval, you obtain

$$\text{Area} = \lim_{n\to\infty} \sum_{i=1}^{n} f(c_i)\,\Delta x = \lim_{n\to\infty} \sum_{i=1}^{n} \left[4 - \left(1 + \frac{i}{n}\right)^2\right]\left(\frac{1}{n}\right)$$

$$= \lim_{n\to\infty} \sum_{i=1}^{n} \left(3 - \frac{2i}{n} - \frac{i^2}{n^2}\right)\left(\frac{1}{n}\right)$$

$$= \lim_{n\to\infty} \left(\frac{1}{n}\sum_{i=1}^{n} 3 - \frac{2}{n^2}\sum_{i=1}^{n} i - \frac{1}{n^3}\sum_{i=1}^{n} i^2\right)$$

$$= \lim_{n\to\infty} \left\{\frac{1}{n}(3n) - \frac{2}{n^2}\left[\frac{n(n+1)}{2}\right] - \frac{1}{n^3}\left[\frac{n(n+1)(2n+1)}{6}\right]\right\}$$

$$= \lim_{n\to\infty} \left[3 - \left(1 + \frac{1}{n}\right) - \left(\frac{1}{3} + \frac{1}{2n} + \frac{1}{6n^2}\right)\right]$$

$$= 3 - 1 - \frac{1}{3}$$

$$= \frac{5}{3}.$$

The area of the region is $\frac{5}{3}$. ■

The last example in this section looks at a region that is bounded by the y-axis (rather than by the x-axis).

EXAMPLE 7 A Region Bounded by the y-axis

Find the area of the region bounded by the graph of $f(y) = y^2$ and the y-axis for $0 \le y \le 1$, as shown in Figure 6.16.

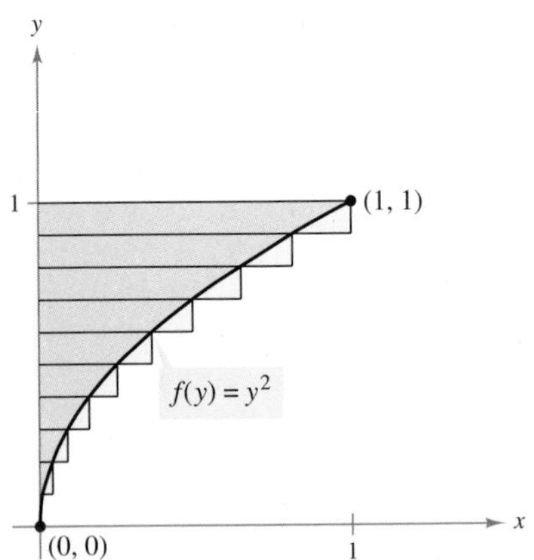

The area of the region bounded by the graph of f and the y-axis for $0 \le y \le 1$ is $\frac{1}{3}$.
Figure 6.16

Solution When f is a continuous, nonnegative function of y, you still can use the same basic procedure shown in Examples 5 and 6. Begin by partitioning the interval $[0, 1]$ into n subintervals, each of width $\Delta y = 1/n$. Then, using the upper endpoints $c_i = i/n$, you obtain

$$\begin{aligned}\text{Area} = \lim_{n\to\infty}\sum_{i=1}^{n} f(c_i)\,\Delta y &= \lim_{n\to\infty}\sum_{i=1}^{n}\left(\frac{i}{n}\right)^2\left(\frac{1}{n}\right) \qquad \text{Upper endpoints: } c_i = \frac{i}{n}\\ &= \lim_{n\to\infty}\frac{1}{n^3}\sum_{i=1}^{n} i^2\\ &= \lim_{n\to\infty}\frac{1}{n^3}\left[\frac{n(n+1)(2n+1)}{6}\right]\\ &= \lim_{n\to\infty}\left(\frac{1}{3}+\frac{1}{2n}+\frac{1}{6n^2}\right)\\ &= \frac{1}{3}.\end{aligned}$$

The area of the region is $\frac{1}{3}$. ■

6.2 Exercises

See www.CalcChat.com for worked-out solutions to odd-numbered exercises.

In Exercises 1–6, find the sum. Use the summation capabilities of a graphing utility to verify your result.

1. $\displaystyle\sum_{i=1}^{6}(3i+2)$
2. $\displaystyle\sum_{k=5}^{8}k(k-4)$
3. $\displaystyle\sum_{k=0}^{4}\frac{1}{k^2+1}$
4. $\displaystyle\sum_{j=4}^{7}\frac{2}{j}$
5. $\displaystyle\sum_{k=1}^{4}c$
6. $\displaystyle\sum_{i=1}^{4}[(i-1)^2+(i+1)^3]$

In Exercises 7–14, use sigma notation to write the sum.

7. $\dfrac{1}{5(1)}+\dfrac{1}{5(2)}+\dfrac{1}{5(3)}+\cdots+\dfrac{1}{5(11)}$
8. $\dfrac{9}{1+1}+\dfrac{9}{1+2}+\dfrac{9}{1+3}+\cdots+\dfrac{9}{1+14}$
9. $\left[7\left(\frac{1}{6}\right)+5\right]+\left[7\left(\frac{2}{6}\right)+5\right]+\cdots+\left[7\left(\frac{6}{6}\right)+5\right]$
10. $\left[1-\left(\frac{1}{4}\right)^2\right]+\left[1-\left(\frac{2}{4}\right)^2\right]+\cdots+\left[1-\left(\frac{4}{4}\right)^2\right]$
11. $\left[\left(\frac{2}{n}\right)^3-\frac{2}{n}\right]\left(\frac{2}{n}\right)+\cdots+\left[\left(\frac{2n}{n}\right)^3-\frac{2n}{n}\right]\left(\frac{2}{n}\right)$
12. $\left[1-\left(\frac{2}{n}-1\right)^2\right]\left(\frac{2}{n}\right)+\cdots+\left[1-\left(\frac{2n}{n}-1\right)^2\right]\left(\frac{2}{n}\right)$
13. $\left[2\left(1+\frac{3}{n}\right)^2\right]\left(\frac{3}{n}\right)+\cdots+\left[2\left(1+\frac{3n}{n}\right)^2\right]\left(\frac{3}{n}\right)$
14. $\left(\frac{1}{n}\right)\sqrt{1-\left(\frac{0}{n}\right)^2}+\cdots+\left(\frac{1}{n}\right)\sqrt{1-\left(\frac{n-1}{n}\right)^2}$

In Exercises 15–22, use the properties of summation and Theorem 6.2 to evaluate the sum. Use the summation capabilities of a graphing utility to verify your result.

15. $\displaystyle\sum_{i=1}^{12}7$
16. $\displaystyle\sum_{i=1}^{30}-18$
17. $\displaystyle\sum_{i=1}^{24}4i$
18. $\displaystyle\sum_{i=1}^{16}(5i-4)$
19. $\displaystyle\sum_{i=1}^{20}(i-1)^2$
20. $\displaystyle\sum_{i=1}^{10}(i^2-1)$
21. $\displaystyle\sum_{i=1}^{15}i(i-1)^2$
22. $\displaystyle\sum_{i=1}^{10}i(i^2+1)$

In Exercises 23 and 24, use the summation capabilities of a graphing utility to evaluate the sum. Then use the properties of summation and Theorem 6.2 to verify the sum.

23. $\displaystyle\sum_{i=1}^{20}(i^2+3)$
24. $\displaystyle\sum_{i=1}^{15}(i^3-2i)$

25. Consider the function $f(x) = 3x + 2$.

(a) Estimate the area between the graph of f and the x-axis between $x = 0$ and $x = 3$ using six rectangles and right endpoints. Sketch the graph and the rectangles.

(b) Repeat part (a) using left endpoints.

26. Consider the function $g(x) = x^2 + x - 4$.

(a) Estimate the area between the graph of g and the x-axis between $x = 2$ and $x = 4$ using four rectangles and right endpoints. Sketch the graph and the rectangles.

(b) Repeat part (a) using left endpoints.

In Exercises 27–30, bound the area of the shaded region by approximating the upper and lower sums. Use rectangles of width 1.

27.

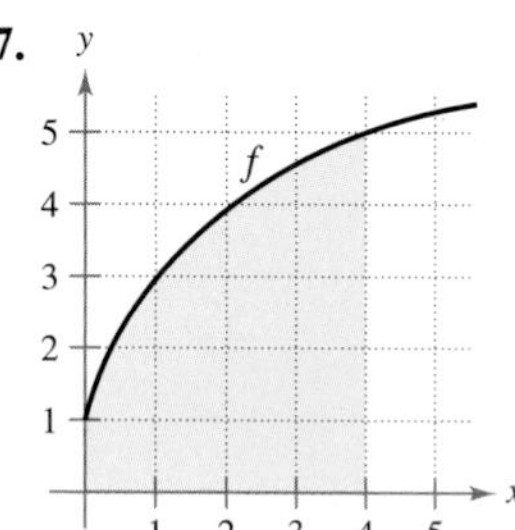

28.

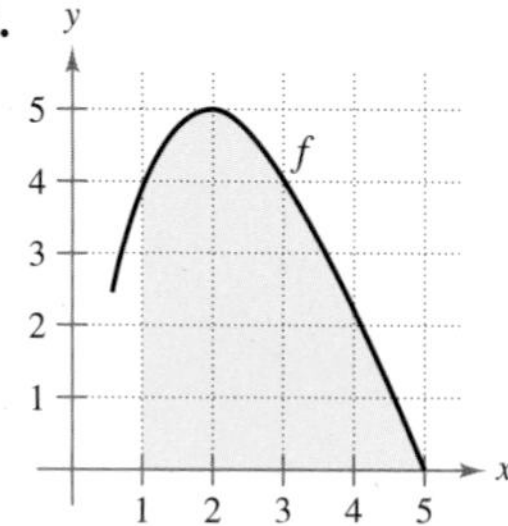

29.

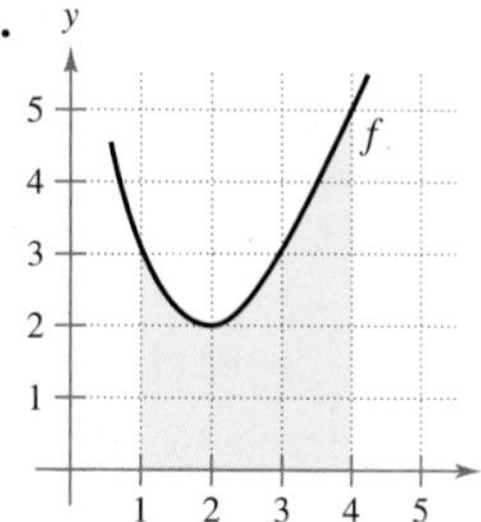

30.

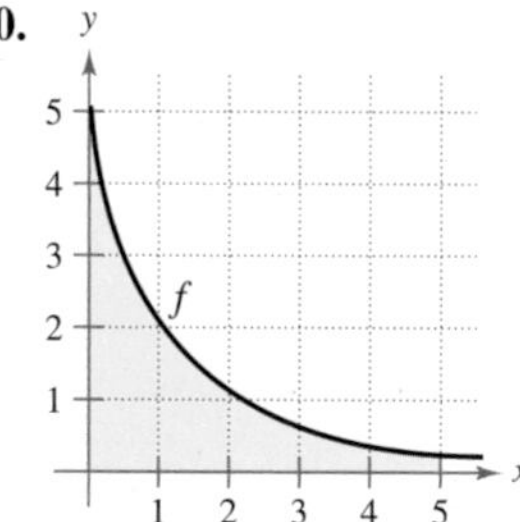

In Exercises 31–34, find the limit of $s(n)$ as $n \to \infty$.

31. $s(n) = \dfrac{81}{n^4}\left[\dfrac{n^2(n+1)^2}{4}\right]$

32. $s(n) = \dfrac{64}{n^3}\left[\dfrac{n(n+1)(2n+1)}{6}\right]$

33. $s(n) = \dfrac{18}{n^2}\left[\dfrac{n(n+1)}{2}\right]$

34. $s(n) = \dfrac{1}{n^2}\left[\dfrac{n(n+1)}{2}\right]$

In Exercises 35–38, use upper and lower sums to approximate the area of the region using the given number of subintervals (of equal width).

35. $y = \sqrt{x}$

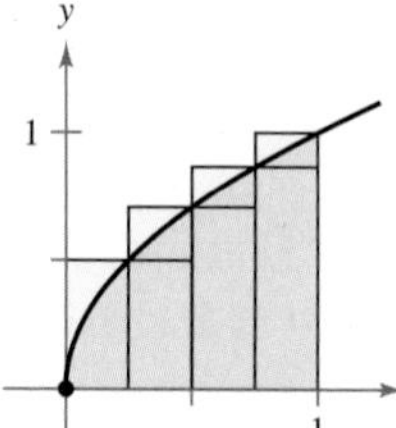

36. $y = \sqrt{x} + 2$

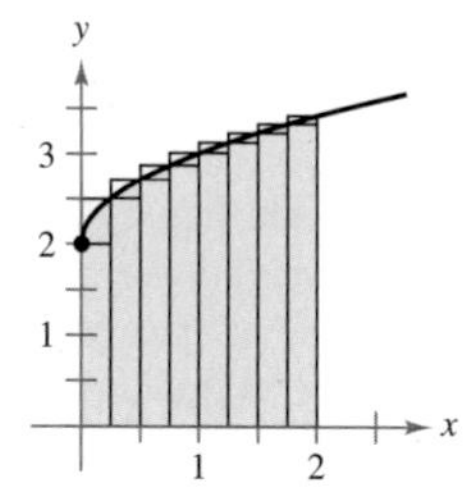

37. $y = \dfrac{1}{x}$

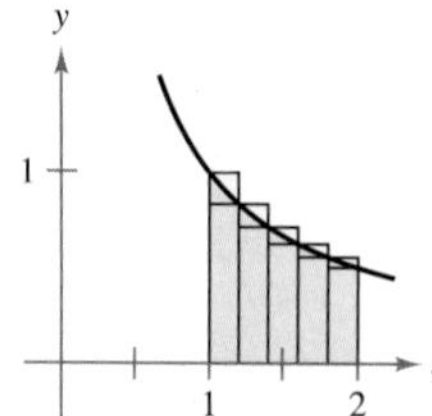

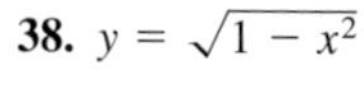

38. $y = \sqrt{1 - x^2}$

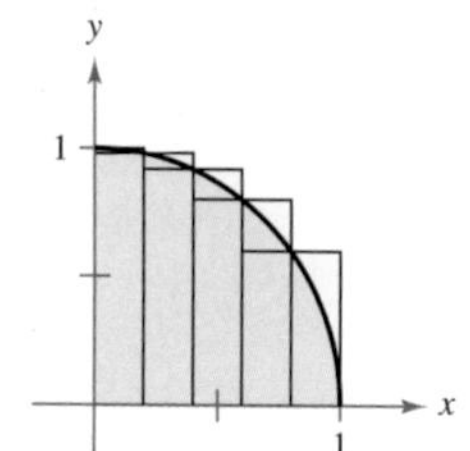

In Exercises 39–42, use the summation formulas to rewrite the expression without the summation notation. Use the result to find the sums for $n = 10, 100, 1000$, and $10{,}000$.

39. $\displaystyle\sum_{i=1}^{n} \frac{2i+1}{n^2}$

40. $\displaystyle\sum_{j=1}^{n} \frac{4j+3}{n^2}$

41. $\displaystyle\sum_{k=1}^{n} \frac{6k(k-1)}{n^3}$

42. $\displaystyle\sum_{i=1}^{n} \frac{4i^2(i-1)}{n^4}$

In Exercises 43–48, find a formula for the sum of n terms. Use the formula to find the limit as $n \to \infty$.

43. $\displaystyle\lim_{n\to\infty} \sum_{i=1}^{n} \frac{24i}{n^2}$

44. $\displaystyle\lim_{n\to\infty} \sum_{i=1}^{n} \left(\frac{2i}{n}\right)\left(\frac{2}{n}\right)$

45. $\displaystyle\lim_{n\to\infty} \sum_{i=1}^{n} \frac{1}{n^3}(i-1)^2$

46. $\displaystyle\lim_{n\to\infty} \sum_{i=1}^{n} \left(1 + \frac{2i}{n}\right)^2\left(\frac{2}{n}\right)$

47. $\displaystyle\lim_{n\to\infty} \sum_{i=1}^{n} \left(1 + \frac{i}{n}\right)\left(\frac{2}{n}\right)$

48. $\displaystyle\lim_{n\to\infty} \sum_{i=1}^{n} \left(1 + \frac{2i}{n}\right)^3\left(\frac{2}{n}\right)$

49. ***Numerical Reasoning*** Consider a triangle of area 2 bounded by the graphs of $y = x$, $y = 0$, and $x = 2$.

(a) Sketch the region.

(b) Divide the interval $[0, 2]$ into n subintervals of equal width and show that the endpoints are

$$0 < 1\left(\frac{2}{n}\right) < \cdots < (n-1)\left(\frac{2}{n}\right) < n\left(\frac{2}{n}\right).$$

(c) Show that $s(n) = \displaystyle\sum_{i=1}^{n} \left[(i-1)\left(\frac{2}{n}\right)\right]\left(\frac{2}{n}\right)$.

(d) Show that $S(n) = \displaystyle\sum_{i=1}^{n} \left[i\left(\frac{2}{n}\right)\right]\left(\frac{2}{n}\right)$.

(e) Complete the table.

n	5	10	50	100
$s(n)$				
$S(n)$				

(f) Show that $\displaystyle\lim_{n\to\infty} s(n) = \lim_{n\to\infty} S(n) = 2$.

50. ***Numerical Reasoning*** Consider a trapezoid of area 4 bounded by the graphs of $y = x$, $y = 0$, $x = 1$, and $x = 3$.

(a) Sketch the region.

(b) Divide the interval $[1, 3]$ into n subintervals of equal width and show that the endpoints are

$$1 < 1 + 1\left(\frac{2}{n}\right) < \cdot \cdot \cdot < 1 + (n-1)\left(\frac{2}{n}\right) < 1 + n\left(\frac{2}{n}\right).$$

(c) Show that $s(n) = \sum_{i=1}^{n}\left[1 + (i-1)\left(\frac{2}{n}\right)\right]\left(\frac{2}{n}\right)$.

(d) Show that $S(n) = \sum_{i=1}^{n}\left[1 + i\left(\frac{2}{n}\right)\right]\left(\frac{2}{n}\right)$.

(e) Complete the table.

n	5	10	50	100
$s(n)$				
$S(n)$				

(f) Show that $\lim_{n\to\infty} s(n) = \lim_{n\to\infty} S(n) = 4$.

In Exercises 51–60, use the limit process to find the area of the region between the graph of the function and the x-axis over the given interval. Sketch the region.

51. $y = -4x + 5, \quad [0, 1]$

52. $y = 3x - 2, \quad [2, 5]$

53. $y = x^2 + 2, \quad [0, 1]$

54. $y = x^2 + 1, \quad [0, 3]$

55. $y = 25 - x^2, \quad [1, 4]$

56. $y = 4 - x^2, \quad [-2, 2]$

57. $y = 27 - x^3, \quad [1, 3]$

58. $y = 2x - x^3, \quad [0, 1]$

59. $y = x^2 - x^3, \quad [-1, 1]$

60. $y = x^2 - x^3, \quad [-1, 0]$

In Exercises 61–66, use the limit process to find the area of the region between the graph of the function and the y-axis over the given y-interval. Sketch the region.

61. $f(y) = 4y, 0 \le y \le 2$

62. $g(y) = \frac{1}{2}y, 2 \le y \le 4$

63. $f(y) = y^2, 0 \le y \le 5$

64. $f(y) = 4y - y^2, 1 \le y \le 2$

65. $g(y) = 4y^2 - y^3, 1 \le y \le 3$

66. $h(y) = y^3 + 1, 1 \le y \le 2$

In Exercises 67–70, use the *Midpoint Rule*

$$\text{Area} \approx \sum_{i=1}^{n} f\left(\frac{x_i + x_{i-1}}{2}\right)\Delta x$$

with $n = 4$ to approximate the area of the region bounded by the graph of the function and the x-axis over the given interval.

67. $f(x) = x^2 + 3, \quad [0, 2]$

68. $f(x) = x^2 + 4x, \quad [0, 4]$

69. $f(x) = \sqrt{x - 1}, \quad [1, 2]$

70. $f(x) = \frac{1}{x^2 + 1}, \quad [0, 2]$

Programming **Write a program for a graphing utility to approximate areas by using the Midpoint Rule. Assume that the function is positive over the given interval and that the subintervals are of equal width. In Exercises 71–74, use the program to approximate the area of the region between the graph of the function and the x-axis over the given interval, and complete the table.**

n	4	8	12	16	20
Approximate Area					

71. $f(x) = \sqrt{x}, \quad [0, 4]$

72. $f(x) = x^{3/2} + 2, \quad [0, 2]$

73. $f(x) = \frac{8}{x^2 + 1}, \quad [2, 6]$

74. $f(x) = \frac{5x}{x^2 + 1}, \quad [1, 3]$

WRITING ABOUT CONCEPTS

75. In your own words and using appropriate figures, describe the methods of upper sums and lower sums in approximating the area of a region.

76. Give the definition of the area of a region in the plane.

77. ***Graphical Reasoning*** Consider the region bounded by the graphs of $f(x) = 8x/(x + 1)$, $x = 0$, $x = 4$, and $y = 0$, as shown in the figure. To print an enlarged copy of the graph, go to the website *www.mathgraphs.com*.

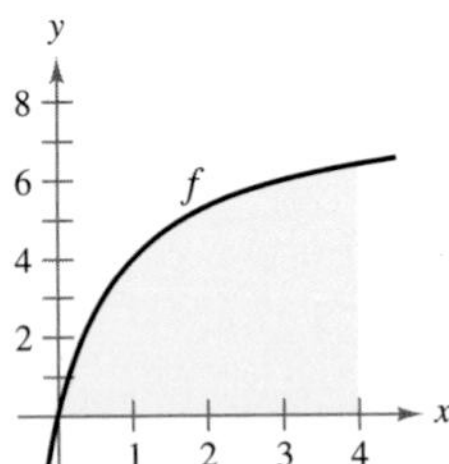

(a) Redraw the figure, and complete and shade the rectangles representing the lower sum when $n = 4$. Find this lower sum.

(b) Redraw the figure, and complete and shade the rectangles representing the upper sum when $n = 4$. Find this upper sum.

(c) Redraw the figure, and complete and shade the rectangles whose heights are determined by the functional values at the midpoint of each subinterval when $n = 4$. Find this sum using the Midpoint Rule.

(d) Verify the following formulas for approximating the area of the region using n subintervals of equal width.

Lower sum: $s(n) = \sum_{i=1}^{n} f\left[(i-1)\frac{4}{n}\right]\left(\frac{4}{n}\right)$

Upper sum: $S(n) = \sum_{i=1}^{n} f\left[(i)\frac{4}{n}\right]\left(\frac{4}{n}\right)$

Midpoint Rule: $M(n) = \sum_{i=1}^{n} f\left[\left(i-\frac{1}{2}\right)\frac{4}{n}\right]\left(\frac{4}{n}\right)$

(e) Use a graphing utility and the formulas in part (d) to complete the table.

n	4	8	20	100	200
$s(n)$					
$S(n)$					
$M(n)$					

(f) Explain why $s(n)$ increases and $S(n)$ decreases for increasing values of n, as shown in the table in part (e).

CAPSTONE

78. Consider a function $f(x)$ that is increasing on the interval $[1, 4]$. The interval $[1, 4]$ is divided into 12 subintervals.

(a) What are the left endpoints of the first and last subintervals?

(b) What are the right endpoints of the first two subintervals?

(c) When using the right endpoints, will the rectangles lie above or below the graph of $f(x)$? Use a graph to explain your answer.

(d) What can you conclude about the heights of the rectangles if a function is constant on the given interval?

Approximation **In Exercises 79 and 80, determine which value best approximates the area of the region between the x-axis and the graph of the function over the indicated interval. (Make your selection on the basis of a sketch of the region and not by performing calculations.)**

79. $f(x) = 4 - x^2, \quad [0, 2]$

(a) -2 (b) 6 (c) 10 (d) 3 (e) 8

80. $f(x) = \dfrac{4}{x^2}, \quad [1, 4]$

(a) 3 (b) 1 (c) -2 (d) 8 (e) 6

True or False? **In Exercises 81 and 82, determine whether the statement is true or false. If it is false, explain why or give an example that shows it is false.**

81. The sum of the first n positive integers is $n(n + 1)/2$.

82. If f is continuous and nonnegative on $[a, b]$, then the limits as $n \to \infty$ of its lower sum $s(n)$ and upper sum $S(n)$ both exist and are equal.

83. ***Writing*** Use the figure to write a short paragraph explaining why the formula

$$1 + 2 + \cdot \cdot \cdot + n = \tfrac{1}{2}n(n + 1)$$

is valid for all positive integers n.

84. ***Building Blocks*** A child places n cubic building blocks in a row to form the base of a triangular design (see figure). Each successive row contains two fewer blocks than the preceding row. Find a formula for the number of blocks used in the design. (*Hint:* The number of building blocks in the design depends on whether n is odd or even.)

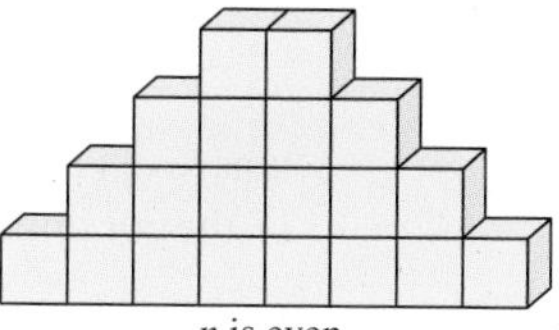

n is even.

85. ***Modeling Data*** The table lists the measurements of a lot bounded by a stream and two straight roads that meet at right angles, where x and y are measured in feet (see figure).

x	0	50	100	150	200	250	300
y	450	362	305	268	245	156	0

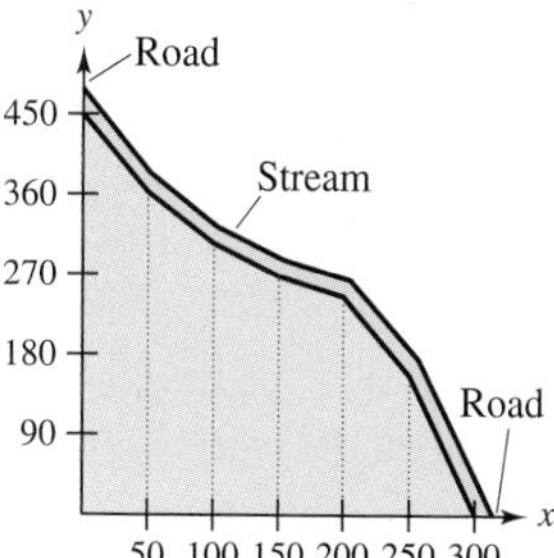

(a) Use the regression capabilities of a graphing utility to find a model of the form $y = ax^3 + bx^2 + cx + d$.

(b) Use a graphing utility to plot the data and graph the model.

(c) Use the model in part (a) to estimate the area of the lot.

86. Prove each formula by mathematical induction. (You may need to review the method of proof by induction from a precalculus text.)

(a) $\sum_{i=1}^{n} 2i = n(n + 1)$

(b) $\sum_{i=1}^{n} i^3 = \dfrac{n^2(n + 1)^2}{4}$

6.3 Riemann Sums and Definite Integrals

- **Understand the definition of a Riemann sum.**
- **Evaluate a definite integral using limits.**
- **Evaluate a definite integral using properties of definite integrals.**

Riemann Sums

In the definition of area given in Section 6.2, the partitions have subintervals of *equal width*. This was done only for computational convenience. The following example shows that it is not necessary to have subintervals of equal width.

EXAMPLE 1 A Partition with Subintervals of Unequal Widths

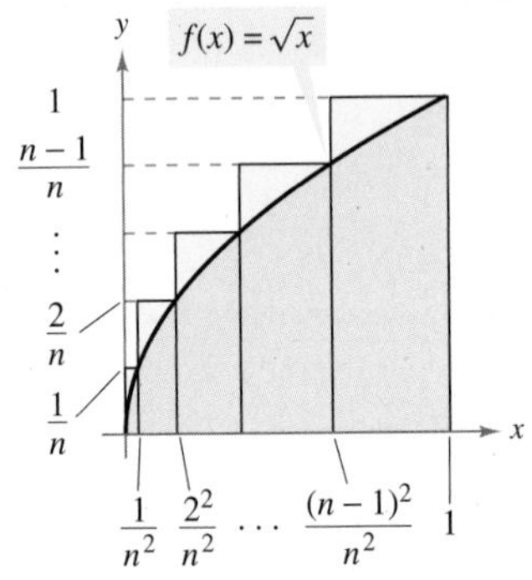

The subintervals do not have equal widths.
Figure 6.17

Consider the region bounded by the graph of $f(x) = \sqrt{x}$ and the x-axis for $0 \le x \le 1$, as shown in Figure 6.17. Evaluate the limit

$$\lim_{n\to\infty} \sum_{i=1}^{n} f(c_i)\,\Delta x_i$$

where c_i is the right endpoint of the partition given by $c_i = i^2/n^2$ and Δx_i is the width of the ith interval.

Solution The width of the ith interval is given by

$$\begin{aligned}\Delta x_i &= \frac{i^2}{n^2} - \frac{(i-1)^2}{n^2}\\ &= \frac{i^2 - i^2 + 2i - 1}{n^2}\\ &= \frac{2i-1}{n^2}.\end{aligned}$$

So, the limit is

$$\begin{aligned}\lim_{n\to\infty} \sum_{i=1}^{n} f(c_i)\,\Delta x_i &= \lim_{n\to\infty} \sum_{i=1}^{n} \sqrt{\frac{i^2}{n^2}}\left(\frac{2i-1}{n^2}\right)\\ &= \lim_{n\to\infty} \frac{1}{n^3} \sum_{i=1}^{n} (2i^2 - i)\\ &= \lim_{n\to\infty} \frac{1}{n^3}\left[2\left(\frac{n(n+1)(2n+1)}{6}\right) - \frac{n(n+1)}{2}\right]\\ &= \lim_{n\to\infty} \frac{4n^3 + 3n^2 - n}{6n^3}\\ &= \frac{2}{3}.\end{aligned}$$

■

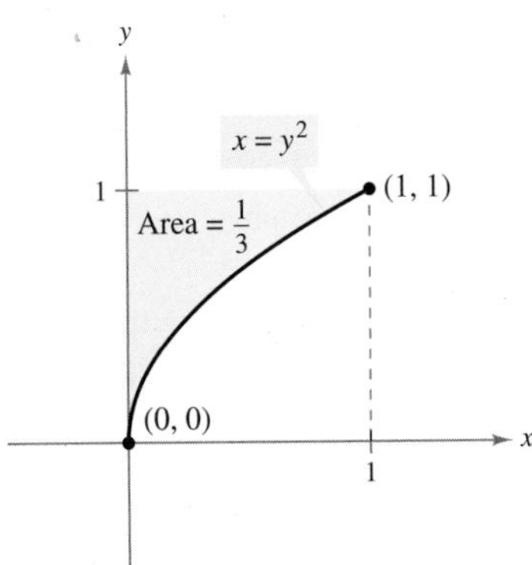

The area of the region bounded by the graph of $x = y^2$ and the y-axis for $0 \le y \le 1$ is $\frac{1}{3}$.
Figure 6.18

From Example 7 in Section 6.2, you know that the region shown in Figure 6.18 has an area of $\frac{1}{3}$. Because the square bounded by $0 \le x \le 1$ and $0 \le y \le 1$ has an area of 1, you can conclude that the area of the region shown in Figure 6.17 has an area of $\frac{2}{3}$. This agrees with the limit found in Example 1, even though that example used a partition having subintervals of unequal widths. The reason this particular partition gave the proper area is that as n increases, the *width of the largest subinterval approaches zero*. This is a key feature of the development of definite integrals.

GEORG FRIEDRICH BERNHARD RIEMANN (1826–1866)

German mathematician Riemann did his most famous work in the areas of non-Euclidean geometry, differential equations, and number theory. It was Riemann's results in physics and mathematics that formed the structure on which Einstein's General Theory of Relativity is based.

In the preceding section, the limit of a sum was used to define the area of a region in the plane. Finding area by this means is only one of *many* applications involving the limit of a sum. A similar approach can be used to determine quantities as diverse as arc lengths, average values, centroids, volumes, work, and surface areas. The following definition is named after Georg Friedrich Bernhard Riemann. Although the definite integral had been defined and used long before the time of Riemann, he generalized the concept to cover a broader category of functions.

In the following definition of a Riemann sum, note that the function f has no restrictions other than being defined on the interval $[a, b]$. (In the preceding section, the function f was assumed to be continuous and nonnegative because we were dealing with the area under a curve.)

DEFINITION OF RIEMANN SUM

Let f be defined on the closed interval $[a, b]$, and let Δ be a partition of $[a, b]$ given by

$$a = x_0 < x_1 < x_2 < \cdot \cdot \cdot < x_{n-1} < x_n = b$$

where Δx_i is the width of the ith subinterval. If c_i is *any* point in the ith subinterval $[x_{i-1}, x_i]$, then the sum

$$\sum_{i=1}^{n} f(c_i)\,\Delta x_i, \quad x_{i-1} \le c_i \le x_i$$

is called a **Riemann sum** of f for the partition Δ.

NOTE The sums in Section 6.2 are examples of Riemann sums, but there are more general Riemann sums than those covered there. ■

The width of the largest subinterval of a partition Δ is the **norm** of the partition and is denoted by $\|\Delta\|$. If every subinterval is of equal width, the partition is **regular** and the norm is denoted by

$$\|\Delta\| = \Delta x = \frac{b - a}{n}. \qquad \text{Regular partition}$$

For a general partition, the norm is related to the number of subintervals of $[a, b]$ in the following way.

$$\frac{b - a}{\|\Delta\|} \le n \qquad \text{General partition}$$

So, the number of subintervals in a partition approaches infinity as the norm of the partition approaches 0. That is, $\|\Delta\| \to 0$ implies that $n \to \infty$.

The converse of this statement is not true. For example, let Δ_n be the partition of the interval $[0, 1]$ given by

$$0 < \frac{1}{2^n} < \frac{1}{2^{n-1}} < \cdot \cdot \cdot < \frac{1}{8} < \frac{1}{4} < \frac{1}{2} < 1.$$

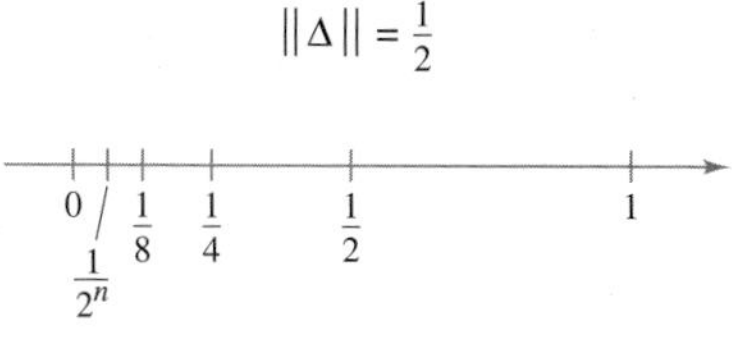

$n \to \infty$ does not imply that $\|\Delta\| \to 0$.
Figure 6.19

As shown in Figure 6.19, for any positive value of n, the norm of the partition Δ_n is $\frac{1}{2}$. So, letting n approach infinity does not force $\|\Delta\|$ to approach 0. In a regular partition, however, the statements $\|\Delta\| \to 0$ and $n \to \infty$ are equivalent.

Definite Integrals

To define the definite integral, consider the following limit.

$$\lim_{\|\Delta\| \to 0} \sum_{i=1}^{n} f(c_i)\,\Delta x_i = L$$

To say that this limit exists means there exists a real number L such that for each $\varepsilon > 0$ there exists a $\delta > 0$ so that for every partition with $\|\Delta\| < \delta$ it follows that

$$\left| L - \sum_{i=1}^{n} f(c_i)\,\Delta x_i \right| < \varepsilon$$

regardless of the choice of c_i in the ith subinterval of each partition Δ.

■ **FOR FURTHER INFORMATION** For insight into the history of the definite integral, see the article "The Evolution of Integration" by A. Shenitzer and J. Steprāns in *The American Mathematical Monthly*. To view this article, go to the website *www.matharticles.com*.

DEFINITION OF DEFINITE INTEGRAL

If f is defined on the closed interval $[a, b]$ and the limit of Riemann sums over partitions Δ

$$\lim_{\|\Delta\| \to 0} \sum_{i=1}^{n} f(c_i)\,\Delta x_i$$

exists (as described above), then f is said to be **integrable** on $[a, b]$ and the limit is denoted by

$$\lim_{\|\Delta\| \to 0} \sum_{i=1}^{n} f(c_i)\,\Delta x_i = \int_a^b f(x)\,dx.$$

The limit is called the **definite integral** of f from a to b. The number a is the **lower limit** of integration, and the number b is the **upper limit** of integration.

It is not a coincidence that the notation for definite integrals is similar to that used for indefinite integrals. You will see why in the next section when the Fundamental Theorem of Calculus is introduced. For now it is important to see that definite integrals and indefinite integrals are different concepts. A definite integral is a *number*, whereas an indefinite integral is a *family of functions*.

Though Riemann sums were defined for functions with very few restrictions, a sufficient condition for a function f to be integrable on $[a, b]$ is that it is continuous on $[a, b]$. A proof of this theorem is beyond the scope of this text.

STUDY TIP Later in this chapter, you will learn convenient methods for calculating $\int_a^b f(x)\,dx$ for continuous functions. For now, you must use the limit definition.

THEOREM 6.4 CONTINUITY IMPLIES INTEGRABILITY

If a function f is continuous on the closed interval $[a, b]$, then f is integrable on $[a, b]$. That is, $\int_a^b f(x)\,dx$ exists.

EXPLORATION

The Converse of Theorem 6.4 Is the converse of Theorem 6.4 true? That is, if a function is integrable, does it have to be continuous? Explain your reasoning and give examples.

Describe the relationships among continuity, differentiability, and integrability. Which is the strongest condition? Which is the weakest? Which conditions imply other conditions?

EXAMPLE 2 Evaluating a Definite Integral as a Limit

Evaluate the definite integral $\int_{-2}^{1} 2x\, dx$.

Solution The function $f(x) = 2x$ is integrable on the interval $[-2, 1]$ because it is continuous on $[-2, 1]$. Moreover, the definition of integrability implies that any partition whose norm approaches 0 can be used to determine the limit. For computational convenience, define Δ by subdividing $[-2, 1]$ into n subintervals of equal width

$$\Delta x_i = \Delta x = \frac{b-a}{n} = \frac{3}{n}.$$

Choosing c_i as the right endpoint of each subinterval produces

$$c_i = a + i(\Delta x) = -2 + \frac{3i}{n}.$$

So, the definite integral is given by

$$\begin{aligned}
\int_{-2}^{1} 2x\, dx &= \lim_{\|\Delta\|\to 0} \sum_{i=1}^{n} f(c_i)\, \Delta x_i \\
&= \lim_{n\to\infty} \sum_{i=1}^{n} f(c_i)\, \Delta x \\
&= \lim_{n\to\infty} \sum_{i=1}^{n} 2\left(-2 + \frac{3i}{n}\right)\left(\frac{3}{n}\right) \\
&= \lim_{n\to\infty} \frac{6}{n} \sum_{i=1}^{n} \left(-2 + \frac{3i}{n}\right) \\
&= \lim_{n\to\infty} \frac{6}{n} \left\{-2n + \frac{3}{n}\left[\frac{n(n+1)}{2}\right]\right\} \\
&= \lim_{n\to\infty} \left(-12 + 9 + \frac{9}{n}\right) \\
&= -3.
\end{aligned}$$

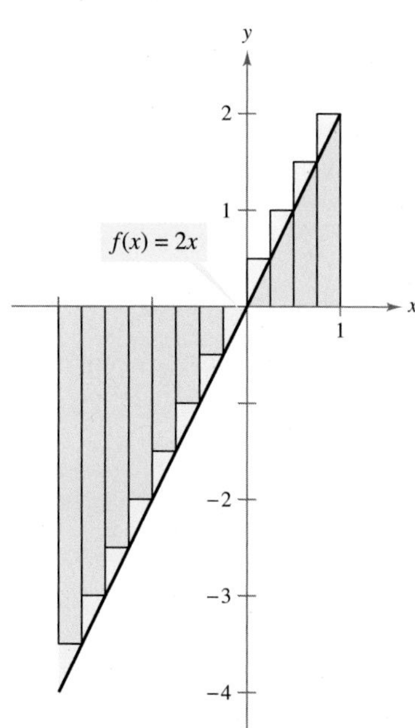

Because the definite integral is negative, it does not represent the area of the region.
Figure 6.20

Because the definite integral in Example 2 is negative, it *does not* represent the area of the region shown in Figure 6.20. Definite integrals can be positive, negative, or zero. For a definite integral to be interpreted as an area (as defined in Section 6.2), the function f must be continuous and nonnegative on $[a, b]$, as stated in the following theorem. The proof of this theorem is straightforward—you simply use the definition of area given in Section 6.2, because it is a Riemann sum.

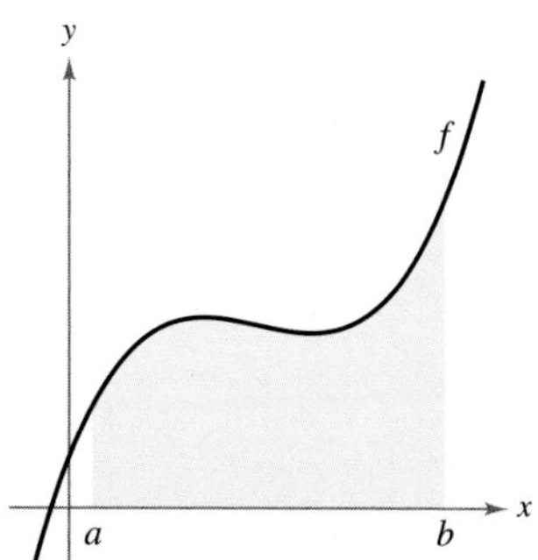

You can use a definite integral to find the area of the region bounded by the graph of f, the x-axis, $x = a$, and $x = b$.
Figure 6.21

THEOREM 6.5 THE DEFINITE INTEGRAL AS THE AREA OF A REGION

If f is continuous and nonnegative on the closed interval $[a, b]$, then the area of the region bounded by the graph of f, the x-axis, and the vertical lines $x = a$ and $x = b$ is given by

$$\text{Area} = \int_a^b f(x)\, dx.$$

(See Figure 6.21.)

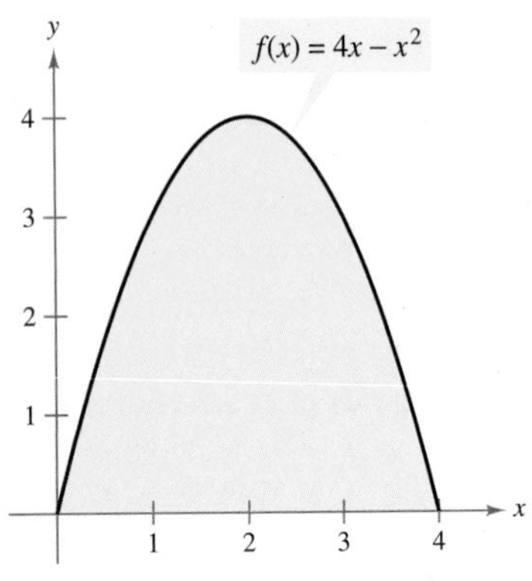

Area $= \int_0^4 (4x - x^2)\, dx$

Figure 6.22

As an example of Theorem 6.5, consider the region bounded by the graph of

$$f(x) = 4x - x^2$$

and the x-axis, as shown in Figure 6.22. Because f is continuous and nonnegative on the closed interval $[0, 4]$, the area of the region is

$$\text{Area} = \int_0^4 (4x - x^2)\, dx.$$

A straightforward technique for evaluating a definite integral such as this will be discussed in Section 6.4. For now, however, you can evaluate a definite integral in two ways—you can use the limit definition *or* you can check to see whether the definite integral represents the area of a common geometric region such as a rectangle, triangle, or semicircle.

EXAMPLE 3 Areas of Common Geometric Figures

Sketch the region corresponding to each definite integral. Then evaluate each integral using a geometric formula.

a. $\displaystyle\int_1^3 4\, dx$

b. $\displaystyle\int_0^3 (x + 2)\, dx$

c. $\displaystyle\int_{-2}^2 \sqrt{4 - x^2}\, dx$

Solution A sketch of each region is shown in Figure 6.23.

a. This region is a rectangle of height 4 and width 2.

$$\int_1^3 4\, dx = (\text{Area of rectangle}) = 4(2) = 8$$

b. This region is a trapezoid with an altitude of 3 and parallel bases of lengths 2 and 5. The formula for the area of a trapezoid is $\frac{1}{2}h(b_1 + b_2)$.

$$\int_0^3 (x + 2)\, dx = (\text{Area of trapezoid}) = \frac{1}{2}(3)(2 + 5) = \frac{21}{2}$$

c. This region is a semicircle of radius 2. The formula for the area of a semicircle is $\frac{1}{2}\pi r^2$.

$$\int_{-2}^2 \sqrt{4 - x^2}\, dx = (\text{Area of semicircle}) = \frac{1}{2}\pi(2^2) = 2\pi$$

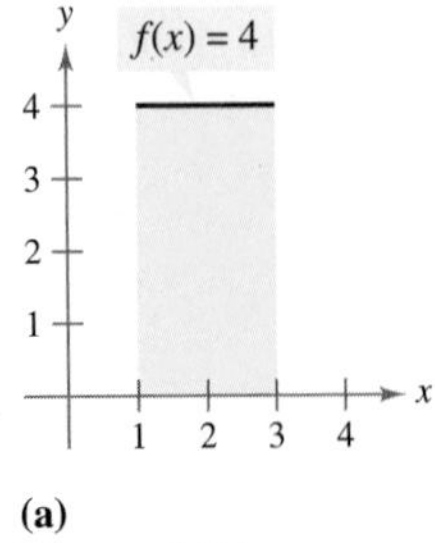

(a)

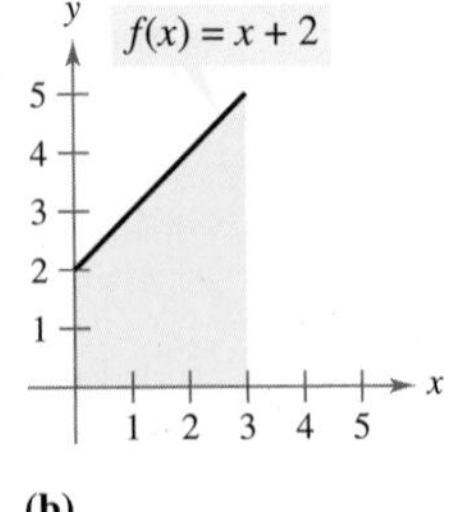

(b)

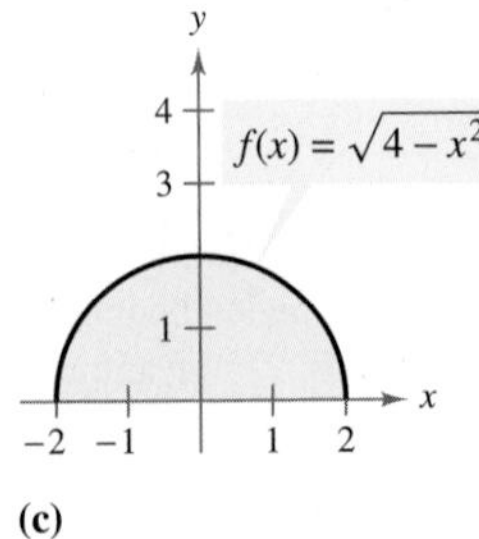

(c)

Figure 6.23

NOTE The variable of integration in a definite integral is sometimes called a *dummy variable* because it can be replaced by any other variable without changing the value of the integral. For instance, the definite integrals

$$\int_0^3 (x + 2)\, dx$$

and

$$\int_0^3 (t + 2)\, dt$$

have the same value.

Properties of Definite Integrals

The definition of the definite integral of f on the interval $[a, b]$ specifies that $a < b$. Now, however, it is convenient to extend the definition to cover cases in which $a = b$ or $a > b$. Geometrically, the following two definitions seem reasonable. For instance, it makes sense to define the area of a region of zero width and finite height to be 0.

DEFINITIONS OF TWO SPECIAL DEFINITE INTEGRALS

1. If f is defined at $x = a$, then we define $\int_a^a f(x)\, dx = 0$.

2. If f is integrable on $[a, b]$, then we define $\int_b^a f(x)\, dx = -\int_a^b f(x)\, dx$.

EXAMPLE 4 Evaluating Definite Integrals

a. Because the integrand is defined at $x = 2$, and the upper and lower limits of integration are equal, you can write

$$\int_2^2 \sqrt{x^2 + 1}\, dx = 0.$$

b. The integral $\int_3^0 (x + 2)\, dx$ is the same as that given in Example 3(b) except that the upper and lower limits are interchanged. Because the integral in Example 3(b) has a value of $\frac{21}{2}$, you can write

$$\int_3^0 (x + 2)\, dx = -\int_0^3 (x + 2)\, dx = -\frac{21}{2}.$$

■

In Figure 6.24, the larger region can be divided at $x = c$ into two subregions whose intersection is a line segment. Because the line segment has zero area, it follows that the area of the larger region is equal to the sum of the areas of the two smaller regions.

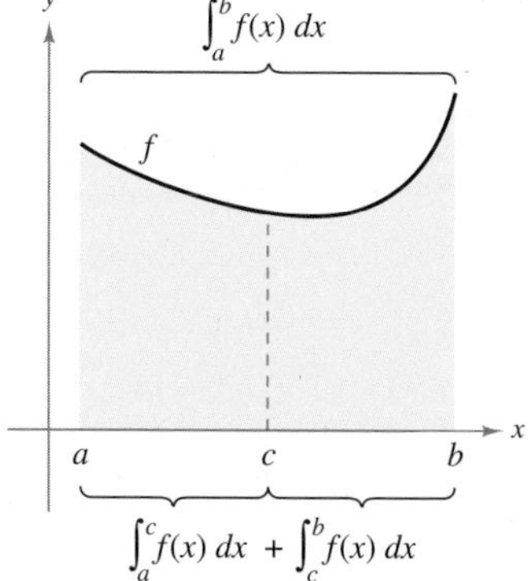

Figure 6.24

THEOREM 6.6 ADDITIVE INTERVAL PROPERTY

If f is integrable on the three closed intervals determined by a, b, and c, then

$$\int_a^b f(x)\, dx = \int_a^c f(x)\, dx + \int_c^b f(x)\, dx.$$

EXAMPLE 5 Using the Additive Interval Property

Evaluate the definite integral $\int_{-1}^1 |x|\, dx$.

Solution

$$\int_{-1}^1 |x|\, dx = \int_{-1}^0 -x\, dx + \int_0^1 x\, dx = \frac{1}{2} + \frac{1}{2} = 1$$

■

Because the definite integral is defined as the limit of a sum, it inherits the properties of summation given at the top of page 409.

THEOREM 6.7 PROPERTIES OF DEFINITE INTEGRALS

If f and g are integrable on $[a, b]$ and k is a constant, then the functions kf and $f \pm g$ are integrable on $[a, b]$, and

1. $\displaystyle\int_a^b kf(x)\,dx = k\int_a^b f(x)\,dx$
2. $\displaystyle\int_a^b [f(x) \pm g(x)]\,dx = \int_a^b f(x)\,dx \pm \int_a^b g(x)\,dx.$

NOTE Property 2 of Theorem 6.7 can be extended to cover any finite number of functions. For example,

$$\int_a^b [f(x) + g(x) + h(x)]\,dx = \int_a^b f(x)\,dx + \int_a^b g(x)\,dx + \int_a^b h(x)\,dx.$$

EXAMPLE 6 Evaluation of a Definite Integral

Evaluate $\displaystyle\int_1^3 (-x^2 + 4x - 3)\,dx$ using each of the following values.

$$\int_1^3 x^2\,dx = \frac{26}{3}, \qquad \int_1^3 x\,dx = 4, \qquad \int_1^3 dx = 2$$

Solution

$$\begin{aligned}\int_1^3 (-x^2 + 4x - 3)\,dx &= \int_1^3 (-x^2)\,dx + \int_1^3 4x\,dx + \int_1^3 (-3)\,dx\\ &= -\int_1^3 x^2\,dx + 4\int_1^3 x\,dx - 3\int_1^3 dx\\ &= -\left(\frac{26}{3}\right) + 4(4) - 3(2) = \frac{4}{3}\end{aligned}$$

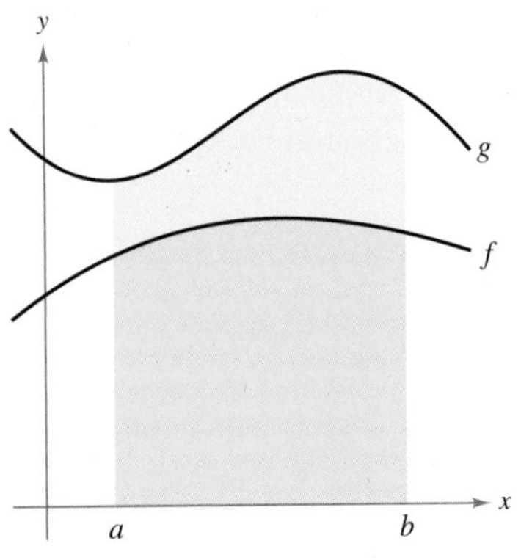

$\displaystyle\int_a^b f(x)\,dx \le \int_a^b g(x)\,dx$

Figure 6.25

If f and g are continuous on the closed interval $[a, b]$ and $0 \le f(x) \le g(x)$ for $a \le x \le b$, the following properties are true.

1. The area of the region bounded by the graph of f and the x-axis (between a and b) must be nonnegative.
2. This area must be less than or equal to the area of the region bounded by the graph of g and the x-axis (between a and b), as shown in Figure 6.25.

These two properties are generalized in Theorem 6.8. (A proof of this theorem is given in Appendix A.)

THEOREM 6.8 PRESERVATION OF INEQUALITY

1. If f is integrable and nonnegative on the closed interval $[a, b]$, then
$$0 \le \int_a^b f(x)\,dx.$$
2. If f and g are integrable on the closed interval $[a, b]$ and $f(x) \le g(x)$ for every x in $[a, b]$, then
$$\int_a^b f(x)\,dx \le \int_a^b g(x)\,dx.$$

6.3 Exercises

See www.CalcChat.com for worked-out solutions to odd-numbered exercises.

In Exercises 1 and 2, use Example 1 as a model to evaluate the limit

$$\lim_{n\to\infty} \sum_{i=1}^{n} f(c_i)\,\Delta x_i$$

over the region bounded by the graphs of the equations.

1. $f(x) = \sqrt{x}, \quad y = 0, \quad x = 0, \quad x = 3$
   (*Hint:* Let $c_i = 3i^2/n^2$.)
2. $f(x) = \sqrt[3]{x}, \quad y = 0, \quad x = 0, \quad x = 1$
   (*Hint:* Let $c_i = i^3/n^3$.)

In Exercises 3–8, evaluate the definite integral by the limit definition.

3. $\int_2^6 8\,dx$
4. $\int_{-2}^3 x\,dx$
5. $\int_{-1}^1 x^3\,dx$
6. $\int_1^4 4x^2\,dx$
7. $\int_1^2 (x^2 + 1)\,dx$
8. $\int_{-2}^1 (2x^2 + 3)\,dx$

In Exercises 9–12, write the limit as a definite integral on the interval $[a, b]$, where c_i is any point in the ith subinterval.

Limit	*Interval*
9. $\lim_{\|\Delta\|\to 0} \sum_{i=1}^{n} (3c_i + 10)\,\Delta x_i$	$[-1, 5]$
10. $\lim_{\|\Delta\|\to 0} \sum_{i=1}^{n} 6c_i(4 - c_i)^2\,\Delta x_i$	$[0, 4]$
11. $\lim_{\|\Delta\|\to 0} \sum_{i=1}^{n} \sqrt{c_i^2 + 4}\,\Delta x_i$	$[0, 3]$
12. $\lim_{\|\Delta\|\to 0} \sum_{i=1}^{n} \left(\frac{3}{c_i^2}\right)\Delta x_i$	$[1, 3]$

In Exercises 13–20, set up a definite integral that yields the area of the region. (Do not evaluate the integral.)

13. $f(x) = 5$

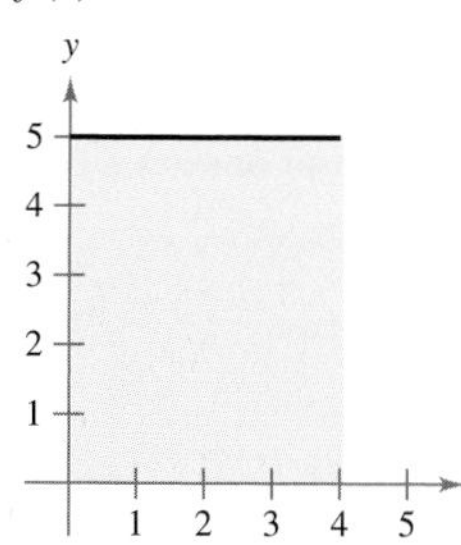

14. $f(x) = 6 - 3x$

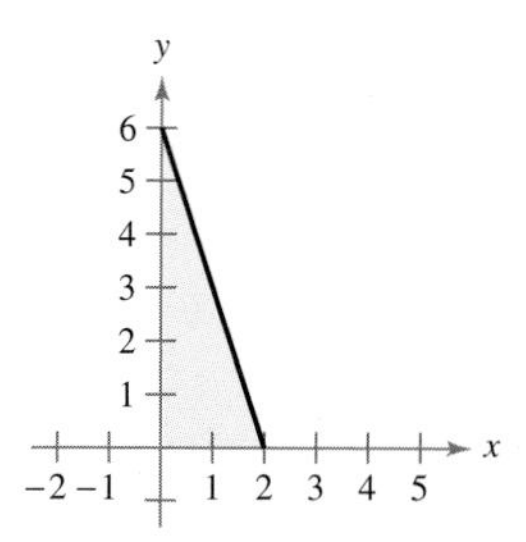

15. $f(x) = 4 - |x|$

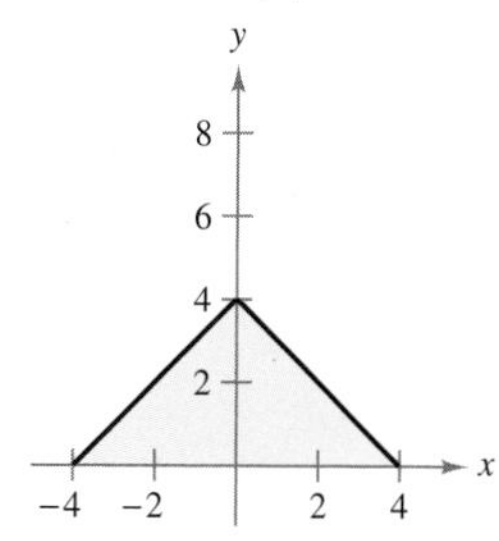

16. $f(x) = x^2$

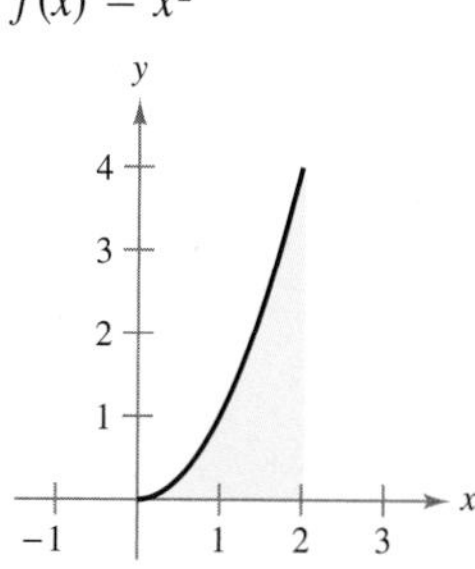

17. $f(x) = 25 - x^2$

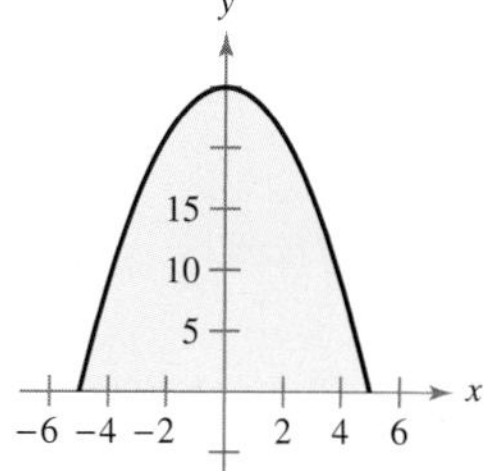

18. 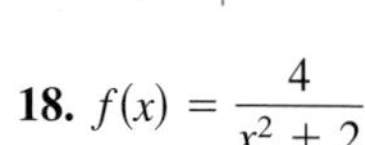 $f(x) = \dfrac{4}{x^2 + 2}$

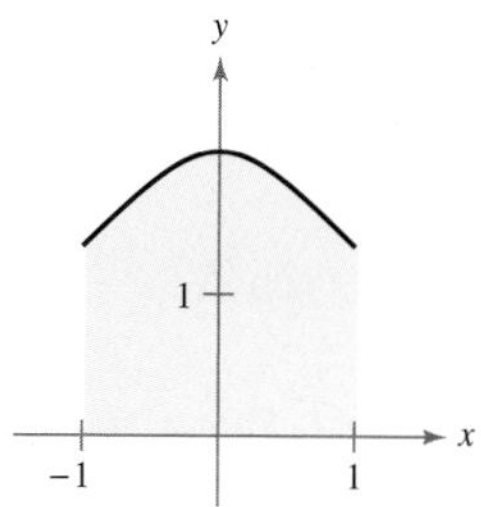

19. $g(y) = y^3$

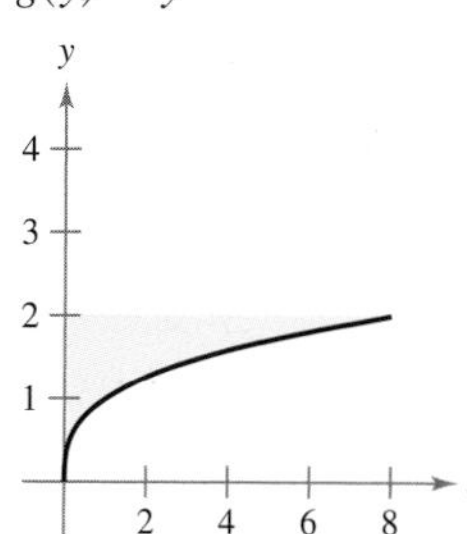

20. $f(y) = (y - 2)^2$

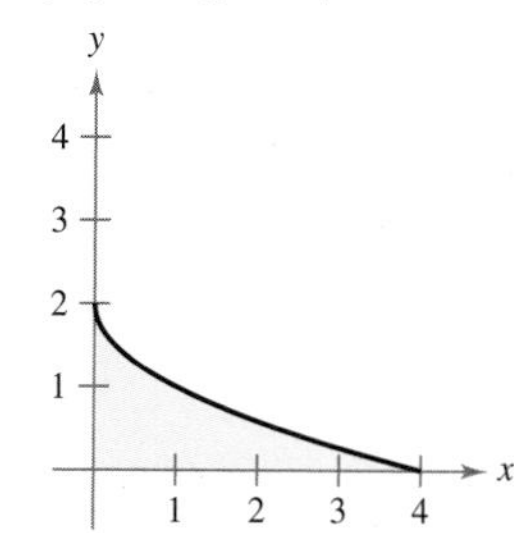

In Exercises 21–30, sketch the region whose area is given by the definite integral. Then use a geometric formula to evaluate the integral ($a > 0, r > 0$).

21. $\int_0^3 4\,dx$
22. $\int_{-a}^a 4\,dx$
23. $\int_0^4 x\,dx$
24. $\int_0^4 \frac{x}{2}\,dx$
25. $\int_0^2 (3x + 4)\,dx$
26. $\int_0^6 (6 - x)\,dx$
27. $\int_{-1}^1 (1 - |x|)\,dx$
28. $\int_{-a}^a (a - |x|)\,dx$
29. $\int_{-7}^7 \sqrt{49 - x^2}\,dx$
30. $\int_{-r}^r \sqrt{r^2 - x^2}\,dx$

In Exercises 31–38, evaluate the integral using the following values.

$$\int_2^4 x^3\,dx = 60, \qquad \int_2^4 x\,dx = 6, \qquad \int_2^4 dx = 2$$

31. $\int_4^2 x\,dx$
32. $\int_2^2 x^3\,dx$

33. $\int_2^4 8x\,dx$

34. $\int_2^4 25\,dx$

35. $\int_2^4 (x - 9)\,dx$

36. $\int_2^4 (x^3 + 4)\,dx$

37. $\int_2^4 \left(\frac{1}{2}x^3 - 3x + 2\right)dx$

38. $\int_2^4 (10 + 4x - 3x^3)\,dx$

39. Given $\int_0^5 f(x)\,dx = 10$ and $\int_5^7 f(x)\,dx = 3$, evaluate

(a) $\int_0^7 f(x)\,dx$. (b) $\int_5^0 f(x)\,dx$.

(c) $\int_5^5 f(x)\,dx$. (d) $\int_0^5 3f(x)\,dx$.

40. Given $\int_0^3 f(x)\,dx = 4$ and $\int_3^6 f(x)\,dx = -1$, evaluate

(a) $\int_0^6 f(x)\,dx$. (b) $\int_6^3 f(x)\,dx$.

(c) $\int_3^3 f(x)\,dx$. (d) $\int_3^6 -5f(x)\,dx$.

41. Given $\int_2^6 f(x)\,dx = 10$ and $\int_2^6 g(x)\,dx = -2$, evaluate

(a) $\int_2^6 [f(x) + g(x)]\,dx$. (b) $\int_2^6 [g(x) - f(x)]\,dx$.

(c) $\int_2^6 2g(x)\,dx$. (d) $\int_2^6 3f(x)\,dx$.

42. Given $\int_{-1}^1 f(x)\,dx = 0$ and $\int_0^1 f(x)\,dx = 5$, evaluate

(a) $\int_{-1}^0 f(x)\,dx$. (b) $\int_0^1 f(x)\,dx - \int_{-1}^0 f(x)\,dx$.

(c) $\int_{-1}^1 3f(x)\,dx$. (d) $\int_0^1 3f(x)\,dx$.

43. ***Think About It*** The graph of f consists of line segments and a semicircle, as shown in the figure. Evaluate each definite integral by using geometric formulas.

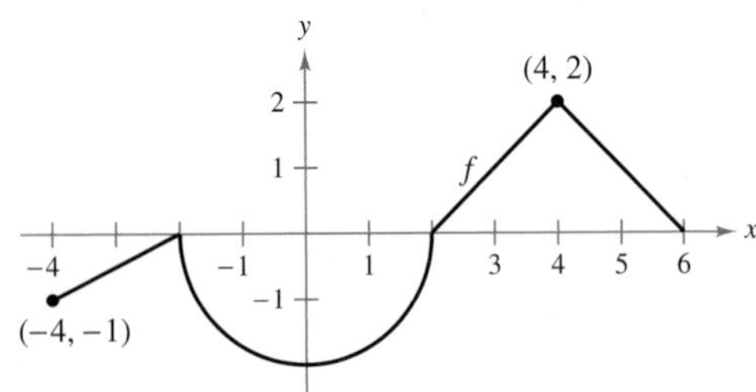

(a) $\int_0^2 f(x)\,dx$ (b) $\int_2^6 f(x)\,dx$

(c) $\int_{-4}^2 f(x)\,dx$ (d) $\int_{-4}^6 f(x)\,dx$

(e) $\int_{-4}^6 |f(x)|\,dx$ (f) $\int_{-4}^6 [f(x) + 2]\,dx$

44. ***Think About It*** The graph of f consists of line segments, as shown in the figure. Evaluate each definite integral by using geometric formulas.

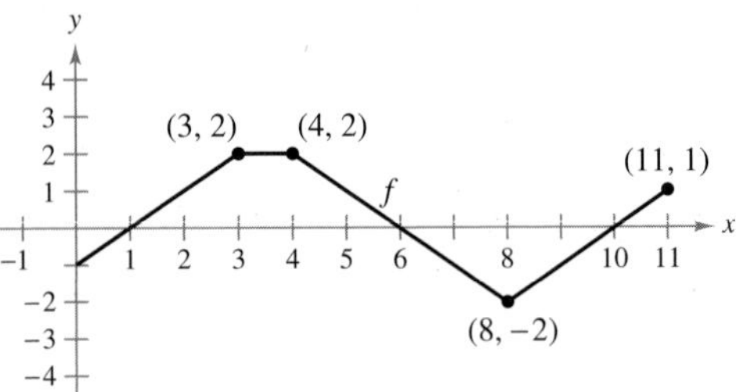

(a) $\int_0^1 -f(x)\,dx$ (b) $\int_3^4 3\,f(x)\,dx$

(c) $\int_0^7 f(x)\,dx$ (d) $\int_5^{11} f(x)\,dx$

(e) $\int_0^{11} f(x)\,dx$ (f) $\int_4^{10} f(x)\,dx$

45. ***Think About It*** Consider the function f that is continuous on the interval $[-5, 5]$ and for which

$$\int_0^5 f(x)\,dx = 4.$$

Evaluate each integral.

(a) $\int_0^5 [f(x) + 2]\,dx$ (b) $\int_{-2}^3 f(x + 2)\,dx$

(c) $\int_{-5}^5 f(x)\,dx$ (f is even.) (d) $\int_{-5}^5 f(x)\,dx$ (f is odd.)

46. ***Think About It*** A function f is defined below. Use geometric formulas to find $\int_0^8 f(x)\,dx$.

$$f(x) = \begin{cases} 4, & x < 4 \\ x, & x \ge 4 \end{cases}$$

47. ***Think About It*** A function f is defined below. Use geometric formulas to find $\int_0^{12} f(x)\,dx$.

$$f(x) = \begin{cases} 6, & x > 6 \\ -\frac{1}{2}x + 9, & x \le 6 \end{cases}$$

CAPSTONE

48. Find possible values of a and b that make the statement true. If possible, use a graph to support your answer. (There may be more than one correct answer.)

(a) $\int_{-2}^1 f(x)\,dx + \int_1^5 f(x)\,dx = \int_a^b f(x)\,dx$

(b) $\int_{-3}^3 f(x)\,dx + \int_3^6 f(x)\,dx - \int_a^b f(x)\,dx = \int_{-1}^6 f(x)\,dx$

WRITING ABOUT CONCEPTS

In Exercises 49 and 50, use the figure to fill in the blank with the symbol <, >, or =.

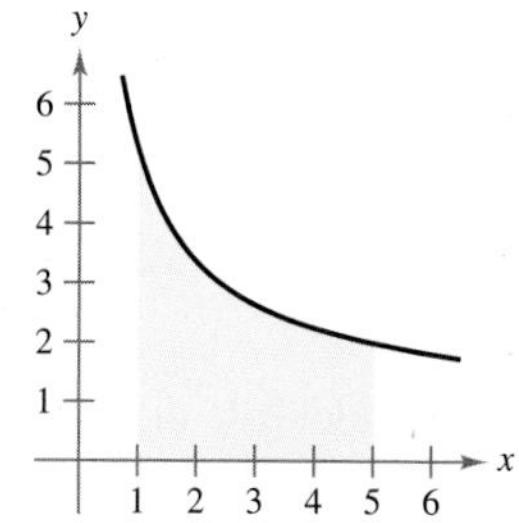

49. The interval $[1, 5]$ is partitioned into n subintervals of equal width Δx, and x_i is the left endpoint of the ith subinterval.

$$\sum_{i=1}^{n} f(x_i)\,\Delta x \quad \blacksquare \quad \int_1^5 f(x)\,dx$$

50. The interval $[1, 5]$ is partitioned into n subintervals of equal width Δx, and x_i is the right endpoint of the ith subinterval.

$$\sum_{i=1}^{n} f(x_i)\,\Delta x \quad \blacksquare \quad \int_1^5 f(x)\,dx$$

51. Determine whether the function $f(x) = \dfrac{1}{x-4}$ is integrable on the interval $[3, 5]$. Explain.

52. Give an example of a function that is integrable on the interval $[-1, 1]$, but not continuous on $[-1, 1]$.

In Exercises 53 and 54, determine which value best approximates the definite integral. Make your selection on the basis of a sketch.

53. $\displaystyle\int_0^4 \sqrt{x}\,dx$

(a) 5 (b) -3 (c) 10 (d) 2 (e) 8

54. $\displaystyle\int_1^3 \frac{x^3+1}{x^2}\,dx$

(a) 2 (b) -2 (c) 16 (d) 5 (e) 10

True or False? **In Exercises 55–60, determine whether the statement is true or false. If it is false, explain why or give an example that shows it is false.**

55. $\displaystyle\int_a^b [f(x) + g(x)]\,dx = \int_a^b f(x)\,dx + \int_a^b g(x)\,dx$

56. $\displaystyle\int_a^b f(x)g(x)\,dx = \left[\int_a^b f(x)\,dx\right]\left[\int_a^b g(x)\,dx\right]$

57. If the norm of a partition approaches zero, then the number of subintervals approaches infinity.

58. If f is increasing on $[a, b]$, then the minimum value of $f(x)$ on $[a, b]$ is $f(a)$.

59. The value of $\int_a^b f(x)\,dx$ must be positive.

60. If $\int_a^b f(x)\,dx > 0$, then f is nonnegative for all x in $[a, b]$.

Programming **Write a program for your graphing utility to approximate a definite integral using the Riemann sum**

$$\sum_{i=1}^{n} f(c_i)\Delta x_i$$

where the subintervals are of equal width. The output should give three approximations of the integral, where c_i is the left-hand endpoint $L(n)$, the midpoint $M(n)$, and the right-hand endpoint $R(n)$ of each subinterval. In Exercises 61 and 62, use the program to approximate the definite integral and complete the table.

n	4	8	12	16	20
$L(n)$					
$M(n)$					
$R(n)$					

61. $\displaystyle\int_0^3 x\sqrt{3-x}\,dx$

62. $\displaystyle\int_0^3 \frac{5}{x^2+1}\,dx$

63. Find the Riemann sum for $f(x) = x^2 + 3x$ over the interval $[0, 8]$, where $x_0 = 0$, $x_1 = 1$, $x_2 = 3$, $x_3 = 7$, and $x_4 = 8$, and where $c_1 = 1$, $c_2 = 2$, $c_3 = 5$, and $c_4 = 8$.

64. ***Think About It*** Determine whether the Dirichlet function

$$f(x) = \begin{cases} 1, & x \text{ is rational} \\ 0, & x \text{ is irrational} \end{cases}$$

is integrable on the interval $[0, 1]$. Explain.

65. Suppose the function f is defined on $[0, 1]$ as shown.

$$f(x) = \begin{cases} 0, & x = 0 \\ \dfrac{1}{x}, & 0 < x \le 1 \end{cases}$$

Show that $\int_0^1 f(x)\,dx$ does not exist. Why doesn't this contradict Theorem 6.4?

66. Find the constants a and b that maximize the value of

$$\int_a^b (1 - x^2)\,dx.$$

Explain your reasoning.

67. Evaluate, if possible, the integral $\displaystyle\int_0^2 [\![x]\!]\,dx$.

68. Determine

$$\lim_{n\to\infty} \frac{1}{n^3}[1^2 + 2^2 + 3^2 + \cdots + n^2]$$

by using an appropriate Riemann sum.

6.4 The Fundamental Theorem of Calculus

- Evaluate a definite integral using the Fundamental Theorem of Calculus.
- Understand and use the Mean Value Theorem for Integrals.
- Find the average value of a function over a closed interval.
- Understand and use the Second Fundamental Theorem of Calculus.
- Understand and use the Net Change Theorem.

> **EXPLORATION**
>
> ***Integration and Antidifferentiation*** Throughout this chapter, you have been using the integral sign to denote an antiderivative (a family of functions) and a definite integral (a number).
>
> 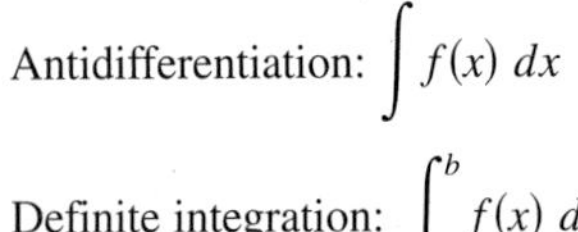
>
> Antidifferentiation: $\int f(x)\,dx$
>
> Definite integration: $\int_a^b f(x)\,dx$
>
> The use of this same symbol for both operations makes it appear that they are related. In the early work with calculus, however, it was not known that the two operations were related. Do you think the symbol $\int$ was first applied to antidifferentiation or to definite integration? Explain your reasoning. (*Hint:* The symbol was first used by Leibniz and was derived from the letter *S*.)

The Fundamental Theorem of Calculus

You have now been introduced to the two major branches of calculus: differential calculus (introduced with the tangent line problem) and integral calculus (introduced with the area problem). At this point, these two problems might seem unrelated—but there is a very close connection. The connection was discovered independently by Isaac Newton and Gottfried Leibniz and is stated in a theorem that is appropriately called the **Fundamental Theorem of Calculus.**

Informally, the theorem states that differentiation and (definite) integration are inverse operations, in the same sense that division and multiplication are inverse operations. To see how Newton and Leibniz might have anticipated this relationship, consider the approximations shown in Figure 6.26. The slope of the tangent line was defined using the *quotient* $\Delta y/\Delta x$ (the slope of the secant line). Similarly, the area of a region under a curve was defined using the *product* $\Delta y \Delta x$ (the area of a rectangle). So, at least in the primitive approximation stage, the operations of differentiation and definite integration appear to have an inverse relationship in the same sense that division and multiplication are inverse operations. The Fundamental Theorem of Calculus states that the limit processes (used to define the derivative and definite integral) preserve this inverse relationship.

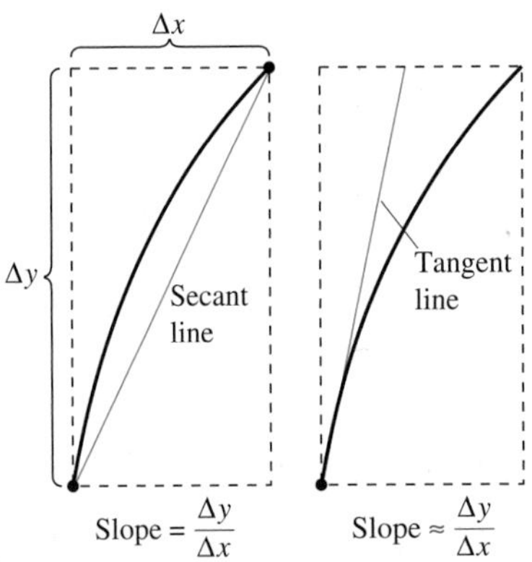

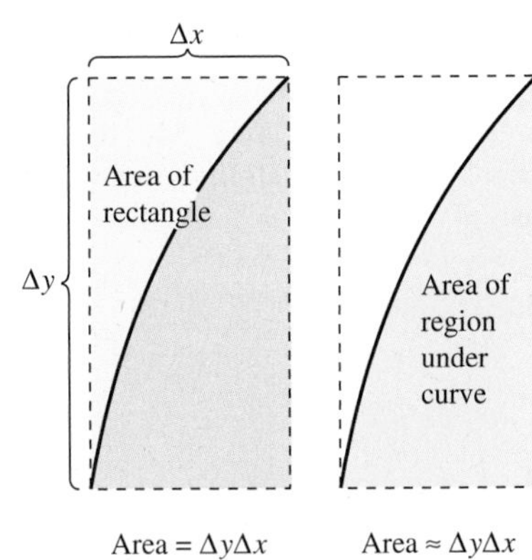

(a) Differentiation (b) Definite integration

Differentiation and definite integration have an "inverse" relationship.

Figure 6.26

> **THEOREM 6.9 THE FUNDAMENTAL THEOREM OF CALCULUS**
>
> If a function f is continuous on the closed interval $[a, b]$ and F is an antiderivative of f on the interval $[a, b]$, then
>
> $$\int_a^b f(x)\,dx = F(b) - F(a).$$

PROOF The key to the proof is in writing the difference $F(b) - F(a)$ in a convenient form. Let Δ be any partition of $[a, b]$.

$$a = x_0 < x_1 < x_2 < \cdot \cdot \cdot < x_{n-1} < x_n = b$$

By pairwise subtraction and addition of like terms, you can write

$$\begin{aligned} F(b) - F(a) &= F(x_n) - F(x_{n-1}) + F(x_{n-1}) - \cdot \cdot \cdot - F(x_1) + F(x_1) - F(x_0) \\ &= \sum_{i=1}^{n} [F(x_i) - F(x_{i-1})]. \end{aligned}$$

By the Mean Value Theorem, you know that there exists a number c_i in the ith subinterval such that

$$F'(c_i) = \frac{F(x_i) - F(x_{i-1})}{x_i - x_{i-1}}.$$

Because $F'(c_i) = f(c_i)$, you can let $\Delta x_i = x_i - x_{i-1}$ and obtain

$$F(b) - F(a) = \sum_{i=1}^{n} f(c_i)\, \Delta x_i.$$

This important equation tells you that by repeatedly applying the Mean Value Theorem, you can always find a collection of c_i's such that the *constant* $F(b) - F(a)$ is a Riemann sum of f on $[a, b]$ for any partition. Theorem 6.4 guarantees that the limit of Riemann sums over the partition with $\|\Delta\| \to 0$ exists. So, taking the limit (as $\|\Delta\| \to 0$) produces

$$F(b) - F(a) = \int_a^b f(x)\, dx.$$ ■

The following guidelines can help you understand the use of the Fundamental Theorem of Calculus.

GUIDELINES FOR USING THE FUNDAMENTAL THEOREM OF CALCULUS

1. *Provided you can find* an antiderivative of f, you now have a way to evaluate a definite integral without having to use the limit of a sum.
2. When applying the Fundamental Theorem of Calculus, the following notation is convenient.

$$\begin{aligned} \int_a^b f(x)\, dx &= F(x)\Big]_a^b \\ &= F(b) - F(a) \end{aligned}$$

For instance, to evaluate $\int_1^3 x^3\, dx$, you can write

$$\int_1^3 x^3\, dx = \frac{x^4}{4}\Big]_1^3 = \frac{3^4}{4} - \frac{1^4}{4} = \frac{81}{4} - \frac{1}{4} = 20.$$

3. It is not necessary to include a constant of integration C in the antiderivative because

$$\begin{aligned} \int_a^b f(x)\, dx &= \Big[F(x) + C\Big]_a^b \\ &= [F(b) + C] - [F(a) + C] \\ &= F(b) - F(a). \end{aligned}$$

EXAMPLE 1 Evaluating a Definite Integral

Evaluate each definite integral.

a. $\int_1^2 (x^2 - 3)\,dx$ **b.** $\int_1^4 3\sqrt{x}\,dx$

Solution

a. $$\int_1^2 (x^2 - 3)\,dx = \underbrace{\left[\frac{x^3}{3} - 3x\right]}_{F(x)}\Big|_1^2 = \underbrace{\left(\frac{8}{3} - 6\right)}_{F(2)} - \underbrace{\left(\frac{1}{3} - 3\right)}_{F(1)} = -\frac{2}{3}$$

b. $$\int_1^4 3\sqrt{x}\,dx = 3\int_1^4 x^{1/2}\,dx = 3\left[\frac{x^{3/2}}{3/2}\right]_1^4 = 2(4)^{3/2} - 2(1)^{3/2} = 14$$

EXAMPLE 2 A Definite Integral Involving Absolute Value

Evaluate $\int_0^2 |2x - 1|\,dx$.

Solution Using Figure 6.27 and the definition of absolute value, you can rewrite the integrand as shown.

$$|2x - 1| = \begin{cases} -(2x - 1), & x < \frac{1}{2} \\ 2x - 1, & x \ge \frac{1}{2} \end{cases}$$

The definite integral of y on $[0, 2]$ is $\frac{5}{2}$.
Figure 6.27

From this, you can rewrite the integral in two parts.

$$\begin{aligned}\int_0^2 |2x - 1|\,dx &= \int_0^{1/2} -(2x - 1)\,dx + \int_{1/2}^2 (2x - 1)\,dx \\ &= \Big[-x^2 + x\Big]_0^{1/2} + \Big[x^2 - x\Big]_{1/2}^2 \\ &= \left(-\frac{1}{4} + \frac{1}{2}\right) - (0 + 0) + (4 - 2) - \left(\frac{1}{4} - \frac{1}{2}\right) \\ &= \frac{5}{2}\end{aligned}$$

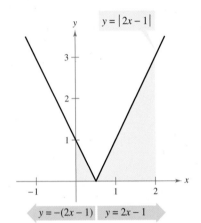

EXAMPLE 3 Using the Fundamental Theorem to Find Area

Find the area of the region bounded by the graph of $y = 2x^2 - 3x + 2$, the x-axis, and the vertical lines $x = 0$ and $x = 2$, as shown in Figure 6.28.

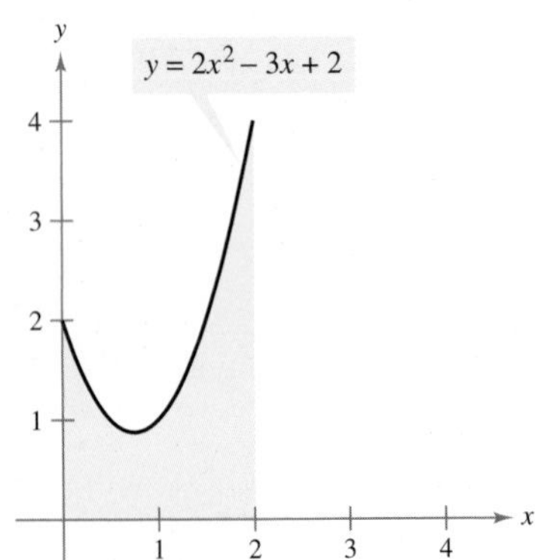

The area of the region bounded by the graph of y, the x-axis, $x = 0$, and $x = 2$ is $\frac{10}{3}$.
Figure 6.28

Solution Note that $y > 0$ on the interval $[0, 2]$.

$$\begin{aligned}\text{Area} &= \int_0^2 (2x^2 - 3x + 2)\,dx && \text{Integrate between } x = 0 \text{ and } x = 2. \\ &= \left[\frac{2x^3}{3} - \frac{3x^2}{2} + 2x\right]_0^2 && \text{Find antiderivative.} \\ &= \left(\frac{16}{3} - 6 + 4\right) - (0 - 0 + 0) && \text{Apply Fundamental Theorem.} \\ &= \frac{10}{3} && \text{Simplify.}\end{aligned}$$

■

The Mean Value Theorem for Integrals

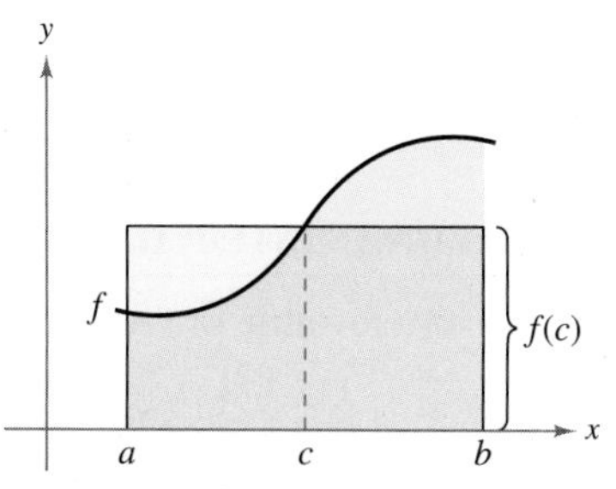

Mean value rectangle:

$f(c)(b-a) = \int_a^b f(x)\, dx$

Figure 6.29

In Section 6.2, you saw that the area of a region under a curve is greater than the area of an inscribed rectangle and less than the area of a circumscribed rectangle. The Mean Value Theorem for Integrals states that somewhere "between" the inscribed and circumscribed rectangles there is a rectangle whose area is precisely equal to the area of the region under the curve, as shown in Figure 6.29.

> **THEOREM 6.10 MEAN VALUE THEOREM FOR INTEGRALS**
>
> If f is continuous on the closed interval $[a, b]$, then there exists a number c in the closed interval $[a, b]$ such that
>
> $$\int_a^b f(x)\, dx = f(c)(b-a).$$

PROOF

Case 1: If f is constant on the interval $[a, b]$, the theorem is clearly valid because c can be any point in $[a, b]$.

Case 2: If f is not constant on $[a, b]$, then, by the Extreme Value Theorem, you can choose $f(m)$ and $f(M)$ to be the minimum and maximum values of f on $[a, b]$. Because $f(m) \le f(x) \le f(M)$ for all x in $[a, b]$, you can apply Theorem 6.8 to write the following.

$$\int_a^b f(m)\, dx \le \int_a^b f(x)\, dx \le \int_a^b f(M)\, dx \qquad \text{See Figure 6.30.}$$

$$f(m)(b-a) \le \int_a^b f(x)\, dx \le f(M)(b-a)$$

$$f(m) \le \frac{1}{b-a}\int_a^b f(x)\, dx \le f(M)$$

From the third inequality, you can apply the Intermediate Value Theorem to conclude that there exists some c in $[a, b]$ such that

$$f(c) = \frac{1}{b-a}\int_a^b f(x)\, dx \qquad \text{or} \qquad f(c)(b-a) = \int_a^b f(x)\, dx.$$

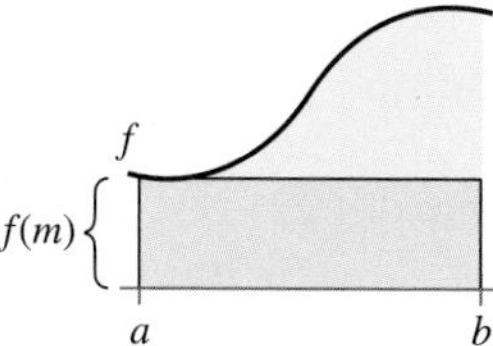

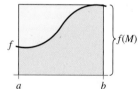

Inscribed rectangle
(less than actual area)

$\int_a^b f(m)\, dx = f(m)(b-a)$

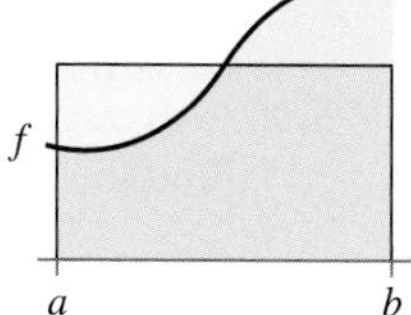

Mean value rectangle
(equal to actual area)

$\int_a^b f(x)\, dx$

Circumscribed rectangle
(greater than actual area)

$\int_a^b f(M)\, dx = f(M)(b-a)$

Figure 6.30

■

NOTE Notice that Theorem 6.10 does not specify how to determine c. It merely guarantees the existence of at least one number c in the interval. ■

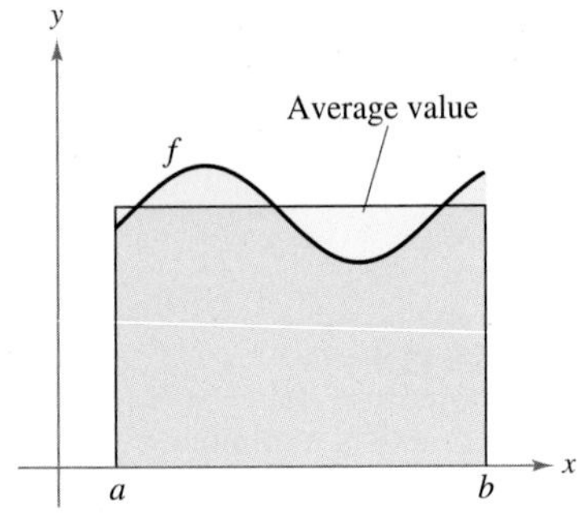

Average value $= \dfrac{1}{b-a}\displaystyle\int_a^b f(x)\,dx$

Figure 6.31

Average Value of a Function

The value of $f(c)$ given in the Mean Value Theorem for Integrals is called the **average value** of f on the interval $[a, b]$.

DEFINITION OF THE AVERAGE VALUE OF A FUNCTION ON AN INTERVAL

If f is integrable on the closed interval $[a, b]$, then the **average value** of f on the interval is

$$\frac{1}{b-a}\int_a^b f(x)\,dx.$$

NOTE Notice in Figure 6.31 that the area of the region under the graph of f is equal to the area of the rectangle whose height is the average value. ■

To see why the average value of f is defined in this way, suppose that you partition $[a, b]$ into n subintervals of equal width $\Delta x = (b-a)/n$. If c_i is any point in the ith subinterval, the arithmetic average (or mean) of the function values at the c_i's is given by

$$a_n = \frac{1}{n}[f(c_1) + f(c_2) + \cdot\cdot\cdot + f(c_n)]. \qquad \text{Average of } f(c_1), \ldots, f(c_n)$$

By multiplying and dividing by $(b-a)$, you can write the average as

$$\begin{aligned} a_n = \frac{1}{n}\sum_{i=1}^{n} f(c_i)\left(\frac{b-a}{b-a}\right) &= \frac{1}{b-a}\sum_{i=1}^{n} f(c_i)\left(\frac{b-a}{n}\right) \\ &= \frac{1}{b-a}\sum_{i=1}^{n} f(c_i)\,\Delta x. \end{aligned}$$

Finally, taking the limit as $n \to \infty$ produces the average value of f on the interval $[a, b]$, as given in the definition above.

This development of the average value of a function on an interval is only one of many practical uses of definite integrals to represent summation processes. In a later course, you will study other applications, such as volume, arc length, centers of mass, and work.

EXAMPLE 4 Finding the Average Value of a Function

Find the average value of $f(x) = 3x^2 - 2x$ on the interval $[1, 4]$.

Solution The average value is given by

$$\begin{aligned} \frac{1}{b-a}\int_a^b f(x)\,dx &= \frac{1}{4-1}\int_1^4 (3x^2 - 2x)\,dx \\ &= \frac{1}{3}\Big[x^3 - x^2\Big]_1^4 \\ &= \frac{1}{3}[64 - 16 - (1 - 1)] \\ &= \frac{48}{3} = 16. \end{aligned}$$

(See Figure 6.32.) ■

$f(x) = 3x^2 - 2x$; (4, 40); (1, 1); Average value = 16

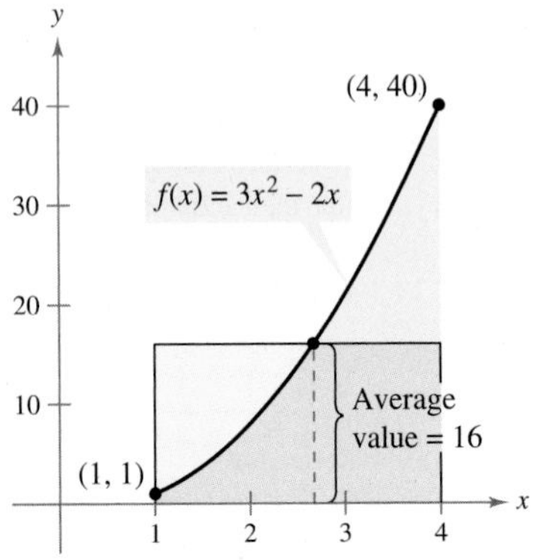

Figure 6.32

George Hall/CORBIS

The first person to fly at a speed greater than the speed of sound was Charles Yeager. On October 14, 1947, Yeager was clocked at 295.9 meters per second at an altitude of 12.2 kilometers. If Yeager had been flying at an altitude below 11.275 kilometers, this speed would not have "broken the sound barrier." The photo above shows an F-14 *Tomcat*, a supersonic, twin-engine strike fighter. Currently, the *Tomcat* can reach heights of 15.24 kilometers and speeds up to 2 mach (707.78 meters per second).

EXAMPLE 5 The Speed of Sound

At different altitudes in Earth's atmosphere, sound travels at different speeds. The speed of sound $s(x)$ (in meters per second) can be modeled by

$$s(x) = \begin{cases} -4x + 341, & 0 \le x < 11.5 \\ 295, & 11.5 \le x < 22 \\ \frac{3}{4}x + 278.5, & 22 \le x < 32 \\ \frac{3}{2}x + 254.5, & 32 \le x < 50 \\ -\frac{3}{2}x + 404.5, & 50 \le x \le 80 \end{cases}$$

where x is the altitude in kilometers (see Figure 6.33). What is the average speed of sound over the interval $[0, 80]$?

Solution Begin by integrating $s(x)$ over the interval $[0, 80]$. To do this, you can break the integral into five parts.

$$\int_0^{11.5} s(x)\,dx = \int_0^{11.5} (-4x + 341)\,dx = \Big[-2x^2 + 341x\Big]_0^{11.5} = 3657$$

$$\int_{11.5}^{22} s(x)\,dx = \int_{11.5}^{22} (295)\,dx = \Big[295x\Big]_{11.5}^{22} = 3097.5$$

$$\int_{22}^{32} s(x)\,dx = \int_{22}^{32} \left(\tfrac{3}{4}x + 278.5\right)dx = \Big[\tfrac{3}{8}x^2 + 278.5x\Big]_{22}^{32} = 2987.5$$

$$\int_{32}^{50} s(x)\,dx = \int_{32}^{50} \left(\tfrac{3}{2}x + 254.5\right)dx = \Big[\tfrac{3}{4}x^2 + 254.5x\Big]_{32}^{50} = 5688$$

$$\int_{50}^{80} s(x)\,dx = \int_{50}^{80} \left(-\tfrac{3}{2}x + 404.5\right)dx = \Big[-\tfrac{3}{4}x^2 + 404.5x\Big]_{50}^{80} = 9210$$

By adding the values of the five integrals, you have

$$\int_0^{80} s(x)\,dx = 24{,}640.$$

So, the average speed of sound from an altitude of 0 kilometers to an altitude of 80 kilometers is

$$\text{Average speed} = \frac{1}{80}\int_0^{80} s(x)\,dx = \frac{24{,}640}{80} = 308 \text{ meters per second.}$$

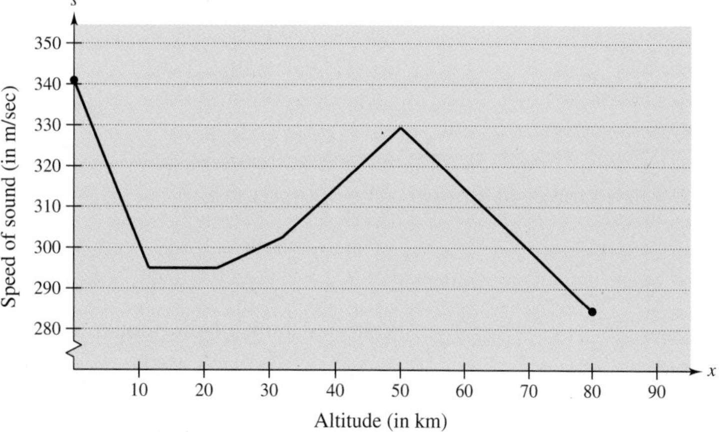

Speed of sound depends on altitude.
Figure 6.33

■

The Second Fundamental Theorem of Calculus

Earlier you saw that the definite integral of f on the interval $[a, b]$ was defined using the constant b as the upper limit of integration and x as the variable of integration. However, a slightly different situation may arise in which the variable x is used in the upper limit of integration. To avoid the confusion of using x in two different ways, t is temporarily used as the variable of integration. (Remember that the definite integral is *not* a function of its variable of integration.)

The Definite Integral as a Number

Constant

$$\int_a^b f(x)\,dx$$

Constant

f is a function of x.

The Definite Integral as a Function of x

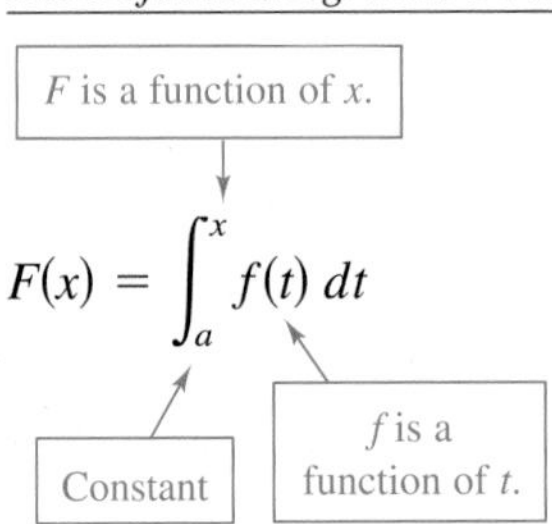

EXAMPLE 6 The Definite Integral as a Function

Evaluate the function

$$F(x) = \int_0^x (3 - 3t^2)\,dt$$

at $x = 0, \frac{1}{4}, \frac{1}{2}, \frac{3}{4}$, and 1.

Solution You could evaluate five different definite integrals, one for each of the given upper limits. However, it is much simpler to fix x (as a constant) temporarily to obtain

$$\int_0^x (3 - 3t^2)\,dt = \left[3t - t^3\right]_0^x = [3x - x^3] - [3(0) - 0^3] = 3x - x^3.$$

Now, using $F(x) = 3x - x^3$, you can obtain the results shown in Figure 6.34.

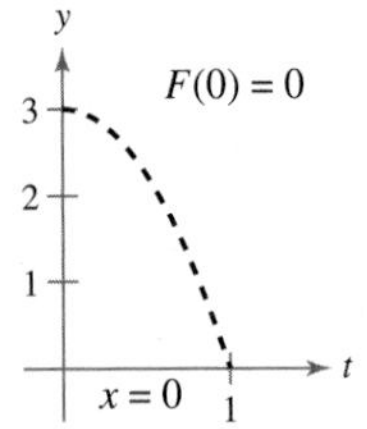

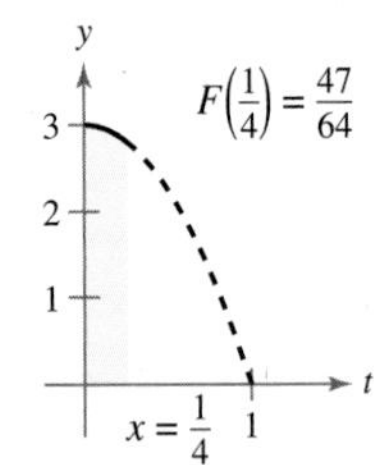

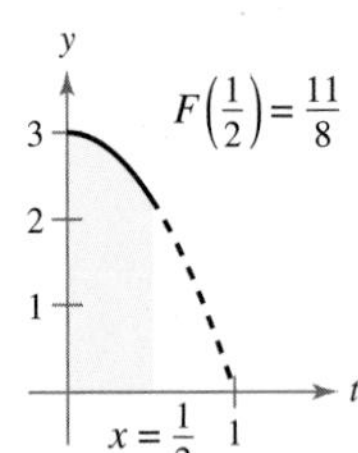

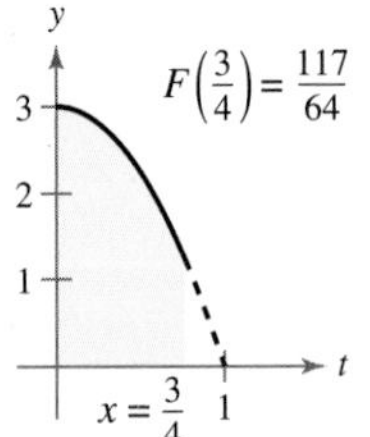

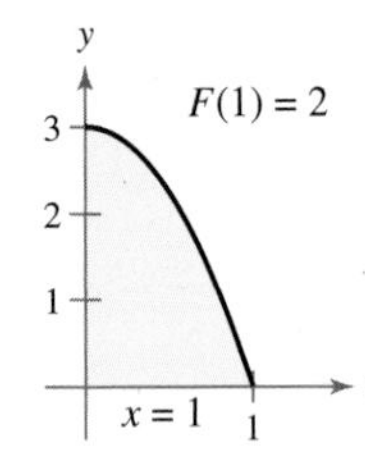

$F(x) = \int_0^x (3 - 3t^2)\,dt$ is the area under the curve $f(t) = 3 - 3t^2$ from 0 to x.

Figure 6.34

■

You can think of the function $F(x)$ as *accumulating* the area under the curve $f(t) = 3 - 3t^2$ from $t = 0$ to $t = x$. For $x = 0$, the area is 0 and $F(0) = 0$. For $x = 1$, $F(1) = 2$ gives the accumulated area under the curve on the entire interval $[0, 1]$. This interpretation of an integral as an **accumulation function** is used often in applications of integration.

In Example 6, note that the derivative of F is the original integrand (with only the variable changed). That is,

$$\frac{d}{dx}[F(x)] = \frac{d}{dx}[3x - x^3] = \frac{d}{dx}\left[\int_0^x (3 - 3t^2)\,dt\right] = 3 - 3x^2.$$

This result is generalized in the following theorem, called the **Second Fundamental Theorem of Calculus.**

THEOREM 6.11 THE SECOND FUNDAMENTAL THEOREM OF CALCULUS

If f is continuous on an open interval I containing a, then, for every x in the interval,

$$\frac{d}{dx}\left[\int_a^x f(t)\,dt\right] = f(x).$$

PROOF Begin by defining F as

$$F(x) = \int_a^x f(t)\,dt.$$

Then, by the definition of the derivative, you can write

$$\begin{aligned}
F'(x) &= \lim_{\Delta x \to 0} \frac{F(x + \Delta x) - F(x)}{\Delta x} \\
&= \lim_{\Delta x \to 0} \frac{1}{\Delta x}\left[\int_a^{x+\Delta x} f(t)\,dt - \int_a^x f(t)\,dt\right] \\
&= \lim_{\Delta x \to 0} \frac{1}{\Delta x}\left[\int_a^{x+\Delta x} f(t)\,dt + \int_x^a f(t)\,dt\right] \\
&= \lim_{\Delta x \to 0} \frac{1}{\Delta x}\left[\int_x^{x+\Delta x} f(t)\,dt\right].
\end{aligned}$$

From the Mean Value Theorem for Integrals (assuming $\Delta x > 0$), you know there exists a number c in the interval $[x, x + \Delta x]$ such that the integral in the expression above is equal to $f(c)\,\Delta x$. Moreover, because $x \le c \le x + \Delta x$, it follows that $c \to x$ as $\Delta x \to 0$. So, you obtain

$$\begin{aligned}
F'(x) &= \lim_{\Delta x \to 0} \left[\frac{1}{\Delta x} f(c)\,\Delta x\right] \\
&= \lim_{\Delta x \to 0} f(c) \\
&= f(x).
\end{aligned}$$

A similar argument can be made for $\Delta x < 0$. ■

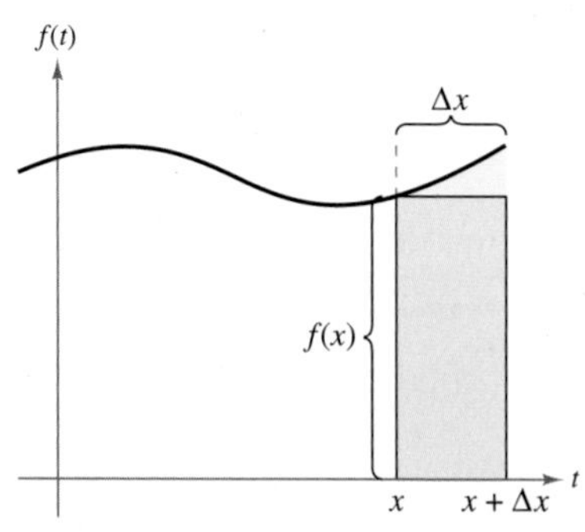

$$f(x)\,\Delta x \approx \int_x^{x+\Delta x} f(t)\,dt$$

Figure 6.35

NOTE Using the area model for definite integrals, you can view the approximation

$$f(x)\,\Delta x \approx \int_x^{x+\Delta x} f(t)\,dt$$

as saying that the area of the rectangle of height $f(x)$ and width Δx is approximately equal to the area of the region lying between the graph of f and the x-axis on the interval $[x, x + \Delta x]$, as shown in Figure 6.35. ■

Note that the Second Fundamental Theorem of Calculus tells you that if a function is continuous, you can be sure that it has an antiderivative. This antiderivative need not, however, be an *elementary* function.

EXAMPLE 7 Using the Second Fundamental Theorem of Calculus

Evaluate $\frac{d}{dx}\left[\int_0^x \sqrt{t^2+1}\,dt\right]$.

Solution Note that $f(t) = \sqrt{t^2+1}$ is continuous on the entire real line. So, using the Second Fundamental Theorem of Calculus, you can write

$$\frac{d}{dx}\left[\int_0^x \sqrt{t^2+1}\,dt\right] = \sqrt{x^2+1}.$$

■

> **CLASSIFICATION OF FUNCTIONS**
>
> By the end of the eighteenth century, mathematicians and scientists had concluded that most real-world phenomena could be represented by mathematical models taken from a collection of functions called **elementary functions.** Elementary functions fall into three categories.
>
> 1. Algebraic functions (polynomial, radical, rational)
> 2. Exponential and logarithmic functions
> 3. Trigonometric functions (sine, cosine, tangent, and so on)
>
> Exponential and logarithmic functions are introduced in Chapter 7 and trigonometric and inverse trigonometric functions are introduced in Chapter 9.

The differentiation shown in Example 7 is a straightforward application of the Second Fundamental Theorem of Calculus. The next example shows how this theorem can be combined with the Chain Rule to find the derivative of a function.

EXAMPLE 8 Using the Second Fundamental Theorem of Calculus

Find the derivative of $F(x) = \int_0^{x^3} t^2\,dt$.

Solution Using $u = x^3$, you can apply the Second Fundamental Theorem of Calculus with the Chain Rule as shown.

$$\begin{aligned}
F'(x) &= \frac{dF}{du}\frac{du}{dx} && \text{Chain Rule}\\
&= \frac{d}{du}[F(x)]\frac{du}{dx} && \text{Definition of } \frac{dF}{du}\\
&= \frac{d}{du}\left[\int_0^{x^3} t^2\,dt\right]\frac{du}{dx} && \text{Substitute } \int_0^{x^3} t^2\,dt \text{ for } F(x).\\
&= \frac{d}{du}\left[\int_0^{u} t^2\,dt\right]\frac{du}{dx} && \text{Substitute } u \text{ for } x^3.\\
&= (u^2)(3x^2) && \text{Apply Second Fundamental Theorem of Calculus.}\\
&= (x^3)^2(3x^2) && \text{Rewrite as function of } x.\\
&= 3x^8 && \text{Simplify.}
\end{aligned}$$

■

Because the integrand in Example 8 is easily integrated, you can verify the derivative as follows.

$$F(x) = \int_0^{x^3} t^2\,dt = \frac{t^3}{3}\bigg]_0^{x^3} = \frac{x^9}{3}$$

In this form, you can apply the Power Rule to verify that the derivative is the same as that obtained in Example 8.

$$F'(x) = 3x^8$$

Net Change Theorem

The Fundamental Theorem of Calculus (Theorem 6.9) states that if f is continuous on the closed interval $[a, b]$ and F is an antiderivative of f on $[a, b]$, then

$$\int_a^b f(x)\,dx = F(b) - F(a).$$

But because $F'(x) = f(x)$, this statement can be rewritten as

$$\int_a^b F'(x)\,dx = F(b) - F(a)$$

where the quantity $F(b) - F(a)$ represents the *net change of F* on the interval $[a, b]$.

THEOREM 6.12 THE NET CHANGE THEOREM

The definite integral of the rate of change of a quantity $F'(x)$ gives the total change, or **net change,** in that quantity on the interval $[a, b]$.

$$\int_a^b F'(x)\,dx = F(b) - F(a) \qquad \text{Net change of } F$$

EXAMPLE 9 Using the Net Change Theorem

A chemical flows into a storage tank at a rate of $180 + 3t$ liters per minute, where $0 \le t \le 60$. Find the amount of the chemical that flows into the tank during the first 20 minutes.

Solution Let $c(t)$ be the amount of the chemical in the tank at time t. Then $c'(t)$ represents the rate at which the chemical flows into the tank at time t. During the first 20 minutes, the amount that flows into the tank is

$$\begin{aligned}\int_0^{20} c'(t)\,dt &= \int_0^{20} (180 + 3t)\,dt \\ &= \left[180t + \frac{3}{2}t^2\right]_0^{20} \\ &= 3600 + 600 \\ &= 4200.\end{aligned}$$

So, the amount that flows into the tank during the first 20 minutes is 4200 liters. ■

Another way to illustrate the Net Change Theorem is to examine the velocity of a particle moving along a straight line where $s(t)$ is the position at time t. Then its velocity is $v(t) = s'(t)$ and

$$\int_a^b v(t)\,dt = s(b) - s(a).$$

This definite integral represents the net change in position, or **displacement,** of the particle.

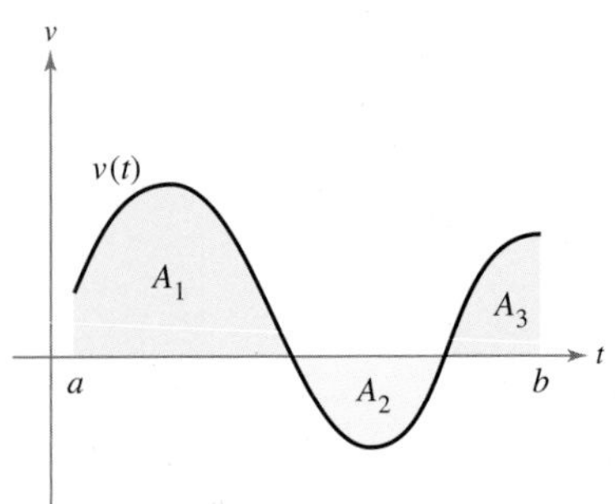

A_1, A_2, and A_3 are the areas of the shaded regions.
Figure 6.36

When calculating the *total* distance traveled by the particle, you must consider the intervals where $v(t) \le 0$ and the intervals where $v(t) \ge 0$. When $v(t) \le 0$, the particle moves to the left, and when $v(t) \ge 0$, the particle moves to the right. To calculate the total distance traveled, integrate the absolute value of velocity $|v(t)|$. So, the displacement of a particle and the total distance traveled by a particle over $[a, b]$ can be written as

$$\textbf{Displacement on } [a, b] = \int_a^b v(t)\,dt = A_1 - A_2 + A_3$$

$$\textbf{Total distance traveled on } [a, b] = \int_a^b |v(t)|\,dt = A_1 + A_2 + A_3$$

(see Figure 6.36).

EXAMPLE 10 Solving a Particle Motion Problem

A particle is moving along a line so that its velocity is $v(t) = t^3 - 10t^2 + 29t - 20$ feet per second at time t.

a. What is the displacement of the particle on the time interval $1 \le t \le 5$?

b. What is the total distance traveled by the particle on the time interval $1 \le t \le 5$?

Solution

a. By definition, you know that the displacement is

$$\begin{aligned}\int_1^5 v(t)\,dt &= \int_1^5 (t^3 - 10t^2 + 29t - 20)\,dt\\ &= \left[\frac{t^4}{4} - \frac{10}{3}t^3 + \frac{29}{2}t^2 - 20t\right]_1^5\\ &= \frac{25}{12} - \left(-\frac{103}{12}\right)\\ &= \frac{128}{12}\\ &= \frac{32}{3}.\end{aligned}$$

So, the particle moves $\frac{32}{3}$ feet to the right.

b. To find the total distance traveled, calculate $\int_1^5 |v(t)|\,dt$. Using Figure 6.37 and the fact that $v(t)$ can be factored as $(t - 1)(t - 4)(t - 5)$, you can determine that $v(t) \ge 0$ on $[1, 4]$ and $v(t) \le 0$ on $[4, 5]$. So, the total distance traveled is

$$\begin{aligned}\int_1^5 |v(t)|\,dt &= \int_1^4 v(t)\,dt - \int_4^5 v(t)\,dt\\ &= \int_1^4 (t^3 - 10t^2 + 29t - 20)\,dt - \int_4^5 (t^3 - 10t^2 + 29t - 20)\,dt\\ &= \left[\frac{t^4}{4} - \frac{10}{3}t^3 + \frac{29}{2}t^2 - 20t\right]_1^4 - \left[\frac{t^4}{4} - \frac{10}{3}t^3 + \frac{29}{2}t^2 - 20t\right]_4^5\\ &= \frac{45}{4} - \left(-\frac{7}{12}\right)\\ &= \frac{71}{6} \text{ feet.}\end{aligned}$$

■

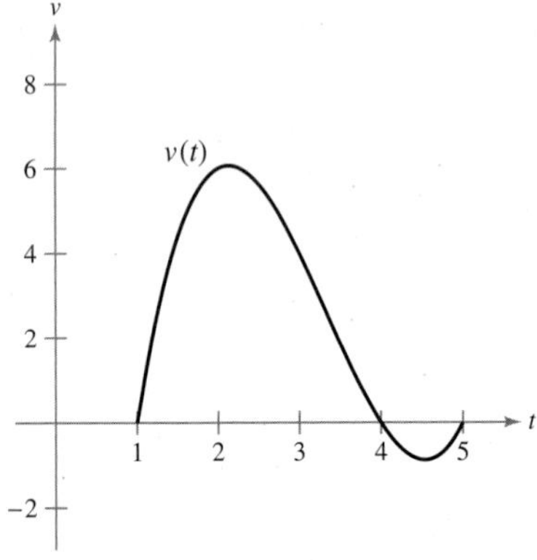

Figure 6.37

6.4 Exercises

See www.CalcChat.com for worked-out solutions to odd-numbered exercises.

Graphical Reasoning **In Exercises 1–4, use a graphing utility to graph the integrand. Use the graph to determine whether the definite integral is positive, negative, or zero.**

1. $\int_0^{\pi} \frac{4}{x^2+1}\,dx$

2. $\int_1^5 \sqrt[3]{x-3}\,dx$

3. $\int_{-2}^2 x\sqrt{x^2+1}\,dx$

4. $\int_{-2}^2 x\sqrt{2-x}\,dx$

In Exercises 5–26, evaluate the definite integral of the algebraic function. Use a graphing utility to verify your result.

5. $\int_0^2 6x\,dx$

6. $\int_4^9 5\,dv$

7. $\int_{-1}^0 (2x-1)\,dx$

8. $\int_2^5 (-3v+4)\,dv$

9. $\int_{-1}^1 (t^2-2)\,dt$

10. $\int_1^7 (6x^2+2x-3)\,dx$

11. $\int_0^1 (2t-1)^2\,dt$

12. $\int_{-1}^1 (t^3-9t)\,dt$

13. $\int_1^2 \left(\frac{3}{x^2}-1\right)dx$

14. $\int_{-2}^{-1} \left(u-\frac{1}{u^2}\right)du$

15. $\int_1^4 \frac{u-2}{\sqrt{u}}\,du$

16. $\int_{-3}^3 v^{1/3}\,dv$

17. $\int_{-1}^1 \left(\sqrt[3]{t}-2\right)dt$

18. $\int_1^8 \sqrt{\frac{2}{x}}\,dx$

19. $\int_0^1 \frac{x-\sqrt{x}}{3}\,dx$

20. $\int_0^2 (2-t)\sqrt{t}\,dt$

21. $\int_{-1}^0 \left(t^{1/3}-t^{2/3}\right)dt$

22. $\int_{-8}^{-1} \frac{x-x^2}{2\sqrt[3]{x}}\,dx$

23. $\int_0^5 |2x-5|\,dx$

24. $\int_1^4 (3-|x-3|)\,dx$

25. $\int_0^4 |x^2-9|\,dx$

26. $\int_0^4 |x^2-4x+3|\,dx$

In Exercises 27–32, determine the area of the given region.

27. $y = x - x^2$

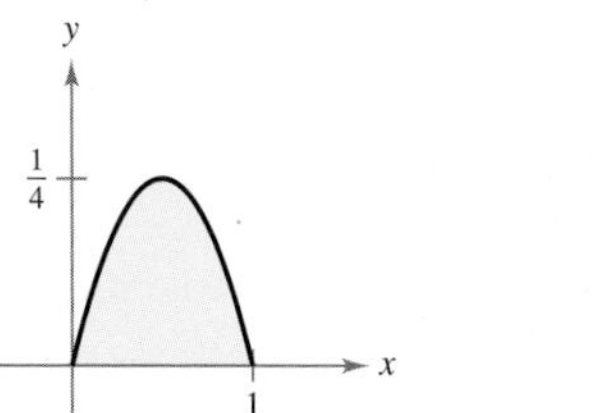

28. $y = -x^2 + 2x + 3$

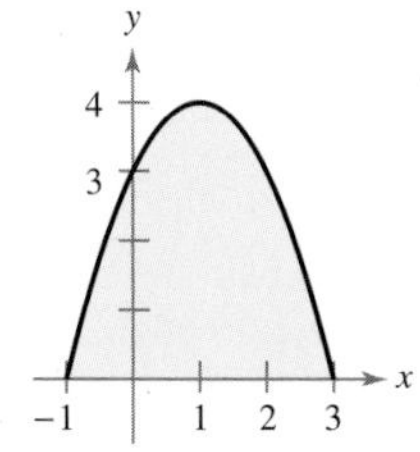

29. $y = 1 - x^4$

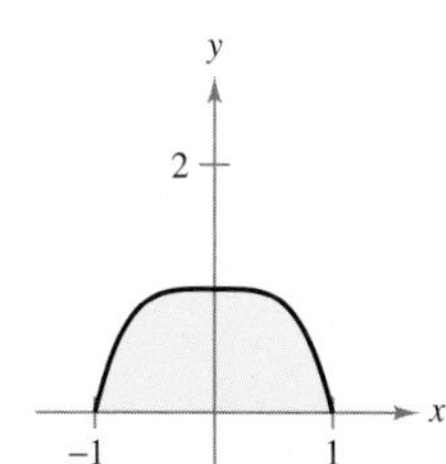

30. $y = \frac{1}{x^2}$

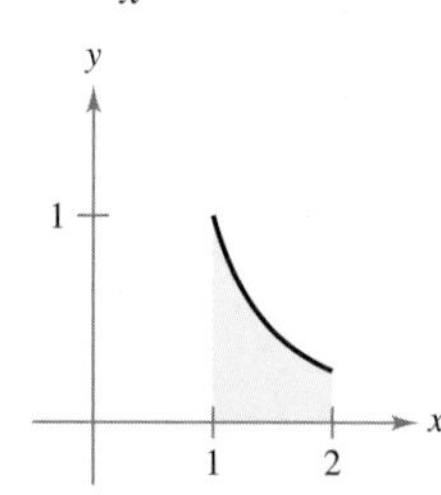

31. $y = \sqrt[3]{2x}$

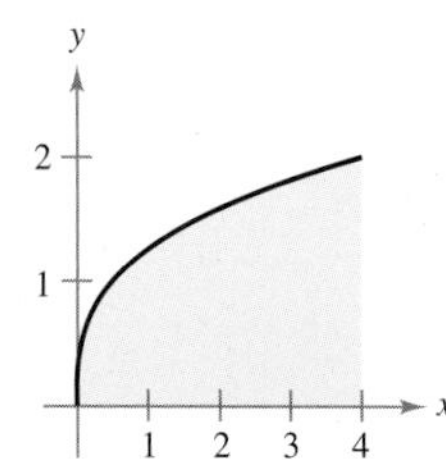

32. $y = (3-x)\sqrt{x}$

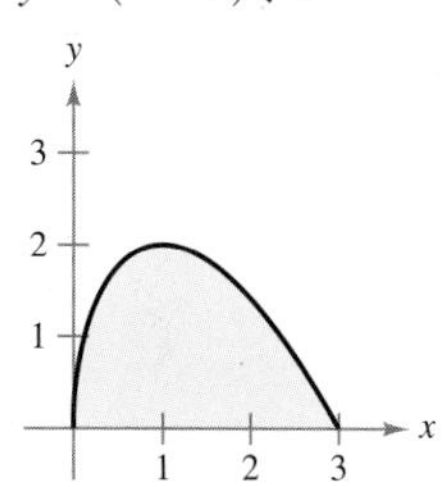

In Exercises 33–38, find the area of the region bounded by the graphs of the equations.

33. $y = 5x^2 + 2,\quad x = 0,\quad x = 2,\quad y = 0$

34. $y = x^3 + x,\quad x = 2,\quad y = 0$

35. $y = 1 + \sqrt[3]{x},\quad x = 0,\quad x = 8,\quad y = 0$

36. $y = (3-x)\sqrt{x},\quad y = 0$

37. $y = -x^2 + 4x,\quad y = 0$

38. $y = 1 - x^4,\quad y = 0$

In Exercises 39–42, find the value(s) of *c* guaranteed by the Mean Value Theorem for Integrals for the function over the given interval.

39. $f(x) = x^3,\quad [0, 3]$

40. $f(x) = \frac{9}{x^3},\quad [1, 3]$

41. $f(x) = -x^2 + 4x,\quad [0, 3]$

42. $f(x) = \sqrt{x},\quad [4, 9]$

In Exercises 43–48, find the average value of the function over the given interval and all values of *x* in the interval for which the function equals its average value.

43. $f(x) = 9 - x^2,\quad [-3, 3]$

44. $f(x) = \frac{4(x^2+1)}{x^2},\quad [1, 3]$

45. $f(x) = x^3,\quad [0, 1]$

46. $f(x) = 4x^3 - 3x^2,\quad [-1, 2]$

47. $f(x) = x - 2\sqrt{x},\quad [0, 4]$

48. $f(x) = \frac{1}{(x-3)^2},\quad [0, 2]$

49. *Velocity* The graph shows the velocity, in feet per second, of a car accelerating from rest. Use the graph to estimate the distance the car travels in 8 seconds.

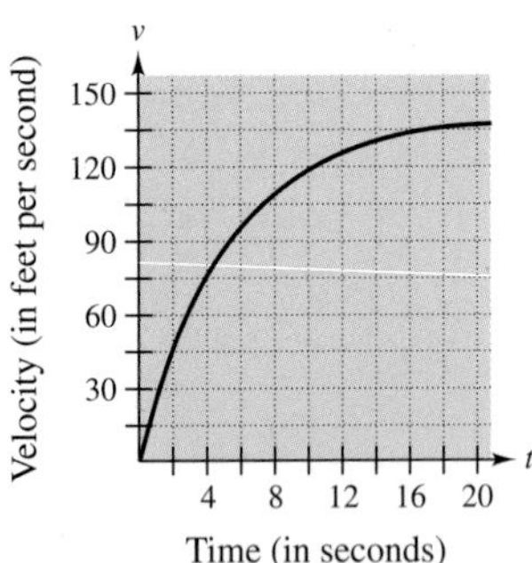

50. *Velocity* The graph shows the velocity, in feet per second, of a decelerating car after the driver applies the brakes. Use the graph to estimate how far the car travels before it comes to a stop.

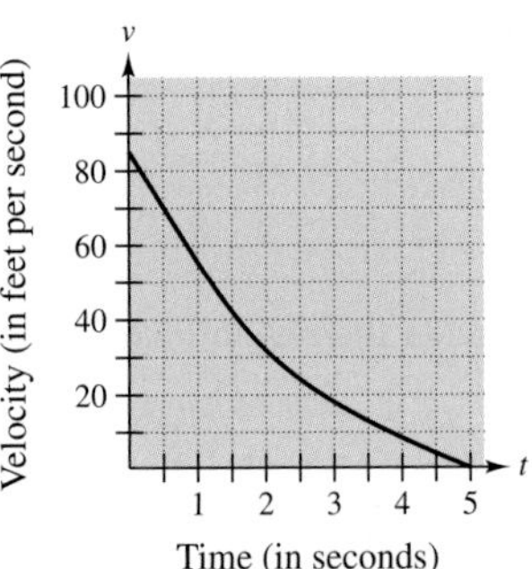

WRITING ABOUT CONCEPTS

51. State the Fundamental Theorem of Calculus.

52. The graph of f is shown in the figure.

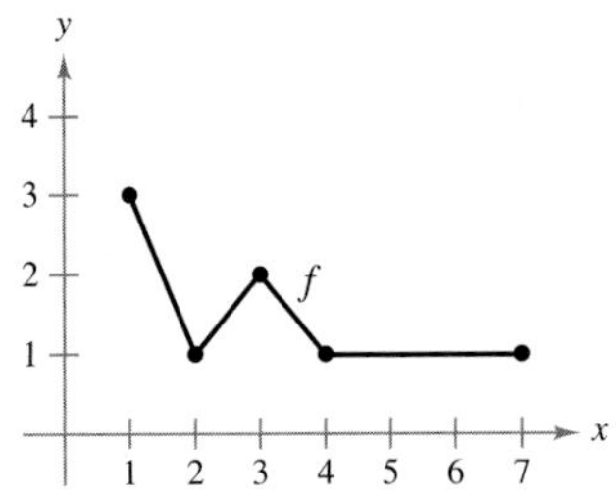

(a) Evaluate $\int_1^7 f(x)\,dx$.

(b) Determine the average value of f on the interval $[1, 7]$.

(c) Determine the answers to parts (a) and (b) if the graph is translated two units upward.

53. If $r'(t)$ represents the rate of growth of a dog in pounds per year, what does $r(t)$ represent? What does

$$\int_2^6 r'(t)\,dx$$

represent about the dog?

CAPSTONE

54. The graph of f is shown in the figure. The shaded region A has an area of 1.5, and $\int_0^6 f(x)\,dx = 3.5$. Use this information to fill in the blanks.

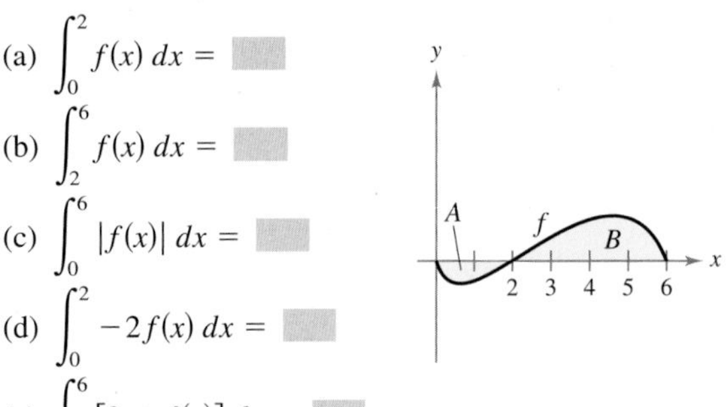

(a) $\int_0^2 f(x)\,dx =$ ▭

(b) $\int_2^6 f(x)\,dx =$ ▭

(c) $\int_0^6 |f(x)|\,dx =$ ▭

(d) $\int_0^2 -2f(x)\,dx =$ ▭

(e) $\int_0^6 [2 + f(x)]\,dx =$ ▭

(f) The average value of f over the interval $[0, 6]$ is ▭.

55. *Respiratory Cycle* The volume V, in liters, of air in the lungs during a five-second respiratory cycle is approximated by the model $V = 0.1729t + 0.1522t^2 - 0.0374t^3$, where t is the time in seconds. Approximate the average volume of air in the lungs during one cycle.

56. *Blood Flow* The velocity v of the flow of blood at a distance r from the central axis of an artery of radius R is $v = k(R^2 - r^2)$, where k is the constant of proportionality. Find the average rate of flow of blood along a radius of the artery. (Use 0 and R as the limits of integration.)

57. *Modeling Data* An experimental vehicle is tested on a straight track. It starts from rest, and its velocity v (in meters per second) is recorded every 10 seconds for 1 minute (see table).

t	0	10	20	30	40	50	60
v	0	5	21	40	62	78	83

(a) Use a graphing utility to find a model of the form $v = at^3 + bt^2 + ct + d$ for the data.

(b) Use a graphing utility to plot the data and graph the model.

(c) Use the Fundamental Theorem of Calculus to approximate the distance traveled by the vehicle during the test.

58. *Modeling Data* A department store manager wants to estimate the number of customers that enter the store from noon until closing at 9 P.M. The table shows the number of customers N entering the store during a randomly selected minute each hour from $t - 1$ to t, with $t = 0$ corresponding to noon.

t	1	2	3	4	5	6	7	8	9
N	6	7	9	12	15	14	11	7	2

(a) Draw a histogram of the data.

(b) Estimate the total number of customers entering the store between noon and 9 P.M.

(c) Use the *regression* feature of a graphing utility to find a model of the form

$$N(t) = at^3 + bt^2 + ct + d$$

for the data.

(d) Use a graphing utility to plot the data and graph the model.

(e) Use a graphing utility to evaluate $\int_0^9 N(t)\,dt$, and use the result to estimate the number of customers entering the store between noon and 9 P.M. Compare this with your answer in part (b).

(f) Estimate the average number of customers entering the store per minute between 3 P.M. and 7 P.M.

In Exercises 59–62, find F as a function of x and evaluate it at $x = 2$, $x = 5$, and $x = 8$.

59. $F(x) = \int_0^x (4t - 7)\,dt$

60. $F(x) = \int_2^x (t^3 + 2t - 2)\,dt$

61. $F(x) = \int_1^x \frac{20}{v^2}\,dv$

62. $F(x) = \int_2^x -\frac{2}{t^3}\,dt$

63. Let $g(x) = \int_0^x f(t)\,dt$, where f is the function whose graph is shown in the figure.

(a) Estimate $g(0)$, $g(2)$, $g(4)$, $g(6)$, and $g(8)$.

(b) Find the largest open interval on which g is increasing. Find the largest open interval on which g is decreasing.

(c) Identify any extrema of g.

(d) Sketch a rough graph of g.

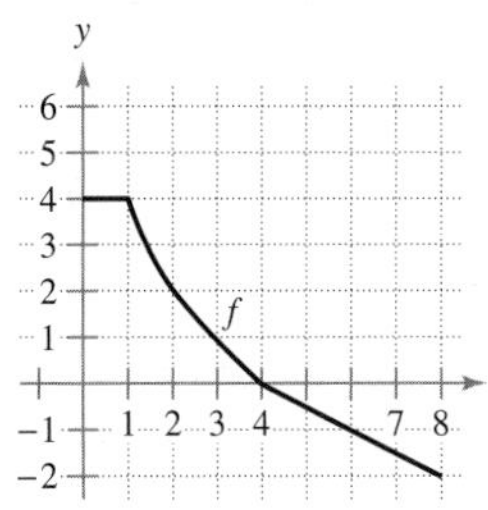

Figure for 63

Figure for 64

64. Let $g(x) = \int_0^x f(t)\,dt$, where f is the function whose graph is shown in the figure.

(a) Estimate $g(0)$, $g(2)$, $g(4)$, $g(6)$, and $g(8)$.

(b) Find the largest open interval on which g is increasing. Find the largest open interval on which g is decreasing.

(c) Identify any extrema of g.

(d) Sketch a rough graph of g.

In Exercises 65–70, (a) integrate to find F as a function of x and (b) demonstrate the Second Fundamental Theorem of Calculus by differentiating the result in part (a).

65. $F(x) = \int_0^x (t + 2)\,dt$

66. $F(x) = \int_0^x t(t^2 + 1)\,dt$

67. $F(x) = \int_8^x \sqrt[3]{t}\,dt$

68. $F(x) = \int_4^x \sqrt{t}\,dt$

69. $F(x) = \int_1^x \frac{1}{t^2}\,dt$

70. $F(x) = \int_0^x t^{3/2}\,dt$

In Exercises 71–74, use the Second Fundamental Theorem of Calculus to find $F'(x)$.

71. $F(x) = \int_{-2}^x (t^2 - 2t)\,dt$

72. $F(x) = \int_1^x \frac{t^2}{t^2 + 1}\,dt$

73. $F(x) = \int_{-1}^x \sqrt{t^4 + 1}\,dt$

74. $F(x) = \int_1^x \sqrt[4]{t}\,dt$

In Exercises 75–78, find $F'(x)$.

75. $F(x) = \int_x^{x+2} (4t + 1)\,dt$

76. $F(x) = \int_{-x}^x t^3\,dt$

77. $F(x) = \int_0^{3x} \sqrt{1 + t^3}\,dt$

78. $F(x) = \int_2^{x^2} \frac{1}{t^3}\,dt$

79. ***Graphical Analysis*** Sketch an approximate graph of g on the interval $0 \le x \le 4$, where $g(x) = \int_0^x f(t)\,dt$. Identify the x-coordinate of an extremum of g. To print an enlarged copy of the graph, go to the website *www.mathgraphs.com*.

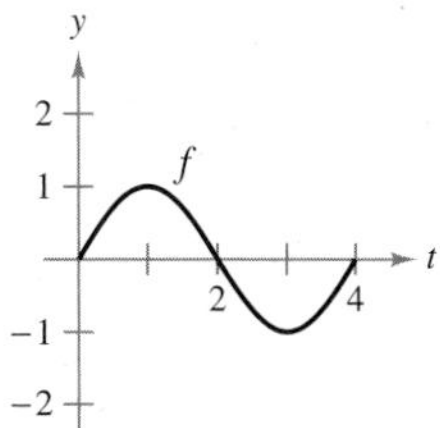

80. Use the graph of the function f shown in the figure and the function g defined by $g(x) = \int_0^x f(t)\,dt$.

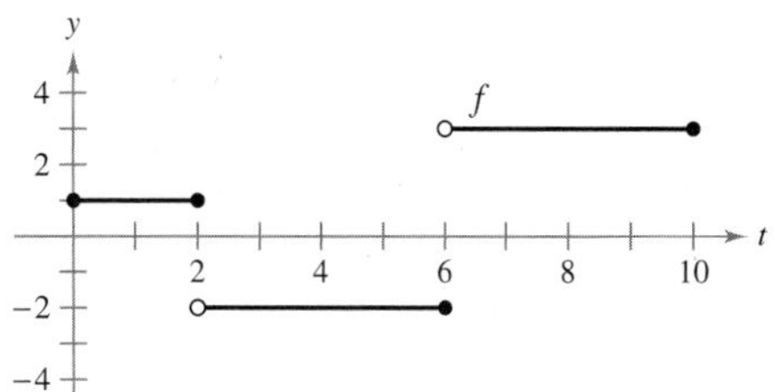

(a) Complete the table.

x	1	2	3	4	5	6	7	8	9	10
$g(x)$										

(b) Plot the points from the table in part (a) and graph g.

(c) Where does g have its minimum? Explain.

(d) Where does g have a maximum? Explain.

(e) On what interval does g increase at the greatest rate? Explain.

(f) Identify the zeros of g.

81. ***Cost*** The total cost C (in dollars) of purchasing and maintaining a piece of equipment for x years is

$$C(x) = 5000\left(25 + 3\int_0^x t^{1/4}\,dt\right).$$

(a) Perform the integration to write C as a function of x.

(b) Find $C(1)$, $C(5)$, and $C(10)$.

82. ***Area*** The area A between the graph of the function $g(t) = 4 - 4/t^2$ and the t-axis over the interval $[1, x]$ is

$$A(x) = \int_1^x \left(4 - \frac{4}{t^2}\right) dt.$$

(a) Find the horizontal asymptote of the graph of g.

(b) Integrate to find A as a function of x. Does the graph of A have a horizontal asymptote? Explain.

In Exercises 83–86, the velocity function, in feet per second, is given for a particle moving along a straight line. Find (a) the displacement and (b) the total distance that the particle travels over the given interval.

83. $v(t) = 5t - 7, \quad 0 \le t \le 3$

84. $v(t) = t^2 - t - 12, \quad 1 \le t \le 5$

85. $v(t) = t^3 - 10t^2 + 27t - 18, \quad 1 \le t \le 7$

86. $v(t) = t^3 - 8t^2 + 15t, \quad 0 \le t \le 5$

Rectilinear Motion **In Exercises 87–89, consider a particle moving along the x-axis where $x(t)$ is the position of the particle at time t, $x'(t)$ is its velocity, and $\int_a^b |x'(t)|\,dt$ is the distance the particle travels in the interval of time.**

87. The position function is given by

$x(t) = t^3 - 6t^2 + 9t - 2, 0 \le t \le 5.$

Find the total distance the particle travels in 5 units of time.

88. Repeat Exercise 87 for the position function given by

$x(t) = (t - 1)(t - 3)^2, \quad 0 \le t \le 5.$

89. A particle moves along the x-axis with velocity

$v(t) = 1/\sqrt{t}, \quad t > 0.$

At time $t = 1$, its position is $x = 4$. Find the total distance traveled by the particle on the interval $1 \le t \le 4$.

90. ***Water Flow*** Water flows from a storage tank at a rate of $500 - 5t$ liters per minute. Find the amount of water that flows out of the tank during the first 18 minutes.

In Exercises 91 and 92, describe why the statement is incorrect.

91. $\displaystyle\int_{-1}^{1} x^{-2}\,dx = \left[-x^{-1}\right]_{-1}^{1} = (-1) - 1 = -2$

92. $\displaystyle\int_{-2}^{1} \frac{2}{x^3}\,dx = \left[-\frac{1}{x^2}\right]_{-2}^{1} = -\frac{3}{4}$

True or False? **In Exercises 93 and 94, determine whether the statement is true or false. If it is false, explain why or give an example that shows it is false.**

93. If $F'(x) = G'(x)$ on the interval $[a, b]$, then $F(b) - F(a) = G(b) - G(a)$.

94. If f is continuous on $[a, b]$, then f is integrable on $[a, b]$.

95. Show that the function

$$f(x) = \int_0^{1/x} \frac{1}{t^2 + 1}\,dt + \int_0^x \frac{1}{t^2 + 1}\,dt$$

is constant for $x > 0$.

96. Prove that $\displaystyle\frac{d}{dx}\left[\int_{u(x)}^{v(x)} f(t)\,dt\right] = f(v(x))v'(x) - f(u(x))u'(x).$

97. Let

$$G(x) = \int_0^x \left[s\int_0^s f(t)\,dt\right] ds$$

where f is continuous for all real t. Find (a) $G(0)$, (b) $G'(0)$, (c) $G''(x)$, and (d) $G''(0)$.

SECTION PROJECT

Demonstrating the Fundamental Theorem

Use a graphing utility to graph the function

$$y_1 = \frac{t}{\sqrt{1 + t}}$$

on the interval $2 \le t \le 5$. Let $F(x)$ be the following function of x.

$$F(x) = \int_2^x \frac{t}{\sqrt{1 + t}}\,dt$$

(a) Complete the table. Explain why the values of F are increasing.

x	2	2.5	3	3.5	4	4.5	5
$F(x)$							

(b) Use the integration capabilities of a graphing utility to graph F.

(c) Use the differentiation capabilities of a graphing utility to graph $F'(x)$. How is this graph related to the graph in part (b)?

(d) Verify that the derivative of

$$y = \frac{2}{3}(t - 2)\sqrt{1 + t}$$

is $t/\sqrt{1 + t}$. Graph y and write a short paragraph about how this graph is related to those in parts (b) and (c).

6.5 Integration by Substitution

- **Use pattern recognition to find an indefinite integral.**
- **Use a change of variables to find an indefinite integral.**
- **Use the General Power Rule for Integration to find an indefinite integral.**
- **Use a change of variables to evaluate a definite integral.**
- **Evaluate a definite integral involving an even or odd function.**

Pattern Recognition

In this section you will study techniques for integrating composite functions. The discussion is split into two parts—*pattern recognition* and *change of variables.* Both techniques involve a ***u*-substitution.** With pattern recognition you perform the substitution mentally, and with change of variables you write the substitution steps.

The role of substitution in integration is comparable to the role of the Chain Rule in differentiation. Recall that for differentiable functions given by $y = F(u)$ and $u = g(x)$, the Chain Rule states that

$$\frac{d}{dx}[F(g(x))] = F'(g(x))g'(x).$$

From the definition of an antiderivative, it follows that

$$\int F'(g(x))g'(x)\,dx = F(g(x)) + C.$$

These results are summarized in the following theorem.

NOTE The statement of Theorem 6.13 doesn't tell how to distinguish between $f(g(x))$ and $g'(x)$ in the integrand. As you become more experienced at integration, your skill in doing this will increase. Of course, part of the key is familiarity with derivatives.

THEOREM 6.13 ANTIDIFFERENTIATION OF A COMPOSITE FUNCTION

Let g be a function whose range is an interval I, and let f be a function that is continuous on I. If g is differentiable on its domain and F is an antiderivative of f on I, then

$$\int f(g(x))g'(x)\,dx = F(g(x)) + C.$$

Letting $u = g(x)$ gives $du = g'(x)\,dx$ and

$$\int f(u)\,du = F(u) + C.$$

Example 1 shows how to apply Theorem 6.13 *directly*, by recognizing the presence of $f(g(x))$ and $g'(x)$. Note that the composite function in the integrand has an *outside function* f and an *inside function* g. Moreover, the derivative $g'(x)$ is present as a factor of the integrand.

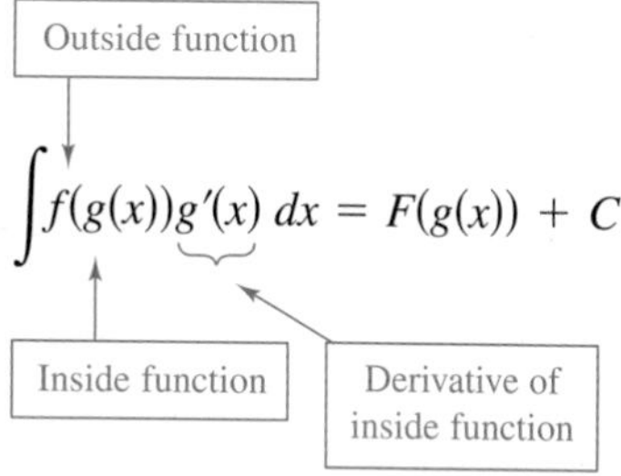

EXAMPLE 1 Recognizing the $f(g(x))g'(x)$ Pattern

Find the integral.

a. $\int (x^2 + 1)^2(2x)\,dx$ **b.** $\int 5\sqrt{5x + 1}\,dx$

Solution

a. Letting $g(x) = x^2 + 1$, you obtain

$$g'(x) = 2x$$

and

$$f(g(x)) = f(x^2 + 1) = (x^2 + 1)^2.$$

From this, you can recognize that the integrand follows the $f(g(x))g'(x)$ pattern. Using the Power Rule for Integration and Theorem 6.13, you can write

$$\int \overbrace{(x^2 + 1)^2}^{f(g(x))}\overbrace{(2x)}^{g'(x)}\,dx = \frac{1}{3}(x^2 + 1)^3 + C.$$

Try using the Chain Rule to check that the derivative of $\frac{1}{3}(x^2 + 1)^3 + C$ is the integrand of the original integral.

b. Letting $g(x) = 5x + 1$, you obtain

$$g'(x) = 5$$

and

$$f(g(x)) = f(5x + 1) = (5x + 1)^{1/2}.$$

From this, you can recognize that the integrand follows the $f(g(x))g'(x)$ pattern. Using the Power Rule for Integration and Theorem 6.13, you can write

$$\int \overbrace{(5x + 1)^{1/2}}^{f(g(x))}\overbrace{(5)}^{g'(x)}\,dx = \frac{2}{3}(5x + 1)^{3/2} + C.$$

You can check this by differentiating $\frac{2}{3}(5x + 1)^{3/2} + C$ to obtain the original integrand. ■

TECHNOLOGY Try using a computer algebra system, such as *Maple*, *Mathematica*, or the *TI-89*, to solve the integrals given in Example 1. Do you obtain the same antiderivatives that are listed in the examples?

STUDY TIP There are several techniques for applying substitution, each differing slightly from the others. However, you should remember that the goal is the same with every technique—*you are trying to find an antiderivative of the integrand.*

EXPLORATION

Recognizing Patterns The integrand in each of the following integrals fits the pattern $f(g(x))g'(x)$. Identify the pattern and use the result to evaluate the integral.

a. $\int 2x(x^2 + 1)^4\,dx$ **b.** $\int 3x^2\sqrt{x^3 + 1}\,dx$

The next two integrals are similar to the first two. Show how you can multiply and divide by a constant to evaluate these integrals.

c. $\int x(x^2 + 1)^4\,dx$ **d.** $\int x^2\sqrt{x^3 + 1}\,dx$

The integrands in Example 1 fit the $f(g(x))g'(x)$ pattern exactly—you only had to recognize the pattern. You can extend this technique considerably with the Constant Multiple Rule

$$\int kf(x)\,dx = k\int f(x)\,dx.$$

Many integrands contain the essential part (the variable part) of $g'(x)$ but are missing a constant multiple. In such cases, you can multiply and divide by the necessary constant multiple, as shown in Example 2.

EXAMPLE 2 Multiplying and Dividing by a Constant

Find

$$\int x(x^2+1)^2\,dx.$$

Solution This is similar to the integral given in Example 1(a), except that the integrand is missing a factor of 2. Recognizing that $2x$ is the derivative of x^2+1, you can let $g(x) = x^2 + 1$ and supply the $2x$ as follows.

$$\begin{aligned}
\int x(x^2+1)^2\,dx &= \int (x^2+1)^2\left(\frac{1}{2}\right)(2x)\,dx && \text{Multiply and divide by 2.}\\
&= \frac{1}{2}\int \overbrace{(x^2+1)^2}^{f(g(x))}\,\overbrace{(2x)}^{g'(x)}\,dx && \text{Constant Multiple Rule}\\
&= \frac{1}{2}\left[\frac{(x^2+1)^3}{3}\right] + C && \text{Integrate.}\\
&= \frac{1}{6}(x^2+1)^3 + C && \text{Simplify.}
\end{aligned}$$

■

In practice, most people would not write as many steps as are shown in Example 2. For instance, you could evaluate the integral by simply writing

$$\begin{aligned}
\int x(x^2+1)^2\,dx &= \frac{1}{2}\int (x^2+1)^2\,2x\,dx\\
&= \frac{1}{2}\left[\frac{(x^2+1)^3}{3}\right] + C\\
&= \frac{1}{6}(x^2+1)^3 + C.
\end{aligned}$$

NOTE Be sure you see that the *Constant* Multiple Rule applies only to *constants*. You cannot multiply and divide by a variable and then move the variable outside the integral sign. For instance,

$$\int (x^2+1)^2\,dx \neq \frac{1}{2x}\int (x^2+1)^2(2x)\,dx.$$

After all, if it were legitimate to move variable quantities outside the integral sign, you could move the entire integrand out and simplify the whole process. But the result would be incorrect.

■

Change of Variables

With a formal **change of variables,** you completely rewrite the integral in terms of u and du (or any other convenient variable). Although this procedure can involve more written steps than the pattern recognition illustrated in Examples 1 and 2, it is useful for complicated integrands. The change of variables technique uses the Leibniz notation for the differential. That is, if $u = g(x)$, then $du = g'(x)\,dx$, and the integral in Theorem 6.13 takes the form

$$\int f(g(x))g'(x)\,dx = \int f(u)\,du = F(u) + C.$$

EXAMPLE 3 Change of Variables

Find $\int \sqrt{2x-1}\,dx$.

Solution First, let u be the inner function, $u = 2x - 1$. Then calculate the differential du to be $du = 2\,dx$. Now, using $\sqrt{2x-1} = \sqrt{u}$ and $dx = du/2$, substitute to obtain

$$\int \sqrt{2x-1}\,dx = \int \sqrt{u}\left(\frac{du}{2}\right) \qquad \text{Integral in terms of } u$$

$$= \frac{1}{2}\int u^{1/2}\,du \qquad \text{Constant Multiple Rule}$$

$$= \frac{1}{2}\left(\frac{u^{3/2}}{3/2}\right) + C \qquad \text{Antiderivative in terms of } u$$

$$= \frac{1}{3}u^{3/2} + C \qquad \text{Simplify.}$$

$$= \frac{1}{3}(2x-1)^{3/2} + C. \qquad \text{Antiderivative in terms of } x$$

STUDY TIP Because integration is usually more difficult than differentiation, you should always check your answer to an integration problem by differentiating. For instance, in Example 3 you should differentiate $\frac{1}{3}(2x-1)^{3/2} + C$ to verify that you obtain the original integrand.

EXAMPLE 4 Change of Variables

Find $\int x\sqrt{2x-1}\,dx$.

Solution As in the previous example, let $u = 2x - 1$ and obtain $dx = du/2$. Because the integrand contains a factor of x, you must also solve for x in terms of u, as shown.

$$u = 2x - 1 \;\Longrightarrow\; x = (u+1)/2 \qquad \text{Solve for } x \text{ in terms of } u.$$

Now, using substitution, you obtain

$$\int x\sqrt{2x-1}\,dx = \int \left(\frac{u+1}{2}\right)u^{1/2}\left(\frac{du}{2}\right)$$

$$= \frac{1}{4}\int (u^{3/2} + u^{1/2})\,du$$

$$= \frac{1}{4}\left(\frac{u^{5/2}}{5/2} + \frac{u^{3/2}}{3/2}\right) + C$$

$$= \frac{1}{10}(2x-1)^{5/2} + \frac{1}{6}(2x-1)^{3/2} + C.$$

■

To complete the change of variables in Example 4, you solved for x in terms of u. Sometimes this is very difficult. Fortunately it is not always necessary, as shown in the next example.

EXAMPLE 5 Change of Variables

Find $\int x\sqrt{x^2 - 1}\, dx$.

Solution Because $\sqrt{x^2 - 1} = (x^2 - 1)^{1/2}$, let $u = x^2 - 1$. Then $du = (2x)\, dx$. Now, because $x\, dx$ is part of the original integral, you can write

$$\frac{du}{2} = x\, dx.$$

STUDY TIP When making a change of variables, be sure that your answer is written using the same variables as in the original integrand. For instance, in Example 5, you should not leave your answer as

$$\tfrac{1}{3}u^{3/2} + C$$

but rather, replace u by $x^2 - 1$.

Substituting u and $du/2$ in the original integral yields

$$\begin{aligned}\int x\sqrt{x^2 - 1}\, dx &= \int u^{1/2}\frac{du}{2}\\ &= \frac{1}{2}\int u^{1/2}\, du\\ &= \frac{1}{2}\left(\frac{u^{3/2}}{3/2}\right) + C\\ &= \frac{1}{3}u^{3/2} + C.\end{aligned}$$

Back-substitution of $u = x^2 - 1$ yields

$$\int x\sqrt{x^2 - 1}\, dx = \frac{1}{3}(x^2 - 1)^{3/2} + C.$$

You can check this by differentiating.

$$\begin{aligned}\frac{d}{dx}\left[\frac{1}{3}(x^2 - 1)^{3/2}\right] &= \left(\frac{1}{3}\right)\left(\frac{3}{2}\right)(x^2 - 1)^{1/2}(2x)\\ &= x\sqrt{x^2 - 1}\end{aligned}$$

Because differentiation produces the original integrand, you know that you have obtained the correct antiderivative. ■

The steps used for integration by substitution are summarized in the following guidelines.

GUIDELINES FOR MAKING A CHANGE OF VARIABLES

1. Choose a substitution $u = g(x)$. Usually, it is best to choose the *inner* part of a composite function, such as a quantity raised to a power.
2. Compute $du = g'(x)\, dx$.
3. Rewrite the integral in terms of the variable u.
4. Find the resulting integral in terms of u.
5. Replace u by $g(x)$ to obtain an antiderivative in terms of x.
6. Check your answer by differentiating.

The General Power Rule for Integration

One of the most common u-substitutions involves quantities in the integrand that are raised to a power. Because of the importance of this type of substitution, it is given a special name—the **General Power Rule for Integration.** A proof of this rule follows directly from the (simple) Power Rule for Integration, together with Theorem 6.13.

THEOREM 6.14 THE GENERAL POWER RULE FOR INTEGRATION

If g is a differentiable function of x, then

$$\int [g(x)]^n g'(x)\,dx = \frac{[g(x)]^{n+1}}{n+1} + C, \quad n \neq -1.$$

Equivalently, if $u = g(x)$, then

$$\int u^n\,du = \frac{u^{n+1}}{n+1} + C, \quad n \neq -1.$$

EXAMPLE 6 Substitution and the General Power Rule

a. $$\int 3(3x-1)^4\,dx = \int \overbrace{(3x-1)^4}^{u^4}\overbrace{(3)\,dx}^{du} = \overbrace{\frac{(3x-1)^5}{5}}^{u^5/5} + C$$

b. $$\int (2x+1)(x^2+x)\,dx = \int \overbrace{(x^2+x)^1}^{u^1}\overbrace{(2x+1)\,dx}^{du} = \overbrace{\frac{(x^2+x)^2}{2}}^{u^2/2} + C$$

c. $$\int x^2\sqrt{x^3-2}\,dx = \frac{1}{3}\int \overbrace{(x^3-2)^{1/2}}^{u^{1/2}}\overbrace{(3x^2)\,dx}^{du} = \frac{1}{3}\overbrace{\left(\frac{(x^3-2)^{3/2}}{3/2}\right)}^{u^{3/2}/(3/2)} + C$$

$$= \frac{2}{9}(x^3-2)^{3/2} + C$$

d. $$\int \frac{x}{(1-2x^2)^2}\,dx = -\frac{1}{4}\int \overbrace{(1-2x^2)^{-2}}^{u^{-2}}\overbrace{(-4x)\,dx}^{du} = -\frac{1}{4}\overbrace{\left(\frac{(1-2x^2)^{-1}}{-1}\right)}^{u^{-1}/(-1)} + C$$

$$= \frac{1}{4(1-2x^2)} + C$$

■

EXPLORATION

Suppose you were asked to find one of the following integrals. Which one would you choose? Explain your reasoning.

$$\int \sqrt{x^3+1}\,dx \quad \text{or}$$

$$\int x^2\sqrt{x^3+1}\,dx$$

Some integrals whose integrands involve quantities raised to powers cannot be found by the General Power Rule. Consider the two integrals

$$\int x(x^2+1)^2\,dx \quad \text{and} \quad \int (x^2+1)^2\,dx.$$

The substitution $u = x^2 + 1$ works in the first integral but not in the second. In the second, the substitution fails because the integrand lacks the factor x needed for du. Fortunately, *for this particular integral*, you can expand the integrand as $(x^2+1)^2 = x^4 + 2x^2 + 1$ and use the (simple) Power Rule to integrate each term.

Change of Variables for Definite Integrals

When using u-substitution with a definite integral, it is often convenient to determine the limits of integration for the variable u rather than to convert the antiderivative back to the variable x and evaluate at the original limits. This change of variables is stated explicitly in the next theorem. The proof follows from Theorem 6.13 combined with the Fundamental Theorem of Calculus.

THEOREM 6.15 CHANGE OF VARIABLES FOR DEFINITE INTEGRALS

If the function $u = g(x)$ has a continuous derivative on the closed interval $[a, b]$ and f is continuous on the range of g, then

$$\int_a^b f(g(x))g'(x)\,dx = \int_{g(a)}^{g(b)} f(u)\,du.$$

EXAMPLE 7 Change of Variables

Evaluate $\displaystyle\int_0^1 x(x^2 + 1)^3\,dx$.

Solution To evaluate this integral, let $u = x^2 + 1$. Then, you obtain

$$u = x^2 + 1 \implies du = 2x\,dx.$$

Before substituting, determine the new upper and lower limits of integration.

Lower Limit	*Upper Limit*
When $x = 0$, $u = 0^2 + 1 = 1$.	When $x = 1$, $u = 1^2 + 1 = 2$.

Now, you can substitute to obtain

$$\int_0^1 x(x^2 + 1)^3\,dx = \frac{1}{2}\int_0^1 (x^2 + 1)^3(2x)\,dx \qquad \text{Integration limits for } x$$

$$= \frac{1}{2}\int_1^2 u^3\,du \qquad \text{Integration limits for } u$$

$$= \frac{1}{2}\left[\frac{u^4}{4}\right]_1^2$$

$$= \frac{1}{2}\left(4 - \frac{1}{4}\right)$$

$$= \frac{15}{8}.$$

STUDY TIP If you are able to use pattern recognition to find the antiderivative, then you do not need to change the limits of integration. The steps for Example 7 would be

$$\int_0^1 x(x^2 + 1)^3\,dx$$

$$= \frac{1}{2}\int_0^1 (x^2 + 1)^3\,(2x)\,dx$$

$$= \frac{1}{2}\left[\frac{(x^2 + 1)^4}{4}\right]_0^1$$

$$= \frac{1}{2}\left[4 - \frac{1}{4}\right]$$

$$= \frac{15}{8}.$$

Try rewriting the antiderivative $\frac{1}{2}(u^4/4)$ in terms of the variable x and evaluate the definite integral at the original limits of integration, as shown.

$$\frac{1}{2}\left[\frac{u^4}{4}\right]_1^2 = \frac{1}{2}\left[\frac{(x^2 + 1)^4}{4}\right]_0^1$$

$$= \frac{1}{2}\left(4 - \frac{1}{4}\right) = \frac{15}{8}$$

Notice that you obtain the same result. ■

EXAMPLE 8 Change of Variables

Evaluate $A = \int_1^5 \frac{x}{\sqrt{2x-1}}\,dx.$

Solution To evaluate this integral, let $u = \sqrt{2x-1}$. Then, you obtain

$$u^2 = 2x - 1$$

$$u^2 + 1 = 2x$$

$$\frac{u^2+1}{2} = x$$

$$u\,du = dx. \qquad \text{Differentiate each side.}$$

Before substituting, determine the new upper and lower limits of integration.

Lower Limit	*Upper Limit*
When $x = 1$, $u = \sqrt{2-1} = 1$.	When $x = 5$, $u = \sqrt{10-1} = 3$.

Now, substitute to obtain

$$\int_1^5 \frac{x}{\sqrt{2x-1}}\,dx = \int_1^3 \frac{1}{u}\left(\frac{u^2+1}{2}\right)u\,du \qquad \text{Rewrite integral in terms of } u.$$

$$= \frac{1}{2}\int_1^3 (u^2+1)\,du \qquad \text{Simplify.}$$

$$= \frac{1}{2}\left[\frac{u^3}{3} + u\right]_1^3 \qquad \text{Integrate.}$$

$$= \frac{1}{2}\left(9 + 3 - \frac{1}{3} - 1\right) \qquad \text{Evaluate.}$$

$$= \frac{16}{3}. \qquad \text{Simplify.}$$

■

The region before substitution has an area of $\frac{16}{3}$.
Figure 6.38

Geometrically, you can interpret the equation

$$\int_1^5 \frac{x}{\sqrt{2x-1}}\,dx = \int_1^3 \frac{u^2+1}{2}\,du$$

to mean that the two *different* regions shown in Figures 6.38 and 6.39 have the *same* area.

When evaluating definite integrals by substitution, it is possible for the upper limit of integration of the u-variable form to be smaller than the lower limit. If this happens, don't rearrange the limits. Simply evaluate as usual. For example, after substituting $u = \sqrt{1-x}$ in the integral

$$\int_0^1 x^2(1-x)^{1/2}\,dx$$

you obtain $u = \sqrt{1-1} = 0$ when $x = 1$, and $u = \sqrt{1-0} = 1$ when $x = 0$. So, the correct u-variable form of this integral is

$$-2\int_1^0 (1-u^2)^2u^2\,du.$$

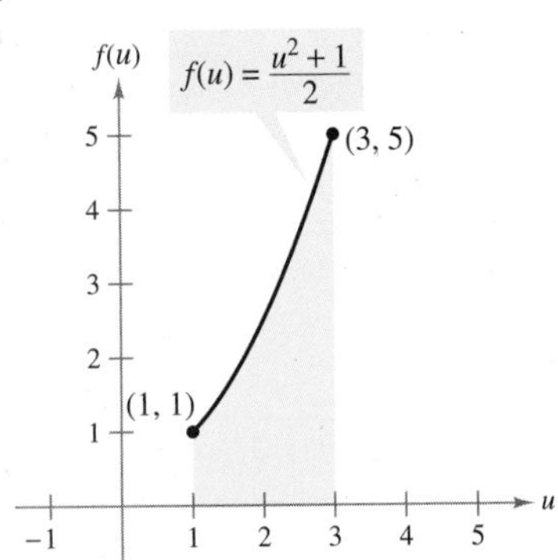

The region after substitution has an area of $\frac{16}{3}$.
Figure 6.39

NOTE In Example 8, you could also let $u = 2x - 1$. The substitution $u = \sqrt{2x-1}$ simply eliminates fractional exponents in the variable u. Let $u = 2x - 1$ in Example 8 to see that you get the same result. ■

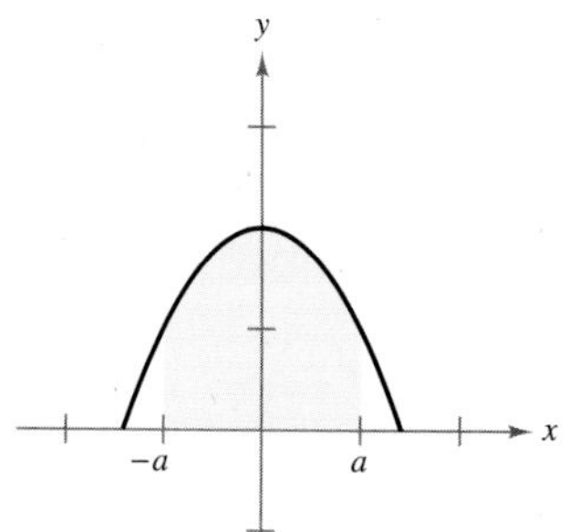

Even function

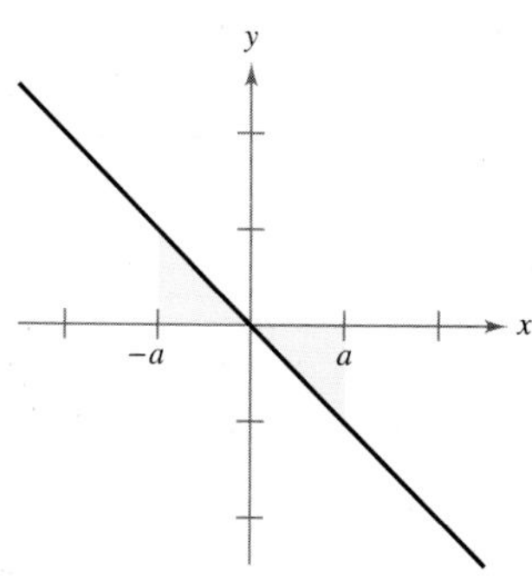

Odd function
Figure 6.40

Integration of Even and Odd Functions

Even with a change of variables, integration can be difficult. Occasionally, you can simplify the evaluation of a definite integral over an interval that is symmetric about the y-axis or about the origin by recognizing the integrand to be an even or odd function (see Figure 6.40).

> **THEOREM 6.16 INTEGRATION OF EVEN AND ODD FUNCTIONS**
>
> Let f be integrable on the closed interval $[-a, a]$.
>
> **1.** If f is an *even* function, then $\int_{-a}^{a} f(x)\,dx = 2\int_{0}^{a} f(x)\,dx.$
>
> **2.** If f is an *odd* function, then $\int_{-a}^{a} f(x)\,dx = 0.$

PROOF Because f is even, you know that $f(x) = f(-x)$. Using Theorem 6.13 with the substitution $u = -x$ produces

$$\int_{-a}^{0} f(x)\,dx = \int_{a}^{0} f(-u)(-du) = -\int_{a}^{0} f(u)\,du = \int_{0}^{a} f(u)\,du = \int_{0}^{a} f(x)\,dx.$$

Finally, using Theorem 6.6, you obtain

$$\begin{aligned}\int_{-a}^{a} f(x)\,dx &= \int_{-a}^{0} f(x)\,dx + \int_{0}^{a} f(x)\,dx\\ &= \int_{0}^{a} f(x)\,dx + \int_{0}^{a} f(x)\,dx = 2\int_{0}^{a} f(x)\,dx.\end{aligned}$$

This proves the first property. The proof of the second property is left to you (see Exercise 90). ■

EXAMPLE 9 Integration of an Odd Function

Evaluate $\int_{-2}^{2} (x^5 - 4x^3 + 6x)\,dx.$

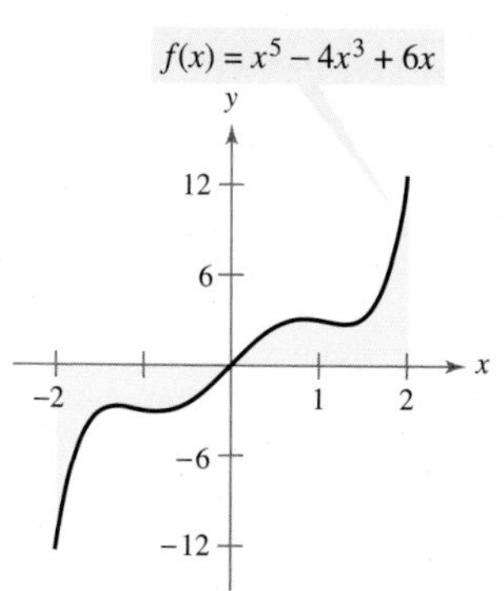

Because f is an odd function,
$\int_{-2}^{2} f(x)\,dx = 0.$
Figure 6.41

Solution Letting $f(x) = x^5 - 4x^3 + 6x$ produces

$$\begin{aligned}f(-x) &= (-x)^5 - 4(-x)^3 + 6(-x)\\ &= -x^5 + 4x^3 - 6x\\ &= -f(x).\end{aligned}$$

So, f is an odd function, and because f is symmetric about the origin over $[-2, 2]$, you can apply Theorem 6.16 to conclude that

$$\int_{-2}^{2} (x^5 - 4x^3 + 6x)\,dx = 0.$$ ■

NOTE From Figure 6.41 you can see that the two regions on either side of the y-axis have the same area. However, because one lies below the x-axis and one lies above it, integration produces a cancellation effect. ■

6.5 Exercises

See www.CalcChat.com for worked-out solutions to odd-numbered exercises.

In Exercises 1–4, complete the table by identifying u and du for the integral.

$\int f(g(x))g'(x)\,dx$	$u = g(x)$	$du = g'(x)\,dx$
1. $\int (8x^2 + 1)^2(16x)\,dx$		
2. $\int (x^3 + 3)3x^2\,dx$		
3. $\int x^2\sqrt{x^3 + 1}\,dx$		
4. $\int \frac{x}{\sqrt{x^2 + 1}}\,dx$		

In Exercises 5 and 6, determine whether it is necessary to use substitution to evaluate the integral. (Do not evaluate the integral.)

5. $\int \sqrt{x}(6 - x)\,dx$

6. $\int x\sqrt{x + 4}\,dx$

In Exercises 7–34, find the indefinite integral and check the result by differentiation.

7. $\int (1 + 6x)^4(6)\,dx$

8. $\int (x^2 - 9)^3(2x)\,dx$

9. $\int \sqrt{25 - x^2}\,(-2x)\,dx$

10. $\int \sqrt[3]{3 - 4x^2}(-8x)\,dx$

11. $\int x^3(x^4 + 3)^2\,dx$

12. $\int x^2(x^3 + 5)^4\,dx$

13. $\int x^2(x^3 - 1)^4\,dx$

14. $\int x(5x^2 + 4)^3\,dx$

15. $\int t\sqrt{t^2 + 2}\,dt$

16. $\int t^3\sqrt{t^4 + 5}\,dt$

17. $\int 5x\sqrt[3]{1 - x^2}\,dx$

18. $\int u^2\sqrt{u^3 + 2}\,du$

19. $\int \frac{x}{(1 - x^2)^3}\,dx$

20. $\int \frac{x^3}{(1 + x^4)^2}\,dx$

21. $\int \frac{x^2}{(1 + x^3)^2}\,dx$

22. $\int \frac{x^2}{(16 - x^3)^2}\,dx$

23. $\int \frac{x}{\sqrt{1 - x^2}}\,dx$

24. $\int \frac{x^3}{\sqrt{1 + x^4}}\,dx$

25. $\int \left(1 + \frac{1}{t}\right)^3\left(\frac{1}{t^2}\right)dt$

26. $\int \left[x^2 + \frac{1}{(3x)^2}\right]dx$

27. $\int \frac{1}{\sqrt{2x}}\,dx$

28. $\int \frac{1}{2\sqrt{x}}\,dx$

29. $\int \frac{x^2 + 5x - 8}{\sqrt{x}}\,dx$

30. $\int \frac{t - 9t^2}{\sqrt{t}}\,dt$

31. $\int t^2\left(t - \frac{8}{t}\right)dt$

32. $\int \left(\frac{t^3}{3} + \frac{1}{4t^2}\right)dt$

33. $\int (9 - y)\sqrt{y}\,dy$

34. $\int 4\pi y(6 + y^{3/2})\,dy$

In Exercises 35–38, solve the differential equation.

35. $\frac{dy}{dx} = 4x + \frac{4x}{\sqrt{16 - x^2}}$

36. $\frac{dy}{dx} = \frac{10x^2}{\sqrt{1 + x^3}}$

37. $\frac{dy}{dx} = \frac{x + 1}{(x^2 + 2x - 3)^2}$

38. $\frac{dy}{dx} = \frac{x - 4}{\sqrt{x^2 - 8x + 1}}$

Slope Fields **In Exercises 39 and 40, a differential equation, a point, and a slope field are given. A *slope field* consists of line segments with slopes given by the differential equation. These line segments give a visual perspective of the directions of the solutions of the differential equation. (a) Sketch two approximate solutions of the differential equation on the slope field, one of which passes through the given point. (To print an enlarged copy of the graph, go to the website *www.mathgraphs.com*.) (b) Use integration to find the particular solution of the differential equation and use a graphing utility to graph the solution. Compare the result with the sketches in part (a).**

39. $\frac{dy}{dx} = x\sqrt{4 - x^2}$

$(2, 2)$

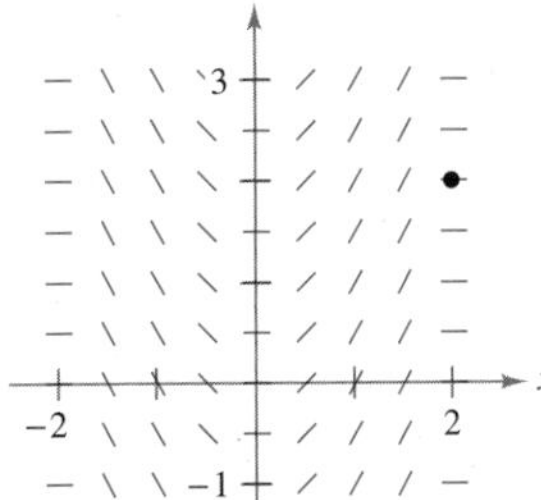

40. $\frac{dy}{dx} = x^2(x^3 - 1)^2$

$(1, 0)$

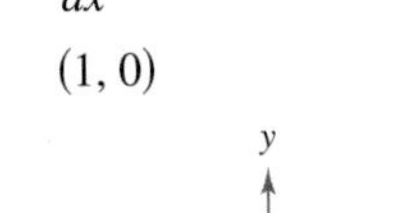

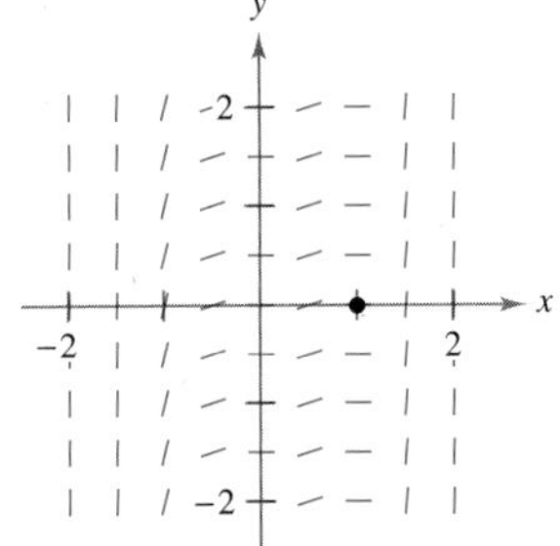

In Exercises 41 and 42, find an equation for the function f that has the given derivative and whose graph passes through the given point.

Derivative	*Point*
41. $f'(x) = 2x(4x^2 - 10)^2$	$(2, 10)$
42. $f'(x) = -2x\sqrt{8 - x^2}$	$(2, 7)$

In Exercises 43–50, find the indefinite integral by the method shown in Example 4.

43. $\int x\sqrt{x + 6}\,dx$

44. $\int x\sqrt{4x + 1}\,dx$

45. $\int x^2\sqrt{1 - x}\,dx$

46. $\int (x + 1)\sqrt{2 - x}\,dx$

47. $\int \frac{x^2 - 1}{\sqrt{2x - 1}}\,dx$

48. $\int \frac{2x + 1}{\sqrt{x + 4}}\,dx$

49. $\int \frac{-x}{(x + 1) - \sqrt{x + 1}}\,dx$

50. $\int t\sqrt[3]{t + 10}\,dt$

In Exercises 51–62, evaluate the definite integral. Use a graphing utility to verify your result.

51. $\int_{-1}^{1} x(x^2 + 1)^3\, dx$

52. $\int_{-2}^{4} x^2(x^3 + 8)^2\, dx$

53. $\int_{1}^{2} 2x^2\sqrt{x^3 + 1}\, dx$

54. $\int_{0}^{1} x\sqrt{1 - x^2}\, dx$

55. $\int_{0}^{4} \frac{1}{\sqrt{2x + 1}}\, dx$

56. $\int_{0}^{2} \frac{x}{\sqrt{1 + 2x^2}}\, dx$

57. $\int_{1}^{9} \frac{1}{\sqrt{x}\left(1 + \sqrt{x}\right)^2}\, dx$

58. $\int_{0}^{2} x\sqrt[3]{4 + x^2}\, dx$

59. $\int_{1}^{2} (x - 1)\sqrt{2 - x}\, dx$

60. $\int_{1}^{5} \frac{x}{\sqrt{2x - 1}}\, dx$

61. $\int_{5}^{14} x\sqrt{x - 5}\, dx$

62. $\int_{0}^{1} \frac{1}{\sqrt{x} + 1}\, dx$

Differential Equations **In Exercises 63–66, the graph of a function f is shown. Use the differential equation and the given point to find an equation of the function.**

63. $\frac{dy}{dx} = 18x^2(2x^3 + 1)^2$

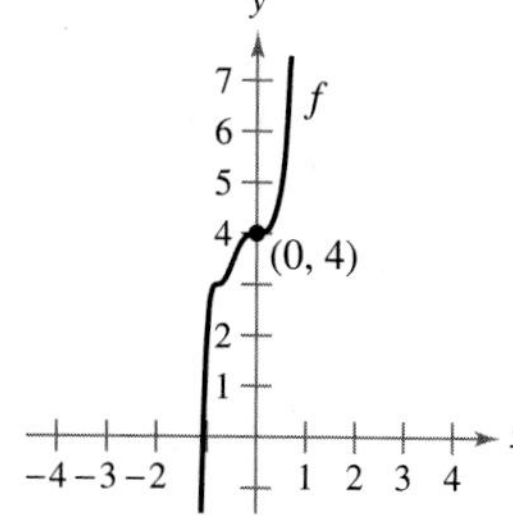

64. $\frac{dy}{dx} = \frac{-48}{(3x + 5)^3}$

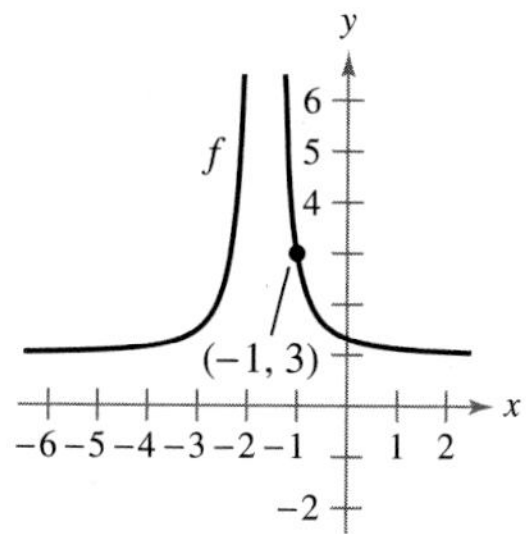

65. $\frac{dy}{dx} = \frac{2x}{\sqrt{2x^2 - 1}}$

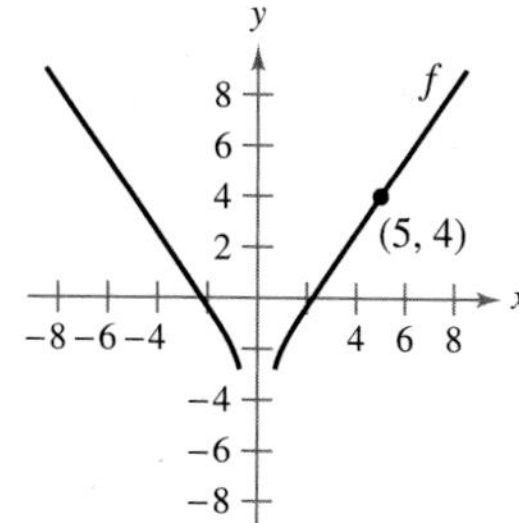

66. 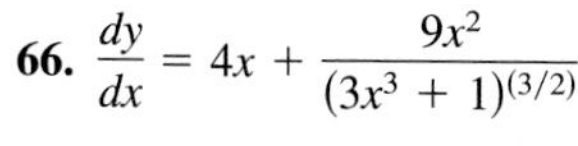 $\frac{dy}{dx} = 4x + \frac{9x^2}{(3x^3 + 1)^{(3/2)}}$

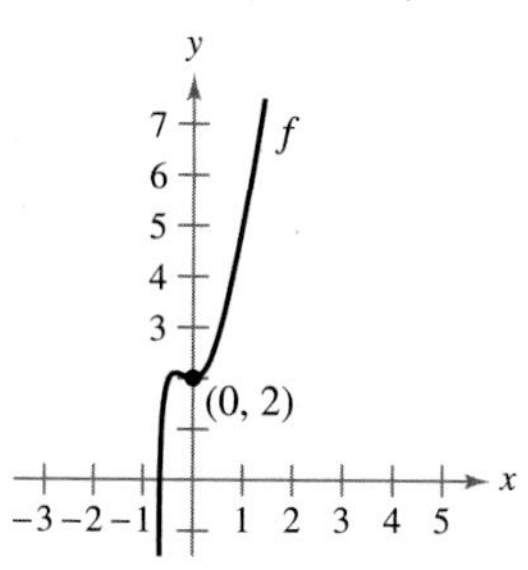

In Exercises 67 and 68, find the area of the region. Use a graphing utility to verify your result.

67. $\int_{0}^{7} x\sqrt[3]{x + 1}\, dx$

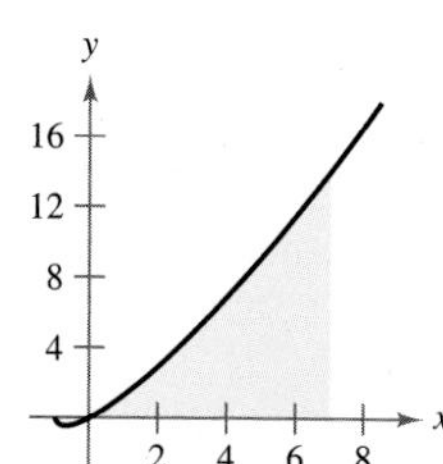

68. 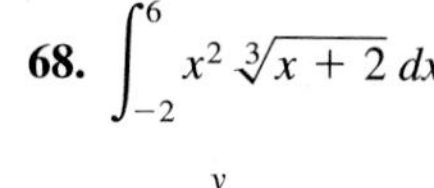 $\int_{-2}^{6} x^2\sqrt[3]{x + 2}\, dx$

In Exercises 69–72, use a graphing utility to evaluate the integral. Graph the region whose area is given by the definite integral.

69. $\int_{0}^{6} \frac{x}{\sqrt{4x + 1}}\, dx$

70. $\int_{0}^{2} x^3\sqrt{2x + 3}\, dx$

71. $\int_{3}^{7} x\sqrt{x - 3}\, dx$

72. $\int_{1}^{5} x^2\sqrt{x - 1}\, dx$

In Exercises 73–76, evaluate the integral using the properties of even and odd functions as an aid.

73. $\int_{-2}^{2} x^2(x^2 + 1)\, dx$

74. $\int_{-2}^{2} x(x^2 + 1)^3\, dx$

75. $\int_{-3}^{3} (9 - x^2)\, dx$

76. $\int_{-3}^{3} x\sqrt{9 - x^2}\, dx$

77. Use $\int_0^4 x^2\, dx = \frac{64}{3}$ to evaluate each definite integral without using the Fundamental Theorem of Calculus.

(a) $\int_{-4}^{0} x^2\, dx$

(b) $\int_{-4}^{4} x^2\, dx$

(c) $\int_{0}^{4} -x^2\, dx$

(d) $\int_{-4}^{0} 3x^2\, dx$

78. Use symmetry as an aid in evaluating each definite integral.

(a) $\int_{-1}^{1} x^2(x^2 + 1)\, dx$

(b) $\int_{-2}^{2} x^3(x^2 + 1)\, dx$

(c) $\int_{-5}^{5} \frac{x}{\sqrt{x^2 + 1}}\, dx$

(d) $\int_{-3}^{3} |x|(x^2 + 1)\, dx$

In Exercises 79 and 80, write the integral as the sum of the integral of an odd function and the integral of an even function. Use this simplification to evaluate the integral.

79. $\int_{-3}^{3} (x^3 + 4x^2 - 3x - 6)\, dx$

80. $\int_{-2}^{2} (x^4 - 3x + 5)\, dx$

WRITING ABOUT CONCEPTS

81. (a) Describe why

$$\int x(5 - x^2)^3\,dx \neq \int u^3\,du$$

where $u = 5 - x^2$.

(b) Without integrating, explain why $\int_{-2}^{2} x(x^2 + 1)^2\,dx = 0$.

CAPSTONE

82. ***Writing*** Find the indefinite integral in two ways. Explain any difference in the forms of the answers.

(a) $\int (2x - 1)^2\,dx$ (b) $\int x(x^2 - 1)^2\,dx$

83. ***Cash Flow*** The rate of disbursement dQ/dt of a 2 million dollar federal grant is proportional to the square of $100 - t$. Time t is measured in days $(0 \le t \le 100)$, and Q is the amount that remains to be disbursed. Find the amount that remains to be disbursed after 50 days. Assume that all the money will be disbursed in 100 days.

84. ***Depreciation*** The rate of depreciation dV/dt of a machine is inversely proportional to the square of $t + 1$, where V is the value of the machine t years after it was purchased. The initial value of the machine was \$500,000, and its value decreased \$100,000 in the first year. Estimate its value after 4 years.

Probability **In Exercises 85 and 86, the function**

$$f(x) = kx^n(1 - x)^m, \quad 0 \le x \le 1$$

where $n > 0$, $m > 0$, and k is a constant, can be used to represent various probability distributions. If k is chosen such that

$$\int_0^1 f(x)\,dx = 1$$

the probability that x will fall between a and b $(0 \le a \le b \le 1)$ is

$$P_{a,b} = \int_a^b f(x)\,dx.$$

85. The probability that a person will remember between $100a\%$ and $100b\%$ of material learned in an experiment is

$$P_{a,b} = \int_a^b \frac{15}{4}x\sqrt{1 - x}\,dx$$

where x represents the proportion remembered. (See figure.)

(a) For a randomly chosen individual, what is the probability that he or she will recall between 50% and 75% of the material?

(b) What is the median percent recall? That is, for what value of b is it true that the probability of recalling 0 to b is 0.5?

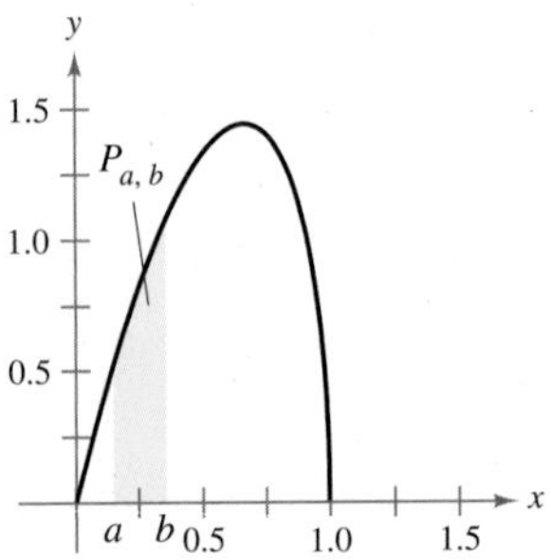

Figure for 85

86. The probability that ore samples taken from a region contain between $100a\%$ and $100b\%$ iron is

$$P_{a,b} = \int_a^b \frac{1155}{32}x^3(1 - x)^{3/2}\,dx$$

where x represents the proportion of iron. (See figure.) What is the probability that a sample will contain between

(a) 0% and 25% iron?

(b) 50% and 100% iron?

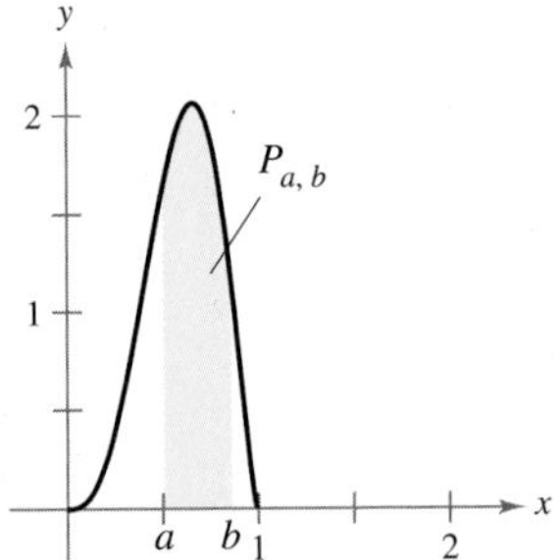

True or False? **In Exercises 87–89, determine whether the statement is true or false. If it is false, explain why or give an example that shows it is false.**

87. $\int (2x + 1)^2\,dx = \frac{1}{3}(2x + 1)^3 + C$

88. $\int x(x^2 + 1)\,dx = \frac{1}{2}x^2\left(\frac{1}{3}x^3 + x\right) + C$

89. $\int_{-10}^{10} (ax^3 + bx^2 + cx + d)\,dx = 2\int_0^{10} (bx^2 + d)\,dx$

90. Complete the proof of Theorem 6.16.

91. (a) Show that $\int_0^1 x^2(1 - x)^5\,dx = \int_0^1 x^5(1 - x)^2\,dx$.

(b) Show that $\int_0^1 x^a(1 - x)^b\,dx = \int_0^1 x^b(1 - x)^a\,dx$.

92. Assume that f is continuous everywhere and that c is a constant. Show that

$$\int_{ca}^{cb} f(x)\,dx = c\int_a^b f(cx)\,dx.$$

93. Show that if f is continuous on the entire real number line, then

$$\int_a^b f(x + h)\,dx = \int_{a+h}^{b+h} f(x)\,dx.$$

6.6 Numerical Integration

- **Approximate a definite integral using the Trapezoidal Rule.**
- **Approximate a definite integral using Simpson's Rule.**
- **Analyze the approximate errors in the Trapezoidal Rule and Simpson's Rule.**

The Trapezoidal Rule

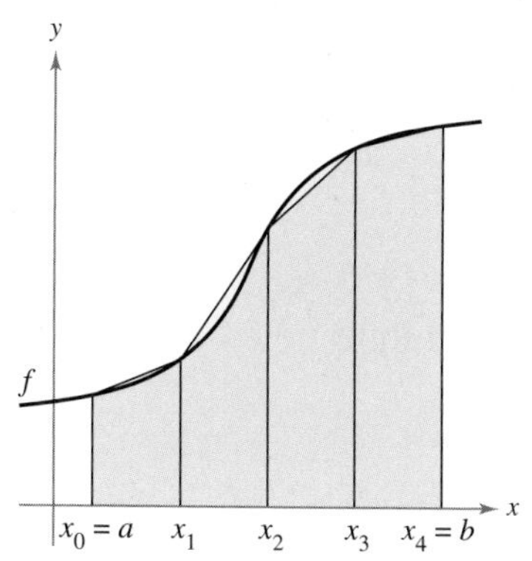

The area of the region can be approximated using four trapezoids.
Figure 6.42

Some elementary functions simply do not have antiderivatives that are elementary functions. For example, there is no elementary function that has any of the following functions as its derivative.

$$\sqrt[3]{x}\sqrt{1-x}, \qquad \sqrt{1+x^3}, \qquad \sqrt{1-x^3}, \qquad \sqrt{\frac{1+4x^2}{81-9x^2}}$$

If you need to evaluate a definite integral involving a function whose antiderivative cannot be found, then while the Fundamental Theorem of Calculus is still true, it cannot be easily applied. In this case, it is easier to resort to an approximation technique. Two such techniques are described in this section.

One way to approximate a definite integral is to use n trapezoids, as shown in Figure 6.42. In the development of this method, assume that f is continuous and positive on the interval $[a, b]$. So, the definite integral

$$\int_a^b f(x)\,dx$$

represents the area of the region bounded by the graph of f and the x-axis, from $x = a$ to $x = b$. First, partition the interval $[a, b]$ into n subintervals, each of width $\Delta x = (b - a)/n$, such that

$$a = x_0 < x_1 < x_2 < \cdot \cdot \cdot < x_n = b.$$

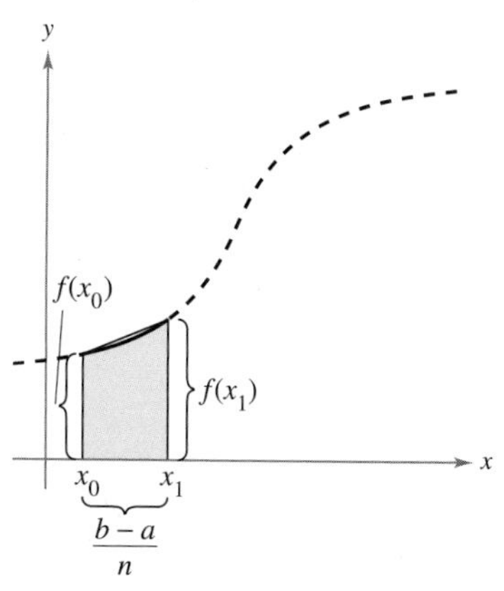

The area of the first trapezoid is $\left[\frac{f(x_0)+f(x_1)}{2}\right]\left(\frac{b-a}{n}\right)$.
Figure 6.43

Then form a trapezoid for each subinterval (see Figure 6.43). The area of the ith trapezoid is

$$\text{Area of } i\text{th trapezoid} = \left[\frac{f(x_{i-1}) + f(x_i)}{2}\right]\left(\frac{b-a}{n}\right).$$

This implies that the sum of the areas of the n trapezoids is

$$\begin{aligned}
\text{Area} &= \left(\frac{b-a}{n}\right)\left[\frac{f(x_0)+f(x_1)}{2} + \cdot \cdot \cdot + \frac{f(x_{n-1})+f(x_n)}{2}\right] \\
&= \left(\frac{b-a}{2n}\right)[f(x_0) + f(x_1) + f(x_1) + f(x_2) + \cdot \cdot \cdot + f(x_{n-1}) + f(x_n)] \\
&= \left(\frac{b-a}{2n}\right)[f(x_0) + 2f(x_1) + 2f(x_2) + \cdot \cdot \cdot + 2f(x_{n-1}) + f(x_n)].
\end{aligned}$$

Letting $\Delta x = (b - a)/n$, you can take the limit as $n \to \infty$ to obtain

$$\begin{aligned}
&\lim_{n\to\infty} \left(\frac{b-a}{2n}\right)[f(x_0) + 2f(x_1) + \cdot \cdot \cdot + 2f(x_{n-1}) + f(x_n)] \\
&= \lim_{n\to\infty} \left[\frac{[f(a) - f(b)]\,\Delta x}{2} + \sum_{i=1}^{n} f(x_i)\,\Delta x\right] \\
&= \lim_{n\to\infty} \frac{[f(a) - f(b)](b-a)}{2n} + \lim_{n\to\infty} \sum_{i=1}^{n} f(x_i)\,\Delta x \\
&= 0 + \int_a^b f(x)\,dx.
\end{aligned}$$

The result is summarized in Theorem 6.17.

THEOREM 6.17 THE TRAPEZOIDAL RULE

Let f be continuous on $[a, b]$. The Trapezoidal Rule for approximating $\int_a^b f(x)\,dx$ is given by

$$\int_a^b f(x)\,dx \approx \frac{b-a}{2n}[f(x_0) + 2f(x_1) + 2f(x_2) + \cdot \cdot \cdot + 2f(x_{n-1}) + f(x_n)].$$

Moreover, as $n \to \infty$, the right-hand side approaches $\int_a^b f(x)\,dx$.

NOTE Observe that the coefficients in the Trapezoidal Rule have the following pattern.

1 2 2 2 . . . 2 2 1 ■

EXAMPLE 1 Approximation with the Trapezoidal Rule

Use the Trapezoidal Rule to approximate

$$\int_0^1 \sqrt{x+1}\,dx.$$

Compare the results for $n = 4$ and $n = 8$, as shown in Figure 6.44.

Solution When $n = 4$, $\Delta x = 1/4$, and you obtain

$$\int_0^1 \sqrt{x+1}\,dx \approx \frac{1}{8}\left(\sqrt{1} + 2\sqrt{\frac{5}{4}} + 2\sqrt{\frac{6}{4}} + 2\sqrt{\frac{7}{4}} + \sqrt{2}\right)$$

$$= \frac{1}{8}\left[1 + 2\left(\frac{\sqrt{5}}{2}\right) + 2\left(\frac{\sqrt{6}}{2}\right) + 2\left(\frac{\sqrt{7}}{2}\right) + \sqrt{2}\right] \approx 1.2182.$$

When $n = 8$, $\Delta x = 1/8$, and you obtain

$$\int_0^1 \sqrt{x+1}\,dx \approx \frac{1}{16}\left[\sqrt{1} + 2\sqrt{\frac{9}{8}} + 2\sqrt{\frac{10}{8}} + 2\sqrt{\frac{11}{8}} + 2\sqrt{\frac{12}{8}} + 2\sqrt{\frac{13}{8}} + 2\sqrt{\frac{14}{8}} + 2\sqrt{\frac{15}{8}} + \sqrt{2}\right]$$

$$\approx 1.2188.$$

For this particular integral, you could have found an antiderivative and determined that the exact area of the region is $\frac{2}{3}(2^{3/2} - 1) \approx 1.2190$. ■

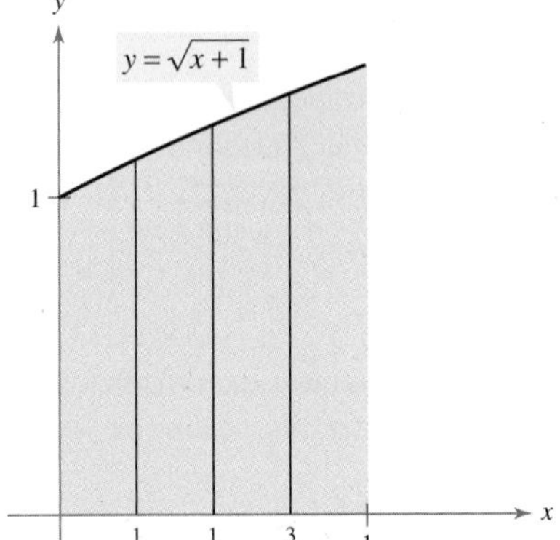

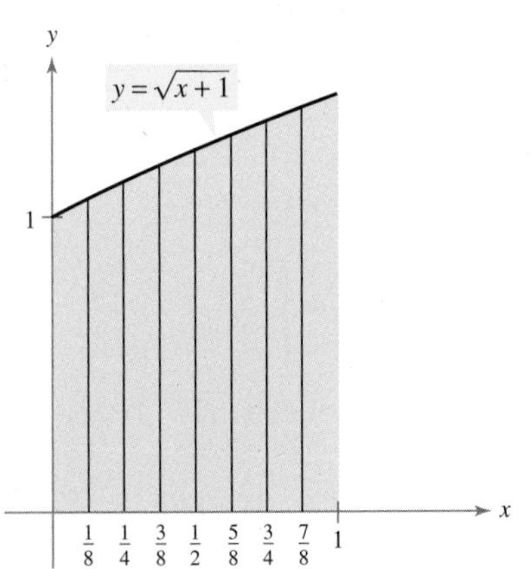

Trapezoidal approximations
Figure 6.44

TECHNOLOGY Most graphing utilities and computer algebra systems have built-in programs that can be used to approximate the value of a definite integral. Try using such a program to approximate the integral in Example 1.

When you use such a program, you need to be aware of its limitations. Often, you are given no indication of the degree of accuracy of the approximation. Other times, you may be given an approximation that is completely wrong. For instance, try using a built-in numerical integration program to evaluate

$$\int_{-1}^2 \frac{1}{x}\,dx.$$

Your calculator should give an error message. Does yours?

It is interesting to compare the Trapezoidal Rule with the Midpoint Rule given in Section 6.2 (Exercises 67–70). For the Trapezoidal Rule, you average the function values at the endpoints of the subintervals, but for the Midpoint Rule you take the function values of the subinterval midpoints.

$$\int_a^b f(x)\,dx \approx \sum_{i=1}^{n} f\left(\frac{x_i + x_{i-1}}{2}\right)\Delta x \qquad \text{Midpoint Rule}$$

$$\int_a^b f(x)\,dx \approx \sum_{i=1}^{n} \left(\frac{f(x_i) + f(x_{i-1})}{2}\right)\Delta x \qquad \text{Trapezoidal Rule}$$

NOTE There are two important points that should be made concerning the Trapezoidal Rule (or the Midpoint Rule). First, the approximation tends to become more accurate as n increases. For instance, in Example 1, if $n = 16$, the Trapezoidal Rule yields an approximation of 1.2189. Second, although you could have used the Fundamental Theorem to evaluate the integral in Example 1, this theorem cannot be used to evaluate an integral as simple as $\int_0^1 \sqrt{x^3 + 1}\,dx$ because $\sqrt{x^3 + 1}$ has no elementary antiderivative. Yet, the Trapezoidal Rule can be applied easily to estimate this integral. ■

Simpson's Rule

One way to view the trapezoidal approximation of a definite integral is to say that on each subinterval you approximate f by a *first*-degree polynomial. In Simpson's Rule, named after the English mathematician Thomas Simpson (1710–1761), you take this procedure one step further and approximate f by *second*-degree polynomials.

Before presenting Simpson's Rule, we list a theorem for evaluating integrals of polynomials of degree 2 (or less).

THEOREM 6.18 INTEGRAL OF $p(x) = Ax^2 + Bx + C$

If $p(x) = Ax^2 + Bx + C$, then

$$\int_a^b p(x)\,dx = \left(\frac{b-a}{6}\right)\left[p(a) + 4p\left(\frac{a+b}{2}\right) + p(b)\right].$$

PROOF

$$\begin{aligned}\int_a^b p(x)\,dx &= \int_a^b (Ax^2 + Bx + C)\,dx\\ &= \left[\frac{Ax^3}{3} + \frac{Bx^2}{2} + Cx\right]_a^b\\ &= \frac{A(b^3 - a^3)}{3} + \frac{B(b^2 - a^2)}{2} + C(b - a)\\ &= \left(\frac{b-a}{6}\right)[2A(a^2 + ab + b^2) + 3B(b + a) + 6C]\end{aligned}$$

By expansion and collection of terms, the expression inside the brackets becomes

$$\underbrace{(Aa^2 + Ba + C)}_{p(a)} + \underbrace{4\left[A\left(\frac{b+a}{2}\right)^2 + B\left(\frac{b+a}{2}\right) + C\right]}_{4p\left(\frac{a+b}{2}\right)} + \underbrace{(Ab^2 + Bb + C)}_{p(b)}$$

and you can write

$$\int_a^b p(x)\,dx = \left(\frac{b-a}{6}\right)\left[p(a) + 4p\left(\frac{a+b}{2}\right) + p(b)\right].$$

■

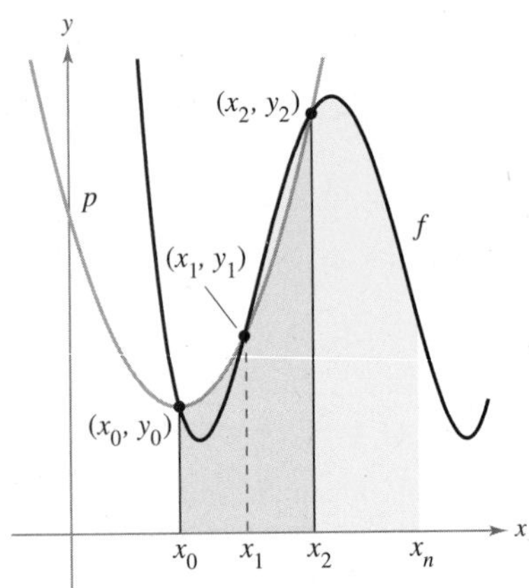

$\int_{x_0}^{x_2} p(x)\,dx \approx \int_{x_0}^{x_2} f(x)\,dx$

Figure 6.45

To develop Simpson's Rule for approximating a definite integral, you again partition the interval $[a, b]$ into n subintervals, each of width $\Delta x = (b - a)/n$. This time, however, n is required to be even, and the subintervals are grouped in pairs such that

$$a = \underbrace{x_0 < x_1 < x_2}_{[x_0,\, x_2]} < \underbrace{x_3 < x_4}_{[x_2,\, x_4]} < \cdots < \underbrace{x_{n-2} < x_{n-1} < x_n}_{[x_{n-2},\, x_n]} = b.$$

On each (double) subinterval $[x_{i-2}, x_i]$, you can approximate f by a polynomial p of degree less than or equal to 2. (See Exercise 48.) For example, on the subinterval $[x_0, x_2]$, choose the polynomial of least degree passing through the points (x_0, y_0), (x_1, y_1), and (x_2, y_2), as shown in Figure 6.45. Now, using p as an approximation of f on this subinterval, you have, by Theorem 6.18,

$$\begin{aligned}\int_{x_0}^{x_2} f(x)\,dx \approx \int_{x_0}^{x_2} p(x)\,dx &= \frac{x_2 - x_0}{6}\left[p(x_0) + 4p\left(\frac{x_0 + x_2}{2}\right) + p(x_2)\right]\\ &= \frac{2[(b-a)/n]}{6}[p(x_0) + 4p(x_1) + p(x_2)]\\ &= \frac{b-a}{3n}[f(x_0) + 4f(x_1) + f(x_2)].\end{aligned}$$

Repeating this procedure on the entire interval $[a, b]$ produces the following theorem.

NOTE Observe that the coefficients in Simpson's Rule have the following pattern.

1 4 2 4 2 4 . . . 4 2 4 1

THEOREM 6.19 SIMPSON'S RULE

Let f be continuous on $[a, b]$ and let n be an even integer. Simpson's Rule for approximating $\int_a^b f(x)\,dx$ is

$$\int_a^b f(x)\,dx \approx \frac{b-a}{3n}[f(x_0) + 4f(x_1) + 2f(x_2) + 4f(x_3) + \cdots + 4f(x_{n-1}) + f(x_n)].$$

Moreover, as $n \to \infty$, the right-hand side approaches $\int_a^b f(x)\,dx$.

In Example 1, the Trapezoidal Rule was used to estimate $\int_0^1 \sqrt{x+1}\,dx$. In the next example, Simpson's Rule is applied to the same integral.

EXAMPLE 2 Approximation with Simpson's Rule

Use Simpson's Rule to approximate

$$\int_0^1 \sqrt{x+1}\,dx.$$

Compare the results for $n = 4$ and $n = 8$.

Solution When $n = 4$, you have

$$\begin{aligned}\int_0^1 \sqrt{x+1}\,dx &\approx \frac{1}{12}\left[\sqrt{1} + 4\sqrt{\frac{5}{4}} + 2\sqrt{\frac{6}{4}} + 4\sqrt{\frac{7}{4}} + \sqrt{2}\right]\\ &\approx 1.218945.\end{aligned}$$

When $n = 8$, you have $\int_0^1 \sqrt{x+1}\,dx \approx 1.218951.$ ■

Error Analysis

If you must use an approximation technique, it is important to know how accurate you can expect the approximation to be. The following theorem, which is listed without proof, gives the formulas for estimating the errors involved in the use of Simpson's Rule and the Trapezoidal Rule. In general, when using an approximation, you can think of the error E as the difference between $\int_a^b f(x)\,dx$ and the approximation.

THEOREM 6.20 ERRORS IN THE TRAPEZOIDAL RULE AND SIMPSON'S RULE

If f has a continuous second derivative on $[a, b]$, then the error E in approximating $\int_a^b f(x)\,dx$ by the Trapezoidal Rule is

$$|E| \le \frac{(b-a)^3}{12n^2}\left[\max|f''(x)|\right], \quad a \le x \le b. \qquad \text{Trapezoidal Rule}$$

Moreover, if f has a continuous fourth derivative on $[a, b]$, then the error E in approximating $\int_a^b f(x)\,dx$ by Simpson's Rule is

$$|E| \le \frac{(b-a)^5}{180n^4}\left[\max|f^{(4)}(x)|\right], \quad a \le x \le b. \qquad \text{Simpson's Rule}$$

Theorem 6.20 states that the errors generated by the Trapezoidal Rule and Simpson's Rule have upper bounds dependent on the extreme values of $f''(x)$ and $f^{(4)}(x)$ in the interval $[a, b]$. Furthermore, these errors can be made arbitrarily small by *increasing n*, provided that f'' and $f^{(4)}$ are continuous and therefore bounded in $[a, b]$.

TECHNOLOGY Use a computer algebra system to evaluate the definite integral in Example 3. You should obtain a value of

$$\int_0^1 \sqrt{1+x^2}\,dx = \tfrac{1}{2}\left[\sqrt{2} + \ln\left(1+\sqrt{2}\right)\right] \approx 1.14779.$$

("ln" represents the natural logarithmic function, which you will study in Chapters 7 and 8.)

EXAMPLE 3 The Approximate Error in the Trapezoidal Rule

Determine a value of n such that the Trapezoidal Rule will approximate the value of $\int_0^1 \sqrt{1+x^2}\,dx$ with an error that is less than or equal to 0.01.

Solution Begin by letting $f(x) = \sqrt{1+x^2}$ and finding the second derivative of f.

$$f'(x) = x(1+x^2)^{-1/2} \quad \text{and} \quad f''(x) = (1+x^2)^{-3/2}$$

The maximum value of $|f''(x)|$ on the interval $[0, 1]$ is $|f''(0)| = 1$. So, by Theorem 6.20, you can write

$$|E| \le \frac{(b-a)^3}{12n^2}|f''(0)| = \frac{1}{12n^2}(1) = \frac{1}{12n^2}.$$

To obtain an error E that is less than 0.01, you must choose n such that $1/(12n^2) \le 1/100$.

$$100 \le 12n^2 \quad \Longrightarrow \quad n \ge \sqrt{\tfrac{100}{12}} \approx 2.89$$

So, you can choose $n = 3$ (because n must be greater than or equal to 2.89) and apply the Trapezoidal Rule, as shown in Figure 6.46, to obtain

$$\int_0^1 \sqrt{1+x^2}\,dx \approx \frac{1}{6}\left[\sqrt{1+0^2} + 2\sqrt{1+\left(\tfrac{1}{3}\right)^2} + 2\sqrt{1+\left(\tfrac{2}{3}\right)^2} + \sqrt{1+1^2}\right] \approx 1.154.$$

So, by adding and subtracting the error from this estimate, you know that

$$1.144 \le \int_0^1 \sqrt{1+x^2}\,dx \le 1.164.$$

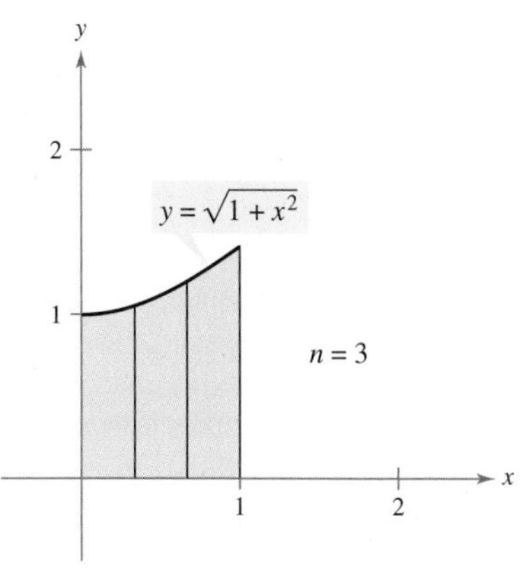

$1.144 \le \int_0^1 \sqrt{1+x^2}\,dx \le 1.164$

Figure 6.46

6.6 Exercises

See www.CalcChat.com for worked-out solutions to odd-numbered exercises.

In Exercises 1–10, use the Trapezoidal Rule and Simpson's Rule to approximate the value of the definite integral for the given value of n. Round your answer to four decimal places and compare the results with the exact value of the definite integral.

1. $\int_0^2 x^2\,dx,\quad n = 4$
2. $\int_1^2 \left(\frac{x^2}{4} + 1\right) dx,\quad n = 4$
3. $\int_0^2 x^3\,dx,\quad n = 4$
4. $\int_2^3 \frac{2}{x^2}\,dx,\quad n = 4$
5. $\int_1^3 x^3\,dx,\quad n = 6$
6. $\int_0^8 \sqrt[3]{x}\,dx,\quad n = 8$
7. $\int_4^9 \sqrt{x}\,dx,\quad n = 8$
8. $\int_1^4 (4 - x^2)\,dx,\quad n = 6$
9. $\int_0^1 \frac{2}{(x + 2)^2}\,dx,\quad n = 4$
10. $\int_0^2 x\sqrt{x^2 + 1}\,dx,\quad n = 4$

In Exercises 11–20, approximate the definite integral using the Trapezoidal Rule and Simpson's Rule. Compare these results with the approximation of the integral using a graphing utility.

11. $\int_0^4 \frac{1}{x + 1}\,dx,\quad n = 4$
12. $\int_0^2 \frac{1}{\sqrt{1 + x^3}}\,dx,\quad n = 4$
13. $\int_0^2 \sqrt{1 + x^3}\,dx,\quad n = 4$
14. $\int_0^4 \sqrt{x^2 + 1}\,dx,\quad n = 4$
15. $\int_0^1 \sqrt{x}\,\sqrt{1 - x}\,dx,\quad n = 4$
16. $\int_0^1 \frac{1}{x^2 + 1}\,dx,\quad n = 2$
17. $\int_{-2}^2 \frac{1}{x^2 + 1}\,dx,\quad n = 8$
18. $\int_{-1}^1 x\sqrt{x + 1}\,dx,\quad n = 4$
19. $\int_1^7 \frac{\sqrt{x - 1}}{x}\,dx,\quad n = 6$
20. $\int_3^6 \frac{1}{1 - \sqrt{x - 1}}\,dx,\quad n = 6$

WRITING ABOUT CONCEPTS

21. The Trapezoidal Rule and Simpson's Rule yield approximations of a definite integral $\int_a^b f(x)\,dx$ based on polynomial approximations of f. What is the degree of the polynomials used for each?
22. Describe the size of the error when the Trapezoidal Rule is used to approximate $\int_a^b f(x)\,dx$ when $f(x)$ is a linear function. Use a graph to explain your answer.

In Exercises 23–26, use the error formulas in Theorem 6.20 to estimate the errors in approximating the integral, with $n = 4$, using (a) the Trapezoidal Rule and (b) Simpson's Rule.

23. $\int_1^3 2x^3\,dx$
24. $\int_3^5 (5x + 2)\,dx$
25. $\int_0^1 \frac{1}{x + 1}\,dx$
26. $\int_2^4 \frac{1}{(x - 1)^2}\,dx$

In Exercises 27–30, use the error formulas in Theorem 6.20 to find n such that the error in the approximation of the definite integral is less than or equal to 0.00001 using (a) the Trapezoidal Rule and (b) Simpson's Rule.

27. $\int_1^3 \frac{1}{x}\,dx$
28. $\int_0^1 \frac{1}{1 + x}\,dx$
29. $\int_0^2 \sqrt{x + 2}\,dx$
30. $\int_1^3 \frac{1}{\sqrt{x}}\,dx$

In Exercises 31–34, use a computer algebra system and the error formulas to find n such that the error in the approximation of the definite integral is less than or equal to 0.00001 using (a) the Trapezoidal Rule and (b) Simpson's Rule.

31. $\int_1^3 \frac{1}{x^2}\,dx$
32. $\int_0^1 \frac{1}{x + 2}\,dx$
33. $\int_0^2 \sqrt{1 + x}\,dx$
34. $\int_0^2 (x + 1)^{2/3}\,dx$

35. Approximate the area of the shaded region using (a) the Trapezoidal Rule and (b) Simpson's Rule with $n = 4$.

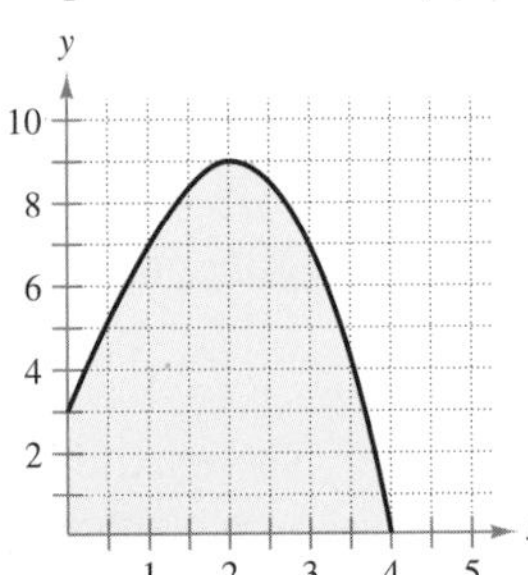

Figure for 35

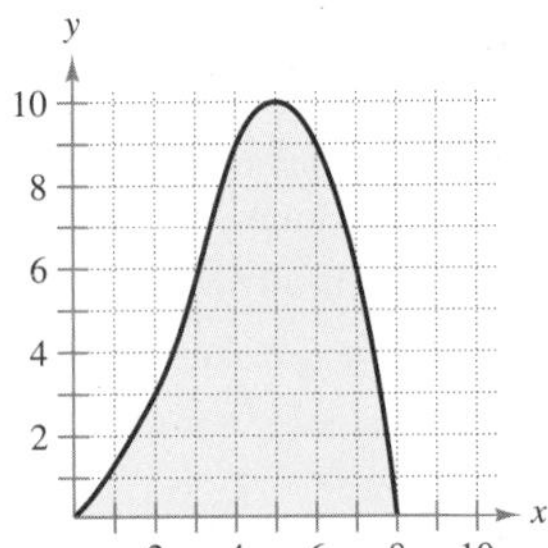

Figure for 36

36. Approximate the area of the shaded region using (a) the Trapezoidal Rule and (b) Simpson's Rule with $n = 8$.
37. ***Programming*** Write a program for a graphing utility to approximate a definite integral using the Trapezoidal Rule and Simpson's Rule. Start with the program written in Section 6.3, Exercises 61 and 62, and note that the Trapezoidal Rule can be written as

$$T(n) = \tfrac{1}{2}[L(n) + R(n)]$$

and Simpson's Rule can be written as

$$S(n) = \tfrac{1}{3}[T(n/2) + 2M(n/2)].$$

[Recall that $L(n)$, $M(n)$, and $R(n)$ represent the Riemann sums using the left-hand endpoints, midpoints, and right-hand endpoints of subintervals of equal width.]

Programming In Exercises 38 and 39, use the program in Exercise 37 to approximate the definite integral and complete the table.

n	$L(n)$	$M(n)$	$R(n)$	$T(n)$	$S(n)$
4					
8					
10					
12					
16					
20					

38. $\int_0^4 \sqrt{2 + 3x^2}\, dx$

39. $\int_0^1 \sqrt{1 - x^2}\, dx$

CAPSTONE

40. Consider a function $f(x)$ that is concave upward on the interval $[0, 2]$ and a function $g(x)$ that is concave downward on $[0, 2]$.

(a) Using the Trapezoidal Rule, which integral would be overestimated? Which integral would be underestimated? Assume $n = 4$. Use graphs to explain your answer.

(b) Which rule would you use for more accurate approximations of $\int_0^2 f(x)\, dx$ and $\int_0^2 g(x)\, dx$, the Trapezoidal Rule or Simpson's Rule? Explain your reasoning.

41. ***Work*** To determine the size of the motor required to operate a press, a company must know the amount of work done when the press moves an object linearly 5 feet. The variable force to move the object is

$$F(x) = 100x\sqrt{125 - x^3}$$

where F is given in pounds and x gives the position of the unit in feet. Use Simpson's Rule with $n = 12$ to approximate the work W (in foot-pounds) done through one cycle if $W = \int_0^5 F(x)\, dx$.

42. The table lists several measurements gathered in an experiment to approximate an unknown continuous function $y = f(x)$.

x	0.00	0.25	0.50	0.75	1.00
y	4.32	4.36	4.58	5.79	6.14

x	1.25	1.50	1.75	2.00
y	7.25	7.64	8.08	8.14

(a) Approximate the integral $\int_0^2 f(x)\, dx$ using the Trapezoidal Rule and Simpson's Rule.

(b) Use a graphing utility to find a model of the form $y = ax^3 + bx^2 + cx + d$ for the data. Integrate the resulting polynomial over $[0, 2]$ and compare the result with the integral from part (a).

Approximation of Pi **In Exercises 43 and 44, use Simpson's Rule with $n = 6$ to approximate π using the given equation. (In Section 11.5, you will be able to evaluate the integral using inverse trigonometric functions.)**

43. $\pi = \int_0^{1/2} \frac{6}{\sqrt{1 - x^2}}\, dx$

44. $\pi = \int_0^1 \frac{4}{1 + x^2}\, dx$

Area **In Exercises 45 and 46, use the Trapezoidal Rule to estimate the number of square meters of land in a lot, where x and y are measured in meters, as shown in the figures. The land is bounded by a stream and two straight roads that meet at right angles.**

45.

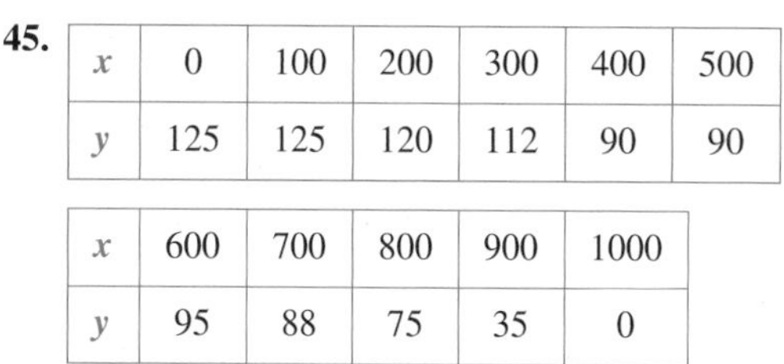

x	0	100	200	300	400	500
y	125	125	120	112	90	90

x	600	700	800	900	1000
y	95	88	75	35	0

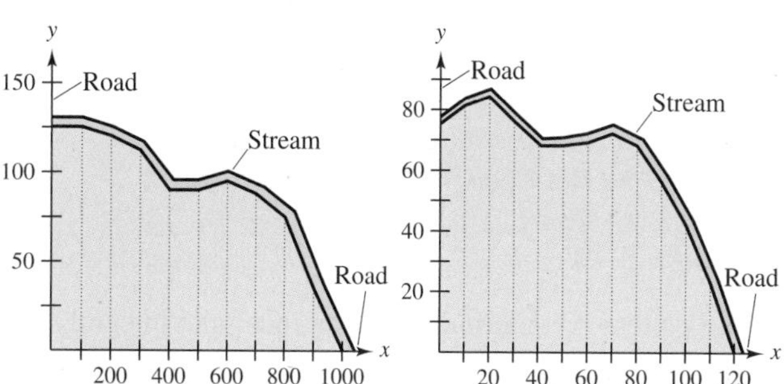

Figure for 45 **Figure for 46**

46.

x	0	10	20	30	40	50	60
y	75	81	84	76	67	68	69

x	70	80	90	100	110	120
y	72	68	56	42	23	0

47. Prove that Simpson's Rule is exact when approximating the integral of a cubic polynomial function, and demonstrate the result for $\int_0^1 x^3\, dx$, $n = 2$.

48. Prove that you can find a polynomial $p(x) = Ax^2 + Bx + C$ that passes through any three points (x_1, y_1), (x_2, y_2), and (x_3, y_3), where the x_i's are distinct.

49. Use Simpson's Rule with $n = 10$ and a computer algebra system to approximate t in the integral equation

$$\int_0^t \frac{1}{x + 1}\, dx = 2.$$

6 CHAPTER SUMMARY

	Review Exercises
Section 6.1	
■ Write the general solution of a differential equation *(p. 398),* use indefinite integral notation for antiderivatives *(p. 399)*, and use basic integration rules to find antiderivatives *(p. 400).*	*1–8*
■ Find a particular solution of a differential equation *(p. 402).*	*9–16*
Section 6.2	
■ Use sigma notation to write and evaluate a sum *(p. 408).*	*17–24*
■ Understand the concept of area *(p. 410),* and use rectangles to approximate the area of a plane region *(p. 411).*	*25, 26*
■ Find the area of a plane region using limits *(p. 412).*	*27–32*
Section 6.3	
■ Understand the definition of a Riemann sum *(p. 420),* and evaluate a definite integral using limits *(p. 422).*	*33, 34*
■ Evaluate a definite integral using properties of definite integrals *(p. 425).*	*35–40*
Section 6.4	
■ Evaluate a definite integral using the Fundamental Theorem of Calculus *(p. 430).*	*41–56*
■ Understand and use the Mean Value Theorem for Integrals *(p. 433),* and find the average value of a function over a closed interval *(p. 434).*	*57, 58*
■ Understand and use the Second Fundamental Theorem of Calculus *(p. 436),* and understand and use the Net Change Theorem *(p. 439).*	*59, 60*
Section 6.5	
■ Use pattern recognition to find an indefinite integral *(p. 445),* use a change of variables to find an indefinite integral *(p. 448),* and use the General Power Rule for Integration to find an indefinite integral *(p. 450).*	*61–70*
■ Use a change of variables to evaluate a definite integral *(p. 451),* and evaluate a definite integral involving an even or odd function *(p. 453).*	*71–79*
Section 6.6	
■ Approximate a definite integral using the Trapezoidal Rule *(p. 457),* approximate a definite integral using Simpson's Rule *(p. 459),* and analyze the approximate errors in the Trapezoidal Rule and Simpson's Rule *(p. 461).*	*80–82*

6 REVIEW EXERCISES

See www.CalcChat.com for worked-out solutions to odd-numbered exercises.

In Exercises 1 and 2, use the graph of f' to sketch a graph of f. To print an enlarged copy of the graph, go to the website *www.mathgraphs.com*.

1. 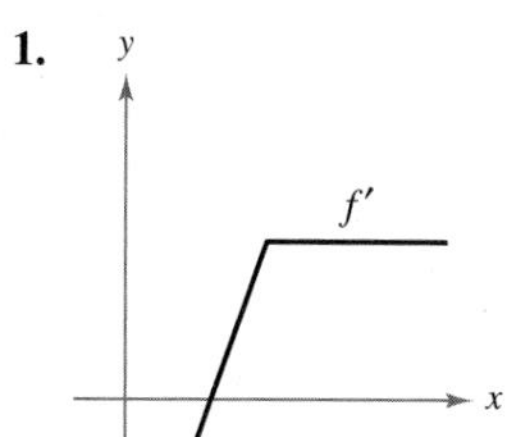

2. 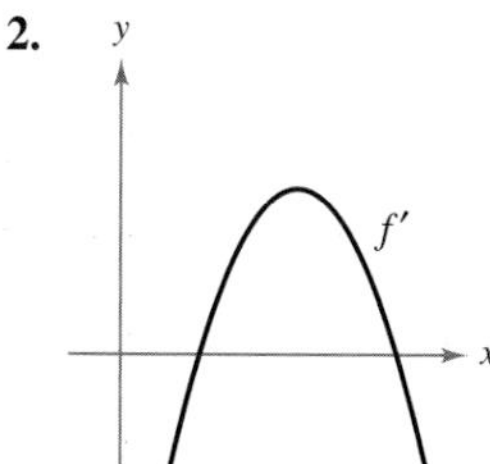

In Exercises 3–8, find the indefinite integral.

3. $\displaystyle\int (4x^2 + x + 3)\, dx$

4. $\displaystyle\int \frac{2}{\sqrt[3]{3x}}\, dx$

5. $\displaystyle\int \frac{x^4 + 8}{x^3}\, dx$

6. $\displaystyle\int \frac{x^4 - 4x^2 + 1}{x^2}\, dx$

7. $\displaystyle\int \sqrt[3]{x}\,(x + 3)\, dx$

8. $\displaystyle\int x^2 (x + 5)^2\, dx$

9. Find the particular solution of the differential equation $f'(x) = -6x$ whose graph passes through the point $(1, -2)$.

10. Find the particular solution of the differential equation $f''(x) = 6(x - 1)$ whose graph passes through the point $(2, 1)$ and is tangent to the line $3x - y - 5 = 0$ at that point.

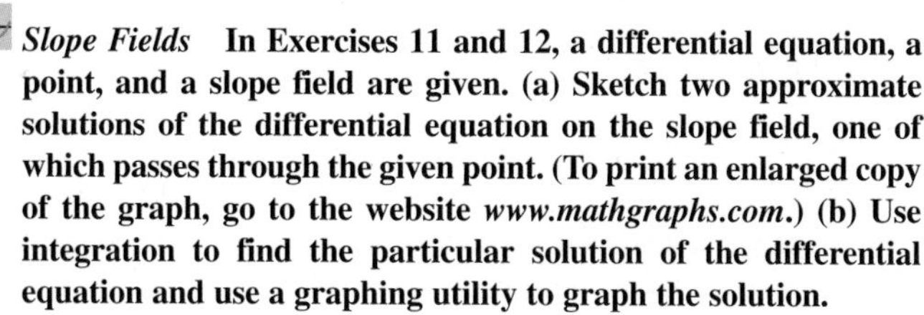

Slope Fields **In Exercises 11 and 12, a differential equation, a point, and a slope field are given. (a) Sketch two approximate solutions of the differential equation on the slope field, one of which passes through the given point. (To print an enlarged copy of the graph, go to the website *www.mathgraphs.com*.) (b) Use integration to find the particular solution of the differential equation and use a graphing utility to graph the solution.**

11. $\dfrac{dy}{dx} = 2x - 4, \quad (4, -2)$

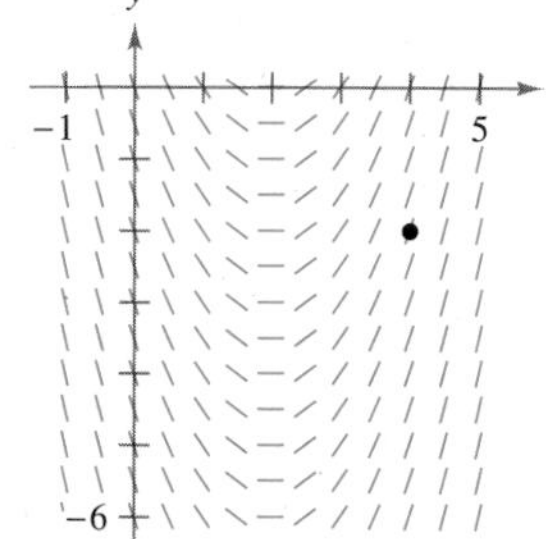

12. $\dfrac{dy}{dx} = \dfrac{1}{2}x^2 - 2x, \quad (6, 2)$

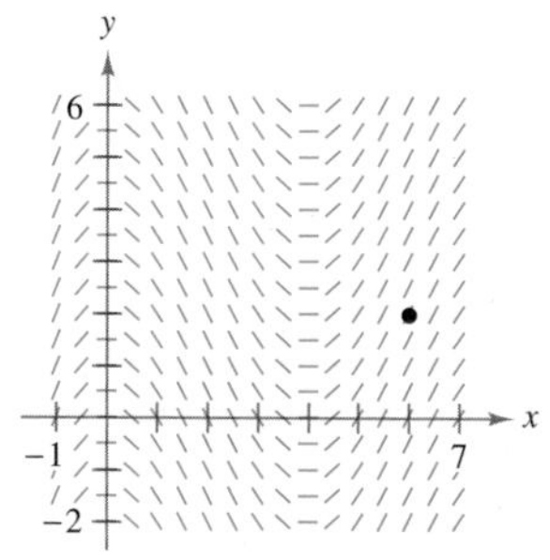

13. ***Velocity and Acceleration*** An airplane taking off from a runway travels 3600 feet before lifting off. The airplane starts from rest, moves with constant acceleration, and makes the run in 30 seconds. With what speed does it lift off?

14. ***Velocity and Acceleration*** The speed of a car traveling in a straight line is reduced from 45 to 30 miles per hour in a distance of 264 feet. Find the distance in which the car can be brought to rest from 30 miles per hour, assuming the same constant deceleration.

15. ***Velocity and Acceleration*** A ball is thrown vertically upward from ground level with an initial velocity of 96 feet per second.
    (a) How long will it take the ball to rise to its maximum height? What is the maximum height?
    (b) After how many seconds is the velocity of the ball one-half the initial velocity?
    (c) What is the height of the ball when its velocity is one-half the initial velocity?

16. ***Modeling Data*** The table shows the velocities (in miles per hour) of two cars on an entrance ramp to an interstate highway. The time t is in seconds.

t	0	5	10	15	20	25	30
v_1	0	2.5	7	16	29	45	65
v_2	0	21	38	51	60	64	65

    (a) Rewrite the velocities in feet per second.
    (b) Use the regression capabilities of a graphing utility to find quadratic models for the data in part (a).
    (c) Approximate the distance traveled by each car during the 30 seconds. Explain the difference in the distances.

In Exercises 17 and 18, use sigma notation to write the sum.

17. $\dfrac{1}{3(1)} + \dfrac{1}{3(2)} + \dfrac{1}{3(3)} + \cdots + \dfrac{1}{3(10)}$

18. $\left(\dfrac{3}{n}\right)\left(\dfrac{1+1}{n}\right)^2 + \left(\dfrac{3}{n}\right)\left(\dfrac{2+1}{n}\right)^2 + \cdots + \left(\dfrac{3}{n}\right)\left(\dfrac{n+1}{n}\right)^2$

In Exercises 19–22, use the properties of summation and Theorem 6.2 to evaluate the sum.

19. $\displaystyle\sum_{i=1}^{20} 2i$

20. $\displaystyle\sum_{i=1}^{20} (4i - 1)$

21. $\displaystyle\sum_{i=1}^{20} (i + 1)^2$

22. $\displaystyle\sum_{i=1}^{12} i(i^2 - 1)$

23. Write in sigma notation (a) the sum of the first ten positive odd integers, (b) the sum of the cubes of the first n positive integers, and (c) $6 + 10 + 14 + 18 + \cdots + 42$.

24. Evaluate each sum for $x_1 = 2$, $x_2 = -1$, $x_3 = 5$, $x_4 = 3$, and $x_5 = 7$.
    (a) $\displaystyle\frac{1}{5}\sum_{i=1}^{5} x_i$
    (b) $\displaystyle\sum_{i=1}^{5} \frac{1}{x_i}$
    (c) $\displaystyle\sum_{i=1}^{5} (2x_i - x_i^2)$
    (d) $\displaystyle\sum_{i=2}^{5} (x_i - x_{i-1})$

In Exercises 25 and 26, use upper and lower sums to approximate the area of the region using the indicated number of subintervals of equal width.

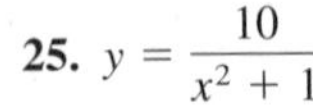

25. $y = \dfrac{10}{x^2 + 1}$

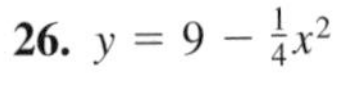

26. $y = 9 - \frac{1}{4}x^2$

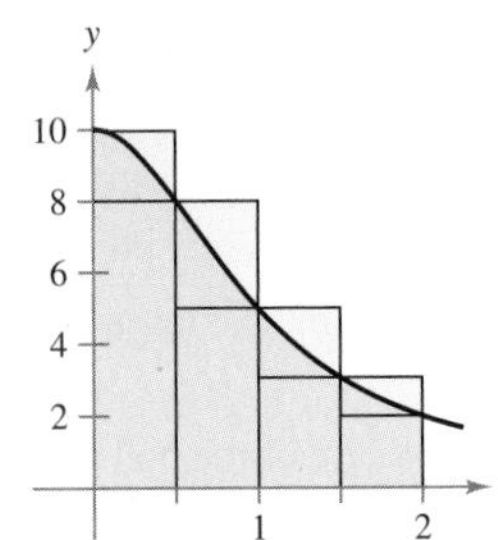

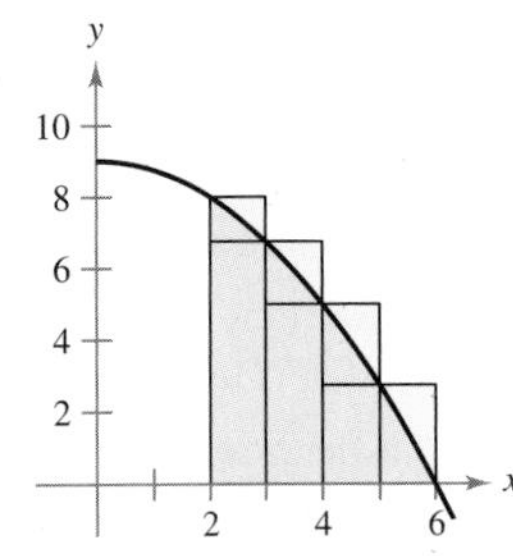

In Exercises 27–30, use the limit process to find the area of the region between the graph of the function and the x-axis over the given interval. Sketch the region.

27. $y = 8 - 2x, \quad [0, 3]$

28. $y = x^2 + 3, \quad [0, 2]$

29. $y = 5 - x^2, \quad [-2, 1]$

30. $y = \frac{1}{4}x^3, \quad [2, 4]$

31. Use the limit process to find the area of the region bounded by $x = 5y - y^2$, $x = 0$, $y = 2$, and $y = 5$.

32. Consider the region bounded by $y = mx$, $y = 0$, $x = 0$, and $x = b$.

(a) Find the upper and lower sums to approximate the area of the region when $\Delta x = b/4$.

(b) Find the upper and lower sums to approximate the area of the region when $\Delta x = b/n$.

(c) Find the area of the region by letting n approach infinity in both sums in part (b). Show that in each case you obtain the formula for the area of a triangle.

In Exercises 33 and 34, write the limit as a definite integral on the interval $[a, b]$, where c_i is any point in the ith subinterval.

	Limit	*Interval*
33.	$\lim\limits_{\|\Delta\| \to 0} \sum\limits_{i=1}^{n} (2c_i - 3)\,\Delta x_i$	$[4, 6]$
34.	$\lim\limits_{\|\Delta\| \to 0} \sum\limits_{i=1}^{n} 3c_i(9 - c_i^2)\,\Delta x_i$	$[1, 3]$

In Exercises 35 and 36, set up a definite integral that yields the area of the region. (Do not evaluate the integral.)

35. $f(x) = 2x + 8$

36. $f(x) = 100 - x^2$

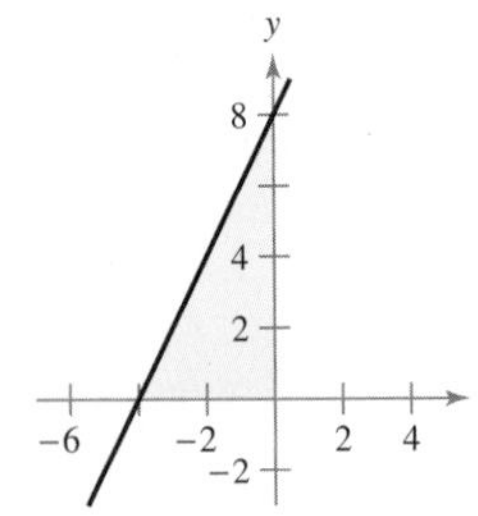

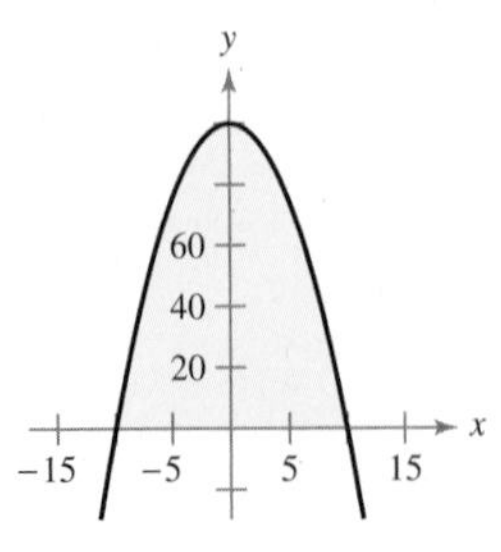

In Exercises 37 and 38, sketch the region whose area is given by the definite integral. Then use a geometric formula to evaluate the integral.

37. $\displaystyle\int_0^5 (5 - |x - 5|)\,dx$

38. $\displaystyle\int_{-6}^6 \sqrt{36 - x^2}\,dx$

39. Given $\displaystyle\int_4^8 f(x)\,dx = 12$ and $\displaystyle\int_4^8 g(x)\,dx = 5$, evaluate

(a) $\displaystyle\int_4^8 [f(x) + g(x)]\,dx.$ (b) $\displaystyle\int_4^8 [f(x) - g(x)]\,dx.$

(c) $\displaystyle\int_4^8 [2f(x) - 3g(x)]\,dx.$ (d) $\displaystyle\int_4^8 7f(x)\,dx.$

40. Given $\displaystyle\int_0^3 f(x)\,dx = 4$ and $\displaystyle\int_3^6 f(x)\,dx = -1$, evaluate

(a) $\displaystyle\int_0^6 f(x)\,dx.$ (b) $\displaystyle\int_6^3 f(x)\,dx.$

(c) $\displaystyle\int_4^4 f(x)\,dx.$ (d) $\displaystyle\int_3^6 -10f(x)\,dx.$

In Exercises 41 and 42, select the correct value of the definite integral.

41. $\displaystyle\int_1^8 \left(\sqrt[3]{x} + 1\right)dx$

(a) $\frac{81}{4}$ (b) $\frac{331}{12}$ (c) $\frac{73}{4}$ (d) $\frac{355}{12}$

42. $\displaystyle\int_1^3 \frac{12}{x^3}\,dx$

(a) $\frac{320}{9}$ (b) $-\frac{16}{3}$ (c) $-\frac{5}{9}$ (d) $\frac{16}{3}$

In Exercises 43–48, use the Fundamental Theorem of Calculus to evaluate the definite integral.

43. $\displaystyle\int_0^8 (3 + x)\,dx$

44. $\displaystyle\int_{-3}^3 (t^2 + 1)\,dt$

45. $\displaystyle\int_{-1}^1 (4t^3 - 2t)\,dt$

46. $\displaystyle\int_{-2}^{-1} (x^4 + 3x^2 - 4)\,dx$

47. $\displaystyle\int_4^9 x\sqrt{x}\,dx$

48. $\displaystyle\int_1^2 \left(\frac{1}{x^2} - \frac{1}{x^3}\right)dx$

In Exercises 49–54, sketch the graph of the region whose area is given by the integral, and find the area.

49. $\displaystyle\int_2^4 (3x - 4)\,dx$

50. $\displaystyle\int_0^6 (8 - x)\,dx$

51. $\displaystyle\int_3^4 (x^2 - 9)\,dx$

52. $\displaystyle\int_{-2}^3 (-x^2 + x + 6)\,dx$

53. $\displaystyle\int_0^1 (x - x^3)\,dx$

54. $\displaystyle\int_0^1 \sqrt{x}(1 - x)\,dx$

In Exercises 55 and 56, sketch the region bounded by the graphs of the equations, and determine its area.

55. $y = \dfrac{4}{\sqrt{x}}, \quad y = 0, \quad x = 1, \quad x = 9$

56. $y = x - x^5, \quad y = 0, \quad x = 0, \quad x = 1$

In Exercises 57 and 58, find the average value of the function over the given interval. Find the values of x at which the function assumes its average value, and graph the function.

57. $f(x) = \dfrac{1}{\sqrt{x}}, \quad [4, 9]$

58. $f(x) = x^3, \quad [0, 2]$

In Exercises 59 and 60, use the Second Fundamental Theorem of Calculus to find $F'(x)$.

59. $F(x) = \displaystyle\int_0^x t^2\sqrt{1 + t^3}\, dt$

60. $F(x) = \displaystyle\int_{-3}^x (t^2 + 3t + 2)\, dt$

In Exercises 61–70, find the indefinite integral.

61. $\displaystyle\int (3 - x^2)^3\, dx$

62. $\displaystyle\int \left(x + \frac{1}{x}\right)^2 dx$

63. $\displaystyle\int x(x^2 + 1)^3\, dx$

64. $\displaystyle\int 3x^2\sqrt{2x^3 - 5}\, dx$

65. $\displaystyle\int \frac{x^2}{\sqrt{x^3 + 3}}\, dx$

66. $\displaystyle\int \frac{x}{\sqrt{25 - 9x^2}}\, dx$

67. $\displaystyle\int x(1 - 3x^2)^4\, dx$

68. $\displaystyle\int \frac{x + 4}{(x^2 + 8x - 7)^2}\, dx$

69. $\displaystyle\int x^2\sqrt{x + 5}\, dx$

70. $\displaystyle\int x\sqrt{x + 5}\, dx$

In Exercises 71–76, evaluate the definite integral. Use a graphing utility to verify your result.

71. $\displaystyle\int_{-2}^1 x(x^2 - 6)\, dx$

72. $\displaystyle\int_0^1 x^2(x^3 - 2)^3\, dx$

73. $\displaystyle\int_0^3 \frac{1}{\sqrt{1 + x}}\, dx$

74. $\displaystyle\int_3^6 \frac{x}{3\sqrt{x^2 - 8}}\, dx$

75. $2\pi\displaystyle\int_0^1 (y + 1)\sqrt{1 - y}\, dy$

76. $2\pi\displaystyle\int_{-1}^0 x^2\sqrt{x + 1}\, dx$

Slope Fields **In Exercises 77 and 78, a differential equation, a point, and a slope field are given. (a) Sketch two approximate solutions of the differential equation on the slope field, one of which passes through the given point. (To print an enlarged copy of the graph, go to the website *www.mathgraphs.com*.) (b) Use integration to find the particular solution of the differential equation and use a graphing utility to graph the solution.**

77. $\dfrac{dy}{dx} = x\sqrt{9 - x^2}, \quad (0, -4)$

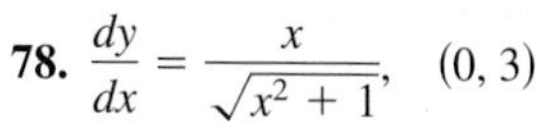

78. $\dfrac{dy}{dx} = \dfrac{x}{\sqrt{x^2 + 1}}, \quad (0, 3)$

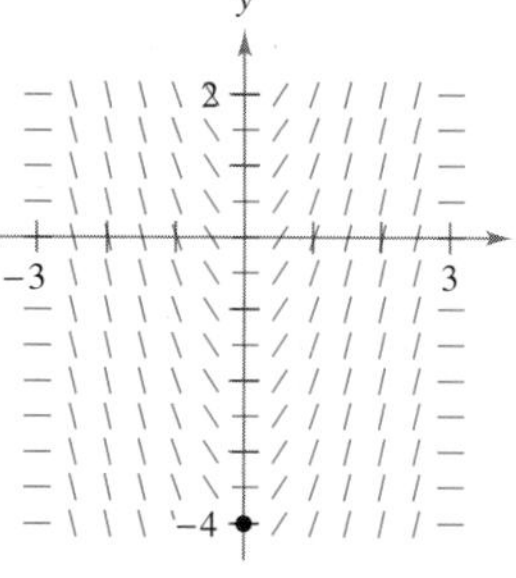

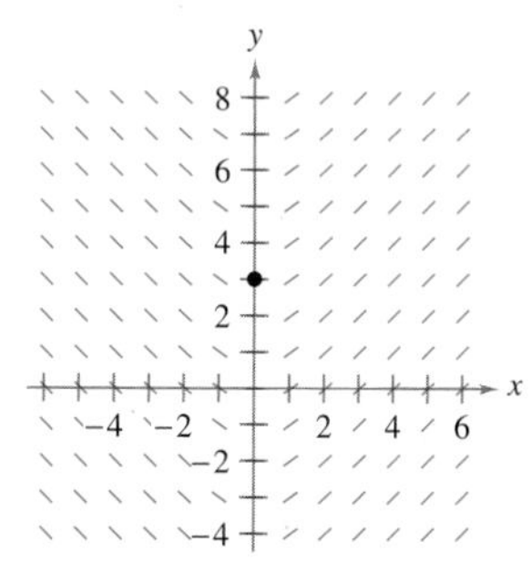

79. Find the area of the region. Use a graphing utility to verify your result.

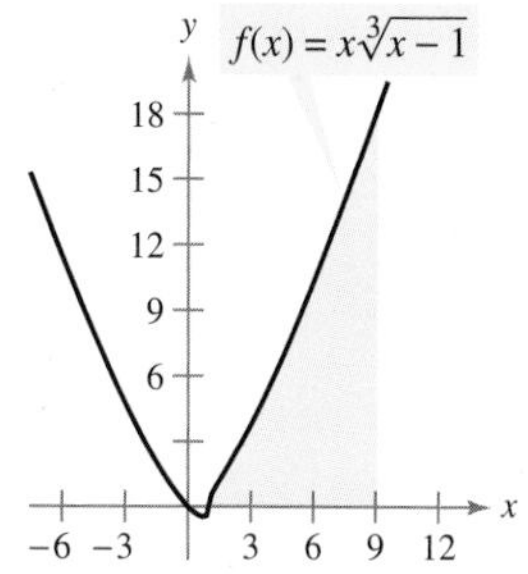

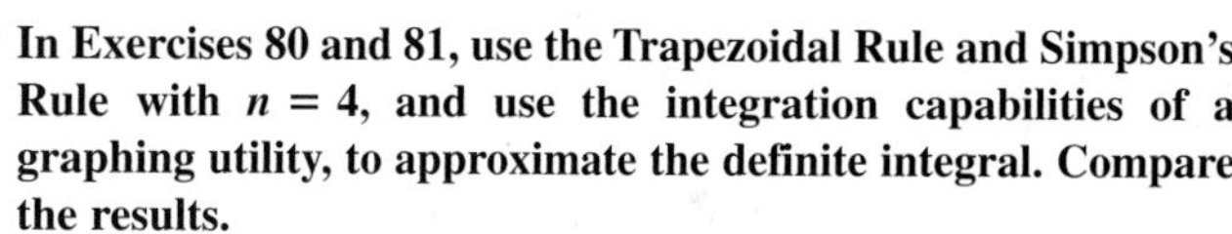

In Exercises 80 and 81, use the Trapezoidal Rule and Simpson's Rule with $n = 4$, and use the integration capabilities of a graphing utility, to approximate the definite integral. Compare the results.

80. $\displaystyle\int_2^3 \frac{2}{1 + x^2}\, dx$

81. $\displaystyle\int_0^1 \frac{x^{3/2}}{3 - x^2}\, dx$

82. Let

$$I = \int_0^4 f(x)\, dx$$

where f is shown in the figure. Let $L(n)$ and $R(n)$ represent the Riemann sums using the left-hand endpoints and right-hand endpoints of n subintervals of equal width. (Assume n is even.) Let $T(n)$ and $S(n)$ be the corresponding values of the Trapezoidal Rule and Simpson's Rule.

(a) For any n, list $L(n)$, $R(n)$, $T(n)$, and I in increasing order.

(b) Approximate $S(4)$.

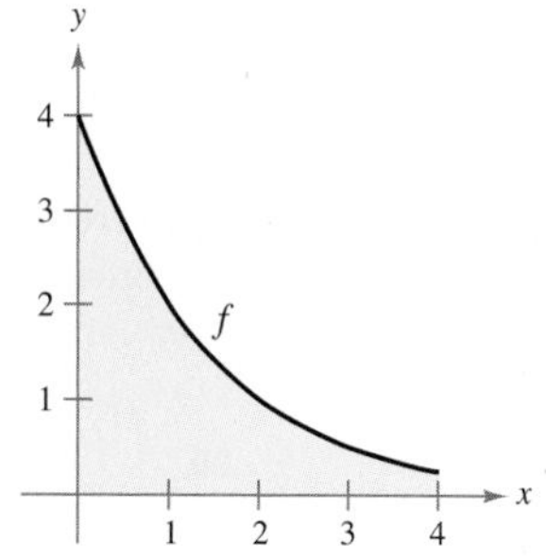

6 CHAPTER TEST

Take this test as you would take a test in class. When you are finished, check your work against the answers given in the back of the book.

In Exercises 1–3, find the indefinite integral and check the result by differentiation.

1. $\int \frac{x^2 + x + 1}{\sqrt{x}}\, dx$ **2.** $\int t\sqrt{t^2 + 7}\, dt$ **3.** $\int x\sqrt{x + 2}\, dx$

$f(x) = 4 - x^2$

Figure for 8

In Exercises 4 and 5, solve the differential equation.

4. $f'(x) = 12x,\ f(0) = -5$

5. $f''(x) = x^3,\ f'(2) = 7,\ f(0) = 5$

6. Use the properties of summation and Theorem 6.2 to evaluate $\sum_{i=1}^{12} (4i^2 - 15i + 12)$.

7. Use the limit process to find the area of the region between the graph of $y = 2x + 3$ and the x-axis over the interval $[0, 3]$. Sketch the region.

8. Set up a definite integral that yields the area of the region shown at the left. (Do not evaluate the integral.)

9. Sketch the region whose area is given by $\int_{-4}^{4} \sqrt{16 - x^2}\, dx$. Then use a geometric formula to evaluate the integral.

10. Given $\int_3^7 f(x)\, dx = 8$ and $\int_3^7 g(x)\, dx = 2$, evaluate

(a) $\int_3^7 [f(x) + g(x)]\, dx$. (b) $\int_3^7 [2f(x) - g(x)]\, dx$. (c) $\int_3^7 6g(x)\, dx$.

In Exercises 11–13, evaluate the definite integral.

11. $\int_2^4 (2x^2 + 4x - 7)\, dx$ **12.** $\int_3^8 \frac{x - \sqrt{x}}{5}\, dx$ **13.** $\int_{-1}^{1} x^2(1 - x^3)^2\, dx$

In Exercises 14 and 15, find the area of the region bounded by the graphs of the equations.

14. $y = 3x^2 + 1, x = 0, x = 2, y = 0$ **15.** $y = -x^2 + 8x, y = 0$

16. Find the average value of $f(x) = 6x^2 - 4x$ over the interval $[1, 3]$ and all value(s) of x in the interval for which the function equals its average value.

17. Find the derivative of $F(x) = \int_{x-3}^{x} (6t + 5)\, dt$.

18. Find an equation for the function f whose graph passes through the point $(1, 8)$ and whose derivative is $f'(x) = 3x(6x^2 - 2)^3$.

19. Evaluate $\int_{-2}^{2} (x^4 + 6x^2 + 2)\, dx$ using the properties of even and odd functions.

In Exercises 20 and 21, use the Trapezoidal Rule and Simpson's Rule with $n = 4$, and use the integration capabilities of a graphing utility, to approximate the definite integral. Compare the results.

20. $\int_0^1 \left(\frac{x^2}{2} + 1\right) dx$ **21.** $\int_1^9 \sqrt{x}\, dx$

P.S. PROBLEM SOLVING

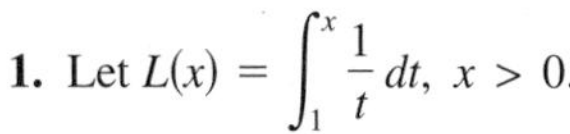

1. Let $L(x) = \int_1^x \frac{1}{t}\,dt,\ x > 0.$

(a) Find $L(1)$.

(b) Find $L'(x)$ and $L'(1)$.

(c) Use a graphing utility to approximate the value of x (to three decimal places) for which $L(x) = 1$.

(d) Prove that $L(x_1x_2) = L(x_1) + L(x_2)$ for all positive values of x_1 and x_2.

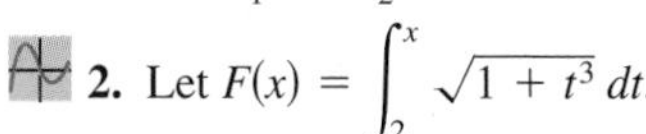

2. Let $F(x) = \int_2^x \sqrt{1 + t^3}\,dt.$

(a) Use a graphing utility to complete the table.

x	0	1.0	1.5	1.9	2.0
$F(x)$					

x	2.1	2.5	3.0	4.0	5.0
$F(x)$					

(b) Let $G(x) = \frac{1}{x-2}F(x) = \frac{1}{x-2}\int_2^x \sqrt{1 + t^3}\,dt.$ Use a graphing utility to complete the table and estimate $\lim_{x\to 2} G(x)$.

x	1.9	1.95	1.99	2.01	2.1
$G(x)$					

(c) Use the definition of the derivative to find the exact value of the limit $\lim_{x\to 2} G(x)$.

In Exercises 3 and 4, (a) write the area under the graph of the given function defined on the given interval as a limit. Then (b) evaluate the sum in part (a), and (c) evaluate the limit using the result of part (b).

3. $y = x^4 - 4x^3 + 4x^2, \quad [0, 2]$

$\left(\textit{Hint: } \sum_{i=1}^{n} i^4 = \frac{n(n+1)(2n+1)(3n^2+3n-1)}{30}\right)$

4. $y = \frac{1}{2}x^5 + 2x^3, \quad [0, 2]$

$\left(\textit{Hint: } \sum_{i=1}^{n} i^5 = \frac{n^2(n+1)^2(2n^2+2n-1)}{12}\right)$

5. The **Two-Point Gaussian Quadrature Approximation** for f is

$$\int_{-1}^{1} f(x)\,dx \approx f\left(-\frac{1}{\sqrt{3}}\right) + f\left(\frac{1}{\sqrt{3}}\right).$$

(a) Use this formula to approximate $\int_{-1}^{1} \sqrt{x+2}\,dx$. Find the error of the approximation.

(b) Use this formula to approximate $\int_{-1}^{1} \frac{1}{1+x^2}\,dx$.

(c) Prove that the Two-Point Gaussian Quadrature Approximation is exact for all polynomials of degree 3 or less.

6. Let $f(x) = x^2$ on the interval $[0, 3]$, as indicated in the figure.

(a) Find the slope of the segment OB.

(b) Find the average value of the slope of the tangent line to the graph of f on the interval $[0, 3]$.

(c) Let f be an arbitrary function having a continuous first derivative on the interval $[a, b]$. Find the average value of the slope of the tangent line to the graph of f on the interval $[a, b]$.

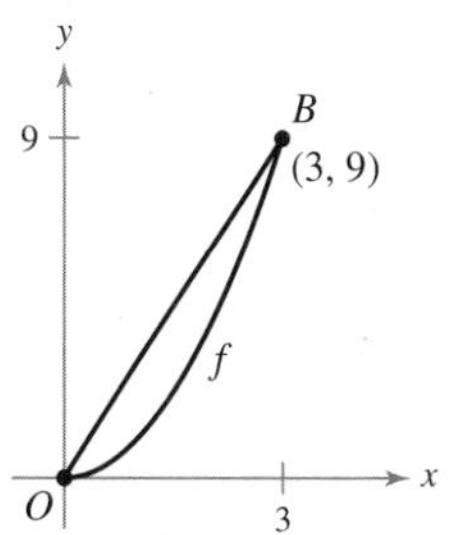

Figure for 6

Figure for 7

7. Archimedes showed that the area of a parabolic arch is equal to $\frac{2}{3}$ the product of the base and the height (see figure).

(a) Graph the parabolic arch bounded by $y = 9 - x^2$ and the x-axis. Use an appropriate integral to find the area A.

(b) Find the base and height of the arch and verify Archimedes' formula.

(c) Prove Archimedes' formula for a general parabola.

8. Galileo Galilei (1564–1642) stated the following proposition concerning falling objects:

The time in which any space is traversed by a uniformly accelerating body is equal to the time in which that same space would be traversed by the same body moving at a uniform speed whose value is the mean of the highest speed of the accelerating body and the speed just before acceleration began.

Use the techniques of this chapter to verify this proposition.

9. The graph of the function f consists of the three line segments joining the points $(0, 0)$, $(2, -2)$, $(6, 2)$, and $(8, 3)$. The function F is defined by the integral

$$F(x) = \int_0^x f(t)\,dt.$$

(a) Sketch the graph of f.

(b) Complete the table.

x	0	1	2	3	4	5	6	7	8
$F(x)$									

(c) Find the extrema of F on the interval $[0, 8]$.

(d) Determine all points of inflection of F on the interval $(0, 8)$.

10. A car travels in a straight line for 1 hour. Its velocity v in miles per hour at six-minute intervals is shown in the table.

t (hours)	0	0.1	0.2	0.3	0.4	0.5
v (mi/h)	0	10	20	40	60	50

t (hours)	0.6	0.7	0.8	0.9	1.0
v (mi/h)	40	35	40	50	65

(a) Produce a reasonable graph of the velocity function v by graphing these points and connecting them with a smooth curve.

(b) Find the open intervals over which the acceleration a is positive.

(c) Find the average acceleration of the car (in miles per hour squared) over the interval $[0, 0.4]$.

(d) What does the integral $\int_0^1 v(t)\,dt$ signify? Approximate this integral using the Trapezoidal Rule with five subintervals.

(e) Approximate the acceleration at $t = 0.8$.

11. Prove $\displaystyle\int_0^x f(t)(x-t)\,dt = \int_0^x\left(\int_0^t f(v)\,dv\right)dt.$

12. Prove $\displaystyle\int_a^b f(x)f'(x)\,dx = \tfrac{1}{2}([f(b)]^2 - [f(a)]^2).$

13. Use an appropriate Riemann sum to evaluate the limit

$$\lim_{n\to\infty}\frac{\sqrt{1}+\sqrt{2}+\sqrt{3}+\cdots+\sqrt{n}}{n^{3/2}}.$$

14. Use an appropriate Riemann sum to evaluate the limit

$$\lim_{n\to\infty}\frac{1^5+2^5+3^5+\cdots+n^5}{n^6}.$$

15. Suppose that f is integrable on $[a, b]$ and $0 < m \le f(x) \le M$ for all x in the interval $[a, b]$. Prove that

$$m(a-b) \le \int_a^b f(x)\,dx \le M(b-a).$$

Use this result to estimate $\int_0^1 \sqrt{1+x^4}\,dx$.

16. Prove that if f is a continuous function on a closed interval $[a, b]$, then

$$\left|\int_a^b f(x)\,dx\right| \le \int_a^b |f(x)|\,dx.$$

17. Verify that

$$\sum_{i=1}^{n} i^2 = \frac{n(n+1)(2n+1)}{6}$$

by showing the following.

(a) $(1+i)^3 - i^3 = 3i^2 + 3i + 1$

(b) $(n+1)^3 = \sum_{i=1}^{n}(3i^2+3i+1) + 1$

(c) $\sum_{i=1}^{n} i^2 = \dfrac{n(n+1)(2n+1)}{6}$

18. In this exercise you will use the formula

$$1+2+3+\cdots+n = \frac{n(n+1)}{2}$$

to derive the formula

$$1^3+2^3+3^3+\cdots+n^3 = \frac{n^2(n+1)^2}{4}.$$

Let $S = 1+2+3+\cdots+n$ be the length of the sides of the square in the figure. Mark off segments of lengths $1, 2, 3, \ldots, n$ along two adjacent sides.

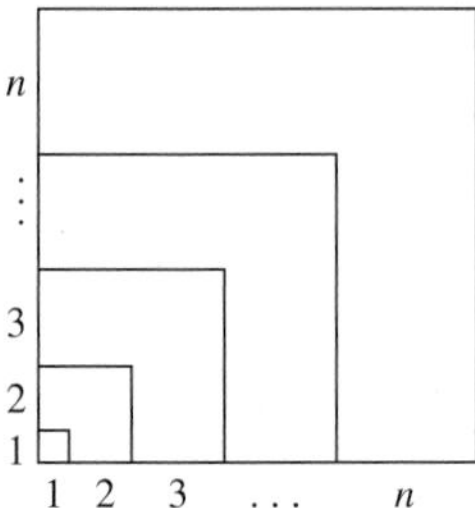

(a) Show that the area A of the square is

$$A = S^2 = (1+2+3+\cdots+n)^2 = \left[\frac{n(n+1)}{2}\right]^2.$$

(b) Show that the area of the shaded region is $A_k = k^3$. (*Hint:* Divide the region into two rectangles as indicated.)

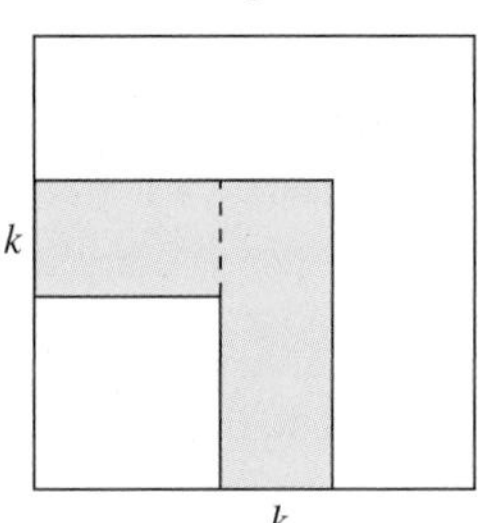

(c) Verify the formula

$$A = \left[\frac{n(n+1)}{2}\right]^2 = 1^3+2^3+3^3+\cdots+n^3.$$

19. ***Oil Leak*** At 1:00 P.M., oil begins leaking from a tank at a rate of $4 + 0.75t$ gallons per hour.

(a) How much oil is lost from 1:00 P.M. to 4:00 P.M.?

(b) How much oil is lost from 4:00 P.M. to 7:00 P.M.?

(c) Compare your answers from parts (a) and (b). What do you notice?

20. Find the function $f(x)$ and all values of c such that

$$\int_c^x f(t)\,dt = x^2 + x - 2.$$

21. Determine the limits of integration where $a \le b$ such that

$$\int_a^b (x^2 - 16)\,dx$$

has minimal value.

7 Exponential and Logarithmic Functions

In this chapter you will study two types of nonalgebraic functions– *exponential functions* and *logarithmic functions.* Exponential and logarithmic functions are widely used in describing economic and physical phenomena such as compound interest, population growth, memory retention, and decay of radioactive material.

In this chapter, you should learn the following.

- How to recognize, evaluate, and graph exponential functions. (7.1)
- How to recognize, evaluate, and graph logarithmic functions. (7.2)
- How to use properties of logarithms to evaluate, rewrite, expand, or condense logarithmic expressions. (7.3)
- How to solve exponential and logarithmic equations. (7.4)
- How to use exponential growth models, exponential decay models, Gaussian models, logistic growth models, and logarithmic models to solve real-life problems. (7.5)

Juniors Bildarchiv / Alamy

Given data about four-legged animals, how can you find a logarithmic function that can be used to relate an animal's weight and its lowest galloping speed? (See Section 7.3, Exercise 96.)

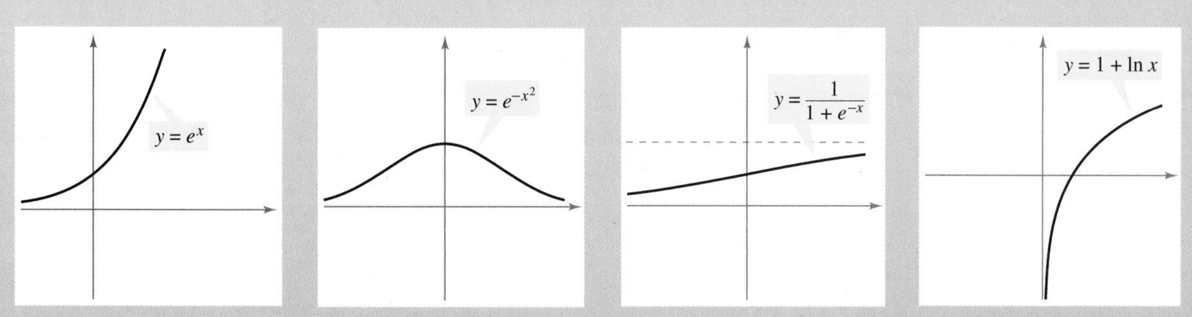

You can use exponential and logarithmic functions to model many real-life situations. You will learn about the types of data that are best represented by the different models. (See Section 7.5.)

7.1 Exponential Functions and Their Graphs

- **Recognize and evaluate exponential functions with base *a*.**
- **Graph exponential functions.**
- **Recognize, evaluate, and graph exponential functions with base *e*.**
- **Use exponential functions to model and solve real-life problems.**

Exponential Functions

So far, this text has dealt mainly with **algebraic functions,** which include polynomial functions and rational functions. In this chapter, you will study two types of nonalgebraic functions—*exponential* functions and *logarithmic* functions. These functions are examples of **transcendental functions.**

> **DEFINITION OF EXPONENTIAL FUNCTION**
>
> The **exponential function f with base a** is denoted by
>
> $$f(x) = a^x$$
>
> where $a > 0$, $a \neq 1$, and x is any real number.

The base $a = 1$ is excluded because it yields $f(x) = 1^x = 1$. This is a constant function, not an exponential function.

You have evaluated a^x for integer and rational values of x. For example, you know that $4^3 = 64$ and $4^{1/2} = 2$. However, to evaluate 4^x for any real number x, you need to interpret forms with *irrational* exponents. For the purposes of this text, it is sufficient to think of

$$a^{\sqrt{2}} \quad (\text{where } \sqrt{2} \approx 1.41421356)$$

as the number that has the successively closer approximations

$$a^{1.4}, a^{1.41}, a^{1.414}, a^{1.4142}, a^{1.41421}, \ldots .$$

Graphs of Exponential Functions

The graphs of all exponential functions have similar characteristics, as shown in Examples 1 through 3.

EXAMPLE 1 Graphs of $y = a^x$

In the same coordinate plane, sketch the graphs of $f(x) = 2^x$ and $g(x) = 4^x$.

Solution The table below lists some values for each function, and Figure 7.1 shows the graphs of the two functions. Note that both graphs are increasing. Moreover, the graph of $g(x) = 4^x$ is increasing more rapidly than the graph of $f(x) = 2^x$.

x	-3	-2	-1	0	1	2
2^x	$\frac{1}{8}$	$\frac{1}{4}$	$\frac{1}{2}$	1	2	4
4^x	$\frac{1}{64}$	$\frac{1}{16}$	$\frac{1}{4}$	1	4	16

The *table* feature of a graphing utility could be used to expand the table. ■

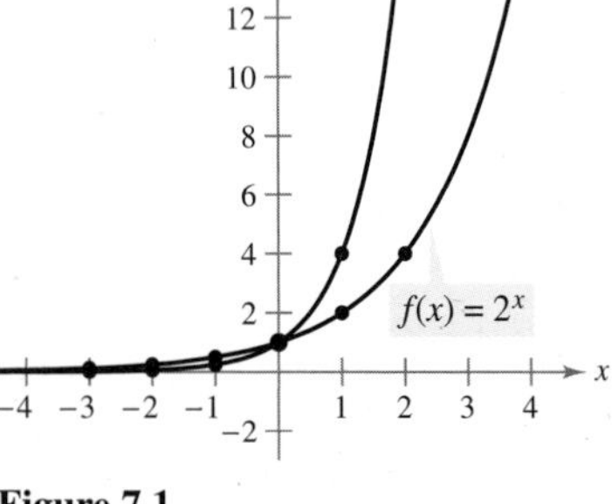

Figure 7.1

EXAMPLE 2 Graphs of $y = a^{-x}$

In the same coordinate plane, sketch the graphs of $F(x) = 2^{-x}$ and $G(x) = 4^{-x}$.

Solution The table below lists some values for each function, and Figure 7.2 shows the graphs of the two functions. Note that both graphs are decreasing. Moreover, the graph of $G(x) = 4^{-x}$ is decreasing more rapidly than the graph of $F(x) = 2^{-x}$.

x	-2	-1	0	1	2	3
2^{-x}	4	2	1	$\frac{1}{2}$	$\frac{1}{4}$	$\frac{1}{8}$
4^{-x}	16	4	1	$\frac{1}{4}$	$\frac{1}{16}$	$\frac{1}{64}$

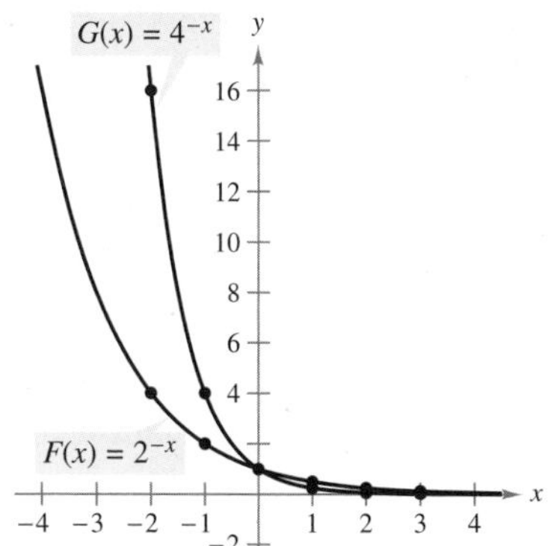

Figure 7.2

In Example 2, note that by using the properties of exponents, the functions $F(x) = 2^{-x}$ and $G(x) = 4^{-x}$ can be rewritten with positive exponents.

$$F(x) = 2^{-x} = \frac{1}{2^x} = \left(\frac{1}{2}\right)^x \quad \text{and} \quad G(x) = 4^{-x} = \frac{1}{4^x} = \left(\frac{1}{4}\right)^x$$

Comparing the functions in Examples 1 and 2, observe that

$$F(x) = 2^{-x} = f(-x) \qquad \text{and} \qquad G(x) = 4^{-x} = g(-x).$$

Consequently, the graph of F is a reflection (in the y-axis) of the graph of f. The graphs of G and g have the same relationship. The graphs in Figures 7.1 and 7.2 are typical of the exponential functions $y = a^x$ and $y = a^{-x}$. They have one y-intercept and one horizontal asymptote (the x-axis), and they are continuous. The basic characteristics of these exponential functions are summarized in Figures 7.3 and 7.4.

STUDY TIP Notice that the range of an exponential function is $(0, \infty)$, which means that $a^x > 0$ for all values of x.

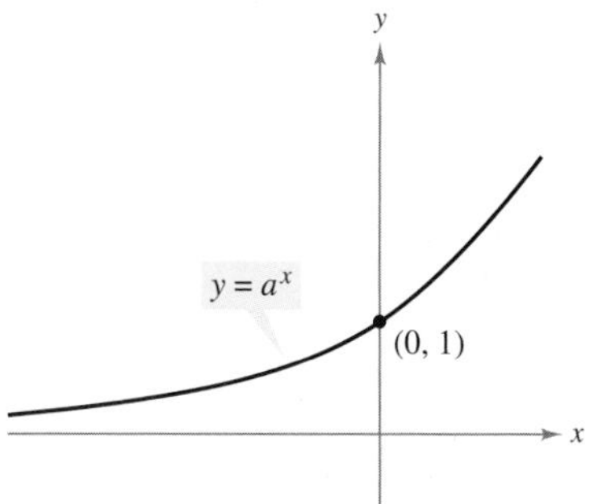

Figure 7.3

Graph of $y = a^x, a > 1$
- Domain: $(-\infty, \infty)$
- Range: $(0, \infty)$
- y-intercept: $(0, 1)$
- Increasing
- x-axis is a horizontal asymptote ($a^x \to 0$ as $x \to -\infty$).
- Continuous

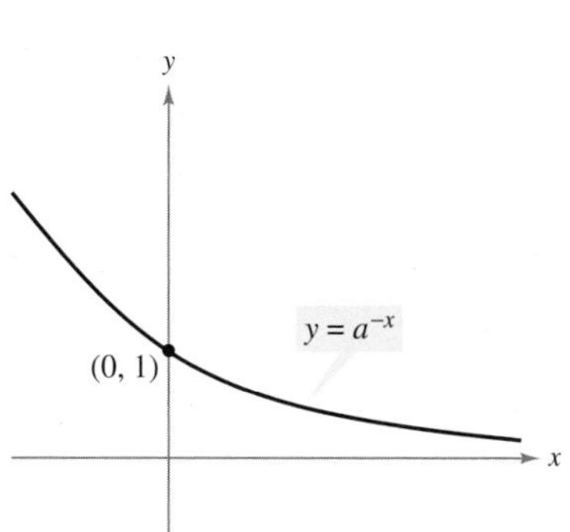

Figure 7.4

Graph of $y = a^{-x}, a > 1$
- Domain: $(-\infty, \infty)$
- Range: $(0, \infty)$
- y-intercept: $(0, 1)$
- Decreasing
- x-axis is a horizontal asymptote ($a^{-x} \to 0$ as $x \to \infty$).
- Continuous

EXPLORATION

Use a graphing utility to graph

$$y = a^x$$

for $a = 3, 5$, and 7 in the same viewing window. (Use a viewing window in which $-2 \le x \le 1$ and $0 \le y \le 2$.)

How do the graphs compare with each other? Which graph is on the top in the interval $(-\infty, 0)$? Which is on the bottom? Which graph is on the top in the interval $(0, \infty)$? Which is on the bottom?

Repeat this experiment with the graphs of $y = b^x$ for $b = \frac{1}{3}, \frac{1}{5}$, and $\frac{1}{7}$. (Use a viewing window in which $-1 \le x \le 2$ and $0 \le y \le 2$.) What can you conclude about the shape of the graph of $y = b^x$ and the value of b?

In the following example, notice how the graph of $y = a^x$ can be used to sketch the graphs of functions of the form $f(x) = b \pm a^{x+c}$.

EXAMPLE 3 Transformations of Graphs of Exponential Functions

Use the graph of $f(x) = 3^x$ to describe the transformation that yields the graph of g.

a. $g(x) = 3^{x+1}$ **b.** $g(x) = 3^x - 2$ **c.** $g(x) = -3^x$ **d.** $g(x) = 3^{-x}$

Solution

a. Because $g(x) = 3^{x+1} = f(x + 1)$, the graph of g can be obtained by shifting the graph of f one unit to the left. See Figure 7.5(a).

b. Because $g(x) = 3^x - 2 = f(x) - 2$, the graph of g can be obtained by shifting the graph of f down two units. See Figure 7.5(b).

c. Because $g(x) = -3^x = -f(x)$, the graph of g can be obtained by reflecting the graph of f in the x-axis. See Figure 7.5(c).

d. Because $g(x) = 3^{-x} = f(-x)$, the graph of g can be obtained by reflecting the graph of f in the y-axis. See Figure 7.5(d).

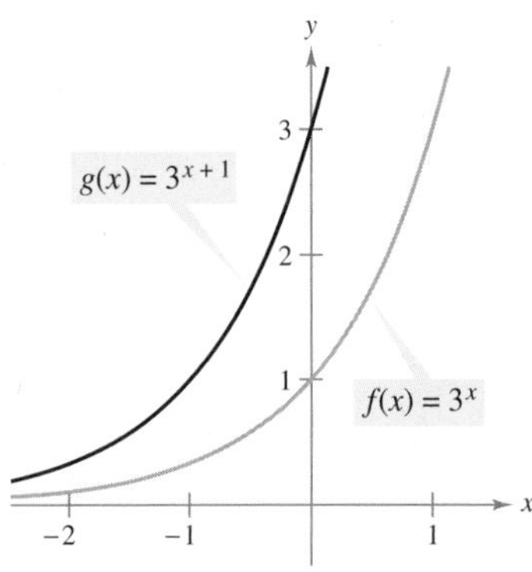

(a) Horizontal shift to the left

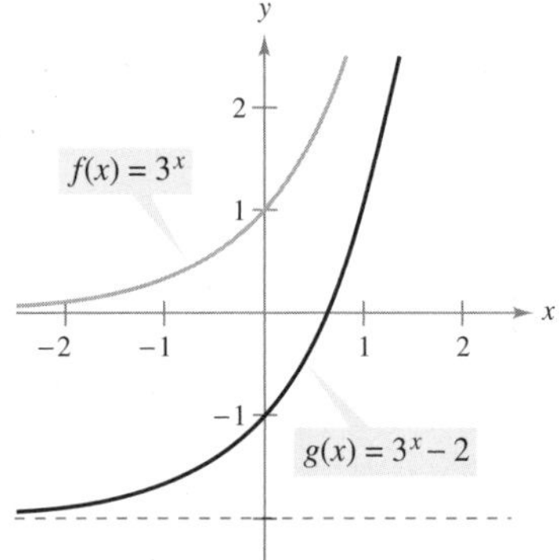

(b) Vertical shift downward

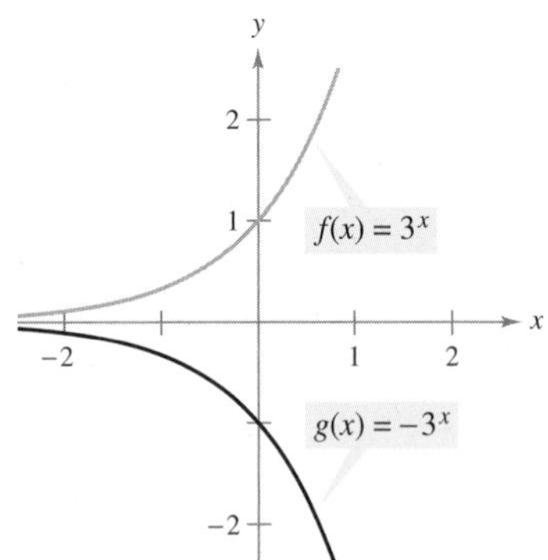

(c) Reflection in the x-axis

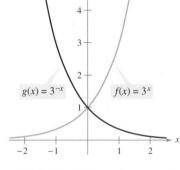

(d) Reflection in the y-axis

Figure 7.5

In Figure 7.5, notice that the transformations in parts (a), (c), and (d) keep the x-axis as a horizontal asymptote, but the transformation in part (b) yields a new horizontal asymptote of $y = -2$. Also, be sure to note how the y-intercept is affected by each transformation.

The Natural Base *e*

In many applications, the most convenient choice for a base is the irrational number $e \approx 2.718281828\ldots$. This number is called the *natural base*. The function given by $f(x) = e^x$ is called the natural exponential function. Its graph is shown in Figure 7.6. Be sure you see that for the exponential function given by $f(x) = e^x$, e is the constant $2.718281828\ldots$, whereas x is the variable.

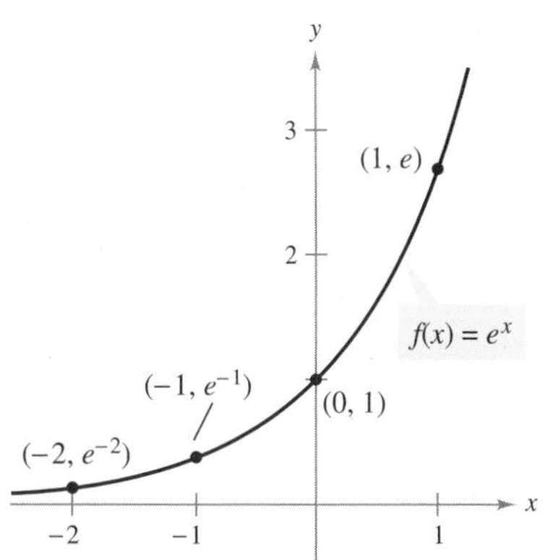

Figure 7.6

THEOREM 7.1 A LIMIT INVOLVING *e*

The following limits exist and are equal. The real number that is the limit is defined to be $e \approx 2.718281828\ldots$.

$$\lim_{x \to \infty} \left(1 + \frac{1}{x}\right)^x = e$$

$$\lim_{x \to 0} (1 + x)^{1/x} = e$$

THE NUMBER *e*

The symbol *e* was first used by mathematician Leonhard Euler to represent the base of natural logarithms in a letter to another mathematician, Christian Goldbach, in 1731.

STUDY TIP The choice of *e* as a base for exponential functions may seem anything but "natural." In Section 8.1, you will see more clearly why *e* is the convenient choice for a base.

EXAMPLE 4 Graphing Natural Exponential Functions

Sketch the graph of each natural exponential function.

a. $f(x) = 2e^{0.24x}$

b. $g(x) = \frac{1}{2}e^{-0.58x}$

Solution To sketch these two graphs, you can use a graphing utility to construct a table of values, as shown below. After constructing the table, plot the points and connect them with smooth curves, as shown in Figure 7.7. Note that the graph in part (a) is increasing whereas the graph in part (b) is decreasing.

x	-3	-2	-1	0	1	2	3
$f(x)$	0.974	1.238	1.573	2.000	2.542	3.232	4.109
$g(x)$	2.849	1.595	0.893	0.500	0.280	0.157	0.088

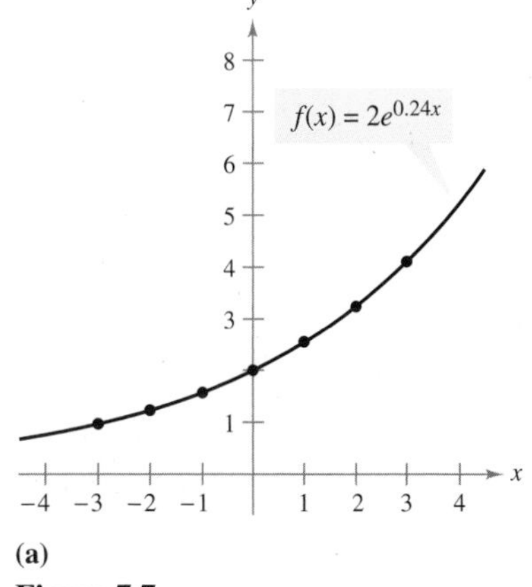

(a)

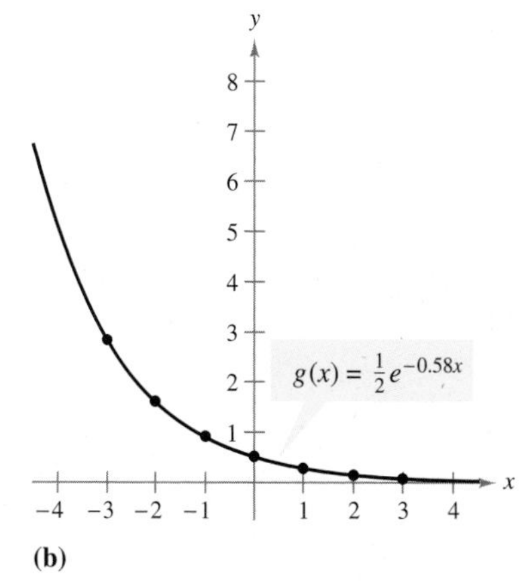

(b)

Figure 7.7

■

Applications

One of the most familiar examples of exponential growth is that of an investment earning *continuously compounded interest.* Using exponential functions, you can develop a formula for the balance in an account that pays compound interest, and show how it leads to continuous compounding.

Suppose a principal P is invested at an annual interest rate r, compounded once a year. If the interest is added to the principal at the end of the year, the new balance P_1 is

$$P_1 = P + Pr = P(1 + r).$$

This pattern of multiplying the previous principal by $1 + r$ is then repeated each successive year, as shown below.

Year	*Balance After Each Compounding*
0	$P = P$
1	$P_1 = P(1 + r)$
2	$P_2 = P_1(1 + r) = P(1 + r)(1 + r) = P(1 + r)^2$
3	$P_3 = P_2(1 + r) = P(1 + r)^2(1 + r) = P(1 + r)^3$
⋮	⋮
t	$P_t = P(1 + r)^t$

To accommodate more frequent (quarterly, monthly, or daily) compounding of interest, let n be the number of compoundings per year and let t be the number of years. Then the rate per compounding is r/n and the account balance after t years is

$$A = P\left(1 + \frac{r}{n}\right)^{nt}. \qquad \text{Amount (balance) with } n \text{ compoundings per year}$$

If you let the number of compoundings n increase without bound, the process approaches what is called **continuous compounding.** In the formula for n compoundings per year, let $m = n/r$. This produces

$$A = P\left(1 + \frac{r}{n}\right)^{nt} \qquad \text{Amount with } n \text{ compoundings per year}$$

$$= P\left(1 + \frac{r}{mr}\right)^{mrt} \qquad \text{Substitute } mr \text{ for } n.$$

$$= P\left(1 + \frac{1}{m}\right)^{mrt} \qquad \text{Simplify.}$$

$$= P\left[\left(1 + \frac{1}{m}\right)^{m}\right]^{rt}. \qquad \text{Property of exponents}$$

As m increases without bound, $[1 + (1/m)]^m$ approaches e. From this, you can conclude that the formula for continuous compounding is

$$A = Pe^{rt}. \qquad \text{Substitute } e \text{ for } (1 + 1/m)^m.$$

FORMULAS FOR COMPOUND INTEREST

After t years, the balance A in an account with principal P and annual interest rate r (in decimal form) is given by the following formulas.

1. For n compoundings per year: $A = P\left(1 + \frac{r}{n}\right)^{nt}$

2. For continuous compounding: $A = Pe^{rt}$

STUDY TIP Be sure you see that the annual interest rate must be expressed in decimal form when using the compound interest formula. For instance, 6% should be expressed as 0.06.

NOTE In Example 5, note that continuous compounding yields more than quarterly, monthly, or daily compounding. This is typical of the two types of compounding. That is, for a given principal, interest rate, and time, continuous compounding will always yield a larger balance than compounding n times a year.

EXAMPLE 5 Compound Interest

A total of \$12,000 is invested at an annual interest rate of 9%. Find the balance after 5 years if it is compounded

a. quarterly. **b.** monthly. **c.** daily. **d.** continuously.

Solution

a. For quarterly compoundings, you have $n = 4$. So, in 5 years at 9%, the balance is

$$A = P\left(1 + \frac{r}{n}\right)^{nt} = 12{,}000\left(1 + \frac{0.09}{4}\right)^{4(5)} \approx \$18{,}726.11.$$

b. For monthly compoundings, you have $n = 12$. So, in 5 years at 9%, the balance is

$$A = P\left(1 + \frac{r}{n}\right)^{nt} = 12{,}000\left(1 + \frac{0.09}{12}\right)^{12(5)} \approx \$18{,}788.17.$$

c. For daily compoundings, you have $n = 365$. So, in 5 years at 9%, the balance is

$$A = P\left(1 + \frac{r}{n}\right)^{nt} = 12{,}000\left(1 + \frac{0.09}{365}\right)^{365(5)} \approx \$18{,}818.70.$$

d. For continuous compounding, the balance is

$$A = Pe^{rt} = 12{,}000e^{0.09(5)} \approx \$18{,}819.75.$$ ■

EXPLORATION

Use a graphing utility to make a table of values that shows the amount of time it would take to *double* the investment in Example 5 using continuous compounding.

EXAMPLE 6 Radioactive Decay

The *half-life* of radioactive radium (^{226}Ra) is about 1599 years. That is, for a given amount of radium, *half* of the original amount will remain after 1599 years. After another 1599 years, one-quarter of the original amount will remain, and so on. Let y represent the mass, in grams, of a quantity of radium. The quantity present after t years, then, is $y = 25\left(\frac{1}{2}\right)^{t/1599}$.

a. What is the initial mass (when $t = 0$)?

b. How much of the initial mass is present after 2500 years?

Algebraic Solution

a. $y = 25\left(\frac{1}{2}\right)^{t/1599}$ Write original equation.

$= 25\left(\frac{1}{2}\right)^{0/1599}$ Substitute 0 for t.

$= 25$ Simplify.

So, the initial mass is 25 grams.

b. $y = 25\left(\frac{1}{2}\right)^{t/1599}$ Write original equation.

$= 25\left(\frac{1}{2}\right)^{2500/1599}$ Substitute 2500 for t.

$\approx 25\left(\frac{1}{2}\right)^{1.563}$ Simplify.

≈ 8.46 Use a calculator.

So, about 8.46 grams is present after 2500 years.

Graphical Solution

Use a graphing utility to graph $y = 25\left(\frac{1}{2}\right)^{t/1599}$.

a. Use the *value* feature or the *zoom* and *trace* features of the graphing utility to determine that when $x = 0$, the value of y is 25, as shown in Figure 7.8(a). So, the initial mass is 25 grams.

b. Use the *value* feature or the *zoom* and *trace* features of the graphing utility to determine that when $x = 2500$, the value of y is about 8.46, as shown in Figure 7.8(b). So, about 8.46 grams is present after 2500 years.

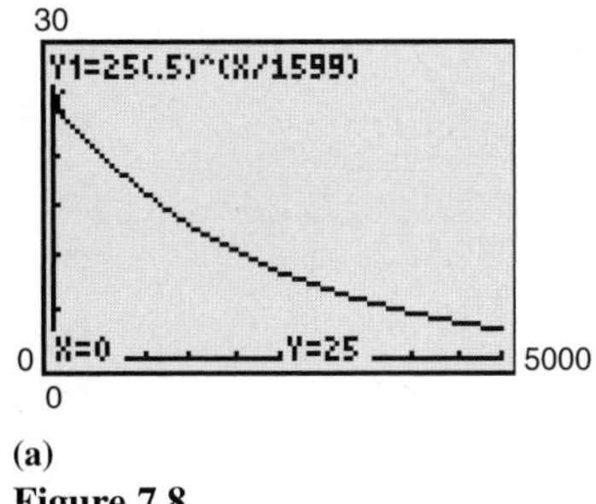

(a)

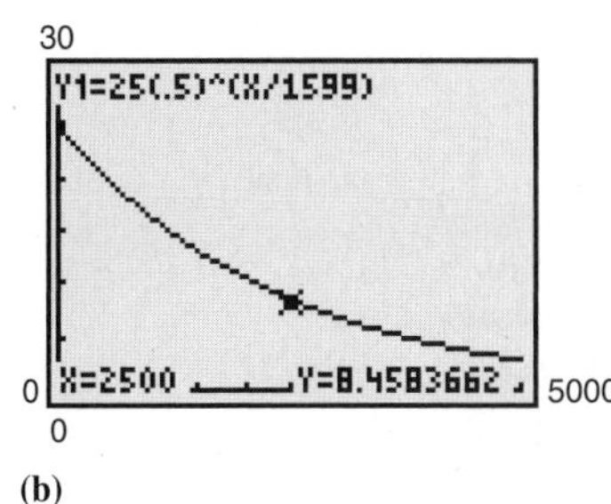

(b)

Figure 7.8

7.1 Exercises

See www.CalcChat.com for worked-out solutions to odd-numbered exercises.

In Exercises 1–6, fill in the blanks.

1. Polynomial and rational functions are examples of ________ functions.
2. Exponential and logarithmic functions are examples of nonalgebraic functions, also called ________ functions.
3. You can use ________ of the graph of $y = a^x$ to sketch the graphs of functions of the form $f(x) = b \pm a^{x+c}$.
4. The exponential function given by $f(x) = e^x$ is called the ________ ________ function, and the base e is called the ________ base.
5. To find the amount A in an account after t years with principal P and an annual interest rate r compounded n times per year, you can use the formula ________.
6. To find the amount A in an account after t years with principal P and an annual interest rate r compounded continuously, you can use the formula ________.

In Exercises 7–10, evaluate the function at the indicated value of x. Round your result to three decimal places.

	Function	*Value*
7.	$f(x) = 0.9^x$	$x = 1.4$
8.	$f(x) = 2.3^x$	$x = \frac{3}{2}$
9.	$f(x) = 5^x$	$x = -\pi$
10.	$f(x) = \left(\frac{2}{3}\right)^{5x}$	$x = \frac{3}{10}$

In Exercises 11–14, match the exponential function with its graph. [The graphs are labeled (a), (b), (c), and (d).]

(a)

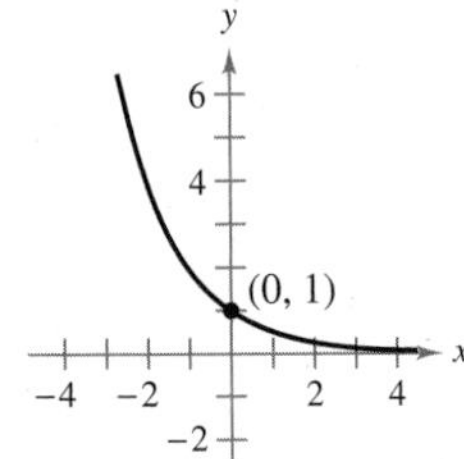

(b)

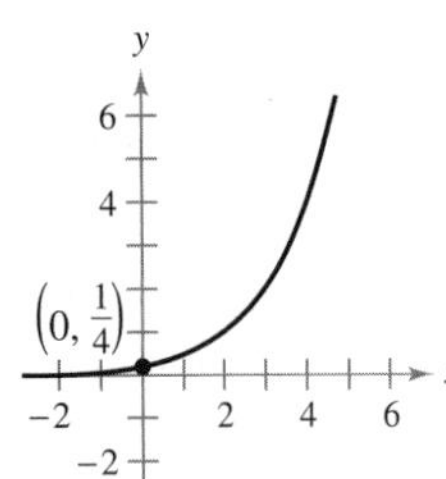

(c)

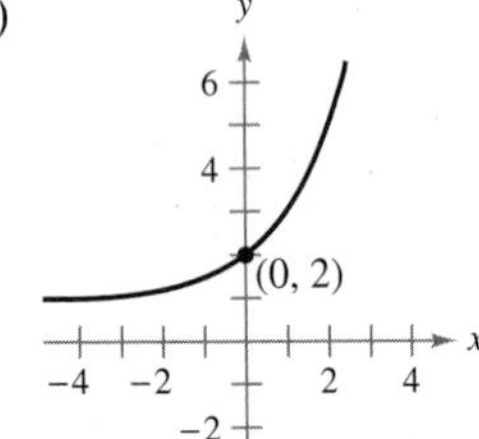

(d)

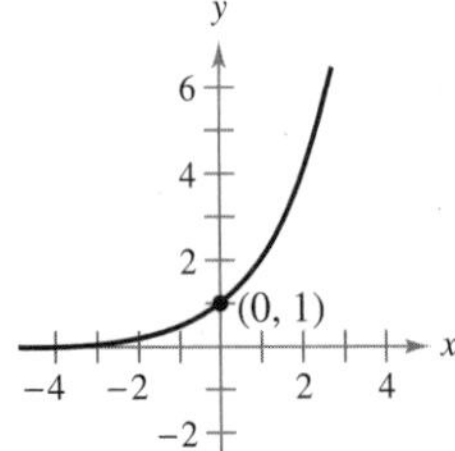

11. $f(x) = 2^x$
12. $f(x) = 2^x + 1$
13. $f(x) = 2^{-x}$
14. $f(x) = 2^{x-2}$

In Exercises 15–20, use a graphing utility to construct a table of values for the function. Then sketch the graph of the function.

15. $f(x) = \left(\frac{1}{2}\right)^x$
16. $f(x) = \left(\frac{1}{2}\right)^{-x}$
17. $f(x) = 6^{-x}$
18. $f(x) = 6^x$
19. $f(x) = 2^{x-1}$
20. $f(x) = 4^{x-3} + 3$

In Exercises 21–26, use the graph of f to describe the transformation that yields the graph of g.

21. $f(x) = 3^x, \quad g(x) = 3^x + 1$
22. $f(x) = 4^x, \quad g(x) = 4^{x-3}$
23. $f(x) = 2^x, \quad g(x) = 3 - 2^x$
24. $f(x) = 10^x, \quad g(x) = 10^{-x+3}$
25. $f(x) = \left(\frac{7}{2}\right)^x, \quad g(x) = -\left(\frac{7}{2}\right)^{-x}$
26. $f(x) = 0.3^x, \quad g(x) = -0.3^x + 5$

In Exercises 27–30, use a graphing utility to graph the exponential function.

27. $y = 2^{-x^2}$
28. $y = 3^{-|x|}$
29. $y = 3^{x-2} + 1$
30. $y = 4^{x+1} - 2$

In Exercises 31–34, evaluate the function at the indicated value of x. Round your result to three decimal places.

	Function	*Value*
31.	$h(x) = e^{-x}$	$x = \frac{3}{4}$
32.	$f(x) = e^x$	$x = 3.2$
33.	$f(x) = 2e^{-5x}$	$x = 10$
34.	$f(x) = 1.5e^{x/2}$	$x = 240$

In Exercises 35–40, use a graphing utility to construct a table of values for the function. Then sketch the graph of the function.

35. $f(x) = e^x$
36. $f(x) = e^{-x}$
37. $f(x) = 3e^{x+4}$
38. $f(x) = 2e^{-0.5x}$
39. $f(x) = 2e^{x-2} + 4$
40. $f(x) = 2 + e^{x-5}$

In Exercises 41–46, use a graphing utility to graph the exponential function.

41. $y = 1.08^{-5x}$

42. $y = 1.08^{5x}$

43. $s(t) = 2e^{0.12t}$

44. $s(t) = 3e^{-0.2t}$

45. $g(x) = 1 + e^{-x}$

46. $h(x) = e^{x-2}$

WRITING ABOUT CONCEPTS

In Exercises 47–50, use properties of exponents to determine which functions (if any) are the same.

47. $f(x) = 3^{x-2}$
$g(x) = 3^x - 9$
$h(x) = \frac{1}{9}(3^x)$

48. $f(x) = 4^x + 12$
$g(x) = 2^{2x+6}$
$h(x) = 64(4^x)$

49. $f(x) = 16(4^{-x})$
$g(x) = \left(\frac{1}{4}\right)^{x-2}$
$h(x) = 16(2^{-2x})$

50. $f(x) = 5^{-x} + 3$
$g(x) = 5^{3-x}$
$h(x) = -5^{x-3}$

51. Graph the functions given by $y = 3^x$ and $y = 4^x$ and use the graphs to solve each inequality.

(a) $4^x < 3^x$

(b) $4^x > 3^x$

52. Graph the functions given by $y = \left(\frac{1}{2}\right)^x$ and $y = \left(\frac{1}{4}\right)^x$ and use the graphs to solve each inequality.

(a) $\left(\frac{1}{4}\right)^x < \left(\frac{1}{2}\right)^x$

(b) $\left(\frac{1}{4}\right)^x > \left(\frac{1}{2}\right)^x$

53. Use a graphing utility to graph $y_1 = e^x$ and each of the functions $y_2 = x^2$, $y_3 = x^3$, $y_4 = \sqrt{x}$, and $y_5 = |x|$. Which function increases at the greatest rate as x approaches $+\infty$?

54. Use the result of Exercise 53 to make a conjecture about the rate of growth of $y_1 = e^x$ and $y = x^n$, where n is a natural number and x approaches $+\infty$.

55. Use the results of Exercises 53 and 54 to describe what is implied when it is stated that a quantity is growing exponentially.

56. Which functions are exponential?

(a) $f(x) = 3x$

(b) $f(x) = 3x^2$

(c) $f(x) = 3^x$

(d) $f(x) = 2^{-x}$

Compound Interest **In Exercises 57–60, complete the table to determine the balance A for P dollars invested at rate r for t years and compounded n times per year.**

n	1	2	4	12	365	Continuous
A						

57. $P = \$1500$, $r = 2\%$, $t = 10$ years

58. $P = \$2500$, $r = 3.5\%$, $t = 10$ years

59. $P = \$2500$, $r = 4\%$, $t = 20$ years

60. $P = \$1000$, $r = 6\%$, $t = 40$ years

Compound Interest **In Exercises 61–64, complete the table to determine the balance A for \$12,000 invested at rate r for t years, compounded continuously.**

t	10	20	30	40	50
A					

61. $r = 4\%$

62. $r = 6\%$

63. $r = 6.5\%$

64. $r = 3.5\%$

65. ***Trust Fund*** On the day of a child's birth, a deposit of \$30,000 is made in a trust fund that pays 5% interest, compounded continuously. Determine the balance in this account on the child's 25th birthday.

66. ***Trust Fund*** A deposit of \$5000 is made in a trust fund that pays 7.5% interest, compounded continuously. It is specified that the balance will be given to the college from which the donor graduated after the money has earned interest for 50 years. How much will the college receive?

67. ***Inflation*** If the annual rate of inflation averages 4% over the next 10 years, the approximate costs C of goods or services during any year in that decade will be modeled by $C(t) = P(1.04)^t$, where t is the time in years and P is the present cost. The price of an oil change for your car is presently \$23.95. Estimate the price 10 years from now.

68. ***Computer Virus*** The number V of computers infected by a computer virus increases according to the model $V(t) = 100e^{4.6052t}$, where t is the time in hours. Find the number of computers infected after (a) 1 hour, (b) 1.5 hours, and (c) 2 hours.

69. ***Population Growth*** The projected populations of California for the years 2015 through 2030 can be modeled by $P = 34.696e^{0.0098t}$, where P is the population (in millions) and t is the time (in years), with $t = 15$ corresponding to 2015. *(Source: U.S. Census Bureau)*

(a) Use a graphing utility to graph the function for the years 2015 through 2030.

(b) Use the *table* feature of a graphing utility to create a table of values for the same time period as in part (a).

(c) According to the model, when will the population of California exceed 50 million?

70. ***Population*** The populations P (in millions) of Italy from 1990 through 2008 can be approximated by the model $P = 56.8e^{0.0015t}$, where t represents the year, with $t = 0$ corresponding to 1990. *(Source: U.S. Census Bureau, International Data Base)*

(a) According to the model, is the population of Italy increasing or decreasing? Explain.

(b) Find the populations of Italy in 2000 and 2008.

(c) Use the model to predict the populations of Italy in 2015 and 2020.

71. ***Radioactive Decay*** Let Q represent a mass of radioactive plutonium (^{239}Pu) (in grams), whose half-life is 24,100 years. The quantity of plutonium present after t years is $Q = 16\left(\frac{1}{2}\right)^{t/24,100}$.

(a) Determine the initial quantity (when $t = 0$).

(b) Determine the quantity present after 75,000 years.

(c) Use a graphing utility to graph the function over the interval $t = 0$ to $t = 150,000$.

72. ***Radioactive Decay*** Let Q represent a mass of carbon 14 (^{14}C) (in grams), whose half-life is 5715 years. The quantity of carbon 14 present after t years is $Q = 10\left(\frac{1}{2}\right)^{t/5715}$.

(a) Determine the initial quantity (when $t = 0$).

(b) Determine the quantity present after 2000 years.

(c) Sketch the graph of this function over the interval $t = 0$ to $t = 10,000$.

73. ***Depreciation*** After t years, the value of a wheelchair conversion van that originally cost \$30,500 depreciates so that each year it is worth $\frac{7}{8}$ of its value for the previous year.

(a) Find a model for $V(t)$, the value of the van after t years.

(b) Determine the value of the van 4 years after it was purchased.

74. ***Drug Concentration*** Immediately following an injection, the concentration of a drug in the bloodstream is 300 milligrams per milliliter. After t hours, the concentration is 75% of the level of the previous hour.

(a) Find a model for $C(t)$, the concentration of the drug after t hours.

(b) Determine the concentration of the drug after 8 hours.

True or False? **In Exercises 75 and 76, determine whether the statement is true or false. Justify your answer.**

75. The line $y = -2$ is an asymptote for the graph of $f(x) = 10^x - 2$.

76. $e = \dfrac{271,801}{99,990}$

77. Use a graphing utility to graph each function. Use the graph to find where the function is increasing and decreasing, and approximate any relative maximum or minimum values.

(a) $f(x) = x^2e^{-x}$

(b) $g(x) = x2^{3-x}$

78. ***Graphical Analysis*** Use a graphing utility to graph $y_1 = (1 + 1/x)^x$ and $y_2 = e$ in the same viewing window. Using the *trace* feature, explain what happens to the graph of y_1 as x increases.

79. ***Graphical Analysis*** Use a graphing utility to graph

$$f(x) = \left(1 + \frac{0.5}{x}\right)^x \quad \text{and} \quad g(x) = e^{0.5}$$

in the same viewing window. What is the relationship between f and g as x increases and decreases without bound?

80. ***Graphical Analysis*** Use a graphing utility to graph each pair of functions in the same viewing window. Describe any similarities and differences in the graphs.

(a) $y_1 = 2^x$, $y_2 = x^2$

(b) $y_1 = 3^x$, $y_2 = x^3$

81. ***Compound Interest*** Use the formula

$$A = P\left(1 + \frac{r}{n}\right)^{nt}$$

to calculate the balance of an account when $P = \$3000$, $r = 6\%$, and $t = 10$ years, and compounding is done (a) by the day, (b) by the hour, (c) by the minute, and (d) by the second. Does increasing the number of compoundings per year result in unlimited growth of the balance of the account? Explain.

CAPSTONE

82. The figure shows the graphs of $y = 2^x$, $y = e^x$, $y = 10^x$, $y = 2^{-x}$, $y = e^{-x}$, and $y = 10^{-x}$. Match each function with its graph. [The graphs are labeled (a) through (f).] Explain your reasoning.

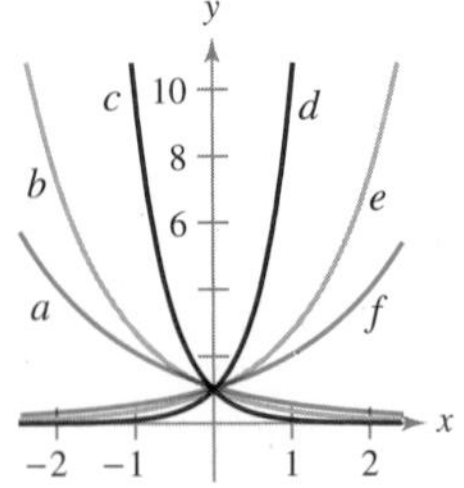

7.2 Logarithmic Functions and Their Graphs

- **Recognize and evaluate logarithmic functions with base *a*.**
- **Graph logarithmic functions.**
- **Recognize, evaluate, and graph natural logarithmic functions.**
- **Use logarithmic functions to model and solve real-life problems.**

Logarithmic Functions

In Section 1.5, you studied the concept of an inverse function. There, you learned that if a function is one-to-one—that is, if the function has the property that no horizontal line intersects the graph of the function more than once—the function must have an inverse function. By looking back at the graphs of the exponential functions introduced in Section 7.1, you will see that every function of the form $f(x) = a^x$ passes the Horizontal Line Test and therefore must have an inverse function. This inverse function is called the **logarithmic function with base *a*.**

> **DEFINITION OF LOGARITHMIC FUNCTION WITH BASE *a***
>
> Let $a > 0$ and $a \neq 1$. For $x > 0$,
>
> $y = \log_a x$ if and only if $x = a^y$.
>
> The function given by
>
> $f(x) = \log_a x$ Read as "log base a of x."
>
> is called the **logarithmic function with base *a*.**

The equations

$y = \log_a x$ and $x = a^y$

are equivalent. The first equation is in logarithmic form and the second is in exponential form. For example, the logarithmic equation $2 = \log_3 9$ can be rewritten in exponential form as $9 = 3^2$. The exponential equation $5^3 = 125$ can be rewritten in logarithmic form as $\log_5 125 = 3$.

When evaluating logarithms, remember that *a logarithm is an exponent.* This means that $\log_a x$ is the exponent to which a must be raised to obtain x. For instance, $\log_2 8 = 3$ because 2 must be raised to the third power to get 8.

EXAMPLE 1 Evaluating Logarithmic Functions

Use the definition of logarithmic function to evaluate each function at the given value of x.

a. $f(x) = \log_2 x, \quad x = 32$ **b.** $f(x) = \log_3 x, \quad x = 1$

c. $f(x) = \log_4 x, \quad x = 2$ **d.** $f(x) = \log_{10} x, \quad x = \frac{1}{100}$

Solution

a. $f(32) = \log_2 32 = 5$ because $2^5 = 32$.

b. $f(1) = \log_3 1 = 0$ because $3^0 = 1$.

c. $f(2) = \log_4 2 = \frac{1}{2}$ because $4^{1/2} = \sqrt{4} = 2$.

d. $f\left(\frac{1}{100}\right) = \log_{10} \frac{1}{100} = -2$ because $10^{-2} = \frac{1}{10^2} = \frac{1}{100}$. ■

The logarithmic function with base 10 is called the **common logarithmic function.** It is denoted by $\log_{10}$ or simply by log. On most calculators, this function is denoted by (LOG). Example 2 shows how to use a calculator to evaluate common logarithmic functions. You will learn how to use a calculator to calculate logarithms to any base in the next section.

EXAMPLE 2 Evaluating the Common Logarithmic Function

Use a calculator to evaluate the function given by $f(x) = \log x$ at each value of x.

a. $x = 10$ **b.** $x = \frac{1}{3}$

c. $x = 2.5$ **d.** $x = -2$

Solution

Function Value	*Graphing Calculator Keystrokes*	*Display*
a. $f(10) = \log 10$	(LOG) 10 (ENTER)	1
b. $f(\frac{1}{3}) = \log \frac{1}{3}$	(LOG) (( ) 1 (÷) 3 ( )) (ENTER)	−0.4771213
c. $f(2.5) = \log 2.5$	(LOG) 2.5 (ENTER)	0.3979400
d. $f(-2) = \log(-2)$	(LOG) ((−)) 2 (ENTER)	ERROR

Note that the calculator displays an error message (or a complex number) when you try to evaluate $\log(-2)$. The reason for this is that there is no real number power to which 10 can be raised to obtain -2. ■

The following properties follow directly from the definition of the logarithmic function with base a.

THEOREM 7.2 PROPERTIES OF LOGARITHMS

1. $\log_a 1 = 0$ because $a^0 = 1$.
2. $\log_a a = 1$ because $a^1 = a$.
3. $\log_a a^x = x$ and $a^{\log_a x} = x$ — Inverse Properties
4. If $\log_a x = \log_a y$, then $x = y$. — One-to-One Property

EXAMPLE 3 Using Properties of Logarithms

a. Solve the equation $\log_2 x = \log_2 3$ for x.

b. Solve the equation $\log_4 4 = x$ for x.

c. Simplify the expression $\log_5 5^x$.

d. Simplify the expression $6^{\log_6 20}$.

Solution

a. Using the One-to-One Property (Property 4), you can conclude that $x = 3$.

b. Using Property 2, you can conclude that $x = 1$.

c. Using the Inverse Property (Property 3), it follows that $\log_5 5^x = x$.

d. Using the Inverse Property (Property 3), it follows that $6^{\log_6 20} = 20$. ■

Graphs of Logarithmic Functions

To sketch the graph of

$$y = \log_a x$$

you can use the fact that the graphs of inverse functions are reflections of each other in the line $y = x$.

EXAMPLE 4 Graphs of Exponential and Logarithmic Functions

In the same coordinate plane, sketch the graph of each function.

a. $f(x) = 2^x$

b. $g(x) = \log_2 x$

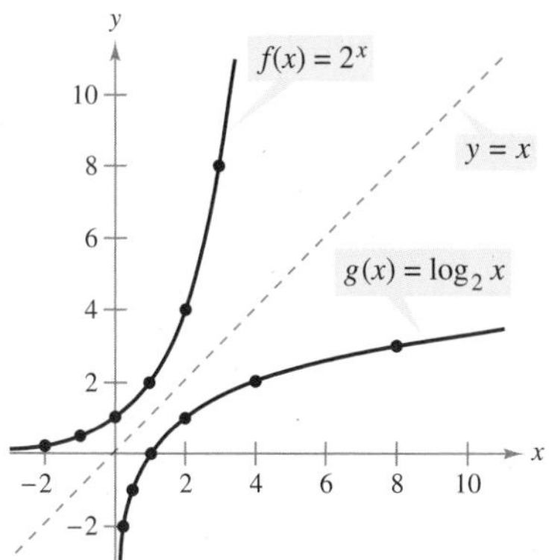

Figure 7.9

Solution

a. For $f(x) = 2^x$, construct a table of values.

x	-2	-1	0	1	2	3
$f(x) = 2^x$	$\frac{1}{4}$	$\frac{1}{2}$	1	2	4	8

By plotting these points and connecting them with a smooth curve, you obtain the graph shown in Figure 7.9.

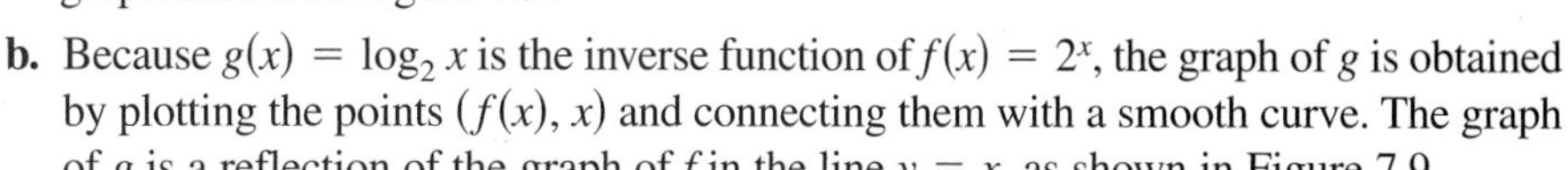

b. Because $g(x) = \log_2 x$ is the inverse function of $f(x) = 2^x$, the graph of g is obtained by plotting the points $(f(x), x)$ and connecting them with a smooth curve. The graph of g is a reflection of the graph of f in the line $y = x$, as shown in Figure 7.9.

EXAMPLE 5 Sketching the Graph of a Logarithmic Function

Sketch the graph of the common logarithmic function $f(x) = \log x$. Identify the x-intercept and vertical asymptote.

Solution Begin by constructing a table of values. Note that some of the values can be obtained without a calculator by using the Inverse Property of Logarithms. Others require a calculator. Next, plot the points and connect them with a smooth curve, as shown in Figure 7.10. The x-intercept of the graph is $(1, 0)$ and the vertical asymptote is $x = 0$ (y-axis).

	Without calculator				With calculator		
x	$\frac{1}{100}$	$\frac{1}{10}$	1	10	2	5	8
$f(x) = \log x$	-2	-1	0	1	0.301	0.699	0.903

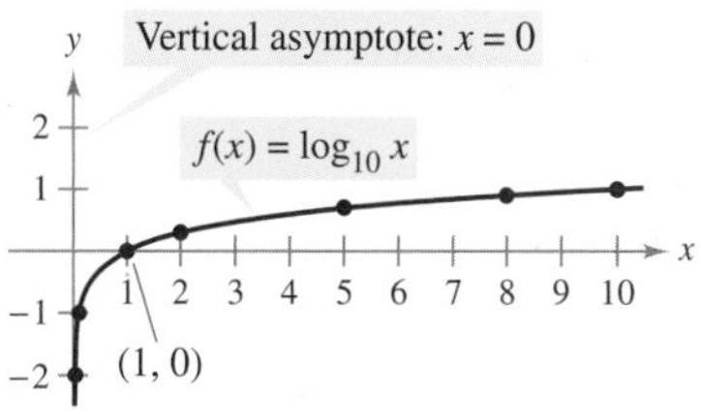

Figure 7.10

■

The nature of the graph in Figure 7.10 is typical of functions of the form $f(x) = \log_a x$, $a > 1$. They have one x-intercept and one vertical asymptote. Notice how slowly the graph rises for $x > 1$. The basic characteristics of logarithmic graphs are summarized in Figure 7.11.

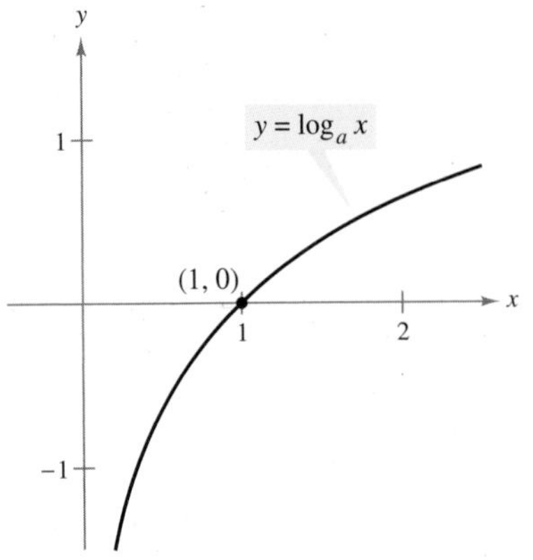

Figure 7.11

Graph of $y = \log_a x$, $a > 1$

- Domain: $(0, \infty)$
- Range: $(-\infty, \infty)$
- x-intercept: $(1, 0)$
- Increasing
- One-to-one, therefore has an inverse function
- y-axis is a vertical asymptote ($\log_a x \to -\infty$ as $x \to 0^+$).
- Continuous
- Reflection of graph of $y = a^x$ about the line $y = x$
- The vertical asymptote occurs at $x = 0$, where $\log_a x$ is undefined.

The basic characteristics of the graph of $f(x) = a^x$ are shown below to illustrate the inverse relation between $f(x) = a^x$ and $g(x) = \log_a x$.

- Domain: $(-\infty, \infty)$
- Range: $(0, \infty)$
- y-intercept: $(0, 1)$
- x-axis is a horizontal asymptote ($a^x \to 0$ as $x \to -\infty$).

In the next example, the graph of $y = \log_a x$ is used to sketch the graphs of functions of the form $f(x) = b \pm \log_a(x + c)$. Notice how a horizontal shift of the graph results in a horizontal shift of the vertical asymptote.

STUDY TIP You can use your understanding of transformations to identify vertical asymptotes of logarithmic functions. For instance, in Example 6(a), the graph of $g(x) = f(x - 1)$ shifts the graph of $f(x)$ one unit to the right. So, the vertical asymptote of $g(x)$ is $x = 1$, one unit to the right of the vertical asymptote of the graph of $f(x)$.

EXAMPLE 6 Shifting Graphs of Logarithmic Functions

The graph of each of the functions is similar to the graph of $f(x) = \log x$.

a. Because $g(x) = \log(x - 1) = f(x - 1)$, the graph of g can be obtained by shifting the graph of f one unit to the right, as shown in Figure 7.12(a).

b. Because $h(x) = 2 + \log x = 2 + f(x)$, the graph of h can be obtained by shifting the graph of f two units upward, as shown in Figure 7.12(b).

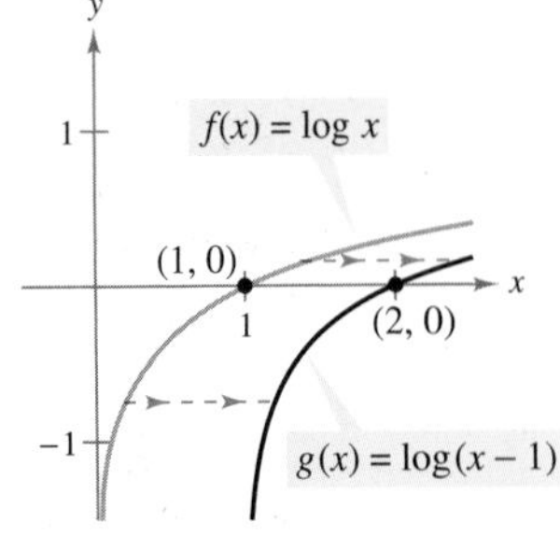

(a)

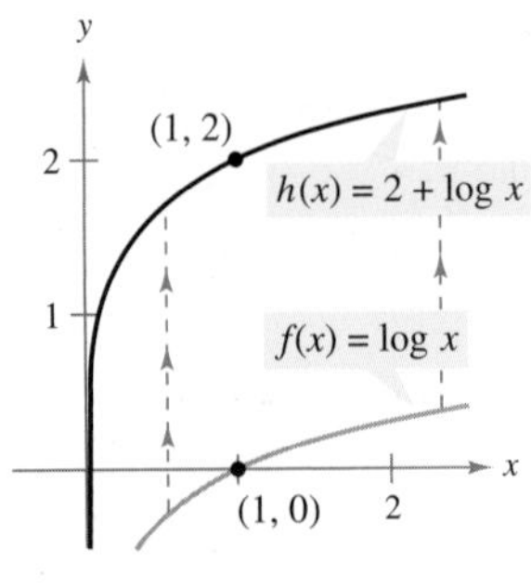

(b)

Figure 7.12

■

The Natural Logarithmic Function

By looking back at the graph of the natural exponential function introduced in Section 7.1, you will see that $f(x) = e^x$ is one-to-one and so has an inverse function. This inverse function is called the **natural logarithmic function** and is denoted by the special symbol $\ln x$, read as "the natural log of x" or "el en of x." Note that the natural logarithm is written without a base. The base is understood to be e.

STUDY TIP Notice that as with every other logarithmic function, the domain of the natural logarithmic function is the set of *positive real numbers*—be sure you see that $\ln x$ is not defined for zero or for negative numbers.

THEOREM 7.3 THE NATURAL LOGARITHMIC FUNCTION

The function defined by

$$f(x) = \log_e x = \ln x, \quad x > 0$$

is called the **natural logarithmic function.**

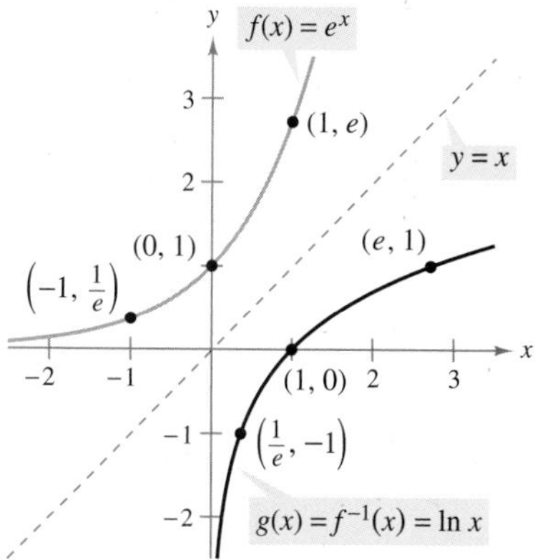

Reflection of graph of $f(x) = e^x$ about the line $y = x$.
Figure 7.13

The definition above implies that the natural logarithmic function and the natural exponential function are inverse functions of each other. So, every logarithmic equation can be written in an equivalent exponential form and every exponential equation can be written in logarithmic form. That is, $y = \ln x$ and $x = e^y$ are equivalent equations.

Because the functions given by $f(x) = e^x$ and $g(x) = \ln x$ are inverse functions of each other, their graphs are reflections of each other in the line $y = x$. This reflective property is illustrated in Figure 7.13.

The four properties of logarithms listed on page 482 are also valid for natural logarithms.

THEOREM 7.4 PROPERTIES OF NATURAL LOGARITHMS

1. $\ln 1 = 0$ because $e^0 = 1$.

2. $\ln e = 1$ because $e^1 = e$.

3. $\ln e^x = x$ and $e^{\ln x} = x$. Inverse Properties

4. If $\ln x = \ln y$, then $x = y$. One-to-One Property

EXAMPLE 7 Using Properties of Natural Logarithms

Use the properties of natural logarithms to simplify each expression.

a. $\ln \frac{1}{e}$ **b.** $e^{\ln 5}$ **c.** $\frac{\ln 1}{3}$ **d.** $2 \ln e$ **e.** $\ln e^2$ **f.** $e^{\ln(x+1)}$

Solution

a. $\ln \frac{1}{e} = \ln e^{-1} = -1$ Inverse Property

b. $e^{\ln 5} = 5$ Inverse Property

c. $\frac{\ln 1}{3} = \frac{0}{3} = 0$ Property 1

d. $2 \ln e = 2(1) = 2$ Property 2

e. $\ln e^2 = 2$ Inverse Property

f. $e^{\ln(x+1)} = x + 1$ Inverse Property ■

On most calculators, the natural logarithm is denoted by (LN), as illustrated in Example 8.

EXAMPLE 8 Evaluating the Natural Logarithmic Function

Use a calculator to evaluate the function given by $f(x) = \ln x$ for each value of x.

a. $x = 2$ **b.** $x = 0.3$ **c.** $x = -1$ **d.** $x = 1 + \sqrt{2}$

Solution

Function Value	*Graphing Calculator Keystrokes*	*Display*
a. $f(2) = \ln 2$	(LN) 2 (ENTER)	0.6931472
b. $f(0.3) = \ln 0.3$	(LN) .3 (ENTER)	−1.2039728
c. $f(-1) = \ln(-1)$	(LN) ((−)) 1 (ENTER)	ERROR
d. $f(1 + \sqrt{2}) = \ln(1 + \sqrt{2})$	(LN) (() 1 (+) (√) 2 ()) (ENTER)	0.8813736

■

In Example 8, be sure you see that $\ln(-1)$ gives an error message on most calculators. This occurs because the domain of $\ln x$ is the set of positive real numbers (see Figure 7.13). So, $\ln(-1)$ is undefined.

NOTE Some graphing utilities display a complex number instead of an ERROR message when evaluating an expression such as $\ln(-1)$. ■

EXAMPLE 9 Finding the Domains of Logarithmic Functions

Find the domain of each function.

a. $f(x) = \ln(x - 2)$

b. $g(x) = \ln(2 - x)$

c. $h(x) = \ln x^2$

Solution

a. Because $\ln(x - 2)$ is defined only if $x - 2 > 0$, it follows that the domain of f is $(2, \infty)$. The graph of f is shown in Figure 7.14(a).

b. Because $\ln(2 - x)$ is defined only if $2 - x > 0$, it follows that the domain of g is $(-\infty, 2)$. The graph of g is shown in Figure 7.14(b).

c. Because $\ln x^2$ is defined only if $x^2 > 0$, it follows that the domain of h is all real numbers except $x = 0$. The graph of h is shown in Figure 7.14(c).

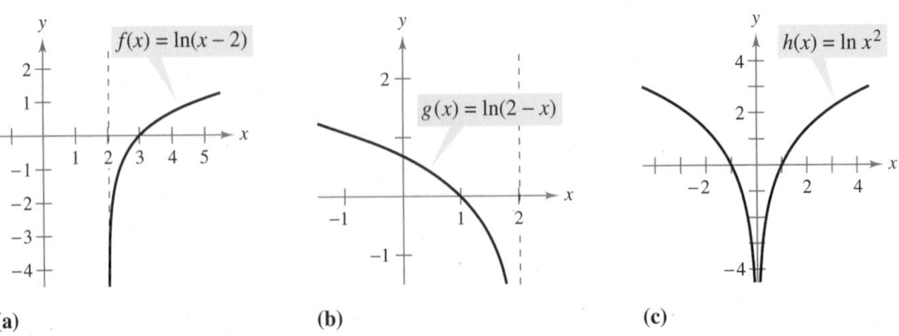

Figure 7.14 ■

Application

EXAMPLE 10 Human Memory Model

Students participating in a psychology experiment attended several lectures on a subject and were given an exam. Every month for a year after the exam, the students were retested to see how much of the material they remembered. The average scores for the group are given by the *human memory model*

$$f(t) = 75 - 6\ln(t + 1),\ 0 \le t \le 12$$

where t is the time in months.

a. What was the average score on the original ($t = 0$) exam?

b. What was the average score at the end of $t = 2$ months?

c. What was the average score at the end of $t = 6$ months?

Algebraic Solution

a. The original average score was

$f(0) = 75 - 6\ln(0 + 1)$	Substitute 0 for t.
$= 75 - 6\ln 1$	Simplify.
$= 75 - 6(0)$	Property of natural logarithms
$= 75.$	Solution

b. After 2 months, the average score was

$f(2) = 75 - 6\ln(2 + 1)$	Substitute 2 for t.
$= 75 - 6\ln 3$	Simplify.
$\approx 75 - 6(1.0986)$	Use a calculator.
$\approx 68.4.$	Solution

c. After 6 months, the average score was

$f(6) = 75 - 6\ln(6 + 1)$	Substitute 6 for t.
$= 75 - 6\ln 7$	Simplify.
$\approx 75 - 6(1.9459)$	Use a calculator.
$\approx 63.3.$	Solution

Graphical Solution

Use a graphing utility to graph the model $y = 75 - 6\ln(x + 1)$. Then use the *value* or *trace* feature to approximate the following.

a. When $x = 0$, $y = 75$ (see Figure 7.15(a)). So, the original average score was 75.

b. When $x = 2$, $y \approx 68.4$ (see Figure 7.15(b)). So, the average score after 2 months was about 68.4.

c. When $x = 6$, $y \approx 63.3$ (see Figure 7.15(c)). So, the average score after 6 months was about 63.3.

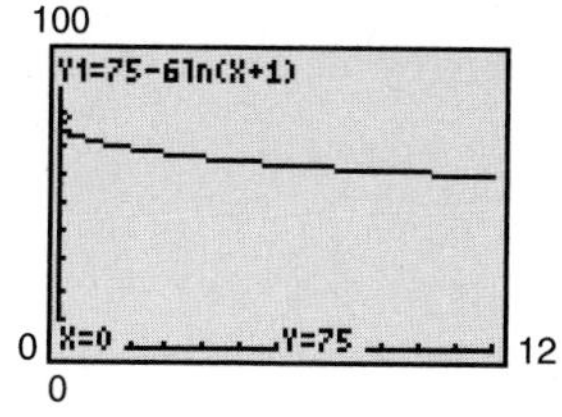

(a)

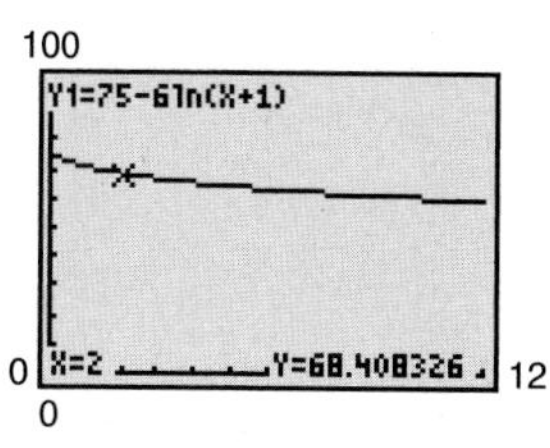

(b)

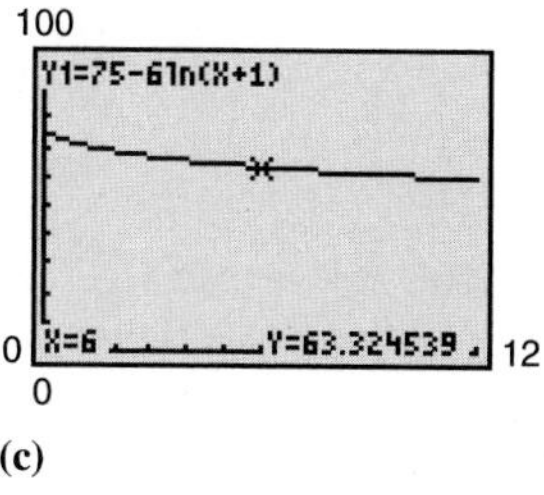

(c)

Figure 7.15

7.2 Exercises

See www.CalcChat.com for worked-out solutions to odd-numbered exercises.

In Exercises 1–6, fill in the blanks.

1. The inverse function of the exponential function given by $f(x) = a^x$ is called the ________ function with base a.
2. The common logarithmic function has base ________ .
3. The logarithmic function given by $f(x) = \ln x$ is called the ________ logarithmic function and has base ________.
4. The Inverse Properties of logarithms and exponentials state that $\log_a a^x = x$ and ________.
5. The One-to-One Property of natural logarithms states that if $\ln x = \ln y$, then ________.
6. The domain of the natural logarithmic function is the set of ________ ________ ________ .

In Exercises 7–14, write the logarithmic equation in exponential form. For example, the exponential form of $\log_5 25 = 2$ is $5^2 = 25$.

7. $\log_4 16 = 2$
8. $\log_7 343 = 3$
9. $\log_9 \frac{1}{81} = -2$
10. $\log \frac{1}{1000} = -3$
11. $\log_{32} 4 = \frac{2}{5}$
12. $\log_{16} 8 = \frac{3}{4}$
13. $\log_{64} 8 = \frac{1}{2}$
14. $\log_8 4 = \frac{2}{3}$

In Exercises 15–22, write the exponential equation in logarithmic form. For example, the logarithmic form of $2^3 = 8$ is $\log_2 8 = 3$.

15. $5^3 = 125$
16. $13^2 = 169$
17. $81^{1/4} = 3$
18. $9^{3/2} = 27$
19. $6^{-2} = \frac{1}{36}$
20. $4^{-3} = \frac{1}{64}$
21. $24^0 = 1$
22. $10^{-3} = 0.001$

In Exercises 23–28, evaluate the function at the given value of x without using a calculator.

	Function	*Value*
23.	$f(x) = \log_2 x$	$x = 64$
24.	$f(x) = \log_{25} x$	$x = 5$
25.	$f(x) = \log_8 x$	$x = 1$
26.	$f(x) = \log x$	$x = 10$
27.	$g(x) = \log_a x$	$x = a^2$
28.	$g(x) = \log_b x$	$x = b^{-3}$

In Exercises 29–32, use a calculator to evaluate $f(x) = \log x$ at the given value of x. Round your result to three decimal places.

29. $x = \frac{7}{8}$
30. $x = \frac{1}{500}$
31. $x = 12.5$
32. $x = 96.75$

In Exercises 33–36, use the properties of logarithms to simplify the expression.

33. $\log_{11} 11^7$
34. $\log_{3.2} 1$
35. $\log_\pi \pi$
36. $9^{\log_9 15}$

In Exercises 37–44, find the domain, x-intercept, and vertical asymptote of the logarithmic function and sketch its graph.

37. $f(x) = \log_4 x$
38. $g(x) = \log_6 x$
39. $y = -\log_3 x + 2$
40. $h(x) = \log_4(x - 3)$
41. $f(x) = -\log_6(x + 2)$
42. $y = \log_5(x - 1) + 4$
43. $y = \log\left(\frac{x}{7}\right)$
44. $y = \log(-x)$

In Exercises 45–50, use the graph of $g(x) = \log_3 x$ to match the given function with its graph. Then describe the relationship between the graphs of f and g. [The graphs are labeled (a), (b), (c), (d), (e), and (f).]

(a)

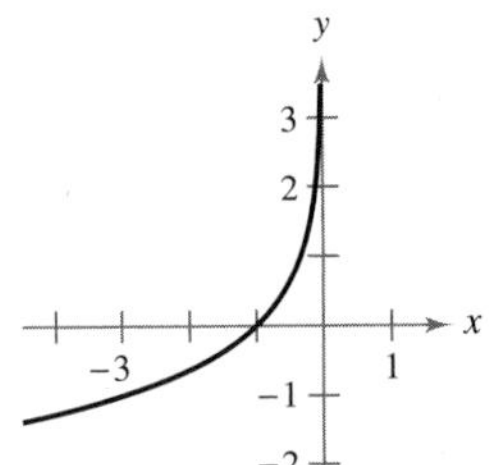

(b)

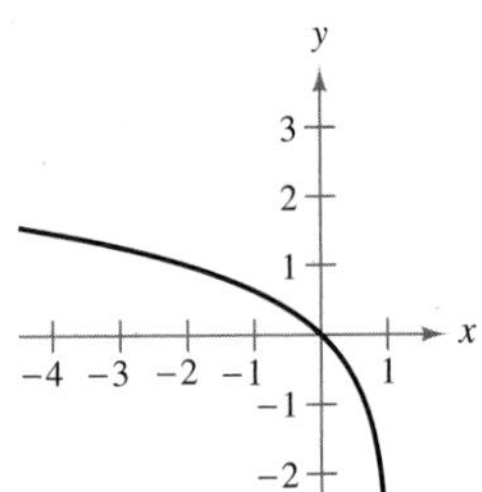

(c)

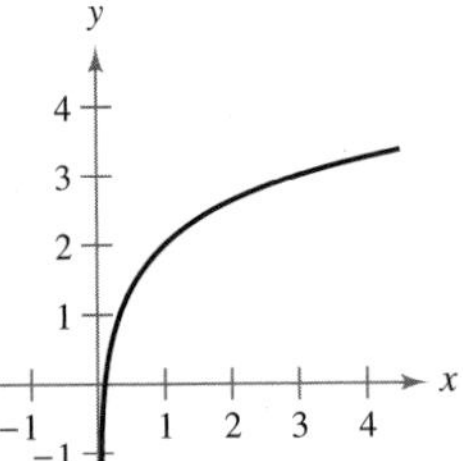

(d)

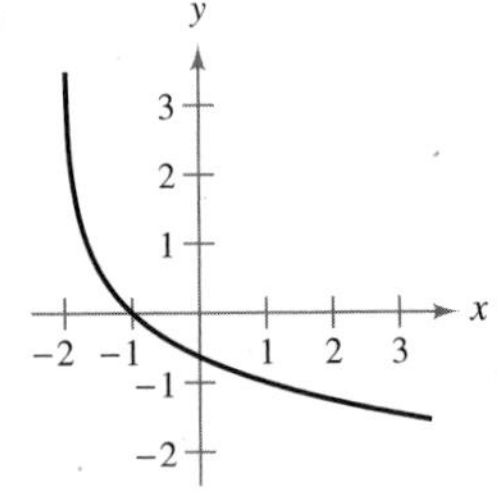

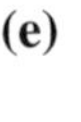

(e)

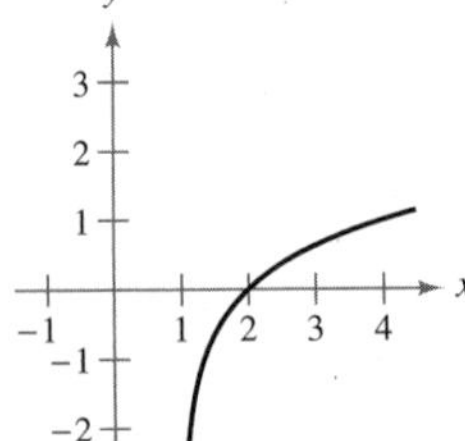

(f)

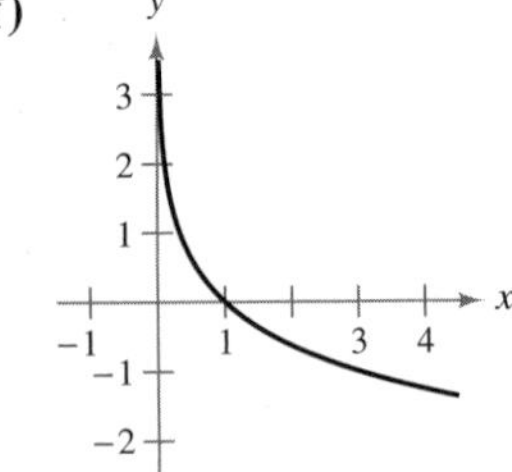

45. $f(x) = \log_3 x + 2$
46. $f(x) = -\log_3 x$
47. $f(x) = -\log_3(x + 2)$
48. $f(x) = \log_3(x - 1)$
49. $f(x) = \log_3(1 - x)$
50. $f(x) = -\log_3(-x)$

In Exercises 51–58, write the logarithmic equation in exponential form.

51. $\ln \frac{1}{2} = -0.693\ldots$ **52.** $\ln \frac{2}{5} = -0.916\ldots$

53. $\ln 7 = 1.945\ldots$ **54.** $\ln 10 = 2.302\ldots$

55. $\ln 250 = 5.521\ldots$ **56.** $\ln 1084 = 6.988\ldots$

57. $\ln 1 = 0$ **58.** $\ln e = 1$

In Exercises 59–66, write the exponential equation in logarithmic form.

59. $e^4 = 54.598\ldots$ **60.** $e^2 = 7.3890\ldots$

61. $e^{1/2} = 1.6487\ldots$ **62.** $e^{1/3} = 1.3956\ldots$

63. $e^{-0.9} = 0.406\ldots$ **64.** $e^{-4.1} = 0.0165\ldots$

65. $e^x = 4$ **66.** $e^{2x} = 3$

In Exercises 67–70, use a calculator to evaluate the function at the given value of x. Round your result to three decimal places.

	Function	*Value*
67.	$f(x) = \ln x$	$x = 18.42$
68.	$f(x) = 3 \ln x$	$x = 0.74$
69.	$g(x) = 8 \ln x$	$x = 0.05$
70.	$g(x) = -\ln x$	$x = \frac{1}{2}$

In Exercises 71–74, evaluate $g(x) = \ln x$ at the given value of x without using a calculator.

71. $x = e^5$ **72.** $x = e^{-4}$

73. $x = e^{-5/6}$ **74.** $x = e^{-5/2}$

In Exercises 75–78, find the domain, x-intercept, and vertical asymptote of the logarithmic function and sketch its graph.

75. $f(x) = \ln(x - 4)$ **76.** $h(x) = \ln(x + 5)$

77. $g(x) = \ln(-x)$ **78.** $f(x) = \ln(3 - x)$

In Exercises 79–84, use a graphing utility to graph the function. Be sure to use an appropriate viewing window.

79. $f(x) = \log(x + 9)$ **80.** $f(x) = \log(x - 6)$

81. $f(x) = \ln(x - 1)$ **82.** $f(x) = \ln(x + 2)$

83. $f(x) = \ln x + 8$ **84.** $f(x) = 3 \ln x - 1$

In Exercises 85–92, use the One-to-One Property to solve the equation for x.

85. $\log_5(x + 1) = \log_5 6$ **86.** $\log_2(x - 3) = \log_2 9$

87. $\log(2x + 1) = \log 15$ **88.** $\log(5x + 3) = \log 12$

89. $\ln(x + 4) = \ln 12$ **90.** $\ln(x - 7) = \ln 7$

91. $\ln(x^2 - 2) = \ln 23$ **92.** $\ln(x^2 - x) = \ln 6$

WRITING ABOUT CONCEPTS

In Exercises 93–96, sketch the graphs of f and g and describe the relationship between the graphs of f and g. What is the relationship between the functions f and g?

93. $f(x) = 3^x, \quad g(x) = \log_3 x$

94. $f(x) = 5^x, \quad g(x) = \log_5 x$

95. $f(x) = e^x, \quad g(x) = \ln x$

96. $f(x) = 8^x, \quad g(x) = \log_8 x$

97. ***Graphical Analysis*** Use a graphing utility to graph f and g in the same viewing window and determine which is increasing at the greater rate as x approaches $+\infty$. What can you conclude about the rate of growth of the natural logarithmic function?

(a) $f(x) = \ln x, \quad g(x) = \sqrt{x}$

(b) $f(x) = \ln x, \quad g(x) = \sqrt[4]{x}$

98. ***Compound Interest*** A principal P, invested at $5\frac{1}{2}\%$ and compounded continuously, increases to an amount K times the original principal after t years, where t is given by $t = (\ln K)/0.055$.

(a) Complete the table and interpret your results.

K	1	2	4	6	8	10	12
t							

(b) Sketch a graph of the function.

99. ***Cable Television*** The numbers of cable television systems C (in thousands) in the United States from 2001 through 2006 can be approximated by the model $C = 10.355 - 0.298t \ln t$, $1 \le t \le 6$, where t represents the year, with $t = 1$ corresponding to 2001. *(Source: Warren Communication News)*

(a) Complete the table.

t	1	2	3	4	5	6
C						

(b) Use a graphing utility to graph the function.

(c) Can the model be used to predict the numbers of cable television systems beyond 2006? Explain.

100. ***Population*** The time t in years for the world population to double if it is increasing at a continuous rate of r is given by $t = (\ln 2)/r$.

(a) Complete the table and interpret your results.

r	0.005	0.010	0.015	0.020	0.025	0.030
t						

(b) Use a graphing utility to graph the function.

101. ***Human Memory Model*** Students in a mathematics class were given an exam and then retested monthly with an equivalent exam. The average scores for the class are given by the human memory model $f(t) = 80 - 17 \log(t + 1)$, $0 \le t \le 12$, where t is the time in months.

(a) Use a graphing utility to graph the model over the specified domain.

(b) What was the average score on the original exam $(t = 0)$?

(c) What was the average score after 4 months?

(d) What was the average score after 10 months?

102. ***Sound Intensity*** The relationship between the number of decibels β and the intensity of a sound I in watts per square meter is

$$\beta = 10 \log\left(\frac{I}{10^{-12}}\right).$$

(a) Determine the number of decibels of a sound with an intensity of 1 watt per square meter.

(b) Determine the number of decibels of a sound with an intensity of 10^{-2} watt per square meter.

(c) The intensity of the sound in part (a) is 100 times as great as that in part (b). Is the number of decibels 100 times as great? Explain.

103. ***Monthly Payment*** The model

$$t = 16.625 \ln\left(\frac{x}{x - 750}\right), \quad x > 750$$

approximates the length of a home mortgage of \$150,000 at 6% in terms of the monthly payment. In the model, t is the length of the mortgage in years and x is the monthly payment in dollars.

(a) Use the model to approximate the lengths of a \$150,000 mortgage at 6% when the monthly payment is \$897.72 and when the monthly payment is \$1659.24.

(b) Approximate the total amounts paid over the term of the mortgage with a monthly payment of \$897.72 and with a monthly payment of \$1659.24.

(c) Approximate the total interest charges for a monthly payment of \$897.72 and for a monthly payment of \$1659.24.

(d) What is the vertical asymptote for the model? Interpret its meaning in the context of the problem.

True or False? **In Exercises 104 and 105, determine whether the statement is true or false. Justify your answer.**

104. You can determine the graph of $f(x) = \log_6 x$ by graphing $g(x) = 6^x$ and reflecting it about the x-axis.

105. The graph of $f(x) = \log_3 x$ contains the point $(27, 3)$.

106. ***Think About It*** Complete the table for $f(x) = 10^x$.

x	-2	-1	0	1	2
$f(x)$					

Complete the table for $f(x) = \log x$.

x	$\frac{1}{100}$	$\frac{1}{10}$	1	10	100
$f(x)$					

Compare the two tables. What is the relationship between $f(x) = 10^x$ and $f(x) = \log x$?

107. (a) Complete the table for the function given by $f(x) = (\ln x)/x$.

x	1	5	10	10^2	10^4	10^6
$f(x)$						

(b) Use the table in part (a) to determine what value $f(x)$ approaches as x increases without bound.

(c) Use a graphing utility to confirm the result of part (b).

CAPSTONE

108. The table of values was obtained by evaluating a function. Determine which of the statements may be true and which must be false.

x	y
1	0
2	1
8	3

(a) y is an exponential function of x.

(b) y is a logarithmic function of x.

(c) x is an exponential function of y.

(d) y is a linear function of x.

109. ***Writing*** Explain why $\log_a x$ is defined only for $0 < a < 1$ and $a > 1$.

In Exercises 110 and 111, (a) use a graphing utility to graph the function, (b) use the graph to determine the intervals in which the function is increasing and decreasing, and (c) approximate any relative maximum or minimum values of the function.

110. $f(x) = |\ln x|$

111. $h(x) = \ln(x^2 + 1)$

7.3 Using Properties of Logarithms

- Use the change-of-base formula to rewrite and evaluate logarithmic expressions.
- Use properties of logarithms to evaluate or rewrite logarithmic expressions.
- Use properties of logarithms to expand or condense logarithmic expressions.
- Use logarithmic functions to model and solve real-life problems.

Change of Base

Most calculators have only two types of log keys, one for common logarithms (base 10) and one for natural logarithms (base e). Although common logarithms and natural logarithms are the most frequently used, you may occasionally need to evaluate logarithms with other bases. To do this, you can use the following **change-of-base formula.**

THEOREM 7.5 CHANGE-OF-BASE FORMULA

Let a, b, and x be positive real numbers such that $a \neq 1$ and $b \neq 1$. Then $\log_a x$ can be converted to a different base as follows.

Base b	*Base 10*	*Base e*
$\log_a x = \dfrac{\log_b x}{\log_b a}$	$\log_a x = \dfrac{\log x}{\log a}$	$\log_a x = \dfrac{\ln x}{\ln a}$

One way to look at the change-of-base formula is that logarithms with base a are simply *constant multiples* of logarithms with base b. The constant multiplier is $1/(\log_b a)$.

EXAMPLE 1 Changing Bases Using Common Logarithms

a. $\log_4 25 = \dfrac{\log 25}{\log 4}$ $\qquad \log_a x = \dfrac{\log x}{\log a}$

$\approx \dfrac{1.39794}{0.60206}$ $\qquad$ Use a calculator.

≈ 2.3219 $\qquad$ Simplify.

b. $\log_2 12 = \dfrac{\log 12}{\log 2} \approx \dfrac{1.07918}{0.30103} \approx 3.5850$

EXAMPLE 2 Changing Bases Using Natural Logarithms

a. $\log_4 25 = \dfrac{\ln 25}{\ln 4}$ $\qquad \log_a x = \dfrac{\ln x}{\ln a}$

$\approx \dfrac{3.21888}{1.38629}$ $\qquad$ Use a calculator.

≈ 2.3219 $\qquad$ Simplify.

b. $\log_2 12 = \dfrac{\ln 12}{\ln 2} \approx \dfrac{2.48491}{0.69315} \approx 3.5850$ ■

Properties of Logarithms

You know from the preceding section that the logarithmic function with base a is the *inverse function* of the exponential function with base a. So, it makes sense that the properties of exponents should have corresponding properties involving logarithms. For instance, the exponential property $a^0 = 1$ has the corresponding logarithmic property $\log_a 1 = 0$.

STUDY TIP There is no general property that can be used to rewrite $\log_a(u \pm v)$. Specifically, $\log_a(u + v)$ is *not* equal to $\log_a u + \log_a v$.

THEOREM 7.6 PROPERTIES OF LOGARITHMS

Let a be a positive number such that $a \neq 1$, and let n be a real number. If u and v are positive real numbers, the following properties are true.

	Logarithm with Base a	*Natural Logarithm*
1. Product Property:	$\log_a(uv) = \log_a u + \log_a v$	$\ln(uv) = \ln u + \ln v$
2. Quotient Property:	$\log_a \frac{u}{v} = \log_a u - \log_a v$	$\ln \frac{u}{v} = \ln u - \ln v$
3. Power Property:	$\log_a u^n = n \log_a u$	$\ln u^n = n \ln u$

NOTE Pay attention to the domain when applying the properties of logarithms to a logarithmic function. For example, the domain of $f(x) = \ln x^2$ is all real $x \neq 0$, whereas the domain of $g(x) = 2 \ln x$ is all real $x > 0$. ■

A proof of the first property listed above is given in Appendix A.

EXAMPLE 3 Using Properties of Logarithms

Write each logarithm in terms of ln 2 and ln 3.

a. $\ln 6$ **b.** $\ln \frac{2}{27}$

Solution

a. $\ln 6 = \ln(2 \cdot 3)$ Rewrite 6 as $2 \cdot 3$.

$= \ln 2 + \ln 3$ Product Property

b. $\ln \frac{2}{27} = \ln 2 - \ln 27$ Quotient Property

$= \ln 2 - \ln 3^3$ Rewrite 27 as 3^3.

$= \ln 2 - 3 \ln 3$ Power Property

The Granger Collection, New York

JOHN NAPIER

John Napier, a Scottish mathematician, developed logarithms as a way to simplify some of the tedious calculations of his day. Beginning in 1594, Napier worked about 20 years on the invention of logarithms. Napier was only partially successful in his quest to simplify tedious calculations. Nonetheless, the development of logarithms was a step forward and received immediate recognition.

EXAMPLE 4 Using Properties of Logarithms

Find the exact value of each expression without using a calculator.

a. $\log_5 \sqrt[3]{5}$ **b.** $\ln e^6 - \ln e^2$

Solution

a. $\log_5 \sqrt[3]{5} = \log_5 5^{1/3} = \frac{1}{3} \log_5 5 = \frac{1}{3}(1) = \frac{1}{3}$

b. $\ln e^6 - \ln e^2 = \ln \frac{e^6}{e^2} = \ln e^4 = 4 \ln e = 4(1) = 4$ ■

Rewriting Logarithmic Expressions

STUDY TIP In Section 8.2, you will see that properties of logarithms can also be used to rewrite logarithmic functions in forms that simplify the operations of calculus.

The properties of logarithms are useful for rewriting logarithmic expressions in forms that simplify the operations of algebra. This is true because these properties convert complicated products, quotients, and exponential forms into simpler sums, differences, and products, respectively.

EXAMPLE 5 Expanding Logarithmic Expressions

Expand each logarithmic expression.

a. $\log_4 5x^3y$

b. $\ln \dfrac{\sqrt{3x-5}}{7}$

Solution

a. $\log_4 5x^3y = \log_4 5 + \log_4 x^3 + \log_4 y$ — Product Property

$= \log_4 5 + 3\log_4 x + \log_4 y$ — Power Property

b. $\ln \dfrac{\sqrt{3x-5}}{7} = \ln \dfrac{(3x-5)^{1/2}}{7}$ — Rewrite using rational exponent.

$= \ln(3x-5)^{1/2} - \ln 7$ — Quotient Property

$= \dfrac{1}{2}\ln(3x-5) - \ln 7$ — Power Property ■

In Example 5, the properties of logarithms were used to *expand* logarithmic expressions. In Example 6, this procedure is reversed and the properties of logarithms are used to *condense* logarithmic expressions.

EXAMPLE 6 Condensing Logarithmic Expressions

Condense each logarithmic expression.

a. $\frac{1}{2}\log x + 3\log(x+1)$

b. $2\ln(x+2) - \ln x$

c. $\frac{1}{3}[\log_2 x + \log_2(x+1)]$

Solution

a. $\frac{1}{2}\log x + 3\log(x+1) = \log x^{1/2} + \log(x+1)^3$ — Power Property

$= \log\left[\sqrt{x}(x+1)^3\right]$ — Product Property

b. $2\ln(x+2) - \ln x = \ln(x+2)^2 - \ln x$ — Power Property

$= \ln \dfrac{(x+2)^2}{x}$ — Quotient Property

c. $\frac{1}{3}[\log_2 x + \log_2(x+1)] = \frac{1}{3}\{\log_2[x(x+1)]\}$ — Product Property

$= \log_2[x(x+1)]^{1/3}$ — Power Property

$= \log_2 \sqrt[3]{x(x+1)}$ — Rewrite with a radical.

■

Application

One method of determining how the x- and y-values for a set of nonlinear data are related is to take the natural logarithm of each of the x- and y-values. If these new points are graphed and fall on a line, then you can determine that the x- and y-values are related by the equation

$$\ln y = m \ln x + b$$

where m is the slope of the line.

EXAMPLE 7 Finding a Mathematical Model

The table shows the mean distance from the sun x and the period y (the time it takes a planet to orbit the sun) for each of the six planets that are closest to the sun. In the table, the mean distance is given in terms of astronomical units (where Earth's mean distance is defined as 1.0), and the period is given in years. Find an equation that relates y and x.

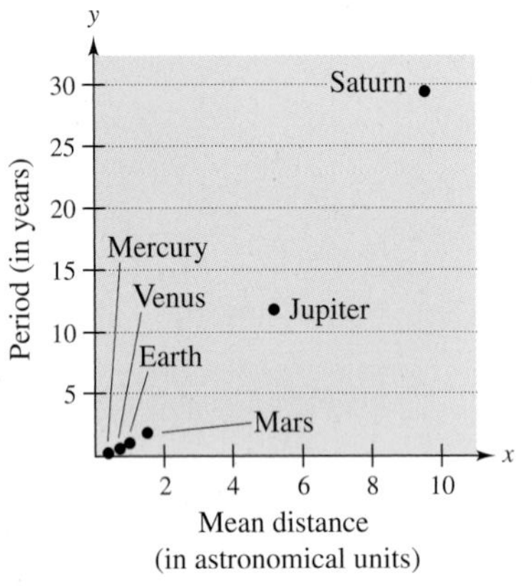

Figure 7.16

Planet	Mercury	Venus	Earth	Mars	Jupiter	Saturn
Mean distance, x	0.387	0.723	1.000	1.524	5.203	9.537
Period, y	0.241	0.615	1.000	1.881	11.860	29.460

Solution The points in the table above are plotted in Figure 7.16. From this figure it is not clear how to find an equation that relates y and x. To solve this problem, take the natural logarithm of each of the x- and y-values in the table. For instance,

$$\ln 0.241 = -1.423$$

and

$$\ln 0.387 = -0.949.$$

Continuing this produces the following results.

Planet	Mercury	Venus	Earth	Mars	Jupiter	Saturn
$\ln x$	-0.949	-0.324	0.000	0.421	1.649	2.255
$\ln y$	-1.423	-0.486	0.000	0.632	2.473	3.383

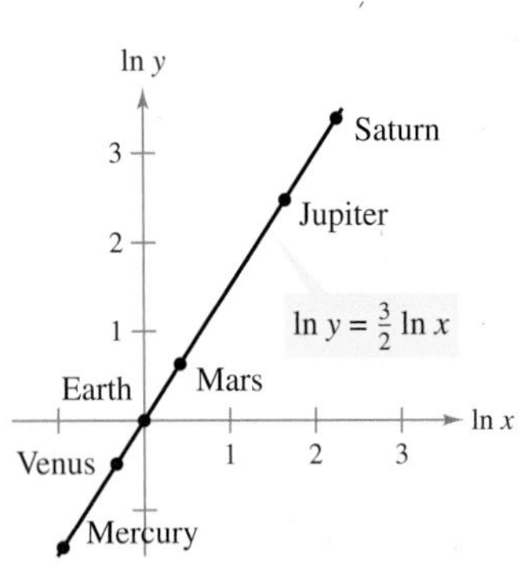

Figure 7.17

Now, by plotting the points in the second table, you can see that all six of the points appear to lie in a line (see Figure 7.17). Using any two points, the slope of this line is found to be $\frac{3}{2}$. You can therefore conclude that

$$\ln y = \frac{3}{2} \ln x.$$

The graph of this equation is shown in Figure 7.17. Using properties of logarithms, you can solve for y as shown below.

$$\ln y = \ln x^{3/2} \quad \text{Power Property}$$
$$y = x^{3/2} \quad \text{One-to-One Property}$$

■

7.3 Exercises

See www.CalcChat.com for worked-out solutions to odd-numbered exercises.

In Exercises 1–4, fill in the blanks.

1. To evaluate a logarithm to any base, you can use the ________ formula.
2. The change-of-base formula for base e is given by $\log_a x =$ ________.
3. You can consider $\log_a x$ to be a constant multiple of $\log_b x$; the constant multiplier is ________.
4. The properties of logarithms are useful for ________ logarithmic expressions in forms that simplify the operations of algebra.

In Exercises 5–12, rewrite the logarithm as a ratio of (a) common logarithms and (b) natural logarithms.

5. $\log_5 16$
6. $\log_3 47$
7. $\log_{1/5} x$
8. $\log_{1/3} x$
9. $\log_x \frac{3}{10}$
10. $\log_x \frac{3}{4}$
11. $\log_{2.6} x$
12. $\log_{7.1} x$

In Exercises 13–20, evaluate the logarithm using the change-of-base formula. Round your result to three decimal places.

13. $\log_3 7$
14. $\log_7 4$
15. $\log_{1/2} 4$
16. $\log_{1/4} 5$
17. $\log_9 0.1$
18. $\log_{20} 0.25$
19. $\log_{15} 1250$
20. $\log_3 0.015$

In Exercises 21–26, use the properties of logarithms to rewrite and simplify the logarithmic expression.

21. $\log_4 8$
22. $\log_2(4^2 \cdot 3^4)$
23. $\log_5 \frac{1}{250}$
24. $\log \frac{9}{300}$
25. $\ln(5e^6)$
26. $\ln \dfrac{6}{e^2}$

In Exercises 27–42, find the exact value of the logarithmic expression without using a calculator. (If this is not possible, state the reason.)

27. $\log_3 9$
28. $\log_5 \frac{1}{125}$
29. $\log_2 \sqrt[4]{8}$
30. $\log_6 \sqrt[3]{6}$
31. $\log_4 16^2$
32. $\log_3 81^{-3}$
33. $\log_2(-2)$
34. $\log_3(-27)$
35. $\ln e^{4.5}$
36. $3 \ln e^4$
37. $\ln \dfrac{1}{\sqrt{e}}$
38. $\ln \sqrt[4]{e^3}$
39. $\ln e^2 + \ln e^5$
40. $2 \ln e^6 - \ln e^5$
41. $\log_5 75 - \log_5 3$
42. $\log_4 2 + \log_4 32$

In Exercises 43–64, use the properties of logarithms to expand the expression as a sum, difference, and/or constant multiple of logarithms. (Assume all variables are positive.)

43. $\ln 4x$
44. $\log_3 10z$
45. $\log_8 x^4$
46. $\log_{10} \dfrac{y}{2}$
47. $\log_5 \dfrac{5}{x}$
48. $\log_6 \dfrac{1}{z^3}$
49. $\ln \sqrt{z}$
50. $\ln \sqrt[3]{t}$
51. $\ln xyz^2$
52. $\log 4x^2 y$
53. $\ln z(z-1)^2,\ z > 1$
54. $\ln\left(\dfrac{x^2-1}{x^3}\right),\ x > 1$
55. $\log_2 \dfrac{\sqrt{a-1}}{9},\ a > 1$
56. $\ln \dfrac{6}{\sqrt{x^2+1}}$
57. $\ln \sqrt[3]{\dfrac{x}{y}}$
58. $\ln \sqrt{\dfrac{x^2}{y^3}}$
59. $\ln x^2 \sqrt{\dfrac{y}{z}}$
60. $\log_2 x^4 \sqrt{\dfrac{y}{z^3}}$
61. $\log_5 \dfrac{x^2}{y^2 z^3}$
62. $\log_{10} \dfrac{xy^4}{z^5}$
63. $\ln \sqrt[4]{x^3(x^2+3)}$
64. $\ln \sqrt{x^2(x+2)}$

In Exercises 65–82, condense the expression to the logarithm of a single quantity.

65. $\ln 2 + \ln x$
66. $\ln y + \ln t$
67. $\log_4 z - \log_4 y$
68. $\log_5 8 - \log_5 t$
69. $2 \log_2 x + 4 \log_2 y$
70. $\frac{2}{3} \log_7(z-2)$
71. $\frac{1}{4} \log_3 5x$
72. $-4 \log_6 2x$
73. $\log x - 2 \log(x+1)$
74. $2 \ln 8 + 5 \ln(z-4)$
75. $\log x - 2 \log y + 3 \log z$
76. $3 \log_3 x + 4 \log_3 y - 4 \log_3 z$
77. $\ln x - [\ln(x+1) + \ln(x-1)]$
78. $4[\ln z + \ln(z+5)] - 2 \ln(z-5)$
79. $\frac{1}{3}[2 \ln(x+3) + \ln x - \ln(x^2-1)]$
80. $2[3 \ln x - \ln(x+1) - \ln(x-1)]$
81. $\frac{1}{3}[\log_8 y + 2 \log_8(y+4)] - \log_8(y-1)$
82. $\frac{1}{2}[\log_4(x+1) + 2 \log_4(x-1)] + 6 \log_4 x$

In Exercises 83 and 84, compare the logarithmic quantities. If two are equal, explain why.

83. $\dfrac{\log_2 32}{\log_2 4}$, $\log_2 \dfrac{32}{4}$, $\log_2 32 - \log_2 4$

84. $\log_7 \sqrt{70}$, $\log_7 35$, $\frac{1}{2} + \log_7 \sqrt{10}$

Sound Intensity **In Exercises 85–88, use the following information. The relationship between the number of decibels β and the intensity of a sound I in watts per square meter is given by**

$$\beta = 10 \log\left(\frac{I}{10^{-12}}\right).$$

85. Use the properties of logarithms to write the formula in simpler form, and determine the number of decibels of a sound with an intensity of 10^{-6} watt per square meter.

86. Find the difference in loudness between an average office with an intensity of 1.26×10^{-7} watt per square meter and a broadcast studio with an intensity of 3.16×10^{-10} watt per square meter.

87. Find the difference in loudness between a vacuum cleaner with an intensity of 10^{-4} watt per square meter and rustling leaves with an intensity of 10^{-11} watt per square meter.

88. You and your roommate are playing your stereos at the same time and at the same intensity. How much louder is the music when both stereos are playing compared with just one stereo playing?

Curve Fitting **In Exercises 89–92, find a logarithmic equation that relates y and x. Explain the steps used to find the equation.**

89.

x	1	2	3	4	5	6
y	1	1.189	1.316	1.414	1.495	1.565

90.

x	1	2	3	4	5	6
y	1	1.587	2.080	2.520	2.924	3.302

91.

x	1	2	3	4	5	6
y	2.5	2.102	1.9	1.768	1.672	1.597

92.

x	1	2	3	4	5	6
y	0.5	2.828	7.794	16	27.951	44.091

WRITING ABOUT CONCEPTS

In Exercises 93 and 94, use a graphing utility to graph the two functions in the same viewing window. Use the graphs to verify that the expressions are equivalent.

93. $f(x) = \log_{10} x$

$g(x) = \dfrac{\ln x}{\ln 10}$

94. $f(x) = \ln x$

$g(x) = \dfrac{\log_{10} x}{\log_{10} e}$

95. Sketch the graphs of

$$f(x) = \ln \frac{x}{2}, \quad g(x) = \frac{\ln x}{\ln 2}, \quad h(x) = \ln x - \ln 2$$

on the same set of axes. Which two functions have identical graphs? Explain your reasoning.

96. ***Galloping Speeds of Animals*** Four-legged animals run with two different types of motion: trotting and galloping. An animal that is trotting has at least one foot on the ground at all times, whereas an animal that is galloping has all four feet off the ground at some point in its stride. The number of strides per minute at which an animal breaks from a trot to a gallop depends on the weight of the animal. Use the table to find a logarithmic equation that relates an animal's weight x (in pounds) and its lowest galloping speed y (in strides per minute).

Weight, x	25	35	50
Galloping speed, y	191.5	182.7	173.8

Weight, x	75	500	1000
Galloping speed, y	164.2	125.9	114.2

97. ***Nail Length*** The approximate lengths and diameters (in inches) of common nails are shown in the table. Find a logarithmic equation that relates the diameter y of a common nail to its length x.

Length, x	1	2	3
Diameter, y	0.072	0.120	0.148

Length, x	4	5	6
Diameter, y	0.203	0.238	0.284

98. ***Comparing Models*** A cup of water at an initial temperature of 78°C is placed in a room at a constant temperature of 21°C. The temperature of the water is measured every 5 minutes during a half-hour period. The results are recorded as ordered pairs of the form (t, T), where t is the time (in minutes) and T is the temperature (in degrees Celsius).

$(0, 78.0°), (5, 66.0°), (10, 57.5°), (15, 51.2°), (20, 46.3°), (25, 42.4°), (30, 39.6°)$

(a) The graph of the model for the data should be asymptotic with the graph of the temperature of the room. Subtract the room temperature from each of the temperatures in the ordered pairs. Use a graphing utility to plot the data points (t, T) and $(t, T - 21)$.

(b) An exponential model for the data $(t, T - 21)$ is given by

$$T - 21 = 54.4(0.964)^t.$$

Solve for T and graph the model. Compare the result with the plot of the original data.

(c) Take the natural logarithms of the revised temperatures. Use a graphing utility to plot the points $(t, \ln(T - 21))$ and observe that the points appear to be linear. Use the *regression* feature of the graphing utility to fit a line to these data. This resulting line has the form

$$\ln(T - 21) = at + b.$$

Solve for T, and verify that the result is equivalent to the model in part (b).

(d) Fit a rational model to the data. Take the reciprocals of the y-coordinates of the revised data points to generate the points

$$\left(t, \frac{1}{T - 21}\right).$$

Use a graphing utility to graph these points and observe that they appear to be linear. Use the *regression* feature of a graphing utility to fit a line to these data. The resulting line has the form

$$\frac{1}{T - 21} = at + b.$$

Solve for T, and use a graphing utility to graph the rational function and the original data points.

(e) Why did taking the logarithms of the temperatures lead to a linear scatter plot? Why did taking the reciprocals of the temperatures lead to a linear scatter plot?

True or False? **In Exercises 99–104, determine whether the statement is true or false given that $f(x) = \ln x$. Justify your answer.**

99. $f(0) = 0$

100. $f(ax) = f(a) + f(x), \quad a > 0, x > 0$

101. $f(x - 2) = f(x) - f(2), \quad x > 2$

102. $\sqrt{f(x)} = \frac{1}{2}f(x)$

103. If $f(u) = 2f(v)$, then $v = u^2$.

104. If $f(x) < 0$, then $0 < x < 1$.

In Exercises 105–110, use the change-of-base formula to rewrite the logarithm as a ratio of logarithms. Then use a graphing utility to graph the ratio.

105. $f(x) = \log_2 x$

106. $f(x) = \log_4 x$

107. $f(x) = \log_{1/2} x$

108. $f(x) = \log_{1/4} x$

109. $f(x) = \log_{11.8} x$

110. $f(x) = \log_{12.4} x$

111. ***Graphical Analysis*** Use a graphing utility to graph the functions given by

$$y_1 = \ln x - \ln(x - 3)$$

and

$$y_2 = \ln \frac{x}{x - 3}$$

in the same viewing window. Does the graphing utility show the functions with the same domain? If so, should it? Explain your reasoning.

CAPSTONE

112. A classmate claims that the following are true.

(a) $\ln(u + v) = \ln u + \ln v = \ln(uv)$

(b) $\ln(u - v) = \ln u - \ln v = \ln \frac{u}{v}$

(c) $(\ln u)^n = n(\ln u) = \ln u^n$

Discuss how you would demonstrate that these claims are not true.

113. ***Proof*** Prove that $\log_b \frac{u}{v} = \log_b u - \log_b v$.

114. ***Proof*** Prove that $\log_b u^n = n \log_b u$.

115. ***Think About It*** For how many integers between 1 and 20 can the natural logarithms be approximated given the values $\ln 2 \approx 0.6931$, $\ln 3 \approx 1.0986$, and $\ln 5 \approx 1.6094$? Approximate these logarithms (do not use a calculator).

7.4 Exponential and Logarithmic Equations

- **Solve simple exponential and logarithmic equations.**
- **Solve more complicated exponential equations.**
- **Solve more complicated logarithmic equations.**
- **Use exponential and logarithmic equations to model and solve real-life problems.**

Introduction

So far in this chapter, you have studied the definitions, graphs, and properties of exponential and logarithmic functions. In this section, you will study procedures for *solving equations* involving these exponential and logarithmic functions.

There are two basic strategies for solving exponential or logarithmic equations. The first is based on the One-to-One Properties and the second is based on the Inverse Properties. For $a > 0$ and $a \neq 1$, the following properties are true for all x and y for which $\log_a x$ and $\log_a y$ are defined.

One-to-One Properties

$a^x = a^y$ if and only if $x = y$.

$\log_a x = \log_a y$ if and only if $x = y$.

Inverse Properties

$a^{\log_a x} = x$

$\log_a a^x = x$

EXAMPLE 1 Solving Simple Equations

	Original Equation	*Rewritten Equation*	*Solution*	*Property*
a.	$2^x = 32$	$2^x = 2^5$	$x = 5$	One-to-One
b.	$\ln x - \ln 3 = 0$	$\ln x = \ln 3$	$x = 3$	One-to-One
c.	$\left(\frac{1}{3}\right)^x = 9$	$3^{-x} = 3^2$	$x = -2$	One-to-One
d.	$e^x = 7$	$\ln e^x = \ln 7$	$x = \ln 7$	Inverse
e.	$\ln x = -3$	$e^{\ln x} = e^{-3}$	$x = e^{-3}$	Inverse
f.	$\log x = -1$	$10^{\log x} = 10^{-1}$	$x = 10^{-1} = \frac{1}{10}$	Inverse
g.	$\log_3 x = 4$	$3^{\log_3 x} = 3^4$	$x = 81$	Inverse

■

The strategies used in Example 1 are summarized as follows.

STRATEGIES FOR SOLVING EXPONENTIAL AND LOGARITHMIC EQUATIONS

1. Rewrite the original equation in a form that allows the use of the One-to-One Properties of exponential or logarithmic functions.
2. Rewrite an *exponential* equation in logarithmic form and apply the Inverse Property of logarithmic functions.
3. Rewrite a *logarithmic* equation in exponential form and apply the Inverse Property of exponential functions.

Solving Exponential Equations

EXAMPLE 2 Solving Exponential Equations

Solve each equation and approximate the result to three decimal places, if necessary.

a. $e^{-x^2} = e^{-3x-4}$

b. $3(2^x) = 42$

Solution

a.

$$e^{-x^2} = e^{-3x-4} \quad \text{Write original equation.}$$
$$-x^2 = -3x - 4 \quad \text{One-to-One Property}$$
$$x^2 - 3x - 4 = 0 \quad \text{Write in general form.}$$
$$(x + 1)(x - 4) = 0 \quad \text{Factor.}$$
$$(x + 1) = 0 \Rightarrow x = -1 \quad \text{Set 1st factor equal to 0.}$$
$$(x - 4) = 0 \Rightarrow x = 4 \quad \text{Set 2nd factor equal to 0.}$$

The solutions are $x = -1$ and $x = 4$. Check these in the original equation.

b.

$$3(2^x) = 42 \quad \text{Write original equation.}$$
$$2^x = 14 \quad \text{Divide each side by 3.}$$
$$\log_2 2^x = \log_2 14 \quad \text{Take log (base 2) of each side.}$$
$$x = \log_2 14 \quad \text{Inverse Property}$$
$$x = \frac{\ln 14}{\ln 2} \quad \text{Change-of-base formula}$$
$$x \approx 3.807 \quad \text{Use a calculator.}$$

The solution is $x = \log_2 14 \approx 3.807$. Check this in the original equation. ■

STUDY TIP Another way to solve Example 2(b) is by taking the natural log of each side and then applying the Power Property, as follows.

$$3(2^x) = 42$$
$$2^x = 14$$
$$\ln 2^x = \ln 14$$
$$x \ln 2 = \ln 14$$
$$x = \frac{\ln 14}{\ln 2}$$
$$x \approx 3.807$$

As you can see, you obtain the same result as in Example 2(b).

In Example 2(b), the exact solution is $x = \log_2 14$ and the approximate solution is $x \approx 3.807$. An exact answer is preferred when the solution is an intermediate step in a larger problem. For a final answer, an approximate solution is easier to comprehend.

EXAMPLE 3 Solving an Exponential Equation

Solve

$$e^x + 5 = 60$$

and approximate the result to three decimal places.

Solution

$$e^x + 5 = 60 \quad \text{Write original equation.}$$
$$e^x = 55 \quad \text{Subtract 5 from each side.}$$
$$\ln e^x = \ln 55 \quad \text{Take natural log of each side.}$$
$$x = \ln 55 \quad \text{Inverse Property}$$
$$x \approx 4.007 \quad \text{Use a calculator.}$$

The solution is $x = \ln 55 \approx 4.007$. Check this in the original equation. ■

STUDY TIP When taking the logarithm of each side of an exponential equation, choose the base for the logarithm to be the same as the base in the exponential equation. In Example 2(b), base 2 was chosen, and in Example 3, base e was chosen for the logarithm.

EXAMPLE 4 Solving an Exponential Equation

Solve

$$2(3^{2t-5}) - 4 = 11$$

and approximate the result to three decimal places.

Solution

$$\begin{aligned} 2(3^{2t-5}) - 4 &= 11 && \text{Write original equation.} \\ 2(3^{2t-5}) &= 15 && \text{Add 4 to each side.} \\ 3^{2t-5} &= \frac{15}{2} && \text{Divide each side by 2.} \\ \log_3 3^{2t-5} &= \log_3 \frac{15}{2} && \text{Take log (base 3) of each side.} \\ 2t - 5 &= \log_3 \frac{15}{2} && \text{Inverse Property} \\ 2t &= 5 + \log_3 7.5 && \text{Add 5 to each side.} \\ t &= \frac{5}{2} + \frac{1}{2}\log_3 7.5 && \text{Divide each side by 2.} \\ t &\approx 3.417 && \text{Use a calculator.} \end{aligned}$$

The solution is $t = \frac{5}{2} + \frac{1}{2}\log_3 7.5 \approx 3.417$. Check this in the original equation.

When an equation involves two or more exponential expressions, you can still use a procedure similar to that demonstrated in Examples 2, 3, and 4. However, the algebra is a bit more complicated. In such cases, remember that a graph can help you check the reasonableness of your solution.

EXAMPLE 5 Solving an Exponential Equation of Quadratic Type

Solve

$$e^{2x} - 3e^x + 2 = 0.$$

Algebraic Solution

$$\begin{aligned} e^{2x} - 3e^x + 2 &= 0 && \text{Write original equation.} \\ (e^x)^2 - 3e^x + 2 &= 0 && \text{Write in quadratic form.} \\ (e^x - 2)(e^x - 1) &= 0 && \text{Factor.} \\ e^x - 2 &= 0 && \text{Set 1st factor equal to 0.} \\ x &= \ln 2 && \text{Solution} \\ e^x - 1 &= 0 && \text{Set 2nd factor equal to 0.} \\ x &= 0 && \text{Solution} \end{aligned}$$

The solutions are $x = \ln 2 \approx 0.693$ and $x = 0$. Check these in the original equation.

Graphical Solution

Use a graphing utility to graph $y = e^{2x} - 3e^x + 2$. Use the *zero* or *root* feature or the *zoom* and *trace* features of the graphing utility to approximate the values of x for which $y = 0$. In Figure 7.18, you can see that the zeros occur at $x = 0$ and at $x \approx 0.693$. So, the solutions are $x = 0$ and $x \approx 0.693$.

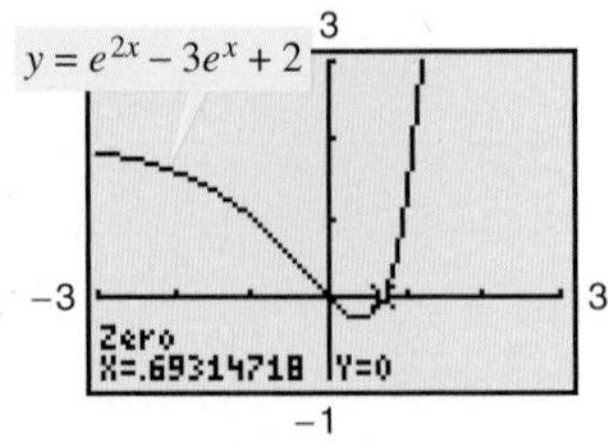

Figure 7.18

Solving Logarithmic Equations

To solve a logarithmic equation, you can write it in exponential form.

$\ln x = 3$	Logarithmic form
$e^{\ln x} = e^3$	Exponentiate each side.
$x = e^3$	Exponential form

This procedure is called *exponentiating* each side of an equation.

EXAMPLE 6 Solving Logarithmic Equations

STUDY TIP Remember to check your solutions in the original equation when solving equations to verify that the answer is correct and to make sure that the answer lies in the domain of the original equation.

a. $\ln x = 2$	Original equation
$e^{\ln x} = e^2$	Exponentiate each side.
$x = e^2$	Inverse Property
b. $\log_3(5x - 1) = \log_3(x + 7)$	Original equation
$5x - 1 = x + 7$	One-to-One Property
$4x = 8$	Add $-x$ and 1 to each side.
$x = 2$	Divide each side by 4.
c. $\log_6(3x + 14) - \log_6 5 = \log_6 2x$	Original equation
$\log_6\left(\frac{3x + 14}{5}\right) = \log_6 2x$	Quotient Property of Logarithms
$\frac{3x + 14}{5} = 2x$	One-to-One Property
$3x + 14 = 10x$	Cross multiply.
$-7x = -14$	Isolate x.
$x = 2$	Divide each side by -7.

EXAMPLE 7 Solving a Logarithmic Equation

Solve

$$5 + 2 \ln x = 4$$

and approximate the result to three decimal places.

Algebraic Solution

$5 + 2 \ln x = 4$	Write original equation.
$2 \ln x = -1$	Subtract 5 from each side.
$\ln x = -\frac{1}{2}$	Divide each side by 2.
$e^{\ln x} = e^{-1/2}$	Exponentiate each side.
$x = e^{-1/2}$	Inverse Property
$x \approx 0.607$	Use a calculator.

The solution is $x = e^{-1/2} \approx 0.607$. Check this in the original equation.

Graphical Solution

Use a graphing utility to graph $y_1 = 5 + 2 \ln x$ and $y_2 = 4$ in the same viewing window. Use the *intersect* feature or the *zoom* and *trace* features to approximate the intersection point, as shown in Figure 7.19. So, the solution is $x \approx 0.607$.

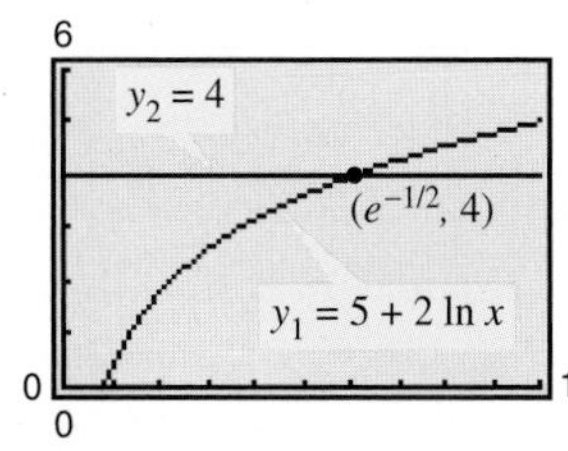

Figure 7.19

EXAMPLE 8 Solving a Logarithmic Equation

Solve

$2 \log_5 3x = 4.$

Solution

$$\begin{aligned} 2 \log_5 3x &= 4 && \text{Write original equation.} \\ \log_5 3x &= 2 && \text{Divide each side by 2.} \\ 5^{\log_5 3x} &= 5^2 && \text{Exponentiate each side (base 5).} \\ 3x &= 25 && \text{Inverse Property} \\ x &= \frac{25}{3} && \text{Divide each side by 3.} \end{aligned}$$

The solution is $x = \frac{25}{3}$. Check this in the original equation. ■

STUDY TIP Notice in Example 9 that the logarithmic part of the equation is condensed into a single logarithm before exponentiating each side of the equation.

Because the domain of a logarithmic function generally does not include all real numbers, you should be sure to check for extraneous solutions of logarithmic equations.

EXAMPLE 9 Checking for Extraneous Solutions

Solve

$\log 5x + \log(x - 1) = 2.$

Algebraic Solution

$$\begin{aligned} \log 5x + \log(x - 1) &= 2 && \text{Write original equation.} \\ \log[5x(x - 1)] &= 2 && \text{Product Property of Logarithms} \\ 10^{\log(5x^2 - 5x)} &= 10^2 && \text{Exponentiate each side (base 10).} \\ 5x^2 - 5x &= 100 && \text{Inverse Property} \\ x^2 - x - 20 &= 0 && \text{Write in general form.} \\ (x - 5)(x + 4) &= 0 && \text{Factor.} \\ x - 5 &= 0 && \text{Set 1st factor equal to 0.} \\ x &= 5 && \text{Solution} \\ x + 4 &= 0 && \text{Set 2nd factor equal to 0.} \\ x &= -4 && \text{Solution} \end{aligned}$$

The solutions appear to be $x = 5$ and $x = -4$. However, when you check these in the original equation, you can see that $x = 5$ is the only solution.

Graphical Solution

Use a graphing utility to graph

$y_1 = \log 5x + \log(x - 1)$

and

$y_2 = 2$

in the same viewing window. From the graph shown in Figure 7.20, it appears that the graphs intersect at one point. Use the *intersect* feature or the *zoom* and *trace* features to determine that the graphs intersect at approximately $(5, 2)$. So, the solution is $x = 5$. Verify that 5 is an exact solution algebraically.

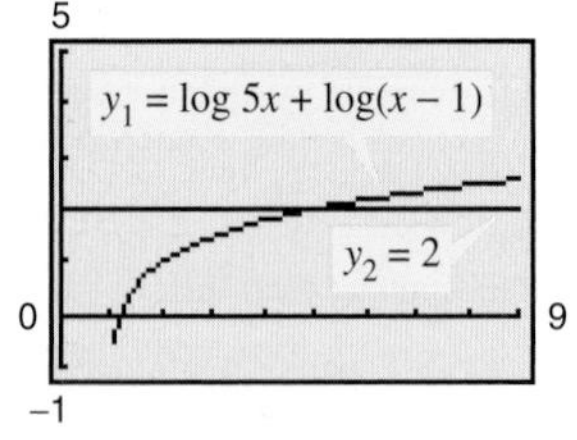

Figure 7.20 ■

In Example 9, the domain of $\log 5x$ is $x > 0$ and the domain of $\log(x - 1)$ is $x > 1$, so the domain of the original equation is $x > 1$. Because the domain is all real numbers greater than 1, the solution $x = -4$ is extraneous. The graph in Figure 7.20 verifies this conclusion.

Applications

EXAMPLE 10 Doubling an Investment

You have deposited \$500 in an account that pays 6.75% interest, compounded continuously. How long will it take your money to double?

Solution Using the formula for continuous compounding, you can find that the balance in the account is

$$A = Pe^{rt}$$
$$A = 500e^{0.0675t}.$$

To find the time required for the balance to double, let $A = 1000$ and solve the resulting equation for t.

$500e^{0.0675t} = 1000$ Substitute 1000 for A.

$e^{0.0675t} = 2$ Divide each side by 500.

$\ln e^{0.0675t} = \ln 2$ Take natural log of each side.

$0.0675t = \ln 2$ Inverse Property

$t = \dfrac{\ln 2}{0.0675}$ Divide each side by 0.0675.

$t \approx 10.27$ Use a calculator.

The balance in the account will double after approximately 10.27 years. This result is demonstrated graphically in Figure 7.21.

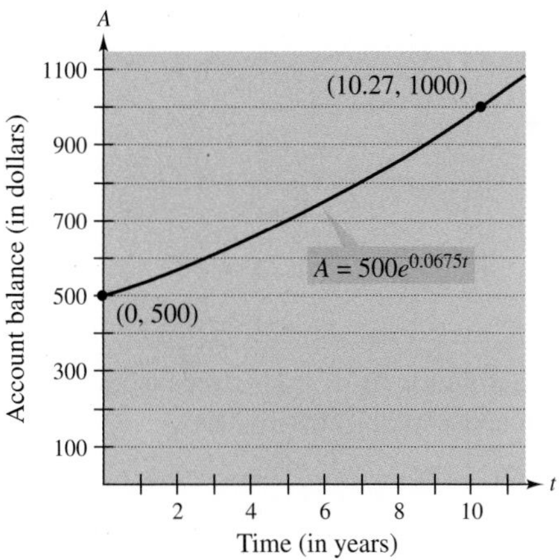

Figure 7.21

In Example 10, an approximate answer of 10.27 years is given. Within the context of the problem, the exact solution,

$$\frac{\ln 2}{0.0675} \text{ years}$$

does not make sense as an answer.

EXAMPLE 11 Retail Sales

The retail sales y (in billions) of e-commerce companies in the United States from 2002 through 2007 can be modeled by

$$y = -549 + 236.7 \ln t, \quad 12 \le t \le 17$$

where t represents the year, with $t = 12$ corresponding to 2002 (see Figure 7.22). During which year did the sales reach \$108 billion? *(Source: U.S. Census Bureau)*

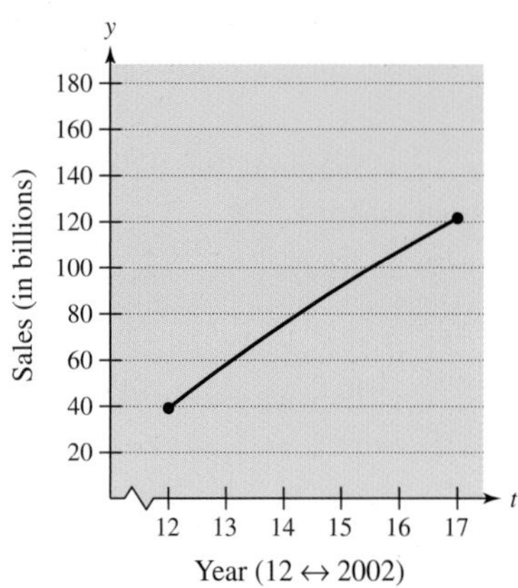

Figure 7.22

Solution

$-549 + 236.7 \ln t = y$	Write original equation.
$-549 + 236.7 \ln t = 108$	Substitute 108 for y.
$236.7 \ln t = 657$	Add 549 to each side.
$\ln t = \dfrac{657}{236.7}$	Divide each side by 236.7.
$e^{\ln t} = e^{657/236.7}$	Exponentiate each side.
$t = e^{657/236.7}$	Inverse Property
$t \approx 16$	Use a calculator.

The solution is $t \approx 16$. Because $t = 12$ represents 2002, it follows that the sales reached \$108 billion in 2006. ■

7.4 Exercises

See www.CalcChat.com for worked-out solutions to odd-numbered exercises.

In Exercises 1–4, fill in the blanks.

1. To ________ an equation in x means to find all values of x for which the equation is true.
2. To solve exponential and logarithmic equations, you can use the following One-to-One and Inverse Properties.
   (a) $a^x = a^y$ if and only if ________.
   (b) $\log_a x = \log_a y$ if and only if ________.
   (c) $a^{\log_a x} =$ ________
   (d) $\log_a a^x =$ ________
3. To solve exponential and logarithmic equations, you can use the following strategies.
   (a) Rewrite the original equation in a form that allows the use of the ________ Properties of exponential or logarithmic functions.
   (b) Rewrite an exponential equation in ________ form and apply the Inverse Property of ________ functions.
   (c) Rewrite a logarithmic equation in ________ form and apply the Inverse Property of ________ functions.
4. An ________ solution does not satisfy the original equation.

In Exercises 5–12, determine whether each x-value is a solution (or an approximate solution) of the equation.

5. $4^{2x-7} = 64$
   (a) $x = 5$
   (b) $x = 2$

6. $2^{3x+1} = 32$
   (a) $x = -1$
   (b) $x = 2$

7. $3e^{x+2} = 75$
   (a) $x = -2 + e^{25}$
   (b) $x = -2 + \ln 25$
   (c) $x \approx 1.219$

8. $4e^{x-1} = 60$
   (a) $x = 1 + \ln 15$
   (b) $x \approx 3.7081$
   (c) $x = \ln 16$

9. $\log_4(3x) = 3$
   (a) $x \approx 21.333$
   (b) $x = -4$
   (c) $x = \frac{64}{3}$

10. $\log_2(x + 3) = 10$
   (a) $x = 1021$
   (b) $x = 17$
   (c) $x = 10^2 - 3$

11. $\ln(2x + 3) = 5.8$
   (a) $x = \frac{1}{2}(-3 + \ln 5.8)$
   (b) $x = \frac{1}{2}(-3 + e^{5.8})$
   (c) $x \approx 163.650$

12. $\ln(x - 1) = 3.8$
   (a) $x = 1 + e^{3.8}$
   (b) $x \approx 45.701$
   (c) $x = 1 + \ln 3.8$

In Exercises 13–24, solve for x.

13. $4^x = 16$
14. $3^x = 243$
15. $\left(\frac{1}{2}\right)^x = 32$
16. $\left(\frac{1}{4}\right)^x = 64$
17. $\ln x - \ln 2 = 0$
18. $\ln x - \ln 5 = 0$
19. $e^x = 2$
20. $e^x = 4$
21. $\ln x = -1$
22. $\log x = -2$
23. $\log_4 x = 3$
24. $\log_5 x = \frac{1}{2}$

In Exercises 25–28, approximate the point of intersection of the graphs of f and g. Then solve the equation $f(x) = g(x)$ algebraically to verify your approximation.

25. $f(x) = 2^x$
    $g(x) = 8$

26. $f(x) = 27^x$
    $g(x) = 9$

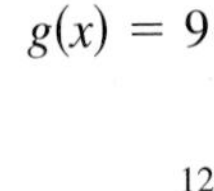

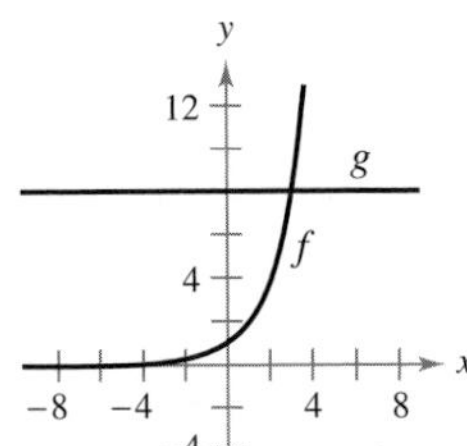

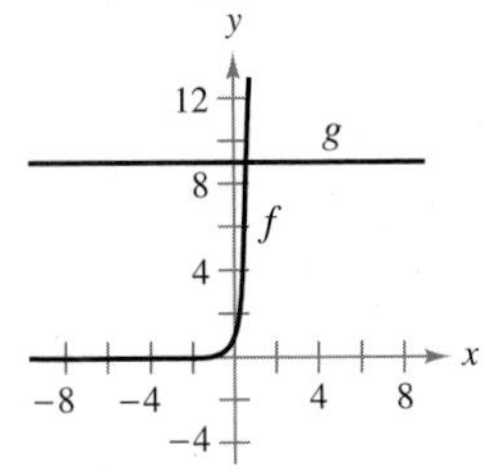

27. $f(x) = \log_3 x$
    $g(x) = 2$

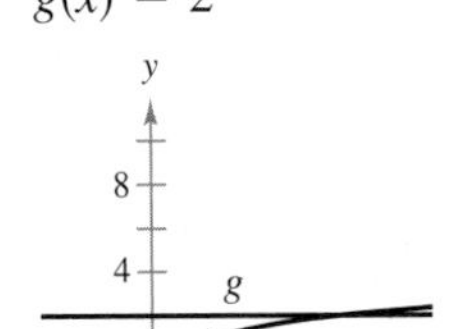

28. $f(x) = \ln(x - 4)$
    $g(x) = 0$

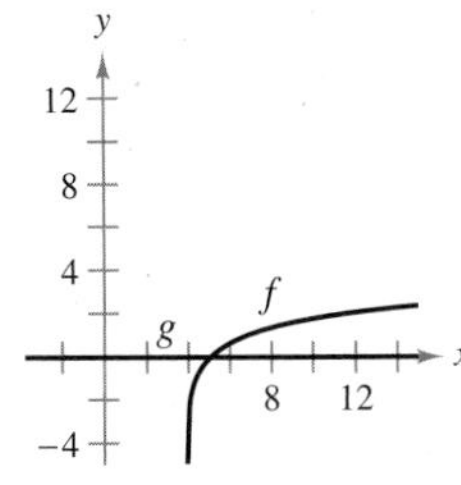

In Exercises 29–70, solve the exponential equation algebraically. Approximate the result to three decimal places.

29. $e^x = e^{x^2-2}$
30. $e^{2x} = e^{x^2-8}$
31. $e^{x^2-3} = e^{x-2}$
32. $e^{-x^2} = e^{x^2-2x}$
33. $4(3^x) = 20$
34. $2(5^x) = 32$
35. $2e^x = 10$
36. $4e^x = 91$
37. $e^x - 9 = 19$
38. $6^x + 10 = 47$
39. $3^{2x} = 80$
40. $6^{5x} = 3000$
41. $5^{-t/2} = 0.20$
42. $4^{-3t} = 0.10$
43. $3^{x-1} = 27$
44. $2^{x-3} = 32$
45. $2^{3-x} = 565$
46. $8^{-2-x} = 431$
47. $8(10^{3x}) = 12$
48. $5(10^{x-6}) = 7$
49. $3(5^{x-1}) = 21$
50. $8(3^{6-x}) = 40$
51. $e^{3x} = 12$
52. $e^{2x} = 50$
53. $500e^{-x} = 300$
54. $1000e^{-4x} = 75$
55. $7 - 2e^x = 5$
56. $-14 + 3e^x = 11$
57. $6(2^{3x-1}) - 7 = 9$
58. $8(4^{6-2x}) + 13 = 41$
59. $e^{2x} - 4e^x - 5 = 0$
60. $e^{2x} - 5e^x + 6 = 0$
61. $e^{2x} - 3e^x - 4 = 0$
62. $e^{2x} + 9e^x + 36 = 0$
63. $\dfrac{500}{100 - e^{x/2}} = 20$
64. $\dfrac{400}{1 + e^{-x}} = 350$
65. $\dfrac{3000}{2 + e^{2x}} = 2$
66. $\dfrac{119}{e^{6x} - 14} = 7$
67. $\left(1 + \dfrac{0.065}{365}\right)^{365t} = 4$
68. $\left(4 - \dfrac{2.471}{40}\right)^{9t} = 21$
69. $\left(1 + \dfrac{0.10}{12}\right)^{12t} = 2$
70. $\left(16 - \dfrac{0.878}{26}\right)^{3t} = 30$

In Exercises 71–80, use a graphing utility to graph and solve the equation. Approximate the result to three decimal places. Verify your result algebraically.

71. $7 = 2^x$
72. $5^x = 212$
73. $6e^{1-x} = 25$
74. $-4e^{-x-1} + 15 = 0$
75. $3e^{3x/2} = 962$
76. $8e^{-2x/3} = 11$
77. $e^{0.09t} = 3$
78. $-e^{1.8x} + 7 = 0$
79. $e^{0.125t} - 8 = 0$
80. $e^{2.724x} = 29$

In Exercises 81–112, solve the logarithmic equation algebraically. Approximate the result to three decimal places.

81. $\ln x = -3$ **82.** $\ln x = 1.6$

83. $\ln x - 7 = 0$ **84.** $\ln x + 1 = 0$

85. $\ln 2x = 2.4$ **86.** $2.1 = \ln 6x$

87. $\log x = 6$ **88.** $\log 3z = 2$

89. $3 \ln 5x = 10$ **90.** $2 \ln x = 7$

91. $\ln \sqrt{x + 2} = 1$ **92.** $\ln \sqrt{x - 8} = 5$

93. $7 + 3 \ln x = 5$

94. $2 - 6 \ln x = 10$

95. $-2 + 2 \ln 3x = 17$

96. $2 + 3 \ln x = 12$

97. $6 \log_3(0.5x) = 11$

98. $4 \log(x - 6) = 11$

99. $\ln x - \ln(x + 1) = 2$

100. $\ln x + \ln(x + 1) = 1$

101. $\ln x + \ln(x - 2) = 1$

102. $\ln x + \ln(x + 3) = 1$

103. $\ln(x + 5) = \ln(x - 1) - \ln(x + 1)$

104. $\ln(x + 1) - \ln(x - 2) = \ln x$

105. $\log_2(2x - 3) = \log_2(x + 4)$

106. $\log(3x + 4) = \log(x - 10)$

107. $\log(x + 4) - \log x = \log(x + 2)$

108. $\log_2 x + \log_2(x + 2) = \log_2(x + 6)$

109. $\log_4 x - \log_4(x - 1) = \frac{1}{2}$

110. $\log_3 x + \log_3(x - 8) = 2$

111. $\log 8x - \log(1 + \sqrt{x}) = 2$

112. $\log 4x - \log(12 + \sqrt{x}) = 2$

In Exercises 113–116, use a graphing utility to graph and solve the equation. Approximate the result to three decimal places. Verify your result algebraically.

113. $3 - \ln x = 0$ **114.** $10 - 4 \ln(x - 2) = 0$

115. $2 \ln(x + 3) = 3$ **116.** $\ln(x + 1) = 2 - \ln x$

In Exercises 117–124, solve the equation algebraically. Round the result to three decimal places. Verify your answer using a graphing utility.

117. $2x^2e^{2x} + 2xe^{2x} = 0$ **118.** $-x^2e^{-x} + 2xe^{-x} = 0$

119. $-xe^{-x} + e^{-x} = 0$ **120.** $e^{-2x} - 2xe^{-2x} = 0$

121. $2x \ln x + x = 0$ **122.** $\dfrac{1 - \ln x}{x^2} = 0$

123. $\dfrac{1 + \ln x}{2} = 0$ **124.** $2x \ln\left(\dfrac{1}{x}\right) - x = 0$

WRITING ABOUT CONCEPTS

125. ***Finance*** You are investing P dollars at an annual interest rate of r, compounded continuously, for t years. Which of the following would result in the highest value of the investment? Explain your reasoning.

(a) Double the amount you invest.

(b) Double your interest rate.

(c) Double the number of years.

126. Write a paragraph explaining whether the time required for an investment to double depends on the size of the investment.

Compound Interest **In Exercises 127–130, \$2500 is invested in an account at interest rate r, compounded continuously. Find the time required for the amount to (a) double and (b) triple.**

127. $r = 0.05$ **128.** $r = 0.045$

129. $r = 0.025$ **130.** $r = 0.0375$

131. ***Think About It*** Are the times required for the investments in Exercises 127–130 to quadruple twice as long as the times for them to double? Give a reason for your answer and verify your answer algebraically.

132. ***Demand*** The demand equation for a limited edition coin set is

$$p = 1000\left(1 - \frac{5}{5 + e^{-0.001x}}\right).$$

Find the demand x for a price of (a) $p = \$139.50$ and (b) $p = \$99.99$.

133. ***Demand*** The demand equation for a hand-held electronic organizer is

$$p = 5000\left(1 - \frac{4}{4 + e^{-0.002x}}\right).$$

Find the demand x for a price of (a) $p = \$600$ and (b) $p = \$400$.

134. ***Forest Yield*** The yield V (in millions of cubic feet per acre) for a forest at age t years is given by $V = 6.7e^{-48.1/t}$.

(a) Use a graphing utility to graph the function.

(b) Determine the horizontal asymptote of the function. Interpret its meaning in the context of the problem.

(c) Find the time necessary to obtain a yield of 1.3 million cubic feet.

135. ***Trees per Acre*** The number N of trees of a given species per acre is approximated by the model $N = 68(10^{-0.04x})$, $5 \le x \le 40$, where x is the average diameter of the trees (in inches) 3 feet above the ground. Use the model to approximate the average diameter of the trees in a test plot when $N = 21$.

136. ***U.S. Currency*** The values y (in billions of dollars) of U.S. currency in circulation in the years 2000 through 2007 can be modeled by $y = -451 + 444 \ln t$, $10 \le t \le 17$, where t represents the year, with $t = 10$ corresponding to 2000. During which year did the value of U.S. currency in circulation exceed \$690 billion? *(Source: Board of Governors of the Federal Reserve System)*

137. ***Medicine*** The numbers y of freestanding ambulatory care surgery centers in the United States from 2000 through 2007 can be modeled by

$$y = 2875 + \frac{2635.11}{1 + 14.215e^{-0.8038t}}, \quad 0 \le t \le 7$$

where t represents the year, with $t = 0$ corresponding to 2000. *(Source: Verispan)*

(a) Use a graphing utility to graph the model.

(b) Use the *trace* feature of the graphing utility to estimate the year in which the number of surgery centers exceeded 3600.

138. ***Average Heights*** The percent m of American males between the ages of 18 and 24 who are no more than x inches tall is modeled by

$$m(x) = \frac{100}{1 + e^{-0.6114(x-69.71)}}$$

and the percent f of American females between the ages of 18 and 24 who are no more than x inches tall is modeled by

$$f(x) = \frac{100}{1 + e^{-0.66607(x-64.51)}}.$$

(Source: U.S. National Center for Health Statistics)

(a) Use the graph to determine any horizontal asymptotes of the graphs of the functions. Interpret the meaning in the context of the problem.

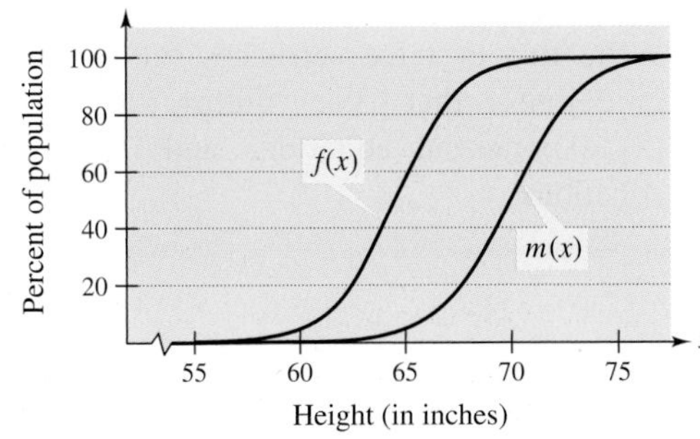

(b) What is the average height of each sex?

139. ***Learning Curve*** In a group project in learning theory, a mathematical model for the proportion P of correct responses after n trials was found to be

$$P = \frac{0.83}{1 + e^{-0.2n}}.$$

(a) Use a graphing utility to graph the function.

(b) Use the graph to determine any horizontal asymptotes of the graph of the function. Interpret the meaning of the upper asymptote in the context of this problem.

(c) After how many trials will 60% of the responses be correct?

140. ***Automobiles*** Automobiles are designed with crumple zones that help protect their occupants in crashes. The crumple zones allow the occupants to move short distances when the automobiles come to abrupt stops. The greater the distance moved, the fewer g's the crash victims experience. (One g is equal to the acceleration due to gravity. For very short periods of time, humans have withstood as much as 40 g's.) In crash tests with vehicles moving at 90 kilometers per hour, analysts measured the numbers of g's experienced during deceleration by crash dummies that were permitted to move x meters during impact. The data are shown in the table. A model for the data is given by

$$y = -3.00 + 11.88 \ln x + \frac{36.94}{x}$$

where y is the number of g's.

x	0.2	0.4	0.6	0.8	1.0
g's	158	80	53	40	32

(a) Complete the table using the model.

x	0.2	0.4	0.6	0.8	1.0
y					

(b) Use a graphing utility to graph the data points and the model in the same viewing window. How do they compare?

(c) Use the model to estimate the distance traveled during impact if the passenger deceleration must not exceed 30 g's.

(d) Do you think it is practical to lower the number of g's experienced during impact to fewer than 23? Explain your reasoning.

141. ***Data Analysis*** An object at a temperature of 160°C was removed from a furnace and placed in a room at 20°C. The temperature T of the object was measured each hour h and recorded in the table. A model for the data is given by $T = 20[1 + 7(2^{-h})]$. The graph of this model is shown in the figure.

Hour, h	0	1	2	3	4	5
Temperature, T	160°	90°	56°	38°	29°	24°

(a) Use the graph to identify the horizontal asymptote of the model and interpret the asymptote in the context of the problem.

(b) Use the model to approximate the time when the temperature of the object was 100°C.

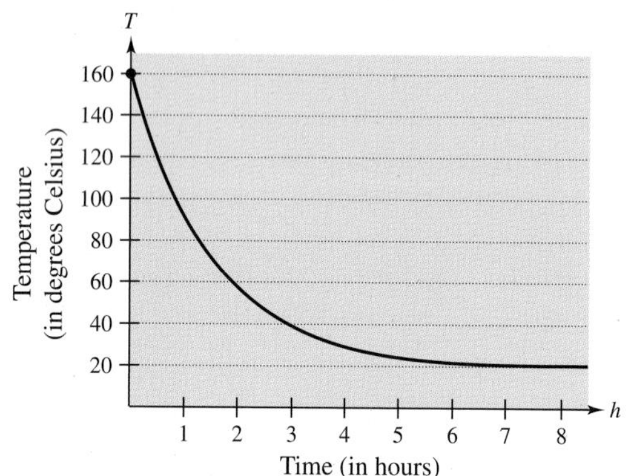

142. ***Data Analysis*** The personal consumption medical care expenditures E (in billions of dollars) for selected years from 1960 to 2000 are shown in the table.

t	1960	1970	1980	1990	2000
E	20.0	49.9	207.2	619.7	1171.1

A model for these data is $E = 20.39e^{0.1066t}$, where t is the time in years, with $t = 0$ corresponding to 1960. *(Source: U.S. Bureau of Economic Analysis)*

(a) Use a graphing utility to graph the data points and the model in the same viewing window. How do they compare?

(b) Use the model to estimate the personal consumption medical expenditures for 2010, 2015, and 2020.

(c) Algebraically find the year, according to the model, when personal consumption medical expenditures exceed 1 trillion dollars.

(d) Do you believe that the future personal consumption medical expenditures can be predicted using the given model? Explain your reasoning.

True or False? **In Exercises 143–146, rewrite each verbal statement as an equation. Then decide whether the statement is true or false. Justify your answer.**

143. The logarithm of the product of two numbers is equal to the sum of the logarithms of the numbers.

144. The logarithm of the sum of two numbers is equal to the product of the logarithms of the numbers.

145. The logarithm of the difference of two numbers is equal to the difference of the logarithms of the numbers.

146. The logarithm of the quotient of two numbers is equal to the difference of the logarithms of the numbers.

147. ***Think About It*** Is it possible for a logarithmic equation to have more than one extraneous solution? Explain.

148. The *effective yield* of a savings plan is the percent increase in the balance after 1 year. Find the effective yield for each savings plan when $1000 is deposited in a savings account. Which savings plan has the greatest effective yield? Which savings plan will have the highest balance after 5 years?

(a) 7% annual interest rate, compounded annually

(b) 7% annual interest rate, compounded continuously

(c) 7% annual interest rate, compounded quarterly

(d) 7.25% annual interest rate, compounded quarterly

149. ***Graphical Analysis*** Let

$f(x) = \log_a x$

and

$g(x) = a^x$

where $a > 1$.

(a) Let $a = 1.2$ and use a graphing utility to graph the two functions in the same viewing window. What do you observe? Approximate any points of intersection of the two graphs.

(b) Determine the value(s) of a for which the two graphs have one point of intersection.

(c) Determine the value(s) of a for which the two graphs have two points of intersection.

CAPSTONE

150. Write two or three sentences stating the general guidelines that you follow when solving (a) exponential equations and (b) logarithmic equations.

7.5 Exponential and Logarithmic Models

- Recognize the five most common types of models involving exponential and logarithmic functions.
- Use exponential growth and decay functions to model and solve real-life problems.
- Use Gaussian functions to model and solve real-life problems.
- Use logistic growth functions to model and solve real-life problems.
- Use logarithmic functions to model and solve real-life problems.

Introduction

The five most common types of mathematical models involving exponential functions and logarithmic functions are as follows.

STUDY TIP The models for exponential growth and decay vary only in the sign of the real number b.

1. **Exponential growth model:** $y = ae^{bx}, \quad a > 0, \quad b > 0$
2. **Exponential decay model:** $y = ae^{-bx}, \quad a > 0, \quad b > 0$
3. **Gaussian model:** $y = ae^{-(x-b)^2/c}, \quad a > 0$
4. **Logistic growth model:** $y = \dfrac{a}{1 + be^{-rx}}, \quad a > 0$
5. **Logarithmic models:** $y = a + b \ln x, \quad y = a + b \log x$

The basic shapes of the graphs of these functions are shown in Figure 7.23.

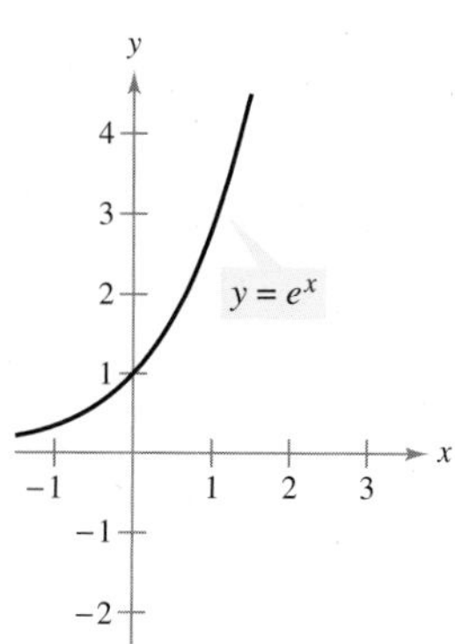

Exponential growth model

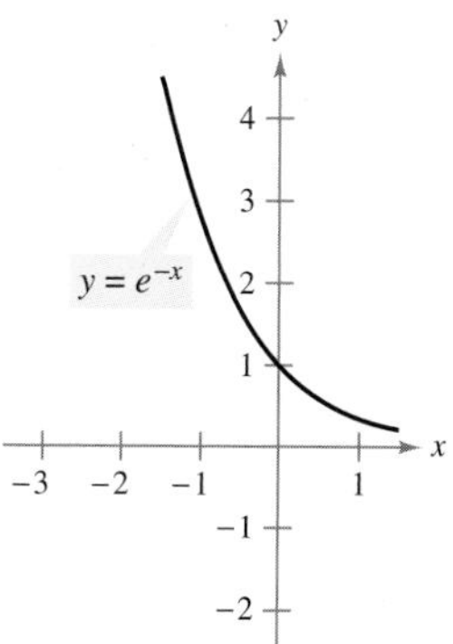

Exponential decay model

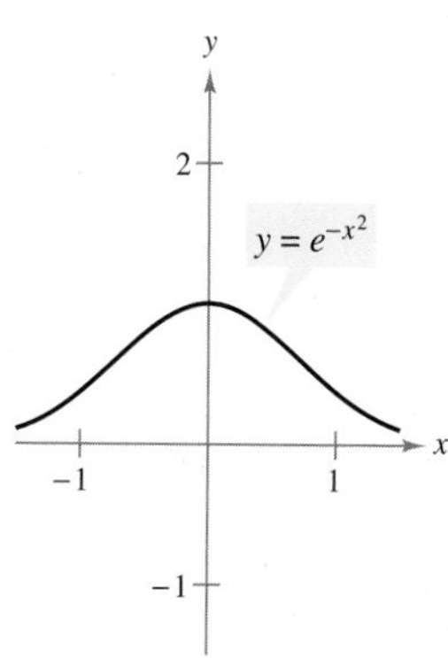

Gaussian model

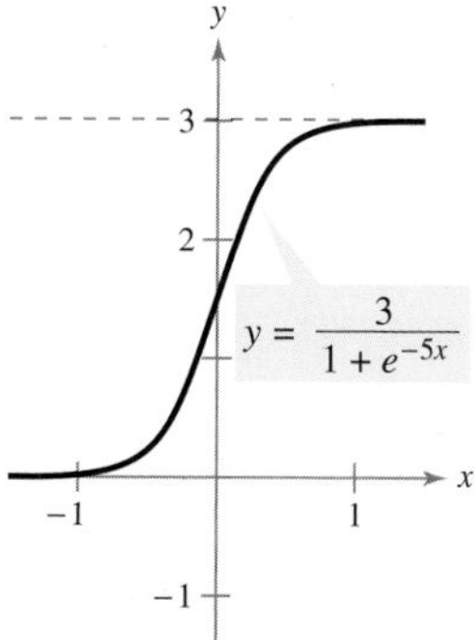

Logistic growth model

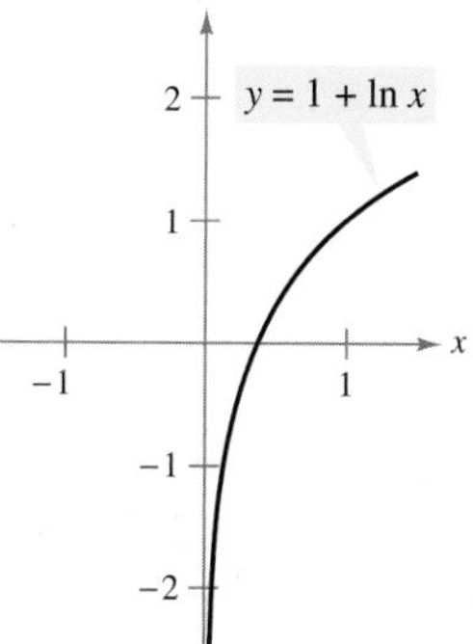

Natural logarithmic model

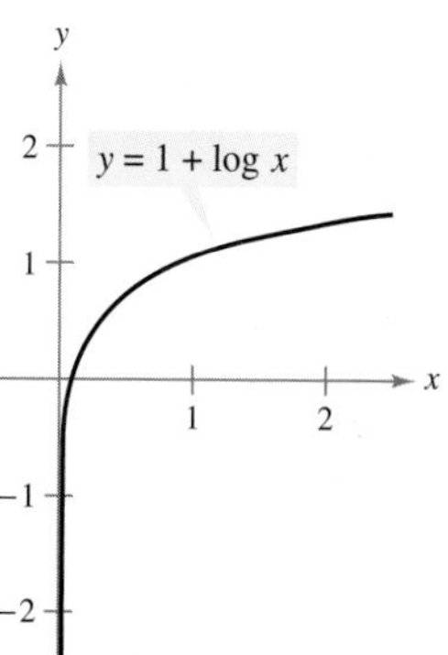

Common logarithmic model

Figure 7.23

You can often gain quite a bit of insight into a situation modeled by an exponential or logarithmic function by identifying and interpreting the function's asymptotes. Use the graphs in Figure 7.23 to identify the asymptotes of the graph of each function.

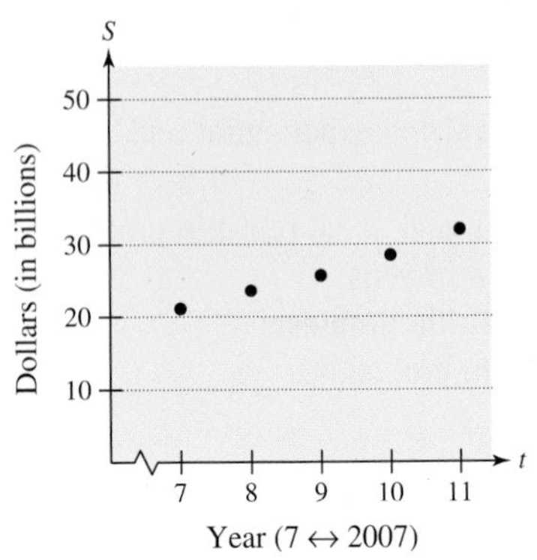

Figure 7.24

Exponential Growth and Decay

EXAMPLE 1 Online Advertising

Estimates of the amounts (in billions of dollars) of U.S. online advertising spending from 2007 through 2011 are shown in the table. A scatter plot of the data is shown in Figure 7.24. *(Source: eMarketer)*

Year	2007	2008	2009	2010	2011
Advertising spending	21.1	23.6	25.7	28.5	32.0

An exponential growth model that approximates these data is given by

$$S = 10.33e^{0.1022t}, \ 7 \le t \le 11$$

where S is the amount of spending (in billions) and $t = 7$ represents 2007. Compare the values given by the model with the estimates shown in the table. According to this model, when will the amount of U.S. online advertising spending reach \$40 billion?

Algebraic Solution

The following table compares the two sets of advertising spending figures.

Year	2007	2008	2009	2010	2011
Advertising spending	21.1	23.6	25.7	28.5	32.0
Model	21.1	23.4	25.9	28.7	31.8

To find when the amount of U.S. online advertising spending will reach \$40 billion, let $S = 40$ in the model and solve for t.

$$10.33e^{0.1022t} = S \quad \text{Write original model.}$$

$$10.33e^{0.1022t} = 40 \quad \text{Substitute 40 for } S.$$

$$e^{0.1022t} \approx 3.8722 \quad \text{Divide each side by 10.33.}$$

$$\ln e^{0.1022t} \approx \ln 3.8722 \quad \text{Take natural log of each side.}$$

$$0.1022t \approx 1.3538 \quad \text{Inverse Property}$$

$$t \approx 13.2 \quad \text{Divide each side by 0.1022.}$$

According to the model, the amount of U.S. online advertising spending will reach \$40 billion in 2013.

Graphical Solution

Use a graphing utility to graph the model $y = 10.33e^{0.1022x}$ and the data in the same viewing window. You can see in Figure 7.25 that the model appears to fit the data closely.

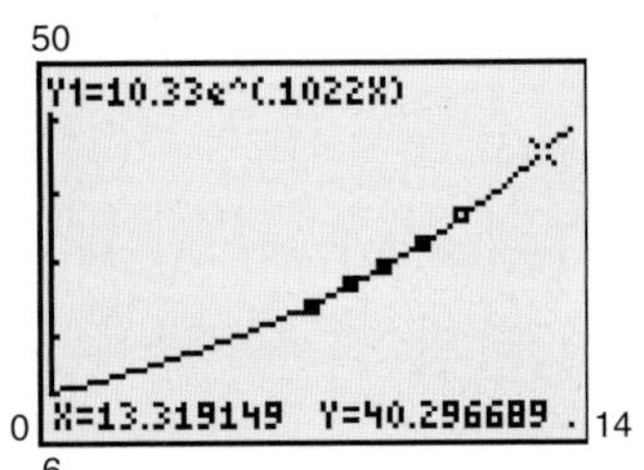

Figure 7.25

Use the *zoom* and *trace* features of the graphing utility to find that the approximate value of x for $y = 40$ is $x \approx 13.2$. So, according to the model, the amount of U.S. online advertising spending will reach \$40 billion in 2013. ■

TECHNOLOGY Some graphing utilities have an *exponential regression* feature that can be used to find exponential models that represent data. If you have such a graphing utility, try using it to find an exponential model for the data given in Example 1. How does your model compare with the model given in Example 1?

In Example 1, you were given the exponential growth model. But suppose this model were not given. How could you find such a model? One technique for doing this is demonstrated in Example 2.

EXAMPLE 2 Modeling Population Growth

In a research experiment, a population of fruit flies is increasing according to the law of exponential growth. After 2 days there are 100 flies, and after 4 days there are 300 flies. How many flies will there be after 5 days?

Solution Let y be the number of flies at time t. From the given information, you know that $y = 100$ when $t = 2$ and $y = 300$ when $t = 4$. Substituting this information into the model $y = ae^{bt}$ produces

$$100 = ae^{2b} \quad \text{and} \quad 300 = ae^{4b}.$$

To solve for b, solve for a in the first equation.

$$100 = ae^{2b} \Longrightarrow a = \frac{100}{e^{2b}} \qquad \text{Solve for } a \text{ in the first equation.}$$

Then substitute the result into the second equation.

$$300 = ae^{4b} \qquad \text{Write second equation.}$$

$$300 = \left(\frac{100}{e^{2b}}\right)e^{4b} \qquad \text{Substitute } \frac{100}{e^{2b}} \text{ for } a.$$

$$\frac{300}{100} = e^{2b} \qquad \text{Divide each side by 100.}$$

$$\ln 3 = 2b \qquad \text{Take natural log of each side.}$$

$$\frac{1}{2}\ln 3 = b \qquad \text{Solve for } b.$$

Using $b = \frac{1}{2}\ln 3$ and the equation you found for a, you can determine that

$$a = \frac{100}{e^{2[(1/2)\ln 3]}} = \frac{100}{e^{\ln 3}} = \frac{100}{3} \approx 33.$$

So, with $a \approx 33$ and $b = \frac{1}{2}\ln 3 \approx 0.5493$, the exponential growth model is

$$y = 33e^{0.5493t}$$

as shown in Figure 7.26. This implies that, after 5 days, the population will be

$$y = 33e^{0.5493(5)} \approx 514 \text{ flies.}$$

■

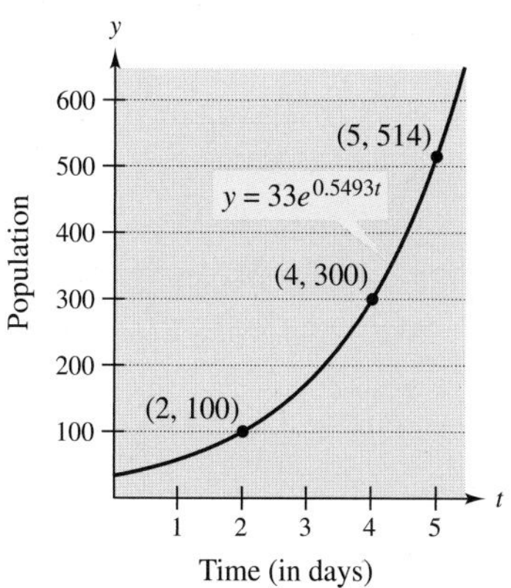

Figure 7.26

In living organic material, the ratio of the number of radioactive carbon isotopes (carbon 14) to the number of nonradioactive carbon isotopes (carbon 12) is about 1 to 10^{12}. When organic material dies, its carbon 12 content remains fixed, whereas its radioactive carbon 14 begins to decay with a half-life of about 5700 years. To estimate the age of dead organic material, scientists use the following formula, which denotes the ratio of carbon 14 to carbon 12 present at any time t (in years).

$$R = \frac{1}{10^{12}}e^{-t/8223} \qquad \text{Carbon dating model}$$

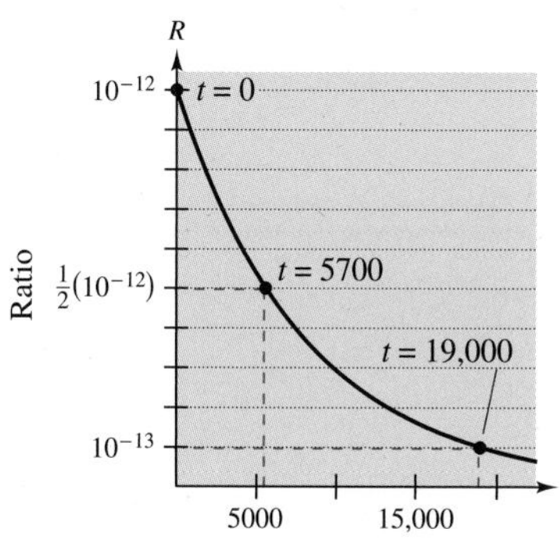

Figure 7.27

The graph of R is shown in Figure 7.27. Note that R decreases as t increases.

Gaussian Models

As mentioned at the beginning of this section, Gaussian models are of the form

$$y = ae^{-(x-b)^2/c}.$$

This type of model is commonly used in probability and statistics to represent populations that are **normally distributed.** For *standard* normal distributions, the model takes the form

$$y = \frac{1}{\sigma\sqrt{2\pi}}e^{-x^2/(2\sigma^2)}$$

where $\sigma = 1$ is the standard deviation (σ is the lowercase Greek letter sigma). The graph of a Gaussian model is called a **bell-shaped curve.** Try to sketch the standard normal distribution curve with a graphing utility. Can you see why it is called a bell-shaped curve?

The average value for a population can be found from the bell-shaped curve by observing where the maximum y-value of the function occurs. The x-value corresponding to the maximum y-value of the function represents the average value of the independent variable—in this case, x.

EXAMPLE 3 SAT Scores

In 2008, the Scholastic Aptitude Test (SAT) math scores for college-bound seniors roughly followed the normal distribution given by

$$y = 0.0034e^{-(x-515)^2/26{,}912}, \quad 200 \le x \le 800$$

where x is the SAT score for mathematics. Sketch the graph of this function. From the graph, estimate the average SAT score. *(Source: College Board)*

Solution The graph of the function is shown in Figure 7.28. On this bell-shaped curve, the maximum value of the curve represents the average score. From the graph, you can estimate that the average mathematics score for college-bound seniors in 2008 was 515.

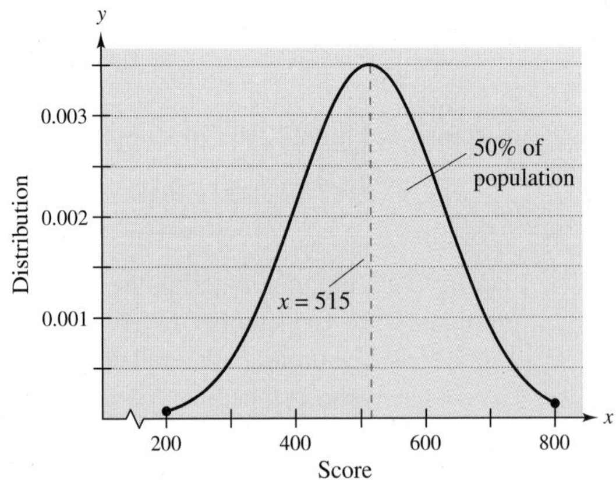

Figure 7.28

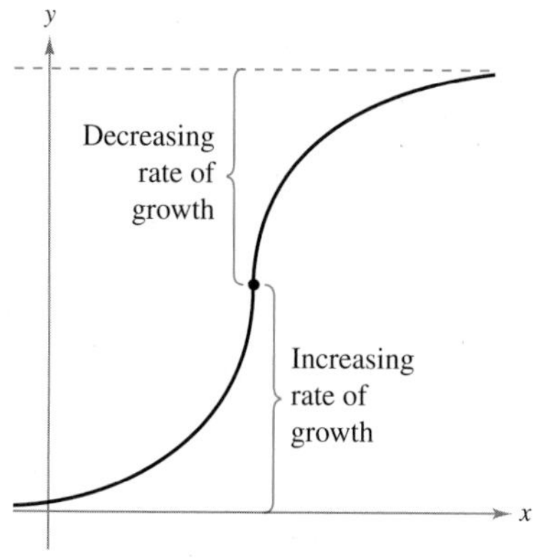

Figure 7.29

Logistic Growth Models

Some populations initially have rapid growth, followed by a declining rate of growth, as indicated by the graph in Figure 7.29. One model for describing this type of growth pattern is the **logistic curve** given by the function

$$y = \frac{a}{1 + be^{-rx}}$$

where y is the population size and x is the time. An example is a bacteria culture that is initially allowed to grow under ideal conditions, and then under less favorable conditions that inhibit growth. A logistic growth curve is also called a **sigmoidal curve.**

EXAMPLE 4 Spread of a Virus

On a college campus of 5000 students, one student returns from vacation with a contagious and long-lasting flu virus. The spread of the virus is modeled by

$$y = \frac{5000}{1 + 4999e^{-0.8t}}, \quad t \geq 0$$

where y is the total number of students infected after t days. The college will cancel classes when 40% or more of the students are infected.

a. How many students are infected after 5 days?

b. After how many days will the college cancel classes?

Algebraic Solution

a. After 5 days, the number of students infected is

$$y = \frac{5000}{1 + 4999e^{-0.8(5)}} = \frac{5000}{1 + 4999e^{-4}} \approx 54.$$

b. Classes are canceled when the number infected is $(0.40)(5000) = 2000$.

$$2000 = \frac{5000}{1 + 4999e^{-0.8t}}$$

$$1 + 4999e^{-0.8t} = 2.5$$

$$e^{-0.8t} = \frac{1.5}{4999}$$

$$\ln e^{-0.8t} = \ln \frac{1.5}{4999}$$

$$-0.8t = \ln \frac{1.5}{4999}$$

$$t = -\frac{1}{0.8} \ln \frac{1.5}{4999}$$

$$t \approx 10.1$$

So, after about 10 days, at least 40% of the students will be infected, and the college will cancel classes.

Graphical Solution

a. Use a graphing utility to graph $y = \dfrac{5000}{1 + 4999e^{-0.8x}}$. Use the *value* feature or the *zoom* and *trace* features of the graphing utility to estimate that $y \approx 54$ when $x = 5$. So, after 5 days, about 54 students will be infected.

b. Classes are canceled when the number of infected students is $(0.40)(5000) = 2000$. Use a graphing utility to graph

$$y_1 = \frac{5000}{1 + 4999e^{-0.8x}} \text{ and } y_2 = 2000$$

in the same viewing window. Use the *intersect* feature or the *zoom* and *trace* features of the graphing utility to find the point of intersection of the graphs. In Figure 7.30, you can see that the point of intersection occurs near $x \approx 10.1$. So, after about 10 days, at least 40% of the students will be infected, and the college will cancel classes.

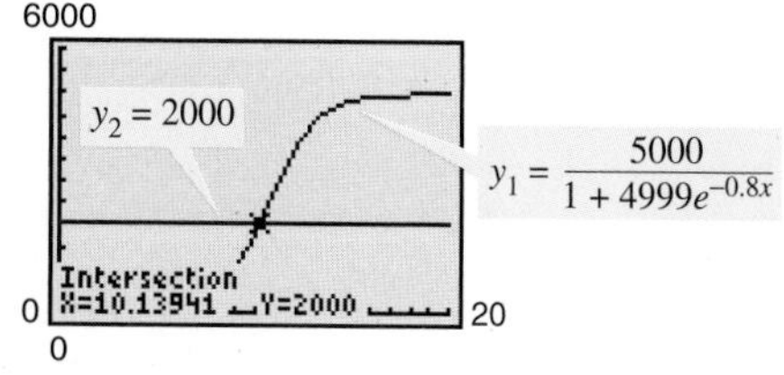

Figure 7.30

■

CLARO CORTES IV/Reuters /Landov

On May 12, 2008, an earthquake of magnitude 7.9 struck Eastern Sichuan Province, China. The total economic loss was estimated at 86 billion U.S. dollars.

Logarithmic Models

EXAMPLE 5 Magnitudes of Earthquakes

On the Richter scale, the magnitude R of an earthquake of intensity I is given by

$$R = \log \frac{I}{I_0}$$

where $I_0 = 1$ is the minimum intensity used for comparison. Find the intensity of each earthquake. (Intensity is a measure of the wave energy of an earthquake.)

a. Nevada in 2008: $R = 6.0$

b. Eastern Sichuan, China in 2008: $R = 7.9$

c. Offshore Maule, Chile in 2010: $R = 8.8$

Solution

a. Because $I_0 = 1$ and $R = 6.0$, you have

$$6.0 = \log \frac{I}{1} \qquad \text{Substitute 1 for } I_0 \text{ and 6.0 for } R.$$

$$10^{6.0} = 10^{\log I} \qquad \text{Exponentiate each side.}$$

$$I = 10^{6.0} \qquad \text{Inverse Property}$$

$$= 1{,}000{,}000.$$

b. For $R = 7.9$, you have

$$7.9 = \log \frac{I}{1} \qquad \text{Substitute 1 for } I_0 \text{ and 7.9 for } R.$$

$$10^{7.9} = 10^{\log I} \qquad \text{Exponentiate each side.}$$

$$I = 10^{7.9} \qquad \text{Inverse Property}$$

$$\approx 79{,}400{,}000.$$

c. For $R = 8.8$, you have

$$8.8 = \log \frac{I}{1} \qquad \text{Substitute 1 for } I_0 \text{ and 8.8 for } R.$$

$$10^{8.8} = 10^{\log I} \qquad \text{Exponentiate each side.}$$

$$I = 10^{8.8} \qquad \text{Inverse Property}$$

$$\approx 631{,}000{,}000.$$

Note that an increase of 1.9 units on the Richter scale (from 6.0 to 7.9) represents an increase in intensity by a factor of

$$\frac{79{,}400{,}000}{1{,}000{,}000} = 79.4.$$

In other words, the intensity of the earthquake in Eastern Sichuan was about 79 times as great as that of the earthquake in Nevada. ■

7.5 Exercises

See www.CalcChat.com for worked-out solutions to odd-numbered exercises.

In Exercises 1–6, fill in the blanks.

1. An exponential growth model has the form ________ and an exponential decay model has the form ________.
2. A logarithmic model has the form ________ or ________.
3. Gaussian models are commonly used in probability and statistics to represent populations that are ________ ________.
4. The graph of a Gaussian model is ________ shaped, where the ________ ________ is the x-value corresponding to the maximum y-value of the graph.
5. A logistic growth model has the form ________.
6. A logistic curve is also called a ________ curve.

In Exercises 7–12, match the function with its graph. [The graphs are labeled (a), (b), (c), (d), (e), and (f).]

(a)

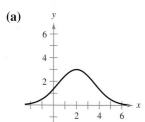

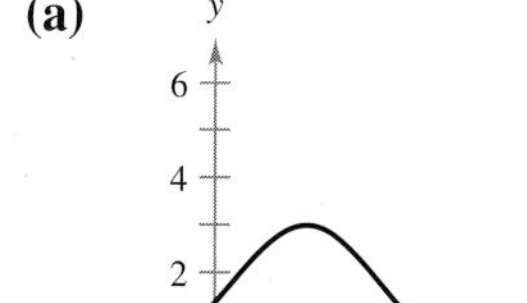

(b)

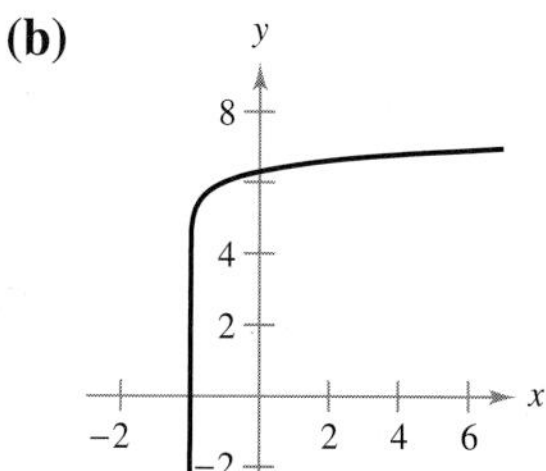

(c)

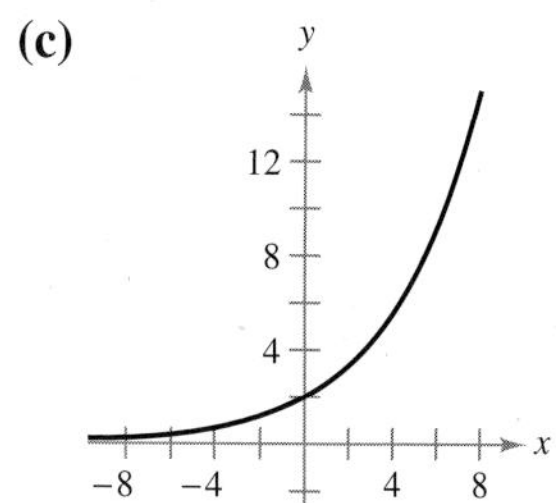

(d)

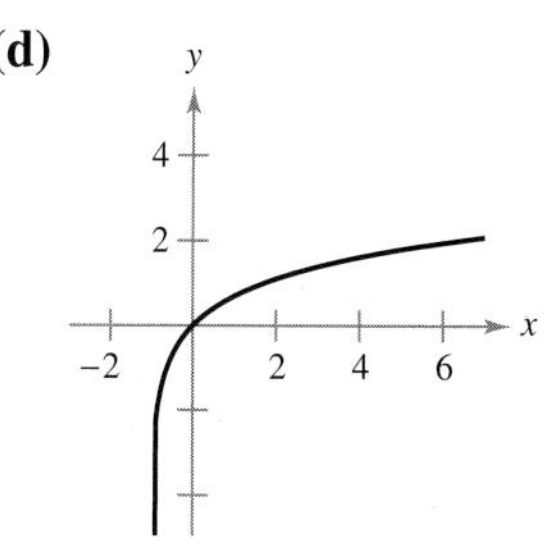

(e)

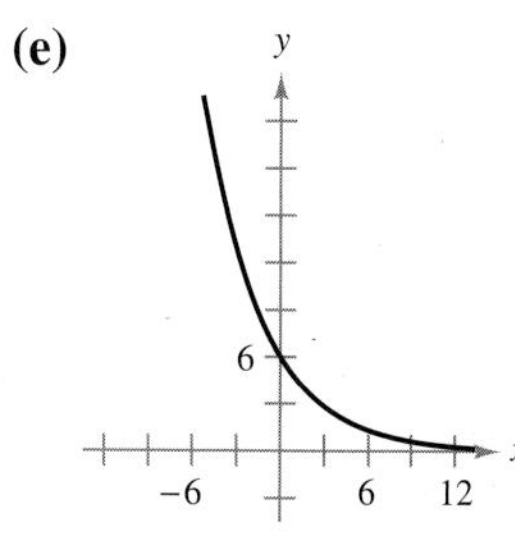

(f)

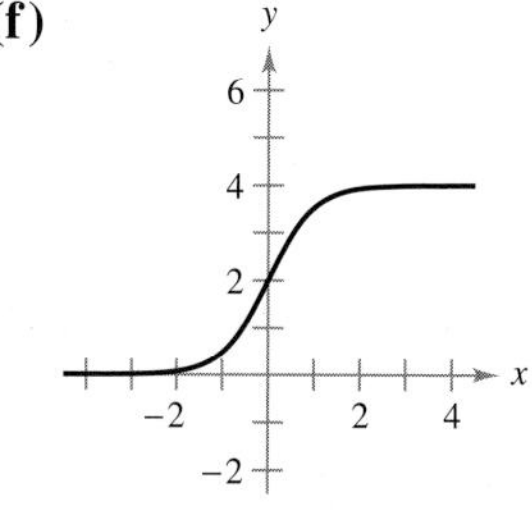

7. $y = 2e^{x/4}$
8. $y = 6e^{-x/4}$
9. $y = 6 + \log(x + 2)$
10. $y = 3e^{-(x-2)^2/5}$
11. $y = \ln(x + 1)$
12. $y = \dfrac{4}{1 + e^{-2x}}$

WRITING ABOUT CONCEPTS

13. Find the values of b such that the logistic curve $y = a/(1 + be^{-xt})$ has a vertical asymptote.
14. Find the values of b such that the logistic curve $y = a/(1 + be^{-xt})$ does not have a vertical asymptote.
15. The height of American men between 18 and 24 years old is normally distributed according to the model

$$y = \frac{1}{3\sqrt{2\pi}}e^{-(x-70)^2/18}$$

where x is the height in inches. Briefly describe the shape of the curve, noting the location of the maximum value of the function and its meaning in this real-life setting. *(Source: U.S. National Center for Health Statistics)*

16. ***Writing*** Use your school's library, the Internet, or some other reference source to write a paper describing John Napier's work with logarithms.

Compound Interest **In Exercises 17–24, complete the table for a savings account in which interest is compounded continuously.**

	Initial Investment	*Annual % Rate*	*Time to Double*	*Amount After 10 years*
17.	\$1000	3.5%		
18.	\$750	$10\frac{1}{2}$%		
19.	\$750		$7\frac{3}{4}$ yr	
20.	\$10,000		12 yr	
21.	\$500			\$1505.00
22.	\$600			\$19,205.00
23.		4.5%		\$10,000.00
24.		2%		\$2000.00

Compound Interest **In Exercises 25 and 26, determine the principal P that must be invested at rate r, compounded monthly, so that \$500,000 will be available for retirement in t years.**

25. $r = 5\%, t = 10$
26. $r = 3\frac{1}{2}\%, t = 15$

Compound Interest **In Exercises 27 and 28, determine the time necessary for \$1000 to double if it is invested at interest rate r compounded (a) annually, (b) monthly, (c) daily, and (d) continuously.**

27. $r = 10\%$
28. $r = 6.5\%$

29. ***Compound Interest*** Complete the table for the time t (in years) necessary for P dollars to triple if interest is compounded continuously at rate r.

r	2%	4%	6%	8%	10%	12%
t						

30. ***Modeling Data*** Draw a scatter plot of the data in Exercise 29. Use the *regression* feature of a graphing utility to find a model for the data.

31. ***Compound Interest*** Complete the table for the time t (in years) necessary for P dollars to triple if interest is compounded annually at rate r.

r	2%	4%	6%	8%	10%	12%
t						

32. ***Modeling Data*** Draw a scatter plot of the data in Exercise 31. Use the *regression* feature of a graphing utility to find a model for the data.

33. ***Comparing Models*** If \$1 is invested in an account over a 10-year period, the amount in the account, where t represents the time in years, is given by $A = 1 + 0.075[\![t]\!]$ or $A = e^{0.07t}$ depending on whether the account pays simple interest at $7\frac{1}{2}\%$ or continuous compound interest at 7%. Graph each function on the same set of axes. Which grows at a higher rate? (Remember that $[\![t]\!]$ is the greatest integer function discussed in Section 1.2.)

34. ***Comparing Models*** If \$1 is invested in an account over a 10-year period, the amount in the account, where t represents the time in years, is given by $A = 1 + 0.06[\![t]\!]$ or $A = [1 + (0.055/365)]^{[\![365t]\!]}$ depending on whether the account pays simple interest at 6% or compound interest at $5\frac{1}{2}\%$ compounded daily. Use a graphing utility to graph each function in the same viewing window. Which grows at a higher rate?

In Exercises 35–38, find the exponential model $y = ae^{bx}$ that fits the points shown in the graph or table.

35.

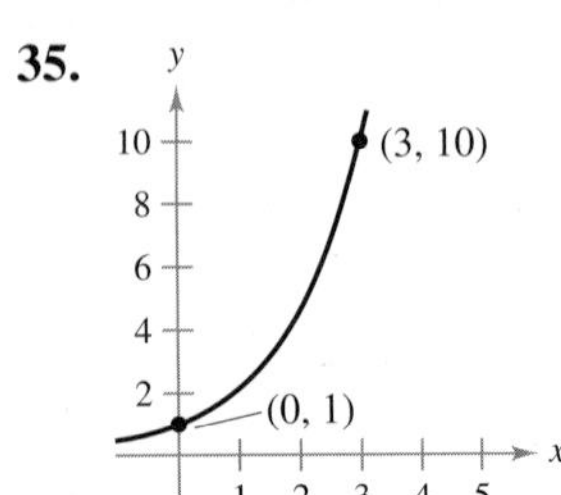

36.

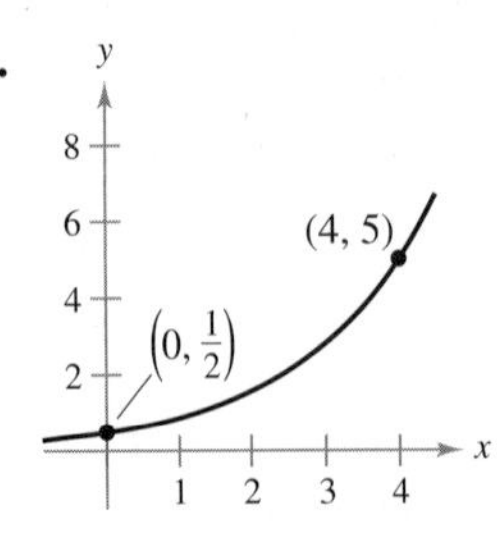

37.

x	0	4
y	5	1

38.

x	0	3
y	1	$\frac{1}{4}$

39. ***Population*** The populations P (in thousands) of Horry County, South Carolina from 1970 through 2007 can be modeled by

$$P = -18.5 + 92.2e^{0.0282t}$$

where t represents the year, with $t = 0$ corresponding to 1970. *(Source: U.S. Census Bureau)*

(a) Use the model to complete the table.

Year	1970	1980	1990	2000	2007
Population					

(b) According to the model, when will the population of Horry County reach 300,000?

(c) Do you think the model is valid for long-term predictions of the population? Explain.

40. ***Population*** The table shows the populations (in millions) of five countries in 2000 and the projected populations (in millions) for the year 2015. *(Source: U.S. Census Bureau)*

Country	2000	2015
Bulgaria	7.8	6.9
Canada	31.1	35.1
China	1268.9	1393.4
United Kingdom	59.5	62.2
United States	282.2	325.5

(a) Find the exponential growth or decay model $y = ae^{bt}$ or $y = ae^{-bt}$ for the population of each country by letting $t = 0$ correspond to 2000. Use the model to predict the population of each country in 2030.

(b) You can see that the populations of the United States and the United Kingdom are growing at different rates. What constant in the equation $y = ae^{bt}$ is determined by these different growth rates? Discuss the relationship between the different growth rates and the magnitude of the constant.

(c) You can see that the population of China is increasing while the population of Bulgaria is decreasing. What constant in the equation $y = ae^{bt}$ reflects this difference? Explain.

41. *Website Growth* The number y of hits a new search-engine website receives each month can be modeled by

$y = 4080e^{kt}$

where t represents the number of months the website has been operating. In the website's third month, there were 10,000 hits. Find the value of k, and use this value to predict the number of hits the website will receive after 24 months.

42. *Value of a Painting* The value V (in millions of dollars) of a famous painting can be modeled by

$V = 10e^{kt}$

where t represents the year, with $t = 0$ corresponding to 2000. In 2008, the same painting was sold for $65 million. Find the value of k, and use this value to predict the value of the painting in 2014.

43. *Population* The populations P (in thousands) of Reno, Nevada from 2000 through 2007 can be modeled by $P = 346.8e^{kt}$, where t represents the year, with $t = 0$ corresponding to 2000. In 2005, the population of Reno was about 395,000. *(Source: U.S. Census Bureau)*

(a) Find the value of k. Is the population increasing or decreasing? Explain.

(b) Use the model to find the populations of Reno in 2010 and 2015. Are the results reasonable? Explain.

(c) According to the model, during what year will the population reach 500,000?

44. *Population* The populations P (in thousands) of Orlando, Florida from 2000 through 2007 can be modeled by $P = 1656.2e^{kt}$, where t represents the year, with $t = 0$ corresponding to 2000. In 2005, the population of Orlando was about 1,940,000. *(Source: U.S. Census Bureau)*

(a) Find the value of k. Is the population increasing or decreasing? Explain.

(b) Use the model to find the populations of Orlando in 2010 and 2015. Are the results reasonable? Explain.

(c) According to the model, during what year will the population reach 2.2 million?

45. *Bacteria Growth* The number of bacteria in a culture is increasing according to the law of exponential growth. After 3 hours, there are 100 bacteria, and after 5 hours, there are 400 bacteria. How many bacteria will there be after 6 hours?

46. *Bacteria Growth* The number of bacteria in a culture is increasing according to the law of exponential growth. The initial population is 250 bacteria, and the population after 10 hours is double the population after 1 hour. How many bacteria will there be after 6 hours?

47. *Carbon Dating*

(a) The ratio of carbon 14 to carbon 12 in a piece of wood discovered in a cave is $R = 1/8^{14}$. Estimate the age of the piece of wood.

(b) The ratio of carbon 14 to carbon 12 in a piece of paper buried in a tomb is $R = 1/13^{11}$. Estimate the age of the piece of paper.

48. *Radioactive Decay* Carbon 14 dating assumes that the carbon dioxide on Earth today has the same radioactive content as it did centuries ago. If this is true, the amount of ^{14}C absorbed by a tree that grew several centuries ago should be the same as the amount of ^{14}C absorbed by a tree growing today. A piece of ancient charcoal contains only 15% as much radioactive carbon as a piece of modern charcoal. How long ago was the tree burned to make the ancient charcoal if the half-life of ^{14}C is 5715 years?

49. *Depreciation* A sport utility vehicle that costs $23,300 new has a book value of $12,500 after 2 years.

(a) Find the linear model $V = mt + b$.

(b) Find the exponential model $V = ae^{kt}$.

(c) Use a graphing utility to graph the two models in the same viewing window. Which model depreciates faster in the first 2 years?

(d) Find the book values of the vehicle after 1 year and after 3 years using each model.

(e) Explain the advantages and disadvantages of using each model to a buyer and a seller.

50. *Depreciation* A laptop computer that costs $1150 new has a book value of $550 after 2 years.

(a) Find the linear model $V = mt + b$.

(b) Find the exponential model $V = ae^{kt}$.

(c) Use a graphing utility to graph the two models in the same viewing window. Which model depreciates faster in the first 2 years?

(d) Find the book values of the computer after 1 year and after 3 years using each model.

(e) Explain the advantages and disadvantages of using each model to a buyer and a seller.

51. *Sales* The sales S (in thousands of units) of a new CD burner after it has been on the market for t years are modeled by

$S(t) = 100(1 - e^{kt})$.

Fifteen thousand units of the new product were sold the first year.

(a) Complete the model by solving for k.

(b) Sketch the graph of the model.

(c) Use the model to estimate the number of units sold after 5 years.

52. ***Learning Curve*** The management at a plastics factory has found that the maximum number of units a worker can produce in a day is 30. The learning curve for the number N of units produced per day after a new employee has worked t days is modeled by $N = 30(1 - e^{kt})$. After 20 days on the job, a new employee produces 19 units.

(a) Find the learning curve for this employee (first, find the value of k).

(b) How many days should pass before this employee is producing 25 units per day?

53. ***IQ Scores*** The IQ scores for a sample of a class of returning adult students at a small northeastern college roughly follow the normal distribution $y = 0.0266e^{-(x-100)^2/450}$, $70 \le x \le 115$, where x is the IQ score.

(a) Use a graphing utility to graph the function.

(b) From the graph in part (a), estimate the average IQ score of an adult student.

54. ***Education*** The amount of time (in hours per week) a student utilizes a math-tutoring center roughly follows the normal distribution $y = 0.7979e^{-(x-5.4)^2/0.5}$, $4 \le x \le 7$, where x is the number of hours.

(a) Use a graphing utility to graph the function.

(b) From the graph in part (a), estimate the average number of hours per week a student uses the tutoring center.

55. ***Cell Sites*** A cell site is a site where electronic communications equipment is placed in a cellular network for the use of mobile phones. The numbers y of cell sites from 1985 through 2008 can be modeled by

$$y = \frac{237{,}101}{1 + 1950e^{-0.355t}}$$

where t represents the year, with $t = 5$ corresponding to 1985. *(Source: CTIA-The Wireless Association)*

(a) Use the model to find the numbers of cell sites in the years 1985, 2000, and 2006.

(b) Use a graphing utility to graph the function.

(c) Use the graph to determine the year in which the number of cell sites will reach 235,000.

(d) Confirm your answer to part (c) algebraically.

56. ***Population*** The populations P (in thousands) of Pittsburgh, Pennsylvania from 2000 through 2007 can be modeled by

$$P = \frac{2632}{1 + 0.083e^{0.0500t}}$$

where t represents the year, with $t = 0$ corresponding to 2000. *(Source: U.S. Census Bureau)*

(a) Use the model to find the populations of Pittsburgh in the years 2000, 2005, and 2007.

(b) Use a graphing utility to graph the function.

(c) Use the graph to determine the year in which the population will reach 2.2 million.

(d) Confirm your answer to part (c) algebraically.

57. ***Population Growth*** A conservation organization releases 100 animals of an endangered species into a game preserve. The organization believes that the preserve has a carrying capacity of 1000 animals and that the growth of the pack will be modeled by the logistic curve

$$p(t) = \frac{1000}{1 + 9e^{-0.1656t}}$$

where t is measured in months (see figure).

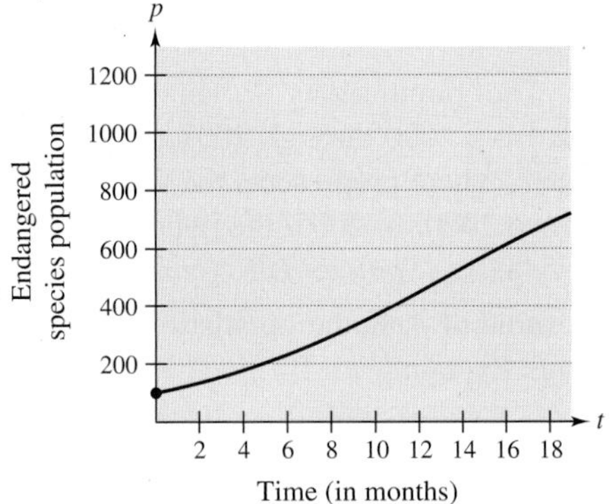

(a) Estimate the population after 5 months.

(b) After how many months will the population be 500?

(c) Use a graphing utility to graph the function. Use the graph to determine the horizontal asymptotes, and interpret the meaning of the asymptotes in the context of the problem.

Geology **In Exercises 58 and 59, use the Richter scale**

$$R = \log \frac{I}{I_0}$$

for measuring the magnitudes of earthquakes.

58. Find the intensity I of an earthquake measuring R on the Richter scale (let $I_0 = 1$).

(a) Southern Sumatra, Indonesia in 2007, $R = 8.5$

(b) Illinois in 2008, $R = 5.4$

(c) Costa Rica in 2009, $R = 6.1$

59. Find the magnitude R of each earthquake of intensity I (let $I_0 = 1$).

(a) $I = 199{,}500{,}000$ (b) $I = 48{,}275{,}000$

(c) $I = 17{,}000$

Intensity of Sound **In Exercises 60–63, use the following information for determining sound intensity. The level of sound β, in decibels, with an intensity of I, is given by $\beta = 10 \log(I/I_0)$, where I_0 is an intensity of 10^{-12} watt per square meter, corresponding roughly to the faintest sound that can be heard by the human ear. In Exercises 60 and 61, find the level of sound β.**

60. (a) $I = 10^{-10}$ watt per m^2 (quiet room)
(b) $I = 10^{-5}$ watt per m^2 (busy street corner)
(c) $I = 10^{-8}$ watt per m^2 (quiet radio)
(d) $I = 10^{0}$ watt per m^2 (threshold of pain)

61. (a) $I = 10^{-11}$ watt per m^2 (rustle of leaves)
(b) $I = 10^{2}$ watt per m^2 (jet at 30 meters)
(c) $I = 10^{-4}$ watt per m^2 (door slamming)
(d) $I = 10^{-2}$ watt per m^2 (siren at 30 meters)

62. Due to the installation of noise suppression materials, the noise level in an auditorium was reduced from 93 to 80 decibels. Find the percent decrease in the intensity level of the noise as a result of the installation of these materials.

63. Due to the installation of a muffler, the noise level of an engine was reduced from 88 to 72 decibels. Find the percent decrease in the intensity level of the noise as a result of the installation of the muffler.

pH Levels **In Exercises 64–69, use the acidity model given by $\text{pH} = -\log[H^+]$, where acidity (pH) is a measure of the hydrogen ion concentration $[H^+]$ (measured in moles of hydrogen per liter) of a solution.**

64. Find the pH if $[H^+] = 2.3 \times 10^{-5}$.

65. Find the pH if $[H^+] = 1.13 \times 10^{-5}$.

66. Compute $[H^+]$ for a solution in which $\text{pH} = 5.8$.

67. Compute $[H^+]$ for a solution in which $\text{pH} = 3.2$.

68. Apple juice has a pH of 2.9 and drinking water has a pH of 8.0. The hydrogen ion concentration of the apple juice is how many times the concentration of drinking water?

69. The pH of a solution is decreased by one unit. The hydrogen ion concentration is increased by what factor?

True or False? **In Exercises 70–73, determine whether the statement is true or false. Justify your answer.**

70. The domain of a logistic growth function cannot be the set of real numbers.

71. A logistic growth function will always have an x-intercept.

72. The graph of $f(x) = \dfrac{4}{1 + 6e^{-2x}} + 5$ is the graph of $g(x) = \dfrac{4}{1 + 6e^{-2x}}$ shifted to the right five units.

73. The graph of a Gaussian model will never have an x-intercept.

CAPSTONE

74. Identify each model as exponential, Gaussian, linear, logarithmic, logistic, or quadratic. Explain your reasoning.

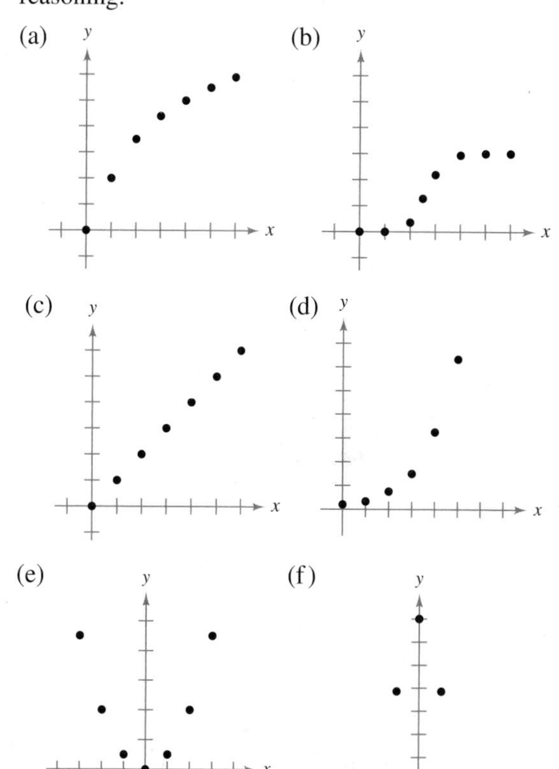

75. ***Forensics*** At 8:30 A.M., a coroner was called to the home of a person who had died during the night. In order to estimate the time of death, the coroner took the person's temperature twice. At 9:00 A.M. the temperature was 85.7°F, and at 11:00 A.M. the temperature was 82.8°F. From these two temperatures, the coroner was able to determine that the time elapsed since death and the body temperature were related by the formula

$$t = -10 \ln \frac{T - 70}{98.6 - 70}$$

where t is the time in hours elapsed since the person died and T is the temperature (in degrees Fahrenheit) of the person's body. (This formula is derived from a general cooling principle called *Newton's Law of Cooling*. It uses the assumptions that the person had a normal body temperature of 98.6°F at death, and that the room temperature was a constant 70°F.) Use the formula to estimate the time of death of the person.

7 CHAPTER SUMMARY

Section 7.1	Review Exercises
■ Recognize and evaluate exponential functions with base a *(p. 472).*	*1–6*
■ Graph exponential functions *(p. 473).*	*7–20, 29, 30*
■ Recognize, evaluate, and graph exponential functions with base e *(p. 475).*	*21–28, 31, 32*
■ Use exponential functions to model and solve real-life problems *(p. 476).*	*33–36*
Section 7.2	
■ Recognize and evaluate logarithmic functions with base a *(p. 481).*	*37–48*
■ Graph logarithmic functions *(p. 483).*	*49–52*
■ Recognize, evaluate, and graph natural logarithmic functions *(p. 485).*	*53–58*
■ Use logarithmic functions to model and solve real-life problems *(p. 487).*	*59, 60*
Section 7.3	
■ Use the change-of-base formula to rewrite and evaluate logarithmic expressions *(p. 491).*	*61–64*
■ Use properties of logarithms to evaluate, rewrite, expand, or condense logarithmic expressions *(p. 492).*	*65–80*
■ Use logarithmic functions to model and solve real-life problems *(p. 494).*	*81, 82*
Section 7.4	
■ Solve simple exponential and logarithmic equations *(p. 498).*	*83–88*
■ Solve more complicated exponential equations *(p. 499)* and logarithmic equations *(p. 501).*	*89–108*
■ Use exponential and logarithmic equations to model and solve real-life problems *(p. 503).*	*109, 110*
Section 7.5	
■ Recognize the five most common types of models involving exponential and logarithmic functions *(p. 509).*	*111–116*
■ Use exponential growth and decay functions to model and solve real-life problems *(p. 510).*	*117–120*
■ Use Gaussian functions *(p. 512),* logistic growth functions *(p. 513),* and logarithmic functions *(p. 514)* to model and solve real-life problems.	*121–123*

7 REVIEW EXERCISES

See www.CalcChat.com for worked-out solutions to odd-numbered exercises.

In Exercises 1–6, evaluate the function at the indicated value of x. Round your result to three decimal places.

1. $f(x) = 0.3^x, \quad x = 1.5$
2. $f(x) = 30^x, \quad x = \sqrt{3}$
3. $f(x) = 2^{-0.5x}, \quad x = \pi$
4. $f(x) = 1278^{x/5}, \quad x = 1$
5. $f(x) = 7(0.2^x), \quad x = -\sqrt{11}$
6. $f(x) = -14(5^x), \quad x = -0.8$

In Exercises 7–14, use the graph of f to describe the transformation that yields the graph of g.

7. $f(x) = 2^x, \quad g(x) = 2^x - 2$
8. $f(x) = 5^x, \quad g(x) = 5^x + 1$
9. $f(x) = 4^x, \quad g(x) = 4^{-x+2}$
10. $f(x) = 6^x, \quad g(x) = 6^{x+1}$
11. $f(x) = 3^x, \quad g(x) = 1 - 3^x$
12. $f(x) = 0.1^x, \quad g(x) = -0.1^x$
13. $f(x) = \left(\frac{1}{2}\right)^x, \quad g(x) = -\left(\frac{1}{2}\right)^{x+2}$
14. $f(x) = \left(\frac{2}{3}\right)^x, \quad g(x) = 8 - \left(\frac{2}{3}\right)^x$

In Exercises 15–20, use a graphing utility to construct a table of values for the function. Then sketch the graph of the function.

15. $f(x) = 4^{-x} + 4$
16. $f(x) = 2.65^{x-1}$
17. $f(x) = 5^{x-2} + 4$
18. $f(x) = 2^{x-6} - 5$
19. $f(x) = \left(\frac{1}{2}\right)^{-x} + 3$
20. $f(x) = \left(\frac{1}{8}\right)^{x+2} - 5$

In Exercises 21–24, evaluate $f(x) = e^x$ at the indicated value of x. Round your result to three decimal places.

21. $x = 8$
22. $x = \frac{5}{8}$
23. $x = -1.7$
24. $x = 0.278$

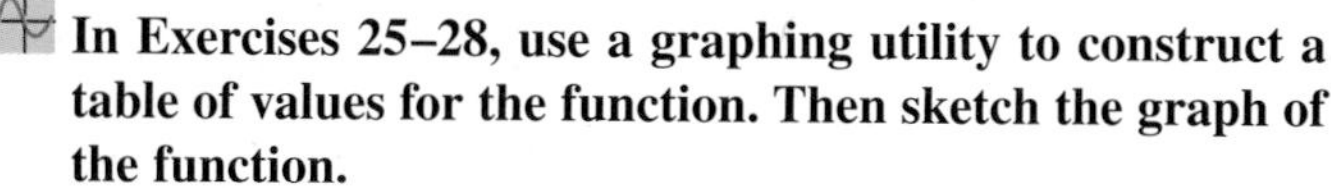

In Exercises 25–28, use a graphing utility to construct a table of values for the function. Then sketch the graph of the function.

25. $h(x) = e^{-x/2}$
26. $h(x) = 2 - e^{-x/2}$
27. $f(x) = e^{x+2}$
28. $s(t) = 4e^{-2/t}, \quad t > 0$

In Exercises 29–32, use a graphing utility to graph the exponential function.

29. $y = 4^{-(x-1)^2}$
30. $y = 2^{-|x+2|}$
31. $g(x) = 2.85e^{-x/4}$
32. $s(t) = 3 - 2e^{-0.25t}$

Compound Interest **In Exercises 33 and 34, complete the table to determine the balance A for P dollars invested at rate r for t years and compounded n times per year.**

n	1	2	4	12	365	Continuous
A						

33. $P = \$5000, \; r = 3\%, \; t = 10$ years
34. $P = \$4500, \; r = 2.5\%, \; t = 30$ years

35. ***Waiting Times*** The average time between incoming calls at a switchboard is 3 minutes. The probability F of waiting less than t minutes until the next incoming call is approximated by the model $F(t) = 1 - e^{-t/3}$. A call has just come in. Find the probability that the next call will be within
 (a) $\frac{1}{2}$ minute. (b) 2 minutes. (c) 5 minutes.

36. ***Depreciation*** After t years, the value V of a car that originally cost \$23,970 is given by $V(t) = 23{,}970\left(\frac{3}{4}\right)^t$.
 (a) Use a graphing utility to graph the function.
 (b) Find the value of the car 2 years after it was purchased.
 (c) According to the model, when does the car depreciate most rapidly? Is this realistic? Explain.
 (d) According to the model, when will the car have no value?

In Exercises 37–40, write the exponential equation in logarithmic form. For example, the logarithmic form of $2^3 = 8$ is $\log_2 8 = 3$.

37. $3^3 = 27$
38. $25^{3/2} = 125$
39. $e^{0.8} = 2.2255\ldots$
40. $e^0 = 1$

In Exercises 41–44, evaluate the function at the indicated value of x without using a calculator.

41. $f(x) = \log x, x = 1000$
42. $g(x) = \log_9 x, x = 3$
43. $g(x) = \log_2 x, x = \frac{1}{4}$
44. $f(x) = \log_3 x, x = \frac{1}{81}$

In Exercises 45–48, use the One-to-One Property to solve the equation for x.

45. $\log_4(x + 7) = \log_4 14$
46. $\log_8(3x - 10) = \log_8 5$
47. $\ln(x + 9) = \ln 4$
48. $\ln(2x - 1) = \ln 11$

In Exercises 49–52, find the domain, x-intercept, and vertical asymptote of the logarithmic function and sketch its graph.

49. $g(x) = \log_7 x$
50. $f(x) = \log\left(\dfrac{x}{3}\right)$
51. $f(x) = 4 - \log(x + 5)$
52. $f(x) = \log(x - 3) + 1$

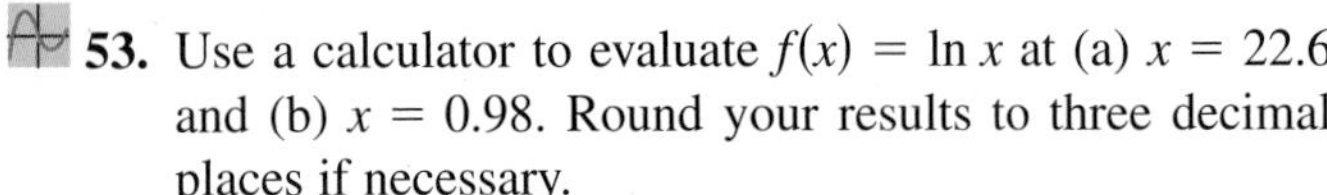

53. Use a calculator to evaluate $f(x) = \ln x$ at (a) $x = 22.6$ and (b) $x = 0.98$. Round your results to three decimal places if necessary.

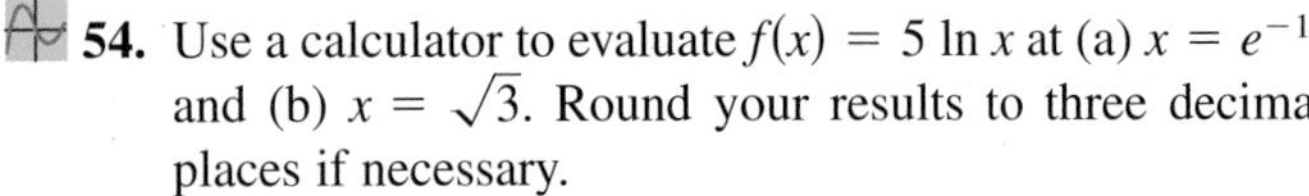

54. Use a calculator to evaluate $f(x) = 5 \ln x$ at (a) $x = e^{-12}$ and (b) $x = \sqrt{3}$. Round your results to three decimal places if necessary.

In Exercises 55–58, find the domain, *x*-intercept, and vertical asymptote of the logarithmic function and sketch its graph.

55. $f(x) = \ln x + 3$ **56.** $f(x) = \ln(x - 3)$

57. $h(x) = \ln(x^2)$ **58.** $f(x) = \frac{1}{4} \ln x$

59. ***Antler Spread*** The antler spread a (in inches) and shoulder height h (in inches) of an adult male American elk are related by the model $h = 116 \log(a + 40) - 176$. Approximate the shoulder height of a male American elk with an antler spread of 55 inches.

60. ***Snow Removal*** The number of miles s of roads cleared of snow is approximated by the model

$$s = 25 - \frac{13 \ln(h/12)}{\ln 3}, \quad 2 \le h \le 15$$

where h is the depth of the snow in inches. Use this model to find s when $h = 10$ inches.

In Exercises 61–64, evaluate the logarithm using the change-of-base formula. Do each exercise twice, once with common logarithms and once with natural logarithms. Round the results to three decimal places.

61. $\log_2 6$ **62.** $\log_{12} 200$

63. $\log_{1/2} 5$ **64.** $\log_3 0.28$

In Exercises 65–68, use the properties of logarithms to rewrite and simplify the logarithmic expression.

65. $\log 18$ **66.** $\log_2\left(\frac{1}{12}\right)$

67. $\ln 20$ **68.** $\ln(3e^{-4})$

In Exercises 69–74, use the properties of logarithms to expand the expression as a sum, difference, and/or constant multiple of logarithms. (Assume all variables are positive.)

69. $\log_5 5x^2$ **70.** $\log 7x^4$

71. $\log_3 \dfrac{9}{\sqrt{x}}$ **72.** $\log_7 \dfrac{\sqrt[3]{x}}{14}$

73. $\ln x^2y^2z$ **74.** $\ln\left(\dfrac{y-1}{4}\right)^2, \quad y > 1$

In Exercises 75–80, condense the expression to the logarithm of a single quantity.

75. $\log_2 5 + \log_2 x$ **76.** $\log_6 y - 2 \log_6 z$

77. $\ln x - \frac{1}{4} \ln y$ **78.** $3 \ln x + 2 \ln(x + 1)$

79. $\frac{1}{2} \log_3 x - 2 \log_3(y + 8)$

80. $5 \ln(x - 2) - \ln(x + 2) - 3 \ln x$

81. ***Climb Rate*** The time t (in minutes) for a small plane to climb to an altitude of h feet is modeled by $t = 50 \log[18{,}000/(18{,}000 - h)]$, where 18,000 feet is the plane's absolute ceiling.

(a) Determine the domain of the function in the context of the problem.

(b) Use a graphing utility to graph the function and identify any asymptotes.

(c) As the plane approaches its absolute ceiling, what can be said about the time required to increase its altitude?

(d) Find the time for the plane to climb to an altitude of 4000 feet.

82. ***Human Memory Model*** Students in a learning theory study were given an exam and then retested monthly for 6 months with an equivalent exam. The data obtained in the study are given as the ordered pairs (t, s), where t is the time in months after the initial exam and s is the average score for the class. Use these data to find a logarithmic equation that relates t and s.

$(1, 84.2), (2, 78.4), (3, 72.1), (4, 68.5), (5, 67.1), (6, 65.3)$

In Exercises 83–88, solve for *x*.

83. $5^x = 125$ **84.** $6^x = \frac{1}{216}$

85. $e^x = 3$ **86.** $\log_6 x = -1$

87. $\ln x = 4$ **88.** $\ln x = -1.6$

In Exercises 89–92, solve the exponential equation algebraically. Approximate your result to three decimal places.

89. $e^{4x} = e^{x^2+3}$ **90.** $e^{3x} = 25$

91. $2^x - 3 = 29$ **92.** $e^{2x} - 6e^x + 8 = 0$

In Exercises 93 and 94, use a graphing utility to graph and solve the equation. Approximate the result to three decimal places.

93. $25e^{-0.3x} = 12$ **94.** $2^x = 3 + x - e^x$

In Exercises 95–104, solve the logarithmic equation algebraically. Approximate the result to three decimal places.

95. $\ln 3x = 8.2$ **96.** $4 \ln 3x = 15$

97. $\ln x - \ln 3 = 2$ **98.** $\ln x - \ln 5 = 4$

99. $\ln\sqrt{x} = 4$ **100.** $\ln\sqrt{x+8} = 3$

101. $\log_8(x-1) = \log_8(x-2) - \log_8(x+2)$

102. $\log_6(x+2) - \log_6 x = \log_6(x+5)$

103. $\log(1-x) = -1$ **104.** $\log(-x-4) = 2$

In Exercises 105–108, use a graphing utility to graph and solve the equation. Approximate the result to three decimal places.

105. $2\ln(x+3) - 3 = 0$ **106.** $x - 2\log(x+4) = 0$

107. $6\log(x^2+1) - x = 0$

108. $3\ln x + 2\log x = e^x - 25$

109. ***Compound Interest*** You deposit \$8500 in an account that pays 3.5% interest, compounded continuously. How long will it take for the money to triple?

110. ***Meteorology*** The speed of the wind S (in miles per hour) near the center of a tornado and the distance d (in miles) the tornado travels are related by the model $S = 93\log d + 65$. On March 18, 1925, a large tornado struck portions of Missouri, Illinois, and Indiana with a wind speed at the center of about 283 miles per hour. Approximate the distance traveled by this tornado.

In Exercises 111–116, match the function with its graph. [The graphs are labeled (a), (b), (c), (d), (e), and (f).]

(a)
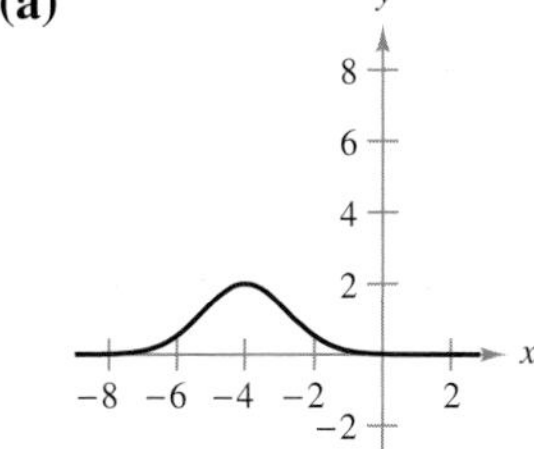

(b)
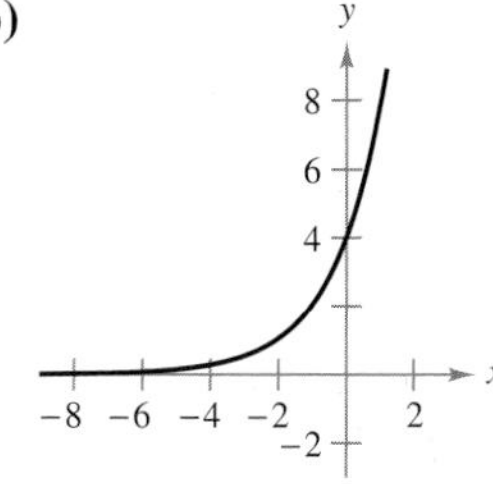

(c)
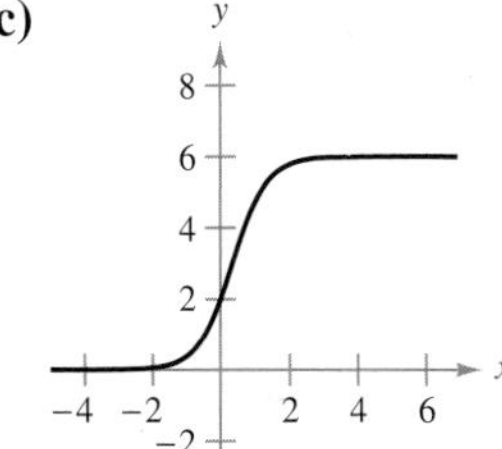

(d)
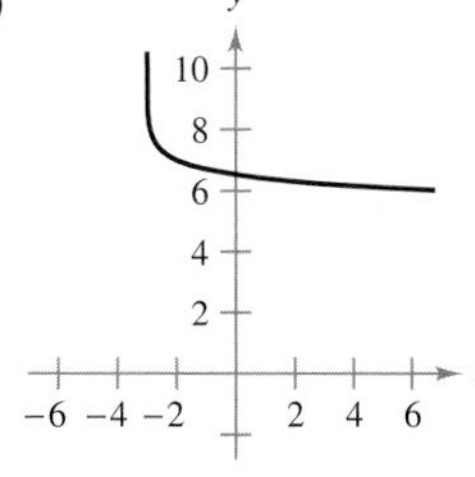

(e)
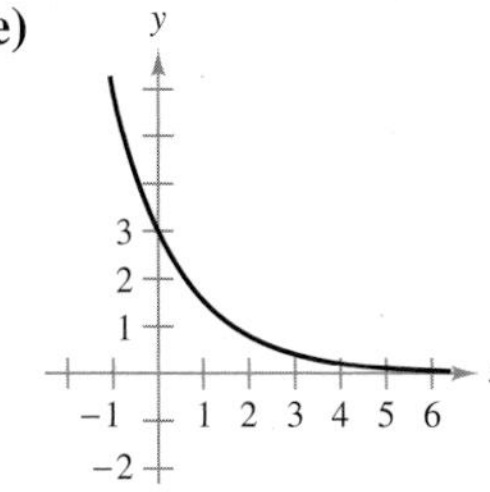

(f)
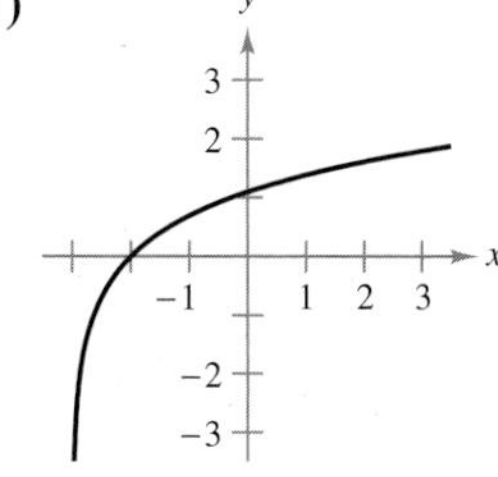

111. $y = 3e^{-2x/3}$ **112.** $y = 4e^{2x/3}$

113. $y = \ln(x+3)$ **114.** $y = 7 - \log(x+3)$

115. $y = 2e^{-(x+4)^2/3}$ **116.** $y = \dfrac{6}{1+2e^{-2x}}$

In Exercises 117 and 118, find the exponential model $y = ae^{bx}$ that passes through the points.

117. $(0, 2), (4, 3)$ **118.** $\left(0, \frac{1}{2}\right), (5, 5)$

119. ***Population*** In 2007, the population of Florida residents aged 65 and over was about 3.10 million. In 2015 and 2020, the populations of Florida residents aged 65 and over are projected to be about 4.13 million and 5.11 million, respectively. An exponential growth model that approximates these data is given by $P = 2.36e^{0.0382t}$, $7 \le t \le 20$, where P is the population (in millions) and $t = 7$ represents 2007. *(Source: U.S. Census Bureau)*

(a) Use a graphing utility to graph the model and the data in the same viewing window. Is the model a good fit for the data? Explain.

(b) According to the model, when will the population of Florida residents aged 65 and over reach 5.5 million? Does your answer seem reasonable? Explain.

120. ***Wildlife Population*** A species of bat is in danger of becoming extinct. Five years ago, the total population of the species was 2000. Two years ago, the total population of the species was 1400. What was the total population of the species one year ago?

121. ***Test Scores*** The test scores for a biology test follow a normal distribution modeled by $y = 0.0499e^{-(x-71)^2/128}$, $40 \le x \le 100$, where x is the test score. Use a graphing utility to graph the equation and estimate the average test score.

122. ***Typing Speed*** In a typing class, the average number N of words per minute typed after t weeks of lessons was found to be $N = 157/(1 + 5.4e^{-0.12t})$. Find the time necessary to type (a) 50 words per minute and (b) 75 words per minute.

123. ***Sound Intensity*** The relationship between the number of decibels β and the intensity of a sound I in watts per square meter is $\beta = 10\log(I/10^{-12})$. Find I for each decibel level β.

(a) $\beta = 60$ (b) $\beta = 135$ (c) $\beta = 1$

124. Consider the graph of $y = e^{kt}$. Describe the characteristics of the graph when k is positive and when k is negative.

True or False? **In Exercises 125 and 126, determine whether the equation is true or false. Justify your answer.**

125. $\log_b b^{2x} = 2x$ **126.** $\ln(x+y) = \ln x + \ln y$

7 CHAPTER TEST

Take this test as you would take a test in class. When you are finished, check your work against the answers given in the back of the book.

In Exercises 1–4, evaluate the expression. Approximate your result to three decimal places.

1. $4.2^{0.6}$ **2.** $4^{3\pi/2}$ **3.** $e^{-7/10}$ **4.** $e^{3.1}$

In Exercises 5–7, construct a table of values. Then sketch the graph of the function.

5. $f(x) = 10^{-x}$ **6.** $f(x) = -6^{x-2}$ **7.** $f(x) = 1 - e^{2x}$

8. Evaluate (a) $\log_7 7^{-0.89}$ and (b) $4.6 \ln e^2$.

In Exercises 9–11, construct a table of values. Then sketch the graph of the function. Identify any asymptotes.

9. $f(x) = -\log x - 6$ **10.** $f(x) = \ln(x - 4)$ **11.** $f(x) = 1 + \ln(x + 6)$

In Exercises 12–14, evaluate the logarithm using the change-of-base formula. Round your result to three decimal places.

12. $\log_7 44$ **13.** $\log_{16} 0.63$ **14.** $\log_{3/4} 24$

In Exercises 15–17, use the properties of logarithms to expand the expression as a sum, difference, and/or constant multiple of logarithms.

15. $\log_2 3a^4$ **16.** $\ln \dfrac{5\sqrt{x}}{6}$ **17.** $\log \dfrac{(x-1)^3}{y^2 z}$

In Exercises 18–20, condense the expression to the logarithm of a single quantity.

18. $\log_3 13 + \log_3 y$ **19.** $4 \ln x - 4 \ln y$

20. $3 \ln x - \ln(x + 3) + 2 \ln y$

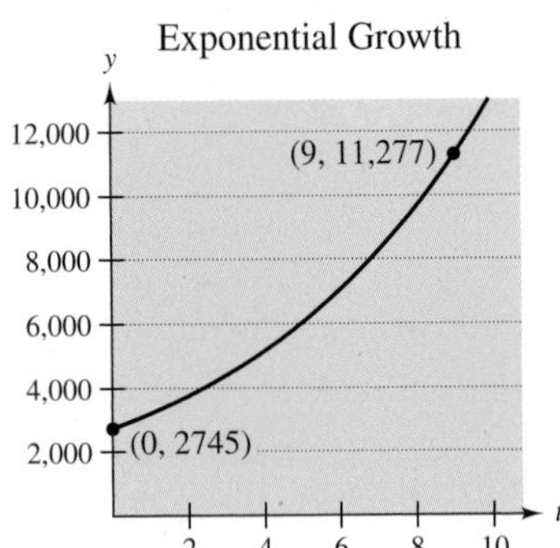

Figure for 27

In Exercises 21–26, solve the equation algebraically. Approximate your result to three decimal places.

21. $5^x = \dfrac{1}{25}$ **22.** $3e^{-5x} = 132$

23. $\dfrac{1025}{8 + e^{4x}} = 5$ **24.** $\ln x = \dfrac{1}{2}$

25. $18 + 4 \ln x = 7$ **26.** $\log x + \log(x - 15) = 2$

27. Find an exponential growth model for the graph shown in the figure.

28. The number of bacteria in a culture is increasing according to the law of exponential growth. After 2 hours, there are 80 bacteria, and after 4 hours, there are 300 bacteria. How many bacteria will there be after 6 hours?

29. A model that can be used for predicting the height H (in centimeters) of a child based on his or her age is $H = 70.228 + 5.104x + 9.222 \ln x$, $\frac{1}{4} \le x \le 6$, where x is the age of the child in years. *(Source: Snapshots of Applications in Mathematics)*

(a) Construct a table of values. Then sketch the graph of the model.

(b) Use the graph from part (a) to estimate the height of a four-year-old child. Then calculate the actual height using the model.

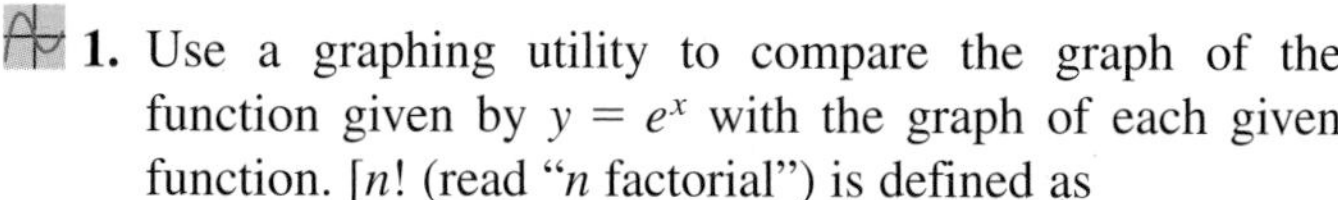

P.S. PROBLEM SOLVING

1. Use a graphing utility to compare the graph of the function given by $y = e^x$ with the graph of each given function. [$n!$ (read "n factorial") is defined as

$n! = 1 \cdot 2 \cdot 3 \cdot \cdot \cdot (n - 1) \cdot n.$]

(a) $y_1 = 1 + \dfrac{x}{1!}$

(b) $y_2 = 1 + \dfrac{x}{1!} + \dfrac{x^2}{2!}$

(c) $y_3 = 1 + \dfrac{x}{1!} + \dfrac{x^2}{2!} + \dfrac{x^3}{3!}$

2. Identify the pattern of successive polynomials given in Exercise 1. Extend the pattern one more term and compare the graph of the resulting polynomial function with the graph of $y = e^x$. What do you think this pattern implies?

3. Given the exponential function

$f(x) = a^x$

show that

(a) $f(u + v) = f(u) \cdot f(v)$.

(b) $f(2x) = [f(x)]^2$.

4. Use a graphing utility to compare the graph of the function given by $y = \ln x$ with the graph of each given function.

(a) $y_1 = x - 1$

(b) $y_2 = (x - 1) - \frac{1}{2}(x - 1)^2$

(c) $y_3 = (x - 1) - \frac{1}{2}(x - 1)^2 + \frac{1}{3}(x - 1)^3$

5. Identify the pattern of successive polynomials given in Exercise 4. Extend the pattern one more term and compare the graph of the resulting polynomial function with the graph of $y = \ln x$. What do you think the pattern implies?

6. Approximate the natural logarithms of as many integers as possible between 1 and 20 given that $\ln 2 \approx 0.6931$, $\ln 3 \approx 1.0986$, and $\ln 5 \approx 1.6094$. (Do not use a calculator.)

7. Use a graphing utility to graph

$y = (1 + x)^{1/x}$.

Describe the behavior of the graph near $x = 0$. Is there a y-intercept? Create a table that shows values of y for values of x near $x = 0$ to verify the behavior of the graph near this point.

8. The table shows the time t (in seconds) required to attain a speed of s miles per hour from a standing start for a car.

s	30	40	50	60	70	80	90
t	3.4	5.0	7.0	9.3	12.0	15.8	20.0

Two models for these data are shown below.

$t_1 = 40.757 + 0.556s - 15.817 \ln s$

$t_2 = 1.2259 + 0.0023s^2$

(a) Use a graphing utility to fit a linear model t_3 and an exponential model t_4 to the data.

(b) Use a graphing utility to graph the data points and each model.

(c) Create a table comparing the data with estimates obtained from each model.

(d) Use the results of part (c) to find the sum of the absolute values of the differences between the data and estimated values given by each model. Based on the four sums, which model do you think better fits the data? Explain your reasoning.

9. A \$120,000 home mortgage for 30 years at $7\frac{1}{2}\%$ has a monthly payment of \$839.06. Part of the monthly payment is paid toward the interest charge on the unpaid balance, and the remainder of the payment is used to reduce the principal. The amount that is paid toward the interest is

$$u = M - \left(M - \frac{Pr}{12}\right)\left(1 + \frac{r}{12}\right)^{12t}$$

and the amount that is paid toward the reduction of the principal is

$$v = \left(M - \frac{Pr}{12}\right)\left(1 + \frac{r}{12}\right)^{12t}.$$

In these formulas, P is the size of the mortgage, r is the interest rate, M is the monthly payment, and t is the time (in years).

(a) Use a graphing utility to graph each function in the same viewing window. (The viewing window should show all 30 years of mortgage payments.)

(b) In the early years of the mortgage, is the larger part of the monthly payment paid toward the interest or the principal? Approximate the time when the monthly payment is evenly divided between interest and principal reduction.

(c) Repeat parts (a) and (b) for a repayment period of 20 years ($M = \$966.71$). What can you conclude?

10. The table shows the colonial population estimates of the American colonies from 1700 to 1780. *(Source: U.S. Census Bureau)*

Year	Population
1700	250,900
1710	331,700
1720	466,200
1730	629,400
1740	905,600
1750	1,170,800
1760	1,593,600
1770	2,148,100
1780	2,780,400

In each of the following, let y represent the population in the year t, with $t = 0$ corresponding to 1700.

(a) Use the *regression* feature of a graphing utility to find an exponential model for the data.

(b) Use the *regression* feature of the graphing utility to find a quadratic model for the data.

(c) Use the graphing utility to plot the data and the models from parts (a) and (b) in the same viewing window.

(d) Which model is a better fit for the data? Would you use this model to predict the population of the United States in 2015? Explain your reasoning.

In Exercises 11 and 12, use the model

$$y = 80.4 - 11 \ln x, \quad 100 \le x \le 1500$$

which approximates the minimum required ventilation rate in terms of the air space per child in a public school classroom. In the model, x is the air space per child in cubic feet and y is the ventilation rate per child in cubic feet per minute.

11. Use a graphing utility to graph the model and approximate the required ventilation rate if there is 300 cubic feet of air space per child.

12. A classroom is designed for 30 students. The air conditioning system in the room has the capacity of moving 450 cubic feet of air per minute.

(a) Determine the ventilation rate per child, assuming that the room is filled to capacity.

(b) Estimate the air space required per child.

(c) Determine the minimum number of square feet of floor space required for the room if the ceiling height is 30 feet.

13. By observation, identify the equation that corresponds to the graph. Explain your reasoning.

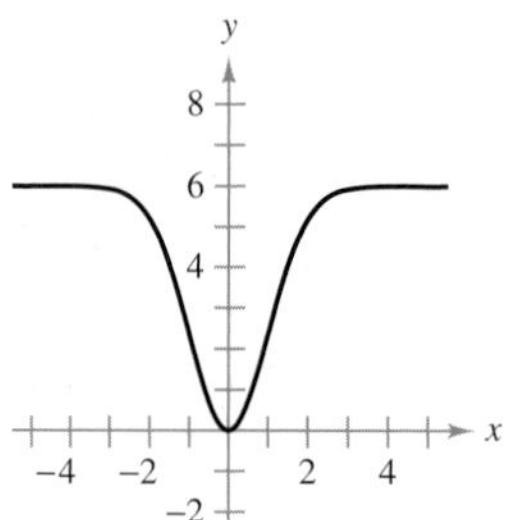

(a) $y = 6e^{-x^2/2}$

(b) $y = \dfrac{6}{1 + e^{-x/2}}$

(c) $y = 6\left(1 - e^{-x^2/2}\right)$

14. Solve the logarithmic equation

$(\ln x)^2 = \ln x^2$.

15. Two different samples of radioactive isotopes are decaying. The isotopes have initial amounts of c_1 and c_2, as well as half-lives of k_1 and k_2, respectively. Find the time required for the samples to decay to equal amounts.

16. Show that

$$\frac{\log_a x}{\log_{a/b} x} = 1 + \log_a \frac{1}{b}.$$

17. Graph the function given by

$f(x) = e^x - e^{-x}$.

From the graph, the function appears to be one-to-one. Assuming that the function has an inverse, find $f^{-1}(x)$.

18. Given that

$$f(x) = \frac{e^x + e^{-x}}{2} \quad \text{and} \quad g(x) = \frac{e^x - e^{-x}}{2}$$

show that

$[f(x)]^2 - [g(x)]^2 = 1$.

19. Find a pattern for $f^{-1}(x)$ if

$$f(x) = \frac{a^x + 1}{a^x - 1}$$

where $a > 0$, $a \neq 1$.

20. A lab culture initially contains 500 bacteria. Two hours later the number of bacteria decreases to 200. Find the exponential decay model of the form

$B = B_0 a^{kt}$

that can be used to approximate the number of bacteria after t hours.

Exponential and Logarithmic Functions and Calculus

In Chapter 7, you were introduced to exponential functions and logarithmic functions. Now, you will study their derivatives and antiderivatives.

In this chapter, you should learn the following.

- How to find the derivative and antiderivative of the natural exponential function. (8.1)
- How to find the derivative of the natural logarithmic function, use logarithms as an aid in differentiating nonlogarithmic functions, and find the derivatives of exponential and logarithmic functions in bases other than e. (8.2)
- How to find the antiderivative of the natural logarithmic function. (8.3)
- How to use an exponential function to model growth and decay. (8.4)

Brian Maslyar/Photolibrary.com

A geyser is a hot spring that erupts periodically when groundwater in a confined space boils and produces steam. The steam forces overlying water up and out through an opening on Earth's surface. The temperature at which water boils is affected by pressure. Do you think an *increase* or a *decrease* in pressure causes water to boil at a lower temperature? Why? (See Section 8.2, Exercise 88.)

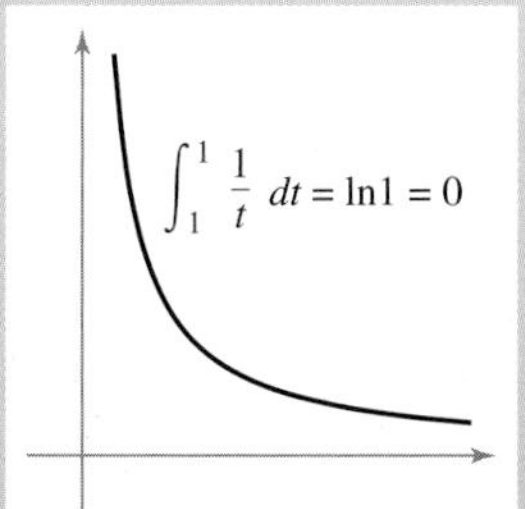

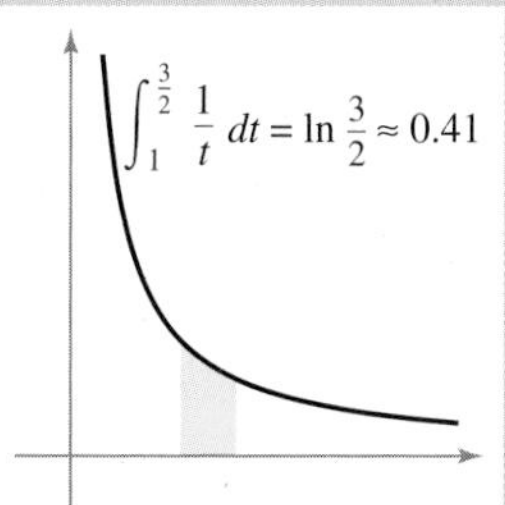

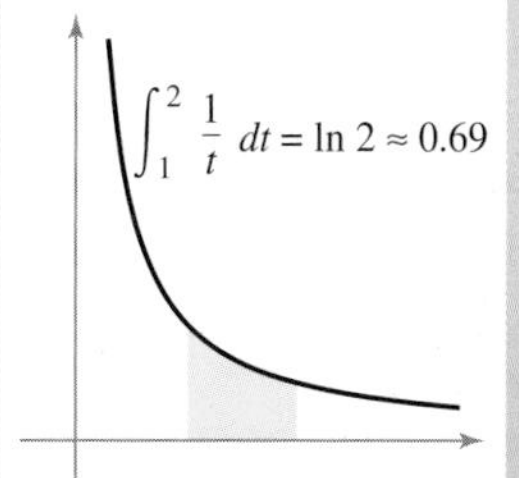

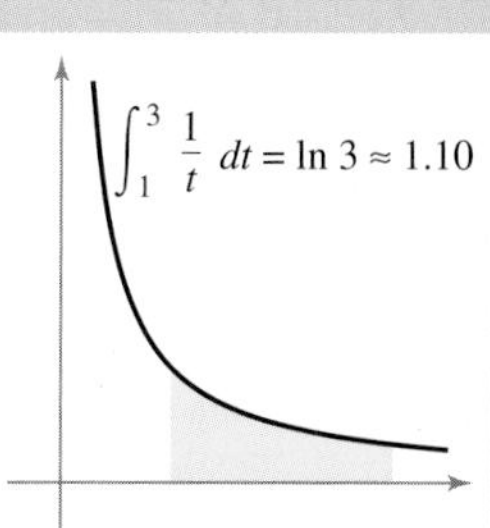

You know how to integrate functions of the form $f(x) = x^n$, provided that $n \neq -1$. In Chapter 8, you will learn how to integrate rational functions of the form $f(x) = 1/x$ using the Log Rule.

8.1 Exponential Functions: Differentiation and Integration

- **Differentiate natural exponential functions.**
- **Integrate natural exponential functions.**

Differentiation of Exponential Functions

In Section 7.1, it was stated that the natural base e is the most convenient base for exponential functions. One reason for this claim is that the natural exponential function $f(x) = e^x$ *is its own derivative*. The proof of this is shown below.

PROOF

$$f'(x) = \lim_{\Delta x \to 0} \frac{f(x + \Delta x) - f(x)}{\Delta x}$$
$$= \lim_{\Delta x \to 0} \frac{e^{x+\Delta x} - e^x}{\Delta x}$$
$$= \lim_{\Delta x \to 0} \frac{e^x[e^{\Delta x} - 1]}{\Delta x}$$

Now, the definition of e

$$e = \lim_{\Delta x \to 0} (1 + \Delta x)^{1/\Delta x}$$

tells you that for small values of Δx, you have $e \approx (1 + \Delta x)^{1/\Delta x}$, which implies that $e^{\Delta x} \approx 1 + \Delta x$. Replacing $e^{\Delta x}$ by this approximation produces

$$f'(x) = \lim_{\Delta x \to 0} \frac{e^x[e^{\Delta x} - 1]}{\Delta x}$$
$$= \lim_{\Delta x \to 0} \frac{e^x[(1 + \Delta x) - 1]}{\Delta x}$$
$$= \lim_{\Delta x \to 0} \frac{e^x(\Delta x)}{\Delta x} = e^x.$$

This result is summarized, along with its "Chain Rule version," in Theorem 8.1. ■

■ **FOR FURTHER INFORMATION** To find out about derivatives of exponential functions of order 1/2, see the article "A Child's Garden of Fractional Derivatives" by Marcia Kleinz and Thomas J. Osler in *The College Mathematics Journal.* To view this article, go to the website *www.matharticles.com.*

THEOREM 8.1 DERIVATIVES OF NATURAL EXPONENTIAL FUNCTION

Let u be a differentiable function of x.

1. $\dfrac{d}{dx}[e^x] = e^x$ **2.** $\dfrac{d}{dx}[e^u] = e^u\dfrac{du}{dx} = e^u \cdot u'$

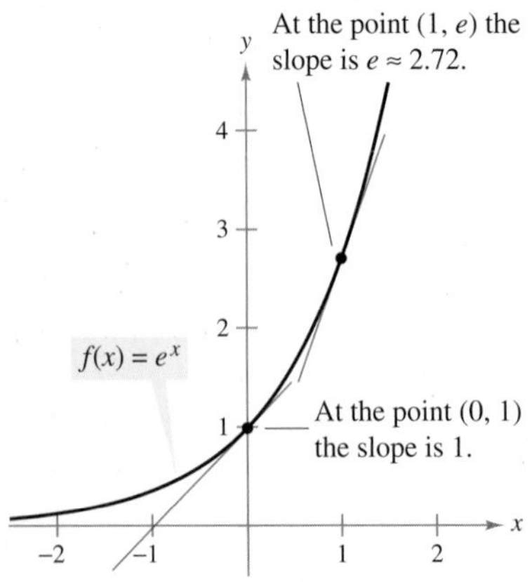

NOTE You can interpret this result geometrically by saying that the slope of the graph of $f(x) = e^x$ at any point (x, e^x) is equal to the y-coordinate of the point, as shown in Figure 8.1. ■

At the point (1, e) the slope is e ≈ 2.72.
f(x) = e^x
At the point (0, 1) the slope is 1.
y
x
4
3
2
1
−2
−1
1
2

Figure 8.1

EXAMPLE 1 Differentiating an Exponential Function

Find the derivative of $f(x) = e^{2x-1}$.

Solution Let $u = 2x - 1$. Then $u' = 2$ and you have

$$f'(x) = e^u \cdot u' = e^{2x-1}(2) = 2e^{2x-1}.$$ ■

EXAMPLE 2 Differentiating an Exponential Function

Find the derivative of

$$f(x) = e^{-3x}.$$

Solution Let $u = -\dfrac{3}{x}$. Then $u' = 3x^{-2} = \dfrac{3}{x^2}$ and you have

$$f'(x) = e^u \cdot u' = e^{-3/x}\left(\frac{3}{x^2}\right) = \frac{3e^{-3/x}}{x^2}.$$

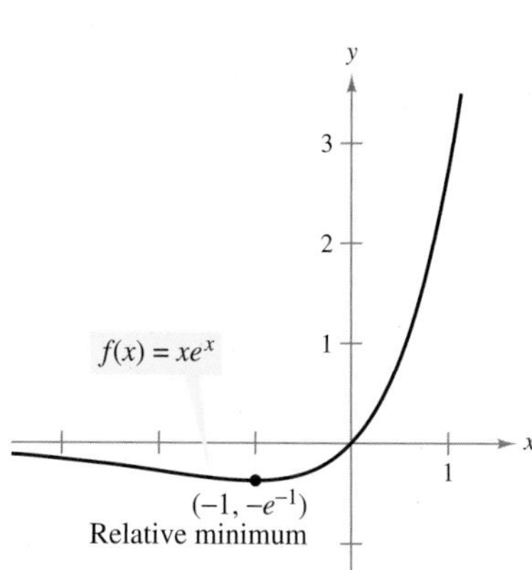

The derivative of f changes from negative to positive at $x = -1$.
Figure 8.2

EXAMPLE 3 Locating Relative Extrema

Find the relative extrema of

$$f(x) = xe^x.$$

Solution The derivative of f is

$$f'(x) = x(e^x) + e^x(1) \qquad \text{Product Rule}$$
$$= e^x(x + 1).$$

Because e^x is never 0, the derivative is 0 only when $x = -1$. Moreover, by the First Derivative Test, you can determine that this corresponds to a relative minimum, as shown in Figure 8.2. Because $f'(x) = e^x(x + 1)$ is defined for all x, there are no other critical points.

EXAMPLE 4 The Standard Normal Probability Density Function

Show that the graph of the *standard normal probability density function*

$$f(x) = \frac{1}{\sqrt{2\pi}}e^{-x^2/2}$$

has points of inflection when $x = \pm 1$.

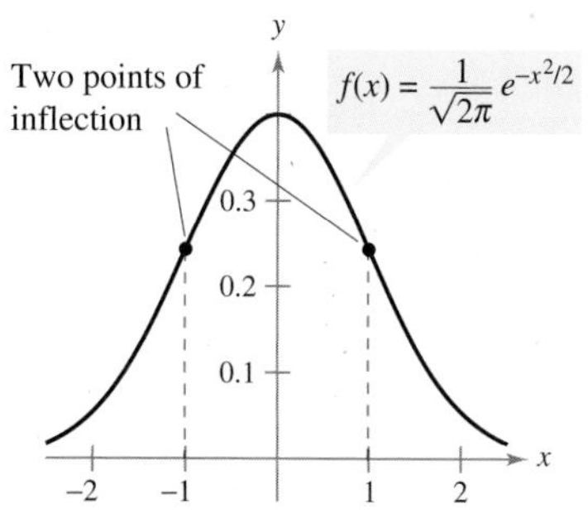

The bell-shaped curve given by a standard normal probability density function
Figure 8.3

Solution To locate possible points of inflection, find the x-values for which the second derivative is 0.

$$f'(x) = \frac{1}{\sqrt{2\pi}}(-x)e^{-x^2/2} \qquad \text{First derivative}$$

$$f''(x) = \frac{1}{\sqrt{2\pi}}[(-x)(-x)e^{-x^2/2} + (-1)e^{-x^2/2}] \qquad \text{Product Rule}$$

$$= \frac{1}{\sqrt{2\pi}}(e^{-x^2/2})(x^2 - 1) \qquad \text{Second derivative}$$

So, $f''(x) = 0$ when $x = \pm 1$, and you can apply the techniques of Chapter 5 to conclude that these values yield the two points of inflection shown in Figure 8.3. ■

NOTE The general form of a normal probability density function is

$$f(x) = \frac{1}{\sigma\sqrt{2\pi}}e^{-x^2/2\sigma^2}$$

where σ is the standard deviation (σ is the lowercase Greek letter sigma). By following the procedure of Example 4, you can show that the bell-shaped curve of this function has points of inflection when $x = \pm\sigma$. ■

Integration of Exponential Functions

Each of the differentiation formulas for exponential functions has a corresponding integration formula, as shown in Theorem 8.2.

THEOREM 8.2 INTEGRATION RULES FOR EXPONENTIAL FUNCTIONS

Let u be a differentiable function of x.

1. $\displaystyle\int e^x\,dx = e^x + C$

2. $\displaystyle\int e^u\,du = e^u + C$

EXAMPLE 5 Integrating an Exponential Function

Find $\displaystyle\int e^{3x+1}\,dx$.

Solution If you let $u = 3x + 1$, then $du = 3\,dx$.

$$\int e^{3x+1}\,dx = \frac{1}{3}\int e^{3x+1}(3)\,dx \qquad \text{Multiply and divide by 3.}$$

$$= \frac{1}{3}\int e^u\,du \qquad \text{Substitute: } u = 3x + 1.$$

$$= \frac{1}{3}e^u + C \qquad \text{Apply Exponential Rule.}$$

$$= \frac{e^{3x+1}}{3} + C \qquad \text{Back-substitute.}$$

NOTE In Example 5, the missing *constant* factor 3 was introduced to create $du = 3\,dx$. However, remember that you cannot introduce a missing *variable* factor in the integrand. For instance,

$$\int e^{-x^2}\,dx \neq \frac{1}{x}\int e^{-x^2}(x\,dx).$$

EXAMPLE 6 Integrating an Exponential Function

Find $\displaystyle\int 5xe^{-x^2}\,dx$.

Solution If you let $u = -x^2$, then $du = -2x\,dx$, which implies that $x\,dx = -du/2$.

$$\int 5xe^{-x^2}\,dx = \int 5e^{-x^2}(x\,dx) \qquad \text{Regroup integrand.}$$

$$= \int 5e^u\left(-\frac{du}{2}\right) \qquad \text{Substitute: } u = -x^2.$$

$$= -\frac{5}{2}\int e^u\,du \qquad \text{Constant Multiple Rule}$$

$$= -\frac{5}{2}e^u + C \qquad \text{Apply Exponential Rule.}$$

$$= -\frac{5}{2}e^{-x^2} + C \qquad \text{Back-substitute.}$$

■

EXAMPLE 7 Integrating Exponential Functions

$$\textbf{a.}\quad \int \frac{e^{1/x}}{x^2}\,dx = -\int \overbrace{e^{1/x}}^{e^u}\overbrace{\left(-\frac{1}{x^2}\right)dx}^{du} \qquad u = \frac{1}{x},\ du = -\frac{1}{x^2}\,dx$$

$$= -e^{1/x} + C$$

$$\textbf{b.}\quad \int (1 + e^x)^2\,dx = \int (1 + 2e^x + e^{2x})\,dx = x + 2e^x + \frac{1}{2}e^{2x} + C$$

EXAMPLE 8 Finding Areas Bounded by Exponential Functions

Find the area of the region bounded by the graph of f and the x-axis, for $0 \le x \le 1$.

a. $f(x) = e^{-x}$ **b.** $f(x) = \dfrac{e^x}{\sqrt{1 + e^x}}$

Solution

a. The region is shown in Figure 8.4(a), and its area is

$$\begin{aligned} \text{Area} = \int_0^1 f(x)\,dx &= \int_0^1 e^{-x}\,dx \\ &= -e^{-x}\Big]_0^1 \\ &= -e^{-1} - (-1) \\ &= 1 - \frac{1}{e} \approx 0.632. \end{aligned}$$

b. The region is shown in Figure 8.4(b), and its area is

$$\begin{aligned} \text{Area} = \int_0^1 f(x)\,dx &= \int_0^1 \frac{e^x}{\sqrt{1 + e^x}}\,dx \qquad u = 1 + e^x,\ du = e^x\,dx \\ &= \int_0^1 (1 + e^x)^{-1/2}(e^x\,dx) \\ &= 2\sqrt{1 + e^x}\Big]_0^1 \\ &= 2\sqrt{1 + e} - 2\sqrt{2} \\ &\approx 1.028. \end{aligned}$$

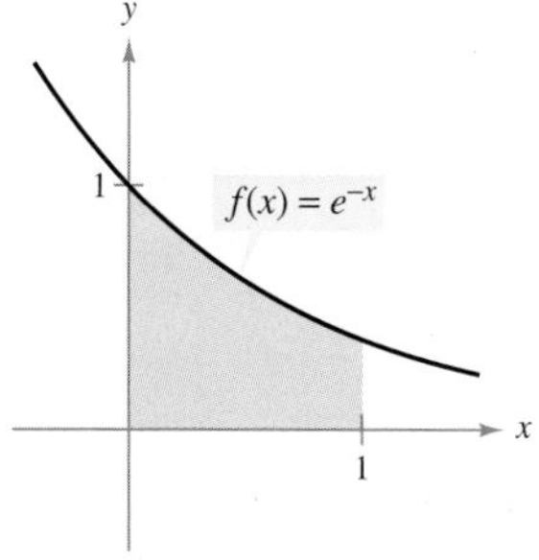

(a)

(b)

Areas bounded by exponential functions

Figure 8.4

8.1 Exercises

See www.CalcChat.com for worked-out solutions to odd-numbered exercises.

In Exercises 1 and 2, find the slope of the tangent line to the graph of each function at the point (0, 1).

1. (a) $y = e^{3x}$ (b) $y = e^{-3x}$

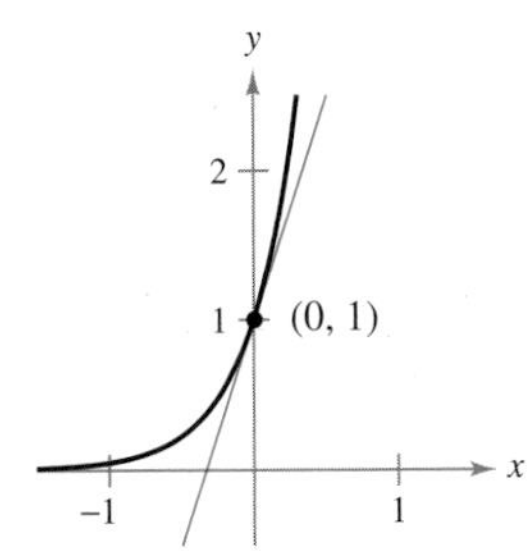

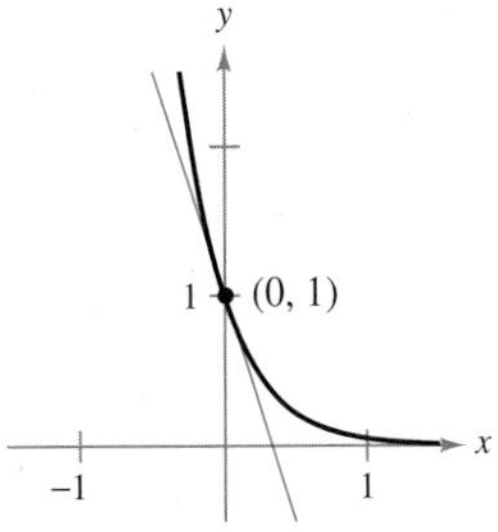

2. (a) $y = e^{2x}$ (b) $y = e^{-2x}$

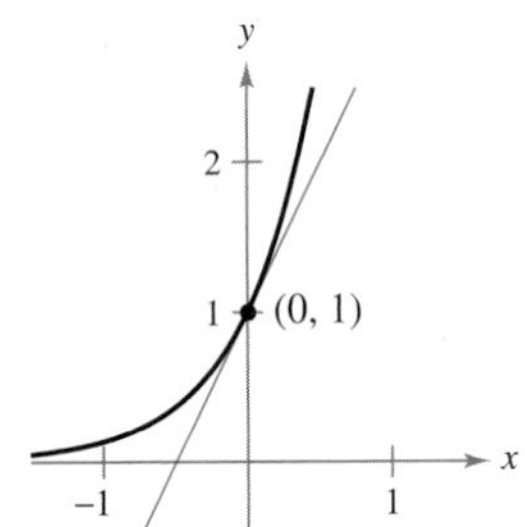

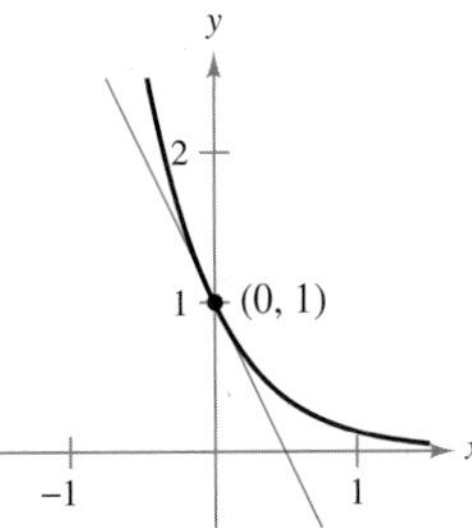

In Exercises 3–20, find the derivative of the function.

3. $f(x) = e^{2x}$
4. $f(x) = e^{1-x}$
5. $y = e^{-2x+x^2}$
6. $y = e^{-x^2}$
7. $f(x) = e^{1/x}$
8. $f(x) = e^{-1/x^2}$
9. $y = e^{\sqrt{x}}$
10. $g(x) = e^{3\sqrt{x}}$
11. $f(x) = (x + 1)e^{3x}$
12. $y = x^2e^{-x}$
13. $f(x) = \dfrac{e^{x^2}}{x}$
14. $f(x) = \dfrac{e^{x/2}}{\sqrt{x}}$
15. $g(t) = (e^{-t} + e^t)^3$
16. $y = (1 - e^{-x})^2$
17. $y = \dfrac{2}{e^x + e^{-x}}$
18. $y = \dfrac{e^x - e^{-x}}{2}$
19. $y = x^2e^x - 2xe^x + 2e^x$
20. $y = xe^x - e^x$

In Exercises 21–24, find an equation of the tangent line to the graph of the function at the given point.

21. $f(x) = e^{1-x}$, $(1, 1)$
22. $y = e^{-2x+x^2}$, $(2, 1)$
23. $y = x^2e^x - 2xe^x + 2e^x$, $(1, e)$
24. $y = xe^x - e^x$, $(1, 0)$

In Exercises 25 and 26, use implicit differentiation to find dy/dx.

25. $xe^y - 10x + 3y = 0$
26. $e^{xy} + x^2 - y^2 = 10$

In Exercises 27–30, find the second derivative of the function.

27. $f(x) = 2e^{3x} + 3e^{-2x}$
28. $f(x) = 5e^{-x} - 2e^{-5x}$
29. $f(x) = (3 + 2x)e^{-3x}$
30. $g(x) = (1 + 2x)e^{4x}$

In Exercises 31–38, find the extrema and the points of inflection (if any exist) of the function. Use a graphing utility to graph the function and confirm your results.

31. $f(x) = \dfrac{e^x + e^{-x}}{2}$
32. $f(x) = \dfrac{e^x - e^{-x}}{2}$
33. $g(x) = \dfrac{1}{\sqrt{2\pi}}e^{-(x-2)^2/2}$
34. $g(x) = \dfrac{1}{\sqrt{2\pi}}e^{-(x-3)^2/2}$
35. $f(x) = x^2e^{-x}$
36. $f(x) = xe^{-x}$
37. $g(t) = 1 + (2 + t)e^{-t}$
38. $f(x) = -2 + e^{3x}(4 - 2x)$

39. ***Area*** Find the area of the largest rectangle that can be inscribed under the curve given by $y = e^{-x^2}$ in the first and second quadrants.

40. ***Area*** Perform the following steps to find the maximum area of the rectangle shown in the figure.

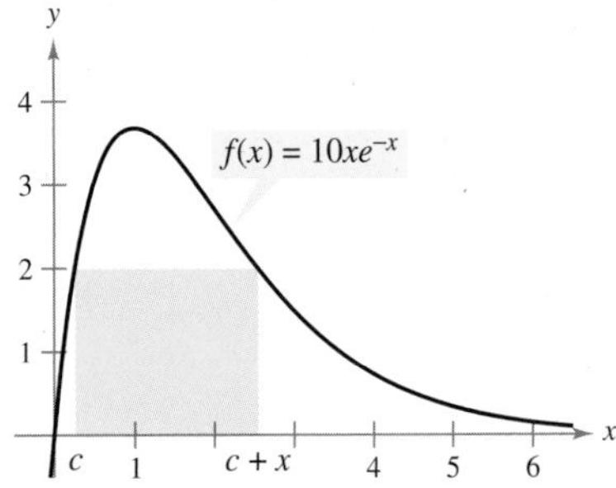

(a) Solve for c in the equation $f(c) = f(c + x)$.

(b) Use the result in part (a) to write the area A as a function of x. [*Hint:* $A = xf(c)$.]

(c) Use a graphing utility to graph the area function. Use the graph to approximate the dimensions of the rectangle of maximum area. Determine the maximum area.

(d) Use a graphing utility to graph the expression for c found in part (a). Use the graph to approximate

$$\lim_{x \to 0^+} c \quad \text{and} \quad \lim_{x \to \infty} c.$$

Use this result to describe the changes in dimensions and position of the rectangle for $0 < x < \infty$.

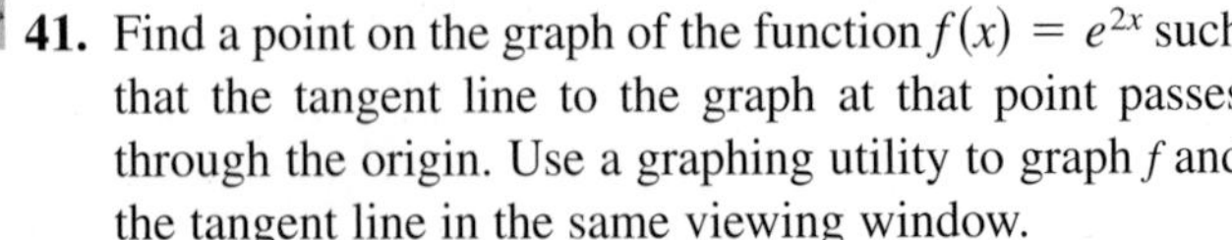

41. Find a point on the graph of the function $f(x) = e^{2x}$ such that the tangent line to the graph at that point passes through the origin. Use a graphing utility to graph f and the tangent line in the same viewing window.

42. Find the point on the graph of $y = e^{-x}$ where the normal line to the curve passes through the origin. (Use the *zero* or *root* feature of a graphing utility.)

43. ***Depreciation*** The value V of an item t years after it is purchased is given by

$$V = 15{,}000e^{-0.6286t}, \quad 0 \le t \le 10.$$

(a) Use a graphing utility to graph the function.

(b) Find the rates of change of V with respect to t when $t = 1$ and $t = 5$.

(c) Use a graphing utility to graph the tangent lines to the function when $t = 1$ and $t = 5$.

44. ***Modeling Data*** The table lists the approximate values V of a mid-sized sedan for the years 2005 through 2011. The variable t represents the time in years, with $t = 5$ corresponding to 2005.

t	5	6	7	8
V	\$23,046	\$20,596	\$18,851	\$17,001

t	9	10	11
V	\$15,226	\$14,101	\$12,841

(a) Use the regression capabilities of a graphing utility to fit linear and quadratic models to the data. Plot the data and graph the models.

(b) What does the slope represent in the linear model in part (a)?

(c) Use the regression capabilities of a graphing utility to fit an exponential model to the data.

(d) Determine the horizontal asymptote of the exponential model found in part (c). Interpret its meaning in the context of the problem.

(e) Find the rate of decrease in the value of the sedan when $t = 6$ and $t = 10$ using the exponential model.

Linear and Quadratic Approximations **In Exercises 45 and 46, use a graphing utility to graph the function. Then graph**

$$P_1(x) = f(0) + f'(0)(x - 0)$$

and

$$P_2(x) = f(0) + f'(0)(x - 0) + \tfrac{1}{2}f''(0)(x - 0)^2$$

in the same viewing window. Compare the values of f, P_1, and P_2, and their first derivatives, at $x = 0$.

45. $f(x) = e^x$

46. $f(x) = e^{x/2}$

In Exercises 47–60, find the indefinite integral.

47. $\int e^{5x}(5)\,dx$

48. $\int e^{-x^4}(-4x^3)\,dx$

49. $\int xe^{-x^2}\,dx$

50. $\int x^2e^{x^3/2}\,dx$

51. $\int \frac{e^{\sqrt{x}}}{\sqrt{x}}\,dx$

52. $\int \frac{e^{1/x^2}}{x^3}\,dx$

53. $\int (1 + e^x)^2\,dx$

54. $\int e^{2x}(1 - 3e^{2x})^2\,dx$

55. $\int e^x\sqrt{1 - e^x}\,dx$

56. $\int e^x(e^x - e^{-x})\,dx$

57. $\int \frac{5 - e^x}{e^{2x}}\,dx$

58. $\int \frac{e^{2x}}{(1 + e^{2x})^2}\,dx$

59. $\int \frac{e^x + e^{-x}}{\sqrt{e^x - e^{-x}}}\,dx$

60. $\int \frac{2e^x - 2e^{-x}}{(e^x + e^{-x})^2}\,dx$

In Exercises 61–68, evaluate the definite integral. Use a graphing utility to verify your result.

61. $\int_0^1 e^{-2x}\,dx$

62. $\int_3^4 e^{3-x}\,dx$

63. $\int_1^3 \frac{e^{3/x}}{x^2}\,dx$

64. $\int_0^{\sqrt{2}} xe^{-(x^2/2)}\,dx$

65. $\int_{-1}^0 e^{-x}(1 + e^{-x})^2\,dx$

66. $\int_{-2}^2 (e^x - e^{-x})^2\,dx$

67. $\int_0^1 \frac{e^{-x}}{(1 + e^{-x})^2}\,dx$

68. $\int_{-3}^3 \frac{e^{2x} + 2e^x + 1}{e^x}\,dx$

Slope Fields **In Exercises 69 and 70, a differential equation, a point, and a slope field are given. (To print an enlarged copy of the graph, go to the website *www.mathgraphs.com*.) (a) Sketch two approximate solutions of the differential equation on the slope field, one of which passes through the indicated point. (b) Use integration to find the particular solution of the differential equation and use a graphing utility to graph the solution. Compare the result with the sketches in part (a).**

69. $\frac{dy}{dx} = 2e^{-x/2}, \quad (0, 1)$

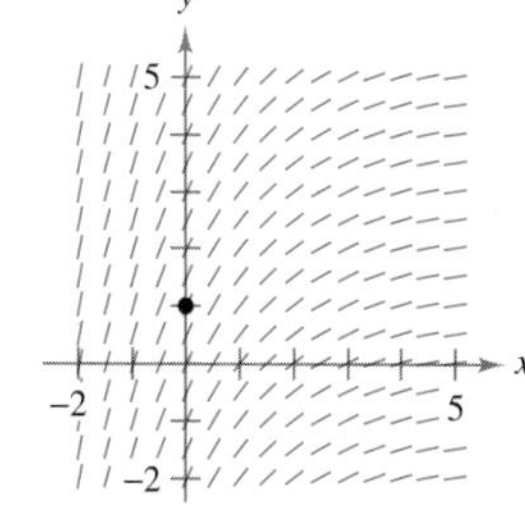

70. $\frac{dy}{dx} = xe^{-0.2x^2}, \quad \left(0, -\frac{3}{2}\right)$

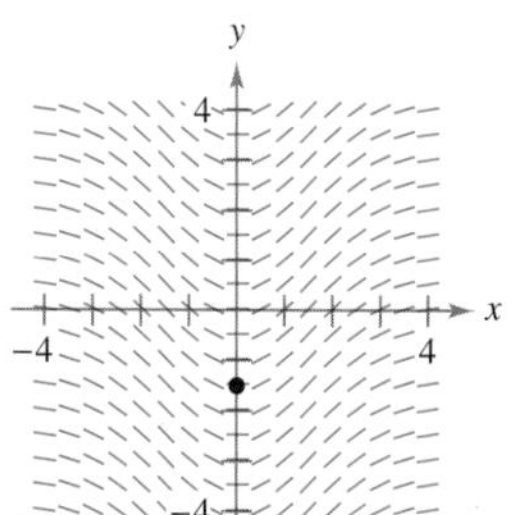

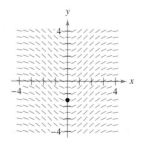

In Exercises 71 and 72, solve the differential equation.

71. $\dfrac{dy}{dx} = xe^{ax^2}$

72. $\dfrac{dy}{dx} = (e^x - e^{-x})^2$

In Exercises 73 and 74, find the particular solution of the differential equation that satisfies the initial conditions.

73. $f''(x) = \frac{1}{2}(e^x + e^{-x}),$
$f(0) = 1, \quad f'(0) = 0$

74. $f''(x) = x + e^{2x},$
$f(0) = \frac{1}{4}, \quad f'(0) = \frac{1}{2}$

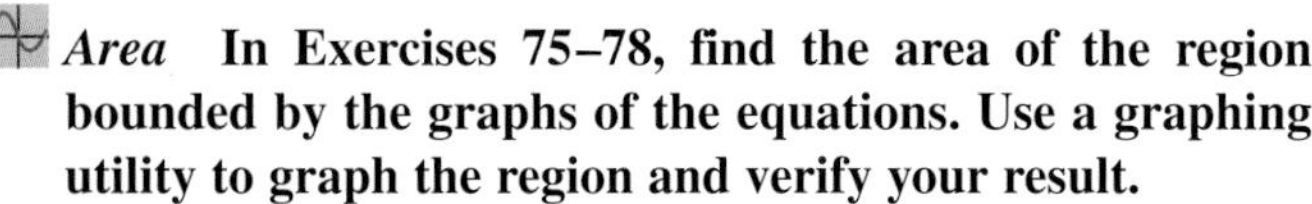

Area **In Exercises 75–78, find the area of the region bounded by the graphs of the equations. Use a graphing utility to graph the region and verify your result.**

75. $y = e^x, \quad y = 0, \quad x = 0, \quad x = 5$

76. $y = e^{-x}, \quad y = 0, \quad x = a, \quad x = b$

77. $y = xe^{-x^2/4}, \quad y = 0, \quad x = 0, \quad x = \sqrt{6}$

78. $y = e^{-2x} + 2, \quad y = 0, \quad x = 0, \quad x = 2$

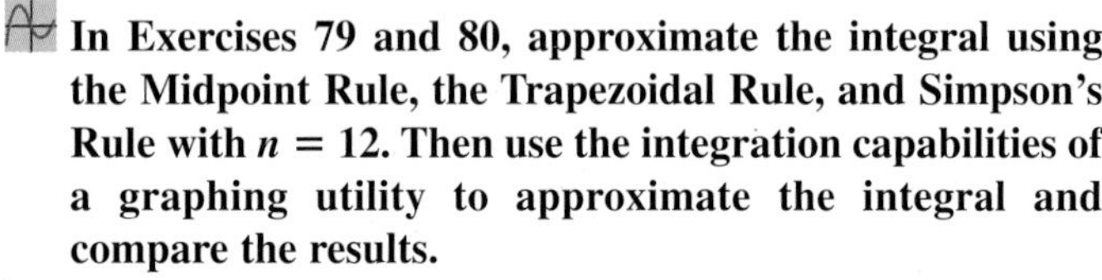

In Exercises 79 and 80, approximate the integral using the Midpoint Rule, the Trapezoidal Rule, and Simpson's Rule with $n = 12$. Then use the integration capabilities of a graphing utility to approximate the integral and compare the results.

79. $\displaystyle\int_0^4 \sqrt{x}\, e^x \, dx$

80. $\displaystyle\int_0^2 2xe^{-x} \, dx$

81. ***Probability*** A car battery has an average lifetime of 48 months with a standard deviation of 6 months. The battery lives are normally distributed. The probability that a given battery will last between 48 months and 60 months is

$$0.0665 \int_{48}^{60} e^{-0.0139(t-48)^2} \, dt.$$

Use the integration capabilities of a graphing utility to approximate the integral. Interpret the resulting probability.

82. ***Probability*** The median waiting time (in minutes) for people waiting for service in a convenience store is given by the solution of the equation

$$\int_0^x 0.3e^{-0.3t} \, dt = \frac{1}{2}.$$

Solve the equation.

83. ***Horizontal Motion*** The position function of a particle moving along the x-axis is $x(t) = Ae^{kt} + Be^{-kt}$, where A, B, and k are positive constants.

(a) During what times t is the particle closest to the origin?

(b) Show that the acceleration of the particle is proportional to the position of the particle. What is the constant of proportionality?

CAPSTONE

84. Is there a function f such that $f(x) = f'(x)$? If so, identify it. How many such functions are there?

85. ***Modeling Data*** A valve on a storage tank is opened for 4 hours to release a chemical in a manufacturing process. The flow rate R (in liters per hour) at time t (in hours) is given in the table.

t	0	1	2	3	4
R	425	240	118	71	36

(a) Use the *regression* feature of a graphing utility to find an exponential model for the data.

(b) Use a graphing utility to plot the data and graph the exponential model.

(c) Use the definite integral to approximate the number of liters of chemical released during the 4 hours.

WRITING ABOUT CONCEPTS

86. Without integrating, state the integration formula you can use to integrate each integral.

(a) $\displaystyle\int \frac{e^x}{(e^x + 1)^2} \, dx$ (b) $\displaystyle\int xe^{x^2} \, dx$

87. ***Writing*** Consider the function given by

$f(x) = 2/(1 + e^{1/x})$.

(a) Use a graphing utility to graph f.

(b) Write a short paragraph explaining why the graph has a horizontal asymptote at $y = 1$ and why the function has a nonremovable discontinuity at $x = 0$.

88. Explain why $\displaystyle\int_0^2 e^{-x} \, dx > 0$.

89. Prove that $\dfrac{e^a}{e^b} = e^{a-b}$.

90. Given $e^x \geq 1$ for $x \geq 0$, it follows that

$$\int_0^x e^t \, dt \geq \int_0^x 1 \, dt.$$

Perform this integration to derive the inequality $e^x \geq 1 + x$ for $x \geq 0$.

91. Find the value of a such that the area bounded by $y = e^{-x}$, the x-axis, $x = -a$, and $x = a$ is $\frac{8}{3}$.

92. Verify that the function given by

$$y = \frac{L}{1 + ae^{-x/b}}, \quad a > 0, b > 0, L > 0$$

increases at a maximum rate when $y = L/2$.

8.2 Logarithmic Functions and Differentiation

- Find derivatives of functions involving the natural logarithmic function.
- Use logarithms as an aid in differentiating nonlogarithmic functions.
- Find derivatives of exponential and logarithmic functions in bases other than *e*.

Differentiation of the Natural Logarithmic Function

The derivative of the natural logarithmic function is given in the following theorem. You are asked to prove the theorem in Exercise 101.

EXPLORATION

Use a graphing utility to graph

$$y_1 = \frac{1}{x}$$

and

$$y_2 = \frac{d}{dx}[\ln x]$$

in the same viewing window in which $0.1 \le x \le 5$ and $-2 \le y \le 8$. Explain why the graphs appear to be identical.

THEOREM 8.3 DERIVATIVES OF NATURAL LOGARITHMIC FUNCTION

Let u be a differentiable function of x.

1. $\dfrac{d}{dx}[\ln x] = \dfrac{1}{x}, \quad x > 0$

2. $\dfrac{d}{dx}[\ln u] = \dfrac{1}{u}\dfrac{du}{dx} = \dfrac{u'}{u}, \quad u > 0$

So far in your development of the natural logarithmic function, it would have been difficult to predict its intimate relationship to the rational function $1/x$. Hidden relationships such as this not only illustrate the joy of mathematical discovery, they also give you logical alternatives in constructing a mathematical system. An alternative that Theorem 8.3 provides is that the natural logarithmic function could have been developed as the *antiderivative* of $1/x$, rather than as the inverse of e^x. If you are interested in pursuing this alternative development of ln x, you can consult *Calculus*, 9th edition, by Larson and Edwards.

EXAMPLE 1 Differentiation of Logarithmic Functions

Find the derivative of each function.

a. $f(x) = \ln(2x)$ **b.** $f(x) = \ln(x^2 + 1)$

c. $f(x) = x \ln x$ **d.** $f(x) = (\ln x)^3$

Solution

a. $\dfrac{d}{dx}[\ln(2x)] = \dfrac{u'}{u} = \dfrac{2}{2x} = \dfrac{1}{x}$ $\qquad u = 2x$

b. $\dfrac{d}{dx}[\ln(x^2 + 1)] = \dfrac{u'}{u} = \dfrac{2x}{x^2 + 1}$ $\qquad u = x^2 + 1$

c. $\dfrac{d}{dx}[x \ln x] = x\left(\dfrac{d}{dx}[\ln x]\right) + (\ln x)\left(\dfrac{d}{dx}[x]\right)$ $\qquad$ Product Rule

$$= x\left(\frac{1}{x}\right) + (\ln x)(1)$$

$$= 1 + \ln x$$

d. $\dfrac{d}{dx}[(\ln x)^3] = 3(\ln x)^2 \dfrac{d}{dx}[\ln x]$ $\qquad$ Chain Rule

$$= 3(\ln x)^2 \frac{1}{x}$$

■

The properties of logarithms can be used to simplify the work involved in differentiating complicated logarithmic functions, as demonstrated in the next three examples.

EXAMPLE 2 Logarithmic Properties as Aids to Differentiation

Differentiate $f(x) = \ln \sqrt{x+1}$.

Solution Because

$$f(x) = \ln \sqrt{x+1} = \ln(x+1)^{1/2} = \frac{1}{2}\ln(x+1)$$

you can write

$$f'(x) = \frac{d}{dx}\left[\frac{1}{2}\ln(x+1)\right] = \frac{1}{2}\left(\frac{1}{x+1}\right) = \frac{1}{2(x+1)}.$$

EXAMPLE 3 Logarithmic Properties as Aids to Differentiation

Differentiate $f(x) = \ln\left[x\sqrt{1-x^2}\right]$.

Solution Because

$$\begin{aligned} f(x) = \ln\left[x\sqrt{1-x^2}\right] &= \ln x + \ln(1-x^2)^{1/2} \\ &= \ln x + \frac{1}{2}\ln(1-x^2) \end{aligned}$$

you can write

$$\begin{aligned} f'(x) = \frac{1}{x} + \frac{1}{2}\left(\frac{-2x}{1-x^2}\right) &= \frac{1}{x} - \frac{x}{1-x^2} \\ &= \frac{1-x^2-x^2}{x(1-x^2)} \\ &= \frac{1-2x^2}{x(1-x^2)}. \end{aligned}$$

EXAMPLE 4 Logarithmic Properties as Aids to Differentiation

Differentiate $f(x) = \ln \dfrac{x(x^2+1)^2}{\sqrt{2x^3-1}}$.

Solution Because

$$\begin{aligned} f(x) &= \ln \frac{x(x^2+1)^2}{\sqrt{2x^3-1}} \\ &= \ln x + 2\ln(x^2+1) - \frac{1}{2}\ln(2x^3-1) \end{aligned}$$

you can write

$$\begin{aligned} f'(x) &= \frac{1}{x} + 2\left(\frac{2x}{x^2+1}\right) - \frac{1}{2}\left(\frac{6x^2}{2x^3-1}\right) \\ &= \frac{1}{x} + \frac{4x}{x^2+1} - \frac{3x^2}{2x^3-1}. \end{aligned}$$

■

NOTE In Examples 2, 3, and 4, be sure you see the benefit in applying logarithmic properties *before* differentiating. For instance, consider the difficulty of direct differentiation of the function given in Example 4.

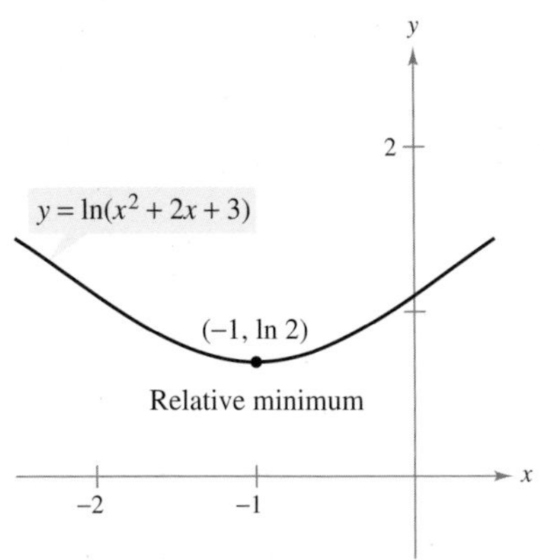

The derivative of y changes from negative to positive at $x = -1$.
Figure 8.5

EXAMPLE 5 Finding Relative Extrema

Find the relative extrema of $y = \ln(x^2 + 2x + 3)$.

Solution Differentiating y, you obtain

$$\frac{dy}{dx} = \frac{2x + 2}{x^2 + 2x + 3}.$$

Because $dy/dx = 0$ when $x = -1$, you can apply the First Derivative Test and conclude that the point $(-1, \ln 2)$ is a relative minimum. Because there are no other critical points, it follows that this is the only relative extremum (see Figure 8.5). ■

Logarithmic Differentiation

On occasion, it is convenient to use logarithms as an aid in differentiating *nonlogarithmic* functions. This procedure is called **logarithmic differentiation** and is illustrated in Examples 6 and 7.

LOGARITHMIC DIFFERENTIATION

To differentiate the function $y = u$, use the following steps.

1. Take the natural logarithm of each side: $\ln y = \ln u$
2. Use logarithmic properties to rid $\ln u$ of as many products, quotients, and exponents as possible.
3. Differentiate *implicitly*: $\frac{y'}{y} = \frac{d}{dx}[\ln u]$
4. Solve for y': $y' = y\frac{d}{dx}[\ln u]$
5. Substitute for y and simplify: $y' = u\frac{d}{dx}[\ln u]$

EXAMPLE 6 Logarithmic Differentiation

Find the derivative of $y = x\sqrt{x^2 + 1}$.

Solution Begin by taking the natural logarithm of each side of the equation. Then, apply logarithmic properties and differentiate implicitly. Finally, solve for y'.

$$y = x\sqrt{x^2 + 1}$$ Write original function.

$$\ln y = \ln\left[x\sqrt{x^2 + 1}\right]$$ Take the natural logarithm of each side.

$$\ln y = \ln x + \frac{1}{2}\ln(x^2 + 1)$$ Rewrite using logarithmic properties.

$$\frac{y'}{y} = \frac{1}{x} + \frac{1}{2}\left(\frac{2x}{x^2 + 1}\right) = \frac{2x^2 + 1}{x(x^2 + 1)}$$ Differentiate implicitly.

$$y' = y\left[\frac{2x^2 + 1}{x(x^2 + 1)}\right]$$ Solve for y'.

$$y' = x\sqrt{x^2 + 1}\left[\frac{2x^2 + 1}{x(x^2 + 1)}\right] = \frac{2x^2 + 1}{\sqrt{x^2 + 1}}$$ Substitute for y. ■

EXAMPLE 7 Logarithmic Differentiation

Find the derivative of $y = \dfrac{(x-2)^2}{\sqrt{x^2+1}}$.

Solution

$$\ln y = \ln \frac{(x-2)^2}{\sqrt{x^2+1}} \qquad \text{Take the natural logarithm of each side.}$$

$$\ln y = 2\ln(x-2) - \frac{1}{2}\ln(x^2+1) \qquad \text{Rewrite using logarithmic properties.}$$

$$\frac{y'}{y} = 2\left(\frac{1}{x-2}\right) - \frac{1}{2}\left(\frac{2x}{x^2+1}\right) \qquad \text{Differentiate implicitly.}$$

$$= \frac{2}{x-2} - \frac{x}{x^2+1}$$

$$= \frac{x^2+2x+2}{(x-2)(x^2+1)}$$

$$y' = y\left(\frac{x^2+2x+2}{(x-2)(x^2+1)}\right) \qquad \text{Solve for } y'.$$

$$= \frac{(x-2)^2}{\sqrt{x^2+1}}\left[\frac{x^2+2x+2}{(x-2)(x^2+1)}\right] \qquad \text{Substitute for } y.$$

$$= \frac{(x-2)(x^2+2x+2)}{(x^2+1)^{3/2}} \qquad \text{Simplify.}$$

■

Because the natural logarithm is undefined for negative numbers, you will encounter expressions of the form $\ln|u|$. The following theorem states that you can differentiate functions of the form $y = \ln|u|$ as if the absolute value sign were not present.

THEOREM 8.4 DERIVATIVES INVOLVING ABSOLUTE VALUE

If u is a differentiable function of x such that $u \neq 0$, then

$$\frac{d}{dx}[\ln|u|] = \frac{u'}{u}.$$

PROOF If $u > 0$, then $|u| = u$, and the result follows from Theorem 8.3. If $u < 0$, then $|u| = -u$, and you have

$$\frac{d}{dx}[\ln|u|] = \frac{d}{dx}[\ln(-u)] = \frac{-u'}{-u} = \frac{u'}{u}.$$

■

EXAMPLE 8 Derivative Involving Absolute Value

Find the derivative of $f(x) = \ln|2x-1|$.

Solution Using Theorem 8.4, let $u = 2x - 1$ and write

$$\frac{d}{dx}[\ln|2x-1|] = \frac{u'}{u} = \frac{2}{2x-1}.$$

■

Bases Other Than e

To differentiate exponential and logarithmic functions to other bases, you have three options: (1) use the properties of a^x and $\log_a x$

$$a^x = e^{\ln a^x} = e^{x \ln a} = e^{(\ln a)x} \quad \text{and} \quad \log_a x = \frac{\ln x}{\ln a}$$

and differentiate using the rules for the natural exponential and logarithmic functions, (2) use logarithmic differentiation, or (3) use the following differentiation rules for bases other than e.

THEOREM 8.5 DERIVATIVES FOR BASES OTHER THAN e

Let a be a positive real number $(a \neq 1)$ and let u be a differentiable function of x.

1. $\dfrac{d}{dx}[a^x] = (\ln a)a^x$
2. $\dfrac{d}{dx}[a^u] = (\ln a)a^u\dfrac{du}{dx}$
3. $\dfrac{d}{dx}[\log_a x] = \dfrac{1}{(\ln a)x}$
4. $\dfrac{d}{dx}[\log_a u] = \dfrac{1}{(\ln a)u}\dfrac{du}{dx}$

PROOF You know that $a^x = e^{(\ln a)x}$. So, you can prove the first rule by letting $u = (\ln a)x$ and differentiating with base e to obtain

$$\frac{d}{dx}[a^x] = \frac{d}{dx}[e^{(\ln a)x}] = e^u\frac{du}{dx} = e^{(\ln a)x}(\ln a) = (\ln a)a^x.$$

To prove the third rule, you can write

$$\frac{d}{dx}[\log_a x] = \frac{d}{dx}\left[\frac{1}{\ln a}\ln x\right] = \frac{1}{\ln a}\left(\frac{1}{x}\right) = \frac{1}{(\ln a)x}.$$

The second and fourth rules are simply the Chain Rule versions of the first and third rules. ■

NOTE These differentiation rules are similar to those for the natural exponential function and natural logarithmic function. In fact, they differ only by the constant factors $\ln a$ and $1/\ln a$. This points out one reason why, for calculus, e is the most *convenient* base. ■

EXAMPLE 9 Differentiating Functions to Other Bases

Find the derivative of each function.

a. $y = 2^x$ **b.** $y = 2^{3x}$ **c.** $y = \log_2(x^2 + 1)$

Solution

a. $y' = \dfrac{d}{dx}[2^x] = (\ln 2)2^x$

b. $y' = \dfrac{d}{dx}[2^{3x}] = (\ln 2)2^{3x}(3) = (3 \ln 2)2^{3x}$

c. $y' = \dfrac{d}{dx}[\log_2(x^2 + 1)] = \dfrac{1}{(\ln 2)(x^2 + 1)}(2x) = \dfrac{1}{\ln 2} \cdot \dfrac{2x}{x^2 + 1}$ ■

NOTE In Example 9(b), try writing 2^{3x} as 8^x and differentiating to see that you obtain the same result.

When the Power Rule, $D_x[x^n] = nx^{n-1}$, was introduced in Chapter 4, the exponent n was required to be a rational number. The rule is now extended to cover any real value of n. Try to prove this theorem using logarithmic differentiation.

THEOREM 8.6 THE POWER RULE FOR REAL EXPONENTS

Let n be any real number and let u be a differentiable function of x.

1. $\dfrac{d}{dx}[x^n] = nx^{n-1}$ **2.** $\dfrac{d}{dx}[u^n] = nu^{n-1}\dfrac{du}{dx}$

The next example compares the derivatives of four types of functions. Each function uses a different differentiation formula, depending on whether the base and exponent are constants or variables.

EXAMPLE 10 Comparing Variables and Constants

a. $\dfrac{d}{dx}[e^e] = 0$ Constant Rule

b. $\dfrac{d}{dx}[e^x] = e^x$ Exponential Rule

c. $\dfrac{d}{dx}[x^e] = ex^{e-1}$ Power Rule

d. $y = x^x$ Logarithmic differentiation

$\ln y = \ln x^x = x \ln x$

$\dfrac{y'}{y} = x\left(\dfrac{1}{x}\right) + (\ln x)(1) = 1 + \ln x$

$y' = y(1 + \ln x) = x^x(1 + \ln x)$ ■

NOTE There is no simple differentiation rule for *directly* calculating a derivative of the form

$$y = u(x)^{v(x)}.$$

In general, logarithmic differentiation is very useful. Another option is to rewrite the function as

$$y = e^{\ln[u(x)^{v(x)}]} = e^{v(x)\ln u(x)}$$

and then differentiate this exponential form. Try this method with $y = x^x$.

8.2 Exercises

See www.CalcChat.com for worked-out solutions to odd-numbered exercises.

In Exercises 1–4, find the limit.

1. $\lim\limits_{x\to 3^+} \ln(x - 3)$ **2.** $\lim\limits_{x\to 6^-} \ln(6 - x)$

3. $\lim\limits_{x\to 2^-} \ln[x^2(3 - x)]$ **4.** $\lim\limits_{x\to 5^+} \ln\left(x/\sqrt{x - 4}\right)$

In Exercises 5–8, find the slope of the tangent line to the logarithmic function at the point (1, 0).

5. $y = \ln x^3$ **6.** $y = \ln x^{3/2}$

7. $y = \ln x^2$ **8.** $y = \ln x^{1/2}$

In Exercises 9–36, find the derivative of the function.

9. $g(x) = \ln x^2$ **10.** $h(x) = \ln(2x^2 + 1)$

11. $y = \ln\sqrt{x^4 - 4x}$ **12.** $y = \ln(1 - x)^{3/2}$

13. $y = (\ln x)^4$ **14.** $y = x \ln x$

15. $y = \ln\left(x\sqrt{x^2 - 1}\right)$ **16.** $y = \ln\sqrt{x^2 - 4}$

17. $f(x) = \ln[x/(x^2 + 1)]$ **18.** $f(x) = \ln[2x/(x + 3)]$

19. $g(t) = \ln(t/t^2)$ **20.** $h(t) = \ln(t/t)$

21. $y = \ln(\ln x^2)$ **22.** $y = \ln(\ln x)$

23. $y = \ln\sqrt{(x + 1)/(x - 1)}$

24. $y = \ln\sqrt[3]{(x - 1)/(x + 1)}$

25. $y = \ln(e^{x^2})$ **26.** $y = \ln e^{-x/2}$

27. $y = \ln\left(\dfrac{1 + e^x}{1 - e^x}\right)$ **28.** $y = \ln\left(\dfrac{e^x + e^{-x}}{2}\right)$

29. $f(x) = e^{-x} \ln x$ **30.** $g(x) = e^3 \ln x$

31. $f(x) = \ln\left(\dfrac{\sqrt{4 + x^2}}{x}\right)$ **32.** $f(x) = \ln\left(x + \sqrt{4 + x^2}\right)$

33. $y = \dfrac{-\sqrt{x^2 + 1}}{x} + \ln\left(x + \sqrt{x^2 + 1}\right)$

34. $y = \dfrac{-\sqrt{x^2 + 4}}{2x^2} - \dfrac{1}{4}\ln\left(\dfrac{2 + \sqrt{x^2 + 4}}{x}\right)$

35. $f(x) = \ln|x^2 - 1|$

36. $f(x) = \ln\left|\dfrac{x + 5}{x}\right|$

In Exercises 37–52, find the derivative of the function.

37. $f(x) = 4^x$

38. $g(x) = 2^{-x}$

39. $y = 5^{x-2}$

40. $y = x(6^{-2x})$

41. $g(t) = t^2 2^t$

42. $f(t) = \dfrac{3^{2t}}{t}$

43. $y = \log_3 x$

44. $y = \log_{10} 2x$

45. $y = \log_4(5x + 1)$

46. $y = \log_3(x^2 - 3x)$

47. $f(x) = \log_2 \dfrac{x^2}{x - 1}$

48. $h(x) = \log_3 \dfrac{x\sqrt{x - 1}}{2}$

49. $y = \log_5 \sqrt{x^2 - 1}$

50. $y = \log_{10} \dfrac{x^2 - 1}{x}$

51. $g(t) = \dfrac{10 \log_4 t}{t}$

52. $f(t) = t^{3/2} \log_2 \sqrt{t + 1}$

In Exercises 53–56, (a) find an equation of the tangent line to the graph of the function at the indicated point, (b) use a graphing utility to graph the function and its tangent line at the point, and (c) use the *derivative* feature of a graphing utility to confirm your results.

53. $y = 3x^2 - \ln x, \quad (1, 3)$

54. $y = 4 - x^2 - \ln\left(\frac{1}{2}x + 1\right), \quad (0, 4)$

55. $f(x) = x^3 \ln x, \quad (1, 0)$

56. $f(x) = \dfrac{1}{2}x \ln x^2, \quad (-1, 0)$

In Exercises 57–60, use implicit differentiation to find dy/dx.

57. $x^2 - 3 \ln y + y^2 = 10$

58. $\ln xy + 5x = 30$

59. $4x^3 + \ln y^2 + 2y = 2x$

60. $4xy + \ln x^2y = 7$

In Exercises 61 and 62, show that the function is a solution of the differential equation.

	Function	*Differential Equation*
61.	$y = 2 \ln x + 3$	$xy'' + y' = 0$
62.	$y = x \ln x - 4x$	$x + y - xy' = 0$

In Exercises 63–70, find any relative extrema and inflection points. Use a graphing utility to confirm your results.

63. $y = \dfrac{x^2}{2} - \ln x$

64. $y = x - \ln x$

65. $y = x \ln x$

66. $y = \dfrac{\ln x}{x}$

67. $y = \dfrac{x}{\ln x}$

68. $y = x^2 \ln \dfrac{x}{4}$

69. $y = x^2 - \ln x$

70. $y = (\ln x)^2$

Linear and Quadratic Approximations **In Exercises 71 and 72, use a graphing utility to graph the function. Then graph**

$$P_1(x) = f(1) + f'(1)(x - 1)$$

and

$$P_2(x) = f(1) + f'(1)(x - 1) + \tfrac{1}{2}f''(1)(x - 1)^2$$

in the same viewing window. Compare the values of f, P_1, and P_2, and their first derivatives, at $x = 1$.

71. $f(x) = \ln x$

72. $f(x) = x \ln x$

In Exercises 73–82, find dy/dx using logarithmic differentiation.

73. $y = x\sqrt{x^2 - 1}$

74. $y = \sqrt{(x - 1)(x - 2)(x - 3)}$

75. $y = \dfrac{x^2\sqrt{3x - 2}}{(x - 1)^2}$

76. $y = \sqrt{\dfrac{x^2 - 1}{x^2 + 1}}$

77. $y = \dfrac{x(x - 1)^{3/2}}{\sqrt{x + 1}}$

78. $y = \dfrac{(x + 1)(x + 2)}{(x - 1)(x - 2)}$

79. $y = x^{2/x}$

80. $y = x^{x-1}$

81. $y = (x - 2)^{x+1}$

82. $y = (1 + x)^{1/x}$

WRITING ABOUT CONCEPTS

83. Let f be a function that is positive and differentiable on the entire real line. Let $g(x) = \ln f(x)$.
    (a) If the graph of g is increasing, must the graph of f be increasing? Explain your reasoning.
    (b) If the graph of f is concave upward, must the graph of g be concave upward? Explain your reasoning.

84. Consider the function given by $f(x) = x - 2 \ln x$ on the interval $[1, 3]$.
    (a) Explain why Rolle's Theorem does not apply.
    (b) Do you think the conclusion of Rolle's Theorem is true for f? Explain your reasoning.

85. ***Ordering Functions*** Order the functions

$$f(x) = \log_2 x, \ g(x) = x^x, \ h(x) = x^2, \text{ and } k(x) = 2^x$$

from the one with the greatest rate of growth to the one with the smallest rate of growth for "large" values of x.

CAPSTONE

86. Find the derivative of each function, given that a is constant.
    (a) $y = \ln x^a$ (b) $y = (\ln x)^a$ (c) $y = x^a$
    (d) $y = a^x$ (e) $y = x^x$ (f) $y = a^a$
    (g) $y = \log_a x$

87. *Home Mortgage* The term t (in years) of a \$200,000 home mortgage at 7.5% interest can be approximated by

$$t = 13.375 \ln\left(\frac{x}{x - 1250}\right), \quad x > 1250$$

where x is the monthly payment in dollars.

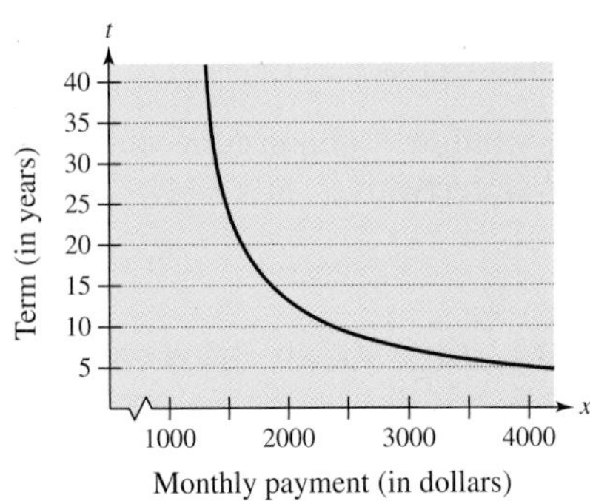

(a) Use a graphing utility to graph the model.

(b) Use the model to approximate the term of a home mortgage for which the monthly payment is \$1398.43. What is the total amount paid?

(c) Use the model to approximate the term of a home mortgage for which the monthly payment is \$1611.19. What is the total amount paid?

(d) Find the instantaneous rates of change of t with respect to x when $x = \$1398.43$ and $x = \$1611.19$.

(e) Write a short paragraph describing the benefit of the higher monthly payment.

88. *Modeling Data* The table shows the temperatures T (in degrees Fahrenheit) at which water boils at selected pressures p (in pounds per square inch). *(Source: Standard Handbook of Mechanical Engineers)*

p	5	10	14.696 (1 atm)	20
T	162.24°	193.21°	212.00°	227.96°

p	30	40	60	80	100
T	250.33°	267.25°	292.71°	312.03°	327.81°

A model that approximates the data is

$$T = 87.97 + 34.96 \ln p + 7.91\sqrt{p}.$$

(a) Use a graphing utility to plot the data and graph the model.

(b) Find the rates of change of T with respect to p when $p = 10$ and $p = 70$.

(c) Use a graphing utility to graph T'. Find $\lim_{p\to\infty} T'(p)$ and interpret the result in the context of the problem.

89. *Modeling Data* The atmospheric pressure decreases with increasing altitude. At sea level, the average air pressure is 1 atmosphere (1.033227 kilograms per square centimeter). The table shows the pressures p (in atmospheres) at various altitudes h (in kilometers).

h	0	5	10	15	20	25
p	1	0.55	0.25	0.12	0.06	0.02

(a) Use a graphing utility to find a model of the form $p = a + b \ln h$ for the data. Explain why the result is an error message.

(b) Use a graphing utility to find the logarithmic model $h = a + b \ln p$ for the data.

(c) Use a graphing utility to plot the data and graph the logarithmic model.

(d) Use the model to estimate the altitude at which the pressure is 0.75 atmosphere.

(e) Use the model to estimate the pressure at an altitude of 13 kilometers.

(f) Use the model to find the rates of change of pressure when $h = 5$ and $h = 20$. Interpret the results in the context of the problem.

90. *Learning Theory* A group of 200 college students was tested every 6 months over a 4-year period. The group was composed of students who took French during the fall semester of their freshman year and did not take subsequent French courses. The average test score p (in percent) is modeled by

$$p = 91.6 - 15.6 \ln(t + 1), \quad 0 \le t \le 48$$

where t is the time in months. At what rate was the average score changing after 1 year?

91. *Minimum Average Cost* The cost of producing x units of a product is given by

$$C = 6000 + 300x + 300x \ln x.$$

Use a graphing utility to find the minimum average cost. Then confirm your result analytically.

92. *Learning Theory* In a group project in learning theory, a mathematical model for the proportion P of correct responses after n trials was found to be

$$P = \frac{0.86}{1 + 0.779^{n}}.$$

(a) Find the limiting proportion of correct responses as n approaches infinity.

(b) Find the rates at which P is changing after $n = 3$ trials and $n = 10$ trials.

93. ***Timber Yield*** The yield V (in millions of cubic feet per acre) for a stand of timber at age t is given by

$$V = 6.7e^{-48.1/t}$$

where t is measured in years.

(a) Find the limiting volume of wood per acre as t approaches infinity.

(b) Find the rates at which the yield is changing when $t = 20$ years and $t = 60$ years.

94. ***Tractrix*** A person walking along a dock drags a boat by a 10-meter rope. The boat travels along a path known as a *tractrix* (see figure). The equation of this path is given by

$$y = 10 \ln\left(\frac{10 + \sqrt{100 - x^2}}{x}\right) - \sqrt{100 - x^2}.$$

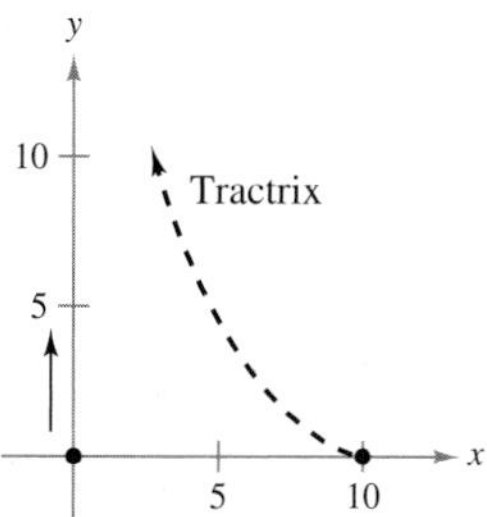

(a) Use a graphing utility to graph the function.

(b) What are the slopes of this path when $x = 5$ and $x = 9$?

(c) What does the slope of the path approach as $x \to 10$?

95. ***Conjecture*** Use a graphing utility to graph f and g in the same viewing window and determine which is increasing at the greater rate for large values of x. What can you conclude about the rate of growth of the natural logarithmic function?

(a) $f(x) = \ln x,\ g(x) = \sqrt{x}$

(b) $f(x) = \ln x,\ g(x) = \sqrt[4]{x}$

96. Prove that the natural logarithmic function is one-to-one.

True or False? **In Exercises 97 and 98, determine whether the statement is true or false. If it is false, explain why or give an example that shows it is false.**

97. $\dfrac{d}{dx}[\ln(x^2 + 5x)] = \dfrac{d}{dx}[\ln x^2] + \dfrac{d}{dx}[\ln(5x)]$

98. If $y = \ln \pi$, then $y' = 1/\pi$.

99. Let $f(x) = \ln x/x$.

(a) Graph f on $(0, \infty)$ and show that f is strictly decreasing on (e, ∞).

(b) Show that if $e \le A < B$, then $A^B > B^A$.

(c) Use part (b) to show that $e^\pi > \pi^e$.

100. To approximate e^x, you can use a function of the form $f(x) = \dfrac{a + bx}{1 + cx}$. (This function is known as a **Padé approximation.**) The values of $f(0)$, $f'(0)$, and $f''(0)$ are equal to the corresponding values of e^x. Show that these values are equal to 1 and find the values of a, b, and c such that $f(0) = f'(0) = f''(0) = 1$. Then use a graphing utility to compare the graphs of f and e^x.

101. Prove that $1/x$ is the derivative of $\ln x$.

SECTION PROJECT

An Alternative Definition of ln *x*

Recall from Section 7.2 that the natural logarithmic function was defined as $f(x) = \log_e x = \ln x, x > 0$. In this project, use the Second Fundamental Theorem of Calculus to define the natural logarithmic function using an integral.

(a) Complete the table below. Use a graphing utility and Simpson's Rule with $n = 10$ to approximate the integral $\int_1^x (1/t)\, dt$.

x	0.5	1.5	2	2.5	3	3.5	4
$\int_1^x (1/t)\, dt$							
$\ln x$							

What can you conclude about the relationship between $\ln x$ and the integral $\int_1^x (1/t)\, dt$?

(b) Use a graphing utility to graph $y = \int_1^x (1/t)\, dt$ for $0 < x \le 4$. Compare the result with the graph of $y = \ln x$. Do the graphs support your conclusion in part (a)?

(c) Use the results of parts (a) and (b) to write an alternative integral definition of the natural logarithmic function. Provide a geometric interpretation of $\ln x$ as an area under a curve.

(d) Use a graphing utility to evaluate each logarithm first by using the natural logarithm key and then by using the graphing utility's integration capabilities to evaluate the integral $y = \int_1^x (1/t)\, dt$.

(i) $\ln 45$ (ii) $\ln 8.3$ (iii) $\ln 0.8$ (iv) $\ln 0.6$

8.3 Logarithmic Functions and Integration

■ **Use the Log Rule for Integration to integrate a rational function.**

Log Rule for Integration

The differentiation rules

$$\frac{d}{dx}[\ln|x|] = \frac{1}{x} \quad \text{and} \quad \frac{d}{dx}[\ln|u|] = \frac{u'}{u}$$

fill the hole in the General Power Rule for Integration. Recall from Section 6.5 that

$$\int u^n\, du = \frac{u^{n+1}}{n+1} + C$$

provided $n \neq -1$. Having the differentiation formulas for logarithmic functions, you are now in a position to evaluate $\int u^n\, du$ for $n = -1$, as stated in the theorem.

THEOREM 8.7 LOG RULE FOR INTEGRATION

Let u be a differentiable function of x.

1. $\displaystyle\int \frac{1}{x}\, dx = \ln|x| + C$ **2.** $\displaystyle\int \frac{1}{u}\, du = \ln|u| + C$

STUDY TIP The alternative form of the Log Rule is useful for integrating by pattern recognition rather than by changing variables. In either case, u is the *denominator* of the rational integrand.

Because $du = u'\, dx$, the second formula can also be written as

$$\int \frac{u'}{u}\, dx = \ln|u| + C. \qquad \text{Alternative form of Log Rule}$$

EXAMPLE 1 Using the Log Rule for Integration

$$\int \frac{2}{x}\, dx = 2\int \frac{1}{x}\, dx = 2\ln|x| + C = \ln x^2 + C$$

Because x^2 cannot be negative, the absolute value notation is unnecessary in the final form of the antiderivative. ■

EXPLORATION

Integrating Rational Functions Each of the following rational functions can be integrated using the Log Rule.

$\dfrac{2}{x}$	Example 1	$\dfrac{1}{2x-1}$	Example 2	$\dfrac{x}{x^2+1}$	Example 3
$\dfrac{3x^2+1}{x^3+x}$	Example 4(a)	$\dfrac{x+1}{x^2+2x}$	Example 4(b)	$\dfrac{1}{3x+2}$	Example 4(c)
$\dfrac{x^2+x+1}{x^2+1}$	Example 5	$\dfrac{2x}{(x+1)^2}$	Example 6		

There are still some rational functions that cannot be integrated using the Log Rule. Give examples of these functions, and explain your reasoning.

EXAMPLE 2 Using the Log Rule with a Change of Variables

Find $\int \frac{1}{2x - 1}\, dx$.

Solution If you let $u = 2x - 1$, then $du = 2\, dx$.

$$\begin{aligned}
\int \frac{1}{2x-1}\, dx &= \frac{1}{2}\int \left(\frac{1}{2x-1}\right) 2\, dx && \text{Multiply and divide by 2.}\\
&= \frac{1}{2}\int \frac{1}{u}\, du && \text{Substitute: } u = 2x - 1.\\
&= \frac{1}{2}\ln|u| + C && \text{Apply Log Rule.}\\
&= \frac{1}{2}\ln|2x - 1| + C && \text{Back-substitute.}
\end{aligned}$$

■

Example 3 uses the alternative form of the Log Rule

$$\int \frac{u'}{u}\, dx = \ln|u| + C.$$

This form of the Log Rule is convenient, especially for simpler integrals. In order to apply this rule, look for quotients in which the numerator is the derivative of the denominator.

EXAMPLE 3 Finding Area with the Log Rule

Find the area of the region bounded by the graph of

$$y = \frac{x}{x^2 + 1}$$

the x-axis, and the line $x = 3$.

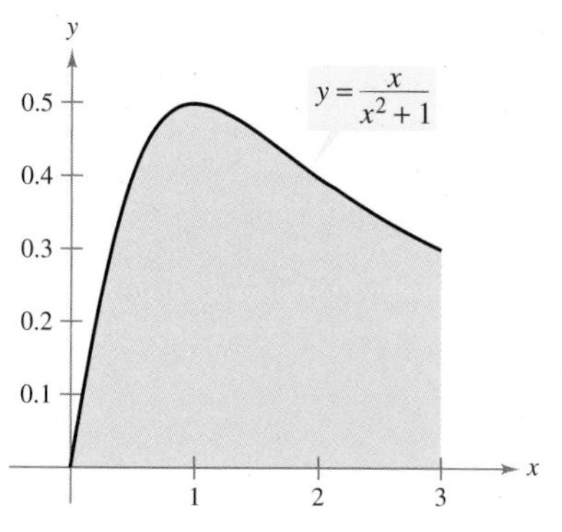

Area $= \int_0^3 \frac{x}{x^2+1}\, dx$

Figure 8.6

Solution From Figure 8.6, you can see that the area of the region is given by the definite integral

$$\int_0^3 \frac{x}{x^2 + 1}\, dx.$$

If you let $u = x^2 + 1$, then $u' = 2x$. To apply the Log Rule, multiply and divide by 2 as shown.

$$\begin{aligned}
\int_0^3 \frac{x}{x^2+1}\, dx &= \frac{1}{2}\int_0^3 \frac{2x}{x^2+1}\, dx && \text{Multiply and divide by 2.}\\
&= \frac{1}{2}\Big[\ln(x^2+1)\Big]_0^3 && \int \frac{u'}{u}\, dx = \ln|u| + C\\
&= \frac{1}{2}(\ln 10 - \ln 1)\\
&= \frac{1}{2}\ln 10 && \ln 1 = 0\\
&\approx 1.151
\end{aligned}$$

The area of the region bounded by the graph of y, the x-axis, and $x = 3$ is $\frac{1}{2}\ln 10$, which is approximately equal to 1.151. ■

EXAMPLE 4 Recognizing Quotient Forms of the Log Rule

a. $\displaystyle\int \frac{3x^2 + 1}{x^3 + x}\,dx = \ln|x^3 + x| + C$ $\qquad u = x^3 + x$

b. $\displaystyle\int \frac{x + 1}{x^2 + 2x}\,dx = \frac{1}{2}\int \frac{2x + 2}{x^2 + 2x}\,dx$ $\qquad u = x^2 + 2x$

$\displaystyle = \frac{1}{2}\ln|x^2 + 2x| + C$

c. $\displaystyle\int \frac{1}{3x + 2}\,dx = \frac{1}{3}\int \frac{3}{3x + 2}\,dx$ $\qquad u = 3x + 2$

$\displaystyle = \frac{1}{3}\ln|3x + 2| + C$ ■

With antiderivatives involving logarithms, it is easy to obtain forms that look quite different but are still equivalent. For instance, both of the following are equivalent to the antiderivative listed in Example 4(c).

$\ln|(3x + 2)^{1/3}| + C$ and $\ln|3x + 2|^{1/3} + C$

Integrals to which the Log Rule can be applied often appear in disguised form. For instance, if a rational function has a *numerator of degree greater than or equal to that of the denominator*, division may reveal a form to which you can apply the Log Rule. This is shown in Example 5.

EXAMPLE 5 Using Long Division Before Integrating

Find $\displaystyle\int \frac{x^2 + x + 1}{x^2 + 1}\,dx.$

Solution Begin by using long division to rewrite the integrand.

$$\frac{x^2 + x + 1}{x^2 + 1} \quad \Rightarrow \quad \begin{array}{r} 1 \\ x^2 + 1\,\overline{)\,x^2 + x + 1} \\ \underline{x^2 \qquad + 1} \\ x \end{array} \quad \Rightarrow \quad 1 + \frac{x}{x^2 + 1}$$

Now, you can integrate to obtain

$$\int \frac{x^2 + x + 1}{x^2 + 1}\,dx = \int \left(1 + \frac{x}{x^2 + 1}\right) dx \qquad \text{Rewrite using long division.}$$

$$= \int dx + \frac{1}{2}\int \frac{2x}{x^2 + 1}\,dx \qquad \text{Rewrite as two integrals.}$$

$$= x + \frac{1}{2}\ln(x^2 + 1) + C. \qquad \text{Integrate.}$$

Check this result by differentiating to obtain the original integrand. ■

TECHNOLOGY If you have access to a computer algebra system, use it to find the indefinite integrals in Examples 5 and 6. How do the forms of the antiderivatives that it gives you compare with those given in Examples 5 and 6?

The next example gives another instance in which the use of the Log Rule is disguised. In this case, a change of variables helps you recognize the Log Rule.

EXAMPLE 6 Change of Variables with the Log Rule

Find $\displaystyle\int \frac{2x}{(x+1)^2}\,dx.$

Solution If you let $u = x + 1$, then $du = dx$ and $x = u - 1$.

$$\begin{aligned}
\int \frac{2x}{(x+1)^2}\,dx &= \int \frac{2(u-1)}{u^2}\,du && \text{Substitute.}\\
&= 2\int \left[\frac{u}{u^2} - \frac{1}{u^2}\right] du && \text{Rewrite as two fractions.}\\
&= 2\int \frac{du}{u} - 2\int u^{-2}\,du && \text{Rewrite as two integrals.}\\
&= 2\ln|u| - 2\left(\frac{u^{-1}}{-1}\right) + C && \text{Integrate.}\\
&= 2\ln|u| + \frac{2}{u} + C && \text{Simplify.}\\
&= 2\ln|x+1| + \frac{2}{x+1} + C && \text{Back-substitute.}
\end{aligned}$$

Check this result by differentiating to obtain the original integrand. ■

As you study the methods shown in Examples 5 and 6, be aware that both methods involve rewriting a disguised integrand so that it fits one or more of the basic integration formulas. In Chapter 11, time will be devoted to integration techniques. To master these techniques, you must recognize the "form-fitting" nature of integration. In this sense, integration is not nearly as straightforward as differentiation. Differentiation takes the form

"Here is the question; what is the answer?"

Integration is more like

"Here is the answer; what is the question?"

You can use the following guidelines for integration.

GUIDELINES FOR INTEGRATION

1. Learn a basic list of integration formulas. At this point our list consists only of the Power Rule, the Exponential Rule, and the Log Rule. By the end of Section 11.5, this list will have expanded to 20 basic rules.
2. Find an integration formula that resembles all or part of the integrand, and, by trial and error, find a choice of u that will make the integrand conform to the formula.
3. If you cannot find a u-substitution that works, try altering the integrand. You might try long division, multiplication and division by the same quantity, or addition and subtraction of the same quantity. Be creative.
4. If you have access to computer software that will find antiderivatives symbolically, use it.

EXAMPLE 7 *u*-Substitution and the Log Rule

Solve the differential equation

$$\frac{dy}{dx} = \frac{1}{x \ln x}.$$

Solution The solution can be written as an indefinite integral.

$$y = \int \frac{1}{x \ln x}\,dx$$

Because the integrand is a quotient whose denominator is raised to the first power, you should try the Log Rule. There are three basic choices for u. The choices $u = x$ and $u = x \ln x$ fail to fit the u'/u form of the Log Rule. However, the third choice does fit. Letting $u = \ln x$ produces $u' = 1/x$, and you obtain

$$\int \frac{1}{x \ln x}\,dx = \int \frac{1/x}{\ln x}\,dx \qquad \text{Divide numerator and denominator by } x.$$

$$= \int \frac{u'}{u}\,dx \qquad \text{Substitute: } u = \ln x.$$

$$= \ln|u| + C \qquad \text{Apply Log Rule.}$$

$$= \ln|\ln x| + C. \qquad \text{Back-substitute.}$$

So, the solution is $y = \ln|\ln x| + C$. ■

Because integration is more difficult than differentiation, keep in mind that you can check your answer to an integration problem by differentiating the answer. For instance, in Example 7, the derivative of

$$y = \ln|\ln x| + C \quad \text{is} \quad y' = \frac{1}{x \ln x}.$$

EXAMPLE 8 *u*-Substitution and the Log Rule

Find $\displaystyle\int \frac{1}{\sqrt{x} + 1}\,dx.$

Solution Because neither the Power Rule, the Exponential Rule, nor the Log Rule applies to the integral as given, consider the substitution $u = \sqrt{x}$. Then

$$u^2 = x \quad \text{and} \quad 2u\,du = dx.$$

Substitution in the original integral yields

$$\int \frac{1}{\sqrt{x} + 1}\,dx = \int \frac{1}{u + 1}(2u\,du)$$

$$= 2\int \frac{u}{u + 1}\,du.$$

Because the degree of the numerator is equal to the degree of the denominator, you can use long division to divide u by $(u + 1)$ to obtain

$$\int \frac{1}{\sqrt{x} + 1}\,dx = 2\int \left(1 - \frac{1}{u + 1}\right) du$$

$$= 2(u - \ln|u + 1|) + C$$

$$= 2\sqrt{x} - 2\ln(\sqrt{x} + 1) + C.$$

■

EXAMPLE 9 An Application

As a volume of gas at pressure p expands from V_0 to V_1, the work done by the gas is

$$W = \int_{V_0}^{V_1} p\, dV.$$

A quantity of gas with an initial volume of 1 cubic foot and a pressure of 500 pounds per square foot expands to a volume of 2 cubic feet. Find the work done by the gas. (Assume the pressure is inversely proportional to the volume.)

Solution Because $p = k/V$ and $p = 500$ when $V = 1$, $k = 500$. So, the work is

$$\begin{aligned} W &= \int_{V_0}^{V_1} \frac{k}{V}\, dV \\ &= \int_1^2 \frac{500}{V}\, dV \\ &= 500\Big[\ln V\Big]_1^2 \\ &= 500(\ln 2 - \ln 1) \approx 346.6 \text{ ft-lb.} \end{aligned}$$

■

Occasionally, an integrand involves an exponential function to a base other than e. When this occurs, there are two options: (1) convert to base e using the formula $a^x = e^{(\ln a)x}$ and then integrate, or (2) integrate directly, using the integration formula

$$\int a^x\, dx = \left(\frac{1}{\ln a}\right)a^x + C$$

(which follows from Theorem 8.5).

NOTE For bases other than e, the integration rule is the same except the antiderivative contains a constant factor $\left(\dfrac{1}{\ln a}\right)$.

EXAMPLE 10 Integrating an Exponential Function to Another Base

Find $\displaystyle\int 2^x\, dx$.

Solution

$$\int 2^x\, dx = \frac{1}{\ln 2}2^x + C \qquad \frac{1}{\ln 2} \text{ is a constant.}$$

■

8.3 Exercises

See www.CalcChat.com for worked-out solutions to odd-numbered exercises.

In Exercises 1–28, find the indefinite integral.

1. $\displaystyle\int \frac{3}{x}\, dx$

2. $\displaystyle\int \frac{10}{x}\, dx$

3. $\displaystyle\int \frac{1}{x+1}\, dx$

4. $\displaystyle\int \frac{1}{x-5}\, dx$

5. $\displaystyle\int \frac{1}{3-2x}\, dx$

6. $\displaystyle\int \frac{1}{3x+2}\, dx$

7. $\displaystyle\int \frac{x}{x^2+1}\, dx$

8. $\displaystyle\int \frac{x^2}{3-x^3}\, dx$

9. $\displaystyle\int \frac{x^2-4}{x}\, dx$

10. $\displaystyle\int \frac{x}{\sqrt{9-x^2}}\, dx$

11. $\displaystyle\int \frac{x^2+2x+3}{x^3+3x^2+9x}\, dx$

12. $\displaystyle\int \frac{x(x+2)}{x^3+3x^2-4}\, dx$

13. $\displaystyle\int \frac{x^2-3x+2}{x+1}\, dx$

14. $\displaystyle\int \frac{2x^2+7x-3}{x-2}\, dx$

15. $\displaystyle\int \frac{x^3 - 3x^2 + 5}{x - 3}\,dx$

16. $\displaystyle\int \frac{x^3 - 6x - 20}{x + 5}\,dx$

17. $\displaystyle\int \frac{x^4 + x - 4}{x^2 + 2}\,dx$

18. $\displaystyle\int \frac{x^3 - 3x^2 + 4x - 9}{x^2 + 3}\,dx$

19. $\displaystyle\int \frac{(\ln x)^2}{x}\,dx$

20. $\displaystyle\int \frac{1}{x\ln(x^3)}\,dx$

21. $\displaystyle\int \frac{e^{-x}}{1 + e^{-x}}\,dx$

22. $\displaystyle\int \frac{e^{2x}}{1 + e^{2x}}\,dx$

23. $\displaystyle\int \frac{e^x + e^{-x}}{e^x - e^{-x}}\,dx$

24. $\displaystyle\int \ln(e^{2x-1})\,dx$

25. $\displaystyle\int \frac{1}{\sqrt{x + 1}}\,dx$

26. $\displaystyle\int \frac{1}{x^{2/3}(1 + x^{1/3})}\,dx$

27. $\displaystyle\int \frac{2x}{(x - 1)^2}\,dx$

28. $\displaystyle\int \frac{x(x - 2)}{(x - 1)^3}\,dx$

In Exercises 29–32, find the indefinite integral by *u*-substitution.

29. $\displaystyle\int \frac{1}{1 + \sqrt{2x}}\,dx$

30. $\displaystyle\int \frac{1}{1 + \sqrt{3x}}\,dx$

31. $\displaystyle\int \frac{\sqrt{x}}{\sqrt{x} - 3}\,dx$

32. $\displaystyle\int \frac{\sqrt[3]{x}}{\sqrt[3]{x} - 1}\,dx$

In Exercises 33–38, find the indefinite integral.

33. $\displaystyle\int 3^x\,dx$

34. $\displaystyle\int 5^{-x}\,dx$

35. $\displaystyle\int x(5^{-x^2})\,dx$

36. $\displaystyle\int (3 - x)7^{(3-x)^2}\,dx$

37. $\displaystyle\int \frac{3^{2x}}{1 + 3^{2x}}\,dx$

38. $\displaystyle\int \frac{2^{-x}}{1 + 2^{-x}}\,dx$

In Exercises 39–46, evaluate the definite integral. Use a graphing utility to verify your result.

39. $\displaystyle\int_{-1}^{2} 2^x\,dx$

40. $\displaystyle\int_{0}^{1} (5^x - 3^x)\,dx$

41. $\displaystyle\int_{0}^{4} \frac{5}{3x + 1}\,dx$

42. $\displaystyle\int_{-1}^{1} \frac{1}{x + 2}\,dx$

43. $\displaystyle\int_{1}^{e} \frac{(1 + \ln x)^2}{x}\,dx$

44. $\displaystyle\int_{e}^{e^2} \frac{1}{x\ln x}\,dx$

45. $\displaystyle\int_{0}^{1} \frac{x^2 - 2}{x + 1}\,dx$

46. $\displaystyle\int_{0}^{1} \frac{x - 1}{x + 1}\,dx$

In Exercises 47 and 48, use a computer algebra system to find the indefinite integral. Graph the integrand.

47. $\displaystyle\int \frac{1}{1 + \sqrt{x}}\,dx$

48. $\displaystyle\int \frac{1 - \sqrt{x}}{1 + \sqrt{x}}\,dx$

In Exercises 49 and 50, solve the differential equation. Use a graphing utility to graph three solutions, one of which passes through the indicated point.

49. $\dfrac{dy}{dx} = \dfrac{3}{2 - x}$, $(1, 0)$

50. $\dfrac{dy}{dx} = \dfrac{2x}{x^2 - 9}$, $(0, 4)$

Slope Fields **In Exercises 51–54, a differential equation, a point, and a slope field are given. (To print an enlarged copy of the graph, go to the website *www.mathgraphs.com*.) (a) Sketch two approximate solutions of the differential equation on the slope field, one of which passes through the indicated point. (b) Use integration to find the particular solution of the differential equation and use a graphing utility to graph the solution. Compare the result with the sketches in part (a).**

51. $\dfrac{dy}{dx} = \dfrac{1}{x + 2}$, $(0, 1)$

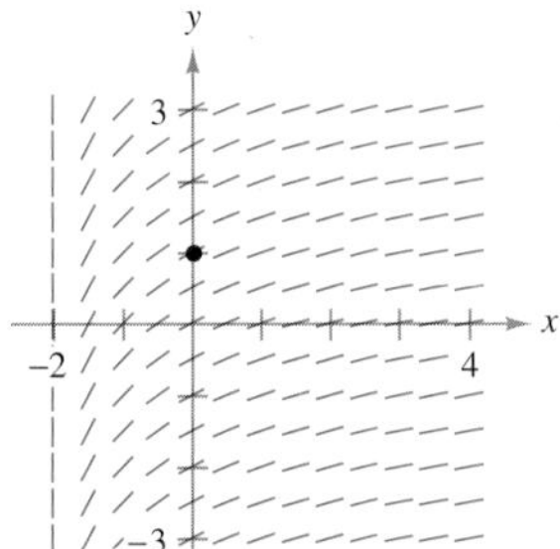

52. $\dfrac{dy}{dx} = \dfrac{\ln x}{x}$, $(1, -2)$

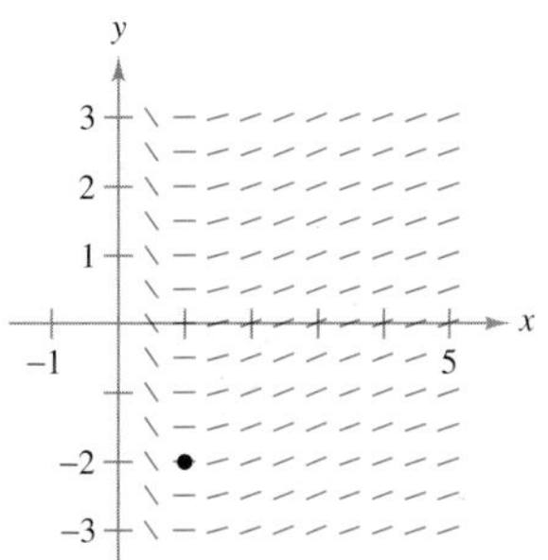

53. $\dfrac{dy}{dx} = 0.4^{x/3}$, $\left(0, \frac{1}{2}\right)$

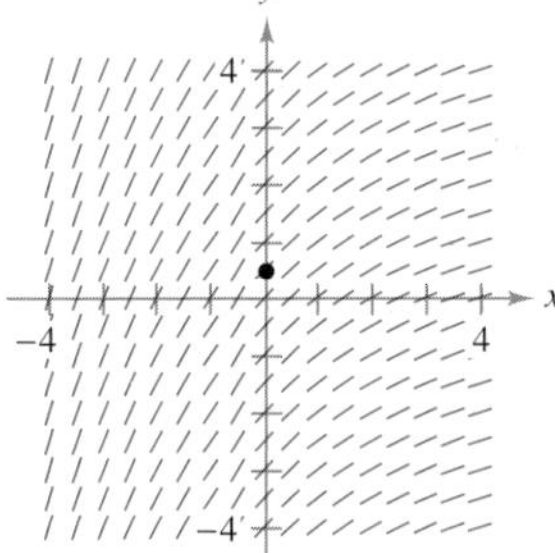

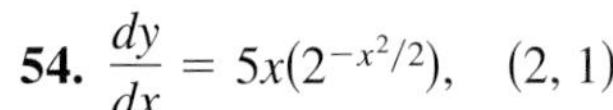

54. $\dfrac{dy}{dx} = 5x(2^{-x^2/2})$, $(2, 1)$

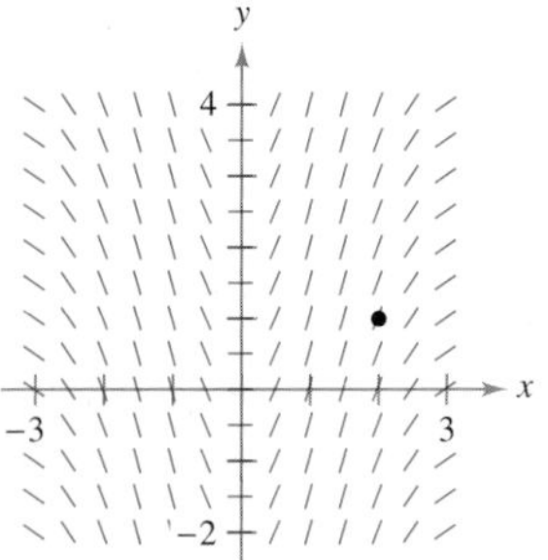

In Exercises 55–58, find $F'(x)$.

55. $F(x) = \displaystyle\int_{1}^{x} \frac{1}{t}\,dt$

56. $F(x) = \displaystyle\int_{0}^{x} \frac{1}{t + 1}\,dt$

57. $F(x) = \displaystyle\int_{x}^{3x} \frac{1}{t}\,dt$

58. $F(x) = \displaystyle\int_{1}^{x^2} \frac{1}{t}\,dt$

Area **In Exercises 59–62, find the area of the region bounded by the graphs of the equations. Use a graphing utility to graph the region and verify your result.**

59. $y = \dfrac{x^2 + 4}{x}$, $x = 1$, $x = 4$, $y = 0$

60. $y = \dfrac{x+6}{x}, x = 1, x = 5, y = 0$

61. $y = 3^x, y = 0, x = 0, x = 3$

62. $y = 2^{-x}, y = 0, x = 0, x = 4$

WRITING ABOUT CONCEPTS

In Exercises 63–66, state the integration formula you would use to perform the integration. (Do not integrate.)

63. $\displaystyle\int \sqrt[3]{x}\,dx$

64. $\displaystyle\int \frac{x}{(x^2+4)^3}\,dx$

65. $\displaystyle\int \frac{x}{x^2+4}\,dx$

66. $\displaystyle\int xe^{-x^2/2}\,dx$

67. What is the first step when integrating

$$\int \frac{x^2}{x+1}\,dx?$$

68. Make a list of the integration formulas studied so far in the course.

69. Find a value of x such that $\displaystyle\int_1^x \frac{3}{t}\,dt = \int_{1/4}^x \frac{1}{t}\,dt.$

CAPSTONE

70. Find a value of x such that

$$\int_1^x \frac{1}{t}\,dt$$

is equal to (a) $\ln 5$ and (b) 1.

In Exercises 71–74, find the average value of the function over the given interval.

71. $f(x) = \dfrac{8}{x^2}, \quad [2, 4]$

72. $f(x) = \dfrac{4(x+1)}{x^2}, \quad [2, 4]$

73. $f(x) = \dfrac{\ln x}{x}, \quad [1, e]$

74. $f(x) = \dfrac{8}{x+2}, \quad [0, 6]$

75. ***Population Growth*** A population of bacteria is changing at a rate of

$$\frac{dP}{dt} = \frac{3000}{1+0.25t}$$

where t is the time (in days). The initial population (when $t = 0$) is 1000. Write an equation that gives the population at any time t, and find the population when $t = 3$ days.

76. ***Heat Transfer*** Find the time required for an object to cool from 300°F to 250°F by evaluating

$$t = \frac{10}{\ln 2}\int_{250}^{300} \frac{1}{T-100}\,dT$$

where t is time (in minutes).

77. ***Average Price*** The demand equation for a product is

$$p = \frac{90{,}000}{400+3x}.$$

Find the *average* price p on the interval $40 \le x \le 50$.

78. ***Sales*** The rate of change in sales S is inversely proportional to time t $(t > 1)$ measured in weeks. Find S as a function of t if sales after 2 and 4 weeks are 200 units and 300 units, respectively.

79. ***Conjecture***

(a) Use a graphing utility to approximate the integrals of the functions given by

$$f(t) = 4\left(\frac{3}{8}\right)^{2t/3}, \quad g(t) = 4\left(\frac{\sqrt[3]{9}}{4}\right)^t,$$

and

$$h(t) = 4e^{-0.653886t}$$

on the interval $[0, 4]$.

(b) Use a graphing utility to graph the three functions.

(c) Use the results in parts (a) and (b) to make a conjecture about the three functions. Could you make the conjecture using only part (a)? Explain your reasoning. Prove your conjecture analytically.

True or False? **In Exercises 80–83, determine whether the statement is true or false. If it is false, explain why or give an example that shows it is false.**

80. $(\ln x)^{1/2} = \frac{1}{2}(\ln x)$

81. $\displaystyle\int \ln x\,dx = (1/x) + C$

82. $\displaystyle\int \frac{1}{x}\,dx = \ln|cx|, \quad c \ne 0$

83. $\displaystyle\int_{-1}^{2} \frac{1}{x}\,dx = \Big[\ln|x|\Big]_{-1}^{2} = \ln 2 - \ln 1 = \ln 2$

84. Graph the function given by

$$f(x) = \frac{x}{1+x^2}$$

on the interval $[0, \infty)$.

(a) Find the area bounded by the graph of f and the line $y = \frac{1}{2}x$.

(b) Determine the values of the slope m such that the line $y = mx$ and the graph of f enclose a finite region.

(c) Calculate the area of this region as a function of m.

85. Prove that the function given by

$$F(x) = \int_x^{2x} \frac{1}{t}\,dt$$

is constant on the interval $(0, \infty)$.

8.4 Differential Equations: Growth and Decay

- **Use separation of variables to solve a simple differential equation.**
- **Use exponential functions to model growth and decay in applied problems.**

Differential Equations

Up to now in the text, you have learned to solve only two types of differential equations—those of the forms

$$y' = f(x) \quad \text{and} \quad y'' = f(x).$$

In this section, you will learn how to solve a more general type of differential equation. The strategy is to rewrite the equation so that each variable occurs on only one side of the equation. This strategy is called *separation of variables.*

EXAMPLE 1 Solving a Differential Equation

$$y' = \frac{2x}{y} \qquad \text{Original equation}$$

$$yy' = 2x \qquad \text{Multiply both sides by } y.$$

$$\int yy'\,dx = \int 2x\,dx \qquad \text{Integrate with respect to } x.$$

$$\int y\,dy = \int 2x\,dx \qquad dy = y'\,dx$$

$$\frac{1}{2}y^2 = x^2 + C_1 \qquad \text{Apply Power Rule.}$$

$$y^2 - 2x^2 = C \qquad \text{Rewrite, letting } C = 2C_1.$$

So, the general solution is given by $y^2 - 2x^2 = C$. ■

STUDY TIP You can use implicit differentiation to check the solution in Example 1.

EXPLORATION

In Example 1, the general solution of the differential equation is

$$y^2 - 2x^2 = C.$$

Use a graphing utility to sketch the particular solutions for $C = \pm 2$, $C = \pm 1$, and $C = 0$. Describe the solutions graphically. Is the following statement true of each solution?

The slope of the graph at the point (x, y) is equal to twice the ratio of x and y.

Explain your reasoning. Are all curves for which this statement is true represented by the general solution?

Notice that when you integrate both sides of the equation in Example 1, you don't need to add a constant of integration to both sides. If you did, you would obtain the same result.

$$\int y\,dy = \int 2x\,dx$$

$$\tfrac{1}{2}y^2 + C_2 = x^2 + C_3$$

$$\tfrac{1}{2}y^2 = x^2 + (C_3 - C_2)$$

$$\tfrac{1}{2}y^2 = x^2 + C_1$$

Some people prefer to use Leibniz notation and differentials when applying separation of variables. The solution of Example 1 is shown below using this notation.

$$\frac{dy}{dx} = \frac{2x}{y}$$

$$y\,dy = 2x\,dx$$

$$\int y\,dy = \int 2x\,dx$$

$$\frac{1}{2}y^2 = x^2 + C_1$$

$$y^2 - 2x^2 = C$$

Growth and Decay Models

In many applications, the rate of change of a variable y is proportional to the value of y. If y is a function of time t, the proportion can be written as follows.

Rate of change of y is proportional to y.

$$\frac{dy}{dt} = ky$$

The general solution of this differential equation is given in the following theorem.

THEOREM 8.8 EXPONENTIAL GROWTH AND DECAY MODEL

If y is a differentiable function of t such that $y > 0$ and $y' = ky$ for some constant k, then

$$y = Ce^{kt}.$$

C is the **initial value** of y, and k is the **proportionality constant. Exponential growth** occurs when $k > 0$, and **exponential decay** occurs when $k < 0$.

PROOF

$$y' = ky \qquad \text{Write original equation.}$$

$$\frac{y'}{y} = k \qquad \text{Separate variables.}$$

$$\int \frac{y'}{y}\,dt = \int k\,dt \qquad \text{Integrate with respect to } t.$$

$$\int \frac{1}{y}\,dy = \int k\,dt \qquad dy = y'\,dt$$

$$\ln y = kt + C_1 \qquad \text{Find antiderivative of each side.}$$

$$y = e^{kt}e^{C_1} \qquad \text{Solve for } y.$$

$$y = Ce^{kt} \qquad \text{Let } C = e^{C_1}.$$

So, all solutions of $y' = ky$ are of the form $y = Ce^{kt}$. Remember that you can differentiate the function $y = Ce^{kt}$ with respect to t to verify that $y' = ky$. ■

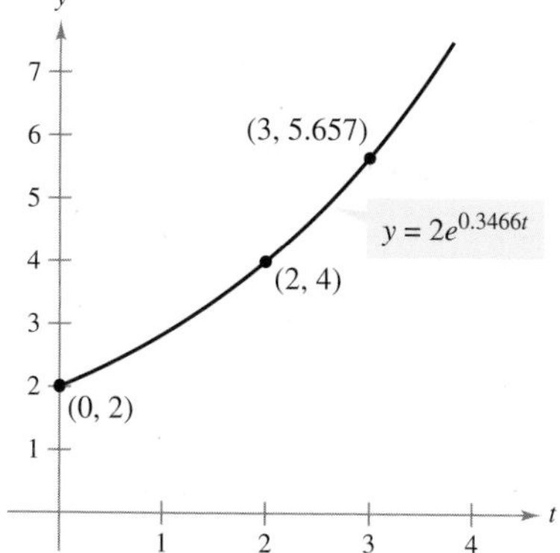

If the rate of change of y is proportional to y, then y follows an exponential model.
Figure 8.7

STUDY TIP Using logarithmic properties, note that the value of k in Example 2 can also be written as $\ln(\sqrt{2})$. So, the model becomes $y = 2e^{(\ln\sqrt{2})t}$, which can then be rewritten as $y = 2(\sqrt{2})^t$.

EXAMPLE 2 Using an Exponential Growth Model

The rate of change of y is proportional to y. When $t = 0$, $y = 2$, and when $t = 2$, $y = 4$. What is the value of y when $t = 3$?

Solution Because $y' = ky$, you know that y and t are related by the equation $y = Ce^{kt}$. You can find the values of the constants C and k by applying the initial conditions.

$$2 = Ce^0 \implies C = 2 \qquad \text{When } t = 0,\ y = 2.$$

$$4 = 2e^{2k} \implies k = \frac{1}{2}\ln 2 \approx 0.3466 \qquad \text{When } t = 2,\ y = 4.$$

So, the model is $y = 2e^{0.3466t}$. When $t = 3$, the value of y is $2e^{0.3466(3)} \approx 5.657$ (see Figure 8.7). ■

TECHNOLOGY Most graphing utilities have curve-fitting capabilities that can be used to find models that represent data. Use the *exponential regression* feature of a graphing utility and the information in Example 2 to find a model for the data. How does your model compare with the given model?

Radioactive decay is measured in terms of *half-life*—the number of years required for half of the atoms in a sample of radioactive material to decay. The rate of decay is proportional to the amount present. The half-lives of some common radioactive isotopes are shown below.

Uranium (^{238}U)	4,470,000,000 years
Plutonium (^{239}Pu)	24,100 years
Carbon (^{14}C)	5715 years
Radium (^{226}Ra)	1599 years
Einsteinium (^{254}Es)	276 days
Nobelium (^{257}No)	25 seconds

EXAMPLE 3 Radioactive Decay

Suppose that 10 grams of the plutonium isotope ^{239}Pu was released in the Chernobyl nuclear accident. How long will it take for the 10 grams to decay to 1 gram?

Solution Let y represent the mass (in grams) of the plutonium. Because the rate of decay is proportional to y, you know that

$$y = Ce^{kt}$$

where t is the time in years. To find the values of the constants C and k, apply the initial conditions. Using the fact that $y = 10$ when $t = 0$, you can write

$$10 = Ce^{k(0)} = Ce^{0}$$

which implies that $C = 10$. Next, using the fact that the half-life of ^{239}Pu is 24,100 years, you have $y = 10/2 = 5$ when $t = 24{,}100$, so you can write

$$5 = 10e^{k(24{,}100)}$$
$$\frac{1}{2} = e^{24{,}100k}$$
$$\ln\frac{1}{2} = \ln(e^{24{,}100k})$$
$$\ln\frac{1}{2} = 24{,}100k$$
$$\frac{1}{24{,}100}\ln\frac{1}{2} = k$$
$$-0.000028761 \approx k.$$

So, the model is

$$y = 10e^{-0.000028761t}. \qquad \text{Half-life model}$$

To find the time it would take for 10 grams to decay to 1 gram, you can solve for t in the equation

$$1 = 10e^{-0.000028761t}.$$

The solution is approximately 80,059 years. ■

LAZARENKO NIKOLAI/ITAR-TASS /Landov

NOTE The exponential decay model in Example 3 could also be written as $y = 10\left(\frac{1}{2}\right)^{t/24{,}100}$. This model is much easier to derive, but for some applications it is not as convenient to use.

From Example 3, notice that in an exponential growth or decay problem, it is easy to solve for C when you are given the value of y at $t = 0$. To determine an exponential model when the values of y are known for two nonzero values of t, review Example 2 in Section 7.5.

In Examples 2 and 3, you did not actually have to solve the differential equation

$$y' = ky.$$

(This was done once in the proof of Theorem 8.8.) The next example demonstrates a problem whose solution involves the separation of variables technique. The example concerns **Newton's Law of Cooling,** which states that the rate of change in the temperature of an object is proportional to the difference between the object's temperature and the temperature of the surrounding medium.

EXAMPLE 4 Newton's Law of Cooling

Let y represent the temperature (in °F) of an object in a room whose temperature is kept at a constant 60°. If the object cools from 100° to 90° in 10 minutes, how much longer will it take for its temperature to decrease to 80°?

Solution From Newton's Law of Cooling, you know that the rate of change in y is proportional to the difference between y and 60. This can be written as

$$y' = k(y - 60), \quad 80 \le y \le 100.$$

To solve this differential equation, use separation of variables, as follows.

$$\frac{dy}{dt} = k(y - 60) \qquad \text{Differential equation}$$

$$\left(\frac{1}{y - 60}\right) dy = k\,dt \qquad \text{Separate variables.}$$

$$\int \frac{1}{y - 60}\,dy = \int k\,dt \qquad \text{Integrate each side.}$$

$$\ln|y - 60| = kt + C_1 \qquad \text{Find antiderivative of each side.}$$

Because $y > 60$, $|y - 60| = y - 60$, and you can omit the absolute value signs. Using exponential notation, you have

$$y - 60 = e^{kt + C_1} \quad \Longrightarrow \quad y = 60 + Ce^{kt}. \qquad C = e^{C_1}$$

Using $y = 100$ when $t = 0$, you obtain $100 = 60 + Ce^{k(0)} = 60 + C$, which implies that $C = 40$. Because $y = 90$ when $t = 10$,

$$90 = 60 + 40e^{k(10)}$$

$$30 = 40e^{10k}$$

$$k = \tfrac{1}{10} \ln \tfrac{3}{4} \approx -0.02877.$$

So, the model is

$$y = 60 + 40e^{-0.02877t} \qquad \text{Cooling model}$$

and finally, when $y = 80$, you obtain

$$80 = 60 + 40e^{-0.02877t}$$

$$20 = 40e^{-0.02877t}$$

$$\tfrac{1}{2} = e^{-0.02877t}$$

$$\ln \tfrac{1}{2} = -0.02877t$$

$$t \approx 24.09 \text{ minutes.}$$

So, it will require about 14.09 *more* minutes for the object to cool to a temperature of 80° (see Figure 8.8). ■

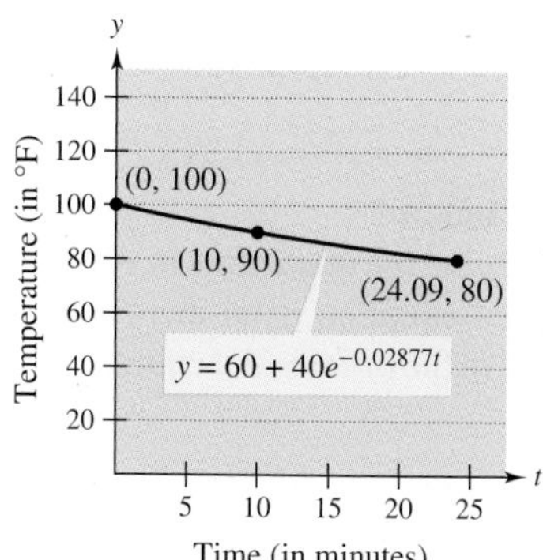

Figure 8.8

8.4 Exercises

See www.CalcChat.com for worked-out solutions to odd-numbered exercises.

In Exercises 1–10, solve the differential equation.

1. $\frac{dy}{dx} = x + 3$
2. $\frac{dy}{dx} = 6 - x$
3. $\frac{dy}{dx} = y + 3$
4. $\frac{dy}{dx} = 6 - y$
5. $y' = \frac{5x}{y}$
6. $y' = \frac{\sqrt{x}}{7y}$
7. $y' = \sqrt{x}\,y$
8. $y' = x(1 + y)$
9. $(1 + x^2)y' - 2xy = 0$
10. $xy + y' = 100x$

In Exercises 11–14, write and solve the differential equation that models the verbal statement.

11. The rate of change of Q with respect to t is inversely proportional to the square of t.
12. The rate of change of P with respect to t is proportional to $25 - t$.
13. The rate of change of N with respect to s is proportional to $500 - s$.
14. The rate of change of y with respect to x varies jointly as x and $L - y$.

Slope Fields **In Exercises 15 and 16, a differential equation, a point, and a slope field are given. (a) Sketch two approximate solutions of the differential equation on the slope field, one of which passes through the given point. (b) Use integration to find the particular solution of the differential equation and use a graphing utility to graph the solution. Compare the result with the sketch in part (a). To print an enlarged copy of the graph, go to the website *www.mathgraphs.com.***

15. $\frac{dy}{dx} = x(6 - y)$, $(0, 0)$
16. $\frac{dy}{dx} = xy$, $\left(0, \frac{1}{2}\right)$

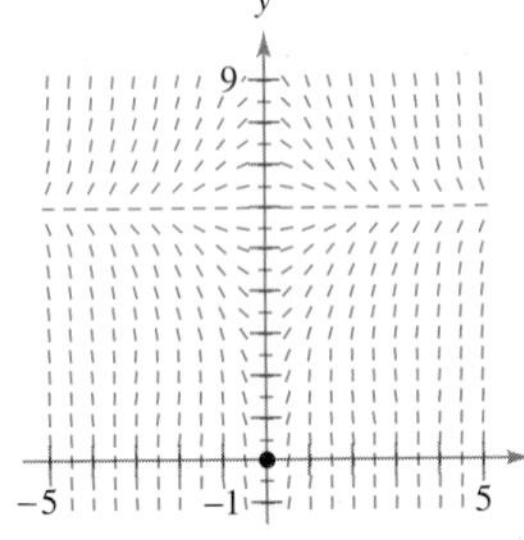

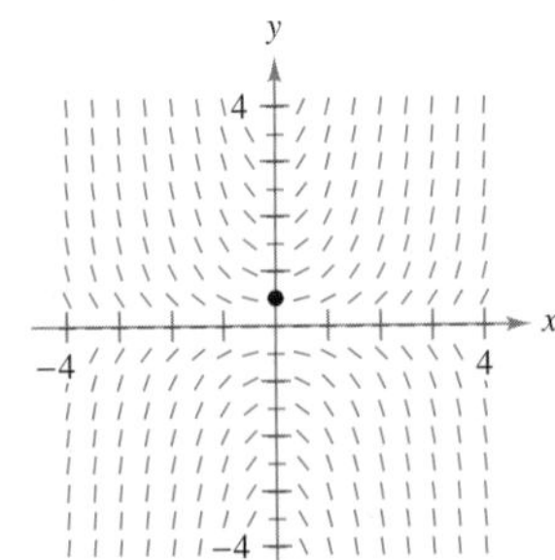

In Exercises 17–20, find the function $y = f(t)$ passing through the point $(0, 10)$ with the given first derivative. Use a graphing utility to graph the solution.

17. $\frac{dy}{dt} = \frac{1}{2}t$
18. $\frac{dy}{dt} = -\frac{3}{4}\sqrt{t}$
19. $\frac{dy}{dt} = -\frac{1}{2}y$
20. $\frac{dy}{dt} = \frac{3}{4}y$

In Exercises 21–24, write and solve the differential equation that models the verbal statement. Evaluate the solution at the specified value of the independent variable.

21. The rate of change of y is proportional to y. When $x = 0$, $y = 6$, and when $x = 4$, $y = 15$. What is the value of y when $x = 8$?
22. The rate of change of N is proportional to N. When $t = 0$, $N = 250$, and when $t = 1$, $N = 400$. What is the value of N when $t = 4$?
23. The rate of change of V is proportional to V. When $t = 0$, $V = 20{,}000$, and when $t = 4$, $V = 12{,}500$. What is the value of V when $t = 6$?
24. The rate of change of P is proportional to P. When $t = 0$, $P = 5000$, and when $t = 1$, $P = 4750$. What is the value of P when $t = 5$?

In Exercises 25–28, find the exponential function $y = Ce^{kt}$ that passes through the two given points.

25.

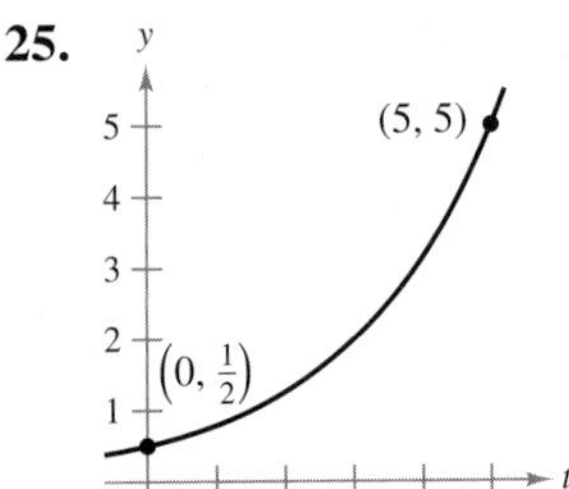

26.

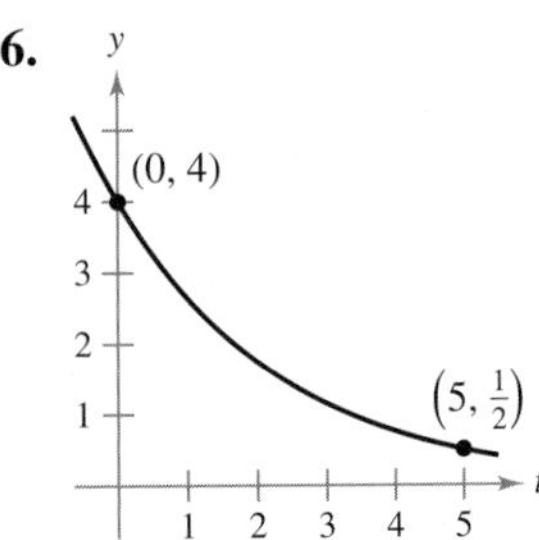

27.

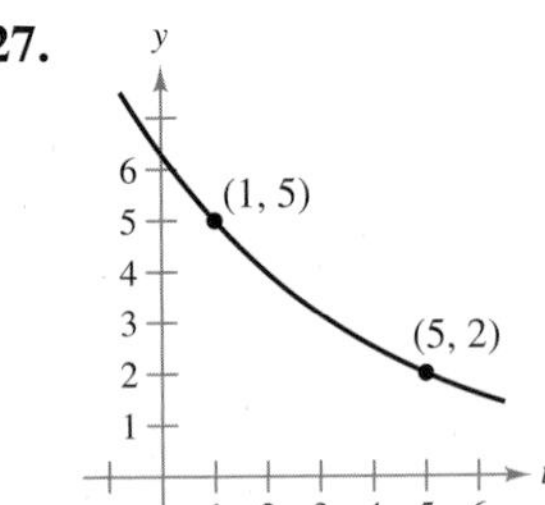

28.

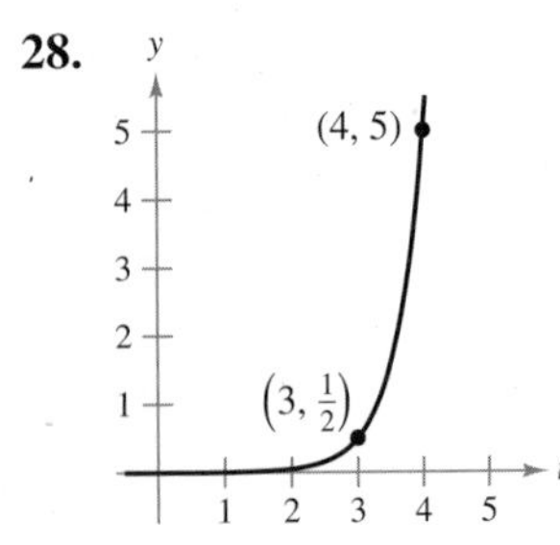

WRITING ABOUT CONCEPTS

In Exercises 29 and 30, determine the quadrants in which the solution of the differential equation is an increasing function. Explain. (Do not solve the differential equation.)

29. $\frac{dy}{dx} = \frac{1}{2}xy$
30. $\frac{dy}{dx} = \frac{1}{2}x^2y$

Radioactive Decay **In Exercises 31–38, complete the table for the radioactive isotope.**

Isotope	Half-life (in years)	Initial Quantity	Amount After 1000 Years	Amount After 10,000 Years
31. ^{226}Ra	1599	20 g		
32. ^{226}Ra	1599		1.5 g	
33. ^{226}Ra	1599			0.1 g
34. ^{14}C	5715			3 g
35. ^{14}C	5715	5 g		
36. ^{14}C	5715		1.6 g	
37. ^{239}Pu	24,100		2.1 g	
38. ^{239}Pu	24,100			0.4 g

39. ***Radioactive Decay*** Radioactive radium has a half-life of approximately 1599 years. What percent of a given amount remains after 100 years?

40. ***Carbon Dating*** Carbon-14 dating assumes that the carbon dioxide on Earth today has the same radioactive content as it did centuries ago. If this is true, the amount of ^{14}C absorbed by a tree that grew several centuries ago should be the same as the amount of ^{14}C absorbed by a tree growing today. A piece of ancient charcoal contains only 15% as much of the radioactive carbon as a piece of modern charcoal. How long ago was the tree burned to make the ancient charcoal? (The half-life of ^{14}C is 5715 years.)

Population **In Exercises 41–45, the population (in millions) of a country in 2007 and the expected continuous annual rate of change k of the population are given.** ***(Source: U.S. Census Bureau, International Data Base)***

(a) Find the exponential growth model $P = Ce^{kt}$ for the population by letting $t = 0$ correspond to 2000.

(b) Use the model to predict the population of the country in 2015.

(c) Discuss the relationship between the sign of k and the change in population for the country.

Country	2007 Population	k
41. Latvia	2.3	−0.006
42. Egypt	80.3	0.017
43. Paraguay	6.7	0.024
44. Hungary	10.0	−0.003
45. Uganda	30.3	0.036

CAPSTONE

46. (a) Suppose an insect population increases by a constant number each month. Explain why the number of insects can be represented by a linear function.

(b) Suppose an insect population increases by a constant percentage each month. Explain why the number of insects can be represented by an exponential function.

47. ***Modeling Data*** One hundred bacteria are started in a culture and the number N of bacteria is counted each hour for 5 hours. The results are shown in the table, where t is the time in hours.

t	0	1	2	3	4	5
N	100	126	151	198	243	297

(a) Use the regression capabilities of a graphing utility to find an exponential model for the data.

(b) Use the model to estimate the time required for the population to quadruple in size.

48. ***Bacteria Growth*** The number of bacteria in a culture is increasing according to the law of exponential growth. There are 125 bacteria in the culture after 2 hours and 350 bacteria after 4 hours.

(a) Find the initial population.

(b) Write an exponential growth model for the bacteria population. Let t represent time in hours.

(c) Use the model to determine the number of bacteria after 8 hours.

(d) After how many hours will the bacteria count be 25,000?

49. ***Learning Curve*** The management at a certain factory has found that a worker can produce at most 30 units in a day. The learning curve for the number of units N produced per day after a new employee has worked t days is $N = 30(1 - e^{kt})$. After 20 days on the job, a particular worker produces 19 units.

(a) Find the learning curve for this worker.

(b) How many days should pass before this worker is producing 25 units per day?

50. ***Learning Curve*** If the management in Exercise 49 requires a new employee to produce at least 20 units per day after 30 days on the job, find (a) the learning curve that describes this minimum requirement and (b) the number of days before a minimal achiever is producing 25 units per day.

51. ***Modeling Data*** The table shows the populations P (in millions) of the United States from 1960 to 2000. *(Source: U.S. Census Bureau)*

Year	1960	1970	1980	1990	2000
Population, P	181	205	228	250	282

(a) Use the 1960 and 1970 data to find an exponential model P_1 for the data. Let $t = 0$ represent 1960.

(b) Use a graphing utility to find an exponential model P_2 for all the data. Let $t = 0$ represent 1960.

(c) Use a graphing utility to plot the data and graph models P_1 and P_2 in the same viewing window. Compare the actual data with the predictions. Which model better fits the data?

(d) Estimate when the population will be 320 million.

52. ***Modeling Data*** The table shows the net receipts and the amounts required to service the national debt (interest on Treasury debt securities) of the United States from 2001 through 2010. The years 2007 through 2010 are estimated, and the monetary amounts are given in billions of dollars. *(Source: U.S. Office of Management and Budget)*

Year	2001	2002	2003	2004	2005
Receipts	1991.4	1853.4	1782.5	1880.3	2153.9
Interest	359.5	332.5	318.1	321.7	352.3

Year	2006	2007	2008	2009	2010
Receipts	2407.3	2540.1	2662.5	2798.3	2954.7
Interest	405.9	433.0	469.9	498.0	523.2

(a) Use the regression capabilities of a graphing utility to find an exponential model R for the receipts and a quartic model I for the amount required to service the debt. Let t represent the time in years, with $t = 1$ corresponding to 2001.

(b) Use a graphing utility to plot the points corresponding to the receipts, and graph the exponential model. Based on the model, what is the continuous rate of growth of the receipts?

(c) Use a graphing utility to plot the points corresponding to the amounts required to service the debt, and graph the quartic model.

(d) Find a function $P(t)$ that approximates the percent of the receipts that is required to service the national debt. Use a graphing utility to graph this function.

53. ***Sound Intensity*** The level of sound β (in decibels) with an intensity of I is $\beta(I) = 10 \log_{10} (I/I_0)$, where I_0 is an intensity of 10^{-16} watt per square centimeter, corresponding roughly to the faintest sound that can be heard. Determine $\beta(I)$ for the following.

(a) $I = 10^{-14}$ watt per square centimeter (whisper)

(b) $I = 10^{-9}$ watt per square centimeter (busy street corner)

(c) $I = 10^{-6.5}$ watt per square centimeter (air hammer)

(d) $I = 10^{-4}$ watt per square centimeter (threshold of pain)

54. ***Earthquake Intensity*** On the Richter scale, the magnitude R of an earthquake of intensity I is

$$R = \frac{\ln I - \ln I_0}{\ln 10}$$

where I_0 is the minimum intensity used for comparison. Assume that $I_0 = 1$.

(a) Find the intensity of the 1906 San Francisco earthquake ($R = 8.3$).

(b) Find the factor by which the intensity is increased if the Richter scale measurement is doubled.

(c) Find dR/dI.

55. ***Forestry*** The value of a tract of timber is

$$V(t) = 100{,}000e^{0.8\sqrt{t}}$$

where t is the time in years, with $t = 0$ corresponding to 2008. If money earns interest continuously at 10%, the present value of the timber at any time t is $A(t) = V(t)e^{-0.10t}$. Find the year in which the timber should be harvested to maximize the present value function.

56. ***Newton's Law of Cooling*** A container of hot liquid is placed in a freezer that is kept at a constant temperature of 20°F. The initial temperature of the liquid is 160°F. After 5 minutes, the liquid's temperature is 60°F. How much longer will it take for its temperature to decrease to 30°F?

True or False? **In Exercises 57–60, determine whether the statement is true or false. If it is false, explain why or give an example that shows it is false.**

57. In exponential growth, the rate of growth is constant.

58. In linear growth, the rate of growth is constant.

59. If prices are rising at a rate of 0.5% per month, then they are rising at a rate of 6% per year.

60. The differential equation modeling exponential growth is $dy/dx = ky$, where k is a constant.

8 CHAPTER SUMMARY

	Review Exercises
Section 8.1	
■ Differentiate natural exponential functions *(p. 528).*	*1–14*
■ Integrate natural exponential functions *(p. 530).*	*15–28*
Section 8.2	
■ Differentiate natural logarithmic functions *(p. 535).*	*29–36, 51–56*
■ Use logarithmic differentiation to differentiate nonlogarithmic functions *(p. 537).*	*47–50*
■ Differentiate exponential and logarithmic functions for bases other than e *(p. 539).*	*37–46, 57, 58*
Section 8.3	
■ Use the Log Rule for Integration to integrate a rational function *(p. 544).*	*59–70, 75–78*
■ Integrate exponential functions for bases other than e *(p. 549).*	*71–74*
Section 8.4	
■ Use separation of variables to solve simple differential equations *(p. 552).*	*79–82*
■ Use exponential functions to model growth and decay *(p. 553).*	*83–90*

8 REVIEW EXERCISES

See www.CalcChat.com for worked-out solutions to odd-numbered exercises.

In Exercises 1–8, find the derivative of the function.

1. $f(x) = e^{-x^3}$
2. $h(z) = e^{-z^2/2}$
3. $g(t) = t^2e^t$
4. $g(x) = xe^x$
5. $y = \sqrt{e^{2x} + e^{-2x}}$
6. $y = 3e^{-3/t}$
7. $g(x) = x^2/e^x$
8. $f(t) = e^t/(1 + e^t)$

In Exercises 9 and 10, find an equation of the tangent line to the graph of the function at the given point.

9. $f(x) = e^{4-x}$, $(4, 1)$
10. $f(x) = e^{x^2-3x}$, $(3, 1)$

In Exercises 11 and 12, use implicit differentiation to find dy/dx.

11. $e^x + y^2 = 0$
12. $x + y = xe^y$

13. Show that $y = 5e^{2x} - 12e^{3x}$ satisfies the differential equation $y'' - 5y' + 6y = 0$.

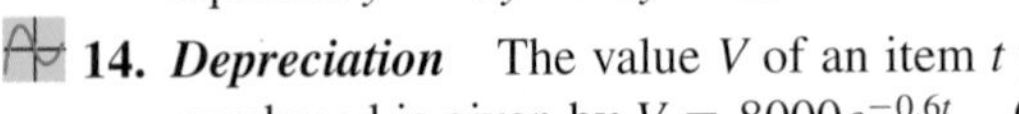

14. ***Depreciation*** The value V of an item t years after it is purchased is given by $V = 8000e^{-0.6t}$, $0 \le t \le 5$.
 (a) Use a graphing utility to graph the function.
 (b) Find the rates of change of V with respect to t when $t = 1$ and $t = 4$.
 (c) Use a graphing utility to sketch the tangent lines to the function when $t = 1$ and $t = 4$.

In Exercises 15–20, find the indefinite integral.

15. $\int xe^{1-x^2}\,dx$
16. $\int x^2e^{x^3+1}\,dx$
17. $\int \frac{e^{4x} - e^{2x} + 1}{e^x}\,dx$
18. $\int \frac{e^{2x} - e^{-2x}}{(e^{2x} + e^{-2x})^2}\,dx$
19. $\int e^{4x}(2 + 5e^{4x})^3\,dx$
20. $\int \frac{e^{2/x^3}}{x^4}\,dx$

In Exercises 21–26, evaluate the definite integral. Use a graphing utility to verify your result.

21. $\int_0^7 e^{7-x}\,dx$
22. $\int_0^1 xe^{-3x^2}\,dx$
23. $\int_{1/2}^2 \frac{e^{1/x}}{x^2}\,dx$
24. $\int_0^2 \frac{e^{2x}}{\sqrt{e^{2x} + 1}}\,dx$
25. $\int_1^3 \frac{e^x}{(e^x - 1)^{3/2}}\,dx$
26. $\int_0^{1/2} (e^{2x} + e^{-x})^2\,dx$

In Exercises 27 and 28, find the area of the region bounded by the graphs of the equations.

27. $y = xe^{-x^2}$, $y = 0$, $x = 0$, $x = 4$
28. $y = 2e^{-x}$, $y = 0$, $x = 0$, $x = 2$

In Exercises 29–36, find the derivative of the function.

29. $f(x) = \ln(5x)$
30. $y = \ln(2x^2 - 3)$
31. $g(x) = \ln\sqrt{x}$
32. $f(x) = \ln\sqrt{x^3 + 6x}$
33. $f(x) = x\sqrt{\ln x}$
34. $y = e^{2x}\ln x$
35. $h(x) = \ln\frac{x(x - 1)}{x - 2}$
36. $f(x) = \ln[x(x^2 - 2)^{2/3}]$

In Exercises 37–46, find the derivative of the function.

37. $f(x) = 3^{x-1}$
38. $y = 5^{3x}$
39. $f(x) = x^3\,3^x$
40. $f(x) = (4e)^x$
41. $y = x^{2x+1}$
42. $y = x(4^{-x})$
43. $f(x) = \log_5 x$
44. $y = \log_{10} 6x$
45. $g(x) = \log_3\sqrt{1 - x}$
46. $h(x) = \log_5\frac{x}{x - 1}$

In Exercises 47–50, find dy/dx using logarithmic differentiation.

47. $y = \sqrt{\frac{6x}{x^2 + 1}}$
48. $y = \sqrt{x(x + 4)(x - 3)}$
49. $y = x\sqrt{(x + 1)(x + 2)}$
50. $y = x^{3x}$

In Exercises 51 and 52, use implicit differentiation to find dy/dx.

51. $\ln x + y^2 = 0$
52. $x\ln y - 3xy = 4$

53. Show that $y\ln|1 - x| = 1$ satisfies the differential equation $\frac{dy}{dx} = \frac{y^2}{1 - x}$.

54. Show that $y = 5\ln x + 6$ satisfies the differential equation $xy'' + y' = 0$.

In Exercises 55 and 56, find any relative extrema and inflection points. Use a graphing utility to confirm your results.

55. $y = \frac{x^3}{3} - \ln x$
56. $y = \ln x - x$

57. ***Inflation*** If the annual rate of inflation averages 5% over the next 10 years, the approximate cost C of goods or services during any year in that decade is

$$C(t) = P(1.05)^t$$

where t is the time in years and P is the present cost.

(a) The price of an oil change for your car is presently \$24.95. Estimate the price 10 years from now.

(b) Find the rates of change of C with respect to t when $t = 1$ and $t = 8$.

(c) Verify that the rate of change of C is proportional to C. What is the constant of proportionality?

58. *Depreciation* After t years, the value of a car purchased for \$25,000 is

$$V(t) = 25{,}000\left(\tfrac{3}{4}\right)^t.$$

(a) Use a graphing utility to graph the function and determine the value of the car 2 years after it was purchased.

(b) Find the rates of change of V with respect to t when $t = 1$ and $t = 4$.

(c) Use a graphing utility to graph $V'(t)$ and determine the horizontal asymptote of $V'(t)$. Interpret its meaning in the context of the problem.

In Exercises 59–64, find the integral.

59. $\displaystyle\int \frac{1}{7x - 2}\,dx$ **60.** $\displaystyle\int \frac{x}{x^2 - 1}\,dx$

61. $\displaystyle\int \frac{x^2 + 4x + 5}{x - 3}\,dx$ **62.** $\displaystyle\int \frac{\ln \sqrt{x}}{x}\,dx$

63. $\displaystyle\int \frac{e^{2x} - e^{-2x}}{e^{2x} + e^{-2x}}\,dx$ **64.** $\displaystyle\int \frac{e^{2x}}{e^{2x} + 1}\,dx$

In Exercises 65–70, evaluate the definite integral. Use a graphing utility to verify your result.

65. $\displaystyle\int_1^5 \frac{7}{x}\,dx$ **66.** $\displaystyle\int_2^4 \frac{1}{4x + 1}\,dx$

67. $\displaystyle\int_1^4 \frac{x + 1}{x}\,dx$ **68.** $\displaystyle\int_0^6 \frac{x^3 - 21x + 30}{x + 5}\,dx$

69. $\displaystyle\int_1^e \frac{\ln x}{x}\,dx$ **70.** $\displaystyle\int_1^2 \frac{e^x}{e^x - 1}\,dx$

In Exercises 71–74, find the indefinite integral.

71. $\displaystyle\int 4^x\,dx$ **72.** $\displaystyle\int 8^{-x}\,dx$

73. $\displaystyle\int (x + 1)5^{(x+1)^2}\,dx$ **74.** $\displaystyle\int \frac{2^{-1/t}}{t^2}\,dt$

In Exercises 75 and 76, solve the differential equation. Use a graphing utility to graph three solutions, one of which passes through the indicated point.

75. $\dfrac{dy}{dx} = \dfrac{2}{5 - x}$, $(4, -2)$ **76.** $\dfrac{dy}{dx} = \dfrac{2x}{x^2 - 4}$, $(0, 3)$

77. *Probability* Two numbers between 0 and 10 are chosen at random. The probability that their product is less than n $(0 < n < 100)$ is

$$P = \frac{1}{100}\left(n + \int_{n/10}^{10} \frac{n}{x}\,dx\right).$$

(a) What is the probability that the product is less than 25?

(b) What is the probability that the product is less than 50?

78. Find the average value of the function given by $f(x) = 1/(x - 1)$ over the interval $[5, 10]$.

In Exercises 79–82, solve the differential equation.

79. $\dfrac{dy}{dx} = 8 - x$ **80.** $\dfrac{dy}{dx} = y + 8$

81. $\dfrac{dy}{dx} = \dfrac{x^2 + 3}{x}$ **82.** $\dfrac{dy}{dx} = \dfrac{e^{-2x}}{1 + e^{-2x}}$

In Exercises 83–86, find the exponential function $y = Ce^{kt}$ that passes through the two points.

83. $\left(0, \frac{3}{4}\right)$, $(5, 5)$ **84.** $\left(2, \frac{3}{2}\right)$, $(4, 5)$

85. $(0, 5)$, $\left(5, \frac{1}{6}\right)$ **86.** $(1, 9)$, $(6, 2)$

87. *Air Pressure* Under ideal conditions, air pressure decreases continuously with height above sea level at a rate proportional to the pressure at that height. If the barometer reads 30 inches at sea level and 15 inches at 18,000 feet, find the barometric pressure at 35,000 feet.

88. *Radioactive Decay* Radioactive radium has a half-life of approximately 1599 years. If the initial quantity is 5 grams, how much remains after 600 years?

89. *Population Growth* A population grows exponentially with a proportionality constant of 1.5%. How long will it take the population to double?

90. *Fuel Economy* An automobile gets 28 miles per gallon of gasoline for speeds up to 50 miles per hour. Over 50 miles per hour, the number of miles per gallon drops at the rate of 12 percent for each 10 miles per hour.

(a) If s is the speed and y is the number of miles per gallon, find y as a function of s by solving the differential equation

$$\frac{dy}{ds} = -0.012y, \quad s > 50.$$

(b) Use the function in part (a) to complete the table.

Speed	50	55	60	65	70
Miles per Gallon					

8 CHAPTER TEST

Take this test as you would take a test in class. When you are finished, check your work against the answers given in the back of the book.

In Exercises 1–6, find the derivative of the function.

1. $g(x) = e^{-5x}$ **2.** $f(x) = e^{4-x}$ **3.** $y = (x^2 + 4x)e^{6x}$

4. $y = \ln(4x^3 - 1)$ **5.** $y = 5^{-4x}$ **6.** $g(x) = \log_5(4 - x)^2$

7. Find an equation of the tangent line to the graph of $f(x) = e^{x+3} + 3$ at the point $(-3, 4)$.

8. Find the second derivative of $f(x) = (3x - 1)e^{-2x}$.

In Exercises 9–11, find the indefinite integral.

9. $\int e^{2x-1} dx$ **10.** $\int \frac{x^2 + 1}{x} dx$ **11.** $\int x4^{3x^2} dx$

In Exercises 12–14, evaluate the definite integral.

12. $\int_2^4 (e^x + 2)^2 dx$ **13.** $\int_3^5 \frac{1}{3x - 8} dx$ **14.** $\int_0^1 9^x dx$

15. Find the area of the region bounded by the graphs of $y = xe^{2x^2} + 3$, $y = 0$, $x = 0$, and $x = 1$. Use a graphing utility to graph the region and verify your result.

16. Find any relative extrema and inflection points of $y = x^2 \ln x$. Use a graphing utility to confirm your result.

In Exercises 17–19, find dy/dx using logarithmic differentiation.

17. $y = x\sqrt{2x^2 + 7x}$ **18.** $y = \frac{(x - 5)(x - 8)}{(x + 5)(x + 8)}$ **19.** $y = x^{x^2}$

20. Solve $\frac{dy}{dx} = \frac{2}{x + 4}$. Use a graphing utility to graph three solutions, one of which passes through the point $(-3, 2)$.

21. Find the average value of the function given by

$$f(x) = \frac{6}{x + 1}$$

over the interval $[0, 3]$.

In Exercises 22–24, solve the differential equation.

22. $\frac{dy}{dx} = 2x + 6$ **23.** $\frac{dy}{dx} = \frac{x}{4y}$ **24.** $\frac{dy}{dx} = x^2y$

In Exercises 25–27, find the exponential function $y = Ce^{kt}$ that passes through the two points.

25. $(0, 6), (3, 1)$ **26.** $\left(0, \frac{3}{2}\right), (4, 3)$ **27.** $(1, 1), (5, 5)$

28. Radioactive carbon has a half-life of approximately 5715 years. If the initial quantity is 20 grams, how much remains after 1000 years?

P.S. PROBLEM SOLVING

1. The tangent line to the curve $y = e^{-x}$ at the point $P(a, b)$ intersects the x- and y-axes at the points Q and R, as indicated in the figure. Find the coordinates of the point P that yield the maximum area of $\triangle OQR$. What is the maximum area?

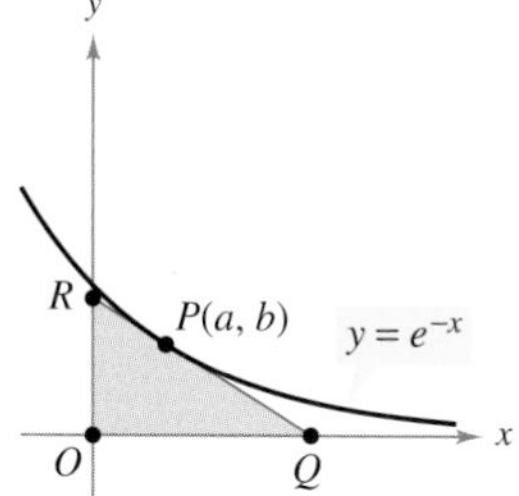

2. (a) Find the polynomial $P_1(x) = a_0 + a_1x$ whose value and slope agree with the value and slope of $f(x) = \ln(1 + x)$ at the point $(0, 0)$.

   (b) Find the polynomial $P_2(x) = a_0 + a_1x + a_2x^2$ whose value and first two derivatives agree with the value and first two derivatives of $f(x) = \ln(1 + x)$ at the point $(0, 0)$. This polynomial is called the second-degree **Taylor polynomial** of $f(x) = \ln(1 + x)$ at $x = 0$.

   (c) Complete the table comparing the values of f and P_2. What do you observe?

x	-1.0	-0.01	-0.0001
$f(x) = \ln(1 + x)$			
$P_2(x)$			

x	0	0.0001	0.01	1.0
$f(x) = \ln(1 + x)$				
$P_2(x)$				

   (d) Use a graphing utility to graph the polynomial $P_2(x)$ together with $f(x) = \ln(1 + x)$ in the same viewing window. What do you observe?

3. (a) Prove that $\displaystyle\int_0^1 [f(x) + f(x + 1)]\, dx = \int_0^2 f(x)\, dx.$

   (b) Use the result of part (a) to evaluate

$$\int_0^1 \left(\sqrt{x} + \sqrt{x + 1}\right) dx.$$

   (c) Use the result of part (a) to evaluate

$$\int_0^1 (e^x + e^{x+1})\, dx.$$

4. Consider the two regions A and B determined by the graph of $f(x) = \ln x$, as indicated in the figure below.

   (a) Calculate the area of region A.

   (b) Use your answer in part (a) to evaluate the integral

$$\int_1^e \ln x\, dx.$$

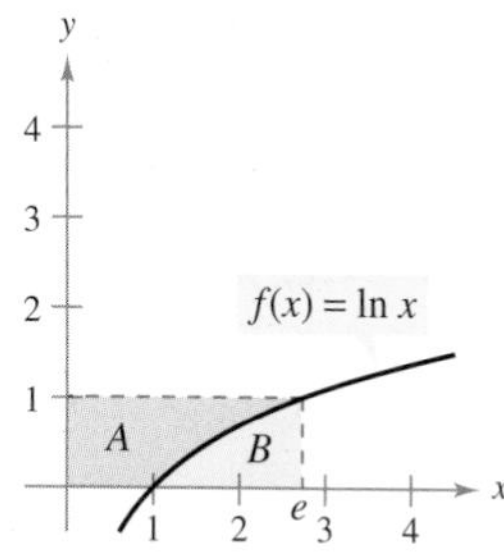

Figure for 4

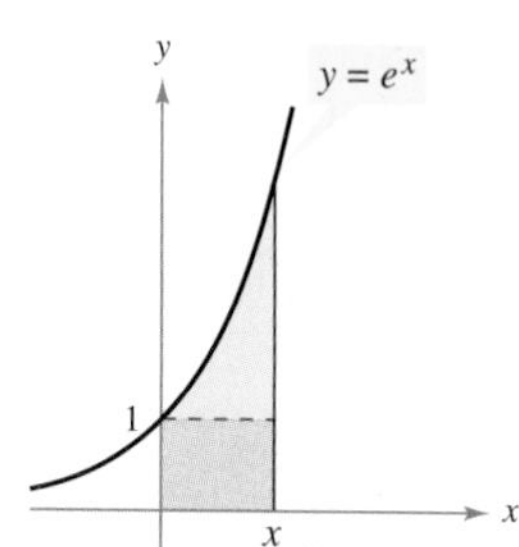

Figure for 5

5. Let x be a positive number.

   (a) Use the figure above to prove that

$$e^x > 1 + x.$$

   (b) Prove that

$$e^x > 1 + x + \frac{x^2}{2}.$$

   (c) Prove in general that for all positive integers n,

$$e^x > 1 + x + \frac{x^2}{2} + \cdot \cdot \cdot + \frac{x^n}{n!}$$

   where $n! = 1 \cdot 2 \cdot 3 \cdot \cdot \cdot (n - 1) \cdot n$.

6. Let L be the tangent line to the graph of the function given by $y = \ln x$ at the point (a, b). (See figure.) Show that the distance between b and c is always equal to 1.

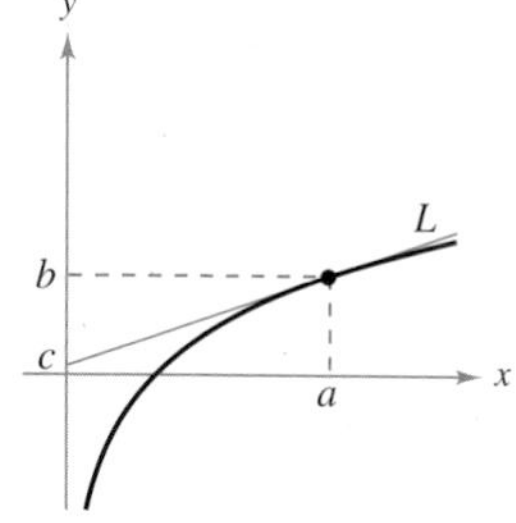

Figure for 6

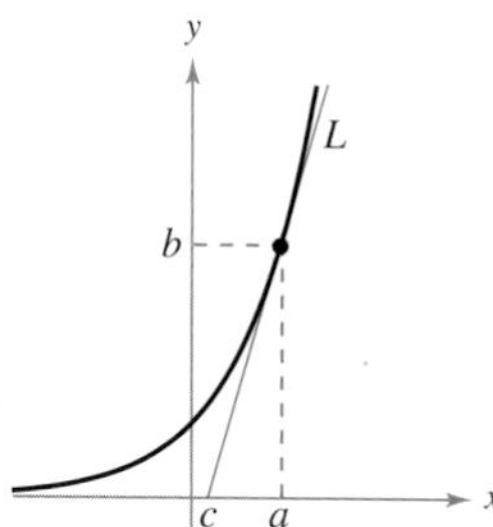

Figure for 7

7. Let L be the tangent line to the graph of the function $y = e^x$ at the point (a, b). (See figure.) Show that the distance between a and c is always equal to 1.

8. Graph the exponential function $y = a^x$ for $a = 0.5, 1.2,$ and 2.0. Which of these curves intersects the line $y = x$? Determine all positive numbers a for which the curve $y = a^x$ intersects the line $y = x$.

9. The differential equation

$$\frac{dy}{dt} = ky^{1+\varepsilon}$$

where k and ε are positive constants, is called the *doomsday equation.*

(a) Solve the doomsday equation $dy/dt = y^{1.01}$ given that $y(0) = 1$. Find the time T at which $\lim_{t \to T^-} y(t) = \infty$.

(b) Solve the doomsday equation $dy/dt = ky^{1+\varepsilon}$ given that $y(0) = y_0$. Explain why this equation is called the doomsday equation.

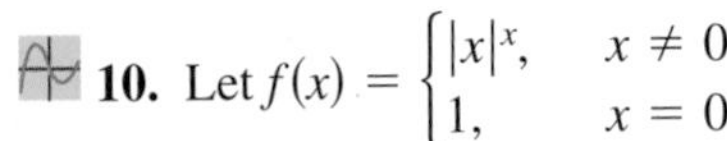

10. Let $f(x) = \begin{cases} |x|^x, & x \neq 0 \\ 1, & x = 0 \end{cases}$.

(a) Use a graphing utility to graph f in the viewing window $-3 \le x \le 3$, $-2 \le y \le 2$. What is the domain of f?

(b) Use the *zoom* and *trace* features of a graphing utility to estimate $\lim_{x \to 0} f(x)$.

(c) Write a short paragraph explaining why the function f is continuous for all real numbers.

(d) Visually estimate the slope of f at the point $(0, 1)$.

(e) Explain why the derivative of a function can be approximated by the formula

$$\frac{f(x + \Delta x) - f(x - \Delta x)}{2\Delta x}$$

for small values of Δx. Use this formula to approximate the slope of f at the point $(0, 1)$.

$$f'(0) \approx \frac{f(0 + \Delta x) - f(0 - \Delta x)}{2\Delta x} = \frac{f(\Delta x) - f(-\Delta x)}{2\Delta x}$$

What do you think the slope of the graph of f is at $(0, 1)$?

(f) Find a formula for the derivative of f and determine $f'(0)$. Write a short paragraph explaining how a graphing utility might lead you to approximate the slope of a graph incorrectly.

(g) Use your formula for the derivative of f to find the relative extrema of f. Verify your answer with a graphing utility.

■ **FOR FURTHER INFORMATION** For more information on using graphing utilities to estimate slope, see the article "Computer-Aided Delusions" by Richard L. Hall in *The College Mathematics Journal.* To view this article, go to the website *www.matharticles.com.* ■

11. Use integration by substitution to find the area under the curve

$$y = \frac{1}{\sqrt{x} + x}$$

between $x = 1$ and $x = 4$.

12. Show that $f(x) = \dfrac{\ln x^n}{x}$ is a decreasing function for $x > e$ and $n > 0$.

13. The differential equation $dy/dt = ky(L - y)$, where k and L are positive constants, is called the **logistic equation.**

(a) Solve the logistic equation $dy/dt = y(1 - y)$ given that $y(0) = \frac{1}{4}$.

$\left(\textit{Hint: } \dfrac{1}{y(1 - y)} = \dfrac{1}{y} + \dfrac{1}{1 - y}.\right)$

(b) Graph the solution on the interval $-6 \le t \le 6$. Show that the rate of growth of the solution is maximum at the point of inflection.

(c) Solve the logistic equation $dy/dt = y(1 - y)$ given that $y(0) = 2$. How does this solution differ from that in part (a)?

14. Let S represent sales of a new product (in thousands of units), let L represent the maximum level of sales (in thousands of units), and let t represent time (in months). The rate of change of S with respect to t varies jointly as the product of S and $L - S$.

(a) Write the differential equation for the sales model if $L = 100$, $S = 10$ when $t = 0$, and $S = 20$ when $t = 1$. Verify that

$$S = \frac{L}{1 + Ce^{-kt}}.$$

(b) At what time is the growth in sales increasing most rapidly?

(c) Use a graphing utility to graph the sales function.

(d) Sketch the solution in part (a) on the slope field shown in the figure. (To print an enlarged copy of the graph, go to the website *www.mathgraphs.com.*)

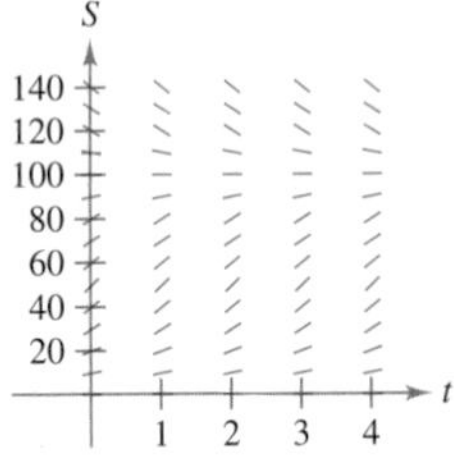

(e) If the estimated maximum level of sales is correct, use the slope field to describe the shape of the solution curves for sales if, at some period of time, sales exceed L.

9 Trigonometric Functions

In this chapter, you will study trigonometric functions. Trigonometry is used to find relationships between the sides and angles of triangles, and to write trigonometric functions as models of real-life quantities. Trigonometric functions are also used to model quantities that are periodic.

In this chapter, you should learn the following.

- How to describe angles, use radian measure, and use degree measure. (9.1)
- How to evaluate trigonometric functions using the unit circle. (9.2)
- How to evaluate trigonometric functions of acute angles and use the fundamental trigonometric identities. (9.3)
- How to use reference angles to evaluate trigonometric functions of any angle. (9.4)
- How to sketch the graphs of sine and cosine functions. (9.5)
- How to sketch the graphs of tangent, cotangent, secant, and cosecant functions. (9.6)
- How to evaluate inverse trigonometric functions. (9.7)
- How to solve real-life problems involving right triangles, directional bearings, and harmonic motion. (9.8)

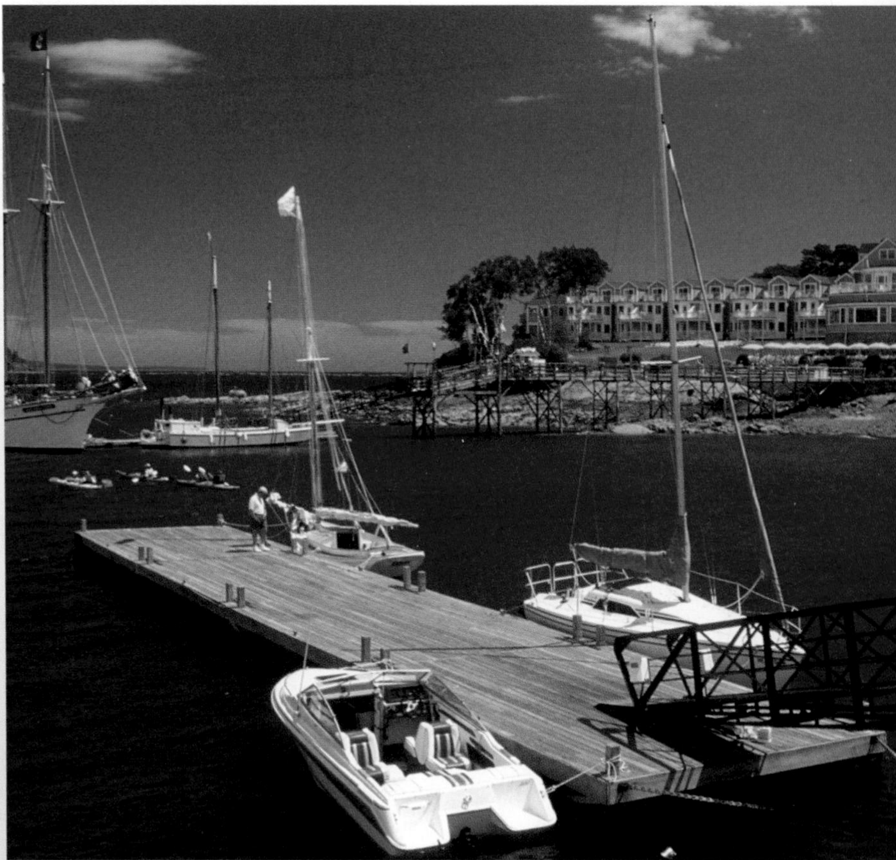

Andre Jenny / Alamy

Throughout the day, the depth of water at the end of a dock in Bar Harbor, Maine varies with the tide. How can the depth be modeled by a trigonometric function? (See Section 9.5, Example 7.)

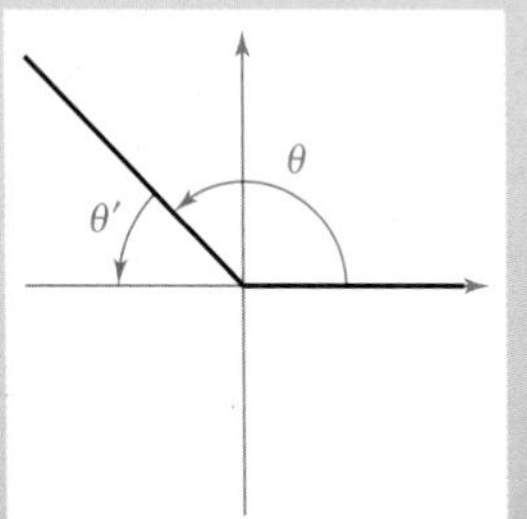

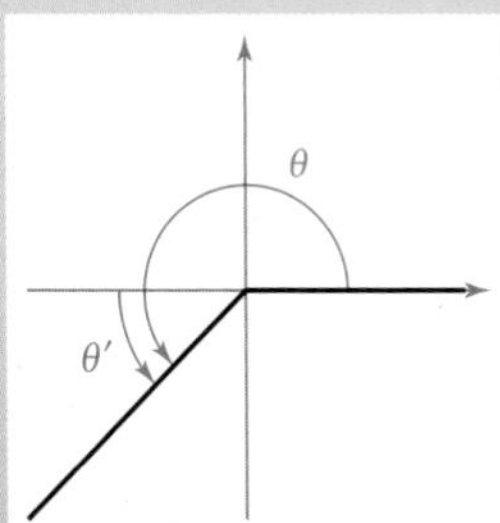

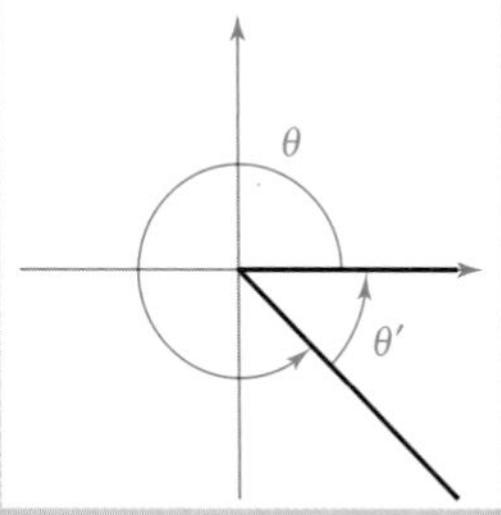

A reference angle is the acute angle θ' formed by the terminal side of θ and the horizontal axis. You will learn how to use reference angles to calculate the values of trigonometric functions of angles greater than 90 degrees. (See Section 9.4.)

9.1 Radian and Degree Measure

- Describe angles.
- Use radian measure.
- Use degree measure.
- Use angles to model and solve real-life problems.

Angles

As derived from the Greek language, the word **trigonometry** means "measurement of triangles." Initially, trigonometry dealt with relationships among the sides and angles of triangles and was used in the development of astronomy, navigation, and surveying. With the development of calculus and the physical sciences in the 17th century, a different perspective arose—one that viewed the classic trigonometric relationships as *functions* with the set of real numbers as their domains. Consequently, the applications of trigonometry expanded to include a vast number of physical phenomena involving rotations and vibrations. These phenomena include sound waves, light rays, planetary orbits, vibrating strings, pendulums, and orbits of atomic particles.

The approach in this text incorporates *both* perspectives, starting with angles and their measure.

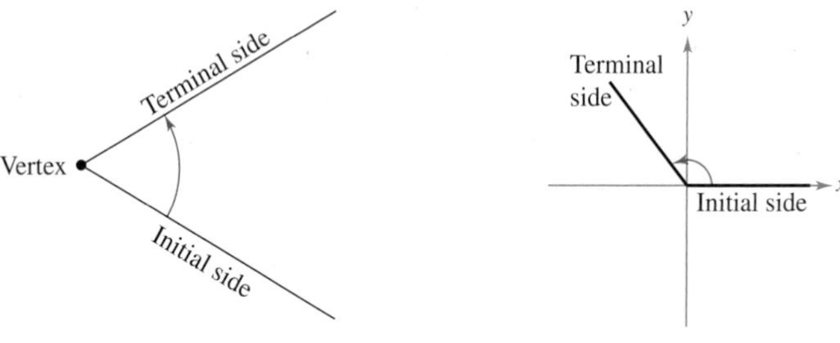

(a) Angle
(b) Angle in standard position
Figure 9.1

An **angle** is determined by rotating a ray (half-line) about its endpoint. The starting position of the ray is the **initial side** of the angle, and the position after rotation is the **terminal side,** as shown in Figure 9.1(a). The endpoint of the ray is the **vertex** of the angle. This perception of an angle fits a coordinate system in which the origin is the vertex and the initial side coincides with the positive x-axis. Such an angle is in **standard position,** as shown in Figure 9.1(b). **Positive angles** are generated by counterclockwise rotation, and **negative angles** by clockwise rotation, as shown in Figure 9.2. Angles are labeled with Greek letters α (alpha), β (beta), and θ (theta), as well as uppercase letters A, B, and C. In Figure 9.3, note that angles α and β have the same initial and terminal sides. Such angles are **coterminal.**

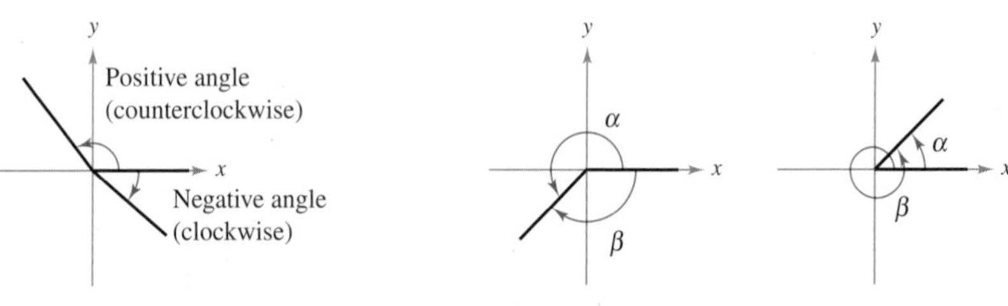

Positive and negative angles
Figure 9.2

Coterminal angles
Figure 9.3

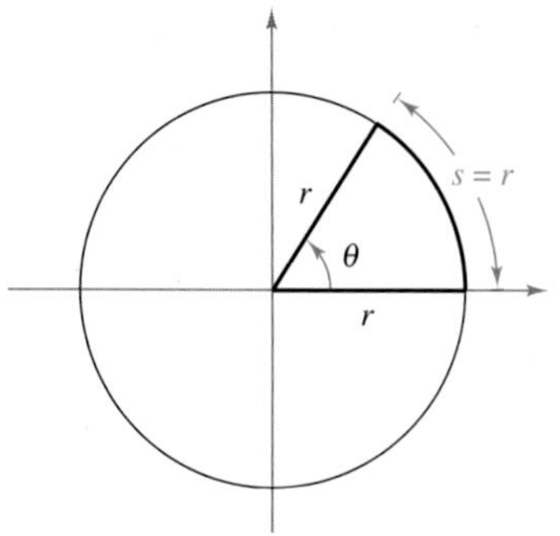

Arc length = radius when $\theta = 1$ radian

Figure 9.4

Radian Measure

The **measure of an angle** is determined by the amount of rotation from the initial side to the terminal side. One way to measure angles is in **radians.** This type of measure is especially useful in calculus. To define a radian, you can use a **central angle** of a circle, one whose vertex is the center of the circle, as shown in Figure 9.4.

> **DEFINITION OF RADIAN**
>
> One **radian** is the measure of a central angle θ that intercepts an arc s equal in length to the radius r of the circle. See Figure 9.4.

Because the circumference of a circle is $2\pi r$ units, it follows that a central angle of one full revolution (counterclockwise) corresponds to an arc length of

$$s = 2\pi r.$$

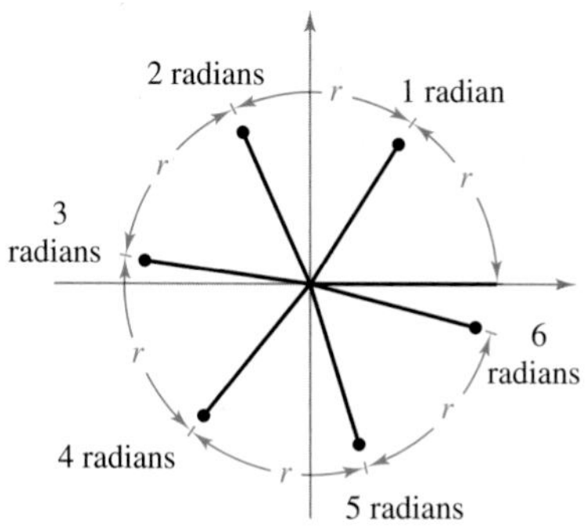

Figure 9.5

Moreover, because $2\pi \approx 6.28$, there are just over six radius lengths in a full circle, as shown in Figure 9.5. In general, the radian measure of a central angle θ is obtained by dividing the arc length s by r. That is, $s/r = \theta$, where θ is *measured in radians.* Because the units of measure for s and r are the same, this ratio is unitless—it is simply a real number.

Because the radian measure of an angle of one full revolution is 2π, you can obtain

$$\frac{1}{2} \text{ revolution} = \frac{2\pi}{2} = \pi \text{ radians}$$

$$\frac{1}{4} \text{ revolution} = \frac{2\pi}{4} = \frac{\pi}{2} \text{ radians}$$

$$\frac{1}{6} \text{ revolution} = \frac{2\pi}{6} = \frac{\pi}{3} \text{ radians.}$$

These and other common angles are shown in Figure 9.6.

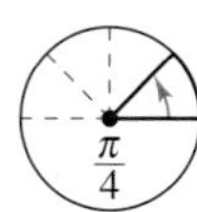

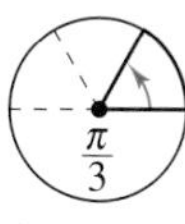

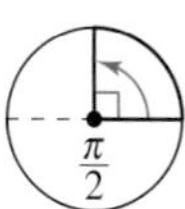

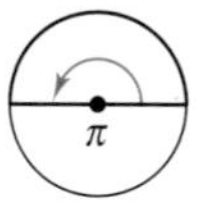

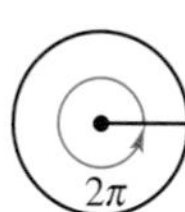

Figure 9.6

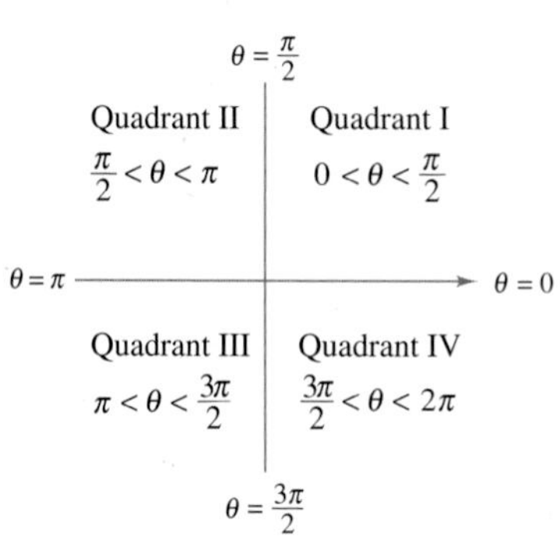

Figure 9.7

Recall that the four quadrants in a coordinate system are numbered I, II, III, and IV. Figure 9.7 shows which angles between 0 and 2π lie in each of the four quadrants. Note that angles between 0 and $\pi/2$ are **acute** and that angles between $\pi/2$ and π are **obtuse.**

Because two angles are coterminal if they have the same initial and terminal sides, the angles 0 and 2π are coterminal, as are the angles $\pi/6$ and $13\pi/6$. You can find an angle that is coterminal to a given angle θ by adding or subtracting 2π (one revolution), as demonstrated in Example 1. A given angle θ has infinitely many coterminal angles. For instance, $\theta = \pi/6$ is coterminal with

$$\frac{\pi}{6} + 2n\pi$$

where n is an integer.

STUDY TIP The phrase "the terminal side of θ lies in a quadrant" is often abbreviated by simply saying that "θ lies in a quadrant." The terminal sides of the "quadrant angles" 0, $\pi/2$, π, and $3\pi/2$ do not lie within quadrants.

EXAMPLE 1 Sketching and Finding Coterminal Angles

a. For the positive angle $13\pi/6$, subtract 2π to obtain a coterminal angle

$$\frac{13\pi}{6} - 2\pi = \frac{\pi}{6}.$$ See Figure 9.8(a).

b. For the positive angle $3\pi/4$, subtract 2π to obtain a coterminal angle

$$\frac{3\pi}{4} - 2\pi = -\frac{5\pi}{4}.$$ See Figure 9.8(b).

c. For the negative angle $-2\pi/3$, add 2π to obtain a coterminal angle

$$-\frac{2\pi}{3} + 2\pi = \frac{4\pi}{3}.$$ See Figure 9.8(c).

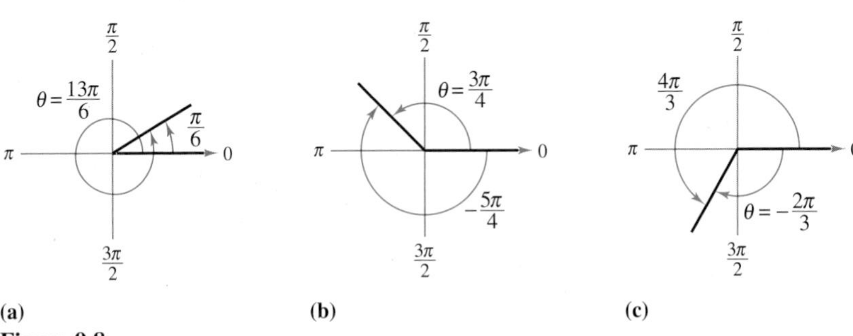

Figure 9.8

Two positive angles α and β are **complementary** (complements of each other) if their sum is $\pi/2$. Two positive angles are **supplementary** (supplements of each other) if their sum is π. See Figure 9.9.

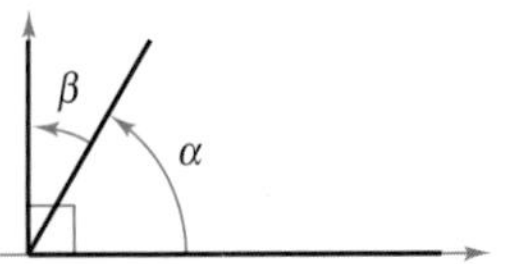

Complementary angles: $\alpha + \beta = \pi/2$

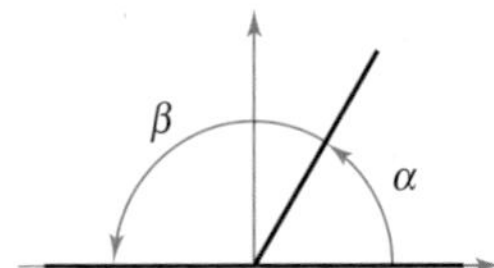

Supplementary angles: $\alpha + \beta = \pi$

Figure 9.9

EXAMPLE 2 Complementary and Supplementary Angles

a. The complement of $2\pi/5$ is

$$\frac{\pi}{2} - \frac{2\pi}{5} = \frac{5\pi}{10} - \frac{4\pi}{10} = \frac{\pi}{10}.$$

The supplement of $2\pi/5$ is

$$\pi - \frac{2\pi}{5} = \frac{5\pi}{5} - \frac{2\pi}{5} = \frac{3\pi}{5}.$$

b. Because $4\pi/5$ is greater than $\pi/2$, it has no complement. (Remember that complements are *positive* angles.) The supplement is

$$\pi - \frac{4\pi}{5} = \frac{5\pi}{5} - \frac{4\pi}{5} = \frac{\pi}{5}.$$

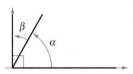

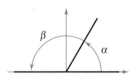

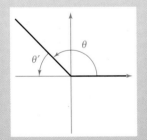

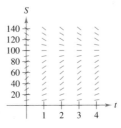

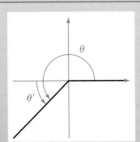

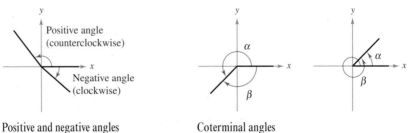

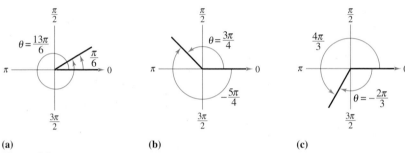

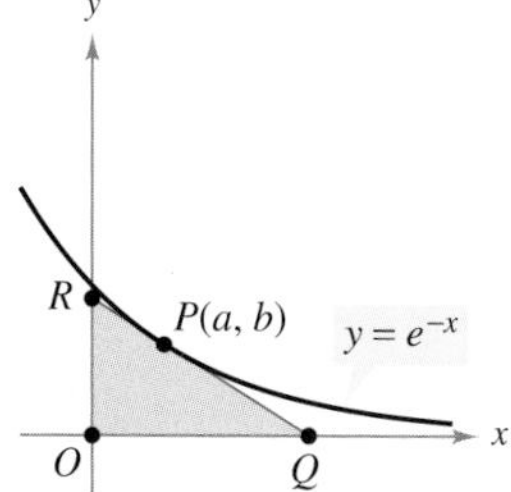

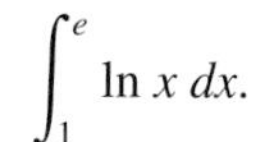

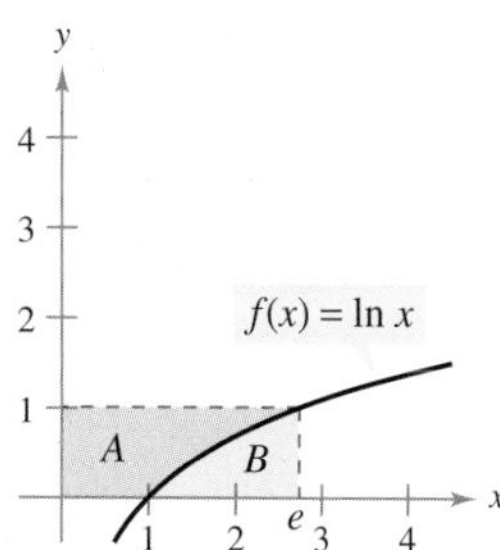

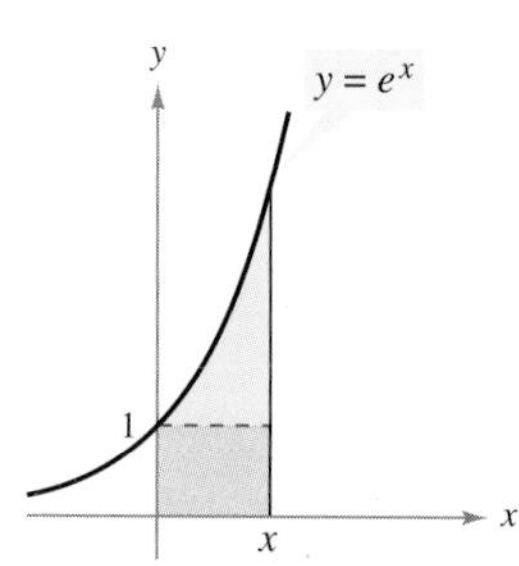

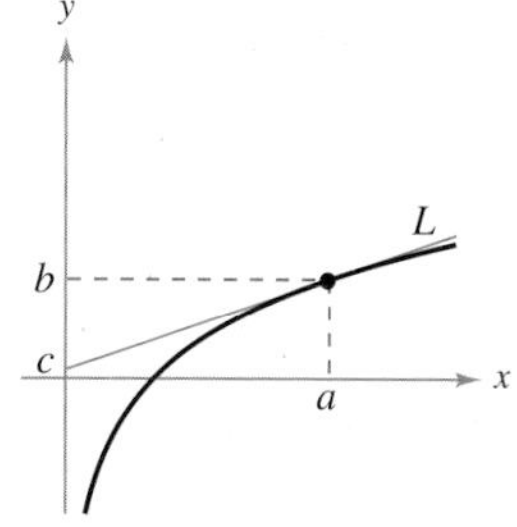

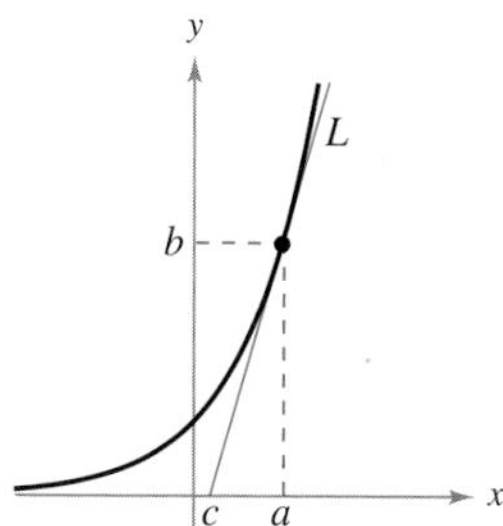

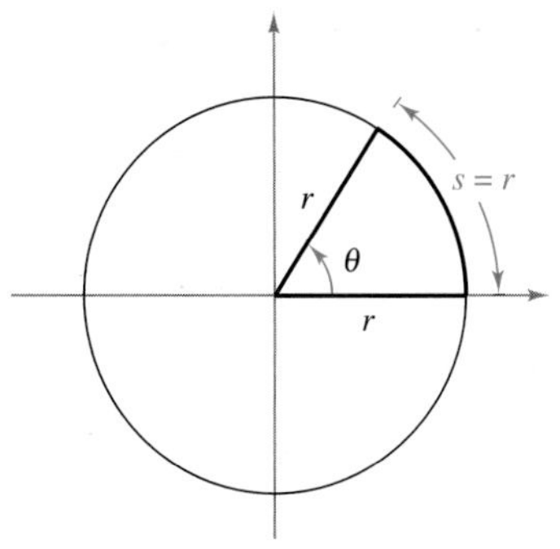

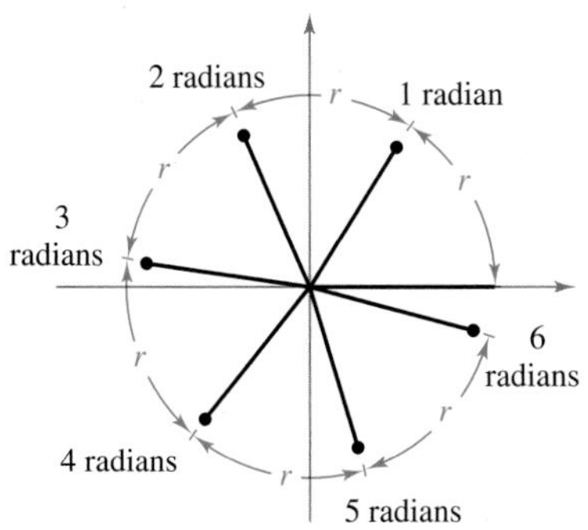

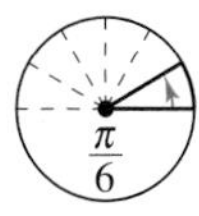

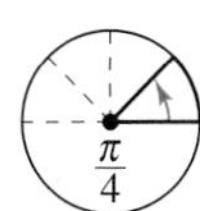

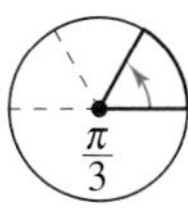

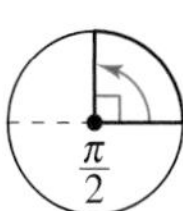

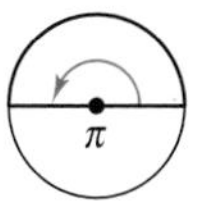

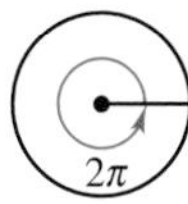

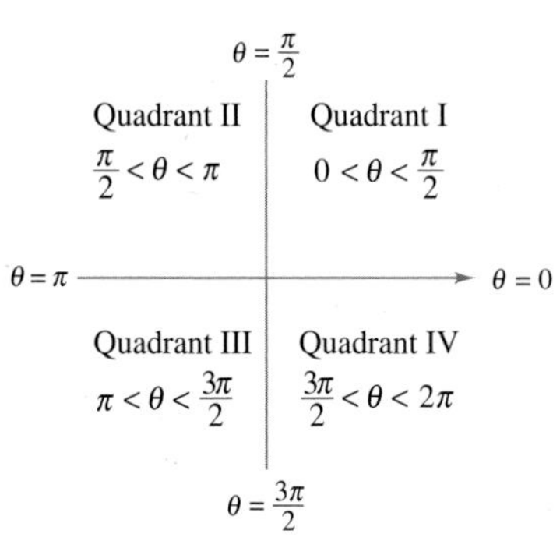

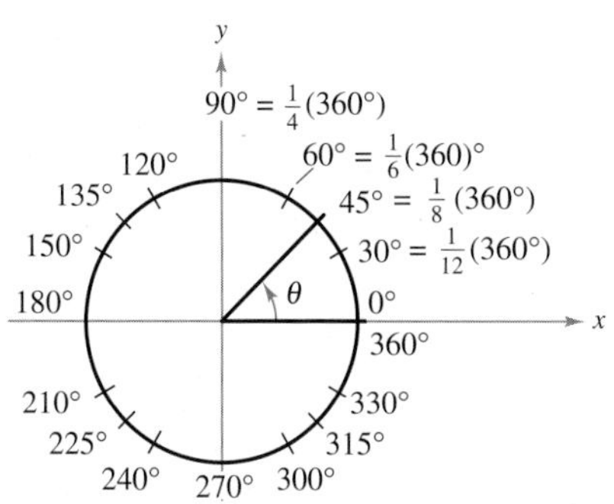

Figure 9.10

Degree Measure

A second way to measure angles is in terms of **degrees,** denoted by the symbol °. A measure of one degree (1°) is equivalent to a rotation of $\frac{1}{360}$ of a complete revolution about the vertex. To measure angles, it is convenient to mark degrees on the circumference of a circle, as shown in Figure 9.10. So, a full revolution (counterclockwise) corresponds to 360°, a half revolution to 180°, a quarter revolution to 90°, and so on.

Because 2π radians corresponds to one complete revolution, degrees and radians are related by the equations

$$360° = 2\pi \text{ rad} \qquad \text{and} \qquad 180° = \pi \text{ rad}.$$

From the latter equation, you obtain

$$1° = \frac{\pi}{180} \text{ rad} \qquad \text{and} \qquad 1 \text{ rad} = \left(\frac{180°}{\pi}\right)$$

which lead to the following conversion rules.

CONVERSIONS BETWEEN DEGREES AND RADIANS

1. To convert degrees to radians, multiply degrees by $\dfrac{\pi \text{ rad}}{180°}$.
2. To convert radians to degrees, multiply radians by $\dfrac{180°}{\pi \text{ rad}}$.

To apply these two conversion rules, use the basic relationship $\pi \text{ rad} = 180°$. (See Figure 9.11.)

STUDY TIP When no units of angle measure are specified, *radian measure is implied.* For instance, if you write $\theta = \pi$ or $\theta = 2$, you imply that $\theta = \pi$ radians or $\theta = 2$ radians.

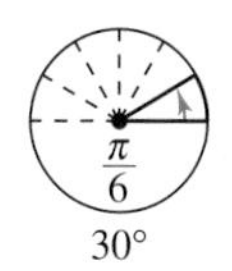

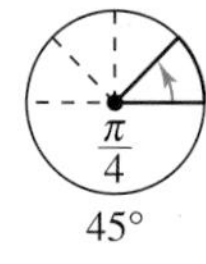

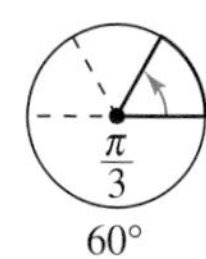

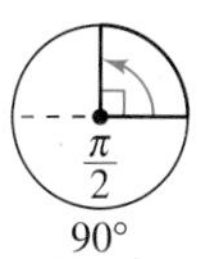

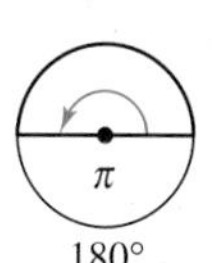

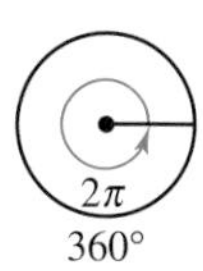

Figure 9.11

TECHNOLOGY With calculators it is convenient to use *decimal* degrees to denote fractional parts of degrees. Historically, however, fractional parts of degrees were expressed in *minutes* and *seconds,* using the prime (′) and double prime (″) notations, respectively. That is,

$$1' = \text{one minute} = \tfrac{1}{60}(1°)$$
$$1'' = \text{one second} = \tfrac{1}{3600}(1°).$$

Consequently, an angle of 64 degrees, 32 minutes, and 47 seconds is represented by $\theta = 64° \, 32' \, 47''$. Many calculators have special keys for converting an angle in degrees, minutes, and seconds (D° M′ S″) to decimal degree form, and vice versa.

EXAMPLE 3 Converting from Degrees to Radians

a. $135° = (135 \text{ deg})\left(\dfrac{\pi \text{ rad}}{180 \text{ deg}}\right) = \dfrac{3\pi}{4} \text{ radians}$ Multiply by $\pi/180$.

b. $540° = (540 \text{ deg})\left(\dfrac{\pi \text{ rad}}{180 \text{ deg}}\right) = 3\pi \text{ radians}$ Multiply by $\pi/180$.

c. $-270° = (-270 \text{ deg})\left(\dfrac{\pi \text{ rad}}{180 \text{ deg}}\right) = -\dfrac{3\pi}{2} \text{ radians}$ Multiply by $\pi/180$.

EXAMPLE 4 Converting from Radians to Degrees

a. $-\dfrac{\pi}{2} \text{ rad} = \left(-\dfrac{\pi}{2} \text{ rad}\right)\left(\dfrac{180 \text{ deg}}{\pi \text{ rad}}\right) = -90°$ Multiply by $180/\pi$.

b. $\dfrac{9\pi}{2} \text{ rad} = \left(\dfrac{9\pi}{2} \text{ rad}\right)\left(\dfrac{180 \text{ deg}}{\pi \text{ rad}}\right) = 810°$ Multiply by $180/\pi$.

c. $2 \text{ rad} = (2 \text{ rad})\left(\dfrac{180 \text{ deg}}{\pi \text{ rad}}\right) = \dfrac{360°}{\pi} \approx 114.59°$ Multiply by $180/\pi$. ■

Applications

The *radian measure* formula, $\theta = s/r$, can be used to measure arc length along a circle.

ARC LENGTH

For a circle of radius r, a central angle θ intercepts an arc of length s given by

$$s = r\theta \qquad \text{Length of circular arc}$$

where θ is measured in radians. Note that if $r = 1$, then $s = \theta$, and the radian measure of θ equals the arc length.

EXAMPLE 5 Finding Arc Length

A circle has a radius of 4 inches. Find the length of the arc intercepted by a central angle of 240°, as shown in Figure 9.12.

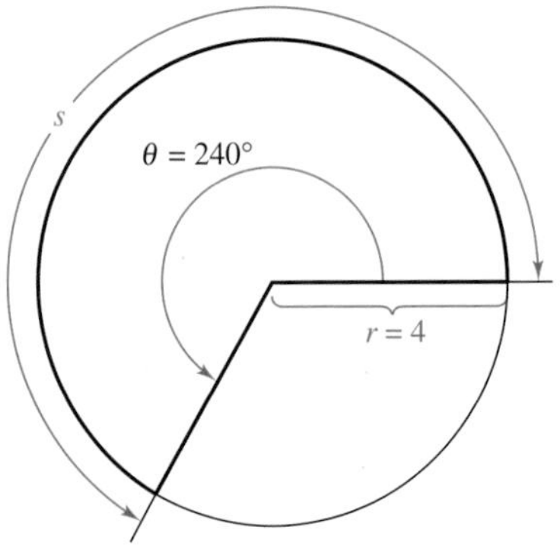

s

$\theta = 240°$

$r = 4$

Figure 9.12

Solution To use the formula $s = r\theta$, first convert 240° to radian measure.

$$240° = (240 \text{ deg})\left(\frac{\pi \text{ rad}}{180 \text{ deg}}\right) = \frac{4\pi}{3} \text{ radians}$$

Then, using a radius of $r = 4$ inches, you can find the arc length to be

$$s = r\theta = 4\left(\frac{4\pi}{3}\right) = \frac{16\pi}{3} \approx 16.76 \text{ inches.}$$

Note that the units for $r\theta$ are determined by the units for r because θ is given in radian measure, which has no units. ■

The formula for the length of a circular arc can be used to analyze the motion of a particle moving at a *constant speed* along a circular path.

LINEAR AND ANGULAR SPEEDS

Consider a particle moving at a constant speed along a circular arc of radius r. If s is the length of the arc traveled in time t, then the **linear speed** v of the particle is

$$\text{Linear speed } v = \frac{\text{arc length}}{\text{time}} = \frac{s}{t}.$$

Moreover, if θ is the angle (in radian measure) corresponding to the arc length s, then the **angular speed** ω (the lowercase Greek letter omega) of the particle is

$$\text{Angular speed } \omega = \frac{\text{central angle}}{\text{time}} = \frac{\theta}{t}.$$

STUDY TIP Linear speed measures how fast the particle moves, and angular speed measures how fast the angle changes. By dividing the formula for arc length by t, you can establish a relationship between linear speed v and angular speed ω, as shown.

$$s = r\theta$$

$$\frac{s}{t} = \frac{r\theta}{t}$$

$$v = r\omega$$

Figure 9.13

EXAMPLE 6 Finding Linear Speed

The second hand of a clock is 10.2 centimeters long, as shown in Figure 9.13. Find the linear speed of the tip of this second hand as it passes around the clock face.

Solution In one revolution, the arc length traveled is

$$\begin{aligned} s &= 2\pi r \\ &= 2\pi(10.2) && \text{Substitute for } r. \\ &= 20.4\pi \text{ centimeters.} \end{aligned}$$

The time required for the second hand to travel this distance is

$$t = 1 \text{ minute} = 60 \text{ seconds.}$$

So, the linear speed of the tip of the second hand is

$$\begin{aligned} \text{Linear speed} &= \frac{s}{t} \\ &= \frac{20.4\pi \text{ centimeters}}{60 \text{ seconds}} \\ &\approx 1.068 \text{ centimeters per second.} \end{aligned}$$

Figure 9.14

EXAMPLE 7 Finding Angular and Linear Speeds

The blades of a wind turbine are 116 feet long (see Figure 9.14). The propeller rotates at 15 revolutions per minute.

a. Find the angular speed of the propeller in radians per minute.

b. Find the linear speed of the tips of the blades.

Solution

a. Because each revolution generates 2π radians, it follows that the propeller turns

$$(15)(2\pi) = 30\pi \text{ radians per minute.}$$

In other words, the angular speed is

$$\begin{aligned} \text{Angular speed} &= \frac{\theta}{t} \\ &= \frac{30\pi \text{ radians}}{1 \text{ minute}} \\ &= 30\pi \text{ radians per minute.} \end{aligned}$$

b. The linear speed is

$$\begin{aligned} \text{Linear speed} &= \frac{s}{t} \\ &= \frac{r\theta}{t} \\ &= \frac{(116)(30\pi) \text{ feet}}{1 \text{ minute}} \\ &\approx 10{,}933 \text{ feet per minute.} \end{aligned}$$

■

9.1 Exercises

See www.CalcChat.com for worked-out solutions to odd-numbered exercises.

In Exercises 1–10, fill in the blanks.

1. ________ means "measurement of triangles."
2. An ________ is determined by rotating a ray about its endpoint.
3. Two angles that have the same initial and terminal sides are ________.
4. One ________ is the measure of a central angle that intercepts an arc equal to the radius of the circle.
5. Angles that measure between 0 and $\pi/2$ are ________ angles, and angles that measure between $\pi/2$ and π are ________ angles.
6. Two positive angles that have a sum of $\pi/2$ are ________ angles, whereas two positive angles that have a sum of π are ________ angles.
7. The angle measure that is equivalent to a rotation of $\frac{1}{360}$ of a complete revolution about an angle's vertex is one ________.
8. 180 degrees = ________ radians.
9. The radian measure formula, $\theta = s/r$, can be used to measure ________ ________ along a circle.
10. The ________ speed of a particle is the ratio of arc length to time traveled, and the ________ speed of a particle is the ratio of central angle to time traveled.

In Exercises 11–16, estimate the angle to the nearest one-half radian.

11. 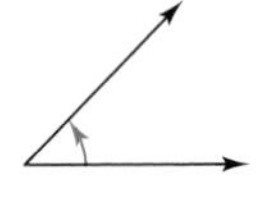

12. 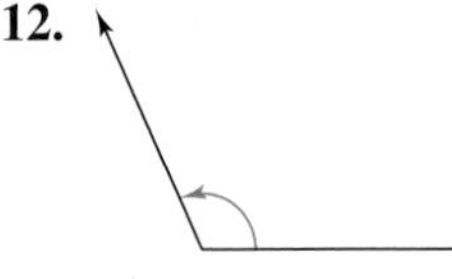

13. 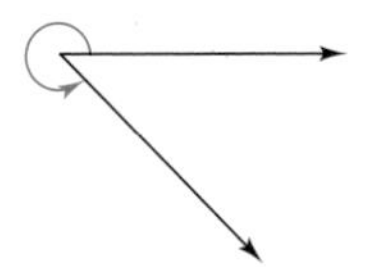

14. 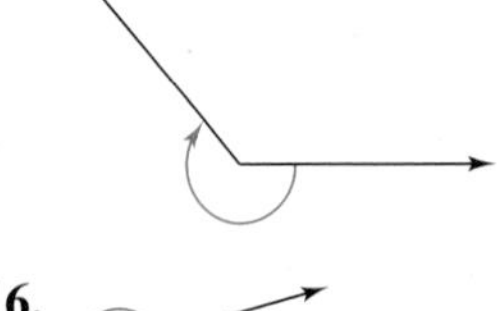

15. 

16. 

In Exercises 17–22, determine the quadrant in which each angle lies. (The angle measure is given in radians.)

17. (a) $\frac{\pi}{4}$ (b) $\frac{5\pi}{4}$
18. (a) $\frac{11\pi}{8}$ (b) $\frac{9\pi}{8}$
19. (a) $-\frac{\pi}{6}$ (b) $-\frac{\pi}{3}$
20. (a) $-\frac{5\pi}{6}$ (b) $-\frac{11\pi}{9}$
21. (a) 3.5 (b) 2.25
22. (a) 6.02 (b) -4.25

In Exercises 23–26, sketch each angle in standard position.

23. (a) $\frac{\pi}{3}$ (b) $-\frac{2\pi}{3}$
24. (a) $-\frac{7\pi}{4}$ (b) $\frac{5\pi}{2}$
25. (a) $\frac{11\pi}{6}$ (b) -3
26. (a) 4 (b) 7π

In Exercises 27–30, determine two coterminal angles (one positive and one negative) for each angle. Give your answers in radians.

27. (a) 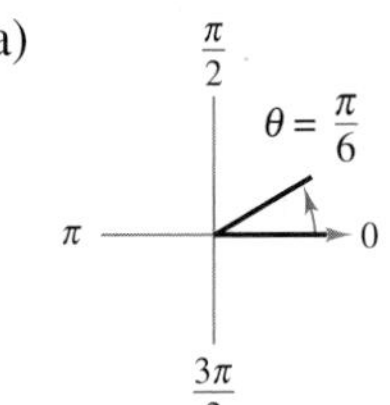

(b) 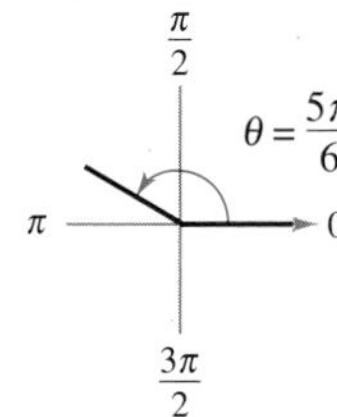

28. (a) 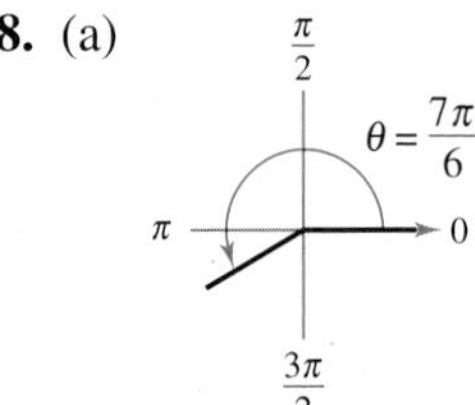

(b) 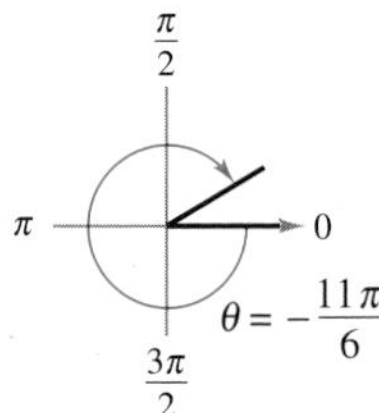

29. (a) $\theta = \frac{2\pi}{3}$ (b) $\theta = \frac{\pi}{12}$
30. (a) $\theta = -\frac{9\pi}{4}$ (b) $\theta = -\frac{2\pi}{15}$

In Exercises 31–34, find (if possible) the complement and supplement of each angle.

31. (a) $\pi/3$ (b) $\pi/4$
32. (a) $\pi/12$ (b) $11\pi/12$
33. (a) 1 (b) 2
34. (a) 3 (b) 1.5

In Exercises 35–40, estimate the number of degrees in the angle. Use a protractor to check your answer.

35. 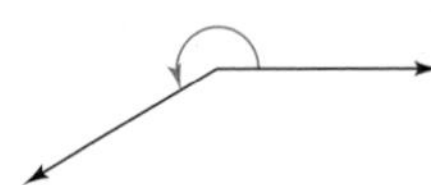

36. 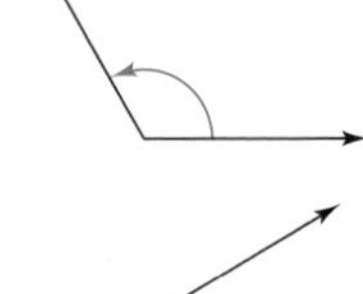

37. 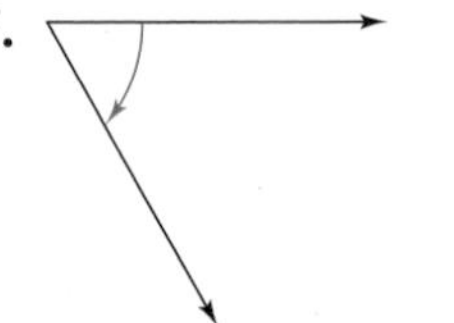

38. 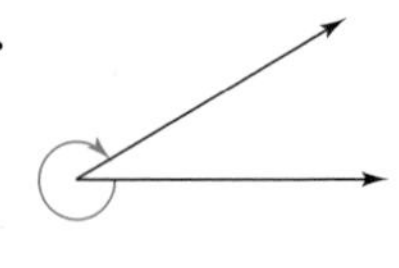

39.

40.

In Exercises 41–44, determine the quadrant in which each angle lies.

41. (a) $130°$ (b) $285°$

42. (a) $8.3°$ (b) $257° 30'$

43. (a) $-132° 50'$ (b) $-336°$

44. (a) $-260°$ (b) $-3.4°$

In Exercises 45–48, sketch each angle in standard position.

45. (a) $90°$ (b) $180°$

46. (a) $270°$ (b) $120°$

47. (a) $-30°$ (b) $-135°$

48. (a) $-750°$ (b) $-600°$

In Exercises 49–52, determine two coterminal angles (one positive and one negative) for each angle. Give your answers in degrees.

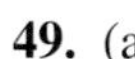

49. (a)

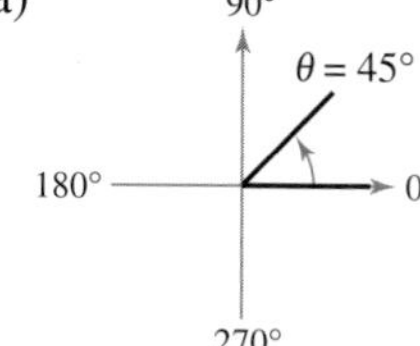

(b)

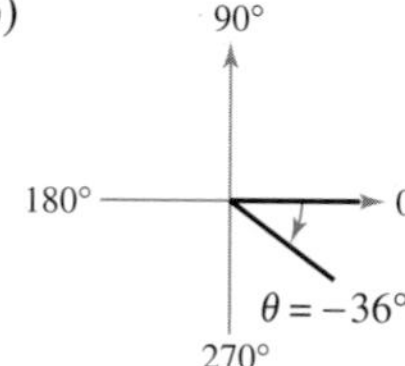

50. (a)

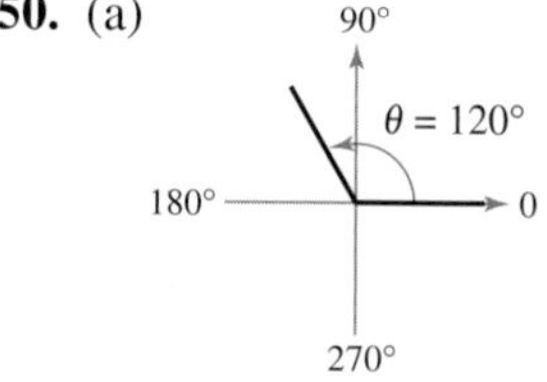

(b)

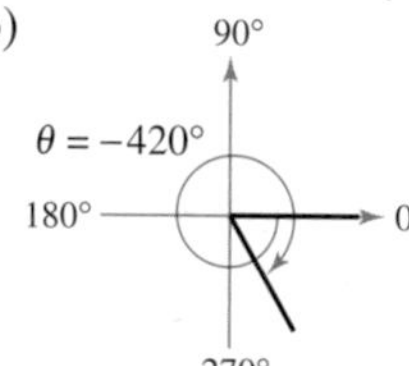

51. (a) $\theta = 240°$ (b) $\theta = -180°$

52. (a) $\theta = -390°$ (b) $\theta = 230°$

In Exercises 53–56, find (if possible) the complement and supplement of each angle.

53. (a) $18°$ (b) $85°$

54. (a) $46°$ (b) $93°$

55. (a) $150°$ (b) $79°$

56. (a) $130°$ (b) $170°$

In Exercises 57–60, rewrite each angle in radian measure as a multiple of π. (Do not use a calculator.)

57. (a) $30°$ (b) $45°$

58. (a) $315°$ (b) $120°$

59. (a) $-20°$ (b) $-60°$

60. (a) $-270°$ (b) $144°$

In Exercises 61–64, rewrite each angle in degree measure. (Do not use a calculator.)

61. (a) $\dfrac{3\pi}{2}$ (b) $\dfrac{7\pi}{6}$

62. (a) $-\dfrac{7\pi}{12}$ (b) $\dfrac{\pi}{9}$

63. (a) $\dfrac{5\pi}{4}$ (b) $-\dfrac{7\pi}{3}$

64. (a) $\dfrac{11\pi}{6}$ (b) $\dfrac{34\pi}{15}$

In Exercises 65–72, convert the angle measure from degrees to radians. Round to three decimal places.

65. $45°$

66. $87.4°$

67. $-216.35°$

68. $-48.27°$

69. $532°$

70. $345°$

71. $-0.83°$

72. $0.54°$

In Exercises 73–80, convert the angle measure from radians to degrees. Round to three decimal places.

73. $\pi/7$

74. $5\pi/11$

75. $15\pi/8$

76. $13\pi/2$

77. -4.2π

78. 4.8π

79. -2

80. -0.57

In Exercises 81–84, convert each angle measure to decimal degree form without using a calculator. Then check your answers using a calculator.

81. (a) $54° 45'$ (b) $-128° 30'$

82. (a) $245° 10'$ (b) $2° 12'$

83. (a) $85° 18' 30''$ (b) $330° 25''$

84. (a) $-135° 36''$ (b) $-408° 16' 20''$

In Exercises 85–88, convert each angle measure to degrees, minutes, and seconds without using a calculator. Then check your answers using a calculator.

85. (a) $240.6°$ (b) $-145.8°$

86. (a) $-345.12°$ (b) $0.45°$

87. (a) $2.5°$ (b) $-3.58°$

88. (a) $-0.36°$ (b) $0.79°$

In Exercises 89–92, find the length of the arc on a circle of radius r intercepted by a central angle θ.

	Radius r	*Central Angle θ*
89.	15 inches	$120°$
90.	9 feet	$60°$
91.	3 meters	$150°$
92.	20 centimeters	$45°$

In Exercises 93–96, find the radian measure of the central angle of a circle of radius r that intercepts an arc of length s.

	Radius r	*Arc Length s*
93.	4 inches	18 inches
94.	14 feet	8 feet
95.	25 centimeters	10.5 centimeters
96.	80 kilometers	150 kilometers

In Exercises 97–100, use the given arc length and radius to find the angle θ (in radians).

97.

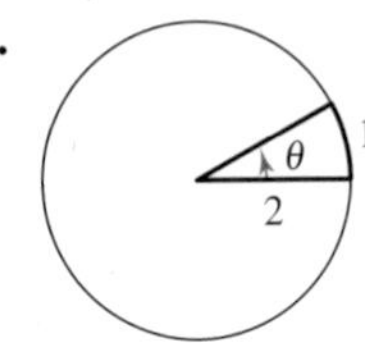

98.

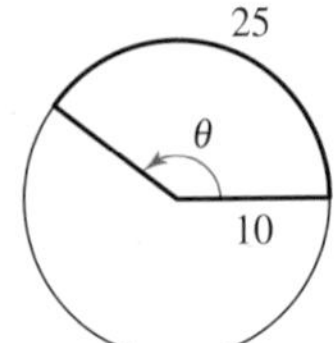

99.

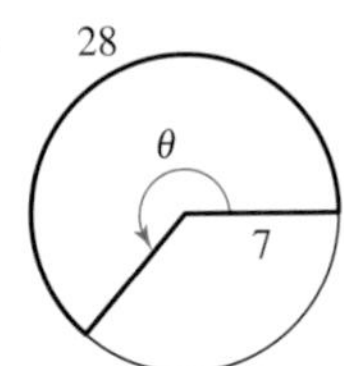

100.

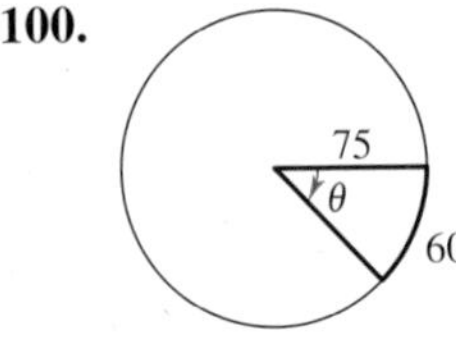

WRITING ABOUT CONCEPTS

101. A fan motor turns at a given angular speed. How does the speed of the tips of the blades change if a fan of greater diameter is installed on the motor? Explain.

102. Is a degree or a radian the larger unit of measure? Explain your reasoning.

103. If the radius of a circle is increasing and the magnitude of a central angle is held constant, how is the length of the intercepted arc changing? Explain your reasoning.

Distance Between Cities **In Exercises 104 and 105, find the distance between the cities. Assume that Earth is a sphere of radius 4000 miles and that the cities are on the same longitude (one city is due north of the other).**

	City	*Latitude*
104.	Dallas, Texas	32° 47′ 39″ N
	Omaha, Nebraska	41° 15′ 50″ N
105.	San Francisco, California	37° 47′ 36″ N
	Seattle, Washington	47° 37′ 18″ N

106. ***Linear and Angular Speeds*** A circular power saw has a $7\frac{1}{4}$-inch-diameter blade that rotates at 5000 revolutions per minute.

(a) Find the angular speed of the saw blade in radians per minute.

(b) Find the linear speed (in feet per minute) of one of the 24 cutting teeth as they contact the wood being cut.

107. ***Linear and Angular Speeds*** A carousel with a 50-foot diameter makes 4 revolutions per minute.

(a) Find the angular speed of the carousel in radians per minute.

(b) Find the linear speed (in feet per minute) of the platform rim of the carousel.

108. ***Angular Speed*** A two-inch-diameter pulley on an electric motor that runs at 1700 revolutions per minute is connected by a belt to a four-inch-diameter pulley on a saw arbor.

(a) Find the angular speed (in radians per minute) of each pulley.

(b) Find the revolutions per minute of the saw.

109. ***Angular Speed*** A car is moving at a rate of 65 miles per hour, and the diameter of its wheels is 2 feet.

(a) Find the number of revolutions per minute the wheels are rotating.

(b) Find the angular speed of the wheels in radians per minute.

CAPSTONE

110. Write a short paper in your own words explaining the meaning of each of the following concepts to a classmate.

(a) an angle in standard position

(b) positive and negative angles

(c) coterminal angles

(d) angle measure in degrees and radians

(e) obtuse and acute angles

(f) complementary and supplementary angles

111. ***Speed of a Bicycle*** The radii of the pedal sprocket, the wheel sprocket, and the wheel of the bicycle in the figure are 4 inches, 2 inches, and 14 inches, respectively. A cyclist is pedaling at a rate of 1 revolution per second. Find the speed of the bicycle in feet per second and miles per hour.

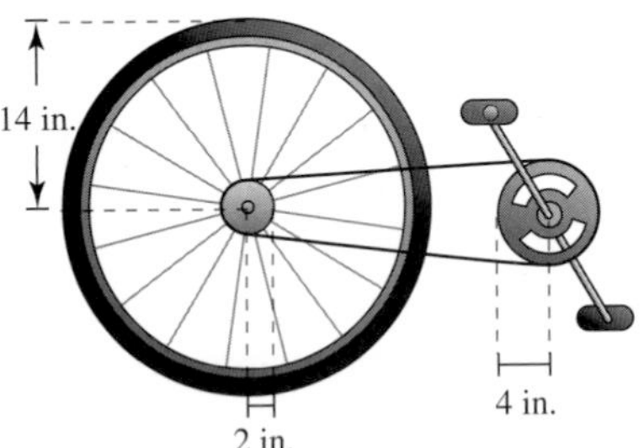

True or False? **In Exercises 112–114, determine whether the statement is true or false. Justify your answer.**

112. A measurement of 4 radians corresponds to two complete revolutions from the initial side to the terminal side of an angle.

113. The difference between the measures of two coterminal angles is always a multiple of 360° if expressed in degrees and is always a multiple of 2π radians if expressed in radians.

114. An angle that measures $-1260°$ lies in Quadrant III.

9.2 Trigonometric Functions: The Unit Circle

- Identify a unit circle and describe its relationship to real numbers.
- Evaluate trigonometric functions using the unit circle.
- Use the domain and period to evaluate sine and cosine functions.
- Use a calculator to evaluate trigonometric functions.

The Unit Circle

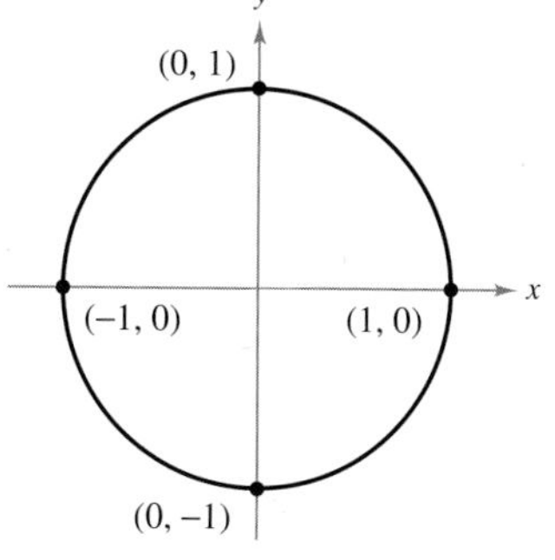

Unit circle
Figure 9.15

The two historical perspectives of trigonometry incorporate different methods for introducing the trigonometric functions. Our first introduction to these functions is based on the unit circle.

Consider the **unit circle** given by

$$x^2 + y^2 = 1 \qquad \text{Unit circle}$$

as shown in Figure 9.15. Imagine that the real number line is wrapped around this circle, with positive numbers corresponding to a counterclockwise wrapping and negative numbers corresponding to a clockwise wrapping, as shown in Figure 9.16.

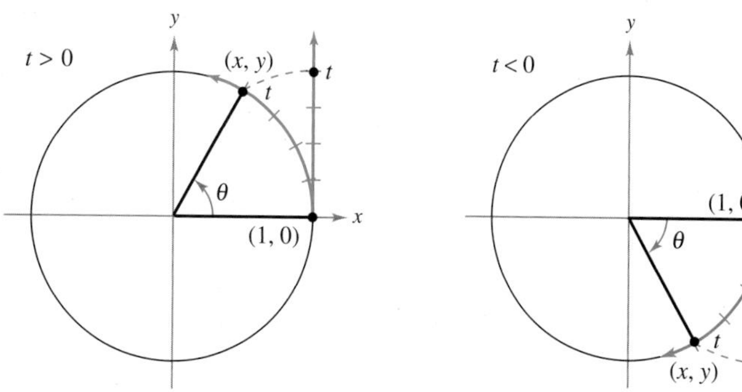

Figure 9.16

As the real number line is wrapped around the unit circle, each real number t corresponds to a point (x, y) on the circle. For example, the real number 0 corresponds to the point $(1, 0)$. Moreover, because the unit circle has a circumference of 2π, the real number 2π also corresponds to the point $(1, 0)$.

Real Number	*Point on Unit Circle*
0	$(1, 0)$
$\frac{\pi}{2}$	$(0, 1)$
π	$(-1, 0)$
$\frac{3\pi}{2}$	$(0, -1)$
2π	$(1, 0)$

In general, each real number t also corresponds to a central angle θ (in standard position) whose radian measure is t. With this interpretation of t, the arc length formula $s = r\theta$ (with $r = 1$) indicates that the real number t is the (directional) length of the arc intercepted by the angle θ, given in radians.

The Trigonometric Functions

From the preceding discussion, it follows that the coordinates x and y are two functions of the real variable t. You can use these coordinates to define the six trigonometric functions of t.

sine **cosine** **tangent** **cotangent** **secant** **cosecant**

These six functions are normally abbreviated sin, cos, tan, cot, sec, and csc, respectively.

NOTE Observe that the functions in the second row are the *reciprocals* of the corresponding functions in the first row.

DEFINITIONS OF TRIGONOMETRIC FUNCTIONS

Let t be a real number and let (x, y) be the point on the unit circle corresponding to t.

$$\sin t = y \qquad \cos t = x \qquad \tan t = \frac{y}{x},\quad x \neq 0$$

$$\csc t = \frac{1}{y},\quad y \neq 0 \qquad \sec t = \frac{1}{x},\quad x \neq 0 \qquad \cot t = \frac{x}{y},\quad y \neq 0$$

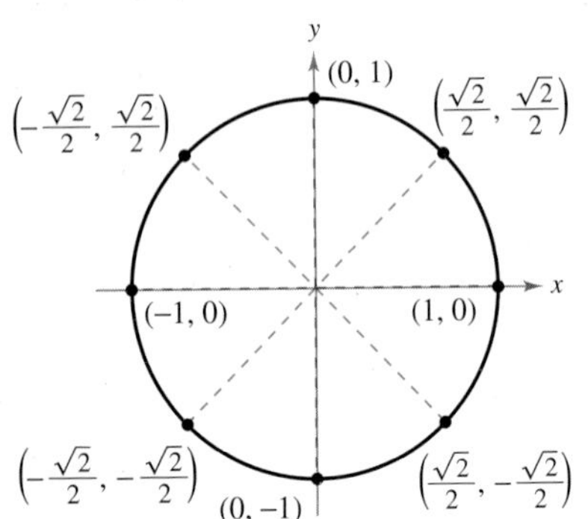

Figure 9.17

In the definitions of the trigonometric functions, note that the tangent and secant are not defined when $x = 0$. For instance, because $t = \pi/2$ corresponds to $(x, y) = (0, 1)$, it follows that $\tan(\pi/2)$ and $\sec(\pi/2)$ are *undefined.* Similarly, the cotangent and cosecant are not defined when $y = 0$. For instance, because $t = 0$ corresponds to $(x, y) = (1, 0)$, cot 0 and csc 0 are *undefined.*

In Figure 9.17, the unit circle has been divided into eight equal arcs, corresponding to t-values of

$$0, \frac{\pi}{4}, \frac{\pi}{2}, \frac{3\pi}{4}, \pi, \frac{5\pi}{4}, \frac{3\pi}{2}, \frac{7\pi}{4}, \text{ and } 2\pi.$$

Similarly, in Figure 9.18, the unit circle has been divided into 12 equal arcs, corresponding to t-values of

$$0, \frac{\pi}{6}, \frac{\pi}{3}, \frac{\pi}{2}, \frac{2\pi}{3}, \frac{5\pi}{6}, \pi, \frac{7\pi}{6}, \frac{4\pi}{3}, \frac{3\pi}{2}, \frac{5\pi}{3}, \frac{11\pi}{6}, \text{ and } 2\pi.$$

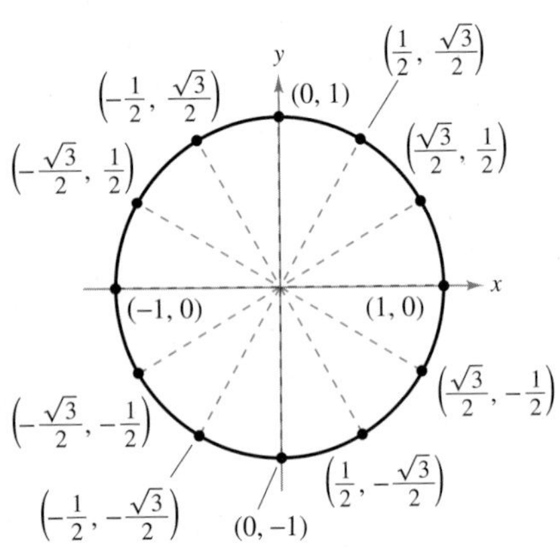

Figure 9.18

To verify the points on the unit circle in Figure 9.17, note that $\left(\frac{\sqrt{2}}{2}, \frac{\sqrt{2}}{2}\right)$ also lies on the line $y = x$. So, substituting x for y in the equation of the unit circle produces the following.

$$x^2 + x^2 = 1 \quad \Rightarrow \quad 2x^2 = 1 \quad \Rightarrow \quad x^2 = \frac{1}{2} \quad \Rightarrow \quad x = \pm\frac{\sqrt{2}}{2}$$

Because the point is in the first quadrant, $x = \frac{\sqrt{2}}{2}$ and because $y = x$, you also have $y = \frac{\sqrt{2}}{2}$. You can use similar reasoning to verify the rest of the points in Figure 9.17 and the points in Figure 9.18.

Using the (x, y) coordinates in Figures 9.17 and 9.18, you can evaluate the trigonometric functions for common t-values. This procedure is demonstrated in Examples 1, 2, and 3. You should study and learn these exact function values for common t-values because they will help you in later sections to perform calculations.

STUDY TIP On the unit circles in Figures 9.17 and 9.18, each point (x, y) can be designated as $(\cos t, \sin t)$.

EXAMPLE 1 Evaluating Trigonometric Functions

Evaluate the six trigonometric functions at each real number.

a. $t = \dfrac{\pi}{6}$

b. $t = \dfrac{5\pi}{4}$

c. $t = 0$

d. $t = \pi$

Solution For each t-value, begin by finding the corresponding point (x, y) on the unit circle. Then use the definitions of trigonometric functions listed on the previous page.

a. $t = \dfrac{\pi}{6}$ corresponds to the point $(x, y) = \left(\dfrac{\sqrt{3}}{2}, \dfrac{1}{2}\right) = (\cos t, \sin t)$.

$$\sin\frac{\pi}{6} = y = \frac{1}{2} \qquad \csc\frac{\pi}{6} = \frac{1}{y} = \frac{1}{1/2} = 2$$

$$\cos\frac{\pi}{6} = x = \frac{\sqrt{3}}{2} \qquad \sec\frac{\pi}{6} = \frac{1}{x} = \frac{2}{\sqrt{3}} = \frac{2\sqrt{3}}{3}$$

$$\tan\frac{\pi}{6} = \frac{y}{x} = \frac{1/2}{\sqrt{3}/2} = \frac{1}{\sqrt{3}} = \frac{\sqrt{3}}{3} \qquad \cot\frac{\pi}{6} = \frac{x}{y} = \frac{\sqrt{3}/2}{1/2} = \sqrt{3}$$

b. $t = \dfrac{5\pi}{4}$ corresponds to the point $(x, y) = \left(-\dfrac{\sqrt{2}}{2}, -\dfrac{\sqrt{2}}{2}\right) = (\cos t, \sin t)$.

$$\sin\frac{5\pi}{4} = y = -\frac{\sqrt{2}}{2} \qquad \csc\frac{5\pi}{4} = \frac{1}{y} = -\frac{2}{\sqrt{2}} = -\sqrt{2}$$

$$\cos\frac{5\pi}{4} = x = -\frac{\sqrt{2}}{2} \qquad \sec\frac{5\pi}{4} = \frac{1}{x} = -\frac{2}{\sqrt{2}} = -\sqrt{2}$$

$$\tan\frac{5\pi}{4} = \frac{y}{x} = \frac{-\sqrt{2}/2}{-\sqrt{2}/2} = 1 \qquad \cot\frac{5\pi}{4} = \frac{x}{y} = \frac{-\sqrt{2}/2}{-\sqrt{2}/2} = 1$$

c. $t = 0$ corresponds to the point $(x, y) = (1, 0)$.

$$\sin 0 = y = 0 \qquad \csc 0 = \frac{1}{y} \text{ is undefined.}$$

$$\cos 0 = x = 1 \qquad \sec 0 = \frac{1}{x} = \frac{1}{1} = 1$$

$$\tan 0 = \frac{y}{x} = \frac{0}{1} = 0 \qquad \cot 0 = \frac{x}{y} \text{ is undefined.}$$

d. $t = \pi$ corresponds to the point $(x, y) = (-1, 0)$.

$$\sin \pi = y = 0 \qquad \csc \pi = \frac{1}{y} \text{ is undefined.}$$

$$\cos \pi = x = -1 \qquad \sec \pi = \frac{1}{x} = \frac{1}{-1} = -1$$

$$\tan \pi = \frac{y}{x} = \frac{0}{-1} = 0 \qquad \cot \pi = \frac{x}{y} \text{ is undefined.}$$

EXAMPLE 2 Evaluating Trigonometric Functions

Evaluate the six trigonometric functions at $t = -\dfrac{\pi}{3}$.

Solution Moving *clockwise* around the unit circle, it follows that $t = -\pi/3$ corresponds to the point $(x, y) = \left(1/2, -\sqrt{3}/2\right)$.

$$\sin\left(-\frac{\pi}{3}\right) = -\frac{\sqrt{3}}{2} \qquad \csc\left(-\frac{\pi}{3}\right) = -\frac{2}{\sqrt{3}} = -\frac{2\sqrt{3}}{3}$$

$$\cos\left(-\frac{\pi}{3}\right) = \frac{1}{2} \qquad \sec\left(-\frac{\pi}{3}\right) = 2$$

$$\tan\left(-\frac{\pi}{3}\right) = \frac{-\sqrt{3}/2}{1/2} = -\sqrt{3} \qquad \cot\left(-\frac{\pi}{3}\right) = \frac{1/2}{-\sqrt{3}/2} = -\frac{1}{\sqrt{3}} = -\frac{\sqrt{3}}{3}$$

■

Domain and Period of Sine and Cosine

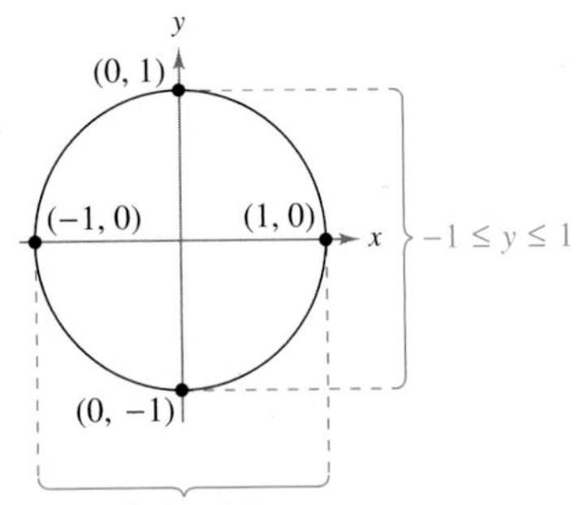

Figure 9.19

The *domain* of the sine and cosine functions is the set of all real numbers. To determine the *range* of these two functions, consider the unit circle shown in Figure 9.19. By definition, $\sin t = y$ and $\cos t = x$. Because (x, y) is on the unit circle, you know that $-1 \le y \le 1$ and $-1 \le x \le 1$. So, the values of sine and cosine also range between -1 and 1.

$$-1 \le y \le 1 \qquad \text{and} \qquad -1 \le x \le 1$$

$$-1 \le \sin t \le 1 \qquad\qquad -1 \le \cos t \le 1$$

Adding 2π to each value of t in the interval $[0, 2\pi]$ completes a second revolution around the unit circle, as shown in Figure 9.20. The values of $\sin(t + 2\pi)$ and $\cos(t + 2\pi)$ correspond to those of $\sin t$ and $\cos t$. Similar results can be obtained for repeated revolutions (positive or negative) on the unit circle. This leads to the general result

$$\sin(t + 2\pi n) = \sin t$$

and

$$\cos(t + 2\pi n) = \cos t$$

for any integer n and real number t. Functions that behave in such a repetitive (or cyclic) manner are called **periodic.**

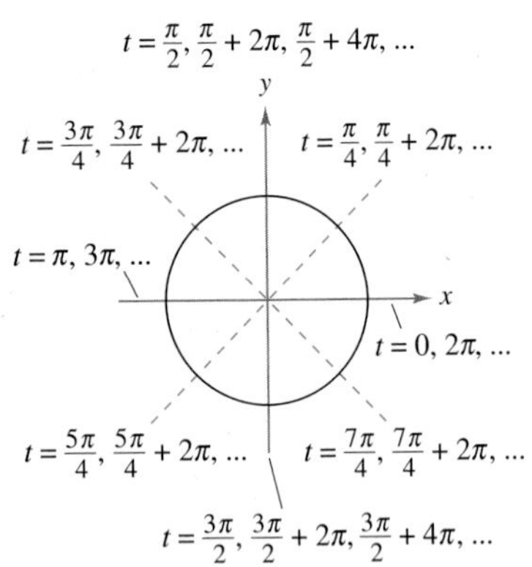

Figure 9.20

NOTE In Figure 9.20, note that *positive* multiples of 2π are added to the t-values. You could just as well have added *negative* multiples. For instance, $\pi/4 - 2\pi$ and $\pi/4 - 4\pi$ are also coterminal with $\pi/4$. ■

DEFINITION OF PERIODIC FUNCTION

A function f is **periodic** if there exists a positive real number c such that

$$f(t + c) = f(t)$$

for all t in the domain of f. The smallest number c for which f is periodic is called the **period** of f.

Recall from Section 1.2 that a function f is *even* if $f(-t) = f(t)$, and is *odd* if $f(-t) = -f(t)$.

EVEN AND ODD TRIGONOMETRIC FUNCTIONS

The cosine and secant functions are *even.*

$$\cos(-t) = \cos t \qquad \sec(-t) = \sec t$$

The sine, cosecant, tangent, and cotangent functions are *odd.*

$$\sin(-t) = -\sin t \qquad \csc(-t) = -\csc t$$

$$\tan(-t) = -\tan t \qquad \cot(-t) = -\cot t$$

EXAMPLE 3 Using the Period to Evaluate the Sine and Cosine

a. Because $\frac{13\pi}{6} = 2\pi + \frac{\pi}{6}$, you have $\sin\frac{13\pi}{6} = \sin\left(2\pi + \frac{\pi}{6}\right) = \sin\frac{\pi}{6} = \frac{1}{2}$.

b. Because $-\frac{7\pi}{2} = -4\pi + \frac{\pi}{2}$, you have

$$\cos\left(-\frac{7\pi}{2}\right) = \cos\left(-4\pi + \frac{\pi}{2}\right) = \cos\frac{\pi}{2} = 0.$$

c. For $\sin t = \frac{4}{5}$, $\sin(-t) = -\frac{4}{5}$ because the sine function is odd. ■

STUDY TIP From the definition of periodic function, it follows that the sine and cosine functions are periodic and have a period of 2π. The other four trigonometric functions are also periodic, and will be discussed further in Section 9.6.

Evaluating Trigonometric Functions with a Calculator

When evaluating a trigonometric function with a calculator, you need to set the calculator to the desired *mode* of measurement (*degree* or *radian*).

Most calculators do not have keys for the cosecant, secant, and cotangent functions. To evaluate these functions, you can use the [x⁻¹] key with their respective reciprocal functions sine, cosine, and tangent. For instance, to evaluate $\csc(\pi/8)$, use the fact that

$$\csc\frac{\pi}{8} = \frac{1}{\sin(\pi/8)}$$

and enter the following keystroke sequence in *radian* mode.

[(] [SIN] [(] [π] [÷] 8 [)] [)] [x^{-1}] [ENTER] Display 2.6131259

TECHNOLOGY When evaluating trigonometric functions with a calculator, remember to enclose all fractional angle measures in parentheses. For instance, if you want to evaluate $\sin t$ for $t = \pi/6$, you should enter

[SIN] [(] [π] [÷] 6 [)] [ENTER].

These keystrokes yield the correct value of 0.5. Note that some calculators automatically place a left parenthesis after trigonometric functions. Check the user's guide for your calculator for specific keystrokes on how to evaluate trigonometric functions.

EXAMPLE 4 Using a Calculator

Function	*Mode*	*Calculator Keystrokes*	*Display*
a. $\sin\frac{2\pi}{3}$	Radian	[SIN] [(] 2 [π] [÷] 3 [)] [ENTER]	0.8660254
b. cot 1.5	Radian	[(] [TAN] [(] 1.5 [)] [)] [x^{-1}] [ENTER]	0.0709148

■

9.2 Exercises

See www.CalcChat.com for worked-out solutions to odd-numbered exercises.

In Exercises 1–4, fill in the blanks.

1. Each real number t corresponds to a point (x, y) on the ________ ________.
2. A function f is ________ if there exists a positive real number c such that $f(t + c) = f(t)$ for all t in the domain of f.
3. The smallest number c for which a function f is periodic is called the ________ of f.
4. A function f is ________ if $f(-t) = -f(t)$ and ________ if $f(-t) = f(t)$.

In Exercises 5–8, determine the exact values of the six trigonometric functions of the real number t.

5.
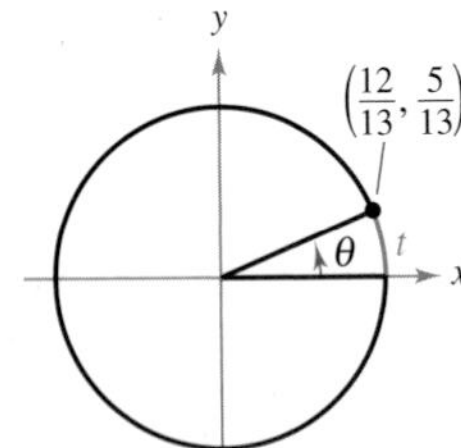

6.
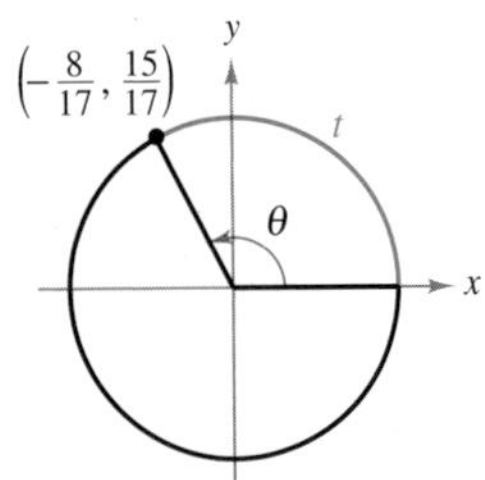

7.
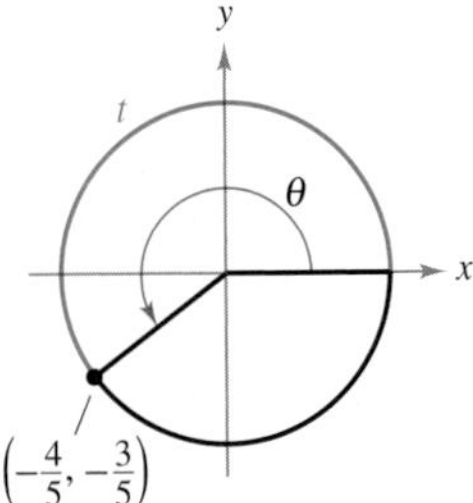

8.
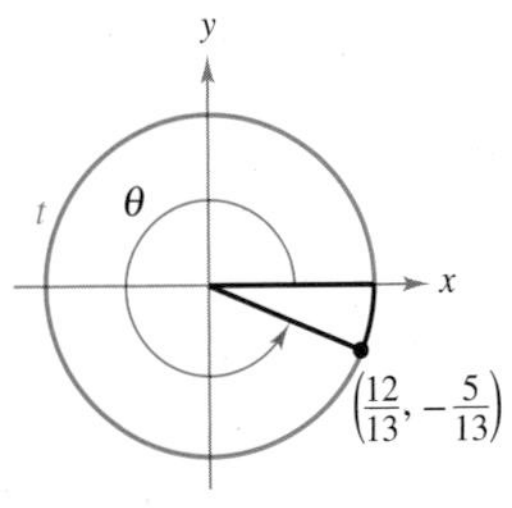

In Exercises 9–16, find the point (x, y) on the unit circle that corresponds to the real number t.

9. $t = \frac{\pi}{2}$
10. $t = \pi$
11. $t = \frac{\pi}{4}$
12. $t = \frac{\pi}{3}$
13. $t = \frac{5\pi}{6}$
14. $t = \frac{3\pi}{4}$
15. $t = \frac{4\pi}{3}$
16. $t = \frac{5\pi}{3}$

In Exercises 17–26, evaluate (if possible) the sine, cosine, and tangent of the real number t.

17. $t = \frac{\pi}{4}$
18. $t = \frac{\pi}{3}$
19. $t = -\frac{\pi}{6}$
20. $t = -\frac{\pi}{4}$
21. $t = -\frac{7\pi}{4}$
22. $t = -\frac{4\pi}{3}$
23. $t = \frac{11\pi}{6}$
24. $t = \frac{5\pi}{3}$
25. $t = -\frac{3\pi}{2}$
26. $t = -2\pi$

In Exercises 27–34, evaluate (if possible) the six trigonometric functions of the real number t.

27. $t = \frac{2\pi}{3}$
28. $t = \frac{5\pi}{6}$
29. $t = \frac{4\pi}{3}$
30. $t = \frac{7\pi}{4}$
31. $t = \frac{3\pi}{4}$
32. $t = \frac{3\pi}{2}$
33. $t = -\frac{\pi}{2}$
34. $t = -\pi$

In Exercises 35–42, evaluate the trigonometric function using its period as an aid.

35. $\sin 4\pi$
36. $\cos 3\pi$
37. $\cos \frac{7\pi}{3}$
38. $\sin \frac{9\pi}{4}$
39. $\cos \frac{17\pi}{4}$
40. $\sin \frac{19\pi}{6}$
41. $\sin\left(-\frac{8\pi}{3}\right)$
42. $\cos\left(-\frac{9\pi}{4}\right)$

In Exercises 43–48, use the value of the trigonometric function to evaluate the indicated functions.

43. $\sin t = \frac{1}{2}$
 (a) $\sin(-t)$
 (b) $\csc(-t)$
44. $\sin(-t) = \frac{3}{8}$
 (a) $\sin t$
 (b) $\csc t$
45. $\cos(-t) = -\frac{1}{5}$
 (a) $\cos t$
 (b) $\sec(-t)$
46. $\cos t = -\frac{3}{4}$
 (a) $\cos(-t)$
 (b) $\sec(-t)$
47. $\sin t = \frac{4}{5}$
 (a) $\sin(\pi - t)$
 (b) $\sin(t + \pi)$
48. $\cos t = \frac{4}{5}$
 (a) $\cos(\pi - t)$
 (b) $\cos(t + \pi)$

In Exercises 49–58, use a calculator to evaluate the trigonometric function. Round your answer to four decimal places. (Be sure the calculator is set in the correct angle mode.)

49. $\sin \frac{\pi}{4}$

50. $\tan \frac{\pi}{3}$

51. $\cot \frac{\pi}{4}$

52. $\csc \frac{2\pi}{3}$

53. $\cos(-1.7)$

54. $\cos(-2.5)$

55. $\csc 0.8$

56. $\sec 1.8$

57. $\sec(-22.8)$

58. $\cot(-0.9)$

59. ***Harmonic Motion*** The displacement from equilibrium of an oscillating weight suspended by a spring is given by $y(t) = \frac{1}{4}\cos 6t$, where y is the displacement (in feet) and t is the time (in seconds). Find the displacements when (a) $t = 0$, (b) $t = \frac{1}{4}$, and (c) $t = \frac{1}{2}$.

60. ***Harmonic Motion*** The displacement from equilibrium of an oscillating weight suspended by a spring and subject to the damping effect of friction is given by $y(t) = \frac{1}{4}e^{-t}\cos 6t$, where y is the displacement (in feet) and t is the time (in seconds).

(a) Complete the table.

t	0	$\frac{1}{4}$	$\frac{1}{2}$	$\frac{3}{4}$	1
y					

(b) Use the *table* feature of a graphing utility to approximate the time when the weight reaches equilibrium.

(c) What appears to happen to the displacement as t increases?

True or False? **In Exercises 61–64, determine whether the statement is true or false. Justify your answer.**

61. Because $\sin(-t) = -\sin t$, it can be said that the sine of a negative angle is a negative number.

62. Because the sine is an odd function, its graph is symmetric to the y-axis.

63. The real number 0 corresponds to the point $(0, 1)$ on the unit circle.

64. $\cos\left(-\frac{7\pi}{2}\right) = \cos\left(\pi + \frac{\pi}{2}\right)$

65. Verify that $\cos 2t \neq 2\cos t$ by approximating $\cos 1.5$ and $2\cos 0.75$.

66. Verify that $\sin(t_1 + t_2) \neq \sin t_1 + \sin t_2$ by approximating $\sin 0.25$, $\sin 0.75$, and $\sin 1$.

WRITING ABOUT CONCEPTS

67. Let (x_1, y_1) and (x_2, y_2) be points on the unit circle corresponding to $t = t_1$ and $t = \pi - t_1$, respectively.

(a) Identify the symmetry of the points (x_1, y_1) and (x_2, y_2).

(b) Make a conjecture about any relationship between $\sin t_1$ and $\sin(\pi - t_1)$.

(c) Make a conjecture about any relationship between $\cos t_1$ and $\cos(\pi - t_1)$.

68. Use the unit circle to verify that the cosine and secant functions are even and that the sine, cosecant, tangent, and cotangent functions are odd.

69. Because $f(t) = \sin t$ is an odd function and $g(t) = \cos t$ is an even function, what can be said about the function $h(t) = f(t)g(t)$?

70. Because $f(t) = \sin t$ and $g(t) = \tan t$ are odd functions, what can be said about the function $h(t) = f(t)g(t)$?

71. ***Graphical Analysis*** With your graphing utility in *radian* and *parametric* modes, enter the equations

$X_{1T} = \cos T$ and $Y_{1T} = \sin T$

and use the following settings.

Tmin = 0, Tmax = 6.3, Tstep = 0.1

Xmin = −1.5, Xmax = 1.5, Xscl = 1

Ymin = −1, Ymax = 1, Yscl = 1

(a) Graph the entered equations and describe the graph.

(b) Use the *trace* feature to move the cursor around the graph. What do the t-values represent? What do the x- and y-values represent?

(c) What are the least and greatest values of x and y?

CAPSTONE

72. A student you are tutoring has used a unit circle divided into 8 equal parts to complete the table for selected values of t. What is wrong?

t	0	$\frac{\pi}{4}$	$\frac{\pi}{2}$	$\frac{3\pi}{4}$	π
x	1	$\frac{\sqrt{2}}{2}$	0	$-\frac{\sqrt{2}}{2}$	-1
y	0	$\frac{\sqrt{2}}{2}$	1	$\frac{\sqrt{2}}{2}$	0
$\sin t$	1	$\frac{\sqrt{2}}{2}$	0	$-\frac{\sqrt{2}}{2}$	-1
$\cos t$	0	$\frac{\sqrt{2}}{2}$	1	$\frac{\sqrt{2}}{2}$	0
$\tan t$	Undef.	1	0	-1	Undef.

9.3 Right Triangle Trigonometry

- Evaluate trigonometric functions of acute angles.
- Use fundamental trigonometric identities.
- Use trigonometric functions to model and solve real-life problems.

The Six Trigonometric Functions

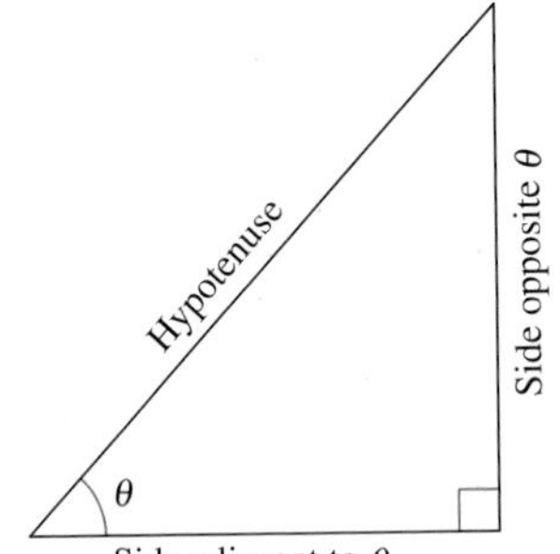

Figure 9.21

Our second look at the trigonometric functions is from a *right triangle* perspective. Consider a right triangle, with one acute angle labeled θ, as shown in Figure 9.21. Relative to the angle θ, the three sides of the triangle are the **hypotenuse,** the **opposite side** (the side opposite the angle θ), and the **adjacent side** (the side adjacent to the angle θ).

Using the lengths of these three sides, you can form six ratios that define the six trigonometric functions of the acute angle θ.

sine cosecant cosine secant tangent cotangent

In the following definitions, it is important to see that $0° < \theta < 90°$ (θ lies in the first quadrant) and that for such angles the value of each trigonometric function is *positive.*

RIGHT TRIANGLE DEFINITIONS OF TRIGONOMETRIC FUNCTIONS

Let θ be an *acute* angle of a right triangle. The six trigonometric functions of the angle θ are defined as follows. (Note that the functions in the second row are the *reciprocals* of the corresponding functions in the first row.)

$$\sin\theta = \frac{\text{opp}}{\text{hyp}} \qquad \cos\theta = \frac{\text{adj}}{\text{hyp}} \qquad \tan\theta = \frac{\text{opp}}{\text{adj}}$$

$$\csc\theta = \frac{\text{hyp}}{\text{opp}} \qquad \sec\theta = \frac{\text{hyp}}{\text{adj}} \qquad \cot\theta = \frac{\text{adj}}{\text{opp}}$$

The abbreviations opp, adj, and hyp represent the lengths of the three sides of a right triangle.

opp = the length of the side *opposite* θ

adj = the length of the side *adjacent to* θ

hyp = the length of the *hypotenuse*

HISTORICAL NOTE

Georg Joachim Rhaeticus (1514–1574) was the leading Teutonic mathematical astronomer of the 16th century. He was the first to define the trigonometric functions as ratios of the sides of a right triangle.

EXAMPLE 1 Evaluating Trigonometric Functions

Use the triangle in Figure 9.22 to find the values of the six trigonometric functions of θ.

Solution By the Pythagorean Theorem, $(\text{hyp})^2 = (\text{opp})^2 + (\text{adj})^2$, it follows that

$$\text{hyp} = \sqrt{4^2 + 3^2} = \sqrt{25} = 5.$$

So, the six trigonometric functions of θ are

$$\sin\theta = \frac{\text{opp}}{\text{hyp}} = \frac{4}{5} \qquad \cos\theta = \frac{\text{adj}}{\text{hyp}} = \frac{3}{5} \qquad \tan\theta = \frac{\text{opp}}{\text{adj}} = \frac{4}{3}$$

$$\csc\theta = \frac{\text{hyp}}{\text{opp}} = \frac{5}{4} \qquad \sec\theta = \frac{\text{hyp}}{\text{adj}} = \frac{5}{3} \qquad \cot\theta = \frac{\text{adj}}{\text{opp}} = \frac{3}{4}.$$

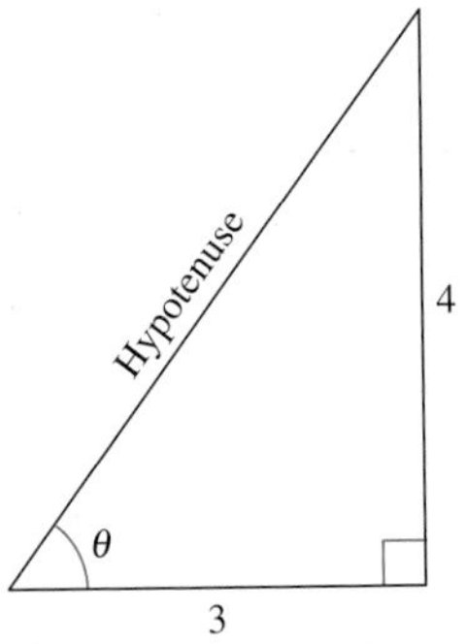

Hypotenuse, 4, 3, θ

Figure 9.22

EXAMPLE 2 Evaluating Trigonometric Functions of 45°

Find the values of sin 45°, cos 45°, and tan 45°.

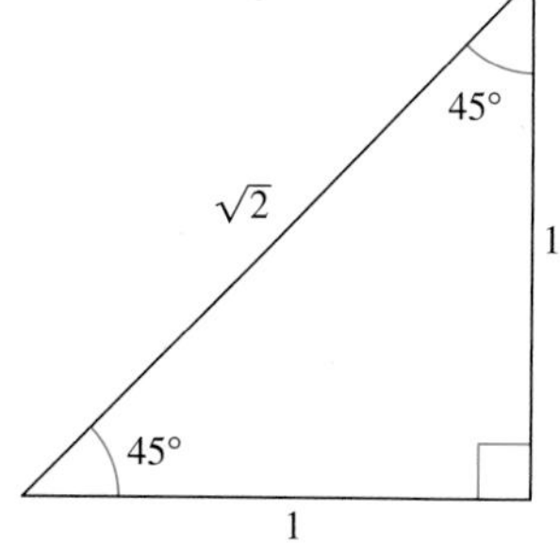

Figure 9.23

Solution Construct a right triangle having 45° as one of its acute angles, as shown in Figure 9.23. Choose the length of the adjacent side to be 1. From geometry, you know that the other acute angle is also 45°. So, the triangle is isosceles and the length of the opposite side is also 1. Using the Pythagorean Theorem, you find the length of the hypotenuse to be $\sqrt{2}$.

$$\sin 45° = \frac{\text{opp}}{\text{hyp}} = \frac{1}{\sqrt{2}} = \frac{\sqrt{2}}{2}$$

$$\cos 45° = \frac{\text{adj}}{\text{hyp}} = \frac{1}{\sqrt{2}} = \frac{\sqrt{2}}{2}$$

$$\tan 45° = \frac{\text{opp}}{\text{adj}} = \frac{1}{1} = 1$$

EXAMPLE 3 Evaluating Trigonometric Functions of 30° and 60°

Use the equilateral triangle shown in Figure 9.24 to find the values of sin 60°, cos 60°, sin 30°, and cos 30°.

Figure 9.24

Solution Use the Pythagorean Theorem and the equilateral triangle in Figure 9.24 to verify the lengths of the sides shown in the figure. For $\theta = 60°$, you have adj $= 1$, opp $= \sqrt{3}$, and hyp $= 2$. So,

$$\sin 60° = \frac{\text{opp}}{\text{hyp}} = \frac{\sqrt{3}}{2} \quad \text{and} \quad \cos 60° = \frac{\text{adj}}{\text{hyp}} = \frac{1}{2}.$$

For $\theta = 30°$, adj $= \sqrt{3}$, opp $= 1$, and hyp $= 2$. So,

$$\sin 30° = \frac{\text{opp}}{\text{hyp}} = \frac{1}{2} \quad \text{and} \quad \cos 30° = \frac{\text{adj}}{\text{hyp}} = \frac{\sqrt{3}}{2}.$$

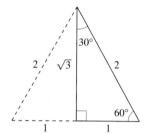

■

TECHNOLOGY You can use a calculator to convert the answers in Example 3 to decimals. However, the radical form is the exact value and in most cases, the exact value is preferred.

STUDY TIP Because the angles 30°, 45°, and 60° ($\pi/6$, $\pi/4$, and $\pi/3$) occur frequently in trigonometry, you should learn to construct the triangles shown in Figures 9.23 and 9.24.

SINES, COSINES, AND TANGENTS OF SPECIAL ANGLES

$$\sin 30° = \sin\frac{\pi}{6} = \frac{1}{2} \qquad \cos 30° = \cos\frac{\pi}{6} = \frac{\sqrt{3}}{2} \qquad \tan 30° = \tan\frac{\pi}{6} = \frac{\sqrt{3}}{3}$$

$$\sin 45° = \sin\frac{\pi}{4} = \frac{\sqrt{2}}{2} \qquad \cos 45° = \cos\frac{\pi}{4} = \frac{\sqrt{2}}{2} \qquad \tan 45° = \tan\frac{\pi}{4} = 1$$

$$\sin 60° = \sin\frac{\pi}{3} = \frac{\sqrt{3}}{2} \qquad \cos 60° = \cos\frac{\pi}{3} = \frac{1}{2} \qquad \tan 60° = \tan\frac{\pi}{3} = \sqrt{3}$$

In the preceding box, note that $\sin 30° = \frac{1}{2} = \cos 60°$. This occurs because 30° and 60° are complementary angles. In general, it can be shown from the right triangle definitions that *cofunctions of complementary angles are equal.* That is, if θ is an acute angle, the following relationships are true.

$$\sin(90° - \theta) = \cos\theta \qquad \cos(90° - \theta) = \sin\theta$$

$$\tan(90° - \theta) = \cot\theta \qquad \cot(90° - \theta) = \tan\theta$$

$$\sec(90° - \theta) = \csc\theta \qquad \csc(90° - \theta) = \sec\theta$$

Trigonometric Identities

In trigonometry, a great deal of time is spent studying relationships between trigonometric functions (identities).

FUNDAMENTAL TRIGONOMETRIC IDENTITIES

Reciprocal Identities

$\sin\theta = \dfrac{1}{\csc\theta}$ $\cos\theta = \dfrac{1}{\sec\theta}$ $\tan\theta = \dfrac{1}{\cot\theta}$

$\csc\theta = \dfrac{1}{\sin\theta}$ $\sec\theta = \dfrac{1}{\cos\theta}$ $\cot\theta = \dfrac{1}{\tan\theta}$

Quotient Identities

$\tan\theta = \dfrac{\sin\theta}{\cos\theta}$ $\cot\theta = \dfrac{\cos\theta}{\sin\theta}$

Pythagorean Identities

$\sin^2\theta + \cos^2\theta = 1$ $1 + \tan^2\theta = \sec^2\theta$

$1 + \cot^2\theta = \csc^2\theta$

Note that $\sin^2\theta$ represents $(\sin\theta)^2$, $\cos^2\theta$ represents $(\cos\theta)^2$, and so on.

EXAMPLE 4 Applying Trigonometric Identities

Let θ be an acute angle such that $\sin\theta = 0.6$. Find the values of (a) $\cos\theta$ and (b) $\tan\theta$ using trigonometric identities.

Solution

a. To find the value of $\cos\theta$, use the Pythagorean identity

$$\sin^2\theta + \cos^2\theta = 1.$$

So, you have

$$(0.6)^2 + \cos^2\theta = 1 \quad \text{Substitute 0.6 for } \sin\theta.$$

$$\cos^2\theta = 1 - (0.6)^2 = 0.64 \quad \text{Subtract } (0.6)^2 \text{ from each side.}$$

$$\cos\theta = \sqrt{0.64} = 0.8. \quad \text{Extract the positive square root.}$$

b. Now, knowing the sine and cosine of θ, you can find the tangent of θ to be

$$\tan\theta = \frac{\sin\theta}{\cos\theta} = \frac{0.6}{0.8} = 0.75.$$

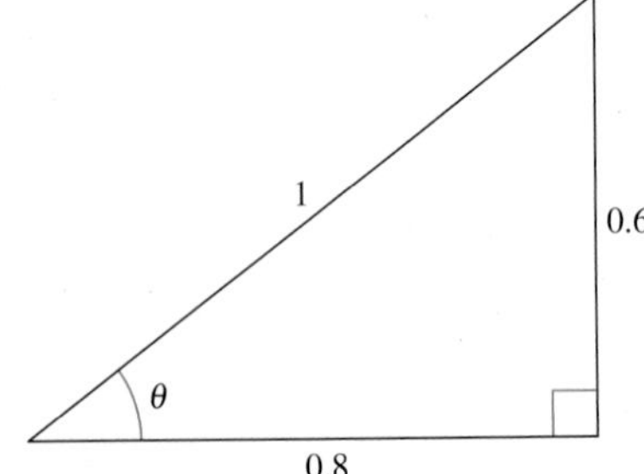

Figure 9.25

Use the definitions of $\cos\theta$ and $\tan\theta$, and the triangle shown in Figure 9.25, to check these results. ■

Applications Involving Right Triangles

To use a calculator to evaluate trigonometric functions of angles measured in degrees, first set the calculator to *degree* mode and then proceed as demonstrated in Section 9.2. For instance, you can find values of cos 28° and sec 28° as shown.

Function	*Calculator Keystrokes*	*Display*
cos 28°	COS 28 ENTER	0.8829476
sec 28°	(COS 28) x^{-1} ENTER	1.1325701

NOTE Throughout this text, angles are assumed to be measured in radians unless noted otherwise. For example, sin 1 means the sine of 1 radian and sin 1° means the sine of 1 degree.

Many applications of trigonometry involve a process called **solving right triangles.** In this type of application, you are usually given one side of a right triangle and one of the acute angles and asked to find one of the other sides, *or* you are given two sides and asked to find one of the acute angles.

In Example 5, the angle you are given is the **angle of elevation,** which represents the angle from the horizontal upward to an object. For objects that lie below the horizontal, it is common to use the term **angle of depression.**

EXAMPLE 5 Using Trigonometry to Solve a Right Triangle

A surveyor is standing 115 feet from the base of the Washington Monument, as shown in Figure 9.26. The surveyor measures the angle of elevation to the top of the monument as 78.3°. How tall is the Washington Monument?

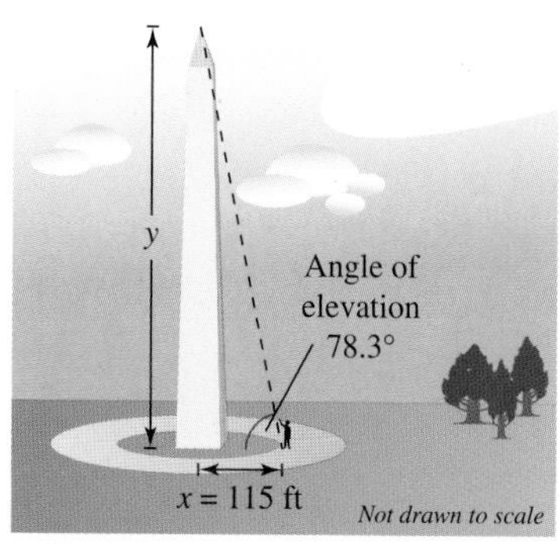

Figure 9.26

Solution From Figure 9.26, you see that

$$\tan 78.3° = \frac{\text{opp}}{\text{adj}} = \frac{y}{x}$$

where $x = 115$ and y is the height of the monument. So, the height of the Washington Monument is

$$y = x \tan 78.3° \approx 115(4.82882) \approx 555 \text{ feet.}$$

EXAMPLE 6 Using Trigonometry to Solve a Right Triangle

You are 200 yards from a river. Rather than walking directly to the river, you walk 400 yards along a straight path to the river's edge. Find the acute angle θ between this path and the river's edge, as illustrated in Figure 9.27.

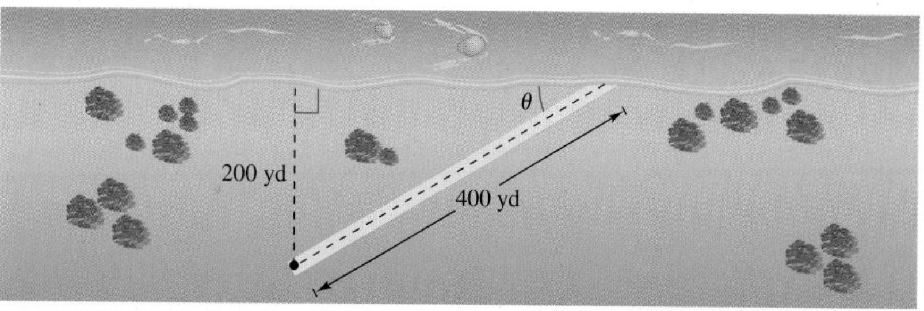

Figure 9.27

Solution From Figure 9.27, you can see that the sine of the angle θ is

$$\sin \theta = \frac{\text{opp}}{\text{hyp}} = \frac{200}{400} = \frac{1}{2}.$$

So, $\theta = 30°$. ■

By now you are able to recognize that $\theta = 30°$ is the acute angle that satisfies the equation $\sin\theta = \frac{1}{2}$. Suppose, however, that you were given the equation $\sin\theta = 0.6$ and were asked to find the acute angle θ. Because

$$\sin 30° = \frac{1}{2}$$
$$= 0.5000$$

and

$$\sin 45° = \frac{1}{\sqrt{2}}$$
$$\approx 0.7071$$

you might guess that θ lies somewhere between 30° and 45°. In a later section, you will study a method by which a more precise value of θ can be determined.

EXAMPLE 7 Solving a Right Triangle

Find the length c of the skateboard ramp shown in Figure 9.28.

Figure 9.28

Solution From Figure 9.28, you can see that

$$\sin 18.4° = \frac{\text{opp}}{\text{hyp}}$$
$$= \frac{4}{c}.$$

So, the length of the skateboard ramp is

$$c = \frac{4}{\sin 18.4°}$$
$$\approx \frac{4}{0.3156}$$
$$\approx 12.7 \text{ feet.}$$

■

9.3 Exercises

See www.CalcChat.com for worked-out solutions to odd-numbered exercises.

1. Match the trigonometric function with its right triangle definition.

(a) Sine	(i) $\dfrac{\text{hypotenuse}}{\text{adjacent}}$
(b) Cosine	(ii) $\dfrac{\text{adjacent}}{\text{opposite}}$
(c) Tangent	(iii) $\dfrac{\text{hypotenuse}}{\text{opposite}}$
(d) Cosecant	(iv) $\dfrac{\text{adjacent}}{\text{hypotenuse}}$
(e) Secant	(v) $\dfrac{\text{opposite}}{\text{hypotenuse}}$
(f) Cotangent	(vi) $\dfrac{\text{opposite}}{\text{adjacent}}$

In Exercises 2–4, fill in the blanks.

2. Relative to the angle θ, the three sides of a right triangle are the ________ side, the ________ side, and the ________.

3. Cofunctions of ________ angles are equal.

4. An angle that measures from the horizontal upward to an object is called the angle of ________, whereas an angle that measures from the horizontal downward to an object is called the angle of ________.

In Exercises 5–8, find the exact values of the six trigonometric functions of the angle θ shown in the figure. (Use the Pythagorean Theorem to find the third side of the triangle.)

5.

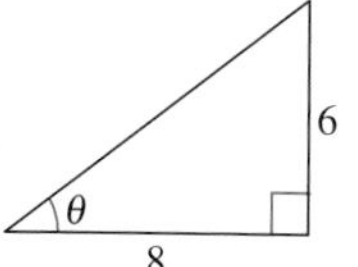

6.

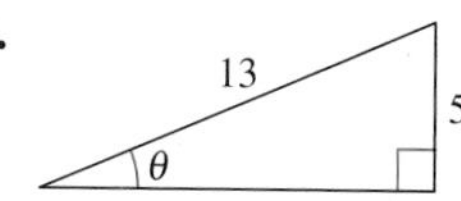

7.

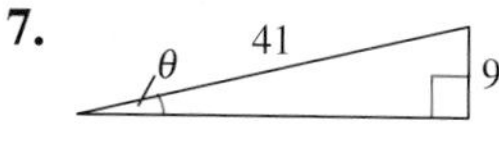

8.

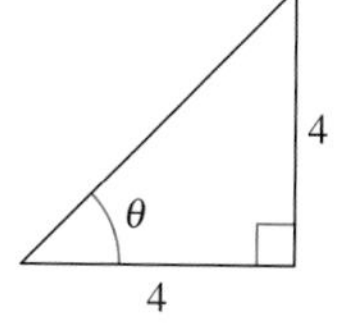

In Exercises 9–12, find the exact values of the six trigonometric functions of the angle θ for each of the two triangles. Explain why the function values are the same.

9.

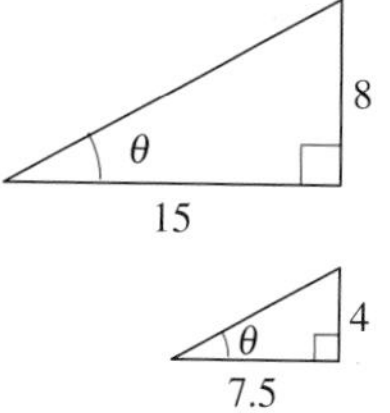

10.

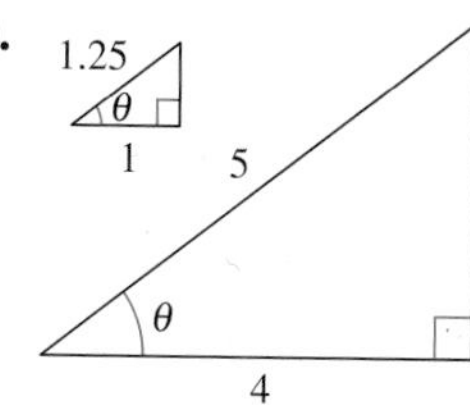

11.

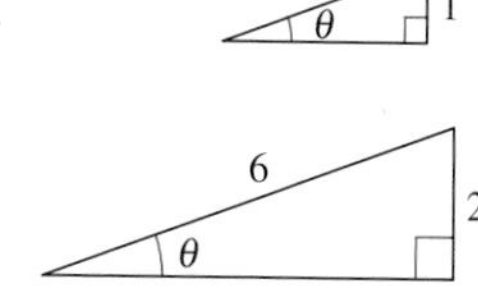

12.

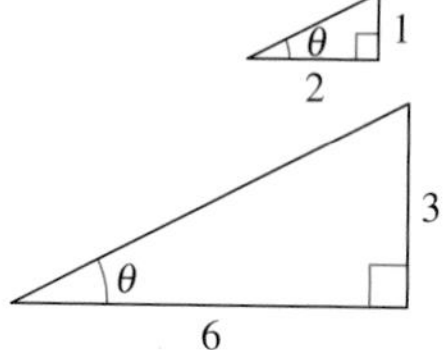

In Exercises 13–20, sketch a right triangle corresponding to the trigonometric function of the acute angle θ. Use the Pythagorean Theorem to determine the third side and then find the other five trigonometric functions of θ.

13. $\tan \theta = \dfrac{3}{4}$

14. $\cos \theta = \dfrac{5}{6}$

15. $\sec \theta = \dfrac{3}{2}$

16. $\tan \theta = \dfrac{4}{5}$

17. $\sin \theta = \dfrac{1}{5}$

18. $\sec \theta = \dfrac{17}{7}$

19. $\cot \theta = 3$

20. $\csc \theta = 9$

In Exercises 21–30, construct an appropriate triangle to complete the table. $(0° \leq \theta \leq 90°, 0 \leq \theta \leq \pi/2)$

	Function	*θ (deg)*	*θ (rad)*	*Function Value*
21.	sin	30°		
22.	cos	45°		
23.	sec		$\dfrac{\pi}{4}$	
24.	tan		$\dfrac{\pi}{3}$	
25.	cot			$\dfrac{\sqrt{3}}{3}$
26.	csc			$\sqrt{2}$
27.	csc		$\dfrac{\pi}{6}$	
28.	sin		$\dfrac{\pi}{4}$	
29.	cot			1
30.	tan			$\dfrac{\sqrt{3}}{3}$

In Exercises 31–36, use the given function value(s), and trigonometric identities (including the cofunction identities), to find the indicated trigonometric functions.

31. $\sin 60° = \dfrac{\sqrt{3}}{2}, \quad \cos 60° = \dfrac{1}{2}$

(a) $\sin 30°$ (b) $\cos 30°$

(c) $\tan 60°$ (d) $\cot 60°$

32. $\sin 30° = \dfrac{1}{2}, \quad \tan 30° = \dfrac{\sqrt{3}}{3}$

(a) $\csc 30°$ (b) $\cot 60°$

(c) $\cos 30°$ (d) $\cot 30°$

33. $\cos \theta = \frac{1}{3}$

(a) $\sin \theta$ (b) $\tan \theta$

(c) $\sec \theta$ (d) $\csc(90° - \theta)$

34. $\sec \theta = 5$

(a) $\cos \theta$ (b) $\cot \theta$

(c) $\cot(90° - \theta)$ (d) $\sin \theta$

35. $\cot \alpha = 5$

(a) $\tan \alpha$ (b) $\csc \alpha$

(c) $\cot(90° - \alpha)$ (d) $\cos \alpha$

36. $\cos \beta = \dfrac{\sqrt{7}}{4}$

(a) $\sec \beta$ (b) $\sin \beta$

(c) $\cot \beta$ (d) $\sin(90° - \beta)$

In Exercises 37–46, use trigonometric identities to transform the left side of the equation into the right side $(0 < \theta < \pi/2)$.

37. $\tan \theta \cot \theta = 1$

38. $\cos \theta \sec \theta = 1$

39. $\tan \alpha \cos \alpha = \sin \alpha$

40. $\cot \alpha \sin \alpha = \cos \alpha$

41. $(1 + \sin \theta)(1 - \sin \theta) = \cos^2 \theta$

42. $(1 + \cos \theta)(1 - \cos \theta) = \sin^2 \theta$

43. $(\sec \theta + \tan \theta)(\sec \theta - \tan \theta) = 1$

44. $\sin^2 \theta - \cos^2 \theta = 2 \sin^2 \theta - 1$

45. $\dfrac{\sin \theta}{\cos \theta} + \dfrac{\cos \theta}{\sin \theta} = \csc \theta \sec \theta$

46. $\dfrac{\tan \beta + \cot \beta}{\tan \beta} = \csc^2 \beta$

In Exercises 47–56, use a calculator to evaluate each function. Round your answers to four decimal places. (Be sure the calculator is in the correct angle mode.)

47. (a) $\sin 10°$ (b) $\cos 80°$

48. (a) $\tan 23.5°$ (b) $\cot 66.5°$

49. (a) $\sin 16.35°$ (b) $\csc 16.35°$

50. (a) $\cot 79.56°$ (b) $\sec 79.56°$

51. (a) $\cos 4° 50' 15''$ (b) $\sec 4° 50' 15''$

52. (a) $\sec 42° 12'$ (b) $\csc 48° 7'$

53. (a) $\cot 11° 15'$ (b) $\tan 11° 15'$

54. (a) $\sec 56° 8' 10''$ (b) $\cos 56° 8' 10''$

55. (a) $\csc 32° 40' 3''$ (b) $\tan 44° 28' 16''$

56. (a) $\sec\left(\frac{9}{5} \cdot 20 + 32\right)°$ (b) $\cot\left(\frac{9}{5} \cdot 30 + 32\right)°$

In Exercises 57–62, find the values of θ in degrees $(0° < \theta < 90°)$ and radians $(0 < \theta < \pi/2)$ without the aid of a calculator.

57. (a) $\sin \theta = \frac{1}{2}$ (b) $\csc \theta = 2$

58. (a) $\cos \theta = \dfrac{\sqrt{2}}{2}$ (b) $\tan \theta = 1$

59. (a) $\sec \theta = 2$ (b) $\cot \theta = 1$

60. (a) $\tan \theta = \sqrt{3}$ (b) $\cos \theta = \frac{1}{2}$

61. (a) $\csc \theta = \dfrac{2\sqrt{3}}{3}$ (b) $\sin \theta = \dfrac{\sqrt{2}}{2}$

62. (a) $\cot \theta = \dfrac{\sqrt{3}}{3}$ (b) $\sec \theta = \sqrt{2}$

In Exercises 63–66, solve for x, y, or r as indicated.

63. Solve for y.

18
y
60°

64. Solve for x.

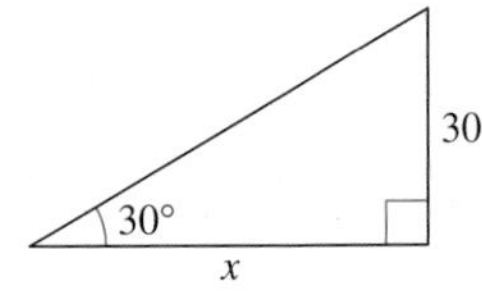

65. Solve for x.

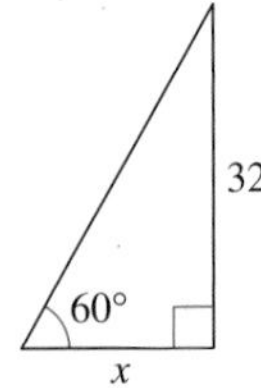

66. Solve for r.

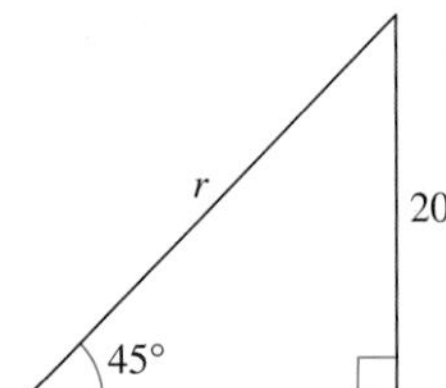

WRITING ABOUT CONCEPTS

67. In right triangle trigonometry, explain why $\sin 30° = \frac{1}{2}$ regardless of the size of the triangle.

68. You are given only the value $\tan \theta$. Is it possible to find the value of $\sec \theta$ without finding the measure of θ? Explain.

WRITING ABOUT CONCEPTS (continued)

69. (a) Complete the table.

θ	0.1	0.2	0.3	0.4	0.5
$\sin \theta$					

(b) Is θ or $\sin \theta$ greater for θ in the interval $(0, 0.5]$?

(c) As θ approaches 0, how do θ and $\sin \theta$ compare? Explain.

70. (a) Complete the table.

θ	0°	18°	36°	54°	72°	90°
$\sin \theta$						
$\cos \theta$						

(b) Discuss the behavior of the sine function for θ in the range from 0° to 90°.

(c) Discuss the behavior of the cosine function for θ in the range from 0° to 90°.

(d) Use the definitions of the sine and cosine functions to explain the results of parts (b) and (c).

71. ***Empire State Building*** You are standing 45 meters from the base of the Empire State Building. You estimate that the angle of elevation to the top of the 86th floor (the observatory) is 82°. If the total height of the building is another 123 meters above the 86th floor, what is the approximate height of the building? One of your friends is on the 86th floor. What is the distance between you and your friend?

72. ***Height*** A six-foot person walks from the base of a broadcasting tower directly toward the tip of the shadow cast by the tower. When the person is 132 feet from the tower and 3 feet from the tip of the shadow, the person's shadow starts to appear beyond the tower's shadow.

(a) Draw a right triangle that gives a visual representation of the problem. Show the known quantities of the triangle and use a variable to indicate the height of the tower.

(b) Use a trigonometric function to write an equation involving the unknown quantity.

(c) What is the height of the tower?

73. ***Angle of Elevation*** You are skiing down a mountain with a vertical height of 1500 feet. The distance from the top of the mountain to the base is 3000 feet. What is the angle of elevation from the base to the top of the mountain?

74. ***Width of a River*** A biologist wants to know the width w of a river so that instruments for studying the pollutants in the water can be set properly. From point A, the biologist walks downstream 100 feet and sights to point C (see figure). From this sighting, it is determined that $\theta = 54°$. How wide is the river?

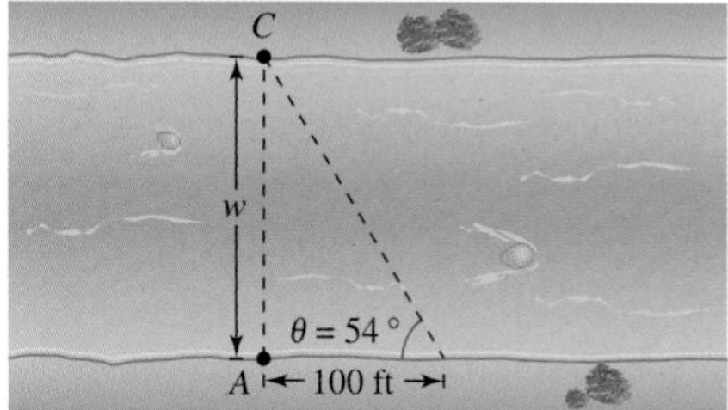

75. ***Machine Shop Calculations*** A steel plate has the form of one-fourth of a circle with a radius of 60 centimeters. Two two-centimeter holes are to be drilled in the plate positioned as shown in the figure. Find the coordinates of the center of each hole.

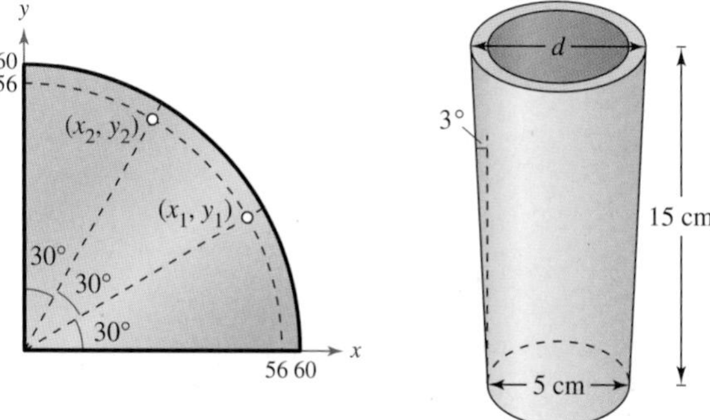

Figure for 75 **Figure for 76**

76. ***Machine Shop Calculations*** A tapered shaft has a diameter of 5 centimeters at the small end and is 15 centimeters long (see figure). The taper is 3°. Find the diameter d of the large end of the shaft.

77. ***Geometry*** Use a compass to sketch a quarter of a circle of radius 10 centimeters. Using a protractor, construct an angle of 20° in standard position (see figure). Drop a perpendicular line from the point of intersection of the terminal side of the angle and the arc of the circle. By actual measurement, calculate the coordinates (x, y) of the point of intersection and use these measurements to approximate the six trigonometric functions of a 20° angle.

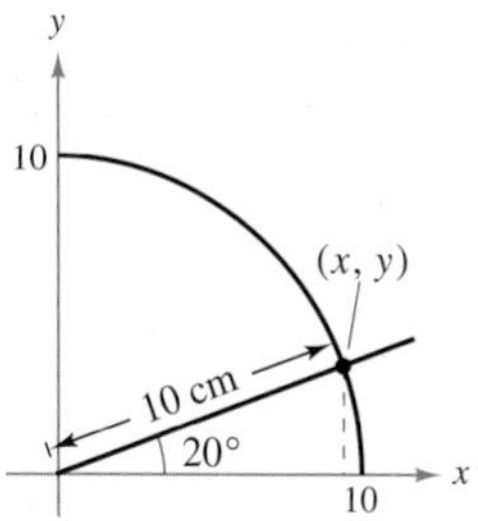

78. ***Height*** A 20-meter line is used to tether a helium-filled balloon. Because of a breeze, the line makes an angle of approximately 85° with the ground.

(a) Draw a right triangle that gives a visual representation of the problem. Show the known quantities of the triangle and use a variable to indicate the height of the balloon.

(b) Use a trigonometric function to write an equation involving the unknown quantity.

(c) What is the height of the balloon?

(d) The breeze becomes stronger and the angle the balloon makes with the ground decreases. How does this affect the triangle you drew in part (a)?

(e) Complete the table, which shows the heights (in meters) of the balloon for decreasing angle measures θ.

Angle, θ	80°	70°	60°	50°
Height				

Angle, θ	40°	30°	20°	10°
Height				

(f) As the angle the balloon makes with the ground approaches 0°, how does this affect the height of the balloon? Draw a right triangle to explain your reasoning.

79. ***Length*** A guy wire runs from the ground to a cell tower. The wire is attached to the cell tower 150 feet above the ground. The angle formed between the wire and the ground is 43° (see figure).

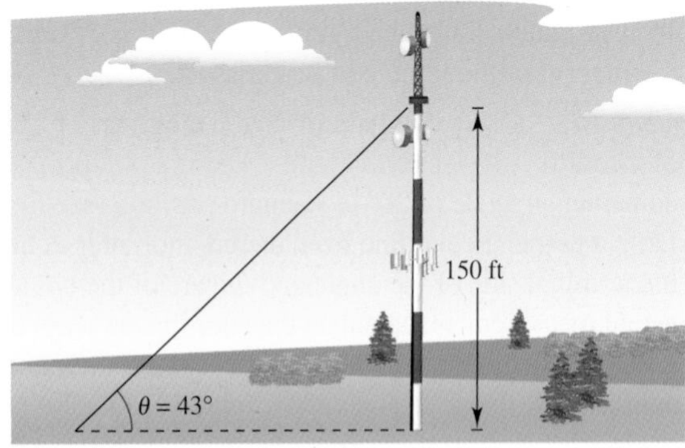

(a) How long is the guy wire?

(b) How far from the base of the tower is the guy wire anchored to the ground?

80. ***Height of a Mountain*** In traveling across flat land, you notice a mountain directly in front of you. Its angle of elevation (to the peak) is 3.5°. After you drive 13 miles closer to the mountain, the angle of elevation is 9°. Approximate the height of the mountain.

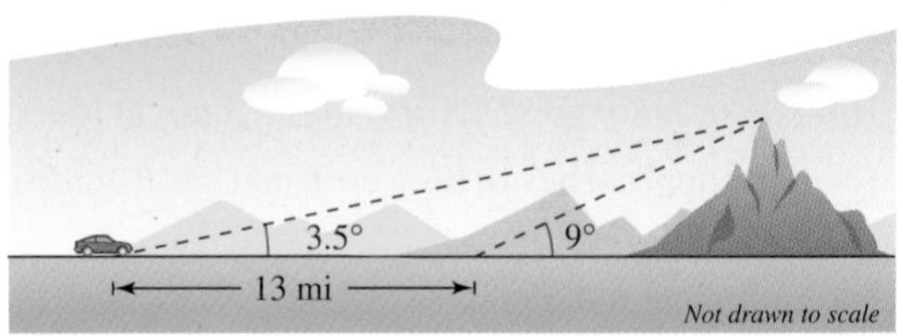

True or False? **In Exercises 81–86, determine whether the statement is true or false. Justify your answer.**

81. $\sin 60° \csc 60° = 1$

82. $\sec 30° = \csc 60°$

83. $\sin 45° + \cos 45° = 1$

84. $\cot^2 10° - \csc^2 10° = -1$

85. $\dfrac{\sin 60°}{\sin 30°} = \sin 2°$

86. $\tan[(5°)^2] = \tan^2 5°$

87. ***Geometry*** Use the equilateral triangle shown in Figure 9.24 and similar triangles to verify the points in Figure 9.18 (in Section 9.2) that do not lie on the axes.

CAPSTONE

88. The Johnstown Inclined Plane in Pennsylvania is one of the longest and steepest hoists in the world. The railway cars travel a distance of 896.5 feet at an angle of approximately 35.4°, rising to a height of 1693.5 feet above sea level.

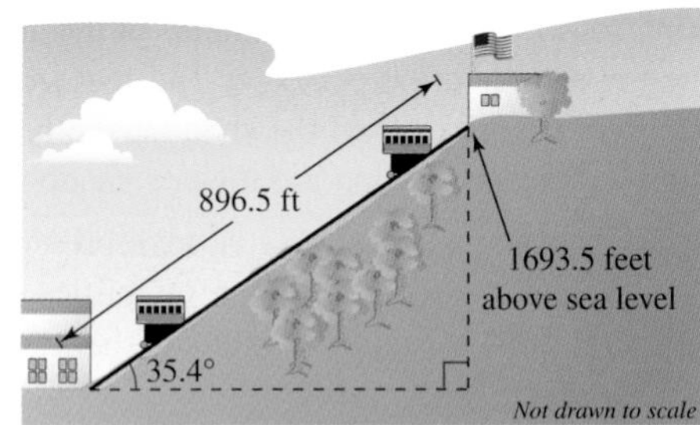

(a) Find the vertical rise of the inclined plane.

(b) Find the elevation of the lower end of the inclined plane.

(c) The cars move up the mountain at a rate of 300 feet per minute. Find the rate at which they rise vertically.

9.4 Trigonometric Functions of Any Angle

- **Evaluate trigonometric functions of any angle.**
- **Use reference angles to evaluate trigonometric functions.**

Introduction

In Section 9.3, the definitions of trigonometric functions were restricted to acute angles. In this section, the definitions are extended to cover *any* angle. If θ is an *acute* angle, these definitions coincide with those given in the preceding section.

DEFINITIONS OF TRIGONOMETRIC FUNCTIONS OF ANY ANGLE

Let θ be an angle in standard position with (x, y) a point on the terminal side of θ and $r = \sqrt{x^2 + y^2} \neq 0$.

$$\sin\theta = \frac{y}{r} \qquad \cos\theta = \frac{x}{r}$$

$$\tan\theta = \frac{y}{x}, \quad x \neq 0 \qquad \cot\theta = \frac{x}{y}, \quad y \neq 0$$

$$\sec\theta = \frac{r}{x}, \quad x \neq 0 \qquad \csc\theta = \frac{r}{y}, \quad y \neq 0$$

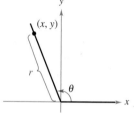

Because $r = \sqrt{x^2 + y^2}$ *cannot* be zero, it follows that the sine and cosine functions are defined for any real value of θ. However, if $x = 0$, the tangent and secant of θ are undefined. For example, the tangent of 90° is undefined. Similarly, if $y = 0$, the cotangent and cosecant of θ are undefined.

EXAMPLE 1 Evaluating Trigonometric Functions

Let $(-3, 4)$ be a point on the terminal side of θ. Find each trigonometric function.

a. $\sin\theta$ **b.** $\csc\theta$

c. $\cos\theta$ **d.** $\sec\theta$

e. $\tan\theta$ **f.** $\cot\theta$

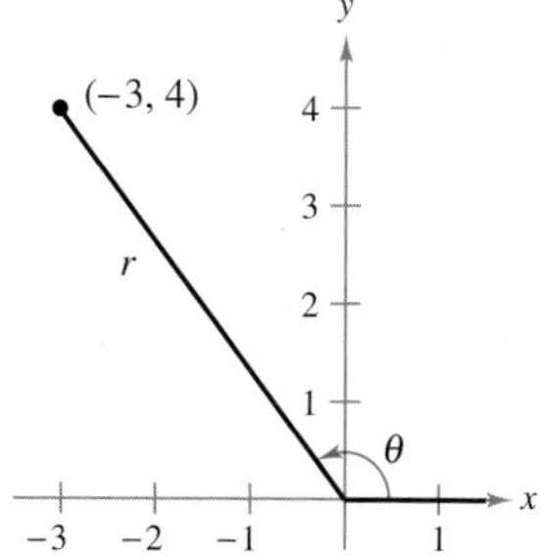

Figure 9.29

Solution Referring to Figure 9.29, you see that $x = -3$, $y = 4$, and

$$r = \sqrt{x^2 + y^2}$$
$$= \sqrt{(-3)^2 + 4^2} = \sqrt{25} = 5.$$

So, you have

a. $\sin\theta = \frac{y}{r} = \frac{4}{5}$ **b.** $\csc\theta = \frac{r}{y} = \frac{5}{4}$ Positive values

c. $\cos\theta = \frac{x}{r} = -\frac{3}{5}$ **d.** $\sec\theta = \frac{r}{x} = -\frac{5}{3}$ Negative values

e. $\tan\theta = \frac{y}{x} = -\frac{4}{3}$ **f.** $\cot\theta = \frac{x}{y} = -\frac{3}{4}.$ Negative values

Notice that the sign of each trigonometric function is the same as the sign of its reciprocal. ■

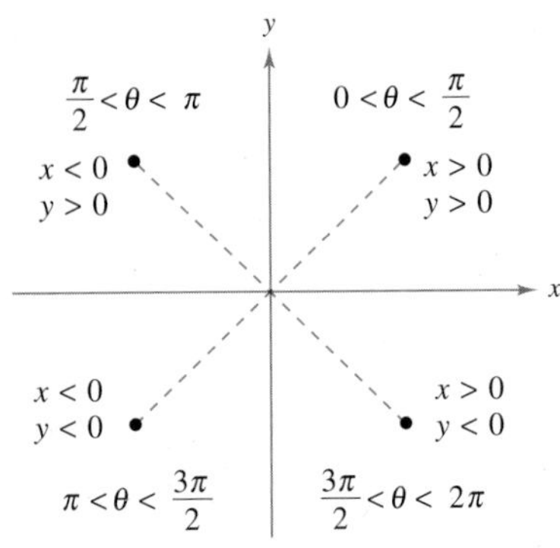

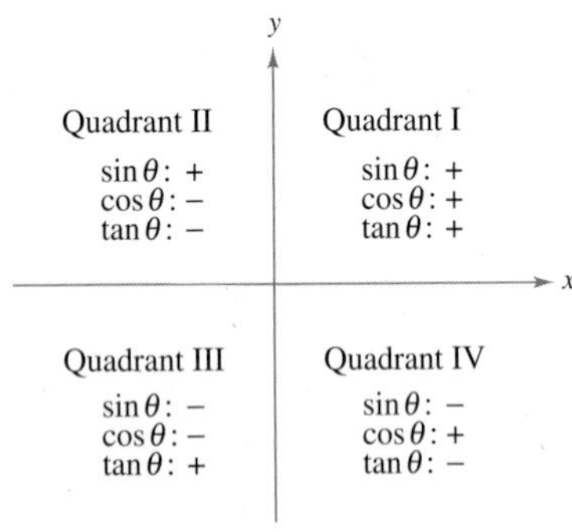

Figure 9.30

The *signs* of the trigonometric functions in the four quadrants can be determined from the definitions of the functions. For instance, because $\cos\theta = x/r$, it follows that $\cos\theta$ is positive wherever $x > 0$, which is in Quadrants I and IV. (Remember, r is always positive.) In a similar manner, you can verify the results shown in Figure 9.30.

EXAMPLE 2 Evaluating Trigonometric Functions

Given $\tan\theta = -\frac{5}{4}$ and $\cos\theta > 0$, find $\sin\theta$ and $\sec\theta$.

Solution Note that θ lies in Quadrant IV because that is the only quadrant in which the tangent is negative and the cosine is positive. Moreover, using

$$\tan\theta = \frac{y}{x} = -\frac{5}{4}$$

and the fact that y is negative in Quadrant IV, you can let $y = -5$ and $x = 4$. So, $r = \sqrt{16 + 25} = \sqrt{41}$ and you have

$$\sin\theta = \frac{y}{r} = \frac{-5}{\sqrt{41}} \approx -0.7809$$

$$\sec\theta = \frac{r}{x} = \frac{\sqrt{41}}{4} \approx 1.6008.$$

EXAMPLE 3 Trigonometric Functions of Quadrant Angles

Evaluate the cosine and tangent functions at the four quadrant angles $0, \frac{\pi}{2}, \pi$, and $\frac{3\pi}{2}$.

Solution To begin, choose a point on the terminal side of each angle, as shown in Figure 9.31. For each of the four points, $r = 1$, and you have the following.

$$\cos 0 = \frac{x}{r} = \frac{1}{1} = 1 \qquad \tan 0 = \frac{y}{x} = \frac{0}{1} = 0 \qquad (x, y) = (1, 0)$$

$$\cos\frac{\pi}{2} = \frac{x}{r} = \frac{0}{1} = 0 \qquad \tan\frac{\pi}{2} = \frac{y}{x} = \frac{1}{0} \Rightarrow \text{undefined} \qquad (x, y) = (0, 1)$$

$$\cos\pi = \frac{x}{r} = \frac{-1}{1} = -1 \qquad \tan\pi = \frac{y}{x} = \frac{0}{-1} = 0 \qquad (x, y) = (-1, 0)$$

$$\cos\frac{3\pi}{2} = \frac{x}{r} = \frac{0}{1} = 0 \qquad \tan\frac{3\pi}{2} = \frac{y}{x} = \frac{-1}{0} \Rightarrow \text{undefined} \qquad (x, y) = (0, -1)$$

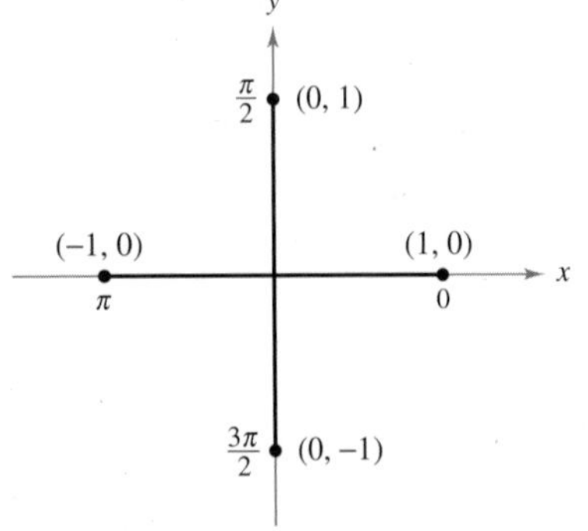

Figure 9.31

■

Reference Angles

The values of the trigonometric functions of angles greater than 90° (or less than 0°) can be determined from their values at corresponding acute angles called **reference angles.**

> **DEFINITION OF REFERENCE ANGLE**
>
> Let θ be an angle in standard position. Its **reference angle** is the acute angle θ' formed by the terminal side of θ and the horizontal axis.

Figure 9.32 shows the reference angles for θ in Quadrants II, III, and IV.

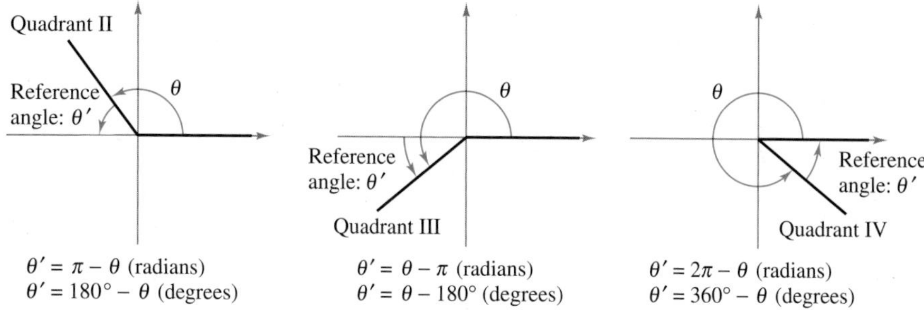

Figure 9.32

EXAMPLE 4 Finding Reference Angles

Find the reference angle θ'.

a. $\theta = 300°$ **b.** $\theta = 2.3$ **c.** $\theta = -135°$

Solution

a. Because 300° lies in Quadrant IV, the angle it makes with the x-axis is

$$\theta' = 360° - 300°$$
$$= 60°. \qquad \text{Degrees}$$

Figure 9.33(a) shows the angle $\theta = 300°$ and its reference angle $\theta' = 60°$.

b. Because 2.3 lies between $\pi/2 \approx 1.5708$ and $\pi \approx 3.1416$, it follows that it is in Quadrant II and its reference angle is

$$\theta' = \pi - 2.3$$
$$\approx 0.8416. \qquad \text{Radians}$$

Figure 9.33(b) shows the angle $\theta = 2.3$ and its reference angle $\theta' = \pi - 2.3$.

c. First, determine that $-135°$ is coterminal with 225°, which lies in Quadrant III. So, the reference angle is

$$\theta' = 225° - 180°$$
$$= 45°. \qquad \text{Degrees}$$

Figure 9.33(c) shows the angle $\theta = -135°$ and its reference angle $\theta' = 45°$. ■

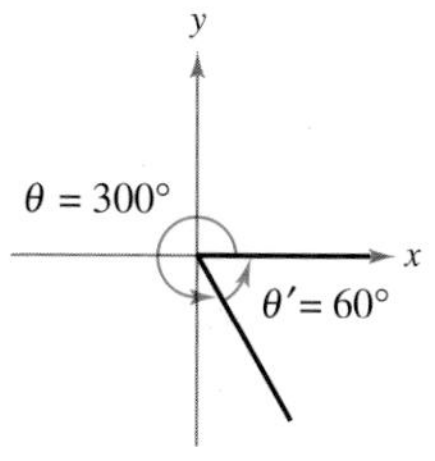

(a)

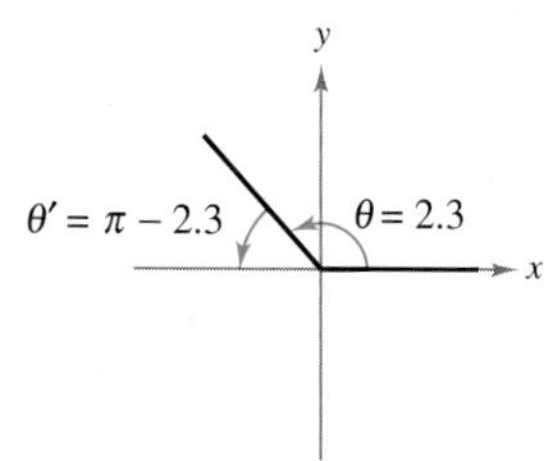

(b)

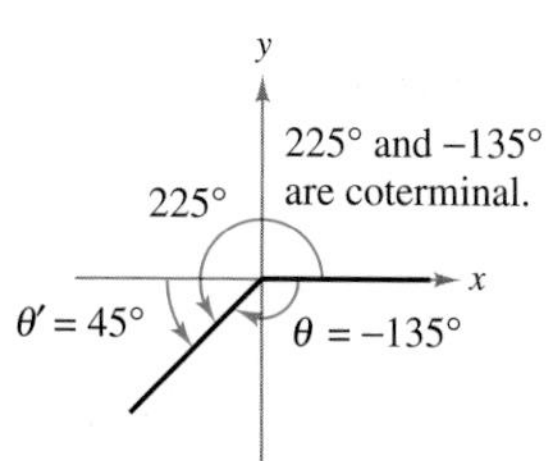

(c)

Figure 9.33

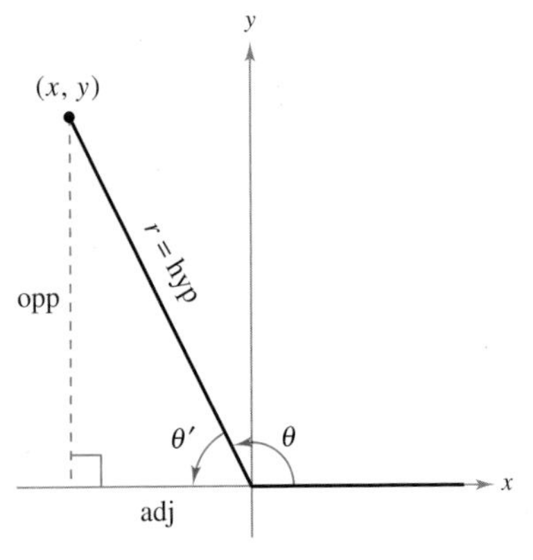

$\text{opp} = |y|, \text{adj} = |x|$

Figure 9.34

Trigonometric Functions of Any Angle

To see how a reference angle is used to evaluate a trigonometric function, consider the point (x, y) on the terminal side of θ, as shown in Figure 9.34. By definition, you know that

$$\sin \theta = \frac{y}{r} \quad \text{and} \quad \tan \theta = \frac{y}{x}.$$

For the right triangle with acute angle θ' and sides of lengths $|x|$ and $|y|$, you have

$$\sin \theta' = \frac{\text{opp}}{\text{hyp}} = \frac{|y|}{r} \quad \text{and} \quad \tan \theta' = \frac{\text{opp}}{\text{adj}} = \frac{|y|}{|x|}.$$

So, it follows that $\sin \theta$ and $\sin \theta'$ are equal, *except possibly in sign.* The same is true for $\tan \theta$ and $\tan \theta'$ *and* for the other four trigonometric functions. In all cases, the sign of the function value can be determined by the quadrant in which θ lies.

EVALUATING TRIGONOMETRIC FUNCTIONS OF ANY ANGLE

To find the value of a trigonometric function of any angle θ:

1. Determine the function value for the associated reference angle θ'.
2. Depending on the quadrant in which θ lies, affix the appropriate sign to the function value.

By using reference angles and the special angles discussed in the preceding section, you can greatly extend the scope of *exact* trigonometric values. For instance, knowing the function values of 30° means that you know the function values of all angles for which 30° is a reference angle. For convenience, the table below shows the exact values of the trigonometric functions of special angles and quadrant angles.

Trigonometric Values of Common Angles

θ (degrees)	0°	30°	45°	60°	90°	180°	270°
θ (radians)	0	$\frac{\pi}{6}$	$\frac{\pi}{4}$	$\frac{\pi}{3}$	$\frac{\pi}{2}$	π	$\frac{3\pi}{2}$
$\sin \theta$	0	$\frac{1}{2}$	$\frac{\sqrt{2}}{2}$	$\frac{\sqrt{3}}{2}$	1	0	-1
$\cos \theta$	1	$\frac{\sqrt{3}}{2}$	$\frac{\sqrt{2}}{2}$	$\frac{1}{2}$	0	-1	0
$\tan \theta$	0	$\frac{\sqrt{3}}{3}$	1	$\sqrt{3}$	Undef.	0	Undef.

STUDY TIP Learning the table of values above is worth the effort. Doing so will increase both your efficiency and your confidence, especially with calculus. Here are patterns for the sine and cosine functions that may help you remember the values.

θ	0°	30°	45°	60°	90°
$\sin \theta$	$\frac{\sqrt{0}}{2}$	$\frac{\sqrt{1}}{2}$	$\frac{\sqrt{2}}{2}$	$\frac{\sqrt{3}}{2}$	$\frac{\sqrt{4}}{2}$

θ	0°	30°	45°	60°	90°
$\cos \theta$	$\frac{\sqrt{4}}{2}$	$\frac{\sqrt{3}}{2}$	$\frac{\sqrt{2}}{2}$	$\frac{\sqrt{1}}{2}$	$\frac{\sqrt{0}}{2}$

EXAMPLE 5 Trigonometric Functions of Nonacute Angles

Evaluate each trigonometric function.

a. $\cos \frac{4\pi}{3}$ **b.** $\tan(-210°)$ **c.** $\csc \frac{11\pi}{4}$

Solution

a. Because $\theta = 4\pi/3$ lies in Quadrant III, the reference angle is $\theta' = (4\pi/3) - \pi = \pi/3$, as shown in Figure 9.35(a). Moreover, the cosine is negative in Quadrant III, so

$$\cos \frac{4\pi}{3} = (-) \cos \frac{\pi}{3} = -\frac{1}{2}.$$

b. Because $-210° + 360° = 150°$, it follows that $-210°$ is coterminal with the second-quadrant angle 150°. So, the reference angle is $\theta' = 180° - 150° = 30°$, as shown in Figure 9.35(b). Finally, because the tangent is negative in Quadrant II, you have

$$\tan(-210°) = (-) \tan 30° = -\frac{\sqrt{3}}{3}.$$

c. Because $(11\pi/4) - 2\pi = 3\pi/4$, it follows that $11\pi/4$ is coterminal with the second-quadrant angle $3\pi/4$. So, the reference angle is $\theta' = \pi - (3\pi/4) = \pi/4$, as shown in Figure 9.35(c). Because the cosecant is positive in Quadrant II, you have

$$\csc \frac{11\pi}{4} = (+) \csc \frac{\pi}{4} = \frac{1}{\sin(\pi/4)} = \sqrt{2}.$$

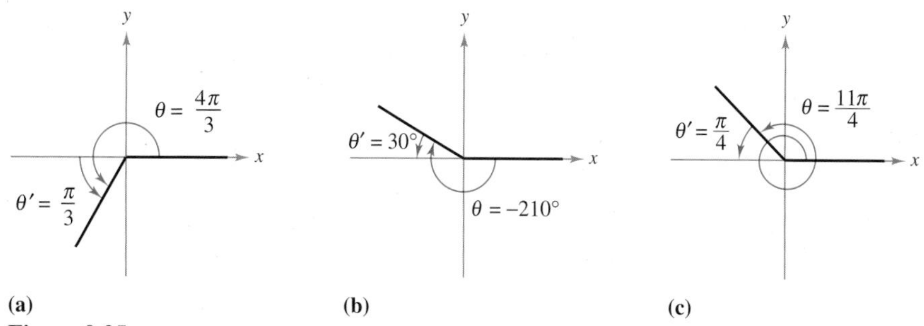

Figure 9.35

EXAMPLE 6 Using Trigonometric Identities

Let θ be an angle in Quadrant II such that $\sin \theta = \frac{1}{3}$. Find $\cos \theta$.

Solution Using the Pythagorean identity $\sin^2 \theta + \cos^2 \theta = 1$, you obtain

$$\left(\frac{1}{3}\right)^2 + \cos^2 \theta = 1 \qquad \text{Substitute } \tfrac{1}{3} \text{ for } \sin \theta.$$

$$\cos^2 \theta = 1 - \frac{1}{9} = \frac{8}{9}.$$

Because $\cos \theta < 0$ in Quadrant II, you can use the negative root to obtain

$$\cos \theta = -\frac{\sqrt{8}}{\sqrt{9}} = -\frac{2\sqrt{2}}{3}.$$

■

9.4 Exercises

See www.CalcChat.com for worked-out solutions to odd-numbered exercises.

In Exercises 1–6, fill in the blanks. Let θ be an angle in standard position, with (x, y) a point on the terminal side of θ and $r = \sqrt{x^2 + y^2} \neq 0$.

1. $\sin \theta =$ ________
2. $\dfrac{r}{y} =$ ________
3. $\tan \theta =$ ________
4. $\sec \theta =$ ________
5. $\dfrac{x}{r} =$ ________
6. $\dfrac{x}{y} =$ ________

In Exercises 7 and 8, fill in the blanks.

7. Because $r = \sqrt{x^2 + y^2}$ cannot be ________, the sine and cosine functions are ________ for any real value of θ.
8. The acute positive angle that is formed by the terminal side of the angle θ and the horizontal axis is called the ________ angle of θ and is denoted by θ'.

In Exercises 9–12, determine the exact values of the six trigonometric functions of the angle θ.

9. (a)

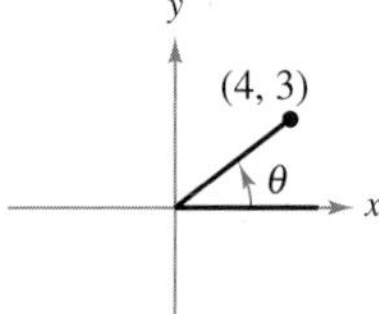

(b)

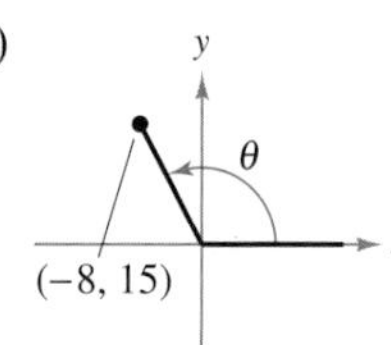

10. (a)

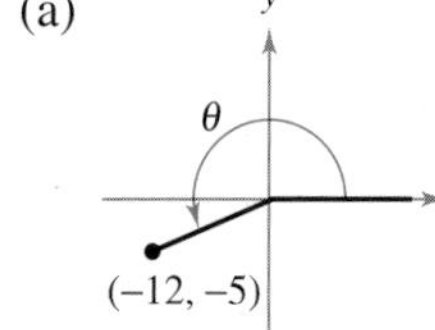

(b)

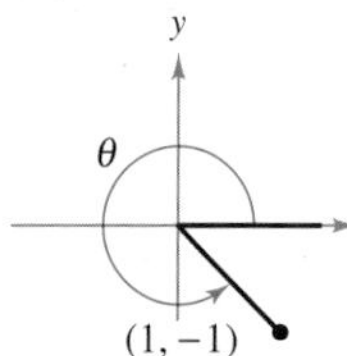

11. (a)

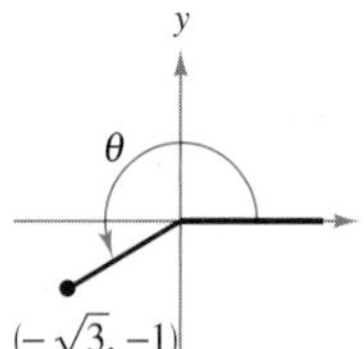

(b)

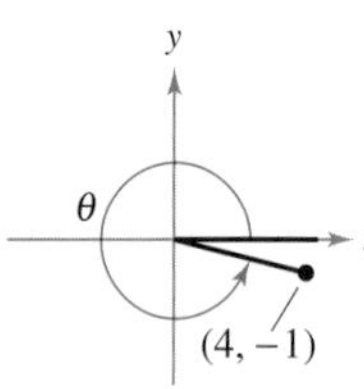

12. (a)

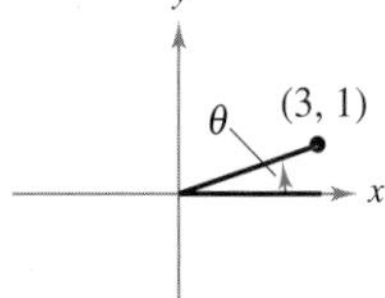

(b)

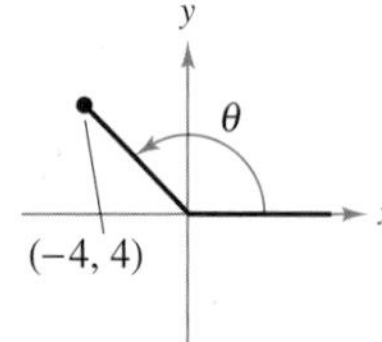

In Exercises 13–18, the point is on the terminal side of an angle in standard position. Determine the exact values of the six trigonometric functions of the angle.

13. $(5, 12)$
14. $(8, 15)$
15. $(-5, -2)$
16. $(-4, 10)$
17. $(-5.4, 7.2)$
18. $\left(3\frac{1}{2}, -7\frac{3}{4}\right)$

In Exercises 19–22, state the quadrant in which θ lies.

19. $\sin \theta > 0$ and $\cos \theta > 0$
20. $\sin \theta < 0$ and $\cos \theta < 0$
21. $\sin \theta > 0$ and $\cos \theta < 0$
22. $\sec \theta > 0$ and $\cot \theta < 0$

In Exercises 23–32, find the values of the six trigonometric functions of θ with the given constraint.

	Function Value	*Constraint*
23.	$\tan \theta = -\frac{15}{8}$	$\sin \theta > 0$
24.	$\cos \theta = \frac{8}{17}$	$\tan \theta < 0$
25.	$\sin \theta = \frac{3}{5}$	θ lies in Quadrant II.
26.	$\cos \theta = -\frac{4}{5}$	θ lies in Quadrant III.
27.	$\cot \theta = -3$	$\cos \theta > 0$
28.	$\csc \theta = 4$	$\cot \theta < 0$
29.	$\sec \theta = -2$	$\sin \theta < 0$
30.	$\sin \theta = 0$	$\sec \theta = -1$
31.	$\cot \theta$ is undefined.	$\pi/2 \le \theta \le 3\pi/2$
32.	$\tan \theta$ is undefined.	$\pi \le \theta \le 2\pi$

In Exercises 33–36, the terminal side of θ lies on the given line in the specified quadrant. Find the values of the six trigonometric functions of θ by finding a point on the line.

	Line	*Quadrant*
33.	$y = -x$	II
34.	$y = \frac{1}{3}x$	III
35.	$2x - y = 0$	III
36.	$4x + 3y = 0$	IV

In Exercises 37–44, evaluate the trigonometric function of the quadrant angle.

37. $\sin \pi$
38. $\csc \dfrac{3\pi}{2}$
39. $\sec \dfrac{3\pi}{2}$
40. $\sec \pi$
41. $\sin \dfrac{\pi}{2}$
42. $\cot \pi$
43. $\csc \pi$
44. $\cot \dfrac{\pi}{2}$

In Exercises 45–52, find the reference angle θ', and sketch θ and θ' in standard position.

45. $\theta = 160°$ **46.** $\theta = 309°$

47. $\theta = -125°$ **48.** $\theta = -215°$

49. $\theta = \dfrac{2\pi}{3}$ **50.** $\theta = \dfrac{7\pi}{6}$

51. $\theta = 4.8$ **52.** $\theta = 11.6$

In Exercises 53–68, evaluate the sine, cosine, and tangent of the angle without using a calculator.

53. $225°$ **54.** $300°$

55. $750°$ **56.** $-405°$

57. $-150°$ **58.** $-840°$

59. $\dfrac{2\pi}{3}$ **60.** $\dfrac{3\pi}{4}$

61. $\dfrac{5\pi}{4}$ **62.** $\dfrac{7\pi}{6}$

63. $-\dfrac{\pi}{6}$ **64.** $-\dfrac{\pi}{2}$

65. $\dfrac{9\pi}{4}$ **66.** $\dfrac{10\pi}{3}$

67. $-\dfrac{3\pi}{2}$ **68.** $-\dfrac{23\pi}{4}$

In Exercises 69–74, find the indicated trigonometric value in the specified quadrant.

	Function	*Quadrant*	*Trigonometric Value*
69.	$\sin\theta = -\frac{3}{5}$	IV	$\cos\theta$
70.	$\cot\theta = -3$	II	$\sin\theta$
71.	$\tan\theta = \frac{3}{2}$	III	$\sec\theta$
72.	$\csc\theta = -2$	IV	$\cot\theta$
73.	$\cos\theta = \frac{5}{8}$	I	$\sec\theta$
74.	$\sec\theta = -\frac{9}{4}$	III	$\tan\theta$

In Exercises 75–86, use a calculator to evaluate the trigonometric function. Round your answer to four decimal places. (Be sure the calculator is set in the correct angle mode.)

75. $\sin 10°$ **76.** $\sec 225°$

77. $\cos(-110°)$ **78.** $\csc(-330°)$

79. $\tan 4.5$ **80.** $\cot 1.35$

81. $\tan(\pi/9)$ **82.** $\tan(-\pi/9)$

83. $\sin(-0.65)$ **84.** $\sec 0.29$

85. $\cot(-11\pi/8)$ **86.** $\csc(-15\pi/14)$

In Exercises 87–92, find two solutions of the equation. Give your answers in degrees ($0° \le \theta < 360°$) and in radians ($0 \le \theta < 2\pi$). Do not use a calculator.

87. (a) $\sin\theta = \frac{1}{2}$ (b) $\sin\theta = -\frac{1}{2}$

88. (a) $\cos\theta = \sqrt{2}/2$ (b) $\cos\theta = -\sqrt{2}/2$

89. (a) $\csc\theta = \dfrac{2\sqrt{3}}{3}$ (b) $\cot\theta = -1$

90. (a) $\sec\theta = 2$ (b) $\sec\theta = -2$

91. (a) $\tan\theta = 1$ (b) $\cot\theta = -\sqrt{3}$

92. (a) $\sin\theta = \sqrt{3}/2$ (b) $\sin\theta = -\sqrt{3}/2$

WRITING ABOUT CONCEPTS

93. Consider an angle in standard position with $r = 12$ centimeters, as shown in the figure. Write a short paragraph describing the changes in the values of x, y, $\sin\theta$, $\cos\theta$, and $\tan\theta$ as θ increases continuously from 0° to 90°.

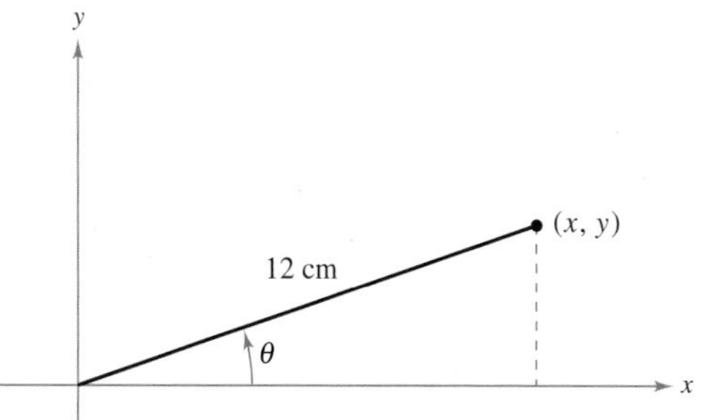

94. The figure shows point $P(x, y)$ on a unit circle and right triangle OAP.

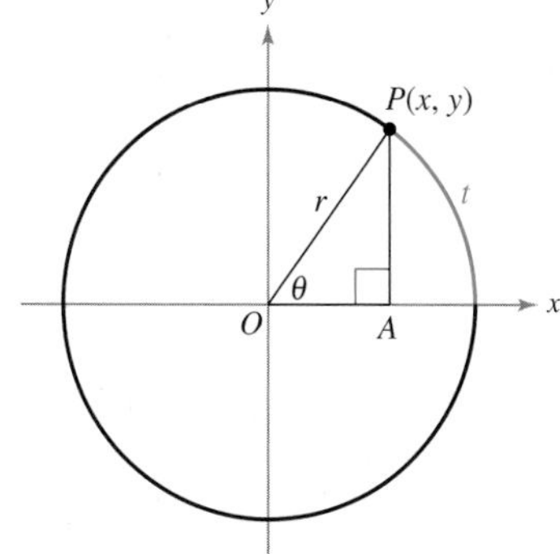

(a) Find $\sin t$ and $\cos t$ using the unit circle definitions of sine and cosine (from Section 9.2).

(b) What is the value of r? Explain.

(c) Use the definitions of sine and cosine given in this section to find $\sin\theta$ and $\cos\theta$. Write your answers in terms of x and y.

(d) Based on your answers to parts (a) and (c), what can you conclude?

95. ***Sales*** A company that produces snowboards, which are seasonal products, forecasts monthly sales over the next 2 years to be $S = 23.1 + 0.442t + 4.3\cos(\pi t/6)$, where S is measured in thousands of units and t is the time in months, with $t = 1$ representing January 2010. Predict sales for each of the following months.

(a) February 2010 (b) February 2011

(c) June 2010 (d) June 2011

96. ***Harmonic Motion*** The displacement from equilibrium of an oscillating weight suspended by a spring is given by $y(t) = 2\cos 6t$, where y is the displacement (in centimeters) and t is the time (in seconds). Find the displacement when (a) $t = 0$, (b) $t = \frac{1}{4}$, and (c) $t = \frac{1}{2}$.

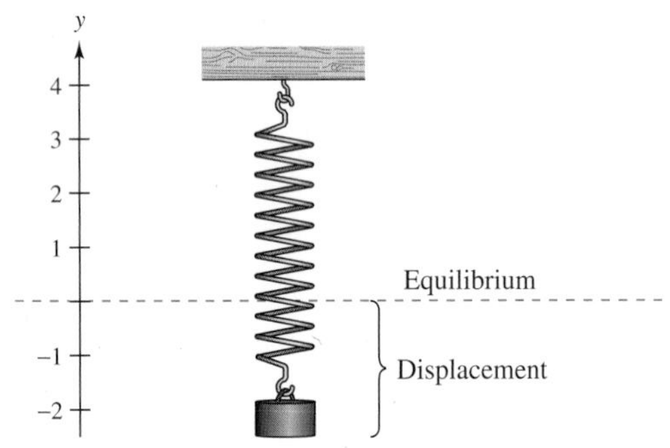

Figure for 96 and 97

97. ***Harmonic Motion*** The displacement from equilibrium of an oscillating weight suspended by a spring and subject to the damping effect of friction is given by $y(t) = 2e^{-t}\cos 6t$, where y is the displacement (in centimeters) and t is the time (in seconds). Find the displacement when (a) $t = 0$, (b) $t = \frac{1}{4}$, and (c) $t = \frac{1}{2}$.

98. ***Electric Circuits*** The current I (in amperes) when 100 volts is applied to a circuit is given by $I = 5e^{-2t}\sin t$, where t is the time (in seconds) after the voltage is applied. Approximate the current at $t = 0.7$ second after the voltage is applied.

99. ***Data Analysis: Meteorology*** The table shows the monthly normal temperatures (in degrees Fahrenheit) for selected months in New York City (N) and Fairbanks, Alaska (F). *(Source: National Climatic Data Center)*

Month	New York City, N	Fairbanks, F
January	33	−10
April	52	32
July	77	62
October	58	24
December	38	−6

(a) Use the *regression* feature of a graphing utility to find a model of the form $y = a\sin(bt + c) + d$ for each city. Let t represent the month, with $t = 1$ corresponding to January.

(b) Use the models from part (a) to find the monthly normal temperatures for the two cities in February, March, May, June, August, September, and November.

(c) Compare the models for the two cities.

100. ***Distance*** An airplane, flying at an altitude of 6 miles, is on a flight path that passes directly over an observer (see figure). If θ is the angle of elevation from the observer to the plane, find the distance d from the observer to the plane when (a) $\theta = 30°$, (b) $\theta = 90°$, and (c) $\theta = 120°$.

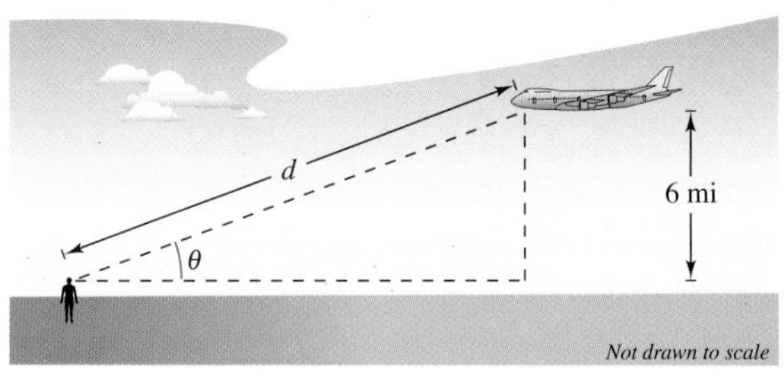

True or False? **In Exercises 101–103, determine whether the statement is true or false. Justify your answer.**

101. In each of the four quadrants, the signs of the secant function and sine function will be the same.

102. To find the reference angle for an angle θ (given in degrees), find the integer n such that $0 \le 360°n - \theta \le 360°$. The difference $360°n - \theta$ is the reference angle.

103. If $\sin\theta = \dfrac{1}{4}$ and $\cos\theta < 0$, then

$$\tan\theta = -\frac{\sqrt{15}}{15}.$$

CAPSTONE

104. Write a short paper in your own words explaining to a classmate how to evaluate the six trigonometric functions of any angle θ in standard position. Include an explanation of reference angles and how to use them, the signs of the functions in each of the four quadrants, and the trigonometric values of common angles. Be sure to include figures or diagrams in your paper.

9.5 Graphs of Sine and Cosine Functions

- **Sketch the graphs of basic sine and cosine functions.**
- **Use amplitude and period to help sketch the graphs of sine and cosine functions.**
- **Sketch translations of the graphs of sine and cosine functions.**
- **Use sine and cosine functions to model real-life data.**

Basic Sine and Cosine Curves

In this section, you will study techniques for sketching the graphs of the sine and cosine functions. The graph of the sine function is a **sine curve.** In Figure 9.36, the black portion of the graph represents one period of the function and is called **one cycle** of the sine curve. The gray portion of the graph indicates that the basic sine curve repeats indefinitely in the positive and negative directions. The graph of the cosine function is shown in Figure 9.37.

Recall from Section 9.2 that the domain of the sine and cosine functions is the set of all real numbers. Moreover, the range of each function is the interval $[-1, 1]$, and each function has a period of 2π. Do you see how this information is consistent with the basic graphs shown in Figures 9.36 and 9.37?

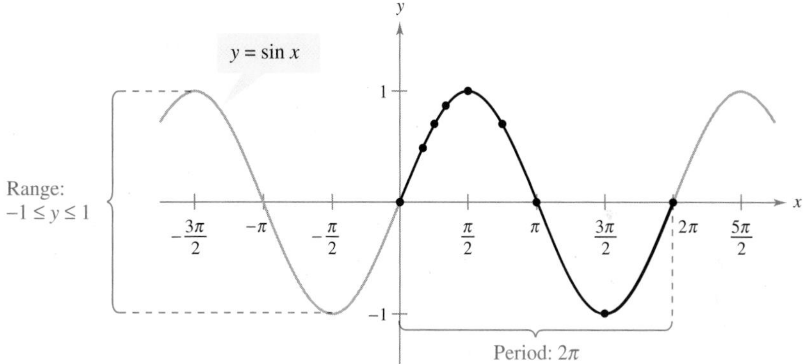

Figure 9.36

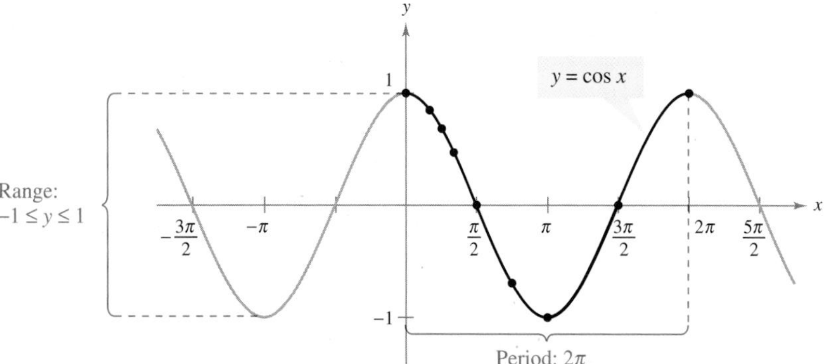

Figure 9.37

Note in Figures 9.36 and 9.37 that the sine curve is symmetric with respect to the *origin,* whereas the cosine curve is symmetric with respect to the *y-axis*. These properties of symmetry follow from the fact that the sine function is odd and the cosine function is even. Note also that the cosine curve appears to be a left shift (of $\pi/2$) of the sine curve. More will be said about this later in the section.

To sketch the graphs of the basic sine and cosine functions by hand, it helps to note five **key points** in one period of each graph: the *intercepts*, *maximum points*, and *minimum points* (see Figure 9.38).

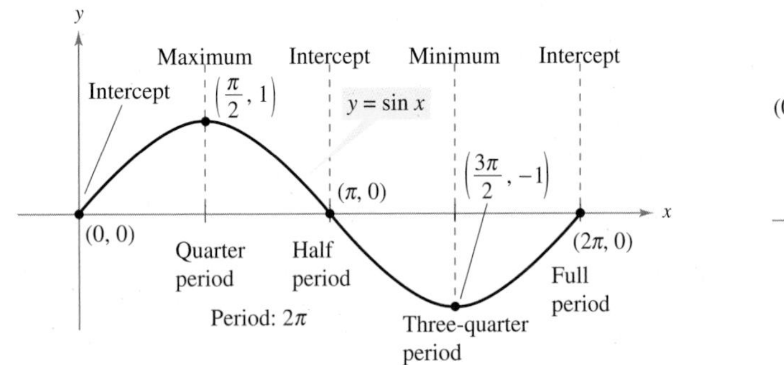

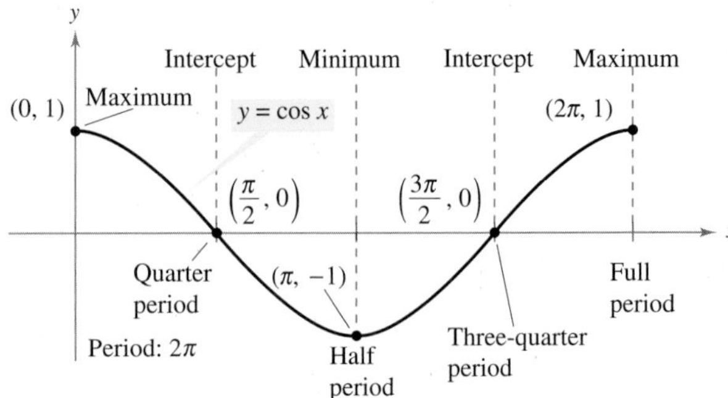

Figure 9.38

EXAMPLE 1 Using Key Points to Sketch a Sine Curve

Sketch the graph of $y = 2 \sin x$ on the interval $[-\pi, 4\pi]$.

Solution Note that

$$y = 2 \sin x = 2(\sin x)$$

indicates that the y-values for the key points will have twice the magnitude of those on the graph of $y = \sin x$. Divide the period 2π into four equal parts to get the key points for $y = 2 \sin x$.

Intercept	*Maximum*	*Intercept*	*Minimum*		*Intercept*
$(0, 0)$,	$\left(\frac{\pi}{2}, 2\right)$,	$(\pi, 0)$,	$\left(\frac{3\pi}{2}, -2\right)$,	and	$(2\pi, 0)$

By connecting these key points with a smooth curve and extending the curve in both directions over the interval $[-\pi, 4\pi]$, you obtain the graph shown in Figure 9.39.

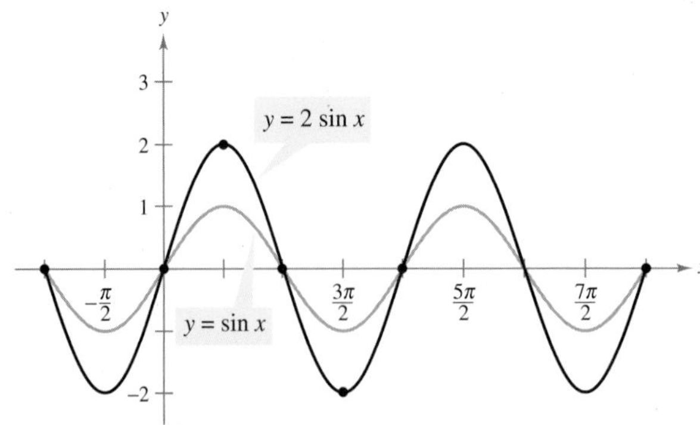

Figure 9.39

TECHNOLOGY When using a graphing utility to graph trigonometric functions, pay special attention to the viewing window you use. For instance, try graphing $y = [\sin(10x)]/10$ in the standard viewing window in *radian* mode. What do you observe? Use the *zoom* feature to find a viewing window that displays a good view of the graph.

Amplitude and Period

In the remainder of this section you will study the graphic effect of each of the constants a, b, c, and d in equations of the forms

$$y = d + a\sin(bx - c)$$

and

$$y = d + a\cos(bx - c).$$

A quick review of the transformations you studied in Section 1.3 should help in this investigation.

The constant factor a in $y = a\sin x$ acts as a *scaling factor—a vertical stretch* or *vertical shrink* of the basic sine curve. If $|a| > 1$, the basic sine curve is stretched, and if $|a| < 1$, the basic sine curve is shrunk. The result is that the graph of $y = a\sin x$ ranges between $-a$ and a instead of between -1 and 1. The absolute value of a is the **amplitude** of the function $y = a\sin x$. The range of the function $y = a\sin x$ for $a > 0$ is $-a \le y \le a$.

DEFINITION OF AMPLITUDE OF SINE AND COSINE CURVES

The **amplitude** of $y = a\sin x$ and $y = a\cos x$ represents half the distance between the maximum and minimum values of the function and is given by

$$\text{Amplitude} = |a|.$$

EXAMPLE 2 Scaling: Vertical Shrinking and Stretching

On the same coordinate axes, sketch the graph of each function.

a. $y = \dfrac{1}{2}\cos x$

b. $y = 3\cos x$

Solution

a. Because the amplitude of $y = \frac{1}{2}\cos x$ is $\frac{1}{2}$, the maximum value is $\frac{1}{2}$ and the minimum value is $-\frac{1}{2}$. Divide one cycle, $0 \le x \le 2\pi$, into four equal parts to get the key points

Maximum	*Intercept*	*Minimum*	*Intercept*		*Maximum*
$\left(0, \frac{1}{2}\right)$,	$\left(\frac{\pi}{2}, 0\right)$,	$\left(\pi, -\frac{1}{2}\right)$,	$\left(\frac{3\pi}{2}, 0\right)$,	and	$\left(2\pi, \frac{1}{2}\right)$.

b. A similar analysis shows that the amplitude of $y = 3\cos x$ is 3, and the key points are

Maximum	*Intercept*	*Minimum*	*Intercept*		*Maximum*
$(0, 3)$,	$\left(\frac{\pi}{2}, 0\right)$,	$(\pi, -3)$,	$\left(\frac{3\pi}{2}, 0\right)$,	and	$(2\pi, 3)$.

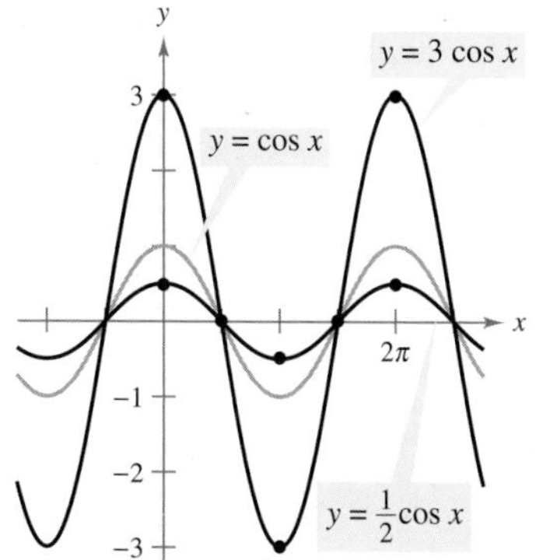

Figure 9.40

The graphs of these two functions are shown in Figure 9.40. Notice that the graph of $y = \frac{1}{2}\cos x$ is a vertical *shrink* of the graph of $y = \cos x$ and the graph of $y = 3\cos x$ is a vertical *stretch* of the graph of $y = \cos x$. ■

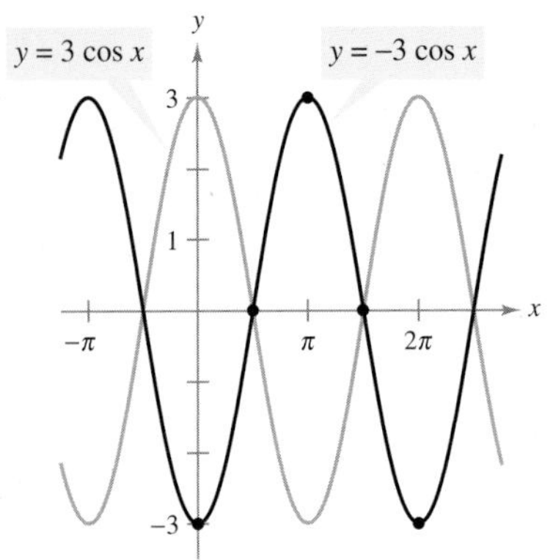

Figure 9.41

You know from Section 1.3 that the graph of $y = -f(x)$ is a **reflection** in the x-axis of the graph of $y = f(x)$. For instance, the graph of $y = -3 \cos x$ is a reflection of the graph of $y = 3 \cos x$, as shown in Figure 9.41.

Because $y = a \sin x$ completes one cycle from $x = 0$ to $x = 2\pi$, it follows that $y = a \sin bx$ completes one cycle from $x = 0$ to $x = 2\pi/b$.

PERIOD OF SINE AND COSINE FUNCTIONS

Let b be a positive real number. The **period** of $y = a \sin bx$ and $y = a \cos bx$ is given by

$$\text{Period} = \frac{2\pi}{b}.$$

Note that if $0 < b < 1$, the period of $y = a \sin bx$ is greater than 2π and represents a *horizontal stretching* of the graph of $y = a \sin x$. Similarly, if $b > 1$, the period of $y = a \sin bx$ is less than 2π and represents a *horizontal shrinking* of the graph of $y = a \sin x$. If b is negative, the identities $\sin(-x) = -\sin x$ and $\cos(-x) = \cos x$ are used to rewrite the function.

EXAMPLE 3 Scaling: Horizontal Stretching

Sketch the graph of

$$y = \sin \frac{x}{2}.$$

Solution The amplitude is 1. Moreover, because $b = \frac{1}{2}$, the period is

$$\frac{2\pi}{b} = \frac{2\pi}{\frac{1}{2}} = 4\pi. \qquad \text{Substitute for } b.$$

Now, divide the period-interval $[0, 4\pi]$ into four equal parts with the values π, 2π, and 3π to obtain the key points on the graph.

Intercept	*Maximum*	*Intercept*	*Minimum*		*Intercept*
$(0, 0)$,	$(\pi, 1)$,	$(2\pi, 0)$,	$(3\pi, -1)$,	and	$(4\pi, 0)$

The graph is shown in Figure 9.42.

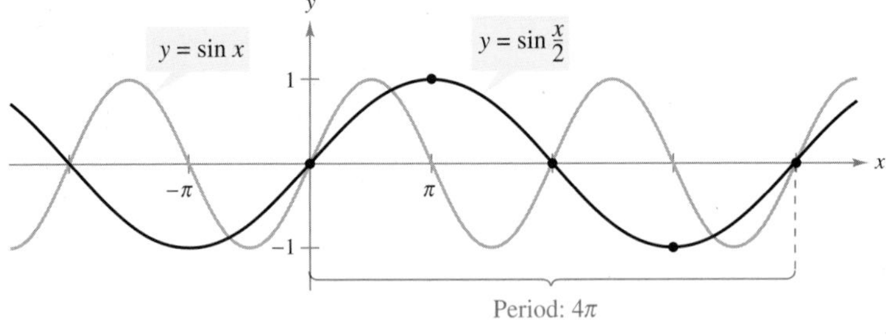

Figure 9.42

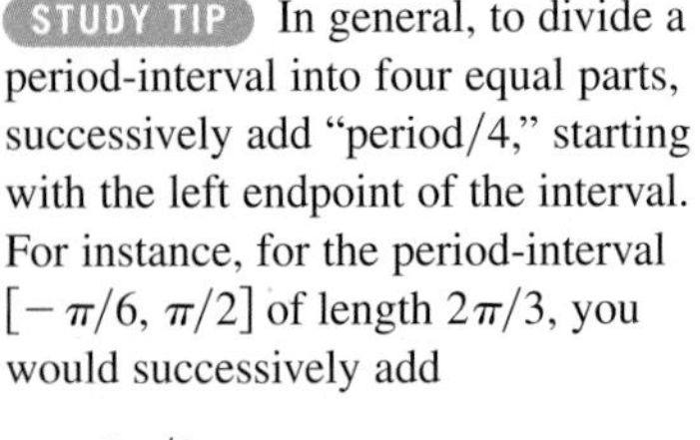

STUDY TIP In general, to divide a period-interval into four equal parts, successively add "period/4," starting with the left endpoint of the interval. For instance, for the period-interval $[-\pi/6, \pi/2]$ of length $2\pi/3$, you would successively add

$$\frac{2\pi/3}{4} = \frac{\pi}{6}$$

to get $-\pi/6, 0, \pi/6, \pi/3$, and $\pi/2$ as the x-values for the key points on the graph.

Translations of Sine and Cosine Curves

The constant c in the general equations

$$y = a\sin(bx - c) \quad \text{and} \quad y = a\cos(bx - c)$$

creates a *horizontal translation* (shift) of the basic sine and cosine curves. Comparing $y = a\sin bx$ with $y = a\sin(bx - c)$, you find that the graph of $y = a\sin(bx - c)$ completes one cycle from $bx - c = 0$ to $bx - c = 2\pi$. By solving for x, you can find the interval for one cycle to be

$$\overbrace{\frac{c}{b}}^{\text{Left endpoint}} \le x \le \overbrace{\frac{c}{b} + \underbrace{\frac{2\pi}{b}}_{\text{Period}}}^{\text{Right endpoint}}.$$

This implies that the period of $y = a\sin(bx - c)$ is $2\pi/b$, and the graph of $y = a\sin bx$ is shifted by an amount c/b. The number c/b is the **phase shift.**

GRAPHS OF SINE AND COSINE FUNCTIONS

The graphs of $y = a\sin(bx - c)$ and $y = a\cos(bx - c)$ have the following characteristics. (Assume $b > 0$.)

$$\text{Amplitude} = |a| \qquad \text{Period} = \frac{2\pi}{b}$$

The left and right endpoints of a one-cycle interval can be determined by solving the equations $bx - c = 0$ and $bx - c = 2\pi$.

EXAMPLE 4 Horizontal Translation

Analyze the graph of

$$y = \frac{1}{2}\sin\left(x - \frac{\pi}{3}\right).$$

Algebraic Solution

The amplitude is $\frac{1}{2}$ and the period is 2π. By solving the equations

$$x - \frac{\pi}{3} = 0 \quad \Rightarrow \quad x = \frac{\pi}{3}$$

and

$$x - \frac{\pi}{3} = 2\pi \quad \Rightarrow \quad x = \frac{7\pi}{3}$$

you see that the interval $[\pi/3, 7\pi/3]$ corresponds to one cycle of the graph. Dividing this interval into four equal parts produces the key points

Intercept	*Maximum*	*Intercept*	*Minimum*		*Intercept*
$\left(\frac{\pi}{3}, 0\right)$,	$\left(\frac{5\pi}{6}, \frac{1}{2}\right)$,	$\left(\frac{4\pi}{3}, 0\right)$,	$\left(\frac{11\pi}{6}, -\frac{1}{2}\right)$,	and	$\left(\frac{7\pi}{3}, 0\right)$.

Graphical Solution

Use a graphing utility set in *radian* mode to graph $y = (1/2)\sin(x - \pi/3)$, as shown in Figure 9.43. Use the *minimum, maximum,* and *zero* or *root* features of the graphing utility to approximate the key points (1.05, 0), (2.62, 0.5), (4.19, 0), (5.76, −0.5), and (7.33, 0).

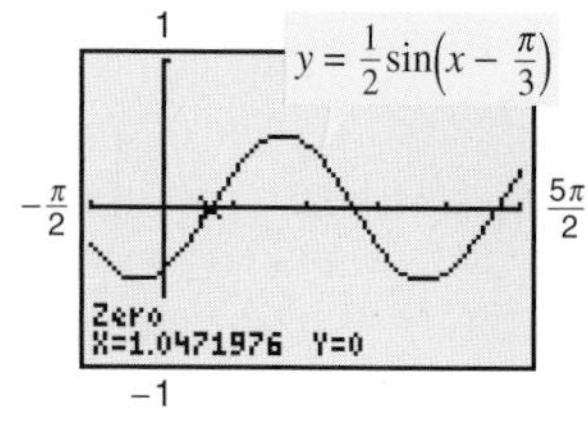

Figure 9.43

■

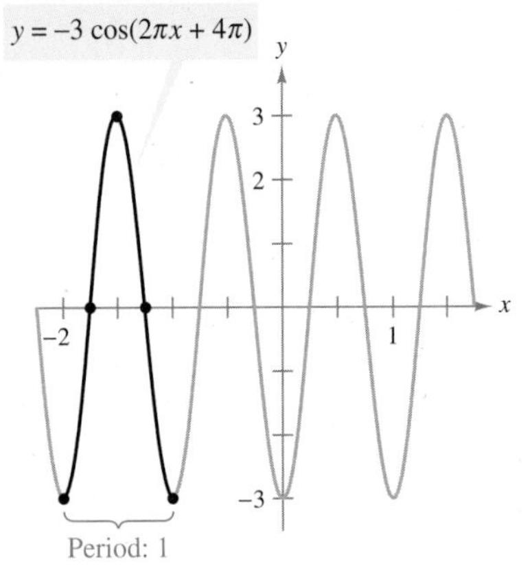

Figure 9.44

EXAMPLE 5 Horizontal Translation

Sketch the graph of

$$y = -3\cos(2\pi x + 4\pi).$$

Solution The amplitude is 3 and the period is $2\pi/2\pi = 1$. By solving the equations

$$2\pi x + 4\pi = 0 \qquad\qquad 2\pi x + 4\pi = 2\pi$$
$$2\pi x = -4\pi \quad \text{and} \quad 2\pi x = -2\pi$$
$$x = -2 \qquad\qquad x = -1$$

you see that the interval $[-2, -1]$ corresponds to one cycle of the graph. Dividing this interval into four equal parts produces the key points

Minimum	*Intercept*	*Maximum*	*Intercept*		*Minimum*
$(-2, -3)$,	$\left(-\frac{7}{4}, 0\right)$,	$\left(-\frac{3}{2}, 3\right)$,	$\left(-\frac{5}{4}, 0\right)$,	and	$(-1, -3)$.

The graph is shown in Figure 9.44. ■

The final type of transformation is the *vertical translation* caused by the constant d in the equations

$$y = d + a\sin(bx - c) \quad \text{and} \quad y = d + a\cos(bx - c).$$

The shift is d units upward for $d > 0$ and d units downward for $d < 0$. In other words, the graph oscillates about the horizontal line $y = d$ instead of about the x-axis.

EXAMPLE 6 Vertical Translation

Sketch the graph of

$$y = 2 + 3\cos 2x.$$

Solution The amplitude is 3 and the period is π. The key points over the interval $[0, \pi]$ are

$$(0, 5), \quad \left(\frac{\pi}{4}, 2\right), \quad \left(\frac{\pi}{2}, -1\right), \quad \left(\frac{3\pi}{4}, 2\right), \quad \text{and} \quad (\pi, 5).$$

The graph is shown in Figure 9.45. Compared with the graph of $f(x) = 3\cos 2x$, the graph of $y = 2 + 3\cos 2x$ is shifted upward two units.

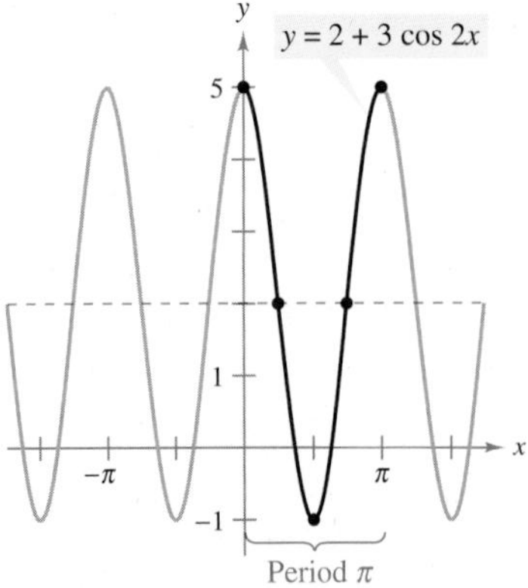

Figure 9.45 ■

Mathematical Modeling

Sine and cosine functions can be used to model many real-life situations, including electric currents, musical tones, radio waves, tides, and weather patterns.

EXAMPLE 7 Finding a Trigonometric Model

Throughout the day, the depth of water at the end of a dock in Bar Harbor, Maine varies with the tides. The table shows the depths (in feet) at various times during the morning. *(Source: Nautical Software, Inc.)*

Time, t	Midnight	2 A.M.	4 A.M.	6 A.M.	8 A.M.	10 A.M.	Noon
Depth, y	3.4	8.7	11.3	9.1	3.8	0.1	1.2

a. Use a trigonometric function to model the data.

b. Find the depths at 9 A.M. and 3 P.M.

c. A boat needs at least 10 feet of water to moor at the dock. During what times in the afternoon can it safely dock?

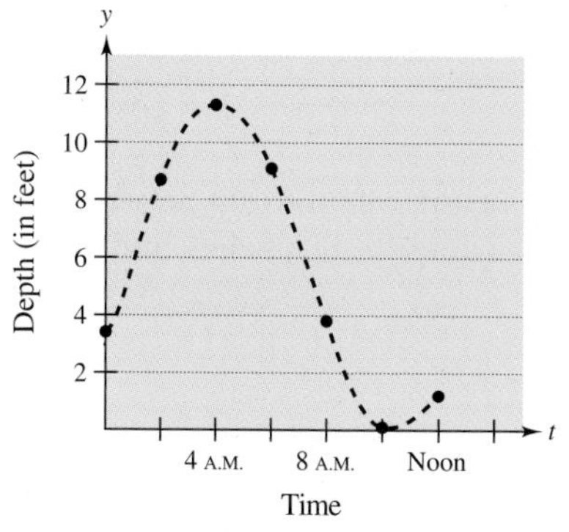

Figure 9.46

Solution

a. Begin by graphing the data, as shown in Figure 9.46. You can use either a sine or a cosine model. Use a cosine model of the form

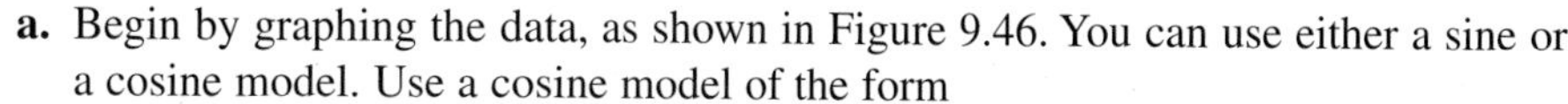

$$y = a\cos(bt - c) + d.$$

The difference between the maximum height and the minimum height of the graph is twice the amplitude of the function. So, the amplitude is

$$a = \frac{1}{2}[(\text{maximum depth}) - (\text{minimum depth})] = \frac{1}{2}(11.3 - 0.1) = 5.6.$$

The cosine function completes one half of a cycle between the times at which the maximum and minimum depths occur. So, the period is

$$p = 2[(\text{time of min. depth}) - (\text{time of max. depth})] = 2(10 - 4) = 12$$

which implies that $b = 2\pi/p \approx 0.524$. Because high tide occurs 4 hours after midnight, consider the left endpoint to be $c/b = 4$, so $c \approx 2.094$. Moreover, because the average depth is $\frac{1}{2}(11.3 + 0.1) = 5.7$, it follows that $d = 5.7$. So, you can model the depth with the function given by

$$y = 5.6\cos(0.524t - 2.094) + 5.7.$$

b. The depths at 9 A.M. and 3 P.M. are as follows.

$$y = 5.6\cos(0.524 \cdot 9 - 2.094) + 5.7$$

$$\approx 0.84 \text{ foot} \qquad \text{9 A.M.}$$

$$y = 5.6\cos(0.524 \cdot 15 - 2.094) + 5.7$$

$$\approx 10.57 \text{ feet} \qquad \text{3 P.M.}$$

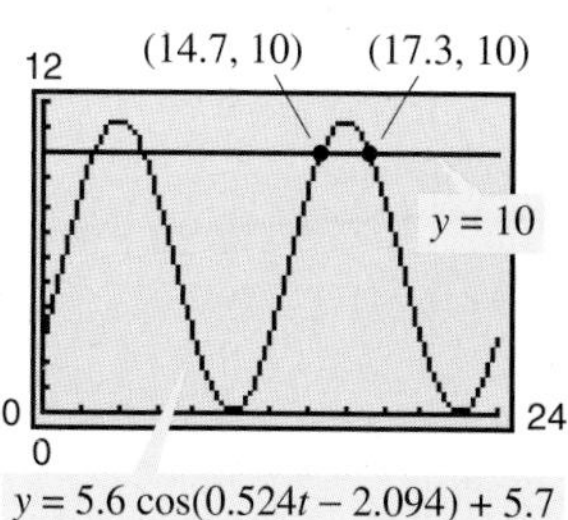

Figure 9.47

c. To find out when the depth y is at least 10 feet, you can graph the model with the line $y = 10$ using a graphing utility, as shown in Figure 9.47. Using the *intersect* feature, you can determine that the depth is at least 10 feet between 2:42 P.M. ($t \approx 14.7$) and 5:18 P.M. ($t \approx 17.3$). ■

9.5 Exercises

See www.CalcChat.com for worked-out solutions to odd-numbered exercises.

In Exercises 1–4, fill in the blanks.

1. One period of a sine or cosine function is called one ________ of the sine or cosine curve.
2. The ________ of a sine or cosine curve represents half the distance between the maximum and minimum values of the function.
3. For the function given by $y = a\sin(bx - c)$, $\frac{c}{b}$ represents the ________ ________ of the graph of the function.
4. For the function given by $y = d + a\cos(bx - c)$, d represents a ________ ________ of the graph of the function.

In Exercises 5–18, find the period and amplitude.

5. $y = 2\sin 5x$

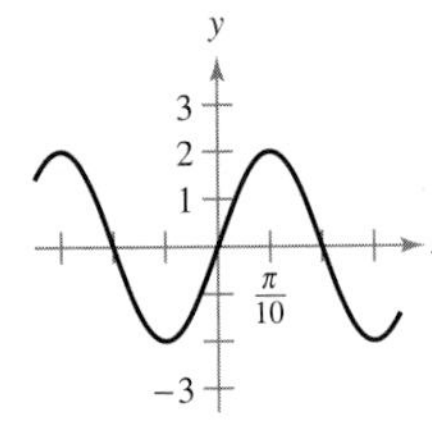

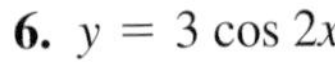

6. $y = 3\cos 2x$

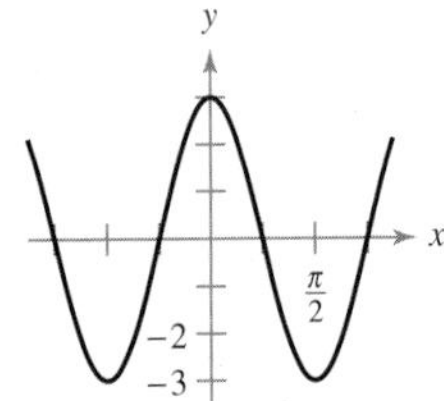

7. $y = \frac{3}{4}\cos\frac{x}{2}$

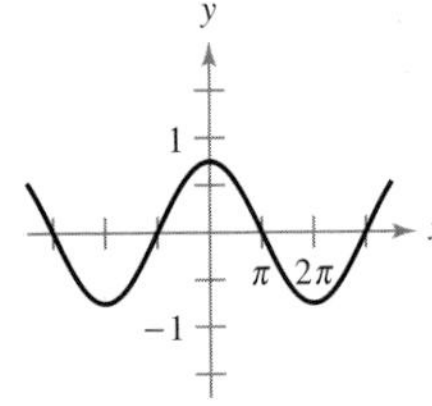

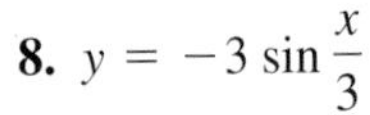

8. $y = -3\sin\frac{x}{3}$

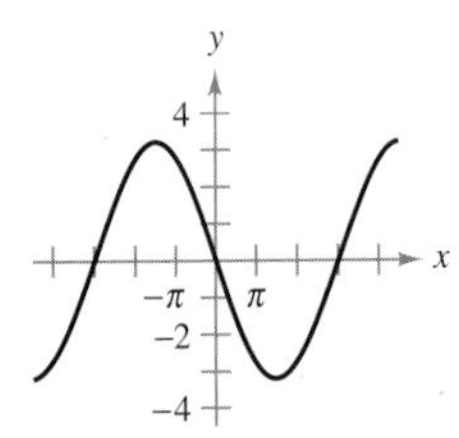

9. $y = \frac{1}{2}\sin\frac{\pi x}{3}$

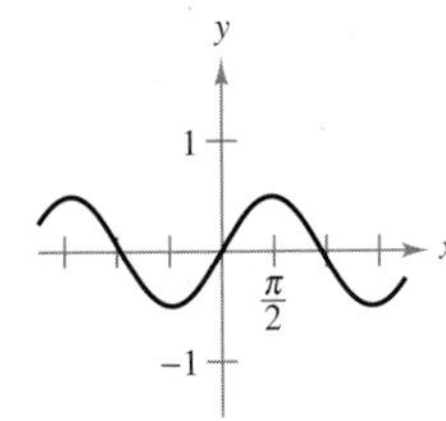

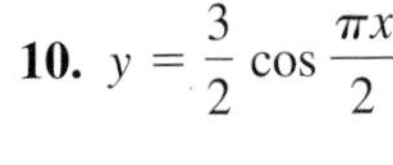

10. $y = \frac{3}{2}\cos\frac{\pi x}{2}$

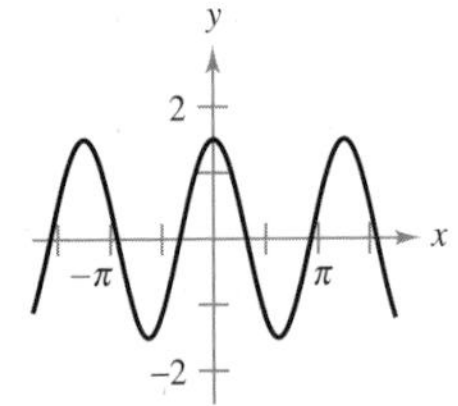

11. $y = -4\sin x$
12. $y = -\cos\frac{2x}{3}$
13. $y = 3\sin 10x$
14. $y = \frac{1}{5}\sin 6x$
15. $y = \frac{5}{3}\cos\frac{4x}{5}$
16. $y = \frac{5}{2}\cos\frac{x}{4}$
17. $y = \frac{1}{4}\sin 2\pi x$
18. $y = \frac{2}{3}\cos\frac{\pi x}{10}$

In Exercises 19–26, describe the relationship between the graphs of *f* and *g*. Consider amplitude, period, and shifts.

19. $f(x) = \sin x$
    $g(x) = \sin(x - \pi)$
20. $f(x) = \cos x$
    $g(x) = \cos(x + \pi)$
21. $f(x) = \cos 2x$
    $g(x) = -\cos 2x$
22. $f(x) = \sin 3x$
    $g(x) = \sin(-3x)$
23. $f(x) = \cos x$
    $g(x) = \cos 2x$
24. $f(x) = \sin x$
    $g(x) = \sin 3x$
25. $f(x) = \sin 2x$
    $g(x) = 3 + \sin 2x$
26. $f(x) = \cos 4x$
    $g(x) = -2 + \cos 4x$

In Exercises 27–30, describe the relationship between the graphs of *f* and *g*. Consider amplitude, period, and shifts.

27.

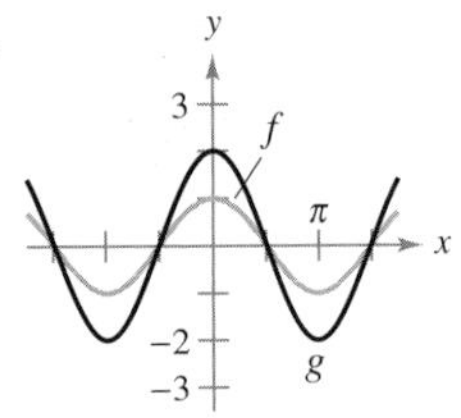

28.

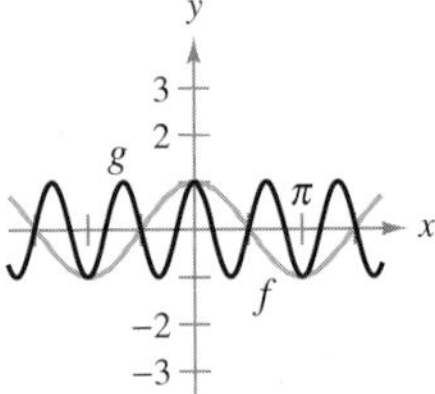

29.

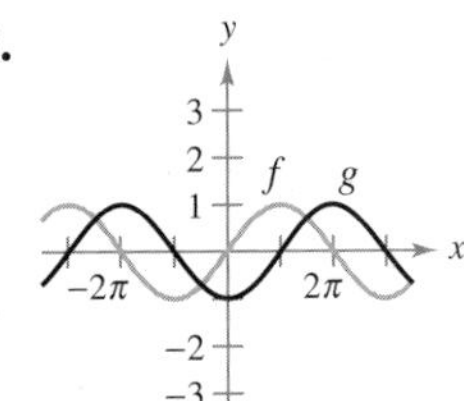

30.

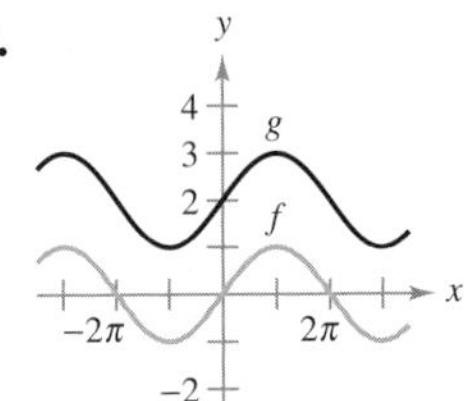

In Exercises 31–38, graph *f* and *g* on the same set of coordinate axes. (Include two full periods.)

31. $f(x) = -2\sin x$
    $g(x) = 4\sin x$
32. $f(x) = \sin x$
    $g(x) = \sin\frac{x}{3}$
33. $f(x) = \cos x$
    $g(x) = 2 + \cos x$
34. $f(x) = 2\cos 2x$
    $g(x) = -\cos 4x$
35. $f(x) = -\frac{1}{2}\sin\frac{x}{2}$
    $g(x) = 3 - \frac{1}{2}\sin\frac{x}{2}$
36. $f(x) = 4\sin \pi x$
    $g(x) = 4\sin \pi x - 3$
37. $f(x) = 2\cos x$
    $g(x) = 2\cos(x + \pi)$
38. $f(x) = -\cos x$
    $g(x) = -\cos(x - \pi)$

In Exercises 39–60, sketch the graph of the function. (Include two full periods.)

39. $y = 5 \sin x$

40. $y = \frac{1}{4} \sin x$

41. $y = \frac{1}{3} \cos x$

42. $y = 4 \cos x$

43. $y = \cos \dfrac{x}{2}$

44. $y = \sin 4x$

45. $y = \cos 2\pi x$

46. $y = \sin \dfrac{\pi x}{4}$

47. $y = -\sin \dfrac{2\pi x}{3}$

48. $y = -10 \cos \dfrac{\pi x}{6}$

49. $y = \sin\left(x - \dfrac{\pi}{2}\right)$

50. $y = \sin(x - 2\pi)$

51. $y = 3 \cos(x + \pi)$

52. $y = 4 \cos\left(x + \dfrac{\pi}{4}\right)$

53. $y = 2 - \sin \dfrac{2\pi x}{3}$

54. $y = -3 + 5 \cos \dfrac{\pi t}{12}$

55. $y = 2 + \frac{1}{10} \cos 60\pi x$

56. $y = 2 \cos x - 3$

57. $y = 3 \cos(x + \pi) - 3$

58. $y = 4 \cos\left(x + \dfrac{\pi}{4}\right) + 4$

59. $y = \dfrac{2}{3} \cos\left(\dfrac{x}{2} - \dfrac{\pi}{4}\right)$

60. $y = -3 \cos(6x + \pi)$

In Exercises 61–66, g is related to a parent function $f(x) = \sin(x)$ or $f(x) = \cos(x)$. (a) Describe the sequence of transformations from f to g. (b) Sketch the graph of g. (c) Use function notation to write g in terms of f.

61. $g(x) = \sin(4x - \pi)$

62. $g(x) = \sin(2x + \pi)$

63. $g(x) = \cos(x - \pi) + 2$

64. $g(x) = 1 + \cos(x + \pi)$

65. $g(x) = 2 \sin(4x - \pi) - 3$

66. $g(x) = 4 - \sin(2x + \pi)$

In Exercises 67–72, use a graphing utility to graph the function. Include two full periods. Be sure to choose an appropriate viewing window.

67. $y = -2 \sin(4x + \pi)$

68. $y = -4 \sin\left(\dfrac{2}{3}x - \dfrac{\pi}{3}\right)$

69. $y = \cos\left(2\pi x - \dfrac{\pi}{2}\right) + 1$

70. $y = 3 \cos\left(\dfrac{\pi x}{2} + \dfrac{\pi}{2}\right) - 2$

71. $y = -0.1 \sin\left(\dfrac{\pi x}{10} + \pi\right)$

72. $y = \dfrac{1}{100} \sin 120\pi t$

Graphical Reasoning **In Exercises 73–76, find a and d for the function $f(x) = a \cos x + d$ such that the graph of f matches the figure.**

73.

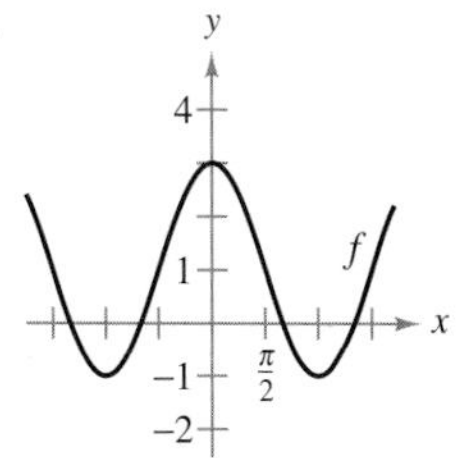

74.

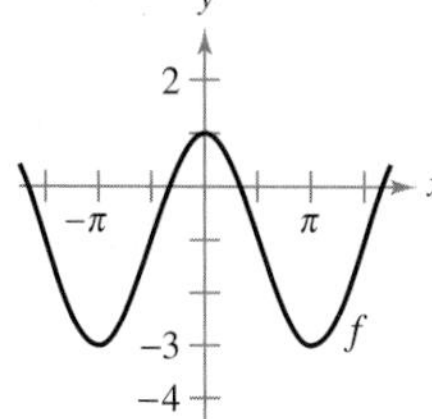

75.

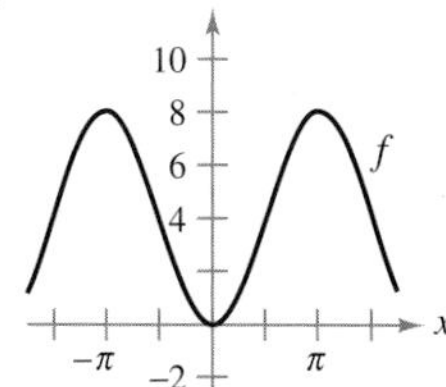

76.

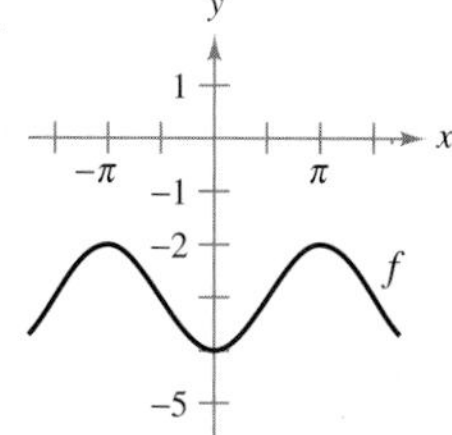

Graphical Reasoning **In Exercises 77–80, find a, b, and c for the function $f(x) = a \sin(bx - c)$ such that the graph of f matches the figure.**

77.

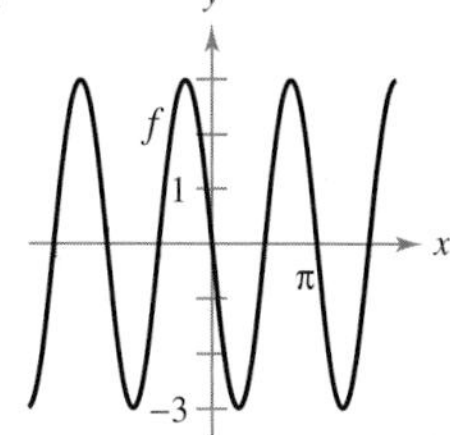

78.

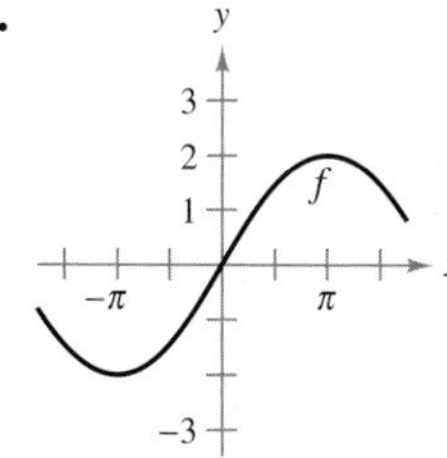

79.

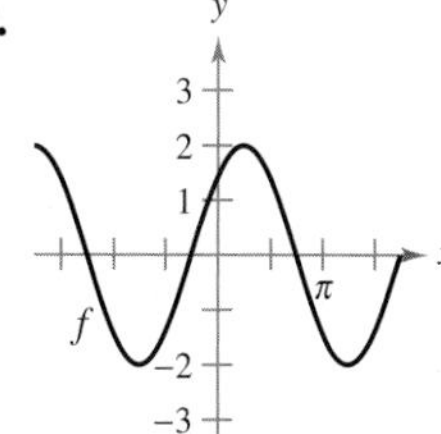

80.

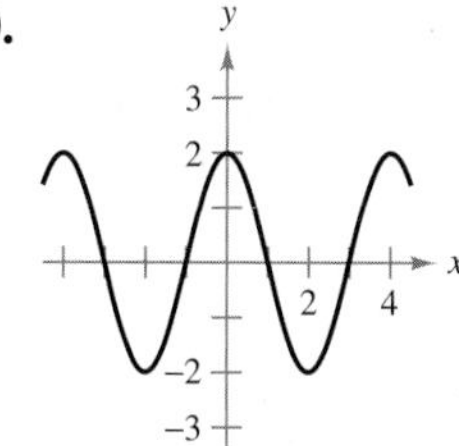

In Exercises 81 and 82, use a graphing utility to graph y_1 and y_2 in the interval $[-2\pi, 2\pi]$. Use the graphs to find real numbers x such that $y_1 = y_2$.

81. $y_1 = \sin x$; $y_2 = -\frac{1}{2}$

82. $y_1 = \cos x$; $y_2 = -1$

In Exercises 83–86, write an equation for the function that is described by the given characteristics.

83. A sine curve with a period of π, an amplitude of 2, a right phase shift of $\pi/2$, and a vertical translation up 1 unit

84. A sine curve with a period of 4π, an amplitude of 3, a left phase shift of $\pi/4$, and a vertical translation down 1 unit

85. A cosine curve with a period of π, an amplitude of 1, a left phase shift of π, and a vertical translation down $\frac{3}{2}$ units

86. A cosine curve with a period of 4π, an amplitude of 3, a right phase shift of $\pi/2$, and a vertical translation up 2 units

WRITING ABOUT CONCEPTS

87. Sketch the graph of $y = \cos bx$ for $b = \frac{1}{2}$, 2, and 3. How does the value of b affect the graph? How many complete cycles occur between 0 and 2π for each value of b?

88. Sketch the graph of $y = \sin(x - c)$ for $c = -\pi/4$, 0, and $\pi/4$. How does the value of c affect the graph?

89. Use a graphing utility to graph h, and use the graph to decide whether h is even, odd, or neither.

(a) $h(x) = \cos^2 x$ (b) $h(x) = \sin^2 x$

90. If f is an even function and g is an odd function, use the results of Exercise 89 to make a conjecture about h, where

(a) $h(x) = [f(x)]^2$. (b) $h(x) = [g(x)]^2$.

91. ***Respiratory Cycle*** For a person at rest, the velocity v (in liters per second) of airflow during a respiratory cycle (the time from the beginning of one breath to the beginning of the next) is given by $v = 0.85 \sin \frac{\pi t}{3}$, where t is the time (in seconds). (Inhalation occurs when $v > 0$, and exhalation occurs when $v < 0$.)

(a) Find the time for one full respiratory cycle.

(b) Find the number of cycles per minute.

(c) Sketch the graph of the velocity function.

92. ***Respiratory Cycle*** After exercising for a few minutes, a person has a respiratory cycle for which the velocity of airflow is approximated by $v = 1.75 \sin \frac{\pi t}{2}$, where t is the time (in seconds). (Inhalation occurs when $v > 0$, and exhalation occurs when $v < 0$.)

(a) Find the time for one full respiratory cycle.

(b) Find the number of cycles per minute.

(c) Sketch the graph of the velocity function.

93. ***Data Analysis: Meteorology*** The table shows the maximum daily high temperatures in Las Vegas L and International Falls I (in degrees Fahrenheit) for month t, with $t = 1$ corresponding to January. *(Source: National Climatic Data Center)*

t	1	2	3	4	5	6
L	57.1	63.0	69.5	78.1	87.8	98.9
I	13.8	22.4	34.9	51.5	66.6	74.2

t	7	8	9	10	11	12
L	104.1	101.8	93.8	80.8	66.0	57.3
I	78.6	76.3	64.7	51.7	32.5	18.1

(a) A model for the temperature in Las Vegas is given by

$$L(t) = 80.60 + 23.50 \cos\left(\frac{\pi t}{6} - 3.67\right).$$

Find a trigonometric model for International Falls.

(b) Use a graphing utility to graph the data points and the model for the temperatures in Las Vegas. How well does the model fit the data?

(c) Use a graphing utility to graph the data points and the model for the temperatures in International Falls. How well does the model fit the data?

(d) Use the models to estimate the average maximum temperature in each city. Which term of the models did you use? Explain.

(e) What is the period of each model? Are the periods what you expected? Explain.

(f) Which city has the greater variability in temperature throughout the year? Which factor of the models determines this variability? Explain.

94. ***Health*** The function given by

$$P = 100 - 20 \cos \frac{5\pi t}{3}$$

approximates the blood pressure P (in millimeters of mercury) at time t (in seconds) for a person at rest.

(a) Find the period of the function.

(b) Find the number of heartbeats per minute.

95. ***Piano Tuning*** When tuning a piano, a technician strikes a tuning fork for the A above middle C and sets up a wave motion that can be approximated by $y = 0.001 \sin 880\pi t$, where t is the time (in seconds).

(a) What is the period of the function?

(b) The frequency f is given by $f = 1/p$. What is the frequency of the note?

96. ***Data Analysis: Astronomy*** The percents y (in decimal form) of the moon's face that was illuminated on day x in the year 2009, where $x = 1$ represents January 1, are shown in the table. *(Source: U.S. Naval Observatory)*

x	4	11	18	26	33	40
y	0.5	1.0	0.5	0.0	0.5	1.0

(a) Create a scatter plot of the data.

(b) Find a trigonometric model that fits the data.

(c) Add the graph of your model in part (b) to the scatter plot. How well does the model fit the data?

(d) What is the period of the model?

(e) Estimate the moon's percent illumination for March 12, 2009.

97. ***Fuel Consumption*** The daily consumption C (in gallons) of diesel fuel on a farm is modeled by

$$C = 30.3 + 21.6 \sin\left(\frac{2\pi t}{365} + 10.9\right)$$

where t is the time (in days), with $t = 1$ corresponding to January 1.

(a) What is the period of the model? Is it what you expected? Explain.

(b) What is the average daily fuel consumption? Which term of the model did you use? Explain.

(c) Use a graphing utility to graph the model. Use the graph to approximate the time of the year when consumption exceeds 40 gallons per day.

98. ***Ferris Wheel*** A Ferris wheel is built such that the height h (in feet) above ground of a seat on the wheel at time t (in seconds) can be modeled by

$$h(t) = 53 + 50 \sin\left(\frac{\pi}{10}t - \frac{\pi}{2}\right).$$

(a) Find the period of the model. What does the period tell you about the ride?

(b) Find the amplitude of the model. What does the amplitude tell you about the ride?

(c) Use a graphing utility to graph one cycle of the model.

True or False? **In Exercises 99–101, determine whether the statement is true or false. Justify your answer.**

99. The graph of the function given by $f(x) = \sin(x + 2\pi)$ translates the graph of $f(x) = \sin x$ exactly one period to the right so that the two graphs look identical.

100. The function given by $y = \frac{1}{2}\cos 2x$ has an amplitude that is twice that of the function given by $y = \cos x$.

101. The graph of $y = -\cos x$ is a reflection of the graph of $y = \sin(x + \pi/2)$ in the x-axis.

CAPSTONE

102. Use a graphing utility to graph the function given by $y = d + a\sin(bx - c)$, for several different values of a, b, c, and d. Write a paragraph describing the changes in the graph corresponding to changes in each constant.

Conjecture **In Exercises 103 and 104, graph f and g on the same set of coordinate axes. Include two full periods. Make a conjecture about the functions.**

103. $f(x) = \sin x, \quad g(x) = \cos\left(x - \frac{\pi}{2}\right)$

104. $f(x) = \sin x, \quad g(x) = -\cos\left(x + \frac{\pi}{2}\right)$

SECTION PROJECT

Approximating Sine and Cosine Functions

Using calculus, it can be shown that the sine and cosine functions can be approximated by the polynomials

$$\sin x \approx x - \frac{x^3}{3!} + \frac{x^5}{5!} \quad \text{and} \quad \cos x \approx 1 - \frac{x^2}{2!} + \frac{x^4}{4!}$$

where x is in radians.

(a) Use a graphing utility to graph the sine function and its polynomial approximation in the same viewing window. How do the graphs compare?

(b) Use a graphing utility to graph the cosine function and its polynomial approximation in the same viewing window. How do the graphs compare?

(c) Study the patterns in the polynomial approximations of the sine and cosine functions and guess the next term in each. Then repeat parts (a) and (b). How did the accuracy of the approximations change when additional terms were added?

(d) Use the polynomial approximation for the sine function to approximate the following functional values. Compare the results with those given by a calculator. Is the error in the approximation the same in each case? Explain your reasoning.

(i) $\sin\frac{1}{2}$ (ii) $\sin 1$ (iii) $\sin\frac{\pi}{6}$

(e) Use the polynomial approximation for the cosine function to approximate the following functional values. Compare the results with those given by a calculator. Is the error in the approximation the same in each case?

(i) $\cos(-0.5)$ (ii) $\cos 1$ (iii) $\cos\frac{\pi}{4}$

9.6 Graphs of Other Trigonometric Functions

- **Sketch the graphs of tangent functions.**
- **Sketch the graphs of cotangent functions.**
- **Sketch the graphs of secant and cosecant functions.**
- **Sketch the graphs of damped trigonometric functions.**

Graph of the Tangent Function

Recall that the tangent function is odd. That is, $\tan(-x) = -\tan x$. Consequently, the graph of $y = \tan x$ is symmetric with respect to the origin. You also know from the identity $\tan x = \sin x/\cos x$ that the tangent is undefined for values at which $\cos x = 0$. Two such values are $x = \pm\pi/2 \approx \pm 1.5708$.

x	$-\frac{\pi}{2}$	-1.57	-1.5	$-\frac{\pi}{4}$	0	$\frac{\pi}{4}$	1.5	1.57	$\frac{\pi}{2}$
$\tan x$	Undef.	-1255.8	-14.1	-1	0	1	14.1	1255.8	Undef.

As indicated in the table, $\tan x$ increases without bound as x approaches $\pi/2$ from the left, and decreases without bound as x approaches $-\pi/2$ from the right. So, the graph of $y = \tan x$ has *vertical asymptotes* at $x = \pi/2$ and $x = -\pi/2$, as shown in Figure 9.48. Moreover, because the period of the tangent function is π, vertical asymptotes also occur when $x = \pi/2 + n\pi$, where n is an integer. The domain of the tangent function is the set of all real numbers other than $x = \pi/2 + n\pi$, and the range is the set of all real numbers.

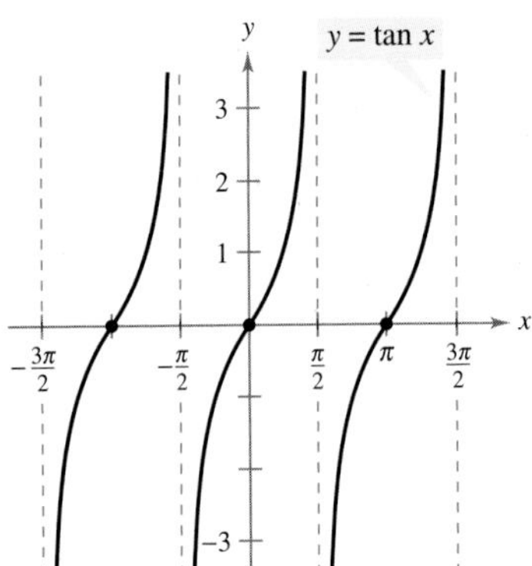

Figure 9.48

Period: π
Domain: all $x \neq \frac{\pi}{2} + n\pi$
Range: $(-\infty, \infty)$
Vertical asymptotes: $x = \frac{\pi}{2} + n\pi$
Symmetry: Origin

Sketching the graph of $y = a\tan(bx - c)$ is similar to sketching the graph of $y = a\sin(bx - c)$ in that you locate key points that identify the intercepts and asymptotes. Two consecutive vertical asymptotes can be found by solving the equations

$$bx - c = -\frac{\pi}{2} \quad \text{and} \quad bx - c = \frac{\pi}{2}.$$

The midpoint between two consecutive vertical asymptotes is an x-intercept of the graph. The period of the function $y = a\tan(bx - c)$ is the distance between two consecutive vertical asymptotes. The amplitude of a tangent function is not defined. After plotting the asymptotes and the x-intercept, plot a few additional points between the two asymptotes and sketch one cycle. Finally, sketch one or two additional cycles to the left and right.

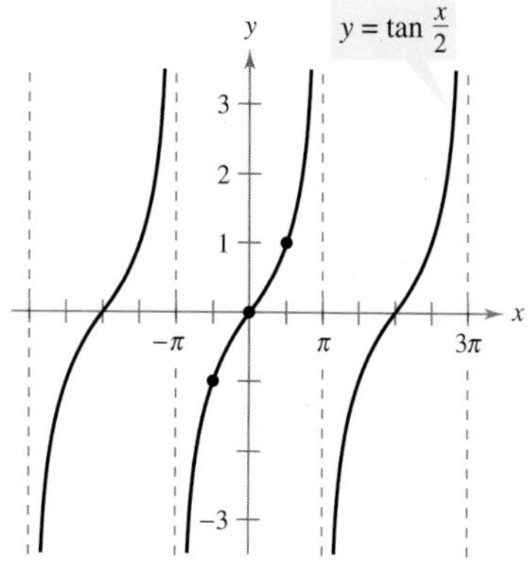

Figure 9.49

EXAMPLE 1 Sketching the Graph of a Tangent Function

Sketch the graph of

$$y = \tan\frac{x}{2}.$$

Solution By solving the equations

$$\frac{x}{2} = -\frac{\pi}{2} \quad \text{and} \quad \frac{x}{2} = \frac{\pi}{2}$$

$$x = -\pi \qquad\qquad x = \pi$$

you can see that two consecutive vertical asymptotes occur at

$$x = -\pi \quad \text{and} \quad x = \pi.$$

Between these two asymptotes, plot a few points, including the x-intercept, as shown in the table. Three cycles of the graph are shown in Figure 9.49.

x	$-\pi$	$-\frac{\pi}{2}$	0	$\frac{\pi}{2}$	π
$\tan\frac{x}{2}$	Undef.	-1	0	1	Undef.

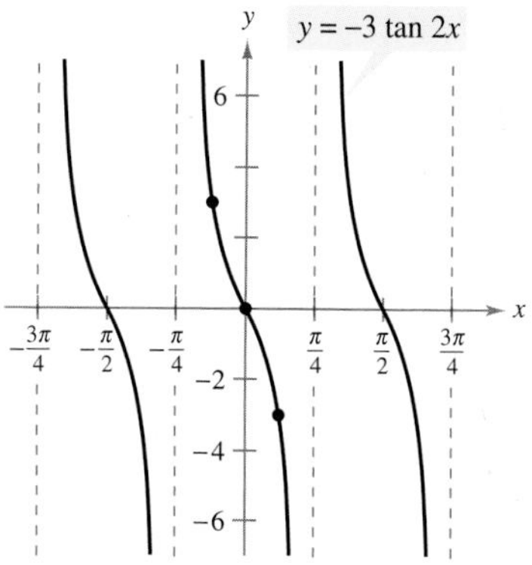

EXAMPLE 2 Sketching the Graph of a Tangent Function

Figure 9.50

Sketch the graph of

$$y = -3\tan 2x.$$

Solution By solving the equations

$$2x = -\frac{\pi}{2} \quad \text{and} \quad 2x = \frac{\pi}{2}$$

$$x = -\frac{\pi}{4} \qquad\qquad x = \frac{\pi}{4}$$

you can see that two consecutive vertical asymptotes occur at

$$x = -\frac{\pi}{4} \quad \text{and} \quad x = \frac{\pi}{4}.$$

Between these two asymptotes, plot a few points, including the x-intercept, as shown in the table. Three cycles of the graph are shown in Figure 9.50.

x	$-\frac{\pi}{4}$	$-\frac{\pi}{8}$	0	$\frac{\pi}{8}$	$\frac{\pi}{4}$
$-3\tan 2x$	Undef.	3	0	-3	Undef.

By comparing the graphs in Examples 1 and 2, you can see that the graph of $y = a\tan(bx - c)$ increases between consecutive vertical asymptotes when $a > 0$, and decreases between consecutive vertical asymptotes when $a < 0$. In other words, the graph for $a < 0$ is a reflection in the x-axis of the graph for $a > 0$. ■

Graph of the Cotangent Function

The graph of the cotangent function is similar to the graph of the tangent function. It also has a period of π. However, from the identity

$$y = \cot x = \frac{\cos x}{\sin x}$$

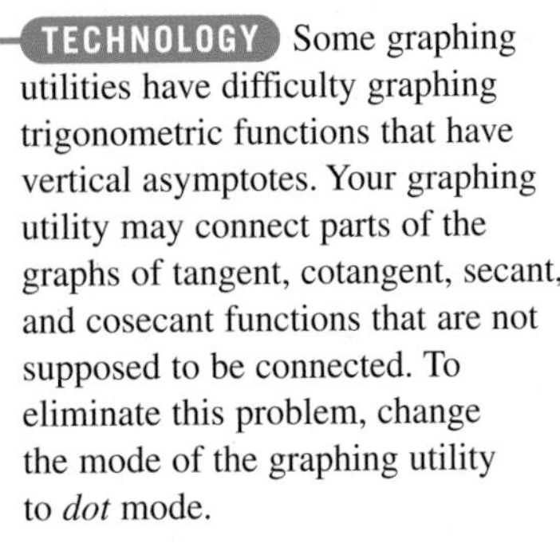

TECHNOLOGY Some graphing utilities have difficulty graphing trigonometric functions that have vertical asymptotes. Your graphing utility may connect parts of the graphs of tangent, cotangent, secant, and cosecant functions that are not supposed to be connected. To eliminate this problem, change the mode of the graphing utility to *dot* mode.

you can see that the cotangent function has vertical asymptotes when $\sin x$ is zero, which occurs at $x = n\pi$, where n is an integer. The graph of the cotangent function is shown in Figure 9.51. Note that two consecutive vertical asymptotes of the graph of $y = a\cot(bx - c)$ can be found by solving the equations $bx - c = 0$ and $bx - c = \pi$.

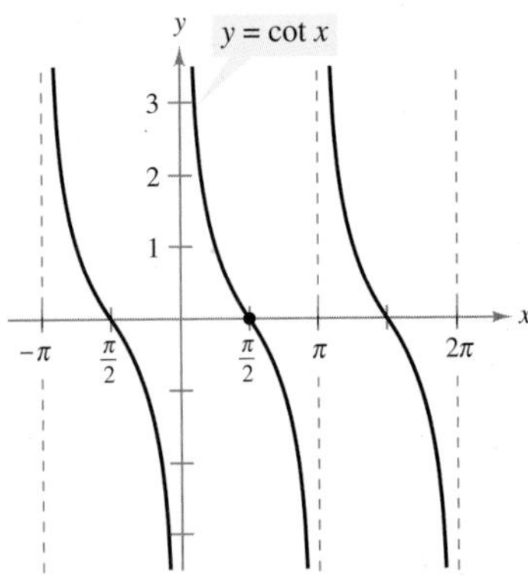

Figure 9.51

Period: π
Domain: all $x \neq n\pi$
Range: $(-\infty, \infty)$
Vertical asymptotes: $x = n\pi$
Symmetry: Origin

EXAMPLE 3 Sketching the Graph of a Cotangent Function

Sketch the graph of

$$y = 2\cot\frac{x}{3}.$$

Solution By solving the equations

$$\frac{x}{3} = 0 \qquad \text{and} \qquad \frac{x}{3} = \pi$$

$$x = 0 \qquad\qquad x = 3\pi$$

you can see that two consecutive vertical asymptotes occur at

$$x = 0 \qquad \text{and} \qquad x = 3\pi.$$

Between these two asymptotes, plot a few points, including the x-intercept, as shown in the table. Three cycles of the graph are shown in Figure 9.52. Note that the period is 3π, the distance between consecutive asymptotes.

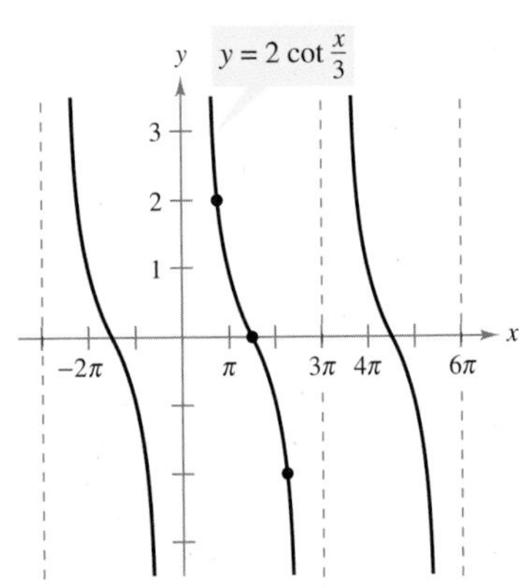

Figure 9.52

x	0	$\frac{3\pi}{4}$	$\frac{3\pi}{2}$	$\frac{9\pi}{4}$	3π
$2\cot\frac{x}{3}$	Undef.	2	0	-2	Undef.

■

Graphs of the Reciprocal Functions

The graphs of the two remaining trigonometric functions can be obtained from the graphs of the sine and cosine functions using the reciprocal identities

$$\csc x = \frac{1}{\sin x} \quad \text{and} \quad \sec x = \frac{1}{\cos x}.$$

For instance, at a given value of x, the y-coordinate of $\sec x$ is the reciprocal of the y-coordinate of $\cos x$. Of course, when $\cos x = 0$, the reciprocal does not exist. Near such values of x, the behavior of the secant function is similar to that of the tangent function. In other words, the graphs of

$$\tan x = \frac{\sin x}{\cos x} \quad \text{and} \quad \sec x = \frac{1}{\cos x}$$

have vertical asymptotes at $x = \pi/2 + n\pi$, where n is an integer, and the cosine is zero at these x-values. Similarly,

$$\cot x = \frac{\cos x}{\sin x} \quad \text{and} \quad \csc x = \frac{1}{\sin x}$$

have vertical asymptotes where $\sin x = 0$—that is, at $x = n\pi$.

To sketch the graph of a secant or cosecant function, you should first make a sketch of its reciprocal function. For instance, to sketch the graph of $y = \csc x$, first sketch the graph of $y = \sin x$. Then take reciprocals of the y-coordinates to obtain points on the graph of $y = \csc x$. This procedure is used to obtain the graphs shown in Figure 9.53.

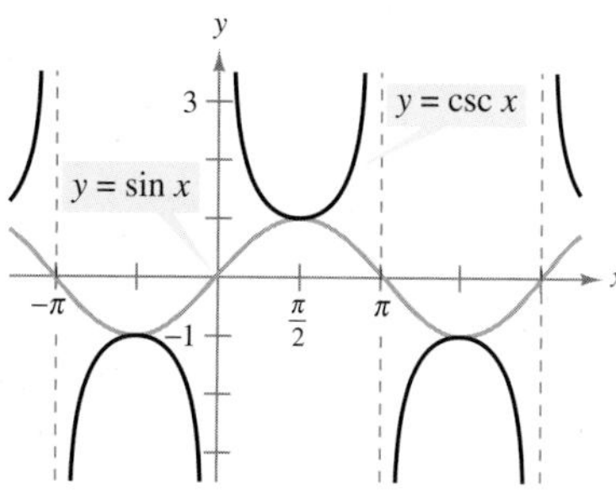

Period: 2π
Domain: All $x \neq n\pi$
Range: $(-\infty, -1] \cup [1, \infty)$
Vertical asymptotes: $x = n\pi$
Symmetry: Origin

Figure 9.53

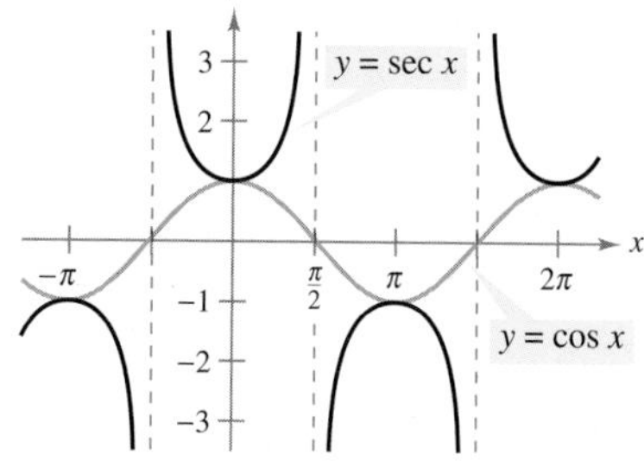

Period: 2π
Domain: All $x \neq \frac{\pi}{2} + n\pi$
Range: $(-\infty, -1] \cup [1, \infty)$
Vertical asymptotes: $x = \frac{\pi}{2} + n\pi$
Symmetry: y-axis

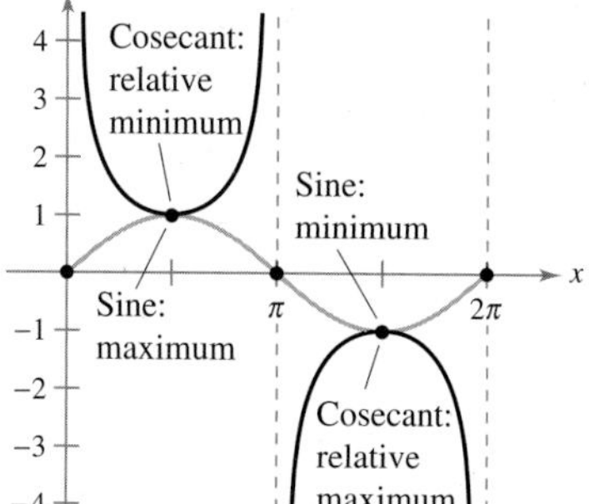

Figure 9.54

In comparing the graphs of the cosecant and secant functions with those of the sine and cosine functions, note that the "hills" and "valleys" are interchanged. For example, a hill (or maximum point) on the sine curve corresponds to a valley (a relative minimum) on the cosecant curve, and a valley (or minimum point) on the sine curve corresponds to a hill (a relative maximum) on the cosecant curve, as shown in Figure 9.54. Additionally, x-intercepts of the sine and cosine functions become vertical asymptotes of the cosecant and secant functions, respectively (see Figure 9.54).

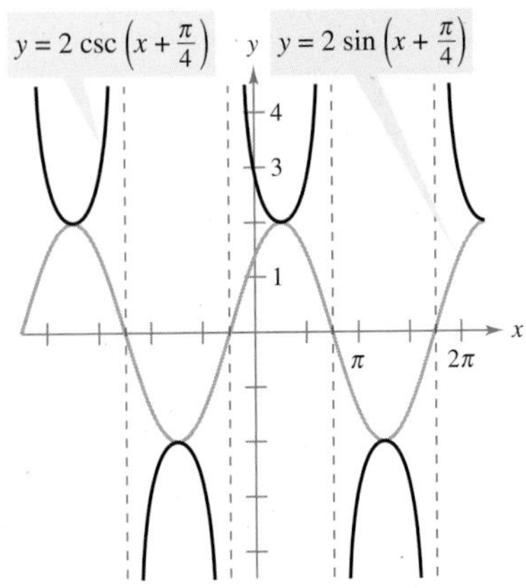

Figure 9.55

EXAMPLE 4 Sketching the Graph of a Cosecant Function

Sketch the graph of

$$y = 2\csc\left(x + \frac{\pi}{4}\right).$$

Solution Begin by sketching the graph of

$$y = 2\sin\left(x + \frac{\pi}{4}\right).$$

For this function, the amplitude is 2 and the period is 2π. By solving the equations

$$x + \frac{\pi}{4} = 0 \qquad \text{and} \qquad x + \frac{\pi}{4} = 2\pi$$

$$x = -\frac{\pi}{4} \qquad\qquad x = \frac{7\pi}{4}$$

you can see that one cycle of the sine function corresponds to the interval from $x = -\pi/4$ to $x = 7\pi/4$. The graph of this sine function is represented by the gray curve in Figure 9.55. Because the sine function is zero at the midpoint and endpoints of this interval, the corresponding cosecant function

$$y = 2\csc\left(x + \frac{\pi}{4}\right)$$

$$= 2\left(\frac{1}{\sin[x + (\pi/4)]}\right)$$

has vertical asymptotes at

$$x = -\frac{\pi}{4}, x = \frac{3\pi}{4}, x = \frac{7\pi}{4}, \text{ etc.}$$

The graph of the cosecant function is represented by the black curve in Figure 9.55.

EXAMPLE 5 Sketching the Graph of a Secant Function

Sketch the graph of

$$y = \sec 2x.$$

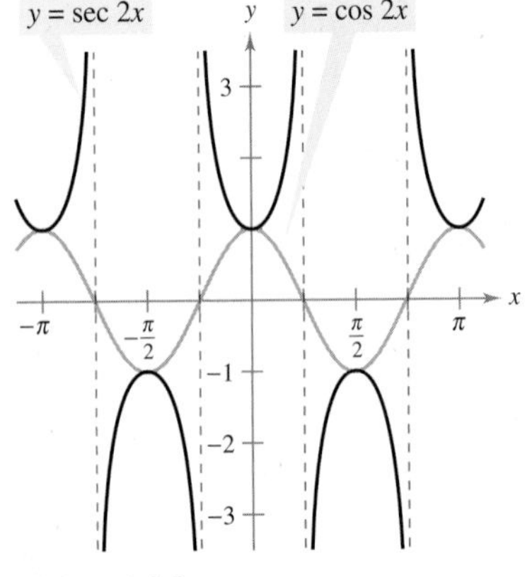

Figure 9.56

Solution Begin by sketching the graph of $y = \cos 2x$, as indicated by the gray curve in Figure 9.56. Then, form the graph of $y = \sec 2x$ as the black curve in the figure. Note that the x-intercepts of $y = \cos 2x$

$$\left(-\frac{\pi}{4}, 0\right), \quad \left(\frac{\pi}{4}, 0\right), \quad \left(\frac{3\pi}{4}, 0\right), \ldots$$

correspond to the vertical asymptotes

$$x = -\frac{\pi}{4}, \quad x = \frac{\pi}{4}, \quad x = \frac{3\pi}{4}, \ldots$$

of the graph of $y = \sec 2x$. Moreover, notice that the period of $y = \cos 2x$ and $y = \sec 2x$ is π. ■

Damped Trigonometric Graphs

A *product* of two functions can be graphed using properties of the individual functions. For instance, consider the function

$$f(x) = x \sin x$$

as the product of the functions $y = x$ and $y = \sin x$. Using properties of absolute value and the fact that $|\sin x| \leq 1$, you have $0 \leq |x||\sin x| \leq |x|$. Consequently,

$$-|x| \leq x \sin x \leq |x|$$

which means that the graph of $f(x) = x \sin x$ lies between the lines $y = -x$ and $y = x$. Furthermore, because

$$f(x) = x \sin x = \pm x \quad \text{at} \quad x = \frac{\pi}{2} + n\pi$$

and

$$f(x) = x \sin x = 0 \quad \text{at} \quad x = n\pi$$

the graph of f touches the line $y = -x$ or the line $y = x$ at $x = \pi/2 + n\pi$ and has x-intercepts at $x = n\pi$. A sketch of f is shown in Figure 9.57. In the function $f(x) = x \sin x$, the factor x is called the **damping factor.**

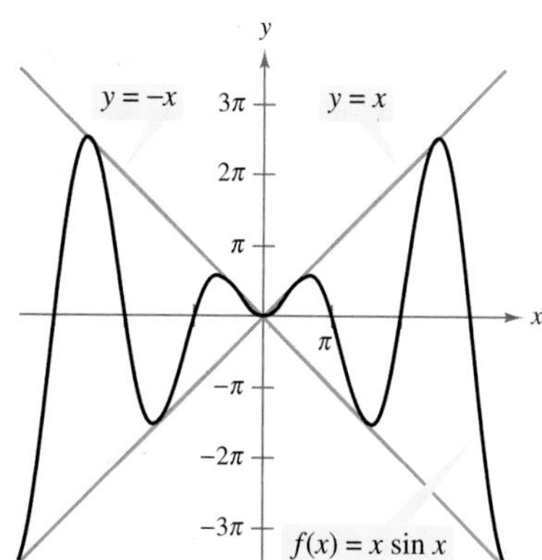

Figure 9.57

EXAMPLE 6 Damped Sine Wave

Sketch the graph of

$$f(x) = e^{-x} \sin 3x.$$

Solution Consider f as the product of the two functions

$$y = e^{-x} \quad \text{and} \quad y = \sin 3x$$

each of which has the set of real numbers as its domain. For any real number x, you know that $e^{-x} \geq 0$ and $|\sin 3x| \leq 1$. So, $e^{-x}|\sin 3x| \leq e^{-x}$, which means that

$$-e^{-x} \leq e^{-x} \sin 3x \leq e^{-x}.$$

Furthermore, because

$$f(x) = e^{-x} \sin 3x = \pm e^{-x} \quad \text{at} \quad x = \frac{\pi}{6} + \frac{n\pi}{3}$$

and

$$f(x) = e^{-x} \sin 3x = 0 \quad \text{at} \quad x = \frac{n\pi}{3}$$

the graph of f touches the curves

$$y = -e^{-x} \text{ and } y = e^{-x}$$

at

$$x = \frac{\pi}{6} + \frac{n\pi}{3}$$

and has intercepts at

$$x = \frac{n\pi}{3}.$$

A sketch is shown in Figure 9.58. ■

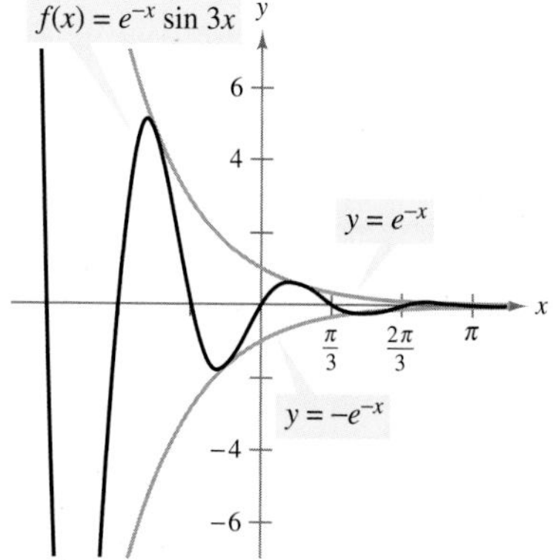

Figure 9.58

9.6 Exercises

See www.CalcChat.com for worked-out solutions to odd-numbered exercises.

In Exercises 1–8, fill in the blanks.

1. The tangent, cotangent, and cosecant functions are ________, so the graphs of these functions have symmetry with respect to the ________.
2. The graphs of the tangent, cotangent, secant, and cosecant functions all have ________ asymptotes.
3. To sketch the graph of a secant or cosecant function, first make a sketch of its corresponding ________ function.
4. For the functions given by $f(x) = g(x) \cdot \sin x$, $g(x)$ is called the ________ factor of the function $f(x)$.
5. The period of $y = \tan x$ is ________.
6. The domain of $y = \cot x$ is all real numbers such that ________.
7. The range of $y = \sec x$ is ________.
8. The period of $y = \csc x$ is ________.

In Exercises 9–14, match the function with its graph. State the period of the function. [The graphs are labeled (a), (b), (c), (d), (e), and (f).]

(a)

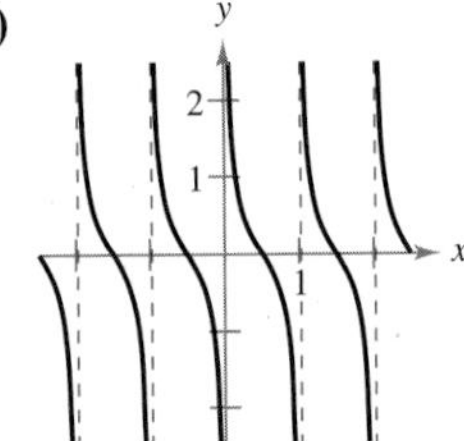

(b)

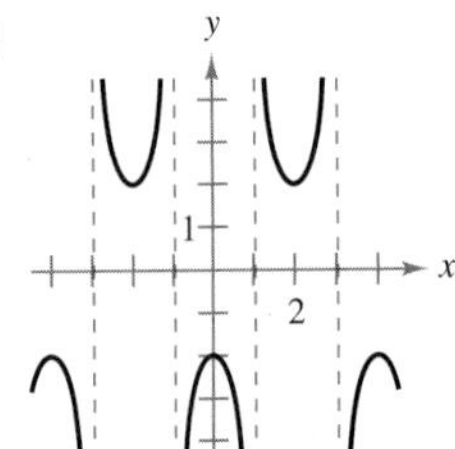

(c)

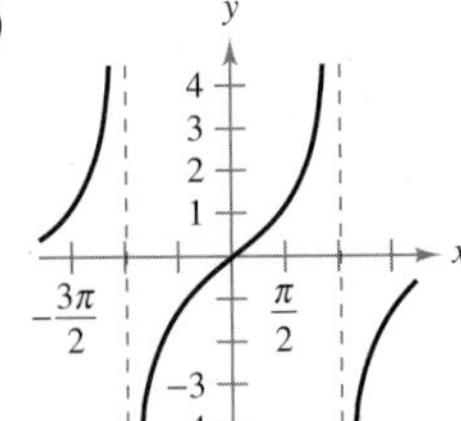

(d)

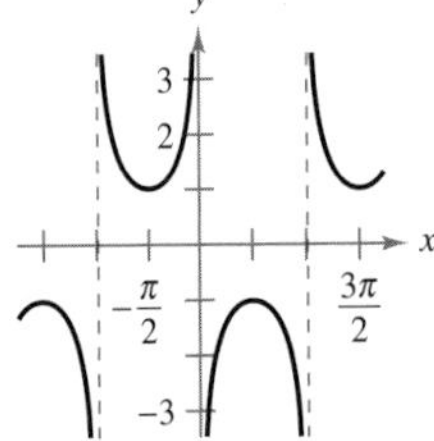

(e)

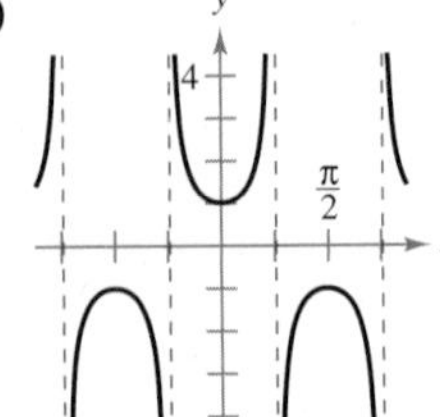

(f)

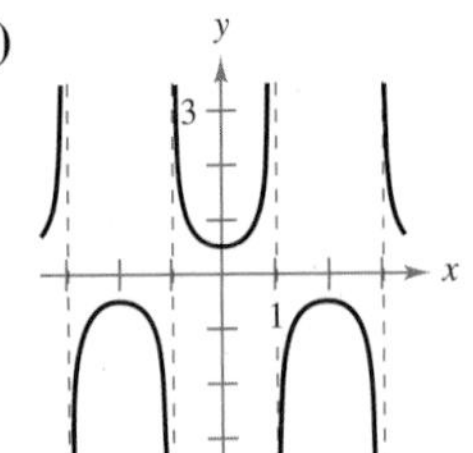

9. $y = \sec 2x$
10. $y = \tan(x/2)$
11. $y = \frac{1}{2} \cot \pi x$
12. $y = -\csc x$
13. $y = \frac{1}{2} \sec(\pi x/2)$
14. $y = -2 \sec(\pi x/2)$

In Exercises 15–38, sketch the graph of the function. Include two full periods.

15. $y = \frac{1}{3} \tan x$
16. $y = \tan 4x$
17. $y = -2 \tan 3x$
18. $y = -3 \tan \pi x$
19. $y = -\frac{1}{2} \sec x$
20. $y = \frac{1}{4} \sec x$
21. $y = \csc \pi x$
22. $y = 3 \csc 4x$
23. $y = \frac{1}{2} \sec \pi x$
24. $y = -2 \sec 4x + 2$
25. $y = \csc \dfrac{x}{2}$
26. $y = \csc \dfrac{x}{3}$
27. $y = 3 \cot 2x$
28. $y = 3 \cot \dfrac{\pi x}{2}$
29. $y = 2 \sec 3x$
30. $y = -\frac{1}{2} \tan x$
31. $y = \tan \dfrac{\pi x}{4}$
32. $y = \tan(x + \pi)$
33. $y = 2 \csc(x - \pi)$
34. $y = \csc(2x - \pi)$
35. $y = 2 \sec(x + \pi)$
36. $y = -\sec \pi x + 1$
37. $y = \dfrac{1}{4} \csc\left(x + \dfrac{\pi}{4}\right)$
38. $y = 2 \cot\left(x + \dfrac{\pi}{2}\right)$

 **In Exercises 39–48, use a graphing utility to graph the function. Include two full periods.**

39. $y = \tan \dfrac{x}{3}$
40. $y = -\tan 2x$
41. $y = -2 \sec 4x$
42. $y = \sec \pi x$
43. $y = \tan\left(x - \dfrac{\pi}{4}\right)$
44. $y = \dfrac{1}{4} \cot\left(x - \dfrac{\pi}{2}\right)$
45. $y = -\csc(4x - \pi)$
46. $y = 2 \sec(2x - \pi)$
47. $y = 0.1 \tan\left(\dfrac{\pi x}{4} + \dfrac{\pi}{4}\right)$
48. $y = \dfrac{1}{3} \sec\left(\dfrac{\pi x}{2} + \dfrac{\pi}{2}\right)$

In Exercises 49–56, use a graph to solve the equation on the interval $[-2\pi, 2\pi]$.

49. $\tan x = 1$
50. $\tan x = \sqrt{3}$
51. $\cot x = -\dfrac{\sqrt{3}}{3}$
52. $\cot x = 1$
53. $\sec x = -2$
54. $\sec x = 2$
55. $\csc x = \sqrt{2}$
56. $\csc x = -\dfrac{2\sqrt{3}}{3}$

In Exercises 57–64, use the graph of the function to determine whether the function is even, odd, or neither. Verify your answer algebraically.

57. $f(x) = \sec x$

58. $f(x) = \tan x$

59. $g(x) = \cot x$

60. $g(x) = \csc x$

61. $f(x) = x + \tan x$

62. $f(x) = x^2 - \sec x$

63. $g(x) = x \csc x$

64. $g(x) = x^2 \cot x$

In Exercises 65–70, use a graphing utility to graph the two equations in the same viewing window. Use the graphs to determine whether the expressions are equivalent. Verify the results algebraically.

65. $y_1 = \sin x \csc x, \quad y_2 = 1$

66. $y_1 = \sin x \sec x, \quad y_2 = \tan x$

67. $y_1 = \dfrac{\cos x}{\sin x}, \quad y_2 = \cot x$

68. $y_1 = \tan x \cot^2 x, \quad y_2 = \cot x$

69. $y_1 = 1 + \cot^2 x, \quad y_2 = \csc^2 x$

70. $y_1 = \sec^2 x - 1, \quad y_2 = \tan^2 x$

In Exercises 71–74, match the function with its graph. Describe the behavior of the function as x approaches zero. [The graphs are labeled (a), (b), (c), and (d).]

(a)

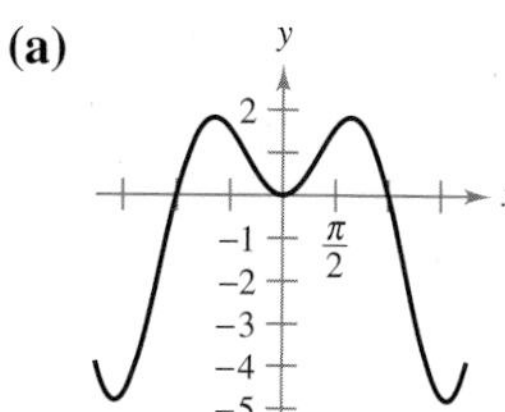

(b)

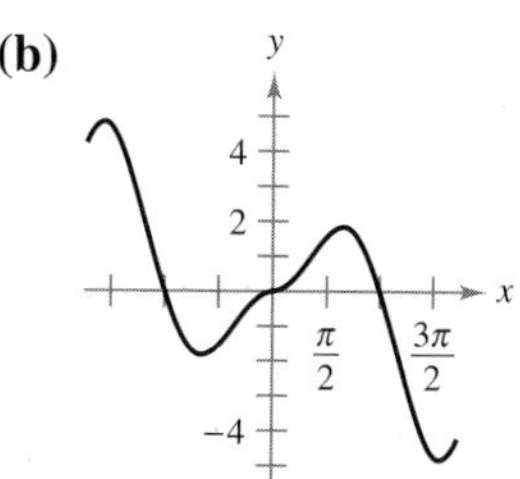

(c)

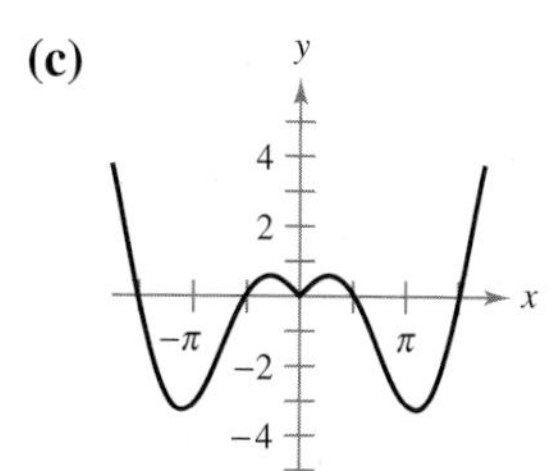

(d)

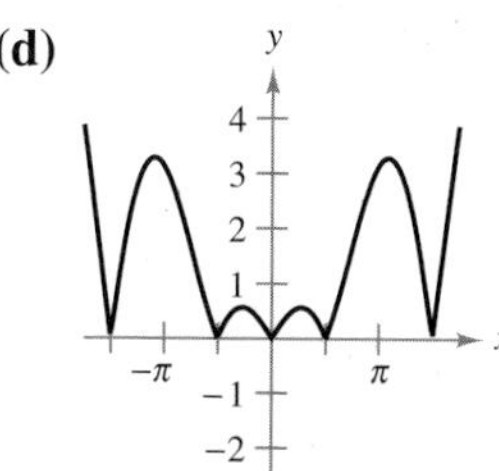

71. $f(x) = |x \cos x|$

72. $f(x) = x \sin x$

73. $g(x) = |x| \sin x$

74. $g(x) = |x| \cos x$

Conjecture **In Exercises 75–78, graph the functions f and g. Use the graphs to make a conjecture about the relationship between the functions.**

75. $f(x) = \sin x + \cos\left(x + \dfrac{\pi}{2}\right), \quad g(x) = 0$

76. $f(x) = \sin x - \cos\left(x + \dfrac{\pi}{2}\right), \quad g(x) = 2 \sin x$

77. $f(x) = \sin^2 x, \quad g(x) = \frac{1}{2}(1 - \cos 2x)$

78. $f(x) = \cos^2 \dfrac{\pi x}{2}, \quad g(x) = \dfrac{1}{2}(1 + \cos \pi x)$

In Exercises 79–82, use a graphing utility to graph the function and the damping factor of the function in the same viewing window. Describe the behavior of the function as x increases without bound.

79. $g(x) = e^{-x^2/2} \sin x$

80. $f(x) = e^{-x} \cos x$

81. $f(x) = 2^{-x/4} \cos \pi x$

82. $h(x) = 2^{-x^2/4} \sin x$

In Exercises 83–88, use a graphing utility to graph the function. Describe the behavior of the function as x approaches zero.

83. $y = \dfrac{6}{x} + \cos x, \quad x > 0$

84. $y = \dfrac{4}{x} + \sin 2x, \quad x > 0$

85. $g(x) = \dfrac{\sin x}{x}$

86. $f(x) = \dfrac{1 - \cos x}{x}$

87. $f(x) = \sin \dfrac{1}{x}$

88. $h(x) = x \sin \dfrac{1}{x}$

WRITING ABOUT CONCEPTS

89. Consider the functions given by

$$f(x) = 2 \sin x$$

and

$$g(x) = \frac{1}{2} \csc x$$

on the interval $(0, \pi)$.

(a) Graph f and g in the same coordinate plane.

(b) Approximate the interval in which $f > g$.

(c) Describe the behavior of each of the functions as x approaches π. How is the behavior of g related to the behavior of f as x approaches π?

90. Consider the functions given by

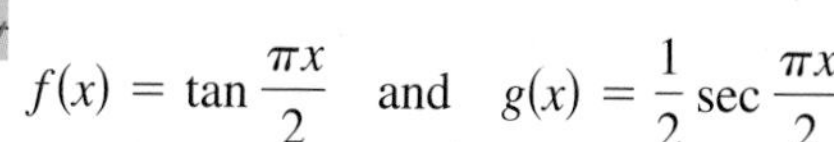

$$f(x) = \tan \frac{\pi x}{2} \quad \text{and} \quad g(x) = \frac{1}{2} \sec \frac{\pi x}{2}$$

on the interval $(-1, 1)$.

(a) Use a graphing utility to graph f and g in the same viewing window.

(b) Approximate the interval in which $f < g$.

(c) Approximate the interval in which $2f < 2g$. How does the result compare with that of part (b)? Explain.

91. ***Distance*** A plane flying at an altitude of 7 miles above a radar antenna will pass directly over the radar antenna (see figure). Let d be the ground distance from the antenna to the point directly under the plane and let x be the angle of elevation to the plane from the antenna. (d is positive as the plane approaches the antenna.) Write d as a function of x and graph the function over the interval $0 < x < \pi$.

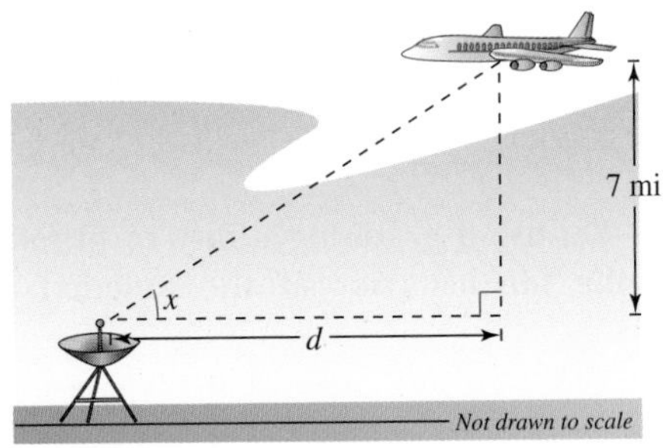

92. ***Television Coverage*** A television camera is on a reviewing platform 27 meters from the street on which a parade will be passing from left to right (see figure). Write the distance d from the camera to a particular unit in the parade as a function of the angle x, and graph the function over the interval $-\pi/2 < x < \pi/2$. (Consider x as negative when a unit in the parade approaches from the left.)

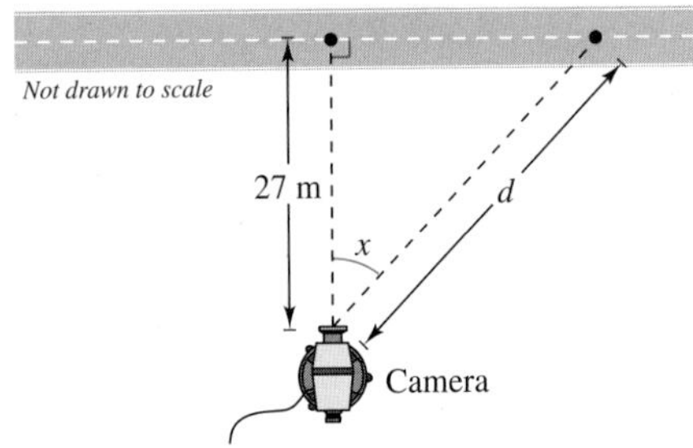

93. ***Meteorology*** The normal monthly high temperatures H (in degrees Fahrenheit) in Erie, Pennsylvania are approximated by

$$H(t) = 56.94 - 20.86 \cos(\pi t/6) - 11.58 \sin(\pi t/6)$$

and the normal monthly low temperatures L are approximated by

$$L(t) = 41.80 - 17.13 \cos(\pi t/6) - 13.39 \sin(\pi t/6)$$

where t is the time (in months), with $t = 1$ corresponding to January (see figure). *(Source: National Climatic Data Center)*

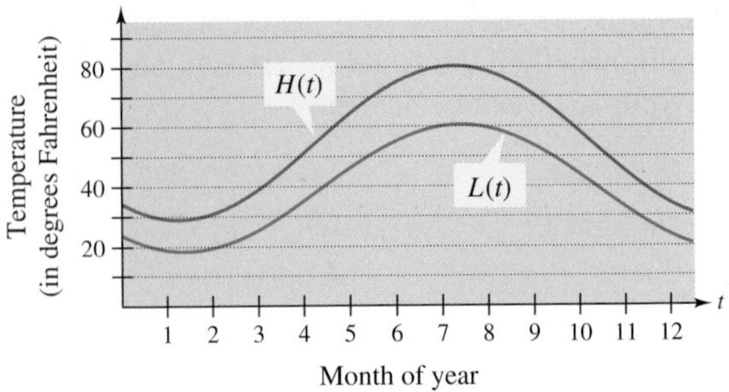

(a) What is the period of each function?

(b) During what part of the year is the difference between the normal high and normal low temperatures greatest? When is it smallest?

(c) The sun is northernmost in the sky around June 21, but the graph shows the warmest temperatures at a later date. Approximate the lag time of the temperatures relative to the position of the sun.

94. ***Sales*** The projected monthly sales S (in thousands of units) of lawn mowers (a seasonal product) are modeled by $S = 74 + 3t - 40 \cos(\pi t/6)$, where t is the time (in months), with $t = 1$ corresponding to January. Graph the sales function over 1 year.

95. ***Harmonic Motion*** An object weighing W pounds is suspended from the ceiling by a steel spring (see figure). The weight is pulled downward (positive direction) from its equilibrium position and released. The resulting motion of the weight is described by the function $y = \frac{1}{2}e^{-t/4} \cos 4t$, $t > 0$, where y is the distance (in feet) and t is the time (in seconds).

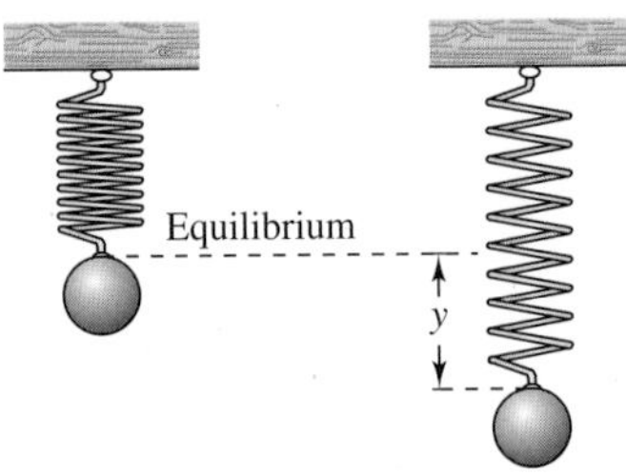

(a) Use a graphing utility to graph the function.

(b) Describe the behavior of the displacement function for increasing values of time t.

True or False? **In Exercises 96 and 97, determine whether the statement is true or false. Justify your answer.**

96. The graph of $y = \csc x$ can be obtained on a calculator by graphing the reciprocal of $y = \sin x$.

97. The graph of $y = \sec x$ can be obtained on a calculator by graphing a translation of the reciprocal of $y = \sin x$.

CAPSTONE

98. Determine which function is represented by the graph. Do not use a calculator. Explain your reasoning.

(a)

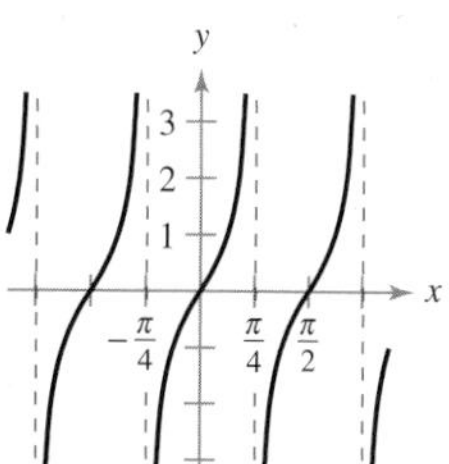

(b)

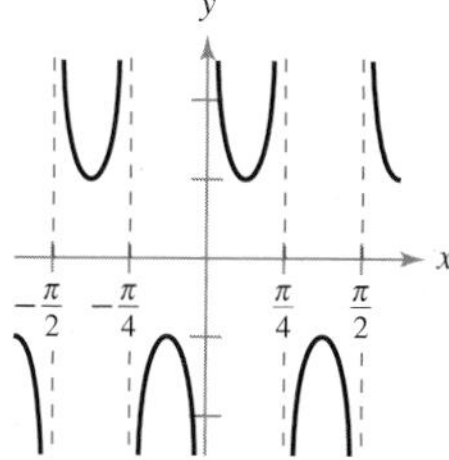

(a)	(b)
(i) $f(x) = \tan 2x$	(i) $f(x) = \sec 4x$
(ii) $f(x) = \tan(x/2)$	(ii) $f(x) = \csc 4x$
(iii) $f(x) = 2 \tan x$	(iii) $f(x) = \csc(x/4)$
(iv) $f(x) = -\tan 2x$	(iv) $f(x) = \sec(x/4)$
(v) $f(x) = -\tan(x/2)$	(v) $f(x) = \csc(4x - \pi)$

In Exercises 99 and 100, use a graphing utility to graph the function. Use the graph to determine the behavior of the function as $x \to c$.

(a) $x \to \frac{\pi}{2}^+$ **(as x approaches $\frac{\pi}{2}$ from the right)**

(b) $x \to \frac{\pi}{2}^-$ **(as x approaches $\frac{\pi}{2}$ from the left)**

(c) $x \to -\frac{\pi}{2}^+$ **(as x approaches $-\frac{\pi}{2}$ from the right)**

(d) $x \to -\frac{\pi}{2}^-$ **(as x approaches $-\frac{\pi}{2}$ from the left)**

99. $f(x) = \tan x$ **100.** $f(x) = \sec x$

In Exercises 101 and 102, use a graphing utility to graph the function. Use the graph to determine the behavior of the function as $x \to c$.

(a) As $x \to 0^+$, the value of $f(x) \to$ ▭.

(b) As $x \to 0^-$, the value of $f(x) \to$ ▭.

(c) As $x \to \pi^+$, the value of $f(x) \to$ ▭.

(d) As $x \to \pi^-$, the value of $f(x) \to$ ▭.

101. $f(x) = \cot x$ **102.** $f(x) = \csc x$

103. ***Think About It*** Consider the function given by $f(x) = x - \cos x$.

(a) Use a graphing utility to graph the function and verify that there exists a zero between 0 and 1. Use the graph to approximate the zero.

(b) Starting with $x_0 = 1$, generate a sequence $x_1, x_2, x_3, \ldots$, where $x_n = \cos(x_{n-1})$. For example,

$$x_0 = 1$$
$$x_1 = \cos(x_0)$$
$$x_2 = \cos(x_1)$$
$$x_3 = \cos(x_2)$$
$$\vdots$$

What value does the sequence approach?

104. ***Approximation*** Using calculus, it can be shown that the tangent function can be approximated by the polynomial

$$\tan x \approx x + \frac{2x^3}{3!} + \frac{16x^5}{5!}$$

where x is in radians. Use a graphing utility to graph the tangent function and its polynomial approximation in the same viewing window. How do the graphs compare?

105. ***Approximation*** Using calculus, it can be shown that the secant function can be approximated by the polynomial

$$\sec x \approx 1 + \frac{x^2}{2!} + \frac{5x^4}{4!}$$

where x is in radians. Use a graphing utility to graph the secant function and its polynomial approximation in the same viewing window. How do the graphs compare?

106. ***Pattern Recognition***

(a) Use a graphing utility to graph each function.

$$y_1 = \frac{4}{\pi}\left(\sin \pi x + \frac{1}{3} \sin 3\pi x\right)$$

$$y_2 = \frac{4}{\pi}\left(\sin \pi x + \frac{1}{3} \sin 3\pi x + \frac{1}{5} \sin 5\pi x\right)$$

(b) Identify the pattern started in part (a) and find a function y_3 that continues the pattern one more term. Use a graphing utility to graph y_3.

(c) The graphs in parts (a) and (b) approximate the periodic function in the figure. Find a function y_4 that is a better approximation.

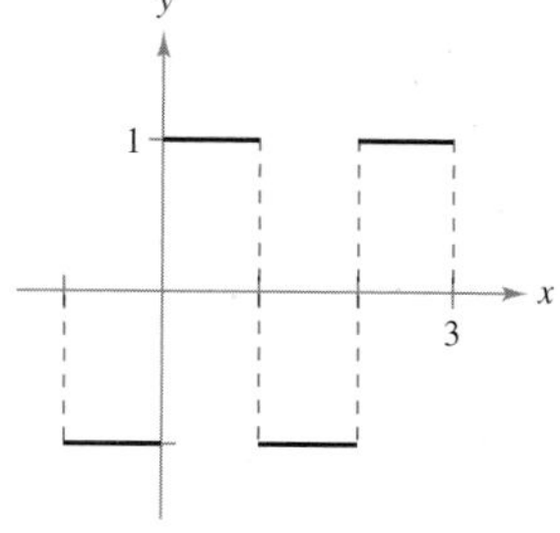

9.7 Inverse Trigonometric Functions

- **Evaluate and graph the inverse sine function.**
- **Evaluate and graph the other inverse trigonometric functions.**
- **Evaluate and graph the compositions of trigonometric functions.**

Inverse Sine Function

Recall from Section 1.5 that, for a function to have an inverse function, it must be one-to-one—that is, it must pass the Horizontal Line Test. From Figure 9.59, you can see that $y = \sin x$ does not pass the test because different values of x yield the same y-value.

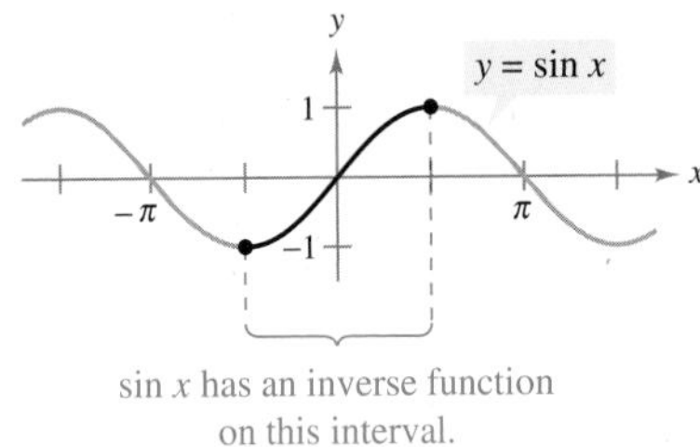

Figure 9.59

However, if you restrict the domain to the interval $-\pi/2 \le x \le \pi/2$ (corresponding to the black portion of the graph in Figure 9.59), the following properties hold.

1. On the interval $[-\pi/2, \pi/2]$, the function $y = \sin x$ is increasing.
2. On the interval $[-\pi/2, \pi/2]$, $y = \sin x$ takes on its full range of values, $-1 \le \sin x \le 1$.
3. On the interval $[-\pi/2, \pi/2]$, $y = \sin x$ is one-to-one.

So, on the restricted domain $-\pi/2 \le x \le \pi/2$, $y = \sin x$ has a unique inverse function called the **inverse sine function.** It is denoted by

$$y = \arcsin x \qquad \text{or} \qquad y = \sin^{-1} x.$$

The notation $\sin^{-1} x$ is consistent with the inverse function notation f^{-1}. The $\arcsin x$ notation (read as "the arcsine of x") comes from the association of a central angle with its intercepted *arc length* on a unit circle. So, $\arcsin x$ means the angle (or arc) whose sine is x. Both notations, $\arcsin x$ and $\sin^{-1} x$, are commonly used in mathematics, so remember that $\sin^{-1} x$ denotes the *inverse* sine function rather than $1/\sin x$. The values of $\arcsin x$ lie in the interval $-\pi/2 \le \arcsin x \le \pi/2$. The graph of $y = \arcsin x$ is shown in Example 2.

DEFINITION OF THE INVERSE SINE FUNCTION

The **inverse sine function** is defined by

$$y = \arcsin x \qquad \text{if and only if} \qquad \sin y = x$$

where $-1 \le x \le 1$ and $-\pi/2 \le y \le \pi/2$. The domain of $y = \arcsin x$ is $[-1, 1]$, and the range is $[-\pi/2, \pi/2]$.

NOTE When evaluating the inverse sine function, it helps to remember the phrase "the arcsine of x is the angle (or number) whose sine is x." ■

EXAMPLE 1 Evaluating the Inverse Sine Function

If possible, find the exact value.

a. $\arcsin\left(-\frac{1}{2}\right)$ **b.** $\sin^{-1}\frac{\sqrt{3}}{2}$ **c.** $\sin^{-1} 2$

Solution

a. Because $\sin\left(-\frac{\pi}{6}\right) = -\frac{1}{2}$ for $-\frac{\pi}{2} \le y \le \frac{\pi}{2}$, it follows that

$$\arcsin\left(-\frac{1}{2}\right) = -\frac{\pi}{6}. \qquad \text{Angle whose sine is } -\tfrac{1}{2}$$

b. Because $\sin\frac{\pi}{3} = \frac{\sqrt{3}}{2}$ for $-\frac{\pi}{2} \le y \le \frac{\pi}{2}$, it follows that

$$\sin^{-1}\frac{\sqrt{3}}{2} = \frac{\pi}{3}. \qquad \text{Angle whose sine is } \sqrt{3}/2$$

c. It is not possible to evaluate $y = \sin^{-1} x$ when $x = 2$ because there is no angle whose sine is 2. Remember that the domain of the inverse sine function is $[-1, 1]$.

STUDY TIP As with the trigonometric functions, much of the work with the inverse trigonometric functions can be done by *exact* calculations rather than by calculator approximations. Exact calculations help to increase your understanding of the inverse functions by relating them to the right triangle definitions of the trigonometric functions.

EXAMPLE 2 Graphing the Arcsine Function

Sketch a graph of

$$y = \arcsin x.$$

Solution By definition, the equations

$$y = \arcsin x \quad \text{and} \quad \sin y = x$$

are equivalent for $-\pi/2 \le y \le \pi/2$. So, their graphs are the same. From the interval $[-\pi/2, \pi/2]$, you can assign values to y in the second equation to make a table of values. Then plot the points and draw a smooth curve through the points.

y	$-\frac{\pi}{2}$	$-\frac{\pi}{4}$	$-\frac{\pi}{6}$	0	$\frac{\pi}{6}$	$\frac{\pi}{4}$	$\frac{\pi}{2}$
$x = \sin y$	-1	$-\frac{\sqrt{2}}{2}$	$-\frac{1}{2}$	0	$\frac{1}{2}$	$\frac{\sqrt{2}}{2}$	1

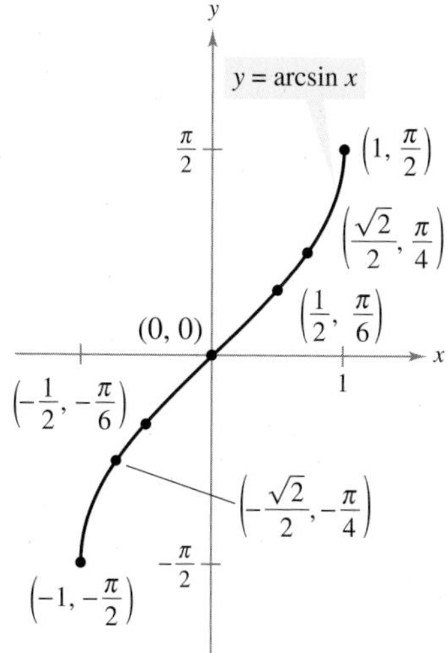

Figure 9.60

The resulting graph for

$$y = \arcsin x$$

is shown in Figure 9.60. Note that it is the reflection (in the line $y = x$) of the black portion of the graph in Figure 9.59. Be sure you see that Figure 9.60 shows the *entire* graph of the inverse sine function. Remember that the domain of $y = \arcsin x$ is the closed interval $[-1, 1]$ and the range is the closed interval $[-\pi/2, \pi/2]$.

Other Inverse Trigonometric Functions

The cosine function is decreasing and one-to-one on the interval $0 \le x \le \pi$, as shown in Figure 9.61.

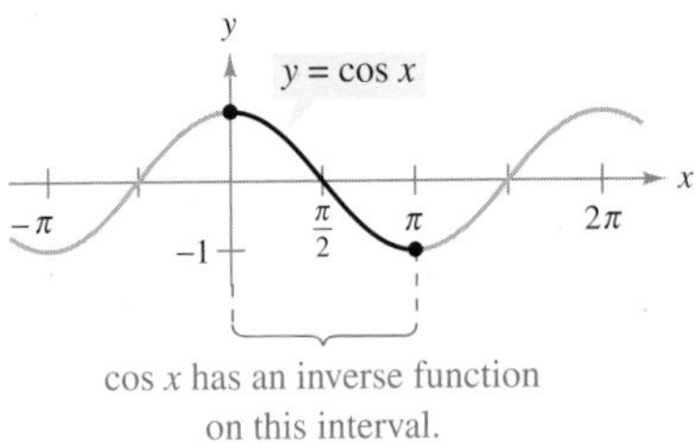

Figure 9.61

Consequently, on this interval the cosine function has an inverse function—the **inverse cosine function**—denoted by

$$y = \arccos x \qquad \text{or} \qquad y = \cos^{-1} x.$$

Similarly, you can define an **inverse tangent function** by restricting the domain of $y = \tan x$ to the interval $(-\pi/2, \pi/2)$. The following list summarizes the definitions of the three most common inverse trigonometric functions. The remaining three are defined in Exercises 93–95.

DEFINITIONS OF THE INVERSE TRIGONOMETRIC FUNCTIONS

Function	*Domain*	*Range*
$y = \arcsin x$ if and only if $\sin y = x$	$-1 \le x \le 1$	$-\frac{\pi}{2} \le y \le \frac{\pi}{2}$
$y = \arccos x$ if and only if $\cos y = x$	$-1 \le x \le 1$	$0 \le y \le \pi$
$y = \arctan x$ if and only if $\tan y = x$	$-\infty < x < \infty$	$-\frac{\pi}{2} < y < \frac{\pi}{2}$

The graphs of these three inverse trigonometric functions are shown in Figure 9.62.

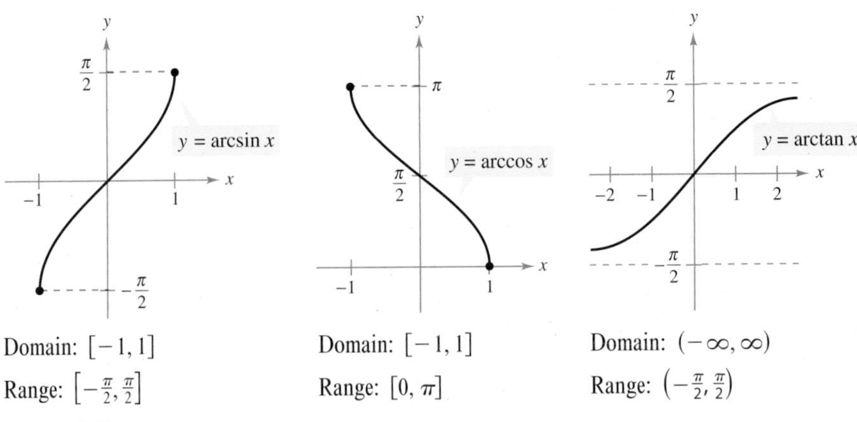

Domain: $[-1, 1]$
Range: $\left[-\frac{\pi}{2}, \frac{\pi}{2}\right]$

Domain: $[-1, 1]$
Range: $[0, \pi]$

Domain: $(-\infty, \infty)$
Range: $\left(-\frac{\pi}{2}, \frac{\pi}{2}\right)$

Figure 9.62

EXAMPLE 3 Evaluating Inverse Trigonometric Functions

Find the exact value.

a. $\arccos \dfrac{\sqrt{2}}{2}$

b. $\cos^{-1}(-1)$

c. $\arctan 0$

d. $\tan^{-1}(-1)$

Solution

a. Because $\cos(\pi/4) = \sqrt{2}/2$, and $\pi/4$ lies in $[0, \pi]$, it follows that

$$\arccos \frac{\sqrt{2}}{2} = \frac{\pi}{4}. \qquad \text{Angle whose cosine is } \sqrt{2}/2$$

b. Because $\cos \pi = -1$, and π lies in $[0, \pi]$, it follows that

$$\cos^{-1}(-1) = \pi. \qquad \text{Angle whose cosine is } -1$$

c. Because $\tan 0 = 0$, and 0 lies in $(-\pi/2, \pi/2)$, it follows that

$$\arctan 0 = 0. \qquad \text{Angle whose tangent is } 0$$

d. Because $\tan(-\pi/4) = -1$, and $-\pi/4$ lies in $(-\pi/2, \pi/2)$, it follows that

$$\tan^{-1}(-1) = -\frac{\pi}{4}. \qquad \text{Angle whose tangent is } -1$$

EXAMPLE 4 Calculators and Inverse Trigonometric Functions

Use a calculator to approximate the value (if possible).

a. $\arctan(-8.45)$

b. $\sin^{-1} 0.2447$

c. $\arccos 2$

Solution

Function	*Mode*	*Calculator Keystrokes*
a. $\arctan(-8.45)$	Radian	TAN⁻¹ ((−) 8.45) ENTER

From the display, it follows that $\arctan(-8.45) \approx -1.453001$.

b. $\sin^{-1} 0.2447$	Radian	SIN⁻¹ (0.2447) ENTER

From the display, it follows that $\sin^{-1} 0.2447 \approx 0.2472103$.

c. $\arccos 2$	Radian	COS⁻¹ (2) ENTER

In *real number* mode, the calculator should display an *error message* because the domain of the inverse cosine function is $[-1, 1]$. ■

STUDY TIP Remember that the domain of the inverse sine function and the inverse cosine function is $[-1, 1]$, as indicated in Example 4(c).

In Example 4, if you had set the calculator to *degree* mode, the displays would have been in degrees rather than radians. This convention is peculiar to calculators. By definition, the values of inverse trigonometric functions are *always in radians*.

Compositions of Functions

Recall from Section 1.5 that for all x in the domains of f and f^{-1}, inverse functions have the properties

$$f(f^{-1}(x)) = x \quad \text{and} \quad f^{-1}(f(x)) = x.$$

INVERSE PROPERTIES OF TRIGONOMETRIC FUNCTIONS

If $-1 \le x \le 1$ and $-\pi/2 \le y \le \pi/2$, then

$$\sin(\arcsin x) = x \quad \text{and} \quad \arcsin(\sin y) = y.$$

If $-1 \le x \le 1$ and $0 \le y \le \pi$, then

$$\cos(\arccos x) = x \quad \text{and} \quad \arccos(\cos y) = y.$$

If x is a real number and $-\pi/2 < y < \pi/2$, then

$$\tan(\arctan x) = x \quad \text{and} \quad \arctan(\tan y) = y.$$

Keep in mind that these inverse properties do not apply for arbitrary values of x and y. For instance,

$$\arcsin\left(\sin \frac{3\pi}{2}\right) = \arcsin(-1) = -\frac{\pi}{2} \neq \frac{3\pi}{2}.$$

In other words, the property

$$\arcsin(\sin y) = y$$

is not valid for values of y outside the interval $[-\pi/2, \pi/2]$.

EXAMPLE 5 Using Inverse Properties

If possible, find the exact value.

a. $\tan[\arctan(-5)]$ **b.** $\arcsin\left(\sin \frac{5\pi}{3}\right)$ **c.** $\cos(\cos^{-1} \pi)$

Solution

a. Because -5 lies in the domain of the arctan function, the inverse property applies, and you have

$$\tan[\arctan(-5)] = -5.$$

b. In this case, $5\pi/3$ does not lie within the range of the arcsine function, $-\pi/2 \le y \le \pi/2$. However, $5\pi/3$ is coterminal with

$$\frac{5\pi}{3} - 2\pi = -\frac{\pi}{3}$$

which does lie in the range of the arcsine function, and you have

$$\arcsin\left(\sin \frac{5\pi}{3}\right) = \arcsin\left[\sin\left(-\frac{\pi}{3}\right)\right] = -\frac{\pi}{3}.$$

c. The expression $\cos(\cos^{-1} \pi)$ is not defined because $\cos^{-1} \pi$ is not defined. Remember that the domain of the inverse cosine function is $[-1, 1]$. ■

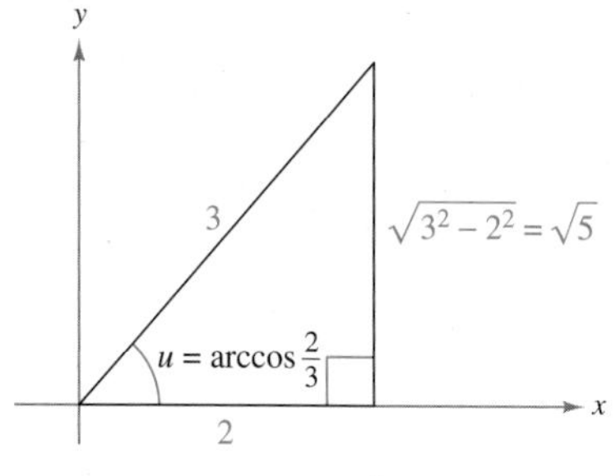

(a) Angle whose cosine is $\frac{2}{3}$

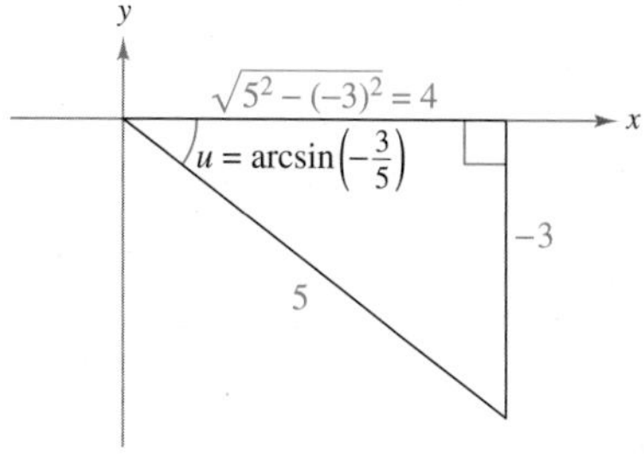

(b) Angle whose sine is $-\frac{3}{5}$
Figure 9.63

Example 6 shows how to use right triangles to find exact values of compositions of inverse functions. Then, Example 7 shows how to use right triangles to convert a trigonometric expression into an algebraic expression. This conversion technique is used frequently in calculus.

EXAMPLE 6 Evaluating Compositions of Functions

Find the exact value.

a. $\tan\left(\arccos \frac{2}{3}\right)$

b. $\cos\left[\arcsin\left(-\frac{3}{5}\right)\right]$

Solution

a. If you let $u = \arccos \frac{2}{3}$, then $\cos u = \frac{2}{3}$. Because $\cos u$ is positive, u is a *first*-quadrant angle. You can sketch and label angle u as shown in Figure 9.63(a). Consequently,

$$\tan\left(\arccos \frac{2}{3}\right) = \tan u = \frac{\text{opp}}{\text{adj}} = \frac{\sqrt{5}}{2}.$$

b. If you let $u = \arcsin\left(-\frac{3}{5}\right)$, then $\sin u = -\frac{3}{5}$. Because $\sin u$ is negative, u is a *fourth*-quadrant angle. You can sketch and label angle u as shown in Figure 9.63(b). Consequently,

$$\cos\left[\arcsin\left(-\frac{3}{5}\right)\right] = \cos u = \frac{\text{adj}}{\text{hyp}} = \frac{4}{5}.$$

EXAMPLE 7 Some Problems from Calculus

Write each of the following as an algebraic expression in x.

a. $\sin(\arccos 3x), \quad 0 \le x \le \frac{1}{3}$

b. $\cot(\arccos 3x), \quad 0 \le x < \frac{1}{3}$

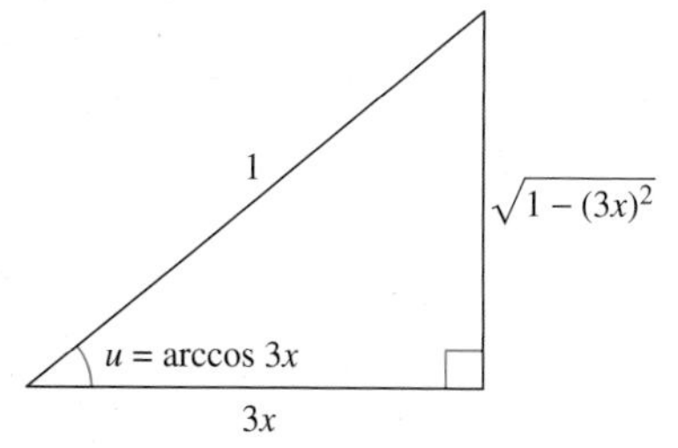

Angle whose cosine is $3x$
Figure 9.64

Solution If you let $u = \arccos 3x$, then $\cos u = 3x$, where $-1 \le 3x \le 1$. Because

$$\cos u = \frac{\text{adj}}{\text{hyp}} = \frac{3x}{1}$$

you can sketch a right triangle with acute angle u, as shown in Figure 9.64. From this triangle, you can easily convert each expression to algebraic form.

a. $\sin(\arccos 3x) = \sin u = \frac{\text{opp}}{\text{hyp}} = \sqrt{1 - 9x^2}, \quad 0 \le x \le \frac{1}{3}$

b. $\cot(\arccos 3x) = \cot u = \frac{\text{adj}}{\text{opp}} = \frac{3x}{\sqrt{1 - 9x^2}}, \quad 0 \le x < \frac{1}{3}$ ■

In Example 7, similar arguments can be made for x-values lying in the interval $\left[-\frac{1}{3}, 0\right]$.

9.7 Exercises

See www.CalcChat.com for worked-out solutions to odd-numbered exercises.

In Exercises 1–4, fill in the blanks.

	Function	*Alternative Notation*	*Domain*	*Range*
1.	$y = \arcsin x$	________	________	$-\frac{\pi}{2} \le y \le \frac{\pi}{2}$
2.	________	$y = \cos^{-1} x$	$-1 \le x \le 1$	________
3.	$y = \arctan x$	________	________	________

4. Without restrictions, no trigonometric function has a(n) ________ function.

In Exercises 5–20, evaluate the expression without using a calculator.

5. $\arcsin \frac{1}{2}$

6. $\arcsin 0$

7. $\arccos \frac{1}{2}$

8. $\arccos 0$

9. $\arctan \dfrac{\sqrt{3}}{3}$

10. $\arctan(1)$

11. $\cos^{-1}\left(-\dfrac{\sqrt{3}}{2}\right)$

12. $\sin^{-1}\left(-\dfrac{\sqrt{2}}{2}\right)$

13. $\arctan(-\sqrt{3})$

14. $\arctan \sqrt{3}$

15. $\arccos\left(-\dfrac{1}{2}\right)$

16. $\arcsin \dfrac{\sqrt{2}}{2}$

17. $\sin^{-1}\left(-\dfrac{\sqrt{3}}{2}\right)$

18. $\tan^{-1}\left(-\dfrac{\sqrt{3}}{3}\right)$

19. $\tan^{-1} 0$

20. $\cos^{-1} 1$

In Exercises 21 and 22, use a graphing utility to graph *f*, *g*, and $y = x$ in the same viewing window to verify geometrically that *g* is the inverse function of *f*. (Be sure to restrict the domain of *f* properly.)

21. $f(x) = \sin x, \quad g(x) = \arcsin x$

22. $f(x) = \tan x, \quad g(x) = \arctan x$

In Exercises 23–40, use a calculator to evaluate the expression. Round your result to two decimal places.

23. $\arccos 0.37$

24. $\arcsin 0.65$

25. $\arcsin(-0.75)$

26. $\arccos(-0.7)$

27. $\arctan(-3)$

28. $\arctan 25$

29. $\sin^{-1} 0.31$

30. $\cos^{-1} 0.26$

31. $\arccos(-0.41)$

32. $\arcsin(-0.125)$

33. $\arctan 0.92$

34. $\arctan 2.8$

35. $\arcsin \frac{7}{8}$

36. $\arccos\left(-\frac{1}{3}\right)$

37. $\tan^{-1} \frac{19}{4}$

38. $\tan^{-1}\left(-\frac{95}{7}\right)$

39. $\tan^{-1}(-\sqrt{372})$

40. $\tan^{-1}(-\sqrt{2165})$

In Exercises 41 and 42, determine the missing coordinates of the points on the graph of the function.

41.

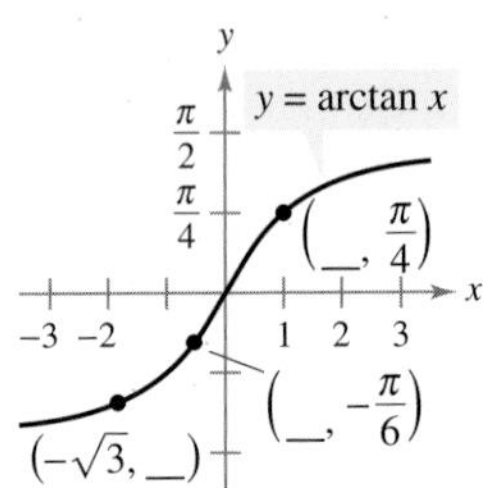

42.

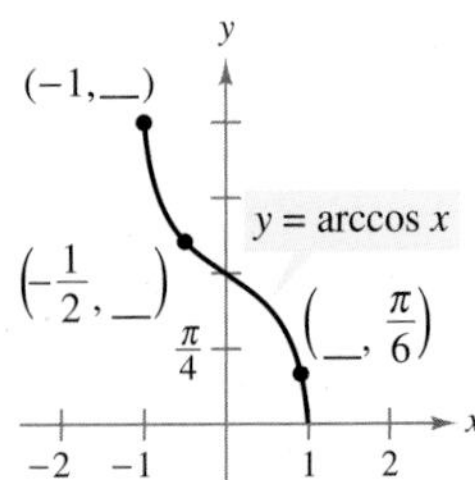

In Exercises 43–48, use an inverse trigonometric function to write θ as a function of *x*.

43.

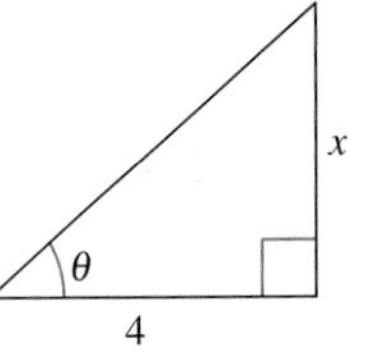

44.

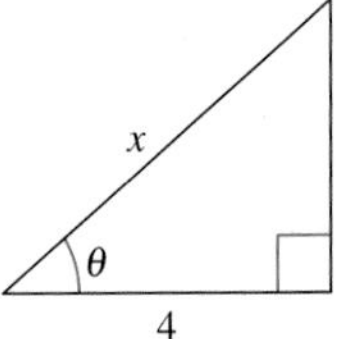

45.

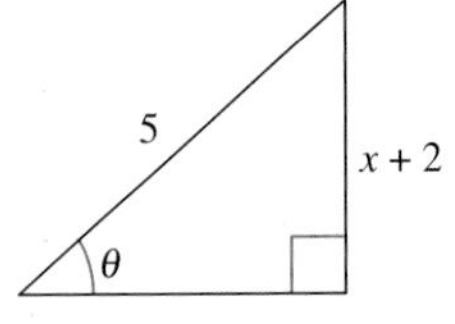

46.

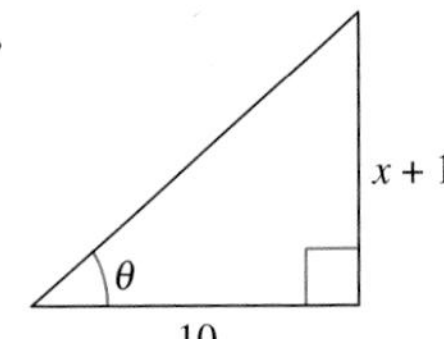

47.

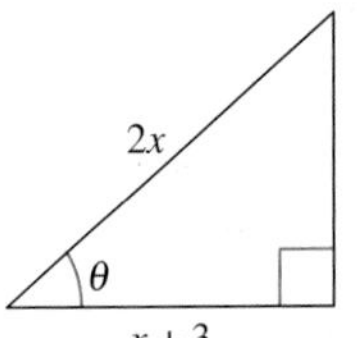

48.

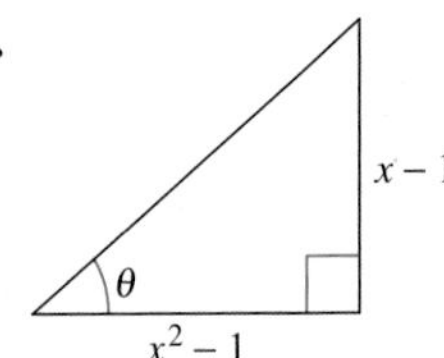

In Exercises 49–54, use the properties of inverse trigonometric functions to evaluate the expression.

49. $\sin(\arcsin 0.3)$

50. $\tan(\arctan 45)$

51. $\cos[\arccos(-0.1)]$

52. $\sin[\arcsin(-0.2)]$

53. $\arcsin(\sin 3\pi)$

54. $\arccos\left(\cos \dfrac{7\pi}{2}\right)$

In Exercises 55–64, find the exact value of the expression. (*Hint:* Sketch a right triangle.)

55. $\sin\left(\arctan \frac{3}{4}\right)$

56. $\sec\left(\arcsin \frac{4}{5}\right)$

57. $\cos(\tan^{-1} 2)$

58. $\sin\left(\cos^{-1} \dfrac{\sqrt{5}}{5}\right)$

59. $\cos\left(\arcsin \frac{5}{13}\right)$

60. $\csc\left[\arctan\left(-\frac{5}{12}\right)\right]$

61. $\sec\left[\arctan\left(-\frac{3}{5}\right)\right]$

62. $\tan\left[\arcsin\left(-\frac{3}{4}\right)\right]$

63. $\sin\left[\arccos\left(-\frac{2}{3}\right)\right]$

64. $\cot\left(\arctan \frac{5}{8}\right)$

In Exercises 65–74, write an algebraic expression that is equivalent to the expression. (*Hint:* Sketch a right triangle, as demonstrated in Example 7.)

65. $\cot(\arctan x)$
66. $\sin(\arctan x)$
67. $\cos(\arcsin 2x)$
68. $\sec(\arctan 3x)$
69. $\sin(\arccos x)$
70. $\sec[\arcsin(x - 1)]$
71. $\tan\left(\arccos \dfrac{x}{3}\right)$
72. $\cot\left(\arctan \dfrac{1}{x}\right)$
73. $\csc\left(\arctan \dfrac{x}{\sqrt{2}}\right)$
74. $\cos\left(\arcsin \dfrac{x - h}{r}\right)$

In Exercises 75 and 76, use a graphing utility to graph *f* and *g* in the same viewing window to verify that the two functions are equal. Explain why they are equal. Identify any asymptotes of the graphs.

75. $f(x) = \sin(\arctan 2x), \quad g(x) = \dfrac{2x}{\sqrt{1 + 4x^2}}$

76. $f(x) = \tan\left(\arccos \dfrac{x}{2}\right), \quad g(x) = \dfrac{\sqrt{4 - x^2}}{x}$

In Exercises 77–80, fill in the blank.

77. $\arctan \dfrac{9}{x} = \arcsin(\quad), \quad x \neq 0$

78. $\arcsin \dfrac{\sqrt{36 - x^2}}{6} = \arccos(\quad), \quad 0 \leq x \leq 6$

79. $\arccos \dfrac{3}{\sqrt{x^2 - 2x + 10}} = \arcsin(\quad)$

80. $\arccos \dfrac{x - 2}{2} = \arctan(\quad), \quad |x - 2| \leq 2$

In Exercises 81–86, sketch a graph of the function.

81. $y = 2 \arccos x$
82. $g(t) = \arccos(t + 2)$
83. $f(x) = \arctan 2x$
84. $f(x) = \dfrac{\pi}{2} + \arctan x$
85. $h(v) = \tan(\arccos v)$
86. $f(x) = \arccos \dfrac{x}{4}$

In Exercises 87–92, use a graphing utility to graph the function.

87. $f(x) = 2 \arccos(2x)$
88. $f(x) = \pi \arcsin(4x)$
89. $f(x) = \arctan(2x - 3)$
90. $f(x) = -3 + \arctan(\pi x)$
91. $f(x) = \pi - \sin^{-1}\left(\dfrac{2}{3}\right)$
92. $f(x) = \dfrac{\pi}{2} + \cos^{-1}\left(\dfrac{1}{\pi}\right)$

WRITING ABOUT CONCEPTS

93. Define the inverse cotangent function by restricting the domain of the cotangent function to the interval $(0, \pi)$, and sketch its graph.

94. Define the inverse secant function by restricting the domain of the secant function to the intervals $[0, \pi/2)$ and $(\pi/2, \pi]$, and sketch its graph.

95. Define the inverse cosecant function by restricting the domain of the cosecant function to the intervals $[-\pi/2, 0)$ and $(0, \pi/2]$, and sketch its graph.

96. Use the results of Exercises 93–95 to evaluate the following without using a calculator.
(a) $\text{arcsec}\,\sqrt{2}$
(b) $\text{arcsec}\,1$
(c) $\text{arccot}(-\sqrt{3})$
(d) $\text{arccsc}\,2$

In Exercises 97–99, prove the identity.

97. $\arcsin(-x) = -\arcsin x$
98. $\arctan(-x) = -\arctan x$
99. $\arccos(-x) = \pi - \arccos x$

100. Consider the functions given by
$f(x) = \sin x \quad \text{and} \quad f^{-1}(x) = \arcsin x.$
(a) Use a graphing utility to graph the composite functions $f \circ f^{-1}$ and $f^{-1} \circ f$.
(b) Explain why the graphs in part (a) are not the graph of the line $y = x$. Why do the graphs of $f \circ f^{-1}$ and $f^{-1} \circ f$ differ?

In Exercises 101 and 102, write the function in terms of the sine function by using the identity

$$A \cos \omega t + B \sin \omega t = \sqrt{A^2 + B^2} \sin\left(\omega t + \arctan \frac{A}{B}\right).$$

Use a graphing utility to graph both forms of the function. What does the graph imply?

101. $f(t) = 3 \cos 2t + 3 \sin 2t$
102. $f(t) = 4 \cos \pi t + 3 \sin \pi t$

In Exercises 103–108, fill in the blank. If not possible, state the reason.

103. As $x \to 1^-$, the value of $\arcsin x \to$ ___.
104. As $x \to 1^-$, the value of $\arccos x \to$ ___.
105. As $x \to \infty$, the value of $\arctan x \to$ ___.
106. As $x \to -1^+$, the value of $\arcsin x \to$ ___.
107. As $x \to -1^+$, the value of $\arccos x \to$ ___.
108. As $x \to -\infty$, the value of $\arctan x \to$ ___.

109. ***Docking a Boat*** A boat is pulled in by means of a winch located on a dock 5 feet above the deck of the boat (see figure). Let θ be the angle of elevation from the boat to the winch and let s be the length of the rope from the winch to the boat.

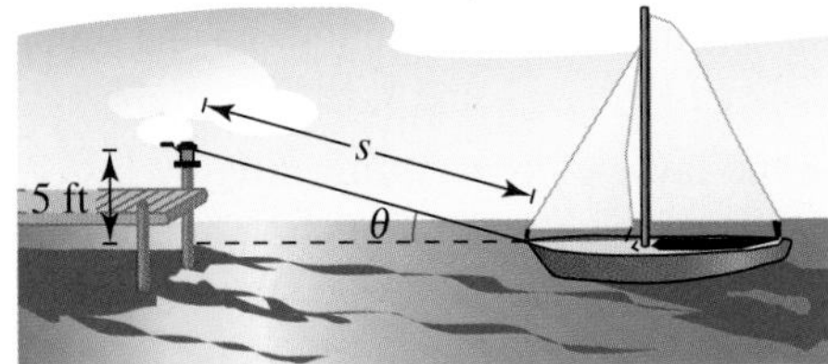

(a) Write θ as a function of s.

(b) Find θ when $s = 40$ feet and $s = 20$ feet.

110. ***Photography*** A television camera at ground level is filming the lift-off of a space shuttle at a point 750 meters from the launch pad (see figure). Let θ be the angle of elevation to the shuttle and let s be the height of the shuttle.

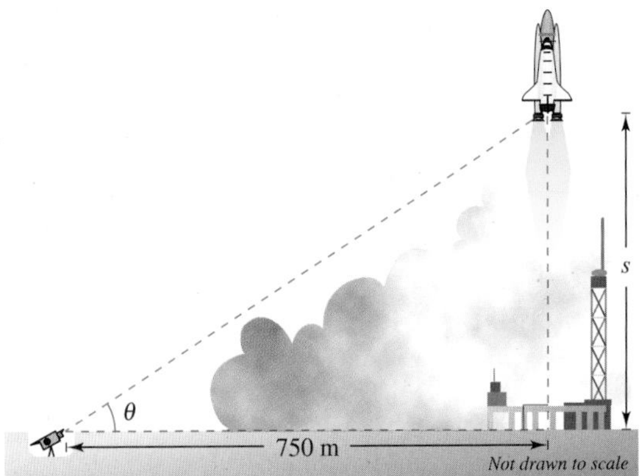

(a) Write θ as a function of s.

(b) Find θ when $s = 300$ meters and $s = 1200$ meters.

111. ***Photography*** A photographer is taking a picture of a three-foot-tall painting hung in an art gallery. The camera lens is 1 foot below the lower edge of the painting (see figure). The angle β subtended by the camera lens x feet from the painting is $\beta = \arctan[3x/(x^2 + 4)], \quad x > 0$.

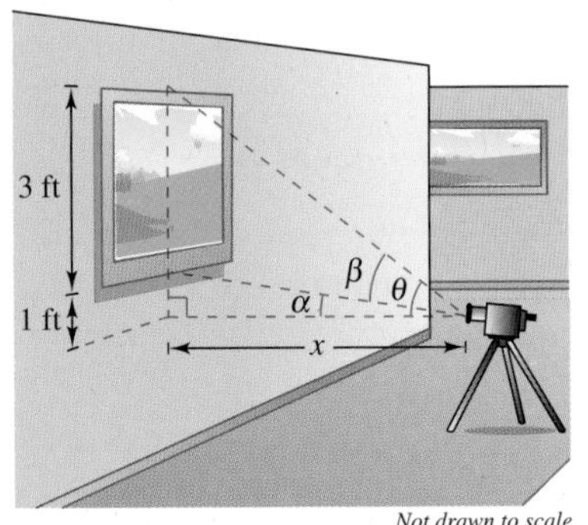

(a) Use a graphing utility to graph β as a function of x.

(b) Move the cursor along the graph to approximate the distance from the picture when β is maximum.

(c) Identify the asymptote of the graph and discuss its meaning in the context of the problem.

112. ***Angle of Elevation*** An airplane flies at an altitude of 6 miles toward a point directly over an observer. Consider θ and x as shown in the figure.

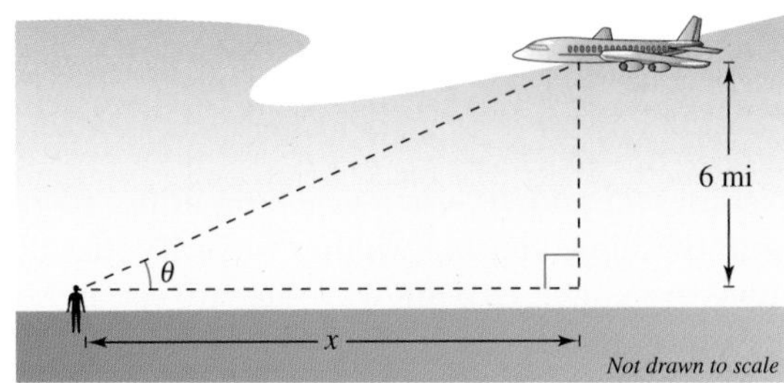

(a) Write θ as a function of x.

(b) Find θ when $x = 7$ miles and $x = 1$ mile.

113. ***Security Patrol*** A security car with its spotlight on is parked 20 meters from a warehouse. Consider θ and x as shown in the figure.

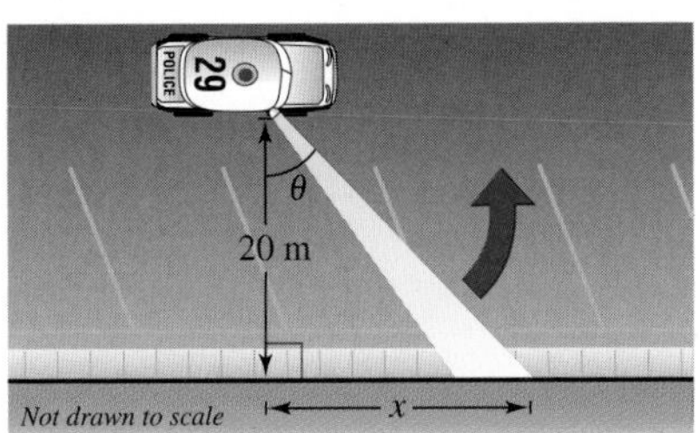

(a) Write θ as a function of x.

(b) Find θ when $x = 5$ meters and $x = 12$ meters.

CAPSTONE

114. Use the results of Exercises 93–95 to explain how to graph (a) the inverse cotangent function, (b) the inverse secant function, and (c) the inverse cosecant function on a graphing utility.

True or False? **In Exercises 115–117, determine whether the statement is true or false. Justify your answer.**

115. $\sin \frac{5\pi}{6} = \frac{1}{2} \quad \Rightarrow \quad \arcsin \frac{1}{2} = \frac{5\pi}{6}$

116. $\tan \frac{5\pi}{4} = 1 \quad \Rightarrow \quad \arctan 1 = \frac{5\pi}{4}$

117. $\arctan x = \dfrac{\arcsin x}{\arccos x}$

9.8 Applications and Models

- Solve real-life problems involving right triangles.
- Solve real-life problems involving directional bearings.
- Solve real-life problems involving harmonic motion.

Applications Involving Right Triangles

In keeping with a twofold perspective of trigonometry, this section includes both right triangle applications and applications that emphasize the periodic nature of the trigonometric functions.

EXAMPLE 1 Solving a Right Triangle

NOTE In this section, the three angles of a right triangle are denoted by the letters A, B, and C (where C is the right angle), and the lengths of the sides opposite these angles by the letters a, b, and c (where c is the hypotenuse).

Solve the right triangle shown in Figure 9.65 for all unknown sides and angles.

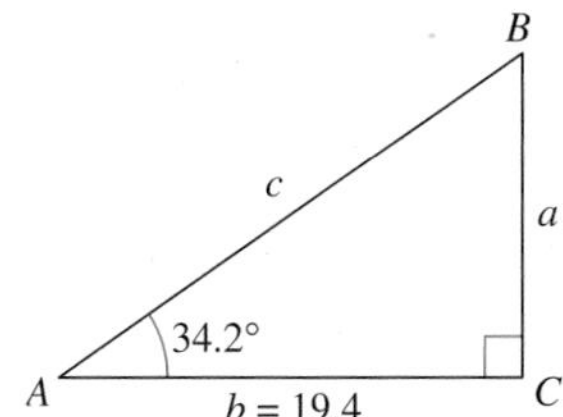

Figure 9.65

Solution Because $C = 90°$, it follows that

$$A + B = 90° \text{ and } B = 90° - 34.2° = 55.8°.$$

To solve for a, use the fact that

$$\tan A = \frac{\text{opp}}{\text{adj}} = \frac{a}{b} \quad \Longrightarrow \quad a = b \tan A.$$

So, $a = 19.4 \tan 34.2° \approx 13.18$. Similarly, to solve for c, use the fact that

$$\cos A = \frac{\text{adj}}{\text{hyp}} = \frac{b}{c} \quad \Longrightarrow \quad c = \frac{b}{\cos A}.$$

So, $c = \dfrac{19.4}{\cos 34.2°} \approx 23.46$.

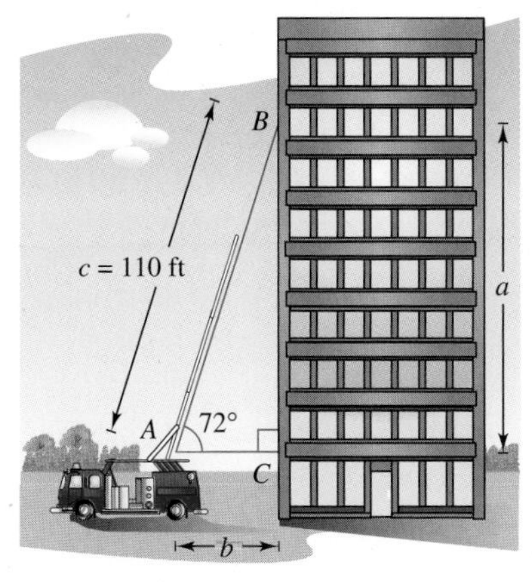

Figure 9.66

EXAMPLE 2 Finding a Side of a Right Triangle

A safety regulation states that the maximum angle of elevation for a rescue ladder is 72°. A fire department's longest ladder is 110 feet. What is the maximum safe rescue height?

Solution A sketch is shown in Figure 9.66. From the equation $\sin A = a/c$, it follows that

$$\begin{aligned} a &= c \sin A \\ &= 110 \sin 72° \\ &\approx 104.6. \end{aligned}$$

So, the maximum safe rescue height is about 104.6 feet above the height of the fire truck. ■

EXAMPLE 3 Finding a Side of a Right Triangle

At a point 200 feet from the base of a building, the angle of elevation to the *bottom* of a smokestack is 35°, whereas the angle of elevation to the *top* is 53°, as shown in Figure 9.67. Find the height s of the smokestack alone.

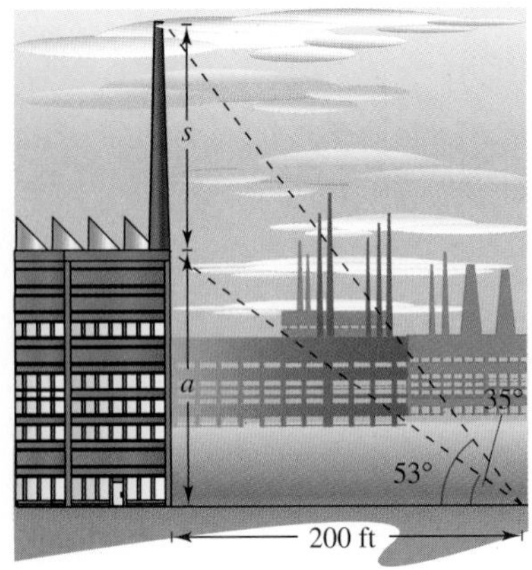

Figure 9.67

Solution Note from Figure 9.67 that this problem involves two right triangles. For the smaller right triangle, use the fact that

$$\tan 35° = \frac{a}{200}$$

to conclude that the height of the building is

$$a = 200 \tan 35°.$$

For the larger right triangle, use the equation

$$\tan 53° = \frac{a + s}{200}$$

to conclude that $a + s = 200 \tan 53°$. So, the height of the smokestack is

$$\begin{aligned} s &= 200 \tan 53° - a \\ &= 200 \tan 53° - 200 \tan 35° \\ &\approx 125.4 \text{ feet.} \end{aligned}$$

EXAMPLE 4 Finding an Acute Angle of a Right Triangle

A swimming pool is 20 meters long and 12 meters wide. The bottom of the pool is slanted so that the water depth is 1.3 meters at the shallow end and 4 meters at the deep end, as shown in Figure 9.68. Find the angle of depression of the bottom of the pool.

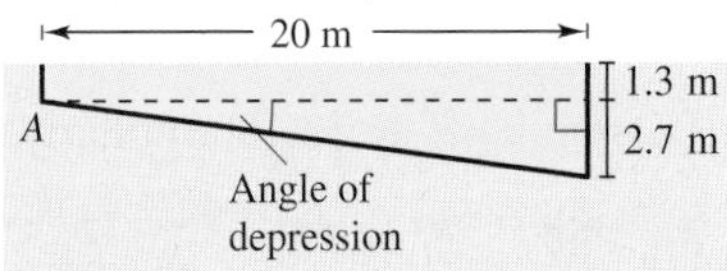

Figure 9.68

Solution Using the tangent function, you can see that

$$\begin{aligned} \tan A &= \frac{\text{opp}}{\text{adj}} \\ &= \frac{2.7}{20} \\ &= 0.135. \end{aligned}$$

So, the angle of depression is

$$\begin{aligned} A &= \arctan 0.135 \\ &\approx 0.13419 \text{ radian} \\ &\approx 7.69°. \end{aligned}$$

■

Trigonometry and Bearings

STUDY TIP In *air navigation,* bearings are measured in degrees *clockwise* from north. Examples of air navigation bearings are shown below.

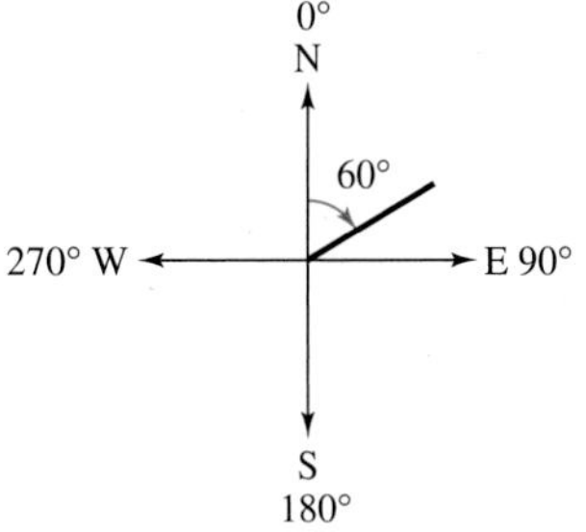

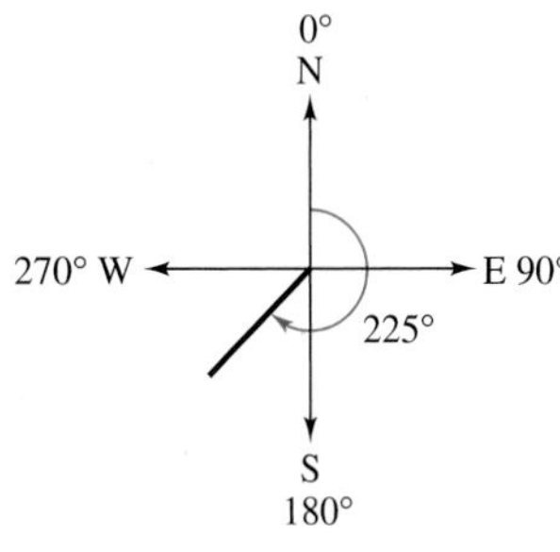

In surveying and navigation, directions can be given in terms of **bearings.** A bearing measures the acute angle that a path or line of sight makes with a fixed north-south line, as shown in Figure 9.69. For instance, the bearing S 35° E in Figure 9.69 means 35 degrees east of south.

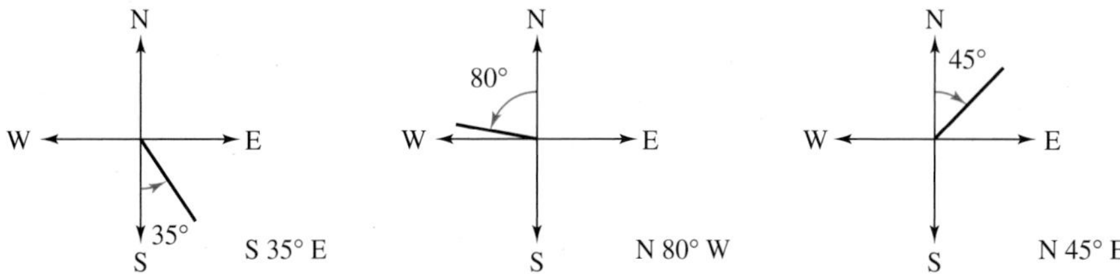

Figure 9.69

EXAMPLE 5 Finding Directions in Terms of Bearings

A ship leaves port at noon and heads due west at 20 knots, or 20 nautical miles (nm) per hour. At 2 P.M. the ship changes course to N 54° W, as shown in Figure 9.70. Find the ship's bearing and distance from the port of departure at 3 P.M.

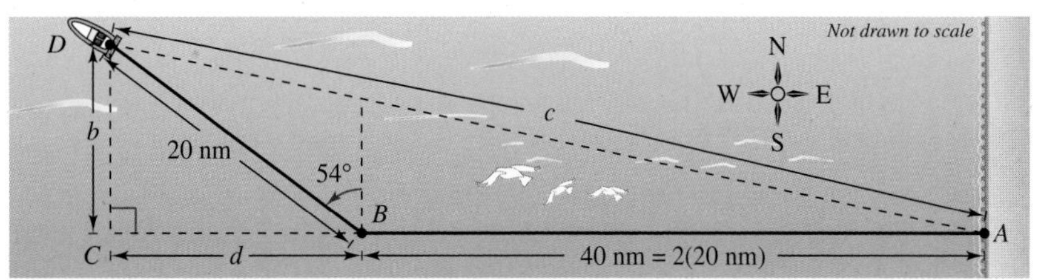

Figure 9.70

Solution For triangle BCD, you have

$$B = 90° - 54° = 36°.$$

The two sides of this triangle can be determined to be

$$b = 20 \sin 36° \quad \text{and} \quad d = 20 \cos 36°.$$

For triangle ACD, you can find angle A as follows.

$$\tan A = \frac{b}{d + 40} = \frac{20 \sin 36°}{20 \cos 36° + 40} \approx 0.2092494$$

$$A \approx \arctan 0.2092494 \approx 11.82°$$

The angle with the north-south line is $90° - 11.82° = 78.18°$. So, the bearing of the ship is N 78.18° W. Finally, from triangle ACD, you have $\sin A = b/c$, which yields

$$c = \frac{b}{\sin A}$$

$$= \frac{20 \sin 36°}{\sin 11.82°}$$

$$\approx 57.4 \text{ nautical miles.}$$ Distance from port

■

Harmonic Motion

The periodic nature of the trigonometric functions is useful for describing the motion of a point on an object that vibrates, oscillates, rotates, or is moved by wave motion.

For example, consider a ball that is bobbing up and down on the end of a spring, as shown in Figure 9.71. Suppose that 10 centimeters is the maximum distance the ball moves vertically upward or downward from its equilibrium (at rest) position. Suppose further that the time it takes for the ball to move from its maximum displacement above zero to its maximum displacement below zero and back again is $t = 4$ seconds. Assuming the ideal conditions of perfect elasticity and no friction or air resistance, the ball would continue to move up and down in a uniform manner.

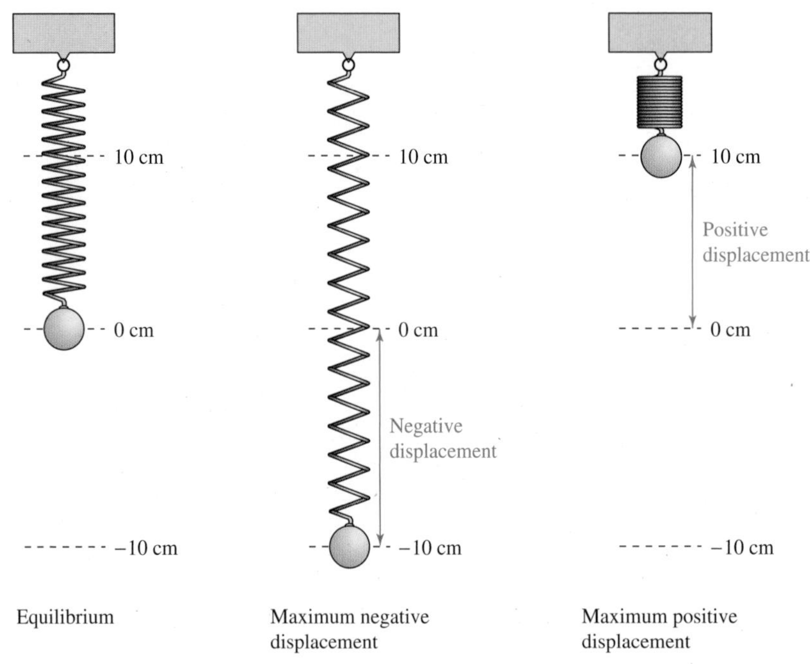

Figure 9.71

From this spring you can conclude that the period (time for one complete cycle) of the motion is

Period $= 4$ seconds

and that its amplitude (maximum displacement from equilibrium) is

Amplitude $= 10$ centimeters.

Motion of this nature can be described by a sine or cosine function, and is called **simple harmonic motion.**

DEFINITION OF SIMPLE HARMONIC MOTION

A point that moves on a coordinate line is said to be in **simple harmonic motion** if its distance d from the origin at time t is given by either

$d = a \sin \omega t$ or $d = a \cos \omega t$

where a and ω are real numbers such that $\omega > 0$. The motion has amplitude $|a|$, period $2\pi/\omega$, and frequency $\omega/(2\pi)$.

EXAMPLE 6 Simple Harmonic Motion

Write the equation for the simple harmonic motion of the ball described in Figure 9.71, where the period is 4 seconds.

a. What is the frequency of this harmonic motion?

b. Find the value of d when $t = 3$.

Solution

a. Assuming that the spring is at equilibrium ($d = 0$) when $t = 0$, you use the equation

$$d = a \sin \omega t.$$

Moreover, because the maximum displacement from zero is 10 and the period is 4, you have

$$\text{Amplitude} = |a| = 10$$

$$\text{Period} = \frac{2\pi}{\omega} = 4 \quad \Longrightarrow \quad \omega = \frac{\pi}{2}.$$

Consequently, the equation of motion is

$$d = 10 \sin \frac{\pi}{2} t.$$

Note that the choice of $a = 10$ or $a = -10$ depends on whether the ball initially moves up or down. The frequency is

$$\text{Frequency} = \frac{\omega}{2\pi} = \frac{\pi/2}{2\pi} = \frac{1}{4} \text{ cycle per second.}$$

b. When $t = 3$,

$$d = 10 \sin \frac{\pi}{2}(3) = 10 \sin \frac{3\pi}{2} = 10(-1) = -10.$$

■

9.8 Exercises

See www.CalcChat.com for worked-out solutions to odd-numbered exercises.

In Exercises 1–4, fill in the blanks.

1. A ________ measures the acute angle a path or line of sight makes with a fixed north-south line.
2. A point that moves on a coordinate line is said to be in simple ________ ________ if its distance d from the origin at time t is given by either $d = a \sin \omega t$ or $d = a \cos \omega t$.
3. The time for one complete cycle of a point in simple harmonic motion is its ________.
4. The number of cycles per second of a point in simple harmonic motion is its ________.

In Exercises 5–14, solve the right triangle shown in the figure for all unknown sides and angles. Round your answers to two decimal places.

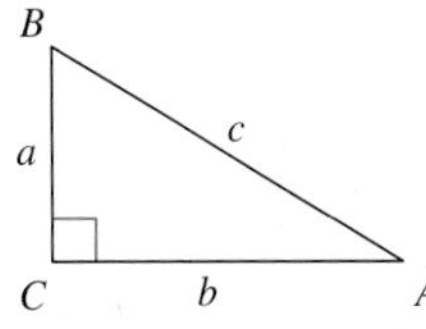

5. $A = 30°, \quad b = 3$

6. $B = 54°, \quad c = 15$

7. $B = 71°, \quad b = 24$

8. $A = 8.4°, \quad a = 40.5$

9. $a = 3, \quad b = 4$

10. $a = 25, \quad c = 35$

11. $b = 16, \quad c = 52$

12. $b = 1.32, \quad c = 9.45$

13. $A = 12°15', \quad c = 430.5$

14. $B = 65°12', \quad a = 14.2$

In Exercises 15–18, find the altitude of the isosceles triangle shown in the figure. Round your answers to two decimal places.

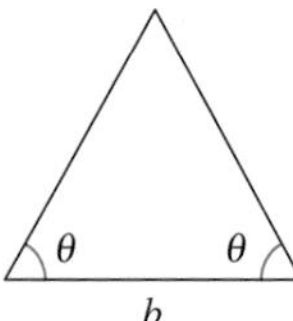

15. $\theta = 45°, \quad b = 6$

16. $\theta = 18°, \quad b = 10$

17. $\theta = 32°, \quad b = 8$

18. $\theta = 27°, \quad b = 11$

WRITING ABOUT CONCEPTS

In Exercises 19–22, determine whether the right triangle can be solved for all of the unknown parts, if the indicated parts of the triangle are known.

19. A, C, a

20. A, B, C

21. C, a, c

22. B, C, a

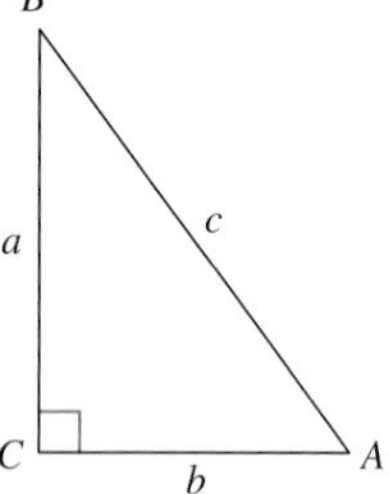

In Exercises 23–26, sketch the bearing.

23. N 45° W

24. S 60° E

25. S 75° W

26. N 30° E

27. Simple harmonic motion can be modeled by $d = a \sin \omega t$. What is the amplitude of this motion?

28. Simple harmonic motion can be modeled by $d = a \cos \omega t$. What is the period of this motion?

29. ***Length*** The sun is 25° above the horizon. Find the length of a shadow cast by a building that is 100 feet tall (see figure).

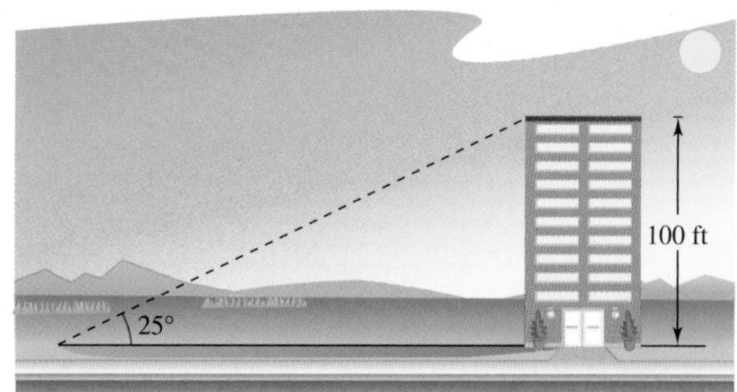

30. ***Length*** The sun is 20° above the horizon. Find the length of a shadow cast by a park statue that is 12 feet tall.

31. ***Height*** A ladder 20 feet long leans against the side of a house. Find the height from the top of the ladder to the ground if the angle of elevation of the ladder is 80°.

32. ***Height*** The length of a shadow of a tree is 125 feet when the angle of elevation of the sun is 33°. Approximate the height of the tree.

33. ***Height*** From a point 50 feet in front of a church, the angles of elevation to the base of the steeple and the top of the steeple are 35° and 47° 40′, respectively. Find the height of the steeple.

34. ***Distance*** An observer in a lighthouse 350 feet above sea level observes two ships directly offshore. The angles of depression to the ships are 4° and 6.5° (see figure). How far apart are the ships?

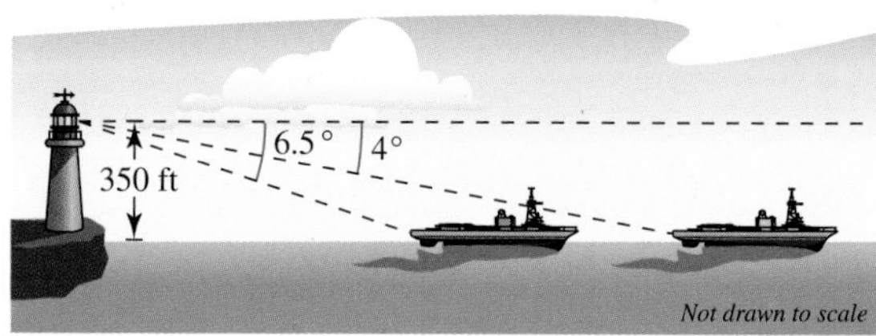

35. ***Distance*** A passenger in an airplane at an altitude of 10 kilometers sees two towns directly to the east of the plane. The angles of depression to the towns are 28° and 55° (see figure). How far apart are the towns?

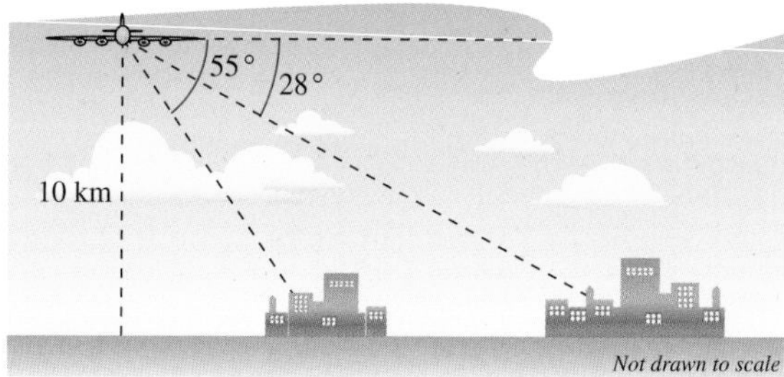

36. ***Altitude*** You observe a plane approaching overhead and assume that its speed is 550 miles per hour. The angle of elevation of the plane is 16° at one time and 57° one minute later. Approximate the altitude of the plane.

37. ***Angle of Elevation*** An engineer erects a 75-foot cellular telephone tower. Find the angle of elevation to the top of the tower at a point on level ground 50 feet from its base.

38. ***Angle of Elevation*** The height of an outdoor basketball backboard is $12\frac{1}{2}$ feet, and the backboard casts a shadow $17\frac{1}{3}$ feet long.

(a) Draw a right triangle that gives a visual representation of the problem. Label the known and unknown quantities.

(b) Use a trigonometric function to write an equation involving the unknown quantity.

(c) Find the angle of elevation of the sun.

39. ***Angle of Depression*** A cellular telephone tower that is 150 feet tall is placed on top of a mountain that is 1200 feet above sea level. What is the angle of depression from the top of the tower to a cell phone user who is 5 horizontal miles away and 400 feet above sea level?

40. ***Angle of Depression*** A Global Positioning System satellite orbits 12,500 miles above Earth's surface (see figure). Find the angle of depression from the satellite to the horizon. Assume the radius of Earth is 4000 miles.

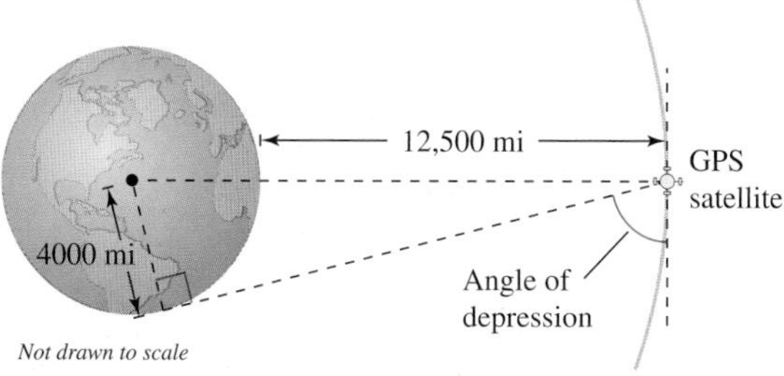

41. ***Height*** You are holding one of the tethers attached to the top of a giant character balloon in a parade. Before the start of the parade the balloon is upright and the bottom is floating approximately 20 feet above ground level. You are standing approximately 100 feet ahead of the balloon (see figure).

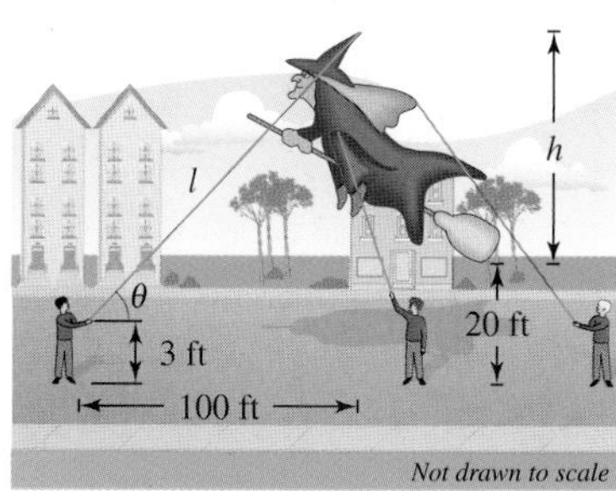

(a) Find the length l of the tether you are holding in terms of h, the height of the balloon from top to bottom.

(b) Find an expression for the angle of elevation θ from you to the top of the balloon.

(c) Find the height h of the balloon if the angle of elevation to the top of the balloon is 35°.

42. ***Height*** The designers of a water park are creating a new slide and have sketched some preliminary drawings. The length of the ladder is 30 feet, and its angle of elevation is 60° (see figure).

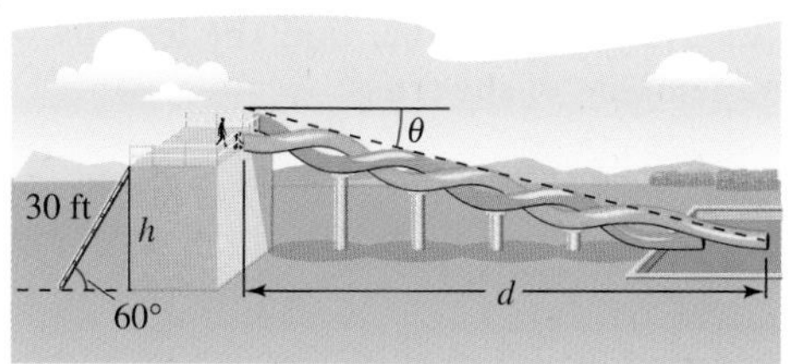

(a) Find the height h of the slide.

(b) Find the angle of depression θ from the top of the slide to the end of the slide at the ground in terms of the horizontal distance d the rider travels.

(c) The angle of depression of the ride is bounded by safety restrictions to be no less than 25° and not more than 30°. Find an interval for how far the rider travels horizontally.

43. ***Speed Enforcement*** A police department has set up a speed enforcement zone on a straight length of highway. A patrol car is parked parallel to the zone, 200 feet from one end and 150 feet from the other end (see figure).

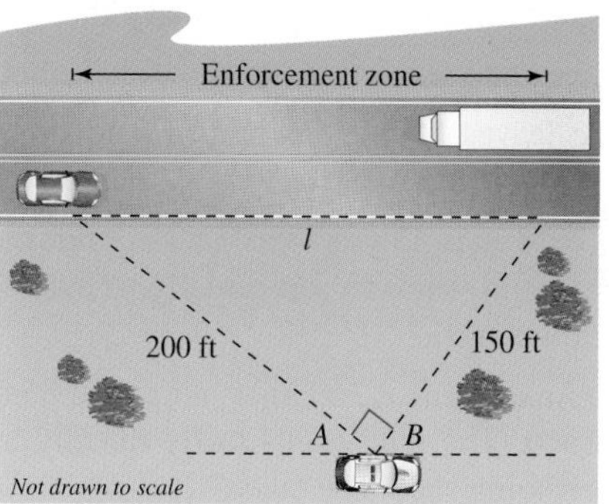

(a) Find the length l of the zone and the measures of the angles A and B (in degrees).

(b) Find the minimum amount of time (in seconds) it takes for a vehicle to pass through the zone without exceeding the posted speed limit of 35 miles per hour.

44. ***Airplane Ascent*** During takeoff, an airplane's angle of ascent is 18° and its speed is 275 feet per second.

(a) Find the plane's altitude after 1 minute.

(b) How long will it take the plane to climb to an altitude of 10,000 feet?

45. ***Navigation*** An airplane flying at 600 miles per hour has a bearing of 52°. After flying for 1.5 hours, how far north and how far east will the plane have traveled from its point of departure?

46. ***Navigation*** A jet leaves Reno, Nevada and is headed toward Miami, Florida at a bearing of 100°. The distance between the two cities is approximately 2472 miles.

(a) How far north and how far west is Reno relative to Miami?

(b) If the jet is to return directly to Reno from Miami, at what bearing should it travel?

47. ***Navigation*** A ship leaves port at noon and has a bearing of S 29° W. The ship sails at 20 knots.

(a) How many nautical miles south and how many nautical miles west will the ship have traveled by 6:00 P.M.?

(b) At 6:00 P.M., the ship changes course to due west. Find the ship's bearing and distance from the port of departure at 7:00 P.M.

48. ***Navigation*** A privately owned yacht leaves a dock in Myrtle Beach, South Carolina and heads toward Freeport in the Bahamas at a bearing of S 1.4° E. The yacht averages a speed of 20 knots over the 428-nautical-mile trip.

(a) How long will it take the yacht to make the trip?

(b) How far east and south is the yacht after 12 hours?

(c) If a plane leaves Myrtle Beach to fly to Freeport, what bearing should be taken?

49. ***Navigation*** A ship is 45 miles east and 30 miles south of port. The captain wants to sail directly to port. What bearing should be taken?

50. ***Navigation*** An airplane is 160 miles north and 85 miles east of an airport. The pilot wants to fly directly to the airport. What bearing should be taken?

51. ***Surveying*** A surveyor wants to find the distance across a swamp (see figure). The bearing from A to B is N 32° W. The surveyor walks 50 meters from A, and at the point C the bearing to B is N 68° W. Find (a) the bearing from A to C and (b) the distance from A to B.

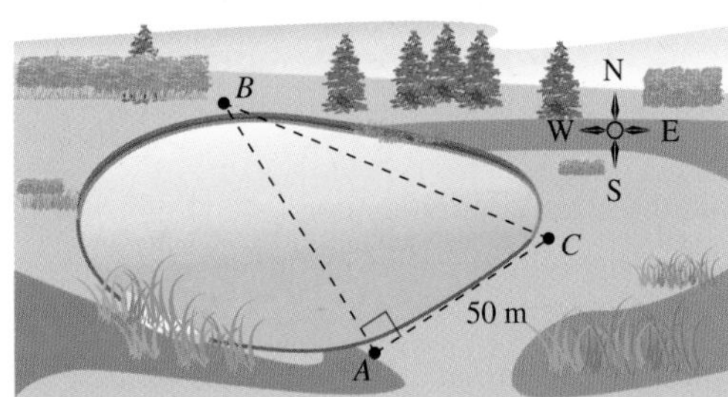

52. ***Location of a Fire*** Two fire towers are 30 kilometers apart, where tower A is due west of tower B. A fire is spotted from the towers, and the bearings from A and B are N 76° E and N 56° W, respectively (see figure). Find the distance d of the fire from the line segment AB.

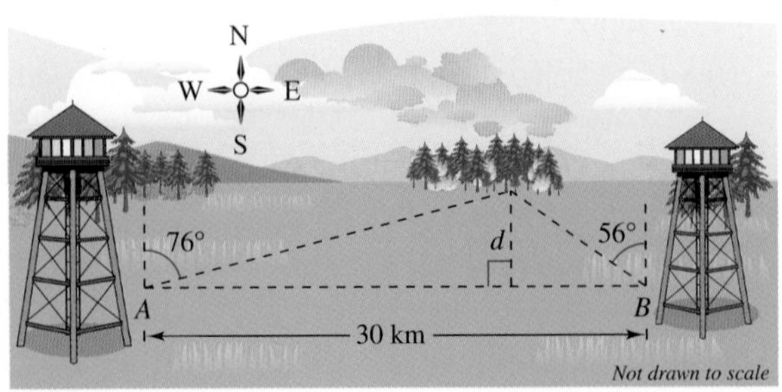

Geometry **In Exercises 53 and 54, find the angle α between two nonvertical lines L_1 and L_2. The angle α satisfies the equation**

$$\tan \alpha = \left| \frac{m_2 - m_1}{1 + m_2 m_1} \right|$$

where m_1 and m_2 are the slopes of L_1 and L_2, respectively. (Assume that $m_1 m_2 \neq -1$.)

53. L_1: $3x - 2y = 5$
L_2: $x + y = 1$

54. L_1: $2x - y = 8$
L_2: $x - 5y = -4$

55. ***Geometry*** Determine the angle between the diagonal of a cube and the diagonal of its base, as shown in the figure.

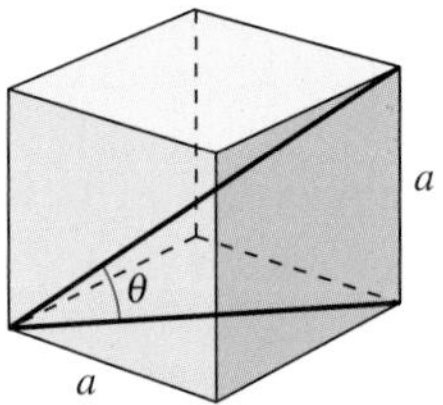

Figure for 55

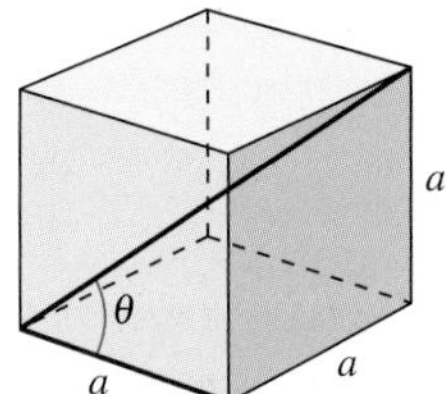

Figure for 56

56. ***Geometry*** Determine the angle between the diagonal of a cube and its edge, as shown in the figure.

57. ***Hardware*** Write the distance y across the flat sides of a hexagonal nut as a function of r (see figure).

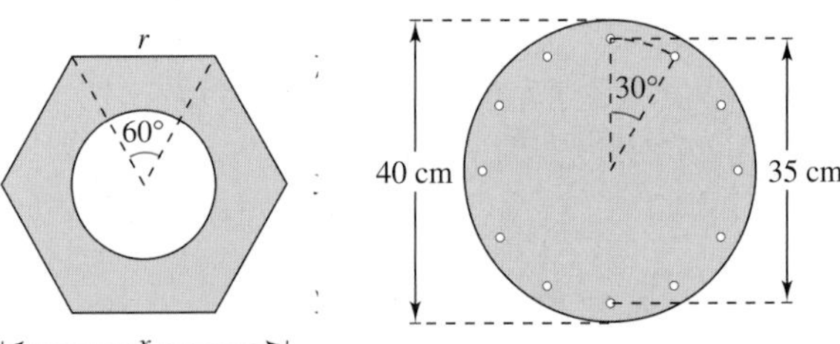

Figure for 57 Figure for 58

58. ***Bolt Holes*** The figure shows a circular piece of sheet metal that has a diameter of 40 centimeters and contains 12 equally-spaced bolt holes. Determine the straight-line distance between the centers of consecutive bolt holes.

59. ***Geometry*** Find the length of the sides of a regular pentagon inscribed in a circle of radius 25 inches.

60. ***Geometry*** Find the length of the sides of a regular hexagon inscribed in a circle of radius 25 inches.

Trusses **In Exercises 61 and 62, find the lengths of all the unknown members of the truss.**

61.

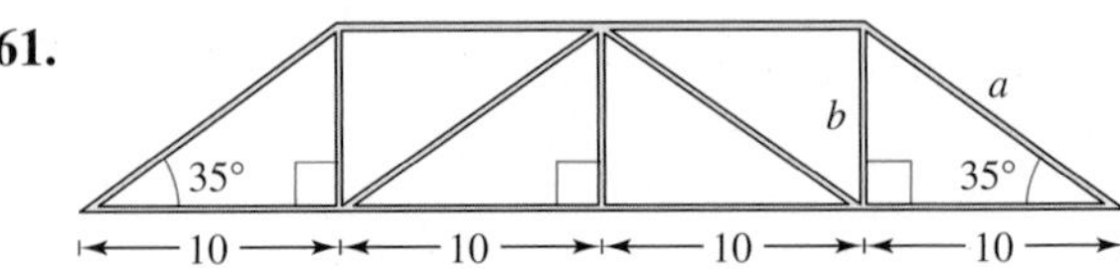

62.

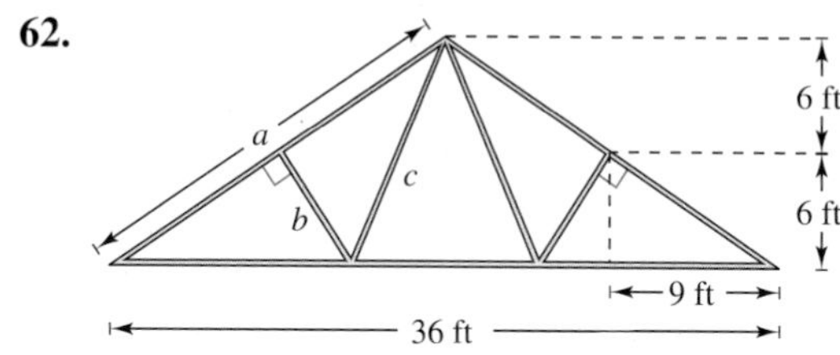

Harmonic Motion **In Exercises 63–66, find a model for simple harmonic motion satisfying the specified conditions.**

	Displacement $(t = 0)$	*Amplitude*	*Period*
63.	0	4 centimeters	2 seconds
64.	0	3 meters	6 seconds
65.	3 inches	3 inches	1.5 seconds
66.	2 feet	2 feet	10 seconds

Harmonic Motion **In Exercises 67–70, for the simple harmonic motion described by the trigonometric function, find (a) the maximum displacement, (b) the frequency, (c) the value of d when $t = 5$, and (d) the least positive value of t for which $d = 0$. Use a graphing utility to verify your results.**

67. $d = 9 \cos \dfrac{6\pi}{5}t$

68. $d = \dfrac{1}{2}\cos 20\pi t$

69. $d = \dfrac{1}{4}\sin 6\pi t$

70. $d = \dfrac{1}{64}\sin 792\pi t$

71. ***Tuning Fork*** A point on the end of a tuning fork moves in simple harmonic motion described by

$d = a \sin \omega t.$

Find ω given that the tuning fork for middle C has a frequency of 264 vibrations per second.

72. ***Wave Motion*** A buoy oscillates in simple harmonic motion as waves go past. It is noted that the buoy moves a total of 3.5 feet from its low point to its high point (see figure), and that it returns to its high point every 10 seconds. Write an equation that describes the motion of the buoy if its high point is at $t = 0$.

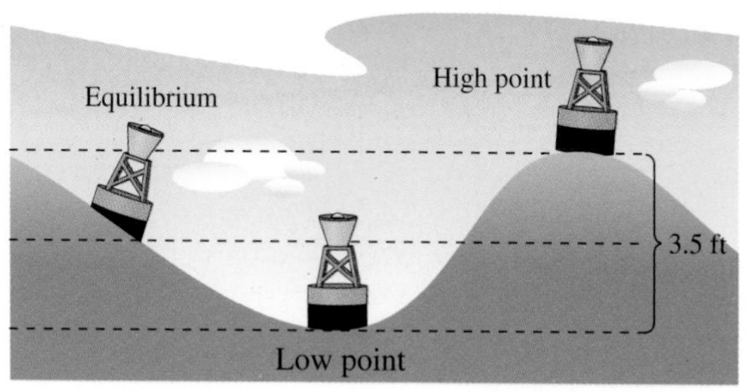

73. ***Oscillation of a Spring*** A ball that is bobbing up and down on the end of a spring has a maximum displacement of 3 inches. Its motion (in ideal conditions) is modeled by

$y = \frac{1}{4}\cos 16t \ (t > 0)$

where y is measured in feet and t is the time in seconds.

(a) Graph the function.

(b) What is the period of the oscillations?

(c) Determine the first time the weight passes the point of equilibrium $(y = 0)$.

CAPSTONE

74. While walking across flat land, you notice a wind turbine tower of height h feet directly in front of you. The angle of elevation to the top of the tower is A degrees. After you walk d feet closer to the tower, the angle of elevation increases to B degrees.

(a) Draw a diagram to represent the situation.

(b) Write an expression for the height h of the tower in terms of the angles A and B and the distance d.

75. ***Data Analysis*** The number of hours H of daylight in Denver, Colorado on the 15th of each month are:

1(9.67), 2(10.72), 3(11.92), 4(13.25), 5(14.37), 6(14.97), 7(14.72), 8(13.77), 9(12.48), 10(11.18), 11(10.00), 12(9.38).

The month is represented by t, with $t = 1$ corresponding to January. A model for the data is given by

$H(t) = 12.13 + 2.77 \sin[(\pi t/6) - 1.60].$

(a) Use a graphing utility to graph the data points and the model in the same viewing window.

(b) What is the period of the model? Is it what you expected? Explain.

(c) What is the amplitude of the model? What does it represent in the context of the problem? Explain.

True or False? **In Exercises 76 and 77, determine whether the statement is true or false. Justify your answer.**

76. The Leaning Tower of Pisa is not vertical, but if you know the angle of elevation θ to the top of the tower when you stand d feet away from it, you can find its height h using the formula $h = d \tan \theta$.

77. N 24° E means 24 degrees north of east.

9 CHAPTER SUMMARY

	Review Exercises
Section 9.1	
■ Describe angles *(p. 566)*.	*1–8*
■ Convert between degrees and radians *(p. 569)*.	*9–20*
■ Use angles to model and solve real-life problems *(p. 570)*.	*21–24*
Section 9.2	
■ Identify a unit circle and describe its relationship to real numbers *(p. 575)*.	*25–28*
■ Evaluate trigonometric functions using the unit circle *(p. 576)*.	*29–32*
■ Use domain and period to evaluate sine and cosine functions *(p. 578)*.	*33–36*
■ Use a calculator to evaluate trigonometric functions *(p. 579)*.	*37–40*
Section 9.3	
■ Evaluate trigonometric functions of acute angles *(p. 582)*.	*41, 42*
■ Use fundamental trigonometric identities *(p. 584)*.	*43–46*
■ Use a calculator to evaluate trigonometric functions *(p. 585)*.	*47–54*
■ Use trigonometric functions to model and solve real-life problems *(p. 585)*.	*55, 56*
Section 9.4	
■ Evaluate trigonometric functions of any angle *(p. 591)*.	*57–70*
■ Use reference angles to evaluate trigonometric functions *(p. 593)*.	*71–84*
Section 9.5	
■ Sketch the graphs of sine and cosine functions using amplitude and period *(p. 599)*.	*85–88*
■ Sketch translations of the graphs of sine and cosine functions *(p. 603)*.	*89–92*
■ Use sine and cosine functions to model real-life data *(p. 605)*.	*93, 94*
Section 9.6	
■ Sketch the graphs of tangent *(p. 611)*, cotangent *(p. 612)*, cosecant *(p. 614)*, and secant *(p. 614)* functions.	*95–102*
■ Sketch the graphs of damped trigonometric functions *(p. 615)*.	*103, 104*
Section 9.7	
■ Evaluate and graph inverse trigonometric functions *(p. 620)*.	*105–122, 131–138*
■ Evaluate and graph compositions of trigonometric functions *(p. 624)*.	*123–130*
Section 9.8	
■ Solve real-life problems involving right triangles *(p. 629)*.	*139, 140*
■ Solve real-life problems involving directional bearings *(p. 631)*.	*141*
■ Solve real-life problems involving harmonic motion *(p. 633)*.	*142*

9 REVIEW EXERCISES

See www.CalcChat.com for worked-out solutions to odd-numbered exercises.

In Exercises 1–8, (a) sketch the angle in standard position, (b) determine the quadrant in which the angle lies, and (c) determine one positive and one negative coterminal angle.

1. $15\pi/4$
2. $2\pi/9$
3. $-4\pi/3$
4. $-23\pi/3$
5. $70°$
6. $280°$
7. $-110°$
8. $-405°$

In Exercises 9–12, convert the angle measure from degrees to radians. Round your answer to three decimal places.

9. $450°$
10. $-112.5°$
11. $-33° 45'$
12. $197° 17'$

In Exercises 13–16, convert the angle measure from radians to degrees. Round your answer to three decimal places.

13. $3\pi/10$
14. $-11\pi/6$
15. -3.5
16. 5.7

In Exercises 17–20, convert each angle measure to degrees, minutes, and seconds without using a calculator.

17. $198.4°$
18. $-70.2°$
19. $0.65°$
20. $-5.96°$

21. ***Arc Length*** Find the length of the arc on a circle with a radius of 20 inches intercepted by a central angle of $138°$.

22. ***Phonograph*** Phonograph records are vinyl discs that rotate on a turntable. A typical record album is 12 inches in diameter and plays at $33\frac{1}{3}$ revolutions per minute.

(a) What is the angular speed of a record album?

(b) What is the linear speed of the outer edge of a record album?

23. ***Bicycle*** At what speed is a bicyclist traveling if his 27-inch-diameter tires are rotating at an angular speed of 5π radians per second?

24. ***Bicycle*** At what speed is a bicyclist traveling if her 26-inch-diameter tires are rotating at an angular speed of 6π radians per second?

In Exercises 25–28, find the point (x, y) on the unit circle that corresponds to the real number t.

25. $t = 2\pi/3$
26. $t = 7\pi/4$
27. $t = 7\pi/6$
28. $t = -4\pi/3$

In Exercises 29–32, evaluate (if possible) the six trigonometric functions of the real number.

29. $t = 7\pi/6$
30. $t = 3\pi/4$
31. $t = -2\pi/3$
32. $t = 2\pi$

In Exercises 33–36, evaluate the trigonometric function using its period as an aid.

33. $\sin(11\pi/4)$
34. $\cos 4\pi$
35. $\sin(-17\pi/6)$
36. $\cos(-13\pi/3)$

In Exercises 37–40, use a calculator to evaluate the trigonometric function. Round your answer to four decimal places.

37. $\tan 33$
38. $\csc 10.5$
39. $\sec(12\pi/5)$
40. $\sin(-\pi/9)$

In Exercises 41 and 42, find the exact values of the six trigonometric functions of the angle θ shown in the figure.

41.

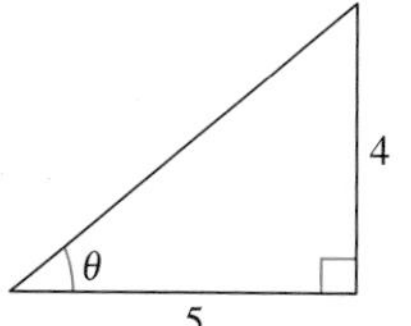

42.

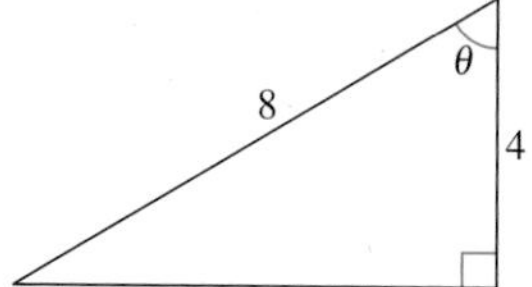

In Exercises 43–46, use the given function value and trigonometric identities (including the cofunction identities) to find the indicated trigonometric functions.

43. $\sin \theta = \frac{1}{3}$ (a) $\csc \theta$ (b) $\cos \theta$ (c) $\sec \theta$ (d) $\tan \theta$

44. $\tan \theta = 4$ (a) $\cot \theta$ (b) $\sec \theta$ (c) $\cos \theta$ (d) $\csc \theta$

45. $\csc \theta = 4$ (a) $\sin \theta$ (b) $\cos \theta$ (c) $\sec \theta$ (d) $\tan \theta$

46. $\csc \theta = 5$ (a) $\sin \theta$ (b) $\cot \theta$ (c) $\tan \theta$ (d) $\sec(90° - \theta)$

In Exercises 47–54, use a calculator to evaluate the trigonometric function. Round your answer to four decimal places.

47. $\tan 33°$
48. $\csc 11°$
49. $\sin 34.2°$
50. $\sec 79.3°$
51. $\cot 15° 14'$
52. $\csc 44° 35'$
53. $\tan 31° 24' 5''$
54. $\cos 78° 11' 58''$

55. ***Railroad Grade*** A train travels 3.5 kilometers on a straight track with a grade of $1° 10'$ (see figure on the next page). What is the vertical rise of the train in that distance?

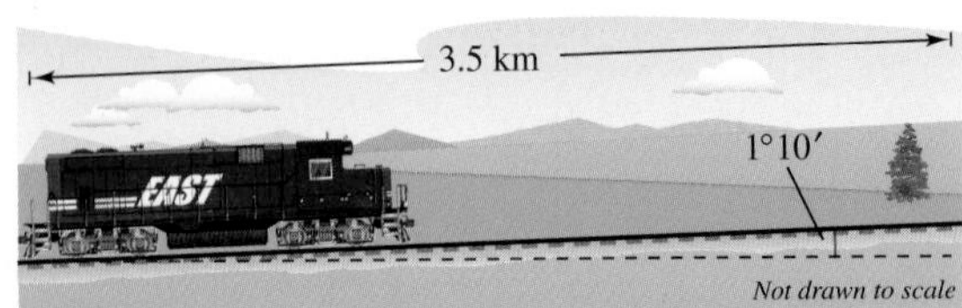

Figure for 55

56. ***Guy Wire*** A guy wire runs from the ground to the top of a 25-foot telephone pole. The angle formed between the wire and the ground is 52°. How far from the base of the pole is the wire attached to the ground?

In Exercises 57–64, the point is on the terminal side of an angle θ in standard position. Determine the exact values of the six trigonometric functions of the angle θ.

57. $(12, 16)$ **58.** $(3, -4)$

59. $\left(\frac{2}{3}, \frac{5}{2}\right)$ **60.** $\left(-\frac{10}{3}, -\frac{2}{3}\right)$

61. $(-0.5, 4.5)$ **62.** $(0.3, 0.4)$

63. $(x, 4x), \quad x > 0$ **64.** $(-2x, -3x), \quad x > 0$

In Exercises 65–70, find the values of the remaining five trigonometric functions of θ.

	Function Value	*Constraint*
65.	$\sec\theta = \frac{6}{5}$	$\tan\theta < 0$
66.	$\csc\theta = \frac{3}{2}$	$\cos\theta < 0$
67.	$\sin\theta = \frac{3}{8}$	$\cos\theta < 0$
68.	$\tan\theta = \frac{5}{4}$	$\cos\theta < 0$
69.	$\cos\theta = -\frac{2}{5}$	$\sin\theta > 0$
70.	$\sin\theta = -\frac{1}{2}$	$\cos\theta > 0$

In Exercises 71–74, find the reference angle θ' and sketch θ and θ' in standard position.

71. $\theta = 264°$ **72.** $\theta = 635°$

73. $\theta = -6\pi/5$ **74.** $\theta = 17\pi/3$

In Exercises 75–80, evaluate the sine, cosine, and tangent of the angle without using a calculator.

75. $\pi/3$ **76.** $\pi/4$

77. $-7\pi/3$ **78.** $-5\pi/4$

79. $495°$ **80.** $-150°$

In Exercises 81–84, use a calculator to evaluate the trigonometric function. Round your answer to four decimal places.

81. $\sin 4$ **82.** $\cot(-4.8)$

83. $\sin(12\pi/5)$ **84.** $\tan(-25\pi/7)$

In Exercises 85–92, sketch the graph of the function. Include two full periods.

85. $y = \sin 6x$ **86.** $y = -\cos 3x$

87. $f(x) = 5\sin(2x/5)$ **88.** $f(x) = 8\cos(-x/4)$

89. $y = 5 + \sin x$ **90.** $y = -4 - \cos\pi x$

91. $g(t) = \frac{5}{2}\sin(t - \pi)$ **92.** $g(t) = 3\cos(t + \pi)$

93. ***Sound Waves*** Sound waves can be modeled by sine functions of the form $y = a\sin bx$, where x is measured in seconds.

(a) Write an equation of a sound wave whose amplitude is 2 and whose period is $\frac{1}{264}$ second.

(b) What is the frequency of the sound wave described in part (a)?

94. ***Data Analysis: Meteorology*** The times S of sunset (Greenwich Mean Time) at 40° north latitude on the 15th of each month are: 1(16:59), 2(17:35), 3(18:06), 4(18:38), 5(19:08), 6(19:30), 7(19:28), 8(18:57), 9(18:09), 10(17:21), 11(16:44), 12(16:36). The month is represented by t, with $t = 1$ corresponding to January. A model (in which minutes have been converted to the decimal parts of an hour) for the data is $S(t) = 18.09 + 1.41\sin[(\pi t/6) + 4.60]$.

(a) Use a graphing utility to graph the data points and the model in the same viewing window.

(b) What is the period of the model? Is it what you expected? Explain.

(c) What is the amplitude of the model? What does it represent in the model? Explain.

In Exercises 95–102, sketch a graph of the function. Include two full periods.

95. $f(x) = 3\tan 2x$ **96.** $f(t) = \tan\left(t + \frac{\pi}{2}\right)$

97. $f(x) = \frac{1}{2}\cot x$ **98.** $g(t) = 2\cot 2t$

99. $f(x) = 3\sec x$ **100.** $h(t) = \sec\left(t - \frac{\pi}{4}\right)$

101. $f(x) = \frac{1}{2}\csc\frac{x}{2}$ **102.** $f(t) = 3\csc\left(2t + \frac{\pi}{4}\right)$

In Exercises 103 and 104, use a graphing utility to graph the function and the damping factor of the function in the same viewing window. Describe the behavior of the function as x increases without bound.

103. $f(x) = x\cos x$ **104.** $g(x) = x^4\cos x$

In Exercises 105–110, evaluate the expression. If necessary, round your answer to two decimal places.

105. $\arcsin\left(-\frac{1}{2}\right)$ **106.** $\arcsin(-1)$

107. $\arcsin 0.4$ **108.** $\arcsin 0.213$

109. $\sin^{-1}(-0.44)$ **110.** $\sin^{-1} 0.89$

In Exercises 111–114, evaluate the expression without using a calculator.

111. $\arccos\left(-\sqrt{2}/2\right)$
112. $\arccos\left(\sqrt{2}/2\right)$
113. $\cos^{-1}(-1)$
114. $\cos^{-1}\left(\sqrt{3}/2\right)$

In Exercises 115–118, use a calculator to evaluate the expression. Round your answer to two decimal places.

115. $\arccos 0.324$
116. $\arccos(-0.888)$
117. $\tan^{-1}(-1.5)$
118. $\tan^{-1} 8.2$

In Exercises 119–122, use a graphing utility to graph the function.

119. $f(x) = 2 \arcsin x$
120. $f(x) = 3 \arccos x$
121. $f(x) = \arctan(x/2)$
122. $f(x) = -\arcsin 2x$

In Exercises 123–128, find the exact value of the expression.

123. $\cos\left(\arctan \frac{3}{4}\right)$
124. $\tan\left(\arccos \frac{3}{5}\right)$
125. $\sec\left(\tan^{-1} \frac{12}{5}\right)$
126. $\sec\left[\sin^{-1}\left(-\frac{1}{4}\right)\right]$
127. $\cot\left(\arctan \frac{7}{10}\right)$
128. $\cot\left[\arcsin\left(-\frac{12}{13}\right)\right]$

In Exercises 129 and 130, write an algebraic expression that is equivalent to the expression.

129. $\tan[\arccos(x/2)]$
130. $\sec[\arcsin(x-1)]$

In Exercises 131–134, evaluate each expression without using a calculator.

131. $\operatorname{arccot} \sqrt{3}$
132. $\operatorname{arcsec}(-1)$
133. $\operatorname{arcsec}\left(-\sqrt{2}\right)$
134. $\operatorname{arccsc} 1$

In Exercises 135–138, use a calculator to approximate the value of the expression. Round your result to two decimal places.

135. $\operatorname{arccot}(10.5)$
136. $\operatorname{arcsec}(-7.5)$
137. $\operatorname{arcsec}\left(-\frac{5}{2}\right)$
138. $\operatorname{arccsc}(-2.01)$

139. ***Angle of Elevation*** The height of a radio transmission tower is 70 meters, and it casts a shadow of length 30 meters. Draw a diagram and find the angle of elevation of the sun.

140. ***Height*** Your football has landed at the edge of the roof of your school building. When you are 25 feet from the base of the building, the angle of elevation to your football is 21°. How high off the ground is your football?

141. ***Distance*** From city A to city B, a plane flies 650 miles at a bearing of 48°. From city B to city C, the plane flies 810 miles at a bearing of 115°. Find the distance from city A to city C and the bearing from city A to city C.

142. ***Wave Motion*** Your fishing bobber oscillates in simple harmonic motion from the waves in the lake where you fish. Your bobber moves a total of 1.5 inches from its high point to its low point and returns to its high point every 3 seconds. Write an equation modeling the motion of your bobber if it is at its high point at time $t = 0$.

True or False? **In Exercises 143 and 144, determine whether the statement is true or false. Justify your answer.**

143. $y = \sin \theta$ is not a function because $\sin 30° = \sin 150°$.

144. Because $\tan(3\pi/4) = -1$, $\arctan(-1) = 3\pi/4$.

145. ***Writing*** Describe the behavior of $f(\theta) = \sec \theta$ at the zeros of $g(\theta) = \cos \theta$. Explain your reasoning.

146. ***Conjecture***

(a) Use a graphing utility to complete the table.

θ	0.1	0.4	0.7	1.0	1.3
$\tan\left(\theta - \frac{\pi}{2}\right)$					
$-\cot \theta$					

(b) Make a conjecture about the relationship between $\tan[\theta - (\pi/2)]$ and $-\cot \theta$.

147. ***Writing*** When graphing the sine and cosine functions, determining the amplitude is part of the analysis. Explain why this is not true for the other four trigonometric functions.

148. ***Oscillation of a Spring*** A weight is suspended from a ceiling by a steel spring. The weight is lifted (positive direction) from the equilibrium position and released. The resulting motion of the weight is modeled by $y = Ae^{-kt} \cos bt = \frac{1}{5}e^{-t/10} \cos 6t$, where y is the distance in feet from equilibrium and t is the time in seconds. The graph of the function is shown in the figure. For each of the following, describe the change in the system without graphing the resulting function.

(a) A is changed from $\frac{1}{5}$ to $\frac{1}{3}$.

(b) k is changed from $\frac{1}{10}$ to $\frac{1}{3}$.

(c) b is changed from 6 to 9.

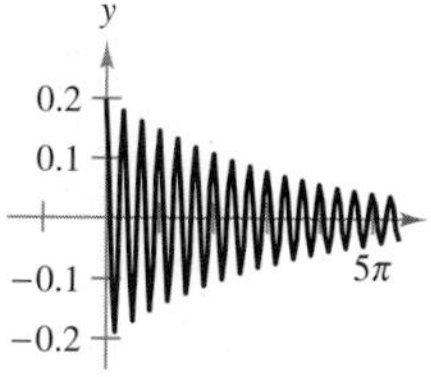

9 CHAPTER TEST

Take this test as you would take a test in class. When you are finished, check your work against the answers given in the back of the book.

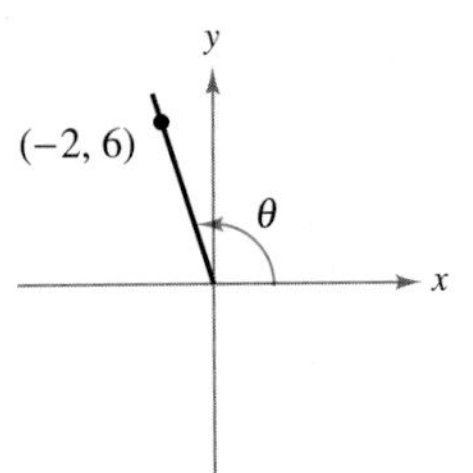

Figure for 3

1. Consider an angle that measures $\frac{5\pi}{4}$ radians.
   (a) Sketch the angle in standard position.
   (b) Determine two coterminal angles (one positive and one negative).
   (c) Convert the angle to degree measure.
2. A truck is moving at a rate of 105 kilometers per hour, and the diameter of its wheels is 1 meter. Find the angular speed of the wheels in radians per minute.
3. Find the exact values of the six trigonometric functions of the angle θ shown in the figure.
4. Given that $\tan \theta = \frac{3}{2}$, find the other five trigonometric functions of θ.
5. Determine the reference angle θ' for the angle $\theta = 205°$ and sketch θ and θ' in standard position.
6. Determine the quadrant in which θ lies if $\sec \theta < 0$ and $\tan \theta > 0$.
7. Find two exact values of θ in degrees $(0 \le \theta < 360°)$ if $\cos \theta = -\sqrt{3}/2$. (Do not use a calculator.)
8. Use a calculator to approximate two values of θ in radians $(0 \le \theta < 2\pi)$ if $\csc \theta = 1.030$. Round the results to two decimal places.

In Exercises 9 and 10, find the remaining five trigonometric functions of θ satisfying the conditions.

9. $\cos \theta = \frac{3}{5}$, $\tan \theta < 0$
10. $\sec \theta = -\frac{29}{20}$, $\sin \theta > 0$

In Exercises 11 and 12, sketch the graph of the function. (Include two full periods.)

11. $g(x) = -2 \sin\left(x - \frac{\pi}{4}\right)$
12. $f(\alpha) = \frac{1}{2} \tan 2\alpha$

In Exercises 13 and 14, use a graphing utility to graph the function. If the function is periodic, find its period.

13. $y = \sin 2\pi x + 2 \cos \pi x$
14. $y = 6e^{-0.12t} \cos(0.25t)$, $0 \le t \le 32$

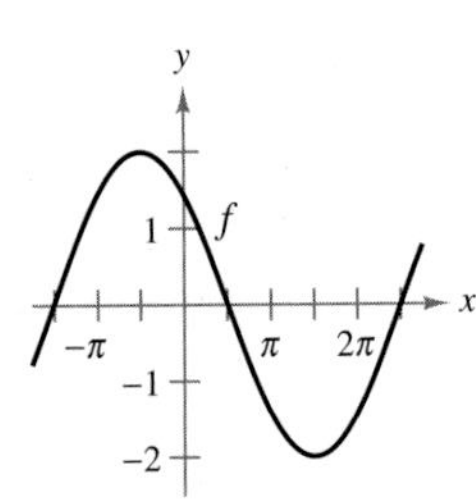

Figure for 15

15. Find a, b, and c for the function $f(x) = a \sin(bx + c)$ such that the graph of f matches the figure.
16. Find the exact value of $\cot\left(\arcsin \frac{3}{8}\right)$ without the aid of a calculator.
17. Graph the function $f(x) = 2 \arcsin\left(\frac{1}{2}x\right)$.
18. A plane is 90 miles south and 110 miles east of London Heathrow Airport. What bearing should be taken to fly directly to the airport?
19. Write the equation for the simple harmonic motion of a ball on a spring that starts at its lowest point of 6 inches below equilibrium, bounces to its maximum height of 6 inches above equilibrium, and returns to its lowest point in a total of 2 seconds.

P.S. PROBLEM SOLVING

1. The pressure P (in millimeters of mercury) against the walls of the blood vessels of a patient is modeled by

$$P = 100 - 20\cos\left(\frac{8\pi}{3}t\right)$$

where t is time (in seconds).

(a) Use a graphing utility to graph the model.

(b) What is the period of the model? What does the period tell you about this situation?

(c) What is the amplitude of the model? What does it tell you about this situation?

(d) If one cycle of this model is equivalent to one heartbeat, what is the pulse of this patient?

(e) If a physician wants this patient's pulse rate to be 64 beats per minute or less, what should the period be? What should the coefficient of t be?

In Exercises 2–4, prove the identity.

2. $\arctan x + \arctan \frac{1}{x} = \frac{\pi}{2}, \quad x > 0$

3. $\arcsin x + \arccos x = \frac{\pi}{2}$

4. $\arcsin x = \arctan \frac{x}{\sqrt{1 - x^2}}$

5. The table shows the average sales S (in millions of dollars) of an outerwear manufacturer for each month t, where $t = 1$ represents January.

t	1	2	3	4	5	6
S	13.46	11.15	7.00	4.85	2.54	1.70

t	7	8	9	10	11	12
S	2.54	4.85	8.00	11.15	13.46	14.30

(a) Create a scatter plot of the data.

(b) Find a trigonometric model that fits the data. Graph the model on your scatter plot. How well does the model fit the data?

(c) What is the period of the model? Do you think it is reasonable given the context? Explain your reasoning.

(d) Interpret the meaning of the model's amplitude in the context of the problem.

6. A two-meter-high fence is 3 meters from the side of a grain storage bin. A grain elevator must reach from ground level outside the fence to the storage bin (see figure). The objective is to determine the shortest elevator that meets the constraints.

(a) Complete two additional rows of the table.

θ	L_1	L_2	$L_1 + L_2$
0.1	$\frac{2}{\sin 0.1}$	$\frac{3}{\cos 0.1}$	23.0
0.2	$\frac{2}{\sin 0.2}$	$\frac{3}{\cos 0.2}$	13.1

(b) Use a graphing utility to generate additional rows of the table. Use the table to estimate the minimum length of the elevator.

(c) Write the length $L_1 + L_2$ as a function of θ.

(d) Use a graphing utility to graph the function. Use the graph to estimate the minimum length. How does your estimate compare with that of part (b)?

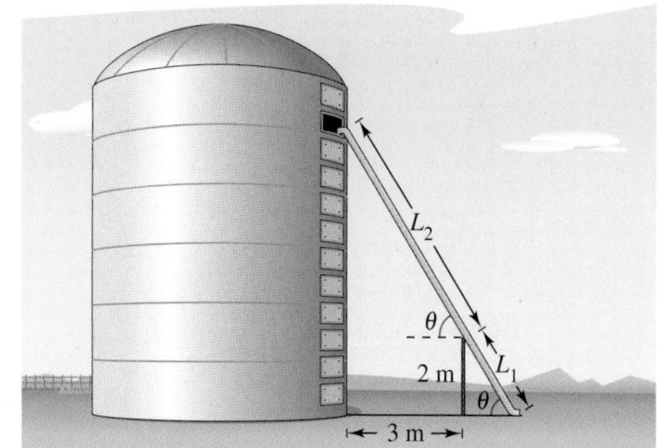

7. In calculus, it can be shown that the arctangent function can be approximated by the polynomial

$$\arctan x \approx x - \frac{x^3}{3} + \frac{x^5}{5} - \frac{x^7}{7}$$

where x is in radians.

(a) Use a graphing utility to graph the arctangent function and its polynomial approximation in the same viewing window. How do the graphs compare?

(b) Study the pattern in the polynomial approximation of the arctangent function and guess the next term. Then repeat part (a). How did the accuracy of the approximation change when additional terms were added?

8. Use a graphing utility to graph the functions given by

$f(x) = \sqrt{x}$ and $g(x) = 6 \arctan x$.

For $x > 0$, it appears that $g > f$. Explain why you know that there exists a positive real number a such that $g < f$ for $x > a$. Approximate the number a.

9. The cross sections of an irrigation canal are isosceles trapezoids, where the length of three of the sides is 8 feet (see figure). The objective is to find the angle θ that maximizes the area of the cross sections. [*Hint:* The area of a trapezoid is $(h/2)(b_1 + b_2)$.]

(a) Complete seven rows of the table.

Base 1	Base 2	Altitude	Area
8	$8 + 16 \cos 10°$	$8 \sin 10°$	22.1
8	$8 + 16 \cos 20°$	$8 \sin 20°$	42.5

(b) Use a graphing utility to generate additional rows of the table. Use the table to estimate the maximum cross-sectional area.

(c) Write the area A as a function of θ.

(d) Use a graphing utility to graph the function. Use the graph to estimate the maximum cross-sectional area. How does your estimate compare with that of part (b)?

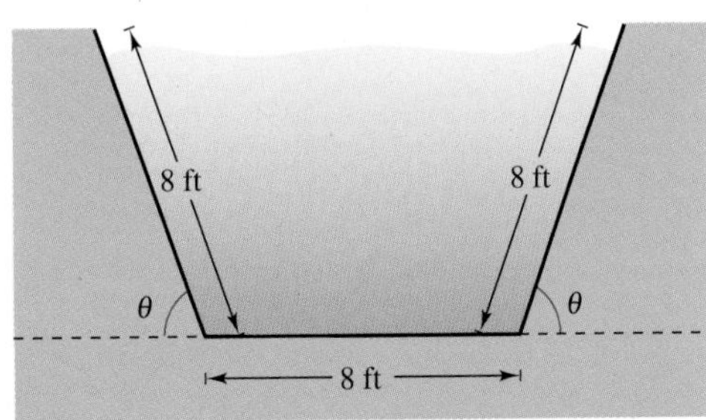

10. The following equation is true for all values of x.

$$d_1 + a_1 \sin(b_1 x + c_1) = d_2 + a_2 \cos(b_2 x + c_2)$$

(a) Describe the relationship between d_1 and d_2.

(b) Describe the relationship between a_1 and a_2.

(c) Describe the relationship between b_1 and b_2.

(d) Describe the relationship between c_1 and c_2.

(e) Give several examples of values of d_1, a_1, b_1, c_1, d_2, a_2, b_2, and c_2 that make the equation true for all values of x.

11. Show that for $f(x) = \sin^k x$, if k is a positive even integer, f is an even function and if k is a positive odd integer, f is an odd function. Is the same true for $f(x) = \cos^k x$? Explain your reasoning.

12. Find the distance in miles that the tip of a six-inch second hand travels in 365 days.

13. The model for the height h (in feet) of a Ferris wheel car is

$$h = 50 + 50 \sin 8\pi t$$

where t is the time (in minutes). (The Ferris wheel has a radius of 50 feet.) This model yields a height of 50 feet when $t = 0$. Alter the model so that the height of the car is 1 foot when $t = 0$.

14. If you stand in shallow water and look at an object below the surface of the water, the object will look farther away from you than it really is. This is because when light rays pass between air and water, the water refracts, or bends, the light rays. The index of refraction for water is 1.333. This is the ratio of the sine of θ_1 and the sine of θ_2 (see figure).

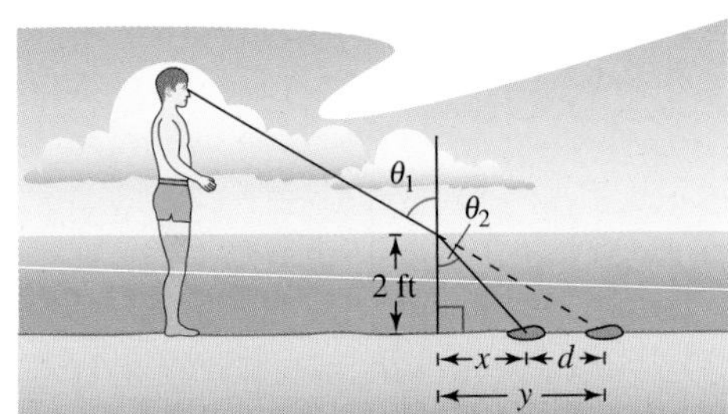

(a) You are standing in water that is 2 feet deep and are looking at a rock at angle $\theta_1 = 60°$ (measured from a line perpendicular to the surface of the water). Find θ_2.

(b) Find the distances x and y.

(c) Find the distance d between where the rock is and where it appears to be.

(d) What happens to d as you move closer to the rock? Explain your reasoning.

15. The function f is periodic, with period c. Therefore, $f(t + c) = f(t)$. Are the following equal? Explain your reasoning.

(a) $f(t - 2c) \stackrel{?}{=} f(t)$

(b) $f\left(t + \frac{1}{2}c\right) \stackrel{?}{=} f\left(\frac{1}{2}t\right)$

(c) $f\left(\frac{1}{2}(t + c)\right) \stackrel{?}{=} f\left(\frac{1}{2}t\right)$

10 Analytic Trigonometry

In this chapter, you will study analytic trigonometry. Analytic trigonometry is used to simplify trigonometric expressions and solve trigonometric equations.

In this chapter, you should learn the following.

- How to use the fundamental trigonometric identities to evaluate, simplify, and rewrite trigonometric expressions. (10.1)
- How to verify trigonometric identities. (10.2)
- How to use standard algebraic techniques to solve trigonometric equations, solve trigonometric equations of quadratic type, solve trigonometric equations involving multiple angles, and use inverse trigonometric functions to solve trigonometric equations. (10.3)
- How to use sum and difference formulas to evaluate trigonometric functions, verify identities, and solve trigonometric equations. (10.4)
- How to use multiple-angle formulas, power-reducing formulas, half-angle formulas, product-to-sum formulas, and sum-to-product formulas to rewrite and evaluate trigonometric functions, and rewrite real-life problems. (10.5)

Steve Chenn/Brand X Pictures/Jupiter Images

Given a function that models the range of a javelin in terms of the velocity and the angle thrown, how can you determine the angle needed to throw a javelin 130 feet at a velocity of 75 feet per second? (See Section 10.5, Exercise 142.)

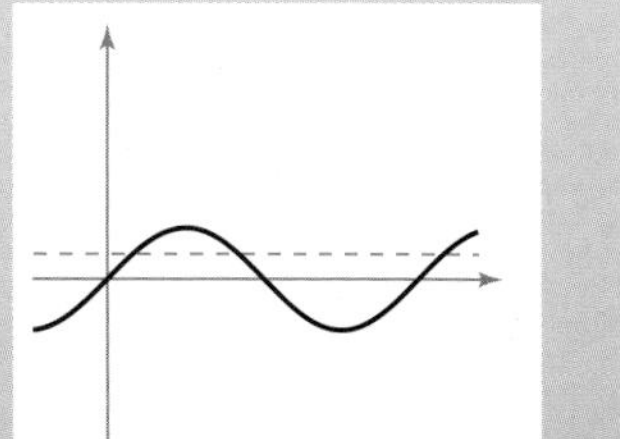
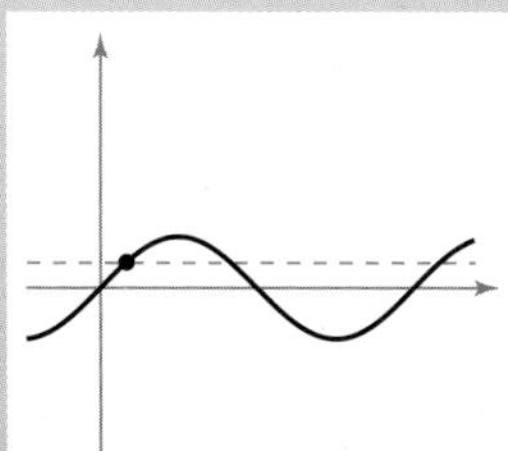
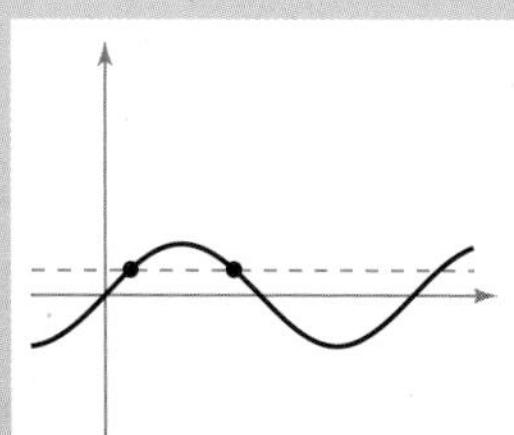
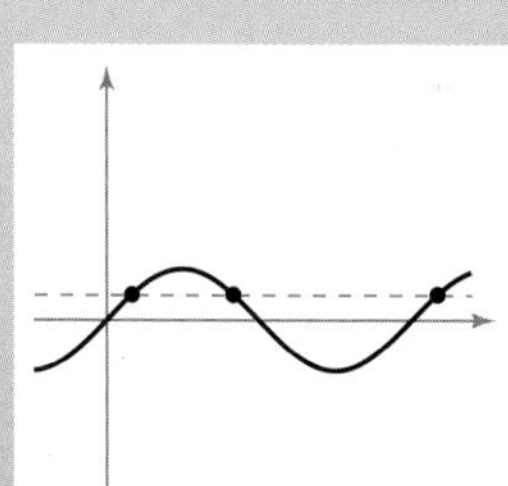

Many trigonometric equations have an infinite number of solutions. You will learn how to use fundamental trigonometric identities and the rules of algebra to find all possible solutions to trigonometric equations. (See Section 10.3.)

10.1 Using Fundamental Trigonometric Identities

- **Recognize and write the fundamental trigonometric identities.**
- **Use the fundamental trigonometric identities to evaluate trigonometric functions, simplify trigonometric expressions, and rewrite trigonometric expressions.**

Introduction

In Chapter 9, you studied the basic definitions, properties, graphs, and applications of the individual trigonometric functions. In this chapter, you will learn how to use the fundamental trigonometric identities to do the following.

1. Evaluate trigonometric functions.
2. Simplify trigonometric expressions.
3. Develop additional trigonometric identities.
4. Solve trigonometric equations.

STUDY TIP Recall that an identity is an equation that is true for every value in the domain of the variable. For instance,

$$\csc u = \frac{1}{\sin u}$$

is an identity because it is true for all values of u for which $\csc u$ is defined.

FUNDAMENTAL TRIGONOMETRIC IDENTITIES

Reciprocal Identities

$$\sin u = \frac{1}{\csc u} \qquad \cos u = \frac{1}{\sec u} \qquad \tan u = \frac{1}{\cot u}$$

$$\csc u = \frac{1}{\sin u} \qquad \sec u = \frac{1}{\cos u} \qquad \cot u = \frac{1}{\tan u}$$

Quotient Identities

$$\tan u = \frac{\sin u}{\cos u} \qquad \cot u = \frac{\cos u}{\sin u}$$

Pythagorean Identities

$$\sin^2 u + \cos^2 u = 1 \qquad 1 + \tan^2 u = \sec^2 u \qquad 1 + \cot^2 u = \csc^2 u$$

Cofunction Identities

$$\sin\left(\frac{\pi}{2} - u\right) = \cos u \qquad \cos\left(\frac{\pi}{2} - u\right) = \sin u$$

$$\tan\left(\frac{\pi}{2} - u\right) = \cot u \qquad \cot\left(\frac{\pi}{2} - u\right) = \tan u$$

$$\sec\left(\frac{\pi}{2} - u\right) = \csc u \qquad \csc\left(\frac{\pi}{2} - u\right) = \sec u$$

Even/Odd Identities

$$\sin(-u) = -\sin u \qquad \cos(-u) = \cos u \qquad \tan(-u) = -\tan u$$

$$\csc(-u) = -\csc u \qquad \sec(-u) = \sec u \qquad \cot(-u) = -\cot u$$

STUDY TIP You should learn the fundamental trigonometric identities well, because they are used frequently in trigonometry and they will also appear later in calculus. Note that u can be an angle, a real number, or a variable.

Pythagorean identities are sometimes used in radical form such as

$$\sin u = \pm\sqrt{1 - \cos^2 u}$$

or

$$\tan u = \pm\sqrt{\sec^2 u - 1}$$

where the sign depends on the choice of u.

Using the Fundamental Trigonometric Identities

One common application of trigonometric identities is to use given values of trigonometric functions to evaluate other trigonometric functions.

EXAMPLE 1 Using Trigonometric Identities to Evaluate a Function

Use the values $\sec u = -\frac{3}{2}$ and $\tan u > 0$ to find the values of all six trigonometric functions.

Solution Using a reciprocal identity, you have

$$\cos u = \frac{1}{\sec u} \qquad \text{Reciprocal identity}$$

$$= \frac{1}{-3/2} \qquad \text{Substitute } -\tfrac{3}{2} \text{ for } \sec u.$$

$$= -\frac{2}{3}. \qquad \text{Simplify.}$$

Using a Pythagorean identity, you have

$$\sin^2 u = 1 - \cos^2 u \qquad \text{Pythagorean identity}$$

$$= 1 - \left(-\frac{2}{3}\right)^2 \qquad \text{Substitute } -\tfrac{2}{3} \text{ for } \cos u.$$

$$= 1 - \frac{4}{9} = \frac{5}{9}. \qquad \text{Simplify.}$$

Because $\sec u < 0$ and $\tan u > 0$, it follows that u lies in Quadrant III. Moreover, because $\sin u$ is negative when u is in Quadrant III, you can choose the negative root and obtain $\sin u = -\sqrt{5}/3$. Now, knowing the values of the sine and cosine, you can find the values of all six trigonometric functions.

$$\sin u = -\frac{\sqrt{5}}{3} \qquad \csc u = \frac{1}{\sin u} = -\frac{3}{\sqrt{5}} = -\frac{3\sqrt{5}}{5}$$

$$\cos u = -\frac{2}{3} \qquad \sec u = \frac{1}{\cos u} = -\frac{3}{2}$$

$$\tan u = \frac{\sin u}{\cos u} = \frac{-\sqrt{5}/3}{-2/3} = \frac{\sqrt{5}}{2} \qquad \cot u = \frac{1}{\tan u} = \frac{2}{\sqrt{5}} = \frac{2\sqrt{5}}{5}$$

TECHNOLOGY You can use a graphing utility to check the result of Example 2. To do this, graph

$$y_1 = \sin x \cos^2 x - \sin x$$

and

$$y_2 = -\sin^3 x$$

in the same viewing window, as shown below. Because Example 2 shows the equivalence algebraically and the two graphs appear to coincide, you can conclude that the expressions are equivalent.

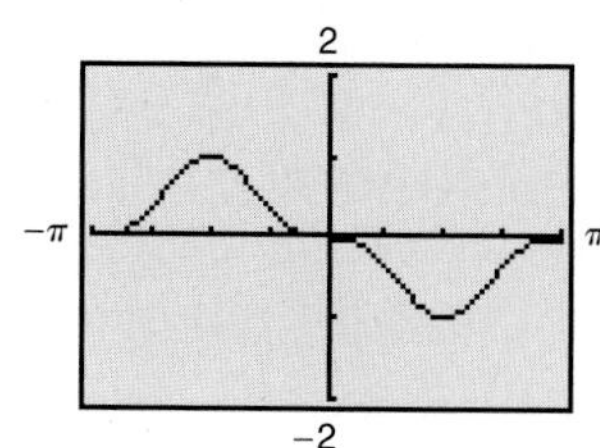

EXAMPLE 2 Simplifying a Trigonometric Expression

Simplify

$$\sin x \cos^2 x - \sin x.$$

Solution First factor out a common monomial factor and then use a fundamental identity.

$$\sin x \cos^2 x - \sin x = \sin x(\cos^2 x - 1) \qquad \text{Factor out common monomial factor.}$$

$$= -\sin x(1 - \cos^2 x) \qquad \text{Factor out } -1.$$

$$= -\sin x(\sin^2 x) \qquad \text{Pythagorean identity}$$

$$= -\sin^3 x \qquad \text{Multiply.}$$

■

When factoring trigonometric expressions, it is helpful to find a special polynomial factoring form that fits the expression, as shown in Example 3.

EXAMPLE 3 Factoring Trigonometric Expressions

Factor each expression.

a. $\sec^2\theta - 1$ **b.** $4\tan^2\theta + \tan\theta - 3$

Solution

a. This expression has the form $u^2 - v^2$, which is the difference of two squares. It factors as

$$\sec^2\theta - 1 = (\sec\theta - 1)(\sec\theta + 1).$$

b. This expression has the polynomial form $ax^2 + bx + c$, and it factors as

$$4\tan^2\theta + \tan\theta - 3 = (4\tan\theta - 3)(\tan\theta + 1).$$

■

On occasion, factoring or simplifying can best be done by first rewriting the expression in terms of just *one* trigonometric function or in terms of *sine and cosine only.* These strategies are shown in Examples 4 and 5, respectively.

EXAMPLE 4 Factoring a Trigonometric Expression

Factor

$$\csc^2 x - \cot x - 3.$$

Solution Use the identity

$$\csc^2 x = 1 + \cot^2 x$$

to rewrite the expression in terms of the cotangent.

$$\begin{aligned}\csc^2 x - \cot x - 3 &= (1 + \cot^2 x) - \cot x - 3 && \text{Pythagorean identity}\\ &= \cot^2 x - \cot x - 2 && \text{Combine like terms.}\\ &= (\cot x - 2)(\cot x + 1) && \text{Factor.}\end{aligned}$$

EXAMPLE 5 Simplifying a Trigonometric Expression

Simplify

$$\sin t + \cot t\cos t.$$

Solution Begin by rewriting $\cot t$ in terms of sine and cosine.

$$\begin{aligned}\sin t + \cot t\cos t &= \sin t + \left(\frac{\cos t}{\sin t}\right)\cos t && \text{Quotient identity}\\ &= \frac{\sin^2 t + \cos^2 t}{\sin t} && \text{Add fractions.}\\ &= \frac{1}{\sin t} && \text{Pythagorean identity}\\ &= \csc t && \text{Reciprocal identity}\end{aligned}$$

■

STUDY TIP Remember that when adding rational expressions, you must first find the least common denominator (LCD). In Example 5, the LCD is $\sin t$.

EXAMPLE 6 Adding Trigonometric Expressions

Perform the addition and simplify.

$$\frac{\sin\theta}{1+\cos\theta}+\frac{\cos\theta}{\sin\theta}$$

Solution

$$\begin{aligned}
\frac{\sin\theta}{1+\cos\theta}+\frac{\cos\theta}{\sin\theta} &= \frac{(\sin\theta)(\sin\theta)+(\cos\theta)(1+\cos\theta)}{(1+\cos\theta)(\sin\theta)} \\
&= \frac{\sin^2\theta+\cos^2\theta+\cos\theta}{(1+\cos\theta)(\sin\theta)} && \text{Multiply.} \\
&= \frac{1+\cos\theta}{(1+\cos\theta)(\sin\theta)} && \text{Pythagorean identity: } \sin^2\theta+\cos^2\theta=1 \\
&= \frac{1}{\sin\theta} && \text{Divide out common factor.} \\
&= \csc\theta && \text{Reciprocal identity}
\end{aligned}$$

The last two examples in this section involve techniques for rewriting expressions in forms that are useful when integrating. In particular, it is often useful to convert a fraction with a binomial denominator into one with a monomial denominator.

EXAMPLE 7 Rewriting a Trigonometric Expression

Rewrite

$$\frac{1}{1+\sin x}$$

so that it is *not* in fractional form.

Solution From the Pythagorean identity

$$\begin{aligned}
\cos^2 x &= 1-\sin^2 x \\
&= (1-\sin x)(1+\sin x)
\end{aligned}$$

you can see that multiplying both the numerator and the denominator by $(1-\sin x)$ will produce a monomial denominator.

$$\begin{aligned}
\frac{1}{1+\sin x} &= \frac{1}{1+\sin x}\cdot\frac{1-\sin x}{1-\sin x} && \text{Multiply numerator and denominator by } (1-\sin x). \\
&= \frac{1-\sin x}{1-\sin^2 x} && \text{Multiply.} \\
&= \frac{1-\sin x}{\cos^2 x} && \text{Pythagorean identity} \\
&= \frac{1}{\cos^2 x}-\frac{\sin x}{\cos^2 x} && \text{Write as separate fractions.} \\
&= \frac{1}{\cos^2 x}-\frac{\sin x}{\cos x}\cdot\frac{1}{\cos x} && \text{Product of fractions} \\
&= \sec^2 x-\tan x\sec x && \text{Reciprocal and quotient identities}
\end{aligned}$$

EXAMPLE 8 Trigonometric Substitution That Removes a Radical

Use the substitution $x = 2 \tan \theta,\ 0 < \theta < \pi/2$, to write

$$\sqrt{4 + x^2}$$

as a trigonometric function of θ.

Solution Begin by letting $x = 2 \tan \theta$. Then, you can obtain

$$\begin{aligned}
\sqrt{4 + x^2} &= \sqrt{4 + (2 \tan \theta)^2} && \text{Substitute } 2 \tan \theta \text{ for } x. \\
&= \sqrt{4 + 4 \tan^2 \theta} && \text{Rule of exponents} \\
&= \sqrt{4(1 + \tan^2 \theta)} && \text{Factor.} \\
&= \sqrt{4 \sec^2 \theta} && \text{Pythagorean identity} \\
&= 2 \sec \theta. && \sec \theta > 0 \text{ for } 0 < \theta < \pi/2
\end{aligned}$$

■

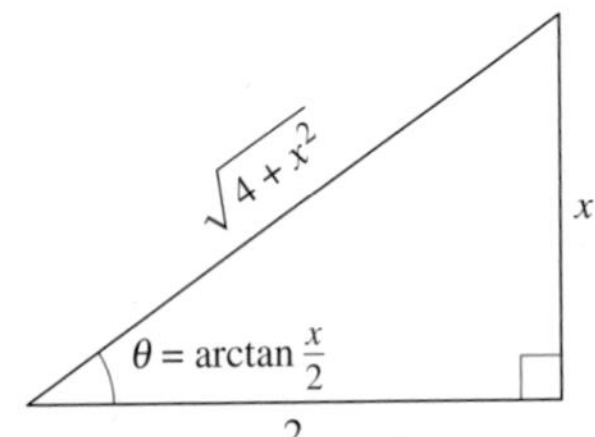

Angle whose tangent is $x/2$.
Figure 10.1

Figure 10.1 shows the right triangle illustration of the trigonometric substitution $x = 2 \tan \theta$ in Example 8. For $0 < \theta < \pi/2$, you have

$$\text{opp} = x, \quad \text{adj} = 2, \quad \text{and} \quad \text{hyp} = \sqrt{4 + x^2}.$$

With these expressions, you can write the following.

$$\sec \theta = \frac{\sqrt{4 + x^2}}{2} \quad \Longrightarrow \quad 2 \sec \theta = \sqrt{4 + x^2}$$

10.1 Exercises

See www.CalcChat.com for worked-out solutions to odd-numbered exercises.

In Exercises 1–10, fill in the blank to complete the trigonometric identity.

1. $\dfrac{\sin u}{\cos u} = $ ________
2. $\dfrac{1}{\csc u} = $ ________
3. $\dfrac{1}{\tan u} = $ ________
4. $\dfrac{1}{\cos u} = $ ________
5. $1 + $ ________ $= \csc^2 u$
6. $1 + \tan^2 u = $ ________
7. $\sin\left(\dfrac{\pi}{2} - u\right) = $ ________
8. $\sec\left(\dfrac{\pi}{2} - u\right) = $ ________
9. $\cos(-u) = $ ________
10. $\tan(-u) = $ ________

In Exercises 11–24, use the given values to evaluate (if possible) all six trigonometric functions.

11. $\sin x = \dfrac{1}{2}, \quad \cos x = \dfrac{\sqrt{3}}{2}$
12. $\tan x = \dfrac{\sqrt{3}}{3}, \quad \cos x = -\dfrac{\sqrt{3}}{2}$
13. $\sec \theta = \sqrt{2}, \quad \sin \theta = -\dfrac{\sqrt{2}}{2}$
14. $\csc \theta = \dfrac{25}{7}, \quad \tan \theta = \dfrac{7}{24}$
15. $\tan x = \dfrac{8}{15}, \quad \sec x = -\dfrac{17}{15}$
16. $\cot \phi = -3, \quad \sin \phi = \dfrac{\sqrt{10}}{10}$
17. $\sec \phi = \dfrac{3}{2}, \quad \csc \phi = -\dfrac{3\sqrt{5}}{5}$
18. $\cos\left(\dfrac{\pi}{2} - x\right) = \dfrac{3}{5}, \quad \cos x = \dfrac{4}{5}$
19. $\sin(-x) = -\dfrac{1}{3}, \quad \tan x = -\dfrac{\sqrt{2}}{4}$
20. $\sec x = 4, \quad \sin x > 0$
21. $\tan \theta = 2, \quad \sin \theta < 0$
22. $\csc \theta = -5, \quad \cos \theta < 0$
23. $\sin \theta = -1, \quad \cot \theta = 0$
24. $\tan \theta$ is undefined, $\quad \sin \theta > 0$

In Exercises 25–30, match the trigonometric expression with one of the following.

(a) $\sec x$ (b) -1 (c) $\cot x$

(d) 1 (e) $-\tan x$ (f) $\sin x$

25. $\sec x \cos x$
26. $\tan x \csc x$
27. $\cot^2 x - \csc^2 x$
28. $(1 - \cos^2 x)(\csc x)$
29. $\dfrac{\sin(-x)}{\cos(-x)}$
30. $\dfrac{\sin[(\pi/2) - x]}{\cos[(\pi/2) - x]}$

In Exercises 31–48, use the fundamental trigonometric identities to simplify the expression. There is more than one correct form of each answer.

31. $\cot \theta \sec \theta$
32. $\cos \beta \tan \beta$
33. $\sin \phi(\csc \phi - \sin \phi)$
34. $\sec^2 x(1 - \sin^2 x)$
35. $\dfrac{\cot x}{\csc x}$
36. $\dfrac{\csc \theta}{\sec \theta}$
37. $\dfrac{1 - \sin^2 x}{\csc^2 x - 1}$
38. $\dfrac{1}{\tan^2 x + 1}$
39. $\sec \alpha \cdot \dfrac{\sin \alpha}{\tan \alpha}$
40. $\dfrac{\tan^2 \theta}{\sec^2 \theta}$
41. $\cos\left(\dfrac{\pi}{2} - x\right) \sec x$
42. $\cot\left(\dfrac{\pi}{2} - x\right) \cos x$
43. $\dfrac{\cos^2 y}{1 - \sin y}$
44. $\cos t(1 + \tan^2 t)$
45. $\sin \beta \tan \beta + \cos \beta$
46. $\csc \phi \tan \phi + \sec \phi$
47. $\cot u \sin u + \tan u \cos u$
48. $\sin \theta \sec \theta + \cos \theta \csc \theta$

In Exercises 49–60, factor the expression and use the fundamental identities to simplify. There is more than one correct form of each answer.

49. $\tan^2 x - \tan^2 x \sin^2 x$
50. $\sin^2 x \csc^2 x - \sin^2 x$
51. $\sin^2 x \sec^2 x - \sin^2 x$
52. $\cos^2 x + \cos^2 x \tan^2 x$
53. $\dfrac{\sec^2 x - 1}{\sec x - 1}$
54. $\dfrac{\cos^2 x - 4}{\cos x - 2}$
55. $\tan^4 x + 2 \tan^2 x + 1$
56. $1 - 2 \cos^2 x + \cos^4 x$
57. $\sin^4 x - \cos^4 x$
58. $\sec^4 x - \tan^4 x$
59. $\csc^3 x - \csc^2 x - \csc x + 1$
60. $\sec^3 x - \sec^2 x - \sec x + 1$

In Exercises 61–64, perform the multiplication and use the fundamental identities to simplify. There is more than one correct form of each answer.

61. $(\sin x + \cos x)^2$
62. $(\cot x + \csc x)(\cot x - \csc x)$
63. $(2 \csc x + 2)(2 \csc x - 2)$
64. $(3 - 3 \sin x)(3 + 3 \sin x)$

In Exercises 65–68, perform the addition or subtraction and use the fundamental identities to simplify. There is more than one correct form of each answer.

65. $\dfrac{1}{1 + \cos x} + \dfrac{1}{1 - \cos x}$
66. $\dfrac{1}{\sec x + 1} - \dfrac{1}{\sec x - 1}$
67. $\dfrac{\cos x}{1 + \sin x} + \dfrac{1 + \sin x}{\cos x}$
68. $\tan x - \dfrac{\sec^2 x}{\tan x}$

In Exercises 69–72, rewrite the expression so that it is not in fractional form. There is more than one correct form of each answer.

69. $\dfrac{\sin^2 y}{1 - \cos y}$
70. $\dfrac{5}{\tan x + \sec x}$
71. $\dfrac{3}{\sec x - \tan x}$
72. $\dfrac{\tan^2 x}{\csc x + 1}$

WRITING ABOUT CONCEPTS

In Exercises 73–78, determine whether the equation is an identity, and give a reason for your answer.

73. $\cos \theta = \sqrt{1 - \sin^2 \theta}$
74. $\cot \theta = \sqrt{\csc^2 \theta + 1}$
75. $(\sin k\theta)/(\cos k\theta) = \tan \theta$, k is a constant.
76. $1/(5 \cos \theta) = 5 \sec \theta$
77. $\sin \theta \csc \theta = 1$
78. $\csc^2 \theta = 1$

79. Express each of the other trigonometric functions of θ in terms of $\sin \theta$.
80. Express each of the other trigonometric functions of θ in terms of $\cos \theta$.

Numerical and Graphical Analysis **In Exercises 81–84, use a graphing utility to complete the table and graph the functions. Make a conjecture about y_1 and y_2.**

x	0.2	0.4	0.6	0.8	1.0	1.2	1.4
y_1							
y_2							

81. $y_1 = \cos\left(\dfrac{\pi}{2} - x\right)$, $y_2 = \sin x$
82. $y_1 = \sec x - \cos x$, $y_2 = \sin x \tan x$
83. $y_1 = \dfrac{\cos x}{1 - \sin x}$, $y_2 = \dfrac{1 + \sin x}{\cos x}$
84. $y_1 = \sec^4 x - \sec^2 x$, $y_2 = \tan^2 x + \tan^4 x$

In Exercises 85–88, use a graphing utility to determine which of the six trigonometric functions is equal to the expression. Verify your answer algebraically.

85. $\cos x \cot x + \sin x$ **86.** $\sec x \csc x - \tan x$

87. $\dfrac{1}{\sin x}\left(\dfrac{1}{\cos x} - \cos x\right)$

88. $\dfrac{1}{2}\left(\dfrac{1 + \sin \theta}{\cos \theta} + \dfrac{\cos \theta}{1 + \sin \theta}\right)$

In Exercises 89–94, use the trigonometric substitution to write the algebraic expression as a trigonometric function of θ, where $0 < \theta < \pi/2$.

89. $\sqrt{9 - x^2}$, $x = 3 \cos \theta$

90. $\sqrt{64 - 16x^2}$, $x = 2 \cos \theta$

91. $\sqrt{49 - x^2}$, $x = 7 \sin \theta$

92. $\sqrt{x^2 - 4}$, $x = 2 \sec \theta$

93. $\sqrt{x^2 + 100}$, $x = 10 \tan \theta$

94. $\sqrt{9x^2 + 25}$, $3x = 5 \tan \theta$

In Exercises 95–98, use the trigonometric substitution to write the algebraic equation as a trigonometric equation of θ, where $-\pi/2 < \theta < \pi/2$. Then find $\sin \theta$ and $\cos \theta$.

95. $3 = \sqrt{9 - x^2}$, $x = 3 \sin \theta$

96. $3 = \sqrt{36 - x^2}$, $x = 6 \sin \theta$

97. $2\sqrt{2} = \sqrt{16 - 4x^2}$, $x = 2 \cos \theta$

98. $-5\sqrt{3} = \sqrt{100 - x^2}$, $x = 10 \cos \theta$

In Exercises 99–102, use a graphing utility to solve the equation for θ, where $0 \le \theta < 2\pi$.

99. $\sin \theta = \sqrt{1 - \cos^2 \theta}$ **100.** $\cos \theta = -\sqrt{1 - \sin^2 \theta}$

101. $\sec \theta = \sqrt{1 + \tan^2 \theta}$ **102.** $\csc \theta = \sqrt{1 + \cot^2 \theta}$

In Exercises 103–106, rewrite the expression as a single logarithm and simplify the result.

103. $\ln|\cos x| - \ln|\sin x|$ **104.** $\ln|\sec x| + \ln|\sin x|$

105. $\ln|\cot t| + \ln(1 + \tan^2 t)$

106. $\ln(\cos^2 t) + \ln(1 + \tan^2 t)$

In Exercises 107–110, use a calculator to demonstrate the identity for each value of θ.

107. $\csc^2 \theta - \cot^2 \theta = 1$ (a) $\theta = 132°$ (b) $\theta = \dfrac{2\pi}{7}$

108. $\tan^2 \theta + 1 = \sec^2 \theta$ (a) $\theta = 346°$ (b) $\theta = 3.1$

109. $\cos\left(\dfrac{\pi}{2} - \theta\right) = \sin \theta$ (a) $\theta = 80°$ (b) $\theta = 0.8$

110. $\sin(-\theta) = -\sin \theta$ (a) $\theta = 250°$ (b) $\theta = \frac{1}{2}$

111. ***Friction*** The forces acting on an object weighing W units on an inclined plane positioned at an angle of θ with the horizontal (see figure) are modeled by

$$\mu W \cos \theta = W \sin \theta$$

where μ is the coefficient of friction. Solve the equation for μ and simplify the result.

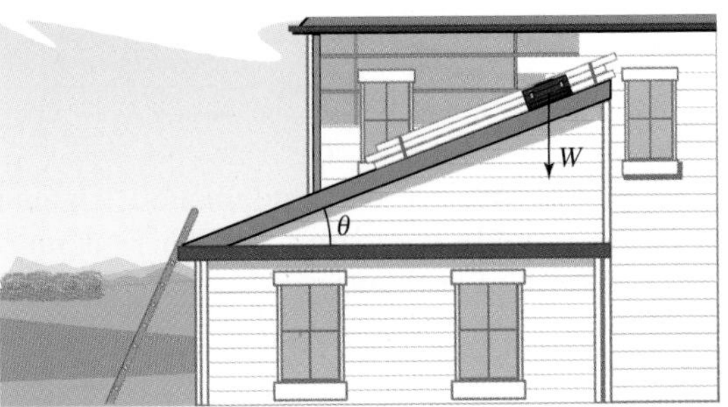

CAPSTONE

112. (a) Use the definitions of sine and cosine to derive the Pythagorean identity $\sin^2 \theta + \cos^2 \theta = 1$.

(b) Use the Pythagorean identity $\sin^2 \theta + \cos^2 \theta = 1$ to derive the other Pythagorean identities, $1 + \tan^2 \theta = \sec^2 \theta$ and $1 + \cot^2 \theta = \csc^2 \theta$. Discuss how to remember these identities and other fundamental identities.

113. ***Rate of Change*** The rate of change of the function $f(x) = \sec x + \cos x$ is given by the expression $\sec x \tan x - \sin x$. Show that this expression can also be written as $\sin x \tan^2 x$.

True or False? **In Exercises 114 and 115, determine whether the statement is true or false. Justify your answer.**

114. The even and odd trigonometric identities are helpful for determining whether the value of a trigonometric function is positive or negative.

115. A cofunction identity can be used to transform a tangent function so that it can be represented by a cosecant function.

In Exercises 116–119, fill in the blanks. (*Note:* The notation $x \to c^+$ indicates that x approaches c from the right and $x \to c^-$ indicates that x approaches c from the left.)

116. As $x \to \dfrac{\pi}{2}^-$, $\sin x \to$ ___ and $\csc x \to$ ___.

117. As $x \to 0^+$, $\cos x \to$ ___ and $\sec x \to$ ___.

118. As $x \to \dfrac{\pi}{2}^-$, $\tan x \to$ ___ and $\cot x \to$ ___.

119. As $x \to \pi^+$, $\sin x \to$ ___ and $\csc x \to$ ___.

10.2 Verifying Trigonometric Identities

- **Verify trigonometric identities.**

Introduction

In this section, you will study techniques for verifying trigonometric identities. In the next section, you will study techniques for solving trigonometric equations. The key to verifying identities *and* solving equations is the ability to use the fundamental identities and the rules of algebra to rewrite trigonometric expressions.

Remember that a *conditional equation* is an equation that is true for only some of the values in its domain. For example, the conditional equation

$$\sin x = 0 \qquad \text{Conditional equation}$$

is true only for $x = n\pi$, where n is an integer. When you find these values, you are *solving* the equation.

On the other hand, an equation that is true for all real values in the domain of the variable is an *identity.* For example, the familiar equation

$$\sin^2 x = 1 - \cos^2 x \qquad \text{Identity}$$

is true for all real numbers x. So, it is an identity.

Verifying Trigonometric Identities

Although there are similarities, verifying that a trigonometric equation is an identity is quite different from solving an equation. There is no well-defined set of rules to follow in verifying trigonometric identities, and the process is best learned by practice.

GUIDELINES FOR VERIFYING TRIGONOMETRIC IDENTITIES

1. Work with one side of the equation at a time. It is often better to work with the more complicated side first.
2. Look for opportunities to factor an expression, add fractions, square a binomial, or create a monomial denominator.
3. Look for opportunities to use the fundamental identities. Note which functions are in the final expression you want. Sines and cosines pair up well, as do secants and tangents, and cosecants and cotangents.
4. If the preceding guidelines do not help, try converting all terms to sines and cosines.
5. Always try *something.* Even paths that lead to dead ends provide insights.

Verifying trigonometric identities is a useful process if you need to convert a trigonometric expression into a form that is more useful algebraically. When you verify an identity, you cannot *assume* that the two sides of the equation are equal because you are trying to verify that they *are* equal. As a result, when verifying identities, you cannot use operations such as adding the same quantity to each side of the equation or cross multiplication.

EXAMPLE 1 Verifying a Trigonometric Identity

Verify the identity

$$\frac{\sec^2 \theta - 1}{\sec^2 \theta} = \sin^2 \theta.$$

STUDY TIP Remember that an identity is only true for all real values in the domain of the variable. For instance, in Example 1 the identity is not true when $\theta = \pi/2$ because $\sec^2 \theta$ is not defined when $\theta = \pi/2$.

Solution The left side is more complicated, so start with it.

$$\begin{aligned}
\frac{\sec^2 \theta - 1}{\sec^2 \theta} &= \frac{(\tan^2 \theta + 1) - 1}{\sec^2 \theta} && \text{Pythagorean identity} \\
&= \frac{\tan^2 \theta}{\sec^2 \theta} && \text{Simplify.} \\
&= \tan^2 \theta(\cos^2 \theta) && \text{Reciprocal identity} \\
&= \frac{\sin^2 \theta}{\cancel{(\cos^2 \theta)}} \cancel{(\cos^2 \theta)} && \text{Quotient identity} \\
&= \sin^2 \theta && \text{Simplify.}
\end{aligned}$$

Notice how the identity is verified. You start with the left side of the equation (the more complicated side) and use the fundamental trigonometric identities to simplify it until you obtain the right side. ■

There can be more than one way to verify an identity. Here is another way to verify the identity in Example 1.

$$\begin{aligned}
\frac{\sec^2 \theta - 1}{\sec^2 \theta} &= \frac{\sec^2 \theta}{\sec^2 \theta} - \frac{1}{\sec^2 \theta} && \text{Rewrite as the difference of fractions.} \\
&= 1 - \cos^2 \theta && \text{Reciprocal identity} \\
&= \sin^2 \theta && \text{Pythagorean identity}
\end{aligned}$$

EXAMPLE 2 Verifying a Trigonometric Identity

Verify the identity $2 \sec^2 \alpha = \dfrac{1}{1 - \sin \alpha} + \dfrac{1}{1 + \sin \alpha}$.

Algebraic Solution

The right side is more complicated, so start with it.

$$\begin{aligned}
\frac{1}{1 - \sin \alpha} + \frac{1}{1 + \sin \alpha} &= \frac{1 + \sin \alpha + 1 - \sin \alpha}{(1 - \sin \alpha)(1 + \sin \alpha)} && \text{Add fractions.} \\
&= \frac{2}{1 - \sin^2 \alpha} && \text{Simplify.} \\
&= \frac{2}{\cos^2 \alpha} && \text{Pythagorean identity} \\
&= 2 \sec^2 \alpha && \text{Reciprocal identity}
\end{aligned}$$

Numerical Solution

Use the *table* feature of a graphing utility set in *radian* mode to create a table that shows the values of $y_1 = 2/\cos^2 x$ and $y_2 = 1/(1 - \sin x) + 1/(1 + \sin x)$ for different values of x, as shown in Figure 10.2. From the table, you can see that the values appear to be identical, so $2 \sec^2 x = 1/(1 - \sin x) + 1/(1 + \sin x)$ appears to be an identity.

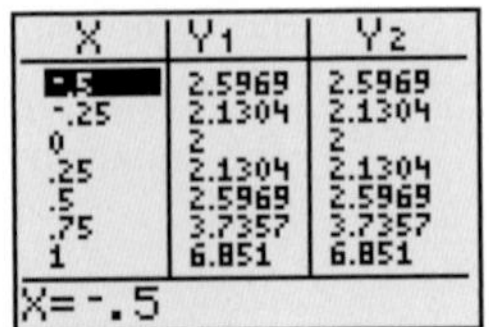

X	Y1	Y2
-.5	2.5969	2.5969
-.25	2.1304	2.1304
0	2	2
.25	2.1304	2.1304
.5	2.5969	2.5969
.75	3.7357	3.7357
1	6.851	6.851

X=-.5

Figure 10.2 ■

EXAMPLE 3 Verifying a Trigonometric Identity

Verify the identity $(\tan^2 x + 1)(\cos^2 x - 1) = -\tan^2 x$.

Algebraic Solution

By applying identities before multiplying, you obtain the following.

$$\begin{aligned}(\tan^2 x + 1)(\cos^2 x - 1) &= (\sec^2 x)(-\sin^2 x) && \text{Pythagorean identities}\\ &= -\frac{\sin^2 x}{\cos^2 x} && \text{Reciprocal identity}\\ &= -\left(\frac{\sin x}{\cos x}\right)^2 && \text{Rule of exponents}\\ &= -\tan^2 x && \text{Quotient identity}\end{aligned}$$

Graphical Solution

Use a graphing utility set in *radian* mode to graph the left side of the identity $y_1 = (\tan^2 x + 1)(\cos^2 x - 1)$ and the right side of the identity $y_2 = -\tan^2 x$ in the same viewing window, as shown in Figure 10.3. (Select the *line* style for y_1 and the *path* style for y_2.) Because the graphs appear to coincide, $(\tan^2 x + 1)(\cos^2 x - 1) = -\tan^2 x$ appears to be an identity.

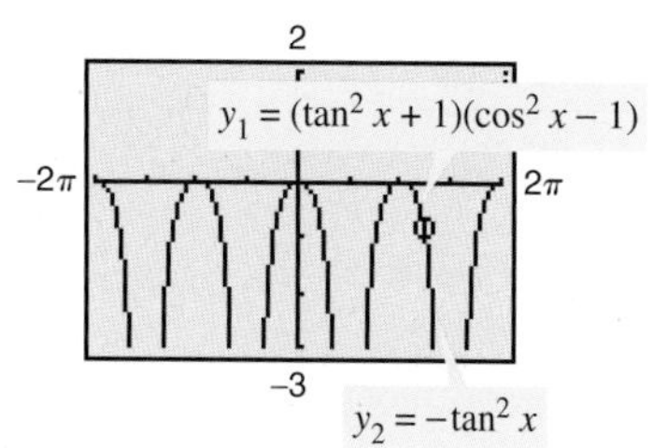

Figure 10.3

■

EXAMPLE 4 Converting to Sines and Cosines

STUDY TIP Although a graphing utility can be useful in helping to verify an identity, you must use algebraic techniques to produce a *valid* proof.

Verify the identity $\tan x + \cot x = \sec x \csc x$.

Solution Try converting the left side into sines and cosines.

$$\begin{aligned}\tan x + \cot x &= \frac{\sin x}{\cos x} + \frac{\cos x}{\sin x} && \text{Quotient identities}\\ &= \frac{\sin^2 x + \cos^2 x}{\cos x \sin x} && \text{Add fractions.}\\ &= \frac{1}{\cos x \sin x} && \text{Pythagorean identity}\\ &= \frac{1}{\cos x} \cdot \frac{1}{\sin x} && \text{Product of fractions}\\ &= \sec x \csc x && \text{Reciprocal identities}\end{aligned}$$

■

Recall from algebra that *rationalizing the denominator* using conjugates is, on occasion, a powerful simplification technique. A related form of this technique, shown below, works for simplifying trigonometric expressions as well.

$$\begin{aligned}\frac{1}{1 - \cos x} &= \frac{1}{1 - \cos x}\left(\frac{1 + \cos x}{1 + \cos x}\right)\\ &= \frac{1 + \cos x}{1 - \cos^2 x}\\ &= \frac{1 + \cos x}{\sin^2 x}\\ &= \csc^2 x(1 + \cos x)\end{aligned}$$

STUDY TIP As shown at the right, $\csc^2 x(1 + \cos x)$ is considered a simplified form of $1/(1 - \cos x)$ because the expression does not contain any fractions.

This technique is demonstrated in the next example.

EXAMPLE 5 Verifying a Trigonometric Identity

Verify the identity

$$\sec x + \tan x = \frac{\cos x}{1 - \sin x}.$$

Algebraic Solution

Begin with the *right* side because you can create a monomial denominator by multiplying the numerator and denominator by $1 + \sin x$.

$$\frac{\cos x}{1 - \sin x} = \frac{\cos x}{1 - \sin x}\left(\frac{1 + \sin x}{1 + \sin x}\right) \qquad \text{Multiply numerator and denominator by } 1 + \sin x.$$

$$= \frac{\cos x + \cos x \sin x}{1 - \sin^2 x} \qquad \text{Multiply.}$$

$$= \frac{\cos x + \cos x \sin x}{\cos^2 x} \qquad \text{Pythagorean identity}$$

$$= \frac{\cos x}{\cos^2 x} + \frac{\cos x \sin x}{\cos^2 x} \qquad \text{Write as separate fractions.}$$

$$= \frac{1}{\cos x} + \frac{\sin x}{\cos x} \qquad \text{Simplify.}$$

$$= \sec x + \tan x \qquad \text{Identities}$$

Graphical Solution

Use a graphing utility set in the *radian* and *dot* modes to graph $y_1 = \sec x + \tan x$ and $y_2 = \cos x/(1 - \sin x)$ in the same viewing window, as shown in Figure 10.4. Because the graphs appear to coincide, $\sec x + \tan x = \cos x/(1 - \sin x)$ appears to be an identity.

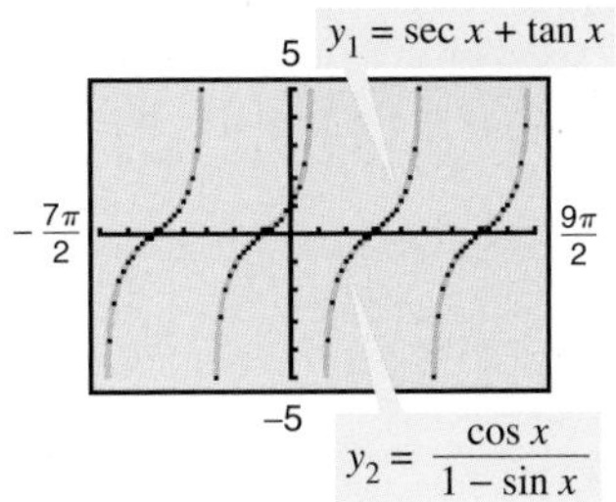

Figure 10.4

In Examples 1 through 5, you have been verifying trigonometric identities by working with one side of the equation and converting to the form given on the other side. On occasion, it is practical to work with each side *separately*, to obtain one common form equivalent to both sides. This is illustrated in Example 6.

EXAMPLE 6 Working with Each Side Separately

Verify the identity $\dfrac{\cot^2 \theta}{1 + \csc \theta} = \dfrac{1 - \sin \theta}{\sin \theta}$.

Algebraic Solution

Working with the left side, you have

$$\frac{\cot^2 \theta}{1 + \csc \theta} = \frac{\csc^2 \theta - 1}{1 + \csc \theta} \qquad \text{Pythagorean identity}$$

$$= \frac{(\csc \theta - 1)\cancel{(\csc \theta + 1)}}{\cancel{1 + \csc \theta}} \qquad \text{Factor.}$$

$$= \csc \theta - 1. \qquad \text{Simplify.}$$

Now, simplifying the right side, you have

$$\frac{1 - \sin \theta}{\sin \theta} = \frac{1}{\sin \theta} - \frac{\sin \theta}{\sin \theta} \qquad \text{Write as separate fractions.}$$

$$= \csc \theta - 1. \qquad \text{Reciprocal identity}$$

The identity is verified because both sides are equal to $\csc \theta - 1$.

Numerical Solution

Use the *table* feature of a graphing utility set in *radian* mode to create a table that shows the values of $y_1 = \cot^2 x/(1 + \csc x)$ and $y_2 = (1 - \sin x)/\sin x$ for different values of x, as shown in Figure 10.5. From the table you can see that the values appear to be identical, so $\cot^2 x/(1 + \csc x) = (1 - \sin x)/\sin x$ appears to be an identity.

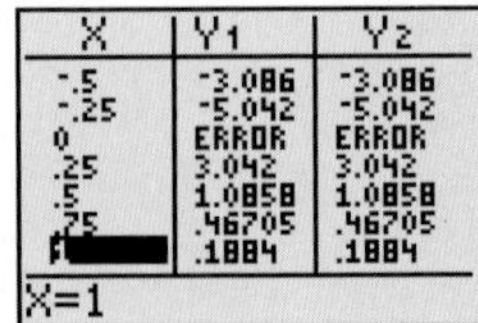

Figure 10.5

In Example 7, powers of trigonometric functions are rewritten as more complicated sums of products of trigonometric functions. This is a common procedure used to integrate trigonometric power functions.

EXAMPLE 7 Verifying Trigonometric Identities

Verify each identity.

a. $\tan^4 x = \tan^2 x \sec^2 x - \tan^2 x$

b. $\sin^3 x \cos^4 x = (\cos^4 x - \cos^6 x) \sin x$

c. $\csc^4 x \cot x = \csc^2 x(\cot x + \cot^3 x)$

Solution

a. $\tan^4 x = (\tan^2 x)(\tan^2 x)$ Write as separate factors.

$= \tan^2 x(\sec^2 x - 1)$ Pythagorean identity

$= \tan^2 x \sec^2 x - \tan^2 x$ Multiply.

b. $\sin^3 x \cos^4 x = \sin^2 x \cos^4 x \sin x$ Write as separate factors.

$= (1 - \cos^2 x) \cos^4 x \sin x$ Pythagorean identity

$= (\cos^4 x - \cos^6 x) \sin x$ Multiply.

c. $\csc^4 x \cot x = \csc^2 x \csc^2 x \cot x$ Write as separate factors.

$= \csc^2 x(1 + \cot^2 x) \cot x$ Pythagorean identity

$= \csc^2 x(\cot x + \cot^3 x)$ Multiply. ■

10.2 Exercises

See www.CalcChat.com for worked-out solutions to odd-numbered exercises.

In Exercises 1 and 2, fill in the blanks.

1. An equation that is true for all real values in its domain is called an ________.
2. An equation that is true for only some values in its domain is called a ________ ________.

In Exercises 3–8, fill in the blank to complete the trigonometric identity.

3. $\dfrac{1}{\cot u} =$ ______
4. $\dfrac{\cos u}{\sin u} =$ ______
5. $\sin^2 u +$ ______ $= 1$
6. $\cos\left(\dfrac{\pi}{2} - u\right) =$ ______
7. $\csc(-u) =$ ______
8. $\sec(-u) =$ ______

In Exercises 9–46, verify the identity.

9. $\tan t \cot t = 1$
10. $\sec y \cos y = 1$
11. $\cot^2 y(\sec^2 y - 1) = 1$
12. $\cos x + \sin x \tan x = \sec x$
13. $(1 + \sin \alpha)(1 - \sin \alpha) = \cos^2 \alpha$
14. $\cos^2 \beta - \sin^2 \beta = 2\cos^2 \beta - 1$
15. $\cos^2 \beta - \sin^2 \beta = 1 - 2\sin^2 \beta$
16. $\sin^2 \alpha - \sin^4 \alpha = \cos^2 \alpha - \cos^4 \alpha$
17. $\dfrac{\tan^2 \theta}{\sec \theta} = \sin \theta \tan \theta$
18. $\dfrac{\cot^3 t}{\csc t} = \cos t(\csc^2 t - 1)$
19. $\dfrac{\cot^2 t}{\csc t} = \dfrac{1 - \sin^2 t}{\sin t}$
20. $\dfrac{1}{\tan \beta} + \tan \beta = \dfrac{\sec^2 \beta}{\tan \beta}$
21. $\sin^{1/2} x \cos x - \sin^{5/2} x \cos x = \cos^3 x\sqrt{\sin x}$
22. $\sec^6 x(\sec x \tan x) - \sec^4 x(\sec x \tan x) = \sec^5 x \tan^3 x$
23. $\dfrac{\cot x}{\sec x} = \csc x - \sin x$
24. $\dfrac{\sec \theta - 1}{1 - \cos \theta} = \sec \theta$
25. $\csc x - \sin x = \cos x \cot x$
26. $\sec x - \cos x = \sin x \tan x$
27. $\dfrac{1}{\tan x} + \dfrac{1}{\cot x} = \tan x + \cot x$
28. $\dfrac{1}{\sin x} - \dfrac{1}{\csc x} = \csc x - \sin x$
29. $\dfrac{1 + \sin \theta}{\cos \theta} + \dfrac{\cos \theta}{1 + \sin \theta} = 2\sec \theta$

30. $\dfrac{\cos\theta\cot\theta}{1-\sin\theta} - 1 = \csc\theta$

31. $\dfrac{1}{\cos x + 1} + \dfrac{1}{\cos x - 1} = -2\csc x\cot x$

32. $\cos x - \dfrac{\cos x}{1-\tan x} = \dfrac{\sin x\cos x}{\sin x - \cos x}$

33. $\tan\left(\dfrac{\pi}{2} - \theta\right)\tan\theta = 1$ **34.** $\dfrac{\cos[(\pi/2) - x]}{\sin[(\pi/2) - x]} = \tan x$

35. $\dfrac{\tan x\cot x}{\cos x} = \sec x$ **36.** $\dfrac{\csc(-x)}{\sec(-x)} = -\cot x$

37. $(1 + \sin y)[1 + \sin(-y)] = \cos^2 y$

38. $\dfrac{\tan x + \tan y}{1 - \tan x\tan y} = \dfrac{\cot x + \cot y}{\cot x\cot y - 1}$

39. $\dfrac{\tan x + \cot y}{\tan x\cot y} = \tan y + \cot x$

40. $\dfrac{\cos x - \cos y}{\sin x + \sin y} + \dfrac{\sin x - \sin y}{\cos x + \cos y} = 0$

41. $\sqrt{\dfrac{1+\sin\theta}{1-\sin\theta}} = \dfrac{1+\sin\theta}{|\cos\theta|}$

42. $\sqrt{\dfrac{1-\cos\theta}{1+\cos\theta}} = \dfrac{1-\cos\theta}{|\sin\theta|}$

43. $\cos^2\beta + \cos^2\left(\dfrac{\pi}{2} - \beta\right) = 1$

44. $\sec^2 y - \cot^2\left(\dfrac{\pi}{2} - y\right) = 1$

45. $\sin t\csc\left(\dfrac{\pi}{2} - t\right) = \tan t$

46. $\sec^2\left(\dfrac{\pi}{2} - x\right) - 1 = \cot^2 x$

In Exercises 47–54, (a) use a graphing utility to graph each side of the equation to determine whether the equation is an identity, (b) use the *table* feature of a graphing utility to determine whether the equation is an identity, and (c) confirm the results of parts (a) and (b) algebraically.

47. $(1 + \cot^2 x)(\cos^2 x) = \cot^2 x$

48. $\csc x(\csc x - \sin x) + \dfrac{\sin x - \cos x}{\sin x} + \cot x = \csc^2 x$

49. $2 + \cos^2 x - 3\cos^4 x = \sin^2 x(3 + 2\cos^2 x)$

50. $\tan^4 x + \tan^2 x - 3 = \sec^2 x(4\tan^2 x - 3)$

51. $\csc^4 x - 2\csc^2 x + 1 = \cot^4 x$

52. $(\sin^4\beta - 2\sin^2\beta + 1)\cos\beta = \cos^5\beta$

53. $\dfrac{1+\cos x}{\sin x} = \dfrac{\sin x}{1 - \cos x}$

54. $\dfrac{\cot\alpha}{\csc\alpha + 1} = \dfrac{\csc\alpha + 1}{\cot\alpha}$

In Exercises 55–58, verify the identity.

55. $\tan^5 x = \tan^3 x\sec^2 x - \tan^3 x$

56. $\sec^4 x\tan^2 x = (\tan^2 x + \tan^4 x)\sec^2 x$

57. $\cos^3 x\sin^2 x = (\sin^2 x - \sin^4 x)\cos x$

58. $\sin^4 x + \cos^4 x = 1 - 2\cos^2 x + 2\cos^4 x$

WRITING ABOUT CONCEPTS

In Exercises 59–64, explain why the equation is *not* an identity and find one value of the variable for which the equation is not true.

59. $\sin\theta = \sqrt{1 - \cos^2\theta}$ **60.** $\tan\theta = \sqrt{\sec^2\theta - 1}$

61. $\sqrt{\tan^2 x} = \tan x$

62. $\sqrt{\sin^2 x + \cos^2 x} = \sin x + \cos x$

63. $1 + \tan\theta = \sec\theta$ **64.** $\csc\theta - 1 = \cot\theta$

65. Verify that for all integers n, $\cos\left[\dfrac{(2n+1)\pi}{2}\right] = 0.$

66. Verify that for all integers n, $\sin\left[\dfrac{(12n+1)\pi}{6}\right] = \dfrac{1}{2}.$

In Exercises 67–70, use the cofunction identities to evaluate the expression without using a calculator.

67. $\sin^2 25° + \sin^2 65°$ **68.** $\cos^2 55° + \cos^2 35°$

69. $\cos^2 20° + \cos^2 52° + \cos^2 38° + \cos^2 70°$

70. $\tan^2 63° + \cot^2 16° - \sec^2 74° - \csc^2 27°$

71. ***Rate of Change*** The rate of change of the function $f(x) = \sin x + \csc x$ with respect to change in the variable x is given by the expression $\cos x - \csc x\cot x$. Show that the expression for the rate of change can also be $-\cos x\cot^2 x$.

CAPSTONE

72. Write a short paper in your own words explaining to a classmate the difference between a trigonometric identity and a conditional equation. Include suggestions on how to verify a trigonometric identity.

True or False? **In Exercises 73 and 74, determine whether the statement is true or false. Justify your answer.**

73. There can be more than one way to verify a trigonometric identity.

74. The equation $\sin^2\theta + \cos^2\theta = 1 + \tan^2\theta$ is an identity because $\sin^2(0) + \cos^2(0) = 1$ and $1 + \tan^2(0) = 1$.

10.3 Solving Trigonometric Equations

- **Use standard algebraic techniques to solve trigonometric equations.**
- **Solve trigonometric equations of quadratic type.**
- **Solve trigonometric equations involving multiple angles.**
- **Use inverse trigonometric functions to solve trigonometric equations.**

Introduction

To solve a trigonometric equation, use standard algebraic techniques such as collecting like terms and factoring. Your preliminary goal in solving a trigonometric equation is to *isolate* the trigonometric function in the equation. For example, to solve the equation $2 \sin x = 1$, divide each side by 2 to obtain $\sin x = \frac{1}{2}$. To solve for x, note in Figure 10.6 that the equation $\sin x = \frac{1}{2}$ has solutions $x = \pi/6$ and $x = 5\pi/6$ in the interval $[0, 2\pi)$. Moreover, because $\sin x$ has a period of 2π, there are infinitely many other solutions, which can be written as

$$x = \frac{\pi}{6} + 2n\pi \quad \text{and} \quad x = \frac{5\pi}{6} + 2n\pi \qquad \text{General solution}$$

where n is an integer, as shown in Figure 10.6.

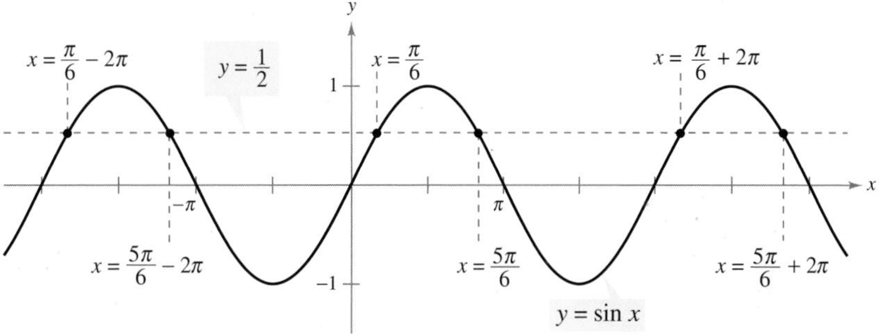

Figure 10.6

Another way to show that the equation $\sin x = \frac{1}{2}$ has infinitely many solutions is indicated in Figure 10.7. Any angles that are coterminal with $\pi/6$ or $5\pi/6$ will also be solutions of the equation.

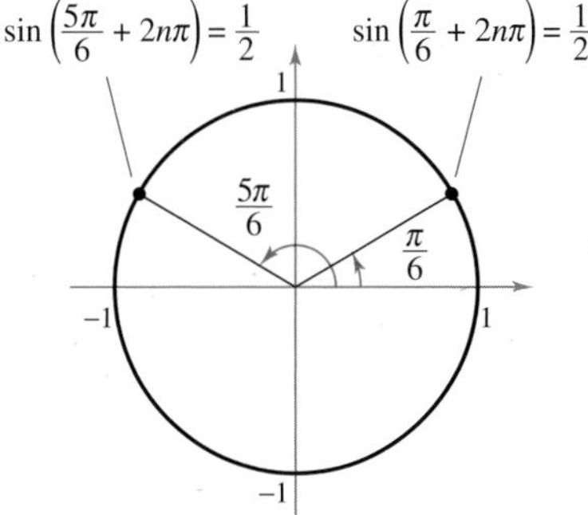

Figure 10.7

When solving trigonometric equations, you should write your answer(s) using exact values rather than decimal approximations.

EXAMPLE 1 Collecting Like Terms

Solve

$$\sin x + \sqrt{2} = -\sin x.$$

Solution Begin by rewriting the equation so that $\sin x$ is isolated on one side of the equation.

$$\sin x + \sqrt{2} = -\sin x \qquad \text{Write original equation.}$$

$$\sin x + \sin x + \sqrt{2} = 0 \qquad \text{Add } \sin x \text{ to each side.}$$

$$\sin x + \sin x = -\sqrt{2} \qquad \text{Subtract } \sqrt{2} \text{ from each side.}$$

$$2 \sin x = -\sqrt{2} \qquad \text{Combine like terms.}$$

$$\sin x = -\frac{\sqrt{2}}{2} \qquad \text{Divide each side by 2.}$$

Because $\sin x$ has a period of 2π, first find all solutions in the interval $[0, 2\pi)$. These solutions are $x = 5\pi/4$ and $x = 7\pi/4$. Finally, add multiples of 2π to each of these solutions to get the general form

$$x = \frac{5\pi}{4} + 2n\pi \quad \text{and} \quad x = \frac{7\pi}{4} + 2n\pi \qquad \text{General solution}$$

where n is an integer.

EXAMPLE 2 Extracting Square Roots

Solve

$$3 \tan^2 x - 1 = 0.$$

STUDY TIP When you extract square roots, make sure you account for both the positive and negative solutions.

Solution Begin by rewriting the equation so that $\tan x$ is isolated on one side of the equation.

$$3 \tan^2 x - 1 = 0 \qquad \text{Write original equation.}$$

$$3 \tan^2 x = 1 \qquad \text{Add 1 to each side.}$$

$$\tan^2 x = \frac{1}{3} \qquad \text{Divide each side by 3.}$$

$$\tan x = \pm\frac{1}{\sqrt{3}} \qquad \text{Extract square roots.}$$

$$= \pm\frac{\sqrt{3}}{3}$$

Because $\tan x$ has a period of π, first find all solutions in the interval $[0, \pi)$. These solutions are $x = \pi/6$ and $x = 5\pi/6$. Finally, add multiples of π to each of these solutions to get the general form

$$x = \frac{\pi}{6} + n\pi \quad \text{and} \quad x = \frac{5\pi}{6} + n\pi \qquad \text{General solution}$$

where n is an integer. ■

The equations in Examples 1 and 2 involved only one trigonometric function. When two or more functions occur in the same equation, collect all terms on one side and try to separate the functions by factoring or by using appropriate identities. This may produce factors that yield no solutions, as illustrated in Example 3.

EXAMPLE 3 Factoring

Solve

$$\cot x \cos^2 x = 2 \cot x.$$

Solution Begin by rewriting the equation so that all terms are collected on one side of the equation.

$$\cot x \cos^2 x = 2 \cot x \qquad \text{Write original equation.}$$

$$\cot x \cos^2 x - 2 \cot x = 0 \qquad \text{Subtract } 2 \cot x \text{ from each side.}$$

$$\cot x(\cos^2 x - 2) = 0 \qquad \text{Factor.}$$

By setting each of these factors equal to zero, you obtain

$$\cot x = 0 \quad \text{and} \quad \cos^2 x - 2 = 0$$

$$x = \frac{\pi}{2} \qquad \cos^2 x = 2$$

$$\cos x = \pm\sqrt{2}.$$

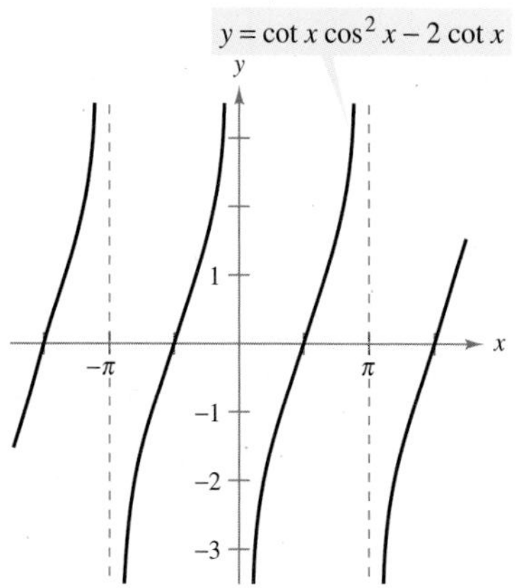

Figure 10.8

The equation $\cot x = 0$ has the solution $x = \pi/2$ [in the interval $(0, \pi)$]. No solution is obtained for $\cos x = \pm\sqrt{2}$ because $\pm\sqrt{2}$ are outside the range of the cosine function. Because $\cot x$ has a period of π, the general form of the solution is obtained by adding multiples of π to $x = \pi/2$, to get

$$x = \frac{\pi}{2} + n\pi \qquad \text{General solution}$$

where n is an integer. You can confirm this graphically by sketching the graph of $y = \cot x \cos^2 x - 2 \cot x$, as shown in Figure 10.8. From the graph you can see that the x-intercepts occur at $-3\pi/2$, $-\pi/2$, $\pi/2$, $3\pi/2$, and so on. These x-intercepts correspond to the solutions of $\cot x \cos^2 x - 2 \cot x = 0$. ■

NOTE In Example 3, don't make the mistake of dividing each side of the equation by $\cot x$. If you do this, you lose the solutions. Can you see why? ■

Equations of Quadratic Type

Many trigonometric equations are of quadratic type

$$ax^2 + bx + c = 0.$$

Here are a couple of examples.

Quadratic in sin x	*Quadratic in sec x*
$2 \sin^2 x - \sin x - 1 = 0$	$\sec^2 x - 3 \sec x - 2 = 0$
$2(\sin x)^2 - \sin x - 1 = 0$	$(\sec x)^2 - 3(\sec x) - 2 = 0$

To solve equations of this type, factor the quadratic or, if this is not possible, use the Quadratic Formula.

EXAMPLE 4 Factoring an Equation of Quadratic Type

Find all solutions of

$$2\sin^2 x - \sin x - 1 = 0$$

in the interval $[0, 2\pi)$.

Algebraic Solution

Begin by treating the equation as a quadratic in $\sin x$ and factoring.

$$2\sin^2 x - \sin x - 1 = 0 \qquad \text{Write original equation.}$$

$$(2\sin x + 1)(\sin x - 1) = 0 \qquad \text{Factor.}$$

Setting each factor equal to zero, you obtain the following solutions in the interval $[0, 2\pi)$.

$$2\sin x + 1 = 0 \qquad \text{and} \qquad \sin x - 1 = 0$$

$$\sin x = -\frac{1}{2} \qquad \sin x = 1$$

$$x = \frac{7\pi}{6}, \frac{11\pi}{6} \qquad x = \frac{\pi}{2}$$

Graphical Solution

Use a graphing utility set in *radian* mode to graph $y = 2\sin^2 x - \sin x - 1$ for $0 \le x < 2\pi$, as shown in Figure 10.9. Use the *zero* or *root* feature or the *zoom* and *trace* features to approximate the x-intercepts to be

$$x \approx 1.571 \approx \frac{\pi}{2}, \quad x \approx 3.665 \approx \frac{7\pi}{6}, \quad \text{and} \quad x \approx 5.760 \approx \frac{11\pi}{6}.$$

These values are the approximate solutions of $2\sin^2 x - \sin x - 1 = 0$ in the interval $[0, 2\pi)$.

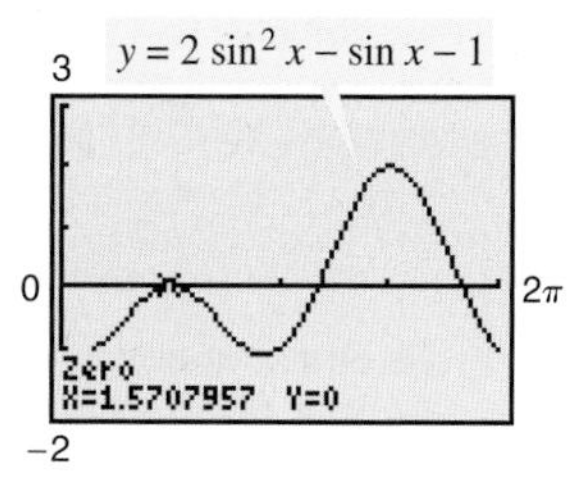

Figure 10.9

EXAMPLE 5 Rewriting with a Single Trigonometric Function

Solve

$$2\sin^2 x + 3\cos x - 3 = 0.$$

Solution This equation contains both sine and cosine functions. You can rewrite the equation so that it has only cosine functions by using the identity $\sin^2 x = 1 - \cos^2 x$.

$$2\sin^2 x + 3\cos x - 3 = 0 \qquad \text{Write original equation.}$$

$$2(1 - \cos^2 x) + 3\cos x - 3 = 0 \qquad \text{Pythagorean identity}$$

$$2\cos^2 x - 3\cos x + 1 = 0 \qquad \text{Multiply each side by } -1.$$

$$(2\cos x - 1)(\cos x - 1) = 0 \qquad \text{Factor.}$$

Set each factor equal to zero to find the solutions in the interval $[0, 2\pi)$.

$$2\cos x - 1 = 0 \qquad \text{and} \qquad \cos x - 1 = 0$$

$$\cos x = \frac{1}{2} \qquad \cos x = 1$$

$$x = \frac{\pi}{3}, \frac{5\pi}{3} \qquad x = 0$$

Because $\cos x$ has a period of 2π, the general form of the solution is obtained by adding multiples of 2π to get

$$x = 2n\pi, \quad x = \frac{\pi}{3} + 2n\pi, \quad x = \frac{5\pi}{3} + 2n\pi \qquad \text{General solution}$$

where n is an integer.

STUDY TIP In Example 5, conversion to cosine was chosen because the identity relating sine and cosine

$$\sin^2\theta + \cos^2\theta = 1$$

involves their squares. When using the Pythagorean identities to convert equations to one function, keep in mind their function pairs and powers.

Sometimes you must square each side of an equation to obtain a quadratic, as demonstrated in the next example. Because this procedure can introduce extraneous solutions, you should check any solutions in the original equation to see whether they are valid or extraneous.

STUDY TIP You square each side of the equation in Example 6 because the squares of the sine and cosine functions are related by a Pythagorean identity. The same is true for the squares of the secant and tangent functions and for the squares of the cosecant and cotangent functions.

EXAMPLE 6 Squaring and Converting to Quadratic Type

Find all solutions of

$$\cos x + 1 = \sin x$$

in the interval $[0, 2\pi)$.

Solution It is not clear how to rewrite this equation in terms of a single trigonometric function. Notice what happens when you square each side of the equation.

$$\cos x + 1 = \sin x \quad \text{Write original equation.}$$

$$\cos^2 x + 2\cos x + 1 = \sin^2 x \quad \text{Square each side.}$$

$$\cos^2 x + 2\cos x + 1 = 1 - \cos^2 x \quad \text{Pythagorean identity}$$

$$\cos^2 x + \cos^2 x + 2\cos x + 1 - 1 = 0 \quad \text{Rewrite equation.}$$

$$2\cos^2 x + 2\cos x = 0 \quad \text{Combine like terms.}$$

$$2\cos x(\cos x + 1) = 0 \quad \text{Factor.}$$

Setting each factor equal to zero produces

$$2\cos x = 0 \qquad \text{and} \qquad \cos x + 1 = 0$$

$$\cos x = 0 \qquad\qquad \cos x = -1$$

$$x = \frac{\pi}{2}, \frac{3\pi}{2} \qquad\qquad x = \pi.$$

Because you squared the original equation, check for extraneous solutions.

Check $x = \pi/2$

$$\cos\frac{\pi}{2} + 1 \stackrel{?}{=} \sin\frac{\pi}{2} \quad \text{Substitute } \pi/2 \text{ for } x.$$

$$0 + 1 = 1 \quad \text{Solution checks. } \checkmark$$

Check $x = 3\pi/2$

$$\cos\frac{3\pi}{2} + 1 \stackrel{?}{=} \sin\frac{3\pi}{2} \quad \text{Substitute } 3\pi/2 \text{ for } x.$$

$$0 + 1 \neq -1 \quad \text{Solution does not check.}$$

Check $x = \pi$

$$\cos\pi + 1 \stackrel{?}{=} \sin\pi \quad \text{Substitute } \pi \text{ for } x.$$

$$-1 + 1 = 0 \quad \text{Solution checks. } \checkmark$$

Of the three possible solutions, $x = 3\pi/2$ is extraneous. So, in the interval $[0, 2\pi)$, the only two solutions are $x = \pi/2$ and $x = \pi$. ■

NOTE In Example 6, the general solution is

$$x = \frac{\pi}{2} + 2n\pi \quad \text{and} \quad x = \pi + 2n\pi$$

where n is an integer. ■

Functions Involving Multiple Angles

The next two examples involve trigonometric functions of multiple angles of the forms $\cos ku$ and $\tan ku$. To solve equations of these forms, first solve the equation for ku, then divide your result by k.

EXAMPLE 7 Functions of Multiple Angles

Solve

$$2 \cos 3t - 1 = 0.$$

Solution

$$2 \cos 3t - 1 = 0 \qquad \text{Write original equation.}$$

$$2 \cos 3t = 1 \qquad \text{Add 1 to each side.}$$

$$\cos 3t = \frac{1}{2} \qquad \text{Divide each side by 2.}$$

In the interval $[0, 2\pi)$, you know that $3t = \pi/3$ and $3t = 5\pi/3$ are the only solutions, so, in general, you have

$$3t = \frac{\pi}{3} + 2n\pi \quad \text{and} \quad 3t = \frac{5\pi}{3} + 2n\pi.$$

Dividing these results by 3, you obtain the general solution

$$t = \frac{\pi}{9} + \frac{2n\pi}{3} \quad \text{and} \quad t = \frac{5\pi}{9} + \frac{2n\pi}{3} \qquad \text{General solution}$$

where n is an integer.

NOTE Two different intervals, $[0, 2\pi)$ and $[0, \pi)$, that correspond to the periods of the cosine and tangent functions are considered in Examples 7 and 8, respectively.

EXAMPLE 8 Functions of Multiple Angles

Solve $3 \tan \frac{x}{2} + 3 = 0.$

Solution

$$3 \tan \frac{x}{2} + 3 = 0 \qquad \text{Write original equation.}$$

$$3 \tan \frac{x}{2} = -3 \qquad \text{Subtract 3 from each side.}$$

$$\tan \frac{x}{2} = -1 \qquad \text{Divide each side by 3.}$$

In the interval $[0, \pi)$, you know that $x/2 = 3\pi/4$ is the only solution, so, in general, you have

$$\frac{x}{2} = \frac{3\pi}{4} + n\pi.$$

Multiplying this result by 2, you obtain the general solution

$$x = \frac{3\pi}{2} + 2n\pi \qquad \text{General solution}$$

where n is an integer. ■

Using Inverse Functions

In the next example, you will see how inverse trigonometric functions can be used to solve an equation.

EXAMPLE 9 Using Inverse Functions

Solve $\sec^2 x - 2\tan x = 4$.

Solution

$$\sec^2 x - 2\tan x = 4 \quad \text{Write original equation.}$$
$$1 + \tan^2 x - 2\tan x - 4 = 0 \quad \text{Pythagorean identity}$$
$$\tan^2 x - 2\tan x - 3 = 0 \quad \text{Combine like terms.}$$
$$(\tan x - 3)(\tan x + 1) = 0 \quad \text{Factor.}$$

Setting each factor equal to zero, you obtain two solutions in the interval $(-\pi/2, \pi/2)$. [Recall that the range of the inverse tangent function is $(-\pi/2, \pi/2)$.]

$$\tan x - 3 = 0 \quad \text{and} \quad \tan x + 1 = 0$$
$$\tan x = 3 \qquad \tan x = -1$$
$$x = \arctan 3 \qquad x = -\frac{\pi}{4}$$

Finally, because $\tan x$ has a period of π, you obtain the general solution by adding multiples of π

$$x = \arctan 3 + n\pi \quad \text{and} \quad x = -\frac{\pi}{4} + n\pi \quad \text{General solution}$$

where n is an integer. You can use a calculator to approximate the value of arctan 3. ■

10.3 Exercises

See www.CalcChat.com for worked-out solutions to odd-numbered exercises.

In Exercises 1–4, fill in the blanks.

1. When solving a trigonometric equation, the preliminary goal is to ________ the trigonometric function involved in the equation.
2. The equation $2\sin\theta + 1 = 0$ has the solutions $\theta = \frac{7\pi}{6} + 2n\pi$ and $\theta = \frac{11\pi}{6} + 2n\pi$, which are called ________ solutions.
3. The equation $2\tan^2 x - 3\tan x + 1 = 0$ is a trigonometric equation that is of ________ type.
4. A solution of an equation that does not satisfy the original equation is called an ________ solution.

In Exercises 5–10, verify that the x-values are solutions of the equation.

5. $2\cos x - 1 = 0$
   (a) $x = \frac{\pi}{3}$ (b) $x = \frac{5\pi}{3}$
6. $\sec x - 2 = 0$
   (a) $x = \frac{\pi}{3}$ (b) $x = \frac{5\pi}{3}$
7. $3\tan^2 2x - 1 = 0$
   (a) $x = \frac{\pi}{12}$ (b) $x = \frac{5\pi}{12}$
8. $2\cos^2 4x - 1 = 0$
   (a) $x = \frac{\pi}{16}$ (b) $x = \frac{3\pi}{16}$
9. $2\sin^2 x - \sin x - 1 = 0$
   (a) $x = \frac{\pi}{2}$ (b) $x = \frac{7\pi}{6}$
10. $\csc^4 x - 4\csc^2 x = 0$
    (a) $x = \frac{\pi}{6}$ (b) $x = \frac{5\pi}{6}$

In Exercises 11–20, solve the equation.

11. $2\cos x + 1 = 0$

12. $2\sin x + 1 = 0$

13. $\sqrt{3}\csc x - 2 = 0$

14. $\tan x + \sqrt{3} = 0$

15. $3\sec^2 x - 4 = 0$

16. $3\cot^2 x - 1 = 0$

17. $\sin x(\sin x + 1) = 0$

18. $(3\tan^2 x - 1)(\tan^2 x - 3) = 0$

19. $4\cos^2 x - 1 = 0$

20. $\sin^2 x = 3\cos^2 x$

In Exercises 21–32, find all solutions of the equation in the interval $[0, 2\pi)$.

21. $\cos^3 x = \cos x$

22. $\sec^2 x - 1 = 0$

23. $3\tan^3 x = \tan x$

24. $2\sin^2 x = 2 + \cos x$

25. $\sec^2 x - \sec x = 2$

26. $\sec x \csc x = 2\csc x$

27. $2\sin x + \csc x = 0$

28. $\sec x + \tan x = 1$

29. $2\cos^2 x + \cos x - 1 = 0$

30. $2\sin^2 x + 3\sin x + 1 = 0$

31. $2\sec^2 x + \tan^2 x - 3 = 0$

32. $\cos x + \sin x \tan x = 2$

In Exercises 33–40, solve the multiple-angle equation.

33. $\cos 2x = \dfrac{1}{2}$

34. $\sin 2x = -\dfrac{\sqrt{3}}{2}$

35. $\tan 3x = 1$

36. $\sec 4x = 2$

37. $\cos \dfrac{x}{2} = \dfrac{\sqrt{2}}{2}$

38. $\sin \dfrac{x}{2} = -\dfrac{\sqrt{3}}{2}$

39. $2\sin^2 2x = 1$

40. $\tan^2 3x = 3$

In Exercises 41–44, find the x-intercepts of the graph.

41. $y = \sin \dfrac{\pi x}{2} + 1$

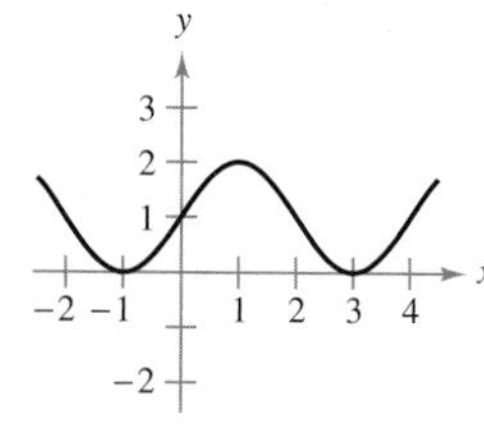

42. $y = \sin \pi x + \cos \pi x$

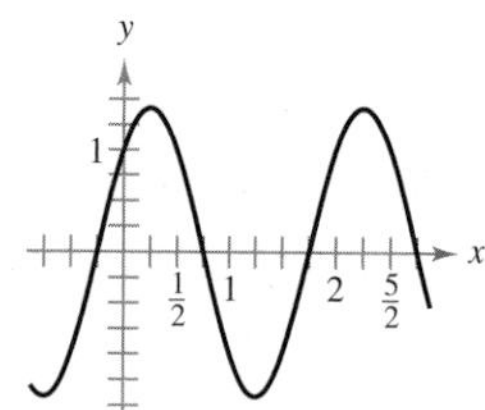

43. $y = \tan^2\left(\dfrac{\pi x}{6}\right) - 3$

44. $y = \sec^4\left(\dfrac{\pi x}{8}\right) - 4$

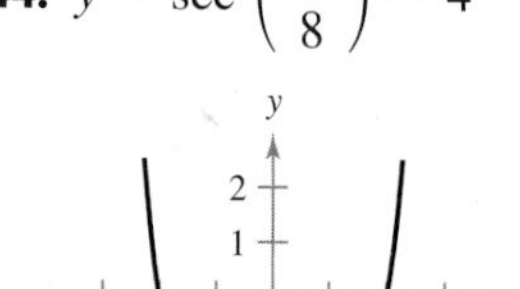

WRITING ABOUT CONCEPTS

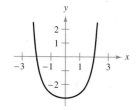

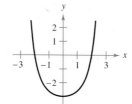

In Exercises 45 and 46, solve both equations. How do the solutions of the algebraic equation compare with the solutions of the trigonometric equation?

45. $6y^2 - 13y + 6 = 0$

$6\cos^2 x - 13\cos x + 6 = 0$

46. $y^2 + y - 20 = 0$

$\sin^2 x + \sin x - 20 = 0$

In Exercises 47–56, use a graphing utility to approximate the solutions (to three decimal places) of the equation in the interval $[0, 2\pi)$.

47. $2\sin x + \cos x = 0$

48. $4\sin^3 x + 2\sin^2 x - 2\sin x - 1 = 0$

49. $\dfrac{1 + \sin x}{\cos x} + \dfrac{\cos x}{1 + \sin x} = 4$

50. $\dfrac{\cos x \cot x}{1 - \sin x} = 3$

51. $x\tan x - 1 = 0$

52. $x\cos x - 1 = 0$

53. $\sec^2 x + 0.5\tan x - 1 = 0$

54. $\csc^2 x + 0.5\cot x - 5 = 0$

55. $2\tan^2 x + 7\tan x - 15 = 0$

56. $6\sin^2 x - 7\sin x + 2 = 0$

In Exercises 57–60, use the Quadratic Formula to solve the equation in the interval $[0, 2\pi)$. Then use a graphing utility to approximate the angle x.

57. $12\sin^2 x - 13\sin x + 3 = 0$

58. $3\tan^2 x + 4\tan x - 4 = 0$

59. $\tan^2 x + 3\tan x + 1 = 0$

60. $4\cos^2 x - 4\cos x - 1 = 0$

In Exercises 61–64, use inverse functions where needed to find all solutions of the equation in the interval $[0, 2\pi)$.

61. $\tan^2 x - 6\tan x + 5 = 0$

62. $\sec^2 x + \tan x - 3 = 0$

63. $2\cos^2 x - 5\cos x + 2 = 0$

64. $2\sin^2 x - 7\sin x + 3 = 0$

In Exercises 65–68, (a) use a graphing utility to graph the function and approximate the maximum and minimum points on the graph in the interval $[0, 2\pi)$, and (b) solve the trigonometric equation and demonstrate that its solutions are the x-coordinates of the maximum and minimum points of f.

	Function	*Trigonometric Equation*
65.	$f(x) = \sin^2 x + \cos x$	$2 \sin x \cos x - \sin x = 0$
66.	$f(x) = \cos^2 x - \sin x$	$-2 \sin x \cos x - \cos x = 0$
67.	$f(x) = \sin x + \cos x$	$\cos x - \sin x = 0$
68.	$f(x) = 2 \sin x + \cos 2x$	$2 \cos x - 4 \sin x \cos x = 0$

69. ***Graphical Reasoning*** Consider the function given by

$$f(x) = \cos \frac{1}{x}$$

and its graph shown in the figure.

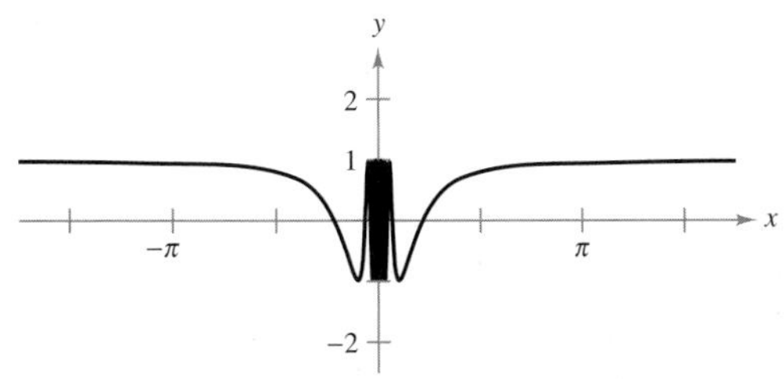

(a) What is the domain of the function?

(b) Identify any symmetry and any asymptotes of the graph.

(c) Describe the behavior of the function as $x \to 0$.

(d) How many solutions does the equation

$$\cos \frac{1}{x} = 0$$

have in the interval $[-1, 1]$? Find the solutions.

(e) Does the equation $\cos(1/x) = 0$ have a greatest solution? If so, approximate the solution. If not, explain why.

70. ***Graphical Reasoning*** Consider the function given by $f(x) = (\sin x)/x$ and its graph shown in the figure.

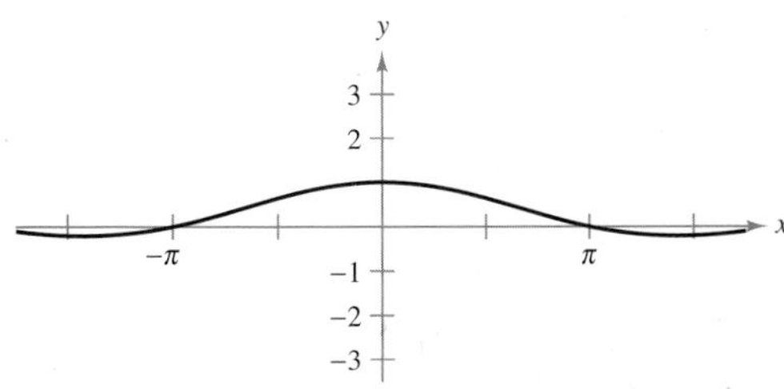

(a) What is the domain of the function?

(b) Identify any symmetry and any asymptotes of the graph.

(c) Describe the behavior of the function as $x \to 0$.

(d) How many solutions does the equation

$$\frac{\sin x}{x} = 0$$

have in the interval $[-8, 8]$? Find the solutions.

71. ***Harmonic Motion*** A weight is oscillating on the end of a spring. The position of the weight relative to the point of equilibrium is given by

$$y = \tfrac{1}{12}(\cos 8t - 3 \sin 8t)$$

where y is the displacement (in meters) and t is the time (in seconds). Find the times when the weight is at the point of equilibrium ($y = 0$) for $0 \le t \le 1$.

72. ***Damped Harmonic Motion*** The displacement from equilibrium of a weight oscillating on the end of a spring is given by $y = 1.56e^{-0.22t} \cos 4.9t$, where y is the displacement (in feet) and t is the time (in seconds). Use a graphing utility to graph the displacement function for $0 \le t \le 10$. Find the time beyond which the displacement does not exceed 1 foot from equilibrium.

73. ***Sales*** The monthly sales S (in thousands of units) of a seasonal product are approximated by

$$S = 74.50 + 43.75 \sin \frac{\pi t}{6}$$

where t is the time (in months), with $t = 1$ corresponding to January. Determine the months in which sales exceed 100,000 units.

74. ***Projectile Motion*** A batted baseball leaves the bat at an angle of θ with the horizontal and an initial velocity of $v_0 = 100$ feet per second. The ball is caught by an outfielder 300 feet from home plate (see figure). Find θ if the range r of a projectile is given by $r = \frac{1}{32} v_0^2 \sin 2\theta$.

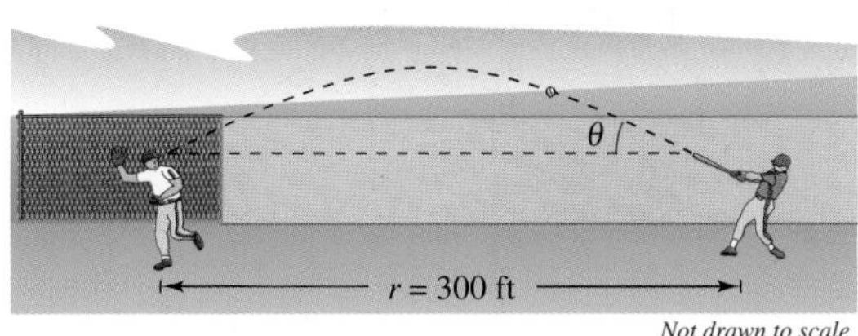

Not drawn to scale

75. ***Projectile Motion*** A sharpshooter intends to hit a target at a distance of 1000 yards with a gun that has a muzzle velocity of 1200 feet per second (see figure). Neglecting air resistance, determine the gun's minimum angle of elevation θ if the range r is given by

$$r = \frac{1}{32} v_0^2 \sin 2\theta.$$

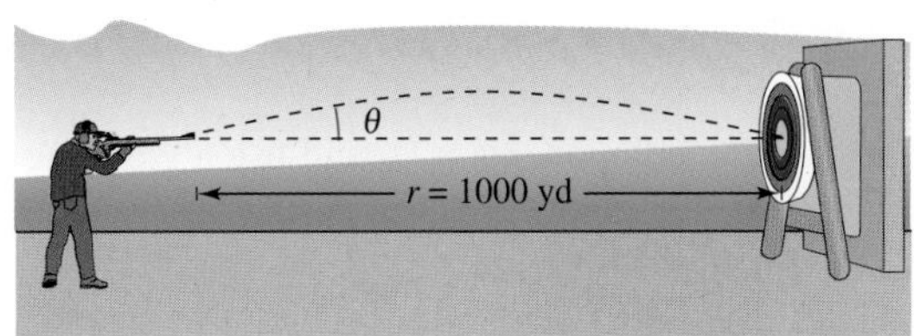

Not drawn to scale

76. ***Data Analysis: Meteorology*** The table shows the average daily high temperatures in Houston H (in degrees Fahrenheit) for month t, with $t = 1$ corresponding to January. *(Source: National Climatic Data Center)*

t	1	2	3	4	5	6
H	62.3	66.5	73.3	79.1	85.5	90.7

t	7	8	9	10	11	12
H	93.6	93.5	89.3	82.0	72.0	64.6

(a) Create a scatter plot of the data.

(b) Find a cosine model for the temperatures in Houston.

(c) Use a graphing utility to graph the data points and the model for the temperatures in Houston. How well does the model fit the data?

(d) What is the overall average daily high temperature in Houston?

(e) Use a graphing utility to describe the months during which the average daily high temperature is above 86°F and below 86°F.

77. ***Geometry*** The area of a rectangle (see figure) inscribed in one arc of the graph of $y = \cos x$ is given by $A = 2x \cos x$, $0 < x < \pi/2$.

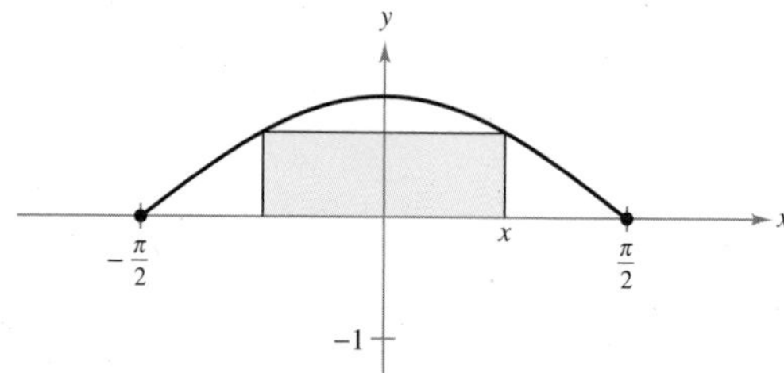

(a) Use a graphing utility to graph the area function, and approximate the area of the largest inscribed rectangle.

(b) Determine the values of x for which $A \geq 1$.

CAPSTONE

78. Consider the equation $2 \sin x - 1 = 0$. Explain the similarities and differences between finding all solutions in the interval $\left[0, \frac{\pi}{2}\right)$, finding all solutions in the interval $[0, 2\pi)$, and finding the general solution.

True or False? **In Exercises 79 and 80, determine whether the statement is true or false. Justify your answer.**

79. The equation $2 \sin 4t - 1 = 0$ has four times the number of solutions in the interval $[0, 2\pi)$ as the equation $2 \sin t - 1 = 0$.

80. If you correctly solve a trigonometric equation to the statement $\sin x = 3.4$, then you can finish solving the equation by using an inverse function.

81. ***Think About It*** Explain what would happen if you divided each side of the equation $\cot x \cos^2 x = 2 \cot x$ by $\cot x$. Is this a correct method to use when solving equations?

82. ***Graphical Reasoning*** Use a graphing utility to confirm the solutions found in Example 6 in two different ways.

(a) Graph both sides of the equation and find the x-coordinates of the points at which the graphs intersect.

Left side: $y = \cos x + 1$ *Right side:* $y = \sin x$

(b) Graph the equation $y = \cos x + 1 - \sin x$ and find the x-intercepts of the graph. Do both methods produce the same x-values? Which method do you prefer? Explain.

SECTION PROJECT

Modeling a Sound Wave

A particular sound wave is modeled by

$$p(t) = \frac{1}{4\pi}(p_1(t) + 30p_2(t) + p_3(t) + p_5(t) + 30p_6(t))$$

where $p_n(t) = \frac{1}{n}\sin(524\,n\pi t)$, and t is the time in seconds.

(a) Find the sine components $p_n(t)$ and use a graphing utility to graph each component. Then verify the graph of p that is shown at the right.

(b) Find the period of each sine component of p. Is p periodic? If so, what is its period?

(c) Use the *zero* or *root* feature or the *zoom* and *trace* features to find the t-intercepts of the graph of p over one cycle.

(d) Use the *maximum* and *minimum* features of a graphing utility to approximate the absolute maximum and absolute minimum values of p over one cycle.

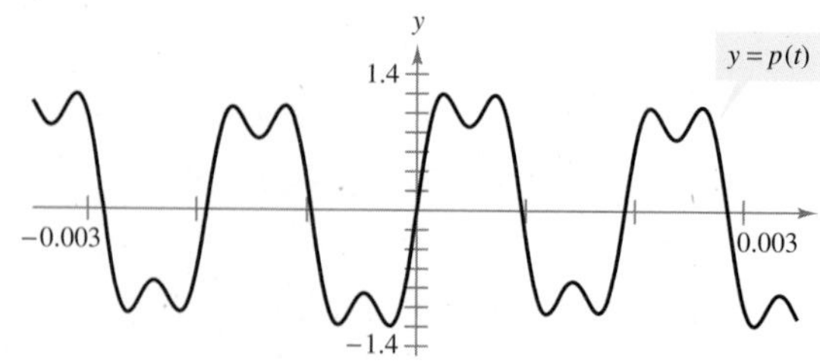

10.4 Sum and Difference Formulas

■ **Use sum and difference formulas to evaluate trigonometric functions, verify identities, and solve trigonometric equations.**

Using Sum and Difference Formulas

In this and the following section, you will study the uses of several trigonometric identities and formulas. (Proofs of these formulas are given in Appendix A.)

SUM AND DIFFERENCE FORMULAS

$$\sin(u + v) = \sin u \cos v + \cos u \sin v$$

$$\sin(u - v) = \sin u \cos v - \cos u \sin v$$

$$\cos(u + v) = \cos u \cos v - \sin u \sin v$$

$$\cos(u - v) = \cos u \cos v + \sin u \sin v$$

$$\tan(u + v) = \frac{\tan u + \tan v}{1 - \tan u \tan v} \qquad \tan(u - v) = \frac{\tan u - \tan v}{1 + \tan u \tan v}$$

NOTE Note that $\sin(u + v) \neq \sin u + \sin v$. Similar statements can be made for $\cos(u + v)$ and $\tan(u + v)$. ■

Examples 1 and 2 show how **sum and difference formulas** can be used to find exact values of trigonometric functions involving sums or differences of special angles.

EXAMPLE 1 Evaluating a Trigonometric Function

Find the exact value of $\sin \frac{\pi}{12}$.

Solution To find the *exact* value of $\sin \frac{\pi}{12}$, use the fact that

$$\frac{\pi}{12} = \frac{\pi}{3} - \frac{\pi}{4}.$$

Consequently, the formula for $\sin(u - v)$ yields

$$\begin{aligned}\sin \frac{\pi}{12} &= \sin\left(\frac{\pi}{3} - \frac{\pi}{4}\right)\\ &= \sin \frac{\pi}{3} \cos \frac{\pi}{4} - \cos \frac{\pi}{3} \sin \frac{\pi}{4}\\ &= \frac{\sqrt{3}}{2}\left(\frac{\sqrt{2}}{2}\right) - \frac{1}{2}\left(\frac{\sqrt{2}}{2}\right)\\ &= \frac{\sqrt{6} - \sqrt{2}}{4}.\end{aligned}$$

Try checking this result on your calculator. You will find that $\sin \frac{\pi}{12} \approx 0.259$. ■

STUDY TIP Another way to solve Example 2 is to use the fact that $75° = 120° - 45°$ together with the formula for $\cos(u - v)$.

EXAMPLE 2 Evaluating a Trigonometric Function

Find the exact value of $\cos 75°$.

Solution Using the fact that $75° = 30° + 45°$, together with the formula for $\cos(u + v)$, you obtain

$$\begin{aligned}\cos 75° &= \cos(30° + 45°)\\ &= \cos 30° \cos 45° - \sin 30° \sin 45°\\ &= \frac{\sqrt{3}}{2}\left(\frac{\sqrt{2}}{2}\right) - \frac{1}{2}\left(\frac{\sqrt{2}}{2}\right)\\ &= \frac{\sqrt{6} - \sqrt{2}}{4}.\end{aligned}$$

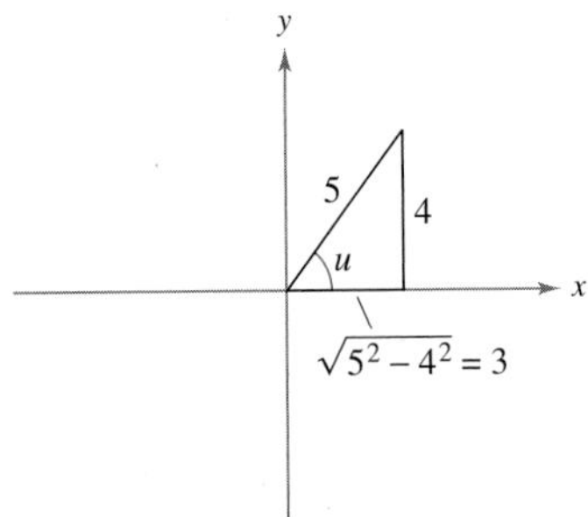

Figure 10.10

EXAMPLE 3 Evaluating a Trigonometric Expression

Find the exact value of $\sin(u + v)$ given

$$\sin u = \frac{4}{5}, \text{ where } 0 < u < \frac{\pi}{2}, \quad \text{and} \quad \cos v = -\frac{12}{13}, \text{ where } \frac{\pi}{2} < v < \pi.$$

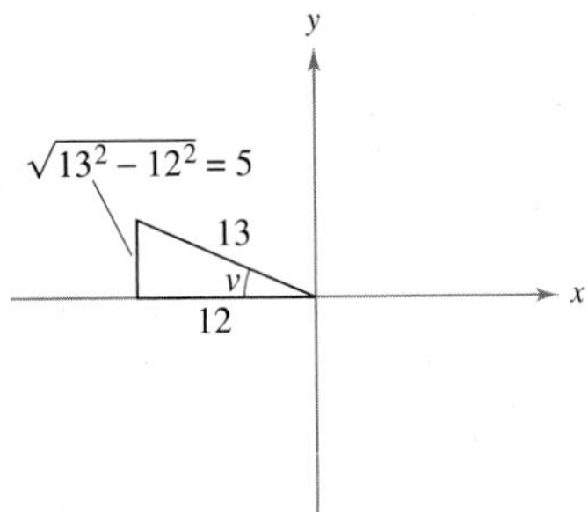

Figure 10.11

Solution Because $\sin u = 4/5$ and u is in Quadrant I, $\cos u = 3/5$, as shown in Figure 10.10. Because $\cos v = -12/13$ and v is in Quadrant II, $\sin v = 5/13$, as shown in Figure 10.11. You can find $\sin(u + v)$ as follows.

$$\begin{aligned}\sin(u + v) &= \sin u \cos v + \cos u \sin v\\ &= \left(\frac{4}{5}\right)\left(-\frac{12}{13}\right) + \left(\frac{3}{5}\right)\left(\frac{5}{13}\right)\\ &= -\frac{48}{65} + \frac{15}{65}\\ &= -\frac{33}{65}\end{aligned}$$

EXAMPLE 4 An Application of a Sum Formula

Write

$$\cos(\arctan 1 + \arccos x)$$

as an algebraic expression.

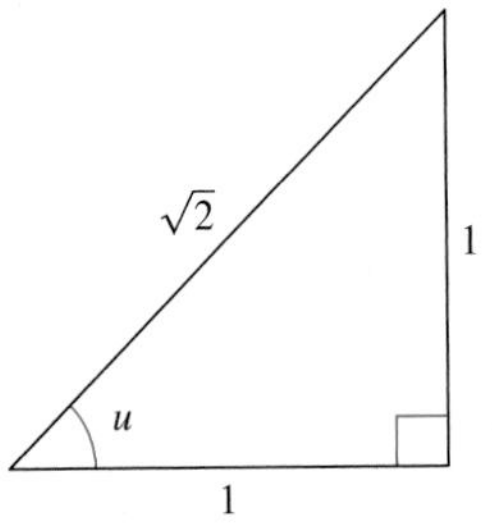

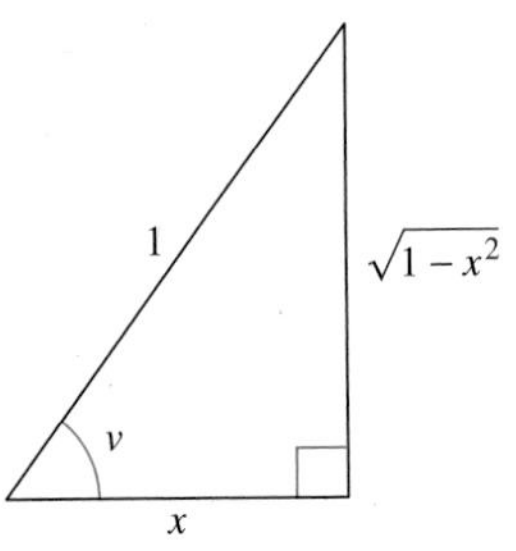

Figure 10.12

Solution This expression fits the formula for $\cos(u + v)$. Angles $u = \arctan 1$ and $v = \arccos x$ are shown in Figure 10.12. So

$$\begin{aligned}\cos(u + v) &= \cos(\arctan 1) \cos(\arccos x) - \sin(\arctan 1) \sin(\arccos x)\\ &= \frac{1}{\sqrt{2}} \cdot x - \frac{1}{\sqrt{2}} \cdot \sqrt{1 - x^2}\\ &= \frac{x - \sqrt{1 - x^2}}{\sqrt{2}}.\end{aligned}$$

■

NOTE In Example 4, you can test the reasonableness of your solution by evaluating both expressions for particular values of x. Try doing this for $x = 0$. ■

The Granger Collection, New York

HIPPARCHUS

Hipparchus, considered the most eminent of Greek astronomers, was born about 190 B.C. in Nicaea. He was credited with the invention of trigonometry. He also derived the sum and difference formulas for $\sin(A \pm B)$ and $\cos(A \pm B)$.

Example 5 shows how to use a difference formula to prove the cofunction identity $\cos\left(\frac{\pi}{2} - x\right) = \sin x$.

EXAMPLE 5 Proving a Cofunction Identity

Prove the cofunction identity

$$\cos\left(\frac{\pi}{2} - x\right) = \sin x.$$

Solution Using the formula for $\cos(u - v)$, you have

$$\begin{aligned}\cos\left(\frac{\pi}{2} - x\right) &= \cos\frac{\pi}{2}\cos x + \sin\frac{\pi}{2}\sin x\\ &= (0)(\cos x) + (1)(\sin x)\\ &= \sin x.\end{aligned}$$

■

Sum and difference formulas can be used to rewrite expressions such as

$$\sin\left(\theta + \frac{n\pi}{2}\right) \quad \text{and} \quad \cos\left(\theta + \frac{n\pi}{2}\right), \quad \text{where } n \text{ is an integer}$$

as expressions involving only $\sin\theta$ or $\cos\theta$. The resulting formulas are called **reduction formulas.**

EXAMPLE 6 Deriving Reduction Formulas

Simplify each expression.

a. $\cos\left(\theta - \frac{3\pi}{2}\right)$

b. $\tan(\theta + 3\pi)$

Solution

a. Using the formula for $\cos(u - v)$, you have

$$\begin{aligned}\cos\left(\theta - \frac{3\pi}{2}\right) &= \cos\theta\cos\frac{3\pi}{2} + \sin\theta\sin\frac{3\pi}{2}\\ &= (\cos\theta)(0) + (\sin\theta)(-1)\\ &= -\sin\theta.\end{aligned}$$

b. Using the formula for $\tan(u + v)$, you have

$$\begin{aligned}\tan(\theta + 3\pi) &= \frac{\tan\theta + \tan 3\pi}{1 - \tan\theta\tan 3\pi}\\ &= \frac{\tan\theta + 0}{1 - (\tan\theta)(0)}\\ &= \tan\theta.\end{aligned}$$

■

EXAMPLE 7 Solving a Trigonometric Equation

Find all solutions of $\sin\left(x+\frac{\pi}{4}\right)+\sin\left(x-\frac{\pi}{4}\right)=-1$ in the interval $[0, 2\pi)$.

Algebraic Solution

Using sum and difference formulas, rewrite the equation as

$$\sin x\cos\frac{\pi}{4}+\cos x\sin\frac{\pi}{4}+\sin x\cos\frac{\pi}{4}-\cos x\sin\frac{\pi}{4}=-1$$

$$2\sin x\cos\frac{\pi}{4}=-1$$

$$2(\sin x)\left(\frac{\sqrt{2}}{2}\right)=-1$$

$$\sin x=-\frac{1}{\sqrt{2}}$$

$$\sin x=-\frac{\sqrt{2}}{2}.$$

So, the only solutions in the interval $[0, 2\pi)$ are

$$x=\frac{5\pi}{4}\quad\text{and}\quad x=\frac{7\pi}{4}.$$

Graphical Solution

Sketch the graph of

$$y=\sin\left(x+\frac{\pi}{4}\right)+\sin\left(x-\frac{\pi}{4}\right)+1\text{ for }0\le x<2\pi$$

as shown in Figure 10.13. From the graph you can see that the x-intercepts are $5\pi/4$ and $7\pi/4$. So, the solutions in the interval $[0, 2\pi)$ are

$$x=\frac{5\pi}{4}\quad\text{and}\quad x=\frac{7\pi}{4}.$$

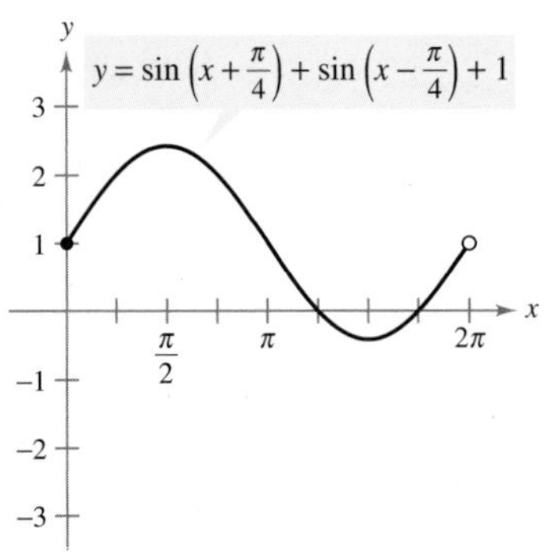

Figure 10.13 ■

The next example will be used to derive the *derivative* of the sine function. (See Section 11.2.)

EXAMPLE 8 An Identity Used to Find a Derivative

Given that $\Delta x \neq 0$, verify that

$$\frac{\sin(x+\Delta x)-\sin x}{\Delta x}=(\cos x)\left(\frac{\sin\Delta x}{\Delta x}\right)-(\sin x)\left(\frac{1-\cos\Delta x}{\Delta x}\right).$$

Solution Using the formula for $\sin(u+v)$, you have

$$\frac{\sin(x+\Delta x)-\sin x}{\Delta x}=\frac{\sin x\cos\Delta x+\cos x\sin\Delta x-\sin x}{\Delta x}$$

$$=\frac{\cos x\sin\Delta x-\sin x(1-\cos\Delta x)}{\Delta x}$$

$$=(\cos x)\left(\frac{\sin\Delta x}{\Delta x}\right)-(\sin x)\left(\frac{1-\cos\Delta x}{\Delta x}\right).$$ ■

10.4 Exercises

See www.CalcChat.com for worked-out solutions to odd-numbered exercises.

In Exercises 1–6, fill in the blank.

1. $\sin(u - v) =$ ________ **2.** $\cos(u + v) =$ ________

3. $\tan(u + v) =$ ________ **4.** $\sin(u + v) =$ ________

5. $\cos(u - v) =$ ________ **6.** $\tan(u - v) =$ ________

In Exercises 7–12, find the exact value of each expression.

7. (a) $\cos\left(\frac{\pi}{4} + \frac{\pi}{3}\right)$ (b) $\cos\frac{\pi}{4} + \cos\frac{\pi}{3}$

8. (a) $\sin\left(\frac{3\pi}{4} + \frac{5\pi}{6}\right)$ (b) $\sin\frac{3\pi}{4} + \sin\frac{5\pi}{6}$

9. (a) $\sin\left(\frac{7\pi}{6} - \frac{\pi}{3}\right)$ (b) $\sin\frac{7\pi}{6} - \sin\frac{\pi}{3}$

10. (a) $\cos(120° + 45°)$ (b) $\cos 120° + \cos 45°$

11. (a) $\sin(135° - 30°)$ (b) $\sin 135° - \cos 30°$

12. (a) $\sin(315° - 60°)$ (b) $\sin 315° - \sin 60°$

In Exercises 13–28, find the exact values of the sine, cosine, and tangent of the angle.

13. $\frac{11\pi}{12} = \frac{3\pi}{4} + \frac{\pi}{6}$ **14.** $\frac{7\pi}{12} = \frac{\pi}{3} + \frac{\pi}{4}$

15. $\frac{17\pi}{12} = \frac{9\pi}{4} - \frac{5\pi}{6}$ **16.** $-\frac{\pi}{12} = \frac{\pi}{6} - \frac{\pi}{4}$

17. $105° = 60° + 45°$

18. $165° = 135° + 30°$

19. $195° = 225° - 30°$

20. $255° = 300° - 45°$

21. $\frac{13\pi}{12}$ **22.** $-\frac{7\pi}{12}$

23. $-\frac{13\pi}{12}$ **24.** $\frac{5\pi}{12}$

25. $285°$ **26.** $-105°$

27. $-165°$ **28.** $15°$

In Exercises 29–36, write the expression as the sine, cosine, or tangent of an angle.

29. $\sin 3 \cos 1.2 - \cos 3 \sin 1.2$

30. $\cos\frac{\pi}{7}\cos\frac{\pi}{5} - \sin\frac{\pi}{7}\sin\frac{\pi}{5}$

31. $\sin 60° \cos 15° + \cos 60° \sin 15°$

32. $\cos 130° \cos 40° - \sin 130° \sin 40°$

33. $\dfrac{\tan 45° - \tan 30°}{1 + \tan 45° \tan 30°}$ **34.** $\dfrac{\tan 140° - \tan 60°}{1 + \tan 140° \tan 60°}$

35. $\dfrac{\tan 2x + \tan x}{1 - \tan 2x \tan x}$

36. $\cos 3x \cos 2y + \sin 3x \sin 2y$

In Exercises 37–42, find the exact value of the expression.

37. $\sin\frac{\pi}{12}\cos\frac{\pi}{4} + \cos\frac{\pi}{12}\sin\frac{\pi}{4}$

38. $\cos\frac{\pi}{16}\cos\frac{3\pi}{16} - \sin\frac{\pi}{16}\sin\frac{3\pi}{16}$

39. $\sin 120° \cos 60° - \cos 120° \sin 60°$

40. $\cos 120° \cos 30° + \sin 120° \sin 30°$

41. $\dfrac{\tan(5\pi/6) - \tan(\pi/6)}{1 + \tan(5\pi/6)\tan(\pi/6)}$

42. $\dfrac{\tan 25° + \tan 110°}{1 - \tan 25° \tan 110°}$

In Exercises 43–50, find the exact value of the trigonometric function given that $\sin u = \frac{5}{13}$ and $\cos v = -\frac{3}{5}$. (Both u and v are in Quadrant II.)

43. $\sin(u + v)$ **44.** $\cos(u - v)$

45. $\cos(u + v)$ **46.** $\sin(v - u)$

47. $\tan(u + v)$ **48.** $\csc(u - v)$

49. $\sec(v - u)$ **50.** $\cot(u + v)$

In Exercises 51–56, find the exact value of the trigonometric function given that $\sin u = -\frac{7}{25}$ and $\cos v = -\frac{4}{5}$. (Both u and v are in Quadrant III.)

51. $\cos(u + v)$ **52.** $\sin(u + v)$

53. $\tan(u - v)$ **54.** $\cot(v - u)$

55. $\csc(u - v)$ **56.** $\sec(v - u)$

In Exercises 57–62, prove the identity.

57. $\sin\left(\frac{\pi}{2} - x\right) = \cos x$ **58.** $\sin\left(\frac{\pi}{2} + x\right) = \cos x$

59. $\sin\left(\frac{\pi}{6} + x\right) = \frac{1}{2}(\cos x + \sqrt{3}\sin x)$

60. $\cos\left(\frac{5\pi}{4} - x\right) = -\frac{\sqrt{2}}{2}(\cos x + \sin x)$

61. $\cos(\pi - \theta) + \sin\left(\frac{\pi}{2} + \theta\right) = 0$

62. $\tan\left(\frac{\pi}{4} - \theta\right) = \dfrac{1 - \tan\theta}{1 + \tan\theta}$

WRITING ABOUT CONCEPTS

In Exercises 63–66, verify the identity.

63. $\cos(x + y)\cos(x - y) = \cos^2 x - \sin^2 y$

64. $\sin(x + y)\sin(x - y) = \sin^2 x - \sin^2 y$

65. $\sin(x + y) + \sin(x - y) = 2 \sin x \cos y$

66. $\cos(x + y) + \cos(x - y) = 2 \cos x \cos y$

In Exercises 67–70, write the trigonometric expression as an algebraic expression.

67. $\sin(\arcsin x + \arccos x)$

68. $\sin(\arctan 2x - \arccos x)$

69. $\cos(\arccos x + \arcsin x)$

70. $\cos(\arccos x - \arctan x)$

In Exercises 71–74, simplify the expression algebraically and use a graphing utility to confirm your answer graphically.

71. $\cos\left(\frac{3\pi}{2} - x\right)$

72. $\cos(\pi + x)$

73. $\sin\left(\frac{3\pi}{2} + \theta\right)$

74. $\tan(\pi + \theta)$

In Exercises 75–80, find all solutions of the equation in the interval $[0, 2\pi)$.

75. $\sin\left(x + \frac{\pi}{6}\right) - \sin\left(x - \frac{\pi}{6}\right) = \frac{1}{2}$

76. $\sin\left(x + \frac{\pi}{3}\right) + \sin\left(x - \frac{\pi}{3}\right) = 1$

77. $\cos\left(x + \frac{\pi}{4}\right) - \cos\left(x - \frac{\pi}{4}\right) = 1$

78. $\tan(x + \pi) + 2 \sin(x + \pi) = 0$

79. $\sin\left(x + \frac{\pi}{2}\right) - \cos^2 x = 0$

80. $\cos\left(x - \frac{\pi}{2}\right) + \sin^2 x = 0$

True or False? **In Exercises 81–84, determine whether the statement is true or false. Justify your answer.**

81. $\sin(u \pm v) = \sin u \cos v \pm \cos u \sin v$

82. $\cos(u \pm v) = \cos u \cos v \pm \sin u \sin v$

83. $\tan\left(x - \frac{\pi}{4}\right) = \frac{\tan x + 1}{1 - \tan x}$

84. $\sin\left(x - \frac{\pi}{2}\right) = -\cos x$

In Exercises 85–88, verify the identity.

85. $\cos(n\pi + \theta) = (-1)^n \cos \theta$, n is an integer.

86. $\sin(n\pi + \theta) = (-1)^n \sin \theta$, n is an integer.

87. $a \sin B\theta + b \cos B\theta = \sqrt{a^2 + b^2} \sin(B\theta + C)$, where $C = \arctan(b/a)$ and $a > 0$

88. $a \sin B\theta + b \cos B\theta = \sqrt{a^2 + b^2} \cos(B\theta - C)$, where $C = \arctan(a/b)$ and $b > 0$

In Exercises 89–92, use the formulas given in Exercises 87 and 88 to write the trigonometric expression in the following forms.

(a) $\sqrt{a^2 + b^2} \sin(B\theta + C)$ **(b)** $\sqrt{a^2 + b^2} \cos(B\theta - C)$

89. $\sin \theta + \cos \theta$

90. $3 \sin 2\theta + 4 \cos 2\theta$

91. $12 \sin 3\theta + 5 \cos 3\theta$

92. $\sin 2\theta + \cos 2\theta$

In Exercises 93 and 94, use the formulas given in Exercises 87 and 88 to write the trigonometric expression in the form $a \sin B\theta + b \cos B\theta$.

93. $2 \sin\left(\theta + \frac{\pi}{4}\right)$

94. $5 \cos\left(\theta - \frac{\pi}{4}\right)$

95. Verify the following identity used in calculus.

$$\frac{\cos(x + \Delta x) - \cos x}{\Delta x} = \frac{\cos x(\cos \Delta x - 1)}{\Delta x} - \frac{\sin x \sin \Delta x}{\Delta x}$$

CAPSTONE

96. Give an example to justify each statement.

(a) $\sin(u + v) \neq \sin u + \sin v$

(b) $\sin(u - v) \neq \sin u - \sin v$

(c) $\cos(u + v) \neq \cos u + \cos v$

(d) $\cos(u - v) \neq \cos u - \cos v$

(e) $\tan(u + v) \neq \tan u + \tan v$

(f) $\tan(u - v) \neq \tan u - \tan v$

In Exercises 97 and 98, use a graphing utility to graph y_1 and y_2 in the same viewing window. Use the graphs to determine whether $y_1 = y_2$. Explain your reasoning.

97. $y_1 = \cos(x + 2)$, $y_2 = \cos x + \cos 2$

98. $y_1 = \sin(x + 4)$, $y_2 = \sin x + \sin 4$

99. ***Proof***

(a) Write a proof of the formula for $\sin(u + v)$.

(b) Write a proof of the formula for $\sin(u - v)$.

10.5 Multiple-Angle and Product-to-Sum Formulas

- **Use multiple-angle formulas to rewrite and evaluate trigonometric functions.**
- **Use power-reducing formulas to rewrite and evaluate trigonometric functions.**
- **Use half-angle formulas to rewrite and evaluate trigonometric functions.**
- **Use product-to-sum and sum-to-product formulas to rewrite and evaluate trigonometric functions.**
- **Use trigonometric formulas to rewrite real-life models.**

Multiple-Angle Formulas

In this section, you will study four other categories of trigonometric identities.

1. The first category involves functions of multiple angles such as $\sin ku$ and $\cos ku$.
2. The second category involves squares of trigonometric functions such as $\sin^2 u$.
3. The third category involves functions of half-angles such as $\sin(u/2)$.
4. The fourth category involves products of trigonometric functions such as $\sin u \cos v$.

You should learn the **double-angle formulas** because they are used most often. (Proofs of the double-angle formulas are given in Appendix A.)

DOUBLE-ANGLE FORMULAS

$$\sin 2u = 2 \sin u \cos u \qquad \cos 2u = \cos^2 u - \sin^2 u$$

$$\tan 2u = \frac{2 \tan u}{1 - \tan^2 u} \qquad = 2 \cos^2 u - 1$$

$$= 1 - 2 \sin^2 u$$

NOTE Remember that $\sin 2u \neq 2 \sin u$. Think of $\sin 2u$ as $\sin(2u)$, where $2u$ is the input of the function. Similar statements can be made for $\cos 2u$ and $\tan 2u$. ■

EXAMPLE 1 Solving a Multiple-Angle Equation

Solve

$$2 \cos x + \sin 2x = 0.$$

Solution Begin by rewriting the equation so that it involves functions of x (rather than $2x$). Then factor and solve.

$$2 \cos x + \sin 2x = 0 \qquad \text{Write original equation.}$$

$$2 \cos x + 2 \sin x \cos x = 0 \qquad \text{Double-angle formula}$$

$$2 \cos x(1 + \sin x) = 0 \qquad \text{Factor.}$$

$$\cos x = 0 \quad \text{and} \quad 1 + \sin x = 0 \qquad \text{Set factors equal to zero.}$$

$$x = \frac{\pi}{2}, \frac{3\pi}{2} \qquad x = \frac{3\pi}{2} \qquad \text{Solutions in } [0, 2\pi)$$

So, the general solution is

$$x = \frac{\pi}{2} + 2n\pi \quad \text{and} \quad x = \frac{3\pi}{2} + 2n\pi$$

where n is an integer. Verify these solutions graphically. ■

EXAMPLE 2 Using Double-Angle Formulas in Sketching Graphs

Use a double-angle formula to rewrite the equation

$$y = 4\cos^2 x - 2.$$

Then sketch the graph of the equation over the interval $[0, 2\pi]$.

Solution Using the double-angle formula for $\cos 2u$, you can rewrite the original equation as

$$y = 4\cos^2 x - 2 \qquad \text{Write original equation.}$$
$$= 2(2\cos^2 x - 1) \qquad \text{Factor.}$$
$$= 2\cos 2x. \qquad \text{Double-angle formula}$$

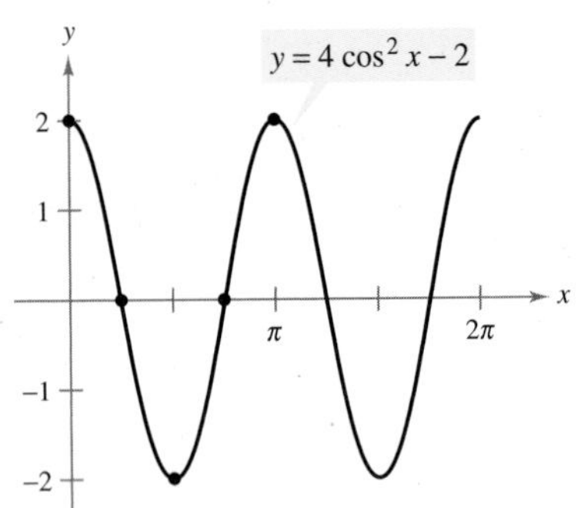

Figure 10.14

Using the techniques discussed in Section 9.5, you can recognize that the graph of this function has an amplitude of 2 and a period of π. The key points in the interval $[0, \pi]$ are as shown.

Maximum	*Intercept*	*Minimum*	*Intercept*	*Maximum*
$(0, 2)$	$\left(\frac{\pi}{4}, 0\right)$	$\left(\frac{\pi}{2}, -2\right)$	$\left(\frac{3\pi}{4}, 0\right)$	$(\pi, 2)$

Two cycles of the graph are shown in Figure 10.14.

EXAMPLE 3 Evaluating Functions Involving Double Angles

Use the following to find $\sin 2\theta$, $\cos 2\theta$, and $\tan 2\theta$.

$$\cos\theta = \frac{5}{13}, \qquad \frac{3\pi}{2} < \theta < 2\pi$$

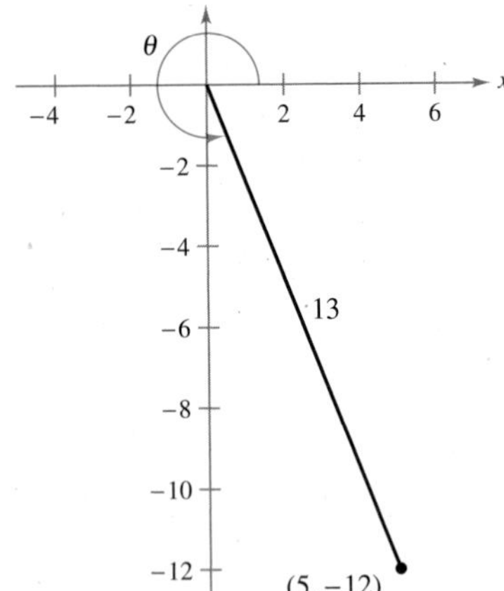

Figure 10.15

Solution From Figure 10.15, you can see that

$$\sin\theta = \frac{y}{r} = -\frac{12}{13}.$$

Consequently, using each of the double-angle formulas, you can write

$$\sin 2\theta = 2\sin\theta\cos\theta = 2\left(\frac{-12}{13}\right)\left(\frac{5}{13}\right) = -\frac{120}{169}$$

$$\cos 2\theta = 2\cos^2\theta - 1 = 2\left(\frac{25}{169}\right) - 1 = -\frac{119}{169}$$

$$\tan 2\theta = \frac{\sin 2\theta}{\cos 2\theta} = \frac{120}{119}.$$

■

The double-angle formulas are not restricted to angles 2θ and θ. Other *double* combinations, such as 4θ and 2θ or 6θ and 3θ, are also valid. Here are two examples.

$$\sin 4\theta = 2\sin 2\theta\cos 2\theta \qquad \text{and} \qquad \cos 6\theta = \cos^2 3\theta - \sin^2 3\theta$$

By using double-angle formulas together with the sum formulas given in the preceding section, you can form other multiple-angle formulas.

EXAMPLE 4 Deriving a Triple-Angle Formula

Express

$\sin 3x$

in terms of $\sin x$.

Solution

$$\begin{aligned}
\sin 3x &= \sin(2x + x) \\
&= \sin 2x \cos x + \cos 2x \sin x \\
&= 2 \sin x \cos x \cos x + (1 - 2 \sin^2 x) \sin x \\
&= 2 \sin x \cos^2 x + \sin x - 2 \sin^3 x \\
&= 2 \sin x(1 - \sin^2 x) + \sin x - 2 \sin^3 x \\
&= 2 \sin x - 2 \sin^3 x + \sin x - 2 \sin^3 x \\
&= 3 \sin x - 4 \sin^3 x
\end{aligned}$$

■

Power-Reducing Formulas

The double-angle formulas can be used to obtain the following **power-reducing formulas.** (Proofs of the power-reducing formulas are given in Appendix A.) Example 5 shows a typical power reduction that is used in calculus.

POWER-REDUCING FORMULAS

$$\sin^2 u = \frac{1 - \cos 2u}{2} \qquad \cos^2 u = \frac{1 + \cos 2u}{2} \qquad \tan^2 u = \frac{1 - \cos 2u}{1 + \cos 2u}$$

EXAMPLE 5 Reducing a Power

Rewrite

$\sin^4 x$

as a sum of first powers of the cosines of multiple angles.

Solution Note the repeated use of power-reducing formulas.

$$\begin{aligned}
\sin^4 x &= (\sin^2 x)^2 && \text{Property of exponents} \\
&= \left(\frac{1 - \cos 2x}{2}\right)^2 && \text{Power-reducing formula} \\
&= \frac{1}{4}(1 - 2 \cos 2x + \cos^2 2x) && \text{Square binomial.} \\
&= \frac{1}{4}\left(1 - 2 \cos 2x + \frac{1 + \cos 4x}{2}\right) && \text{Power-reducing formula} \\
&= \frac{1}{4} - \frac{1}{2} \cos 2x + \frac{1}{8} + \frac{1}{8} \cos 4x && \text{Simplify.} \\
&= \frac{1}{8}(3 - 4 \cos 2x + \cos 4x) && \text{Factor out common factor.}
\end{aligned}$$

■

NOTE Example 5 illustrates techniques used to integrate sine and cosine functions raised to powers greater than 1. ■

Half-Angle Formulas

You can derive some useful alternative forms of the power-reducing formulas by replacing u with $u/2$. The results are called **half-angle formulas.**

> **HALF-ANGLE FORMULAS**
>
> $$\sin\frac{u}{2} = \pm\sqrt{\frac{1-\cos u}{2}} \qquad \cos\frac{u}{2} = \pm\sqrt{\frac{1+\cos u}{2}}$$
>
> $$\tan\frac{u}{2} = \frac{1-\cos u}{\sin u} = \frac{\sin u}{1+\cos u}$$
>
> The signs of $\sin(u/2)$ and $\cos(u/2)$ depend on the quadrant in which $u/2$ lies.

EXAMPLE 6 Using a Half-Angle Formula

Find the exact value of $\sin 105°$.

Solution Begin by noting that $105°$ is half of $210°$. Then, using the half-angle formula for $\sin(u/2)$ and the fact that $105°$ lies in Quadrant II, you have

$$\sin 105° = \sqrt{\frac{1-\cos 210°}{2}} = \sqrt{\frac{1-(-\cos 30°)}{2}} = \sqrt{\frac{1+(\sqrt{3}/2)}{2}}$$

$$= \frac{\sqrt{2+\sqrt{3}}}{2}.$$

The positive square root is chosen because $\sin\theta$ is positive in Quadrant II.

EXAMPLE 7 Solving a Trigonometric Equation

Find all solutions of $2 - \sin^2 x = 2\cos^2\frac{x}{2}$ in the interval $[0, 2\pi)$.

Algebraic Solution

$2 - \sin^2 x = 2\cos^2\frac{x}{2}$ — Write original equation.

$2 - \sin^2 x = 2\left(\pm\sqrt{\frac{1+\cos x}{2}}\right)^2$ — Half-angle formula

$2 - \sin^2 x = 2\left(\frac{1+\cos x}{2}\right)$ — Simplify.

$2 - \sin^2 x = 1 + \cos x$ — Simplify.

$2 - (1 - \cos^2 x) = 1 + \cos x$ — Pythagorean identity

$\cos^2 x - \cos x = 0$ — Simplify.

$\cos x(\cos x - 1) = 0$ — Factor.

By setting the factors $\cos x$ and $\cos x - 1$ equal to zero, you find that the solutions in the interval $[0, 2\pi)$ are

$$x = \frac{\pi}{2}, \quad x = \frac{3\pi}{2}, \quad \text{and} \quad x = 0.$$

Graphical Solution

Use a graphing utility set in *radian* mode to graph $y = 2 - \sin^2 x - 2\cos^2(x/2)$, as shown in Figure 10.16. Use the *zero* or *root* feature or the *zoom* and *trace* features to approximate the x-intercepts in the interval $[0, 2\pi)$ to be

$$x = 0,\ x \approx 1.571 \approx \frac{\pi}{2},\ \text{and } x \approx 4.712 \approx \frac{3\pi}{2}.$$

These values are the approximate solutions of $2 - \sin^2 x - 2\cos^2(x/2) = 0$ in the interval $[0, 2\pi)$.

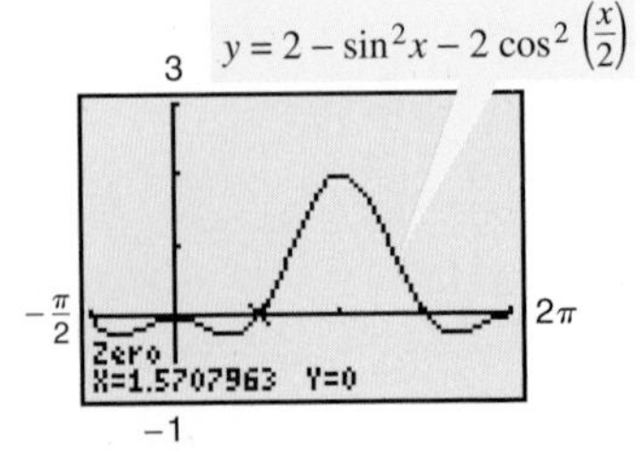

Figure 10.16

■

Product-to-Sum Formulas

Each of the following **product-to-sum formulas** can be verified using the sum and difference formulas discussed in the preceding section. These formulas are used when integrating a trigonometric product in which the angles are different.

PRODUCT-TO-SUM FORMULAS

$$\sin u \sin v = \frac{1}{2}[\cos(u - v) - \cos(u + v)]$$

$$\cos u \cos v = \frac{1}{2}[\cos(u - v) + \cos(u + v)]$$

$$\sin u \cos v = \frac{1}{2}[\sin(u + v) + \sin(u - v)]$$

$$\cos u \sin v = \frac{1}{2}[\sin(u + v) - \sin(u - v)]$$

EXAMPLE 8 Writing Products as Sums

$$\cos 5x \sin 4x = \frac{1}{2}[\sin(5x + 4x) - \sin(5x - 4x)] = \frac{1}{2}\sin 9x - \frac{1}{2}\sin x$$ ■

Occasionally, it is useful to reverse the procedure and write a sum of trigonometric functions as a product. This can be accomplished with the following **sum-to-product formulas.** (A proof of the first formula is given in Appendix A.)

SUM-TO-PRODUCT FORMULAS

$$\sin x + \sin y = 2 \sin\left(\frac{x + y}{2}\right) \cos\left(\frac{x - y}{2}\right)$$

$$\sin x - \sin y = 2 \cos\left(\frac{x + y}{2}\right) \sin\left(\frac{x - y}{2}\right)$$

$$\cos x + \cos y = 2 \cos\left(\frac{x + y}{2}\right) \cos\left(\frac{x - y}{2}\right)$$

$$\cos x - \cos y = -2 \sin\left(\frac{x + y}{2}\right) \sin\left(\frac{x - y}{2}\right)$$

EXAMPLE 9 Using a Sum-to-Product Formula

$$\cos 195° + \cos 105° = 2 \cos\left(\frac{195° + 105°}{2}\right) \cos\left(\frac{195° - 105°}{2}\right)$$

$$= 2 \cos 150° \cos 45°$$

$$= 2\left(-\frac{\sqrt{3}}{2}\right)\left(\frac{\sqrt{2}}{2}\right) = -\frac{\sqrt{6}}{2}$$ ■

EXAMPLE 10 Solving a Trigonometric Equation

Solve

$$\sin 5x + \sin 3x = 0.$$

Algebraic Solution

$$\sin 5x + \sin 3x = 0 \qquad \text{Write original equation.}$$

$$2\sin\left(\frac{5x + 3x}{2}\right)\cos\left(\frac{5x - 3x}{2}\right) = 0 \qquad \text{Sum-to-product formula}$$

$$2\sin 4x \cos x = 0 \qquad \text{Simplify.}$$

By setting the factor $2\sin 4x$ equal to zero, you can find that the solutions in the interval $[0, 2\pi)$ are

$$x = 0, \frac{\pi}{4}, \frac{\pi}{2}, \frac{3\pi}{4}, \pi, \frac{5\pi}{4}, \frac{3\pi}{2}, \frac{7\pi}{4}.$$

The equation $\cos x = 0$ yields no additional solutions, so you can conclude that the solutions are of the form

$$x = \frac{n\pi}{4}$$

where n is an integer.

Graphical Solution

Sketch the graph of

$$y = \sin 5x + \sin 3x$$

as shown in Figure 10.17. From the graph you can see that the x-intercepts occur at multiples of $\pi/4$. So, you can conclude that the solutions are of the form

$$x = \frac{n\pi}{4}$$

where n is an integer.

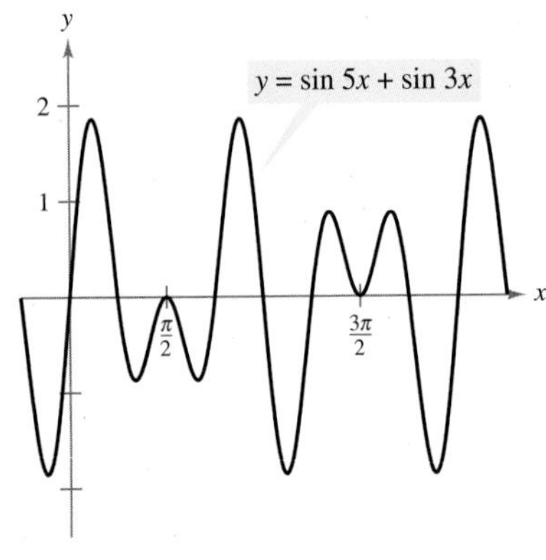

Figure 10.17

■

EXAMPLE 11 Verifying a Trigonometric Identity

Verify the identity

$$\frac{\sin 3x - \sin x}{\cos x + \cos 3x} = \tan x.$$

Solution Using appropriate sum-to-product formulas, you have

$$\frac{\sin 3x - \sin x}{\cos x + \cos 3x} = \frac{2\cos\left(\frac{3x + x}{2}\right)\sin\left(\frac{3x - x}{2}\right)}{2\cos\left(\frac{x + 3x}{2}\right)\cos\left(\frac{x - 3x}{2}\right)}$$

$$= \frac{2\cos(2x)\sin x}{2\cos(2x)\cos(-x)}$$

$$= \frac{\sin x}{\cos(-x)}$$

$$= \frac{\sin x}{\cos x}$$

$$= \tan x.$$

■

Application

EXAMPLE 12 Projectile Motion

Ignoring air resistance, the range of a projectile fired at an angle θ with the horizontal and with an initial velocity of v_0 feet per second is given by

$$r = \frac{1}{16}v_0^2 \sin \theta \cos \theta$$

where r is the horizontal distance (in feet) that the projectile will travel. A place kicker for a football team can kick a football from ground level with an initial velocity of 80 feet per second (see Figure 10.18).

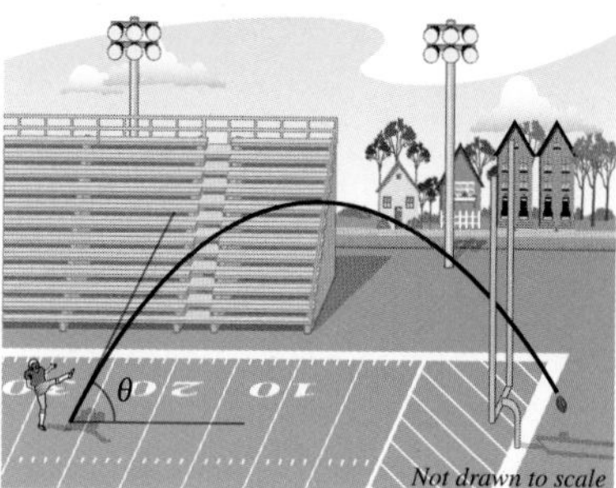

Figure 10.18

a. Write the projectile motion model in a simpler form.

b. At what angle must the player kick the football so that the football travels 200 feet?

c. For what angle is the horizontal distance the football travels a maximum?

Solution

a. You can use a double-angle formula to rewrite the projectile motion model as

$$r = \frac{1}{32}v_0^2(2 \sin \theta \cos \theta) \qquad \text{Rewrite original projectile motion model.}$$

$$= \frac{1}{32}v_0^2 \sin 2\theta. \qquad \text{Rewrite model using a double-angle formula.}$$

b.

$$r = \frac{1}{32}v_0^2 \sin 2\theta \qquad \text{Write projectile motion model.}$$

$$200 = \frac{1}{32}(80)^2 \sin 2\theta \qquad \text{Substitute 200 for } r \text{ and 80 for } v_0.$$

$$200 = 200 \sin 2\theta \qquad \text{Simplify.}$$

$$1 = \sin 2\theta \qquad \text{Divide each side by 200.}$$

You know that $2\theta = \pi/2$, so dividing this result by 2 produces $\theta = \pi/4$. Because $\pi/4 = 45°$, you can conclude that the player must kick the football at an angle of 45° so that the football will travel 200 feet.

c. From the model $r = 200 \sin 2\theta$, you can see that the amplitude is 200. So the maximum range is $r = 200$ feet. From part (b), you know that this corresponds to an angle of 45°. Therefore, kicking the football at an angle of 45° will produce a maximum horizontal distance of 200 feet. ■

10.5 Exercises

See www.CalcChat.com for worked-out solutions to odd-numbered exercises.

In Exercises 1–10, fill in the blank to complete the trigonometric formula.

1. $\sin 2u =$ ______
2. $\dfrac{1 + \cos 2u}{2} =$ ______
3. $\cos 2u =$ ______
4. $\dfrac{1 - \cos 2u}{1 + \cos 2u} =$ ______
5. $\sin \dfrac{u}{2} =$ ______
6. $\tan \dfrac{u}{2} =$ ______
7. $\cos u \cos v =$ ______
8. $\sin u \cos v =$ ______
9. $\sin u + \sin v =$ ______
10. $\cos u - \cos v =$ ______

In Exercises 11–18, use the figure to find the exact value of the trigonometric function.

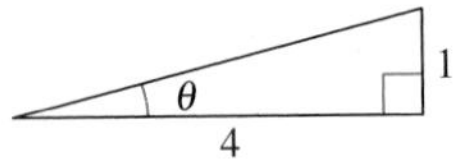

11. $\cos 2\theta$
12. $\sin 2\theta$
13. $\tan 2\theta$
14. $\sec 2\theta$
15. $\csc 2\theta$
16. $\cot 2\theta$
17. $\sin 4\theta$
18. $\tan 4\theta$

In Exercises 19–28, find the exact solutions of the equation in the interval $[0, 2\pi)$.

19. $\sin 2x - \sin x = 0$
20. $\sin 2x + \cos x = 0$
21. $4 \sin x \cos x = 1$
22. $\sin 2x \sin x = \cos x$
23. $\cos 2x - \cos x = 0$
24. $\cos 2x + \sin x = 0$
25. $\sin 4x = -2 \sin 2x$
26. $(\sin 2x + \cos 2x)^2 = 1$
27. $\tan 2x - \cot x = 0$
28. $\tan 2x - 2 \cos x = 0$

In Exercises 29–36, use a double-angle formula to rewrite the expression.

29. $6 \sin x \cos x$
30. $\sin x \cos x$
31. $6 \cos^2 x - 3$
32. $\cos^2 x - \frac{1}{2}$
33. $4 - 8 \sin^2 x$
34. $10 \sin^2 x - 5$
35. $(\cos x + \sin x)(\cos x - \sin x)$
36. $(\sin x - \cos x)(\sin x + \cos x)$

In Exercises 37–42, find the exact values of $\sin 2u$, $\cos 2u$, and $\tan 2u$ using the double-angle formulas.

37. $\sin u = -\dfrac{3}{5}, \quad \dfrac{3\pi}{2} < u < 2\pi$
38. $\cos u = -\dfrac{4}{5}, \quad \dfrac{\pi}{2} < u < \pi$
39. $\tan u = \dfrac{3}{5}, \quad 0 < u < \dfrac{\pi}{2}$
40. $\cot u = \sqrt{2}, \quad \pi < u < \dfrac{3\pi}{2}$
41. $\sec u = -2, \quad \dfrac{\pi}{2} < u < \pi$
42. $\csc u = 3, \quad \dfrac{\pi}{2} < u < \pi$

In Exercises 43–52, use the power-reducing formulas to rewrite the expression in terms of the first power of the cosine.

43. $\cos^4 x$
44. $\sin^4 2x$
45. $\cos^4 2x$
46. $\sin^8 x$
47. $\tan^4 2x$
48. $\sin^2 x \cos^4 x$
49. $\sin^2 2x \cos^2 2x$
50. $\tan^2 2x \cos^4 2x$
51. $\sin^4 x \cos^2 x$
52. $\sin^4 x \cos^4 x$

In Exercises 53–58, use the figure to find the exact value of the trigonometric function.

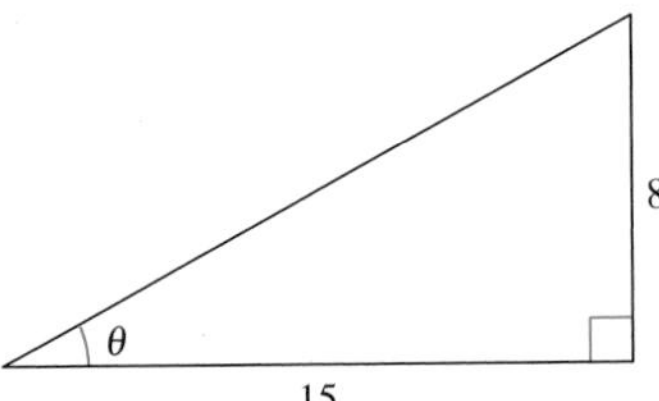

53. $\cos \dfrac{\theta}{2}$
54. $\sin \dfrac{\theta}{2}$
55. $\tan \dfrac{\theta}{2}$
56. $\sec \dfrac{\theta}{2}$
57. $\csc \dfrac{\theta}{2}$
58. $\cot \dfrac{\theta}{2}$

In Exercises 59–66, use the half-angle formulas to determine the exact values of the sine, cosine, and tangent of the angle.

59. $75°$
60. $165°$
61. $112° \, 30'$
62. $67° \, 30'$
63. $\pi/8$
64. $\pi/12$
65. $3\pi/8$
66. $7\pi/12$

In Exercises 67–72, (a) determine the quadrant in which $u/2$ lies, and (b) find the exact values of $\sin(u/2)$, $\cos(u/2)$, and $\tan(u/2)$ using the half-angle formulas.

67. $\cos u = \frac{7}{25}, \quad 0 < u < \frac{\pi}{2}$

68. $\sin u = \frac{5}{13}, \quad \frac{\pi}{2} < u < \pi$

69. $\tan u = -\frac{5}{12}, \quad \frac{3\pi}{2} < u < 2\pi$

70. $\cot u = 3, \quad \pi < u < \frac{3\pi}{2}$

71. $\csc u = -\frac{5}{3}, \quad \pi < u < \frac{3\pi}{2}$

72. $\sec u = \frac{7}{2}, \quad \frac{3\pi}{2} < u < 2\pi$

In Exercises 73–76, use the half-angle formulas to simplify the expression.

73. $\sqrt{\frac{1 - \cos 6x}{2}}$ **74.** $\sqrt{\frac{1 + \cos 4x}{2}}$

75. $-\sqrt{\frac{1 - \cos 8x}{1 + \cos 8x}}$ **76.** $-\sqrt{\frac{1 - \cos(x - 1)}{2}}$

In Exercises 77–80, find all solutions of the equation in the interval $[0, 2\pi)$. Use a graphing utility to graph the equation and verify the solutions.

77. $\sin \frac{x}{2} + \cos x = 0$ **78.** $\sin \frac{x}{2} + \cos x - 1 = 0$

79. $\cos \frac{x}{2} - \sin x = 0$ **80.** $\tan \frac{x}{2} - \sin x = 0$

In Exercises 81–90, use the product-to-sum formulas to write the product as a sum or difference.

81. $\sin \frac{\pi}{3} \cos \frac{\pi}{6}$ **82.** $4 \cos \frac{\pi}{3} \sin \frac{5\pi}{6}$

83. $10 \cos 75° \cos 15°$ **84.** $6 \sin 45° \cos 15°$

85. $\sin 5\theta \sin 3\theta$ **86.** $3 \sin(-4\alpha) \sin 6\alpha$

87. $7 \cos(-5\beta) \sin 3\beta$ **88.** $\cos 2\theta \cos 4\theta$

89. $\sin(x + y) \sin(x - y)$ **90.** $\sin(x + y) \cos(x - y)$

In Exercises 91–98, use the sum-to-product formulas to write the sum or difference as a product.

91. $\sin 3\theta + \sin \theta$ **92.** $\sin 5\theta - \sin 3\theta$

93. $\cos 6x + \cos 2x$ **94.** $\cos x + \cos 4x$

95. $\sin(\alpha + \beta) - \sin(\alpha - \beta)$

96. $\cos(\phi + 2\pi) + \cos \phi$

97. $\cos\left(\theta + \frac{\pi}{2}\right) - \cos\left(\theta - \frac{\pi}{2}\right)$

98. $\sin\left(x + \frac{\pi}{2}\right) + \sin\left(x - \frac{\pi}{2}\right)$

In Exercises 99–102, use the sum-to-product formulas to find the exact value of the expression.

99. $\sin 75° + \sin 15°$ **100.** $\cos 120° + \cos 60°$

101. $\cos \frac{3\pi}{4} - \cos \frac{\pi}{4}$ **102.** $\sin \frac{5\pi}{4} - \sin \frac{3\pi}{4}$

In Exercises 103–106, find all solutions of the equation in the interval $[0, 2\pi)$. Use a graphing utility to graph the equation and verify the solutions.

103. $\sin 6x + \sin 2x = 0$ **104.** $\cos 2x - \cos 6x = 0$

105. $\frac{\cos 2x}{\sin 3x - \sin x} - 1 = 0$ **106.** $\sin^2 3x - \sin^2 x = 0$

In Exercises 107–116, use the figure to find the exact value of the trigonometric function.

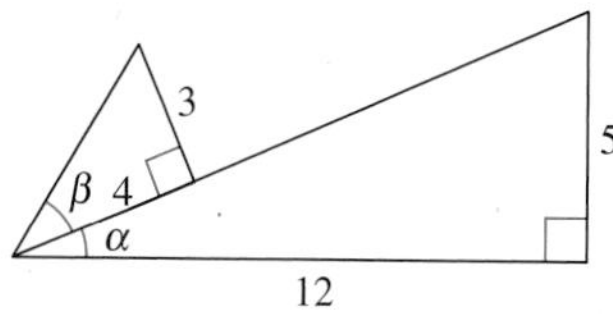

107. $\sin 2\alpha$ **108.** $\cos 2\beta$

109. $\tan 2\alpha$ **110.** $\sin^2 \alpha$

111. $\cos(\beta/2)$ **112.** $\tan(\beta/2)$

113. $\cos \alpha \sin \beta$ **114.** $\sin \alpha + \sin \beta$

115. $\cos \alpha - \cos \beta$ **116.** $\sin(\alpha + \beta)$

In Exercises 117–130, verify the identity.

117. $\csc 2\theta = \frac{\csc \theta}{2 \cos \theta}$

118. $\sec 2\theta = \frac{\sec^2 \theta}{2 - \sec^2 \theta}$

119. $\sin \frac{\alpha}{3} \cos \frac{\alpha}{3} = \frac{1}{2} \sin \frac{2\alpha}{3}$

120. $\frac{\cos 3\beta}{\cos \beta} = 1 - 4 \sin^2 \beta$

121. $1 + \cos 10y = 2 \cos^2 5y$

122. $\cos^4 x - \sin^4 x = \cos 2x$

123. $\cos 4\alpha = \cos^2 2\alpha - \sin^2 2\alpha$

124. $(\sin x + \cos x)^2 = 1 + \sin 2x$

125. $\tan \frac{u}{2} = \csc u - \cot u$

126. $\sec \frac{u}{2} = \pm\sqrt{\frac{2 \tan u}{\tan u + \sin u}}$

127. $\frac{\cos 4x + \cos 2x}{\sin 4x + \sin 2x} = \cot 3x$

128. $\frac{\sin x \pm \sin y}{\cos x + \cos y} = \tan \frac{x \pm y}{2}$

129. $\sin\left(\frac{\pi}{6} + x\right) + \sin\left(\frac{\pi}{6} - x\right) = \cos x$

130. $\cos\left(\frac{\pi}{3} + x\right) + \cos\left(\frac{\pi}{3} - x\right) = \cos x$

In Exercises 131–134, use a graphing utility to verify the identity. Confirm that it is an identity algebraically.

131. $\cos 3\beta = \cos^3 \beta - 3 \sin^2 \beta \cos \beta$

132. $\sin 4\beta = 4 \sin \beta \cos \beta(1 - 2 \sin^2 \beta)$

133. $(\cos 4x - \cos 2x)/(2 \sin 3x) = -\sin x$

134. $(\cos 3x - \cos x)/(\sin 3x - \sin x) = -\tan 2x$

In Exercises 135 and 136, graph the function by hand in the interval $[0, 2\pi]$ by using the power-reducing formulas.

135. $f(x) = \sin^2 x$

136. $f(x) = \cos^2 x$

In Exercises 137–140, write the trigonometric expression as an algebraic expression.

137. $\sin(2 \arcsin x)$

138. $\cos(2 \arccos x)$

139. $\cos(2 \arcsin x)$

140. $\sin(2 \arccos x)$

WRITING ABOUT CONCEPTS

141. Consider the function given by

$$f(x) = 2 \sin x\left[2 \cos^2\left(\frac{x}{2}\right) - 1\right].$$

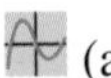
(a) Use a graphing utility to graph the function.

(b) Make a conjecture about the function that is an identity with *f*.

(c) Verify your conjecture analytically.

142. ***Projectile Motion*** The range of a projectile fired at an angle θ with the horizontal and with an initial velocity of v_0 feet per second is

$$r = \frac{1}{32} v_0^2 \sin 2\theta$$

where *r* is measured in feet. An athlete throws a javelin at 75 feet per second. At what angle must the athlete throw the javelin so that the javelin travels 130 feet?

143. ***Mach Number*** The mach number *M* of an airplane is the ratio of its speed to the speed of sound. When an airplane travels faster than the speed of sound, the sound waves form a cone behind the airplane (see figure). The mach number is related to the apex angle θ of the cone by $\sin(\theta/2) = 1/M$.

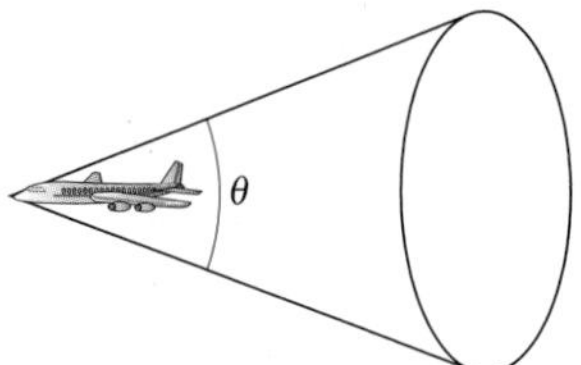

(a) Find the angle θ that corresponds to a mach number of 1.

(b) Find the angle θ that corresponds to a mach number of 4.5.

(c) The speed of sound is about 760 miles per hour. Determine the speed of an object with the mach numbers from parts (a) and (b).

(d) Rewrite the equation in terms of θ.

CAPSTONE

144. Consider the function given by

$$f(x) = \sin^4 x + \cos^4 x.$$

(a) Use the power-reducing formulas to write the function in terms of cosine to the first power.

(b) Determine another way of rewriting the function. Use a graphing utility to rule out incorrectly rewritten functions.

(c) Add a trigonometric term to the function so that it becomes a perfect square trinomial. Rewrite the function as a perfect square trinomial minus the term that you added. Use a graphing utility to rule out incorrectly rewritten functions.

(d) Rewrite the result of part (c) in terms of the sine of a double angle. Use a graphing utility to rule out incorrectly rewritten functions.

(e) When you rewrite a trigonometric expression, the result may not be the same as a friend's. Does this mean that one of you is wrong? Explain.

True or False? **In Exercises 145 and 146, determine whether the statement is true or false. Justify your answer.**

145. Because the sine function is an odd function, for a negative number *u*, $\sin 2u = -2 \sin u \cos u$.

146. $\sin \frac{u}{2} = -\sqrt{\frac{1 - \cos u}{2}}$ when *u* is in the second quadrant.

10 CHAPTER SUMMARY

	Review Exercises
Section 10.1	
■ Recognize and write the fundamental trigonometric identities *(p. 646).*	*1–6*
■ Use the fundamental trigonometric identities to evaluate trigonometric functions, and simplify and rewrite trigonometric expressions *(p. 647).*	*7–24*
Section 10.2	
■ Verify trigonometric identities *(p. 653).*	*25–32*
Section 10.3	
■ Use standard algebraic techniques to solve trigonometric equations *(p. 659).*	*33–38*
■ Solve trigonometric equations of quadratic type *(p. 661).*	*39–42*
■ Solve trigonometric equations involving multiple angles *(p. 664).*	*43–46*
■ Use inverse trigonometric functions to solve trigonometric equations *(p. 665).*	*47–50*
Section 10.4	
■ Use sum and difference formulas to evaluate trigonometric functions, verify identities, and solve trigonometric equations *(p. 669).*	*51–66*
Section 10.5	
■ Use multiple-angle formulas to rewrite and evaluate trigonometric functions *(p. 675).*	*67–72*
■ Use power-reducing formulas to rewrite and evaluate trigonometric functions *(p. 677).*	*73–76*
■ Use half-angle formulas to rewrite and evaluate trigonometric functions *(p. 678).*	*77–82*
■ Use product-to-sum formulas and sum-to-product formulas to rewrite and evaluate trigonometric functions *(p. 679).*	*83–90*
■ Use trigonometric formulas to rewrite real-life models *(p. 681).*	*91–96*

10 REVIEW EXERCISES

See www.CalcChat.com for worked-out solutions to odd-numbered exercises.

In Exercises 1–6, name the trigonometric function that is equivalent to the expression.

1. $\dfrac{\sin x}{\cos x}$
2. $\dfrac{1}{\sin x}$
3. $\dfrac{1}{\sec x}$
4. $\dfrac{1}{\tan x}$
5. $\sqrt{\cot^2 x + 1}$
6. $\sqrt{1 + \tan^2 x}$

In Exercises 7–10, use the given values and trigonometric identities to evaluate (if possible) all six trigonometric functions.

7. $\sin x = \frac{5}{13}, \quad \cos x = \frac{12}{13}$
8. $\tan \theta = \dfrac{2}{3}, \quad \sec \theta = \dfrac{\sqrt{13}}{3}$
9. $\sin\left(\dfrac{\pi}{2} - x\right) = \dfrac{\sqrt{2}}{2}, \quad \sin x = -\dfrac{\sqrt{2}}{2}$
10. $\csc\left(\dfrac{\pi}{2} - \theta\right) = 9, \quad \sin \theta = \dfrac{4\sqrt{5}}{9}$

In Exercises 11–24, use the fundamental trigonometric identities to simplify the expression. There is more than one correct form of each answer.

11. $\dfrac{1}{\cot^2 x + 1}$
12. $\dfrac{\tan \theta}{1 - \cos^2 \theta}$
13. $\tan^2 x(\csc^2 x - 1)$
14. $\cot^2 x(\sin^2 x)$
15. $\dfrac{\sin\left(\dfrac{\pi}{2} - \theta\right)}{\sin \theta}$
16. $\dfrac{\cot\left(\dfrac{\pi}{2} - u\right)}{\cos u}$
17. $\dfrac{\sin^2 \theta + \cos^2 \theta}{\sin \theta}$
18. $\dfrac{\sec^2(-\theta)}{\csc^2 \theta}$
19. $\cos^2 x + \cos^2 x \cot^2 x$
20. $\tan^2 \theta \csc^2 \theta - \tan^2 \theta$
21. $(\tan x + 1)^2 \cos x$
22. $(\sec x - \tan x)^2$
23. $\dfrac{1}{\csc \theta + 1} - \dfrac{1}{\csc \theta - 1}$
24. $\dfrac{\tan^2 x}{1 + \sec x}$

In Exercises 25–32, verify the identity.

25. $\cos x(\tan^2 x + 1) = \sec x$
26. $\sec^2 x \cot x - \cot x = \tan x$
27. $\sec\left(\dfrac{\pi}{2} - \theta\right) = \csc \theta$
28. $\cot\left(\dfrac{\pi}{2} - x\right) = \tan x$
29. $\dfrac{1}{\tan \theta \csc \theta} = \cos \theta$
30. $\dfrac{1}{\tan x \csc x \sin x} = \cot x$
31. $\sin^5 x \cos^2 x = (\cos^2 x - 2\cos^4 x + \cos^6 x)\sin x$
32. $\cos^3 x \sin^2 x = (\sin^2 x - \sin^4 x)\cos x$

In Exercises 33–38, solve the equation.

33. $\sin x = \sqrt{3} - \sin x$
34. $4\cos \theta = 1 + 2\cos \theta$
35. $3\sqrt{3}\tan u = 3$
36. $\frac{1}{2}\sec x - 1 = 0$
37. $3\csc^2 x = 4$
38. $4\tan^2 u - 1 = \tan^2 u$

In Exercises 39–46, find all solutions of the equation in the interval $[0, 2\pi)$.

39. $2\cos^2 x - \cos x = 1$
40. $2\sin^2 x - 3\sin x = -1$
41. $\cos^2 x + \sin x = 1$
42. $\sin^2 x + 2\cos x = 2$
43. $2\sin 2x - \sqrt{2} = 0$
44. $\sqrt{3}\tan 3x = 0$
45. $\cos 4x(\cos x - 1) = 0$
46. $3\csc^2 5x = -4$

In Exercises 47–50, use inverse functions where needed to find all solutions of the equation in the interval $[0, 2\pi)$.

47. $\sin^2 x - 2\sin x = 0$
48. $2\cos^2 x + 3\cos x = 0$
49. $\tan^2 \theta + \tan \theta - 6 = 0$
50. $\sec^2 x + 6\tan x + 4 = 0$

In Exercises 51–54, find the exact values of the sine, cosine, and tangent of the angle by using a sum or difference formula.

51. $285° = 315° - 30°$
52. $345° = 300° + 45°$
53. $\dfrac{25\pi}{12} = \dfrac{11\pi}{6} + \dfrac{\pi}{4}$
54. $\dfrac{19\pi}{12} = \dfrac{11\pi}{6} - \dfrac{\pi}{4}$

In Exercises 55–58, write the expression as the sine, cosine, or tangent of an angle.

55. $\sin 60° \cos 45° - \cos 60° \sin 45°$
56. $\cos 45° \cos 120° - \sin 45° \sin 120°$
57. $\dfrac{\tan 25° + \tan 10°}{1 - \tan 25° \tan 10°}$
58. $\dfrac{\tan 68° - \tan 115°}{1 + \tan 68° \tan 115°}$

In Exercises 59–64, find the exact value of the trigonometric function given that $\tan u = \frac{3}{4}$ and $\cos v = -\frac{4}{5}$. (u is in Quadrant I and v is in Quadrant III.)

59. $\sin(u + v)$
60. $\tan(u + v)$
61. $\cos(u - v)$
62. $\sin(u - v)$
63. $\cos(u + v)$
64. $\tan(u - v)$

In Exercises 65 and 66, find all solutions of the equation in the interval $[0, 2\pi)$.

65. $\sin\left(x + \dfrac{\pi}{4}\right) - \sin\left(x - \dfrac{\pi}{4}\right) = 1$
66. $\cos\left(x + \dfrac{\pi}{6}\right) - \cos\left(x - \dfrac{\pi}{6}\right) = 1$

In Exercises 67–70, find the exact values of sin 2*u*, cos 2*u*, and tan 2*u* using the double-angle formulas.

67. $\sin u = -\frac{4}{5}, \quad \pi < u < \frac{3\pi}{2}$

68. $\cos u = -\frac{2}{\sqrt{5}}, \quad \frac{\pi}{2} < u < \pi$

69. $\sec u = -3, \quad \frac{\pi}{2} < u < \pi$

70. $\cot u = 2, \quad \pi < u < \frac{3\pi}{2}$

In Exercises 71 and 72, use double-angle formulas to verify the identity algebraically and use a graphing utility to confirm your result graphically.

71. $\sin 4x = 8 \cos^3 x \sin x - 4 \cos x \sin x$

72. $\tan^2 x = \frac{1 - \cos 2x}{1 + \cos 2x}$

In Exercises 73–76, use the power-reducing formulas to rewrite the expression in terms of the first power of the cosine.

73. $\tan^2 2x$

74. $\cos^2 3x$

75. $\sin^2 x \tan^2 x$

76. $\cos^2 x \tan^2 x$

In Exercises 77–80, use the half-angle formulas to determine the exact values of the sine, cosine, and tangent of the angle.

77. $-75°$

78. $15°$

79. $\frac{19\pi}{12}$

80. $-\frac{17\pi}{12}$

In Exercises 81 and 82, use the half-angle formulas to simplify the expression.

81. $-\sqrt{\frac{1 + \cos 10x}{2}}$

82. $\frac{\sin 6x}{1 + \cos 6x}$

In Exercises 83–86, use the product-to-sum formulas to write the product as a sum or difference.

83. $\cos \frac{\pi}{6} \sin \frac{\pi}{6}$

84. $6 \sin 15° \sin 45°$

85. $\cos 4\theta \sin 6\theta$

86. $2 \sin 7\theta \cos 3\theta$

In Exercises 87–90, use the sum-to-product formulas to write the sum or difference as a product.

87. $\sin 4\theta - \sin 8\theta$

88. $\cos 6\theta + \cos 5\theta$

89. $\cos\left(x + \frac{\pi}{6}\right) - \cos\left(x - \frac{\pi}{6}\right)$

90. $\sin\left(x + \frac{\pi}{4}\right) - \sin\left(x - \frac{\pi}{4}\right)$

91. ***Projectile Motion*** A baseball leaves the hand of the player at first base at an angle of θ with the horizontal and at an initial velocity of $v_0 = 80$ feet per second. The ball is caught by the player at second base 100 feet away. Find θ if the range r of a projectile is

$$r = \frac{1}{32} v_0^2 \sin 2\theta.$$

92. ***Geometry*** A trough for feeding cattle is 4 meters long and its cross sections are isosceles triangles with the two equal sides being $\frac{1}{2}$ meter (see figure). The angle between the two sides is θ.

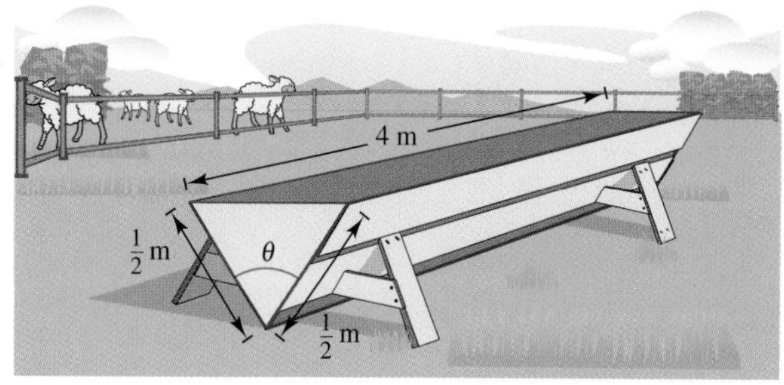

(a) Write the trough's volume as a function of $\theta/2$.

(b) Write the volume of the trough as a function of θ and determine the value of θ such that the volume is maximum.

Harmonic Motion **In Exercises 93–96, use the following information. A weight is attached to a spring suspended vertically from a ceiling. When a driving force is applied to the system, the weight moves vertically from its equilibrium position, and this motion is described by the model $y = 1.5 \sin 8t - 0.5 \cos 8t$, where y is the distance from equilibrium (in feet) and t is the time (in seconds).**

93. Use a graphing utility to graph the model.

94. Write the model in the form

$$y = \sqrt{a^2 + b^2} \sin(Bt + C).$$

95. Find the amplitude of the oscillations of the weight.

96. Find the frequency of the oscillations of the weight.

In Exercises 97 and 98, use the *zero* or *root* feature of a graphing utility to approximate the zeros of the function.

97. $y = \sqrt{x + 3} + 4 \cos x$

98. $y = 2 - \frac{1}{2}x^2 + 3 \sin \frac{\pi x}{2}$

10 CHAPTER TEST

Take this test as you would take a test in class. When you are finished, check your work against the answers given in the back of the book.

1. If $\tan \theta = \frac{6}{5}$ and $\cos \theta < 0$, use the fundamental identities to evaluate all six trigonometric functions of θ.
2. Use the fundamental identities to simplify $\csc^2 \beta(1 - \cos^2 \beta)$.
3. Factor and simplify $\dfrac{\sec^4 x - \tan^4 x}{\sec^2 x + \tan^2 x}$.
4. Add and simplify $\dfrac{\cos \theta}{\sin \theta} + \dfrac{\sin \theta}{\cos \theta}$.
5. Determine the values of $\theta, 0 \le \theta < 2\pi$, for which $\tan \theta = -\sqrt{\sec^2 \theta - 1}$ is true.
6. Use a graphing utility to graph the functions $y_1 = \cos x + \sin x \tan x$ and $y_2 = \sec x$. Make a conjecture about y_1 and y_2. Verify the result algebraically.

In Exercises 7–12, verify the identity.

7. $\sin \theta \sec \theta = \tan \theta$
8. $\sec^2 x \tan^2 x + \sec^2 x = \sec^4 x$
9. $\dfrac{\csc \alpha + \sec \alpha}{\sin \alpha + \cos \alpha} = \cot \alpha + \tan \alpha$
10. $\tan\left(x + \dfrac{\pi}{2}\right) = -\cot x$
11. $\sin(n\pi + \theta) = (-1)^n \sin \theta$, n is an integer.
12. $(\sin x + \cos x)^2 = 1 + \sin 2x$

13. Rewrite $\sin^4 \dfrac{x}{2}$ in terms of the first power of the cosine.
14. Use a half-angle formula to simplify the expression $\sin 4\theta/(1 + \cos 4\theta)$.
15. Write $4 \sin 3\theta \cos 2\theta$ as a sum or difference.
16. Write $\cos 3\theta - \cos \theta$ as a product.

In Exercises 17–20, find all solutions of the equation in the interval $[0, 2\pi)$.

17. $\tan^2 x + \tan x = 0$
18. $\sin 2\alpha - \cos \alpha = 0$
19. $4 \cos^2 x - 3 = 0$
20. $\csc^2 x - \csc x - 2 = 0$

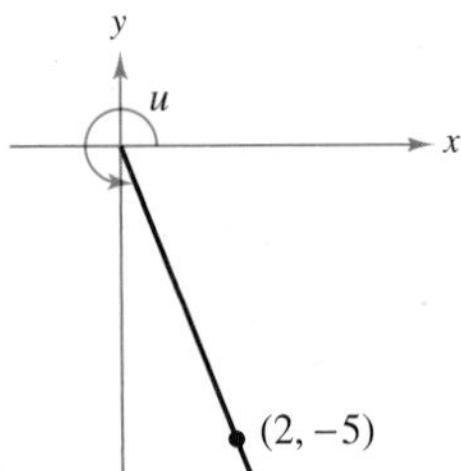

Figure for 23

21. Use a graphing utility to approximate the solutions of the equation $5 \sin x - x = 0$ accurate to three decimal places.
22. Find the exact value of $\cos 105°$ using the fact that $105° = 135° - 30°$.
23. Use the figure to find the exact values of $\sin 2u$, $\cos 2u$, and $\tan 2u$.
24. Cheyenne, Wyoming has a latitude of 41°N. At this latitude, the position of the sun at sunrise can be modeled by

$$D = 31 \sin\left(\frac{2\pi}{365}t - 1.4\right)$$

where t is the time (in days) and $t = 1$ represents January 1. In this model, D represents the number of degrees north or south of due east that the sun rises. Use a graphing utility to determine the days on which the sun is more than 20° north of due east at sunrise.

25. The heights h (in feet) of two people in different seats on a Ferris wheel can be modeled by

$$h_1 = 28 \cos 10t + 38 \quad \text{and} \quad h_2 = 28 \cos\left[10\left(t - \frac{\pi}{6}\right)\right] + 38,\ 0 \le t \le 2$$

where t is the time (in minutes). When are the two people at the same height?

P.S. PROBLEM SOLVING

In Exercises 1 and 2, find the smallest positive fixed point of the function f. A fixed point of a function f is a real number c such that $f(c) = c$.

1. $f(x) = \tan \dfrac{\pi x}{4}$ **2.** $f(x) = \cos x$

3. Use the figure to derive the formulas for

$$\sin \frac{\theta}{2}, \cos \frac{\theta}{2}, \text{ and } \tan \frac{\theta}{2}$$

where θ is an acute angle.

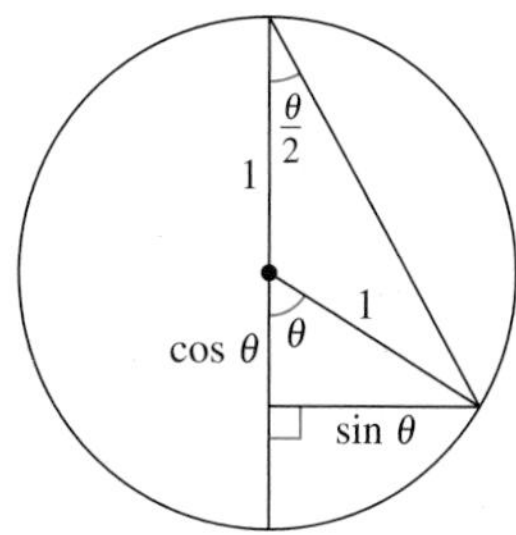

4. The force F (in pounds) on a person's back when he or she bends over at an angle θ is modeled by

$$F = \frac{0.6W \sin(\theta + 90°)}{\sin 12°}$$

where W is the person's weight (in pounds).

(a) Simplify the model.

(b) Use a graphing utility to graph the model, where $W = 185$ and $0° < \theta < 90°$.

(c) At what angle is the force a maximum? At what angle is the force a minimum?

5. A weight is attached to a spring suspended vertically from a ceiling. When a driving force is applied to the system, the weight moves vertically from its equilibrium position, and this motion is modeled by

$$y = \frac{1}{3} \sin 2t + \frac{1}{4} \cos 2t$$

where y is the distance from equilibrium measured in feet and t is the time in seconds.

(a) Use the identity

$$a \sin B\theta + b \cos B\theta = \sqrt{a^2 + b^2} \sin(B\theta + C)$$

where $C = \arctan(b/a)$, $a > 0$, to write the model in the form

$$y = \sqrt{a^2 + b^2} \sin(Bt + C).$$

(b) Find the amplitude of the oscillations of the weight.

(c) Find the frequency of the oscillations of the weight.

6. Consider the function given by

$$f(x) = 3 \sin(0.6x - 2).$$

(a) Approximate the zero of the function in the interval $[0, 6]$.

(b) A quadratic approximation agreeing with f at $x = 5$ is

$$g(x) = -0.45x^2 + 5.52x - 13.70.$$

Use a graphing utility to graph f and g in the same viewing window. Describe the result.

(c) Use the Quadratic Formula to find the zeros of g. Compare the zero in the interval $[0, 6]$ with the result of part (a).

7. The equation of a standing wave is obtained by adding the displacements of two waves traveling in opposite directions (see figure). Assume that each of the waves has amplitude A, period T, and wavelength λ. If the models for these waves are

$$y_1 = A \cos 2\pi\left(\frac{t}{T} - \frac{x}{\lambda}\right) \quad \text{and}$$

$$y_2 = A \cos 2\pi\left(\frac{t}{T} + \frac{x}{\lambda}\right)$$

show that

$$y_1 + y_2 = 2A \cos \frac{2\pi t}{T} \cos \frac{2\pi x}{\lambda}.$$

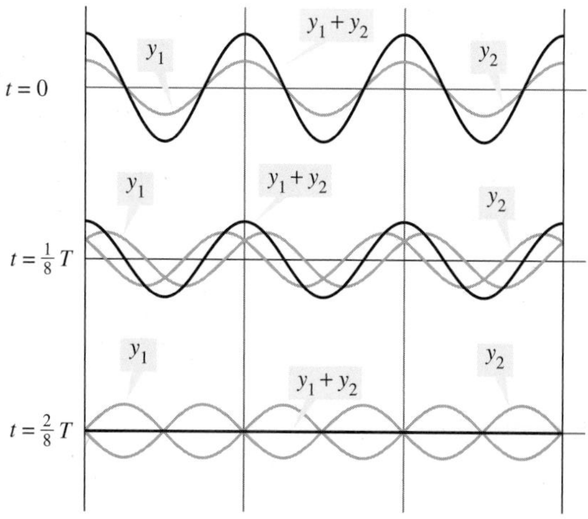

8. Find the solution of each inequality in the interval $[0, 2\pi]$.

(a) $\sin x \geq 0.5$

(b) $\cos x \leq -0.5$

(c) $\tan x < \sin x$

(d) $\cos x \geq \sin x$

In Exercises 9 and 10, use the figure, which shows two lines whose equations are

$$y_1 = m_1x + b_1 \quad \text{and} \quad y_2 = m_2x + b_2.$$

Assume that both lines have positive slopes.

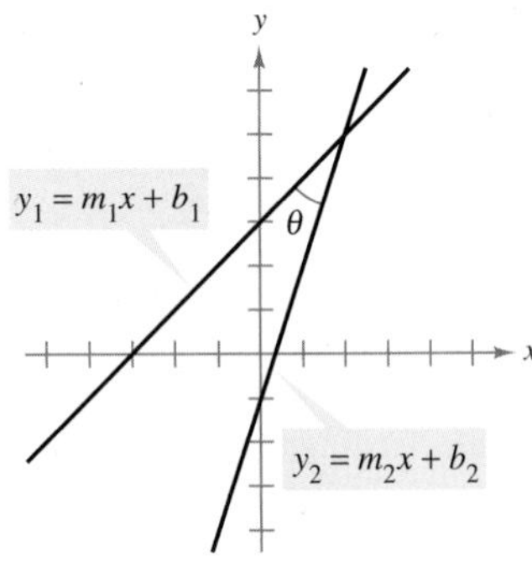

9. Derive a formula for the angle θ between the two lines.

10. Use your formula from Exercise 9 to find the angle between the given pair of lines.

(a) $y = x$ and $y = \sqrt{3}x$

(b) $y = x$ and $y = \dfrac{1}{\sqrt{3}}x$

11. Consider the function given by

$$f(\theta) = \sin^2\left(\theta + \frac{\pi}{4}\right) + \sin^2\left(\theta - \frac{\pi}{4}\right).$$

Use a graphing utility to graph the function and use the graph to write an identity. Prove your conjecture.

12. Three squares of side s are placed side by side (see figure). Make a conjecture about the relationship between $u + v$ and w. Prove your conjecture by using the identity for the tangent of the sum of two angles.

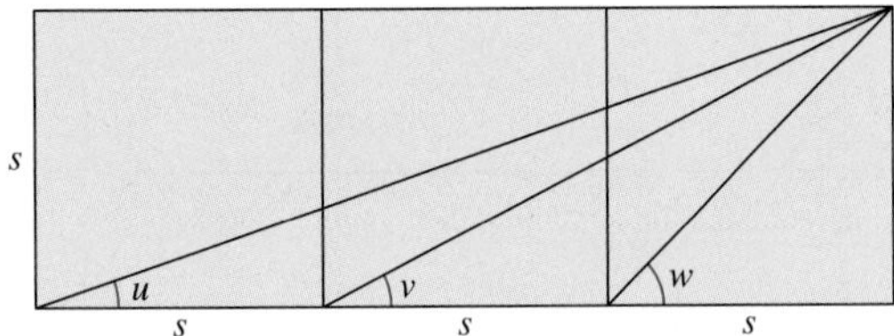

13. (a) Write a sum formula for $\sin(u + v + w)$.

(b) Write a sum formula for $\tan(u + v + w)$.

14. (a) Derive a formula for $\cos 3\theta$.

(b) Derive a formula for $\cos 4\theta$.

15. The length of each of the two equal sides of an isosceles triangle is 10 meters (see figure). The angle between the two sides is θ.

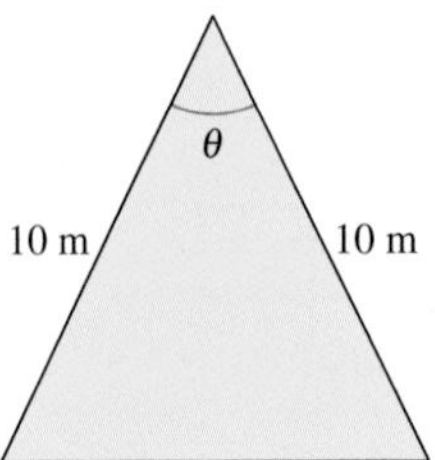

(a) Express the area of the triangle as a function of $\theta/2$.

(b) Express the area of the triangle as a function of θ. Determine the value of θ such that the area is a maximum.

16. When two railroad tracks merge, the overlapping portions of the tracks are in the shapes of circular arcs (see figure). The radius of each arc r (in feet) and the angle θ are related by

$$\frac{x}{2} = 2r \sin^2 \frac{\theta}{2}.$$

Write a formula for x in terms of $\cos \theta$.

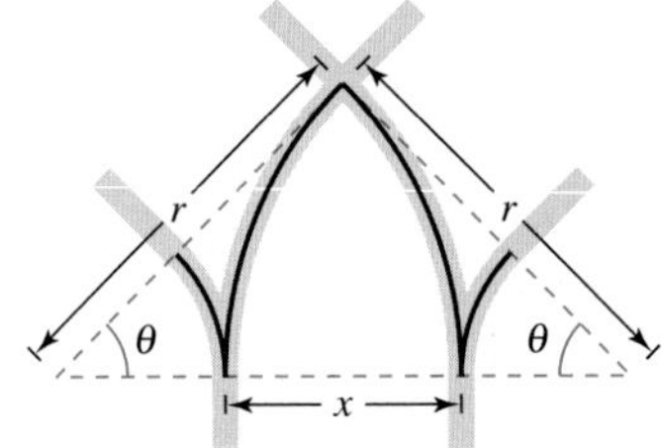

17. Determine all values of the constant a such that the following function is continuous for all real numbers.

$$f(x) = \begin{cases} \dfrac{ax}{\tan x}, & x \geq 0 \\ a^2 - 2, & x < 0 \end{cases}$$

18. For $x > 0$, show that $A + B = \dfrac{\pi}{4}$. See figure.

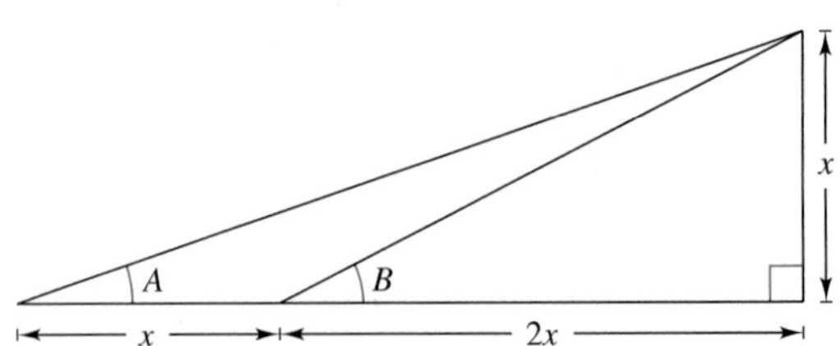

11 Trigonometric Functions and Calculus

So far in your study of calculus, you have learned to find limits, differentiate, and integrate several types of functions, including algebraic, exponential, and logarithmic functions. In this chapter, the first five sections discuss calculus of trigonometric functions, and the last section briefly discusses a class of functions called *hyperbolic functions.*

In this chapter, you should learn the following.

- How to determine limits of trigonometric functions. (11.1)
- How to find derivatives of trigonometric functions. (11.2)
- How to find antiderivatives of trigonometric functions. (11.3)
- How to find derivatives of inverse trigonometric functions. (11.4)
- How to find antiderivatives involving inverse trigonometric functions. (11.5)
- The properties of hyperbolic functions. How to find derivatives and antiderivatives of hyperbolic functions. (11.6)

Owaki - Kulla/Photolibrary

The Gateway Arch in St. Louis, Missouri is over 600 feet high and covered with 886 tons of quarter-inch stainless steel. A mathematical equation used to construct the arch involves which function? (See Section 11.6, Section Project.)

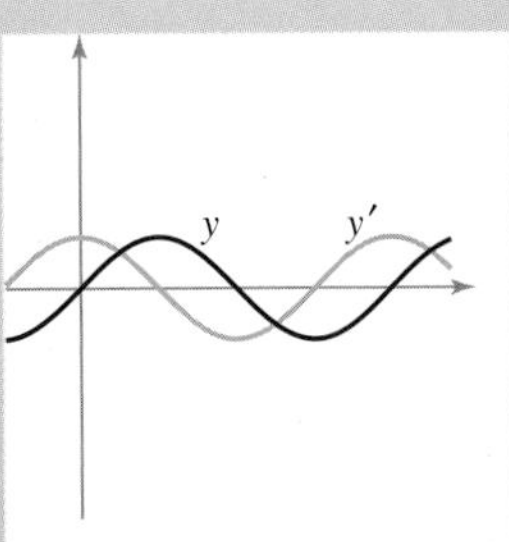

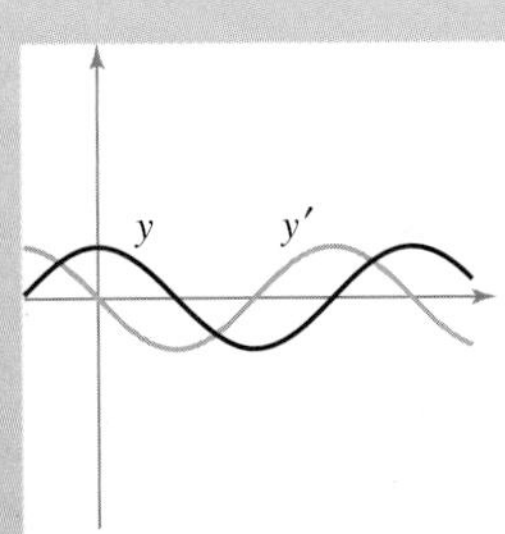

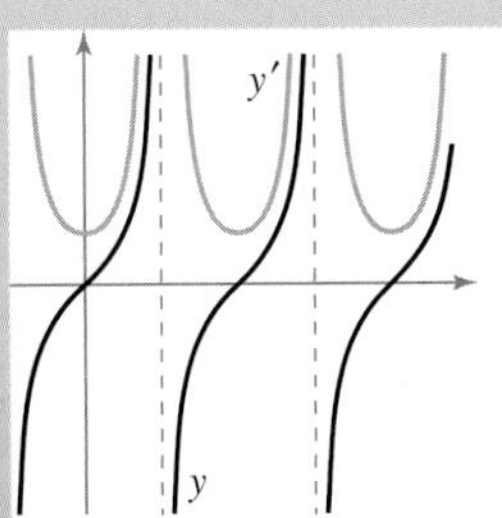

You can use derivatives of trigonometric functions to find slopes and help you sketch the graphs of trigonometric functions. (See Section 11.2.)

11.1 Limits of Trigonometric Functions

- **Determine the limits of trigonometric functions.**

Limits of Trigonometric Functions

In the next section, you will see how the derivatives of the trigonometric functions can assist you in sketching graphs of trigonometric functions. However, before discussing these derivatives, you need to obtain some results regarding the limits of trigonometric functions. In Chapter 3, you saw that the limits of many algebraic functions can be evaluated by direct substitution. Each of the six basic trigonometric functions also exhibit this desirable quality, as shown in the next theorem (presented without proof).

THEOREM 11.1 LIMITS OF TRIGONOMETRIC FUNCTIONS

Let c be a real number in the domain of the given trigonometric function.

1. $\lim_{x \to c} \sin x = \sin c$ **2.** $\lim_{x \to c} \cos x = \cos c$

3. $\lim_{x \to c} \tan x = \tan c$ **4.** $\lim_{x \to c} \cot x = \cot c$

5. $\lim_{x \to c} \sec x = \sec c$ **6.** $\lim_{x \to c} \csc x = \csc c$

NOTE From Theorem 11.1, it follows that each of the six trigonometric functions is *continuous* at every point in its domain. ■

EXAMPLE 1 Limits Involving Trigonometric Functions

a. By Theorem 11.1, you have

$$\lim_{x \to 0} \sin x = \sin(0) = 0.$$

b. By Theorem 11.1 and Property 3 of Theorem 3.2, you have

$$\lim_{x \to \pi} (x \cos x) = \left[\lim_{x \to \pi} x\right]\left[\lim_{x \to \pi} \cos x\right] = \pi \cos(\pi) = -\pi.$$

EXAMPLE 2 A Limit Involving Trigonometric Functions

Find the following limit.

$$\lim_{x \to 0} \frac{\tan x}{\sin x}$$

Solution Direct substitution yields the indeterminate form $0/0$. However, by using the fact that $\tan x = (\sin x)/(\cos x)$, you can rewrite the function as

$$\frac{\tan x}{\sin x} = \frac{(\sin x)/(\cos x)}{\sin x} = \frac{\cancel{\sin x}}{(\cos x)(\cancel{\sin x})} = \frac{1}{\cos x}.$$

So, you have

$$\lim_{x \to 0} \frac{\tan x}{\sin x} = \lim_{x \to 0} \frac{1}{\cos x} = \frac{1}{1} = 1.$$

■

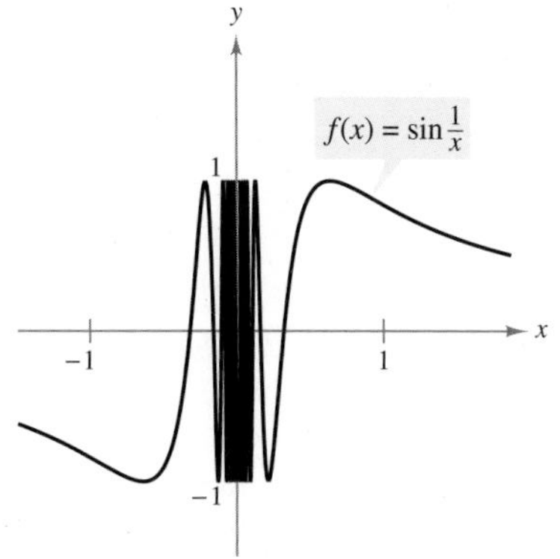

$\lim_{x \to 0} f(x)$ does not exist.

Figure 11.1

EXAMPLE 3 A Limit That Does Not Exist

Discuss the existence of the limit

$$\lim_{x \to 0} \sin(1/x).$$

Solution Let $f(x) = \sin(1/x)$. In Figure 11.1, it appears that as x approaches 0, $f(x)$ oscillates between -1 and 1. So, the limit does not exist because no matter how small you choose δ, it is possible to choose x_1 and x_2 within δ units of 0 such that $\sin(1/x_1) = 1$ and $\sin(1/x_2) = -1$, as shown in the table.

x	$2/\pi$	$2/(3\pi)$	$2/(5\pi)$	$2/(7\pi)$	$2/(9\pi)$	$x \to 0$
$\sin(1/x)$	1	-1	1	-1	1	Limit does not exist.

EXAMPLE 4 Testing for Continuity

Describe the interval(s) on which each function is continuous.

a. $f(x) = \tan x$ **b.** $h(x) = \begin{cases} x\sin(1/x), & x \neq 0 \\ 0, & x = 0 \end{cases}$

Solution

a. The tangent function is undefined at $x = (\pi/2) + n\pi$, where n is an integer. At all other points it is continuous. So, $f(x) = \tan x$ is continuous on the open intervals

$$\ldots, \left(-\frac{3\pi}{2}, -\frac{\pi}{2}\right), \left(-\frac{\pi}{2}, \frac{\pi}{2}\right), \left(\frac{\pi}{2}, \frac{3\pi}{2}\right), \ldots$$

as shown in Figure 11.2(a).

b. This function is similar to the function in Example 3 except that the oscillations are damped by the factor x. Using the Squeeze Theorem, you obtain

$$-|x| \le x\sin(1/x) \le |x|, \quad x \neq 0$$

and you can conclude that the limit as $x \to 0$ is zero. So, h is continuous on the entire real line, as shown in Figure 11.2(b).

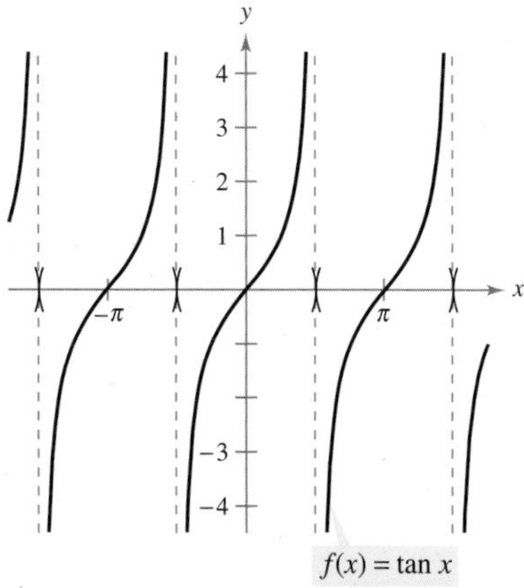

(a) f is continuous on each open interval in its domain.

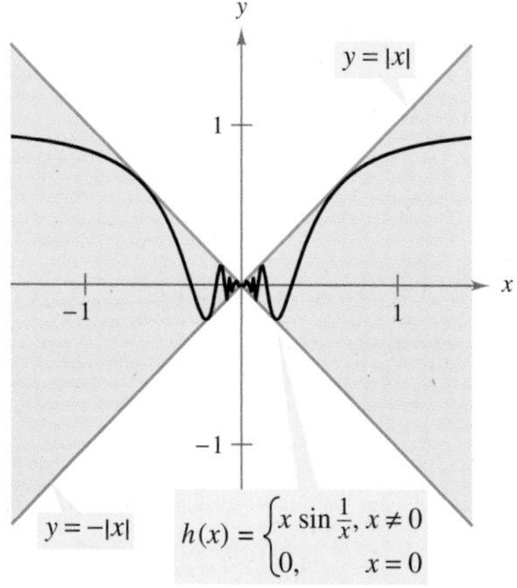

(b) h is continuous on the entire real line.

Figure 11.2

■

In the next section, you will see that the following two important limits are useful in determining the derivatives of trigonometric functions.

THEOREM 11.2 TWO SPECIAL TRIGONOMETRIC LIMITS

1. $\lim_{x \to 0} \frac{\sin x}{x} = 1$ **2.** $\lim_{x \to 0} \frac{1 - \cos x}{x} = 0$

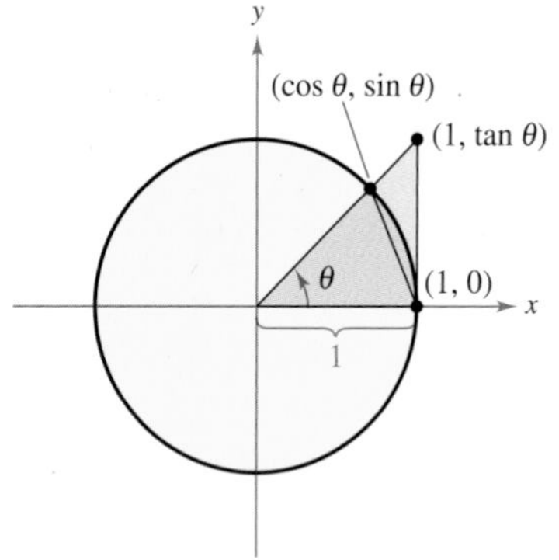

A circular sector is used to prove Theorem 11.2.
Figure 11.3

PROOF To avoid the confusion of two different uses of x, the proof is presented using the variable θ, where θ is an acute positive angle *measured in radians*. Figure 11.3 shows a circular section that is squeezed between two triangles. The areas of the three figures below satisfy the inequality that follows.

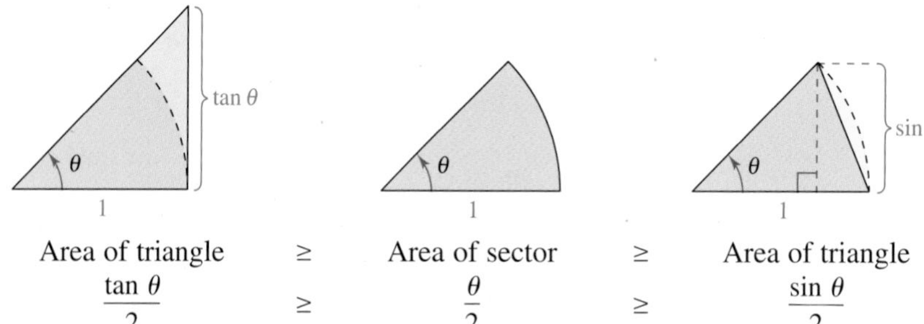

Area of triangle $\geq$ Area of sector $\geq$ Area of triangle

$$\frac{\tan \theta}{2} \geq \frac{\theta}{2} \geq \frac{\sin \theta}{2}$$

Multiplying each expression by $2/\sin \theta$ produces

$$\frac{1}{\cos \theta} \geq \frac{\theta}{\sin \theta} \geq 1$$

and taking reciprocals and reversing the inequalities yields

$$\cos \theta \leq \frac{\sin \theta}{\theta} \leq 1.$$

Because

$$\cos \theta = \cos(-\theta)$$

and

$$\frac{\sin \theta}{\theta} = \frac{\sin(-\theta)}{-\theta}$$

you can conclude that this inequality is valid for all nonzero θ in the open interval $(-\pi/2, \pi/2)$. Finally, because

$$\lim_{\theta \to 0} \cos \theta = 1$$

and

$$\lim_{\theta \to 0} 1 = 1$$

you can apply the Squeeze Theorem to conclude that

$$\lim_{\theta \to 0} \frac{\sin \theta}{\theta} = 1.$$

■

The proof of the second limit is left as an exercise. (See Exercise 73.)

TECHNOLOGY Use a graphing utility to confirm the limits in the examples and exercise set. For instance, Figure 11.4 shows the graph of

$$f(x) = \frac{\tan x}{x}.$$

Note that the graph appears to contain the point (0, 1), which lends support to the conclusions obtained in Example 5.

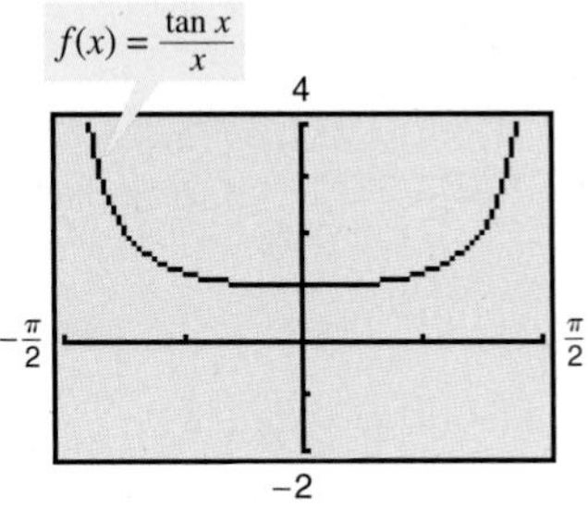

The limit of $f(x)$ as x approaches 0 is 1.
Figure 11.4

EXAMPLE 5 A Limit Involving a Trigonometric Function

Find the limit: $\lim_{x \to 0} \frac{\tan x}{x}$.

Solution Direct substitution yields the indeterminate form 0/0. To solve this problem, you can write $\tan x = (\sin x)/(\cos x)$ and obtain

$$\lim_{x \to 0} \frac{\tan x}{x} = \lim_{x \to 0} \left(\frac{\sin x}{x}\right)\left(\frac{1}{\cos x}\right).$$

Now, because

$$\lim_{x \to 0} \frac{\sin x}{x} = 1 \quad \text{and} \quad \lim_{x \to 0} \frac{1}{\cos x} = 1$$

you can obtain

$$\lim_{x \to 0} \frac{\tan x}{x} = \left(\lim_{x \to 0} \frac{\sin x}{x}\right)\left(\lim_{x \to 0} \sec x\right) = (1)(1) = 1.$$

EXAMPLE 6 A Limit Involving a Trigonometric Function

Find the limit: $\lim_{x \to 0} \frac{\sin 2x}{x}$.

Solution Direct substitution gives the indeterminate form 0/0. To solve this problem, you can rewrite the limit as

$$\lim_{x \to 0} \frac{\sin 2x}{x} = 2\left(\lim_{x \to 0} \frac{\sin 2x}{2x}\right). \qquad \text{Multiply and divide by 2.}$$

Now, by letting $y = 2x$ and observing that $x \to 0$ if and only if $y \to 0$, you can write

$$\lim_{x \to 0} \frac{\sin 2x}{x} = 2\left(\lim_{x \to 0} \frac{\sin 2x}{2x}\right) = 2\left(\lim_{y \to 0} \frac{\sin y}{y}\right) = 2(1) = 2.$$ ■

11.1 Exercises

See www.CalcChat.com for worked-out solutions to odd-numbered exercises.

In Exercises 1 and 2, complete the table and use the result to estimate the limit. Use a graphing utility to graph the function to confirm your result.

1. $\lim_{x \to 0} \frac{\sin x}{x}$

x	−0.1	−0.01	−0.001	0.001	0.01	0.1
$f(x)$						

2. $\lim_{x \to 0} \frac{\cos x - 1}{x}$

x	−0.1	−0.01	−0.001	0.001	0.01	0.1
$f(x)$						

In Exercises 3–6, use the graph to find the limit (if it exists). If the limit does not exist, explain why.

3. $\lim_{x \to \pi/2} \tan x$

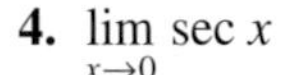

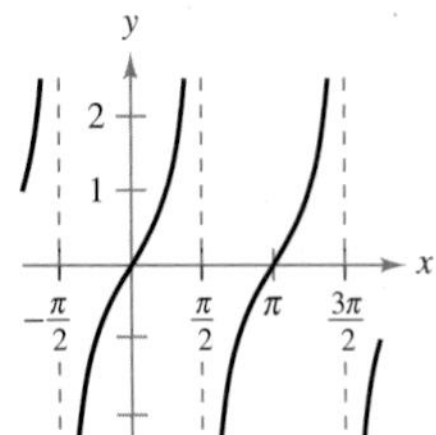

4. $\lim_{x \to 0} \sec x$

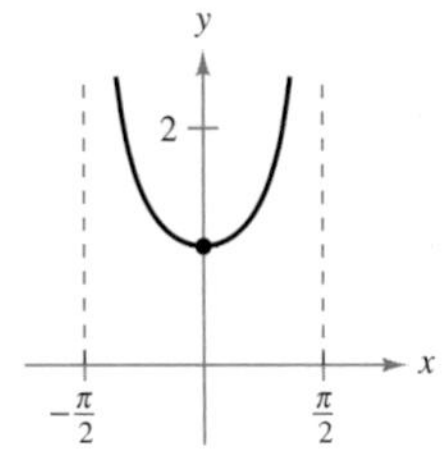

5. $\lim_{x \to 0} \cos \frac{1}{x}$

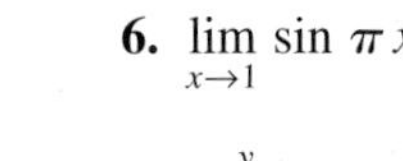

6. $\lim_{x \to 1} \sin \pi x$

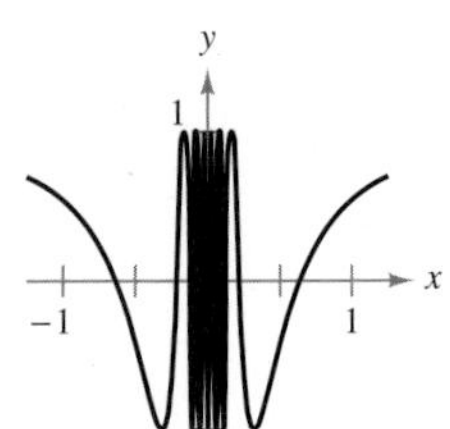

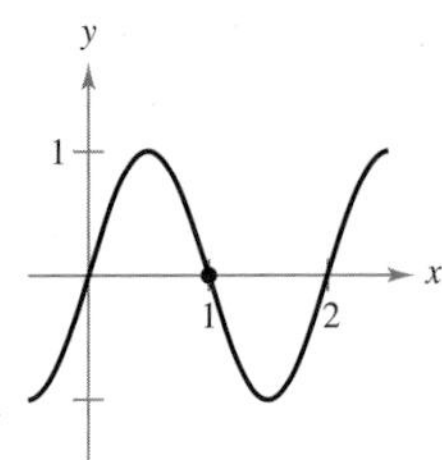

In Exercises 7–16, find the limit of the trigonometric function.

7. $\lim_{x \to \pi/2} \sin x$

8. $\lim_{x \to \pi} \tan x$

9. $\lim_{x \to 1} \cos \frac{\pi x}{3}$

10. $\lim_{x \to 2} \sin \frac{\pi x}{2}$

11. $\lim_{x \to 0} \sec 2x$

12. $\lim_{x \to \pi} \cos 3x$

13. $\lim_{x \to 5\pi/6} \sin x$

14. $\lim_{x \to 5\pi/3} \cos x$

15. $\lim_{x \to 3} \tan\left(\frac{\pi x}{4}\right)$

16. $\lim_{x \to 7} \sec\left(\frac{\pi x}{6}\right)$

In Exercises 17–22, find the *x*-values (if any) at which *f* is not continuous. Which of the discontinuities are removable?

17. $f(x) = 3x - \cos x$

18. $f(x) = \cos \frac{\pi x}{2}$

19. $f(x) = \begin{cases} \tan \frac{\pi x}{4}, & |x| < 1 \\ x, & |x| \geq 1 \end{cases}$

20. $f(x) = \begin{cases} \csc \frac{\pi x}{6}, & |x - 3| \leq 2 \\ 2, & |x - 3| > 2 \end{cases}$

21. $f(x) = \csc 2x$

22. $f(x) = \tan \frac{\pi x}{2}$

Writing **In Exercises 23 and 24, use a graphing utility to graph the function on the interval $[-4, 4]$. Does the graph of the function appear to be continuous on this interval? Is the function continuous on $[-4, 4]$? Write a short paragraph about the importance of examining a function analytically as well as graphically.**

23. $f(x) = \frac{\sin x}{x}$

24. $f(x) = \frac{\tan x}{x}$

In Exercises 25–28, find the vertical asymptotes (if any) of the graph of the function.

25. $f(x) = \tan \pi x$

26. $f(x) = \sec \pi x$

27. $s(t) = \frac{t}{\sin t}$

28. $g(\theta) = \frac{\tan \theta}{\theta}$

Graphical, Numerical, and Analytic Analysis **In Exercises 29–32, use a graphing utility to graph the function and estimate the limit. Use a table to reinforce your conclusion. Then find the limit by analytic methods.**

29. $\lim_{t \to 0} \frac{\sin 3t}{t}$

30. $\lim_{x \to 0} \frac{\cos x - 1}{2x^2}$

31. $\lim_{x \to 0} \frac{\sin x^2}{x}$

32. $\lim_{x \to 0} \frac{\sin x}{\sqrt[3]{x}}$

In Exercises 33–38, use a graphing utility to graph the given function and the equations $y = |x|$ and $y = -|x|$ in the same viewing window. Using the graphs to observe the Squeeze Theorem visually, find $\lim_{x \to 0} f(x)$.

33. $f(x) = x \cos x$

34. $f(x) = |x \sin x|$

35. $f(x) = |x| \sin x$

36. $f(x) = |x| \cos x$

37. $f(x) = x \sin \frac{1}{x}$

38. $h(x) = x \cos \frac{1}{x}$

In Exercises 39–58, determine the limit of the trigonometric function (if it exists).

39. $\lim_{x \to 0} \frac{\sin x}{5x}$

40. $\lim_{x \to 0} \frac{3(1 - \cos x)}{x}$

41. $\lim_{x \to 0} \frac{\sin x(1 - \cos x)}{x^2}$

42. $\lim_{\theta \to 0} \frac{\cos \theta \tan \theta}{\theta}$

43. $\lim_{x \to 0} \frac{\sin^2 x}{x}$

44. $\lim_{x \to 0} \frac{\tan^2 x}{x}$

45. $\lim_{h \to 0} \frac{(1 - \cos h)^2}{h}$

46. $\lim_{\phi \to \pi} \phi \sec \phi$

47. $\lim_{x \to \pi/2} \frac{\cos x}{\cot x}$

48. $\lim_{x \to \pi/4} \frac{1 - \tan x}{\sin x - \cos x}$

49. $\lim_{t \to 0} \frac{\sin 3t}{2t}$

50. $\lim_{x \to 0} \frac{\sin 2x}{\sin 3x}$ $\left[\textit{Hint:} \text{ Find } \lim_{x \to 0} \left(\frac{2 \sin 2x}{2x}\right)\left(\frac{3x}{3 \sin 3x}\right).\right]$

51. $\lim_{x \to \pi} \cot x$

52. $\lim_{x \to \pi/2} \sec x$

53. $\lim_{x \to 0^+} \frac{2}{\sin x}$

54. $\lim_{x \to (\pi/2)^+} \frac{-2}{\cos x}$

55. $\lim_{x \to \pi} \frac{\sqrt{x}}{\csc x}$

56. $\lim_{x \to 0} \frac{x + 2}{\cot x}$

57. $\lim_{x \to 1/2} x \sec \pi x$

58. $\lim_{x \to 1/2} x^2 \tan \pi x$

WRITING ABOUT CONCEPTS

59. Write a brief description of the meaning of the notation $\lim_{x \to \pi/6} \sin x = \frac{1}{2}$.

60. Write a trigonometric function with a vertical asymptote at $x = 2n$, where n is an integer.

61. Use a graphing utility to evaluate the limit $\lim_{x \to 0} \frac{\tan nx}{x}$ for several values of n. What do you notice?

CAPSTONE

62. (a) If $f(3) = 8$, can you conclude anything about $\lim_{x \to 3} f(x)$? Explain.

(b) If the limit of $f(x)$ as x approaches π is 4, can you conclude anything about $f(\pi)$? Explain.

63. ***Writing*** Use a graphing utility to graph

$$f(x) = x, \quad g(x) = \sin x, \quad \text{and} \quad h(x) = \frac{\sin x}{x}$$

in the same viewing window. Compare the magnitudes of $f(x)$ and $g(x)$ when x is close to 0. Use your comparison to write a short paragraph explaining why

$$\lim_{x \to 0} h(x) = 1.$$

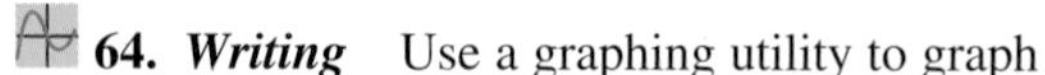

64. ***Writing*** Use a graphing utility to graph

$$f(x) = x, \quad g(x) = \sin^2 x, \quad \text{and} \quad h(x) = \frac{\sin^2 x}{x}$$

in the same viewing window. Compare the magnitudes of $f(x)$ and $g(x)$ when x is close to 0. Use your comparison to write a short paragraph explaining why

$$\lim_{x \to 0} h(x) = 0.$$

65. ***Numerical and Graphical Analysis*** Use a graphing utility to complete the table for each function and graph each function to approximate the limit. Describe the changes in the limits with increasing powers of x.

x	1	0.5	0.2	0.1	0.01	0.001	0.0001
$f(x)$							

(a) $\lim_{x \to 0^+} \frac{x - \sin x}{x}$

(b) $\lim_{x \to 0^+} \frac{x - \sin x}{x^2}$

(c) $\lim_{x \to 0^+} \frac{x - \sin x}{x^3}$

(d) $\lim_{x \to 0^+} \frac{x - \sin x}{x^4}$

66. ***Graphical Reasoning*** Consider $f(x) = \frac{\sec x - 1}{x^2}$.

(a) Find the domain of f.

(b) Use a graphing utility to graph f. Is the domain of f obvious from the graph? If not, explain.

(c) Use the graph of f to approximate $\lim_{x \to 0} f(x)$.

(d) Confirm your answer in part (c) analytically.

67. ***Approximation***

(a) Find $\lim_{x \to 0} \frac{1 - \cos x}{x^2}$.

(b) Use your answer to part (a) to derive the approximation $\cos x \approx 1 - \frac{1}{2}x^2$ for x near 0.

(c) Use your answer to part (b) to approximate $\cos(0.1)$.

(d) Use a calculator to approximate $\cos(0.1)$ to four decimal places. Compare the result with part (c).

68. ***Rate of Change*** A patrol car is parked 50 feet from a long warehouse (see figure). The revolving light on top of the car turns at a rate of $\frac{1}{2}$ revolution per second. The rate at which the light beam moves along the wall is $r = 50\pi \sec^2 \theta$ ft/sec.

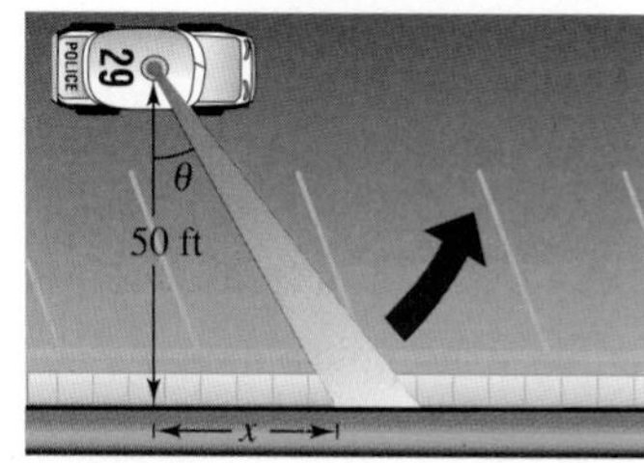

(a) Find the rate r when θ is $\pi/6$.

(b) Find the rate r when θ is $\pi/3$.

(c) Find the limit of r as $\theta \to (\pi/2)^-$.

69. ***Numerical and Graphical Analysis*** Consider the shaded region outside the sector of a circle of radius 10 meters and inside a right triangle (see figure).

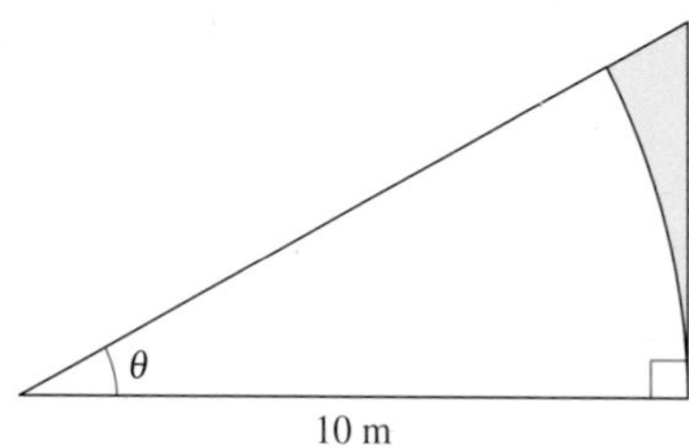

(a) Write the area $A = f(\theta)$ of the region as a function of θ. Determine the domain of the function.

(b) Use a graphing utility to complete the table and graph the function over the appropriate domain.

θ	0.3	0.6	0.9	1.2	1.5
$f(\theta)$					

(c) Find the limit of A as $\theta \to (\pi/2)^-$.

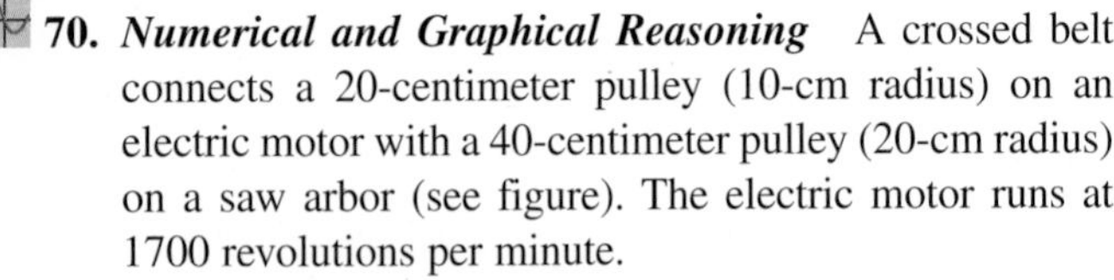

70. ***Numerical and Graphical Reasoning*** A crossed belt connects a 20-centimeter pulley (10-cm radius) on an electric motor with a 40-centimeter pulley (20-cm radius) on a saw arbor (see figure). The electric motor runs at 1700 revolutions per minute.

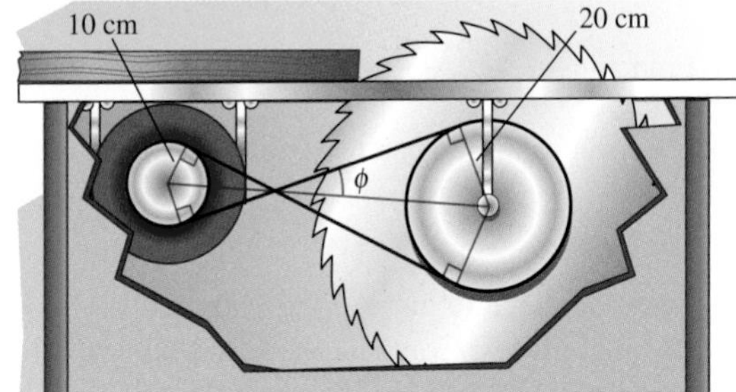

(a) Determine the number of revolutions per minute of the saw.

(b) How does crossing the belt affect the saw in relation to the motor?

(c) Let L be the total length of the belt. Write L as a function of the angle ϕ (see figure), where ϕ is measured in radians. What is the domain of the function? (*Hint:* Add the lengths of the straight sections of the belt and the length of the belt around each pulley.)

(d) Use a graphing utility to complete the table.

ϕ	0.3	0.6	0.9	1.2	1.5
L					

(e) Use a graphing utility to graph the function over the appropriate domain.

(f) Find $\lim_{\phi \to (\pi/2)^-} L$. Use a geometric argument as the basis of a second method of finding this limit.

(g) Find $\lim_{\phi \to 0^+} L$.

71. ***Think About It*** When using a graphing utility to generate a table to approximate $\lim_{x \to 0} [(\tan 2x)/x]$, a student concluded that the limit was 0.03491 rather than 2. Determine the probable cause of the error.

72. ***Think About It*** When using a graphing utility to generate a table to approximate $\lim_{x \to 0} [(\sin x)/x]$, a student concluded that the limit was 0.01745 rather than 1. Determine the probable cause of the error.

73. Prove the second part of Theorem 11.2 by proving that

$$\lim_{x \to 0} \frac{1 - \cos x}{x} = 0.$$

74. Prove that for any real number y there exists x in $(-\pi/2, \pi/2)$ such that $\tan x = y$.

Writing **In Exercises 75 and 76, explain why the function has a zero in the given interval.**

	Function	*Interval*
75.	$f(x) = x^2 - 2 - \cos x$	$[0, \pi]$
76.	$f(x) = -\dfrac{5}{x} + \tan\left(\dfrac{\pi x}{10}\right)$	$[1, 4]$

True or False? **In Exercises 77–79, determine whether the statement is true or false. If it is false, explain why or give an example that shows it is false.**

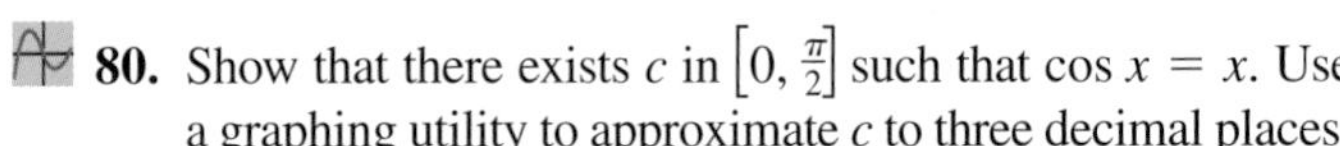

77. Each of the six trigonometric functions is continuous on the set of real numbers.

78. If c is any real number, $\lim_{x \to c} \tan x = \tan c$.

79. The graphs of trigonometric functions have no vertical asymptotes.

80. Show that there exists c in $\left[0, \frac{\pi}{2}\right]$ such that $\cos x = x$. Use a graphing utility to approximate c to three decimal places.

SECTION PROJECT

Graphs and Limits of Trigonometric Functions

Recall from Theorem 11.2 that $\lim_{x \to 0} [(\sin x)/x] = 1$.

(a) Use a graphing utility to graph the function f on the interval $-\pi \le 0 \le \pi$. Explain how this graph helps confirm this theorem.

(b) Explain how you could use a table of values to confirm the value of this limit numerically.

(c) Graph $g(x) = \sin x$ by hand. Sketch a tangent line at the point $(0, 0)$ and visually estimate the slope of this tangent line.

(d) Let $(x, \sin x)$ be a point on the graph of g near $(0, 0)$ and write a formula for the slope of the secant line joining $(x, \sin x)$ and $(0, 0)$. Evaluate this formula at $x = 0.1$ and $x = 0.01$. Then find the exact slope of the tangent line to g at the point $(0, 0)$.

(e) Sketch the graph of the cosine function $h(x) = \cos x$. What is the slope of the tangent line at the point $(0, 1)$? Use limits to find this slope analytically.

(f) Find the slope of the tangent line to $k(x) = \tan x$ at $(0, 0)$.

11.2 Trigonometric Functions: Differentiation

- **Find and use the derivatives of the sine and cosine functions.**
- **Find and use the derivatives of other trigonometric functions.**
- **Apply the First Derivative Test to find the minima and maxima of a function.**

Derivatives of Sine and Cosine Functions

In the preceding section, you studied the following limits.

$$\lim_{\Delta x \to 0} \frac{\sin \Delta x}{\Delta x} = 1 \quad \text{and} \quad \lim_{\Delta x \to 0} \frac{1 - \cos \Delta x}{\Delta x} = 0$$

These two limits are crucial in the proofs of the derivatives of the sine and cosine functions. The derivatives of the other four trigonometric functions follow easily from these two.

THEOREM 11.3 DERIVATIVES OF SINE AND COSINE

$$\frac{d}{dx}[\sin x] = \cos x \qquad \frac{d}{dx}[\cos x] = -\sin x$$

FOR FURTHER INFORMATION For the outline of a geometric proof of the derivatives of the sine and cosine functions, see the article "The Spider's Spacewalk Derivation of sin′ and cos′" by Tim Hesterberg in *The College Mathematics Journal*. To view this article, go to the website *www.matharticles.com.*

PROOF

$$\begin{aligned}
\frac{d}{dx}[\sin x] &= \lim_{\Delta x \to 0} \frac{\sin(x + \Delta x) - \sin x}{\Delta x} && \text{Definition of derivative} \\
&= \lim_{\Delta x \to 0} \frac{\sin x \cos \Delta x + \cos x \sin \Delta x - \sin x}{\Delta x} && \text{Trigonometric identity} \\
&= \lim_{\Delta x \to 0} \frac{\cos x \sin \Delta x - \sin x(1 - \cos \Delta x)}{\Delta x} \\
&= \lim_{\Delta x \to 0} \left[\cos x \left(\frac{\sin \Delta x}{\Delta x}\right) - \sin x \left(\frac{1 - \cos \Delta x}{\Delta x}\right)\right] \\
&= \cos x \left(\lim_{\Delta x \to 0} \frac{\sin \Delta x}{\Delta x}\right) - \sin x \left(\lim_{\Delta x \to 0} \frac{1 - \cos \Delta x}{\Delta x}\right) \\
&= (\cos x)(1) - (\sin x)(0) \\
&= \cos x
\end{aligned}$$

This differentiation formula is shown graphically in Figure 11.5. Note that for each x, the *slope* of the sine curve is equal to the *value* of the cosine. Moreover, when y is increasing, y' is positive, and when y is decreasing, y' is negative.

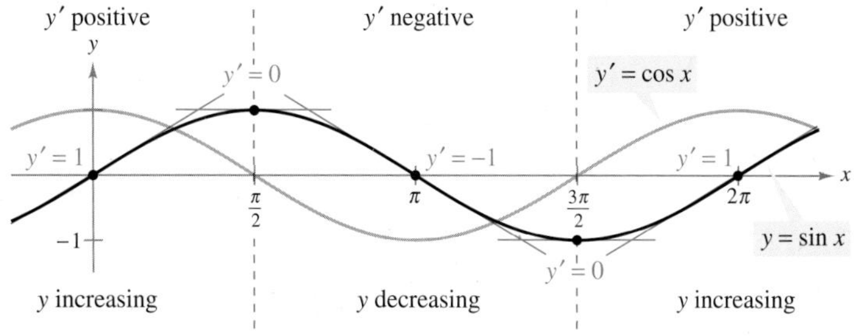

The derivative of the sine function is the cosine function.
Figure 11.5

NOTE The proof of the second rule is left to you.

TECHNOLOGY A graphing utility can provide insight into the interpretation of a derivative. For instance, Figure 11.6 shows the graphs of $y = a \sin x$ for $a = \frac{1}{2}, 1, \frac{3}{2}$, and 2. Estimate the slope of each graph at the point $(0, 0)$. Then verify your estimates analytically by evaluating the derivative of each function when $x = 0$.

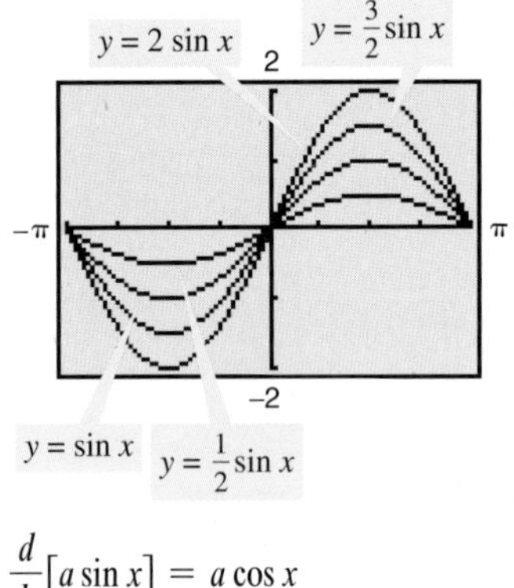

$$\frac{d}{dx}[a \sin x] = a \cos x$$

Figure 11.6

When taking the derivatives of trigonometric functions, the standard differentiation rules still apply—the Sum Rule, the Constant Multiple Rule, the Product Rule, and so on.

EXAMPLE 1 Derivatives Involving Sines and Cosines

Function	*Derivative*
a. $y = 3 \sin x$	$y' = 3 \cos x$
b. $y = \dfrac{\sin x}{2} = \dfrac{1}{2} \sin x$	$y' = \dfrac{1}{2} \cos x = \dfrac{\cos x}{2}$
c. $y = x + \cos x$	$y' = 1 - \sin x$

EXAMPLE 2 A Derivative Involving the Product Rule

Find the derivative of $y = 2x \cos x - 2 \sin x$.

Solution

$$\frac{dy}{dx} = \overbrace{(2x)\left(\frac{d}{dx}[\cos x]\right) + (\cos x)\left(\frac{d}{dx}[2x]\right)}^{\text{Product Rule}} - \overbrace{2\frac{d}{dx}[\sin x]}^{\text{Constant Multiple Rule}}$$

$$= (2x)(-\sin x) + (\cos x)(2) - 2(\cos x)$$

$$= -2x \sin x$$

EXAMPLE 3 Using a Derivative to Find the Slope of a Curve

Find the slope of the graph of $f(x) = 2 \cos x$ at the following points.

a. $\left(-\dfrac{\pi}{2}, 0\right)$

b. $\left(\dfrac{\pi}{3}, 1\right)$

c. $(\pi, -2)$

Solution The derivative of f is $f'(x) = -2 \sin x$. So, the slopes at the indicated points are shown below.

a. At $x = -\dfrac{\pi}{2}$, the slope is

$$f'\left(-\frac{\pi}{2}\right) = -2 \sin\left(-\frac{\pi}{2}\right) = -2(-1) = 2.$$

b. At $x = \dfrac{\pi}{3}$, the slope is

$$f'\left(\frac{\pi}{3}\right) = -2 \sin \frac{\pi}{3} = -2\left(\frac{\sqrt{3}}{2}\right) = -\sqrt{3}.$$

c. At $x = \pi$, the slope is

$$f'(\pi) = -2 \sin \pi = -2(0) = 0.$$

See Figure 11.7. ■

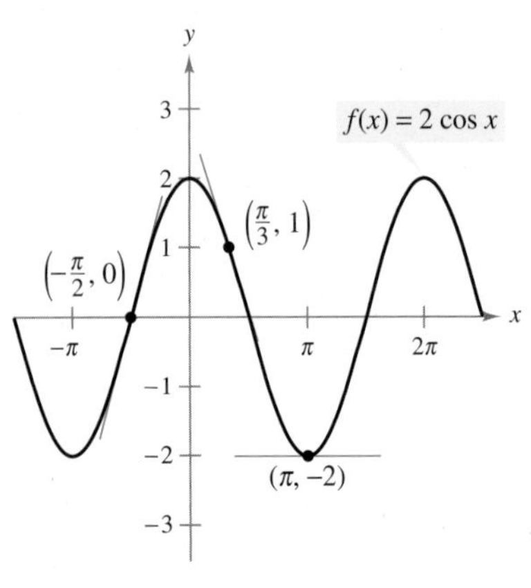

Figure 11.7

Derivatives of Other Trigonometric Functions

Knowing the derivatives of the sine and cosine functions, you can use the Quotient Rule to find the derivatives of the four remaining trigonometric functions.

THEOREM 11.4 DERIVATIVES OF TRIGONOMETRIC FUNCTIONS

$$\frac{d}{dx}[\tan x] = \sec^2 x \qquad \frac{d}{dx}[\cot x] = -\csc^2 x$$

$$\frac{d}{dx}[\sec x] = \sec x \tan x \qquad \frac{d}{dx}[\csc x] = -\csc x \cot x$$

NOTE Because of trigonometric identities, the derivative of a trigonometric function can take many forms. This presents a challenge when you are trying to match your answer to one given in the back of the text.

PROOF Consider $\tan x = (\sin x)/(\cos x)$ and apply the Quotient Rule.

$$\frac{d}{dx}[\tan x] = \frac{(\cos x)(\cos x) - (\sin x)(-\sin x)}{\cos^2 x} = \frac{\cos^2 x + \sin^2 x}{\cos^2 x} = \frac{1}{\cos^2 x} = \sec^2 x$$

■

You are asked to prove the other three derivatives in Theorem 11.4 in Exercise 121.

EXAMPLE 4 Differentiating Trigonometric Functions

Function	Derivative	
a. $y = x - \tan x$	$\dfrac{dy}{dx} = 1 - \sec^2 x$	Difference Rule
b. $y = x \sec x$	$y' = x(\sec x \tan x) + (\sec x)(1)$	Product Rule
	$= (\sec x)(1 + x \tan x)$	Factor.

EXAMPLE 5 Different Forms of a Derivative

Differentiate both forms of $y = \dfrac{1 - \cos x}{\sin x} = \csc x - \cot x$.

Solution

First form: $y = \dfrac{1 - \cos x}{\sin x}$

$$y' = \frac{(\sin x)(\sin x) - (1 - \cos x)(\cos x)}{\sin^2 x} \qquad \text{Quotient Rule}$$

$$= \frac{\sin^2 x + \cos^2 x - \cos x}{\sin^2 x} = \frac{1 - \cos x}{\sin^2 x} \qquad \text{Trigonometric identity}$$

Second form: $y = \csc x - \cot x$

$$y' = -\csc x \cot x + \csc^2 x$$

To show that the two derivatives are equal, you can write

$$\frac{1 - \cos x}{\sin^2 x} = \frac{1}{\sin^2 x} - \frac{\cos x}{\sin^2 x}$$

$$= \frac{1}{\sin^2 x} - \left(\frac{1}{\sin x}\right)\left(\frac{\cos x}{\sin x}\right) = \csc^2 x - \csc x \cot x.$$

■

With the Chain Rule, you can extend the six trigonometric differentiation rules to cover composite functions. The "Chain Rule Versions" of the six basic formulas are summarized below.

DERIVATIVES OF TRIGONOMETRIC FUNCTIONS

$$\frac{d}{dx}[\sin u] = \cos u \frac{du}{dx} \qquad \frac{d}{dx}[\cos u] = -\sin u \frac{du}{dx}$$

$$\frac{d}{dx}[\tan u] = \sec^2 u \frac{du}{dx} \qquad \frac{d}{dx}[\cot u] = -\csc^2 u \frac{du}{dx}$$

$$\frac{d}{dx}[\sec u] = \sec u \tan u \frac{du}{dx} \qquad \frac{d}{dx}[\csc u] = -\csc u \cot u \frac{du}{dx}$$

EXAMPLE 6 Applying the Chain Rule to Trigonometric Functions

Function	*Derivative*	
a. $y = \sin \overbrace{2x}^{u}$	$y' = \overbrace{\cos 2x}^{\cos u} \overbrace{\frac{d}{dx}[2x]}^{u'} = (\cos 2x)(2) = 2 \cos 2x$	
b. $y = \cos(x - 1)$	$y' = -\sin(x - 1)$	$u = x - 1$
c. $y = \tan e^x$	$y' = e^x \sec^2 e^x$	$u = e^x$

■

Be sure that you understand the mathematical conventions regarding parentheses and trigonometric functions. For instance, in part (a) of Example 6, $\sin 2x$ means $\sin(2x)$. The next example shows the effects of different placements of parentheses.

STUDY TIP If you are having difficulty getting the correct answer to a calculus problem, be sure to check that your algebra is correct. Frequent algebraic errors can make calculus seem confusing and difficult.

EXAMPLE 7 Parentheses and Trigonometric Functions

Function	*Derivative*
a. $y = \cos 3x^2 = \cos(3x^2)$	$y' = (-\sin 3x^2)(6x) = -6x \sin 3x^2$
b. $y = (\cos 3)x^2$	$y' = (\cos 3)(2x) = 2x \cos 3$
c. $y = \cos(3x)^2 = \cos(9x^2)$	$y' = (-\sin 9x^2)(18x) = -18x \sin 9x^2$
d. $y = \cos^2 3x = (\cos 3x)^2$	$y' = (2 \cos 3x)D_x[\cos 3x]$
	$= 2(\cos 3x)(-\sin 3x)(3)$
	$= -6 \cos 3x \sin 3x$

NOTE For composite functions such as the one in Example 8, the chain rules are *nested*. In this case, $u = \sin 4t$ for the power function, and $u = 4t$ for the trigonometric function.

EXAMPLE 8 Differentiating a Composite Function

Differentiate $f(t) = \sqrt{\sin 4t}$.

Solution First you can write

$$f(t) = (\sin 4t)^{1/2}.$$

Then, by the Power Rule, you have

$$f'(t) = \frac{1}{2}(\sin 4t)^{-1/2} \frac{d}{dt}[\sin 4t] = \frac{1}{2}(\sin 4t)^{-1/2}(\cos 4t)(4) = \frac{2 \cos 4t}{\sqrt{\sin 4t}}.$$

■

Applications

In the remainder of this section, some applications of the derivative in the context of trigonometric functions will be reviewed. An application to minimum and maximum values of a function on a closed interval is shown in Example 9.

EXAMPLE 9 Finding Extrema on a Closed Interval

Find the extrema of $f(x) = 2 \sin x - \cos 2x$ on the interval $[0, 2\pi]$.

Solution This function is differentiable for all real x, so you can find all critical numbers by setting $f'(x)$ equal to zero, as shown below.

$$f'(x) = 2 \cos x + 2 \sin 2x = 0$$

$$2 \cos x + 4 \sin x \cos x = 0 \qquad \sin 2x = 2 \sin x \cos x$$

$$2(\cos x)(1 + 2 \sin x) = 0 \qquad \text{Factor.}$$

By setting the two factors equal to zero and solving for x in the interval $[0, 2\pi]$, you have the following.

$$\cos x = 0 \implies x = \frac{\pi}{2}, \frac{3\pi}{2} \qquad \text{Critical numbers}$$

$$\sin x = -\frac{1}{2} \implies x = \frac{7\pi}{6}, \frac{11\pi}{6} \qquad \text{Critical numbers}$$

Finally, by evaluating f at these four critical numbers and at the endpoints of the interval, you can conclude that the maximum is $f(\pi/2) = 3$ and the minimum occurs at *two* points, $f(7\pi/6) = -3/2$ and $f(11\pi/6) = -3/2$, as shown in the table.

Left Endpoint	Critical Number	Critical Number	Critical Number	Critical Number	Right Endpoint
$f(x) = -1$	$f\left(\frac{\pi}{2}\right) = 3$ Maximum	$f\left(\frac{7\pi}{6}\right) = -\frac{3}{2}$ Minimum	$f\left(\frac{3\pi}{2}\right) = -1$	$f\left(\frac{11\pi}{6}\right) = -\frac{3}{2}$ Minimum	$f(2\pi) = -1$

The graph is shown in Figure 11.8.

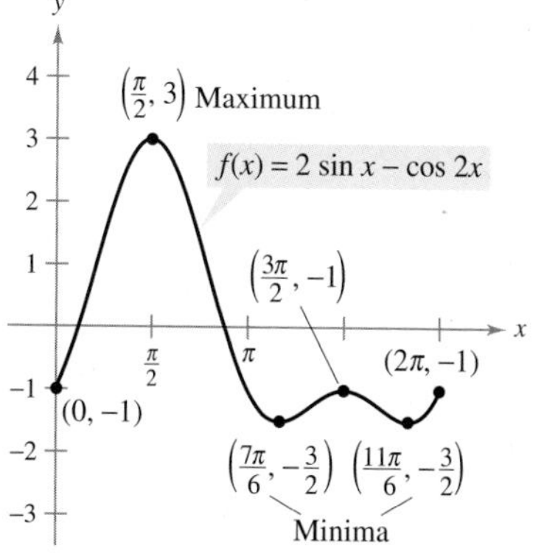

On the closed interval $[0, 2\pi]$, f has two minima at $(7\pi/6, -3/2)$ and $(11\pi/6, -3/2)$ and a maximum at $(\pi/2, 3)$.
Figure 11.8

■

EXAMPLE 10 Modeling Seasonal Sales

A fertilizer manufacturer finds that the sales of one of its fertilizer brands follows a seasonal pattern that can be modeled by

$$F = 100{,}000\left[1 + \sin\frac{2\pi(t-60)}{365}\right], \quad t \geq 0$$

where F is the amount sold (in pounds) and t is the time (in days), with $t = 1$ representing January 1. On which day of the year is the maximum amount of fertilizer sold?

Solution You can find the derivative of the model as shown below.

$$F = 100{,}000\left[1 + \sin\frac{2\pi(t-60)}{365}\right] \qquad \text{Write original function.}$$

$$= 100{,}000 + 100{,}000\sin\frac{2\pi(t-60)}{365} \qquad \text{Distributive Property}$$

$$\frac{dF}{dt} = 100{,}000\cos\frac{2\pi(t-60)}{365}\left[\frac{d}{dt}\left(\frac{2\pi(t-60)}{365}\right)\right] \qquad \text{Apply Chain Rule.}$$

$$= 100{,}000\left(\frac{2\pi}{365}\right)\cos\frac{2\pi(t-60)}{365} \qquad \text{Simplify.}$$

Setting this derivative equal to zero produces

$$\cos\frac{2\pi(t-60)}{365} = 0.$$

Because the cosine is zero at $\frac{\pi}{2}$ and $\frac{3\pi}{2}$, you obtain the following.

$$\frac{2\pi(t-60)}{365} = \frac{\pi}{2} \qquad\qquad \frac{2\pi(t-60)}{365} = \frac{3\pi}{2}$$

$$t - 60 = \frac{365}{4} \qquad\qquad t - 60 = \frac{3(365)}{4}$$

$$t = \frac{365}{4} + 60 \qquad\qquad t = \frac{3(365)}{4} + 60$$

$$t \approx 151 \qquad\qquad t \approx 334$$

The 151st day of the year is May 31 and the 334th day of the year is November 30. From the graph in Figure 11.9, you can see that according to the model, the maximum sales occur on May 31.

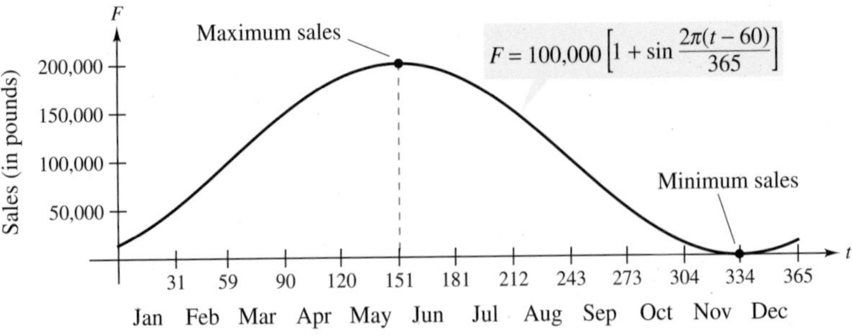

Figure 11.9

■

11.2 Exercises

See www.CalcChat.com for worked-out solutions to odd-numbered exercises.

In Exercises 1–24, use the rules of differentiation to find the derivative of the function.

1. $f(x) = 2\sin x + 3\cos x$
2. $g(t) = \pi\cos t$
3. $f(x) = 6\sqrt{x} + 5\cos x$
4. $g(t) = \sqrt[4]{t} + 6\csc t$
5. $f(x) = -x + \tan x$
6. $y = x + \cot x$
7. $y = \dfrac{1}{x} - 3\sin x$
8. $h(x) = \dfrac{1}{x} - 12\sec x$
9. $f(x) = x^3\cos x$
10. $g(x) = \sqrt{x}\sin x$
11. $f(t) = t^2\sin t$
12. $f(\theta) = (\theta + 1)\cos\theta$
13. $f(t) = \dfrac{\cos t}{t}$
14. $f(x) = \dfrac{\sin x}{x}$
15. $g(x) = \dfrac{\sin x}{x^2}$
16. $f(t) = \dfrac{\cos t}{t^3}$
17. $g(\theta) = \dfrac{\theta}{1 - \sin\theta}$
18. $f(\theta) = \dfrac{\sin\theta}{1 - \cos\theta}$
19. $y = \dfrac{3(1 - \sin x)}{2\cos x}$
20. $y = \dfrac{\sec x}{x}$
21. $y = -\csc x - \sin x$
22. $f(x) = x^2\tan x$
23. $y = 2x\sin x + x^2\cos x$
24. $h(\theta) = 5\theta\sec\theta + \theta\tan\theta$

In Exercises 25 and 26, find the slope of the tangent line to the sine function at the origin. Compare this value with the number of complete cycles in the interval $[0, 2\pi]$. What can you conclude about the slope of the sine function $\sin ax$ at the origin?

25. (a) $y = \sin x$ (b) $y = \sin 2x$

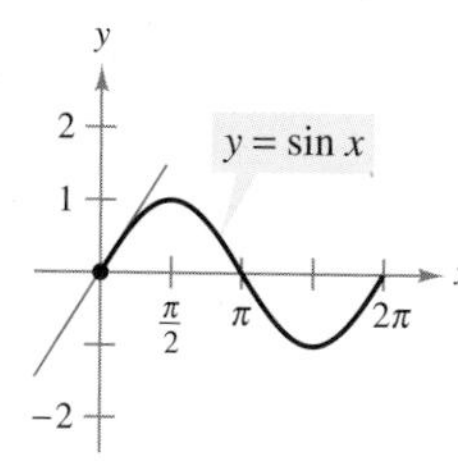

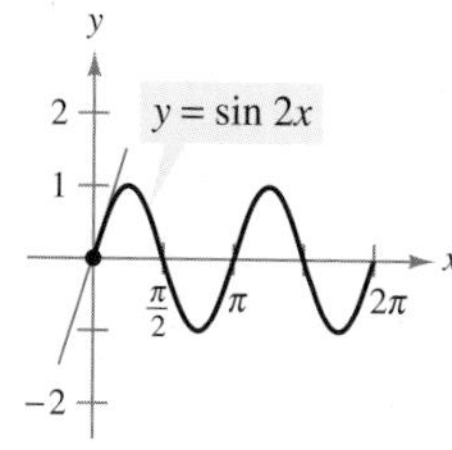

26. (a) $y = \sin 3x$ (b) $y = \sin\dfrac{x}{2}$

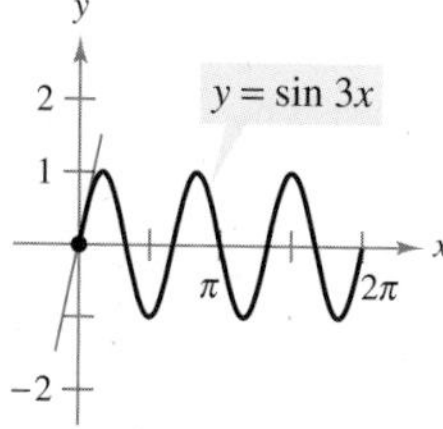

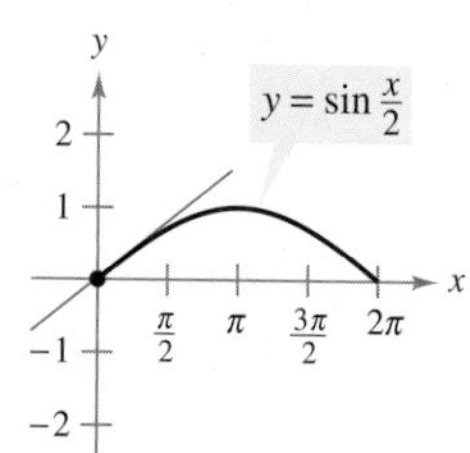

In Exercises 27–30, find an equation of the tangent line to the graph of f at the given point.

27. $f(x) = \sin x\cos x, \quad \left(\dfrac{\pi}{2}, 0\right)$
28. $f(x) = x\sin x + \cos x, \quad (\pi, -1)$
29. $f(x) = e^{\sin x}, \quad (0, 1)$
30. $f(x) = \sin(\cos x), \quad \left(\dfrac{3\pi}{2}, 0\right)$

In Exercises 31–60, find the derivative of the function.

31. $y = \cos 4x$
32. $y = \sin\pi x$
33. $g(x) = 5\tan 3x$
34. $h(x) = \sec x^2$
35. $y = \sin(\pi x)^2$
36. $y = \cos(1 - 2x)^2$
37. $h(x) = \sin 2x\cos 2x$
38. $g(\theta) = \sec\left(\frac{1}{2}\theta\right)\tan\left(\frac{1}{2}\theta\right)$
39. $f(x) = \dfrac{\cot x}{\sin x}$
40. $g(v) = \dfrac{\cos v}{\csc v}$
41. $y = 4\sec^2 x$
42. $y = 2\tan^3 x$
43. $f(\theta) = \frac{1}{4}\sin^2 2\theta$
44. $g(t) = 5\cos^2 \pi t$
45. $f(\theta) = \tan^2 5\theta$
46. $g(\theta) = \cos^2 8\theta$
47. $f(t) = 3\sec^2(\pi t - 1)$
48. $h(t) = 2\cot^2(\pi t + 2)$
49. $y = \sqrt{x} + \frac{1}{4}\sin(2x)^2$
50. $y = 3x - 5\cos(\pi x)^2$
51. $y = \sin\sqrt[3]{x} + \sqrt[3]{\sin x}$
52. $y = e^x(\sin x + \cos x)$
53. $y = \ln|\sin x|$
54. $y = \ln|\csc x|$
55. $y = \ln|\csc x - \cot x|$
56. $y = \ln|\sec x + \tan x|$
57. $y = \ln\left|\dfrac{\cos x}{1 - \sin x}\right|$
58. $y = \ln\sqrt{2 + \cos^2 x}$
59. $y = \sin(\tan 2x)$
60. $y = \cos\sqrt{\sin(\tan\pi x)}$

In Exercises 61 and 62, determine whether the Mean Value Theorem can be applied to f on the closed interval $[a, b]$. If the Mean Value Theorem can be applied, find all values of c in the open interval (a, b) such that $f'(c) = \dfrac{f(b) - f(a)}{b - a}$. If the Mean Value Theorem cannot be applied, explain why not.

61. $f(x) = \sin x, \quad [0, \pi]$
62. $f(x) = \cos x + \tan x, \quad [0, \pi]$

In Exercises 63–66, evaluate the derivative of the function at the given point. Use a graphing utility to verify your result.

Function	*Point*
63. $y = \dfrac{1 + \csc x}{1 - \csc x}$	$\left(\dfrac{\pi}{6}, -3\right)$
64. $f(x) = \tan x \cot x$	$(1, 1)$
65. $h(t) = \dfrac{\sec t}{t}$	$\left(\pi, -\dfrac{1}{\pi}\right)$
66. $f(x) = \sin x(\sin x + \cos x)$	$\left(\dfrac{\pi}{4}, 1\right)$

In Exercises 67 and 68, show that the function $y = f(x)$ is a solution of the differential equation.

67. $y = e^x(\cos\sqrt{2}x + \sin\sqrt{2}x)$

$y'' - 2y' + 3y = 0$

68. $y = e^x(3\cos 2x - 4\sin 2x)$

$y'' - 2y' + 5y = 0$

In Exercises 69 and 70, find the derivatives of the function f for $n = 1, 2, 3$, and 4. Use the results to write a general rule for $f'(x)$ in terms of n.

69. $f(x) = x^n \sin x$

70. $f(x) = \cos x/x^n$

In Exercises 71–76, find dy/dx by implicit differentiation.

71. $\sin x + 2\cos 2y = 1$

72. $(\sin \pi x + \cos \pi y)^2 = 2$

73. $\sin x = x(1 + \tan y)$

74. $\cot y = x - y$

75. $y = \sin xy$

76. $x = \sec 1/y$

Linear and Quadratic Approximations **The linear and quadratic approximations of a function f at $x = a$ are**

$P_1(x) = f'(a)(x - a) + f(a)$ **and**

$P_2(x) = \frac{1}{2}f''(a)(x - a)^2 + f'(a)(x - a) + f(a).$

In Exercises 77–80, (a) find the specified linear and quadratic approximations of f, (b) use a graphing utility to graph f and the approximations, (c) determine whether P_1 or P_2 is the better approximation, and (d) state how the accuracy changes as you move farther from $x = a$.

77. $f(x) = \cos x, \quad a = \dfrac{\pi}{3}$

78. $f(x) = \sin x, \quad a = \dfrac{\pi}{2}$

79. $f(x) = \tan x, \quad a = \dfrac{\pi}{4}$

80. $f(x) = \sec x, \quad a = \dfrac{\pi}{6}$

In Exercises 81 and 82, a point is moving along the graph of the given function such that dx/dt is 2 centimeters per second. Find dy/dt for the given values of x.

81. $y = \tan x$ (a) $x = -\dfrac{\pi}{3}$ (b) $x = -\dfrac{\pi}{4}$ (c) $x = 0$

82. $y = \cos x$ (a) $x = \dfrac{\pi}{6}$ (b) $x = \dfrac{\pi}{4}$ (c) $x = \dfrac{\pi}{3}$

In Exercises 83–86, locate the absolute extrema of the function on the closed interval.

83. $f(x) = \cos \pi x, \quad \left[0, \dfrac{1}{6}\right]$

84. $g(x) = \sec x, \quad \left[-\dfrac{\pi}{6}, \dfrac{\pi}{3}\right]$

85. $y = 3\cos x, \quad [0, 2\pi]$

86. $y = \tan\left(\dfrac{\pi x}{8}\right), \quad [0, 2]$

In Exercises 87–92, determine whether Rolle's Theorem can be applied to f on the closed interval $[a, b]$. If Rolle's Theorem can be applied, find all values of c on the open interval (a, b) such that $f'(c) = 0$. If Rolle's Theorem cannot be applied, explain why not.

87. $f(x) = \sin x, \quad [0, 2\pi]$

88. $f(x) = \cos x, \quad [0, 2\pi]$

89. $f(x) = \cos 2x, \quad [-\pi, \pi]$

90. $f(x) = \dfrac{6x}{\pi} - 4\sin^2 x, \quad \left[0, \dfrac{\pi}{6}\right]$

91. $f(x) = \tan x, \quad [0, \pi]$

92. $f(x) = \sec x, \quad [\pi, 2\pi]$

In Exercises 93–100, sketch a graph of the function over the given interval. Use a graphing utility to verify your graph.

93. $f(x) = 2\sin x + \sin 2x, \quad 0 \le x \le 2\pi$

94. $f(x) = 2\sin x + \cos 2x, \quad 0 \le x \le 2\pi$

95. $y = \sin x - \frac{1}{18}\sin 3x, \quad 0 \le x \le 2\pi$

96. $y = \cos x - \frac{1}{4}\cos 2x, \quad 0 \le x \le 2\pi$

97. $f(x) = x - \sin x, \quad 0 \le x \le 4\pi$

98. $f(x) = \cos x - x, \quad 0 \le x \le 4\pi$

99. $f(x) = 2(\csc x + \sec x), \quad 0 < x < \dfrac{\pi}{2}$

100. $g(x) = x\tan x, \quad -\dfrac{3\pi}{2} < x < \dfrac{3\pi}{2}$

WRITING ABOUT CONCEPTS

101. The derivative of $f(x) = \sin x$ is negative in the interval $(\pi/2, 3\pi/2)$. Explain why $f(2) > f(4)$.

102. Suppose $f(0) = 5$ and $4 \le f'(x) \le 6$ for all x in the interval $[-4, 4]$. Determine the greatest and least possible values of $f(3)$.

103. Sketch the graph of a function f such that $f' < 0$ and $f'' > 0$ for all x.

CAPSTONE

104. Determine if the statement is true. If it is false, explain why and correct it.

(a) $\frac{d}{dx}[\sin x^2] = 2x \cos x^2$ (b) $\frac{d}{dx}[\cos x^2] = 2x \sin x^2$

(c) $\frac{d}{dx}[\tan x^2] = 2x \sec x^2$ (d) $\frac{d}{dx}[\cot x^2] = -2x \csc^2 x^2$

(e) $\frac{d}{dx}[\sec x^2] = 2x \sec x \tan x$

(f) $\frac{d}{dx}[\csc x^2] = -2x \csc x^2 \cot x^2$

105. ***Wave Motion*** A buoy oscillates in simple harmonic motion

$$y = A \cos \omega t$$

as waves move past it. The buoy moves a total of 3.5 feet (vertically) from its low point to its high point. It returns to its high point every 10 seconds.

(a) Write an equation describing the motion of the buoy if it is at its high point at $t = 0$.

(b) Determine the velocity of the buoy as a function of t.

106. ***Modeling Data*** The normal daily maximum temperatures T (in degrees Fahrenheit) for Chicago, Illinois are shown in the table. *(Source: National Oceanic and Atmospheric Administration)*

Month	Jan	Feb	Mar	Apr	May	Jun
Temperature	29.6	34.7	46.1	58.0	69.9	79.2

Month	Jul	Aug	Sep	Oct	Nov	Dec
Temperature	83.5	81.2	73.9	62.1	47.1	34.4

(a) Use a graphing utility to plot the data and find a model for the data of the form

$$T(t) = a + b \sin(ct - d)$$

where T is the temperature and t is the time in months, with $t = 1$ corresponding to January.

(b) Use a graphing utility to graph the model. How well does the model fit the data?

(c) Find T' and use a graphing utility to graph the derivative.

(d) Based on the graph of the derivative, during what times does the temperature change most rapidly? Most slowly? Do your answers agree with your observations of the temperature changes? Explain.

107. ***Angle of Elevation*** An airplane flies at an altitude of 5 miles toward a point directly over an observer (see figure). The speed of the plane is 600 miles per hour. Find the rate at which the angle of elevation θ is changing when the angle is (a) $\theta = 30°$, (b) $\theta = 60°$, and (c) $\theta = 75°$.

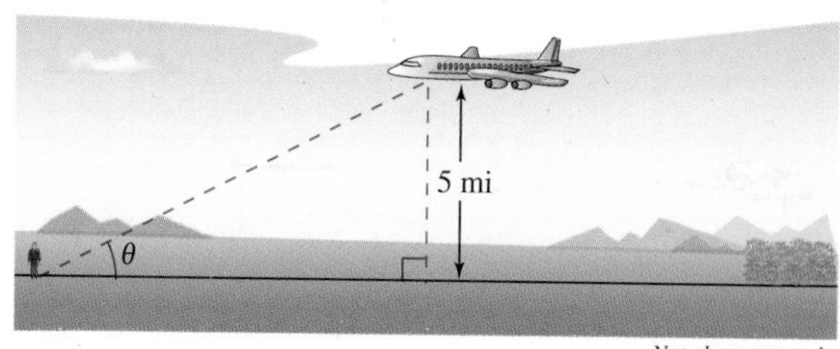

Not drawn to scale

108. ***Linear vs. Angular Speed*** A patrol car is parked 50 feet from a long warehouse (see figure). The revolving light on top of the car turns at a rate of 30 revolutions per minute. How fast is the light beam moving along the wall when the beam makes angles of (a) $\theta = 30°$, (b) $\theta = 60°$, and (c) $\theta = 70°$ with the perpendicular line from the light to the wall?

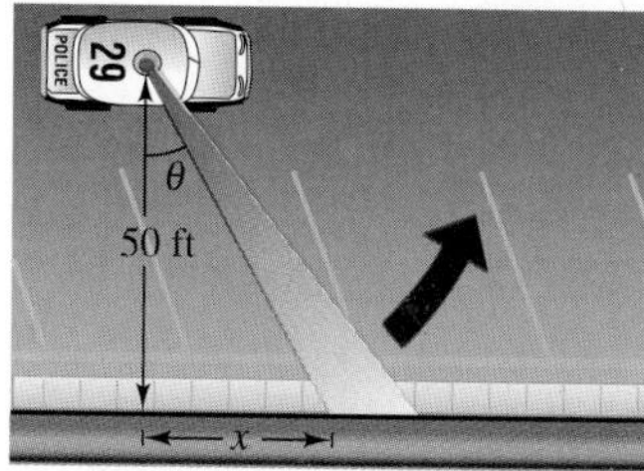

109. ***Projectile Motion*** The range R of a projectile is

$$R = \frac{v_0^2}{32}(\sin 2\theta)$$

where v_0 is the initial velocity in feet per second and θ is the angle of elevation. If $v_0 = 2500$ feet per second and θ is changed from 10° to 11°, use differentials to approximate the change in range.

110. ***Surveying*** A surveyor standing 50 feet from the base of a large tree measures the angle of elevation to the top of the tree as 71.5°. How accurately must the angle be measured if the error in estimating the height of the tree is to be less than 6%?

111. ***Minimum Force*** A component is designed to slide a block of steel with weight W across a table and into a chute (see figure on next page). The motion of the block is resisted by a frictional force proportional to its apparent weight. (Let k be the constant of proportionality.) Find the minimum force F needed to slide the block, and find the corresponding value of θ. (*Hint:* $F \cos \theta$ is the force in the direction of motion, and $F \sin \theta$ is the amount of force tending to lift the block. Therefore, the apparent weight of the block is $W - F \sin \theta$.)

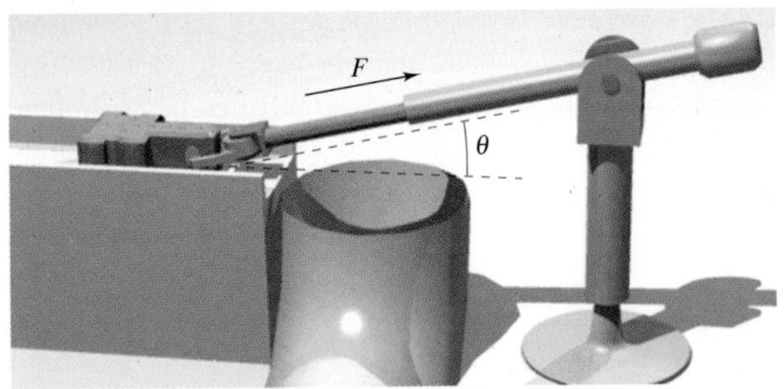

Figure for 111

112. ***Maximum Volume*** A sector with central angle θ is cut from a circle of radius 12 inches (see figure), and the edges of the sector are brought together to form a cone. Find the magnitude of θ such that the volume of the cone is a maximum.

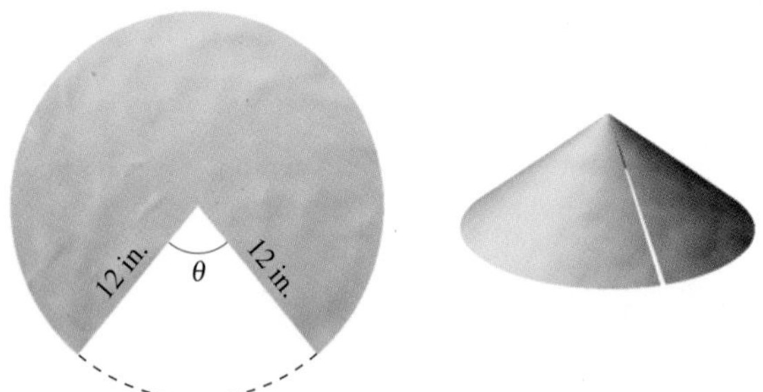

113. ***Numerical, Graphical, and Analytic Analysis*** The cross sections of an irrigation canal are isosceles trapezoids of which three sides are 8 feet long (see figure). Determine the angle of elevation θ of the sides such that the area of the cross sections is a maximum by completing the following.

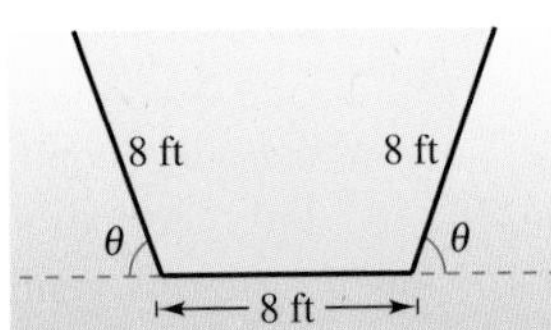

(a) Analytically complete six rows of a table such as the one below. (The first two rows are shown.)

Base 1	Base 2	Altitude	Area
8	$8 + 16 \cos 10°$	$8 \sin 10°$	≈ 22.1
8	$8 + 16 \cos 20°$	$8 \sin 20°$	≈ 42.5

(b) Use a graphing utility to generate additional rows of the table and estimate the maximum cross-sectional area. (*Hint*: Use the *table* feature of the graphing utility.)

(c) Write the cross-sectional area A as a function of θ.

(d) Find the critical number of the function in part (c) and find the angle that will yield the maximum cross-sectional area.

(e) Use a graphing utility to graph the function in part (c) and verify the maximum cross-sectional area.

114. ***Conjecture*** Let f be a differentiable function of period p.

(a) Is the function f' periodic? Verify your answer.

(b) Consider the function $g(x) = f(2x)$. Is the function $g'(x)$ periodic? Verify your answer.

115. ***Graphical Reasoning*** Consider the function

$$f(x) = \frac{\cos^2 \pi x}{\sqrt{x^2 + 1}}, \quad 0 < x < 4.$$

(a) Use a computer algebra system to graph the function and use the graph to approximate the critical numbers visually.

(b) Use a computer algebra system to find f' and approximate the critical numbers. Are the results the same as the visual approximation in part (a)? Explain.

116. ***Graphical Reasoning*** Consider the function

$$f(x) = \tan(\sin \pi x).$$

(a) Use a graphing utility to graph the function.

(b) Identify any symmetry of the graph.

(c) Is the function periodic? If so, what is the period?

(d) Identify any extrema on $(-1, 1)$.

(e) Use a graphing utility to determine the concavity of the graph on $(0, 1)$.

True or False? **In Exercises 117–120, determine whether the statement is true or false. If it is false, explain why or give an example that shows it is false.**

117. If $f(x) = \sin^2 2x$, then $f'(x) = 2(\sin 2x)(\cos 2x)$.

118. You would first apply the General Power Rule to find the derivative of $y = x \sin^3 x$.

119. The maximum value of $y = 3 \sin x + 2 \cos x$ is 5.

120. The maximum slope of the graph of $y = \sin(bx)$ is b.

121. Prove the following differentiation rules.

(a) $\dfrac{d}{dx}[\sec x] = \sec x \tan x$

(b) $\dfrac{d}{dx}[\csc x] = -\csc x \cot x$

(c) $\dfrac{d}{dx}[\cot x] = -\csc^2 x$

122. ***Writing*** If $g(x) = f(1 - 2x)$, what is the relationship between f' and g'? Explain.

11.3 Trigonometric Functions: Integration

- **Integrate trigonometric functions using trigonometric identities and *u*-substitution.**
- **Use integrals to find the average value of a function.**

Integrals of Trigonometric Functions

Corresponding to each trigonometric differentiation formula is an integration formula. For instance, the differentiation formula

$$\frac{d}{dx}[\cos u] = -\sin u \frac{du}{dx}$$

corresponds to the integration formula

$$\int \sin u \, du = -\cos u + C.$$

The following list summarizes all six integration formulas corresponding to the derivatives of the basic trigonometric functions. Keep in mind that if u is a differentiable function of x, then $du = (u')\,dx$.

THEOREM 11.5 BASIC TRIGONOMETRIC INTEGRATION FORMULAS

Let u be a differentiable function of x.

Integration Formula	*Differentiation Formula*
$\int \cos u \, du = \sin u + C$	$\frac{d}{dx}[\sin u] = \cos u \frac{du}{dx}$
$\int \sin u \, du = -\cos u + C$	$\frac{d}{dx}[\cos u] = -\sin u \frac{du}{dx}$
$\int \sec^2 u \, du = \tan u + C$	$\frac{d}{dx}[\tan u] = \sec^2 u \frac{du}{dx}$
$\int \sec u \tan u \, du = \sec u + C$	$\frac{d}{dx}[\sec u] = \sec u \tan u \frac{du}{dx}$
$\int \csc^2 u \, du = -\cot u + C$	$\frac{d}{dx}[\cot u] = -\csc^2 u \frac{du}{dx}$
$\int \csc u \cot u \, du = -\csc u + C$	$\frac{d}{dx}[\csc u] = -\csc u \cot u \frac{du}{dx}$

EXAMPLE 1 Integration of Trigonometric Functions

a. $\int 2 \cos x \, dx = 2 \int \cos x \, dx = 2 \sin x + C$ $\quad u = x$

b. $\int 3x^2 \sin x^3 \, dx = \int \underbrace{\sin x^3}_{\sin u} \underbrace{(3x^2) \, dx}_{du} = -\cos x^3 + C$ $\quad u = x^3$

c. $\int \sec^2 3x \, dx = \frac{1}{3} \int \underbrace{(\sec^2 3x)}_{\sec^2 u} \underbrace{(3) \, dx}_{du} = \frac{1}{3} \tan 3x + C$ $\quad u = 3x$

■

Each integral in Example 1 is easily recognized as fitting one of the basic integration formulas in Theorem 11.5. However, because of the variety of trigonometric identities, it often happens that an integrand that fits one of the basic formulas will come in a disguised form. This is shown in the next two examples.

EXAMPLE 2 Using a Trigonometric Identity

Find $\int \tan^2 x\, dx$.

Solution

$$\int \tan^2 x\, dx = \int (-1 + \sec^2 x)\, dx \qquad \text{Pythagorean identity}$$
$$= -x + \tan x + C$$

EXAMPLE 3 Using a Trigonometric Identity

Find $\int (\csc x + \sin x)(\csc x)\, dx$.

Solution

$$\int (\csc x + \sin x)(\csc x)\, dx = \int (\csc^2 x + 1)\, dx \qquad \text{Reciprocal identity}$$
$$= -\cot x + x + C$$

■

In addition to using trigonometric identities, another useful technique in evaluating trigonometric integrals is u-substitution, as shown in the next example.

EXAMPLE 4 Integration by u-Substitution

Find $\int \frac{\sec^2 \sqrt{x}}{\sqrt{x}}\, dx$.

Solution Let $u = \sqrt{x}$. Then you have

$$u = \sqrt{x} \quad \Rightarrow \quad du = \frac{1}{2\sqrt{x}}\, dx \quad \Rightarrow \quad 2\, du = \frac{1}{\sqrt{x}}\, dx.$$

So,

$$\int \frac{\sec^2 \sqrt{x}}{\sqrt{x}}\, dx = \int \sec^2 \sqrt{x} \left(\frac{1}{\sqrt{x}}\, dx\right)$$
$$= \int \sec^2 u(2\, du)$$
$$= 2\int \sec^2 u\, du$$
$$= 2 \tan u + C$$
$$= 2 \tan \sqrt{x} + C.$$

■

One of the most common u-substitutions involves quantities in the integrand that are raised to a power, as shown in the next two examples.

EXAMPLE 5 Integration by u-Substitution and the Power Rule

Find $\int \sin^2 3x \cos 3x\, dx$.

Solution Because $\sin^2 3x = (\sin 3x)^2$, you can let $u = \sin 3x$. Then

$$du = (\cos 3x)(3)\, dx \quad \Longrightarrow \quad \frac{du}{3} = \cos 3x\, dx.$$

Substituting u and $du/3$ in the given integral yields

$$\begin{aligned}\int \sin^2 3x \cos 3x\, dx &= \int u^2 \frac{du}{3} \\ &= \frac{1}{3}\int u^2\, du \\ &= \frac{1}{3}\left(\frac{u^3}{3}\right) + C \\ &= \frac{1}{9}\sin^3 3x + C.\end{aligned}$$

NOTE In Examples 4 and 5, u-substitution is used with a *change of variables* to find the antiderivative. In Example 6, u-substitution is used with *pattern recognition* to find the antiderivative. These procedures are equivalent, and you can use either one.

EXAMPLE 6 Substitution and the Power Rule

Find each integral.

a. $\displaystyle\int \frac{\sin x}{\cos^2 x}\, dx$ **b.** $\displaystyle\int 4\cos^2 4x \sin 4x\, dx$ **c.** $\displaystyle\int \frac{\sec^2 x}{\sqrt{\tan x}}\, dx$

Solution

a. $$\int \frac{\sin x}{\cos^2 x}\, dx = -\int \overbrace{(\cos x)^{-2}}^{u^{-2}}\overbrace{(-\sin x)\, dx}^{du}$$

$$= -\frac{\overbrace{(\cos x)^{-1}}^{u^{-1}/(-1)}}{-1} + C = \sec x + C$$

b. $$\int 4\cos^2 4x \sin 4x\, dx = -\int \overbrace{(\cos 4x)^2}^{u^2}\overbrace{(-4\sin 4x)\, dx}^{du}$$

$$= -\overbrace{\frac{(\cos 4x)^3}{3}}^{u^3/3} + C$$

c. $$\int \frac{\sec^2 x}{\sqrt{\tan x}}\, dx = \int \overbrace{(\tan x)^{-1/2}}^{u^{-1/2}}\overbrace{(\sec^2 x)\, dx}^{du} = \overbrace{\frac{(\tan x)^{1/2}}{1/2}}^{u^{1/2}/(1/2)} + C$$

$$= 2\sqrt{\tan x} + C$$

■

In Theorem 11.5, six trigonometric integration formulas are listed—the six that correspond directly to differentiation rules. With the Log Rule, you can now complete the set of basic trigonometric integration formulas.

EXAMPLE 7 The Antiderivative of the Tangent

Find $\int \tan x\,dx$.

Solution This integral does not seem to fit any formulas on our basic list. However, by using a trigonometric identity, you obtain the following quotient form.

$$\int \tan x\,dx = \int \frac{\sin x}{\cos x}\,dx$$

Now, knowing that $D_x[\cos x] = -\sin x$, you can let $u = \cos x$ and write

$$\begin{aligned}
\int \tan x\,dx &= -\int \frac{(-\sin x)}{\cos x}\,dx && \text{Trigonometric identity}\\
&= -\int \frac{u'}{u}\,dx && \text{Substitute: } u = \cos x.\\
&= -\ln|u| + C && \text{Apply Log Rule.}\\
&= -\ln|\cos x| + C. && \text{Back-substitute.}
\end{aligned}$$

■

Example 7 uses a trigonometric identity together with the Log Rule to derive an integration formula for the tangent function. The next example takes a rather unusual step (multiplying and dividing by the same quantity) to derive an integration formula for the secant function.

EXAMPLE 8 Antiderivative of the Secant

Find $\int \sec x\,dx$.

Solution Consider the following procedure.

$$\begin{aligned}
\int \sec x\,dx &= \int \sec x\left(\frac{\sec x + \tan x}{\sec x + \tan x}\right)dx\\
&= \int \frac{\sec^2 x + \sec x \tan x}{\sec x + \tan x}\,dx
\end{aligned}$$

Now, letting u be the denominator of this quotient produces

$$u = \sec x + \tan x \quad \Rightarrow \quad u' = \sec x \tan x + \sec^2 x.$$

So, you can conclude that

$$\begin{aligned}
\int \sec x\,dx &= \int \frac{\sec^2 x + \sec x \tan x}{\sec x + \tan x}\,dx\\
&= \int \frac{u'}{u}\,dx\\
&= \ln|u| + C\\
&= \ln|\sec x + \tan x| + C.
\end{aligned}$$

■

NOTE Using trigonometric identities and properties of logarithms, you could rewrite these six integration rules in other forms. For instance, you could write

$$\int \csc u\, du = \ln|\csc u - \cot u| + C.$$

(See Exercises 89–92.)

With the results of Examples 7 and 8, you now have integration formulas for $\sin x$, $\cos x$, $\tan x$, and $\sec x$. All six trigonometric functions are summarized below.

INTEGRALS OF THE SIX BASIC TRIGONOMETRIC FUNCTIONS

$$\int \sin u\, du = -\cos u + C \qquad \int \cos u\, du = \sin u + C$$

$$\int \tan u\, du = -\ln|\cos u| + C \qquad \int \cot u\, du = \ln|\sin u| + C$$

$$\int \sec u\, du = \ln|\sec u + \tan u| + C \qquad \int \csc u\, du = -\ln|\csc u + \cot u| + C$$

EXAMPLE 9 Integrating Trigonometric Functions

Evaluate $\displaystyle\int_0^{\pi/4} \sqrt{1 + \tan^2 x}\, dx$.

Solution Because $1 + \tan^2 x = \sec^2 x$, you can write

$$\begin{aligned}
\int_0^{\pi/4} \sqrt{1 + \tan^2 x}\, dx &= \int_0^{\pi/4} \sqrt{\sec^2 x}\, dx \\
&= \int_0^{\pi/4} \sec x\, dx \qquad \sec x \ge 0 \text{ for } 0 \le x \le \frac{\pi}{4}. \\
&= \Big[\ln|\sec x + \tan x|\Big]_0^{\pi/4} \\
&= \ln\left(\sqrt{2} + 1\right) - \ln(1) \approx 0.881.
\end{aligned}$$

■

Application

EXAMPLE 10 Finding the Average Value

Find the average value of $f(x) = \tan x$ on the interval $[0, \pi/4]$.

Solution

$$\begin{aligned}
\text{Average value} &= \frac{1}{(\pi/4) - 0}\int_0^{\pi/4} \tan x\, dx \qquad \text{Average value} = \frac{1}{b-a}\int_a^b f(x)\, dx \\
&= \frac{4}{\pi}\int_0^{\pi/4} \tan x\, dx \qquad \text{Simplify.} \\
&= \frac{4}{\pi}\Big[-\ln|\cos x|\Big]_0^{\pi/4} \qquad \text{Integrate.} \\
&= -\frac{4}{\pi}\left[\ln\left(\frac{\sqrt{2}}{2}\right) - \ln(1)\right] \\
&= -\frac{4}{\pi}\ln\left(\frac{\sqrt{2}}{2}\right) \approx 0.441
\end{aligned}$$

The average value is about 0.441, as shown in Figure 11.10. ■

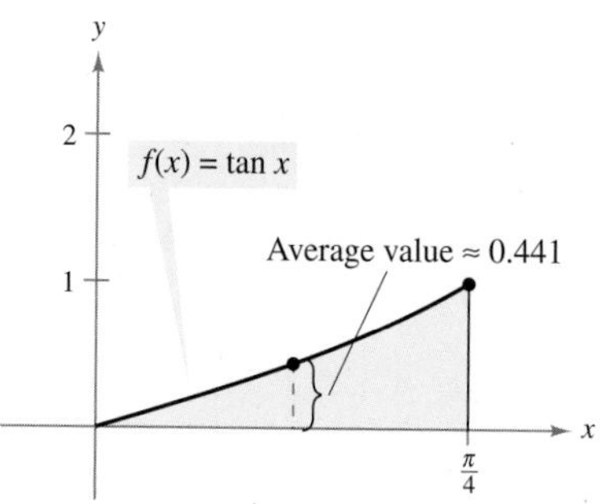

Figure 11.10

11.3 Exercises

See www.CalcChat.com for worked-out solutions to odd-numbered exercises.

In Exercises 1–32, find the indefinite integral.

1. $\int (5 \cos x + 4 \sin x)\, dx$
2. $\int (t^2 - \cos t)\, dt$
3. $\int (1 - \csc t \cot t)\, dt$
4. $\int (\theta^2 + \sec^2 \theta)\, d\theta$
5. $\int (\sec^2 \theta - \sin \theta)\, d\theta$
6. $\int \sec y (\tan y - \sec y)\, dy$
7. $\int (\tan^2 y + 1)\, dy$
8. $\int \frac{\cos x}{1 - \cos^2 x}\, dx$
9. $\int \pi \sin \pi x\, dx$
10. $\int 4x^3 \sin x^4\, dx$
11. $\int \sin 4x\, dx$
12. $\int \cos 8x\, dx$
13. $\int \frac{1}{\theta^2} \cos \frac{1}{\theta}\, d\theta$
14. $\int x \sin x^2\, dx$
15. $\int \sin 2x \cos 2x\, dx$
16. $\int \sec(1 - x) \tan(1 - x)\, dx$
17. $\int \tan^4 x \sec^2 x\, dx$
18. $\int \sqrt{\tan x} \sec^2 x\, dx$
19. $\int \frac{\csc^2 x}{\cot^3 x}\, dx$
20. $\int \frac{\sin x}{\cos^3 x}\, dx$
21. $\int \cot^2 x\, dx$
22. $\int \csc^2 \left(\frac{x}{2}\right) dx$
23. $\int e^x \cos e^x\, dx$
24. $\int e^{\sin x} \cos x\, dx$
25. $\int \cot \frac{\theta}{3}\, d\theta$
26. $\int \tan 5\theta\, d\theta$
27. $\int \csc 2x\, dx$
28. $\int \sec \frac{x}{2}\, dx$
29. $\int \frac{\cos t}{1 + \sin t}\, dt$
30. $\int \frac{\csc^2 t}{\cot t}\, dt$
31. $\int \frac{\sec x \tan x}{\sec x - 1}\, dx$
32. $\int (\sec 2x + \tan 2x)\, dx$

In Exercises 33–38, evaluate the definite integral. Use a graphing utility to verify your result.

33. $\int_0^{\pi} (1 + \sin x)\, dx$
34. $\int_0^{\pi/4} \frac{1 - \sin^2 \theta}{\cos^2 \theta}\, d\theta$
35. $\int_{-\pi/6}^{\pi/6} \sec^2 x\, dx$
36. $\int_{\pi/4}^{\pi/2} (2 - \csc^2 x)\, dx$
37. $\int_{-\pi/3}^{\pi/3} 4 \sec \theta \tan \theta\, d\theta$
38. $\int_{-\pi/2}^{\pi/2} (2t + \cos t)\, dt$

In Exercises 39 and 40, determine which value best approximates the definite integral. Make your selection on the basis of a sketch.

39. $\int_0^{1/2} 4 \cos \pi x\, dx$

(a) 4 (b) $\frac{4}{3}$ (c) 16 (d) 2π (e) -6

40. $\int_0^{1} 2 \sin \pi x\, dx$

(a) 6 (b) $\frac{1}{2}$ (c) 4 (d) $\frac{5}{4}$

Slope Fields **In Exercises 41 and 42, a differential equation, a point, and a slope field are given. (a) Sketch two approximate solutions of the differential equation on the slope field, one of which passes through the indicated point. (b) Use integration to find the particular solution of the differential equation and use a graphing utility to graph the solution. Compare the result with the sketches in part (a). To print an enlarged copy of the graph, go to the website *www.mathgraphs.com*.**

41. $\frac{dy}{dx} = \cos x, \quad (0, 4)$
42. $\frac{dy}{dx} = x \cos x^2, \quad (0, 1)$

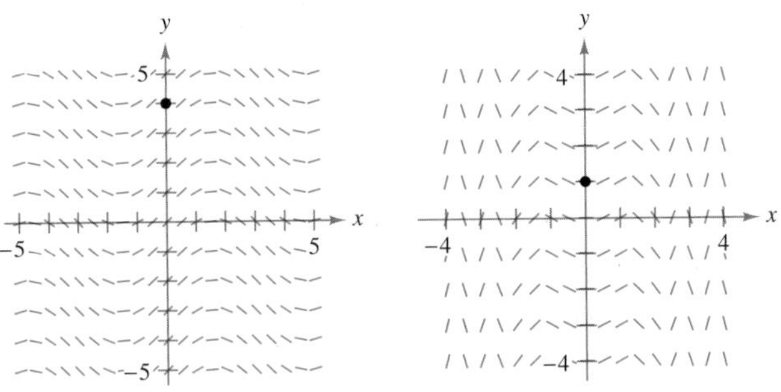

In Exercises 43 and 44, solve the differential equation. Use a graphing utility to graph three solutions, one of which passes through the given point.

43. $\frac{ds}{d\theta} = \tan 2\theta, \quad (0, 2)$
44. $\frac{dr}{dt} = \frac{\sec^2 t}{\tan t + 1}, \quad (\pi, 4)$

In Exercises 45–48, use a computer algebra system to find or evaluate the integral. Graph the integrand.

45. $\int \cos(1 - x)\, dx$
46. $\int \frac{\tan^2 2x}{\sec 2x}\, dx$
47. $\int_{\pi/4}^{\pi/2} (\csc x - \sin x)\, dx$
48. $\int_{-\pi/4}^{\pi/4} \frac{\sin^2 x - \cos^2 x}{\cos x}\, dx$

In Exercises 49–52, use a graphing utility to approximate the definite integral.

49. $\int_0^{\pi/2} \sin^2 x\,dx$

50. $\int_0^3 x \sin x\,dx$

51. $\int_0^4 \sin \sqrt{x}\,dx$

52. $\int_1^2 \frac{\sin x}{x}\,dx$

In Exercises 53–58, approximate the definite integral using the Trapezoidal Rule and Simpson's Rule with $n = 4$. Compare these results with the approximation of the integral using a graphing utility.

53. $\int_0^{\sqrt{\pi/2}} \sin x^2\,dx$

54. $\int_0^{\sqrt{\pi/4}} \tan x^2\,dx$

55. $\int_3^{3.1} \cos x^2\,dx$

56. $\int_0^{\pi/2} \sqrt{1 + \sin^2 x}\,dx$

57. $\int_0^{\pi/4} x \tan x\,dx$

58. $\int_0^{\pi} f(x)\,dx, \quad f(x) = \begin{cases} \dfrac{\sin x}{x}, & x > 0 \\ 1, & x = 0 \end{cases}$

In Exercises 59–64, determine the area of the given region.

59. $y = \cos x$

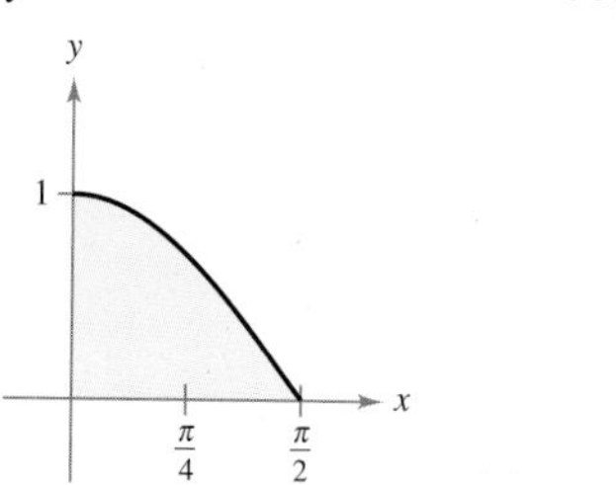

60. $y = x + \sin x$

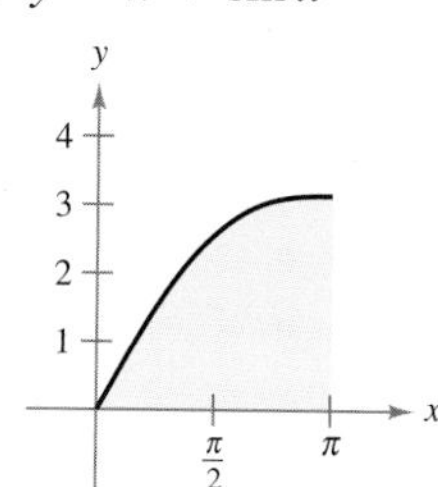

61. $y = 2 \sin x + \sin 2x$

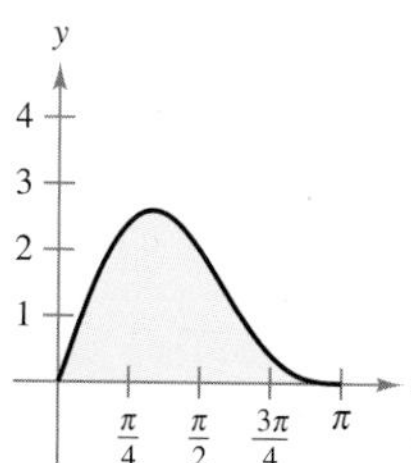

62. $y = \sin x + \cos 2x$

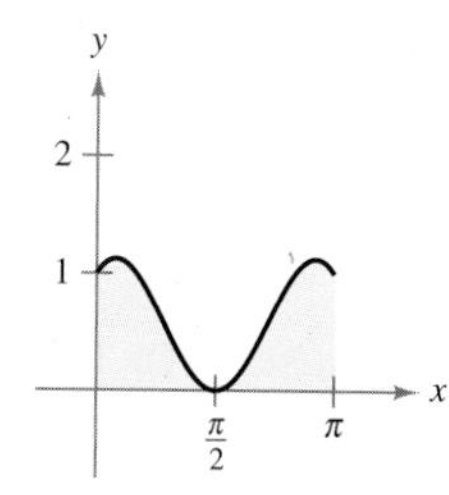

63. $\int_{\pi/2}^{2\pi/3} \sec^2\left(\frac{x}{2}\right) dx$

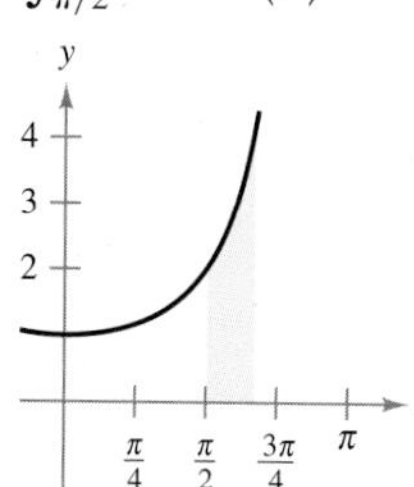

64. $\int_{\pi/12}^{\pi/4} \csc 2x \cot 2x\,dx$

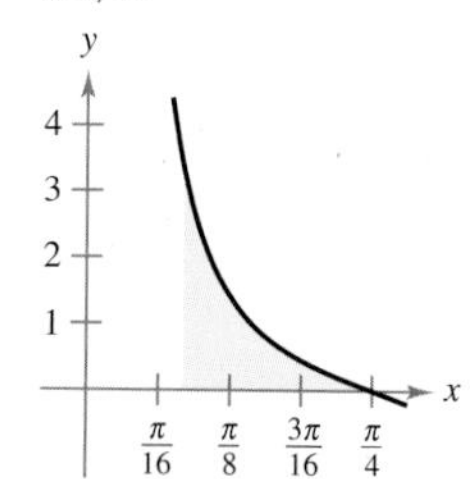

In Exercises 65–68, evaluate the integral using the properties of even and odd functions as an aid.

65. $\int_{-\pi/4}^{\pi/4} \sin x\,dx$

66. $\int_{-\pi/4}^{\pi/4} \cos x\,dx$

67. $\int_{-\pi/2}^{\pi/2} \sin^2 x \cos x\,dx$

68. $\int_{-\pi/2}^{\pi/2} \sin x \cos x\,dx$

In Exercises 69 and 70, find the value(s) of c guaranteed by the Mean Value Theorem for Integrals for the function over the given interval.

69. $f(x) = 2 \sec^2 x, \quad [-\pi/4, \pi/4]$

70. $f(x) = \cos x, \quad [-\pi/3, \pi/3]$

In Exercises 71 and 72, use a graphing utility to graph the function over the given interval. Find the average value of the function over the interval and all values of x in the interval for which the function equals its average value.

71. $f(x) = \sin x, \quad [0, \pi]$

72. $f(x) = \cos x, \quad [0, \pi/2]$

In Exercises 73–76, use the Second Fundamental Theorem of Calculus to find $F'(x)$.

73. $F(x) = \int_0^x t \cos t\,dt$

74. $F(x) = \int_0^x \sec^3 t\,dt$

75. $F(x) = \int_{\pi/4}^x \sec^2 t\,dt$

76. $F(x) = \int_{\pi/3}^x \sec t \tan t\,dt$

WRITING ABOUT CONCEPTS

77. Determine whether the function $f(x) = \tan x$ is integrable on the interval $[0, \pi]$.

78. Describe why $\int \sec^2(\pi x)\,dx \neq \int \sec^2 u\,du$, where $u = \pi x$.

79. Without integrating, explain why $\int_{-\pi}^{\pi} x \cos x\,dx = 0$.

80. Will the Trapezoidal Rule yield a result greater than or less than $\int_0^{\pi} \sin x\,dx$? Explain your reasoning.

81. Find possible values of a and b that make the statement true. If possible, use a graph to support your answer. (There may be more than one correct answer.)

(a) $\int_a^b \sin x\,dx < 0$ (b) $\int_a^b \cos x\,dx = 0$

CAPSTONE

82. ***Writing*** Find the indefinite integral in two ways. Explain any difference in the forms of the answers.

(a) $\int \sin x \cos x\,dx$ (b) $\int \tan x \sec^2 x\,dx$

83. ***Average Sales*** A company fits a model to the monthly sales data for a seasonal product. The model is

$$S(t) = \frac{t}{4} + 1.8 + 0.5 \sin\left(\frac{\pi t}{6}\right), \quad 0 \le t \le 24$$

where S is sales (in thousands) and t is time in months.

(a) Use a graphing utility to graph $f(t) = 0.5 \sin(\pi t/6)$ for $0 \le t \le 24$. Use the graph to explain why the average value of $f(t)$ is 0 over the interval.

(b) Use a graphing utility to graph $S(t)$ and the line $g(t) = t/4 + 1.8$ in the same viewing window. Use the graph and the result of part (a) to explain why g is called the *trend line*.

84. ***Precipitation*** The normal monthly precipitation at the Seattle-Tacoma airport can be approximated by the model

$$R = 2.876 + 2.202 \sin(0.576t + 0.847)$$

where R is measured in inches and t is the time in months, with $t = 0$ corresponding to January 1. *(Source: U.S. National Oceanic and Atmospheric Administration)*

(a) Determine the extrema of the function over a one-year period.

(b) Use integration to approximate the normal annual precipitation. (*Hint:* Integrate over the interval [0, 12].)

(c) Approximate the average monthly precipitation during the months of October, November, and December.

85. ***Sales*** The sales S (in thousands of units) of a seasonal product are given by the model

$$S = 74.50 + 43.75 \sin \frac{\pi t}{6}$$

where t is the time in months, with $t = 1$ corresponding to January. Find the average monthly sales for each time period.

(a) The first quarter $(0 \le t \le 3)$

(b) The second quarter $(3 \le t \le 6)$

(c) The entire year $(0 \le t \le 12)$

86. ***Water Supply*** A model for the flow rate of water at a pumping station on a given day is

$$R(t) = 53 + 7 \sin\left(\frac{\pi t}{6} + 3.6\right) + 9 \cos\left(\frac{\pi t}{12} + 8.9\right)$$

where $0 \le t \le 24$. R is the flow rate in thousands of gallons per hour, and t is the time in hours.

(a) Use a graphing utility to graph the rate function and approximate the maximum flow rate at the pumping station.

(b) Approximate the total volume of water pumped in 1 day.

87. ***Electricity*** The oscillating current in an electrical circuit is

$$I = 2 \sin(60\pi t) + \cos(120\pi t)$$

where I is measured in amperes and t is measured in seconds. Find the average current for each time interval.

(a) $0 \le t \le \frac{1}{60}$

(b) $0 \le t \le \frac{1}{240}$

(c) $0 \le t \le \frac{1}{30}$

88. Use Simpson's Rule with $n = 10$ and a computer algebra system to approximate t to three decimal places in the integral equation

$$\int_0^t \sin\sqrt{x}\, dx = 2.$$

In Exercises 89–92, verify that the two formulas are equivalent.

89. $\int \tan x\, dx = -\ln|\cos x| + C$

$\int \tan x\, dx = \ln|\sec x| + C$

90. $\int \cot x\, dx = \ln|\sin x| + C$

$\int \cot x\, dx = -\ln|\csc x| + C$

91. $\int \sec x\, dx = \ln|\sec x + \tan x| + C$

$\int \sec x\, dx = -\ln|\sec x - \tan x| + C$

92. $\int \csc x\, dx = -\ln|\csc x + \cot x| + C$

$\int \csc x\, dx = \ln|\csc x - \cot x| + C$

True or False? **In Exercises 93–95, determine whether the statement is true or false. If it is false, explain why or give an example that shows it is false.**

93. $\int_a^b \sin x\, dx = \int_a^{b+2\pi} \sin x\, dx$

94. $4\int \sin x \cos x\, dx = -\cos 2x + C$

95. $\int \sin^2 2x \cos 2x\, dx = \frac{1}{3} \sin^3 2x + C$

11.4 Inverse Trigonometric Functions: Differentiation

- **Differentiate an inverse trigonometric function.**
- **Review the basic differentiation rules for elementary functions.**

Derivatives of Inverse Trigonometric Functions

In Section 8.2, you saw that the derivative of the *transcendental* function $f(x) = \ln x$ is the *algebraic* function $f'(x) = 1/x$. You will now see that the derivatives of the inverse trigonometric functions also are algebraic (even though the inverse trigonometric functions are themselves transcendental).

Theorem 11.6 lists the derivatives of the six inverse trigonometric functions.

NOTE Observe that the derivatives of $\operatorname{arccos} u$, $\operatorname{arccot} u$, and $\operatorname{arccsc} u$ are the *negatives* of the derivatives of $\arcsin u$, $\arctan u$, and $\operatorname{arcsec} u$, respectively.

THEOREM 11.6 DERIVATIVES OF INVERSE TRIGONOMETRIC FUNCTIONS

Let u be a differentiable function of x.

$$\frac{d}{dx}[\arcsin u] = \frac{u'}{\sqrt{1-u^2}} \qquad \frac{d}{dx}[\arccos u] = \frac{-u'}{\sqrt{1-u^2}}$$

$$\frac{d}{dx}[\arctan u] = \frac{u'}{1+u^2} \qquad \frac{d}{dx}[\operatorname{arccot} u] = \frac{-u'}{1+u^2}$$

$$\frac{d}{dx}[\operatorname{arcsec} u] = \frac{u'}{|u|\sqrt{u^2-1}} \qquad \frac{d}{dx}[\operatorname{arccsc} u] = \frac{-u'}{|u|\sqrt{u^2-1}}$$

To derive these formulas, you can use implicit differentiation. For instance, if $y = \arcsin x$, then $\sin y = x$ and $(\cos y)y' = 1$. (See Exercise 54.)

TECHNOLOGY If your graphing utility does not have the arcsecant function, you can obtain its graph using

$$f(x) = \operatorname{arcsec} x = \arccos \frac{1}{x}.$$

EXAMPLE 1 Differentiating Inverse Trigonometric Functions

a. $\dfrac{d}{dx}[\arcsin(2x)] = \dfrac{2}{\sqrt{1-(2x)^2}} = \dfrac{2}{\sqrt{1-4x^2}}$

b. $\dfrac{d}{dx}[\arctan(3x)] = \dfrac{3}{1+(3x)^2} = \dfrac{3}{1+9x^2}$

c. $\dfrac{d}{dx}\left[\arcsin\sqrt{x}\right] = \dfrac{(1/2)\,x^{-1/2}}{\sqrt{1-x}} = \dfrac{1}{2\sqrt{x}\sqrt{1-x}} = \dfrac{1}{2\sqrt{x-x^2}}$

d. $\dfrac{d}{dx}[\operatorname{arcsec} e^{2x}] = \dfrac{2e^{2x}}{e^{2x}\sqrt{(e^{2x})^2-1}} = \dfrac{2e^{2x}}{e^{2x}\sqrt{e^{4x}-1}} = \dfrac{2}{\sqrt{e^{4x}-1}}$

The absolute value sign is not necessary because $e^{2x} > 0$.

NOTE From Example 2, you can see one of the benefits of inverse trigonometric functions—they can be used to integrate common algebraic functions. For instance, from the result shown in the example, it follows that

$$\int \sqrt{1-x^2}\,dx = \frac{1}{2}\left(\arcsin x + x\sqrt{1-x^2}\right).$$

EXAMPLE 2 A Derivative That Can Be Simplified

Differentiate $y = \arcsin x + x\sqrt{1-x^2}$.

Solution

$$\begin{aligned} y' &= \frac{1}{\sqrt{1-x^2}} + x\left(\frac{1}{2}\right)(-2x)(1-x^2)^{-1/2} + \sqrt{1-x^2} \\ &= \frac{1}{\sqrt{1-x^2}} - \frac{x^2}{\sqrt{1-x^2}} + \sqrt{1-x^2} \\ &= \sqrt{1-x^2} + \sqrt{1-x^2} = 2\sqrt{1-x^2} \end{aligned}$$

■

EXAMPLE 3 Analyzing an Inverse Trigonometric Graph

Analyze the graph of

$$y = (\arctan x)^2.$$

Solution From the derivative

$$y' = 2(\arctan x)\frac{d}{dx}(\arctan x) \qquad \text{Power Rule}$$

$$= \frac{2 \arctan x}{1 + x^2}$$

you can see that the only critical number is $x = 0$. By the First Derivative Test, this value corresponds to a relative minimum. From the second derivative

$$y'' = \frac{(1 + x^2)\left(\frac{2}{1 + x^2}\right) - (2 \arctan x)(2x)}{(1 + x^2)^2}$$

$$= \frac{2(1 - 2x \arctan x)}{(1 + x^2)^2}$$

it follows that points of inflection occur when $2x \arctan x = 1$. By using a graphing utility, you can see that these points occur when $x \approx \pm 0.765$. Finally, because

$$\lim_{x \to \pm\infty} (\arctan x)^2 = \left(\frac{\pi}{2}\right)^2 = \frac{\pi^2}{4}$$

it follows that the graph has a horizontal asymptote at $y = \pi^2/4$. The graph is shown in Figure 11.11.

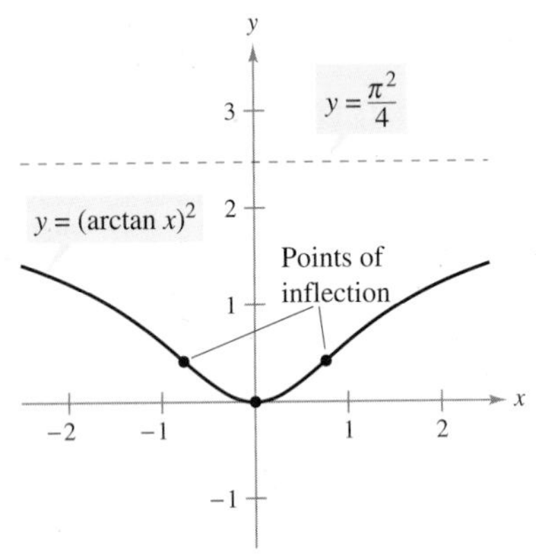

The graph of $y = (\arctan x)^2$ has a horizontal asymptote at $y = \pi^2/4$.
Figure 11.11

EXAMPLE 4 Maximizing an Angle

A photographer is taking a picture of a painting hung in an art gallery. The height of the painting is 4 feet. The camera lens is 1 foot below the lower edge of the painting, as shown in Figure 11.12. How far should the camera be from the painting to maximize the angle subtended by the camera lens?

Solution In Figure 11.12, let β be the angle to be maximized.

$$\beta = \theta - \alpha$$

$$= \operatorname{arccot}\frac{x}{5} - \operatorname{arccot} x$$

Differentiating produces

$$\frac{d\beta}{dx} = \frac{-1/5}{1 + (x^2/25)} - \frac{-1}{1 + x^2}$$

$$= \frac{-5}{25 + x^2} + \frac{1}{1 + x^2}$$

$$= \frac{4(5 - x^2)}{(25 + x^2)(1 + x^2)}.$$

Because $d\beta/dx = 0$ when $x = \sqrt{5}$, you can conclude from the First Derivative Test that this distance yields a maximum value of β. So, the distance is $x \approx 2.236$ feet and the angle is $\beta \approx 0.7297$ radian $\approx 41.81°$. ■

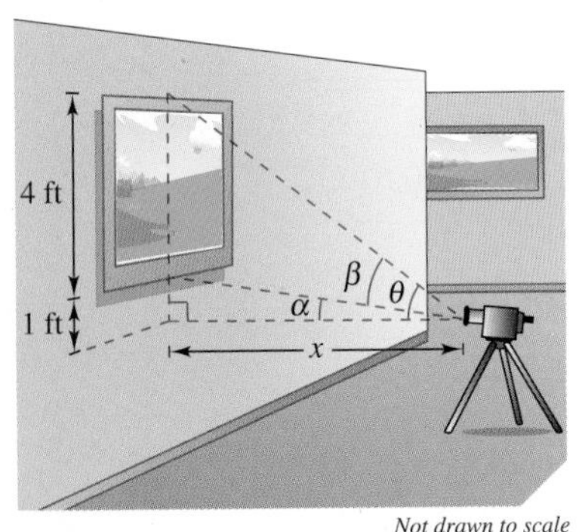

The camera should be 2.236 feet from the painting to maximize the angle β.
Figure 11.12

GALILEO GALILEI (1564–1642)

Galileo's approach to science departed from the accepted Aristotelian view that nature had describable *qualities*, such as "fluidity" and "potentiality." He chose to describe the physical world in terms of measurable *quantities*, such as time, distance, force, and mass.

Review of Basic Differentiation Rules

In the 1600s, Europe was ushered into the scientific age by such great thinkers as Descartes, Galileo, Huygens, Newton, and Kepler. These men believed that nature is governed by basic laws—laws that can, for the most part, be written in terms of mathematical equations. One of the most influential publications of this period—*Dialogue on the Great World Systems*, by Galileo Galilei—has become a classic description of modern scientific thought.

As mathematics has developed during the past few hundred years, a small number of elementary functions have proven sufficient for modeling most* phenomena in physics, chemistry, biology, engineering, economics, and a variety of other fields. An **elementary function** is a function from the following list or one that can be formed as the sum, product, quotient, or composition of functions in the list.

Algebraic Functions	*Transcendental Functions*
Polynomial functions	Logarithmic functions
Rational functions	Exponential functions
Functions involving radicals	Trigonometric functions
	Inverse trigonometric functions

With the differentiation rules introduced so far in the text, you can differentiate *any* elementary function. For convenience, these differentiation rules are summarized below.

BASIC DIFFERENTIATION RULES FOR ELEMENTARY FUNCTIONS

1. $\frac{d}{dx}[cu] = cu'$
2. $\frac{d}{dx}[u \pm v] = u' \pm v'$
3. $\frac{d}{dx}[uv] = uv' + vu'$
4. $\frac{d}{dx}\left[\frac{u}{v}\right] = \frac{vu' - uv'}{v^2}$
5. $\frac{d}{dx}[c] = 0$
6. $\frac{d}{dx}[u^n] = nu^{n-1}u'$
7. $\frac{d}{dx}[x] = 1$
8. $\frac{d}{dx}[|u|] = \frac{u}{|u|}(u'), \quad u \neq 0$
9. $\frac{d}{dx}[\ln u] = \frac{u'}{u}$
10. $\frac{d}{dx}[e^u] = e^u u'$
11. $\frac{d}{dx}[\log_a u] = \frac{u'}{(\ln a)u}$
12. $\frac{d}{dx}[a^u] = (\ln a)a^u u'$
13. $\frac{d}{dx}[\sin u] = (\cos u)u'$
14. $\frac{d}{dx}[\cos u] = -(\sin u)u'$
15. $\frac{d}{dx}[\tan u] = (\sec^2 u)u'$
16. $\frac{d}{dx}[\cot u] = -(\csc^2 u)u'$
17. $\frac{d}{dx}[\sec u] = (\sec u \tan u)u'$
18. $\frac{d}{dx}[\csc u] = -(\csc u \cot u)u'$
19. $\frac{d}{dx}[\arcsin u] = \frac{u'}{\sqrt{1 - u^2}}$
20. $\frac{d}{dx}[\arccos u] = \frac{-u'}{\sqrt{1 - u^2}}$
21. $\frac{d}{dx}[\arctan u] = \frac{u'}{1 + u^2}$
22. $\frac{d}{dx}[\text{arccot } u] = \frac{-u'}{1 + u^2}$
23. $\frac{d}{dx}[\text{arcsec } u] = \frac{u'}{|u|\sqrt{u^2 - 1}}$
24. $\frac{d}{dx}[\text{arccsc } u] = \frac{-u'}{|u|\sqrt{u^2 - 1}}$

* *Some important functions used in engineering and science (such as Bessel functions and gamma functions) are not elementary functions.*

11.4 Exercises

See www.CalcChat.com for worked-out solutions to odd-numbered exercises.

In Exercises 1–4, find the slope of the tangent line to the arcsine function at the origin.

1. $y = \arcsin x$

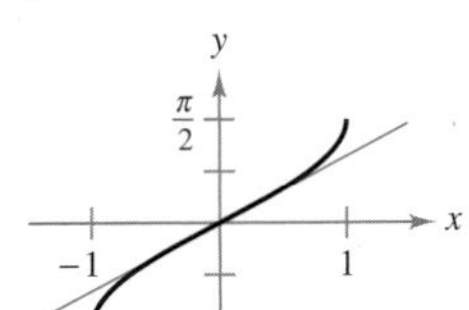

2. $y = \arcsin 2x$

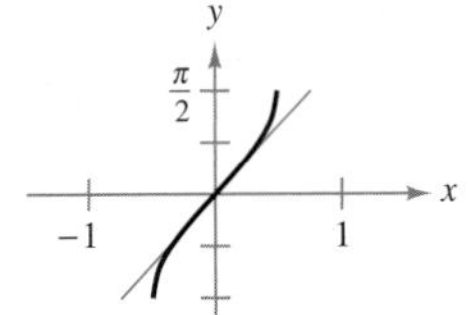

3. $y = \arcsin \dfrac{x}{2}$

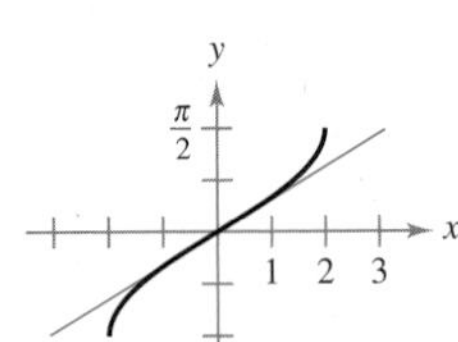

4. $y = \arcsin \dfrac{x}{3}$

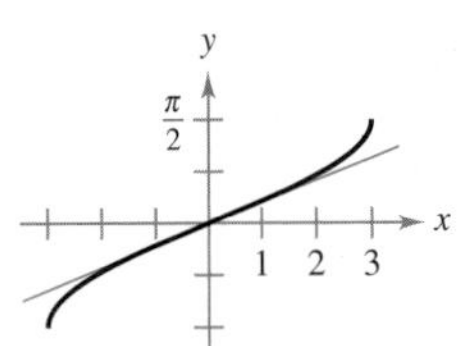

In Exercises 5–22, find the derivative of the function.

5. $f(x) = 2 \arcsin(x - 1)$

6. $f(t) = \arcsin t^2$

7. $g(x) = 3 \arccos \dfrac{x}{2}$

8. $f(x) = \operatorname{arcsec} 2x$

9. $f(x) = \arctan e^x$

10. $f(x) = \arctan \sqrt{x}$

11. $g(x) = \dfrac{\arcsin 3x}{x}$

12. $h(x) = x^2 \arctan 5x$

13. $h(t) = \sin(\arccos t)$

14. $f(x) = \arcsin x + \arccos x$

15. $y = 2x \arccos x - 2\sqrt{1 - x^2}$

16. $y = \ln(t^2 + 4) - \dfrac{1}{2} \arctan \dfrac{t}{2}$

17. $y = 25 \arcsin \dfrac{x}{5} - x\sqrt{25 - x^2}$

18. $y = \dfrac{1}{2}\left[x\sqrt{4 - x^2} + 4 \arcsin\left(\dfrac{x}{2}\right)\right]$

19. $y = x \arcsin x + \sqrt{1 - x^2}$

20. $y = x \arctan 2x - \frac{1}{4} \ln(1 + 4x^2)$

21. $y = \arctan x + \dfrac{x}{1 + x^2}$

22. $y = \arctan \dfrac{x}{2} - \dfrac{1}{2(x^2 + 4)}$

In Exercises 23–26, find an equation of the tangent line to the graph of the function at the given point.

23. $y = 2 \arcsin x, \quad \left(\dfrac{1}{2}, \dfrac{\pi}{3}\right)$

24. $y = \dfrac{1}{2} \arccos x, \quad \left(-\dfrac{\sqrt{2}}{2}, \dfrac{3\pi}{8}\right)$

25. $y = \arctan \dfrac{x}{2}, \quad \left(2, \dfrac{\pi}{4}\right)$

26. $y = \operatorname{arcsec} 4x, \quad \left(\dfrac{\sqrt{2}}{4}, \dfrac{\pi}{4}\right)$

Linear and Quadratic Approximations **In Exercises 27–30, use a computer algebra system to find the linear approximation $P_1(x) = f(a) + f'(a)(x - a)$ and the quadratic approximation $P_2(x) = f(a) + f'(a)(x - a) + \frac{1}{2}f''(a)(x - a)^2$ of the function f at $x = a$. Sketch the graph of the function and its linear and quadratic approximations.**

27. $f(x) = \arctan x, \quad a = 0$

28. $f(x) = \arccos x, \quad a = 0$

29. $f(x) = \arcsin x, \quad a = \frac{1}{2}$

30. $f(x) = \arctan x, \quad a = 1$

In Exercises 31–34, find any relative extrema of the function.

31. $f(x) = \operatorname{arcsec} x - x$

32. $f(x) = \arcsin x - 2x$

33. $f(x) = \arctan x - \arctan(x - 4)$

34. $h(x) = \arcsin x - 2 \arctan x$

Implicit Differentiation **In Exercises 35–38, find an equation of the tangent line to the graph of the equation at the given point.**

35. $x^2 + x \arctan y = y - 1, \quad \left(-\dfrac{\pi}{4}, 1\right)$

36. $\arctan(xy) = \arcsin(x + y), \quad (0, 0)$

37. $\arcsin x + \arcsin y = \dfrac{\pi}{2}, \quad \left(\dfrac{\sqrt{2}}{2}, \dfrac{\sqrt{2}}{2}\right)$

38. $\arctan(x + y) = y^2 + \dfrac{\pi}{4}, \quad (1, 0)$

WRITING ABOUT CONCEPTS

39. Explain why the domains of the trigonometric functions are restricted when finding the inverse trigonometric functions.

40. Explain why $\dfrac{d}{dx}[\tan(\arctan x)] = 1$.

41. Give a geometric argument explaining why the derivative of $y = \arcsin x$ is positive.

42. Give a geometric argument explaining why the derivative of $y = \operatorname{arccot} x$ is negative.

43. Are the derivatives of the inverse trigonometric functions algebraic or transcendental functions? List the derivatives of the inverse trigonometric functions.

CAPSTONE

44. The point $\left(\frac{3\pi}{2}, 0\right)$ is on the graph of $y = \cos x$. Does $\left(0, \frac{3\pi}{2}\right)$ lie on the graph of $y = \arccos x$? If not, does this contradict the definition of inverse function?

True or False? **In Exercises 45–48, determine whether the statement is true or false. If it is false, explain why or give an example that shows it is false.**

45. Because $\cos\left(-\frac{\pi}{3}\right) = \frac{1}{2}$, it follows that $\arccos \frac{1}{2} = -\frac{\pi}{3}$.

46. $\arcsin \frac{\pi}{4} = \frac{\sqrt{2}}{2}$

47. The slope of the graph of the inverse tangent function is positive for all x.

48. The range of $y = \arcsin x$ is $[0, \pi]$.

49. ***Angular Rate of Change*** An airplane flies at an altitude of 5 miles toward a point directly over an observer. Consider θ and x as shown in the figure.

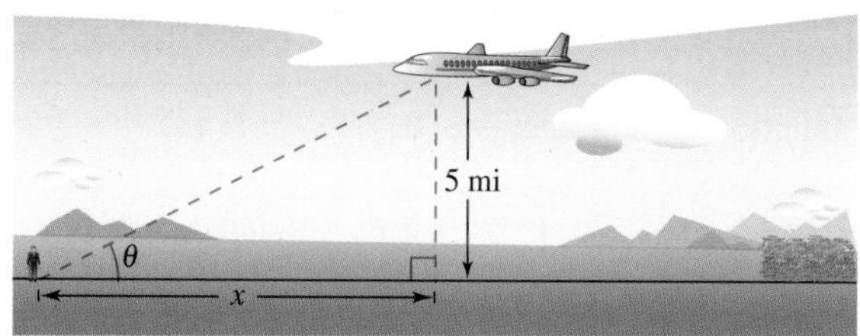

(a) Write θ as a function of x.

(b) The speed of the plane is 400 miles per hour. Find $d\theta/dt$ when $x = 10$ miles and $x = 3$ miles.

50. ***Writing*** Repeat Exercise 49 for an altitude of 3 miles and describe how the altitude affects the rate of change of θ.

51. ***Angular Rate of Change*** In a free-fall experiment, an object is dropped from a height of 256 feet. A camera on the ground 500 feet from the point of impact records the fall of the object (see figure).

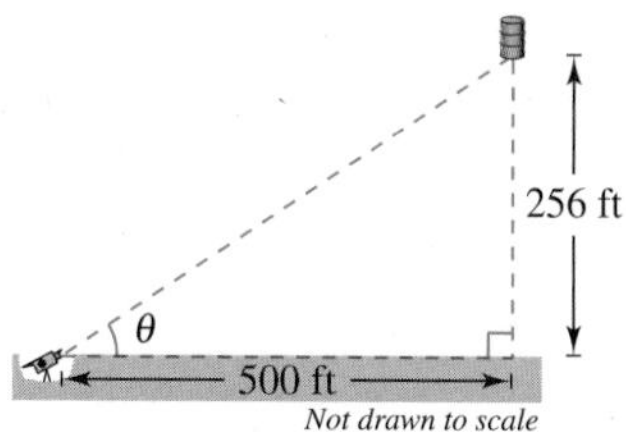

(a) Find the position function that yields the height of the object at time t assuming the object is released at time $t = 0$. At what time will the object reach ground level?

(b) Find the rates of change of the angle of elevation of the camera when $t = 1$ and $t = 2$.

52. ***Angular Rate of Change*** A television camera at ground level is filming the lift-off of a space shuttle at a point 800 meters from the launch pad. Let θ be the angle of elevation of the shuttle and let s be the distance between the camera and the shuttle (see figure). Write θ as a function of s for the period of time when the shuttle is moving vertically. Differentiate the result to find $d\theta/dt$ in terms of s and ds/dt.

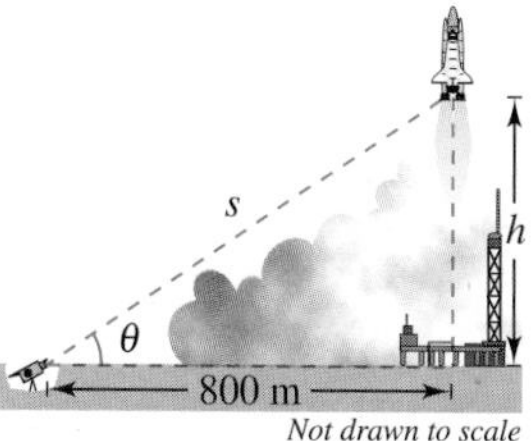

53. (a) Prove that

$$\arctan x + \arctan y = \arctan \frac{x+y}{1-xy}, \quad xy \neq 1.$$

(b) Use the formula in part (a) to show that

$$\arctan \frac{1}{2} + \arctan \frac{1}{3} = \frac{\pi}{4}.$$

54. Verify each differentiation formula.

(a) $\frac{d}{dx}[\arcsin u] = \frac{u'}{\sqrt{1-u^2}}$

(b) $\frac{d}{dx}[\arctan u] = \frac{u'}{1+u^2}$

(c) $\frac{d}{dx}[\operatorname{arcsec} u] = \frac{u'}{|u|\sqrt{u^2-1}}$

(d) $\frac{d}{dx}[\arccos u] = \frac{-u'}{\sqrt{1-u^2}}$

(e) $\frac{d}{dx}[\operatorname{arccot} u] = \frac{-u'}{1+u^2}$

(f) $\frac{d}{dx}[\operatorname{arccsc} u] = \frac{-u'}{|u|\sqrt{u^2-1}}$

55. Show that the function given by

$$f(x) = \arcsin\left(\frac{x-2}{2}\right) - 2\arcsin\frac{\sqrt{x}}{2}$$

is constant for $0 \leq x \leq 4$.

56. ***Think About It*** Use a graphing utility to graph

$f(x) = \sin x$ and $g(x) = \arcsin(\sin x)$.

(a) Why isn't the graph of g the line $y = x$?

(b) Determine the extrema of g.

11.5 Inverse Trigonometric Functions: Integration

- **Integrate functions whose antiderivatives involve inverse trigonometric functions.**
- **Use the method of completing the square to integrate a function.**
- **Review the basic integration rules involving elementary functions.**

Integrals Involving Inverse Trigonometric Functions

The derivatives of the six inverse trigonometric functions fall into three pairs. In each pair, the derivative of one function is the negative of the other. For example,

$$\frac{d}{dx}[\arcsin x] = \frac{1}{\sqrt{1-x^2}}$$

and

$$\frac{d}{dx}[\arccos x] = -\frac{1}{\sqrt{1-x^2}}.$$

When listing the *antiderivative* that corresponds to each of the inverse trigonometric functions, you need to use only one member from each pair. It is conventional to use $\arcsin x$ as the antiderivative of $1/\sqrt{1-x^2}$, rather than $-\arccos x$. The next theorem gives one antiderivative formula for each of the three pairs. The proofs of these integration rules are left to you (see Exercises 59–61).

FOR FURTHER INFORMATION For a detailed proof of rule 2 of Theorem 11.7, see the article "A Direct Proof of the Integral Formula for Arctangent" by Arnold J. Insel in *The College Mathematics Journal*. To view this article, go to the website *www.matharticles.com*.

THEOREM 11.7 INTEGRALS INVOLVING INVERSE TRIGONOMETRIC FUNCTIONS

Let u be a differentiable function of x, and let $a > 0$.

1. $\displaystyle\int \frac{du}{\sqrt{a^2-u^2}} = \arcsin\frac{u}{a} + C$

2. $\displaystyle\int \frac{du}{a^2+u^2} = \frac{1}{a}\arctan\frac{u}{a} + C$

3. $\displaystyle\int \frac{du}{u\sqrt{u^2-a^2}} = \frac{1}{a}\operatorname{arcsec}\frac{|u|}{a} + C$

EXAMPLE 1 Integration with Inverse Trigonometric Functions

a. $\displaystyle\int \frac{dx}{\sqrt{4-x^2}} = \arcsin\frac{x}{2} + C$

b. $\displaystyle\int \frac{dx}{2+9x^2} = \frac{1}{3}\int \frac{3\,dx}{(\sqrt{2})^2+(3x)^2}$ $\quad u = 3x,\ a = \sqrt{2}$

$$= \frac{1}{3\sqrt{2}}\arctan\frac{3x}{\sqrt{2}} + C$$

c. $\displaystyle\int \frac{dx}{x\sqrt{4x^2-9}} = \int \frac{2\,dx}{2x\sqrt{(2x)^2-3^2}}$ $\quad u = 2x,\ a = 3$

$$= \frac{1}{3}\operatorname{arcsec}\frac{|2x|}{3} + C$$

■

The integrals in Example 1 are fairly straightforward applications of integration formulas. Unfortunately, this is not typical. The integration formulas for inverse trigonometric functions can be disguised in many ways.

TECHNOLOGY PITFALL

Computer software that can perform symbolic integration is useful for integrating functions such as the one in Example 2. When using such software, however, you must remember that it can fail to find an antiderivative for two reasons. First, some elementary functions simply do not have antiderivatives that are elementary functions. Second, every symbolic integration utility has limitations—you might have entered a function that the software was not programmed to handle. You should also remember that antiderivatives involving trigonometric functions or logarithmic functions can be written in many different forms. For instance, one symbolic integration utility found the integral in Example 2 to be

$$\int \frac{dx}{\sqrt{e^{2x}-1}} = \arctan \sqrt{e^{2x}-1} + C.$$

Try showing that this antiderivative is equivalent to that obtained in Example 2.

EXAMPLE 2 Integration by Substitution

Find

$$\int \frac{dx}{\sqrt{e^{2x}-1}}.$$

Solution As it stands, this integral doesn't fit any of the three inverse trigonometric formulas. Using the substitution $u = e^x$, however, produces

$$u = e^x \quad \Longrightarrow \quad du = e^x\,dx \quad \Longrightarrow \quad dx = \frac{du}{e^x} = \frac{du}{u}.$$

With this substitution, you can integrate as follows.

$$\int \frac{dx}{\sqrt{e^{2x}-1}} = \int \frac{dx}{\sqrt{(e^x)^2-1}} \qquad \text{Write } e^{2x} \text{ as } (e^x)^2.$$

$$= \int \frac{du/u}{\sqrt{u^2-1}} \qquad \text{Substitute.}$$

$$= \int \frac{du}{u\sqrt{u^2-1}} \qquad \text{Rewrite to fit Arcsecant Rule.}$$

$$= \operatorname{arcsec} \frac{|u|}{1} + C \qquad \text{Apply Arcsecant Rule.}$$

$$= \operatorname{arcsec} e^x + C \qquad \text{Back-substitute.}$$

EXAMPLE 3 Rewriting as the Sum of Two Quotients

Find $\displaystyle\int \frac{x+2}{\sqrt{4-x^2}}\,dx.$

Solution This integral does not appear to fit any of the basic integration formulas. By splitting the integrand into two parts, however, you can see that the first part can be found with the Power Rule and the second part yields an inverse sine function.

$$\int \frac{x+2}{\sqrt{4-x^2}}\,dx = \int \frac{x}{\sqrt{4-x^2}}\,dx + \int \frac{2}{\sqrt{4-x^2}}\,dx$$

$$= -\frac{1}{2}\int (4-x^2)^{-1/2}(-2x)\,dx + 2\int \frac{1}{\sqrt{4-x^2}}\,dx$$

$$= -\frac{1}{2}\left[\frac{(4-x^2)^{1/2}}{1/2}\right] + 2\arcsin\frac{x}{2} + C$$

$$= -\sqrt{4-x^2} + 2\arcsin\frac{x}{2} + C$$

■

Completing the Square

Completing the square helps when quadratic functions are involved in the integrand. For example, in Section P.1, you learned that the quadratic $x^2 + bx + c$ can be written as the difference of two squares by adding and subtracting $(b/2)^2$.

$$x^2 + bx + c = x^2 + bx + \left(\frac{b}{2}\right)^2 - \left(\frac{b}{2}\right)^2 + c$$

$$= \left(x + \frac{b}{2}\right)^2 - \left(\frac{b}{2}\right)^2 + c$$

EXAMPLE 4 Completing the Square

Find $\displaystyle\int \frac{dx}{x^2 - 4x + 7}$.

Solution You can write the denominator as the sum of two squares, as follows.

$$\begin{aligned} x^2 - 4x + 7 &= (x^2 - 4x + 4) - 4 + 7 \\ &= (x - 2)^2 + 3 = u^2 + a^2 \end{aligned}$$

Now, in this completed square form, let $u = x - 2$ and $a = \sqrt{3}$.

$$\int \frac{dx}{x^2 - 4x + 7} = \int \frac{dx}{(x - 2)^2 + 3} = \frac{1}{\sqrt{3}} \arctan \frac{x - 2}{\sqrt{3}} + C$$

If the leading coefficient is not 1, it helps to factor before completing the square. For instance, you can complete the square of $2x^2 - 8x + 10$ by factoring first.

$$\begin{aligned} 2x^2 - 8x + 10 &= 2(x^2 - 4x + 5) \\ &= 2(x^2 - 4x + 4 - 4 + 5) \\ &= 2[(x - 2)^2 + 1] \end{aligned}$$

To complete the square when the coefficient of x^2 is negative, use the same factoring process shown above. For instance, you can complete the square for $3x - x^2$ as shown.

$$\begin{aligned} 3x - x^2 &= -(x^2 - 3x) \\ &= -\left[x^2 - 3x + \left(\tfrac{3}{2}\right)^2 - \left(\tfrac{3}{2}\right)^2\right] \\ &= \left(\tfrac{3}{2}\right)^2 - \left(x - \tfrac{3}{2}\right)^2 \end{aligned}$$

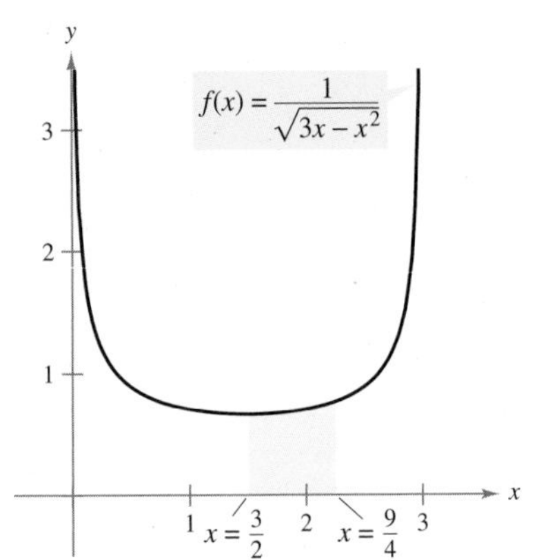

The area of the region bounded by the graph of f, the x-axis, $x = \frac{3}{2}$, and $x = \frac{9}{4}$ is $\pi/6$.
Figure 11.13

EXAMPLE 5 Completing the Square (Negative Leading Coefficient)

Find the area of the region bounded by the graph of

$$f(x) = \frac{1}{\sqrt{3x - x^2}}$$

the x-axis, and the lines $x = \frac{3}{2}$ and $x = \frac{9}{4}$.

Solution In Figure 11.13, you can see that the area is given by

$$\text{Area} = \int_{3/2}^{9/4} \frac{1}{\sqrt{3x - x^2}}\, dx.$$

Using the completed square form derived above, you can integrate as shown.

$$\begin{aligned} \int_{3/2}^{9/4} \frac{dx}{\sqrt{3x - x^2}} &= \int_{3/2}^{9/4} \frac{dx}{\sqrt{(3/2)^2 - [x - (3/2)]^2}} \qquad u = \left(x - \tfrac{3}{2}\right) \\ &= \arcsin \frac{x - (3/2)}{3/2} \Bigg]_{3/2}^{9/4} \\ &= \arcsin \frac{1}{2} - \arcsin 0 \\ &= \frac{\pi}{6} \\ &\approx 0.524 \end{aligned}$$

TECHNOLOGY With definite integrals such as the one given in Example 5, remember that you can resort to a numerical solution. For instance, applying Simpson's Rule (with $n = 12$) to the integral in the example, you obtain

$$\int_{3/2}^{9/4} \frac{1}{\sqrt{3x - x^2}}\, dx \approx 0.523599.$$

This differs from the exact value of the integral ($\pi/6 \approx 0.5235988$) by less than one millionth.

Review of Basic Integration Rules

You have now completed the introduction of the **basic integration rules.** To be efficient at applying these rules, you should have practiced enough so that each rule is committed to memory.

BASIC INTEGRATION RULES ($a > 0$)

1. $\int kf(u)\,du = k\int f(u)\,du$

2. $\int [f(u) \pm g(u)]\,du = \int f(u)\,du \pm \int g(u)\,du$

3. $\int du = u + C$

4. $\int u^n\,du = \dfrac{u^{n+1}}{n+1} + C, \quad n \neq -1$

5. $\int \dfrac{du}{u} = \ln|u| + C$

6. $\int e^u\,du = e^u + C$

7. $\int a^u\,du = \left(\dfrac{1}{\ln a}\right)a^u + C$

8. $\int \sin u\,du = -\cos u + C$

9. $\int \cos u\,du = \sin u + C$

10. $\int \tan u\,du = -\ln|\cos u| + C$

11. $\int \cot u\,du = \ln|\sin u| + C$

12. $\int \sec u\,du = \ln|\sec u + \tan u| + C$

13. $\int \csc u\,du = -\ln|\csc u + \cot u| + C$

14. $\int \sec^2 u\,du = \tan u + C$

15. $\int \csc^2 u\,du = -\cot u + C$

16. $\int \sec u \tan u\,du = \sec u + C$

17. $\int \csc u \cot u\,du = -\csc u + C$

18. $\int \dfrac{du}{\sqrt{a^2 - u^2}} = \arcsin\dfrac{u}{a} + C$

19. $\int \dfrac{du}{a^2 + u^2} = \dfrac{1}{a}\arctan\dfrac{u}{a} + C$

20. $\int \dfrac{du}{u\sqrt{u^2 - a^2}} = \dfrac{1}{a}\operatorname{arcsec}\dfrac{|u|}{a} + C$

You can learn a lot about the nature of integration by comparing this list with the summary of differentiation rules given in the preceding section. For differentiation, you now have rules that allow you to differentiate *any* elementary function. For integration, this is far from true.

The integration rules listed above are primarily those that were happened on during the development of differentiation rules. So far, you have not learned any rules or techniques for finding the antiderivative of a general product or quotient, the natural logarithmic function, or the inverse trigonometric functions. More importantly, you cannot apply any of the rules in this list unless you can create the proper *du* corresponding to the *u* in the formula.

The next two examples should give you a better feeling for the integration problems that you *can* and *cannot* do with the techniques and rules you now know.

EXAMPLE 6 Comparing Integration Problems

Find as many of the following integrals as you can using the formulas and techniques you have studied so far in the text.

a. $\displaystyle\int \frac{dx}{x\sqrt{x^2-1}}$ **b.** $\displaystyle\int \frac{x\,dx}{\sqrt{x^2-1}}$ **c.** $\displaystyle\int \frac{dx}{\sqrt{x^2-1}}$

Solution

a. You *can* find this integral (it fits the Arcsecant Rule with $u = x$).

$$\int \frac{dx}{x\sqrt{x^2-1}} = \operatorname{arcsec}|x| + C$$

b. You *can* find this integral (it fits the Power Rule with $u = x^2 - 1$).

$$\begin{aligned}\int \frac{x\,dx}{\sqrt{x^2-1}} &= \frac{1}{2}\int (x^2-1)^{-1/2}(2x)\,dx\\ &= \frac{1}{2}\left[\frac{(x^2-1)^{1/2}}{1/2}\right] + C\\ &= \sqrt{x^2-1} + C\end{aligned}$$

c. You *cannot* find this integral using the techniques you have studied so far. (You should scan the list of basic integration rules to verify this conclusion.)

EXAMPLE 7 Comparing Integration Problems

Find as many of the following integrals as you can using the formulas and techniques you have studied so far in the text.

a. $\displaystyle\int \frac{dx}{x\ln x}$ **b.** $\displaystyle\int \frac{\ln x\,dx}{x}$ **c.** $\displaystyle\int \ln x\,dx$

Solution

a. You *can* find this integral (it fits the Log Rule with $u = \ln x$).

$$\begin{aligned}\int \frac{dx}{x\ln x} &= \int \frac{1/x}{\ln x}\,dx\\ &= \ln|\ln x| + C\end{aligned}$$

b. You *can* find this integral (it fits the Power Rule with $u = \ln x$).

$$\begin{aligned}\int \frac{\ln x\,dx}{x} &= \int (\ln x)^1\left(\frac{1}{x}\right)dx\\ &= \frac{(\ln x)^2}{2} + C\end{aligned}$$

c. You *cannot* find this integral using the techniques you have studied so far. ■

NOTE Examples 6 and 7 illustrate that the *simplest* functions are often the ones that you cannot yet integrate. ■

11.5 Exercises

See www.CalcChat.com for worked-out solutions to odd-numbered exercises.

In Exercises 1–20, find the integral.

1. $\displaystyle\int \frac{dx}{\sqrt{9-x^2}}$

2. $\displaystyle\int \frac{dx}{\sqrt{1-4x^2}}$

3. $\displaystyle\int \frac{7}{16+x^2}\,dx$

4. $\displaystyle\int \frac{12}{1+9x^2}\,dx$

5. $\displaystyle\int \frac{1}{x\sqrt{4x^2-1}}\,dx$

6. $\displaystyle\int \frac{1}{4+(x-3)^2}\,dx$

7. $\displaystyle\int \frac{t}{\sqrt{1-t^4}}\,dt$

8. $\displaystyle\int \frac{1}{x\sqrt{x^4-4}}\,dx$

9. $\displaystyle\int \frac{t}{t^4+25}\,dt$

10. $\displaystyle\int \frac{1}{x\sqrt{1-(\ln x)^2}}\,dx$

11. $\displaystyle\int \frac{\sec^2 x}{\sqrt{25-\tan^2 x}}\,dx$

12. $\displaystyle\int \frac{\sin x}{7+\cos^2 x}\,dx$

13. $\displaystyle\int \frac{x^3}{x^2+1}\,dx$

14. $\displaystyle\int \frac{x^4-1}{x^2+1}\,dx$

15. $\displaystyle\int \frac{1}{\sqrt{x}\sqrt{1-x}}\,dx$

16. $\displaystyle\int \frac{3}{2\sqrt{x}(1+x)}\,dx$

17. $\displaystyle\int \frac{x-3}{x^2+1}\,dx$

18. $\displaystyle\int \frac{4x+3}{\sqrt{1-x^2}}\,dx$

19. $\displaystyle\int \frac{x+5}{\sqrt{9-(x-3)^2}}\,dx$

20. $\displaystyle\int \frac{x-2}{(x+1)^2+4}\,dx$

In Exercises 21–32, evaluate the integral.

21. $\displaystyle\int_0^{1/6} \frac{3}{\sqrt{1-9x^2}}\,dx$

22. $\displaystyle\int_0^1 \frac{dx}{\sqrt{4-x^2}}$

23. $\displaystyle\int_0^{\sqrt{3}/2} \frac{1}{1+4x^2}\,dx$

24. $\displaystyle\int_{\sqrt{3}}^3 \frac{6}{9+x^2}\,dx$

25. $\displaystyle\int_{-1/2}^0 \frac{x}{\sqrt{1-x^2}}\,dx$

26. $\displaystyle\int_{-\sqrt{3}}^0 \frac{x}{1+x^2}\,dx$

27. $\displaystyle\int_3^6 \frac{1}{25+(x-3)^2}\,dx$

28. $\displaystyle\int_1^4 \frac{1}{x\sqrt{16x^2-5}}\,dx$

29. $\displaystyle\int_{\pi/2}^{\pi} \frac{\sin x}{1+\cos^2 x}\,dx$

30. $\displaystyle\int_0^{\pi/2} \frac{\cos x}{1+\sin^2 x}\,dx$

31. $\displaystyle\int_0^{1/\sqrt{2}} \frac{\arcsin x}{\sqrt{1-x^2}}\,dx$

32. $\displaystyle\int_0^{1/\sqrt{2}} \frac{\arccos x}{\sqrt{1-x^2}}\,dx$

In Exercises 33–44, find or evaluate the integral. (Complete the square, if necessary.)

33. $\displaystyle\int_0^2 \frac{dx}{x^2-2x+2}$

34. $\displaystyle\int_{-2}^2 \frac{dx}{x^2+4x+13}$

35. $\displaystyle\int \frac{2x}{x^2+6x+13}\,dx$

36. $\displaystyle\int \frac{2x-5}{x^2+2x+2}\,dx$

37. $\displaystyle\int \frac{1}{\sqrt{-x^2-4x}}\,dx$

38. $\displaystyle\int \frac{2}{\sqrt{-x^2+4x}}\,dx$

39. $\displaystyle\int \frac{x+2}{\sqrt{-x^2-4x}}\,dx$

40. $\displaystyle\int \frac{x-1}{\sqrt{x^2-2x}}\,dx$

41. $\displaystyle\int_2^3 \frac{2x-3}{\sqrt{4x-x^2}}\,dx$

42. $\displaystyle\int \frac{1}{(x-1)\sqrt{x^2-2x}}\,dx$

43. $\displaystyle\int \frac{x}{x^4+2x^2+2}\,dx$

44. $\displaystyle\int \frac{x}{\sqrt{9+8x^2-x^4}}\,dx$

In Exercises 45 and 46, use the specified substitution to find or evaluate the integral.

45. $\displaystyle\int \sqrt{e^t-3}\,dt$
$u=\sqrt{e^t-3}$

46. $\displaystyle\int \frac{\sqrt{x-2}}{x+1}\,dx$
$u=\sqrt{x-2}$

Slope Fields **In Exercises 47 and 48, a differential equation, a point, and a slope field are given. (a) Sketch two approximate solutions of the differential equation on the slope field, one of which passes through the given point. (b) Use integration to find the particular solution of the differential equation and use a graphing utility to graph the solution. Compare the result with the sketches in part (a). To print an enlarged copy of the graph, go to the website *www.mathgraphs.com*.**

47. $\dfrac{dy}{dx}=\dfrac{3}{1+x^2},\quad (0,0)$

48. 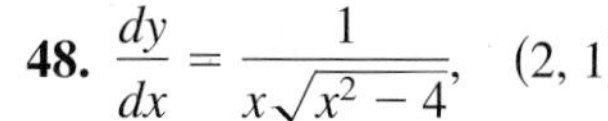
$\dfrac{dy}{dx}=\dfrac{1}{x\sqrt{x^2-4}},\quad (2,1)$

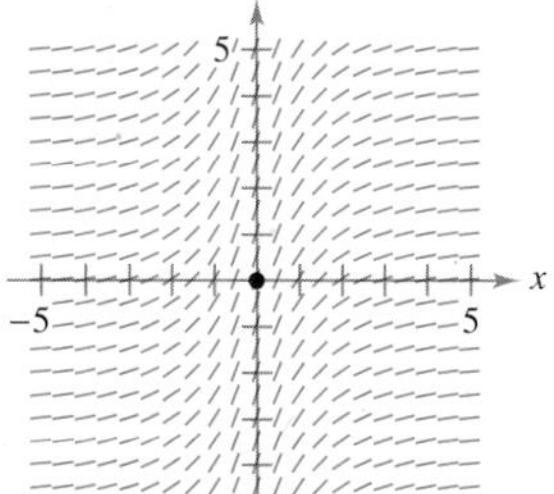

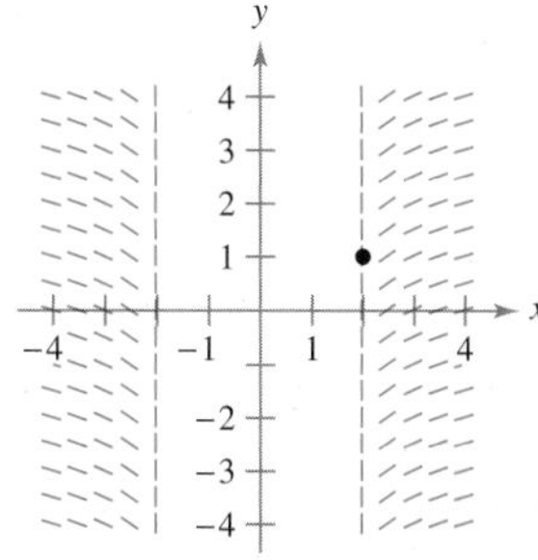

Slope Fields **In Exercises 49 and 50, use a computer algebra system to graph the slope field for the differential equation and graph the solution satisfying the specified initial condition.**

49. $\dfrac{dy}{dx}=\dfrac{10}{x\sqrt{x^2-1}},\quad y(3)=0$

50. $\dfrac{dy}{dx}=\dfrac{2y}{\sqrt{16-x^2}},\quad y(0)=2$

Area **In Exercises 51 and 52, find the area of the region.**

51. $y = \dfrac{2}{\sqrt{4 - x^2}}$

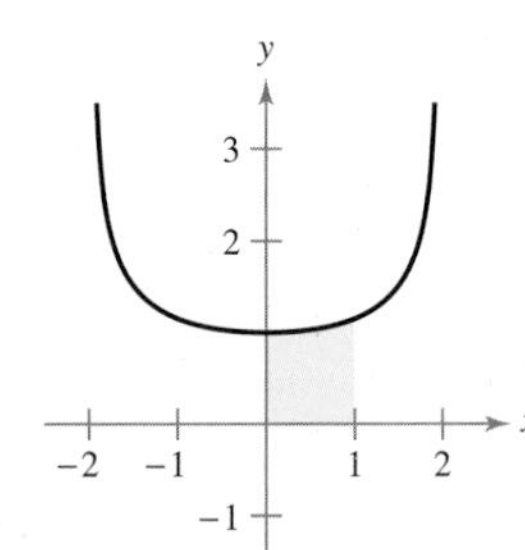

52. $y = \dfrac{1}{x^2 - 2x + 5}$

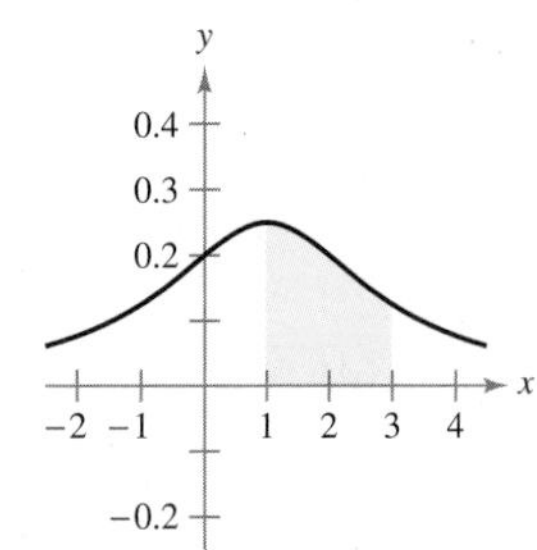

WRITING ABOUT CONCEPTS

In Exercises 53–55, determine which of the integrals can be found using the basic integration formulas you have studied so far in the text.

53. (a) $\displaystyle\int \frac{1}{\sqrt{1 - x^2}}\,dx$ (b) $\displaystyle\int \frac{x}{\sqrt{1 - x^2}}\,dx$

(c) $\displaystyle\int \frac{1}{x\sqrt{1 - x^2}}\,dx$

54. (a) $\displaystyle\int e^{x^2}\,dx$ (b) $\displaystyle\int xe^{x^2}\,dx$ (c) $\displaystyle\int \frac{1}{x^2}e^{1/x}\,dx$

55. (a) $\displaystyle\int \sqrt{x - 1}\,dx$ (b) $\displaystyle\int x\sqrt{x - 1}\,dx$

(c) $\displaystyle\int \frac{x}{\sqrt{x - 1}}\,dx$

CAPSTONE

56. Determine which of the integrals can be found using the basic integration formulas you have studied so far in the text.

(a) $\displaystyle\int \frac{1}{1 + x^4}\,dx$ (b) $\displaystyle\int \frac{x}{1 + x^4}\,dx$ (c) $\displaystyle\int \frac{x^3}{1 + x^4}\,dx$

57. (a) Show that $\displaystyle\int_0^1 \frac{4}{1 + x^2}\,dx = \pi$.

(b) Approximate the number π using Simpson's Rule (with $n = 6$) and the integral in part (a).

(c) Approximate the number π by using the integration capabilities of a graphing utility.

58. Graph $y_1 = \dfrac{x}{1 + x^2}$, $y_2 = \arctan x$, and $y_3 = x$ on $[0, 10]$.

Prove that $\dfrac{x}{1 + x^2} < \arctan x < x$ for $x > 0$.

Verifying Integration Rules **In Exercises 59–61, verify each rule by differentiating. Let $a > 0$.**

59. $\displaystyle\int \frac{du}{\sqrt{a^2 - u^2}} = \arcsin \frac{u}{a} + C$

60. $\displaystyle\int \frac{du}{a^2 + u^2} = \frac{1}{a} \arctan \frac{u}{a} + C$

61. $\displaystyle\int \frac{du}{u\sqrt{u^2 - a^2}} = \frac{1}{a} \operatorname{arcsec} \frac{|u|}{a} + C$

62. *Vertical Motion* An object is projected upward from ground level with an initial velocity of 500 feet per second. (In this exercise, the goal is to analyze the motion of the object during its upward flight.)

(a) If air resistance is neglected, find the velocity of the object as a function of time. Use a graphing utility to graph this function.

(b) Use the result of part (a) to find the position function and determine the maximum height attained by the object.

(c) If the air resistance is proportional to the square of the velocity, you obtain the equation

$$\frac{dv}{dt} = -(32 + kv^2)$$

where -32 feet per second per second is the acceleration due to gravity and k is a constant. Find the velocity as a function of time by solving the equation

$$\int \frac{dv}{32 + kv^2} = -\int dt.$$

(d) Use a graphing utility to graph the velocity function $v(t)$ in part (c) for $k = 0.001$. Use the graph to approximate the time t_0 at which the object reaches its maximum height.

(e) Use the integration capabilities of a graphing utility to approximate the integral

$$\int_0^{t_0} v(t)\,dt$$

where $v(t)$ and t_0 are those found in part (d). This is the approximation of the maximum height of the object.

(f) Explain the difference between the results in parts (b) and (e).

FOR FURTHER INFORMATION For more information on this topic, see "What Goes Up Must Come Down; Will Air Resistance Make It Return Sooner, or Later?" by John Lekner in *Mathematics Magazine*. To view this article, go to the website *www.matharticles.com*.

11.6 Hyperbolic Functions

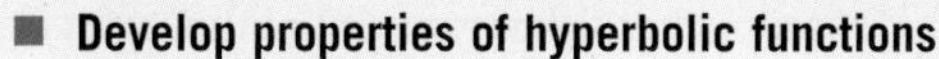

- Develop properties of hyperbolic functions.
- Differentiate and integrate hyperbolic functions.
- Develop properties of inverse hyperbolic functions.
- Differentiate and integrate functions involving inverse hyperbolic functions.

Hyperbolic Functions

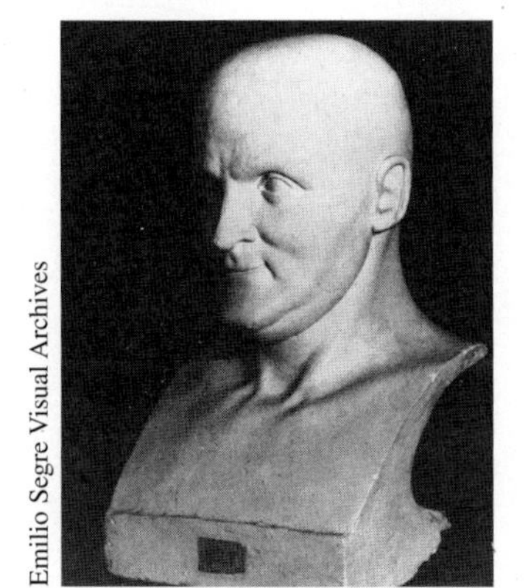

Emilio Segre Visual Archives

JOHANN HEINRICH LAMBERT (1728–1777)

The first person to publish a comprehensive study on hyperbolic functions was Johann Heinrich Lambert, a Swiss-German mathematician and colleague of Euler.

In this section, you will look briefly at a special class of exponential functions called **hyperbolic functions.** The name *hyperbolic function* arose from comparison of the area of a semicircular region, as shown in Figure 11.14, with the area of a region under a hyperbola, as shown in Figure 11.15. The integral for the semicircular region involves an inverse trigonometric (circular) function:

$$\int_{-1}^{1} \sqrt{1 - x^2}\, dx = \frac{1}{2}\left[x\sqrt{1 - x^2} + \arcsin x\right]_{-1}^{1} = \frac{\pi}{2} \approx 1.571.$$

The integral for the hyperbolic region involves an inverse hyperbolic function:

$$\int_{-1}^{1} \sqrt{1 + x^2}\, dx = \frac{1}{2}\left[x\sqrt{1 + x^2} + \sinh^{-1} x\right]_{-1}^{1} \approx 2.296.$$

This is only one of many ways in which the hyperbolic functions are similar to the trigonometric functions.

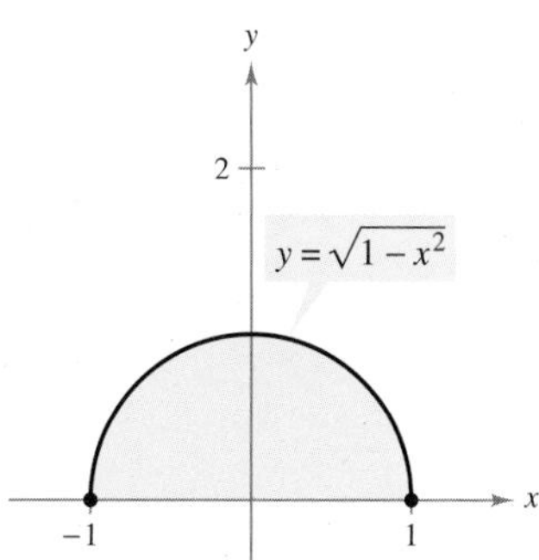

Circle: $x^2 + y^2 = 1$
Figure 11.14

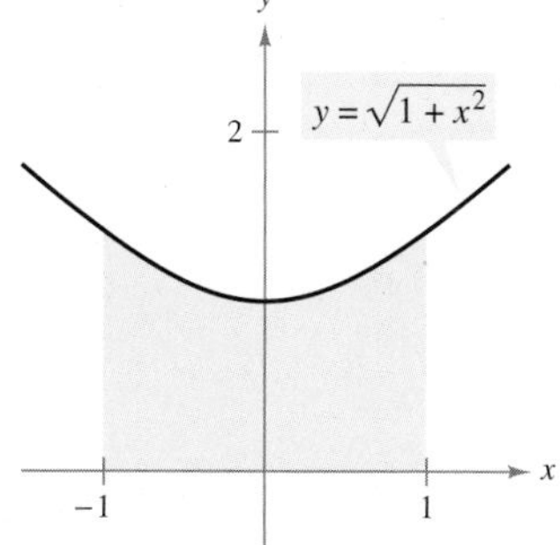

Hyperbola: $-x^2 + y^2 = 1$
Figure 11.15

■ **FOR FURTHER INFORMATION** For more information on the development of hyperbolic functions, see the article "An Introduction to Hyperbolic Functions in Elementary Calculus" by Jerome Rosenthal in *Mathematics Teacher*. To view this article, go to the website *www.matharticles.com*.

DEFINITIONS OF THE HYPERBOLIC FUNCTIONS

$$\sinh x = \frac{e^x - e^{-x}}{2} \qquad \operatorname{csch} x = \frac{1}{\sinh x}, \quad x \neq 0$$

$$\cosh x = \frac{e^x + e^{-x}}{2} \qquad \operatorname{sech} x = \frac{1}{\cosh x}$$

$$\tanh x = \frac{\sinh x}{\cosh x} \qquad \coth x = \frac{1}{\tanh x}, \quad x \neq 0$$

NOTE $\sinh x$ is read as "the hyperbolic sine of x," $\cosh x$ as "the hyperbolic cosine of x," and so on. ■

The graphs of the six hyperbolic functions and their domains and ranges are shown in Figure 11.16. Note that the graph of $\sinh x$ can be obtained by adding the corresponding y-coordinates of the exponential functions $f(x) = \frac{1}{2}e^x$ and $g(x) = -\frac{1}{2}e^{-x}$. Likewise, the graph of $\cosh x$ can be obtained by adding the corresponding y-coordinates of the exponential functions $f(x) = \frac{1}{2}e^x$ and $h(x) = \frac{1}{2}e^{-x}$.

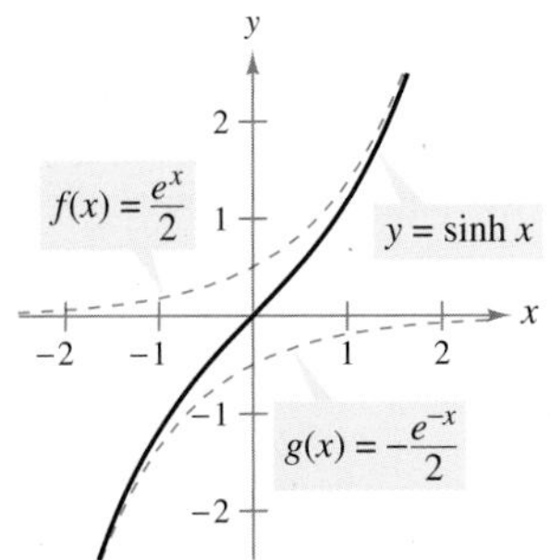

Domain: $(-\infty, \infty)$
Range: $(-\infty, \infty)$

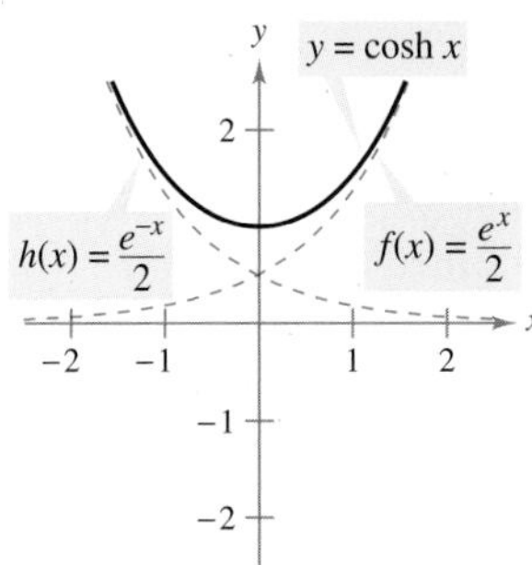

Domain: $(-\infty, \infty)$
Range: $[1, \infty)$

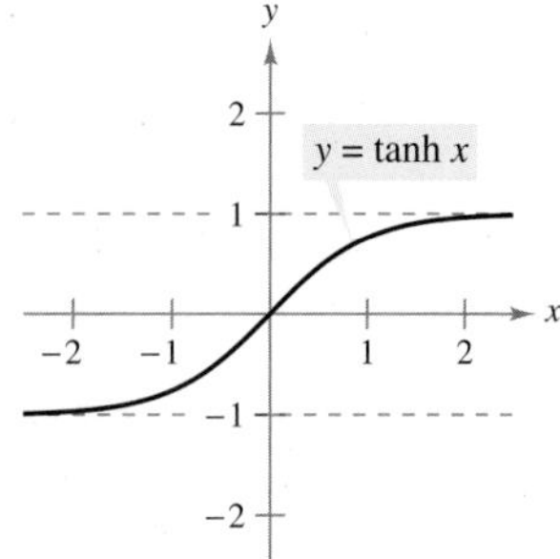

Domain: $(-\infty, \infty)$
Range: $(-1, 1)$

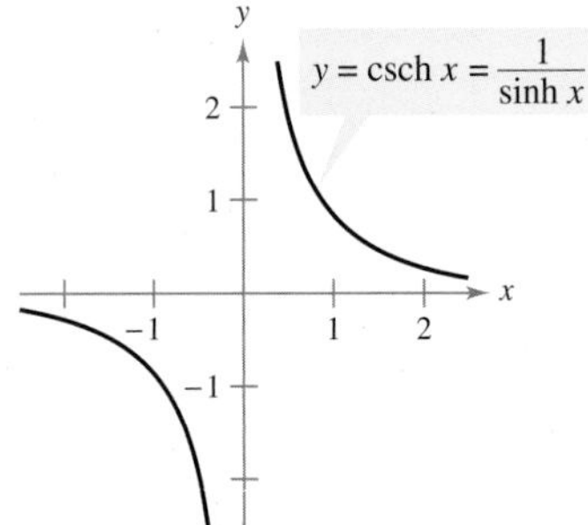

Domain: $(-\infty, 0) \cup (0, \infty)$
Range: $(-\infty, 0) \cup (0, \infty)$

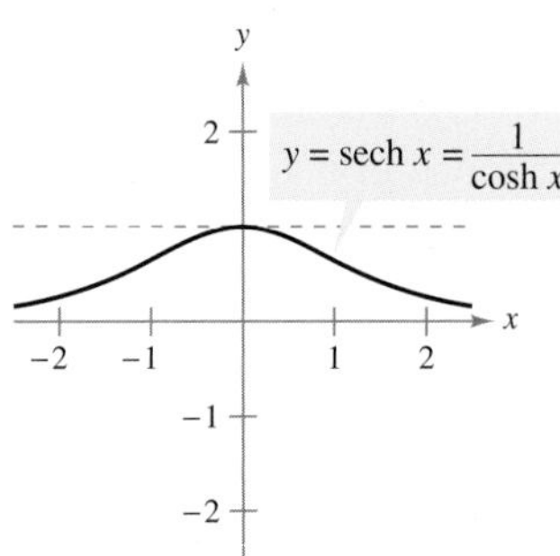

Domain: $(-\infty, \infty)$
Range: $(0, 1]$

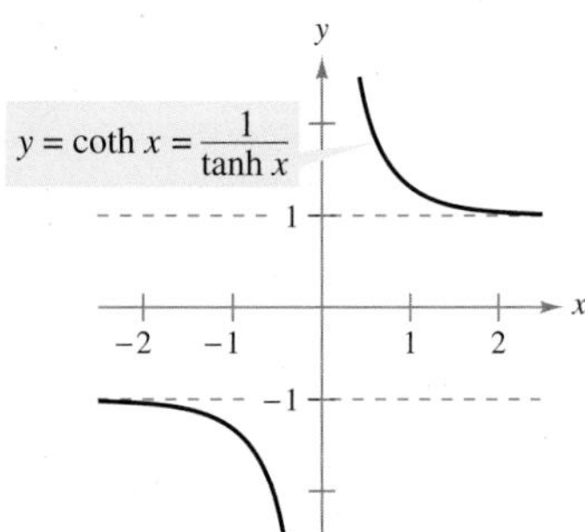

Domain: $(-\infty, 0) \cup (0, \infty)$
Range: $(-\infty, -1) \cup (1, \infty)$

Figure 11.16

Many of the trigonometric identities have corresponding *hyperbolic identities.* For instance,

$$\begin{aligned}
\cosh^2 x - \sinh^2 x &= \left(\frac{e^x + e^{-x}}{2}\right)^2 - \left(\frac{e^x - e^{-x}}{2}\right)^2 \\
&= \frac{e^{2x} + 2 + e^{-2x}}{4} - \frac{e^{2x} - 2 + e^{-2x}}{4} \\
&= \frac{4}{4} \\
&= 1
\end{aligned}$$

and

$$\begin{aligned}
2 \sinh x \cosh x &= 2\left(\frac{e^x - e^{-x}}{2}\right)\left(\frac{e^x + e^{-x}}{2}\right) \\
&= \frac{e^{2x} - e^{-2x}}{2} \\
&= \sinh 2x.
\end{aligned}$$

HYPERBOLIC IDENTITIES

$\cosh^2 x - \sinh^2 x = 1$

$\tanh^2 x + \operatorname{sech}^2 x = 1$

$\coth^2 x - \operatorname{csch}^2 x = 1$

$\sinh(x + y) = \sinh x \cosh y + \cosh x \sinh y$

$\sinh(x - y) = \sinh x \cosh y - \cosh x \sinh y$

$\cosh(x + y) = \cosh x \cosh y + \sinh x \sinh y$

$\cosh(x - y) = \cosh x \cosh y - \sinh x \sinh y$

$\sinh^2 x = \dfrac{-1 + \cosh 2x}{2}$

$\cosh^2 x = \dfrac{1 + \cosh 2x}{2}$

$\sinh 2x = 2 \sinh x \cosh x$

$\cosh 2x = \cosh^2 x + \sinh^2 x$

Differentiation and Integration of Hyperbolic Functions

Because the hyperbolic functions are written in terms of e^x and e^{-x}, you can easily derive rules for their derivatives. The following theorem lists these derivatives with the corresponding integration rules.

THEOREM 11.8 DERIVATIVES AND INTEGRALS OF HYPERBOLIC FUNCTIONS

Let u be a differentiable function of x.

$$\frac{d}{dx}[\sinh u] = (\cosh u)u' \qquad \int \cosh u\, du = \sinh u + C$$

$$\frac{d}{dx}[\cosh u] = (\sinh u)u' \qquad \int \sinh u\, du = \cosh u + C$$

$$\frac{d}{dx}[\tanh u] = (\operatorname{sech}^2 u)u' \qquad \int \operatorname{sech}^2 u\, du = \tanh u + C$$

$$\frac{d}{dx}[\coth u] = -(\operatorname{csch}^2 u)u' \qquad \int \operatorname{csch}^2 u\, du = -\coth u + C$$

$$\frac{d}{dx}[\operatorname{sech} u] = -(\operatorname{sech} u \tanh u)u' \qquad \int \operatorname{sech} u \tanh u\, du = -\operatorname{sech} u + C$$

$$\frac{d}{dx}[\operatorname{csch} u] = -(\operatorname{csch} u \coth u)u' \qquad \int \operatorname{csch} u \coth u\, du = -\operatorname{csch} u + C$$

PROOF

$$\frac{d}{dx}[\sinh x] = \frac{d}{dx}\left[\frac{e^x - e^{-x}}{2}\right] = \frac{e^x + e^{-x}}{2} = \cosh x$$

$$\frac{d}{dx}[\tanh x] = \frac{d}{dx}\left[\frac{\sinh x}{\cosh x}\right] = \frac{\cosh x(\cosh x) - \sinh x(\sinh x)}{\cosh^2 x} = \frac{1}{\cosh^2 x} = \operatorname{sech}^2 x$$

■

In Exercises 76–78, you are asked to prove some of the other differentiation rules.

EXAMPLE 1 Differentiation of Hyperbolic Functions

a. $\dfrac{d}{dx}[\sinh(x^2 - 3)] = 2x \cosh(x^2 - 3)$ $\qquad u = x^2 - 3$

b. $\dfrac{d}{dx}[\ln(\cosh x)] = \dfrac{\sinh x}{\cosh x} = \tanh x$ $\qquad u = \cosh x$

c. $\dfrac{d}{dx}[x \sinh x - \cosh x] = x \cosh x + \sinh x - \sinh x = x \cosh x$

EXAMPLE 2 Finding Relative Extrema

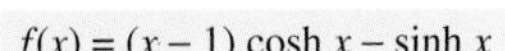

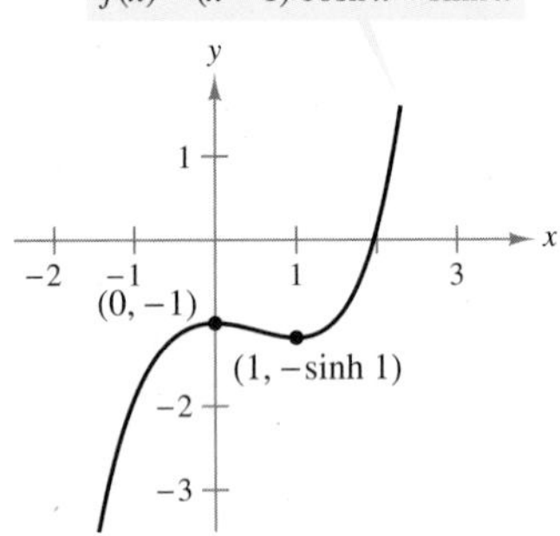

$f''(0) < 0$, so $(0, -1)$ is a relative maximum. $f''(1) > 0$, so $(1, -\sinh 1)$ is a relative minimum.
Figure 11.17

Find the relative extrema of $f(x) = (x - 1) \cosh x - \sinh x$.

Solution Begin by setting the first derivative of f equal to 0.

$$f'(x) = (x - 1) \sinh x + \cosh x - \cosh x = 0$$
$$(x - 1) \sinh x = 0$$

So, the critical numbers are $x = 1$ and $x = 0$. Using the Second Derivative Test, you can verify that the point $(0, -1)$ yields a relative maximum and the point $(1, -\sinh 1)$ yields a relative minimum, as shown in Figure 11.17. Try using a graphing utility to confirm this result. If your graphing utility does not have hyperbolic functions, you can use exponential functions, as follows.

$$\begin{aligned} f(x) &= (x - 1)\left(\tfrac{1}{2}\right)(e^x + e^{-x}) - \tfrac{1}{2}(e^x - e^{-x}) \\ &= \tfrac{1}{2}(xe^x + xe^{-x} - e^x - e^{-x} - e^x + e^{-x}) \\ &= \tfrac{1}{2}(xe^x + xe^{-x} - 2e^x) \end{aligned}$$

■

When a uniform flexible cable, such as a telephone wire, is suspended from two points, it takes the shape of a *catenary*, as discussed in Example 3.

EXAMPLE 3 Hanging Power Cables

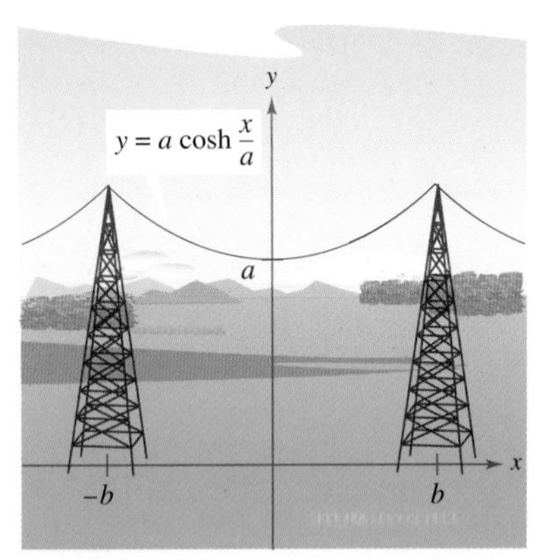

Catenary
Figure 11.18

Power cables are suspended between two towers, forming the catenary shown in Figure 11.18. The equation for this catenary is

$$y = a \cosh \frac{x}{a}.$$

The distance between the two towers is $2b$. Find the slope of the catenary at the point where the cable meets the right-hand tower.

Solution Differentiating produces

$$y' = a\left(\frac{1}{a}\right) \sinh \frac{x}{a} = \sinh \frac{x}{a}.$$

At the point $(b, a \cosh(b/a))$, the slope (from the left) is given by $m = \sinh \dfrac{b}{a}$.

■

■ **FOR FURTHER INFORMATION** In Example 3, the cable is a catenary between two supports at the same height. To learn about the shape of a cable hanging between supports of different heights, see the article "Reexamining the Catenary" by Paul Cella in *The College Mathematics Journal*. To view this article, go to the website *www.matharticles.com*. ■

EXAMPLE 4 Integrating a Hyperbolic Function

Find $\int \cosh 2x \sinh^2 2x\, dx$.

Solution

$$\begin{aligned}\int \cosh 2x \sinh^2 2x\, dx &= \frac{1}{2}\int (\sinh 2x)^2(2\cosh 2x)\, dx \qquad u = \sinh 2x\\ &= \frac{1}{2}\left[\frac{(\sinh 2x)^3}{3}\right] + C\\ &= \frac{\sinh^3 2x}{6} + C\end{aligned}$$

■

Inverse Hyperbolic Functions

Unlike trigonometric functions, hyperbolic functions are not periodic. In fact, by looking back at Figure 11.16, you can see that four of the six hyperbolic functions are actually one-to-one (the hyperbolic sine, tangent, cosecant, and cotangent). So, you can conclude that these four functions have inverse functions. The other two (the hyperbolic cosine and secant) are one-to-one if their domains are restricted to the positive real numbers, and for this restricted domain they also have inverse functions. Because the hyperbolic functions are defined in terms of exponential functions, it is not surprising to find that the inverse hyperbolic functions can be written in terms of logarithmic functions, as shown in Theorem 11.9.

THEOREM 11.9 INVERSE HYPERBOLIC FUNCTIONS

Function	*Domain*
$\sinh^{-1} x = \ln\left(x + \sqrt{x^2 + 1}\right)$	$(-\infty, \infty)$
$\cosh^{-1} x = \ln\left(x + \sqrt{x^2 - 1}\right)$	$[1, \infty)$
$\tanh^{-1} x = \frac{1}{2}\ln\frac{1 + x}{1 - x}$	$(-1, 1)$
$\coth^{-1} x = \frac{1}{2}\ln\frac{x + 1}{x - 1}$	$(-\infty, -1) \cup (1, \infty)$
$\text{sech}^{-1} x = \ln\frac{1 + \sqrt{1 - x^2}}{x}$	$(0, 1]$
$\text{csch}^{-1} x = \ln\left(\frac{1}{x} + \frac{\sqrt{1 + x^2}}{\lvert x\rvert}\right)$	$(-\infty, 0) \cup (0, \infty)$

PROOF The proof of this theorem is a straightforward application of the properties of the exponential and logarithmic functions. For example, if

$$f(x) = \sinh x = \frac{e^x - e^{-x}}{2}$$

and

$$g(x) = \ln\left(x + \sqrt{x^2 + 1}\right)$$

you can show that $f(g(x)) = x$ and $g(f(x)) = x$, which implies that g is the inverse function of f. ■

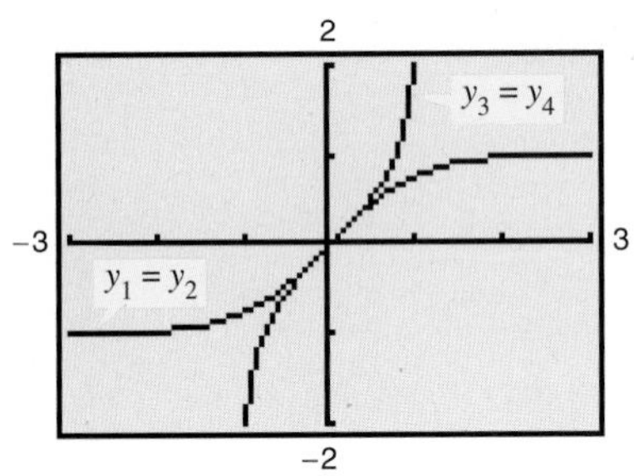

Graphs of the hyperbolic tangent function and the inverse hyperbolic tangent function
Figure 11.19

TECHNOLOGY You can use a graphing utility to confirm graphically the results of Theorem 11.9. For instance, graph the following functions.

$$y_1 = \tanh x \qquad \text{Hyperbolic tangent}$$

$$y_2 = \frac{e^x - e^{-x}}{e^x + e^{-x}} \qquad \text{Definition of hyperbolic tangent}$$

$$y_3 = \tanh^{-1} x \qquad \text{Inverse hyperbolic tangent}$$

$$y_4 = \frac{1}{2}\ln\frac{1 + x}{1 - x} \qquad \text{Definition of inverse hyperbolic tangent}$$

The resulting display is shown in Figure 11.19. As you watch the graphs being traced out, notice that $y_1 = y_2$ and $y_3 = y_4$. Also notice that the graph of y_1 is the reflection of the graph of y_3 in the line $y = x$.

The graphs of the inverse hyperbolic functions are shown in Figure 11.20.

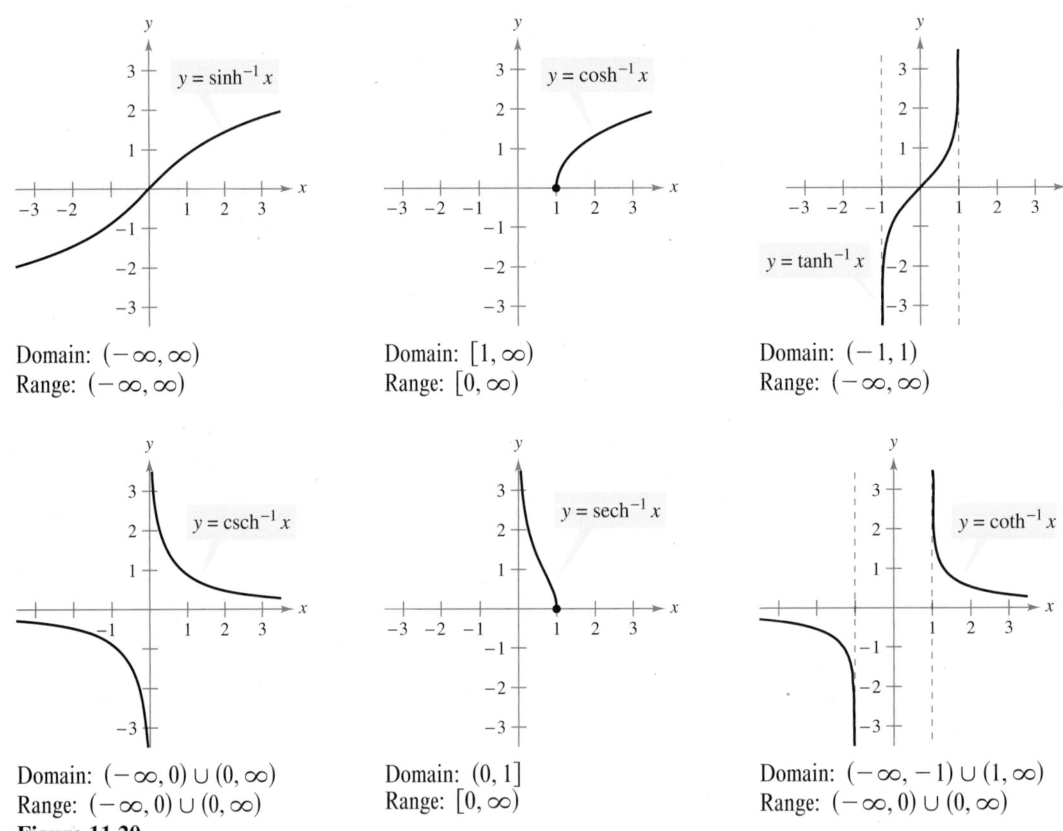

Figure 11.20

The inverse hyperbolic secant can be used to define a curve called a *tractrix* or *pursuit curve*, as discussed in Example 5.

EXAMPLE 5 A Tractrix

A person is holding a rope that is tied to a boat, as shown in Figure 11.21. As the person walks along the dock, the boat travels along a **tractrix,** given by the equation

$$y = a \operatorname{sech}^{-1} \frac{x}{a} - \sqrt{a^2 - x^2}$$

where a is the length of the rope. If $a = 20$ feet, find the distance the person must walk to bring the boat to a position 5 feet from the dock.

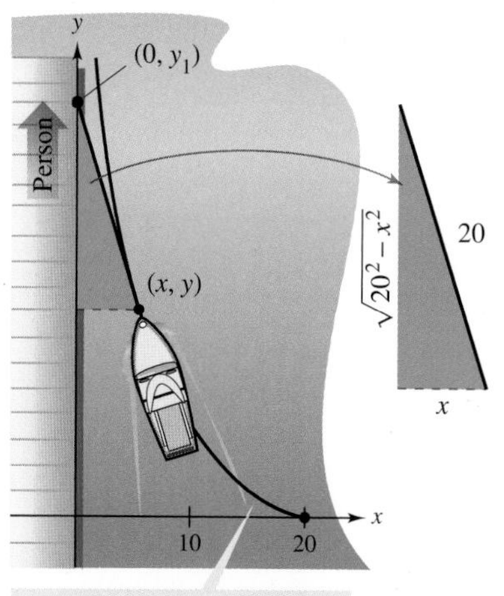

A person must walk 41.27 feet to bring the boat to a position 5 feet from the dock.
Figure 11.21

Solution In Figure 11.21, notice that the distance the person has walked is given by

$$\begin{aligned} y_1 = y + \sqrt{20^2 - x^2} &= \left(20 \operatorname{sech}^{-1} \frac{x}{20} - \sqrt{20^2 - x^2}\right) + \sqrt{20^2 - x^2} \\ &= 20 \operatorname{sech}^{-1} \frac{x}{20}. \end{aligned}$$

When $x = 5$, this distance is

$$\begin{aligned} y_1 = 20 \operatorname{sech}^{-1} \frac{5}{20} &= 20 \ln \frac{1 + \sqrt{1 - (1/4)^2}}{1/4} \\ &= 20 \ln\left(4 + \sqrt{15}\right) \\ &\approx 41.27 \text{ feet.} \end{aligned}$$

■

Differentiation and Integration of Inverse Hyperbolic Functions

The derivatives of the inverse hyperbolic functions, which resemble the derivatives of the inverse trigonometric functions, are listed in Theorem 11.10 with the corresponding integration formulas (in logarithmic form). You can verify each of these formulas by applying the logarithmic definitions of the inverse hyperbolic functions. (See Exercises 73–75.)

THEOREM 11.10 DIFFERENTIATION AND INTEGRATION INVOLVING INVERSE HYPERBOLIC FUNCTIONS

Let u be a differentiable function of x.

$$\frac{d}{dx}[\sinh^{-1} u] = \frac{u'}{\sqrt{u^2 + 1}} \qquad \frac{d}{dx}[\cosh^{-1} u] = \frac{u'}{\sqrt{u^2 - 1}}$$

$$\frac{d}{dx}[\tanh^{-1} u] = \frac{u'}{1 - u^2} \qquad \frac{d}{dx}[\coth^{-1} u] = \frac{u'}{1 - u^2}$$

$$\frac{d}{dx}[\operatorname{sech}^{-1} u] = \frac{-u'}{u\sqrt{1 - u^2}} \qquad \frac{d}{dx}[\operatorname{csch}^{-1} u] = \frac{-u'}{|u|\sqrt{1 + u^2}}$$

$$\int \frac{du}{\sqrt{u^2 \pm a^2}} = \ln\left(u + \sqrt{u^2 \pm a^2}\right) + C$$

$$\int \frac{du}{a^2 - u^2} = \frac{1}{2a} \ln\left|\frac{a + u}{a - u}\right| + C$$

$$\int \frac{du}{u\sqrt{a^2 \pm u^2}} = -\frac{1}{a} \ln \frac{a + \sqrt{a^2 \pm u^2}}{|u|} + C$$

EXAMPLE 6 More About a Tractrix

For the tractrix given in Example 5, show that the boat is always pointing toward the person.

Solution For a point (x, y) on a tractrix, the slope of the graph gives the direction of the boat, as shown in Figure 11.21.

$$\begin{aligned} y' &= \frac{d}{dx}\left[20 \operatorname{sech}^{-1}\frac{x}{20} - \sqrt{20^2 - x^2}\right] \\ &= -20\left(\frac{1}{20}\right)\left[\frac{1}{(x/20)\sqrt{1-(x/20)^2}}\right] - \left(\frac{1}{2}\right)\left(\frac{-2x}{\sqrt{20^2 - x^2}}\right) \\ &= \frac{-20^2}{x\sqrt{20^2 - x^2}} + \frac{x}{\sqrt{20^2 - x^2}} \\ &= -\frac{\sqrt{20^2 - x^2}}{x} \end{aligned}$$

However, from Figure 11.21, you can see that the slope of the line segment connecting the point $(0, y_1)$ with the point (x, y) is also

$$m = -\frac{\sqrt{20^2 - x^2}}{x}.$$

So, the boat is always pointing toward the person. (It is because of this property that a tractrix is called a *pursuit curve*.)

EXAMPLE 7 Integration Using Inverse Hyperbolic Functions

Find $\displaystyle\int \frac{dx}{x\sqrt{4 - 9x^2}}$.

Solution Let $a = 2$ and $u = 3x$.

$$\begin{aligned} \int \frac{dx}{x\sqrt{4 - 9x^2}} &= \int \frac{3\,dx}{(3x)\sqrt{4 - 9x^2}} & \int \frac{du}{u\sqrt{a^2 - u^2}} \\ &= -\frac{1}{2}\ln\frac{2 + \sqrt{4 - 9x^2}}{|3x|} + C & -\frac{1}{a}\ln\frac{a + \sqrt{a^2 - u^2}}{|u|} + C \end{aligned}$$

EXAMPLE 8 Integration Using Inverse Hyperbolic Functions

Find $\displaystyle\int \frac{dx}{5 - 4x^2}$.

Solution Let $a = \sqrt{5}$ and $u = 2x$.

$$\begin{aligned} \int \frac{dx}{5 - 4x^2} &= \frac{1}{2}\int \frac{2\,dx}{(\sqrt{5})^2 - (2x)^2} & \int \frac{du}{a^2 - u^2} \\ &= \frac{1}{2}\left(\frac{1}{2\sqrt{5}}\ln\left|\frac{\sqrt{5} + 2x}{\sqrt{5} - 2x}\right|\right) + C & \frac{1}{2a}\ln\left|\frac{a + u}{a - u}\right| + C \\ &= \frac{1}{4\sqrt{5}}\ln\left|\frac{\sqrt{5} + 2x}{\sqrt{5} - 2x}\right| + C \end{aligned}$$

■

11.6 Exercises

See www.CalcChat.com for worked-out solutions to odd-numbered exercises.

In Exercises 1–4, evaluate the function. If the value is not a rational number, give the answer to three-decimal-place accuracy.

1. (a) $\sinh 3$ (b) $\tanh(-2)$
2. (a) $\cosh 0$ (b) $\operatorname{sech} 1$
3. (a) $\cosh^{-1} 2$ (b) $\operatorname{sech}^{-1} \frac{2}{3}$
4. (a) $\sinh^{-1} 0$ (b) $\tanh^{-1} 0$

In Exercises 5–12, verify the identity.

5. $e^x = \sinh x + \cosh x$
6. $e^{2x} = \sinh 2x + \cosh 2x$
7. $\tanh^2 x + \operatorname{sech}^2 x = 1$
8. $\coth^2 x - \operatorname{csch}^2 x = 1$
9. $\cosh^2 x = \dfrac{1 + \cosh 2x}{2}$
10. $\sinh^2 x = \dfrac{-1 + \cosh 2x}{2}$
11. $\sinh(x + y) = \sinh x \cosh y + \cosh x \sinh y$
12. $\cosh x + \cosh y = 2 \cosh \dfrac{x + y}{2} \cosh \dfrac{x - y}{2}$

In Exercises 13 and 14, use the value of the given hyperbolic function to find the values of the other hyperbolic functions at x.

13. $\sinh x = \dfrac{3}{2}$
14. $\tanh x = \dfrac{1}{2}$

In Exercises 15–24, find the derivative of the function.

15. $f(x) = \sinh 3x$
16. $f(x) = \cosh(x - 2)$
17. $f(x) = \ln(\sinh x)$
18. $g(x) = \ln(\cosh x)$
19. $y = \ln\left(\tanh \dfrac{x}{2}\right)$
20. $y = x \cosh x - \sinh x$
21. $h(x) = \dfrac{1}{4} \sinh 2x - \dfrac{x}{2}$
22. $h(t) = t - \coth t$
23. $f(t) = \arctan(\sinh t)$
24. $g(x) = \operatorname{sech}^2 3x$

In Exercises 25–28, find any relative extrema of the function. Use a graphing utility to confirm your result.

25. $f(x) = \sin x \sinh x - \cos x \cosh x, \quad -4 \le x \le 4$
26. $f(x) = x \sinh(x - 1) - \cosh(x - 1)$
27. $g(x) = x \operatorname{sech} x$
28. $h(x) = 2 \tanh x - x$

In Exercises 29 and 30, show that the function satisfies the differential equation.

	Function	*Differential Equation*
29.	$y = a \sinh x$	$y''' - y' = 0$
30.	$y = a \cosh x$	$y'' - y = 0$

Catenary **In Exercises 31 and 32, a model for a power cable suspended between two towers is given. (a) Graph the model, (b) find the heights of the cable at the towers and at the midpoint between the towers, and (c) find the slope of the model at the point where the cable meets the right-hand tower.**

31. $y = 10 + 15 \cosh \dfrac{x}{15}, \quad -15 \le x \le 15$
32. $y = 18 + 25 \cosh \dfrac{x}{25}, \quad -25 \le x \le 25$

In Exercises 33–44, find the integral.

33. $\displaystyle\int \cosh 2x \, dx$
34. $\displaystyle\int \operatorname{sech}^2 (-x) \, dx$
35. $\displaystyle\int \sinh(1 - 2x) \, dx$
36. $\displaystyle\int \frac{\cosh \sqrt{x}}{\sqrt{x}} \, dx$
37. $\displaystyle\int \frac{\cosh x}{\sinh x} \, dx$
38. $\displaystyle\int \operatorname{sech}^2(2x - 1) \, dx$
39. $\displaystyle\int x \operatorname{csch}^2 \frac{x^2}{2} \, dx$
40. $\displaystyle\int \operatorname{sech}^3 x \tanh x \, dx$
41. $\displaystyle\int \frac{\operatorname{csch}(1/x) \coth(1/x)}{x^2} \, dx$
42. $\displaystyle\int \frac{\cosh x}{\sqrt{9 - \sinh^2 x}} \, dx$
43. $\displaystyle\int \frac{x}{x^4 + 1} \, dx$
44. $\displaystyle\int \frac{2}{x\sqrt{1 + 4x^2}} \, dx$

In Exercises 45–48, evaluate the integral.

45. $\displaystyle\int_0^4 \frac{1}{25 - x^2} \, dx$
46. $\displaystyle\int_0^1 \cosh^2 x \, dx$
47. $\displaystyle\int_0^{\sqrt{2}/4} \frac{2}{\sqrt{1 - 4x^2}} \, dx$
48. $\displaystyle\int_0^4 \frac{1}{\sqrt{25 - x^2}} \, dx$

In Exercises 49–58, find the derivative of the function.

49. $y = \cosh^{-1}(3x)$
50. $y = \tanh^{-1} \dfrac{x}{2}$
51. $y = \tanh^{-1} \sqrt{x}$
52. $f(x) = \coth^{-1}(x^2)$
53. $y = \sinh^{-1}(\tan x)$
54. $y = \tanh^{-1}(\sin 2x)$
55. $y = (\operatorname{csch}^{-1} x)^2$
56. $y = \operatorname{sech}^{-1}(\cos 2x), \quad 0 < x < \pi/4$
57. $y = 2x \sinh^{-1}(2x) - \sqrt{1 + 4x^2}$
58. $y = x \tanh^{-1} x + \ln \sqrt{1 - x^2}$

WRITING ABOUT CONCEPTS

59. Sketch the graph of each hyperbolic function. Then identify the domain and range of each function.

CAPSTONE

60. Which hyperbolic functions take on only positive values? Which hyperbolic functions are increasing on their domains?

In Exercises 61–66, find the indefinite integral using the formulas from Theorem 11.10.

61. $\displaystyle\int \frac{1}{3 - 9x^2}\,dx$

62. $\displaystyle\int \frac{1}{2x\sqrt{1 - 4x^2}}\,dx$

63. $\displaystyle\int \frac{1}{\sqrt{1 + e^{2x}}}\,dx$

64. $\displaystyle\int \frac{1}{\sqrt{x}\sqrt{1 + x}}\,dx$

65. $\displaystyle\int \frac{-1}{4x - x^2}\,dx$

66. $\displaystyle\int \frac{dx}{(x + 1)\sqrt{2x^2 + 4x + 8}}$

In Exercises 67 and 68, solve the differential equation.

67. $\dfrac{dy}{dx} = \dfrac{x^3 - 21x}{5 + 4x - x^2}$

68. $\dfrac{dy}{dx} = \dfrac{1 - 2x}{4x - x^2}$

In Exercises 69 and 70, find the area of the region bounded by the graphs of the equations.

69. $y = \tanh 2x,\quad y = 0,\quad x = 2$

70. $y = \dfrac{6}{\sqrt{x^2 - 4}},\quad y = 0,\quad x = 3,\quad x = 5$

Tractrix **In Exercises 71 and 72, use the equation of the tractrix $y = a\,\text{sech}^{-1}(x/a) - \sqrt{a^2 - x^2},\ a > 0$.**

71. Find dy/dx.

72. Let L be the tangent line to the tractrix at the point P. If L intersects the y-axis at the point Q, show that the distance between P and Q is a.

In Exercises 73–78, verify the differentiation formula.

73. $\dfrac{d}{dx}[\text{sech}^{-1} x] = \dfrac{-1}{x\sqrt{1 - x^2}}$

74. $\dfrac{d}{dx}[\cosh^{-1} x] = \dfrac{1}{\sqrt{x^2 - 1}}$

75. $\dfrac{d}{dx}[\sinh^{-1} x] = \dfrac{1}{\sqrt{x^2 + 1}}$

76. $\dfrac{d}{dx}[\text{sech}\, x] = -\text{sech}\, x \tanh x$

77. $\dfrac{d}{dx}[\cosh x] = \sinh x$

78. $\dfrac{d}{dx}[\coth x] = -\text{csch}^2 x$

SECTION PROJECT

St. Louis Arch

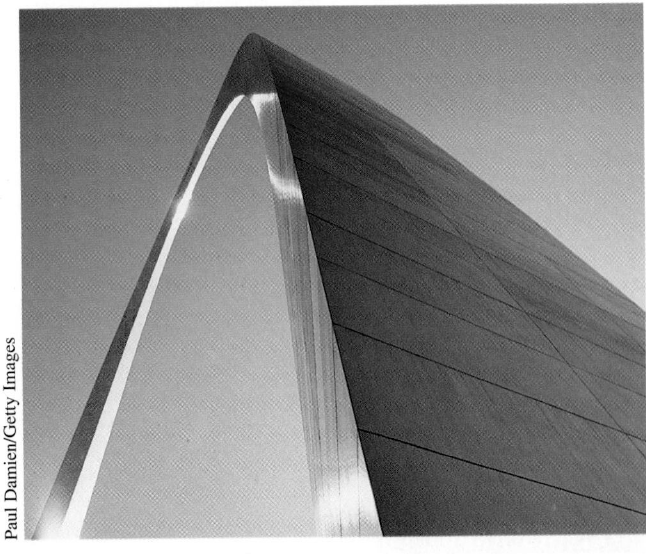
Paul Damien/Getty Images

The Gateway Arch in St. Louis, Missouri was constructed using the hyperbolic cosine function. The equation used for construction was

$y = 693.8597 - 68.7672 \cosh 0.0100333x,$

$-299.2239 \le x \le 299.2239$

where x and y are measured in feet. Cross sections of the arch are equilateral triangles, and (x, y) traces the path of the centers of mass of the cross-sectional triangles. For each value of x, the area of the cross-sectional triangle is $A = 125.1406 \cosh 0.0100333x$. *(Source: Owner's Manual for the Gateway Arch, Saint Louis, MO, by William Thayer)*

(a) How high above the ground is the center of the highest triangle? (At ground level, $y = 0$.)

(b) What is the height of the arch? (*Hint:* For an equilateral triangle, $A = \sqrt{3}c^2$, where c is one-half the base of the triangle, and the center of mass of the triangle is located at two-thirds the height of the triangle.)

(c) How wide is the arch at ground level?

11 CHAPTER SUMMARY

	Review Exercises
Section 11.1	
■ Determine the limits of trigonometric functions *(p. 692)*.	*1–16*
Section 11.2	
■ Find and use the derivatives of the sine and cosine functions *(p. 699)*, and find and use the derivatives of other trigonometric functions *(p. 701)*.	*17–44*
■ Apply the First Derivative Test to find the minima and maxima of a function *(p. 703)*.	*45–48*
Section 11.3	
■ Integrate trigonometric functions using trigonometric identities and u-substitution *(p. 709)*.	*49–72*
■ Use integrals to find the average value of a function *(p. 713)*.	*73, 74*
Section 11.4	
■ Differentiate an inverse trigonometric function *(p. 717)*, and review the basic differentiation rules for elementary functions *(p. 719)*.	*75–80*
Section 11.5	
■ Integrate functions whose antiderivatives involve inverse trigonometric functions *(p. 722)*, use the method of completing the square to integrate a function *(p. 723)*, and review the basic integration rules involving elementary functions *(p. 725)*.	*81–87*
Section 11.6	
■ Develop properties of hyperbolic functions *(p. 729)*, differentiate and integrate hyperbolic functions *(p. 731)*, develop properties of inverse hyperbolic functions *(p. 733)*, and differentiate and integrate functions involving inverse hyperbolic functions *(p. 735)*.	*88–91*

11 REVIEW EXERCISES

See www.CalcChat.com for worked-out solutions to odd-numbered exercises.

In Exercises 1–12, find the limit (if it exists).

1. $\lim_{x\to 3} \sec(\pi x)$
2. $\lim_{\theta\to -1/2} \tan(2\pi\,\theta)$
3. $\lim_{x\to \pi/2} \cot x$
4. $\lim_{x\to \pi/3} \sec x$
5. $\lim_{\alpha\to 0} \dfrac{\sin 5\alpha}{3\alpha}$
6. $\lim_{t\to 0} \dfrac{2(1-\cos t)}{10t}$
7. $\lim_{\Delta x\to 0} \dfrac{\sin[(\pi/6)+\Delta x]-(1/2)}{\Delta x}$

   [*Hint:* $\sin(\theta+\phi) = \sin\theta\cos\phi + \cos\theta\sin\phi$]
8. $\lim_{\Delta x\to 0} \dfrac{\cos(\pi+\Delta x)+1}{\Delta x}$

   [*Hint:* $\cos(\theta+\phi) = \cos\theta\cos\phi - \sin\theta\sin\phi$]
9. $\lim_{x\to 0^+} \dfrac{\sin 4x}{5x}$
10. $\lim_{x\to 0^+} \dfrac{\sec x}{x}$
11. $\lim_{x\to 0^+} \dfrac{\csc 2x}{x}$
12. $\lim_{x\to 0^-} \dfrac{\cos^2 x}{x}$

In Exercises 13 and 14, determine the intervals on which the function is continuous.

13. $f(x) = \csc \dfrac{\pi x}{2}$
14. $f(x) = \tan 2x$

15. ***Writing*** Give a written explanation of why the function

$$f(x) = -\frac{4}{x} + \tan\left(\frac{\pi x}{8}\right)$$

has a zero in the interval $[1, 3]$.

16. The function f is defined as follows.

$$f(x) = \frac{\tan 2x}{x}, \quad x \neq 0$$

(a) Find $\lim_{x\to 0} \dfrac{\tan 2x}{x}$ (if it exists).

(b) Can the function f be defined such that it is continuous at $x = 0$?

In Exercises 17–36, find the derivative of the function.

17. $f(\theta) = 4\theta - 5\sin\theta$
18. $g(\alpha) = 4\cos\alpha + 6$
19. $h(x) = \sqrt{x}\sin x$
20. $f(t) = 2t^5\cos t$
21. $y = \dfrac{x^4}{\cos x}$
22. $y = \dfrac{\sin x}{x^4}$
23. $y = 3\sec x$
24. $y = 2x - \tan x$
25. $y = \frac{1}{2}\csc 2x$
26. $y = \csc 3x + \cot 3x$
27. $y = \dfrac{x}{2} - \dfrac{\sin 2x}{4}$
28. $y = \dfrac{1+\sin x}{1-\sin x}$
29. $y = \frac{2}{3}\sin^{3/2} x - \frac{2}{7}\sin^{7/2} x$
30. $y = \dfrac{\sec^7 x}{7} - \dfrac{\sec^5 x}{5}$
31. $y = -x\tan x$
32. $y = x\cos x - \sin x$
33. $y = \dfrac{\sin \pi x}{x+2}$
34. $y = \dfrac{\cos(x-1)}{x-1}$
35. $y = \ln|\tan\theta|$
36. $y = e^{-x}\cos \pi x$

In Exercises 37–40, find the second derivative of the function.

37. $f(x) = \cot x$
38. $y = \sin^2 x$
39. $f(\theta) = 3\tan\theta$
40. $h(t) = 4\sin t - 5\cos t$

In Exercises 41 and 42, show that the function satisfies the equation.

	Function	*Equation*
41.	$y = 2\sin x + 3\cos x$	$y'' + y = 0$
42.	$y = \dfrac{10 - \cos x}{x}$	$xy' + y = \sin x$

In Exercises 43 and 44, find dy/dx by implicit differentiation and evaluate the derivative at the given point.

43. $\tan(x+y) = x, \quad (0, 0)$
44. $x\cos y = 1, \quad \left(2, \dfrac{\pi}{3}\right)$

In Exercises 45 and 46, determine the absolute extrema of the function in the closed interval and the x-values where they occur.

45. $g(x) = \csc x, \quad \left[\dfrac{\pi}{6}, \dfrac{\pi}{3}\right]$
46. $f(x) = 2x - \tan x, \quad \left[-\dfrac{\pi}{2}, \dfrac{\pi}{4}\right]$

47. ***Distance*** A hallway of width 6 feet meets a hallway of width 9 feet at right angles. Find the length of the longest pipe that can be carried level around this corner. [*Hint:* If L is the length of the pipe, show that

$$L = 6\csc\theta + 9\csc\left(\frac{\pi}{2} - \theta\right)$$

where θ is the angle between the pipe and the wall of the narrower hallway.]

48. ***Length*** Rework Exercise 47, given that one hallway is of width a meters and the other is of width b meters.

In Exercises 49–60, find the indefinite integral.

49. $\displaystyle\int (2x - 9\sin x)\,dx$
50. $\displaystyle\int (5\cos x - 2\sec^2 x)\,dx$

51. $\int \sin^3 x \cos x\, dx$

52. $\int x \sin 3x^2\, dx$

53. $\int \frac{\cos \theta}{\sqrt{1 - \sin \theta}}\, d\theta$

54. $\int \frac{\sin x}{\sqrt{\cos x}}\, dx$

55. $\int \tan^n x \sec^2 x\, dx, \quad n \neq -1$

56. $\int (1 + \sec \pi x)^2 \sec \pi x \tan \pi x\, dx$

57. $\int \sec 2x \tan 2x\, dx$

58. $\int \cot^4 \alpha \csc^2 \alpha\, d\alpha$

59. $\int \frac{\sin x}{1 + \cos x}\, dx$

60. $\int \sec^2 x e^{\tan x}\, dx$

In Exercises 61 and 62, find an equation for the function f that has the given derivative and whose graph passes through the given point.

61. $f'(x) = \cos \frac{x}{2}, \quad (0, 3)$

62. $f'(x) = \pi \sec \pi x \tan \pi x, \quad \left(\frac{1}{3}, 1\right)$

In Exercises 63–66, evaluate the definite integral. Use a graphing utility to verify your result.

63. $\int_0^{\pi} \cos \frac{x}{2}\, dx$

64. $\int_{-\pi/4}^{\pi/4} \sin 2x\, dx$

65. $\int_0^{\pi/3} \sec \theta\, d\theta$

66. $\int_0^{\pi/4} \tan\left(\frac{\pi}{4} - x\right) dx$

In Exercises 67 and 68, sketch the region bounded by the graphs of the equations, and determine its area.

67. $y = \sec^2 x, \quad y = 0, \quad x = 0, \quad x = \frac{\pi}{3}$

68. $y = \cos x, \quad y = 0, \quad x = -\frac{\pi}{4}, \quad x = \frac{\pi}{4}$

In Exercises 69 and 70, use the Second Fundamental Theorem of Calculus to find $F'(x)$.

69. $F(x) = \int_0^x \tan^4 t\, dt$

70. $F(x) = \int_{\pi/2}^{x^3} \cos t\, dt$

In Exercises 71 and 72, approximate the definite integral using (a) the Trapezoidal Rule and (b) Simpson's Rule.

71. $\int_0^{\pi} \sqrt{x} \sin x\, dx, \quad n = 4$

72. $\int_0^{\pi/2} \sqrt{1 + \cos^2 x}\, dx, \quad n = 2$

In Exercises 73 and 74, use a graphing utility to graph the function over the given interval. Find the average value of the function over the interval and all values of x in the interval for which the function equals its average value.

73. $f(x) = \tan x, \quad \left[0, \frac{\pi}{4}\right]$

74. $f(x) = \sec x, \quad \left[-\frac{\pi}{3}, \frac{\pi}{3}\right]$

In Exercises 75–80, find the derivative of the function.

75. $y = \tan(\arcsin x)$

76. $y = \arctan(x^2 - 1)$

77. $y = x \operatorname{arcsec} x$

78. $y = \frac{1}{2} \arctan e^{2x}$

79. $y = x(\arcsin x)^2 - 2x + 2\sqrt{1 - x^2} \arcsin x$

80. $y = \sqrt{x^2 - 4} - 2 \operatorname{arcsec} \frac{x}{2}, \quad 2 < x < 4$

In Exercises 81–86, find the indefinite integral.

81. $\int \frac{1}{e^{2x} + e^{-2x}}\, dx$

82. $\int \frac{1}{3 + 25x^2}\, dx$

83. $\int \frac{x}{16 + x^2}\, dx$

84. $\int \frac{1}{\sqrt{2x - x^2}}\, dx$

85. $\int \frac{\arctan(x/2)}{4 + x^2}\, dx$

86. $\int \frac{\arcsin 2x}{\sqrt{1 - 4x^2}}\, dx$

87. ***Harmonic Motion*** A weight of mass m is attached to a spring and oscillates with simple harmonic motion. By Hooke's Law, you can determine that

$$\int \frac{dy}{\sqrt{A^2 - y^2}} = \int \sqrt{\frac{k}{m}}\, dt$$

where A is the maximum displacement, t is the time, and k is a constant. Find y as a function of t, given that $y = 0$ when $t = 0$.

In Exercises 88 and 89, find the derivative of the function.

88. $y = 2x - \cosh \sqrt{x}$

89. $y = x \tanh^{-1} 2x$

In Exercises 90 and 91, find the indefinite integral.

90. $\int \frac{x}{\sqrt{x^4 - 1}}\, dx$

91. $\int x^2 \operatorname{sech}^2 x^3\, dx$

11 CHAPTER TEST

Take this test as you would take a test in class. When you are finished, check your work against the answers given in the back of the book.

In Exercises 1 and 2, find the limit (if it exists). If the limit does not exist, explain why.

1. $\lim_{x\to 0} \frac{\sin x}{3x}$

2. $\lim_{x\to 0} \sin \frac{1}{2x}$

In Exercises 3–6, find the derivative of the function.

3. $f(x) = x^4 \cos x$

4. $f(\theta) = \frac{1}{3}\sin^2 3\theta$

5. $h(t) = \csc(\operatorname{arccot} t)$

6. $y = \cosh(1 + x^2)$

In Exercises 7 and 8, find an equation of the tangent line to the graph of *f* at the given point.

7. $f(x) = e^{\tan x}$, $(0, 1)$

8. $f(x) = 3 \arccos x$, $\left(\frac{1}{2}, \pi\right)$

9. Find $\frac{dy}{dx}$ by implicit differentiation given that $y = \sin(x + y)$.

10. Determine whether Rolle's Theorem can be applied to $f(x) = \cot x$ on the closed interval $\left[\frac{\pi}{4}, \frac{5\pi}{4}\right]$. If Rolle's Theorem can be applied, find all values of c on the open interval $\left(\frac{\pi}{4}, \frac{5\pi}{4}\right)$ such that $f'(c) = 0$. If Rolle's Theorem cannot be applied, explain why not.

In Exercises 11–14, find the indefinite integral.

11. $\int \sin 4x \cos 4x \, dx$

12. $\int \frac{t}{\sqrt{4 - t^4}} \, dt$

13. $\int \frac{2x}{x^2 + 8x + 17} \, dx$

14. $\int \sinh(1 + 3x) \, dx$

In Exercises 15 and 16, evaluate the definite integral.

15. $\int_{-\pi/3}^{\pi/3} \sec^2 x \, dx$

16. $\int_{-\pi/6}^{\pi/6} \sin x \, dx$

In Exercises 17 and 18, find the area of the region bounded by the graphs of the equations.

17. $y = \sin x, y = 0, x = 0, x = \frac{\pi}{2}$

18. $y = \frac{1}{\sqrt{16 - x^2}}, y = 0, x = 0, x = 2$

19. Given $F(x) = \int_0^x t \tan t \, dt$, use the Second Fundamental Theorem of Calculus to find $F'(x)$.

20. Find any relative extrema of $f(x) = \arcsin x - 3x$.

21. Evaluate each hyperbolic function. Give the answers to three-decimal-place accuracy.

(a) $\sinh 2$ (b) $\tanh(-3)$

22. Verify the identity: $\cosh 2x = \cosh^2 x + \sinh^2 x$.

P.S. PROBLEM SOLVING

1. Find a function of the form $f(x) = a + b \cos cx$ that is tangent to the line $y = 1$ at the point $(0, 1)$, and tangent to the line

$$y = x + \frac{3}{2} - \frac{\pi}{4}$$

at the point $\left(\frac{\pi}{4}, \frac{3}{2}\right)$.

2. The fundamental limit $\lim_{x \to 0} \frac{\sin x}{x} = 1$ assumes that x is measured in radians. What happens if you assume that x is measured in degrees instead of radians?

(a) Set your calculator to *degree* mode and complete the table.

z (in degrees)	0.1	0.01	0.0001
$\frac{\sin z}{z}$			

(b) Use the table to estimate

$$\lim_{z \to 0} \frac{\sin z}{z}$$

for z in degrees. What is the exact value of this limit? (*Hint:* $180° = \pi$ radians)

(c) Use the limit definition of the derivative to find

$$\frac{d}{dz} \sin z$$

for z in degrees.

(d) Define the new functions $S(z) = \sin(cz)$ and $C(z) = \cos(cz)$, where $c = \pi/180$. Find $S(90)$ and $C(180)$. Use the Chain Rule to calculate

$$\frac{d}{dz} S(z).$$

(e) Explain why differentiation is made easier by using radians instead of degrees.

3. The efficiency E of a screw with square threads is

$$E = \frac{\tan \phi(1 - \mu \tan \phi)}{\mu + \tan \phi}$$

where μ is the coefficient of sliding friction and ϕ is the angle of inclination of the threads to a plane perpendicular to the axis of the screw. Find the angle ϕ that yields maximum efficiency when $\mu = 0.1$.

4. (a) Let x be a positive number. Use the *table* feature of a graphing utility to verify that $\sin x < x$.

(b) Use the Mean Value Theorem to prove that $\sin x < x$ for all positive real numbers x.

5. The amount of illumination of a surface is proportional to the intensity of the light source, inversely proportional to the square of the distance from the light source, and proportional to $\sin \theta$, where θ is the angle at which the light strikes the surface. A rectangular room measures 10 feet by 24 feet, with a 10-foot ceiling (see figure). Determine the height at which the light should be placed to allow the corners of the floor to receive as much light as possible.

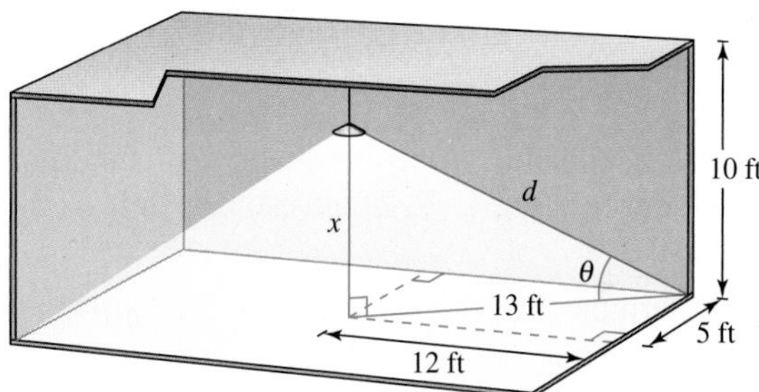

6. Let f be continuous on the interval $[0, b]$, where $f(x) + f(b - x) \neq 0$ on $[0, b]$.

(a) Show that $\displaystyle\int_0^b \frac{f(x)}{f(x) + f(b - x)}\, dx = \frac{b}{2}$.

(b) Use the result in part (a) to evaluate

$$\int_0^1 \frac{\sin x}{\sin(1 - x) + \sin x}\, dx.$$

(c) Use the result in part (a) to evaluate

$$\int_0^3 \frac{\sqrt{x}}{\sqrt{x} + \sqrt{3 - x}}\, dx.$$

7. (a) Let $P(\cos t, \sin t)$ be a point on the unit circle $x^2 + y^2 = 1$ in the first quadrant (see figure). Show that t is equal to twice the area of the shaded circular sector AOP.

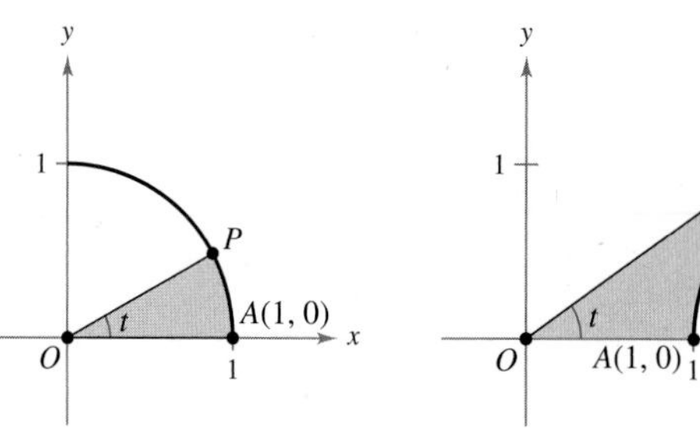

Figure for 7(a) Figure for 7(b)

(b) Let $P(\cosh t, \sinh t)$ be a point on the unit hyperbola $x^2 - y^2 = 1$ in the first quadrant (see figure). Show that t is equal to twice the area of the shaded region AOP. [*Hint:* Begin by showing that the area of the shaded region AOP is given by the formula

$$A(t) = \frac{1}{2}\cosh t \sinh t - \int_1^{\cosh t} \sqrt{x^2 - 1}\, dx.]$$

8. Let $f(x) = \sin(\ln x)$.

(a) Determine the domain of the function f.

(b) Find two values of x satisfying $f(x) = 1$.

(c) Find two values of x satisfying $f(x) = -1$.

(d) What is the range of the function f?

(e) Calculate $f'(x)$ and find the maximum value of f on the interval $[1, 10]$.

(f) Use a graphing utility to graph f in the viewing window $[0, 5] \times [-2, 2]$ and estimate $\lim_{x\to 0^+} f(x)$, if it exists.

(g) Determine $\lim_{x\to 0^+} f(x)$ analytically, if it exists.

9. Find the value of a that maximizes the angle θ shown in the figure. What is the approximate measure of this angle?

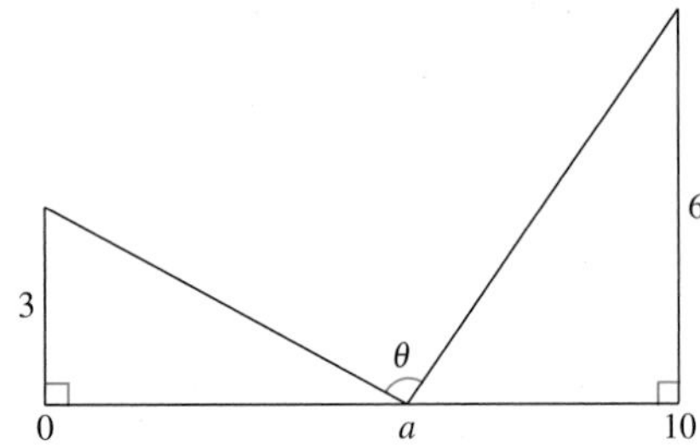

10. Use integration by substitution to find the area under the curve

$$y = \frac{1}{\sin^2 x + 4\cos^2 x}$$

between $x = 0$ and $x = \pi/4$.

11. Recall that the graph of a function $y = f(x)$ is symmetric with respect to the origin if, whenever (x, y) is a point on the graph, $(-x, -y)$ is also a point on the graph. The graph of the function $y = f(x)$ is **symmetric with respect to the point (a, b)** if, whenever $(a - x, b - y)$ is a point on the graph, $(a + x, b + y)$ is also a point on the graph, as shown in the figure.

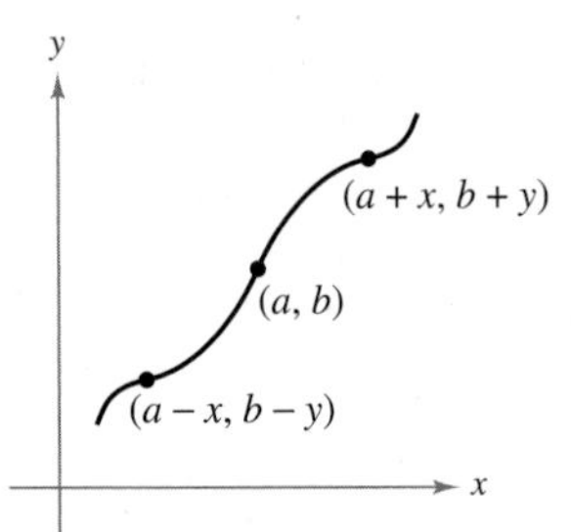

(a) Sketch the graph of $y = \sin x$ on the interval $[0, 2\pi]$. Write a short paragraph explaining how the symmetry of the graph with respect to the point $(0, \pi)$ allows you to conclude that

$$\int_0^{2\pi} \sin x\, dx = 0.$$

(b) Sketch the graph of $y = \sin x + 2$ on the interval $[0, 2\pi]$. Use the symmetry of the graph with respect to the point $(\pi, 2)$ to evaluate the integral

$$\int_0^{2\pi} (\sin x + 2)\, dx.$$

(c) Sketch the graph of $y = \arccos x$ on the interval $[-1, 1]$. Use the symmetry of the graph to evaluate the integral

$$\int_{-1}^{1} \arccos x\, dx.$$

(d) Evaluate the integral $\displaystyle\int_0^{\pi/2} \frac{1}{1 + (\tan x)^{\sqrt{2}}}\, dx.$

12. An object is dropped from a height of 400 feet.

(a) Find the velocity of the object as a function of time (neglect air resistance on the object).

(b) Use the result in part (a) to find the position function.

(c) If the air resistance is proportional to the square of the velocity, then

$$dv/dt = -32 + kv^2$$

where -32 feet per second per second is the acceleration due to gravity and k is a constant. Show that the velocity v as a function of time is

$$v(t) = -\sqrt{32/k}\,\tanh\left(\sqrt{32k}\, t\right)$$

by performing $\int dv/(32 - kv^2) = -\int dt$ and simplifying the result.

(d) Use the result of part (c) to find $\lim_{t\to\infty} v(t)$ and give its interpretation.

(e) Integrate the velocity function in part (c) and find the position s of the object as a function of t. Use a graphing utility to graph the position function when $k = 0.01$ and the position function in part (b) in the same viewing window. Estimate the additional time required for the object to reach ground level when air resistance is not neglected.

(f) Give a written description of what you believe would happen if k were increased. Then test your assertion with a particular value of k.

12 Topics in Analytic Geometry

In this chapter, you will analyze and write equations of conics using their properties. You will also learn how to write and graph parametric equations and polar equations. In addition to the rectangular equations of conics, you will also study polar equations of conics.

In this chapter, you should learn the following.

- How to write equations of parabolas in standard form and solve real-life problems. (12.1)
- How to write equations of ellipses in standard form and solve real-life problems. (12.2)
- How to write equations of hyperbolas in standard form, solve real-life problems, and classify conics from their general equations. (12.3)
- How to sketch curves that are represented by sets of parametric equations. (12.4)
- How to understand the polar coordinate system and rewrite rectangular coordinates and equations in polar form and vice versa. (12.5)
- How to sketch graphs of polar equations and recognize special polar graphs. (12.6)
- How to write and graph equations of conics in polar form. (12.7)

Chuck Savage/Corbis Edge/Corbis

The path of a baseball hit at a particular height at an angle with the horizontal can be modeled using parametric equations. How can a set of parametric equations be used to find the minimum angle at which the ball must leave the bat in order for the hit to be a home run? (See Section 12.4, Exercise 103.)

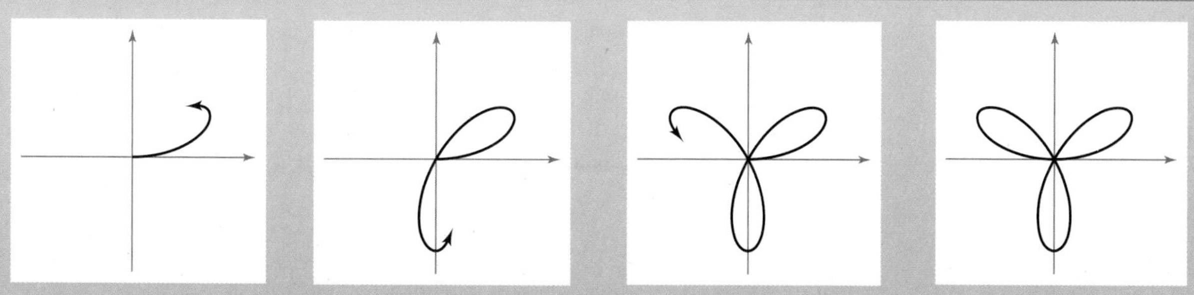

Graphing an equation in the polar coordinate system involves tracing a curve about a fixed point called the pole. One special type of polar graph is called a *rose curve* because each loop on the graph forms a *petal*. You will learn how to sketch polar equations such as rose curves by plotting points and using your knowledge of symmetry, zeros, and maximum values. (See Section 12.6.)

12.1 Introduction to Conics: Parabolas

- **Recognize a conic as the intersection of a plane and a double-napped cone.**
- **Write equations of parabolas in standard form and graph parabolas.**
- **Use the reflective property of parabolas to solve real-life problems.**

Conics

Conic sections were discovered during the classical Greek period, 600 to 300 B.C. The early Greeks were concerned largely with the geometric properties of conics. It was not until the 17th century that the broad applicability of conics became apparent and played a prominent role in the early development of calculus.

A **conic section** (or simply **conic**) is the intersection of a plane and a double-napped cone. Notice in Figure 12.1 that in the formation of the four basic conics, the intersecting plane does not pass through the vertex of the cone. When the plane does pass through the vertex, the resulting figure is a **degenerate conic,** as shown in Figure 12.2.

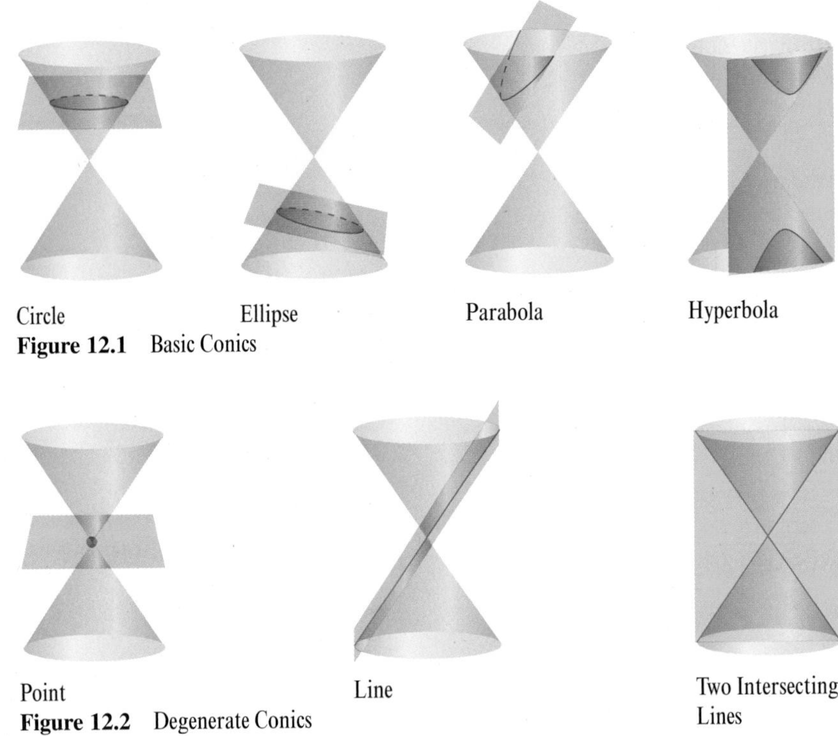

Figure 12.1 Basic Conics

Figure 12.2 Degenerate Conics

There are several ways to approach the study of conics. You could begin by defining conics in terms of the intersections of planes and cones, as the Greeks did, or you could define them algebraically, in terms of the general second-degree equation

$$Ax^2 + Bxy + Cy^2 + Dx + Ey + F = 0. \qquad \text{General second-degree equation}$$

However, you will study a third approach, in which each of the conics is defined as a **locus** (collection) of points satisfying a geometric property. For example, in Section P.4, you learned that a circle is defined as the collection of all points (x, y) that are equidistant from a fixed point (h, k). This leads to the standard form of the equation of a circle

$$(x - h)^2 + (y - k)^2 = r^2. \qquad \text{Equation of circle}$$

Parabolas

The first type of conic is called a **parabola** and is defined below.

DEFINITION OF PARABOLA

A **parabola** is the set of all points (x, y) in a plane that are equidistant from a fixed line (**directrix**) and a fixed point (**focus**) not on the line.

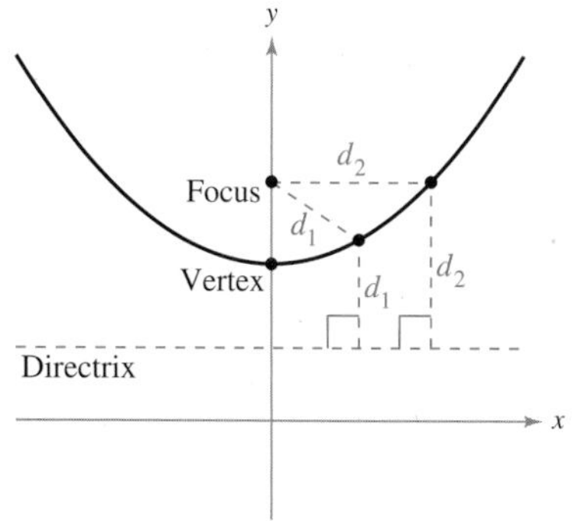

Figure 12.3 Parabola

The midpoint between the focus and the directrix is called the **vertex,** and the line passing through the focus and the vertex is called the **axis** of the parabola. Note in Figure 12.3 that a parabola is symmetric with respect to its axis. Using the definition of a parabola, you can derive the following **standard form** of the equation of a parabola whose directrix is parallel to the x-axis or to the y-axis.

THEOREM 12.1 STANDARD EQUATION OF A PARABOLA

The **standard form of the equation of a parabola** with vertex at (h, k) is as follows.

$$(x - h)^2 = 4p(y - k),\ p \neq 0 \qquad \text{Vertical axis, directrix: } y = k - p$$

$$(y - k)^2 = 4p(x - h),\ p \neq 0 \qquad \text{Horizontal axis, directrix: } x = h - p$$

The focus lies on the axis p units (*directed distance*) from the vertex. If the vertex is at the origin $(0, 0)$, the equation takes one of the following forms.

$$x^2 = 4py \quad \text{Vertical axis} \qquad y^2 = 4px \quad \text{Horizontal axis}$$

See Figure 12.4.

PROOF The case for which the directrix is parallel to the x-axis and the focus lies above the vertex, as shown in Figure 12.4(a), is proven here. If (x, y) is any point on the parabola, then, by definition, it is equidistant from the focus $(h, k + p)$ and the directrix $y = k - p$, and you have

$$\sqrt{(x - h)^2 + [y - (k + p)]^2} = y - (k - p)$$

$$(x - h)^2 + [y - (k + p)]^2 = [y - (k - p)]^2$$

$$(x - h)^2 + y^2 - 2y(k + p) + (k + p)^2 = y^2 - 2y(k - p) + (k - p)^2$$

$$(x - h)^2 - 2py + 2pk = 2py - 2pk$$

$$(x - h)^2 = 4p(y - k).$$

■

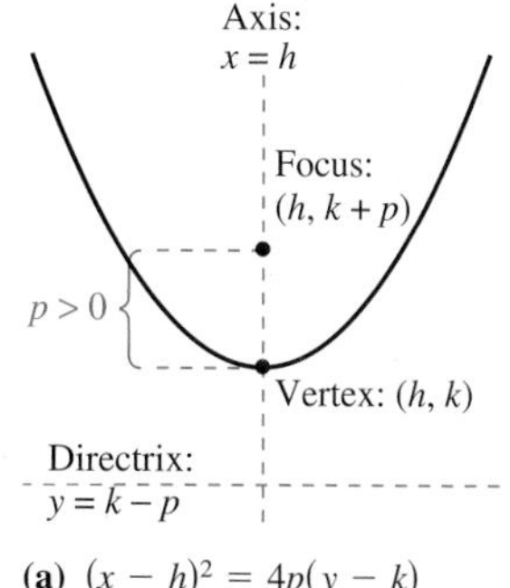

(a) $(x - h)^2 = 4p(y - k)$
Vertical axis: $p > 0$

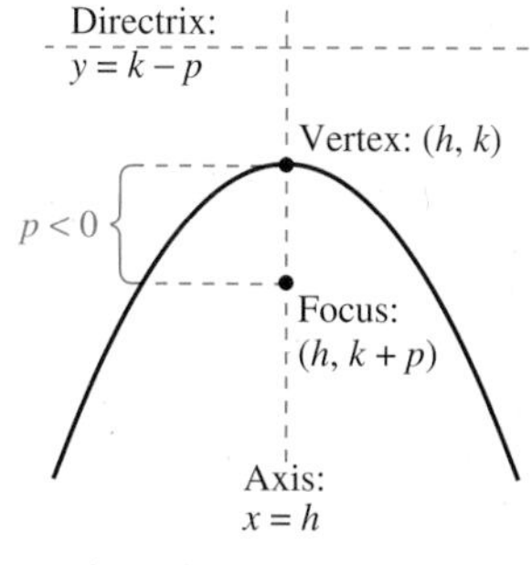

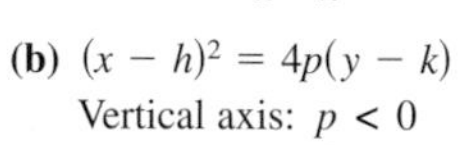

(b) $(x - h)^2 = 4p(y - k)$
Vertical axis: $p < 0$

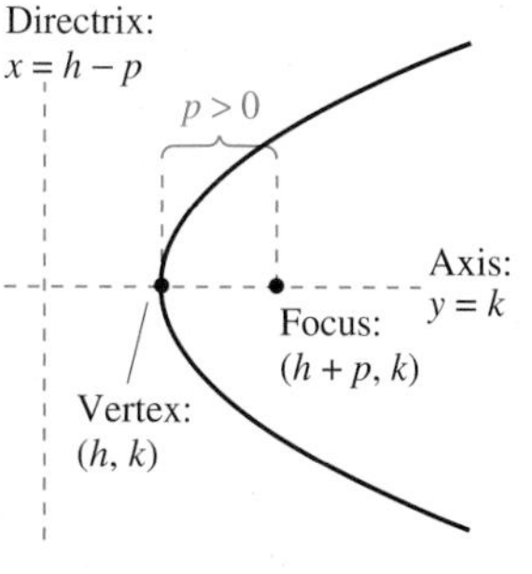

(c) $(y - k)^2 = 4p(x - h)$
Horizontal axis: $p > 0$

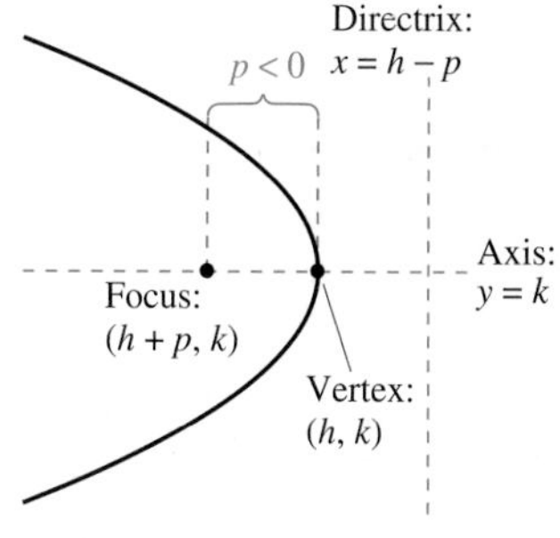

(d) $(y - k)^2 = 4p(x - h)$
Horizontal axis: $p < 0$

Figure 12.4

TECHNOLOGY Use a graphing utility to confirm the equation found in Example 1. In order to graph the equation, you may have to use two separate equations:

$$y_1 = \sqrt{8x} \qquad \text{Upper part}$$

and

$$y_2 = -\sqrt{8x}. \qquad \text{Lower part}$$

EXAMPLE 1 Vertex at the Origin

Find the standard equation of the parabola with vertex at the origin and focus $(2, 0)$.

Solution The axis of the parabola is horizontal, passing through $(0, 0)$ and $(2, 0)$, as shown in Figure 12.5.

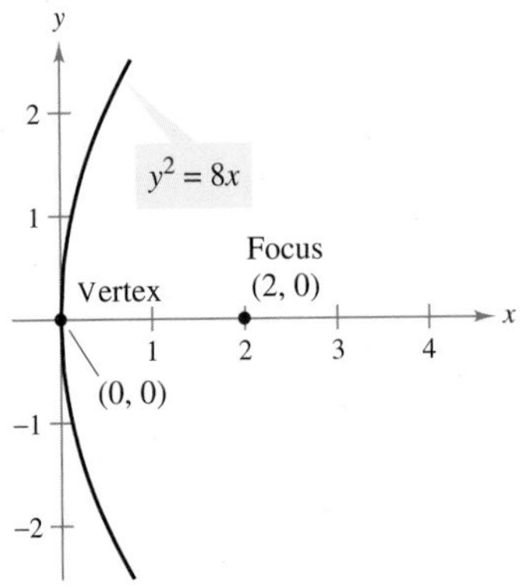

Figure 12.5

The standard form is $y^2 = 4px$, where $h = 0$, $k = 0$, and $p = 2$. So, the equation is

$$y^2 = 8x.$$

EXAMPLE 2 Finding the Focus of a Parabola

Find the focus of the parabola given by

$$y = -\frac{1}{2}x^2 - x + \frac{1}{2}.$$

Solution To find the focus, convert to standard form by completing the square.

$y = -\frac{1}{2}x^2 - x + \frac{1}{2}$	Write original equation.
$-2y = x^2 + 2x - 1$	Multiply each side by −2.
$1 - 2y = x^2 + 2x$	Add 1 to each side.
$1 + 1 - 2y = x^2 + 2x + 1$	Complete the square.
$2 - 2y = x^2 + 2x + 1$	Combine like terms.
$-2(y - 1) = (x + 1)^2$	Standard form

Comparing this equation with

$$(x - h)^2 = 4p(y - k)$$

you can conclude that $h = -1$, $k = 1$, and $p = -\frac{1}{2}$. Because p is negative, the parabola opens downward, as shown in Figure 12.6. So, the focus of the parabola is

$$(h, k + p) = \left(-1, \frac{1}{2}\right). \qquad \text{Focus}$$

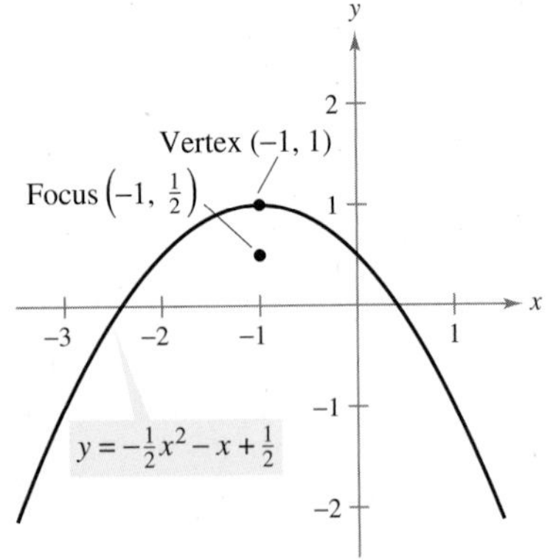

Figure 12.6

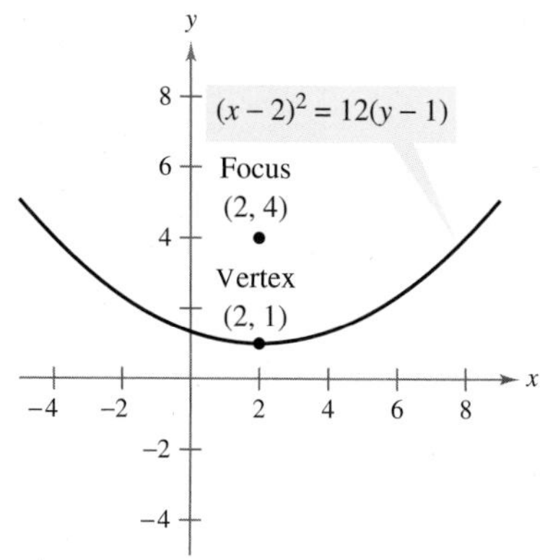

Figure 12.7

EXAMPLE 3 Finding the Standard Equation of a Parabola

Find the standard form of the equation of the parabola with vertex (2, 1) and focus (2, 4). Then write the quadratic form of the equation.

Solution Because the axis of the parabola is vertical, passing through (2, 1) and (2, 4), consider the equation

$$(x - h)^2 = 4p(y - k)$$

where $h = 2$, $k = 1$, and $p = 4 - 1 = 3$. So, the standard form is

$$(x - 2)^2 = 12(y - 1).$$

You can obtain the more common quadratic form as follows.

$$(x - 2)^2 = 12(y - 1) \qquad \text{Write original equation.}$$
$$x^2 - 4x + 4 = 12y - 12 \qquad \text{Multiply.}$$
$$x^2 - 4x + 16 = 12y \qquad \text{Add 12 to each side.}$$
$$\frac{1}{12}(x^2 - 4x + 16) = y \qquad \text{Divide each side by 12.}$$

The graph of this parabola is shown in Figure 12.7. ■

NOTE You may want to review the technique of completing the square found in Section P.1, which will be used to rewrite each of the conics in standard form. ■

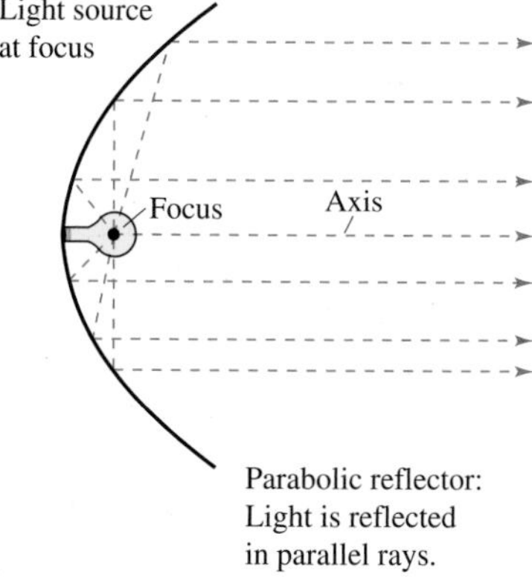

Figure 12.8

Application

A line segment that passes through the focus of a parabola and has endpoints on the parabola is called a **focal chord.** The specific focal chord perpendicular to the axis of the parabola is called the **latus rectum.**

Parabolas occur in a wide variety of applications. For instance, a parabolic reflector can be formed by revolving a parabola around its axis. The resulting surface has the property that all incoming rays parallel to the axis are reflected through the focus of the parabola. This is the principle behind the construction of the parabolic mirrors used in reflecting telescopes. Conversely, the light rays emanating from the focus of a parabolic reflector used in a flashlight are all parallel to one another, as shown in Figure 12.8.

Tangent lines to parabolas have special properties related to the use of parabolas in constructing reflective surfaces.

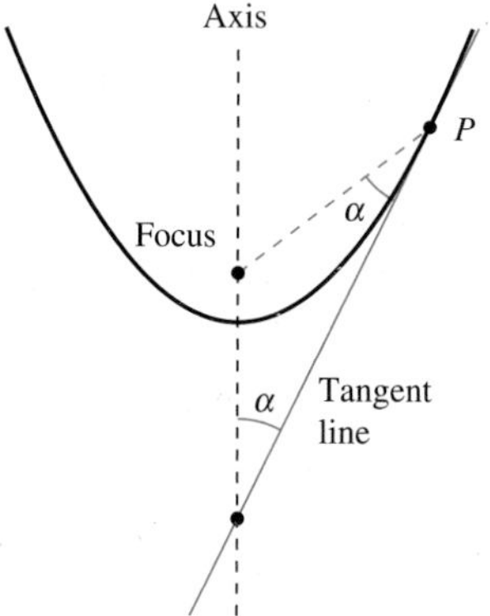

Figure 12.9

THEOREM 12.2 REFLECTIVE PROPERTY OF A PARABOLA

The tangent line to a parabola at a point P makes equal angles with the following two lines (see Figure 12.9).

1. The line passing through P and the focus
2. The axis of the parabola

12.1 Exercises

See www.CalcChat.com for worked-out solutions to odd-numbered exercises.

In Exercises 1–8, fill in the blanks.

1. A ________ is the intersection of a plane and a double-napped cone.
2. When a plane passes through the vertex of a double-napped cone, the intersection is a ________ ________.
3. A collection of points satisfying a geometric property can also be referred to as a ________ of points.
4. A ________ is defined as the set of all points (x, y) in a plane that are equidistant from a fixed line, called the ________, and a fixed point, called the ________, not on the line.
5. The line that passes through the focus and the vertex of a parabola is called the ________ of the parabola.
6. The ________ of a parabola is the midpoint between the focus and the directrix.
7. A line segment that passes through the focus of a parabola and has endpoints on the parabola is called a ________ ________.
8. A line is ________ to a parabola at a point on the parabola if the line intersects, but does not cross, the parabola at the point.

In Exercises 9–14, match the equation with its graph. [The graphs are labeled (a), (b), (c), (d), (e), and (f).]

(a)

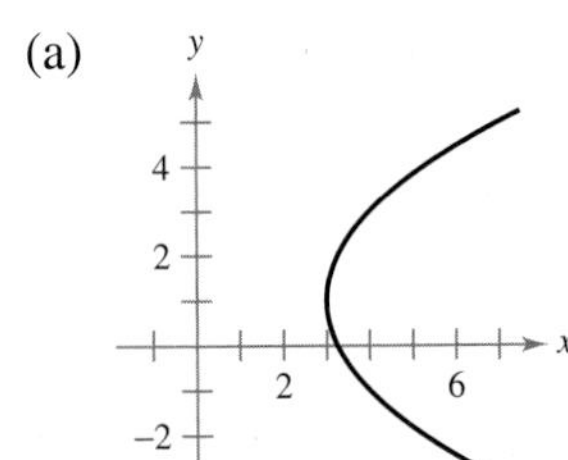

(b)

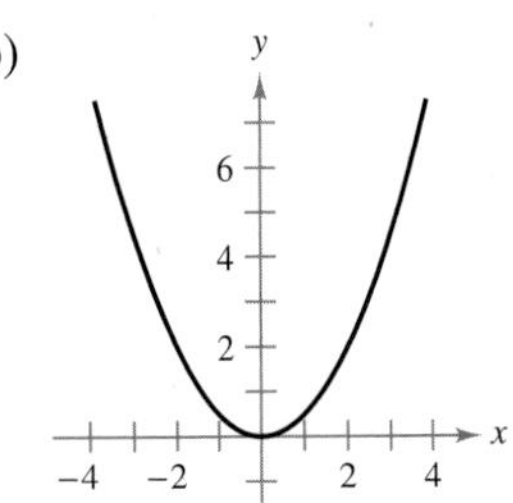

(c)

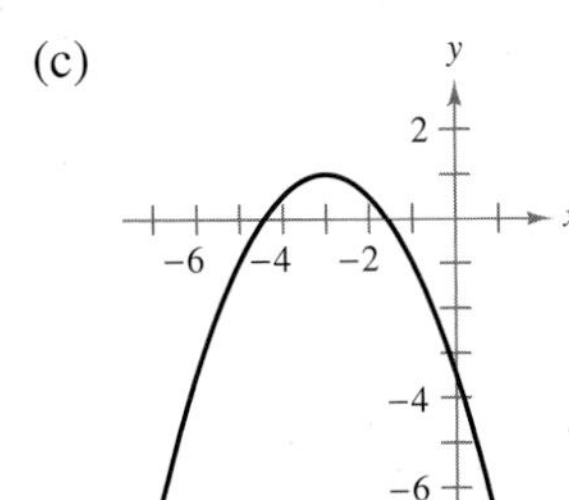

(d)

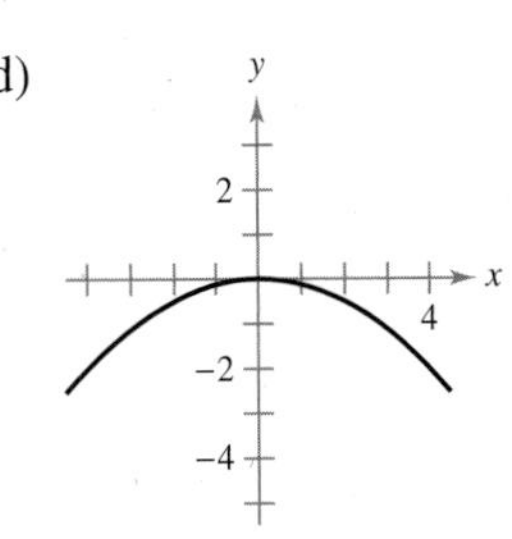

(e)

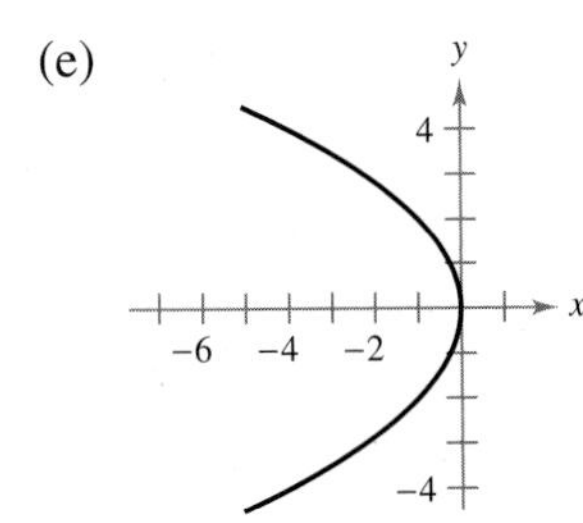

(f)

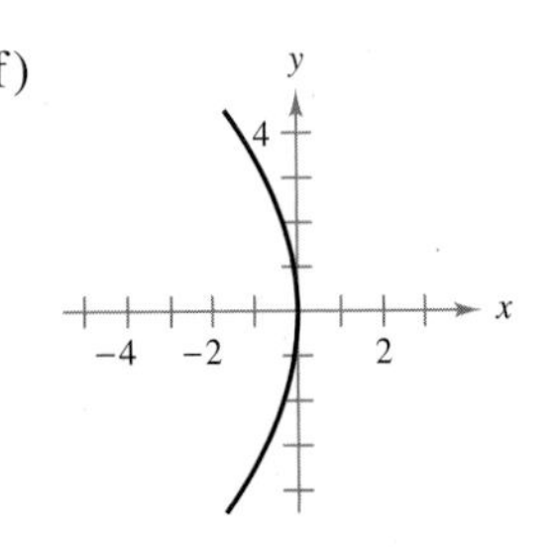

9. $y^2 = -4x$
10. $x^2 = 2y$
11. $x^2 = -8y$
12. $y^2 = -12x$
13. $(y - 1)^2 = 4(x - 3)$
14. $(x + 3)^2 = -2(y - 1)$

In Exercises 15–28, find the standard form of the equation of the parabola with the given characteristic(s) and vertex at the origin.

15.

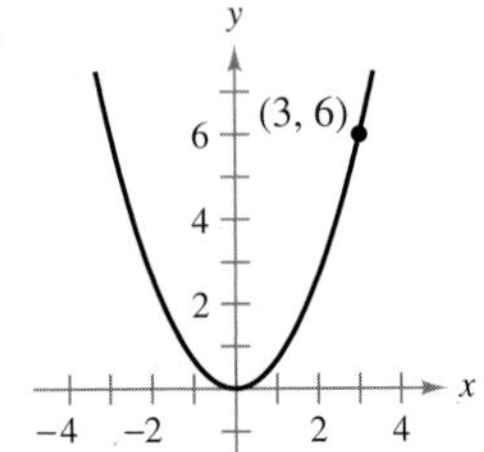

16.

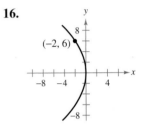

17. Focus: $\left(0, \frac{1}{2}\right)$
18. Focus: $\left(-\frac{3}{2}, 0\right)$
19. Focus: $(-2, 0)$
20. Focus: $(0, -2)$
21. Directrix: $y = 1$
22. Directrix: $y = -2$
23. Directrix: $x = -1$
24. Directrix: $x = 3$
25. Vertical axis and passes through the point $(4, 6)$
26. Vertical axis and passes through the point $(-3, -3)$
27. Horizontal axis and passes through the point $(-2, 5)$
28. Horizontal axis and passes through the point $(3, -2)$

In Exercises 29–42, find the vertex, focus, and directrix of the parabola, and sketch its graph.

29. $y = \frac{1}{2}x^2$
30. $y = -2x^2$
31. $y^2 = -6x$
32. $y^2 = 3x$
33. $x^2 + 6y = 0$
34. $x + y^2 = 0$
35. $(x - 1)^2 + 8(y + 2) = 0$
36. $(x + 5) + (y - 1)^2 = 0$
37. $(x + 3)^2 = 4\left(y - \frac{3}{2}\right)$
38. $\left(x + \frac{1}{2}\right)^2 = 4(y - 1)$
39. $y = \frac{1}{4}(x^2 - 2x + 5)$
40. $x = \frac{1}{4}(y^2 + 2y + 33)$
41. $y^2 + 6y + 8x + 25 = 0$
42. $y^2 - 4y - 4x = 0$

In Exercises 43–46, find the vertex, focus, and directrix of the parabola. Use a graphing utility to graph the parabola.

43. $x^2 + 4x + 6y - 2 = 0$
44. $x^2 - 2x + 8y + 9 = 0$
45. $y^2 + x + y = 0$
46. $y^2 - 4x - 4 = 0$

In Exercises 47–56, find the standard form of the equation of the parabola with the given characteristics.

47.

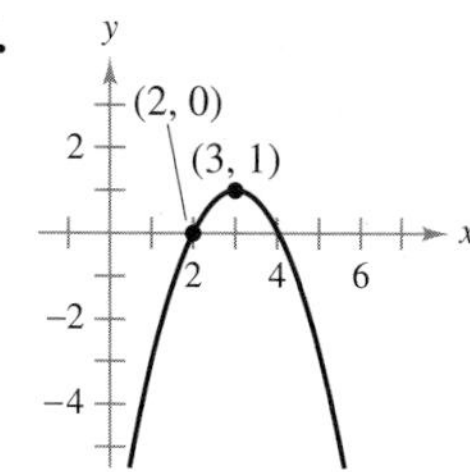

48.

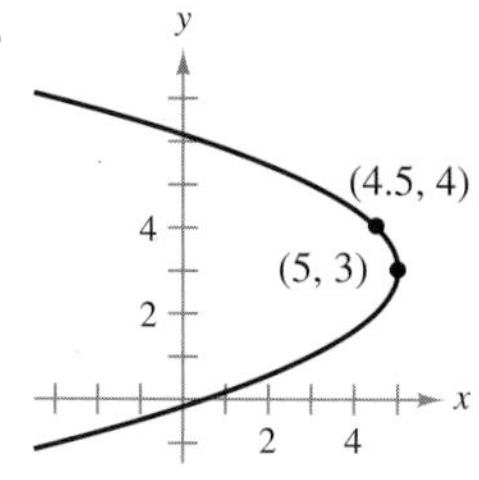

49.

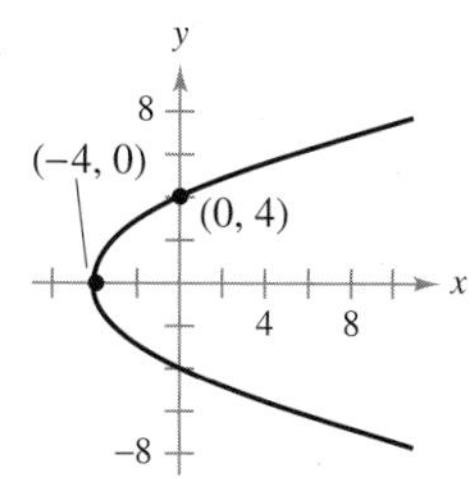

50.

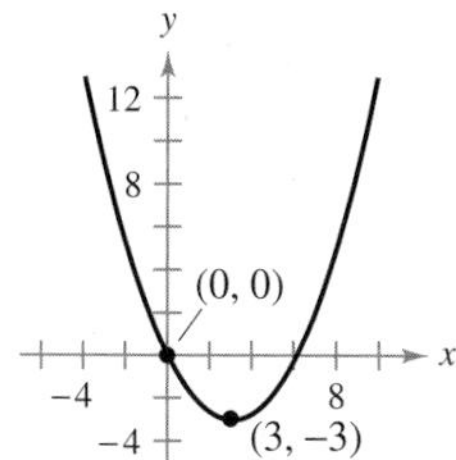

51. Vertex: $(4, 3)$; focus: $(6, 3)$

52. Vertex: $(-1, 2)$; focus: $(-1, 0)$

53. Vertex: $(0, 2)$; directrix: $y = 4$

54. Vertex: $(1, 2)$; directrix: $y = -1$

55. Focus: $(2, 2)$; directrix: $x = -2$

56. Focus: $(0, 0)$; directrix: $y = 8$

In Exercises 57 and 58, change the equation of the parabola so that its graph matches the description.

57. $(y - 3)^2 = 6(x + 1)$; upper half of parabola

58. $(y + 1)^2 = 2(x - 4)$; lower half of parabola

In Exercises 59 and 60, the equations of a parabola and a tangent line to the parabola are given. Use a graphing utility to graph both equations in the same viewing window. Determine the coordinates of the point of tangency.

	Parabola	*Tangent Line*
59.	$y^2 - 8x = 0$	$x - y + 2 = 0$
60.	$x^2 + 12y = 0$	$x + y - 3 = 0$

In Exercises 61–68, find dy/dx.

61. $x^2 = 4y$

62. $x^2 = \frac{1}{4}y$

63. $y^2 = 6x$

64. $y^2 = -8x$

65. $(x - 2)^2 = 6(y + 3)$

66. $(x + 4)^2 = -3(y - 1)$

67. $(y + 3)^2 = -8(x - 2)$

68. $\left(y - \frac{3}{2}\right)^2 = 4(x + 4)$

In Exercises 69–76, find an equation of the tangent line to the parabola at the given point.

69. $x^2 = 2y,\ (4, 8)$

70. $x^2 = 2y,\ \left(-3, \frac{9}{2}\right)$

71. $y = -2x^2,\ (-1, -2)$

72. $y = -2x^2,\ (2, -8)$

73. $y^2 = 2(x - 3),\ (5, 2)$

74. $y^2 = 2(x - 3),\ (11, 4)$

75. $(x - 1)^2 = 6(y + 2),\ (-5, 4)$

76. $(x - 1)^2 = 6(y + 2),\ (10, 11.5)$

77. ***Revenue*** The revenue R (in dollars) generated by the sale of x units of a patio furniture set is given by

$$(x - 106)^2 = -\frac{4}{5}(R - 14{,}045).$$

Use a graphing utility to graph the function and approximate the number of sales that will maximize revenue.

78. ***Revenue*** The revenue R (in dollars) generated by the sale of x units of a digital camera is given by

$$(x - 135)^2 = -\frac{5}{7}(R - 25{,}515).$$

Use a graphing utility to graph the function and approximate the number of sales that will maximize revenue.

WRITING ABOUT CONCEPTS

In Exercises 79–82, describe in words how a plane could intersect with the double-napped cone shown to form the conic section.

79. Circle

80. Ellipse

81. Parabola

82. Hyperbola

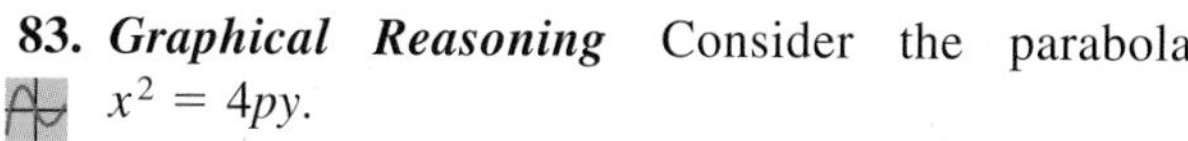

83. ***Graphical Reasoning*** Consider the parabola $x^2 = 4py$.

(a) Use a graphing utility to graph the parabola for $p = 1$, $p = 2$, $p = 3$, and $p = 4$. Describe the effect on the graph when p increases.

(b) Locate the focus for each parabola in part (a).

(c) For each parabola in part (a), find the length of the latus rectum (see figure). How can the length of the latus rectum be determined directly from the standard form of the equation of the parabola?

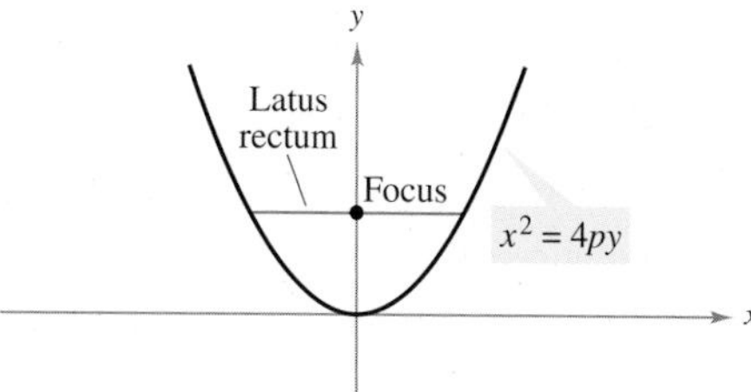

(d) Explain how the result of part (c) can be used as a sketching aid when graphing parabolas.

WRITING ABOUT CONCEPTS (continued)

84. Let (x_1, y_1) be the coordinates of a point on the parabola $x^2 = 4py$. The equation of the line tangent to the parabola at the point is $y - y_1 = \frac{x_1}{2p}(x - x_1)$. What is the slope of the tangent line?

85. ***Suspension Bridge*** Each cable of the Golden Gate Bridge is suspended (in the shape of a parabola) between two towers that are 1280 meters apart. The top of each tower is 152 meters above the roadway. The cables touch the roadway midway between the towers.

(a) Draw a sketch of the bridge. Locate the origin of a rectangular coordinate system at the center of the roadway. Label the coordinates of the known points.

(b) Write an equation that models the cables.

(c) Complete the table by finding the height y of the suspension cables over the roadway at a distance of x meters from the center of the bridge.

x	0	100	250	400	500
y					

86. ***Satellite Dish*** The receiver in a parabolic satellite dish is 4.5 feet from the vertex and is located at the focus (see figure). Write an equation for a cross section of the reflector. (Assume that the dish is directed upward and the vertex is at the origin.)

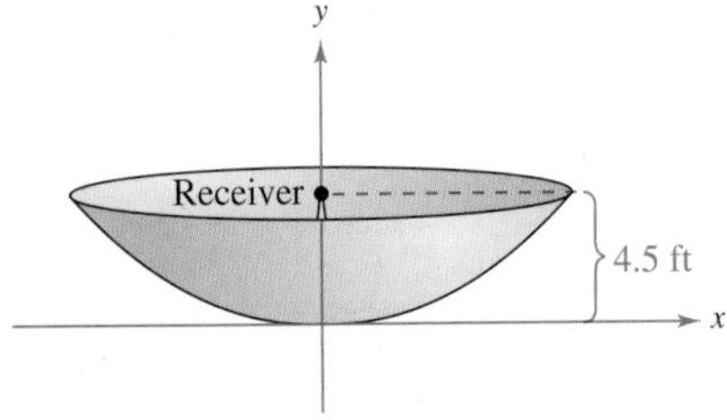

87. ***Road Design*** Roads are often designed with parabolic surfaces to allow rain to drain off. A particular road that is 32 feet wide is 0.4 foot higher in the center than it is on the sides (see figure).

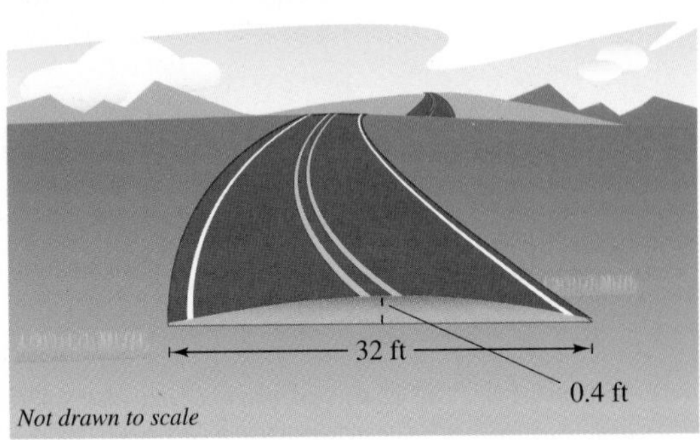

Cross section of road surface

(a) Find an equation of the parabola that models the road surface. (Assume that the origin is at the center of the road.)

(b) How far from the center of the road is the road surface 0.1 foot lower than in the middle?

88. ***Highway Design*** Highway engineers design a parabolic curve for an entrance ramp from a straight street to an interstate highway (see figure). Find an equation of the parabola.

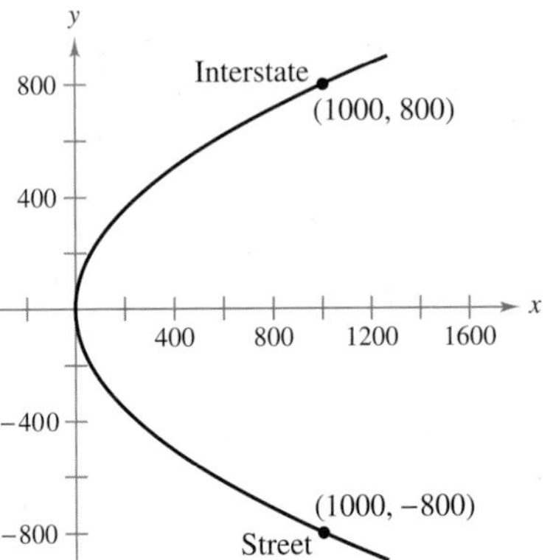

89. ***Beam Deflection*** A simply supported beam is 12 meters long and has a load at the center (see figure). The deflection of the beam at its center is 2 centimeters. Assume that the shape of the deflected beam is parabolic.

(a) Write an equation of the parabola. (Assume that the origin is at the center of the deflected beam.)

(b) How far from the center of the beam is the deflection equal to 1 centimeter?

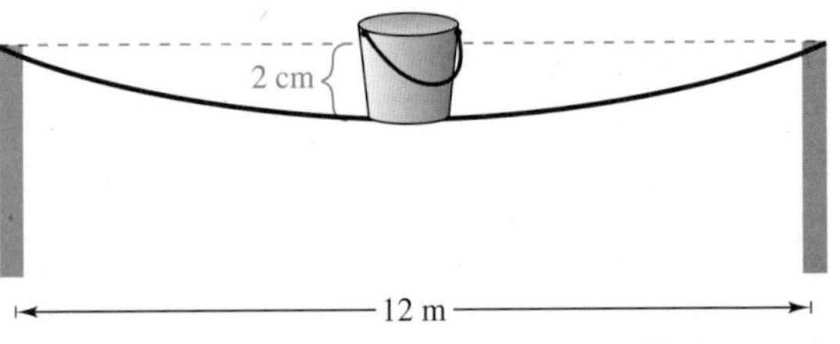

90. ***Beam Deflection*** Repeat Exercise 89 if the length of the beam is 16 meters and the deflection of the beam at the center is 3 centimeters.

91. ***Fluid Flow*** Water is flowing from a horizontal pipe 48 feet above the ground. The falling stream of water has the shape of a parabola whose vertex $(0, 48)$ is at the end of the pipe (see figure on the next page). The stream of water strikes the ground at the point $(10\sqrt{3}, 0)$. Find the equation of the path taken by the water.

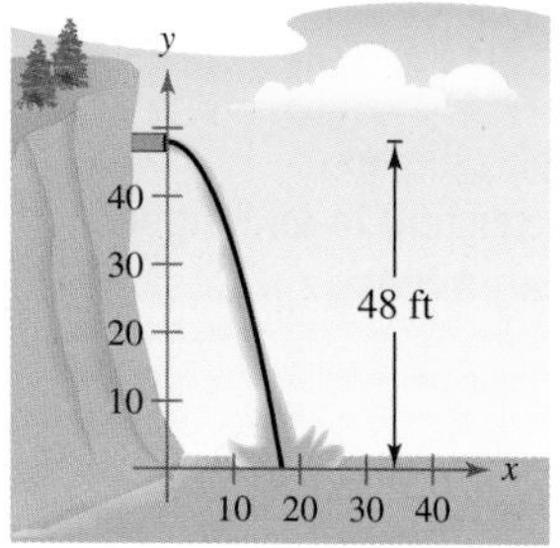

Figure for 91

92. ***Lattice Arch*** A parabolic lattice arch is 16 feet high at the vertex. At a height of 6 feet, the width of the lattice arch is 4 feet (see figure). How wide is the lattice arch at ground level?

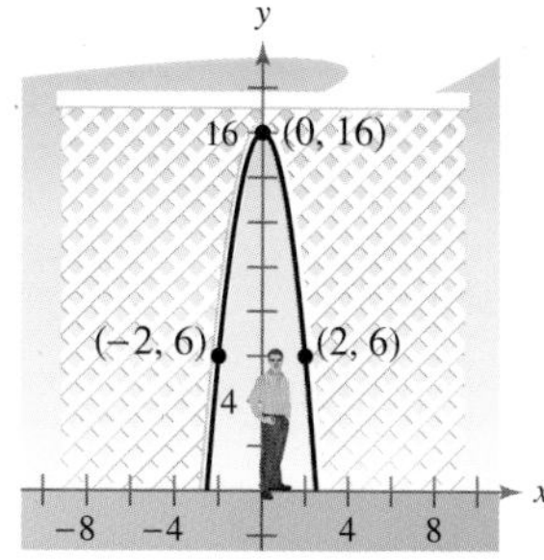

93. ***Satellite Orbit*** A satellite in a 100-mile-high circular orbit around Earth has a velocity of approximately 17,500 miles per hour. If this velocity is multiplied by $\sqrt{2}$, the satellite will have the minimum velocity necessary to escape Earth's gravity and it will follow a parabolic path with the center of Earth as the focus.

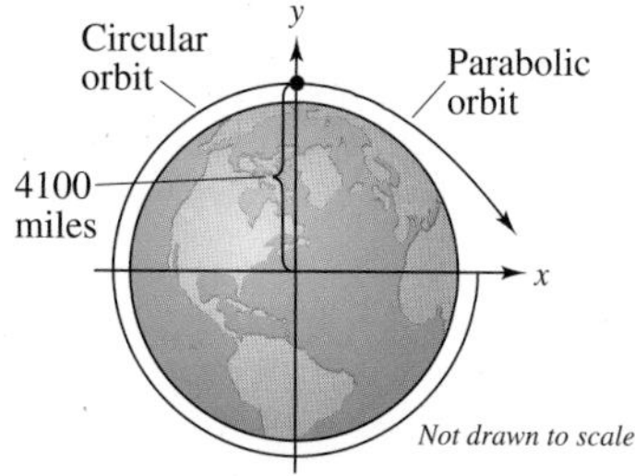

(a) Find the escape velocity of the satellite.

(b) Find an equation of the parabolic path of the satellite (assume that the radius of Earth is 4000 miles).

94. ***Path of a Projectile*** The path of a softball is given by the equation

$$y = -0.08x^2 + x + 4.$$

The coordinates x and y are measured in feet, with $x = 0$ corresponding to the position from which the ball was thrown.

(a) Use a graphing utility to graph the trajectory of the softball.

(b) Move the cursor along the path to approximate the highest point. Approximate the range of the trajectory.

(c) Analytically find the maximum height of the softball.

95. ***Projectile Motion*** A bomber is flying at an altitude of 30,000 feet and a speed of 540 miles per hour. When should a bomb be dropped so that it will hit the target if the path of the bomb is modeled by

$$y = 30{,}000 - \frac{x^2}{39{,}204}$$

where x is measured in feet?

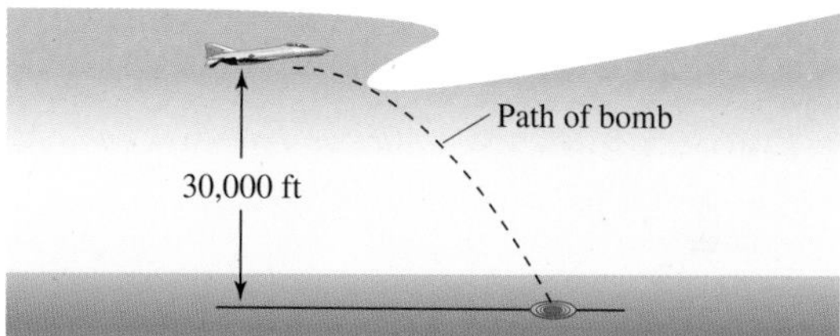

CAPSTONE

96. Explain what each of the following equations represents, and how equations (a) and (b) are equivalent.

(a) $y = a(x - h)^2 + k, \quad a \neq 0$

(b) $(x - h)^2 = 4p(y - k), \quad p \neq 0$

(c) $(y - k)^2 = 4p(x - h), \quad p \neq 0$

True or False? **In Exercises 97 and 98, determine whether the statement is true or false. Justify your answer.**

97. It is possible for a parabola to intersect its directrix.

98. If the vertex and focus of a parabola are on a horizontal line, then the directrix of the parabola is vertical.

99. ***Distance*** Find the point on the graph of $y^2 = 6x$ that is closest to the focus of the parabola.

Area **In Exercises 100–105, find the area of the region bounded by the graphs of the given equations.**

100. $x^2 = 2y, \quad y = 3$

101. $x^2 = 4(y - 1), \quad y = 10$

102. $y^2 = 4x, \quad x = 5$

103. $y^2 = -4(x + 1), \quad x = -5$

104. $(x - 2)^2 = 4y, \quad x = 0, \ x = 4, \ y = 0$

105. $(x + 1)^2 = -8(y - 2), \quad y = 0$

12.2 Ellipses and Implicit Differentiation

- **Write equations of ellipses in standard form and graph ellipses.**
- **Use implicit differentiation to find the slope of a line tangent to an ellipse.**
- **Use properties of ellipses to model and solve real-life problems.**
- **Find eccentricities of ellipses.**

Introduction

The second type of conic is called an **ellipse,** and is defined as follows.

DEFINITION OF ELLIPSE
An **ellipse** is the set of all points (x, y) in a plane, the sum of whose distances from two distinct fixed points (**foci**) is constant. See Figure 12.10.

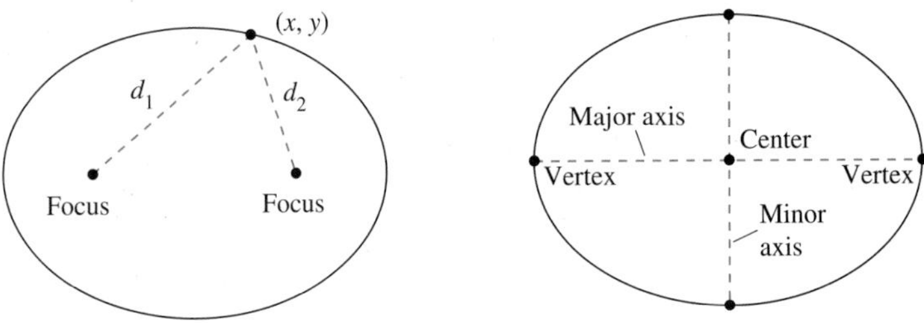

$d_1 + d_2$ is constant.
Figure 12.10

Figure 12.11

The line through the foci intersects the ellipse at two points called **vertices.** The chord joining the vertices is the **major axis,** and its midpoint is the **center** of the ellipse. The chord perpendicular to the major axis at the center is the **minor axis** of the ellipse. See Figure 12.11.

You can visualize the definition of an ellipse by imagining two thumbtacks placed at the foci, as shown in Figure 12.12. If the ends of a fixed length of string are fastened to the thumbtacks and the string is *drawn taut* with a pencil, the path traced by the pencil will be an ellipse.

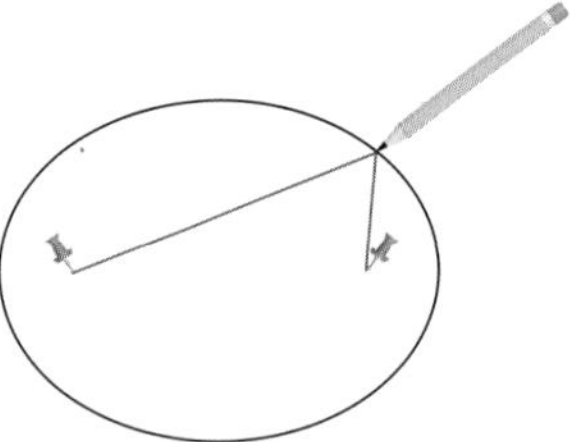

Figure 12.12

To derive the standard form of the equation of an ellipse, consider the ellipse in Figure 12.13 with the following points: center, (h, k); vertices, $(h \pm a, k)$; foci, $(h \pm c, k)$. Note that the center is the midpoint of the segment joining the foci. The sum of the distances from any point on the ellipse to the two foci is constant. Using a vertex point, this constant sum is $(a + c) + (a - c) = 2a$ or simply the length of the major axis. Now, if you let (x, y) be any point on the ellipse, the sum of the distances between (x, y) and the two foci must also be $2a$.

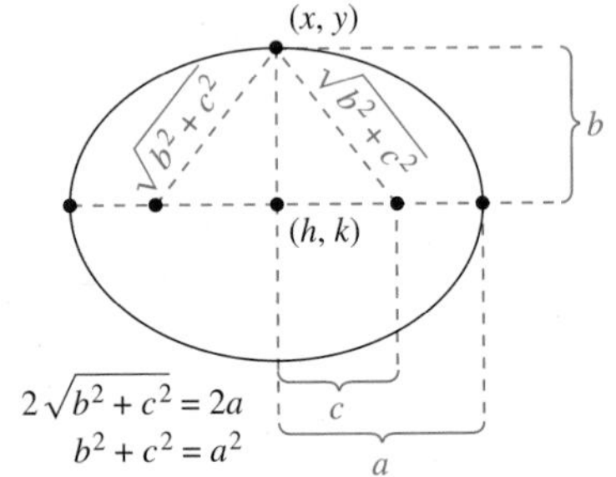

Figure 12.13

That is,

$$\sqrt{[x-(h-c)]^2+(y-k)^2}+\sqrt{[x-(h+c)]^2+(y-k)^2}=2a$$

which, after expanding and regrouping, reduces to

$$(a^2-c^2)(x-h)^2+a^2(y-k)^2=a^2(a^2-c^2).$$

Finally, in Figure 12.13, you can see that

$$b^2=a^2-c^2$$

which implies that the equation of the ellipse is

$$b^2(x-h)^2+a^2(y-k)^2=a^2b^2$$
$$\frac{(x-h)^2}{a^2}+\frac{(y-k)^2}{b^2}=1.$$

FOR FURTHER INFORMATION To learn about how an ellipse may be "exploded" into a parabola, see the article "Exploding the Ellipse" by Arnold Good in *The Mathematics Teacher.* To view this article, go to the website *www.matharticles.com.*

You would obtain a similar equation in the derivation by starting with a vertical major axis. Both results are summarized as follows.

THEOREM 12.3 STANDARD EQUATION OF AN ELLIPSE

The **standard form of the equation of an ellipse,** with center (h, k) and major and minor axes of lengths $2a$ and $2b$, respectively, where $0 < b < a$, is

$$\frac{(x-h)^2}{a^2}+\frac{(y-k)^2}{b^2}=1 \qquad \text{Major axis is horizontal.}$$

$$\frac{(x-h)^2}{b^2}+\frac{(y-k)^2}{a^2}=1. \qquad \text{Major axis is vertical.}$$

The foci lie on the major axis, c units from the center, with $c^2=a^2-b^2$. If the center is at the origin $(0, 0)$, the equation takes one of the following forms.

$$\frac{x^2}{a^2}+\frac{y^2}{b^2}=1 \qquad \text{Major axis is horizontal.}$$

$$\frac{x^2}{b^2}+\frac{y^2}{a^2}=1 \qquad \text{Major axis is vertical.}$$

Figure 12.14 shows both the horizontal and vertical orientations for an ellipse.

STUDY TIP Consider the equation of the ellipse

$$\frac{(x-h)^2}{a^2}+\frac{(y-k)^2}{b^2}=1.$$

If you let $a=b$, then the equation can be rewritten as

$$(x-h)^2+(y-k)^2=a^2$$

which is the standard form of the equation of a circle with radius $r=a$ (see Section P.4). Geometrically, when $a=b$ for an ellipse, the major and minor axes are of equal length, and so the graph is a circle.

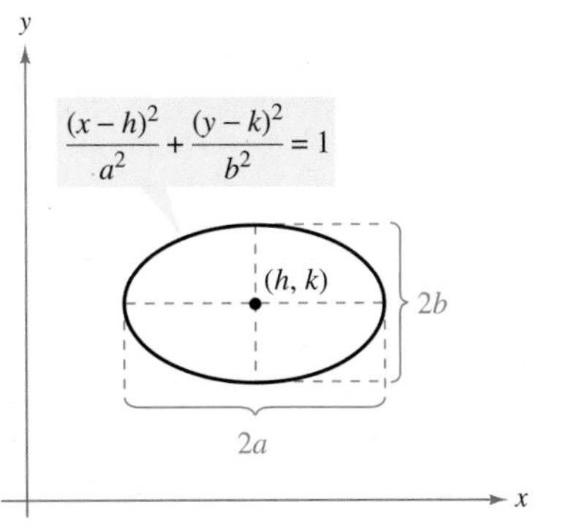

Major axis is horizontal.

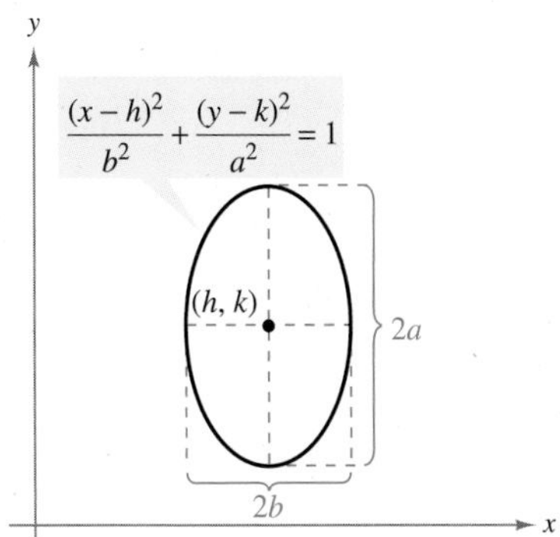

Major axis is vertical.

Figure 12.14

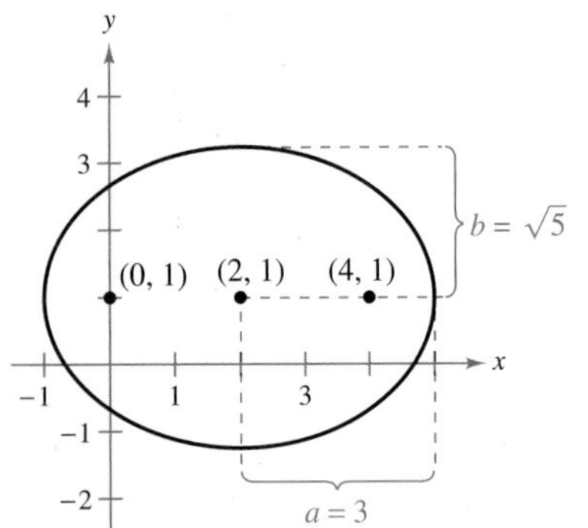

Figure 12.15

NOTE In Example 1, note the use of the equation $c^2 = a^2 - b^2$. Don't confuse this equation with the Pythagorean Theorem—there is a difference in sign.

EXAMPLE 1 Finding the Standard Equation of an Ellipse

Find the standard form of the equation of the ellipse having foci at (0, 1) and (4, 1) and a major axis of length 6, as shown in Figure 12.15.

Solution Because the foci occur at (0, 1) and (4, 1), the center of the ellipse is (2, 1) and the distance from the center to one of the foci is $c = 2$. Because $2a = 6$, you know that $a = 3$. Now, from $c^2 = a^2 - b^2$, you have

$$b = \sqrt{a^2 - c^2} = \sqrt{3^2 - 2^2} = \sqrt{5}.$$

Because the major axis is horizontal, the standard equation is

$$\frac{(x-2)^2}{3^2} + \frac{(y-1)^2}{(\sqrt{5})^2} = 1.$$

This equation simplifies to

$$\frac{(x-2)^2}{9} + \frac{(y-1)^2}{5} = 1.$$

EXAMPLE 2 Sketching an Ellipse

Sketch the ellipse given by

$$x^2 + 4y^2 + 6x - 8y + 9 = 0.$$

Solution Begin by writing the original equation in standard form. In the fourth step, note that 9 and 4 are added to *both* sides of the equation when completing the squares.

$$x^2 + 4y^2 + 6x - 8y + 9 = 0 \quad \text{Write original equation.}$$

$$(x^2 + 6x + \quad) + (4y^2 - 8y + \quad) = -9 \quad \text{Group terms.}$$

$$(x^2 + 6x + \quad) + 4(y^2 - 2y + \quad) = -9 \quad \text{Factor 4 out of } y\text{-terms.}$$

$$(x^2 + 6x + 9) + 4(y^2 - 2y + 1) = -9 + 9 + 4(1)$$

$$(x+3)^2 + 4(y-1)^2 = 4 \quad \text{Write in completed square form.}$$

$$\frac{(x+3)^2}{4} + \frac{(y-1)^2}{1} = 1 \quad \text{Divide each side by 4.}$$

$$\frac{(x+3)^2}{2^2} + \frac{(y-1)^2}{1^2} = 1 \quad \text{Write in standard form.}$$

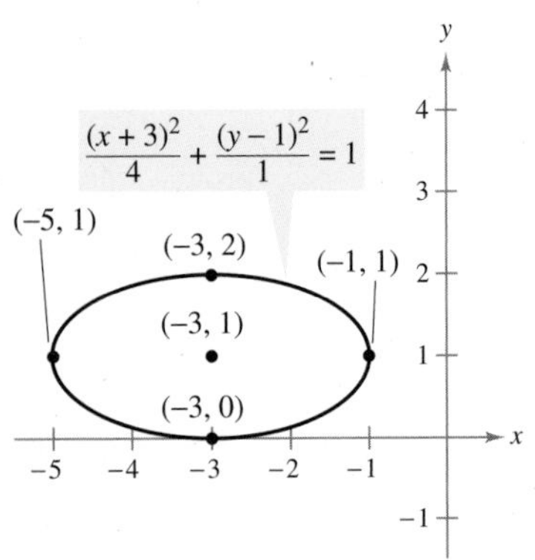

Figure 12.16

From this standard form, it follows that the center is $(h, k) = (-3, 1)$. Because the denominator of the x-term is $a^2 = 2^2$, the endpoints of the major axis lie two units to the right and left of the center. Similarly, because the denominator of the y-term is $b^2 = 1^2$, the endpoints of the minor axis lie one unit up and down from the center. Now, from $c^2 = a^2 - b^2$, you have $c = \sqrt{2^2 - 1^2} = \sqrt{3}$. So, the foci of the ellipse are $(-3 - \sqrt{3}, 1)$ and $(-3 + \sqrt{3}, 1)$. The ellipse is shown in Figure 12.16. ■

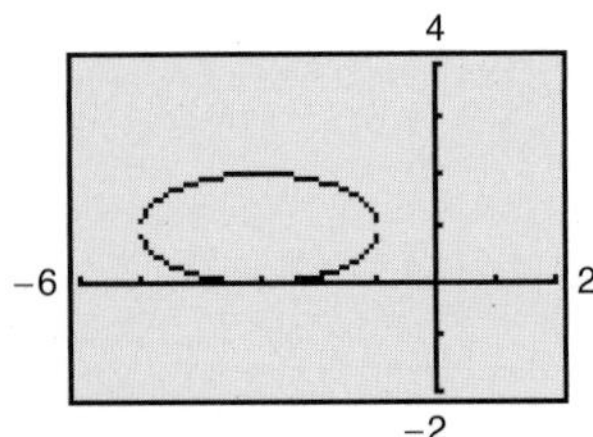

Figure 12.17

TECHNOLOGY You can use a graphing utility to graph an ellipse by graphing the upper and lower portions in the same viewing window. For instance, to graph the ellipse in Example 2, first solve for y to get

$$y_1 = 1 + \sqrt{1 - \frac{1}{4}(x+3)^2} \quad \text{and} \quad y_2 = 1 - \sqrt{1 - \frac{1}{4}(x+3)^2}.$$

The graph is shown in Figure 12.17.

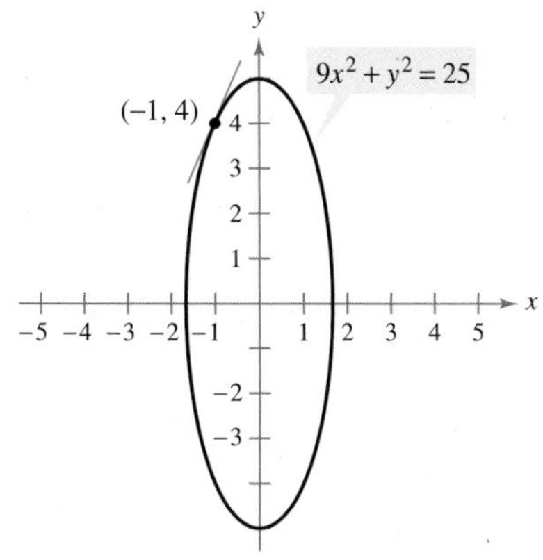

Figure 12.18

EXAMPLE 3 Finding the Slope of a Graph Implicitly

Determine the slope of the tangent line to the graph of $9x^2 + y^2 = 25$ at the point $(-1, 4)$. See Figure 12.18.

Solution Implicit differentiation of the equation $9x^2 + y^2 = 25$ with respect to x yields

$$18x + 2y\frac{dy}{dx} = 0$$

$$\frac{dy}{dx} = \frac{-18x}{2y} = \frac{-9x}{y}.$$

So, at $(-1, 4)$, the slope is

$$\frac{dy}{dx} = \frac{-9(-1)}{4} = \frac{9}{4}.$$

■

NOTE To see the benefit of implicit differentiation, try doing Example 3 using the explicit function $y = \sqrt{25 - 9x^2}$. ■

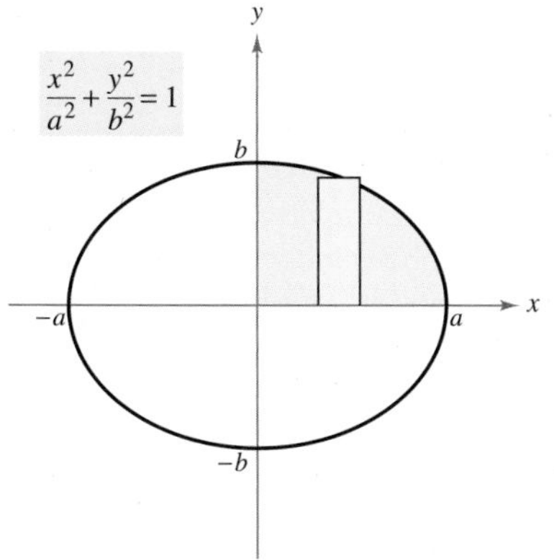

Figure 12.19

EXAMPLE 4 Finding the Area of an Ellipse

Find the area of an ellipse whose major and minor axes have lengths of $2a$ and $2b$, respectively.

Solution For simplicity, choose an ellipse centered at the origin

$$\frac{x^2}{a^2} + \frac{y^2}{b^2} = 1.$$

Then, using symmetry, you can find the area of the entire region lying within the ellipse by finding the area of the region in the first quadrant and multiplying by 4, as indicated in Figure 12.19. In the first quadrant, you have

$$y = \frac{b}{a}\sqrt{a^2 - x^2}$$

which implies that the entire area is

$$A = 4\int_0^a \frac{b}{a}\sqrt{a^2 - x^2}\,dx.$$

NOTE For a review of trigonometric integration, see Section 11.3.

Using the trigonometric substitution $x = a\sin\theta$ where $dx = a\cos\theta\,d\theta$, you have

$$A = \frac{4b}{a}\int_0^{\pi/2} a^2\cos^2\theta\,d\theta$$

$$= 4ab\int_0^{\pi/2} \frac{1 + \cos 2\theta}{2}\,d\theta$$

$$= 2ab\int_0^{\pi/2} 1 + \cos 2\theta\,d\theta$$

$$= 2ab\left[\theta + \frac{\sin 2\theta}{2}\right]_0^{\pi/2} = 2ab\left(\frac{\pi}{2}\right) = \pi ab.$$

■

NOTE Observe that if $a = b$, then the formula for the area of an ellipse reduces to the formula for the area of a circle. ■

Application

Ellipses have many practical and aesthetic uses. For instance, machine gears, supporting arches, and acoustic designs often involve elliptical shapes. The orbits of satellites and planets are also ellipses. Example 5 investigates the elliptical orbit of the moon about Earth.

EXAMPLE 5 An Application Involving an Elliptical Orbit

The moon travels about Earth in an elliptical orbit with Earth at one focus, as shown in Figure 12.20. The major and minor axes of the orbit have lengths of 768,800 kilometers and 767,640 kilometers, respectively. Find the greatest and smallest distances (the *apogee* and *perigee,* respectively) from Earth's center to the moon's center.

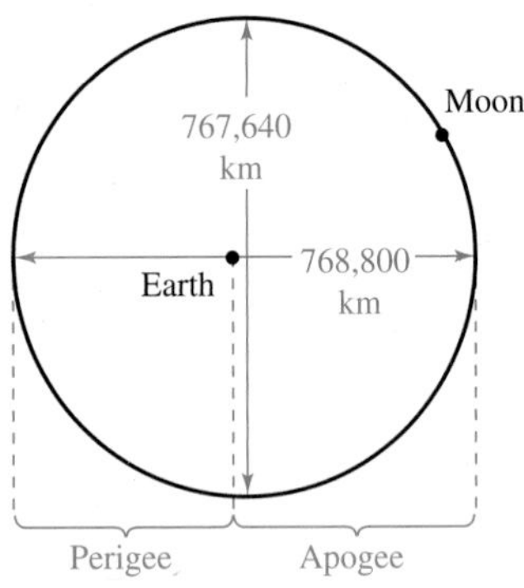

Figure 12.20

Solution Because $2a = 768{,}800$ and $2b = 767{,}640$, you have $a = 384{,}400$ and $b = 383{,}820$ which implies that $c = \sqrt{a^2 - b^2} = \sqrt{384{,}400^2 - 383{,}820^2} \approx 21{,}108$. So, the greatest distance between the center of Earth and the center of the moon is $a + c \approx 405{,}508$ kilometers and the smallest distance is $a - c \approx 363{,}292$ kilometers. ■

Eccentricity

One of the reasons it was difficult for early astronomers to detect that the orbits of the planets are ellipses is that the foci of the planetary orbits are relatively close to their centers, and so the orbits are nearly circular. To measure the ovalness of an ellipse, you can use the concept of **eccentricity.**

NOTE $0 < e < 1$ for *every* ellipse.

DEFINITION OF ECCENTRICITY OF AN ELLIPSE

The **eccentricity** e of an ellipse is given by the ratio $e = \dfrac{c}{a}$.

To see how this ratio is used to describe the shape of an ellipse, note that because the foci of an ellipse are located along the major axis between the vertices and the center, it follows that $0 < c < a$. For an ellipse that is nearly circular, the foci are close to the center and the ratio c/a is small, as shown in Figure 12.21. On the other hand, for an elongated ellipse, the foci are close to the vertices and the ratio c/a is close to 1, as shown in Figure 12.22.

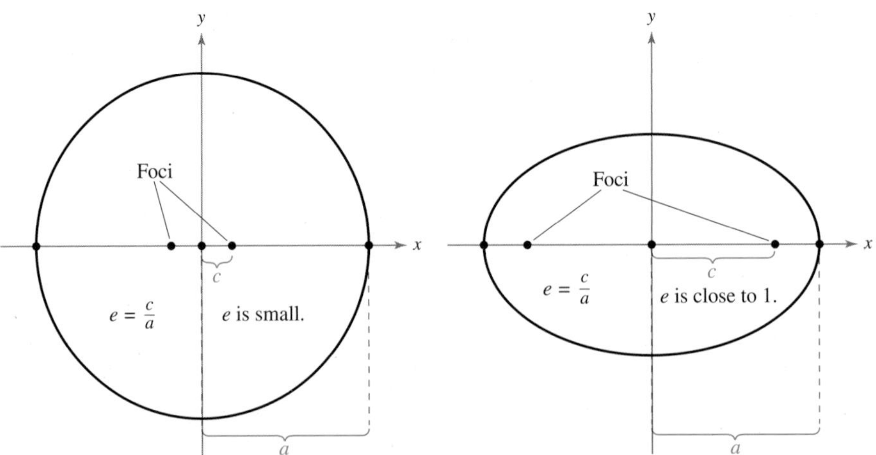

Figure 12.21 **Figure 12.22**

12.2 Exercises

See www.CalcChat.com for worked-out solutions to odd-numbered exercises.

In Exercises 1–4, fill in the blanks.

1. An ________ is the set of all points (x, y) in a plane, the sum of whose distances from two distinct fixed points, called ________, is constant.
2. The chord joining the vertices of an ellipse is called the ________ ________, and its midpoint is the ________ of the ellipse.
3. The chord perpendicular to the major axis at the center of the ellipse is called the ________ ________ of the ellipse.
4. The concept of ________ is used to measure the ovalness of an ellipse.

In Exercises 5–10, match the equation with its graph. [The graphs are labeled (a), (b), (c), (d), (e), and (f).]

(a)

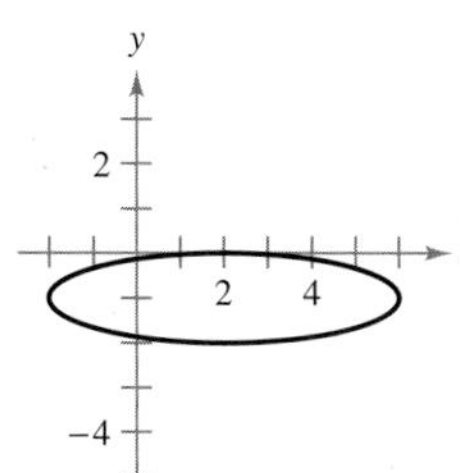

(b)

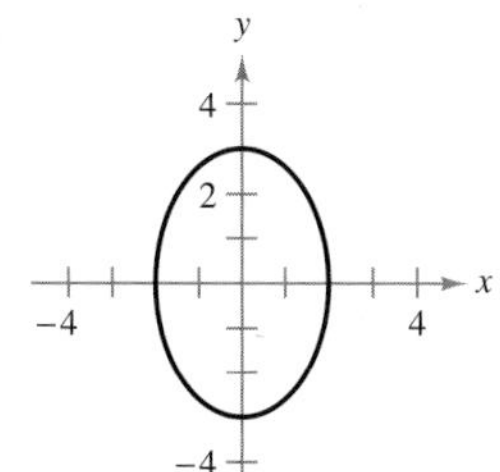

(c)

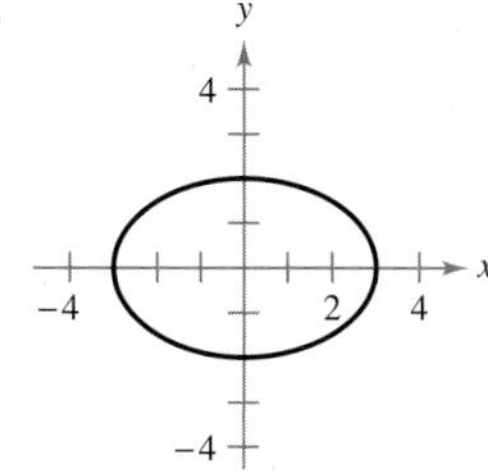

(d)

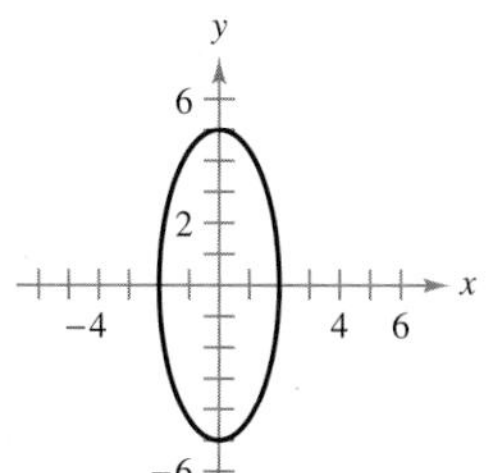

(e)

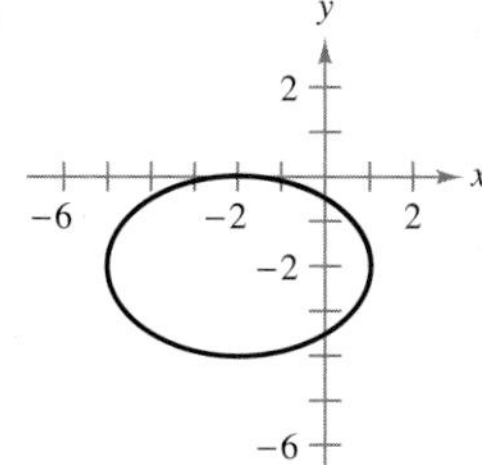

(f)

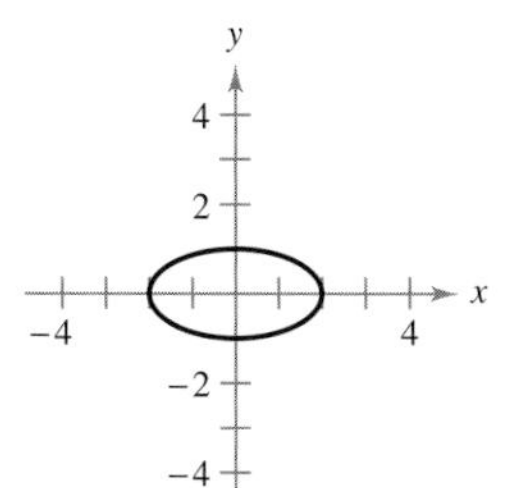

5. $\dfrac{x^2}{4} + \dfrac{y^2}{9} = 1$
6. $\dfrac{x^2}{9} + \dfrac{y^2}{4} = 1$
7. $\dfrac{x^2}{4} + \dfrac{y^2}{25} = 1$
8. $\dfrac{x^2}{4} + y^2 = 1$
9. $\dfrac{(x-2)^2}{16} + (y+1)^2 = 1$
10. $\dfrac{(x+2)^2}{9} + \dfrac{(y+2)^2}{4} = 1$

In Exercises 11–18, find the standard form of the equation of the ellipse with the given characteristics and center at the origin.

11.

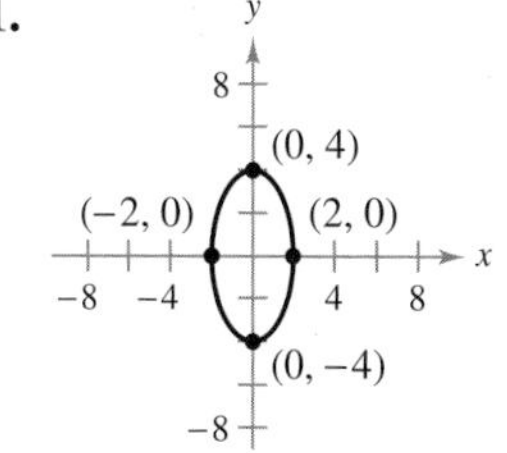

12.

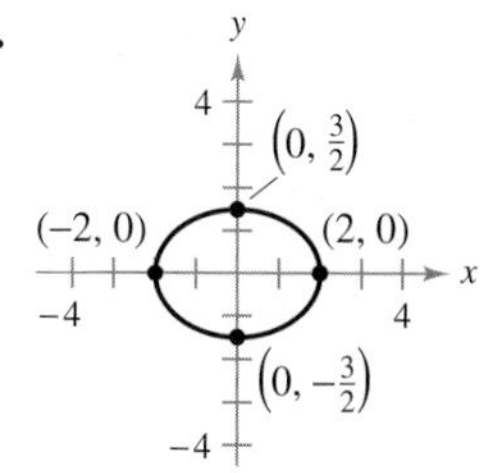

13. Vertices: $(\pm 7, 0)$; foci: $(\pm 2, 0)$
14. Vertices: $(0, \pm 8)$; foci: $(0, \pm 4)$
15. Foci: $(\pm 5, 0)$; major axis of length 14
16. Foci: $(\pm 2, 0)$; major axis of length 10
17. Vertices: $(0, \pm 5)$; passes through the point $(4, 2)$
18. Vertical major axis; passes through the points $(0, 6)$ and $(3, 0)$

In Exercises 19–28, find the standard form of the equation of the ellipse with the given characteristics.

19.

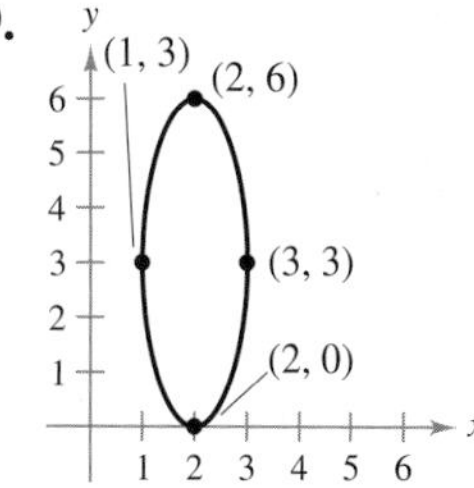

20.

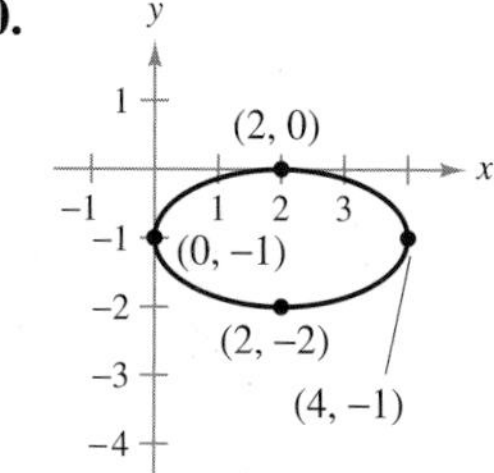

21. Vertices: $(0, 2)$, $(8, 2)$; minor axis of length 2
22. Foci: $(0, 0)$, $(4, 0)$; major axis of length 6
23. Foci: $(0, 0)$, $(0, 8)$; major axis of length 16
24. Center: $(2, -1)$; vertex: $\left(2, \frac{1}{2}\right)$; minor axis of length 2
25. Center: $(0, 4)$; $a = 2c$; vertices: $(-4, 4)$, $(4, 4)$
26. Center: $(3, 2)$; $a = 3c$; foci: $(1, 2)$, $(5, 2)$
27. Vertices: $(0, 2)$, $(4, 2)$; endpoints of the minor axis: $(2, 3)$, $(2, 1)$
28. Vertices: $(5, 0)$, $(5, 12)$; endpoints of the minor axis: $(1, 6)$, $(9, 6)$

In Exercises 29–52, identify the conic as a circle or an ellipse. Then find the center, radius, vertices, foci, and eccentricity of the conic (if applicable), and sketch its graph.

29. $\dfrac{x^2}{25} + \dfrac{y^2}{16} = 1$
30. $\dfrac{x^2}{16} + \dfrac{y^2}{81} = 1$

31. $\frac{x^2}{25} + \frac{y^2}{25} = 1$

32. $\frac{x^2}{9} + \frac{y^2}{9} = 1$

33. $\frac{x^2}{5} + \frac{y^2}{9} = 1$

34. $\frac{x^2}{64} + \frac{y^2}{28} = 1$

35. $\frac{(x-4)^2}{16} + \frac{(y+1)^2}{25} = 1$

36. $\frac{(x+3)^2}{12} + \frac{(y-2)^2}{16} = 1$

37. $\frac{x^2}{4/9} + \frac{(y+1)^2}{4/9} = 1$

38. $\frac{(x+5)^2}{9/4} + (y-1)^2 = 1$

39. $(x+2)^2 + \frac{(y+4)^2}{1/4} = 1$

40. $\frac{(x-3)^2}{25/4} + \frac{(y-1)^2}{25/4} = 1$

41. $9x^2 + 4y^2 + 36x - 24y + 36 = 0$

42. $9x^2 + 4y^2 - 54x + 40y + 37 = 0$

43. $x^2 + y^2 - 2x + 4y - 31 = 0$

44. $x^2 + 5y^2 - 8x - 30y - 39 = 0$

45. $3x^2 + y^2 + 18x - 2y - 8 = 0$

46. $6x^2 + 2y^2 + 18x - 10y + 2 = 0$

47. $x^2 + 4y^2 - 6x + 20y - 2 = 0$

48. $x^2 + y^2 - 4x + 6y - 3 = 0$

49. $9x^2 + 9y^2 + 18x - 18y + 14 = 0$

50. $16x^2 + 25y^2 - 32x + 50y + 16 = 0$

51. $9x^2 + 25y^2 - 36x - 50y + 60 = 0$

52. $16x^2 + 16y^2 - 64x + 32y + 55 = 0$

In Exercises 53–56, use a graphing utility to graph the ellipse. Find the center, foci, and vertices. (Recall that it may be necessary to solve the equation for *y* and obtain two equations.)

53. $5x^2 + 3y^2 = 15$

54. $3x^2 + 4y^2 = 12$

55. $12x^2 + 20y^2 - 12x + 40y - 37 = 0$

56. $36x^2 + 9y^2 + 48x - 36y - 72 = 0$

In Exercises 57–60, find the eccentricity of the ellipse.

57. $\frac{x^2}{4} + \frac{y^2}{9} = 1$

58. $\frac{x^2}{25} + \frac{y^2}{36} = 1$

59. $x^2 + 9y^2 - 10x + 36y + 52 = 0$

60. $4x^2 + 3y^2 - 8x + 18y + 19 = 0$

61. Find an equation of the ellipse with vertices $(\pm 5, 0)$ and eccentricity $e = \frac{3}{5}$.

62. Find an equation of the ellipse with vertices $(0, \pm 8)$ and eccentricity $e = \frac{1}{2}$.

WRITING ABOUT CONCEPTS

63. At the beginning of this section it was noted that an ellipse can be drawn using two thumbtacks, a string of fixed length (greater than the distance between the two tacks), and a pencil. If the ends of the string are fastened at the tacks and the string is drawn taut with a pencil, the path traced by the pencil is an ellipse.

(a) What is the length of the string in terms of a?

(b) Explain why the path is an ellipse.

64. Consider an ellipse with the major axis horizontal and 10 units in length. The number b in the standard form of the equation of the ellipse must be less than what real number? Explain the change in the shape of the ellipse as b approaches this number.

In Exercises 65–70, find dy/dx.

65. $\frac{x^2}{9} + \frac{y^2}{4} = 1$

66. $\frac{x^2}{25} + \frac{y^2}{64} = 1$

67. $\frac{(x-4)^2}{4} + \frac{(y+2)^2}{16} = 1$

68. $\frac{(x+5)^2}{36} + \frac{(y-3)^2}{9} = 1$

69. $9x^2 + 4y^2 - 36x + 8y + 31 = 0$

70. $3x^2 + 25y^2 - 216x - 300y + 324 = 0$

In Exercises 71 and 72, (a) find an equation of the tangent line to the ellipse at the specified point, (b) use the symmetry of the ellipse to write the equation of a tangent line parallel to the one found in part (a), and (c) use a graphing utility to graph the ellipse and the tangent lines found in parts (a) and (b).

71. $\frac{(x-2)^2}{16} + \frac{y^2}{12} = 1, \quad (0, 3)$

72. $\frac{(x-2)^2}{4} + (y+1)^2 = 1, \quad \left(3, -\frac{2+\sqrt{3}}{2}\right)$

In Exercises 73 and 74, determine the points at which dy/dx is zero or does not exist to locate the endpoints of the major and minor axes of the ellipse.

73. $x^2 + 4y^2 + 6x - 16y + 9 = 0$

74. $9x^2 + y^2 - 90x + 2y + 190 = 0$

In Exercises 75–78, find the area of the region bounded by the ellipse.

75. $\frac{x^2}{4} + \frac{y^2}{1} = 1$

76. $\frac{x^2}{16} + \frac{y^2}{9} = 1$

77. $3x^2 + 2y^2 = 6$

78. $5x^2 + 7y^2 = 70$

79. *Architecture* A semielliptical arch over a tunnel for a one-way road through a mountain has a major axis of 50 feet and a height at the center of 10 feet.

(a) Draw a rectangular coordinate system on a sketch of the tunnel with the center of the road entering the tunnel at the origin. Identify the coordinates of the known points.

(b) Find an equation of the semielliptical arch.

(c) You are driving a moving truck that has a width of 8 feet and a height of 9 feet. Will the moving truck clear the opening of the arch?

80. *Architecture* A fireplace arch is to be constructed in the shape of a semiellipse. The opening is to have a height of 2 feet at the center and a width of 6 feet along the base (see figure). The contractor draws the outline of the ellipse using tacks as described at the beginning of this section. Determine the required positions of the tacks and the length of the string.

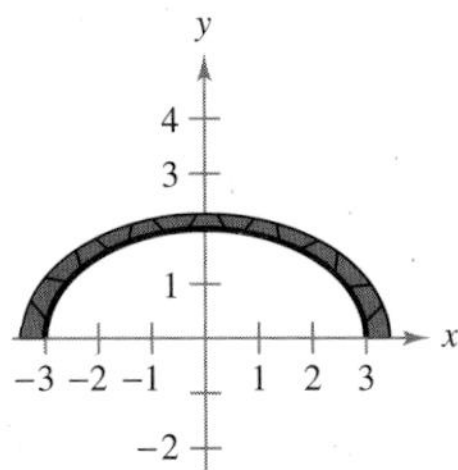

81. *Geometry* The area of the ellipse in the figure is twice the area of the circle. What is the length of the major axis?

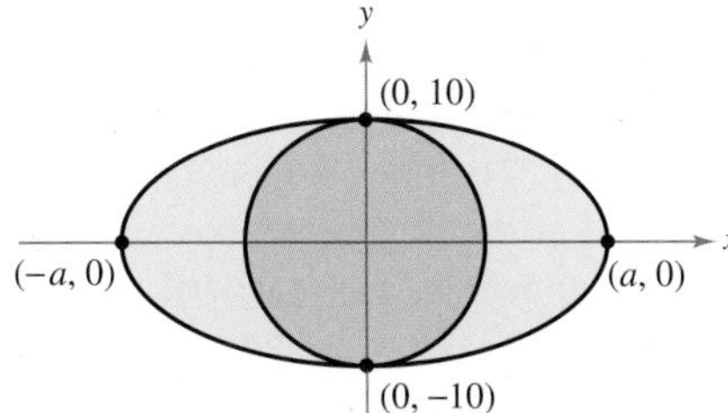

82. *Satellite Orbit* The first artificial satellite to orbit Earth was Sputnik I (launched by the former Soviet Union in 1957). Its highest point above Earth's surface was 947 kilometers, and its lowest point was 228 kilometers (see figure). The center of Earth was at one focus of the elliptical orbit, and the radius of Earth is 6378 kilometers. Find the eccentricity of the orbit.

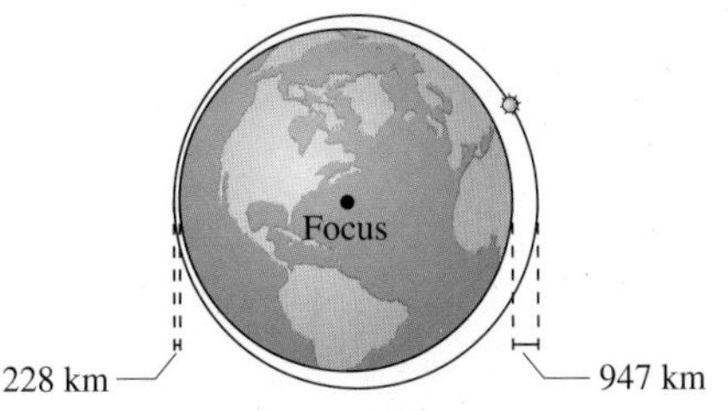

83. *Comet Orbit* Halley's comet has an elliptical orbit, with the sun at one focus. The eccentricity of the orbit is approximately 0.967. The length of the major axis of the orbit is approximately 35.88 astronomical units. (An astronomical unit is about 93 million miles.)

(a) Find an equation of the orbit. Place the center of the orbit at the origin, and place the major axis on the x-axis.

(b) Use a graphing utility to graph the equation of the orbit.

(c) Find the greatest (aphelion) and smallest (perihelion) distances from the sun's center to the comet's center.

84. *Area* Find the dimensions of the rectangle (with sides parallel to the coordinate axes) of maximum area that can be inscribed in the ellipse $x^2/25 + y^2/16 = 1$.

True or False? **In Exercises 85 and 86, determine whether the statement is true or false. Justify your answer.**

85. The graph of $x^2 + 4y^4 - 4 = 0$ is an ellipse.

86. It is easier to distinguish the graph of an ellipse from the graph of a circle if the eccentricity of the ellipse is large (close to 1).

87. Consider the ellipse $x^2/a^2 + y^2/b^2 = 1, \quad a + b = 20$.

(a) The area of the ellipse is given by $A = \pi ab$. Write the area of the ellipse as a function of a.

(b) Find the equation of an ellipse with an area of 264 square centimeters.

(c) Complete the table using your equation from part (a), and make a conjecture about the shape of the ellipse with maximum area.

a	8	9	10	11	12	13
A						

(d) Use a graphing utility to graph the area function and use the graph to support your conjecture in part (c).

CAPSTONE

88. Describe the relationship between circles and ellipses. How are they similar? How do they differ?

89. *Think About It* Find the equation of an ellipse such that for any point on the ellipse, the sum of the distances from the points (2, 2) and (10, 2) is 36.

90. *Proof* Show that $a^2 = b^2 + c^2$ for the ellipse

$$\frac{x^2}{a^2} + \frac{y^2}{b^2} = 1$$

where $a > 0$, $b > 0$, and the distance from the center of the ellipse (0, 0) to a focus is c.

12.3 Hyperbolas and Implicit Differentiation

- **Write equations of hyperbolas in standard form.**
- **Find asymptotes of and graph hyperbolas.**
- **Use implicit differentiation to find the slope of a line tangent to a hyperbola.**
- **Use properties of hyperbolas to solve real-life problems.**
- **Classify conics from their general equations.**

Introduction

The third type of conic is called a **hyperbola.** The definition of a hyperbola is similar to that of an ellipse. The difference is that for an ellipse the *sum* of the distances between the foci and a point on the ellipse is fixed, whereas for a hyperbola the *difference* of the distances between the foci and a point on the hyperbola is fixed.

DEFINITION OF HYPERBOLA

A **hyperbola** is the set of all points (x, y) in a plane, the difference of whose distances from two distinct fixed points (**foci**) is a positive constant. See Figure 12.23.

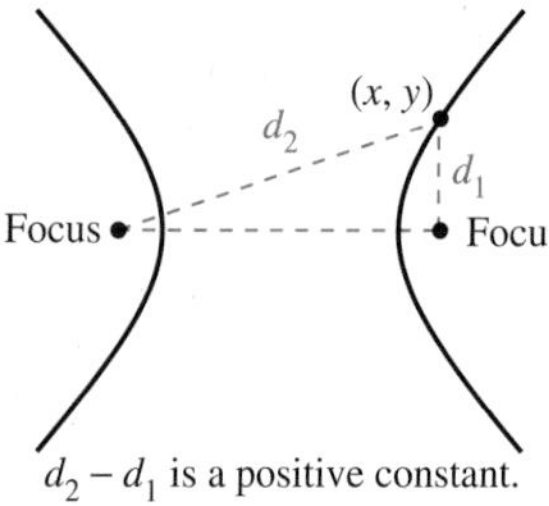

Figure 12.23

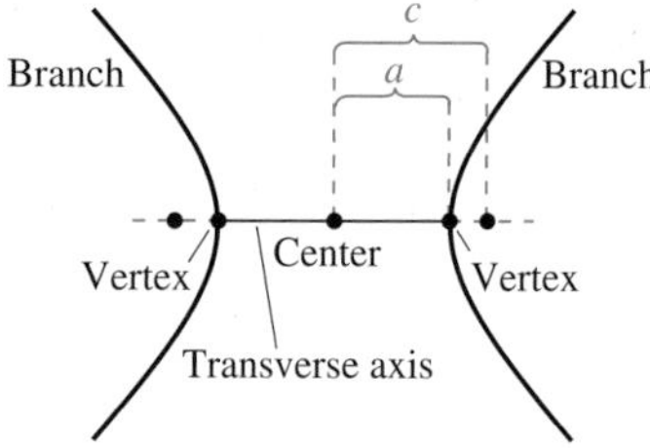

Figure 12.24

The graph of a hyperbola has two disconnected **branches.** The line through the two foci intersects the hyperbola at its two **vertices.** The line segment connecting the vertices is the **transverse axis,** and the midpoint of the transverse axis is the **center** of the hyperbola. See Figure 12.24. The development of the standard form of the equation of a hyperbola is similar to that of an ellipse. Note in the definition below that a, b, and c are related differently for hyperbolas than for ellipses.

THEOREM 12.4 STANDARD EQUATION OF A HYPERBOLA

The **standard form of the equation of a hyperbola** with center (h, k) is

$$\frac{(x-h)^2}{a^2} - \frac{(y-k)^2}{b^2} = 1 \qquad \text{Transverse axis is horizontal.}$$

$$\frac{(y-k)^2}{a^2} - \frac{(x-h)^2}{b^2} = 1. \qquad \text{Transverse axis is vertical.}$$

The vertices are a units from the center, and the foci are c units from the center. Moreover, $c^2 = a^2 + b^2$. If the center of the hyperbola is at the origin $(0, 0)$, the equation takes one of the following forms.

$$\frac{x^2}{a^2} - \frac{y^2}{b^2} = 1 \qquad \text{Transverse axis is horizontal.} \qquad \frac{y^2}{a^2} - \frac{x^2}{b^2} = 1 \qquad \text{Transverse axis is vertical.}$$

Figure 12.25 shows both the horizontal and vertical orientations for a hyperbola.

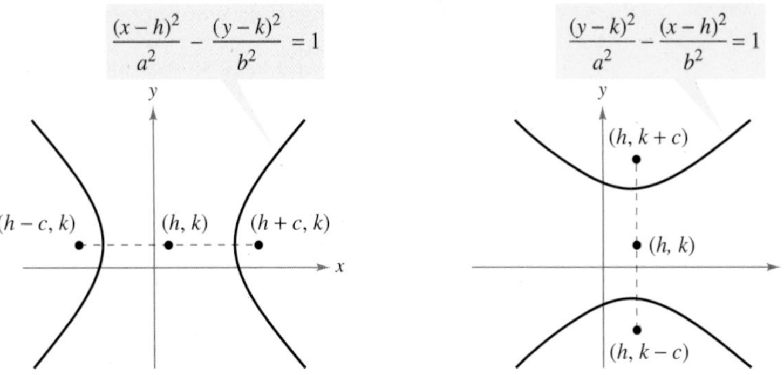

Transverse axis is horizontal.
Figure 12.25

Transverse axis is vertical.

EXAMPLE 1 Finding the Standard Equation of a Hyperbola

Find the standard form of the equation of the hyperbola with foci $(-1, 2)$ and $(5, 2)$ and vertices $(0, 2)$ and $(4, 2)$.

Solution By the Midpoint Formula, the center of the hyperbola occurs at the point $(2, 2)$. Furthermore, $c = 5 - 2 = 3$ and $a = 4 - 2 = 2$, and it follows that

$$b = \sqrt{c^2 - a^2} = \sqrt{3^2 - 2^2} = \sqrt{9 - 4} = \sqrt{5}.$$

So, the hyperbola has a horizontal transverse axis and the standard form of the equation is

$$\frac{(x-2)^2}{2^2} - \frac{(y-2)^2}{(\sqrt{5})^2} = 1.$$ See Figure 12.26.

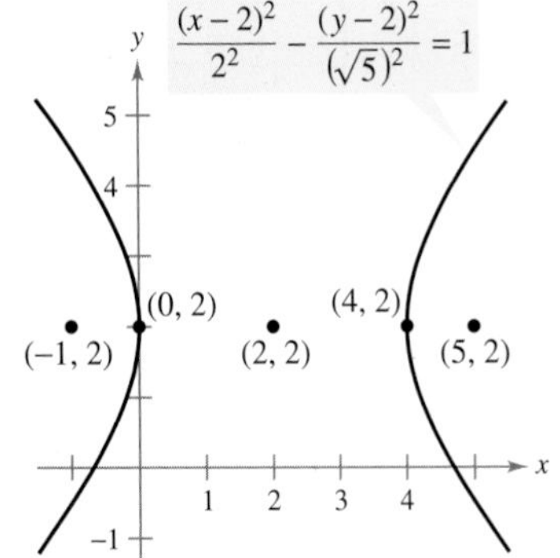

Figure 12.26

This equation simplifies to

$$\frac{(x-2)^2}{4} - \frac{(y-2)^2}{5} = 1.$$ ■

Asymptotes of a Hyperbola

Each hyperbola has two **asymptotes** that intersect at the center of the hyperbola, as shown in Figure 12.27. The asymptotes pass through the vertices of a rectangle of dimensions $2a$ by $2b$, with its center at (h, k). The line segment of length $2b$ joining $(h, k + b)$ and $(h, k - b)$ $[$or $(h + b, k)$ and $(h - b, k)]$ is the **conjugate axis** of the hyperbola.

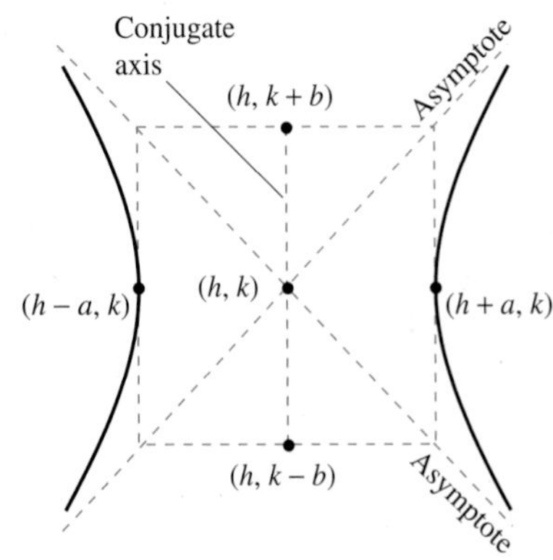

Figure 12.27

THEOREM 12.5 ASYMPTOTES OF A HYPERBOLA

The equations of the asymptotes of a hyperbola are

$$y = k \pm \frac{b}{a}(x - h)$$ Transverse axis is horizontal.

$$y = k \pm \frac{a}{b}(x - h).$$ Transverse axis is vertical.

EXAMPLE 2 Using Asymptotes to Sketch a Hyperbola

Sketch the hyperbola whose equation is $4x^2 - y^2 = 16$.

Solution Divide each side of the original equation by 16, and rewrite the equation in standard form.

$$\frac{x^2}{2^2} - \frac{y^2}{4^2} = 1 \qquad \text{Write in standard form.}$$

From this, you can conclude that $a = 2$, $b = 4$, and the transverse axis is horizontal. So, the vertices occur at $(-2, 0)$ and $(2, 0)$, and the endpoints of the conjugate axis occur at $(0, -4)$ and $(0, 4)$. Using these four points, you are able to sketch the rectangle shown in Figure 12.28(a). Finally, after drawing the asymptotes through the corners of this rectangle, you can complete the sketch, as shown in Figure 12.28(b).

STUDY TIP A convenient way to remember the equation of the asymptotes is to use the point-slope form from Section P.5,

$$y - k = m(x - h)$$

where (h, k) is the center and

$$m = \frac{\text{vertical change}}{\text{horizontal change}}$$

$$= \pm\frac{b}{a} \quad \text{or} \quad \pm\frac{a}{b}$$

depending on the orientation of the hyperbola.

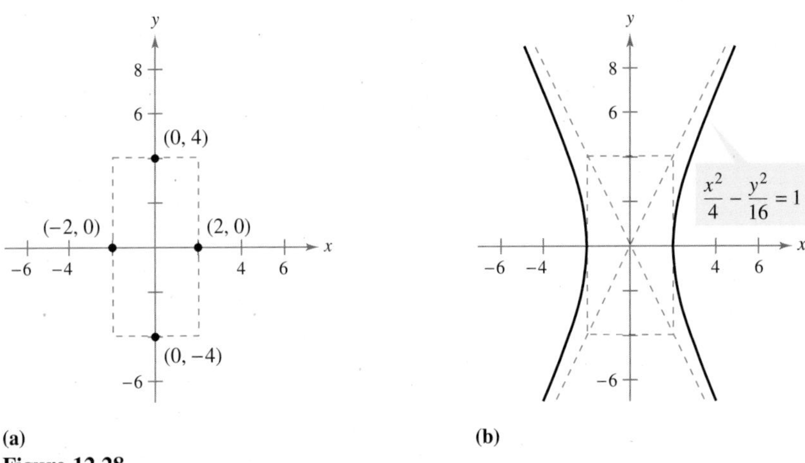

(a) (b)
Figure 12.28

EXAMPLE 3 Finding the Asymptotes of a Hyperbola

Sketch the hyperbola given by $4x^2 - 3y^2 + 8x + 16 = 0$ and find the equations of its asymptotes.

Solution

$$4x^2 - 3y^2 + 8x + 16 = 0 \qquad \text{Write original equation.}$$

$$4(x^2 + 2x) - 3y^2 = -16 \qquad \text{Subtract 16 from each side and factor.}$$

$$4(x^2 + 2x + 1) - 3y^2 = -16 + 4 \qquad \text{Add 4 to each side.}$$

$$4(x + 1)^2 - 3y^2 = -12 \qquad \text{Write in completed square form.}$$

$$\frac{y^2}{2^2} - \frac{(x + 1)^2}{\left(\sqrt{3}\right)^2} = 1 \qquad \text{Write in standard form.}$$

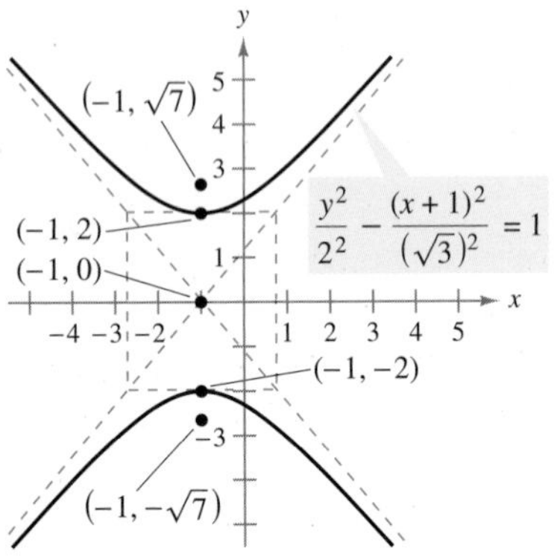

From this equation you can conclude that the hyperbola has a vertical transverse axis, centered at $(-1, 0)$, has vertices $(-1, 2)$ and $(-1, -2)$, and has a conjugate axis with endpoints $\left(-1 - \sqrt{3}, 0\right)$ and $\left(-1 + \sqrt{3}, 0\right)$. To sketch the hyperbola, draw a rectangle through these four points. The asymptotes are the lines passing through the corners of the rectangle, as shown in Figure 12.29. Finally, using $a = 2$ and $b = \sqrt{3}$, you can conclude that the equations of the asymptotes are

$$y = \frac{2}{\sqrt{3}}(x + 1) \quad \text{and} \quad y = -\frac{2}{\sqrt{3}}(x + 1).$$

■

Figure 12.29

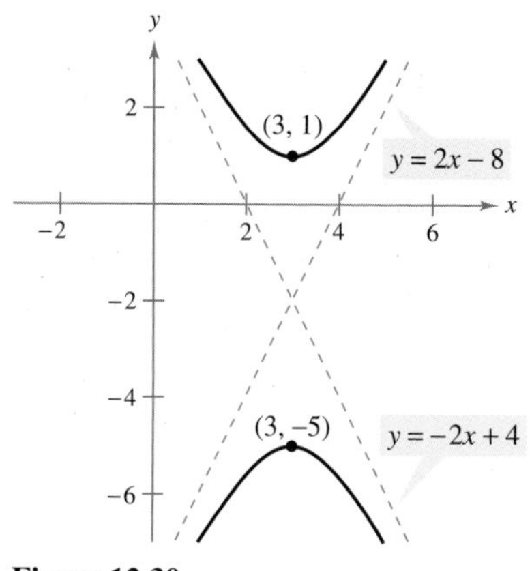

Figure 12.30

EXAMPLE 4 Using Asymptotes to Find the Standard Equation

Find the standard form of the equation of the hyperbola having vertices $(3, -5)$ and $(3, 1)$ and having asymptotes

$$y = 2x - 8 \quad \text{and} \quad y = -2x + 4$$

as shown in Figure 12.30.

Solution By the Midpoint Formula, the center of the hyperbola is $(3, -2)$. Furthermore, the hyperbola has a vertical transverse axis with $a = 3$. From the original equations, you can determine the slopes of the asymptotes to be

$$m_1 = 2 = \frac{a}{b} \quad \text{and} \quad m_2 = -2 = -\frac{a}{b}$$

and, because $a = 3$, you can conclude

$$2 = \frac{a}{b} \implies 2 = \frac{3}{b} \implies b = \frac{3}{2}.$$

So, the standard form of the equation is

$$\frac{(y+2)^2}{3^2} - \frac{(x-3)^2}{\left(\frac{3}{2}\right)^2} = 1.$$

■

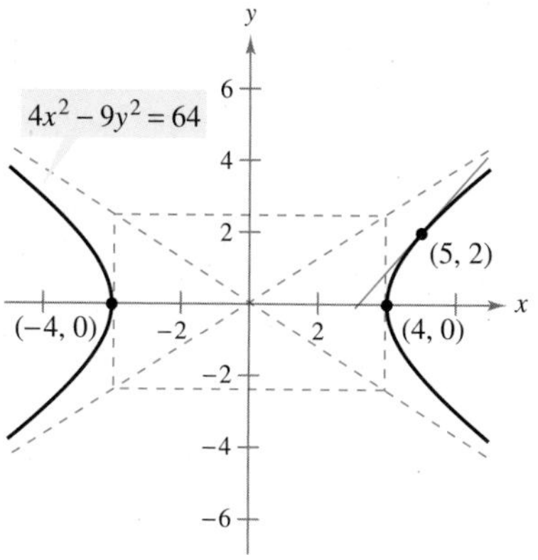

EXAMPLE 5 Finding the Slope of a Graph Implicitly

Determine the slope of the tangent line to the graph of

$$4x^2 - 9y^2 = 64$$

at the point $(5, 2)$. See Figure 12.31.

Figure 12.31

Solution Implicit differentiation of the equation $4x^2 - 9y^2 = 64$ with respect to x yields

$$8x - 18y\frac{dy}{dx} = 0$$

$$\frac{dy}{dx} = \frac{8x}{18y}$$

$$= \frac{4x}{9y}.$$

So, at $(5, 2)$, the slope is

$$\frac{dy}{dx} = \frac{4(5)}{9(2)}$$

$$= \frac{10}{9}.$$

■

DEFINITION OF ECCENTRICITY OF A HYPERBOLA

The **eccentricity** e of a hyperbola is given by the ratio

$$e = \frac{c}{a}.$$

Because $c > a$ for a hyperbola, it follows that $e > 1$. If the eccentricity is large, the branches of the hyperbola are nearly flat, as shown in Figure 12.32. If the eccentricity is close to 1, the branches of the hyperbola are more narrow, as shown in Figure 12.33.

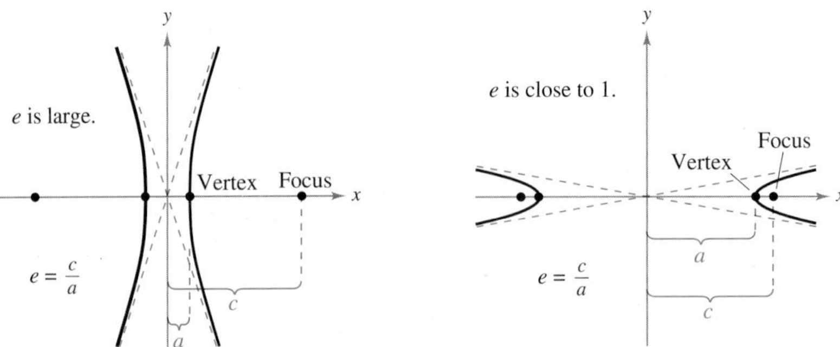

Figure 12.32 **Figure 12.33**

Applications

The following application was developed during World War II. It shows how the properties of hyperbolas can be used in radar and other detection systems.

EXAMPLE 6 An Application Involving Hyperbolas

Two microphones, 1 mile apart, record an explosion. Microphone A receives the sound 2 seconds before microphone B. Where did the explosion occur? (Assume sound travels at 1100 feet per second.)

Solution Assuming sound travels at 1100 feet per second, you know that the explosion took place 2200 feet farther from B than from A, as shown in Figure 12.34. The locus of all points that are 2200 feet closer to A than to B is one branch of the hyperbola $x^2/a^2 - y^2/b^2 = 1$, where $c = 5280/2 = 2640$ and $a = 2200/2 = 1100$. So, $b^2 = c^2 - a^2 = 2640^2 - 1100^2 = 5{,}759{,}600$, and you can conclude that the explosion occurred somewhere on the right branch of the hyperbola

$$\frac{x^2}{1{,}210{,}000} - \frac{y^2}{5{,}759{,}600} = 1.$$

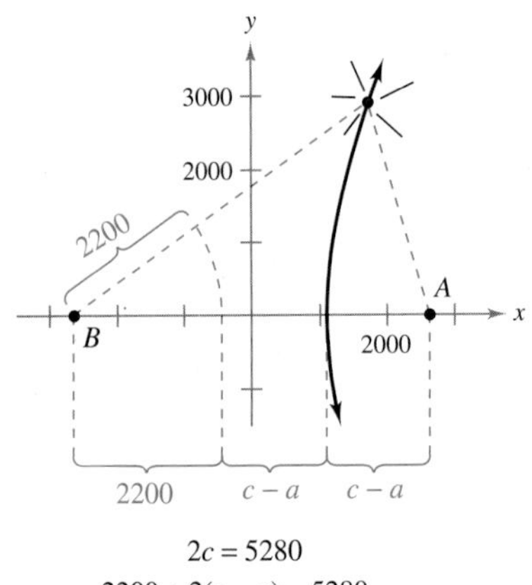

Figure 12.34

Another interesting application of conic sections involves the orbits of comets in our solar system. Of the 610 comets identified prior to 1970, 245 have elliptical orbits, 295 have parabolic orbits, and 70 have hyperbolic orbits. The center of the sun is a focus of each of these orbits, and each orbit has a vertex at the point where the comet is closest to the sun, as shown in Figure 12.35. Undoubtedly, there have been many comets with parabolic or hyperbolic orbits that were not identified. We only get to see such comets *once*. Comets with elliptical orbits, such as Halley's comet, are the only ones that remain in our solar system.

If p is the distance between the vertex and the focus (in meters), and v is the velocity of the comet at the vertex (in meters per second), then the type of orbit is determined as follows.

Ellipse: $v < \sqrt{2GM/p}$ Parabola: $v = \sqrt{2GM/p}$ Hyperbola: $v > \sqrt{2GM/p}$

In each of these relations, $M = 1.989 \times 10^{30}$ kilograms (the mass of the sun) and $G \approx 6.67 \times 10^{-11}$ cubic meter per kilogram-second squared (the universal gravitational constant).

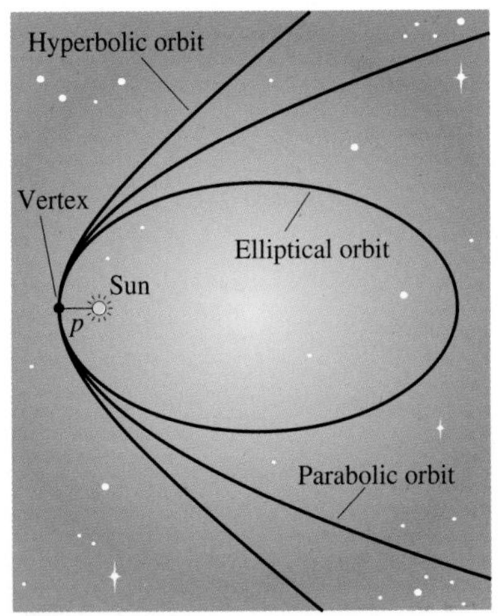

Figure 12.35

NOTE The test at the right is valid if the graph is a conic. The test does not apply to equations such as $x^2 + y^2 = -1$, which has no real graph.

General Equations of Conics

CLASSIFYING A CONIC FROM ITS GENERAL EQUATION

The graph of $Ax^2 + Cy^2 + Dx + Ey + F = 0$ is one of the following.

1. *Circle:* $A = C$
2. *Parabola:* $AC = 0$ — $A = 0$ or $C = 0$, but not both.
3. *Ellipse:* $AC > 0$ — A and C have like signs.
4. *Hyperbola:* $AC < 0$ — A and C have unlike signs.

EXAMPLE 7 Classifying Conics from General Equations

Classify the graph of each equation.

a. $4x^2 - 9x + y - 5 = 0$

b. $4x^2 - y^2 + 8x - 6y + 4 = 0$

c. $2x^2 + 4y^2 - 4x + 12y = 0$

d. $2x^2 + 2y^2 - 8x + 12y + 2 = 0$

Solution

a. For the equation $4x^2 - 9x + y - 5 = 0$, you have

$$AC = 4(0) = 0. \qquad \text{Parabola}$$

So, the graph is a parabola, as shown in Figure 12.36(a).

b. For the equation $4x^2 - y^2 + 8x - 6y + 4 = 0$, you have

$$AC = 4(-1) < 0. \qquad \text{Hyperbola}$$

So, the graph is a hyperbola, as shown in Figure 12.36(b).

c. For the equation $2x^2 + 4y^2 - 4x + 12y = 0$, you have

$$AC = 2(4) > 0. \qquad \text{Ellipse}$$

So, the graph is an ellipse, as shown in Figure 12.36(c).

d. For the equation $2x^2 + 2y^2 - 8x + 12y + 2 = 0$, you have

$$A = C = 2. \qquad \text{Circle}$$

So, the graph is a circle, as shown in Figure 12.36(d).

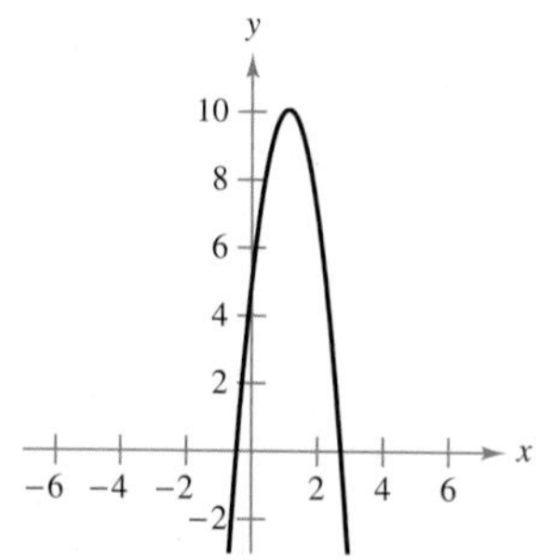

(a)

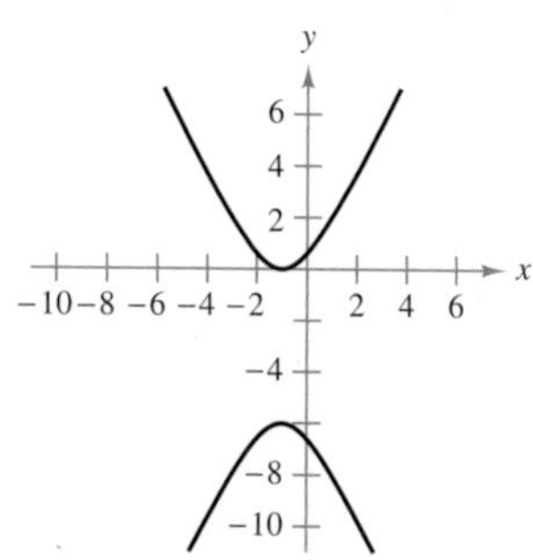

(b)

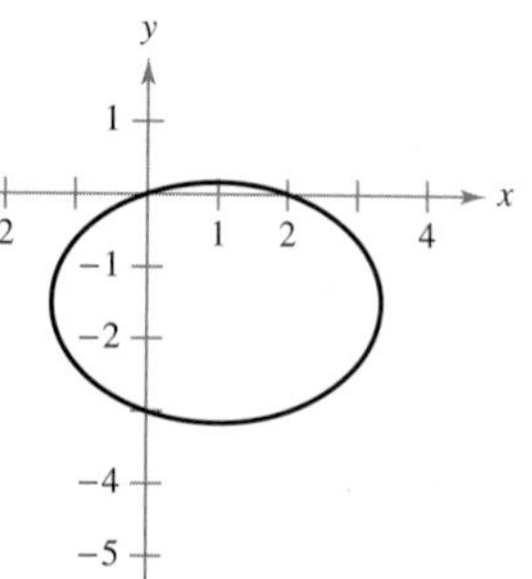

(c)

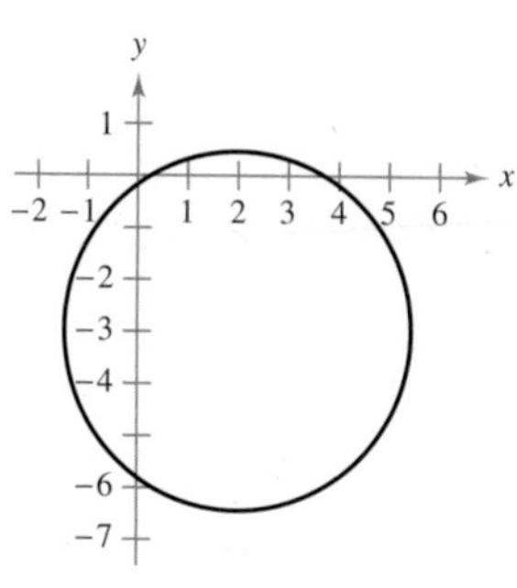

(d)

Figure 12.36

■

12.3 Exercises

See www.CalcChat.com for worked-out solutions to odd-numbered exercises.

In Exercises 1–4, fill in the blanks.

1. A ________ is the set of all points (x, y) in a plane, the difference of whose distances from two distinct fixed points, called ________, is a positive constant.
2. The graph of a hyperbola has two disconnected parts called ________.
3. The line segment connecting the vertices of a hyperbola is called the ________ ________, and the midpoint of the line segment is the ________ of the hyperbola.
4. Each hyperbola has two ________ that intersect at the center of the hyperbola.

In Exercises 5–8, match the equation with its graph. [The graphs are labeled (a), (b), (c), and (d).]

(a)

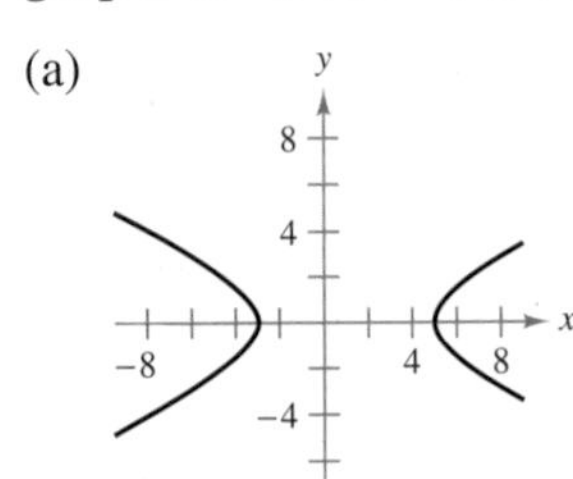

(b)

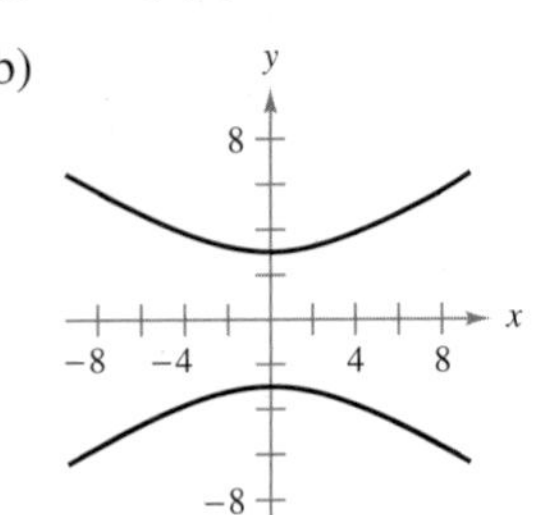

(c)

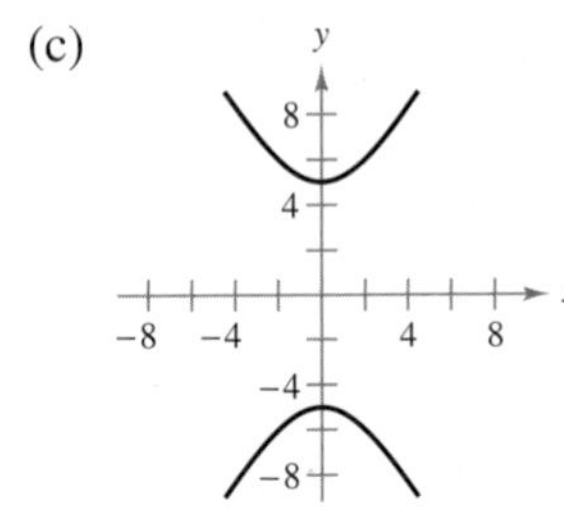

(d)

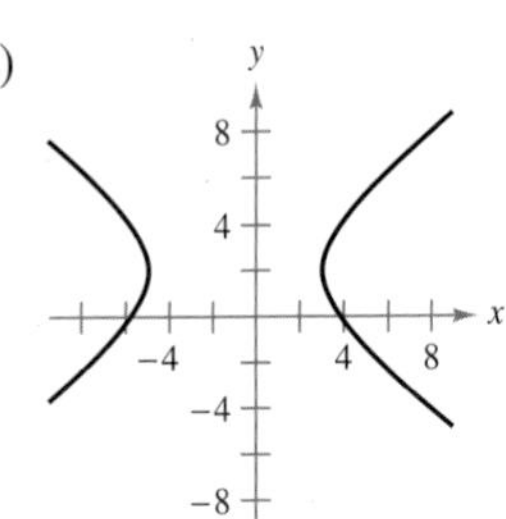

5. $\frac{y^2}{9} - \frac{x^2}{25} = 1$
6. $\frac{y^2}{25} - \frac{x^2}{9} = 1$
7. $\frac{(x-1)^2}{16} - \frac{y^2}{4} = 1$
8. $\frac{(x+1)^2}{16} - \frac{(y-2)^2}{9} = 1$

In Exercises 9–22, find the center, vertices, foci, and the equations of the asymptotes of the hyperbola, and sketch its graph using the asymptotes as an aid.

9. $x^2 - y^2 = 1$
10. $\frac{x^2}{9} - \frac{y^2}{25} = 1$
11. $\frac{y^2}{25} - \frac{x^2}{81} = 1$
12. $\frac{x^2}{36} - \frac{y^2}{4} = 1$
13. $\frac{y^2}{1} - \frac{x^2}{4} = 1$
14. $\frac{y^2}{9} - \frac{x^2}{1} = 1$
15. $\frac{(x-1)^2}{4} - \frac{(y+2)^2}{1} = 1$
16. $\frac{(x+3)^2}{144} - \frac{(y-2)^2}{25} = 1$
17. $\frac{(y+6)^2}{1/9} - \frac{(x-2)^2}{1/4} = 1$
18. $\frac{(y-1)^2}{1/4} - \frac{(x+3)^2}{1/16} = 1$
19. $9x^2 - y^2 - 36x - 6y + 18 = 0$
20. $x^2 - 9y^2 + 36y - 72 = 0$
21. $x^2 - 9y^2 + 2x - 54y - 80 = 0$
22. $16y^2 - x^2 + 2x + 64y + 63 = 0$

In Exercises 23–28, find the center, vertices, foci, and the equations of the asymptotes of the hyperbola. Use a graphing utility to graph the hyperbola and its asymptotes.

23. $2x^2 - 3y^2 = 6$
24. $6y^2 - 3x^2 = 18$
25. $4x^2 - 9y^2 = 36$
26. $25x^2 - 4y^2 = 100$
27. $9y^2 - x^2 + 2x + 54y + 62 = 0$
28. $9x^2 - y^2 + 54x + 10y + 55 = 0$

In Exercises 29–34, find the standard form of the equation of the hyperbola with the given characteristics and center at the origin.

29. Vertices: $(0, \pm 2)$; foci: $(0, \pm 4)$
30. Vertices: $(\pm 4, 0)$; foci: $(\pm 6, 0)$
31. Vertices: $(\pm 1, 0)$; asymptotes: $y = \pm 5x$
32. Vertices: $(0, \pm 3)$; asymptotes: $y = \pm 3x$
33. Foci: $(0, \pm 8)$; asymptotes: $y = \pm 4x$
34. Foci: $(\pm 10, 0)$; asymptotes: $y = \pm \frac{3}{4}x$

In Exercises 35–46, find the standard form of the equation of the hyperbola with the given characteristics.

35. Vertices: $(2, 0), (6, 0)$; foci: $(0, 0), (8, 0)$
36. Vertices: $(2, 3), (2, -3)$; foci: $(2, 6), (2, -6)$
37. Vertices: $(4, 1), (4, 9)$; foci: $(4, 0), (4, 10)$
38. Vertices: $(-2, 1), (2, 1)$; foci: $(-3, 1), (3, 1)$
39. Vertices: $(2, 3), (2, -3)$;
    passes through the point $(0, 5)$
40. Vertices: $(-2, 1), (2, 1)$;
    passes through the point $(5, 4)$

41. Vertices: $(0, 4), (0, 0)$;
passes through the point $(\sqrt{5}, -1)$

42. Vertices: $(1, 2), (1, -2)$;
passes through the point $(0, \sqrt{5})$

43. Vertices: $(1, 2), (3, 2)$; asymptotes: $y = x,\ y = 4 - x$

44. Vertices: $(3, 0), (3, 6)$; asymptotes: $y = 6 - x,\ y = x$

45. Vertices: $(0, 2), (6, 2)$; asymptotes: $y = \frac{2}{3}x,\ y = 4 - \frac{2}{3}x$

46. Vertices: $(3, 0), (3, 4)$; asymptotes: $y = \frac{2}{3}x,\ y = 4 - \frac{2}{3}x$

In Exercises 47–50, write the standard form of the equation of the hyperbola.

47.

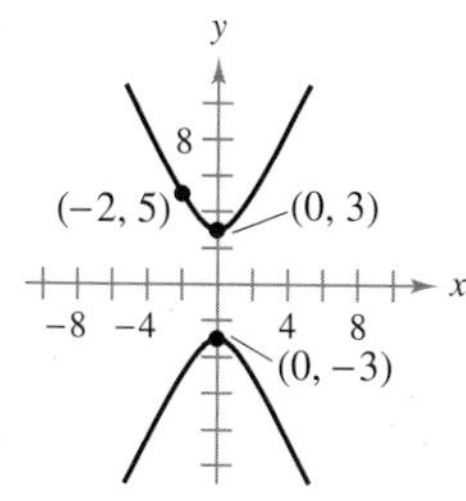

48.

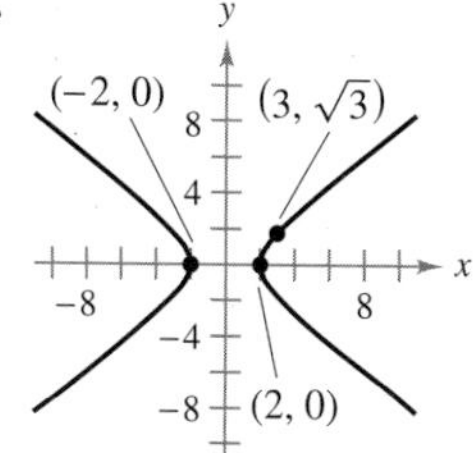

49.

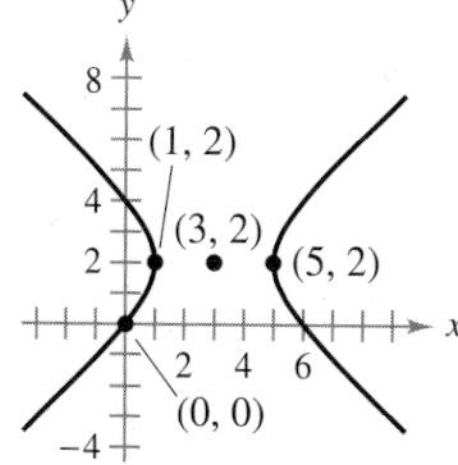

50.

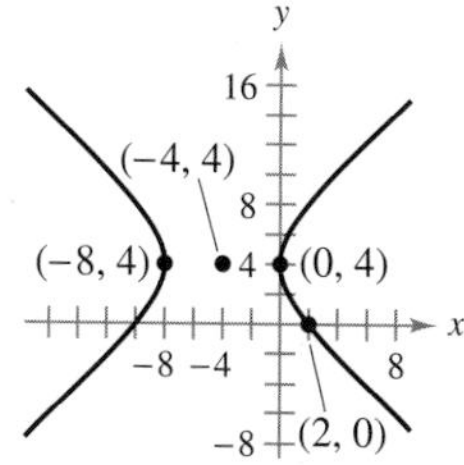

WRITING ABOUT CONCEPTS

51. Find an equation of the hyperbola such that for any point on the hyperbola, the difference between its distances from the points $(2, 2)$ and $(10, 2)$ is 6.

52. Find an equation of the hyperbola such that for any point on the hyperbola, the difference between its distances from the points $(-3, 0)$ and $(-3, 3)$ is 2.

53. Consider a hyperbola centered at the origin with a horizontal transverse axis. Use the definition of a hyperbola to derive its standard form.

54. Explain how the central rectangle of a hyperbola can be used to sketch its asymptotes.

In Exercises 55–62, find dy/dx.

55. $\dfrac{x^2}{64} - \dfrac{y^2}{36} = 1$

56. $\dfrac{y^2}{16} - \dfrac{x^2}{25} = 1$

57. $\dfrac{(y-3)^2}{9} - \dfrac{(x+1)^2}{9} = 1$

58. $(x+1)^2 - \dfrac{(y-2)^2}{9} = 1$

59. $x^2 - 2y^2 + 8y - 17 = 0$

60. $x^2 - 5y^2 + 20x + 2y - 35 = 0$

61. $x^2 - 4y^2 + 2x + 16y - 19 = 0$

62. $4y^2 - x^2 + 4x - 5 = 0$

In Exercises 63 and 64, (a) find an equation of the tangent line to the hyperbola at the specified point, (b) use the symmetry of the hyperbola to write the equation of the tangent line parallel to the one found in part (a), and (c) use a graphing utility to graph the hyperbola and the tangent lines found in parts (a) and (b).

63. $\dfrac{(x-2)^2}{16} - \dfrac{y^2}{12} = 1, \quad (10, 6)$

64. $\dfrac{(y-3)^2}{4} - \dfrac{(x-1)^2}{9} = 1, \quad \left(5, \dfrac{19}{3}\right)$

In Exercises 65 and 66, determine the points at which dy/dx is zero or does not exist as an aid in locating the vertices of the hyperbola.

65. $4y^2 - x^2 + 6x + 40y + 75 = 0$

66. $16x^2 - 9y^2 + 64x + 18y + 19 = 0$

67. ***Art*** A sculpture has a hyperbolic cross section (see figure).

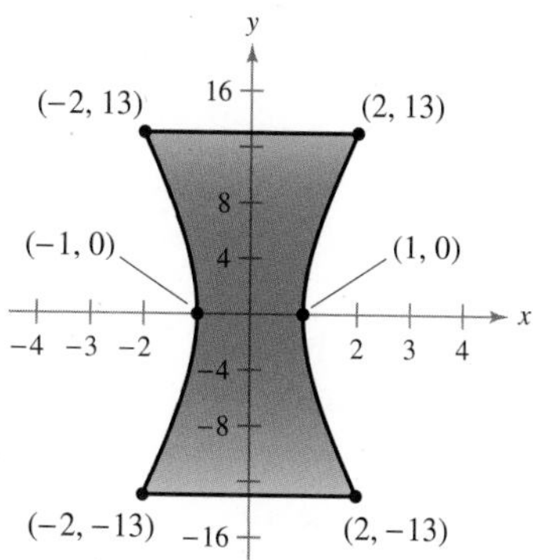

(a) Write an equation that models the curved sides of the sculpture.

(b) Each unit in the coordinate plane represents 1 foot. Find the width of the sculpture at a height of 5 feet.

68. ***Sound Location*** You and a friend live 4 miles apart (on the same "east-west" street) and are talking on the phone. You hear a clap of thunder from lightning in a storm, and 18 seconds later your friend hears the thunder. Find an equation that gives the possible places where the lightning could have occurred. (Assume that the coordinate system is measured in feet and that sound travels at 1100 feet per second.)

69. ***Sound Location*** Three listening stations located at $(3300, 0)$, $(3300, 1100)$, and $(-3300, 0)$ monitor an explosion. The last two stations detect the explosion 1 second and 4 seconds after the first, respectively. Determine the coordinates of the explosion. (Assume that the coordinate system is measured in feet and that sound travels at 1100 feet per second.)

70. ***LORAN*** Long distance radio navigation for aircraft and ships uses synchronized pulses transmitted by widely separated transmitting stations. These pulses travel at the speed of light (186,000 miles per second). The difference in the times of arrival of these pulses at an aircraft or ship is constant on a hyperbola having the transmitting stations as foci. Assume that two stations, 300 miles apart, are positioned on the rectangular coordinate system at points with coordinates $(-150, 0)$ and $(150, 0)$, and that a ship is traveling on a hyperbolic path with coordinates $(x, 75)$ (see figure).

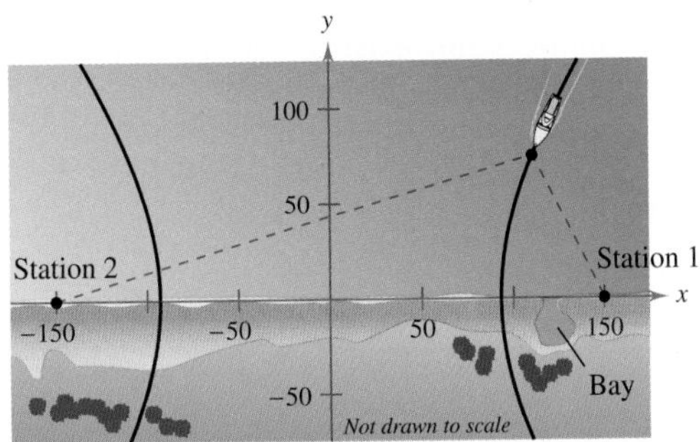

(a) Find the x-coordinate of the position of the ship if the time difference between the pulses from the transmitting stations is 1000 microseconds (0.001 second).

(b) Determine the distance between the ship and station 1 when the ship reaches the shore.

(c) The ship wants to enter a bay located between the two stations. The bay is 30 miles from station 1. What should be the time difference between the pulses?

(d) The ship is 60 miles offshore when the time difference in part (c) is obtained. What is the position of the ship?

In Exercises 71–86, classify the graph of the equation as a circle, a parabola, an ellipse, or a hyperbola.

71. $9x^2 + 4y^2 - 18x + 16y - 119 = 0$

72. $x^2 + y^2 - 4x - 6y - 23 = 0$

73. $4x^2 - y^2 - 4x - 3 = 0$

74. $y^2 - 6y - 4x + 21 = 0$

75. $y^2 - 4x^2 + 4x - 2y - 4 = 0$

76. $x^2 + y^2 - 4x + 6y - 3 = 0$

77. $y^2 + 12x + 4y + 28 = 0$

78. $4x^2 + 25y^2 + 16x + 250y + 541 = 0$

79. $4x^2 + 3y^2 + 8x - 24y + 51 = 0$

80. $4y^2 - 2x^2 - 4y - 8x - 15 = 0$

81. $25x^2 - 10x - 200y - 119 = 0$

82. $4y^2 + 4x^2 - 24x + 35 = 0$

83. $x^2 - 6x - 2y + 7 = 0$

84. $9x^2 + 4y^2 - 90x + 8y + 228 = 0$

85. $100x^2 + 100y^2 - 100x + 400y + 409 = 0$

86. $4x^2 - y^2 + 4x + 2y - 1 = 0$

True or False? **In Exercises 87–90, determine whether the statement is true or false. Justify your answer.**

87. In the standard form of the equation of a hyperbola, the larger the ratio of b to a, the larger the eccentricity of the hyperbola.

88. In the standard form of the equation of a hyperbola, the trivial solution of two intersecting lines occurs when $b = 0$.

89. If $D \neq 0$ and $E \neq 0$, then the graph of $x^2 - y^2 + Dx + Ey = 0$ is a hyperbola.

90. If the asymptotes of the hyperbola $\frac{x^2}{a^2} - \frac{y^2}{b^2} = 1$, where $a, b > 0$, intersect at right angles, then $a = b$.

91. ***Think About It*** Change the equation of the hyperbola so that its graph is the bottom half of the hyperbola.

$9x^2 - 54x - 4y^2 + 8y + 41 = 0$

CAPSTONE

92. Given the hyperbolas

$$\frac{x^2}{16} - \frac{y^2}{9} = 1 \quad \text{and} \quad \frac{y^2}{9} - \frac{x^2}{16} = 1$$

describe any common characteristics that the hyperbolas share, as well as any differences in the graphs of the hyperbolas. Verify your results by using a graphing utility to graph each of the hyperbolas in the same viewing window.

93. A circle and a parabola can have 0, 1, 2, 3, or 4 points of intersection. Sketch the circle given by $x^2 + y^2 = 4$. Discuss how this circle could intersect a parabola with an equation of the form $y = x^2 + C$. Then find the values of C for each of the five cases described below. Use a graphing utility to verify your results.

(a) No points of intersection

(b) One point of intersection

(c) Two points of intersection

(d) Three points of intersection

(e) Four points of intersection

12.4 Parametric Equations and Calculus

- **Evaluate sets of parametric equations for given values of the parameter.**
- **Sketch curves that are represented by sets of parametric equations.**
- **Rewrite sets of parametric equations as single rectangular equations by eliminating the parameter.**
- **Find sets of parametric equations for graphs.**
- **Find the slope of a tangent line to a curve given by a set of parametric equations.**

Plane Curves

Up to this point you have been representing a graph by a single equation involving the *two* variables x and y. In this section, you will study situations in which it is useful to introduce a *third* variable to represent a curve in the plane.

To see the usefulness of this procedure, consider the path followed by an object that is propelled into the air at an angle of 45°. If the initial velocity of the object is 48 feet per second, it can be shown that the object follows the parabolic path

$$y = -\frac{x^2}{72} + x \qquad \text{Rectangular equation}$$

as shown in Figure 12.37. However, this equation does not tell the whole story. Although it does tell you *where* the object has been, it does not tell you *when* the object was at a given point (x, y) on the path. To determine this time, you can introduce a third variable t, called a **parameter.** It is possible to write both x and y as functions of t to obtain the **parametric equations**

$$x = 24\sqrt{2}t \qquad \text{Parametric equation for } x$$

$$y = -16t^2 + 24\sqrt{2}t. \qquad \text{Parametric equation for } y$$

From this set of equations you can determine that at time $t = 0$, the object is at the point $(0, 0)$. Similarly, at time $t = 1$, the object is at the point $\left(24\sqrt{2}, 24\sqrt{2} - 16\right)$, and so on, as shown in Figure 12.37.

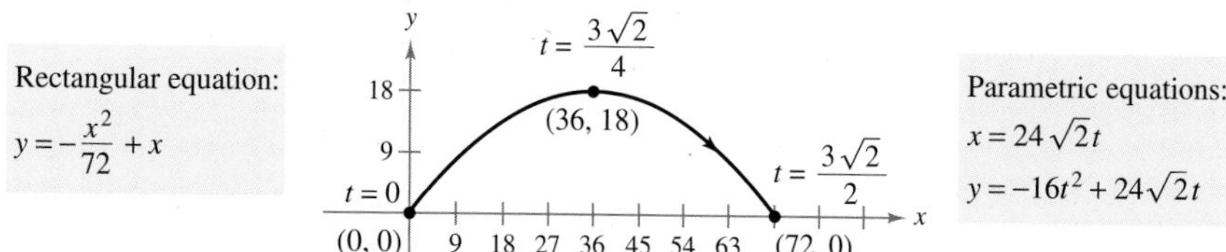

Curvilinear Motion: Two Variables for Position, One Variable for Time
Figure 12.37

For this particular motion problem, x and y are continuous functions of t, and the resulting path is a **plane curve.** (Informally, you might say that a function is continuous if its graph can be traced without lifting the pencil from the paper.)

DEFINITION OF A PLANE CURVE

If f and g are continuous functions of t on an interval I, the set of ordered pairs $(f(t), g(t))$ is a **plane curve** C. The equations

$$x = f(t) \qquad \text{and} \qquad y = g(t)$$

are **parametric equations** for C, and t is the **parameter.**

Sketching a Plane Curve

When sketching a curve represented by a pair of parametric equations, you still plot points in the xy-plane. Each set of coordinates (x, y) is determined from a value chosen for the parameter t. Plotting the resulting points in the order of *increasing* values of t traces the curve in a specific direction. This is called the **orientation** of the curve.

EXAMPLE 1 Sketching a Curve

STUDY TIP When using a value of t to find x, be sure to use the same value of t to find the corresponding value of y. Organizing your results in a table, as shown in Example 1, can be helpful.

Sketch the curve given by the parametric equations

$$x = t^2 - 4 \quad \text{and} \quad y = \frac{t}{2}, \quad -2 \le t \le 3.$$

Solution Using values of t in the specified interval, the parametric equations yield the points (x, y) shown in the table.

t	-2	-1	0	1	2	3
x	0	-3	-4	-3	0	5
y	-1	$-\frac{1}{2}$	0	$\frac{1}{2}$	1	$\frac{3}{2}$

By plotting these points in the order of increasing t, you obtain the curve C shown in Figure 12.38. Note that the arrows on the curve indicate its orientation as t increases from -2 to 3. So, if a particle were moving on this curve, it would start at $(0, -1)$ and then move along the curve to the point $\left(5, \frac{3}{2}\right)$.

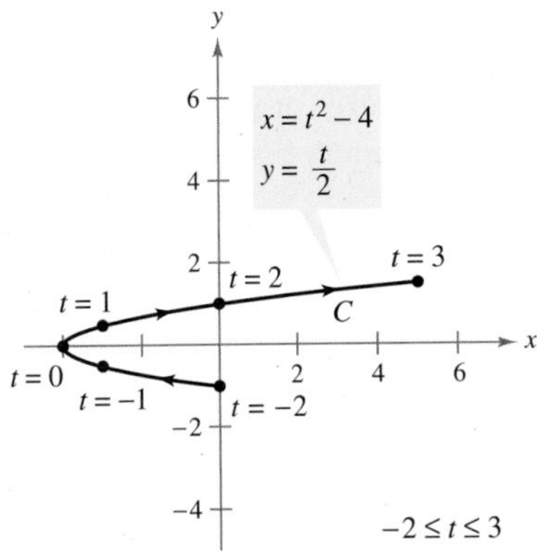

Figure 12.38

Note that the graph shown in Figure 12.38 does not define y as a function of x. This points out one benefit of parametric equations—they can be used to represent graphs that are more general than graphs of functions. It often happens that two different sets of parametric equations have the same graph. For example, the set of parametric equations

$$x = 4t^2 - 4 \quad \text{and} \quad y = t, \quad -1 \le t \le \frac{3}{2}$$

has the same graph as the set given in Example 1. However, by comparing the values of t in Figures 12.38 and 12.39, you can see that this second graph is traced out more *rapidly* (considering t as time) than the first graph. So, in applications, different parametric representations can be used to represent various *speeds* at which objects travel along a given path.

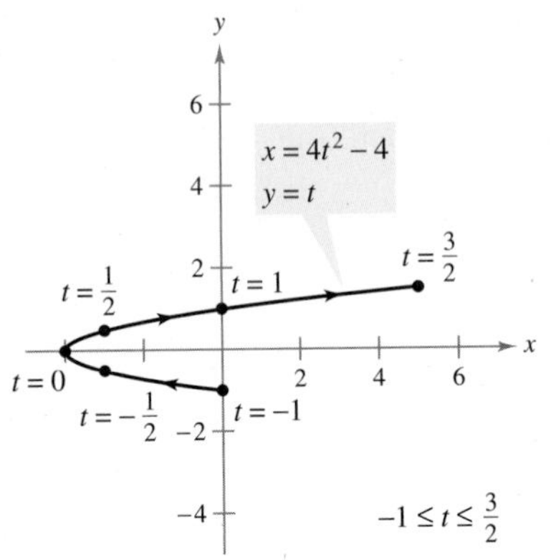

Figure 12.39

Eliminating the Parameter

Example 1 uses simple point plotting to sketch the curve. This tedious process can sometimes be simplified by finding a rectangular equation (in x and y) that has the same graph. This process is called **eliminating the parameter.**

Parametric equations	⇨	Solve for t in one equation.	⇨	Substitute in other equation.	⇨	Rectangular equation
$x = t^2 - 4$ $y = \dfrac{t}{2}$		$t = 2y$		$x = (2y)^2 - 4$		$x = 4y^2 - 4$

Now you can recognize that the equation $x = 4y^2 - 4$ represents a parabola with a horizontal axis and vertex at $(-4, 0)$.

When converting equations from parametric to rectangular form, you may need to alter the domain of the rectangular equation so that its graph matches the graph of the parametric equations. Such a situation is demonstrated in Example 2.

EXAMPLE 2 Eliminating the Parameter

Sketch the curve represented by the equations

$$x = \frac{1}{\sqrt{t+1}} \quad \text{and} \quad y = \frac{t}{t+1}$$

by eliminating the parameter and adjusting the domain of the resulting rectangular equation.

Solution Solving for t in the equation for x produces

$$x = \frac{1}{\sqrt{t+1}} \quad \Rightarrow \quad x^2 = \frac{1}{t+1}$$

which implies that

$$t = \frac{1 - x^2}{x^2}.$$

Now, substituting in the equation for y, you obtain the rectangular equation

$$y = \frac{t}{t+1} = \frac{\dfrac{1-x^2}{x^2}}{\left(\dfrac{1-x^2}{x^2}\right) + 1} = \frac{\dfrac{1-x^2}{x^2}}{\dfrac{1-x^2}{x^2} + 1} \cdot \frac{x^2}{x^2} = 1 - x^2.$$

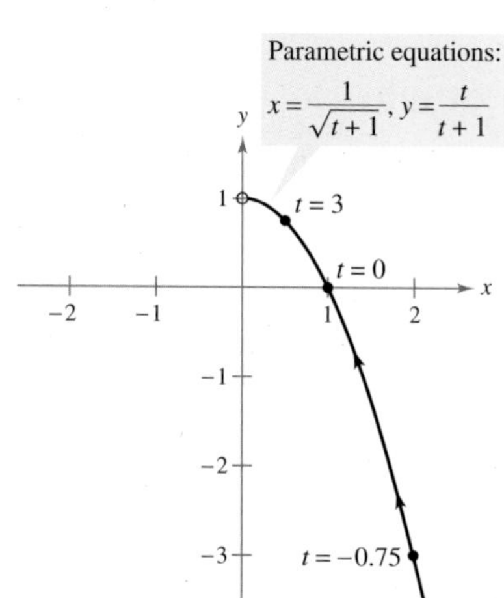

Figure 12.40

From this rectangular equation, you can recognize that the curve is a parabola that opens downward and has its vertex at $(0, 1)$. Also, this rectangular equation is defined for all values of x, but from the parametric equation for x you can see that the curve is defined only when $t > -1$. This implies that you should restrict the domain of x to positive values, as shown in Figure 12.40. ■

It is not necessary for the parameter in a set of parametric equations to represent time. The next example uses an *angle* as the parameter.

EXAMPLE 3 Eliminating an Angle Parameter

Sketch the curve represented by

$$x = 3\cos\theta \quad \text{and} \quad y = 4\sin\theta, \quad 0 \le \theta \le 2\pi$$

by eliminating the parameter.

Solution Begin by solving for $\cos\theta$ and $\sin\theta$ in the equations.

$$\cos\theta = \frac{x}{3} \quad \text{and} \quad \sin\theta = \frac{y}{4}$$ Solve for $\cos\theta$ and $\sin\theta$.

Use the identity $\sin^2\theta + \cos^2\theta = 1$ to form an equation involving only x and y.

$$\cos^2\theta + \sin^2\theta = 1$$ Pythagorean identity

$$\left(\frac{x}{3}\right)^2 + \left(\frac{y}{4}\right)^2 = 1$$ Substitute $\frac{x}{3}$ for $\cos\theta$ and $\frac{y}{4}$ for $\sin\theta$.

$$\frac{x^2}{9} + \frac{y^2}{16} = 1$$ Rectangular equation

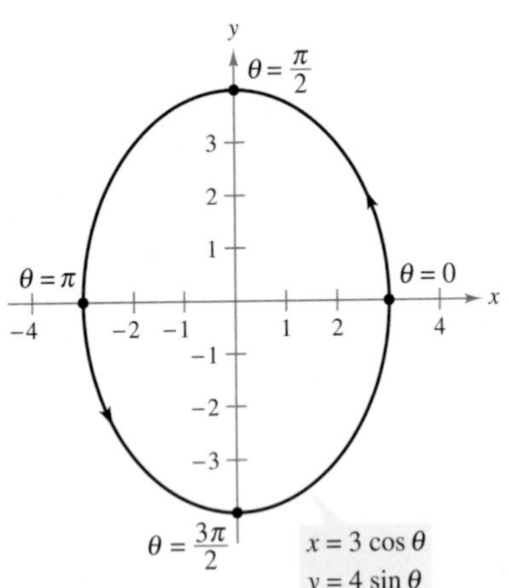

Figure 12.41

From this rectangular equation, you can see that the graph is an ellipse centered at $(0, 0)$, with vertices $(0, 4)$ and $(0, -4)$ and minor axis of length $2b = 6$, as shown in Figure 12.41. Note that the elliptic curve is traced out *counterclockwise* as θ varies from 0 to 2π. ■

STUDY TIP To eliminate the parameter in equations involving trigonometric functions, try using identities such as

$$\sin^2\theta + \cos^2\theta = 1 \quad \text{or} \quad \sec^2\theta - \tan^2\theta = 1$$

as shown in Example 3. ■

In Examples 2 and 3, it is important to realize that eliminating the parameter is primarily an *aid to curve sketching.* If the parametric equations represent the path of a moving object, the graph alone is not sufficient to describe the object's motion. You still need the parametric equations to tell you the *position, direction,* and *speed* at a given time.

Finding Parametric Equations for a Graph

You have been studying techniques for sketching the graph represented by a set of parametric equations. Now consider the *reverse* problem—that is, how can you find a set of parametric equations for a given graph or a given physical description? From the discussion following Example 1, you know that such a representation is not unique. That is, the equations

$$x = 4t^2 - 4 \quad \text{and} \quad y = t, \ -1 \le t \le \frac{3}{2}$$

produced the same graph as the equations

$$x = t^2 - 4 \quad \text{and} \quad y = \frac{t}{2}, \ -2 \le t \le 3.$$

This is further demonstrated in Example 4.

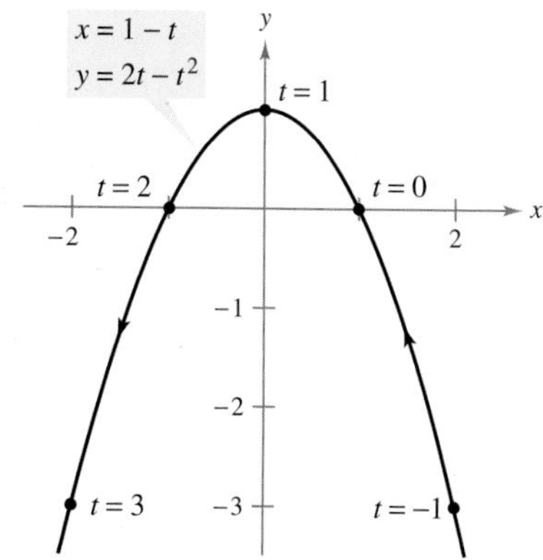

Figure 12.42

EXAMPLE 4 Finding Parametric Equations for a Graph

Find a set of parametric equations to represent the graph of

$$y = 1 - x^2$$

using the following parameters.

a. $t = x$

b. $t = 1 - x$

Solution

a. Letting $t = x$, you obtain the parametric equations

$$x = t \qquad \text{and} \qquad y = 1 - x^2 = 1 - t^2.$$

b. Letting $t = 1 - x$, you obtain the parametric equations

$$x = 1 - t \qquad \text{and} \qquad y = 1 - x^2 = 1 - (1 - t)^2 = 2t - t^2.$$

In Figure 12.42, note how the resulting curve is oriented by the increasing values of t. For part (a), the curve would have the opposite orientation.

EXAMPLE 5 Parametric Equations for a Cycloid

Describe the **cycloid** traced out by a point P on the circumference of a circle of radius a as the circle rolls along a straight line in a plane.

Solution As the parameter, let θ be the measure of the circle's rotation, and let the point $P = (x, y)$ begin at the origin. When $\theta = 0$, P is at the origin; when $\theta = \pi$, P is at a maximum point $(\pi a, 2a)$; and when $\theta = 2\pi$, P is back on the x-axis at $(2\pi a, 0)$. From Figure 12.43, you can see that $\angle APC = 180° - \theta$. So, you have

$$\sin \theta = \sin(180° - \theta) = \sin(\angle APC) = \frac{AC}{a} = \frac{BD}{a}$$

$$\cos \theta = -\cos(180° - \theta) = -\cos(\angle APC) = -\frac{AP}{a}$$

STUDY TIP In Example 5, $\widehat{PD}$ represents the arc of the circle between points P and D.

which implies that $BD = a \sin \theta$ and $AP = -a \cos \theta$. Because the circle rolls along the x-axis, you know that $OD = \widehat{PD} = a\theta$. Furthermore, because $BA = DC = a$, you have

$$x = OD - BD = a\theta - a \sin \theta \qquad \text{and} \qquad y = BA + AP = a - a \cos \theta.$$

So, the parametric equations are $x = a(\theta - \sin \theta)$ and $y = a(1 - \cos \theta)$.

TECHNOLOGY You can use a graphing utility in *parametric* mode to obtain a graph similar to Figure 12.43 by graphing the following equations.

$$X_{1T} = T - \sin T$$

$$Y_{1T} = 1 - \cos T$$

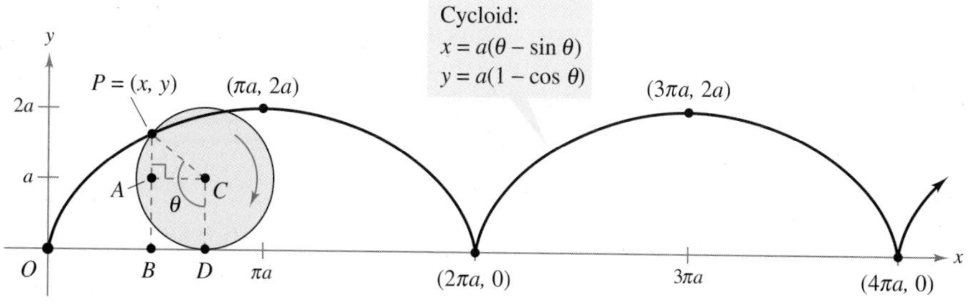

Figure 12.43

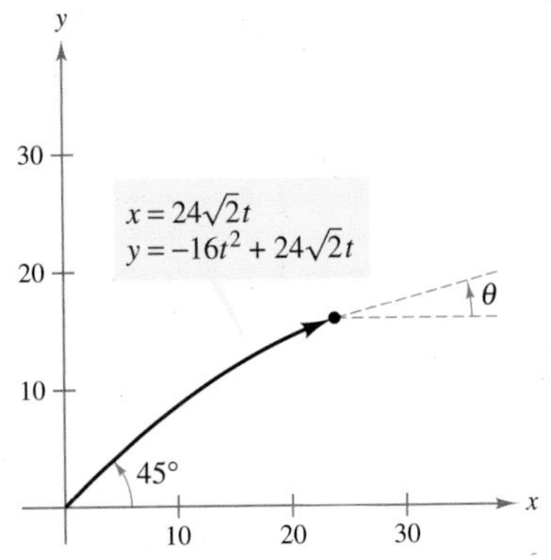

At time t, the angle of elevation of the projectile is θ, the slope of the tangent line at that point.

Figure 12.44

Slope and Tangent Lines

Now that you can represent a graph in the plane by a set of parametric equations, it is natural to ask how to use calculus to study plane curves. To begin, let's take another look at the projectile represented by the parametric equations

$$x = 24\sqrt{2}t \quad \text{and} \quad y = -16t^2 + 24\sqrt{2}t$$

as shown in Figure 12.44. You know that these equations enable you to locate the position of the projectile at a given time. You also know that the object is initially projected at an angle of 45°. But how can you find the angle θ representing the object's direction at some other time t? The following theorem answers this question by giving a formula for the slope of the tangent line as a function of t.

THEOREM 12.6 PARAMETRIC FORM OF THE DERIVATIVE

If a smooth curve C is given by the continuous functions $x = f(t)$ and $y = g(t)$, then the slope of C at (x, y) is

$$\frac{dy}{dx} = \frac{dy/dt}{dx/dt}, \quad \frac{dx}{dt} \neq 0.$$

Because dy/dx is a function of t, you can use Theorem 12.6 repeatedly to find *higher-order* derivatives. For instance,

$$\frac{d^2y}{dx^2} = \frac{d}{dx}\left[\frac{dy}{dx}\right] = \frac{\frac{d}{dt}\left[\frac{dy}{dx}\right]}{dx/dt}. \qquad \text{Second derivative}$$

EXAMPLE 6 Finding Slope and Concavity

For the curve given by

$$x = \sqrt{t} \quad \text{and} \quad y = \frac{1}{4}(t^2 - 4), \quad t \geq 0$$

find the slope and concavity at the point (2, 3).

Solution Because

$$\frac{dy}{dx} = \frac{dy/dt}{dx/dt} = \frac{(1/2)t}{(1/2)t^{-1/2}} = t^{3/2} \qquad \text{Parametric form of first derivative}$$

you can find the second derivative to be

$$\frac{d^2y}{dx^2} = \frac{\frac{d}{dt}\left[\frac{dy}{dx}\right]}{dx/dt} = \frac{\frac{d}{dt}\left[t^{3/2}\right]}{dx/dt} = \frac{(3/2)t^{1/2}}{(1/2)t^{-1/2}} = 3t. \qquad \text{Parametric form of second derivative}$$

At $(x, y) = (2, 3)$, it follows that $t = 4$, and the slope is $\frac{dy}{dx} = (4)^{3/2} = 8$. Moreover, when $t = 4$, the second derivative is $\frac{d^2y}{dx^2} = 3(4) = 12 > 0$ and you can conclude that the graph is concave upward at (2, 3), as shown in Figure 12.45. ■

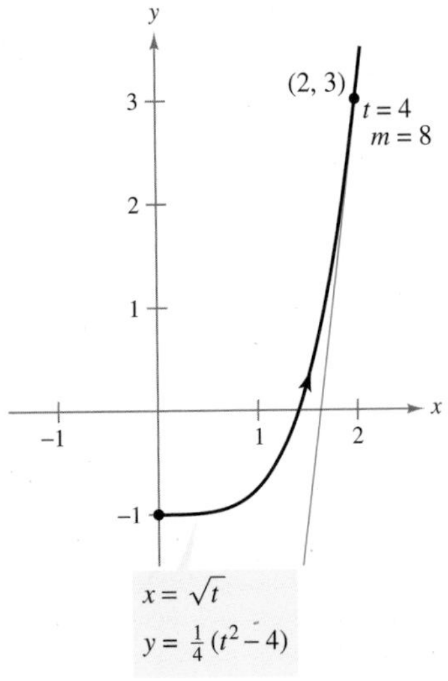

$x = \sqrt{t}$
$y = \frac{1}{4}(t^2 - 4)$

(2, 3) $t = 4$ $m = 8$

The graph is concave upward at (2, 3), when $t = 4$.

Figure 12.45

EXAMPLE 7 Finding Horizontal and Vertical Tangent Lines

For the curve given by $x = \sin t$ and $y = \cos t$, find all horizontal and vertical tangent lines on the interval $0 \le t < 2\pi$.

Solution Because

$$\frac{dy}{dx} = \frac{dy/dt}{dx/dt} = \frac{-\sin t}{\cos t} = -\tan t$$

the horizontal tangent lines occur when $\tan t = 0$. So, the given curve has a horizontal tangent line at the point $(0, 1)$ when $t = 0$ and at the point $(0, -1)$ when $t = \pi$. The vertical tangent lines occur when $\tan t$ is undefined. So, the given curve has a vertical tangent line at the point $(1, 0)$ when $t = \pi/2$ and at the point $(-1, 0)$ when $t = 3\pi/2$. ■

12.4 Exercises

See www.CalcChat.com for worked-out solutions to odd-numbered exercises.

In Exercises 1–4, fill in the blanks.

1. If f and g are continuous functions of t on an interval I, the set of ordered pairs $(f(t), g(t))$ is a ________ ________ C.
2. The ________ of a curve is the direction in which the curve is traced out for increasing values of the parameter.
3. The process of converting a set of parametric equations to a corresponding rectangular equation is called ________ the ________.
4. A curve traced by a point on the circumference of a circle as the circle rolls along a straight line in a plane is called a ________.
5. Consider the parametric equations $x = \sqrt{t}$ and $y = 3 - t$.
   (a) Create a table of x- and y-values using $t = 0, 1, 2, 3,$ and 4.
   (b) Plot the points (x, y) generated in part (a), and sketch a graph of the parametric equations.
   (c) Find the rectangular equation by eliminating the parameter. Sketch its graph. How do the graphs differ?
6. Consider the parametric equations $x = 4\cos^2\theta$ and $y = 2\sin\theta$.
   (a) Create a table of x- and y-values using $\theta = -\pi/2, -\pi/4, 0, \pi/4,$ and $\pi/2$.
   (b) Plot the points (x, y) generated in part (a), and sketch a graph of the parametric equations.
   (c) Find the rectangular equation by eliminating the parameter. Sketch its graph. How do the graphs differ?

In Exercises 7–26, (a) sketch the curve represented by the parametric equations (indicate the orientation of the curve) and (b) eliminate the parameter and write the corresponding rectangular equation whose graph represents the curve. Adjust the domain of the resulting rectangular equation if necessary.

7. $x = t - 1$, $y = 3t + 1$
8. $x = 3 - 2t$, $y = 2 + 3t$
9. $x = \frac{1}{4}t$, $y = t^2$
10. $x = t$, $y = t^3$
11. $x = t + 2$, $y = t^2$
12. $x = \sqrt{t}$, $y = 1 - t$
13. $x = t + 1$, $y = \dfrac{t}{t + 1}$
14. $x = t - 1$, $y = \dfrac{t}{t - 1}$
15. $x = 2(t + 1)$, $y = |t - 2|$
16. $x = |t - 1|$, $y = t + 2$
17. $x = 4\cos\theta$, $y = 2\sin\theta$
18. $x = 2\cos\theta$, $y = 3\sin\theta$
19. $x = 6\sin 2\theta$, $y = 6\cos 2\theta$
20. $x = \cos\theta$, $y = 2\sin 2\theta$
21. $x = 1 + \cos\theta$, $y = 1 + 2\sin\theta$
22. $x = 2 + 5\cos\theta$, $y = -6 + 4\sin\theta$
23. $x = e^{-t}$, $y = e^{3t}$
24. $x = e^{2t}$, $y = e^{t}$
25. $x = t^3$, $y = 3\ln t$
26. $x = \ln 2t$, $y = 2t^2$

In Exercises 27 and 28, determine how the plane curves differ from each other.

27. (a) $x = t$, $y = 2t + 1$
    (b) $x = \cos\theta$, $y = 2\cos\theta + 1$
    (c) $x = e^{-t}$, $y = 2e^{-t} + 1$
    (d) $x = e^{t}$, $y = 2e^{t} + 1$

28. (a) $x = t$
$y = t^2 - 1$
(b) $x = t^2$
$y = t^4 - 1$
(c) $x = \sin t$
$y = \sin^2 t - 1$
(d) $x = e^t$
$y = e^{2t} - 1$

In Exercises 29–32, eliminate the parameter and obtain the standard form of the rectangular equation.

29. Line: $x = x_1 + t(x_2 - x_1),\ y = y_1 + t(y_2 - y_1)$

30. Circle: $x = h + r\cos\theta,\ y = k + r\sin\theta$

31. Ellipse: $x = h + a\cos\theta,\ y = k + b\sin\theta$

32. Hyperbola: $x = h + a\sec\theta,\ y = k + b\tan\theta$

In Exercises 33–40, use the results of Exercises 29–32 to find a set of parametric equations for the line or conic.

33. Line: passes through $(0, 0)$ and $(3, 6)$

34. Line: passes through $(3, 2)$ and $(-6, 3)$

35. Circle: center: $(3, 2)$; radius: 4

36. Circle: center: $(5, -3)$; radius: 4

37. Ellipse: vertices: $(\pm 5, 0)$; foci: $(\pm 4, 0)$

38. Ellipse: vertices: $(3, 7), (3, -1)$; foci: $(3, 5), (3, 1)$

39. Hyperbola: vertices: $(\pm 4, 0)$; foci: $(\pm 5, 0)$

40. Hyperbola: vertices: $(\pm 2, 0)$; foci: $(\pm 4, 0)$

In Exercises 41–48, find a set of parametric equations for the rectangular equation using (a) $t = x$ and (b) $t = 2 - x$.

41. $y = 3x - 2$

42. $x = 3y - 2$

43. $y = 2 - x$

44. $y = x^2 + 1$

45. $y = x^2 - 3$

46. $y = 1 - 2x^2$

47. $y = \dfrac{1}{x}$

48. $y = \dfrac{1}{2x}$

49. The graph of the parametric equations $x = 2\sec t$ and $y = 3\tan t$ is given in the figure. Would the graph change for the equations $x = 2\sec(-t)$ and $y = 3\tan(-t)$? If so, how would it change?

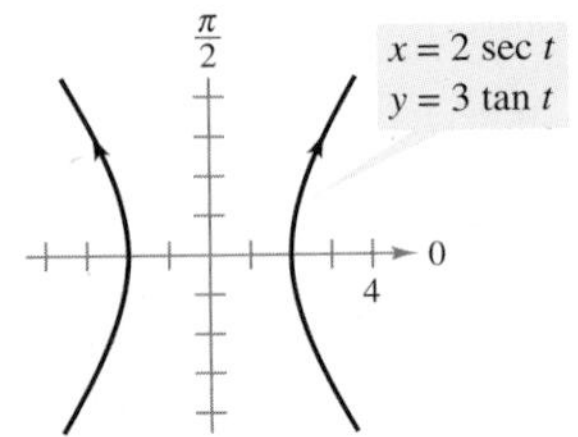

50. A moving object is modeled by the parametric equations $x = 4\cos t$ and $y = 3\sin t$, where t is time (see figure). How would the orbit change for the following?

(a) $x = 4\cos 2t,\quad y = 3\sin 2t$

(b) $x = 5\cos t,\quad y = 3\sin t$

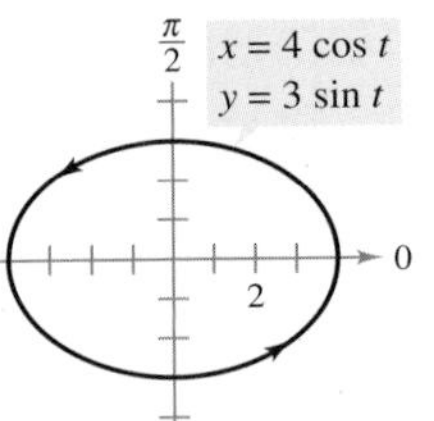

Figure for 50

In Exercises 51–58, use a graphing utility to graph the curve represented by the parametric equations.

51. Cycloid: $x = 4(\theta - \sin\theta),\ y = 4(1 - \cos\theta)$

52. Cycloid: $x = \theta + \sin\theta,\ y = 1 - \cos\theta$

53. Prolate cycloid: $x = \theta - \frac{3}{2}\sin\theta,\ y = 1 - \frac{3}{2}\cos\theta$

54. Prolate cycloid: $x = 2\theta - 4\sin\theta,\ y = 2 - 4\cos\theta$

55. Hypocycloid: $x = 3\cos^3\theta,\ y = 3\sin^3\theta$

56. Curtate cycloid: $x = 8\theta - 4\sin\theta,\ y = 8 - 4\cos\theta$

57. Witch of Agnesi: $x = 2\cot\theta,\ y = 2\sin^2\theta$

58. Folium of Descartes: $x = \dfrac{3t}{1 + t^3},\ y = \dfrac{3t^2}{1 + t^3}$

In Exercises 59–62, match the parametric equations with the correct graph and describe the domain and range. [The graphs are labeled (a), (b), (c), and (d).]

(a)

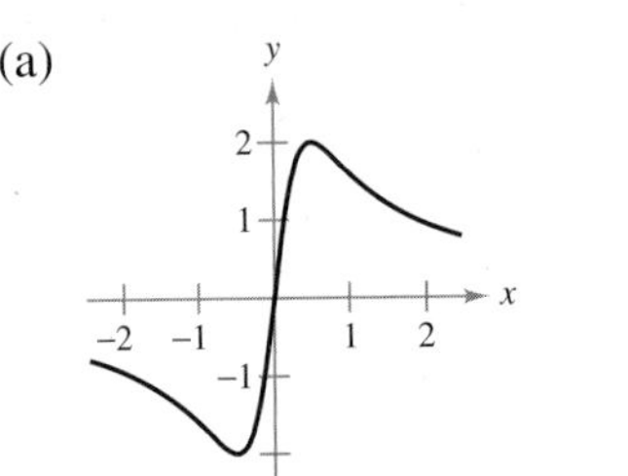

(b)

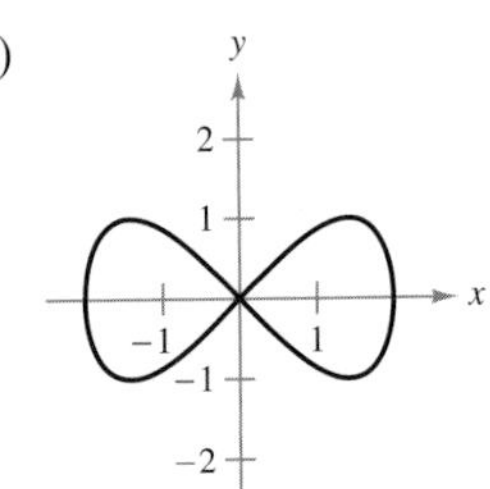

(c)

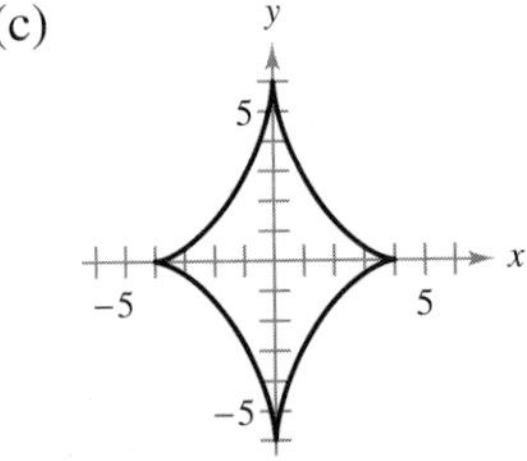

(d)

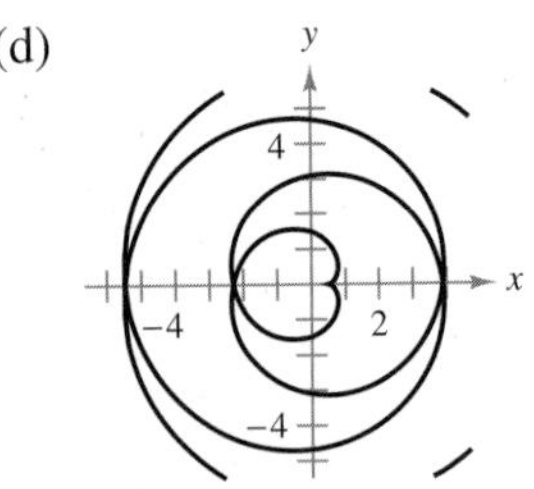

59. Lissajous curve: $x = 2\cos\theta,\ y = \sin 2\theta$

60. Evolute of ellipse: $x = 4\cos^3\theta,\ y = 6\sin^3\theta$

61. Involute of circle: $x = \frac{1}{2}(\cos\theta + \theta\sin\theta)$
$y = \frac{1}{2}(\sin\theta - \theta\cos\theta)$

62. Serpentine curve: $x = \frac{1}{2}\cot\theta,\ y = 4\sin\theta\cos\theta$

63. ***Writing*** Review Exercises 27 and 28 and write a short paragraph describing how the graphs of curves represented by different sets of parametric equations can differ even though eliminating the parameter from each yields the same rectangular equation.

64. *Conjecture*

(a) Use a graphing utility to graph the curves represented by the two sets of parametric equations.

$x = 4 \cos t$ $\qquad$ $x = 4 \cos(-t)$

$y = 3 \sin t$ $\qquad$ $y = 3 \sin(-t)$

(b) Describe the change in the graph when the sign of the parameter is changed.

(c) Make a conjecture about the change in the graph of parametric equations when the sign of the parameter is changed.

(d) Test your conjecture with another set of parametric equations.

WRITING ABOUT CONCEPTS

65. State the definition of a plane curve given by parametric equations.

66. Explain the process of sketching a plane curve given by parametric equations. What is meant by the orientation of the curve?

Projectile Motion **A projectile is launched at a height of h feet above the ground at an angle of θ with the horizontal. The initial velocity is v_0 feet per second, and the path of the projectile is modeled by the parametric equations**

$$x = (v_0 \cos \theta)t \quad \text{and} \quad y = h + (v_0 \sin \theta)t - 16t^2.$$

In Exercises 67 and 68, use a graphing utility to graph the paths of a projectile launched from ground level at each value of θ and v_0. For each case, use the graph to approximate the maximum height and the range of the projectile.

67. (a) $\theta = 60°$, $v_0 = 88$ feet per second
(b) $\theta = 60°$, $v_0 = 132$ feet per second
(c) $\theta = 45°$, $v_0 = 88$ feet per second
(d) $\theta = 45°$, $v_0 = 132$ feet per second

68. (a) $\theta = 15°$, $v_0 = 50$ feet per second
(b) $\theta = 15°$, $v_0 = 120$ feet per second
(c) $\theta = 10°$, $v_0 = 50$ feet per second
(d) $\theta = 10°$, $v_0 = 120$ feet per second

In Exercises 69–78, given that

$$\frac{dy}{dx} = \frac{dy/dt}{dx/dt}, \quad \frac{dx}{dt} \neq 0$$

(a) find dy/dx using this formula.

(b) Eliminate the parameter and find dy/dx. Then compare your result with that of part (a).

69. $x = 2t$, $y = 3t - 1$

70. $x = \sqrt{t}$, $y = 3t - 1$

71. $x = t + 1$, $y = t^2 + 3t$

72. $x = t^2 + 3t + 2$, $y = 2t$

73. $x = 2 \cos t$, $y = 2 \sin t$

74. $x = \cos t$, $y = 3 \sin t$

75. $x = 2 + \sec t$, $y = 1 + 2 \tan t$

76. $x = \sqrt{t}$, $y = \sqrt{t - 1}$

77. $x = \cos^3 t$, $y = \sin^3 t$

78. $x = t - \sin t$, $y = 1 - \cos t$

In Exercises 79–82, find dy/dx.

79. $x = t^2, \quad y = 5 - 4t$

80. $x = \sqrt[3]{t}, \quad y = 4 - t$

81. $x = \sin^2 \theta, \quad y = \cos^2 \theta$

82. $x = 2e^{\theta}, \quad y = e^{-\theta/2}$

In Exercises 83–92, find dy/dx and d^2y/dx^2, and find the slope and concavity (if possible) at the given value of the parameter.

	Parametric Equations	*Point*
83.	$x = 2t, \ y = 3t - 1$	$t = 3$
84.	$x = \sqrt{t}, \ y = 3t - 1$	$t = 1$
85.	$x = t + 1, \ y = t^2 + 3t$	$t = -1$
86.	$x = t^2 + 3t + 2, \quad y = 2t$	$t = 0$
87.	$x = 2 \cos \theta, \quad y = 2 \sin \theta$	$\theta = \frac{\pi}{4}$
88.	$x = \cos \theta, \quad y = 3 \sin \theta$	$\theta = 0$
89.	$x = 2 + \sec \theta, \quad y = 1 + 2 \tan \theta$	$\theta = \frac{\pi}{6}$
90.	$x = \sqrt{t}, \quad y = \sqrt{t - 1}$	$t = 2$
91.	$x = \cos^3 \theta, \quad y = \sin^3 \theta$	$\theta = \frac{\pi}{4}$
92.	$x = \theta - \sin \theta, \quad y = 1 - \cos \theta$	$\theta = \pi$

In Exercises 93–102, find all points (if any) of horizontal and vertical tangency to the curve. Use a graphing utility to confirm your results.

93. $x = 1 - t, \quad y = t^2$

94. $x = t + 1, \quad y = t^2 + 3t$

95. $x = 1 - t, \quad y = t^3 - 3t$

96. $x = t^2 - t + 2, \quad y = t^3 - 3t$

97. $x = 3 \cos \theta, \quad y = 3 \sin \theta$

98. $x = \cos \theta, \quad y = 2 \sin 2\theta$

99. $x = 4 + 2 \cos \theta, \quad y = -1 + \sin \theta$

100. $x = 4 \cos^2 \theta, \quad y = 2 \sin \theta$

101. $x = \sec \theta, \quad y = \tan \theta$

102. $x = \cos^2 \theta, \quad y = \cos \theta$

103. ***Sports*** The center field fence in Yankee Stadium is 7 feet high and 408 feet from home plate. A baseball is hit at a point 3 feet above the ground. It leaves the bat at an angle of θ degrees with the horizontal at a speed of 100 miles per hour (see figure).

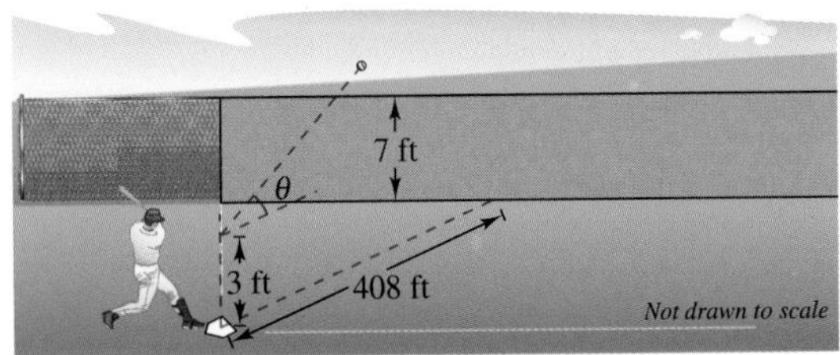

(a) Write a set of parametric equations that model the path of the baseball.

(b) Use a graphing utility to graph the path of the baseball when $\theta = 15°$. Is the hit a home run?

(c) Use the graphing utility to graph the path of the baseball when $\theta = 23°$. Is the hit a home run?

(d) Find the minimum angle required for the hit to be a home run.

104. ***Sports*** An archer releases an arrow from a bow at a point 5 feet above the ground. The arrow leaves the bow at an angle of 15° with the horizontal and at an initial speed of 225 feet per second.

(a) Write a set of parametric equations that model the path of the arrow.

(b) Assuming the ground is level, find the distance the arrow travels before it hits the ground. (Ignore air resistance.)

(c) Use a graphing utility to graph the path of the arrow and approximate its maximum height. Verify your result analytically.

(d) Find the total time the arrow is in the air.

105. ***Projectile Motion*** Eliminate the parameter t from the parametric equations $x = (v_0 \cos \theta)t$ and $y = h + (v_0 \sin \theta)t - 16t^2$ for the motion of a projectile to show that the rectangular equation is

$$y = -\frac{16 \sec^2 \theta}{v_0^2}x^2 + (\tan \theta)x + h.$$

106. ***Path of a Projectile*** The path of a projectile is given by the rectangular equation $y = 7 + x - 0.02x^2$.

(a) Use the result of Exercise 105 to find h, v_0, and θ. Find the parametric equations of the path.

(b) Use a graphing utility to graph the rectangular equation for the path of the projectile. Confirm your answer in part (a) by sketching the curve represented by the parametric equations.

(c) Use the graphing utility to approximate the maximum height of the projectile and its range.

107. ***Curtate Cycloid*** A wheel of radius a units rolls along a straight line without slipping. The curve traced by a point P that is b units from the center $(b < a)$ is called a **curtate cycloid** (see figure). Use the angle θ shown in the figure to find a set of parametric equations for the curve.

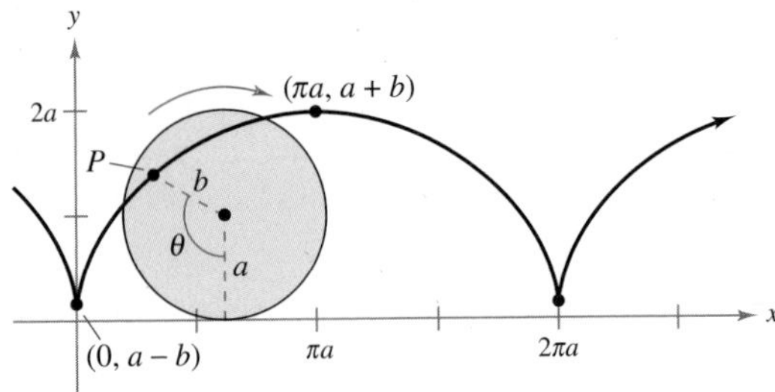

108. ***Epicycloid*** A circle of radius one unit rolls around the outside of a circle of radius two units without slipping. The curve traced by a point on the circumference of the smaller circle is called an **epicycloid** (see figure). Use the angle θ shown in the figure to find a set of parametric equations for the curve.

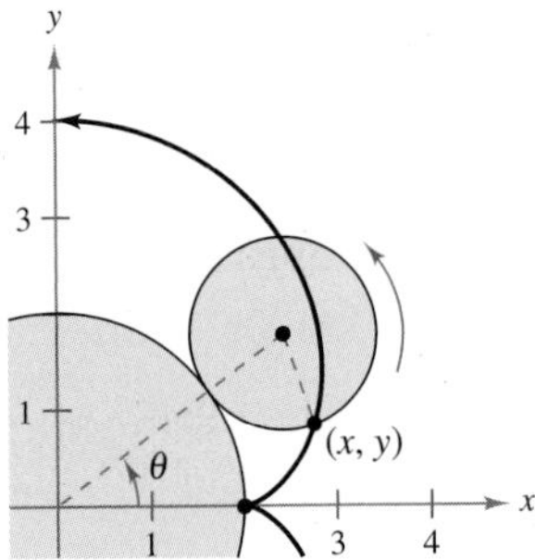

True or False? **In Exercises 109 and 110, determine whether the statement is true or false. Justify your answer.**

109. The two sets of parametric equations $x = t$, $y = t^2 + 1$ and $x = 3t$, $y = 9t^2 + 1$ have the same rectangular equation.

110. If y is a function of t and x is a function of t, then y must be a function of x.

111. Use a graphing utility set in *parametric* mode to enter the parametric equations from Example 2. Over what values should you let t vary to obtain the graph shown in Figure 12.40?

CAPSTONE

112. (a) Describe the curve represented by the parametric equations $x = 8 \cos t$ and $y = 8 \sin t$.

(b) How does the curve represented by the parametric equations $x = 8 \cos t + 3$ and $y = 8 \sin t + 6$ compare with the curve described in part (a)?

(c) How does the original curve change when cosine and sine are interchanged?

12.5 Polar Coordinates and Calculus

- Understand the polar coordinate system.
- Rewrite rectangular coordinates and equations in polar form and vice versa.
- Find the slope of a tangent line to a polar graph.

Polar Coordinates

So far, you have been representing graphs as collections of points (x, y) on the rectangular coordinate system. The corresponding equations for these graphs have been in either rectangular or parametric form. In this section you will study a coordinate system called the **polar coordinate system.**

To form the polar coordinate system in the plane, fix a point O, called the **pole** (or **origin**), and construct from O an initial ray called the **polar axis,** as shown in Figure 12.46. Then each point P in the plane can be assigned **polar coordinates** (r, θ), as follows.

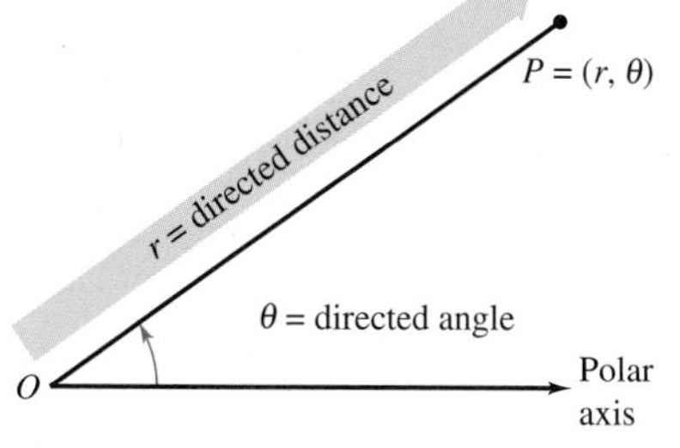

Polar coordinates
Figure 12.46

$r =$ *directed distance* from O to P

$\theta =$ *directed angle*, counterclockwise from polar axis to segment $\overline{OP}$

Figure 12.47 shows three points on the polar coordinate system. Notice that in this system, it is convenient to locate points with respect to a grid of concentric circles intersected by **radial lines** through the pole.

POLAR COORDINATES

The mathematician credited with first using polar coordinates was James Bernoulli, who introduced them in 1691. However, there is some evidence that it may have been Isaac Newton who first used them.

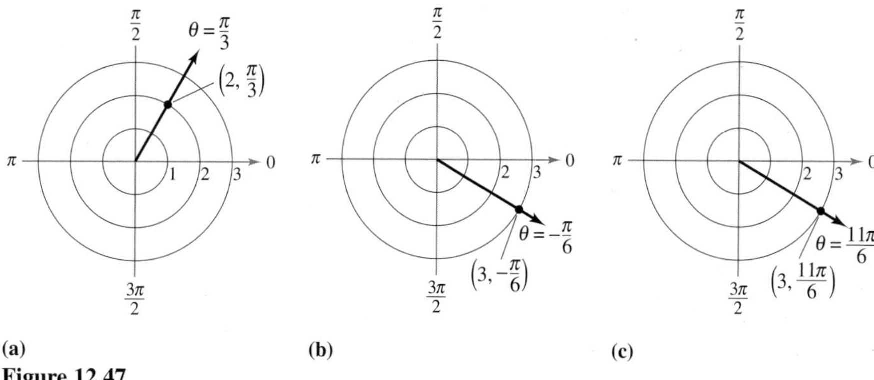

Figure 12.47

With rectangular coordinates, each point (x, y) has a unique representation. This is not true with polar coordinates. For instance, the coordinates (r, θ) and $(r, \theta + 2\pi)$ represent the same point [see parts (b) and (c) in Figure 12.47]. Also, because r is a *directed distance,* the coordinates (r, θ) and $(-r, \theta + \pi)$ represent the same point. In general, the point (r, θ) can be written as

$$(r, \theta) = (r, \theta + 2n\pi) \qquad \text{or} \qquad (r, \theta) = (-r, \theta + (2n + 1)\pi)$$

where n is any integer. Moreover, the pole is represented by $(0, \theta)$, where θ is any angle.

Coordinate Conversion

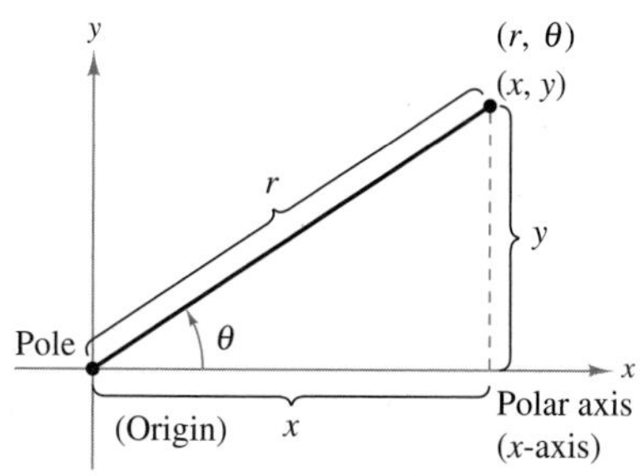

Relating polar and rectangular coordinates
Figure 12.48

To establish the relationship between polar and rectangular coordinates, let the polar axis coincide with the positive x-axis and the pole with the origin, as shown in Figure 12.48. Because (x, y) lies on a circle of radius r, it follows that $r^2 = x^2 + y^2$. Moreover, for $r > 0$, the definitions of the trigonometric functions imply that $\tan \theta = y/x$, $\cos \theta = x/r$, and $\sin \theta = y/r$. If $r < 0$, you can show that the same relationships hold.

THEOREM 12.7 COORDINATE CONVERSION

The polar coordinates (r, θ) of a point are related to the rectangular coordinates (x, y) of the point as follows.

1. $x = r\cos\theta$ **2.** $\tan\theta = \dfrac{y}{x}$

$y = r\sin\theta$ $r^2 = x^2 + y^2$

EXAMPLE 1 Polar-to-Rectangular Conversion

Convert each point from polar coordinates to rectangular coordinates.

a. $(2, \pi)$ **b.** $\left(\sqrt{3}, \dfrac{\pi}{6}\right)$

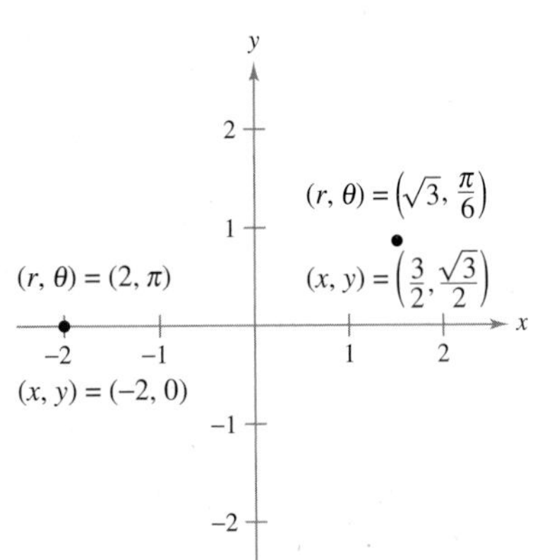

To convert from polar to rectangular coordinates, let $x = r\cos\theta$ and $y = r\sin\theta$.
Figure 12.49

Solution

a. For the point $(r, \theta) = (2, \pi)$,

$$x = r\cos\theta = 2\cos\pi = -2 \quad \text{and} \quad y = r\sin\theta = 2\sin\pi = 0.$$

So, the rectangular coordinates are $(x, y) = (-2, 0)$. See Figure 12.49.

b. For the point $(r, \theta) = \left(\sqrt{3}, \dfrac{\pi}{6}\right)$,

$$x = \sqrt{3}\cos\frac{\pi}{6} = \frac{3}{2} \quad \text{and} \quad y = \sqrt{3}\sin\frac{\pi}{6} = \frac{\sqrt{3}}{2}.$$

So, the rectangular coordinates are $(x, y) = \left(\dfrac{3}{2}, \dfrac{\sqrt{3}}{2}\right)$. See Figure 12.49.

EXAMPLE 2 Rectangular-to-Polar Conversion

Convert each point from rectangular coordinates to polar coordinates.

a. $(-1, 1)$

b. $(0, 2)$

Solution

a. For the second quadrant point $(x, y) = (-1, 1)$,

$$\tan\theta = \frac{y}{x} = -1 \quad \Longrightarrow \quad \theta = \frac{3\pi}{4}.$$

Because θ was chosen to be in the same quadrant as (x, y), you should use a positive value of r.

$$\begin{aligned} r &= \sqrt{x^2 + y^2} \\ &= \sqrt{(-1)^2 + (1)^2} \\ &= \sqrt{2} \end{aligned}$$

This implies that *one* set of polar coordinates is $(r, \theta) = \left(\sqrt{2}, 3\pi/4\right)$. See Figure 12.50.

b. Because the point $(x, y) = (0, 2)$ lies on the positive y-axis, choose $\theta = \pi/2$ and $r = 2$, and *one* set of polar coordinates is $(r, \theta) = (2, \pi/2)$. See Figure 12.50.

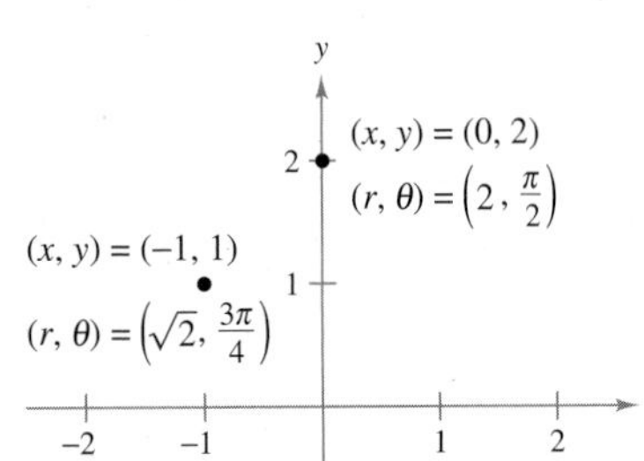

To convert from rectangular to polar coordinates, let $\tan\theta = y/x$ and $r = \sqrt{x^2 + y^2}$.
Figure 12.50

Equation Conversion

By comparing Examples 1 and 2, you can see that point conversion from the polar to the rectangular system is straightforward, whereas point conversion from the rectangular to the polar system is more involved. For equations, the opposite is true. To convert a rectangular equation to polar form, you simply replace x by $r\cos\theta$ and y by $r\sin\theta$. For instance, the rectangular equation $y = x^2$ can be written in polar form as follows.

$$y = x^2 \qquad \text{Rectangular equation}$$
$$r\sin\theta = (r\cos\theta)^2 \qquad \text{Polar equation}$$
$$r = \sec\theta\tan\theta \qquad \text{Simplest form}$$

On the other hand, converting a polar equation to rectangular form requires considerable ingenuity.

Example 3 demonstrates several polar-to-rectangular conversions that enable you to sketch the graphs of some polar equations.

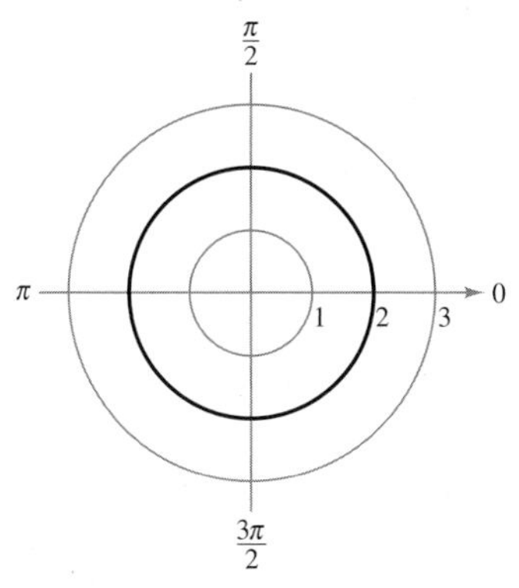

Figure 12.51

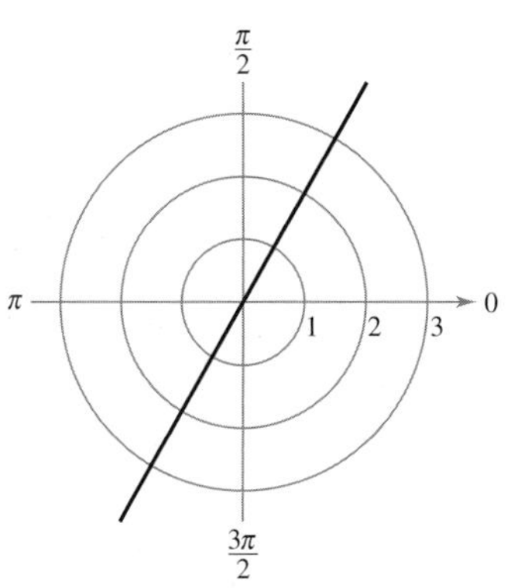

Figure 12.52

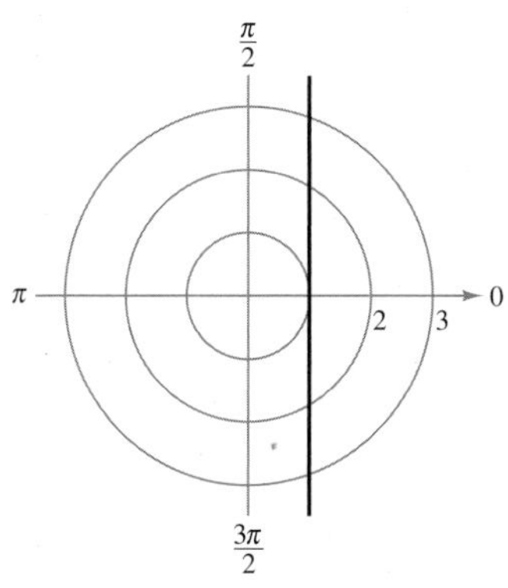

Figure 12.53

EXAMPLE 3 Converting Polar Equations to Rectangular Form

Describe the graph of each polar equation and find the corresponding rectangular equation.

a. $r = 2$

b. $\theta = \dfrac{\pi}{3}$

c. $r = \sec\theta$

Solution

a. The graph of the polar equation $r = 2$ consists of all points that are two units from the pole. In other words, this graph is a circle centered at the origin with a radius of 2, as shown in Figure 12.51. You can confirm this by converting to rectangular form, using the relationship $r^2 = x^2 + y^2$.

$$\underbrace{r = 2}_{\text{Polar equation}} \quad \Longrightarrow \quad r^2 = 2^2 \quad \Longrightarrow \quad \underbrace{x^2 + y^2 = 2^2}_{\text{Rectangular equation}}$$

b. The graph of the polar equation $\theta = \pi/3$ consists of all points on the line that makes an angle of $\pi/3$ with the positive polar axis, as shown in Figure 12.52. To convert to rectangular form, make use of the relationship $\tan\theta = y/x$.

$$\underbrace{\theta = \frac{\pi}{3}}_{\text{Polar equation}} \quad \Longrightarrow \quad \tan\theta = \sqrt{3} \quad \Longrightarrow \quad \underbrace{y = \sqrt{3}x}_{\text{Rectangular equation}}$$

c. The graph of the polar equation $r = \sec\theta$ is not evident by simple inspection, so convert to rectangular form by using the relationship $r\cos\theta = x$.

$$\underbrace{r = \sec\theta}_{\text{Polar equation}} \quad \Longrightarrow \quad r\cos\theta = 1 \quad \Longrightarrow \quad \underbrace{x = 1}_{\text{Rectangular equation}}$$

Now you see that the graph is a vertical line, as shown in Figure 12.53. ■

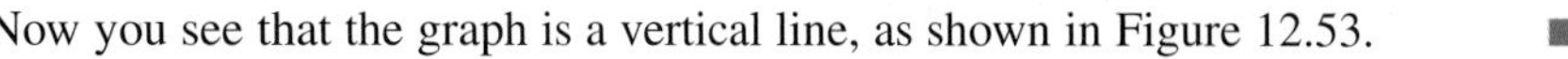

Slope and Tangent Lines

To find the slope of a tangent line to a polar graph, consider a differentiable function given by $r = f(\theta)$. To find the slope in polar form, use the parametric equations

$$x = r\cos\theta = f(\theta)\cos\theta \quad \text{and} \quad y = r\sin\theta = f(\theta)\sin\theta.$$

Using the parametric form of dy/dx given in Theorem 12.6, you have

$$\begin{aligned}\frac{dy}{dx} &= \frac{dy/d\theta}{dx/d\theta}\\ &= \frac{f(\theta)\cos\theta + f'(\theta)\sin\theta}{-f(\theta)\sin\theta + f'(\theta)\cos\theta}\end{aligned}$$

which establishes the following theorem.

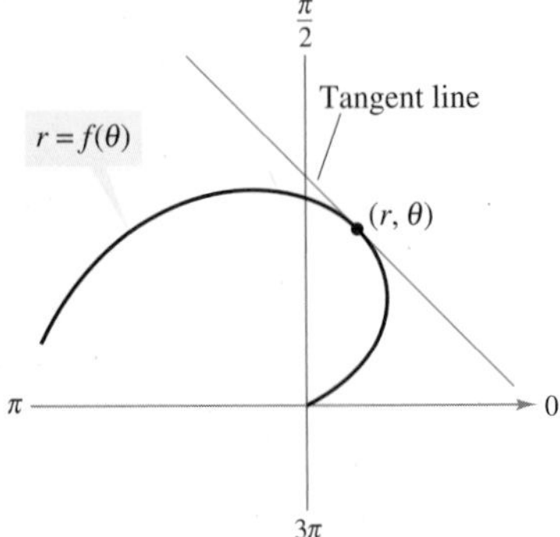

Tangent line to polar curve
Figure 12.54

THEOREM 12.8 SLOPE IN POLAR FORM

If f is a differentiable function of θ, then the *slope* of the tangent line to the graph of $r = f(\theta)$ at the point (r, θ) is given by

$$\frac{dy}{dx} = \frac{dy/d\theta}{dx/d\theta} = \frac{f(\theta)\cos\theta + f'(\theta)\sin\theta}{-f(\theta)\sin\theta + f'(\theta)\cos\theta}$$

provided that $dx/d\theta \neq 0$ at (r, θ). See Figure 12.54.

From Theorem 12.8, you can make the following observations.

1. Solutions to $\dfrac{dy}{d\theta} = 0$ yield horizontal tangents, provided that $\dfrac{dx}{d\theta} \neq 0$.
2. Solutions to $\dfrac{dx}{d\theta} = 0$ yield vertical tangents, provided that $\dfrac{dy}{d\theta} \neq 0$.

If $dy/d\theta$ and $dx/d\theta$ are *simultaneously* 0, no conclusion can be drawn about tangent lines.

EXAMPLE 4 Finding Horizontal and Vertical Tangent Lines

Find the horizontal and vertical tangent lines of $r = \sin\theta$, $0 \le \theta \le \pi$.

Solution Begin by writing the equation in parametric form.

$$x = r\cos\theta = \sin\theta\cos\theta$$

and

$$y = r\sin\theta = \sin\theta\sin\theta = \sin^2\theta$$

Next, differentiate x and y with respect to θ and set each derivative equal to 0.

$$\frac{dx}{d\theta} = \cos^2\theta - \sin^2\theta = \cos 2\theta = 0 \quad \Longrightarrow \quad \theta = \frac{\pi}{4}, \frac{3\pi}{4}$$

$$\frac{dy}{d\theta} = 2\sin\theta\cos\theta = \sin 2\theta = 0 \quad \Longrightarrow \quad \theta = 0, \frac{\pi}{2}$$

So, the graph has vertical tangent lines at $(\sqrt{2}/2, \pi/4)$ and $(\sqrt{2}/2, 3\pi/4)$, and it has horizontal tangent lines at $(0, 0)$ and $(1, \pi/2)$, as shown in Figure 12.55. ■

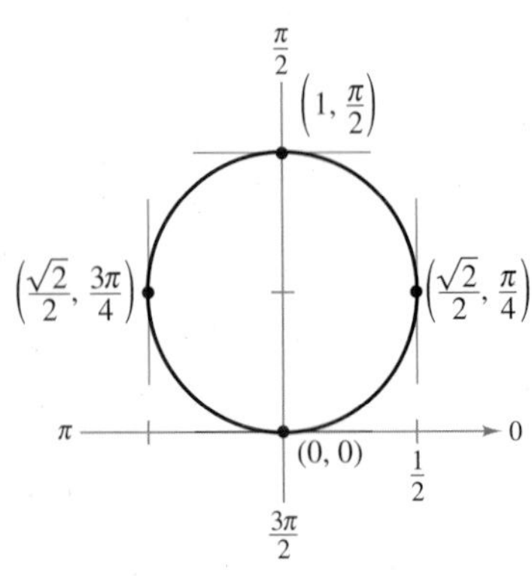

Horizontal and vertical tangent lines of $r = \sin\theta$
Figure 12.55

12.5 Exercises

See www.CalcChat.com for worked-out solutions to odd-numbered exercises.

In Exercises 1–4, fill in the blanks.

1. The origin of the polar coordinate system is called the _______.
2. For the point (r, θ), r is the _______ _______ from O to P and θ is the _______ _______, counterclockwise from the polar axis to the line segment $\overline{OP}$.
3. To plot the point (r, θ), use the _______ coordinate system.
4. The polar coordinates (r, θ) are related to the rectangular coordinates (x, y) as follows:

$x =$ _______ $y =$ _______

$\tan \theta =$ _______ $r^2 =$ _______

In Exercises 5–18, plot the point given in polar coordinates and find two additional polar representations of the point, using $-2\pi < \theta < 2\pi$.

5. $\left(2, \dfrac{5\pi}{6}\right)$
6. $\left(3, \dfrac{5\pi}{4}\right)$
7. $\left(4, -\dfrac{\pi}{3}\right)$
8. $\left(-1, -\dfrac{3\pi}{4}\right)$
9. $(2, 3\pi)$
10. $\left(4, \dfrac{5\pi}{2}\right)$
11. $\left(-2, \dfrac{2\pi}{3}\right)$
12. $\left(-3, \dfrac{11\pi}{6}\right)$
13. $\left(0, -\dfrac{7\pi}{6}\right)$
14. $\left(0, -\dfrac{7\pi}{2}\right)$
15. $(\sqrt{2}, 2.36)$
16. $(2\sqrt{2}, 4.71)$
17. $(-3, -1.57)$
18. $(-5, -2.36)$

In Exercises 19–28, a point in polar coordinates is given. Convert the point to rectangular coordinates.

19. $(3, \pi/2)$
20. $(3, 3\pi/2)$
21. $(-1, 5\pi/4)$
22. $(0, -\pi)$
23. $(2, 3\pi/4)$
24. $(1, 5\pi/4)$
25. $(-2, 7\pi/6)$
26. $(-3, 5\pi/6)$
27. $(-2.5, 1.1)$
28. $(-2, 5.76)$

In Exercises 29–36, use a graphing utility to find the rectangular coordinates of the point given in polar coordinates. Round your results to two decimal places.

29. $(2, 2\pi/9)$
30. $(4, 11\pi/9)$
31. $(-4.5, 1.3)$
32. $(8.25, 3.5)$
33. $(2.5, 1.58)$
34. $(5.4, 2.85)$
35. $(-4.1, -0.5)$
36. $(8.2, -3.2)$

In Exercises 37–54, a point in rectangular coordinates is given. Convert the point to polar coordinates.

37. $(1, 1)$
38. $(2, 2)$
39. $(-3, -3)$
40. $(-4, -4)$
41. $(-6, 0)$
42. $(3, 0)$
43. $(0, -5)$
44. $(0, 5)$
45. $(-3, 4)$
46. $(-4, -3)$
47. $(-\sqrt{3}, -\sqrt{3})$
48. $(-\sqrt{3}, \sqrt{3})$
49. $(\sqrt{3}, -1)$
50. $(-1, \sqrt{3})$
51. $(6, 9)$
52. $(6, 2)$
53. $(5, 12)$
54. $(7, 15)$

In Exercises 55–64, use a graphing utility to find one set of polar coordinates for the point given in rectangular coordinates.

55. $(3, -2)$
56. $(-4, -2)$
57. $(-5, 2)$
58. $(7, -2)$
59. $(\sqrt{3}, 2)$
60. $(5, -\sqrt{2})$
61. $\left(\frac{5}{2}, \frac{4}{3}\right)$
62. $\left(\frac{9}{5}, \frac{11}{2}\right)$
63. $\left(\frac{7}{4}, \frac{3}{2}\right)$
64. $\left(-\frac{7}{9}, -\frac{3}{4}\right)$

In Exercises 65–84, convert the rectangular equation to polar form. Assume $a > 0$.

65. $x^2 + y^2 = 9$
66. $x^2 + y^2 = 16$
67. $y = 4$
68. $y = x$
69. $x = 10$
70. $x = 4a$
71. $y = -2$
72. $y = 1$
73. $3x - y + 2 = 0$
74. $3x + 5y - 2 = 0$
75. $xy = 16$
76. $2xy = 1$
77. $y^2 - 8x - 16 = 0$
78. $(x^2 + y^2)^2 = 9(x^2 - y^2)$
79. $x^2 + y^2 = a^2$
80. $x^2 + y^2 = 9a^2$
81. $x^2 + y^2 - 2ax = 0$
82. $x^2 + y^2 - 2ay = 0$
83. $y^3 = x^2$
84. $y^2 = x^3$

In Exercises 85–108, convert the polar equation to rectangular form.

85. $r = 4 \sin \theta$
86. $r = 2 \cos \theta$
87. $r = -2 \cos \theta$
88. $r = -5 \sin \theta$
89. $\theta = 2\pi/3$
90. $\theta = 5\pi/3$
91. $\theta = 11\pi/6$
92. $\theta = 5\pi/6$
93. $r = 4$
94. $r = 10$

95. $r = 4 \csc \theta$

96. $r = 2 \csc \theta$

97. $r = -3 \sec \theta$

98. $r = -\sec \theta$

99. $r^2 = \cos \theta$

100. $r^2 = 2 \sin \theta$

101. $r^2 = \sin 2\theta$

102. $r^2 = \cos 2\theta$

103. $r = 2 \sin 3\theta$

104. $r = 3 \cos 2\theta$

105. $r = \dfrac{2}{1 + \sin \theta}$

106. $r = \dfrac{1}{1 - \cos \theta}$

107. $r = \dfrac{6}{2 - 3 \sin \theta}$

108. $r = \dfrac{6}{2 \cos \theta - 3 \sin \theta}$

In Exercises 109–118, describe the graph of the polar equation and find the corresponding rectangular equation. Sketch its graph.

109. $r = 6$

110. $r = 8$

111. $\theta = \pi/6$

112. $\theta = 3\pi/4$

113. $r = 2 \sin \theta$

114. $r = 4 \cos \theta$

115. $r = -6 \cos \theta$

116. $r = -3 \sin \theta$

117. $r = 3 \sec \theta$

118. $r = 2 \csc \theta$

In Exercises 119–122, find the points of horizontal and vertical tangency (if any) to the polar curve for $0 \le \theta < 2\pi$. Then use a graphing utility to verify your results.

119. $r = 1 + \sin \theta$

120. $r = 1 - 2 \cos \theta$

121. $r = 2 \csc \theta + 3$

122. $r = \sec \theta + 2$

WRITING ABOUT CONCEPTS

123. Convert the polar equation $r = 2(h \cos \theta + k \sin \theta)$ to rectangular form and verify that it is the equation of a circle. Find the radius and the rectangular coordinates of the center of the circle.

124. Convert the polar equation $r = \cos \theta + 3 \sin \theta$ to rectangular form and identify the graph.

125. Identify the type of symmetry each of the following polar points has with the point $(4, \pi/6)$.

(a) $(-4, \pi/6)$ (b) $(4, -\pi/6)$ (c) $(-4, -\pi/6)$

126. What is the relationship between the graphs of the rectangular and polar equations?

(a) $x^2 + y^2 = 25,\ r = 5$

(b) $x - y = 0,\ \theta = \pi/4$

True or False? **In Exercises 127 and 128, determine whether the statement is true or false. Justify your answer.**

127. If $\theta_1 = \theta_2 + 2\pi n$ for some integer n, then (r, θ_1) and (r, θ_2) represent the same point on the polar coordinate system.

128. If $|r_1| = |r_2|$, then (r_1, θ) and (r_2, θ) represent the same point on the polar coordinate system.

129. ***Think About It***

(a) Show that the distance between the points (r_1, θ_1) and (r_2, θ_2) is

$$\sqrt{r_1^2 + r_2^2 - 2r_1 r_2 \cos(\theta_1 - \theta_2)}.$$

(b) Describe the positions of the points relative to each other for $\theta_1 = \theta_2$. Simplify the Distance Formula for this case. Is the simplification what you expected? Explain.

(c) Simplify the Distance Formula for $\theta_1 - \theta_2 = 90°$. Is the simplification what you expected? Explain.

(d) Choose two points on the polar coordinate system and find the distance between them. Then choose different polar representations of the same two points and apply the Distance Formula again. Discuss the result.

130. ***Graphical Reasoning***

(a) Set the window format of your graphing utility on rectangular coordinates and locate the cursor at any position off the coordinate axes. Move the cursor horizontally and observe any changes in the displayed coordinates of the points. Explain the changes in the coordinates. Now repeat the process moving the cursor vertically.

(b) Set the window format of your graphing utility on polar coordinates and locate the cursor at any position off the coordinate axes. Move the cursor horizontally and observe any changes in the displayed coordinates of the points. Explain the changes in the coordinates. Now repeat the process moving the cursor vertically.

(c) Explain why the results of parts (a) and (b) are not the same.

131. ***Graphical Reasoning***

(a) Use a graphing utility in *polar* mode to graph the equation $r = 3$.

(b) Use the *trace* feature to move the cursor around the circle. Can you locate the point $(3, 5\pi/4)$?

(c) Can you find other polar representations of the point $(3, 5\pi/4)$? If so, explain how you did it.

CAPSTONE

132. In the rectangular coordinate system, each point (x, y) has a unique representation. Explain why this is not true for a point (r, θ) in the polar coordinate system.

12.6 Graphs of Polar Equations

- **Graph polar equations by point plotting.**
- **Use symmetry, zeros, and maximum *r*-values to sketch graphs of polar equations.**
- **Recognize special polar graphs.**

Introduction

In previous chapters, you learned how to sketch graphs on rectangular coordinate systems. You began with the basic point-plotting method, which was then enhanced by sketching aids such as symmetry, intercepts, asymptotes, relative extrema, concavity, periods, and shifts. This section approaches curve sketching on the polar coordinate system similarly, beginning with a demonstration of point plotting.

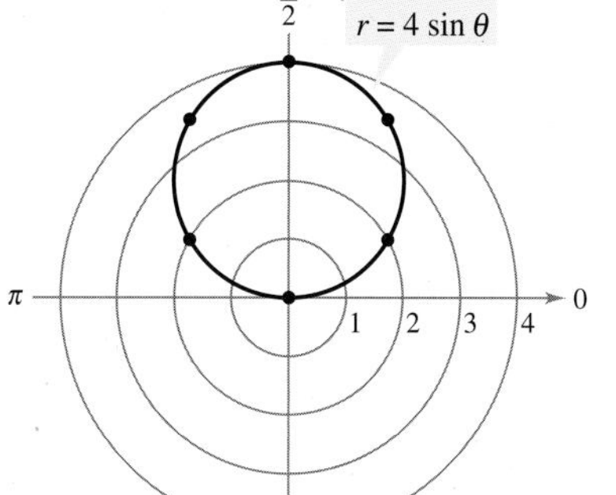

Figure 12.56

EXAMPLE 1 Graphing a Polar Equation by Point Plotting

Sketch the graph of the polar equation $r = 4 \sin \theta$.

Solution The sine function is periodic, so you can get a full range of r-values by considering values of θ in the interval $0 \le \theta \le 2\pi$, as shown in the following table.

θ	0	$\frac{\pi}{6}$	$\frac{\pi}{3}$	$\frac{\pi}{2}$	$\frac{2\pi}{3}$	$\frac{5\pi}{6}$	π	$\frac{7\pi}{6}$	$\frac{3\pi}{2}$	$\frac{11\pi}{6}$	2π
r	0	2	$2\sqrt{3}$	4	$2\sqrt{3}$	2	0	-2	-4	-2	0

If you plot these points, as shown in Figure 12.56, it appears that the graph is a circle of radius 2 whose center is at the point $(x, y) = (0, 2)$. Try confirming this by letting $\sin \theta = y/r$ in the polar equation and converting the result to rectangular form. ■

Symmetry

In Figure 12.56, note that as θ increases from 0 to 2π, the graph is traced out twice. Moreover, note that the graph is *symmetric with respect to the line* $\theta = \pi/2$. Had you known about this symmetry and retracing ahead of time, you could have used fewer points.

Symmetry with respect to the line $\theta = \pi/2$ is one of three important types of symmetry to consider in polar curve sketching, as shown in Figure 12.57.

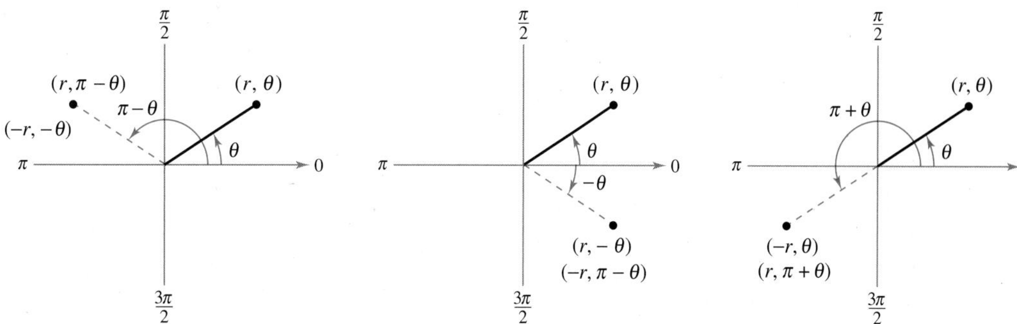

Symmetry with Respect to the Line $\theta = \pi/2$ | Symmetry with Respect to the Polar Axis | Symmetry with Respect to the Pole

Figure 12.57

STUDY TIP Note in Example 2 that $\cos(-\theta) = \cos\theta$. This is because the cosine function is *even*. Recall from Section 9.2 that the cosine function is even and the sine function is odd. That is, $\sin(-\theta) = -\sin\theta$.

TESTS FOR SYMMETRY IN POLAR COORDINATES

The graph of a polar equation is symmetric with respect to the following if the given substitution yields an equivalent equation.

1. *The line $\theta = \pi/2$:* Replace (r, θ) by $(r, \pi - \theta)$ or $(-r, -\theta)$.
2. *The polar axis:* Replace (r, θ) by $(r, -\theta)$ or $(-r, \pi - \theta)$.
3. *The pole:* Replace (r, θ) by $(r, \pi + \theta)$ or $(-r, \theta)$.

EXAMPLE 2 Using Symmetry to Sketch a Polar Graph

Use symmetry to sketch the graph of

$$r = 3 + 2\cos\theta.$$

Solution Replacing (r, θ) by $(r, -\theta)$ produces

$$r = 3 + 2\cos(-\theta)$$
$$= 3 + 2\cos\theta.$$

So, you can conclude that the curve is symmetric with respect to the polar axis. Plotting the points in the table and using polar axis symmetry, you obtain the graph of a **limaçon,** as shown in Figure 12.58.

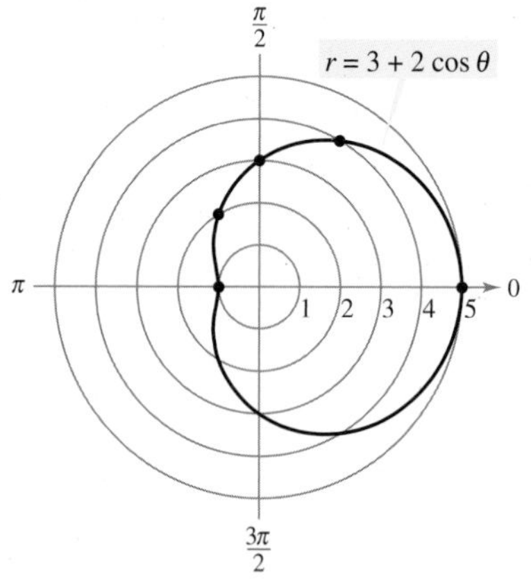

Figure 12.58

θ	0	$\frac{\pi}{3}$	$\frac{\pi}{2}$	$\frac{2\pi}{3}$	π
r	5	4	3	2	1

■

The three tests for symmetry in polar coordinates are sufficient to guarantee symmetry, but they are not necessary. For instance, Figure 12.59 shows the graph of $r = \theta + 2\pi$ to be symmetric with respect to the line $\theta = \pi/2$, and yet the tests fail to indicate symmetry.

The equations discussed in Examples 1 and 2 are of the form

$$r = 4\sin\theta = f(\sin\theta) \qquad \text{and} \qquad r = 3 + 2\cos\theta = g(\cos\theta).$$

The graph of the first equation is symmetric with respect to the line $\theta = \pi/2$, and the graph of the second equation is symmetric with respect to the polar axis. This observation can be generalized to yield the following tests.

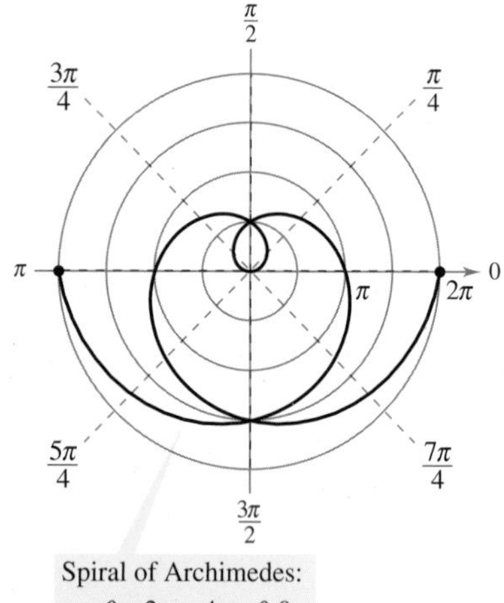

Spiral of Archimedes:
$r = \theta + 2\pi, -4\pi \quad \theta \; 0$

Figure 12.59

QUICK TESTS FOR SYMMETRY IN POLAR COORDINATES

1. The graph of $r = f(\sin\theta)$ is symmetric with respect to the line $\theta = \dfrac{\pi}{2}$.
2. The graph of $r = g(\cos\theta)$ is symmetric with respect to the polar axis.

Zeros and Maximum *r*-Values

Two additional aids to sketching graphs of polar equations involve knowing the θ-values for which $|r|$ is maximum and knowing the θ-values for which $r = 0$. For instance, in Example 1, the maximum value of $|r|$ for $r = 4\sin\theta$ is $|r| = 4$, and this occurs when $\theta = \pi/2$, as shown in Figure 12.56. Moreover, $r = 0$ when $\theta = 0$.

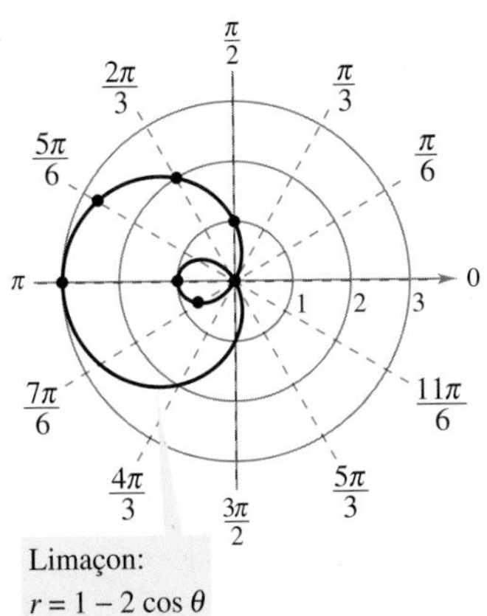

Figure 12.60

EXAMPLE 3 Sketching a Polar Graph

Sketch the graph of $r = 1 - 2\cos\theta$.

Solution From the equation $r = 1 - 2\cos\theta$, you can obtain the following.

Symmetry: With respect to the polar axis

Maximum value of $|r|$*:* $r = 3$ when $\theta = \pi$

Zero of r: $r = 0$ when $\theta = \pi/3$

The table shows several θ-values in the interval $[0, \pi]$. By plotting the corresponding points, you can sketch the graph shown in Figure 12.60.

θ	0	$\frac{\pi}{6}$	$\frac{\pi}{3}$	$\frac{\pi}{2}$	$\frac{2\pi}{3}$	$\frac{5\pi}{6}$	π
r	-1	-0.73	0	1	2	2.73	3

Note how the negative r-values determine the *inner loop* of the graph in Figure 12.60. ■

Some curves reach their zeros and maximum r-values at more than one point. Example 4 shows how to handle this situation.

EXAMPLE 4 Sketching a Polar Graph

Sketch the graph of $r = 2\cos 3\theta$.

Solution

Symmetry: With respect to the polar axis

Maximum value of $|r|$*:* $|r| = 2$ when $3\theta = 0, \pi, 2\pi, 3\pi$ or $\theta = 0, \pi/3, 2\pi/3, \pi$

Zero of r: $r = 0$ when $3\theta = \pi/2, 3\pi/2, 5\pi/2$ or $\theta = \pi/6, \pi/2, 5\pi/6$

θ	0	$\frac{\pi}{12}$	$\frac{\pi}{6}$	$\frac{\pi}{4}$	$\frac{\pi}{3}$	$\frac{5\pi}{12}$	$\frac{\pi}{2}$
r	2	$\sqrt{2}$	0	$-\sqrt{2}$	-2	$-\sqrt{2}$	0

NOTE In Example 4, note how the entire curve is generated as θ increases from 0 to π in increments of $\pi/6$.

By plotting these points and using the specified symmetry, zeros, and maximum values, you can obtain the graph shown in Figure 12.61. This graph is called a **rose curve,** and each of the loops on the graph is called a *petal* of the rose curve.

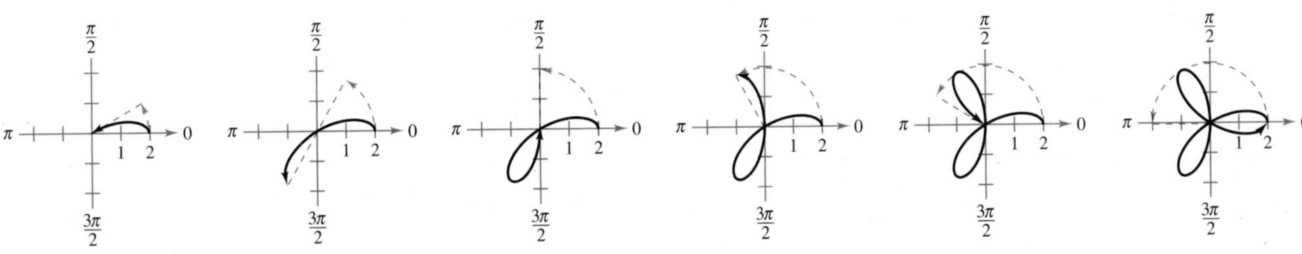

Figure 12.61

■

Special Polar Graphs

Several important types of graphs have equations that are simpler in polar form than in rectangular form. For example, the circle

$$r = 4 \sin \theta$$

in Example 1 has the more complicated rectangular equation

$$x^2 + (y - 2)^2 = 4.$$

Several other types of graphs that have simple polar equations are shown below.

Limaçons

$r = a \pm b \cos \theta$

$r = a \pm b \sin \theta$

$(a > 0, b > 0)$

$\frac{a}{b} < 1$

Limaçon with inner loop

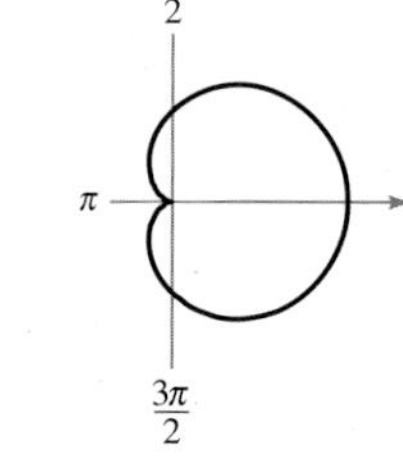

$\frac{a}{b} = 1$

Cardioid (heart-shaped)

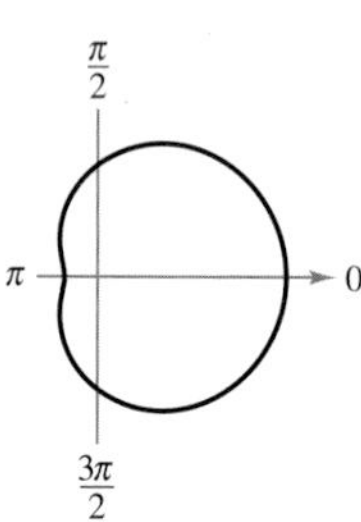

$1 < \frac{a}{b} < 2$

Dimpled limaçon

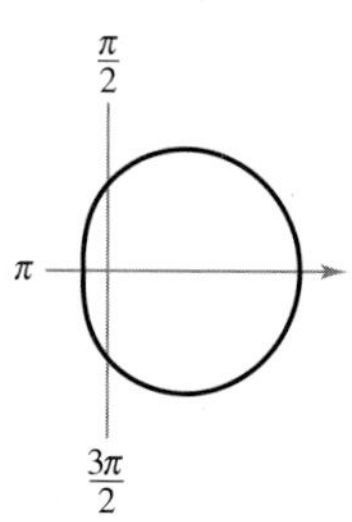

$\frac{a}{b} \geq 2$

Convex limaçon

Rose Curves

n petals if n is odd,

$2n$ petals if n is even

$(n \geq 2)$

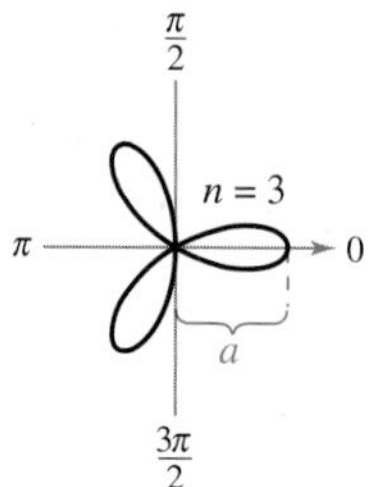

$r = a \cos n\theta$
Rose curve

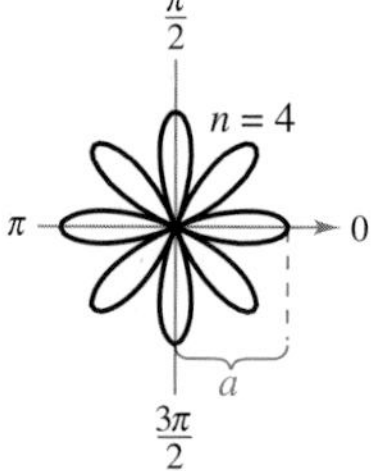

$r = a \cos n\theta$
Rose curve

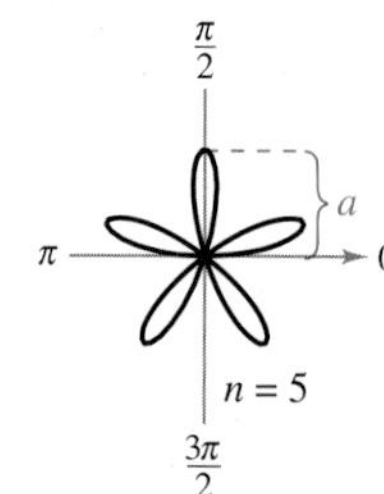

$r = a \sin n\theta$
Rose curve

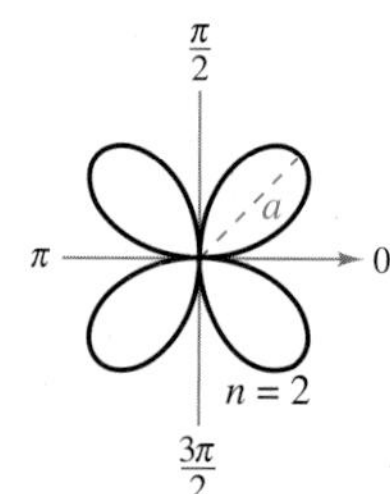

$r = a \sin n\theta$
Rose curve

Circles and Lemniscates

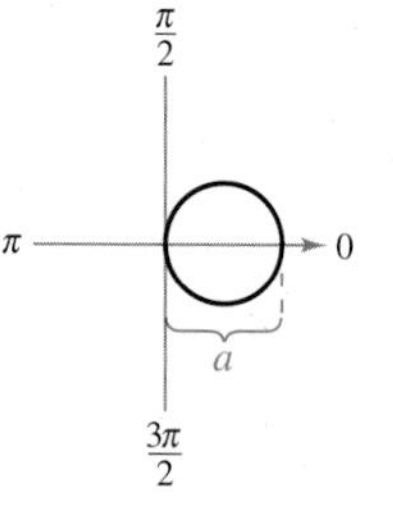

$r = a \cos \theta$
Circle

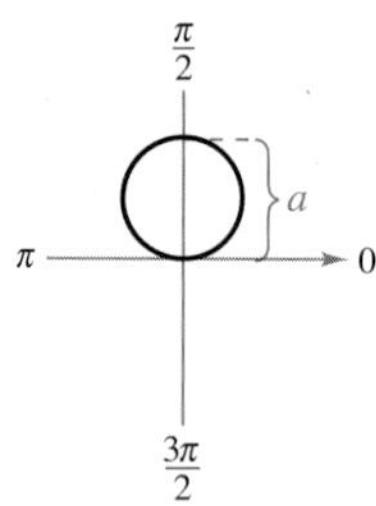

$r = a \sin \theta$
Circle

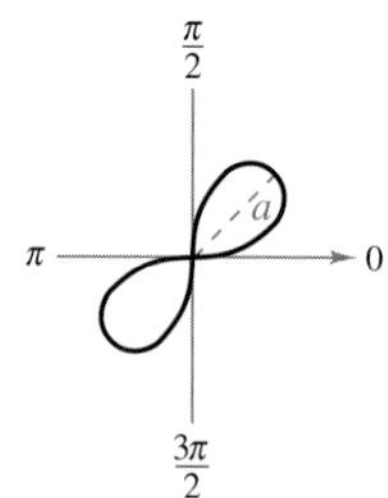

$r^2 = a^2 \sin 2\theta$
Lemniscate

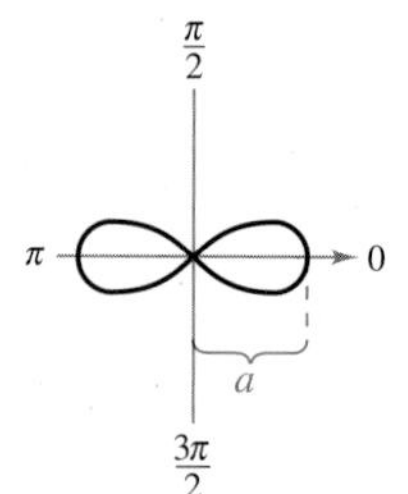

$r^2 = a^2 \cos 2\theta$
Lemniscate

TECHNOLOGY The rose curves described above are of the form $r = a \cos n\theta$ or $r = a \sin n\theta$, where n is a positive integer that is greater than or equal to 2. Try using a graphing utility to sketch the graph of $r = a \cos n\theta$ or $r = a \sin n\theta$ for some noninteger values of n. Are these graphs also rose curves? For example, try sketching the graph of $r = \cos \frac{2}{3}\theta$, $0 \leq \theta \leq 6\pi$.

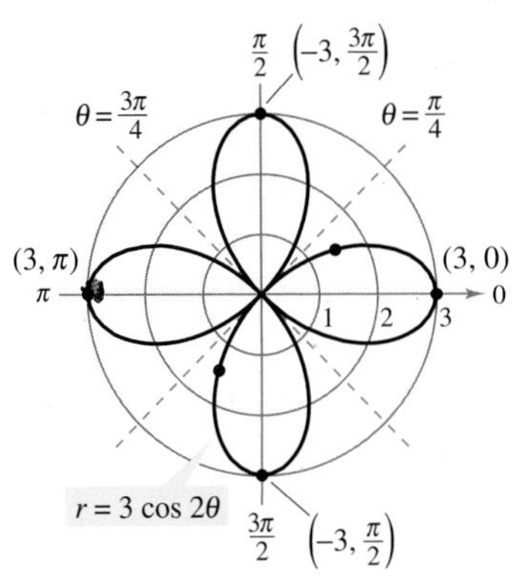

Figure 12.62

EXAMPLE 5 Sketching a Rose Curve

Sketch the graph of $r = 3\cos 2\theta$.

Solution

Type of curve: Rose curve with $2n = 4$ petals

Symmetry: With respect to polar axis, the line $\theta = \pi/2$, and the pole

Maximum value of $|r|$: $|r| = 3$ when $\theta = 0, \pi/2, \pi, 3\pi/2$

Zeros of r: $r = 0$ when $\theta = \pi/4, 3\pi/4$

Using this information together with the additional points shown in the following table, you obtain the graph shown in Figure 12.62.

θ	0	$\frac{\pi}{6}$	$\frac{\pi}{4}$	$\frac{\pi}{3}$
r	3	$\frac{3}{2}$	0	$-\frac{3}{2}$

■

12.6 Exercises

See www.CalcChat.com for worked-out solutions to odd-numbered exercises.

In Exercises 1–6, fill in the blanks.

1. The graph of $r = f(\sin\theta)$ is symmetric with respect to the line ________.
2. The graph of $r = g(\cos\theta)$ is symmetric with respect to the ________ ________.
3. The equation $r = 2 + \cos\theta$ represents a ________ ________.
4. The equation $r = 2\cos\theta$ represents a ________.
5. The equation $r^2 = 4\sin 2\theta$ represents a ________.
6. The equation $r = 1 + \sin\theta$ represents a ________.

In Exercises 7–12, identify the type of polar graph.

7.

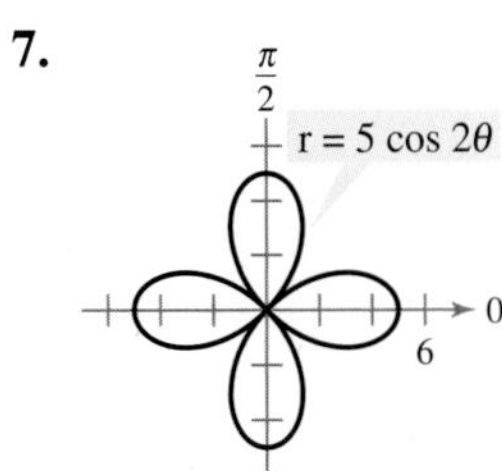

8.

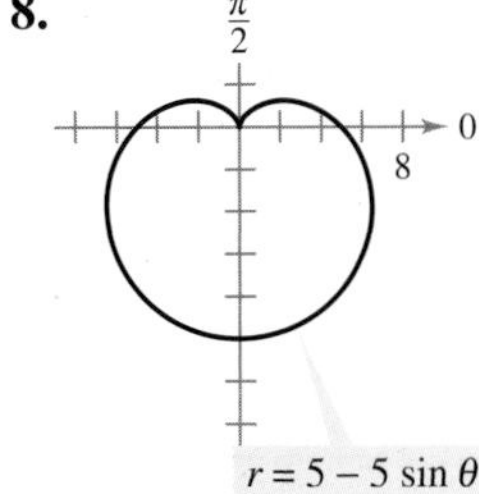

9.

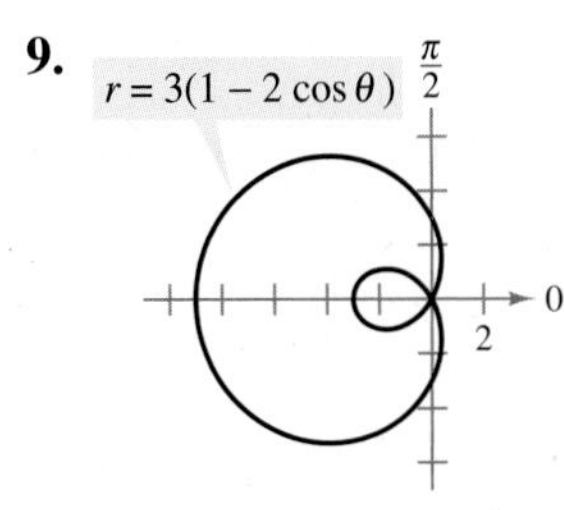

10.

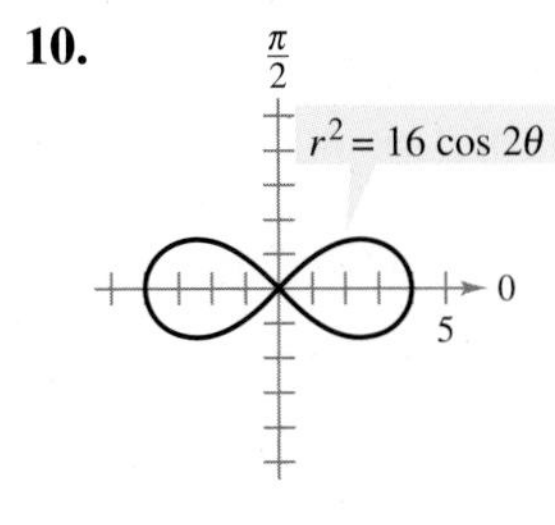

11.

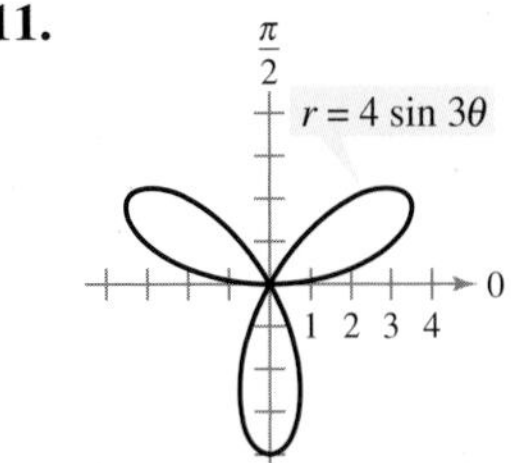

12.

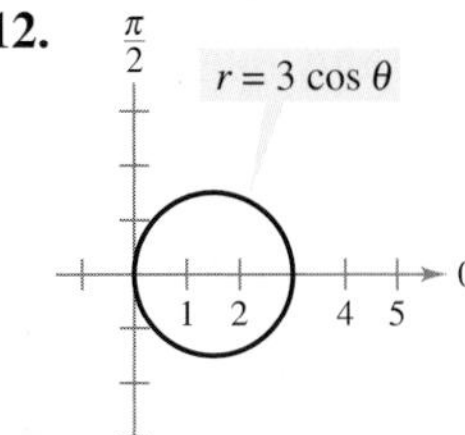

In Exercises 13–18, test for symmetry with respect to $\theta = \pi/2$, the polar axis, and the pole.

13. $r = 4 + 3\cos\theta$
14. $r = 9\cos 3\theta$
15. $r = \dfrac{2}{1 + \sin\theta}$
16. $r = \dfrac{3}{2 + \cos\theta}$
17. $r^2 = 36\cos 2\theta$
18. $r^2 = 25\sin 2\theta$

In Exercises 19–22, find the maximum value of $|r|$ and any zeros of r.

19. $r = 10 - 10\sin\theta$
20. $r = 6 + 12\cos\theta$
21. $r = 4\cos 3\theta$
22. $r = 3\sin 2\theta$

In Exercises 23–48, sketch the graph of the polar equation using symmetry, zeros, maximum r-values, and any other additional points.

23. $r = 4$
24. $r = -7$
25. $r = \dfrac{\pi}{3}$
26. $r = -\dfrac{3\pi}{4}$
27. $r = \sin\theta$
28. $r = 4\cos\theta$

29. $r = 3(1 - \cos\theta)$
30. $r = 4(1 - \sin\theta)$
31. $r = 4(1 + \sin\theta)$
32. $r = 2(1 + \cos\theta)$
33. $r = 3 + 6\sin\theta$
34. $r = 4 - 3\sin\theta$
35. $r = 1 - 2\sin\theta$
36. $r = 2 - 4\cos\theta$
37. $r = 3 - 4\cos\theta$
38. $r = 4 + 3\cos\theta$
39. $r = 5\sin 2\theta$
40. $r = 2\cos 2\theta$
41. $r = 6\cos 3\theta$
42. $r = 3\sin 3\theta$
43. $r = 2\sec\theta$
44. $r = 5\csc\theta$
45. $r = \dfrac{3}{\sin\theta - 2\cos\theta}$
46. $r = \dfrac{6}{2\sin\theta - 3\cos\theta}$
47. $r^2 = 9\cos 2\theta$
48. $r^2 = 4\sin\theta$

In Exercises 49–54, use a graphing utility to graph the polar equation. Describe your viewing window.

49. $r = 8\cos\theta$
50. $r = \cos 2\theta$
51. $r = 3(2 - \sin\theta)$
52. $r = 2\cos(3\theta - 2)$
53. $r = 8\sin\theta\cos^2\theta$
54. $r = 2\csc\theta + 5$

In Exercises 55–60, use a graphing utility to graph the polar equation. Find an interval for θ for which the graph is traced *only once*.

55. $r = 3 - 8\cos\theta$
56. $r = 5 + 4\cos\theta$
57. $r = 2\cos\left(\dfrac{3\theta}{2}\right)$
58. $r = 3\sin\left(\dfrac{5\theta}{2}\right)$
59. $r^2 = 16\sin 2\theta$
60. $r^2 = \dfrac{1}{\theta}$

WRITING ABOUT CONCEPTS

61. Sketch the graph of $r = 6\cos\theta$ over each interval. Describe the part of the graph obtained in each case.

(a) $0 \le \theta \le \dfrac{\pi}{2}$
(b) $\dfrac{\pi}{2} \le \theta \le \pi$
(c) $-\dfrac{\pi}{2} \le \theta \le \dfrac{\pi}{2}$
(d) $\dfrac{\pi}{4} \le \theta \le \dfrac{3\pi}{4}$

62. Graph and identify $r = 2 + k\sin\theta$ for $k = 0, 1, 2,$ and 3.

In Exercises 63–66, use a graphing utility to graph the polar equation and show that the indicated line is an asymptote of the graph.

	Name of Graph	*Polar Equation*	*Asymptote*
63.	Conchoid	$r = 2 - \sec\theta$	$x = -1$
64.	Conchoid	$r = 2 + \csc\theta$	$y = 1$
65.	Hyperbolic spiral	$r = \dfrac{3}{\theta}$	$y = 3$
66.	Strophoid	$r = 2\cos 2\theta\sec\theta$	$x = -2$

***True or False?* In Exercises 67–72, determine whether the statement is true or false. Justify your answer.**

67. In the polar coordinate system, if a graph that has symmetry with respect to the polar axis were folded on the line $\theta = 0$, the portion of the graph above the polar axis would coincide with the portion of the graph below the polar axis.

68. In the polar coordinate system, if a graph that has symmetry with respect to the pole were folded on the line $\theta = 3\pi/4$, the portion of the graph on one side of the fold would coincide with the portion of the graph on the other side of the fold.

69. The graph of $r = 1 + 2\cos\theta$ is a limaçon with an inner loop.

70. The graph of $r = 3 - 2\sin\theta$ is a cardioid.

71. The graph of $r = 2\sin 3\theta$ is a three-petal rose.

72. The graph of $r = \cos 2\theta$ is a four-petal rose.

73. Sketch the graph of each equation.

(a) $r = 1 - \sin\theta$
(b) $r = 1 - \sin\left(\theta - \dfrac{\pi}{4}\right)$

74. Sketch the graph of each equation.

(a) $r = 3\sec\theta$
(b) $r = 3\sec\left(\theta - \dfrac{\pi}{4}\right)$
(c) $r = 3\sec\left(\theta + \dfrac{\pi}{3}\right)$
(d) $r = 3\sec\left(\theta - \dfrac{\pi}{2}\right)$

75. ***Think About It*** How many petals do the rose curves given by $r = 2\cos 4\theta$ and $r = 2\sin 3\theta$ have? Determine the numbers of petals for the curves given by $r = 2\cos n\theta$ and $r = 2\sin n\theta$, where n is a positive integer.

CAPSTONE

76. Write a brief paragraph that describes why some polar curves have equations that are simpler in polar form than in rectangular form. Besides a circle, give an example of a curve that is simpler in polar form than in rectangular form. Give an example of a curve that is simpler in rectangular form than in polar form.

77. Consider the equation $r = 3\sin k\theta$.

(a) Use a graphing utility to graph the equation for $k = 1.5$. Find the interval for θ over which the graph is traced only once.

(b) Use a graphing utility to graph the equation for $k = 2.5$. Find the interval for θ over which the graph is traced only once.

(c) Is it possible to find an interval for θ over which the graph is traced only once for any rational number k? Explain.

12.7 Polar Equations of Conics

- **Define conics in terms of eccentricity.**
- **Write and graph equations of conics in polar form.**
- **Use equations of conics in polar form to model real-life problems.**

Alternative Definition of Conic

In Sections 12.2 and 12.3, you learned that the rectangular equations of ellipses and hyperbolas take simple forms when the origin lies at their *centers*. As it happens, there are many important applications of conics in which it is more convenient to use one of the *foci* as the origin. In this section, you will learn that polar equations of conics take simple forms if one of the foci lies at the pole.

To begin, consider the following alternative definition of conic that uses the concept of eccentricity.

ALTERNATIVE DEFINITION OF CONIC

The locus of a point in the plane that moves so that its distance from a fixed point (focus) is in a constant ratio to its distance from a fixed line (directrix) is a **conic.** The constant ratio is the **eccentricity** of the conic and is denoted by e. Moreover, the conic is an **ellipse** if $e < 1$, a **parabola** if $e = 1$, and a **hyperbola** if $e > 1$. (See Figure 12.63.)

In Figure 12.63, note that for each type of conic, the focus is at the pole.

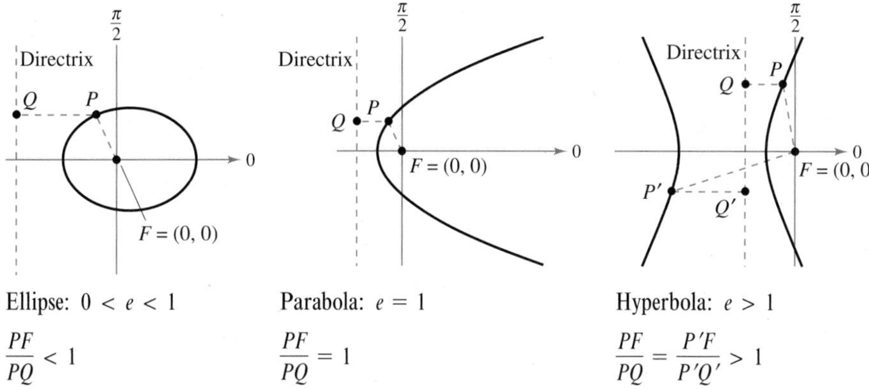

Ellipse: $0 < e < 1$ Parabola: $e = 1$ Hyperbola: $e > 1$

$\frac{PF}{PQ} < 1$ $\frac{PF}{PQ} = 1$ $\frac{PF}{PQ} = \frac{P'F}{P'Q'} > 1$

Figure 12.63

Polar Equations of Conics

The benefit of locating a focus of a conic at the pole is that the equation of the conic takes on a simpler form. A proof of the polar form is given in Appendix A.

THEOREM 12.9 POLAR EQUATIONS OF CONICS

The graph of a polar equation of the form

1. $r = \dfrac{ep}{1 \pm e \cos \theta}$ or   **2.** $r = \dfrac{ep}{1 \pm e \sin \theta}$

is a conic, where $e > 0$ is the eccentricity and $|p|$ is the distance between the focus (pole) and the directrix.

Equations of the form

$$r = \frac{ep}{1 \pm e \cos \theta} = g(\cos \theta) \qquad \text{Vertical directrix}$$

correspond to conics with a vertical directrix and symmetry with respect to the polar axis. Equations of the form

$$r = \frac{ep}{1 \pm e \sin \theta} = g(\sin \theta) \qquad \text{Horizontal directrix}$$

correspond to conics with a horizontal directrix and symmetry with respect to the line $\theta = \pi/2$. Moreover, the converse is also true—that is, any conic with a focus at the pole and having a horizontal or vertical directrix can be represented by one of these equations.

EXAMPLE 1 Identifying a Conic from Its Equation

Identify the type of conic represented by the equation $r = \dfrac{15}{3 - 2\cos\theta}$.

Algebraic Solution

To identify the type of conic, rewrite the equation in the form $r = (ep)/(1 \pm e \cos \theta)$.

$$r = \frac{15}{3 - 2\cos\theta} \qquad \text{Write original equation.}$$

$$= \frac{5}{1 - (2/3)\cos\theta} \qquad \text{Divide numerator and denominator by 3.}$$

Because $e = \frac{2}{3} < 1$, you can conclude that the graph is an ellipse.

Graphical Solution

You can start sketching the graph by plotting points from $\theta = 0$ to $\theta = \pi$. Because the equation is of the form $r = g(\cos \theta)$, the graph of r is symmetric with respect to the polar axis. So, you can complete the sketch, as shown in Figure 12.64. From this, you can conclude that the graph is an ellipse.

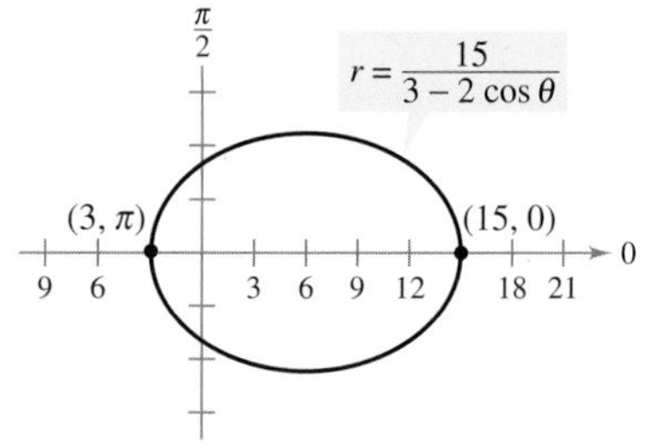

Figure 12.64

For the ellipse in Figure 12.64, the major axis is horizontal and the vertices lie at $(15, 0)$ and $(3, \pi)$. So, the length of the *major* axis is $2a = 18$. To find the length of the *minor* axis, you can use the equations $e = c/a$ and $b^2 = a^2 - c^2$ to conclude that

$$\begin{aligned} b^2 &= a^2 - c^2 \\ &= a^2 - (ea)^2 \\ &= a^2(1 - e^2). \qquad \text{Ellipse} \end{aligned}$$

Because $e = \frac{2}{3}$, you have $b^2 = 9^2\left[1 - \left(\frac{2}{3}\right)^2\right] = 45$, which implies that $b = \sqrt{45} = 3\sqrt{5}$. So, the length of the minor axis is $2b = 6\sqrt{5}$. A similar analysis for hyperbolas yields

$$\begin{aligned} b^2 &= c^2 - a^2 \\ &= (ea)^2 - a^2 \\ &= a^2(e^2 - 1). \qquad \text{Hyperbola} \end{aligned}$$

EXAMPLE 2 Sketching a Conic from Its Polar Equation

Identify the conic

$$r = \frac{32}{3 + 5 \sin \theta}$$

and sketch its graph.

Solution Dividing the numerator and denominator by 3, you have

$$r = \frac{32/3}{1 + (5/3) \sin \theta}.$$

Because $e = \frac{5}{3} > 1$, the graph is a hyperbola. The transverse axis of the hyperbola lies on the line $\theta = \pi/2$, and the vertices occur at $(4, \pi/2)$ and $(-16, 3\pi/2)$. Because the length of the transverse axis is 12, you can see that $a = 6$. To find b, write

$$b^2 = a^2(e^2 - 1) = 6^2\left[\left(\frac{5}{3}\right)^2 - 1\right] = 64.$$

So, $b = 8$. Finally, you can use a and b to convert the equation to a rectangular equation to find the center $(0, 10)$ and determine that the asymptotes of the hyperbola are $y = 10 \pm \frac{3}{4}x$. The graph is shown in Figure 12.65. ■

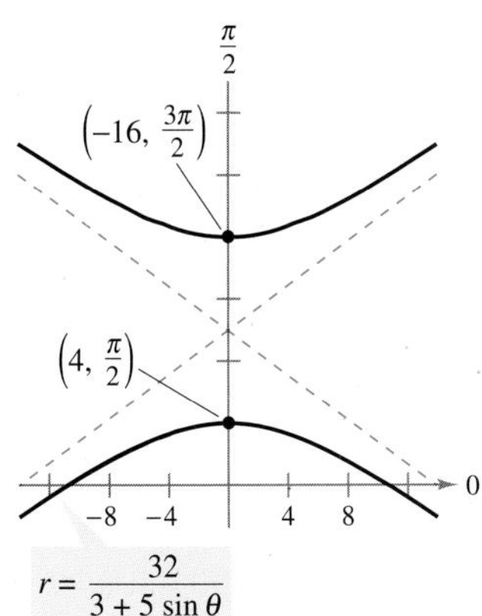

Figure 12.65

In the next example, you are asked to find a polar equation of a specified conic. To do this, let p be the distance between the pole and the directrix.

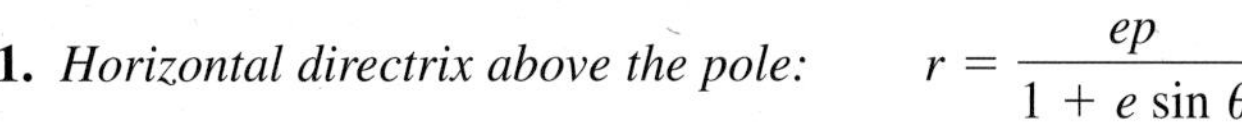

TECHNOLOGY Use a graphing utility set in *polar* mode to verify the four orientations shown at the right.

1. *Horizontal directrix above the pole:* $r = \dfrac{ep}{1 + e \sin \theta}$
2. *Horizontal directrix below the pole:* $r = \dfrac{ep}{1 - e \sin \theta}$
3. *Vertical directrix to the right of the pole:* $r = \dfrac{ep}{1 + e \cos \theta}$
4. *Vertical directrix to the left of the pole:* $r = \dfrac{ep}{1 - e \cos \theta}$

EXAMPLE 3 Finding the Polar Equation of a Conic

Find the polar equation of the parabola whose focus is the pole and whose directrix is the line $y = 3$.

Solution From Figure 12.66, you can see that the directrix is horizontal and above the pole, so you can choose an equation of the form

$$r = \frac{ep}{1 + e \sin \theta}.$$

Moreover, because the eccentricity of a parabola is $e = 1$ and the distance between the pole and the directrix is $p = 3$, you have the equation

$$r = \frac{3}{1 + \sin \theta}.$$ ■

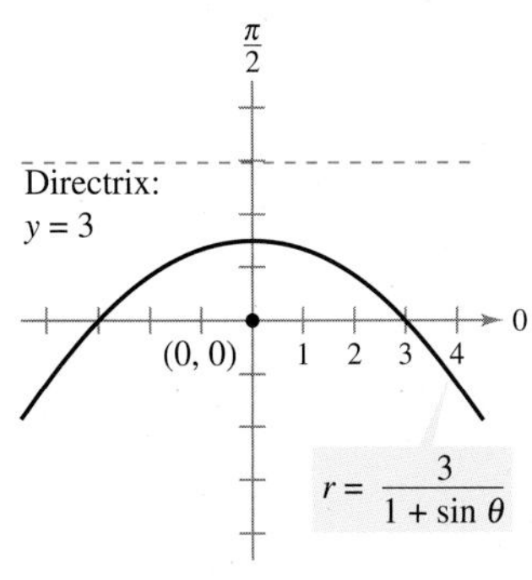

Figure 12.66

Applications

Kepler's Laws (listed below), named after the German astronomer Johannes Kepler (1571–1630), can be used to describe the orbits of the planets about the sun.

1. Each planet moves in an elliptical orbit with the sun at one focus.
2. A ray from the sun to the planet sweeps out equal areas of the ellipse in equal times.
3. The square of the period (the time it takes for a planet to orbit the sun) is proportional to the cube of the mean distance between the planet and the sun.

Although Kepler simply stated these laws on the basis of observation, they were later validated by Isaac Newton (1642–1727). In fact, Newton was able to show that each law can be deduced from a set of universal laws of motion and gravitation that govern the movement of all heavenly bodies, including comets and satellites. This is illustrated in the next example, which involves the comet named after the English mathematician and physicist Edmund Halley (1656–1742).

If you use Earth as a reference with a period of 1 year and a distance of 1 astronomical unit (an *astronomical unit* is defined as the mean distance between Earth and the sun, or about 93 million miles), the proportionality constant in Kepler's third law is 1. For example, because Mars has a mean distance to the sun of $d = 1.524$ astronomical units, its period P is given by $d^3 = P^2$. So, the period of Mars is $P \approx 1.88$ years.

EXAMPLE 4 Halley's Comet

Halley's comet has an elliptical orbit with an eccentricity of $e \approx 0.967$. The length of the major axis of the orbit is approximately 35.88 astronomical units. Find a polar equation for the orbit. How close does Halley's comet come to the sun?

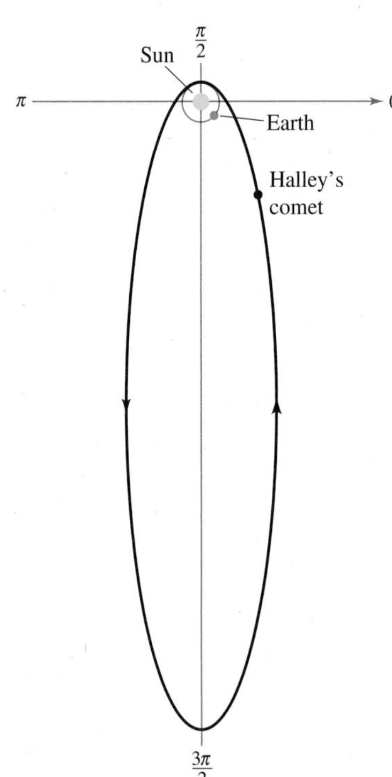

Figure 12.67

Solution Using a vertical axis, as shown in Figure 12.67, choose an equation of the form

$$r = \frac{ep}{1 + e \sin \theta}.$$

Because the vertices of the ellipse occur when $\theta = \pi/2$ and $\theta = 3\pi/2$, you can determine the length of the major axis to be the sum of the r-values of the vertices. That is,

$$2a = \frac{0.967p}{1 + 0.967} + \frac{0.967p}{1 - 0.967} \approx 29.79p \approx 35.88.$$

So, $p \approx 1.204$ and $ep \approx (0.967)(1.204) \approx 1.164$. Using this value of ep in the equation, you have

$$r = \frac{1.164}{1 + 0.967 \sin \theta}$$

where r is measured in astronomical units. To find the closest point to the sun (the focus), substitute $\theta = \pi/2$ in this equation to obtain

$$r = \frac{1.164}{1 + 0.967 \sin(\pi/2)}$$
$$\approx 0.59 \text{ astronomical unit}$$
$$\approx 55{,}000{,}000 \text{ miles.}$$

■

12.7 Exercises

See www.CalcChat.com for worked-out solutions to odd-numbered exercises.

In Exercises 1–3, fill in the blanks.

1. The locus of a point in the plane that moves so that its distance from a fixed point (focus) is in a constant ratio to its distance from a fixed line (directrix) is a ________.
2. The constant ratio is the ________ of the conic and is denoted by ________.
3. An equation of the form $r = \dfrac{ep}{1 + e\cos\theta}$ has a ________ directrix to the ________ of the pole.
4. Match the conic with its eccentricity.

(a) $e < 1$ (b) $e = 1$ (c) $e > 1$

(i) parabola (ii) hyperbola (iii) ellipse

In Exercises 5–8, write the polar equation of the conic for $e = 1$, $e = 0.5$, and $e = 1.5$. Identify the conic for each equation. Verify your answers with a graphing utility.

5. $r = \dfrac{2e}{1 + e\cos\theta}$
6. $r = \dfrac{2e}{1 - e\cos\theta}$
7. $r = \dfrac{2e}{1 - e\sin\theta}$
8. $r = \dfrac{2e}{1 + e\sin\theta}$

In Exercises 9–14, match the polar equation with its graph. [The graphs are labeled (a), (b), (c), (d), (e), and (f).]

(a)

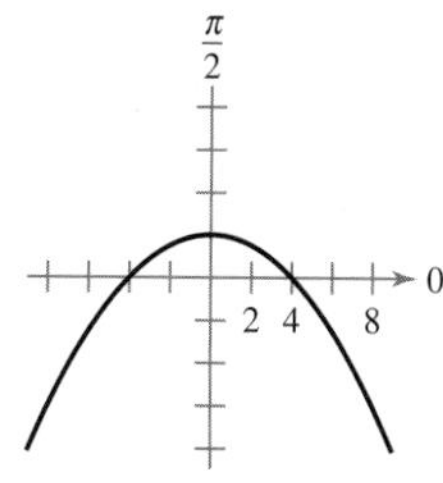

(b)

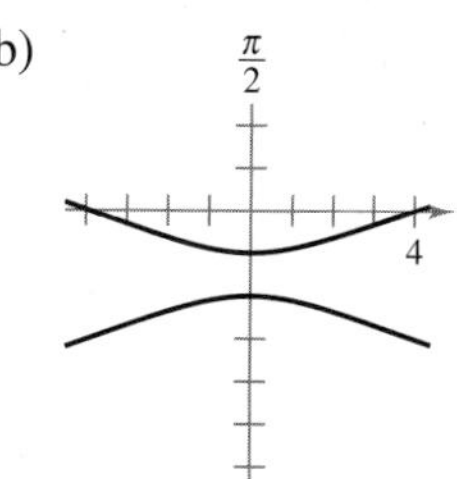

(c)

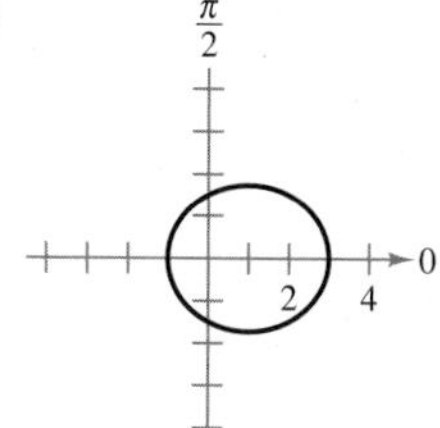

(d)

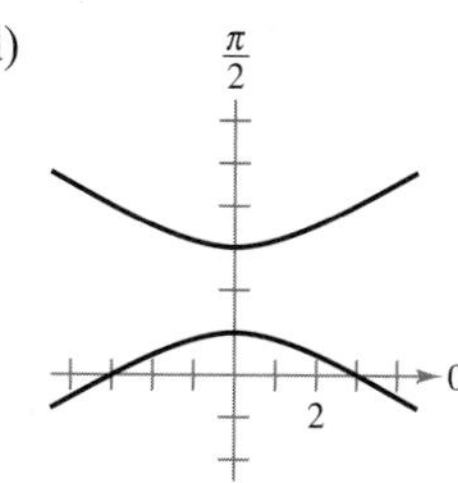

(e)

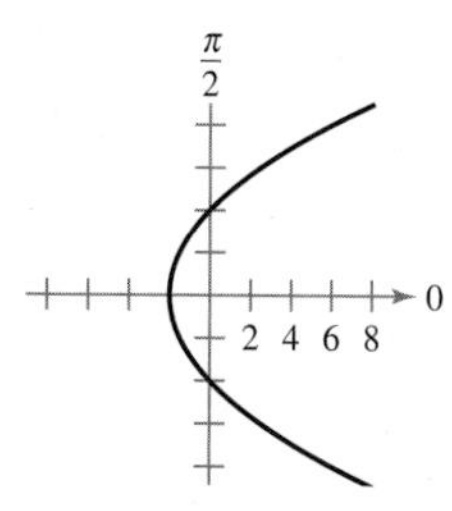

(f)

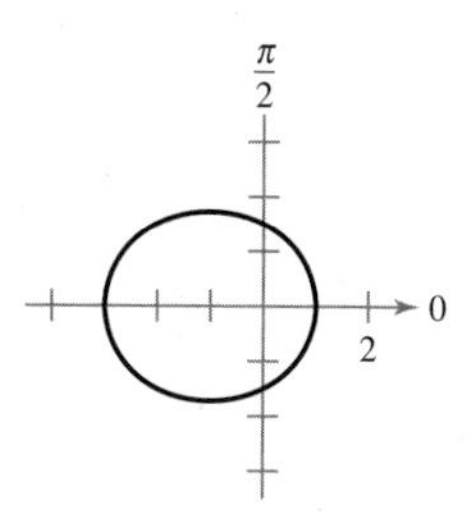

9. $r = \dfrac{4}{1 - \cos\theta}$
10. $r = \dfrac{3}{2 - \cos\theta}$
11. $r = \dfrac{3}{1 + 2\sin\theta}$
12. $r = \dfrac{3}{2 + \cos\theta}$
13. $r = \dfrac{4}{1 + \sin\theta}$
14. $r = \dfrac{4}{1 - 3\sin\theta}$

In Exercises 15–28, identify the conic and sketch its graph.

15. $r = \dfrac{3}{1 - \cos\theta}$
16. $r = \dfrac{7}{1 + \sin\theta}$
17. $r = \dfrac{5}{1 + \sin\theta}$
18. $r = \dfrac{6}{1 + \cos\theta}$
19. $r = \dfrac{2}{2 - \cos\theta}$
20. $r = \dfrac{4}{4 + \sin\theta}$
21. $r = \dfrac{6}{2 + \sin\theta}$
22. $r = \dfrac{9}{3 - 2\cos\theta}$
23. $r = \dfrac{3}{2 + 4\sin\theta}$
24. $r = \dfrac{5}{-1 + 2\cos\theta}$
25. $r = \dfrac{3}{2 - 6\cos\theta}$
26. $r = \dfrac{3}{2 + 6\sin\theta}$
27. $r = \dfrac{4}{2 - \cos\theta}$
28. $r = \dfrac{2}{2 + 3\sin\theta}$

In Exercises 29–38, use a graphing utility to graph the polar equation. Identify the graph.

29. $r = \dfrac{-1}{1 - \sin\theta}$
30. $r = \dfrac{-5}{2 + 4\sin\theta}$
31. $r = \dfrac{3}{-4 + 2\cos\theta}$
32. $r = \dfrac{4}{1 - 2\cos\theta}$
33. $r = \dfrac{14}{14 + 17\sin\theta}$
34. $r = \dfrac{12}{2 - \cos\theta}$
35. $r = \dfrac{2}{1 + \sin\theta}$
36. $r = \dfrac{6}{3 - 2\cos\theta}$
37. $r = \dfrac{5}{-1 + 2\cos\theta}$
38. $r = \dfrac{42}{2 - 3\sin\theta}$

In Exercises 39–54, find a polar equation of the conic with its focus at the pole.

Conic	Eccentricity	Directrix
39. Parabola	$e = 1$	$x = -1$
40. Parabola	$e = 1$	$y = -4$
41. Ellipse	$e = \frac{1}{2}$	$y = 1$
42. Ellipse	$e = \frac{3}{4}$	$y = -2$
43. Hyperbola	$e = 2$	$x = 1$
44. Hyperbola	$e = \frac{3}{2}$	$x = -1$

Conic	Vertex or Vertices
45. Parabola	$(1, -\pi/2)$
46. Parabola	$(8, 0)$
47. Parabola	$(5, \pi)$
48. Parabola	$(10, \pi/2)$
49. Ellipse	$(2, 0), (10, \pi)$
50. Ellipse	$(2, \pi/2), (4, 3\pi/2)$
51. Ellipse	$(20, 0), (4, \pi)$
52. Hyperbola	$(2, 0), (8, 0)$
53. Hyperbola	$(1, 3\pi/2), (9, 3\pi/2)$
54. Hyperbola	$(4, \pi/2), (1, \pi/2)$

WRITING ABOUT CONCEPTS

55. Verify that the polar equation of the ellipse

$$\frac{x^2}{a^2} + \frac{y^2}{b^2} = 1 \quad \text{is} \quad r^2 = \frac{b^2}{1 - e^2 \cos^2 \theta}.$$

56. Verify that the polar equation of the hyperbola

$$\frac{x^2}{a^2} - \frac{y^2}{b^2} = 1 \quad \text{is} \quad r^2 = \frac{-b^2}{1 - e^2 \cos^2 \theta}.$$

In Exercises 57–62, use the results of Exercises 55 and 56 to write the polar form of the equation of the conic.

57. $x^2/169 + y^2/144 = 1$

58. $x^2/25 + y^2/16 = 1$

59. $x^2/9 - y^2/16 = 1$

60. $x^2/36 - y^2/4 = 1$

61. Hyperbola One focus: $(5, 0)$
Vertices: $(4, 0), (4, \pi)$

62. Ellipse One focus: $(4, 0)$
Vertices: $(5, 0), (5, \pi)$

63. ***Astronomy*** The comet Encke has an elliptical orbit with an eccentricity of $e \approx 0.847$. The length of the major axis of the orbit is approximately 4.42 astronomical units. Find a polar equation for the orbit. How close does the comet come to the sun?

64. ***Astronomy*** The comet Hale-Bopp has an elliptical orbit with an eccentricity of $e \approx 0.995$. The length of the major axis of the orbit is approximately 500 astronomical units. Find a polar equation for the orbit. How close does the comet come to the sun?

65. ***Satellite Tracking*** A satellite in a 100-mile-high circular orbit around Earth has a velocity of approximately 17,500 miles per hour. If this velocity is multiplied by $\sqrt{2}$, the satellite will have the minimum velocity necessary to escape Earth's gravity and will follow a parabolic path with the center of Earth as the focus (see figure).

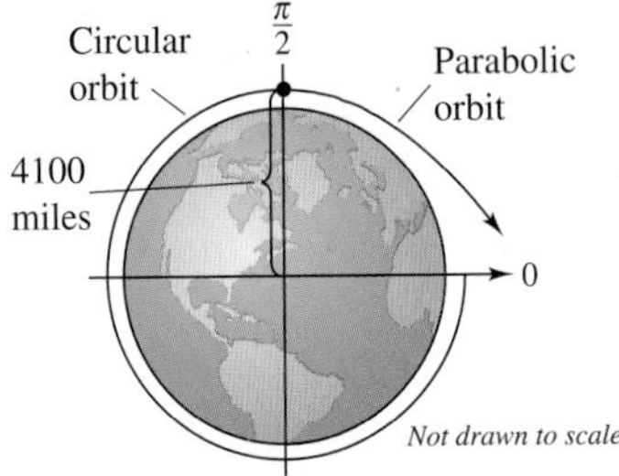

(a) Find a polar equation of the parabolic path of the satellite (assume the radius of Earth is 4000 miles).

(b) Use a graphing utility to graph the equation you found in part (a).

(c) Find the distance between the surface of the Earth and the satellite when $\theta = 30°$.

(d) Find the distance between the surface of Earth and the satellite when $\theta = 60°$.

66. ***Roman Coliseum*** The Roman Coliseum is an elliptical amphitheater measuring approximately 188 meters long and 156 meters wide.

(a) Find an equation to model the coliseum that is of the form

$$\frac{x^2}{a^2} + \frac{y^2}{b^2} = 1.$$

(b) Find a polar equation to model the coliseum. (Assume $e \approx 0.5581$ and $p \approx 115.98$.)

(c) Use a graphing utility to graph the equations you found in parts (a) and (b). Are the graphs the same? Why or why not?

(d) In part (c), did you prefer graphing the rectangular equation or the polar equation? Explain.

True or False? **In Exercises 67–72, determine whether the statement is true or false. Justify your answer.**

67. The graph of $r = \dfrac{4}{-3 - 3 \sin \theta}$ has a horizontal directrix above the pole.

68. For a given value of $e > 1$ over the interval $\theta = 0$ to $\theta = 2\pi$, the graph of

$$r = \frac{ex}{1 - e\cos\theta}$$

is the same as the graph of

$$r = \frac{e(-x)}{1 + e\cos\theta}.$$

69. The conic represented by the following equation is an ellipse.

$$r^2 = \frac{16}{9 - 4\cos\theta}$$

70. The conic represented by the following equation is a parabola.

$$r = \frac{6}{3 - 2\cos\theta}$$

71. The polar equation of the ellipse

$$\frac{(x+4)^2}{36} + \frac{y^2}{20} = 1 \text{ is } r = \frac{10}{3 + 2\cos\theta}.$$

72. The polar equation of the hyperbola

$$\frac{y^2}{16} - \frac{(x-6)^2}{20} = 1 \text{ is } r = \frac{10}{2 + \sin\theta}.$$

73. ***Writing*** Explain how the graph of each conic differs from the graph of $r = \dfrac{5}{1 + \sin\theta}$. (See Exercise 17.)

(a) $r = \dfrac{5}{1 - \cos\theta}$ (b) $r = \dfrac{5}{1 - \sin\theta}$

(c) $r = \dfrac{5}{1 + \cos\theta}$ (d) $r = \dfrac{5}{1 - \sin[\theta - (\pi/4)]}$

CAPSTONE

74. In your own words, define the term *eccentricity* and explain how it can be used to classify conics.

75. The equation

$$r = \frac{ep}{1 \pm e\sin\theta}$$

is the equation of an ellipse with $e < 1$. What happens to the lengths of both the major axis and the minor axis when the value of e remains fixed and the value of p changes? Use an example to explain your reasoning.

76. Consider the polar equation

$$r = \frac{4}{1 - 0.4\cos\theta}.$$

(a) Identify the conic without graphing the equation.

(b) Without graphing the following polar equations, describe how each differs from the given polar equation.

$$r_1 = \frac{4}{1 + 0.4\cos\theta} \qquad r_2 = \frac{4}{1 - 0.4\sin\theta}$$

(c) Use a graphing utility to verify your results in part (b).

SECTION PROJECT

Polar Equations of Planetary Orbits

The polar equation of the orbit of a planet is

$$r = \frac{(1 - e^2)a}{1 - e\cos\theta}$$

where e is the eccentricity. The perihelion distance (minimum distance) from the sun to the planet is $r = a(1 - e)$ and the aphelion distance (maximum distance) is $r = a(1 + e)$. Find the polar equation of the planet's orbit and the perihelion and aphelion distances.

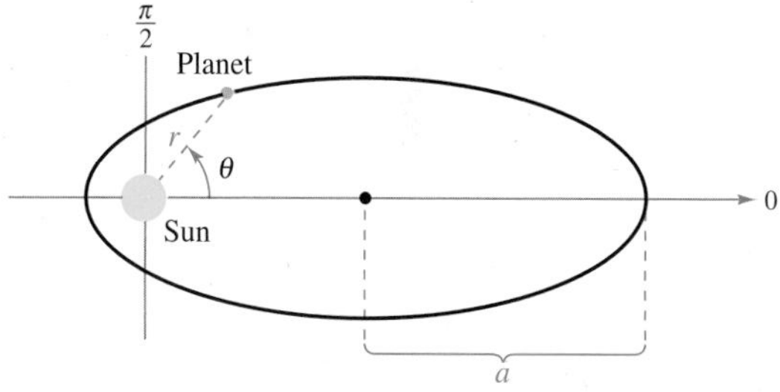

(a) Earth $a = 1.496 \times 10^8$ kilometers
$e = 0.0167$

(b) Saturn $a = 1.427 \times 10^9$ kilometers
$e = 0.0542$

(c) Venus $a = 1.082 \times 10^8$ kilometers
$e = 0.0068$

(d) Mercury $a = 5.791 \times 10^7$ kilometers
$e = 0.2056$

(e) Mars $a = 2.279 \times 10^8$ kilometers
$e = 0.0934$

(f) Jupiter $a = 7.784 \times 10^8$ kilometers
$e = 0.0484$

12 CHAPTER SUMMARY

	Review Exercises
Section 12.1	
■ Recognize a conic as the intersection of a plane and a double-napped cone *(p. 746).*	*1–2*
■ Write equations of parabolas in standard form and graph parabolas *(p. 747).*	*3–6*
■ Use the reflexive property of parabolas to solve real-life problems *(p. 749).*	*7–10*
Section 12.2	
■ Write equations of ellipses in standard form and graph ellipses *(p. 755).*	*11–14*
■ Use properties of ellipses to model and solve real-life problems *(p. 758).*	*15, 16*
■ Find eccentricities of ellipses *(p. 758).*	*17–20*
■ Find the equation of the tangent line to an ellipse at a given point *(p. 757).*	*21, 22*
■ Find the area of a region bounded by an ellipse *(p. 757).*	*23–26*
Section 12.3	
■ Write equations of hyperbolas in standard form *(p. 762)* and find asymptotes of and graph hyperbolas *(p. 763).*	*27–34*
■ Use properties of hyperbolas to solve real-life problems *(p. 766).*	*35, 36*
■ Classify conics from their general equations *(p. 767).*	*37–40*
Section 12.4	
■ Sketch or graph curves that are represented by sets of parametric equations *(p. 772).*	*41–48, 54*
■ Rewrite sets of parametric equations as single rectangular equations by eliminating the parameter *(p. 773).*	*43–48*
■ Find sets of parametric equations for graphs *(p. 774).*	*49–53*
■ Find the derivative of a set of parametric equations *(p. 776).*	*55–58*
■ Find all points of horizontal and vertical tangency for a set of parametric equations *(p. 777).*	*59–62*
Section 12.5	
■ Plot points on the polar coordinate system and find additional polar representations of the point *(p. 781).*	*63–66*
■ Convert points *(p. 781)* and equations *(p. 783)* from rectangular to polar form and vice versa.	*67–86*
■ Find all points of horizontal and vertical tangency to the graph of a polar curve *(p. 784).*	*87–90*
Section 12.6	
■ Use point plotting *(p. 787)*, symmetry *(p. 787)*, and zeros and maximum r-values *(p. 788)* to sketch graphs of polar equations.	*91–100*
■ Recognize special polar graphs *(p. 790).*	*101–104*
Section 12.7	
■ Define conics in polar form in terms of eccentricity and sketch their graphs *(p. 793).*	*105–108*
■ Write equations of conics in polar form *(p. 795).*	*109–112*
■ Use equations of conics in polar form to model real-life problems *(p. 796).*	*113, 114*

12 REVIEW EXERCISES

See www.CalcChat.com for worked-out solutions to odd-numbered exercises.

In Exercises 1 and 2, state what type of conic is formed by the intersection of the plane and the double-napped cone.

1. 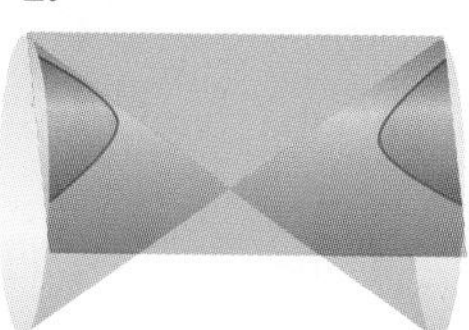

2. 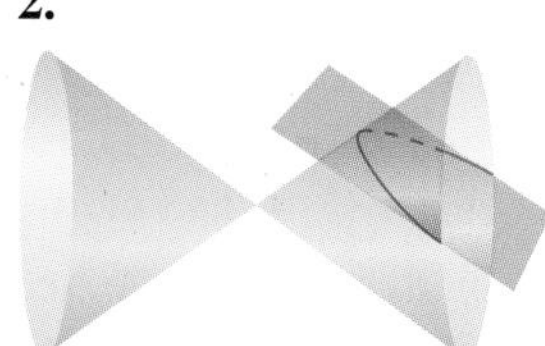

In Exercises 3–6, find the standard form of the equation of the parabola with the given characteristics. Then graph the parabola.

3. Vertex: $(0, 0)$
   Focus: $(4, 0)$

4. Vertex: $(2, 0)$
   Focus: $(0, 0)$

5. Vertex: $(0, 2)$
   Directrix: $x = -3$

6. Vertex: $(-3, -3)$
   Directrix: $y = 0$

In Exercises 7 and 8, find an equation of the tangent line to the parabola at the given point.

7. $y = 2x^2$, $(-1, 2)$

8. $x^2 = -2y$, $(-4, -8)$

9. ***Architecture*** A parabolic archway is 12 meters high at the vertex. At a height of 10 meters, the width of the archway is 8 meters (see figure). How wide is the archway at ground level?

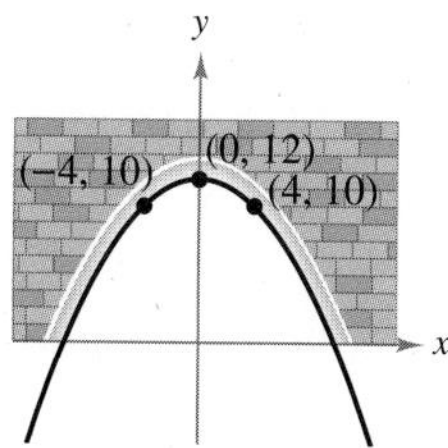

Figure for 9

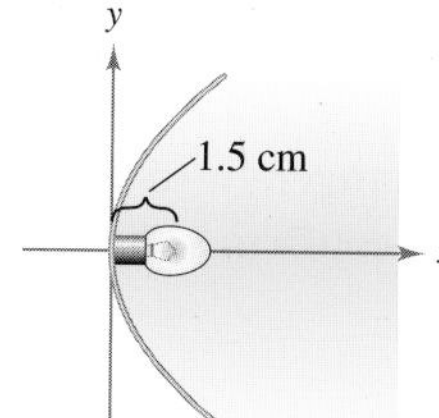

Figure for 10

10. ***Flashlight*** The light bulb in a flashlight is at the focus of its parabolic reflector, 1.5 centimeters from the vertex of the reflector (see figure). Write an equation of a cross section of the flashlight's reflector with its focus on the positive x-axis and its vertex at the origin.

In Exercises 11–14, find the standard form of the equation of the ellipse with the given characteristics. Then graph the ellipse.

11. Vertices: $(-2, 0), (8, 0)$; foci: $(0, 0), (6, 0)$
12. Vertices: $(4, 3), (4, 7)$; foci: $(4, 4), (4, 6)$
13. Vertices: $(0, 1), (4, 1)$; endpoints of the minor axis: $(2, 0), (2, 2)$
14. Vertices: $(-4, -1), (-4, 11)$; endpoints of the minor axis: $(-6, 5), (-2, 5)$
15. ***Architecture*** A semielliptical archway is to be formed over the entrance to an estate. The arch is to be set on pillars that are 10 feet apart and is to have a height (atop the pillars) of 4 feet. Where should the foci be placed in order to sketch the arch?
16. ***Wading Pool*** You are building a wading pool that is in the shape of an ellipse. Your plans give an equation for the elliptical shape of the pool measured in feet as

$$\frac{x^2}{324} + \frac{y^2}{196} = 1.$$

Find the longest distance across the pool, the shortest distance, and the distance between the foci.

In Exercises 17–20, find the center, vertices, foci, and eccentricity of the ellipse.

17. $\dfrac{(x + 1)^2}{25} + \dfrac{(y - 2)^2}{49} = 1$
18. $\dfrac{(x - 5)^2}{1} + \dfrac{(y + 3)^2}{36} = 1$
19. $16x^2 + 9y^2 - 32x + 72y + 16 = 0$
20. $4x^2 + 25y^2 + 16x - 150y + 141 = 0$

In Exercises 21 and 22, find an equation of the tangent line to the ellipse at the given point.

21. $\dfrac{(x - 1)^2}{9} + \dfrac{(y - 5)^2}{25} = 1, \quad (4, 5)$
22. $\dfrac{(x + 2)^2}{4} + \dfrac{(y + 2)^2}{25} = 1, \quad (-2, 3)$

In Exercises 23–26, find the area of the region bounded by the ellipse.

23. $x^2 + 5y^2 = 10$
24. $4x^2 + y^2 = 4$
25. $x^2 + 4y^2 - 2x - 16y + 13 = 0$
26. $16x^2 + 4y^2 - 32x - 8y - 44 = 0$

In Exercises 27–30, find the standard form of the equation of the hyperbola with the given characteristics.

27. Vertices: $(0, \pm 1)$; foci: $(0, \pm 2)$
28. Vertices: $(3, 3), (-3, 3)$; foci: $(4, 3), (-4, 3)$
29. Foci: $(0, 0), (8, 0)$; asymptotes: $y = \pm 2(x - 4)$
30. Foci: $(3, \pm 2)$; asymptotes: $y = \pm 2(x - 3)$

In Exercises 31–34, find the center, vertices, foci, and the equations of the asymptotes of the hyperbola, and sketch its graph using the asymptotes as an aid.

31. $\dfrac{(x-5)^2}{36} - \dfrac{(y+3)^2}{16} = 1$

32. $\dfrac{(y-1)^2}{4} - x^2 = 1$

33. $9x^2 - 16y^2 - 18x - 32y - 151 = 0$

34. $-4x^2 + 25y^2 - 8x + 150y + 121 = 0$

35. ***LORAN*** Radio transmitting station A is located 200 miles east of transmitting station B. A ship is in an area to the north and 40 miles west of station A. Synchronized radio pulses transmitted at 186,000 miles per second by the two stations are received 0.0005 second sooner from station A than from station B. How far north is the ship?

36. ***Locating an Explosion*** Two of your friends live 4 miles apart and on the same "east-west" street, and you live halfway between them. You are having a three-way phone conversation when you hear an explosion. Six seconds later, your friend to the east hears the explosion, and your friend to the west hears it 8 seconds after you do. Find equations of two hyperbolas that would locate the explosion. (Assume that the coordinate system is measured in feet and that sound travels at 1100 feet per second.)

In Exercises 37–40, classify the graph of the equation as a circle, a parabola, an ellipse, or a hyperbola.

37. $5x^2 - 2y^2 + 10x - 4y + 17 = 0$

38. $-4y^2 + 5x + 3y + 7 = 0$

39. $3x^2 + 2y^2 - 12x + 12y + 29 = 0$

40. $4x^2 + 4y^2 - 4x + 8y - 11 = 0$

In Exercises 41 and 42, (a) create a table of x- and y-values for the parametric equations using $t = -2, -1, 0, 1,$ and 2, and (b) plot the points (x, y) generated in part (a) and sketch a graph of the parametric equations.

41. $x = 3t - 2$ and $y = 7 - 4t$

42. $x = \dfrac{1}{4}t$ and $y = \dfrac{6}{t+3}$

In Exercises 43–48, (a) sketch the curve represented by the parametric equations (indicate the orientation of the curve) and (b) eliminate the parameter and write the corresponding rectangular equation whose graph represents the curve. Adjust the domain of the resulting rectangular equation, if necessary. (c) Verify your result with a graphing utility.

43. $x = 2t$
$y = 4t$

44. $x = 1 + 4t$
$y = 2 - 3t$

45. $x = t^2$
$y = \sqrt{t}$

46. $x = t + 4$
$y = t^2$

47. $x = 3\cos\theta$
$y = 3\sin\theta$

48. $x = 3 + 3\cos\theta$
$y = 2 + 5\sin\theta$

49. Find a parametric representation of the line that passes through the points $(-4, 4)$ and $(9, -10)$.

50. Find a parametric representation of the circle with center $(5, 4)$ and radius 6.

51. Find a parametric representation of the ellipse with center $(-3, 4)$, major axis horizontal and eight units in length, and minor axis six units in length.

52. Find a parametric representation of the hyperbola with vertices $(0, \pm 4)$ and foci $(0, \pm 5)$.

53. Find a parametric representation of the hyperbola with asymptotes $y = 3 \pm \frac{1}{2}(x + 1)$.

54. ***Rotary Engine*** The rotary engine was developed by Felix Wankel in the 1950s. The engine features a rotor that is basically a modified equilateral triangle. The rotor moves in a chamber that, in two dimensions, is an epitrochoid. Use a graphing utility to graph the chamber modeled by the parametric equations $x = \cos 3\theta + 5\cos\theta$ and $y = \sin 3\theta + 5\sin\theta$.

In Exercises 55–58, find dy/dx.

55. $x = 4t - 3$
$y = 7t$

56. $x = t^2 + t - 6$
$y = t^3 - 2t$

57. $x = 6\cos t$
$y = 6\sin t$

58. $x = 3\cos t$
$y = 2\sin^2 t$

In Exercises 59–62, find all points of horizontal and vertical tangency (if any) to the curve. Use a graphing utility to confirm your results.

59. $x = t^2 - 4t$
$y = t^3$

60. $x = 2t^3 - 3t^2$
$y = t^2 + 4t$

61. $x = 2\sin t$
$y = -\cos t$

62. $x = 1 + \cos t$
$y = 1 - 2\sin t$

In Exercises 63–66, plot the point given in polar coordinates and find two additional polar representations of the point, using $-2\pi < \theta < 2\pi$.

63. $\left(2, \dfrac{\pi}{4}\right)$

64. $\left(-5, -\dfrac{\pi}{3}\right)$

65. $(-7, 4.19)$

66. $(\sqrt{3}, 2.62)$

In Exercises 67–70, a point in polar coordinates is given. Convert the point to rectangular coordinates.

67. $\left(-1, \dfrac{\pi}{3}\right)$

68. $\left(2, \dfrac{5\pi}{4}\right)$

69. $\left(3, \frac{3\pi}{4}\right)$

70. $\left(0, \frac{\pi}{2}\right)$

In Exercises 71–74, a point in rectangular coordinates is given. Convert the point to polar coordinates.

71. $(0, 1)$

72. $(-\sqrt{5}, \sqrt{5})$

73. $(4, 6)$

74. $(3, -4)$

In Exercises 75–80, convert the rectangular equation to polar form.

75. $x^2 + y^2 = 81$

76. $x^2 + y^2 = 48$

77. $x^2 + y^2 - 6y = 0$

78. $x^2 + y^2 - 4x = 0$

79. $xy = 5$

80. $xy = -2$

In Exercises 81–86, convert the polar equation to rectangular form.

81. $r = 5$

82. $r = 12$

83. $r = 3 \cos \theta$

84. $r = 8 \sin \theta$

85. $r^2 = \sin \theta$

86. $r^2 = 4 \cos 2\theta$

In Exercises 87–90, find the points of horizontal and vertical tangency (if any) to the graph of the polar curve for $0 \le \theta < 2\pi$.

87. $r = 1 - 2 \sin \theta$

88. $r = 1 + 2 \cos \theta$

89. $r = \cos \theta$

90. $r = 1 + \cos \theta$

In Exercises 91–100, determine the symmetry of r, the maximum value of $|r|$, and any zeros of r. Then sketch the graph of the polar equation (plot additional points if necessary).

91. $r = 6$

92. $r = 11$

93. $r = 4 \sin 2\theta$

94. $r = \cos 5\theta$

95. $r = -2(1 + \cos \theta)$

96. $r = 1 - 4 \cos \theta$

97. $r = 2 + 6 \sin \theta$

98. $r = 5 - 5 \cos \theta$

99. $r = -3 \cos 2\theta$

100. $r^2 = \cos 2\theta$

In Exercises 101–104, identify the type of polar graph and use a graphing utility to graph the equation.

101. $r = 3(2 - \cos \theta)$

102. $r = 5(1 - 2 \cos \theta)$

103. $r = 8 \cos 3\theta$

104. $r^2 = 2 \sin 2\theta$

In Exercises 105–108, identify the conic and sketch its graph.

105. $r = \dfrac{1}{1 + 2 \sin \theta}$

106. $r = \dfrac{6}{1 + \sin \theta}$

107. $r = \dfrac{4}{5 - 3 \cos \theta}$

108. $r = \dfrac{16}{4 + 5 \cos \theta}$

In Exercises 109–112, find a polar equation of the conic with its focus at the pole.

109. Parabola Vertex: $(2, \pi)$

110. Parabola Vertex: $(2, \pi/2)$

111. Ellipse Vertices: $(5, 0)$, $(1, \pi)$

112. Hyperbola Vertices: $(1, 0)$, $(7, 0)$

113. ***Explorer 18*** On November 27, 1963, the United States launched Explorer 18. Its low and high points above the surface of Earth were 119 miles and 122,800 miles, respectively. The center of Earth was at one focus of the orbit (see figure). Find the polar equation of the orbit and find the distance between the surface of Earth (assume Earth has a radius of 4000 miles) and the satellite when $\theta = \pi/3$.

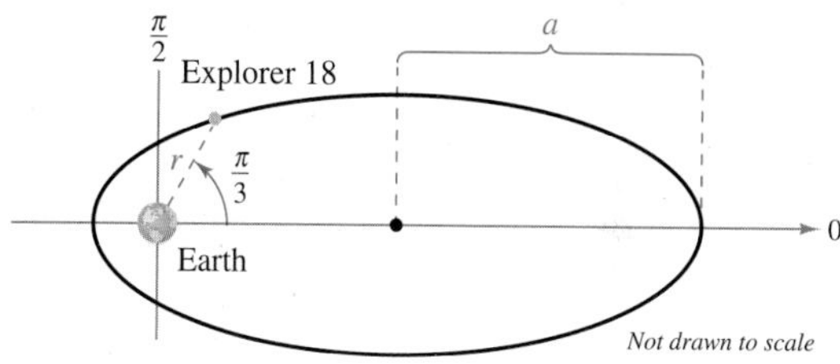

114. ***Asteroid*** An asteroid takes a parabolic path with Earth as its focus. It is about 6,000,000 miles from Earth at its closest approach. Write the polar equation of the path of the asteroid with its vertex at $\theta = \pi/2$. Find the distance between the asteroid and Earth when $\theta = -\pi/3$.

True or False? **In Exercises 115–117, determine whether the statement is true or false. Justify your answer.**

115. The graph of

$$\frac{1}{4}x^2 - y^4 = 1$$

is a hyperbola.

116. Only one set of parametric equations can represent the line $y = 3 - 2x$.

117. There is a unique polar coordinate representation of each point in the plane.

118. Consider an ellipse with the major axis horizontal and 10 units in length. The number b in the standard form of the equation of the ellipse must be less than what real number? Explain the change in the shape of the ellipse as b approaches this number.

119. What is the relationship between the graphs of the rectangular and polar equations?

(a) $x^2 + y^2 = 25, \quad r = 5$

(b) $x - y = 0, \quad \theta = \dfrac{\pi}{4}$

12 CHAPTER TEST

Take this test as you would take a test in class. When you are finished, check your work against the answers given in the back of the book.

In Exercises 1–4, classify the conic and write the equation in standard form. Identify the center, vertices, foci, and asymptotes (if applicable). Then sketch the graph of the conic.

1. $y^2 - 2x + 2 = 0$

2. $x^2 - 4y^2 - 4x = 0$

3. $9x^2 + 16y^2 + 54x - 32y - 47 = 0$

4. $2x^2 + 2y^2 - 8x - 4y + 9 = 0$

5. Find the standard form of the equation of the parabola with vertex $(2, -3)$, with a vertical axis, and passing through the point $(4, 0)$.

6. Find the standard form of the equation of the hyperbola with foci $(0, 0)$ and $(0, 4)$ and asymptotes $y = \pm\frac{1}{2}x + 2$.

7. Sketch the curve represented by the parametric equations $x = 2 + 3\cos\theta$ and $y = 2\sin\theta$. Eliminate the parameter and write the corresponding rectangular equation.

8. Find a set of parametric equations of the line passing through the points $(2, -3)$ and $(6, 4)$. (There are many correct answers.)

In Exercises 9 and 10, find *dy/dx*.

9. $\dfrac{(x-5)^2}{9} + \dfrac{(y-1)^2}{16} = 1$

10. $x = 2t + 3,\ y = 4t^2 - 10t$

In Exercises 11 and 12, find an equation of the tangent line to the conic at the given point.

11. $(x + 2)^2 = 4(y - 3),\quad (0, 4)$

12. $y^2 - \dfrac{(x+2)^2}{4} = 1,\quad \left(-6, \sqrt{5}\right)$

13. Convert the polar coordinate $\left(-2, \dfrac{5\pi}{6}\right)$ to rectangular form.

14. Convert the rectangular coordinate $(2, -2)$ to polar form and find two additional polar representations of this point.

15. Convert the rectangular equation $x^2 + y^2 - 3x = 0$ to polar form.

In Exercises 16–19, sketch the graph of the polar equation. Identify the type of graph.

16. $r = \dfrac{4}{1 + \cos\theta}$

17. $r = \dfrac{4}{2 + \sin\theta}$

18. $r = 2 + 3\sin\theta$

19. $r = 2\sin 4\theta$

20. Find a polar equation of the ellipse with focus at the pole, eccentricity $e = \frac{1}{4}$, and directrix $y = 4$.

21. Find the area of the region bounded by the ellipse $25x^2 + 4y^2 = 100$.

22. A baseball is hit at a point 3 feet above the ground toward the left field fence. The fence is 10 feet high and 375 feet from home plate. The path of the baseball can be modeled by the parametric equations $x = (115\cos\theta)t$ and $y = 3 + (115\sin\theta)t - 16t^2$. Will the baseball go over the fence if it is hit at an angle of $\theta = 30°$? Will the baseball go over the fence if $\theta = 35°$?

P.S. PROBLEM SOLVING

1. A hyperbolic mirror (used in some telescopes) has the property that a light ray directed at a focus will be reflected to the other focus (see figure). The focus of a hyperbolic mirror has coordinates (24, 0). Find the vertex of the mirror if its mount has coordinates (24, 24).

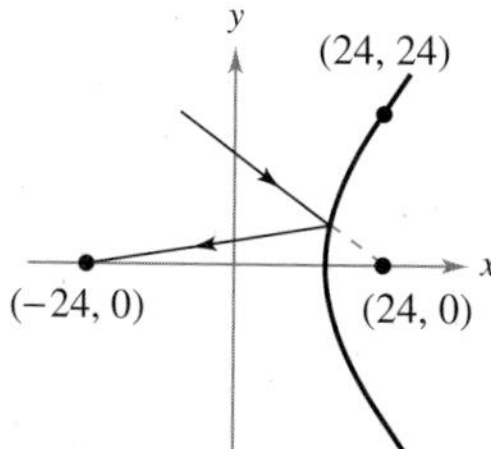

2. A line segment through a focus of an ellipse with endpoints on the ellipse and perpendicular to the major axis is called a **latus rectum** of the ellipse. Therefore, an ellipse has two latera recta. Knowing the length of the latera recta is helpful in sketching an ellipse because it yields other points on the curve (see figure). Show that the length of each latus rectum is $2b^2/a$.

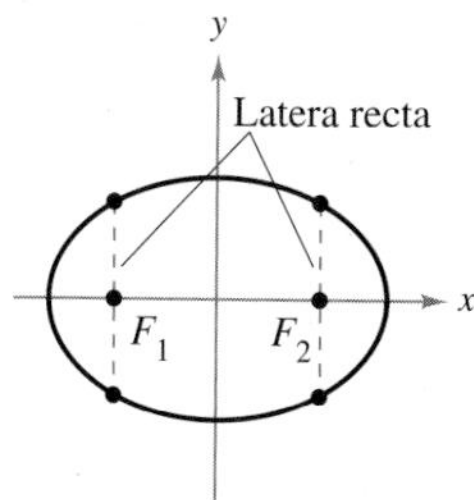

3. Find the equation(s) of all parabolas that have the x-axis as the axis of symmetry and focus at the origin.

4. Find the area of the square inscribed in the ellipse below.

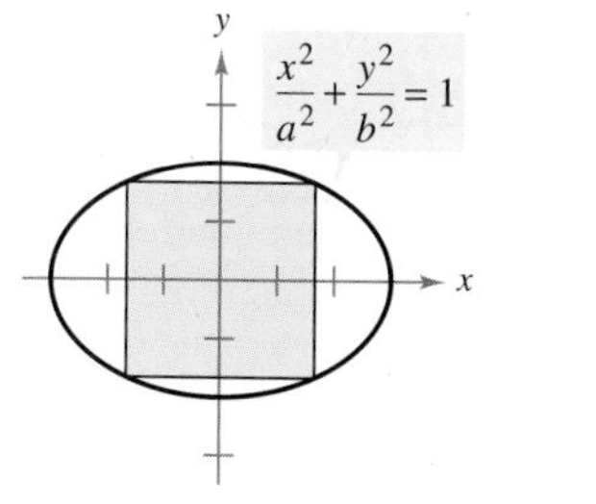

Figure for 4

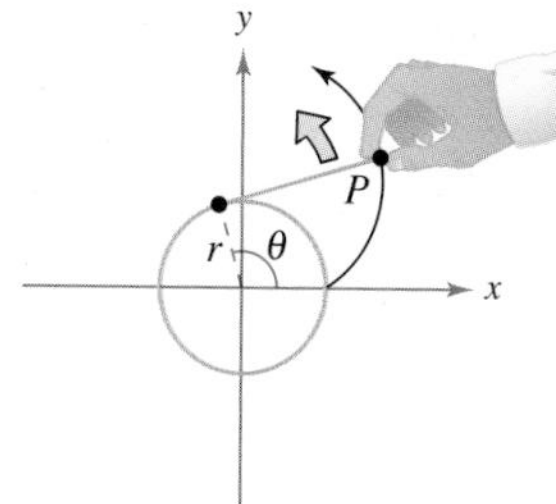

Figure for 5

5. The *involute* of a circle is described by the endpoint P of a string that is held taut as it is unwound from a spool (see figure). The spool does not rotate. Show that

$x = r(\cos\theta + \theta\sin\theta) \qquad y = r(\sin\theta - \theta\cos\theta)$

is a parametric representation of the involute of a circle.

6. A tour boat travels between two islands that are 12 miles apart (see figure). For a trip between the islands, there is enough fuel for a 20-mile trip.

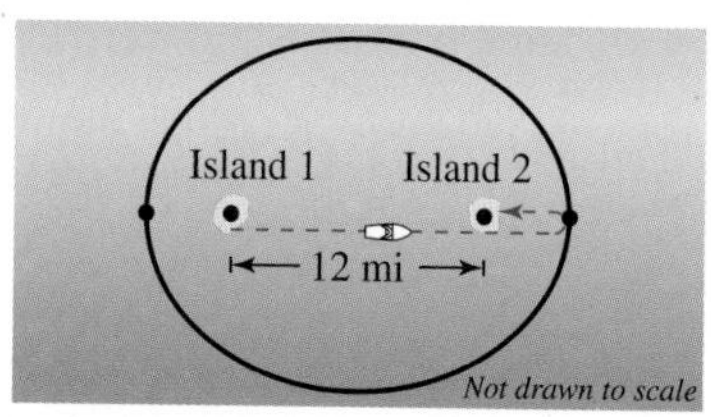

(a) Explain why the region in which the boat can travel is bounded by an ellipse.

(b) Let (0, 0) represent the center of the ellipse. Find the coordinates of each island.

(c) The boat travels from one island, straight past the other island to the vertex of the ellipse, and back to the second island. How many miles does the boat travel? Use your answer to find the coordinates of the vertex.

(d) Use the results from parts (b) and (c) to write an equation of the ellipse that bounds the region in which the boat can travel.

7. Find an equation of the hyperbola such that for any point on the hyperbola, the difference between its distances from the points (2, 2) and (10, 2) is 6.

8. Prove that the graph of the equation

$Ax^2 + Cy^2 + Dx + Ey + F = 0$

is one of the following (except in degenerate cases).

Conic	*Condition*
(a) Circle	$A = C$
(b) Parabola	$A = 0$ or $C = 0$ (but not both)
(c) Ellipse	$AC > 0$
(d) Hyperbola	$AC < 0$

9. The following sets of parametric equations model projectile motion.

$x = (v_0\cos\theta)t \qquad x = (v_0\cos\theta)t$

$y = (v_0\sin\theta)t \qquad y = h + (v_0\sin\theta)t - 16t^2$

(a) Under what circumstances would you use each model?

(b) Eliminate the parameter for each set of equations.

(c) In which case is the path of the moving object not affected by a change in the velocity v? Explain.

10. The area of the shaded region in the figure is

$$A = \frac{8}{3}p^{1/2}b^{3/2}.$$

(a) Use integration to verify the formula for the area of the shaded region in the figure.

(b) Find the area if $p = 2$ and $b = 4$.

(c) Give a geometric explanation of why the area approaches 0 as p approaches 0.

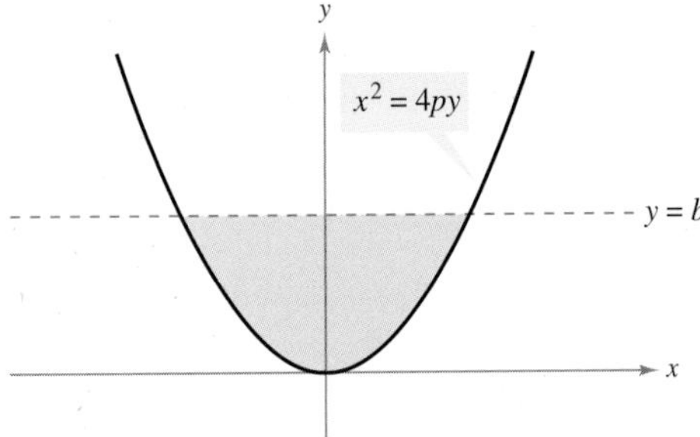

11. The rose curves described in this chapter are of the form

$$r = a \cos n\theta \qquad \text{or} \qquad r = a \sin n\theta$$

where n is a positive integer that is greater than or equal to 2. Use a graphing utility to graph $r = a \cos n\theta$ and $r = a \sin n\theta$ for some noninteger values of n. Describe the graphs.

12. What conic section is represented by the polar equation

$$r = a \sin \theta + b \cos \theta?$$

13. The graph of the polar equation

$$r = e^{\cos \theta} - 2 \cos 4\theta + \sin^5\left(\frac{\theta}{12}\right)$$

is called the *butterfly curve*, as shown in the figure.

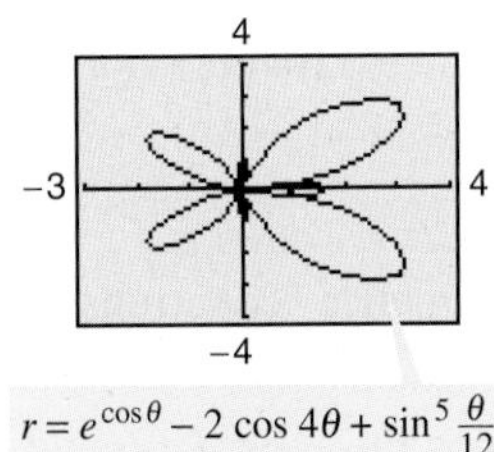

$r = e^{\cos\theta} - 2\cos 4\theta + \sin^5 \frac{\theta}{12}$

(a) The graph shown was produced using $0 \le \theta \le 2\pi$. Does this show the entire graph? Explain your reasoning.

(b) Approximate the maximum r-value of the graph. Does this value change if you use $0 \le \theta \le 4\pi$ instead of $0 \le \theta \le 2\pi$? Explain.

14. As t increases, the ellipse given by the parametric equations $x = \cos t$ and $y = 2 \sin t$ is traced out *counterclockwise*. Find a parametric representation for which the same ellipse is traced out *clockwise*.

15. Use a graphing utility to graph the polar equation

$$r = \cos 5\theta + n \cos \theta$$

for $0 \le \theta \le \pi$ for the integers $n = -5$ to $n = 5$. As you graph these equations, you should see the graph change shape from a heart to a bell. Write a short paragraph explaining what values of n produce the heart portion of the curve and what values of n produce the bell portion.

16. The planets travel in elliptical orbits with the sun at one focus. The polar equation of the orbit of a planet with one focus at the pole and major axis of length $2a$ (see figure) is

$$r = \frac{(1 - e^2)a}{1 - e \cos \theta}$$

where e is the eccentricity. The minimum distance (perihelion) from the sun to a planet is $r = a(1 - e)$ and the maximum distance (aphelion) is $r = a(1 + e)$. For the planet Neptune, $a = 4.495 \times 10^9$ kilometers and $e = 0.0086$. For the dwarf planet Pluto, $a = 5.906 \times 10^9$ kilometers and $e = 0.2488$.

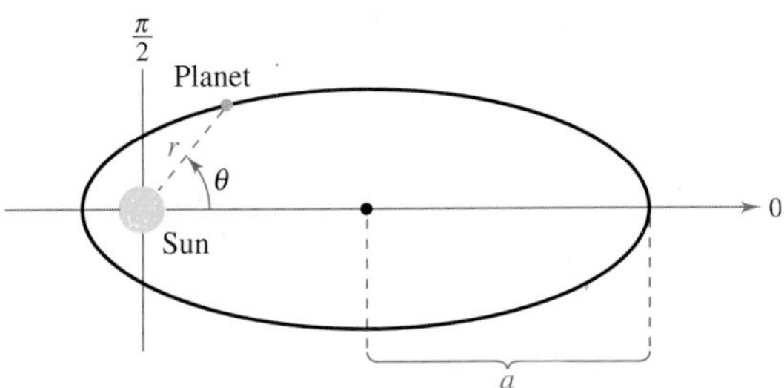

(a) Find the polar equation of the orbit of each planet.

(b) Find the perihelion and aphelion distances for each planet.

(c) Use a graphing utility to graph the equations of the orbits of Neptune and Pluto in the same viewing window.

(d) Is Pluto ever closer to the sun than Neptune? Until recently, Pluto was considered the ninth planet. Why was Pluto called the ninth planet and Neptune the eighth planet?

(e) Do the orbits of Neptune and Pluto intersect? Will Neptune and Pluto ever collide? Why or why not?

13 Additional Topics in Trigonometry

In Chapter 9, you studied trigonometric functions. In this chapter, you will use your knowledge of trigonometric functions to study additional topics, including solving triangles, finding areas, estimating heights, representing vectors, and writing complex numbers.

In this chapter, you should learn the following.

- How to use the Law of Sines to solve oblique triangles and how to find areas of oblique triangles. (13.1)
- How to use the Law of Cosines to solve oblique triangles and how to use Heron's Area Formula to find areas of triangles. (13.2)
- How to write vectors, perform basic vector operations, and represent vectors graphically. (13.3)
- How to find the dot product of two vectors in the plane. (13.4)
- How to write trigonometric forms of complex numbers and how to multiply, divide, find powers of, and find roots of complex numbers. (13.5)

A 30,000-pound truck is parked on a hill of slope $d°$. What force is required to keep the truck from rolling down the hill for varying values of d? (See Section 13.4, Exercise 58.)

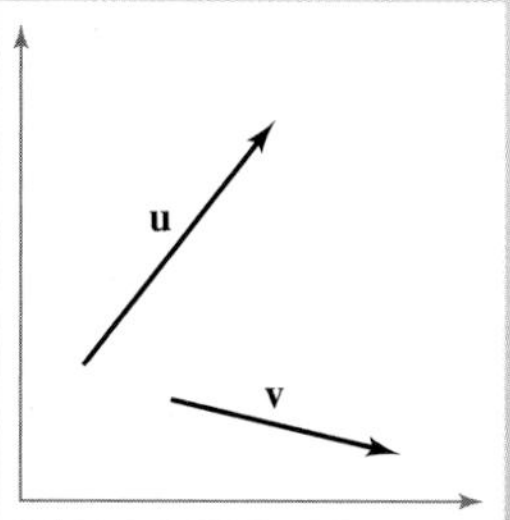

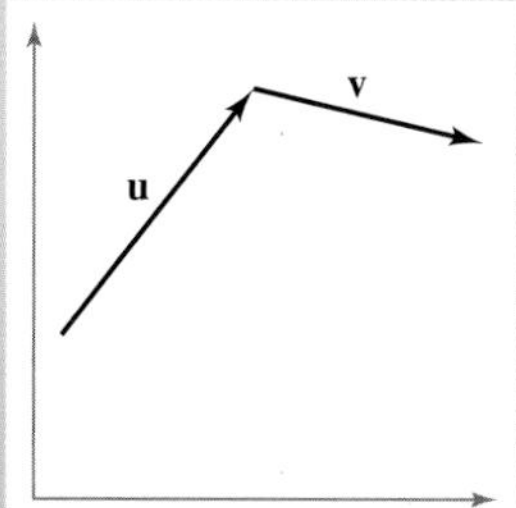

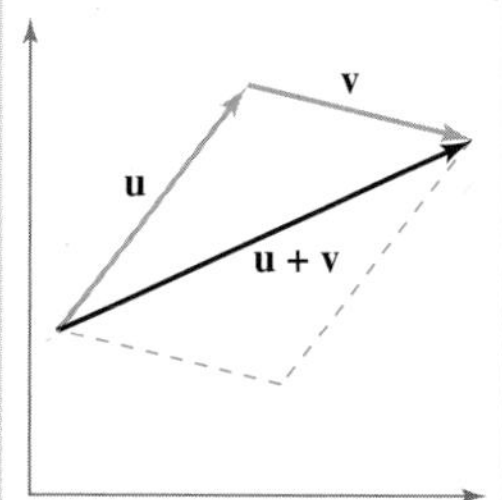

Vectors indicate quantities that involve both magnitude and direction. You can represent vector operations geometrically. For example, the graphs shown above represent vector addition in the plane. (See Section 13.3.)

13.1 Law of Sines

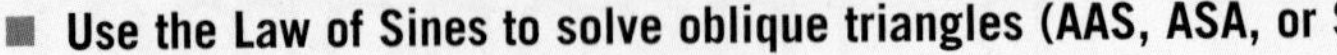

- **Use the Law of Sines to solve oblique triangles (AAS, ASA, or SSA).**
- **Find the areas of oblique triangles.**
- **Use the Law of Sines to model and solve real-life problems.**

Introduction

Figure 13.1

In Chapter 9, you studied techniques for solving right triangles. In this section and the next, you will solve **oblique triangles**—triangles that have no right angles. As standard notation, the angles of a triangle are labeled A, B, and C, and their opposite sides are labeled a, b, and c, as shown in Figure 13.1.

To solve an oblique triangle, you need to know the measure of at least one side and any two other measures of the triangle—either two sides, two angles, or one angle and one side. This breaks down into the following four cases.

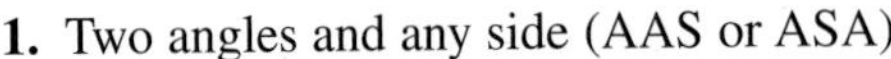

1. Two angles and any side (AAS or ASA)
2. Two sides and an angle opposite one of them (SSA)
3. Three sides (SSS)
4. Two sides and their included angle (SAS)

The first two cases can be solved using the **Law of Sines,** whereas the last two cases require the Law of Cosines (see Section 13.2).

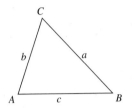

THEOREM 13.1 LAW OF SINES

If ABC is a triangle with sides a, b, and c, then

$$\frac{a}{\sin A} = \frac{b}{\sin B} = \frac{c}{\sin C}.$$

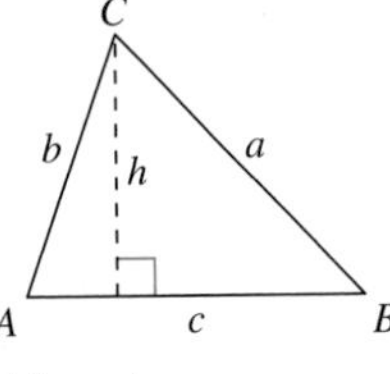

A is acute.

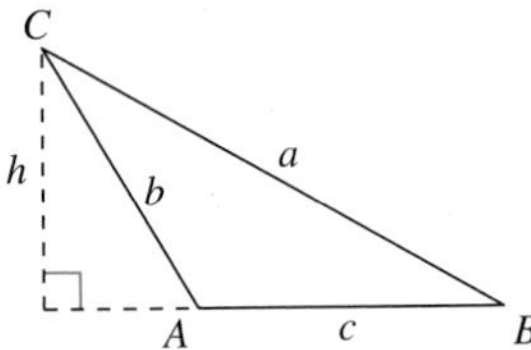

A is obtuse.

NOTE The Law of Sines can also be written in the reciprocal form

$$\frac{\sin A}{a} = \frac{\sin B}{b} = \frac{\sin C}{c}.$$

PROOF Let h be the altitude of either triangle. Then you have $h = b \sin A$ and $h = a \sin B$. Equating these two values of h, you have

$$a \sin B = b \sin A \quad \text{or} \quad \frac{a}{\sin A} = \frac{b}{\sin B}.$$

Note that $\sin A \neq 0$ and $\sin B \neq 0$ because no angle of a triangle can have a measure of 0° or 180°. In a similar manner, by constructing an altitude from vertex B to side AC (extended), you can show that

$$\frac{a}{\sin A} = \frac{c}{\sin C}.$$

So, the Law of Sines is established. ■

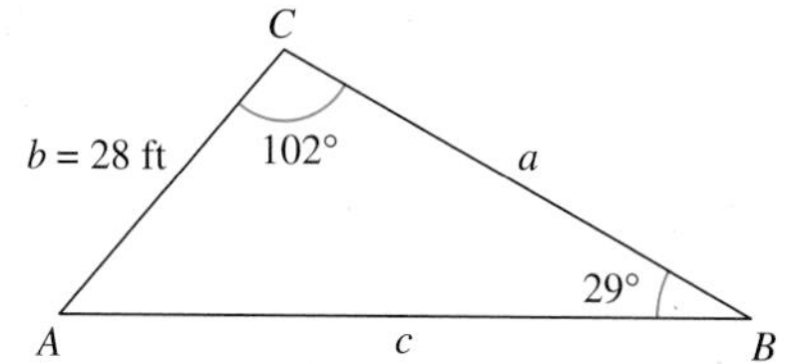

Figure 13.2

EXAMPLE 1 Given Two Angles and One Side—AAS

For the triangle in Figure 13.2, $C = 102°$, $B = 29°$, and $b = 28$ feet. Find the remaining angle and sides.

Solution The third angle of the triangle is

$$\begin{aligned} A &= 180° - B - C \\ &= 180° - 29° - 102° \\ &= 49°. \end{aligned}$$

By the Law of Sines, you have

$$\frac{a}{\sin A} = \frac{b}{\sin B} = \frac{c}{\sin C}.$$

Using $b = 28$ produces

$$\begin{aligned} a &= \frac{b}{\sin B}(\sin A) \\ &= \frac{28}{\sin 29°}(\sin 49°) \approx 43.59 \text{ feet} \end{aligned}$$

and

$$\begin{aligned} c &= \frac{b}{\sin B}(\sin C) \\ &= \frac{28}{\sin 29°}(\sin 102°) \approx 56.49 \text{ feet.} \end{aligned}$$

STUDY TIP When solving triangles, a careful sketch is useful as a quick test for the feasibility of an answer. Remember that the longest side lies opposite the largest angle, and the shortest side lies opposite the smallest angle.

EXAMPLE 2 Given Two Angles and One Side—ASA

A pole tilts *toward* the sun at an 8° angle from the vertical, and it casts a 22-foot shadow. The angle of elevation from the tip of the shadow to the top of the pole is 43°. How tall is the pole?

Solution From Figure 13.3, note that $A = 43°$ and $B = 90° + 8° = 98°$. So, the third angle is

$$\begin{aligned} C &= 180° - A - B \\ &= 180° - 43° - 98° \\ &= 39°. \end{aligned}$$

By the Law of Sines, you have

$$\frac{a}{\sin A} = \frac{c}{\sin C}.$$

Because $c = 22$ feet, the length of the pole is

$$\begin{aligned} a &= \frac{c}{\sin C}(\sin A) \\ &= \frac{22}{\sin 39°}(\sin 43°) \approx 23.84 \text{ feet.} \end{aligned}$$

■

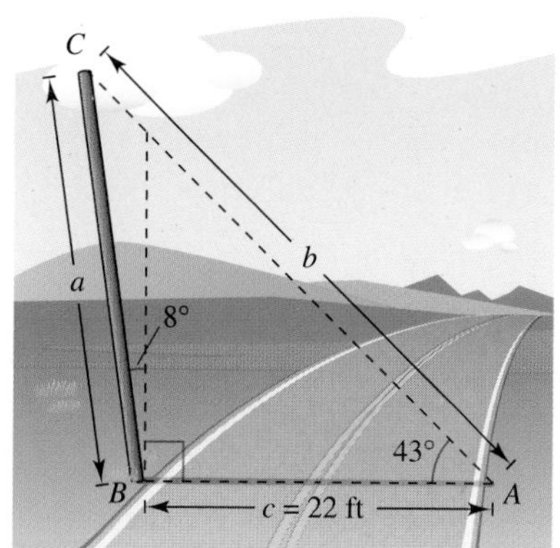

Figure 13.3

For practice, try reworking Example 2 for a pole that tilts *away from* the sun under the same conditions.

The Ambiguous Case (SSA)

In Examples 1 and 2, you saw that two angles and one side determine a unique triangle. However, if two sides and one opposite angle are given, three possible situations can occur: (1) no such triangle exists, (2) one such triangle exists, or (3) two distinct triangles may satisfy the conditions.

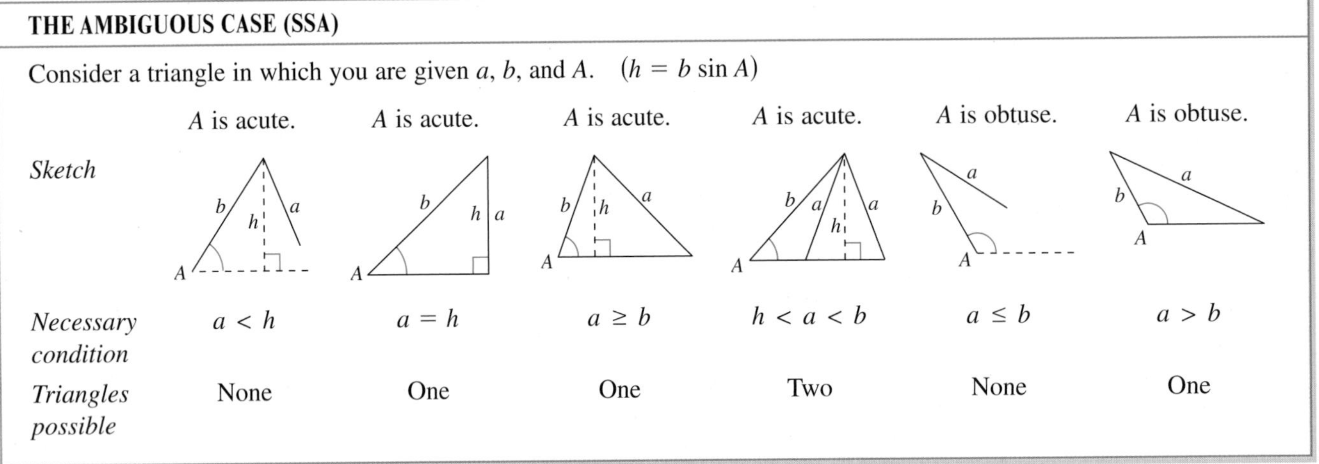

THE AMBIGUOUS CASE (SSA)

Consider a triangle in which you are given a, b, and A. $(h = b \sin A)$

	A is acute.	A is acute.	A is acute.	A is acute.	A is obtuse.	A is obtuse.
Sketch						
Necessary condition	$a < h$	$a = h$	$a \geq b$	$h < a < b$	$a \leq b$	$a > b$
Triangles possible	None	One	One	Two	None	One

EXAMPLE 3 Single-Solution Case—SSA

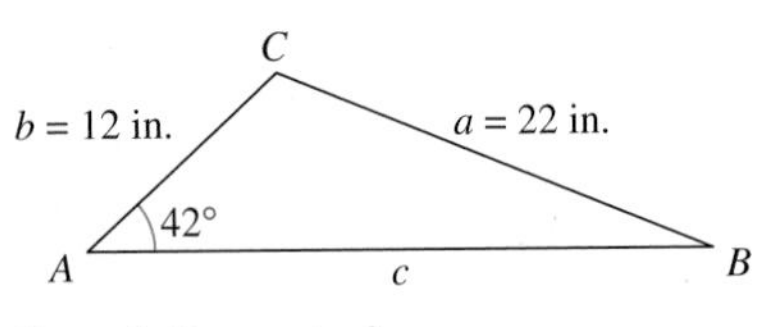

One solution: $a \geq b$

Figure 13.4

For the triangle in Figure 13.4, $a = 22$ inches, $b = 12$ inches, and $A = 42°$. Find the remaining side and angles.

Solution By the Law of Sines, you have

$$\frac{\sin B}{b} = \frac{\sin A}{a} \qquad \text{Reciprocal form}$$

$$\sin B = b\left(\frac{\sin A}{a}\right) \qquad \text{Multiply each side by } b.$$

$$\sin B = 12\left(\frac{\sin 42°}{22}\right) \qquad \text{Substitute for } A, a, \text{ and } b.$$

$$B \approx 21.41°. \qquad B \text{ is acute.}$$

Now, you can determine that

$$C \approx 180° - 42° - 21.41°$$
$$= 116.59°.$$

Then, the remaining side is

$$\frac{c}{\sin C} = \frac{a}{\sin A}$$

$$c = \frac{a}{\sin A}(\sin C)$$

$$= \frac{22}{\sin 42°}(\sin 116.59°)$$

$$\approx 29.40 \text{ inches.}$$

■

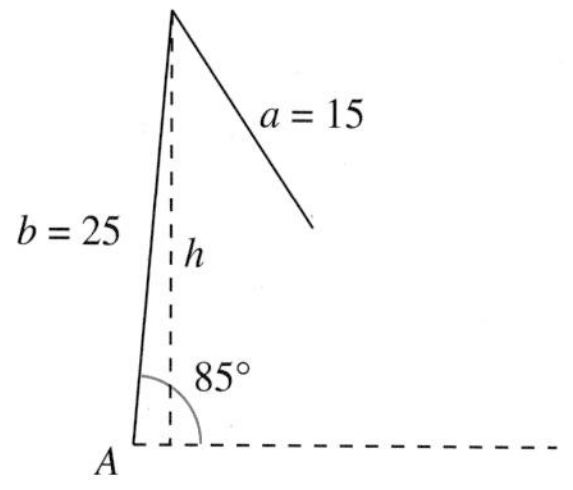

No solution: $a < h$
Figure 13.5

EXAMPLE 4 No-Solution Case—SSA

Show that there is no triangle for which $a = 15$, $b = 25$, and $A = 85°$.

Solution Begin by making the sketch shown in Figure 13.5. From this figure it appears that no triangle is formed. You can verify this using the Law of Sines.

$$\frac{\sin B}{b} = \frac{\sin A}{a} \qquad \text{Reciprocal form}$$

$$\sin B = b\left(\frac{\sin A}{a}\right) \qquad \text{Multiply each side by } b.$$

$$\sin B = 25\left(\frac{\sin 85°}{15}\right) \approx 1.660 > 1$$

This contradicts the fact that $|\sin B| \leq 1$. So, no triangle can be formed having sides $a = 15$ and $b = 25$ and an angle of $A = 85°$.

EXAMPLE 5 Two-Solution Case—SSA

Find two triangles for which $a = 12$ meters, $b = 31$ meters, and $A = 20.5°$.

Solution By the Law of Sines, you have

$$\frac{\sin B}{b} = \frac{\sin A}{a} \qquad \text{Reciprocal form}$$

$$\sin B = b\left(\frac{\sin A}{a}\right)$$

$$= 31\left(\frac{\sin 20.5°}{12}\right) \approx 0.9047.$$

There are two angles, $B_1 \approx 64.8°$ and $B_2 \approx 180° - 64.8° = 115.2°$, between 0° and 180° whose sine is 0.9047. For $B_1 \approx 64.8°$, you obtain

$$C \approx 180° - 20.5° - 64.8° = 94.7°$$

$$c = \frac{a}{\sin A}(\sin C)$$

$$= \frac{12}{\sin 20.5°}(\sin 94.7°) \approx 34.15 \text{ meters.}$$

For $B_2 \approx 115.2°$, you obtain

$$C \approx 180° - 20.5° - 115.2° = 44.3°$$

$$c = \frac{a}{\sin A}(\sin C)$$

$$= \frac{12}{\sin 20.5°}(\sin 44.3°) \approx 23.93 \text{ meters.}$$

The resulting triangles are shown in Figure 13.6.

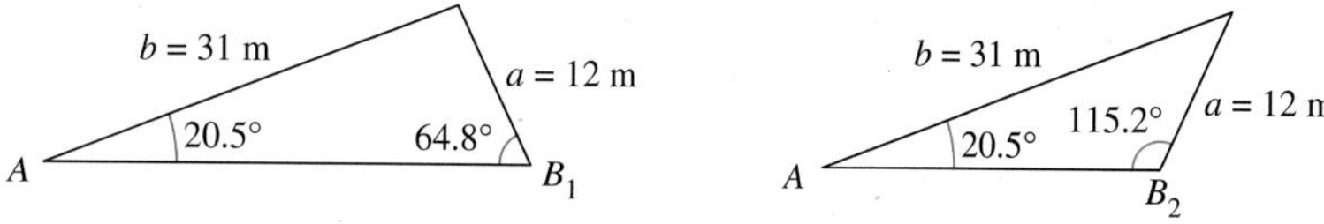

Figure 13.6

Area of an Oblique Triangle

STUDY TIP To see how to obtain the height of the obtuse triangle in Figure 13.7, notice the use of the reference angle $180° - A$ and the difference formula for sine, as follows.

$$\begin{aligned} h &= b \sin(180° - A) \\ &= b(\sin 180° \cos A \\ &\quad - \cos 180° \sin A) \\ &= b[0 \cdot \cos A - (-1) \cdot \sin A] \\ &= b \sin A \end{aligned}$$

The procedure used to prove the Law of Sines leads to a simple formula for the area of an oblique triangle. Referring to Figure 13.7, note that each triangle has a height of $h = b \sin A$. Consequently, the area of each triangle is

$$\text{Area} = \frac{1}{2}(\text{base})(\text{height}) = \frac{1}{2}(c)(b \sin A) = \frac{1}{2}bc \sin A.$$

By similar arguments, you can develop the formulas

$$\text{Area} = \frac{1}{2}ab \sin C = \frac{1}{2}ac \sin B.$$

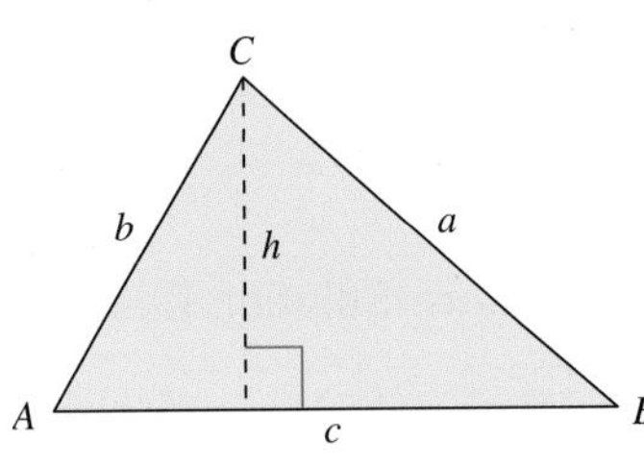

A is acute.

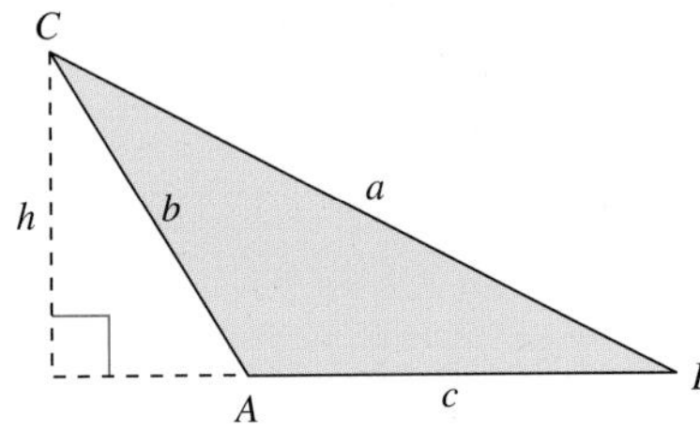

A is obtuse.

Figure 13.7

AREA OF AN OBLIQUE TRIANGLE

The area of any triangle is one-half the product of the lengths of two sides times the sine of their included angle. That is,

$$\text{Area} = \frac{1}{2}bc \sin A = \frac{1}{2}ab \sin C = \frac{1}{2}ac \sin B.$$

Note that if angle A is 90°, the formula gives the area for a right triangle:

$$\text{Area} = \frac{1}{2}bc \sin 90° = \frac{1}{2}bc = \frac{1}{2}(\text{base})(\text{height}). \qquad \sin 90° = 1$$

Similar results are obtained for angles C and B equal to 90°.

EXAMPLE 6 Finding the Area of a Triangular Lot

Find the area of a triangular lot having two sides of lengths 90 meters and 52 meters and an included angle of 102°.

Solution Consider $a = 90$ meters, $b = 52$ meters, and angle $C = 102°$, as shown in Figure 13.8. Then, the area of the triangle is

$$\begin{aligned} \text{Area} &= \frac{1}{2}ab \sin C \\ &= \frac{1}{2}(90)(52)(\sin 102°) \\ &\approx 2289 \text{ square meters.} \end{aligned}$$

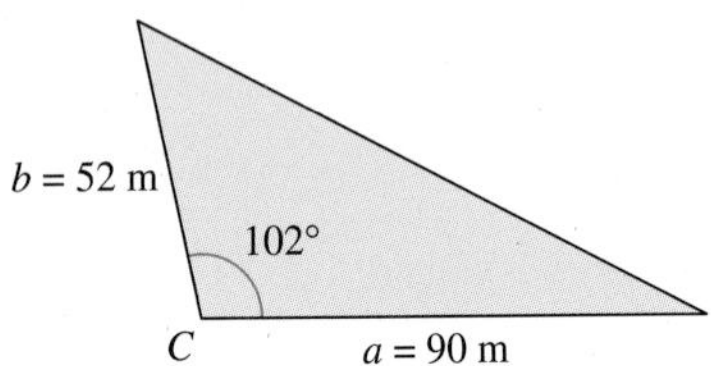

Figure 13.8

Application

EXAMPLE 7 An Application of the Law of Sines

The course for a boat race starts at point A in Figure 13.9 and proceeds in the direction S 52° W to point B, then in the direction S 40° E to point C, and finally back to A. Point C lies 8 kilometers directly south of point A. Approximate the total distance of the race course.

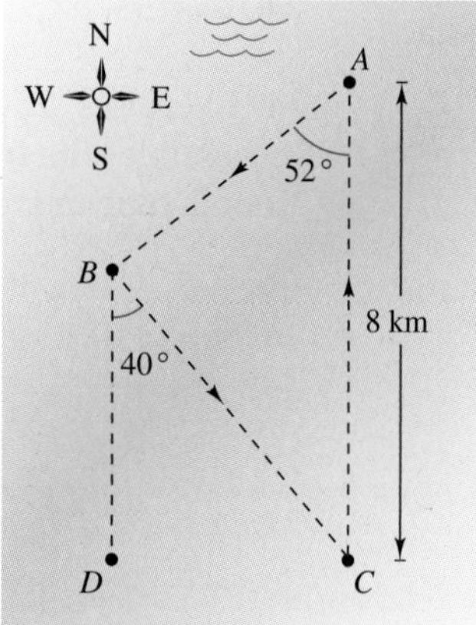

Figure 13.9

Solution Because lines BD and AC are parallel, it follows that $\angle BCA \cong \angle CBD$. Consequently, triangle ABC has the measures shown in Figure 13.10. The measure of angle B is $180° - 52° - 40° = 88°$. Using the Law of Sines,

$$\frac{a}{\sin 52°} = \frac{b}{\sin 88°} = \frac{c}{\sin 40°}.$$

Because $b = 8$,

$$a = \frac{8}{\sin 88°}(\sin 52°) \approx 6.308$$

and

$$c = \frac{8}{\sin 88°}(\sin 40°) \approx 5.145.$$

The total length of the course is approximately

$$\begin{aligned}\text{Length} &\approx 8 + 6.308 + 5.145 \\ &= 19.453 \text{ kilometers.}\end{aligned}$$

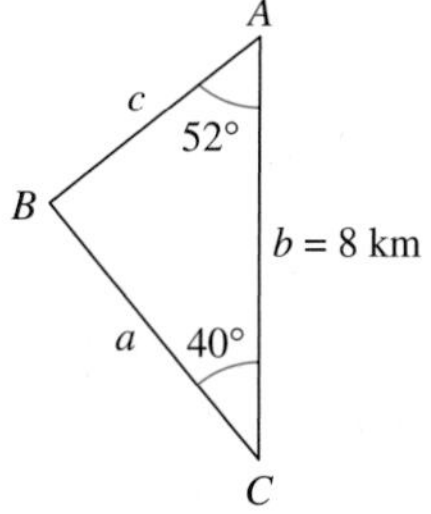

Figure 13.10

■

13.1 Exercises

See www.CalcChat.com for worked-out solutions to odd-numbered exercises.

In Exercises 1–4, fill in the blanks.

1. An ________ triangle is a triangle that has no right angle.
2. For triangle ABC, the Law of Sines is given by $\dfrac{a}{\sin A} = $ ________ $= \dfrac{c}{\sin C}$.
3. Two ________ and one ________ determine a unique triangle.
4. The area of an oblique triangle is given by $\frac{1}{2}bc \sin A = \frac{1}{2}ab \sin C =$ ________ .

In Exercises 5–24, use the Law of Sines to solve the triangle. Round your answers to two decimal places.

5.

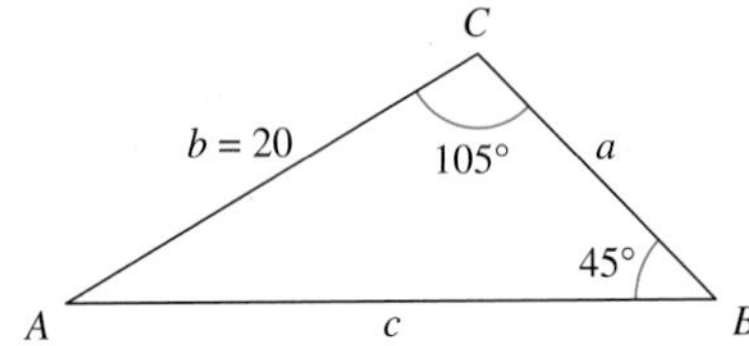

6.

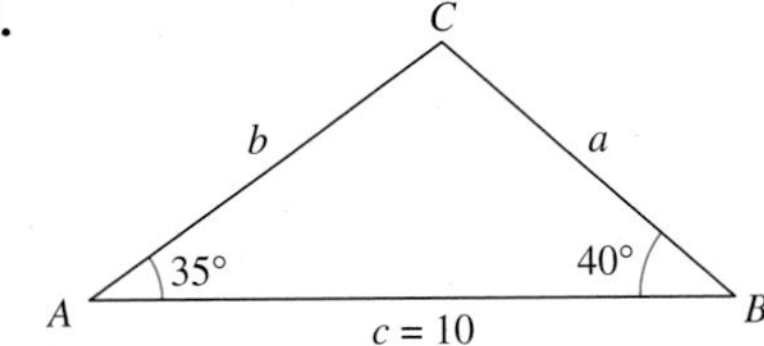

7.

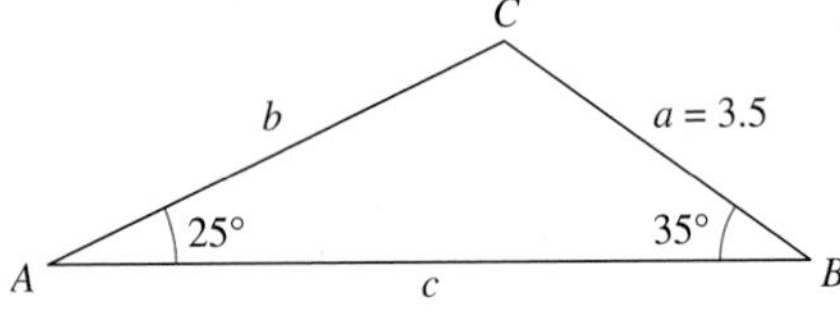

8.

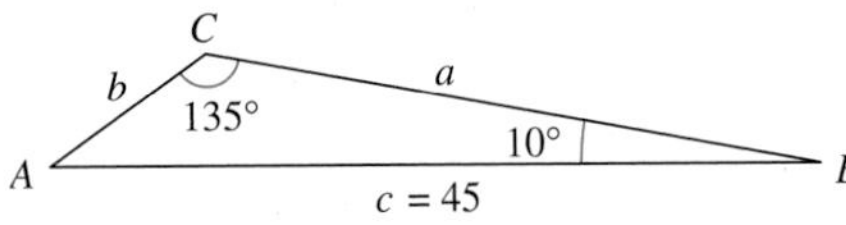

9. $A = 102.4°,\quad C = 16.7°,\quad a = 21.6$
10. $A = 24.3°,\quad C = 54.6°,\quad c = 2.68$
11. $A = 83° 20',\quad C = 54.6°,\quad c = 18.1$
12. $A = 5° 40',\quad B = 8° 15',\quad b = 4.8$
13. $A = 35°,\quad B = 65°,\quad c = 10$
14. $A = 120°,\quad B = 45°,\quad c = 16$
15. $A = 55°,\quad B = 42°,\quad c = \frac{3}{4}$
16. $B = 28°,\quad C = 104°,\quad a = 3\frac{5}{8}$
17. $A = 36°,\quad a = 8,\quad b = 5$
18. $A = 60°,\quad a = 9,\quad c = 10$
19. $B = 15° 30',\quad a = 4.5,\quad b = 6.8$
20. $B = 2° 45',\quad b = 6.2,\quad c = 5.8$
21. $A = 145°,\quad a = 14,\quad b = 4$
22. $A = 100°,\quad a = 125,\quad c = 10$
23. $A = 110° 15',\quad a = 48,\quad b = 16$
24. $C = 95.20°,\quad a = 35,\quad c = 50$

In Exercises 25–34, use the Law of Sines to solve (if possible) the triangle. If two solutions exist, find both. Round your answers to two decimal places.

25. $A = 110°,\quad a = 125,\quad b = 100$
26. $A = 110°,\quad a = 125,\quad b = 200$
27. $A = 76°,\quad a = 18,\quad b = 20$
28. $A = 76°,\quad a = 34,\quad b = 21$
29. $A = 58°,\quad a = 11.4,\quad b = 12.8$
30. $A = 58°,\quad a = 4.5,\quad b = 12.8$
31. $A = 120°,\quad a = b = 25$
32. $A = 120°,\quad a = 25,\quad b = 24$
33. $A = 45°,\quad a = b = 1$
34. $A = 25° 4',\quad a = 9.5,\quad b = 22$

WRITING ABOUT CONCEPTS

In Exercises 35–38, find a value for b such that the triangle has (a) one solution, (b) two solutions, and (c) no solution.

35. $A = 36°,\quad a = 5$
36. $A = 60°,\quad a = 10$
37. $A = 10°,\quad a = 10.8$
38. $A = 88°,\quad a = 315.6$

39. State the Law of Sines.
40. Write a short paragraph explaining how the Law of Sines can be used to solve a right triangle.

In Exercises 41–46, find the area of the triangle having the indicated angle and sides.

41. $C = 120°,\quad a = 4,\quad b = 6$
42. $B = 130°,\quad a = 62,\quad c = 20$
43. $A = 43° 45',\quad b = 57,\quad c = 85$
44. $A = 5° 15',\quad b = 4.5,\quad c = 22$
45. $B = 72° 30',\quad a = 105,\quad c = 64$
46. $C = 84° 30',\quad a = 16,\quad b = 20$

47. ***Height*** Because of prevailing winds, a tree grew so that it was leaning 4° from the vertical. At a point 40 meters from the tree, the angle of elevation to the top of the tree is 30° (see figure). Find the height h of the tree.

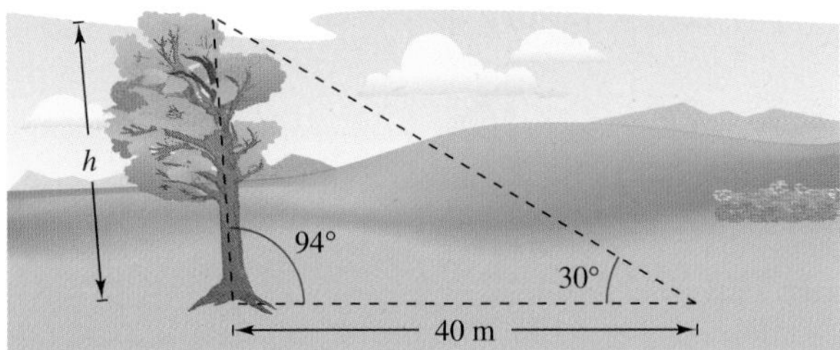

48. ***Height*** A flagpole at a right angle to the horizontal is located on a slope that makes an angle of 12° with the horizontal. The flagpole's shadow is 16 meters long and points directly up the slope. The angle of elevation from the tip of the shadow to the sun is 20°.
    (a) Draw a triangle to represent the situation. Show the known quantities on the triangle and use a variable to indicate the height of the flagpole.
    (b) Write an equation that can be used to find the height of the flagpole.
    (c) Find the height of the flagpole.

49. ***Angle of Elevation*** A 10-meter utility pole casts a 17-meter shadow directly down a slope when the angle of elevation of the sun is 42° (see figure). Find θ, the angle of elevation of the ground.

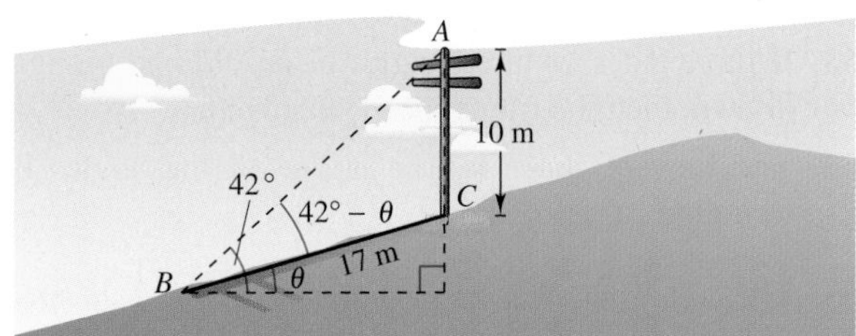

50. ***Flight Path*** A plane flies 500 kilometers with a bearing of 316° from Naples to Elgin (see figure). The plane then flies 720 kilometers from Elgin to Canton (Canton is due west of Naples). Find the bearing of the flight from Elgin to Canton.

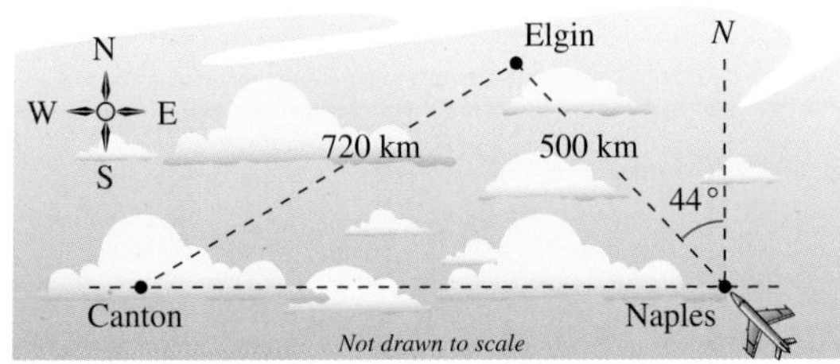

51. ***Bridge Design*** A bridge is to be built across a small lake from a gazebo to a dock (see figure). The bearing from the gazebo to the dock is S 41° W. From a tree 100 meters from the gazebo, the bearings to the gazebo and the dock are S 74° E and S 28° E, respectively. Find the distance from the gazebo to the dock.

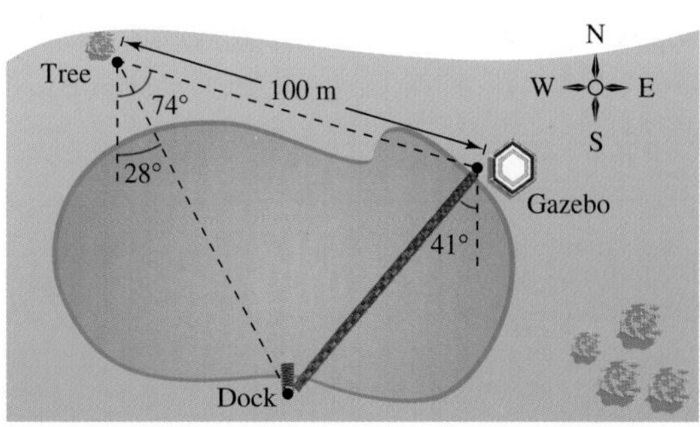

52. ***Railroad Track Design*** The circular arc of a railroad curve has a chord of length 3000 feet corresponding to a central angle of 40°.
    (a) Draw a diagram that visually represents the situation. Show the known quantities on the diagram and use the variables r and s to represent the radius of the arc and the length of the arc, respectively.
    (b) Find the radius r of the circular arc.
    (c) Find the length s of the circular arc.

53. ***Glide Path*** A pilot has just started on the glide path for landing at an airport with a runway of length 9000 feet. The angles of depression from the plane to the ends of the runway are 17.5° and 18.8°.
    (a) Draw a diagram that visually represents the situation.
    (b) Find the air distance the plane must travel until touching down on the near end of the runway.
    (c) Find the ground distance the plane must travel until touching down.
    (d) Find the altitude of the plane when the pilot begins the descent.

54. ***Locating a Fire*** The bearing from the Pine Knob fire tower to the Colt Station fire tower is N 65° E, and the two towers are 30 kilometers apart. A fire spotted by rangers in each tower has a bearing of N 80° E from Pine Knob and S 70° E from Colt Station (see figure). Find the distance of the fire from each tower.

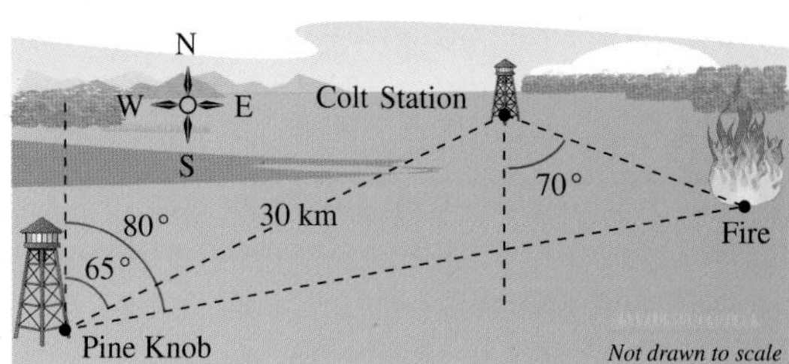

55. ***Distance*** A boat is sailing due east parallel to the shoreline at a speed of 10 miles per hour. At a given time, the bearing to the lighthouse is S 70° E, and 15 minutes later the bearing is S 63° E (see figure). The lighthouse is located at the shoreline. What is the distance from the boat to the shoreline?

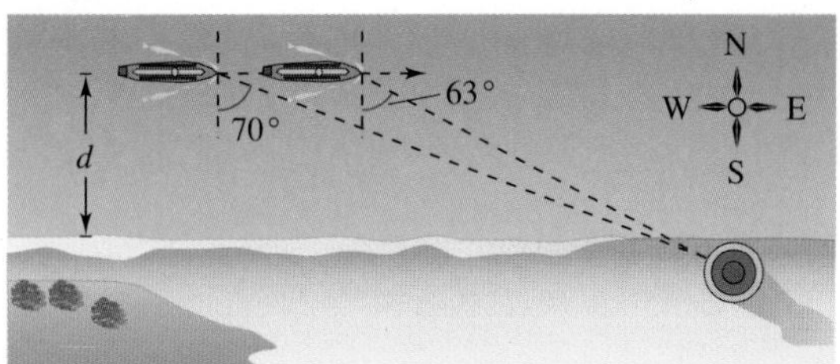

56. ***Distance*** A family is traveling due west on a road that passes a famous landmark. At a given time the bearing to the landmark is N 62° W, and after the family travels 5 miles farther the bearing is N 38° W. What is the closest the family will come to the landmark while on the road?

57. ***Altitude*** The angles of elevation to an airplane from two points A and B on level ground are 55° and 72°, respectively. The points A and B are 2.2 miles apart, and the airplane is east of both points in the same vertical plane. Find the altitude of the plane.

58. ***Distance*** The angles of elevation θ and ϕ to an airplane from the airport control tower and from an observation post 2 miles away are being continuously monitored (see figure). Write an equation giving the distance d between the plane and observation post in terms of θ and ϕ.

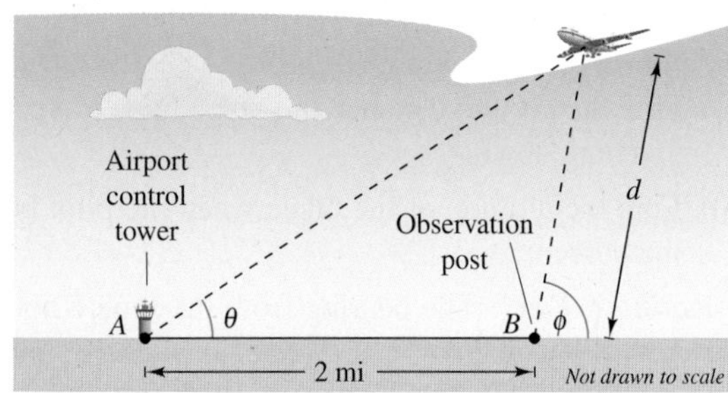

59. ***Area*** You are seeding a triangular courtyard. One side of the courtyard is 52 feet long and another side is 46 feet long. The angle opposite the 52-foot side is 65°.

(a) Draw a diagram that gives a visual representation of the situation.

(b) How long is the third side of the courtyard?

(c) One bag of grass seed covers an area of 50 square feet. How many bags of grass seed will you need to cover the courtyard?

60. ***Graphical Analysis***

(a) Write the area A of the shaded region in the figure as a function of θ.

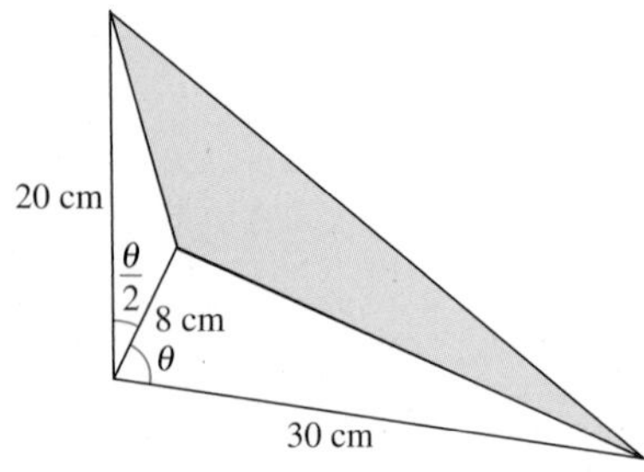

(b) Use a graphing utility to graph the function.

(c) Determine the domain of the function. Explain how the area of the region and the domain of the function would change if the 8-centimeter line segment were decreased in length.

(d) Differentiate the function and use the *zero* or *root* feature of a graphing utility to approximate the critical number.

True or False? **In Exercises 61–65, determine whether the statement is true or false. If it is false, explain why or give an example that shows it is false.**

61. It is not possible to create an obtuse triangle whose longest side is one of the sides that forms its obtuse angle.

62. Two angles and one side of a triangle do not necessarily determine a unique triangle.

63. If three sides or three angles of an oblique triangle are known, then the triangle can be solved.

64. The Law of Sines is true if one of the angles in the triangle is a right angle.

65. The area of an oblique triangle is Area $= \frac{1}{2}ab \sin A$.

CAPSTONE

66. In the figure, a triangle is to be formed by drawing a line segment of length a from $(4, 3)$ to the positive x-axis. For what value(s) of a can you form (a) one triangle, (b) two triangles, and (c) no triangles? Explain your reasoning.

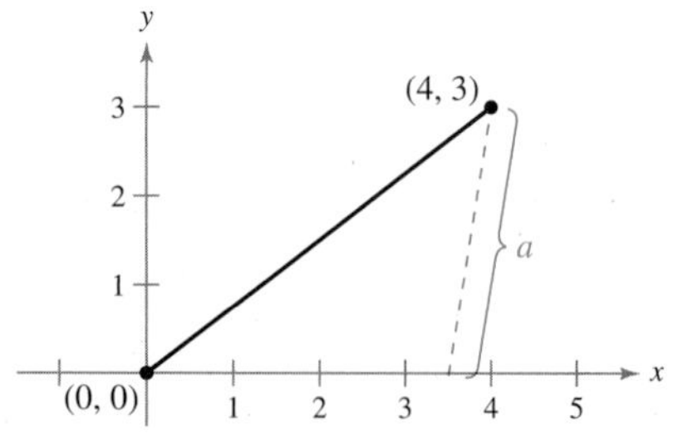

13.2 Law of Cosines

- **Use the Law of Cosines to solve oblique triangles (SSS or SAS).**
- **Use Heron's Area Formula to find the area of a triangle.**
- **Use the Law of Cosines to model and solve real-life problems.**

Introduction

Two cases remain in the list of conditions needed to solve an oblique triangle—SSS and SAS. If you are given three sides (SSS), or two sides and their included angle (SAS), none of the ratios in the Law of Sines would be complete. In such cases, you can use the **Law of Cosines.** See Appendix A for a proof of the Law of Cosines.

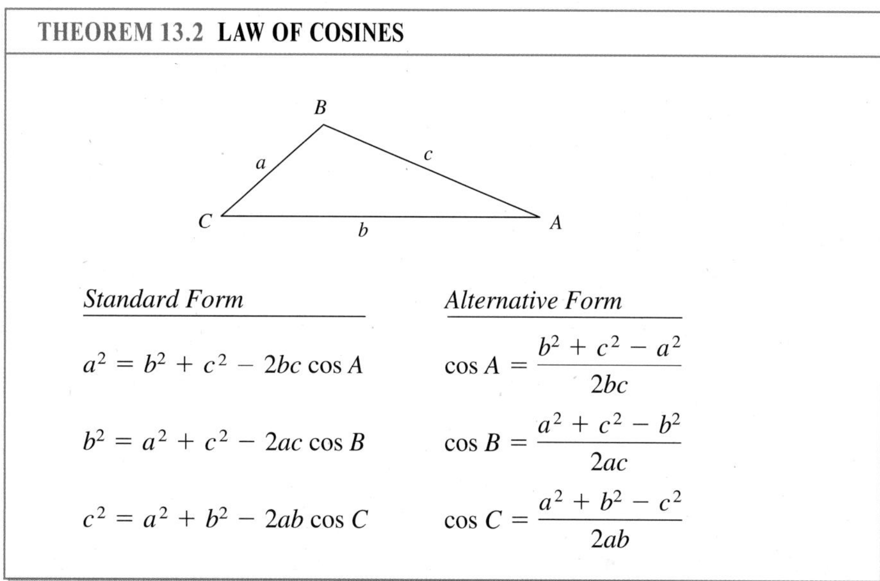

THEOREM 13.2 LAW OF COSINES

Standard Form	*Alternative Form*
$a^2 = b^2 + c^2 - 2bc \cos A$	$\cos A = \dfrac{b^2 + c^2 - a^2}{2bc}$
$b^2 = a^2 + c^2 - 2ac \cos B$	$\cos B = \dfrac{a^2 + c^2 - b^2}{2ac}$
$c^2 = a^2 + b^2 - 2ab \cos C$	$\cos C = \dfrac{a^2 + b^2 - c^2}{2ab}$

EXAMPLE 1 Three Sides of a Triangle—SSS

Find the three angles of the triangle in Figure 13.11.

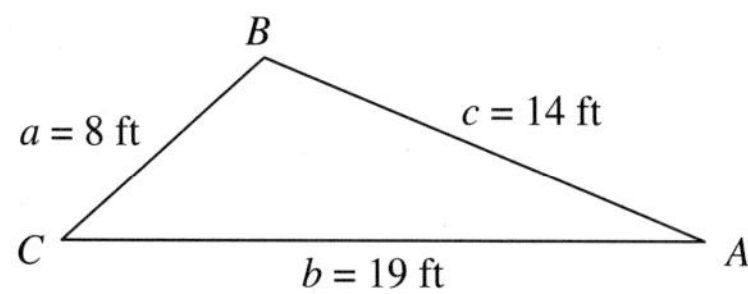

Figure 13.11

Solution It is a good idea first to find the angle opposite the longest side—side b in this case. Using the alternative form of the Law of Cosines, you find that

$$\cos B = \frac{a^2 + c^2 - b^2}{2ac} = \frac{8^2 + 14^2 - 19^2}{2(8)(14)} \approx -0.45089. \qquad \text{Alternative form}$$

Because $\cos B$ is negative, you know that B is an *obtuse* angle given by $B \approx 116.80°$. At this point, it is simpler to use the Law of Sines to determine A.

$$\sin A = a\left(\frac{\sin B}{b}\right)$$

$$\approx 8\left(\frac{\sin 116.80°}{19}\right) \approx 0.37583$$

You know that A must be acute because B is obtuse, and a triangle can have, at most, one obtuse angle. So, $A \approx 22.08°$ and

$$C \approx 180° - 22.08° - 116.80° = 41.12°.$$

■

EXPLORATION

What familiar formula do you obtain when you use the third form of the Law of Cosines

$$c^2 = a^2 + b^2 - 2ab \cos C$$

and you let $C = 90°$? What is the relationship between the Law of Cosines and this formula?

Do you see why it was wise to find the largest angle *first* in Example 1? Knowing the cosine of an angle, you can determine whether the angle is acute or obtuse. That is,

$$\cos \theta > 0 \quad \text{for} \quad 0° < \theta < 90° \qquad \text{Acute}$$

$$\cos \theta < 0 \quad \text{for} \quad 90° < \theta < 180°. \qquad \text{Obtuse}$$

So, in Example 1, once you found that angle B was obtuse, you knew that angles A and C were both acute. If the largest angle is acute, the remaining two angles are acute also.

EXAMPLE 2 Two Sides and the Included Angle—SAS

Find the remaining angles and side of the triangle in Figure 13.12.

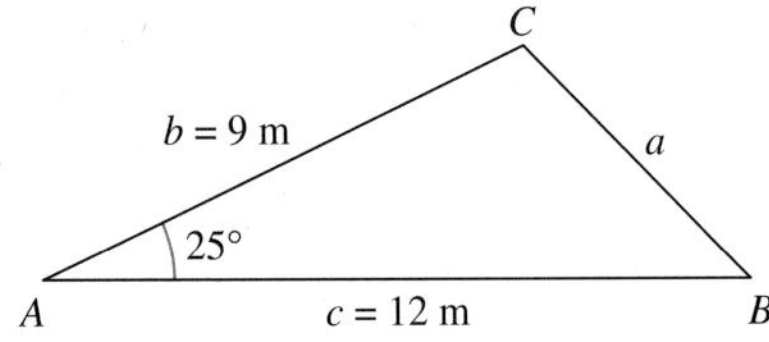

Figure 13.12

Solution Use the Law of Cosines to find the unknown side a in the figure.

$$a^2 = b^2 + c^2 - 2bc \cos A$$

$$a^2 = 9^2 + 12^2 - 2(9)(12) \cos 25° \qquad \text{Standard form}$$

$$a^2 \approx 29.2375$$

$$a \approx 5.4072$$

Because $a \approx 5.4072$ meters, you now know the ratio $(\sin A)/a$ and you can use the Law of Sines $(\sin B)/b = (\sin A)/a$ to solve for B.

$$\sin B = b\left(\frac{\sin A}{a}\right) = 9\left(\frac{\sin 25°}{5.4072}\right) \approx 0.7034$$

So, $B = \arcsin 0.7034 \approx 44.7°$ and $C \approx 180° - 25° - 44.7° = 110.3°$. ■

Heron's Area Formula

HERON OF ALEXANDRIA

Heron of Alexandria (c. 100 B.C.) was a Greek geometer and inventor. His works describe how to find the areas of triangles, quadrilaterals, regular polygons having 3 to 12 sides, and circles as well as the surface areas and volumes of three-dimensional objects.

The Law of Cosines can be used to establish the following formula for the area of a triangle. This formula is called **Heron's Area Formula** after the Greek mathematician Heron (c. 100 B.C.). A proof of this formula is given in Appendix A.

THEOREM 13.3 HERON'S AREA FORMULA

Given any triangle with sides of lengths a, b, and c, the area of the triangle is

$$\text{Area} = \sqrt{s(s-a)(s-b)(s-c)}$$

where $s = (a + b + c)/2$.

EXAMPLE 3 Using Heron's Area Formula

Find the area of a triangle having sides of lengths $a = 43$ meters, $b = 53$ meters, and $c = 72$ meters.

Solution Because $s = (a + b + c)/2 = 168/2 = 84$, Heron's Area Formula yields

$$\text{Area} = \sqrt{s(s-a)(s-b)(s-c)}$$

$$= \sqrt{84(41)(31)(12)}$$

$$\approx 1131.89 \text{ square meters.}$$ ■

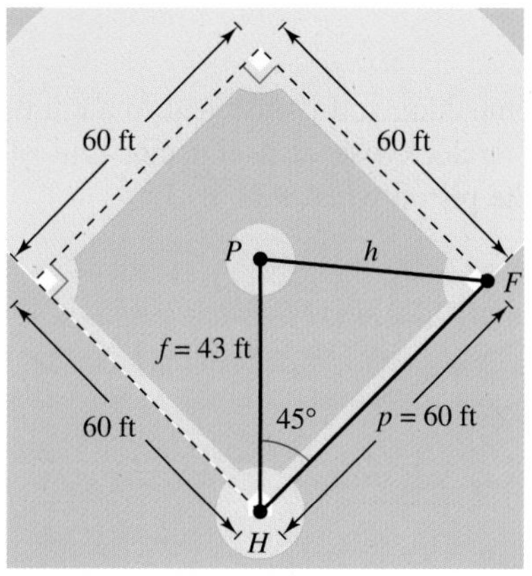

Figure 13.13

Applications

EXAMPLE 4 An Application of the Law of Cosines

The pitcher's mound on a women's softball field is 43 feet from home plate and the distance between the bases is 60 feet, as shown in Figure 13.13. (The pitcher's mound is not halfway between home plate and second base.) How far is the pitcher's mound from first base?

Solution In triangle HPF, $H = 45°$ (line HP bisects the right angle at H), $f = 43$, and $p = 60$. Using the Law of Cosines for this SAS case, you have

$$\begin{aligned} h^2 &= f^2 + p^2 - 2fp \cos H \\ &= 43^2 + 60^2 - 2(43)(60) \cos 45° \\ &\approx 1800.3. \end{aligned}$$

So, the approximate distance from the pitcher's mound to first base is

$$\begin{aligned} h &\approx \sqrt{1800.3} \\ &\approx 42.43 \text{ feet.} \end{aligned}$$

EXAMPLE 5 An Application of the Law of Cosines

A ship travels 60 miles due east, then adjusts its course northward, as shown in Figure 13.14. After traveling 80 miles in that direction, the ship is 139 miles from its point of departure. Describe the bearing from point B to point C.

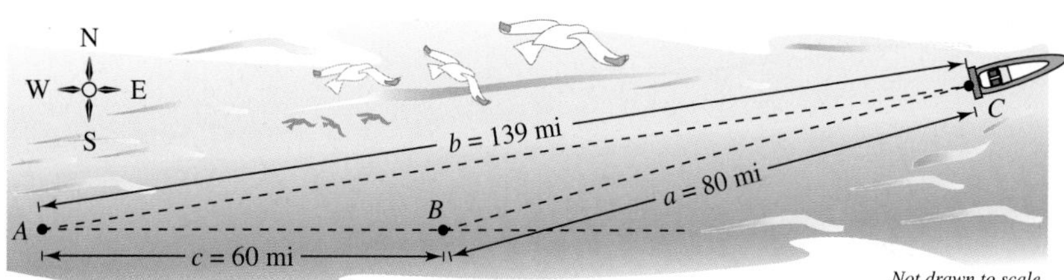

Figure 13.14

Solution You have $a = 80$, $b = 139$, and $c = 60$. So, using the alternative form of the Law of Cosines, you have

$$\begin{aligned} \cos B &= \frac{a^2 + c^2 - b^2}{2ac} \\ &= \frac{80^2 + 60^2 - 139^2}{2(80)(60)} \\ &\approx -0.97094. \end{aligned}$$

So,

$$\begin{aligned} B &\approx \arccos(-0.97094) \\ &\approx 166.15° \end{aligned}$$

and thus the bearing measured from due north from point B to point C is

$$166.15° - 90° = 76.15°, \text{ or N } 76.15° \text{ E.}$$

■

EXAMPLE 6 The Velocity of a Piston

In the engine shown in Figure 13.15, a 7-inch connecting rod is fastened to a crank of radius 3 inches. The crankshaft rotates counterclockwise at a constant rate of 200 revolutions per minute. Find the velocity of the piston when $\theta = \pi/3$.

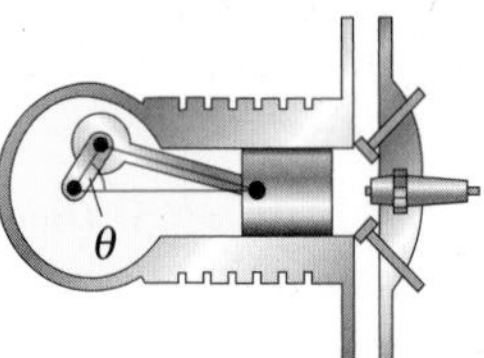

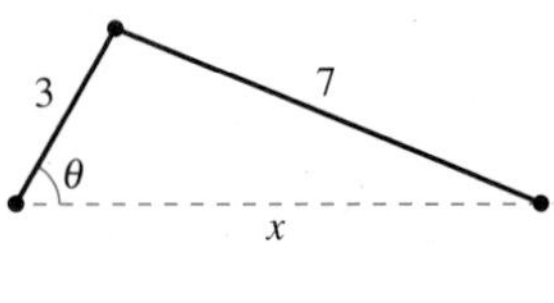

The velocity of a piston is related to the angle of the crankshaft.
Figure 13.15

Solution Label the distances as shown in Figure 13.15. Because a complete revolution corresponds to 2π radians, it follows that $d\theta/dt = 200(2\pi) = 400\pi$ radians per minute.

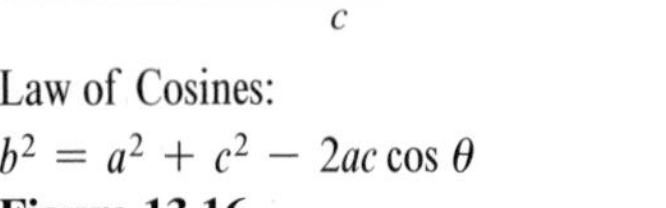

Law of Cosines:
$b^2 = a^2 + c^2 - 2ac\cos\theta$
Figure 13.16

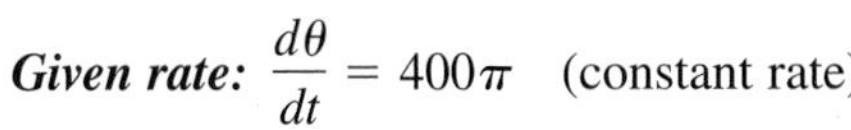

Given rate: $\frac{d\theta}{dt} = 400\pi$ (constant rate)

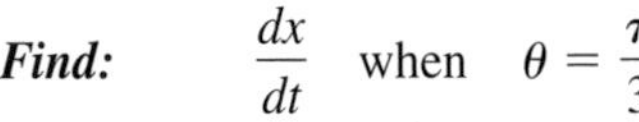

Find: $\frac{dx}{dt}$ when $\theta = \frac{\pi}{3}$

You can use the Law of Cosines (Figure 13.16) to find an equation that relates x and θ.

Equation:

$$7^2 = 3^2 + x^2 - 2(3)(x)\cos\theta$$

$$0 = 2x\frac{dx}{dt} - 6\left(-x\sin\theta\frac{d\theta}{dt} + \cos\theta\frac{dx}{dt}\right)$$

$$(6\cos\theta - 2x)\frac{dx}{dt} = 6x\sin\theta\frac{d\theta}{dt}$$

$$\frac{dx}{dt} = \frac{6x\sin\theta}{6\cos\theta - 2x}\left(\frac{d\theta}{dt}\right)$$

When $\theta = \pi/3$, you can solve for x as shown.

$$7^2 = 3^2 + x^2 - 2(3)(x)\cos\frac{\pi}{3}$$

$$49 = 9 + x^2 - 6x\left(\frac{1}{2}\right)$$

$$0 = x^2 - 3x - 40$$

$$0 = (x - 8)(x + 5)$$

$$x = 8$$ Choose positive solution.

So, when $x = 8$ and $\theta = \pi/3$, the velocity of the piston is

$$\frac{dx}{dt} = \frac{6(8)(\sqrt{3}/2)}{6(1/2) - 16}(400\pi)$$

$$= \frac{9600\pi\sqrt{3}}{-13}$$

$$\approx -4018 \text{ inches per minute.}$$

NOTE The velocity in Example 6 is negative because x represents a distance that is decreasing.

13.2 Exercises

See www.CalcChat.com for worked-out solutions to odd-numbered exercises.

In Exercises 1–4, fill in the blanks.

1. If you are given three sides of a triangle, you would use the Law of ________ to find the three angles of the triangle.
2. If you are given two angles and any side of a triangle, you would use the Law of ________ to solve the triangle.
3. The standard form of the Law of Cosines for $\cos B = \dfrac{a^2 + c^2 - b^2}{2ac}$ is ________ .
4. The Law of Cosines can be used to establish a formula for finding the area of a triangle called ________ ________ Formula.

In Exercises 5–20, use the Law of Cosines to solve the triangle. Round your answers to two decimal places.

5. $a = 10, \quad b = 12, \quad c = 16$
6. $a = 7, \quad b = 3, \quad c = 8$
7. $A = 30°, \quad b = 15, \quad c = 30$
8. $C = 105°, \quad a = 9, \quad b = 4.5$
9. $a = 11, \quad b = 15, \quad c = 21$
10. $a = 55, \quad b = 25, \quad c = 72$
11. $a = 75.4, \quad b = 52, \quad c = 52$
12. $a = 1.42, \quad b = 0.75, \quad c = 1.25$
13. $A = 120°, \quad b = 6, \quad c = 7$
14. $A = 48°, \quad b = 3, \quad c = 14$
15. $B = 10°\,35', \quad a = 40, \quad c = 30$
16. $B = 75°\,20', \quad a = 6.2, \quad c = 9.5$
17. $B = 125°\,40', \quad a = 37, \quad c = 37$
18. $C = 15°\,15', \quad a = 7.45, \quad b = 2.15$
19. $C = 43°, \quad a = \frac{4}{9}, \quad b = \frac{7}{9}$
20. $C = 101°, \quad a = \frac{3}{8}, \quad b = \frac{3}{4}$

In Exercises 21–26, use Heron's Area Formula to find the area of the triangle.

21. $a = 8, \quad b = 12, \quad c = 17$
22. $a = 33, \quad b = 36, \quad c = 25$
23. $a = 2.5, \quad b = 10.2, \quad c = 9$
24. $a = 75.4, \quad b = 52, \quad c = 52$
25. $a = 12.32, \quad b = 8.46, \quad c = 15.05$
26. $a = 3.05, \quad b = 0.75, \quad c = 2.45$

WRITING ABOUT CONCEPTS

27. State the Law of Cosines.

WRITING ABOUT CONCEPTS (continued)

28. List the four cases for solving an oblique triangle. Explain when to use the Law of Sines and when to use the Law of Cosines.

29. ***Navigation*** A boat race runs along a triangular course marked by buoys A, B, and C. The race starts with the boats headed west for 3700 meters. The other two sides of the course lie to the north of the first side, and their lengths are 1700 meters and 3000 meters. Draw a figure that gives a visual representation of the situation, and find the bearings for the last two legs of the race.

30. ***Navigation*** A plane flies 810 miles from Franklin to Centerville with a bearing of 75°. Then it flies 648 miles from Centerville to Rosemount with a bearing of 32°. Draw a figure that visually represents the situation, and find the straight-line distance and bearing from Franklin to Rosemount.

31. ***Surveying*** To approximate the length of a marsh, a surveyor walks 250 meters from point A to point B, then turns 75° and walks 220 meters to point C (see figure). Approximate the length AC of the marsh.

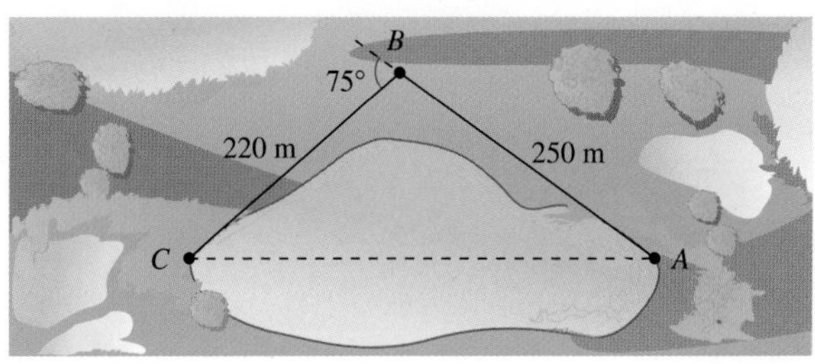

32. ***Surveying*** A triangular parcel of land has 115 meters of frontage, and the other boundaries have lengths of 76 meters and 92 meters. What angles does the frontage make with the two other boundaries?

33. ***Surveying*** A triangular parcel of ground has sides of lengths 725 feet, 650 feet, and 575 feet. Find the measure of the largest angle.

34. ***Streetlight Design*** Determine the angle θ in the design of the streetlight shown in the figure.

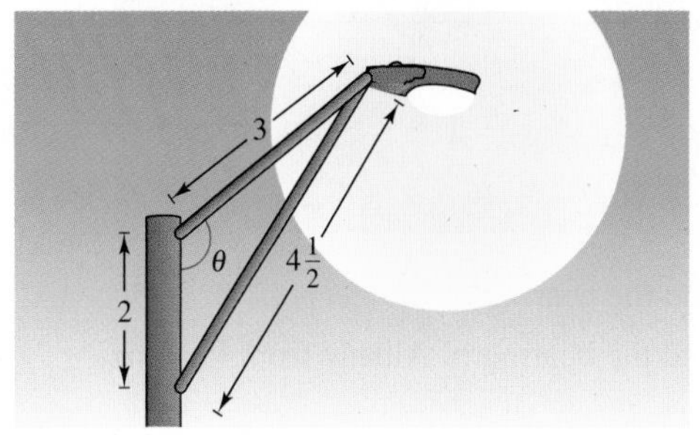

35. ***Distance*** Two ships leave a port at 9 A.M. One travels at a bearing of N 53° W at 12 miles per hour, and the other travels at a bearing of S 67° W at 16 miles per hour. Approximate how far apart they are at noon that day.

36. ***Length*** A 100-foot vertical tower is to be erected on the side of a hill that makes a 6° angle with the horizontal (see figure). Find the length of each of the two guy wires that will be anchored 75 feet uphill and downhill from the base of the tower.

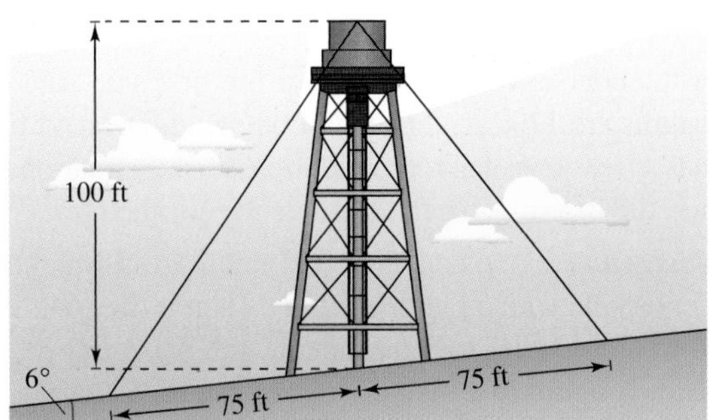

37. ***Navigation*** On a map, Orlando is 178 millimeters due south of Niagara Falls, Denver is 273 millimeters from Orlando, and Denver is 235 millimeters from Niagara Falls (see figure).

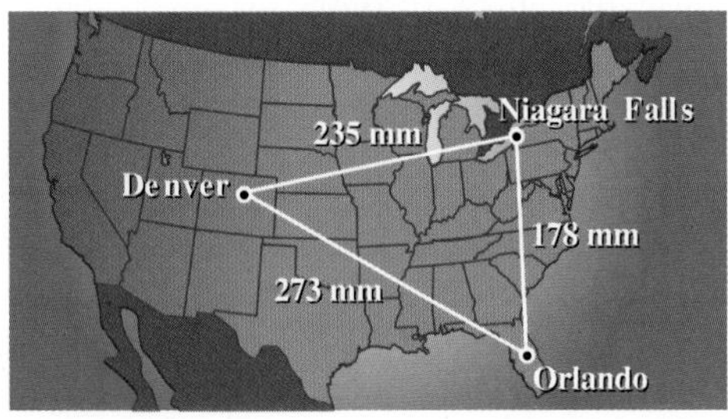

(a) Find the bearing of Denver from Orlando.

(b) Find the bearing of Denver from Niagara Falls.

38. ***Navigation*** On a map, Minneapolis is 165 millimeters due west of Albany, Phoenix is 216 millimeters from Minneapolis, and Phoenix is 368 millimeters from Albany (see figure).

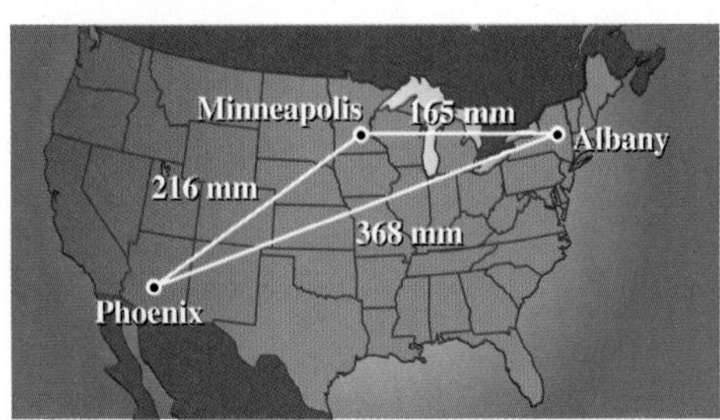

(a) Find the bearing of Minneapolis from Phoenix.

(b) Find the bearing of Albany from Phoenix.

39. ***Baseball*** On a baseball diamond with 90-foot sides, the pitcher's mound is 60.5 feet from home plate. How far is it from the pitcher's mound to third base?

40. ***Baseball*** The baseball player in center field is playing approximately 330 feet from the television camera that is behind home plate. A batter hits a fly ball that goes to the wall 420 feet from the camera (see figure). (a) The camera turns 8° to follow the play. Approximately how far does the center fielder have to run to make the catch? (b) When $\theta = 3°$, the camera was turning at the rate of 1.4° per second. Find the speed of the center fielder.

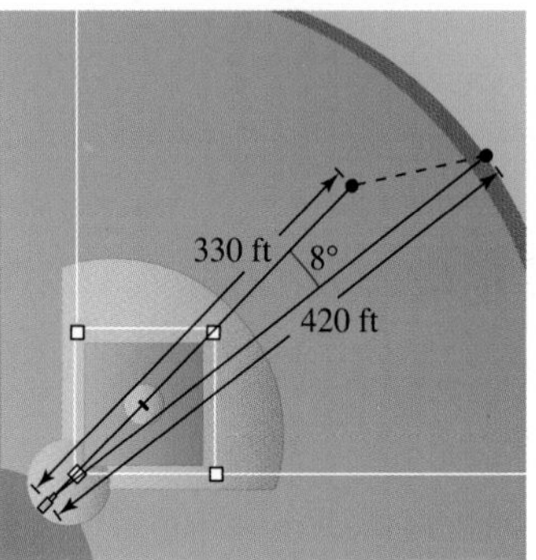

41. ***Aircraft Tracking*** To determine the distance between two aircraft, a tracking station continuously determines the distance to each aircraft and the angle A between them (see figure). (a) Determine the distance a between the planes when $A = 42°$, $b = 35$ miles, and $c = 20$ miles. (b) The plane at angle B is flying at 300 miles per hour and the plane at angle C is flying at 375 miles per hour. What is the rate of separation of the planes at the time of the conditions of part (a)?

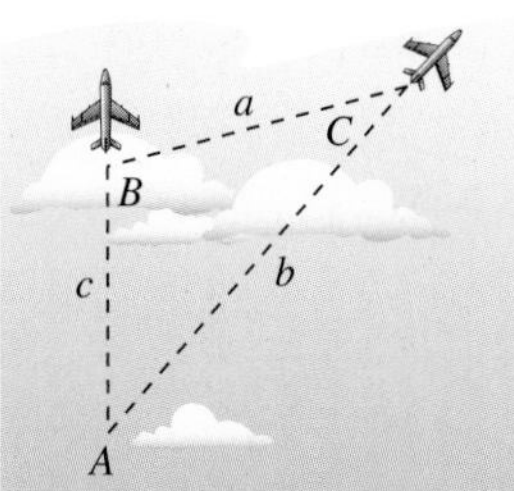

42. ***Aircraft Tracking*** Use the figure for Exercise 41 to determine the distance a between the planes when $A = 11°$, $b = 20$ miles, and $c = 20$ miles.

43. ***Trusses*** Q is the midpoint of the line segment $\overline{PR}$ in the truss rafter shown in the figure. What are the lengths of the line segments $\overline{PQ}$, $\overline{QS}$, and $\overline{RS}$?

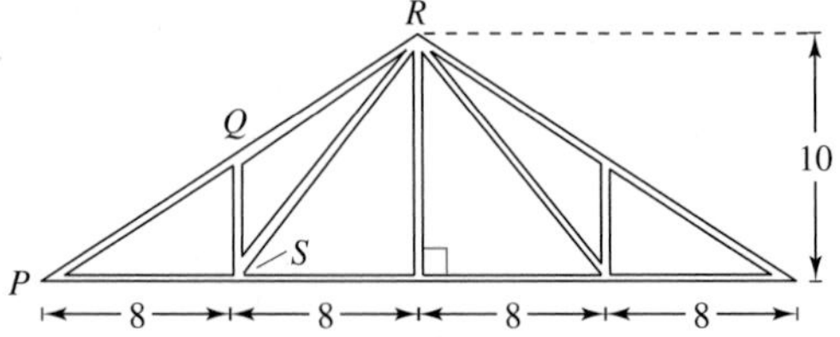

44. ***Velocity of a Piston*** An engine has a 7-inch connecting rod fastened to a crank (see figure). Let d be the distance the piston is from the top of its stroke for an angle θ.

(a) Use the Law of Cosines to write a relationship between x and θ. Use the Quadratic Formula to write x as a function of θ. (Select the sign that yields positive values of x.)

(b) Use the result of part (a) to write d as a function of θ.

(c) Complete the table.

θ	0°	45°	90°	135°	180°
d					

(d) The spark plug fires at $\theta = 5°$ before top dead center. How far is the piston from the top of its stroke?

(e) Use a graphing utility to find the first and second derivatives of the function d. For what values of θ is the speed of the piston 0? For what value in the interval $[0, \pi]$ is it moving at the greatest speed?

(f) If the engine is running at 2500 revolutions per minute, find the speed of the piston when $\theta = 0°$, $\theta = 30°$, $\theta = 90°$, and $\theta = 150°$.

(g) Use a graphing utility to graph the second derivative. The speed of the piston is the same when $\theta = 0°$ and $\theta = 180°$. Is the acceleration on the piston the same for these two values of θ?

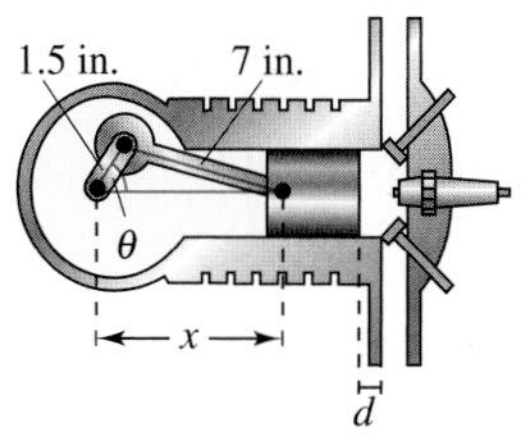

Figure for 44

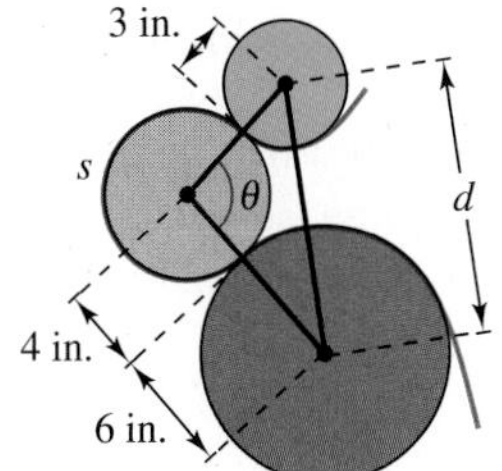

Figure for 45

45. ***Paper Manufacturing*** In a certain process with continuous paper, the paper passes across three rollers of radii 3 inches, 4 inches, and 6 inches (see figure). The centers of the three-inch and six-inch rollers are d inches apart, and the length of the arc in contact with the paper on the four-inch roller is s inches. Complete the following table.

d (inches)	9	10	12	13	14	15	16
θ (degrees)							
s (inches)							

46. ***Awning Design*** A retractable awning above a patio door lowers at an angle of 50° from the exterior wall at a height of 10 feet above the ground (see figure). No direct sunlight is to enter the door when the angle of elevation of the sun is greater than 70°. What is the length x of the awning?

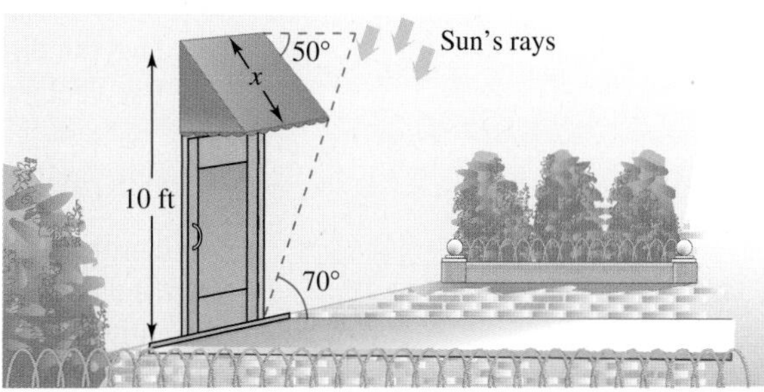

47. ***Geometry*** The lengths of the sides of a triangular parcel of land are approximately 200 feet, 500 feet, and 600 feet. Approximate the area of the parcel.

48. ***Geometry*** A parking lot has the shape of a parallelogram (see figure). The lengths of two adjacent sides are 70 meters and 100 meters. The angle between the two sides is 70°. What is the area of the parking lot?

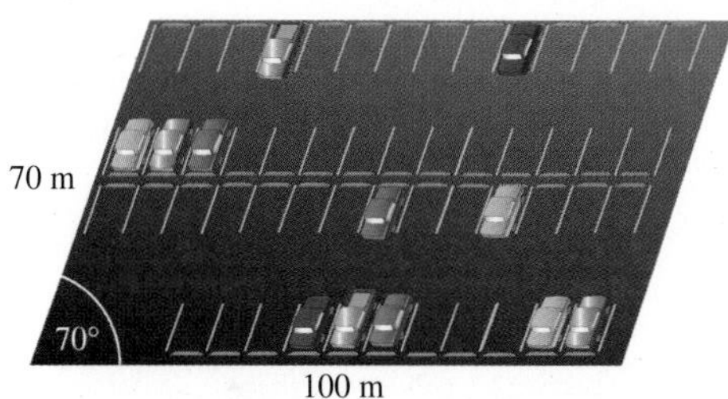

True or False? **In Exercises 49–51, determine whether the statement is true or false. If it is false, explain why or give an example that shows it is false.**

49. In Heron's Area Formula, s is the average of the lengths of the three sides of the triangle.

50. In addition to SSS and SAS, the Law of Cosines can be used to solve triangles with SSA conditions.

51. If the cosine of the largest angle in a triangle is negative, then all the angles in a triangle are acute angles.

CAPSTONE

52. Determine whether the Law of Sines or the Law of Cosines is needed to solve the triangle.

(a) A, C, and a
(b) a, c, and C
(c) b, c, and A
(d) A, B, and c
(e) b, c, and C
(f) a, b, and c

13.3 Vectors in the Plane

- **Represent vectors as directed line segments.**
- **Write the component forms of vectors.**
- **Perform basic vector operations and represent them graphically.**
- **Write vectors as linear combinations of unit vectors.**
- **Find the direction angles of vectors.**
- **Use vectors to model and solve real-life problems.**

Introduction

Quantities such as force and velocity cannot be completely characterized by a single real number because they involve both *magnitude* and *direction*. To represent such a quantity, you can use a **directed line segment,** as shown in Figure 13.17. The directed line segment $\overrightarrow{PQ}$ has **initial point** P and **terminal point** Q. Its **magnitude** (or length) is denoted by $\|\overrightarrow{PQ}\|$ and can be found using the Distance Formula.

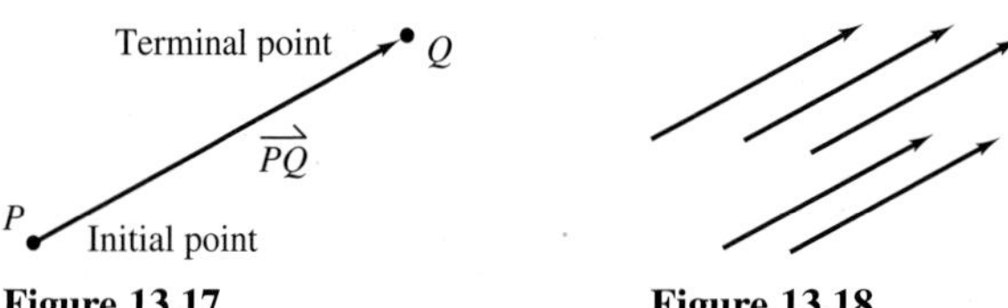

Figure 13.17 **Figure 13.18**

Two directed line segments that have the same magnitude and direction are equivalent. For example, the directed line segments in Figure 13.18 are all equivalent. The set of all directed line segments that are equivalent to the directed line segment $\overrightarrow{PQ}$ is a **vector v in the plane,** written $\mathbf{v} = \overrightarrow{PQ}$. Vectors are denoted by lowercase, boldface letters such as **u**, **v**, and **w**.

EXAMPLE 1 Vector Representation by Directed Line Segments

Let **u** be represented by the directed line segment from $P(0, 3)$ to $Q(4, 5)$, and let **v** be represented by the directed line segment from $R(2, 1)$ to $S(6, 3)$, as shown in Figure 13.19. Show that **u** and **v** are equivalent.

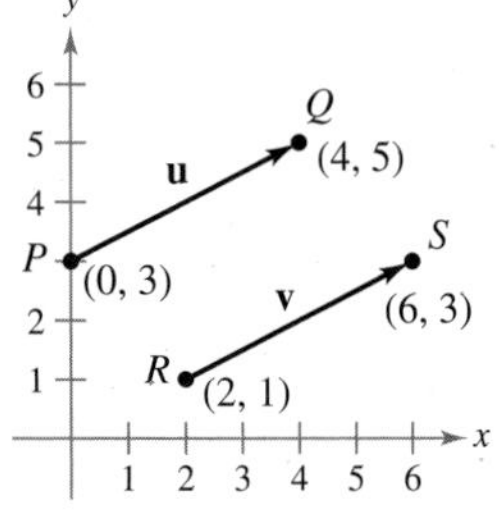

Figure 13.19

Solution From the Distance Formula, it follows that $\overrightarrow{PQ}$ and $\overrightarrow{RS}$ have the *same magnitude.*

$$\begin{aligned}\|\overrightarrow{PQ}\| &= \sqrt{(4-0)^2 + (5-3)^2}\\ &= \sqrt{16+4}\\ &= \sqrt{20}\\ &= 2\sqrt{5}\\ \|\overrightarrow{RS}\| &= \sqrt{(6-2)^2 + (3-1)^2}\\ &= \sqrt{16+4}\\ &= \sqrt{20}\\ &= 2\sqrt{5}\end{aligned}$$

Moreover, both line segments have the *same direction* because they are both directed toward the upper right on lines having a slope of

$$\frac{5-3}{4-0} = \frac{3-1}{6-2} = \frac{2}{4} = \frac{1}{2}.$$

Because $\overrightarrow{PQ}$ and $\overrightarrow{RS}$ have the same magnitude and direction, **u** and **v** are equivalent. ■

TECHNOLOGY You can graph vectors with a graphing utility by graphing directed line segments. Consult the user's guide for your graphing utility for specific instructions.

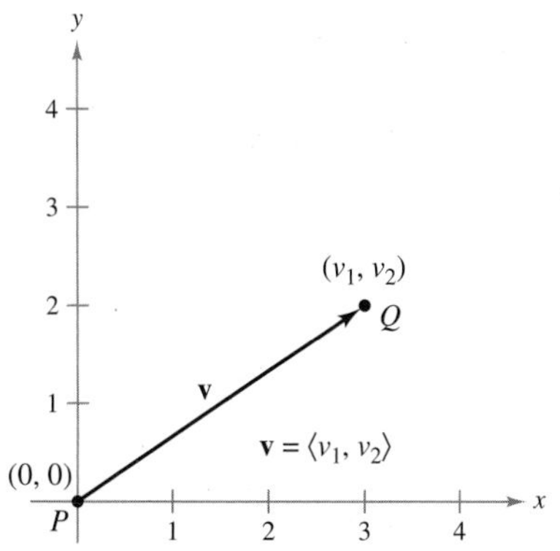

The standard position of a vector
Figure 13.20

Component Form of a Vector

The directed line segment whose initial point is the origin is often the most convenient representative of a set of equivalent directed line segments. This representative of the vector **v** is in **standard position,** as shown in Figure 13.20.

A vector whose initial point is the origin $(0, 0)$ can be uniquely represented by the coordinates of its terminal point (v_1, v_2). This is the **component form of a vector v,** written as $\mathbf{v} = \langle v_1, v_2 \rangle$. The coordinates v_1 and v_2 are the components of **v**. If both the initial point and the terminal point lie at the origin, **v** is the **zero vector** and is denoted by $\mathbf{0} = \langle 0, 0 \rangle$.

COMPONENT FORM OF A VECTOR

The component form of the vector with initial point $P(p_1, p_2)$ and terminal point $Q(q_1, q_2)$ is given by $\overrightarrow{PQ} = \langle q_1 - p_1, q_2 - p_2 \rangle = \langle v_1, v_2 \rangle = \mathbf{v}$.
The **magnitude** (or length) of **v** is given by

$$\|\mathbf{v}\| = \sqrt{(q_1 - p_1)^2 + (q_2 - p_2)^2} = \sqrt{v_1^2 + v_2^2}.$$

If $\|\mathbf{v}\| = 1$, **v** is a **unit vector.** Moreover, $\|\mathbf{v}\| = 0$ if and only if **v** is the zero vector **0**.

Two vectors $\mathbf{u} = \langle u_1, u_2 \rangle$ and $\mathbf{v} = \langle v_1, v_2 \rangle$ are *equal* if and only if $u_1 = v_1$ and $u_2 = v_2$. For instance, in Example 1, the vector **u** from $P(0, 3)$ to $Q(4, 5)$ is $\mathbf{u} = \overrightarrow{PQ} = \langle 4 - 0, 5 - 3 \rangle = \langle 4, 2 \rangle$, and the vector **v** from $R(2, 1)$ to $S(6, 3)$ is $\mathbf{v} = \overrightarrow{RS} = \langle 6 - 2, 3 - 1 \rangle = \langle 4, 2 \rangle$.

EXAMPLE 2 Finding the Component Form of a Vector

Find the component form and magnitude of the vector **v** that has initial point $(4, -7)$ and terminal point $(-1, 5)$.

Algebraic Solution

Let

$$P(4, -7) = (p_1, p_2)$$

and

$$Q(-1, 5) = (q_1, q_2).$$

Then, the components of $\mathbf{v} = \langle v_1, v_2 \rangle$ are

$$v_1 = q_1 - p_1 = -1 - 4 = -5$$
$$v_2 = q_2 - p_2 = 5 - (-7) = 12.$$

So, $\mathbf{v} = \langle -5, 12 \rangle$ and the magnitude of **v** is

$$\|\mathbf{v}\| = \sqrt{(-5)^2 + 12^2} = \sqrt{169} = 13.$$

Graphical Solution

Use centimeter graph paper to plot the points $P(4, -7)$ and $Q(-1, 5)$. Carefully sketch the vector **v**. Use the sketch to find the components of $\mathbf{v} = \langle v_1, v_2 \rangle$. Then use a centimeter ruler to find the magnitude of **v**.

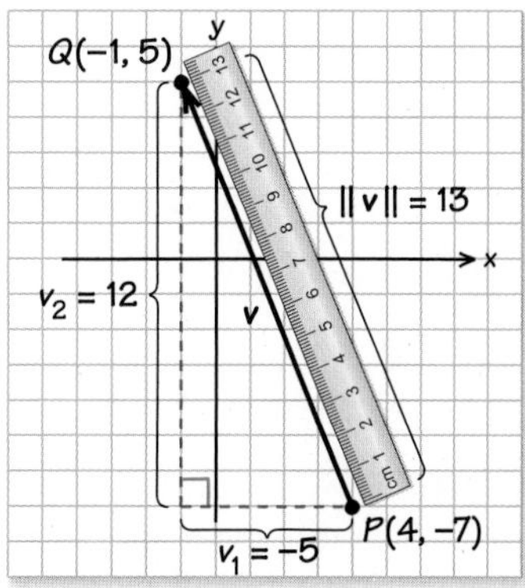

Figure 13.21

Figure 13.21 shows that the components of **v** are $v_1 = -5$ and $v_2 = 12$, so $\mathbf{v} = \langle -5, 12 \rangle$. Figure 13.21 also shows that the magnitude of **v** is $\|\mathbf{v}\| = 13$. ■

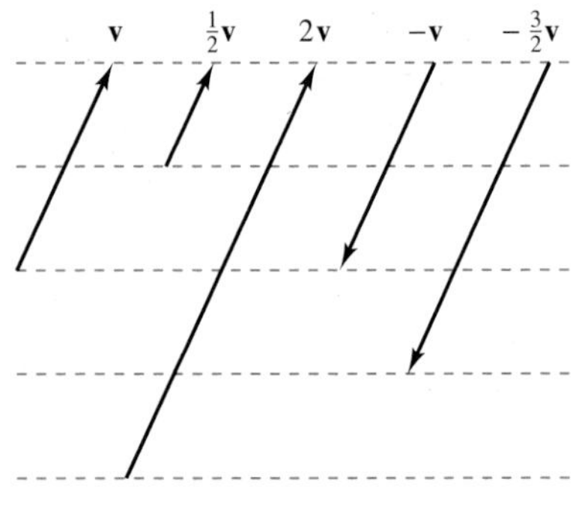

Figure 13.22

Vector Operations

The two basic vector operations are **scalar multiplication** and **vector addition.** In operations with vectors, numbers are usually referred to as **scalars.** In this text, scalars will always be real numbers. Geometrically, the product of a vector $\mathbf{v}$ and a scalar k is the vector that is $|k|$ times as long as $\mathbf{v}$. If k is positive, $k\mathbf{v}$ has the same direction as $\mathbf{v}$, and if k is negative, $k\mathbf{v}$ has the direction *opposite* that of $\mathbf{v}$, as shown in Figure 13.22.

To add two vectors $\mathbf{u}$ and $\mathbf{v}$ geometrically, first position them (without changing their lengths or directions) so that the initial point of the second vector $\mathbf{v}$ coincides with the terminal point of the first vector $\mathbf{u}$. The sum $\mathbf{u} + \mathbf{v}$ is the vector formed by joining the initial point of the first vector $\mathbf{u}$ with the terminal point of the second vector $\mathbf{v}$, as shown in Figure 13.23. This technique is called the **parallelogram law** for vector addition because the vector $\mathbf{u} + \mathbf{v}$, often called the **resultant** of vector addition, is the diagonal of a parallelogram having adjacent sides $\mathbf{u}$ and $\mathbf{v}$.

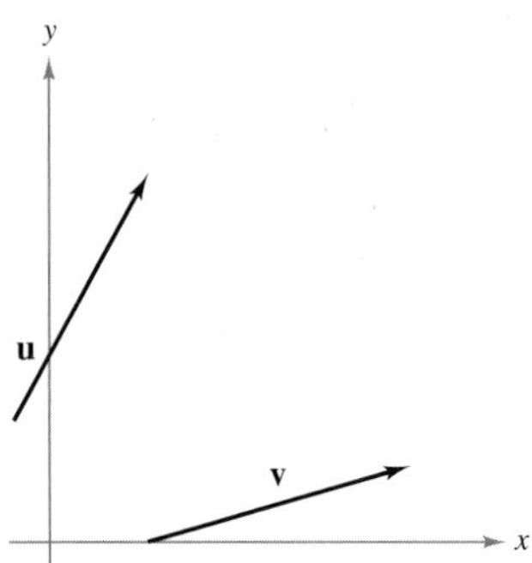

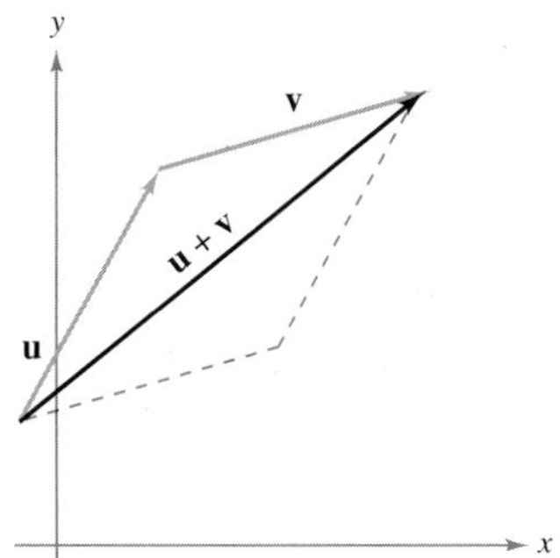

Figure 13.23

DEFINITIONS OF VECTOR ADDITION AND SCALAR MULTIPLICATION

Let $\mathbf{u} = \langle u_1, u_2\rangle$ and $\mathbf{v} = \langle v_1, v_2\rangle$ be vectors and let k be a scalar (a real number). Then the *sum* of $\mathbf{u}$ and $\mathbf{v}$ is the vector

$$\mathbf{u} + \mathbf{v} = \langle u_1 + v_1, u_2 + v_2\rangle \qquad \text{Sum}$$

and the *scalar multiple* of k times $\mathbf{u}$ is the vector

$$k\mathbf{u} = k\langle u_1, u_2\rangle = \langle ku_1, ku_2\rangle. \qquad \text{Scalar multiple}$$

The **negative** of $\mathbf{v} = \langle v_1, v_2\rangle$ is

$$-\mathbf{v} = (-1)\mathbf{v}$$
$$= \langle -v_1, -v_2\rangle \qquad \text{Negative}$$

and the **difference** of $\mathbf{u}$ and $\mathbf{v}$ is

$$\mathbf{u} - \mathbf{v} = \mathbf{u} + (-\mathbf{v})$$
$$= \langle u_1 - v_1, u_2 - v_2\rangle. \qquad \text{Difference}$$

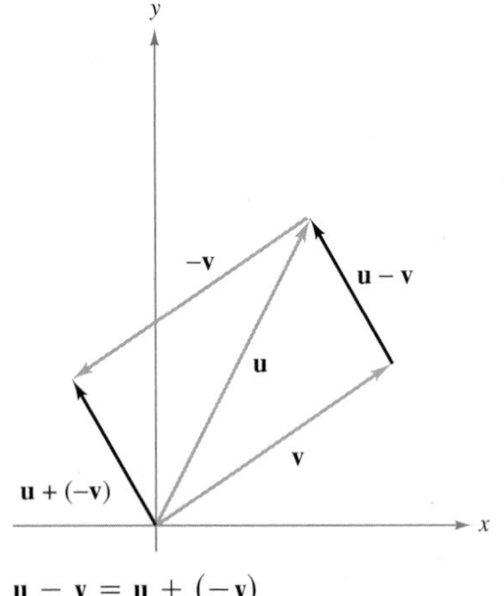

$\mathbf{u} - \mathbf{v} = \mathbf{u} + (-\mathbf{v})$
Figure 13.24

To represent $\mathbf{u} - \mathbf{v}$ geometrically, you can use directed line segments with the *same* initial point. The difference $\mathbf{u} - \mathbf{v}$ is the vector from the terminal point of $\mathbf{v}$ to the terminal point of $\mathbf{u}$, which is equal to $\mathbf{u} + (-\mathbf{v})$, as shown in Figure 13.24.

The component definitions of vector addition and scalar multiplication are illustrated in Example 3. In this example, notice that each of the vector operations can be interpreted geometrically.

EXAMPLE 3 Vector Operations

Let $\mathbf{v} = \langle 3, -1 \rangle$ and $\mathbf{w} = \langle -4, 4 \rangle$, and find (a) $2\mathbf{v}$, (b) $\mathbf{v} - \mathbf{w}$, and (c) $2\mathbf{v} + 3\mathbf{w}$.

Solution

a. Because $\mathbf{v} = \langle 3, -1 \rangle$, you have

$$2\mathbf{v} = 2\langle 3, -1 \rangle = \langle 2(3), 2(-1) \rangle = \langle 6, -2 \rangle.$$

A sketch of $2\mathbf{v}$ is shown in Figure 13.25(a).

b. The difference of $\mathbf{v}$ and $\mathbf{w}$ is

$$\mathbf{v} - \mathbf{w} = \langle 3, -1 \rangle - \langle -4, 4 \rangle = \langle 3 - (-4), -1 - 4 \rangle = \langle 7, -5 \rangle.$$

A sketch of $\mathbf{v} - \mathbf{w}$ is shown in Figure 13.25(b).

c. The sum of $2\mathbf{v}$ and $3\mathbf{w}$ is

$$\begin{aligned} 2\mathbf{v} + 3\mathbf{w} &= 2\langle 3, -1 \rangle + 3\langle -4, 4 \rangle = \langle 2(3), 2(-1) \rangle + \langle 3(-4), 3(4) \rangle \\ &= \langle 6, -2 \rangle + \langle -12, 12 \rangle = \langle 6 - 12, -2 + 12 \rangle = \langle -6, 10 \rangle. \end{aligned}$$

A sketch of $2\mathbf{v} + 3\mathbf{w}$ is shown in Figure 13.25(c).

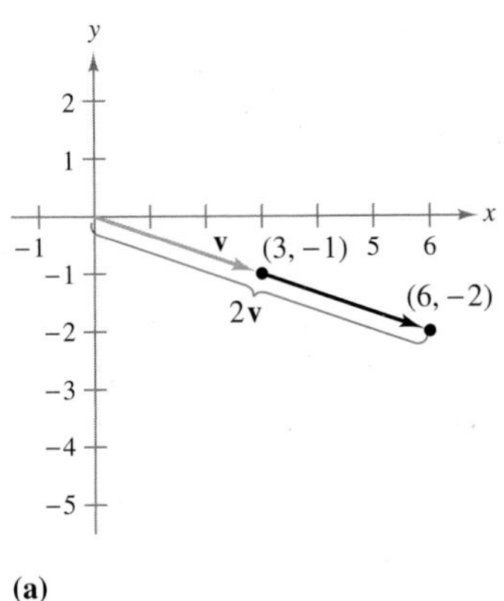

(a)

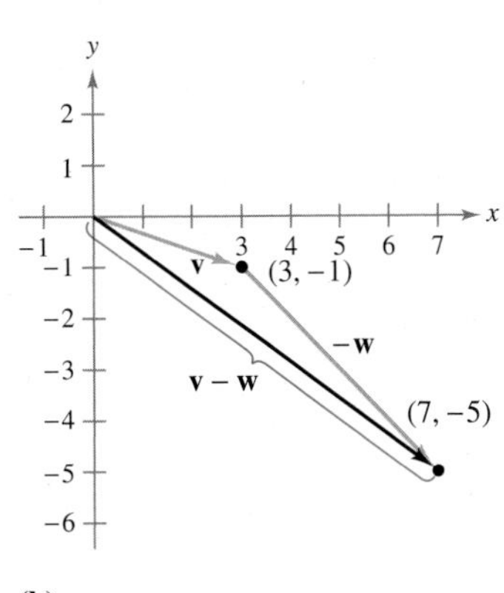

(b)

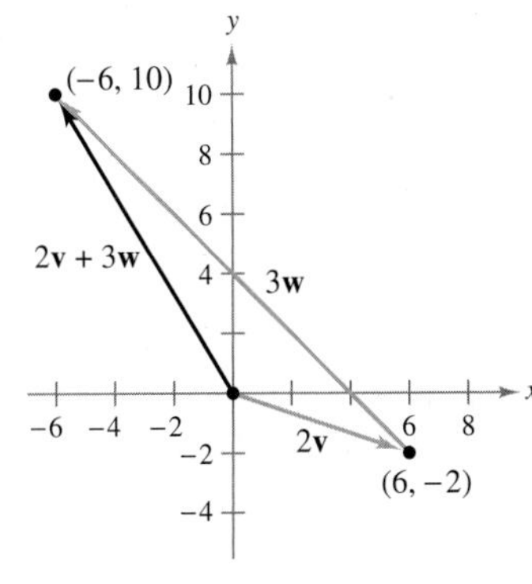

(c)

Figure 13.25

Vector addition and scalar multiplication share many of the properties of ordinary arithmetic.

NOTE Property 9 can be stated as follows: The magnitude of the vector $c\mathbf{v}$ is the absolute value of c times the magnitude of $\mathbf{v}$.

THEOREM 13.4 PROPERTIES OF VECTOR ADDITION AND SCALAR MULTIPLICATION

Let $\mathbf{u}$, $\mathbf{v}$, and $\mathbf{w}$ be vectors and let c and d be scalars. Then the following properties are true.

1. $\mathbf{u} + \mathbf{v} = \mathbf{v} + \mathbf{u}$

2. $(\mathbf{u} + \mathbf{v}) + \mathbf{w} = \mathbf{u} + (\mathbf{v} + \mathbf{w})$

3. $\mathbf{u} + \mathbf{0} = \mathbf{u}$

4. $\mathbf{u} + (-\mathbf{u}) = \mathbf{0}$

5. $c(d\mathbf{u}) = (cd)\mathbf{u}$

6. $(c + d)\mathbf{u} = c\mathbf{u} + d\mathbf{u}$

7. $c(\mathbf{u} + \mathbf{v}) = c\mathbf{u} + c\mathbf{v}$

8. $1(\mathbf{u}) = \mathbf{u}, \quad 0(\mathbf{u}) = \mathbf{0}$

9. $\|c\mathbf{v}\| = |c|\,\|\mathbf{v}\|$

Unit Vectors

In many applications of vectors, it is useful to find a unit vector that has the same direction as a given nonzero vector **v**. To do this, you can divide **v** by its magnitude.

THEOREM 13.5 UNIT VECTOR IN THE DIRECTION OF v

If **v** is a nonzero vector in the plane, then the vector

$$\mathbf{u} = \frac{\mathbf{v}}{\|\mathbf{v}\|} = \frac{1}{\|\mathbf{v}\|}\mathbf{v}$$

has length 1 and the same direction as **v**. The vector **u** is called a unit vector in the direction of **v**.

PROOF Because $1/\|\mathbf{v}\|$ is positive and $\mathbf{u} = (1/\|\mathbf{v}\|)\mathbf{v}$, you can conclude that **u** has the same direction as **v**. To see that $\|\mathbf{u}\| = 1$, note that

$$\|\mathbf{u}\| = \left\|\left(\frac{1}{\|\mathbf{v}\|}\right)\mathbf{v}\right\| = \left|\frac{1}{\|\mathbf{v}\|}\right|\|\mathbf{v}\| = \frac{1}{\|\mathbf{v}\|}\|\mathbf{v}\| = 1.$$

So, **u** has length 1 and the same direction as **v**. ■

EXAMPLE 4 Finding a Unit Vector

Find a unit vector in the direction of $\mathbf{v} = \langle -4, 5\rangle$ and verify that the result has a magnitude of 1.

Solution The unit vector in the direction of **v** is

$$\frac{\mathbf{v}}{\|\mathbf{v}\|} = \frac{\langle -4, 5\rangle}{\sqrt{(-4)^2 + (5)^2}} = \frac{1}{\sqrt{41}}\langle -4, 5\rangle = \left\langle \frac{-4}{\sqrt{41}}, \frac{5}{\sqrt{41}}\right\rangle.$$

This vector has a magnitude of 1 because

$$\sqrt{\left(\frac{-4}{\sqrt{41}}\right)^2 + \left(\frac{5}{\sqrt{41}}\right)^2} = \sqrt{\frac{16}{41} + \frac{25}{41}} = \sqrt{\frac{41}{41}} = 1.$$ ■

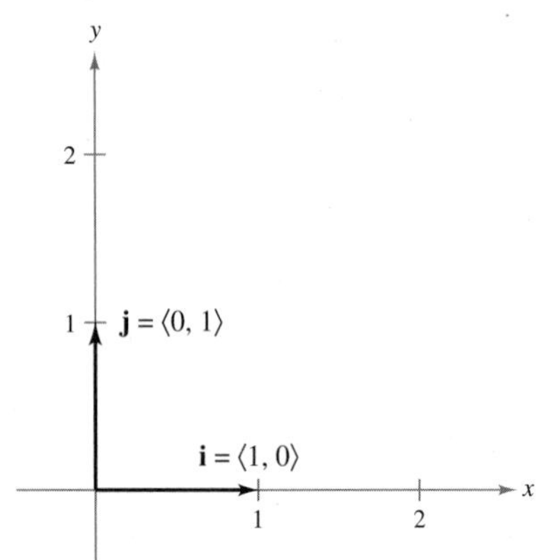

Figure 13.26

Unit vectors $\langle 1, 0\rangle$ and $\langle 0, 1\rangle$, called the **standard unit vectors,** are denoted by

$$\mathbf{i} = \langle 1, 0\rangle \quad \text{and} \quad \mathbf{j} = \langle 0, 1\rangle \qquad \text{Standard unit vectors}$$

as shown in Figure 13.26. (Note that the lowercase letters **i** and **j** are written in boldface to distinguish them from scalars, variables, or the imaginary number $i = \sqrt{-1}$.) These vectors can be used to represent any vector $\mathbf{v} = \langle v_1, v_2\rangle$ as shown.

$$\mathbf{v} = \langle v_1, v_2\rangle = v_1\langle 1, 0\rangle + v_2\langle 0, 1\rangle = v_1\mathbf{i} + v_2\mathbf{j}$$

The scalars v_1 and v_2 are called the **horizontal** and **vertical components of v**, respectively. The vector sum

$$v_1\mathbf{i} + v_2\mathbf{j}$$

is called a **linear combination** of the vectors **i** and **j**. Any vector in the plane can be written as a linear combination of the standard unit vectors **i** and **j**. For instance, the vector in Figure 13.27 can be written as

$$\mathbf{u} = \langle -1 - 2, 3 - (-5)\rangle = \langle -3, 8\rangle = -3\mathbf{i} + 8\mathbf{j}.$$

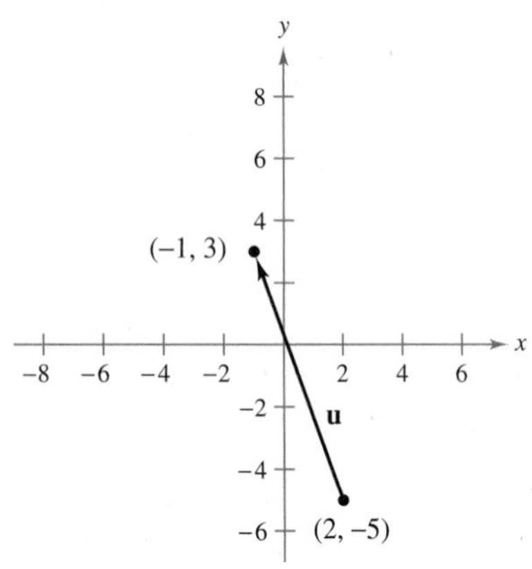

Figure 13.27

EXAMPLE 5 Vector Operations

Let $\mathbf{u} = -2\mathbf{i} + 5\mathbf{j}$ and let $\mathbf{v} = 3\mathbf{i} - \mathbf{j}$. Find $4\mathbf{u} - 3\mathbf{v}$.

Solution You could solve this problem by converting **u** and **v** to component form. This, however, is not necessary. It is just as easy to perform the operations in unit vector form.

$$4\mathbf{u} - 3\mathbf{v} = 4(-2\mathbf{i} + 5\mathbf{j}) - 3(3\mathbf{i} - \mathbf{j}) = -8\mathbf{i} + 20\mathbf{j} - 9\mathbf{i} + 3\mathbf{j} = -17\mathbf{i} + 23\mathbf{j}$$

■

Direction Angles

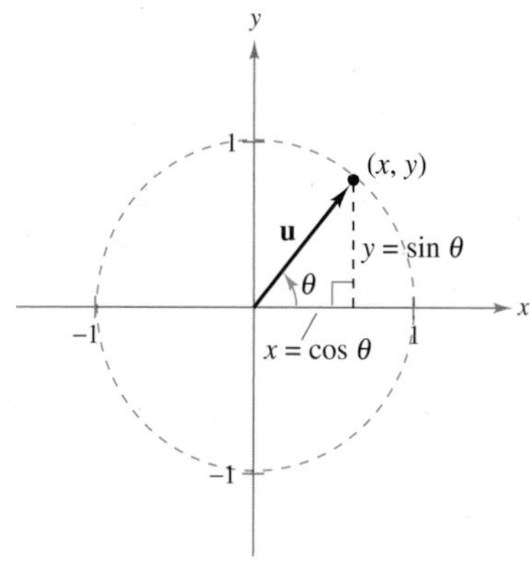

$\|\mathbf{u}\| = 1$
Figure 13.28

If **u** is a *unit vector* such that θ is the angle (measured counterclockwise) from the positive x-axis to **u**, the terminal point of **u** lies on the unit circle and you have

$$\mathbf{u} = \langle x, y \rangle = \langle \cos\theta, \sin\theta \rangle = (\cos\theta)\mathbf{i} + (\sin\theta)\mathbf{j}$$

as shown in Figure 13.28. The angle θ is the **direction angle** of the vector **u**.

Suppose that **u** is a unit vector with direction angle θ. If $\mathbf{v} = a\mathbf{i} + b\mathbf{j}$ is any vector that makes an angle θ with the positive x-axis, it has the same direction as **u** and you can write

$$\begin{aligned}\mathbf{v} &= \|\mathbf{v}\|\langle \cos\theta, \sin\theta \rangle \\ &= \|\mathbf{v}\|(\cos\theta)\mathbf{i} + \|\mathbf{v}\|(\sin\theta)\mathbf{j}.\end{aligned}$$

Because $\mathbf{v} = a\mathbf{i} + b\mathbf{j} = \|\mathbf{v}\|(\cos\theta)\mathbf{i} + \|\mathbf{v}\|(\sin\theta)\mathbf{j}$, it follows that the direction angle θ for **v** is determined from

$$\tan\theta = \frac{\sin\theta}{\cos\theta} = \frac{\|\mathbf{v}\|\sin\theta}{\|\mathbf{v}\|\cos\theta} = \frac{b}{a}.$$

EXAMPLE 6 Finding Direction Angles of Vectors

Find the direction angle of each vector.

a. $\mathbf{u} = 3\mathbf{i} + 3\mathbf{j}$

b. $\mathbf{v} = 3\mathbf{i} - 4\mathbf{j}$

Solution

a. The direction angle is

$$\tan\theta = \frac{b}{a} = \frac{3}{3} = 1.$$

So, $\theta = 45°$, as shown in Figure 13.29(a).

b. The direction angle is

$$\tan\theta = \frac{b}{a} = \frac{-4}{3}.$$

Moreover, because $\mathbf{v} = 3\mathbf{i} - 4\mathbf{j}$ lies in Quadrant IV, θ lies in Quadrant IV and its reference angle is

$$\theta = \left|\arctan\left(-\frac{4}{3}\right)\right| \approx |-53.13°| = 53.13°.$$

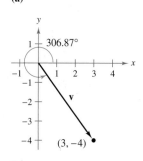

So, it follows that $\theta \approx 360° - 53.13° = 306.87°$, as shown in Figure 13.29(b).

■

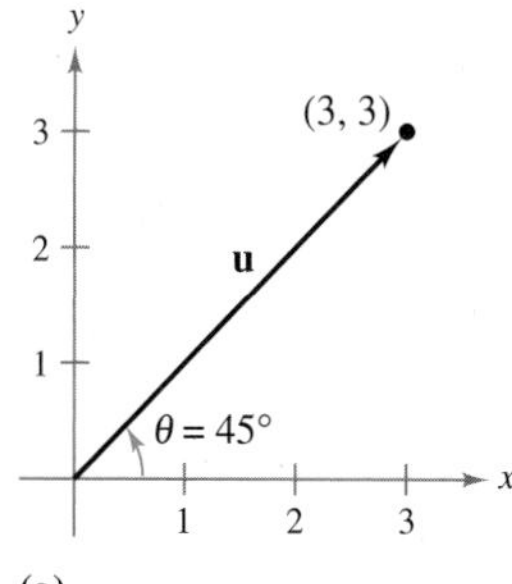

(a)

(b)
Figure 13.29

Applications of Vectors

EXAMPLE 7 Finding the Component Form of a Vector

Find the component form of the vector that represents the velocity of an airplane descending at a speed of 150 miles per hour at an angle 20° below the horizontal, as shown in Figure 13.30.

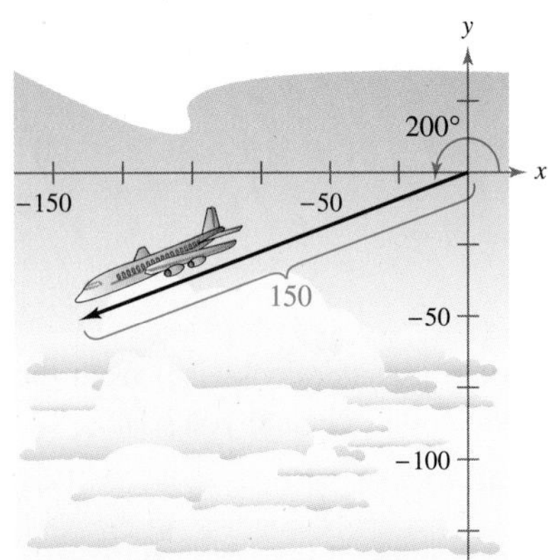

Figure 13.30

Solution The velocity vector **v** has a magnitude of 150 and a direction angle of $\theta = 180° + 20° = 200°$. So,

$$\begin{aligned}\mathbf{v} &= \|\mathbf{v}\|(\cos\theta)\mathbf{i} + \|\mathbf{v}\|(\sin\theta)\mathbf{j}\\ &= 150(\cos 200°)\mathbf{i} + 150(\sin 200°)\mathbf{j}\\ &\approx 150(-0.9397)\mathbf{i} + 150(-0.3420)\mathbf{j}\\ &\approx -140.96\mathbf{i} - 51.30\mathbf{j}\\ &= \langle -140.96, -51.30\rangle.\end{aligned}$$

You can check that **v** has a magnitude of 150, as follows.

$$\begin{aligned}\|\mathbf{v}\| &\approx \sqrt{(-140.96)^2 + (-51.30)^2}\\ &\approx \sqrt{19{,}869.72 + 2631.69}\\ &= \sqrt{22{,}501.41} \approx 150\end{aligned}$$

EXAMPLE 8 Using Vectors to Determine Weight

A force of 600 pounds is required to pull a boat and trailer up a ramp inclined at 15° from the horizontal. Find the combined weight of the boat and trailer.

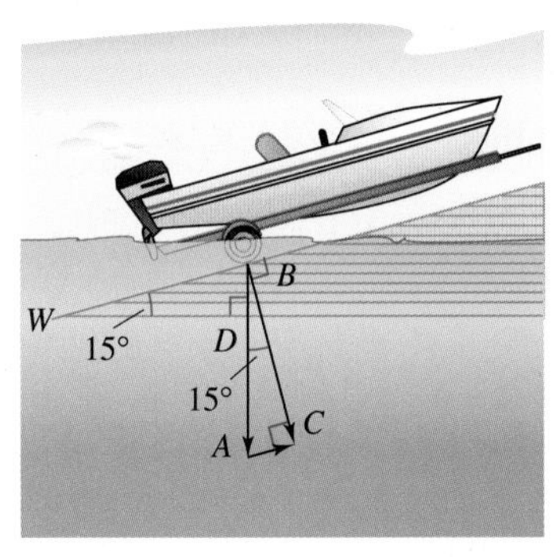

Figure 13.31

Solution Based on Figure 13.31, you can make the following observations.

$\|\overrightarrow{BA}\|$ = force of gravity = combined weight of boat and trailer

$\|\overrightarrow{BC}\|$ = force against ramp

$\|\overrightarrow{AC}\|$ = force required to move boat up ramp = 600 pounds

By construction, triangles *BWD* and *ABC* are similar. Therefore, angle *ABC* is 15°. So, in triangle *ABC* you have

$$\sin 15° = \frac{\|\overrightarrow{AC}\|}{\|\overrightarrow{BA}\|}$$

$$\sin 15° = \frac{600}{\|\overrightarrow{BA}\|}$$

$$\|\overrightarrow{BA}\| = \frac{600}{\sin 15°}$$

$$\|\overrightarrow{BA}\| \approx 2318.$$

Consequently, the combined weight is approximately 2318 pounds. ■

NOTE In Figure 13.31, note that $\overrightarrow{AC}$ is parallel to the ramp. ■

STUDY TIP Recall from Section 9.8 that in air navigation, bearings can be measured in degrees clockwise from north.

EXAMPLE 9 Using Vectors to Find Speed and Direction

An airplane is traveling at a speed of 500 miles per hour with a bearing of 330° at a fixed altitude with a negligible wind velocity as shown in Figure 13.32(a). When the airplane reaches a certain point, it encounters a wind with a velocity of 70 miles per hour in the direction N 45° E, as shown in Figure 13.32(b). What are the resultant speed and direction of the airplane?

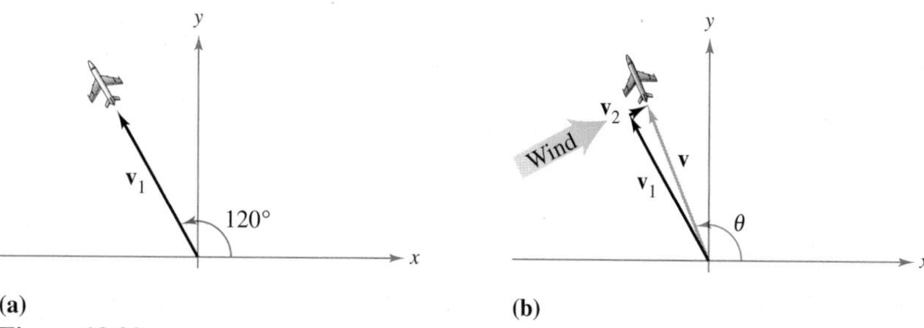

Figure 13.32

Solution Using Figure 13.32, the velocity of the airplane (alone) is

$$\begin{aligned}\mathbf{v}_1 &= 500\langle\cos 120°, \sin 120°\rangle\\ &= \langle -250, 250\sqrt{3}\rangle\end{aligned}$$

and the velocity of the wind is

$$\begin{aligned}\mathbf{v}_2 &= 70\langle\cos 45°, \sin 45°\rangle\\ &= \langle 35\sqrt{2}, 35\sqrt{2}\rangle.\end{aligned}$$

So, the velocity of the airplane (in the wind) is

$$\begin{aligned}\mathbf{v} &= \mathbf{v}_1 + \mathbf{v}_2\\ &= \langle -250 + 35\sqrt{2}, 250\sqrt{3} + 35\sqrt{2}\rangle\\ &\approx \langle -200.5, 482.5\rangle\end{aligned}$$

and the resultant speed of the airplane is

$$\begin{aligned}\|\mathbf{v}\| &\approx \sqrt{(-200.5)^2 + (482.5)^2}\\ &\approx 522.5 \text{ miles per hour.}\end{aligned}$$

Finally, if θ is the direction angle of the flight path, you have

$$\begin{aligned}\tan\theta &\approx \frac{482.5}{-200.5}\\ &\approx -2.4065\end{aligned}$$

which implies that

$$\begin{aligned}\theta &\approx 180° + \arctan(-2.4065)\\ &\approx 180° - 67.4°\\ &= 112.6°.\end{aligned}$$

So, the true direction of the airplane is approximately

$$270° + (180° - 112.6°) = 337.4°.$$

■

13.3 Exercises

See www.CalcChat.com for worked-out solutions to odd-numbered exercises.

In Exercises 1–10, fill in the blanks.

1. A ________ ________ ________ can be used to represent a quantity that involves both magnitude and direction.
2. The directed line segment $\overrightarrow{PQ}$ has ________ point P and ________ point Q.
3. The ________ of the directed line segment $\overrightarrow{PQ}$ is denoted by $\|\overrightarrow{PQ}\|$.
4. The set of all directed line segments that are equivalent to a given directed line segment $\overrightarrow{PQ}$ is a ________ **v** in the plane.
5. In order to show that two vectors are equivalent, you must show that they have the same ________ and the same ________ .
6. The directed line segment whose initial point is the origin is said to be in ________ ________ .
7. A vector that has a length of 1 is called a ________ ________ .
8. The two basic vector operations are scalar ________ and vector ________ .
9. The vector $\mathbf{u} + \mathbf{v}$ is called the ________ of vector addition.
10. The vector sum $v_1\mathbf{i} + v_2\mathbf{j}$ is called a ________ ________ of the vectors **i** and **j**, and the scalars v_1 and v_2 are called the ________ and ________ components of **v**, respectively.

In Exercises 11–22, find the component form and the magnitude of the vector v.

11.

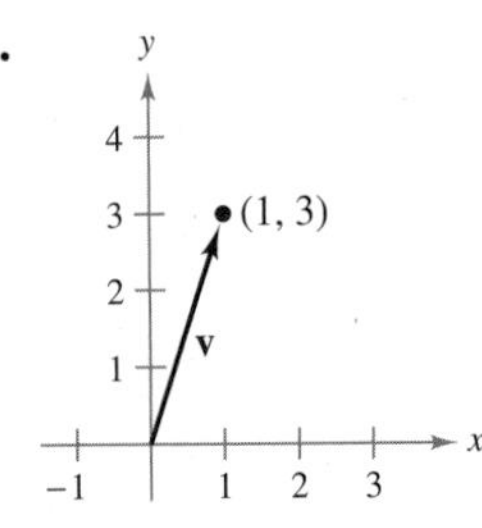

12.

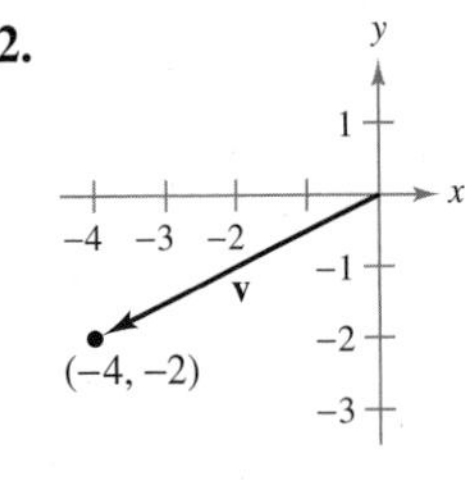

13.

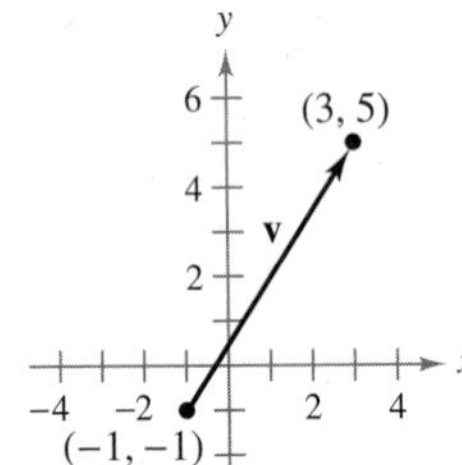

14.

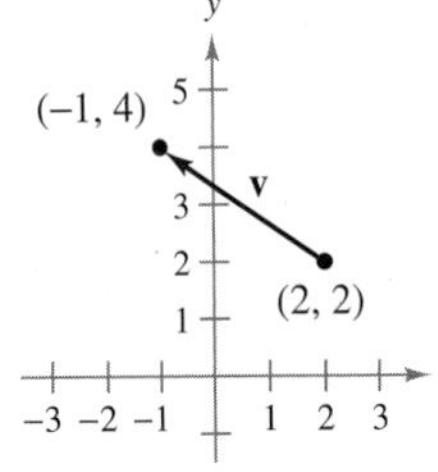

	Initial Point	*Terminal Point*
15.	$(3, -2)$	$(3, 3)$
16.	$(-4, -1)$	$(3, -1)$
17.	$(-3, -5)$	$(5, 1)$
18.	$(-2, 7)$	$(5, -17)$
19.	$(1, 3)$	$(-8, -9)$
20.	$(1, 11)$	$(9, 3)$
21.	$(-1, 5)$	$(15, 12)$
22.	$(-3, 11)$	$(9, 40)$

In Exercises 23–28, use the figure to sketch a graph of the specified vector. To print an enlarged copy of the graph, go to the website *www.mathgraphs.com*.

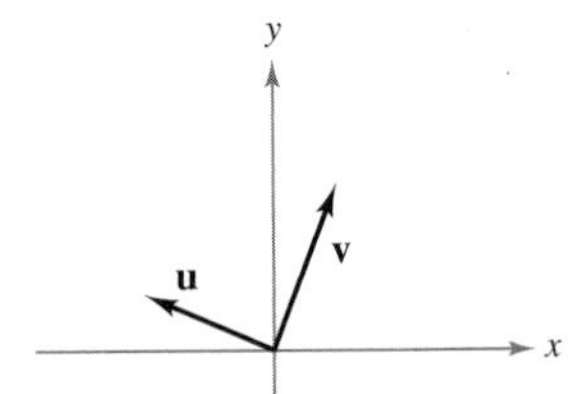

23. $-\mathbf{v}$

24. $5\mathbf{v}$

25. $\mathbf{u} + \mathbf{v}$

26. $\mathbf{u} + 2\mathbf{v}$

27. $\mathbf{u} - \mathbf{v}$

28. $\mathbf{v} - \frac{1}{2}\mathbf{u}$

In Exercises 29–36, find (a) u + v, (b) u − v, and (c) 2u − 3v. Then sketch each resultant vector.

29. $\mathbf{u} = \langle 2, 1 \rangle,\ \mathbf{v} = \langle 1, 3 \rangle$ **30.** $\mathbf{u} = \langle 2, 3 \rangle,\ \mathbf{v} = \langle 4, 0 \rangle$

31. $\mathbf{u} = \langle -5, 3 \rangle,\ \mathbf{v} = \langle 0, 0 \rangle$ **32.** $\mathbf{u} = \langle 0, 0 \rangle,\ \mathbf{v} = \langle 2, 1 \rangle$

33. $\mathbf{u} = \mathbf{i} + \mathbf{j},\ \mathbf{v} = 2\mathbf{i} - 3\mathbf{j}$ **34.** $\mathbf{u} = -2\mathbf{i} + \mathbf{j},\ \mathbf{v} = 3\mathbf{j}$

35. $\mathbf{u} = 2\mathbf{i},\ \mathbf{v} = \mathbf{j}$ **36.** $\mathbf{u} = 2\mathbf{j},\ \mathbf{v} = 3\mathbf{i}$

In Exercises 37–46, find a unit vector in the direction of the given vector. Verify that the result has a magnitude of 1.

37. $\mathbf{u} = \langle 3, 0 \rangle$ **38.** $\mathbf{u} = \langle 0, -2 \rangle$

39. $\mathbf{v} = \langle -2, 2 \rangle$ **40.** $\mathbf{v} = \langle 5, -12 \rangle$

41. $\mathbf{v} = \mathbf{i} + \mathbf{j}$ **42.** $\mathbf{v} = 6\mathbf{i} - 2\mathbf{j}$

43. $\mathbf{w} = 4\mathbf{j}$ **44.** $\mathbf{w} = -6\mathbf{i}$

45. $\mathbf{w} = \mathbf{i} - 2\mathbf{j}$ **46.** $\mathbf{w} = 7\mathbf{j} - 3\mathbf{i}$

In Exercises 47–50, find the vector v with the given magnitude and the same direction as u.

	Magnitude	*Direction*
47.	$\|\mathbf{v}\| = 10$	$\mathbf{u} = \langle -3, 4 \rangle$
48.	$\|\mathbf{v}\| = 3$	$\mathbf{u} = \langle -12, -5 \rangle$
49.	$\|\mathbf{v}\| = 9$	$\mathbf{u} = \langle 2, 5 \rangle$
50.	$\|\mathbf{v}\| = 8$	$\mathbf{u} = \langle 3, 3 \rangle$

In Exercises 51–56, find the component form of v and sketch the specified vector operations geometrically, where $\mathbf{u} = 2\mathbf{i} - \mathbf{j}$, and $\mathbf{w} = \mathbf{i} + 2\mathbf{j}$.

51. $\mathbf{v} = \frac{3}{2}\mathbf{u}$ **52.** $\mathbf{v} = \frac{3}{4}\mathbf{w}$

53. $\mathbf{v} = \mathbf{u} + 2\mathbf{w}$ **54.** $\mathbf{v} = -\mathbf{u} + \mathbf{w}$

55. $\mathbf{v} = \frac{1}{2}(3\mathbf{u} + \mathbf{w})$ **56.** $\mathbf{v} = \mathbf{u} - 2\mathbf{w}$

In Exercises 57–60, find the magnitude and direction angle of the vector v.

57. $\mathbf{v} = 6\mathbf{i} - 6\mathbf{j}$ **58.** $\mathbf{v} = -5\mathbf{i} + 4\mathbf{j}$

59. $\mathbf{v} = 3(\cos 60°\mathbf{i} + \sin 60°\mathbf{j})$

60. $\mathbf{v} = 8(\cos 135°\mathbf{i} + \sin 135°\mathbf{j})$

In Exercises 61–68, find the component form of v given its magnitude and the angle it makes with the positive x-axis. Sketch v.

	Magnitude	*Angle*
61.	$\|\mathbf{v}\| = 3$	$\theta = 0°$
62.	$\|\mathbf{v}\| = 1$	$\theta = 45°$
63.	$\|\mathbf{v}\| = \frac{7}{2}$	$\theta = 150°$
64.	$\|\mathbf{v}\| = \frac{3}{4}$	$\theta = 150°$
65.	$\|\mathbf{v}\| = 2\sqrt{3}$	$\theta = 45°$
66.	$\|\mathbf{v}\| = 4\sqrt{3}$	$\theta = 90°$
67.	$\|\mathbf{v}\| = 3$	$\mathbf{v}$ in the direction $3\mathbf{i} + 4\mathbf{j}$
68.	$\|\mathbf{v}\| = 2$	$\mathbf{v}$ in the direction $\mathbf{i} + 3\mathbf{j}$

In Exercises 69–72, find the component form of the sum of u and v with the given magnitudes and direction angles $\theta_{\mathbf{u}}$ and $\theta_{\mathbf{v}}$.

	Magnitude	*Angle*
69.	$\|\mathbf{u}\| = 5$	$\theta_{\mathbf{u}} = 0°$
	$\|\mathbf{v}\| = 5$	$\theta_{\mathbf{v}} = 90°$
70.	$\|\mathbf{u}\| = 4$	$\theta_{\mathbf{u}} = 60°$
	$\|\mathbf{v}\| = 4$	$\theta_{\mathbf{v}} = 90°$
71.	$\|\mathbf{u}\| = 20$	$\theta_{\mathbf{u}} = 45°$
	$\|\mathbf{v}\| = 50$	$\theta_{\mathbf{v}} = 180°$
72.	$\|\mathbf{u}\| = 50$	$\theta_{\mathbf{u}} = 30°$
	$\|\mathbf{v}\| = 30$	$\theta_{\mathbf{v}} = 110°$

In Exercises 73–76, use the Law of Cosines to find the angle α between the vectors. (Assume $0° \le \alpha \le 180°$.)

73. $\mathbf{v} = \mathbf{i} + \mathbf{j}$, $\mathbf{w} = 2\mathbf{i} - 2\mathbf{j}$

74. $\mathbf{v} = 3\mathbf{i} - 2\mathbf{j}$, $\mathbf{w} = 2\mathbf{i} + 2\mathbf{j}$

75. $\mathbf{v} = \mathbf{i} + \mathbf{j}$, $\mathbf{w} = 3\mathbf{i} - \mathbf{j}$

76. $\mathbf{v} = \mathbf{i} + 2\mathbf{j}$, $\mathbf{w} = 2\mathbf{i} - \mathbf{j}$

WRITING ABOUT CONCEPTS

77. What conditions must be met in order for two vectors to be equivalent? Which vectors in the figure appear to be equivalent?

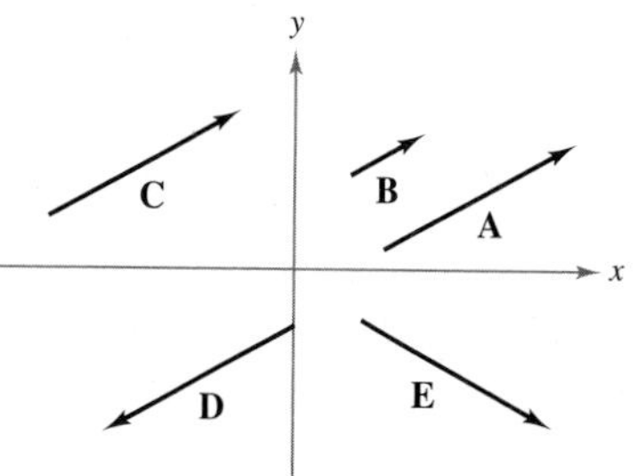

78. The vectors **u** and **v** have the same magnitudes in the two figures. In which figure will the magnitude of the sum be greater? Give a reason for your answer.

(a)

(b)

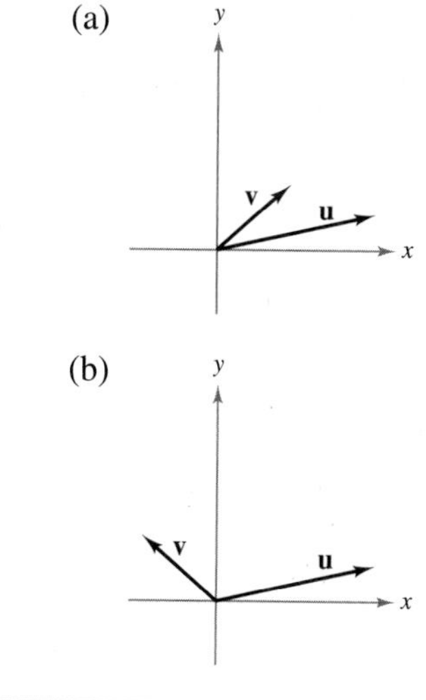

Resultant Force **In Exercises 79 and 80, find the angle between the forces given the magnitude of their resultant. (*Hint:* Write force 1 as a vector in the direction of the positive x-axis and force 2 as a vector at an angle θ with the positive x-axis.)**

	Force 1	*Force 2*	*Resultant Force*
79.	45 pounds	60 pounds	90 pounds
80.	3000 pounds	1000 pounds	3750 pounds

81. ***Velocity*** A gun with a muzzle velocity of 1200 feet per second is fired at an angle of 6° above the horizontal. Find the vertical and horizontal components of the velocity.

82. ***Velocity*** Detroit Tigers pitcher Joel Zumaya was recorded throwing a pitch at a velocity of 104 miles per hour. If he threw the pitch at an angle of 35° below the horizontal, find the vertical and horizontal components of the velocity. *(Source: Damon Lichtenwalner, Baseball Info Solutions)*

83. ***Resultant Force*** Forces with magnitudes of 125 newtons and 300 newtons act on a hook (see figure). The angle between the two forces is 45°. Find the direction and magnitude of the resultant of these forces.

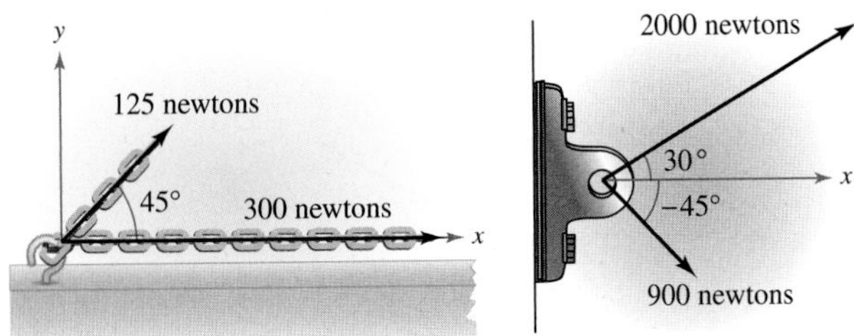

Figure for 83 **Figure for 84**

84. ***Resultant Force*** Forces with magnitudes of 2000 newtons and 900 newtons act on a machine part at angles of 30° and −45°, respectively, with the x-axis (see figure). Find the direction and magnitude of the resultant of these forces.

85. ***Resultant Force*** Three forces with magnitudes of 75 pounds, 100 pounds, and 125 pounds act on an object at angles of 30°, 45°, and 120°, respectively, with the positive x-axis. Find the direction and magnitude of the resultant of these forces.

86. ***Resultant Force*** Three forces with magnitudes of 70 pounds, 40 pounds, and 60 pounds act on an object at angles of −30°, 45°, and 135°, respectively, with the positive x-axis. Find the direction and magnitude of the resultant of these forces.

Cable Tension **In Exercises 87 and 88, use the figure to determine the tension in each cable supporting the load.**

87.

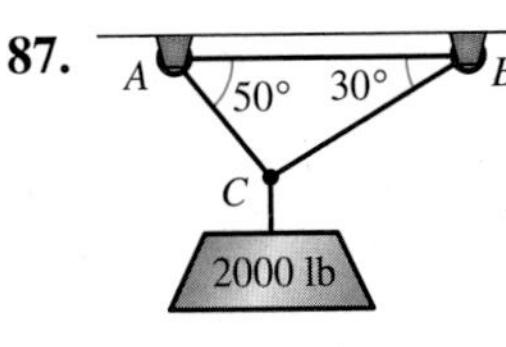

88.

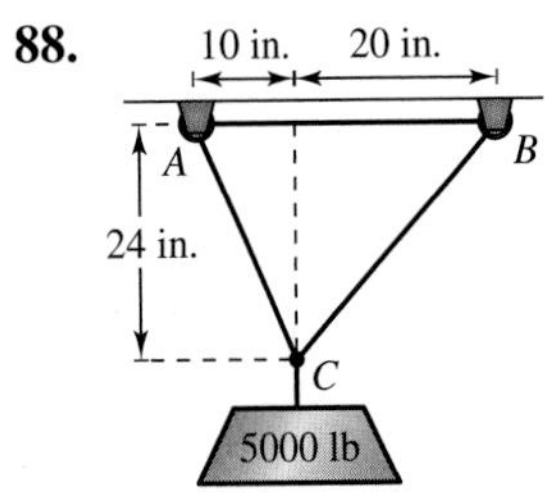

89. ***Tow Line Tension*** A loaded barge is being towed by two tugboats, and the magnitude of the resultant is 6000 pounds directed along the axis of the barge (see figure). Find the tension in the tow lines if they each make an 18° angle with the axis of the barge.

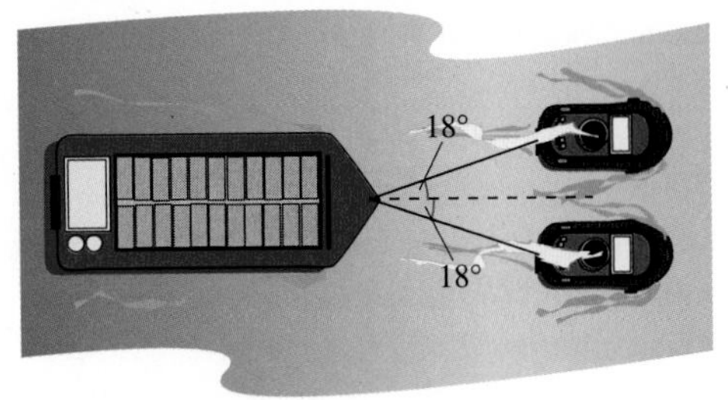

90. ***Rope Tension*** To carry a 100-pound cylindrical weight, two people lift on the ends of short ropes that are tied to an eyelet on the top center of the cylinder. Each rope makes a 20° angle with the vertical. Draw a figure that gives a visual representation of the situation, and find the tension in the ropes.

91. ***Work*** A heavy object is pulled 30 feet across a floor, using a force of 100 pounds. The force is exerted at an angle of 50° above the horizontal (see figure). Find the work done. (Use the formula for work, $W = FD$, where F is the component of the force in the direction of motion and D is the distance.)

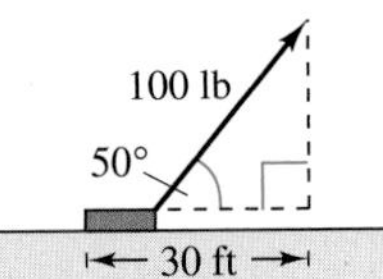

92. ***Rope Tension*** A tetherball weighing 1 pound is pulled outward from the pole by a horizontal force **u** until the rope makes a 45° angle with the pole (see figure). Determine the resulting tension in the rope and the magnitude of **u**.

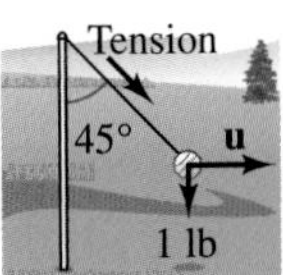

93. ***Navigation*** An airplane is flying in the direction of 148°, with an airspeed of 875 kilometers per hour. Because of the wind, its groundspeed and direction are 800 kilometers per hour and 140°, respectively (see figure). Find the direction and speed of the wind.

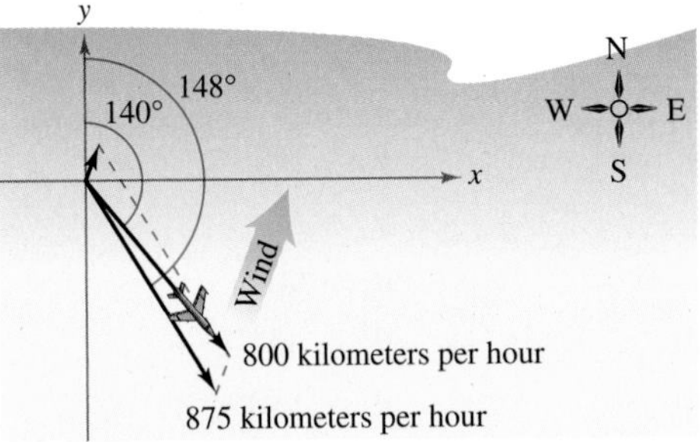

94. ***Navigation*** An airplane's velocity with respect to the air is 580 miles per hour, and its bearing is 332°. The wind, at the altitude of the plane, is from the southwest and has a velocity of 60 miles per hour.

(a) Draw a figure that gives a visual representation of the problem.

(b) What is the true direction of the plane, and what is its speed with respect to the ground?

True or False? **In Exercises 95 and 96, decide whether the statement is true or false. If it is false, explain why or give an example that shows it is false.**

95. If $\mathbf{u}$ and $\mathbf{v}$ have the same magnitude and direction, then $\mathbf{u} = \mathbf{v}$.

96. If $\mathbf{u} = a\mathbf{i} + b\mathbf{j}$ is a unit vector, then $a^2 + b^2 = 1$.

97. *Proof* Prove that $(\cos\theta)\mathbf{i} + (\sin\theta)\mathbf{j}$ is a unit vector for any value of θ.

CAPSTONE

98. The initial and terminal points of vector $\mathbf{v}$ are $(3, -4)$ and $(9, 1)$, respectively.

(a) Write $\mathbf{v}$ in component form.

(b) Write $\mathbf{v}$ as the linear combination of the standard unit vectors $\mathbf{i}$ and $\mathbf{j}$.

(c) Sketch $\mathbf{v}$ with its initial point at the origin.

(d) Find the magnitude of $\mathbf{v}$.

99. *Writing* In your own words, state the difference between a scalar and a vector. Give examples of each.

100. *Writing* Give geometric descriptions of the operations of addition of vectors and multiplication of a vector by a scalar.

101. *Writing* Identify the quantity as a scalar or as a vector. Explain your reasoning.

(a) The muzzle velocity of a bullet

(b) The price of a company's stock

(c) The air temperature in a room

(d) The weight of an automobile

102. *Technology* Write a program for your graphing utility that graphs two vectors and their difference given the vectors in component form.

In Exercises 103 and 104, use the program in Exercise 102 to find the difference of the vectors shown in the figure.

103. **104.**

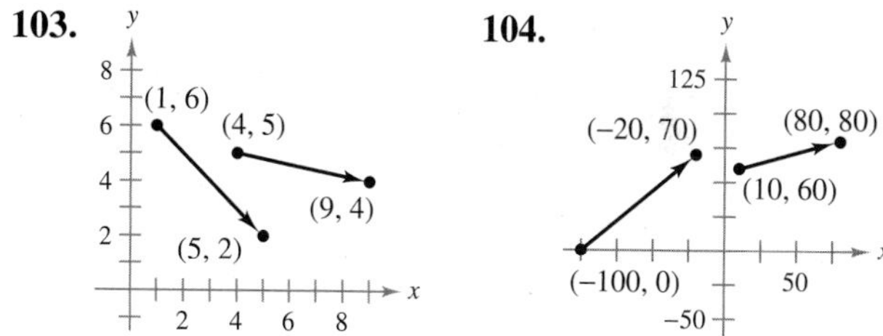

SECTION PROJECT

Adding Vectors Graphically

The pseudo code below can be translated into a program for a graphing utility.

Program

- Input a.
- Input b.
- Input c.
- Input d.
- Draw a line from $(0, 0)$ to (a, b).
- Draw a line from $(0, 0)$ to (c, d).
- Add $a + c$ and store in e.
- Add $b + d$ and store in f.
- Draw a line from $(0, 0)$ to (e, f).
- Draw a line from (a, b) to (c, d).
- Draw a line from (c, d) to (e, f).
- Pause to view graph.
- End program.

The program sketches two vectors $\mathbf{u} = a\mathbf{i} + b\mathbf{j}$ and $\mathbf{v} = c\mathbf{i} + d\mathbf{j}$ in standard position. Then, using the parallelogram law for vector addition, the program also sketches the vector sum $\mathbf{u} + \mathbf{v}$. *Before* running the program, you should set values that produce an appropriate viewing window.

(a) An airplane is flying at a heading of 300° and a speed of 400 miles per hour. The airplane encounters wind of velocity 75 miles per hour in the direction 40°. Use the program to find the resultant speed and direction of the airplane.

(b) After encountering the wind, is the airplane in part (a) traveling at a higher speed or a lower speed? Explain.

(c) Consider the airplane described in part (a), at a heading of 300° and a speed of 400 miles per hour. Use the program to find the wind velocity in the direction of 40° that will produce a resultant direction of 310°.

(d) Consider the airplane described in part (a), at a heading of 300° and a speed of 400 miles per hour. Use the program to find the wind direction at a speed of 75 miles per hour that will produce a resultant direction of 310°.

13.4 Vectors and Dot Products

- **Find the dot product of two vectors and use the properties of the dot product.**
- **Find the angle between two vectors and determine whether two vectors are orthogonal.**
- **Write a vector as the sum of two vector components.**
- **Use vectors to find the work done by a force.**

The Dot Product of Two Vectors

So far you have studied two vector operations—vector addition and multiplication by a scalar—each of which yields another vector. In this section, you will study a third vector operation, the **dot product.** This product yields a scalar, rather than a vector.

DEFINITION OF THE DOT PRODUCT

The **dot product** of $\mathbf{u} = \langle u_1, u_2 \rangle$ and $\mathbf{v} = \langle v_1, v_2 \rangle$ is

$\mathbf{u} \cdot \mathbf{v} = u_1 v_1 + u_2 v_2.$

THEOREM 13.6 PROPERTIES OF THE DOT PRODUCT

Let **u**, **v**, and **w** be vectors in the plane or in space and let c be a scalar.

1. $\mathbf{u} \cdot \mathbf{v} = \mathbf{v} \cdot \mathbf{u}$
2. $\mathbf{0} \cdot \mathbf{v} = 0$
3. $\mathbf{u} \cdot (\mathbf{v} + \mathbf{w}) = \mathbf{u} \cdot \mathbf{v} + \mathbf{u} \cdot \mathbf{w}$
4. $\mathbf{v} \cdot \mathbf{v} = \|\mathbf{v}\|^2$
5. $c(\mathbf{u} \cdot \mathbf{v}) = c\mathbf{u} \cdot \mathbf{v} = \mathbf{u} \cdot c\mathbf{v}$

Proofs of Properties 1 and 4 are given in Appendix A.

EXAMPLE 1 Finding Dot Products

Find each dot product.

a. $\langle 4, 5 \rangle \cdot \langle 2, 3 \rangle$ **b.** $\langle 2, -1 \rangle \cdot \langle 1, 2 \rangle$ **c.** $\langle 0, 3 \rangle \cdot \langle 4, -2 \rangle$

Solution

a. $\langle 4, 5 \rangle \cdot \langle 2, 3 \rangle = 4(2) + 5(3)$
$= 8 + 15$
$= 23$

b. $\langle 2, -1 \rangle \cdot \langle 1, 2 \rangle = 2(1) + (-1)(2)$
$= 2 - 2 = 0$

c. $\langle 0, 3 \rangle \cdot \langle 4, -2 \rangle = 0(4) + 3(-2)$
$= 0 - 6 = -6$

■

NOTE In Example 1, be sure you see that the dot product of two vectors is a scalar (a real number), not a vector. Moreover, notice that the dot product can be positive, zero, or negative. ■

EXAMPLE 2 Using Properties of Dot Products

Let $\mathbf{u} = \langle -1, 3\rangle$, $\mathbf{v} = \langle 2, -4\rangle$, and $\mathbf{w} = \langle 1, -2\rangle$. Find each dot product.

a. $(\mathbf{u} \cdot \mathbf{v})\mathbf{w}$ **b.** $\mathbf{u} \cdot 2\mathbf{v}$ **c.** $\mathbf{u} \cdot (\mathbf{v} + \mathbf{w})$

Solution Begin by finding the dot product of $\mathbf{u}$ and $\mathbf{v}$.

$$\begin{aligned}\mathbf{u} \cdot \mathbf{v} &= \langle -1, 3\rangle \cdot \langle 2, -4\rangle \\ &= (-1)(2) + 3(-4) \\ &= -14\end{aligned}$$

a. $$\begin{aligned}(\mathbf{u} \cdot \mathbf{v})\mathbf{w} &= -14\langle 1, -2\rangle \\ &= \langle -14, 28\rangle\end{aligned}$$

b. $$\begin{aligned}\mathbf{u} \cdot 2\mathbf{v} &= 2(\mathbf{u} \cdot \mathbf{v}) \\ &= 2(-14) \\ &= -28\end{aligned}$$

c. $$\begin{aligned}\mathbf{u} \cdot (\mathbf{v} + \mathbf{w}) &= -14 + \langle -1, 3\rangle \cdot \langle 1, -2\rangle \\ &= -14 + (-1) + (-6) = -21\end{aligned}$$

Notice that the first product is a vector, whereas the second and third are scalars.

EXAMPLE 3 Dot Product and Magnitude

The dot product of $\mathbf{u}$ with itself is 5. What is the magnitude of $\mathbf{u}$?

Solution Because $\|\mathbf{u}\|^2 = \mathbf{u} \cdot \mathbf{u}$ and $\mathbf{u} \cdot \mathbf{u} = 5$, it follows that

$$\|\mathbf{u}\| = \sqrt{\mathbf{u} \cdot \mathbf{u}} = \sqrt{5}.$$

■

The Angle Between Two Vectors

The **angle between two nonzero vectors** is the angle θ, $0 \le \theta \le \pi$, between their respective standard position vectors, as shown in Figure 13.33. This angle can be found using the dot product.

THEOREM 13.7 ANGLE BETWEEN TWO VECTORS

If θ is the angle between two nonzero vectors $\mathbf{u}$ and $\mathbf{v}$, then

$$\cos\theta = \frac{\mathbf{u} \cdot \mathbf{v}}{\|\mathbf{u}\|\,\|\mathbf{v}\|}.$$

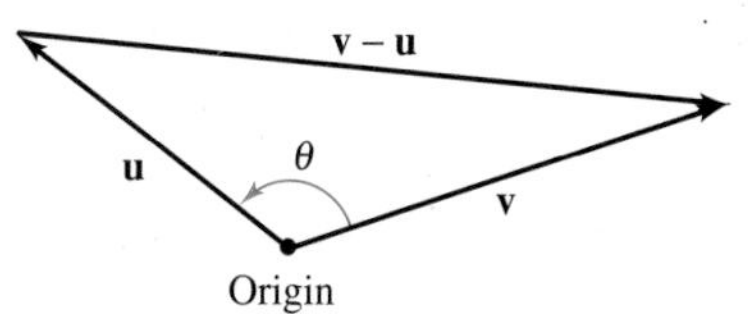

Figure 13.33

PROOF Consider the triangle determined by vectors $\mathbf{u}$, $\mathbf{v}$, and $\mathbf{v} - \mathbf{u}$, as shown in Figure 13.33. By the Law of Cosines, you can write

$$\begin{aligned}\|\mathbf{v} - \mathbf{u}\|^2 &= \|\mathbf{u}\|^2 + \|\mathbf{v}\|^2 - 2\|\mathbf{u}\|\,\|\mathbf{v}\|\cos\theta \\ (\mathbf{v} - \mathbf{u}) \cdot (\mathbf{v} - \mathbf{u}) &= \|\mathbf{u}\|^2 + \|\mathbf{v}\|^2 - 2\|\mathbf{u}\|\,\|\mathbf{v}\|\cos\theta \\ (\mathbf{v} - \mathbf{u}) \cdot \mathbf{v} - (\mathbf{v} - \mathbf{u}) \cdot \mathbf{u} &= \|\mathbf{u}\|^2 + \|\mathbf{v}\|^2 - 2\|\mathbf{u}\|\,\|\mathbf{v}\|\cos\theta \\ \mathbf{v} \cdot \mathbf{v} - \mathbf{u} \cdot \mathbf{v} - \mathbf{v} \cdot \mathbf{u} + \mathbf{u} \cdot \mathbf{u} &= \|\mathbf{u}\|^2 + \|\mathbf{v}\|^2 - 2\|\mathbf{u}\|\,\|\mathbf{v}\|\cos\theta \\ \|\mathbf{v}\|^2 - 2\mathbf{u} \cdot \mathbf{v} + \|\mathbf{u}\|^2 &= \|\mathbf{u}\|^2 + \|\mathbf{v}\|^2 - 2\|\mathbf{u}\|\,\|\mathbf{v}\|\cos\theta \\ -2\mathbf{u} \cdot \mathbf{v} &= -2\|\mathbf{u}\|\,\|\mathbf{v}\|\cos\theta \\ \cos\theta &= \frac{\mathbf{u} \cdot \mathbf{v}}{\|\mathbf{u}\|\,\|\mathbf{v}\|}.\end{aligned}$$

■

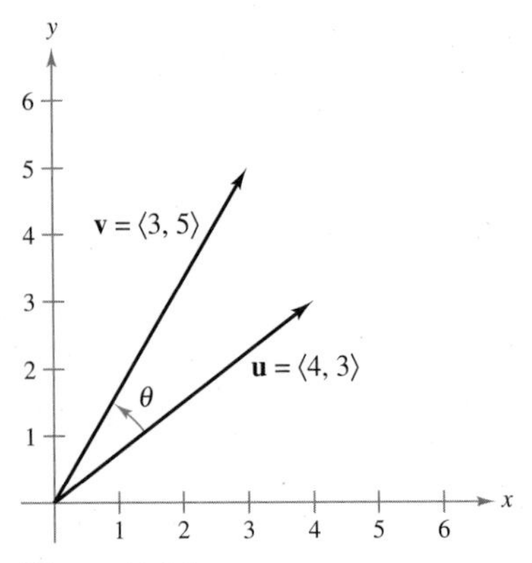

Figure 13.34

EXAMPLE 4 Finding the Angle Between Two Vectors

Find the angle θ between

$$\mathbf{u} = \langle 4, 3 \rangle \text{ and } \mathbf{v} = \langle 3, 5 \rangle.$$

Solution The two vectors and θ are shown in Figure 13.34.

$$\cos\theta = \frac{\mathbf{u} \cdot \mathbf{v}}{\|\mathbf{u}\|\,\|\mathbf{v}\|}$$

$$= \frac{\langle 4, 3 \rangle \cdot \langle 3, 5 \rangle}{\|\langle 4, 3 \rangle\|\,\|\langle 3, 5 \rangle\|}$$

$$= \frac{27}{5\sqrt{34}}$$

This implies that the angle between the two vectors is

$$\theta = \arccos\frac{27}{5\sqrt{34}}$$

$$\approx 22.2^\circ.$$

■

Rewriting the expression for the angle between two vectors in the form

$$\mathbf{u} \cdot \mathbf{v} = \|\mathbf{u}\|\,\|\mathbf{v}\| \cos\theta \qquad \text{Alternative form of dot product}$$

produces an alternative way to calculate the dot product. From this form, you can see that because $\|\mathbf{u}\|$ and $\|\mathbf{v}\|$ are always positive, $\mathbf{u} \cdot \mathbf{v}$ and $\cos\theta$ will always have the same sign. Figure 13.35 shows the five possible orientations of two vectors.

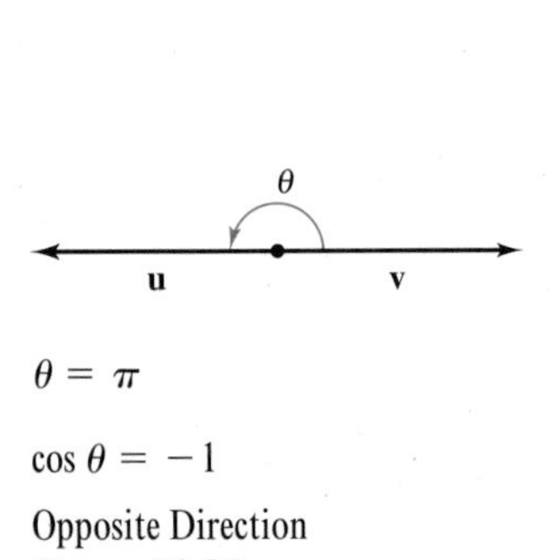

$\theta = \pi$

$\cos\theta = -1$

Opposite Direction

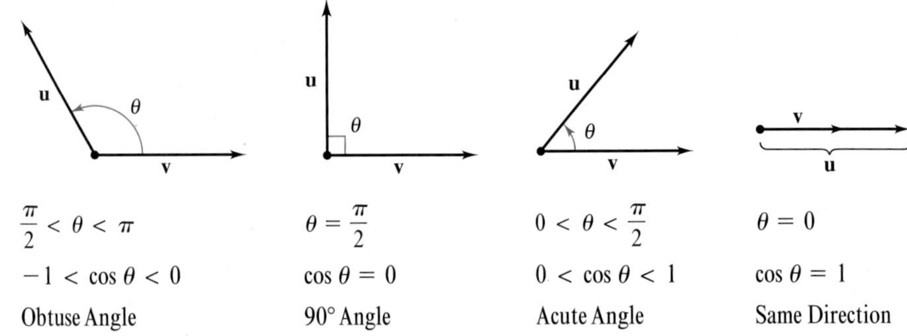

$\frac{\pi}{2} < \theta < \pi$ | $-1 < \cos\theta < 0$ | Obtuse Angle

$\theta = \frac{\pi}{2}$ | $\cos\theta = 0$ | 90° Angle

$0 < \theta < \frac{\pi}{2}$ | $0 < \cos\theta < 1$ | Acute Angle

$\theta = 0$ | $\cos\theta = 1$ | Same Direction

Figure 13.35

DEFINITION OF ORTHOGONAL VECTORS

The vectors **u** and **v** are **orthogonal** if $\mathbf{u} \cdot \mathbf{v} = 0$.

The terms *orthogonal* and *perpendicular* mean essentially the same thing—meeting at right angles. Note that the zero vector is orthogonal to every vector **u**, because $\mathbf{0} \cdot \mathbf{u} = 0$.

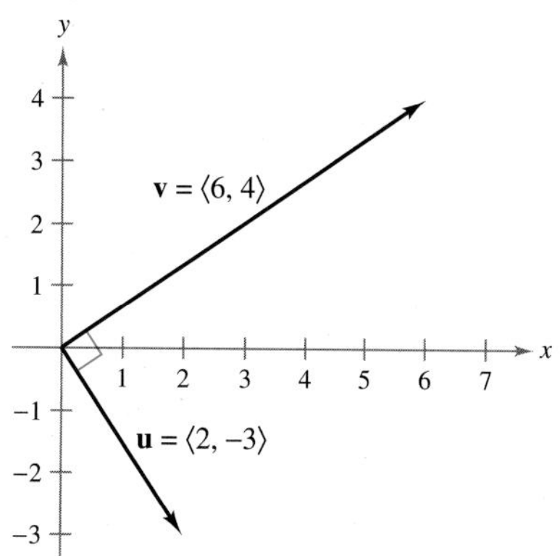

Figure 13.36

EXAMPLE 5 Determining Orthogonal Vectors

Are the vectors $\mathbf{u} = \langle 2, -3\rangle$ and $\mathbf{v} = \langle 6, 4\rangle$ orthogonal?

Solution Find the dot product of the two vectors.

$$\begin{aligned}\mathbf{u} \cdot \mathbf{v} &= \langle 2, -3\rangle \cdot \langle 6, 4\rangle \\ &= 2(6) + (-3)(4) \\ &= 0\end{aligned}$$

Because the dot product is 0, the two vectors are by definition orthogonal (see Figure 13.36). ■

Finding Vector Components

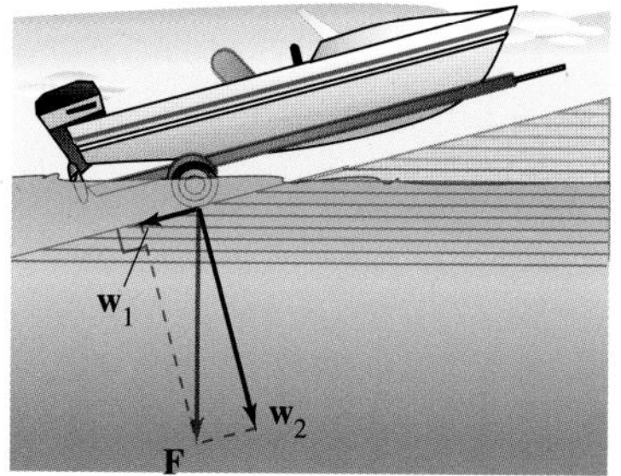

Figure 13.37

You have already seen applications in which two vectors are added to produce a resultant vector. Many applications in physics and engineering pose the *reverse* problem—decomposing a given vector into the sum of two **vector components.**

Consider a boat on an inclined ramp, as shown in Figure 13.37. The force $\mathbf{F}$ due to gravity pulls the boat *down* the ramp and *against* the ramp. These two orthogonal forces, $\mathbf{w}_1$ and $\mathbf{w}_2$, are vector components of $\mathbf{F}$. That is,

$$\mathbf{F} = \mathbf{w}_1 + \mathbf{w}_2. \qquad \text{Vector components of } \mathbf{F}$$

The negative of component $\mathbf{w}_1$ represents the force needed to keep the boat from rolling down the ramp, whereas $\mathbf{w}_2$ represents the force that the tires must withstand against the ramp.

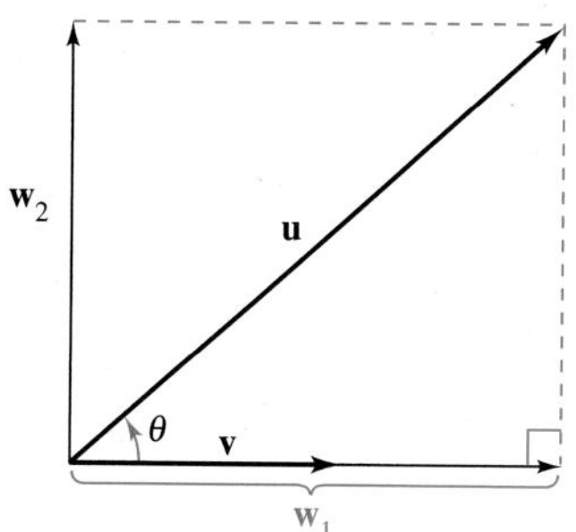

θ is acute.

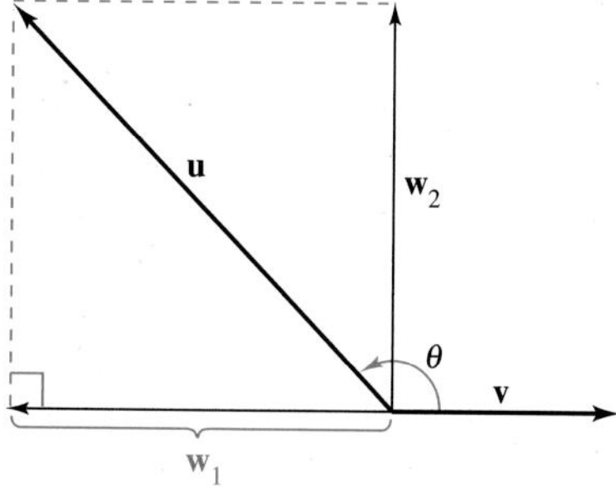

θ is obtuse.
Figure 13.38

DEFINITION OF VECTOR COMPONENTS

Let $\mathbf{u}$ and $\mathbf{v}$ be nonzero vectors such that

$$\mathbf{u} = \mathbf{w}_1 + \mathbf{w}_2$$

where $\mathbf{w}_1$ and $\mathbf{w}_2$ are orthogonal and $\mathbf{w}_1$ is parallel to (or a scalar multiple of) $\mathbf{v}$, as shown in Figure 13.38. The vectors $\mathbf{w}_1$ and $\mathbf{w}_2$ are called **vector components** of $\mathbf{u}$. The vector $\mathbf{w}_1$ is the **projection** of $\mathbf{u}$ onto $\mathbf{v}$ and is denoted by

$$\mathbf{w}_1 = \text{proj}_{\mathbf{v}}\mathbf{u}.$$

The vector $\mathbf{w}_2$ is called the vector component of $\mathbf{u}$ orthogonal to $\mathbf{v}$ and is given by $\mathbf{w}_2 = \mathbf{u} - \mathbf{w}_1$.

From the definition of vector components, you can see that it is easy to find the component $\mathbf{w}_2$ once you have found the projection of $\mathbf{u}$ onto $\mathbf{v}$. To find the projection, you can use the dot product.

THEOREM 13.8 PROJECTION OF u ONTO v

Let $\mathbf{u}$ and $\mathbf{v}$ be nonzero vectors. The projection of $\mathbf{u}$ onto $\mathbf{v}$ is

$$\text{proj}_{\mathbf{v}}\mathbf{u} = \left(\frac{\mathbf{u} \cdot \mathbf{v}}{\|\mathbf{v}\|^2}\right)\mathbf{v}.$$

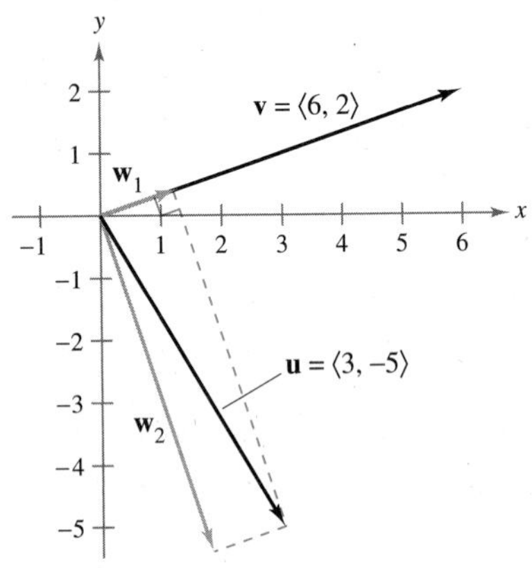

Figure 13.39

EXAMPLE 6 Decomposing a Vector into Orthogonal Components

Find the projection of $\mathbf{u} = \langle 3, -5\rangle$ onto $\mathbf{v} = \langle 6, 2\rangle$. Then write $\mathbf{u}$ as the sum of two orthogonal vectors, one of which is $\text{proj}_{\mathbf{v}}\mathbf{u}$.

Solution The projection of $\mathbf{u}$ onto $\mathbf{v}$ is

$$\mathbf{w}_1 = \text{proj}_{\mathbf{v}}\mathbf{u} = \left(\frac{\mathbf{u}\cdot\mathbf{v}}{\|\mathbf{v}\|^2}\right)\mathbf{v} = \left(\frac{8}{40}\right)\langle 6, 2\rangle = \left\langle\frac{6}{5}, \frac{2}{5}\right\rangle$$

as shown in Figure 13.39. The other component, $\mathbf{w}_2$, is

$$\begin{aligned}\mathbf{w}_2 &= \mathbf{u} - \mathbf{w}_1\\ &= \langle 3, -5\rangle - \left\langle\frac{6}{5}, \frac{2}{5}\right\rangle\\ &= \left\langle\frac{9}{5}, -\frac{27}{5}\right\rangle.\end{aligned}$$

So,

$$\begin{aligned}\mathbf{u} &= \mathbf{w}_1 + \mathbf{w}_2\\ &= \left\langle\frac{6}{5}, \frac{2}{5}\right\rangle + \left\langle\frac{9}{5}, -\frac{27}{5}\right\rangle\\ &= \langle 3, -5\rangle.\end{aligned}$$

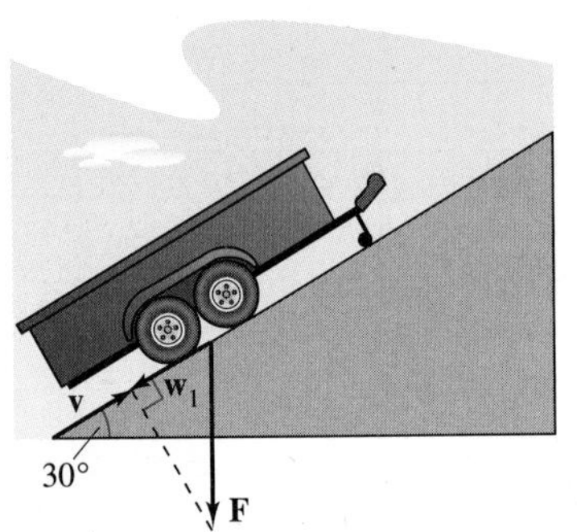

Figure 13.40

EXAMPLE 7 Finding a Force

A 200-pound cart sits on a ramp inclined at 30°, as shown in Figure 13.40. What force is required to keep the cart from rolling down the ramp?

Solution Because the force due to gravity is vertical and downward, you can represent the gravitational force by the vector

$$\mathbf{F} = -200\mathbf{j}. \qquad \text{Force due to gravity}$$

To find the force required to keep the cart from rolling down the ramp, project $\mathbf{F}$ onto a unit vector $\mathbf{v}$ in the direction of the ramp, as follows.

$$\mathbf{v} = (\cos 30°)\mathbf{i} + (\sin 30°)\mathbf{j} = \frac{\sqrt{3}}{2}\mathbf{i} + \frac{1}{2}\mathbf{j} \qquad \text{Unit vector along ramp}$$

Therefore, the projection of $\mathbf{F}$ onto $\mathbf{v}$ is

$$\begin{aligned}\mathbf{w}_1 &= \text{proj}_{\mathbf{v}}\mathbf{F}\\ &= \left(\frac{\mathbf{F}\cdot\mathbf{v}}{\|\mathbf{v}\|^2}\right)\mathbf{v}\\ &= (\mathbf{F}\cdot\mathbf{v})\mathbf{v} && \|\mathbf{v}\|^2 = 1\\ &= \left[(0)\left(\frac{\sqrt{3}}{2}\right) + (-200)\left(\frac{1}{2}\right)\right]\mathbf{v} && F = 0\mathbf{i} - 200\mathbf{j}\\ &= (-200)\left(\frac{1}{2}\right)\mathbf{v}\\ &= -100\left(\frac{\sqrt{3}}{2}\mathbf{i} + \frac{1}{2}\mathbf{j}\right).\end{aligned}$$

The magnitude of this force is 100, and so a force of 100 pounds is required to keep the cart from rolling down the ramp. ■

Work

The work W done by a *constant* force $\mathbf{F}$ acting along the line of motion of an object is given by

$$\begin{aligned} W &= \text{(magnitude of force)(distance)} \\ &= \|\mathbf{F}\|\,\|\overrightarrow{PQ}\| \end{aligned}$$

as shown in Figure 13.41(a). If the constant force $\mathbf{F}$ is not directed along the line of motion, as shown in Figure 13.41(b), the work W done by the force is given by

$$\begin{aligned} W &= \|\text{proj}_{\overrightarrow{PQ}}\,\mathbf{F}\|\,\|\overrightarrow{PQ}\| \\ &= (\cos\theta)\|\mathbf{F}\|\,\|\overrightarrow{PQ}\| \\ &= \mathbf{F}\cdot\overrightarrow{PQ}. \end{aligned}$$

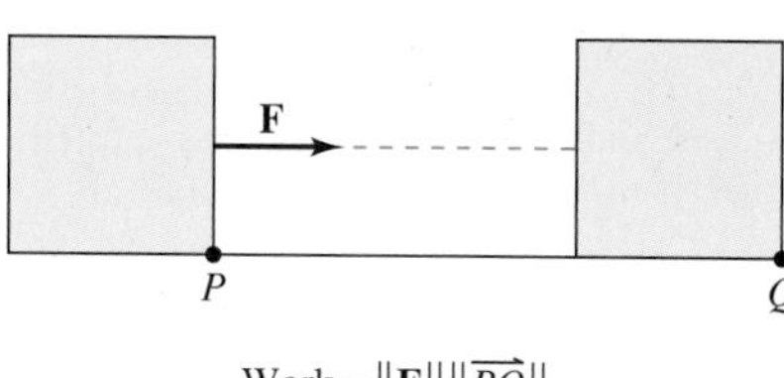

Work = $\|\mathbf{F}\|\|\overrightarrow{PQ}\|$

(a) Force acts along the line of motion.

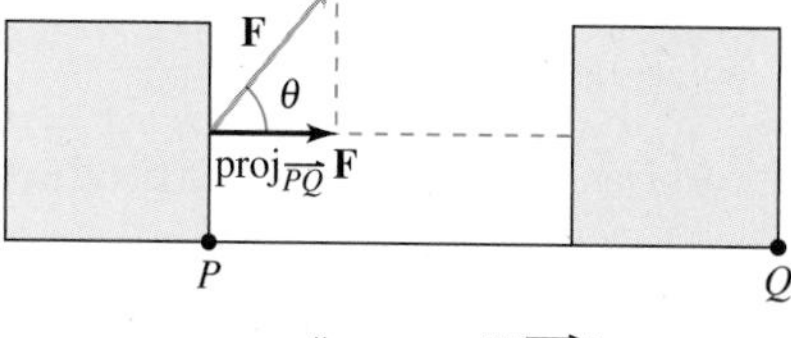

Work = $\|\text{proj}_{\overrightarrow{PQ}}\,\mathbf{F}\|\|\overrightarrow{PQ}\|$

(b) Force acts at an angle θ with the line of motion.

Figure 13.41

This notion of work is summarized in the following definition.

DEFINITION OF WORK

The **work** W done by a constant force $\mathbf{F}$ as its point of application moves along the vector $\overrightarrow{PQ}$ is given by either of the following.

1. $W = \|\text{proj}_{\overrightarrow{PQ}}\mathbf{F}\|\,\|\overrightarrow{PQ}\|$ Projection form

2. $W = \mathbf{F}\cdot\overrightarrow{PQ}$ Dot product form

EXAMPLE 8 Finding Work

Figure 13.42

To close a sliding barn door, a person pulls on a rope with a constant force of 50 pounds at a constant angle of 60°, as shown in Figure 13.42. Find the work done in moving the barn door 12 feet to its closed position.

Solution Using a projection, you can calculate the work as follows.

$$\begin{aligned} W &= \|\text{proj}_{\overrightarrow{PQ}}\,\mathbf{F}\|\,\|\overrightarrow{PQ}\| && \text{Projection form for work} \\ &= (\cos 60^\circ)\|\mathbf{F}\|\,\|\overrightarrow{PQ}\| \\ &= \frac{1}{2}(50)(12) \\ &= 300 \text{ foot-pounds} \end{aligned}$$

So, the work done is 300 foot-pounds. You can verify this result by finding the vectors $\mathbf{F}$ and $\overrightarrow{PQ}$ and calculating their dot product. ■

13.4 Exercises

See www.CalcChat.com for worked-out solutions to odd-numbered exercises.

In Exercises 1–6, fill in the blanks.

1. The ________ ________ of two vectors yields a scalar, rather than a vector.
2. The dot product of $\mathbf{u} = \langle u_1, u_2 \rangle$ and $\mathbf{v} = \langle v_1, v_2 \rangle$ is $\mathbf{u} \cdot \mathbf{v} =$ ________ .
3. If θ is the angle between two nonzero vectors $\mathbf{u}$ and $\mathbf{v}$, then $\cos\theta =$ ________ .
4. The vectors $\mathbf{u}$ and $\mathbf{v}$ are ________ if $\mathbf{u} \cdot \mathbf{v} = 0$.
5. The projection of $\mathbf{u}$ onto $\mathbf{v}$ is given by $\text{proj}_{\mathbf{v}}\mathbf{u} =$ ________ .
6. The work W done by a constant force $\mathbf{F}$ as its point of application moves along the vector $\overrightarrow{PQ}$ is given by $W =$ ________ or $W =$ ________ .

In Exercises 7–10, find the dot product of u and v.

7. $\mathbf{u} = \langle 7, 1 \rangle$, $\mathbf{v} = \langle -3, 2 \rangle$
8. $\mathbf{u} = \langle 6, 10 \rangle$, $\mathbf{v} = \langle -2, 3 \rangle$
9. $\mathbf{u} = 4\mathbf{i} - 2\mathbf{j}$, $\mathbf{v} = \mathbf{i} - \mathbf{j}$
10. $\mathbf{u} = 3\mathbf{i} + 4\mathbf{j}$, $\mathbf{v} = 7\mathbf{i} - 2\mathbf{j}$

In Exercises 11–14, use the vectors $\mathbf{u} = \langle 3, 3 \rangle$ and $\mathbf{v} = \langle -4, 2 \rangle$ to find the indicated quantity. State whether the result is a vector or a scalar.

11. $\mathbf{u} \cdot \mathbf{u}$
12. $3\mathbf{u} \cdot \mathbf{v}$
13. $(\mathbf{u} \cdot \mathbf{v})\mathbf{v}$
14. $2 - \|\mathbf{u}\|$

In Exercises 15–20, use the dot product to find the magnitude of u.

15. $\mathbf{u} = \langle -8, 15 \rangle$
16. $\mathbf{u} = \langle 4, -6 \rangle$
17. $\mathbf{u} = 20\mathbf{i} + 25\mathbf{j}$
18. $\mathbf{u} = 12\mathbf{i} - 16\mathbf{j}$
19. $\mathbf{u} = 6\mathbf{j}$
20. $\mathbf{u} = -21\mathbf{i}$

In Exercises 21–30, find the angle θ between the vectors.

21. $\mathbf{u} = \langle 1, 0 \rangle$, $\mathbf{v} = \langle 0, -2 \rangle$
22. $\mathbf{u} = \langle 3, 2 \rangle$, $\mathbf{v} = \langle 4, 0 \rangle$
23. $\mathbf{u} = 3\mathbf{i} + 4\mathbf{j}$, $\mathbf{v} = -2\mathbf{j}$
24. $\mathbf{u} = 2\mathbf{i} - 3\mathbf{j}$, $\mathbf{v} = \mathbf{i} - 2\mathbf{j}$
25. $\mathbf{u} = 2\mathbf{i} - \mathbf{j}$, $\mathbf{v} = 6\mathbf{i} + 4\mathbf{j}$
26. $\mathbf{u} = -6\mathbf{i} - 3\mathbf{j}$, $\mathbf{v} = -8\mathbf{i} + 4\mathbf{j}$
27. $\mathbf{u} = 5\mathbf{i} + 5\mathbf{j}$, $\mathbf{v} = -6\mathbf{i} + 6\mathbf{j}$
28. $\mathbf{u} = 2\mathbf{i} - 3\mathbf{j}$, $\mathbf{v} = 4\mathbf{i} + 3\mathbf{j}$
29. $\mathbf{u} = \cos\left(\frac{\pi}{3}\right)\mathbf{i} + \sin\left(\frac{\pi}{3}\right)\mathbf{j}$, $\mathbf{v} = \cos\left(\frac{3\pi}{4}\right)\mathbf{i} + \sin\left(\frac{3\pi}{4}\right)\mathbf{j}$
30. $\mathbf{u} = \cos\left(\frac{\pi}{4}\right)\mathbf{i} + \sin\left(\frac{\pi}{4}\right)\mathbf{j}$, $\mathbf{v} = \cos\left(\frac{\pi}{2}\right)\mathbf{i} + \sin\left(\frac{\pi}{2}\right)\mathbf{j}$

In Exercises 31–34, use vectors to find the interior angles of the triangle with the given vertices.

31. $(1, 2), (3, 4), (2, 5)$
32. $(-3, -4), (1, 7), (8, 2)$
33. $(-3, 0), (2, 2), (0, 6)$
34. $(-3, 5), (-1, 9), (7, 9)$

In Exercises 35–38, find $\mathbf{u} \cdot \mathbf{v}$, where θ is the angle between u and v.

35. $\|\mathbf{u}\| = 4,\ \|\mathbf{v}\| = 10,\ \theta = \frac{2\pi}{3}$
36. $\|\mathbf{u}\| = 100,\ \|\mathbf{v}\| = 250,\ \theta = \frac{\pi}{6}$
37. $\|\mathbf{u}\| = 9,\ \|\mathbf{v}\| = 36,\ \theta = \frac{3\pi}{4}$
38. $\|\mathbf{u}\| = 4,\ \|\mathbf{v}\| = 12,\ \theta = \frac{\pi}{3}$

In Exercises 39–44, determine whether u and v are orthogonal, parallel, or neither.

39. $\mathbf{u} = \langle -12, 30 \rangle$, $\mathbf{v} = \left\langle \frac{1}{2}, -\frac{5}{4} \right\rangle$
40. $\mathbf{u} = \langle 3, 15 \rangle$, $\mathbf{v} = \langle -1, 5 \rangle$
41. $\mathbf{u} = \frac{1}{4}(3\mathbf{i} - \mathbf{j})$, $\mathbf{v} = 5\mathbf{i} + 6\mathbf{j}$
42. $\mathbf{u} = \mathbf{i}$, $\mathbf{v} = -2\mathbf{i} + 2\mathbf{j}$
43. $\mathbf{u} = 2\mathbf{i} - 2\mathbf{j}$, $\mathbf{v} = -\mathbf{i} - \mathbf{j}$
44. $\mathbf{u} = \langle \cos\theta, \sin\theta \rangle$, $\mathbf{v} = \langle \sin\theta, -\cos\theta \rangle$

In Exercises 45–48, find the projection of u onto v. Then write u as the sum of two orthogonal vectors, one of which is $\text{proj}_{\mathbf{v}}\mathbf{u}$.

45. $\mathbf{u} = \langle 2, 2 \rangle$, $\mathbf{v} = \langle 6, 1 \rangle$
46. $\mathbf{u} = \langle 4, 2 \rangle$, $\mathbf{v} = \langle 1, -2 \rangle$
47. $\mathbf{u} = \langle 0, 3 \rangle$, $\mathbf{v} = \langle 2, 15 \rangle$
48. $\mathbf{u} = \langle -3, -2 \rangle$, $\mathbf{v} = \langle -4, -1 \rangle$

In Exercises 49–52, find two vectors in opposite directions that are orthogonal to the vector u. (There are many correct answers.)

49. $\mathbf{u} = \langle 3, 5 \rangle$
50. $\mathbf{u} = \langle -8, 3 \rangle$
51. $\mathbf{u} = \frac{1}{2}\mathbf{i} - \frac{2}{3}\mathbf{j}$
52. $\mathbf{u} = -\frac{5}{2}\mathbf{i} - 3\mathbf{j}$

Work **In Exercises 53 and 54, find the work done in moving a particle from P to Q if the magnitude and direction of the force are given by v.**

53. $P(0, 0)$, $Q(4, 7)$, $\mathbf{v} = \langle 1, 4 \rangle$

54. $P(1, 3)$, $Q(-3, 5)$, $\mathbf{v} = -2\mathbf{i} + 3\mathbf{j}$

WRITING ABOUT CONCEPTS

55. Under what conditions is the dot product of two vectors equal to the product of the lengths of the vectors?

56. Two forces of the same magnitude F_1 and F_2 act at angles θ_1 and θ_2, respectively. Compare the work done by F_1 with the work done by F_2 in moving along the vector $\overrightarrow{PQ}$ if

(a) $\theta_1 = -\theta_2$.

(b) $\theta_1 = 60°$ and $\theta_2 = 30°$.

57. ***Revenue*** The vector $\mathbf{u} = \langle 4600, 5250 \rangle$ gives the numbers of units of two models of cellular phones produced by a telecommunications company. The vector $\mathbf{v} = \langle 79.99, 99.99 \rangle$ gives the prices (in dollars) of the two models of cellular phones, respectively.

(a) Find the dot product $\mathbf{u} \cdot \mathbf{v}$ and interpret the result in the context of the problem.

(b) Identify the vector operation used to increase the prices by 5%.

58. ***Braking Load*** A truck with a gross weight of 30,000 pounds is parked on a slope of $d°$ (see figure). Assume that the only force to overcome is the force of gravity.

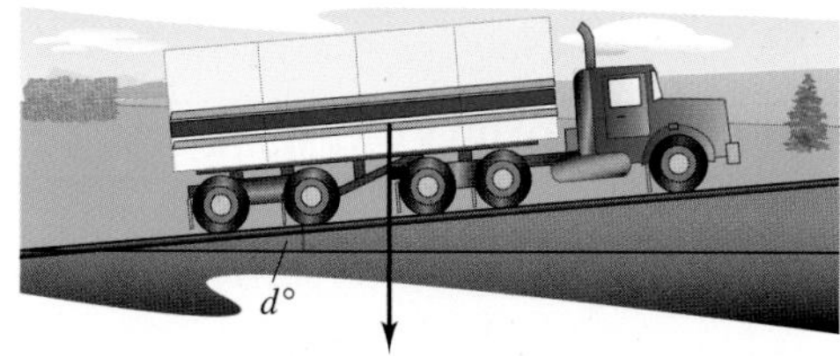

(a) Find the force required to keep the truck from rolling down the hill in terms of the slope d.

(b) Use a graphing utility to complete the table.

d	0°	1°	2°	3°	4°	5°
Force						

d	6°	7°	8°	9°	10°
Force					

(c) Find the force perpendicular to the hill when $d = 5°$.

59. ***Work*** Determine the work done by a person lifting a 245-newton bag of sugar 3 meters.

60. ***Work*** Determine the work done by a crane lifting a 2400-pound car 5 feet.

61. ***Work*** A force of 45 pounds exerted at an angle of 30° above the horizontal is required to slide a table across a floor (see figure). The table is dragged 20 feet. Determine the work done in sliding the table.

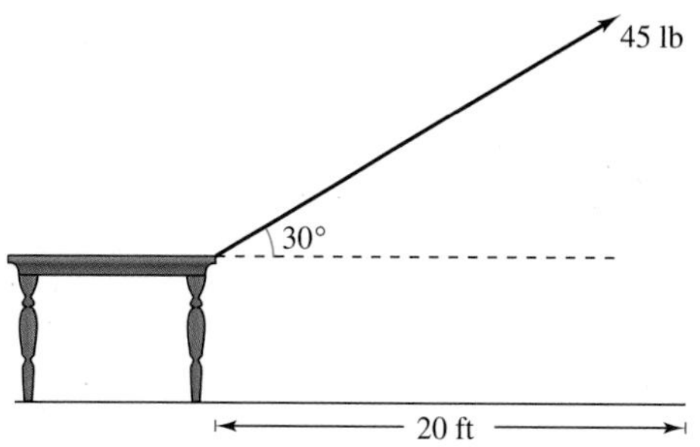

62. ***Work*** A tractor pulls a log 800 meters, and the tension in the cable connecting the tractor and log is approximately 15,691 newtons. The direction of the force is 35° above the horizontal. Approximate the work done in pulling the log.

True or False? **In Exercises 63–65, determine whether the statement is true or false. If it is false, explain why or give an example that shows it is false.**

63. The work W done by a constant force $\mathbf{F}$ acting along the line of motion of an object is represented by a vector.

64. A sliding door moves along the line of vector $\overrightarrow{PQ}$. If a force is applied to the door along a vector that is orthogonal to $\overrightarrow{PQ}$, then no work is done.

65. The dot product of two vectors is a scalar that is always nonnegative.

CAPSTONE

66. What is known about θ, the angle between two nonzero vectors $\mathbf{u}$ and $\mathbf{v}$, under each condition (see figure)?

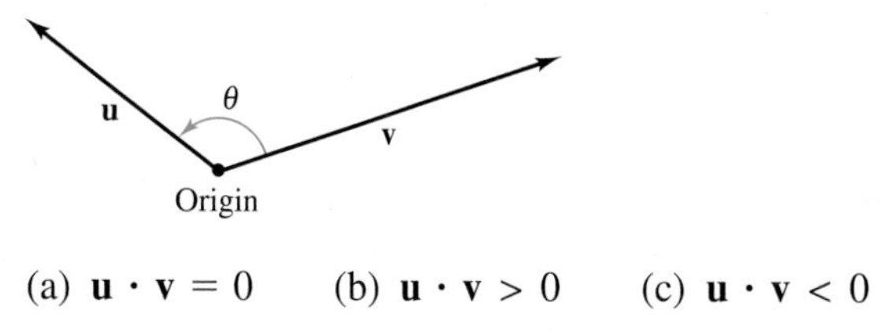

(a) $\mathbf{u} \cdot \mathbf{v} = 0$ (b) $\mathbf{u} \cdot \mathbf{v} > 0$ (c) $\mathbf{u} \cdot \mathbf{v} < 0$

67. Prove the following Properties of the Dot Product.

(a) $\mathbf{0} \cdot \mathbf{v} = 0$ (b) $\mathbf{u} \cdot (\mathbf{v} + \mathbf{w}) = \mathbf{u} \cdot \mathbf{v} + \mathbf{u} \cdot \mathbf{w}$

(c) $c(\mathbf{u} \cdot \mathbf{v}) = \mathbf{u} \cdot c\mathbf{v}$

68. Prove that $4(\mathbf{u} \cdot \mathbf{v}) = \|\mathbf{u} + \mathbf{v}\|^2 - \|\mathbf{u} - \mathbf{v}\|^2$.

13.5 Trigonometric Form of a Complex Number

- Plot complex numbers in the complex plane and find absolute values of complex numbers.
- Write the trigonometric forms of complex numbers.
- Multiply and divide complex numbers written in trigonometric form.
- Use DeMoivre's Theorem to find powers of complex numbers.
- Find nth roots of complex numbers.

The Complex Plane

Just as real numbers can be represented by points on the real number line, you can represent a complex number

$$z = a + bi$$

as the point (a, b) in a coordinate plane (the **complex plane**). The horizontal axis is called the **real axis** and the vertical axis is called the **imaginary axis,** as shown in Figure 13.43.

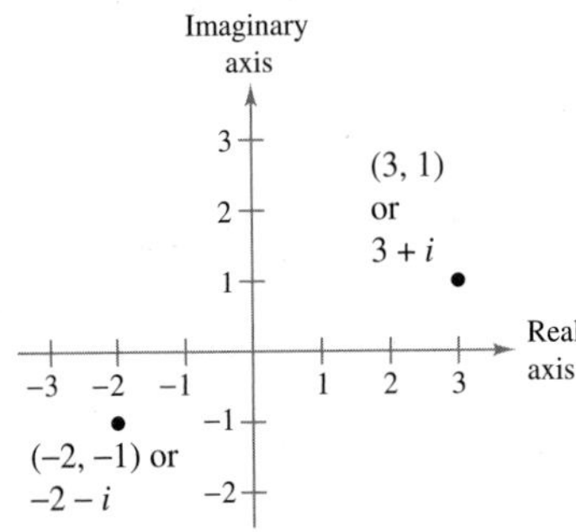

Figure 13.43

The **absolute value** of the complex number $a + bi$ is defined as the distance between the origin $(0, 0)$ and the point (a, b).

DEFINITION OF THE ABSOLUTE VALUE OF A COMPLEX NUMBER

The **absolute value** of the complex number $z = a + bi$ is

$$|a + bi| = \sqrt{a^2 + b^2}.$$

NOTE If the complex number $a + bi$ is a real number (that is, if $b = 0$), then this definition agrees with that given for the absolute value of a real number

$$|a + 0i| = \sqrt{a^2 + 0^2} = |a|.$$

EXAMPLE 1 Finding the Absolute Value of a Complex Number

Plot $z = -2 + 5i$ and find its absolute value.

Solution The number is plotted in Figure 13.44. It has an absolute value of

$$|z| = \sqrt{(-2)^2 + 5^2}$$
$$= \sqrt{29}.$$

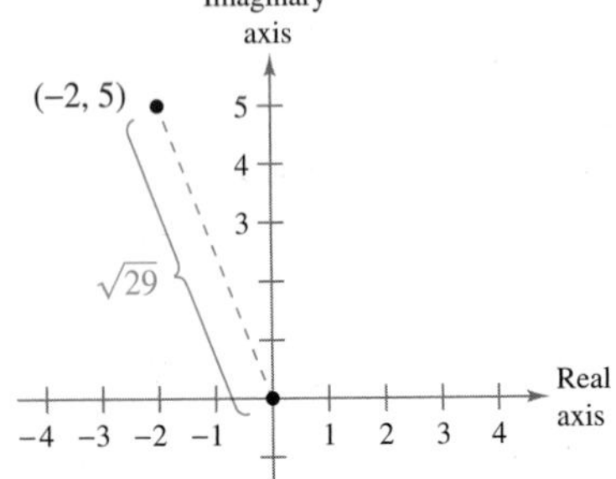

Figure 13.44

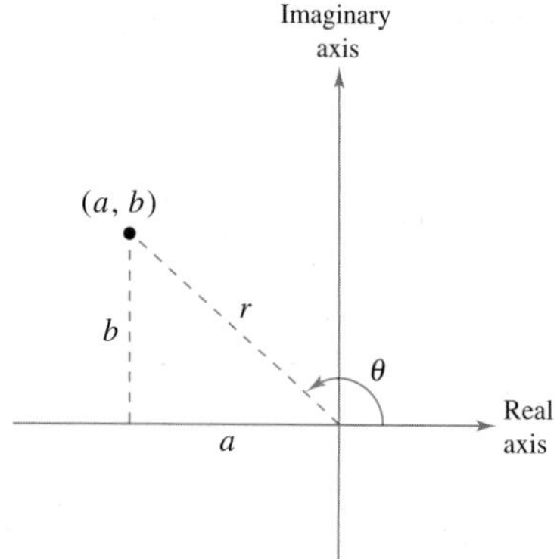

Figure 13.45

Trigonometric Form of a Complex Number

In Section 2.4, you learned how to add, subtract, multiply, and divide complex numbers. To work effectively with *powers* and *roots* of complex numbers, it is helpful to write complex numbers in trigonometric form. In Figure 13.45, consider the nonzero complex number $a + bi$. By letting θ be the angle from the positive real axis (measured counterclockwise) to the line segment connecting the origin and the point (a, b), you can write

$$a = r\cos\theta$$

and

$$b = r\sin\theta$$

where $r = \sqrt{a^2 + b^2}$. Consequently, you have

$$a + bi = (r\cos\theta) + (r\sin\theta)i$$

from which you can obtain the **trigonometric form of a complex number.**

> **TRIGONOMETRIC FORM OF A COMPLEX NUMBER**
>
> The **trigonometric form** of the complex number $z = a + bi$ is
>
> $$z = r(\cos\theta + i\sin\theta)$$
>
> where $a = r\cos\theta$, $b = r\sin\theta$, $r = \sqrt{a^2 + b^2}$, and $\tan\theta = b/a$. The number r is the **modulus** of z, and θ is called an **argument** of z.

The trigonometric form of a complex number is also called the *polar form.* Because there are infinitely many choices for θ, the trigonometric form of a complex number is not unique. Normally, θ is restricted to the interval $0 \le \theta < 2\pi$, although on occasion it is convenient to use $\theta < 0$.

EXAMPLE 2 Writing a Complex Number in Trigonometric Form

Write the complex number $z = -2 - 2\sqrt{3}i$ in trigonometric form.

Solution The absolute value of z is

$$\begin{aligned} r &= \left|-2 - 2\sqrt{3}i\right| \\ &= \sqrt{(-2)^2 + \left(-2\sqrt{3}\right)^2} \\ &= \sqrt{16} = 4 \end{aligned}$$

and the reference angle θ' is given by

$$\tan\theta' = \frac{b}{a} = \frac{-2\sqrt{3}}{-2} = \sqrt{3}.$$

Because $\tan(\pi/3) = \sqrt{3}$ and because $z = -2 - 2\sqrt{3}i$ lies in Quadrant III, you choose θ to be $\theta = \pi + \pi/3 = 4\pi/3$. So, the trigonometric form is

$$\begin{aligned} z &= r(\cos\theta + i\sin\theta) \\ &= 4\left(\cos\frac{4\pi}{3} + i\sin\frac{4\pi}{3}\right). \end{aligned}$$

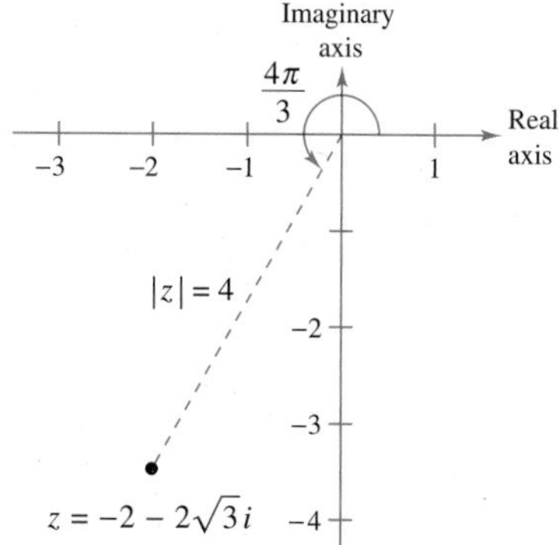

Figure 13.46

See Figure 13.46. ■

EXAMPLE 3 Writing a Complex Number in Standard Form

Write the complex number in standard form $a + bi$.

$$z = \sqrt{8}\left[\cos\left(-\frac{\pi}{3}\right) + i\sin\left(-\frac{\pi}{3}\right)\right]$$

Solution Because

$$\cos\left(-\frac{\pi}{3}\right) = \frac{1}{2} \quad \text{and} \quad \sin\left(-\frac{\pi}{3}\right) = -\frac{\sqrt{3}}{2}$$

you can write

$$\begin{aligned} z &= \sqrt{8}\left[\cos\left(-\frac{\pi}{3}\right) + i\sin\left(-\frac{\pi}{3}\right)\right] \\ &= 2\sqrt{2}\left(\frac{1}{2} - \frac{\sqrt{3}}{2}i\right) \\ &= \sqrt{2} - \sqrt{6}i. \end{aligned}$$

■

TECHNOLOGY You can use a graphing utility to convert a complex number in trigonometric (or polar) form to standard form. For specific keystrokes, see the user's manual for your graphing utility.

Multiplication and Division of Complex Numbers

The trigonometric form adapts nicely to multiplication and division of complex numbers. Suppose you are given two complex numbers

$$z_1 = r_1(\cos\theta_1 + i\sin\theta_1) \quad \text{and} \quad z_2 = r_2(\cos\theta_2 + i\sin\theta_2).$$

The product of z_1 and z_2 is given by

$$\begin{aligned} z_1z_2 &= r_1r_2(\cos\theta_1 + i\sin\theta_1)(\cos\theta_2 + i\sin\theta_2) \\ &= r_1r_2[(\cos\theta_1\cos\theta_2 - \sin\theta_1\sin\theta_2) + i(\sin\theta_1\cos\theta_2 + \cos\theta_1\sin\theta_2)]. \end{aligned}$$

Using the sum and difference formulas for cosine and sine, you can rewrite this equation as

$$z_1z_2 = r_1r_2[\cos(\theta_1 + \theta_2) + i\sin(\theta_1 + \theta_2)].$$

This establishes the first part of the following rule. The second part is left for you to verify (see Exercise 125).

PRODUCT AND QUOTIENT OF TWO COMPLEX NUMBERS

Let $z_1 = r_1(\cos\theta_1 + i\sin\theta_1)$ and $z_2 = r_2(\cos\theta_2 + i\sin\theta_2)$ be complex numbers.

$$z_1z_2 = r_1r_2[\cos(\theta_1 + \theta_2) + i\sin(\theta_1 + \theta_2)] \qquad \text{Product}$$

$$\frac{z_1}{z_2} = \frac{r_1}{r_2}[\cos(\theta_1 - \theta_2) + i\sin(\theta_1 - \theta_2)], \quad z_2 \neq 0 \qquad \text{Quotient}$$

Note that this rule says that to *multiply* two complex numbers you multiply moduli and add arguments, whereas to *divide* two complex numbers you divide moduli and subtract arguments.

EXAMPLE 4 Multiplying Complex Numbers

Find the product z_1z_2 of the complex numbers.

$$z_1 = 2\left(\cos\frac{2\pi}{3} + i\sin\frac{2\pi}{3}\right)$$

$$z_2 = 8\left(\cos\frac{11\pi}{6} + i\sin\frac{11\pi}{6}\right)$$

Solution Use the formula for multiplying complex numbers.

$$\begin{aligned} z_1z_2 &= r_1r_2[\cos(\theta_1 + \theta_2) + i\sin(\theta_1 + \theta_2)] \\ &= (2)(8)\left[\cos\left(\frac{2\pi}{3} + \frac{11\pi}{6}\right) + i\sin\left(\frac{2\pi}{3} + \frac{11\pi}{6}\right)\right] \\ &= 16\left(\cos\frac{5\pi}{2} + i\sin\frac{5\pi}{2}\right) \\ &= 16\left(\cos\frac{\pi}{2} + i\sin\frac{\pi}{2}\right) \qquad \frac{5\pi}{2} \text{ and } \frac{\pi}{2} \text{ are coterminal.} \\ &= 16[0 + i(1)] \\ &= 16i \end{aligned}$$

TECHNOLOGY Some graphing utilities can multiply and divide complex numbers in trigonometric form. If you have access to such a graphing utility, use it to find z_1z_2 and z_1/z_2 in Examples 4 and 5.

You can check this result by first converting the complex numbers to the standard forms

$$z_1 = -1 + \sqrt{3}i \qquad \text{and} \qquad z_2 = 4\sqrt{3} - 4i$$

and then multiplying, as in Section 2.4.

$$\begin{aligned} z_1z_2 &= \left(-1 + \sqrt{3}i\right)\left(4\sqrt{3} - 4i\right) \\ &= -4\sqrt{3} + 4i + 12i + 4\sqrt{3} \\ &= 16i \end{aligned}$$

■

EXAMPLE 5 Dividing Complex Numbers

Find the quotient z_1/z_2 of the complex numbers.

$$z_1 = 24(\cos 300° + i\sin 300°)$$

$$z_2 = 8(\cos 75° + i\sin 75°)$$

Solution Use the formula for dividing complex numbers.

$$\begin{aligned} \frac{z_1}{z_2} &= \frac{r_1}{r_2}[\cos(\theta_1 - \theta_2) + i\sin(\theta_1 - \theta_2)] \\ &= \frac{24}{8}[\cos(300° - 75°) + i\sin(300° - 75°)] \\ &= 3(\cos 225° + i\sin 225°) \\ &= 3\left[\left(-\frac{\sqrt{2}}{2}\right) + i\left(-\frac{\sqrt{2}}{2}\right)\right] \\ &= -\frac{3\sqrt{2}}{2} - \frac{3\sqrt{2}}{2}i \end{aligned}$$

■

Powers of Complex Numbers

The trigonometric form of a complex number is used to raise a complex number to a power. To accomplish this, consider repeated use of the multiplication rule.

$$\begin{aligned} z &= r(\cos\theta + i\sin\theta) \\ z^2 &= r(\cos\theta + i\sin\theta)r(\cos\theta + i\sin\theta) = r^2(\cos 2\theta + i\sin 2\theta) \\ z^3 &= r^2(\cos 2\theta + i\sin 2\theta)r(\cos\theta + i\sin\theta) = r^3(\cos 3\theta + i\sin 3\theta) \\ z^4 &= r^4(\cos 4\theta + i\sin 4\theta) \\ z^5 &= r^5(\cos 5\theta + i\sin 5\theta) \\ &\vdots \end{aligned}$$

This pattern leads to DeMoivre's Theorem, which is named after the French mathematician Abraham DeMoivre (1667–1754).

The Granger Collection, New York

ABRAHAM DEMOIVRE (1667–1754)

DeMoivre is remembered for his work in probability theory and DeMoivre's Theorem. His book *The Doctrine of Chances* (published in 1718) includes the theory of recurring series and the theory of partial fractions.

THEOREM 13.9 DEMOIVRE'S THEOREM

If $z = r(\cos\theta + i\sin\theta)$ is a complex number and n is a positive integer, then

$$\begin{aligned} z^n &= [r(\cos\theta + i\sin\theta)]^n \\ &= r^n(\cos n\theta + i\sin n\theta). \end{aligned}$$

EXAMPLE 6 Finding Powers of a Complex Number

$$\begin{aligned} (i)^6 &= \left[1\left(\cos\frac{\pi}{2} + i\sin\frac{\pi}{2}\right)\right]^6 \\ &= 1^6(\cos 3\pi + i\sin 3\pi) && r = 1,\ n = 6 \\ &= -1 && \cos 3\pi = -1,\ \sin 3\pi = 0 \end{aligned}$$

EXAMPLE 7 Finding Powers of a Complex Number

Use DeMoivre's Theorem to find $\left(-1 + \sqrt{3}i\right)^{12}$.

Solution First convert the complex number to trigonometric form using

$$r = \sqrt{(-1)^2 + \left(\sqrt{3}\right)^2} = 2 \quad \text{and} \quad \theta = \arctan\frac{\sqrt{3}}{-1} = \frac{2\pi}{3}.$$

So, the trigonometric form is

$$z = -1 + \sqrt{3}i = 2\left(\cos\frac{2\pi}{3} + i\sin\frac{2\pi}{3}\right).$$

Then, by DeMoivre's Theorem, you have

$$\begin{aligned} \left(-1 + \sqrt{3}i\right)^{12} &= \left[2\left(\cos\frac{2\pi}{3} + i\sin\frac{2\pi}{3}\right)\right]^{12} \\ &= 2^{12}\left[\cos\frac{12(2\pi)}{3} + i\sin\frac{12(2\pi)}{3}\right] \\ &= 4096(\cos 8\pi + i\sin 8\pi) \\ &= 4096(1 + 0) \\ &= 4096. \end{aligned}$$

■

Roots of Complex Numbers

Recall that a consequence of the Fundamental Theorem of Algebra is that a polynomial equation of degree n has n solutions in the complex number system. So, the equation $x^6 = 1$ has six solutions, and in this particular case you can find the six solutions by factoring and using the Quadratic Formula.

$$\begin{aligned} x^6 - 1 &= (x^3 - 1)(x^3 + 1) \\ &= (x - 1)(x^2 + x + 1)(x + 1)(x^2 - x + 1) = 0 \end{aligned}$$

Consequently, the solutions are

$$x = \pm 1, \qquad x = \frac{-1 \pm \sqrt{3}i}{2}, \qquad \text{and} \qquad x = \frac{1 \pm \sqrt{3}i}{2}.$$

Each of these numbers is a sixth root of 1. In general, an ***n*th root of a complex number** is defined as follows.

DEFINITION OF AN *n*TH ROOT OF A COMPLEX NUMBER

The complex number $u = a + bi$ is an ***n*th root** of the complex number z if

$$z = u^n = (a + bi)^n.$$

EXPLORATION

The nth roots of a complex number are useful for solving some polynomial equations. For instance, explain how you can use DeMoivre's Theorem to solve the polynomial equation

$$x^4 + 16 = 0.$$

[*Hint*: Write -16 as $16(\cos \pi + i \sin \pi)$.]

To find a formula for an nth root of a complex number, let u be an nth root of z, where

$$u = s(\cos \beta + i \sin \beta)$$

and

$$z = r(\cos \theta + i \sin \theta).$$

By DeMoivre's Theorem and the fact that $u^n = z$, you have

$$s^n(\cos n\beta + i \sin n\beta) = r(\cos \theta + i \sin \theta).$$

Taking the absolute value of each side of this equation, it follows that $s^n = r$. Substituting back into the previous equation and dividing by r, you get

$$\cos n\beta + i \sin n\beta = \cos \theta + i \sin \theta.$$

So, it follows that

$$\cos n\beta = \cos \theta$$

and

$$\sin n\beta = \sin \theta.$$

Because both sine and cosine have a period of 2π, these last two equations have solutions if and only if the angles differ by a multiple of 2π. Consequently, there must exist an integer k such that

$$\begin{aligned} n\beta &= \theta + 2\pi k \\ \beta &= \frac{\theta + 2\pi k}{n}. \end{aligned}$$

By substituting this value of β into the trigonometric form of u, you get the result stated in Theorem 13.10 on the following page.

THEOREM 13.10 *n*TH ROOTS OF A COMPLEX NUMBER

For a positive integer n, the complex number $z = r(\cos\theta + i\sin\theta)$ has exactly n distinct nth roots given by

$$\sqrt[n]{r}\left(\cos\frac{\theta + 2\pi k}{n} + i\sin\frac{\theta + 2\pi k}{n}\right)$$

where $k = 0, 1, 2, \ldots, n - 1$.

NOTE When k exceeds $n - 1$, the roots begin to repeat. For instance, if $k = n$, the angle

$$\frac{\theta + 2\pi n}{n} = \frac{\theta}{n} + 2\pi$$

is coterminal with θ/n, which is also obtained when $k = 0$. ■

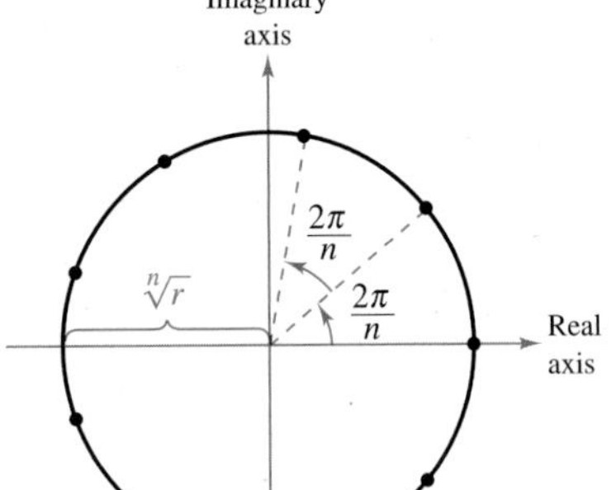

Figure 13.47

The formula for the nth roots of a complex number z has a nice geometrical interpretation, as shown in Figure 13.47. Note that because the nth roots of z all have the same magnitude $\sqrt[n]{r}$, they all lie on a circle of radius $\sqrt[n]{r}$ with center at the origin. Furthermore, because successive nth roots have arguments that differ by $2\pi/n$, the n roots are equally spaced around the circle.

You have already found the sixth roots of 1 by factoring and by using the Quadratic Formula. Example 8 shows how you can solve the same problem with the formula for nth roots.

EXAMPLE 8 Find the *n*th Roots of a Real Number

Find all sixth roots of 1.

Solution First write 1 in the trigonometric form $1 = 1(\cos 0 + i\sin 0)$. Then, by Theorem 13.10, with $n = 6$ and $r = 1$, the roots have the form

$$\sqrt[6]{1}\left(\cos\frac{0 + 2\pi k}{6} + i\sin\frac{0 + 2\pi k}{6}\right) = \cos\frac{\pi k}{3} + i\sin\frac{\pi k}{3}.$$

So, for $k = 0, 1, 2, 3, 4$, and 5, the sixth roots are as follows. (See Figure 13.48.)

$$\cos 0 + i\sin 0 = 1$$

$$\cos\frac{\pi}{3} + i\sin\frac{\pi}{3} = \frac{1}{2} + \frac{\sqrt{3}}{2}i \qquad \text{Increment by } \frac{2\pi}{n} = \frac{2\pi}{6} = \frac{\pi}{3}$$

$$\cos\frac{2\pi}{3} + i\sin\frac{2\pi}{3} = -\frac{1}{2} + \frac{\sqrt{3}}{2}i$$

$$\cos\pi + i\sin\pi = -1$$

$$\cos\frac{4\pi}{3} + i\sin\frac{4\pi}{3} = -\frac{1}{2} - \frac{\sqrt{3}}{2}i$$

$$\cos\frac{5\pi}{3} + i\sin\frac{5\pi}{3} = \frac{1}{2} - \frac{\sqrt{3}}{2}i$$

■

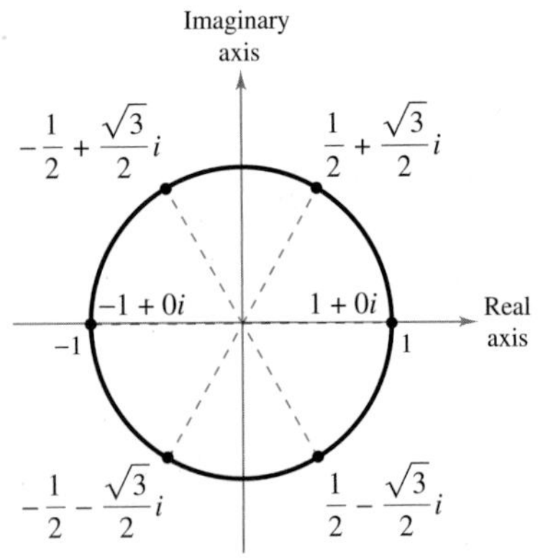

Figure 13.48

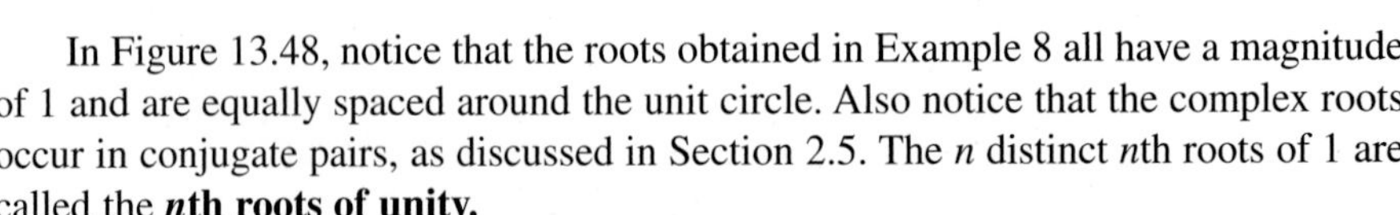

In Figure 13.48, notice that the roots obtained in Example 8 all have a magnitude of 1 and are equally spaced around the unit circle. Also notice that the complex roots occur in conjugate pairs, as discussed in Section 2.5. The n distinct nth roots of 1 are called the ***n*th roots of unity.**

EXAMPLE 9 Finding the *n*th Roots of a Complex Number

Find the three cube roots of $z = -2 + 2i$.

Solution Because z lies in Quadrant II, the trigonometric form of z is

$$z = -2 + 2i = \sqrt{8}\,(\cos 135° + i \sin 135°). \qquad \theta = \arctan\left(\frac{2}{-2}\right) = 135°$$

By Theorem 13.10, the cube roots have the form

$$\sqrt[6]{8}\left(\cos \frac{135° + 360°k}{3} + i \sin \frac{135° + 360°k}{3}\right).$$

Finally, for $k = 0, 1$, and 2, you obtain the roots

$$\sqrt[6]{8}\left(\cos \frac{135° + 360°(0)}{3} + i \sin \frac{135° + 360°(0)}{3}\right) = \sqrt{2}(\cos 45° + i \sin 45°)$$
$$= 1 + i$$
$$\sqrt[6]{8}\left(\cos \frac{135° + 360°(1)}{3} + i \sin \frac{135° + 360°(1)}{3}\right) = \sqrt{2}(\cos 165° + i \sin 165°)$$
$$\approx -1.3660 + 0.3660i$$
$$\sqrt[6]{8}\left(\cos \frac{135° + 360°(2)}{3} + i \sin \frac{135° + 360°(2)}{3}\right) = \sqrt{2}(\cos 285° + i \sin 285°)$$
$$\approx 0.3660 - 1.3660i.$$

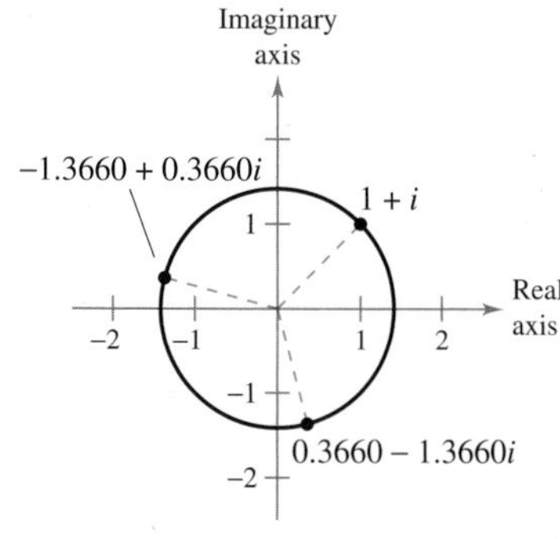

Figure 13.49

See Figure 13.49. ■

13.5 Exercises

See www.CalcChat.com for worked-out solutions to odd-numbered exercises.

In Exercises 1–4, fill in the blanks.

1. The ________ ________ of a complex number $a + bi$ is the distance between the origin $(0, 0)$ and the point (a, b).
2. The ________ ________ of a complex number $z = a + bi$ is given by $z = r(\cos \theta + i \sin \theta)$, where r is the ________ of z and θ is the ________ of z.
3. ________ Theorem states that if $z = r(\cos \theta + i \sin \theta)$ is a complex number and n is a positive integer, then $z^n = r^n(\cos n\theta + i \sin n\theta)$.
4. The complex number $u = a + bi$ is an ________ ________ of the complex number z if $z = u^n = (a + bi)^n$.

In Exercises 5–10, plot the complex number and find its absolute value.

5. $-6 + 8i$
6. $5 - 12i$
7. $-7i$
8. -7
9. $4 - 6i$
10. $-8 + 3i$

In Exercises 11–14, write the complex number in trigonometric form.

11.
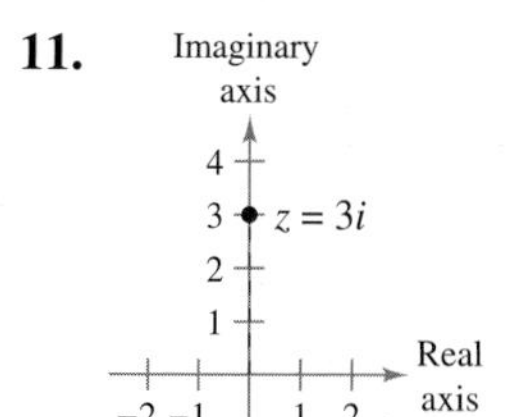

12.
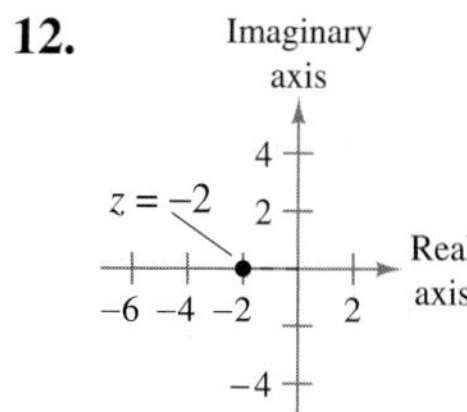

13.
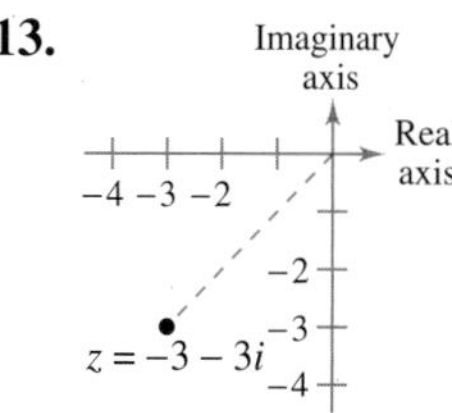

14.
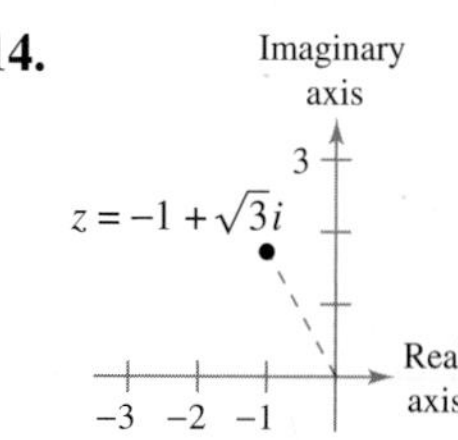

In Exercises 15–34, represent the complex number graphically, and find the trigonometric form of the number.

15. $1 + i$
16. $5 - 5i$
17. $1 - \sqrt{3}i$
18. $4 - 4\sqrt{3}i$

19. $-2(1 + \sqrt{3}i)$
20. $\frac{5}{2}(\sqrt{3} - i)$
21. $-5i$
22. $12i$
23. $-7 + 4i$
24. $3 - i$
25. 2
26. 4
27. $3 + \sqrt{3}i$
28. $2\sqrt{2} - i$
29. $-3 - i$
30. $1 + 3i$
31. $5 + 2i$
32. $8 + 3i$
33. $-8 - 5\sqrt{3}i$
34. $-9 - 2\sqrt{10}i$

In Exercises 35–44, find the standard form of the complex number. Then represent the complex number graphically.

35. $2(\cos 60° + i \sin 60°)$
36. $5(\cos 135° + i \sin 135°)$
37. $\sqrt{48}[\cos(-30°) + i \sin(-30°)]$
38. $\sqrt{8}(\cos 225° + i \sin 225°)$
39. $\frac{9}{4}\left(\cos \frac{3\pi}{4} + i \sin \frac{3\pi}{4}\right)$
40. $6\left(\cos \frac{5\pi}{12} + i \sin \frac{5\pi}{12}\right)$
41. $7(\cos 0 + i \sin 0)$
42. $8\left(\cos \frac{\pi}{2} + i \sin \frac{\pi}{2}\right)$
43. $5[\cos(198° 45') + i \sin(198° 45')]$
44. $9.75[\cos(280° 30') + i \sin(280° 30')]$

In Exercises 45–48, use a graphing utility to represent the complex number in standard form.

45. $5\left(\cos \frac{\pi}{9} + i \sin \frac{\pi}{9}\right)$
46. $10\left(\cos \frac{2\pi}{5} + i \sin \frac{2\pi}{5}\right)$
47. $2(\cos 155° + i \sin 155°)$
48. $9(\cos 58° + i \sin 58°)$

In Exercises 49–60, perform the operation and leave the result in trigonometric form.

49. $\left[2\left(\cos \frac{\pi}{4} + i \sin \frac{\pi}{4}\right)\right]\left[6\left(\cos \frac{\pi}{12} + i \sin \frac{\pi}{12}\right)\right]$
50. $\left[\frac{3}{4}\left(\cos \frac{\pi}{3} + i \sin \frac{\pi}{3}\right)\right]\left[4\left(\cos \frac{3\pi}{4} + i \sin \frac{3\pi}{4}\right)\right]$
51. $\left[\frac{5}{3}(\cos 120° + i \sin 120°)\right]\left[\frac{2}{3}(\cos 30° + i \sin 30°)\right]$
52. $\left[\frac{1}{2}(\cos 100° + i \sin 100°)\right]\left[\frac{4}{5}(\cos 300° + i \sin 300°)\right]$
53. $(\cos 80° + i \sin 80°)(\cos 330° + i \sin 330°)$
54. $(\cos 5° + i \sin 5°)(\cos 20° + i \sin 20°)$
55. $\dfrac{3(\cos 50° + i \sin 50°)}{9(\cos 20° + i \sin 20°)}$
56. $\dfrac{\cos 120° + i \sin 120°}{2(\cos 40° + i \sin 40°)}$
57. $\dfrac{\cos \pi + i \sin \pi}{\cos(\pi/3) + i \sin(\pi/3)}$
58. $\dfrac{5(\cos 4.3 + i \sin 4.3)}{4(\cos 2.1 + i \sin 2.1)}$
59. $\dfrac{12(\cos 92° + i \sin 92°)}{2(\cos 122° + i \sin 122°)}$
60. $\dfrac{6(\cos 40° + i \sin 40°)}{7(\cos 100° + i \sin 100°)}$

In Exercises 61–68, (a) write the trigonometric forms of the complex numbers, (b) perform the indicated operation using the trigonometric forms, and (c) perform the indicated operation using the standard forms, and check your result with that of part (b).

61. $(2 + 2i)(1 - i)$
62. $(\sqrt{3} + i)(1 + i)$
63. $-2i(1 + i)$
64. $3i(1 - \sqrt{2}i)$
65. $\dfrac{3 + 4i}{1 - \sqrt{3}i}$
66. $\dfrac{1 + \sqrt{3}i}{6 - 3i}$
67. $\dfrac{5}{2 + 3i}$
68. $\dfrac{4i}{-4 + 2i}$

In Exercises 69–72, sketch the graphs of all complex numbers z satisfying the given condition.

69. $|z| = 2$
70. $|z| = 3$
71. $\theta = \frac{\pi}{6}$
72. $\theta = \frac{5\pi}{4}$

In Exercises 73 and 74, represent the powers z, z^2, z^3, and z^4 graphically. Describe the pattern.

73. $z = \frac{\sqrt{2}}{2}(1 + i)$
74. $z = \frac{1}{2}(1 + \sqrt{3}i)$

In Exercises 75–92, use DeMoivre's Theorem to find the indicated power of the complex number. Write the result in standard form.

75. $(1 + i)^5$
76. $(2 + 2i)^6$
77. $(-1 + i)^6$
78. $(3 - 2i)^8$
79. $2(\sqrt{3} + i)^{10}$
80. $4(1 - \sqrt{3}i)^3$
81. $[5(\cos 20° + i \sin 20°)]^3$
82. $[3(\cos 60° + i \sin 60°)]^4$

83. $\left(\cos\frac{\pi}{4} + i\sin\frac{\pi}{4}\right)^{12}$ **84.** $\left[2\left(\cos\frac{\pi}{2} + i\sin\frac{\pi}{2}\right)\right]^8$

85. $[5(\cos 3.2 + i\sin 3.2)]^4$ **86.** $(\cos 0 + i\sin 0)^{20}$

87. $(3 - 2i)^5$ **88.** $(\sqrt{5} - 4i)^3$

89. $[3(\cos 15° + i\sin 15°)]^4$ **90.** $[2(\cos 10° + i\sin 10°)]^8$

91. $\left[2\left(\cos\frac{\pi}{10} + i\sin\frac{\pi}{10}\right)\right]^5$ **92.** $\left[2\left(\cos\frac{\pi}{8} + i\sin\frac{\pi}{8}\right)\right]^6$

In Exercises 93–108, (a) use Theorem 13.10 to find the indicated roots of the complex number, (b) represent each of the roots graphically, and (c) write each of the roots in standard form.

93. Square roots of $5(\cos 120° + i\sin 120°)$

94. Square roots of $16(\cos 60° + i\sin 60°)$

95. Cube roots of $8\left(\cos\frac{2\pi}{3} + i\sin\frac{2\pi}{3}\right)$

96. Fifth roots of $32\left(\cos\frac{5\pi}{6} + i\sin\frac{5\pi}{6}\right)$

97. Cube roots of $-\frac{125}{2}(1 + \sqrt{3}i)$

98. Cube roots of $-4\sqrt{2}(-1 + i)$

99. Square roots of $-25i$ **100.** Fourth roots of $625i$

101. Fourth roots of 16 **102.** Fourth roots of i

103. Fifth roots of 1 **104.** Cube roots of 1000

105. Cube roots of -125 **106.** Fourth roots of -4

107. Fifth roots of $4(1 - i)$ **108.** Sixth roots of $64i$

WRITING ABOUT CONCEPTS

In Exercises 109 and 110, use the figure. One of the fourth roots of a complex number z is shown.

109. How many roots are not shown?

110. Describe the other roots.

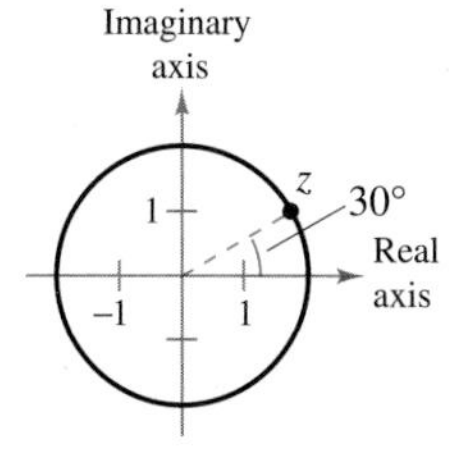

In Exercises 111–118, use Theorem 13.10 to find all the solutions of the equation and represent the solutions graphically.

111. $x^4 + i = 0$ **112.** $x^3 + 1 = 0$

113. $x^5 + 243 = 0$ **114.** $x^3 - 27 = 0$

115. $x^4 + 16i = 0$ **116.** $x^6 + 64i = 0$

117. $x^3 - (1 - i) = 0$ **118.** $x^4 + (1 + i) = 0$

True or False? **In Exercises 119–123, determine whether the statement is true or false. If it is false, explain why or give an example that shows it is false.**

119. Although the square of the complex number bi is given by $(bi)^2 = -b^2$, the absolute value of the complex number $z = a + bi$ is defined as

$$|a + bi| = \sqrt{a^2 + b^2}.$$

120. Geometrically, the nth roots of any complex number z are all equally spaced around the unit circle centered at the origin.

121. The product of the two complex numbers $z_1 = r_1(\cos\theta_1 + i\sin\theta_1)$ and $z_2 = r_2(\cos\theta_2 + i\sin\theta_2)$ is zero only when $r_1 = 0$ and/or $r_2 = 0$.

122. By DeMoivre's Theorem,

$$(4 + \sqrt{6}i)^8 = \cos(32) + i\sin(8\sqrt{6}).$$

123. By DeMoivre's Theorem,

$$(2 - 2\sqrt{3}i)^3 = 64(\cos\pi + i\sin\pi).$$

CAPSTONE

124. Use the graph of the roots of a complex number.

(a) Write each of the roots in trigonometric form.

(b) Identify the complex number whose roots are given. Use a graphing utility to verify your results.

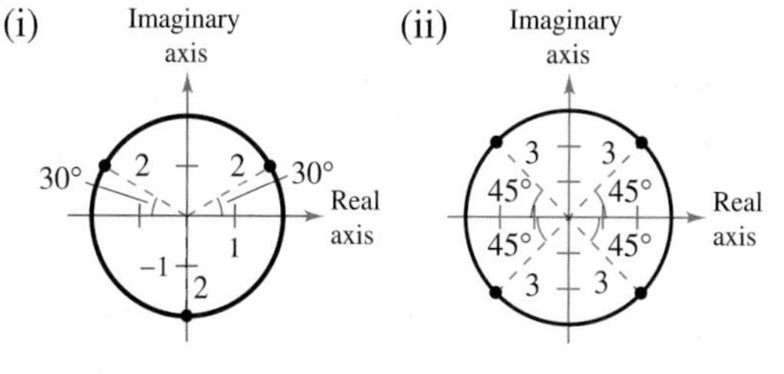

125. Given two complex numbers $z_1 = r_1(\cos\theta_1 + i\sin\theta_1)$ and $z_2 = r_2(\cos\theta_2 + i\sin\theta_2)$, $z_2 \neq 0$, show that

$$\frac{z_1}{z_2} = \frac{r_1}{r_2}[\cos(\theta_1 - \theta_2) + i\sin(\theta_1 - \theta_2)].$$

126. Show that $\bar{z} = r[\cos(-\theta) + i\sin(-\theta)]$ is the complex conjugate of $z = r(\cos\theta + i\sin\theta)$.

127. Use the trigonometric forms of z and $\bar{z}$ in Exercise 126 to find (a) $z\bar{z}$ and (b) $z/\bar{z}$, $\bar{z} \neq 0$.

128. Show that the negative of $z = r(\cos\theta + i\sin\theta)$ is $-z = r[\cos(\theta + \pi) + i\sin(\theta + \pi)]$.

129. Show that $\frac{1}{2}(1 - \sqrt{3}i)$ is a ninth root of -1.

130. Show that $2^{-1/4}(1 - i)$ is a fourth root of -2.

13 CHAPTER SUMMARY

Section 13.1	Review Exercises
■ Use the Law of Sines to solve oblique triangles (AAS, ASA, or SSA) *(p. 808).*	*1–12*
■ Find the areas of oblique triangles *(p. 812).*	*13–16*
■ Use the Law of Sines to model and solve real-life problems *(p. 813).*	*17–20*
Section 13.2	
■ Use the Law of Cosines to solve oblique triangles (SSS or SAS) *(p. 817).*	*21–30*
■ Use Heron's Area Formula to find the area of a triangle *(p. 818).*	*35–38*
■ Use the Law of Cosines to model and solve real-life problems *(p. 819).*	*39, 40*
Section 13.3	
■ Represent vectors as directed line segments *(p. 824).*	*41–44*
■ Write the component forms of vectors *(p. 825).*	*45–50*
■ Perform basic vector operations and represent them graphically *(p. 826).*	*51–62*
■ Write vectors as linear combinations of unit vectors *(p. 828).*	*63–68*
■ Find the direction angles of vectors *(p. 829).*	*69–74*
■ Use vectors to model and solve real-life problems *(p. 830).*	*75–78*
Section 13.4	
■ Find the dot product of two vectors and use the properties of the dot product *(p. 836).*	*79–86*
■ Find the angle between two vectors and determine whether two vectors are orthogonal *(p. 837).*	*87–94*
■ Write a vector as the sum of two vector components *(p. 839).*	*95–98*
■ Use vectors to find the work done by a force *(p. 841).*	*99, 100*
Section 13.5	
■ Plot complex numbers in the complex plane and find absolute values of complex numbers *(p. 844).*	*101–104*
■ Write the trigonometric forms of complex numbers *(p. 845).*	*105–110*
■ Multiply and divide complex numbers written in trigonometric form *(p. 846).*	*111, 112*
■ Use DeMoivre's Theorem to find powers of complex numbers *(p. 848).*	*113–116*
■ Find nth roots of complex numbers *(p. 849).*	*117–128*

13 REVIEW EXERCISES

See www.CalcChat.com for worked-out solutions to odd-numbered exercises.

In Exercises 1–12, use the Law of Sines to solve (if possible) the triangle. If two solutions exist, find both. Round your answers to two decimal places.

1. 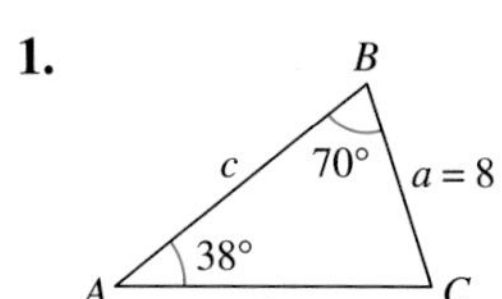

2. 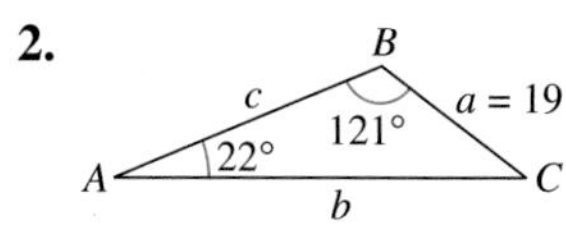

3. $B = 72°,\ C = 82°,\ b = 54$
4. $B = 10°,\ C = 20°,\ c = 33$
5. $A = 16°,\ B = 98°,\ c = 8.4$
6. $A = 95°,\ B = 45°,\ c = 104.8$
7. $A = 24°,\ C = 48°,\ b = 27.5$
8. $B = 64°,\ C = 36°,\ a = 367$
9. $B = 150°,\ b = 30,\ c = 10$
10. $B = 150°,\ a = 10,\ b = 3$
11. $A = 75°,\ a = 51.2,\ b = 33.7$
12. $B = 25°,\ a = 6.2,\ b = 4$

In Exercises 13–16, find the area of the triangle having the indicated angle and sides.

13. $A = 33°,\ b = 7,\ c = 10$
14. $B = 80°,\ a = 4,\ c = 8$
15. $C = 119°,\ a = 18,\ b = 6$
16. $A = 11°,\ b = 22,\ c = 21$

17. ***Height*** From a certain distance, the angle of elevation to the top of a building is 17°. At a point 50 meters closer to the building, the angle of elevation is 31°. Approximate the height of the building.

18. ***Geometry*** Find the length of the side w of the parallelogram.

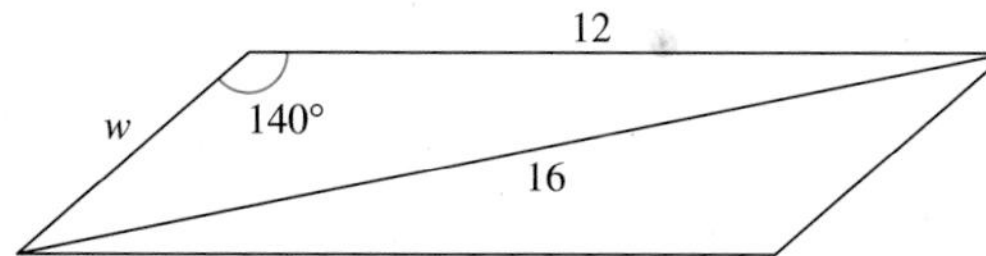

19. ***Height*** A tree stands on a hillside of slope 28° from the horizontal. From a point 75 feet down the hill, the angle of elevation to the top of the tree is 45° (see figure). Find the height of the tree.

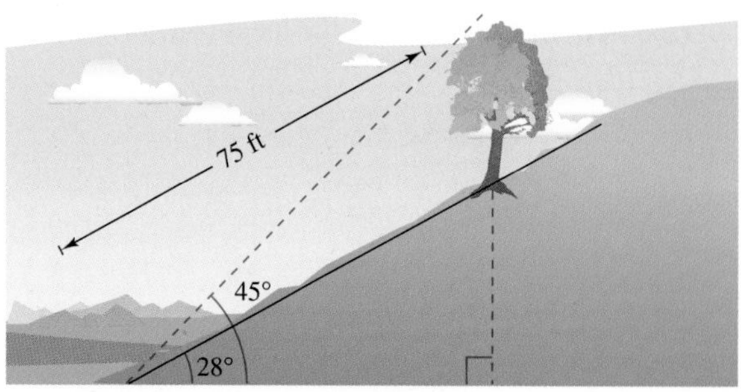

Figure for 19

20. ***River Width*** A surveyor finds that a tree on the opposite bank of a river flowing due east has a bearing of N 22° 30′ E from a certain point and a bearing of N 15° W from a point 400 feet downstream. Find the width of the river.

In Exercises 21–30, use the Law of Cosines to solve the triangle. Round your answers to two decimal places.

21. 22.

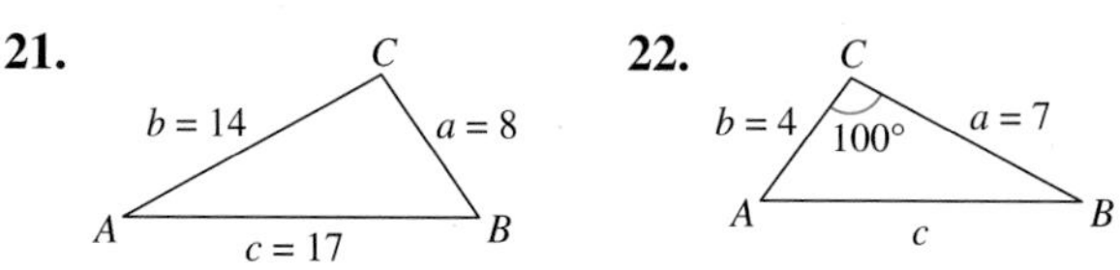

23. $a = 6,\ b = 9,\ c = 14$
24. $a = 75,\ b = 50,\ c = 110$
25. $a = 2.5,\ b = 5.0,\ c = 4.5$
26. $a = 16.4,\ b = 8.8,\ c = 12.2$
27. $B = 108°,\ a = 11,\ c = 11$
28. $B = 150°,\ a = 10,\ c = 20$
29. $C = 43°,\ a = 22.5,\ b = 31.4$
30. $A = 62°,\ b = 11.34,\ c = 19.52$

In Exercises 31–34, determine whether the Law of Sines or the Law of Cosines is needed to solve the triangle. Then solve the triangle.

31. $b = 9,\ c = 13,\ C = 64°$
32. $a = 4,\ c = 5,\ B = 52°$
33. $a = 13,\ b = 15,\ c = 24$
34. $A = 44°,\ B = 31°,\ c = 2.8$

In Exercises 35–38, use Heron's Area Formula to find the area of the triangle.

35. $a = 3,\ b = 6,\ c = 8$
36. $a = 15,\ b = 8,\ c = 10$
37. $a = 12.3,\ b = 15.8,\ c = 3.7$
38. $a = \frac{4}{5},\ b = \frac{3}{4},\ c = \frac{5}{8}$

39. ***Surveying*** To approximate the length of a marsh, a surveyor walks 425 meters from point A to point B. Then the surveyor turns 65° and walks 300 meters to point C (see figure). Approximate the length AC of the marsh.

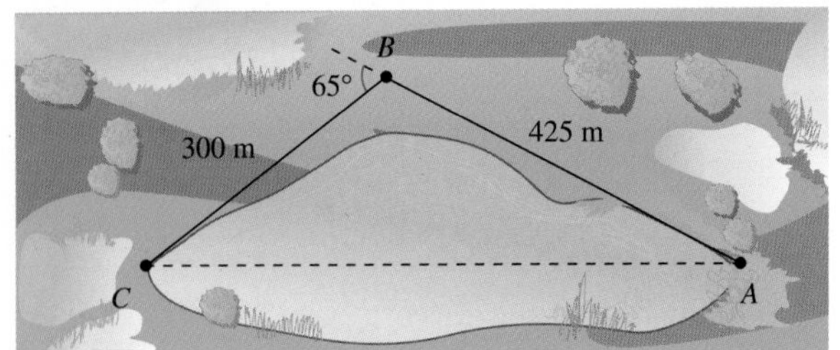

40. ***Navigation*** Two planes leave an airport at approximately the same time. One is flying 425 miles per hour at a bearing of 355°, and the other is flying 530 miles per hour at a bearing of 67°. Draw a figure that gives a visual representation of the situation and determine the distance between the planes after they have flown for 2 hours.

In Exercises 41–44, graph the vector with the given initial point and terminal point.

	Initial Point	*Terminal Point*
41.	(0, 0)	(8, 7)
42.	(3, 4)	(−5, −7)
43.	(−3, 9)	(8, −4)
44.	(−6, −8)	(8, 3)

In Exercises 45–50, find the component form of the vector v satisfying the conditions.

45.

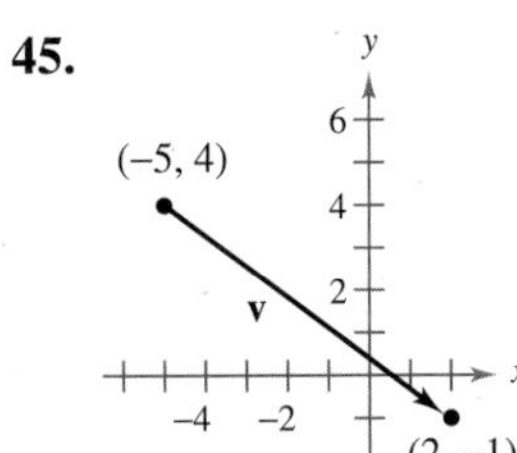

46.

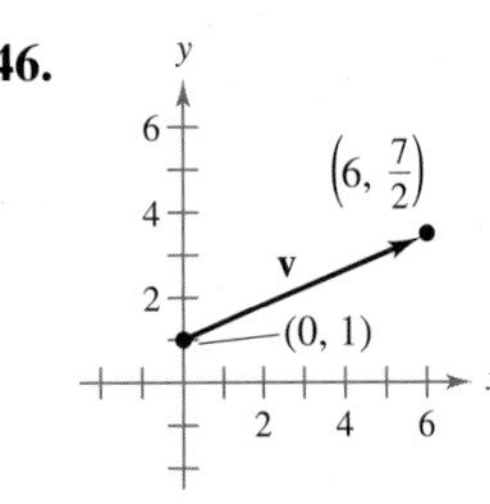

47. Initial point: (0, 10); terminal point: (7, 3)

48. Initial point: (1, 5); terminal point: (15, 9)

49. $\|\mathbf{v}\| = 8, \quad \theta = 120°$

50. $\|\mathbf{v}\| = \frac{1}{2}, \quad \theta = 225°$

In Exercises 51–58, find (a) u + v, (b) u − v, (c) 4u, and (d) 3v + 5u.

51. $\mathbf{u} = \langle -1, -3\rangle, \mathbf{v} = \langle -3, 6\rangle$

52. $\mathbf{u} = \langle 4, 5\rangle, \mathbf{v} = \langle 0, -1\rangle$

53. $\mathbf{u} = \langle -5, 2\rangle, \mathbf{v} = \langle 4, 4\rangle$

54. $\mathbf{u} = \langle 1, -8\rangle, \mathbf{v} = \langle 3, -2\rangle$

55. $\mathbf{u} = 2\mathbf{i} - \mathbf{j}, \mathbf{v} = 5\mathbf{i} + 3\mathbf{j}$

56. $\mathbf{u} = -7\mathbf{i} - 3\mathbf{j}, \mathbf{v} = 4\mathbf{i} - \mathbf{j}$

57. $\mathbf{u} = 4\mathbf{i}, \mathbf{v} = -\mathbf{i} + 6\mathbf{j}$

58. $\mathbf{u} = -6\mathbf{j}, \mathbf{v} = \mathbf{i} + \mathbf{j}$

In Exercises 59–62, find the component form of w and sketch the specified vector operations geometrically, where u = 6i − 5j and v = 10i + 3j.

59. $\mathbf{w} = 2\mathbf{u} + \mathbf{v}$

60. $\mathbf{w} = 4\mathbf{u} - 5\mathbf{v}$

61. $\mathbf{w} = 3\mathbf{v}$

62. $\mathbf{w} = \frac{1}{2}\mathbf{v}$

In Exercises 63–66, write vector u as a linear combination of the standard unit vectors i and j.

63. $\mathbf{u} = \langle -1, 5\rangle$

64. $\mathbf{u} = \langle -6, -8\rangle$

65. **u** has initial point (3, 4) and terminal point (9, 8).

66. **u** has initial point (−2, 7) and terminal point (5, −9).

In Exercises 67 and 68, write the vector v in the form $\|\mathbf{v}\|(\cos\theta)\mathbf{i} + \|\mathbf{v}\|(\sin\theta)\mathbf{j}$.

67. $\mathbf{v} = -10\mathbf{i} + 10\mathbf{j}$

68. $\mathbf{v} = 4\mathbf{i} - \mathbf{j}$

In Exercises 69–74, find the magnitude and the direction angle of the vector v.

69. $\mathbf{v} = 7(\cos 60°\mathbf{i} + \sin 60°\mathbf{j})$

70. $\mathbf{v} = 3(\cos 150°\mathbf{i} + \sin 150°\mathbf{j})$

71. $\mathbf{v} = 5\mathbf{i} + 4\mathbf{j}$

72. $\mathbf{v} = -4\mathbf{i} + 7\mathbf{j}$

73. $\mathbf{v} = -3\mathbf{i} - 3\mathbf{j}$

74. $\mathbf{v} = 8\mathbf{i} - \mathbf{j}$

75. ***Resultant Force*** Forces with magnitudes of 85 pounds and 50 pounds act on a single point. The angle between the forces is 15°. Describe the resultant force.

76. ***Rope Tension*** A 180-pound weight is supported by two ropes, as shown in the figure. Find the tension in each rope.

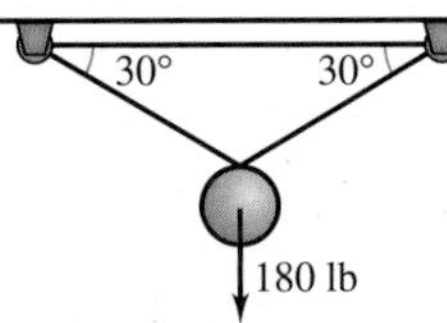

77. ***Navigation*** An airplane has an airspeed of 430 miles per hour at a bearing of 135°. The wind velocity is 35 miles per hour in the direction of N 30° E. Find the resultant speed and direction of the airplane.

78. ***Navigation*** An airplane has an airspeed of 724 kilometers per hour at a bearing of 30°. The wind velocity is 32 kilometers per hour from the west. Find the resultant speed and direction of the airplane.

In Exercises 79–82, find the dot product of u and v.

79. $\mathbf{u} = \langle 6, 7 \rangle$
$\mathbf{v} = \langle -3, 9 \rangle$

80. $\mathbf{u} = \langle -7, 12 \rangle$
$\mathbf{v} = \langle -4, -14 \rangle$

81. $\mathbf{u} = 3\mathbf{i} + 7\mathbf{j}$
$\mathbf{v} = 11\mathbf{i} - 5\mathbf{j}$

82. $\mathbf{u} = -7\mathbf{i} + 2\mathbf{j}$
$\mathbf{v} = 16\mathbf{i} - 12\mathbf{j}$

In Exercises 83–86, use the vectors $\mathbf{u} = \langle -4, 2 \rangle$ and $\mathbf{v} = \langle 5, 1 \rangle$ to find the indicated quantity. State whether the result is a vector or a scalar.

83. $2\mathbf{u} \cdot \mathbf{u}$

84. $3\mathbf{u} \cdot \mathbf{v}$

85. $\mathbf{u}(\mathbf{u} \cdot \mathbf{v})$

86. $(\mathbf{v} \cdot \mathbf{v}) - (\mathbf{v} \cdot \mathbf{u})$

In Exercises 87–90, find the angle θ between the vectors.

87. $\mathbf{u} = \cos \frac{7\pi}{4}\mathbf{i} + \sin \frac{7\pi}{4}\mathbf{j}$
$\mathbf{v} = \cos \frac{5\pi}{6}\mathbf{i} + \sin \frac{5\pi}{6}\mathbf{j}$

88. $\mathbf{u} = \cos 45°\mathbf{i} + \sin 45°\mathbf{j}$
$\mathbf{v} = \cos 300°\mathbf{i} + \sin 300°\mathbf{j}$

89. $\mathbf{u} = \langle 2\sqrt{2}, -4 \rangle, \quad \mathbf{v} = \langle -\sqrt{2}, 1 \rangle$

90. $\mathbf{u} = \langle 3, \sqrt{3} \rangle, \quad \mathbf{v} = \langle 4, 3\sqrt{3} \rangle$

In Exercises 91–94, determine whether u and v are orthogonal, parallel, or neither.

91. $\mathbf{u} = \langle -3, 8 \rangle$
$\mathbf{v} = \langle 8, 3 \rangle$

92. $\mathbf{u} = \langle \frac{1}{4}, -\frac{1}{2} \rangle$
$\mathbf{v} = \langle -2, 4 \rangle$

93. $\mathbf{u} = -\mathbf{i}$
$\mathbf{v} = \mathbf{i} + 2\mathbf{j}$

94. $\mathbf{u} = -2\mathbf{i} + \mathbf{j}$
$\mathbf{v} = 3\mathbf{i} + 6\mathbf{j}$

In Exercises 95–98, find the projection of u onto v. Then write u as the sum of two orthogonal vectors, one of which is $\text{proj}_{\mathbf{v}}\mathbf{u}$.

95. $\mathbf{u} = \langle -4, 3 \rangle, \ \mathbf{v} = \langle -8, -2 \rangle$

96. $\mathbf{u} = \langle 5, 6 \rangle, \ \mathbf{v} = \langle 10, 0 \rangle$

97. $\mathbf{u} = \langle 2, 7 \rangle, \ \mathbf{v} = \langle 1, -1 \rangle$

98. $\mathbf{u} = \langle -3, 5 \rangle, \ \mathbf{v} = \langle -5, 2 \rangle$

Work **In Exercises 99 and 100, find the work done in moving a particle from P to Q if the magnitude and direction of the force are given by v.**

99. $P(5, 3), Q(8, 9), \mathbf{v} = \langle 2, 7 \rangle$

100. $P(-2, -9), Q(-12, 8), \mathbf{v} = 3\mathbf{i} - 6\mathbf{j}$

In Exercises 101–104, plot the complex number and find its absolute value.

101. $7i$

102. $-6i$

103. $5 + 3i$

104. $\sqrt{2} - \sqrt{2}i$

In Exercises 105–110, write the complex number in trigonometric form.

105. $4i$

106. -7

107. $5 - 5i$

108. $5 + 12i$

109. $-5 - 12i$

110. $-3\sqrt{3} + 3i$

In Exercises 111 and 112, (a) write the two complex numbers in trigonometric form, and (b) use the trigonometric forms to find z_1z_2 and z_1/z_2, where $z_2 \neq 0$.

111. $z_1 = 2\sqrt{3} - 2i, \quad z_2 = -10i$

112. $z_1 = -3(1 + i), \quad z_2 = 2(\sqrt{3} + i)$

In Exercises 113–116, use DeMoivre's Theorem to find the indicated power of the complex number. Write the result in standard form.

113. $\left[5\left(\cos \frac{\pi}{12} + i \sin \frac{\pi}{12}\right)\right]^4$

114. $\left[2\left(\cos \frac{4\pi}{15} + i \sin \frac{4\pi}{15}\right)\right]^5$

115. $(2 + 3i)^6$

116. $(1 - i)^8$

Graphical Reasoning **In Exercises 117 and 118, use the graph of the roots of a complex number.**

(a) Write each of the roots in trigonometric form.

(b) Identify the complex number whose roots are given. Use a graphing utility to verify your results.

117. 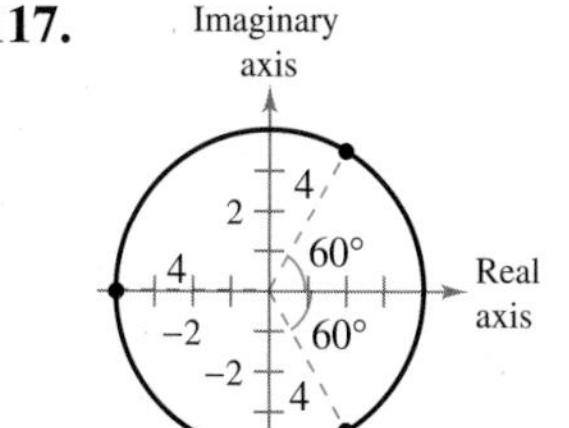

118. 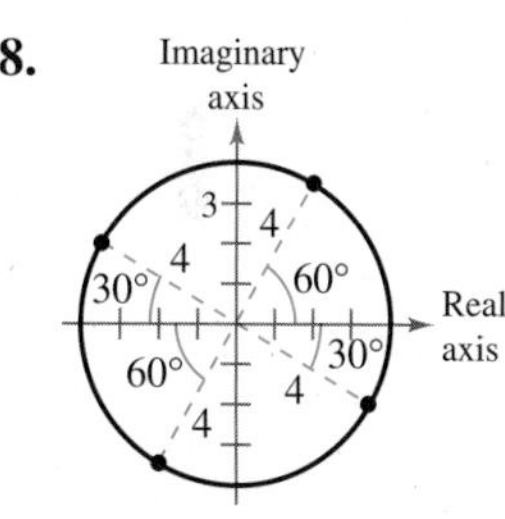

In Exercises 119–122, (a) use Theorem 13.10 to find the indicated roots of the complex number, (b) represent each of the roots graphically, and (c) write each of the roots in standard form.

119. Sixth roots of $-729i$

120. Fourth roots of $256i$

121. Cube roots of 8

122. Fifth roots of -1024

In Exercises 123–128, use Theorem 13.10 to find all solutions of the equation and represent the solutions graphically.

123. $x^4 + 81 = 0$

124. $x^5 - 32 = 0$

125. $x^3 + 8i = 0$

126. $x^4 - 64i = 0$

127. $x^5 + x^3 - x^2 - 1 = 0$

128. $x^5 + 4x^3 - 8x^2 - 32 = 0$

13 CHAPTER TEST

Take this test as you would take a test in class. When you are finished, check your work against the answers given in the back of the book.

In Exercises 1–6, use the information to solve (if possible) the triangle. If two solutions exist, find both solutions. Round your answers to two decimal places.

1. $A = 24°,\ B = 68°,\ a = 12.2$

2. $B = 110°,\ C = 28°,\ a = 15.6$

3. $A = 24°,\ a = 11.2,\ b = 13.4$

4. $a = 4.0,\ b = 7.3,\ c = 12.4$

5. $B = 100°,\ a = 15,\ b = 23$

6. $C = 121°,\ a = 34,\ b = 55$

7. A triangular parcel of land has borders of lengths 60 meters, 70 meters, and 82 meters. Find the area of the parcel of land.

8. An airplane flies 370 miles from point A to point B with a bearing of 24°. It then flies 240 miles from point B to point C with a bearing of 37° (see figure). Find the distance and bearing from point A to point C.

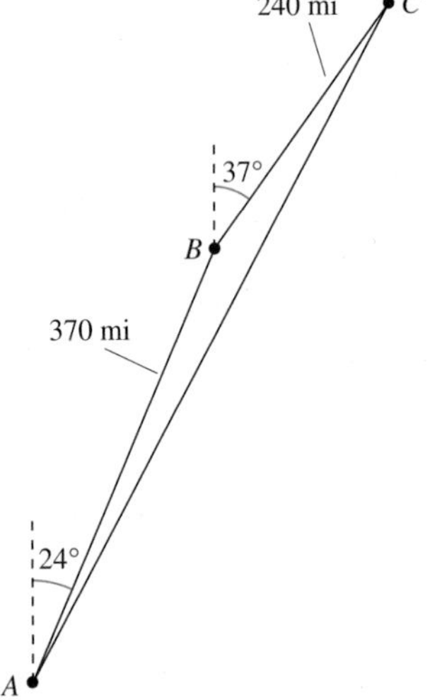

Figure for 8

In Exercises 9 and 10, find the component form of the vector v satisfying the given conditions.

9. Initial point of **v**: $(-3, 7)$; terminal point of **v**: $(11, -16)$

10. Magnitude of **v**: $\|\mathbf{v}\| = 12$; direction of **v**: $\mathbf{u} = \langle 3, -5\rangle$

In Exercises 11–14, $\mathbf{u} = \langle 2, 7\rangle$ and $\mathbf{v} = \langle -6, 5\rangle$. Find the resultant vector and sketch its graph.

11. $\mathbf{u} + \mathbf{v}$

12. $\mathbf{u} - \mathbf{v}$

13. $5\mathbf{u} - 3\mathbf{v}$

14. $4\mathbf{u} + 2\mathbf{v}$

15. Find a unit vector in the direction of $\mathbf{u} = \langle 24, -7\rangle$.

16. Forces with magnitudes of 250 pounds and 130 pounds act on an object at angles of 45° and $-60°$, respectively, with the x-axis. Find the direction and magnitude of the resultant of these forces.

17. Find the angle between the vectors $\mathbf{u} = \langle -1, 5\rangle$ and $\mathbf{v} = \langle 3, -2\rangle$.

18. Are the vectors $\mathbf{u} = \langle 6, -10\rangle$ and $\mathbf{v} = \langle 5, 3\rangle$ orthogonal?

19. Find the projection of $\mathbf{u} = \langle 6, 7\rangle$ onto $\mathbf{v} = \langle -5, -1\rangle$. Then write **u** as the sum of two orthogonal vectors.

20. A 500-pound motorcycle is headed up a hill inclined at 12°. What force is required to keep the motorcycle from rolling down the hill when stopped at a red light?

21. Write the complex number $z = 4 - 4i$ in trigonometric form.

22. Write the complex number $z = 6(\cos 120° + i \sin 120°)$ in standard form.

In Exercises 23 and 24, use DeMoivre's Theorem to find the indicated power of the complex number. Write the result in standard form.

23. $\left[3\left(\cos\frac{7\pi}{6} + i\sin\frac{7\pi}{6}\right)\right]^8$

24. $(3 - 3i)^6$

25. Find the fourth roots of $256(1 + \sqrt{3}i)$.

26. Find all solutions of the equation $x^3 - 27i = 0$ and represent the solutions graphically.

P.S. PROBLEM SOLVING

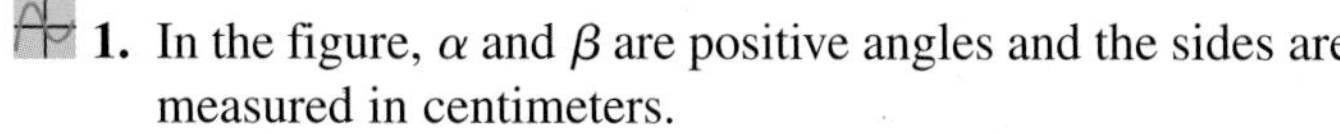

1. In the figure, α and β are positive angles and the sides are measured in centimeters.

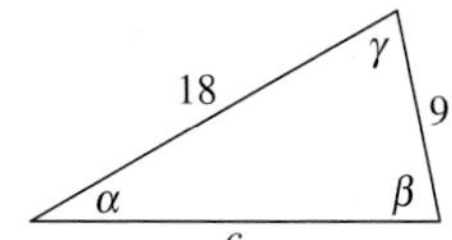

(a) Write α as a function of β and determine its domain.

(b) Differentiate the function and use the derivative to find the maximum of the function. What is the range of the function?

(c) Use a graphing utility to graph the function.

(d) If $d\beta/dt = 0.2$ radian per second, find $d\alpha/dt$ when $\beta = \pi/4$.

(e) Write c as a function of β and determine its domain.

(f) Use a graphing utility to graph the function in part (e). What is the range of the function?

(g) If $d\beta/dt = 0.2$ radian per second, find dc/dt when $\beta = \pi/4$.

(h) Use a graphing utility to complete the table.

β	0	0.4	0.8	1.2	1.6	2.0	2.4	2.8
α								
c								

(i) Explain the value for c in the table when $\beta = 0$.

2. Consider two forces

$$\mathbf{F}_1 = \langle 10, 0 \rangle \text{ and } \mathbf{F}_2 = 5\langle \cos\theta, \sin\theta \rangle.$$

(a) Find $\|\mathbf{F}_1 + \mathbf{F}_2\|$ as a function of θ.

(b) Use a graphing utility to graph the function in part (a) for $0 \le \theta < 2\pi$.

(c) Use the graph in part (b) to determine the range of the function. What is its maximum, and for what value of θ does it occur? What is its minimum, and for what value of θ does it occur?

(d) Explain why the magnitude of the resultant is never 0.

3. Write the vector $\mathbf{w}$ in terms of $\mathbf{u}$ and $\mathbf{v}$, given that the terminal point of $\mathbf{w}$ bisects the line segment.

(a)

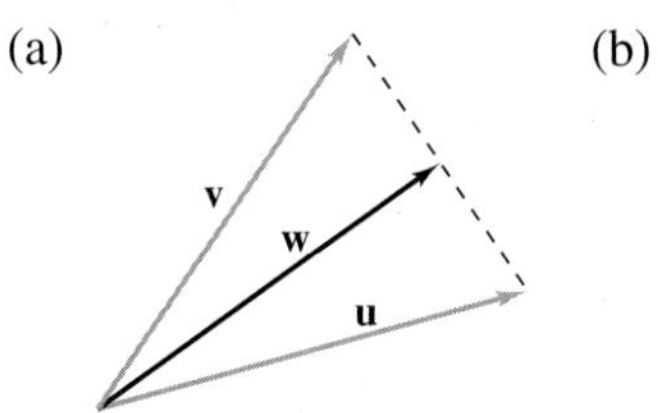

(b)

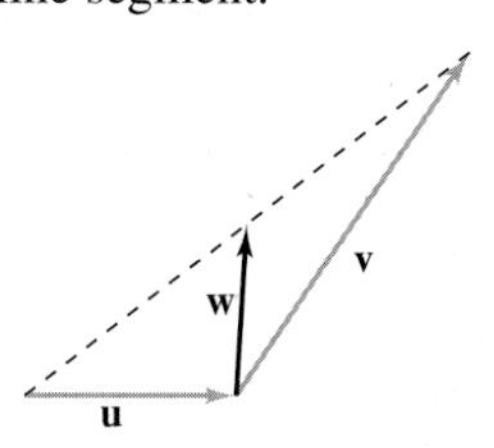

4. Use the Law of Cosines to prove that

$$\frac{1}{2}bc(1 + \cos A) = \frac{a+b+c}{2} \cdot \frac{-a+b+c}{2}.$$

5. Use the Law of Cosines to prove that

$$\frac{1}{2}bc(1 - \cos A) = \frac{a-b+c}{2} \cdot \frac{a+b-c}{2}.$$

6. Let R and r be the radii of the circumscribed and inscribed circles of a triangle ABC, respectively (see figure), and let $s = (a + b + c)/2$.

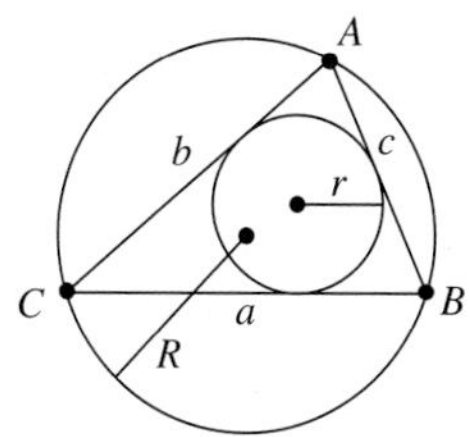

(a) Prove that $2R = \dfrac{a}{\sin A} = \dfrac{b}{\sin B} = \dfrac{c}{\sin C}$.

(b) Prove that $r = \sqrt{\dfrac{(s-a)(s-b)(s-c)}{s}}$.

(c) Given a triangle with $a = 25$, $b = 55$, and $c = 72$, find the areas of (i) the triangle, (ii) the circumscribed circle, and (iii) the inscribed circle.

(d) Find the length of the largest circular track that can be built on a triangular piece of property with sides of lengths 200 feet, 250 feet, and 325 feet.

7. (a) Use an area formula for oblique triangles to find the area of the triangle in the figure.

(b) Find the equations of the two nonvertical lines and use integration to find the area of the triangle.

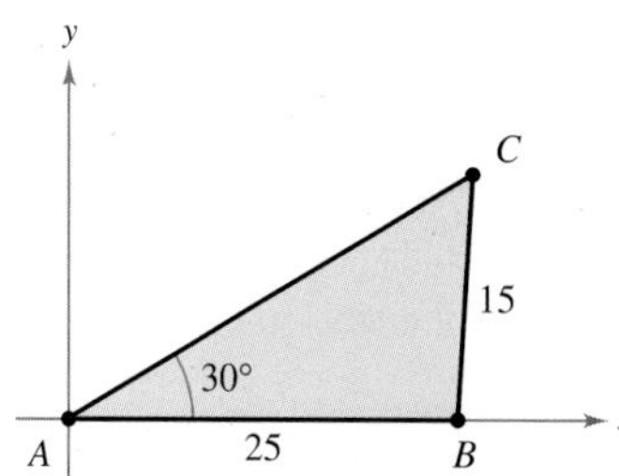

8. Prove that if $\mathbf{u}$ is orthogonal to $\mathbf{v}$ and $\mathbf{w}$, then $\mathbf{u}$ is orthogonal to

$$c\mathbf{v} + d\mathbf{w}$$

for any scalars c and d.

9. Given two vectors **u** and **v**

(a) prove that

$$\|\mathbf{u}+\mathbf{v}\|^2 + \|\mathbf{u}-\mathbf{v}\|^2 = 2\|\mathbf{u}\|^2 + 2\|\mathbf{v}\|^2.$$

(b) The equation in part (a) is called the Parallelogram Law. Use the figure to write a geometric interpretation of the Parallelogram Law.

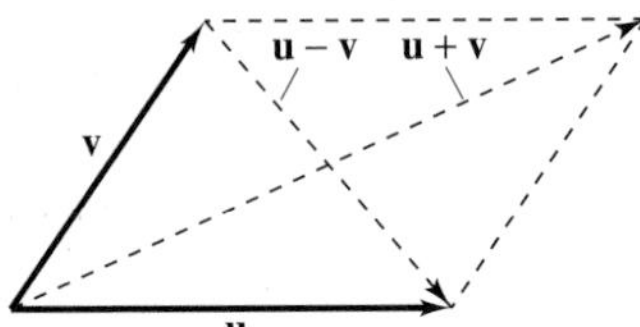

10. In the figure, a beam of light is directed at the blue mirror, reflected to the red mirror, and then reflected back to the blue mirror. Find the distance PT that the light travels from the red mirror back to the blue mirror.

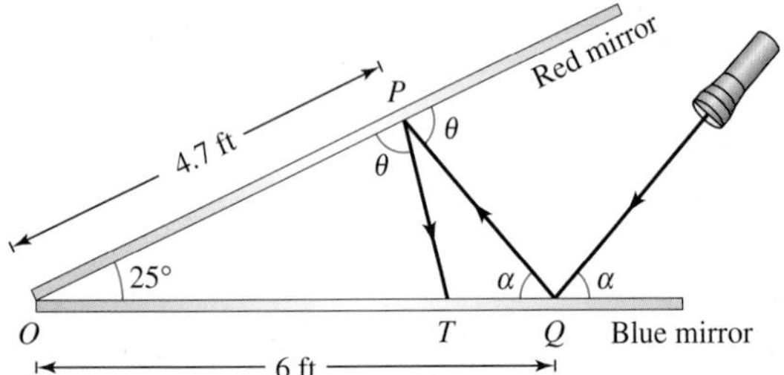

11. For each pair of vectors, find the following.

(i) $\|\mathbf{u}\|$ (ii) $\|\mathbf{v}\|$ (iii) $\|\mathbf{u}+\mathbf{v}\|$

(iv) $\left\|\frac{\mathbf{u}}{\|\mathbf{u}\|}\right\|$ (v) $\left\|\frac{\mathbf{v}}{\|\mathbf{v}\|}\right\|$ (vi) $\left\|\frac{\mathbf{u}+\mathbf{v}}{\|\mathbf{u}+\mathbf{v}\|}\right\|$

(a) $\mathbf{u} = \langle 1, -1\rangle$
$\mathbf{v} = \langle -1, 2\rangle$

(b) $\mathbf{u} = \langle 0, 1\rangle$
$\mathbf{v} = \langle 3, -3\rangle$

(c) $\mathbf{u} = \left\langle 1, \frac{1}{2}\right\rangle$
$\mathbf{v} = \langle 2, 3\rangle$

(d) $\mathbf{u} = \langle 2, -4\rangle$
$\mathbf{v} = \langle 5, 5\rangle$

12. The famous formula

$$e^{a+bi} = e^a(\cos b + i \sin b)$$

is called Euler's Formula, after the Swiss mathematician Leonhard Euler (1707–1783). This formula gives rise to one of the most wonderful equations in mathematics.

$$e^{\pi i} + 1 = 0$$

This elegant equation relates the five most famous numbers in mathematics

$0, 1, \pi, e,$ and i

in a single equation. Show how Euler's Formula can be used to derive this equation.

13. A hiking party is lost in a national park. Two ranger stations have received an emergency SOS signal from the party. Station B is 75 miles due east of Station A. The bearing from Station A to the signal is S 60° E and the bearing from Station B to the signal is S 75° W.

(a) Find the distance from each station to the SOS signal.

(b) A rescue party is in the park 20 miles from Station A at a bearing of S 80° E. Find the distance and the bearing the rescue party must travel to reach the lost hiking party.

14. The figure shows z_1 and z_2. Describe z_1z_2 and z_1/z_2.

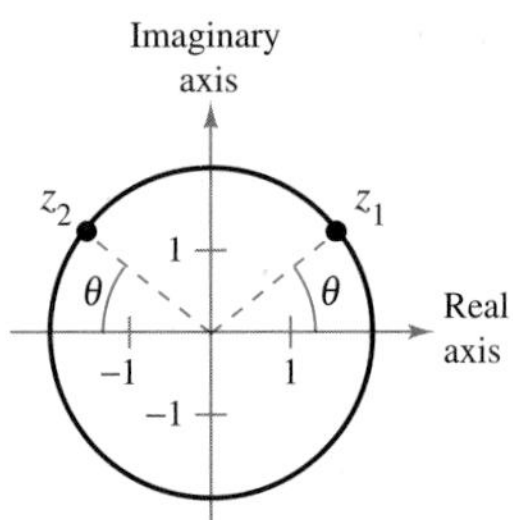

15. A triathlete sets a course to swim S 25° E from a point on shore to a buoy $\frac{3}{4}$ mile away. After swimming 300 yards through a strong current, the triathlete is off course at a bearing of S 35° E. Find the bearing and distance the triathlete needs to swim to correct her course.

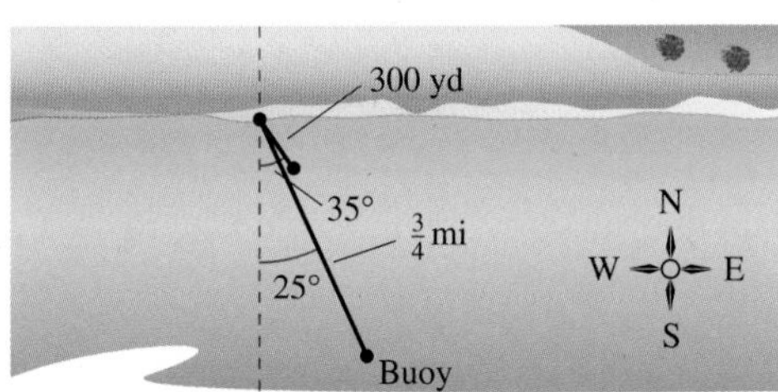

16. Find the volume of the right triangular prism in terms of x, where $V = \frac{1}{3}Bh$. B is the area of the base and h is the height of the prism.

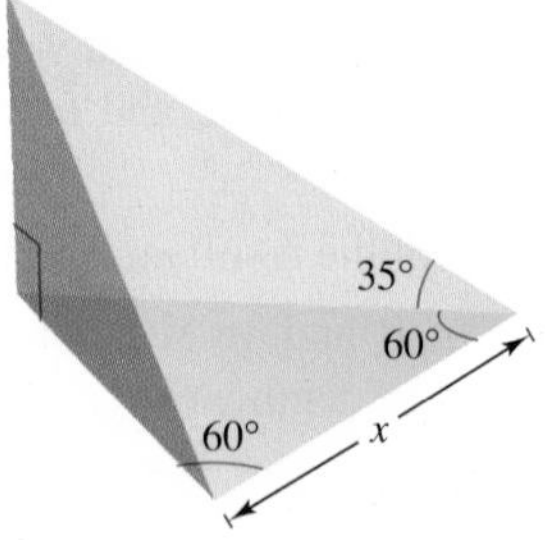

Appendices

A Proofs of Selected Theorems

THEOREM 2.4 LINEAR FACTORIZATION THEOREM (PAGE 181)

If $f(x)$ is a polynomial of degree n, where $n > 0$, then f has precisely n linear factors

$$f(x) = a_n(x - c_1)(x - c_2) \cdot \cdot \cdot (x - c_n)$$

where $c_1, c_2, \ldots, c_n$ are complex numbers.

PROOF Using the Fundamental Theorem of Algebra, you know that f must have at least one zero, c_1. Consequently, $(x - c_1)$ is a factor of $f(x)$, and you have

$$f(x) = (x - c_1)f_1(x).$$

If the degree of $f_1(x)$ is greater than zero, you again apply the Fundamental Theorem to conclude that f_1 must have a zero c_2, which implies that

$$f(x) = (x - c_1)(x - c_2)f_2(x).$$

It is clear that the degree of $f_1(x)$ is $n - 1$, that the degree of $f_2(x)$ is $n - 2$, and that you can repeatedly apply the Fundamental Theorem n times until you obtain

$$f(x) = a_n(x - c_1)(x - c_2) \cdot \cdot \cdot (x - c_n)$$

where a_n is the leading coefficient of the polynomial $f(x)$. ■

THEOREM 3.2 PROPERTIES OF LIMITS (PROPERTIES 2, 3, 4, AND 5) (PAGE 228)

Let b and c be real numbers, let n be a positive integer, and let f and g be functions with the following limits.

$$\lim_{x \to c} f(x) = L \quad \text{and} \quad \lim_{x \to c} g(x) = K$$

2. Sum or difference: $\lim_{x \to c} [f(x) \pm g(x)] = L \pm K$

3. Product: $\lim_{x \to c} [f(x)g(x)] = LK$

4. Quotient: $\lim_{x \to c} \dfrac{f(x)}{g(x)} = \dfrac{L}{K}$, provided $K \neq 0$

5. Power: $\lim_{x \to c} [f(x)]^n = L^n$

PROOF To prove Property 2, choose $\varepsilon > 0$. Because $\varepsilon/2 > 0$, you know that there exists $\delta_1 > 0$ such that $0 < |x - c| < \delta_1$ implies $|f(x) - L| < \varepsilon/2$. You also know that there exists $\delta_2 > 0$ such that $0 < |x - c| < \delta_2$ implies $|g(x) - K| < \varepsilon/2$. Let δ be the smaller of δ_1 and δ_2; then $0 < |x - c| < \delta$ implies that

$$|f(x) - L| < \frac{\varepsilon}{2} \quad \text{and} \quad |g(x) - K| < \frac{\varepsilon}{2}.$$

So, you can apply the Triangle Inequality to conclude that

$$|[f(x) + g(x)] - (L + K)| \le |f(x) - L| + |g(x) - K| < \frac{\varepsilon}{2} + \frac{\varepsilon}{2} = \varepsilon$$

which implies that

$$\lim_{x \to c} [f(x) + g(x)] = L + K = \lim_{x \to c} f(x) + \lim_{x \to c} g(x).$$

The proof that

$$\lim_{x \to c} [f(x) - g(x)] = L - K$$

is similar.

To prove Property 3, given that

$$\lim_{x \to c} f(x) = L \quad \text{and} \quad \lim_{x \to c} g(x) = K$$

you can write

$$f(x)g(x) = [f(x) - L][g(x) - K] + [Lg(x) + Kf(x)] - LK.$$

Because the limit of $f(x)$ is L, and the limit of $g(x)$ is K, you have

$$\lim_{x \to c} [f(x) - L] = 0 \quad \text{and} \quad \lim_{x \to c} [g(x) - K] = 0.$$

Let $0 < \varepsilon < 1$. Then there exists $\delta > 0$ such that if $0 < |x - c| < \delta$, then

$$|f(x) - L - 0| < \varepsilon \quad \text{and} \quad |g(x) - K - 0| < \varepsilon$$

which implies that

$$|[f(x) - L][g(x) - K] - 0| = |f(x) - L|\,|g(x) - K| < \varepsilon\varepsilon < \varepsilon.$$

So,

$$\lim_{x \to c} [f(x) - L][g(x) - K] = 0.$$

Furthermore, by Property 1, you have

$$\lim_{x \to c} Lg(x) = LK \quad \text{and} \quad \lim_{x \to c} Kf(x) = KL.$$

Finally, by Property 2, you obtain

$$\begin{aligned}\lim_{x \to c} f(x)g(x) &= \lim_{x \to c} [f(x) - L][g(x) - K] + \lim_{x \to c} Lg(x) + \lim_{x \to c} Kf(x) - \lim_{x \to c} LK \\ &= 0 + LK + KL - LK \\ &= LK.\end{aligned}$$

To prove Property 4, note that it is sufficient to prove that

$$\lim_{x \to c} \frac{1}{g(x)} = \frac{1}{K}.$$

Then you can use Property 3 to write

$$\lim_{x\to c}\frac{f(x)}{g(x)}=\lim_{x\to c}f(x)\frac{1}{g(x)}=\lim_{x\to c}f(x)\cdot\lim_{x\to c}\frac{1}{g(x)}=\frac{L}{K}.$$

Let $\varepsilon > 0$. Because $\lim_{x\to c} g(x) = K$, there exists $\delta_1 > 0$ such that if

$$0<|x-c|<\delta_1,\text{ then }|g(x)-K|<\frac{|K|}{2}$$

which implies that

$$|K|=|g(x)+[|K|-g(x)]|\le|g(x)|+||K|-g(x)|<|g(x)|+\frac{|K|}{2}.$$

That is, for $0 < |x - c| < \delta_1$,

$$\frac{|K|}{2}<|g(x)|\quad\text{or}\quad\frac{1}{|g(x)|}<\frac{2}{|K|}.$$

Similarly, there exists $\delta_2 > 0$ such that if $0 < |x - c| < \delta_2$, then

$$|g(x)-K|<\frac{|K|^2}{2}\varepsilon.$$

Let δ be the smaller of δ_1 and δ_2. For $0 < |x - c| < \delta$, you have

$$\left|\frac{1}{g(x)}-\frac{1}{K}\right|=\left|\frac{K-g(x)}{g(x)K}\right|=\frac{1}{|K|}\cdot\frac{1}{|g(x)|}|K-g(x)|<\frac{1}{|K|}\cdot\frac{2}{|K|}\frac{|K|^2}{2}\varepsilon=\varepsilon.$$

So, $\lim_{x\to c}\frac{1}{g(x)}=\frac{1}{K}$.

Finally, the proof of Property 5 can be obtained by a straightforward application of mathematical induction coupled with Property 3. ■

THEOREM 3.4 THE LIMIT OF A FUNCTION INVOLVING A RADICAL (PAGE 230)

Let n be a positive integer. The following limit is valid for all c if n is odd, and is valid for $c > 0$ if n is even.

$$\lim_{x\to c}\sqrt[n]{x}=\sqrt[n]{c}$$

PROOF Consider the case for which $c > 0$ and n is any positive integer. For a given $\varepsilon > 0$, you need to find $\delta > 0$ such that

$$\left|\sqrt[n]{x}-\sqrt[n]{c}\right|<\varepsilon\quad\text{whenever}\quad 0<|x-c|<\delta$$

which is the same as saying

$$-\varepsilon<\sqrt[n]{x}-\sqrt[n]{c}<\varepsilon\quad\text{whenever}\quad -\delta<x-c<\delta.$$

Assume that $\varepsilon < \sqrt[n]{c}$, which implies that $0 < \sqrt[n]{c} - \varepsilon < \sqrt[n]{c}$. Now, let δ be the smaller of the two numbers

$$c-\left(\sqrt[n]{c}-\varepsilon\right)^n\quad\text{and}\quad\left(\sqrt[n]{c}+\varepsilon\right)^n-c.$$

Then you have

$$\begin{aligned}
-\delta < x - c &< \delta \\
-\left[c - \left(\sqrt[n]{c} - \varepsilon\right)^n\right] < x - c &< \left(\sqrt[n]{c} + \varepsilon\right)^n - c \\
\left(\sqrt[n]{c} - \varepsilon\right)^n - c < x - c &< \left(\sqrt[n]{c} + \varepsilon\right)^n - c \\
\left(\sqrt[n]{c} - \varepsilon\right)^n < x &< \left(\sqrt[n]{c} + \varepsilon\right)^n \\
\sqrt[n]{c} - \varepsilon < \sqrt[n]{x} &< \sqrt[n]{c} + \varepsilon \\
-\varepsilon < \sqrt[n]{x} - \sqrt[n]{c} &< \varepsilon.
\end{aligned}$$

■

THEOREM 3.5 THE LIMIT OF A COMPOSITE FUNCTION (PAGE 230)

If f and g are functions such that $\lim_{x \to c} g(x) = L$ and $\lim_{x \to L} f(x) = f(L)$, then

$$\lim_{x \to c} f(g(x)) = f\left(\lim_{x \to c} g(x)\right) = f(L).$$

PROOF For a given $\varepsilon > 0$, you must find $\delta > 0$ such that

$$|f(g(x)) - f(L)| < \varepsilon \quad \text{whenever} \quad 0 < |x - c| < \delta.$$

Because the limit of $f(x)$ as $x \to L$ is $f(L)$, you know there exists $\delta_1 > 0$ such that

$$|f(u) - f(L)| < \varepsilon \quad \text{whenever} \quad |u - L| < \delta_1.$$

Moreover, because the limit of $g(x)$ as $x \to c$ is L, you know there exists $\delta > 0$ such that

$$|g(x) - L| < \delta_1 \quad \text{whenever} \quad 0 < |x - c| < \delta.$$

Finally, letting $u = g(x)$, you have

$$|f(g(x)) - f(L)| < \varepsilon \quad \text{whenever} \quad 0 < |x - c| < \delta.$$

■

THEOREM 3.6 FUNCTIONS THAT AGREE AT ALL BUT ONE POINT (PAGE 231)

Let c be a real number and let $f(x) = g(x)$ for all $x \neq c$ in an open interval containing c. If the limit of $g(x)$ as x approaches c exists, then the limit of $f(x)$ also exists and

$$\lim_{x \to c} f(x) = \lim_{x \to c} g(x).$$

PROOF Let L be the limit of $g(x)$ as $x \to c$. Then, for each $\varepsilon > 0$ there exists a $\delta > 0$ such that $f(x) = g(x)$ in the open intervals $(c - \delta, c)$ and $(c, c + \delta)$, and

$$|g(x) - L| < \varepsilon \quad \text{whenever} \quad 0 < |x - c| < \delta.$$

Because $f(x) = g(x)$ for all x in the open interval other than $x = c$, it follows that

$$|f(x) - L| < \varepsilon \quad \text{whenever} \quad 0 < |x - c| < \delta.$$

So, the limit of $f(x)$ as $x \to c$ is also L.

■

THEOREM 3.7 THE SQUEEZE THEOREM (PAGE 233)

If $h(x) \le f(x) \le g(x)$ for all x in an open interval containing c, except possibly at c itself, and if $\lim_{x\to c} h(x) = L = \lim_{x\to c} g(x)$, then $\lim_{x\to c} f(x)$ exists and is equal to L.

PROOF For $\varepsilon > 0$ there exist $\delta_1 > 0$ and $\delta_2 > 0$ such that

$$|h(x) - L| < \varepsilon \quad \text{whenever} \quad 0 < |x - c| < \delta_1$$

and

$$|g(x) - L| < \varepsilon \quad \text{whenever} \quad 0 < |x - c| < \delta_2.$$

Because $h(x) \le f(x) \le g(x)$ for all x in an open interval containing c, except possibly at c itself, there exists $\delta_3 > 0$ such that $h(x) \le f(x) \le g(x)$ for $0 < |x - c| < \delta_3$. Let δ be the smallest of δ_1, δ_2, and δ_3. Then, if $0 < |x - c| < \delta$, it follows that $|h(x) - L| < \varepsilon$ and $|g(x) - L| < \varepsilon$, which implies that

$$-\varepsilon < h(x) - L < \varepsilon \quad \text{and} \quad -\varepsilon < g(x) - L < \varepsilon$$
$$L - \varepsilon < h(x) \quad \text{and} \quad g(x) < L + \varepsilon.$$

Now, because $h(x) \le f(x) \le g(x)$, it follows that $L - \varepsilon < f(x) < L + \varepsilon$, which implies that $|f(x) - L| < \varepsilon$. Therefore,

$$\lim_{x\to c} f(x) = L. \qquad \blacksquare$$

THEOREM 3.9 PROPERTIES OF CONTINUITY (PAGE 241)

If b is a real number and f and g are continuous at $x = c$, then the following functions are also continuous at c.

1. Scalar multiple: bf
2. Sum or difference: $f \pm g$
3. Product: fg
4. Quotient: $\dfrac{f}{g}$, if $g(c) \neq 0$

PROOF Because f and g are continuous at $x = c$, you can write

$$\lim_{x\to c} f(x) = f(c) \quad \text{and} \quad \lim_{x\to c} g(x) = g(c).$$

For Property 1, when b is a real number, it follows from Theorem 3.2 that

$$\lim_{x\to c} [(bf)(x)] = \lim_{x\to c} [bf(x)] = b \lim_{x\to c} [f(x)] = b\,f(c) = (bf)(c).$$

Thus, bf is continuous at $x = c$.

For Property 2, it follows from Theorem 3.2 that

$$\begin{aligned}\lim_{x\to c} (f \pm g)(x) &= \lim_{x\to c} [f(x) \pm g(x)]\\ &= \lim_{x\to c} [f(x)] \pm \lim_{x\to c} [g(x)]\\ &= f(c) \pm g(c)\\ &= (f \pm g)(c).\end{aligned}$$

Thus, $f \pm g$ is continuous at $x = c$.

For Property 3, it follows from Theorem 3.2 that

$$\begin{aligned}\lim_{x\to c}(fg)(x) &= \lim_{x\to c}[f(x)g(x)]\\ &= \lim_{x\to c}[f(x)]\lim_{x\to c}[g(x)]\\ &= f(c)g(c)\\ &= (fg)(c).\end{aligned}$$

Thus, fg is continuous at $x = c$.

For Property 4, when $g(c) \neq 0$, it follows from Theorem 3.2 that

$$\begin{aligned}\lim_{x\to c}\frac{f}{g}(x) &= \lim_{x\to c}\frac{f(x)}{g(x)}\\ &= \frac{\displaystyle\lim_{x\to c} f(x)}{\displaystyle\lim_{x\to c} g(x)}\\ &= \frac{f(c)}{g(c)}\\ &= \frac{f}{g}(c).\end{aligned}$$

Thus, $\dfrac{f}{g}$ is continuous at $x = c$. ■

THEOREM 3.12 VERTICAL ASYMPTOTES (PAGE 249)

Let f and g be continuous on an open interval containing c. If $f(c) \neq 0$, $g(c) = 0$, and there exists an open interval containing c such that $g(x) \neq 0$ for all $x \neq c$ in the interval, then the graph of the function given by

$$h(x) = \frac{f(x)}{g(x)}$$

has a vertical asymptote at $x = c$.

PROOF Consider the case for which $f(c) > 0$, and there exists $b > c$ such that $c < x < b$ implies $g(x) > 0$. Then for $M > 0$, choose δ_1 such that

$$0 < x - c < \delta_1 \quad \text{implies that} \quad \frac{f(c)}{2} < f(x) < \frac{3f(c)}{2}$$

and δ_2 such that

$$0 < x - c < \delta_2 \quad \text{implies that} \quad 0 < g(x) < \frac{f(c)}{2M}.$$

Now let δ be the smaller of δ_1 and δ_2. Then it follows that

$$0 < x - c < \delta \quad \text{implies that} \quad \frac{f(x)}{g(x)} > \frac{f(c)}{2}\left[\frac{2M}{f(c)}\right] = M.$$

So, it follows that

$$\lim_{x\to c^+}\frac{f(x)}{g(x)} = \infty$$

and the line $x = c$ is a vertical asymptote of the graph of h. ■

ALTERNATIVE FORM OF THE DERIVATIVE (PAGE 267)

The derivative of f at c is given by

$$f'(c) = \lim_{x \to c} \frac{f(x) - f(c)}{x - c}$$

provided this limit exists.

PROOF The derivative of f at c is given by

$$f'(c) = \lim_{\Delta x \to 0} \frac{f(c + \Delta x) - f(c)}{\Delta x}.$$

Let $x = c + \Delta x$. Then $x \to c$ as $\Delta x \to 0$. So, replacing $c + \Delta x$ by x, you have

$$f'(c) = \lim_{\Delta x \to 0} \frac{f(c + \Delta x) - f(c)}{\Delta x}$$

$$= \lim_{x \to c} \frac{f(x) - f(c)}{x - c}.$$

■

THEOREM 4.8 THE CHAIN RULE (PAGE 294)

If $y = f(u)$ is a differentiable function of u and $u = g(x)$ is a differentiable function of x, then $y = f(g(x))$ is a differentiable function of x and

$$\frac{dy}{dx} = \frac{dy}{du} \cdot \frac{du}{dx}$$

or, equivalently,

$$\frac{d}{dx}[f(g(x))] = f'(g(x))g'(x).$$

PROOF In Section 4.4, we let $h(x) = f(g(x))$ and used the alternative form of the derivative to show that $h'(c) = f'(g(c))g'(c)$, provided $g(x) \neq g(c)$ for values of x other than c. Now consider a more general proof. Begin by considering the derivative of f.

$$f'(x) = \lim_{\Delta x \to 0} \frac{f(x + \Delta x) - f(x)}{\Delta x} = \lim_{\Delta x \to 0} \frac{\Delta y}{\Delta x}$$

For a fixed value of x, define a function η such that

$$\eta(\Delta x) = \begin{cases} 0, & \Delta x = 0 \\ \dfrac{\Delta y}{\Delta x} - f'(x), & \Delta x \neq 0. \end{cases}$$

Because the limit of $\eta(\Delta x)$ as $\Delta x \to 0$ doesn't depend on the value of $\eta(0)$, you have

$$\lim_{\Delta x \to 0} \eta(\Delta x) = \lim_{\Delta x \to 0} \left[\frac{\Delta y}{\Delta x} - f'(x)\right] = 0$$

and you can conclude that η is continuous at 0. Moreover, because $\Delta y = 0$ when $\Delta x = 0$, the equation

$$\Delta y = \Delta x \eta(\Delta x) + \Delta x f'(x)$$

is valid whether Δx is zero or not. Now, by letting $\Delta u = g(x + \Delta x) - g(x)$, you can use the continuity of g to conclude that

$$\lim_{\Delta x \to 0} \Delta u = \lim_{\Delta x \to 0} [g(x + \Delta x) - g(x)] = 0$$

which implies that

$$\lim_{\Delta x \to 0} \eta(\Delta u) = 0.$$

Finally,

$$\Delta y = \Delta u \eta(\Delta u) + \Delta u f'(u) \rightarrow \frac{\Delta y}{\Delta x} = \frac{\Delta u}{\Delta x}\eta(\Delta u) + \frac{\Delta u}{\Delta x} f'(u), \quad \Delta x \neq 0$$

and taking the limit as $\Delta x \to 0$, you have

$$\begin{aligned} \frac{dy}{dx} = \frac{du}{dx}\left[\lim_{\Delta x \to 0} \eta(\Delta u)\right] + \frac{du}{dx} f'(u) &= \frac{dy}{dx}(0) + \frac{du}{dx} f'(u) \\ &= \frac{du}{dx} f'(u) \\ &= \frac{du}{dx} \cdot \frac{dy}{du}. \end{aligned}$$

■

GRAPHICAL INTERPRETATION OF CONCAVITY (PAGE 348)

1. Let f be differentiable on an open interval I. If the graph of f is concave *upward* on I, then the graph of f lies *above* all of its tangent lines on I.
2. Let f be differentiable on an open interval I. If the graph of f is concave *downward* on I, then the graph of f lies *below* all of its tangent lines on I.

PROOF Assume that f is concave upward on $I = (a, b)$. Then, f' is increasing on (a, b). Let c be a point in the interval $I = (a, b)$. The equation of the tangent line to the graph of f at c is given by

$$g(x) = f(c) + f'(c)(x - c).$$

If x is in the open interval (c, b), then the directed distance from point $(x, f(x))$ (on the graph of f) to the point $(x, g(x))$ (on the tangent line) is given by

$$\begin{aligned} d &= f(x) - [f(c) + f'(c)(x - c)] \\ &= f(x) - f(c) - f'(c)(x - c). \end{aligned}$$

Moreover, by the Mean Value Theorem there exists a number z in (c, x) such that

$$f'(z) = \frac{f(x) - f(c)}{x - c}.$$

So, you have

$$\begin{aligned} d &= f(x) - f(c) - f'(c)(x - c) \\ &= f'(z)(x - c) - f'(c)(x - c) \\ &= [f'(z) - f'(c)](x - c). \end{aligned}$$

The second factor $(x - c)$ is positive because $c < x$. Moreover, because f' is increasing, it follows that the first factor $[f'(z) - f'(c)]$ is also positive. Therefore, $d > 0$ and you can conclude that the graph of f lies above the tangent line at x. If x is in the open interval (a, c), a similar argument can be given. This proves the first statement. The proof of the second statement is similar. ■

THEOREM 5.7 TEST FOR CONCAVITY (PAGE 349)

Let f be a function whose second derivative exists on an open interval I.

1. If $f''(x) > 0$ for all x in I, then the graph of f is concave upward on I.
2. If $f''(x) < 0$ for all x in I, then the graph of f is concave downward on I.

PROOF For Property 1, assume $f''(x) > 0$ for all x in (a, b). Then, by Theorem 5.5, f' is increasing on $[a, b]$. Thus, by the definition of concavity, the graph of f is concave upward on (a, b).

For Property 2, assume $f''(x) < 0$ for all x in (a, b). Then, by Theorem 5.5, f' is decreasing on $[a, b]$. Thus, by the definition of concavity, the graph of f is concave downward on (a, b). ■

THEOREM 5.10 LIMITS AT INFINITY (PAGE 357)

If r is a positive rational number and c is any real number, then

$$\lim_{x\to\infty} \frac{c}{x^r} = 0.$$

Furthermore, if x^r is defined when $x < 0$, then $\displaystyle\lim_{x\to-\infty} \frac{c}{x^r} = 0.$

PROOF Begin by proving that

$$\lim_{x\to\infty} \frac{1}{x} = 0.$$

For $\varepsilon > 0$, let $M = 1/\varepsilon$. Then, for $x > M$, you have

$$x > M = \frac{1}{\varepsilon} \quad \Rightarrow \quad \frac{1}{x} < \varepsilon \quad \Rightarrow \quad \left|\frac{1}{x} - 0\right| < \varepsilon.$$

So, by the definition of a limit at infinity, you can conclude that the limit of $1/x$ as $x \to \infty$ is 0. Now, using this result, and letting $r = m/n$, you can write

$$\begin{aligned}
\lim_{x\to\infty} \frac{c}{x^r} &= \lim_{x\to\infty} \frac{c}{x^{m/n}} \\
&= c\left[\lim_{x\to\infty} \left(\frac{1}{\sqrt[n]{x}}\right)^m\right] \\
&= c\left(\lim_{x\to\infty} \sqrt[n]{\frac{1}{x}}\right)^m \\
&= c\left(\sqrt[n]{\lim_{x\to\infty} \frac{1}{x}}\right)^m \\
&= c\left(\sqrt[n]{0}\right)^m \\
&= 0
\end{aligned}$$

The proof of the second part of the theorem is similar. ■

THEOREM 6.2 SUMMATION FORMULAS (PAGE 409)

1. $\displaystyle\sum_{i=1}^{n} c = cn$

2. $\displaystyle\sum_{i=1}^{n} i = \frac{n(n+1)}{2}$

3. $\displaystyle\sum_{i=1}^{n} i^2 = \frac{n(n+1)(2n+1)}{6}$

4. $\displaystyle\sum_{i=1}^{n} i^3 = \frac{n^2(n+1)^2}{4}$

PROOF The proof of Property 1 is straightforward. By adding c to itself n times, you obtain a sum of cn.

To prove Property 2, write the sum in increasing and decreasing order and add corresponding terms, as follows.

$$\sum_{i=1}^{n} i = 1 + 2 + 3 + \cdots + (n-1) + n$$

$$\sum_{i=1}^{n} i = n + (n-1) + (n-2) + \cdots + 2 + 1$$

$$2\sum_{i=1}^{n} i = \underbrace{(n+1) + (n+1) + (n+1) + \cdots + (n+1) + (n+1)}_{n \text{ terms}}$$

So,

$$\sum_{i=1}^{n} i = \frac{n(n+1)}{2}.$$

To prove Property 3, use mathematical induction. First, if $n = 1$, the result is true because

$$\sum_{i=1}^{1} i^2 = 1^2 = 1 = \frac{1(1+1)(2+1)}{6}.$$

Now, assuming the result is true for $n = k$, you can show that it is true for $n = k + 1$, as shown below.

$$\begin{aligned}\sum_{i=1}^{k+1} i^2 &= \sum_{i=1}^{k} i^2 + (k+1)^2\\ &= \frac{k(k+1)(2k+1)}{6} + (k+1)^2\\ &= \frac{k+1}{6}(2k^2 + k + 6k + 6)\\ &= \frac{k+1}{6}[(2k+3)(k+2)]\\ &= \frac{(k+1)(k+2)[2(k+1)+1]}{6}\end{aligned}$$

Property 4 can be proved using a similar argument with mathematical induction. ■

THEOREM 6.8 PRESERVATION OF INEQUALITY (PAGE 426)

1. If f is integrable and nonnegative on the closed interval $[a, b]$, then

$$0 \le \int_a^b f(x)\,dx.$$

2. If f and g are integrable on the closed interval $[a, b]$ and $f(x) \le g(x)$ for every x in $[a, b]$, then

$$\int_a^b f(x)\,dx \le \int_a^b g(x)\,dx.$$

PROOF To prove Property 1, suppose, on the contrary, that

$$\int_a^b f(x)\,dx = I < 0.$$

Then, let $a = x_0 < x_1 < x_2 < \cdot \cdot \cdot < x_n = b$ be a partition of $[a, b]$, and let

$$R = \sum_{i=1}^{n} f(c_i)\,\Delta x_i$$

be a Riemann sum. Because $f(x) \ge 0$, it follows that $R \ge 0$. Now, for $\|\Delta\|$ sufficiently small, you have $|R - I| < -I/2$, which implies that

$$\sum_{i=1}^{n} f(c_i)\,\Delta x_i = R < I - \frac{I}{2} < 0$$

which is not possible. From this contradiction, you can conclude that

$$0 \le \int_a^b f(x)\,dx.$$

To prove Property 2 of the theorem, note that $f(x) \le g(x)$ implies that $g(x) - f(x) \ge 0$. So, you can apply the result of Property 1 to conclude that

$$0 \le \int_a^b [g(x) - f(x)]\,dx$$

$$0 \le \int_a^b g(x)\,dx - \int_a^b f(x)\,dx$$

$$\int_a^b f(x)\,dx \le \int_a^b g(x)\,dx.$$

■

THEOREM 7.6 PROPERTIES OF LOGARITHMS (PAGE 492)

Let a be a positive number such that $a \neq 1$, and let n be a real number. If u and v are positive real numbers, the following properties are true.

	Logarithm with Base a	*Natural Logarithm*
1. Product Property:	$\log_a(uv) = \log_a u + \log_a v$	$\ln(uv) = \ln u + \ln v$
2. Quotient Property:	$\log_a \frac{u}{v} = \log_a u - \log_a v$	$\ln \frac{u}{v} = \ln u - \ln v$
3. Power Property:	$\log_a u^n = n \log_a u$	$\ln u^n = n \ln u$

PROOF To prove Property 1, let

$$x = \log_a u \quad \text{and} \quad y = \log_a v.$$

The corresponding exponential forms of these two equations are

$$a^x = u \quad \text{and} \quad a^y = v.$$

Multiplying u and v produces

$$uv = a^x a^y = a^{x+y}.$$

The corresponding logarithmic form of $uv = a^{x+y}$ is $\log_a(uv) = x + y$. So,

$$\log_a(uv) = \log_a u + \log_a v.$$

■

SUM AND DIFFERENCE FORMULAS (PAGE 669)

$$\sin(u + v) = \sin u \cos v + \cos u \sin v$$
$$\sin(u - v) = \sin u \cos v - \cos u \sin v$$
$$\cos(u + v) = \cos u \cos v - \sin u \sin v$$
$$\cos(u - v) = \cos u \cos v + \sin u \sin v$$
$$\tan(u + v) = \frac{\tan u + \tan v}{1 - \tan u \tan v}$$
$$\tan(u - v) = \frac{\tan u - \tan v}{1 + \tan u \tan v}$$

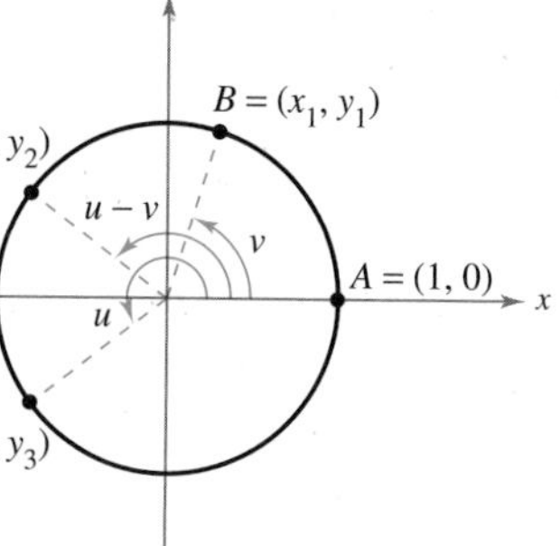

Figure A.1

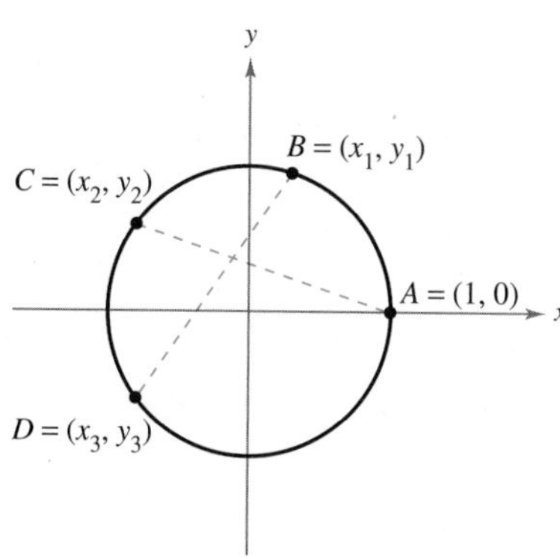

Figure A.2

PROOF Here are proofs for the formulas for $\cos(u \pm v)$. In Figure A.1, let A be the point $(1, 0)$ and then use u and v to locate the points $B = (x_1, y_1)$, $C = (x_2, y_2)$, and $D = (x_3, y_3)$ on the unit circle. So, $x_i^2 + y_i^2 = 1$ for $i = 1, 2,$ and 3. For convenience, assume that $0 < v < u < 2\pi$. In Figure A.2, note that arcs AC and BD have the same length. So, line segments AC and BD are also equal in length, which implies that

$$\sqrt{(x_2 - 1)^2 + (y_2 - 0)^2} = \sqrt{(x_3 - x_1)^2 + (y_3 - y_1)^2}$$
$$x_2^2 - 2x_2 + 1 + y_2^2 = x_3^2 - 2x_1x_3 + x_1^2 + y_3^2 - 2y_1y_3 + y_1^2$$
$$(x_2^2 + y_2^2) + 1 - 2x_2 = (x_3^2 + y_3^2) + (x_1^2 + y_1^2) - 2x_1x_3 - 2y_1y_3$$
$$1 + 1 - 2x_2 = 1 + 1 - 2x_1x_3 - 2y_1y_3$$
$$x_2 = x_3x_1 + y_3y_1.$$

Finally, by substituting the values $x_2 = \cos(u - v)$, $x_3 = \cos u$, $x_1 = \cos v$, $y_3 = \sin u$, and $y_1 = \sin v$, you obtain $\cos(u - v) = \cos u \cos v + \sin u \sin v$.

The formula for $\cos(u + v)$ can be established by considering $u + v = u - (-v)$ and using the formula just derived to obtain

$$\cos(u + v) = \cos[u - (-v)]$$
$$= \cos u \cos(-v) + \sin u \sin(-v)$$
$$= \cos u \cos v - \sin u \sin v.$$

■

DOUBLE-ANGLE FORMULAS (PAGE 675)

$$\sin 2u = 2 \sin u \cos u \qquad \cos 2u = \cos^2 u - \sin^2 u$$
$$\tan 2u = \frac{2 \tan u}{1 - \tan^2 u} \qquad = 2\cos^2 u - 1$$
$$= 1 - 2\sin^2 u$$

PROOF To prove all three formulas, let $v = u$ in the corresponding sum formulas.

$$\sin 2u = \sin(u + u) = \sin u \cos u + \cos u \sin u = 2 \sin u \cos u$$
$$\cos 2u = \cos(u + u) = \cos u \cos u - \sin u \sin u = \cos^2 u - \sin^2 u$$
$$\tan 2u = \tan(u + u) = \frac{\tan u + \tan u}{1 - \tan u \tan u} = \frac{2 \tan u}{1 - \tan^2 u}$$ ■

POWER-REDUCING FORMULAS (PAGE 677)

$$\sin^2 u = \frac{1 - \cos 2u}{2} \qquad \cos^2 u = \frac{1 + \cos 2u}{2} \qquad \tan^2 u = \frac{1 - \cos 2u}{1 + \cos 2u}$$

PROOF The first two formulas can be verified by solving for $\sin^2 u$ and $\cos^2 u$, respectively, in the double-angle formulas

$$\cos 2u = 1 - 2\sin^2 u \quad \text{and} \quad \cos 2u = 2\cos^2 u - 1.$$

The third formula can be verified using the fact that

$$\tan^2 u = \frac{\sin^2 u}{\cos^2 u}.$$ ■

SUM-TO-PRODUCT FORMULAS (PAGE 679)

$$\sin x + \sin y = 2 \sin\left(\frac{x + y}{2}\right) \cos\left(\frac{x - y}{2}\right)$$
$$\sin x - \sin y = 2 \cos\left(\frac{x + y}{2}\right) \sin\left(\frac{x - y}{2}\right)$$
$$\cos x + \cos y = 2 \cos\left(\frac{x + y}{2}\right) \cos\left(\frac{x - y}{2}\right)$$
$$\cos x - \cos y = -2 \sin\left(\frac{x + y}{2}\right) \sin\left(\frac{x - y}{2}\right)$$

PROOF To prove the first formula, let $x = u + v$ and $y = u - v$. Then substitute $u = (x + y)/2$ and $v = (x - y)/2$ in the product-to-sum formula.

$$\sin u \cos v = \frac{1}{2}[\sin(u + v) + \sin(u - v)]$$
$$\sin\left(\frac{x + y}{2}\right) \cos\left(\frac{x - y}{2}\right) = \frac{1}{2}(\sin x + \sin y)$$
$$2 \sin\left(\frac{x + y}{2}\right) \cos\left(\frac{x - y}{2}\right) = \sin x + \sin y$$

The other sum-to-product formulas can be proved in a similar manner. ■

THEOREM 12.9 POLAR EQUATIONS OF CONICS (PAGE 797)

The graph of a polar equation of the form

1. $r = \dfrac{ep}{1 \pm e\cos\theta}$ or **2.** $r = \dfrac{ep}{1 \pm e\sin\theta}$

is a conic, where $e > 0$ is the eccentricity and $|p|$ is the distance between the focus (pole) and the directrix.

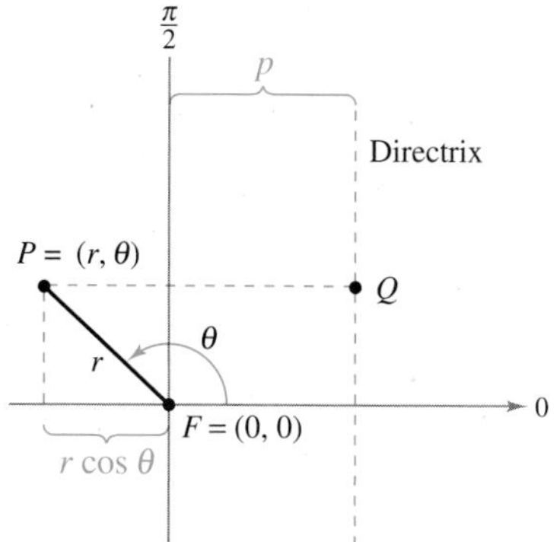

Figure A.3

PROOF A proof for $r = ep/(1 + e\cos\theta)$ with $p > 0$ is given here. The proofs of the other cases are similar. In Figure A.3, consider a vertical directrix, p units to the right of the focus $F = (0, 0)$. If $P = (r, \theta)$ is a point on the graph of

$$r = \frac{ep}{1 + e\cos\theta}$$

the distance between P and the directrix is

$$\begin{aligned} PQ &= |p - x| \\ &= |p - r\cos\theta| \\ &= \left|p - \left(\frac{ep}{1 + e\cos\theta}\right)\cos\theta\right| \\ &= \left|p\left(1 - \frac{e\cos\theta}{1 + e\cos\theta}\right)\right| = \left|\frac{p}{1 + e\cos\theta}\right| = \left|\frac{r}{e}\right|. \end{aligned}$$

Moreover, because the distance between P and the pole is simply $PF = |r|$, the ratio of PF to PQ is

$$\frac{PF}{PQ} = \frac{|r|}{|r/e|} = |e| = e$$

and, by definition, the graph of the equation must be a conic. ■

THEOREM 13.2 LAW OF COSINES (PAGE 821)

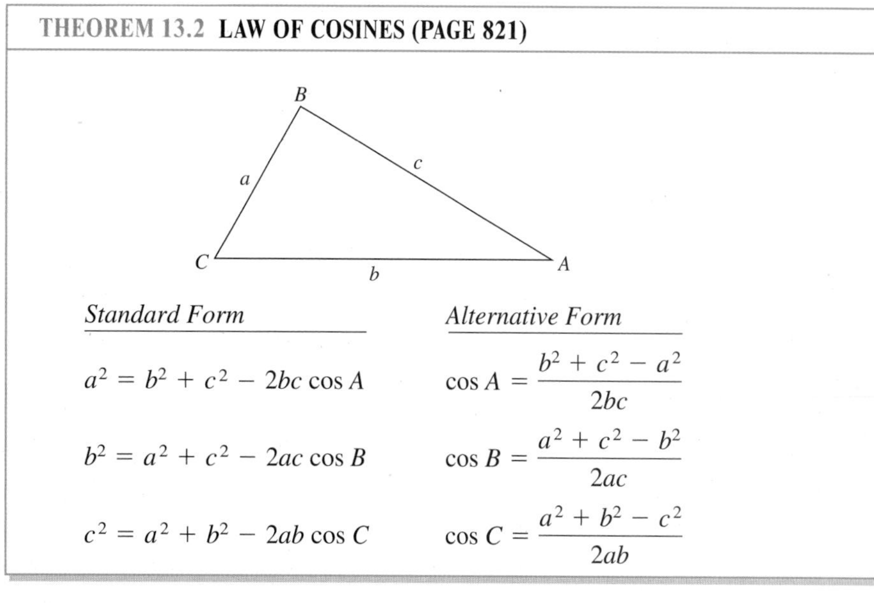

Standard Form	*Alternative Form*
$a^2 = b^2 + c^2 - 2bc\cos A$	$\cos A = \dfrac{b^2 + c^2 - a^2}{2bc}$
$b^2 = a^2 + c^2 - 2ac\cos B$	$\cos B = \dfrac{a^2 + c^2 - b^2}{2ac}$
$c^2 = a^2 + b^2 - 2ab\cos C$	$\cos C = \dfrac{a^2 + b^2 - c^2}{2ab}$

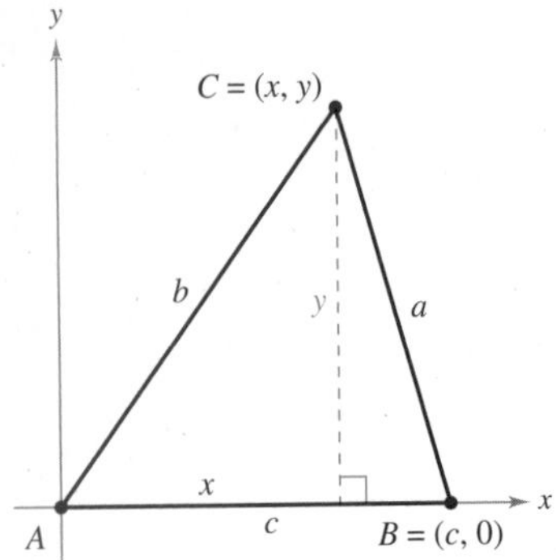

Figure A.4

PROOF Consider a triangle that has three acute angles, as shown in Figure A.4. Note that vertex B has coordinates $(c, 0)$. Furthermore, C has coordinates (x, y), where $x = b \cos A$ and $y = b \sin A$. Because a is the distance from vertex C to vertex B, it follows that

$$
\begin{aligned}
a &= \sqrt{(x - c)^2 + (y - 0)^2} \\
a^2 &= (x - c)^2 + (y - 0)^2 \\
a^2 &= (b \cos A - c)^2 + (b \sin A)^2 \\
a^2 &= b^2 \cos^2 A - 2bc \cos A + c^2 + b^2 \sin^2 A \\
a^2 &= b^2(\sin^2 A + \cos^2 A) + c^2 - 2bc \cos A \\
a^2 &= b^2 + c^2 - 2bc \cos A. && \sin^2 A + \cos^2 A = 1
\end{aligned}
$$

Similar arguments can be used to establish the other two equations. ■

THEOREM 13.3 HERON'S AREA FORMULA (PAGE 822)

Given any triangle with sides of lengths a, b, and c, the area of the triangle is

$$\text{Area} = \sqrt{s(s - a)(s - b)(s - c)}$$

where $s = \dfrac{a + b + c}{2}$.

PROOF From Section 13.1, you know that

$$
\begin{aligned}
\text{Area} &= \frac{1}{2} bc \sin A \\
(\text{Area})^2 &= \frac{1}{4} b^2c^2 \sin^2 A \\
\text{Area} &= \sqrt{\frac{1}{4} b^2c^2 \sin^2 A} \\
&= \sqrt{\frac{1}{4} b^2c^2(1 - \cos^2 A)} \\
&= \sqrt{\left[\frac{1}{2} bc(1 + \cos A)\right]\left[\frac{1}{2} bc(1 - \cos A)\right]}.
\end{aligned}
$$

Using the Law of Cosines, you can show that

$$\frac{1}{2} bc(1 + \cos A) = \frac{a + b + c}{2} \cdot \frac{-a + b + c}{2}$$

and

$$\frac{1}{2} bc(1 - \cos A) = \frac{a - b + c}{2} \cdot \frac{a + b - c}{2}.$$

Letting $s = \dfrac{a + b + c}{2}$, these two equations can be rewritten as

$$\frac{1}{2}bc(1 + \cos A) = s(s - a)$$

and

$$\frac{1}{2}bc(1 - \cos A) = (s - b)(s - c).$$

By substituting into the last formula for area, you can conclude that

$$\text{Area} = \sqrt{s(s - a)(s - b)(s - c)}.$$

■

THEOREM 13.6 PROPERTIES OF THE DOT PRODUCT (PAGE 836)

Let **u**, **v**, and **w** be vectors in the plane or in space and let c be a scalar.

1. $\mathbf{u} \cdot \mathbf{v} = \mathbf{v} \cdot \mathbf{u}$
2. $\mathbf{0} \cdot \mathbf{v} = 0$
3. $\mathbf{u} \cdot (\mathbf{v} + \mathbf{w}) = \mathbf{u} \cdot \mathbf{v} + \mathbf{u} \cdot \mathbf{w}$
4. $\mathbf{v} \cdot \mathbf{v} = \|\mathbf{v}\|^2$
5. $c(\mathbf{u} \cdot \mathbf{v}) = c\mathbf{u} \cdot \mathbf{v} = \mathbf{u} \cdot c\mathbf{v}$

PROOF To prove Property 1, let $\mathbf{u} = \langle u_1, u_2 \rangle$ and $\mathbf{v} = \langle v_1, v_2 \rangle$. Then

$$\begin{aligned}
\mathbf{u} \cdot \mathbf{v} &= u_1v_1 + u_2v_2 \\
&= v_1u_1 + v_2u_2 \\
&= \mathbf{v} \cdot \mathbf{u}.
\end{aligned}$$

To prove Property 4, let $\mathbf{v} = \langle v_1, v_2 \rangle$. Then

$$\begin{aligned}
\mathbf{v} \cdot \mathbf{v} &= v_1^2 + v_2^2 \\
&= \left(\sqrt{v_1^2 + v_2^2}\right)^2 \\
&= \|\mathbf{v}\|^2.
\end{aligned}$$

■

B Additional Topics

B.1 L'Hôpital's Rule

GUILLAUME L'HÔPITAL (1661–1704)

L'Hôpital's Rule is named after the French mathematician Guillaume François Antoine de L'Hôpital. L'Hôpital is credited with writing the first text on differential calculus (in 1696) in which the rule publicly appeared. It was recently discovered that the rule and its proof were written in a letter from John Bernoulli to L'Hôpital. "... I acknowledge that I owe very much to the bright minds of the Bernoulli brothers. ... I have made free use of their discoveries ...," said L'Hôpital.

L'Hôpital's Rule

L'Hôpital's Rule states that under certain conditions the limit of the quotient $f(x)/g(x)$ is determined by the limit of the quotient of the derivatives

$$\frac{f'(x)}{g'(x)}.$$

To prove this theorem, you can use a more general result called the **Extended Mean Value Theorem.**

THEOREM B.1 THE EXTENDED MEAN VALUE THEOREM

If f and g are differentiable on an open interval (a, b) and continuous on $[a, b]$ such that $g'(x) \neq 0$ for any x in (a, b), then there exists a point c in (a, b) such that

$$\frac{f'(c)}{g'(c)} = \frac{f(b) - f(a)}{g(b) - g(a)}.$$

NOTE To see why this is called the Extended Mean Value Theorem, consider the special case in which $g(x) = x$. For this case, you obtain the "standard" Mean Value Theorem as presented in Section 5.2. ■

THEOREM B.2 L'HÔPITAL'S RULE

Let f and g be functions that are differentiable on an open interval (a, b) containing c, except possibly at c itself. Assume that $g'(x) \neq 0$ for all x in (a, b), except possibly at c itself. If the limit of $f(x)/g(x)$ as x approaches c produces the indeterminate form $0/0$, then

$$\lim_{x \to c} \frac{f(x)}{g(x)} = \lim_{x \to c} \frac{f'(x)}{g'(x)}$$

provided the limit on the right exists (or is infinite). This result also applies if the limit of $f(x)/g(x)$ as x approaches c produces any one of the indeterminate forms ∞/∞, $(-\infty)/\infty$, $\infty/(-\infty)$, or $(-\infty)/(-\infty)$.

NOTE People occasionally use L'Hôpital's Rule incorrectly by applying the Quotient Rule to $f(x)/g(x)$. Be sure you see that the rule involves $f'(x)/g'(x)$, not the derivative of $f(x)/g(x)$. ■

TECHNOLOGY *Numerical and Graphical Approaches* Use a numerical or a graphical approach to approximate each limit.

a. $\lim_{x \to 0} \frac{2^{2x} - 1}{x}$

b. $\lim_{x \to 0} \frac{3^{2x} - 1}{x}$

c. $\lim_{x \to 0} \frac{4^{2x} - 1}{x}$

d. $\lim_{x \to 0} \frac{5^{2x} - 1}{x}$

What pattern do you observe? Does an analytic approach have an advantage for these limits? If so, explain your reasoning.

L'Hôpital's Rule can also be applied to one-sided limits. For instance, if the limit of $f(x)/g(x)$ as x approaches c *from the right* produces the indeterminate form $0/0$, then

$$\lim_{x \to c^+} \frac{f(x)}{g(x)} = \lim_{x \to c^+} \frac{f'(x)}{g'(x)}$$

provided the limit exists (or is infinite).

EXAMPLE 1 Indeterminate Form 0/0

Evaluate $\lim_{x \to 0} \frac{e^{2x} - 1}{x}$.

Solution Because direct substitution results in the indeterminate form $0/0$,

$$\lim_{x \to 0} \frac{e^{2x} - 1}{x} \quad \begin{array}{l} \nearrow \lim_{x \to 0} (e^{2x} - 1) = 0 \\ \searrow \lim_{x \to 0} x = 0 \end{array}$$

you can apply L'Hôpital's Rule, as shown below.

$$\lim_{x \to 0} \frac{e^{2x} - 1}{x} = \lim_{x \to 0} \frac{\frac{d}{dx}[e^{2x} - 1]}{\frac{d}{dx}[x]}$$ Apply L'Hôpital's Rule.

$$= \lim_{x \to 0} \frac{2e^{2x}}{1}$$ Differentiate numerator and denominator.

$$= 2$$ Evaluate the limit. ■

NOTE In writing the string of equations in Example 1, you actually do not know that the first limit is equal to the second until you have shown that the second limit exists. In other words, if the second limit had not existed, it would not have been permissible to apply L'Hôpital's Rule.

Another form of L'Hôpital's Rule states that if the limit of $f(x)/g(x)$ as x approaches ∞ (or $-\infty$) produces the indeterminate form $0/0$ or ∞/∞, then

$$\lim_{x \to \infty} \frac{f(x)}{g(x)} = \lim_{x \to \infty} \frac{f'(x)}{g'(x)}$$

provided the limit on the right exists.

EXAMPLE 2 Indeterminate Form ∞/∞

Evaluate $\lim_{x \to \infty} \frac{\ln x}{x}$.

Solution Because direct substitution results in the indeterminate form ∞/∞, you can apply L'Hôpital's Rule to obtain

$$\lim_{x \to \infty} \frac{\ln x}{x} = \lim_{x \to \infty} \frac{\frac{d}{dx}[\ln x]}{\frac{d}{dx}[x]}$$ Apply L'Hôpital's Rule.

$$= \lim_{x \to \infty} \frac{1}{x}$$ Differentiate numerator and denominator.

$$= 0.$$ Evaluate the limit. ■

NOTE Try graphing $y_1 = \ln x$ and $y_2 = x$ in the same viewing window. Which function grows faster as x approaches ∞? How is this observation related to Example 2?

Occasionally it is necessary to apply L'Hôpital's Rule more than once to remove an indeterminate form, as shown in Example 3.

EXAMPLE 3 Applying L'Hôpital's Rule More Than Once

Evaluate $\lim_{x\to-\infty} \dfrac{x^2}{e^{-x}}$.

Solution Because direct substitution results in the indeterminate form ∞/∞, you can apply L'Hôpital's Rule.

$$\lim_{x\to-\infty} \frac{x^2}{e^{-x}} = \lim_{x\to-\infty} \frac{\frac{d}{dx}[x^2]}{\frac{d}{dx}[e^{-x}]} = \lim_{x\to-\infty} \frac{2x}{-e^{-x}}$$

This limit yields the indeterminate form $(-\infty)/(-\infty)$, so you can apply L'Hôpital's Rule again to obtain

$$\lim_{x\to-\infty} \frac{2x}{-e^{-x}} = \lim_{x\to-\infty} \frac{\frac{d}{dx}[2x]}{\frac{d}{dx}[-e^{-x}]} = \lim_{x\to-\infty} \frac{2}{e^{-x}} = 0.$$

■

In addition to the forms $0/0$ and ∞/∞, there are other indeterminate forms such as $0 \cdot \infty$, 1^∞, ∞^0, 0^0, and $\infty - \infty$. For example, consider the following four limits that lead to the indeterminate form $0 \cdot \infty$.

$$\underbrace{\lim_{x\to0} (x)\left(\frac{1}{x}\right)}_{\text{Limit is 1.}}, \quad \underbrace{\lim_{x\to0} (x)\left(\frac{2}{x}\right)}_{\text{Limit is 2.}}, \quad \underbrace{\lim_{x\to\infty} (x)\left(\frac{1}{e^x}\right)}_{\text{Limit is 0.}}, \quad \underbrace{\lim_{x\to\infty} (e^x)\left(\frac{1}{x}\right)}_{\text{Limit is } \infty.}$$

Because each limit is different, it is clear that the form $0 \cdot \infty$ is indeterminate in the sense that it does not determine the value (or even the existence) of the limit. The following examples indicate methods for evaluating these forms. Basically, you attempt to convert each of these forms to $0/0$ or ∞/∞ so that L'Hôpital's Rule can be applied.

EXAMPLE 4 Indeterminate Form $0 \cdot \infty$

Evaluate $\lim_{x\to\infty} e^{-x}\sqrt{x}$.

Solution Because direct substitution produces the indeterminate form $0 \cdot \infty$, you should try to rewrite the limit to fit the form $0/0$ or ∞/∞. In this case, you can rewrite the limit to fit the second form.

$$\lim_{x\to\infty} e^{-x}\sqrt{x} = \lim_{x\to\infty} \frac{\sqrt{x}}{e^x}$$

Now, by L'Hôpital's Rule, you have

$$\lim_{x\to\infty} \frac{\sqrt{x}}{e^x} = \lim_{x\to\infty} \frac{1/(2\sqrt{x})}{e^x}$$

$$= \lim_{x\to\infty} \frac{1}{2\sqrt{x}\,e^x} = 0.$$

■

If rewriting a limit in one of the forms $0/0$ or ∞/∞ does not seem to work, try the other form. For instance, in Example 4 you can write the limit as

$$\lim_{x\to\infty} e^{-x}\sqrt{x} = \lim_{x\to\infty} \frac{e^{-x}}{x^{-1/2}}$$

which yields the indeterminate form $0/0$. As it happens, applying L'Hôpital's Rule to this limit produces

$$\lim_{x\to\infty} \frac{e^{-x}}{x^{-1/2}} = \lim_{x\to\infty} \frac{-e^{-x}}{-1/(2x^{3/2})}$$

which also yields the indeterminate form $0/0$.

The indeterminate forms 1^∞, ∞^0, and 0^0 arise from limits of functions that have variable bases and variable exponents. When you previously encountered this type of function, you used logarithmic differentiation to find the derivative. You can use a similar procedure when taking limits, as shown in the next example.

EXAMPLE 5 Indeterminate Form 1^∞

Evaluate

$$\lim_{x\to\infty} \left(1 + \frac{1}{x}\right)^x.$$

Solution Because direct substitution yields the indeterminate form 1^∞, you can proceed as follows. To begin, assume that the limit exists and is equal to y.

$$y = \lim_{x\to\infty} \left(1 + \frac{1}{x}\right)^x$$

Taking the natural logarithm of each side produces

$$\ln y = \ln\left[\lim_{x\to\infty} \left(1 + \frac{1}{x}\right)^x\right].$$

Because the natural logarithmic function is continuous, you can write

$$\ln y = \lim_{x\to\infty} \left[x \ln\left(1 + \frac{1}{x}\right)\right] \qquad \text{Indeterminate form } \infty \cdot 0$$

$$= \lim_{x\to\infty} \left(\frac{\ln[1 + (1/x)]}{1/x}\right) \qquad \text{Indeterminate form } 0/0$$

$$= \lim_{x\to\infty} \left(\frac{(-1/x^2)\{1/[1 + (1/x)]\}}{-1/x^2}\right) \qquad \text{L'Hôpital's Rule}$$

$$= \lim_{x\to\infty} \frac{1}{1 + (1/x)}$$

$$= 1.$$

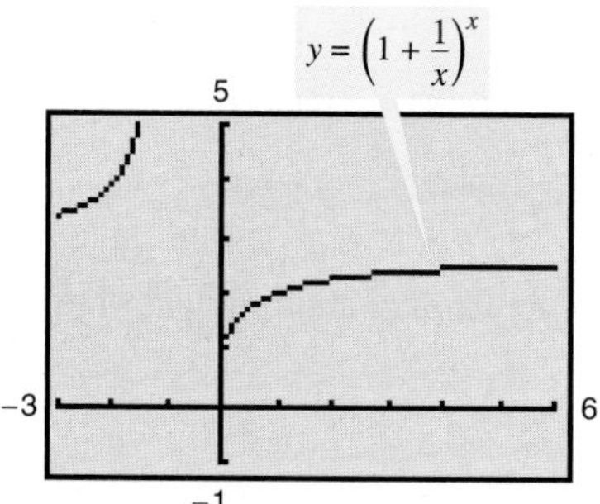

The limit of $[1 + (1/x)]^x$ as x approaches infinity is e.
Figure B.1

Now, because you have shown that $\ln y = 1$, you can conclude that $y = e$ and obtain

$$\lim_{x\to\infty} \left(1 + \frac{1}{x}\right)^x = e.$$

You can use a graphing utility to confirm this result, as shown in Figure B.1. ■

L'Hôpital's Rule can also be applied to one-sided limits, as demonstrated in Examples 6 and 7.

EXAMPLE 6 Indeterminate Form 0^0

Find

$$\lim_{x \to 0^+} (\sin x)^x.$$

Solution Because direct substitution produces the indeterminate form 0^0, you can proceed as shown below. To begin, assume that the limit exists and is equal to y.

$$y = \lim_{x \to 0^+} (\sin x)^x \qquad \text{Indeterminate form } 0^0$$

$$\ln y = \ln\left[\lim_{x \to 0^+} (\sin x)^x\right] \qquad \text{Take natural log of each side.}$$

$$= \lim_{x \to 0^+} \left[\ln(\sin x)^x\right] \qquad \text{Continuity}$$

$$= \lim_{x \to 0^+} \left[x \ln(\sin x)\right] \qquad \text{Indeterminate form } 0 \cdot (-\infty)$$

$$= \lim_{x \to 0^+} \frac{\ln(\sin x)}{1/x} \qquad \text{Indeterminate form } -\infty/\infty$$

$$= \lim_{x \to 0^+} \frac{\cot x}{-1/x^2} \qquad \text{L'Hôpital's Rule}$$

$$= \lim_{x \to 0^+} \frac{-x^2}{\tan x} \qquad \text{Indeterminate form } 0/0$$

$$= \lim_{x \to 0^+} \frac{-2x}{\sec^2 x} = 0 \qquad \text{L'Hôpital's Rule}$$

Now, because $\ln y = 0$, you can conclude that $y = e^0 = 1$, and it follows that

$$\lim_{x \to 0^+} (\sin x)^x = 1.$$

■

TECHNOLOGY When evaluating complicated limits such as the one in Example 6, it is helpful to check the reasonableness of the solution with a computer or with a graphing utility. For instance, the calculations in the following table and the graph in Figure B.2 are consistent with the conclusion that $(\sin x)^x$ approaches 1 as x approaches 0 from the right.

x	1.0	0.1	0.01	0.001	0.0001	0.00001
$(\sin x)^x$	0.8415	0.7942	0.9550	0.9931	0.9991	0.9999

Use a computer algebra system or graphing utility to estimate the following limits:

$$\lim_{x \to 0} (1 - \cos x)^x$$

and

$$\lim_{x \to 0^+} (\tan x)^x.$$

Then see if you can verify your estimates analytically.

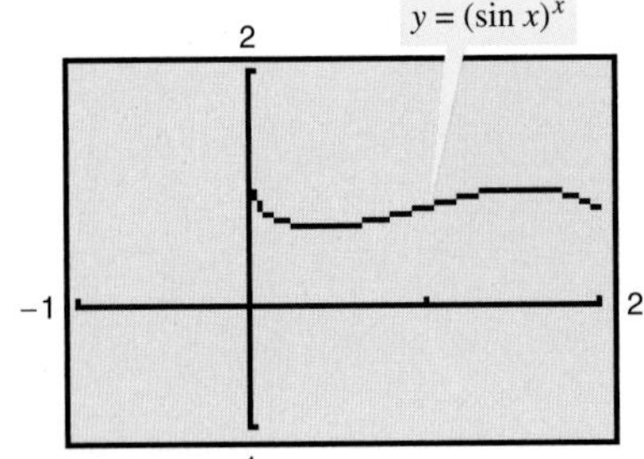

The limit of $(\sin x)^x$ is 1 as x approaches 0 from the right.
Figure B.2

STUDY TIP In each of the examples presented in this section, L'Hôpital's Rule is used to find a limit that exists. It can also be used to conclude that a limit is infinite. For instance, try using L'Hôpital's Rule to show that

$$\lim_{x\to\infty} \frac{e^x}{x} = \infty.$$

EXAMPLE 7 Indeterminate Form $\infty - \infty$

Evaluate

$$\lim_{x\to 1^+} \left(\frac{1}{\ln x} - \frac{1}{x-1}\right).$$

Solution Because direct substitution yields the indeterminate form $\infty - \infty$, you should try to rewrite the expression to produce a form to which you can apply L'Hôpital's Rule. In this case, you can combine the two fractions to obtain

$$\lim_{x\to 1^+} \left(\frac{1}{\ln x} - \frac{1}{x-1}\right) = \lim_{x\to 1^+} \left[\frac{x-1-\ln x}{(x-1)\ln x}\right].$$

Now, because direct substitution produces the indeterminate form $0/0$, you can apply L'Hôpital's Rule to obtain

$$\begin{aligned}\lim_{x\to 1^+} \left(\frac{1}{\ln x} - \frac{1}{x-1}\right) &= \lim_{x\to 1^+} \frac{\frac{d}{dx}[x-1-\ln x]}{\frac{d}{dx}[(x-1)\ln x]}\\ &= \lim_{x\to 1^+} \left[\frac{1-(1/x)}{(x-1)(1/x)+\ln x}\right]\\ &= \lim_{x\to 1^+} \left(\frac{x-1}{x-1+x\ln x}\right).\end{aligned}$$

This limit also yields the indeterminate form $0/0$, so you can apply L'Hôpital's Rule again to obtain

$$\begin{aligned}\lim_{x\to 1^+} \left(\frac{1}{\ln x} - \frac{1}{x-1}\right) &= \lim_{x\to 1^+} \left[\frac{1}{1+x(1/x)+\ln x}\right]\\ &= \frac{1}{2}.\end{aligned}$$

■

The forms $0/0$, ∞/∞, $\infty - \infty$, $0 \cdot \infty$, 0^0, 1^∞, and ∞^0 have been identified as *indeterminate*. There are similar forms that you should recognize as "determinate."

$\infty + \infty \to \infty$	Limit is positive infinity.
$-\infty - \infty \to -\infty$	Limit is negative infinity.
$0^\infty \to 0$	Limit is zero.
$0^{-\infty} \to \infty$	Limit is positive infinity.

As a final comment, remember that L'Hôpital's Rule can be applied only to quotients leading to the indeterminate forms $0/0$ and ∞/∞. For instance, the following application of L'Hôpital's Rule is *incorrect*.

$\lim_{x\to 0} \frac{e^x}{x} = \lim_{x\to 0} \frac{e^x}{1} = 1$
Incorrect use of L'Hôpital's Rule

The reason this application is incorrect is that, even though the limit of the denominator is 0, the limit of the numerator is 1, which means that the hypotheses of L'Hôpital's Rule have not been satisfied.

B.2 Applications of Integration

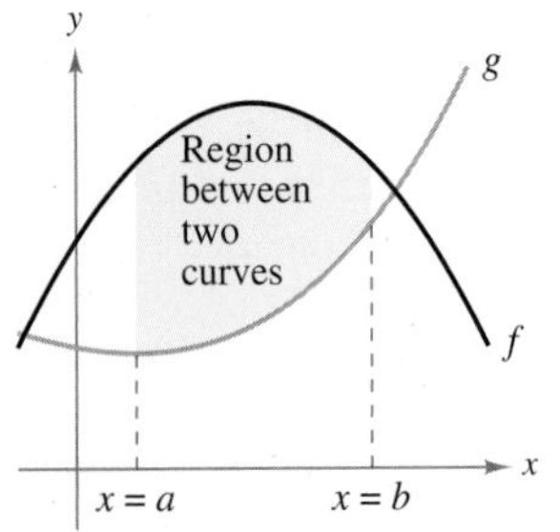

Figure B.3

Area of a Region Between Two Curves

With a few modifications, you can extend the application of definite integrals from the area of a region *under* a curve to the area of a region *between* two curves. Consider two functions f and g that are continuous on the interval $[a, b]$. If, as in Figure B.3, the graphs of both f and g lie above the x-axis, and the graph of g lies below the graph of f, you can geometrically interpret the area of the region between the graphs as the area of the region under the graph of g subtracted from the area of the region under the graph of f, as shown in Figure B.4.

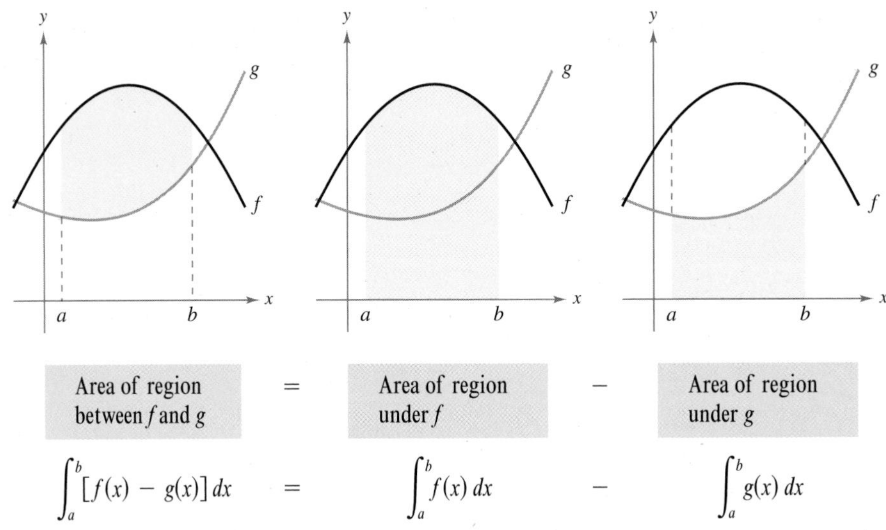

Figure B.4

To verify the reasonableness of the result shown in Figure B.4, you can partition the interval $[a, b]$ into n subintervals, each of width Δx. Then, as shown in Figure B.5, sketch a **representative rectangle** of width Δx and height $f(x_i) - g(x_i)$, where x_i is in the ith subinterval. The area of this representative rectangle is

$$\Delta A_i = (\text{height})(\text{width}) = [f(x_i) - g(x_i)]\,\Delta x.$$

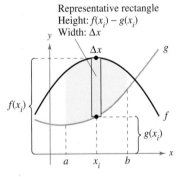

By adding the areas of the n rectangles and taking the limit as $\|\Delta\| \to 0$ $(n \to \infty)$, you obtain

$$\lim_{n \to \infty} \sum_{i=1}^{n} [f(x_i) - g(x_i)]\,\Delta x.$$

Representative rectangle
Height: $f(x_i) - g(x_i)$
Width: Δx

Figure B.5

Because f and g are continuous on $[a, b]$, $f - g$ is also continuous on $[a, b]$ and the limit exists. So, the area of the given region is

$$\text{Area} = \lim_{n \to \infty} \sum_{i=1}^{n} [f(x_i) - g(x_i)]\,\Delta x$$

$$= \int_a^b [f(x) - g(x)]\,dx.$$

AREA OF A REGION BETWEEN TWO CURVES

If f and g are continuous on $[a, b]$ and $g(x) \leq f(x)$ for all x in $[a, b]$, then the area of the region bounded by the graphs of f and g and the vertical lines $x = a$ and $x = b$ is

$$A = \int_a^b [f(x) - g(x)]\, dx.$$

In Figure B.3, the graphs of f and g are shown above the x-axis. This, however, is not necessary. The same integrand $[f(x) - g(x)]$ can be used as long as f and g are continuous and $g(x) \leq f(x)$ for all x in the interval $[a, b]$. This is summarized graphically in Figure B.6. Notice in Figure B.6 that the height of a representative rectangle is $f(x) - g(x)$ regardless of the relative position of the x-axis.

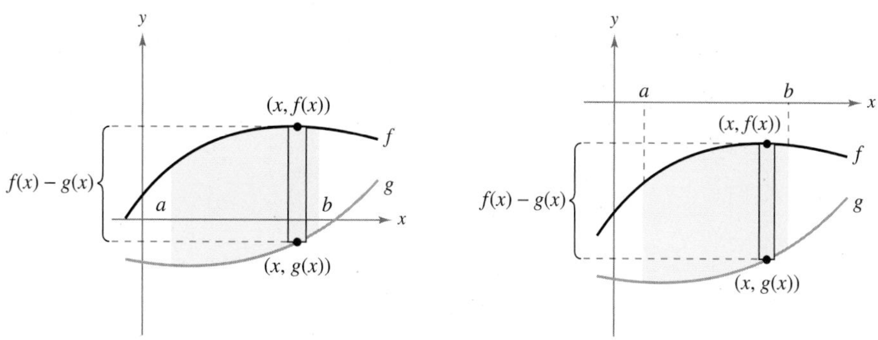

Figure B.6

Representative rectangles are used throughout this section in various applications of integration. A vertical rectangle (of width Δx) implies integration with respect to x, whereas a horizontal rectangle (of width Δy) implies integration with respect to y.

EXAMPLE 1 Finding the Area of a Region Between Two Curves

Find the area of the region bounded by the graphs of $y = x^2 + 2$, $y = -x$, $x = 0$, and $x = 1$.

Solution Let $g(x) = -x$ and $f(x) = x^2 + 2$. Then $g(x) \leq f(x)$ for all x in $[0, 1]$, as shown in Figure B.7. So, the area of the representative rectangle is

$$\begin{aligned}\Delta A &= [f(x) - g(x)]\,\Delta x \\ &= [(x^2 + 2) - (-x)]\,\Delta x\end{aligned}$$

and the area of the region is

$$\begin{aligned}A = \int_a^b [f(x) - g(x)]\, dx &= \int_0^1 [(x^2 + 2) - (-x)]\, dx \\ &= \left[\frac{x^3}{3} + \frac{x^2}{2} + 2x\right]_0^1 \\ &= \frac{1}{3} + \frac{1}{2} + 2 \\ &= \frac{17}{6}.\end{aligned}$$

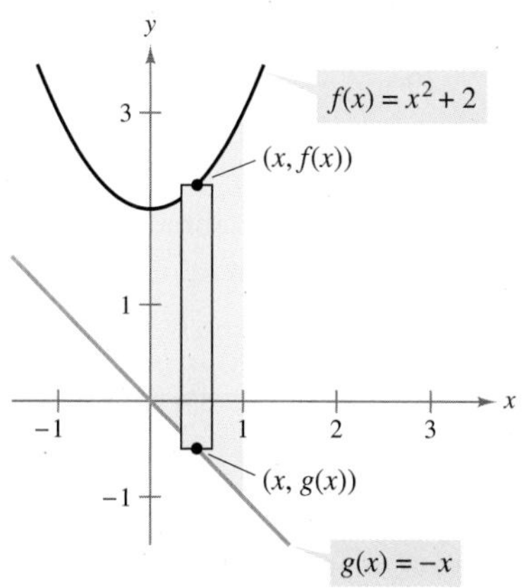

Region bounded by the graph of f, the graph of g, $x = 0$, and $x = 1$
Figure B.7

The Disk Method

You have already learned that area is only one of the *many* applications of the definite integral. Another important application is its use in finding the volume of a three-dimensional solid. In this section you will study a particular type of three-dimensional solid—one whose cross sections are similar. Solids of revolution are used commonly in engineering and manufacturing. Some examples are axles, funnels, pills, bottles, and pistons, as shown in Figure B.8.

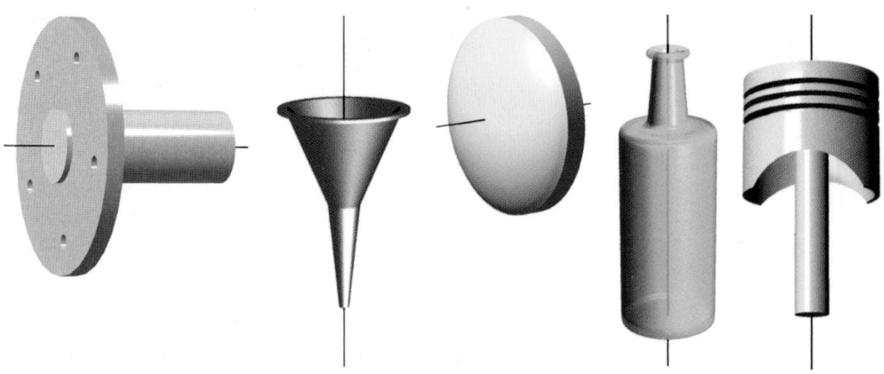

Solids of revolution
Figure B.8

If a region in the plane is revolved about a line, the resulting solid is a **solid of revolution,** and the line is called the **axis of revolution.** The simplest such solid is a right circular cylinder or **disk,** which is formed by revolving a rectangle about an axis adjacent to one side of the rectangle, as shown in Figure B.9. The volume of such a disk is

$$\begin{aligned}\text{Volume of disk} &= (\text{area of disk})(\text{width of disk})\\ &= \pi R^2 w\end{aligned}$$

where R is the radius of the disk and w is the width.

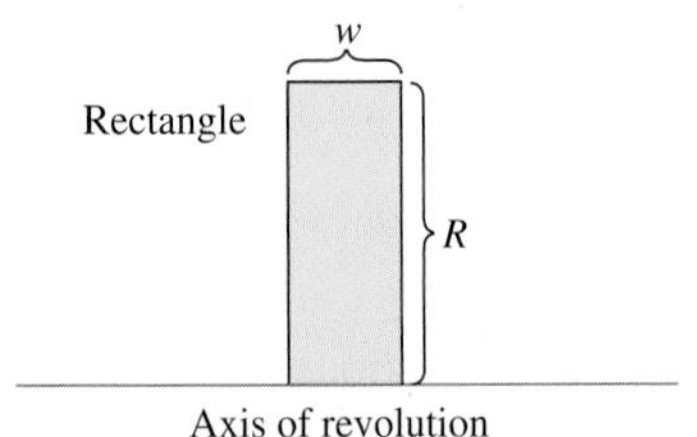

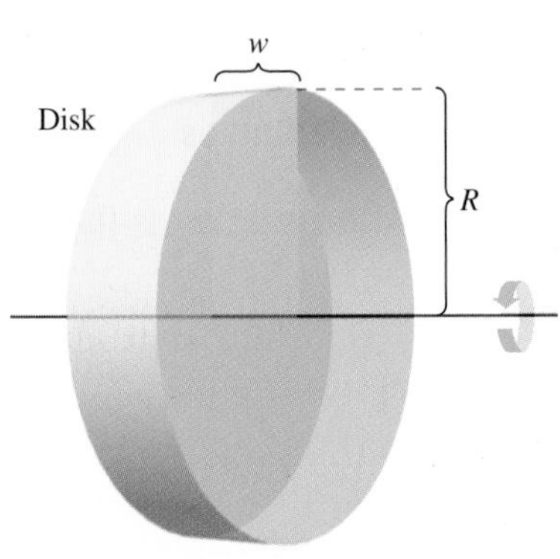

Volume of a disk: $\pi R^2 w$
Figure B.9

To see how to use the volume of a disk to find the volume of a general solid of revolution, consider a solid of revolution formed by revolving the plane region in Figure B.10 about the indicated axis. To determine the volume of this solid, consider a representative rectangle in the plane region. When this rectangle is revolved about the axis of revolution, it generates a representative disk whose volume is

$$\Delta V = \pi R^2 \Delta x.$$

Approximating the volume of the solid by n such disks of width Δx and radius $R(x_i)$ produces

$$\begin{aligned}\text{Volume of solid} &\approx \sum_{i=1}^{n} \pi[R(x_i)]^2 \, \Delta x\\ &= \pi \sum_{i=1}^{n} [R(x_i)]^2 \, \Delta x.\end{aligned}$$

Disk method
Figure B.10

This approximation appears to become better and better as $\|\Delta\| \to 0$ $(n \to \infty)$. So, you can define the volume of the solid as

$$\text{Volume of solid} = \lim_{\|\Delta\| \to 0} \pi \sum_{i=1}^{n} [R(x_i)]^2 \, \Delta x = \pi \int_a^b [R(x)]^2 \, dx.$$

Schematically, the disk method looks like this.

Known Precalculus Formula		*Representative Element*		*New Integration Formula*
Volume of disk $V = \pi R^2 w$	⇨	$\Delta V = \pi[R(x_i)]^2 \, \Delta x$	⇨	Solid of revolution $V = \pi \int_a^b [R(x)]^2 \, dx$

A similar formula can be derived if the axis of revolution is vertical.

THE DISK METHOD

To find the volume of a solid of revolution with the **disk method,** use one of the following, as shown in Figure B.11.

Horizontal Axis of Revolution	*Vertical Axis of Revolution*
$\text{Volume} = V = \pi \int_a^b [R(x)]^2 \, dx$	$\text{Volume} = V = \pi \int_c^d [R(y)]^2 \, dy$

NOTE In Figure B.11, note that you can determine the variable of integration by placing a representative rectangle in the plane region "perpendicular" to the axis of revolution. If the width of the rectangle is Δx, integrate with respect to x, and if the width of the rectangle is Δy, integrate with respect to y.

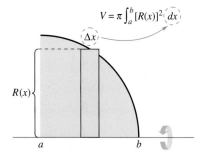

Horizontal axis of revolution
Figure B.11

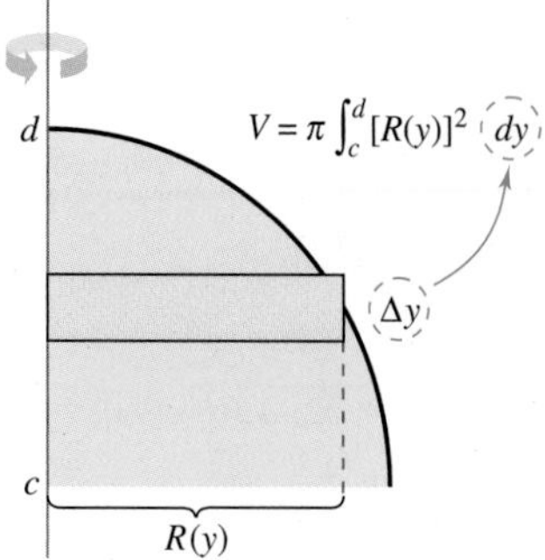

Vertical axis of revolution

The simplest application of the disk method involves a plane region bounded by the graph of f and the x-axis. If the axis of revolution is the x-axis, the radius $R(x)$ is simply $f(x)$.

EXAMPLE 2 Using the Disk Method

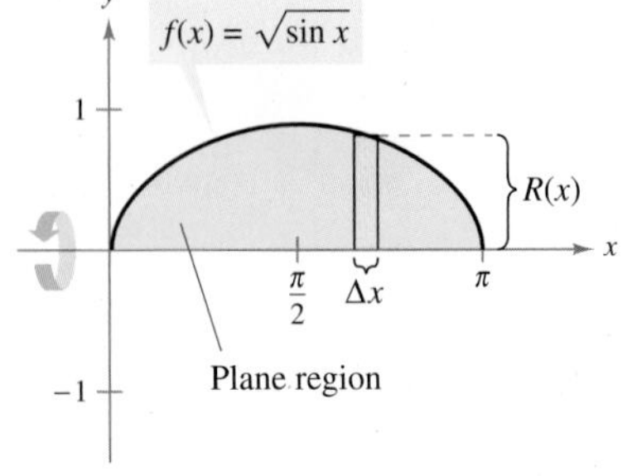

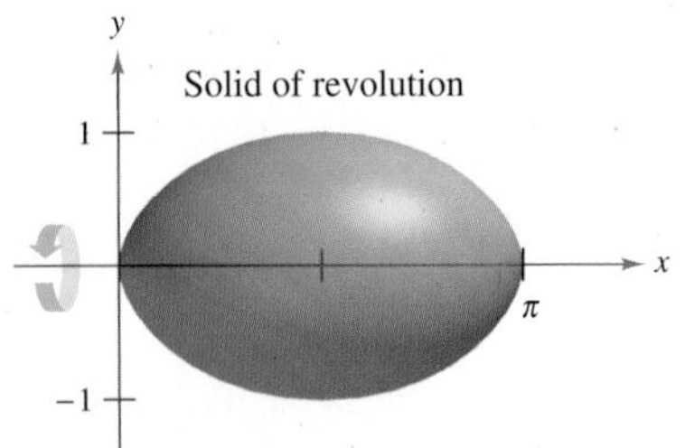

Figure B.12

Find the volume of the solid formed by revolving the region bounded by the graph of

$$f(x) = \sqrt{\sin x}$$

and the x-axis $(0 \le x \le \pi)$ about the x-axis.

Solution From the representative rectangle in the upper graph in Figure B.12, you can see that the radius of this solid is

$$\begin{aligned} R(x) &= f(x) \\ &= \sqrt{\sin x}. \end{aligned}$$

So, the volume of the solid of revolution is

$$\begin{aligned} V = \pi \int_a^b [R(x)]^2\,dx &= \pi \int_0^{\pi} \left(\sqrt{\sin x}\right)^2 dx && \text{Apply disk method.} \\ &= \pi \int_0^{\pi} \sin x\,dx && \text{Simplify.} \\ &= \pi \Big[-\cos x\Big]_0^{\pi} && \text{Integrate.} \\ &= \pi(1 + 1) \\ &= 2\pi. \end{aligned}$$

EXAMPLE 3 Revolving About a Line That Is Not a Coordinate Axis

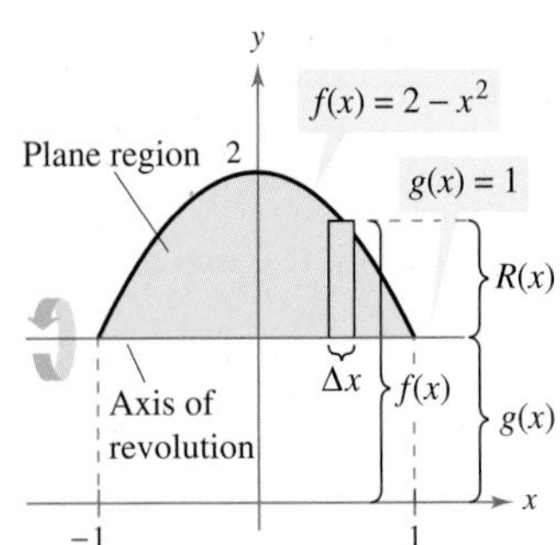

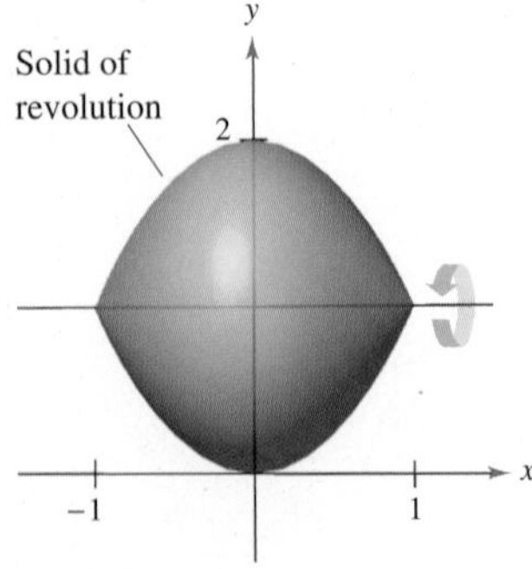

Figure B.13

Find the volume of the solid formed by revolving the region bounded by

$$f(x) = 2 - x^2$$

and $g(x) = 1$ about the line $y = 1$, as shown in Figure B.13.

Solution By equating $f(x)$ and $g(x)$, you can determine that the two graphs intersect when $x = \pm 1$. To find the radius, subtract $g(x)$ from $f(x)$.

$$\begin{aligned} R(x) &= f(x) - g(x) \\ &= (2 - x^2) - 1 \\ &= 1 - x^2 \end{aligned}$$

Finally, integrate between -1 and 1 to find the volume.

$$\begin{aligned} V = \pi \int_a^b [R(x)]^2\,dx &= \pi \int_{-1}^{1} (1 - x^2)^2\,dx && \text{Apply disk method.} \\ &= \pi \int_{-1}^{1} (1 - 2x^2 + x^4)\,dx && \text{Simplify.} \\ &= \pi \left[x - \frac{2x^3}{3} + \frac{x^5}{5}\right]_{-1}^{1} && \text{Integrate.} \\ &= \frac{16\pi}{15} \end{aligned}$$

■

The Washer Method

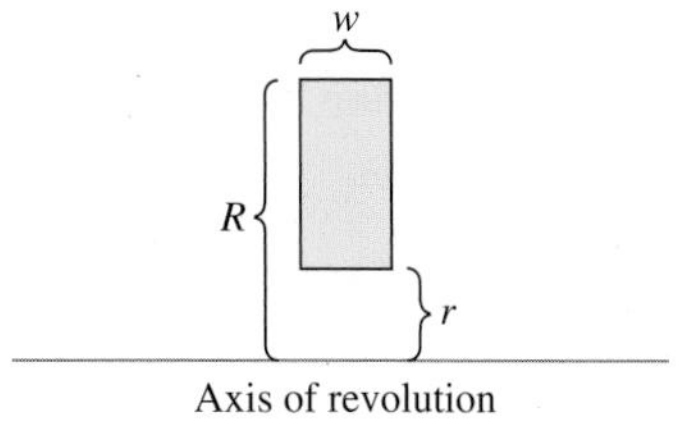

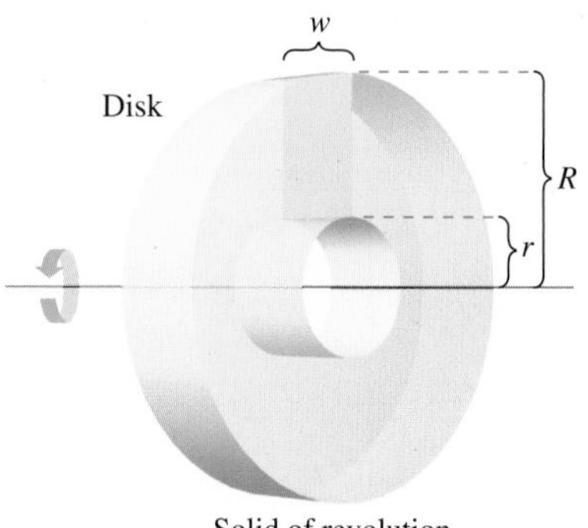

Figure B.14

The disk method can be extended to cover solids of revolution with holes by replacing the representative disk with a representative **washer.** The washer is formed by revolving a rectangle about an axis, as shown in Figure B.14. If r and R are the inner and outer radii of the washer and w is the width of the washer, the volume is given by

$$\text{Volume of washer} = \pi(R^2 - r^2)w.$$

To see how this concept can be used to find the volume of a solid of revolution, consider a region bounded by an **outer radius** $R(x)$ and an **inner radius** $r(x)$, as shown in Figure B.15. If the region is revolved about its axis of revolution, the volume of the resulting solid is given by

$$V = \pi\int_a^b ([R(x)]^2 - [r(x)]^2)\, dx. \qquad \text{Washer method}$$

Note that the integral involving the inner radius represents the volume of the hole and is *subtracted* from the integral involving the outer radius.

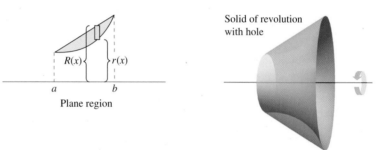

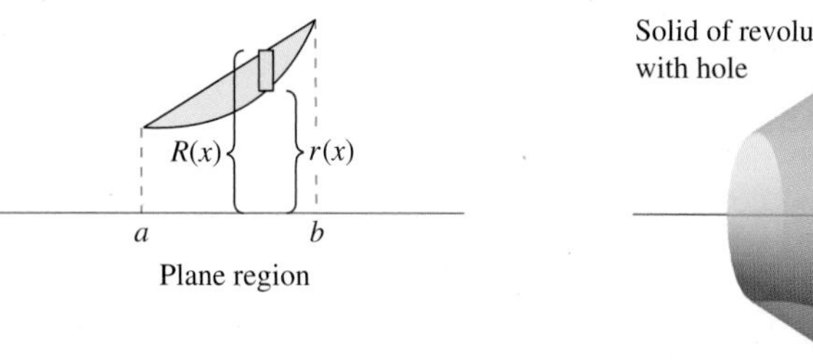

Figure B.15

EXAMPLE 4 Using the Washer Method

Find the volume of the solid formed by revolving the region bounded by the graphs of $y = \sqrt{x}$ and $y = x^2$ about the x-axis, as shown in Figure B.16.

Solution In Figure B.16, you can see that the outer and inner radii are as follows.

$$R(x) = \sqrt{x} \qquad \text{Outer radius}$$
$$r(x) = x^2 \qquad \text{Inner radius}$$

Integrating between 0 and 1 produces

$$V = \pi\int_a^b ([R(x)]^2 - [r(x)]^2)\, dx \qquad \text{Apply washer method.}$$
$$= \pi\int_0^1 \left[(\sqrt{x})^2 - (x^2)^2\right] dx$$
$$= \pi\int_0^1 (x - x^4)\, dx \qquad \text{Simplify.}$$
$$= \pi\left[\frac{x^2}{2} - \frac{x^5}{5}\right]_0^1 \qquad \text{Integrate.}$$
$$= \frac{3\pi}{10}.$$

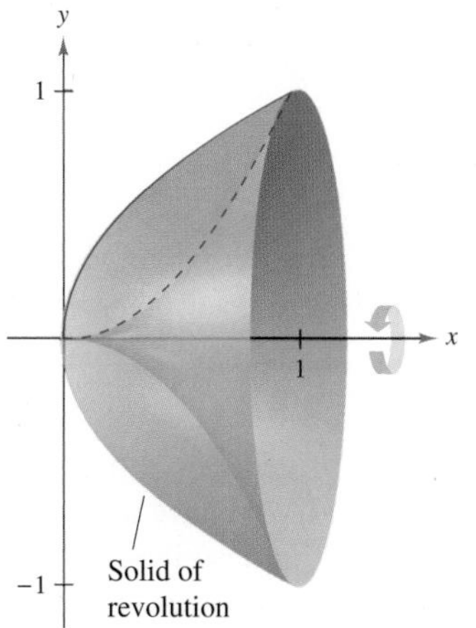

Solid of revolution
Figure B.16

In each example so far, the axis of revolution has been *horizontal* and you have integrated with respect to x. In the next example, the axis of revolution is *vertical* and you integrate with respect to y. In this example, you need two separate integrals to compute the volume.

EXAMPLE 5 Integrating with Respect to y, Two-Integral Case

Find the volume of the solid formed by revolving the region bounded by the graphs of $y = x^2 + 1$, $y = 0$, $x = 0$, and $x = 1$ about the y-axis, as shown in Figure B.17.

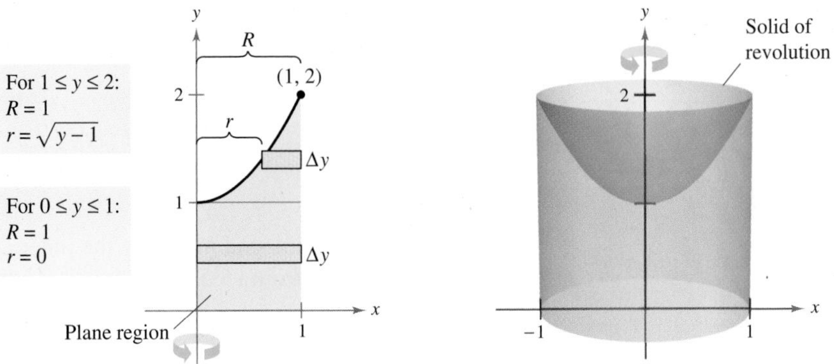

Figure B.17

Solution For the region shown in Figure B.17, the outer radius is simply $R = 1$. There is, however, no convenient formula that represents the inner radius. When $0 \le y \le 1$, $r = 0$, but when $1 \le y \le 2$, r is determined by the equation $y = x^2 + 1$, which implies that $r = \sqrt{y-1}$.

$$r(y) = \begin{cases} 0, & 0 \le y \le 1 \\ \sqrt{y-1}, & 1 \le y \le 2 \end{cases}$$

Using this definition of the inner radius, you can use two integrals to find the volume.

$$V = \pi\int_0^1 (1^2 - 0^2)\,dy + \pi\int_1^2 \left[1^2 - \left(\sqrt{y-1}\right)^2\right]dy \qquad \text{Apply washer method.}$$

$$= \pi\int_0^1 1\,dy + \pi\int_1^2 (2-y)\,dy \qquad \text{Simplify.}$$

$$= \pi\Big[y\Big]_0^1 + \pi\left[2y - \frac{y^2}{2}\right]_1^2 \qquad \text{Integrate.}$$

$$= \pi + \pi\left(4 - 2 - 2 + \frac{1}{2}\right) = \frac{3\pi}{2}$$

Note that the first integral $\pi\int_0^1 1\,dy$ represents the volume of a right circular cylinder of radius 1 and height 1. This portion of the volume could have been determined without using calculus. ■

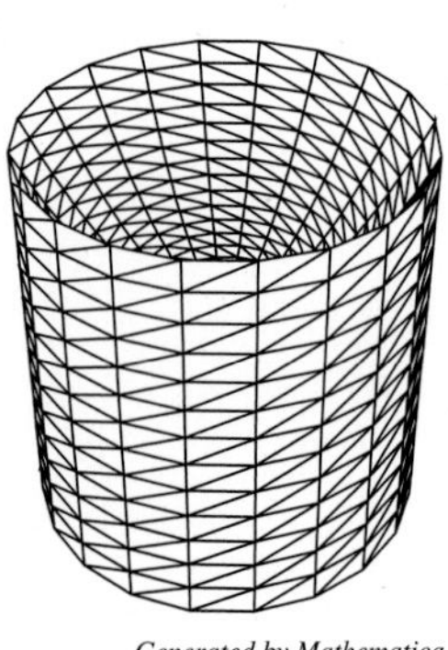

Figure B.18

TECHNOLOGY Some graphing utilities have the capability of generating (or have built-in software capable of generating) a solid of revolution. If you have access to such a utility, use it to graph some of the solids of revolution described in this section. For instance, the solid in Example 5 might appear like that shown in Figure B.18.

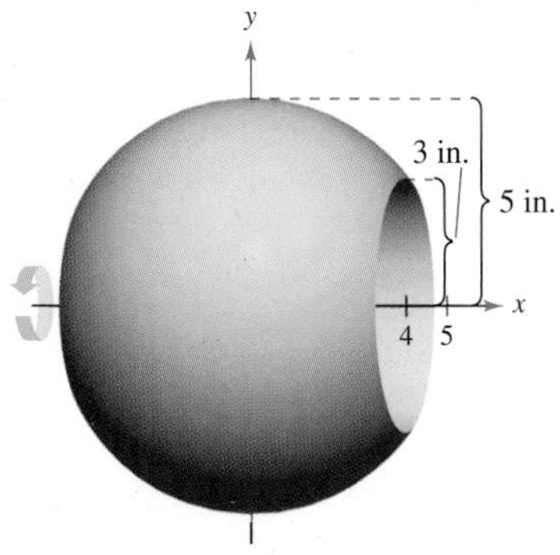

Solid of revolution

(a)

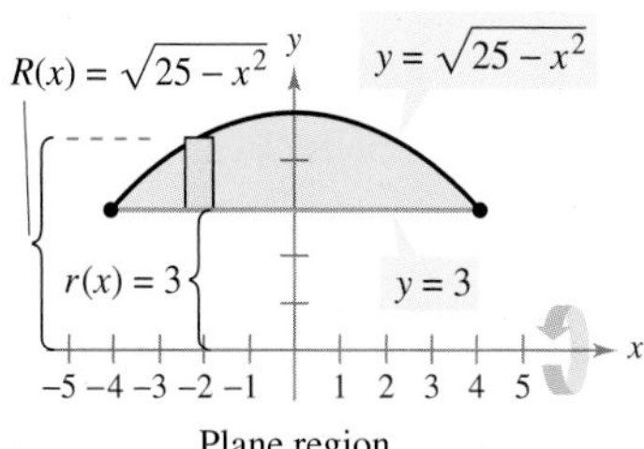

Plane region

(b)

Figure B.19

EXAMPLE 6 Manufacturing

A manufacturer drills a hole through the center of a metal sphere of radius 5 inches, as shown in Figure B.19(a). The hole has a radius of 3 inches. What is the volume of the resulting metal ring?

Solution You can imagine the ring to be generated by a segment of the circle whose equation is

$$x^2 + y^2 = 25$$

as shown in Figure B.19(b). Because the radius of the hole is 3 inches, you can let $y = 3$ and solve the equation $x^2 + y^2 = 25$ to determine that the limits of integration are $x = \pm 4$. So, the inner and outer radii are

$$r(x) = 3 \quad \text{and} \quad R(x) = \sqrt{25 - x^2}$$

and the volume is given by

$$\begin{aligned} V &= \pi \int_a^b \left([R(x)]^2 - [r(x)]^2\right) dx \\ &= \pi \int_{-4}^{4} \left[\left(\sqrt{25 - x^2}\right)^2 - (3)^2\right] dx \\ &= \pi \int_{-4}^{4} (16 - x^2)\, dx \\ &= \pi \left[16x - \frac{x^3}{3}\right]_{-4}^{4} \\ &= \frac{256\pi}{3} \text{ cubic inches.} \end{aligned}$$

■

The Shell Method

Now you will study an alternative method for finding the volume of a solid of revolution. This method is called the **shell method** because it uses cylindrical shells.

To begin, consider a representative rectangle as shown in Figure B.20, where w is the width of the rectangle, h is the height of the rectangle, and p is the distance between the axis of revolution and the *center* of the rectangle. When this rectangle is revolved about its axis of revolution, it forms a cylindrical shell (or tube) of thickness w. To find the volume of this shell, consider two cylinders. The radius of the larger cylinder corresponds to the outer radius of the shell, and the radius of the smaller cylinder corresponds to the inner radius of the shell. Because p is the average radius of the shell, you know the outer radius is $p + (w/2)$ and the inner radius is $p - (w/2)$.

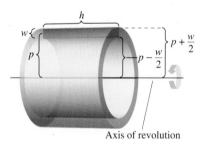

Figure B.20

$$p + \frac{w}{2} \qquad \text{Outer radius}$$

$$p - \frac{w}{2} \qquad \text{Inner radius}$$

So, the volume of the shell is

$$\begin{aligned} \text{Volume of shell} &= (\text{volume of cylinder}) - (\text{volume of hole}) \\ &= \pi\left(p + \frac{w}{2}\right)^2 h - \pi\left(p - \frac{w}{2}\right)^2 h \\ &= 2\pi phw \\ &= 2\pi(\text{average radius})(\text{height})(\text{thickness}). \end{aligned}$$

You can use this formula to find the volume of a solid of revolution. Assume that the plane region in Figure B.21 is revolved about a line to form the indicated solid. If you consider a horizontal rectangle of width Δy, then, as the plane region is revolved about a line parallel to the x-axis, the rectangle generates a representative shell whose volume is

$$\Delta V = 2\pi[p(y)h(y)]\,\Delta y.$$

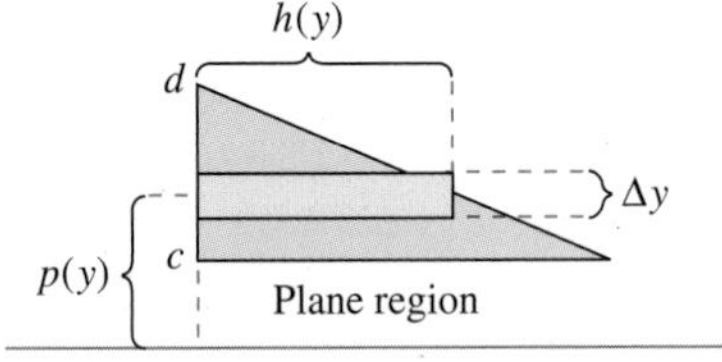

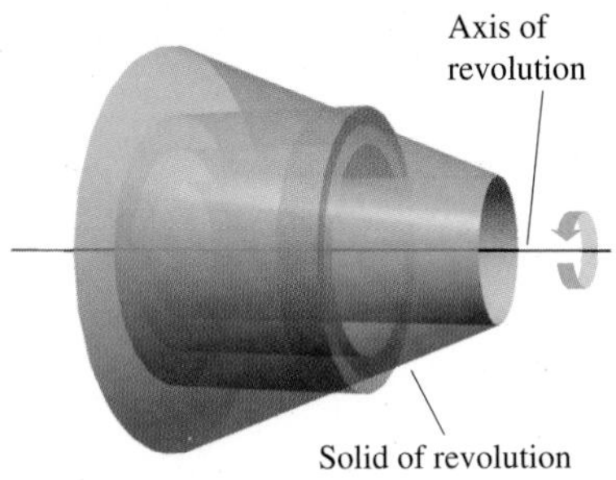

Figure B.21

You can approximate the volume of the solid by n such shells of thickness Δy, height $h(y_i)$, and average radius $p(y_i)$.

$$\text{Volume of solid} \approx \sum_{i=1}^{n} 2\pi[p(y_i)h(y_i)]\,\Delta y$$

$$= 2\pi\sum_{i=1}^{n}[p(y_i)h(y_i)]\,\Delta y$$

This approximation appears to become better and better as $\|\Delta\| \to 0$ $(n \to \infty)$. So, the volume of the solid is

$$\text{Volume of solid} = \lim_{\|\Delta\|\to 0} 2\pi\sum_{i=1}^{n}[p(y_i)h(y_i)]\,\Delta y$$

$$= 2\pi\int_c^d [p(y)h(y)]\,dy.$$

THE SHELL METHOD

To find the volume of a solid of revolution with the **shell method,** use one of the following, as shown in Figure B.22.

Horizontal Axis of Revolution

$$\text{Volume} = V = 2\pi\int_c^d p(y)h(y)\,dy$$

Vertical Axis of Revolution

$$\text{Volume} = V = 2\pi\int_a^b p(x)h(x)\,dx$$

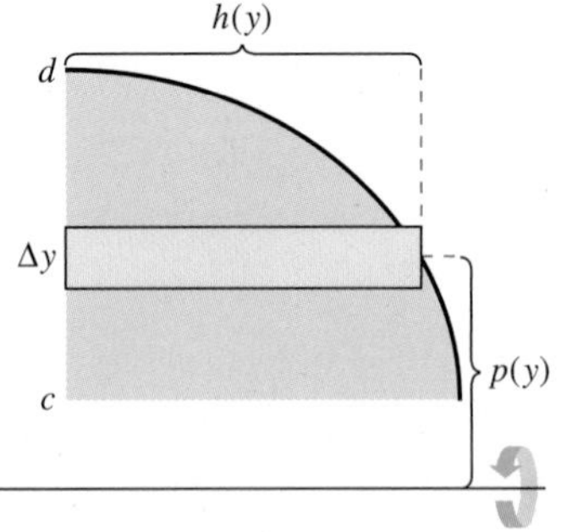

Horizontal axis of revolution

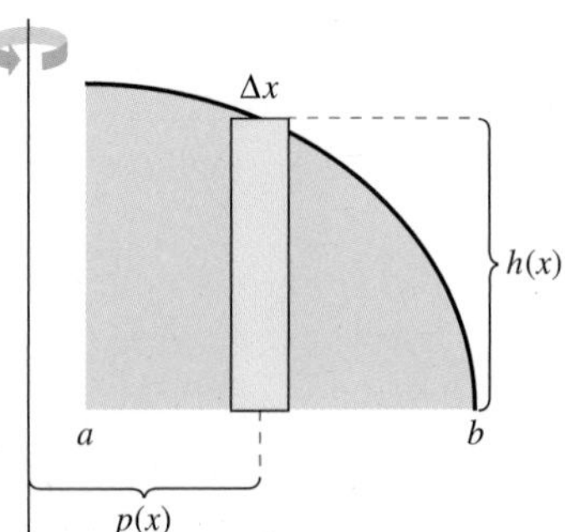

Vertical axis of revolution

Figure B.22

EXAMPLE 7 Using the Shell Method to Find Volume

Find the volume of the solid of revolution formed by revolving the region bounded by

$$y = x - x^3$$

and the x-axis $(0 \leq x \leq 1)$ about the y-axis.

Solution Because the axis of revolution is vertical, use a vertical representative rectangle, as shown in Figure B.23. The width Δx indicates that x is the variable of integration. The distance from the center of the rectangle to the axis of revolution is $p(x) = x$, and the height of the rectangle is

$$h(x) = x - x^3.$$

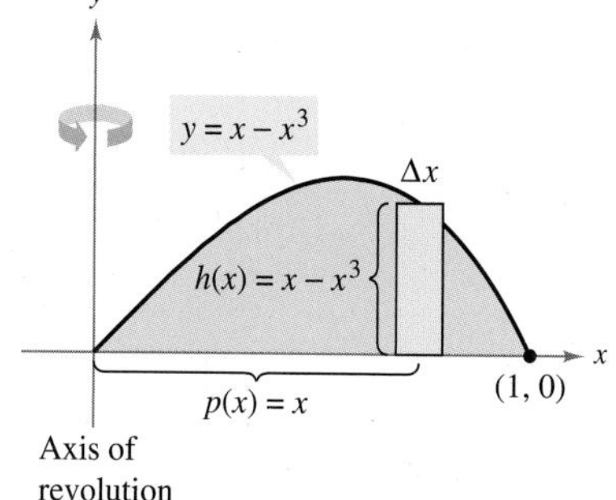

Figure B.23

Because x ranges from 0 to 1, the volume of the solid is

$$\begin{aligned} V = 2\pi \int_a^b p(x)h(x)\,dx &= 2\pi \int_0^1 x(x - x^3)\,dx && \text{Apply shell method.} \\ &= 2\pi \int_0^1 (-x^4 + x^2)\,dx && \text{Simplify.} \\ &= 2\pi \left[-\frac{x^5}{5} + \frac{x^3}{3} \right]_0^1 && \text{Integrate.} \\ &= 2\pi \left(-\frac{1}{5} + \frac{1}{3} \right) = \frac{4\pi}{15}. \end{aligned}$$

EXAMPLE 8 Using the Shell Method to Find Volume

Find the volume of the solid of revolution formed by revolving the region bounded by the graph of

$$x = e^{-y^2}$$

and the y-axis $(0 \leq y \leq 1)$ about the x-axis.

Solution Because the axis of revolution is horizontal, use a horizontal representative rectangle, as shown in Figure B.24. The width Δy indicates that y is the variable of integration. The distance from the center of the rectangle to the axis of revolution is $p(y) = y$, and the height of the rectangle is $h(y) = e^{-y^2}$. Because y ranges from 0 to 1, the volume of the solid is

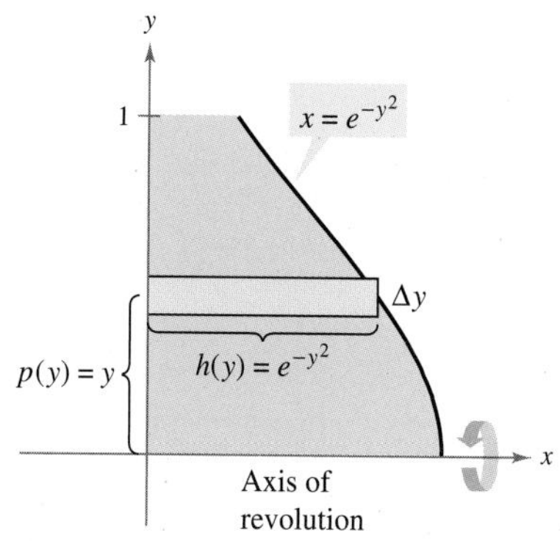

Figure B.24

$$\begin{aligned} V = 2\pi \int_c^d p(y)h(y)\,dy &= 2\pi \int_0^1 ye^{-y^2}\,dy && \text{Apply shell method.} \\ &= -\pi \left[e^{-y^2} \right]_0^1 && \text{Integrate.} \\ &= \pi \left(1 - \frac{1}{e} \right) \approx 1.986. \end{aligned}$$

■

NOTE To see the advantage of using the shell method in Example 8, solve the equation $x = e^{-y^2}$ for y.

$$y = \begin{cases} 1, & 0 \leq x \leq 1/e \\ \sqrt{-\ln x}, & 1/e < x \leq 1 \end{cases}$$

Then use this equation to find the volume using the disk method. ■

Work Done by a Variable Force

If a variable force is applied to an object, calculus is needed to determine the work done, because the amount of force changes as the object changes position. For instance, the force required to compress a spring increases as the spring is compressed.

Suppose that an object is moved along a straight line from $x = a$ to $x = b$ by a continuously varying force $F(x)$. Let Δ be a partition that divides the interval $[a, b]$ into n subintervals determined by

$$a = x_0 < x_1 < x_2 < \cdot \cdot \cdot < x_n = b$$

and let $\Delta x_i = x_i - x_{i-1}$. For each i, choose c_i such that $x_{i-1} \le c_i \le x_i$. Then at c_i the force is given by $F(c_i)$. Because F is continuous, you can approximate the work done in moving the object through the ith subinterval by the increment

$$\Delta W_i = F(c_i)\,\Delta x_i$$

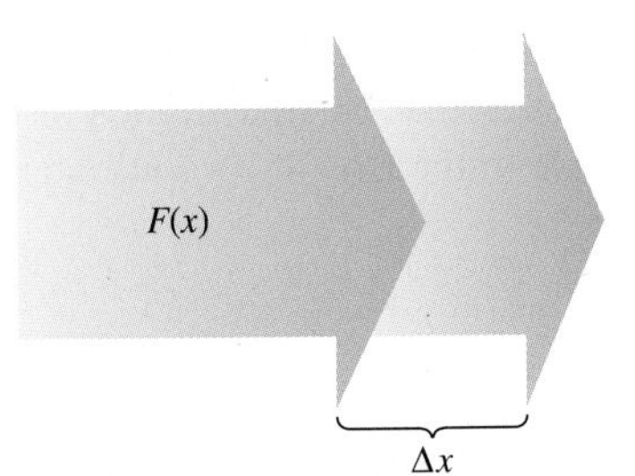

The amount of force changes as an object changes position (Δx).
Figure B.25

as shown in Figure B.25. So, the total work done as the object moves from a to b is approximated by

$$W \approx \sum_{i=1}^{n} \Delta W_i = \sum_{i=1}^{n} F(c_i)\,\Delta x_i.$$

This approximation appears to become better and better as $\|\Delta\| \rightarrow 0$ $(n \rightarrow \infty)$. So, the work done is

$$W = \lim_{\|\Delta\| \rightarrow 0} \sum_{i=1}^{n} F(c_i)\,\Delta x_i = \int_a^b F(x)\,dx.$$

DEFINITION OF WORK DONE BY A VARIABLE FORCE

If an object is moved along a straight line by a continuously varying force $F(x)$, then the **work** W done by the force as the object is moved from $x = a$ to $x = b$ is

$$W = \lim_{\|\Delta\| \rightarrow 0} \sum_{i=1}^{n} \Delta W_i = \int_a^b F(x)\,dx.$$

Bettmann/Corbis

EMILIE DE BRETEUIL (1706–1749)

Another major work by Breteuil was the translation of Newton's "Philosophiae Naturalis Principia Mathematica" into French. Her translation and commentary greatly contributed to the acceptance of Newtonian science in Europe.

The remaining examples in this section use some well-known physical laws. The discoveries of many of these laws occurred during the same period in which calculus was being developed. In fact, during the seventeenth and eighteenth centuries, there was little difference between physicists and mathematicians. One such physicist-mathematician was Emilie de Breteuil. Breteuil was instrumental in synthesizing the work of many other scientists, including Newton, Leibniz, Huygens, Kepler, and Descartes. Her physics text *Institutions* was widely used for many years.

The following three laws of physics were developed by Robert Hooke (1635–1703), Isaac Newton (1642–1727), and Charles Coulomb (1736–1806).

1. **Hooke's Law:** The force F required to compress or stretch a spring (within its elastic limits) is proportional to the distance d that the spring is compressed or stretched from its original length. That is,

$$F = kd$$

where the constant of proportionality k (the spring constant) depends on the specific nature of the spring.

2. **Newton's Law of Universal Gravitation:** The force F of attraction between two particles of masses m_1 and m_2 is proportional to the product of the masses and inversely proportional to the square of the distance d between the two particles. That is,

$$F = k\frac{m_1 m_2}{d^2}.$$

If m_1 and m_2 are given in grams and d in centimeters, F will be in dynes for a value of $k = 6.670 \times 10^{-8}$ cubic centimeter per gram-second squared.

3. **Coulomb's Law:** The force F between two charges q_1 and q_2 in a vacuum is proportional to the product of the charges and inversely proportional to the square of the distance d between the two charges. That is,

$$F = k\frac{q_1 q_2}{d^2}.$$

If q_1 and q_2 are given in electrostatic units and d in centimeters, F will be in dynes for a value of $k = 1$.

EXPLORATION

The work done in compressing the spring in Example 9 from $x = 3$ inches to $x = 6$ inches is 3375 inch-pounds. Should the work done in compressing the spring from $x = 0$ inches to $x = 3$ inches be more than, the same as, or less than this? Explain.

EXAMPLE 9 Compressing a Spring

A force of 750 pounds compresses a spring 3 inches from its natural length of 15 inches. Find the work done in compressing the spring an additional 3 inches.

Solution By Hooke's Law, the force $F(x)$ required to compress the spring x units (from its natural length) is $F(x) = kx$. Using the given data, it follows that $F(3) = 750 = (k)(3)$ and so $k = 250$ and $F(x) = 250x$, as shown in Figure B.26. To find the increment of work, assume that the force required to compress the spring over a small increment Δx is nearly constant. So, the increment of work is

$$\Delta W = (\text{force})(\text{distance increment}) = (250x)\,\Delta x.$$

Because the spring is compressed from $x = 3$ to $x = 6$ inches less than its natural length, the work required is

$$W = \int_a^b F(x)\,dx = \int_3^6 250x\,dx \qquad \text{Formula for work}$$

$$= 125x^2\Big]_3^6 = 4500 - 1125 = 3375 \text{ inch-pounds.}$$

Note that you do *not* integrate from $x = 0$ to $x = 6$ because you were asked to determine the work done in compressing the spring an *additional* 3 inches (not including the first 3 inches). ■

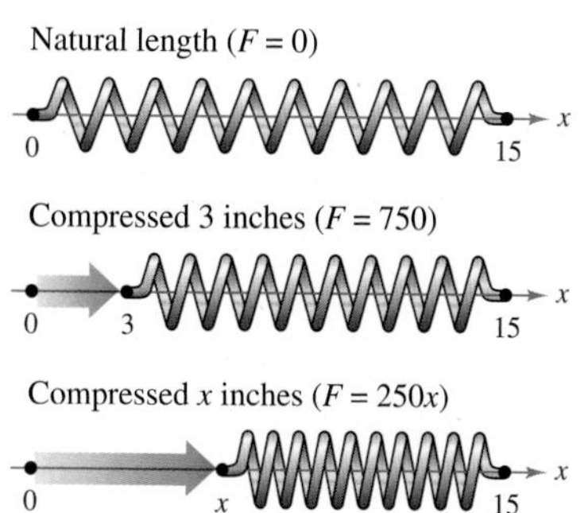

Figure B.26

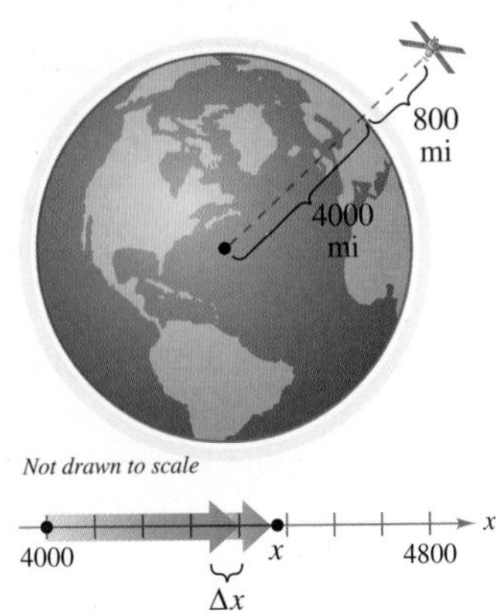

Figure B.27

EXAMPLE 10 Moving a Space Module into Orbit

A space module weighs 15 metric tons on the surface of Earth. How much work is done in propelling the module to a height of 800 miles above Earth, as shown in Figure B.27? (Use 4000 miles as the radius of Earth. Do not consider the effect of air resistance or the weight of the propellant.)

Solution Because the weight of a body varies inversely as the square of its distance from the center of Earth, the force $F(x)$ exerted by gravity is

$$F(x) = \frac{C}{x^2}. \qquad C \text{ is the constant of proportionality.}$$

Because the module weighs 15 metric tons on the surface of Earth and the radius of Earth is approximately 4000 miles, you have

$$15 = \frac{C}{(4000)^2}$$

$$240{,}000{,}000 = C.$$

So, the increment of work is

$$\begin{aligned}\Delta W &= \text{(force)(distance increment)}\\ &= \frac{240{,}000{,}000}{x^2}\Delta x.\end{aligned}$$

Finally, because the module is propelled from $x = 4000$ to $x = 4800$ miles, the total work done is

$$\begin{aligned}W = \int_a^b F(x)\,dx &= \int_{4000}^{4800} \frac{240{,}000{,}000}{x^2}\,dx && \text{Formula for work}\\ &= \frac{-240{,}000{,}000}{x}\Bigg]_{4000}^{4800} && \text{Integrate.}\\ &= -50{,}000 + 60{,}000\\ &= 10{,}000 \text{ mile-tons}\\ &\approx 1.164 \times 10^{11} \text{ foot-pounds.}\end{aligned}$$

In the C-G-S system, using a conversion factor of 1 foot-pound $\approx$ 1.35582 joules, the work done is

$$W \approx 1.578 \times 10^{11} \text{ joules.}$$

■

The solutions to Examples 9 and 10 conform to our development of work as the summation of increments in the form

$$\Delta W = \text{(force)(distance increment)} = (F)(\Delta x).$$

Another way to formulate the increment of work is

$$\Delta W = \text{(force increment)(distance)} = (\Delta F)(x).$$

This second interpretation of ΔW is useful in problems involving the movement of nonrigid substances such as fluids and chains.

EXAMPLE 11 Emptying a Tank of Oil

A spherical tank of radius 8 feet is half full of oil that weighs 50 pounds per cubic foot. Find the work required to pump oil out through a hole in the top of the tank.

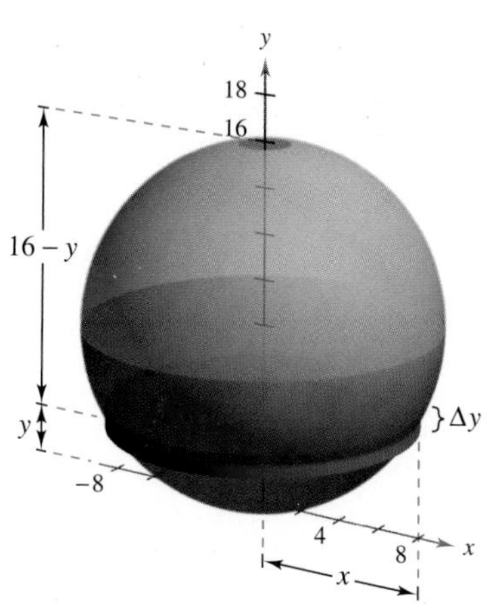

Figure B.28

Solution Consider the oil to be subdivided into disks of thickness Δy and radius x, as shown in Figure B.28. Because the increment of force for each disk is given by its weight, you have

$$\begin{aligned}\Delta F &= \text{weight}\\ &= \left(\frac{50 \text{ pounds}}{\text{cubic foot}}\right)(\text{volume})\\ &= 50(\pi x^2 \Delta y) \text{ pounds.}\end{aligned}$$

For a circle of radius 8 and center at (0, 8), you have

$$\begin{aligned}x^2 + (y-8)^2 &= 8^2\\ x^2 &= 16y - y^2\end{aligned}$$

and you can write the force increment as

$$\begin{aligned}\Delta F &= 50(\pi x^2 \Delta y)\\ &= 50\pi(16y - y^2)\,\Delta y.\end{aligned}$$

In Figure B.28, note that a disk y feet from the bottom of the tank must be moved a distance of $(16 - y)$ feet. So, the increment of work is

$$\begin{aligned}\Delta W &= \Delta F(16 - y)\\ &= 50\pi(16y - y^2)\,\Delta y(16 - y)\\ &= 50\pi(256y - 32y^2 + y^3)\,\Delta y.\end{aligned}$$

Because the tank is half full, y ranges from 0 to 8, and the work required to empty the tank is

$$\begin{aligned}W &= \int_0^8 50\pi(256y - 32y^2 + y^3)\,dy\\ &= 50\pi\left[128y^2 - \frac{32}{3}y^3 + \frac{y^4}{4}\right]_0^8\\ &= 50\pi\left(\frac{11{,}264}{3}\right)\\ &\approx 589{,}782 \text{ foot-pounds.}\end{aligned}$$

■

To estimate the reasonableness of the result in Example 11, consider that the weight of the oil in the tank is

$$\begin{aligned}\left(\frac{1}{2}\right)(\text{volume})(\text{density}) &= \frac{1}{2}\left(\frac{4}{3}\pi 8^3\right)(50)\\ &\approx 53{,}616.5 \text{ pounds.}\end{aligned}$$

Lifting the entire half-tank of oil 8 feet would involve work of $8(53{,}616.5) \approx 428{,}932$ foot-pounds. Because the oil is actually lifted between 8 and 16 feet, it seems reasonable that the work done is 589,782 foot-pounds.

Work required to raise one end of the chain
Figure B.29

EXAMPLE 12 Lifting a Chain

A 20-foot chain weighing 5 pounds per foot is lying coiled on the ground. How much work is required to raise one end of the chain to a height of 20 feet so that it is fully extended, as shown in Figure B.29?

Solution Imagine that the chain is divided into small sections, each of length Δy. Then the weight of each section is the increment of force

$$\Delta F = (\text{weight}) = \left(\frac{5 \text{ pounds}}{\text{foot}}\right)(\text{length}) = 5\Delta y.$$

Because a typical section (initially on the ground) is raised to a height of y, the increment of work is

$$\Delta W = (\text{force increment})(\text{distance}) = (5\,\Delta y)y = 5y\,\Delta y.$$

Because y ranges from 0 to 20, the total work is

$$W = \int_0^{20} 5y\,dy = \frac{5y^2}{2}\bigg]_0^{20} = \frac{5(400)}{2} = 1000 \text{ foot-pounds.}$$ ■

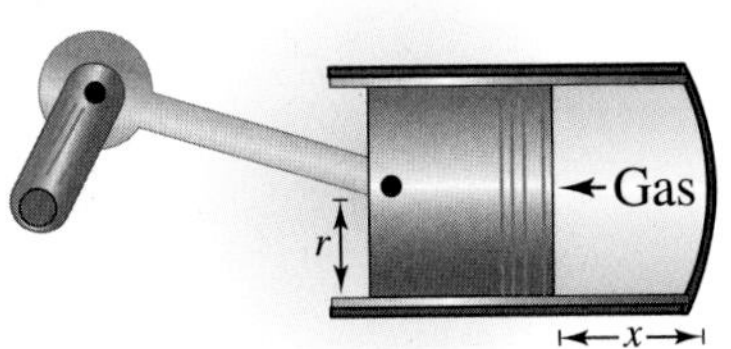

Work done by expanding gas
Figure B.30

In the next example you will consider a piston of radius r in a cylindrical casing, as shown in Figure B.30. As the gas in the cylinder expands, the piston moves and work is done. If p represents the pressure of the gas (in pounds per square foot) against the piston head and V represents the volume of the gas (in cubic feet), the work increment involved in moving the piston Δx feet is

$$\Delta W = (\text{force})(\text{distance increment}) = F(\Delta x) = p(\pi r^2)\,\Delta x = p\,\Delta V.$$

So, as the volume of the gas expands from V_0 to V_1, the work done in moving the piston is

$$W = \int_{V_0}^{V_1} p\,dV.$$

Assuming the pressure of the gas to be inversely proportional to its volume, you have $p = k/V$ and the integral for work becomes

$$W = \int_{V_0}^{V_1} \frac{k}{V}\,dV.$$

EXAMPLE 13 Work Done by an Expanding Gas

A quantity of gas with an initial volume of 1 cubic foot and a pressure of 500 pounds per square foot expands to a volume of 2 cubic feet. Find the work done by the gas. (Assume that the pressure is inversely proportional to the volume.)

Solution Because $p = k/V$ and $p = 500$ when $V = 1$, you have $k = 500$. So, the work is

$$\begin{aligned} W &= \int_{V_0}^{V_1} \frac{k}{V}\,dV \\ &= \int_1^2 \frac{500}{V}\,dV \\ &= 500 \ln|V|\bigg]_1^2 \approx 346.6 \text{ foot-pounds.} \end{aligned}$$ ■

Answers to Odd-Numbered Exercises

Chapter P

Section P.1 (page 12)

1. equation **3.** extraneous **5.** Identity **7.** Identity
9. Conditional equation **11.** 4 **13.** -9 **15.** 5
17. 1 **19.** No solution **21.** $-\frac{96}{23}$ **23.** $-\frac{6}{5}$
25. No solution. The x-terms sum to zero, but the constant terms do not.
27. 10 **29.** 4 **31.** 0
33. No solution. The solution is extraneous.
35. No solution. The solution is extraneous.
37. 0 **39.** $2x^2 + 8x - 3 = 0$ **41.** $3x^2 - 90x - 10 = 0$
43. $0, -\frac{1}{2}$ **45.** $4, -2$ **47.** $5, 7$ **49.** $3, -\frac{1}{2}$
51. $2, -6$ **53.** $-a$ **55.** ± 7 **57.** $\pm 3\sqrt{3}$
59. 8, 16 **61.** $-2 \pm \sqrt{14}$ **63.** $\dfrac{1 \pm 3\sqrt{2}}{2}$
65. 2 **67.** $4, -8$ **69.** $\sqrt{11} - 6, -\sqrt{11} - 6$
71. $2 \pm 2\sqrt{3}$ **73.** $\dfrac{-5 \pm \sqrt{89}}{4}$ **75.** $\dfrac{15 \pm \sqrt{85}}{10}$
77. $\frac{1}{2}, -1$ **79.** $1 \pm \sqrt{3}$ **81.** $-7 \pm \sqrt{5}$
83. $\dfrac{2}{3} \pm \dfrac{\sqrt{7}}{3}$ **85.** $-\dfrac{4}{3}$ **87.** $\dfrac{2}{7}$ **89.** $2 \pm \dfrac{\sqrt{6}}{2}$
91. $6 \pm \sqrt{11}$ **93.** $-3.449, 1.449$
95. $1.687, -0.488$ **97.** $1 \pm \sqrt{2}$
99. $6, -12$ **101.** $\frac{1}{2} \pm \sqrt{3}$ **103.** ± 1 **105.** $0, \pm 5$
107. ± 3 **109.** -6 **111.** $3, 1, -1$ **113.** ± 1
115. $\pm\sqrt{3}, \pm 1$ **117.** $1, -2$ **119.** 50 **121.** 26
123. No solution **125.** $-\frac{513}{2}$ **127.** 6, 7
129. 10 **131.** $-3 \pm 5\sqrt{5}$ **133.** 1 **135.** $2, -\frac{3}{2}$
137. $4, -5$ **139.** $3, -2$ **141.** $\sqrt{3}, -3$
143. $3, \dfrac{-1 - \sqrt{17}}{2}$ **145.** Answers will vary.
147. Equivalent equations have the same solution set, and one is derived from the other by steps for generating equivalent equations.
$2x = 5,\ 2x + 3 = 8$
149. 61.2 in.
151. (a) 1998 (b) 2011; Answers will vary.
153. False. See Example 14 on page 11.
155. $x^2 - 3x - 18 = 0$ **157.** $x^2 - 2x - 1 = 0$
159. Sample answer: $a = 9, b = 9$
161. (a) $x = 0, -\dfrac{b}{a}$ (b) $x = 0, 1$

Section P.2 (page 23)

1. solution set **3.** negative **5.** key; test intervals
7. (a) $0 \le x < 9$ (b) Bounded
9. (a) $-1 \le x \le 5$ (b) Bounded
11. (a) $x > 11$ (b) Unbounded
13. (a) $x < -2$ (b) Unbounded
15. b **16.** h **17.** e **18.** d
19. f **20.** a **21.** g **22.** c
23. (a) Yes (b) No (c) Yes (d) No
25. (a) Yes (b) No (c) No (d) Yes
27. (a) Yes (b) Yes (c) Yes (d) No
29. $x < 3$

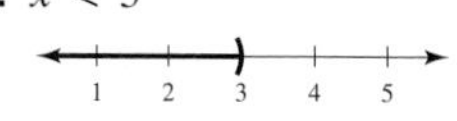

31. $x < \frac{3}{2}$

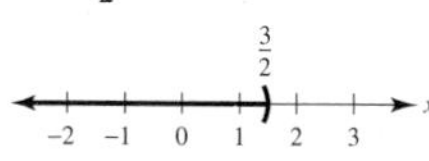

33. $x \ge 12$ 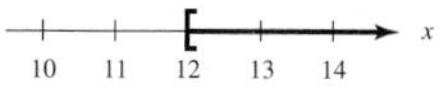
35. $x > 2$ 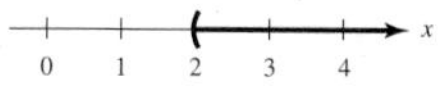
37. $x \ge \frac{2}{7}$ 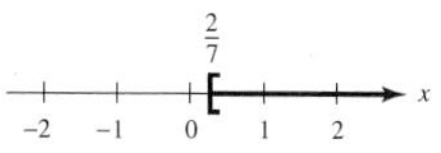
39. $x < 5$
41. $x \ge 4$ 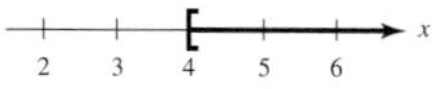
43. $x \ge 2$ 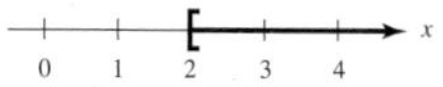
45. $x \ge -4$ 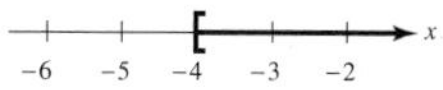
47. $-1 < x < 3$
49. $-2 < x \le 5$
51. $-\frac{9}{2} < x < \frac{15}{2}$
53. $-\frac{3}{4} < x < -\frac{1}{4}$
55. $10.5 \le x \le 13.5$ 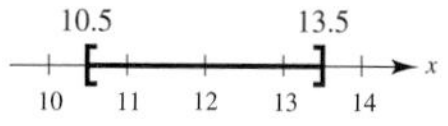
57. $-5 < x < 5$
59. $x < -2, x > 2$ 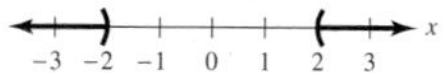
61. No solution
63. $14 \le x \le 26$ 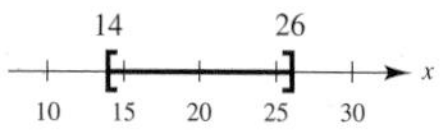
65. $x \le -\frac{3}{2}, x \ge 3$ 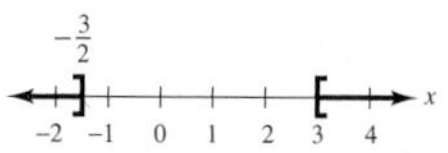
67. $x \le -5,\ x \ge 11$ 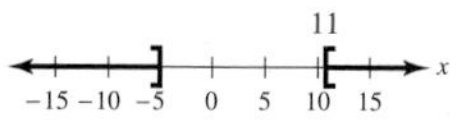
69. $4 < x < 5$
71. $x \le -\frac{29}{2},\ x \ge -\frac{11}{2}$ 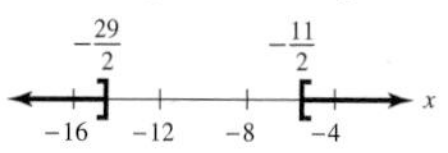
73. 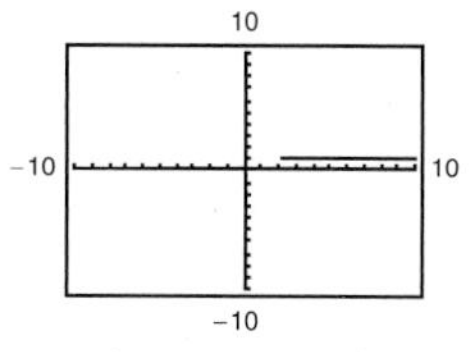
$x > 2$
75.

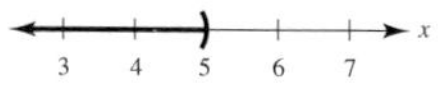

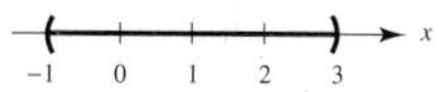

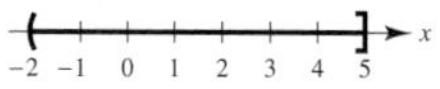

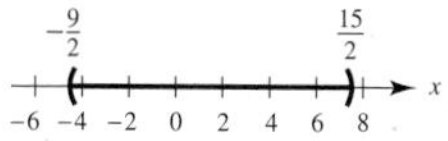

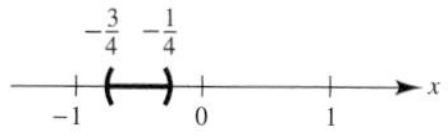

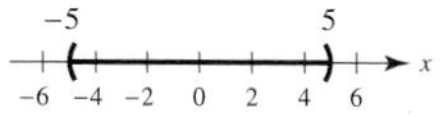

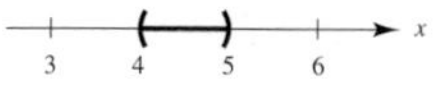

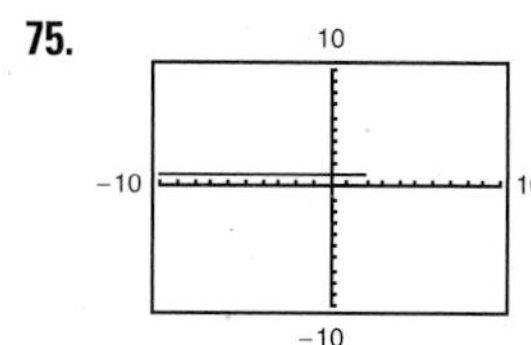

$x \le 2$

77.

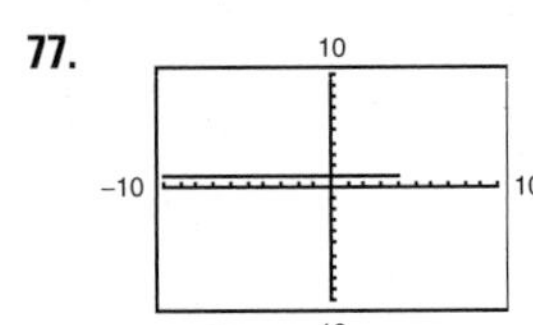

$x \le 4$

79.

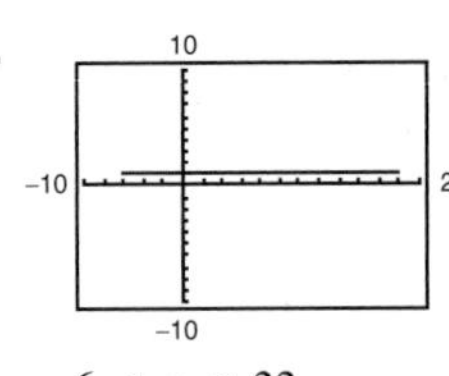

$-6 \le x \le 22$

81.

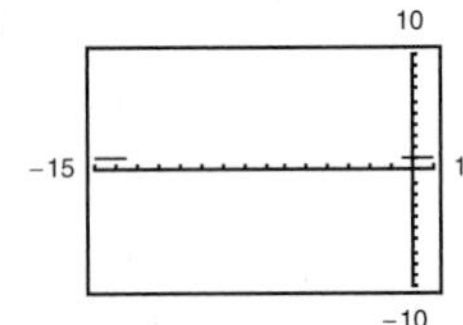

$x \le -\frac{27}{2}, x \ge -\frac{1}{2}$

83. $[5, \infty)$ **85.** $[-3, \infty)$ **87.** $\left(-\infty, \frac{7}{2}\right]$

89. All real numbers within eight units of 10

91. $|x| \le 3$ **93.** $|x - 7| \ge 3$ **95.** $|x - 12| < 10$

97. $|x + 3| > 4$

99. (a) No (b) Yes (c) Yes (d) No

101. (a) Yes (b) No (c) No (d) Yes

103. $-\frac{2}{3}, 1$ **105.** 4, 5

107. $(-3, 3)$

109. $[-7, 3]$

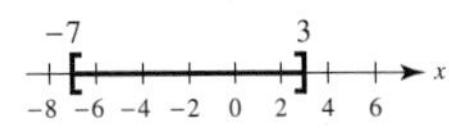

111. $(-\infty, -5] \cup [1, \infty)$

113. $(-3, 2)$

115. $(-3, 1)$

117. $\left(-\infty, -\frac{4}{3}\right) \cup (5, \infty)$

119. $(-\infty, -3) \cup (6, \infty)$

121. $(-1, 1) \cup (3, \infty)$

123. $x = \frac{1}{2}$

125. $(-\infty, 0) \cup \left(0, \frac{3}{2}\right)$ **127.** $[-2, 0] \cup [2, \infty)$

129. $[-2, \infty)$

131. $(-\infty, 0) \cup \left(\frac{1}{4}, \infty\right)$

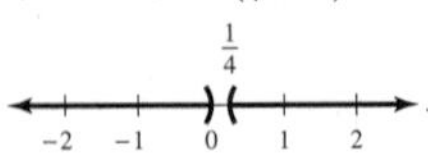

133. $\left(-\infty, \frac{5}{3}\right] \cup (5, \infty)$

135. $(-\infty, -1) \cup (4, \infty)$

137. $(-5, 3) \cup (11, \infty)$

139. $\left(-\frac{3}{4}, 3\right) \cup [6, \infty)$

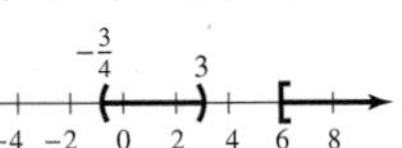

141. $(-3, -2] \cup [0, 3)$

143. $(-\infty, -1) \cup (1, \infty)$

145. $[-2, 2]$ **147.** $(-\infty, 4] \cup [5, \infty)$

149. $(-5, 0] \cup (7, \infty)$ **151.** $(-3.51, 3.51)$

153. $(-0.13, 25.13)$ **155.** $(2.26, 2.39)$

157. b **159.** $a = k, b = 5k, c = 5k, k \ge 0$

161. $9.00 + 0.75x > 13.50;\ x > 6$ **163.** $r > 3.125\%$

165. (a) $1.32 \le t \le 7.89$ (Between 1991 and 1997)
(b) $t > 21.05$ (2011)

167. $65.8 \le h \le 71.2$

169. $13.8 \text{ m} \le L \le 36.2 \text{ m}$ **171.** $r > 4.88\%$

173. (a) $t = 10$ sec (b) 4 sec $< t <$ 6 sec

175. $R_1 \ge 2$ ohms

177. False. c has to be greater than zero.

179. True. The y-values are greater than zero for all values of x.

Section P.3 (page 33)

1. (a) v (b) vi (c) i (d) iv (e) iii (f) ii

3. Distance Formula

5. A: (2, 6), B: $(-6, -2)$, C: $(4, -4)$, D: $(-3, 2)$

7.

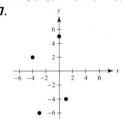

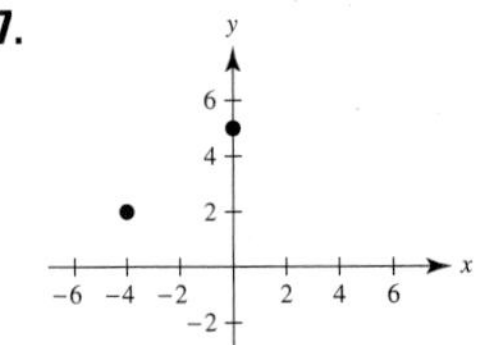

9.

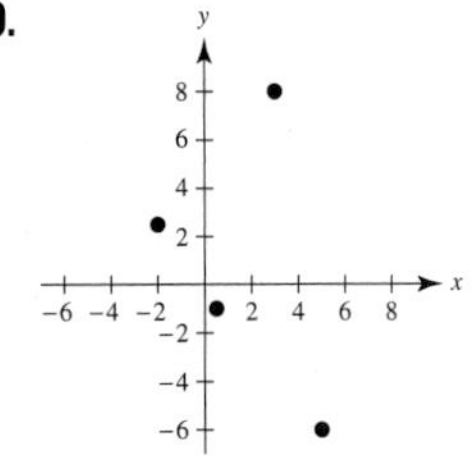

11. $(-3, 4)$ **13.** $(-5, -5)$ **15.** Quadrant IV

17. Quadrant II **19.** Quadrant III or IV **21.** Quadrant III

23. Quadrant I or III

25.

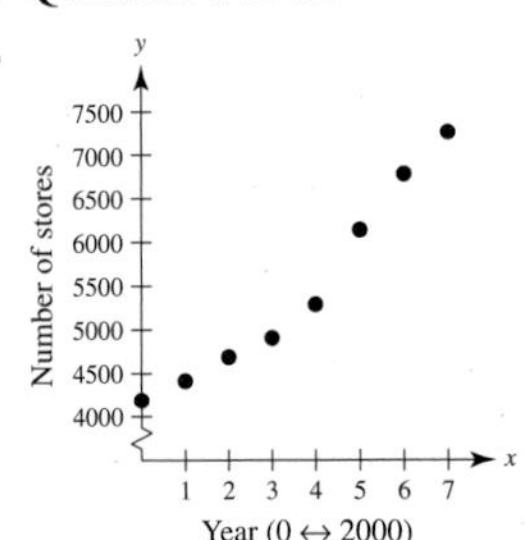

27. 8 **29.** 5 **31.** 13 **33.** $\sqrt{61}$ **35.** $\frac{\sqrt{277}}{6}$

37. 8.47 **39.** (a) 4, 3, 5 (b) $4^2 + 3^2 = 5^2$

41. (a) 10, 3, $\sqrt{109}$ (b) $10^2 + 3^2 = \left(\sqrt{109}\right)^2$

43. $\left(\sqrt{5}\right)^2 + \left(\sqrt{45}\right)^2 = \left(\sqrt{50}\right)^2$

45. Distances between the points: $\sqrt{29}, \sqrt{58}, \sqrt{29}$

47. (a)

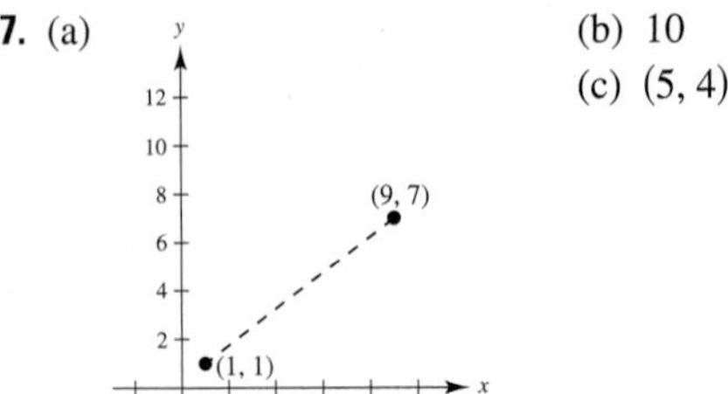

(b) 10
(c) (5, 4)

49. (a)

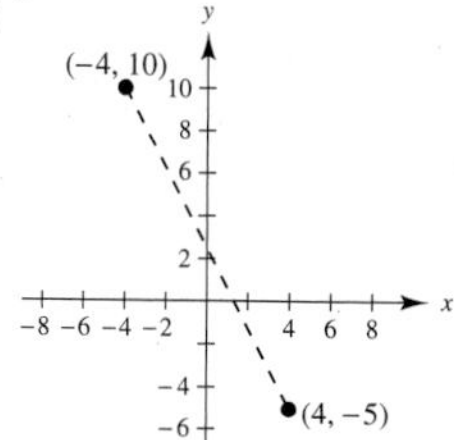

(b) 17
(c) $\left(0, \frac{5}{2}\right)$

51. (a)

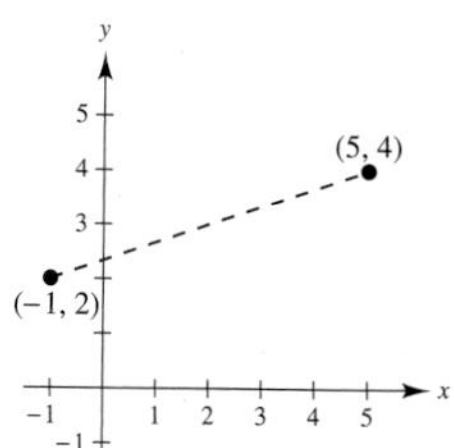

(b) $2\sqrt{10}$
(c) (2, 3)

53. (a)

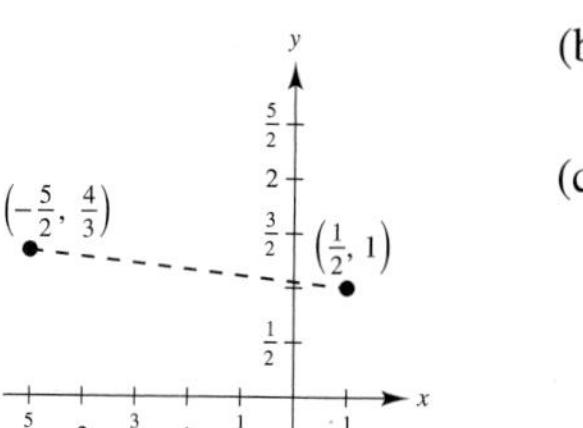

(b) $\dfrac{\sqrt{82}}{3}$
(c) $\left(-1, \dfrac{7}{6}\right)$

55. (a)

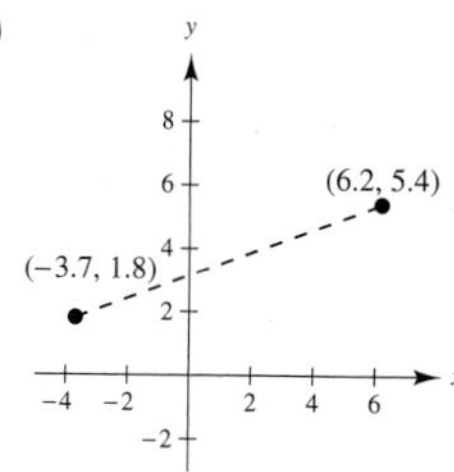

(b) $\sqrt{110.97}$
(c) (1.25, 3.6)

57. $30\sqrt{41} \approx 192$ km **59.** \$4415 million
61. (0, 1), (4, 2), (1, 4) **63.** (−3, 6), (2, 10), (2, 4), (−3, 4)
65. (−1, 5), (−5, 4), (−2, 2)
67. (0, 3), (−3, −2), (−6, 3), (−3, 8)
69. $2\sqrt{5}, 3\sqrt{5}, \sqrt{65}; (2\sqrt{5})^2 + (3\sqrt{5})^2 = (\sqrt{65})^2$
71. The y-coordinate of any point on the x-axis is 0. The x-coordinate of any point on the y-axis is 0.
73. \$3.87/gal; 2007
75. (a) About 9.6% (b) About 28.6%
77. The number of performers elected each year seems to be nearly steady except for the middle years. Five performers will be elected in 2010.
79. \$24,331 million
81. (a)

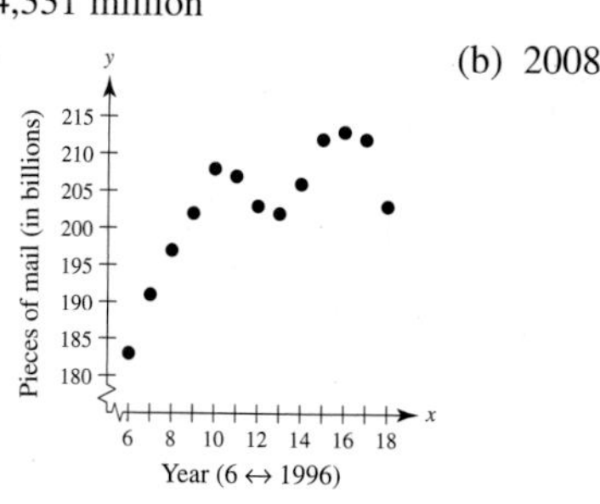

(b) 2008

(c) Answers will vary. Sample answer: Technology now enables us to transport information in many ways other than by mail. The Internet is one example.
83. $(2x_m - x_1, 2y_m - y_1)$
85. $\left(\dfrac{3x_1 + x_2}{4}, \dfrac{3y_1 + y_2}{4}\right), \left(\dfrac{x_1 + x_2}{2}, \dfrac{y_1 + y_2}{2}\right), \left(\dfrac{x_1 + 3x_2}{4}, \dfrac{y_1 + 3y_2}{4}\right)$
87. No. It depends on the magnitudes of the quantities measured.
89. False. The Midpoint Formula would be used 15 times.
91. Use the Midpoint Formula to prove that the diagonals of the parallelogram bisect each other.

$$\left(\frac{b + a}{2}, \frac{c + 0}{2}\right) = \left(\frac{a + b}{2}, \frac{c}{2}\right)$$
$$\left(\frac{a + b + 0}{2}, \frac{c + 0}{2}\right) = \left(\frac{a + b}{2}, \frac{c}{2}\right)$$

Section P.4 (page 46)

1. solution or solution point **3.** intercepts
5. circle; (h, k); r
7. (a) Yes (b) Yes **9.** (a) Yes (b) No
11. (a) Yes (b) No **13.** (a) No (b) Yes
15.

x	−1	0	1	2	$\frac{5}{2}$
y	7	5	3	1	0
(x, y)	(−1, 7)	(0, 5)	(1, 3)	(2, 1)	$\left(\frac{5}{2}, 0\right)$

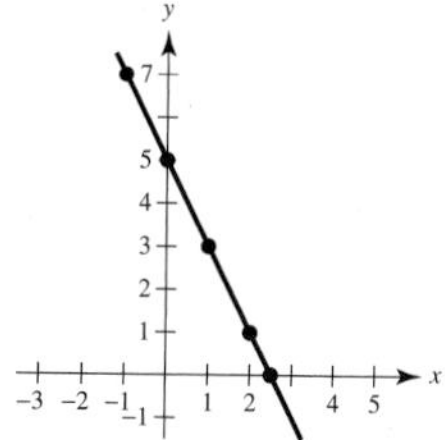

17.

x	−1	0	1	2	3
y	4	0	−2	−2	0
(x, y)	(−1, 4)	(0, 0)	(1, −2)	(2, −2)	(3, 0)

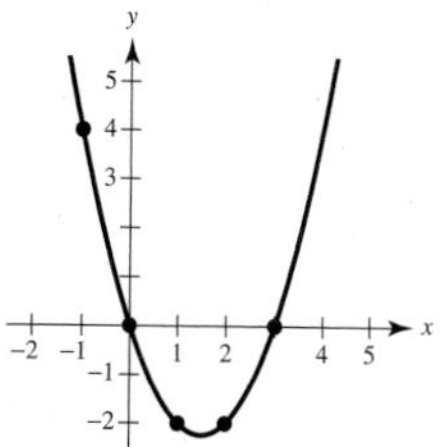

19. x-intercept: (3, 0)
y-intercept: (0, 9)
21. x-intercept: (−2, 0)
y-intercept: (0, 2)
23. x-intercept: $\left(\frac{6}{5}, 0\right)$
y-intercept: (0, −6)
25. x-intercept: (−4, 0)
y-intercept: (0, 2)
27. x-intercept: $\left(\frac{7}{3}, 0\right)$
y-intercept: (0, 7)
29. x-intercepts: (0, 0), (2, 0)
y-intercept: (0, 0)

31. x-intercept: $(6, 0)$
y-intercepts: $(0, \pm\sqrt{6})$

33. y-axis symmetry

35. Origin symmetry

37. Origin symmetry

39. x-axis symmetry

41. x-intercept: $\left(\frac{1}{3}, 0\right)$
y-intercept: $(0, 1)$
No symmetry

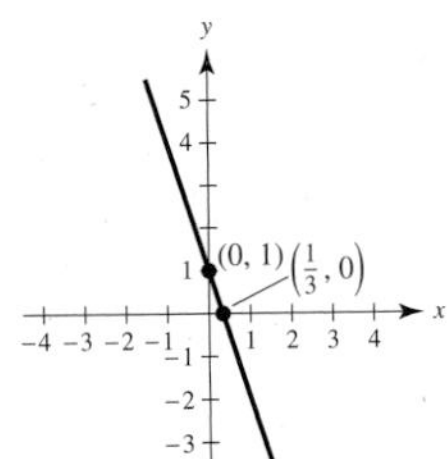

43. x-intercepts: $(0, 0)$, $(2, 0)$
y-intercept: $(0, 0)$
No symmetry

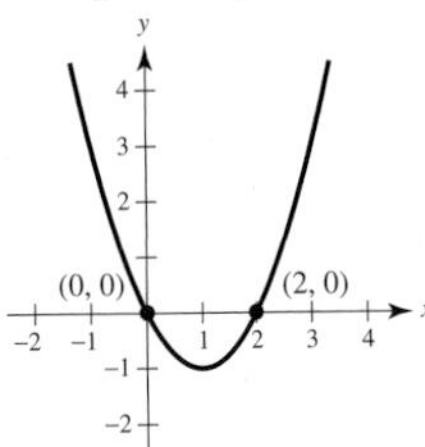

45. x-intercept: $\left(\sqrt[3]{-3}, 0\right)$
y-intercept: $(0, 3)$
No symmetry

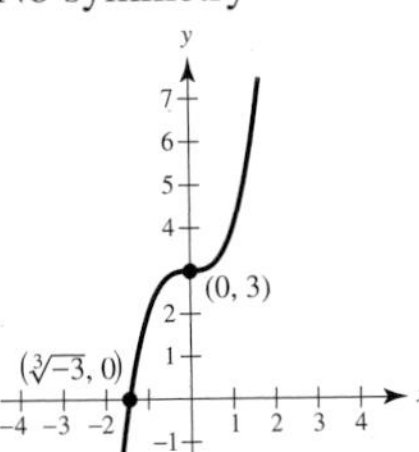

47. x-intercept: $(3, 0)$
y-intercept: None
No symmetry

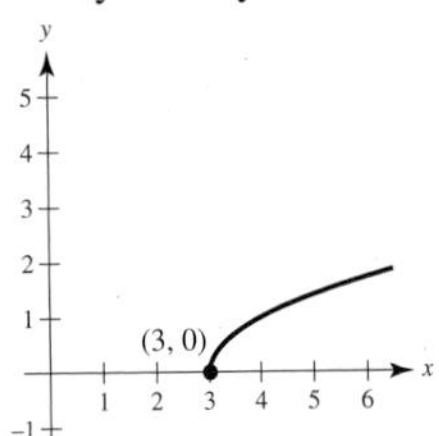

49. x-intercept: $(6, 0)$
y-intercept: $(0, 6)$
No symmetry

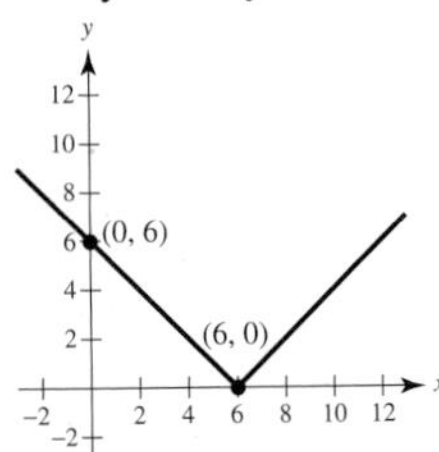

51. x-intercept: $(-1, 0)$
y-intercepts: $(0, \pm 1)$
x-axis symmetry

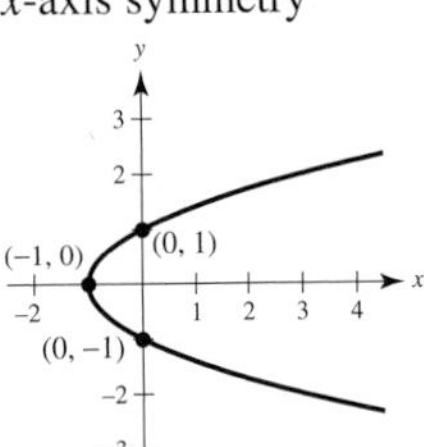

53.

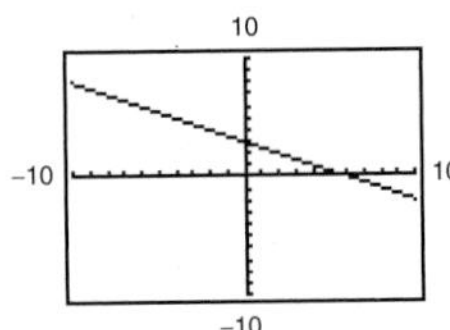

Intercepts: $(6, 0)$, $(0, 3)$

55.

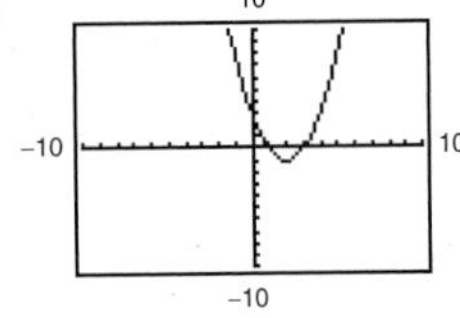

Intercepts: $(3, 0)$, $(1, 0)$, $(0, 3)$

57.

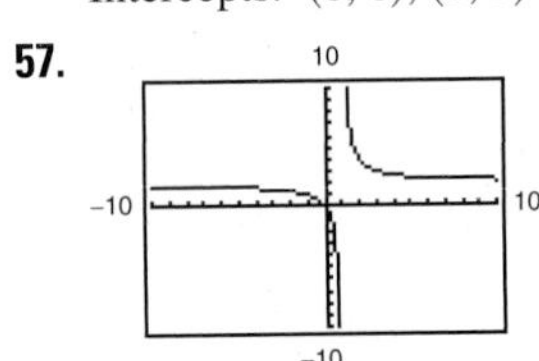

Intercept: $(0, 0)$

59.

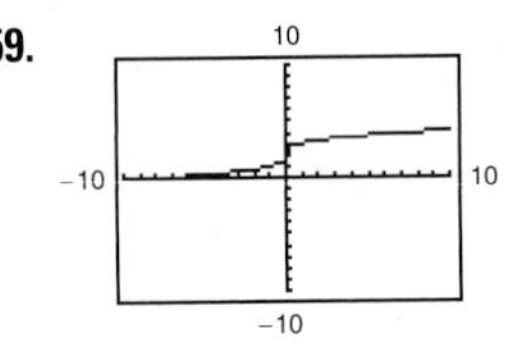

Intercepts: $(-8, 0)$, $(0, 2)$

61.

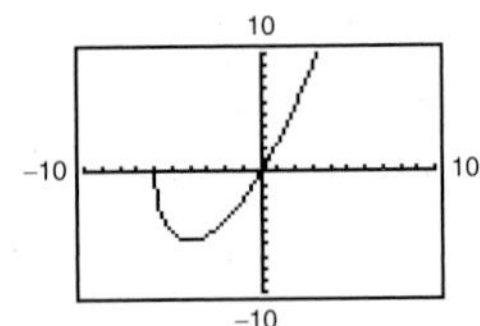

Intercepts: $(0, 0)$, $(-6, 0)$

63.

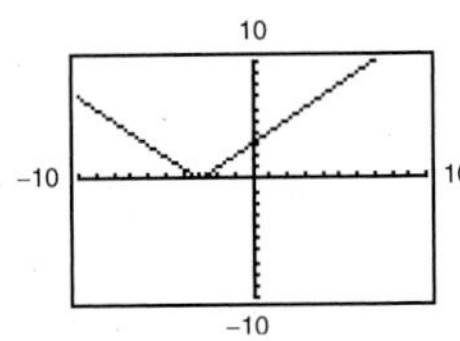

Intercepts: $(-3, 0)$, $(0, 3)$

65. $x^2 + y^2 = 16$

67. $(x - 2)^2 + (y + 1)^2 = 16$

69. $(x + 1)^2 + (y - 2)^2 = 5$

71. $(x - 3)^2 + (y - 4)^2 = 25$

73. Center: $(0, 0)$; Radius: 5

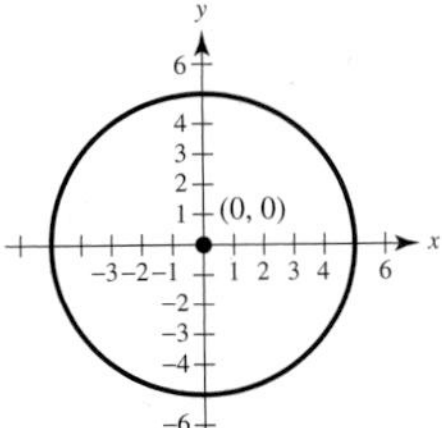

75. Center: $(1, -3)$; Radius: 3

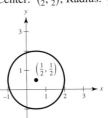

77. Center: $\left(\frac{1}{2}, \frac{1}{2}\right)$; Radius: $\frac{3}{2}$

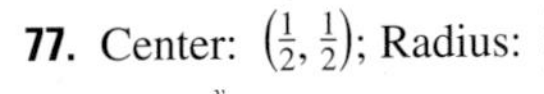

79.

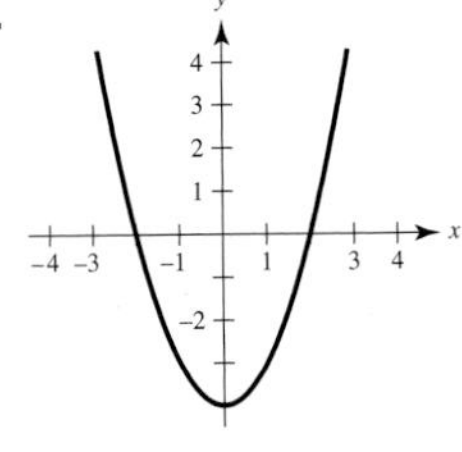

81.

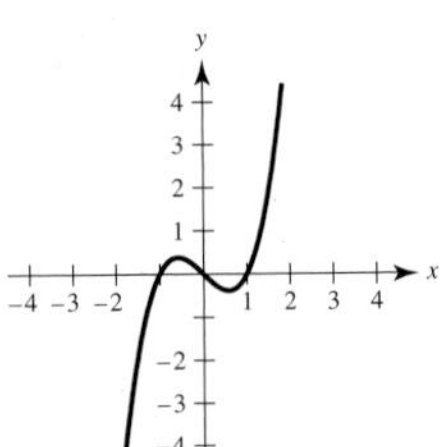

83. Answers will vary.
Sample answer:
$y = x^3 - 8x^2 + 4x + 48$

85. (a)

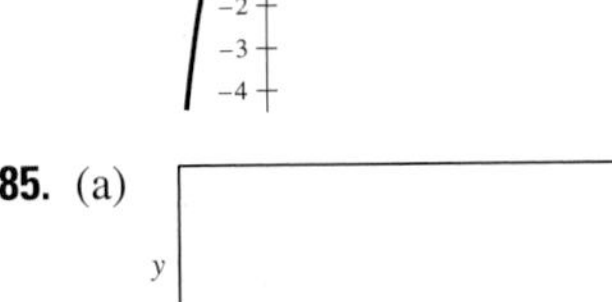

(b) Answers will vary.

(c)

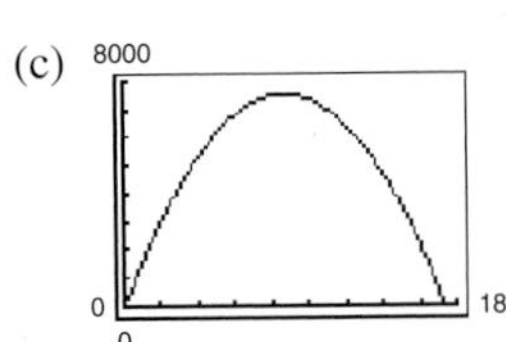

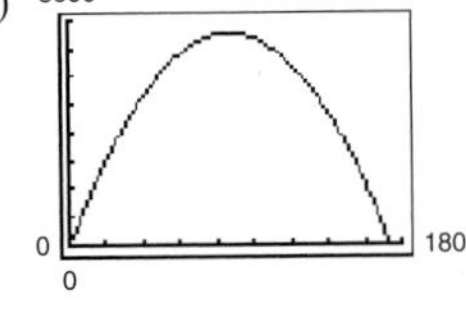

(d) $x = 86\frac{2}{3}$, $y = 86\frac{2}{3}$

(e) A regulation NFL playing field is 120 yards long and $53\frac{1}{3}$ yards wide. The actual area is 6400 square yards.

87. (a)

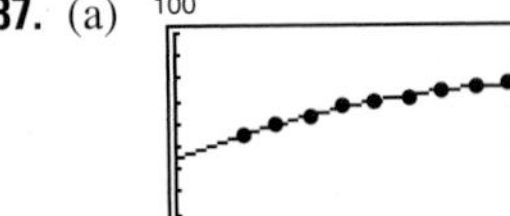

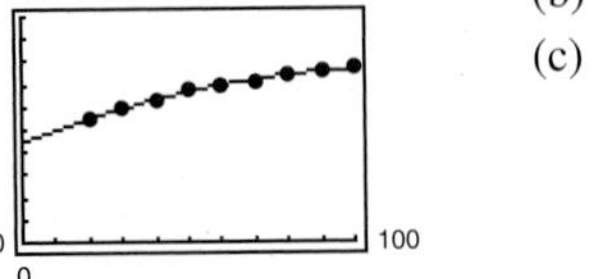

The model fits the data very well.

(b) 75.66 yr

(c) 1993

(d) The projection given by the model, 77.2 years, is less.

(e) Answers will vary.

89. (a)

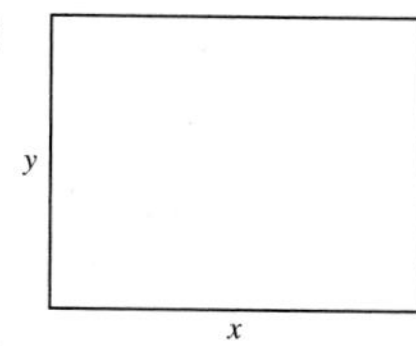

(b) Answers will vary.

(c)

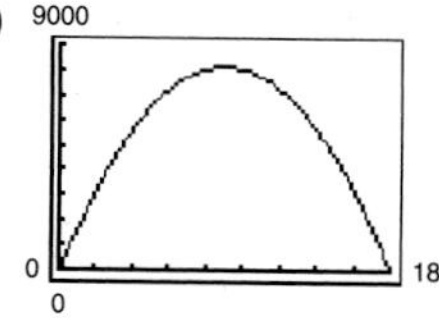

(d) $x = 90, y = 90$

(e) Answers will vary. Sample answer: There are no fixed dimensions for a regulation Major League Soccer field, but they are generally 110 to 115 yards (100.6 to 105.2 meters) long and 75 yards (68.6 meters) wide. This makes the area about 8250 square yards (6901.2 square meters).

91. The viewing window is incorrect. Change the viewing rectangle. Answers will vary.

Section P.5 (page 58)

1. linear **3.** parallel **5.** rate or rate of change

7. general

9.

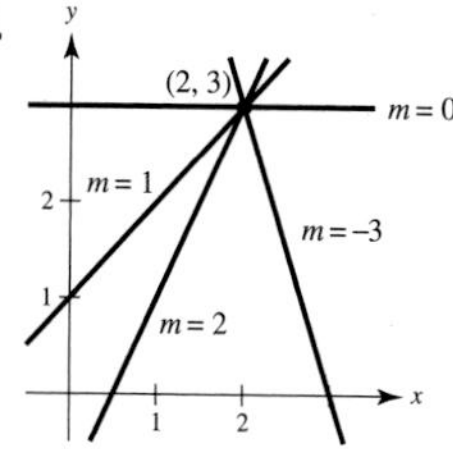

11. $\frac{3}{2}$ **13.** -4

15. $m = 5$

y-intercept: $(0, 3)$

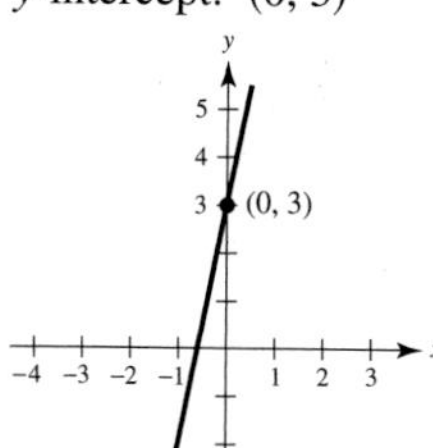

17. $m = -\frac{1}{2}$

y-intercept: $(0, 4)$

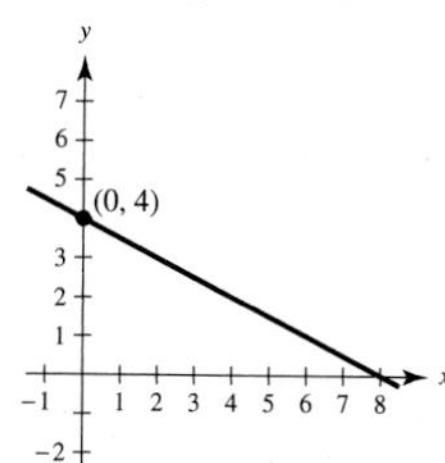

19. m is undefined.

There is no y-intercept.

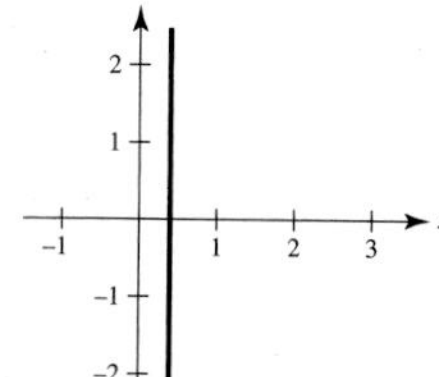

21. $m = -\frac{7}{6}$

y-intercept: $(0, 5)$

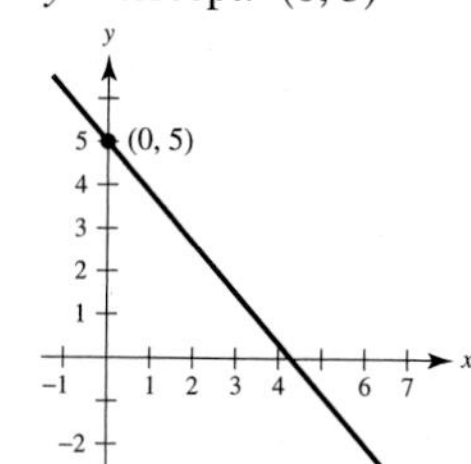

23. $m = 0$

y-intercept: $(0, 3)$

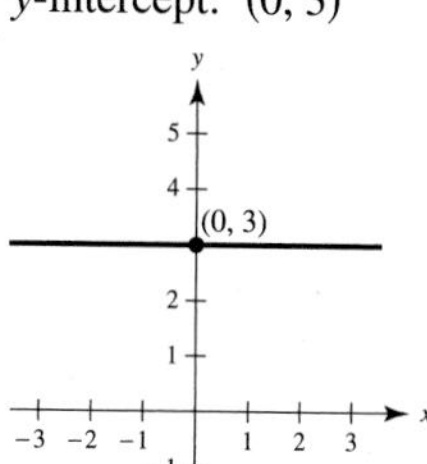

25. m is undefined.

There is no y-intercept.

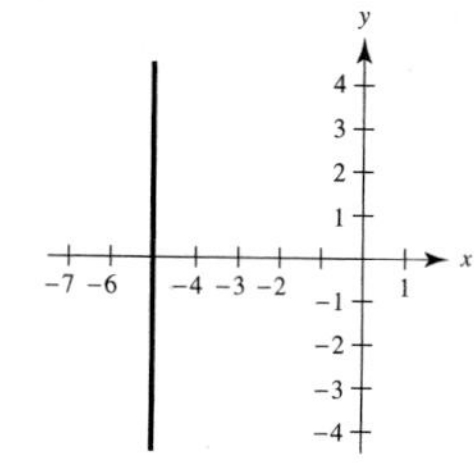

27.

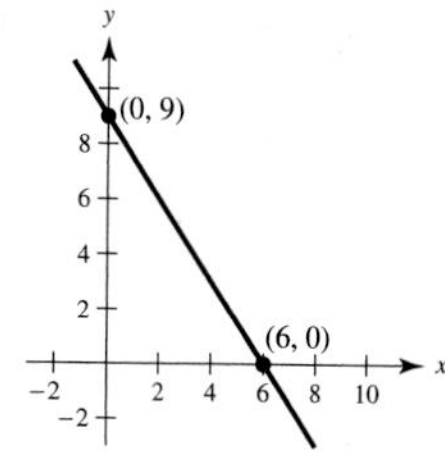

$m = -\frac{3}{2}$

29.

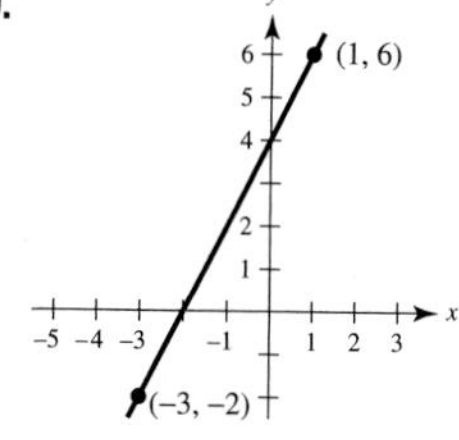

$m = 2$

31.

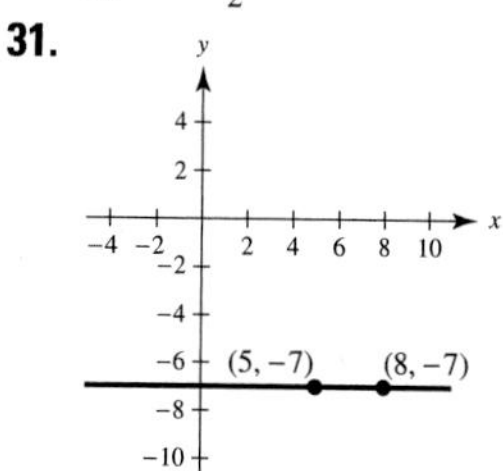

$m = 0$

33.

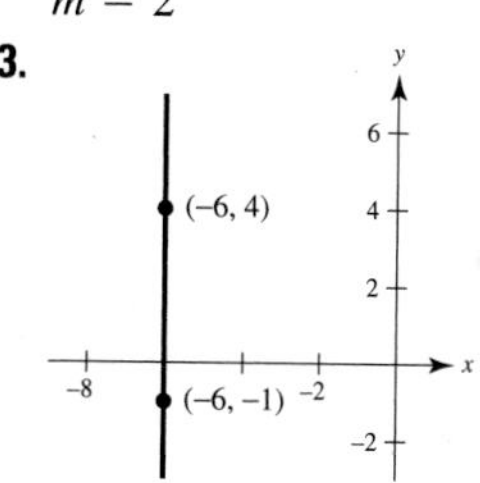

m is undefined.

35.

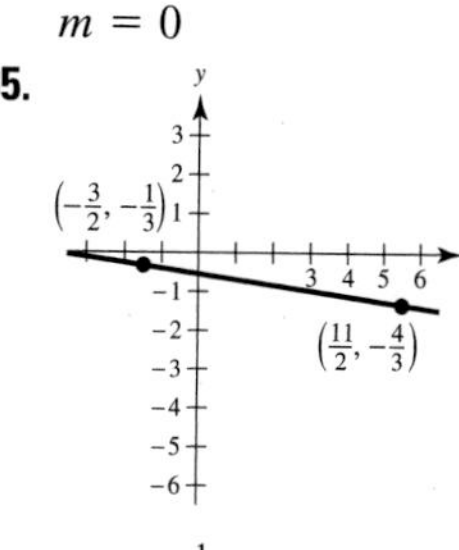

$m = -\frac{1}{7}$

37.

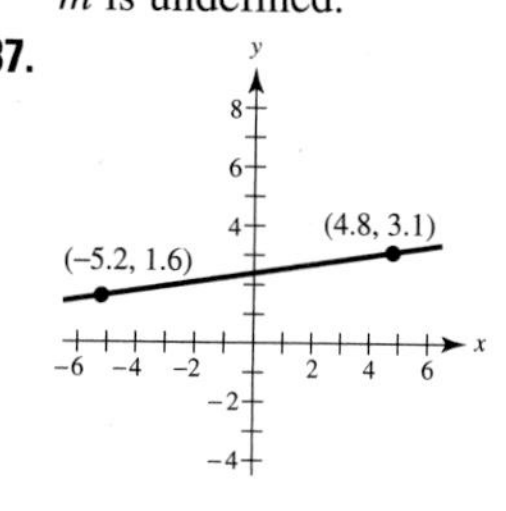

$m = 0.15$

39. (a) L_2 (b) L_3 (c) L_1

41. $(0, 1), (3, 1), (-1, 1)$ **43.** $(6, -5), (7, -4), (8, -3)$

45. $(-8, 0), (-8, 2), (-8, 3)$ **47.** $(-4, 6), (-3, 8), (-2, 10)$

49. $(9, -1), (11, 0), (13, 1)$

51. $y = 3x - 2$

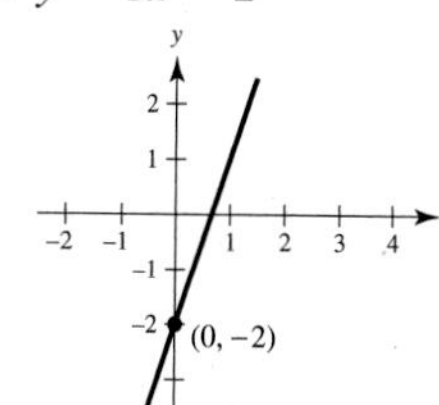

53. $y = -2x$

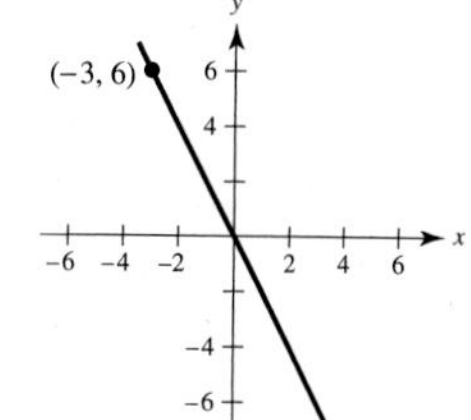

55. $y = -\frac{1}{3}x + \frac{4}{3}$

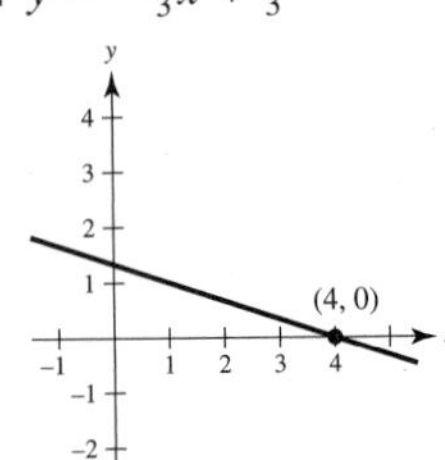

57. $y = -\frac{1}{2}x - 2$

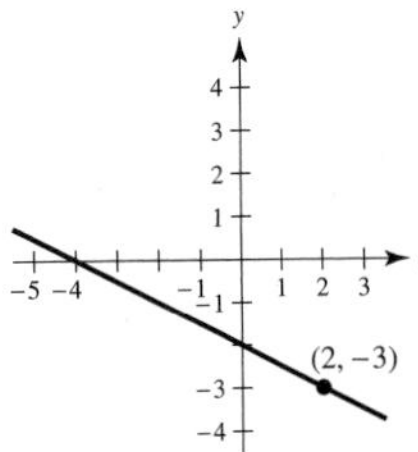

59. $x = 6$

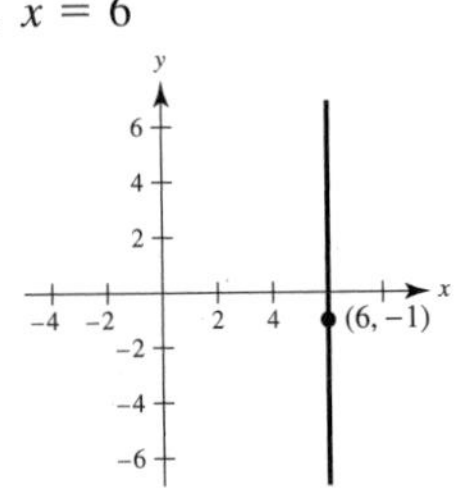

61. $y = \frac{5}{2}$

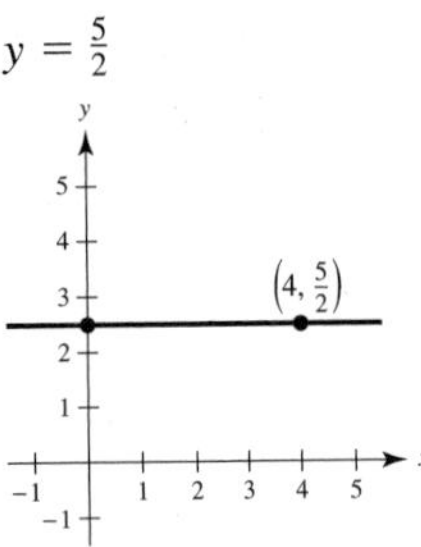

63. $y = 5x + 27.3$

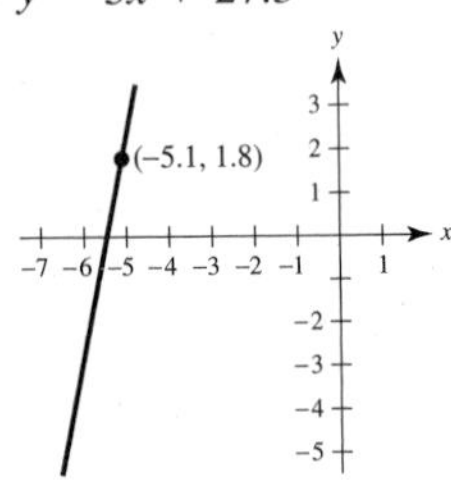

65. $y = -\frac{3}{5}x + 2$

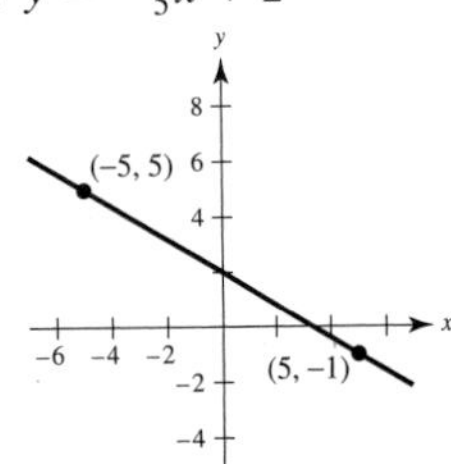

67. $x = -8$

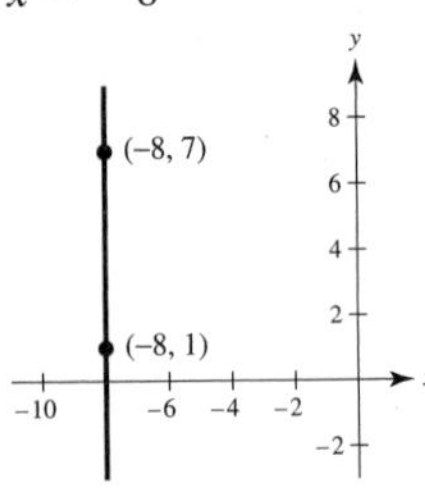

69. $y = -\frac{1}{2}x + \frac{3}{2}$

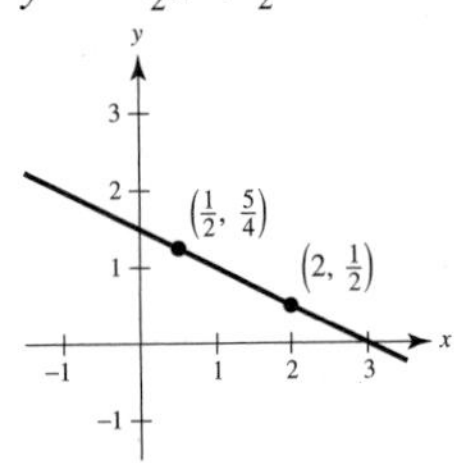

71. $y = -\frac{6}{5}x - \frac{18}{25}$

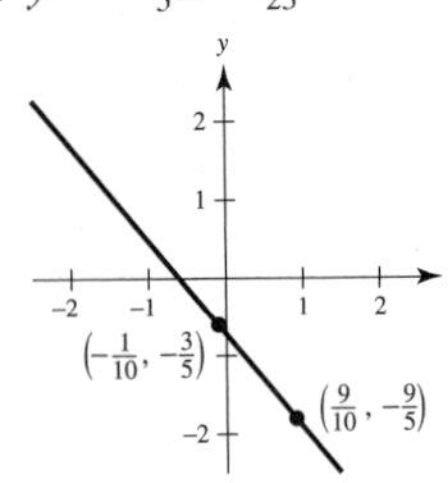

73. $y = 0.4x + 0.2$

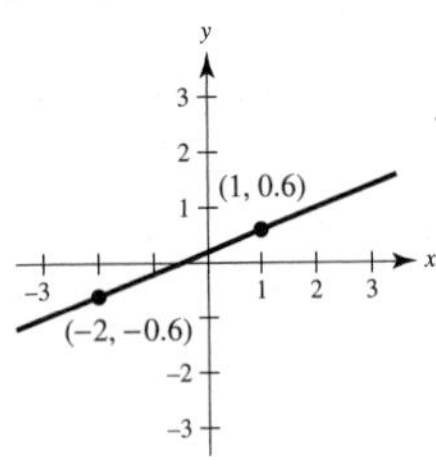

75. $y = -1$

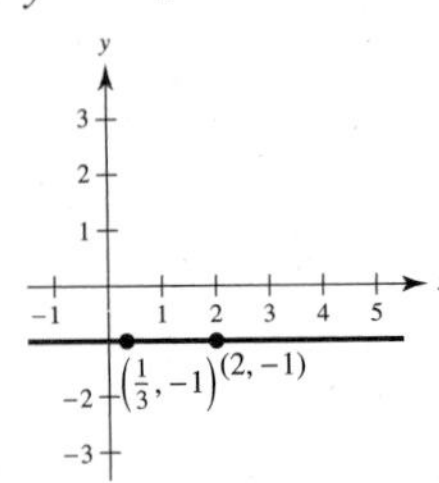

77. $x = \frac{7}{3}$

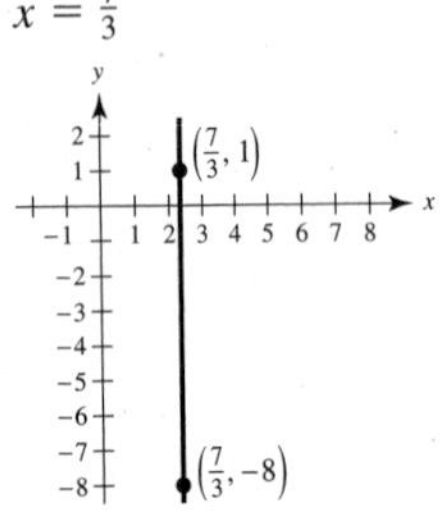

79. Parallel **81.** Neither **83.** Perpendicular

85. Parallel **87.** (a) $y = 2x - 3$ (b) $y = -\frac{1}{2}x + 2$

89. (a) $y = -\frac{3}{4}x + \frac{3}{8}$ (b) $y = \frac{4}{3}x + \frac{127}{72}$

91. (a) $y = 0$ (b) $x = -1$

93. (a) $x = 3$ (b) $y = -2$

95. (a) $y = x + 4.3$ (b) $y = -x + 9.3$

97. $3x + 2y - 6 = 0$ **99.** $12x + 3y + 2 = 0$

101. $x + y - 3 = 0$

103. Line (b) is perpendicular to line (c).

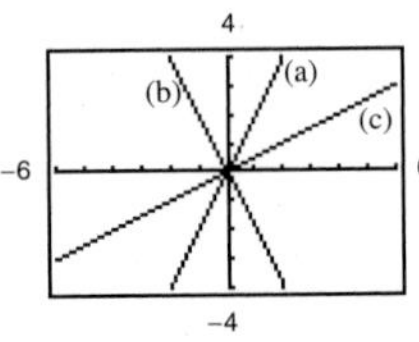

105. Line (a) is parallel to line (b).
Line (c) is perpendicular to line (a) and line (b).

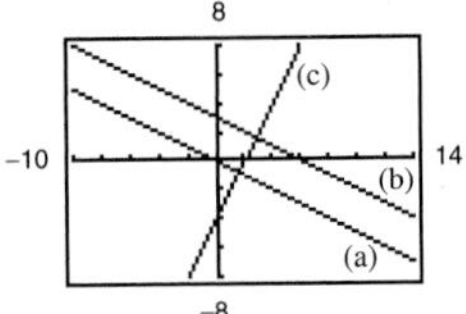

107. $3x - 2y - 1 = 0$ **109.** $80x + 12y + 139 = 0$

111. (a) Sales increasing 135 units/yr
(b) No change in sales
(c) Sales decreasing 40 units/yr

113. (a) The average salary increased the greatest from 2006 to 2008 and increased the least from 2002 to 2004.
(b) $m = 2350.75$
(c) The average salary increased \$2350.75 per year over the 12 years between 1996 and 2008.

115. 12 ft **117.** $V(t) = 3790 - 125t$

119. V-intercept: initial cost; Slope: annual depreciation

121. $V = -175t + 875$ **123.** $S = 0.8L$

125. $W = 0.07S + 2500$

127. $y = 0.03125t + 0.92875$; $y(22) \approx \$1.62$; $y(24) \approx \$1.68$

129. (a) $y(t) = 442.625t + 40{,}571$
(b) $y(10) = 44{,}997$; $y(15) = 47{,}210$
(c) $m = 442.625$; Each year, enrollment increases by about 443 students.

131. (a) $C = 18t + 42{,}000$ (b) $R = 30t$
(c) $P = 12t - 42{,}000$ (d) $t = 3500$ h

133. (a)

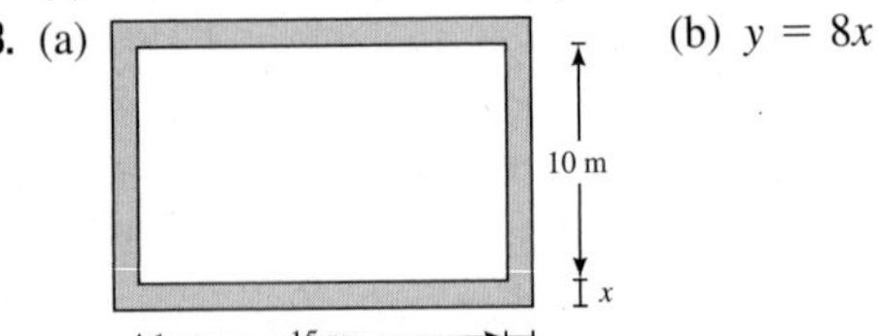

(b) $y = 8x + 50$

(c)

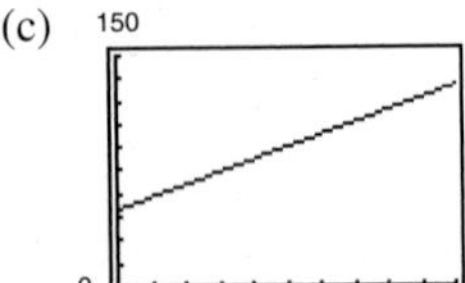

(d) $m = 8$, 8 m

135. (a) and (b)

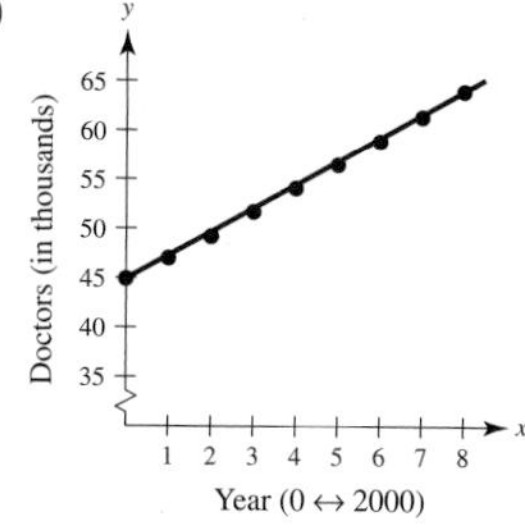

(c) Answers will vary. Sample answer: $y = 2.39x + 44.9$

(d) Answers will vary. Sample answer: The y-intercept indicates that in 2000 there were 44.9 thousand doctors of osteopathic medicine. The slope means that the number of doctors increases by 2.39 thousand each year.

(e) The model is accurate.

(f) Answers will vary. Sample answer: 73.6 thousand

137. False. The slope with the greatest magnitude corresponds to the steepest line.

139. Find the distance between each two points and use the Pythagorean Theorem.

141. No. The slope cannot be determined without knowing the scale on the y-axis. The slopes could be the same.

143. c

145. The line $y = 4x$ rises most quickly, and the line $y = -4x$ falls most quickly. The greater the magnitude of the slope (the absolute value of the slope), the faster the line rises or falls.

147. No. The slopes of two perpendicular lines have opposite signs (assume that neither line is vertical or horizontal).

Review Exercises (page 66)

1. Identity **3.** Identity **5.** 20 **7.** $-\frac{1}{2}$ **9.** $-\frac{7}{2}, 4$
11. $\pm\frac{5}{4}$ **13.** $8 \pm \sqrt{15}$ **15.** $-3 \pm 2\sqrt{3}$
17. $1/2 \pm \sqrt{249}/6$ **19.** $0, \frac{3}{2}$ **21.** $0, -3, \pm\frac{2}{3}$ **23.** 66
25. 2 **27.** 79 **29.** $\pm 2, \pm\frac{2}{3}$ **31.** $2, -5$ **33.** 2, 3
35. $2\frac{6}{7}$ **37.** (a) Yes (b) No **39.** $(-\infty, 12]$
41. $(-2, \infty)$ **43.** $\left(-7, 28\frac{1}{3}\right]$ **45.** $[-6, 4]$
47. $(-\infty, -1), (7, \infty)$ **49.** $x = 37$ units **51.** $(-3, 9)$
53. $\left(-\frac{4}{3}, \frac{1}{2}\right)$ **55.** $[-5, -1), (1, \infty)$ **57.** $[-4, -3], (0, \infty)$
59. $r > 4.88\%$

61.

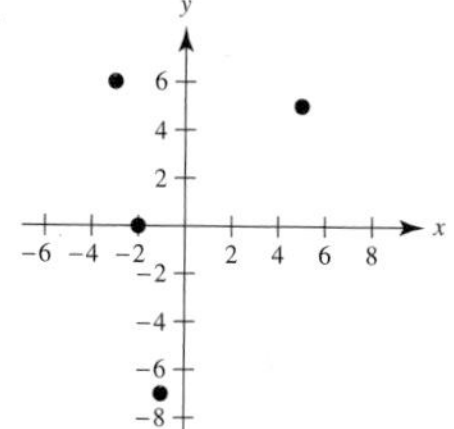

63. Quadrant IV

65. (a)

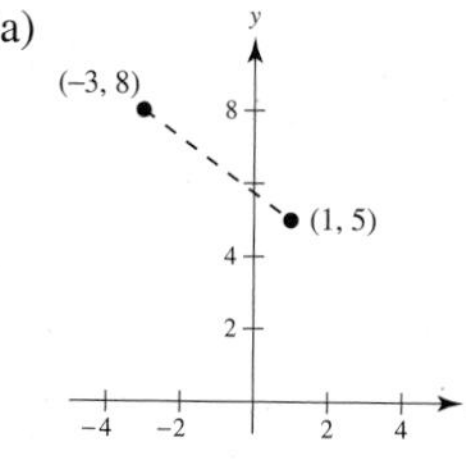

(b) 5
(c) $\left(-1, \frac{13}{2}\right)$

67. (a)

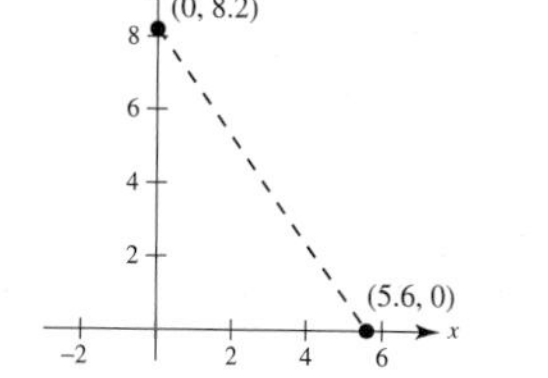

(b) $\sqrt{98.6}$
(c) $(2.8, 4.1)$

69. $(0, 0), (2, 0), (0, -5), (2, -5)$

71. $6.275 billion

73.

x	-2	-1	0	1	2
y	-11	-8	-5	-2	1

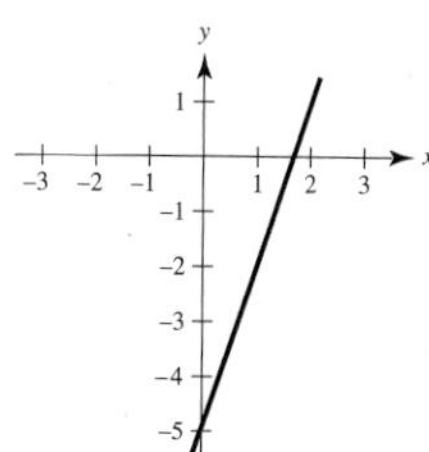

75.

x	-1	0	1	2	3	4
y	4	0	-2	-2	0	4

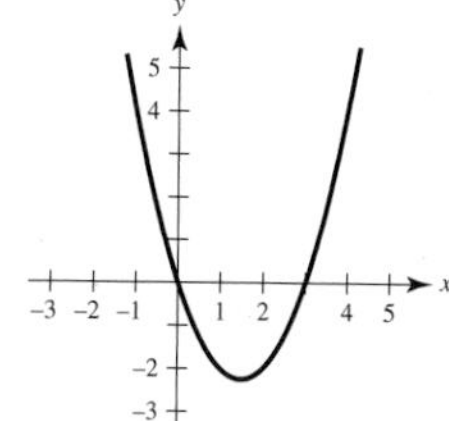

77. **79.**

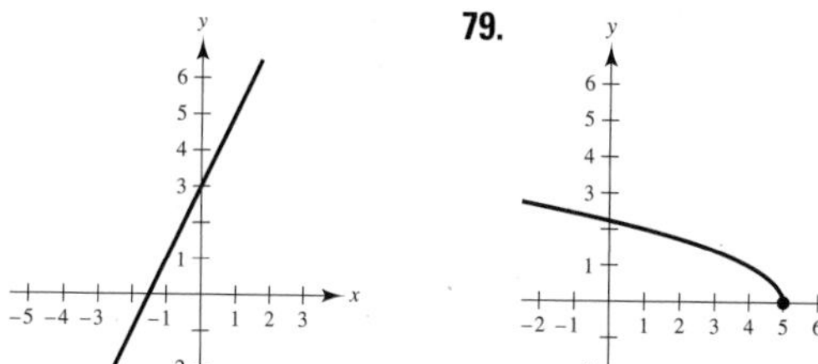

81.

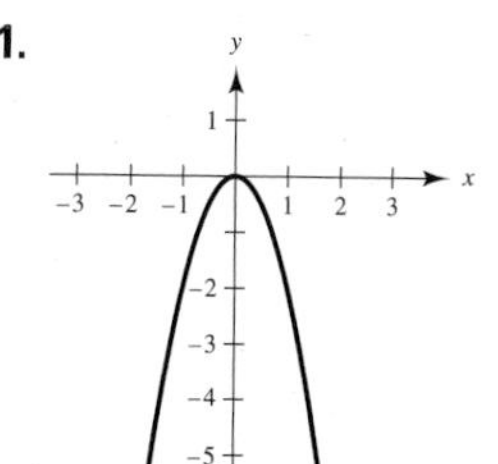

83. x-intercept: $\left(-\frac{7}{2}, 0\right)$
y-intercept: $(0, 7)$

85. x-intercepts: $(1, 0), (5, 0)$
y-intercept: $(0, 5)$

87. x-intercept: $\left(\frac{1}{4}, 0\right)$
y-intercept: $(0, 1)$
No symmetry

89. x-intercepts: $\left(\pm\sqrt{5}, 0\right)$
y-intercept: $(0, 5)$
y-axis symmetry

91. x-intercept: $\left(\sqrt[3]{-3}, 0\right)$
y-intercept: $(0, 3)$
No symmetry

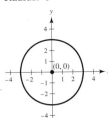

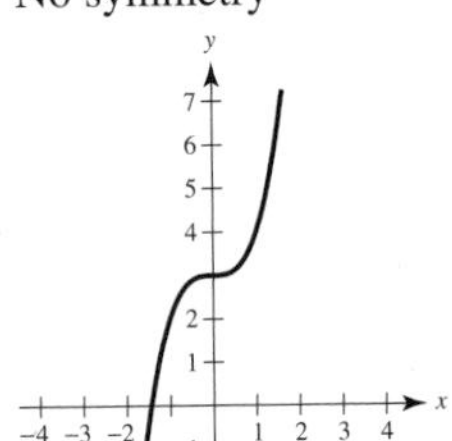

93. x-intercept: $(-5, 0)$
y-intercept: $\left(0, \sqrt{5}\right)$
No symmetry

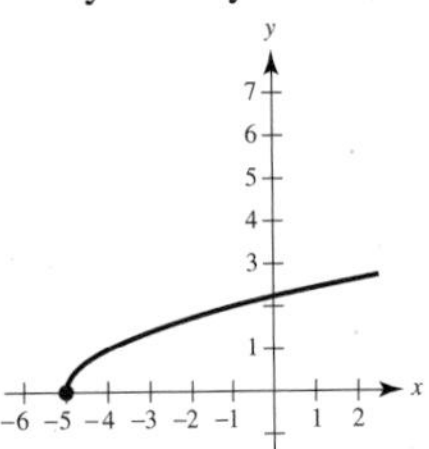

95. Center: $(0, 0)$
Radius: 3

97. Center: $(-2, 0)$
Radius: 4

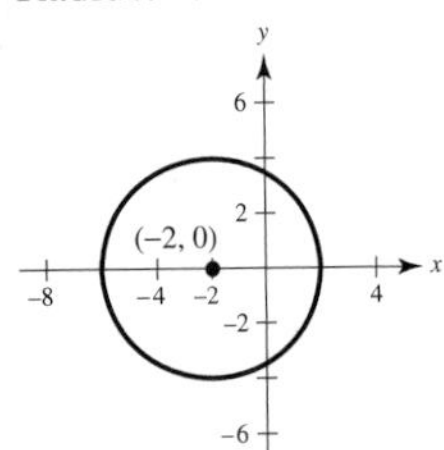

99. Center: $\left(\frac{1}{2}, -1\right)$
Radius: 6

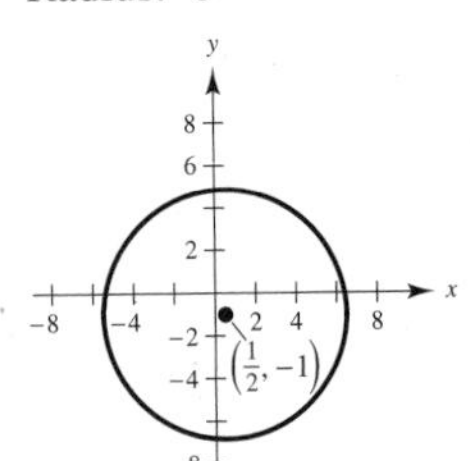

101. $(x - 2)^2 + (y + 3)^2 = 13$

103. (a)

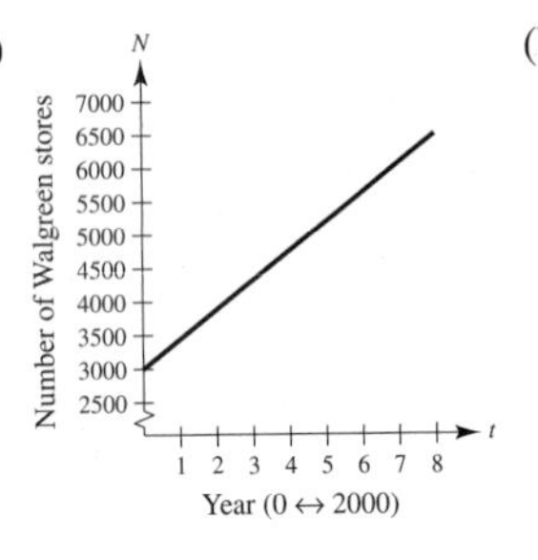

(b) 2008

105. Slope: 0
y-intercept: 6

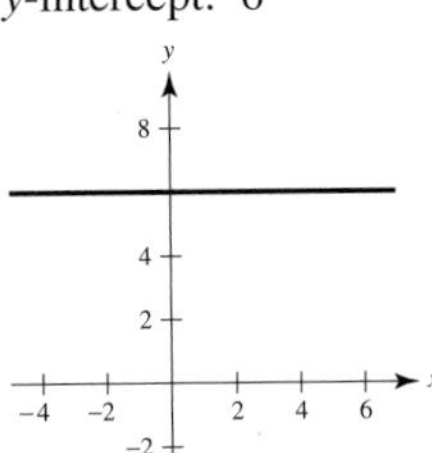

107. Slope: 3
y-intercept: 13

109. (a) L_2 (b) L_3 (c) L_1 (d) L_4

111.

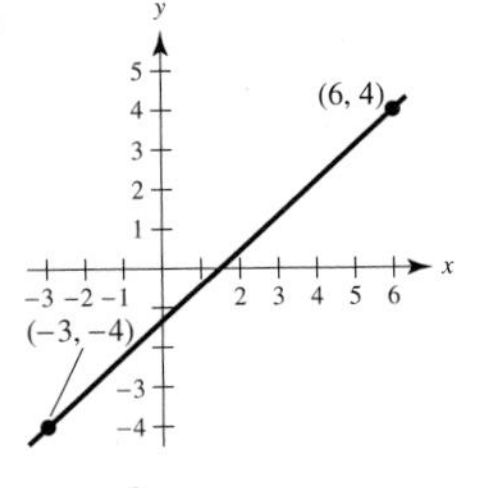

$m = \frac{8}{9}$

113.

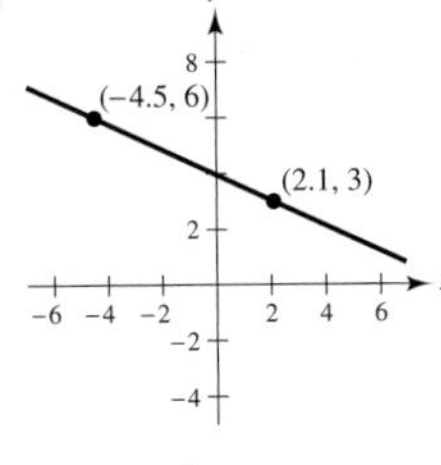

$m = -\frac{5}{11}$

115. $y = \frac{2}{3}x - 2$

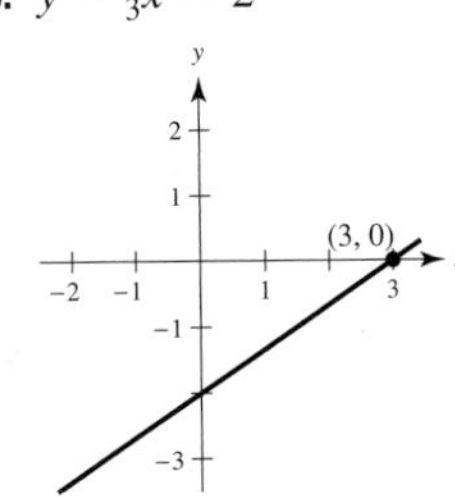

117. $y - 6 = 0$

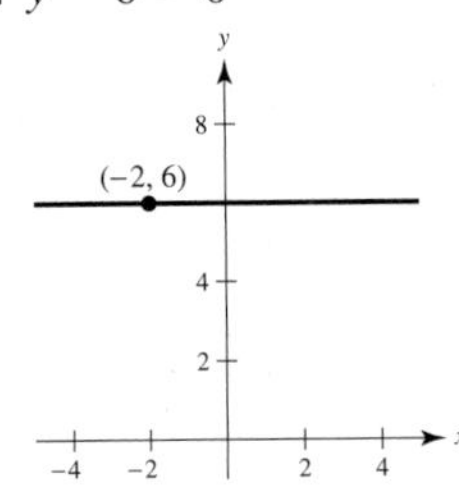

119. $x + 8 = 0$

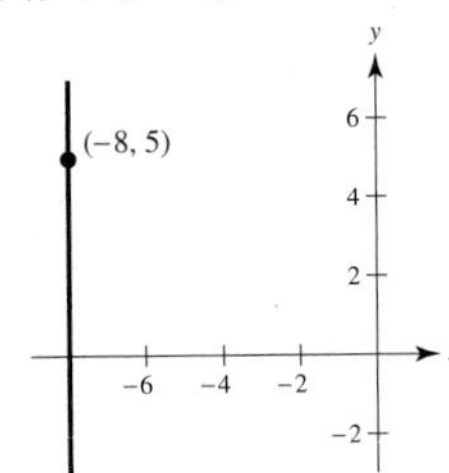

121. $x = 0$ **123.** $y = \frac{2}{7}x + \frac{2}{7}$

125. (a) $x - 2 = 0$ (b) $y + 1 = 0$

127. (a) $y - 1 = 0$ (b) $x + 2 = 0$

129. (a) $y = \frac{5}{4}x - \frac{23}{4}$ (b) $y = -\frac{4}{5}x + \frac{2}{5}$

131. $V = -850t + 21{,}000, \quad 10 \le t \le 15$

133. Sample answer: $a = 20, b = 20$

Chapter Test (page 121)

1. $\frac{128}{11}$ **2.** No solution **3.** $-4, 5$

4. $\pm\sqrt{2}$ **5.** 4 **6.** $-2, \frac{8}{3}$

7. $-\frac{11}{2} \le x < 3$

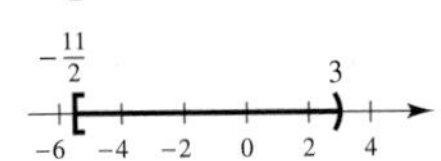

8. $x \le 10$ or $x \ge 20$

9. $x < -4$ or $x > \frac{3}{2}$

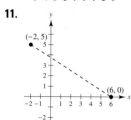

10. $x < -6$ or $0 < x < 4$

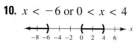

11.

Midpoint: $\left(2, \frac{5}{2}\right)$; Distance: $\sqrt{89}$

12. $(-4, -2), (2, -4), (3, -1)$

13. No symmetry

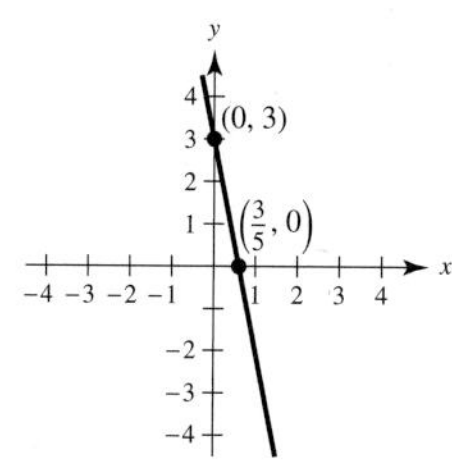

14. y-axis symmetry

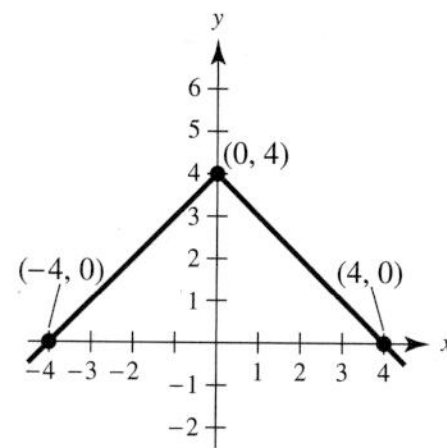

15. y-axis symmetry

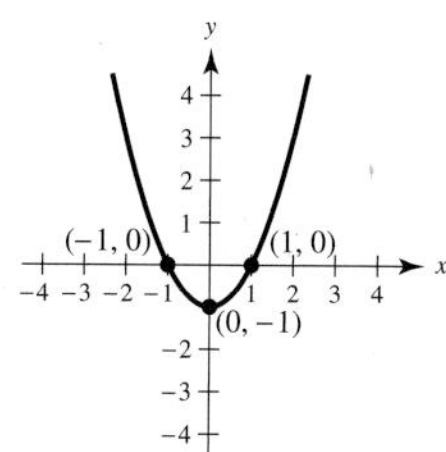

16. $(x - 1)^2 + (y - 3)^2 = 16$

17. $y = -2x + 1$ **18.** $y = -1.7x + 5.9$

19. (a) $5x + 2y - 8 = 0$ (b) $-2x + 5y - 20 = 0$

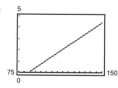

20. (a)

(b) $x \geq 129$

21. $100 \leq r \leq 170$

P.S. Problem Solving (page 69)

1. (a) and (b) $x = -5, -\frac{10}{3}$

(c) The method of part (a) reduces the number of algebraic steps.

3. (a) $(-\infty, -4] \cup [4, \infty)$

(b) $(-\infty, \infty)$

(c) $\left(-\infty, -2\sqrt{30}\right] \cup \left[2\sqrt{30}, \infty\right)$

(d) $\left(-\infty, -2\sqrt{10}\right] \cup \left[2\sqrt{10}, \infty\right)$

(e) If $a > 0$ and $c \leq 0$, b can be any real number. If $a > 0$ and $c > 0$, $b < -2\sqrt{ac}$ or $b > 2\sqrt{ac}$.

(f) 0

5. (a) Neither (b) Both (c) Quadratic (d) Neither

7. Answers will vary.

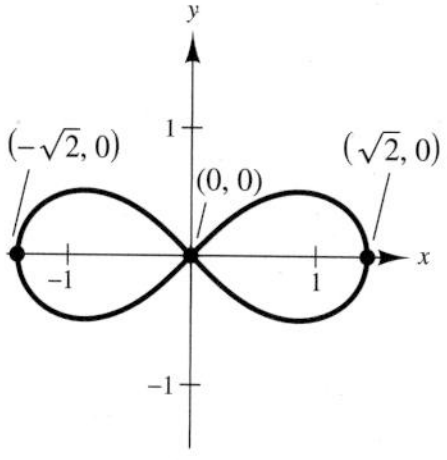

9. (a) 150 students per year

(b) 5950 students in 2003; 6550 in 2007; 6850 in 2009

(c) Letting $x = 0$ represent 2000, the equation of the line is $y = 150x + 5500$. The slope of the line is 150, which means that enrollment increases by approximately 150 students per year.

11. (a) and (b)

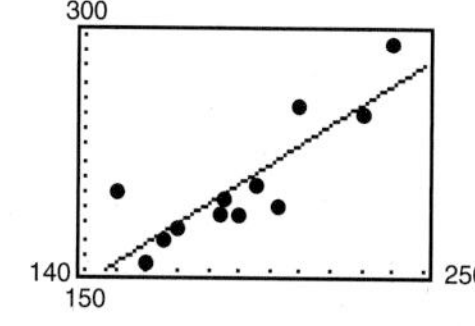

(c) 186.23 lb

(d) Answers will vary.

13. (a) Choice 1: $W = 3000 + 0.07s$

Choice 2: $W = 3400 + 0.05s$

(b) The salaries are the same ($4400 per month) when sales equal $20,000.

(c) Graph the equations representing each wage scale to determine when they balance. Use this information in your decision.

Chapter 1

Section 1.1 (page 79)

1. domain; range; function **3.** independent; dependent

5. implied domain **7.** Yes **9.** No **11.** No

13. Yes, each input value has exactly one output value.

15. No, the input values 7 and 10 each have two different output values.

17. (a) Function

(b) Not a function, because the element 1 in A corresponds to two elements, -2 and 1, in B.

(c) Function

(d) Not a function, because not every element in A is matched with an element in B.

19. Each is a function. For each year there corresponds one and only one circulation.

21. Not a function **23.** Function **25.** Function

27. Not a function **29.** Not a function **31.** Function

33. Function **35.** Not a function **37.** Function

39. (a) -1 (b) -9 (c) $2x - 5$

41. (a) 36π (b) $\frac{9}{2}\pi$ (c) $\frac{32}{3}\pi r^3$

43. (a) 15 (b) $4t^2 - 19t + 27$ (c) $4t^2 - 3t - 10$

45. (a) 1 (b) 2.5 (c) $3 - 2|x|$

47. (a) $-\dfrac{1}{9}$ (b) Undefined (c) $\dfrac{1}{y^2 + 6y}$

49. (a) 1 (b) -1 (c) $\dfrac{|x - 1|}{x - 1}$

51. (a) -1 (b) 2 (c) 6 **53.** (a) -7 (b) 4 (c) 9

55.

x	-2	-1	0	1	2
$f(x)$	1	-2	-3	-2	1

57.

t	-5	-4	-3	-2	-1
$h(t)$	1	$\frac{1}{2}$	0	$\frac{1}{2}$	1

59.

x	-2	-1	0	1	2
$f(x)$	5	$\frac{9}{2}$	4	1	0

61. 5 **63.** $\frac{4}{3}$ **65.** ± 3 **67.** $0, \pm 1$ **69.** $-1, 2$

71. $0, \pm 2$ **73.** All real numbers x

75. All real numbers t except $t = 0$

77. All real numbers y such that $y \geq 10$

79. All real numbers x except $x = 0, -2$

81. All real numbers s such that $s \geq 1$ except $s = 4$

83. All real numbers x such that $x > 0$

85. $\{(-2, 4), (-1, 1), (0, 0), (1, 1), (2, 4)\}$

87. $\{(-2, 4), (-1, 3), (0, 2), (1, 3), (2, 4)\}$

89. No. The element 3 in Set A corresponds to two elements in Set B.

91. $A = \dfrac{P^2}{16}$

93. $A = 8\sqrt{s^2 - 64}$

95. (a) The maximum volume is 1024 cubic centimeters.

(b)

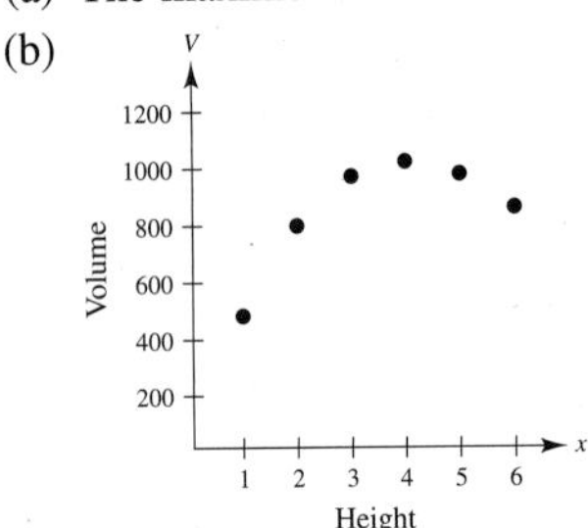

Yes, V is a function of x.

(c) $V = x(24 - 2x)^2, \quad 0 < x < 12$

97. $A = \dfrac{x^2}{2(x - 2)}, \quad x > 2$

99. Yes, the ball will be at a height of 6 feet.

101. 1998: \$136,164 2003: \$180,419
1999: \$140,971 2004: \$195,900
2000: \$147,800 2005: \$216,900
2001: \$156,651 2006: \$224,000
2002: \$167,524 2007: \$217,200

103. (a) $C = 12.30x + 98{,}000$ (b) $R = 17.98x$
(c) $P = 5.68x - 98{,}000$

105. (a) $R = \dfrac{240n - n^2}{20}, \quad n \geq 80$

(b)

n	90	100	110	120	130	140	150
$R(n)$	\$675	\$700	\$715	\$720	\$715	\$700	\$675

The revenue is maximum when 120 people take the trip.

107. (a) About 6.37; Approximately 6.37 million more tax returns were made through e-file each year from 2000 to 2007.

(b)

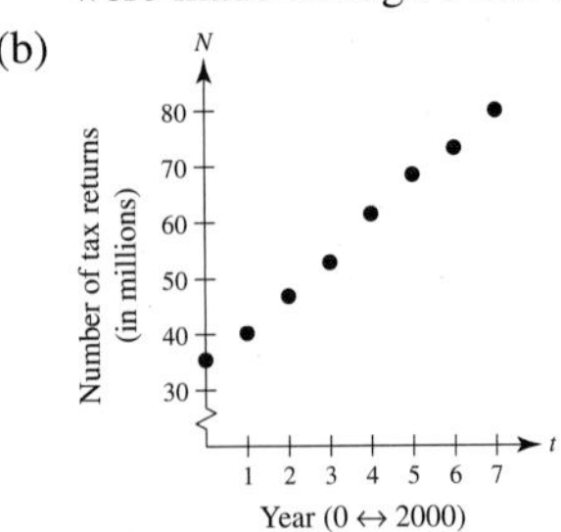

(c) $N = 6.37t + 35.4$

(d)

t	0	1	2	3	4	5	6	7
N	35.4	41.8	48.1	54.5	60.9	67.3	73.6	80.0

The algebraic model is a good fit to the actual data.

(e) $N = 6.56t + 34.4$; Both models are similar, but the model found in part (c) is a slightly better model.

109. $3 + h, \ h \neq 0$ **111.** $3x^2 + 3xc + c^2 + 2, \ c \neq 0$

113. $3, \ x \neq 3$ **115.** $\dfrac{\sqrt{5x} - 5}{x - 5}$

117. $g(x) = cx^2; \ c = -2$ **119.** $r(x) = \dfrac{c}{x}; \ c = 32$

121. False. A function is a special type of relation.

123. True

125. False. The range is $[-1, \infty)$.

127. Domain of $f(x)$: all real numbers $x \geq 1$
Domain of $g(x)$: all real numbers $x > 1$
Notice that the domain of $f(x)$ includes $x = 1$ and the domain of $g(x)$ does not because you cannot divide by 0.

129. No; x is the independent variable, f is the name of the function.

131. (a) Yes. The amount you pay in sales tax will increase as the price of the item purchased increases.
(b) No. The length of time that you study will not necessarily determine how well you do on an exam.

Section 1.2 (page 93)

1. ordered pairs **3.** zeros **5.** maximum **7.** odd

9. Domain: $(-\infty, -1] \cup [1, \infty)$; Range: $[0, \infty)$

11. Domain: $[-4, 4]$; Range: $[0, 4]$

13. Domain: $(-\infty, \infty)$; Range: $[-4, \infty)$
(a) 0 (b) -1 (c) 0 (d) -2

15. Domain: $(-\infty, \infty)$; Range: $(-2, \infty)$
(a) 0 (b) 1 (c) 2 (d) 3

17. Function **19.** Not a function **21.** Function

23. $-\frac{5}{2}, 6$ **25.** 0 **27.** $0, \pm\sqrt{2}$ **29.** $\pm\frac{1}{2}, 6$ **31.** $\frac{1}{2}$

33. **35.**

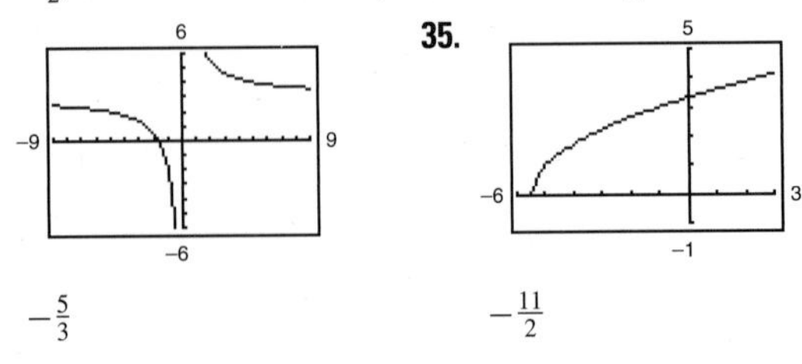

$-\frac{5}{3}$ $-\frac{11}{2}$

37. 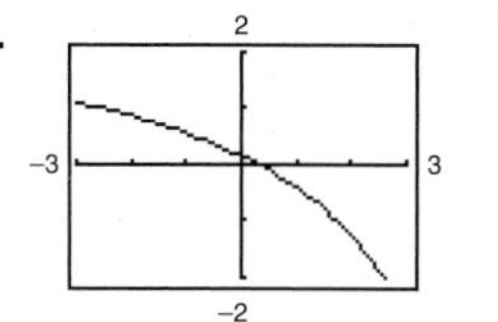

$\frac{1}{3}$

39. Increasing on $(-\infty, \infty)$

41. Increasing on $(-\infty, 0)$ and $(2, \infty)$
Decreasing on $(0, 2)$

43. 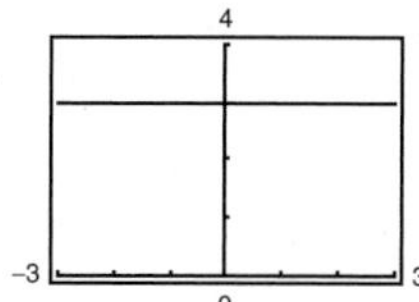

Constant on $(-\infty, \infty)$

45. 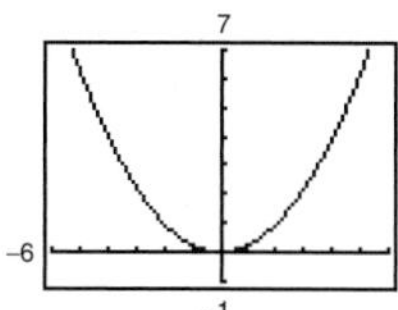

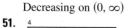

Decreasing on $(-\infty, 0)$
Increasing on $(0, \infty)$

47. 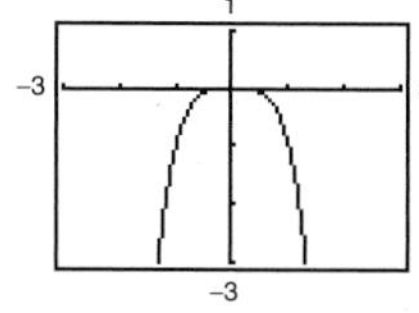

Increasing on $(-\infty, 0)$
Decreasing on $(0, \infty)$

49. 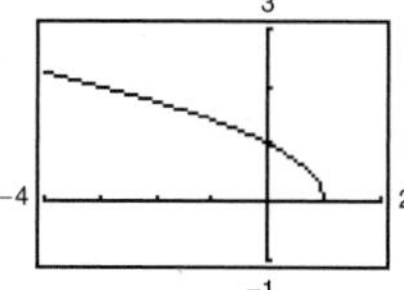

Decreasing on $(-\infty, 1)$

51. 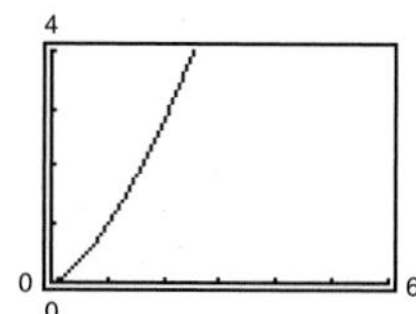

Increasing on $(0, \infty)$

53. 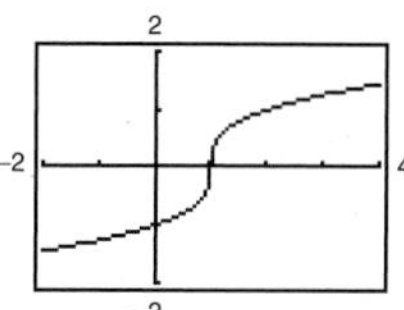

Increasing on $(-\infty, \infty)$

55. 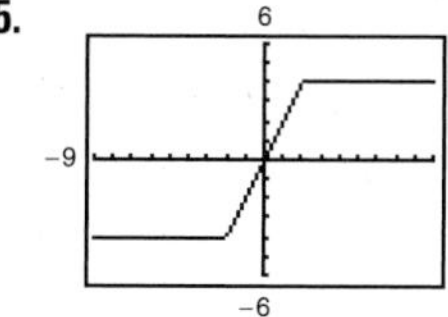

Increasing on $(-2, 2)$

57. 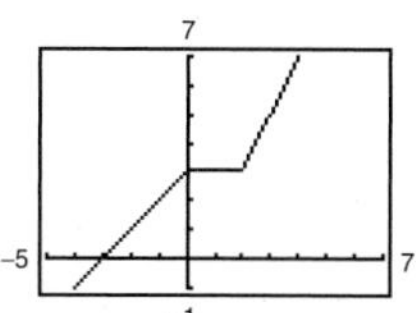

Increasing on $(-\infty, 0)$ and $(2, \infty)$
Constant on $(0, 2)$

59. 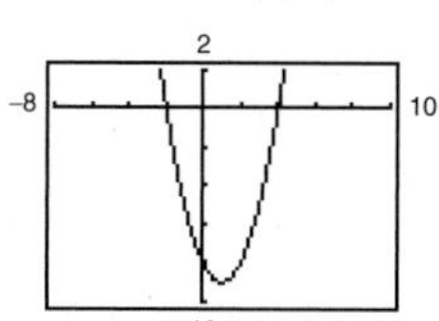

Relative minimum:
$(1, -9)$

61. 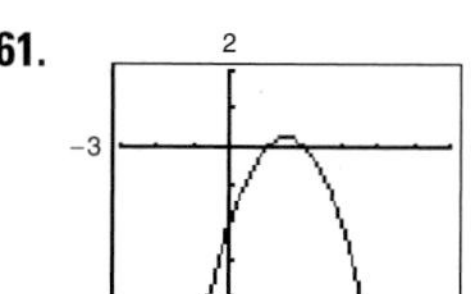

Relative maximum:
$(1.5, 0.25)$

63. 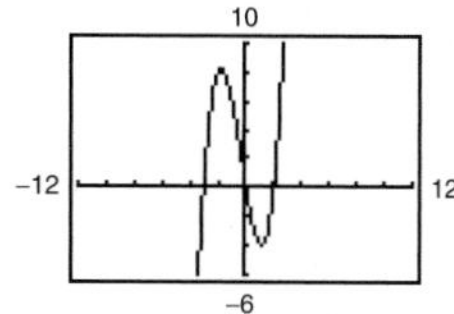

Relative maximum:
$(-1.79, 8.21)$
Relative minimum:
$(1.12, -4.06)$

65. 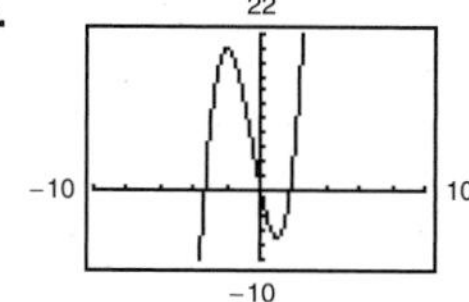

Relative maximum:
$(-2, 20)$
Relative minimum:
$(1, -7)$

67. 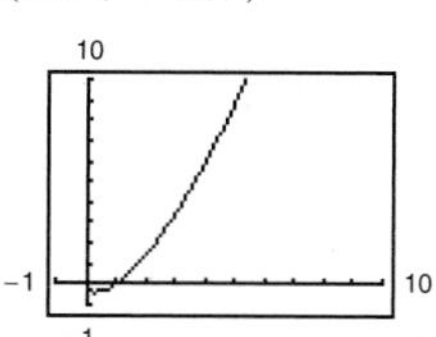

Relative minimum: $(0.33, -0.38)$

69. 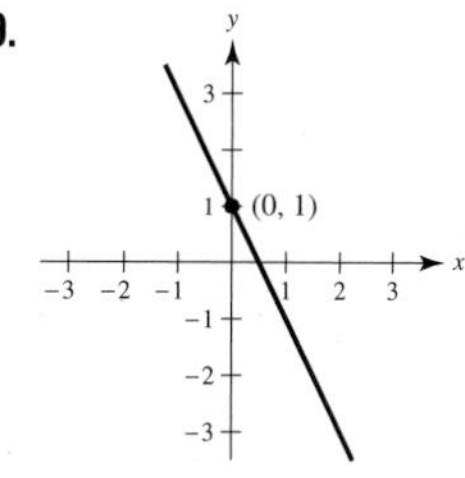

71. 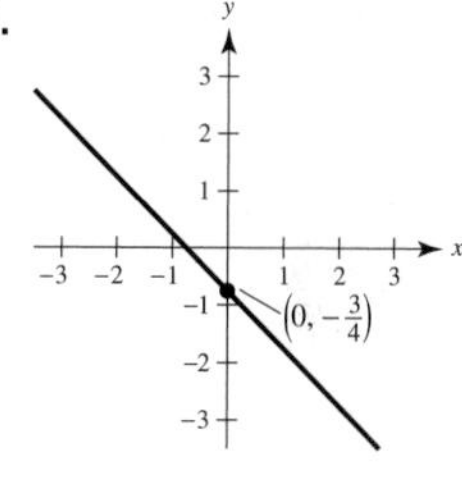

73. 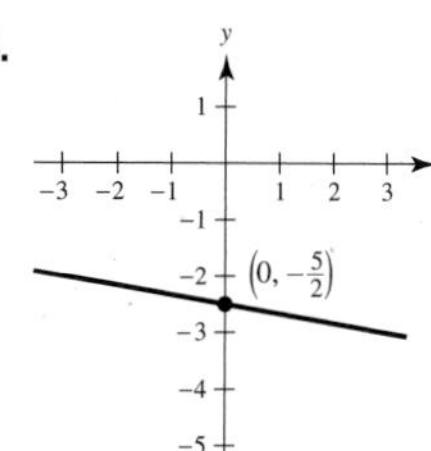

75. 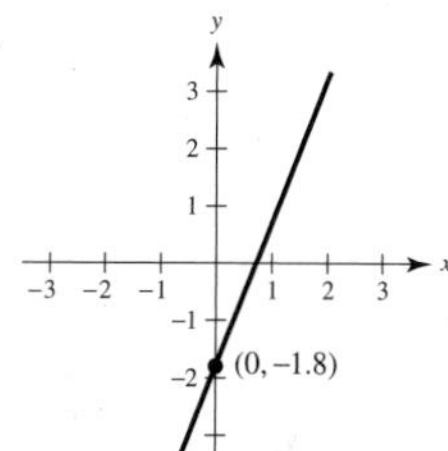

77. (a) $f(x) = -2x + 6$
(b) 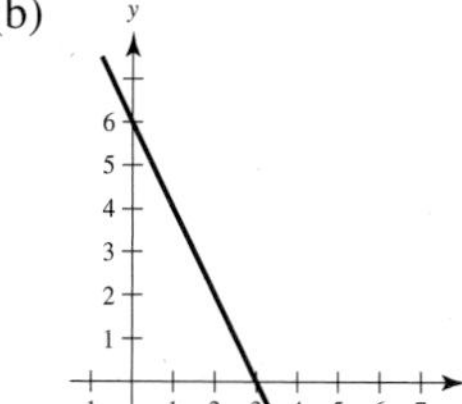

79. (a) $f(x) = -3x + 11$
(b) 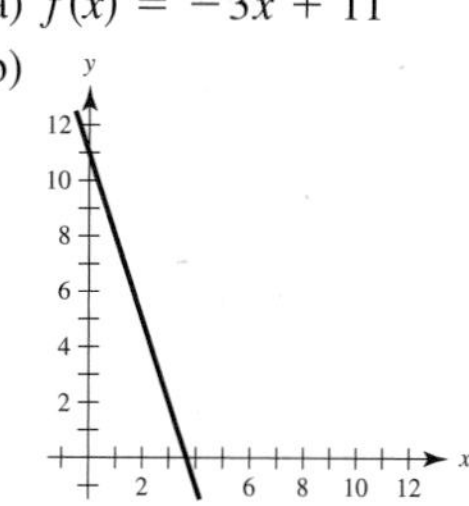

81. (a) $f(x) = -1$
(b) 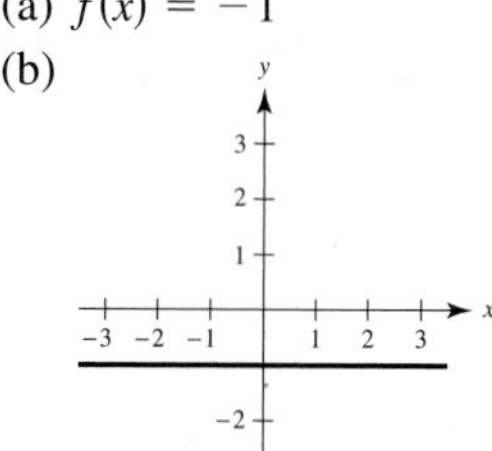

83.

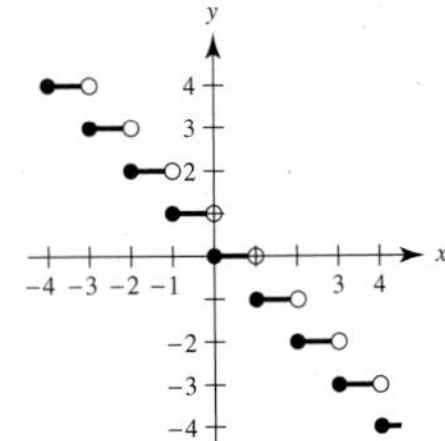

85.

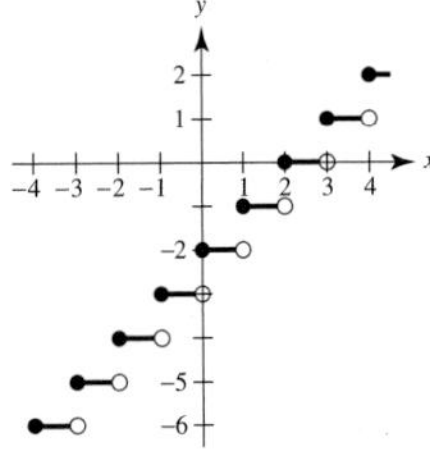

87.

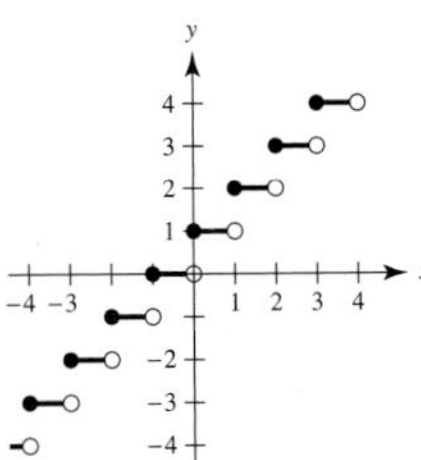

89.

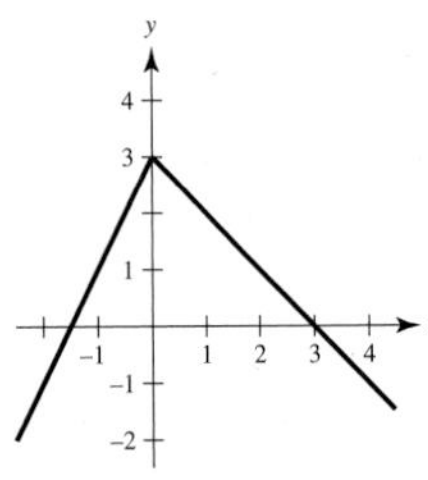

91.

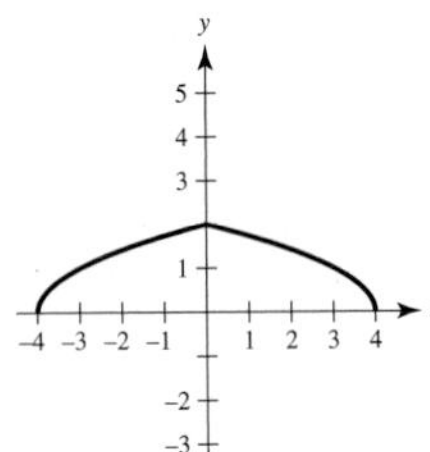

93.

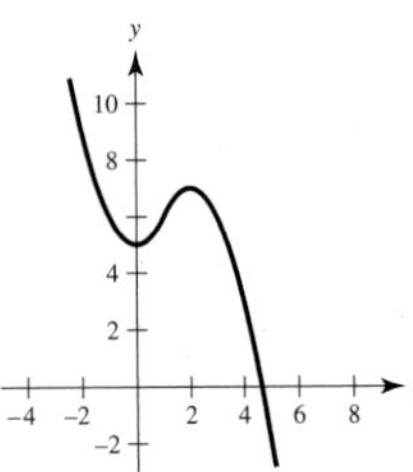

95.

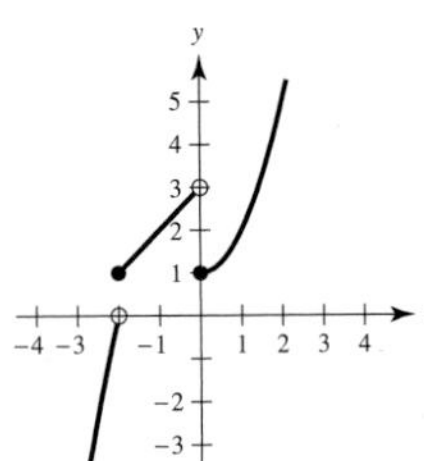

97.

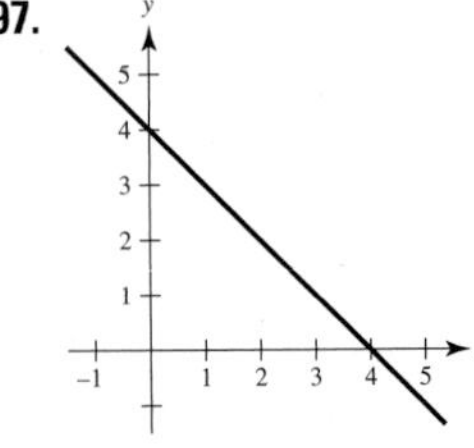

$(-\infty, 4]$

99.

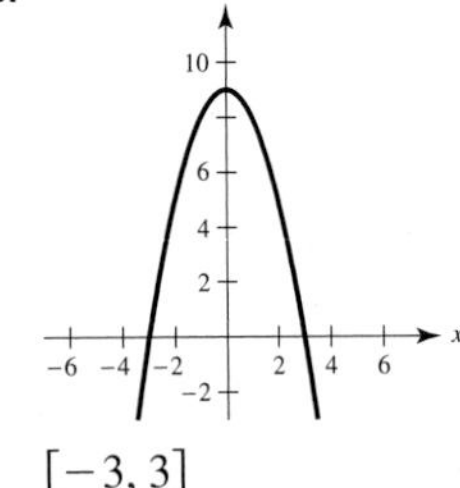

$[-3, 3]$

101.

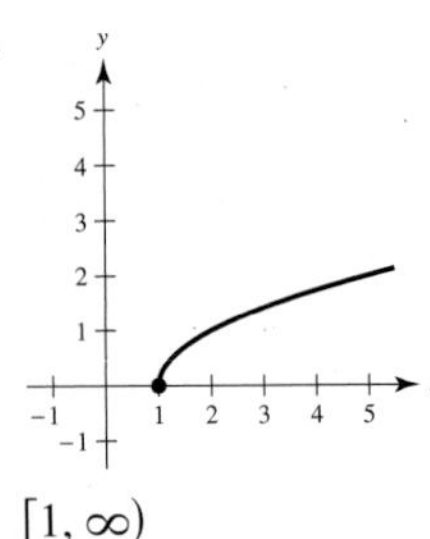

$[1, \infty)$

103.

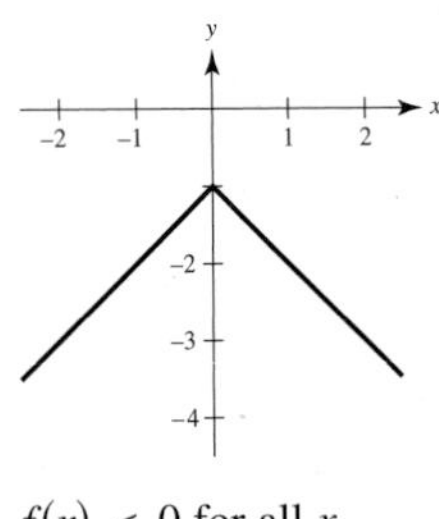

$f(x) < 0$ for all x

105.

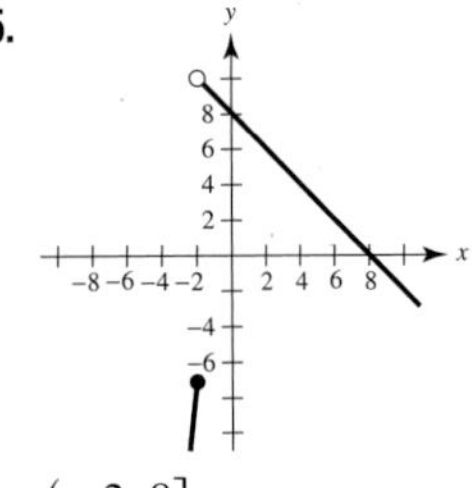

$(-2, 8]$

107. (a)

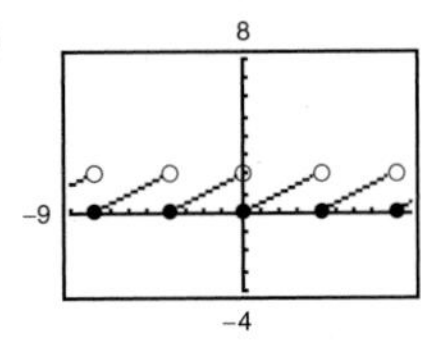

(b) Domain: $(-\infty, \infty)$
Range: $[0, 2)$
(c) Sawtooth pattern

109. (a)

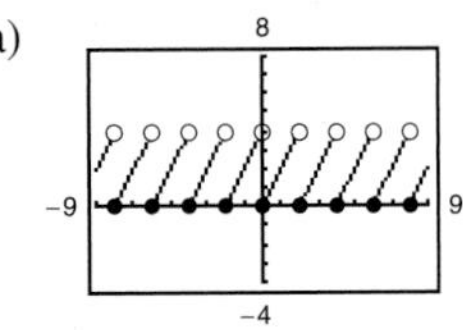

(b) Domain: $(-\infty, \infty)$
Range: $[0, 4)$
(c) Sawtooth pattern

111. (a) $(-\infty, \infty)$ (b) $(-\infty, 4]$
(c) Increasing on $(-\infty, 0)$
Decreasing on $(0, \infty)$

113. (a) $[-2, 2]$ (b) $[-2, 0]$
(c) Increasing on $(0, 2)$
Decreasing on $(-2, 0)$

115. (a) 1 (b) $\sqrt{3}$
(c) Increasing on $(-2, -1.6)$ and $(0, \infty)$
Decreasing on $(-1.6, 0)$

117. Even; y-axis symmetry

119. Odd; origin symmetry

121. Neither; no symmetry

123. Neither; no symmetry

125.

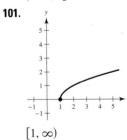

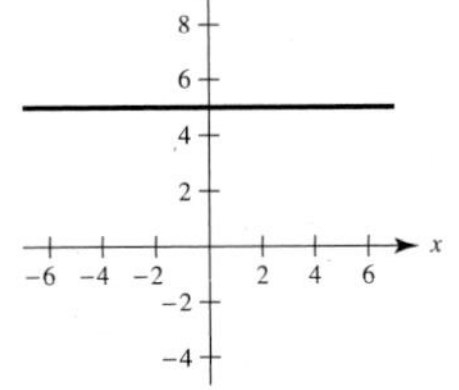

Even

127.

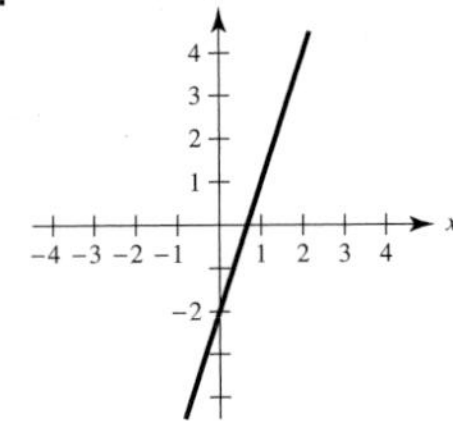

Neither

129.

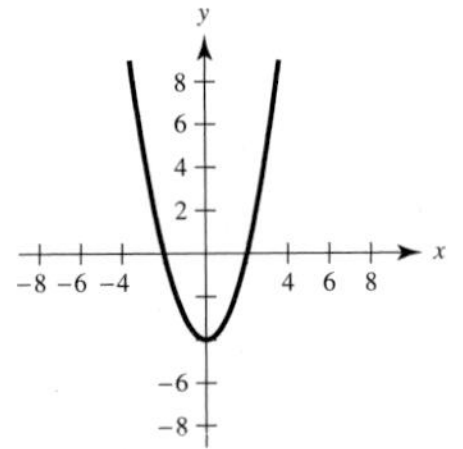

Even

131.

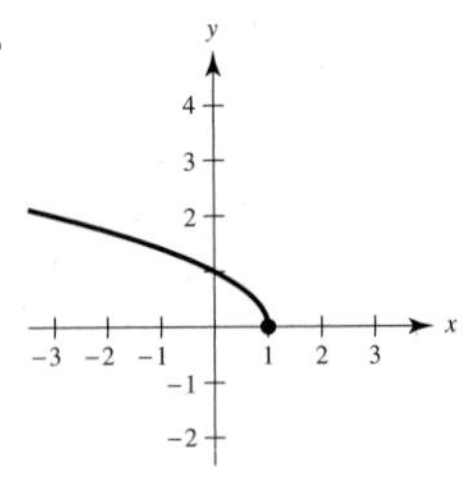

Neither

133.

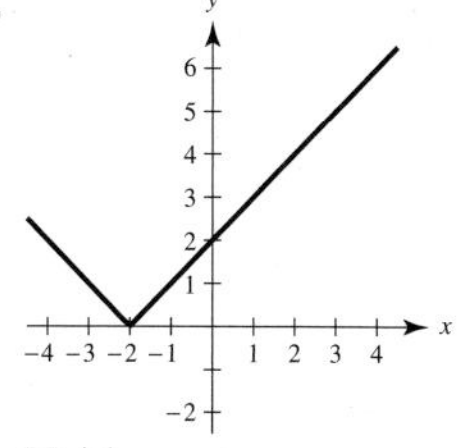

Neither

135. $h = -x^2 + 4x - 3$ **137.** $h = 2x - x^2$

139. $L = \frac{1}{2}y^2$ **141.** $L = 4 - y^2$

143. (a)

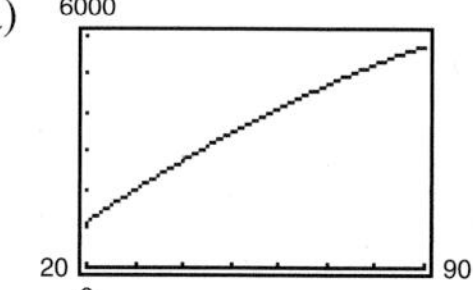

(b) 30 W

145. (a)

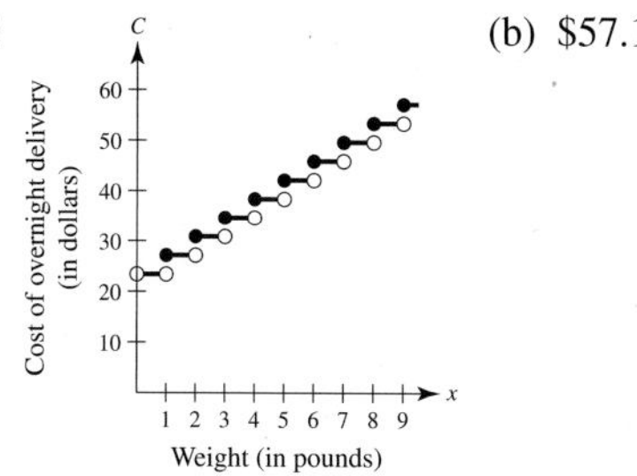

(b) \$57.15

147. (a) Ten thousands (b) Ten millions (c) Percents

149. False. The function $f(x) = \sqrt{x^2 + 1}$ has a domain of all real numbers.

151. (a) Even. The graph is a reflection in the x-axis.
(b) Even. The graph is a reflection in the y-axis.
(c) Even. The graph is a vertical translation of f.
(d) Neither. The graph is a horizontal translation of f.

153. (a) $\left(\frac{3}{2}, 4\right)$ (b) $\left(\frac{3}{2}, -4\right)$

155. (a) $(-4, 9)$ (b) $(-4, -9)$

157. (a) $(-x, -y)$ (b) $(-x, y)$

159. (a) $a = 1, b = -2$ (b) $a = -1, b = 2$

Section 1.3 (page 103)

1. rigid **3.** nonrigid **5.** vertical stretch; vertical shrink

7. (a)

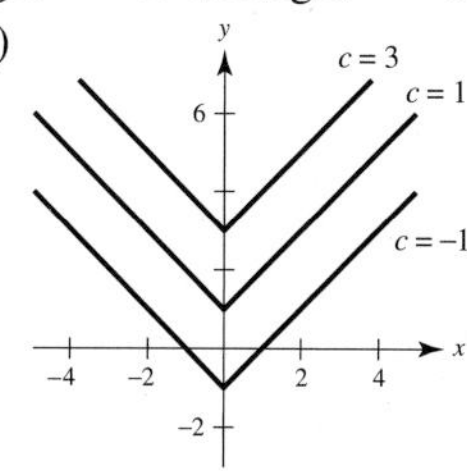

(b)

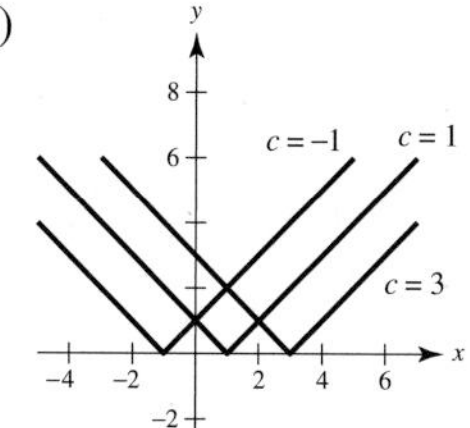

(c)

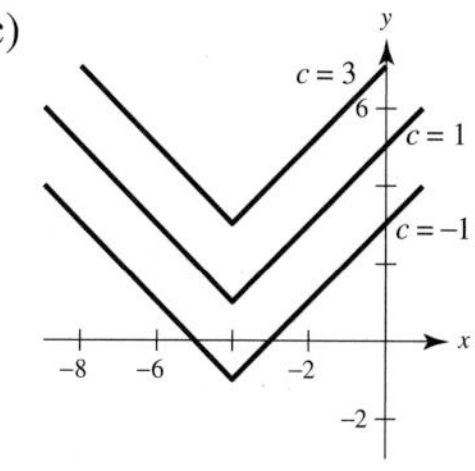

9. (a)

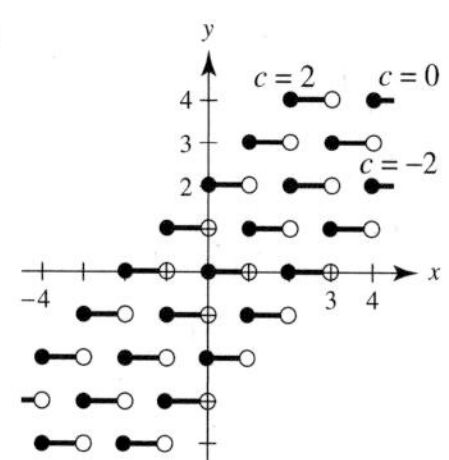

(b)

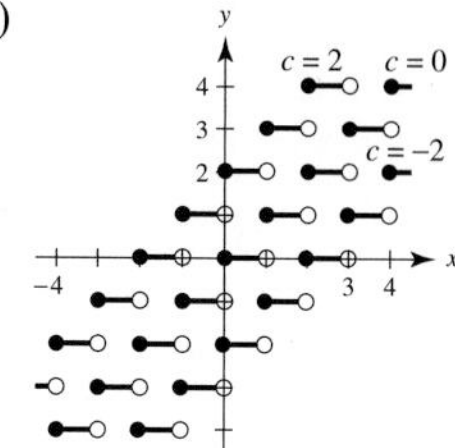

(c)

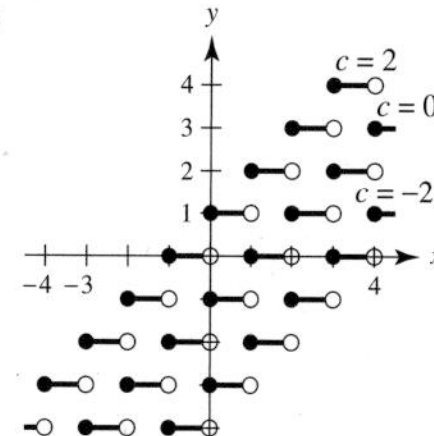

11. (a) $y = x^2 - 1$ (b) $y = 1 - (x + 1)^2$
(c) $y = -(x - 2)^2 + 6$ (d) $y = (x - 5)^2 - 3$

13. (a) $y = |x| + 5$ (b) $y = -|x + 3|$
(c) $y = |x - 2| - 4$ (d) $y = -|x - 6| - 1$

15. Horizontal shift of $y = x^3$; $y = (x - 2)^3$

17. Reflection in the x-axis of $y = x^2$; $y = -x^2$

19. Reflection in the x-axis and vertical shift of $y = \sqrt{x}$; $y = 1 - \sqrt{x}$

21. (a) $f(x) = x^2$
(b) Reflection in the x-axis and vertical shift 12 units upward
(c)

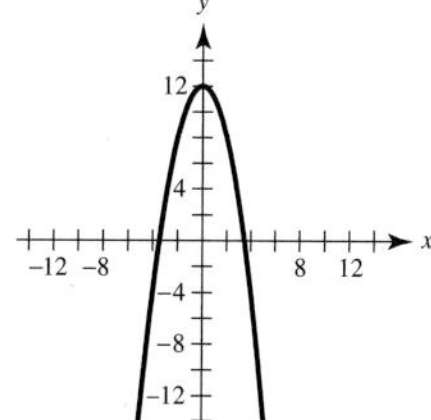

(d) $g(x) = 12 - f(x)$

23. (a) $f(x) = x^3$
(b) Vertical shift seven units upward
(c)

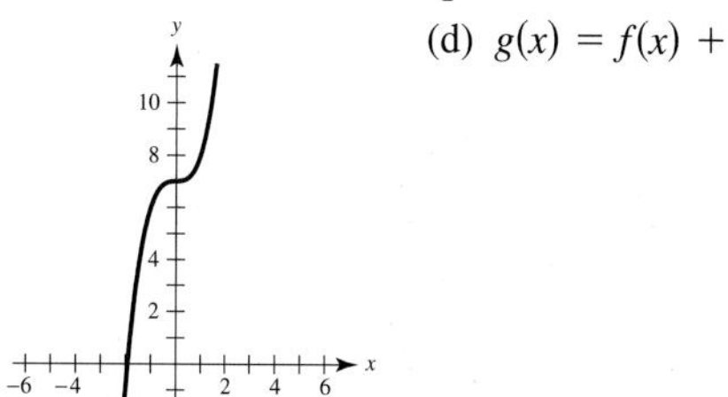

(d) $g(x) = f(x) + 7$

25. (a) $f(x) = x^2$
(b) Reflection in the x-axis, horizontal shift five units to the left, and vertical shift two units upward

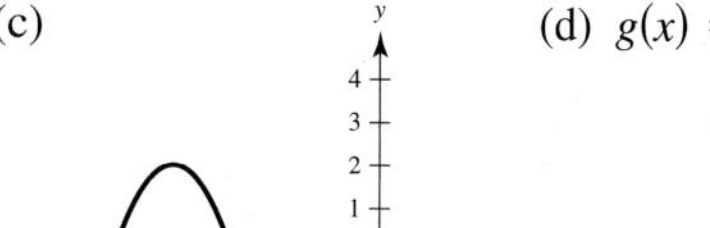

(c)

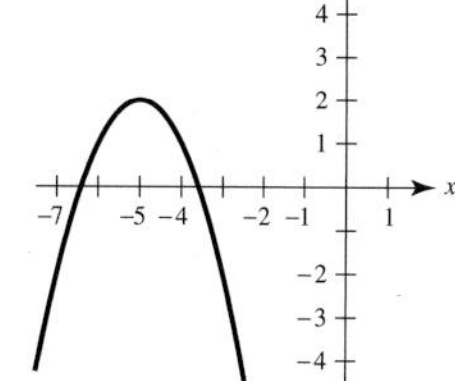

(d) $g(x) = 2 - f(x + 5)$

27. (a) $f(x) = x^3$
(b) Vertical shift two units upward and horizontal shift one unit to the right
(c)

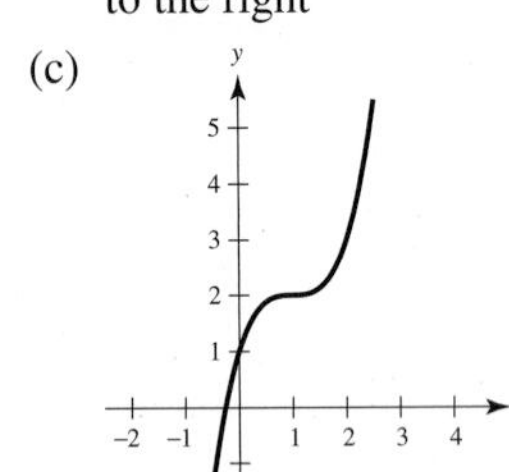
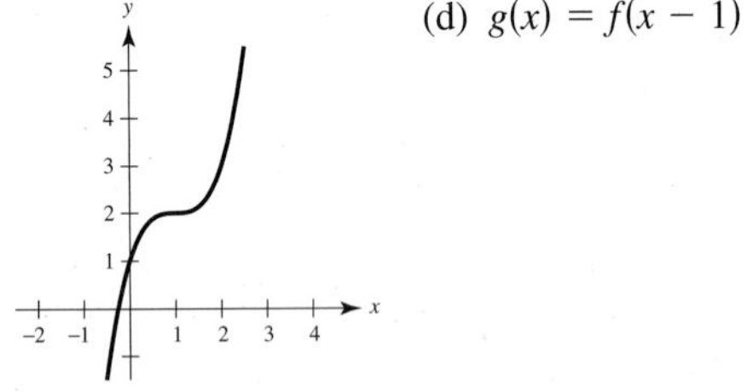

(d) $g(x) = f(x - 1) + 2$

29. (a) $f(x) = |x|$
(b) Reflection in the x-axis and vertical shift two units downward
(c)

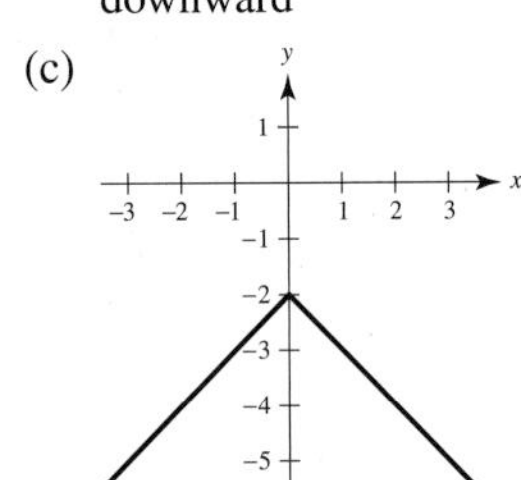

(d) $g(x) = -f(x) - 2$

31. (a) $f(x) = |x|$
(b) Reflection in the x-axis, horizontal shift four units to the left, and vertical shift eight units upward
(c)

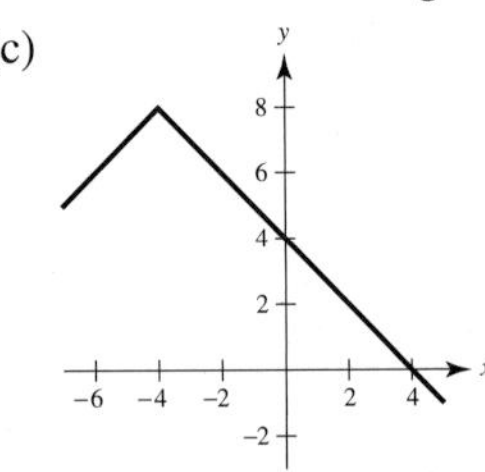

(d) $g(x) = -f(x + 4) + 8$

33. (a) $f(x) = \sqrt{x}$

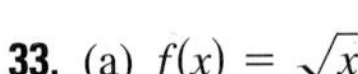

(b) Horizontal shift nine units to the right
(c)

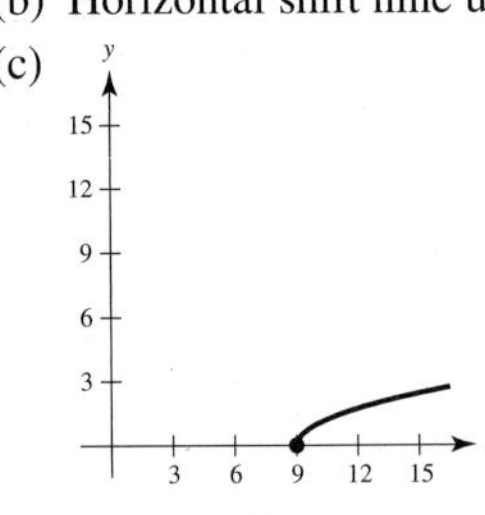

(d) $g(x) = f(x - 9)$

35. (a) $f(x) = \sqrt{x}$
(b) Reflection in the y-axis, horizontal shift seven units to the right, and vertical shift two units downward
(c)

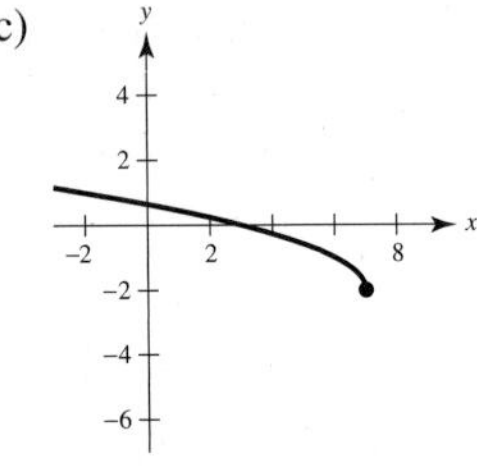

(d) $g(x) = f(7 - x) - 2$

37. $g(x) = (x - 3)^2 - 7$ **39.** $g(x) = (x - 13)^3$
41. $g(x) = -|x| + 12$ **43.** $g(x) = -\sqrt{-x + 6}$
45. (a) $y = -3x^2$ (b) $y = 4x^2 + 3$
47. (a) $y = -\frac{1}{2}|x|$ (b) $y = 3|x| - 3$
49. Vertical stretch of $y = x^3$; $y = 2x^3$
51. Reflection in the x-axis and vertical shrink of $y = x^2$; $y = -\frac{1}{2}x^2$
53. $y = -(x - 2)^3 + 2$ **55.** $y = -\sqrt{x} - 3$
57. (a)

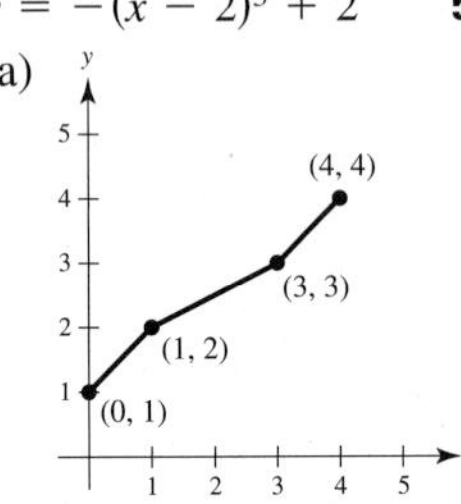

(b)

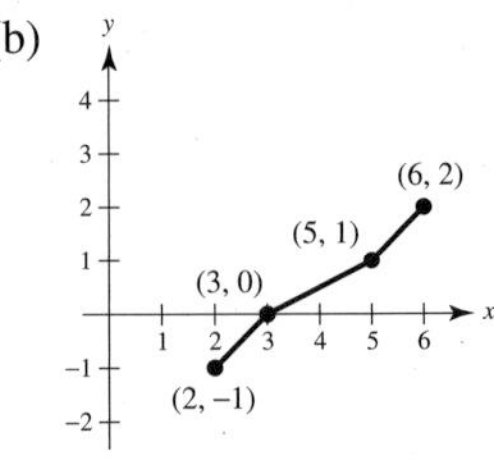

(c)

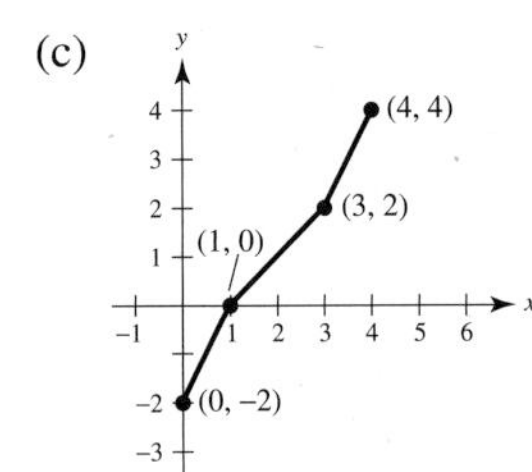

(d)

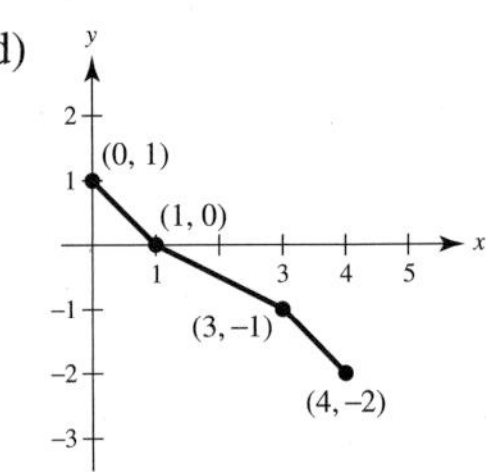

(e)

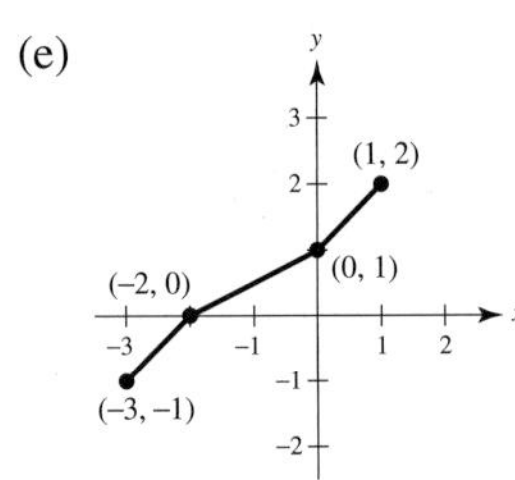

(f)

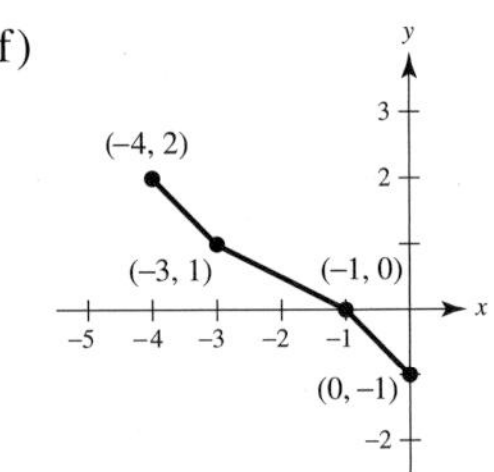

(g)

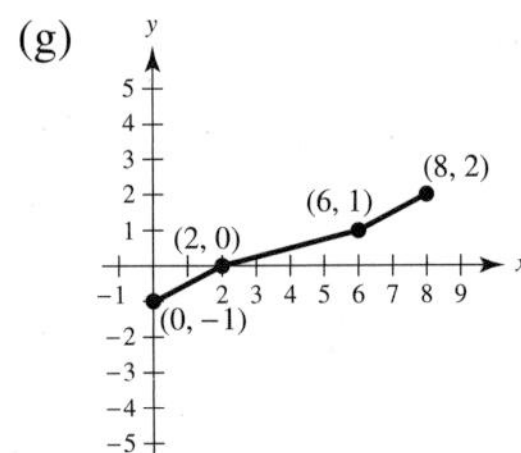

59. (a)

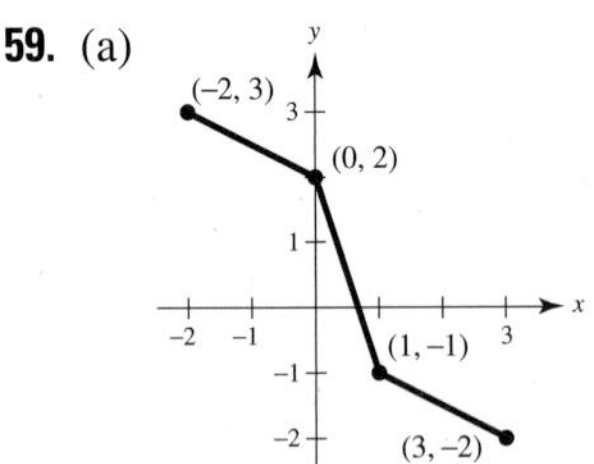

(b)

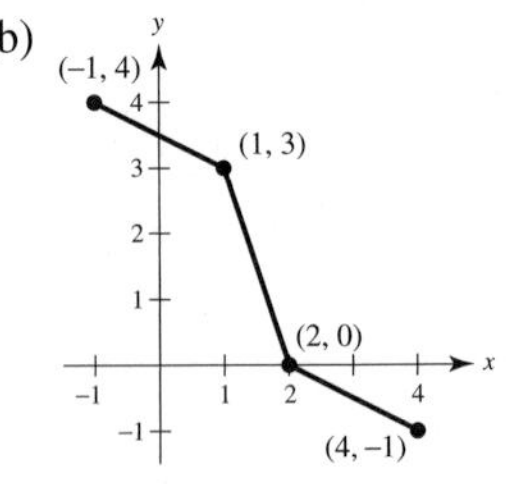

(c)

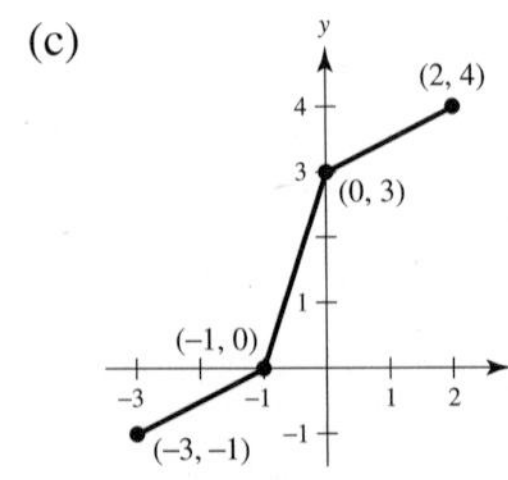

(d)

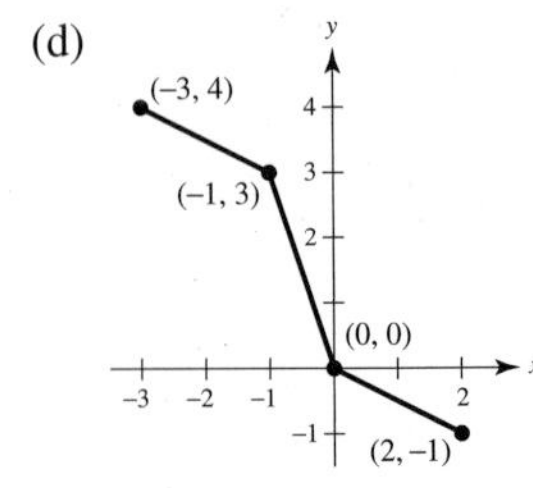

(e)

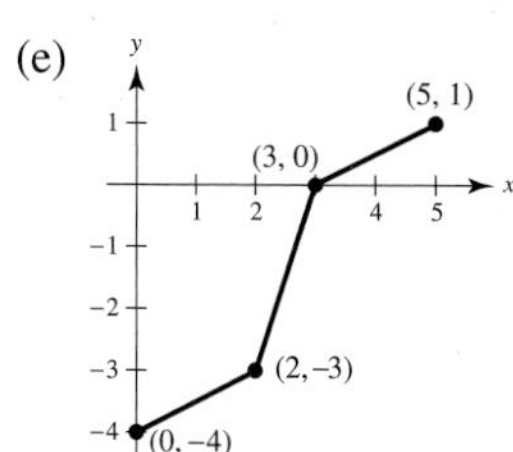

(f)

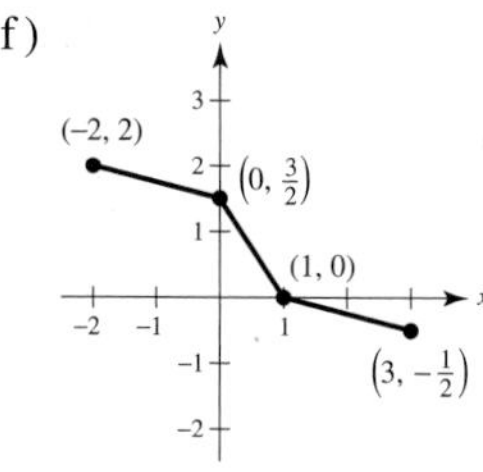

(g)

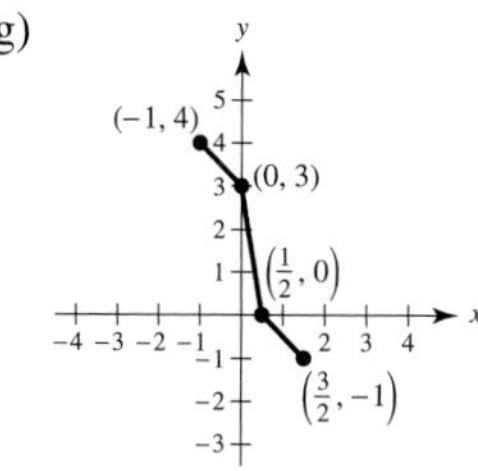

61. (a)

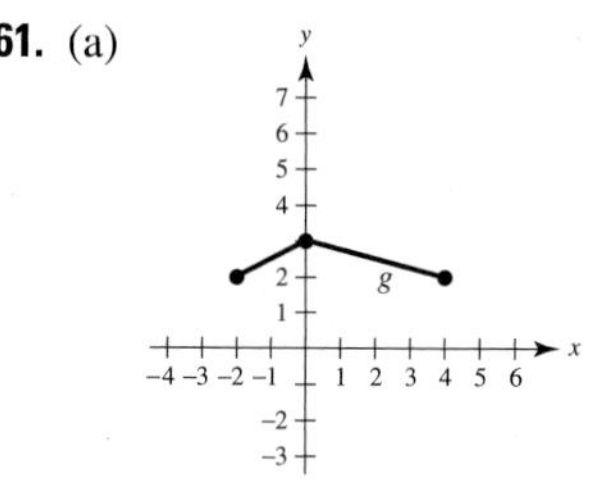

(b)

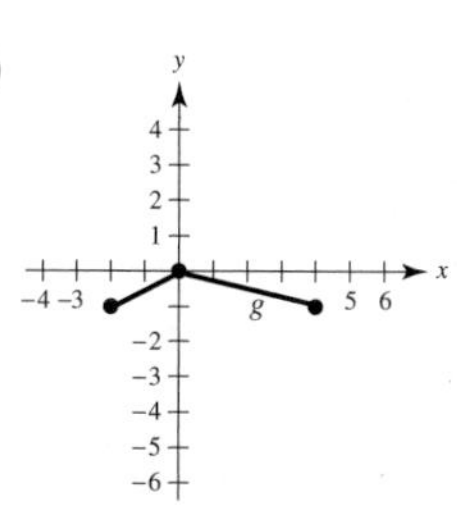

(c)

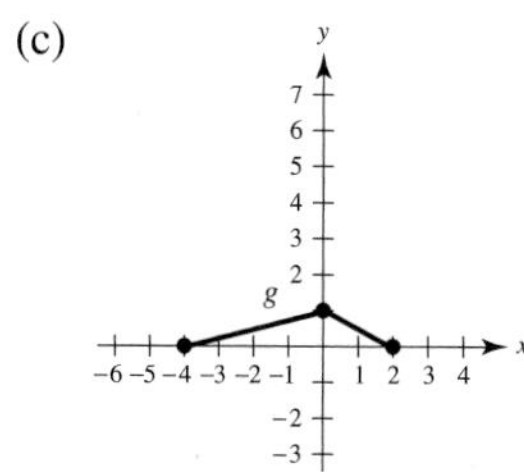

(d)

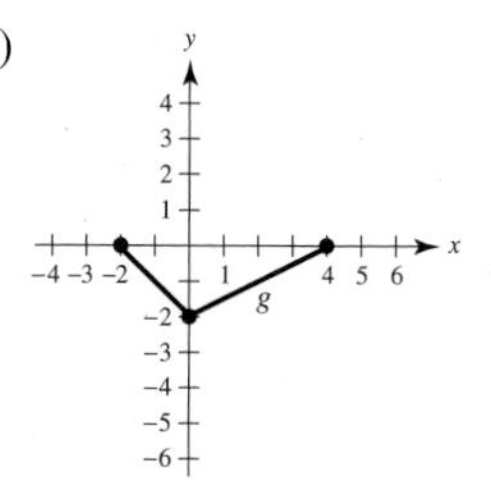

(e)

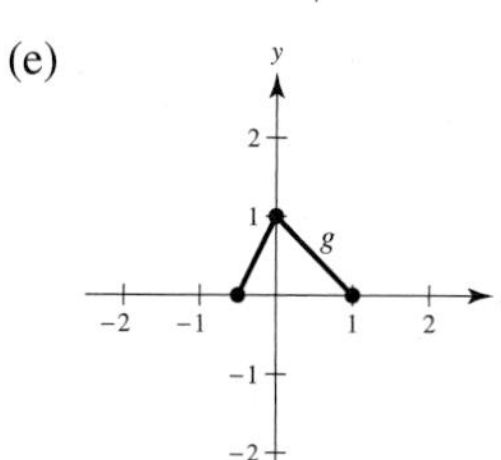

(f)

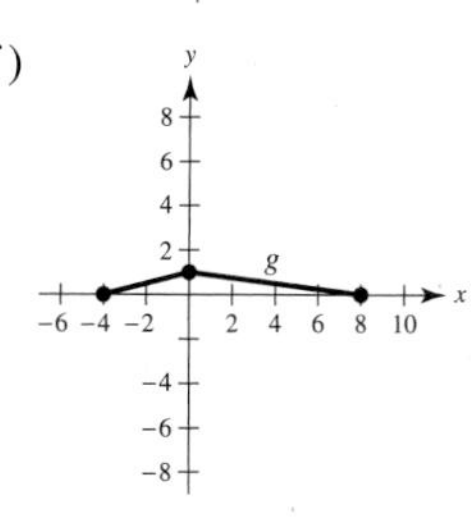

63. (a) Vertical stretch of 128.0 and a vertical shift of 527 units upward

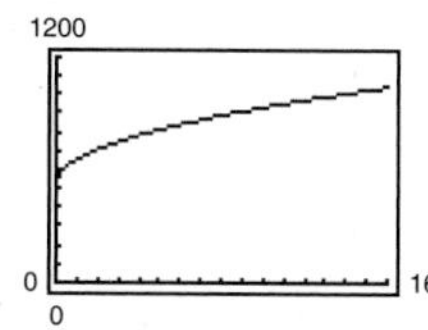

(b) $f(t) = 527 + 128\sqrt{t + 10}$; The graph is shifted 10 units to the left.

65. True. $|-x| = |x|$ **67.** True

Section 1.4 (page 112)

1. addition; subtraction; multiplication; division **3.** $g(x)$

5. (a) $2x$ (b) 4 (c) $x^2 - 4$

(d) $\dfrac{x + 2}{x - 2}$; all real numbers x except $x = 2$

7. (a) $x^2 + 4x - 5$ (b) $x^2 - 4x + 5$ (c) $4x^3 - 5x^2$

(d) $\dfrac{x^2}{4x - 5}$; all real numbers x except $x = \dfrac{5}{4}$

9. (a) $x^2 + 6 + \sqrt{1 - x}$ (b) $x^2 + 6 - \sqrt{1 - x}$

(c) $(x^2 + 6)\sqrt{1 - x}$

(d) $\dfrac{(x^2 + 6)\sqrt{1 - x}}{1 - x}$; all real numbers x such that $x < 1$

11. (a) $\dfrac{x + 1}{x^2}$ (b) $\dfrac{x - 1}{x^2}$ (c) $\dfrac{1}{x^3}$

(d) x; all real numbers x except $x = 0$

13. 3 **15.** 5 **17.** $9t^2 - 3t + 5$ **19.** 74

21. 26 **23.** $\frac{3}{5}$

25.

27.

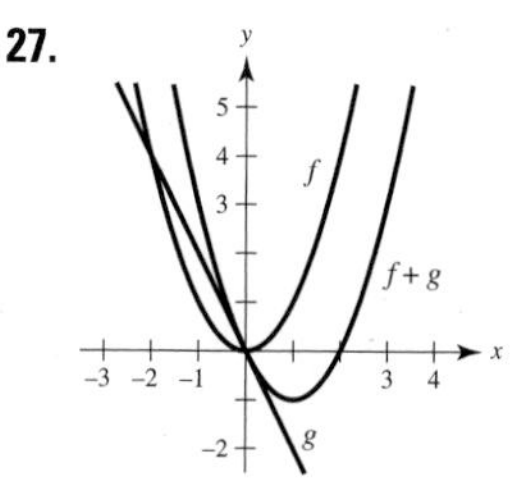

29.

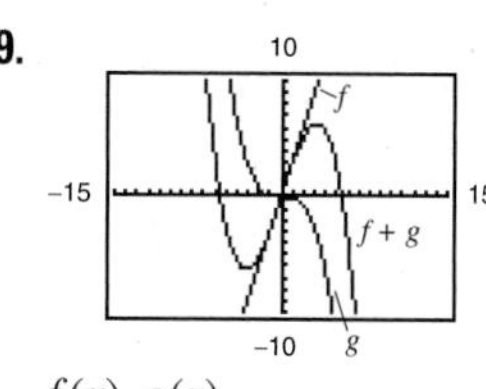

$f(x), g(x)$

31.

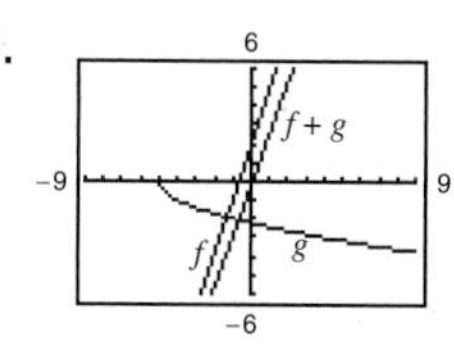

$f(x), f(x)$

33. (a) $(x - 1)^2$ (b) $x^2 - 1$ (c) $x - 2$

35. (a) x (b) x (c) $x^9 + 3x^6 + 3x^3 + 2$

37. (a) $\sqrt{x^2 + 4}$ (b) $x + 4$

Domains of f and $g \circ f$: all real numbers x such that $x \geq -4$

Domains of g and $f \circ g$: all real numbers x

39. (a) $x + 1$ (b) $\sqrt{x^2 + 1}$

Domains of f and $g \circ f$: all real numbers x

Domains of g and $f \circ g$: all real numbers x such that $x \geq 0$

41. (a) $|x + 6|$ (b) $|x| + 6$

Domains of $f, g, f \circ g$, and $g \circ f$: all real numbers x

43. (a) $\dfrac{1}{x + 3}$ (b) $\dfrac{1}{x} + 3$

Domains of f and $g \circ f$: all real numbers x except $x = 0$

Domain of g: all real numbers x

Domain of $f \circ g$: all real numbers x except $x = -3$

45. $f(x) = x^2,\ g(x) = 2x + 1$

47. $f(x) = \sqrt[3]{x},\ g(x) = x^2 - 4$

49. $f(x) = \dfrac{1}{x},\ g(x) = x + 2$ **51.** $f(x) = \dfrac{x + 3}{4 + x},\ g(x) = -x^2$

53.

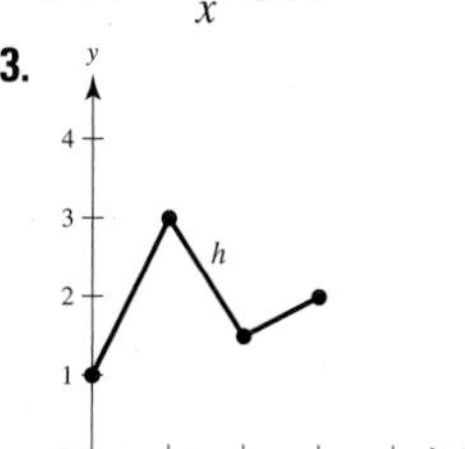

55.

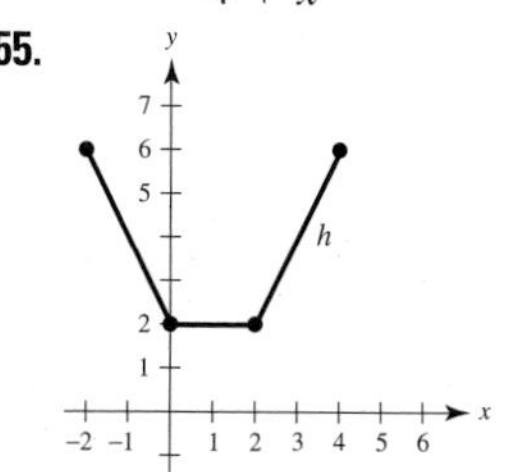

57. (a) 3 (b) 0 **59.** (a) 0 (b) 4

61. (a) $T = \frac{3}{4}x + \frac{1}{15}x^2$

(b)

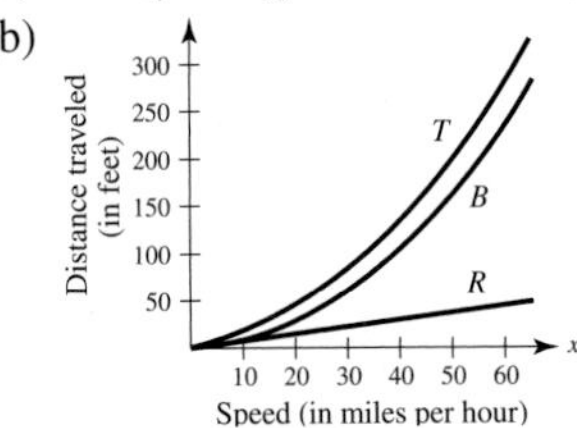

(c) The braking function $B(x)$. As x increases, $B(x)$ increases at a faster rate than $R(x)$.

63. $(B - D)(t) = -0.197t^3 + 10.17t^2 - 128.0t + 2043$, which represents the change in the United States population.

65. (a) $h(t) = \dfrac{0.0233t^4 - 0.3408t^3 + 1.556t^2 - 1.86t + 22.8}{2.78t + 282.5}$, which represents the ratio of the number of people playing tennis in the United States to the U.S. population.

(b) $h(0) = 0.0807$; $h(3) = 0.0822$; $h(6) = 0.0810$

67. (a) $r(x) = \dfrac{x}{2}$ (b) $A(r) = \pi r^2$

(c) $(A \circ r)(x) = \pi\left(\dfrac{x}{2}\right)^2$; $(A \circ r)(x)$ represents the area of the circular base of the tank on the square foundation with side length x.

69. (a) $(C \circ x)(t) = 3000t + 750$; This represents the cost of t hours of production.

(b) \$12,750 (c) 4.75 h

71. False. $(f \circ g)(x) = 6x + 1$ and $(g \circ f)(x) = 6x + 6$

73. (a) Proof

(b) $\frac{1}{2}[f(x) + f(-x)] + \frac{1}{2}[f(x) - f(-x)]$
$= \frac{1}{2}[f(x) + f(-x) + f(x) - f(-x)]$
$= \frac{1}{2}[2f(x)]$
$= f(x)$

(c) $f(x) = (x^2 + 1) + (-2x)$

$$k(x) = \frac{-1}{(x + 1)(x - 1)} + \frac{x}{(x + 1)(x - 1)}$$

Section 1.5 (page 120)

1. inverse **3.** range; domain **5.** one-to-one

7. $f^{-1}(x) = \frac{1}{6}x$ **9.** $f^{-1}(x) = x - 9$ **11.** $f^{-1}(x) = \dfrac{x - 1}{3}$

13. $f^{-1}(x) = 5x + 1$ **15.** $f^{-1}(x) = x^3$

17. (a) $f(g(x)) = f\left(\dfrac{x}{2}\right) = 2\left(\dfrac{x}{2}\right) = x$

$g(f(x)) = g(2x) = \dfrac{(2x)}{2} = x$

(b)

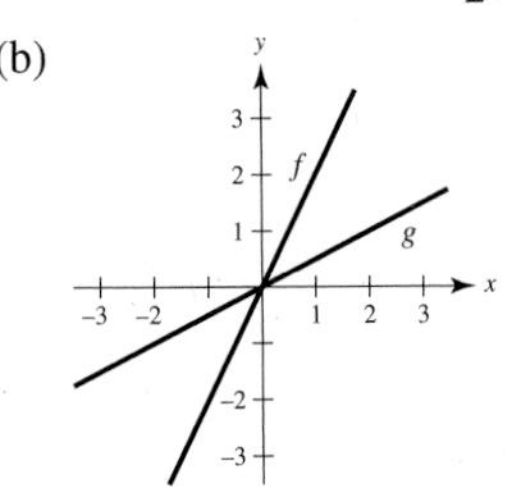

19. (a) $f(g(x)) = f\left(\dfrac{x - 1}{7}\right) = 7\left(\dfrac{x - 1}{7}\right) + 1 = x$

$g(f(x)) = g(7x + 1) = \dfrac{(7x + 1) - 1}{7} = x$

(b)

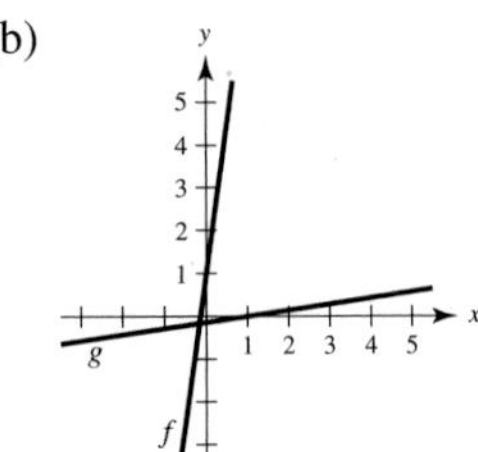

21. (a) $f(g(x)) = f\left(\sqrt[3]{8x}\right) = \dfrac{\left(\sqrt[3]{8x}\right)^3}{8} = x$

$g(f(x)) = g\left(\dfrac{x^3}{8}\right) = \sqrt[3]{8\left(\dfrac{x^3}{8}\right)} = x$

(b)

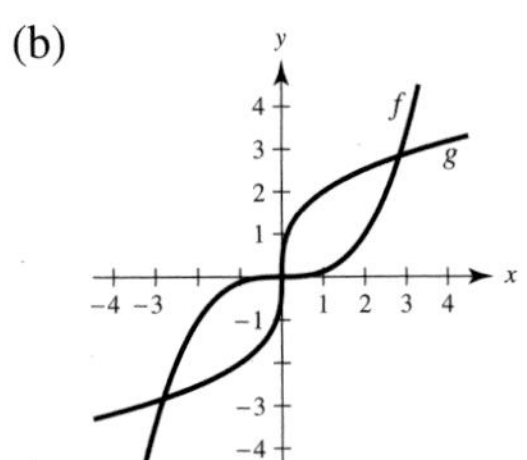

23. (a) $f(g(x)) = f(x^2 + 4),\ x \ge 0$
$= \sqrt{(x^2 + 4) - 4} = x$
$g(f(x)) = g\left(\sqrt{x - 4}\right)$
$= \left(\sqrt{x - 4}\right)^2 + 4 = x$

(b)

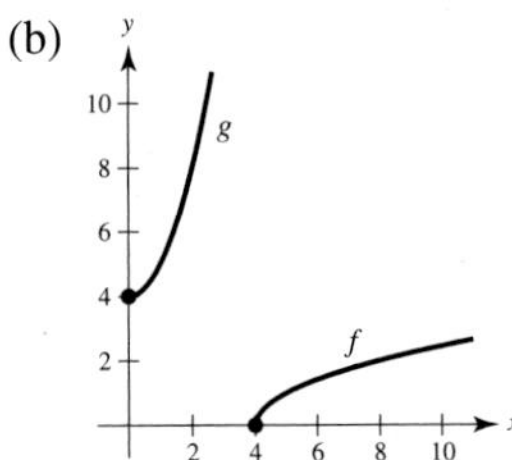

25. (a) $f(g(x)) = f\left(\sqrt{9 - x}\right),\ x \le 9$
$= 9 - \left(\sqrt{9 - x}\right)^2 = x$
$g(f(x)) = g(9 - x^2),\ x \ge 0$
$= \sqrt{9 - (9 - x^2)} = x$

(b)

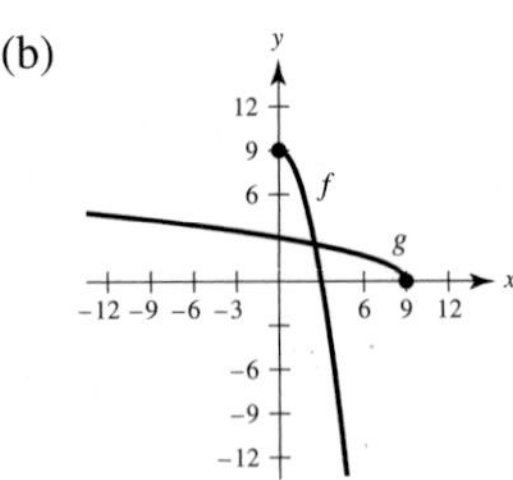

27. (a) $f(g(x)) = f\left(-\dfrac{5x+1}{x-1}\right) = \dfrac{-\left(\dfrac{5x+1}{x-1}\right)-1}{-\left(\dfrac{5x+1}{x-1}\right)+5}$

$$= \frac{-5x-1-x+1}{-5x-1+5x-5} = x$$

$g(f(x)) = g\left(\dfrac{x-1}{x+5}\right) = \dfrac{-5\left(\dfrac{x-1}{x+5}\right)-1}{\dfrac{x-1}{x+5}-1}$

$$= \frac{-5x+5-x-5}{x-1-x-5} = x$$

(b)

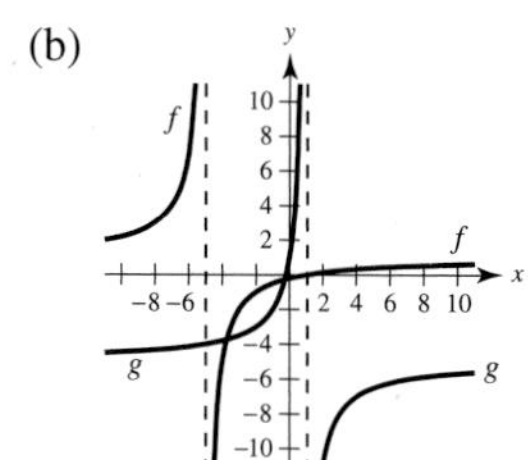

29. No **31.** Yes **33.** No

35.

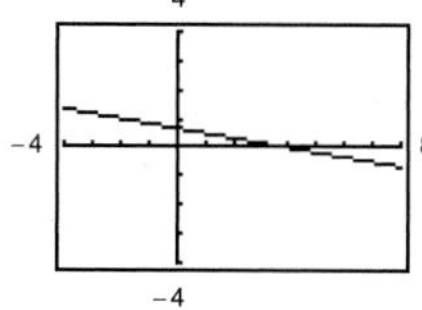

The function has an inverse.

37.

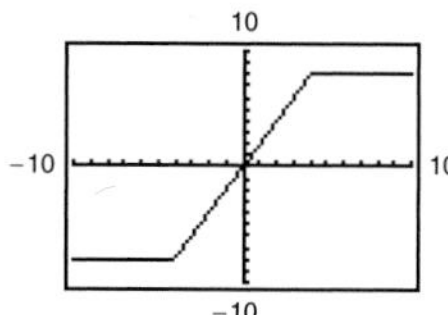

The function does not have an inverse.

39.

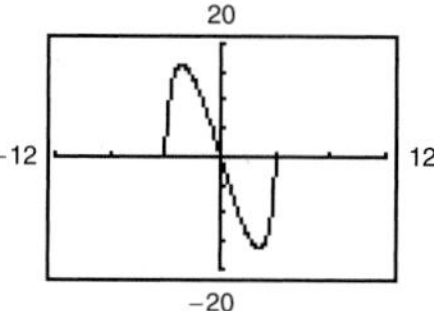

The function does not have an inverse.

41. (a) $f^{-1}(x) = \dfrac{x+3}{2}$

(b)

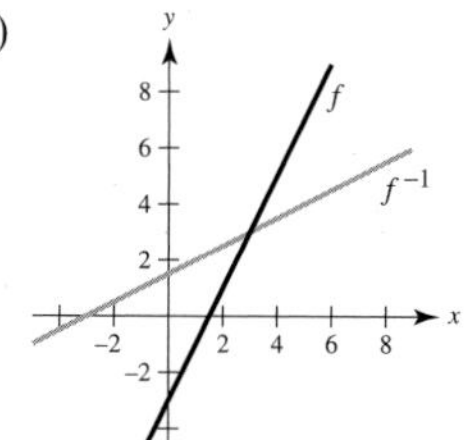

(c) The graph of f^{-1} is the reflection of the graph of f in the line $y = x$.

(d) The domains and ranges of f and f^{-1} are all real numbers.

43. (a) $f^{-1}(x) = \sqrt[5]{x+2}$

(b)

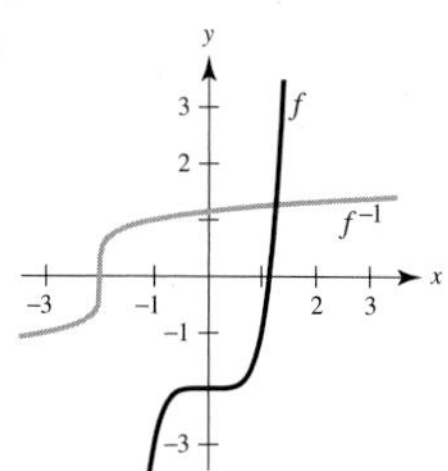

(c) The graph of f^{-1} is the reflection of the graph of f in the line $y = x$.

(d) The domains and ranges of f and f^{-1} are all real numbers.

45. (a) $f^{-1}(x) = \sqrt{4-x^2},\ 0 \le x \le 2$

(b)

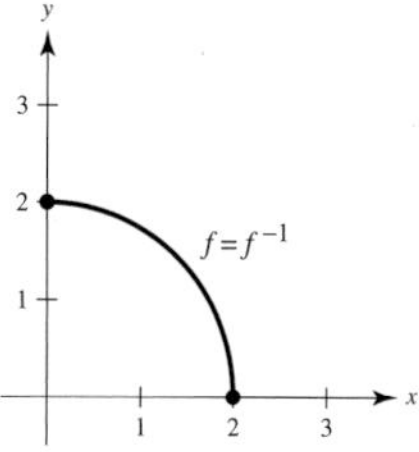

(c) The graph of f^{-1} is the same as the graph of f.

(d) The domains and ranges of f and f^{-1} are all real numbers x such that $0 \le x \le 2$.

47. (a) $f^{-1}(x) = \dfrac{4}{x}$

(b)

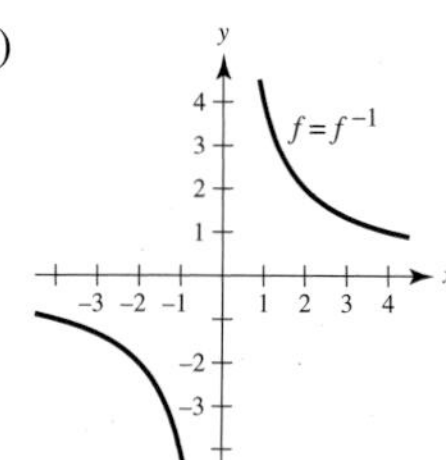

(c) The graph of f^{-1} is the same as the graph of f.

(d) The domains and ranges of f and f^{-1} are all real numbers x except $x = 0$.

49. (a) $f^{-1}(x) = \dfrac{2x+1}{x-1}$

(b)

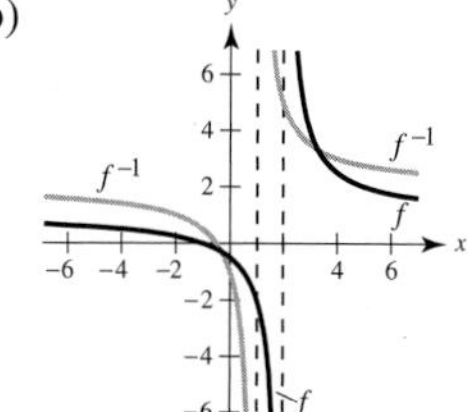

(c) The graph of f^{-1} is the reflection of the graph of f in the line $y = x$.

(d) The domain of f and the range of f^{-1} are all real numbers x except $x = 2$. The domain of f^{-1} and the range of f are all real numbers x except $x = 1$.

51. (a) $f^{-1}(x) = x^3 + 1$
(b)
(c) The graph of f^{-1} is the reflection of the graph of f in the line $y = x$.
(d) The domains and ranges of f and f^{-1} are all real numbers.

53. (a) $f^{-1}(x) = \dfrac{5x - 4}{6 - 4x}$
(b)
(c) The graph of f^{-1} is the reflection of the graph of f in the line $y = x$.
(d) The domain of f and the range of f^{-1} are all real numbers x except $x = -\frac{5}{4}$. The domain of f^{-1} and the range of f are all real numbers x except $x = \frac{3}{2}$.

55. No inverse **57.** $g^{-1}(x) = 8x$ **59.** No inverse

61. $f^{-1}(x) = \sqrt{x} - 3$ **63.** No inverse **65.** No inverse

67. $f^{-1}(x) = \dfrac{x^2 - 3}{2}$, $x \geq 0$ **69.** 32 **71.** 600

73. $2\sqrt[3]{x + 3}$ **75.** $\dfrac{x + 1}{2}$ **77.** $\dfrac{x + 1}{2}$ **79.** c **81.** a

83.

x	-2	0	2	4	6	8
$f^{-1}(x)$	-2	-1	0	1	2	3

85. (a) $y = \dfrac{x - 10}{0.75}$
x = hourly wage; y = number of units produced
(b) 19 units

87. (a) $y = \sqrt{\dfrac{x - 245.50}{0.03}}$, $245.5 < x < 545.5$
x = degrees Fahrenheit; y = % load
(b)
(c) $0 < x \leq 92.11$

89. (a) 5
(b) f^{-1} yields the year for a given amount.
(c) $f(t) = 11.4t + 55.4$
(d) $f^{-1}(t) = (t - 55.4)/11.4$
(e) 15

91. False. $f(x) = x^2$ has no inverse. **93.** Proof

95. This situation could be represented by a one-to-one function if the runner does not stop to rest. The inverse function would represent the time in hours for a given number of miles completed.

97. This situation could be represented by a one-to-one function if the population continues to increase. The inverse function would represent the year for a given population.

Section 1.6 (page 130)

1. variation; regression **3.** least squares regression
5. directly proportional **7.** directly proportional
9. combined

11.

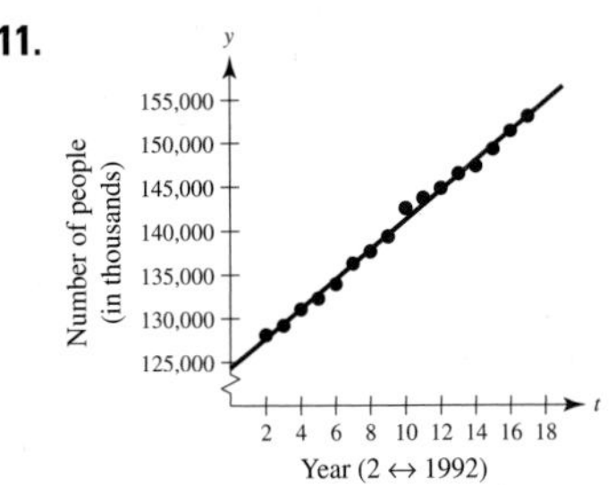

The model is a good fit for the actual data.

13. (a) and (b)

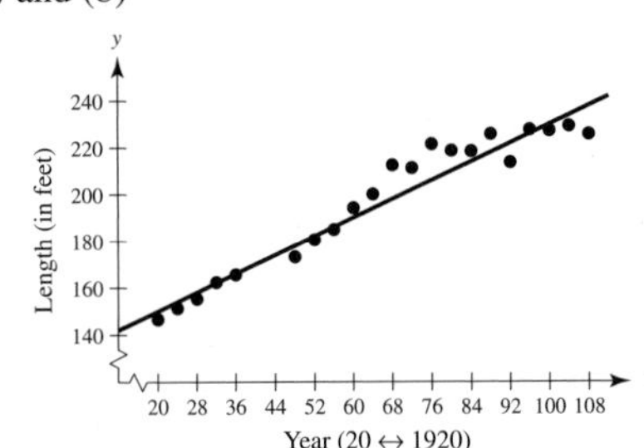

$y \approx t + 130$
(c) $y = 1.01t + 130.82$ (d) The models are similar.
(e) Part (b): 242 ft; Part (c): 243.94 ft
(f) Answers will vary.

15. (a)
(b) $S = 38.3t + 224$
(c)

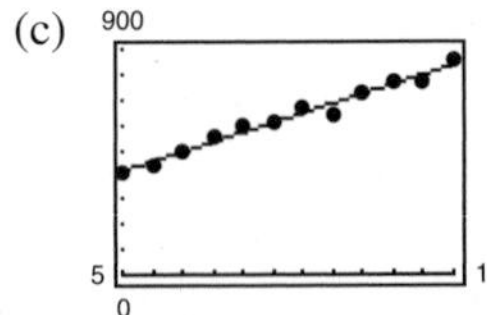

The model is a good fit.
(d) 2007: \$875.1 million; 2009: \$951.7 million
(e) Each year the annual gross ticket sales for Broadway shows in New York City increase by \$38.3 million.

17. Inversely

19.

x	2	4	6	8	10
$y = k/x^2$	$\frac{5}{2}$	$\frac{5}{8}$	$\frac{5}{18}$	$\frac{5}{32}$	$\frac{1}{10}$

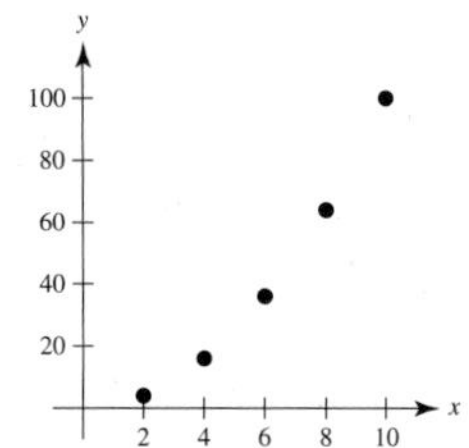

21.

x	2	4	6	8	10
$y = kx^2$	4	16	36	64	100

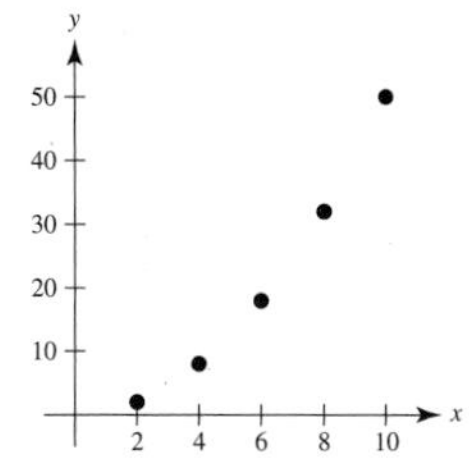

23.

x	2	4	6	8	10
$y = kx^2$	2	8	18	32	50

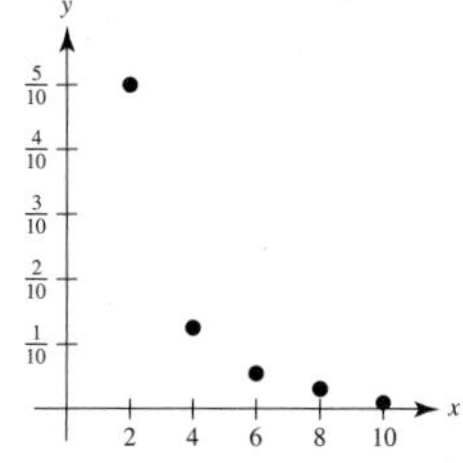

25.

x	2	4	6	8	10
$y = k/x^2$	$\frac{1}{2}$	$\frac{1}{8}$	$\frac{1}{18}$	$\frac{1}{32}$	$\frac{1}{50}$

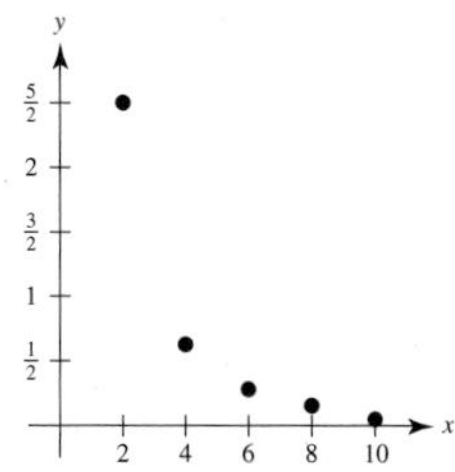

27. $y = \dfrac{5}{x}$ **29.** $y = -\dfrac{7}{10}x$ **31.** $y = \dfrac{12}{5}x$

33. $y = 205x$ **35.** $A = kr^2$ **37.** $y = k/x^2$

39. $F - kg/r^2$

41. The area of a triangle is jointly proportional to its base and its height.

43. The volume of a sphere varies directly as the cube of its radius.

45. Average speed is directly proportional to the distance and inversely proportional to the time.

47. Good approximation **49.** Poor approximation

51.

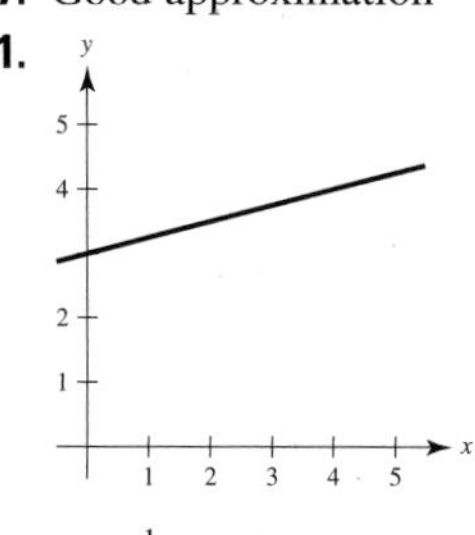

$y = \frac{1}{4}x + 3$

53.

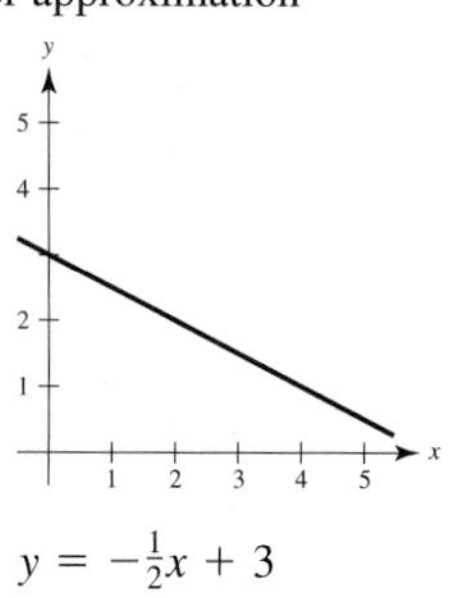

$y = -\frac{1}{2}x + 3$

55. $I = 0.035P$

57. Model: $y = \frac{33}{13}x$; 25.4 cm, 50.8 cm

59. $y = 0.0368x$; \$8280 **61.** $P = \dfrac{k}{V}$ **63.** $F = \dfrac{km_1m_2}{r^2}$

65. (a) 0.05 m (b) $176\frac{2}{3}$ N **67.** 39.47 lb

69. $A = \pi r^2$ **71.** $y = \dfrac{28}{x}$ **73.** $F = 14rs^3$ **75.** $z = \dfrac{2x^2}{3y}$

77. About 0.61 mi/h **79.** 506 ft **81.** 1470 J

83. (a) The velocity is increased by one-third.
(b) The velocity is decreased by one-fourth.

85. (a)

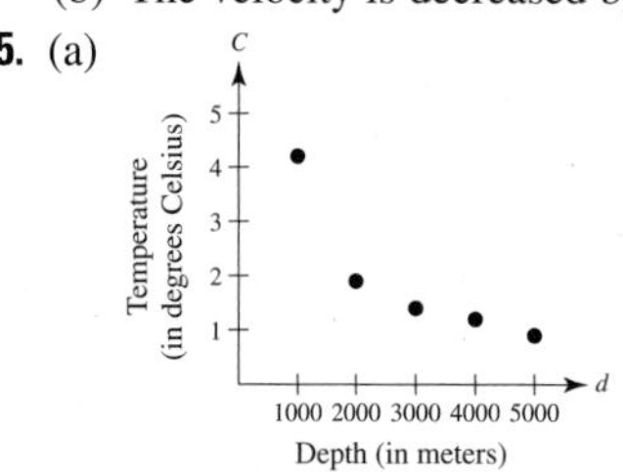

(b) Yes. $k_1 = 4200, k_2 = 3800, k_3 = 4200,$ $k_4 = 4800, k_5 = 4500$

(c) $C = \dfrac{4300}{d}$

(d)

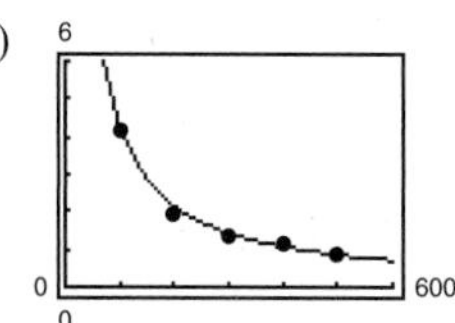

(e) About 1433 m

87. (a)

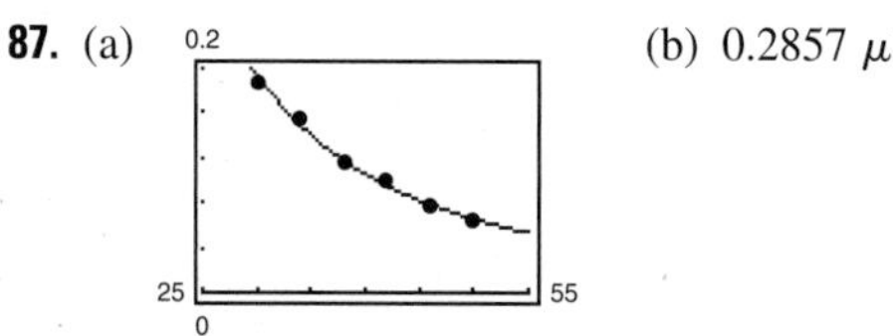

(b) $0.2857\ \mu W/cm^2$

89. False. "y varies directly as x" is equivalent to "y is directly proportional to x" or $y = kx$. "y is inversely proportional to x" is equivalent to "y varies inversely as x" or $y = k/x$.

91. False. E is jointly proportional to the mass of an object and the square of its velocity.

93. The accuracy is questionable when based on such limited data.

95. (a) y will change by a factor of one-fourth.
(b) y will change by a factor of four.

Review Exercises (page 137)

1. No **3.** Yes

5. (a) 5 (b) 17 (c) $t^4 + 1$ (d) $t^2 + 2t + 2$

7. All real numbers x such that $-5 \le x \le 5$

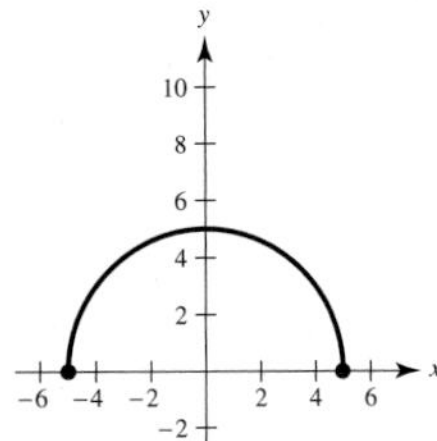

9. All real numbers x except $x = 3, -2$

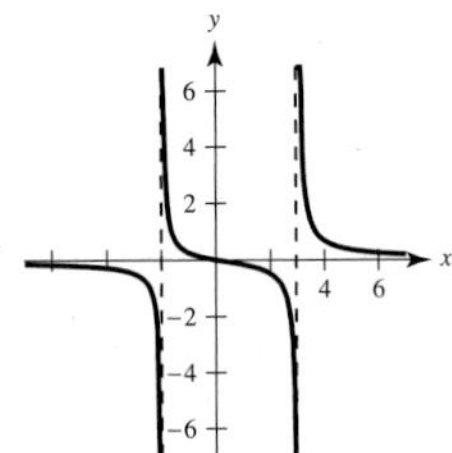

11. (a) 16 ft/sec (b) 1.5 sec (c) -16 ft/sec

13. $4x + 2h + 3,\ h \neq 0$ **15.** Function

17. $\frac{7}{3}, 3$ **19.** $-\frac{3}{8}$

21. Increasing on $(0, \infty)$
Decreasing on $(-\infty, -1)$
Constant on $(-1, 0)$

23.

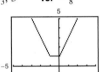

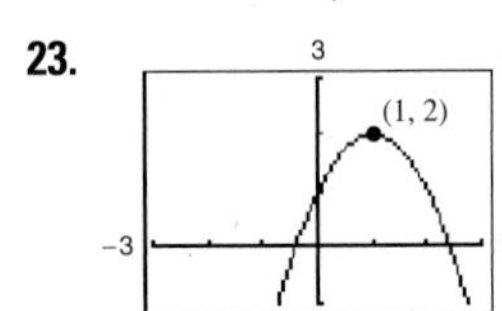

25.

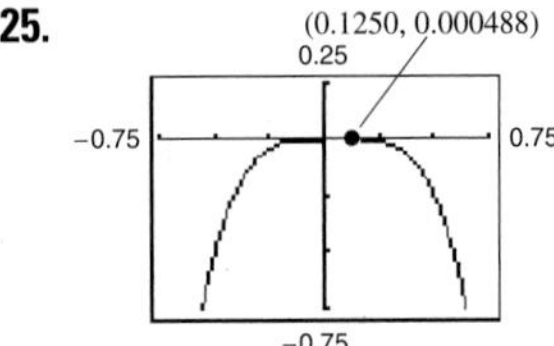

27. $f(x) = -3x$

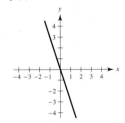

29.

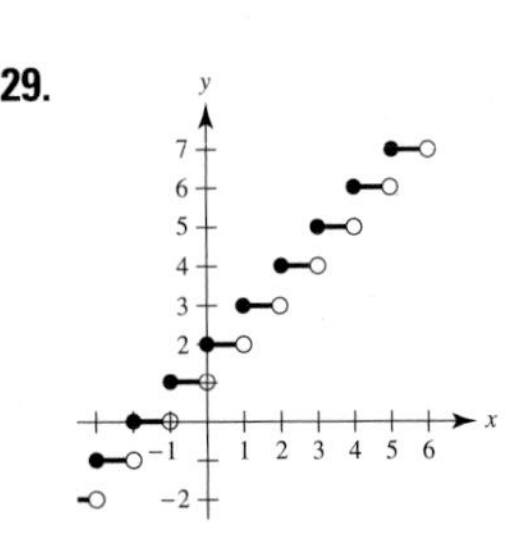

31.

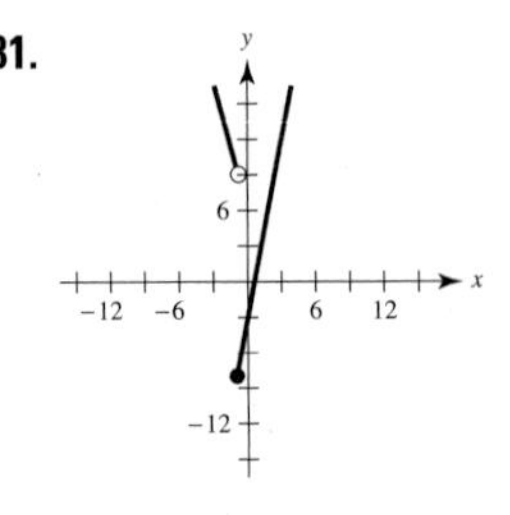

33. Neither **35.** Odd

37. $y = x^3$; Horizontal shift four units to the left and vertical shift four units upward

39. (a) $f(x) = x^2$
(b) Vertical shift nine units downward
(c)

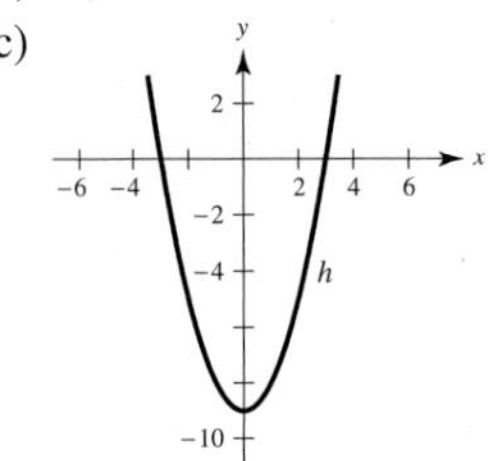

(d) $h(x) = f(x) - 9$

41. (a) $f(x) = \sqrt{x}$
(b) Reflection in the x-axis and vertical shift four units upward
(c)

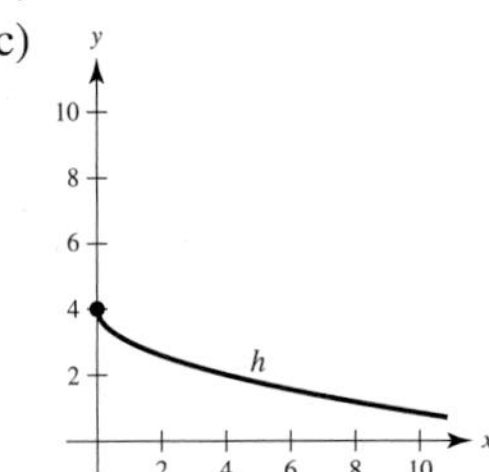

(d) $h(x) = -f(x) + 4$

43. (a) $f(x) = x^2$
(b) Reflection in the x-axis, horizontal shift two units to the left, and vertical shift three units upward
(c)

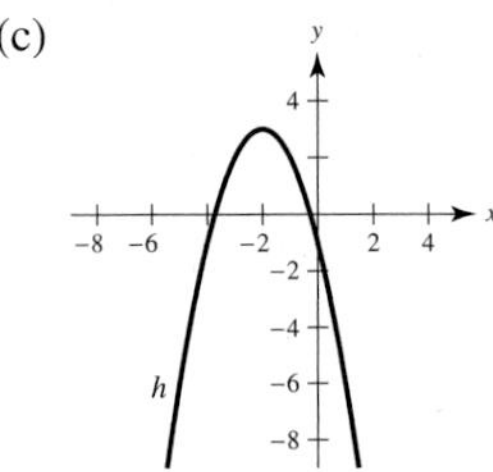

(d) $h(x) = -f(x + 2) + 3$

45. (a) $f(x) = [\![x]\!]$
(b) Reflection in the x-axis and vertical shift six units upward
(c)

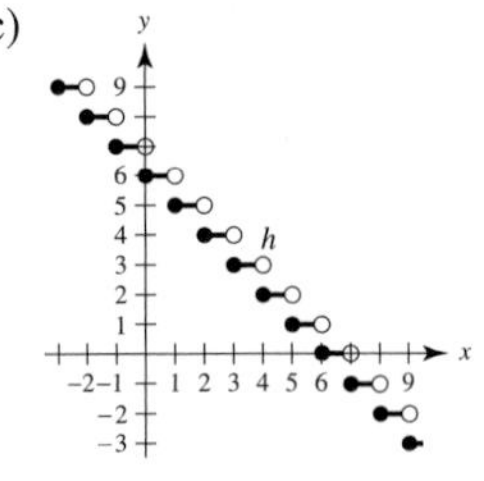

(d) $h(x) = -f(x) + 6$

47. (a) $f(x) = |x|$
(b) Reflections in the x-axis and the y-axis, horizontal shift four units to the right, and vertical shift six units upward
(c)
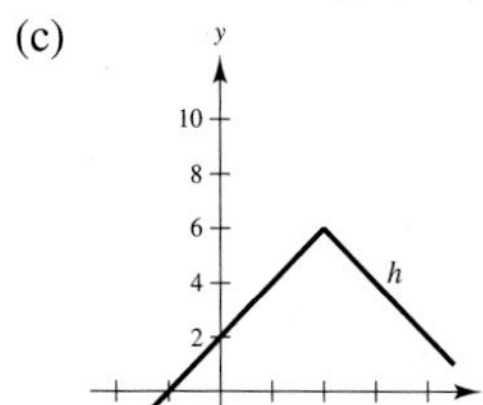
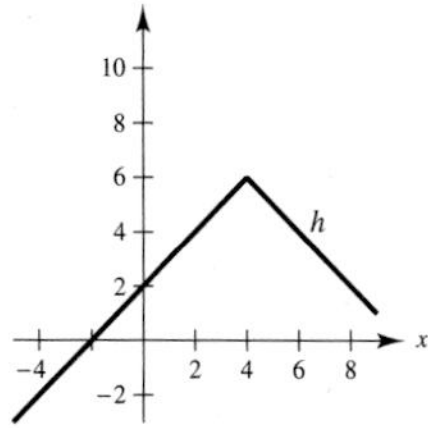
(d) $h(x) = -f(-x + 4) + 6$

49. (a) $f(x) = [\![x]\!]$
(b) Horizontal shift nine units to the right and vertical stretch
(c)
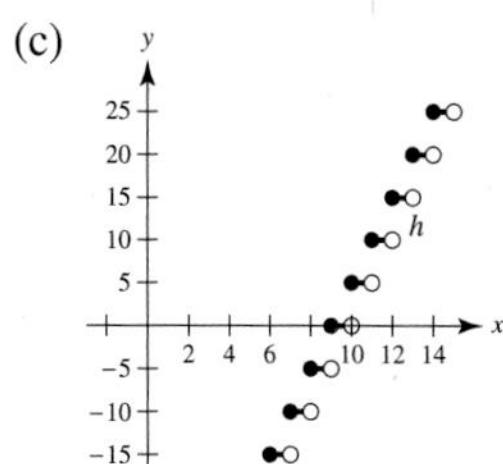
(d) $h(x) = 5\,f(x - 9)$

51. (a) $f(x) = \sqrt{x}$
(b) Reflection in the x-axis, vertical stretch, and horizontal shift four units to the right
(c)
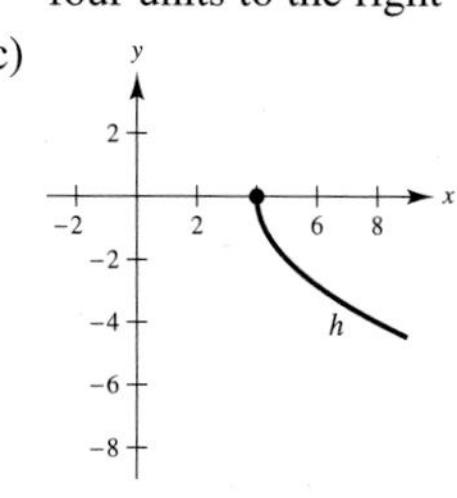
(d) $h(x) = -2f(x - 4)$

53. (a) $x^2 + 2x + 2$ (b) $x^2 - 2x + 4$
(c) $2x^3 - x^2 + 6x - 3$
(d) $\dfrac{x^2 + 3}{2x - 1}$; all real numbers x except $x = \dfrac{1}{2}$

55. (a) $x - \frac{8}{3}$ (b) $x - 8$
Domains of $f, g, f \circ g$, and $g \circ f$: all real numbers

57. $f(x) = x^3, g(x) = 1 - 2x$

59. (a) $(r + c)(t) = 178.8t + 856$; This represents the average annual expenditures for both residential and cellular phone services.
(b)
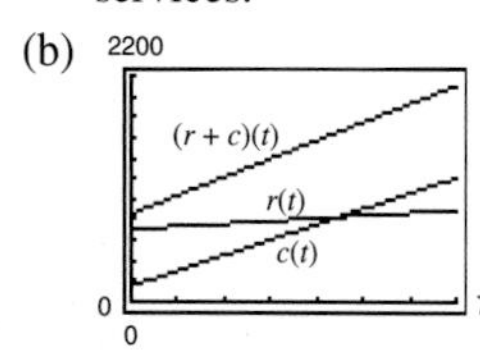

(c) $(r + c)(13) = 3180.4$

61. $f^{-1}(x) = \frac{1}{3}(x - 8)$ **63.** The function has an inverse.

65.
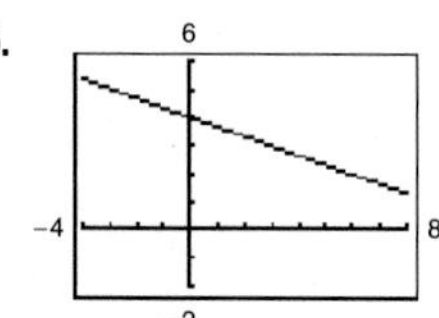

The function has an inverse.

67.
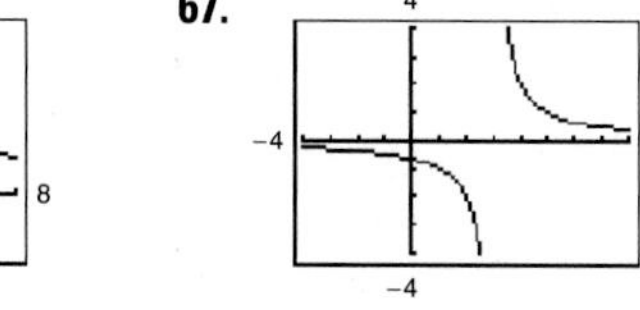

The function has an inverse.

69. (a) $f^{-1}(x) = 2x + 6$
(b)
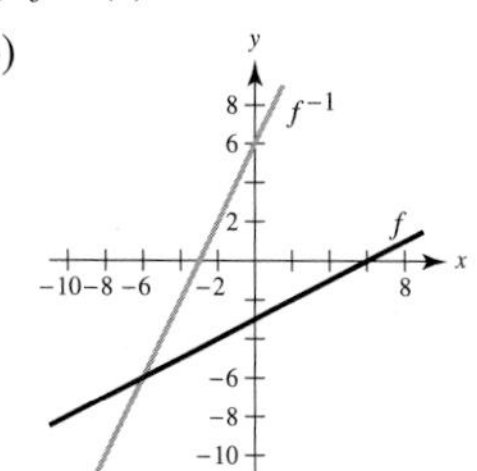
(c) The graph of f^{-1} is the reflection of the graph of f in the line $y = x$.
(d) Both f and f^{-1} have domains and ranges that are all real numbers.

71. (a) $f^{-1}(x) = x^2 - 1,\ x \geq 0$
(b)
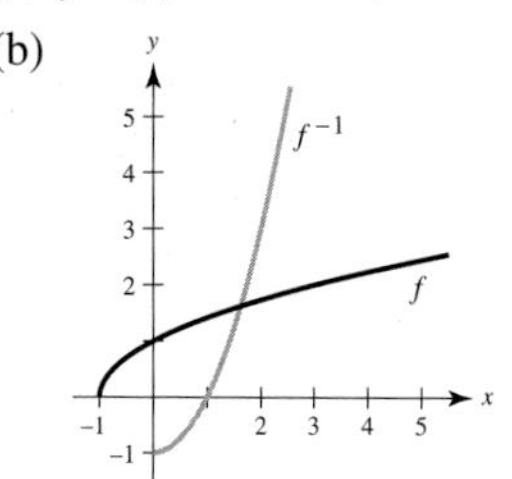
(c) The graph of f^{-1} is the reflection of the graph of f in the line $y = x$.
(d) f has a domain of $[-1, \infty)$ and a range of $[0, \infty)$; f^{-1} has a domain of $[0, \infty)$ and a range of $[-1, \infty)$.

73. $x > 4$; $f^{-1}(x) = \sqrt{\dfrac{x}{2}} + 4, x \neq 0$

75. (a)
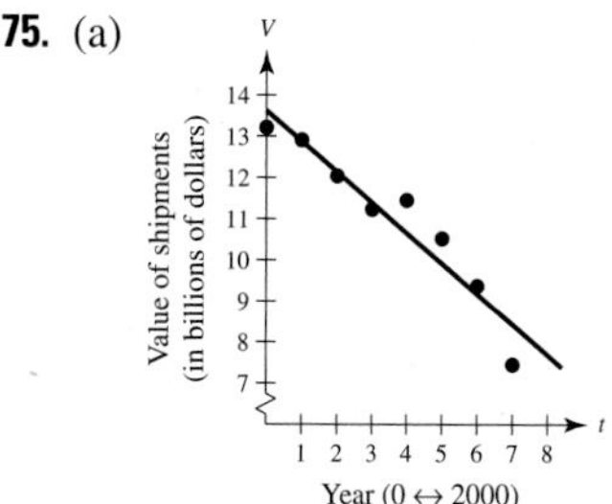

(b) The model is a good fit for the actual data.

77. Model: $k = \frac{8}{5}m$; 3.2 km, 16 km

79. A factor of 4 **81.** About 2 h, 26 min

83. $y = 49.5/x$

85. False. The graph is reflected in the x-axis, shifted 9 units to the left, and then shifted 13 units downward.
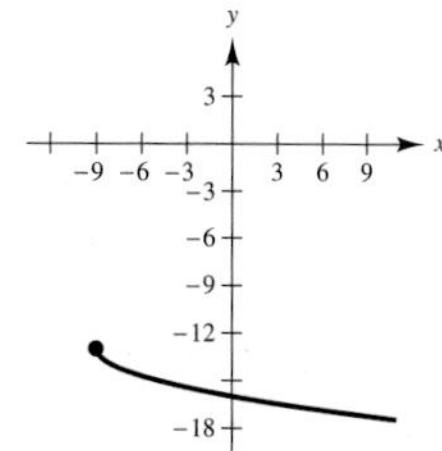

87. The Vertical Line Test is used to determine if the graph of y is a function of x. The Horizontal Line Test is used to determine if a function has an inverse function.

89. The y-intercept is 0.

Chapter Test (page 140)

1. (a) $-\frac{1}{8}$ (b) $-\frac{1}{28}$ (c) $\frac{\sqrt{x}}{x^2 - 18x}$

2. $x \le 3$

3. (a) $0, \pm 0.4314$

(b)

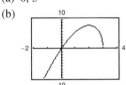

(c) Increasing on $(-0.31, 0), (0.31, \infty)$
Decreasing on $(-\infty, -0.31), (0, 0.31)$

(d) Even

4. (a) 0, 3

(b)

(c) Increasing on $(-\infty, 2)$
Decreasing on $(2, 3)$

(d) Neither

5. (a) -5

(b)

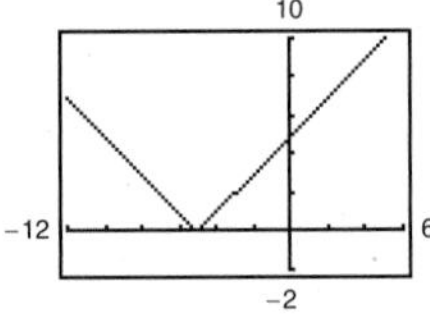

(c) Increasing on $(-5, \infty)$
Decreasing on $(-\infty, -5)$

(d) Neither

6. (a) $f(x) = -\frac{1}{2}x + 7$

(b)

7. (a) $f(x) = \frac{6}{7}x - \frac{45}{7}$

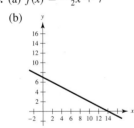

(b)

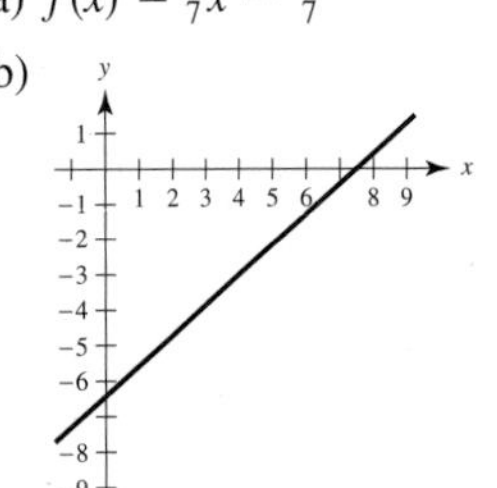

8.

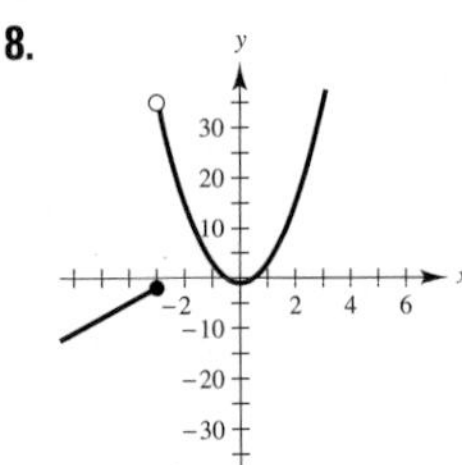

9. Reflection in the x-axis of $y = [\![x]\!]$

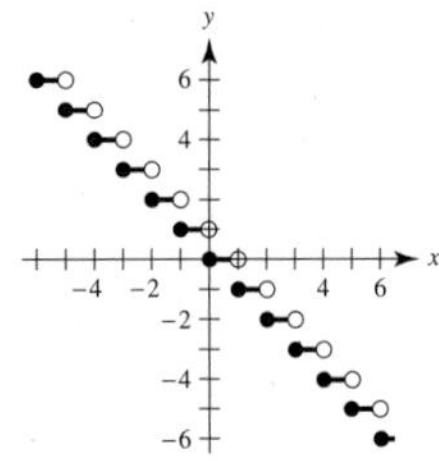

10. Reflection in the x-axis, horizontal shift, and vertical shift of $y = \sqrt{x}$

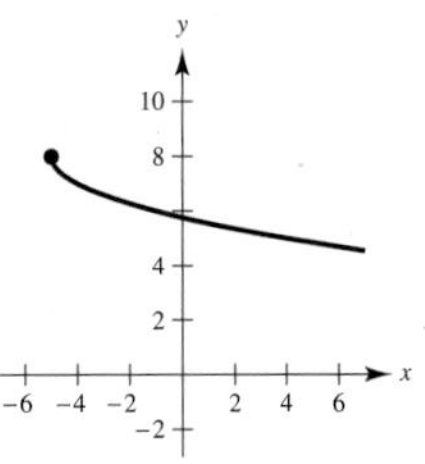

11. Reflection in the x-axis, vertical stretch, horizontal shift, and vertical shift of $y = x^3$

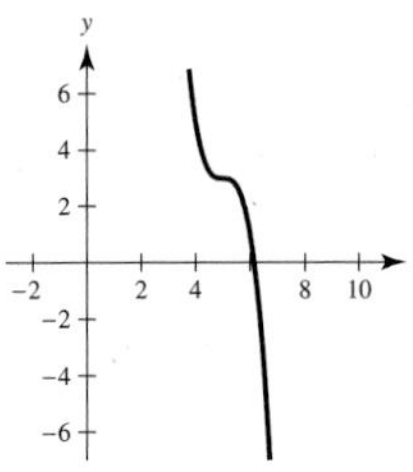

12. (a) $2x^2 - 4x - 2$ (b) $4x^2 + 4x - 12$

(c) $-3x^4 - 12x^3 + 22x^2 + 28x - 35$

(d) $\frac{3x^2 - 7}{-x^2 - 4x + 5}, \quad x \ne -5, 1$

(e) $3x^4 + 24x^3 + 18x^2 - 120x + 68$

(f) $-9x^4 + 30x^2 - 16$

13. (a) $\frac{1 + 2x^{3/2}}{x}, \quad x > 0$ (b) $\frac{1 - 2x^{3/2}}{x}, \quad x > 0$

(c) $\frac{2\sqrt{x}}{x}, \quad x > 0$ (d) $\frac{1}{2x^{3/2}}, \quad x > 0$

(e) $\frac{\sqrt{x}}{2x}, \quad x > 0$ (f) $\frac{2\sqrt{x}}{x}, \quad x > 0$

14. $f^{-1}(x) = \sqrt[3]{x - 8}$ **15.** No inverse

16. $f^{-1}(x) = \left(\frac{1}{3}x\right)^{2/3}, \ x \ge 0$ **17.** $v = 6\sqrt{s}$

18. $A = \frac{25}{6}xy$ **19.** $b = \frac{48}{a}$

20. (a)

(b) $S = 2.3t + 37$

(c)

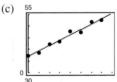

The model is a good fit for the data.

(d) \$71.5 billion; Answers will vary. Sample answer: Yes, this seems reasonable because the model increases steadily.

P.S. Problem Solving (page 141)

1. Mapping numbers onto letters is not a function because each number corresponds to three letters.

Mapping letters onto numbers is a function because every letter is assigned exactly one number.

3. Proof

5. (a)

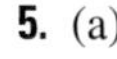

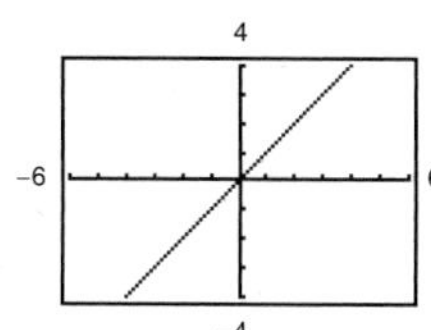

(b)

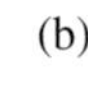

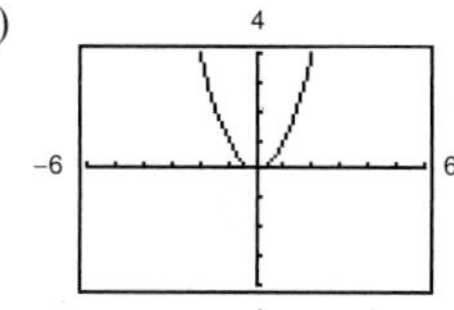

(c)

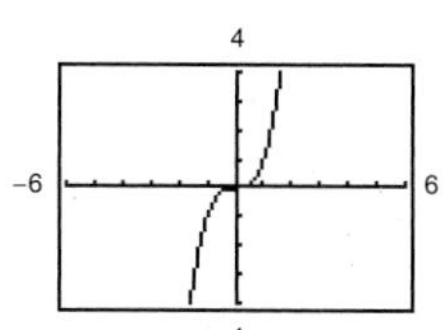

(d)

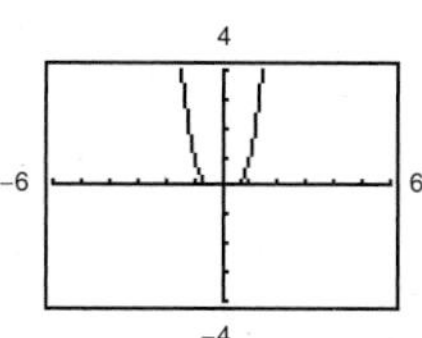

(e)

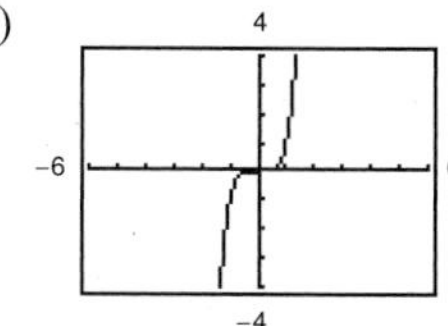

(f)

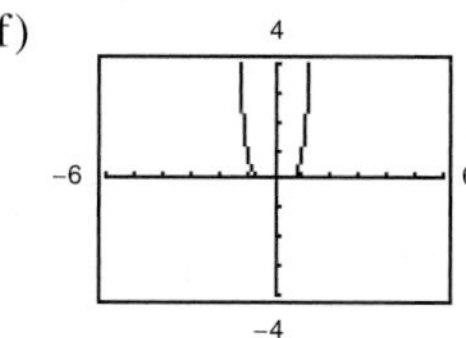

All the graphs pass through the origin. The graphs of the odd powers of x are symmetric with respect to the origin, and the graphs of the even powers are symmetric with respect to the y-axis. As the powers increase, the graphs become flatter in the interval $-1 < x < 1$.

(g) Both graphs will pass through the origin. $y = x^7$ will be symmetric with respect to the origin, and $y = x^8$ will be symmetric with respect to the y-axis.

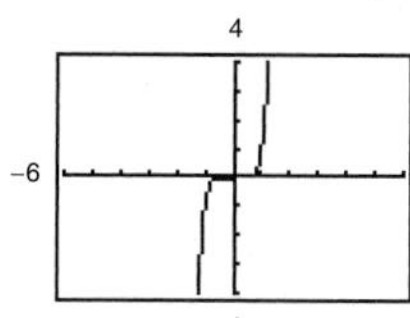

7. $(-2, 0), (-1, 1), (0, 2)$

9. Answers will vary.

11.

x	1	3	4	6
y	1	2	6	7

x	1	2	6	7
$f^{-1}(x)$	1	3	4	6

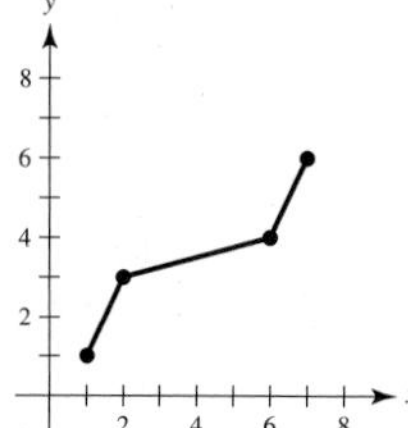

13.

x	-2	-1	3	4
y	6	0	-2	-3

x	-3	-2	0	6
$f^{-1}(x)$	4	3	-1	-2

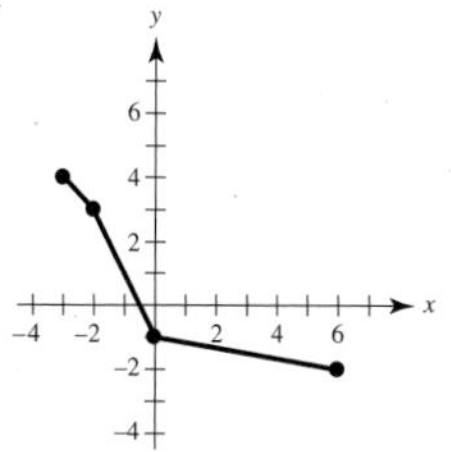

15. $k = \frac{1}{4}$

17. (a)

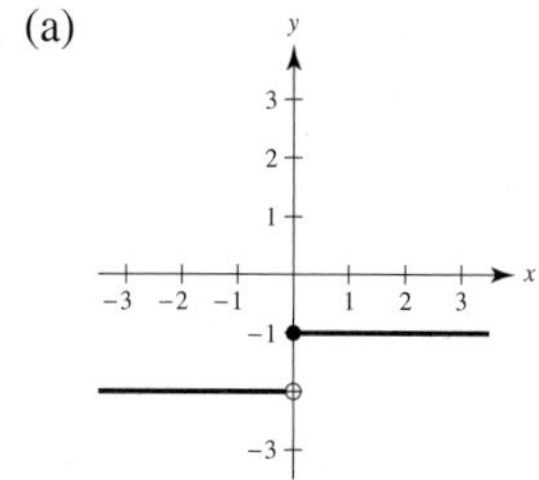

(b)

(c)

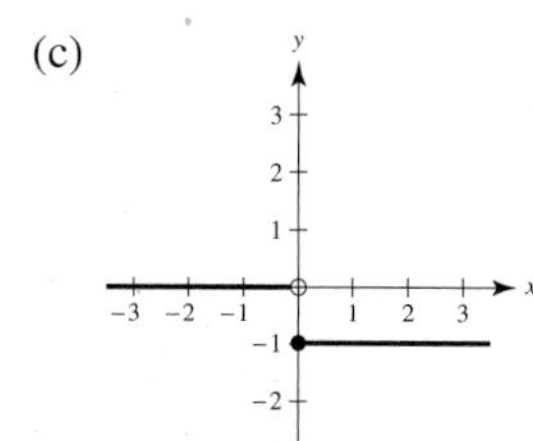

(d)

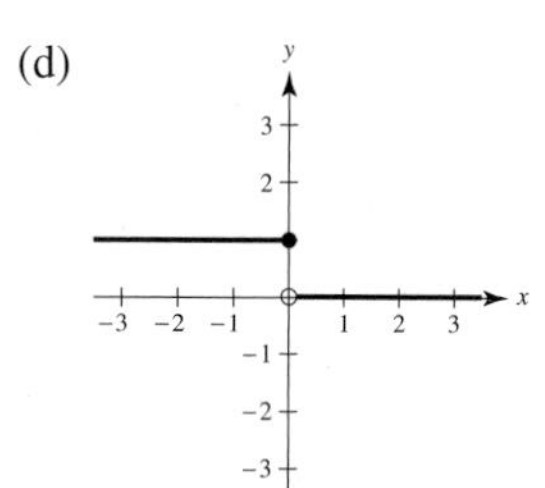

(e)

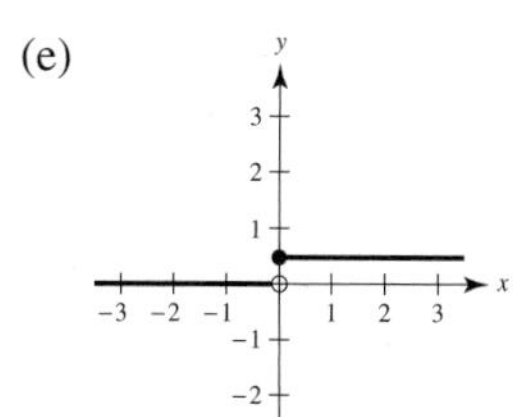

(f)

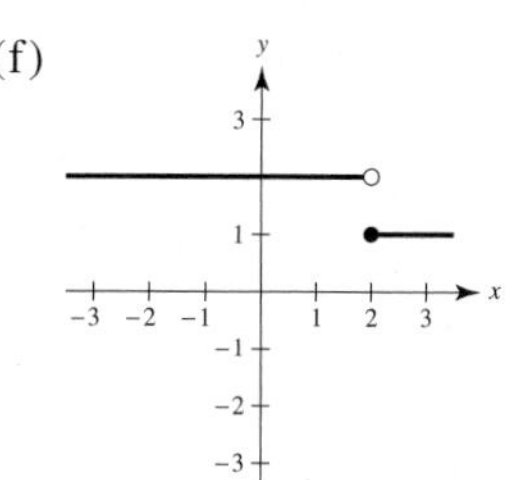

19. Proof

21. (a)

x	-4	-2	0	4
$f(f^{-1}(x))$	-4	-2	0	4

(b)

x	-3	-2	0	1
$(f + f^{-1})(x)$	5	1	-3	-5

(c)

x	-3	-2	0	1
$(f \cdot f^{-1})(x)$	4	0	2	6

(d)

x	-4	-3	0	4
$\lvert f^{-1}(x)\rvert$	2	1	1	3

Chapter 2

Section 2.1 (page 150)

1. polynomial **3.** quadratic; parabola

5. positive; minimum

7. e **8.** c **9.** b **10.** a **11.** f **12.** d

13. (a)

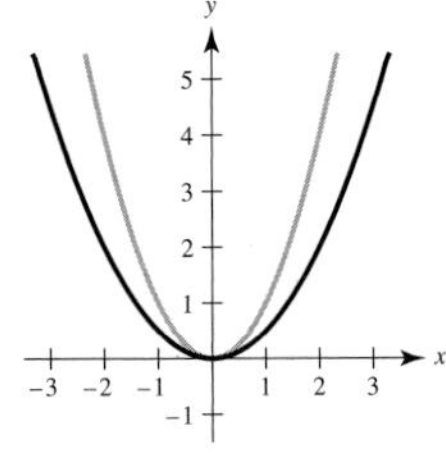

Vertical shrink

(b)

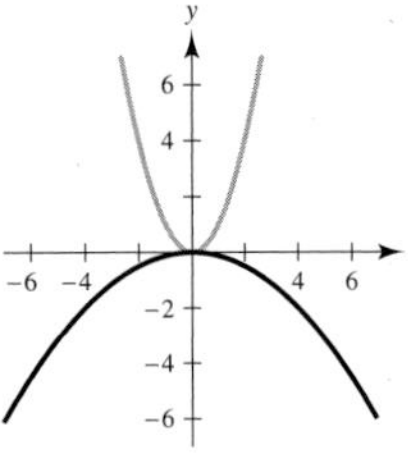

Vertical shrink and reflection in the x-axis

(c)

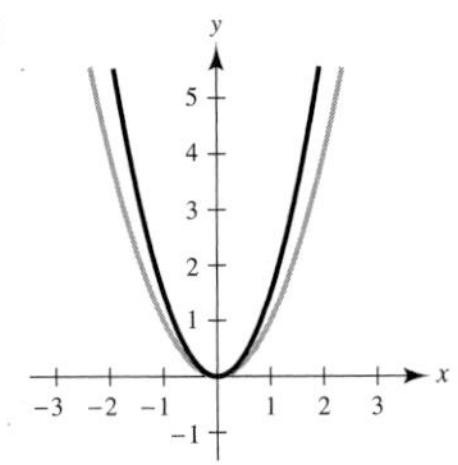

Vertical stretch

(d)

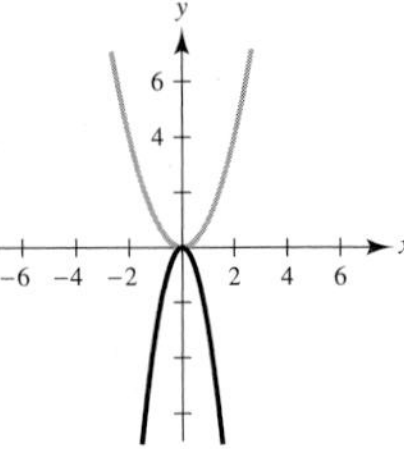

Vertical stretch and reflection in the x-axis

15. (a)

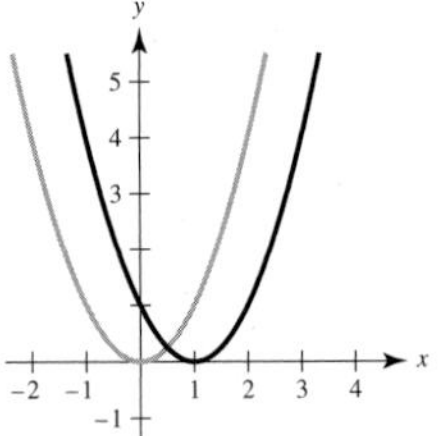

Horizontal shift one unit to the right

(b)

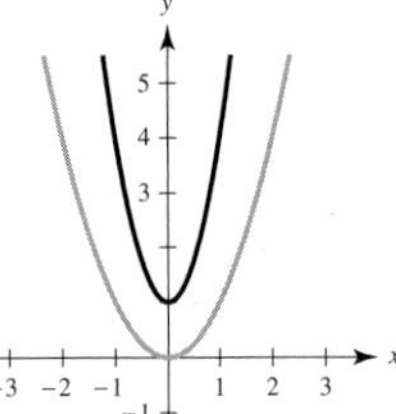

Horizontal shrink and vertical shift one unit upward

(c)

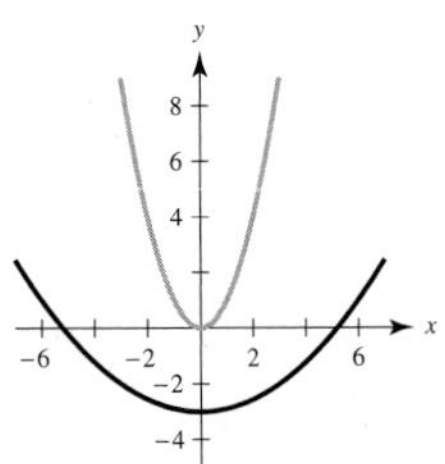

Horizontal stretch and vertical shift three units downward

(d)

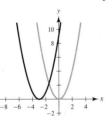

Horizontal shift three units to the left

17.

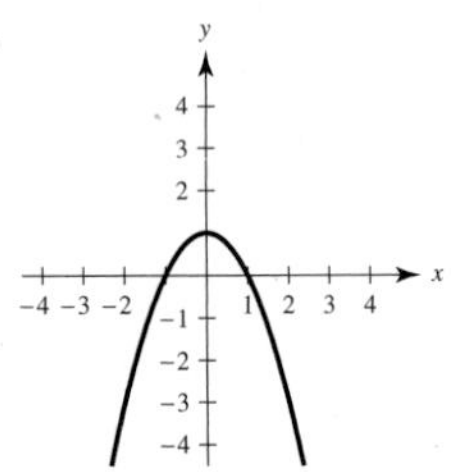

Vertex: $(0, 1)$

Axis of symmetry: y-axis

x-intercepts: $(-1, 0)$ $(1, 0)$

19.

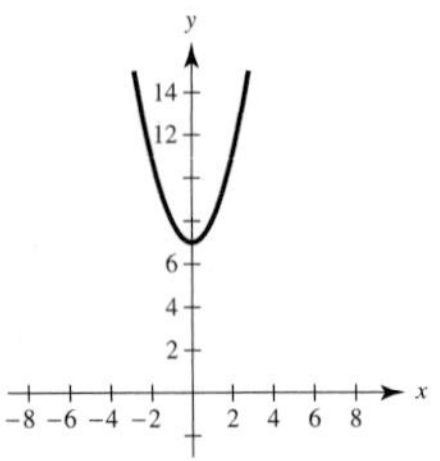

Vertex: $(0, 7)$

Axis of symmetry: y-axis

No x-intercept

21.

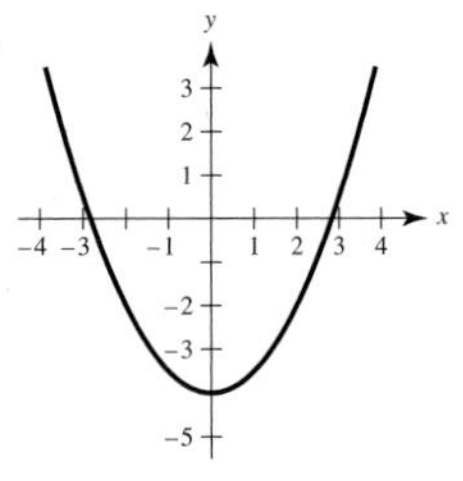

Vertex: $(0, -4)$

Axis of symmetry: y-axis

x-intercepts: $(\pm 2\sqrt{2}, 0)$

23.

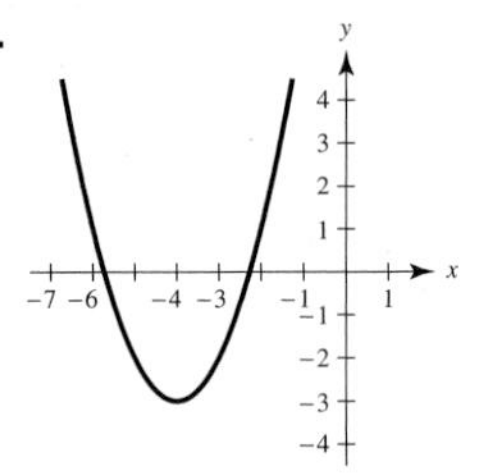

Vertex: $(-4, -3)$

Axis of symmetry: $x = -4$

x-intercepts: $(-4 \pm \sqrt{3}, 0)$

25.

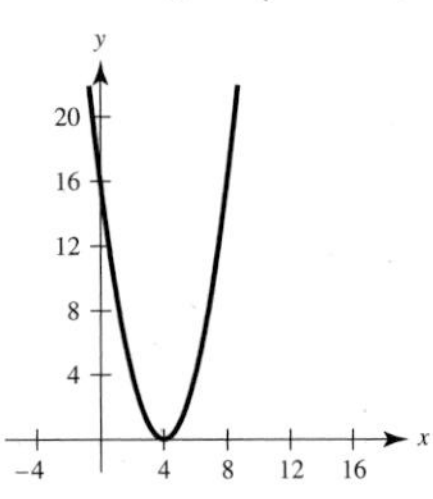

Vertex: $(4, 0)$

Axis of symmetry: $x = 4$

x-intercept: $(4, 0)$

27.

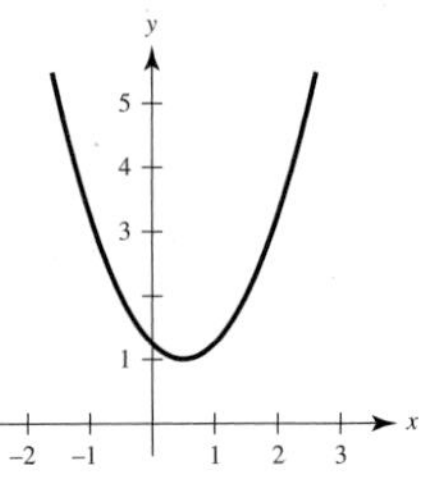

Vertex: $\left(\frac{1}{2}, 1\right)$

Axis of symmetry: $x = \frac{1}{2}$

No x-intercept

29.

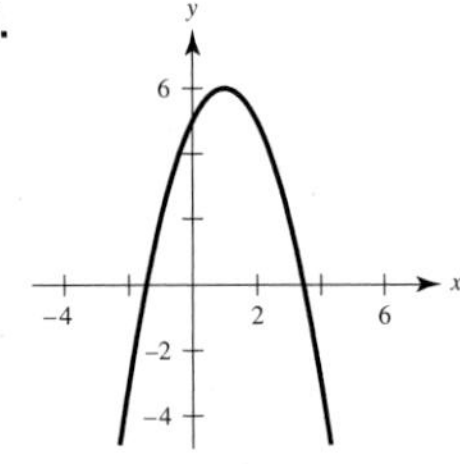

Vertex: $(1, 6)$

Axis of symmetry: $x = 1$

x-intercepts: $(1 \pm \sqrt{6}, 0)$

31.

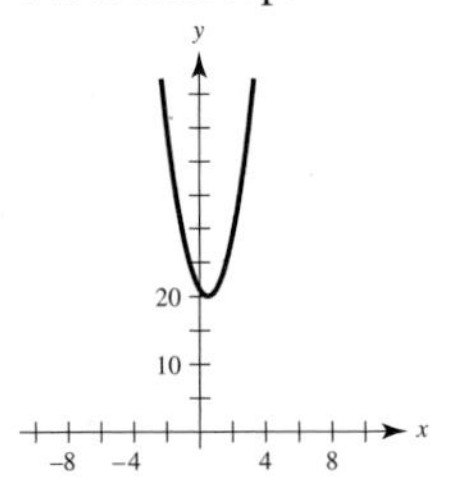

Vertex: $\left(\frac{1}{2}, 20\right)$

Axis of symmetry: $x = \frac{1}{2}$

No x-intercept

33.

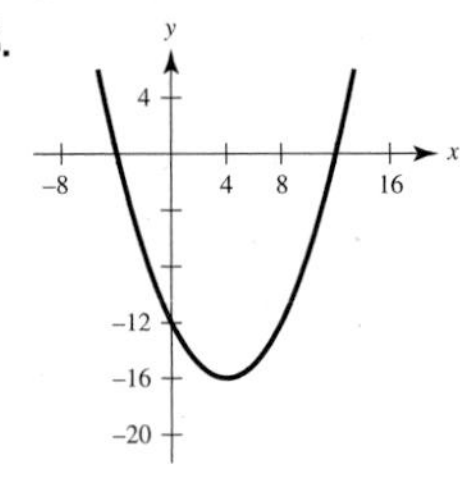

Vertex: $(4, -16)$

Axis of symmetry: $x = 4$

x-intercepts: $(-4, 0)$, $(12, 0)$

35.

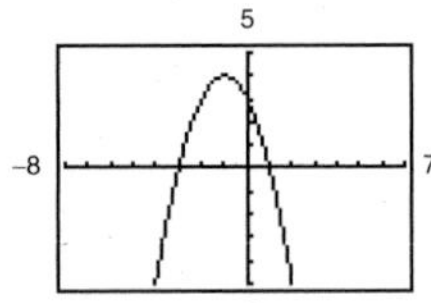

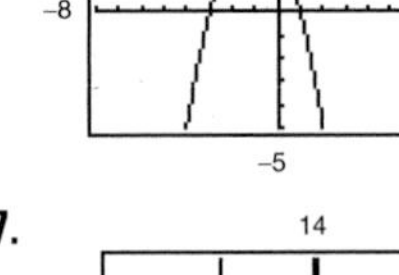

Vertex: $(-1, 4)$

Axis of symmetry: $x = -1$

x-intercepts: $(1, 0)$, $(-3, 0)$

37.

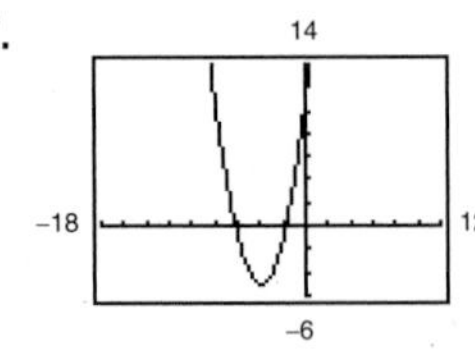

Vertex: $(-4, -5)$

Axis of symmetry: $x = -4$

x-intercepts: $(-4 \pm \sqrt{5}, 0)$

39. 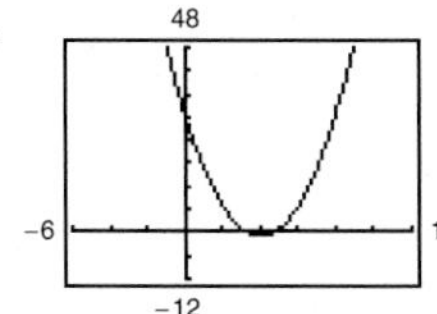
Vertex: $(4, -1)$
Axis of symmetry: $x = 4$
x-intercepts: $\left(4 \pm \frac{1}{2}\sqrt{2}, 0\right)$

41. 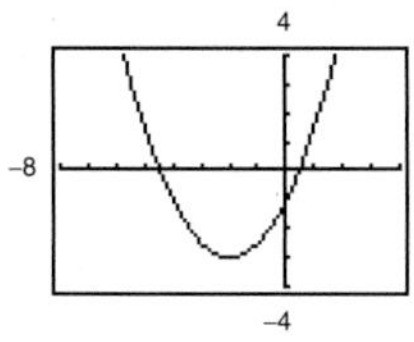
Vertex: $(-2, -3)$
Axis of symmetry: $x = -2$
x-intercepts: $\left(-2 \pm \sqrt{6}, 0\right)$

43. $y = -(x + 1)^2 + 4$
45. $y = -2(x + 2)^2 + 2$
47. $f(x) = (x + 2)^2 + 5$
49. $f(x) = 4(x - 1)^2 - 2$
51. $f(x) = \frac{3}{4}(x - 5)^2 + 12$
53. $f(x) = -\frac{24}{49}\left(x + \frac{1}{4}\right)^2 + \frac{3}{2}$
55. $f(x) = -\frac{16}{3}\left(x + \frac{5}{2}\right)^2$

57. 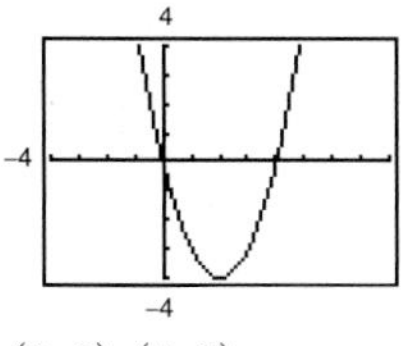
(0, 0), (4, 0)

59. 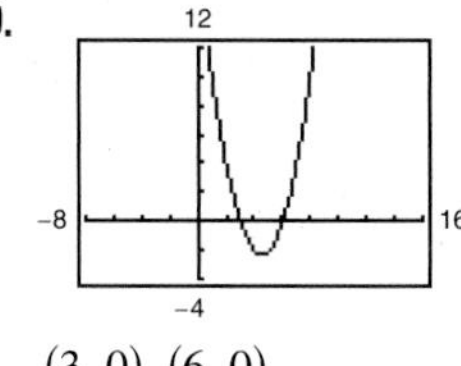
(3, 0), (6, 0)

61. 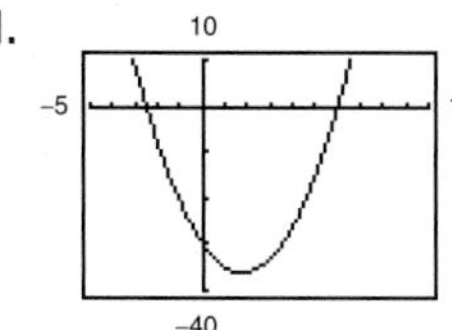
$\left(-\frac{5}{2}, 0\right)$, (6, 0)

63. $f(x) = x^2 - 2x - 3$
$g(x) = -x^2 + 2x + 3$
65. $f(x) = x^2 - 10x$
$g(x) = -x^2 + 10x$
67. $f(x) = 2x^2 + 7x + 3$
$g(x) = -2x^2 - 7x - 3$
69. (a) and (c) $(5, 0), (-1, 0)$
(b) The x-intercepts and solutions of the equation are the same.
71. (a) and (c) $(-1, 0)$
(b) The x-intercepts and solutions of the equation are the same.
73. $f(x) = a\left(x + \frac{b}{2a}\right)^2 + \frac{4ac - b^2}{4a}$
75. 55, 55
77. 12, 6
79. (a) $A = x(50 - x)$, $0 < x < 50$
(b) 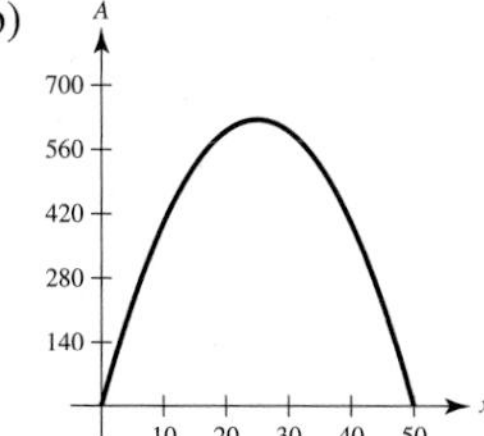
(c) 25 ft × 25 ft 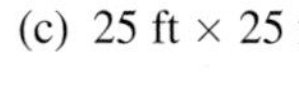
81. 16 ft
83. 20 fixtures
85. (a) \$14,000,000; \$14,375,000; \$13,500,000
(b) \$24; \$14,400,000
Answers will vary.
87. (a) $A = \frac{8x(50 - x)}{3}$
(b)

x	5	10	15	20	25	30
a	600	$1066\frac{2}{3}$	1400	1600	$1666\frac{2}{3}$	1600

$x = 25$ ft, $y = 33\frac{1}{3}$ ft
(c) 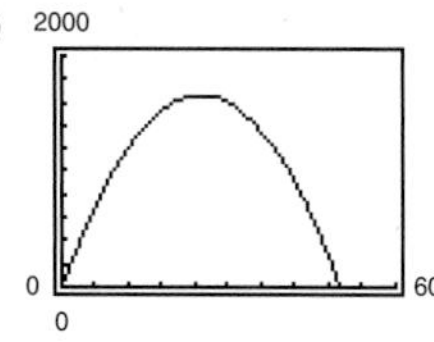
$x = 25$ ft, $y = 33\frac{1}{3}$ ft
(d) $A = -\frac{8}{3}(x - 25)^2 + \frac{5000}{3}$
(e) They are identical.
89. (a) $R = -100x^2 + 3500x$, $15 \le x \le 20$
(b) \$17.50; \$30,625
91. (a) 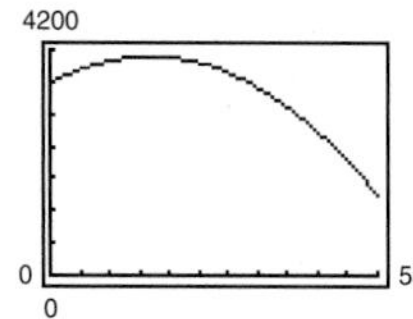
(b) 4075 cigarettes; Yes, the warning had an effect because the maximum consumption occurred in 1966.
(c) 7366 cigarettes per year; 20 cigarettes per day
93. True. The equation has no real solutions, so the graph has no x-intercepts.
95. True. The graph of a quadratic function with a negative leading coefficient will be a downward-opening parabola.
97. $b = \pm 20$
99. $b = \pm 8$
101. The vertex shifts horizontally to the right h units. The parabola will then become narrower $(a > 1)$ or wider $(0 < a < 1)$. The vertex will shift vertically k units upward.
103. Answers will vary.

Section 2.2 (page 161)

1. continuous
3. x^n
5. (a) solution; (b) $(x - a)$; (c) x-intercept
7. touches; crosses
9. c
10. g
11. h
12. f
13. a
14. e
15. d
16. b
17. (a) 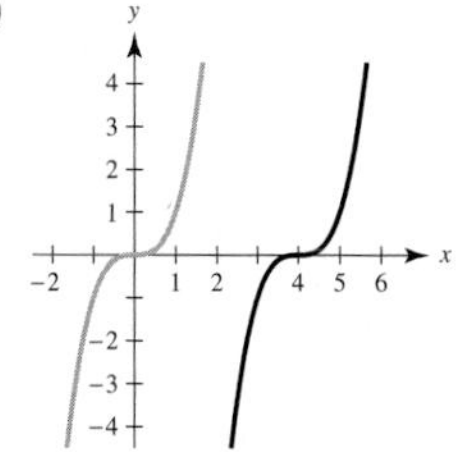
(b) 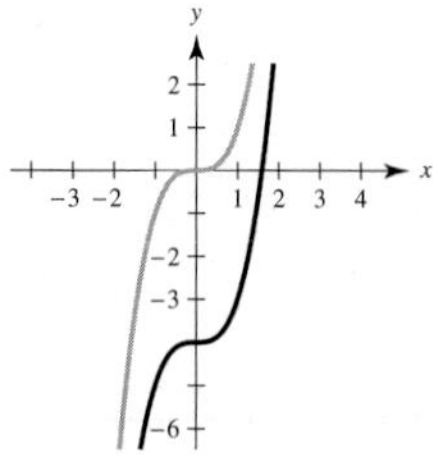
(c) 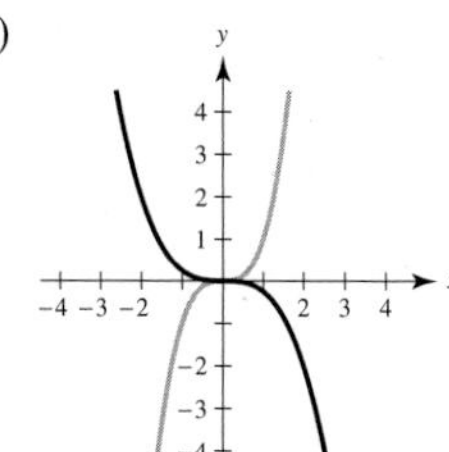
(d) 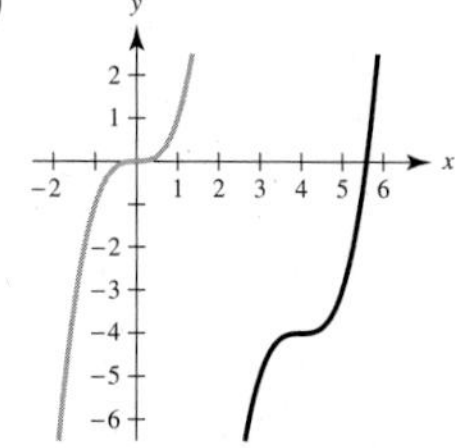

19. (a)

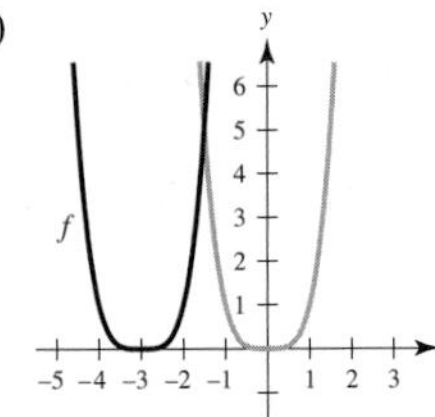

(b)

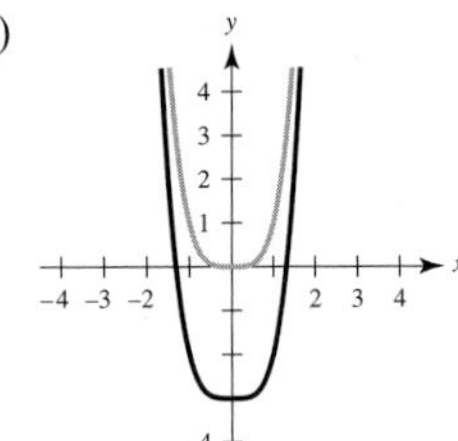

(c)

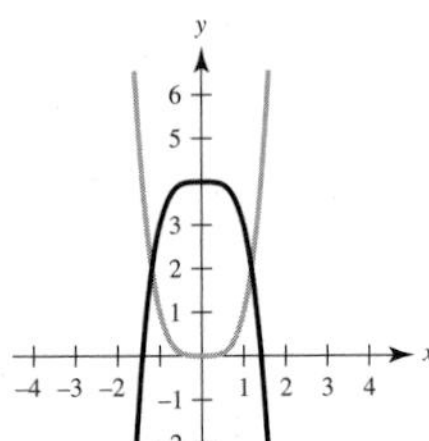

(d)

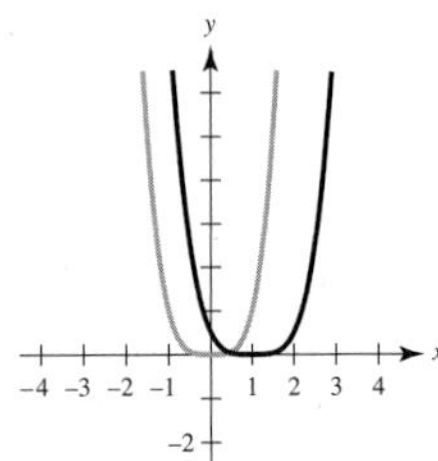

(e)

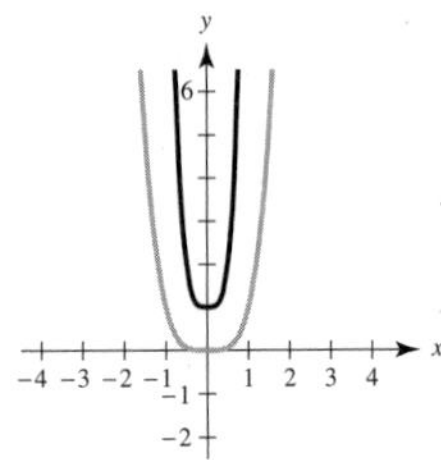

(f)

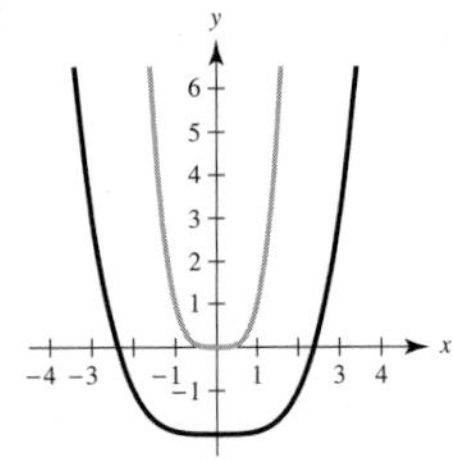

21. Falls to the left, rises to the right
23. Falls to the left, falls to the right
25. Rises to the left, falls to the right
27. Rises to the left, falls to the right
29. Falls to the left, falls to the right

31.

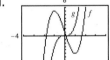

33.

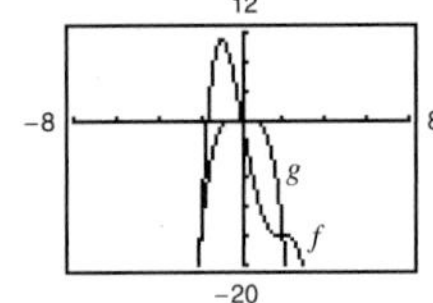

35. (a) ± 6 (b) Odd multiplicity; number of turning points: 1
(c)

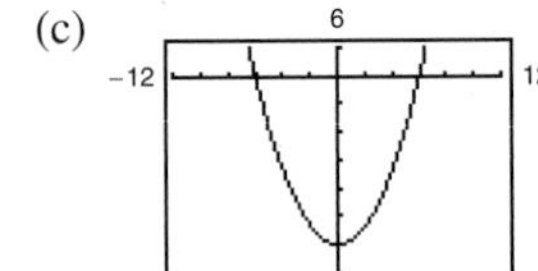

37. (a) 3 (b) Even multiplicity; number of turning points: 1
(c)

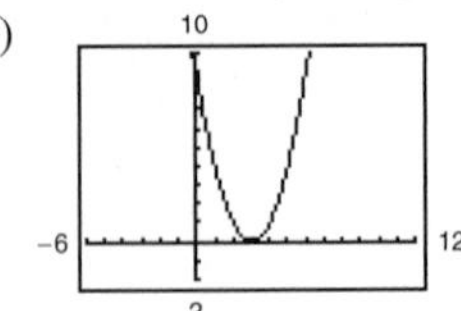

39. (a) $-2, 1$ (b) Odd multiplicity; number of turning points: 1
(c)

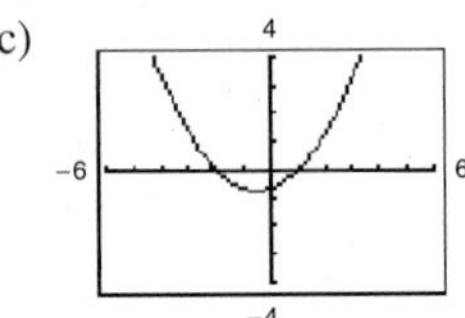

41. (a) $0, 2 \pm \sqrt{3}$
(b) Odd multiplicity; number of turning points: 2
(c)

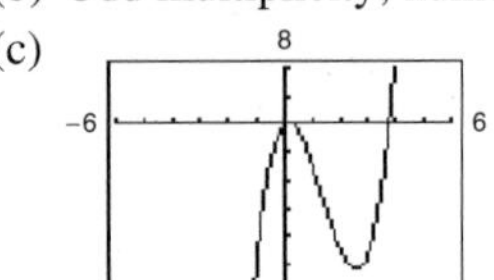

43. (a) 0, 4
(b) 0, odd multiplicity; 4, even multiplicity; number of turning points: 2
(c)

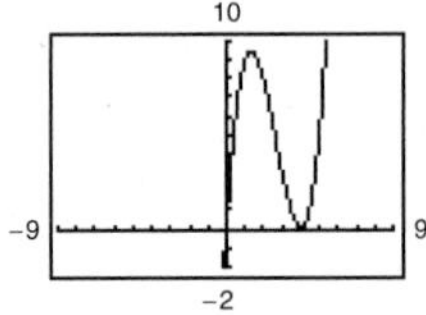

45. (a) $0, \pm\sqrt{3}$
(b) 0, odd multiplicity; $\pm\sqrt{3}$, even multiplicity; number of turning points: 4
(c)

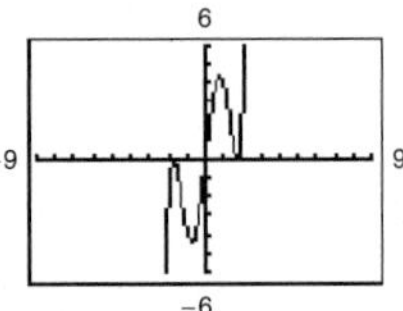

47. (a) No real zeros (b) Number of turning points: 1
(c)

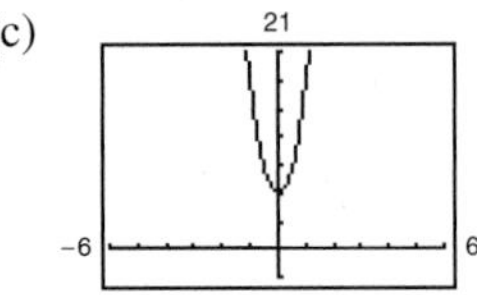

49. (a) $\pm 2, -3$
(b) Odd multiplicity; number of turning points: 2
(c)

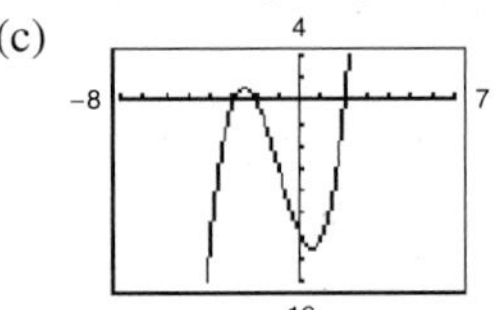

51. (a)

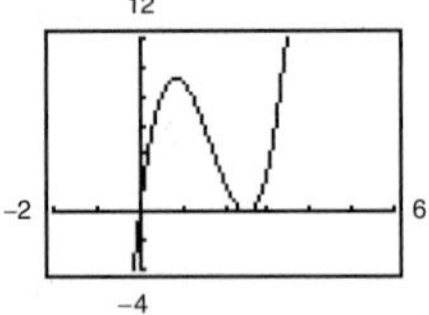

(b) x-intercepts: $(0, 0), \left(\frac{5}{2}, 0\right)$ (c) $x = 0, \frac{5}{2}$
(d) The answers in part (c) match the x-intercepts.

53. (a)

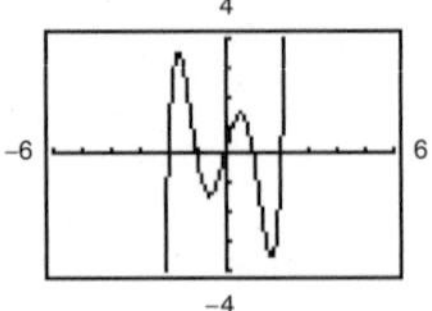

(b) x-intercepts: $(0, 0), (\pm 1, 0), (\pm 2, 0)$
(c) $x = 0, 1, -1, 2, -2$
(d) The answers in part (c) match the x-intercepts.

55. $f(x) = x^2 - 8x$ **57.** $f(x) = x^2 + 4x - 12$
59. $f(x) = x^3 + 9x^2 + 20x$
61. $f(x) = x^4 - 4x^3 - 9x^2 + 36x$ **63.** $f(x) = x^2 - 2x - 2$
65. $f(x) = x^2 + 6x + 9$ **67.** $f(x) = x^3 + 4x^2 - 5x$
69. $f(x) = x^3 - 3x$ **71.** $f(x) = x^4 - x^3 - 6x^2 + 2x + 4$
73. $f(x) = x^5 + 16x^4 + 96x^3 + 256x^2 + 256x$

75. (a) Falls to the left, rises to the right
(b) $0, 5, -5$ (c) Answers will vary.
(d)

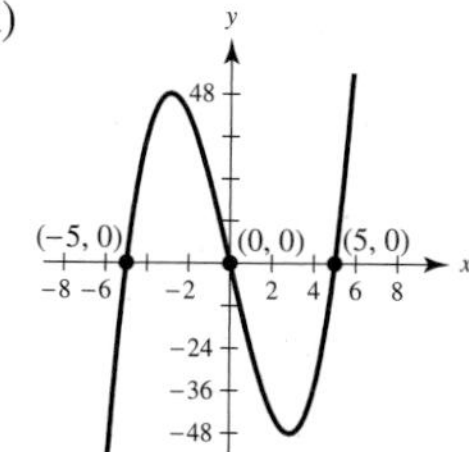

77. (a) Rises to the left, rises to the right
(b) No zeros (c) Answers will vary.
(d)

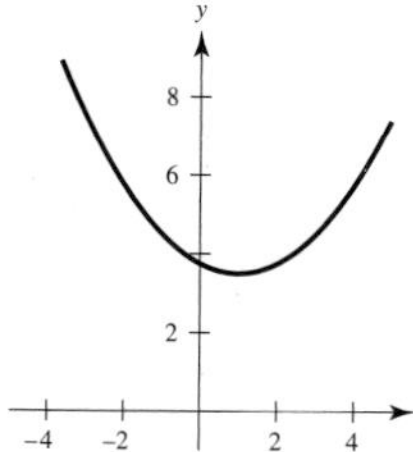

79. (a) Falls to the left, rises to the right
(b) $0, 2$ (c) Answers will vary.
(d)

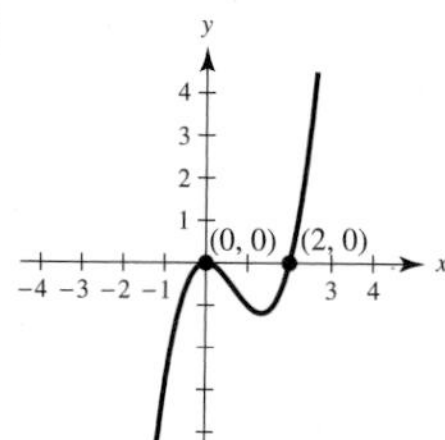

81. (a) Falls to the left, rises to the right
(b) $0, 2, 3$ (c) Answers will vary.
(d)

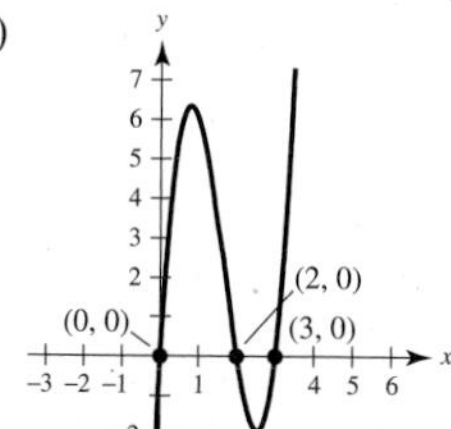

83. (a) Rises to the left, falls to the right
(b) $-5, 0$ (c) Answers will vary.
(d)

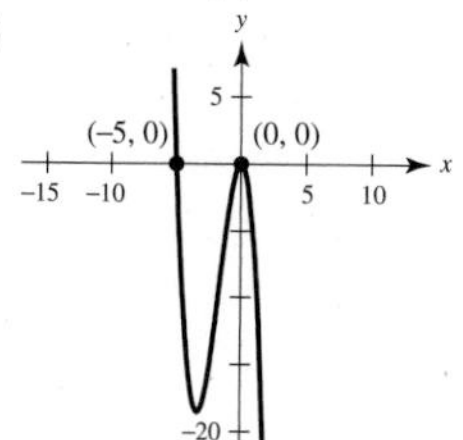

85. (a) Falls to the left, rises to the right
(b) $0, 4$ (c) Answers will vary.
(d)

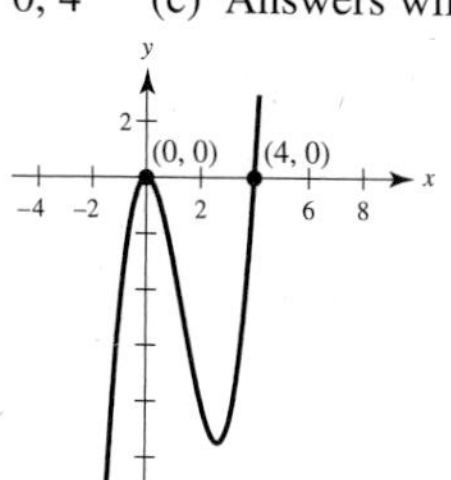

87. (a) Falls to the left, falls to the right
(b) ± 2 (c) Answers will vary.
(d)

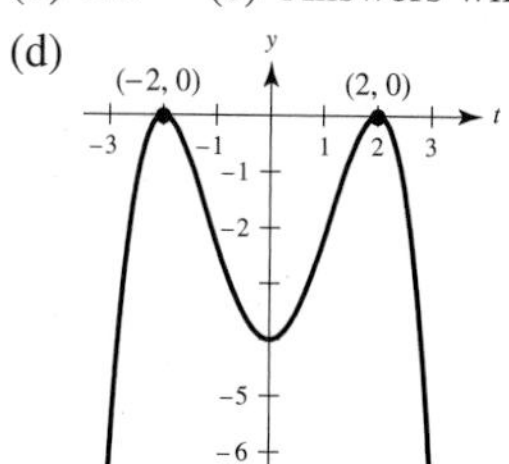

89.

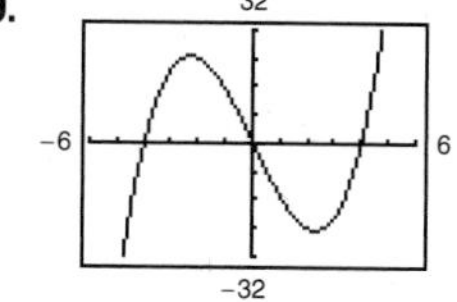

Zeros: $0, \pm 4$, odd multiplicity

91.

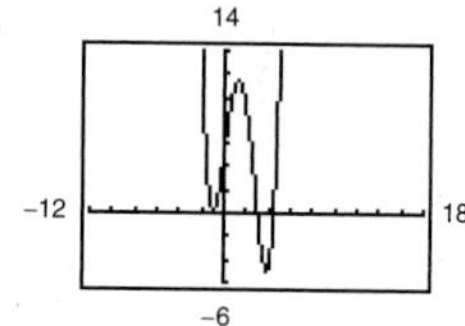

Zeros: -1, even multiplicity; $3, \frac{9}{2}$, odd multiplicity

93.

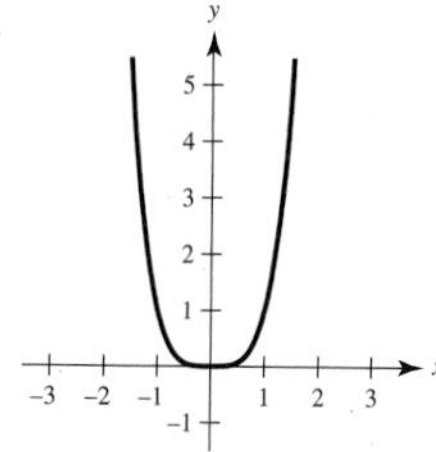

(a) Vertical shift two units upward; Even
(b) Horizontal shift two units to the left; Neither
(c) Reflection in the y-axis; Even
(d) Reflection in the x-axis; Even
(e) Horizontal stretch; Even
(f) Vertical shrink; Even
(g) $g(x) = x^3, x \geq 0$; Neither
(h) $g(x) = x^{16}$; Even

95. (a) $V(x) = x(36 - 2x)^2$ (b) Domain: $0 < x < 18$
(c)

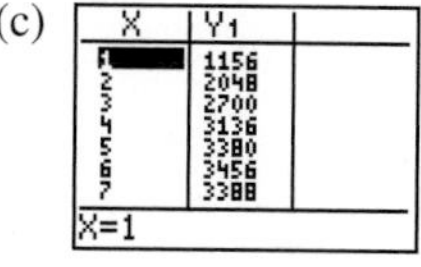

X	Y1
1	1156
2	2048
3	2700
4	3136
5	3380
6	3456
7	3388

X=1

6 in. × 24 in. × 24 in.

(d)

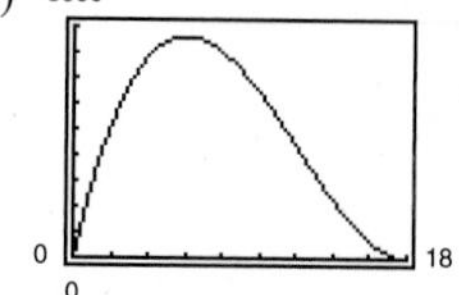

$x = 6$; The results are the same.

97. False. A fifth-degree polynomial can have at most four turning points.

99. True. The degree of the function is odd and its leading coefficient is negative, so the graph rises to the left and falls to the right.

Section 2.3 (page 171)

1. $f(x)$: dividend; $d(x)$: divisor; $q(x)$: quotient; $r(x)$: remainder

3. improper **5.** Factor **7.** Answers will vary.

9. (a) and (b)

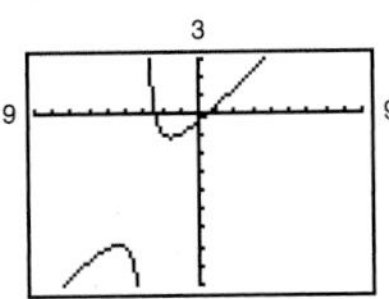
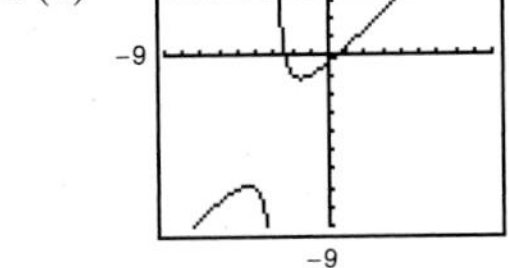

(c) Answers will vary.

11. $2x + 4,\quad x \neq -3$ **13.** $x^2 - 3x + 1,\quad x \neq -\frac{5}{4}$

15. $x^3 + 3x^2 - 1,\quad x \neq -2$ **17.** $x^2 + 3x + 9,\quad x \neq 3$

19. $7 - \dfrac{11}{x+2}$ **21.** $x - \dfrac{x+9}{x^2+1}$ **23.** $2x - 8 + \dfrac{x-1}{x^2+1}$

25. $x + 3 + \dfrac{6x^2 - 8x + 3}{(x-1)^3}$ **27.** $3x^2 - 2x + 5,\quad x \neq 5$

29. $6x^2 + 25x + 74 + \dfrac{248}{x-3}$ **31.** $4x^2 - 9,\quad x \neq -2$

33. $-x^2 + 10x - 25,\quad x \neq -10$

35. $5x^2 + 14x + 56 + \dfrac{232}{x-4}$

37. $10x^3 + 10x^2 + 60x + 360 + \dfrac{1360}{x-6}$

39. $x^2 - 8x + 64,\quad x \neq -8$

41. $-3x^3 - 6x^2 - 12x - 24 - \dfrac{48}{x-2}$

43. $-x^3 - 6x^2 - 36x - 36 - \dfrac{216}{x-6}$

45. $4x^2 + 14x - 30,\quad x \neq -\frac{1}{2}$

47. $f(x) = (x-4)(x^2 + 3x - 2) + 3,\quad f(4) = 3$

49. $f(x) = \left(x + \frac{2}{3}\right)(15x^3 - 6x + 4) + \frac{34}{3},\quad f\left(-\frac{2}{3}\right) = \frac{34}{3}$

51. $f(x) = \left(x - \sqrt{2}\right)\left[x^2 + \left(3 + \sqrt{2}\right)x + 3\sqrt{2}\right] - 8,$
$f\left(\sqrt{2}\right) = -8$

53. $f(x) = \left(x - 1 + \sqrt{3}\right)\left[-4x^2 + \left(2 + 4\sqrt{3}\right)x + \left(2 + 2\sqrt{3}\right)\right],$
$f\left(1 - \sqrt{3}\right) = 0$

55. (a) -2 (b) 1 (c) $-\frac{1}{4}$ (d) 5

57. (a) -35 (b) -22 (c) -10 (d) -211

59. $(x-2)(x+3)(x-1)$; Solutions: $2, -3, 1$

61. $(2x-1)(x-5)(x-2)$; Solutions: $\frac{1}{2}, 5, 2$

63. $\left(x + \sqrt{3}\right)\left(x - \sqrt{3}\right)(x+2)$; Solutions: $-\sqrt{3}, \sqrt{3}, -2$

65. $(x-1)\left(x - 1 - \sqrt{3}\right)\left(x - 1 + \sqrt{3}\right)$;
Solutions: $1,\ 1 + \sqrt{3},\ 1 - \sqrt{3}$

67. (a) Answers will vary. (b) $2x - 1$
(c) $f(x) = (2x-1)(x+2)(x-1)$
(d) $\frac{1}{2}, -2, 1$ (e)

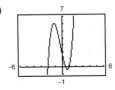

69. (a) Answers will vary. (b) $(x-1), (x-2)$
(c) $f(x) = (x-1)(x-2)(x-5)(x+4)$
(d) $1, 2, 5, -4$ (e)

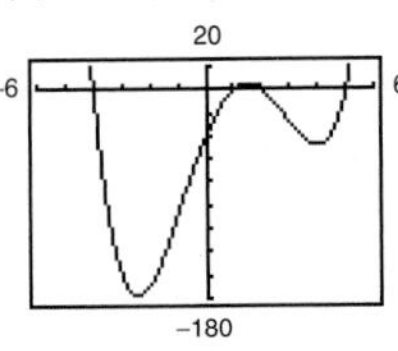

71. (a) Answers will vary. (b) $x + 7$
(c) $f(x) = (x+7)(2x+1)(3x-2)$
(d) $-7, -\frac{1}{2}, \frac{2}{3}$ (e)

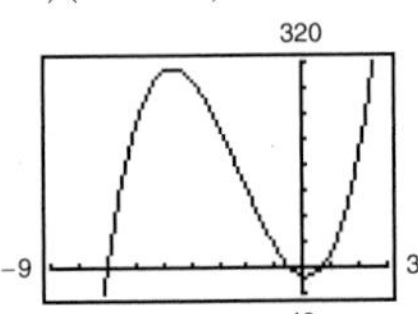

73. (a) Answers will vary. (b) $x - \sqrt{5}$
(c) $f(x) = \left(x - \sqrt{5}\right)\left(x + \sqrt{5}\right)(2x - 1)$
(d) $\pm\sqrt{5}, \frac{1}{2}$ (e)

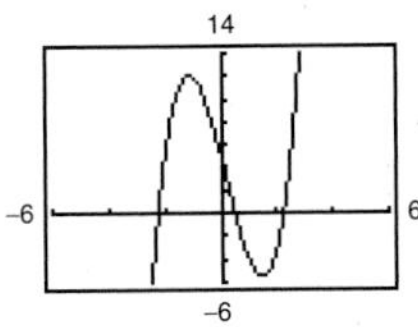

75. (a) Zeros are 2 and about ± 2.236.
(b) $x = 2$ (c) $f(x) = (x-2)\left(x - \sqrt{5}\right)\left(x + \sqrt{5}\right)$

77. (a) Zeros are -2, about 0.268, and about 3.732.
(b) $t = -2$
(c) $h(t) = (t+2)\left[t - \left(2 + \sqrt{3}\right)\right]\left[t - \left(2 - \sqrt{3}\right)\right]$

79. (a) Zeros are 0, 3, 4, and about ± 1.414.
(b) $x = 0$
(c) $h(x) = x(x-4)(x-3)\left(x + \sqrt{2}\right)\left(x - \sqrt{2}\right)$

81. $2x^2 - x - 1,\quad x \neq \frac{3}{2}$ **83.** $x^2 + 3x,\quad x \neq -2, -1$

85. $x^{2n} + 6x^n + 9,\quad x^n \neq -3$ **87.** The remainder is 0.

89. $c = -210$

91. (a) and (b)

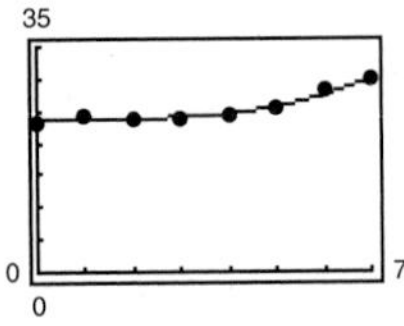

$A = 0.0349t^3 - 0.168t^2 + 0.42t + 23.4$

(c)

t	0	1	2	3
$A(t)$	23.4	23.7	23.8	24.1

t	4	5	6	7
$A(t)$	24.6	25.7	27.4	30.1

(d) \$45.7 billion; No, because the model will approach infinity quickly.

93. False. $-\frac{4}{7}$ is a zero of f.

95. True. The degree of the numerator is greater than the degree of the denominator.

97. False.
To divide $x^4 - 3x^2 + 4x - 1$ by $x + 2$ using synthetic division, the set up would be:

$$\begin{array}{r|rrrrr} -2 & 1 & 0 & -3 & 4 & -1 \\ \hline & & & & & \end{array}$$

A zero must be included for the missing x^3 term.

99. $k = 7$

101. (a) $x + 1,\ \ x \neq 1$ (b) $x^2 + x + 1,\ \ x \neq 1$
(c) $x^3 + x^2 + x + 1,\ \ x \neq 1$
In general, $\dfrac{x^n - 1}{x - 1} = x^{n-1} + x^{n-2} + \cdots + x + 1,\ \ x \neq 1$

Section 2.4 (page 179)

1. (a) iii (b) i (c) ii **3.** principal square
5. $a = -12,\ b = 7$ **7.** $a = 6,\ b = 5$ **9.** $8 + 5i$
11. $4\sqrt{5}i$ **13.** $0.3i$ **15.** $-1 - 10i$ **17.** $10 - 3i$
19. 1 **21.** $3 - 3\sqrt{2}i$ **23.** $-14 + 20i$ **25.** $-5\sqrt{2}$
27. $5 + i$ **29.** $108 + 12i$ **31.** 24 **33.** $-13 + 84i$
35. $9 - 2i,\ 85$ **37.** $-2\sqrt{5}i,\ 20$ **39.** $-3i$
41. $\frac{13}{2} + \frac{13}{2}i$ **43.** $8 - 4i$ **45.** $-\frac{1}{2} - \frac{5}{2}i$ **47.** $\frac{62}{949} + \frac{297}{949}i$
49. $1 \pm i$ **51.** $\dfrac{1}{3} \pm 2i$ **53.** $\dfrac{5}{7} \pm \dfrac{5\sqrt{15}}{7}i$
55. $-1 + 6i$ **57.** i **59.** i
61.
$$\begin{aligned}(a + bi)(a - bi) &= a^2 + abi - abi - b^2i^2 \\ &= a^2 - b^2(-1) \\ &= a^2 + b^2\end{aligned}$$
which is a real number because a and b are real numbers. So, the product of a complex number and its conjugate is a real number.
63. Proof
65. (a) 1 (b) i (c) -1 (d) $-i$
67. False. If the complex number is real, the number equals its conjugate.
69. False.
$i^{44} + i^{150} - i^{74} - i^{109} + i^{61} = 1 - 1 + 1 - i + i = 1$

Section 2.5 (page 188)

1. Fundamental Theorem of Algebra **3.** Rational Zero
5. linear; quadratic; quadratic
7. 0, 6 **9.** $2, -4$ **11.** $-6, \pm i$ **13.** $\pm 1, \pm 2$
15. $\pm 1, \pm 3, \pm 5, \pm 9, \pm 15, \pm 45, \pm\frac{1}{2}, \pm\frac{3}{2}, \pm\frac{5}{2}, \pm\frac{9}{2}, \pm\frac{15}{2}, \pm\frac{45}{2}$
17. 1, 2, 3 **19.** $1, -1, 4$ **21.** $-6, -1$ **23.** $\frac{1}{2}, -1$
25. $-2, 3, \pm\frac{2}{3}$ **27.** $-2, 1$ **29.** $-4, \frac{1}{2}, 1, 1$
31. (a) $\pm 1, \pm 2, \pm 4$
(b)

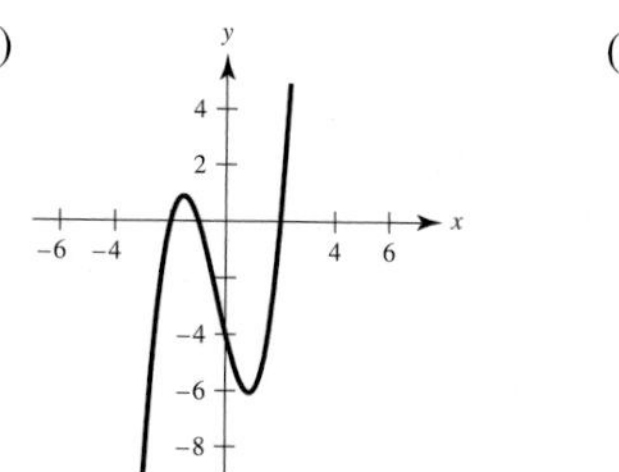

(c) $-2, -1, 2$

33. (a) $\pm 1, \pm 3, \pm\frac{1}{2}, \pm\frac{3}{2}, \pm\frac{1}{4}, \pm\frac{3}{4}$
(b)

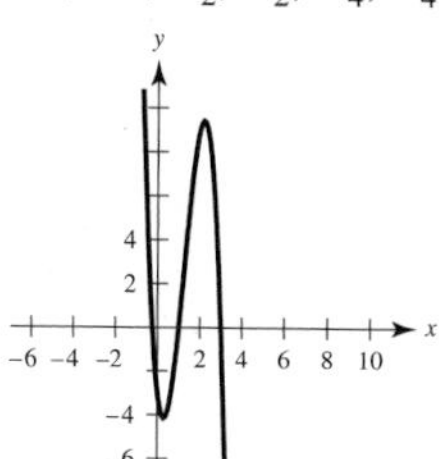

(c) $-\frac{1}{4}, 1, 3$

35. (a) $\pm 1, \pm 2, \pm 4, \pm 8, \pm\frac{1}{2}$
(b)

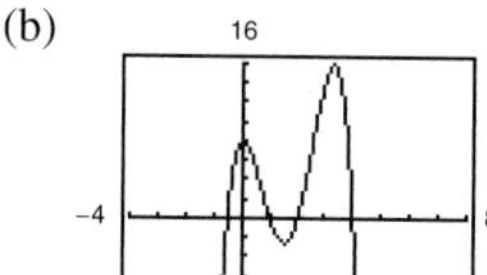

(c) $-\frac{1}{2}, 1, 2, 4$

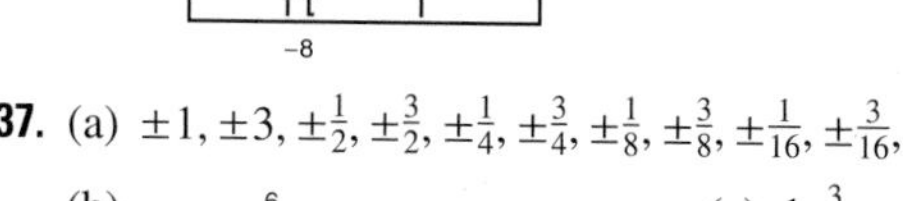

37. (a) $\pm 1, \pm 3, \pm\frac{1}{2}, \pm\frac{3}{2}, \pm\frac{1}{4}, \pm\frac{3}{4}, \pm\frac{1}{8}, \pm\frac{3}{8}, \pm\frac{1}{16}, \pm\frac{3}{16}, \pm\frac{1}{32}, \pm\frac{3}{32}$
(b)

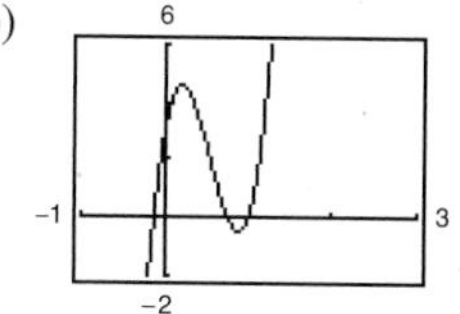

(c) $1, \frac{3}{4}, -\frac{1}{8}$

39. (a) ± 1, about ± 1.414 (b) $\pm 1, \pm\sqrt{2}$
(c) $f(x) = (x + 1)(x - 1)(x + \sqrt{2})(x - \sqrt{2})$
41. (a) 0, 3, 4, about ± 1.414 (b) $0, 3, 4, \pm\sqrt{2}$
(c) $h(x) = x(x - 3)(x - 4)(x + \sqrt{2})(x - \sqrt{2})$
43. $x^3 - x^2 + 25x - 25$ **45.** $x^3 - 12x^2 + 46x - 52$
47. $3x^4 - 17x^3 + 25x^2 + 23x - 22$
49. (a) $(x^2 + 9)(x^2 - 3)$ (b) $(x^2 + 9)(x + \sqrt{3})(x - \sqrt{3})$
(c) $(x + 3i)(x - 3i)(x + \sqrt{3})(x - \sqrt{3})$
51. (a) $(x^2 - 2x - 2)(x^2 - 2x + 3)$
(b) $(x - 1 + \sqrt{3})(x - 1 - \sqrt{3})(x^2 - 2x + 3)$
(c) $(x - 1 + \sqrt{3})(x - 1 - \sqrt{3})(x - 1 + \sqrt{2}i)(x - 1 - \sqrt{2}i)$
53. $\pm 2i, 1$ **55.** $\pm 5i, -\frac{1}{2}, 1$ **57.** $-3 \pm i, \frac{1}{4}$
59. $2, -3 \pm \sqrt{2}i, 1$ **61.** $\pm 6i;\ (x + 6i)(x - 6i)$
63. $1 \pm 4i;\ (x - 1 - 4i)(x - 1 + 4i)$
65. $\pm 2, \pm 2i;\ (x - 2)(x + 2)(x - 2i)(x + 2i)$
67. $1 \pm i;\ (z - 1 + i)(z - 1 - i)$
69. $-1, 2 \pm i;\ (x + 1)(x - 2 + i)(x - 2 - i)$
71. $-2, 1 \pm \sqrt{2}i;\ (x + 2)(x - 1 + \sqrt{2}i)(x - 1 - \sqrt{2}i)$
73. $-\frac{1}{5}, 1 \pm \sqrt{5}i;\ (5x + 1)(x - 1 + \sqrt{5}i)(x - 1 - \sqrt{5}i)$
75. $2, \pm 2i;\ (x - 2)^2(x + 2i)(x - 2i)$
77. $\pm i, \pm 3i;\ (x + i)(x - i)(x + 3i)(x - 3i)$
79. $-10, -7 \pm 5i$ **81.** $-\frac{3}{4}, 1 \pm \frac{1}{2}i$ **83.** $-2, -\frac{1}{2}, \pm i$
85. $1, -\frac{1}{2}$ **87.** $-\frac{3}{4}$ **89.** $\pm 2, \pm\frac{3}{2}$ **91.** $\pm 1, \frac{1}{4}$
93. d **94.** a **95.** b **96.** c

97. Answers will vary. There are infinitely many possible functions for f. Sample equation and graph: $f(x) = -2x^3 + 3x^2 + 11x - 6$

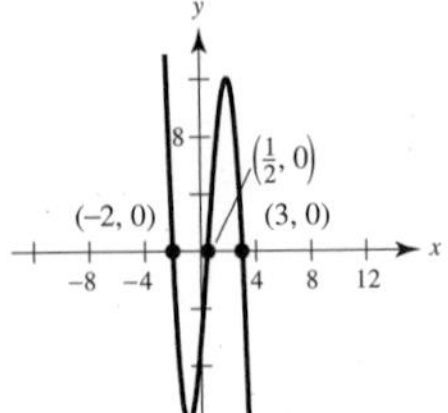

99. (a) $-2, 1, 4$
(b) The graph touches the x-axis at $x = 1$.
(c) The least possible degree of the function is 4, because there are at least four real zeros (1 is repeated) and a function can have at most the number of real zeros equal to the degree of the function. The degree cannot be odd by the definition of multiplicity.
(d) Positive. From the information in the table, it can be concluded that the graph will eventually rise to the left and rise to the right.
(e) $f(x) = x^4 - 4x^3 - 3x^2 + 14x - 8$
(f)

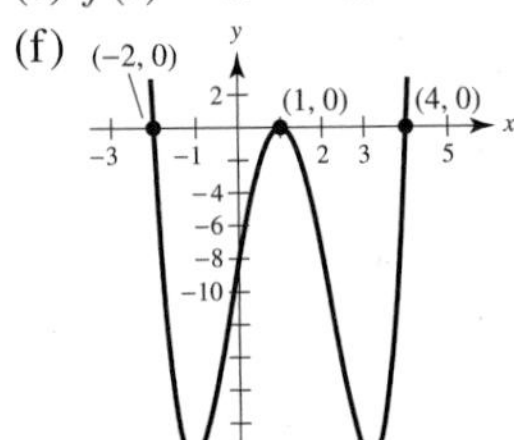

101. (a) $V(x) = 4x^2(30 - x)$
(b) 18,000

0 30
0

20 in. × 20 in. × 40 in.

(c) $15, \dfrac{15}{2} \pm \dfrac{15\sqrt{5}}{2}$; The value $\dfrac{15}{2} - \dfrac{15\sqrt{5}}{2}$ is physically impossible because x is negative.

103. $x \approx 38.4$, or \$384,000

105. (a) $V(x) = x^3 + 9x^2 + 26x + 24 = 120$
(b) 4 ft × 5 ft × 6 ft

107. $x \approx 40$, or 4000 units

109. No. Setting $h = 64$ and solving the resulting equation yields imaginary roots.

111. False. The most complex zeros it can have is two, and the Linear Factorization Theorem guarantees that there are three linear factors, so one zero must be real.

113. r_1, r_2, r_3 **115.** $5 + r_1, 5 + r_2, 5 + r_3$

117. The zeros cannot be determined.

119. $f(x) = x^3 - 3x^2 + 4x - 2$ **121.** $f(x) = x^4 + 5x^2 + 4$

123. Answers will vary.

125. (a) $x^2 + b$ (b) $x^2 - 2ax + a^2 + b^2$

Section 2.6 (page 202)

1. rational functions **3.** horizontal asymptote

5. (a)

x	$f(x)$
0.5	−2
0.9	−10
0.99	−100
0.999	−1000

x	$f(x)$
1.5	2
1.1	10
1.01	100
1.001	1000

x	$f(x)$
5	0.25
10	$0.\overline{1}$
100	$0.\overline{01}$
1000	$0.\overline{001}$

(b) Vertical asymptote: $x = 1$
Horizontal asymptote: $y = 0$
(c) Domain: all real numbers x except $x = 1$

7. (a)

x	$f(x)$
0.5	−1
0.9	−12.79
0.99	−147.8
0.999	−1498

x	$f(x)$
1.5	5.4
1.1	17.29
1.01	152.3
1.001	1502

x	$f(x)$
5	3.125
10	$3.\overline{03}$
100	$3.\overline{0003}$
1000	3

(b) Vertical asymptotes: $x = \pm 1$
Horizontal asymptote: $y = 3$
(c) Domain: all real numbers x except $x = \pm 1$

9. Domain: all real numbers x except $x = 0$
Vertical asymptote: $x = 0$
Horizontal asymptote: $y = 0$

11. Domain: all real numbers x except $x = 5$
Vertical asymptote: $x = 5$
Horizontal asymptote: $y = -1$

13. Domain: all real numbers x except $x = \pm 1$
Vertical asymptotes: $x = \pm 1$

15. Domain: all real numbers x
Horizontal asymptote: $y = 3$

17. d **18.** a **19.** c **20.** b **21.** 3 **23.** 9

25. Domain: all real numbers x except $x = \pm 4$;
Vertical asymptote: $x = -4$; horizontal asymptote: $y = 0$

27. Domain: all real numbers x except $x = -1, 5$;
Vertical asymptote: $x = -1$; horizontal asymptote: $y = 1$

29. Domain: all real numbers x except $x = -1, \frac{1}{2}$;
Vertical asymptote: $x = \frac{1}{2}$; horizontal asymptote: $y = \frac{1}{2}$

31. (a) Domain: all real numbers x except $x = -2$
(b) y-intercept: $\left(0, \frac{1}{2}\right)$
(c) Vertical asymptote: $x = -2$
Horizontal asymptote: $y = 0$
(d)

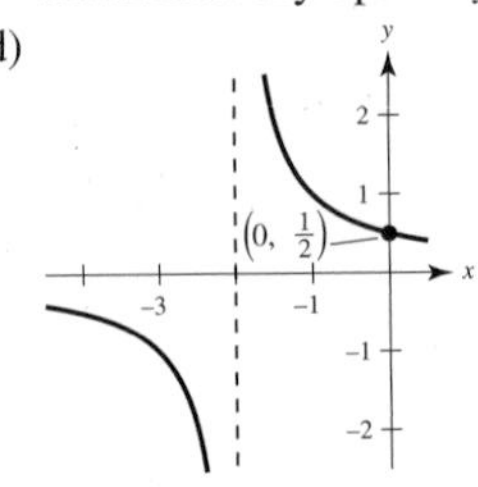

33. (a) Domain: all real numbers x except $x = -4$
(b) y-intercept: $\left(0, -\frac{1}{4}\right)$
(c) Vertical asymptote: $x = -4$
Horizontal asymptote: $y = 0$
(d)

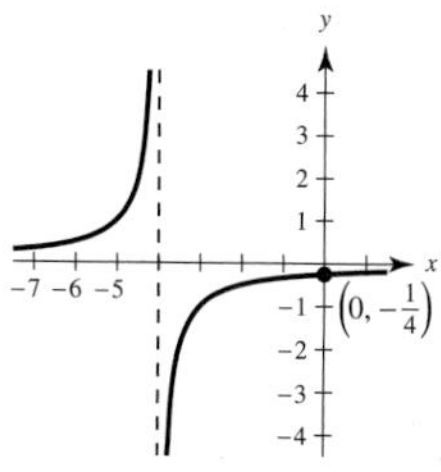

35. (a) Domain: all real numbers x except $x = -2$
(b) x-intercept: $\left(-\frac{7}{2}, 0\right)$
y-intercept: $\left(0, \frac{7}{2}\right)$
(c) Vertical asymptote: $x = -2$
Horizontal asymptote: $y = 2$
(d)

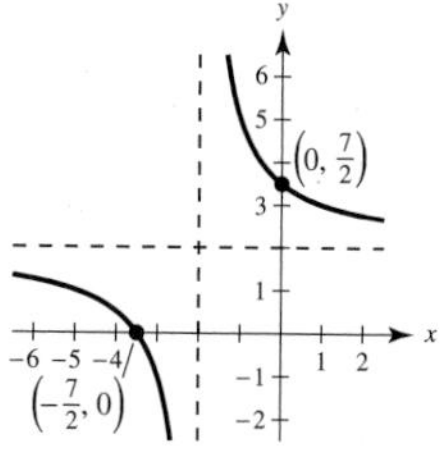

37. (a) Domain: all real numbers x (b) Intercept: $(0, 0)$
(c) Horizontal asymptote: $y = 1$
(d)

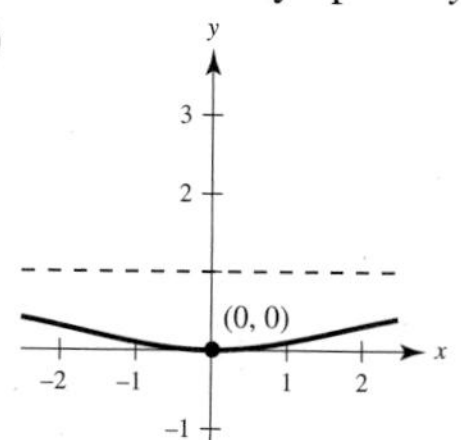

39. (a) Domain: all real numbers s (b) Intercept: $(0, 0)$
(c) Horizontal asymptote: $y = 0$
(d)

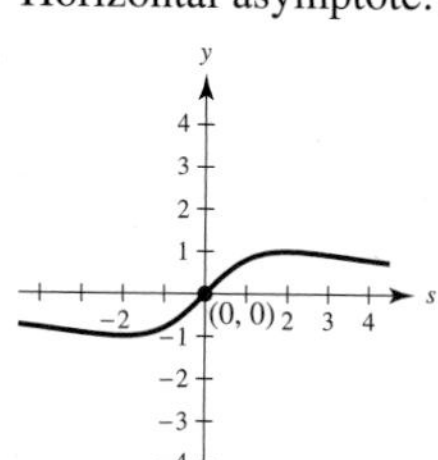

41. (a) Domain: all real numbers x except $x = \pm 2$
(b) x-intercepts: $(1, 0)$ and $(4, 0)$
y-intercept: $(0, -1)$
(c) Vertical asymptotes: $x = \pm 2$
Horizontal asymptote: $y = 1$
(d)

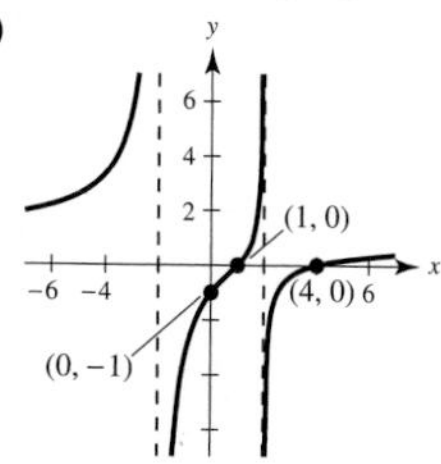

43. (a) Domain: all real numbers x except $x = \pm 1, 2$
(b) x-intercepts: $(3, 0), \left(-\frac{1}{2}, 0\right)$
y-intercept: $\left(0, -\frac{3}{2}\right)$
(c) Vertical asymptotes: $x = 2, x = \pm 1$
Horizontal asymptote: $y = 0$
(d)

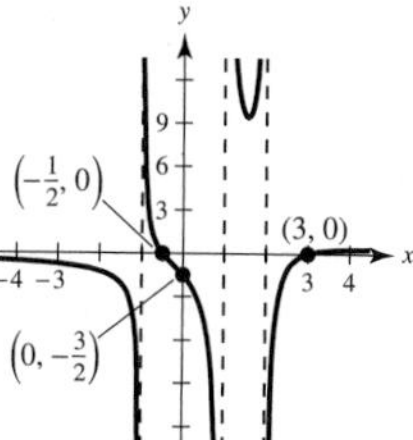

45. (a) Domain: all real numbers x except $x = 2, -3$
(b) Intercept: $(0, 0)$
(c) Vertical asymptote: $x = 2$
Horizontal asymptote: $y = 1$
(d)

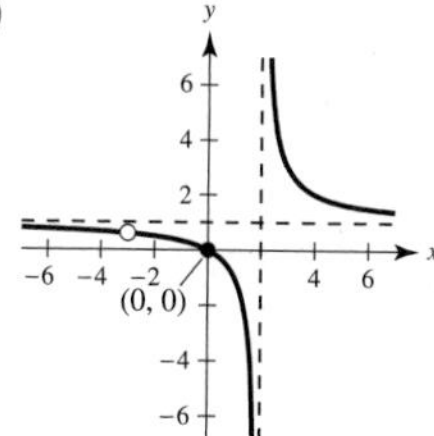

47. (a) Domain: all real numbers x except $x = -\frac{3}{2}, 2$
(b) x-intercept: $\left(\frac{1}{2}, 0\right)$
y-intercept: $\left(0, -\frac{1}{3}\right)$
(c) Vertical asymptote: $x = -\frac{3}{2}$
Horizontal asymptote: $y = 1$
(d)

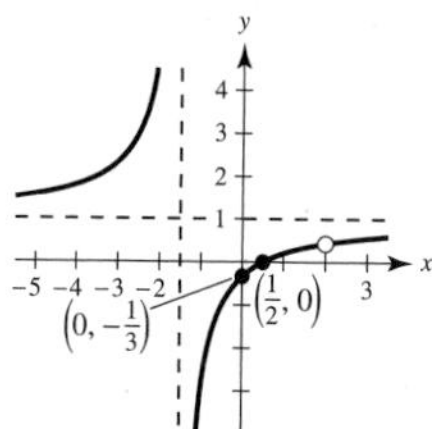

49. (a) Domain: all real numbers t except $t = 1$
(b) t-intercept: $(-1, 0)$
y-intercept: $(0, 1)$
(c) Vertical asymptote: None
Horizontal asymptote: None
(d)

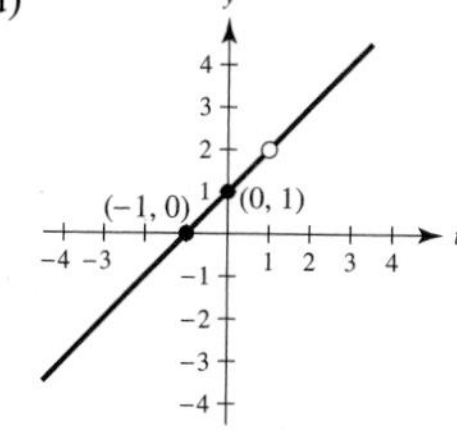

51. (a) Domain of f: all real numbers x except $x = -1$
Domain of g: all real numbers x
(b) $x - 1$; Vertical asymptotes: None
(c)

x	-3	-2	-1.5	-1	-0.5	0	1
$f(x)$	-4	-3	-2.5	Undef.	-1.5	-1	0
$g(x)$	-4	-3	-2.5	-2	-1.5	-1	0

(d)
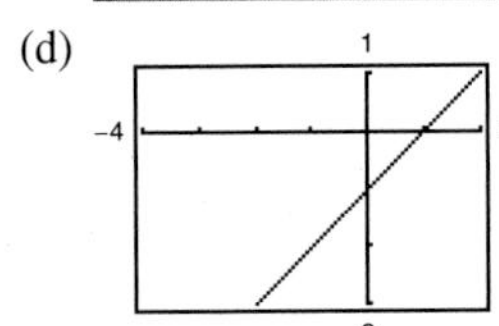

(e) Because there are only a finite number of pixels, the graphing utility may not attempt to evaluate the function where it does not exist.

53. (a) Domain of f: all real numbers x except $x = 0, 2$
Domain of g: all real numbers x except $x = 0$
(b) $\frac{1}{x}$; Vertical asymptote: $x = 0$
(c)

x	-0.5	0	0.5	1	1.5	2	3
$f(x)$	-2	Undef.	2	1	$\frac{2}{3}$	Undef.	$\frac{1}{3}$
$g(x)$	-2	Undef.	2	1	$\frac{2}{3}$	$\frac{1}{2}$	$\frac{1}{3}$

(d)
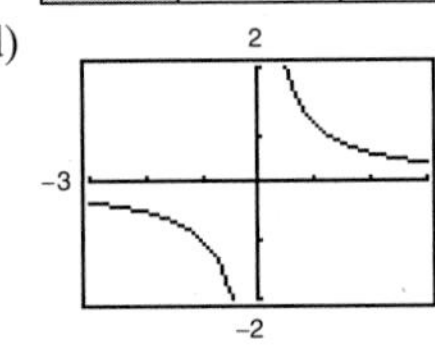

(e) Because there are only a finite number of pixels, the graphing utility may not attempt to evaluate the function where it does not exist.

55. (a) Domain: all real numbers x except $x = 0$
(b) x-intercepts: $(-3, 0)$, $(3, 0)$
(c) Vertical asymptote: $x = 0$
Slant asymptote: $y = x$
(d)
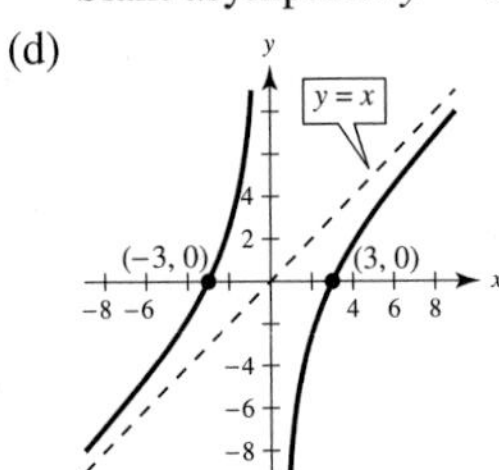

57. (a) Domain: all real numbers x except $x = 0$
(b) No intercepts
(c) Vertical asymptote: $x = 0$
Slant asymptote: $y = 2x$
(d)
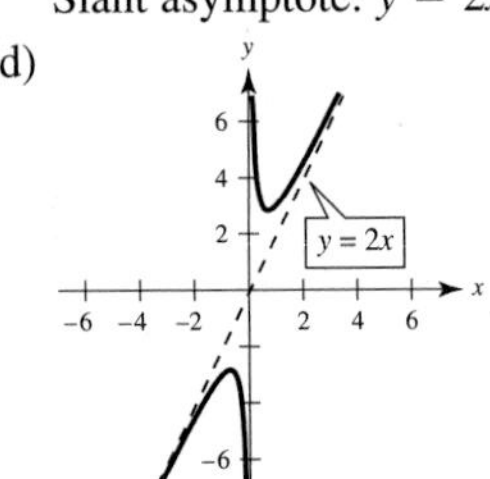

59. (a) Domain: all real numbers x except $x = 0$
(b) No intercepts
(c) Vertical asymptote: $x = 0$
Slant asymptote: $y = x$
(d)
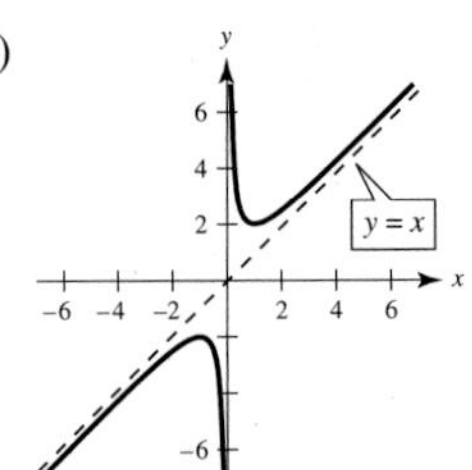

61. (a) Domain: all real numbers t except $t = -5$
(b) y-intercept: $\left(0, -\frac{1}{5}\right)$
(c) Vertical asymptote: $t = -5$
Slant asymptote: $y = -t + 5$
(d)
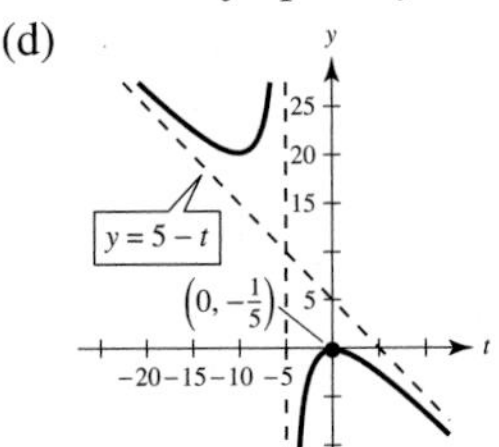

63. (a) Domain: all real numbers x except $x = \pm 2$
(b) Intercept: $(0, 0)$
(c) Vertical asymptotes: $x = \pm 2$
Slant asymptote: $y = x$
(d)
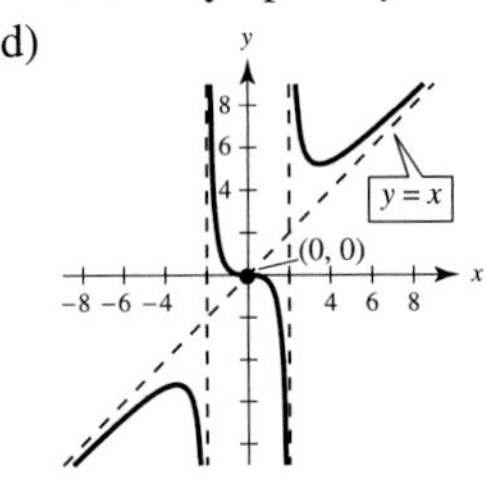

65. (a) Domain: all real numbers x except $x = 1$
(b) y-intercept: $(0, -1)$
(c) Vertical asymptote: $x = 1$
Slant asymptote: $y = x$
(d)
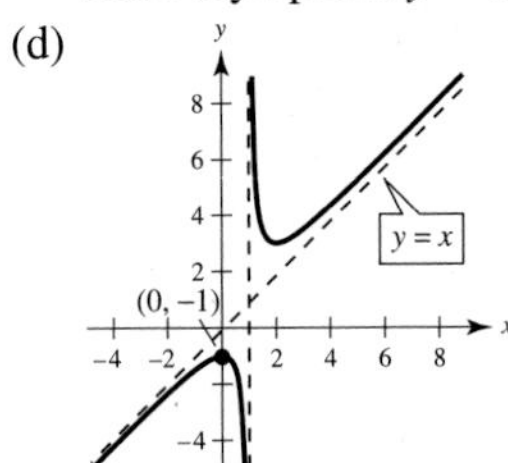

67. (a) Domain: all real numbers x except $x = -1, -2$
(b) y-intercept: $\left(0, \frac{1}{2}\right)$
x-intercepts: $\left(\frac{1}{2}, 0\right)$, $(1, 0)$
(c) Vertical asymptote: $x = -2$
Slant asymptote: $y = 2x - 7$
(d)
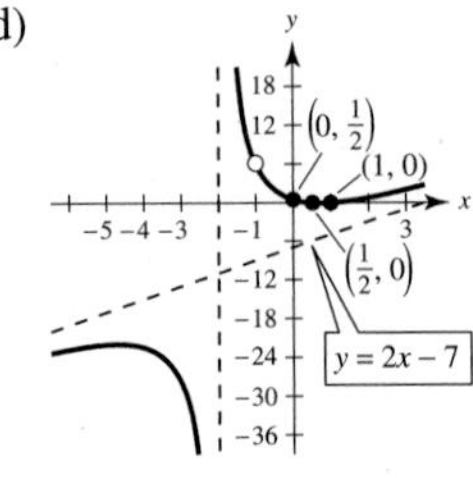

69. $f(x) = \dfrac{1}{x^2 + 2}$; $f(x) = \dfrac{1}{x - 2}$ (Answers are not unique.)

71.

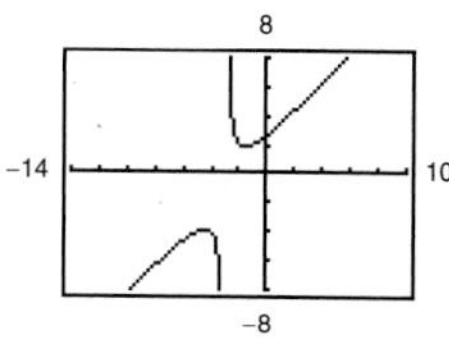

Domain: all real numbers x except $x = -3$
Vertical asymptote: $x = -3$
Slant asymptote: $y = x + 2$
$y = x + 2$

73.

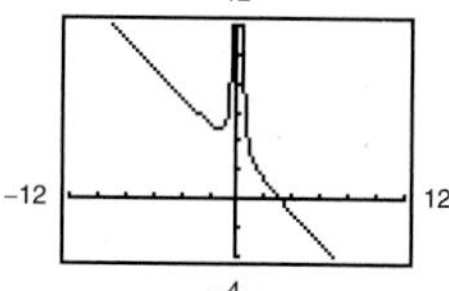

Domain: all real numbers x except $x = 0$
Vertical asymptote: $x = 0$
Slant asymptote: $y = -x + 3$
$y = -x + 3$

75. (a) $(-1, 0)$ (b) -1

77. (a) $(1, 0), (-1, 0)$ (b) ± 1

79. (a)

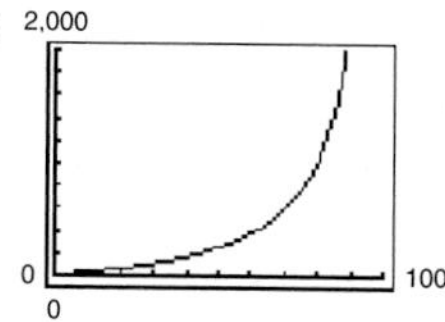

(b) \$28.33 million; \$170 million; \$765 million
(c) No. The function is undefined at $p = 100$.

81. (a) 333 deer, 500 deer, 800 deer (b) 1500 deer

83. (a) Answers will vary.
(b) Vertical asymptote: $x = 25$
Horizontal asymptote: $y = 25$
(c)

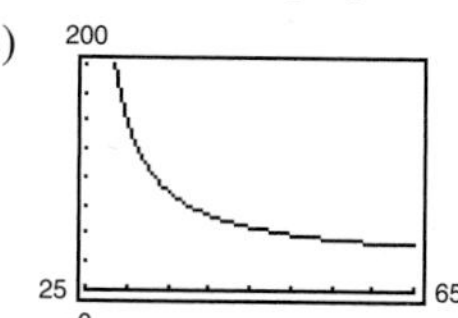

(d)

x	30	35	40	45	50	55	60
y	150	87.5	66.7	56.3	50	45.8	42.9

(e) Sample answer: No. You might expect the average speed for the round trip to be the average of the average speeds for the two parts of the trip.
(f) No. At 20 miles per hour you would use more time in one direction than is required for the round trip at an average speed of 50 miles per hour.

85. False. Polynomials do not have vertical asymptotes.

87. False. If the degree of the numerator is greater than the degree of the denominator, no horizontal asymptote exists. However, a slant asymptote exists only if the degree of the numerator is one greater than the degree of the denominator.

Review Exercises (page 207)

1. (a)

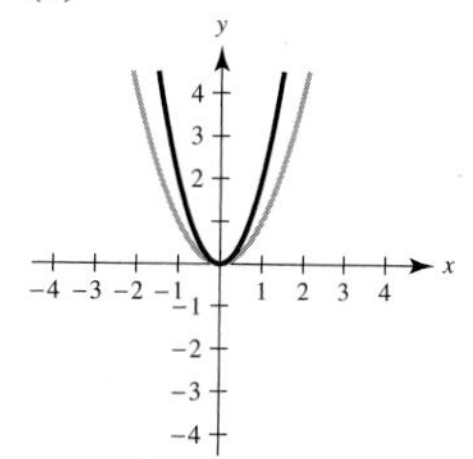

Vertical stretch

(b)

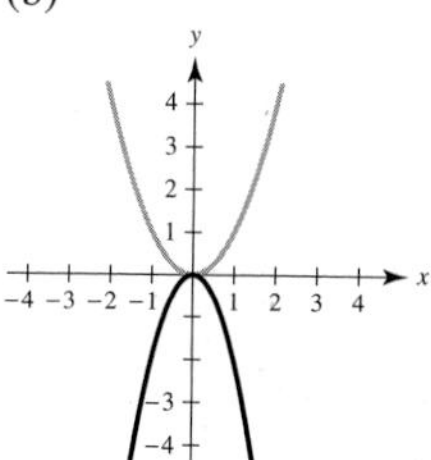

Vertical stretch and reflection in the x-axis

(c)

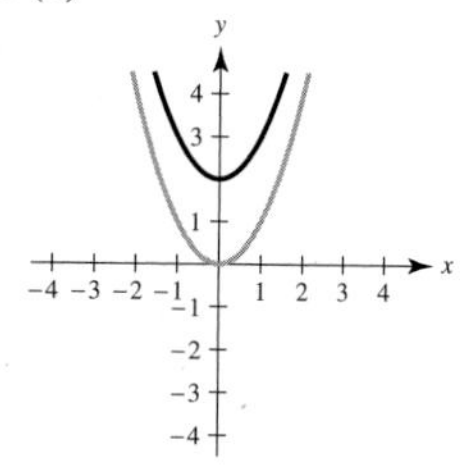

Vertical shift two units upward

(d)

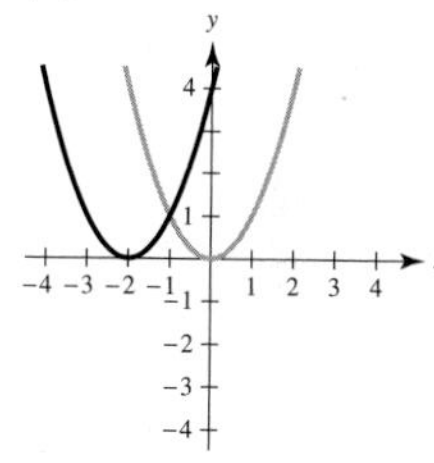

Horizontal shift two units to the left

3. $g(x) = (x - 1)^2 - 1$

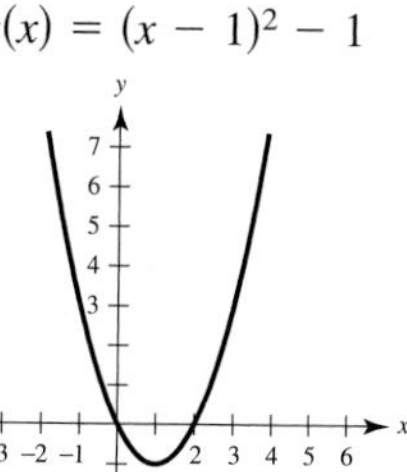

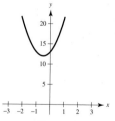

Vertex: $(1, -1)$
Axis of symmetry: $x = 1$
x-intercepts: $(0, 0), (2, 0)$

5. $f(x) = (x + 4)^2 - 6$

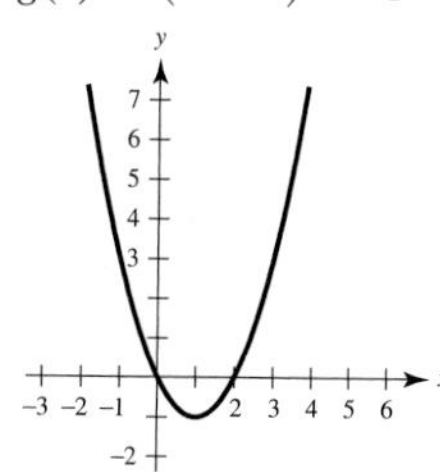

Vertex: $(-4, -6)$
Axis of symmetry: $x = -4$
x-intercepts: $\left(-4 \pm \sqrt{6}, 0\right)$

7. $f(t) = -2(t - 1)^2 + 3$

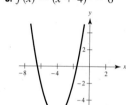

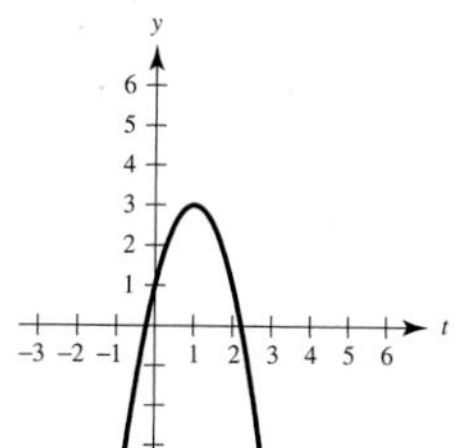

Vertex: $(1, 3)$
Axis of symmetry: $t = 1$
t-intercepts: $\left(1 \pm \dfrac{\sqrt{6}}{2}, 0\right)$

9. $h(x) = 4\left(x + \frac{1}{2}\right)^2 + 12$

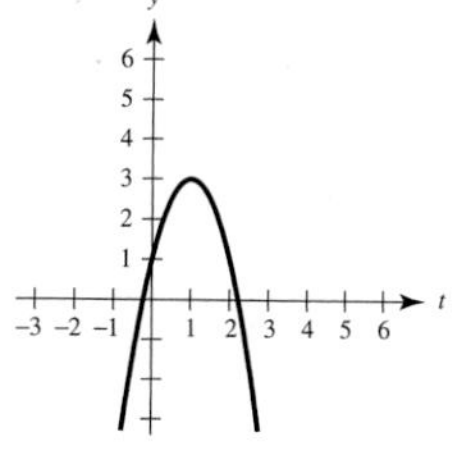

Vertex: $\left(-\frac{1}{2}, 12\right)$
Axis of symmetry: $x = -\frac{1}{2}$
No x-intercept

11. $h(x) = \left(x + \frac{5}{2}\right)^2 - \frac{41}{4}$

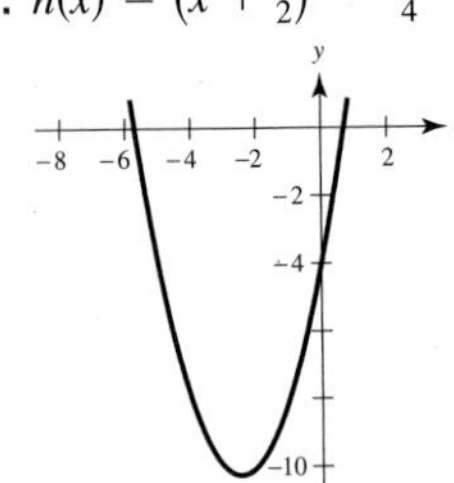

Vertex: $\left(-\frac{5}{2}, -\frac{41}{4}\right)$

Axis of symmetry: $x = -\frac{5}{2}$

x-intercepts: $\left(\frac{\pm\sqrt{41} - 5}{2}, 0\right)$

13. $f(x) = \frac{1}{3}\left(x + \frac{5}{2}\right)^2 - \frac{41}{12}$

Vertex: $\left(-\frac{5}{2}, -\frac{41}{12}\right)$

Axis of symmetry: $x = -\frac{5}{2}$

x-intercepts: $\left(\frac{\pm\sqrt{41} - 5}{2}, 0\right)$

15. $f(x) = -\frac{1}{2}(x - 4)^2 + 1$ **17.** $f(x) = (x - 1)^2 - 4$

19. (a)

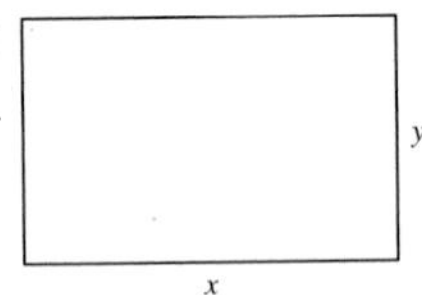

(b) $y = 500 - x$

$A(x) = 500x - x^2$

(c) $x = 250,\ y = 250$

21. 1091 units

23.

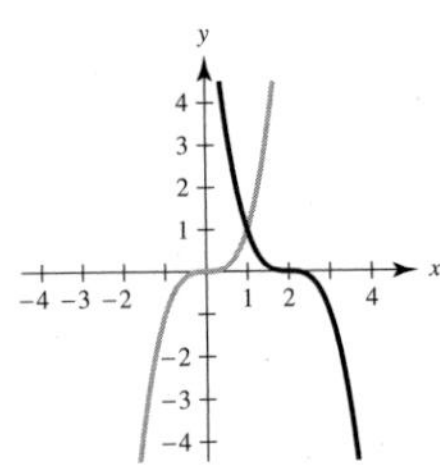

25.

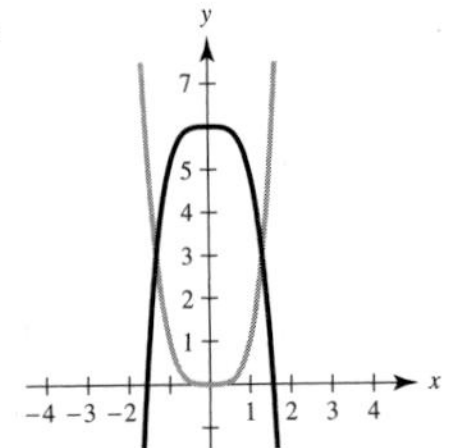

27.

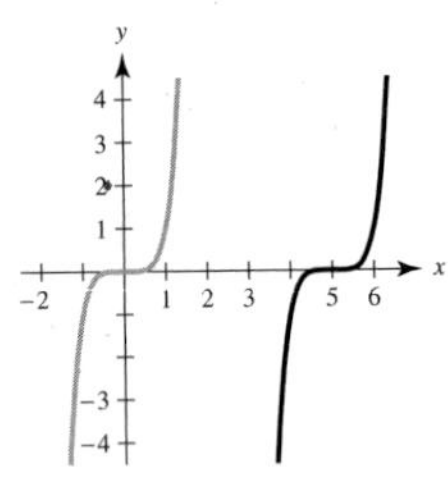

29. Falls to the left, falls to the right

31. Rises to the left, rises to the right

33. $-8, \frac{4}{3}$, odd multiplicity; turning points: 1

35. $0, \pm\sqrt{3}$, odd multiplicity; turning points: 2

37. 0, even multiplicity; $\frac{2}{3}$, odd multiplicity; turning points: 2

39. (a) Rises to the left, falls to the right (b) -1

(c) Answers will vary.

(d)

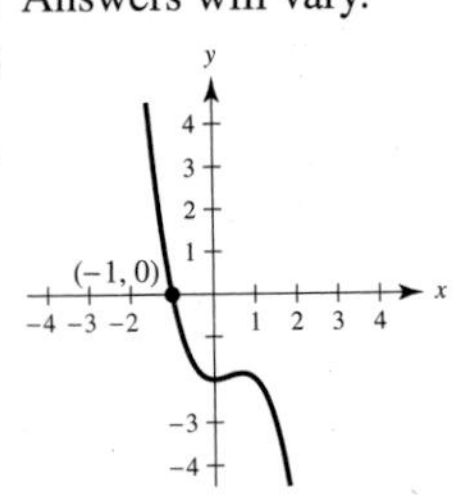

41. (a) Rises to the left, rises to the right (b) $-3, 0, 1$

(c) Answers will vary.

(d)

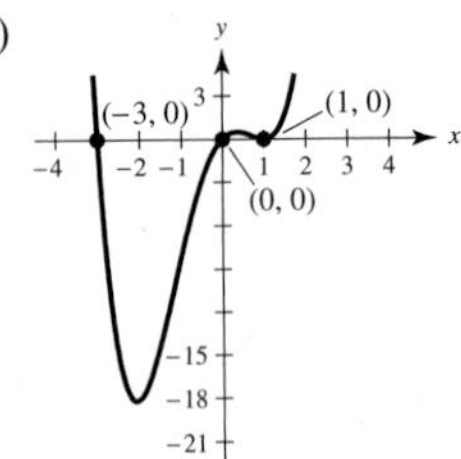

43. $6x + 3 + \dfrac{17}{5x - 3}$ **45.** $5x + 4,\quad x \neq \dfrac{5}{2} \pm \dfrac{\sqrt{29}}{2}$

47. $x^2 - 3x + 2 - \dfrac{1}{x^2 + 2}$

49. $6x^3 + 8x^2 - 11x - 4 - \dfrac{8}{x - 2}$

51. $2x^2 - 9x - 6,\quad x \neq 8$

53. (a) Yes (b) Yes (c) Yes (d) No

55. (a) -421 (b) -9

57. (a) Answers will vary.

(b) $(x + 7), (x + 1)$

(c) $f(x) = (x + 7)(x + 1)(x - 4)$

(d) $-7, -1, 4$

(e)

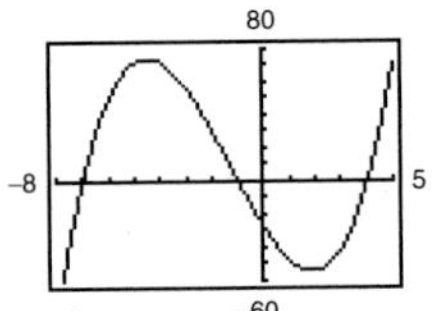

59. (a) Answers will vary. (b) $(x + 1), (x - 4)$

(c) $f(x) = (x + 1)(x - 4)(x + 2)(x - 3)$

(d) $-2, -1, 3, 4$

(e)

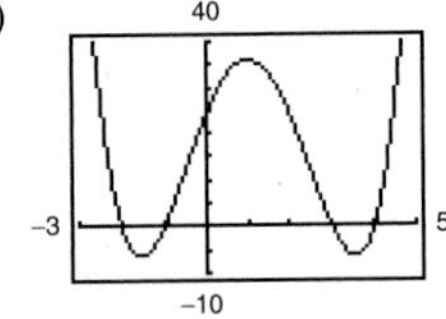

61. $A = 0.0031t^3 - 0.135t^2 + 2.12t - 4.4$

63.

Year, t	7	8	9	10	11	12	13
Attendance, A	4.9	5.5	6.0	6.4	6.7	7.0	7.2

Year, t	14	15	16	17	18	19
Attendance, A	7.3	7.5	7.7	7.9	8.1	8.4

65. $8 + 10i$ **67.** $-1 + 3i$ **69.** $3 + 7i$

71. $63 + 77i$ **73.** $-4 - 46i$ **75.** $39 - 80i$

77. $\dfrac{23}{17} + \dfrac{10}{17}i$ **79.** $\dfrac{21}{13} - \dfrac{1}{13}i$ **81.** $\pm\dfrac{\sqrt{10}}{5}i$

83. $1 \pm 3i$ **85.** $0, 3$ **87.** $2, 9$ **89.** $-4, 6, \pm 2i$

91. $\pm1, \pm3, \pm5, \pm15, \pm\frac{1}{2}, \pm\frac{3}{2}, \pm\frac{5}{2}, \pm\frac{15}{2}, \pm\frac{1}{4}, \pm\frac{3}{4}, \pm\frac{5}{4}, \pm\frac{15}{4}$

93. $-6, -2, 5$ **95.** $1, 8$ **97.** $-4, 3$ **99.** $4, \pm i$

101. $-3, \frac{1}{2}, 2 \pm i$ **103.** $0, 1, -5; f(x) = x(x - 1)(x + 5)$

105. $-4, 2 \pm 3i$; $g(x) = (x + 4)^2(x - 2 - 3i)(x - 2 + 3i)$

107. $f(x) = 3x^4 - 14x^3 + 17x^2 - 42x + 24$

109. Domain: all real numbers x except $x = -10$

111. Domain: all real numbers x except $x = 6, 4$

113. Vertical asymptote: $x = -3$
Horizontal asymptote: $y = 0$

115. Vertical asymptote: $x = 6$
Horizontal asymptote: $y = 0$

117. (a) Domain: all real numbers x except $x = 0$
(b) No intercepts
(c) Vertical asymptote: $x = 0$
Horizontal asymptote: $y = 0$
(d)

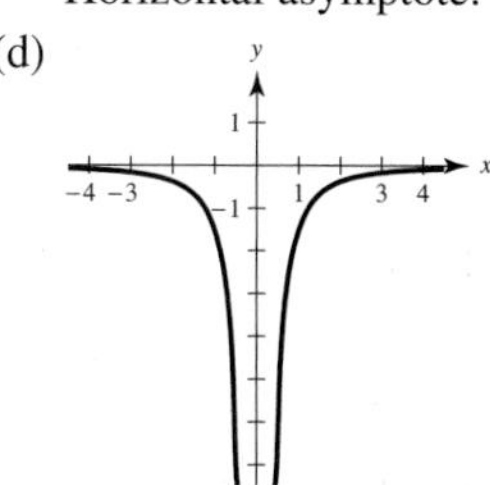

119. (a) Domain: all real numbers x except $x = 1$
(b) x-intercept: $(-2, 0)$
y-intercept: $(0, 2)$
(c) Vertical asymptote: $x = 1$
Horizontal asymptote: $y = -1$
(d)

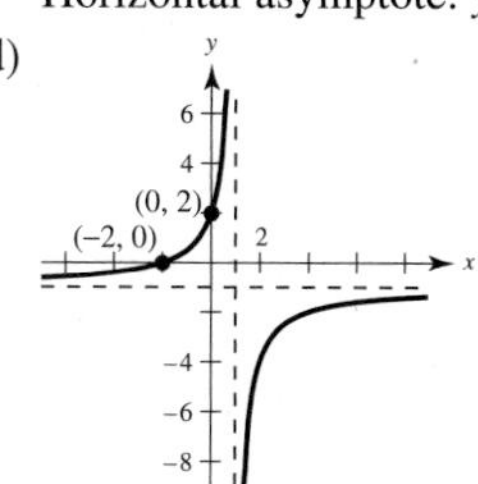

121. (a) Domain: all real numbers x
(b) Intercept: $(0, 0)$
(c) Horizontal asymptote: $y = \frac{5}{4}$
(d)

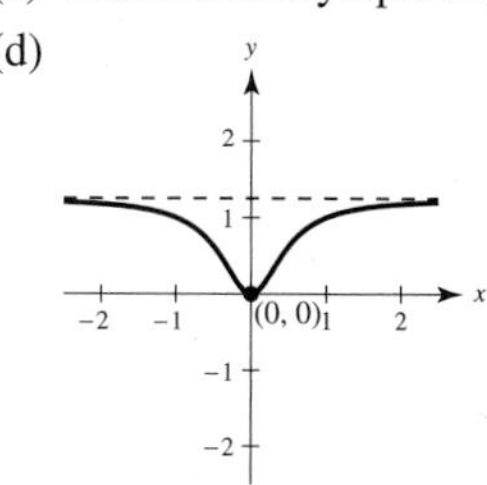

123. (a) Domain: all real numbers x
(b) Intercept: $(0, 0)$
(c) Horizontal asymptote: $y = 0$
(d)

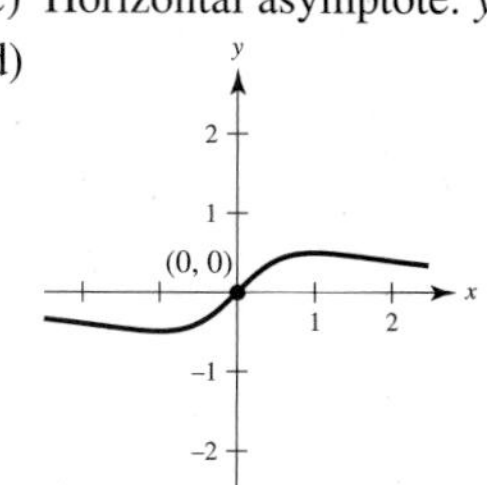

125. (a) Domain: all real numbers x
(b) Intercept: $(0, 0)$
(c) Horizontal asymptote: $y = -6$
(d)

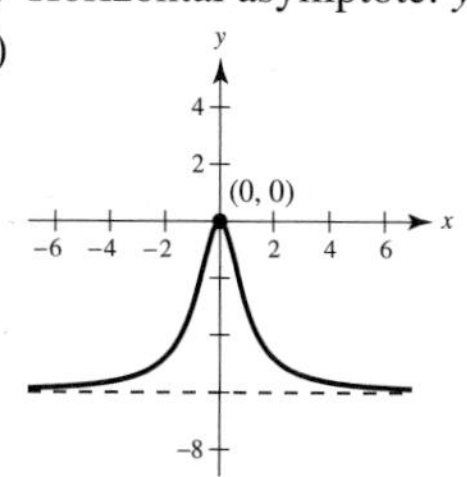

127. (a) Domain: all real numbers x except $x = 0, \frac{1}{3}$
(b) x-intercept: $\left(\frac{3}{2}, 0\right)$
(c) Vertical asymptote: $x = 0$
Horizontal asymptote: $y = 2$
(d)

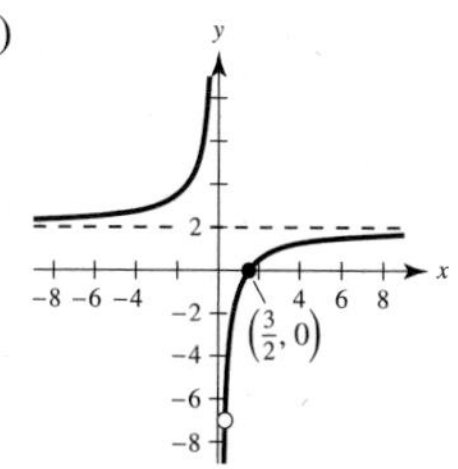

129. (a) Domain: all real numbers x
(b) Intercept: $(0, 0)$ (c) Slant asymptote: $y = 2x$
(d)

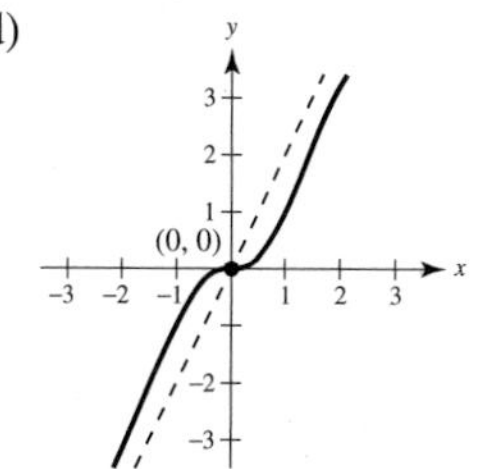

131. (a) Domain: all real numbers x except $x = \frac{4}{3}, -1$
(b) y-intercept: $\left(0, -\frac{1}{2}\right)$
x-intercepts: $\left(\frac{2}{3}, 0\right), (1, 0)$
(c) Vertical asymptote: $x = \frac{4}{3}$
Slant asymptote: $y = x - \frac{1}{3}$
(d)

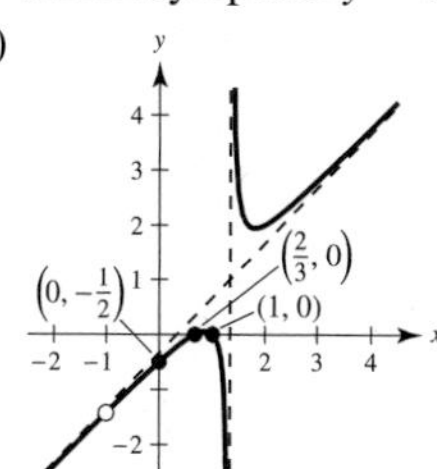

133. $\overline{C} = 0.5 = \$0.50$

Chapter Test (page 210)

1. (a) Reflection in the x-axis followed by a vertical shift two units upward
(b) Horizontal shift $\frac{3}{2}$ units to the right

2. $y = (x - 3)^2 - 6$

3. (a) 50 ft

(b) 5. Yes, changing the constant term results in a vertical translation of the graph and therefore changes the maximum height.

4. Rises to the left, falls to the right

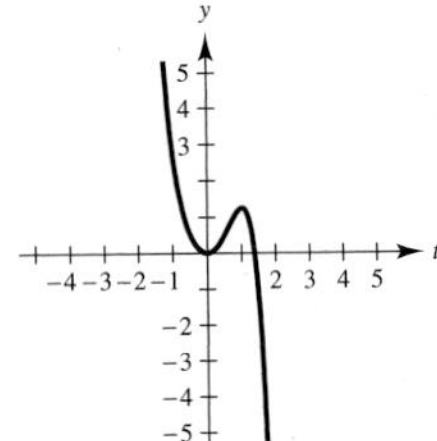

5. $3x + \dfrac{x - 1}{x^2 + 1}$

6. $2x^3 + 4x^2 + 3x + 6 + \dfrac{9}{x - 2}$

7. $(2x - 5)(x + \sqrt{3})(x - \sqrt{3})$;
Zeros: $\frac{5}{2}, \pm\sqrt{3}$

8. (a) $-3 + 5i$ (b) 7 **9.** $2 - i$

10. $f(x) = x^4 - 7x^3 + 17x^2 - 15x$

11. $f(x) = x^4 - 6x^3 + 16x^2 - 24x + 16$

12. $-5, -\frac{2}{3}, 1$ **13.** $-2, 4, -1 \pm \sqrt{2}i$

14. x-intercepts: $(-2, 0), (2, 0)$
Vertical asymptote: $x = 0$
Horizontal asymptote: $y = -1$

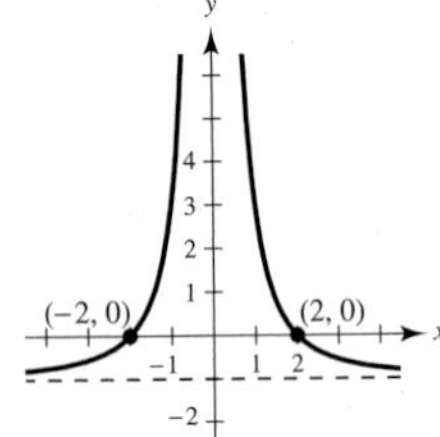

15. x-intercept: $\left(-\frac{3}{2}, 0\right)$
y-intercept: $\left(0, \frac{3}{4}\right)$
Vertical asymptote: $x = -4$
Horizontal asymptote: $y = 2$

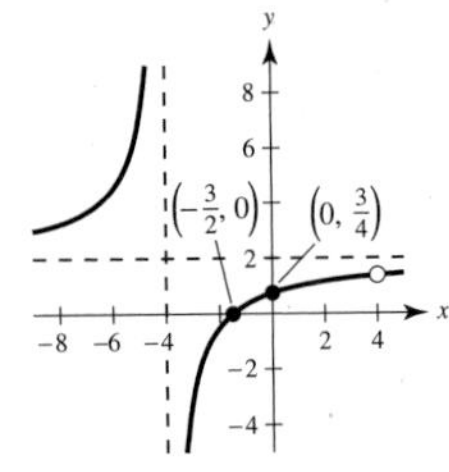

16. y-intercept: $(0, -2)$
Vertical asymptote: $x = 1$
Slant asymptote: $y = x + 1$

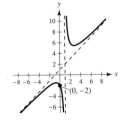

17.

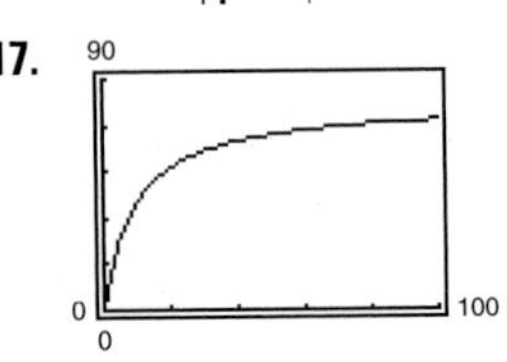

80.3 mg/dm²/h

P.S. Problem Solving (page 211)

1. 2 in. × 2 in. × 5 in.

3. $h = 0, k = 0$, and $a < -1$ produce a stretch that is reflected in the x-axis; $h = 0, k = 0$, and $-1 < a < 0$ produce a shrink that is reflected in the x-axis.

5. (a) As $|a|$ increases, the graph stretches vertically. For $a < 0$, the graph is reflected in the x-axis.

(b) As $|b|$ increases, the vertical asymptote is translated. For $b > 0$, the graph is translated to the right. For $b < 0$, the graph is reflected in the x-axis and is translated to the left.

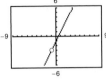

7. (a) No, there is a *hole* at $x = -1$.

(b) $2x - 1, x \neq 1$

(c) As $x \to -1$, $(2x^2 + x - 1)/(x + 1) \to -3$.

9. (a) $\frac{1}{2} - \frac{1}{2}i$ (b) $\frac{3}{10} + \frac{1}{10}i$ (c) $-\frac{1}{34} - \frac{2}{17}i$

11. (a) Slope $= 5$; less than (b) Slope $= 3$; greater than

(c) Slope $= 4.1$; less than (d) $4 + h, h \neq 0$

(e) $h = -1$, slope $= 3$; $h = 1$, slope $= 5$;
$h = 0.1$, slope $= 4.1$

(f) As h approaches 0, the slope approaches 4. So, the slope at $(2, 4)$ is 4.

13. (a) iii (b) ii (c) iv (d) i

Chapter 3

Section 3.1 (page 219)

1. Precalculus: 300 ft

3. Calculus: Slope of the tangent line at $x = 2$ is 0.16.

5. (a) Precalculus: 10 square units

(b) Precalculus: 2π square units

7. (a)

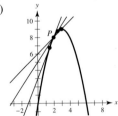

(b) $1; \frac{3}{2}; \frac{5}{2}$ (c) 2. Use points closer to P.

9. Area ≈ 10.417; Area ≈ 9.145

11. (a) 5.66 (b) 6.11 (c) Increase the number of line segments.

Section 3.2 (page 225)

1.

x	3.9	3.99	3.999	4.001	4.01	4.1
$f(x)$	0.2041	0.2004	0.2000	0.2000	0.1996	0.1961

$\displaystyle\lim_{x\to 4} \frac{x - 4}{x^2 - 3x - 4} \approx 0.2000 \left(\text{Actual limit is } \frac{1}{5}.\right)$

3.

x	-0.1	-0.01	-0.001	0.001	0.01	0.1
$f(x)$	0.2050	0.2042	0.2041	0.2041	0.2040	0.2033

$\lim_{x\to 0} \dfrac{\sqrt{x+6}-\sqrt{6}}{x} \approx 0.2041 \left(\text{Actual limit is } \dfrac{1}{2\sqrt{6}}.\right)$

5.

x	2.9	2.99	2.999
$f(x)$	-0.0641	-0.0627	-0.0625

x	3.001	3.01	3.1
$f(x)$	-0.0625	-0.0623	-0.0610

$\lim_{x\to 3} \dfrac{[1/(x+1)]-(1/4)}{x-3} \approx -0.0625 \left(\text{Actual limit is } -\dfrac{1}{16}.\right)$

7.

x	0.9	0.99	0.999	1.001	1.01	1.1
$f(x)$	0.2564	0.2506	0.2501	0.2499	0.2494	0.2439

$\lim_{x\to 1} \dfrac{x-2}{x^2+x-6} \approx 0.2500 \left(\text{Actual limit is } \dfrac{1}{4}.\right)$

9.

x	0.9	0.99	0.999	1.001	1.01	1.1
$f(x)$	0.7340	0.6733	0.6673	0.6660	0.6600	0.6015

$\lim_{x\to 1} \dfrac{x^4-1}{x^6-1} \approx 0.6666 \left(\text{Actual limit is } \dfrac{2}{3}.\right)$

11. 1 **13.** 2

15. Limit does not exist. The function approaches 1 from the right side of 2 but it approaches -1 from the left side of 2.

17. (a) 2
(b) Limit does not exist. The function approaches 1 from the right side of 1 but it approaches 3.5 from the left side of 1.
(c) Value does not exist. The function is undefined at $x = 4$.
(d) 2

19. $\lim_{x\to c} f(x)$ exists for all points on the graph except where $c = -3$.

21. $\delta = \frac{1}{11} \approx 0.091$ **23.** $L = 8$. Let $\delta = 0.01/3 \approx 0.0033$.

25. $L = 1$. Let $\delta = 0.01/5 = 0.002$. **27.** 6

29. -3 **31.** 3 **33.** 0 **35.** 10 **37.** 2 **39.** 4

41.

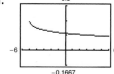

$\lim_{x\to 4} f(x) = \frac{1}{6}$

Domain: $[-5, 4) \cup (4, \infty)$

The graph has a hole at $x = 4$.

43.

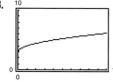

$\lim_{x\to 9} f(x) = 6$

Domain: $[0, 9) \cup (9, \infty)$

The graph has a hole at $x = 9$.

45. Answers will vary. Sample answer: As x approaches 8 from either side, $f(x)$ becomes arbitrarily close to 25.

47. $\lim_{x\to 0^+} p(x) = 14.7 \text{ lb/in.}^2$

49. (a) $r = \dfrac{3}{\pi} \approx 0.9549$ cm
(b) $\dfrac{5.5}{2\pi} \le r \le \dfrac{6.5}{2\pi}$, or approximately $0.8754 < r < 1.0345$
(c) $\lim_{r\to 3/\pi} 2\pi r = 6$; $\varepsilon = 0.5$; $\delta \approx 0.0796$

51.

x	-0.001	-0.0001	-0.00001
$f(x)$	2.7196	2.7184	2.7183

x	0.00001	0.0001	0.001
$f(x)$	2.7183	2.7181	2.7169

$\lim_{x\to 0} f(x) \approx 2.7183$

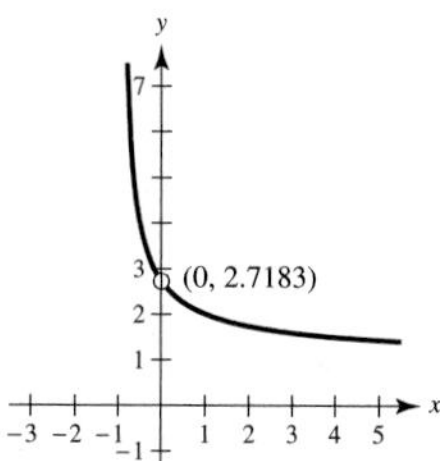

53. False. The existence or nonexistence of $f(x)$ at $x = c$ has no bearing on the existence of the limit of $f(x)$ as $x \to c$.

55. False. See Exercise 13.

57. Yes. As x approaches 0.25 from either side, $\sqrt{x}$ becomes arbitrarily close to 0.5.

59–61. Proofs

63. The value of f at c has no bearing on the limit as x approaches 0.

Section 3.3 (page 234)

1.

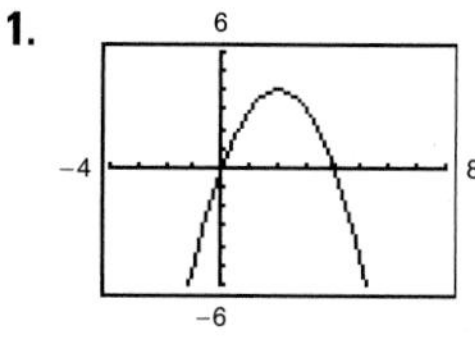

(a) 0 (b) -5

3.

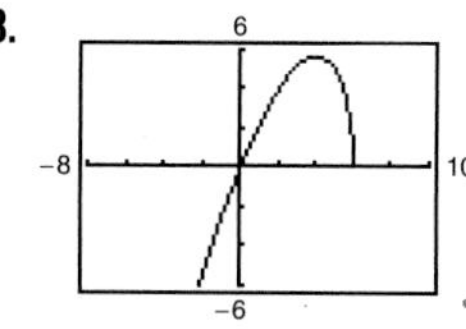

(a) 0 (b) 4

5. 8 **7.** -1 **9.** 0 **11.** 7 **13.** 2 **15.** 1

17. 1/2 **19.** 1/5 **21.** 7 **23.** (a) 4 (b) 64 (c) 64

25. (a) 3 (b) 2 (c) 2

27. (a) 10 (b) 5 (c) 6 (d) 3/2

29. (a) 64 (b) 2 (c) 12 (d) 8

31. (a) -1 (b) -2

$g(x) = \dfrac{x^2 - x}{x}$ and $f(x) = x - 1$ agree except at $x = 0$.

33. (a) 2 (b) 0

$g(x) = \dfrac{x^3 - x}{x - 1}$ and $f(x) = x^2 + x$ agree except at $x = 1$.

35. -2

$f(x) = \dfrac{x^2 - 1}{x + 1}$ and $g(x) = x - 1$ agree except at $x = -1$.

37. 12

$f(x) = \dfrac{x^3 - 8}{x - 2}$ and $g(x) = x^2 + 2x + 4$ agree except at $x = 2$.

39. -1 **41.** $1/8$ **43.** $5/6$ **45.** $1/6$

47. $-1/9$ **49.** 2 **51.** $2x - 2$

53.

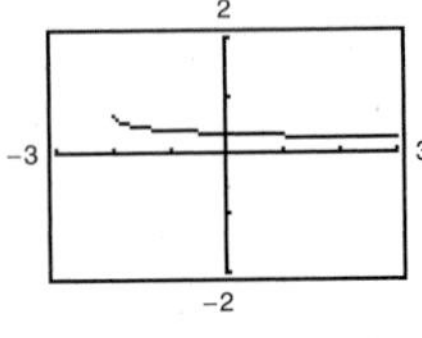

The graph has a hole at $x = 0$.

Answers will vary. Example:

x	-0.1	-0.01	-0.001	0.001	0.01	0.1
$f(x)$	0.358	0.354	0.354	0.354	0.353	0.349

$\lim_{x \to 0} \dfrac{\sqrt{x+2} - \sqrt{2}}{x} \approx 0.354 \left(\text{Actual limit is } \dfrac{1}{2\sqrt{2}} = \dfrac{\sqrt{2}}{4}. \right)$

55.

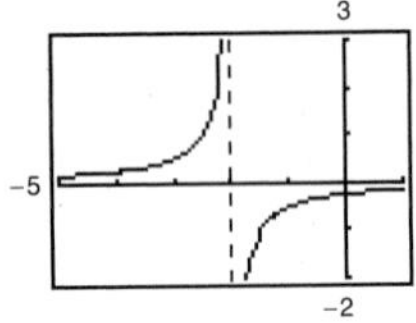

The graph has a hole at $x = 0$.

Answers will vary. Example:

x	-0.1	-0.01	-0.001
$f(x)$	-0.263	-0.251	-0.250

x	0.001	0.01	0.1
$f(x)$	-0.250	-0.249	-0.238

$\lim_{x \to 0} \dfrac{[1/(2+x)] - (1/2)}{x} \approx -0.250 \left(\text{Actual limit is } -\dfrac{1}{4}. \right)$

57. 3 **59.** $-1/(x+3)^2$ **61.** 4

63. f and g agree at all but one point if c is a real number such that $f(x) = g(x)$ for all $x \neq c$.

65. An indeterminate form is obtained when evaluating a limit using direct substitution produces a meaningless fractional form, such as $\frac{0}{0}$.

67. -64 ft/sec (speed $= 64$ ft/sec) **69.** -29.4 m/sec

71. Let $f(x) = 4/x$ and $g(x) = 2/x$.

$\lim_{x \to 0} f(x)$ and $\lim_{x \to 0} g(x)$ do not exist. However,

$\lim_{x \to 0} \left(\dfrac{f(x)}{g(x)} \right) = \lim_{x \to 0} (2) = 2$, and therefore does exist.

73–75. Proofs

77. False. The limit does not exist because the function approaches 1 from the right side of 0 and approaches -1 from the left side of 0. (See graph below.)

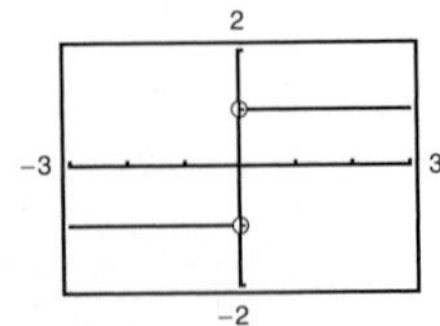

79. True

81. False. The limit does not exist because $f(x)$ approaches 3 from the left side of 2 and approaches 0 from the right side of 2. (See graph below.)

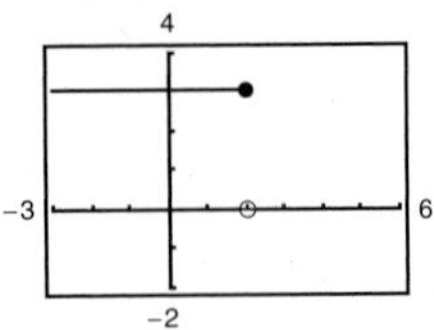

83. $\lim_{x \to 0} f(x)$ does not exist; $\lim_{x \to 0} g(x) = 0$

Section 3.4 (page 243)

1. (a) 3 (b) 3 (c) 3; $f(x)$ is continuous on $(-\infty, \infty)$.

3. (a) 0 (b) 0 (c) 0; Discontinuity at $x = 3$

5. (a) -3 (b) 3 (c) Limit does not exist.
Discontinuity at $x = 2$

7. $\frac{1}{10}$

9. Limit does not exist. The function decreases without bound as x approaches -3 from the left.

11. -1 **13.** $-1/x^2$ **15.** $5/2$ **17.** 2 **19.** 8

21. Limit does not exist. The function approaches 5 from the left side of 3 but approaches 6 from the right side of 3.

23. Discontinuous at $x = -2$ and $x = 2$

25. Discontinuous at every integer

27. Continuous on $[-7, 7]$ **29.** Continuous on $[-1, 4]$

31. Continuous for all real x

33. Nonremovable discontinuity at $x = 1$
Removable discontinuity at $x = 0$

35. Continuous for all real x

37. Removable discontinuity at $x = -2$
Nonremovable discontinuity at $x = 5$

39. Nonremovable discontinuity at $x = -7$

41. Continuous for all real x

43. Nonremovable discontinuity at $x = 2$

45. Nonremovable discontinuities at each integer

47.

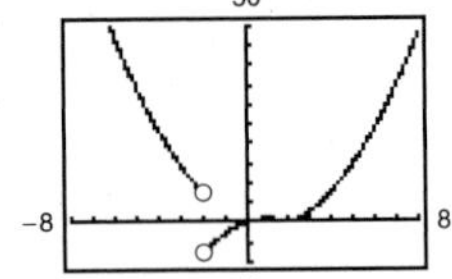

$\lim_{x \to 0^+} f(x) = 0$

$\lim_{x \to 0^-} f(x) = 0$

Discontinuity at $x = -2$

49. $a = 7$ **51.** $a = -1$, $b = 1$

53. Continuous for all real x **55.** Continuous on $(0, \infty)$

57. Nonremovable discontinuities at $x = 1$ and $x = -1$

59.

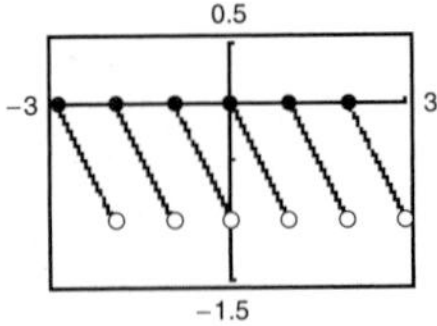

Nonremovable discontinuity at each integer

61.

Nonremovable discontinuity at $x = 4$

63. Continuous on $(-\infty, \infty)$

65. Continuous on $(-\infty, -6) \cup (-6, 6) \cup (6, \infty)$

67. 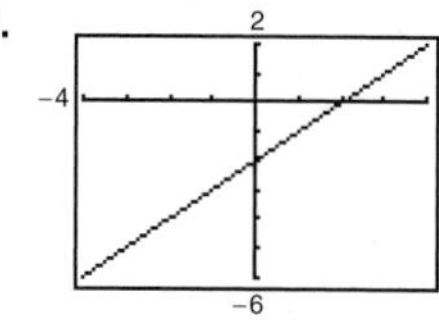
The graph has a hole at $x = -1$. The graph appears to be continuous, but the function is not continuous. It is not obvious from the graph that the function has a discontinuity at $x = -1$.

69. Because $f(x)$ is continuous on the interval $[1, 2]$ and $f(1) = 37/12$ and $f(2) = -8/3$, by the Intermediate Value Theorem there exists a real number c in $[1, 2]$ such that $f(c) = 0$.

71. 0.68, 0.6823 **73.** 0.88, 0.8819

75. $f(3) = 11$ **77.** $f(2) = 4$

79. (a) The limit does not exist at $x = c$.
(b) The function is not defined at $x = c$.
(c) The limit exists, but it is not equal to the value of the function at $x = c$.
(d) The limit does not exist at $x = c$.

81. If f and g are continuous for all real x, then so is $f + g$ (Theorem 3.9, part 2). However, f/g might not be continuous if $g(x) = 0$. For example, let $f(x) = x$ and $g(x) = x^2 - 1$. Then f and g are continuous for all real x, but f/g is not continuous at $x = \pm 1$.

83. True

85. False. A rational function can be written as $P(x)/Q(x)$, where P and Q are polynomials of degree m and n, respectively. It can have, at most, n discontinuities.

87. For non-integer values of x, the functions differ by 1.

89. 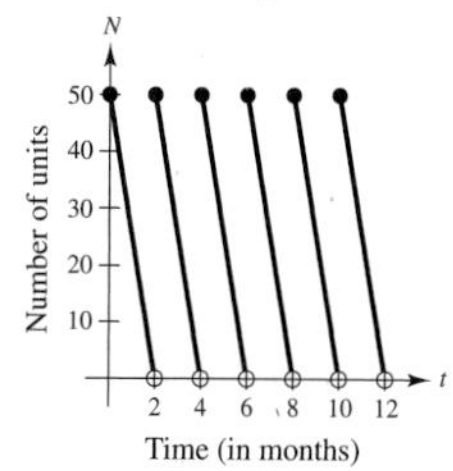
Discontinuous at every positive even integer. The company replenishes its inventory every two months.

91–93. Proofs

95. (a) $f(x) = \begin{cases} 0, & 0 \le x < b \\ b, & b < x \le 2b \end{cases}$

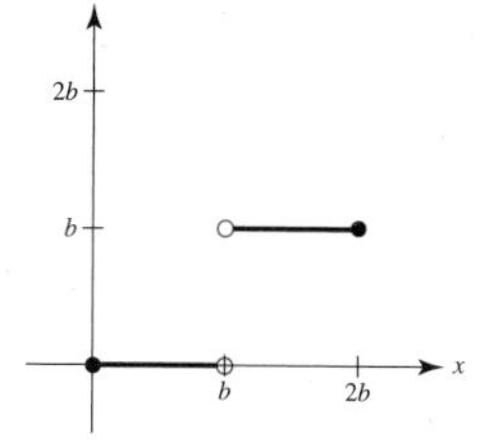
Not continuous at $x = b$

(b) $f(x) = \begin{cases} \dfrac{x}{2}, & 0 \le x < b \\ b - \dfrac{x}{2}, & b < x \le 2b \end{cases}$

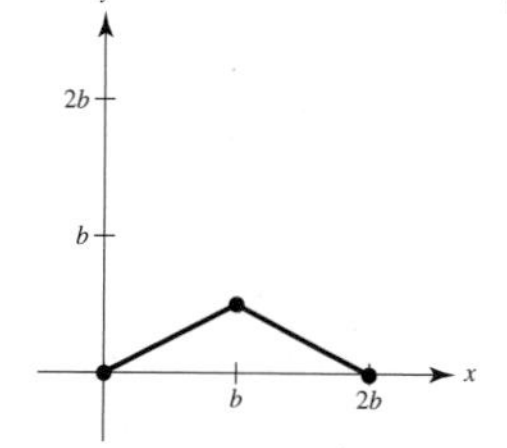
Continuous on $[0, 2b]$

97. Domain: $[-c^2, 0) \cup (0, \infty)$; Let $f(0) = 1/(2c)$.

Section 3.5 (page 252)

1. $\lim_{x\to 4^+} \dfrac{1}{x-4} = \infty$, $\lim_{x\to 4^-} \dfrac{1}{x-4} = -\infty$

3. $\lim_{x\to 4^+} \dfrac{1}{(x-4)^2} = \infty$, $\lim_{x\to 4^-} \dfrac{1}{(x-4)^2} = \infty$

5. $\lim_{x\to -2^+} 2\left|\dfrac{x}{x^2-4}\right| = \infty$, $\lim_{x\to -2^-} 2\left|\dfrac{x}{x^2-4}\right| = \infty$

7.

x	-3.5	-3.1	-3.01	-3.001
$f(x)$	0.31	1.64	16.6	167

x	-2.999	-2.99	-2.9	-2.5
$f(x)$	-167	-16.7	-1.69	-0.36

$\lim_{x\to -3^+} f(x) = -\infty$, $\lim_{x\to -3^-} f(x) = \infty$

9.

x	-3.5	-3.1	-3.01	-3.001
$f(x)$	3.8	16	151	1501

x	-2.999	-2.99	-2.9	-2.5
$f(x)$	-1499	-149	-14	-2.3

$\lim_{x\to -3^+} f(x) = -\infty$, $\lim_{x\to -3^-} f(x) = \infty$

11. $x = -1$, $x = 2$

13. $x = 0$ **15.** $x = \pm 2$ **17.** No vertical asymptote

19. $x = -2$, $x = 3$ **21.** $t = 0$ **23.** $x = -2$, $x = 1$

25. No vertical asymptote **27.** No vertical asymptote

29. Removable discontinuity at $x = -1$

31. Vertical asymptote at $x = -1$ **33.** ∞ **35.** ∞

37. ∞ **39.** $-\frac{1}{5}$ **41.** $\frac{1}{2}$ **43.** $-\infty$

45. ∞ **47.** ∞ **49.** $-\infty$

51. 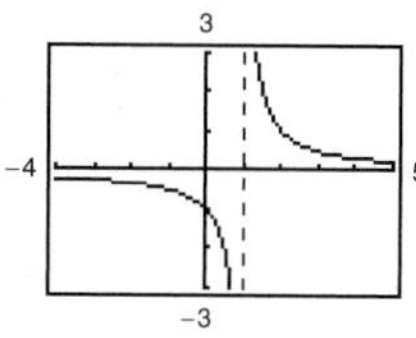
$\lim_{x\to 1^+} f(x) = \infty$

53. 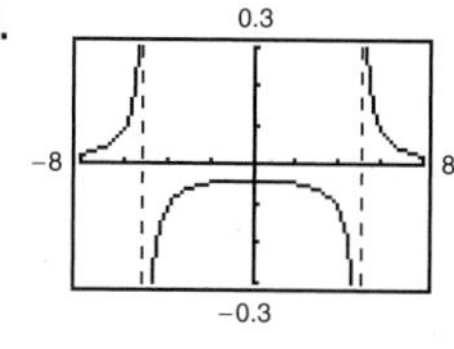
$\lim_{x\to 5^-} f(x) = -\infty$

55. Answers will vary.

57. Answers will vary. Example: $f(x) = \dfrac{x-3}{x^2 - 4x - 12}$

59. 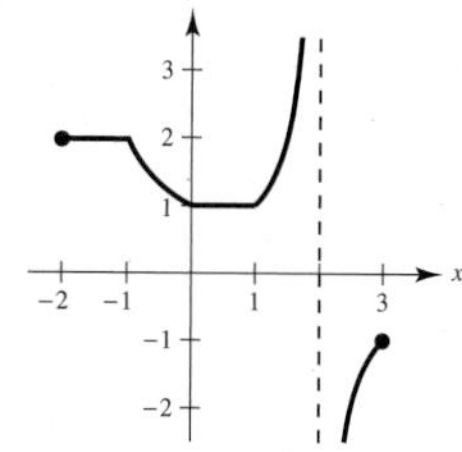

61. $\lim_{V\to 0^+} P = \infty$

63. (a) \$14,117.65 (b) \$80,000 (c) \$720,000
(d) ∞; The cost of removing all the pollution is unbounded.

65. ∞

67. (a) Domain: $x > 25$

(b)

x	30	40	50	60
y	150	66.667	50	42.857

(c) $\lim_{x\to 25^+} \dfrac{25x}{x-25} = \infty$

As x gets closer and closer to 25 mi/h, y becomes larger and larger.

69. False. Let $f(x) = (x^2 - 1)/(x - 1)$.

71. False. Let

$$f(x) = \begin{cases} \dfrac{1}{x}, & x \neq 0 \\ 3, & x = 0. \end{cases}$$

The graph of f has a vertical asymptote at $x = 0$, but $f(0) = 3$.

73. Given $\lim_{x\to c} f(x) = \infty$, let $g(x) = 1$. Then $\lim_{x\to c} \dfrac{g(x)}{f(x)} = 0$ by Theorem 3.13.

75. Answers will vary.

Review Exercises (page 256)

1. Calculus

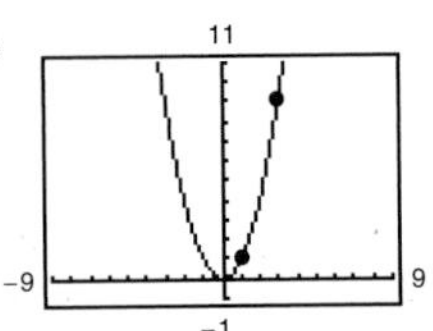

Estimate: 8.3

3.

x	-0.1	-0.01	-0.001
$f(x)$	-1.0526	-1.0050	-1.0005

x	0.001	0.01	0.1
$f(x)$	-0.9995	-0.9950	-0.9524

The estimate of the limit of $f(x)$, as x approaches zero, is -1.00.

5. (a) 4 (b) 5 **7.** 5; Proof **9.** -3; Proof **11.** $-\frac{1}{2}$

13. $\frac{7}{12}$ **15.** 16 **17.** $\sqrt{6} \approx 2.45$ **19.** $-\frac{1}{4}$

21. $\frac{1}{2}$ **23.** -1 **25.** 75

27. (a)

x	1.1	1.01	1.001	1.0001
$f(x)$	0.5680	0.5764	0.5773	0.5773

$\lim_{x\to 1^+} f(x) \approx 0.5773$

(b)

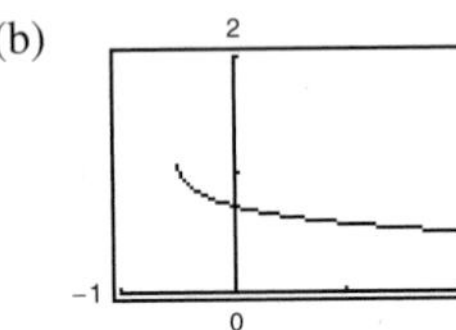

The graph has a hole at $x = 1$.
$\lim_{x\to 1^+} f(x) \approx 0.5774$

(c) $\sqrt{3}/3$

29. -39.2 m/sec **31.** 3 **33.** -1 **35.** 0

37. Limit does not exist. The limit as t approaches 1 from the left is 2, whereas the limit as t approaches 1 from the right is 1.

39. Continuous for all real x

41. Nonremovable discontinuity at each integer
Continuous on $(k, k + 1)$ for all integers k

43. Removable discontinuity at $x = 1$
Continuous on $(-\infty, 1) \cup (1, \infty)$

45. Nonremovable discontinuity at $x = 2$
Continuous on $(-\infty, 2) \cup (2, \infty)$

47. Nonremovable discontinuity at $x = -1$
Continuous on $(-\infty, -1) \cup (-1, \infty)$

49. $c = -\frac{1}{2}$

51. (a) -4 (b) 4 (c) Limit does not exist.

53. $C(x) = 12.80 + 2.50[-[\![-x]\!] - 1], \quad x > 0$
$= 12.80 - 2.50[[\![-x]\!] + 1], \quad x > 0$

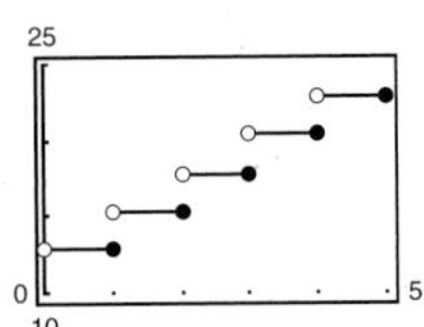

C has a nonremovable discontinuity at each integer 1, 2, 3, ...

55. $-\infty$ **57.** $\frac{1}{3}$ **59.** $-\infty$ **61.** $-\infty$

63. $x = 0$ **65.** $x = 10$

67. (a) $50\sqrt{481}/481 \approx 2.28$ ft/sec (b) $\frac{26}{5} = 5.2$ ft/sec
(c) ∞

Chapter Test (page 258)

1.

x	1.9	1.99	1.999	2.001	2.01	2.1
$f(x)$	0.3448	0.3344	0.3334	0.3332	0.3322	0.3226

$\lim_{x\to 2} [(x - 2)/(x^2 - x - 2)] \approx 0.3333$ (Actual limit is $\frac{1}{3}$.)

2. Limit does not exist. The function approaches 1 from the right side of 5 but it approaches -1 from the left side of 5.

3.

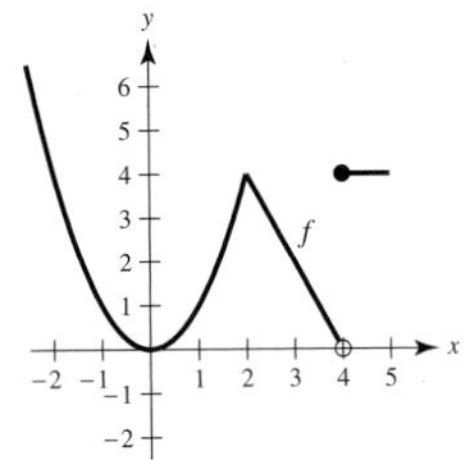

$\lim_{x\to c} f(x)$ exists for all points on the graph except where $c = 4$.

4. 5 **5.** 3 **6.** $\frac{1}{8}$ **7.** $\frac{1}{8}$ **8.** $\frac{1}{8}$

9. (a) 3 (b) 27 (c) 27

10. (a) 15 (b) -3 (c) 10 (d) $\frac{2}{5}$

11. (a) 0 (b) 0 (c) 0; Discontinuity at $x = -3$

12. $f(x)$ has a discontinuity at $x = 1$ because $f(1) = 0 \neq \lim_{x\to 1} f(x) = 1$.

13. Nonremovable discontinuity at $x = -2$

14. Nonremovable discontinuities at $x = 2$ and $x = -2$

15. $f(x)$ is continuous on the interval $[1, 2]$, $f(1) = 2.0625$, and $f(2) = -4$, so by the Intermediate Value Theorem, there exists a real number c in $[1, 2]$ such that $f(c) = 0$.

16. $\lim_{x\to 2^+} \dfrac{1}{(x-2)^2} = \infty, \quad \lim_{x\to 2^-} \dfrac{1}{(x-2)^2} = \infty$

17. $x = 2, x = -3$

18. Removable discontinuity at $x = -3$

19. 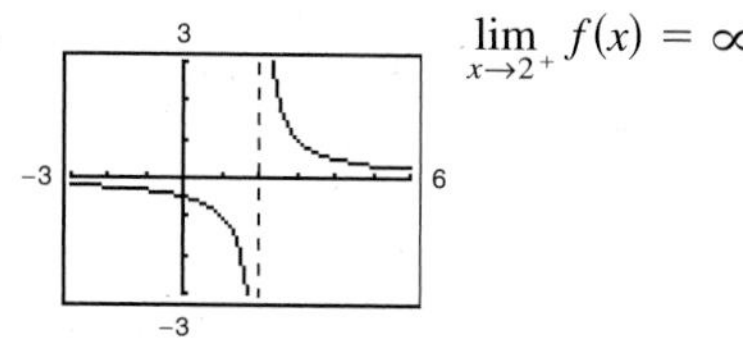

$\lim_{x\to2^+} f(x) = \infty$

P.S. Problem Solving (page 259)

1. (a) Perimeter $\triangle PAO = 1 + \sqrt{(x^2-1)^2 + x^2} + \sqrt{x^4 + x^2}$

Perimeter $\triangle PBO = 1 + \sqrt{x^4 + (x-1)^2} + \sqrt{x^4 + x^2}$

(b)

x	4	2	1
Perimeter $\triangle PAO$	33.0166	9.0777	3.4142
Perimeter $\triangle PBO$	33.7712	9.5952	3.4142
$r(x)$	0.9777	0.9461	1.0000

x	0.1	0.01
Perimeter $\triangle PAO$	2.0955	2.0100
Perimeter $\triangle PBO$	2.0006	2.0000
$r(x)$	1.0475	1.0050

(c) 1

3. (a) $\frac{4}{3}$ (b) $y = -\frac{3}{4}x + \frac{25}{4}$ (c) $m_x = \frac{\sqrt{25 - x^2} - 4}{x - 3}$

(d) $-\frac{3}{4}$; This is the slope of the tangent line at P.

5. $a = 3, b = 6$

7. (a) g_1, g_4 (b) g_1 (c) g_1, g_3, g_4

9. 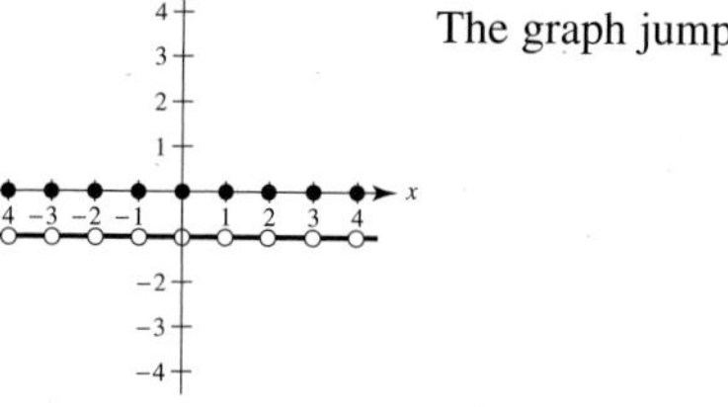

The graph jumps at every integer.

(a) $f(1) = 0,\ f(0) = 0,\ f(\frac{1}{2}) = -1,\ f(-2.7) = -1$

(b) $\lim_{x\to1^-} f(x) = -1,\ \lim_{x\to1^+} f(x) = -1,\ \lim_{x\to1/2} f(x) = -1$

(c) There is a discontinuity at each integer.

11. (a) 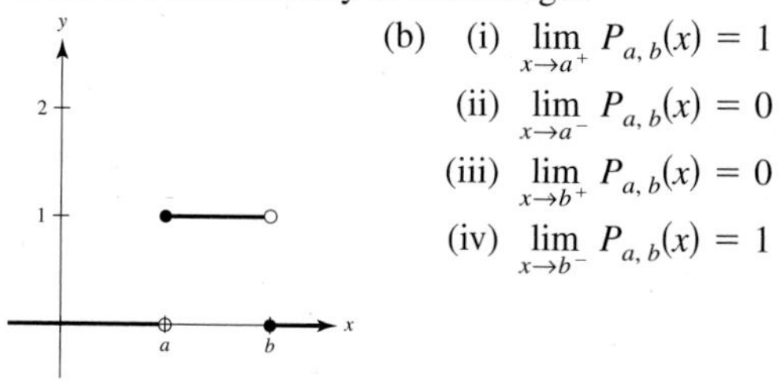

(b) (i) $\lim_{x\to a^+} P_{a,b}(x) = 1$

(ii) $\lim_{x\to a^-} P_{a,b}(x) = 0$

(iii) $\lim_{x\to b^+} P_{a,b}(x) = 0$

(iv) $\lim_{x\to b^-} P_{a,b}(x) = 1$

(c) Continuous for all positive real numbers except a and b

(d) The area under the graph of U and above the x-axis is 1.

Chapter 4

Section 4.1 (page 269)

1. (a) $m_1 = 0, m_2 = 5/2$ (b) $m_1 = -5/2, m_2 = 2$

3. 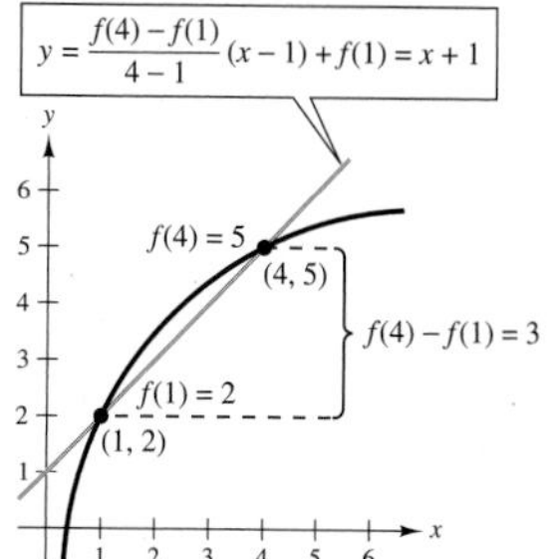

5. $m = -5$ **7.** $m = 4$

9. $m = 3$ **11.** $f'(x) = 0$ **13.** $f'(x) = -10$

15. $h'(s) = \frac{2}{3}$ **17.** $f'(x) = 2x + 1$ **19.** $f'(x) = 3x^2 - 12$

21. $f'(x) = \frac{-1}{(x-1)^2}$ **23.** $f'(x) = \frac{1}{2\sqrt{x+4}}$

25. (a) Tangent line: $y = 2x + 2$

(b) 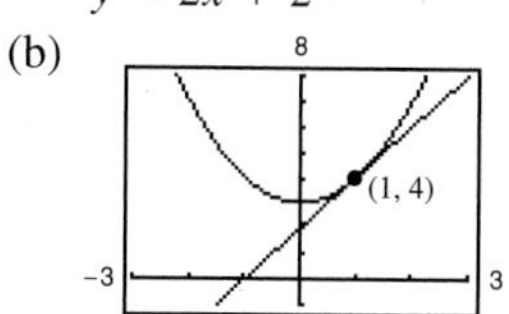

27. (a) Tangent line: $y = 12x - 16$

(b) 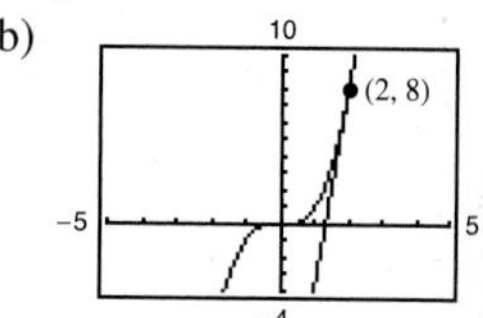

29. (a) Tangent line: $y = \frac{1}{2}x + \frac{1}{2}$

(b) 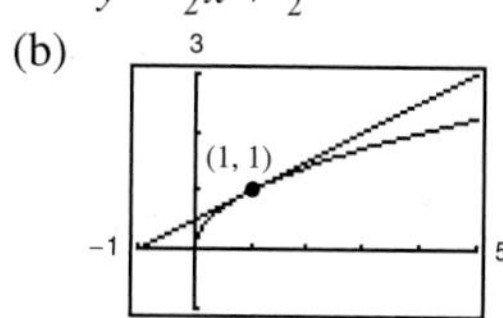

31. (a) Tangent line: $y = \frac{3}{4}x + 2$

(b) 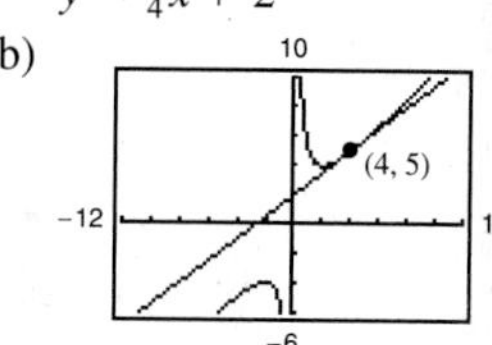

33. $y = 2x - 1$ **35.** $y = 3x - 2;\ y = 3x + 2$

37. $y = -\frac{1}{2}x + \frac{3}{2}$ **39.** b **40.** d **41.** a **42.** c

43. $g(4) = 5;\ g'(4) = -\frac{5}{3}$

45. 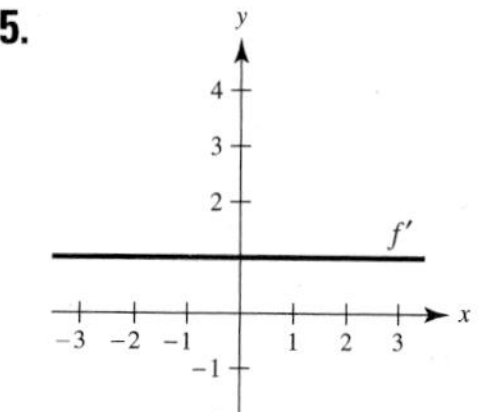

47. 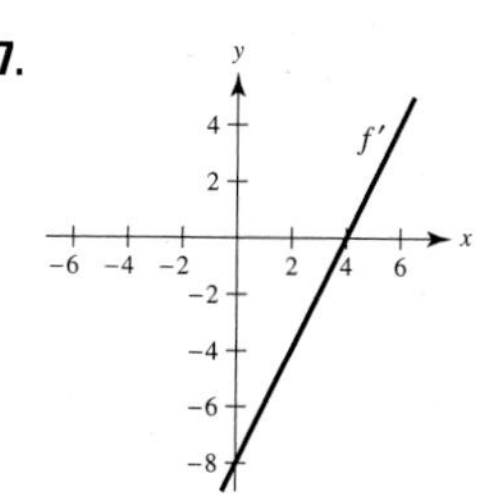

49. 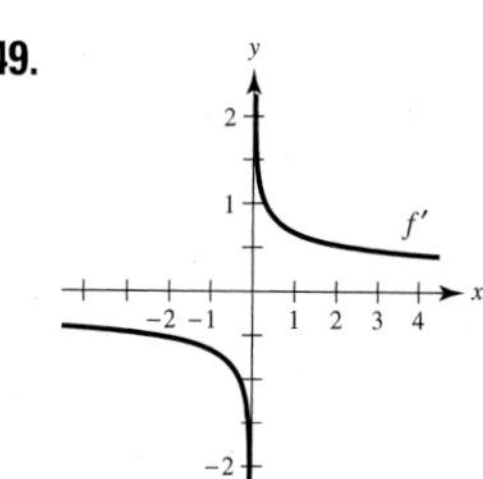

51. Answers will vary.
Sample answer: $y = -x$

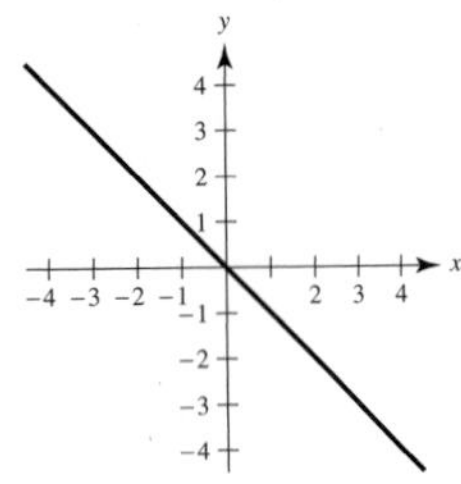

53. $f(x) = 5 - 3x$
$c = 1$

55. $f(x) = -x^2$
$c = 6$

57. $f(x) = -3x + 2$

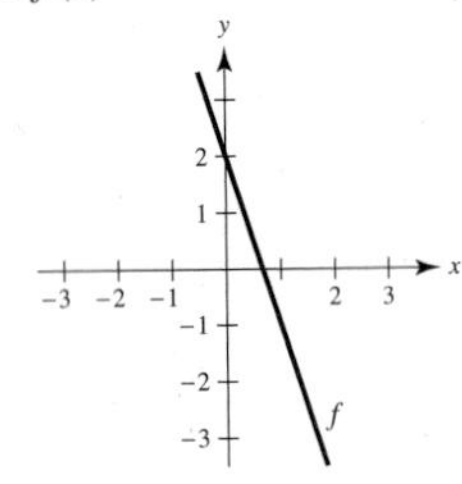

59. Answers will vary.
Sample answer: $f(x) = x^3$

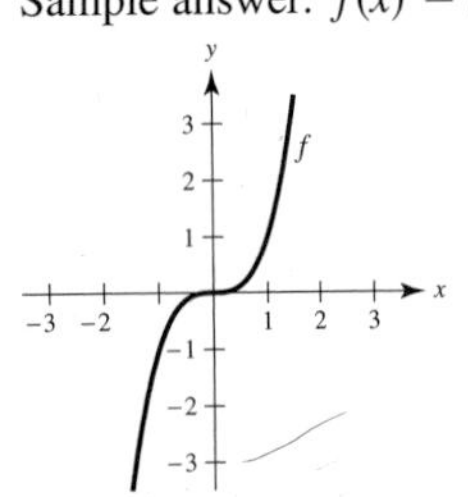

61. $y = 2x + 1$; $y = -2x + 9$

63. (a)

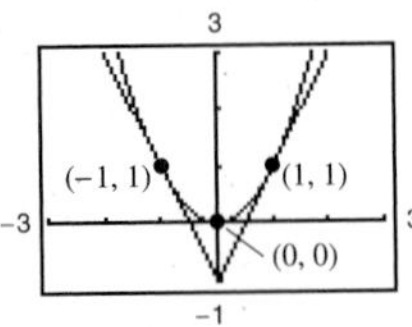

For this function, the slopes of the tangent lines are always distinct for different values of x.

(b)

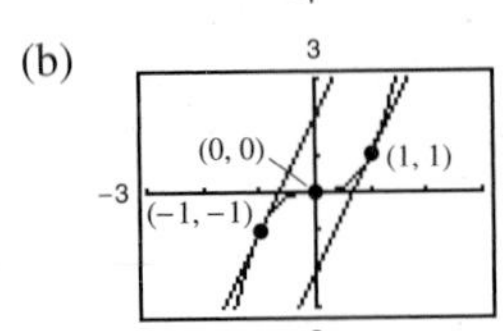

For this function, the slopes of the tangent lines are sometimes the same.

65. (a)

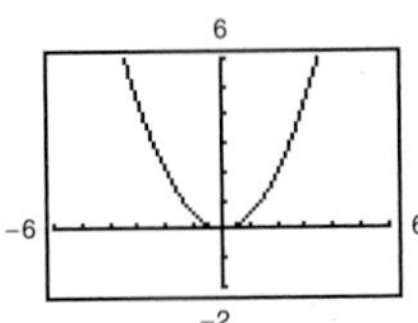

$f'(0) = 0, f'(\frac{1}{2}) = \frac{1}{2}, f'(1) = 1, f'(2) = 2$

(b) $f'(-\frac{1}{2}) = -\frac{1}{2}, f'(-1) = -1, f'(-2) = -2$

(c)

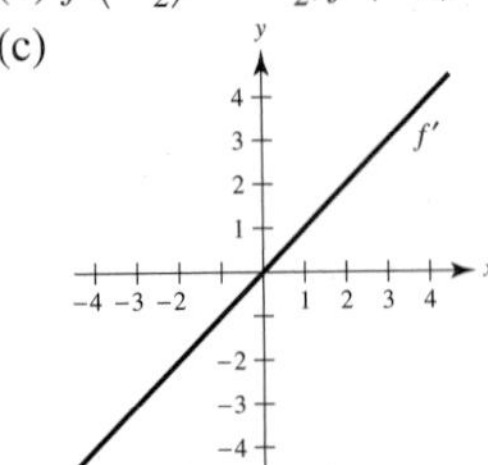

(d) $f'(x) = x$

67.

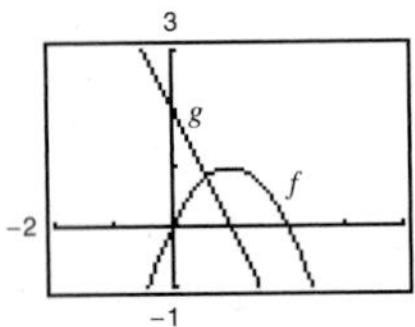

$g(x) \approx f'(x)$

69. $f(2) = 4$; $f(2.1) = 3.99$; $f'(2) \approx -0.1$

71.

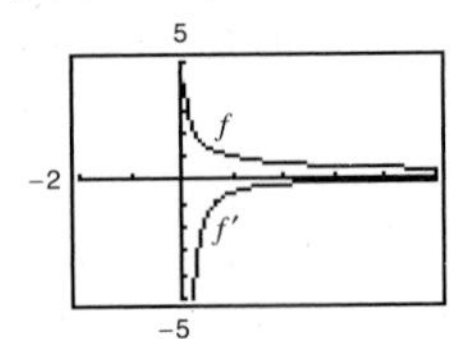

As x approaches infinity, the graph of f approaches a line of slope 0. Thus $f'(x)$ approaches 0.

73. 6 **75.** 4 **77.** $g(x)$ is not differentiable at $x = 0$.

79. $f(x)$ is not differentiable at $x = 6$.

81. $h(x)$ is not differentiable at $x = -7$.

83. $(-\infty, 3) \cup (3, \infty)$ **85.** $(-\infty, -4) \cup (-4, \infty)$

87. $(1, \infty)$

89.

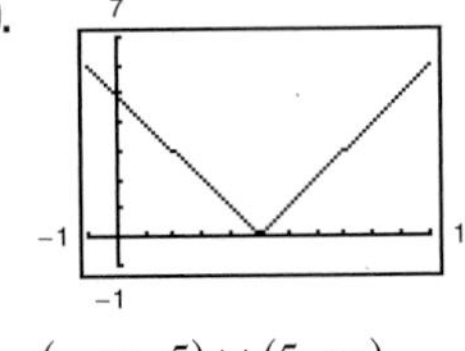

$(-\infty, 5) \cup (5, \infty)$

91.

$(-\infty, 0) \cup (0, \infty)$

93. The derivative from the left is -1 and the derivative from the right is 1, so f is not differentiable at $x = 1$.

95. The derivatives from both the right and the left are 0, so $f'(1) = 0$.

97. f is differentiable at $x = 2$.

99. (a) $d = \left(3|m + 1|\right)/\sqrt{m^2 + 1}$

(b)

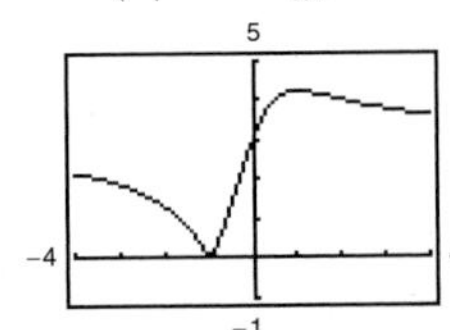

Not differentiable at $m = -1$

101. False. The slope is $\lim_{\Delta x \to 0} \dfrac{f(2 + \Delta x) - f(2)}{\Delta x}$.

103. False. For example: $f(x) = |x|$. The derivative from the left and the derivative from the right both exist but are not equal.

105. (a) Yes (b) No

Section 4.2 (page 280)

1. (a) $\frac{1}{2}$ (b) 3 **3.** 0 **5.** $7x^6$ **7.** $-5/x^6$

9. $1/(5x^{4/5})$ **11.** 1 **13.** $-4t + 3$ **15.** $-3 - x$

17. $2x + 12x^2$ **19.** $3t^2 + 10t - 3$

	Function	*Rewrite*	*Differentiate*	*Simplify*
21.	$y = \dfrac{5}{2x^2}$	$y = \dfrac{5}{2}x^{-2}$	$y' = -5x^{-3}$	$y' = -\dfrac{5}{x^3}$
23.	$y = \dfrac{6}{(5x)^3}$	$y = \dfrac{6}{125}x^{-3}$	$y' = -\dfrac{18}{125}x^{-4}$	$y' = -\dfrac{18}{125x^4}$
25.	$y = \dfrac{\sqrt{x}}{x}$	$y = x^{-1/2}$	$y' = -\dfrac{1}{2}x^{-3/2}$	$y' = -\dfrac{1}{2x^{3/2}}$

27. -2 **29.** 0 **31.** 8 **33.** $2x + 6/x^3$

35. $2t + 12/t^4$ **37.** $8x + 3$ **39.** $(x^3 - 8)/x^3$

41. $3x^2 + 1$ **43.** $\dfrac{1}{2\sqrt{x}} - \dfrac{2}{x^{2/3}}$ **45.** $\dfrac{4}{5s^{1/5}} - \dfrac{2}{3s^{1/3}}$

47. (a) $2x + y - 2 = 0$
(b)
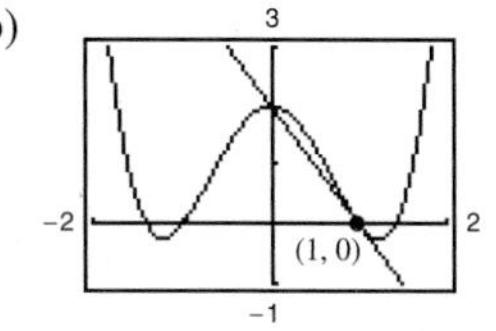

49. (a) $3x + 2y - 7 = 0$
(b)
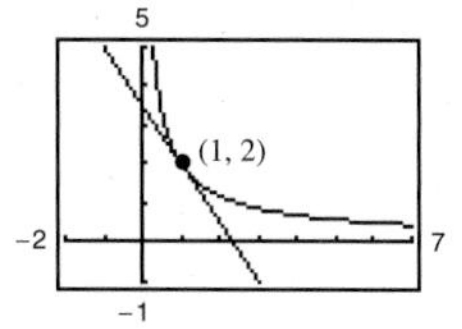

51. $(-1, 2), (0, 3), (1, 2)$ **53.** No horizontal tangents

55. $k = -1,\ k = -9$ **57.** $k = 3$ **59.** $k = 4/27$

61.
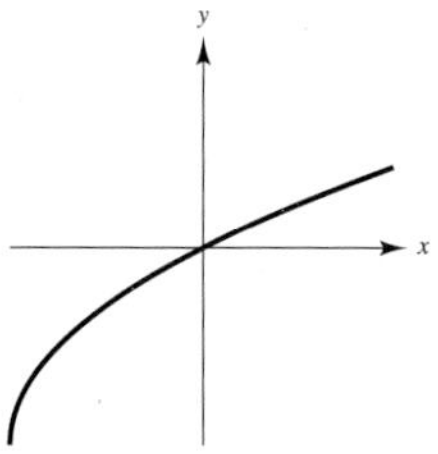

63. $g'(x) = f'(x)$

65.
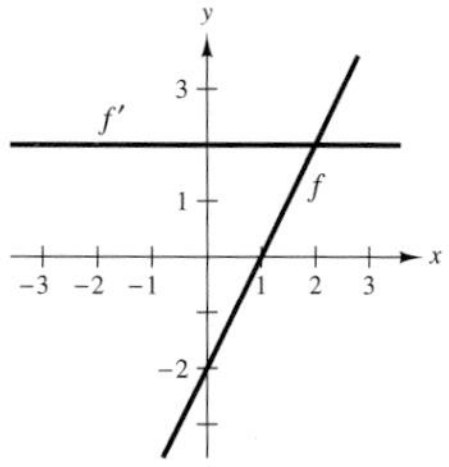

The rate of change of f is constant and therefore f' is a constant function.

67. $y = 2x - 1$
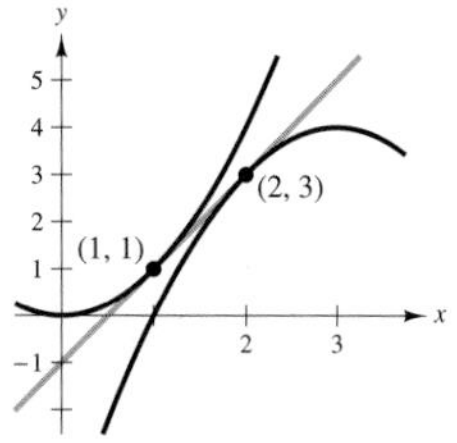

$y = 4x - 4$
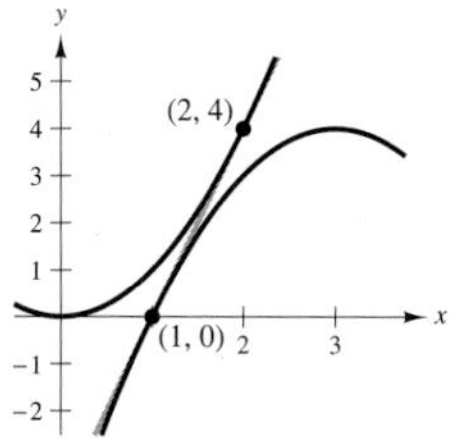

69. $x - 4y + 4 = 0$

71.
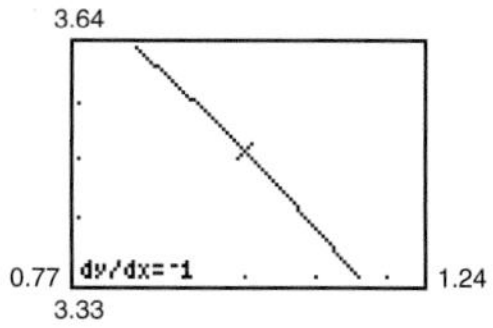

$f'(1)$ appears to be close to -1.

$f'(1) = -1$

73. (a)
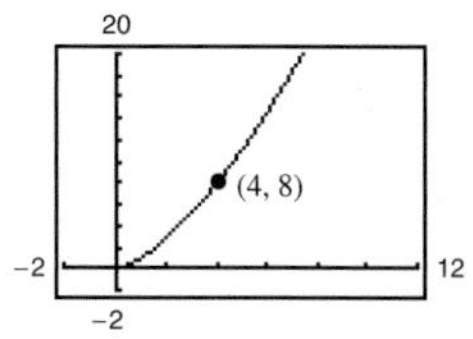

$(3.9, 7.7019)$,
$S(x) = 2.981x - 3.924$

(b) $T(x) = 3(x - 4) + 8 = 3x - 4$
The slope (and equation) of the secant line approaches that of the tangent line at (4, 8) as you choose points closer and closer to (4, 8).

(c)
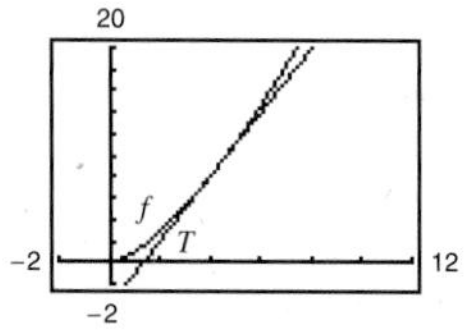

The approximation becomes less accurate.

(d)

Δx	-3	-2	-1	-0.5	-0.1	0
$f(4 + \Delta x)$	1	2.828	5.196	6.548	7.702	8
$T(4 + \Delta x)$	-1	2	5	6.5	7.7	8

Δx	0.1	0.5	1	2	3
$f(4 + \Delta x)$	8.302	9.546	11.180	14.697	18.520
$T(4 + \Delta x)$	8.3	9.5	11	14	17

75. False. Let $f(x) = x$ and $g(x) = x + 1$.

77. False. $dy/dx = 0$ **79.** True

81. Average rate: 4
Instantaneous rates:
$f'(1) = 4; f'(2) = 4$

83. Average rate: $\frac{1}{2}$
Instantaneous rates:
$f'(1) = 1; f'(2) = \frac{1}{4}$

85. (a) $s(t) = -16t^2 + 1362$; $v(t) = -32t$ (b) -48 ft/sec
(c) $s'(1) = -32$ ft/sec; $s'(2) = -64$ ft/sec
(d) $t = \dfrac{\sqrt{1362}}{4} \approx 9.226$ sec (e) -295.242 ft/sec

87. $v(5) = 71$ m/sec; $v(10) = 22$ m/sec

89.
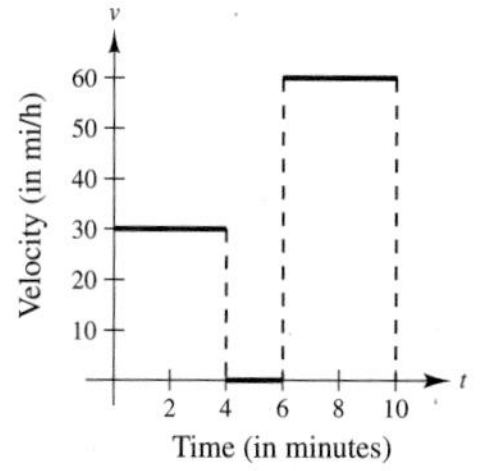

91.
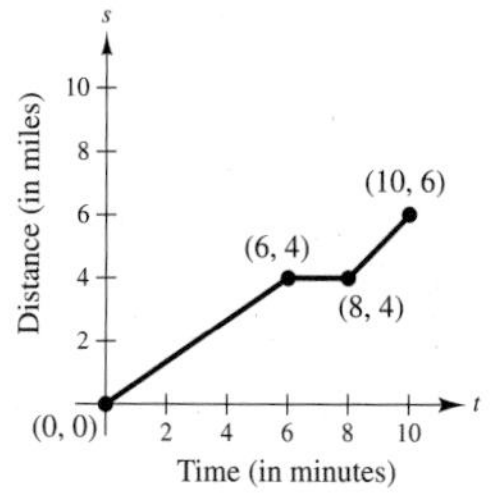

93. (a) $R(v) = 0.417v - 0.02$
(b) $B(v) = 0.0056v^2 + 0.001v + 0.04$
(c) $T(v) = 0.0056v^2 + 0.418v + 0.02$
(d)
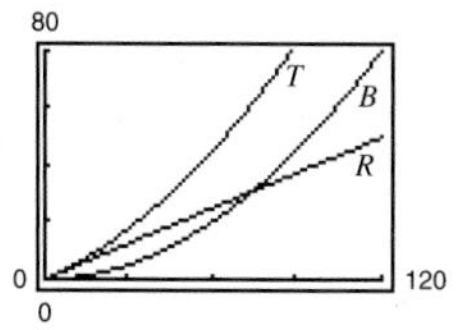

(e) $T'(v) = 0.0112v + 0.418$
$T'(40) = 0.866$
$T'(80) = 1.314$
$T'(100) = 1.538$

(f) Stopping distance increases at an increasing rate.

95. $V'(6) = 108\ \text{cm}^3/\text{cm}$ **97.** Proof

99. (a) The rate of change of the number of gallons of gasoline sold when the price is \$2.979
(b) In general, the rate of change when $p = 2.979$ should be negative. As prices go up, sales go down.

101. $y = 2x^2 - 3x + 1$ **103.** $9x + y = 0,\ 9x + 4y + 27 = 0$

105. $a = \frac{1}{3},\ b = -\frac{4}{3}$

Section 4.3 (page 290)

1. $2(2x^3 - 6x^2 + 3x - 6)$ **3.** $(1 - 5t^2)/(2\sqrt{t})$
5. $-2t(6t^4 - 2t^2 - 11)$ **7.** $(1 - x^2)/(x^2 + 1)^2$
9. $(1 - 5x^3)/[2\sqrt{x}(x^3 + 1)^2]$ **11.** $\dfrac{x^4 - 6x^2 - 4x - 3}{(x^2 - 1)^2}$
13. $f'(x) = -5(x + 6)/x^3; f'(x) = -35$
15. $f'(x) = (x^3 + 4x)(6x + 2) + (3x^2 + 2x - 5)(3x^2 + 4)$
$= 15x^4 + 8x^3 + 21x^2 + 16x - 20$
$f'(0) = -20$
17. $f'(x) = \dfrac{x^2 - 6x + 4}{(x - 3)^2}$ **19.** $f'(x) = 3x^2 - 8x + 5$
$f'(1) = -\dfrac{1}{4}$ $f'(0) = 5$

	Function	*Rewrite*	*Differentiate*	*Simplify*
21.	$y = \dfrac{x^2 + 3x}{7}$	$y = \dfrac{1}{7}x^2 + \dfrac{3}{7}x$	$y' = \dfrac{2}{7}x + \dfrac{3}{7}$	$y' = \dfrac{2x + 3}{7}$
23.	$y = \dfrac{6}{7x^2}$	$y = \dfrac{6}{7}x^{-2}$	$y' = -\dfrac{12}{7}x^{-3}$	$y' = -\dfrac{12}{7x^3}$
25.	$y = \dfrac{4x^{3/2}}{x}$	$y = 4x^{1/2}$, $x > 0$	$y' = 2x^{-1/2}$	$y' = \dfrac{2}{\sqrt{x}}$, $x > 0$

27. $\dfrac{(x^2 - 1)(-3 - 2x) - (4 - 3x - x^2)(2x)}{(x^2 - 1)^2} = \dfrac{3}{(x + 1)^2}$,
$x \neq 1$
29. $1 - 12/(x + 3)^2 = (x^2 + 6x - 3)/(x + 3)^2$
31. $\dfrac{3}{2}x^{-1/2} + \dfrac{1}{2}x^{-3/2} = (3x + 1)/2x^{3/2}$
33. $6s^2(s^3 - 2)$ **35.** $-(2x^2 - 2x + 3)/[x^2(x - 3)^2]$
37. $(6x^2 + 5)(x - 3)(x + 2) + (2x^3 + 5x)(1)(x + 2)$
$+ (2x^3 + 5x)(x - 3)(1)$
$= 10x^4 - 8x^3 - 21x^2 - 10x - 30$
39. $\dfrac{(x^2 - c^2)(2x) - (x^2 + c^2)(2x)}{(x^2 - c^2)^2} = -\dfrac{4xc^2}{(x^2 - c^2)^2}$
41. $\left(\dfrac{x + 1}{x + 2}\right)(2) + (2x - 5)\left[\dfrac{(x + 2)(1) - (x + 1)(1)}{(x + 2)^2}\right]$
$= \dfrac{2x^2 + 8x - 1}{(x + 2)^2}$
43. (a) $y = -3x - 1$
(b)

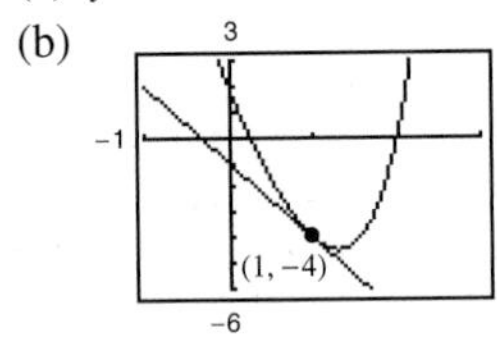

45. (a) $y = 4x + 25$
(b)

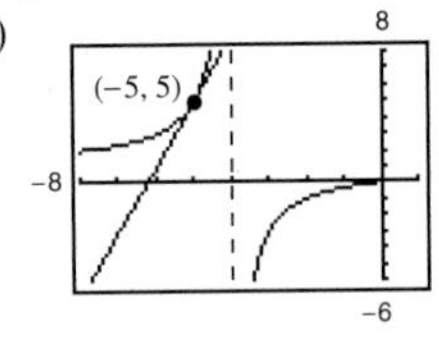

47. $2y + x - 4 = 0$
49. $25y - 12x + 16 = 0$ **51.** $(1, 1)$ **53.** $(0, 0), (2, 4)$
55. Tangent lines: $2y + x = 7$; $2y + x = -1$

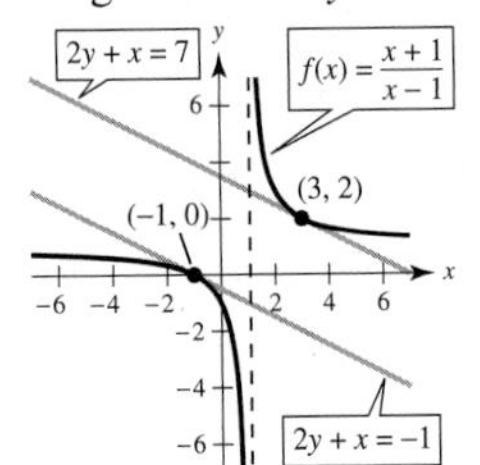

57. $f(x) + 2 = g(x)$
59. (a) $p'(1) = 1$ (b) $q'(4) = -1/3$
61. $(18t + 5)/(2\sqrt{t})$ cm²/sec **63.** 31.55 bacteria/h
65. $12x^2 + 12x - 6$ **67.** $3/\sqrt{x}$ **69.** $2/(x - 1)^3$
71. $2x$ **73.** $1/\sqrt{x}$ **75.** 0 **77.** -10
79. Answers will vary. For example: $f(x) = (x - 2)^2$

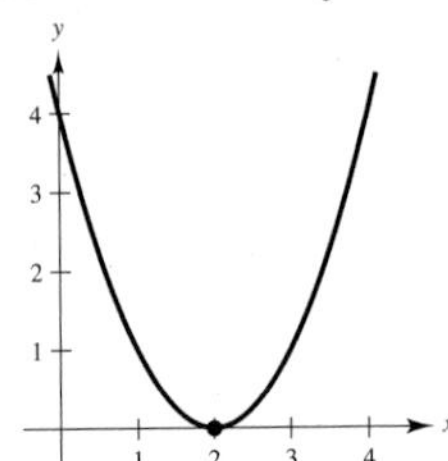

81.

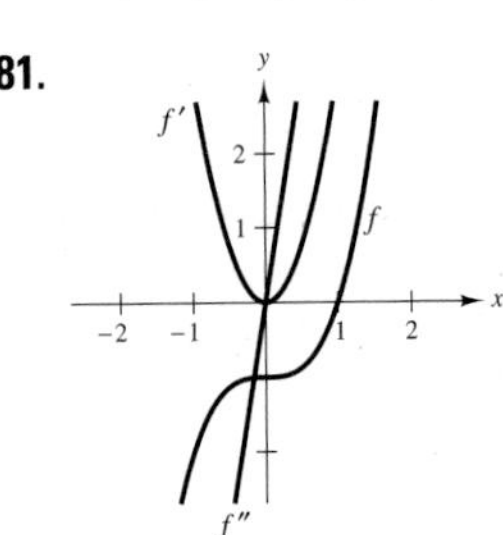

83.

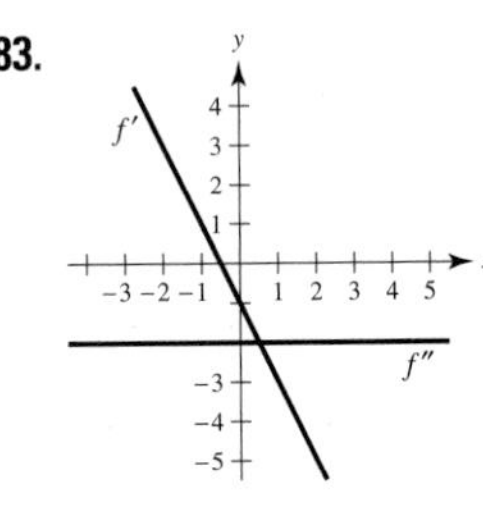

85. $v(3) = 27$ m/sec
$a(3) = -6$ m/sec²
The speed of the object is decreasing.

87.

t	0	1	2	3	4
$s(t)$	0	57.75	99	123.75	132
$v(t)$	66	49.5	33	16.5	0
$a(t)$	-16.5	-16.5	-16.5	-16.5	-16.5

The average velocity on $[0, 1]$ is 57.75, on $[1, 2]$ is 41.25, on $[2, 3]$ is 24.75, and on $[3, 4]$ is 8.25.
89. $f^{(n)}(x) = n(n - 1)(n - 2) \cdots (2)(1) = n!$
91. (a) $f''(x) = g(x)h''(x) + 2g'(x)h'(x) + g''(x)h(x)$
$f'''(x) = g(x)h'''(x) + 3g'(x)h''(x) + 3g''(x)h'(x) + g'''(x)h(x)$
$f^{(4)}(x) = g(x)h^{(4)}(x) + 4g'(x)h'''(x) + 6g''(x)h''(x) + 4g'''(x)h'(x) + g^{(4)}(x)h(x)$
(b) $f^{(n)}(x) = g(x)h^{(n)}(x) + \dfrac{n!}{1!(n - 1)!}g'(x)h^{(n-1)}(x) + \dfrac{n!}{2!(n - 2)!}g''(x)h^{(n-2)}(x) + \cdots + \dfrac{n!}{(n - 1)!1!}g^{(n-1)}(x)h'(x) + g^{(n)}(x)h(x)$
93. $y' = -1/x^2, y'' = 2/x^3$,
$x^3y'' + 2x^2y' = x^3(2/x^3) + 2x^2(-1/x^2)$
$= 2 - 2 = 0$
95. False. $dy/dx = f(x)g'(x) + g(x)f'(x)$ **97.** True
99. True **101.** $f'(x) = 2|x|; f''(0)$ does not exist.

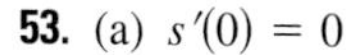

Section 4.4 (page 298)

	$y = f(g(x))$	$u = g(x)$	$y = f(u)$
1.	$y = (5x - 8)^4$	$u = 5x - 8$	$y = u^4$
3.	$y = (x^2 - 3x + 4)^6$	$u = x^2 - 3x + 4$	$y = u^6$
5.	$y = \dfrac{1}{\sqrt{x+1}}$	$u = x + 1$	$y = u^{-1/2}$

7. $12(4x - 1)^2$ **9.** $-108(4 - 9x)^3$

11. $\frac{1}{2}(5 - t)^{-1/2}(-1) = -1/\left(2\sqrt{5 - t}\right)$

13. $\frac{1}{3}(6x^2 + 1)^{-2/3}(12x) = 4x/\sqrt[3]{(6x^2 + 1)^2}$

15. $\frac{1}{2}(9 - x^2)^{-3/4}(-2x) = -x/\sqrt[4]{(9 - x^2)^3}$

17. $-1/(x - 2)^2$ **19.** $-2(t - 3)^{-3}(1) = -2/(t - 3)^3$

21. $-1/\left[2\sqrt{(x + 2)^3}\right]$

23. $x^2[4(x - 2)^3(1)] + (x - 2)^4(2x) = 2x(x - 2)^3(3x - 2)$

25. $x\left(\dfrac{1}{2}\right)(1 - x^2)^{-1/2}(-2x) + (1 - x^2)^{1/2}(1) = \dfrac{1 - 2x^2}{\sqrt{1 - x^2}}$

27. $\dfrac{(x^2 + 1)^{1/2}(1) - x(1/2)(x^2 + 1)^{-1/2}(2x)}{x^2 + 1} = \dfrac{1}{\sqrt{(x^2 + 1)^3}}$

29. $\dfrac{-2(x + 5)(x^2 + 10x - 2)}{(x^2 + 2)^3}$ **31.** $\dfrac{-9(1 - 2v)^2}{(v + 1)^4}$

33. $2((x^2 + 3)^5 + x)(5(x^2 + 3)^4(2x) + 1)$
$= 20x(x^2 + 3)^9 + 2(x^2 + 3)^5 + 20x^2(x^2 + 3)^4 + 2x$

35. $\dfrac{1}{2}(2 + (2 + x^{1/2})^{1/2})^{-1/2}\left(\dfrac{1}{2}(2 + x^{1/2})^{-1/2}\right)\left(\dfrac{1}{2}x^{-1/2}\right)$
$= \dfrac{1}{8\sqrt{x}\left(\sqrt{2 + \sqrt{x}}\right)\left(\sqrt{2 + \sqrt{2 + \sqrt{x}}}\right)}$

37. $(1 - 3x^2 - 4x^{3/2})/[2\sqrt{x}(x^2 + 1)^2]$

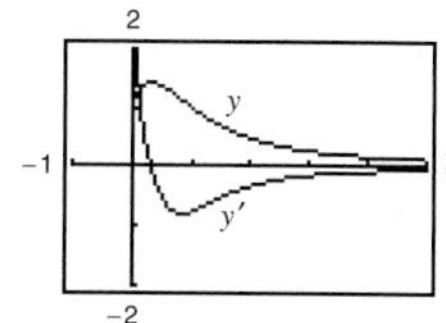

The zero of y' corresponds to the point on the graph of the function where the tangent line is horizontal.

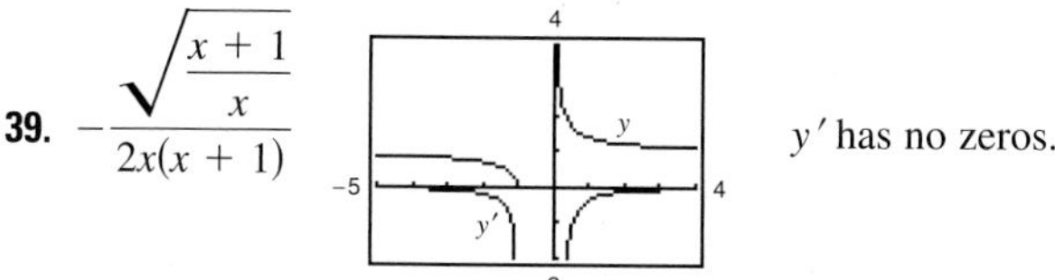

39. $-\dfrac{\sqrt{\dfrac{x + 1}{x}}}{2x(x + 1)}$

y' has no zeros.

41. $s'(t) = \dfrac{t + 3}{\sqrt{t^2 + 6t - 2}}, \dfrac{6}{5}$ **43.** $f'(x) = \dfrac{-15x^2}{(x^3 - 2)^2}, -\dfrac{3}{5}$

45. $f'(t) = \dfrac{-5}{(t - 1)^2}, -5$

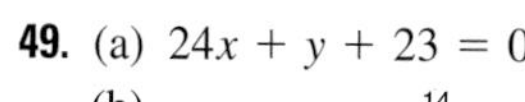

47. (a) $8x - 5y - 7 = 0$
(b)

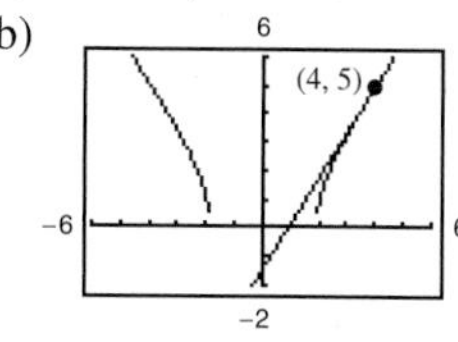

49. (a) $24x + y + 23 = 0$
(b)

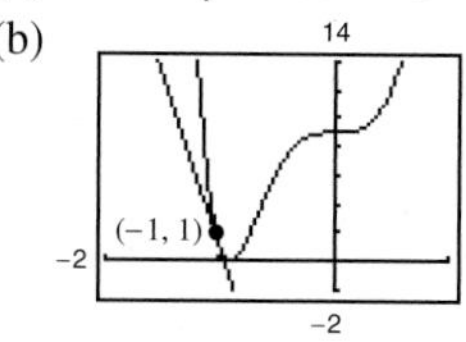

51. (a) $g'(1/2) = -3$
(b) $3x + y - 3 = 0$
(c)

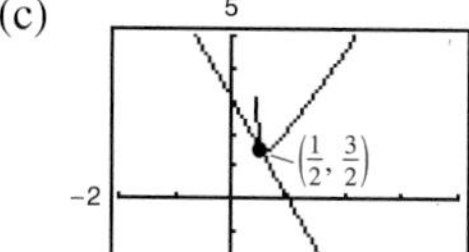

53. (a) $s'(0) = 0$
(b) $y = \frac{4}{3}$
(c)

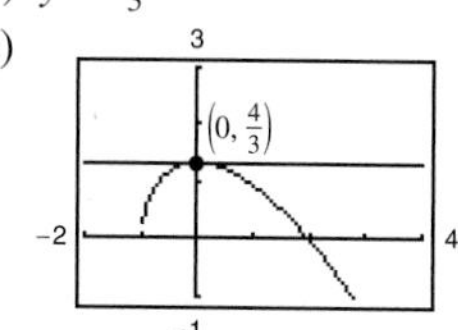

55. $3x + 4y - 25 = 0$

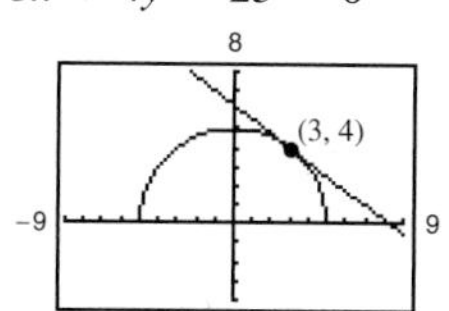

57. $24(x^2 - 2)(5x^2 - 2)$

59. $\dfrac{3}{4(x^2 + x + 1)^{3/2}}$ **61.** $h''(x) = 18x + 6, 24$

63.

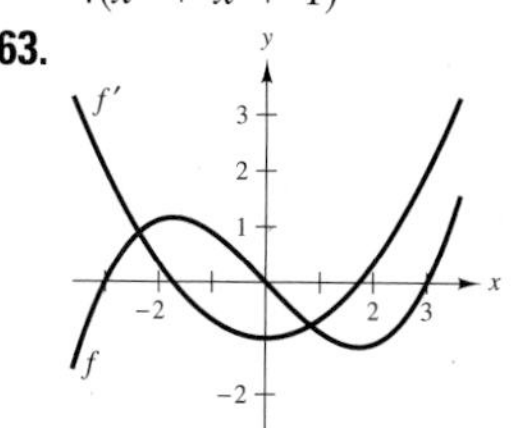

The zeros of f' correspond to the points where the graph of f has horizontal tangents.

65. The rate of change of g is three times as fast as the rate of change of f.

67. (a) $g'(x) = f'(x)$ (b) $h'(x) = 2f'(x)$
(c) $r'(x) = -3f'(-3x)$ (d) $s'(x) = f'(x + 2)$

x	-2	-1	0	1	2	3
$f'(x)$	4	$\frac{2}{3}$	$-\frac{1}{3}$	-1	-2	-4
$g'(x)$	4	$\frac{2}{3}$	$-\frac{1}{3}$	-1	-2	-4
$h'(x)$	8	$\frac{4}{3}$	$-\frac{2}{3}$	-2	-4	-8
$r'(x)$		12	1			
$s'(x)$	$-\frac{1}{3}$	-1	-2	-4		

69. (a) $\frac{1}{2}$
(b) $s'(5)$ does not exist because g is not differentiable at 6.

71. (a) 1.461 (b) −1.016

73. (a) $x = -1.637t^3 + 19.31t^2 - 0.5t - 1$
(b) $\dfrac{dC}{dt} = -294.66t^2 + 2317.2t - 30$
(c) Because x, the number of units produced in t hours, is not a linear function, and therefore the cost with respect to time t is not linear

75. (a) If $f(-x) = -f(x)$, then
$\dfrac{d}{dx}[f(-x)] = \dfrac{d}{dx}[-f(x)]$
$f'(-x)(-1) = -f'(x)$
$f'(-x) = f'(x).$
So, $f'(x)$ is even.

(b) If $f(-x) = f(x)$, then
$\dfrac{d}{dx}[f(-x)] = \dfrac{d}{dx}[f(x)]$
$f'(-x)(-1) = f'(x)$
$f'(-x) = -f'(x).$
So, f' is odd.

77. $g(x) = |3x - 5|$

$g'(x) = 3\left(\frac{3x - 5}{|3x - 5|}\right), \quad x \neq \frac{5}{3}$

79. (a) $P_1(x) = -2(x - 2) + 1$

$P_2(x) = \frac{11}{2}(x - 2)^2 - 2(x - 2) + 1$

(b)

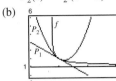

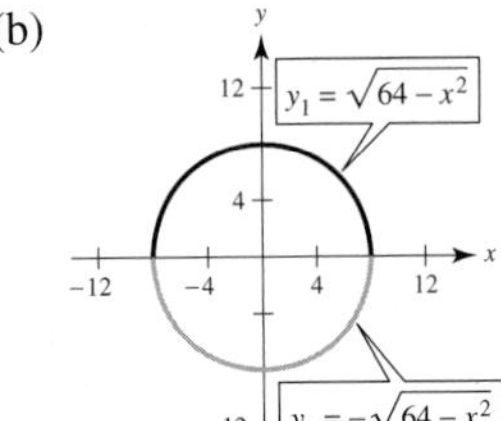

(c) P_2

(d) P_1 and P_2 become less accurate as you move farther from $x = 2$.

81. False. If $y = (1 - x)^{1/2}$, then $y' = \frac{1}{2}(1 - x)^{-1/2}(-1)$.

Section 4.5 (page 306)

1. $-x/y$ **3.** $-\sqrt{y/x}$ **5.** $(y - 3x^2)/(2y - x)$

7. $(1 - 3x^2y^3)/(3x^3y^2 - 1)$

9. $(6xy - 3x^2 - 2y^2)/(4xy - 3x^2)$

11. (a) $y_1 = \sqrt{64 - x^2}$; $y_2 = -\sqrt{64 - x^2}$

(b)

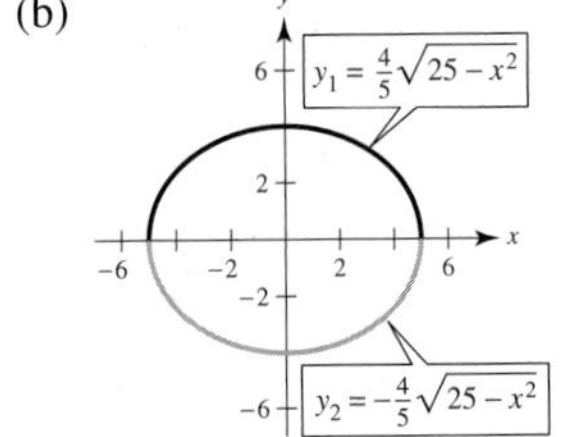

(c) $y' = \mp \frac{x}{\sqrt{64 - x^2}} = -\frac{x}{y}$

(d) $y' = -\frac{x}{y}$

13. (a) $y_1 = \frac{4}{5}\sqrt{25 - x^2}$; $y_2 = -\frac{4}{5}\sqrt{25 - x^2}$

(b)

(c) $y' = \mp \frac{4x}{5\sqrt{25 - x^2}} = -\frac{16x}{25y}$

(d) $y' = -\frac{16x}{25y}$

15. $-\frac{y}{x}, -\frac{1}{6}$ **17.** $\frac{98x}{y(x^2 + 49)^2}$, Undefined

19. $-\sqrt[3]{\frac{y}{x}}, -\frac{1}{2}$ **21.** $-\frac{1}{2}$ **23.** 0

25. $y = -x + 7$ **27.** $y = -x + 2$

29. $y = \sqrt{3}x/6 + 8\sqrt{3}/3$ **31.** $y = -\frac{2}{11}x + \frac{30}{11}$

33. (a) $y = -2x + 4$ (b) Answers will vary.

35. $-4/y^3$ **37.** $-36/y^3$ **39.** $(3x)/(4y)$

41. $2x + 3y - 30 = 0$

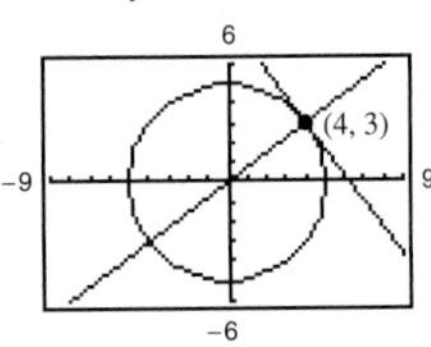

43. At (4, 3):

Tangent line: $4x + 3y - 25 = 0$

Normal line: $3x - 4y = 0$

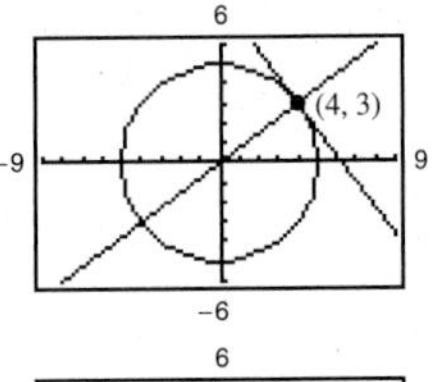

At $(-3, 4)$:

Tangent line: $3x - 4y + 25 = 0$

Normal line: $4x + 3y = 0$

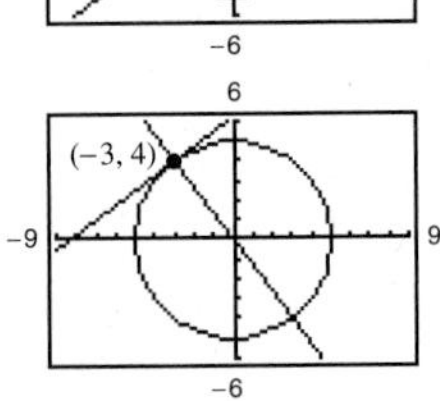

45. $x^2 + y^2 = r^2 \Rightarrow y' = -x/y \Rightarrow y/x =$ slope of normal line. Then for (x_0, y_0) on the circle, $x_0 \neq 0$, an equation of the normal line is $y = (y_0/x_0)x$, which passes through the origin. If $x_0 = 0$, the normal line is vertical and passes through the origin.

47. Horizontal tangents: $(-4, 0), (-4, 10)$

Vertical tangents: $(0, 5), (-8, 5)$

49.

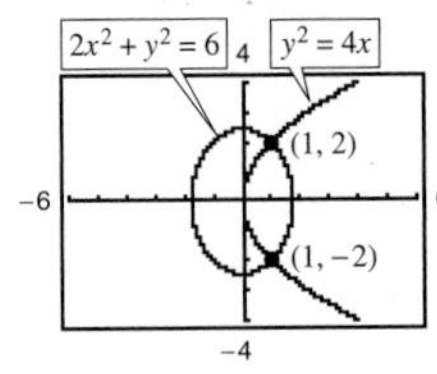

At (1, 2):

Slope of ellipse: −1

Slope of parabola: 1

At $(1, -2)$:

Slope of ellipse: 1

Slope of parabola: −1

51.

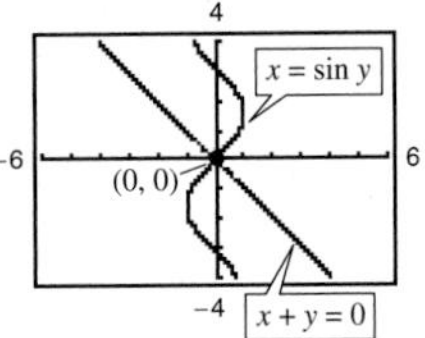

At (0, 0):

Slope of line: −1

Slope of sine curve: 1

53. Derivatives: $\frac{dy}{dx} = -\frac{y}{x}, \frac{dy}{dx} = \frac{x}{y}$

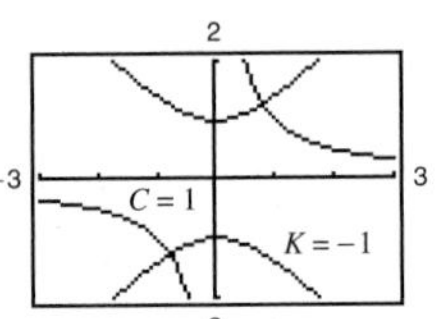

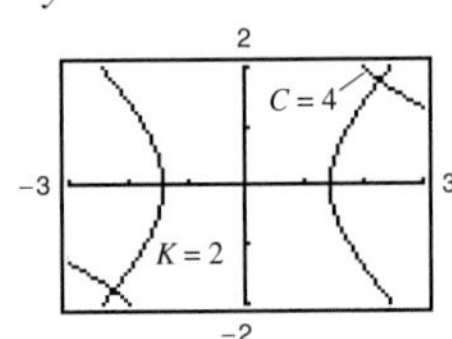

55. (a) $\frac{dy}{dx} = \frac{3x^3}{y}$ (b) $y\frac{dy}{dt} = 3x^3\frac{dx}{dt}$

57. Answers will vary. In the explicit form of a function, the variable is explicitly written as a function of x. In an implicit equation, the function is only implied by an equation. An example of an implicit function is $x^2 + xy = 5$. In explicit form it would be $y = (5 - x^2)/x$.

59.

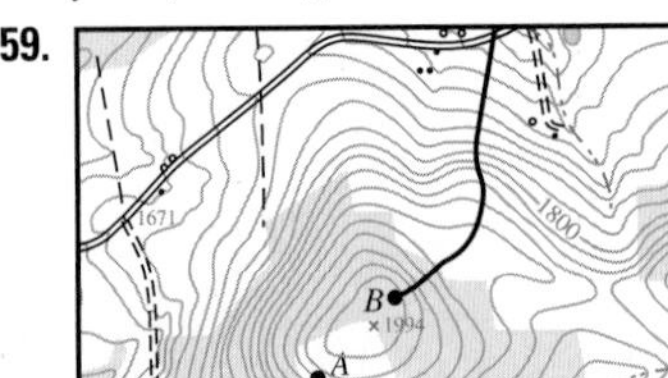

Use starting point B.

61. (a)

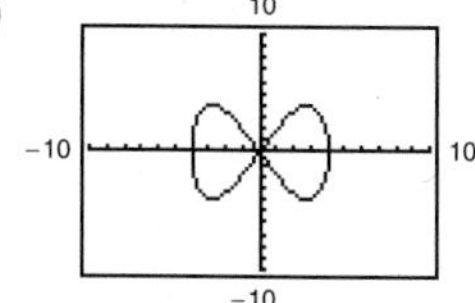

(b)

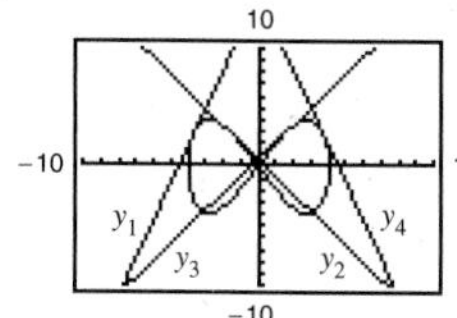

(c) $\left(\dfrac{8\sqrt{7}}{7}, 5\right)$

$y_1 = \frac{1}{3}\left[\left(\sqrt{7} + 7\right)x + \left(8\sqrt{7} + 23\right)\right]$

$y_2 = -\frac{1}{3}\left[\left(-\sqrt{7} + 7\right)x - \left(23 - 8\sqrt{7}\right)\right]$

$y_3 = -\frac{1}{3}\left[\left(\sqrt{7} - 7\right)x - \left(23 - 8\sqrt{7}\right)\right]$

$y_4 = -\frac{1}{3}\left[\left(\sqrt{7} + 7\right)x - \left(8\sqrt{7} + 23\right)\right]$

63. Proof **65.** (0, 1) and (0, −1)

67. (a) 1 (b) 1 (c) 3

$x_0 = \frac{3}{4}$

Section 4.6 (page 313)

1. (a) $\frac{3}{4}$ (b) 20 **3.** (a) $-\frac{5}{8}$ (b) $\frac{3}{2}$

5. (a) −8 cm/sec (b) 0 cm/sec (c) 8 cm/sec

7. (a) −6 cm/sec (b) −2 cm/sec (c) $-\frac{2}{3}$ cm/sec

9. In a linear function, if x changes at a constant rate, so does y. However, unless $a = 1$, y does not change at the same rate as x.

11. $(4x^3 + 6x)/\sqrt{x^4 + 3x^2 + 1}$

13. (a) 64π cm²/min (b) 256π cm²/min

15. (a) $A(b) = b\sqrt{300 - b^2}/4$

(b) When $b = 20$, $dA/dt = 105\sqrt{2}/4 \approx 37.1$ cm/sec.

When $b = 56$, $dA/dt = -501\sqrt{29}/29 \approx -93.0$ cm/sec.

(c) If db/dt are constant, dA/dt is a nonconstant function of b.

17. (a) $2/(9\pi)$ cm/min (b) $1/(18\pi)$ cm/min

19. (a) 144 cm²/sec (b) 720 cm²/sec **21.** $8/(405\pi)$ ft/min

23. (a) 12.5% (b) $\frac{1}{144}$ m/min

25. (a) $-\frac{7}{12}$ ft/sec; $-\frac{3}{2}$ ft/sec; $-\frac{48}{7}$ ft/sec (b) $\frac{527}{24}$ ft²/sec

27. Rate of vertical change: $\frac{1}{5}$ m/sec

Rate of horizontal change: $-\sqrt{3}/15$ m/sec

29. (a) −750 mi/h (b) 30 min

31. $-50/\sqrt{85} \approx -5.42$ ft/sec

33. (a) $\frac{25}{3}$ ft/sec (b) $\frac{10}{3}$ ft/sec

35. Evaporation rate proportional to $S \Rightarrow \dfrac{dV}{dt} = k(4\pi r^2)$

$V = \left(\dfrac{4}{3}\right)\pi r^3 \Rightarrow \dfrac{dV}{dt} = 4\pi r^2 \dfrac{dr}{dt}$. So $k = \dfrac{dr}{dt}$.

37. 0.6 ohm/sec **39.** About −0.1808 ft/sec²

41. (a) $\dfrac{dy}{dt} = 3\dfrac{dx}{dt}$ means that y changes three times as fast as x changes.

(b) y changes slowly when $x \approx 0$ or $x \approx L$. y changes more rapidly when x is near the middle of the interval.

43. About −97.96 m/sec

Review Exercises (page 317)

1. $-\frac{3}{2}$

3. (a) $y = 3x + 1$

(b)

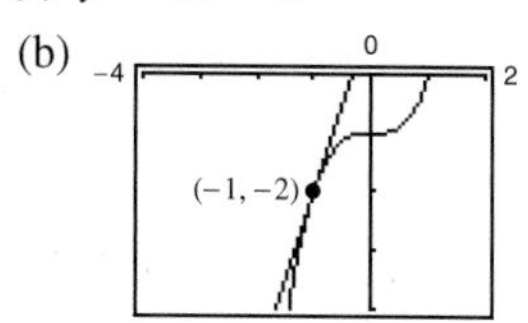

5.

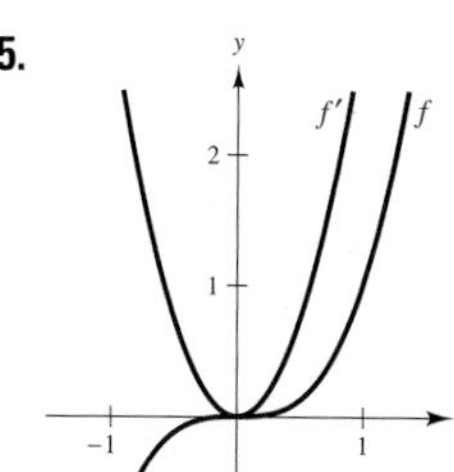

$f' > 0$ where the slopes of tangent lines to the graph of f are positive.

7. $f'(x) = 2x - 4$ **9.** $f'(x) = -2/(x - 1)^2$ **11.** 8

13. f is differentiable at all $x \neq 3$.

15.

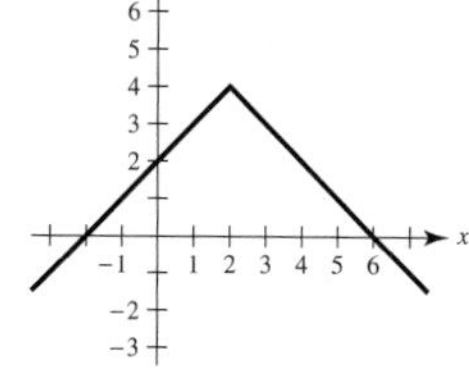

(a) Yes

(b) No, because the derivatives from the left and right are not equal.

17. 0 **19.** $8x^7$ **21.** $52t^3$ **23.** $-4/(3t^3)$

25. $3x^2 - 22x$ **27.** $\dfrac{3}{\sqrt{x}} + \dfrac{1}{\sqrt[3]{x^2}}$

29. (a) 50 vibrations/sec/lb

(b) 33.33 vibrations/sec/lb

31. 1354.24 ft or 412.77 m

33. (a)

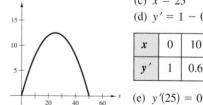

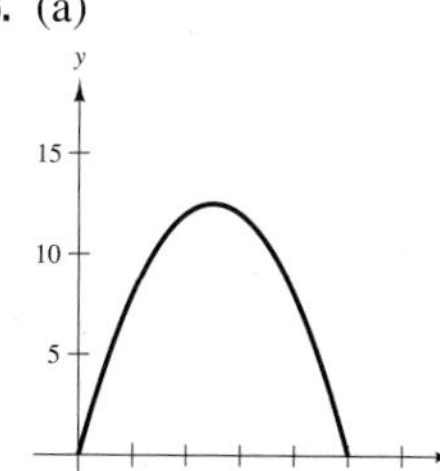

(b) 50

(c) $x = 25$

(d) $y' = 1 - 0.04x$

x	0	10	25	30	50
y'	1	0.6	0	−0.2	−1

(e) $y'(25) = 0$

35. (a) $x'(t) = 2t - 3$ (b) $(-\infty, 1.5)$ (c) $x = -\frac{1}{4}$ (d) 1

37. $4(5x^3 - 15x^2 - 11x - 8)$ **39.** $(7t^3 + 12t - 1)/\left(2\sqrt{t}\right)$

41. $-(x^2 + 1)/(x^2 - 1)^2$ **43.** $(8x)/(9 - 4x^2)^2$

45. $y = 4x - 3$

47. $v(4) = 20$ m/sec; $a(4) = -8$ m/sec²

49. $-48t$ **51.** $\frac{225}{4}\sqrt{x}$ **53.** $4(9x^2 - 9x + 8)$

55. $6x^2(x - 2)(x - 4)^2$ **57.** $32x(1 - 4x^2)$

59. $\dfrac{2(x + 5)(-x^2 - 10x + 3)}{(x^2 + 3)^3}$

61. $s(s^2 - 1)^{3/2}(8s^3 - 3s + 25)$

63. −2

65. $t(t-1)^4(7t-2)$

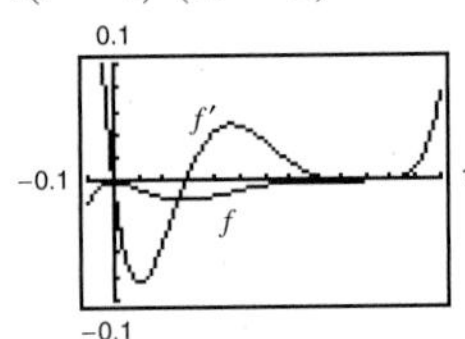

The zeros of f' correspond to the points on the graph of the function where the tangent line is horizontal.

67. $(x+2)/(x+1)^{3/2}$

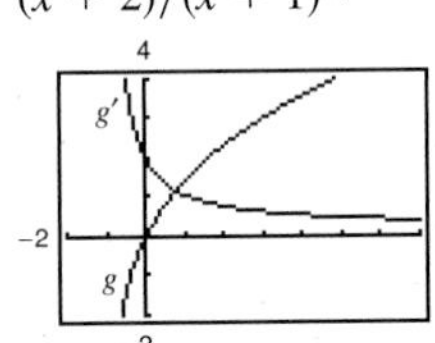

g' is not equal to zero for any x.

69. $5/[6(t+1)^{1/6}]$

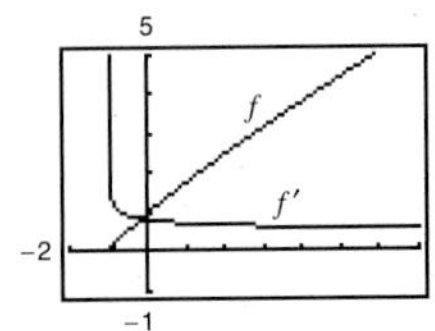

f' has no zeros.

71. (a) $f'(2) = 24$ (b) $y = 24t - 44$

(c)

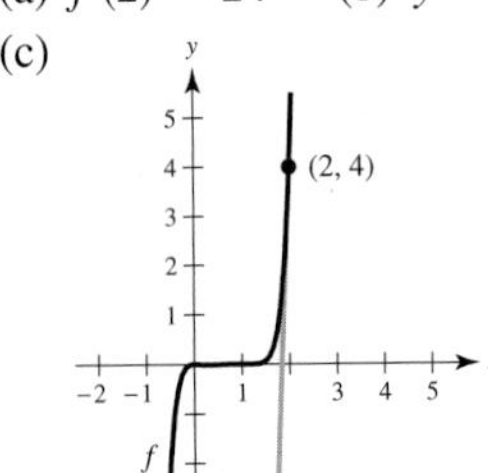

73. $9/(x^2+9)^{3/2}$ **75.** $24/(x-2)^4$

77. $[8(2t+1)]/(1-t)^4$ **79.** $-(x+12)/[4(x+3)^{5/2}]$

81. (a) $3x - y + 7 = 0$

(b)

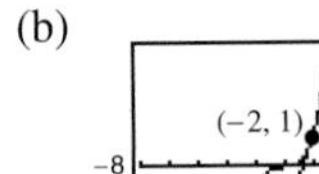

83. (a) $2x - 3y - 3 = 0$

(b)

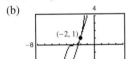

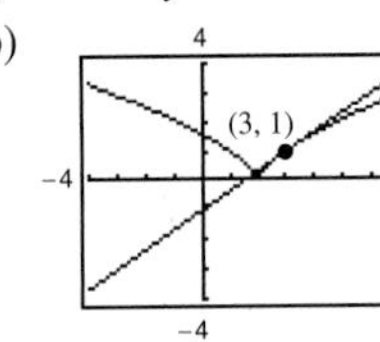

85. (a) $-18.667°/\text{h}$ (b) $-7.284°/\text{h}$

(c) $-3.240°/\text{h}$ (d) $-0.747°/\text{h}$

87. $-\dfrac{2x+3y}{3(x+y^2)}$

89. $\dfrac{\sqrt{y}(2\sqrt{x}-\sqrt{y})}{\sqrt{x}(\sqrt{x}+8\sqrt{y})} = \dfrac{2x-9y}{9x-32y}$

91. Tangent line: $3x + y - 10 = 0$

Normal line: $x - 3y = 0$

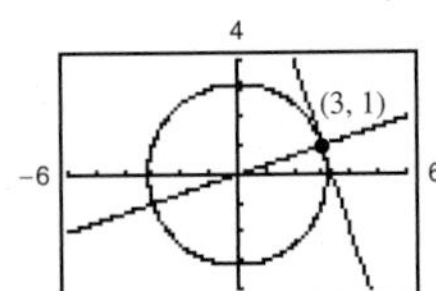

93. (a) $2\sqrt{2}$ units/sec (b) 4 units/sec (c) 8 units/sec

95. 624 cm²/sec **97.** 82 mi/h **99.** $\frac{1}{64}$ ft/min

Chapter Test (page 320)

1. $f'(x) = 2x - 1$

2.

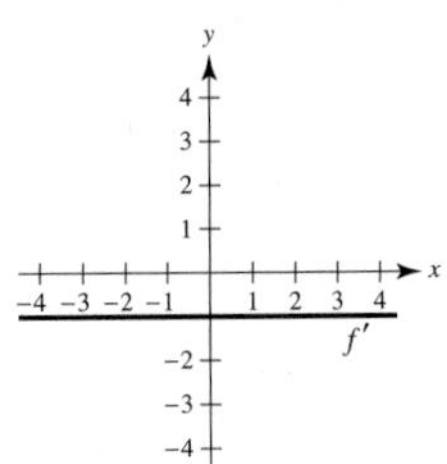

3. $3t^2 - 6t + 2$ **4.** $2(2x^3 - 3x^2 + x - 1)$

5. $(1 - 5x^2)/[3x^{2/3}(x^2+1)^2]$

6. $\dfrac{2}{(x+1)^2},\ x \neq 1$ **7.** $\dfrac{-3}{(t+2)^4}$ **8.** $\dfrac{3}{(x^2+3)^{3/2}}$

9. 0 **10.** $f'(x) = \dfrac{-8}{(3x-2)^2};\ -\dfrac{1}{2}$

11. (a) $y = \frac{1}{4}x + 1$

(b)

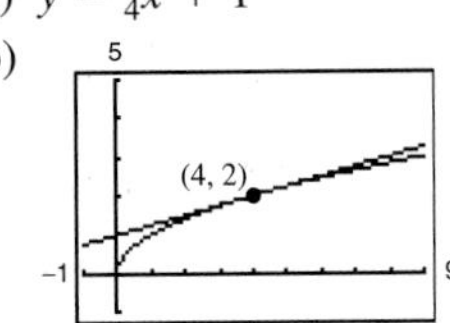

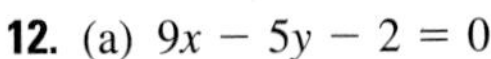

12. (a) $9x - 5y - 2 = 0$

(b)

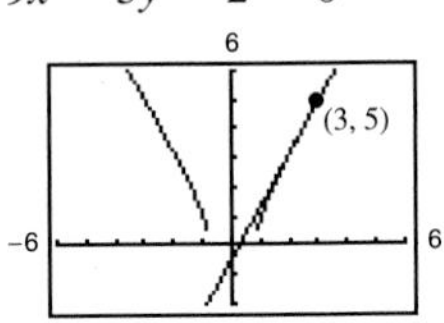

13. $y = -\frac{1}{2}$ **14.** $\left(-1, \frac{2}{3}\right), \left(1, -\frac{2}{3}\right)$

15. $6/(x-3)^3$ **16.** $-x^2/y^2$ **17.** $[y(y-2x)]/[x(x-2y)]$, 1

18. $y = \frac{3}{4}x + 2$ **19.** (a) 36π cm²/min (b) 144π cm²/min

20. (a) $-\dfrac{3}{4}$ ft/sec; $-\dfrac{4}{3}$ ft/sec; $-\dfrac{14}{\sqrt{29}}$ ft/sec (b) $\dfrac{21}{8}$ ft²/sec

P.S. Problem Solving (page 321)

1. (a) $r = \frac{1}{2};\ x^2 + \left(y - \frac{1}{2}\right)^2 = \frac{1}{4}$

(b) Center: $\left(0, \frac{5}{4}\right);\ x^2 + \left(y - \frac{5}{4}\right)^2 = 1$

3. (a) $P_1(x) = 1 + \frac{1}{2}x$ (b) $P_2(x) = 1 + \frac{1}{2}x - \frac{1}{8}x^2$

(c)

x	-1.0	-0.1	-0.001	0
$\sqrt{x+1}$	0	0.9487	0.9995	1
$P_2(x)$	0.375	0.9488	0.9995	1

x	0.001	0.1	1.0
$\sqrt{x+1}$	1.0005	1.0488	1.4142
$P_2(x)$	1.0005	1.0488	1.375

$P_2(x)$ is a good approximation of $f(x) = \sqrt{x+1}$ when x is very close to 0.

(d)

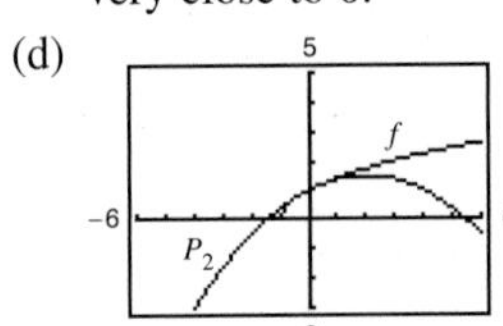

The graphs appear identical in the interval $\left[-\frac{1}{2}, \frac{1}{2}\right]$.

5. $p(x) = 2x^3 + 4x^2 - 5$

7. (a) Graph $\begin{cases} y_1 = \dfrac{1}{a}\sqrt{x^2(a^2 - x^2)} \\ y_2 = -\dfrac{1}{a}\sqrt{x^2(a^2 - x^2)} \end{cases}$ as separate equations.

(b) Answers will vary. Sample answer:

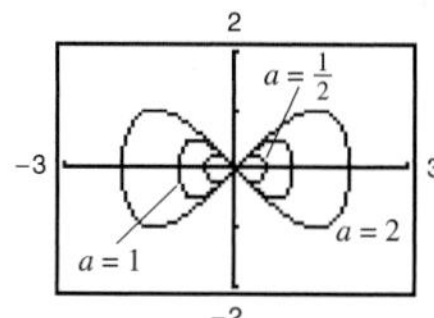

The intercepts will always be $(0, 0)$, $(a, 0)$, and $(-a, 0)$, and the maximum and minimum y-values appear to be $\pm\frac{1}{2}a$.

(c) $\left(\frac{a\sqrt{2}}{2}, \frac{a}{2}\right), \left(\frac{a\sqrt{2}}{2}, -\frac{a}{2}\right), \left(-\frac{a\sqrt{2}}{2}, \frac{a}{2}\right), \left(-\frac{a\sqrt{2}}{2}, -\frac{a}{2}\right)$

9. (a) When the man is 90 ft from the light, the tip of his shadow is $112\frac{1}{2}$ ft from the light. The tip of the child's shadow is $111\frac{1}{9}$ ft from the light, so the man's shadow extends $1\frac{7}{18}$ ft beyond the child's shadow.

(b) When the man is 60 ft from the light, the tip of his shadow is 75 ft from the light. The tip of the child's shadow is $77\frac{7}{9}$ ft from the light, so the child's shadow extends $2\frac{7}{9}$ ft beyond the man's shadow.

(c) $d = 80$ ft

(d) Let x be the distance of the man from the light and let s be the distance from the light to the tip of the shadow.
If $0 < x < 80$, $ds/dt = -50/9$.
If $x > 80$, $ds/dt = -25/4$.
There is a discontinuity at $x = 80$.

11. Proof. The graph of L is a line passing through the origin $(0, 0)$.

13. (a) Proof (b) 2 square units (c) Proof (d) Proof

15. (a) j would be the rate of change of acceleration.
(b) $j = 0$. Acceleration is constant, so there is no change in acceleration.
(c) a: position function, d: velocity function, b: acceleration function, c: jerk function

Chapter 5

Section 5.1 (page 329)

1. $f'(0) = 0$ **3.** $f'(2) = 0$ **5.** $f'(-2)$ is undefined.

7. 2, absolute maximum (and relative maximum)

9. 1, absolute maximum (and relative maximum);
2, absolute minimum (and relative minimum);
3, absolute maximum (and relative maximum)

11. $x = 0, x = 2$ **13.** $t = 8/3$

15. Minimum: $(2, 1)$
Maximum: $(-1, 4)$

17. Minimum: $(1, -1)$
Maximum: $(4, 8)$

19. Minimum: $\left(-1, -\frac{5}{2}\right)$
Maximum: $(2, 2)$

21. Minimum: $(0, 0)$
Maximum: $(-1, 5)$

23. Minimum: $(0, 0)$
Maxima: $\left(-1, \frac{1}{4}\right)$ and $\left(1, \frac{1}{4}\right)$

25. Minimum: $(1, -1)$
Maximum: $\left(0, -\frac{1}{2}\right)$

27. Minimum: $(-1, -1)$
Maximum: $(3, 3)$

29. Minimum value is -2 for $-2 \le x < -1$.
Maximum: $(2, 2)$

31. (a) Minimum: $(0, -3)$;
Maximum: $(2, 1)$
(b) Minimum: $(0, -3)$
(c) Maximum: $(2, 1)$
(d) No extrema

33. (a) Minimum: $(1, -1)$;
Maximum: $(-1, 3)$
(b) Maximum: $(3, 3)$
(c) Minimum: $(1, -1)$
(d) Minimum: $(1, -1)$

35.

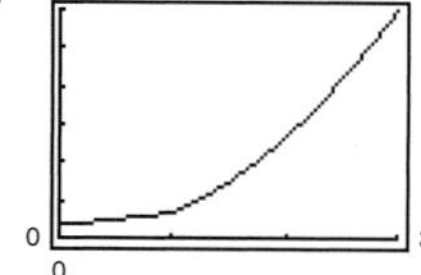

Minimum: $(0, 2)$
Maximum: $(3, 36)$

37.

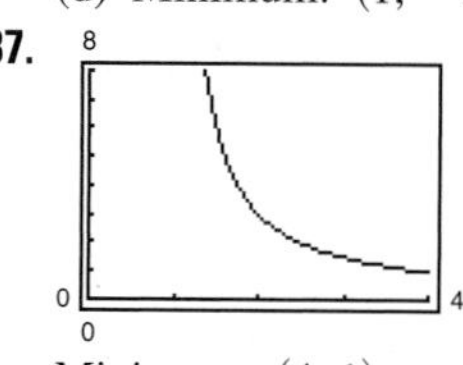

Minimum: $(4, 1)$

39. (a)

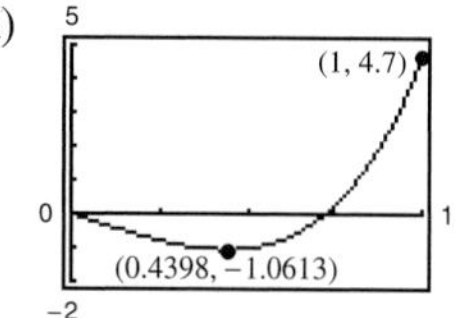

(b) Minimum: $(0.4398, -1.0613)$

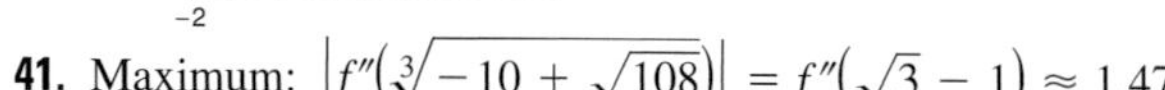

41. Maximum: $\left|f''\left(\sqrt[3]{-10 + \sqrt{108}}\right)\right| = f''\left(\sqrt{3} - 1\right) \approx 1.47$

43. Maximum: $|f^{(4)}(0)| = \frac{56}{81}$

45. Answers will vary. Let $f(x) = 1/x$. f is continuous on $(0, 1)$ but does not have a maximum or minimum.

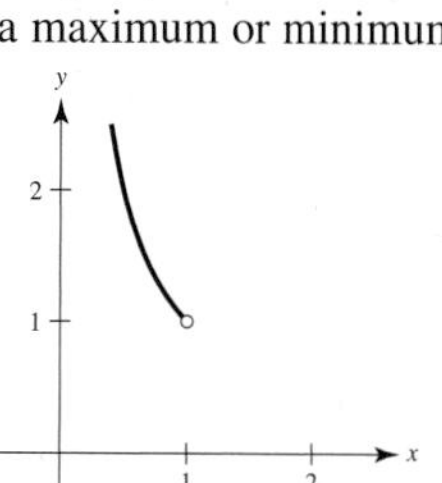

47. Answers will vary. Example:

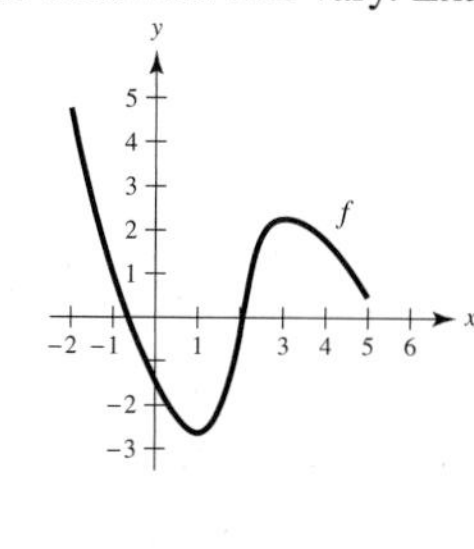

49. (a) Yes (b) No **51.** (a) No (b) Yes

53. Maximum: $P(12) = 72$; No. P is decreasing for $I > 12$.

55. (a) 1970: 2500 per 1000 women
(b) 2005–2006 most rapidly
1975–1980 most slowly
(c) 1970–1975 most rapidly
1980–1985 most slowly
(d) Answers will vary.

57. (a) $A(-500, 45)$, $B(500, 30)$
(b) $y = \dfrac{3}{40{,}000}x^2 - \dfrac{3}{200}x + \dfrac{75}{4}$
(c)

x	-500	-400	-300	-200	-100
d	0	0.75	3	6.75	12

x	0	100	200	300	400	500
d	18.75	12	6.75	3	0.75	0

(d) $(100, 18)$; no

59. True **61.** True **63.** Proof

65. Continuous on $[3, 5]$
Not continuous on $[1, 3]$

Section 5.2 (page 336)

1. $f(-1) = f(1) = 1$; f is not continuous on $[-1, 1]$.

3. $f(0) = f(2) = 0$; f is not differentiable on $(0, 2)$.

5. $(2, 0), (-1, 0); f'\left(\frac{1}{2}\right) = 0$ **7.** $(0, 0), (-4, 0); f'\left(-\frac{8}{3}\right) = 0$

9. $f'\left(\frac{3}{2}\right) = 0$

11. $f'\left(\frac{6 - \sqrt{3}}{3}\right) = 0;\ f'\left(\frac{6 + \sqrt{3}}{3}\right) = 0$

13. Not differentiable at $x = 0$ **15.** $f'\left(-2 + \sqrt{5}\right) = 0$

17. Rolle's Theorem does not apply.

19. (a) $f(1) = f(2) = 38$
(b) Velocity $= 0$ for some t in $(1, 2)$; $t = \frac{3}{2}$ sec

21.

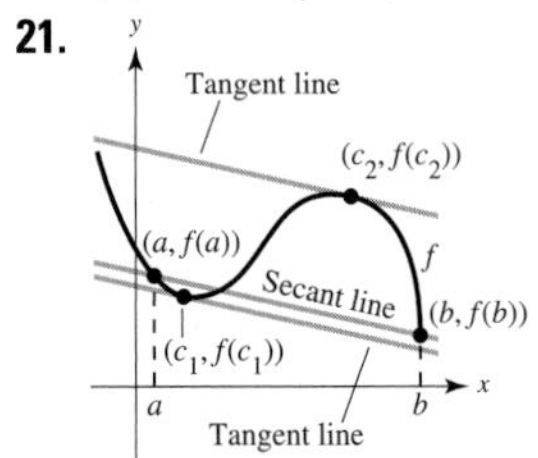

23. The function is not continuous on $[0, 6]$.

25. The function is not continuous on $[0, 6]$.

27. (a) Secant line: $x + y - 3 = 0$ (b) $c = \frac{1}{2}$
(c) Tangent line: $4x + 4y - 21 = 0$
(d)

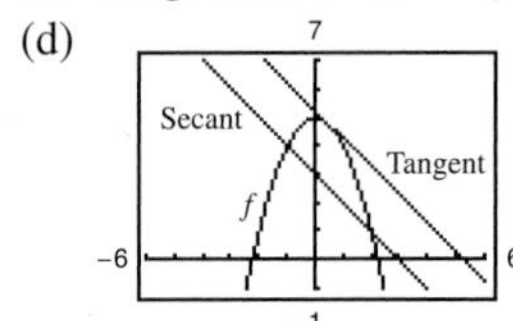

29. $f'(-1/2) = -1$ **31.** $f'(1/\sqrt{3}) = 3, f'(-1/\sqrt{3}) = 3$

33. $f'(8/27) = 1$ **35.** f is not differentiable at $x = -\frac{1}{2}$.

37. (a)–(c)

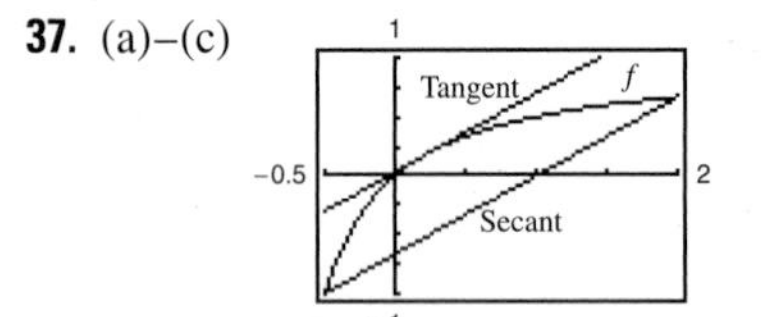

(b) $y = \frac{2}{3}(x - 1)$
(c) $y = \frac{1}{3}\left(2x + 5 - 2\sqrt{6}\right)$

39. (a)–(c)
(b) $y = 0$
(c) $y = -\frac{4}{27}$

41. (a) -14.7 m/sec (b) 1.5 sec

43. No. Let $f(x) = x^2$ on $[-1, 2]$.

45. No. $f(x)$ is not continuous on $[0, 1]$. So it does not satisfy the hypothesis of Rolle's Theorem.

47. By the Mean Value Theorem, there is a time when the speed of the plane must equal the average speed of 454.5 miles/hour. The speed was 400 miles/hour when the plane was accelerating to 454.5 miles/hour and decelerating from 454.5 miles/hour.

49. Proof

51. (a)

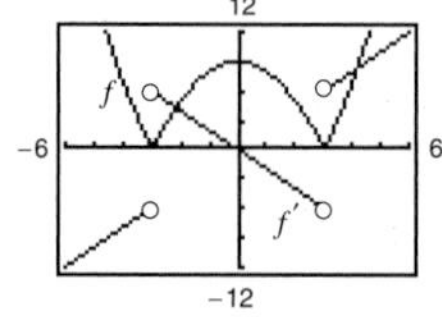

(b) f is continuous and f' is not continuous.
(c) Because $f(-1) = f(1) = 8$ and f is differentiable on $(-1, 1)$, Rolle's Theorem applies on $[-1, 1]$. Because $f(2) = 5$, $f(4) = 7$, and f is not differentiable at $x = 3$, Rolle's Theorem does not apply on $[2, 4]$.
(d) $\lim_{x \to 3^-} f'(x) = -6,\ \lim_{x \to 3^+} f'(x) = 6$

53.

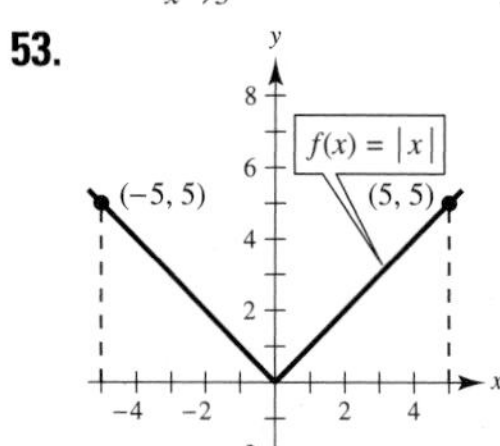

55. $a = 6, b = 1, c = 2$ **57.** $f(x) = 5$ **59.** $f(x) = x^2 - 1$

61. False. f is not continuous on $[-1, 1]$. **63.** True

65–69. Proofs

Section 5.3 (page 345)

1. (a) $(0, 6)$ (b) $(6, 8)$

3. Increasing on $(3, \infty)$; Decreasing on $(-\infty, 3)$

5. Increasing on $(-\infty, -2)$ and $(2, \infty)$; Decreasing on $(-2, 2)$

7. Increasing on $(-\infty, -1)$; Decreasing on $(-1, \infty)$

9. Increasing on $(1, \infty)$; Decreasing on $(-\infty, 1)$

11. Increasing on $\left(-2\sqrt{2}, 2\sqrt{2}\right)$
Decreasing on $\left(-4, -2\sqrt{2}\right)$ and $\left(2\sqrt{2}, 4\right)$

13. (a) Critical number: $x = 2$
(b) Increasing on $(2, \infty)$; Decreasing on $(-\infty, 2)$
(c) Relative minimum: $(2, -4)$

15. (a) Critical number: $x = 1$
(b) Increasing on $(-\infty, 1)$; Decreasing on $(1, \infty)$
(c) Relative maximum: $(1, 5)$

17. (a) Critical numbers: $x = -2, 1$
(b) Increasing on $(-\infty, -2)$ and $(1, \infty)$; Decreasing on $(-2, 1)$
(c) Relative maximum: $(-2, 20)$; Relative minimum: $(1, -7)$

19. (a) Critical numbers: $x = -\frac{5}{3}, 1$
(b) Increasing on $\left(-\infty, -\frac{5}{3}\right), (1, \infty)$
Decreasing on $\left(-\frac{5}{3}, 1\right)$
(c) Relative maximum: $\left(-\frac{5}{3}, \frac{256}{27}\right)$
Relative minimum: $(1, 0)$

21. (a) Critical numbers: $x = \pm 1$
(b) Increasing on $(-\infty, -1)$ and $(1, \infty)$; Decreasing on $(-1, 1)$
(c) Relative maximum: $\left(-1, \frac{4}{5}\right)$; Relative minimum: $\left(1, -\frac{4}{5}\right)$

23. (a) Critical number: $x = 0$
(b) Increasing on $(-\infty, \infty)$
(c) No relative extrema

25. (a) Critical number: $x = -2$
(b) Increasing on $(-2, \infty)$; Decreasing on $(-\infty, -2)$
(c) Relative minimum: $(-2, 0)$

27. (a) Critical number: $x = 5$
(b) Increasing on $(-\infty, 5)$; Decreasing on $(5, \infty)$
(c) Relative maximum: $(5, 5)$

29. (a) Critical numbers: $x = \pm\sqrt{2}/2$; Discontinuity: $x = 0$
(b) Increasing on $(-\infty, -\sqrt{2}/2)$ and $(\sqrt{2}/2, \infty)$
Decreasing on $\left(-\sqrt{2}/2, 0\right)$ and $\left(0, \sqrt{2}/2\right)$
(c) Relative maximum: $\left(-\sqrt{2}/2, -2\sqrt{2}\right)$
Relative minimum: $\left(\sqrt{2}/2, 2\sqrt{2}\right)$

31. (a) Critical number: $x = 0$; Discontinuities: $x = \pm 3$
(b) Increasing on $(-\infty, -3)$ and $(-3, 0)$
Decreasing on $(0, 3)$ and $(3, \infty)$
(c) Relative maximum: $(0, 0)$

33. (a) Critical numbers: $x = -3, 1$; Discontinuity: $x = -1$
(b) Increasing on $(-\infty, -3)$ and $(1, \infty)$
Decreasing on $(-3, -1)$ and $(-1, 1)$
(c) Relative maximum: $(-3, -8)$; Relative minimum: $(1, 0)$

35. (a) Critical number: $x = 0$
(b) Increasing on $(-\infty, 0)$; Decreasing on $(0, \infty)$
(c) No relative extrema

37. (a) Critical number: $x = 1$
(b) Increasing on $(-\infty, 1)$; Decreasing on $(1, \infty)$
(c) Relative maximum: $(1, 4)$

39. (a) $f'(x) = 2(9 - 2x^2)/\sqrt{9 - x^2}$
(b)

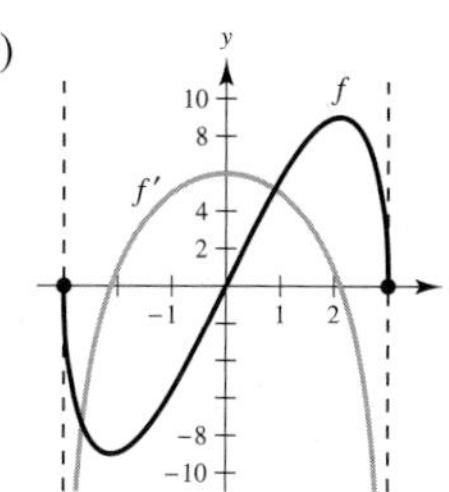
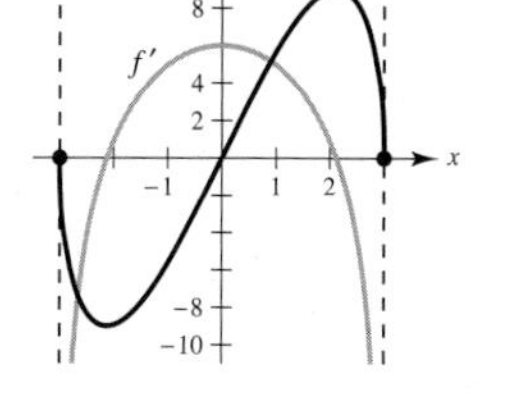

(c) Critical numbers: $x = \pm 3\sqrt{2}/2$
(d) $f' > 0$ on $\left(-3\sqrt{2}/2, 3\sqrt{2}/2\right)$
$f' < 0$ on $\left(-3, -3\sqrt{2}/2\right), \left(3\sqrt{2}/2, 3\right)$
f is increasing when f' is positive and decreasing when f' is negative.

41. $f(x)$ is symmetric with respect to the origin.
Zeros: $(0, 0), \left(\pm\sqrt{3}, 0\right)$

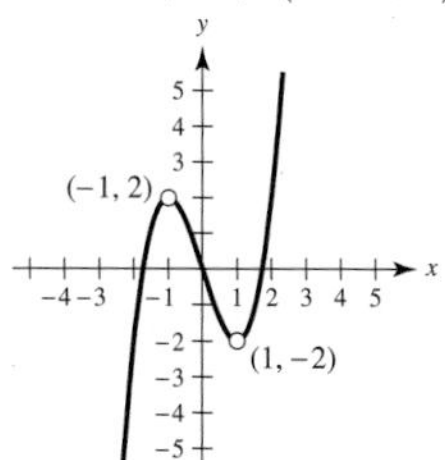

$g(x)$ is continuous on $(-\infty, \infty)$ and $f(x)$ has holes at $x = 1$ and $x = -1$.

43.

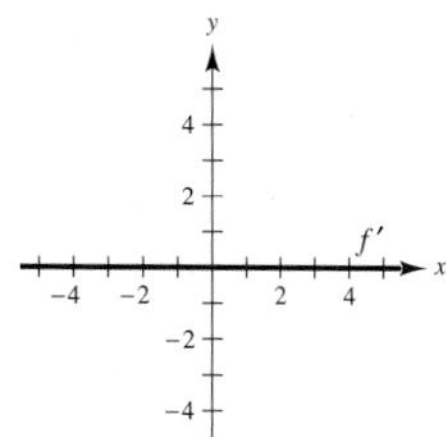

45.

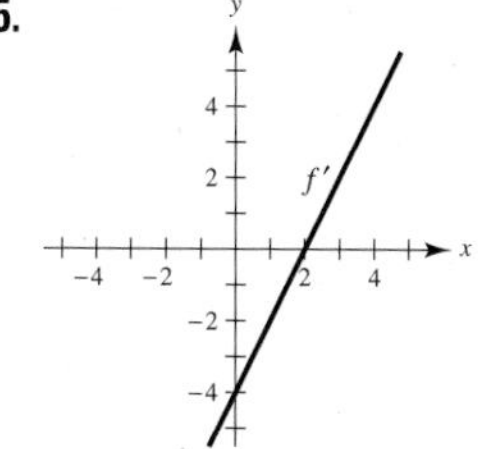

47.

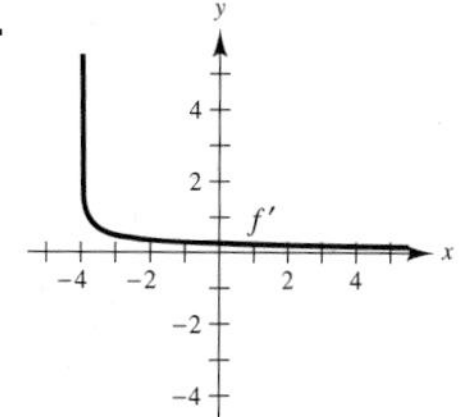

49. (a) Increasing on $(2, \infty)$;
Decreasing on $(-\infty, 2)$
(b) Relative minimum: $x = 2$

51. (a) Increasing on $(-\infty, -1)$ and $(0, 1)$;
Decreasing on $(-1, 0)$ and $(1, \infty)$
(b) Relative maxima: $x = -1$ and $x = 1$
Relative minimum: $x = 0$

53. $g'(0) < 0$ **55.** $g'(-6) < 0$ **57.** $g'(0) > 0$

59. Answers will vary. Sample answer:

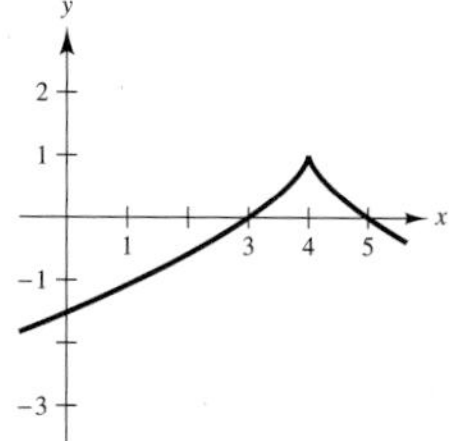

61.

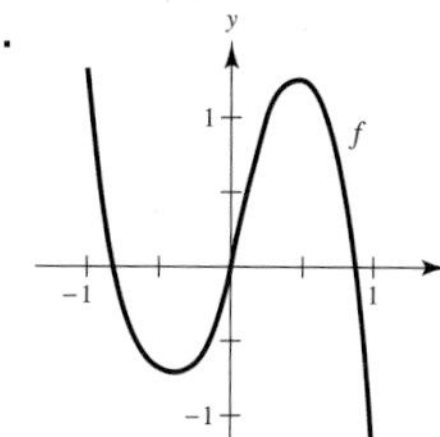

Minimum at the approximate critical number $x = -0.40$
Maximum at the approximate critical number $x = 0.48$

63. $r = 2R/3$ **65.** $x = \sqrt{\dfrac{2Qs}{r}}$

67. (a) $M = 0.03723t^4 - 1.9931t^3 + 37.986t^2 - 282.74t + 825.7$
(b)

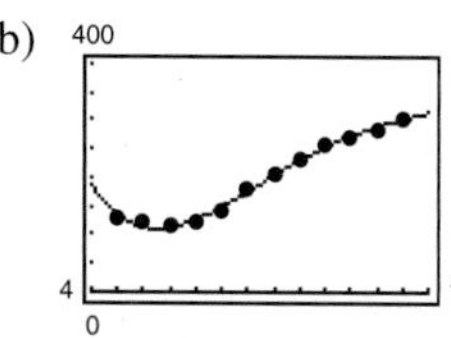

(c) Using a graphing utility, the minimum is (6.5, 111.9) which compares well with the minimum (7, 115.6).

69–71. Proofs

73. (a) $v(t) = 6 - 2t$ (b) $[0, 3)$ (c) $(3, \infty)$ (d) $t = 3$

75. (a) $v(t) = 3t^2 - 10t + 4$
(b) $\left[0, (5 - \sqrt{13})/3\right)$ and $\left((5 + \sqrt{13})/3, \infty\right)$
(c) $\left(\dfrac{5 - \sqrt{13}}{3}, \dfrac{5 + \sqrt{13}}{3}\right)$ (d) $t = \dfrac{5 \pm \sqrt{13}}{3}$

77. (a) Minimum degree: 3
(b) $a_3(0)^3 + a_2(0)^2 + a_1(0) + a_0 = 0$
$a_3(2)^3 + a_2(2)^2 + a_1(2) + a_0 = 2$
$3a_3(0)^2 + 2a_2(0) + a_1 = 0$
$3a_3(2)^2 + 2a_2(2) + a_1 = 0$
(c) $f(x) = -\frac{1}{2}x^3 + \frac{3}{2}x^2$

79. (a) Minimum degree: 4
(b) $a_4(0)^4 + a_3(0)^3 + a_2(0)^2 + a_1(0) + a_0 = 0$
$a_4(2)^4 + a_3(2)^3 + a_2(2)^2 + a_1(2) + a_0 = 4$
$a_4(4)^4 + a_3(4)^3 + a_2(4)^2 + a_1(4) + a_0 = 0$
$4a_4(0)^3 + 3a_3(0)^2 + 2a_2(0) + a_1 = 0$
$4a_4(2)^3 + 3a_3(2)^2 + 2a_2(2) + a_1 = 0$
$4a_4(4)^3 + 3a_3(4)^2 + 2a_2(4) + a_1 = 0$
(c) $f(x) = \frac{1}{4}x^4 - 2x^3 + 4x^2$

81. True **83.** False. Let $f(x) = x^3$.

85. False. Let $f(x) = x^3$. There is a critical number at $x = 0$, but not a relative extremum.

Section 5.4 (page 353)

1. $f' > 0, f'' > 0$ **3.** $f' < 0, f'' < 0$

5. Concave upward: $(-\infty, \infty)$

7. Concave upward: $(-\infty, 1)$; Concave downward: $(1, \infty)$

9. Concave upward: $(-\infty, 2)$; Concave downward: $(2, \infty)$

11. Concave upward: $(-\infty, -2), (2, \infty)$
Concave downward: $(-2, 2)$

13. Concave upward: $(-\infty, -1), (1, \infty)$
Concave downward: $(-1, 1)$

15. Concave upward: $(-2, 2)$
Concave downward: $(-\infty, -2), (2, \infty)$

17. No concavity

19. Points of inflection: $(-2, -8), (0, 0)$
Concave upward: $(-\infty, -2), (0, \infty)$
Concave downward: $(-2, 0)$

21. Point of inflection: $(2, 8)$; Concave downward: $(-\infty, 2)$
Concave upward: $(2, \infty)$

23. Points of inflection: $(\pm 2\sqrt{3}/3, -20/9)$
Concave upward: $(-\infty, -2\sqrt{3}/3), (2\sqrt{3}/3, \infty)$
Concave downward: $(-2\sqrt{3}/3, 2\sqrt{3}/3)$

25. Points of inflection: $(2, -16), (4, 0)$
Concave upward: $(-\infty, 2), (4, \infty)$; Concave downward: $(2, 4)$

27. Concave upward: $(-3, \infty)$

29. Points of inflection: $(-\sqrt{3}/3, 3), (\sqrt{3}/3, 3)$
Concave upward: $(-\infty, -\sqrt{3}/3), (\sqrt{3}/3, \infty)$
Concave downward: $(-\sqrt{3}/3, \sqrt{3}/3)$

31. Relative minimum: $(5, 0)$ **33.** Relative maximum: $(3, 9)$

35. Relative maximum: $(0, 3)$; Relative minimum: $(2, -1)$

37. Relative minimum: $(3, -25)$

39. Relative maximum: $(2.4, 268.74)$; Relative minimum: $(0, 0)$

41. Relative minimum: $(0, -3)$

43. Relative maximum: $(-2, -4)$; Relative minimum: $(2, 4)$

45. (a) $f'(x) = 0.2x(x - 3)^2(5x - 6)$
$f''(x) = 0.4(x - 3)(10x^2 - 24x + 9)$
(b) Relative maximum: $(0, 0)$
Relative minimum: $(1.2, -1.6796)$
Points of inflection: $(0.4652, -0.7048)$, $(1.9348, -0.9048)$, $(3, 0)$
(c)

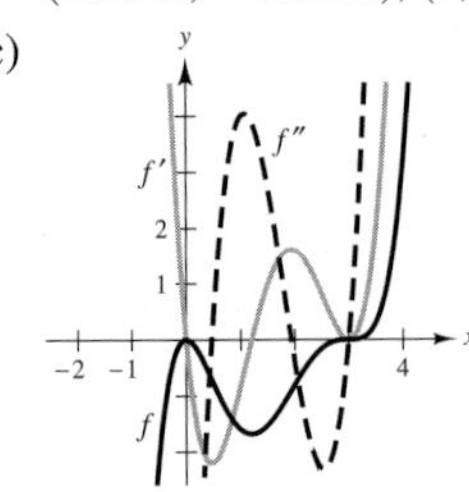

f is increasing when f' is positive, and decreasing when f' is negative. f is concave upward when f'' is positive, and concave downward when f'' is negative.

47. (a)

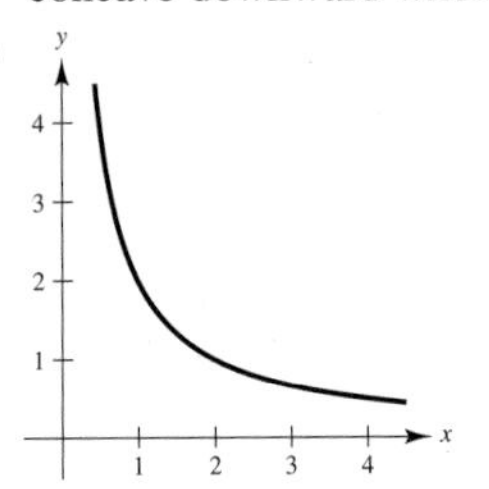

(b)

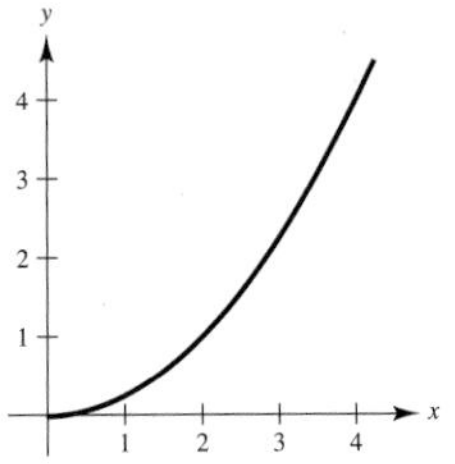

49. Answers will vary. Example:
$f(x) = x^4$; $f''(0) = 0$, but $(0, 0)$ is not a point of inflection.

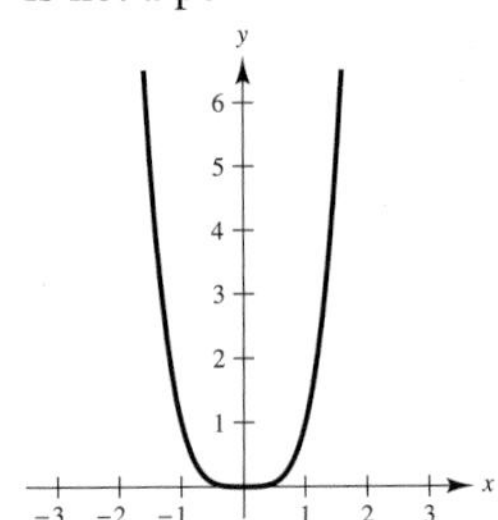

51.

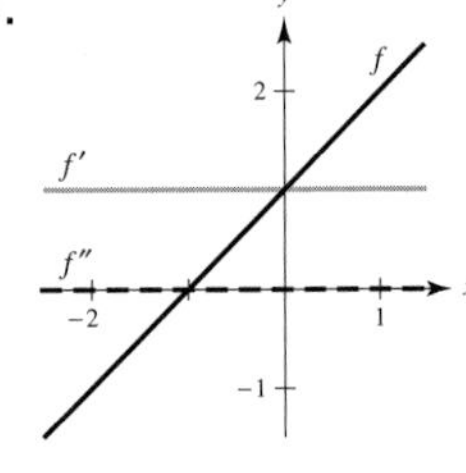

53.

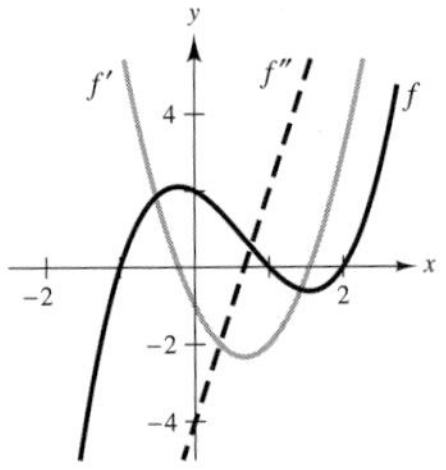

55.

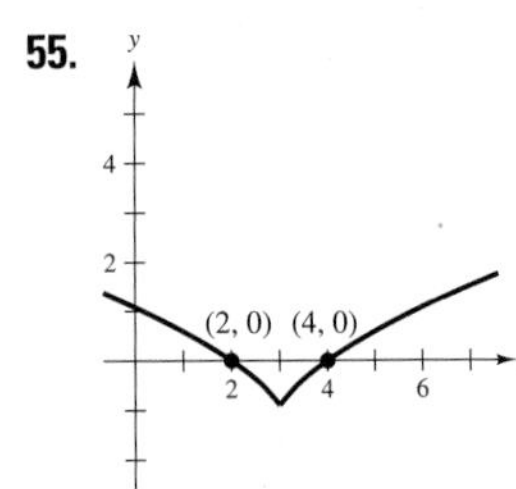

57.

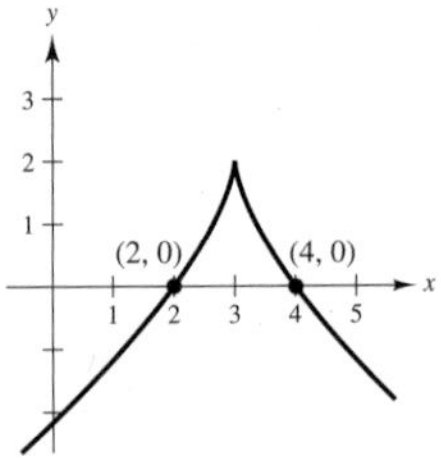

59. Example:

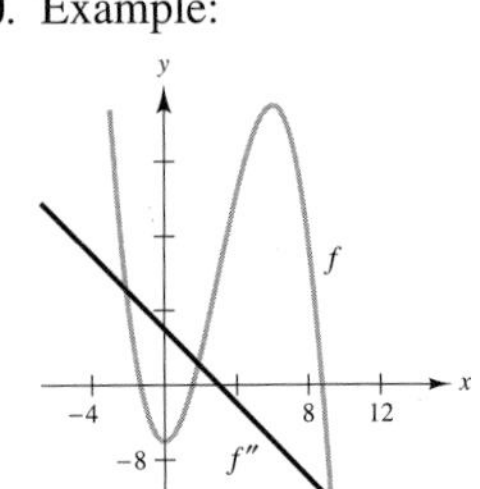

61. (a) $f(x) = (x - 2)^n$ has a point of inflection at $(2, 0)$ if n is odd and $n \geq 3$.

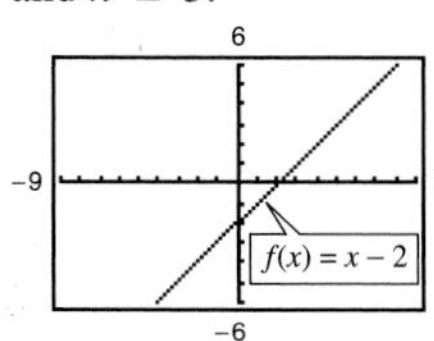

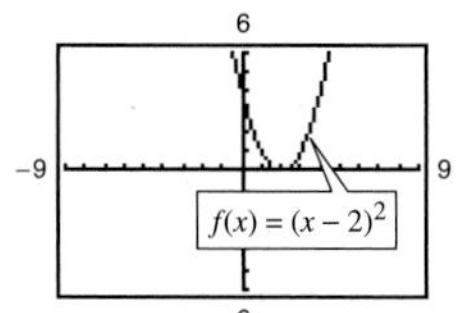

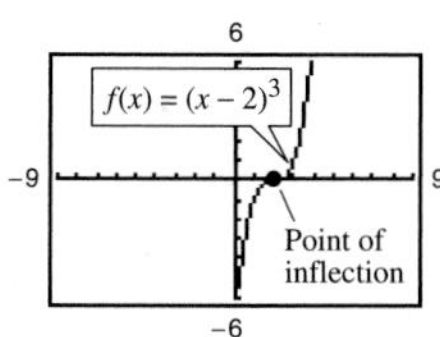

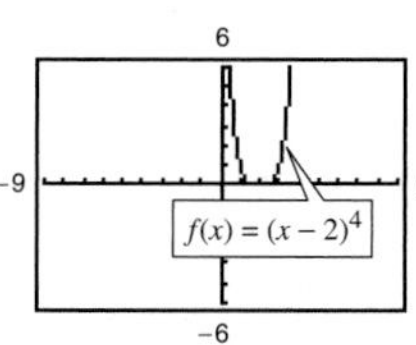

(b) Proof

63. $f(x) = \frac{1}{2}x^3 - 6x^2 + \frac{45}{2}x - 24$

65. (a) $f(x) = \frac{1}{32}x^3 + \frac{3}{16}x^2$ (b) Two miles from touchdown

67. $x = \left(\dfrac{15 - \sqrt{33}}{16}\right)L \approx 0.578L$ **69.** $x = 100$ units

71. (a)

t	0.5	1	1.5	2	2.5	3
S	151.5	555.6	1097.6	1666.7	2193.0	2647.1

$1.5 < t < 2$

(b)

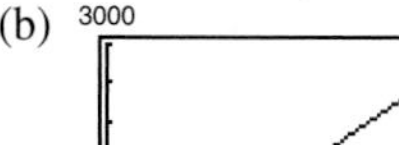

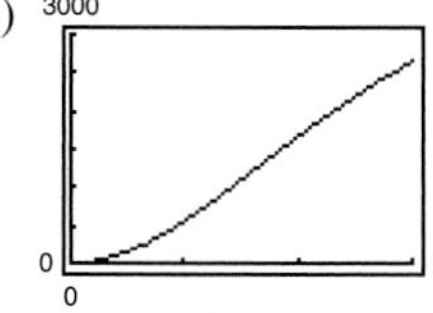

(c) About 1.633 yr

$t \approx 1.5$

73. $P_1(x) = 1 - x/2$
$P_2(x) = 1 - x/2 - x^2/8$

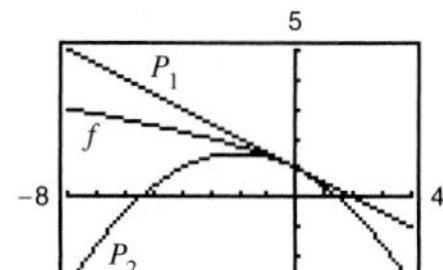

The values of f, P_1, and P_2 and their first derivatives are equal when $x = 0$. The approximations worsen as you move away from $x = 0$.

75. True

77. False. f is concave upward at $x = c$ if $f''(c) > 0$.

79–83. Proofs

Section 5.5 (page 363)

1. f **2.** c **3.** d **4.** a **5.** b **6.** e

7.

x	10^0	10^1	10^2	10^3
$f(x)$	7	2.2632	2.0251	2.0025

x	10^4	10^5	10^6
$f(x)$	2.0003	2.0000	2.0000

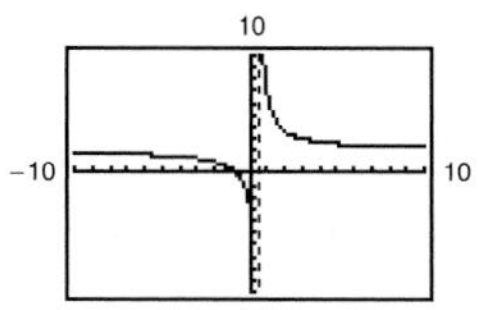

$\lim_{x\to\infty} \dfrac{4x + 3}{2x - 1} = 2$

9.

x	10^0	10^1	10^2	10^3
$f(x)$	−2	−2.9814	−2.9998	−3.0000

x	10^4	10^5	10^6
$f(x)$	−3.0000	−3.0000	−3.0000

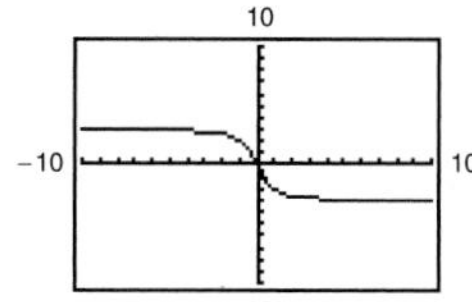

$\lim_{x\to\infty} \dfrac{-6x}{\sqrt{4x^2 + 5}} = -3$

11.

x	10^0	10^1	10^2	10^3
$f(x)$	4.5000	4.9901	4.9999	5.0000

x	10^4	10^5	10^6
$f(x)$	5.0000	5.0000	5.0000

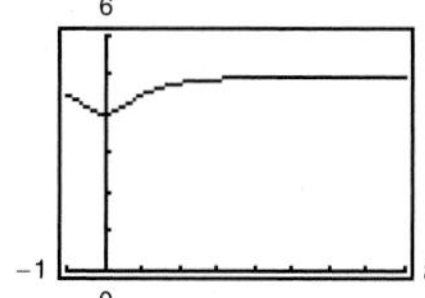

$\lim_{x\to\infty} \left(5 - \dfrac{1}{x^2 + 1}\right) = 5$

13. (a) ∞ (b) 5 (c) 0 **15.** (a) 0 (b) 1 (c) ∞

17. (a) 0 (b) $-\frac{2}{3}$ (c) $-\infty$ **19.** 4 **21.** $\frac{2}{3}$ **23.** 0

25. $-\infty$ **27.** -1 **29.** -2 **31.** $\frac{1}{2}$ **33.** ∞

35.

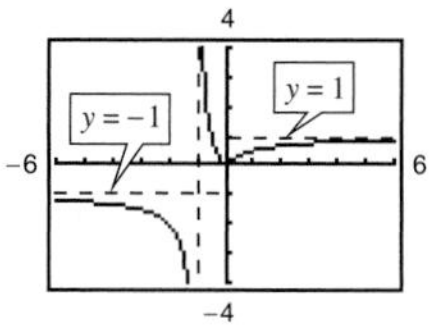

37.

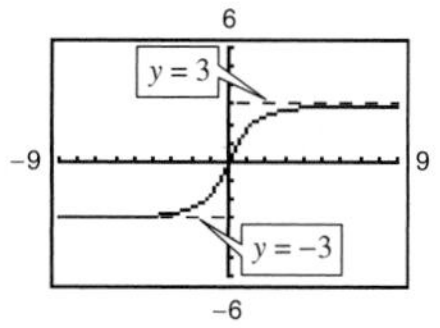

39. 0 **41.** $\frac{1}{6}$

43.

x	10^0	10^1	10^2	10^3	10^4	10^5	10^6
$f(x)$	1.000	0.513	0.501	0.500	0.500	0.500	0.500

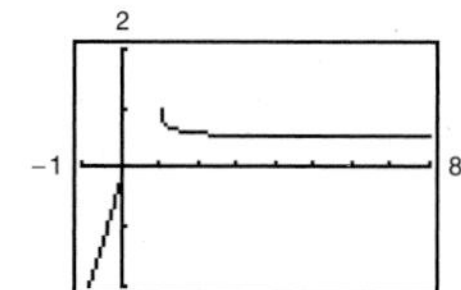

$\lim_{x\to\infty} \left[x - \sqrt{x(x - 1)}\right] = \frac{1}{2}$

45.

x	10^0	10^1	10^2	10^3
$f(x)$	-0.236	-0.025	-0.002	-2.5×10^{-4}

x	10^4	10^5	10^6
$f(x)$	-2.5×10^{-5}	-2.5×10^{-6}	0

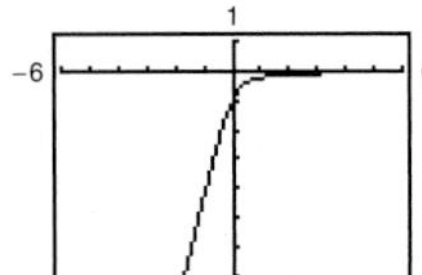

$\lim_{x\to\infty} \left(2x - \sqrt{4x^2 + 1}\right) = 0$

47. Answers will vary. Example: Let $f(x) = \dfrac{-6}{0.1(x-2)^2 + 1} + 6$.

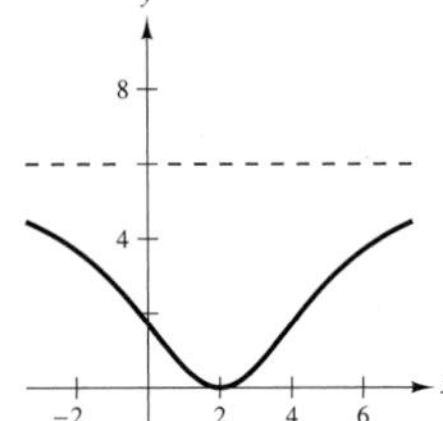

49. (a) 5 (b) -5

51.

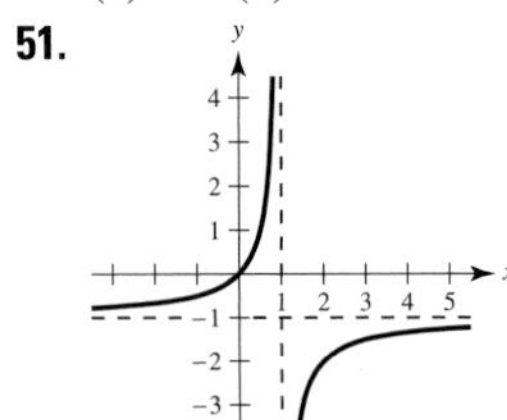

53.

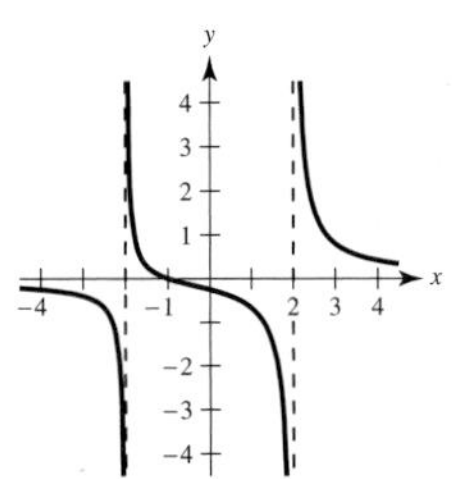

55.

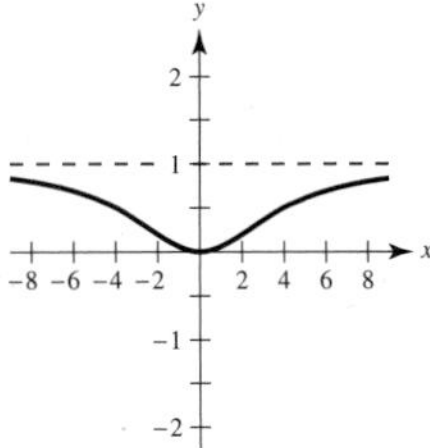

57.

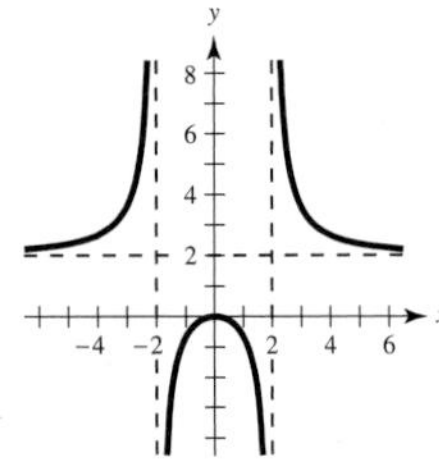

59.

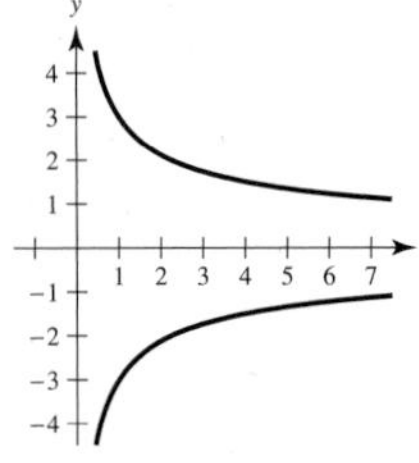

61.

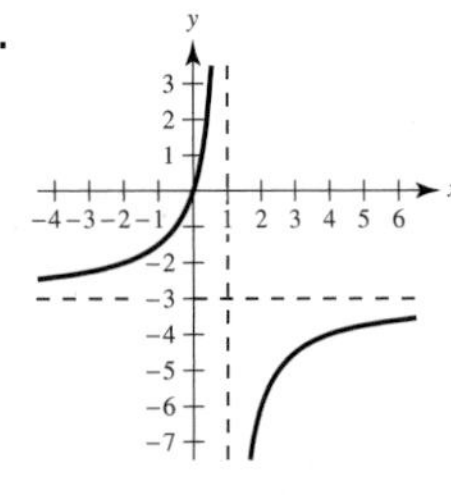

63.

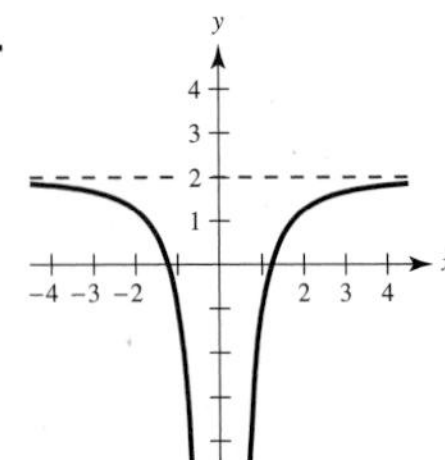

65.

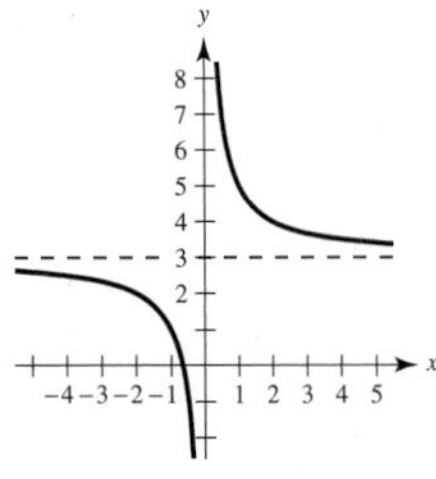

67.

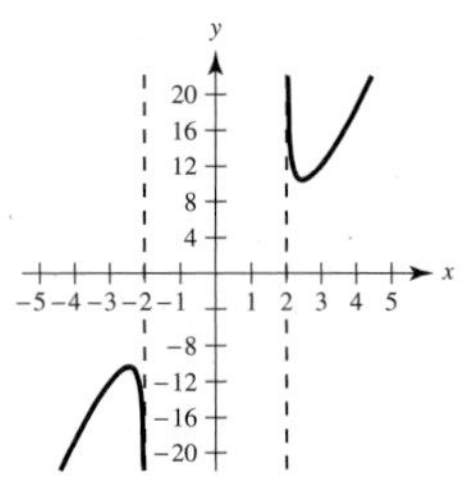

69.

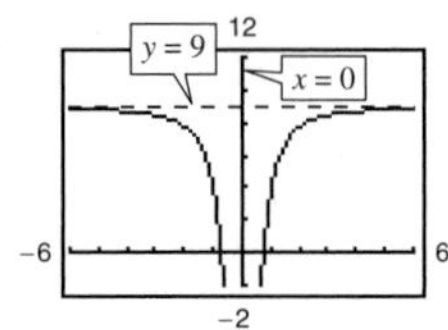

71.

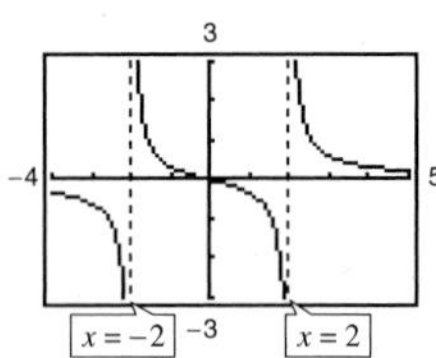

73.

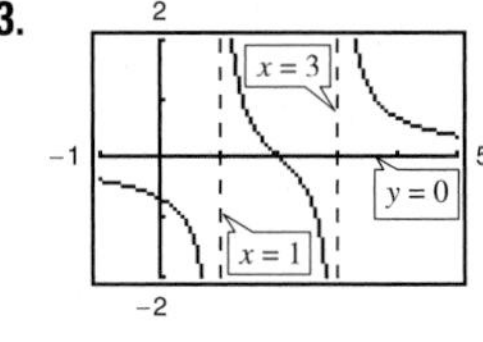

75.

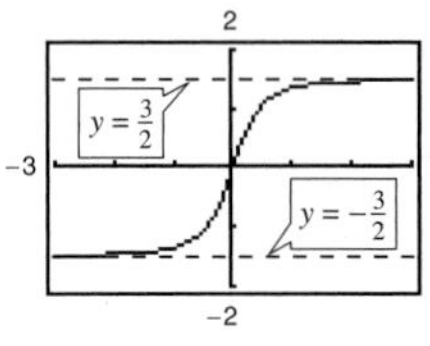

77. (a)

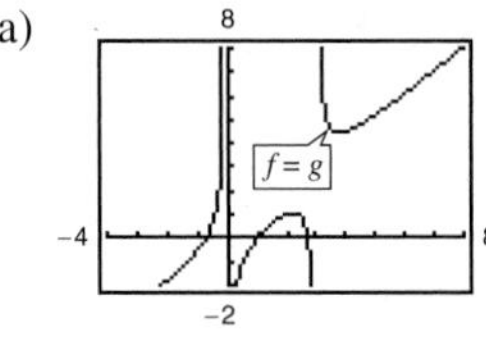

(b) Proof

(c)

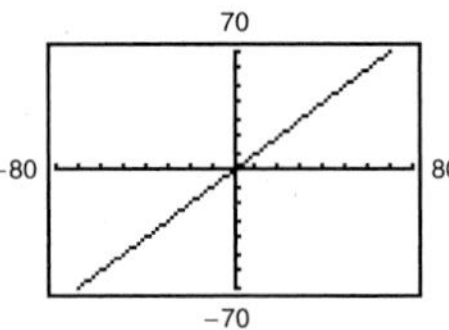

The slant asymptote $y = x$

79. 100%

81. (a) $\lim_{t\to 0^+} T = 1700°$; This is the temperature of the kiln.
(b) $\lim_{t\to\infty} T = 72°$; This is the temperature of the room.
(c) No. $y = 72$ is the horizontal asymptote.

83. (a)

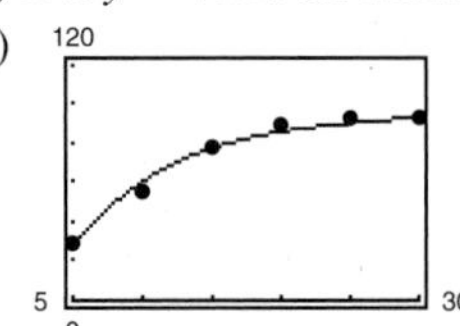

(b) Yes. $\lim_{t\to\infty} S = \frac{100}{1} = 100$

85. (a) $d(m) = \dfrac{|3m + 3|}{\sqrt{m^2 + 1}}$

(b)

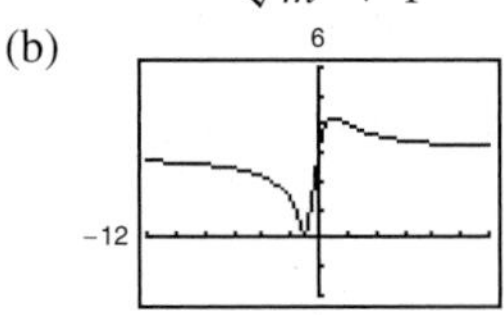

(c) $\lim_{m\to\infty} d(m) = 3$

$\lim_{m\to-\infty} d(m) = 3$

As m approaches $\pm\infty$, the distance approaches 3.

87. False. Let $f(x) = \dfrac{2x}{\sqrt{x^2 + 2}}$. $f'(x) > 0$ for all real numbers.

89. Proof

Section 5.6 (page 371)

1. d **2.** c **3.** a **4.** b

5.

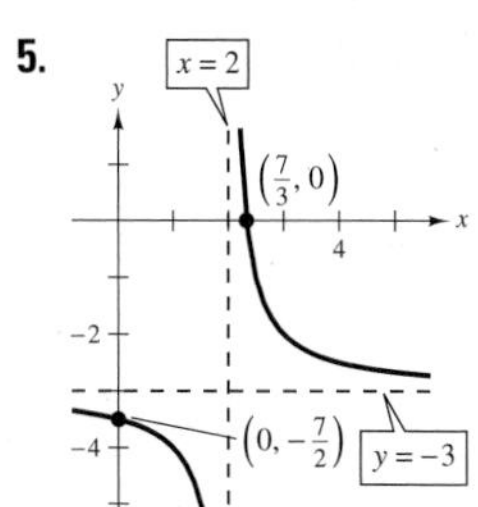

7.

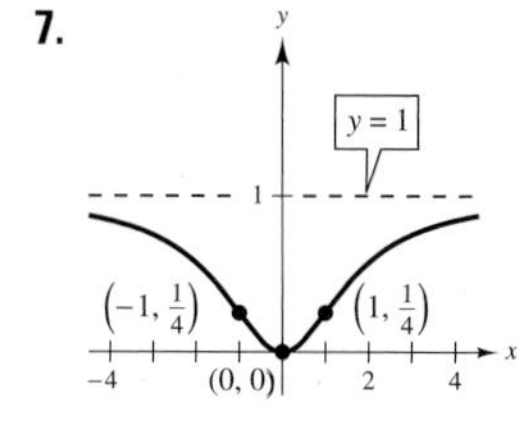

9.

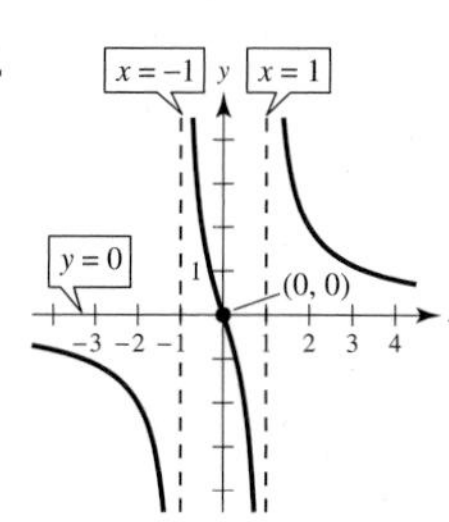

11.

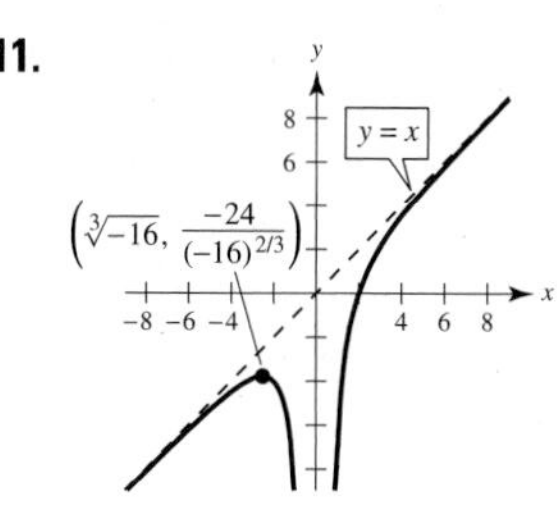

13.

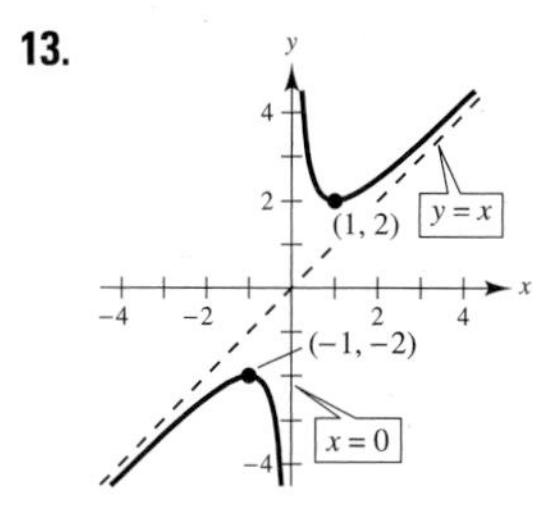

15.

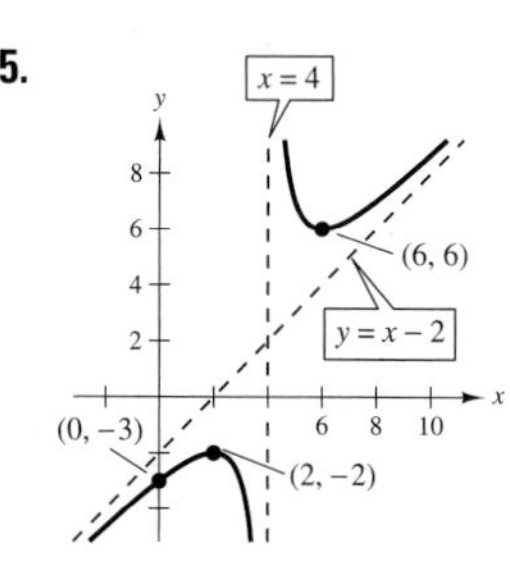

17.

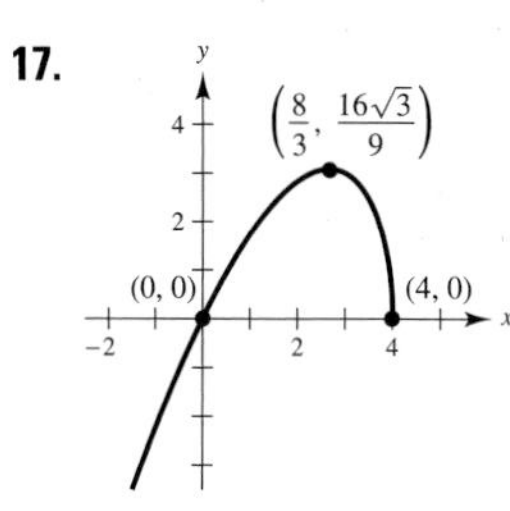

19.

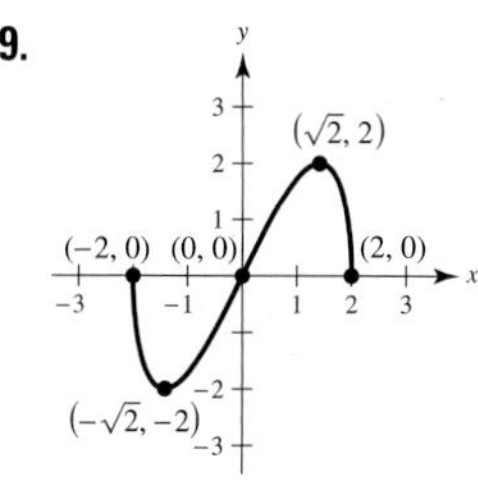

21.

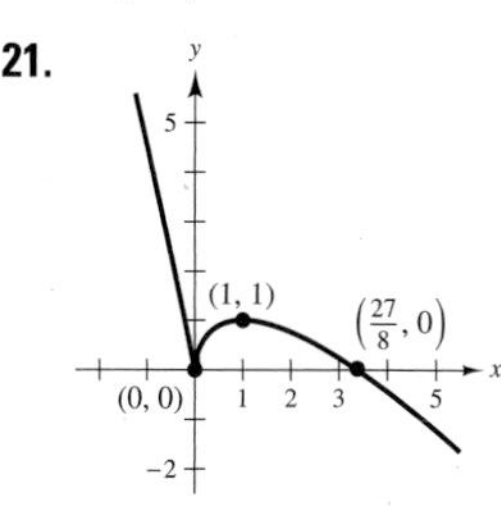

23.

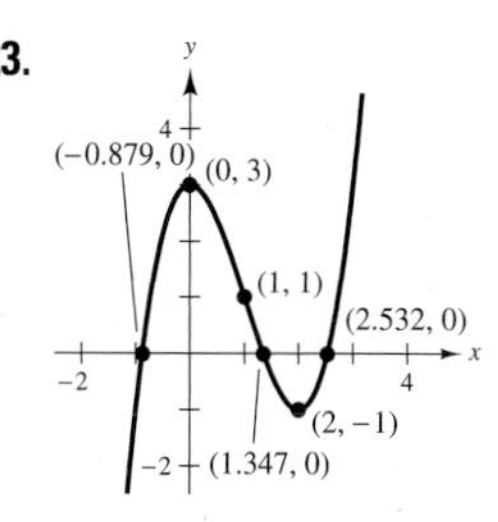

25.

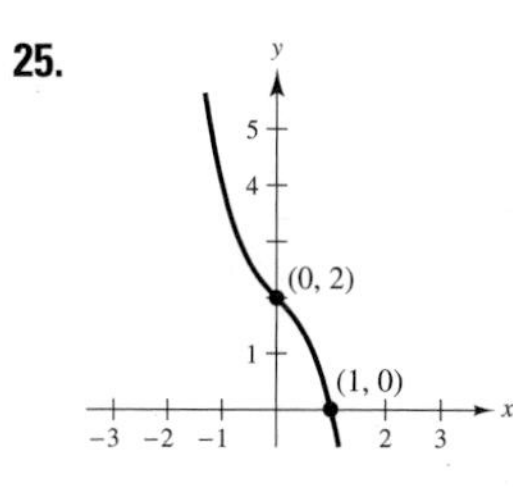

27.

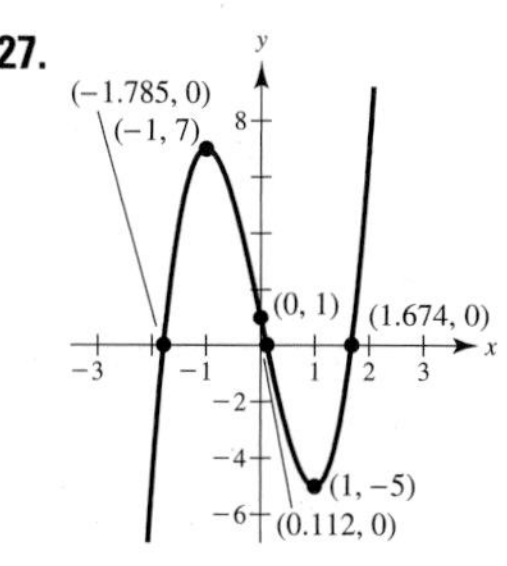

29.

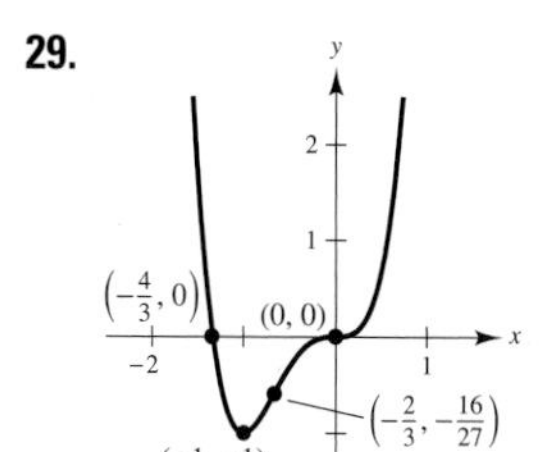

31.

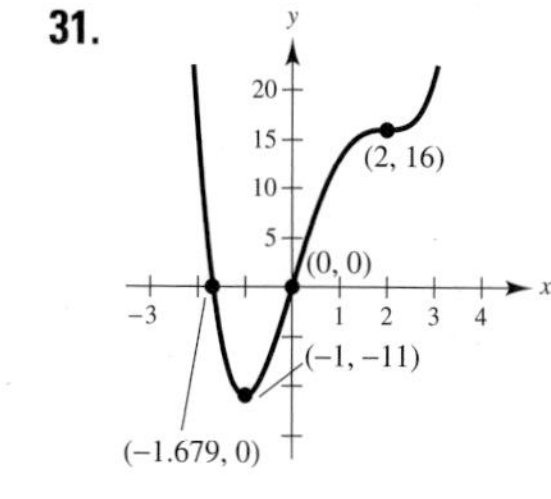

33.

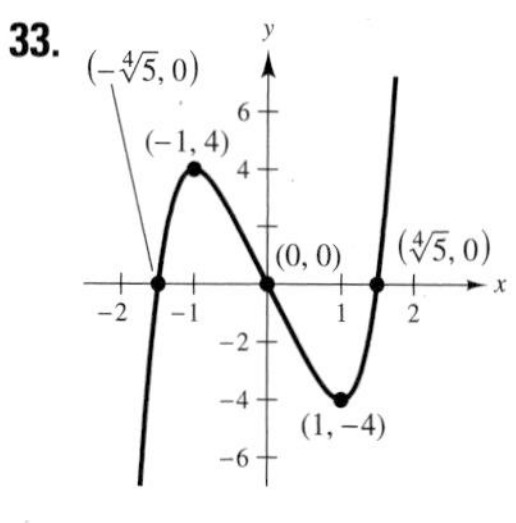

35.

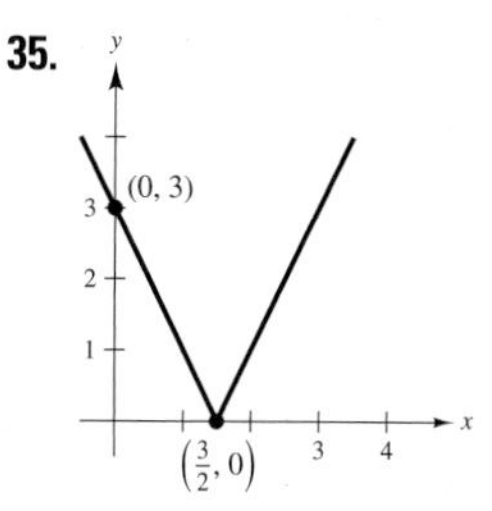

37.

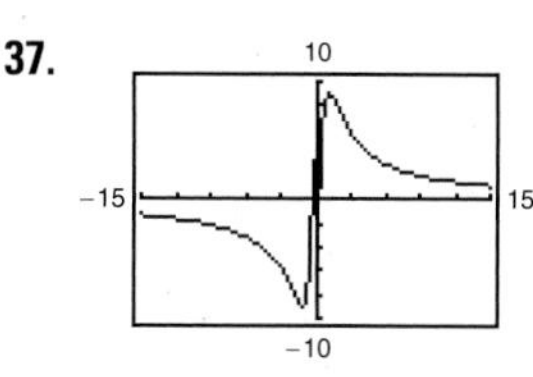

Minimum: $(-1.10, -9.05)$
Maximum: $(1.10, 9.05)$
Points of inflection:
$(-1.84, -7.86), (1.84, 7.86)$
Vertical asymptote: $x = 0$
Horizontal asymptote: $y = 0$

39.

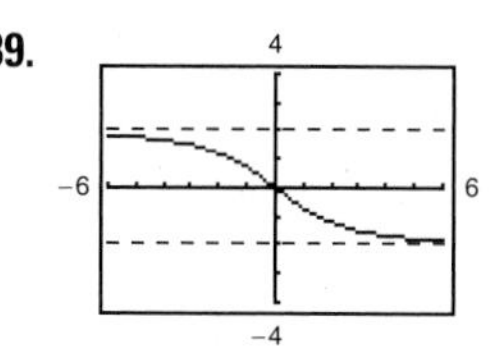

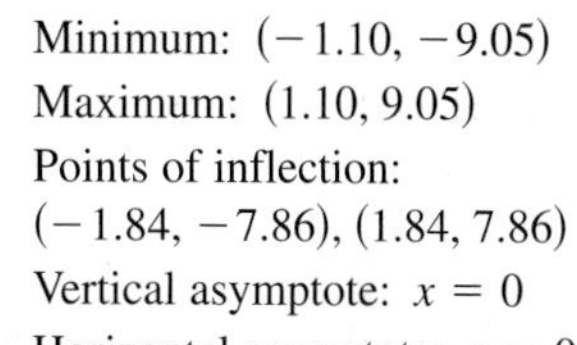

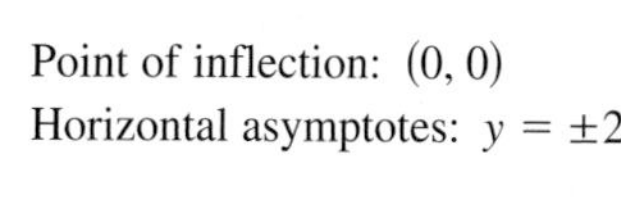

Point of inflection: $(0, 0)$
Horizontal asymptotes: $y = \pm 2$

41. f is decreasing on $(2, 8)$ and therefore $f(3) > f(5)$.

43.

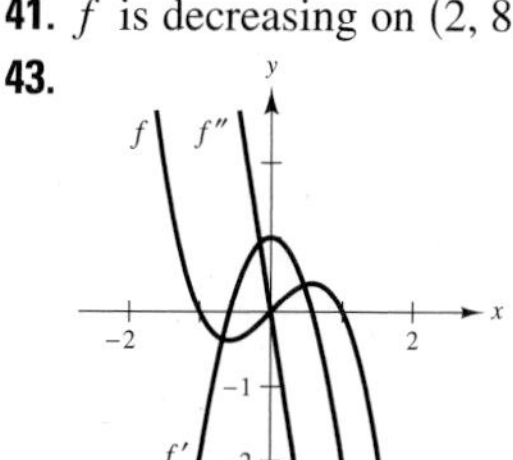

The zeros of f' correspond to the points where the graph of f has horizontal tangents. The zero of f'' corresponds to the point where the graph of f' has a horizontal tangent.

45.

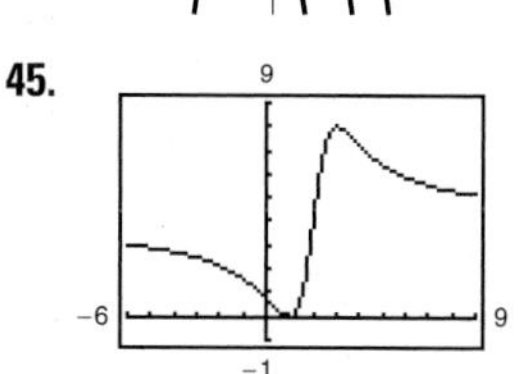

The graph crosses the horizontal asymptote $y = 4$.

The graph of a function f does not cross its vertical asymptote $x = c$ because $f(c)$ does not exist.

47.

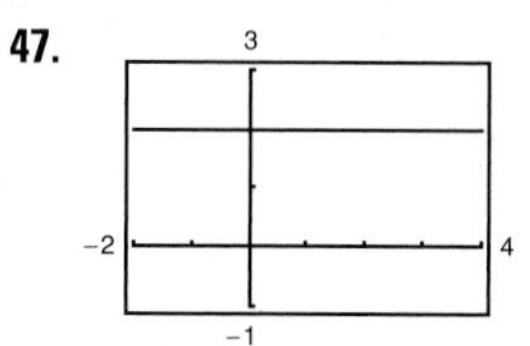

The graph has a hole at $x = 3$. The rational function is not reduced to lowest terms.

49.

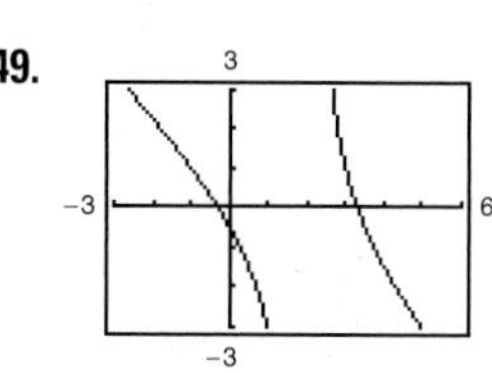

The graph appears to approach the line $y = -x + 1$, which is the slant asymptote.

51. 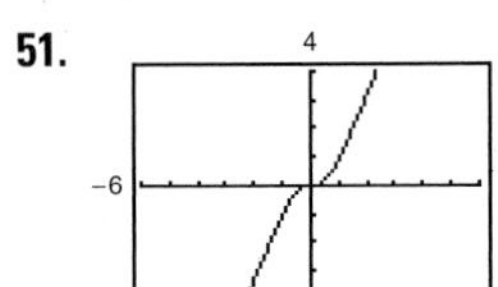
The graph appears to approach the line $y = 2x$, which is the slant asymptote.

53. 

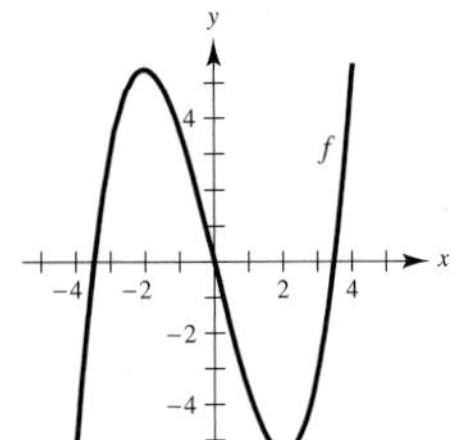

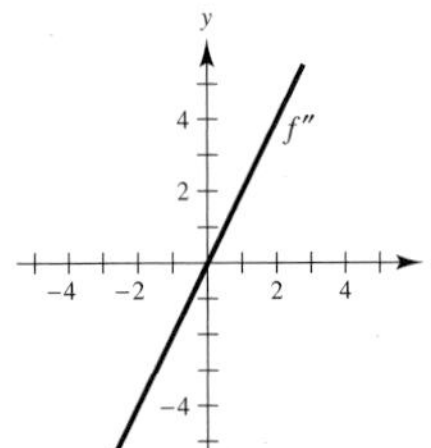

55.

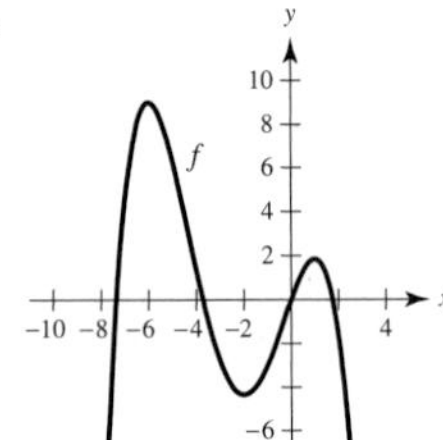

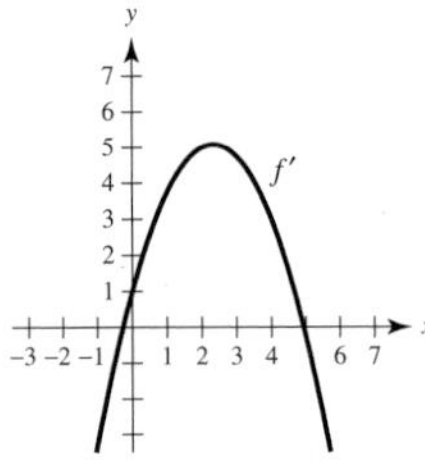

57. Answers will vary. Example: $y = 1/(x - 3)$

59. Answers will vary. Example: $y = (3x^2 - 7x - 5)/(x - 3)$

61. (a) $f'(x) = 0$ for $x = \pm 2$; $f'(x) > 0$ for $(-\infty, -2)$, $(2, \infty)$
$f'(x) < 0$ for $(-2, 2)$
(b) $f''(x) = 0$ for $x = 0$; $f''(x) > 0$ for $(0, \infty)$
$f''(x) < 0$ for $(-\infty, 0)$
(c) $(0, \infty)$
(d) f' is minimum for $x = 0$.
f is decreasing at the greatest rate at $x = 0$.

63. Answers will vary. Sample answer: The graph has a vertical asymptote at $x = b$. If a and b are both positive, or both negative, then the graph of f approaches ∞ as x approaches b, and the graph has a minimum at $x = -b$. If a and b have opposite signs, then the graph of f approaches $-\infty$ as x approaches b, and the graph has a maximum at $x = -b$.

65. (a) If n is even, f is symmetric with respect to the y-axis.
If n is odd, f is symmetric with respect to the origin.
(b) $n = 0, 1, 2, 3$ (c) $n = 4$
(d) When $n = 5$, the slant asymptote is $y = 2x$.
(e)

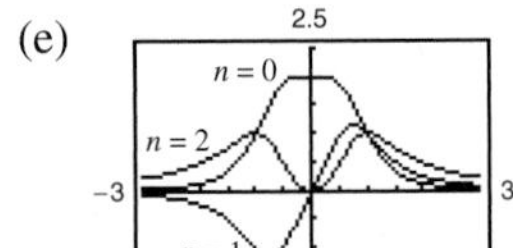

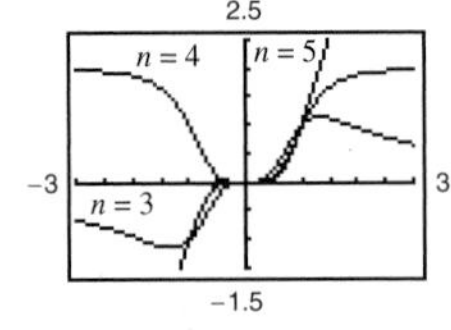

n	0	1	2	3	4	5
M	1	2	3	2	1	0
N	2	3	4	5	2	3

67. (a)

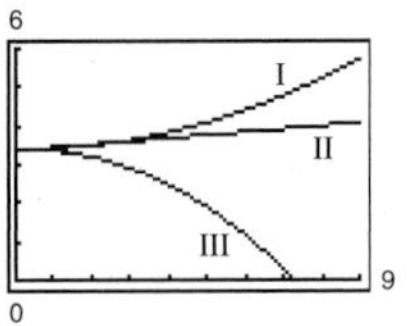

(b) Models I and II
(c) I most optimistic, III most pessimistic; Answers will vary.

69. $y = 4x$, $y = -4x$

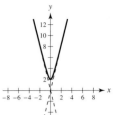

Section 5.7 (page 379)

1. (a) and (b)

First Number x	Second Number	Product P
10	$110 - 10$	$10(110 - 10) = 1000$
20	$110 - 20$	$20(110 - 20) = 1800$
30	$110 - 30$	$30(110 - 30) = 2400$
40	$110 - 40$	$40(110 - 40) = 2800$
50	$110 - 50$	$50(110 - 50) = 3000$
60	$110 - 60$	$60(110 - 60) = 3000$
70	$110 - 70$	$70(110 - 70) = 2800$
80	$110 - 80$	$80(110 - 80) = 2400$
90	$110 - 90$	$90(110 - 90) = 1800$
100	$110 - 100$	$100(110 - 100) = 1000$

The maximum is attained near $x = 50$ and 60.
(c) $P = x(110 - x)$
(d)

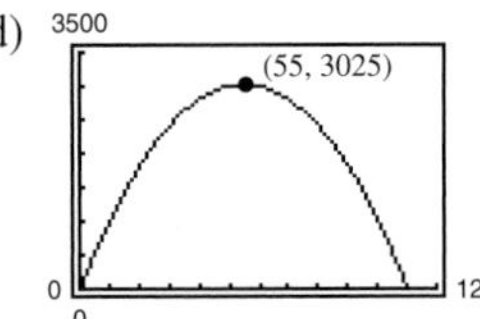

(e) 55 and 55

3. $S/2$ and $S/2$ **5.** 21 and 7 **7.** 54 and 27

9. $l = w = 20$ m **11.** $l = w = 4\sqrt{2}$ ft **13.** $(1, 1)$

15. $\left(\frac{7}{2}, \sqrt{\frac{7}{2}}\right)$

17. Dimensions of page: $(2 + \sqrt{30})$ in. $\times$ $(2 + \sqrt{30})$ in.

19. $x = Q_0/2$ **21.** 700×350 m

23. (a) Proof (b) $V_1 = 99$ in.^{3}, $V_2 = 125$ in.^{3}, $V_3 = 117$ in.^{3}
(c) $5 \times 5 \times 5$ in.

25. Rectangular portion: $16/(\pi + 4) \times 32/(\pi + 4)$ ft

27. (a) $L = \sqrt{x^2 + 4 + \dfrac{8}{x - 1} + \dfrac{4}{(x - 1)^2}}$, $x > 1$
(b)

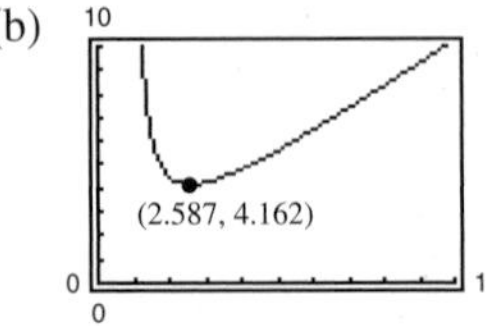

Minimum when $x \approx 2.587$

(c) $(0, 0)$, $(2, 0)$, $(0, 4)$

29. Width: $5\sqrt{2}/2$; Length: $5\sqrt{2}$

31. (a)

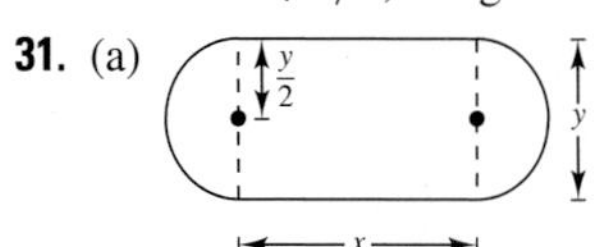

(b)

Length x	Width y	Area xy
10	$(2/\pi)(100 - 10)$	$(10)(2/\pi)(100 - 10) \approx 573$
20	$(2/\pi)(100 - 20)$	$(20)(2/\pi)(100 - 20) \approx 1019$
30	$(2/\pi)(100 - 30)$	$(30)(2/\pi)(100 - 30) \approx 1337$
40	$(2/\pi)(100 - 40)$	$(40)(2/\pi)(100 - 40) \approx 1528$
50	$(2/\pi)(100 - 50)$	$(50)(2/\pi)(100 - 50) \approx 1592$
60	$(2/\pi)(100 - 60)$	$(60)(2/\pi)(100 - 60) \approx 1528$

The maximum area of the rectangle is approximately 1592 m².

(c) $A = (2/\pi)(100x - x^2), \quad 0 < x < 100$

(d) $\dfrac{dA}{dx} = \dfrac{2}{\pi}(100 - 2x)$

$= 0$ when $x = 50$

The maximum value is approximately 1592 when $x = 50$.

(e)

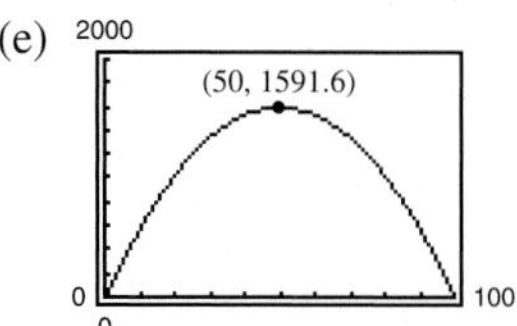

33. $18 \times 18 \times 36$ in. **35.** $32\pi r^3/81$

37. No. The volume changes because the shape of the container changes when squeezed.

39. $r = \sqrt[3]{21/(2\pi)} \approx 1.50$ ($h = 0$, so the solid is a sphere.)

41. Side of square: $\dfrac{10\sqrt{3}}{9 + 4\sqrt{3}}$; Side of triangle: $\dfrac{30}{9 + 4\sqrt{3}}$

43. $w = (20\sqrt{3})/3$ in., $h = (20\sqrt{6})/3$ in. **45.** $\theta = \pi/4$

47. 18 trees; 1296 apples

49. One mile from the nearest point on the coast **51.** 8%

53. $y = \frac{64}{141}x$; $S \approx 6.1$ mi **55.** $y = \frac{3}{10}x$; $S_3 \approx 4.50$ mi

57. (a) $40,000 (b) $s = 20$

Section 5.8 (page 389)

1. $T(x) = 4x - 4$

x	1.9	1.99	2	2.01	2.1
$f(x)$	3.610	3.960	4	4.040	4.410
$T(x)$	3.600	3.960	4	4.040	4.400

3. $T(x) = 80x - 128$

x	1.9	1.99	2	2.01	2.1
$f(x)$	24.761	31.208	32	32.808	40.841
$T(x)$	24.000	31.200	32	32.800	40.000

5. $\Delta y = 0.331$; $dy = 0.3$ **7.** $\Delta y = -0.039$; $dy = -0.040$

9. $6x\,dx$ **11.** $12x^2\,dx$ **13.** $-\dfrac{3}{(2x - 1)^2}\,dx$

15. $1/(2\sqrt{x})\,dx$ **17.** $\dfrac{1 - 2x^2}{\sqrt{1 - x^2}}\,dx$

19. (a) 0.9 (b) 1.04 **21.** (a) 1.05 (b) 0.98

23. (a) 8.035 (b) 7.95 **25.** $\pm\frac{5}{8}$ in.² **27.** $\pm 8\pi$ in.²

29. (a) $\frac{5}{6}$% (b) 1.25% **31.** (a) 2.8% (b) 1.5%

33. $1160; About 2.6%

35. (a) $\Delta p = -0.25 = dp$ (b) $\Delta p = -0.25 = dp$

37. 80π cm³ **39.** (a) $\frac{1}{4}$% (b) 216 sec = 3.6 min

41. $f(x) = \sqrt{x}$, $dy = \dfrac{1}{2\sqrt{x}}\,dx$

$f(99.4) \approx \sqrt{100} + \dfrac{1}{2\sqrt{100}}(-0.6) = 9.97$

Calculator: 9.97

43. $f(x) = \sqrt[4]{x}$, $dy = \dfrac{1}{4x^{3/4}}\,dx$

$f(624) \approx \sqrt[4]{625} + \dfrac{1}{4(625)^{3/4}}(-1) = 4.998$

Calculator: 4.998

45. $y - f(0) = f'(0)(x - 0)$

$y - 2 = \frac{1}{4}x$

$y = 2 + x/4$

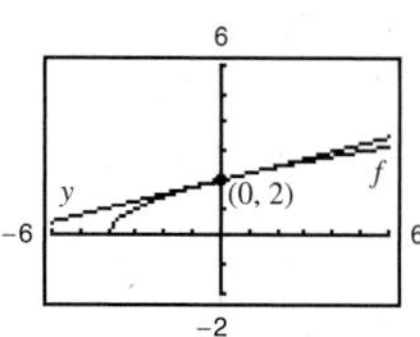

47. The value of dy becomes closer to the value of Δy as Δx decreases.

49. (a) $f(x) = \sqrt{x}$; $dy = \dfrac{1}{2\sqrt{x}}\,dx$

$f(4.02) \approx \sqrt{4} + \dfrac{1}{2\sqrt{4}}(0.02) = 2 + \dfrac{1}{4}(0.02)$

(b) $f(x) = \tan x$; $dy = \sec^2 x\,dx$

$f(0.05) \approx \tan 0 + \sec^2(0)\,(0.05) = 0 + 1(0.05)$

51. True **53.** True

Review Exercises (page 392)

1. Let f be defined at c. If $f'(c) = 0$ or if f' is undefined at c, then c is a critical number of f.

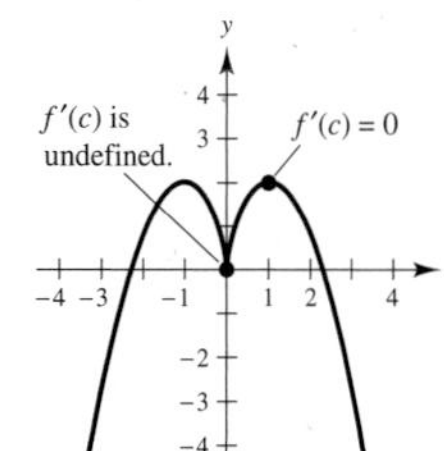

3. Maximum: $(0, 0)$

Minimum: $\left(-\dfrac{5}{2}, -\dfrac{25}{4}\right)$

5. Maximum: $(0, 0)$

Minimum: $\left(\dfrac{10}{3}, \dfrac{10\sqrt{15}}{9}\right)$

7. $f(0) \neq f(4)$ **9.** Not continuous on $[-2, 2]$

11. (a)

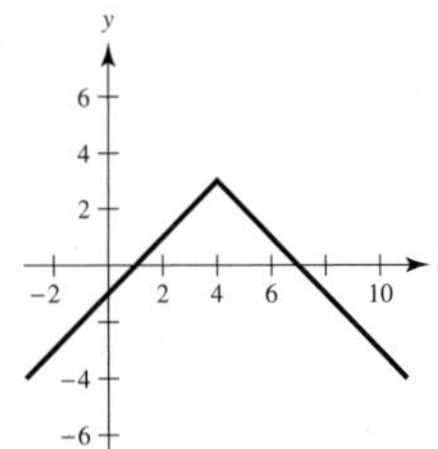

(b) f is not differentiable at $x = 4$.

13. $f'\left(\frac{2744}{729}\right) = \frac{3}{7}$ **15.** f is not differentiable at $x = 5$.

17. f is not defined for $x < 0$. **19.** $c = \dfrac{x_1 + x_2}{2}$

21. Critical number: $x = -\frac{3}{2}$
Increasing on $\left(-\frac{3}{2}, \infty\right)$; Decreasing on $\left(-\infty, -\frac{3}{2}\right)$

23. Critical numbers: $x = 1, \frac{7}{3}$
Increasing on $(-\infty, 1), \left(\frac{7}{3}, \infty\right)$; Decreasing on $\left(1, \frac{7}{3}\right)$

25. Critical number: $x = 1$
Increasing on $(1, \infty)$; Decreasing on $(0, 1)$

27. Relative maximum: $\left(-\dfrac{\sqrt{15}}{6}, \dfrac{5\sqrt{15}}{9}\right)$
Relative minimum: $\left(\dfrac{\sqrt{15}}{6}, -\dfrac{5\sqrt{15}}{9}\right)$

29. Relative minimum: $(2, -12)$

31. $(3, -54)$; Concave upward: $(3, \infty)$;
Concave downward: $(-\infty, 3)$

33. $(0, -16)$; Concave upward: $(0, \infty)$;
Concave downward: $(-\infty, 0)$

35. Relative maxima: $\left(\sqrt{2}/2, 1/2\right), \left(-\sqrt{2}/2, 1/2\right)$
Relative minimum: $(0, 0)$

37.

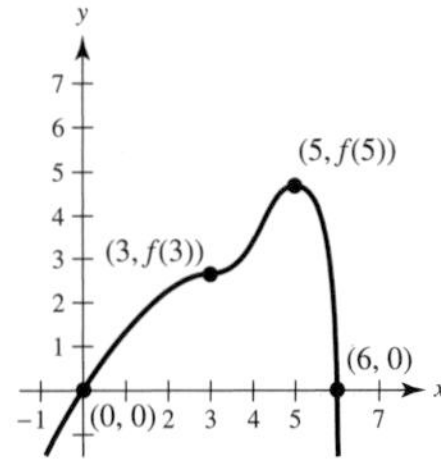

39. Increasing and concave down

41. 8 **43.** $\frac{2}{3}$ **45.** $\frac{1}{2}$

47. Vertical asymptote: $x = 0$; Horizontal asymptote: $y = -2$

49. Vertical asymptote: $x = 4$; Horizontal asymptote: $y = 2$

51.

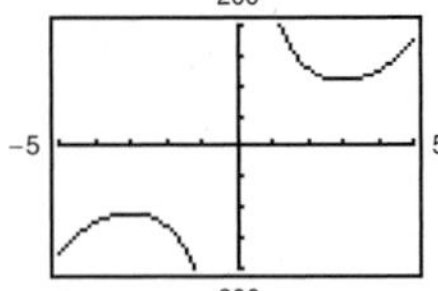

Vertical asymptote: $x = 0$
Relative minimum: $(3, 108)$
Relative maximum: $(-3, -108)$

53.

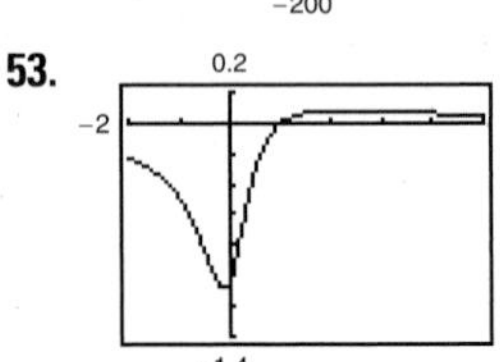

Horizontal asymptote: $y = 0$
Relative minimum:
$(-0.155, -1.077)$
Relative maximum:
$(2.155, 0.077)$

55.

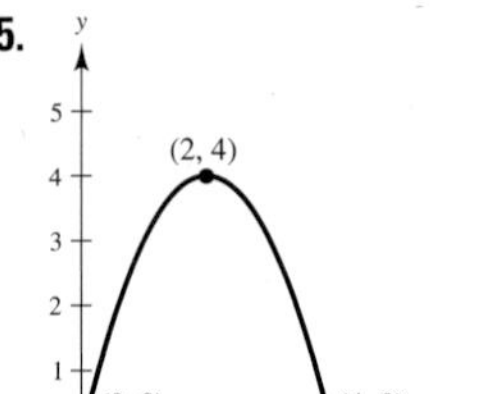

57.

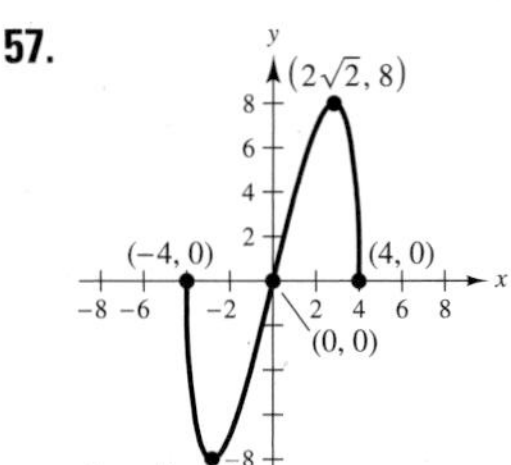

59.

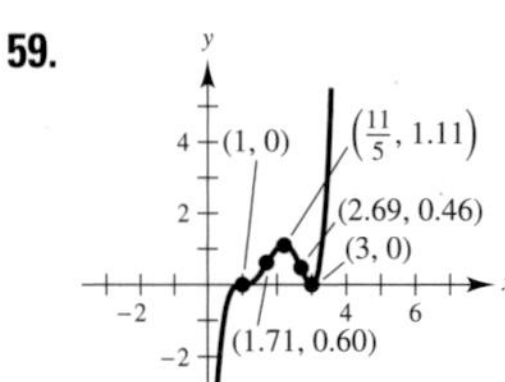

61.

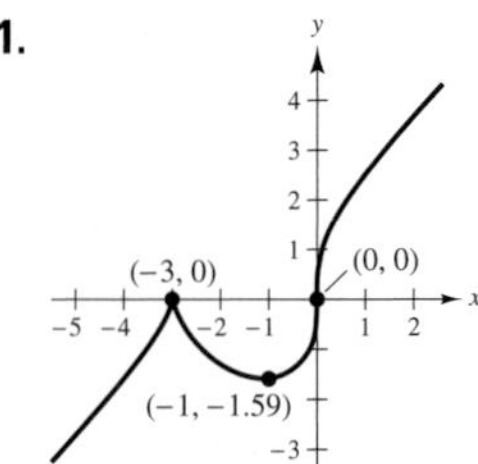

63.

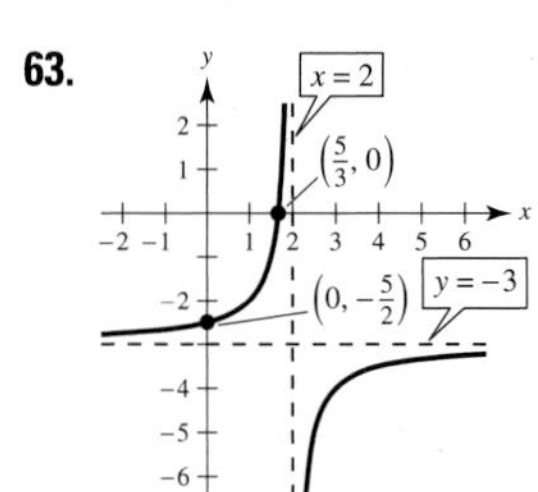

65.

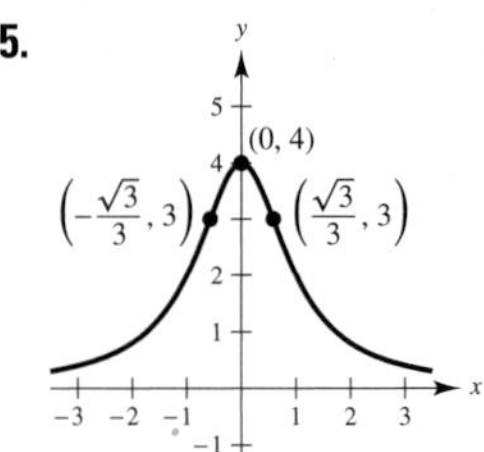

67.

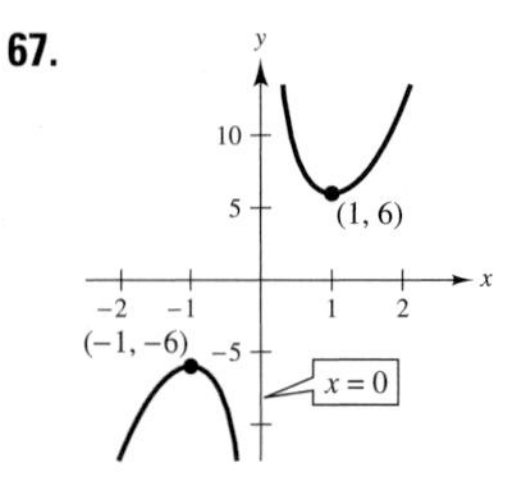

69.

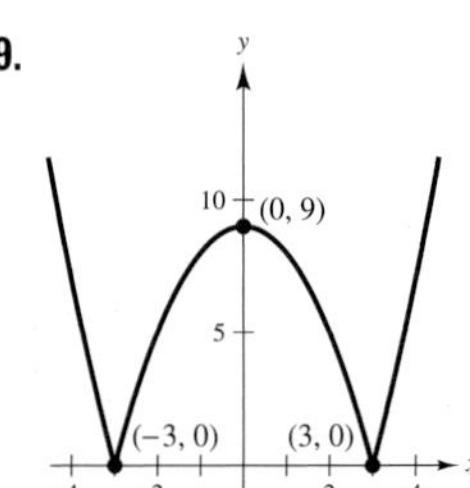

71. (a) and (b) Maximum: $(1, 3)$
Minimum: $(1, 1)$

73. $t \approx 4.92 \approx$ 4:55 P.M.; $d \approx 64$ km **75.** $(0, 0), (5, 0), (0, 10)$

77. Proof **79.** $v \approx 54.77$ mi/h **81.** $dy = 18x(3x^2 - 2)^2\,dx$

83. $dS = \pm 1.8\pi\text{ cm}^2, \dfrac{dS}{S} \times 100 \approx \pm 0.56\%$
$dV = \pm 8.1\pi\text{ cm}^3, \dfrac{dV}{V} \times 100 \approx \pm 0.83\%$

Chapter Test (page 394)

1. Minimum: $(-3, -54)$; Maximum: $(3, 18)$

2. (a) Minimum: $(0, -4)$; Maximum: $(3, 5)$
(b) Minimum: $(0, -4)$
(c) Maximum: $(3, 5)$
(d) No extrema

3. $f'(-1) = 0$ **4.** $f'\left(-\sqrt[3]{2}\right) = 0$

5. Increasing on $\left(-\sqrt{2}, \sqrt{2}\right)$
Decreasing on $\left(-2, -\sqrt{2}\right)$ and $\left(\sqrt{2}, 2\right)$

6. (a) Critical numbers: $x = -1, 3$
(b) Increasing on $(-\infty, -1)$ and $(3, \infty)$; Decreasing on $(-1, 3)$
(c) Relative maximum: $(-1, 5)$; Relative minimum: $(3, -27)$

7. Concave upward: $\left(-\infty, -\frac{\sqrt{3}}{3}\right), \left(\frac{\sqrt{3}}{3}, \infty\right)$

Concave downward: $\left(-\frac{\sqrt{3}}{3}, \frac{\sqrt{3}}{3}\right)$

8. Point of inflection: $(2, -64)$

Concave downward: $(-\infty, 2)$

Concave upward: $(2, \infty)$

9. 0 **10.** 1 **11.** ∞ **12.** 1

13. **14.**

(−8, 8) (−27, 0) (0, 0)

15. (2, 4) **16.** 600 m × 300 m **17.** $x = 27.5$ in., $y = 55$ in.

18. $\Delta y = -0.03940399$, $dy = -0.04$

$\Delta y \approx dy$

19. $\pm 7\pi$ in.^{2}

P.S. Problem Solving (page 395)

1. Choices of a may vary.

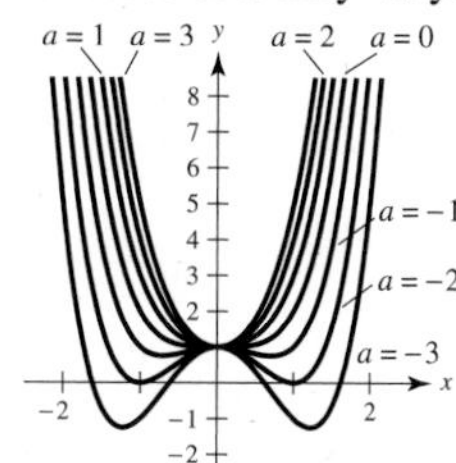

(a) One relative minimum at (0, 1) for $a \geq 0$

(b) One relative maximum at (0, 1) for $a < 0$

(c) Two relative minima for $a < 0$ when $x = \pm\sqrt{-a/2}$

(d) If $a < 0$, there are three critical points; if $a \geq 0$, there is only one critical point.

3. All c, where c is a real number **5–7.** Proofs

9. (a) $p + 2\sqrt{pq} + q$ (b) $4pq$ (c) $\left(p^{2/3} + q^{2/3}\right)^{3/2}$

11. Minimum: $\left(\sqrt{2} - 1\right)d$; There is no maximum.

13. (a)–(c) Proofs

15. (a)

x	0	0.5	1	2
$\sqrt{1 + x}$	1	1.2247	1.4142	1.7321
$\frac{1}{2}x + 1$	1	1.25	1.5	2

(b) Proof

17. Maximum area $= \frac{1}{2}(\text{Area } \triangle ABC)$

Yes. To solve the problem without calculus, divide $\triangle ABC$ into four congruent triangles by joining the midpoints. The parallelogram will consist of two of the triangles.

19. (a) Proof

(b) Domain: $4.25 < x < 8.5$

(c) $x = \frac{51}{8} = 6.375$

(d) For $x = 6.375$, $C \approx 11.0418$ in.

Chapter 6

Section 6.1 (page 404)

1–3. Proofs **5.** $y = 3t^3 + C$ **7.** $y = \frac{2}{5}x^{5/2} + C$

	Original Integral	Rewrite	Integrate	Simplify
9.	$\int \sqrt[3]{x}\,dx$	$\int x^{1/3}\,dx$	$\frac{x^{4/3}}{4/3} + C$	$\frac{3}{4}x^{4/3} + C$
11.	$\int \frac{1}{x\sqrt{x}}\,dx$	$\int x^{-3/2}\,dx$	$\frac{x^{-1/2}}{-1/2} + C$	$-\frac{2}{\sqrt{x}} + C$
13.	$\int \frac{1}{2x^3}\,dx$	$\frac{1}{2}\int x^{-3}\,dx$	$\frac{1}{2}\left(\frac{x^{-2}}{-2}\right) + C$	$-\frac{1}{4x^2} + C$

15. $\frac{1}{2}x^2 + 7x + C$ **17.** $x^2 - x^3 + C$ **19.** $\frac{1}{6}x^6 + x + C$

21. $\frac{2}{5}x^{5/2} + x^2 + x + C$ **23.** $\frac{3}{5}x^{5/3} + C$

25. $-1/(4x^4) + C$

27. $\frac{2}{3}x^{3/2} + 12x^{1/2} + C = \frac{2}{3}x^{1/2}(x + 18) + C$

29. $x^3 + \frac{1}{2}x^2 - 2x + C$ **31.** $\frac{2}{7}y^{7/2} + C$ **33.** $x + C$

35.

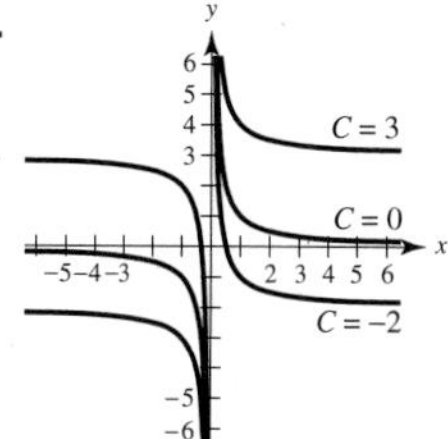

37. Answers will vary.

Example:

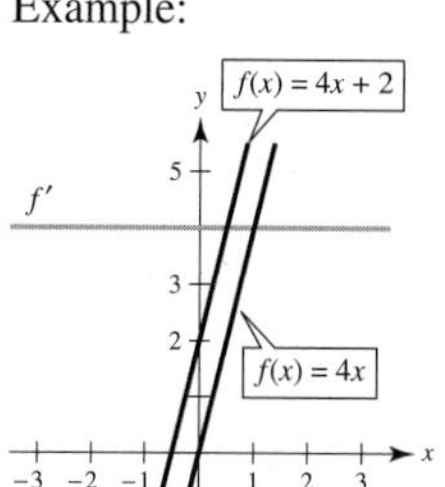

39. Answers will vary.

Example:

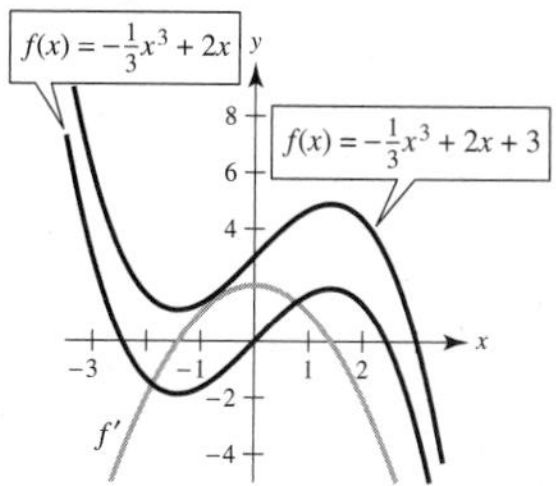

41. $y = x^2 - x + 1$ **43.** $y = x^3 - x + 2$

45. (a) Answers will vary.

Example:

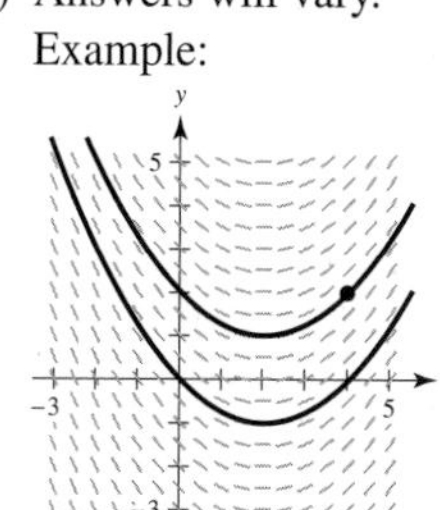

(b) $y = \frac{1}{4}x^2 - x + 2$

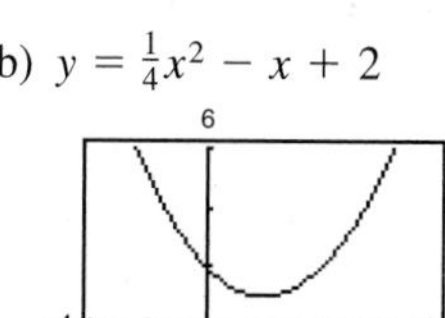

47. (a)

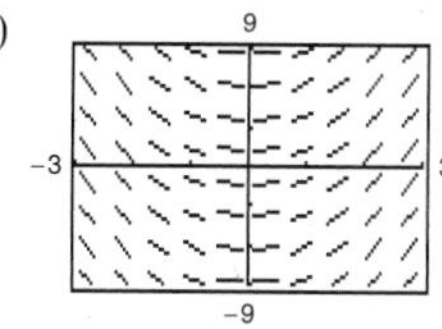

(b) $y = x^2 - 6$

(c)

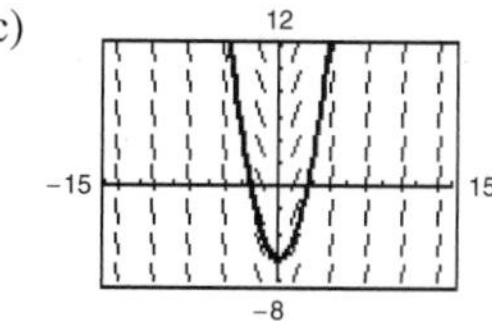

49. $f(x) = 3x^2 + 8$ **51.** $h(t) = 2t^4 + 5t - 11$

53. $f(x) = x^2 + x + 4$

55. (a) $h(t) = \frac{3}{4}t^2 + 5t + 12$ (b) 69 cm

57.

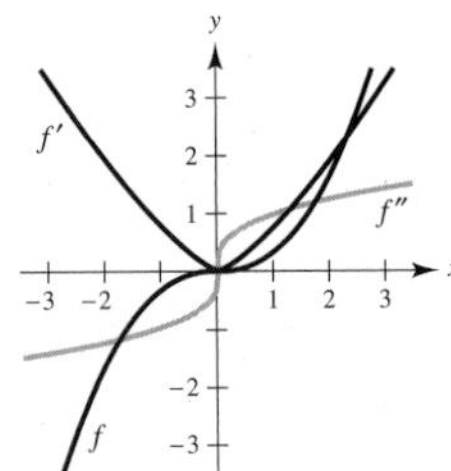

59. 62.25 ft **61.** $v_0 \approx 187.617$ ft/sec

63. $v(t) = -9.8t + C_1 = -9.8t + v_0$
$f(t) = -4.9t^2 + v_0 t + C_2 = -4.9t^2 + v_0 t + s_0$

65. 7.1 m **67.** 320 m; -32 m/sec

69. (a) $v(t) = 3t^2 - 12t + 9$; $a(t) = 6t - 12$
(b) (0, 1), (3, 5) (c) -3

71. $a(t) = -1/(2t^{3/2})$; $x(t) = 2\sqrt{t} + 2$

73. (a) About 73.33 ft
(b) About 117.33 ft
(c)

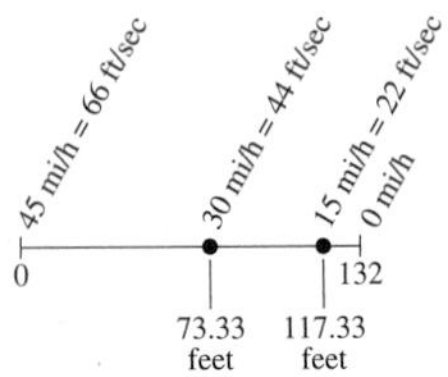

It takes 1.333 seconds to reduce the speed from 45 mi/h to 30 mi/h, 1.333 seconds to reduce the speed from 30 mi/h to 15 mi/h, and 1.333 seconds to reduce the speed from 15 mi/h to 0 mi/h. Each time, less distance is needed to reach the next speed reduction.

75. About 7.45 ft/sec^2

77. (a) $v(t) = 0.6139t^3 - 5.525t^2 + 0.05t + 66.0$ (b) 198 ft

79. True **81.** True

83. False. For example,
$\int x \cdot x\,dx \neq \int x\,dx \cdot \int x\,dx$ because
$\frac{x^3}{3} + C \neq \left(\frac{x^2}{2} + C_1\right)\left(\frac{x^2}{2} + C_2\right)$.

85.

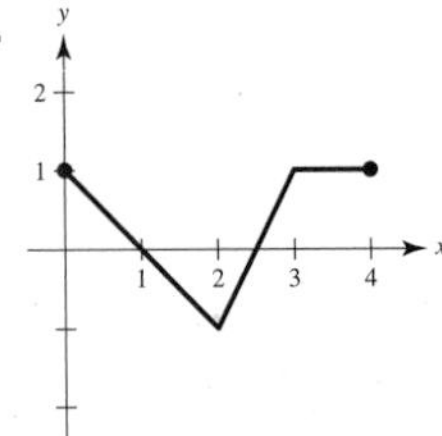

87. Proof

Section 6.2 (page 416)

1. 75 **3.** $\frac{158}{85}$ **5.** $4c$ **7.** $\sum_{i=1}^{11} \frac{1}{5i}$ **9.** $\sum_{j=1}^{6} \left[7\left(\frac{j}{6}\right) + 5\right]$

11. $\frac{2}{n}\sum_{i=1}^{n}\left[\left(\frac{2i}{n}\right)^3 - \left(\frac{2i}{n}\right)\right]$ **13.** $\frac{3}{n}\sum_{i=1}^{n}\left[2\left(1 + \frac{3i}{n}\right)^2\right]$ **15.** 84

17. 1200 **19.** 2470 **21.** 12,040 **23.** 2930

25. (a) (b)

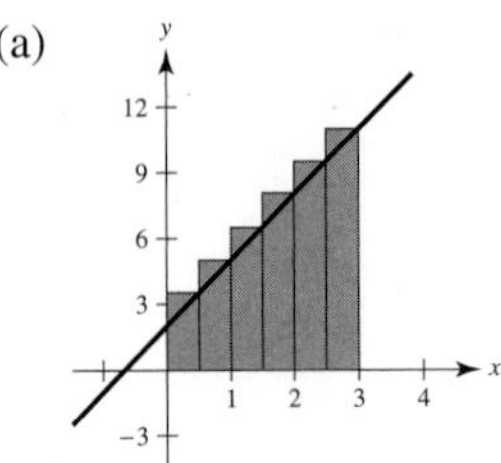

Area ≈ 21.75 Area ≈ 17.25

27. The area of the shaded region falls between 12.5 square units and 16.5 square units.

29. The area of the shaded region falls between 7 square units and 11 square units.

31. $\frac{81}{4}$ **33.** 9 **35.** $A \approx S \approx 0.768$
$A \approx s \approx 0.518$
37. $A \approx S \approx 0.746$
$A \approx s \approx 0.646$

39. $(n + 2)/n$
$n = 10$: $S = 1.2$
$n = 100$: $S = 1.02$
$n = 1000$: $S = 1.002$
$n = 10{,}000$: $S = 1.0002$

41. $[2(n + 1)(n - 1)]/n^2$
$n = 10$: $S = 1.98$
$n = 100$: $S = 1.9998$
$n = 1000$: $S = 1.999998$
$n = 10{,}000$: $S = 1.99999998$

43. $\lim_{n\to\infty}\left[\frac{12(n + 1)}{n}\right] = 12$ **45.** $\lim_{n\to\infty}\frac{1}{6}\left(\frac{2n^3 - 3n^2 + n}{n^3}\right) = \frac{1}{3}$

47. $\lim_{n\to\infty}[(3n + 1)/n] = 3$

49. (a)

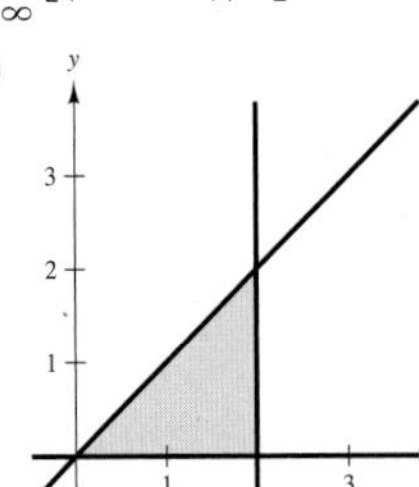

(b) $\Delta x = (2 - 0)/n = 2/n$

(c) $s(n) = \sum_{i=1}^{n} f(x_{i-1})\,\Delta x$
$= \sum_{i=1}^{n}[(i - 1)(2/n)](2/n)$

(d) $S(n) = \sum_{i=1}^{n} f(x_i)\,\Delta x$
$= \sum_{i=1}^{n}[i(2/n)](2/n)$

(e)

n	5	10	50	100
$s(n)$	1.6	1.8	1.96	1.98
$S(n)$	2.4	2.2	2.04	2.02

(f) $\lim_{n\to\infty}\sum_{i=1}^{n}[(i - 1)(2/n)](2/n) = 2$; $\lim_{n\to\infty}\sum_{i=1}^{n}[i(2/n)](2/n) = 2$

51. $A = 3$ **53.** $A = \frac{7}{3}$

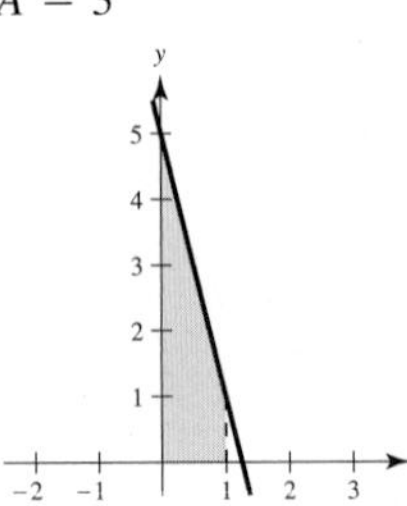

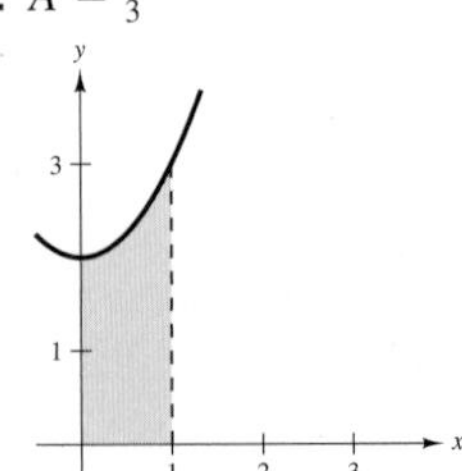

55. $A = 54$

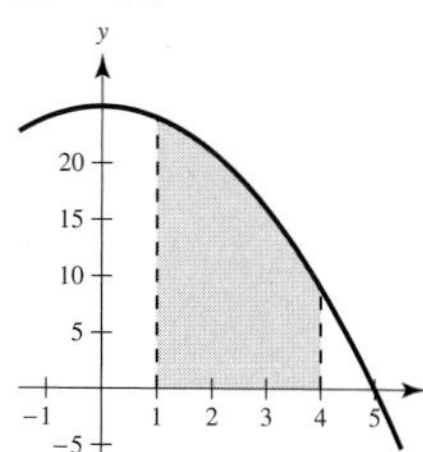

57. $A = 34$

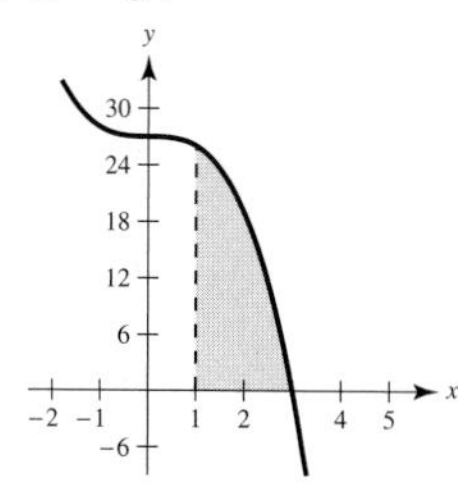

59. $A = \frac{2}{3}$

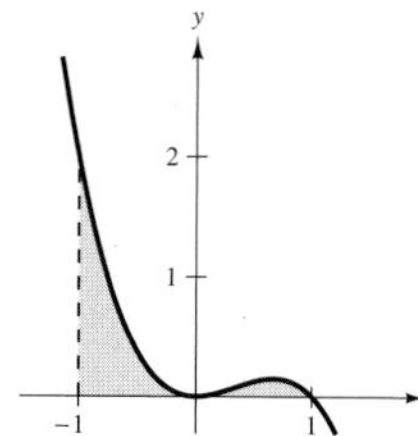

61. $A = 8$

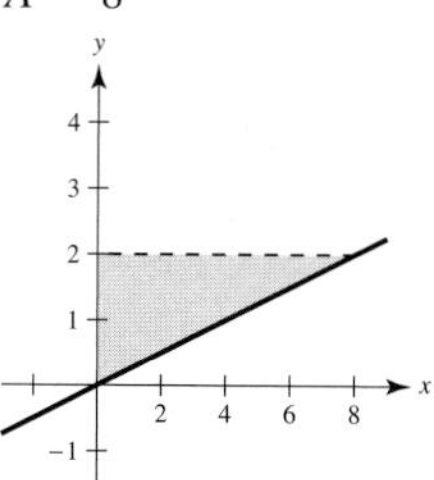

63. $A = \frac{125}{3}$

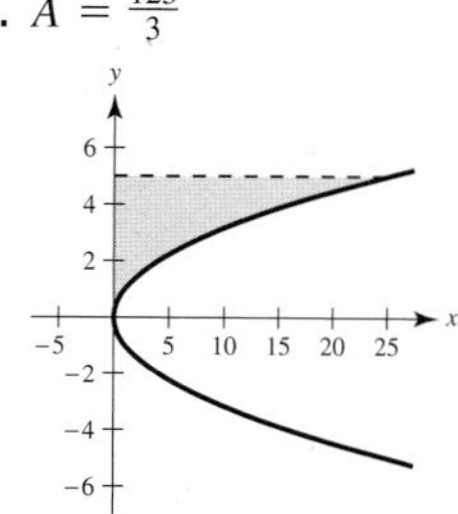

65. $A = \frac{44}{3}$

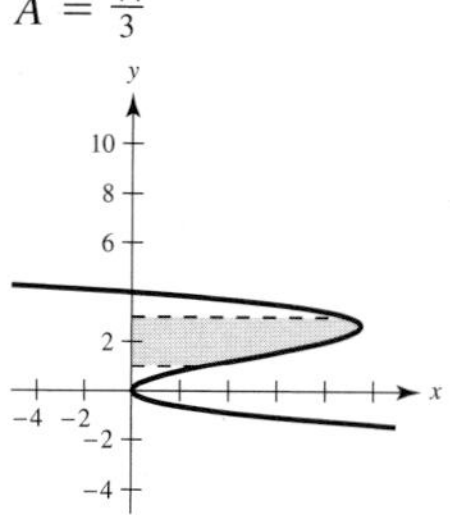

67. $\frac{69}{8}$ **69.** 0.673

71.

n	4	8	12	16	20
Approximate Area	5.3838	5.3523	5.3439	5.3403	5.3384

73.

n	4	8	12	16	20
Approximate Area	2.3397	2.3755	2.3824	2.3848	2.3860

75. You can use the line $y = x$ bounded by $x = a$ and $x = b$. The sum of the areas of the inscribed rectangles in the figure below is the lower sum.

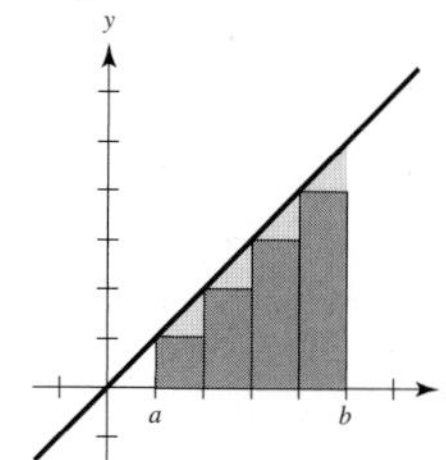

The sum of the areas of the circumscribed rectangles in the figure below is the upper sum.

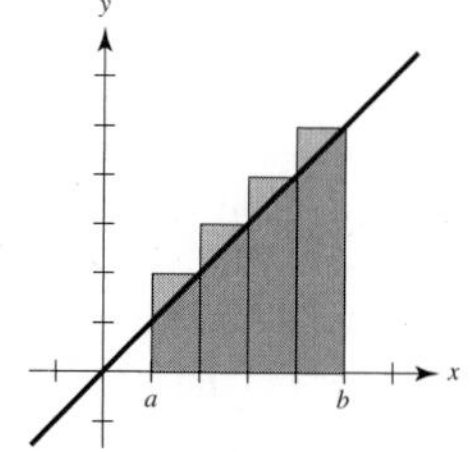

The rectangles in the first graph do not contain all of the area of the region, and the rectangles in the second graph cover more than the area of the region. The exact value of the area lies between these two sums.

77. (a)

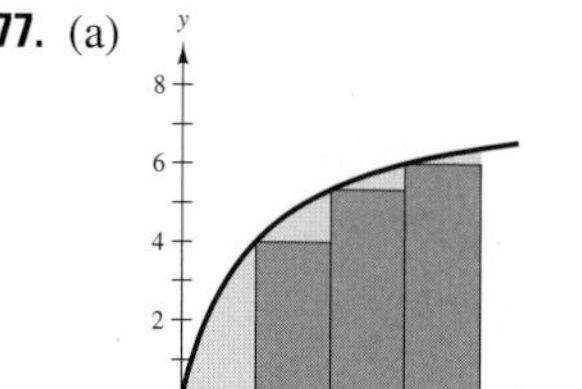

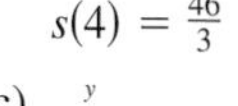

$s(4) = \frac{46}{3}$

(b)

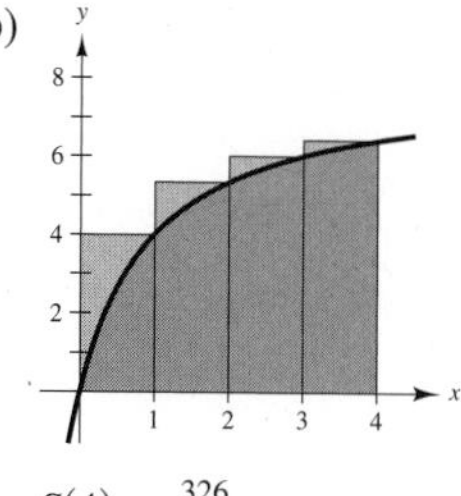

$S(4) = \frac{326}{15}$

(c)

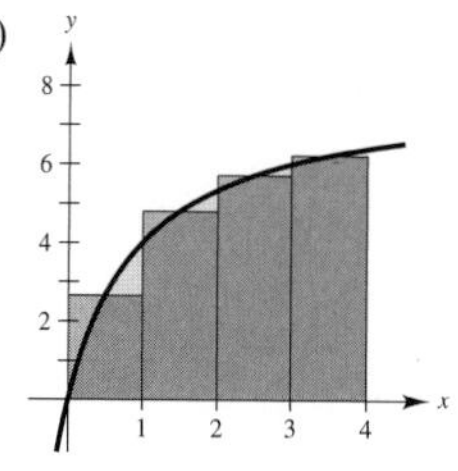

$M(4) = \frac{6112}{315}$

(d) Proof

(e)

n	4	8	20	100	200
$s(n)$	15.333	17.368	18.459	18.995	19.060
$S(n)$	21.733	20.568	19.739	19.251	19.188
$M(n)$	19.403	19.201	19.137	19.125	19.125

(f) Because f is an increasing function, $s(n)$ is always increasing and $S(n)$ is always decreasing.

79. b **81.** True

83. Suppose there are n rows and $n + 1$ columns. The stars on the left total $1 + 2 + \cdots + n$, as do the stars on the right. There are $n(n + 1)$ stars in total. So, $2[1 + 2 + \cdots + n] = n(n + 1)$ and $1 + 2 + \cdots + n = [n(n + 1)]/2$.

85. (a) $y = (-4.09 \times 10^{-5})x^3 + 0.016x^2 - 2.67x + 452.9$

(b)

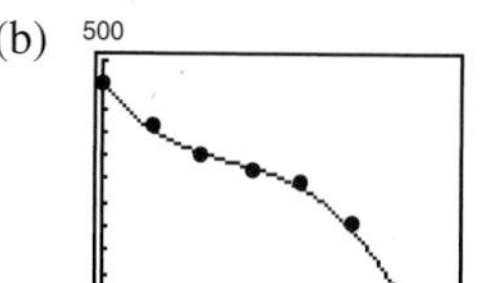

(c) 76,897.5 ft^2

Section 6.3 (page 427)

1. $2\sqrt{3} \approx 3.464$ **3.** 32 **5.** 0 **7.** $\frac{10}{3}$

9. $\int_{-1}^{5} (3x + 10)\, dx$ **11.** $\int_{0}^{3} \sqrt{x^2 + 4}\, dx$ **13.** $\int_{0}^{4} 5\, dx$

15. $\int_{-4}^{4} (4 - |x|)\, dx$ **17.** $\int_{-5}^{5} (25 - x^2)\, dx$ **19.** $\int_{0}^{2} y^3\, dy$

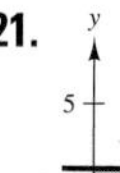

21.

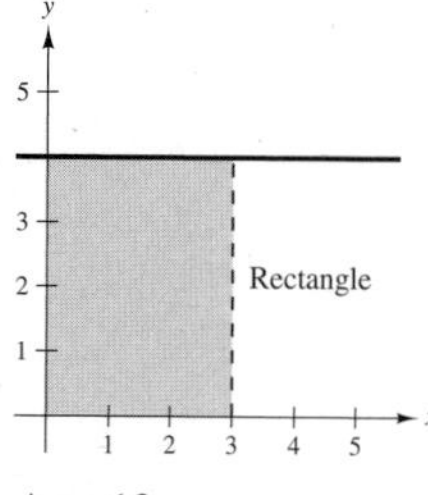

$A = 12$

23.

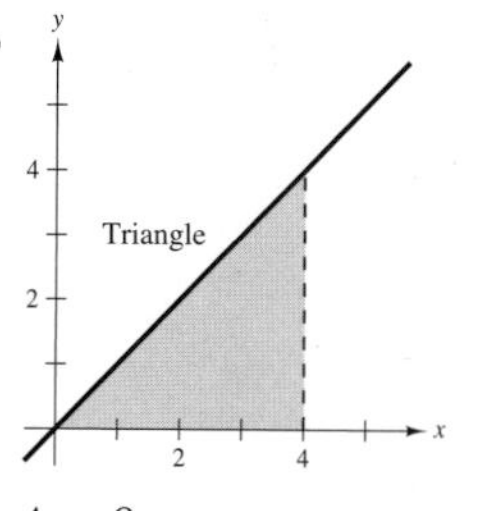

$A = 8$

25.

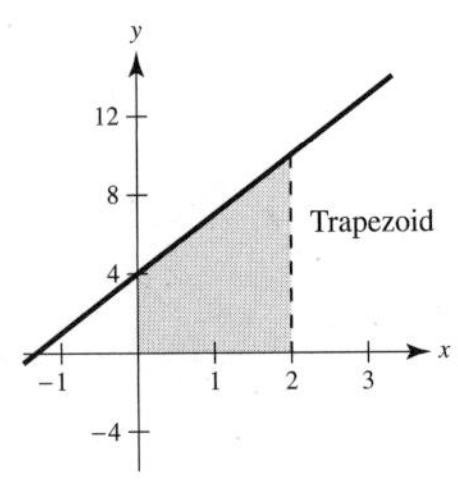

$A = 14$

27.

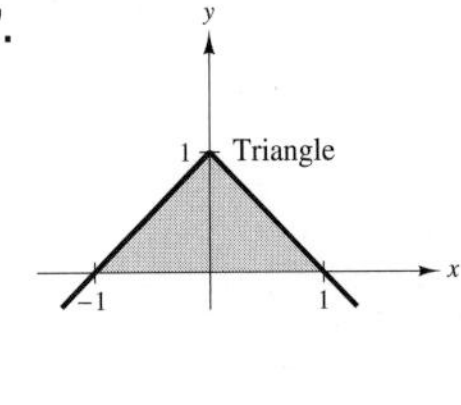

$A = 1$

29.

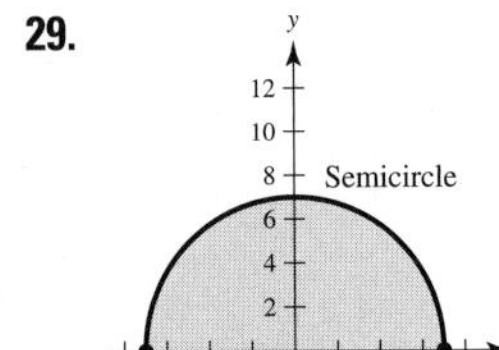

$A = 49\pi/2$

31. -6 **33.** 48 **35.** -12

37. 16 **39.** (a) 13 (b) -10 (c) 0 (d) 30

41. (a) 8 (b) -12 (c) -4 (d) 30

43. (a) $-\pi$ (b) 4 (c) $-(1 + 2\pi)$ (d) $3 - 2\pi$ (e) $5 + 2\pi$ (f) $23 - 2\pi$

45. (a) 14 (b) 4 (c) 8 (d) 0 **47.** 81

49. $\sum_{i=1}^{n} f(x_i)\,\Delta x > \int_1^5 f(x)\,dx$

51. No. There is a discontinuity at $x = 4$. **53.** a

55. True **57.** True **59.** False: $\int_0^2 (-x)\,dx = -2$

61.

n	4	8	12	16	20
$L(n)$	3.6830	3.9956	4.0707	4.1016	4.1177
$M(n)$	4.3082	4.2076	4.1838	4.1740	4.1690
$R(n)$	3.6830	3.9956	4.0707	4.1016	4.1177

63. 272

65. The limit $\lim_{\|\Delta\| \to 0} \sum_{i=1}^{n} f(c_i)\Delta x_i$ does not exist.
This does not contradict Theorem 6.4 because f is not continuous on $[0, 1]$.

67. 1

Section 6.4 (page 441)

1.

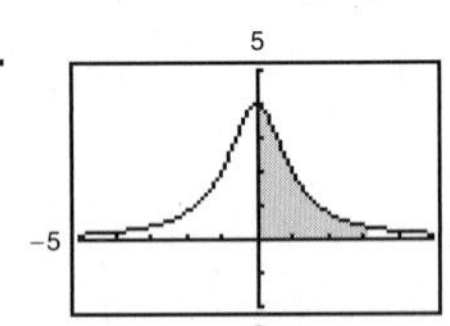

Positive

3.

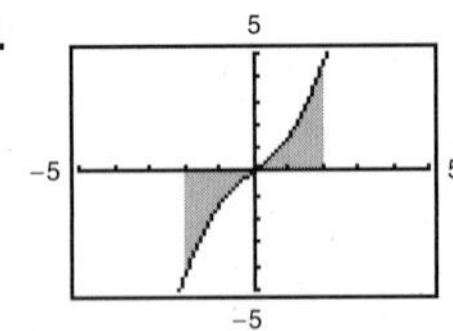

Zero

5. 12 **7.** -2 **9.** $-\frac{10}{3}$ **11.** $\frac{1}{3}$ **13.** $\frac{1}{2}$ **15.** $\frac{2}{3}$

17. -4 **19.** $-\frac{1}{18}$ **21.** $-\frac{27}{20}$ **23.** $\frac{25}{2}$ **25.** $\frac{64}{3}$ **27.** $\frac{1}{6}$

29. $\frac{8}{5}$ **31.** 6 **33.** $\frac{52}{3}$ **35.** 20 **37.** $\frac{32}{3}$

39. $3\sqrt[3]{2}/2 \approx 1.8899$ **41.** 1, 3

43. Average value $= 6$
$x = \pm\sqrt{3} \approx \pm 1.7321$

45. Average value $= \frac{1}{4}$
$x = \sqrt[3]{2}/2 \approx 0.6300$

47. Average value $= -\frac{2}{3}$
$x = 2(2 \pm \sqrt{3})/3;\ x \approx 0.179,\ x \approx 2.488$

49. About 540 ft

51. The Fundamental Theorem of Calculus states that if a function f is continuous on $[a, b]$ and F is an antiderivative of f on $[a, b]$, then $\int_a^b f(x)\,dx = F(b) - F(a)$.

53. $r(t)$ represents the weight in pounds of the dog at time t.
$\int_2^6 r'(t)\,dt$ represents the net change in the weight of the dog from year 2 to year 6.

55. About 0.5318 L

57. (a) $v = -0.00086t^3 + 0.0782t^2 - 0.208t + 0.10$
(b)

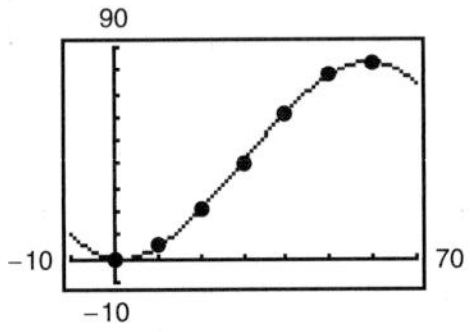

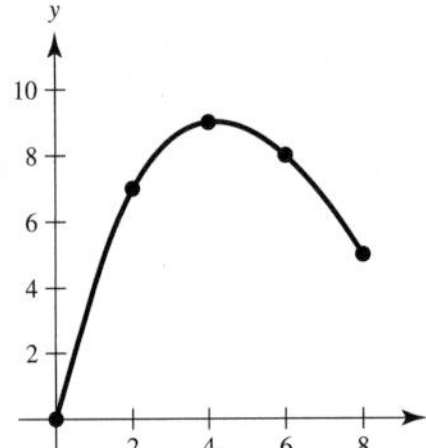

(c) 2475.6 m

59. $F(x) = 2x^2 - 7x$
$F(2) = -6$
$F(5) = 15$
$F(8) = 72$

61. $F(x) = -20/x + 20$
$F(2) = 10$
$F(5) = 16$
$F(8) = \frac{35}{2}$

63. (a) $g(0) = 0,\ g(2) \approx 7,\ g(4) \approx 9,\ g(6) \approx 8,\ g(8) \approx 5$
(b) Increasing: $(0, 4)$; Decreasing: $(4, 8)$
(c) A maximum occurs at $x = 4$.
(d)

65. $\frac{1}{2}x^2 + 2x$ **67.** $\frac{3}{4}x^{4/3} - 12$ **69.** (a) $1 - \frac{1}{x}$ (b) $\frac{1}{x^2}$

71. $x^2 - 2x$ **73.** $\sqrt{x^4 + 1}$ **75.** 8 **77.** $3\sqrt{1 + 27x^3}$

79.

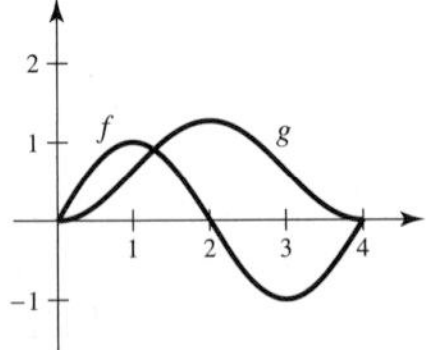

An extremum of g occurs at $x = 2$.

81. (a) $C(x) = 1000(12x^{5/4} + 125)$
(b) $C(1) = \$137{,}000$
$C(5) \approx \$214{,}721$
$C(10) \approx \$338{,}394$

83. (a) $\frac{3}{2}$ ft to the right (b) $\frac{113}{10}$ ft **85.** (a) 0 ft (b) $\frac{63}{2}$ ft

87. 28 units **89.** 2 units

91. $f(x) = x^{-2}$ has a nonremovable discontinuity at $x = 0$.

93. True

95. $f'(x) = \frac{1}{(1/x)^2 + 1}\left(-\frac{1}{x^2}\right) + \frac{1}{x^2 + 1} = 0$
Because $f'(x) = 0$, $f(x)$ is constant.

97. (a) 0 (b) 0 (c) $xf(x) + \int_0^x f(t)\,dt$ (d) 0

Section 6.5 (page 454)

$\int f(g(x))g'(x)\,dx$	$u = g(x)$	$du = g'(x)\,dx$
1. $\int (8x^2 + 1)^2(16x)\,dx$	$8x^2 + 1$	$16x\,dx$
3. $\int x^2\sqrt{x^3 + 1}\,dx$	$x^3 + 1$	$3x^2\,dx$

5. No **7.** $\frac{1}{5}(1 + 6x)^5 + C$
9. $\frac{2}{3}(25 - x^2)^{3/2} + C$ **11.** $\frac{1}{12}(x^4 + 3)^3 + C$
13. $\frac{1}{15}(x^3 - 1)^5 + C$ **15.** $\frac{1}{3}(t^2 + 2)^{3/2} + C$
17. $-\frac{15}{8}(1 - x^2)^{4/3} + C$ **19.** $1/[4(1 - x^2)^2] + C$
21. $-1/[3(1 + x^3)] + C$ **23.** $-\sqrt{1 - x^2} + C$
25. $-\frac{1}{4}(1 + 1/t)^4 + C$ **27.** $\sqrt{2x} + C$
29. $\frac{2}{5}x^{5/2} + \frac{10}{3}x^{3/2} - 16x^{1/2} + C = \frac{1}{15}\sqrt{x}(6x^2 + 50x - 240) + C$
31. $\frac{1}{4}t^4 - 4t^2 + C$
33. $6y^{3/2} - \frac{2}{5}y^{5/2} + C = \frac{2}{5}y^{3/2}(15 - y) + C$
35. $2x^2 - 4\sqrt{16 - x^2} + C$ **37.** $-1/[2(x^2 + 2x - 3)] + C$
39. (a) Answers will vary. (b) $y = -\frac{1}{3}(4 - x^2)^{3/2} + 2$
Example:

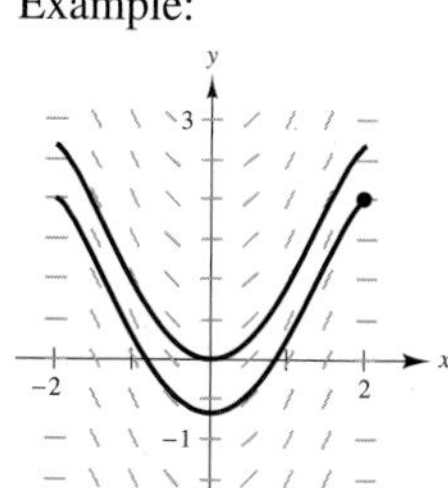

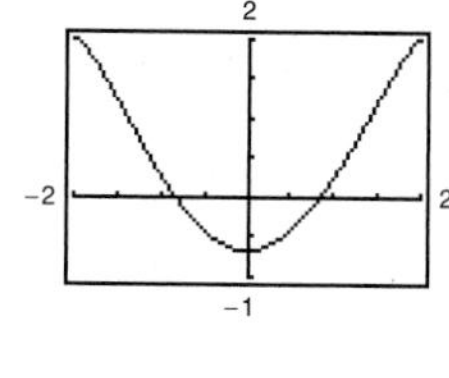

41. $f(x) = \frac{1}{12}(4x^2 - 10)^3 - 8$
43. $\frac{2}{5}(x + 6)^{5/2} - 4(x + 6)^{3/2} + C = \frac{2}{5}(x + 6)^{3/2}(x - 4) + C$
45. $-\left[\frac{2}{3}(1 - x)^{3/2} - \frac{4}{5}(1 - x)^{5/2} + \frac{2}{7}(1 - x)^{7/2}\right] + C = -\frac{2}{105}(1 - x)^{3/2}(15x^2 + 12x + 8) + C$
47. $\frac{1}{8}\left[\frac{2}{5}(2x - 1)^{5/2} + \frac{4}{3}(2x - 1)^{3/2} - 6(2x - 1)^{1/2}\right] + C = \left(\sqrt{2x - 1}/15\right)(3x^2 + 2x - 13) + C$
49. $-x - 1 - 2\sqrt{x + 1} + C$ or $-\left(x + 2\sqrt{x + 1}\right) + C_1$
51. 0 **53.** $12 - \frac{8}{9}\sqrt{2}$ **55.** 2 **57.** $\frac{1}{2}$ **59.** $\frac{4}{15}$ **61.** $\frac{936}{5}$
63. $f(x) = (2x^3 + 1)^3 + 3$ **65.** $f(x) = \sqrt{2x^2 - 1} - 3$
67. 1209/28
69. $\frac{14}{3}$

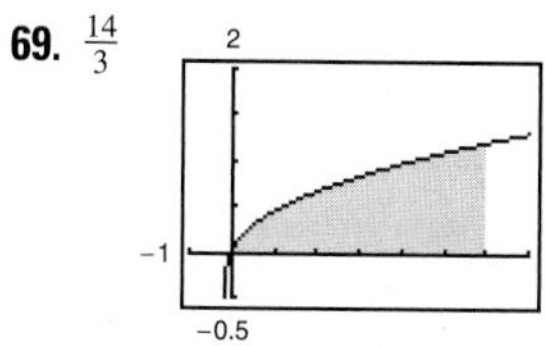

71. $\frac{144}{5}$

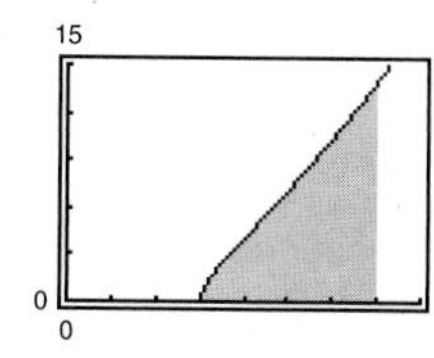

73. $\frac{272}{15}$ **75.** 36 **77.** (a) $\frac{64}{3}$ (b) $\frac{128}{3}$ (c) $-\frac{64}{3}$ (d) 64
79. $2\int_0^3 (4x^2 - 6)\,dx = 36$

81. (a) If $u = 5 - x^2$, then $du = -2x\,dx$ and $\int x(5 - x^2)^3\,dx = -\frac{1}{2}\int(5 - x^2)^3(-2x)\,dx = -\frac{1}{2}\int u^3\,du.$
(b) $f(x) = x(x^2 + 1)^2$ is odd. So, $\int_{-2}^{2} x(x^2 + 1)^2\,dx = 0.$
83. \$250,000
85. (a) $P_{0.50,\,0.75} \approx 35.3\%$ (b) $b \approx 58.6\%$
87. False. $\int (2x + 1)^2\,dx = \frac{1}{6}(2x + 1)^3 + C$
89. True **91.** (a) Proof (b) Proof **93.** Proof

Section 6.6 (page 462)

	Trapezoidal	*Simpson's*	*Exact*
1.	2.7500	2.6667	2.6667
3.	4.2500	4.0000	4.0000
5.	20.2222	20.0000	20.0000
7.	12.6640	12.6667	12.6667
9.	0.3352	0.3334	0.3333

	Trapezoidal	*Simpson's*	*Graphing Utility*
11.	1.6833	1.6222	1.6094
13.	3.2833	3.2396	3.2413
15.	0.3415	0.3720	0.3927
17.	2.2077	2.2103	2.2143
19.	2.3521	2.4385	2.5326

21. Trapezoidal: Linear (1st-degree) polynomials
Simpson's: Quadratic (2nd-degree) polynomials
23. (a) 1.500 (b) 0.000 **25.** (a) 0.01 (b) 0.0005
27. (a) $n = 366$ (b) $n = 26$ **29.** (a) $n = 77$ (b) $n = 8$
31. (a) $n = 633$ (b) $n = 40$ **33.** (a) $n = 130$ (b) $n = 12$
35. (a) 24.5 (b) 25.67 **37.** Answers will vary.
39.

n	$L(n)$	$M(n)$	$R(n)$	$T(n)$	$S(n)$
4	0.8739	0.7960	0.6239	0.7489	0.7709
8	0.8350	0.7892	0.7100	0.7725	0.7803
10	0.8261	0.7881	0.7261	0.7761	0.7818
12	0.8200	0.7875	0.7367	0.7783	0.7826
16	0.8121	0.7867	0.7496	0.7808	0.7836
20	0.8071	0.7864	0.7571	0.7821	0.7841

41. About 10,233.58 ft-lb **43.** 3.1416 **45.** 89,250 m^2
47. Proof **49.** 2.477

Review Exercises (page 465)

1.

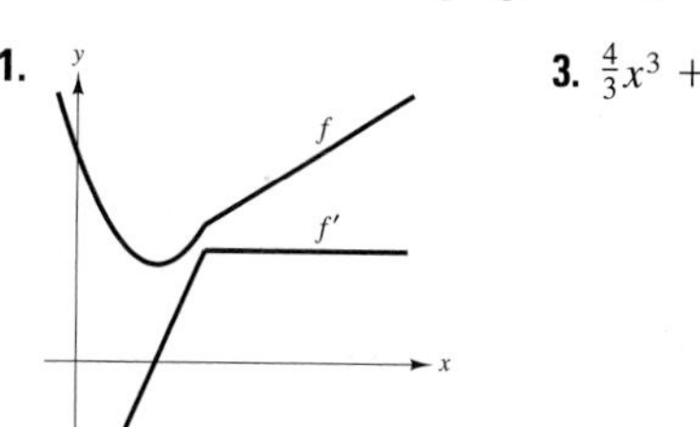

3. $\frac{4}{3}x^3 + \frac{1}{2}x^2 + 3x + C$

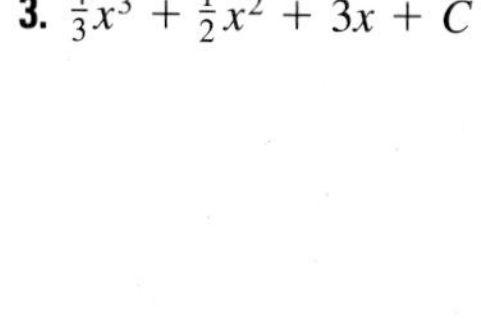

5. $x^2/2 - 4/x^2 + C$ **7.** $\frac{3}{7}x^{7/3} + \frac{9}{4}x^{4/3} + C$
9. $y = 1 - 3x^2$

11. (a) Answers will vary. Example:

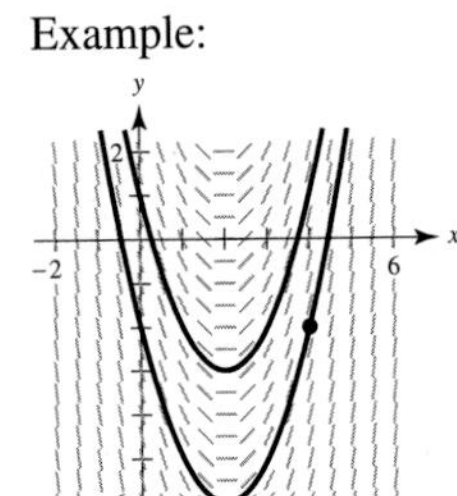

(b) $y = x^2 - 4x - 2$

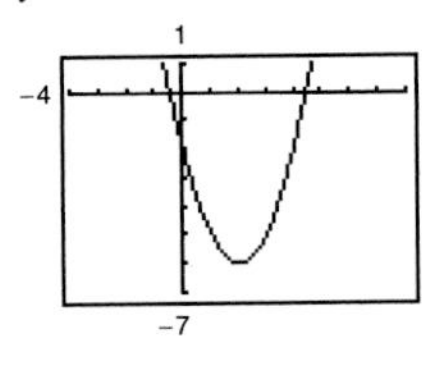

13. 240 ft/sec **15.** (a) 3 sec; 144 ft (b) $\frac{3}{2}$ sec (c) 108 ft

17. $\sum_{n=1}^{10} \frac{1}{3n}$ **19.** 420 **21.** 3310

23. (a) $\sum_{i=1}^{10} (2i - 1)$ (b) $\sum_{i=1}^{n} i^3$ (c) $\sum_{i=1}^{10} (4i + 2)$

25. $9.038 < (\text{Area of region}) < 13.038$

27. $A = 15$ **29.** $A = 12$

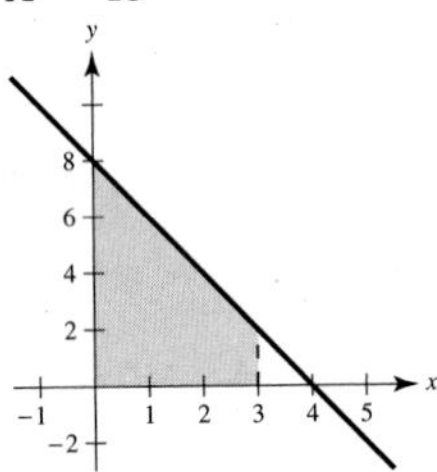

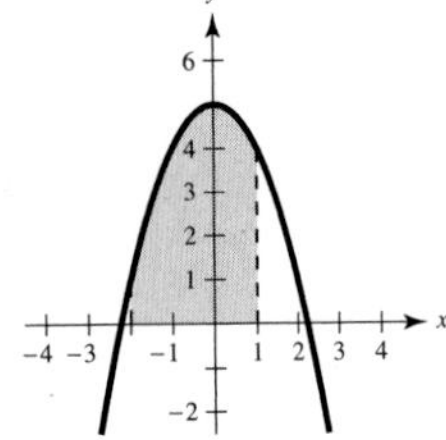

31. $\frac{27}{2}$ **33.** $\int_4^6 (2x - 3)\, dx$ **35.** $\int_{-4}^{0} (2x + 8)\, dx$

37.

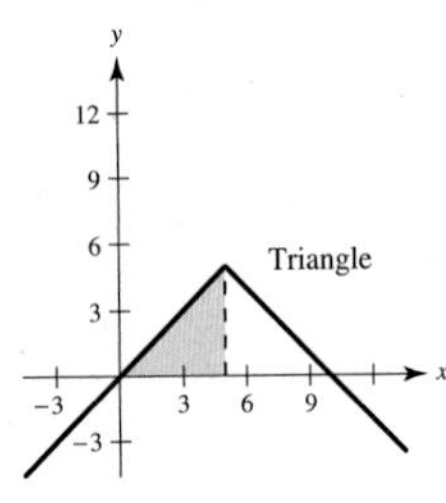

$A = \frac{25}{2}$

39. (a) 17 (b) 7 (c) 9 (d) 84

41. c **43.** 56 **45.** 0 **47.** $\frac{422}{5}$

49.

$A = 10$

51.

$A = \frac{10}{3}$

53.

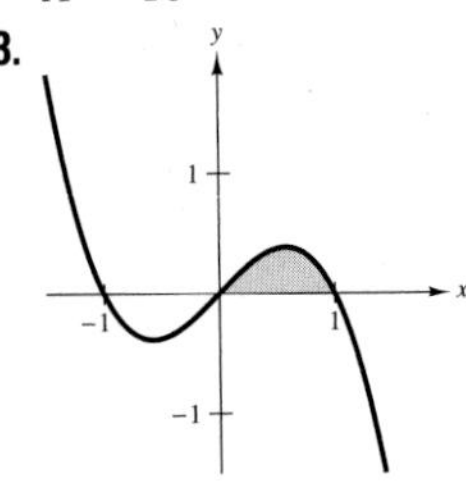

$A = \frac{1}{4}$

55.

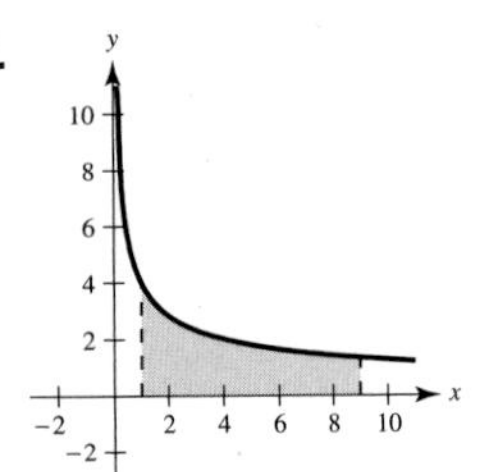

$A = 16$

57. Average value $= \frac{2}{5}$, $x = \frac{25}{4}$

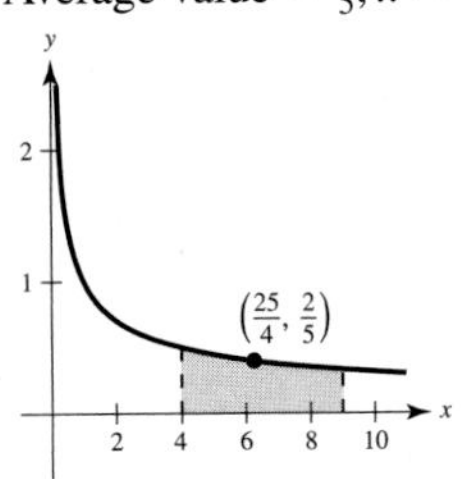

59. $x^2\sqrt{1 + x^3}$ **61.** $-\frac{1}{7}x^7 + \frac{9}{5}x^5 - 9x^3 + 27x + C$

63. $\frac{1}{8}(x^2 + 1)^4 + C$ **65.** $\frac{2}{3}\sqrt{x^3 + 3} + C$

67. $-\frac{1}{30}(1 - 3x^2)^5 + C = \frac{1}{30}(3x^2 - 1)^5 + C$

69. $\frac{2}{21}(x + 5)^{3/2}(3x^2 - 12x + 40) + C$

71. 21/4 **73.** 2 **75.** $28\pi/15$

77. (a) Answers will vary. Example:

(b) $y = -\frac{1}{3}(9 - x^2)^{3/2} + 5$

79. $\frac{468}{7}$

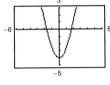

81. Trapezoidal Rule: 0.172; Simpson's Rule: 0.166; Graphing Utility: 0.166

Chapter Test (page 468)

1. $\frac{2}{5}x^{5/2} + \frac{2}{3}x^{3/2} + 2\sqrt{x} + C$ **2.** $\frac{1}{3}(t^2 + 7)^{3/2} + C$

3. $\frac{2}{5}(x + 2)^{5/2} - \frac{4}{3}(x + 2)^{3/2} + C$

4. $f(x) = 6x^2 - 5$ **5.** $f(x) = \frac{1}{20}x^5 + 3x + 5$

6. 1574 **7.** 18 **8.** $\int_{-2}^{2} (4 - x^2)\, dx$

9.

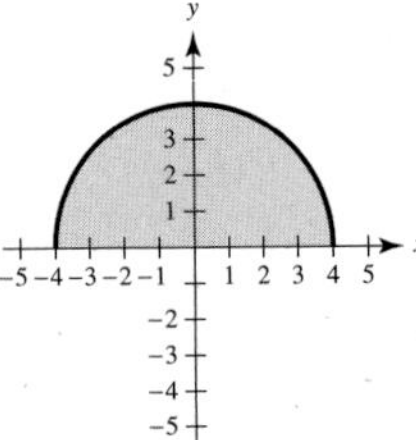

; 8π

10. (a) 10 (b) 14 (c) 12 **11.** $47\frac{1}{3}$

12. $\frac{1}{5}\left(\frac{55}{2} - \frac{32\sqrt{2}}{3} + 2\sqrt{3}\right) \approx 3.1758$

13. $\frac{8}{9}$ **14.** 10 **15.** $85\frac{1}{3}$ **16.** Average value: 18; $\frac{1 + 2\sqrt{7}}{3}$

17. 18 **18.** $f(x) = \frac{1}{16}(6x^2 - 2)^4 - 8$ **19.** 52.8

20. Exact: $\frac{7}{6} \approx 1.1667$; Trapezoidal: $\frac{75}{64} \approx 1.1719$; Simpson's: $\frac{7}{6} \approx 1.1667$

21. Exact: $\frac{52}{3} \approx 17.3333$; Trapezoidal: About 17.2277; Simpson's: About 17.3222

P.S. Problem Solving (page 469)

1. (a) $L(1) = 0$ (b) $L'(x) = 1/x, L'(1) = 1$
(c) $x \approx 2.718$ (d) Proof

3. (a) $\lim_{n\to\infty}\left[\frac{32}{n^5}\sum_{i=1}^{n} i^4 - \frac{64}{n^4}\sum_{i=1}^{n} i^3 + \frac{32}{n^3}\sum_{i=1}^{n} i^2\right]$
(b) $(16n^4 - 16)/(15n^4)$ (c) $16/15$

5. (a) 2.7981; About 0.0007 (b) $\frac{3}{2}$ (c) Proof

7. (a)

Area = 36

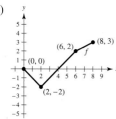

(b) Base = 6, height = 9
Area $= \frac{2}{3}bh = \frac{2}{3}(6)(9) = 36$
(c) Proof

9. (a)

(b)

x	0	1	2	3	4	5	6	7	8
$F(x)$	0	$-\frac{1}{2}$	-2	$-\frac{7}{2}$	-4	$-\frac{7}{2}$	-2	$\frac{1}{4}$	3

(c) $x = 4, 8$ (d) $x = 2$

11. Proof **13.** $\frac{2}{3}$ **15.** $1 \le \int_0^1 \sqrt{1 + x^4}\,dx \le \sqrt{2}$

17. (a) Proof (b) Proof (c) Proof

19. (a) 15.375 gal (b) 22.125 gal
(c) The second answer is larger because the rate of flow is increasing.

21. $a = -4, b = 4$

Chapter 7

Section 7.1 (page 478)

1. algebraic **3.** transformations **5.** $A = P\left(1 + \frac{r}{n}\right)^{nt}$

7. 0.863 **9.** 0.006 **11.** d **12.** c **13.** a **14.** b

15.

x	-2	-1	0	1	2
$f(x)$	4	2	1	0.5	0.25

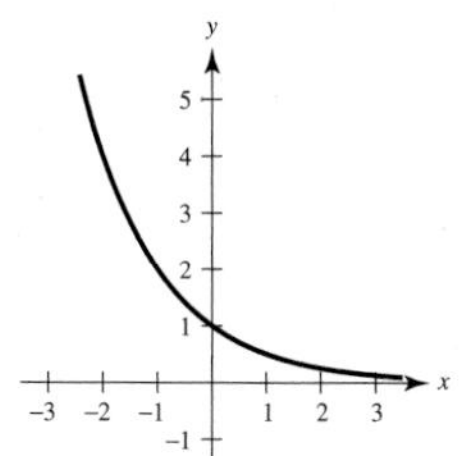

17.

x	-2	-1	0	1	2
$f(x)$	36	6	1	0.167	0.028

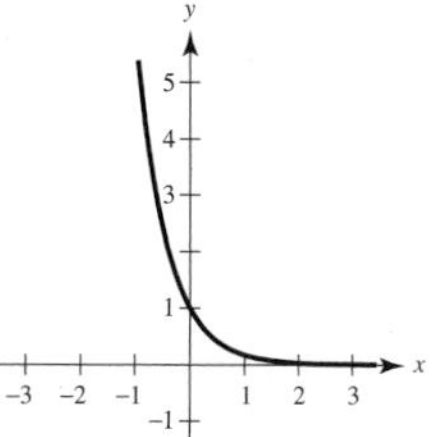

19.

x	-2	-1	0	1	2
$f(x)$	0.125	0.25	0.5	1	2

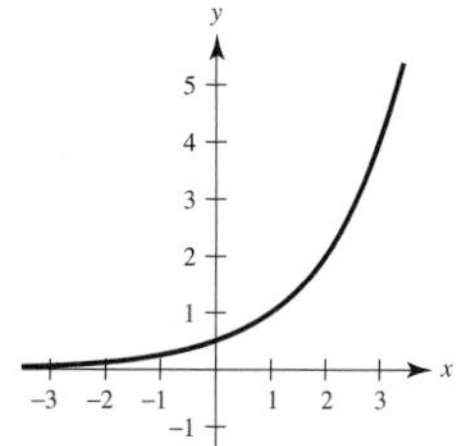

21. Shift the graph of f one unit upward.

23. Reflect the graph of f in the x-axis and shift three units upward.

25. Reflect the graph of f in the origin.

27.

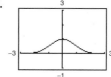

29.

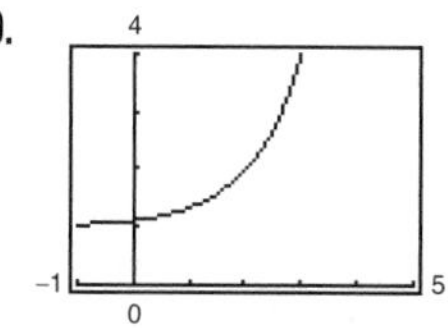

31. 0.472 **33.** 3.857×10^{-22}

35.

x	-2	-1	0	1	2
$f(x)$	0.135	0.368	1	2.718	7.389

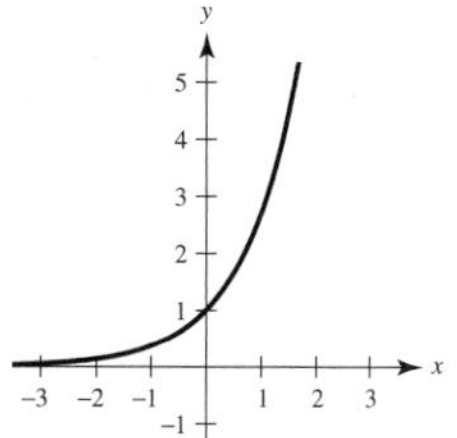

37.

x	-8	-7	-6	-5	-4
$f(x)$	0.055	0.149	0.406	1.104	3

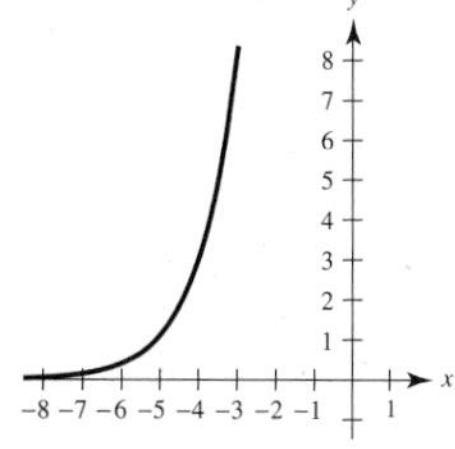

39.

x	-2	-1	0	1	2
$f(x)$	4.037	4.100	4.271	4.736	6

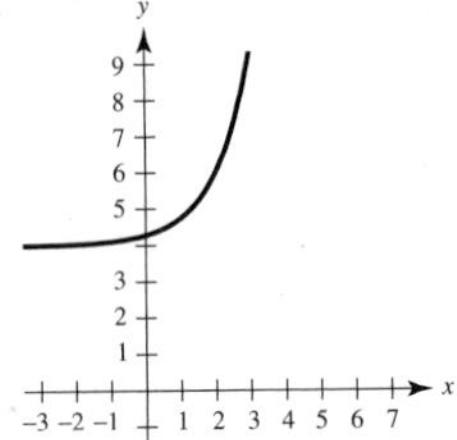

41.

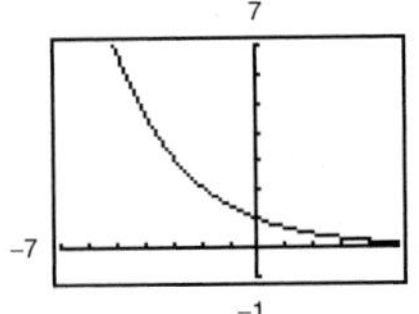

43.

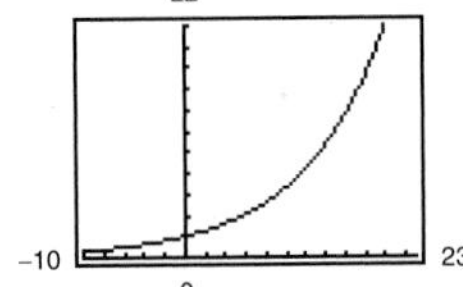

45.

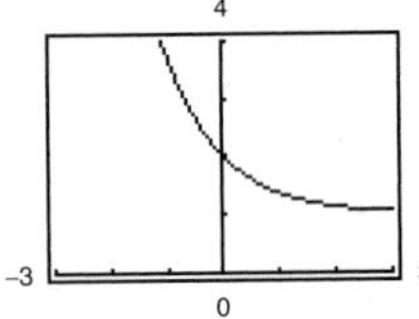

47. $f(x) = h(x)$ **49.** $f(x) = g(x) = h(x)$

51. (a) $x < 0$ (b) $x > 0$

53. $y_1 = e^x$

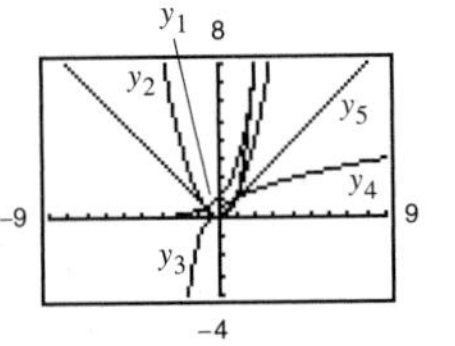

55. It usually implies rapid growth.

57.

n	1	2	4	12	365	Continuous
A	$1828.49	$1830.29	$1831.19	$1831.80	$1832.09	$1832.10

59.

n	1	2	4	12	365	Continuous
A	$5477.81	$5520.10	$5541.79	$5556.46	$5563.61	$5563.85

61.

t	10	20	30	40	50
A	$17,901.90	$26,706.49	$39,841.40	$59,436.39	$88,668.67

63.

t	10	20	30	40	50
A	$22,986.49	$44,031.56	$84,344.25	$161,564.86	$309,484.08

65. $104,710.29 **67.** $35.45

69. (a)

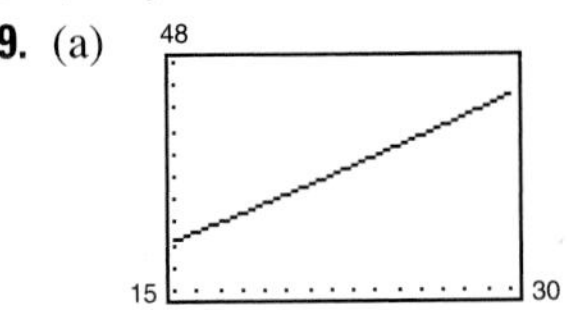

(b)

t	15	16	17	18	19	20
P (in millions)	40.19	40.59	40.99	41.39	41.80	42.21

t	21	22	23	24	25	26
P (in millions)	42.62	43.04	43.47	43.90	44.33	44.77

t	27	28	29	30
P (in millions)	45.21	45.65	46.10	46.56

(c) 2038

71. (a) 16 g (b) 1.85 g

(c)

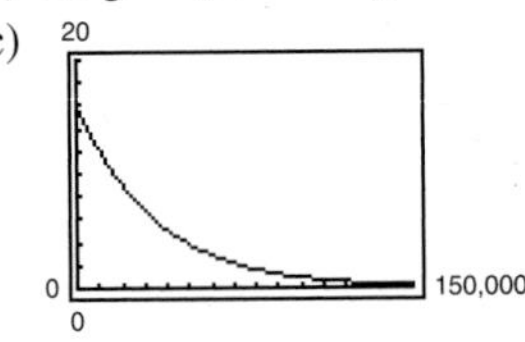

73. (a) $V(t) = 30{,}500\left(\frac{7}{8}\right)^t$ (b) $17,878.54

75. True. As $x \to -\infty$, $f(x) \to -2$ but never reaches -2.

77. (a)

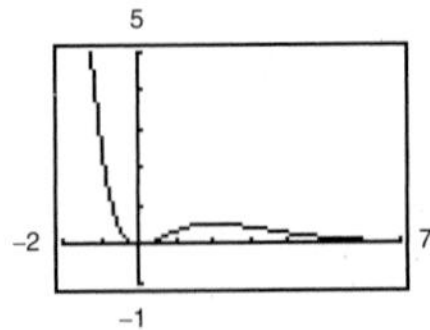

Decreasing: $(-\infty, 0)$, $(2, \infty)$

Increasing: $(0, 2)$

Relative maximum: $(2, 4e^{-2})$

Relative minimum: $(0, 0)$

(b)

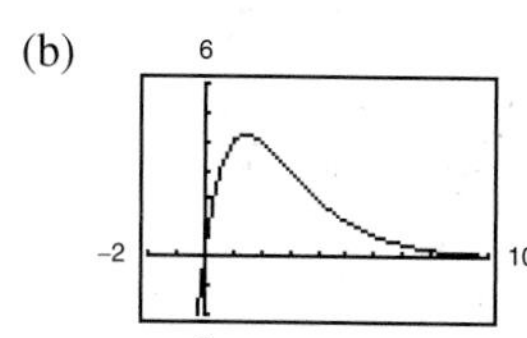

Decreasing: $(1.44, \infty)$
Increasing: $(-\infty, 1.44)$
Relative maximum: $(1.44, 4.25)$

79.

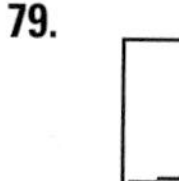

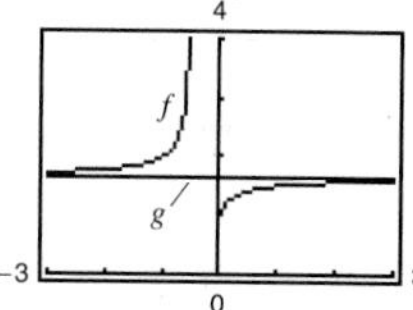

As $x \to \infty, f(x) \to g(x)$.
As $x \to -\infty, f(x) \to g(x)$.

81. (a) $A = \$5466.09$ (b) $A = \$5466.35$
(c) $A = \$5466.36$ (d) $A = \$5466.38$
No. Answers will vary.

Section 7.2 (page 488)

1. logarithmic **3.** natural; e **5.** $x = y$ **7.** $4^2 = 16$
9. $9^{-2} = \frac{1}{81}$ **11.** $32^{2/5} = 4$ **13.** $64^{1/2} = 8$
15. $\log_5 125 = 3$ **17.** $\log_{81} 3 = \frac{1}{4}$ **19.** $\log_6 \frac{1}{36} = -2$
21. $\log_{24} 1 = 0$ **23.** 6 **25.** 0 **27.** 2
29. -0.058 **31.** 1.097 **33.** 7 **35.** 1

37.

Domain: $(0, \infty)$
x-intercept: $(1, 0)$
Vertical asymptote: $x = 0$

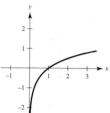

39.

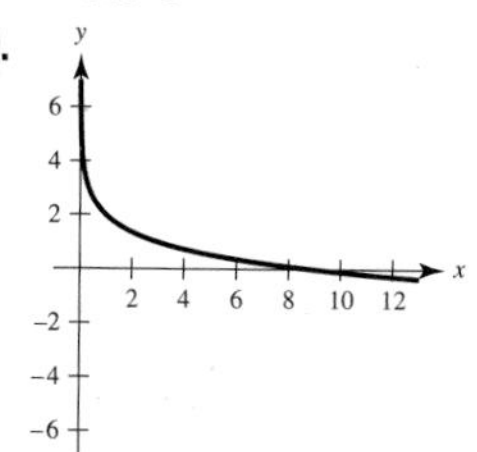

Domain: $(0, \infty)$
x-intercept: $(9, 0)$
Vertical asymptote: $x = 0$

41.

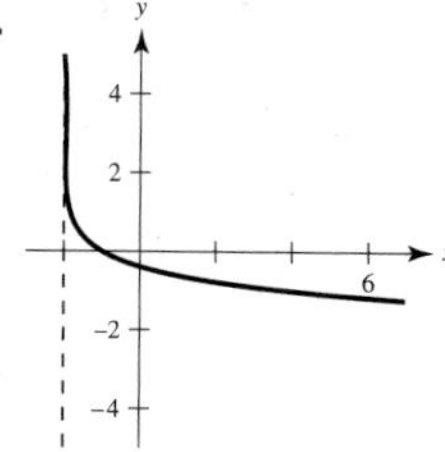

Domain: $(-2, \infty)$
x-intercept: $(-1, 0)$
Vertical asymptote: $x = -2$

43.

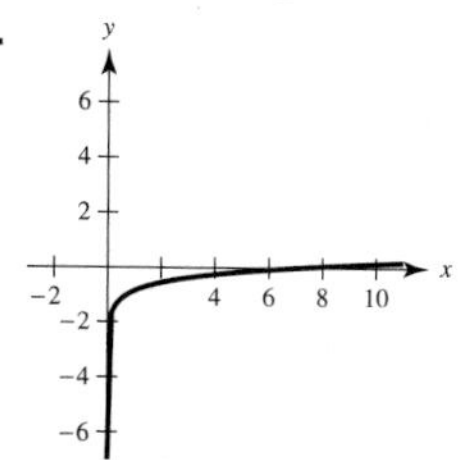

Domain: $(0, \infty)$
x-intercept: $(7, 0)$
Vertical asymptote: $x = 0$

45. c **46.** f **47.** d **48.** e **49.** b **50.** a
51. $e^{-0.693\ldots} = \frac{1}{2}$ **53.** $e^{1.945\ldots} = 7$ **55.** $e^{5.521\ldots} = 250$
57. $e^0 = 1$ **59.** $\ln 54.598\ldots = 4$
61. $\ln 1.6487\ldots = \frac{1}{2}$ **63.** $\ln 0.406\ldots = -0.9$
65. $\ln 4 = x$ **67.** 2.913 **69.** -23.966
71. 5 **73.** $-\frac{5}{6}$

75.

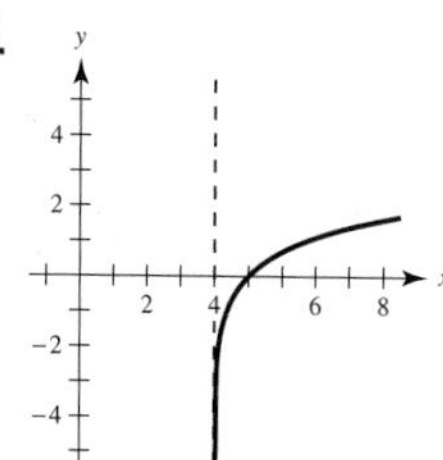

Domain: $(4, \infty)$
x-intercept: $(5, 0)$
Vertical asymptote: $x = 4$

77.

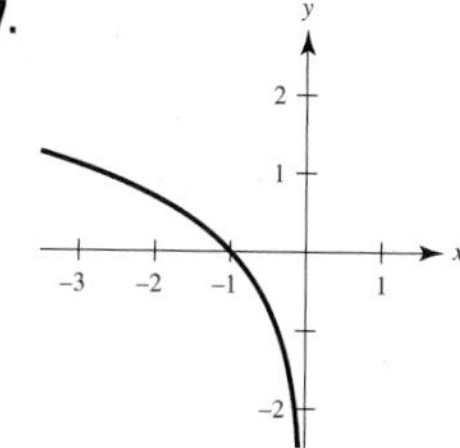

Domain: $(-\infty, 0)$
x-intercept: $(-1, 0)$
Vertical asymptote: $x = 0$

79.

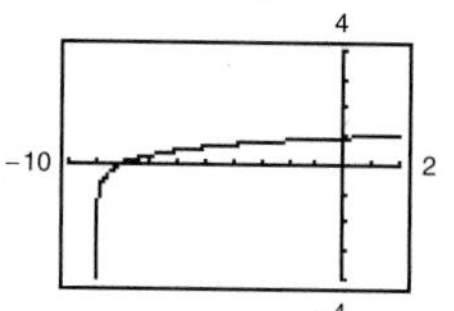

81.

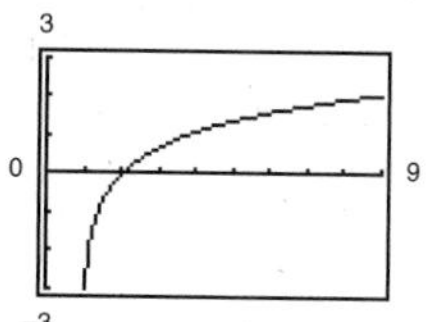

83.

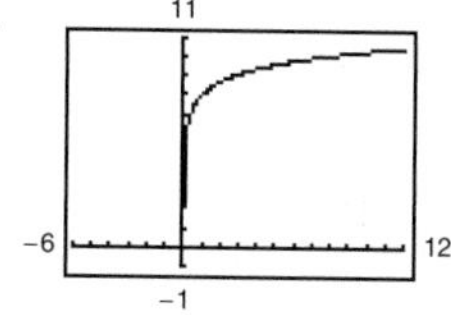

85. $x = 5$ **87.** $x = 7$ **89.** $x = 8$ **91.** $x = -5, 5$

93.

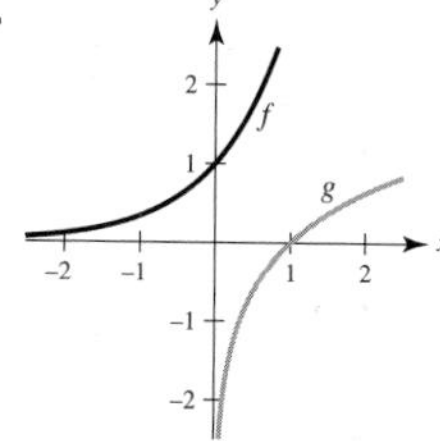

The functions f and g are inverses.

95.

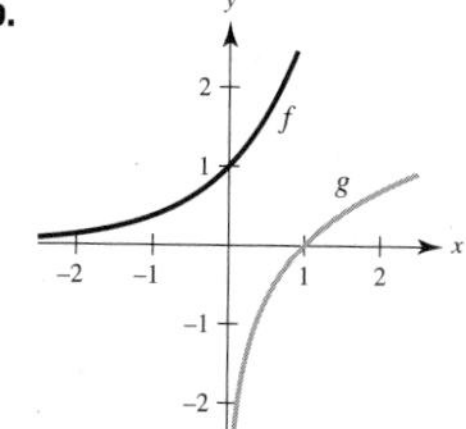

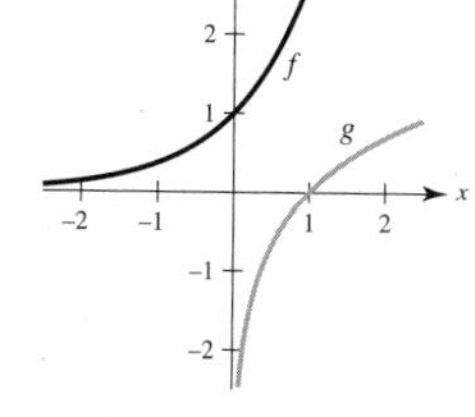

The functions f and g are inverses.

97. (a)

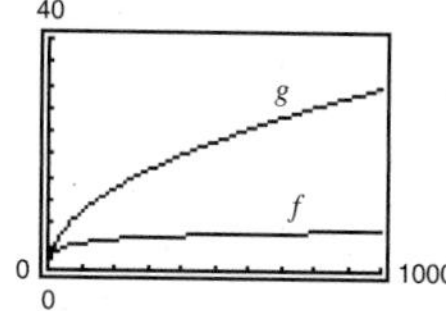

$g(x)$; The natural log function grows at a slower rate than the square root function.

(b)

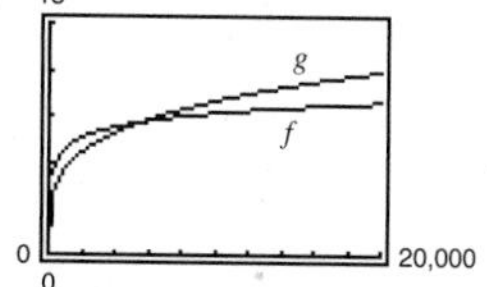

$g(x)$; The natural log function grows at a slower rate than the fourth root function.

99. (a)

t	1	2	3	4	5	6
C	10.36	9.94	9.37	8.70	7.96	7.15

(b)

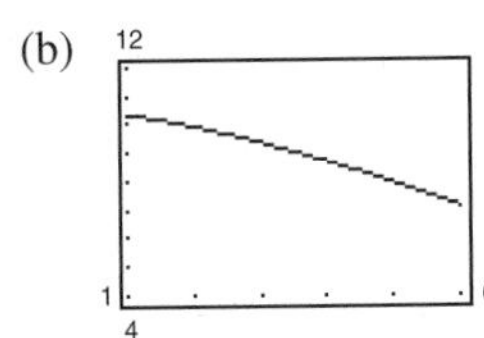

(c) No, the model begins to decrease rapidly, eventually producing negative values.

101. (a)

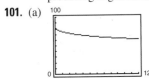

(b) 80 (c) 68.1 (d) 62.3

103. (a) 30 yr; 10 yr (b) \$323,179; \$199,109
(c) \$173,179; \$49,109
(d) $x = 750$; The monthly payment must be greater than \$750.

105. True. $\log_3 27 = 3 \Rightarrow 3^3 = 27$

107. (a)

x	1	5	10	10^2
$f(x)$	0	0.322	0.230	0.046

x	10^4	10^6
$f(x)$	0.00092	0.0000138

(b) 0

(c)

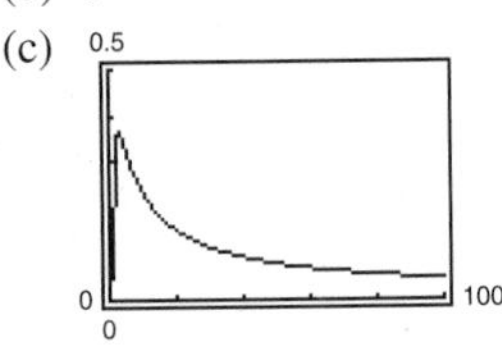

109. Answers will vary.

111. (a)

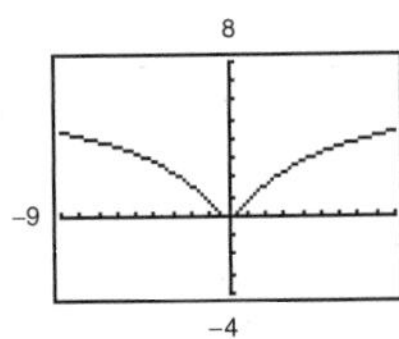

(b) Increasing: $(0, \infty)$
Decreasing: $(-\infty, 0)$
(c) Relative minimum: $(0, 0)$

Section 7.3 (page 495)

1. change-of-base **3.** $\dfrac{1}{\log_b a}$

5. (a) $\dfrac{\log 16}{\log 5}$ (b) $\dfrac{\ln 16}{\ln 5}$ **7.** (a) $\dfrac{\log x}{\log \frac{1}{5}}$ (b) $\dfrac{\ln x}{\ln \frac{1}{5}}$

9. (a) $\dfrac{\log \frac{3}{10}}{\log x}$ (b) $\dfrac{\ln \frac{3}{10}}{\ln x}$ **11.** (a) $\dfrac{\log x}{\log 2.6}$ (b) $\dfrac{\ln x}{\ln 2.6}$

13. 1.771 **15.** -2.000 **17.** -1.048 **19.** 2.633

21. $\frac{3}{2}$ **23.** $-3 - \log_5 2$ **25.** $6 + \ln 5$ **27.** 2

29. $\frac{3}{4}$ **31.** 4 **33.** -2 is not in the domain of $\log_2 x$.

35. 4.5 **37.** $-\frac{1}{2}$ **39.** 7 **41.** 2 **43.** $\ln 4 + \ln x$

45. $4 \log_8 x$ **47.** $1 - \log_5 x$ **49.** $\frac{1}{2} \ln z$

51. $\ln x + \ln y + 2 \ln z$ **53.** $\ln z + 2 \ln(z - 1)$

55. $\frac{1}{2} \log_2(a - 1) - 2 \log_2 3$ **57.** $\frac{1}{3} \ln x - \frac{1}{3} \ln y$

59. $2 \ln x + \frac{1}{2} \ln y - \frac{1}{2} \ln z$

61. $2 \log_5 x - 2 \log_5 y - 3 \log_5 z$

63. $\dfrac{3}{4} \ln x + \dfrac{1}{4} \ln(x^2 + 3)$ **65.** $\ln 2x$ **67.** $\log_4 \dfrac{z}{y}$

69. $\log_2 x^2 y^4$ **71.** $\log_3 \sqrt[4]{5x}$ **73.** $\log \dfrac{x}{(x + 1)^2}$

75. $\log \dfrac{xz^3}{y^2}$ **77.** $\ln \dfrac{x}{(x + 1)(x - 1)}$

79. $\ln \sqrt[3]{\dfrac{x(x + 3)^2}{x^2 - 1}}$ **81.** $\log_8 \dfrac{\sqrt[3]{y(y + 4)^2}}{y - 1}$

83. $\log_2 \frac{32}{4} = \log_2 32 - \log_2 4$; Property 2

85. $\beta = 10(\log I + 12)$; 60 dB **87.** 70 dB

89. $\ln y = \frac{1}{4} \ln x$ **91.** $\ln y = -\frac{1}{4} \ln x + \ln \frac{5}{2}$

93.

95.

$f(x) = h(x)$; Property 2

97. $\ln y = \frac{3}{4} \ln x + \ln 0.072$

99. False; $\ln 1 = 0$ **101.** False; $\ln(x - 2) \neq \ln x - \ln 2$

103. False; $u = v^2$

105. $f(x) = \dfrac{\log x}{\log 2} = \dfrac{\ln x}{\ln 2}$ **107.** $f(x) = \dfrac{\log x}{\log \frac{1}{2}} = \dfrac{\ln x}{\ln \frac{1}{2}}$

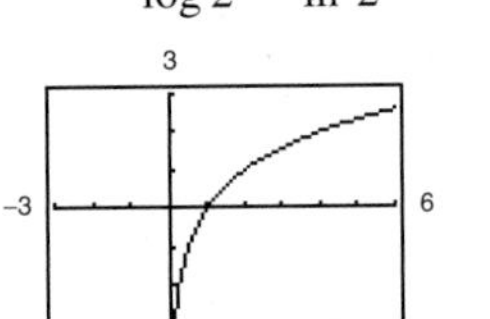

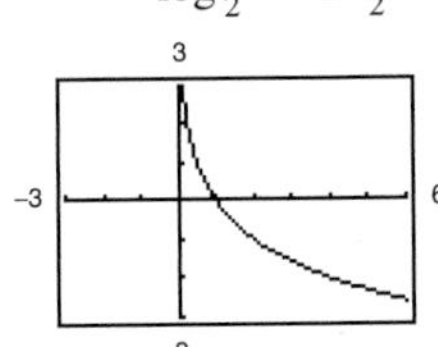

109. $f(x) = \dfrac{\log x}{\log 11.8} = \dfrac{\ln x}{\ln 11.8}$

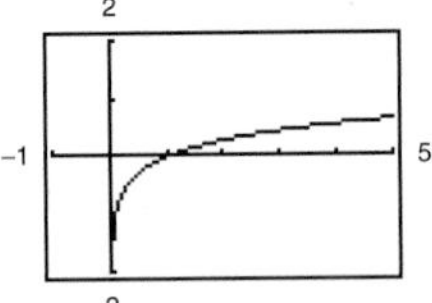

111.

The graphing utility does not show the functions with the same domain. The domain of $y_1 = \ln x - \ln(x - 3)$ is $(3, \infty)$ and the domain of $y_2 = \ln \dfrac{x}{x - 3}$ is $(-\infty, 0) \cup (3, \infty)$.

113. Proof

115. $\ln 1 = 0$ $\ln 9 \approx 2.1972$
$\ln 2 \approx 0.6931$ $\ln 10 \approx 2.3025$
$\ln 3 \approx 1.0986$ $\ln 12 \approx 2.4848$
$\ln 4 \approx 1.3862$ $\ln 15 \approx 2.7080$
$\ln 5 \approx 1.6094$ $\ln 16 \approx 2.7724$
$\ln 6 \approx 1.7917$ $\ln 18 \approx 2.8903$
$\ln 8 \approx 2.0793$ $\ln 20 \approx 2.9956$

Section 7.4 (page 505)

1. solve

3. (a) One-to-One (b) logarithmic; logarithmic (c) exponential; exponential

5. (a) Yes (b) No

7. (a) No (b) Yes (c) Yes, approximate

9. (a) Yes, approximate (b) No (c) Yes

11. (a) No (b) Yes (c) Yes, approximate

13. 2 **15.** -5 **17.** 2 **19.** $\ln 2 \approx 0.693$

21. $e^{-1} \approx 0.368$ **23.** 64 **25.** (3, 8) **27.** (9, 2)

29. 2, -1 **31.** About 1.618, about -0.618

33. $\dfrac{\ln 5}{\ln 3} \approx 1.465$ **35.** $\ln 5 \approx 1.609$ **37.** $\ln 28 \approx 3.332$

39. $\dfrac{\ln 80}{2 \ln 3} \approx 1.994$ **41.** 2 **43.** 4

45. $3 - \dfrac{\ln 565}{\ln 2} \approx -6.142$ **47.** $\dfrac{1}{3}\log\left(\dfrac{3}{2}\right) \approx 0.059$

49. $1 + \dfrac{\ln 7}{\ln 5} \approx 2.209$ **51.** $\dfrac{\ln 12}{3} \approx 0.828$

53. $-\ln\dfrac{3}{5} \approx 0.511$ **55.** 0 **57.** $\dfrac{\ln \frac{8}{3}}{3 \ln 2} + \dfrac{1}{3} \approx 0.805$

59. $\ln 5 \approx 1.609$ **61.** $\ln 4 \approx 1.386$

63. $2 \ln 75 \approx 8.635$ **65.** $\frac{1}{2}\ln 1498 \approx 3.656$

67. $\dfrac{\ln 4}{365 \ln\left(1 + \frac{0.065}{365}\right)} \approx 21.330$ **69.** $\dfrac{\ln 2}{12 \ln\left(1 + \frac{0.10}{12}\right)} \approx 6.960$

71.
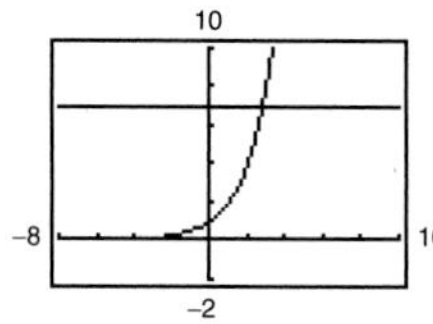

2.807

73.
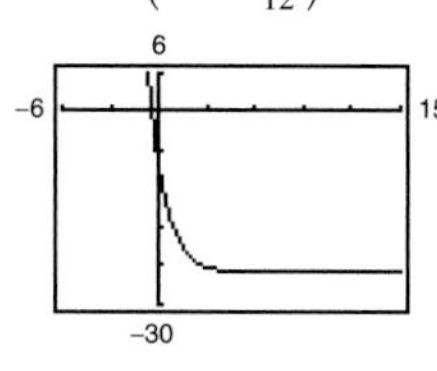

-0.427

75.
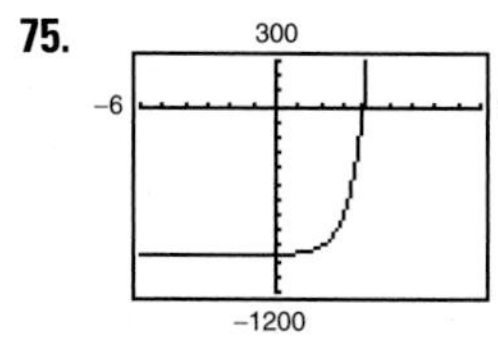

3.847

77.
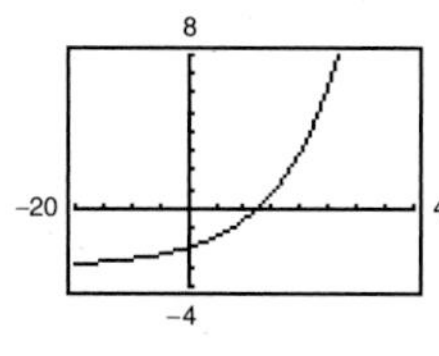

12.207

79.
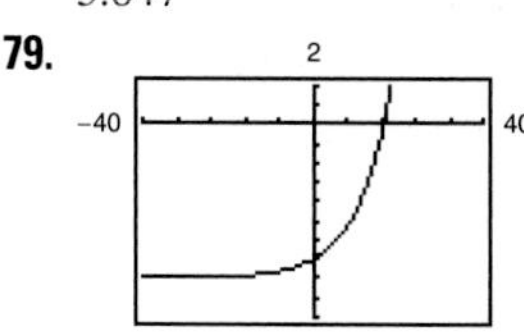

16.636

81. $e^{-3} \approx 0.050$ **83.** $e^7 \approx 1096.633$ **85.** $\dfrac{e^{2.4}}{2} \approx 5.512$

87. 1,000,000 **89.** $\dfrac{e^{10/3}}{5} \approx 5.606$

91. $e^2 - 2 \approx 5.389$ **93.** $e^{-2/3} \approx 0.513$

95. $\dfrac{e^{19/2}}{3} \approx 4453.242$ **97.** $2(3^{11/6}) \approx 14.988$

99. No solution **101.** $1 + \sqrt{1 + e} \approx 2.928$

103. No solution **105.** 7 **107.** $\dfrac{-1 + \sqrt{17}}{2} \approx 1.562$

109. 2 **111.** $\dfrac{725 + 125\sqrt{33}}{8} \approx 180.384$

113.
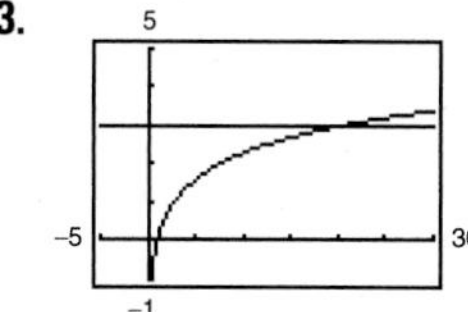

20.086

115.
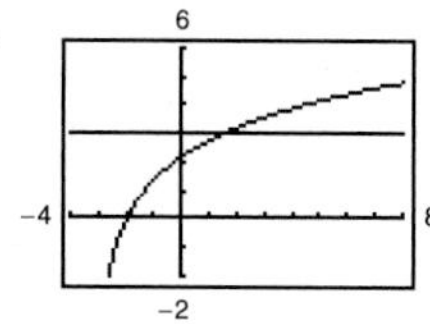

1.482

117. $-1, 0$ **119.** 1 **121.** $e^{-1/2} \approx 0.607$

123. $e^{-1} \approx 0.368$

125. For $rt < \ln 2$ years, double the amount you invest. For $rt > \ln 2$ years, double your interest rate or double the number of years, because either of these will double the exponent in the exponential function.

127. (a) 13.86 yr (b) 21.97 yr

129. (a) 27.73 yr (b) 43.94 yr

131. Yes. Time to double: $t = \dfrac{\ln 2}{r}$;
Time to quadruple: $t = \dfrac{\ln 4}{r} = 2\left(\dfrac{\ln 2}{r}\right)$

133. (a) 303 units (b) 528 units **135.** 12.76 in.

137. (a)
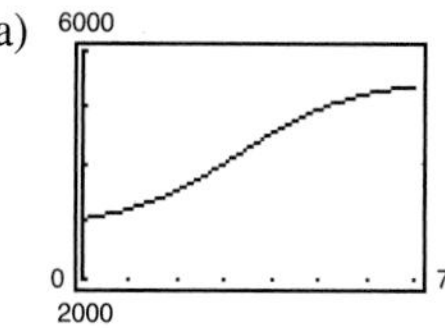

(b) 2002

139. (a)
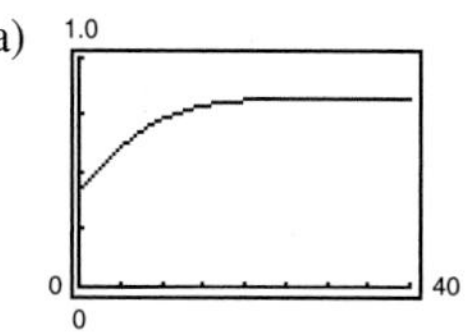

(b) Horizontal asymptotes: $P = 0$, $P = 0.83$
The proportion of correct responses will approach 0.83 as the number of trials increases.
(c) About 5 trials

141. (a) $T = 20$; Room temperature (b) About 0.81 h

143. $\log_b uv = \log_b u + \log_b v$
True by Property 1 in Section 7.3.

145. $\log_b(u - v) = \log_b u - \log_b v$
False
$1.95 \approx \log(100 - 10) \neq \log 100 - \log 10 = 1$

147. Yes. See Exercise 103.

149. (a)

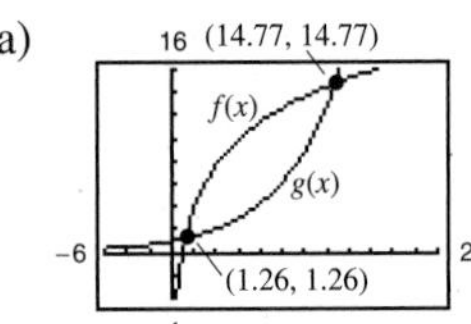

(b) $a = e^{1/e}$
(c) $1 < a < e^{1/e}$

Section 7.5 (page 515)

1. $y = ae^{bx}$; $y = ae^{-bx}$ **3.** normally distributed

5. $y = \dfrac{a}{1 + be^{-rx}}$ **7.** c **8.** e **9.** b

10. a **11.** d **12.** f **13.** $b < 0$

15. This is a bell-shaped curve with a maximum when $x = 70$. The average height of American men between 18 and 24 years old is 70 inches.

	Initial Investment	*Annual % Rate*	*Time to Double*	*Amount After 10 years*
17.	\$1000	3.5%	19.8 yr	\$1419.07
19.	\$750	8.9438%	7.75 yr	\$1834.37
21.	\$500	11.0%	6.3 yr	\$1505.00
23.	\$6376.28	4.5%	15.4 yr	\$10,000.00

25. \$303,580.52

27. (a) 7.27 yr (b) 6.96 yr (c) 6.93 yr (d) 6.93 yr

29.

r	2%	4%	6%	8%	10%	12%
t	54.93	27.47	18.31	13.73	10.99	9.16

31.

r	2%	4%	6%	8%	10%	12%
t	55.48	28.01	18.85	14.27	11.53	9.69

33.

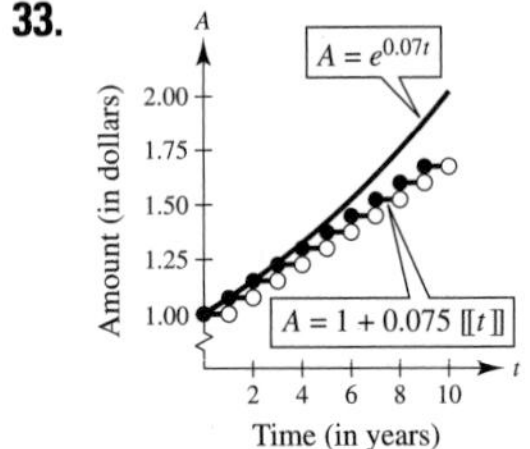

Continuous compounding

35. $y = e^{0.7675x}$ **37.** $y = 5e^{-0.4024x}$

39. (a)

Year	1970	1980	1990	2000	2007
Population	73.7	103.74	143.56	196.35	243.24

(b) 2014
(c) No; The population will not continue to grow at such a quick rate.

41. $k = 0.2988$; About 5,309,734 hits

43. (a) $k = 0.02603$; The population is increasing because $k > 0$.
(b) 449,910; 512,447 (c) 2014

45. About 800 bacteria

47. (a) About 12,180 yr old (b) About 4797 yr old

49. (a) $V = -5400t + 23{,}300$ (b) $V = 23{,}300e^{-0.311t}$

(c)

The exponential model depreciates faster.

(d)

t	1 yr	3 yr
$V = -5400t + 23{,}300$	17,900	7100
$V = 23{,}300e^{-0.311t}$	17,072	9166

(e) Answers will vary.

51. (a) $S(t) = 100(1 - e^{-0.1625t})$
(b)

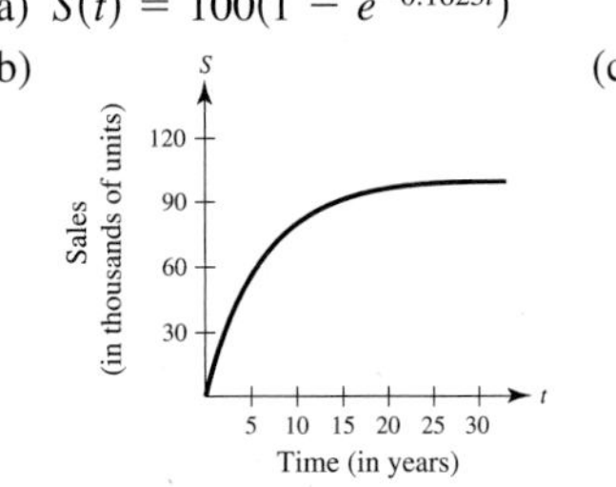

(c) 55,625

53. (a)

(b) 100

55. (a) 715; 90,880; 199,043
(b)

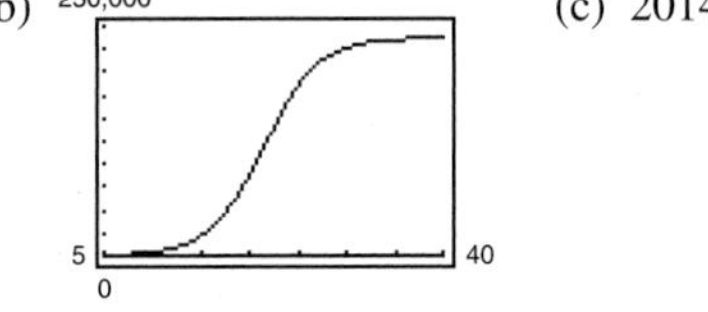

(c) 2014

(d) $235{,}000 = \dfrac{237{,}101}{1 + 1950e^{-0.355t}}$
$t \approx 34.63$

57. (a) 203 animals (b) 13 mo
(c)

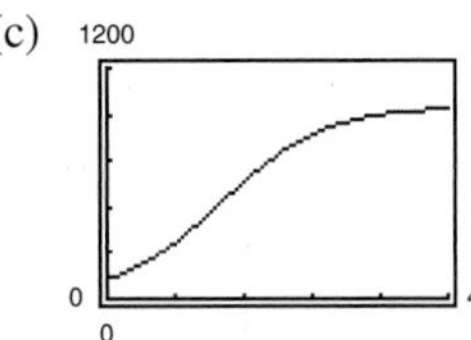

Horizontal asymptotes: $p = 0$, $p = 1000$. The population size will approach 1000 as time increases.

59. (a) 8.30 (b) 7.68 (c) 4.23

61. (a) 10 dB (b) 140 dB (c) 80 dB (d) 100 dB

63. 97% **65.** 4.95

67. $10^{-3.2} \approx 6.3 \times 10^{-4}$ moles/L **69.** 10

71. False. A logistic growth function never has an x-intercept.

73. True. The graph of a Gaussian model will never have an x-intercept.

Review Exercises (page 520)

1. 0.164 **3.** 0.337 **5.** 1456.529

7. Shift the graph of f two units downward.

9. Reflect f in the y-axis and shift two units to the right.
11. Reflect f in the x-axis and shift one unit upward.
13. Reflect f in the x-axis and shift two units to the left.

15.

x	-1	0	1	2	3
$f(x)$	8	5	4.25	4.063	4.016

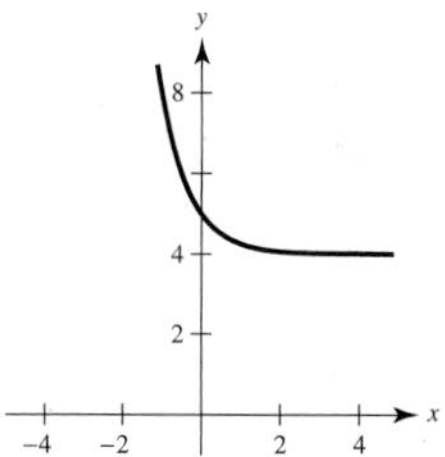

17.

x	-1	0	1	2	3
$f(x)$	4.008	4.04	4.2	5	9

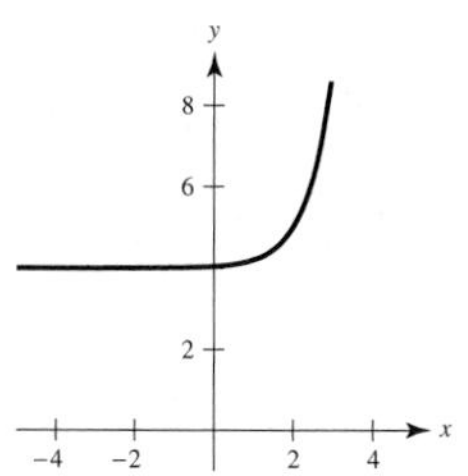

19.

x	-2	-1	0	1	2
$f(x)$	3.25	3.5	4	5	7

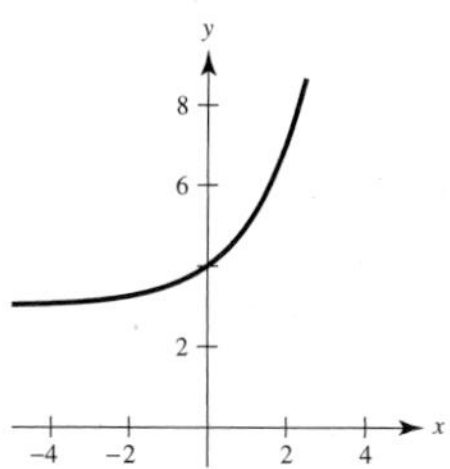

21. 2980.958 **23.** 0.183

25.

x	-2	-1	0	1	2
$h(x)$	2.72	1.65	1	0.61	0.37

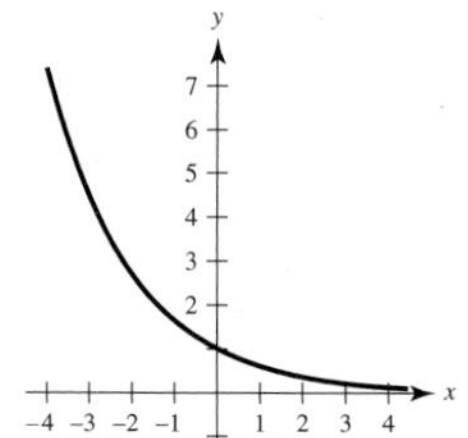

27.

x	-3	-2	-1	0	1
$f(x)$	0.37	1	2.72	7.39	20.09

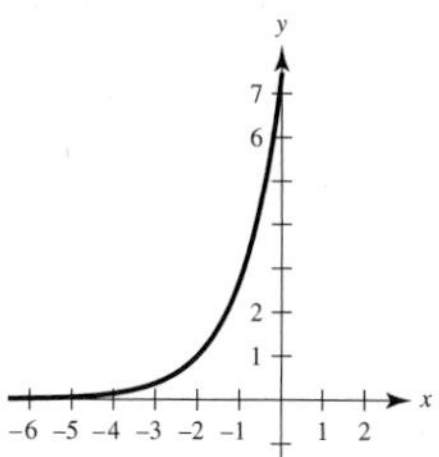

29.

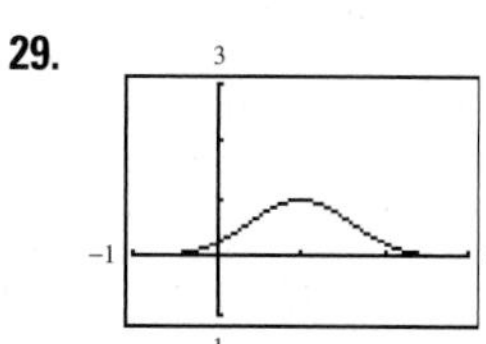

31.

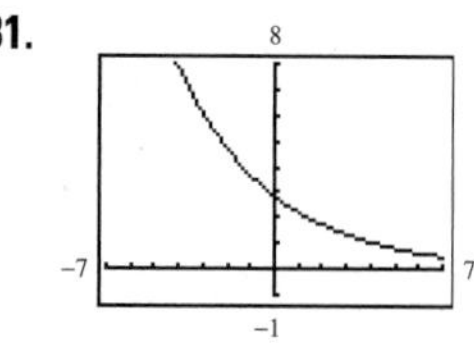

33.

n	1	2	4	12
A	\$6719.58	\$6734.28	\$6741.74	\$6746.77

n	365	Continuous
A	\$6749.21	\$6749.29

35. (a) 0.154 (b) 0.487 (c) 0.811
37. $\log_3 27 = 3$ **39.** $\ln 2.2255 \ldots = 0.8$
41. 3 **43.** -2 **45.** $x = 7$ **47.** $x = -5$

49. Domain: $(0, \infty)$
x-intercept: $(1, 0)$
Vertical asymptote: $x = 0$

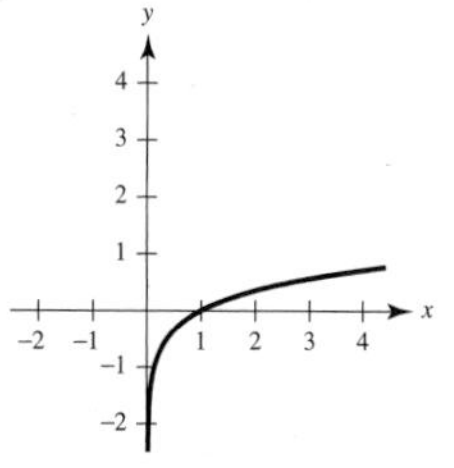

51. Domain: $(-5, \infty)$
x-intercept: $(9995, 0)$
Vertical asymptote: $x = -5$

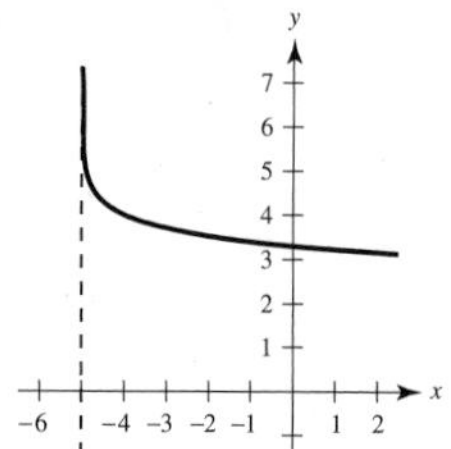

53. (a) 3.118 (b) -0.020

55. Domain: $(0, \infty)$
x-intercept: $(e^{-3}, 0)$
Vertical asymptote: $x = 0$

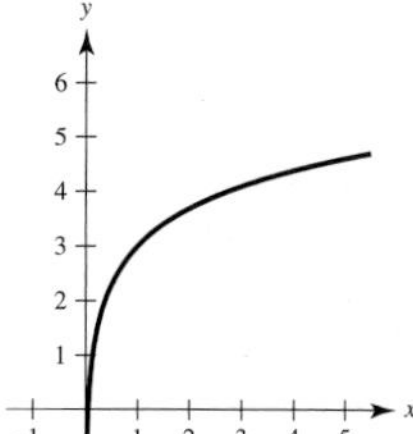

57. Domain: $(-\infty, 0), (0, \infty)$
x-intercept: $(\pm 1, 0)$
Vertical asymptote: $x = 0$

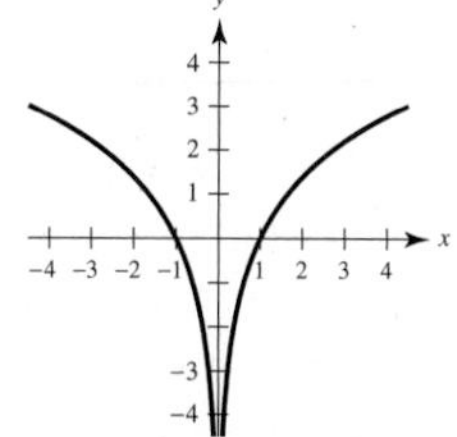

59. 53.4 in. **61.** 2.585 **63.** -2.322
65. $\log 2 + 2 \log 3 \approx 1.255$ **67.** $2 \ln 2 + \ln 5 \approx 2.996$

69. $1 + 2\log_5 x$ **71.** $2 - \frac{1}{2}\log_3 x$

73. $2\ln x + 2\ln y + \ln z$ **75.** $\log_2 5x$ **77.** $\ln \dfrac{x}{\sqrt[4]{y}}$

79. $\log_3 \dfrac{\sqrt{x}}{(y+8)^2}$

81. (a) $0 \le h < 18{,}000$

(b) 100, 0, 20,000, 0

Vertical asymptote: $h = 18{,}000$

(c) The plane is climbing at a slower rate, so the time required increases.

(d) 5.46 min

83. 3 **85.** $\ln 3 \approx 1.099$ **87.** $e^4 \approx 54.598$

89. $x = 1, 3$ **91.** $\dfrac{\ln 32}{\ln 2} = 5$

93.

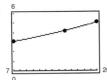

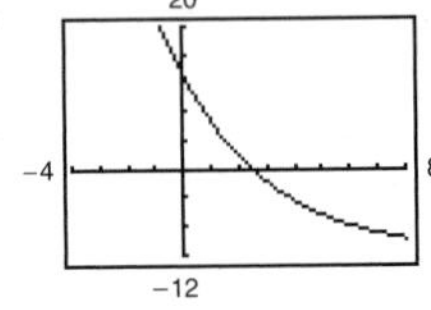

2.447

95. $\frac{1}{3}e^{8.2} \approx 1213.650$ **97.** $3e^2 \approx 22.167$

99. $e^8 \approx 2980.958$ **101.** No solution **103.** 0.900

105.

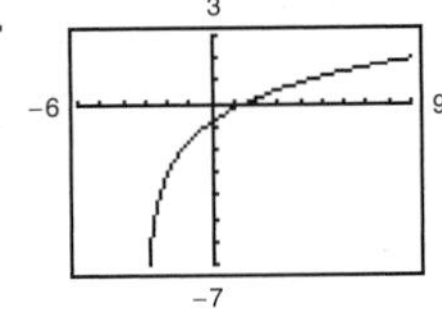

1.482

107.

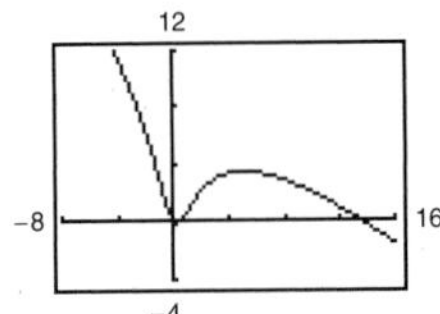

0, 0.416, 13.627

109. 31.4 yr **111.** e **112.** b **113.** f **114.** d

115. a **116.** c **117.** $y = 2e^{0.1014x}$

119. (a) 6, 0, 20, 7

The model fits the data well.

(b) 2022; Answers will vary.

121. (a)

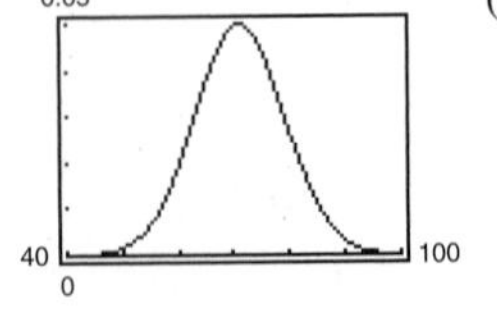

(b) 71

123. (a) $10^{-6}\ \text{W/m}^2$ (b) $10\sqrt{10}\ \text{W/m}^2$

(c) $1.259 \times 10^{-12}\ \text{W/m}^2$

125. True by the inverse properties

Chapter Test (page 523)

1. 2.366 **2.** 687.291 **3.** 0.497 **4.** 22.198

5.

x	-1	$-\frac{1}{2}$	0	$\frac{1}{2}$	1
$f(x)$	10	3.162	1	0.316	0.1

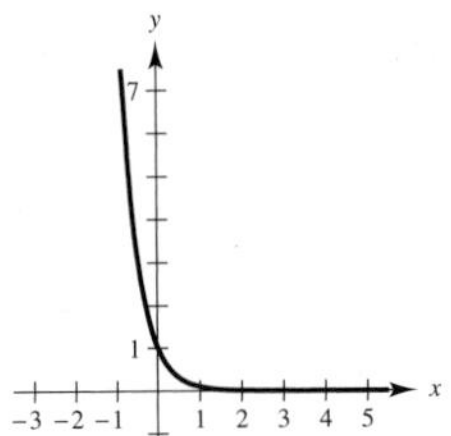

6.

x	-1	0	1	2	3
$f(x)$	-0.005	-0.028	-0.167	-1	-6

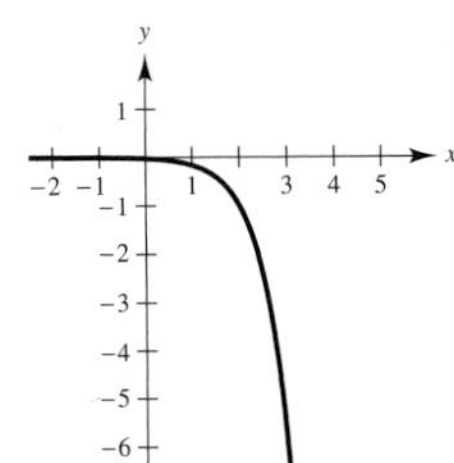

7.

x	-1	$-\frac{1}{2}$	0	$\frac{1}{2}$	1
$f(x)$	0.865	0.632	0	-1.718	-6.389

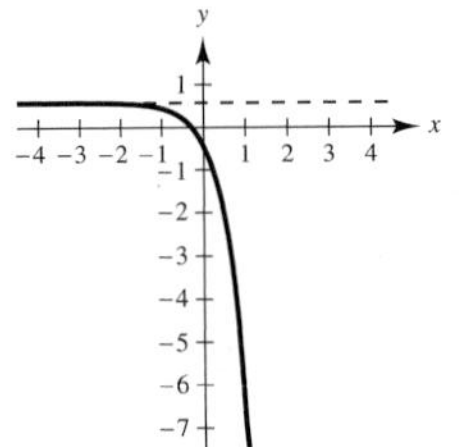

8. (a) -0.89 (b) 9.2

9.

x	$\frac{1}{2}$	1	$\frac{3}{2}$	2	4
$f(x)$	-5.699	-6	-6.176	-6.301	-6.602

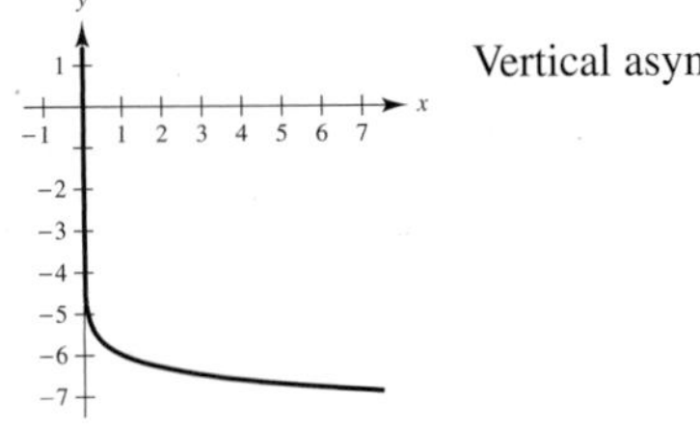

Vertical asymptote: $x = 0$

10.

x	5	7	9	11	13
$f(x)$	0	1.099	1.609	1.946	2.197

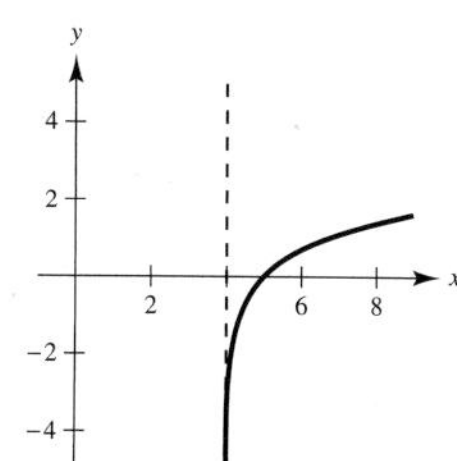

Vertical asymptote: $x = 4$

11.

x	-5	-3	-1	0	1
$f(x)$	1	2.099	2.609	2.792	2.946

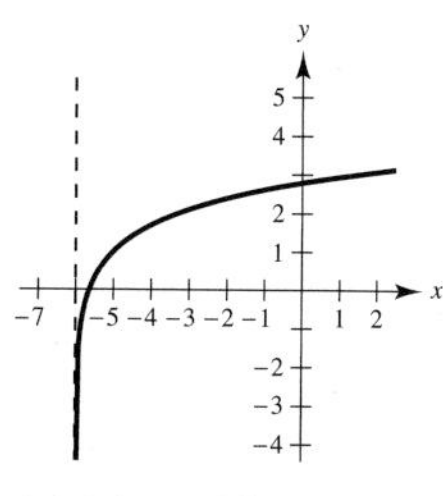

Vertical asymptote: $x = -6$

12. 1.945 **13.** -0.167 **14.** -11.047

15. $\log_2 3 + 4 \log_2 |a|$ **16.** $\ln 5 + \frac{1}{2}\ln x - \ln 6$

17. $3 \log(x - 1) - 2 \log y - \log z$ **18.** $\log_3 13y$

19. $\ln \dfrac{x^4}{y^4}$ **20.** $\ln\left(\dfrac{x^3y^2}{x + 3}\right)$ **21.** $x = -2$

22. $x = \dfrac{\ln 44}{-5} \approx -0.757$ **23.** $\dfrac{\ln 197}{4} \approx 1.321$

24. $e^{1/2} \approx 1.649$ **25.** $e^{-11/4} \approx 0.0639$ **26.** 20

27. $y = 2745e^{0.1570t}$ **28.** About 1125 bacteria

29. (a)

x	$\frac{1}{4}$	1	2	4	5	6
H	58.720	75.332	86.828	103.43	110.59	117.38

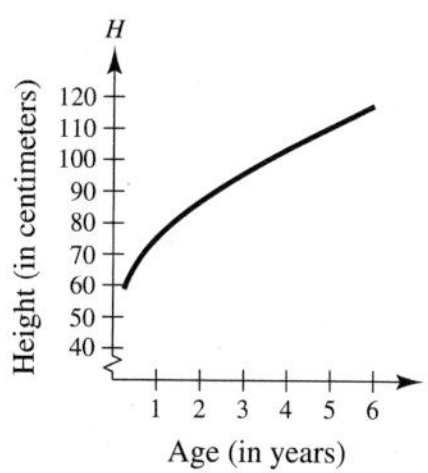

(b) 103 cm; 103.43 cm

P.S. Problem Solving (page 524)

1.

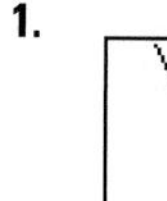

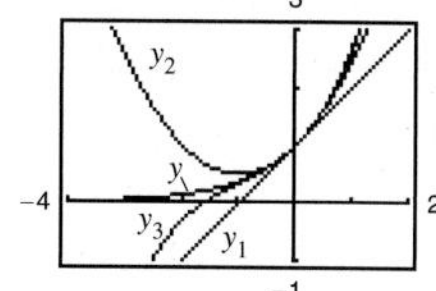

3. (a) $f(u + v) = a^{u+v} = a^u \cdot a^v = f(u) \cdot f(v)$

(b) $f(2x) = a^{2x} = (a^x)^2 = [f(x)]^2$

5. $y_4 = (x - 1) - \frac{1}{2}(x - 1)^2 + \frac{1}{3}(x - 1)^3 - \frac{1}{4}(x - 1)^4$

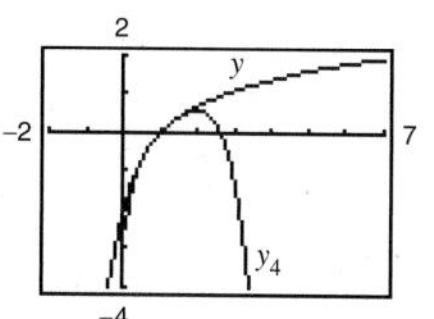

The pattern implies that as we take more terms, the graph will more closely resemble that of $\ln x$ on the interval $(0, 2)$.

7.

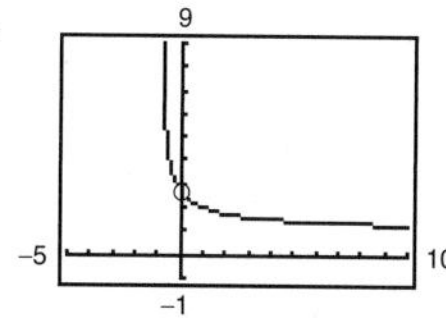

Near $x = 0$, the graph approaches e. There is no y-intercept.

x	0.1	0.01	0.001	0.0001	0.00001
y	2.5937	2.7048	2.7169	2.7181	2.7183

9. (a)

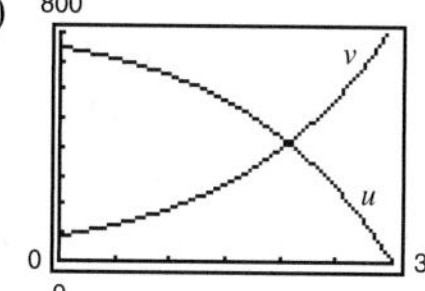

(b) Interest; $t \approx 21$ yr

(c)

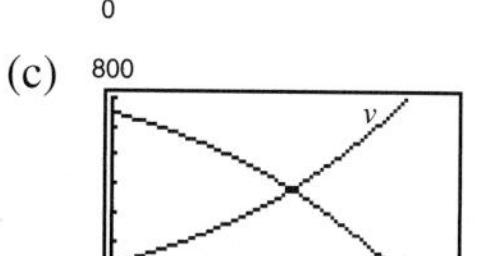

Interest; $t \approx 11$ yr

The interest is still the majority of the monthly payment in the early years, but now the principal and interest are nearly equal when $t \approx 11$ years.

11.

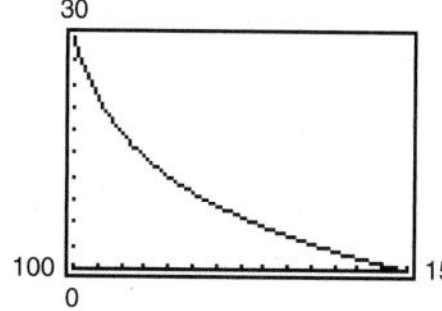

17.7 ft^3/min

13. (c); it passes through the point $(0, 0)$, it is symmetric to the y-axis, and $y = 6$ is a horizontal asymptote.

15. $t = k_1k_2 \ln (c_1/c_2)/[(k_2 - k_1) \ln 2]$

17.

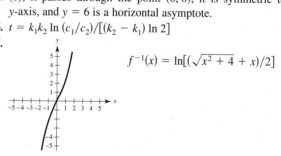

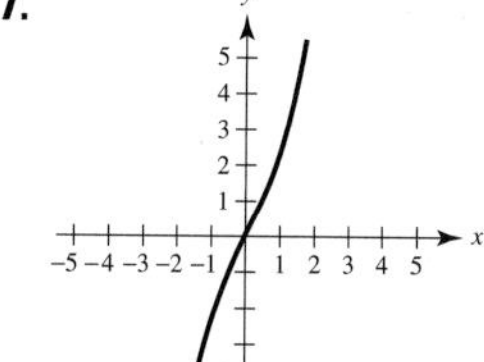

$f^{-1}(x) = \ln[(\sqrt{x^2 + 4} + x)/2]$

19. $f^{-1}(x) = \dfrac{\ln[(x + 1)/(x - 1)]}{\ln a}, a > 0, a \neq 1$

Chapter 8

Section 8.1 (page 532)

1. (a) 3 (b) -3 **3.** $2e^{2x}$ **5.** $2(x-1)e^{-2x+x^2}$
7. $-e^{1/x}/x^2$ **9.** $e^{\sqrt{x}}/(2\sqrt{x})$ **11.** $e^{3x}(3x+4)$
13. $e^{x^2}(2x^2-1)/x^2$ **15.** $3(e^{-t}+e^{t})^2(e^{t}-e^{-t})$
17. $-2(e^x-e^{-x})/(e^x+e^{-x})^2$ **19.** x^2e^x **21.** $y=-x+2$
23. $y=ex$ **25.** $(10-e^y)/(xe^y+3)$ **27.** $6(3e^{3x}+2e^{-2x})$
29. $3(6x+5)e^{-3x}$
31. Relative minimum: $(0, 1)$

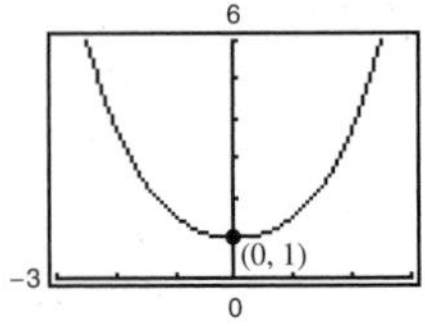

33. Relative maximum: $(2, 1/\sqrt{2\pi})$
Points of inflection: $(1, e^{-1/2}/\sqrt{2\pi})$, $(3, e^{-1/2}/\sqrt{2\pi})$

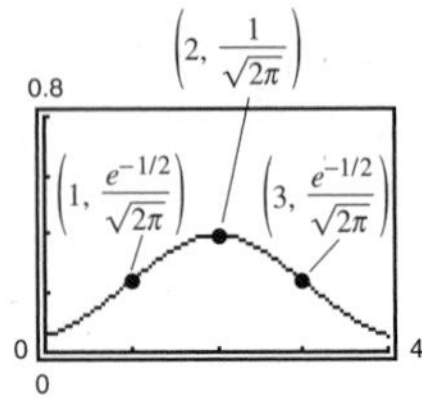

35. Relative minimum: $(0, 0)$
Relative maximum: $(2, 4e^{-2})$
Points of inflection: $(2 \pm \sqrt{2}, (6 \pm 4\sqrt{2})e^{-(2\pm\sqrt{2})})$

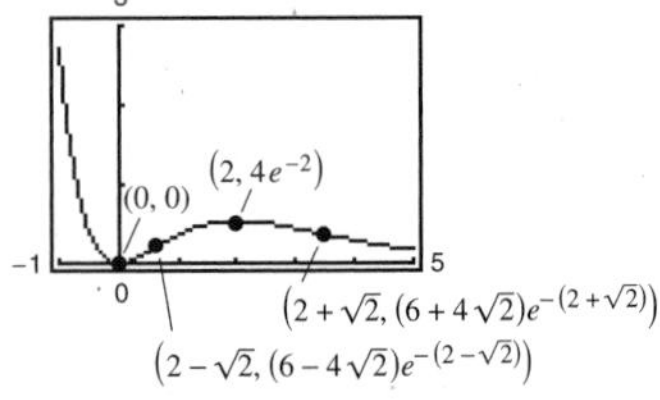

37. Relative maximum: $(-1, 1+e)$
Point of inflection: $(0, 3)$

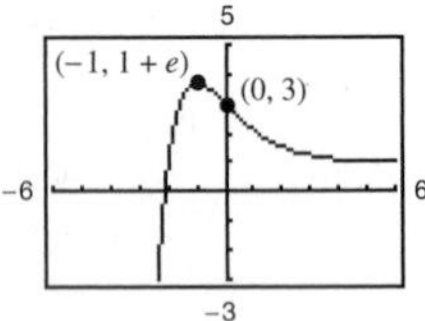

39. $A=\sqrt{2}e^{-1/2}$
41. $(\frac{1}{2}, e)$

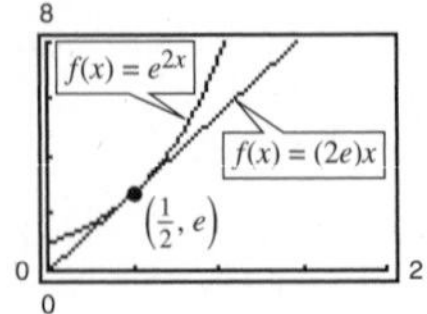

43. (a)

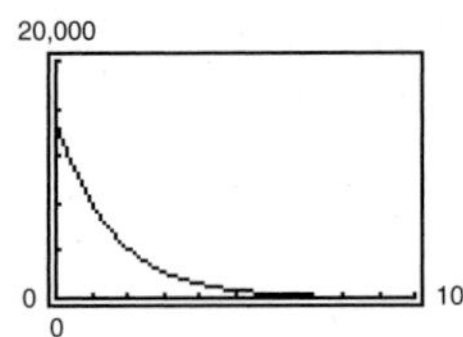

(b) -5028.84; -406.89

(c)

45.

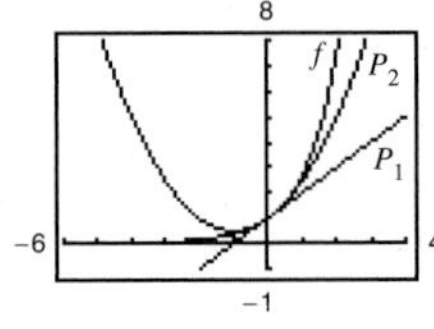

The values of f, P_1, and P_2 and their first derivatives agree at $x=0$.

47. $e^{5x}+C$ **49.** $-\frac{1}{2}e^{-x^2}+C$ **51.** $2e^{\sqrt{x}}+C$
53. $x+2e^x+\frac{1}{2}e^{2x}+C$ **55.** $-\frac{2}{3}(1-e^x)^{3/2}+C$
57. $-\frac{5}{2}e^{-2x}+e^{-x}+C$ **59.** $2\sqrt{e^x-e^{-x}}+C$
61. $(e^2-1)/(2e^2)$ **63.** $e(e^2-1)/3$
65. $-\frac{1}{3}(1+e^{-x})^3+C$ **67.** $1/(1+e^{-x})+C$
69. (a)

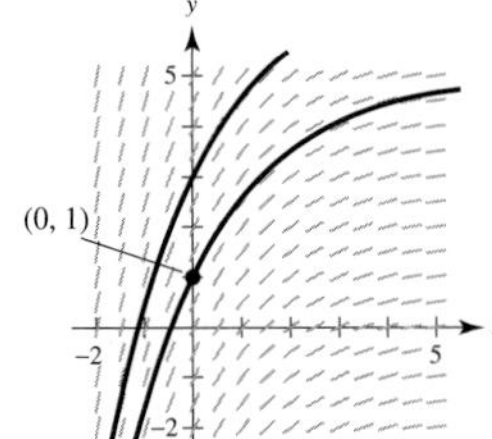

(b) $y=-4e^{-x/2}+5$

71. $e^{ax^2}/(2a)+C$ **73.** $f(x)=\frac{1}{2}(e^x+e^{-x})$
75. $e^5-1\approx 147.413$ **77.** $2(1-e^{-3/2})\approx 1.554$

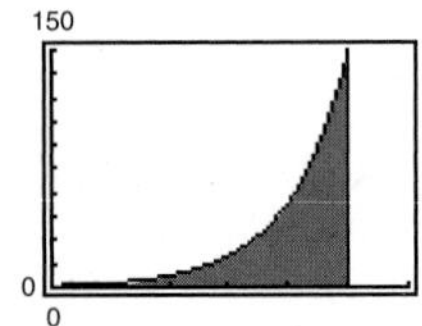

79. Midpoint Rule: 92.1898
Trapezoidal Rule: 93.8371
Simpson's Rule: 92.7385
Graphing utility: 92.7437

81. The probability that a given battery will last between 48 months and 60 months is approximately 47.72%.

83. (a) $t=\dfrac{1}{2k}\ln\dfrac{B}{A}$
(b) $x''(t)=k^2(Ae^{kt}+Be^{-kt})$, k^2 is the constant of proportionality.

85. (a) $R = 428.78e^{-0.6155t}$

(b)

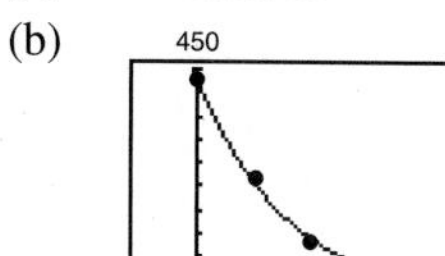

(c) About 637.2 liters

87. (a)

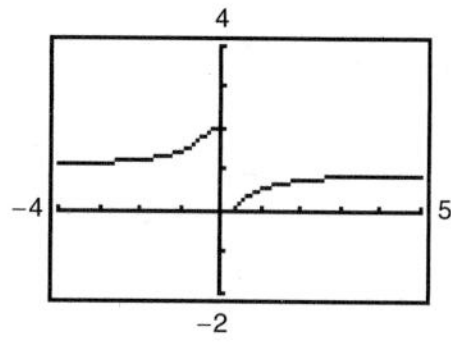

(b) Answers will vary.

89. Proof **91.** $a = \ln 3$

Section 8.2 (page 540)

1. $-\infty$ **3.** $\ln 4$ **5.** 3 **7.** 2 **9.** $2/x$
11. $2(x^3 - 1)/[x(x^3 - 4)]$ **13.** $4(\ln x)^3/x$
15. $(2x^2 - 1)/[x(x^2 - 1)]$ **17.** $(1 - x^2)/[x(x^2 + 1)]$
19. $(1 - 2\ln t)/t^3$ **21.** $2/(x \ln x^2) = 1/(x \ln x)$
23. $1/(1 - x^2)$ **25.** $2x$ **27.** $2e^x/(1 - e^{2x})$
29. $e^{-x}(1/x - \ln x)$ **31.** $-4/[x(x^2 + 4)]$ **33.** $\sqrt{x^2 + 1}/x^2$
35. $(2x)/(x^2 - 1)$ **37.** $(\ln 4)4^x$ **39.** $(\ln 5)5^{x-2}$
41. $t2^t(t \ln 2 + 2)$ **43.** $1/[x(\ln 3)]$ **45.** $5/[(\ln 4)(5x + 1)]$
47. $(x - 2)/[(\ln 2)x(x - 1)]$ **49.** $x/[(\ln 5)(x^2 - 1)]$
51. $5(1 - \ln t)/(t^2 \ln 2)$

53. (a) $5x - y - 2 = 0$

(b)

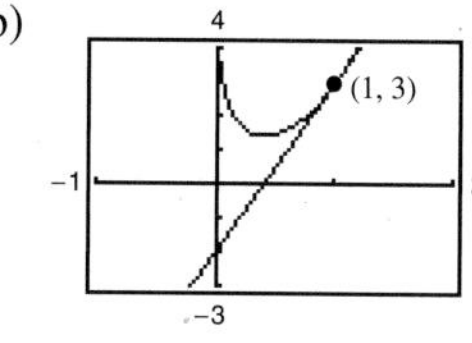

55. (a) $y = x - 1$

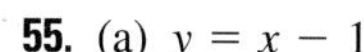

(b)

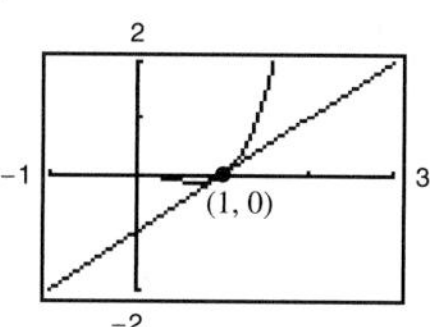

57. $\dfrac{2xy}{3 - 2y^2}$ **59.** $\dfrac{y(1 - 6x^2)}{1 + y}$

61. $xy'' + y' = x(-2/x^2) + 2/x = 0$

63. Relative minimum: $\left(1, \frac{1}{2}\right)$

65. Relative minimum: $(e^{-1}, -e^{-1})$

67. Relative minimum: (e, e)
Point of inflection: $(e^2, e^2/2)$

69. Relative minimum: $\left(\dfrac{\sqrt{2}}{2}, \dfrac{1}{2} - \ln\dfrac{\sqrt{2}}{2}\right) = \left(\dfrac{\sqrt{2}}{2}, \dfrac{1}{2} + \dfrac{1}{2}\ln 2\right)$

71. $P_1 = x - 1$; $P_2 = x - 1 - \frac{1}{2}(x - 1)^2$

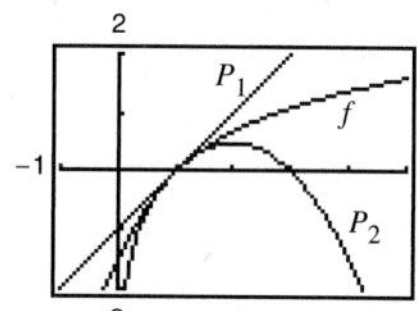

The values of f, P_1, and P_2, and their first derivatives, agree at $x = 1$.

73. $(2x^2 - 1)/\sqrt{x^2 - 1}$
75. $(3x^3 - 15x^2 + 8x)/\left[2(x - 1)^3\sqrt{3x - 2}\right]$
77. $\left[(2x^2 + 2x - 1)\sqrt{x - 1}\right]/(x + 1)^{3/2}$

79. $2(1 - \ln x)x^{(2/x)-2}$ **81.** $(x - 2)^{x+1}\left[\dfrac{x + 1}{x - 2} + \ln(x - 2)\right]$

83. (a) Yes. If the graph of g is increasing, then $g'(x) > 0$. Because $f(x) > 0$, you know that $f'(x) = g'(x)f(x)$ and thus $f'(x) > 0$. Therefore, the graph of f is increasing.
(b) No. Let $f(x) = x^2 + 1$ (positive and concave upward). $g(x) = \ln(x^2 + 1)$ is not concave upward.

85. $g(x), k(x), h(x), f(x)$

87. (a)

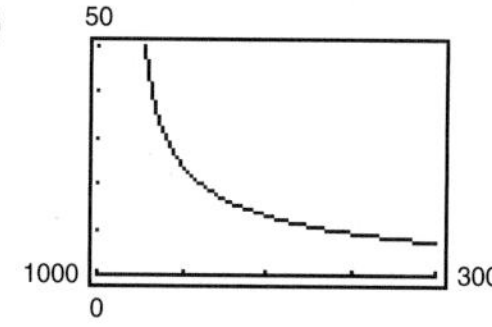

(b) 30 yr; \$503,434.80
(c) 20 yr; \$386,685.60
(d) When $x = 1398.43$, $dt/dx \approx -0.0805$.
When $x = 1611.19$, $dt/dx \approx -0.0287$.
(e) Two benefits of a higher monthly payment are a shorter term and the total amount paid is lower.

89. (a) You get an error message because $\ln h$ does not exist for $h = 0$.
(b) $h = 0.8627 - 6.4474 \ln p$
(c)

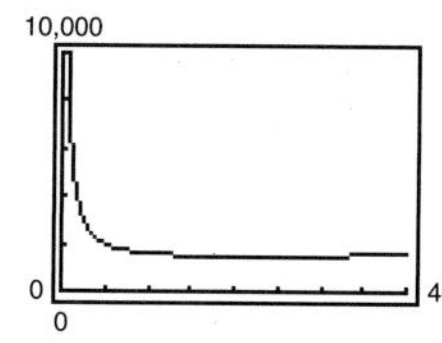

(d) $h \approx 2.72$ km
(e) $p \approx 0.15$ atmosphere
(f) $dp/dh = -0.0853$ atmos/km;
$dp/dh = -0.00931$ atmos/km
As the altitude increases, the rate of change of pressure decreases.

91.

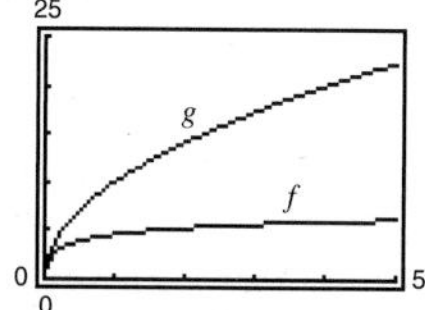

Minimum average cost: \$1498.72

93. (a) 6.7 million ft^3/acre
(b) When $t = 20$, $dV/dt = 0.073$.
When $t = 60$, $dV/dt = 0.040$.

95. (a)

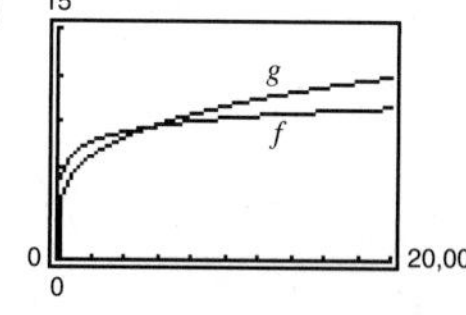

(b)

For large values of x, g increases at a higher rate than f in both cases. The natural logarithmic function increases very slowly for large values of x.

97. False.
$$\frac{d}{dx}[\ln(x^2 + 5x)] = \frac{1}{x^2 + 5x}(2x + 5) \neq \frac{2}{x} + \frac{1}{x}$$

99–101. Proofs

Section 8.3 (page 549)

1. $3 \ln|x| + C$ **3.** $\ln|x + 1| + C$
5. $-\frac{1}{2} \ln|3 - 2x| + C$ **7.** $\ln\sqrt{x^2 + 1} + C$
9. $x^2/2 - \ln(x^4) + C$ **11.** $\frac{1}{3} \ln|x^3 + 3x^2 + 9x| + C$
13. $x^2/2 - 4x + 6 \ln|x + 1| + C$
15. $x^3/3 + 5 \ln|x - 3| + C$ **17.** $x^3/3 - 2x + \ln\sqrt{x^2 + 2} + C$
19. $\frac{1}{3}(\ln x)^3 + C$ **21.** $-\ln(1 + e^{-x}) + C$
23. $\ln|e^x - e^{-x}| + C$ **25.** $2\sqrt{x + 1} + C$
27. $2 \ln|x - 1| - 2/(x - 1) + C$
29. $\sqrt{2x} - \ln(1 + \sqrt{2x}) + C$
31. $x + 6\sqrt{x} + 18 \ln|\sqrt{x} - 3| + C$ **33.** $3^x/\ln 3 + C$
35. $-5^{-x^2}/(2 \ln 5) + C$ **37.** $\ln(1 + 3^{2x})/(2 \ln 3) + C$
39. $7/\ln 4$ **41.** $\frac{5}{3} \ln 13 \approx 4.275$ **43.** $\frac{7}{3}$
45. $-\frac{1}{2} - \ln 2 \approx -1.193$
47. $2[\sqrt{x} - \ln(1 + \sqrt{x})] + C$

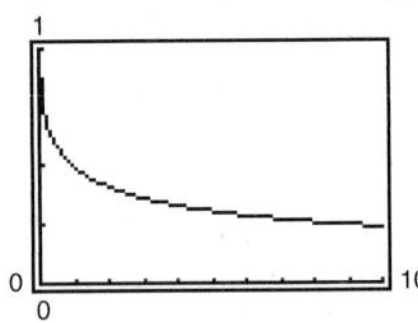

49. $y = -3 \ln|2 - x| + C$

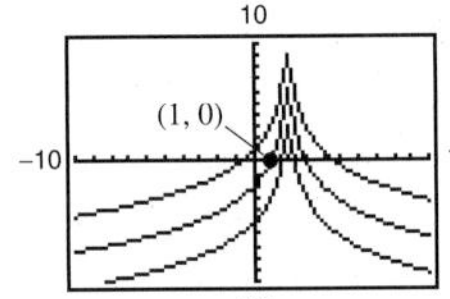

The graph has a hole at $x = 2$.

51. (a)

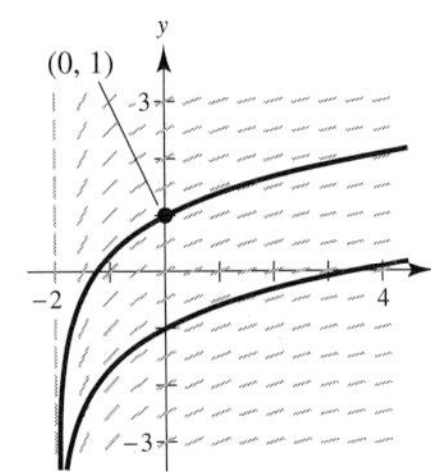

(b) $y = \ln\left|\frac{x + 2}{2}\right| + 1$

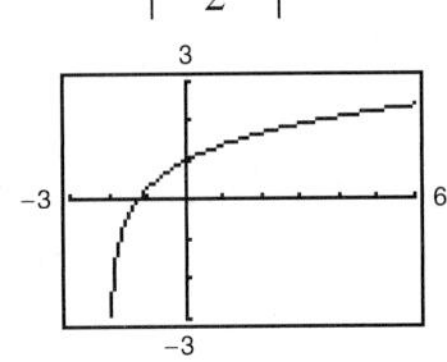

53. (a)

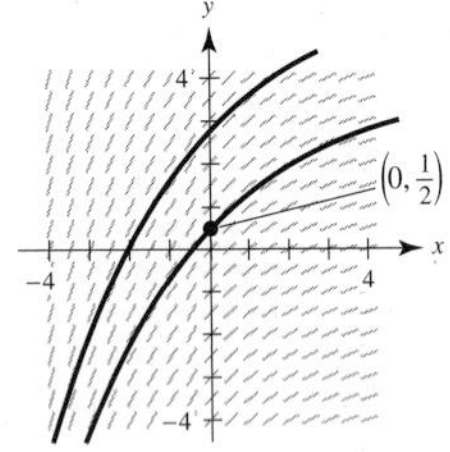

(b) $y = 3(1 - 0.4^{x/3})/\ln 2.5 + 1/2$

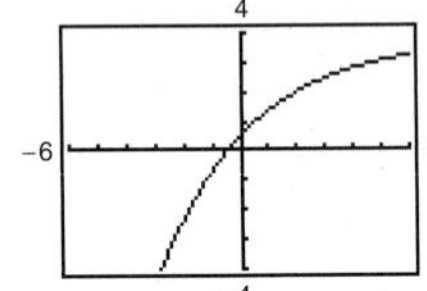

55. $1/x$ **57.** 0

59. $\frac{15}{2} + 8 \ln 2 \approx 13.045$ **61.** $26/\ln 3 \approx 23.666$

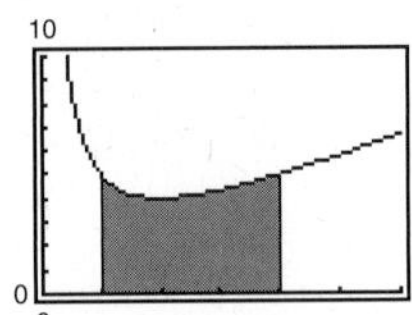

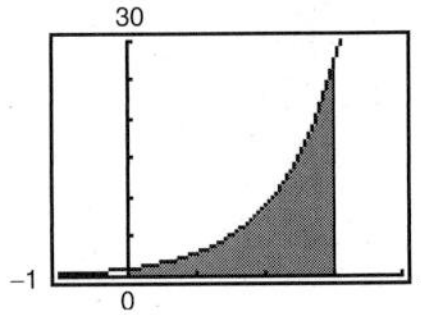

63. Power Rule **65.** u-substitution
67. Use long division to rewrite the integrand. **69.** $x = 2$
71. 1 **73.** $1/[2(e - 1)] \approx 0.291$
75. $P(t) = 1000(12 \ln|1 + 0.25t| + 1)$; $P(3) \approx 7715$
77. \$168.27
79. (a) $\int_0^4 4\left(\frac{3}{8}\right)^{2t/3} \approx 5.66993$

$\int_0^4 4\left(\sqrt[3]{9}/4\right)^t \approx 5.66993$

$\int_0^4 4e^{-0.653886t} \approx 5.66993$

(b)

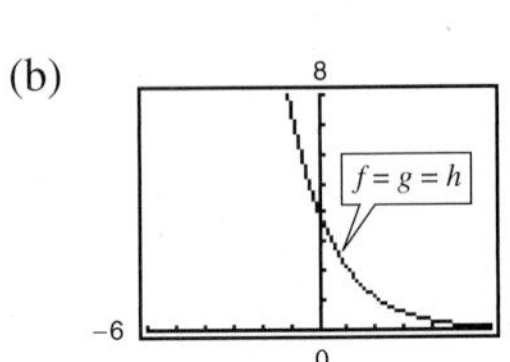

(c) All three functions are equivalent. You cannot make the conjecture using only part (a) because the definite integrals of two functions over a given interval may be equal when the functions are not equal.

81. False. $\frac{d}{dx}[\ln|x|] = \frac{1}{x}$
83. False; the integrand has a nonremovable discontinuity at $x = 0$.
85. Proof

Section 8.4 (page 556)

1. $y = \frac{1}{2}x^2 + 3x + C$ **3.** $y = Ce^x - 3$ **5.** $y^2 - 5x^2 = C$
7. $y = Ce^{(2x^{3/2})/3}$ **9.** $y = C(1 + x^2)$
11. $dQ/dt = k/t^2$
$Q = -k/t + C$
13. $dN/ds = k(500 - s)$
$N = -(k/2)(500 - s)^2 + C$
15. (a)

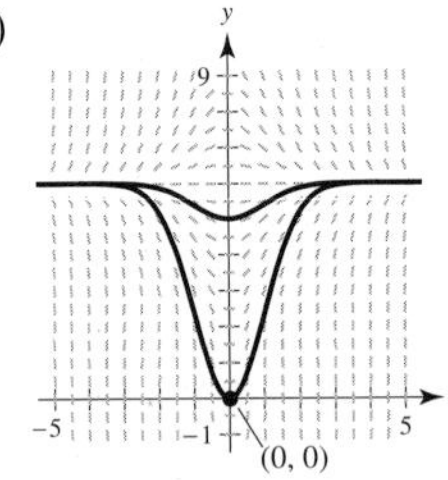

(b) $y = 6 - 6e^{-x^2/2}$

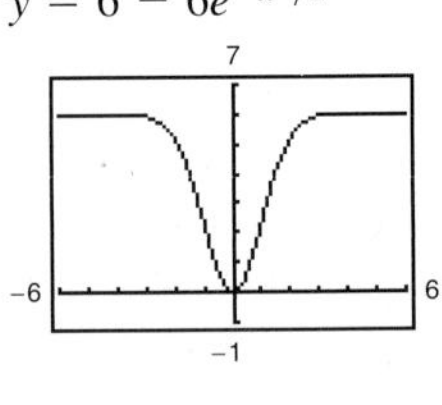

17. $y = \frac{1}{4}t^2 + 10$

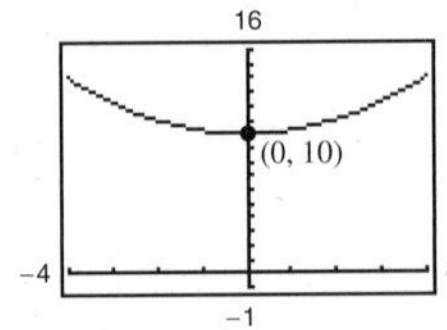

19. $y = 10e^{-t/2}$

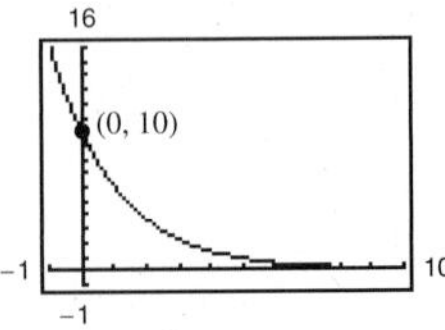

21. $dy/dx = ky$
$y = 6e^{(1/4)\ln(5/2)x} \approx 6e^{0.2291x}$
$y(8) \approx 37.5$
23. $dV/dt = kV$
$V = 20{,}000e^{(1/4)\ln(5/8)t} \approx 20{,}000e^{-0.1175t}$
$V(6) \approx 9882$

25. $y = (1/2)e^{[(\ln 10)/5]t} \approx (1/2)e^{0.4605t}$

27. $y = 5(5/2)^{1/4}e^{[\ln(2/5)/4]t} \approx 6.2872e^{-0.2291t}$

29. Quadrants I and III; dy/dx is positive when both x and y are positive (Quadrant I) or when both x and y are negative (Quadrant III).

31. Amount after 1000 yr: 12.96 g;
Amount after 10,000 yr: 0.26 g

33. Initial quantity: 7.63 g;
Amount after 1000 yr: 4.95 g

35. Amount after 1000 yr: 4.43 g;
Amount after 10,000 yr: 1.49 g

37. Initial quantity: 2.16 g;
Amount after 10,000 yr: 1.62 g

39. 95.76%

41. (a) $P = 2.40e^{-0.006t}$ (b) 2.19 million
(c) Because $k < 0$, the population is decreasing.

43. (a) $P = 5.66e^{0.024t}$ (b) 8.11 million
(c) Because $k > 0$, the population is increasing.

45. (a) $P = 23.55e^{0.036t}$ (b) 40.41 million
(c) Because $k > 0$, the population is increasing.

47. (a) $N = 100.1596(1.2455)^t$ (b) 6.3 h

49. (a) $N \approx 30(1 - e^{-0.0502t})$ (b) 36 days

51. (a) $P_1 \approx 181e^{0.01245t} \approx 181(1.01253)^t$
(b) $P_2 = 182.3248(1.01091)^t$
(c) (d) 2011

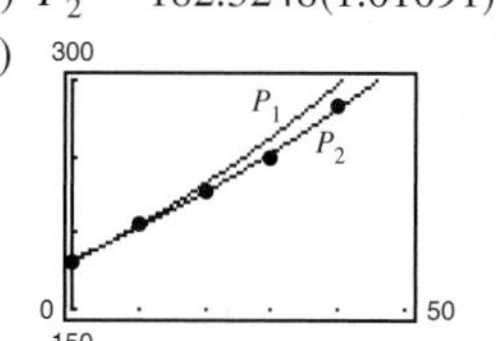

P_2 is a better approximation.

53. (a) 20 dB (b) 70 dB (c) 95 dB (d) 120 dB

55. 2024 ($t = 16$)

57. False. The rate of growth dy/dx is proportional to y.

59. False. The prices are rising at a rate of 6.2% per year.

Review Exercises (page 560)

1. $-3x^2e^{-x^3}$ **3.** $te^t(t + 2)$ **5.** $(e^{2x} - e^{-2x})/\sqrt{e^{2x} + e^{-2x}}$

7. $x(2 - x)/e^x$ **9.** $y = -x + 5$ **11.** $-e^x/(2y)$

13. $y'' - 5y' + 6y = 20e^{2x} - 108e^{3x} - 50e^{2x} + 180e^{3x} + 30e^{2x} - 72e^{3x} = 0$

15. $-\frac{1}{2}e^{1-x^2} + C$ **17.** $(e^{4x} - 3e^{2x} - 3)/(3e^x) + C$

19. $\dfrac{(2 + 5e^{4x})^4}{80} + C$ **21.** $e^7 - 1$ **23.** $e^2 - e^{1/2}$

25. $2\left(\dfrac{1}{\sqrt{e} - 1} - \dfrac{1}{\sqrt{e^3} - 1}\right)$ **27.** $-\frac{1}{2}(e^{-16} - 1) \approx 0.500$

29. $1/x$ **31.** $1/(2x)$ **33.** $(1 + 2\ln x)/(2\sqrt{\ln x})$

35. $(x^2 - 4x + 2)/(x^3 - 3x^2 + 2x)$ **37.** $3^{x-1}\ln 3$

39. $x^2 3^x(x\ln 3 + 3)$ **41.** $x^{2x}(2x\ln x + 2x + 1)$

43. $\dfrac{1}{(\ln 5)x}$ **45.** $\dfrac{-1}{(2 - 2x)\ln 3}$

47. $\sqrt{6}(1 - x^2)/\left[2\sqrt{x}(x^2 + 1)^{3/2}\right]$

49. $\dfrac{4x^2 + 9x + 4}{2\sqrt{(x + 1)(x + 2)}}$ **51.** $\dfrac{-1}{2xy}$ **53.** Proof

55. Relative minimum: $\left(1, \frac{1}{3}\right)$

57. (a) \$40.64 (b) $C'(1) \approx 0.051P, C'(8) \approx 0.072P$
(c) $\ln 1.05$

59. $\frac{1}{7}\ln|7x - 2| + C$ **61.** $\frac{1}{2}x^2 + 7x + 26\ln|x - 3| + C$

63. $\frac{1}{2}\ln(e^{2x} + e^{-2x}) + C$ **65.** $7\ln 5$ **67.** $3 + \ln 4$

69. $\dfrac{1}{2}$ **71.** $\dfrac{4^x}{\ln 4} + C$ **73.** $\dfrac{5^{(x+1)^2}}{2\ln 5} + C$

75. $y = -2\ln|5 - x| + C$ **77.** (a) $P \approx 0.5966$
(b) $P \approx 0.8466$

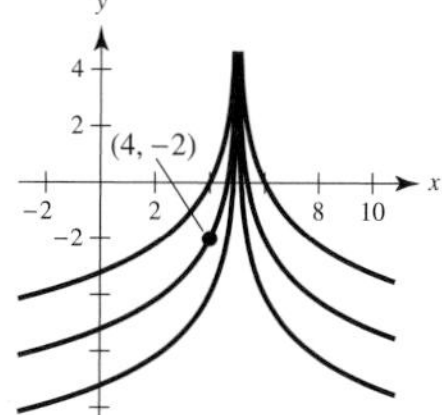

79. $y = 8x - \dfrac{x^2}{2} + C$ **81.** $\dfrac{1}{2}x^2 + 3\ln|x| + C$

83. $y \approx \frac{3}{4}e^{0.379t}$ **85.** $y \approx 5e^{-0.680t}$

87. About 7.79 in. **89.** About 46.2 yr

Chapter Test (page 562)

1. $-5e^{-5x}$ **2.** $-e^{4-x}$ **3.** $e^{6x}(6x^2 + 26x + 4)$

4. $\dfrac{12x^2}{4x^3 - 1}$ **5.** $(-4\ln 5)5^{-4x}$ **6.** $\dfrac{-2}{(\ln 5)(4 - x)}$

7. $y = x + 7$ **8.** $4e^{-2x}(3x - 4)$ **9.** $\frac{1}{2}e^{2x-1} + C$

10. $\dfrac{x^2}{2} + \ln|x| + C$ **11.** $\dfrac{1}{6\ln 4}(4^{3x^2}) + C$

12. $\dfrac{e^8}{2} + \dfrac{7e^4}{2} - 4e^2 + 8$ **13.** $\dfrac{\ln 7}{3}$ **14.** $\dfrac{8}{\ln 9}$ **15.** $\dfrac{e^2 + 11}{4}$

16. Relative minimum: $\left(\dfrac{1}{e^{1/2}}, -\dfrac{1}{2e}\right)$
Point of inflection: $\left(\dfrac{1}{e^{3/2}}, -\dfrac{3}{2e^3}\right)$

17. $\dfrac{8x^2 + 21x}{2\sqrt{2x^2 + 7x}}$ **18.** $\dfrac{26x^2 - 1040}{(x + 5)^2(x + 8)^2}$

19. $x^{x^2+1}(1 + 2\ln x)$

20. $y = 2\ln|x + 4| + C$

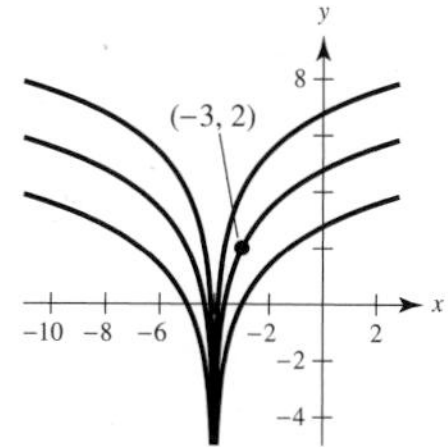

21. $2\ln 4$ **22.** $y = x^2 + 6x + C$ **23.** $4y^2 - x^2 = C$

24. $y = Ce^{x^3/3}$ **25.** $y = 6e^{-0.5973t}$ **26.** $y = \frac{3}{2}e^{0.1733t}$

27. $y = 0.6687e^{0.4024t}$ **28.** 17.72 g

P.S. Problem Solving (page 563)

1. $(1, e^{-1})$; Maximum area $= 2e^{-1} \approx 0.7358$

3. (a) Proof (b) $4\sqrt{2}/3$ (c) $e^2 - 1$

5. (a)–(c) Proofs

7. Proof

9. (a) $y = 1/(1 - 0.01t)^{100}$; $T = 100$
(b) $y = 1/[(1/y_0)^{\varepsilon} - k\varepsilon t]^{1/\varepsilon}$; Answers will vary.

11. $2\ln\left(\frac{3}{2}\right) \approx 0.8109$

13. (a) $y = \dfrac{1}{1 + 3e^{-t}}$
(b) Proof

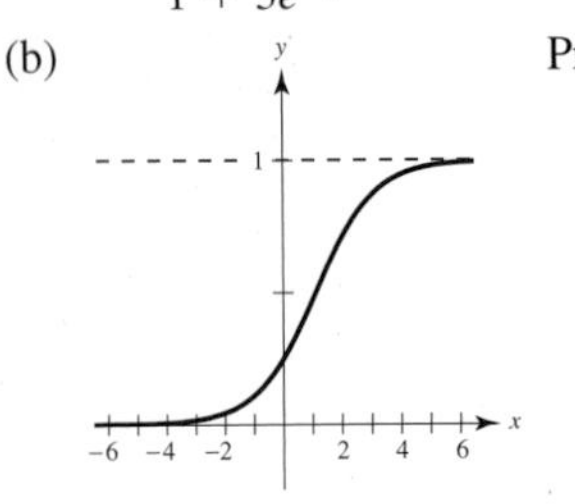

(c) $y = \dfrac{2}{2 - e^{-t}}$;
The graph is different:

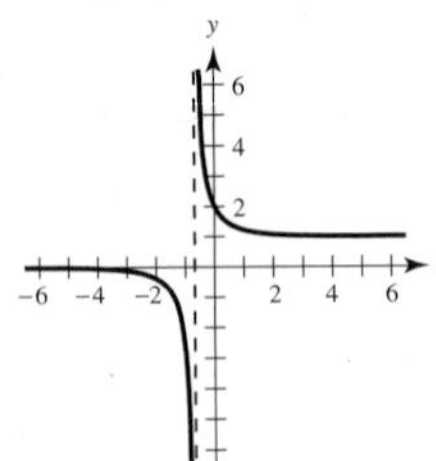

Chapter 9

Section 9.1 (page 572)

1. Trigonometry **3.** coterminal **5.** acute; obtuse
7. degree **9.** arc length **11.** 1 rad **13.** 5.5 rad
15. −3 rad
17. (a) Quadrant I (b) Quadrant III
19. (a) Quadrant IV (b) Quadrant IV
21. (a) Quadrant III (b) Quadrant II
23. (a) (b)

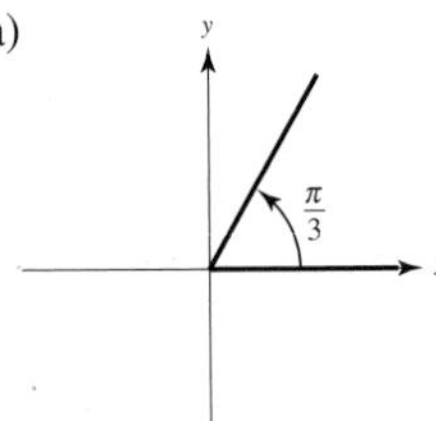

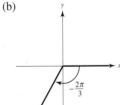

25. (a) (b)

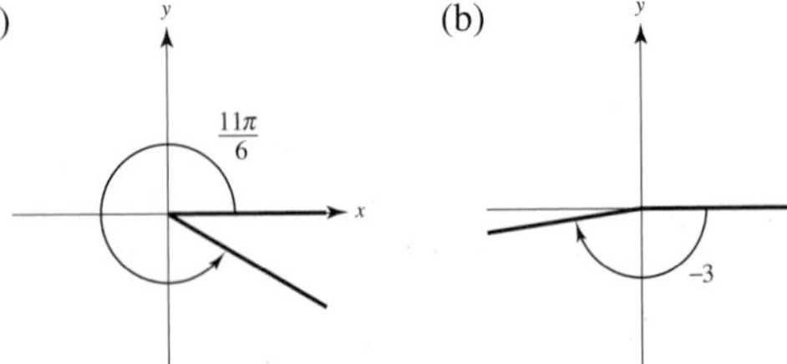

27. Sample answers: (a) $\dfrac{13\pi}{6}, -\dfrac{11\pi}{6}$ (b) $\dfrac{17\pi}{6}, -\dfrac{7\pi}{6}$

29. Sample answers: (a) $\dfrac{8\pi}{3}, -\dfrac{4\pi}{3}$ (b) $\dfrac{25\pi}{12}, -\dfrac{23\pi}{12}$

31. (a) Complement: $\dfrac{\pi}{6}$; Supplement: $\dfrac{2\pi}{3}$
(b) Complement: $\dfrac{\pi}{4}$; Supplement: $\dfrac{3\pi}{4}$

33. (a) Complement: $\dfrac{\pi}{2} - 1 \approx 0.57$;
Supplement: $\pi - 1 \approx 2.14$
(b) Complement: none; Supplement: $\pi - 2 \approx 1.14$

35. 210° **37.** −60° **39.** 165°
41. (a) Quadrant II (b) Quadrant IV
43. (a) Quadrant III (b) Quadrant I
45. (a) (b)

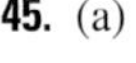

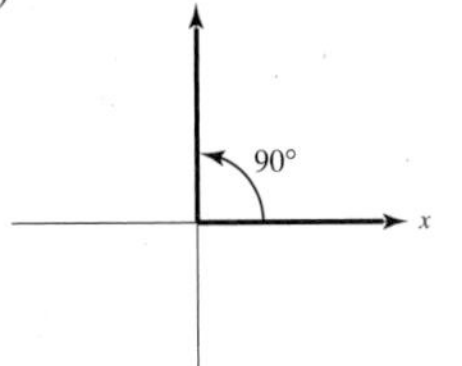

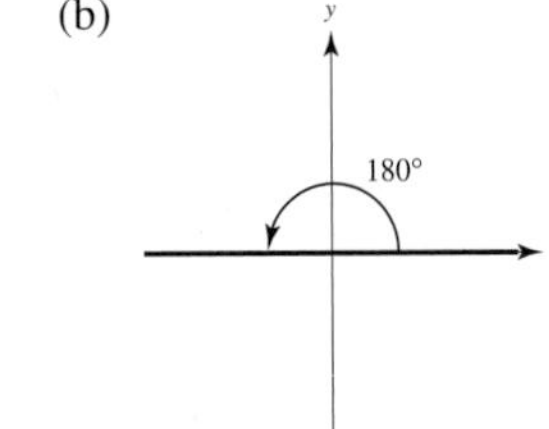

47. (a) (b)

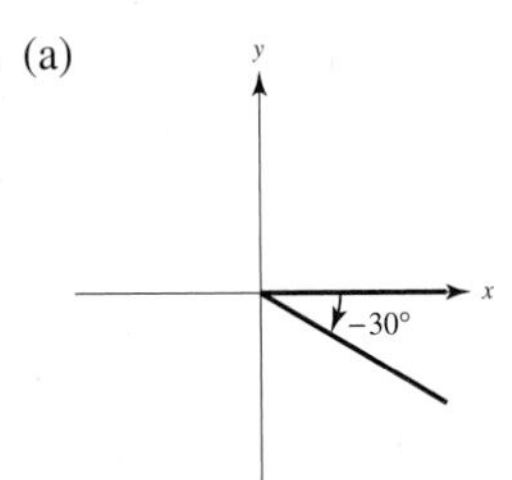

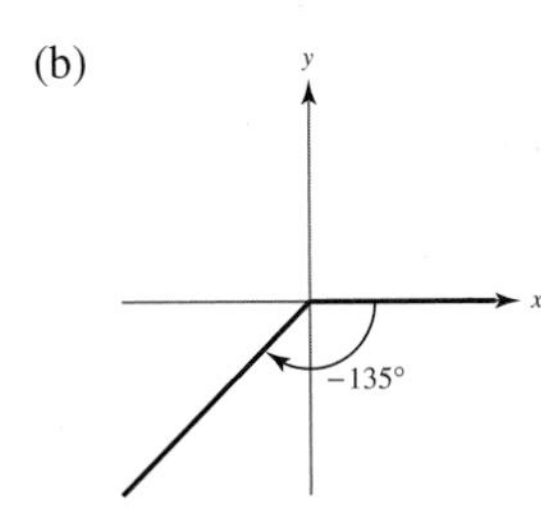

49. Sample answers: (a) 405°, −315° (b) 324°, −396°
51. Sample answers: (a) 600°, −120° (b) 180°, −540°
53. (a) Complement: 72°; Supplement: 162°
(b) Complement: 5°; Supplement: 95°
55. (a) Complement: none; Supplement: 30°
(b) Complement: 11°; Supplement: 101°

57. (a) $\dfrac{\pi}{6}$ (b) $\dfrac{\pi}{4}$ **59.** (a) $-\dfrac{\pi}{9}$ (b) $-\dfrac{\pi}{3}$

61. (a) 270° (b) 210° **63.** (a) 225° (b) −420°
65. 0.785 **67.** −3.776 **69.** 9.285 **71.** −0.014
73. 25.714° **75.** 337.500° **77.** −756.000°
79. −114.592° **81.** (a) 54.75° (b) −128.5°
83. (a) 85.308° (b) 330.007°
85. (a) 240° 36′ (b) −145° 48′
87. (a) 2° 30′ (b) −3° 34′ 48″ **89.** 10π in. ≈ 31.42 in.
91. 2.5π m ≈ 7.85 m **93.** $\frac{9}{2}$ rad **95.** $\frac{21}{50}$ rad **97.** $\frac{1}{2}$ rad
99. 4 rad
101. Increases; because the linear speed is proportional to the radius.
103. The arc length is increasing. In order for the angle θ to remain constant as the radius r increases, the arc length s must increase in proportion to r, as can be seen from the formula $s = r\theta$.
105. 686.2 mi
107. (a) 8π rad/min ≈ 25.13 rad/min
(b) 200π ft/min ≈ 628.3 ft/min
109. (a) 910.37 revolutions/min (b) 5720 rad/min
111. $\dfrac{14\pi}{3}$ ft/sec ≈ 10 mi/h

113. True. Let α and β represent coterminal angles, and let n represent an integer.

$\alpha = \beta + n(360°)$

$\alpha - \beta = n(360°)$

Section 9.2 (page 580)

1. unit circle **3.** period

5. $\sin t = \frac{5}{13}$ $\csc t = \frac{13}{5}$

$\cos t = \frac{12}{13}$ $\sec t = \frac{13}{12}$

$\tan t = \frac{5}{12}$ $\cot t = \frac{12}{5}$

7. $\sin t = -\frac{3}{5}$ $\csc t = -\frac{5}{3}$

$\cos t = -\frac{4}{5}$ $\sec t = -\frac{5}{4}$

$\tan t = \frac{3}{4}$ $\cot t = \frac{4}{3}$

9. $(0, 1)$ **11.** $\left(\frac{\sqrt{2}}{2}, \frac{\sqrt{2}}{2}\right)$

13. $\left(-\frac{\sqrt{3}}{2}, \frac{1}{2}\right)$ **15.** $\left(-\frac{1}{2}, -\frac{\sqrt{3}}{2}\right)$

17. $\sin \frac{\pi}{4} = \frac{\sqrt{2}}{2}$

$\cos \frac{\pi}{4} = \frac{\sqrt{2}}{2}$

$\tan \frac{\pi}{4} = 1$

19. $\sin\left(-\frac{\pi}{6}\right) = -\frac{1}{2}$

$\cos\left(-\frac{\pi}{6}\right) = \frac{\sqrt{3}}{2}$

$\tan\left(-\frac{\pi}{6}\right) = -\frac{\sqrt{3}}{3}$

21. $\sin\left(-\frac{7\pi}{4}\right) = \frac{\sqrt{2}}{2}$

$\cos\left(-\frac{7\pi}{4}\right) = \frac{\sqrt{2}}{2}$

$\tan\left(-\frac{7\pi}{4}\right) = 1$

23. $\sin \frac{11\pi}{6} = -\frac{1}{2}$

$\cos \frac{11\pi}{6} = \frac{\sqrt{3}}{2}$

$\tan \frac{11\pi}{6} = -\frac{\sqrt{3}}{3}$

25. $\sin\left(-\frac{3\pi}{2}\right) = 1$

$\cos\left(-\frac{3\pi}{2}\right) = 0$

$\tan\left(-\frac{3\pi}{2}\right)$ is undefined.

27. $\sin \frac{2\pi}{3} = \frac{\sqrt{3}}{2}$ $\csc \frac{2\pi}{3} = \frac{2\sqrt{3}}{3}$

$\cos \frac{2\pi}{3} = -\frac{1}{2}$ $\sec \frac{2\pi}{3} = -2$

$\tan \frac{2\pi}{3} = -\sqrt{3}$ $\cot \frac{2\pi}{3} = -\frac{\sqrt{3}}{3}$

29. $\sin \frac{4\pi}{3} = -\frac{\sqrt{3}}{2}$ $\csc \frac{4\pi}{3} = -\frac{2\sqrt{3}}{3}$

$\cos \frac{4\pi}{3} = -\frac{1}{2}$ $\sec \frac{4\pi}{3} = -2$

$\tan \frac{4\pi}{3} = \sqrt{3}$ $\cot \frac{4\pi}{3} = \frac{\sqrt{3}}{3}$

31. $\sin \frac{3\pi}{4} = \frac{\sqrt{2}}{2}$ $\csc \frac{3\pi}{4} = \sqrt{2}$

$\cos \frac{3\pi}{4} = -\frac{\sqrt{2}}{2}$ $\sec \frac{3\pi}{4} = -\sqrt{2}$

$\tan \frac{3\pi}{4} = -1$ $\cot \frac{3\pi}{4} = -1$

33. $\sin\left(-\frac{\pi}{2}\right) = -1$ $\csc\left(-\frac{\pi}{2}\right) = -1$

$\cos\left(-\frac{\pi}{2}\right) = 0$ $\sec\left(-\frac{\pi}{2}\right)$ is undefined.

$\tan\left(-\frac{\pi}{2}\right)$ is undefined. $\cot\left(-\frac{\pi}{2}\right) = 0$

35. $\sin 4\pi = \sin 0 = 0$ **37.** $\cos \frac{7\pi}{3} = \cos \frac{\pi}{3} = \frac{1}{2}$

39. $\cos \frac{17\pi}{4} = \cos \frac{\pi}{4} = \frac{\sqrt{2}}{2}$

41. $\sin\left(-\frac{8\pi}{3}\right) = \sin \frac{4\pi}{3} = -\frac{\sqrt{3}}{2}$

43. (a) $-\frac{1}{2}$ (b) -2 **45.** (a) $-\frac{1}{5}$ (b) -5

47. (a) $\frac{4}{5}$ (b) $-\frac{4}{5}$ **49.** 0.7071 **51.** 1.0000

53. -0.1288 **55.** 1.3940 **57.** -1.4486

59. (a) 0.25 ft (b) 0.02 ft (c) -0.25 ft

61. False. $\sin(-t) = -\sin(t)$ means that the function is odd, not that the sine of a negative angle is a negative number.

63. False. The real number 0 corresponds to the point $(1, 0)$.

65. Answers will vary.

67. (a) y-axis symmetry (b) $\sin t_1 = \sin(\pi - t_1)$

(c) $\cos(\pi - t_1) = -\cos t_1$

69. It is an odd function.

71. (a)

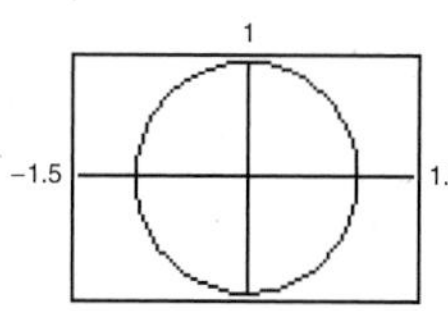

Circle of radius 1 centered at $(0, 0)$

(b) The t-values represent the central angle in radians. The x- and y-values represent the location in the coordinate plane.

(c) $-1 \le x \le 1, -1 \le y \le 1$

Section 9.3 (page 587)

1. (a) v (b) iv (c) vi (d) iii (e) i (f) ii

3. complementary

5. $\sin \theta = \frac{3}{5}$ $\csc \theta = \frac{5}{3}$

$\cos \theta = \frac{4}{5}$ $\sec \theta = \frac{5}{4}$

$\tan \theta = \frac{3}{4}$ $\cot \theta = \frac{4}{3}$

7. $\sin \theta = \frac{9}{41}$ $\csc \theta = \frac{41}{9}$

$\cos \theta = \frac{40}{41}$ $\sec \theta = \frac{41}{40}$

$\tan \theta = \frac{9}{40}$ $\cot \theta = \frac{40}{9}$

9. $\sin \theta = \frac{8}{17}$ $\csc \theta = \frac{17}{8}$

$\cos \theta = \frac{15}{17}$ $\sec \theta = \frac{17}{15}$

$\tan \theta = \frac{8}{15}$ $\cot \theta = \frac{15}{8}$

The triangles are similar, and corresponding sides are proportional.

11. $\sin \theta = \frac{1}{3}$ $\csc \theta = 3$

$\cos \theta = \frac{2\sqrt{2}}{3}$ $\sec \theta = \frac{3\sqrt{2}}{4}$

$\tan \theta = \frac{\sqrt{2}}{4}$ $\cot \theta = 2\sqrt{2}$

The triangles are similar, and corresponding sides are proportional.

13. 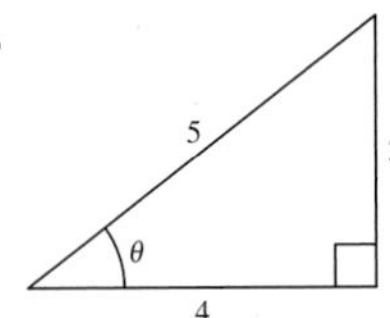

$\sin\theta = \frac{3}{5}$ $\csc\theta = \frac{5}{3}$
$\cos\theta = \frac{4}{5}$ $\sec\theta = \frac{5}{4}$
$\cot\theta = \frac{4}{3}$

15. 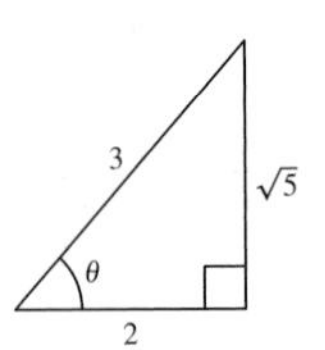

$\sin\theta = \dfrac{\sqrt{5}}{3}$ $\csc\theta = \dfrac{3\sqrt{5}}{5}$
$\cos\theta = \dfrac{2}{3}$
$\tan\theta = \dfrac{\sqrt{5}}{2}$ $\cot\theta = \dfrac{2\sqrt{5}}{5}$

17. $\csc\theta = 5$
$\cos\theta = \dfrac{2\sqrt{6}}{5}$ $\sec\theta = \dfrac{5\sqrt{6}}{12}$
$\tan\theta = \dfrac{\sqrt{6}}{12}$ $\cot\theta = 2\sqrt{6}$

19. $\sin\theta = \dfrac{\sqrt{10}}{10}$ $\csc\theta = \sqrt{10}$
$\cos\theta = \dfrac{3\sqrt{10}}{10}$ $\sec\theta = \dfrac{\sqrt{10}}{3}$
$\tan\theta = \dfrac{1}{3}$

21. $\dfrac{\pi}{6}; \dfrac{1}{2}$ **23.** $45°; \sqrt{2}$ **25.** $60°; \dfrac{\pi}{3}$ **27.** 30°; 2

29. $45°; \dfrac{\pi}{4}$ **31.** (a) $\dfrac{1}{2}$ (b) $\dfrac{\sqrt{3}}{2}$ (c) $\sqrt{3}$ (d) $\dfrac{\sqrt{3}}{3}$

33. (a) $\dfrac{2\sqrt{2}}{3}$ (b) $2\sqrt{2}$ (c) 3 (d) 3

35. (a) $\dfrac{1}{5}$ (b) $\sqrt{26}$ (c) $\dfrac{1}{5}$ (d) $\dfrac{5\sqrt{26}}{26}$

37–45. Answers will vary. **47.** (a) 0.1736 (b) 0.1736

49. (a) 0.2815 (b) 3.5523 **51.** (a) 0.9964 (b) 1.0036

53. (a) 5.0273 (b) 0.1989 **55.** (a) 1.8527 (b) 0.9817

57. (a) $30° = \dfrac{\pi}{6}$ (b) $30° = \dfrac{\pi}{6}$

59. (a) $60° = \dfrac{\pi}{3}$ (b) $45° = \dfrac{\pi}{4}$

61. (a) $60° = \dfrac{\pi}{3}$ (b) $45° = \dfrac{\pi}{4}$ **63.** $9\sqrt{3}$ **65.** $\dfrac{32\sqrt{3}}{3}$

67. Corresponding sides of similar triangles are proportional.

69. (a)

θ	0.1	0.2	0.3	0.4	0.5
$\sin\theta$	0.0998	0.1987	0.2955	0.3894	0.4794

(b) θ (c) As $\theta \to 0$, $\sin\theta \to 0$ and $\dfrac{\theta}{\sin\theta} \to 1$.

71. 443.2 m; 323.3 m **73.** $30° = \pi/6$

75. $(x_1, y_1) = (28\sqrt{3}, 28)$
$(x_2, y_2) = (28, 28\sqrt{3})$

77. $\sin 20° \approx 0.34$, $\cos 20° \approx 0.94$, $\tan 20° \approx 0.36$, $\csc 20° \approx 2.92$, $\sec 20° \approx 1.06$, $\cot 20° \approx 2.75$

79. (a) 219.9 ft (b) 160.9 ft

81. True, $\csc x = \dfrac{1}{\sin x}$. **83.** False, $\dfrac{\sqrt{2}}{2} + \dfrac{\sqrt{2}}{2} \neq 1$.

85. False, $1.7321 \neq 0.0349$.

87. Sample answer:

$x = \cos 30° = \dfrac{\sqrt{3}}{2}$ $y = \sin 30° = \dfrac{1}{2} \Rightarrow \left(\dfrac{\sqrt{3}}{2}, \dfrac{1}{2}\right)$

$x = \cos 60° = \dfrac{1}{2}$ $y = \sin 60° = \dfrac{\sqrt{3}}{2} \Rightarrow \left(\dfrac{1}{2}, \dfrac{\sqrt{3}}{2}\right)$

Signs change depending on what quadrant the point lies in.

Section 9.4 (page 596)

1. $\dfrac{y}{r}$ **3.** $\dfrac{y}{x}$ **5.** $\cos\theta$ **7.** zero; defined

9. (a) $\sin\theta = \frac{3}{5}$ $\csc\theta = \frac{5}{3}$
$\cos\theta = \frac{4}{5}$ $\sec\theta = \frac{5}{4}$
$\tan\theta = \frac{3}{4}$ $\cot\theta = \frac{4}{3}$
(b) $\sin\theta = \frac{15}{17}$ $\csc\theta = \frac{17}{15}$
$\cos\theta = -\frac{8}{17}$ $\sec\theta = -\frac{17}{8}$
$\tan\theta = -\frac{15}{8}$ $\cot\theta = -\frac{8}{15}$

11. (a) $\sin\theta = -\dfrac{1}{2}$ $\csc\theta = -2$
$\cos\theta = -\dfrac{\sqrt{3}}{2}$ $\sec\theta = -\dfrac{2\sqrt{3}}{3}$
$\tan\theta = \dfrac{\sqrt{3}}{3}$ $\cot\theta = \sqrt{3}$
(b) $\sin\theta = -\dfrac{\sqrt{17}}{17}$ $\csc\theta = -\sqrt{17}$
$\cos\theta = \dfrac{4\sqrt{17}}{17}$ $\sec\theta = \dfrac{\sqrt{17}}{4}$
$\tan\theta = -\dfrac{1}{4}$ $\cot\theta = -4$

13. $\sin\theta = \frac{12}{13}$ $\csc\theta = \frac{13}{12}$
$\cos\theta = \frac{5}{13}$ $\sec\theta = \frac{13}{5}$
$\tan\theta = \frac{12}{5}$ $\cot\theta = \frac{5}{12}$

15. $\sin\theta = -\dfrac{2\sqrt{29}}{29}$ $\csc\theta = -\dfrac{\sqrt{29}}{2}$
$\cos\theta = -\dfrac{5\sqrt{29}}{29}$ $\sec\theta = -\dfrac{\sqrt{29}}{5}$
$\tan\theta = \dfrac{2}{5}$ $\cot\theta = \dfrac{5}{2}$

17. $\sin\theta = \frac{4}{5}$ $\csc\theta = \frac{5}{4}$
$\cos\theta = -\frac{3}{5}$ $\sec\theta = -\frac{5}{3}$
$\tan\theta = -\frac{4}{3}$ $\cot\theta = -\frac{3}{4}$

19. Quadrant I **21.** Quadrant II

23. $\sin\theta = \frac{15}{17}$ $\csc\theta = \frac{17}{15}$
$\cos\theta = -\frac{8}{17}$ $\sec\theta = -\frac{17}{8}$
$\tan\theta = -\frac{15}{8}$ $\cot\theta = -\frac{8}{15}$

25. $\sin\theta = \frac{3}{5}$ $\csc\theta = \frac{5}{3}$
$\cos\theta = -\frac{4}{5}$ $\sec\theta = -\frac{5}{4}$
$\tan\theta = -\frac{3}{4}$ $\cot\theta = -\frac{4}{3}$

27. $\sin\theta = -\frac{\sqrt{10}}{10}$ $\csc\theta = -\sqrt{10}$
$\cos\theta = \frac{3\sqrt{10}}{10}$ $\sec\theta = \frac{\sqrt{10}}{3}$
$\tan\theta = -\frac{1}{3}$ $\cot\theta = -3$

29. $\sin\theta = -\frac{\sqrt{3}}{2}$ $\csc\theta = -\frac{2\sqrt{3}}{3}$
$\cos\theta = -\frac{1}{2}$ $\sec\theta = -2$
$\tan\theta = \sqrt{3}$ $\cot\theta = \frac{\sqrt{3}}{3}$

31. $\sin\theta = 0$ $\csc\theta$ is undefined.
$\cos\theta = -1$ $\sec\theta = -1$
$\tan\theta = 0$ $\cot\theta$ is undefined.

33. $\sin\theta = \frac{\sqrt{2}}{2}$ $\csc\theta = \sqrt{2}$
$\cos\theta = -\frac{\sqrt{2}}{2}$ $\sec\theta = -\sqrt{2}$
$\tan\theta = -1$ $\cot\theta = -1$

35. $\sin\theta = -\frac{2\sqrt{5}}{5}$ $\csc\theta = -\frac{\sqrt{5}}{2}$
$\cos\theta = -\frac{\sqrt{5}}{5}$ $\sec\theta = -\sqrt{5}$
$\tan\theta = 2$ $\cot\theta = \frac{1}{2}$

37. 0 **39.** Undefined **41.** 1 **43.** Undefined

45. $\theta' = 20°$

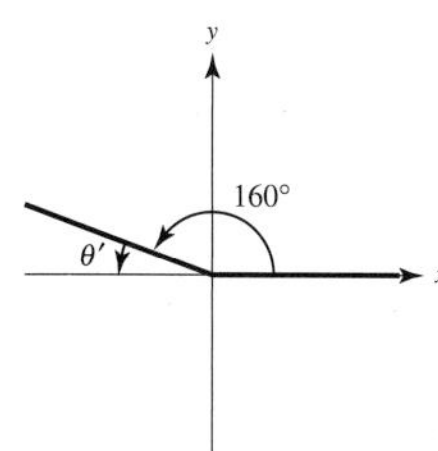

47. $\theta' = 55°$

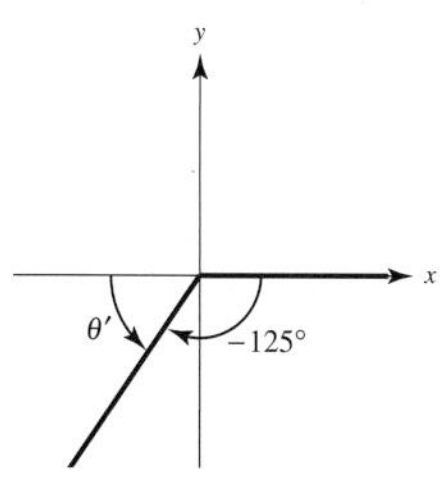

49. $\theta' = \frac{\pi}{3}$

51. $\theta' = 2\pi - 4.8$

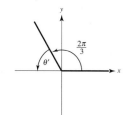

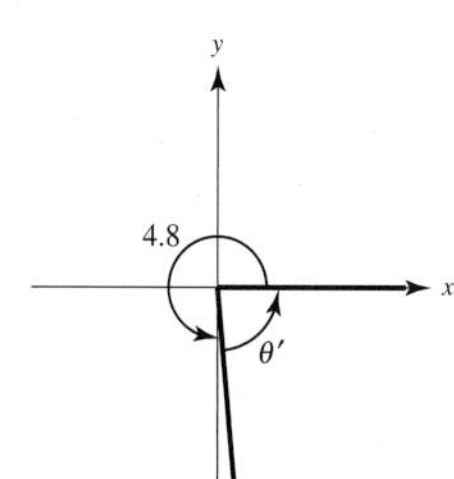

53. $\sin 225° = -\frac{\sqrt{2}}{2}$
$\cos 225° = -\frac{\sqrt{2}}{2}$
$\tan 225° = 1$

55. $\sin 750° = \frac{1}{2}$
$\cos 750° = \frac{\sqrt{3}}{2}$
$\tan 750° = \frac{\sqrt{3}}{3}$

57. $\sin(-150°) = -\frac{1}{2}$
$\cos(-150°) = -\frac{\sqrt{3}}{2}$
$\tan(-150°) = \frac{\sqrt{3}}{3}$

59. $\sin\frac{2\pi}{3} = \frac{\sqrt{3}}{2}$
$\cos\frac{2\pi}{3} = -\frac{1}{2}$
$\tan\frac{2\pi}{3} = -\sqrt{3}$

61. $\sin\frac{5\pi}{4} = -\frac{\sqrt{2}}{2}$
$\cos\frac{5\pi}{4} = -\frac{\sqrt{2}}{2}$
$\tan\frac{5\pi}{4} = 1$

63. $\sin\left(-\frac{\pi}{6}\right) = -\frac{1}{2}$
$\cos\left(-\frac{\pi}{6}\right) = \frac{\sqrt{3}}{2}$
$\tan\left(-\frac{\pi}{6}\right) = -\frac{\sqrt{3}}{3}$

65. $\sin\frac{9\pi}{4} = \frac{\sqrt{2}}{2}$
$\cos\frac{9\pi}{4} = \frac{\sqrt{2}}{2}$
$\tan\frac{9\pi}{4} = 1$

67. $\sin\left(-\frac{3\pi}{2}\right) = 1$
$\cos\left(-\frac{3\pi}{2}\right) = 0$
$\tan\left(-\frac{3\pi}{2}\right)$ is undefined.

69. $\frac{4}{5}$ **71.** $-\frac{\sqrt{13}}{2}$ **73.** $\frac{8}{5}$

75. 0.1736 **77.** -0.3420 **79.** 4.6373

81. 0.3640 **83.** -0.6052 **85.** -0.4142

87. (a) $30° = \frac{\pi}{6}$, $150° = \frac{5\pi}{6}$ (b) $210° = \frac{7\pi}{6}$, $330° = \frac{11\pi}{6}$

89. (a) $60° = \frac{\pi}{3}$, $120° = \frac{2\pi}{3}$ (b) $135° = \frac{3\pi}{4}$, $315° = \frac{7\pi}{4}$

91. (a) $45° = \frac{\pi}{4}$, $225° = \frac{5\pi}{4}$ (b) $150° = \frac{5\pi}{6}$, $330° = \frac{11\pi}{6}$

93. As θ increases from 0° to 90°, x decreases from 12 cm to 0 cm and y increases from 0 cm to 12 cm. Therefore, $\sin\theta = y/12$ increases from 0 to 1 and $\cos\theta = x/12$ decreases from 1 to 0. Thus, $\tan\theta = y/x$ increases without bound. When $\theta = 90°$, the tangent is undefined.

95. (a) 26,134 units (b) 31,438 units
(c) 21,452 units (d) 26,756 units

97. (a) 2 cm (b) 0.11 cm (c) -1.2 cm

99. (a) $N = 22.099\sin(0.522t - 2.219) + 55.008$
$F = 36.641\sin(0.502t - 1.831) + 25.610$
(b) February: $N = 34.6°$, $F = -1.4°$
March: $N = 41.6°$, $F = 13.9°$
May: $N = 63.4°$, $F = 48.6°$
June: $N = 72.5°$, $F = 59.5°$
August: $N = 75.5°$, $F = 55.6°$
September: $N = 68.6°$, $F = 41.7°$
November: $N = 46.8°$, $F = 6.5°$
(c) Answers will vary.

101. False. In each of the four quadrants, the signs of the secant function and the cosine function will be the same, because these functions are reciprocals of each other.

103. True

Section 9.5 (page 606)

1. cycle **3.** phase shift **5.** Period: $\frac{2\pi}{5}$; Amplitude: 2

7. Period: 4π; Amplitude: $\frac{3}{4}$ **9.** Period: 6; Amplitude: $\frac{1}{2}$

11. Period: 2π; Amplitude: 4

13. Period: $\frac{\pi}{5}$; Amplitude: 3 **15.** Period: $\frac{5\pi}{2}$; Amplitude: $\frac{5}{3}$

17. Period: 1; Amplitude: $\frac{1}{4}$

19. g is a shift of f π units to the right.

21. g is a reflection of f in the x-axis.

23. The period of f is twice the period of g.

25. g is a shift of f three units upward.

27. The graph of g has twice the amplitude of the graph of f.

29. The graph of g is a horizontal shift of the graph of f π units to the right.

31.

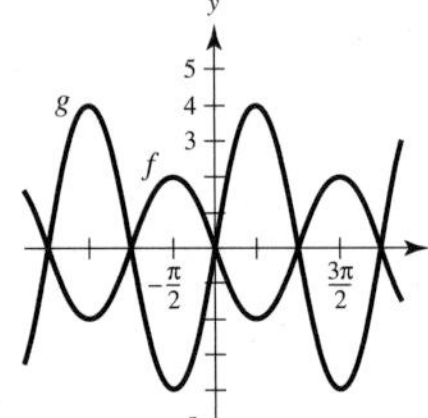

33.

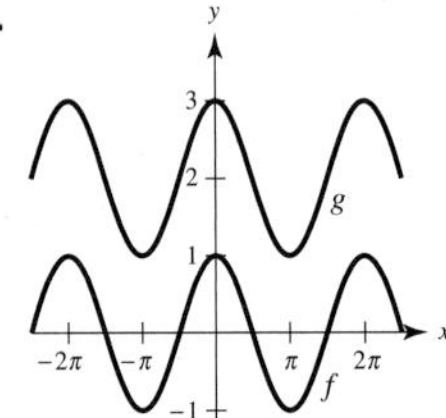

35.

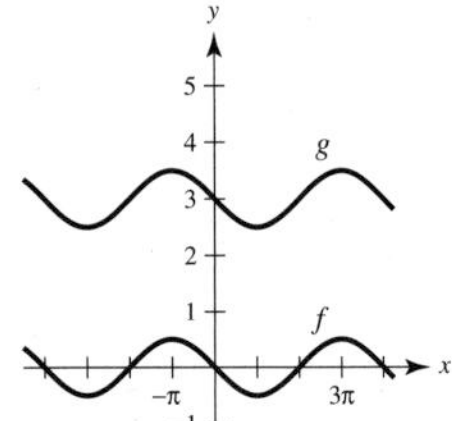

37.

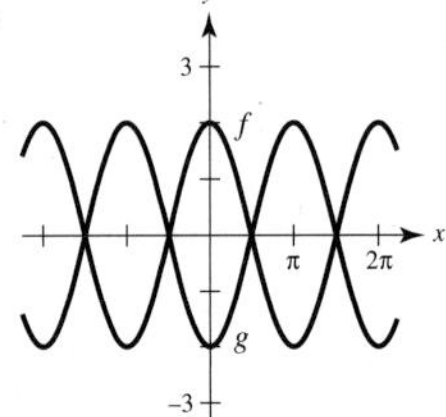

39.

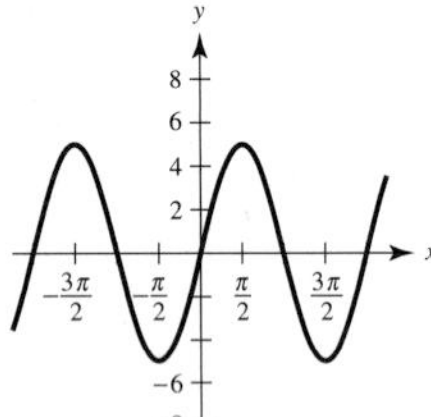

41.

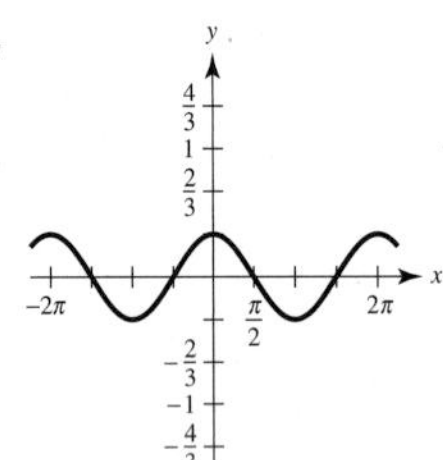

43.

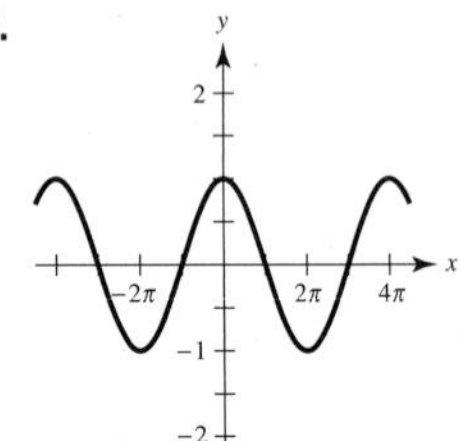

45.

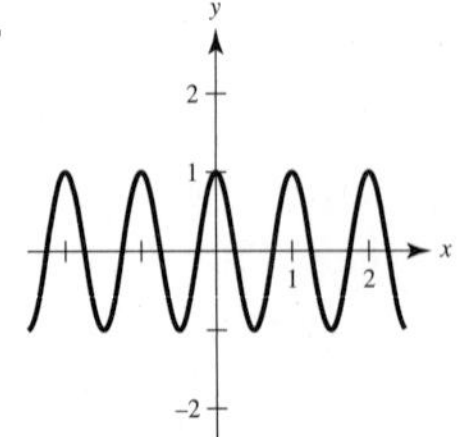

47.

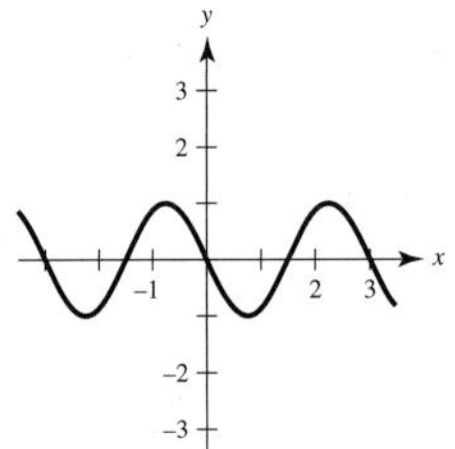

49.

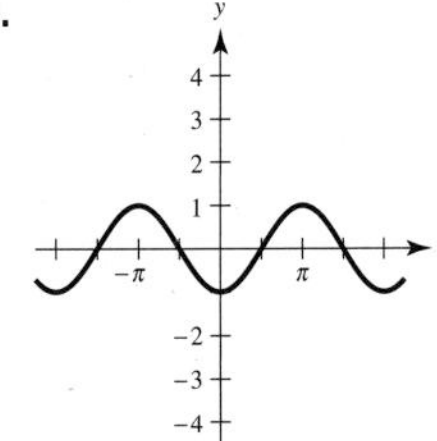

51.

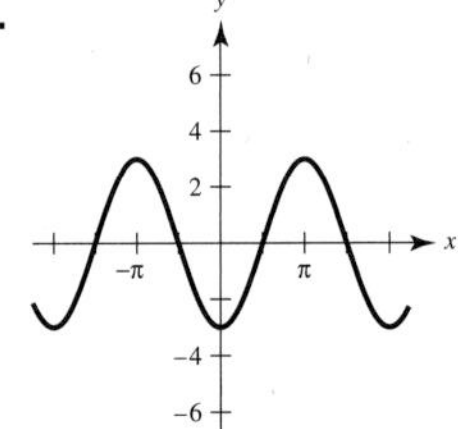

53.

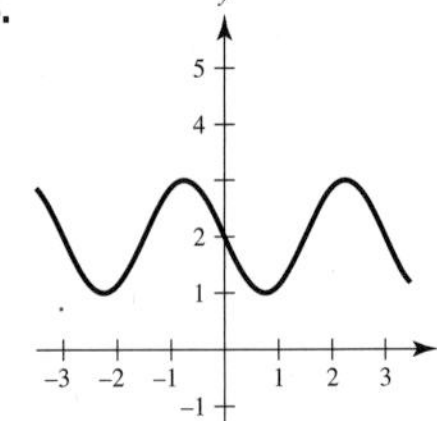

55.

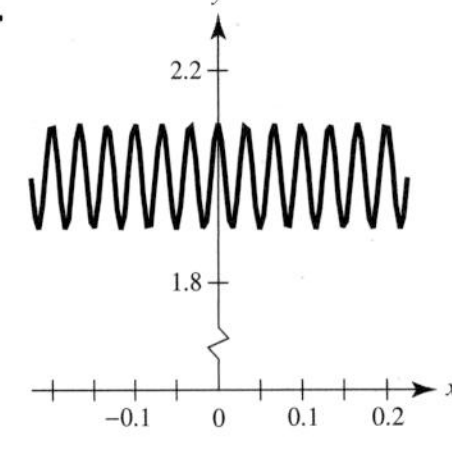

57.

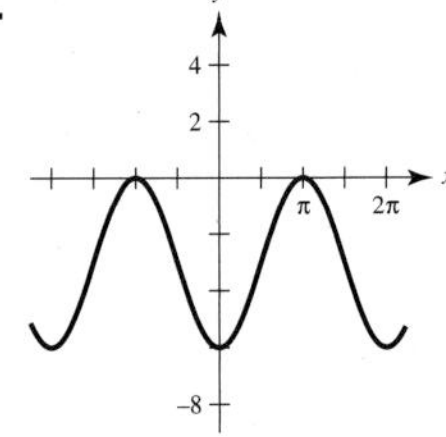

59.

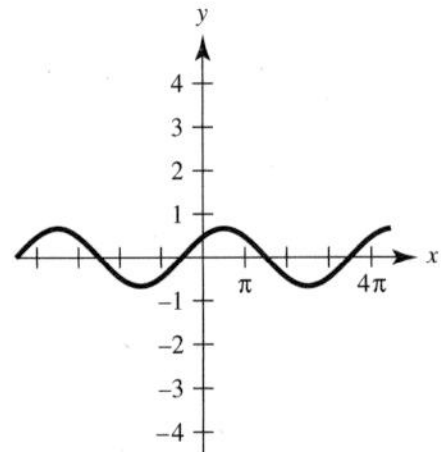

61. (a) $g(x)$ is obtained by a horizontal shrink of four, and one cycle of $g(x)$ corresponds to the interval $[\pi/4, 3\pi/4]$.

(b)

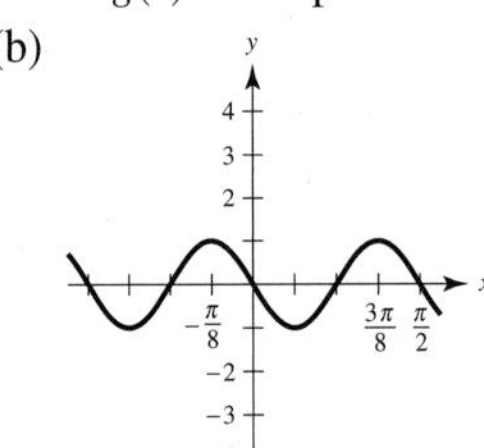

(c) $g(x) = f(4x - \pi)$

63. (a) One cycle of $g(x)$ corresponds to the interval $[\pi, 3\pi]$, and $g(x)$ is obtained by shifting $f(x)$ upward two units.

(b)

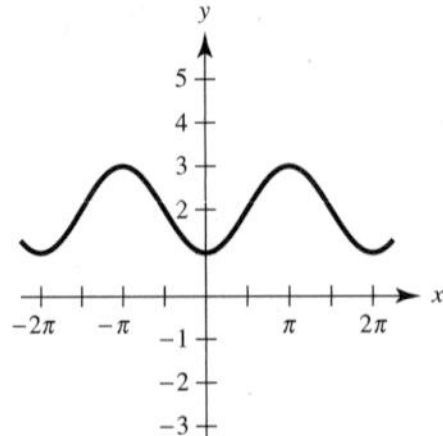

(c) $g(x) = f(x - \pi) + 2$

65. (a) One cycle of $g(x)$ is $[\pi/4, 3\pi/4]$. $g(x)$ is also shifted down three units and has an amplitude of two.

(b)

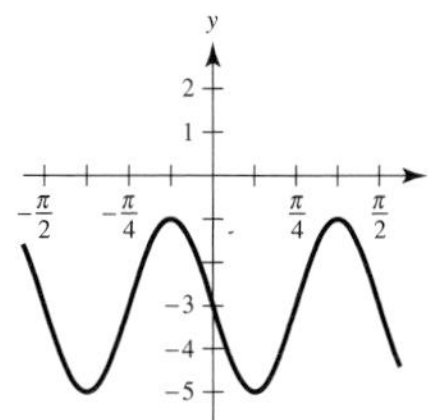

(c) $g(x) = 2f(4x - \pi) - 3$

67.

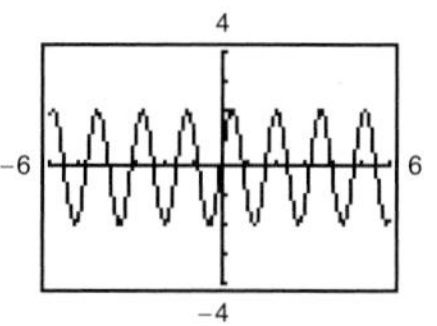

69.

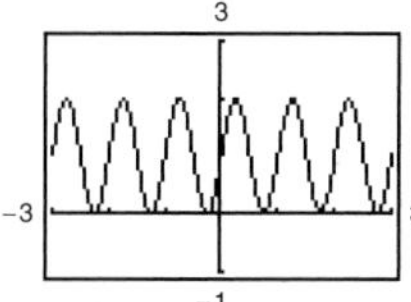

71.

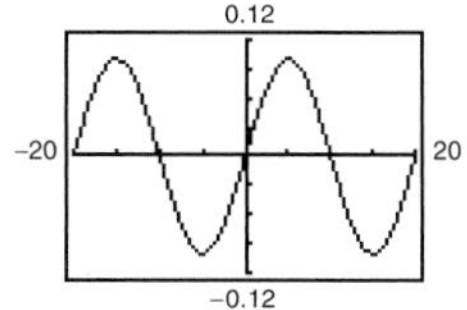

73. $a = 2, d = 1$ **75.** $a = -4, d = 4$

77. $a = -3, b = 2, c = 0$ **79.** $a = 2, b = 1, c = -\dfrac{\pi}{4}$

81.

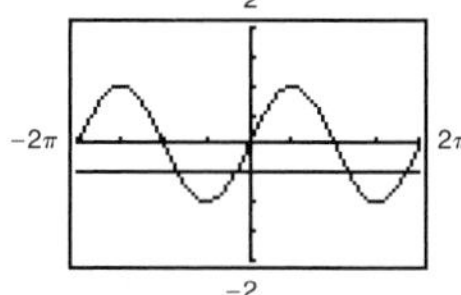

$x = -\dfrac{\pi}{6}, -\dfrac{5\pi}{6}, \dfrac{7\pi}{6}, \dfrac{11\pi}{6}$

83. $y = 1 + 2\sin(2x - \pi)$ **85.** $y = \cos(2x + 2\pi) - \frac{3}{2}$

87.

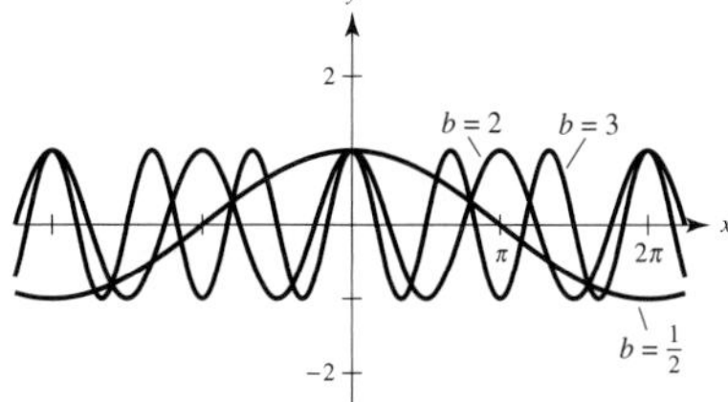

The value of b affects the period of the graph.

$b = \frac{1}{2} \rightarrow \frac{1}{2}$ cycle

$b = 2 \rightarrow 2$ cycles

$b = 3 \rightarrow 3$ cycles

89. (a) Even (b) Even

91. (a) 6 sec (b) 10 cycles/min

(c)

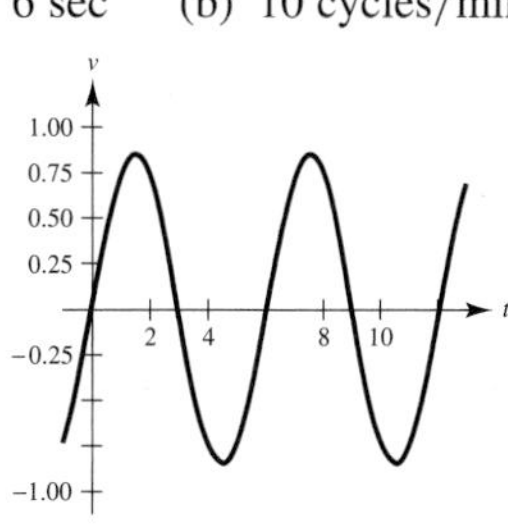

93. (a) $I(t) = 46.2 + 32.4\cos\left(\dfrac{\pi t}{6} - 3.67\right)$

(b)

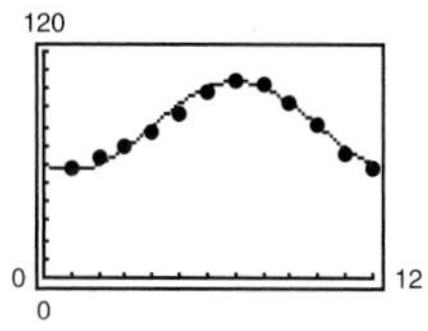

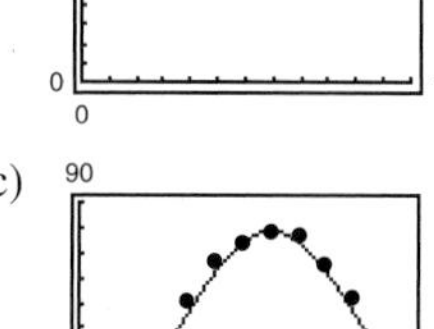

The model fits the data well.

(c)

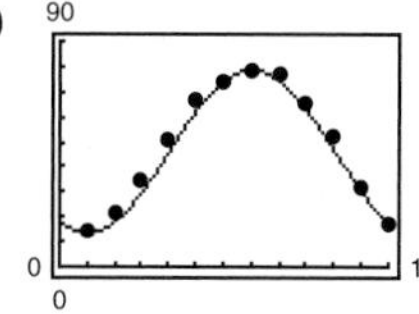

The model fits the data well.

(d) Las Vegas: 80.6°; International Falls: 46.2°
The constant term gives the annual average temperature.

(e) 12; yes; One full period is one year.

(f) International Falls; amplitude; The greater the amplitude, the greater the variability in temperature.

95. (a) $\frac{1}{440}$ sec (b) 440 cycles/sec

97. (a) 365; Yes, because there are 365 days in a year.

(b) 30.3 gal; the constant term

(c)

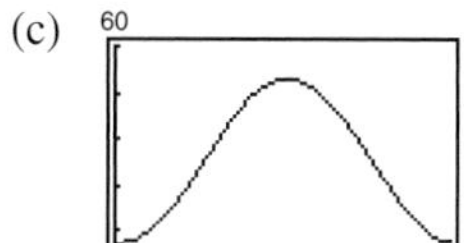

$124 < t < 252$

99. False. The graph of $f(x) = \sin(x + 2\pi)$ translates the graph of $f(x) = \sin x$ exactly one period to the left so that the two graphs look identical.

101. True. Because $\cos x = \sin\left(x + \dfrac{\pi}{2}\right)$, $y = -\cos x$ is a reflection in the x-axis of $y = \sin\left(x + \dfrac{\pi}{2}\right)$.

103.

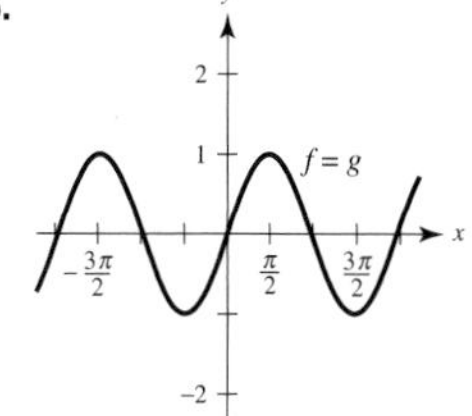

Conjecture:

$\sin x = \cos\left(x - \dfrac{\pi}{2}\right)$

Section 9.6 (page 616)

1. odd; origin **3.** reciprocal **5.** π

7. $(-\infty, -1] \cup [1, \infty)$ **9.** e, π **10.** c, 2π

11. a, 1 **12.** d, 2π **13.** f, 4 **14.** b, 4

15.

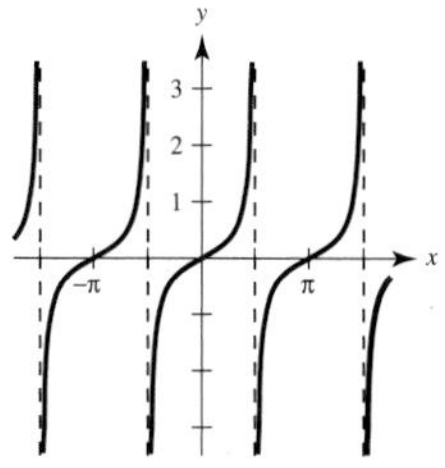

17.

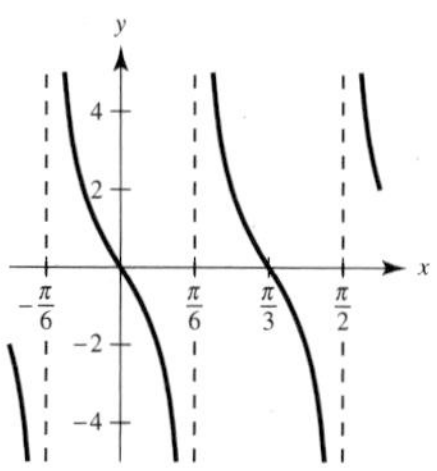

93. Domain: $(-\infty, \infty)$
Range: $(0, \pi)$

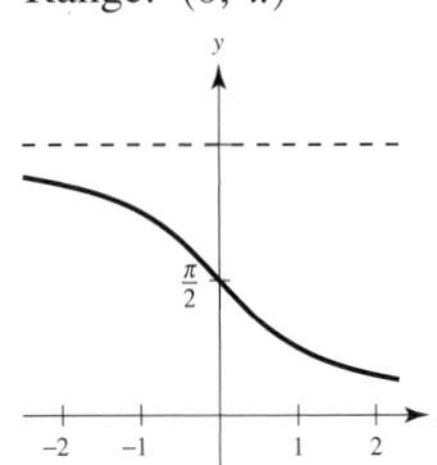

95. Domain: $(-\infty, -1] \cup [1, \infty)$
Range: $[-\pi/2, 0) \cup (0, \pi/2]$

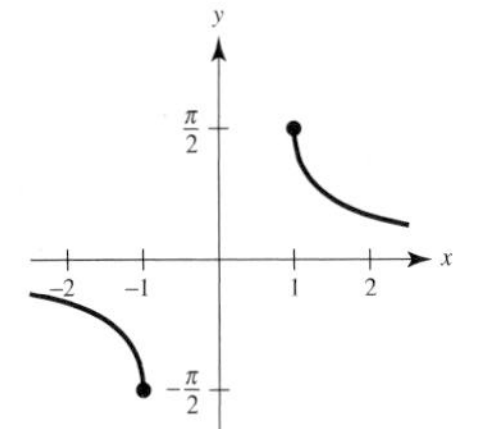

97–99. Proofs

101. 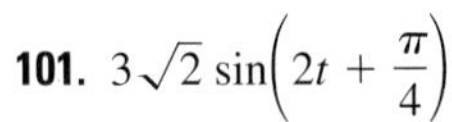
$3\sqrt{2}\sin\left(2t + \frac{\pi}{4}\right)$

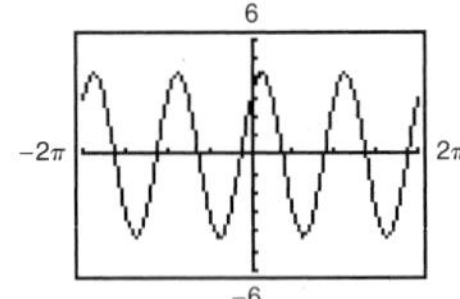

The graph implies that the identity is true.

103. $\frac{\pi}{2}$ **105.** $\frac{\pi}{2}$ **107.** π

109. (a) $\theta = \arcsin \frac{5}{s}$ (b) 0.13, 0.25

111. (a)

(b) 2 ft
(c) $\beta = 0$; As x increases, β approaches 0.

113. (a) $\theta = \arctan \frac{x}{20}$ (b) 14.0°, 31.0°

115. False. $\frac{5\pi}{6}$ is not in the range of the arcsine.

117. False. The graphs are not the same.

Section 9.8 (page 634)

1. bearing **3.** period

5. $a \approx 1.73$
$c \approx 3.46$
$B = 60°$

7. $a \approx 8.26$
$c \approx 25.38$
$A = 19°$

9. $c = 5$
$A \approx 36.87°$
$B \approx 53.13°$

11. $a \approx 49.48$
$A \approx 72.08°$
$B \approx 17.92°$

13. $a \approx 91.34$
$b \approx 420.70$
$B = 77°45'$

15. 3.00 **17.** 2.50 **19.** Yes **21.** Yes

23.

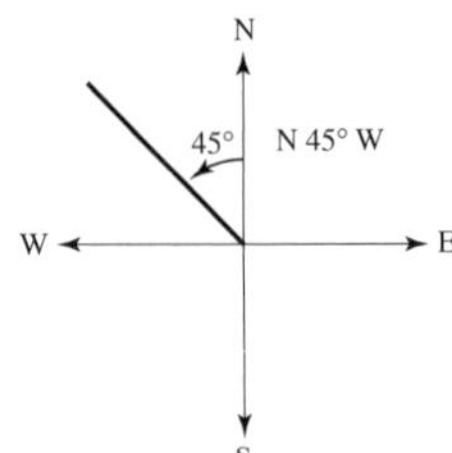

25.

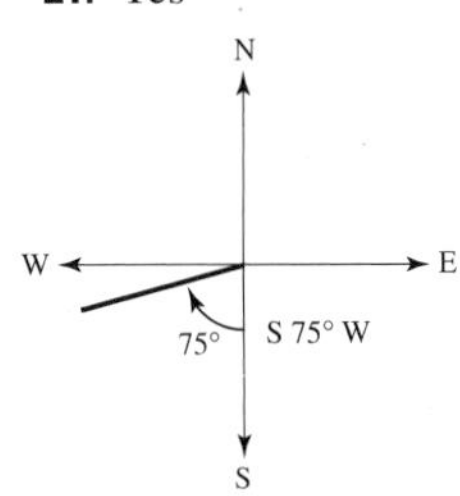

27. $|a|$ **29.** 214.45 ft **31.** 19.7 ft

33. 19.9 ft **35.** 11.8 km **37.** 56.3° **39.** 2.06°

41. (a) $\sqrt{h^2 + 34h + 10{,}289}$ (b) $\theta = \arccos\left(\frac{100}{l}\right)$
(c) 53.02 ft

43. (a) $l = 250$ ft, $A \approx 36.87°$, $B \approx 53.13°$ (b) 4.87 sec

45. 554 mi north; 709 mi east

47. (a) 104.95 nautical mi south; 58.18 nautical mi west
(b) S 36.7° W; distance = 130.9 nautical mi

49. N 56.31° W **51.** (a) N 58° E (b) 68.82 m

53. 78.7° **55.** 35.3° **57.** $y = \sqrt{3}r$ **59.** 29.4 in

61. $a \approx 12.2$, $b \approx 7$ **63.** $d = 4\sin(\pi t)$

65. $d = 3\cos\left(\frac{4\pi t}{3}\right)$

67. (a) 9 (b) $\frac{3}{5}$ (c) 9 (d) $\frac{5}{12}$

69. (a) $\frac{1}{4}$ (b) 3 (c) 0 (d) $\frac{1}{6}$ **71.** $\omega = 528\pi$

73. (a)

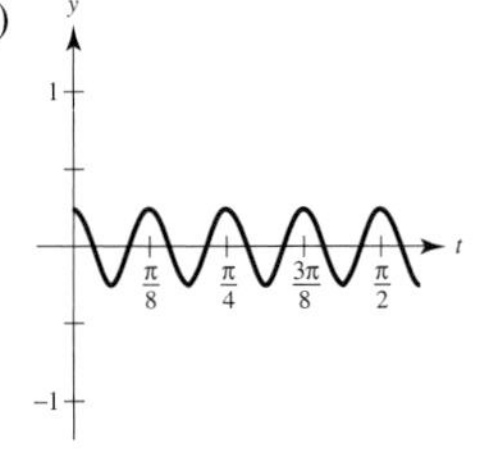

(b) $\frac{\pi}{8}$ (c) $\frac{\pi}{32}$

75. (a)

(b) 12; Yes, there are 12 months in a year.
(c) 2.77; The maximum change in the number of hours of daylight

77. False. N 24° E means 24 degrees east of north.

Review Exercises (page 639)

1. (a)

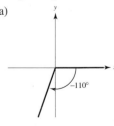

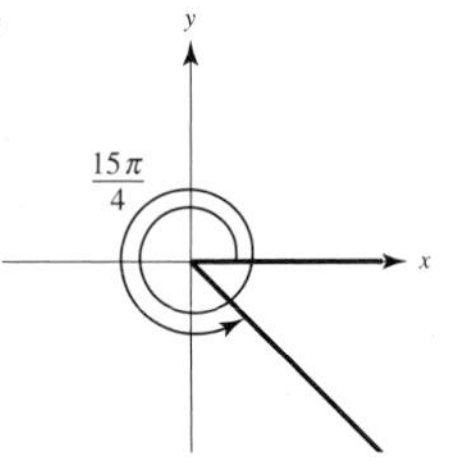

(b) Quadrant IV
(c) $\frac{23\pi}{4}, -\frac{\pi}{4}$

3. (a)

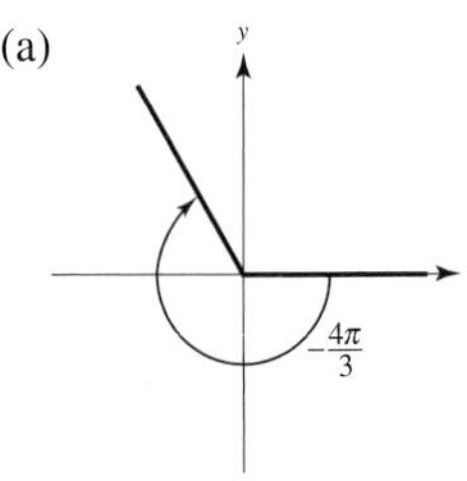

(b) Quadrant II
(c) $\frac{2\pi}{3}, -\frac{10\pi}{3}$

5. (a)

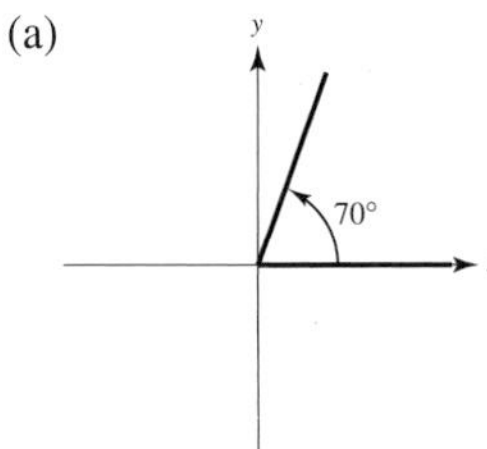

(b) Quadrant I
(c) 430°, −290°

7. (a)

(b) Quadrant III
(c) 250°, −470°

65. (a) One cycle of $g(x)$ is $[\pi/4, 3\pi/4]$. $g(x)$ is also shifted down three units and has an amplitude of two.

(b)

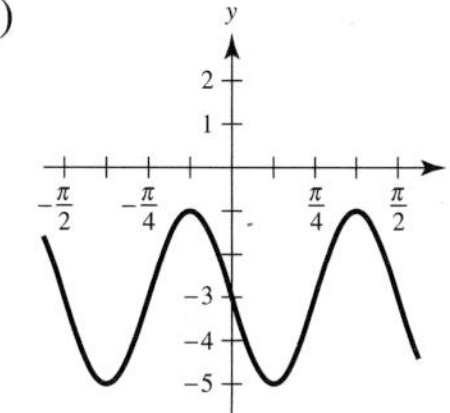

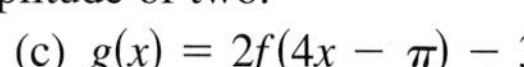
(c) $g(x) = 2f(4x - \pi) - 3$

67.

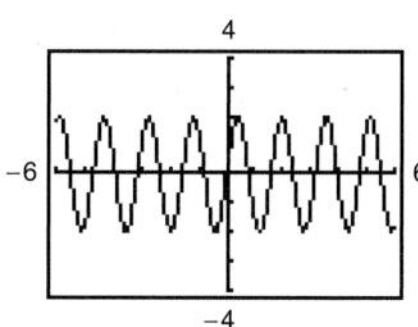

69.

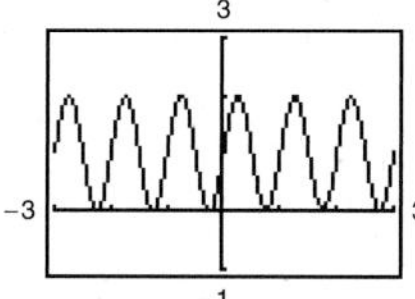

71.

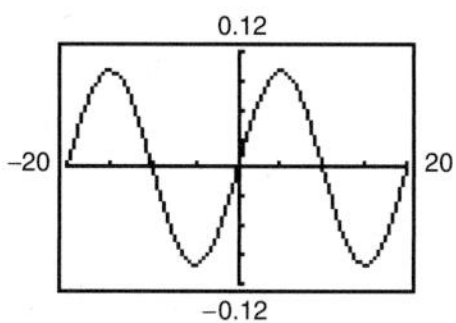

73. $a = 2, d = 1$ **75.** $a = -4, d = 4$

77. $a = -3, b = 2, c = 0$ **79.** $a = 2, b = 1, c = -\dfrac{\pi}{4}$

81.

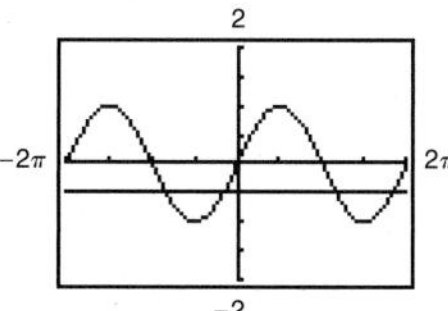

$x = -\dfrac{\pi}{6}, -\dfrac{5\pi}{6}, \dfrac{7\pi}{6}, \dfrac{11\pi}{6}$

83. $y = 1 + 2\sin(2x - \pi)$ **85.** $y = \cos(2x + 2\pi) - \frac{3}{2}$

87.

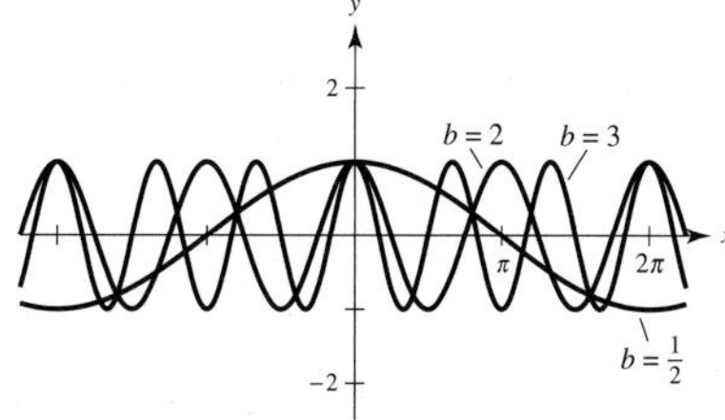

The value of b affects the period of the graph.

$b = \frac{1}{2} \rightarrow \frac{1}{2}$ cycle

$b = 2 \rightarrow 2$ cycles

$b = 3 \rightarrow 3$ cycles

89. (a) Even (b) Even

91. (a) 6 sec (b) 10 cycles/min

(c)

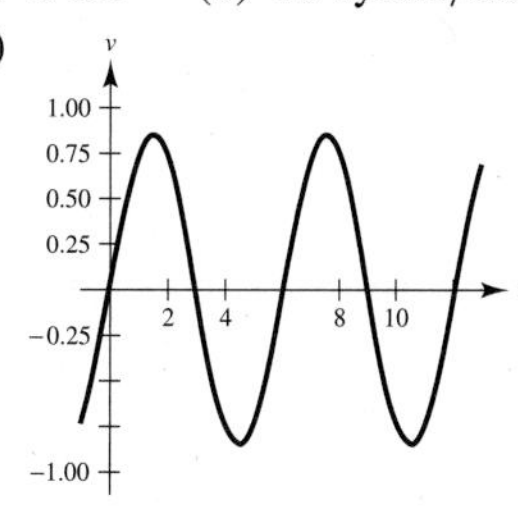

93. (a) $I(t) = 46.2 + 32.4\cos\left(\dfrac{\pi t}{6} - 3.67\right)$

(b)

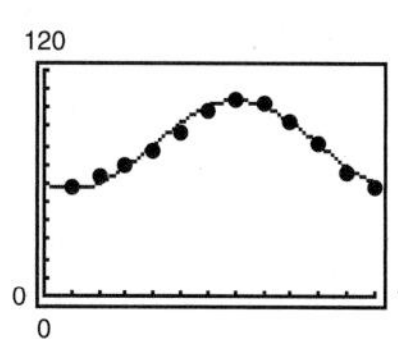

The model fits the data well.

(c)

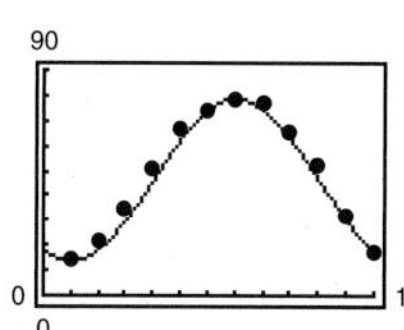

The model fits the data well.

(d) Las Vegas: 80.6°; International Falls: 46.2°

The constant term gives the annual average temperature.

(e) 12; yes; One full period is one year.

(f) International Falls; amplitude; The greater the amplitude, the greater the variability in temperature.

95. (a) $\frac{1}{440}$ sec (b) 440 cycles/sec

97. (a) 365; Yes, because there are 365 days in a year.

(b) 30.3 gal; the constant term

(c)

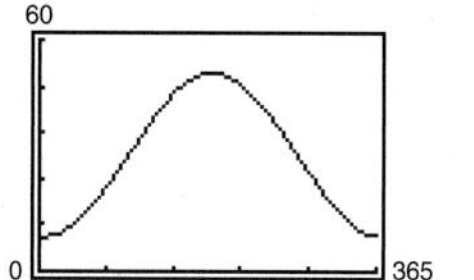

$124 < t < 252$

99. False. The graph of $f(x) = \sin(x + 2\pi)$ translates the graph of $f(x) = \sin x$ exactly one period to the left so that the two graphs look identical.

101. True. Because $\cos x = \sin\left(x + \dfrac{\pi}{2}\right)$, $y = -\cos x$ is a reflection in the x-axis of $y = \sin\left(x + \dfrac{\pi}{2}\right)$.

103.

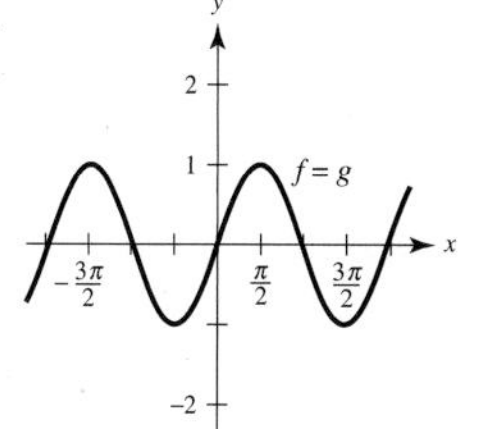

Conjecture:

$\sin x = \cos\left(x - \dfrac{\pi}{2}\right)$

Section 9.6 (page 616)

1. odd; origin **3.** reciprocal **5.** π

7. $(-\infty, -1] \cup [1, \infty)$ **9.** e, π **10.** c, 2π

11. a, 1 **12.** d, 2π **13.** f, 4 **14.** b, 4

15.

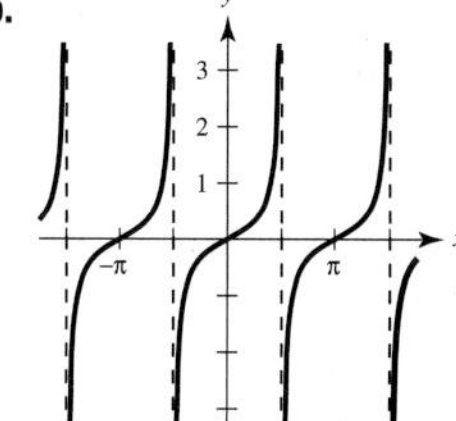

17.

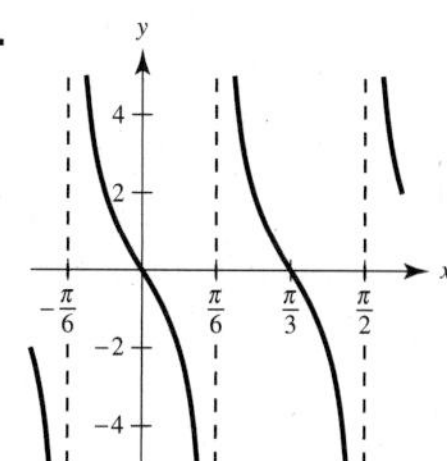

19.

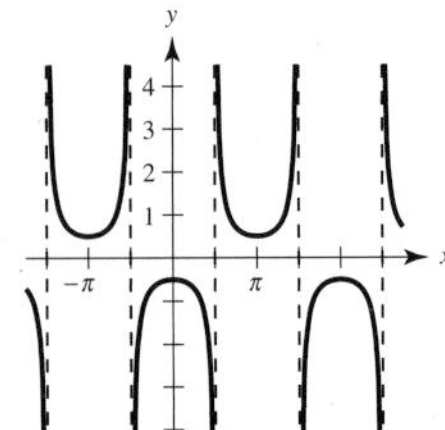

21.

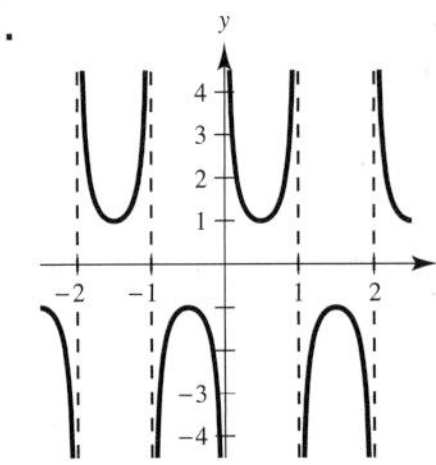

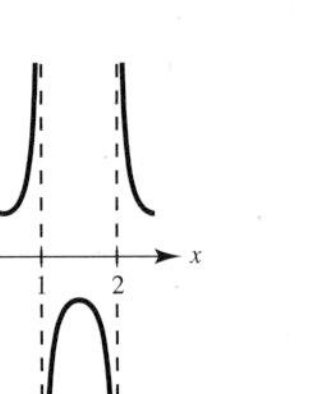

23.

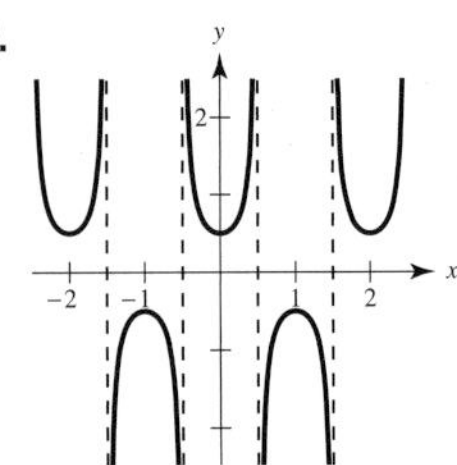

25.

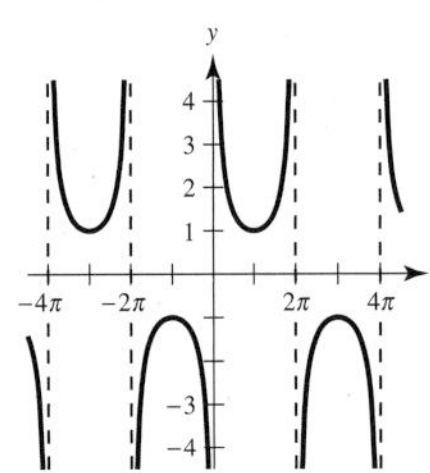

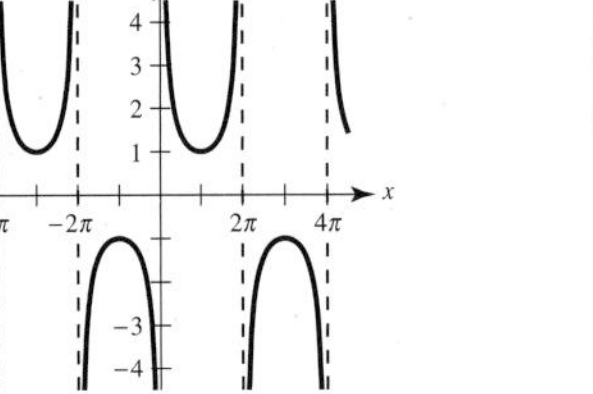

27.

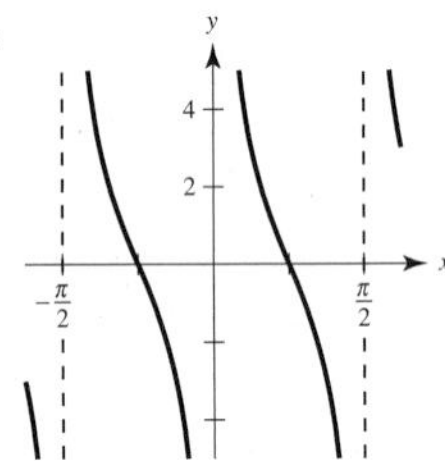

29.

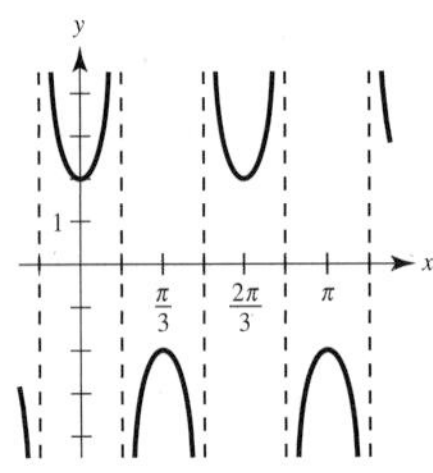

31.

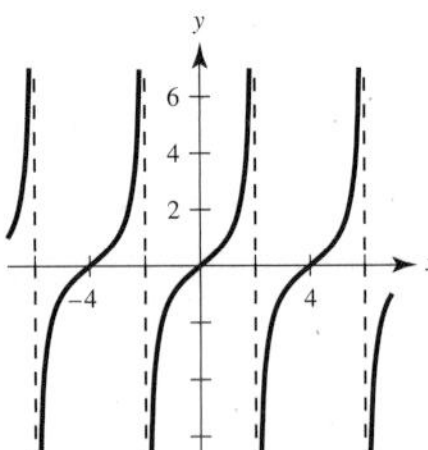

33.

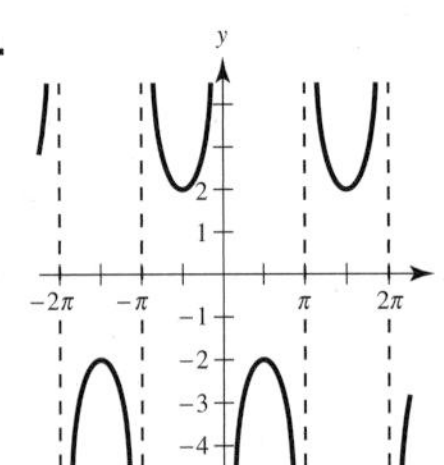

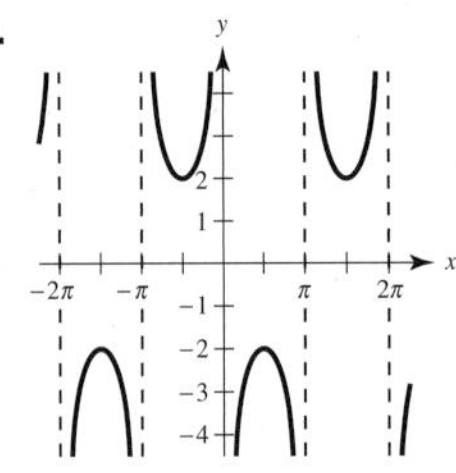

35.

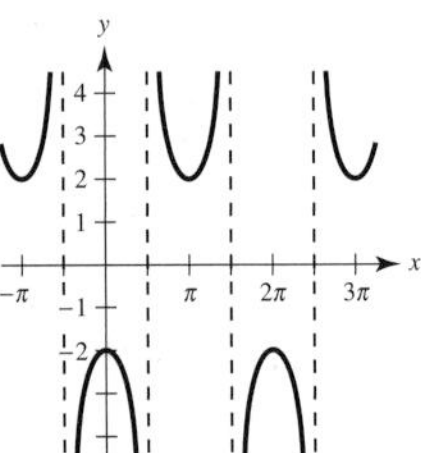

37.

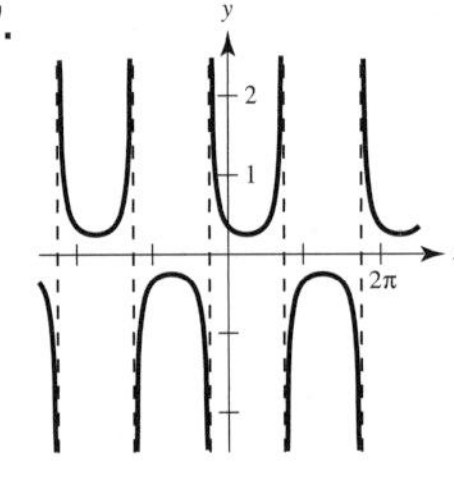

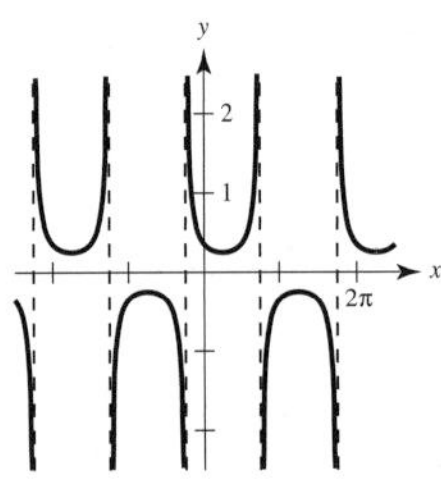

39.

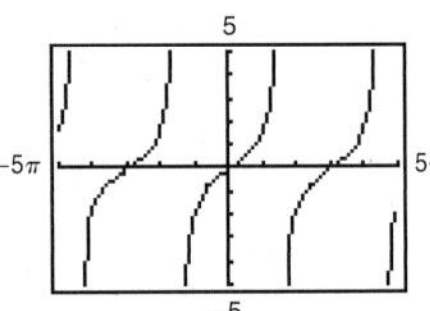

41.

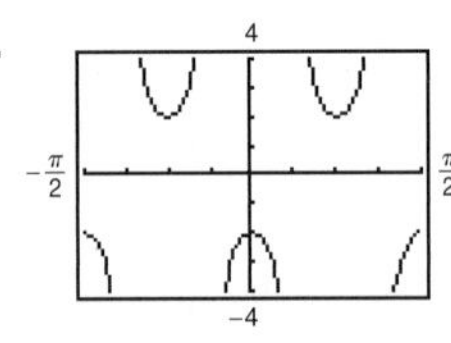

43.

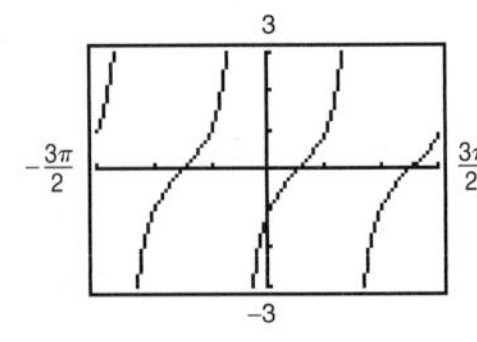

45.

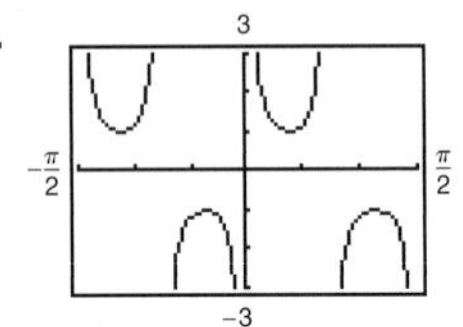

47.

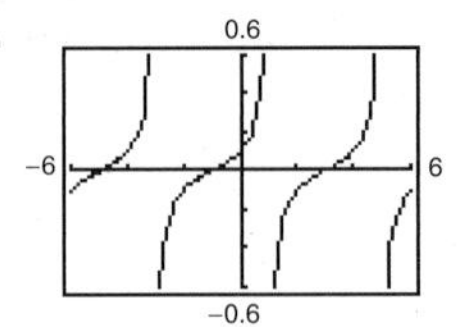

49. $-\frac{7\pi}{4}, -\frac{3\pi}{4}, \frac{\pi}{4}, \frac{5\pi}{4}$ **51.** $-\frac{4\pi}{3}, -\frac{\pi}{3}, \frac{2\pi}{3}, \frac{5\pi}{3}$

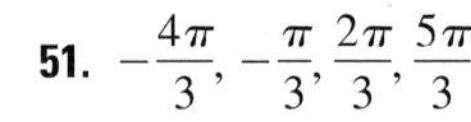

53. $-\frac{4\pi}{3}, -\frac{2\pi}{3}, \frac{2\pi}{3}, \frac{4\pi}{3}$ **55.** $-\frac{7\pi}{4}, -\frac{5\pi}{4}, \frac{\pi}{4}, \frac{3\pi}{4}$

57. Even **59.** Odd **61.** Odd **63.** Even

65. The expressions are equivalent except when $\sin x = 0$. Then, y_1 is undefined.

67.

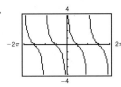

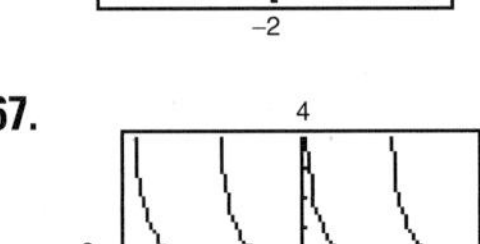

The expressions are equivalent.

69.

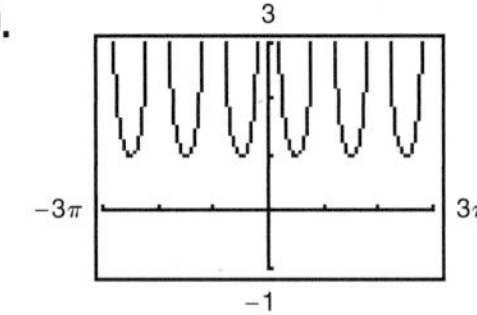

The expressions are equivalent.

71. d, $f \to 0$ as $x \to 0$. **72.** a, $f \to 0$ as $x \to 0$.

73. b, $g \to 0$ as $x \to 0$. **74.** c, $g \to 0$ as $x \to 0$.

75.

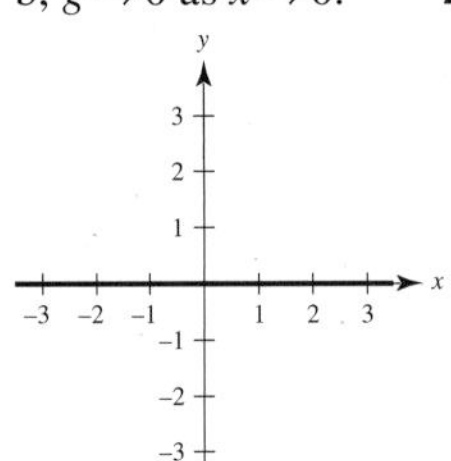

The functions are equal.

77.

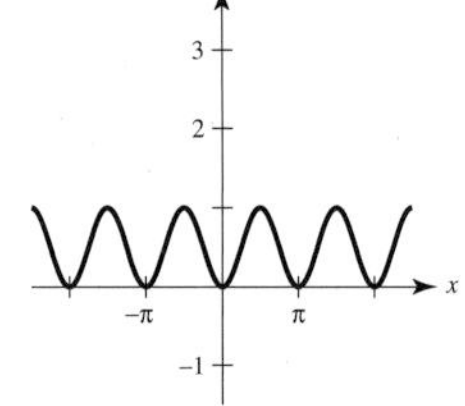

The functions are equal.

79.

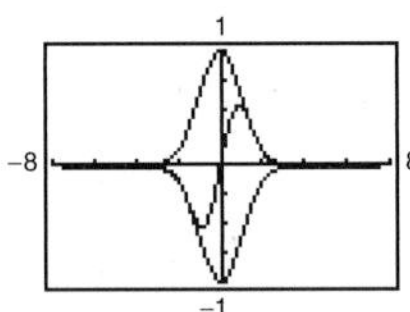

As $x \to \infty$, $g(x) \to 0$.

81.

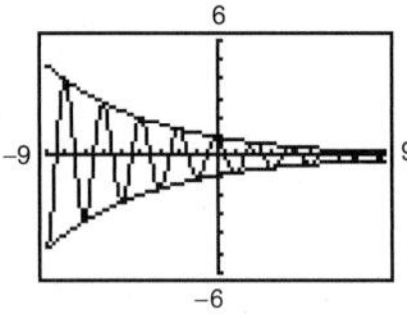

As $x \to \infty$, $f(x) \to 0$.

83.

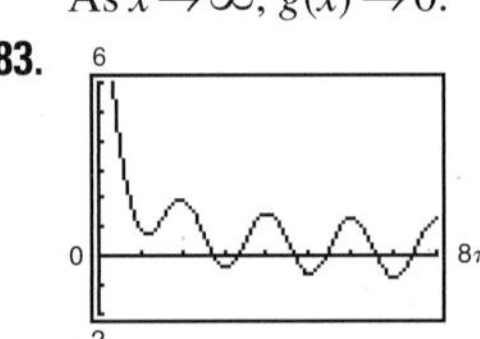

As $x \to 0$, $y \to \infty$.

85.

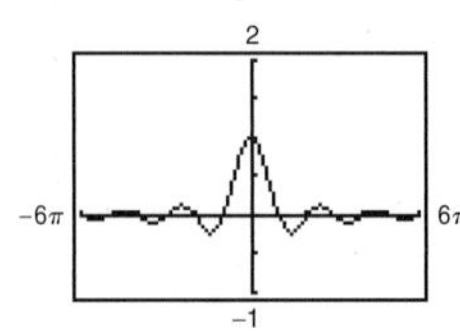

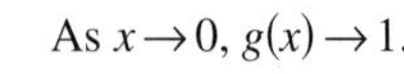

As $x \to 0$, $g(x) \to 1$.

87.

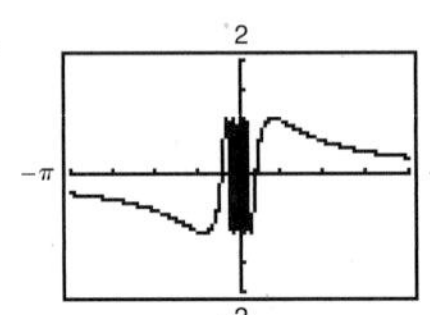

As $x \to 0$, $f(x)$ oscillates between 1 and -1.

89. (a)

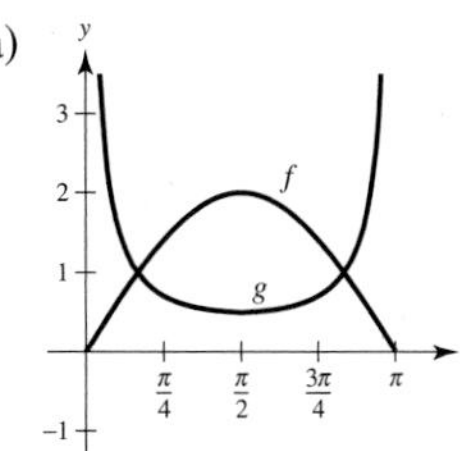

(b) $\frac{\pi}{6} < x < \frac{5\pi}{6}$

(c) f approaches 0 and g approaches $+\infty$ because the cosecant is the reciprocal of the sine.

91. $d = 7 \cot x$

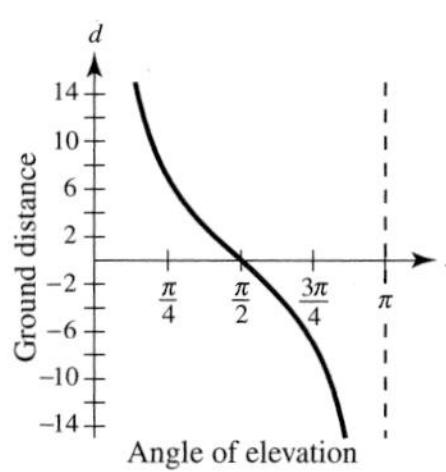

93. (a) Period of $H(t)$: 12 mo
Period of $L(t)$: 12 mo
(b) Summer; winter
(c) About 0.5 mo

95. (a)

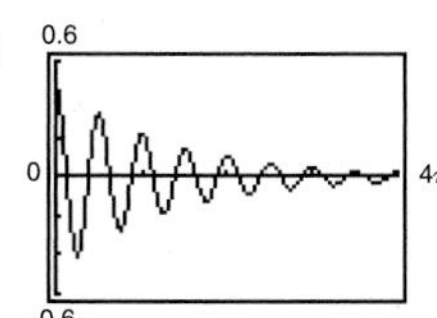

(b) y approaches 0 as t increases.

97. True. $y = \sec x$ is equal to $y = 1/\cos x$, and if the reciprocal of $y = \sin x$ is translated $\pi/2$ units to the left, then

$$\frac{1}{\sin\left(x + \frac{\pi}{2}\right)} = \frac{1}{\cos x} = \sec x.$$

99. (a) As $x \to \frac{\pi}{2}^+$, $f(x) \to -\infty$.
(b) As $x \to \frac{\pi}{2}^-$, $f(x) \to \infty$.
(c) As $x \to -\frac{\pi}{2}^+$, $f(x) \to -\infty$.
(d) As $x \to -\frac{\pi}{2}^-$, $f(x) \to \infty$.

101. (a) As $x \to 0^+$, $f(x) \to \infty$. (b) As $x \to 0^-$, $f(x) \to -\infty$.
(c) As $x \to \pi^+$, $f(x) \to \infty$. (d) As $x \to \pi^-$, $f(x) \to -\infty$.

103. (a)

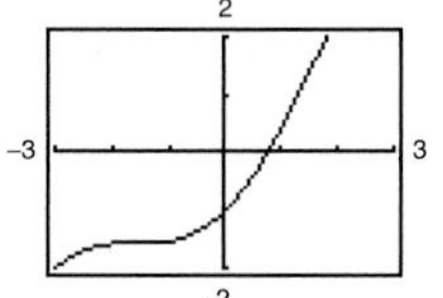

0.7391

(b) 1, 0.5403, 0.8576, 0.6543, 0.7935, 0.7014, 0.7640, 0.7221, 0.7504, 0.7314, . . . ; 0.7391

105.

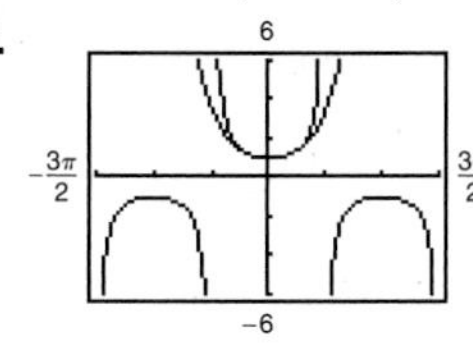

The graphs appear to coincide on the interval $-1.1 \le x \le 1.1$.

Section 9.7 (page 626)

1. $y = \sin^{-1} x$; $-1 \le x \le 1$

3. $y = \tan^{-1} x$; $-\infty < x < \infty$; $-\frac{\pi}{2} < y < \frac{\pi}{2}$ **5.** $\frac{\pi}{6}$ **7.** $\frac{\pi}{3}$

9. $\frac{\pi}{6}$ **11.** $\frac{5\pi}{6}$ **13.** $-\frac{\pi}{3}$ **15.** $\frac{2\pi}{3}$ **17.** $-\frac{\pi}{3}$ **19.** 0

21.

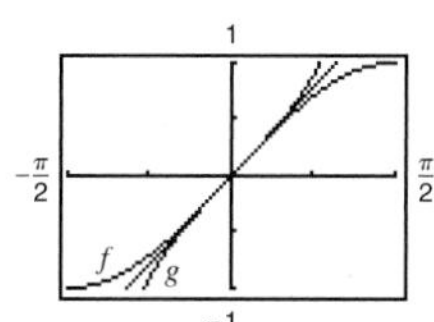

23. 1.19 **25.** -0.85 **27.** -1.25 **29.** 0.32

31. 1.99 **33.** 0.74 **35.** 1.07 **37.** 1.36

39. -1.52 **41.** $-\frac{\pi}{3}, -\frac{\sqrt{3}}{3}, 1$ **43.** $\theta = \arctan \frac{x}{4}$

45. $\theta = \arcsin \frac{x + 2}{5}$ **47.** $\theta = \arccos \frac{x + 3}{2x}$

49. 0.3 **51.** -0.1 **53.** 0 **55.** $\frac{3}{5}$ **57.** $\frac{\sqrt{5}}{5}$

59. $\frac{12}{13}$ **61.** $\frac{\sqrt{34}}{5}$ **63.** $\frac{\sqrt{5}}{3}$ **65.** $\frac{1}{x}$ **67.** $\sqrt{1 - 4x^2}$

69. $\sqrt{1 - x^2}$ **71.** $\frac{\sqrt{9 - x^2}}{x}$ **73.** $\frac{\sqrt{x^2 + 2}}{x}$

75.

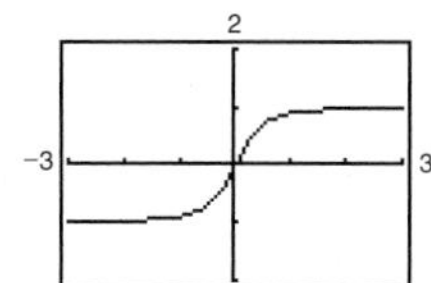

Asymptotes: $y = \pm 1$

77. $\frac{9}{\sqrt{x^2 + 81}}, x > 0; \frac{-9}{\sqrt{x^2 + 81}}, x < 0$ **79.** $\frac{|x - 1|}{\sqrt{x^2 - 2x + 10}}$

81.

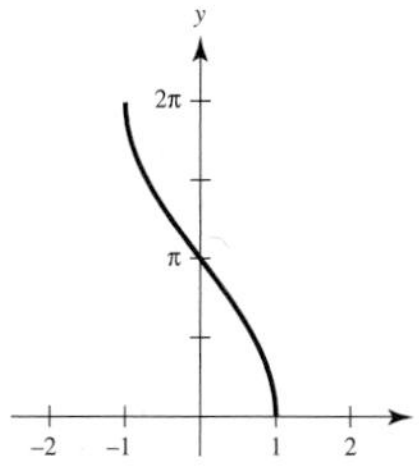

83.

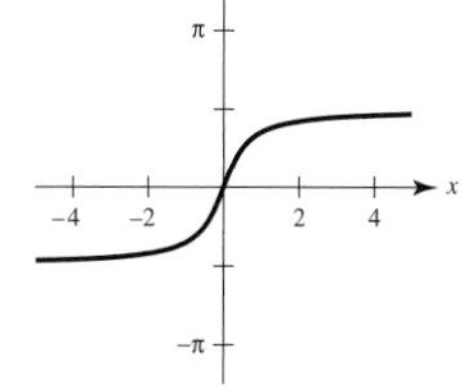

85.

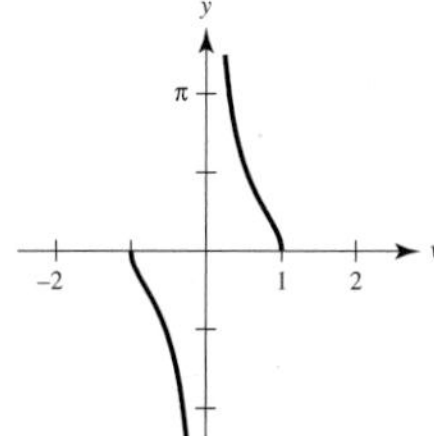

87.

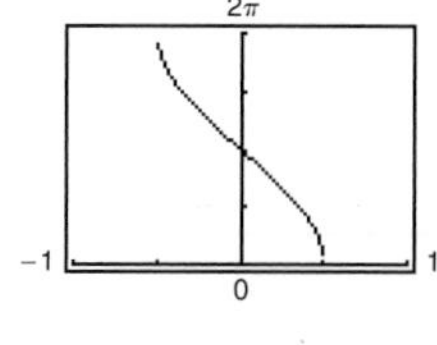

89.

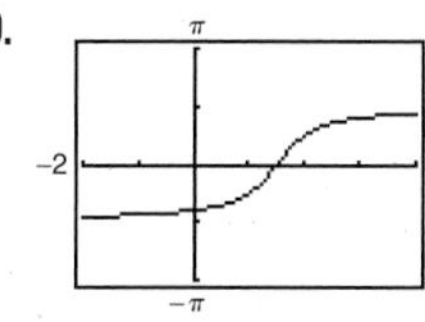

91.

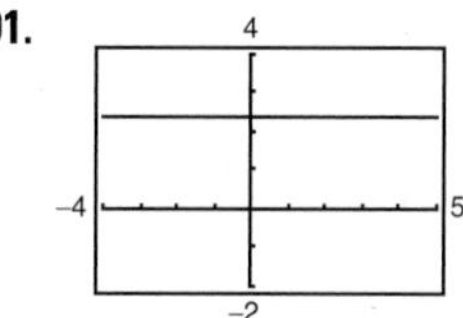

93. Domain: $(-\infty, \infty)$
Range: $(0, \pi)$

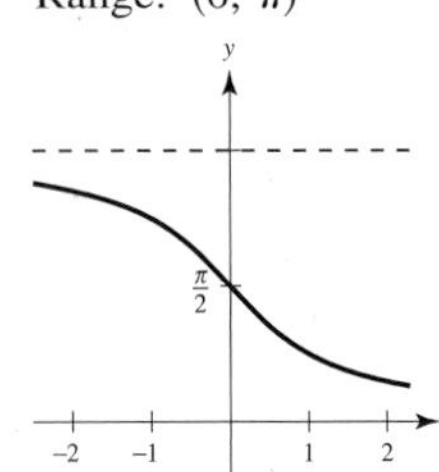

95. Domain: $(-\infty, -1] \cup [1, \infty)$
Range: $[-\pi/2, 0) \cup (0, \pi/2]$

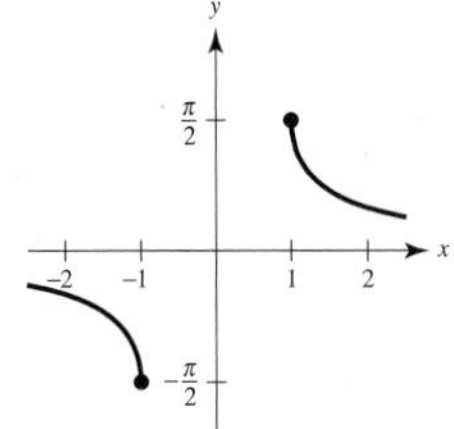

97–99. Proofs

101. $3\sqrt{2}\sin\left(2t + \frac{\pi}{4}\right)$

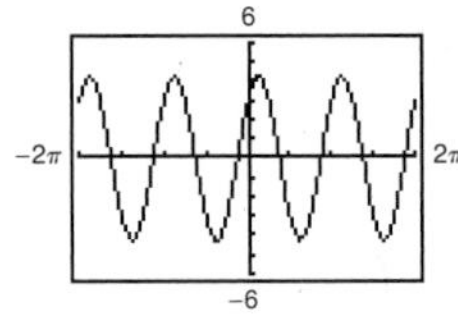

The graph implies that the identity is true.

103. $\frac{\pi}{2}$ **105.** $\frac{\pi}{2}$ **107.** π

109. (a) $\theta = \arcsin \frac{5}{s}$ (b) 0.13, 0.25

111. (a) (b) 2 ft
(c) $\beta = 0$; As x increases, β approaches 0.

113. (a) $\theta = \arctan \frac{x}{20}$ (b) 14.0°, 31.0°

115. False. $\frac{5\pi}{6}$ is not in the range of the arcsine.

117. False. The graphs are not the same.

Section 9.8 (page 634)

1. bearing **3.** period

5. $a \approx 1.73$
$c \approx 3.46$
$B = 60°$

7. $a \approx 8.26$
$c \approx 25.38$
$A = 19°$

9. $c = 5$
$A \approx 36.87°$
$B \approx 53.13°$

11. $a \approx 49.48$
$A \approx 72.08°$
$B \approx 17.92°$

13. $a \approx 91.34$
$b \approx 420.70$
$B = 77°45'$

15. 3.00 **17.** 2.50 **19.** Yes **21.** Yes

23.

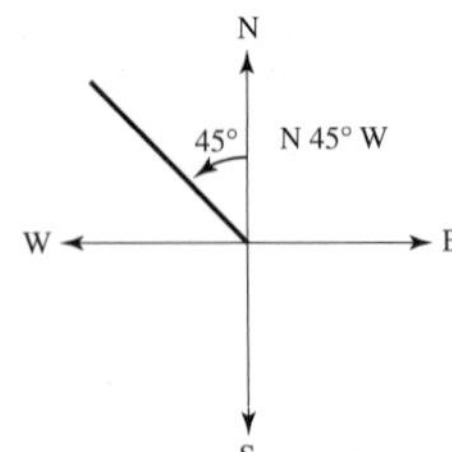

25.

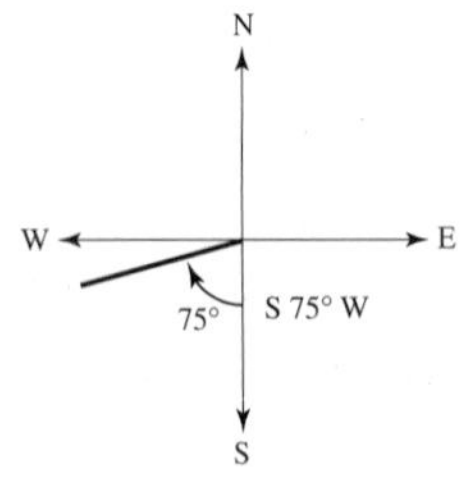

27. $|a|$ **29.** 214.45 ft **31.** 19.7 ft
33. 19.9 ft **35.** 11.8 km **37.** 56.3° **39.** 2.06°

41. (a) $\sqrt{h^2 + 34h + 10{,}289}$ (b) $\theta = \arccos\left(\frac{100}{l}\right)$
(c) 53.02 ft

43. (a) $l = 250$ ft, $A \approx 36.87°$, $B \approx 53.13°$ (b) 4.87 sec

45. 554 mi north; 709 mi east

47. (a) 104.95 nautical mi south; 58.18 nautical mi west
(b) S 36.7° W; distance = 130.9 nautical mi

49. N 56.31° W **51.** (a) N 58° E (b) 68.82 m

53. 78.7° **55.** 35.3° **57.** $y = \sqrt{3}r$ **59.** 29.4 in

61. $a \approx 12.2, b \approx 7$ **63.** $d = 4\sin(\pi t)$

65. $d = 3\cos\left(\frac{4\pi t}{3}\right)$

67. (a) 9 (b) $\frac{3}{5}$ (c) 9 (d) $\frac{5}{12}$

69. (a) $\frac{1}{4}$ (b) 3 (c) 0 (d) $\frac{1}{6}$ **71.** $\omega = 528\pi$

73. (a) (b) $\frac{\pi}{8}$ (c) $\frac{\pi}{32}$

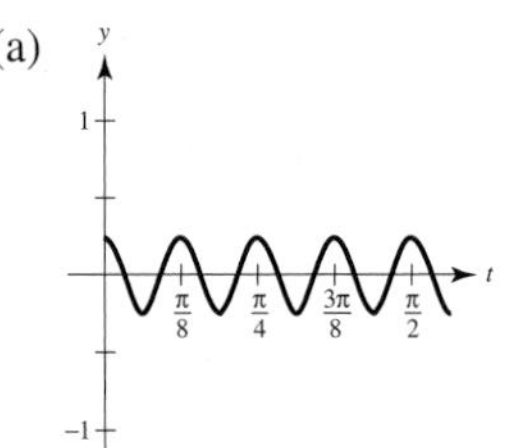

75. (a)

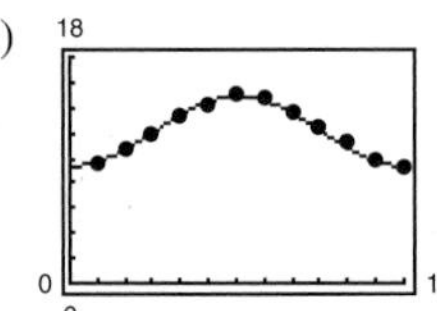

(b) 12; Yes, there are 12 months in a year.
(c) 2.77; The maximum change in the number of hours of daylight

77. False. N 24° E means 24 degrees east of north.

Review Exercises (page 639)

1. (a)

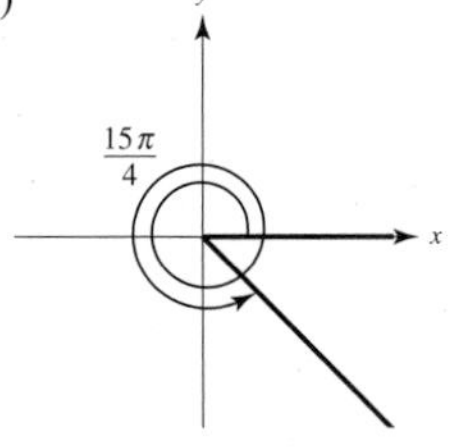

(b) Quadrant IV
(c) $\frac{23\pi}{4}, -\frac{\pi}{4}$

3. (a)

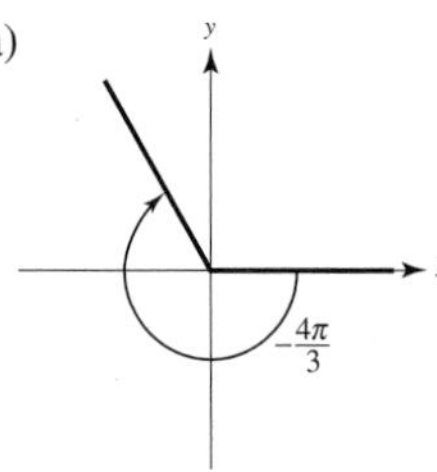

(b) Quadrant II
(c) $\frac{2\pi}{3}, -\frac{10\pi}{3}$

5. (a)

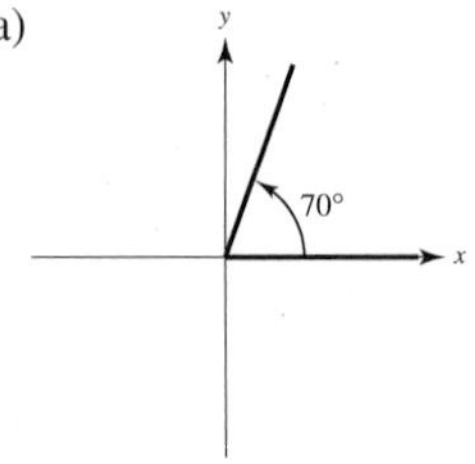

(b) Quadrant I
(c) 430°, −290°

7. (a)

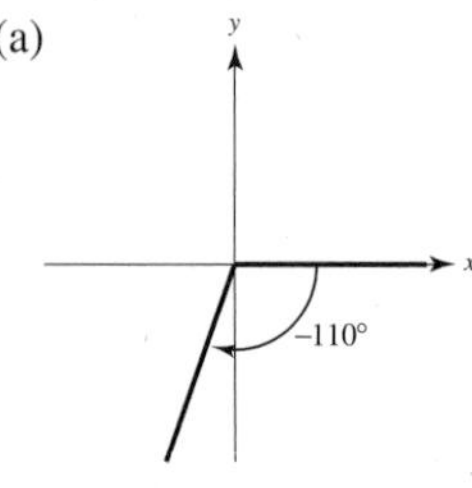

(b) Quadrant III
(c) 250°, −470°

9. 7.854 **11.** -0.589 **13.** $54.000°$ **15.** $-200.535°$

17. $198° \, 24'$ **19.** $0° \, 39'$ **21.** 48.17 in.

23. About 12.05 mi/h **25.** $\left(-\frac{1}{2}, \frac{\sqrt{3}}{2}\right)$ **27.** $\left(-\frac{\sqrt{3}}{2}, -\frac{1}{2}\right)$

29. $\sin\frac{7\pi}{6} = -\frac{1}{2}$ $\csc\frac{7\pi}{6} = -2$

$\cos\frac{7\pi}{6} = -\frac{\sqrt{3}}{2}$ $\sec\frac{7\pi}{6} = -\frac{2\sqrt{3}}{3}$

$\tan\frac{7\pi}{6} = \frac{\sqrt{3}}{3}$ $\cot\frac{7\pi}{6} = \sqrt{3}$

31. $\sin\left(-\frac{2\pi}{3}\right) = -\frac{\sqrt{3}}{2}$ $\csc\left(-\frac{2\pi}{3}\right) = -\frac{2\sqrt{3}}{3}$

$\cos\left(-\frac{2\pi}{3}\right) = -\frac{1}{2}$ $\sec\left(-\frac{2\pi}{3}\right) = -2$

$\tan\left(-\frac{2\pi}{3}\right) = \sqrt{3}$ $\cot\left(-\frac{2\pi}{3}\right) = \frac{\sqrt{3}}{3}$

33. $\sin\frac{11\pi}{4} = \sin\frac{3\pi}{4} = \frac{\sqrt{2}}{2}$ **35.** $\sin\left(-\frac{17\pi}{6}\right) = \sin\frac{7\pi}{6} = -\frac{1}{2}$

37. -75.3130 **39.** 3.2361

41. $\sin\theta = \frac{4\sqrt{41}}{41}$ $\csc\theta = \frac{\sqrt{41}}{4}$

$\cos\theta = \frac{5\sqrt{41}}{41}$ $\sec\theta = \frac{\sqrt{41}}{5}$

$\tan\theta = \frac{4}{5}$ $\cot\theta = \frac{5}{4}$

43. (a) 3 (b) $\frac{2\sqrt{2}}{3}$ (c) $\frac{3\sqrt{2}}{4}$ (d) $\frac{\sqrt{2}}{4}$

45. (a) $\frac{1}{4}$ (b) $\frac{\sqrt{15}}{4}$ (c) $\frac{4\sqrt{15}}{15}$ (d) $\frac{\sqrt{15}}{15}$

47. 0.6494 **49.** 0.5621 **51.** 3.6722

53. 0.6104 **55.** 71.3 m

57. $\sin\theta = \frac{4}{5}$ $\csc\theta = \frac{5}{4}$

$\cos\theta = \frac{3}{5}$ $\sec\theta = \frac{5}{3}$

$\tan\theta = \frac{4}{3}$ $\cot\theta = \frac{3}{4}$

59. $\sin\theta = \frac{15\sqrt{241}}{241}$ $\csc\theta = \frac{\sqrt{241}}{15}$

$\cos\theta = \frac{4\sqrt{241}}{241}$ $\sec\theta = \frac{\sqrt{241}}{4}$

$\tan\theta = \frac{15}{4}$ $\cot\theta = \frac{4}{15}$

61. $\sin\theta = \frac{9\sqrt{82}}{82}$ $\csc\theta = \frac{\sqrt{82}}{9}$

$\cos\theta = \frac{-\sqrt{82}}{82}$ $\sec\theta = -\sqrt{82}$

$\tan\theta = -9$ $\cot\theta = -\frac{1}{9}$

63. $\sin\theta = \frac{4\sqrt{17}}{17}$ $\csc\theta = \frac{\sqrt{17}}{4}$

$\cos\theta = \frac{\sqrt{17}}{17}$ $\sec\theta = \sqrt{17}$

$\tan\theta = 4$ $\cot\theta = \frac{1}{4}$

65. $\sin\theta = -\frac{\sqrt{11}}{6}$ $\csc\theta = -\frac{6\sqrt{11}}{11}$

$\cos\theta = \frac{5}{6}$ $\cot\theta = -\frac{5\sqrt{11}}{11}$

$\tan\theta = -\frac{\sqrt{11}}{5}$

67. $\cos\theta = -\frac{\sqrt{55}}{8}$ $\sec\theta = -\frac{8\sqrt{55}}{55}$

$\tan\theta = -\frac{3\sqrt{55}}{55}$ $\cot\theta = -\frac{\sqrt{55}}{3}$

$\csc\theta = \frac{8}{3}$

69. $\sin\theta = \frac{\sqrt{21}}{5}$ $\sec\theta = -\frac{5}{2}$

$\tan\theta = -\frac{\sqrt{21}}{2}$ $\cot\theta = -\frac{2\sqrt{21}}{21}$

$\csc\theta = \frac{5\sqrt{21}}{21}$

71. $\theta' = 84°$

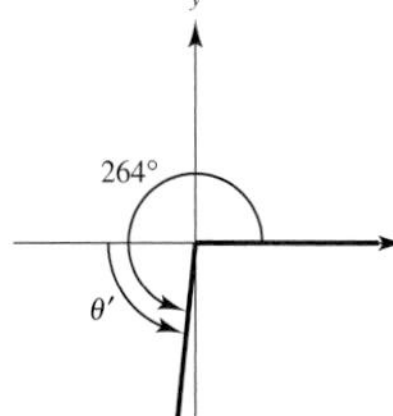

73. $\theta' = \frac{\pi}{5}$

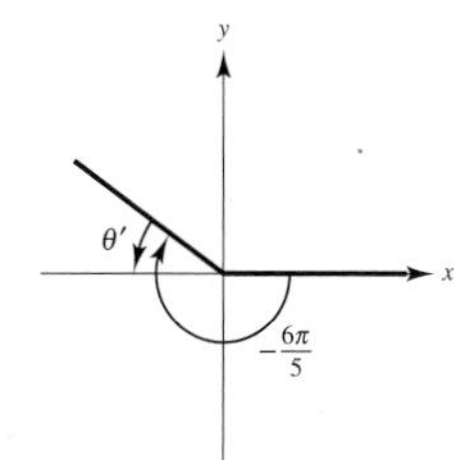

75. $\sin\frac{\pi}{3} = \frac{\sqrt{3}}{2}$; $\cos\frac{\pi}{3} = \frac{1}{2}$; $\tan\frac{\pi}{3} = \sqrt{3}$

77. $\sin\left(-\frac{7\pi}{3}\right) = -\frac{\sqrt{3}}{2}$; $\cos\left(-\frac{7\pi}{3}\right) = \frac{1}{2}$;

$\tan\left(-\frac{7\pi}{3}\right) = -\sqrt{3}$

79. $\sin 495° = \frac{\sqrt{2}}{2}$; $\cos 495° = -\frac{\sqrt{2}}{2}$; $\tan 495° = -1$

81. -0.7568 **83.** 0.9511

85.

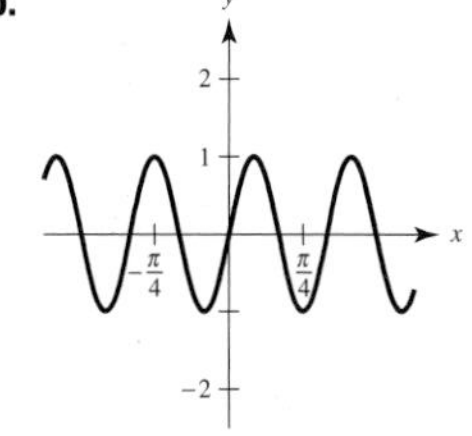

87.

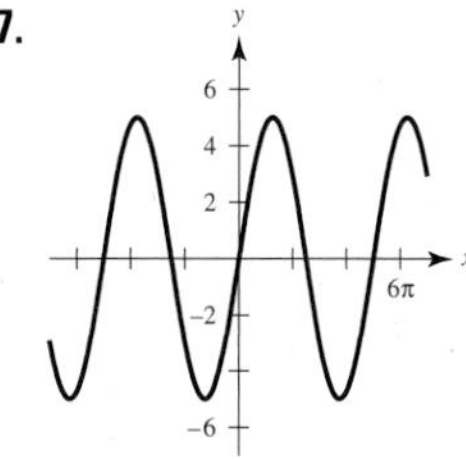

89.

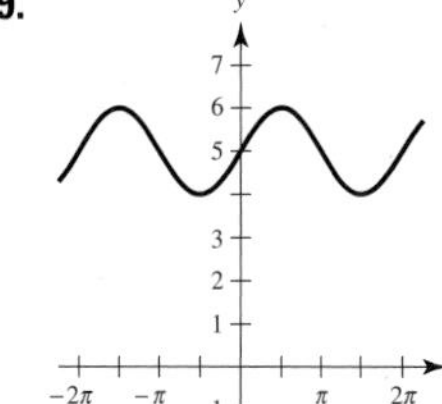

91.

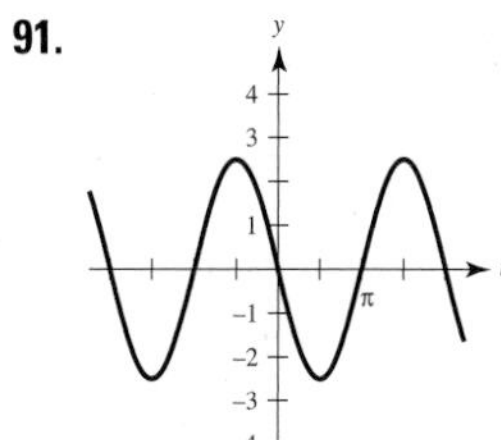

93. (a) $y = 2\sin 528\pi x$ (b) 264 cycles/sec

95.

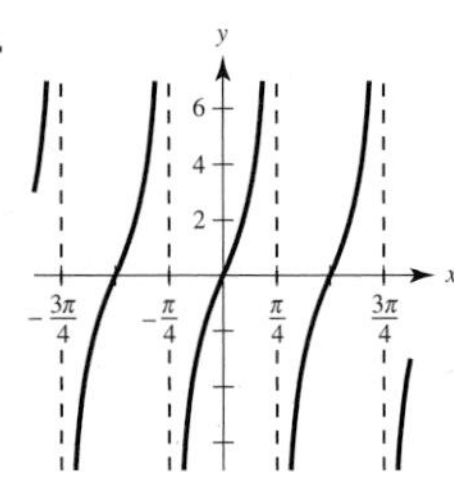

97.

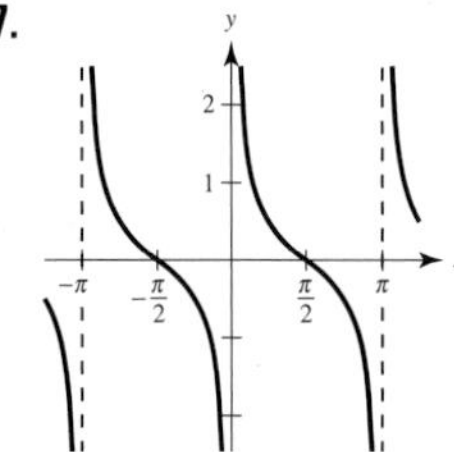

99.

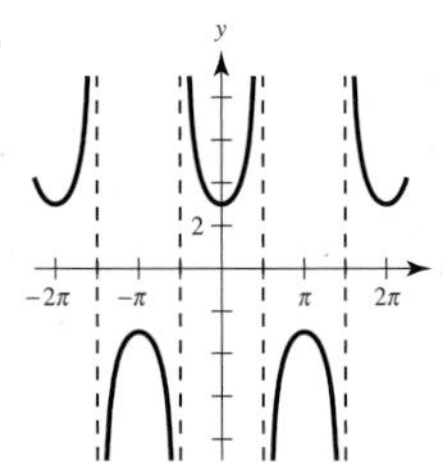

101.

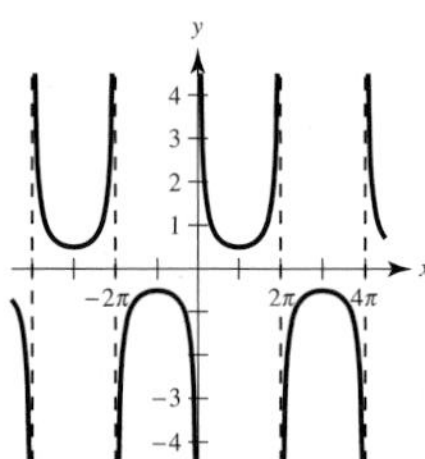

103.

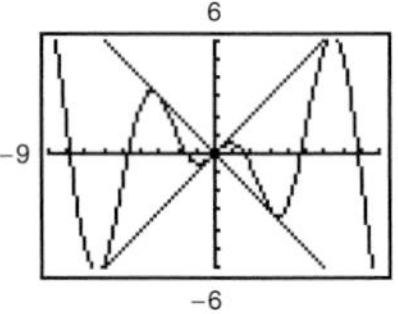

As $x \to +\infty$, $f(x)$

105. $-\frac{\pi}{6}$ **107.** 0.41 **109.** -0.46 **111.** $\frac{3\pi}{4}$

113. π **115.** 1.24 **117.** -0.98

119.

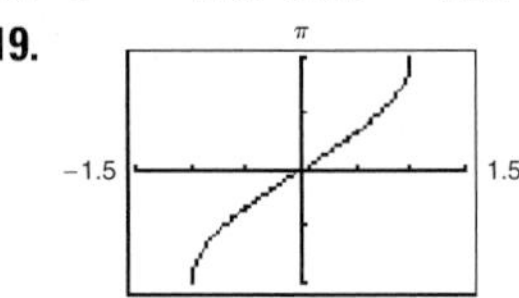

121.

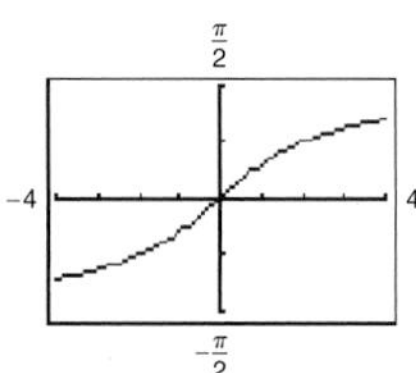

123. $\frac{4}{5}$ **125.** $\frac{13}{5}$ **127.** $\frac{10}{7}$ **129.** $\frac{\sqrt{4-x^2}}{x}$

131. $\frac{\pi}{6}$ **133.** $\frac{3\pi}{4}$ **135.** 0.09 **137.** 1.98

139.

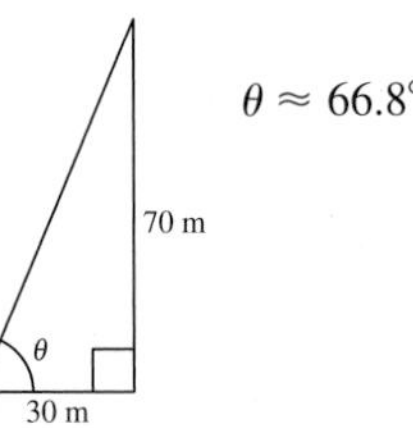

$\theta \approx 66.8°$

141. 1221 mi, 85.6°

143. False. For each θ there corresponds exactly one value of y.

145. The function is undefined because $\sec\theta = 1/\cos\theta$.

147. The ranges of the other four trigonometric functions are $(-\infty, \infty)$ or $(-\infty, -1] \cup [1, \infty)$.

Chapter Test (page 642)

1. (a)

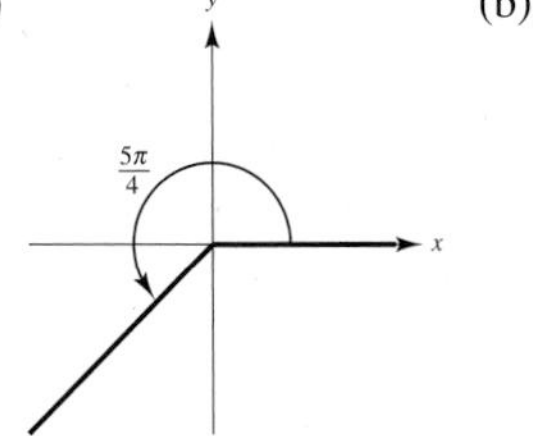

(b) $\frac{13\pi}{4}, -\frac{3\pi}{4}$ (c) 225°

2. 3500 rad/min

3. $\sin\theta = \frac{3\sqrt{10}}{10}$ $\csc\theta = \frac{\sqrt{10}}{3}$

$\cos\theta = -\frac{\sqrt{10}}{10}$ $\sec\theta = -\sqrt{10}$

$\tan\theta = -3$ $\cot\theta = -\frac{1}{3}$

4. For $0 \le \theta < \frac{\pi}{2}$:

$\sin\theta = \frac{3\sqrt{13}}{13}$

$\cos\theta = \frac{2\sqrt{13}}{13}$

$\csc\theta = \frac{\sqrt{13}}{3}$

$\sec\theta = \frac{\sqrt{13}}{2}$

$\cot\theta = \frac{2}{3}$

For $\pi \le \theta < \frac{3\pi}{2}$:

$\sin\theta = -\frac{3\sqrt{13}}{13}$

$\cos\theta = -\frac{2\sqrt{13}}{13}$

$\csc\theta = -\frac{\sqrt{13}}{3}$

$\sec\theta = -\frac{\sqrt{13}}{2}$

$\cot\theta = \frac{2}{3}$

5. $\theta' = 25°$

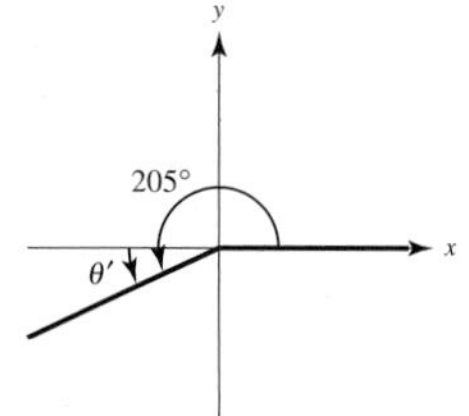

6. Quadrant III **7.** 150°, 210° **8.** 1.33, 1.81

9. $\sin\theta = -\frac{4}{5}$

$\tan\theta = -\frac{4}{3}$

$\csc\theta = -\frac{5}{4}$

$\sec\theta = \frac{5}{3}$

$\cot\theta = -\frac{3}{4}$

10. $\sin\theta = \frac{21}{29}$

$\cos\theta = -\frac{20}{29}$

$\tan\theta = -\frac{21}{20}$

$\csc\theta = \frac{29}{21}$

$\cot\theta = -\frac{20}{21}$

11.

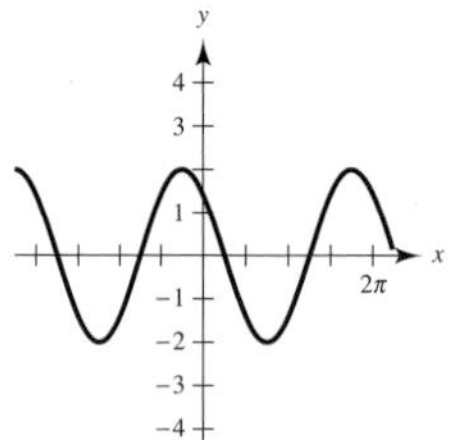

12.

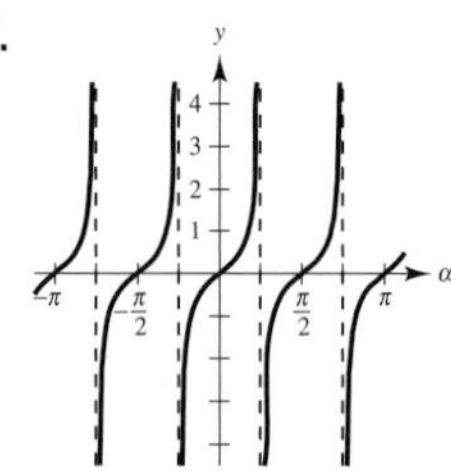

13.

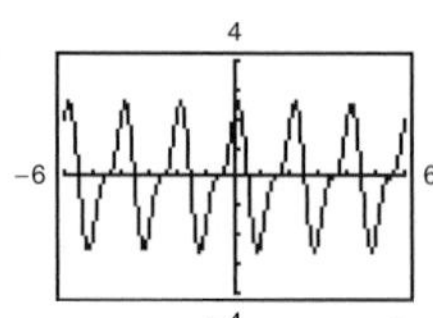

Period: 2

14.

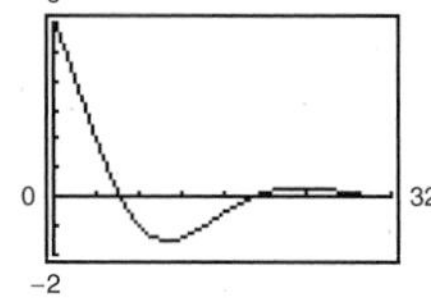

Not periodic

15. $a = -2, b = \frac{1}{2}, c = -\frac{\pi}{4}$ **16.** $\frac{\sqrt{55}}{3}$

17.

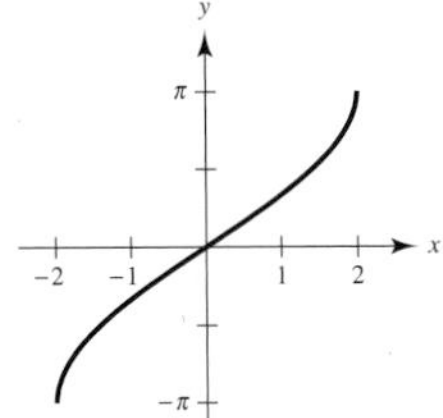

18. 309.3° **19.** $d = -6 \cos \pi t$

P.S. Problem Solving (page 643)

1. (a)

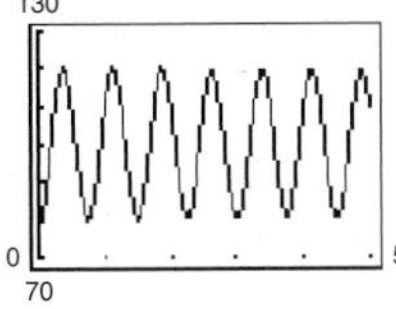

(b) Period $= \frac{3}{4}$ sec; Answers will vary.

(c) 20 mm; Answers will vary.

(d) 80 beats/min

(e) Period $= \dfrac{15}{16}$ sec; $\dfrac{32\pi}{15}$

3. Proof

5. (a)

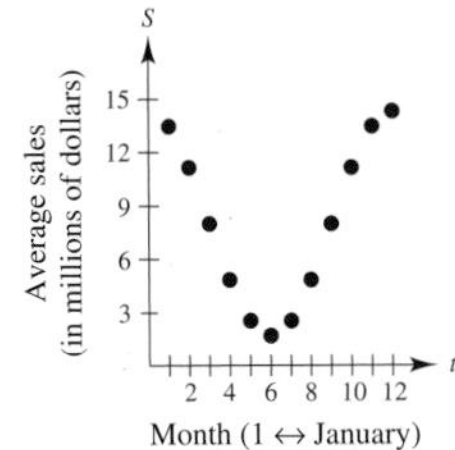

(b)

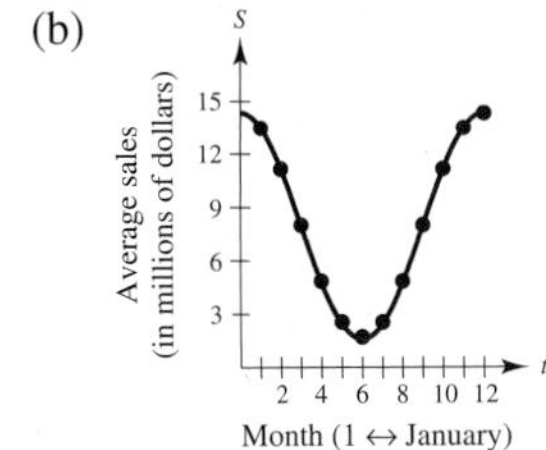

$S = 8 + 6.3 \cos\left(\dfrac{\pi t}{6}\right)$

The model is a good fit.

(c) Period: $\dfrac{2\pi}{\pi/6} = 12$

This corresponds to the 12 months in a year. Because the sales of outerwear is seasonal, this is reasonable.

(d) The amplitude represents the maximum displacement from average sales of 8 million dollars. Sales are greatest in December (cold weather + Christmas) and least in June.

7. (a)

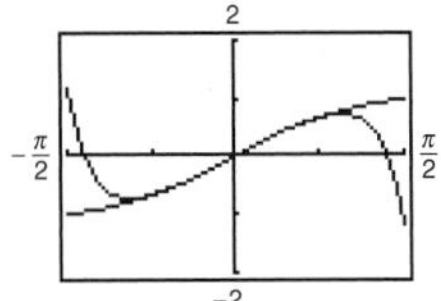

The approximation appears to be accurate over the interval $-0.5 < x < 0.5$.

(b) $\arctan x \approx x - \dfrac{x^3}{3} + \dfrac{x^5}{5} - \dfrac{x^7}{7} + \dfrac{x^9}{9}$

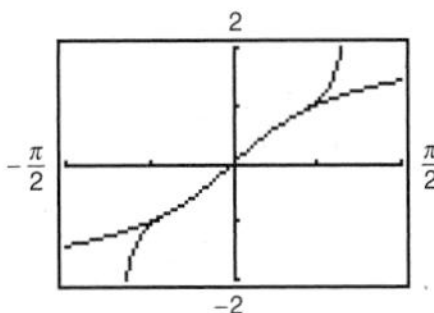

The accuracy improved.

9. (a) & (b)

Base 1	Base 2	Altitude	Area
8	8 + 16 cos 10°	8 sin 10°	22.1
8	8 + 16 cos 20°	8 sin 20°	42.5
8	8 + 16 cos 30°	8 sin 30°	59.7
8	8 + 16 cos 40°	8 sin 40°	72.7
8	8 + 16 cos 50°	8 sin 50°	80.5
8	8 + 16 cos 60°	8 sin 60°	83.1
8	8 + 16 cos 70°	8 sin 70°	80.7

The maximum occurs when $\theta = 60°$ and is approximately 83.1 square feet.

(c) $A(\theta) = 64(1 + \cos \theta)(\sin \theta)$

(d)

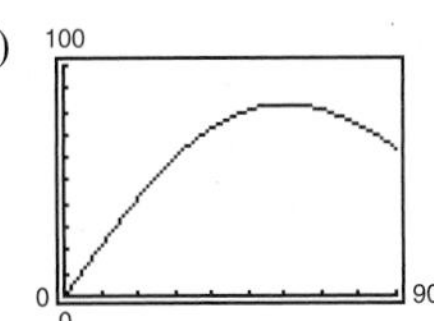

The maximum of 83.1 square feet occurs when $\theta = \dfrac{\pi}{3} = 60°$.

11. Proof

13. $h = 51 - 50 \sin\left(8\pi t + \dfrac{\pi}{2}\right)$

15. (a) $f(t - 2c) = f(t)$, t and $t - 2c$ differ by an integer multiple of the period.

(b) $f\left(t + \frac{1}{2}c\right) \neq f\left(\frac{1}{2}t\right)$, $\left(t + \frac{1}{2}c\right) - \left(\frac{1}{2}t\right) = \frac{1}{2}t + \frac{1}{2}c$, which is not an integer multiple of the period.

(c) $f\left(\frac{1}{2}[t + c]\right) \neq f\left(\frac{1}{2}t\right)$, $\left(\frac{1}{2}[t + c]\right) - \left(\frac{1}{2}t\right) = \frac{1}{2}c$, which is not an integer multiple of the period.

Chapter 10

Section 10.1 (page 651)

1. $\tan u$ **3.** $\cot u$ **5.** $\cot^2 u$ **7.** $\cos u$ **9.** $\cos u$

11. $\sin x = \frac{1}{2}$

$\cos x = \frac{\sqrt{3}}{2}$

$\tan x = \frac{\sqrt{3}}{3}$

$\csc x = 2$

$\sec x = \frac{2\sqrt{3}}{3}$

$\cot x = \sqrt{3}$

13. $\sin \theta = -\frac{\sqrt{2}}{2}$

$\cos \theta = \frac{\sqrt{2}}{2}$

$\tan \theta = -1$

$\csc \theta = -\sqrt{2}$

$\sec \theta = \sqrt{2}$

$\cot \theta = -1$

15. $\sin x = -\frac{8}{17}$

$\cos x = -\frac{15}{17}$

$\tan x = \frac{8}{15}$

$\csc x = -\frac{17}{8}$

$\sec x = -\frac{17}{15}$

$\cot x = \frac{15}{8}$

17. $\sin \phi = -\frac{\sqrt{5}}{3}$

$\cos \phi = \frac{2}{3}$

$\tan \phi = -\frac{\sqrt{5}}{2}$

$\csc \phi = -\frac{3\sqrt{5}}{5}$

$\sec \phi = \frac{3}{2}$

$\cot \phi = -\frac{2\sqrt{5}}{5}$

19. $\sin x = \frac{1}{3}$

$\cos x = -\frac{2\sqrt{2}}{3}$

$\tan x = -\frac{\sqrt{2}}{4}$

$\csc x = 3$

$\sec x = -\frac{3\sqrt{2}}{4}$

$\cot x = -2\sqrt{2}$

21. $\sin \theta = -\frac{2\sqrt{5}}{5}$

$\cos \theta = -\frac{\sqrt{5}}{5}$

$\tan \theta = 2$

$\csc \theta = -\frac{\sqrt{5}}{2}$

$\sec \theta = -\sqrt{5}$

$\cot \theta = \frac{1}{2}$

23. $\sin \theta = -1$ $\csc \theta = -1$

$\cos \theta = 0$ $\sec \theta$ is undefined.

$\tan \theta$ is undefined. $\cot \theta = 0$

25. d **26.** a **27.** b **28.** f **29.** e **30.** c

31. $\csc \theta$ **33.** $\cos^2 \phi$ **35.** $\cos x$ **37.** $\sin^2 x$

39. 1 **41.** $\tan x$ **43.** $1 + \sin y$ **45.** $\sec \beta$

47. $\cos u + \sin u$ **49.** $\sin^2 x$ **51.** $\sin^2 x \tan^2 x$

53. $\sec x + 1$ **55.** $\sec^4 x$ **57.** $\sin^2 x - \cos^2 x$

59. $\cot^2 x(\csc x - 1)$ **61.** $1 + 2 \sin x \cos x$

63. $4 \cot^2 x$ **65.** $2 \csc^2 x$ **67.** $2 \sec x$ **69.** $1 + \cos y$

71. $3(\sec x + \tan x)$

73. Not an identity because $\cos \theta = \pm\sqrt{1 - \sin^2 \theta}$

75. Not an identity because $\sin k\theta/\cos k\theta = \tan k\theta$

77. Identity because $\sin \theta \cdot 1/\sin \theta = 1$

79. $\cos \theta = \pm\sqrt{1 - \sin^2 \theta}$

$\tan \theta = \pm\sin \theta/\sqrt{1 - \sin^2 \theta}$

$\csc \theta = 1/\sin \theta$

$\sec \theta = \pm 1/\sqrt{1 - \sin^2 \theta}$

$\cot \theta = \pm\sqrt{1 - \sin^2 \theta}/\sin \theta$

81.

x	0.2	0.4	0.6	0.8	1.0
y_1	0.1987	0.3894	0.5646	0.7174	0.8415
y_2	0.1987	0.3894	0.5646	0.7174	0.8415

x	1.2	1.4
y_1	0.9320	0.9854
y_2	0.9320	0.9854

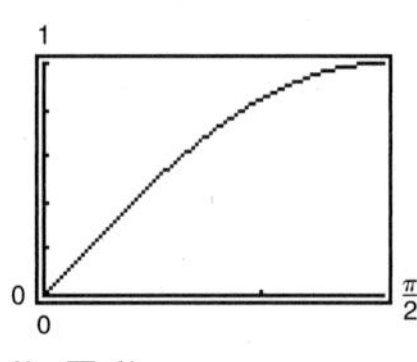

$y_1 = y_2$

83.

x	0.2	0.4	0.6	0.8	1.0
y_1	1.2230	1.5085	1.8958	2.4650	3.4082
y_2	1.2230	1.5085	1.8958	2.4650	3.4082

x	1.2	1.4
y_1	5.3319	11.6814
y_2	5.3319	11.6814

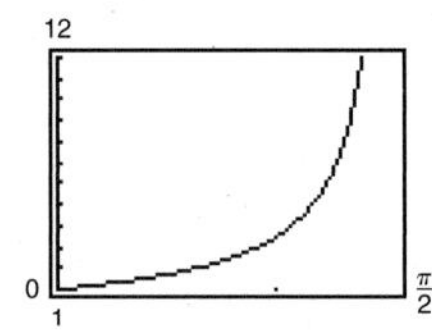

$y_1 = y_2$

85. $\csc x$ **87.** $\tan x$ **89.** $3 \sin \theta$ **91.** $7 \cos \theta$

93. $10 \sec \theta$ **95.** $3 \cos \theta = 3$; $\sin \theta = 0$; $\cos \theta = 1$

97. $4 \sin \theta = 2\sqrt{2}$; $\sin \theta = \frac{\sqrt{2}}{2}$; $\cos \theta = \frac{\sqrt{2}}{2}$

99. $0 \le \theta \le \pi$ **101.** $0 \le \theta < \frac{\pi}{2}, \frac{3\pi}{2} < \theta < 2\pi$

103. $\ln|\cot x|$ **105.** $\ln|\csc t \sec t|$

107. (a) $\csc^2 132° - \cot^2 132° \approx 1.8107 - 0.8107 = 1$

(b) $\csc^2 \frac{2\pi}{7} - \cot^2 \frac{2\pi}{7} \approx 1.6360 - 0.6360 = 1$

109. (a) $\cos(90° - 80°) = \sin 80° \approx 0.9848$

(b) $\cos\left(\frac{\pi}{2} - 0.8\right) = \sin 0.8 \approx 0.7174$

111. $\mu = \tan \theta$ **113.** Answers will vary.

115. False. A cofunction identity can be used to transform a tangent function so that it can be represented by a cotangent function.

117. 1, 1 **119.** $0, -\infty$

Section 10.2 (page 658)

1. identity **3.** $\tan u$ **5.** $\cos^2 u$ **7.** $-\csc u$

9–45. Answers will vary.

47. (a)

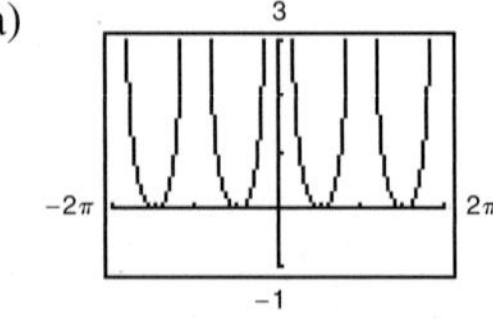

(b)

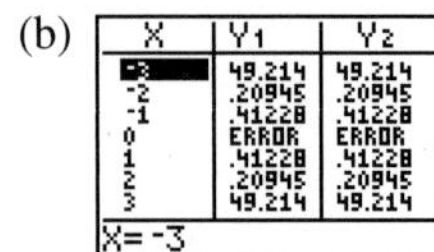

X	Y1	Y2
-3	49.214	49.214
-2	.20945	.20945
-1	.41228	.41228
0	ERROR	ERROR
1	.41228	.41228
2	.20945	.20945
3	49.214	49.214

X=-3

Identity

(c) Answers will vary.

49. (a)

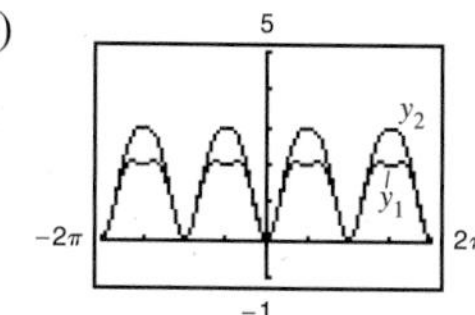

(b)

X	Y_1	Y_2
-4.712	2	3
-3.142	0	0
-1.571	2	3
0	0	0
1.5708	2	3
3.1416	0	0
4.7124	2	3

X=-4.71238898038

Not an identity

(c) Answers will vary.

51. (a)

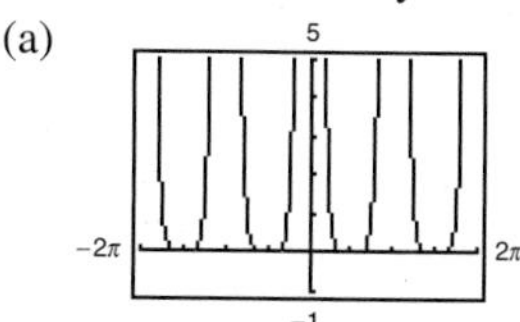

(b)

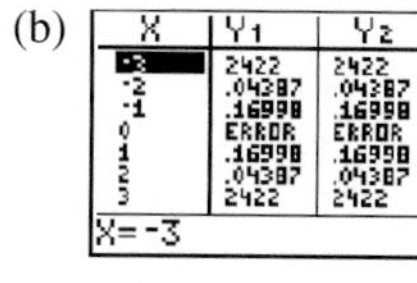

X	Y_1	Y_2
-3	2422	2422
-2	.04387	.04387
-1	.16998	.16998
0	ERROR	ERROR
1	.16998	.16998
2	.04387	.04387
3	2422	2422

X=-3

Identity

(c) Answers will vary.

53. (a)

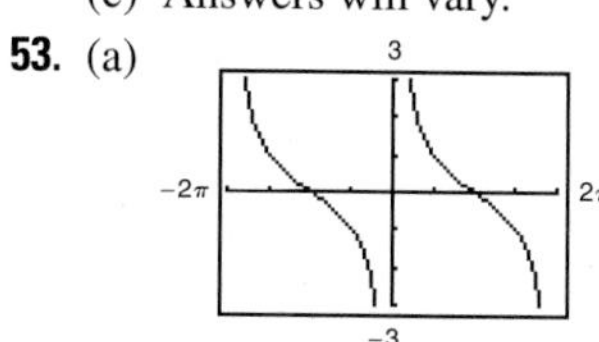

(b)

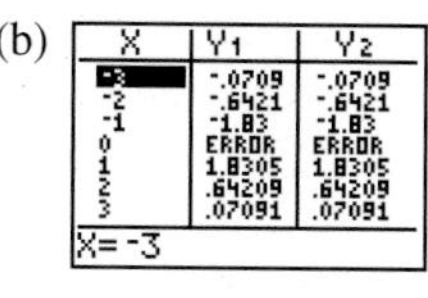

X	Y_1	Y_2
-3	-.0709	-.0709
-2	-.6421	-.6421
-1	-1.83	-1.83
0	ERROR	ERROR
1	1.8305	1.8305
2	.64209	.64209
3	.07091	.07091

X=-3

Identity

(c) Answers will vary.

55–57. Answers will vary.

59. Not an identity because $\sin\theta = \pm\sqrt{1-\cos^2\theta}$
Possible answer: $7\pi/4$

61. Not an identity because $\sqrt{\tan^2 x} = |\tan x|$
Possible answer: $3\pi/4$

63. The equation is not an identity because $1 + \tan^2\theta = \sec^2\theta$.
Possible answer: $\pi/6$

65. Answers will vary. **67.** 1 **69.** 2

71. Answers will vary.

73. True. Many different techniques can be used to verify identities.

Section 10.3 (page 666)

1. isolate **3.** quadratic **5–9.** Answers will vary.

11. $\frac{2\pi}{3} + 2n\pi, \frac{4\pi}{3} + 2n\pi$ **13.** $\frac{\pi}{3} + 2n\pi, \frac{2\pi}{3} + 2n\pi$

15. $\frac{\pi}{6} + n\pi, \frac{5\pi}{6} + n\pi$ **17.** $n\pi, \frac{3\pi}{2} + 2n\pi$

19. $\frac{\pi}{3} + n\pi, \frac{2\pi}{3} + n\pi$ **21.** $0, \frac{\pi}{2}, \pi, \frac{3\pi}{2}$

23. $0, \pi, \frac{\pi}{6}, \frac{5\pi}{6}, \frac{7\pi}{6}, \frac{11\pi}{6}$ **25.** $\frac{\pi}{3}, \frac{5\pi}{3}, \pi$ **27.** No solution

29. $\pi, \frac{\pi}{3}, \frac{5\pi}{3}$ **31.** $\frac{\pi}{6}, \frac{5\pi}{6}, \frac{7\pi}{6}, \frac{11\pi}{6}$ **33.** $\frac{\pi}{6} + n\pi, \frac{5\pi}{6} + n\pi$

35. $\frac{\pi}{12} + \frac{n\pi}{3}$ **37.** $\frac{\pi}{2} + 4n\pi, \frac{7\pi}{2} + 4n\pi$

39. $\frac{\pi}{8} + n\pi, \frac{3\pi}{8} + n\pi, \frac{5\pi}{8} + n\pi, \frac{7\pi}{8} + n\pi$

41. $3 + 4n$ **43.** $-2 + 6n, 2 + 6n$

45. $\frac{2}{3}, \frac{3}{2}$; $\cos^{-1}\left(\frac{2}{3}\right) + 2n\pi \approx 0.8411 + 2n\pi$, $5.4421 + 2n\pi$ and $\cos^{-1}\left(\frac{3}{2}\right) + 2n\pi$

47. 2.678, 5.820 **49.** 1.047, 5.236 **51.** 0.860, 3.426

53. 0, 2.678, 3.142, 5.820 **55.** 0.983, 1.768, 4.124, 4.910

57. 0.3398, 0.8481, 2.2935, 2.8018

59. 1.9357, 2.7767, 5.0773, 5.9183

61. $\frac{\pi}{4}, \frac{5\pi}{4}$, arctan 5, arctan 5 + π **63.** $\frac{\pi}{3}, \frac{5\pi}{3}$

65. (a)

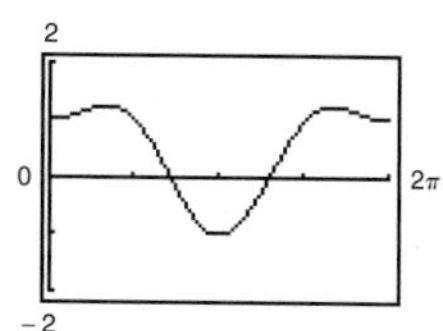

(b) $\frac{\pi}{3} \approx 1.0472$

$\frac{5\pi}{3} \approx 5.2360$

0

$\pi \approx 3.1416$

Maximum: (1.0472, 1.25)
Maximum: (5.2360, 1.25)
Minimum: (0, 1)
Minimum: (3.1416, −1)

67. (a)

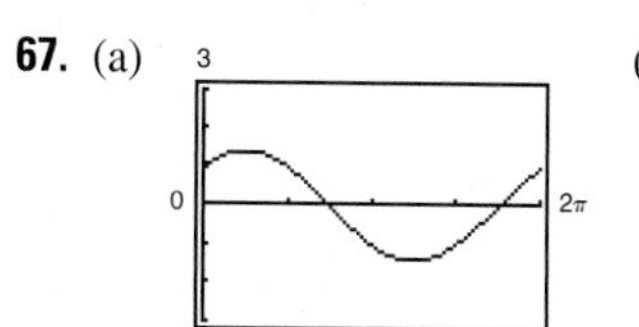

(b) $\frac{\pi}{4} \approx 0.7854$

$\frac{5\pi}{4} \approx 3.9270$

Maximum: (0.7854, 1.4142)
Minimum: (3.9270, −1.4142)

69. (a) All real numbers x except $x = 0$
(b) y-axis symmetry; Horizontal asymptote: $y = 1$
(c) Oscillates (d) Infinitely many solutions; $\frac{2}{2n\pi + \pi}$
(e) Yes, 0.6366

71. 0.04 sec, 0.43 sec, 0.83 sec

73. February, March, and April **75.** 1.9°

77. (a)

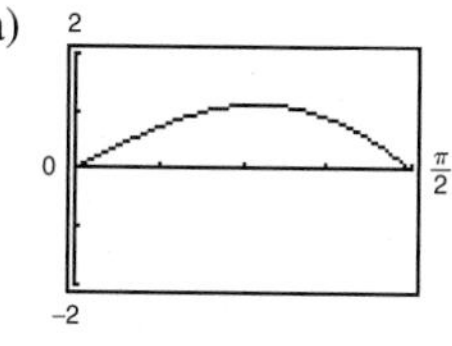

$A \approx 1.12$

(b) $0.6 < x < 1.1$

79. True. The first equation has a smaller period than the second equation, so it will have more solutions in the interval $[0, 2\pi)$.

81. The equation would become $\cos^2 x = 2$; this is not the correct method to use when solving equations.

Section 10.4 (page 673)

1. $\sin u \cos v - \cos u \sin v$ **3.** $\frac{\tan u + \tan v}{1 - \tan u \tan v}$

5. $\cos u \cos v + \sin u \sin v$

7. (a) $\frac{\sqrt{2} - \sqrt{6}}{4}$ $\frac{\sqrt{3} - 1}{2}$ (b) $\frac{\sqrt{2} + 1}{2}$

9. (a) $\frac{1}{2}$ (b)

11. (a) $\frac{\sqrt{6} + \sqrt{2}}{4}$ (b) $\frac{\sqrt{2} - \sqrt{3}}{2}$

13. $\sin\frac{11\pi}{12} = \frac{\sqrt{2}}{4}(\sqrt{3} - 1)$

$\cos\frac{11\pi}{12} = -\frac{\sqrt{2}}{4}(\sqrt{3} + 1)$

$\tan\frac{11\pi}{12} = -2 + \sqrt{3}$

15. $\sin \frac{17\pi}{12} = -\frac{\sqrt{2}}{4}(\sqrt{3} + 1)$

$\cos \frac{17\pi}{12} = \frac{\sqrt{2}}{4}(1 - \sqrt{3})$

$\tan \frac{17\pi}{12} = 2 + \sqrt{3}$

17. $\sin 105° = \frac{\sqrt{2}}{4}(\sqrt{3} + 1)$

$\cos 105° = \frac{\sqrt{2}}{4}(1 - \sqrt{3})$

$\tan 105° = -2 - \sqrt{3}$

19. $\sin 195° = \frac{\sqrt{2}}{4}(1 - \sqrt{3})$

$\cos 195° = -\frac{\sqrt{2}}{4}(\sqrt{3} + 1)$

$\tan 195° = 2 - \sqrt{3}$

21. $\sin \frac{13\pi}{12} = \frac{\sqrt{2}}{4}(1 - \sqrt{3})$

$\cos \frac{13\pi}{12} = -\frac{\sqrt{2}}{4}(1 + \sqrt{3})$

$\tan \frac{13\pi}{12} = 2 - \sqrt{3}$

23. $\sin\left(-\frac{13\pi}{12}\right) = \frac{\sqrt{2}}{4}(\sqrt{3} - 1)$

$\cos\left(-\frac{13\pi}{12}\right) = -\frac{\sqrt{2}}{4}(\sqrt{3} + 1)$

$\tan\left(-\frac{13\pi}{12}\right) = -2 + \sqrt{3}$

25. $\sin 285° = -\frac{\sqrt{2}}{4}(\sqrt{3} + 1)$

$\cos 285° = \frac{\sqrt{2}}{4}(\sqrt{3} - 1)$

$\tan 285° = -(2 + \sqrt{3})$

27. $\sin(-165°) = -\frac{\sqrt{2}}{4}(\sqrt{3} - 1)$

$\cos(-165°) = -\frac{\sqrt{2}}{4}(1 + \sqrt{3})$

$\tan(-165°) = 2 - \sqrt{3}$

29. $\sin 1.8$ **31.** $\sin 75°$ **33.** $\tan 15°$ **35.** $\tan 3x$

37. $\frac{\sqrt{3}}{2}$ **39.** $\frac{\sqrt{3}}{2}$ **41.** $-\sqrt{3}$ **43.** $-\frac{63}{65}$ **45.** $\frac{16}{65}$

47. $-\frac{63}{16}$ **49.** $\frac{65}{56}$ **51.** $\frac{3}{5}$ **53.** $-\frac{44}{117}$ **55.** $-\frac{125}{44}$

57–61. Proofs **63–65.** Answers will vary. **67.** 1

69. 0 **71.** $-\sin x$ **73.** $-\cos \theta$ **75.** $\frac{\pi}{3}, \frac{5\pi}{3}$

77. $\frac{5\pi}{4}, \frac{7\pi}{4}$ **79.** $0, \frac{\pi}{2}, \frac{3\pi}{2}$

81. True. $\sin(u \pm v) = \sin u \cos v \pm \cos u \sin v$

83. False. $\tan\left(x - \frac{\pi}{4}\right) = \frac{\tan x - 1}{1 + \tan x}$

85–87. Answers will vary.

89. (a) $\sqrt{2} \sin\left(\theta + \frac{\pi}{4}\right)$ (b) $\sqrt{2} \cos\left(\theta - \frac{\pi}{4}\right)$

91. (a) $13 \sin(3\theta + 0.3948)$ (b) $13 \cos(3\theta - 1.1760)$

93. $\sqrt{2} \sin \theta + \sqrt{2} \cos \theta$ **95.** Answers will vary.

97.

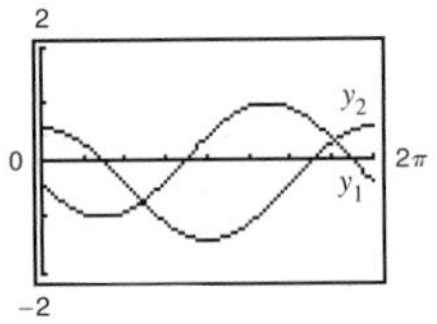

No, $y_1 \neq y_2$ because their graphs are different.

99. (a) and (b) Proofs

Section 10.5 (page 682)

1. $2 \sin u \cos u$

3. $\cos^2 u - \sin^2 u = 2 \cos^2 u - 1 = 1 - 2 \sin^2 u$

5. $\pm\sqrt{\frac{1 - \cos u}{2}}$ **7.** $\frac{1}{2}[\cos(u - v) + \cos(u + v)]$

9. $2 \sin\left(\frac{u + v}{2}\right) \cos\left(\frac{u - v}{2}\right)$ **11.** $\frac{15}{17}$ **13.** $\frac{8}{15}$

15. $\frac{17}{8}$ **17.** $\frac{240}{289}$ **19.** $0, \frac{\pi}{3}, \pi, \frac{5\pi}{3}$

21. $\frac{\pi}{12}, \frac{5\pi}{12}, \frac{13\pi}{12}, \frac{17\pi}{12}$ **23.** $0, \frac{2\pi}{3}, \frac{4\pi}{3}$ **25.** $0, \frac{\pi}{2}, \pi, \frac{3\pi}{2}$

27. $\frac{\pi}{2}, \frac{\pi}{6}, \frac{5\pi}{6}, \frac{7\pi}{6}, \frac{3\pi}{2}, \frac{11\pi}{6}$ **29.** $3 \sin 2x$ **31.** $3 \cos 2x$

33. $4 \cos 2x$ **35.** $\cos 2x$

37. $\sin 2u = -\frac{24}{25}, \cos 2u = \frac{7}{25}, \tan 2u = -\frac{24}{7}$

39. $\sin 2u = \frac{15}{17}, \cos 2u = \frac{8}{17}, \tan 2u = \frac{15}{8}$

41. $\sin 2u = -\frac{\sqrt{3}}{2}, \cos 2u = -\frac{1}{2}, \tan 2u = \sqrt{3}$

43. $\frac{1}{8}(3 + 4 \cos 2x + \cos 4x)$ **45.** $\frac{1}{8}(3 + 4 \cos 4x + \cos 8x)$

47. $\frac{(3 - 4 \cos 4x + \cos 8x)}{(3 + 4 \cos 4x + \cos 8x)}$ **49.** $\frac{1}{8}(1 - \cos 8x)$

51. $\frac{1}{16}(1 - \cos 2x - \cos 4x + \cos 2x \cos 4x)$

53. $\frac{4\sqrt{17}}{17}$ **55.** $\frac{1}{4}$ **57.** $\sqrt{17}$

59. $\sin 75° = \frac{1}{2}\sqrt{2 + \sqrt{3}}$

$\cos 75° = \frac{1}{2}\sqrt{2 - \sqrt{3}}$

$\tan 75° = 2 + \sqrt{3}$

61. $\sin 112° 30' = \frac{1}{2}\sqrt{2 + \sqrt{2}}$

$\cos 112° 30' = -\frac{1}{2}\sqrt{2 - \sqrt{2}}$

$\tan 112° 30' = -1 - \sqrt{2}$

63. $\sin \frac{\pi}{8} = \frac{1}{2}\sqrt{2 - \sqrt{2}}$

$\cos \frac{\pi}{8} = \frac{1}{2}\sqrt{2 + \sqrt{2}}$

$\tan \frac{\pi}{8} = \sqrt{2} - 1$

65. $\sin \frac{3\pi}{8} = \frac{1}{2}\sqrt{2 + \sqrt{2}}$

$\cos \frac{3\pi}{8} = \frac{1}{2}\sqrt{2 - \sqrt{2}}$

$\tan \frac{3\pi}{8} = \sqrt{2} + 1$

67. (a) Quadrant I

(b) $\sin \frac{u}{2} = \frac{3}{5}$, $\cos \frac{u}{2} = \frac{4}{5}$, $\tan \frac{u}{2} = \frac{3}{4}$

69. (a) Quadrant II

(b) $\sin \frac{u}{2} = \frac{\sqrt{26}}{26}$, $\cos \frac{u}{2} = -\frac{5\sqrt{26}}{26}$, $\tan \frac{u}{2} = -\frac{1}{5}$

71. (a) Quadrant II

(b) $\sin \frac{u}{2} = \frac{3\sqrt{10}}{10}$, $\cos \frac{u}{2} = -\frac{\sqrt{10}}{10}$, $\tan \frac{u}{2} = -3$

73. $|\sin 3x|$ **75.** $-|\tan 4x|$

77. π

79. $\frac{\pi}{3}, \pi, \frac{5\pi}{3}$

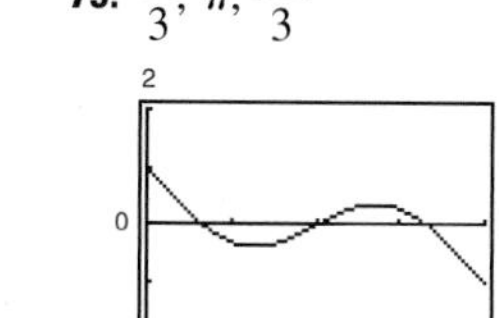

81. $\frac{1}{2}\left(\sin \frac{\pi}{2} + \sin \frac{\pi}{6}\right)$ **83.** $5(\cos 60° + \cos 90°)$

85. $\frac{1}{2}(\cos 2\theta - \cos 8\theta)$ **87.** $\frac{7}{2}(\sin(-2\beta) - \sin(-8\beta))$

89. $\frac{1}{2}(\cos 2y - \cos 2x)$ **91.** $2 \sin 2\theta \cos \theta$

93. $2 \cos 4x \cos 2x$ **95.** $2 \cos \alpha \sin \beta$

97. $-2 \sin \theta \sin \frac{\pi}{2} = -2 \sin \theta$ **99.** $\frac{\sqrt{6}}{2}$ **101.** $-\sqrt{2}$

103. $0, \frac{\pi}{4}, \frac{\pi}{2}, \frac{3\pi}{4}, \pi, \frac{5\pi}{4}, \frac{3\pi}{2}, \frac{7\pi}{4}$

105. $\frac{\pi}{6}, \frac{5\pi}{6}$

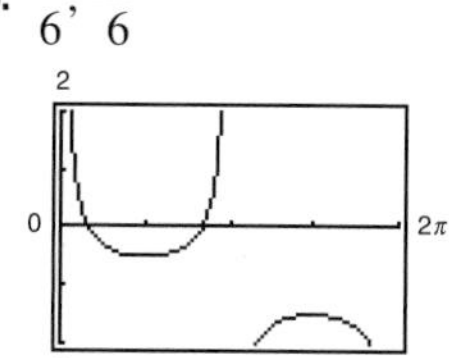

107. $\frac{120}{169}$ **109.** $\frac{120}{119}$ **111.** $\frac{3\sqrt{10}}{10}$ **113.** $\frac{36}{65}$ **115.** $\frac{8}{65}$

117–129. Answers will vary.

131.

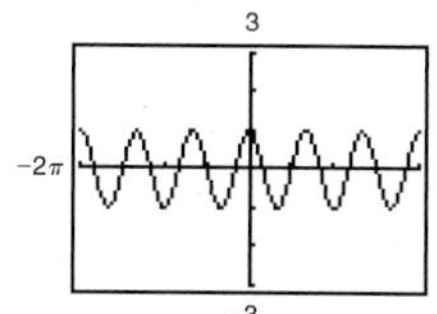

133.

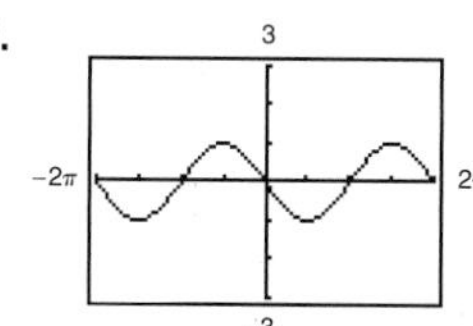

135.

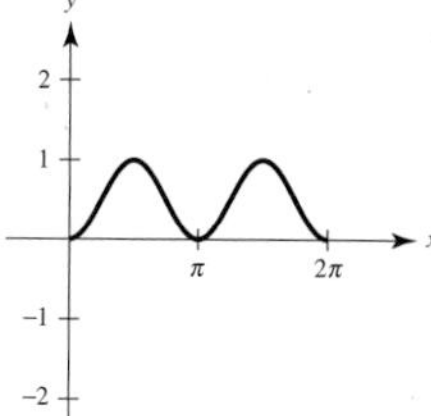

137. $2x\sqrt{1 - x^2}$ **139.** $1 - 2x^2$

141. (a)

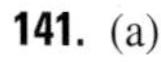

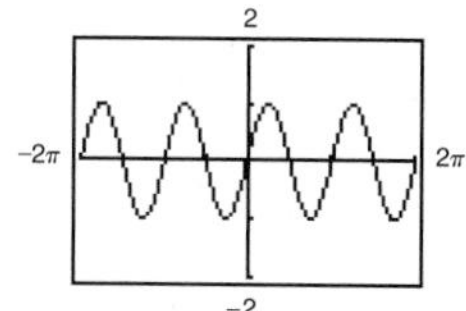

(b) The graph appears to be that of $\sin 2x$.

(c) $2 \sin x\left[2 \cos^2\left(\frac{x}{2}\right) - 1\right] = 2 \sin x \cos x$

$= \sin 2x$

143. (a) π (b) 0.4482 (c) 760 mi/h; 3420 mi/h

(d) $\theta = 2 \sin^{-1}\left(\frac{1}{M}\right)$

145. False. For $u < 0$,

$\sin 2u = -\sin(-2u)$
$= -2 \sin(-u) \cos(-u)$
$= -2(-\sin u) \cos u$
$= 2 \sin u \cos u.$

Review Exercises (page 686)

1. $\tan x$ **3.** $\cos x$ **5.** $|\csc x|$

7. $\tan x = \frac{5}{12}$, $\csc x = \frac{13}{5}$, $\sec x = \frac{13}{12}$, $\cot x = \frac{12}{5}$

9. $\cos x = \frac{\sqrt{2}}{2}$, $\tan x = -1$, $\csc x = -\sqrt{2}$, $\sec x = \sqrt{2}$, $\cot x = -1$

11. $\sin^2 x$ **13.** 1 **15.** $\cot \theta$ **17.** $\csc \theta$ **19.** $\cot^2 x$

21. $\sec x + 2 \sin x$ **23.** $-2 \tan^2 \theta$

25–31. Answers will vary.

33. $\frac{\pi}{3} + 2n\pi, \frac{2\pi}{3} + 2n\pi$ **35.** $\frac{\pi}{6} + n\pi$

37. $\frac{\pi}{3} + n\pi, \frac{2\pi}{3} + n\pi$ **39.** $0, \frac{2\pi}{3}, \frac{4\pi}{3}$ **41.** $0, \frac{\pi}{2}, \pi$

43. $\frac{\pi}{8}, \frac{3\pi}{8}, \frac{9\pi}{8}, \frac{11\pi}{8}$

45. $0, \frac{\pi}{8}, \frac{3\pi}{8}, \frac{5\pi}{8}, \frac{7\pi}{8}, \frac{9\pi}{8}, \frac{11\pi}{8}, \frac{13\pi}{8}, \frac{15\pi}{8}$ **47.** $0, \pi$

49. $\arctan(-3) + \pi$, $\arctan(-3) + 2\pi$, $\arctan 2$, $\arctan 2 + \pi$

51. $\sin 285° = -\frac{\sqrt{2}}{4}(\sqrt{3} + 1)$

$\cos 285° = \frac{\sqrt{2}}{4}(\sqrt{3} - 1)$

$\tan 285° = -2 - \sqrt{3}$

53. $\sin \frac{25\pi}{12} = \frac{\sqrt{2}}{4}(\sqrt{3} - 1)$

$\cos \frac{25\pi}{12} = \frac{\sqrt{2}}{4}(\sqrt{3} + 1)$

$\tan \frac{25\pi}{12} = 2 - \sqrt{3}$

55. $\sin 15°$ **57.** $\tan 35°$ **59.** $-\frac{24}{25}$ **61.** -1

63. $-\frac{7}{25}$ **65.** $\frac{\pi}{4}, \frac{7\pi}{4}$

67. $\sin 2u = \frac{24}{25}$

$\cos 2u = -\frac{7}{25}$

$\tan 2u = -\frac{24}{7}$

69. $\sin 2u = -\frac{4\sqrt{2}}{9}$, $\cos 2u = -\frac{7}{9}$, $\tan 2u = \frac{4\sqrt{2}}{7}$

71.

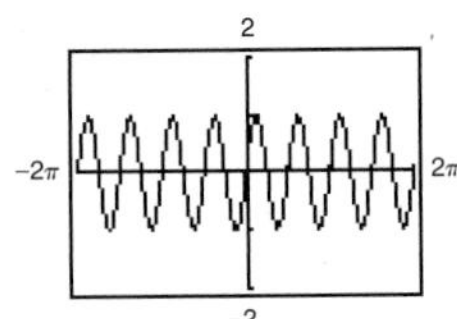

73. $\frac{1 - \cos 4x}{1 + \cos 4x}$ **75.** $\frac{3 - 4\cos 2x + \cos 4x}{4(1 + \cos 2x)}$

77. $\sin(-75°) = -\frac{1}{2}\sqrt{2 + \sqrt{3}}$

$\cos(-75°) = \frac{1}{2}\sqrt{2 - \sqrt{3}}$

$\tan(-75°) = -2 - \sqrt{3}$

79. $\sin\frac{19\pi}{12} = -\frac{1}{2}\sqrt{2 + \sqrt{3}}$

$\cos\frac{19\pi}{12} = \frac{1}{2}\sqrt{2 - \sqrt{3}}$

$\tan\frac{19\pi}{12} = -2 - \sqrt{3}$

81. $-|\cos 5x|$ **83.** $\frac{1}{2}\left(\sin\frac{\pi}{3} - \sin 0\right) = \frac{1}{2}\sin\frac{\pi}{3}$

85. $\frac{1}{2}[\sin 10\theta - \sin(-2\theta)]$ **87.** $2\cos 6\theta \sin(-2\theta)$

89. $-2\sin x \sin\frac{\pi}{6}$ **91.** $\theta = 15°$ or $\frac{\pi}{12}$

93.

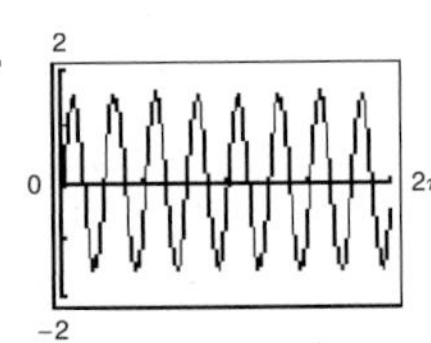

95. $\frac{1}{2}\sqrt{10}$ ft

97. $-1.8431, 2.1758, 3.9903, 8.8935, 9.8820$

Chapter Test (page 688)

1. $\sin\theta = -\frac{6\sqrt{61}}{61}$ $\csc\theta = -\frac{\sqrt{61}}{6}$

$\cos\theta = -\frac{5\sqrt{61}}{61}$ $\sec\theta = -\frac{\sqrt{61}}{5}$

$\tan\theta = \frac{6}{5}$ $\cot\theta = \frac{5}{6}$

2. 1 **3.** 1 **4.** $\csc\theta\sec\theta$

5. $\theta = 0, \frac{\pi}{2} < \theta \le \pi, \frac{3\pi}{2} < \theta < 2\pi$

6.

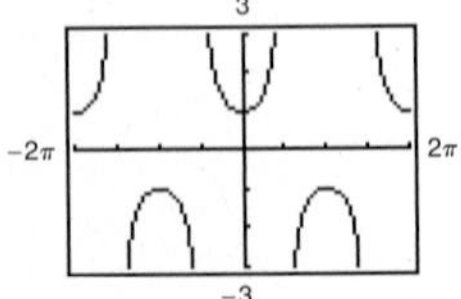

$y_1 = y_2$

7–12. Answers will vary. **13.** $\frac{1}{8}(3 - 4\cos x + \cos 2x)$

14. $\tan 2\theta$ **15.** $2(\sin 5\theta + \sin\theta)$

16. $-2\sin 2\theta \sin\theta$ **17.** $0, \frac{3\pi}{4}, \pi, \frac{7\pi}{4}$

18. $\frac{\pi}{6}, \frac{\pi}{2}, \frac{5\pi}{6}, \frac{3\pi}{2}$ **19.** $\frac{\pi}{6}, \frac{5\pi}{6}, \frac{7\pi}{6}, \frac{11\pi}{6}$

20. $\frac{\pi}{6}, \frac{5\pi}{6}, \frac{3\pi}{2}$ **21.** $-2.596, 0, 2.596$

22. $\frac{\sqrt{2} - \sqrt{6}}{4}$

23. $\sin 2u = -\frac{20}{29}$, $\cos 2u = -\frac{21}{29}$, $\tan 2u = \frac{20}{21}$

24. Day 123 to day 223

25. $t = 0.26$ min

0.58 min

0.89 min

1.20 min

1.52 min

1.83 min

P.S. Problem Solving (page 689)

1. 1

3. $\sin\frac{\theta}{2} = \sqrt{\frac{1 - \cos\theta}{2}}$

$\cos\frac{\theta}{2} = \sqrt{\frac{1 + \cos\theta}{2}}$

$\tan\frac{\theta}{2} = \frac{\sin\theta}{1 + \cos\theta}$

5. (a) $y = \frac{5}{12}\sin(2t + 0.6435)$

(b) $\frac{5}{12}$ foot

(c) $\frac{1}{\pi}$ cycle per second

7. Proof **9.** $\theta = \tan^{-1} m_2 - \tan^{-1} m_1$

11.

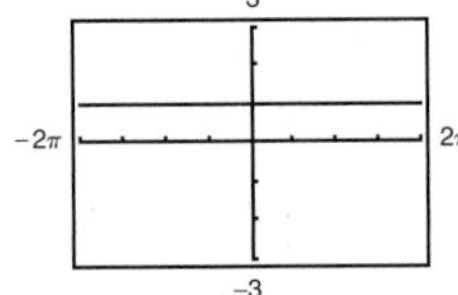

$\sin^2(\theta + \pi/4) + \sin^2(\theta - \pi/4) = 1$

13. (a) $\sin(u + v + w)$

$= \sin u \cos v \cos w - \sin u \sin v \sin w$

$+ \cos u \sin v \cos w + \cos u \cos v \sin w$

(b) $\tan(u + v + w)$

$= \frac{\tan u + \tan v + \tan w - \tan u \tan v \tan w}{1 - \tan u \tan v - \tan u \tan w - \tan v \tan w}$

15. (a) $A = 100\sin(\theta/2)\cos(\theta/2)$

(b) $A = 50\sin\theta$

The area is maximum when $\theta = \pi/2$.

17. $a = -1, 2$

Chapter 11

Section 11.1 (page 695)

1.

x	-0.1	-0.01	-0.001	0.001
$f(x)$	0.9983	0.99998	1.0000	1.0000

x	0.01	0.1
$f(x)$	0.99998	0.9983

$\lim_{x\to 0} (\sin x)/x \approx 1.0000$ (Actual limit is 1.)

3. Limit does not exist because $\lim_{x\to\pi/2^-} = \infty$ and $\lim_{x\to\pi/2^+} = -\infty$.

5. Limit does not exist because as x approaches 0, $\cos(1/x)$ oscillates between -1 and 1.

7. 1 **9.** $\frac{1}{2}$ **11.** 1 **13.** $\frac{1}{2}$ **15.** -1

17. Continuous for all real x **19.** Continuous for all real x

21. Nonremovable discontinuities at integer multiples of $\pi/2$

23.

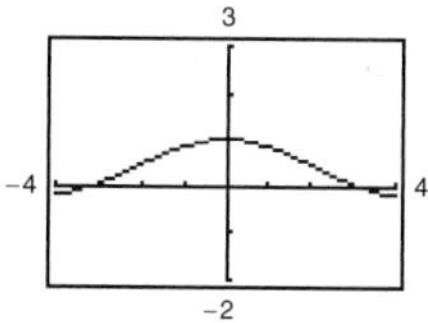

Although the graph appears continuous on $[-4, 4]$, there is a discontinuity at $x = 0$. Examining a function graphically and analytically ensures that you will find all points where the function is not defined.

25. $x = \frac{1}{2} + n$, n is an integer.

27. $t = n\pi$, n is a nonzero integer.

29.

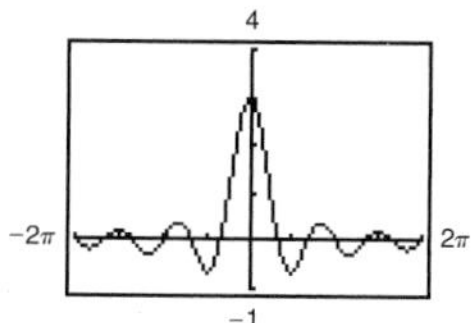

The graph has a hole at $t = 0$.

t	-0.1	-0.01	0	0.01	0.1
$f(t)$	2.96	2.9996	?	2.9996	2.96

$\lim_{t\to 0} (\sin 3t)/t = 3$

31.

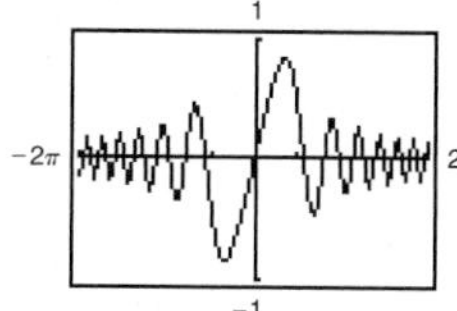

The graph has a hole at $x = 0$.

x	-0.1	-0.01	-0.001	0	0.001	0.01	0.1
$f(x)$	-0.1	-0.01	-0.001	?	0.001	0.01	0.1

$\lim_{x\to 0} (\sin x^2)/x = 0$

33.

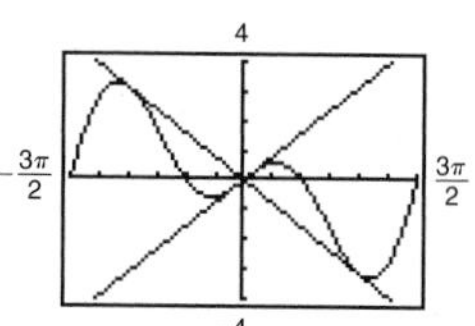

0

35.

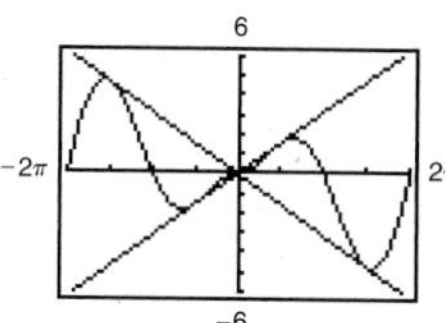

0

37.

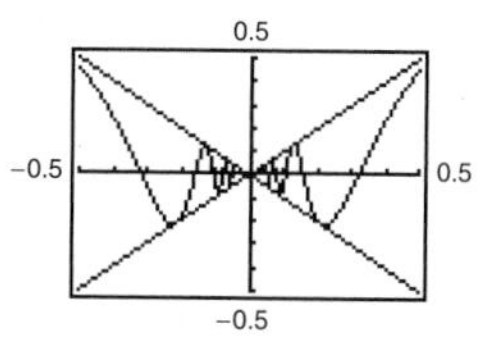

The graph has a hole at $x = 0$.

0

39. $\frac{1}{5}$ **41.** 0 **43.** 0 **45.** 0 **47.** 1 **49.** $\frac{3}{2}$

51. Limit does not exist. **53.** ∞ **55.** 0

57. Limit does not exist.

59. The value of $\sin x$ becomes arbitrarily close to $\frac{1}{2}$ as x approaches $\pi/6$ from either side.

61. $\lim_{x\to 0} \frac{\tan(nx)}{x} = n$

63.

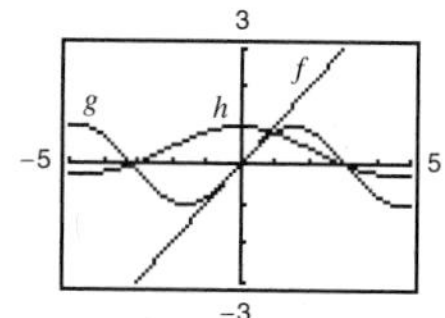

The magnitudes of $f(x)$ and $g(x)$ are approximately equal when x is close to 0. Therefore, their ratio is approximately 1.

65. (a)

x	1	0.5	0.2	0.1
$f(x)$	0.1585	0.0411	0.0067	0.0017

x	0.01	0.001	0.0001
$f(x)$	1.7×10^{-5}	1.7×10^{-7}	1.7×10^{-9}

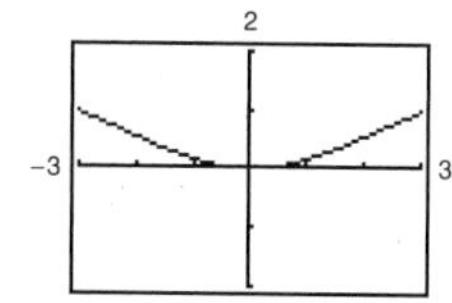

The graph has a hole at $x = 0$.

$\lim_{x\to 0^+} f(x) = 0$

(b)

x	1	0.5	0.2	0.1
$f(x)$	0.1585	0.0823	0.0333	0.0167

x	0.01	0.001	0.0001
$f(x)$	0.0017	1.7×10^{-4}	1.7×10^{-5}

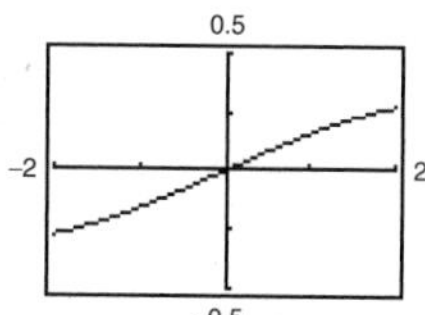

The graph has a hole at $x = 0$.

$\lim_{x\to 0^+} f(x) = 0$

(c)

x	1	0.5	0.2	0.1
$f(x)$	0.1585	0.1646	0.1663	0.1666

x	0.01	0.001	0.0001
$f(x)$	0.1667	0.1667	0.1667

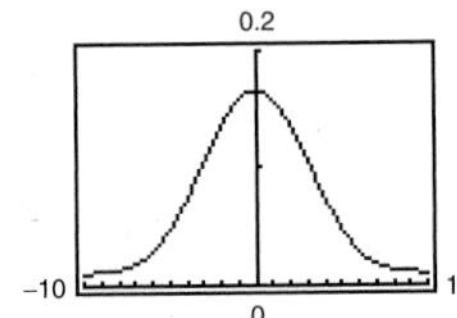

The graph has a hole at $x = 0$.
$\lim_{x\to 0^+} f(x) = \frac{1}{6}$

(d)

x	1	0.5	0.2	0.1
$f(x)$	0.1585	0.3292	0.8317	1.6658

x	0.01	0.001	0.0001
$f(x)$	16.667	166.67	1666.67

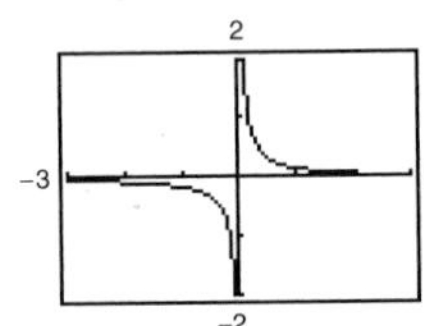

$\lim_{x\to 0^+} f(x) = \infty$
When the power of x is at least 3, the value of the limit is ∞.

67. (a) $\frac{1}{2}$
(b) Because $(1 - \cos x)/x^2 \approx \frac{1}{2}$, it follows that
$1 - \cos x \approx \frac{1}{2}x^2$
$\cos x \approx 1 - \frac{1}{2}x^2$ when $x \approx 0$.
(c) 0.995
(d) Calculator: $\cos(0.1) \approx 0.9950$

69. (a) $A = 50 \tan \theta - 50\theta$
Domain: $(0, \pi/2)$
(b)

θ	0.3	0.6	0.9	1.2	1.5
$f(\theta)$	0.47	4.21	18.0	68.6	630.1

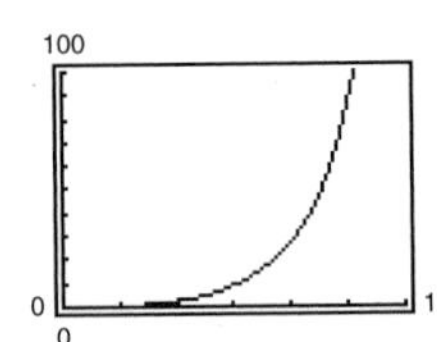

(c) ∞

71. The calculator was set in degree mode. **73.** Proof

75. Because $f(x)$ is continuous on the interval $[0, \pi]$ and $f(0) = -3$ and $f(\pi) \approx 8.87$, by the Intermediate Value Theorem there exists a real number c in $[0, \pi]$ such that $f(c) = 0$.

77. False. $\tan x$, $\cot x$, $\csc x$, and $\sec x$ are not continuous.

79. False. Let $f(x) = \tan x$.

Section 11.2 (page 705)

1. $2 \cos x - 3 \sin x$ **3.** $3/\sqrt{x} - 5 \sin x$

5. $-1 + \sec^2 x = \tan^2 x$ **7.** $-1/x^2 - 3 \cos x$

9. $x^2(3 \cos x - x \sin x)$ **11.** $t(t \cos t + 2 \sin t)$

13. $-(t \sin + \cos t)/t^2$ **15.** $(x \cos x - 2 \sin x)/x^3$

17. $(1 - \sin \theta + \theta \cos \theta)/(1 - \sin \theta)^2$

19. $3(\sin x - 1)/(2 \cos^2 x) = (3/2)(\sec x \tan x - \sec^2 x)$

21. $\csc x \cot x - \cos x$ **23.** $4x \cos x + 2 \sin x - x^2 \sin x$

25. (a) 1; 1 cycle in $[0, 2\pi]$
(b) 2; 2 cycles in $[0, 2\pi]$
The slope of $\sin ax$ at the origin is a.

27. $y = -x + \pi/2$ **29.** $y = x + 1$ **31.** $-4 \sin 4x$

33. $15 \sec^2 3x$ **35.** $2\pi^2 x \cos(\pi^2 x^2)$ **37.** $2 \cos 4x$

39. $(-\cos^2 x - 1)/\sin^3 x$ **41.** $8 \sec^2 x \tan x$

43. $\sin 2\theta \cos 2\theta$ **45.** $10 \tan 5\theta \sec^2 5\theta$

47. $6\pi \sin(\pi t - 1)/\cos^3(\pi t - 1)$

49. $1/(2\sqrt{x}) + 2x \cos(4x^2)$

51. $\cos \sqrt[3]{x}/(3x^{2/3}) + \cos x/(3 \sin^{2/3} x)$

53. $\cot x$ **55.** $\csc x$ **57.** $\sec x$

59. $2 \sec^2 2x \cos(\tan 2x)$ **61.** $\pi/2$

63. $y' = -2 \csc x \cot x/(1 - \csc x)^2, -4\sqrt{3}$

65. $h'(t) = \sec t(t \tan t - 1)/t^2, 1/\pi^2$

67. Answers will vary.

69. $n = 1$: $f'(x) = x \cos x + \sin x$
$n = 2$: $f'(x) = x^2 \cos x + 2x \sin x$
$n = 3$: $f'(x) = x^3 \cos x + 3x^2 \sin x$
$n = 4$: $f'(x) = x^4 \cos x + 4x^3 \sin x$
$f'(x) = x^n \cos x + nx^{n-1} \sin x$

71. $(\cos x)/(4 \sin 2y)$ **73.** $(\cos x - \tan y - 1)/(x \sec^2 y)$

75. $y \cos(xy)/[1 - x \cos(xy)]$

77. (a) $P_1(x) = (-\sqrt{3}/2)(x - \pi/3) + \frac{1}{2}$
$P_2(x) = -\frac{1}{4}(x - \pi/3)^2 - (\sqrt{3}/2)(x - \pi/3) + \frac{1}{2}$
(b)

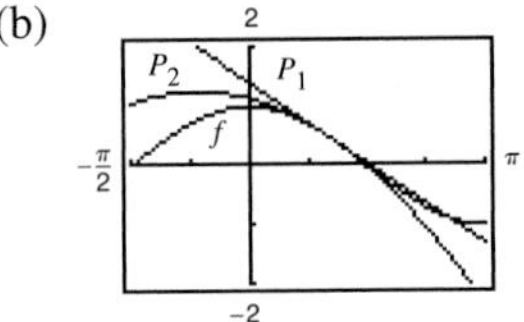

(c) P_2
(d) P_1 and P_2 become less accurate as you move farther from $x = a$.

79. (a) $P_1(x) = 2(x - \pi/4) + 1$
$P_2(x) = 2(x - \pi/4)^2 + 2(x - \pi/4) + 1$
(b)

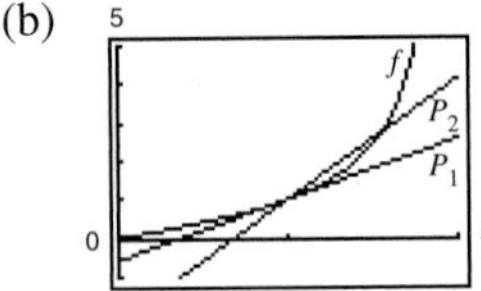

(c) P_2
(d) The accuracy worsens as you move away from $x = \pi/4$.

81. (a) 8 cm/sec (b) 4 cm/sec (c) 2 cm/sec

83. Maximum: $(0, 1)$
Minimum: $\left(\frac{1}{6}, \frac{\sqrt{3}}{2}\right)$

85. Minimum: $(\pi, -3)$
Maxima: $(0, 3)$ and $(2\pi, 3)$

87. $f'(\pi/2) = 0; f'(3\pi/2) = 0$

89. $f'(-\pi/2) = 0; f'(0) = 0, f'(\pi/2) = 0$

91. Not continuous on $[0, \pi]$

93.

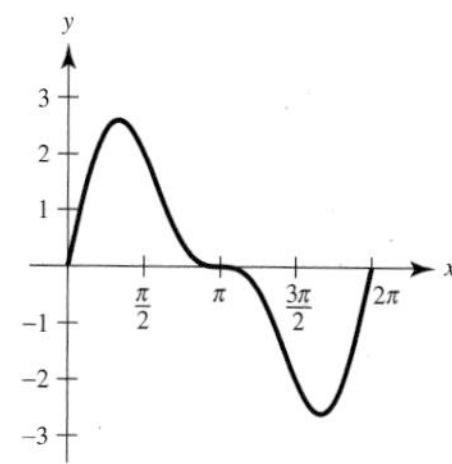

95.

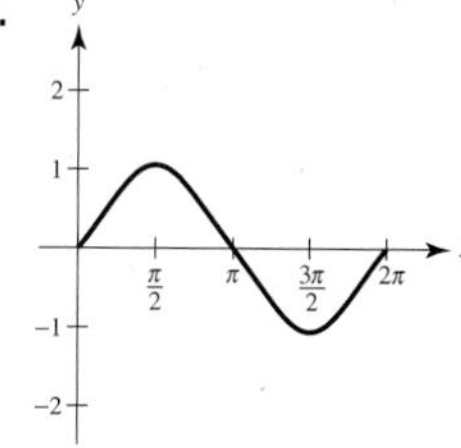

97.

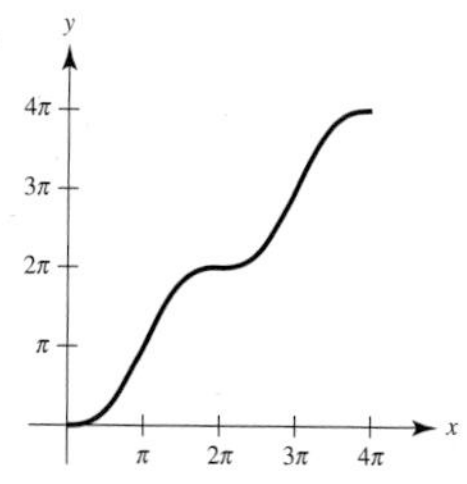

99.

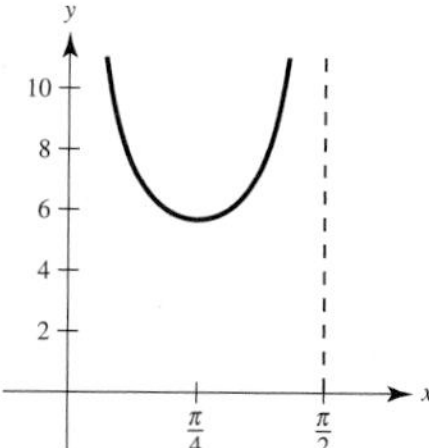

101. f is decreasing on $(\pi/2, 3\pi/2)$ and therefore $f(2) > f(4)$.

103. Answers will vary. One example is:

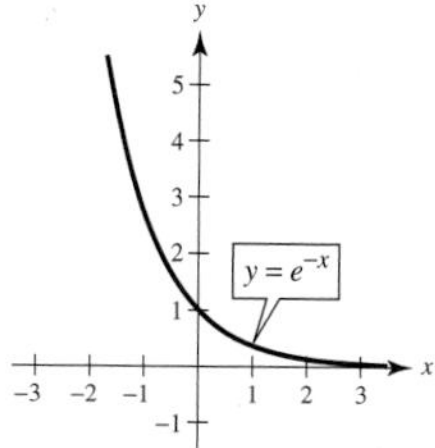

105. (a) $y = 1.75 \cos \pi t/5$ (b) $v = -0.35\pi \sin \pi t/5$

107. (a) $\frac{1}{2}$ rad/min (b) $\frac{3}{2}$ rad/min (c) 1.87 rad/min

109. 6407 ft

111. $F = kW/\sqrt{k^2 + 1}$, $\theta = \arctan k$ or $F = kW/(\cos \theta + k \sin \theta)$

113. (a)

Base 1	Base 2	Altitude	Area
8	8 + 16 cos 10°	8 sin 10°	≈ 22.1
8	8 + 16 cos 20°	8 sin 20°	≈ 42.5
8	8 + 16 cos 30°	8 sin 30°	≈ 59.7
8	8 + 16 cos 40°	8 sin 40°	≈ 72.7
8	8 + 16 cos 50°	8 sin 50°	≈ 80.5
8	8 + 16 cos 60°	8 sin 60°	≈ 83.1

(b)

Base 1	Base 2	Altitude	Area
8	8 + 16 cos 10°	8 sin 10°	≈ 22.1
8	8 + 16 cos 20°	8 sin 20°	≈ 42.5
8	8 + 16 cos 30°	8 sin 30°	≈ 59.7
8	8 + 16 cos 40°	8 sin 40°	≈ 72.7
8	8 + 16 cos 50°	8 sin 50°	≈ 80.5
8	8 + 16 cos 60°	8 sin 60°	≈ 83.1
8	8 + 16 cos 70°	8 sin 70°	≈ 80.7
8	8 + 16 cos 80°	8 sin 80°	≈ 74.0
8	8 + 16 cos 90°	8 sin 90°	≈ 64.0

The maximum cross-sectional area is approximately 83.1 ft^2.

(c) $A = 64(1 + \cos \theta)\sin \theta$, $0 < \theta < 90$

(d) $dA/d\theta = 64(1 + \cos \theta) \cos \theta + (-64 \sin \theta) \sin \theta$
$= 0$ when $\theta = 60, 180, 300$

The maximum occurs when $\theta = 60$.

(e)

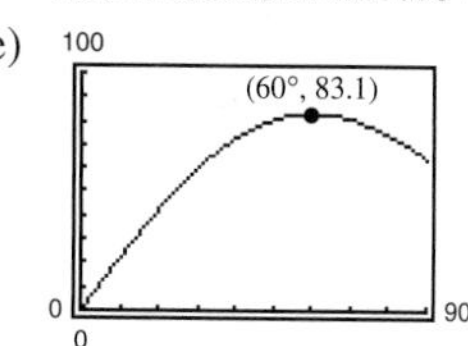

115. (a)

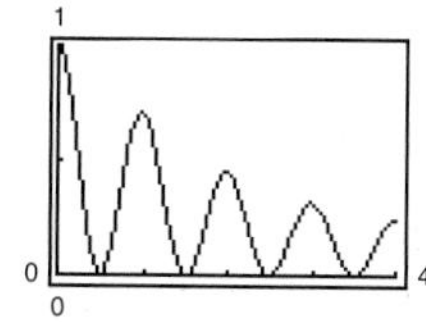

$\{0.5, 0.9747, 1.5, 1.9796, 2.5, 2.9848, 3.5\}$

(b) $\{0.5, 0.9747, 1.5, 1.9796, 2.5, 2.9848, 3.5\}$

The results are the same.

117. False. If $f(x) = \sin^2 2x$, then $f'(x) = 2(\sin 2x)(2 \cos 2x)$.

119. False. $f\left(\tan^{-1} \frac{3}{2}\right) \approx 3.60555$ is the maximum value of y.

121. (a)–(c) Proofs

Section 11.3 (page 714)

1. $5 \sin x - 4 \cos x + C$ **3.** $t + \csc t + C$

5. $\tan \theta + \cos \theta + C$ **7.** $\tan y + C$

9. $-\cos \pi x + C$ **11.** $-\frac{1}{4} \cos 4x + C$

13. $-\sin(1/\theta) + C$

15. $\frac{1}{4} \sin^2 2x + C$ or $-\frac{1}{4} \cos^2 2x + C$ or $-\frac{1}{8} \cos 4x + C$

17. $\frac{1}{5} \tan^5 x + C$ **19.** $\frac{1}{2} \tan^2 x + C$ or $\frac{1}{2} \sec^2 x + C_1$

21. $-\cot x - x + C$ **23.** $\sin e^x + C$

25. $3 \ln\left|\sin \frac{\theta}{3}\right| + C$ **27.** $-\frac{1}{2} \ln|\csc 2x + \cot 2x| + C$

29. $\ln|1 + \sin t| + C$ **31.** $\ln|\sec x - 1| + C$ **33.** $\pi + 2$

35. $2\sqrt{3}/3$ **37.** 0 **39.** b

41. (a) Answers will vary. Sample answer:

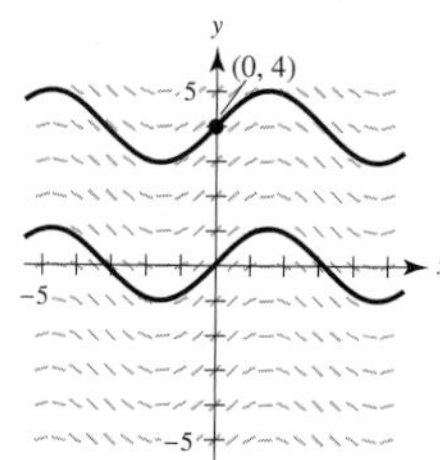

(b) $y = \sin x + 4$

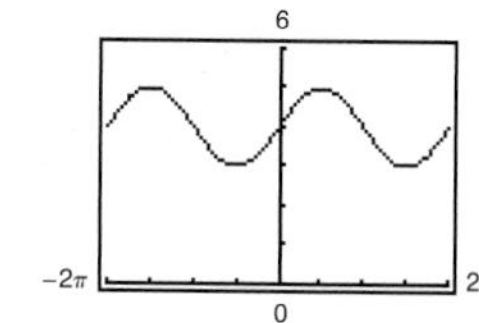

43. $s = -\frac{1}{2}\ln|\cos 2\theta| + C$

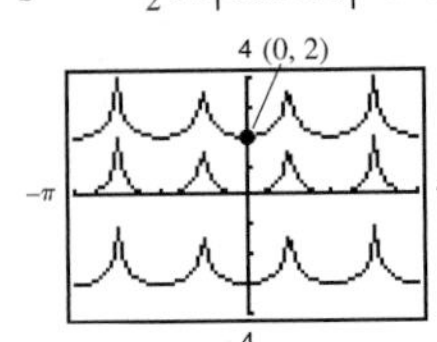

45. $-\sin(1 - x) + C$

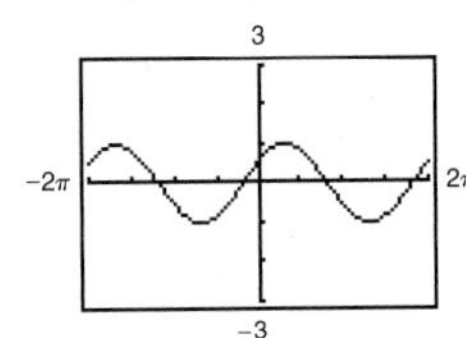

47. $-\ln\left(\sqrt{2} - 1\right) - \sqrt{2}/2$ **49.** 0.7854 **51.** 3.4832

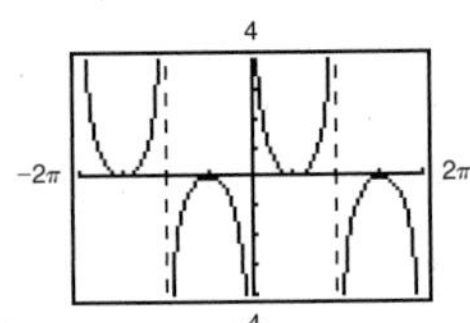
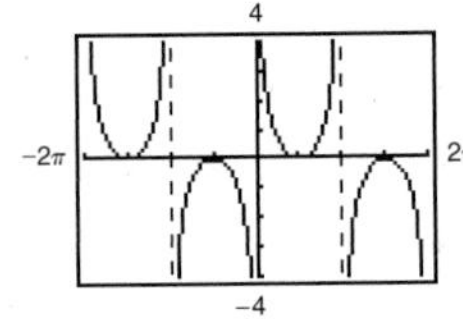

	Trapezoidal	*Simpson's*	*Graphing Utility*
53.	0.5495	0.5483	0.5493
55.	-0.0975	-0.0977	-0.0977
57.	0.1940	0.1860	0.1858

59. 1 **61.** 4 **63.** $2\sqrt{3} - 2$ **65.** 0

67. $\frac{2}{3}$ **69.** $\pm\arccos\left(\sqrt{\pi}/2\right) \approx \pm 0.4817$

71. Average value $= 2/\pi$
$x \approx 0.690, x \approx 2.451$

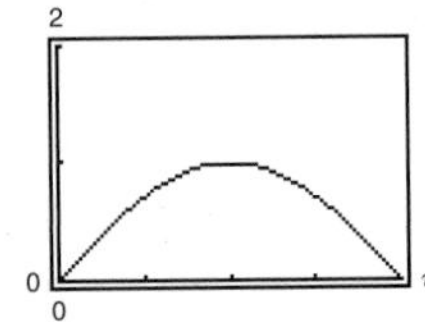

73. $x\cos x$ **75.** $\sec^2 x$

77. The function is not integrable on the interval $[0, \pi]$ because there is a nonremovable discontinuity at $x = \pi/2$.

79. f is an odd function and the interval of integration is centered at $x = 0$.

81. (a) Answers will vary. Sample answer: $a = \pi, b = 2\pi$

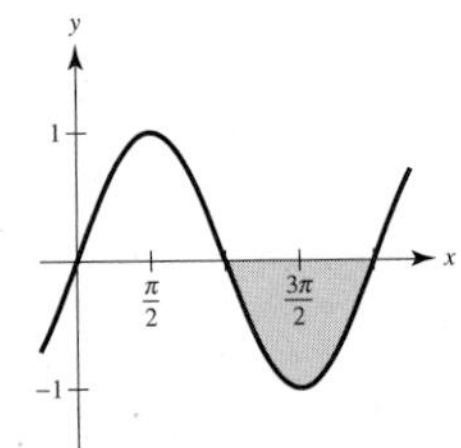

(b) Answers will vary. Sample answer: $a = 0, b = \pi$

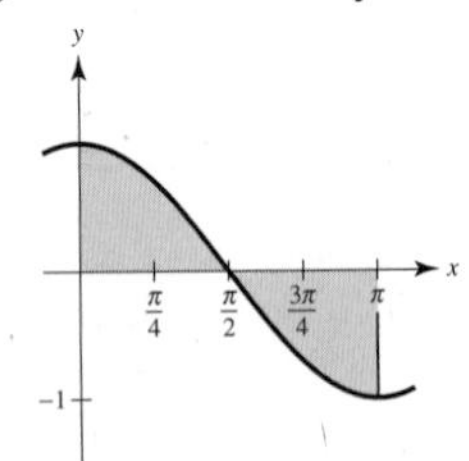

83. (a)

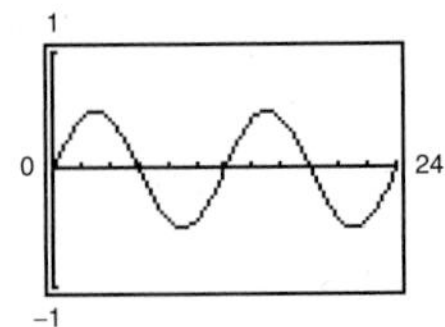

The average value of $f(t)$ over the interval $0 \le t \le 24$ is represented by $\int_0^{24} 0.5\sin(\pi t/6) = 0$.

(b)

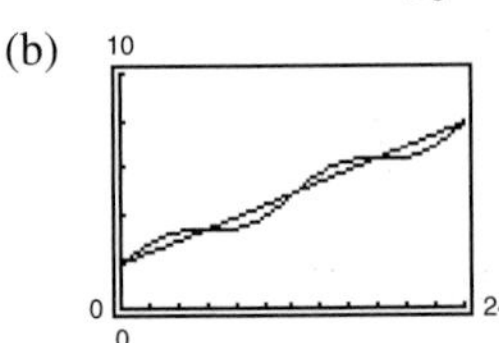

Even though the average value of $f(t) = 0$, the trend represented by g increases over the interval $0 \le t \le 24$ as does $S(t)$.

85. (a) 102.352 thousand units (b) 102.352 thousand units (c) 74.5 thousand units

87. (a) 1.273 amps (b) 1.382 amps (c) 0 amp

89–91. Answers will vary. **93.** True

95. False. $\int \sin^2 2x \cos 2x\,dx = \frac{1}{6}\sin^3 2x + C$

Section 11.4 (page 720)

1. 1 **3.** $\frac{1}{2}$ **5.** $2/\sqrt{2x - x^2}$

7. $-3/\sqrt{4 - x^2}$ **9.** $e^x/(1 + e^{2x})$

11. $\left(3x - \sqrt{1 - 9x^2}\arcsin 3x\right)/\left(x^2\sqrt{1 - 9x^2}\right)$

13. $-t/\sqrt{1 - t^2}$ **15.** $2\arccos x$ **17.** $2x^2/\sqrt{25 - x^2}$

19. $\arcsin x$ **21.** $2/(1 + x^2)^2$

23. $y = \frac{1}{3}\left(4\sqrt{3}x - 2\sqrt{3} + \pi\right)$ **25.** $y = \frac{1}{4}x + (\pi - 2)/4$

27. $P_1(x) = x$; $P_2(x) = x$

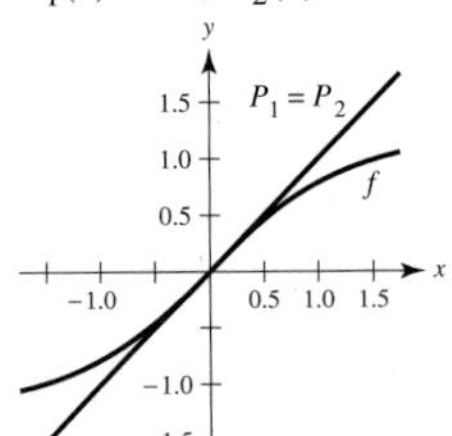

29. $P_1(x) = \frac{\pi}{6} + \frac{2\sqrt{3}}{3}\left(x - \frac{1}{2}\right)$

$P_2(x) = \frac{\pi}{6} + \frac{2\sqrt{3}}{3}\left(x - \frac{1}{2}\right) + \frac{2\sqrt{3}}{9}\left(x - \frac{1}{2}\right)^2$

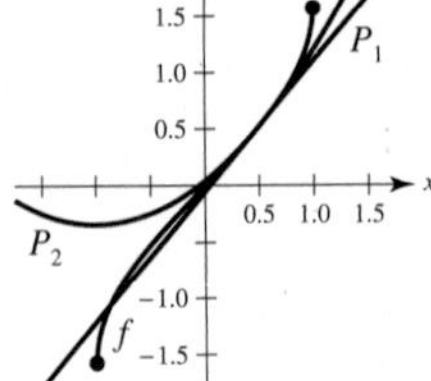

31. Relative maximum: $(1.272, -0.606)$
Relative minimum: $(-1.272, 3.747)$

33. Relative maximum: $(2, 2.214)$

35. $y = -2\pi x/(\pi + 8) + 1 - \pi^2/(2\pi + 16)$

37. $y = -x + \sqrt{2}$

39. If the domains were not restricted, the trigonometric functions would have no inverses because they would not be one-to-one.

41. $f(x) = \arcsin x$ is an increasing function.

43. The derivatives are algebraic. See Theorem 11.8.

45. False. The range of arccos is $[0, \pi]$. **47.** True

49. (a) $\theta = \operatorname{arccot}(x/5)$
(b) $x = 10$: 16 rad/h; $x = 3$: 58.824 rad/h

51. (a) $h(t) = -16t^2 + 256$; $t = 4$ sec
(b) $t = 1$: -0.0520 rad/sec; $t = 2$: -0.1116 rad/sec

53. (a) and (b) Proofs

55. $f'(x) = 0$; therefore $f(x)$ is constant.

Section 11.5 (page 727)

1. $\arcsin\dfrac{x}{3} + C$ **3.** $\dfrac{7}{4}\arctan\dfrac{x}{4} + C$ **5.** $\operatorname{arcsec}|2x| + C$

7. $\dfrac{1}{2}\arcsin t^2 + C$ **9.** $\dfrac{1}{10}\arctan\dfrac{t^2}{5} + C$

11. $\arcsin\left(\dfrac{\tan x}{5}\right) + C$ **13.** $\dfrac{1}{2}x^2 - \dfrac{1}{2}\ln(x^2 + 1) + C$

15. $2\arcsin\sqrt{x} + C$ **17.** $\frac{1}{2}\ln(x^2 + 1) - 3\arctan x + C$

19. $8\arcsin[(x - 3)/3] - \sqrt{6x - x^2} + C$ **21.** $\pi/6$

23. $\pi/6$ **25.** $\frac{1}{2}(\sqrt{3} - 2) \approx -0.134$

27. $\dfrac{1}{5}\arctan\dfrac{3}{5} \approx 0.108$ **29.** $\dfrac{\pi}{4}$ **31.** $\dfrac{1}{32}\pi^2 \approx 0.308$

33. $\dfrac{\pi}{2}$ **35.** $\ln|x^2 + 6x + 13| - 3\arctan[(x + 3)/2] + C$

37. $\arcsin[(x + 2)/2] + C$ **39.** $-\sqrt{-x^2 - 4x} + C$

41. $4 - 2\sqrt{3} + \frac{1}{6}\pi \approx 1.059$ **43.** $\frac{1}{2}\arctan(x^2 + 1) + C$

45. $2\sqrt{e^t - 3} - 2\sqrt{3}\arctan\left(\sqrt{e^t - 3}/\sqrt{3}\right) + C$

47. (a) (b) $y = 3\arctan x$

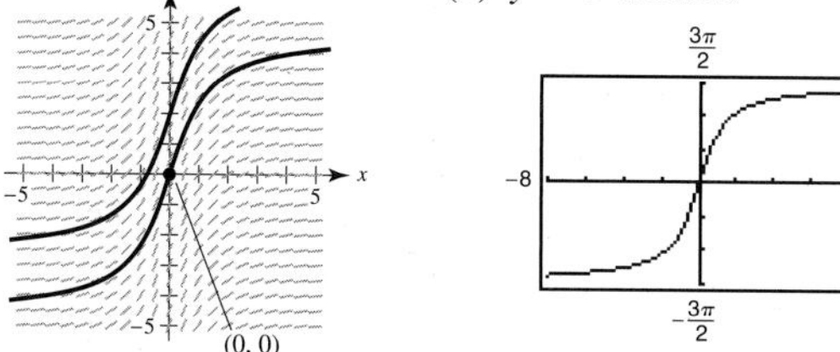

49.

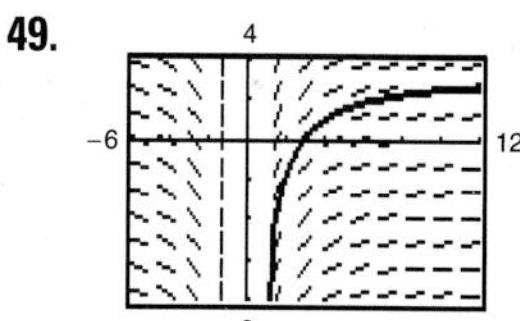

51. $\pi/3$ **53.** a and b **55.** a, b, and c

57. (a) Proof (b) 3.1415918 (c) 3.1415927

59–61. Proofs

Section 11.6 (page 737)

1. (a) 10.018 (b) -0.964

3. (a) 1.317 (b) 0.962 **5–11.** Proofs

13. $\cosh x = \sqrt{13}/2$; $\tanh x = 3\sqrt{13}/13$; $\operatorname{csch} x = 2/3$; $\operatorname{sech} x = 2\sqrt{13}/13$; $\coth x = \sqrt{13}/3$

15. $3\cosh 3x$ **17.** $\coth x$ **19.** $\operatorname{csch} x$

21. $\sinh^2 x$ **23.** $\operatorname{sech} t$

25. Relative maxima: $(\pm\pi, \cosh\pi)$; Relative minimum: $(0, -1)$

27. Relative maximum: $(1.20, 0.66)$
Relative minimum: $(-1.20, -0.66)$

29. $y = a\sinh x$; $y' = a\cosh x$; $y'' = a\sinh x$; $y''' = a\cosh x$; Therefore, $y''' - y' = 0$.

31. (a) (b) 33.146 units; 25 units
(c) $m = \sinh(1) \approx 1.175$

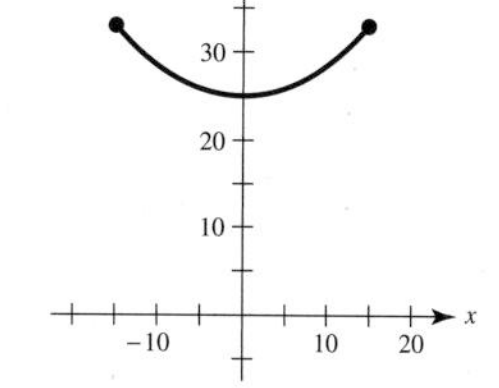

33. $\frac{1}{2}\sinh 2x + C$ **35.** $-\frac{1}{2}\cosh(1 - 2x) + C$

37. $\ln|\sinh x| + C$ **39.** $-\coth(x^2/2) + C$

41. $\operatorname{csch}(1/x) + C$ **43.** $\frac{1}{2}\arctan x^2 + C$ **45.** $\frac{1}{5}\ln 3$

47. $\dfrac{\pi}{4}$ **49.** $\dfrac{3}{\sqrt{9x^2 - 1}}$ **51.** $\dfrac{1}{2\sqrt{x}(1 - x)}$ **53.** $|\sec x|$

55. $\dfrac{-2\operatorname{csch}^{-1} x}{|x|\sqrt{1 + x^2}}$ **57.** $2\sinh^{-1}(2x)$

59. See the definitions and graphs in the textbook.

61. $\dfrac{\sqrt{3}}{18}\ln\left|\dfrac{1 + \sqrt{3}x}{1 - \sqrt{3}x}\right| + C$

63. $\ln\left(\sqrt{e^{2x} + 1} - 1\right) - x + C$ **65.** $\dfrac{1}{4}\ln\left|\dfrac{x - 4}{x}\right| + C$

67. $-\dfrac{x^2}{2} - 4x - \dfrac{10}{3}\ln\left|\dfrac{x - 5}{x + 1}\right| + C$

69. $\ln\sqrt{(e^4 + e^{-4})/2} \approx 1.654$

71. $-\sqrt{a^2 - x^2}/x$ **73–77.** Proofs

Review Exercises (page 740)

1. -1 **3.** 0 **5.** $\dfrac{5}{3}$ **7.** $\dfrac{\sqrt{3}}{2}$ **9.** $\dfrac{4}{5}$ **11.** ∞

13. Nonremovable discontinuity at each even integer
Continuous on $(2k, 2k + 2)$ for all integers k

15. f is continuous and crosses the x-axis in the interval $[1, 3]$.

17. $4 - 5\cos\theta$ **19.** $\sqrt{x}\cos x + \sin x/(2\sqrt{x})$

21. $\dfrac{4x^3\cos x + x^4\sin x}{\cos^2 x}$ **23.** $3\sec x\tan x$

25. $-\csc 2x\cot 2x$ **27.** $\frac{1}{2}(1 - \cos 2x) = \sin^2 x$

29. $\sin^{1/2} x\cos x - \sin^{5/2} x\cos x = \cos^3 x\sqrt{\sin x}$

31. $-x\sec^2 x - \tan x$ **33.** $\dfrac{(x + 2)(\pi\cos\pi x) - \sin\pi x}{(x + 2)^2}$

35. $\cot\theta + \tan\theta = \csc\theta\sec\theta$

37. $2\csc^2 x\cot x$ **39.** $6\sec^2\theta\tan\theta$

41. $y'' + y = -(2\sin x + 3\cos x) + (2\sin x + 3\cos x) = 0$

43. $-\sin^2(x + y)$ or $-\dfrac{x^2}{x^2 + 1}$, 0

45. Minimum: $\frac{2\sqrt{3}}{3} \approx 1.1547$ at $x = \frac{\pi}{3}$; Maximum: 2 at $x = \frac{\pi}{6}$

47. $3(3^{2/3} + 2^{2/3})^{3/2} \approx 21.07$ ft **49.** $x^2 + 9\cos x + C$

51. $\frac{1}{4}\sin^4 x + C$ **53.** $-2\sqrt{1 - \sin\theta} + C$

55. $\tan^{n+1}[x/(n+1)] + C$ **57.** $\frac{1}{2}\sec 2x + C$

59. $-\ln|1 + \cos x| + C$ **61.** $f(x) = 2\sin(x/2) + 3$

63. 2 **65.** $\ln(2 + \sqrt{3})$

67. $A = \sqrt{3}$

69. $F'(x) = \tan^4 x$ **71.** (a) 2.3290 (b) 2.4491

73.

Average value $= (2\ln 2)/\pi \approx 0.441$

$x \approx 0.416$

75. $(1 - x^2)^{-3/2}$ **77.** $\frac{x}{|x|\sqrt{x^2 - 1}} + \operatorname{arcsec} x$

79. $(\arcsin x)^2$ **81.** $\frac{1}{2}\arctan(e^{2x}) + C$ **83.** $\ln\sqrt{16 + x^2} + C$

85. $\frac{1}{4}[\arctan(x/2)]^2 + C$ **87.** $y = A\sin(t\sqrt{k/m})$

89. $y' = x\left(\frac{2}{1 - 4x^2}\right) + \tanh^{-1} 2x = \frac{2x}{1 - 4x^2} + \tanh^{-1} 2x$

91. $\frac{1}{3}\tanh x^3 + C$

Chapter Test (page 742)

1. $\frac{1}{3}$

2. The limit does not exist because the function oscillates between -1 and 1 as x approaches 0.

3. $-x^4\sin x + 4x^3\cos x$ **4.** $2\sin 3\theta\cos 3\theta$

5. $\frac{t}{\sqrt{t^2 + 1}}$ **6.** $2x\sinh(1 + x^2)$ **7.** $y = x + 1$

8. $y = -2x\sqrt{3} + \sqrt{3} + \pi$ **9.** $y' = \frac{\cos(x + y)}{1 - \cos(x + y)}$

10. $f\left(\frac{\pi}{4}\right) = f\left(\frac{5\pi}{4}\right) = 1$, but f is not continuous on $\left(\frac{\pi}{4}, \frac{5\pi}{4}\right)$ because $f(\pi)$ does not exist. Rolle's Theorem does not apply.

11. $\frac{1}{8}\sin^2 4x + C$ or $-\frac{1}{8}\cos^2 4x + C_1$ or $-\frac{1}{16}\cos 8x + C_2$

12. $\frac{1}{2}\arcsin\frac{t^2}{2} + C$

13. $\ln|x^2 + 8x + 17| - 8\arctan(x + 4) + C$

14. $\frac{1}{3}\cosh(1 + 3x) + C$ **15.** $2\sqrt{3}$ **16.** 0 **17.** 1

18. $\frac{\pi}{6}$ **19.** $x\tan x$

20. Relative minimum: $\left(\frac{2\sqrt{2}}{3}, -1.60\right)$;

Relative maximum: $\left(\frac{-2\sqrt{2}}{3}, 1.60\right)$

21. (a) 3.627 (b) -0.995 **22.** Proof

P.S. Problem Solving (page 743)

1. Sample answer: $f(x) = \frac{3}{2} - \frac{1}{2}\cos 2x$

3. $\phi \approx 42.1°$ or 0.736 rad **5.** About 9.19 ft

7. (a)–(b) Proofs **9.** $a \approx 4.7648$; $\theta \approx 1.7263$ or $98.9°$

11. (a)

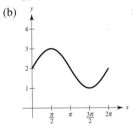

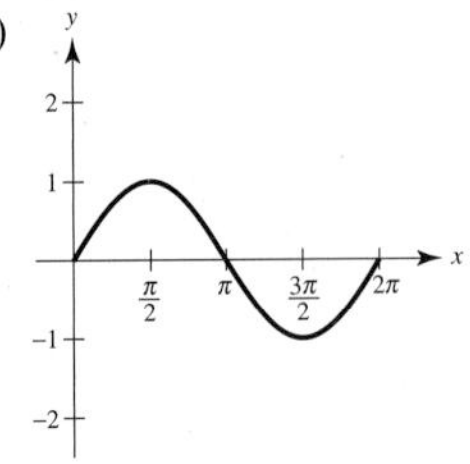

$\int_0^{\pi} \sin x\,dx = -\int_{\pi}^{2\pi} \sin x\,dx \Rightarrow \int_0^{2\pi} \sin x\,dx = 0$

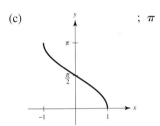

(b) ; 4π

(c) ; π

(d) $\frac{\pi}{4}$

Chapter 12

Section 12.1 (page 750)

1. conic **3.** locus **5.** axis **7.** focal chord

9. e **10.** b **11.** d **12.** f **13.** a **14.** c

15. $x^2 = \frac{3}{2}y$ **17.** $x^2 = 2y$ **19.** $y^2 = -8x$ **21.** $x^2 = -4y$

23. $y^2 = 4x$ **25.** $x^2 = \frac{8}{3}y$ **27.** $y^2 = -\frac{25}{2}x$

29. Vertex: $(0, 0)$

Focus: $\left(0, \frac{1}{2}\right)$

Directrix: $y = -\frac{1}{2}$

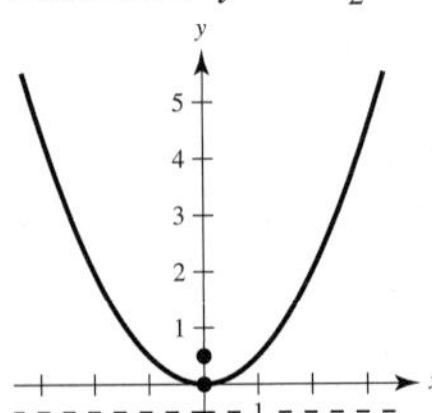

31. Vertex: $(0, 0)$

Focus: $\left(-\frac{3}{2}, 0\right)$

Directrix: $x = \frac{3}{2}$

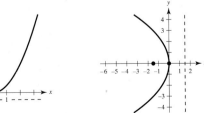

33. Vertex: $(0, 0)$
Focus: $\left(0, -\frac{3}{2}\right)$
Directrix: $y = \frac{3}{2}$

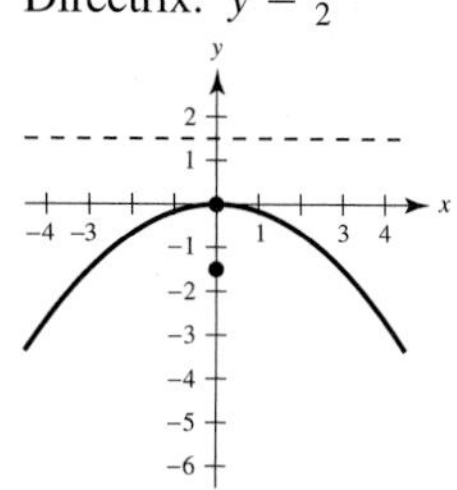

35. Vertex: $(1, -2)$
Focus: $(1, -4)$
Directrix: $y = 0$

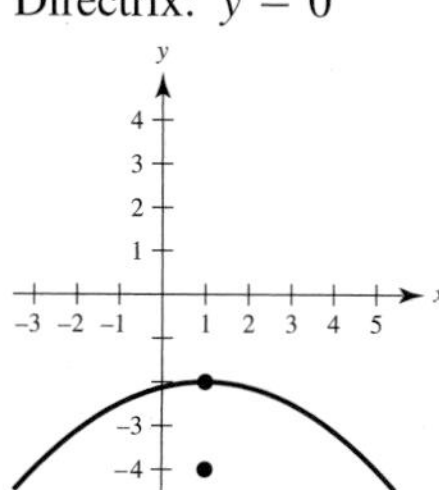

37. Vertex: $\left(-3, \frac{3}{2}\right)$
Focus: $\left(-3, \frac{5}{2}\right)$
Directrix: $y = \frac{1}{2}$

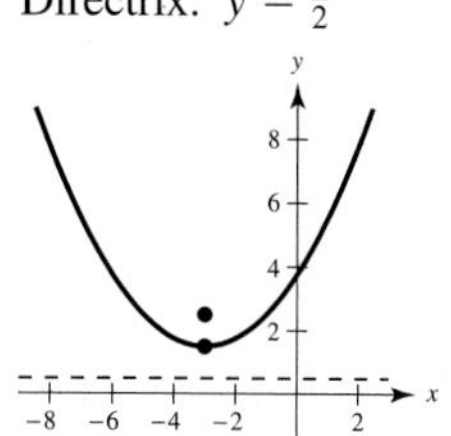

39. Vertex: $(1, 1)$
Focus: $(1, 2)$
Directrix: $y = 0$

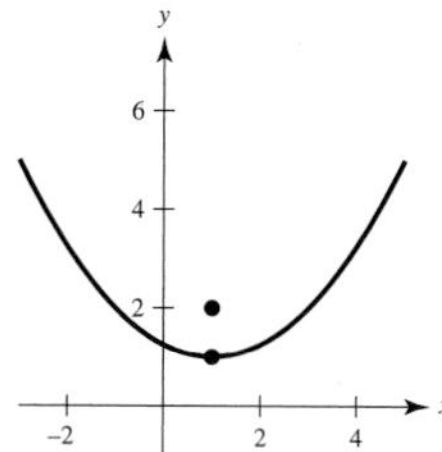

41. Vertex: $(-2, -3)$
Focus: $(-4, -3)$
Directrix: $x = 0$

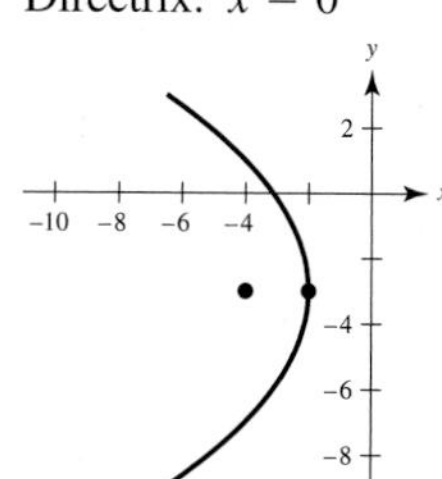

43. Vertex: $(-2, 1)$
Focus: $\left(-2, -\frac{1}{2}\right)$
Directrix: $y = \frac{5}{2}$

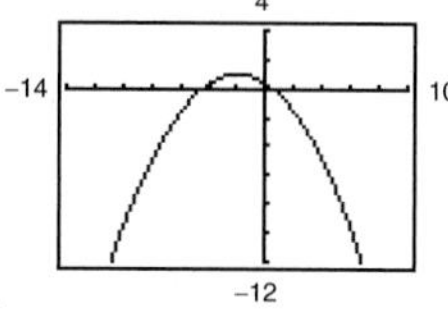

45. Vertex: $\left(\frac{1}{4}, -\frac{1}{2}\right)$
Focus: $\left(0, -\frac{1}{2}\right)$
Directrix: $x = \frac{1}{2}$

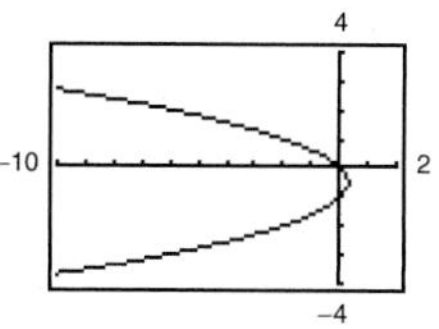

47. $(x - 3)^2 = -(y - 1)$ **49.** $y^2 = 4(x + 4)$
51. $(y - 3)^2 = 8(x - 4)$ **53.** $x^2 = -8(y - 2)$
55. $(y - 2)^2 = 8x$ **57.** $y = \sqrt{6(x + 1)} + 3$
59. $(2, 4)$

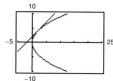

61. $\frac{1}{2}x$ **63.** $3/y$ **65.** $(x - 2)/3$ **67.** $-4/(y + 3)$
69. $4x - y - 8 = 0$ **71.** $4x - y + 2 = 0$
73. $x - 2y - 1 = 0$ **75.** $2x + y + 6 = 0$

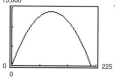

77. $x = 106$ units

79. A circle is formed when a plane intersects the top or bottom half of a double-napped cone and is perpendicular to the axis of the cone.
81. A parabola is formed when a plane intersects the top or bottom half of a double-napped cone, is parallel to the side of the cone, and does not intersect the vertex.
83. (a)

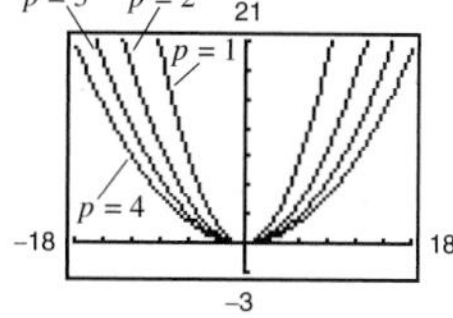

As p increases, the graph becomes wider.
(b) $(0, 1), (0, 2), (0, 3), (0, 4)$ (c) $4, 8, 12, 16;\ 4|p|$
(d) It is an easy way to determine two additional points on the graph.

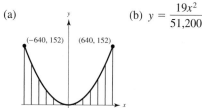

85. (a) (b) $y = \dfrac{19x^2}{51{,}200}$

(c)

Distance, x	0	100	250	400	500
Height, y	0	3.71	23.19	59.38	92.77

87. (a) $y = -\frac{1}{640}x^2$ (b) 8 ft
89. (a) $x^2 = 180{,}000y$ (b) $300\sqrt{2}$ cm ≈ 424.26 cm
91. $x^2 = -\frac{25}{4}(y - 48)$
93. (a) $17{,}500\sqrt{2}$ mi/h $\approx 24{,}750$ mi/h
(b) $x^2 = -16{,}400(y - 4100)$
95. 43.3 sec prior to being over the target
97. False. If the graph crossed the directrix, there would exist points closer to the directrix than the focus.
99. $(0, 0)$ **101.** 72 square units **103.** $\frac{64}{3}$ square units
105. $\frac{32}{3}$ square units

Section 12.2 (page 759)

1. ellipse; foci **3.** minor axis
5. b **6.** c **7.** d **8.** f **9.** a **10.** e
11. $\dfrac{x^2}{4} + \dfrac{y^2}{16} = 1$ **13.** $\dfrac{x^2}{49} + \dfrac{y^2}{45} = 1$ **15.** $\dfrac{x^2}{49} + \dfrac{y^2}{24} = 1$
17. $\dfrac{21x^2}{400} + \dfrac{y^2}{25} = 1$ **19.** $\dfrac{(x - 2)^2}{1} + \dfrac{(y - 3)^2}{9} = 1$
21. $\dfrac{(x - 4)^2}{16} + \dfrac{(y - 2)^2}{1} = 1$ **23.** $\dfrac{x^2}{48} + \dfrac{(y - 4)^2}{64} = 1$

25. $\dfrac{x^2}{16} + \dfrac{(y-4)^2}{12} = 1$

27. $\dfrac{(x-2)^2}{4} + \dfrac{(y-2)^2}{1} = 1$

29. Ellipse
Center: $(0, 0)$
Vertices: $(\pm 5, 0)$
Foci: $(\pm 3, 0)$
Eccentricity: $\frac{3}{5}$

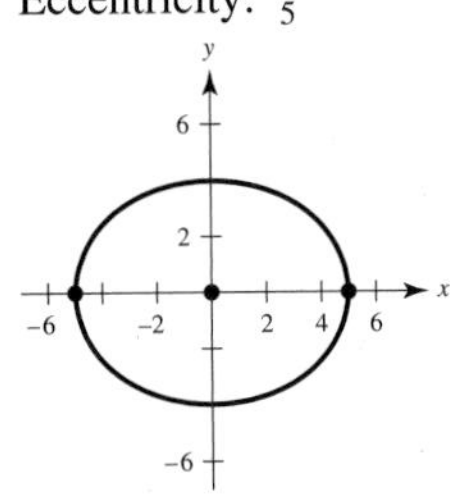

31. Circle
Center: $(0, 0)$
Radius: 5

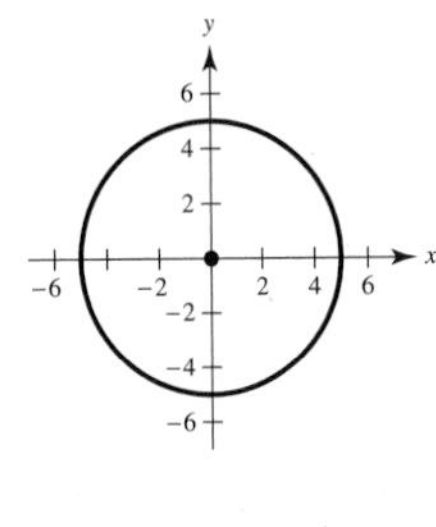

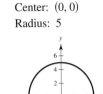

33. Ellipse
Center: $(0, 0)$
Vertices: $(0, \pm 3)$
Foci: $(0, \pm 2)$
Eccentricity: $\frac{2}{3}$

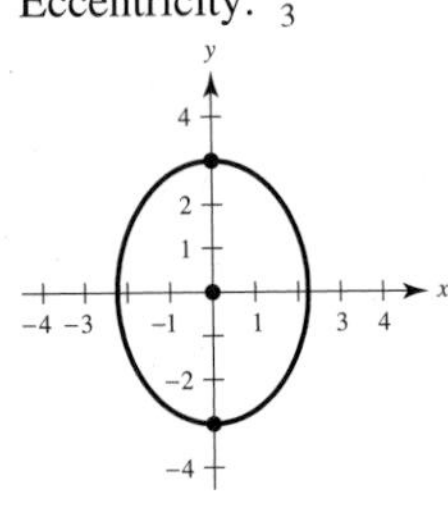

35. Ellipse
Center: $(4, -1)$
Vertices: $(4, -6), (4, 4)$
Foci: $(4, 2), (4, -4)$
Eccentricity: $\frac{3}{5}$

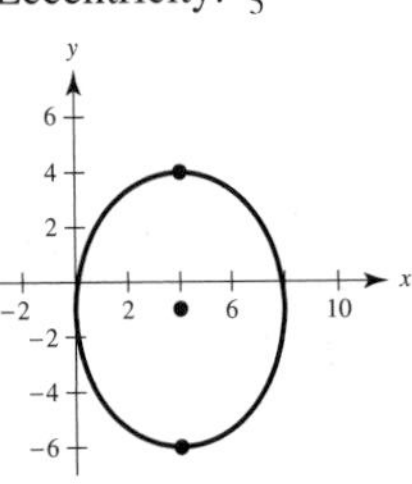

37. Circle
Center: $(0, -1)$
Radius: $\frac{2}{3}$

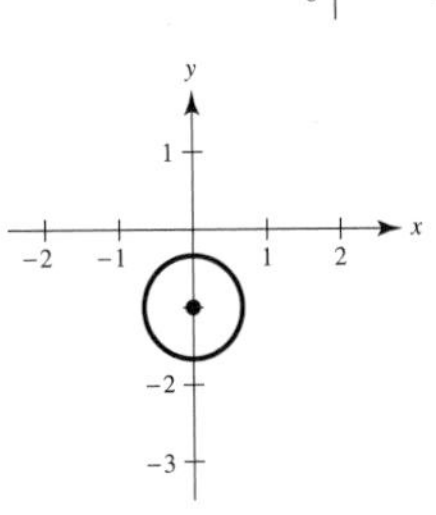

39. Ellipse
Center: $(-2, -4)$
Vertices: $(-3, -4), (-1, -4)$
Foci: $\left(\dfrac{-4 \pm \sqrt{3}}{2}, -4\right)$
Eccentricity: $\dfrac{\sqrt{3}}{2}$

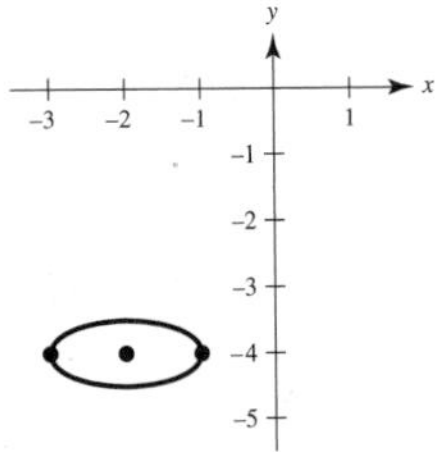

41. Ellipse
Center: $(-2, 3)$
Vertices: $(-2, 6), (-2, 0)$
Foci: $\left(-2, 3 \pm \sqrt{5}\right)$
Eccentricity: $\dfrac{\sqrt{5}}{3}$

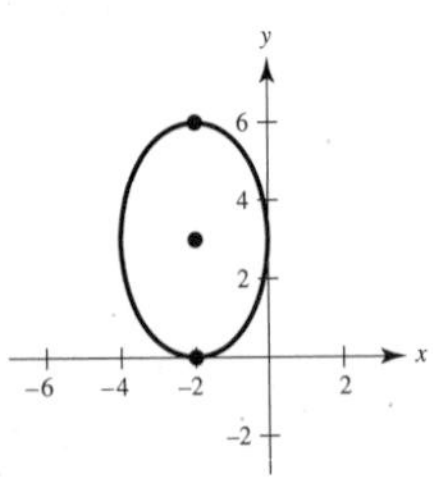

43. Circle
Center: $(1, -2)$
Radius: 6

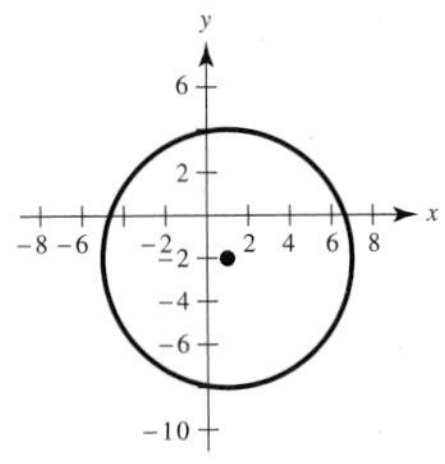

45. Ellipse
Center: $(-3, 1)$
Vertices: $(-3, 7), (-3, -5)$
Foci: $\left(-3, 1 \pm 2\sqrt{6}\right)$
Eccentricity: $\dfrac{\sqrt{6}}{3}$

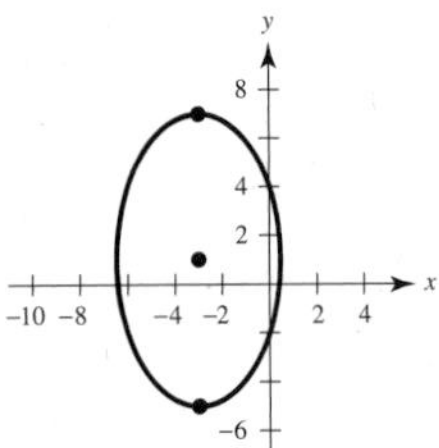

47. Ellipse
Center: $\left(3, -\dfrac{5}{2}\right)$
Vertices: $\left(9, -\dfrac{5}{2}\right), \left(-3, -\dfrac{5}{2}\right)$
Foci: $\left(3 \pm 3\sqrt{3}, -\dfrac{5}{2}\right)$
Eccentricity: $\dfrac{\sqrt{3}}{2}$

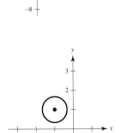

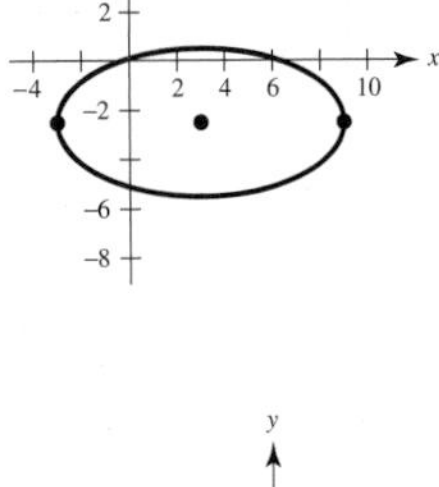

49. Circle
Center: $(-1, 1)$
Radius: $\frac{2}{3}$

51. Ellipse
Center: $(2, 1)$
Vertices: $\left(\frac{7}{3}, 1\right), \left(\frac{5}{3}, 1\right)$
Foci: $\left(\frac{34}{15}, 1\right), \left(\frac{26}{15}, 1\right)$
Eccentricity: $\frac{4}{5}$

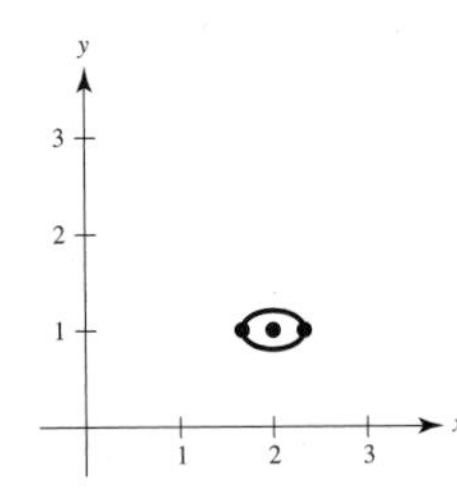

53.

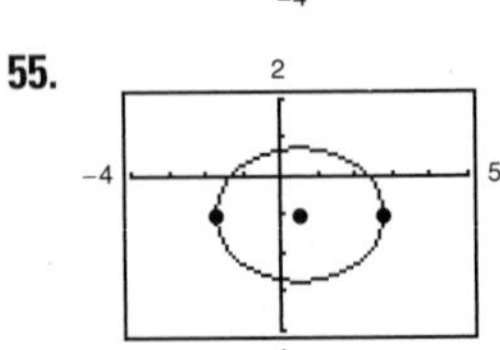

Center: $(0, 0)$
Vertices: $\left(0, \pm\sqrt{5}\right)$
Foci: $\left(0, \pm\sqrt{2}\right)$

55.

Center: $\left(\frac{1}{2}, -1\right)$
Vertices: $\left(\frac{1}{2} \pm \sqrt{5}, -1\right)$
Foci: $\left(\frac{1}{2} \pm \sqrt{2}, -1\right)$

57. $\frac{\sqrt{5}}{3}$ **59.** $\frac{2\sqrt{2}}{3}$ **61.** $\frac{x^2}{25} + \frac{y^2}{16} = 1$

63. (a) $2a$

(b) The sum of the distances from the two fixed points is constant.

65. $-4x/(9y)$ **67.** $-4(x - 4)/(y + 2)$

69. $-\frac{9}{4}(x - 2)/(y + 1)$

71. (a) $\frac{1}{2}x - y + 3 = 0$ (b) $x - 2y - 10 = 0$

(c)

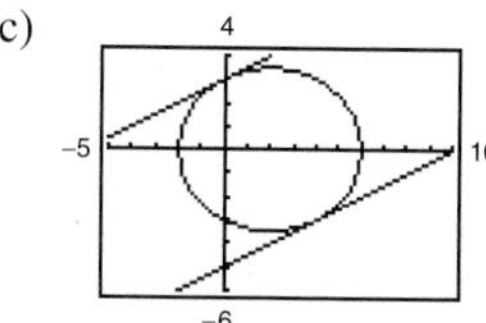

73. $x = -3, y = 2$

Endpoints of major axis: $(1, 2), (-7, 2)$

Endpoints of minor axis: $(-3, 0), (-3, 4)$

75. 2π **77.** $\sqrt{6}\pi$

79. (a)

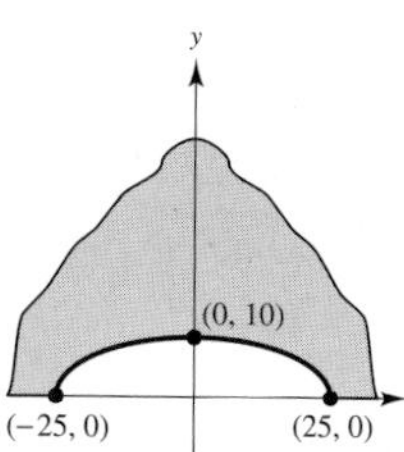

(b) $\frac{x^2}{625} + \frac{y^2}{100} = 1$

(c) Yes

81. 40

83. (a) $\frac{x^2}{321.84} + \frac{y^2}{20.89} = 1$ (b)

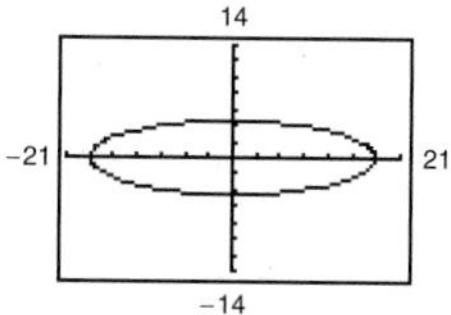

(c) Aphelion:

35.29 astronomical units

Perihelion:

0.59 astronomical unit

85. False. The graph of $(x^2/4) + y^4 = 1$ is not an ellipse. The degree of y is 4, not 2.

87. (a) $A = \pi a(20 - a)$ (b) $\frac{x^2}{196} + \frac{y^2}{36} = 1$

(c)

a	8	9	10	11	12	13
A	301.6	311.0	314.2	311.0	301.6	285.9

$a = 10$, circle

(d)

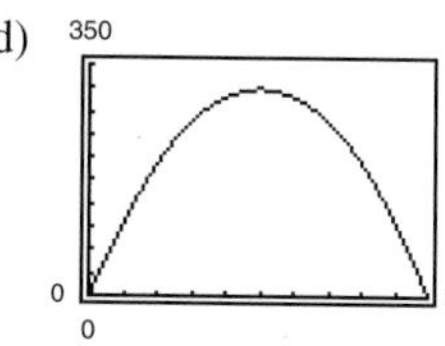

The shape of an ellipse with a maximum area is a circle. The maximum area is found when $a = 10$ (verified in part c) and therefore $b = 10$, so the equation produces a circle.

89. $\frac{(x - 6)^2}{324} + \frac{(y - 2)^2}{308} = 1$

Section 12.3 (page 768)

1. hyperbola; foci **3.** transverse axis; center

5. b **6.** c **7.** a **8.** d

9. Center: $(0, 0)$

Vertices: $(\pm 1, 0)$

Foci: $(\pm\sqrt{2}, 0)$

Asymptotes: $y = \pm x$

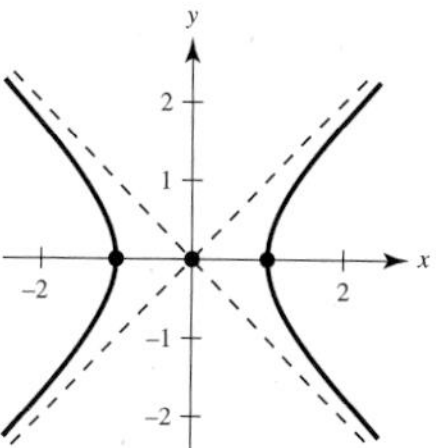

11. Center: $(0, 0)$

Vertices: $(0, \pm 5)$

Foci: $(0, \pm\sqrt{106})$

Asymptotes: $y = \pm\frac{5}{9}x$

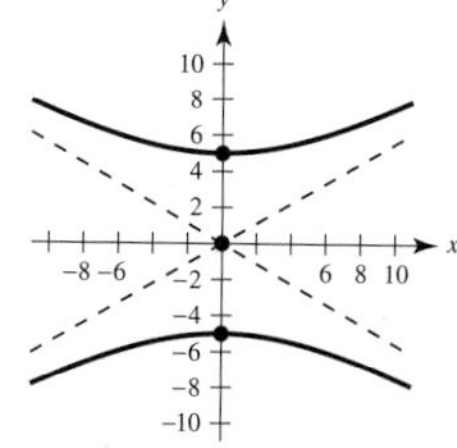

13. Center: $(0, 0)$

Vertices: $(0, \pm 1)$

Foci: $(0, \pm\sqrt{5})$

Asymptotes: $y = \pm\frac{1}{2}x$

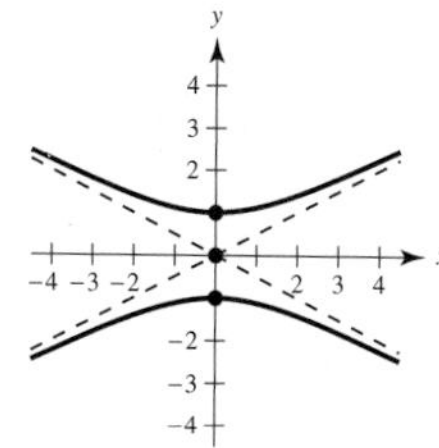

15. Center: $(1, -2)$

Vertices: $(3, -2), (-1, -2)$

Foci: $(1 \pm \sqrt{5}, -2)$

Asymptotes:

$y = -2 \pm \frac{1}{2}(x - 1)$

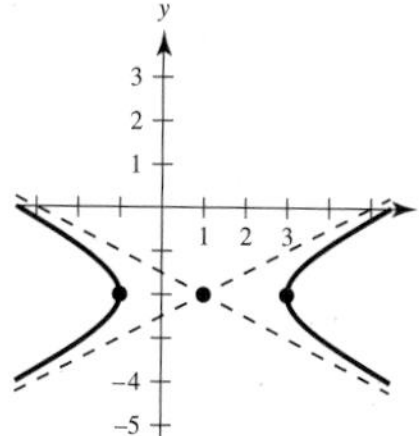

17. Center: $(2, -6)$

Vertices:

$\left(2, -\frac{17}{3}\right), \left(2, -\frac{19}{3}\right)$

Foci: $\left(2, -6 \pm \frac{\sqrt{13}}{6}\right)$

Asymptotes:

$y = -6 \pm \frac{2}{3}(x - 2)$

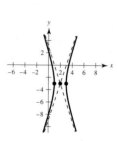

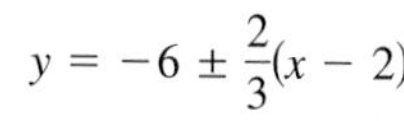

19. Center: $(2, -3)$

Vertices: $(3, -3), (1, -3)$

Foci: $(2 \pm \sqrt{10}, -3)$

Asymptotes:

$y = -3 \pm 3(x - 2)$

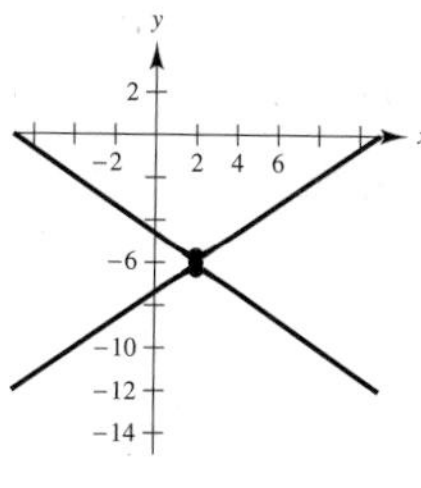

21. The graph of this equation is two lines intersecting at $(-1, -3)$.

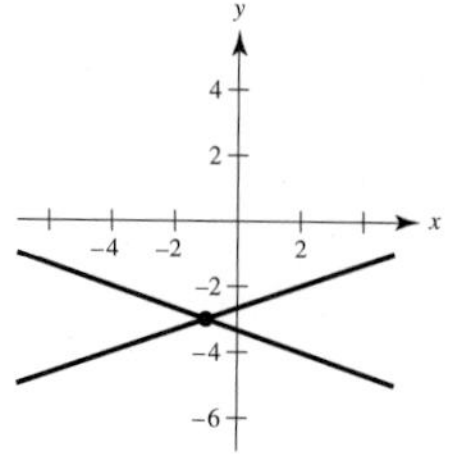

23. Center: $(0, 0)$
Vertices: $(\pm\sqrt{3}, 0)$
Foci: $(\pm\sqrt{5}, 0)$
Asymptotes: $y = \pm\frac{\sqrt{6}}{3}x$

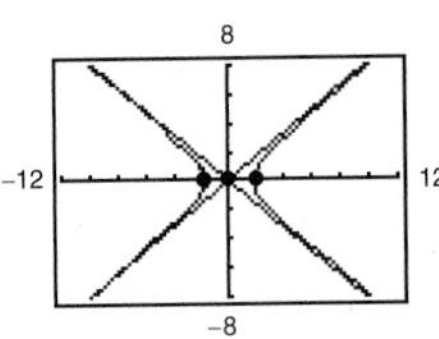

25. Center: $(0, 0)$
Vertices: $(\pm 3, 0)$
Foci: $(\pm\sqrt{13}, 0)$
Asymptotes: $y = \pm\frac{2}{3}x$

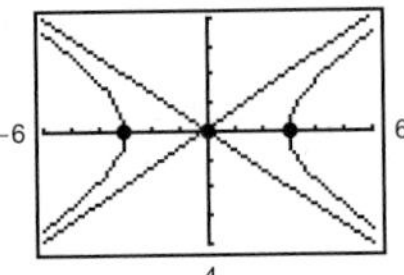

27. Center: $(1, -3)$
Vertices: $(1, -3 \pm \sqrt{2})$
Foci: $(1, -3 \pm 2\sqrt{5})$
Asymptotes:
$y = -3 \pm \frac{1}{3}(x - 1)$

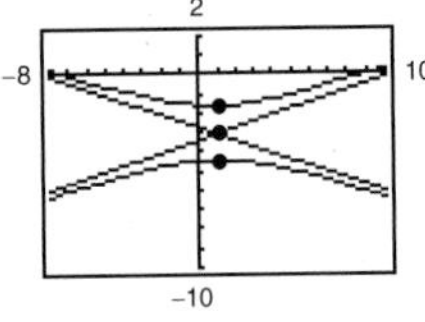

29. $\frac{y^2}{4} - \frac{x^2}{12} = 1$ **31.** $\frac{x^2}{1} - \frac{y^2}{25} = 1$

33. $\frac{17y^2}{1024} - \frac{17x^2}{64} = 1$ **35.** $\frac{(x-4)^2}{4} - \frac{y^2}{12} = 1$

37. $\frac{(y-5)^2}{16} - \frac{(x-4)^2}{9} = 1$ **39.** $\frac{y^2}{9} - \frac{4(x-2)^2}{9} = 1$

41. $\frac{(y-2)^2}{4} - \frac{x^2}{4} = 1$ **43.** $\frac{(x-2)^2}{1} - \frac{(y-2)^2}{1} = 1$

45. $\frac{(x-3)^2}{9} - \frac{(y-2)^2}{4} = 1$ **47.** $\frac{y^2}{9} - \frac{x^2}{9/4} = 1$

49. $\frac{(x-3)^2}{4} - \frac{(y-2)^2}{16/5} = 1$ **51.** $\frac{(x-6)^2}{9} - \frac{(y-2)^2}{7} = 1$

53. Answers will vary. **55.** $9x/(16y)$ **57.** $(x + 1)/(y - 3)$

59. $\frac{1}{2}x/(y - 2)$ **61.** $(x + 1)/(4y - 8)$

63. (a) $x - y - 4 = 0$ (b) $x - y = 0$
(c)

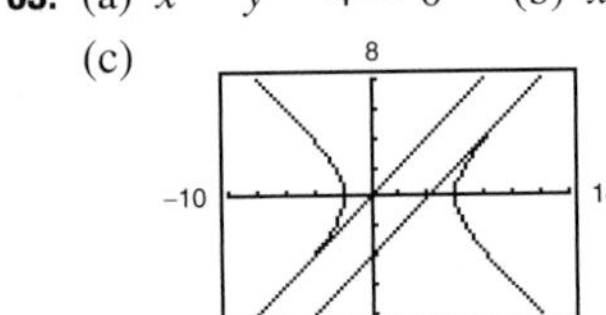

65. $x = 3, y = -5$
Vertices: $(3, -3), (3, -7)$

67. (a) $\frac{x^2}{1} - \frac{y^2}{169/3} = 1$ (b) About 2.403 ft

69. $(3300, -2750)$

71. Ellipse **73.** Hyperbola **75.** Hyperbola

77. Parabola **79.** Ellipse **81.** Parabola

83. Parabola **85.** Circle

87. True. For a hyperbola, $c^2 = a^2 + b^2$. The larger the ratio of b to a, the larger the eccentricity of the hyperbola, $e = c/a$.

89. False. When $D = -E$, the graph is two intersecting lines.

91. $y = 1 - 3\sqrt{\frac{(x-3)^2}{4} - 1}$

93.

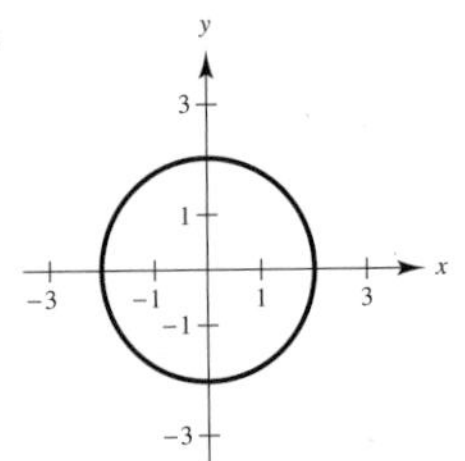

The equation $y = x^2 + C$ is a parabola that could intersect the circle in zero, one, two, three, or four places depending on its location on the y-axis.

(a) $C > 2$ and $C < -\frac{17}{4}$ (b) $C = 2$
(c) $-2 < C < 2$, $C = -\frac{17}{4}$ (d) $C = -2$
(e) $-\frac{17}{4} < C < -2$

Section 12.4 (page 777)

1. plane curve **3.** eliminating; parameter

5. (a)

t	0	1	2	3	4
x	0	1	$\sqrt{2}$	$\sqrt{3}$	2
y	3	2	1	0	-1

(b)

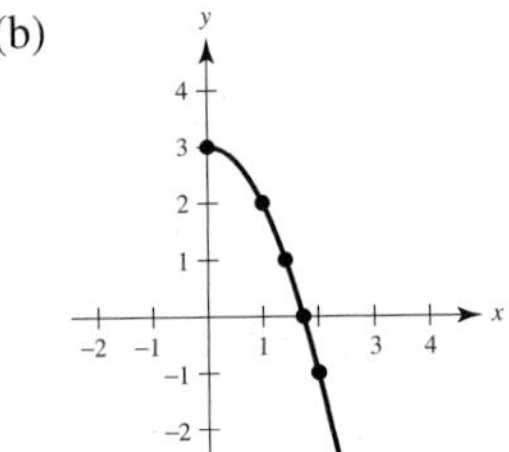

(c) $y = 3 - x^2$

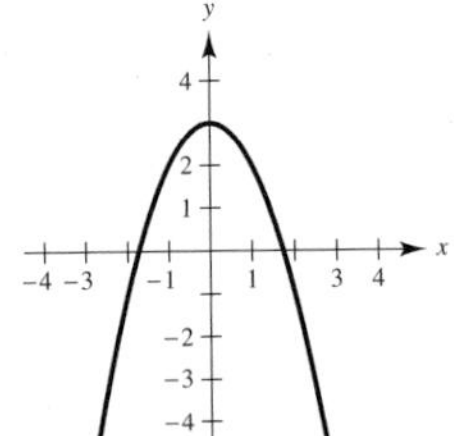

The graph of the rectangular equation shows the entire parabola rather than just the right half.

7. (a)

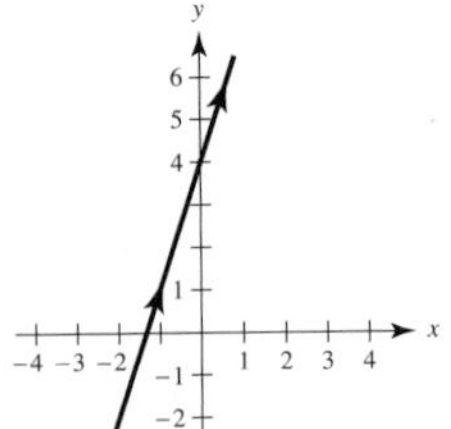

(b) $y = 3x + 4$

9. (a)

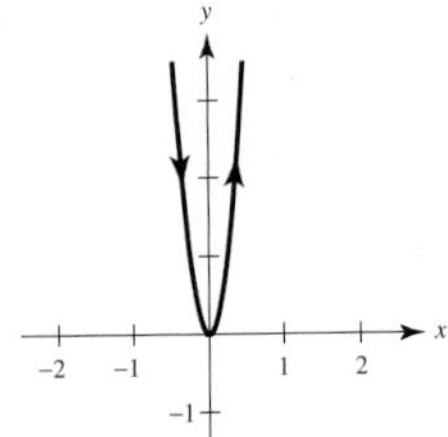

(b) $y = 16x^2$

11. (a)

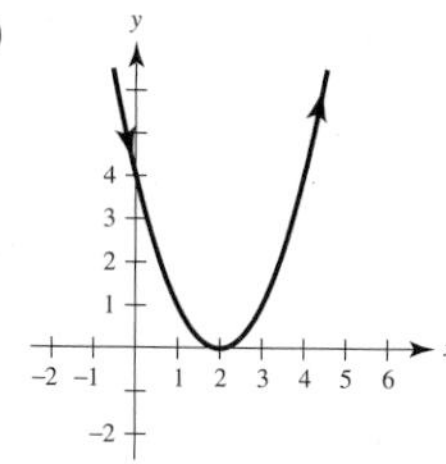

(b) $y = x^2 - 4x + 4$

13. (a)

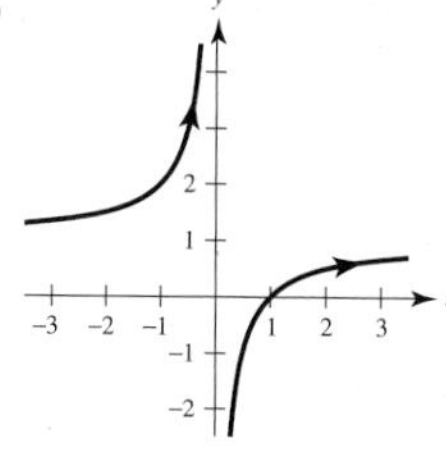

(b) $y = \dfrac{(x - 1)}{x}$

15. (a)

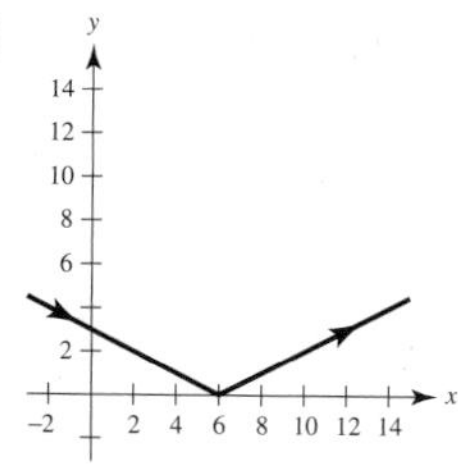

(b) $y = \left|\dfrac{x}{2} - 3\right|$

17. (a)

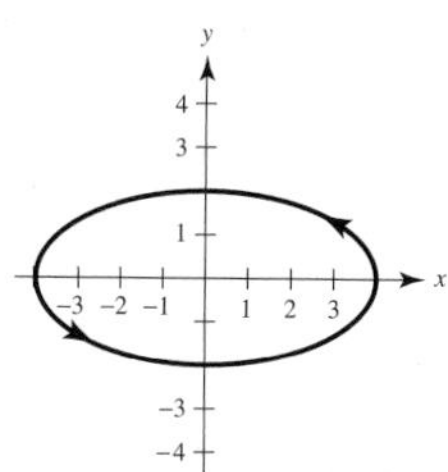

(b) $\dfrac{x^2}{16} + \dfrac{y^2}{4} = 1$

19. (a)

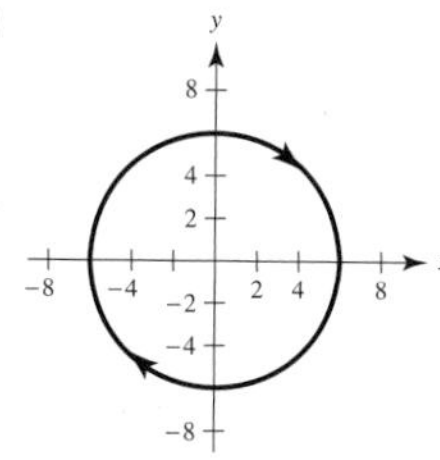

(b) $\dfrac{x^2}{36} + \dfrac{y^2}{36} = 1$

21. (a)

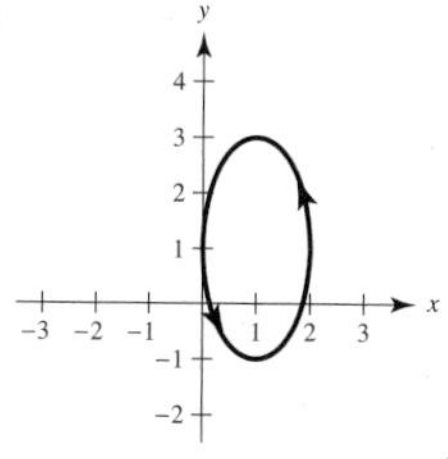

(b) $\dfrac{(x - 1)^2}{1} + \dfrac{(y - 1)^2}{4} = 1$

23. (a)

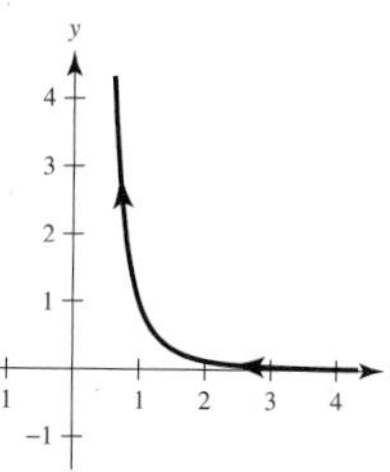

(b) $y = \dfrac{1}{x^3}, \quad x > 0$

25. (a)

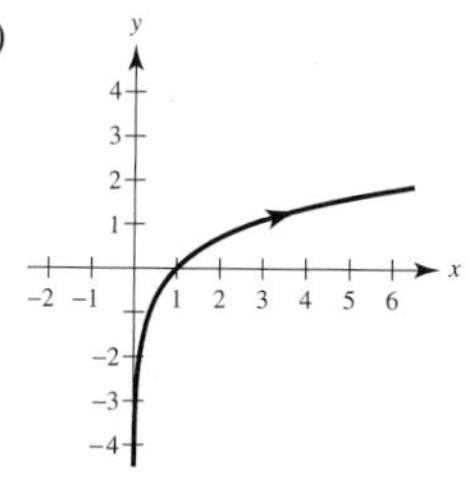

(b) $y = \ln x$

27. Each curve represents a portion of the line $y = 2x + 1$.

Domain	*Orientation*
(a) $(-\infty, \infty)$	Left to right
(b) $[-1, 1]$	Depends on θ
(c) $(0, \infty)$	Right to left
(d) $(0, \infty)$	Left to right

29. $y - y_1 = m(x - x_1)$ **31.** $\dfrac{(x - h)^2}{a^2} + \dfrac{(y - k)^2}{b^2} = 1$

33. $x = 3t$
$y = 6t$

35. $x = 3 + 4 \cos \theta$
$y = 2 + 4 \sin \theta$

37. $x = 5 \cos \theta$
$y = 3 \sin \theta$

39. $x = 4 \sec \theta$
$y = 3 \tan \theta$

41. (a) $x = t,\ y = 3t - 2$ (b) $x = -t + 2,\ y = -3t + 4$

43. (a) $x = t, y = 2 - t$ (b) $x = -t + 2, y = t$

45. (a) $x = t, y = t^2 - 3$ (b) $x = 2 - t, y = t^2 - 4t + 1$

47. (a) $x = t, y = \dfrac{1}{t}$ (b) $x = -t + 2, y = -\dfrac{1}{t - 2}$

49. Yes, the orientation of the curve would be reversed.

51.

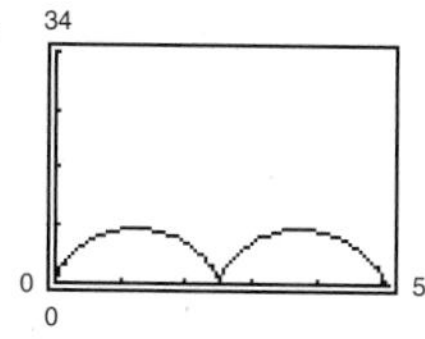

53.

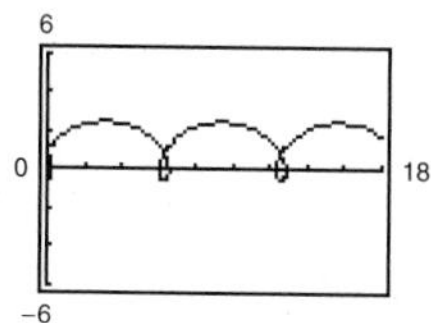

55.

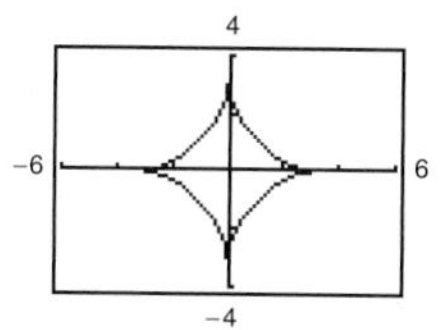

57.

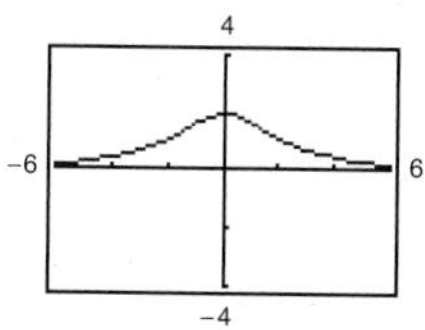

59. b
Domain: $[-2, 2]$
Range: $[-1, 1]$

60. c
Domain: $[-4, 4]$
Range: $[-6, 6]$

61. d
Domain: $(-\infty, \infty)$
Range: $(-\infty, \infty)$

62. a
Domain: $(-\infty, \infty)$
Range: $[-2, 2]$

63. The set of points (x, y) corresponding to the rectangular equation of a set of parametric equations does not show the orientation of the curve nor any restriction on the domain of the original parametric equations.

65. See Definition of a Plane Curve on page 771.

67. (a)

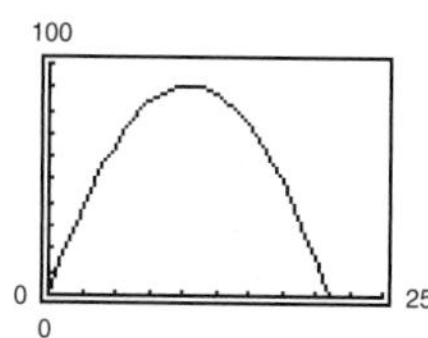

Maximum height: 90.7 ft
Range: 209.6 ft

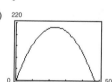

(b) Maximum height: 204.2 ft
Range: 471.6 ft

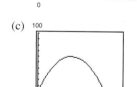

(c) Maximum height: 60.5 ft
Range: 242.0 ft

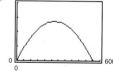

(d) Maximum height: 136.1 ft
Range: 544.5 ft

69. (a) $\frac{3}{2}$ (b) $y = \frac{3}{2}x - 1 \Rightarrow dy/dx = \frac{3}{2}$

71. (a) $2t + 3$
(b) $y = x^2 + x - 2 \Rightarrow dy/dx = 2x + 1 = 2t + 3$

73. (a) $-\cot t$
(b) $x^2 + y^2 = 4 \Rightarrow dy/dx = -x/y = -\cot t$

75. (a) $2 \csc t$
(b) $(x - 2)^2 - \dfrac{(y - 1)^2}{4} = 1 \Rightarrow \dfrac{dy}{dx} = \dfrac{4(x - 2)}{y - 1} = 2 \csc t$

77. (a) $-\tan t$
(b) $x^{2/3} + y^{2/3} = 1 \Rightarrow dy/dx = -y^{1/3}/x^{1/3} = -\tan t$

79. $dy/dx = -2/t$ **81.** $dy/dx = -1$

83. $dy/dx = 3/2$
$d^2y/dx^2 = 0$
Slope: $3/2$
Concavity: none

85. $dy/dx = 2t + 3$
$d^2y/dx^2 = 2$
Slope: 1
Concave upward

87. $dy/dx = -\cot \theta$
$d^2y/dx^2 = -\frac{1}{2}\csc^3 \theta$
Slope: -1
Concave downward

89. $dy/dx = 2 \csc \theta$
$d^2y/dx^2 = -2 \cot^3 \theta$
Slope: 4
Concave downward

91. $dy/dx = -\tan \theta$
$d^2y/dx^2 = \frac{1}{3} \csc \theta \sec^4 \theta$
Slope: -1
Concave upward

93. Horizontal: $(1, 0)$ when $t = 0$
Vertical: none

95. Horizontal: $(0, -2)$ when $t = 1$
$(2, 2)$ when $t = -1$
Vertical: none

97. Horizontal: $(0, -3)$ when $t = 3\pi/2$
$(0, 3)$ when $t = \pi/2$
Vertical: $(3, 0)$ when $t = 0$
$(-3, 0)$ when $t = \pi$

99. Horizontal: $(4, 0)$ when $t = \pi/2$
$(4, -2)$ when $t = 3\pi/2$
Vertical: $(6, -1)$ when $t = 0$
$(2, -1)$ when $t = \pi$

101. Horizontal: none
Vertical: $(-1, 0)$ when $t = \pi$
$(1, 0)$ when $t = 0$

103. (a) $x = (146.67 \cos \theta)t$
$y = 3 + (146.67 \sin \theta)t - 16t^2$

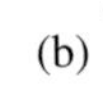

(b) No

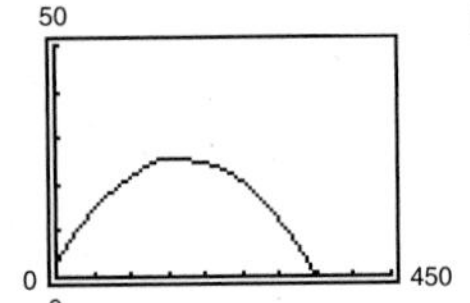

(c) Yes

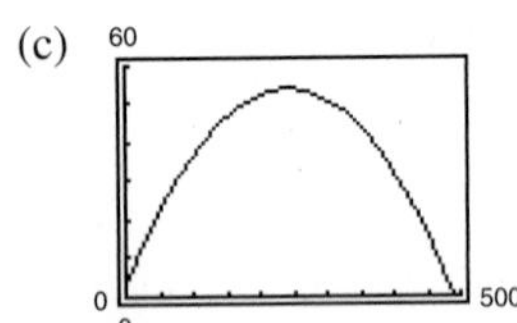

(d) 19.3°

105. Answers will vary.

107. $x = a\theta - b \sin \theta$
$y = a - b \cos \theta$

109. True
$x = t$
$y = t^2 + 1 \Rightarrow y = x^2 + 1$
$x = 3t$
$y = 9t^2 + 1 \Rightarrow y = x^2 + 1$

111. $-1 < t < \infty$

Section 12.5 (page 785)

1. pole **3.** polar

5.

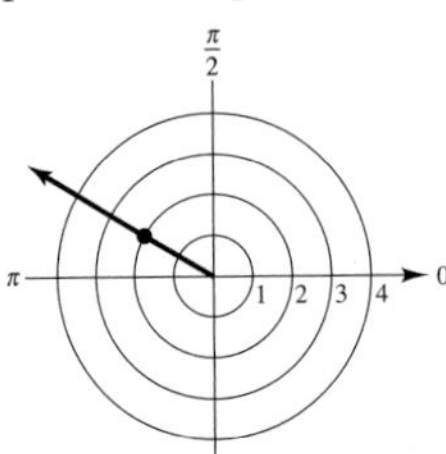

$\left(2, -\dfrac{7\pi}{6}\right), \left(-2, -\dfrac{\pi}{6}\right)$

7.

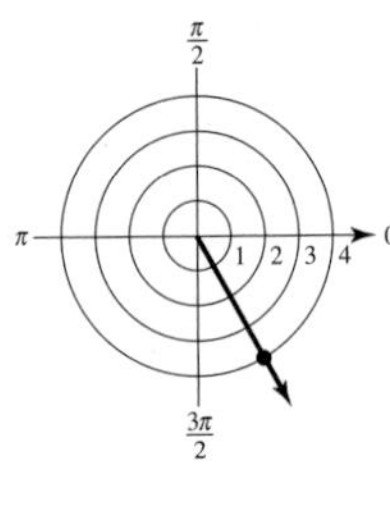

$\left(4, \dfrac{5\pi}{3}\right), \left(-4, -\dfrac{4\pi}{3}\right)$

9.

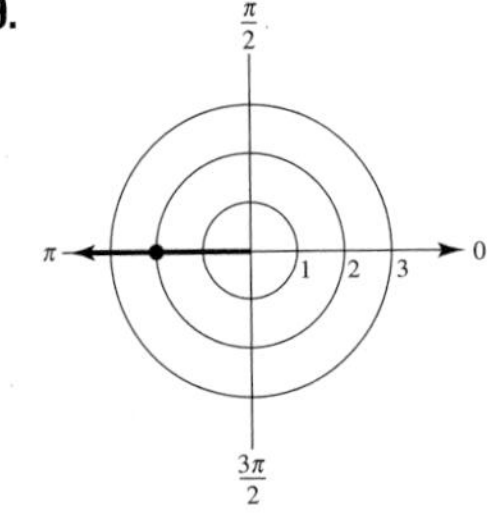

$(2, \pi), (-2, 0)$

11.

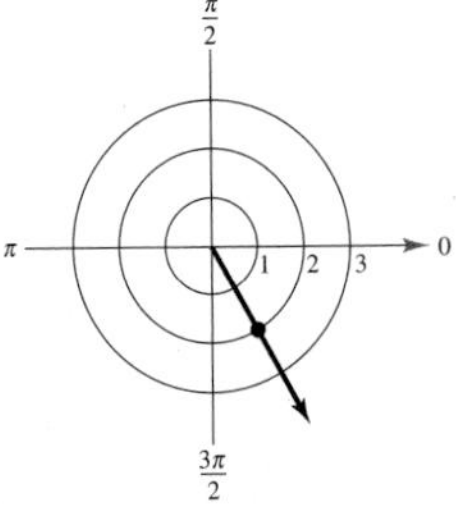

$\left(-2, -\dfrac{4\pi}{3}\right), \left(2, \dfrac{5\pi}{3}\right)$

13.

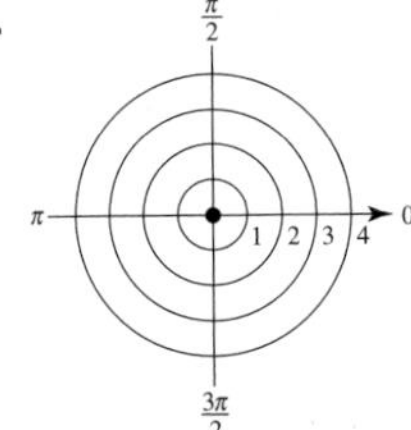

$\left(0, \dfrac{5\pi}{6}\right), \left(0, -\dfrac{\pi}{6}\right)$

15.

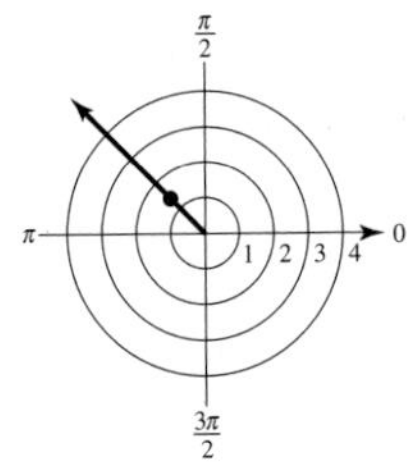

$(\sqrt{2}, -3.92), (-\sqrt{2}, -0.78)$

17.

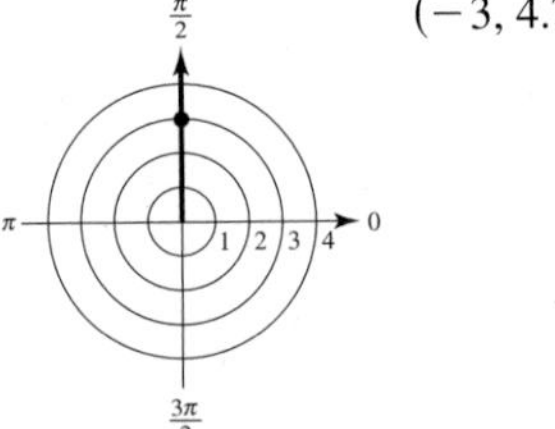

$(-3, 4.71), (3, 1.57)$

19. $(0, 3)$ **21.** $\left(\dfrac{\sqrt{2}}{2}, \dfrac{\sqrt{2}}{2}\right)$ **23.** $(-\sqrt{2}, \sqrt{2})$

25. $(\sqrt{3}, 1)$ **27.** $(-1.1, -2.2)$ **29.** $(1.53, 1.29)$

31. $(-1.20, -4.34)$ **33.** $(-0.02, 2.50)$

35. $(-3.60, 1.97)$ **37.** $\left(\sqrt{2}, \frac{\pi}{4}\right)$ **39.** $\left(3\sqrt{2}, \frac{5\pi}{4}\right)$

41. $(6, \pi)$ **43.** $\left(5, \frac{3\pi}{2}\right)$ **45.** $(5, 2.21)$

47. $\left(\sqrt{6}, \frac{5\pi}{4}\right)$ **49.** $\left(2, \frac{11\pi}{6}\right)$ **51.** $(3\sqrt{13}, 0.98)$

53. $(13, 1.18)$ **55.** $(\sqrt{13}, 5.70)$ **57.** $(\sqrt{29}, 2.76)$

59. $(\sqrt{7}, 0.86)$ **61.** $\left(\frac{17}{6}, 0.49\right)$ **63.** $\left(\frac{\sqrt{85}}{4}, 0.71\right)$

65. $r = 3$ **67.** $r = 4 \csc \theta$ **69.** $r = 10 \sec \theta$

71. $r = -2 \csc \theta$ **73.** $r = \frac{-2}{3 \cos \theta - \sin \theta}$

75. $r^2 = 16 \sec \theta \csc \theta = 32 \csc 2\theta$

77. $r = \frac{4}{1 - \cos \theta}$ or $-\frac{4}{1 + \cos \theta}$ **79.** $r = a$

81. $r = 2a \cos \theta$ **83.** $r = \cot^2 \theta \csc \theta$

85. $x^2 + y^2 - 4y = 0$ **87.** $x^2 + y^2 + 2x = 0$

89. $\sqrt{3}x + y = 0$ **91.** $\frac{\sqrt{3}}{3}x + y = 0$

93. $x^2 + y^2 = 16$ **95.** $y = 4$ **97.** $x = -3$

99. $x^2 + y^2 - x^{2/3} = 0$ **101.** $(x^2 + y^2)^2 = 2xy$

103. $(x^2 + y^2)^2 = 6x^2y - 2y^3$ **105.** $x^2 + 4y - 4 = 0$

107. $4x^2 - 5y^2 - 36y - 36 = 0$

109. The graph of the polar equation consists of all points that are six units from the pole.
$x^2 + y^2 = 36$

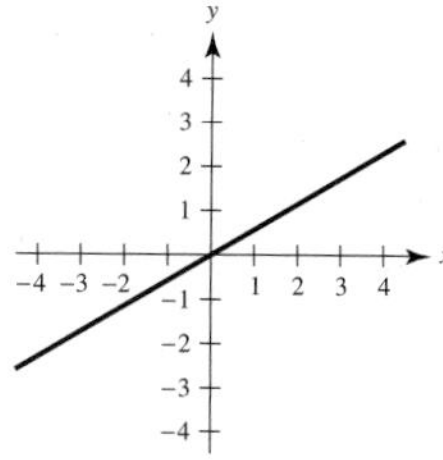

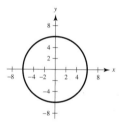

111. The graph of the polar equation consists of all points on the line that makes an angle of $\pi/6$ with the positive polar axis.
$-\sqrt{3}x + 3y = 0$

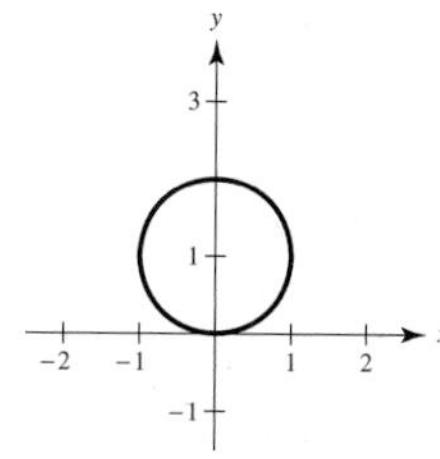

113. The graph of the polar equation is not evident by simple inspection, so convert to rectangular form.
$x^2 + (y - 1)^2 = 1$

115. The graph of the polar equation is not evident by simple inspection, so convert to rectangular form.
$(x + 3)^2 + y^2 = 9$

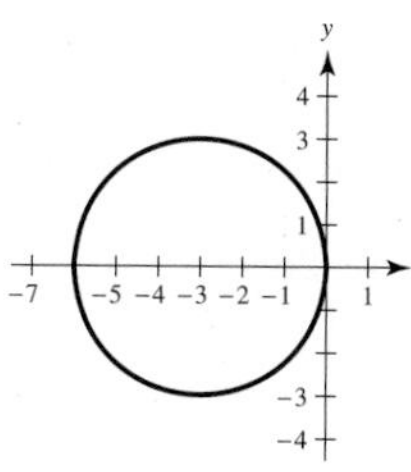

117. The graph of the polar equation is not evident by simple inspection, so convert to rectangular form.
$x - 3 = 0$

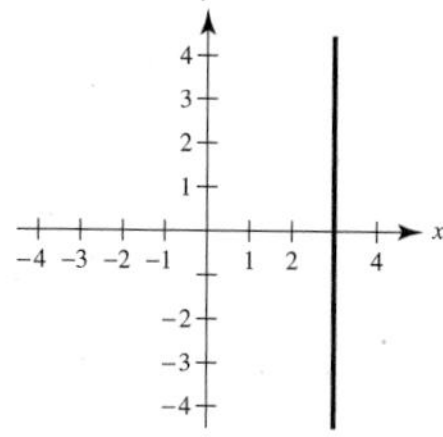

119. Horizontal: $(2, \pi/2), (\frac{1}{2}, 7\pi/6), (\frac{1}{2}, 11\pi/6)$
Vertical: $(\frac{3}{2}, \pi/6), (\frac{3}{2}, 5\pi/6)$

121. Horizontal: $(5, \pi/2), (1, 3\pi/2)$
Vertical: $(0.7105, 4.2041), (0.7105, 5.2207)$

123. $(x - h)^2 + (y - k)^2 = h^2 + k^2$
Radius: $\sqrt{h^2 + k^2}$
Center: (h, k)

125. (a) Symmetric to the pole
(b) Symmetric to the polar axis
(c) Symmetric to the line $\theta = \pi/2$

127. True. Because r is a directed distance, the point (r, θ) can be represented as $(r, \theta \pm 2\pi n)$.

129. (a) Answers will vary.
(b) $(r_1, \theta_1), (r_2, \theta_2)$, and the pole are collinear.
$d = \sqrt{r_1^2 + r_2^2 - 2r_1r_2} = |r_1 - r_2|$
This represents the distance between two points on the line $\theta = \theta_1 = \theta_2$.
(c) $d = \sqrt{r_1^2 + r_2^2}$
This is the result of the Pythagorean Theorem.
(d) Answers will vary. For example:
Points: $(3, \pi/6), (4, \pi/3)$
Distance: 2.053
Points: $(-3, 7\pi/6), (-4, 4\pi/3)$
Distance: 2.053

131. (a)

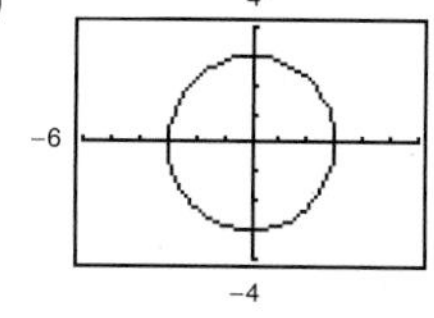

(b) Yes. $\theta \approx 3.927, x \approx -2.121, y \approx -2.121$

(c) Yes. Answers will vary.

Section 12.6 (page 791)

1. $\theta = \frac{\pi}{2}$ **3.** convex limaçon **5.** lemniscate

7. Rose curve with 4 petals **9.** Limaçon with inner loop

11. Rose curve with 3 petals **13.** Polar axis

15. $\theta = \frac{\pi}{2}$ **17.** $\theta = \frac{\pi}{2}$, polar axis, pole

19. Maximum: $|r| = 20$ when $\theta = \frac{3\pi}{2}$
Zero: $r = 0$ when $\theta = \frac{\pi}{2}$

21. Maximum: $|r| = 4$ when $\theta = 0, \frac{\pi}{3}, \frac{2\pi}{3}$
Zeros: $r = 0$ when $\theta = \frac{\pi}{6}, \frac{\pi}{2}, \frac{5\pi}{6}$

23.

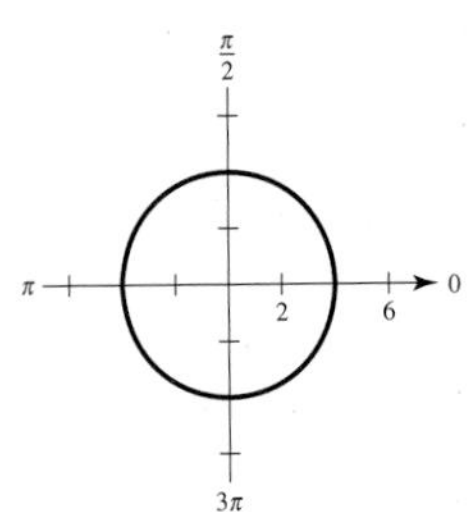

25.

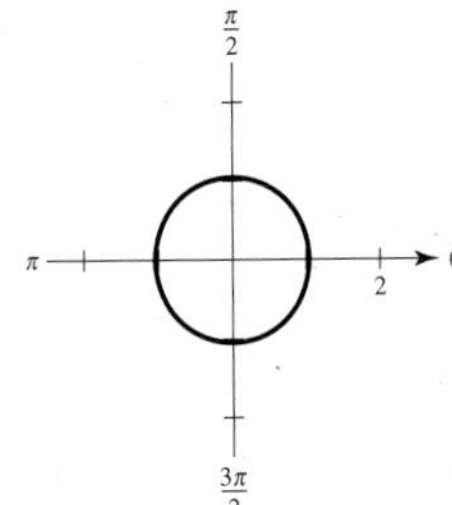

27.

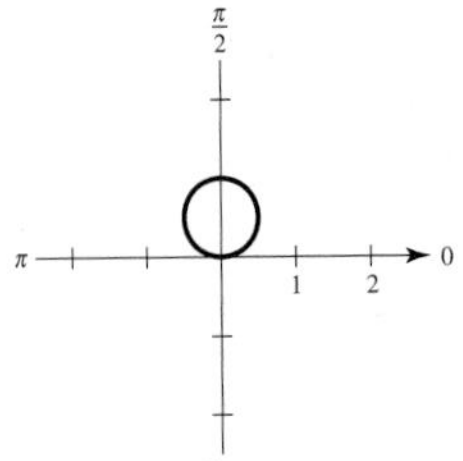

29.

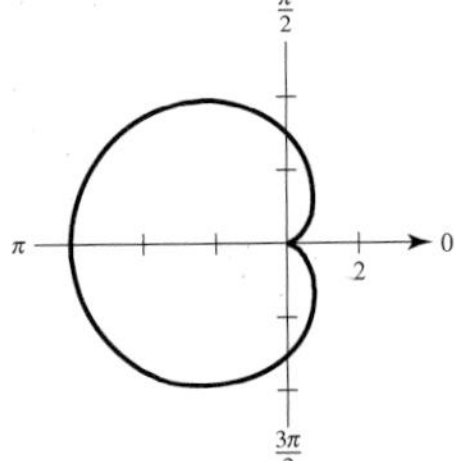

31.

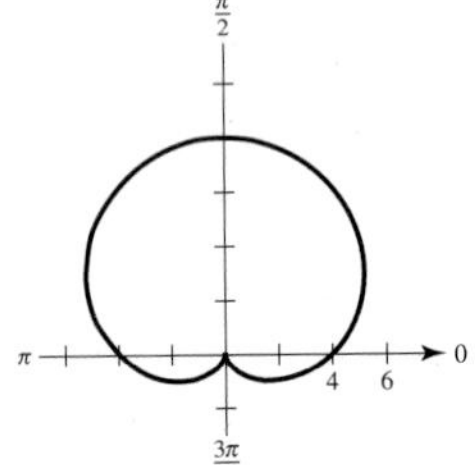

33.

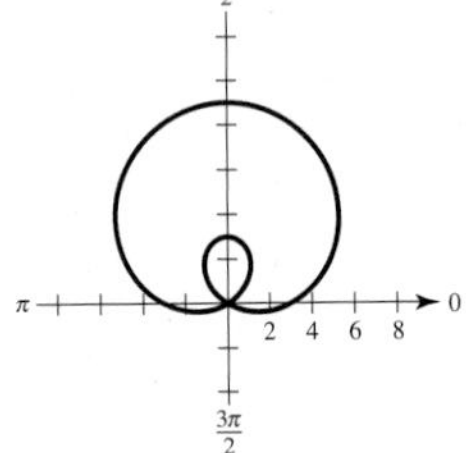

35.

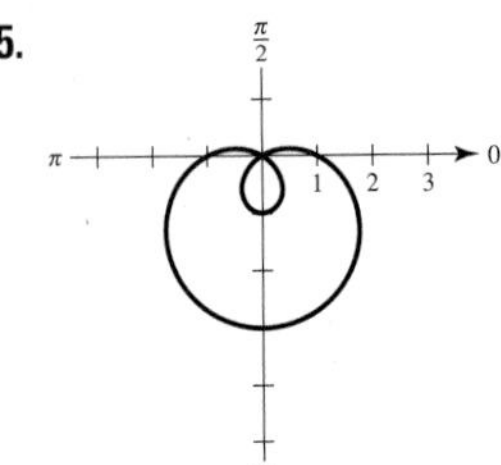

37.

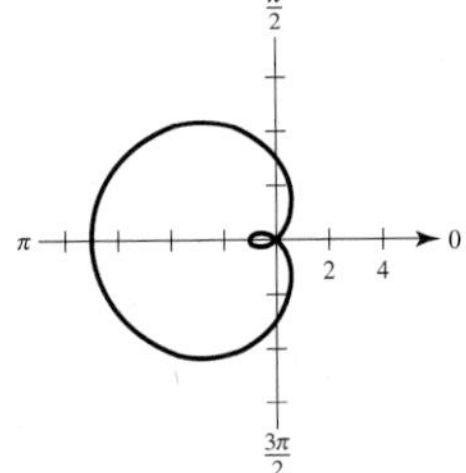

39.

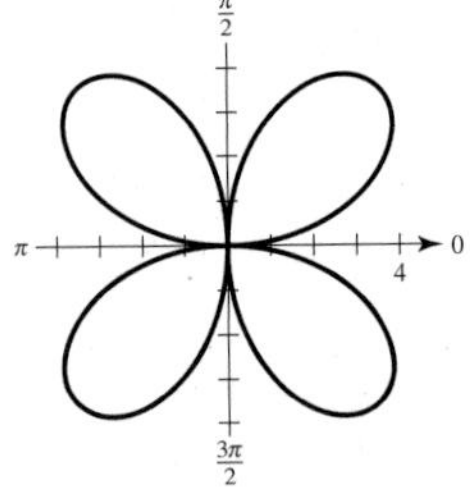

41.

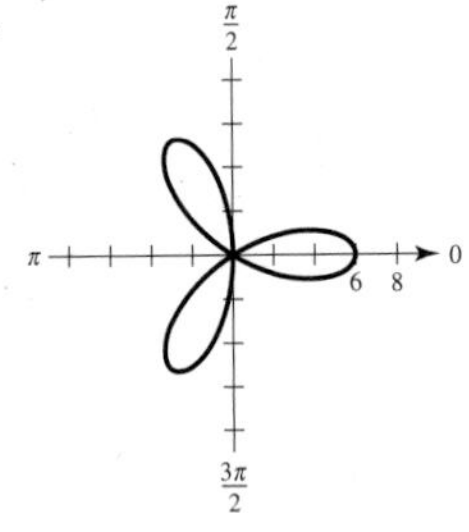

43.

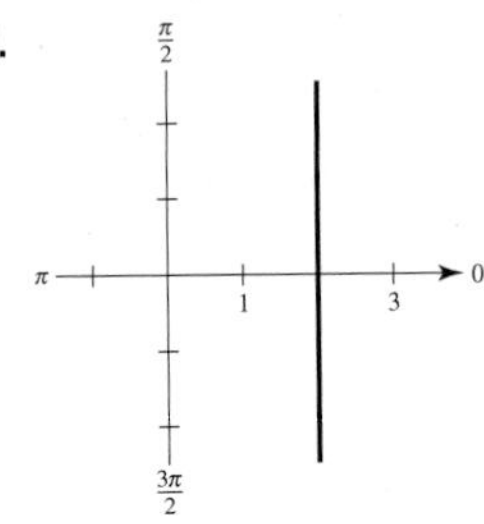

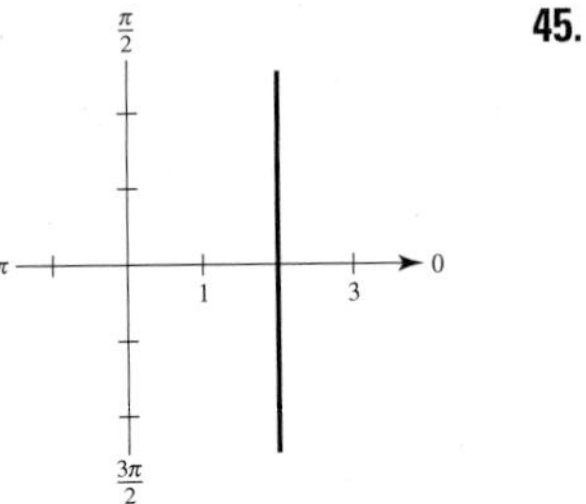

45.

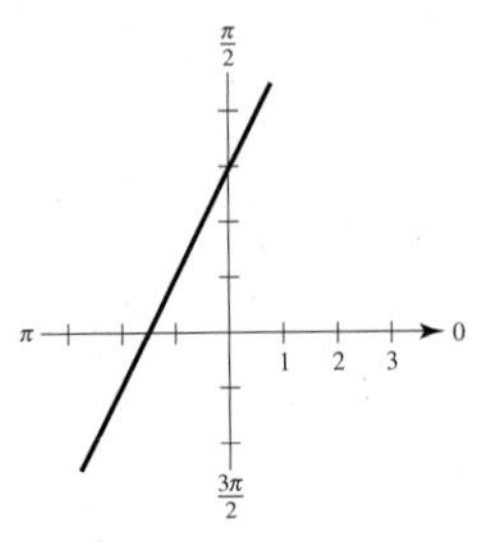

47.

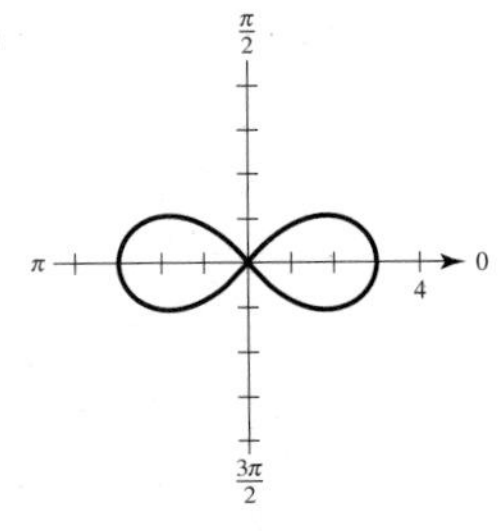

49.

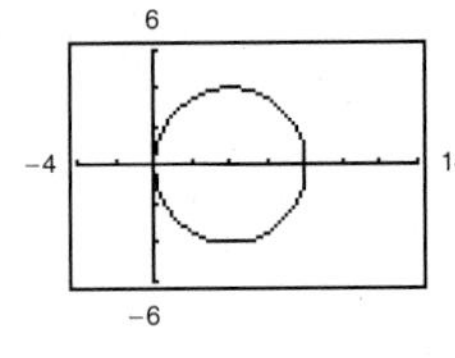

51.

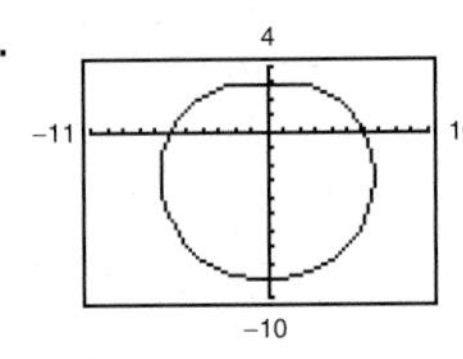

53.

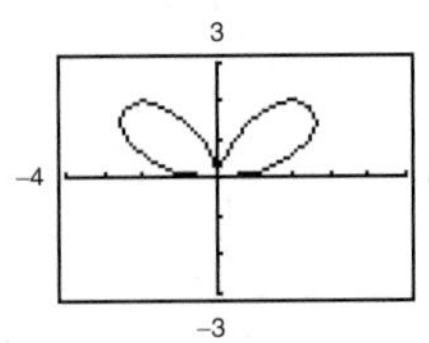

55.

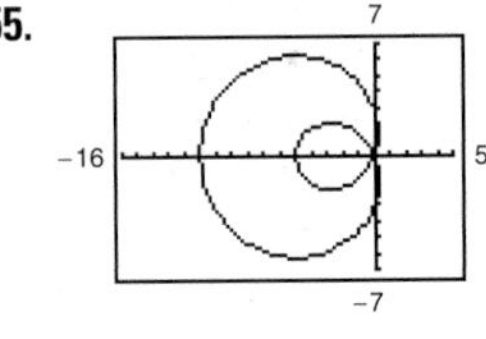

$0 \le \theta < 2\pi$

57.

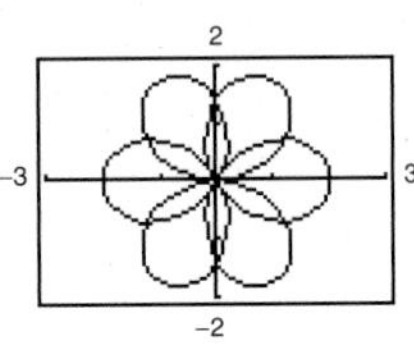

$0 \le \theta < 4\pi$

59.

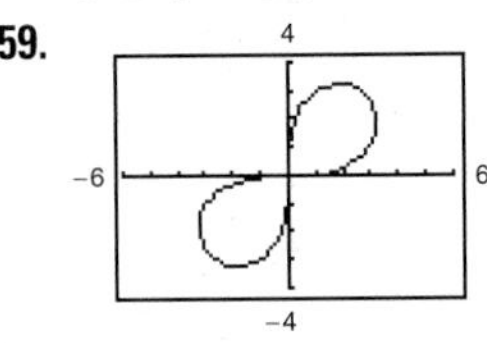

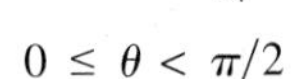

$0 \le \theta < \pi/2$

61. (a)

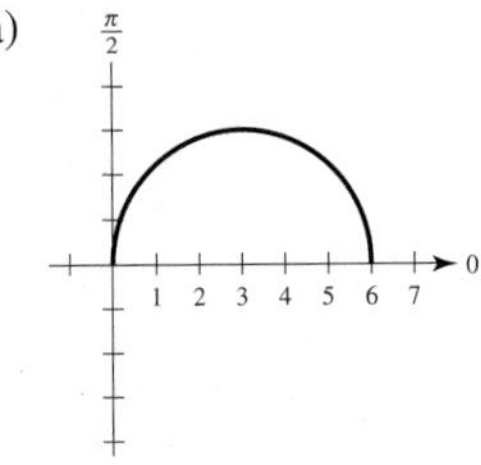

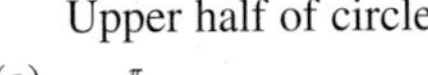

Upper half of circle

(b)

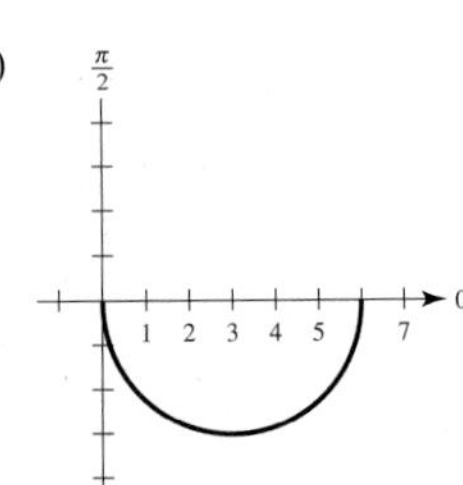

Lower half of circle

(c)

Full circle

(d)

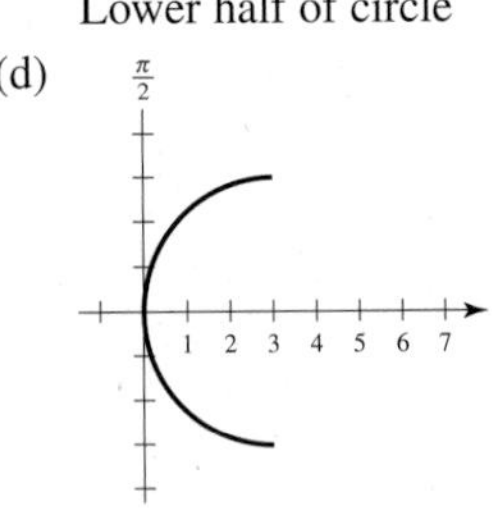

Left half of circle

63.

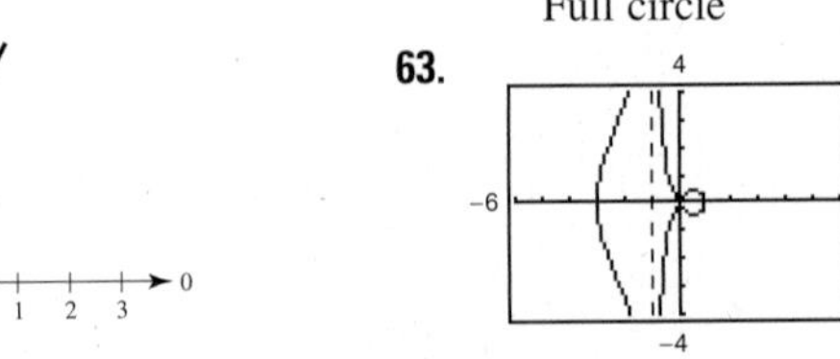

65.

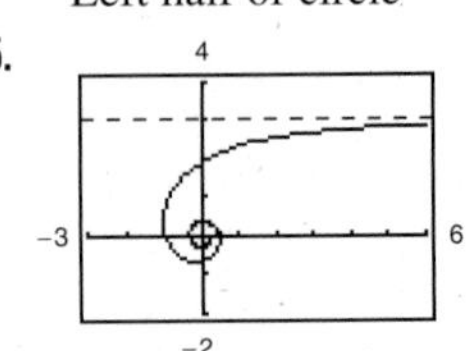

67. True. For a graph to have polar axis symmetry, replace (r, θ) by $(r, -\theta)$ or $(-r, \pi - \theta)$.

69. True **71.** True

73. (a)

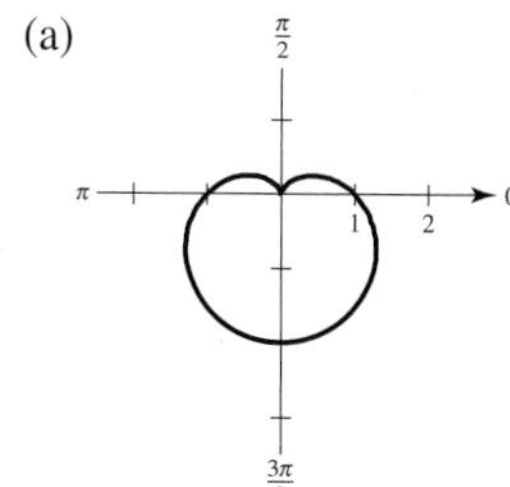

(b)

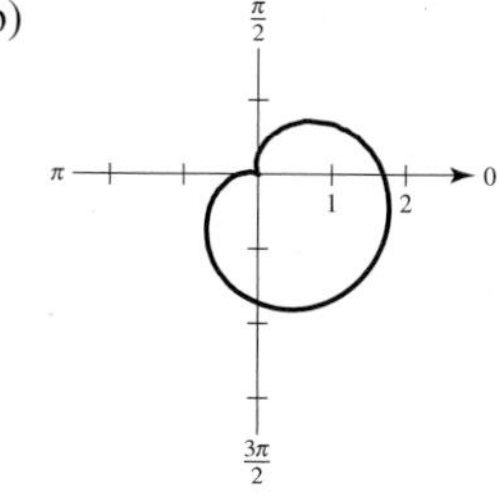

75. 8 petals; 3 petals; For $r = 2 \cos n\theta$ and $r = 2 \sin n\theta$, there are n petals if n is odd, $2n$ petals if n is even.

77. (a)

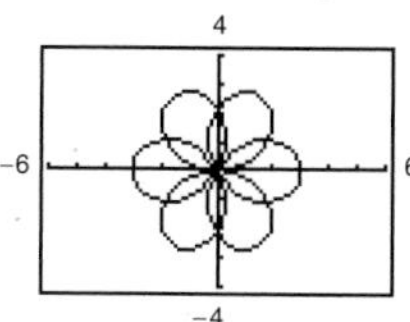

$0 \leq \theta < 4\pi$

(b)

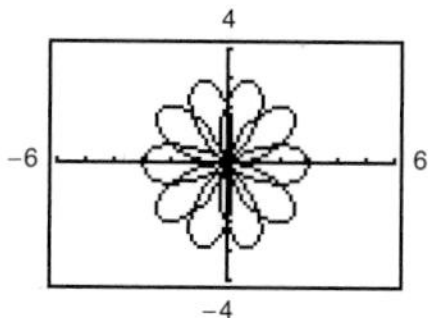

$0 \leq \theta < 4\pi$

(c) Yes. Explanations will vary.

Section 12.7 (page 797)

1. conic **3.** vertical; right

5. $e = 1$: $r = \dfrac{2}{1 + \cos \theta}$, parabola

$e = 0.5$: $r = \dfrac{1}{1 + 0.5 \cos \theta}$, ellipse

$e = 1.5$: $r = \dfrac{3}{1 + 1.5 \cos \theta}$, hyperbola

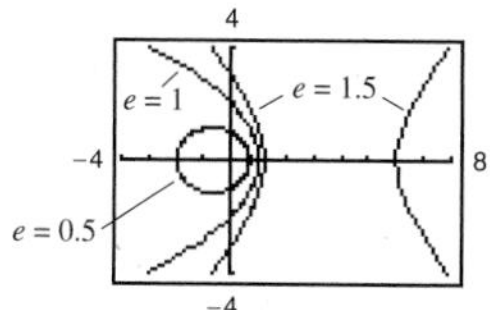

7. $e = 1$: $r = \dfrac{2}{1 - \sin \theta}$, parabola

$e = 0.5$: $r = \dfrac{1}{1 - 0.5 \sin \theta}$, ellipse

$e = 1.5$: $r = \dfrac{3}{1 - 1.5 \sin \theta}$, hyperbola

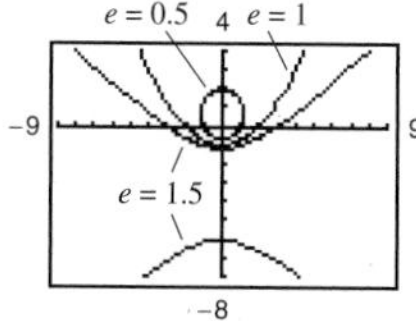

9. e **10.** c **11.** d **12.** f **13.** a **14.** b

15. Parabola

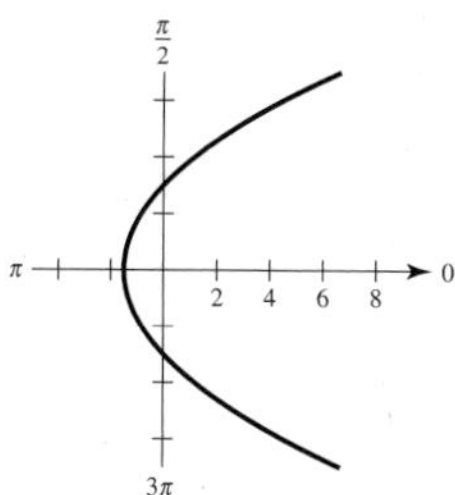

17. Parabola

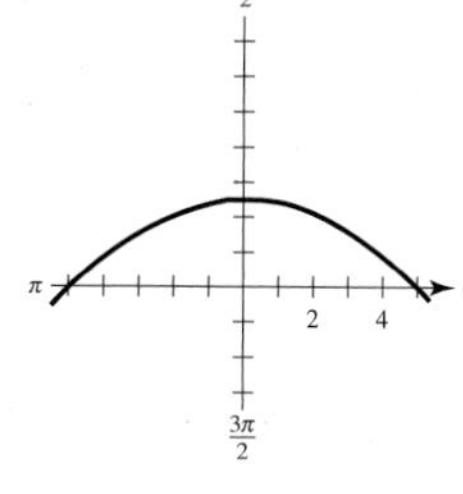

19. Ellipse

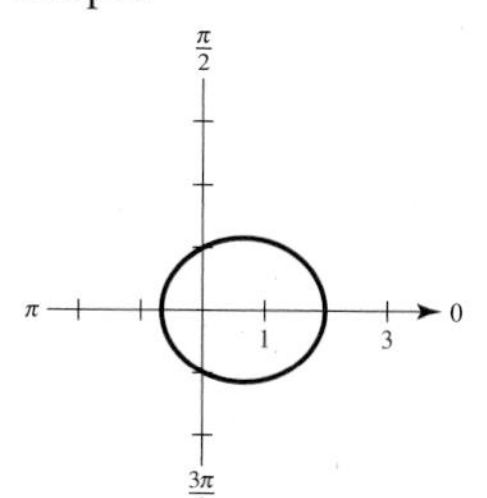

21. Ellipse

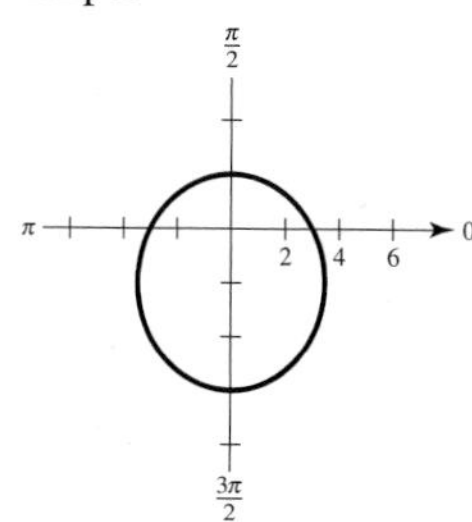

23. Hyperbola

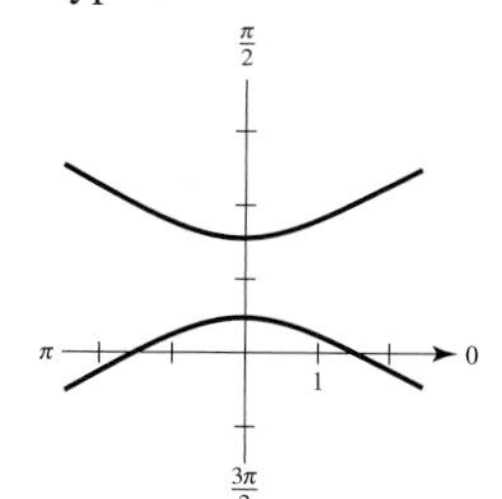

25. Hyperbola

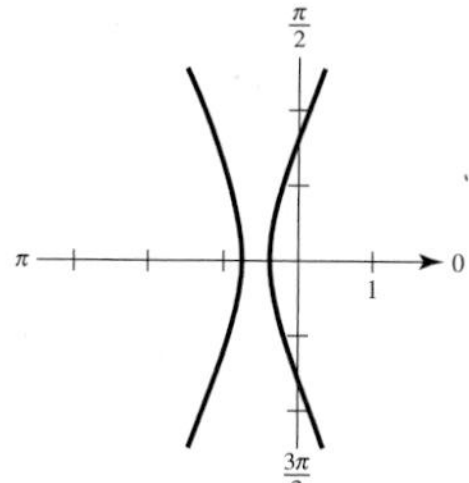

27. Ellipse

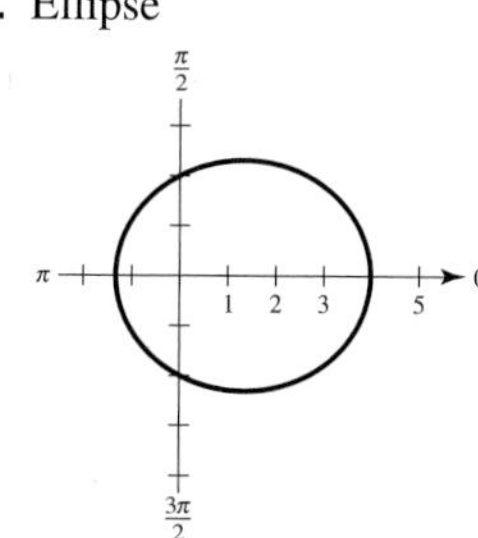

29.

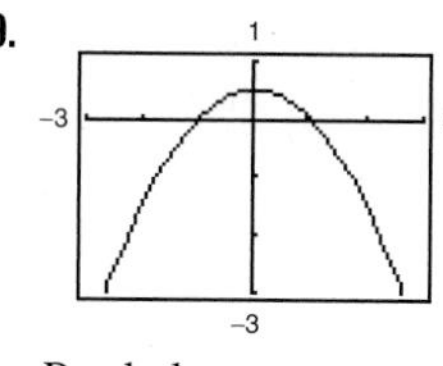

Parabola

31.

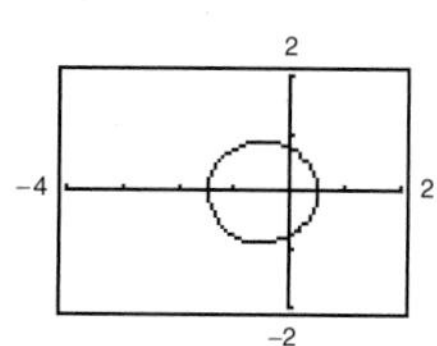

Ellipse

33.

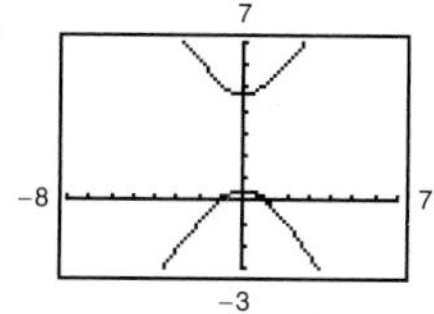

Hyperbola

35.

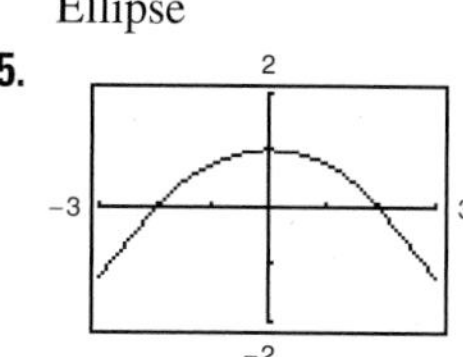

Parabola

37.

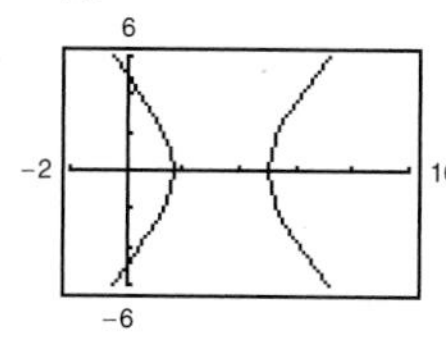

Hyperbola

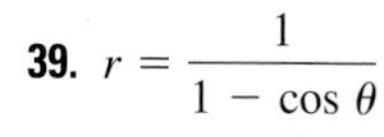

39. $r = \dfrac{1}{1 - \cos \theta}$ **41.** $r = \dfrac{1}{2 + \sin \theta}$

43. $r = \dfrac{2}{1 + 2 \cos \theta}$ **45.** $r = \dfrac{2}{1 - \sin \theta}$

47. $r = \dfrac{10}{1 - \cos \theta}$ **49.** $r = \dfrac{10}{3 + 2 \cos \theta}$

51. $r = \dfrac{20}{3 - 2 \cos \theta}$ **53.** $r = \dfrac{9}{4 - 5 \sin \theta}$

55. Proof

57. $r^2 = 24{,}336/(169 - 25 \cos^2 \theta)$

59. $r^2 = 144/(25 \cos^2 \theta - 9)$

61. $r^2 = -144/(16 - 25 \cos^2 \theta)$

63. $r = \dfrac{0.624}{1 + 0.847 \sin \theta}$; $r \approx 0.338$ astronomical unit

65. (a) $r = \dfrac{8200}{1 + \sin \theta}$

(b)

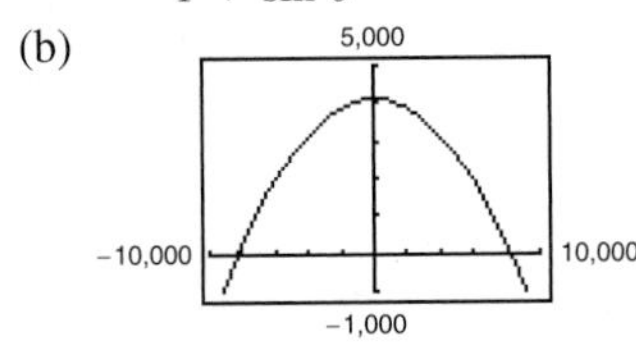

(c) 1467 mi (d) 394 mi

67. False. The graph has a horizontal directrix below the pole.

69. True. The conic is an ellipse because the eccentricity is less than 1.

71. True

73. The original equation graphs as a parabola that opens downward.
(a) The parabola opens to the right.
(b) The parabola opens up.
(c) The parabola opens to the left.
(d) The parabola has been rotated.

75. If e remains fixed and p changes, then the lengths of both the major axis and the minor axis change. For example, graph

$$r = \frac{5}{1 - \frac{2}{3} \sin \theta}$$

with $e = \frac{2}{3}$ and $p = \frac{15}{2}$, and graph

$$r = \frac{6}{1 - \frac{2}{3} \sin \theta}$$

with $e = \frac{2}{3}$ and $p = 9$, on the same set of coordinate axes.

Review Exercises (page 801)

1. Hyperbola

3. $y^2 = 16x$

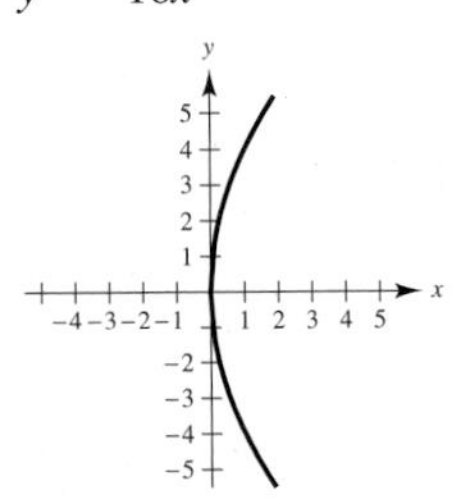

5. $(y - 2)^2 = 12x$

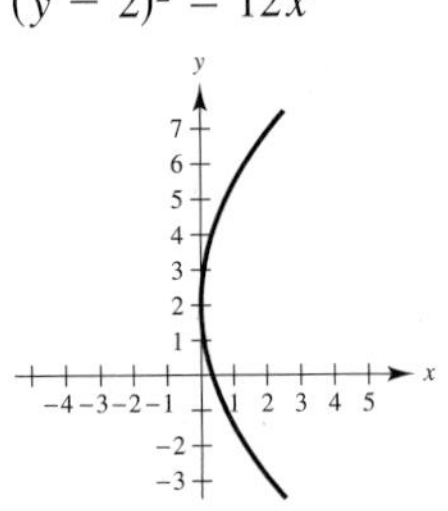

7. $y = -4x - 2$; $\left(-\frac{1}{2}, 0\right)$

9. $8\sqrt{6}$ m

11. $\dfrac{(x - 3)^2}{25} + \dfrac{y^2}{16} = 1$

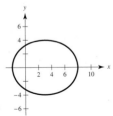

13. $\dfrac{(x - 2)^2}{4} + \dfrac{(y - 1)^2}{1} = 1$

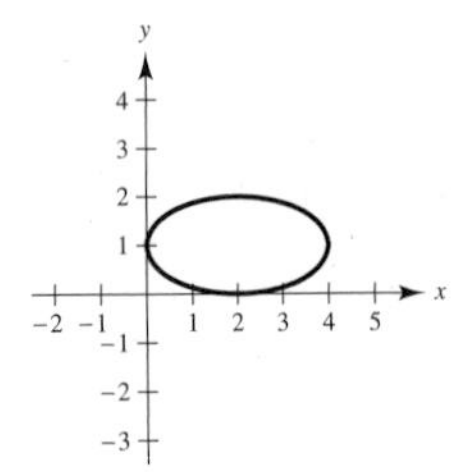

15. The foci occur 3 feet from the center of the arch on a line connecting the tops of the pillars.

17. Center: $(-1, 2)$
Vertices: $(-1, 9), (-1, -5)$
Foci: $\left(-1, 2 \pm 2\sqrt{6}\right)$
Eccentricity: $\dfrac{2\sqrt{6}}{7}$

19. Center:
Vertices: $(1, 0), (1, -8)$
Foci: $\left(1, -4 \pm \sqrt{7}\right)$
Eccentricity: $\dfrac{\sqrt{7}}{4}$

21. $x - 4 = 0$ **23.** $2\sqrt{5}\,\pi$ **25.** 2π

27. $\dfrac{y^2}{1} - \dfrac{x^2}{3} = 1$ **29.** $\dfrac{5(x - 4)^2}{16} - \dfrac{5y^2}{64} = 1$

31. Center: $(5, -3)$
Vertices: $(11, -3), (-1, -3)$
Foci: $\left(5 \pm 2\sqrt{13}, -3\right)$
Asymptotes:
$y = -3 \pm \frac{2}{3}(x - 5)$

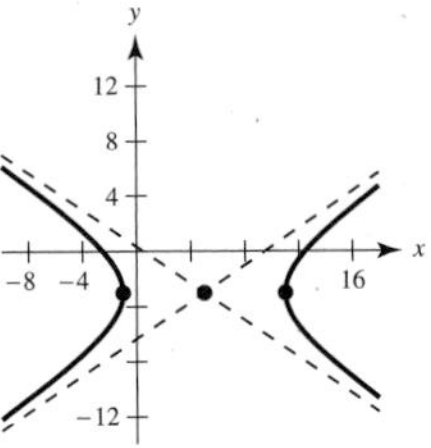

33. Center: $(1, -1)$
Vertices: $(5, -1), (-3, -1)$
Foci: $(6, -1), (-4, -1)$
Asymptotes:
$y = -1 \pm \frac{3}{4}(x - 1)$

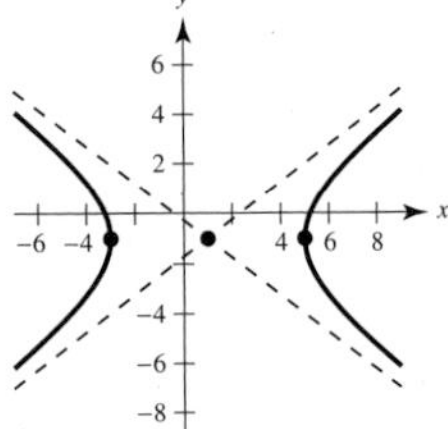

35. 72 mi **37.** Hyperbola **39.** Ellipse

41. (a)

t	−2	−1	0	1	2
x	−8	−5	−2	1	4
y	15	11	7	3	−1

(b)

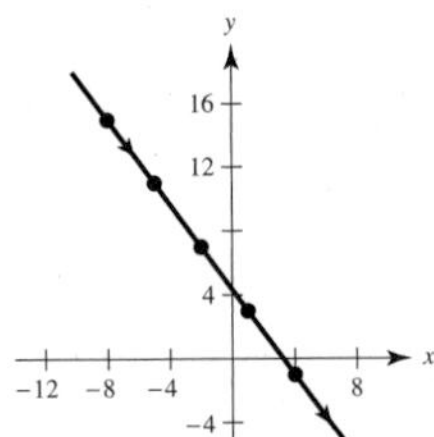

43. (a)

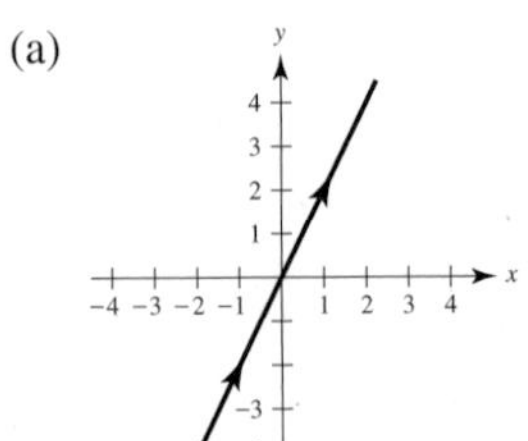

(b) $y = 2x$

45. (a)

(b) $y = \sqrt[4]{x}$

47. (a)

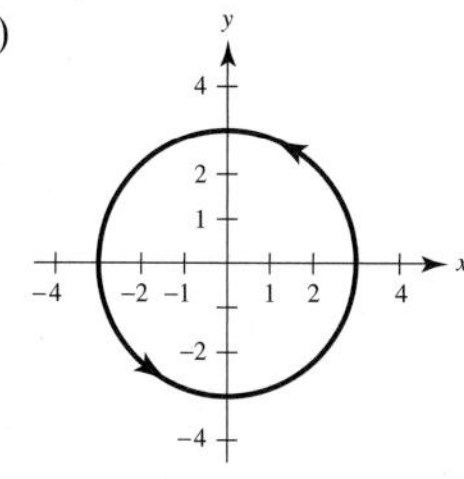

(b) $x^2 + y^2 = 9$

49. $x = -4 + 13t$
$y = 4 - 14t$

51. $x = -3 + 4\cos\theta$
$y = 4 + 3\sin\theta$

53. Vertical transverse axis: $x = -1 + 2\tan\theta$
$y = 3 + \sec\theta$
Horizontal transverse axis: $x = -1 + 2\sec\theta$
$y = 3 + \tan\theta$

55. $\frac{7}{4}$ **57.** $-\cot t$

59. Horizontal: $(0, 0)$ when $t = 0$
Vertical: $(-4, 8)$ when $t = 2$

61. Horizontal: $(0, -1)$ when $t = 0$; $(0, 1)$ when $t = \pi$
Vertical: $(2, 0)$ when $t = \pi/2$; $(-2, 0)$ when $t = 3\pi/2$

63.

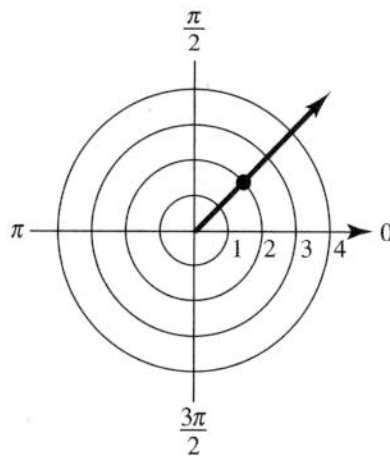

$(2, -7\pi/4), (-2, 5\pi/4)$

65.

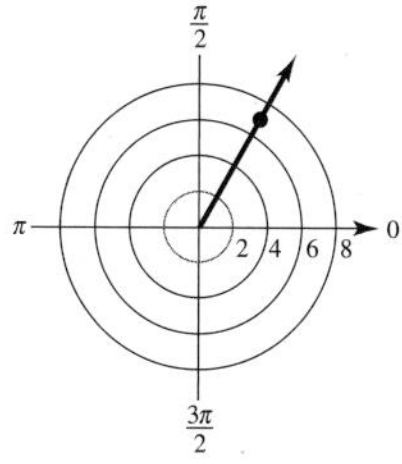

$(7, 1.05), (-7, -2.09)$

67. $\left(-\frac{1}{2}, -\frac{\sqrt{3}}{2}\right)$ **69.** $\left(-\frac{3\sqrt{2}}{2}, \frac{3\sqrt{2}}{2}\right)$ **71.** $\left(1, \frac{\pi}{2}\right)$

73. $\left(2\sqrt{13}, 0.9828\right)$ **75.** $r = 9$ **77.** $r = 6\sin\theta$

79. $r^2 = 10\csc 2\theta$ **81.** $x^2 + y^2 = 25$ **83.** $x^2 + y^2 = 3x$

85. $x^2 + y^2 = y^{2/3}$

87. Horizontal: $(-1, \pi/2), (3, 3\pi/2), \left(-\frac{1}{2}, 0.2527\right), \left(-\frac{1}{2}, 2.8889\right)$
Vertical: $(-0.6862, 1.0030), (-0.6862, 2.1386),$
$(2.1862, 3.7765), (2.1862, 5.6483)$

89. Horizontal: $\left(\sqrt{2}/2, \pi/4\right), \left(-\sqrt{2}/2, 3\pi/4\right)$
Vertical: $(1, 0), (0, \pi/2)$

91. Symmetry: $\theta = \frac{\pi}{2}$, polar axis, pole
Maximum value of $|r|$: $|r| = 6$ for all values of θ
No zeros of r

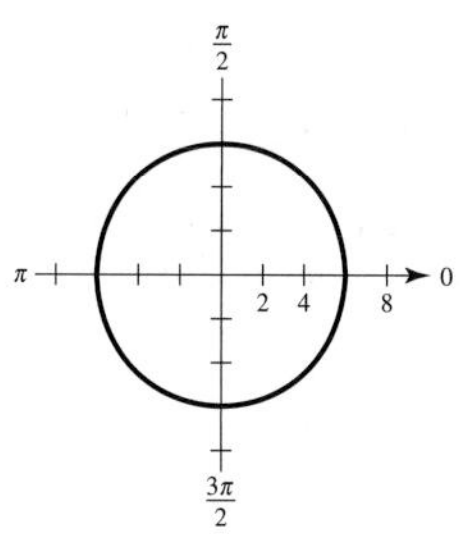

93. Symmetry: $\theta = \frac{\pi}{2}$, polar axis, pole
Maximum value of $|r|$: $|r| = 4$ when $\theta = \frac{\pi}{4}, \frac{3\pi}{4}, \frac{5\pi}{4}, \frac{7\pi}{4}$
Zeros of r: $r = 0$ when
$\theta = 0, \frac{\pi}{2}, \pi, \frac{3\pi}{2}$

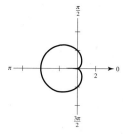

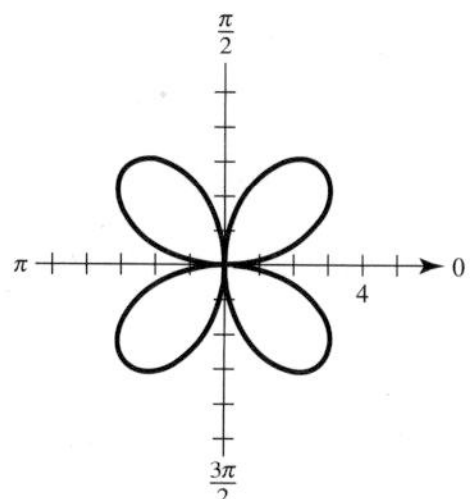

95. Symmetry: polar axis
Maximum value of $|r|$:
$|r| = 4$ when $\theta = 0$
Zeros of r: $r = 0$
when $\theta = \pi$

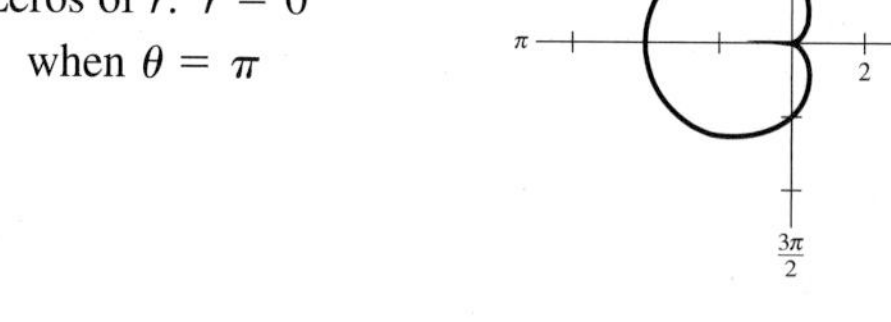

97. Symmetry: $\theta = \frac{\pi}{2}$
Maximum value of $|r|$: $|r| = 8$ when $\theta = \frac{\pi}{2}$
Zeros of r: $r = 0$ when $\theta = 3.4814, 5.9433$

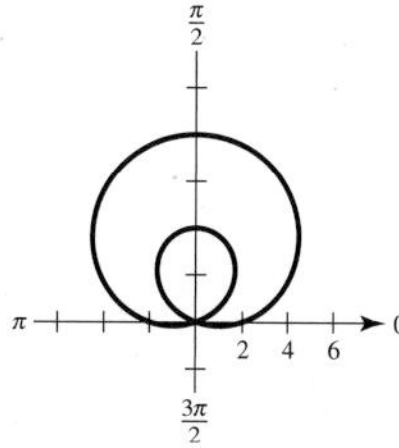

99. Symmetry: $\theta = \frac{\pi}{2}$, polar axis, pole
Maximum value of $|r|$: $|r| = 3$ when $\theta = 0, \frac{\pi}{2}, \pi, \frac{3\pi}{2}$
Zeros of r: $r = 0$ when $\theta = \frac{\pi}{4}, \frac{3\pi}{4}, \frac{5\pi}{4}, \frac{7\pi}{4}$

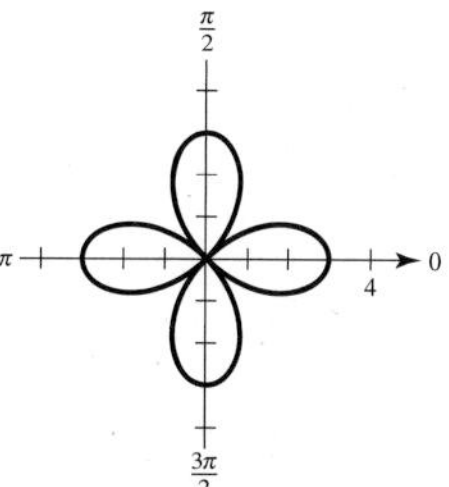

101. Limaçon

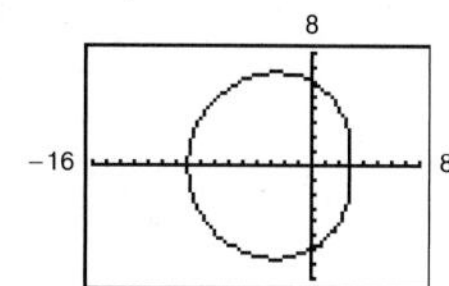

103. Rose curve

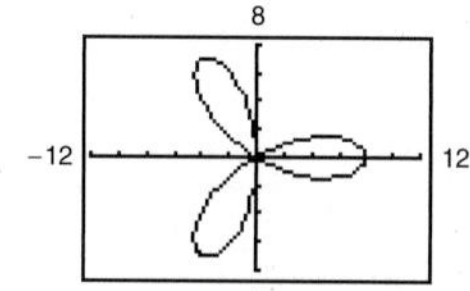

105. Hyperbola

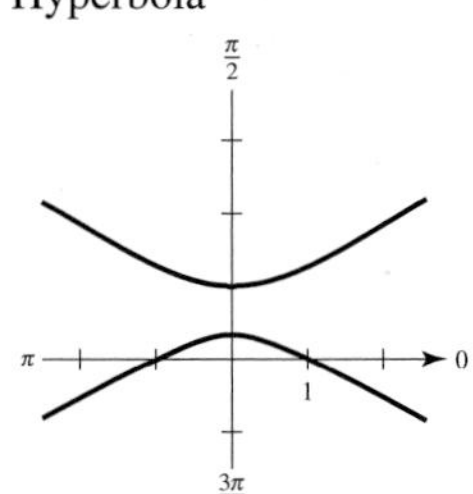

107. Ellipse

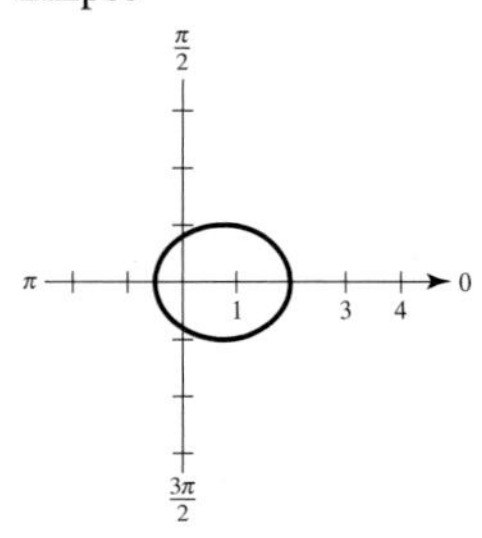

109. $r = \dfrac{4}{1 - \cos\theta}$ **111.** $r = \dfrac{5}{3 - 2\cos\theta}$

113. $r = \dfrac{7978.81}{1 - 0.937\cos\theta}$; 11,011.87 mi

115. False. The equation of a hyperbola is a second-degree equation.

117. False. $(2, \pi/4)$, $(-2, 5\pi/4)$, and $(2, 9\pi/4)$ all represent the same point.

119. (a) The graphs are the same.
(b) The graphs are the same.

Chapter Test (page 804)

1. Parabola: $y^2 = 2(x - 1)$
Vertex: $(1, 0)$
Focus: $\left(\frac{3}{2}, 0\right)$

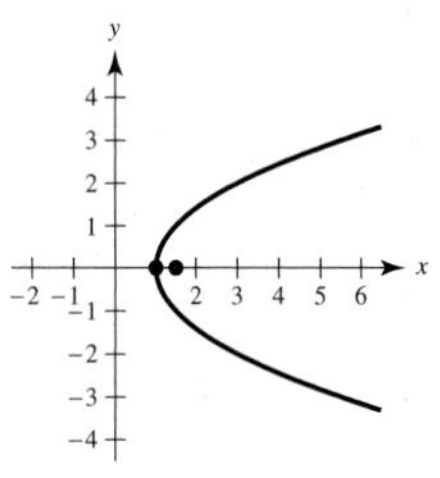

2. Hyperbola: $\dfrac{(x-2)^2}{4} - y^2 = 1$
Center: $(2, 0)$
Vertices: $(0, 0)$, $(4, 0)$
Foci: $\left(2 \pm \sqrt{5}, 0\right)$
Asymptotes: $y = \pm\dfrac{1}{2}(x - 2)$

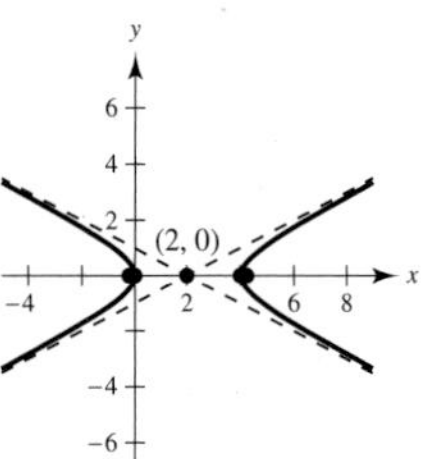

3. Ellipse: $\dfrac{(x+3)^2}{16} + \dfrac{(y-1)^2}{9} = 1$
Center: $(-3, 1)$
Vertices: $(1, 1)$, $(-7, 1)$
Foci: $\left(-3 \pm \sqrt{7}, 1\right)$

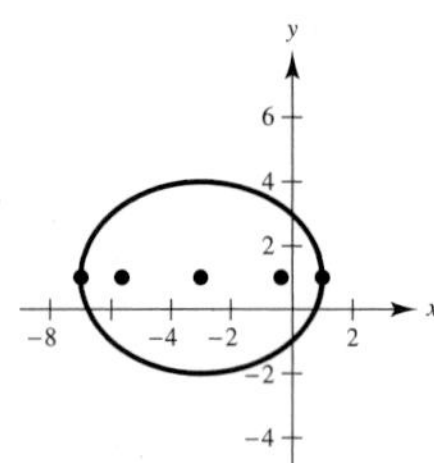

4. Circle: $(x - 2)^2 + (y - 1)^2 = \frac{1}{2}$
Center: $(2, 1)$

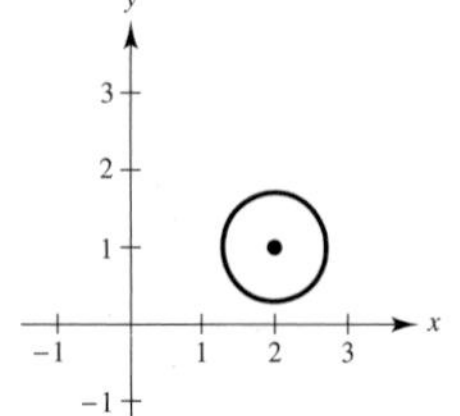

5. $(x - 2)^2 = \dfrac{4}{3}(y + 3)$ **6.** $\dfrac{5(y-2)^2}{4} - \dfrac{5x^2}{16} = 1$

7.

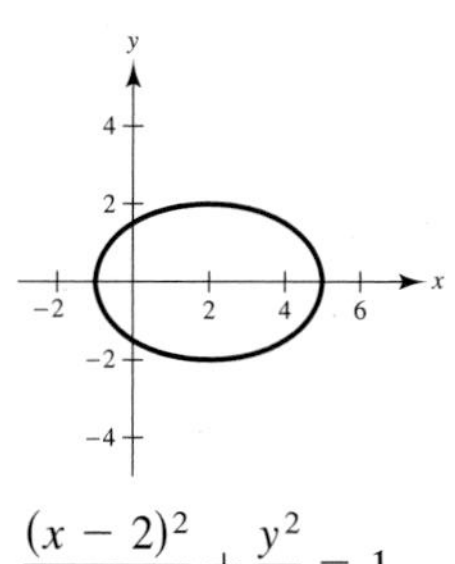

$$\frac{(x-2)^2}{9} + \frac{y^2}{4} = 1$$

8. $x = 6 + 4t$
$y = 4 + 7t$

9. $-\dfrac{16(x-5)}{9(y-1)}$ **10.** $4t - 5$ **11.** $y = x + 4$

12. $y = -\frac{2}{3}x - 1$ **13.** $\left(\sqrt{3}, -1\right)$

14. $\left(2\sqrt{2}, \dfrac{7\pi}{4}\right), \left(-2\sqrt{2}, \dfrac{3\pi}{4}\right), \left(2\sqrt{2}, -\dfrac{\pi}{4}\right)$ **15.** $r = 3\cos\theta$

16.

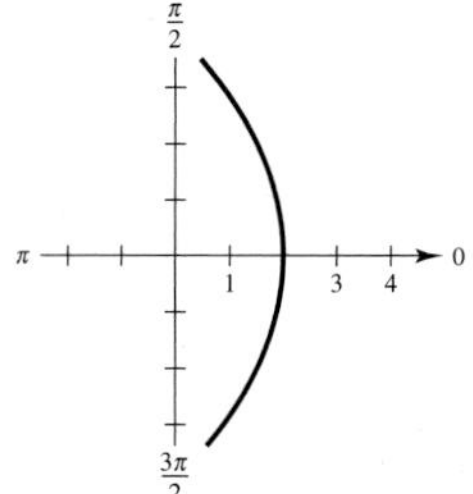

Parabola

17.

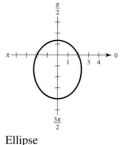

Ellipse

18.

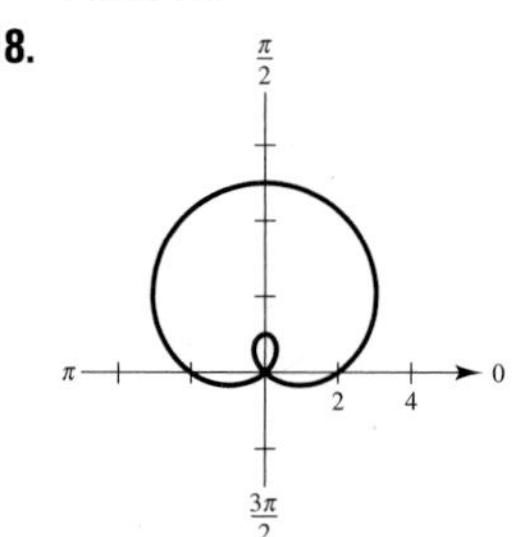

Limaçon with inner loop

19.

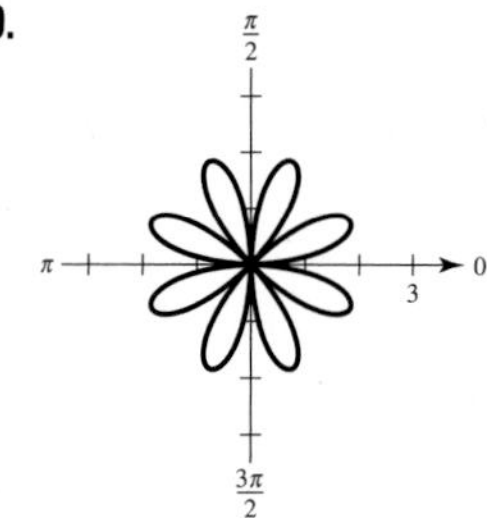

Rose curve

20. Answers will vary. For example: $r = \dfrac{1}{1 + 0.25\sin\theta}$

21. 10π square units **22.** No; Yes

P.S. Problem Solving (page 805)

1. $(14.83, 0)$

3. $y^2 = 4p(x + p)$ **5.** Answers will vary.

7. $\dfrac{(x-6)^2}{9} - \dfrac{(y-2)^2}{7} = 1$

9. (a) The first set of parametric equations models projectile motion along a straight line. The second set of parametric equations models projectile motion of an object launched at a height of h units above the ground that will eventually fall back to the ground.

(b) $y = (\tan\theta)x$; $y = h + x\tan\theta - \dfrac{16x^2\sec^2\theta}{v_0^2}$

(c) In the first case, the path of the moving object is not affected by a change in the velocity because eliminating the parameter removes v_0.

11. 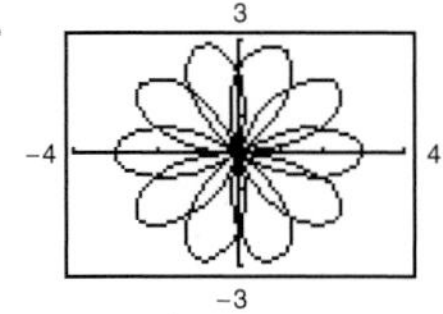 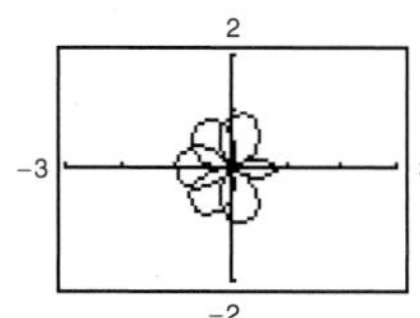

$r = 3\sin\left(\frac{5\theta}{2}\right)$ $r = -\cos(\sqrt{2}\theta),$
$-2\pi \le \theta \le 2\pi$

Sample answer: If n is a rational number, then the curve has a finite number of petals. If n is an irrational number, then the curve has an infinite number of petals.

13. (a) No. Because of the exponential, the graph will continue to trace the butterfly curve at larger values of r.
(b) $r \approx 4.1$. This value will increase if θ is increased.

15. 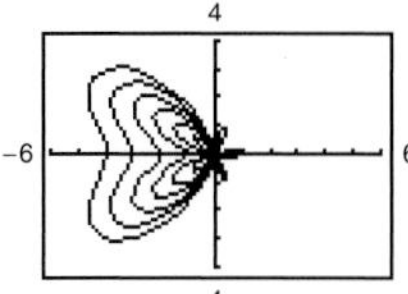 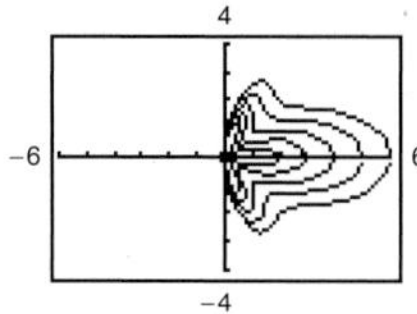

For $n \ge 1$, a bell is produced.
For $n \le -1$, a heart is produced.
For $n = 0$, a rose curve is produced.

Chapter 13

Section 13.1 (page 814)

1. oblique **3.** angles; side
5. $A = 30°, a \approx 14.14, c \approx 27.32$
7. $C = 120°, b \approx 4.75, c \approx 7.17$
9. $B = 60.9°, b \approx 19.32, c \approx 6.36$
11. $B = 42°4', a \approx 22.05, b \approx 14.88$
13. $C = 80°, a \approx 5.82, b \approx 9.20$
15. $C = 83°, a \approx 0.62, b \approx 0.51$
17. $B \approx 21.55°, C \approx 122.45°, c \approx 11.49$
19. $A \approx 10°11', C \approx 154°19', c \approx 11.03$
21. $B \approx 9.43°, C = 25.57°, c \approx 10.53$
23. $B \approx 18°13', C \approx 51°32', c \approx 40.06$
25. $B \approx 48.74°, C \approx 21.26°, c \approx 48.23$
27. No solution
29. Two solutions:
$B \approx 72.21°, C \approx 49.79°, c \approx 10.27$
$B \approx 107.79°, C \approx 14.21°, c \approx 3.30$
31. No solution **33.** $B = 45°, C = 90°, c \approx 1.41$
35. (a) $b \le 5,\ b = \frac{5}{\sin 36°}$ (b) $5 < b < \frac{5}{\sin 36°}$
(c) $b > \frac{5}{\sin 36°}$
37. (a) $b \le 10.8,\ b = \frac{10.8}{\sin 10°}$ (b) $10.8 < b < \frac{10.8}{\sin 10°}$
(c) $b > \frac{10.8}{\sin 10°}$
39. If ABC is a triangle with sides a, b, and c, then $a/\sin A = b/\sin B = c/\sin C$.
41. 10.4 **43.** 1675.2 **45.** 3204.5 **47.** 24.1 m
49. 16.1° **51.** 77 m
53. (a) 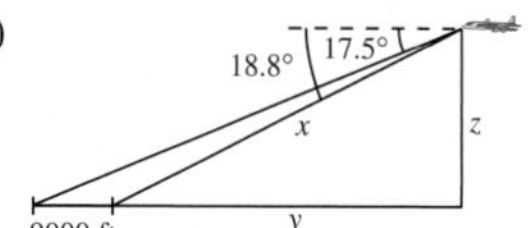

(b) 22.6 mi
(c) 21.4 mi
(d) 7.3 mi
55. 3.2 mi **57.** 5.86 mi
59. (a) 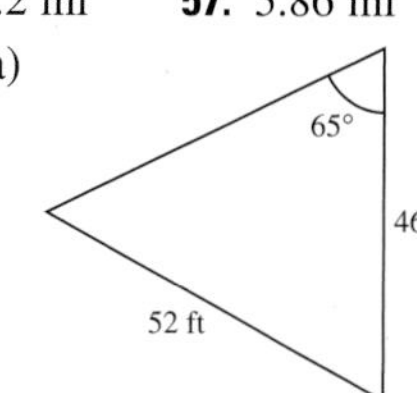

(b) 50.5 ft
(c) 22 bags
61. True
63. False. If just three angles are known, the triangle cannot be solved.
65. False. The area of an oblique triangle is Area $= \frac{1}{2}ab \sin C$, *not* $\frac{1}{2}ab \sin A$.

Section 13.2 (page 821)

1. Cosines **3.** $b^2 = a^2 + c^2 - 2ac\cos B$
5. $A \approx 38.62°, B \approx 48.51°, C \approx 92.87°$
7. $B \approx 23.79°, C \approx 126.21°, a \approx 18.59$
9. $A \approx 30.11°, B \approx 43.16°, C \approx 106.73°$
11. $A \approx 92.94°, B \approx 43.53°, C \approx 43.53°$
13. $B \approx 27.46°, C \approx 32.54°, a \approx 11.27$
15. $A \approx 141°45', C \approx 27°40', b \approx 11.87$
17. $A = 27°10', C = 27°10', b \approx 65.84$
19. $A \approx 33.80°, B \approx 103.20°, c \approx 0.54$
21. 43.52 **23.** 10.4 **25.** 52.11
27. If ABC is a triangle with sides a, b, and c, then
$a^2 = b^2 + c^2 - 2bc\cos A$
$b^2 = a^2 + c^2 - 2ac\cos B$
$c^2 = a^2 + b^2 - 2ab\cos C.$
29. 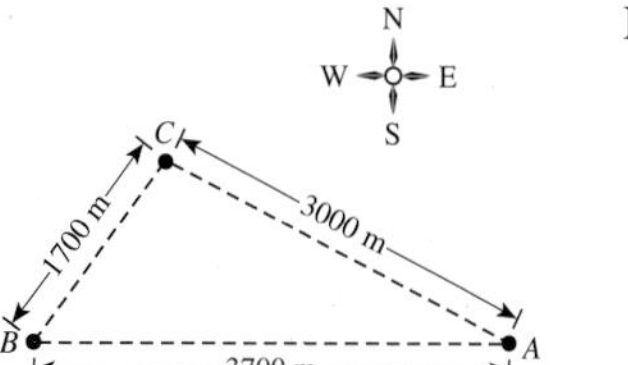

N 37.1° E, S 63.1° E
31. 373.3 m **33.** 72.3° **35.** 43.3 mi
37. (a) N 58.4° W (b) S 81.5° W **39.** 63.7 ft
41. (a) 24.2 mi (b) About 238 mi/h
43. $\overline{PQ} \approx 9.4, \overline{QS} = 5, \overline{RS} \approx 12.8$
45.

d (inches)	9	10	12	13	14
θ (degrees)	60.9°	69.5°	88.0°	98.2°	109.6°
s (inches)	20.88	20.28	18.99	18.28	17.48

d (inches)	15	16
θ (degrees)	122.9°	139.8°
s (inches)	16.55	15.37

47. 46,837.5 ft^2

49. False. For s to be the average of the lengths of the three sides of the triangle, s would be equal to $(a + b + c)/3$.

51. False. Assume $\cos C$ is negative. Then $\frac{\pi}{2} < C < \pi$ and C is obtuse.

Section 13.3 (page 832)

1. directed line segment **3.** magnitude
5. magnitude; direction **7.** unit vector **9.** resultant
11. $\mathbf{v} = \langle 1, 3 \rangle$, $\|\mathbf{v}\| = \sqrt{10}$ **13.** $\mathbf{v} = \langle 4, 6 \rangle$; $\|\mathbf{v}\| = 2\sqrt{13}$
15. $\mathbf{v} = \langle 0, 5 \rangle$; $\|\mathbf{v}\| = 5$ **17.** $\mathbf{v} = \langle 8, 6 \rangle$; $\|\mathbf{v}\| = 10$
19. $\mathbf{v} = \langle -9, -12 \rangle$; $\|\mathbf{v}\| = 15$ **21.** $\mathbf{v} = \langle 16, 7 \rangle$; $\|\mathbf{v}\| = \sqrt{305}$

23.

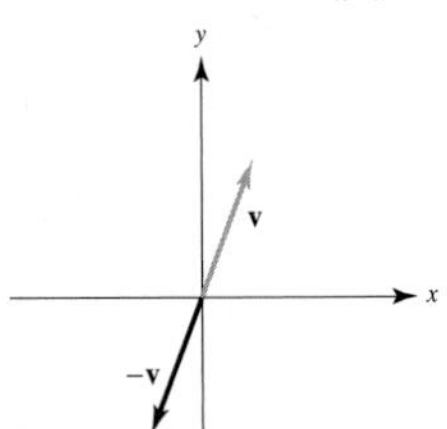

25.

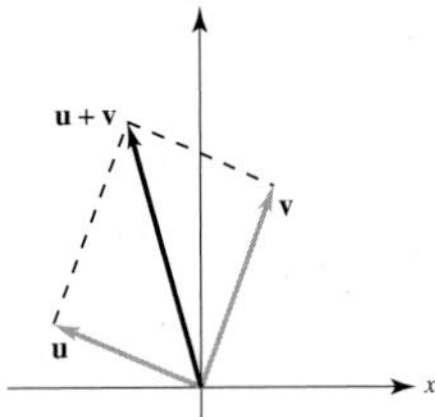

27.

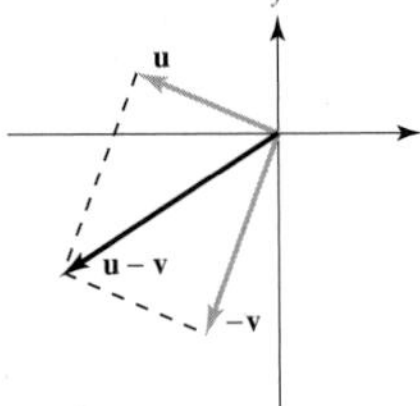

29. (a) $\langle 3, 4 \rangle$

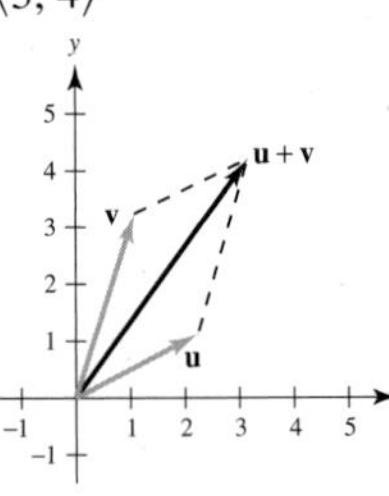

(b) $\langle 1, -2 \rangle$

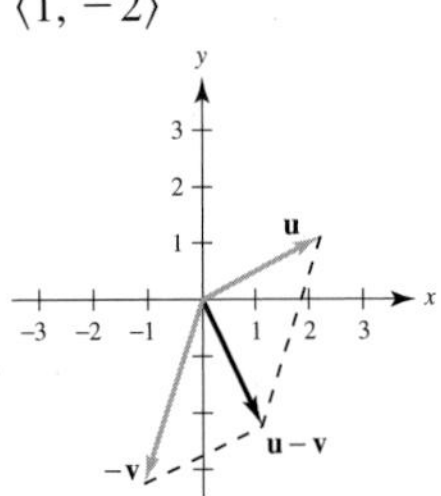

(c) $\langle 1, -7 \rangle$

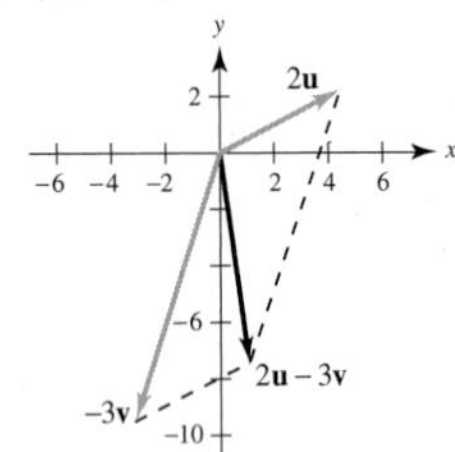

31. (a) $\langle -5, 3 \rangle$

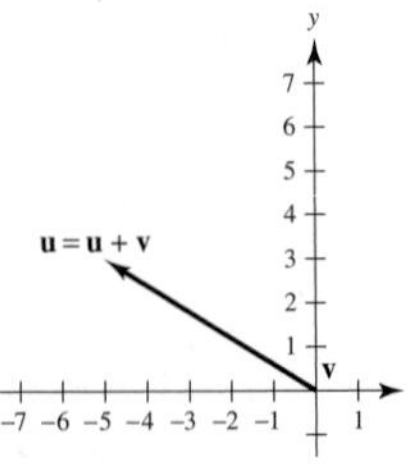

(b) $\langle -5, 3 \rangle$

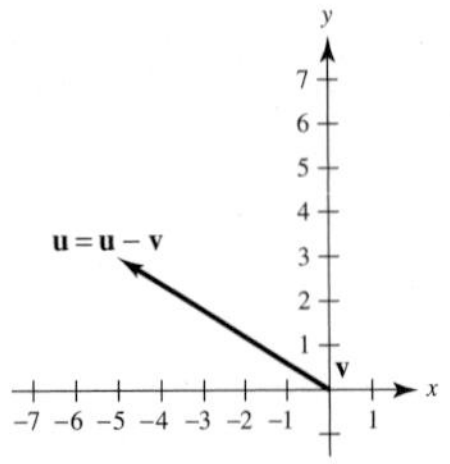

(c) $\langle -10, 6 \rangle$

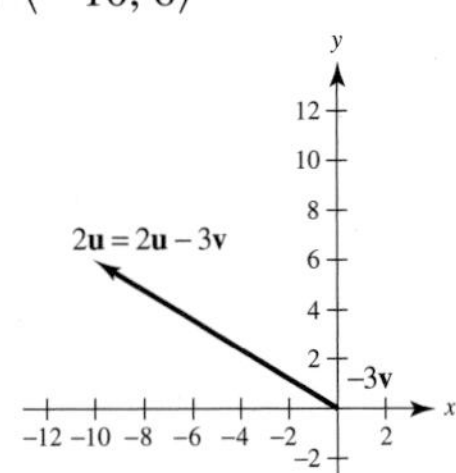

33. (a) $3\mathbf{i} - 2\mathbf{j}$

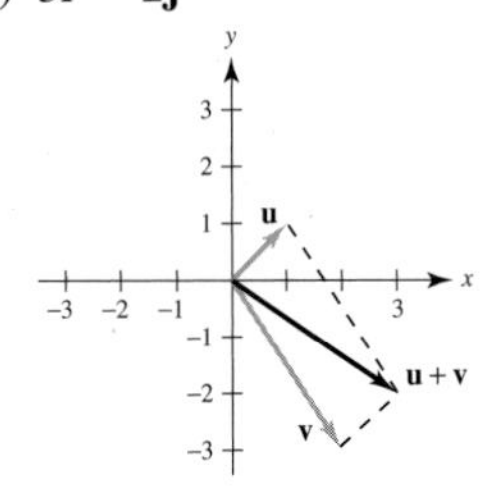

(b) $-\mathbf{i} + 4\mathbf{j}$

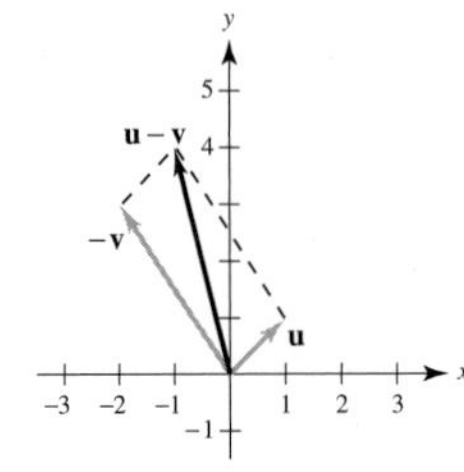

(c) $-4\mathbf{i} + 11\mathbf{j}$

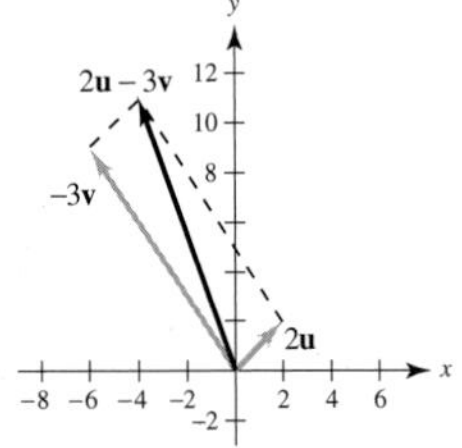

35. (a) $2\mathbf{i} + \mathbf{j}$

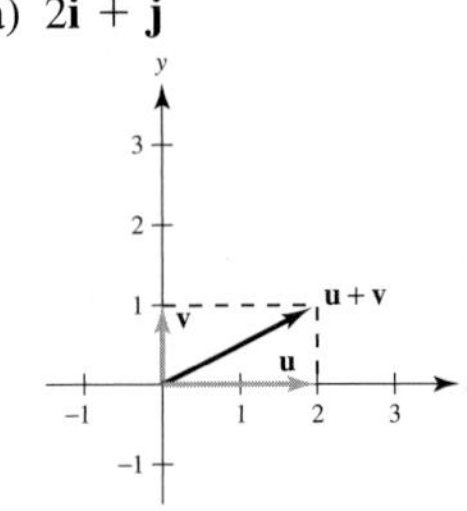

(b) $2\mathbf{i} - \mathbf{j}$

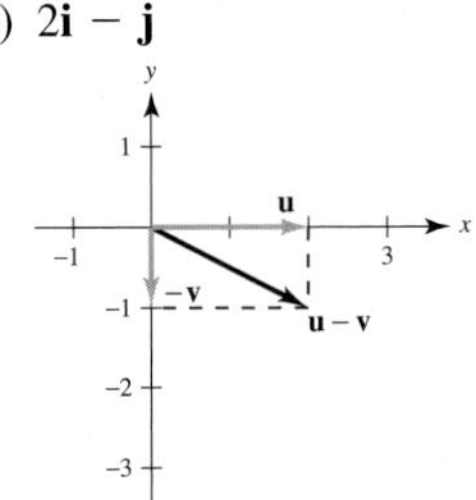

(c) $4\mathbf{i} - 3\mathbf{j}$

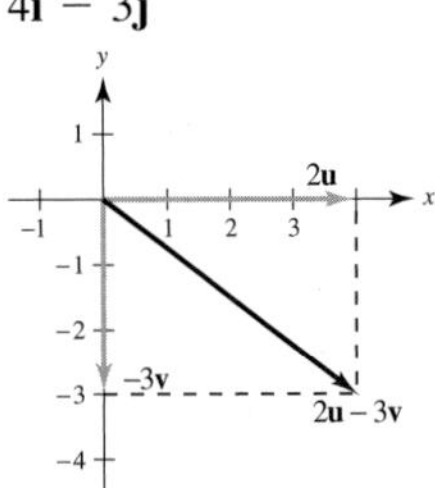

37. $\langle 1, 0 \rangle$ **39.** $\left\langle -\frac{\sqrt{2}}{2}, \frac{\sqrt{2}}{2} \right\rangle$ **41.** $\frac{\sqrt{2}}{2}\mathbf{i} + \frac{\sqrt{2}}{2}\mathbf{j}$

43. $\mathbf{j}$ **45.** $\frac{\sqrt{5}}{5}\mathbf{i} - \frac{2\sqrt{5}}{5}\mathbf{j}$ **47.** $\mathbf{v} = \langle -6, 8 \rangle$

49. $\mathbf{v} = \left\langle \frac{18\sqrt{29}}{29}, \frac{45\sqrt{29}}{29} \right\rangle$

51. $\mathbf{v} = \langle 3, -\frac{3}{2} \rangle$

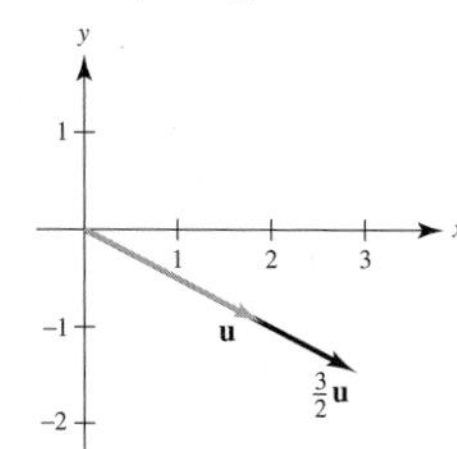

53. $\mathbf{v} = \langle 4, 3 \rangle$

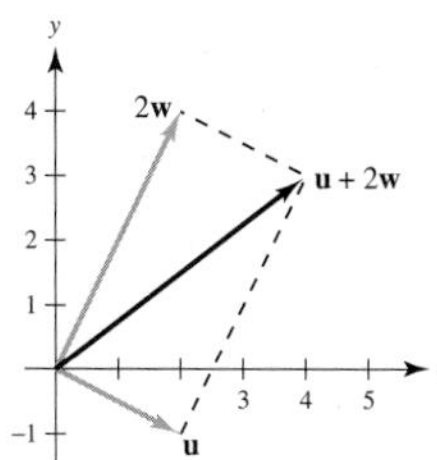

55. $\mathbf{v} = \langle \frac{7}{2}, -\frac{1}{2} \rangle$

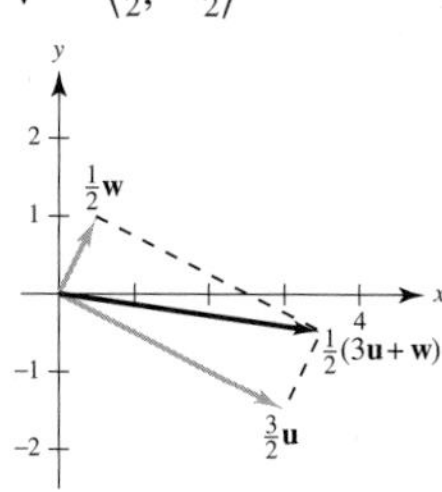

57. $\|\mathbf{v}\| = 6\sqrt{2};\ \theta = 315°$

59. $\|\mathbf{v}\| = 3;\ \theta = 60°$

61. $\mathbf{v} = \langle 3, 0 \rangle$

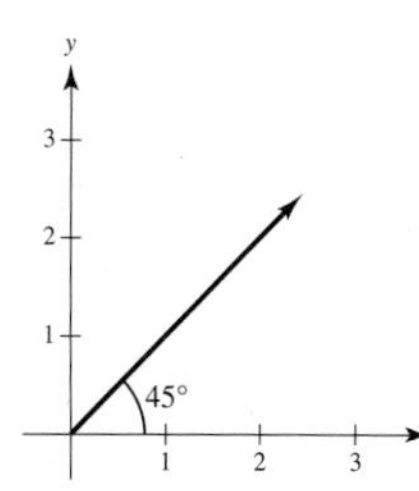

63. $\mathbf{v} = \left\langle -\frac{7\sqrt{3}}{4}, \frac{7}{4} \right\rangle$

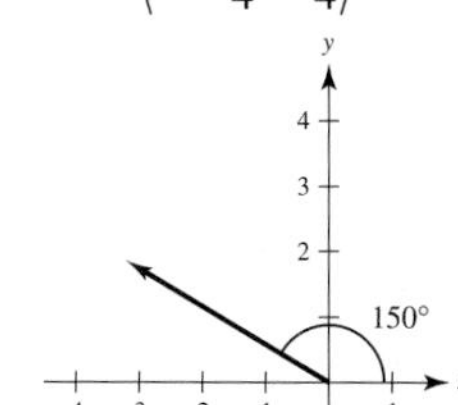

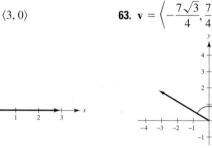

65. $\mathbf{v} = \langle \sqrt{6}, \sqrt{6} \rangle$

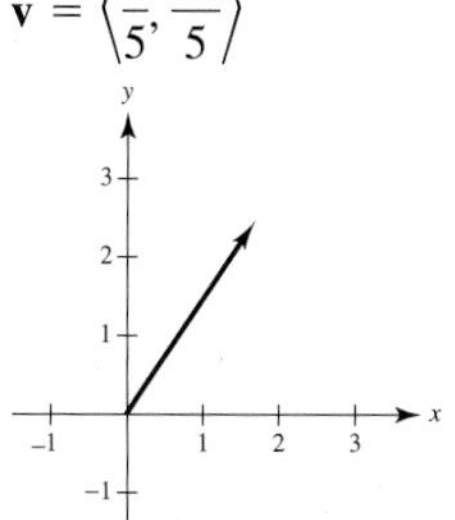

67. $\mathbf{v} = \left\langle \frac{9}{5}, \frac{12}{5} \right\rangle$

69. $\langle 5, 5 \rangle$ **71.** $\langle 10\sqrt{2} - 50, 10\sqrt{2} \rangle$ **73.** 90° **75.** 63.4°

77. Same magnitude and direction; A and C

79. 62.7°

81. Vertical $\approx$ 125.4 ft/sec, Horizontal $\approx$ 1193.4 ft/sec

83. 12.8°; 398.32 N **85.** 71.3°; 228.5 lb

87. $T_{AC} \approx 1758.8$ lb; $T_{BC} \approx 1305.4$ lb

89. 3154.4 lb **91.** 1928.4 ft-lb **93.** N 21.4° E; 138.7 km/h

95. True **97.** Proof **99.** Answers will vary.

101. (a) Vector. Velocity has both magnitude and direction.
(b) Scalar. Price has only magnitude.
(c) Scalar. Temperature has only magnitude.
(d) Vector. Weight has both magnitude and direction.

103. $\langle 1, 3 \rangle$ or $\langle -1, -3 \rangle$

Section 13.4 (page 842)

1. dot product **3.** $\dfrac{\mathbf{u} \cdot \mathbf{v}}{\|\mathbf{u}\|\,\|\mathbf{v}\|}$ **5.** $\left(\dfrac{\mathbf{u} \cdot \mathbf{v}}{\|\mathbf{v}\|^2}\right)\mathbf{v}$

7. -19 **9.** 6 **11.** 18; scalar

13. $\langle 24, -12 \rangle$; vector **15.** 17 **17.** $5\sqrt{41}$ **19.** 6

21. 90° **23.** 143.13° **25.** 60.26° **27.** 90° **29.** $\dfrac{5\pi}{12}$

31. 26.57°, 63.43°, 90° **33.** 41.63°, 53.13°, 85.24°

35. -20 **37.** -229.1 **39.** Parallel **41.** Neither

43. Orthogonal **45.** $\frac{1}{37}\langle 84, 14 \rangle, \frac{1}{37}\langle -10, 60 \rangle$

47. $\frac{45}{229}\langle 2, 15 \rangle, \frac{6}{229}\langle -15, 2 \rangle$ **49.** $\langle -5, 3 \rangle, \langle 5, -3 \rangle$

51. $\frac{2}{3}\mathbf{i} + \frac{1}{2}\mathbf{j}, -\frac{2}{3}\mathbf{i} - \frac{1}{2}\mathbf{j}$ **53.** 32

55. The dot product equals the product of the lengths of two vectors when the angle θ between the vectors is 0.

57. (a) \$892,901.50
This value gives the total revenue that can be earned by selling all of the cellular phones.
(b) $1.05\mathbf{v}$

59. 735 N-m **61.** 779.4 ft-lb

63. False. Work is represented by a scalar.

65. False. The dot product of two vectors is a scalar that can be positive, zero, or negative.

67. (a)–(c) Proofs

Section 13.5 (page 852)

1. absolute value **3.** DeMoivre's

5.

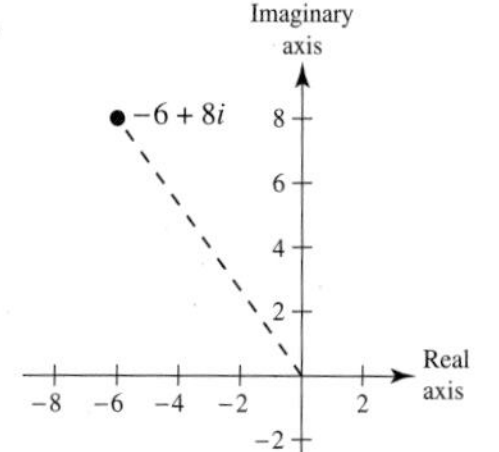

10

7.

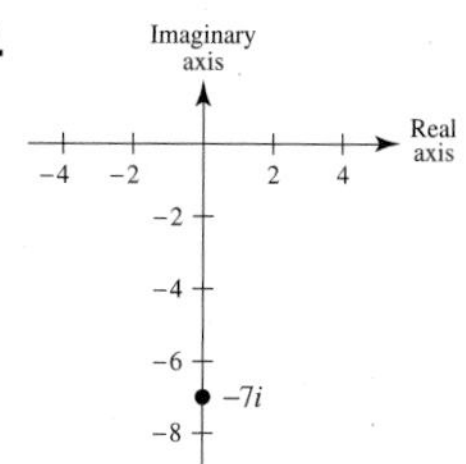

7

9.

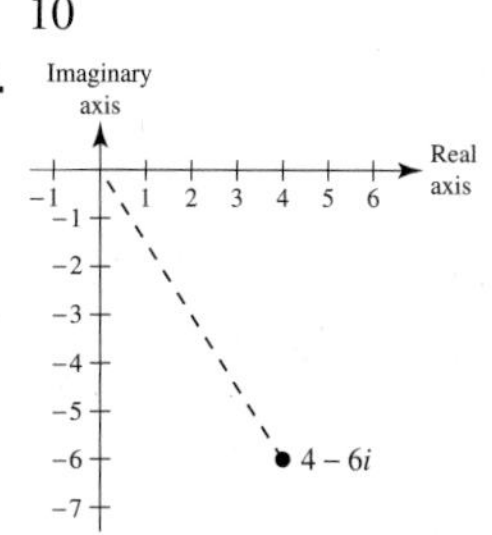

$2\sqrt{13}$

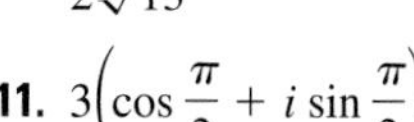

11. $3\left(\cos\frac{\pi}{2} + i\sin\frac{\pi}{2}\right)$

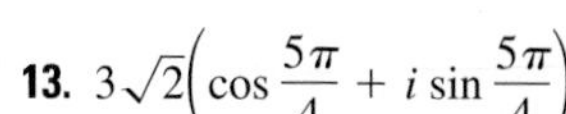

13. $3\sqrt{2}\left(\cos\frac{5\pi}{4} + i\sin\frac{5\pi}{4}\right)$

15. 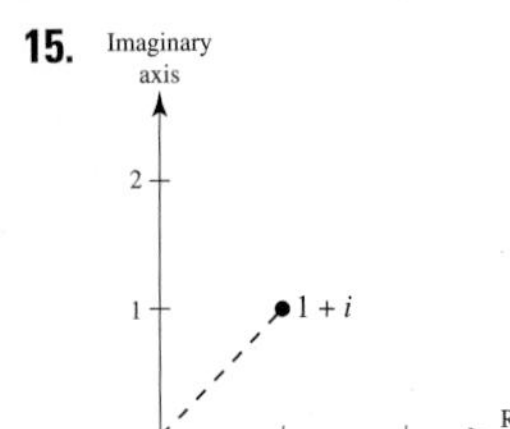

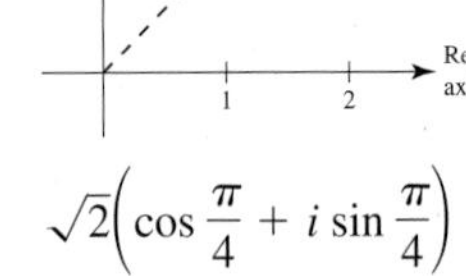
$\sqrt{2}\left(\cos\frac{\pi}{4} + i\sin\frac{\pi}{4}\right)$

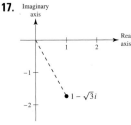

17.
$2\left(\cos\frac{5\pi}{3} + i\sin\frac{5\pi}{3}\right)$

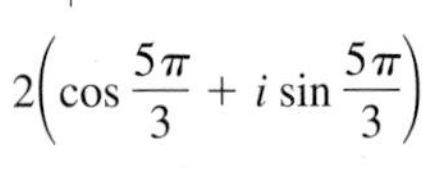

19. 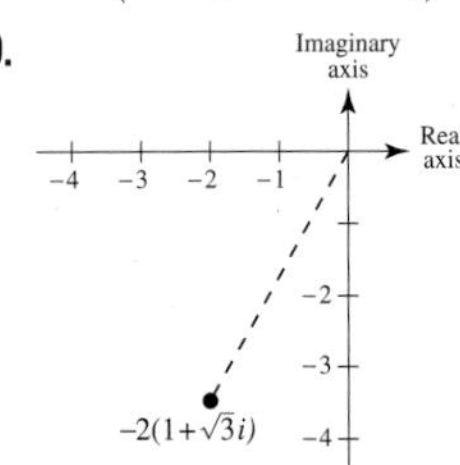

$4\left(\cos\frac{4\pi}{3} + i\sin\frac{4\pi}{3}\right)$

21. 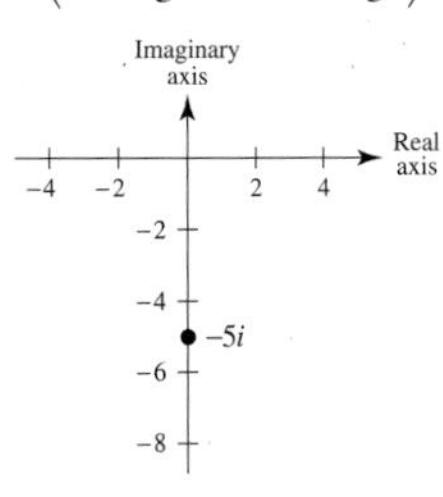

$5\left(\cos\frac{3\pi}{2} + i\sin\frac{3\pi}{2}\right)$

23. 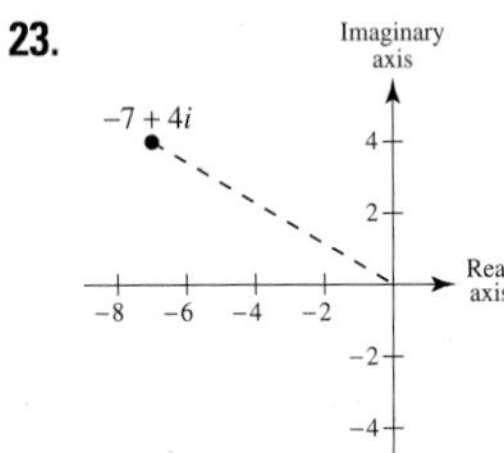

$\sqrt{65}(\cos 2.62 + i\sin 2.62)$

25. 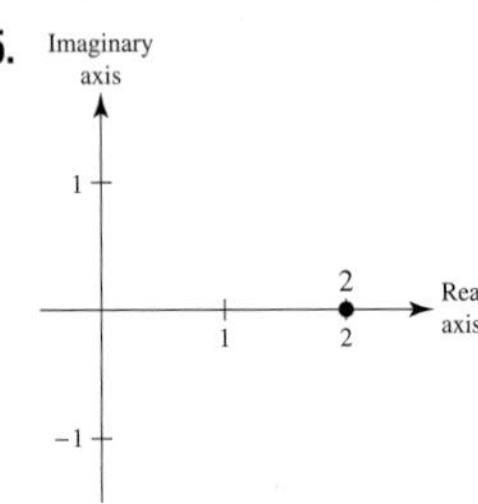

$2(\cos 0 + i\sin 0)$

27.

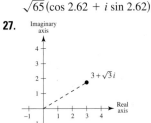

$2\sqrt{3}(\cos \pi/6 + i\sin \pi/6)$

29. 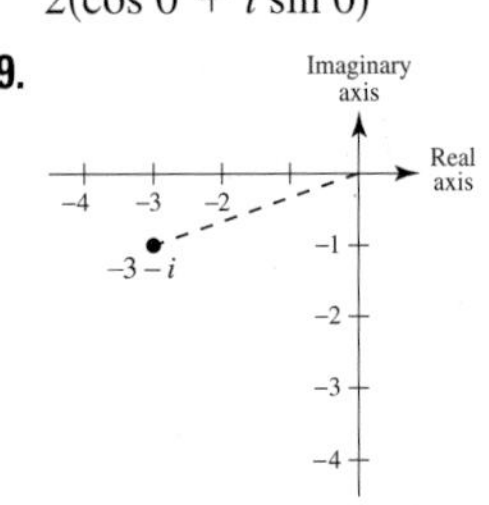

$\sqrt{10}(\cos 3.46 + i\sin 3.46)$

31. 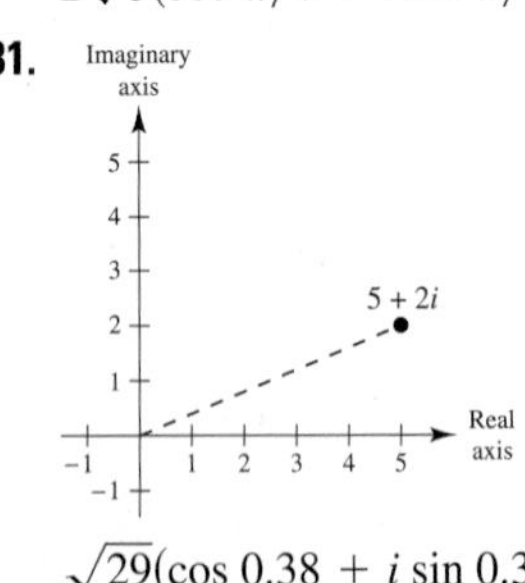

$\sqrt{29}(\cos 0.38 + i\sin 0.38)$

33. 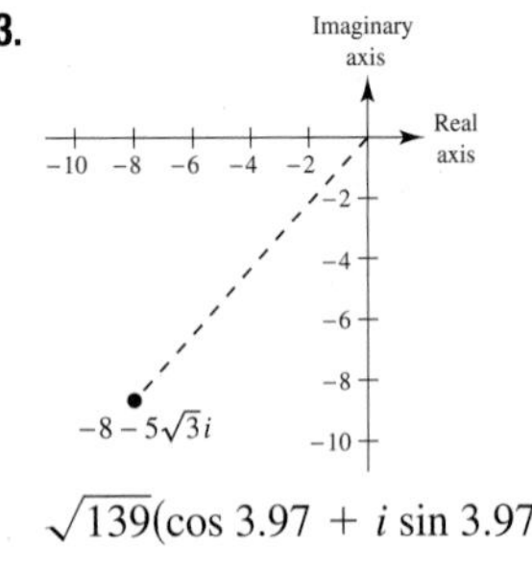

$\sqrt{139}(\cos 3.97 + i\sin 3.97)$

35. 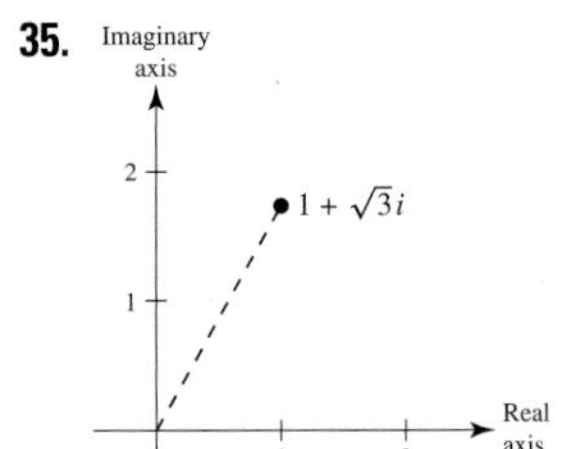

$1 + \sqrt{3}i$

37. 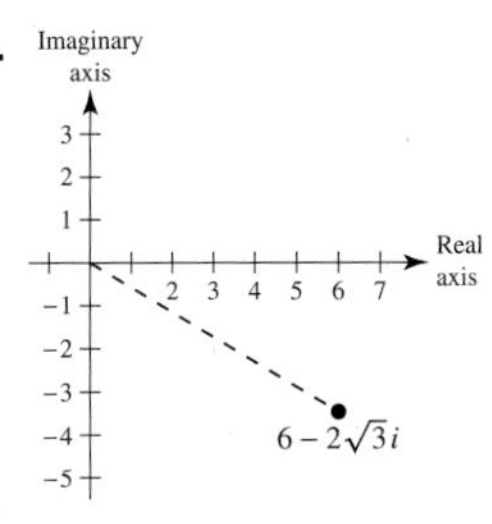

$6 - 2\sqrt{3}i$

39. 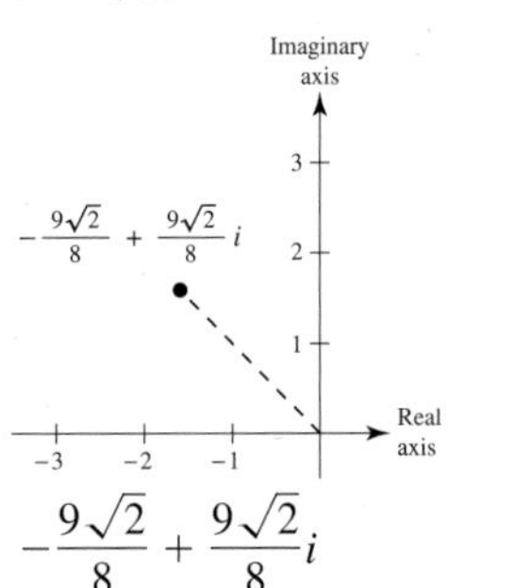

$-\frac{9\sqrt{2}}{8} + \frac{9\sqrt{2}}{8}i$

41. 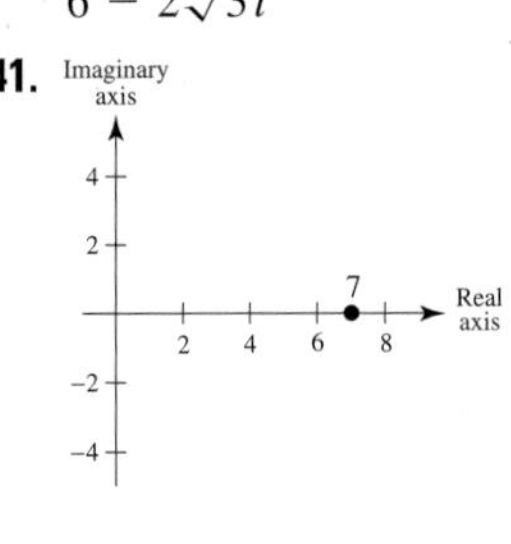

7

43.

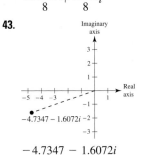

$-4.7347 - 1.6072i$

45. $4.6985 + 1.7101i$ **47.** $-1.8126 + 0.8452i$

49. $12\left(\cos\frac{\pi}{3} + i\sin\frac{\pi}{3}\right)$ **51.** $\frac{10}{9}(\cos 150° + i\sin 150°)$

53. $\cos 50° + i\sin 50°$ **55.** $\frac{1}{3}(\cos 30° + i\sin 30°)$

57. $\cos\frac{2\pi}{3} + i\sin\frac{2\pi}{3}$ **59.** $6(\cos 330° + i\sin 330°)$

61. (a) $\left[2\sqrt{2}\left(\cos\frac{\pi}{4} + i\sin\frac{\pi}{4}\right)\right]\left[\sqrt{2}\left(\cos\frac{7\pi}{4} + i\sin\frac{7\pi}{4}\right)\right]$
(b) $4(\cos 0 + i\sin 0) = 4$ (c) 4

63. (a) $\left[2\left(\cos\frac{3\pi}{2} + i\sin\frac{3\pi}{2}\right)\right]\left[\sqrt{2}\left(\cos\frac{\pi}{4} + i\sin\frac{\pi}{4}\right)\right]$
(b) $2\sqrt{2}\left(\cos\frac{7\pi}{4} + i\sin\frac{7\pi}{4}\right) = 2 - 2i$
(c) $-2i - 2i^2 = -2i + 2 = 2 - 2i$

65. (a) $[5(\cos 0.93 + i\sin 0.93)] \div \left[2\left(\cos\frac{5\pi}{3} + i\sin\frac{5\pi}{3}\right)\right]$
(b) $\frac{5}{2}(\cos 1.97 + i\sin 1.97) \approx -0.982 + 2.299i$
(c) About $-0.982 + 2.299i$

67. (a) $[5(\cos 0 + i\sin 0)] \div \left[\sqrt{13}(\cos 0.98 + i\sin 0.98)\right]$
(b) $(5\sqrt{13}/3)(\cos 5.30 + i\sin 5.30) \approx 0.769 - 1.154i$
(c) $\frac{10}{13} - \frac{15}{13}i \approx 0.769 - 1.154i$

69.

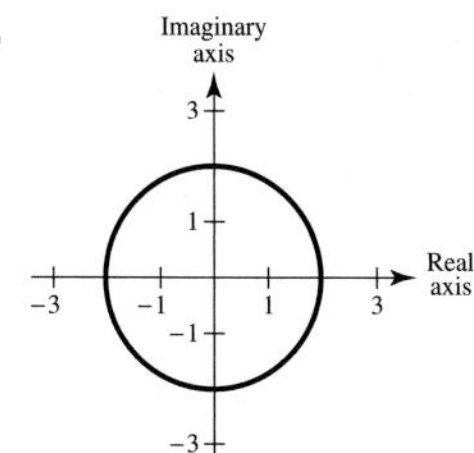

71.

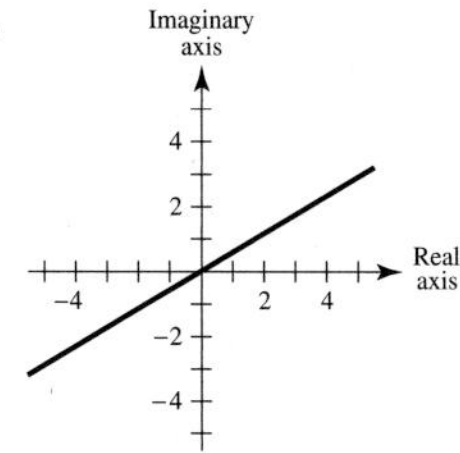
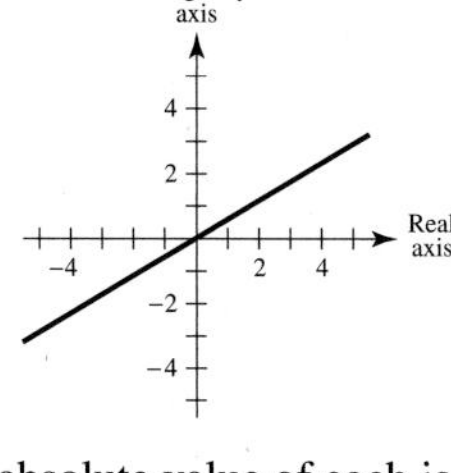

73.

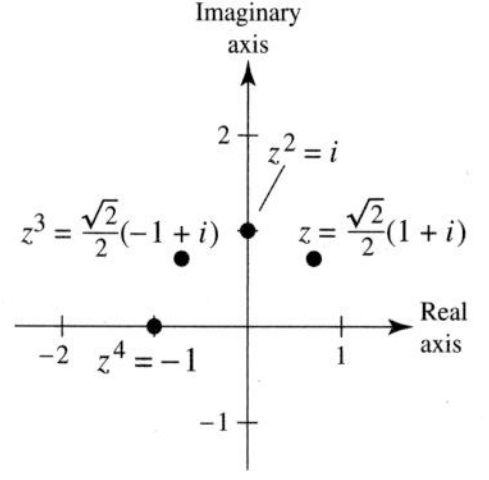

The absolute value of each is 1, and the consecutive powers of z are each 45° apart.

75. $-4 - 4i$ **77.** $8i$ **79.** $1024 - 1024\sqrt{3}i$

81. $\frac{125}{2} + \frac{125\sqrt{3}}{2}i$ **83.** -1 **85.** $608.0 + 144.7i$

87. $-597 - 122i$ **89.** $\frac{81}{2} + \frac{81\sqrt{3}}{2}i$ **91.** $32i$

93. (a) $\sqrt{5}(\cos 60° + i \sin 60°)$
$\sqrt{5}(\cos 240° + i \sin 240°)$

(b)

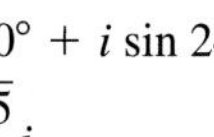
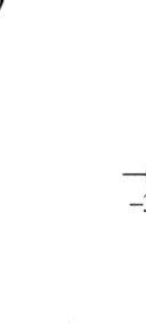

(c) $\frac{\sqrt{5}}{2} + \frac{\sqrt{15}}{2}i$,
$-\frac{\sqrt{5}}{2} - \frac{\sqrt{15}}{2}i$

95. (a) $2\left(\cos \frac{2\pi}{9} + i \sin \frac{2\pi}{9}\right)$
$2\left(\cos \frac{8\pi}{9} + i \sin \frac{8\pi}{9}\right)$
$2\left(\cos \frac{14\pi}{9} + i \sin \frac{14\pi}{9}\right)$

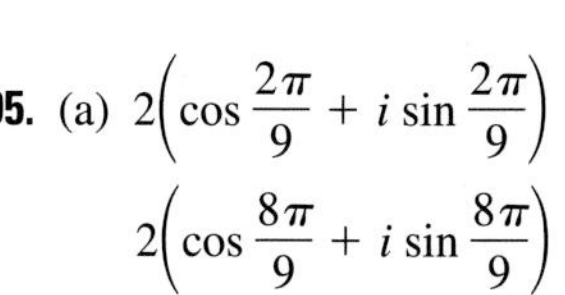

(b)

(c) $1.5321 + 1.2856i$,
$-1.8794 + 0.6840i$,
$0.3473 - 1.9696i$

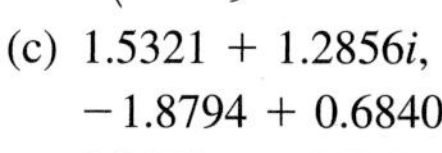

97. (a) $5\left(\cos \frac{4\pi}{9} + i \sin \frac{4\pi}{9}\right)$
$5\left(\cos \frac{10\pi}{9} + i \sin \frac{10\pi}{9}\right)$
$5\left(\cos \frac{16\pi}{9} + i \sin \frac{16\pi}{9}\right)$

(b)

(c) $0.8682 + 4.9240i$,
$-4.6985 - 1.7101i$,
$3.8302 - 3.2140i$

99. (a) $5\left(\cos \frac{3\pi}{4} + i \sin \frac{3\pi}{4}\right)$
$5\left(\cos \frac{7\pi}{4} + i \sin \frac{7\pi}{4}\right)$

(b)

(c) $-\frac{5\sqrt{2}}{2} + \frac{5\sqrt{2}}{2}i$,
$\frac{5\sqrt{2}}{2} - \frac{5\sqrt{2}}{2}i$

101. (a) $2(\cos 0 + i \sin 0)$
$2\left(\cos \frac{\pi}{2} + i \sin \frac{\pi}{2}\right)$
$2(\cos \pi + i \sin \pi)$
$2\left(\cos \frac{3\pi}{2} + i \sin \frac{3\pi}{2}\right)$

(b)

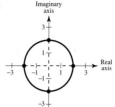

(c) $2, 2i, -2, -2i$

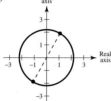

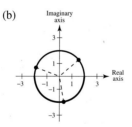

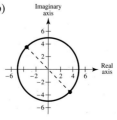

103. (a) $\cos 0 + i \sin 0$
$\cos \frac{2\pi}{5} + i \sin \frac{2\pi}{5}$
$\cos \frac{4\pi}{5} + i \sin \frac{4\pi}{5}$
$\cos \frac{6\pi}{5} + i \sin \frac{6\pi}{5}$
$\cos \frac{8\pi}{5} + i \sin \frac{8\pi}{5}$

(b)

(c) $1, 0.3090 + 0.9511i$,
$-0.8090 + 0.5878i, -0.8090 - 0.5878i$,
$0.3090 - 0.9511i$

105. (a) $5\left(\cos \frac{\pi}{3} + i \sin \frac{\pi}{3}\right)$
$5(\cos \pi + i \sin \pi)$
$5\left(\cos \frac{5\pi}{3} + i \sin \frac{5\pi}{3}\right)$

(b)

(c) $\frac{5}{2} + \frac{5\sqrt{3}}{2}i, -5$,
$\frac{5}{2} - \frac{5\sqrt{3}}{2}i$

107. (a) $\sqrt{2}\left(\cos \frac{7\pi}{20} + i \sin \frac{7\pi}{20}\right)$
$\sqrt{2}\left(\cos \frac{3\pi}{4} + i \sin \frac{3\pi}{4}\right)$
$\sqrt{2}\left(\cos \frac{23\pi}{20} + i \sin \frac{23\pi}{20}\right)$
$\sqrt{2}\left(\cos \frac{31\pi}{20} + i \sin \frac{31\pi}{20}\right)$
$\sqrt{2}\left(\cos \frac{39\pi}{20} + i \sin \frac{39\pi}{20}\right)$

(b)

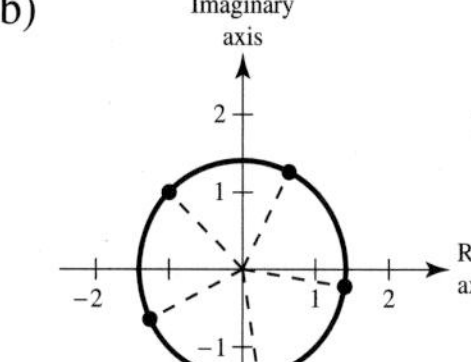

(c) $0.6420 + 1.2601i$,
$-1 + i$,
$-1.2601 - 0.6420i$,
$0.2212 - 1.3968i$,
$1.3968 - 0.2212i$

109. Three roots are not shown.

111. $\cos\frac{3\pi}{8} + i\sin\frac{3\pi}{8}$

$\cos\frac{7\pi}{8} + i\sin\frac{7\pi}{8}$

$\cos\frac{11\pi}{8} + i\sin\frac{11\pi}{8}$

$\cos\frac{15\pi}{8} + i\sin\frac{15\pi}{8}$

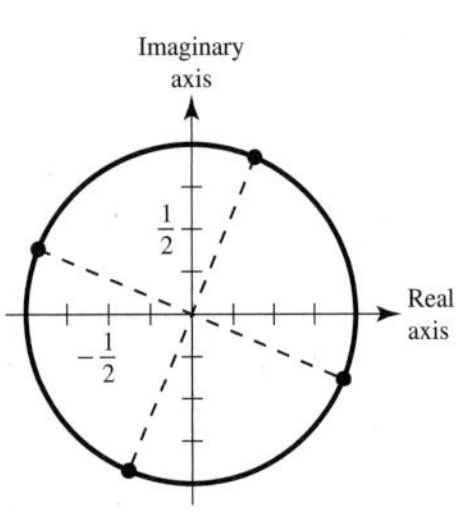

113. $3\left(\cos\frac{\pi}{5} + i\sin\frac{\pi}{5}\right)$

$3\left(\cos\frac{3\pi}{5} + i\sin\frac{3\pi}{5}\right)$

$3(\cos\pi + i\sin\pi)$

$3\left(\cos\frac{7\pi}{5} + i\sin\frac{7\pi}{5}\right)$

$3\left(\cos\frac{9\pi}{5} + i\sin\frac{9\pi}{5}\right)$

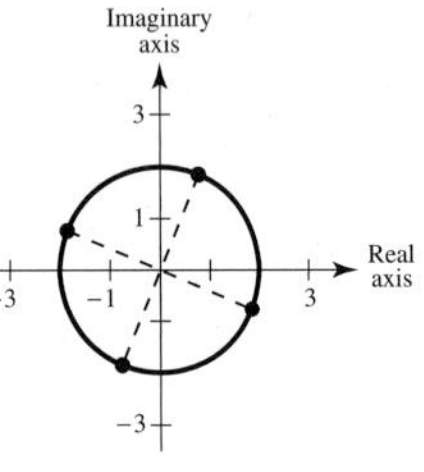

115. $2\left(\cos\frac{3\pi}{8} + i\sin\frac{3\pi}{8}\right)$

$2\left(\cos\frac{7\pi}{8} + i\sin\frac{7\pi}{8}\right)$

$2\left(\cos\frac{11\pi}{8} + i\sin\frac{11\pi}{8}\right)$

$2\left(\cos\frac{15\pi}{8} + i\sin\frac{15\pi}{8}\right)$

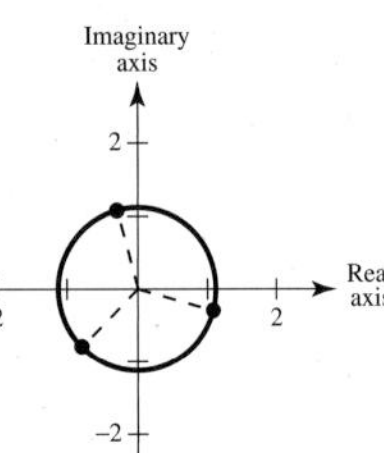

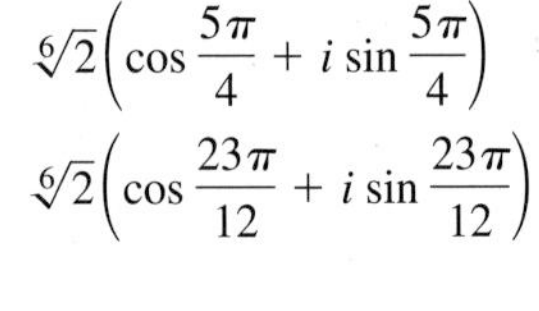

117. $\sqrt[6]{2}\left(\cos\frac{7\pi}{12} + i\sin\frac{7\pi}{12}\right)$

$\sqrt[6]{2}\left(\cos\frac{5\pi}{4} + i\sin\frac{5\pi}{4}\right)$

$\sqrt[6]{2}\left(\cos\frac{23\pi}{12} + i\sin\frac{23\pi}{12}\right)$

119. True **121.** True

123. False. By DeMoivre's Theorem,
$(2 - 2\sqrt{3}i)^3 = 64[\cos(-\pi) + i\sin(-\pi)]$
$= 64(\cos\pi - i\sin\pi)$.

125. Answers will vary. **127.** (a) r^2 (b) $\cos 2\theta + i\sin 2\theta$

129. Answers will vary.

Review Exercises (page 855)

1. $C = 72°, b \approx 12.21, c \approx 12.36$

3. $A = 26°, a \approx 24.89, c \approx 56.23$

5. $C = 66°, a \approx 2.53, b \approx 9.11$

7. $B = 108°, a \approx 11.76, c \approx 21.49$

9. $A \approx 20.41°, C \approx 9.59°, a \approx 20.92$

11. $B \approx 39.48°, C \approx 65.52°, c \approx 48.24$

13. 19.06 **15.** 47.23 **17.** 31.1 m **19.** 31.01 ft

21. $A \approx 27.81°, B \approx 54.75°, C \approx 97.44°$

23. $A \approx 16.99°, B \approx 26.00°, C \approx 137.01°$

25. $A \approx 29.92°, B \approx 86.18°, C \approx 63.90°$

27. $A = 36°, C = 36°, b \approx 17.80$

29. $A \approx 45.76°, B \approx 91.24°, c \approx 21.42$

31. Law of Sines; $A \approx 77.52°, B \approx 38.48°, a \approx 14.12$

33. Law of Cosines; $A \approx 28.62°, B \approx 33.56°, C \approx 117.82°$

35. 7.64 **37.** 8.36 **39.** 615.1 m

41.

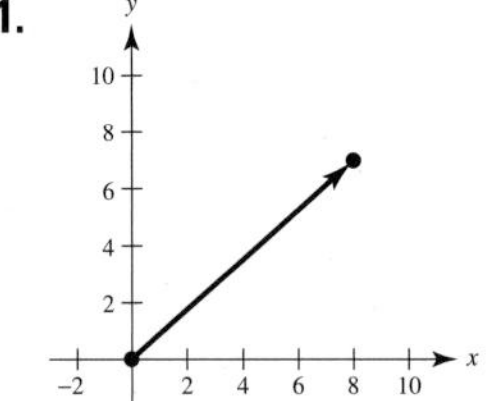

43.

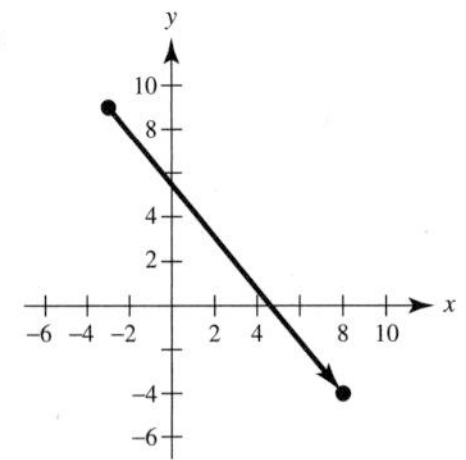

45. $\langle 7, -5\rangle$ **47.** $\langle 7, -7\rangle$ **49.** $\langle -4, 4\sqrt{3}\rangle$

51. (a) $\langle -4, 3\rangle$ (b) $\langle 2, -9\rangle$
(c) $\langle -4, -12\rangle$ (d) $\langle -14, 3\rangle$

53. (a) $\langle -1, 6\rangle$ (b) $\langle -9, -2\rangle$
(c) $\langle -20, 8\rangle$ (d) $\langle -13, 22\rangle$

55. (a) $7\mathbf{i} + 2\mathbf{j}$ (b) $-3\mathbf{i} - 4\mathbf{j}$ (c) $8\mathbf{i} - 4\mathbf{j}$ (d) $25\mathbf{i} + 4\mathbf{j}$

57. (a) $3\mathbf{i} + 6\mathbf{j}$ (b) $5\mathbf{i} - 6\mathbf{j}$ (c) $16\mathbf{i}$ (d) $17\mathbf{i} + 18\mathbf{j}$

59. $\langle 22, -7\rangle$ **61.** $\langle 30, 9\rangle$

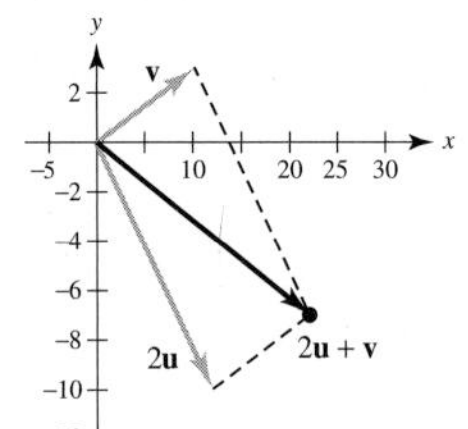

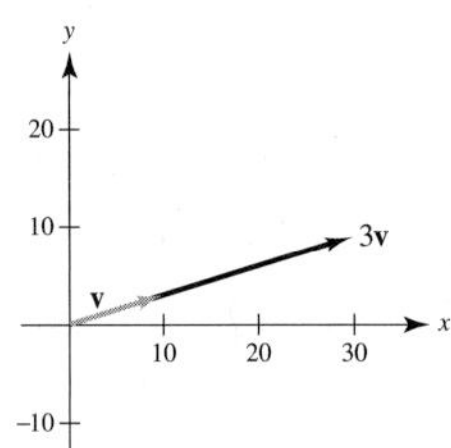

63. $-\mathbf{i} + 5\mathbf{j}$ **65.** $6\mathbf{i} + 4\mathbf{j}$

67. $10\sqrt{2}(\cos 135°\mathbf{i} + \sin 135°\mathbf{j})$ **69.** $\|\mathbf{v}\| = 7$; $\theta = 60°$

71. $\|\mathbf{v}\| = \sqrt{41}$; $\theta = 38.7°$ **73.** $\|\mathbf{v}\| = 3\sqrt{2}$; $\theta = 225°$

75. The resultant force is 133.92 pounds and 5.6° from the 85-pound force.

77. 422.30 mi/h; 130.4° **79.** 45 **81.** −2 **83.** 40; scalar

85. $\langle 72, -36\rangle$; vector **87.** $\frac{11\pi}{12}$

89. 160.5° **91.** Orthogonal **93.** Neither

95. $-\frac{13}{17}\langle 4, 1\rangle, \frac{16}{17}\langle -1, 4\rangle$ **97.** $\frac{5}{2}\langle -1, 1\rangle, \frac{9}{2}\langle 1, 1\rangle$ **99.** 48

101.

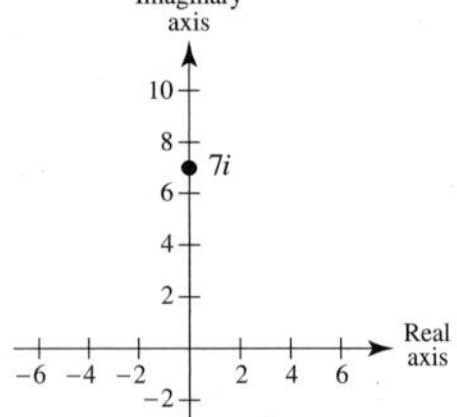

7

103.

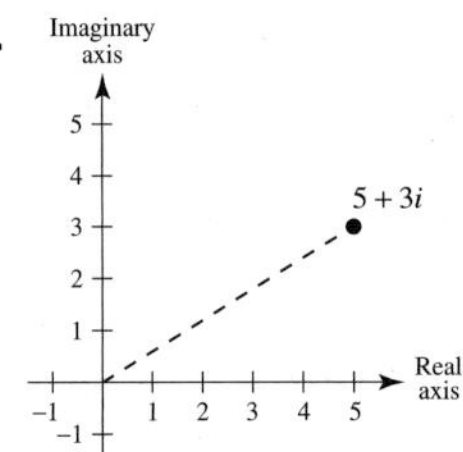

$\sqrt{34}$

105. $4\left(\cos\frac{\pi}{2} + i\sin\frac{\pi}{2}\right)$ **107.** $5\sqrt{2}\left(\cos\frac{7\pi}{4} + i\sin\frac{7\pi}{4}\right)$

109. $13(\cos 4.32 + i\sin 4.32)$

111. (a) $z_1 = 4\left(\cos\frac{11\pi}{6} + i\sin\frac{11\pi}{6}\right)$

$z_2 = 10\left(\cos\frac{3\pi}{2} + i\sin\frac{3\pi}{2}\right)$

(b) $z_1z_2 = 40\left(\cos\frac{10\pi}{3} + i\sin\frac{10\pi}{3}\right)$

$\frac{z_1}{z_2} = \frac{2}{5}\left(\cos\frac{\pi}{3} + i\sin\frac{\pi}{3}\right)$

113. $\frac{625}{2} + \frac{625\sqrt{3}}{2}i$ **115.** $2035 - 828i$

117. (a) $4(\cos 60° + i\sin 60°)$ (b) -64

$4(\cos 180° + i\sin 180°)$

$4(\cos 300° + i\sin 300°)$

119. (a) $3\left(\cos\frac{\pi}{4} + i\sin\frac{\pi}{4}\right)$ (b)

$3\left(\cos\frac{7\pi}{12} + i\sin\frac{7\pi}{12}\right)$

$3\left(\cos\frac{11\pi}{12} + i\sin\frac{11\pi}{12}\right)$

$3\left(\cos\frac{5\pi}{4} + i\sin\frac{5\pi}{4}\right)$

$3\left(\cos\frac{19\pi}{12} + i\sin\frac{19\pi}{12}\right)$

$3\left(\cos\frac{23\pi}{12} + i\sin\frac{23\pi}{12}\right)$

(c) $\frac{3\sqrt{2}}{2} + \frac{3\sqrt{2}}{2}i, -0.776 + 2.898i,$

$-2.898 + 0.776i, -\frac{3\sqrt{2}}{2} - \frac{3\sqrt{2}}{2}i,$

$0.776 - 2.898i, 2.898 - 0.776i$

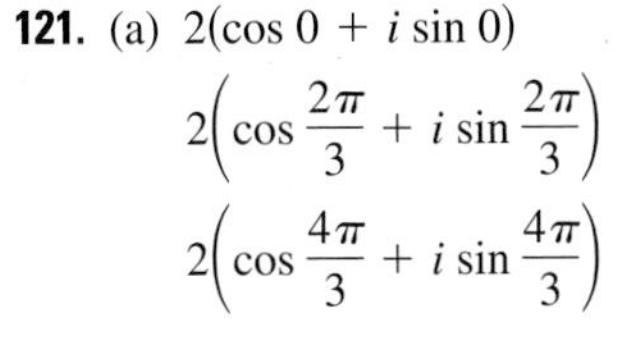

121. (a) $2(\cos 0 + i\sin 0)$ (b)

$2\left(\cos\frac{2\pi}{3} + i\sin\frac{2\pi}{3}\right)$

$2\left(\cos\frac{4\pi}{3} + i\sin\frac{4\pi}{3}\right)$

(c) $2, -1 + \sqrt{3}i, -1 - \sqrt{3}i$

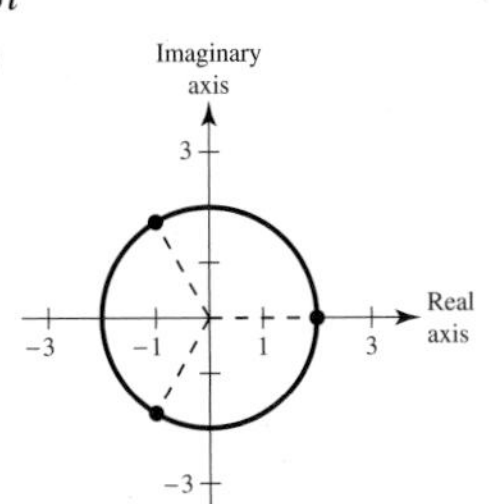

123. $3\left(\cos\frac{\pi}{4} + i\sin\frac{\pi}{4}\right) = \frac{3\sqrt{2}}{2} + \frac{3\sqrt{2}}{2}i$

$3\left(\cos\frac{3\pi}{4} + i\sin\frac{3\pi}{4}\right) = -\frac{3\sqrt{2}}{2} + \frac{3\sqrt{2}}{2}i$

$3\left(\cos\frac{5\pi}{4} + i\sin\frac{5\pi}{4}\right) = -\frac{3\sqrt{2}}{2} - \frac{3\sqrt{2}}{2}i$

$3\left(\cos\frac{7\pi}{4} + i\sin\frac{7\pi}{4}\right) = \frac{3\sqrt{2}}{2} - \frac{3\sqrt{2}}{2}i$

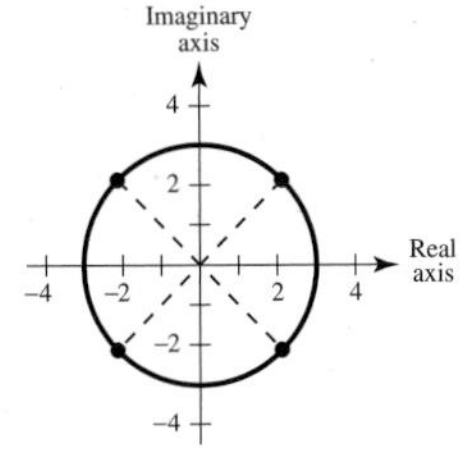

125. $2\left(\cos\frac{\pi}{2} + i\sin\frac{\pi}{2}\right) = 2i$

$2\left(\cos\frac{7\pi}{6} + i\sin\frac{7\pi}{6}\right) = -\sqrt{3} - i$

$2\left(\cos\frac{11\pi}{6} + i\sin\frac{11\pi}{6}\right) = \sqrt{3} - i$

127. $\cos 0 + i\sin 0 = 1$

$\cos\frac{\pi}{2} + i\sin\frac{\pi}{2} = i$

$\cos\frac{2\pi}{3} + i\sin\frac{2\pi}{3} = -\frac{1}{2} + \frac{\sqrt{3}}{2}i$

$\cos\frac{4\pi}{3} + i\sin\frac{4\pi}{3} = -\frac{1}{2} - \frac{\sqrt{3}}{2}i$

$\cos\frac{3\pi}{2} + i\sin\frac{3\pi}{2} = -i$

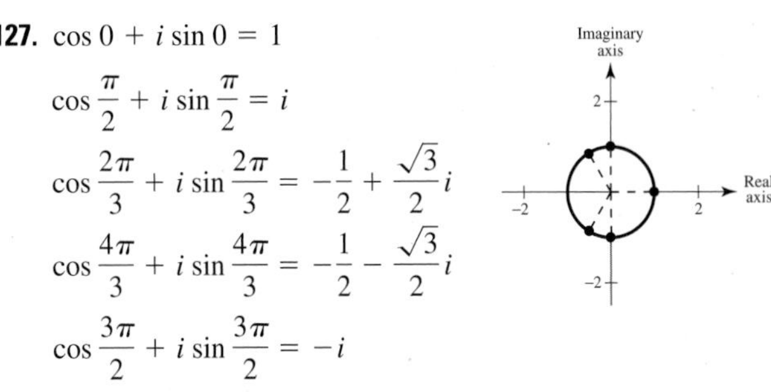

Chapter Test (page 858)

1. $C = 88°, b \approx 27.81, c \approx 29.98$

2. $A = 42°, b \approx 21.91, c \approx 10.95$

3. Two solutions:

$B \approx 29.12°, C \approx 126.88°, c \approx 22.03$

$B \approx 150.88°, C \approx 5.12°, c \approx 2.46$

4. No solution **5.** $A \approx 39.96°, C \approx 40.04°, c \approx 15.02$

6. $A \approx 21.90°, B \approx 37.10°, c \approx 78.15$ **7.** 2052.5 m^2

8. 606.3 mi; 29.1° **9.** $\langle 14, -23\rangle$

10. $\left\langle \frac{18\sqrt{34}}{17}, -\frac{30\sqrt{34}}{17}\right\rangle$

11. $\langle -4, 12\rangle$ **12.** $\langle 8, 2\rangle$

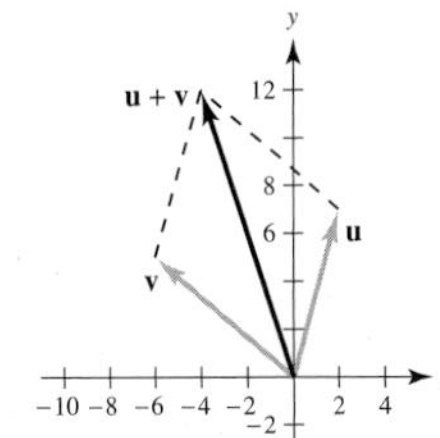

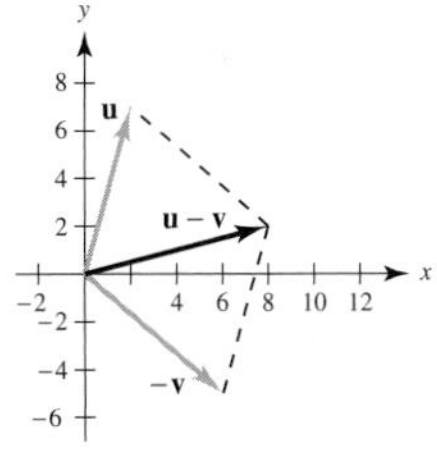

13. $\langle 28, 20\rangle$

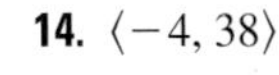

14. $\langle -4, 38\rangle$

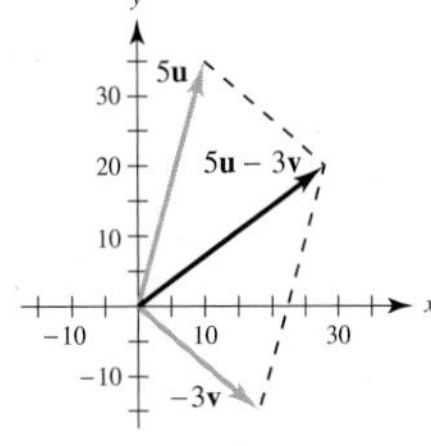

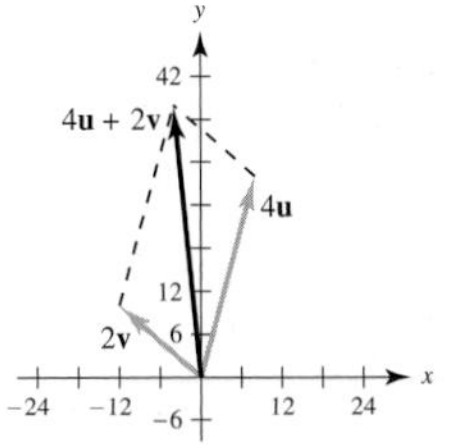

15. $\left\langle \frac{24}{25}, -\frac{7}{25}\right\rangle$ **16.** 14.9°; 250.15 lb **17.** 135° **18.** Yes

19. $\frac{37}{26}\langle 5, 1\rangle; \frac{29}{26}\langle -1, 5\rangle$ **20.** About 104 lb

21. $4\sqrt{2}\left(\cos\frac{7\pi}{4} + i\sin\frac{7\pi}{4}\right)$ **22.** $-3 + 3\sqrt{3}i$

23. $-\frac{6561}{2} - \frac{6561\sqrt{3}}{2}i$ **24.** $5832i$

25. $4\sqrt[4]{2}\left(\cos\frac{\pi}{12} + i\sin\frac{\pi}{12}\right)$

$4\sqrt[4]{2}\left(\cos\frac{7\pi}{12} + i\sin\frac{7\pi}{12}\right)$

$4\sqrt[4]{2}\left(\cos\frac{13\pi}{12} + i\sin\frac{13\pi}{12}\right)$

$4\sqrt[4]{2}\left(\cos\frac{19\pi}{12} + i\sin\frac{19\pi}{12}\right)$

26. $3\left(\cos\frac{\pi}{6} + i\sin\frac{\pi}{6}\right)$

$3\left(\cos\frac{5\pi}{6} + i\sin\frac{5\pi}{6}\right)$

$3\left(\cos\frac{3\pi}{2} + i\sin\frac{3\pi}{2}\right)$

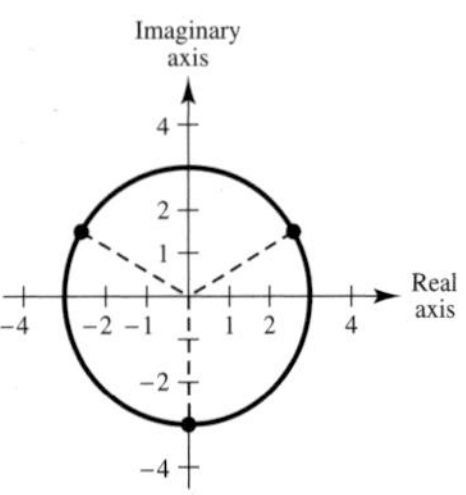

P.S. Problem Solving (page 859)

1. (a) $\alpha = \arcsin(0.5\sin\beta)$
Domain: $0 < \beta < \pi$

(b) $d\alpha/d\beta = \cos\beta/\sqrt{4 - \sin^2\beta}$
Maximum: $(\pi/2, \pi/6)$
Range: $0 < \alpha \leq \pi/6$

(c)

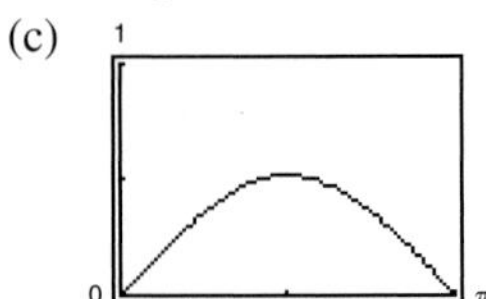

(d) $\sqrt{7}/35 \approx 0.0756$ rad/sec

(e) $c = 18\sin[\pi - \beta - \arcsin(0.5\sin\beta)]/\sin\beta$

(f)

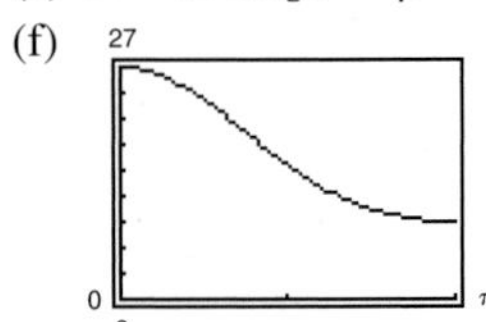

(g) -1.7539 rad/sec

Range: $9 < c < 27$

(h)

β	0	0.4	0.8	1.2	1.6
α	0	0.1960	0.3669	0.4848	0.52534
c	Undef.	25.95	23.07	19.19	15.33

β	2.0	2.4	2.8
α	0.4720	0.3445	0.1683
c	12.29	10.31	9.27

(i) When $\beta = 0$, we can see geometrically that c should be 27, but on a graphing utility we find the function to be undefined. The function obtained using the Law of Sines is not valid when $\beta = 0$ because the figure is no longer a triangle.

3. (a) $\frac{1}{2}(\mathbf{u} + \mathbf{v})$ (b) $\frac{1}{2}(\mathbf{v} - \mathbf{u})$ **5.** Proof

7. (a) About 187.2

(b) $y = \sqrt{3}x/3$, $y = 16.1(x - 25)$, ≈ 187.2

9. (a) Proof

(b) The sum of the squares of the lengths of the diagonals of a parallelogram is equal to the sum of the squares of the lengths of all four sides.

11. (a) (i) $\sqrt{2}$ (ii) $\sqrt{5}$ (iii) 1
(iv) 1 (v) 1 (vi) 1

(b) (i) 1 (ii) $3\sqrt{2}$ (iii) $\sqrt{13}$
(iv) 1 (v) 1 (vi) 1

(c) (i) $\frac{\sqrt{5}}{2}$ (ii) $\sqrt{13}$ (iii) $\frac{\sqrt{85}}{2}$
(iv) 1 (v) 1 (vi) 1

(d) (i) $2\sqrt{5}$ (ii) $5\sqrt{2}$ (iii) $5\sqrt{2}$
(iv) 1 (v) 1 (vi) 1

13. (a) $x \approx 27.452$ mi, $y \approx 53.033$ mi

(b) $z \approx 11.034$ mi; S 21.69° E

15. S 22.09° E

Index Applications

Biology and Life Sciences

Business and Economics

Construction

Consumer

Engineering and Physical Sciences

General

Geometry

INDEX OF APPLICATIONS

Interest Rate

Time and Distance

U.S. Demographics

Index

A

B

C

E

F

INDEX

J

K

L

M

N

O

P

Q

R

S